Department of Economic and Social
Information and Policy Analysis
Statistical Division

Département de l'information économique
et sociale et de l'analyse des politiques
Division de statistique

Statistical Yearbook
Thirty-eighth issue

1990/91
Data available as of
1 September 1992

Annuaire statistique
Trente-huitième édition

Données disponibles
au 1er septembre 1992

United Nations / Nations Unies   New York, 1993

Note

The designations employed and the presentation of material in this publication do not imply the expression of any opinion whatsoever on the part of the Secretariat of the United Nations concerning the legal status of any country, territory, city or area or of its authorities, or concerning the delimitation of its frontiers or boundaries.

In general, statistics contained in the present publication are those available to the United Nations Secretariat up to mid-1992 and refer to mid-1991 or earlier. They therefore reflect country nomenclature in use in mid-1991.

The term "country" as used in this publication also refers, as appropriate, to territories or areas.

The designations "developed" and "developing" are intended for statistical convenience and do not necessarily express a judgment about the stage reached by a particular country or area in the development process.

Symbols of United Nations documents are composed of capital letters combined with figures.

Note

Les appellations employées dans la présente publication et la présentation des données qui y figurent n'impliquent de la part du Secrétariat de l'Organisation des Nations Unies aucune prise de position quant au statut juridique des pays, territoires, villes ou zones ou de leurs autorités, ni quant au tracé de leurs frontières ou limites.

En règle générale, les statistiques contenues dans la présente publication sont celles dont disposait le Secrétariat de l'Organisation des Nations Unies jusqu'à la moitié de 1992 et portent sur la période finissant à la moitié de 1991. Elles reflètent donc la nomenclature des pays en vigueur à l'époque.

Le terme "pays", tel qu'il est utilisé ci-après, peut également désigner des territoires ou des zones.

Les appellations "développées" et "en développement" sont employées à des fins exclusivement statistiques et n'expriment pas nécessairement un jugement quant au niveau de développement atteint par tel pays ou telle région.

Les cotes des documents de l'Organisation des Nations Unies se composent de lettres majuscules et de chiffres.

ST/ESA/STAT/SER.S/14

UNITED NATIONS PUBLICATION
Sales No. E/F.93.XVII.1

PUBLICATION DES NATIONS UNIES
Numero de vente : E/F.93.XVII.1

ISBN 92-1-061152-7
ISSN 0082-8459

Inquiries should be directed to:

SALES SECTION
PUBLISHING DIVISION
UNITED NATIONS
NEW YORK 10017
USA

Adresser toutes demandes de renseignements à la :

SECTION DES VENTES
DIVISION DES PUBLICATIONS
NATIONS UNIES
NEW YORK 10017
USA

HA
12
.5
.U63
1990-91

Ref.

Preface

This is the thirty-eighth issue of the United Nations *Statistical Yearbook*. The *Yearbook* has been prepared by the Statistical Division, Department of Economic and Social Information and Policy Analysis of the United Nations Secretariat, since 1948. The present issue contains series covering, in general, 1981–1990 or 1982–1991, using statistics available to the Statistical Division up to 1 September 1992.

The *Yearbook* is based on data compiled by the Statistical Division in the fields of demographic statistics, national accounts, industry, energy and international trade, data provided by over 20 offices of the United Nations system and international organizations in other specialized fields; and data provided directly to the Statistical Division by countries on railways, motor vehicles in use and international maritime transport.

Acknowledgement is gratefully made to national and international statistical services throughout the world, which have been extremely cooperative in providing data. United Nations agencies and other international organizations which furnished data are listed under "Statistical sources" at the back of this publication.

The present issue of the *Yearbook* is the second to reflect a phased programme of major changes in its organization and presentation undertaken in 1990, the first in the long history of the *Yearbook*. These were adopted in response to the continuing long-term expansion of international data available and demanded by users, in terms of country and subject-matter coverage and level of detail, and the more recent impact of new electronic data processing technologies on data compilation and typesetting.

The new organization of the *Yearbook*, described in the Introduction below in more detail, consists of four parts. Part One, World and Region Summary, is essentially unchanged from previous issues, consisting of key world and regional aggregates and totals. Following Part One, the main subject-matter is presented according to countries or areas and in some cases also world and regions, and is divided into Parts Two, Three and Four: population and social topics, national economic activity and international economic relations. The organization of the population and social topics generally follows the arrangement of subject-matter in the United Nations framework for integration of demographic and social statistics [42]*; economic activity is taken up according to the classes of the United Nations International Standard Industrial Classification of All Economic Activities [44]; and tables on international economic relations cover merchandise trade, international tourism (a major factor in international trade in services) and financial transactions including development assistance. Each chapter includes brief technical notes on statistical sources and methods for the tables in that

Préface

La présente édition est la trente-huitième de l'*Annuaire statistique* des Nations Unies. Depuis 1948, l'*Annuaire* est établi par la Division de statistique du Département de l'information économique et social et de l'analyse des politiques du Secrétariat de l'Organisation des Nations Unies. Elle contient des séries qui portent d'une manière générale sur la période 1981–1990 ou 1982–1991 et pour lesquelles ont été utilisées les informations dont disposait la Division de statistique au 1er septembre 1992.

L'*Annuaire* est établi à partir des données que la Division de statistique a rassemblées dans les domaines suivants : statistiques démographiques, comptabilité nationale, industrie, énergie et commerce international; des données fournies par plus de 20 bureaux du système des Nations Unies et d'organisations internationales dans d'autres domaines spécialisés; enfin, des données que les pays ont communiquées directement à la Division de statistique en ce qui concerne les chemins de fer, les véhicules automobiles en circulation et les transports maritimes internationaux.

Les auteurs de l'*Annuaire statistique* remercient les offices nationaux et internationaux de statistique du monde entier pour leur plus haut degré de coopération. Les institutions spécialisées des Nations Unies et autres organisations internationales qui ont fourni des données sont énumérées dans les "Sources statistiques" figurant à la fin de l'ouvrage.

La présente édition de l'*Annuaire* est la deuxième à tenir compte des importantes transformations par étapes successives à compter de 1990 à son organisation et à sa présentation, la première dans la longue histoire de l'*Annuaire*. Ces modifications furent adoptées face à l'expansion continue et à long terme des données internationales disponibles et demandées par les utilisateurs, par pays et par sujet ainsi qu'avec une précision de plus en plus grande, et compte tenu de l'impact plus récent des nouvelles techniques informatiques sur la compilation des données et la composition automatique des textes.

Le nouveau plan de l'*Annuaire*, qui est décrit ci-après dans l'introduction de manière plus détaillée, comprend quatre parties. La première partie, "Aperçu mondial et régional", est reprise à peu près sans changement des éditions précédentes et se compose des principaux agrégats et totaux aux niveaux mondial et régional. Après la première partie, la matière principale de l'*Annuaire* est présentée par pays ou par région et, dans certains cas, aux niveaux mondial et régional, et s'articule en trois parties (deuxième, troisième et quatrième parties : Population et questions sociales, Activité économique nationale et Relations économiques internationales). L'organisation de la deuxième partie, "Population et questions sociales", suit généralement le plan adopté à l'ONU pour intégrer les statistiques démographiques et sociales [42]*; l'activité économique est présentée conformément aux

---

* Numbers in brackets refer to numbered entries in the section "Statistical sources and references" at the end of this book.

* Les chiffres figurant entre crochets se réfèrent aux entrées numérotées dans la liste des sources et références statistiques à la fin de l'ouvrage.

chapter. Complete references to sources and related methodological publications are provided at the end of the *Yearbook*.

The present issue also incorporates some additional important changes. These include the addition of two detailed tables of statistics and indicators on environment and pollution, prepared in cooperation with the United Nations Environment Programme and the Monitoring and Assessment Research Centre, and the deletion of certain commodities from the commodity production tables in Part Three. The list of tables added and omitted from the last two issues of the *Yearbook* is given in annex III. These changes were reviewed and endorsed by a meeting of experts convened by the United Nations Secretariat and held in New York from 1-5 February 1993. The expert group also recommended a much more detailed programme of changes in the *Yearbook* which will be implemented in phases over the next several issues. The full report of this expert group meeting (in English only) is available upon request at no charge from the Director of the Statistical Division.

The new format adopted for typesetting tables in the *Yearbook* was developed in cooperation with the Graphic Presentation Unit of the United Nations Secretariat, to meet new challenges and design opportunities offered by microcomputer "desktop publishing" technologies. Since the disappearance in the 1970s of mechanical typesetting, photocomposition techniques and equipment and in turn table design for statistics have become much more variable. As a result, it is no longer possible to expect data suppliers for the *Yearbook* to follow a standard design format, nor economically feasible to reset the tables, which are provided in a wide variety of tabulation and database formats, except using microcomputer techniques. This issue of the *Yearbook* continues to make extensive use of microcomputer database, spreadsheet and typesetting technologies, and an "open" table design, which is easier to photocompose than enclosed designs using horizontal and vertical rules and boxes common to the mechanical era.

As described more fully in the Introduction below, every attempt has been made to ensure that the series contained in the *Yearbook* are sufficiently comparable to provide a reliable general description of economic and social topics throughout the world. The reader should nevertheless carefully consult the footnotes and technical notes for any given table for explanations of general limitations of series presented and specific limitations affecting particular data items. Complete information concerning the definitions and concepts used and limitations of the data are provided in the sources and references cited at the end of the *Yearbook*. Readers interested in more detailed figures than those shown in the present publication, and in further information on the full range of internationally assembled statistics on specialized fields, should also consult the specialized publications listed in the statistical sources.

catégories adoptées par l'ONU dans sa Classification internationale type de toutes les activités économiques [44]; les tableaux consacrés aux relations économiques internationales recouvrent le commerce des marchandises, le tourisme international (élément essentiel du secteur international des services) et les opérations financières, y compris l'aide au développement. Chaque chapitre comprend une brève note technique sur les sources et méthodes statistiques utilisées pour les tableaux du chapitre. On trouvera à la fin de l'*Annuaire* des références complètes aux sources et publications méthodologiques connexes.

La présente édition comporte également quelques modifications importantes supplémentaires. Celles-ci comprennent notamment l'addition de deux tableaux détaillés de statistiques et d'indicateurs de l'environnement et de la pollution, établis en coopération avec le Programme des Nations Unies pour l'environnement et le Centre de recherche pour la surveillance et l'évaluation et l'élimination de certains produits de base des tableaux des produits de base de la troisième partie. La liste des tableaux ajoutés et supprimés dans les deux dernières éditions de l'*Annuaire* figure dans l'annexe III. Ces modifications ont été examinées et approuvées par un groupe d'experts réuni par le Secrétariat de l'Organisation des Nations Unies à New York du 1er au 5 février 1993. Le groupe d'experts a également recommandé un programme de modifications beaucoup plus détaillé de l'*Annuaire* qui sera progressivement en oeuvre dans les éditions suivantes. On peut se procurer gratuitement le rapport intégral de la réunion de ce groupe d'experts (en anglais uniquement) auprès du Directeur de la Division de statistique.

La nouvelle présentation adoptée pour la composition automatique des tableaux de l'*Annuaire* a été mise au point en coopération avec le Groupe de la présentation graphique du Secrétariat de l'ONU, afin de satisfaire à de nouvelles tâches et de répondre aux possibilités offertes par les techniques de publication assistée par micro-ordinateur. Depuis la disparition de la composition mécanique dans les années 70, les techniques et le matériel de photocomposition, donc la conception des tableaux statistiques, peuvent varier beaucoup plus librement. En conséquence, on ne peut plus s'attendre à ce que les fournisseurs de données suivent un mode de présentation type et il n'est plus rentable de recomposer les tableaux qui utilisent des présentations et des modèles de bases de données très diverses, si ce n'est en utilisant la micro-informatique. La présente édition de l'*Annuaire* continue à utiliser largement une base de données, un tableur et des techniques de composition empruntés à la micro-informatique, et à adopter un mode de tabulation "ouvert" qui convient mieux à la photocomposition que les modes fermés faisant appel à des lignes horizontales et verticales et à des encadrés, courants à l'époque de la composition mécanique.

Comme il est précisé ci-après dans l'introduction, aucun effort n'a été épargné afin que les séries figurant dans l'*Annuaire* soient suffisamment comparables pour fournir une

Of course much remains to be done to bring the *Yearbook* fully up to date in its scope, coverage, timeliness and design, and in its technical notes. The process is inevitably an evolutionary one. Comments on the present *Yearbook* and its future evolution are welcome and should be addressed to the Director, United Nations Statistical Division, New York 10017 USA.

description générale fiable des problèmes économiques et sociaux dans le monde entier. Le lecteur devra néanmoins consulter avec soin les notes de bas de page et les notes techniques de chaque tableau pour y trouver l'explication des limites générales imposées aux séries présentées et des limites particulières propres à certains types de données. On trouvera dans la liste des sources et références citées à la fin de l'*Annuaire* des renseignements complets concernant les définitions et concepts utilisés et les limites des données.

Les lecteurs qui souhaitent avoir des chiffres plus détaillés que ceux qui figurent dans le présent volume ou qui désirent se procurer des renseignements sur la gamme complète des statistiques relatives à tel ou tel domaine particulier qui ont été compilées à l'échelon international devraient consulter les publications spécialisées énumérées dans les sources statistiques.

Il reste sans aucun doute beaucoup à faire pour mettre l'*Annuaire* pleinement à jour en ce qui concerne sa nature et sa portée, son actualité et sa conception générale, ainsi que ses notes techniques. Il s'agit là inévitablement d'un processus évolutif. Les observations sur la présente édition de l'*Annuaire* et les modifications suggérées pour l'avenir seront reçues avec intérêt et doivent être adressées au Directeur de la Division de statistique de l'ONU, New York, N. Y. 10017 (Etats-Unis d'Amérique).

Contents

Table des matières

Part Four
International Economic Relations

Quatrième partie
Relations économiques internationales

List of tables

Liste des tableaux

Part Three
Economic Activity

Troisième partie
Activité économique

Explanatory notes

The metric system of weights and measures has been employed throughout the *Statistical Yearbook*. For conversion coefficients and factors, see annex II.

Certain tables contain global aggregates designated variously as "total" or "world". Where a figure represents the summation of the country series shown in the table but is not considered comprehensive for the world, it is labelled "total". Where, however, an aggregate is considered to represent substantially complete world coverage, it is labelled "world". As a rule, allowance has been made in the "world" figures for any gaps that may exist in the country series shown.

In some cases, the comparability of the statistics is affected by geographical changes. As a general rule, the data relate to a given country or area within its present de facto boundaries. Where statistically important, attention is called to changes in territory by means of a footnote. The reader is referred to annex I, concerning country and area nomenclature, where changes in designation are listed.

Numbers in brackets refer to numbered entries in the section "Statistical sources and references" at the end of this book.

In general, statistics presented in the present publication are based on information available to the Statistical Division of the United Nations Secretariat up to 1 September 1992.

Symbols and conventions used in the tables

A point (.) is used to indicate decimals.

A hyphen (-) between years, e.g., 1984-1985, indicates the full period involved, including the beginning and end years; a slash (/) indicates a financial year, school year or crop year, e.g., 1984/85.

"Δ p.a." (change per annum) is used to indicate annual rate of change.

Not applicable or not separately reported	..
Data not available	...
Magnitude zero	-
Magnitude zero, or less than half of unit employed	0 or 0.0
Provisional or estimated figure	*
United Nations estimate	x
Marked break in series	#

Details and percentages in tables do not necessarily add to totals because of rounding.

Notes explicatives

Le système métrique de poids et mesures a été utilisé dans tout l'*Annuaire statistique*. On trouvera à l'annexe II les coefficients et facteurs de conversion.

Certains tableaux contiennent des agrégats globaux désignés par la mention "total" ou "monde", selon les cas. Quand un chiffre représente l'addition des chiffres correspondant à chacun des pays qui figurent dans le tableau, mais ne paraît pas recouvrir le monde entier, il porte la mention "total". Si un agrégat semble correspondre au total mondial, ou très peu s'en faut, il porte la mention "monde". En règle générale, les chiffres portant la mention "monde" s'entendent compte tenu des lacunes qui peuvent exister dans la série de pays indiqués.

Dans certains cas, les changements géographiques intervenus influent sur la comparabilité des statistiques. En règle générale, les données renvoient au pays ou zone en question dans ses frontières actuelles effectives. Une note appelle l'attention sur les changements territoriaux, si cela importe du point de vue statistique. Le lecteur est renvoyé à l'annexe A (nomenclature des pays et zones et groupements régionaux) où il trouvera une liste des changements de désignation.

Les chiffres figurant entre crochets se réfèrent aux entrées numérotées dans la liste des sources et références statistiques à la fin de l'ouvrage.

En général, les statistiques qui figurent dans la présente publication sont fondées sur les informations dont disposait la Division de statistique du Secrétariat de l'ONU au 1er septembre 1992.

Signes et conventions employés dans les tableaux

Les décimales sont précédées d'un point (.).

Un tiret (-) entre des années, par exemple "1984-1985", indique que la période est embrassée dans sa totalité, y compris la première et la dernière année; une barre oblique (/) renvoie à un exercice financier, à une année scolaire ou à une campagne agricole, par exemple "1984/85".

Le symbole "Δ p.a." signifie qu'il s'agit du taux annuel de variation.

Non applicable ou non communiqué séparément	..
Données non disponibles	...
Néant	-
Valeur nulle, ou inférieure à la moitié de la dernière unité retenue	0 ou 0.0
Chiffre provisoire ou estimatif	*
Estimation des Nations Unies	x
Discontinuité notable dans la série	#

Les chiffres étant arrondis, les totaux ne correspondent pas toujours à la somme exacte des éléments ou pourcentages figurant dans les tableaux.

Introduction

This is the thirty-eighth edition of the United Nations *Statistical Yearbook*, prepared by the Statistical Division, Department of Economic and Social Information and Policy Analysis, of the United Nations Secretariat. It contains series covering, in general, 1981-1990 or 1982-1991, based on statistics available to the Statistical Division up to 1 September 1992.

The major purpose of the *Statistical Yearbook* is to provide in a single volume a comprehensive compilation of internationally-available statistics on social and economic conditions and activities in the world, at world, regional and national levels, covering roughly a ten-year period.

Most of the statistics presented in the *Yearbook* are extracted from more detailed and specialized publications prepared by the Statistical Division and by many other international statistical services. Thus, while the specialized publications concentrate on monitoring development topics and trends in particular social and economic fields, the *Statistical Yearbook* tables provide data for a more comprehensive, overall description of social and economic structures, conditions, changes and activities.

Towards this end, the objective has been to collect, systematize and coordinate the most essential components of comparable statistical information which can give a broad and, where feasible, a consistent picture of social and economic processes at world, regional and national levels.

More specifically, the *Statistical Yearbook* provides systematic information on a wide range of social and economic issues which are of particular current concern in the United Nations system. A particular challenge facing the *Yearbook* is that these issues are extensively interrelated and meaningful social and economic analysis require systematization and coordination of the data across many fields. Some of these issues include:

— General economic growth and related economic conditions;

— Economic situation in developing countries and progress towards the objectives adopted for the United Nations development decades;

— Population and urbanization, and their growth and impact;

— Employment, inflation and wages;

— Production of energy and development of new energy sources;

— Expansion of trade;

— Supply of food and alleviation of hunger;

— Financial situation and external payments and receipts;

— Education, training and eradication of illiteracy;

— Improvement in general living conditions;

— Pollution and protection of environment;

— Assistance provided to developing countries for social and economic development purposes.

Introduction

La présente édition est la trente-huitième de l'*Annuaire statistique* des Nations Unies, établi par la Division de statistique du Département de l'information économique et sociale et de l'analyse des politiques du Secrétariat de l'Organisation des Nations Unies. Elle contient des séries de données qui portent d'une manière générale sur les années 1981 à 1990 ou 1982 à 1991, et pour lesquelles ont été utilisées les informations dont disposait la Division de statistique au 1er septembre 1992.

L'*Annuaire statistique* a principalement pour objet de présenter en un seul volume un inventaire complet de statistiques disponibles sur le plan international et concernant la situation et les activités sociales et économiques dans le monde, aux échelons mondial, régional et national, pour une période d'environ 10 ans.

Une bonne partie des données qui figurent dans l'*Annuaire* existent sous une forme plus détaillée dans les publications spécialisées établies par la Division de statistique et par bien d'autres services statistiques internationaux. Alors que les publications spécialisées suivent essentiellement l'évolution dans certains domaines socio-économiques précis, l'*Annuaire statistique* présente les données d'une description globale et exhaustive des structures, conditions, changements et activités socio-économiques du monde.

C'est pourquoi on a cherché à recueillir, systématiser et coordonner les principaux éléments de renseignements statistiques comparables, de manière à donner un tableau général et autant que possible cohérent des processus socio-économiques en cours aux échelons mondial, régional et national.

Plus précisément, l'*Annuaire statistique* a pour objet de présenter des renseignements systématiques sur toutes sortes de questions socio-économiques qui sont liées aux préoccupations actuelles du système des Nations Unies. L'une des difficultés tient à ce que ces questions ont d'étroits liens entre elles, et que pour faire une analyse économique et sociale utile il faut systématiser et coordonner des données se rapportant à de nombreux domaines différents. Il s'agit notamment des questions suivantes :

— La croissance économique générale et les aspects connexes de l'économie;

— La situation économique dans les pays en développement et les progrès accomplis vers la réalisation des objectifs des décennies des Nations Unies pour le développement;

— Population et urbanisation, leur croissance et leur impact;

— L'emploi, l'inflation et les salaires;

— La production énergétique et la mise en valeur des nouvelles sources d'énergie;

— L'expansion du commerce;

— Les approvisionnements alimentaires et la lutte contre la faim;

— La situation financière et paiements extérieurs et recettes;

— L'éducation, la formation et l'élimination de l'analphabétisme;

— L'amélioration des conditions de vie en général;

— La pollution et la protection de l'environnement;

— L'assistance fournie aux pays en développement à des fins socio-économiques.

Organization of the *Yearbook*

The contents of the *Statistical Yearbook* are planned to serve a general readership. The *Yearbook* endeavours to provide information for various bodies of the United Nations system as well as for other international organizations, for Governments and non-governmental organizations, for national statistical, economic and social policy bodies, for scientific and educational institutions, for libraries and for the public. Data published in the *Statistical Yearbook* are also of interest to companies and enterprises and to agencies engaged in marketing research.

The range of topics in the *Yearbook* allows comparative analysis of many essential aspects of economic and social development. The *Yearbook* tables are grouped into four broad parts:

— World and Region Summary (tables 1-10);
— Population and Social Statistics (tables 11-28);
— Economic Activity (tables 29-115);
— International Economic Relations (tables 116-131).

These four parts present two levels of aggregation and presentation. The more general type of information shown in Part One provides an overall picture of development at the world and region levels. It is followed and supplemented by more specific and detailed information for analysis concerning individual countries or areas and topics in each of the three following parts. Each of these three parts is divided into more specific chapters, by topic, and each chapter is accompanied by a section entitled "Technical notes". These notes provide brief descriptions of major statistical concepts, definitions and classifications required for interpretation and analysis of the data. Systematic information on the methodology used for computation of figures can also be found in the *1977 Supplement to the Statistical Yearbook and the Monthly Bulletin of Statistics*[47] and in the relevant publications referring to methodology of the United Nations and its agencies, listed in the section "Statistical sources and references" at the end of the *Yearbook*. Additional general information on statistical methodology is provided in the section below on "Comparability of statistics" and in the explanatory notes following the Introduction.

More specifically, Part One, World and Region Summary, comprises 10 tables highlighting the principal trends in the world as a whole as well as in regions and in the major economic and social sectors. It contains global totals of important aggregate statistics needed for the analysis of economic growth, the structure of the world economy, major changes in world population, expansion of external merchandise trade, world production and consumption of energy. The global totals are, as a rule, subdivided into major geographical areas.

Présentation de l'*Annuaire*

Le contenu de l'*Annuaire statistique* a été préparé à l'intention des catégories les plus diverses de lecteurs. Les renseignements fournis devraient pouvoir être utilisés par les divers organismes du système des Nations Unies ainsi que par d'autres organisations internationales, par les gouvernements et les organisations non gouvernementales, par les organismes nationaux chargés des politiques en matière statistique, économique et sociale, par les institutions scientifiques et les établissements d'enseignement, les bibliothèques et le grand public en général. Les données publiées dans l'*Annuaire statistique* présentent également de l'intérêt pour les sociétés et les entreprises, et les organismes spécialisés dans les études de marché.

L'*Annuaire* couvre une gamme de sujets suffisamment large pour permettre une analyse comparative de tous les aspects essentiels du développement économique et social. Les tableaux sont groupés en quatre parties :

— Aperçu mondial et régional (tableaux 1 à 10);
— Statistiques démographiques et sociales (tableaux 11 à 28);
— Activité économique (tableaux 29 à 115);
— Relations économiques internationales (tableaux 116 à 131).

Ces quatre parties représentent deux niveaux de groupement et de présentation de renseignements statistiques. Les renseignements plus généraux apparaissent dans la première partie qui donne un tableau global du développement à l'échelon mondial et régional. Il est suivi et complété par des renseignements plus précis convenant à un type d'analyse qui porte sur des pays individuels ou des domaines et des sujets relevant des trois autres parties. Chacune de ces trois parties est divisée en chapitres portant sur un sujet donné, et chaque chapitre est accompagné de notes techniques. Ces notes fournissent une brève description des principales notions, définitions et classifications statistiques nécessaires pour interpréter et analyser les données. Les méthodes de calcul utilisées sont également décrites de façon systématique dans le *Supplément de 1977 à l'Annuaire statistique et au Bulletin mensuel de statistique*[47], ainsi que dans les publications pertinentes se référant à la méthodologie des Nations Unies et de leurs organismes, énumérées dans la section "Sources et références statistiques", à la fin de l'*Annuaire*. On trouvera un supplément d'informations générales dans la section intitulée "Comparabilité des statistiques", ainsi que dans les notes explicatives qui suivent l'introduction.

Plus spécialement, la première partie, intitulée "Aperçu mondial et régional", comprend 10 tableaux présentant les principales tendances dans le monde et dans les régions ainsi que dans les principaux secteurs économiques et sociaux. Elle fournit des chiffres mondiaux pour les principaux agrégats de statistiques économiques nécessaires pour analyser la croissance économique, la structure de l'économie mondiale, les principaux changements dans la population mondiale, l'expansion du commerce extérieur de la marchandise, la production et la consommation mondiales d'énergie. En général, les chiffres mondiaux sont ventilés par grandes régions géographiques.

Part Two, Population and Social Statistics, comprises 18 tables which contain more detailed statistical series for the appraisal of social conditions, levels of living and the impact of government policies on these. For example, it includes data on education and the development of cultural activities.

Part Three, Economic Activity, provides data in 87 tables on major branches of the economy, that is, agriculture, industry, transport, communications, and so forth. Figures included relate to the output of major industries and commodities, intermediate consumption of selected goods by industries, number of employees engaged in production, wages and salaries paid to employees.

Part Four, International Economic Relations, comprises 16 tables relating to external trade and finance. It focuses on the growth and structure of exports and imports by countries or areas, international tourism, balance of payments and development assistance provided by multilateral and bilateral agencies to developing countries.

Annexes and regional groupings of countries or areas

The annexes to the *Statistical Yearbook* provide additional important information on the *Yearbook*'s contents and presentation of data.

Annex I provides information on countries or areas covered in the *Yearbook* tables and on their grouping into geographical regions. The geographical groupings shown in the *Yearbook* are generally based on continental regions unless otherwise indicated. However, strict consistency in this regard is impossible, due to the wide range of classifications used for different purposes in the various international agencies and other sources of statistics for the *Yearbook*.

Neither is there a common agreement in the United Nations system concerning the terms "developed" and "developing", when referring to the stage of development reached by any given country or area and its corresponding classification in one or the other grouping. Thus, the *Yearbook* refers more generally to "developed" or "developing" regions on the basis of conventional practice. Following this practice, "developed regions" comprises northern America, Europe and the former USSR, Australia, Japan and New Zealand, while all of Africa and the remainder of the Americas, Asia and Oceania comprise the "developing regions". These designations are intended for statistical convenience and do not necessarily express a judgement about the stage reached by a particular country or area in the development process.

Annex II provides detailed information on conversion coeffients and factors used in various tables, and annexe III provides listings of tables added and omitted in the present edition of the *Yearbook*.

La deuxième partie, intitulée "Statistiques démographiques et sociales", comporte 18 tableaux où figurent des séries statistiques plus détaillées servant à l'évaluation des conditions sociales, des niveaux de vie et de l'impact des politiques gouvernementales sur les processus sociaux. Elle comprend notamment des données sur l'éducation et le développement des activités culturelles.

La troisième partie, intitulée "Activité économique", présente, en 87 tableaux, des données sur les principales branches d'activité économique (agriculture, industrie, transports, communications, etc.). Les chiffres cités ont trait à la production des principales branches d'activité industrielle et des principaux produits de base, à la consommation intermédiaire de certains biens par les entreprises, au nombre de personnes occupées à la production, aux salaires et traitements versés aux travailleurs.

La quatrième partie, intitulée "Relations économiques internationales", comprend 16 tableaux relatifs au commerce extérieur et aux finances. Elle est consacrée essentiellement à la croissance et à la structure des exportations et des importations par pays ou zone, au tourisme international, à la balance des paiements et à l'aide au développement fournie aux pays en développement par les organismes multilatéraux et bilatéraux.

Annexes et groupements régionaux des pays ou zones

Les annexes à l'*Annuaire statistique* offrent d'importantes informations supplémentaires sur la teneur et la présentation des données du présent ouvrage.

L'annexe I donne des renseignements sur les pays et les domaines couverts par les tableaux de l'*Annuaire* et sur leur regroupement en régions géographiques. Les groupements géographiques figurant dans l'*Annuaire* sont généralement fondés sur les régions continentales, sauf indication contraire. Toutefois, une présentation absolument systématique est impossible à cet égard en raison de la grande variété de classifications utilisées à différentes fins par les diverses institutions internationales et dans d'autres sources de statistiques employées pour l'*Annuaire*.

Il n'y a pas non plus dans le système des Nations Unies d'accord au sujet des termes "développé" ou "en développement" quand on parle du degré de développement atteint par un pays ou une zone donnés et lorsqu'il s'agit de les classifier dans l'un ou l'autre de ces groupes. Ainsi, dans l'*Annuaire* on parle plus généralement des régions développées et des régions en développement sur la base de l'usage. Selon ce dernier, les régions développées sont l'Amérique du Nord, l'Europe, et l'ancienne URSS, l'Australie, le Japon et la Nouvelle-Zélande, alors que toute l'Afrique et le reste des Amériques, l'Asie et l'Océanie constituent les régions en développement. Ces appellations sont utilisées pour plus de commodité dans la présentation des statistiques et n'impliquent pas nécessairement un jugement quant au stade de développement auquel sont parvenus un pays ou une zone donnés.

L'annexe II fournit des renseignements sur les coefficients et facteurs de conversion utilisés dans les différents tableaux, et l'annexe III contient des listes de tableaux ajoutés ou omis dans la présente édition de l'*Annuaire*

Comparability of statistics

One of the major aims of the *Statistical Yearbook* is to present country series which are as nearly comparable as the available statistics permit. Special efforts have been made to ensure the compatibility of various series by coordinating time periods, base years, prices chosen for valuation and so on. This is indispensable in relating various bodies of data to each other in order to facilitate analysis across different sectors. Thus, for example, relating data on economic output to those on employment makes it possible to derive some trends in the field of productivity; relating data on exports and imports to those on national product allows an evaluation of the relative importance of external trade in different countries and reveals changes in the role of trade over time.

In general, the data presented reflect the methodological recommendations of the United Nations Statistical Commission, issued in various United Nations publications, and of other international bodies concerned with statistics. This promotes not only the international comparability of the data but also ensures a degree of compatibility regarding the underlying concepts, definitions and classifications relating to different series. However, much work remains to be done in this area and, for this reason, some tables serve only as a first source of data, which require further adjustment before being used for in-depth analytical studies. Although, on the whole, a significant degree of comparability has been achieved in international statistics, there are certain limitations, for a variety of reasons, of which the reader should be aware. For example, direct comparison between the official national product data based on the United Nations System of National Accounts (SNA) and those based on the System of Material Product Balances (MPS) would be misleading owing to differences in the underlying concepts. Lack of comparability can also result from the use of different base years in compiling index numbers or from differences in formulas for index numbers.

The most common cause of non-comparability of data is different valuations of statistical aggregates such as national income, wages and salaries, output of industries and so forth. Conversion of these and similar series originally expressed in national prices into a common currency, for example into United States dollars, through the use of exchange rates is not always satisfactory owing to frequent wide fluctuations in market rates and differences between official rates and rates which would be indicated by unofficial markets or purchasing power parities. For this reason, data on national income in United States dollars which are published in the *Yearbook* are subject to certain distortions and can be used as only a rough approximation of the relative magnitudes involved.

Comparabilité des statistiques

L'*Annuaire statistique* a principalement pour objet de présenter des statistiques nationales aussi comparables sur le plan international que les données disponibles le permettent. On s'est efforcé tout particulièrement de faire en sorte que les diverses séries statitiques soient compatibles en harmonisant les périodes de référence, les années de base, les prix utilisés pour les évaluations, etc. Cette démarche est indispensable si l'on veut rapprocher divers ensembles de données pour faciliter l'analyse intersectorielle de l'économie. Lier par exemple les données relatives à la production à celles qui ont trait à l'emploi permet de dégager certaines tendances dans le domaine de la productivité; de même, lier les données relatives aux exportations et aux importations à celles qui ont trait au produit national permet d'évaluer l'importance relative des échanges extérieurs dans différents pays et révèle également les variations du rôle des échanges extérieurs.

De façon générale, les données sont présentées selon les recommandations méthodologiques formulées par la Commission de statistique de l'ONU et par les autres organisations internationales qui s'intéressent aux statistiques. La présentation adoptée tend non seulement à promouvoir la comparabilité des données à l'échelon international, mais elle assure également une certaine compatibilité entre les concepts, les définitions et classifications utilisés. Cependant, comme il reste encore beaucoup à faire dans ce domaine, les données présentées dans un certain nombre de tableaux n'ont qu'une valeur indicative et nécessiteront d'autres ajustements avant de pouvoir servir à des analyses en profondeur.

Bien que dans l'ensemble l'on soit parvenu à un degré de comparabilité appréciable en matière de statistiques internationales, le lecteur doit garder présentes à l'esprit certaines limites qui peuvent être dues à plusieurs facteurs. Par exemple, comparer directement les statistiques officielles du produit national basées sur le Système de comptabilité nationale (SNA) des Nations Unies et celles qui sont fondées sur le Système de comptabilité du produit matériel (MPS), serait une source d'erreur, car les deux systèmes ont une conception différente du produit national.

L'utilisation d'années de base différentes pour le calcul des indices ou l'existence de formules d'indices différentes peut également compromettre la comparabilité des données.

La cause la plus commmune de non-comparabilité des données est l'évaluation différente donnée d'agrégats statistiques tels que le revenu national, les salaires et traitements, la production des différentes branches d'activité industrielle et ainsi de suite. Il n'est pas toujours satisfaisant de ramener la valeur des séries de ce type, exprimée à l'origine en prix nationaux, à une devise commune (par exemple le dollar des Etats-Unis), en appliquant les taux de change officiels, ces taux ne coïncidant pas avec les parités réelles de pouvoir d'achat. C'est pourquoi les données relatives au revenu national exprimé en dollars des Etats-Unis (calculé par application des taux de change officiels) publiées dans l'*Annuaire* ne sont pas exemptes de certaines distorsions et ne peuvent donc servir qu'à donner une idée approximative des ordres de grandeur relatifs.

Non-comparability may also be due to the fact that country practices do not follow international recommendations on methodology.

The use of different kinds of sources for obtaining data is another cause of incomparability. This is true, for example, in the case of employment and unemployment, where data are collected from such non-comparable sources as sample surveys, social insurance statistics and establishment surveys.

Non-comparability of data may also result from differences in the institutional patterns of countries. Certain variations in social and economic organization and institutions may have an impact on the comparability of the data even if the underlying concepts and definitions are identical.

These and other causes of non-comparability of the data are briefly explained in the technical notes to each chapter.

La non-comparabilité peut également être due au fait que les pays n'appliquent pas les méthodes recommandées à l'échelon international.

Le recours à des sources diverses pour la collecte des données est un autre facteur qui limite la comparabilité, en particulier dans les secteurs de l'emploi et du chômage où les statistiques sont obtenues par des moyens aussi peu comparables que les sondages, les fichiers des assurances sociales et les enquêtes auprès des entreprises.

Dans certains cas, les données ne sont pas comparables du fait des différences de structure institutionnelle des pays. Autrement dit, des organisations différentes de l'économie peuvent affecter la comparabilité des données même si les concepts et définitions sont identiques.

Ces causes de non-comparabilité des données, et d'autres encore, sont brièvement expliquées dans les notes techniques concernant les chapitres.

Part One
World and Region Summary

I
World and region summary (tables 1-10)

This part of the *Statistical Yearbook* presents selected aggregate series on principal economic and social topics for the world as a whole and major regions. The topics include population and surface area, agricultural and industrial production, industrial employment, energy, commodity prices, motor vehicles in use, external trade and government financial reserves. More detailed data on individual countries and areas are provided in the subsequent parts of the present *Yearbook*. These comprise Part Two, Population and Social Statistics; Part Three, Economic Activity; and Part Four, International Economic Relations.

Regional totals may contain incomparabilities between series owing to differences in definitions of regions and lack of data for particular regional components. General information on regional groupings is provided in annex I of the *Yearbook*. Supplementary information on regional groupings used in specific series is provided as necessary in table footnotes and in the technical notes at the end of chapter I.

Première partie
Aperçu mondial et régional

I
Aperçu mondial et régional (tableaux 1 à 10)

Cette partie de l'*Annuaire statistique* présente, pour le monde entier et ses principales subdivisions, un choix d'agrégats ayant trait à des questions économiques et sociales essentielles : population et superficie, production agricole et industrielle, emploi industriel, énergie, prix des produits de base, véhicules automobiles en circulation, commerce extérieur et réserves financières publiques. Des statistiques plus détaillées pour divers pays ou zones figurent dans les parties ultérieures de l'*Annuaire*, c'est-à-dire dans les deuxième, troisième et quatrième parties intitulées respectivement : population et statistiques sociales, activités économiques et relations économiques internationales.

Les totaux régionaux peuvent présenter des incomparabilités entre les séries en raison de différences dans la définition des régions et de l'absence de données sur tel ou tel élément régional. A l'annexe I de l'*Annuaire*, on trouvera des renseignements généraux sur les groupements régionaux. Des informations complémentaires sur les groupements régionaux pour certaines séries bien précises sont fournies, lorsqu'il y a lieu, dans les notes figurant au bas des tableaux et dans les notes techniques à la fin du chapitre I.

1
Selected series of world statistics
Séries principales de statistiques mondiales
Population, production, external trade and finance
Population, production, commerce extérieur et finances

Series Séries	Unit or base Unité ou base	1983	1984	1985	1986	1987	1988	1989	1990	1991
World population [1] **Population mondiale** [1]	million	4688	4770	4853	4938	5024	5112	5201	5292	5385

Agriculture, forestry and fishing production • Production agricole, forestière et de la pêche
Index numbers • Indices

All commodities Tous produits	1979-81	106	112	115	116	117	119	123	126	125
Food Produits alimentaires	1979-81	106	112	114	116	117	119	123	126	125
Crops Culture	1979-81	105	113	115	116	116	115	...	...	...
Cereals Céréales	1979-81	105	115	117	119	115	112	...	...	...
Livestock products Produits de l'élevage	1979-81	106	108	112	115	117	119	...	...	...

Quantities • Quantités

Wheat Froment	million t.	494	517	506	534	511	507	543	601	553
Maize (corn) Maïs	million t.	348	453	489	478	451	401	474	480	474
Rice (paddy) Riz (paddy)	million t.	451	470	473	471	464	491	517	524	518
Cotton (lint) Coton (fibre)	million t.	14.8	18.2	17.3	15.2	16.6	18.3	17.1	18.8	20.3
Groundnuts (in shell) Arachides (non décortiquées)	million t.	18.4	18.8	20	21	18.9	20.6	23.1	23.4	23.4
Wool, greasy Laine, en suint	1000 t.	2870	2846	2937	3002	3083	3155	3241	3330	3266
Meat Viande	million t.	142	146	151	157	163	169	172	176	176
Coffee, green Café vert	1000 t.	5564	5128	5800	5275	6470	5787	6096	6250	5868
Tea Thé	1000 t.	2043	2185	2310	2288	2391	2461	2425	2507	2559
Cocoa (beans) Fèves de cacao	1000 t.	1598	1758	1961	2069	2062	2497	2457	2584	2473
Tobacco Tabac	1000 t.	5944	6492	7004	6069	6170	6874	7047	7122	6977
Natural rubber Caoutchouc naturel	1000 t.	4030	4250	4335	4450	4850	5130	5250	5210	5380
Roundwood Bois rond	million m³	3048	3163	3200	3305	3372	3426	3456	3450	...
Fish catches Prises de poissons	million t.	77.5	83.9	86.4	92.8	94.4	99.1	100.3	97.3	...

Industrial production • Production industrielle
Index numbers [2] • Indices [2]

All commodities Tous produits	1980	99.4	105.4	108.4	111.6	115.7	122.2	126.9	127.0	127.5
Mining Mines	1980	85.0	88.2	87.1	89.9	91.0	95.7	100.0	99.6	100.8
Manufacturing Manufactures	1980	101.6	108.3	111.9	115.3	119.9	126.9	131.7	131.6	131.6

Quantities • Quantités

Coal Houille	million t.	2831	2999	3161	3249	3411	3488	3562	3564	3421
Lignite and brown coal [3] Lignite et charbon brun [3]	million t.	1130	1153	1202	1219	1250	1261	1274	1219	1049
Crude petroleum [4] Petrole brut [4]	million t.	2649	2679	2654	2782	2787	2910	2957	3019	2960
Natural gas Gaz naturel	pétajoules	54383	59260	61486	62252	65262	68254	70801	73083	73352
Iron ore Minerai de fer	million t.	740	829	862	868	890	916	920	921	922

1
Selected series of world statistics
Population, production, external trade and finance [*cont.*]
Séries principales de statistiques mondiales
Population, production, commerce extérieur et finaces [*suite*]

Series Séries	Unit or base Unité ou base	1983	1984	1985	1986	1987	1988	1989	1990	1991
Pig-iron and ferro-alloys Fonte et ferro-alliages	million t.	466	500	507	496	503	504	506	506	505
Crude steel Acier brut	million t.	638	679	685	671	682	685	676	685	687
Smelter copper Cuivre de fonderie	1000 t.	8413	8603	8587	8690	8806	8726	8729	8734	8738
Zinc Zinc	1000 t.	6135	6425	6625	6606	6909	6816	6820	6825	6834
Lead Plomb	1000 t.	4943	5036	5241	5152	5197	5283	5285	5273	5276
Tin [5] Etain [5]	1000 t.	195	196	196	191	194	210	210	211	214
Aluminium Aluminium	1000 t.	17990	19916	19649	19699	21219	22361	22390	22378	22365
Cement Ciment	million t.	906	933	949	997	1045	1109	1137	1140	1135
Electricity [6] Electricité [6]	billion milliard kWh	8826	9326	9747	10055	10587	11059	11450	11734	11535
Fertilizers [7] Engrais [7]	million t.	130.9	139.7	136.2	143.5	152.4	152.2	158.3	152.7	147.9
Sugar, raw Sucre, brut	million t.	97.6	99.3	98.8	100.9	101.1	104.1	105.6	110.7	111.8
Wheat flour Farine de froment	million t.	197.3	202.8	206.7	211.3	200.7	205.4	203.2	203.4	...
Wine Vin	million t.	34.9	32.4	29.3	31.9	31.9	27.6	28.6	28.6	...
Cellulosic fibres Fibres cellulosiques	1000 t.	3053	3157	3136	3121	3104	3137	3120	2960	2945
Non-cellulosic fibres Fibres non cellulosiques	1000 t.	10395	11014	11542	11738	12347	12997	13175	11835	11598
Sawnwood Sciages	million m³	438	458	465	480	502	503	499	483	...
Woodpulp · Pâte de bois										
Chemical Chimique	million t.	96.5	101.2	100.9	105.4	109.6	113.9	114.6	114.0	...
Mechanical Mécanique	million t.	27.5	30.1	30.2	31.2	32.4	34.3	35.5	36.3	...
Newsprint Papier journal	million t.	26.0	27.8	28.3	29.3	30.6	32.0	32.3	33.1	...
Merchant vessels · Navires marchands										
Tonnage launched Tonnage lancé	million grt	14.89	17.73	17.25	14.91	9.62	11.80	12.72	14.68	16.68
Tonnage under construction [8] Tonnage en construction [8]	million grt	14.86	15.73	14.73	11.05	9.69	11.62	12.45	13.53	15.90
Motor vehicles · Véhicules automobiles										
Passenger Tourisme	million	29.21	30.57	32.29	32.86	33.01	34.26	35.46	35.70	34.43
Commercial Utilitaires	million	9.79	11.33	12.52	12.44	12.90	13.86	13.44	12.44	12.01

Transport · Transports

Motor vehicles in use · Véhicules automobiles en service

		1983	1984	1985	1986	1987	1988	1989	1990	1991
Passenger cars Voitures de tourisme	1000	352316	364814	374727	386308	394210	412907	424366	444900	...
Commercial vehicles Véhicules utilitaires	1000	103716	108464	112816	113423	121176	126882	132566	138082	...

External trade · Commerce extérieur

Value, billion US$ · Valeur, milliard $E-U

		1983	1984	1985	1986	1987	1988	1989	1990	1991
Imports, c.i.f. Importations c.i.f.		1875.0	1984.9	2028.2	2198.1	2544.4	2908.4	3132.5	3547.0	3613.2
Exports, f.o.b. Exportations f.o.b.		1807.1	1903.2	1926.7	2120.6	2464.7	2814.3	3013.3	3394.2	3482.2

1
Selected series of world statistics
Population, production, external trade and finance [*cont.*]
Séries principales de statistiques mondiales
Population, production, commerce extérieur et finaces [*suite*]

Series Séries	Unit or base Unité ou base	1983	1984	1985	1986	1987	1988	1989	1990	1991
Quantum: index of exports · Quantum : indice des exportations										
All commodities										
Tous produits	1980	100	109	112	120	127	137	146	154	162
Manufactures										
Produits manufacturés	1980	107	119	124	129	139	151	164	172	...
Unit value: index of exports [9] · Valeur unitaire : indice des exportations [9]										
All commodities										
Tous produits	1980	88	87	85	88	98	104	105	114	111
Manufactures										
Produits manufacturés	1980	88	87	87	101	113	121	121	132	...
Primary commodities: price indexes [9][10] · Produits de base : indices des prix [9][10]										
All commodities										
Tous produits	1980	91	90	85	63	66	70	72	79	69
Food										
Produits alimentaires	1980	77	74	66	74	75	87	87	88	85
Non-food: of agricultural origin										
Non alimentaires: d'origine agricole	1980	86	91	74	75	86	100	98	100	93
Minerals										
Minéraux	1980	97	95	93	57	59	59	62	71	59
Machinery [10]										
Machines [10]	1980	99	98	99	118	132	140	138	148	153

Finance [11] • Finances [11]

	Unit or base	1983	1984	1985	1986	1987	1988	1989	1990	1991
Gold reserves [8]	10^6 ounces									
Réserves d'or [8]	10^6 onces	948	947	953	953	947	948	943	941	940
International reserves minus gold, billion SDR [8] · Réserves internationales moins l'or, milliard DTS [8]										
All countries										
Tous les pays		361.9	407.0	405.2	419.0	507.8	542.9	591.1	637.8	672.1
Position in IMF										
Disponibilité au FMI		39.1	41.6	38.7	35.3	31.5	28.3	25.5	21.3	25.5
Foreign exchange										
Devise étrangères		308.3	349.0	348.3	364.1	456.1	494.5	545.2	593.7	625.7
SDR	billion									
DTS	milliard	14.4	16.5	18.2	19.5	20.2	20.2	20.5	20.5	20.4

Source:
Statistical Division of the United Nations Secretariat (New York), Food and Agriculture Oganization of the United Nations (Rome), International Rubber Study Group (London), Lloyd's Register of Shipping (London), Motor Vehicle Manufacturers' Association (Detroit) and International Monetary Fund (Washington, DC).

1 Annual data: mid-year estimates.
2 Excluding Albania, China, Democratic People's Republic of Korea and Viet Nam.
3 Excluding peat for fuel.
4 Excluding natural gas liquids.
5 Primary metal production only.
6 Electricity (hydro, geothermal, thermal, nuclear) generated by establishments for public or private use.
7 1 July - 30 June.
8 End of period.
9 Indexes computer in US dollars.
10 Export price indexes.
11 Excluding Bulgaria, former German Democratic Republic, Mongolia and former USSR.

Source:
Division de statistique du Sécretariat des Nations Unies (New York), Organisation des Nations Unies pour l'alimentation et l'agriculture (Rome), Groupe international d'études du caoutchouc (Londres), "Lloyd's Register of Shipping" (Londres), "Motor Vehicles Manufacturers' Association" (Detroit) et Fonds monétaire internationale (Washington, DC).

1 Données annuelles : estimations au milieu de l'année.
2 Non compris l'Albanie, la Chine, la République populaire démocratique de Corée et le Viet Nam.
3 Non compris la tourbe combustible.
4 Non compris le gaz naturel liquéfié.
5 Production du métal de première fusion seulement.
6 L'électricité (hydraulique, géothermique, thermique, nucléaire) produite par des entreprises d'utilisation publique ou privée.
7 1 juillet - 30 juin.
8 Fin de la période.
9 Indice calculé en dollars des Etats-Unis.
10 Indice des prix à l'exportation.
11 Non compris la Bulgarie, l'ancienne République démocratique allemande, la Mongolie et l'ancienne URSS.

2
Population, rate of increase, birth and death rates, surface area and density
Population, taux d'accroissement, taux de natalité et taux de mortalité, superficie et densité

Macro regions and regions / Grandes régions et régions	Mid-year estimates (millions) / Estimations au milieu de l'année (millions)								Annual rate of increase / Taux d'accrois-sement annuel %	Birth rate / Taux de natalité (0/00)	Death rate / Taux de mortalité (0/00)	Surface area (km²) / Superficie (km²) (000's)	Density[1] / Densité[1]
	1950	1960	1965	1970	1975	1980	1985	1990	1985-90	1985-90	1985-90	1990	1990
World / Monde	**2516**	**3020**	**3336**	**3698**	**4079**	**4448**	**4851**	**5292**	**1.7**	**27**	**10**	**136255**	**39**
Africa / Afrique	**222**	**279**	**317**	**362**	**413**	**477**	**553**	**642**	**3.0**	**45**	**15**	**30305**	**21**
Eastern Africa / Afrique orientale	65	82	94	108	124	144	168	197	3.2	49	16	6354	31
Middle Africa / Afrique centrale	26	32	35	40	45	52	60	70	3.0	45	16	6613	11
Northern Africa / Afrique septentrionale	52	65	73	83	94	107	123	141	2.6	38	11	8525	16
Southern Africa / Afrique méridionale	16	20	23	26	29	32	36	41	2.4	34	10	2675	15
Western Africa / Afrique occidentale	63	81	92	105	122	141	165	194	3.2	48	17	6138	32
Northern America[2] / Amérique septentrionale[2]	**166**	**199**	**214**	**226**	**239**	**252**	**265**	**276**	**0.8**	**15**	**9**	**21962**	**13**
Latin America / Amérique latine	**166**	**218**	**251**	**286**	**323**	**363**	**404**	**448**	**2.1**	**29**	**7**	**20535**	**22**
Caribbean / Caraïbes	17	20	23	25	27	29	31	34	1.5	25	8	235	145
Central America / Amérique centrale	37	50	59	70	81	93	105	118	2.3	31	6	2481	48
South America / Amérique du Sud	112	147	169	191	215	241	268	297	2.0	28	8	17819	17
Asia[3][4] / Asie[3][4]	**1377**	**1668**	**1861**	**2102**	**2354**	**2583**	**2835**	**3113**	**1.9**	**28**	**9**	**27582**	**113**
Eastern Asia[3] / Asie orientale[3]	671	792	874	987	1097	1176	1249	1336	1.3	20	7	11763	114
Southern Asia / Asie méridionale	481	596	670	754	849	948	1070	1201	2.3	35	12	6781	177
South Eastern Asia / Asie mériodionale orientale	182	225	253	287	324	360	401	445	2.0	30	9	4493	99
Western Asia[4] / Asie occidentale[4]	42	56	64	74	85	99	115	132	2.8	36	9	4545	29
Europe[3][4] / Europe[3][4]	**393**	**425**	**445**	**460**	**474**	**484**	**492**	**498**	**0.2**	**13**	**11**	**4933**	**101**
Oceania[2] / Océanie[2]	**12.6**	**15.8**	**17.5**	**19.3**	**21.2**	**22.8**	**24.6**	**26.5**	**1.5**	**19**	**8**	**8536**	**3**
Australia and New Zealand / Australie et Nouvelle Zélande	10.1	12.7	14.0	15.4	16.7	17.8	19.0	20.3	1.3	15	8	7985	2
Melanesia / Mélanésie	2.1	2.6	2.9	3.3	3.7	4.2	4.7	5.3	2.3	34	11	541	10
Micronesia / Micronésie	0.2	0.2	0.2	0.3	0.3	0.3	0.3	0.4	1.6	27	7	3	124
Polynesia / Polynésie	0.2	0.3	0.4	0.4	0.4	0.5	0.5	0.5	1.5	34	5	7	78
former USSR / ancienne URSS	**180**	**214**	**231**	**243**	**254**	**266**	**278**	**289**	**0.8**	**18**	**11**	**22402**	**13**

2
Population, rate of increase, birth and death rates, surface area and density [*cont.*]
Population, taux d'accroissement, taux de natalité et taux de mortalité, superficie et densité [*suite*]

Source:
Demographic statistics database of the Statistical Office of the United Nations Secretariat.

1 Population per square kilometre of surface area. Figures are merely the quotients of population divided by surface area and are not to be considered as either reflecting density in the urban sense or as indicating the supporting power of a territory's land and resources.
2 Hawaii, a state of the United States of America, is included in Northern America rather than Oceania.
3 Excluding the former USSR, shown separately.
4 The European portion of Turkey is included in Western Asia rather than Europe.

Source:
Base de données pour les statistiques demographique du Bureau de statistique du Secrétariat de l'ONU.

1 Habitants per kilomètre carré. Il s'agit simplement du quotient calculé en divisasnt la population par la superficie et n'est pas considéré comme indiquant la densité au sens urbain du mot ni l'effectif de population que les terres et les ressources du territoire sont capables de nourrir.
2 Hawaii, un Etat des Etats-Unis d'Amérique, est compris en Amérique. septentrionale plutôt qu'en Océanie.
3 Non compris l'ancienne URSS, qui fait l'objet d'une rubrique distincte.
4 La partie européenne de la Turquie est comprise en Asie Occidentale plutôt qu'en Europe.

3
Index numbers of total agricultural and food production
Indices de la production agricole totale et de la production alimentaire

1979-1981 = 100

Region Région	1982	1983	1984	1985	1986	1987	1988	1989	1990	1991
A. Total agricultural production · Production agricole totale										
World *Monde*	**106**	**106**	**112**	**115**	**116**	**117**	**119**	**123**	**125**	**125**
Developed regions **Régions développées**	**104**	**100**	**107**	**108**	**109**	**108**	**105**	**110**	**111**	**107**
North America Amérique du Nord	105	88	102	108	103	101	94	103	108	106
Europe Europe	105	103	110	107	109	109	108	110	108	108
Oceania Océanie	96	108	106	109	110	109	114	113	116	115
former USSR ancienne URSS	104	109	109	110	118	118	116	120	120	104
Other Autres	101	96	101	104	103	105	104	107	105	103
Developing regions **Régions en voie de dévelop.**	**108**	**112**	**117**	**121**	**122**	**125**	**131**	**135**	**139**	**143**
Africa Afrique	105	104	105	113	120	118	127	130	132	137
Latin America Amérique latine	106	107	110	115	113	118	123	124	125	126
Near East Proche-Orient	108	109	109	116	122	122	128	121	130	130
Far East Extrême-Orient	109	116	122	125	127	130	137	142	148	152
Other Autres	103	97	111	109	111	107	109	116	117	113
B. Food production · Production alimentaire										
World *Monde*	**106**	**106**	**112**	**114**	**116**	**117**	**119**	**123**	**126**	**125**
Developed regions **Régions développées**	**105**	**100**	**107**	**108**	**109**	**108**	**105**	**110**	**111**	**106**
North America Amérique du Nord	106	90	103	109	104	101	94	104	108	106
Europe Europe	105	103	110	107	109	109	108	110	108	108
Oceania Océanie	94	110	106	107	108	106	111	109	111	109
former USSR ancienne URSS	104	110	110	111	120	119	118	122	122	105
Other Autres	101	96	102	105	105	107	105	109	107	105
Developing regions **Régions en voie de dévelop.**	**108**	**112**	**116**	**120**	**123**	**125**	**132**	**135**	**140**	**143**
Africa Afrique	105	104	105	113	120	118	127	130	132	138
Latin America Amérique latine	108	108	111	116	115	119	124	127	128	129

3
Index numbers of total agricultural and food production
[cont.]
Indices de la production agricole totale et de la production alimentaire
1979-1981 = 100 [suite]

Region Région	1982	1983	1984	1985	1986	1987	1988	1989	1990	1991
Near East Proche-Orient	109	109	109	117	124	124	130	122	133	132
Far East Extrême-Orient	108	116	121	123	127	129	136	142	147	150
Other Autres	105	96	112	109	113	106	109	115	116	113

Source:
Food and Agriculture Organization of the United Nations (Rome).

Source:
Organisation des Nations Unies pour l'alimentation et l'agriculture (Rome).

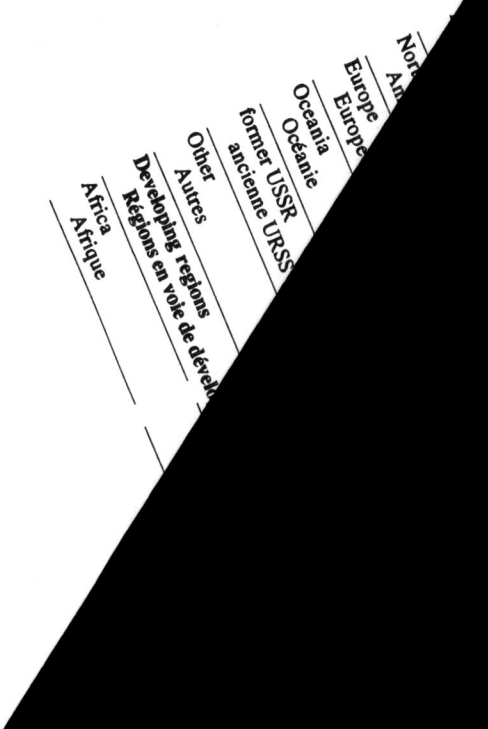

4
Index numbers of per capita total agricultural and food production
Indices de la production agricole totale et de la production alimentaire par habitant
1979-1981 = 100

Region Région	1982	1983	1984	1985	1986	1987	1988	1989	1990	1991
A. Per capita total agricultural production • Production agricole totale par habitant										
World *Monde*	103	101	104	105	104	103	103	105	105	103
Developed regions **Régions développées**	103	98	104	104	104	103	100	103	104	99
North America Amérique du Nord	103	86	98	103	97	94	87	95	99	96
Europe Europe	104	102	109	106	107	106	106	107	105	104
Oceania Océanie	93	104	100	102	102	99	102	100	101	99
former USSR ancienne URSS	103	106	105	106	112	110	108	111	110	95
Other Autres	99	93	97	99	97	98	96	98	95	93
Developing regions **Régions en voie de dévelop.**	104	105	107	109	108	108	111	112	113	113
Africa Afrique	99	95	93	97	100	96	99	99	97	98
Latin America Amérique latine	101	100	101	103	99	102	103	103	102	100
Near East Proche-Orient	101	99	96	99	101	98	101	92	97	94
Far East Extrême-Orient	105	110	114	114	113	114	118	120	123	124
Other Autres	99	91	101	97	97	91	91	95	93	89
B. Per capita food production • Production alimentaire par habitant										
World *Monde*	103	101	104	105	105	103	103	105	106	103
Developed regions **Régions développées**	103	98	104	105	105	103	100	104	104	99
North America Amérique du Nord	104	87	99	103	98	95	87	96	99	96
	104	102	109	106	107	106	106	107	105	104
	92	105	100	101	100	97	100	96	97	94
	103	107	106	106	113	112	110	112	112	96
	99	93	98	99	99	100	98	100	97	95
p.	103	105	107	108	109	108	111	112	113	113
	99	95	93	97	100	96	99	99	97	98

4

Index numbers of per capita total agricultural and food production
[*cont.*]

Indices de la production agricole totale et de la production alimentaire par habitant
1979-1981 = 100 [*suite*]

Region Région	1982	1983	1984	1985	1986	1987	1988	1989	1990	1991
Latin America Amérique latine	103	10!	101	104	101	103	105	105	103	102
Near East Proche-Orient	102	99	96	100	103	100	102	93	99	96
Far East Extrême-Orient	104	110	112	113	114	113	117	120	122	123
Other Autres	100	90	102	97	99	91	91	94	93	88

Source:
Food and Agriculture Organization of the United Nations
(Rome).

Source:
Organisation des Nations Unies pour l'alimentation et
l'agriculture (Rome).

5
# Index numbers of industrial production: world and regions
# Indices de la production industrielle : monde et régions
1980=100

Region and industry [ISIC] Région et industrie [CITI]	Weight(%) Pond.(%)	1983	1984	1985	1986	1987	1988	1989	1990	1991 [1]
World • Monde										
Total industry [2-4]										
Total, industrie [2-4]	**100.0**	**99**	**105**	**108**	**112**	**116**	**122**	**127**	**127**	**101**
Total mining [2]										
Total, industries extractives [2]	**15.7**	**85**	**88**	**87**	**90**	**91**	**96**	**100**	**100**	**100**
Coal										
Houille	1.9	101	101	106	107	107	108	107	99	97
Petroleum, gas										
Pétrole, gaz	11.3	80	83	80	83	83	88	93	93	100
Metal										
Minerais métalliques	1.3	98	104	108	109	113	122	129	136	104
Total manufacturing [3]										
Total, industries manufacturières [3]	**77.8**	**102**	**108**	**112**	**115**	**120**	**127**	**132**	**132**	**101**
Light industry										
Industrie légère	28.2	102	106	108	111	115	119	123	122	101
Heavy industry										
Industrie lourde	49.6	101	110	114	118	123	131	137	137	100
Selected manufacturing										
Industries manufacturières déterminées										
Food, beverages, tobacco										
Industries alimentaires, boissons, tabac	10.7	106	110	111	113	116	120	124	127	104
Textiles										
Textiles	4.4	99	100	103	106	109	110	111	107	98
Apparel, leather, footwear										
Articles d'habillement, cuir et chaussures	3.4	99	102	102	103	104	105	106	103	98
Wood products, furniture										
Bois, meubles	3.2	99	104	104	109	114	119	122	120	97
Paper, printing, publishing										
Papier, imprimerie, édition	4.9	104	111	114	120	127	134	138	142	102
Chemicals and related products										
Produits chimiques et alliés	11.2	104	111	114	119	126	134	138	138	102
Non-metallic mineral products										
Produits minéraux non métalliques	3.5	97	101	102	105	108	114	118	118	99
Basic metals										
Métallurgie de base	6.0	92	99	100	99	103	111	113	112	100
Metal products										
Ouvrages en métaux	29.1	102	112	118	123	128	137	144	144	100
Electricity, gas, water [4]										
Electricité, gaz et eau [4]	**6.5**	**108**	**113**	**118**	**120**	**125**	**130**	**135**	**139**	**105**
A. Developed regions • Régions developpées										
Total industry [2-4]										
Total, industrie [2-4]	**100.0**	**99**	**106**	**109**	**110**	**114**	**120**	**124**	**124**	**100**
Total mining [2]										
Total, industries extractives [2]	**8.8**	**99**	**103**	**106**	**103**	**104**	**106**	**106**	**107**	**101**
Coal										
Houille	1.9	100	98	104	104	103	103	103	96	97
Petroleum, gas										
Pétrole, gaz	4.7	101	106	108	102	103	101	98	99	102
Metal										
Minerais métalliques	1.2	96	106	109	109	116	128	138	148	104
Total manufacturing [3]										
Total, industries manufacturières [3]	**83.3**	**99**	**106**	**109**	**111**	**114**	**121**	**126**	**126**	**99**
Light industry										
Industrie légère	27.0	100	104	105	108	111	115	117	116	99
Heavy industry										
Industrie lourde	56.3	98	107	111	112	116	125	130	130	100
Selected manufacturing										
Industries manufacturières déterminées										
Food, beverages, tobacco										
Industries alimentaires, boissons, tabac	9.1	103	105	107	109	111	114	116	117	103

5

Index numbers of industrial production: world and regions [*cont.*]

Indices de la production industrielle : monde et régions [*suite*]

1980=100

Region g and industry [ISIC] Région g et industrie [CITI]	Weight(%) Pond.(%)	1983	1984	1985	1986	1987	1988	1989	1990	1991[1]
Textiles										
Textiles	3.4	96	97	98	100	102	102	103	98	97
Apparel, leather, footwear										
Articles d'habillement, cuir et chaussures	3.1	95	96	94	94	93	91	90	85	94
Wood products, furniture										
Bois, meubles	3.7	97	102	102	106	112	116	118	116	95
Paper, printing, publishing										
Papier, imprimerie, édition	6.5	103	111	112	117	124	130	134	136	101
Chemicals and related products										
Produits chimiques et alliés	11.6	101	108	110	113	118	125	130	128	100
Non-metallic mineral products										
Produits minéraux non métalliques	3.6	92	96	96	97	99	105	108	108	97
Basic metals										
Métallurgie de base	6.9	88	95	96	92	95	104	106	104	98
Metal products										
Ouvrages en métaux	34.3	100	110	116	118	121	131	139	139	100
Electricity, gas, water [4]										
Electricité, gaz et eau [4]	**7.8**	**105**	**110**	**114**	**116**	**120**	**124**	**128**	**130**	**104**

North America · Amérique du Nord

	Weight(%) Pond.(%)	1983	1984	1985	1986	1987	1988	1989	1990	1991[1]
Total industry [2-4]										
Total, industrie [2-4]	**100.0**	**98**	**109**	**110**	**111**	**116**	**123**	**126**	**126**	**97**
Total mining [2]										
Total, industries extractives [2]	**12.4**	**90**	**98**	**97**	**90**	**92**	**96**	**95**	**97**	**100**
Coal										
Houille	1.6	94	110	110	109	114	120	121	129	96
Petroleum, gas										
Pétrole, gaz	8.8	89	93	92	82	82	83	81	81	102
Metal										
Minerais métalliques	1.2	85	107	107	107	119	132	143	145	98
Total manufacturing [3]										
Total, industries manufacturières [3]	**79.4**	**99**	**110**	**113**	**115**	**120**	**128**	**131**	**132**	**96**
Light industry										
Industrie légère	24.1	105	112	114	118	124	128	131	132	96
Heavy industry										
Industrie lourde	55.3	96	110	112	114	119	127	131	132	96
Selected manufacturing										
Industries manufacturières déterminées										
Food, beverages, tobacco										
Industries alimentaires, boissons, tabac	7.5	103	107	110	112	115	118	120	122	100
Textiles										
Textiles	2.4	102	98	99	106	112	112	114	112	95
Apparel, leather, footwear										
Articles d'habillement, cuir et chaussures	2.6	91	96	91	94	97	98	100	94	91
Wood products, furniture										
Bois, meubles	3.9	112	123	128	136	145	150	150	147	90
Paper, printing, publishing										
Papier, imprimerie, édition	8.0	107	117	118	126	134	138	142	145	98
Chemicals and related products										
Produits chimiques et alliés	10.8	103	111	114	119	126	132	136	138	98
Non-metallic mineral products										
Produits minéraux non métalliques	2.6	96	105	108	111	113	116	116	113	90
Basic metals										
Métallurgie de base	6.3	82	92	91	85	92	102	101	99	91
Metal products										
Ouvrages en métaux	34.1	97	114	117	118	122	133	138	139	96
Electricity, gas, water [4]										
Electricité, gaz et eau [4]	**8.2**	**103**	**108**	**109**	**107**	**109**	**115**	**118**	**117**	**101**

Eastern Europe and former USSR · Europe de l'Est et l'ancienne URSS

	Weight(%) Pond.(%)	1983	1984	1985	1986	1987	1988	1989	1990	1991[1]
Total industry [2-4]										
Total, industrie [2-4]	**100.0**	**108**	**113**	**117**	**122**	**126**	**132**	**133**	**127**	**98**

5
Index numbers of industrial production: world and regions [*cont.*]
Indices de la production industrielle : monde et régions [*suite*]
1980=100

Region ʃ and industry [ISIC] Région ʃ et industrie [CITI]	Weight(%) Pond.(%)	1983	1984	1985	1986	1987	1988	1989	1990	1991[1]
Total mining [2]										
Total, industries extractives [2]	**7.6**	**104**	**105**	**106**	**110**	**112**	**115**	**113**	**105**	**96**
Coal										
Houille	3.8	102	103	104	107	108	110	106	95	96
Petroleum, gas										
Pétrole, gaz	2.0	106	107	108	111	115	116	115	110	97
Metal										
Minerais métalliques	0.4	97	98	100	102	105	107	108	103	90
Total manufacturing [3]										
Total, industries manufacturières [3]	**88.4**	**109**	**114**	**118**	**123**	**127**	**133**	**135**	**128**	**98**
Light industry										
Industrie légère	38	105	109	110	111	115	119	122	116	99
Heavy industry										
Industrie lourde	50.4	112	118	124	131	137	143	144	138	97
Selected manufacturing										
Industries manufacturières déterminées										
Food, beverages, tobacco										
Industries alimentaires, boissons, tabac	17.8	108	112	110	110	113	117	120	117	100
Textiles										
Textiles	7.8	101	101	104	106	108	112	115	107	97
Apparel, leather, footwear										
Articles d'habillement, cuir et chaussures	5.6	104	109	112	113	115	120	123	121	102
Wood products, furniture										
Bois, meubles	2.3	110	114	118	124	127	134	137	130	100
Paper, printing, publishing [2]										
Papier, imprimerie, édition [2]	1.4	110	115	119	126	129	135	137	132	98
Chemicals and related products										
Produits chimiques et alliés	11.2	108	112	115	120	123	129	130	123	98
Non-metallic mineral products										
Produits minéraux non métalliques	4.1	108	111	113	119	123	128	131	124	97
Basic metals										
Métallurgie de base	6.0	104	108	110	115	116	120	119	110	96
Metal products										
Ouvrages en métaux	29.7	115	123	132	140	148	156	157	150	98
Electricity, gas, water [4]										
Électricité, gaz et eau [4]	**4.0**	**109**	**116**	**120**	**124**	**129**	**132**	**132**	**133**	**101**

Western Europe · Europe de l'Ouest

	Weight(%) Pond.(%)	1983	1984	1985	1986	1987	1988	1989	1990	1991[1]
Total industry [2-4]										
Total, industrie [2-4]	**100.0**	**99**	**102**	**105**	**108**	**110**	**115**	**119**	**116**	**100**
Total mining [2]										
Total, industries extractives [2]	**6.9**	**109**	**110**	**116**	**116**	**115**	**112**	**110**	**106**	**99**
Coal										
Houille	2.4	102	91	100	101	96	93	92	78	95
Petroleum, gas										
Pétrole, gaz	3.5	119	128	132	133	135	129	126	128	101
Metal										
Minerais métalliques	0.3	86	92	90	81	76	76	78	78	98
Total manufacturing [3]										
Total, industries manufacturières [3]	**85.8**	**98**	**101**	**104**	**106**	**109**	**114**	**119**	**115**	**99**
Light industry										
Industrie légère	29.9	98	100	101	104	106	109	112	109	100
Heavy industry										
Industrie lourde	55.9	98	101	105	108	110	117	122	118	99
Selected manufacturing										
Industries manufacturières déterminées										
Food, beverages, tobacco										
Industries alimentaires, boissons, tabac	10.6	103	105	106	108	110	113	115	117	104
Textiles										
Textiles	4.1	93	96	98	100	101	101	102	94	96
Apparel, leather, footwear										
Articles d'habillement, cuir et chaussures	3.9	96	96	96	96	92	89	88	82	93

5

Index numbers of industrial production: world and regions [*cont.*]

Indices de la production industrielle : monde et régions [*suite*]

1980=100

Region ș and industry [ISIC] Région ș et industrie [CITI]	Weight(%) Pond.(%)	1983	1984	1985	1986	1987	1988	1989	1990	1991 [1]
Wood products, furniture Bois, meubles	3.8	90	106	89	92	96	101	105	102	99
Paper, printing, publishing Papier, imprimerie, édition	5.6	99	106	105	109	115	122	126	127	101
Chemicals and related products Produits chimiques et alliés	12.6	101	106	109	112	115	123	127	120	99
Non-metallic mineral products Produits minéraux non métalliques	4.2	91	92	91	93	95	100	105	103	97
Basic metals Métallurgie de base	5.8	91	96	98	95	96	104	106	101	98
Metal products Ouvrages en métaux	34.0	99	101	107	111	113	120	127	123	99
Electricity, gas, water [4] **Electricité, gaz et eau [4]**	**7.3**	**105**	**107**	**115**	**118**	**123**	**126**	**129**	**131**	**106**

European Economic Community [+] • Communauté economique européenne [+]

	Weight(%) Pond.(%)	1983	1984	1985	1986	1987	1988	1989	1990	1991 [1]
Total industry [2-4] **Total, industrie [2-4]**	**100.0**	**99**	**101**	**104**	**107**	**109**	**114**	**118**	**114**	**100**
Total mining [2] **Total, industries extractives [2]**	**6.9**	**109**	**108**	**113**	**112**	**110**	**104**	**96**	**91**	**95**
Coal Houille	2.7	102	90	99	100	95	92	90	77	95
Petroleum, gas Pétrole, gaz	3.2	121	127	132	131	129	117	102	102	95
Metal Minerais métalliques	0.2	80	86	81	69	62	63	66	68	104
Total manufacturing [3] **Total, industries manufacturières [3]**	**86.0**	**97**	**100**	**103**	**106**	**108**	**114**	**118**	**114**	**100**
Light industry Industrie légère	29.6	98	100	101	103	106	110	112	109	101
Heavy industry Industrie lourde	56.5	97	100	104	107	109	116	122	117	100
Selected manufacturing **Industries manufacturières déterminées**										
Food, beverages, tobacco Industries alimentaires, boissons, tabac	10.6	103	105	107	109	111	114	117	118	105
Textiles Textiles	4.2	93	96	98	100	101	101	103	95	97
Apparel, leather, footwear Articles d'habillement, cuir et chaussures	3.9	96	96	95	94	91	88	87	81	94
Wood products, furniture Bois, meubles	3.4	89	90	86	89	94	100	101	103	102
Paper, printing, publishing Papier, imprimerie, édition	4.9	97	104	102	106	112	120	124	125	102
Chemicals and related products Produits chimiques et alliés	13.0	100	106	108	111	114	122	125	118	100
Non-metallic mineral products Produits minéraux non métalliques	4.2	90	91	89	91	93	99	104	102	98
Basic metals Métallurgie de base	5.9	89	94	95	92	93	101	104	98	99
Metal products Ouvrages en métaux	34.6	99	101	106	110	112	119	127	122	99
Electricity, gas, water [4] **Electricité, gaz et eau [4]**	**7.1**	**104**	**105**	**113**	**117**	**121**	**124**	**127**	**128**	**106**

European Free Trade Association [+] Association européenne de libre échange [+]

	Weight(%) Pond.(%)	1983	1984	1985	1986	1987	1988	1989	1990	1991 [1]
Total industry [2-4] **Total, industrie [2-4]**	**100.0**	**102**	**107**	**112**	**114**	**118**	**123**	**130**	**133**	**100**
Total mining [2] **Total, industries extractives [2]**	**7.7**	**112**	**126**	**133**	**141**	**155**	**169**	**211**	**221**	**110**
Coal Houille	0.1	165	159	200	190	171	122	165	154	**96**

5
Index numbers of industrial production: world and regions [*cont.*]
Indices de la production industrielle : monde et régions [*suite*]
1980=100

Region & and industry [ISIC] Région & et industrie [CITI]	Weight(%) Pond.(%)	1983	1984	1985	1986	1987	1988	1989	1990	1991 [1]
Petroleum, gas Pétrole, gaz	6.5	113	130	136	146	162	179	228	241	112
Metal Minerais métalliques	0.5	94	102	106	104	106	103	101	99	93
Total manufacturing [3] **Total, industries manufacturières [3]**	**83.9**	**100**	**104**	**108**	**110**	**112**	**117**	**121**	**124**	**98**
Light industry Industrie légère	30.8	97	100	102	104	105	107	109	111	97
Heavy industry Industrie lourde	53.1	101	106	112	114	117	123	128	131	98
Selected manufacturing **Industries manufacturières déterminées**										
Food, beverages, tobacco Industries alimentaires, boissons, tabac	10.4	100	101	102	103	104	105	108	110	102
Textiles Textiles	2.7	88	92	94	93	91	91	90	88	96
Apparel, leather, footwear Articles d'habillement, cuir et chaussures	2.9	93	93	93	92	87	79	75	73	97
Wood products, furniture Bois, meubles	6.8	92	95	94	97	100	103	109	112	92
Paper, printing, publishing Papier, imprimerie, édition	11.4	106	113	116	120	126	132	134	134	99
Chemicals and related products Produits chimiques et alliés	9.8	107	115	119	118	125	135	143	147	96
Non-metallic mineral products Produits minéraux non métalliques	3.5	97	99	104	104	109	109	112	118	93
Basic metals Métallurgie de base	5.0	103	111	112	109	112	121	124	122	97
Metal products Ouvrages en métaux	30.4	99	103	111	114	114	120	126	129	98
Electricity, gas, water [4] **Electricité, gaz et eau [4]**	**8.4**	**114**	**117**	**125**	**125**	**135**	**139**	**142**	**147**	**104**

B. Developing regions • Régions en voie de développement

Total industry [2-4] **Total, industrie [2-4]**	**100.0**	**91**	**97**	**99**	**107**	**113**	**121**	**130**	**136**	**106**
Total mining [2] **Total, industries extractives [2]**	**46.8**	**73**	**76**	**72**	**78**	**79**	**86**	**94**	**94**	**101**
Coal Houille	0.5	118	129	139	147	149	153	163	170	106
Petroleum, gas Pétrole, gaz	42.1	70	72	68	74	74	82	90	90	100
Metal Minerais métalliques	2.5	101	102	108	110	109	114	117	120	103
Total manufacturing [3] **Total, industries manufacturières [3]**	**49.2**	**106**	**114**	**121**	**131**	**141**	**150**	**160**	**171**	**109**
Light industry Industrie légère	23.6	107	113	119	125	130	136	144	153	107
Heavy industry Industrie lourde	25.6	105	116	123	136	151	162	173	187	110
Selected manufacturing **Industries manufacturières déterminées**										
Food, beverages, tobacco Industries alimentaires, boissons, tabac	10.3	114	120	127	131	135	142	155	169	110
Textiles Textiles	4.7	105	108	113	121	126	126	129	132	103
Apparel, leather, footwear Articles d'habillement, cuir et chaussures	2.6	105	113	112	122	129	134	143	144	102
Wood products, furniture Bois, meubles	2.0	99	105	107	112	117	124	127	133	107
Paper, printing, publishing Papier, imprimerie, édition	2.4	107	115	132	141	153	163	177	197	115

5
Index numbers of industrial production: world and regions [*cont.*]
Indices de la production industrielle : monde et régions [*suite*]
1980=100

Region s and industry [ISIC] Région s et industrie [CITI]	Weight(%) Pond.(%)	1983	1984	1985	1986	1987	1988	1989	1990	1991 [1]
Chemicals and related products										
Produits chimiques et alliés	9.7	114	123	131	142	158	173	183	200	111
Non-metallic mineral products										
Produits minéraux non métalliques	2.9	102	109	113	120	129	134	145	151	106
Basic metals										
Métallurgie de base	3.1	102	114	118	128	140	147	160	174	114
Metal products										
Ouvrages en métaux	10.4	99	112	119	137	153	164	173	184	107
Electricity, gas, water [4]										
Electricité, gaz et eau [4]	**4.0**	**120**	**131**	**140**	**149**	**161**	**173**	**186**	**203**	**110**

Latin America and Caribbean · Amérique latine et Caraïbes

Total industry [2-4]										
Total, industrie [2-4]	**100.0**	**94**	**99**	**103**	**110**	**117**	**121**	**130**	**140**	**112**
Total mining [2]										
Total, industries extractives [2]	**16.8**	**98**	**101**	**100**	**103**	**103**	**107**	**109**	**112**	**105**
Coal										
Houille	0.2	124	134	138	135	134	157	150	151	99
Petroleum, gas										
Pétrole, gaz	12.1	96	99	97	100	98	102	103	105	104
Metal										
Minerais métalliques	3.4	107	111	116	117	118	124	134	141	109
Total manufacturing [3]										
Total, industries manufacturières [3]	**77.9**	**92**	**97**	**102**	**110**	**118**	**122**	**132**	**143**	**114**
Light industry										
Industrie légère	36.8	97	101	107	112	116	119	130	141	111
Heavy industry										
Industrie lourde	41.1	87	93	98	108	119	125	134	145	116
Selected manufacturing										
Industries manufacturières déterminées										
Food, beverages, tobacco										
Industries alimentaires, boissons, tabac	18.2	106	109	114	118	122	125	144	160	113
Textiles										
Textiles	5.2	88	91	95	103	106	108	112	114	112
Apparel, leather, footwear										
Articles d'habillement, cuir et chaussures	3.7	87	91	92	94	94	96	101	102	106
Wood products, furniture										
Bois, meubles	3.0	89	93	94	99	104	106	105	112	108
Paper, printing, publishing										
Papier, imprimerie, édition	4.4	98	105	129	136	145	156	169	192	120
Chemicals and related products										
Produits chimiques et alliés	14.0	102	109	117	128	145	158	171	195	118
Non-metallic mineral products										
Produits minéraux non métalliques	4.7	86	87	89	97	106	104	111	114	107
Basic metals										
Métallurgie de base	5.7	89	97	98	106	116	122	139	152	123
Metal products										
Ouvrages en métaux	17.2	73	79	84	95	100	100	102	101	106
Electricity, gas, water [4]										
Electricité, gaz et eau [4]	**5.3**	**113**	**121**	**128**	**137**	**148**	**157**	**166**	**182**	**112**

Asia · Asie

Total industry [2-4]										
Total, industrie [2-4]	**100.0**	**89**	**95**	**95**	**106**	**114**	**126**	**137**	**142**	**103**
Total mining [2]										
Total, industries extractives [2]	**62.9**	**68**	**68**	**64**	**72**	**74**	**84**	**94**	**91**	**98**
Coal										
Houille	0.8	117	129	141	152	161	162	178	187	112
Petroleum, gas										
Pétrole, gaz	59.7	66	66	61	70	71	80	90	87	97
Metal										
Minerais métalliques	1.3	81	77	78	77	82	90	98	91	91

5

Index numbers of industrial production: world and regions [*cont.*]

Indices de la production industrielle : monde et régions [*suite*]

1980=100

Region and industry [ISIC] Région et industrie [CITI]	Weight(%) Pond.(%)	1983	1984	1985	1986	1987	1988	1989	1990	1991 [1]
Total manufacturing [3] **Total, industries manufacturières [3]**	**34.1**	**126**	**141**	**148**	**164**	**182**	**198**	**208**	**226**	**107**
Light industry Industrie légère	16.1	120	130	136	145	154	165	172	180	106
Heavy industry Industrie lourde	18.0	131	150	159	180	206	227	240	267	108
Selected manufacturing **Industries manufacturières déterminées**										
Food, beverages, tobacco Industries alimentaires, boissons, tabac	5.6	132	147	157	157	166	182	192	205	109
Textiles Textiles	4.7	115	117	122	131	140	138	141	148	100
Apparel, leather, footwear Articles d'habillement, cuir et chaussures	1.8	126	139	136	156	179	185	198	198	105
Wood products, furniture Bois, meubles	1.2	118	124	124	130	138	153	165	178	110
Paper, printing, publishing [2] Papier, imprimerie, édition [2]	1.2	126	152	157	175	198	207	224	246	105
Chemicals and related products Produits chimiques et alliés	8.3	126	139	147	160	176	193	200	213	106
Non-metallic mineral products Produits minéraux non métalliques	2.0	124	142	149	153	165	180	197	208	107
Basic metals Métallurgie de base	1.8	128	147	157	174	194	204	212	238	107
Metal products Ouvrages en métaux	6.5	139	166	177	214	256	290	308	354	111
Electricity, gas, water [4] **Electricité, gaz et eau [4]**	**3.0**	**132**	**145**	**158**	**167**	**184**	**202**	**222**	**248**	**110**

Oceania [3] · Océanie [3]

Region and industry [ISIC] Région et industrie [CITI]	Weight(%) Pond.(%)	1983	1984	1985	1986	1987	1988	1989	1990	1991 [1]
Total industry [2-4] **Total, industrie [2-4]**	**100.0**	**99**	**105**	**113**	**114**	**120**	**129**	**137**	**140**	**102**
Total mining [2] **Total, industries extractives [2]**	**16.0**	**115**	**124**	**150**	**156**	**181**	**206**	**237**	**271**	**111**
Coal Houille	4.7	104	107	112	114	136	132	166	145	106
Petroleum, gas Pétrole, gaz	2.5	106	128	146	130	140	132	124	147	97
Metal Minerais métalliques	7.4	131	142	187	206	243	301	350	429	114
Total manufacturing [3] **Total, industries manufacturières [3]**	**73.1**	**94**	**99**	**103**	**102**	**104**	**110**	**112**	**110**	**98**
Light industry Industrie légère	31.8	105	108	110	111	112	116	115	114	100
Heavy industry Industrie lourde	41.3	85	91	97	94	98	105	111	107	96
Selected manufacturing **Industries manufacturières déterminées**										
Food, beverages, tobacco Industries alimentaires, boissons, tabac	13.7	100	103	105	107	109	109	112	113	100
Textiles Textiles	2.7	97	101	105	104	99	103	98	97	103
Apparel, leather, footwear Articles d'habillement, cuir et chaussures	3.0	88	92	95	89	82	82	75	67	96
Wood products, furniture Bois, meubles	4.3	96	101	107	103	108	115	120	116	100
Paper, printing, publishing Papier, imprimerie, édition	6.8	143	146	147	150	155	161	153	156	101
Chemicals and related products Produits chimiques et alliés	8.8	91	92	92	96	102	116	118	115	100
Non-metallic mineral products Produits minéraux non métalliques	3.5	90	97	104	99	104	130	127	121	94

5
Index numbers of industrial production: world and regions [cont.]
Indices de la production industrielle : monde et régions [suite]
1980=100

Region ₰ and industry [ISIC] Région ₰ et industrie [CITI]	Weight(%) Pond.(%)	1983	1984	1985	1986	1987	1988	1989	1990	1991 [1]
Basic metals										
Métallurgie de base	8.3	85	91	97	95	94	103	111	112	98
Metal products										
Ouvrages en métaux	21.3	79	86	94	89	92	96	102	96	93
Electricity, gas, water [4]										
Electricité, gaz et eau [4]	**10.9**	**113**	**120**	**127**	**134**	**139**	**146**	**152**	**154**	**100**

Source:
Industrial statistics database of the Statistical Office of the
United Nations Secretariat.

+   For Member States of this grouping, see Annex I -
     Other groupings.

₰   All series exclude Albania, China, Democratic
     People's Republic of Korea and Viet Nam.
     Series for "Developed regions" include North America
     (Canada and the United States), Western Europe, Australia,
     Israel, Japan, New Zealand and South Africa. Eastern
     Europe and the former USSR are not included in
     "Developed regions" totals but are shown separately
     and include Bulgaria, Czechoslovakia, Hungary, Poland,
     Romania and the former USSR. Series for "Developing
     regions" exclude Australia, Israel, Japan, New Zealand
     and South Africa, and include Yugoslavia.

1   Percentage ratio of the index for January-June 1991
     to the index for January-June 1990.
2   Excluding printing and publishing.
3   Including Australia and New Zealand.

Source :
Base de données de statistiques industrielles du Bureau de statistique
du Secrétariat de l'ONU.

+   Les Etats membres de ce groupement, voir annexe I -
     Autres groupements.

₰   Aucune série ne comprend l'Albanie, la Chine,
     la République populaire démocratique de Corée et le Viet Nam.
     Les régions developpées comprennent l'Amérique du nord
     (le Canada et les Etats-Unis), l'Europe de l'ouest, l'Australie,
     Israël, le Japon, la Nouvelle-Zélande et l'Afrique du sud.
     L'Europe de l'est et l'ancienne URSS ne sont pas compris dans les
     totaux des "Régions developpées" mais sont indiqués séparément et
     comprennent la Bulgarie, la Tchécoslovaquie, la Hongrie, la Pologne,
     la Roumanie et l'ancienne URSS. Les séries pour les "Régions en voie
     de développement" excluent l'Australie, Israël, le Japon,
     la Nouvelle-Zélande et l'Afrique du sud, et incluent la Yougoslavie.

1   Rapport en pourcentage entre l'indice de janvier à juin 1991 et celui
     de janvier à juin 1990.
2   Non compris l'imprimerie et l'édition.
3   Y compris l'Australie et la Nouvelle-Zélande.

6
Index numbers of industrial employment: world and regions
Indices de l'emploi industriel : monde et régions
1980=100

Region s and industry [ISIC] Région s et industrie [CITI]	Weight(%) Pond.(%)	1983	1984	1985	1986	1987	1988	1989	1990
World · Monde									
Total industry [2-4]									
Total, industrie [2-4]	**100.0**	**98**	**98**	**100**	**100**	**102**	**102**	**103**	...
Total mining [2]									
Total, industries extractives [2]	**4.5**	**102**	**101**	**101**	**100**	**98**	**97**	**96**	...
Coal									
Houille	1.8	106	107	107	106	104	102	100	...
Petroleum, gas									
Pétrole, gaz	0.5	114	116	117	116	114	117	117	...
Metal									
Minerais métalliques	1.0	94	91	91	88	86	84	84	...
Total manufacturing [3]									
Total, industries manufacturières [3]	**92.5**	**98**	**98**	**100**	**100**	**102**	**102**	**103**	...
Light industry									
Industrie légère	46.2	98	98	100	101	105	105	107	...
Heavy industry									
Industrie lourde	46.3	98	98	99	98	99	99	99	...
Selected manufacturing									
Industries manufacturières déterminées									
Food, beverages, tobacco									
Industries alimentaires, boissons, tabac	13.9	99	97	103	104	108	106	107	...
Textiles									
Textiles	9.9	90	89	86	86	86	83	82	...
Apparel, leather, footwear									
Articles d'habillement, cuir et chaussures	8.8	101	104	106	107	114	113	115	...
Wood products, furniture									
Bois, meubles	6.2	100	104	108	107	112	124	132	...
Paper, printing, publishing									
Papier, imprimerie, édition	4.4	99	99	100	100	103	104	105	...
Chemicals and related products									
Produits chimiques et alliés	7.4	99	100	104	105	107	107	109	...
Non-metallic mineral products									
Produits minéraux non métalliques	5.5	101	101	103	102	103	104	104	...
Basic metals									
Métallurgie de base	4.1	91	91	89	88	87	86	86	...
Metal products									
Ouvrages en métaux	30.0	98	99	99	99	99	99	100	...
Electricity, gas, water [4]									
Electricité, gaz et eau [4]	**3.0**	**109**	**112**	**110**	**111**	**114**	**116**	**117**	...
A. Developed regions · Régions developpées									
Total industry [2-4]									
Total, industrie [2-4]	**100.0**	**92**	**92**	**91**	**91**	**91**	**91**	**92**	**92**
Total mining [2]									
Total, industries extractives [2]	**3.5**	**94**	**92**	**89**	**87**	**83**	**80**	**77**	**75**
Coal									
Houille	1.4	93	89	84	80	74	68	63	60
Petroleum, gas									
Pétrole, gaz	0.4	114	114	114	108	100	105	98	97
Metal									
Minerais métalliques	1.0	94	94	92	95	93	91	89	85
Total manufacturing [3]									
Total, industries manufacturières [3]	**93.2**	**92**	**92**	**91**	**90**	**91**	**91**	**92**	**93**
Light industry									
Industrie légère	38.0	92	92	90	90	91	92	92	93
Heavy industry									
Industrie lourde	55.2	92	92	92	91	90	91	91	92

6
Index numbers of industrial employment: world and regions [*cont.*]
Indices de l'emploi industriel: monde et régions [*suite*]
1980=100

Regions and industry [ISIC] Régions et industrie [CITI]	Weight(%) Pond.(%)	1983	1984	1985	1986	1987	1988	1989	1990
Selected manufacturing									
Industries manufacturières déterminées									
Food, beverages, tobacco									
Industries alimentaires, boissons, tabac	9.9	96	95	95	95	95	96	96	97
Textiles									
Textiles	6.4	85	84	80	80	80	78	78	78
Apparel, leather, footwear									
Articles d'habillement, cuir et chaussures	6.4	91	89	86	84	85	84	82	82
Wood products, furniture									
Bois, meubles	5.6	87	86	84	84	85	86	86	87
Paper, printing, publishing									
Papier, imprimerie, édition	7.0	96	97	97	98	100	101	102	104
Chemicals and related products									
Produits chimiques et alliés	8.9	95	96	96	96	98	99	101	103
Non-metallic mineral products									
Produits minéraux non métalliques	4.1	88	87	84	83	84	84	85	86
Basic metals									
Métallurgie de base	5.5	79	78	74	72	70	70	69	69
Metal products									
Ouvrages en métaux	37.6	94	95	95	94	94	95	95	96
Electricity, gas, water [4]									
Electricité, gaz et eau [4]	**3.3**	**104**	**104**	**104**	**103**	**104**	**104**	**103**	**104**

North America · Amérique du Nord

	Weight(%) Pond.(%)	1983	1984	1985	1986	1987	1988	1989	1990
Total industry [2-4]									
Total, industrie [2-4]	**100.0**	**91**	**93**	**92**	**90**	**93**	**94**	**94**	**95**
Total mining [2]									
Total, industries extractives [2]	**3.5**	**92**	**91**	**88**	**83**	**77**	**79**	**76**	**74**
Coal									
Houille	1.1	86	83	79	74	69	68	62	59
Petroleum, gas									
Pétrole, gaz	1.1	115	113	113	106	95	101	92	88
Metal									
Minerais métalliques	0.7	69	67	60	56	53	58	61	64
Total manufacturing [3]									
Total, industries manufacturières [3]	**92.7**	**90**	**93**	**91**	**89**	**92**	**94**	**94**	**96**
Light industry									
Industrie légère	35.7	94	94	92	91	97	98	98	101
Heavy industry									
Industrie lourde	57.0	89	92	91	88	90	91	91	92
Selected manufacturing									
Industries manufacturières déterminées									
Food, beverages, tobacco									
Industries alimentaires, boissons, tabac	8.1	94	93	92	92	94	95	96	98
Textiles									
Textiles	4.7	91	90	85	84	88	88	86	87
Apparel, leather, footwear									
Articles d'habillement, cuir et chaussures	6.6	90	85	77	73	76	75	72	72
Wood products, furniture									
Bois, meubles	4.9	89	93	92	93	102	102	102	106
Paper, printing, publishing									
Papier, imprimerie, édition	9.4	99	101	102	103	108	109	110	112
Chemicals and related products									
Produits chimiques et alliés	8.7	96	98	98	97	102	104	107	111
Non-metallic mineral products									
Produits minéraux non métalliques	3.0	86	87	85	85	91	91	93	96
Basic metals									
Métallurgie de base	5.1	70	74	68	64	64	66	66	67
Metal products									
Ouvrages en métaux	40.2	90	94	94	91	93	94	93	95
Electricity, gas, water [4]									
Electricité, gaz et eau [4]	**3.8**	**106**	**106**	**109**	**107**	**109**	**107**	**108**	**108**

6
Index numbers of industrial employment: world and regions [*cont.*]
Indices de l'emploi industriel: monde et régions [*suite*]
1980=100

Region ʄ and industry [ISIC] Région ʄ et industrie [CITI]	Weight(%) Pond.(%)	1983	1984	1985	1986	1987	1988	1989	1990
Eastern Europe and former USSR • Europe de l'Est et l'ancienne URSS									
Total industry [2-4] **Total, industrie [2-4]**	**100.0**	**103**	**100**	**101**	**101**	**101**	**100**	**98**	**97**
Total mining [2] **Total, industries extractives [2]**	**6.6**	**112**	**112**	**114**	**114**	**114**	**113**	**112**	**111**
Coal Houille	3.7	115	116	118	119	120	119	118	117
Petroleum, gas Pétrole, gaz	0.5	126	131	136	138	137	144	145	153
Metal Minerais métalliques	0.5	118	118	120	119	116	112	111	113
Total manufacturing [3] **Total, industries manufacturières [3]**	**91.2**	**102**	**99**	**100**	**100**	**100**	**98**	**96**	**95**
Light industry Industrie légère	34.3	99	94	95	94	94	93	92	92
Heavy industry Industrie lourde	56.9	104	102	103	104	103	102	99	98
Selected manufacturing **Industries manufacturières déterminées**									
Food, beverages, tobacco Industries alimentaires, boissons, tabac	9.2	102	103	103	100	100	100	99	100
Textiles Textiles	7.4	83	82	81	81	80	77	75	74
Apparel, leather, footwear Articles d'habillement, cuir et chaussures	8.4	99	97	98	98	99	96	96	98
Wood products, furniture Bois, meubles	4.3	100	100	100	100	101	99	98	95
Paper, printing, publishing [1] Papier, imprimerie, édition [1]	2.1	101	101	101	101	99	96	92	90
Chemicals and related products Produits chimiques et alliés	6.4	103	103	110	110	109	107	106	106
Non-metallic mineral products Produits minéraux non métalliques	6.6	101	101	102	101	101	99	95	93
Basic metals Métallurgie de base	3.7	101	101	102	102	100	97	93	90
Metal products Ouvrages en métaux	40.6	106	102	103	104	103	102	99	98
Electricity, gas, water [4] **Electricité, gaz et eau [4]**	**2.2**	**105**	**107**	**110**	**112**	**114**	**114**	**115**	**116**
Western Europe • Europe de l'Ouest									
Total industry [2-4] **Total, industrie [2-4]**	**100.0**	**91**	**91**	**89**	**89**	**89**	**88**	**89**	**90**
Total mining [2] **Total, industries extractives [2]**	**3.0**	**94**	**91**	**86**	**83**	**79**	**75**	**72**	**70**
Coal Houille	1.7	95	91	84	81	76	70	66	63
Petroleum, gas Pétrole, gaz	0.2	115	122	122	114	122	121	131	140
Metal Minerais métalliques	0.3	94	92	87	84	77	74	72	71
Total manufacturing [3] **Total, industries manufacturières [3]**	**93.6**	**91**	**90**	**89**	**89**	**89**	**88**	**89**	**90**
Light industry Industrie légère	37.7	92	92	90	89	90	89	90	91
Heavy industry Industrie lourde	55.8	90	89	88	89	88	88	88	89
Selected manufacturing **Industries manufacturières déterminées**									
Food, beverages, tobacco Industries alimentaires, boissons, tabac	10.3	96	95	94	94	94	94	94	95

6

Index numbers of industrial employment: world and regions [*cont.*]

Indices de l'emploi industriel: monde et régions [*suite*]

1980=100

Region ʃ and industry [ISIC] Région ʃ et industrie [CITI]	Weight(%) Pond.(%)	1983	1984	1985	1986	1987	1988	1989	1990
Textiles									
Textiles	6.5	86	87	83	83	83	81	82	82
Apparel, leather, footwear									
Articles d'habillement, cuir et chaussures	6.8	92	91	89	88	89	87	86	85
Wood products, furniture									
Bois, meubles	5.8	90	89	86	84	84	84	85	86
Paper, printing, publishing									
Papier, imprimerie, édition	5.8	93	93	92	91	93	93	95	97
Chemicals and related products									
Produits chimiques et alliés	9.5	92	93	92	93	94	95	97	99
Non-metallic mineral products									
Produits minéraux non métalliques	4.6	89	87	83	81	81	81	81	82
Basic metals									
Métallurgie de base	5.8	79	76	74	72	70	69	68	67
Metal products									
Ouvrages en métaux	37.0	92	91	91	92	91	90	92	92
Electricity, gas, water [4]									
Electricité, gaz et eau [4]	**3.4**	**102**	**103**	**103**	**103**	**104**	**104**	**104**	**105**

European Economic Community [+]

Communauté économique éuropéenne [+]

	Weight(%) Pond.(%)	1983	1984	1985	1986	1987	1988	1989	1990
Total industry [2-4]									
Total, industrie [2-4]	**100.0**	**90**	**89**	**87**	**87**	**86**	**86**	**87**	**88**
Total mining [2]									
Total, industries extractives [2]	**3.0**	**92**	**88**	**82**	**78**	**74**	**70**	**67**	**64**
Coal									
Houille	1.8	94	89	81	78	72	66	61	58
Petroleum, gas									
Pétrole, gaz	0.1	111	117	114	102	113	112	116	122
Metal									
Minerais métalliques	0.2	88	84	76	72	65	64	62	62
Total manufacturing [3]									
Total, industries manufacturières [3]	**93.6**	**90**	**89**	**87**	**86**	**86**	**86**	**87**	**88**
Light industry									
Industrie légère	37.3	90	90	87	86	87	87	88	89
Heavy industry									
Industrie lourde	56.3	89	88	87	87	86	86	86	87
Selected manufacturing									
Industries manufacturières déterminées									
Food, beverages, tobacco									
Industries alimentaires, boissons, tabac	10.3	94	94	93	92	92	92	92	94
Textiles									
Textiles	6.6	85	85	80	80	80	78	79	80
Apparel, leather, footwear									
Articles d'habillement, cuir et chaussures	6.8	90	89	86	84	85	83	81	81
Wood products, furniture									
Bois, meubles	5.4	88	88	84	81	81	82	83	84
Paper, printing, publishing									
Papier, imprimerie, édition	5.4	92	92	91	90	91	92	94	97
Chemicals and related products									
Produits chimiques et alliés	9.8	91	92	91	91	92	93	95	98
Non-metallic mineral products									
Produits minéraux non métalliques	4.7	87	85	81	79	79	79	79	80
Basic metals									
Métallurgie de base	5.8	77	73	70	69	66	64	63	62
Metal products									
Ouvrages en métaux	37.5	91	90	89	90	89	89	90	91
Electricity, gas, water [4]									
Electricité, gaz et eau [4]	**3.3**	**101**	**101**	**101**	**100**	**101**	**101**	**101**	**102**

6

Index numbers of industrial employment: world and regions [*cont.*]

Indices de l'emploi industriel: monde et régions [*suite*]

1980=100

Region s and industry [ISIC] Région s et industrie [CITI]	Weight(%) Pond.(%)	1983	1984	1985	1986	1987	1988	1989	1990
European Free Trade Association [+]									
Association européenne de libre échange [+]									
Total industry [2-4]									
Total, industrie [2-4]	**100.0**	**94**	**92**	**93**	**92**	**91**	**90**	**90**	**88**
Total mining [2]									
Total, industries extractives [2]	**1.8**	**95**	**93**	**94**	**91**	**85**	**81**	**79**	**78**
Coal									
Houille	0.2	93	87	87	87	78	67	51	43
Petroleum, gas									
Pétrole, gaz	0.2	132	150	166	172	168	70	176	178
Metal									
Minerais métalliques	0.7	87	83	81	71	60	52	45	41
Total manufacturing [3]									
Total, industries manufacturières [3]	**94.7**	**94**	**92**	**93**	**92**	**91**	**90**	**90**	**88**
Light industry									
Industrie légère	38.3	94	92	92	91	90	88	87	84
Heavy industry									
Industrie lourde	56.4	93	92	93	92	91	91	91	90
Selected manufacturing									
Industries manufacturières déterminées									
Food, beverages, tobacco									
Industries alimentaires, boissons, tabac	10.8	104	100	100	101	98	98	96	94
Textiles									
Textiles	4.1	80	80	78	77	73	69	67	63
Apparel, leather, footwear									
Articles d'habillement, cuir et chaussures	5.0	89	85	83	80	77	68	64	58
Wood products, furniture									
Bois, meubles	8.4	92	89	88	86	87	86	87	86
Paper, printing, publishing									
Papier, imprimerie, édition	10.3	93	93	93	93	94	92	94	92
Chemicals and related products									
Produits chimiques et alliés	8.0	97	97	97	98	99	99	98	98
Non-metallic mineral products									
Produits minéraux non métalliques	3.5	93	92	92	91	92	92	93	92
Basic metals									
Métallurgie de base	6.2	88	84	84	81	81	80	80	82
Metal products									
Ouvrages en métaux	36.7	94	93	94	94	92	92	92	90
Electricity, gas, water [4]									
Electricité, gaz et eau [4]	**3.5**	**104**	**105**	**105**	**106**	**106**	**107**	**105**	**105**
B. Developing regions · Régions en voie de développement									
Total industry [2-4]									
Total, industrie [2-4]	**100.0**	**103**	**104**	**109**	**110**	**116**	**117**	**121**	**...**
Total mining [2]									
Total, industries extractives [2]	**4.4**	**100**	**97**	**100**	**97**	**96**	**96**	**98**	**...**
Coal									
Houille	1.1	107	110	117	117	114	114	115	...
Petroleum, gas									
Pétrole, gaz	0.7	109	110	111	111	113	114	118	...
Metal									
Minerais métalliques	1.5	88	82	83	75	74	72	73	...
Total manufacturing [3]									
Total, industries manufacturières [3]	**92.5**	**102**	**104**	**110**	**111**	**116**	**118**	**121**	**...**
Light industry									
Industrie légère	63.8	102	103	110	112	118	120	124	...
Heavy industry									
Industrie lourde	28.7	103	106	109	109	113	114	116	...
Selected manufacturing									
Industries manufacturières déterminées									
Food, beverages, tobacco									
Industries alimentaires, boissons, tabac	21.9	101	96	107	110	116	113	115	...

6
Index numbers of industrial employment: world and regions [*cont.*]
Indices de l'emploi industriel: monde et régions [*suite*]
1980=100

Region s and industry [ISIC] Région s et industrie [CITI]	Weight(%) Pond.(%)	1983	1984	1985	1986	1987	1988	1989	1990
Textiles									
Textiles	15.7	94	94	91	91	91	87	86	...
Apparel, leather, footwear									
Articles d'habillement, cuir et chaussures	12.0	108	118	123	126	140	138	145	...
Wood products, furniture									
Bois, meubles	7.8	110	122	131	130	139	167	183	...
Paper, printing, publishing									
Papier, imprimerie, édition	3.0	107	105	108	108	113	115	118	...
Chemicals and related products									
Produits chimiques et alliés	6.3	104	106	115	117	122	121	124	...
Non-metallic mineral products									
Produits minéraux non métalliques	6.5	111	110	119	116	118	121	125	...
Basic metals									
Métallurgie de base	2.9	109	114	112	115	113	116	118	...
Metal products									
Ouvrages en métaux	13.9	99	103	103	103	109	110	112	...
Electricity, gas, water [4]									
Electricité, gaz et eau [4]	3.1	...	...	...	...	...	...	...	...

Latin America and Caribbean · Amérique latine et Caraïbes

Total industry [2-4]									
Total, industrie [2-4]	**100.0**	**92**	**97**	**98**	**98**	**98**	**100**	**102**	...
Total mining [2]									
Total, industries extractives [2]	**5.7**	...	...	...	...	...	...	...	...
Coal									
Houille	0.4	...	...	...	...	...	...	...	...
Petroleum, gas									
Pétrole, gaz	1.2	116	118	116	113	114	114	118	...
Metal									
Minerais métalliques	2.5	...	...	...	...	...	...	...	...
Total manufacturing [3]									
Total, industries manufacturières [3]	**90.5**	**90**	**96**	**96**	**97**	**97**	**99**	**101**	...
Light industry									
Industrie légère	51.8	94	100	101	102	102	105	107	...
Heavy industry									
Industrie lourde	38.7	85	91	90	90	90	92	92	...
Selected manufacturing									
Industries manufacturières déterminées									
Food, beverages, tobacco									
Industries alimentaires, boissons, tabac	21.5	100	102	104	104	103	105	108	...
Textiles									
Textiles	7.2	84	85	84	84	83	83	82	...
Apparel, leather, footwear									
Articles d'habillement, cuir et chaussures	9.3	92	109	109	115	112	116	121	...
Wood products, furniture									
Bois, meubles	6.4	...	...	...	...	...	...	...	...
Paper, printing, publishing									
Papier, imprimerie, édition	4.8	98	94	94	95	96	98	101	...
Chemicals and related products									
Produits chimiques et alliés	8.9	95	102	104	105	106	110	112	...
Non-metallic mineral products									
Produits minéraux non métalliques	6.3	83	91	90	92	90	91	92	...
Basic metals									
Métallurgie de base	3.9	90	91	90	87	86	85	83	...
Metal products									
Ouvrages en métaux	20.7	81	88	88	88	88	90	91	...
Electricity, gas, water [4]									
Electricité, gaz et eau [4]	3.8	...	...	...	...	...	...	...	...

Asia · Asie

Total industry [2-4]									
Total, industrie [2-4]	**100.0**	**106**	**107**	**114**	**114**	**122**	**123**	**127**	...

6
Index numbers of industrial employment: world and regions [cont.]
Indices de l'emploi industriel: monde et régions [suite]
1980=100

Regions and industry [ISIC] Régions et industrie [CITI]	Weight(%) Pond.(%)	1983	1984	1985	1986	1987	1988	1989	1990
Total mining [2]									
Total, industries extractives [2]	**3.0**	**106**	**102**	**105**	**104**	**101**	**100**	**102**	...
Coal									
Houille	1.3	109	112	118	118	115	115	114	...
Petroleum, gas									
Pétrole, gaz	0.5	104	102	103	104	105	106	109	...
Metal									
Minerais métalliques	0.6	97	89	89	85	82	83	83	...
Total manufacturing [3]									
Total, industries manufacturières [3]	**94.4**	**105**	**107**	**114**	**115**	**122**	**123**	**127**	...
Light industry									
Industrie légère	68.9	104	104	112	114	123	124	128	...
Heavy industry									
Industrie lourde	25.5	110	113	117	116	122	122	125	...
Selected manufacturing									
Industries manufacturières déterminées									
Food, beverages, tobacco									
Industries alimentaires, boissons, tabac	22.4	101	93	109	113	121	116	117	...
Textiles									
Textiles	18.2	95	94	91	92	91	87	86	...
Apparel, leather, footwear									
Articles d'habillement, cuir et chaussures	13.6	111	119	125	127	145	141	148	...
Wood products, furniture									
Bois, meubles	8.4	116	131	144	142	154	191	211	...
Paper, printing, publishing									
Papier, imprimerie, édition	1.0	113	112	113	115	124	125	129	...
Chemicals and related products									
Produits chimiques et alliés	5.5	108	109	121	124	131	128	132	...
Non-metallic mineral products									
Produits minéraux non métalliques	6.7	117	115	127	123	126	130	135	...
Basic metals									
Métallurgie de base	2.4	120	130	123	128	126	130	133	...
Metal products									
Ouvrages en métaux	11.8	105	110	110	108	117	117	118	...
Electricity, gas, water [4]									
Electricité, gaz et eau [4]	**2.6**	**115**	**123**	**114**	**114**	**126**	**130**	**133**	...

Oceania [2] • Océanie [2]

	Weight(%) Pond.(%)	1983	1984	1985	1986	1987	1988	1989	1990
Total industry [2-4]									
Total, industrie [2-4]	**100.0**	**97**	**95**	**95**	**95**	**94**	**96**	**95**	**96**
Total mining [2]									
Total, industries extractives [2]	**4.8**	**109**	**109**	**109**	**110**	**108**	**106**	**105**	**104**
Coal									
Houille	1.6	128	125	126	129	126	112	109	104
Petroleum, gas									
Pétrole, gaz	0.2	...	...	...	...	...	...	...	...
Metal									
Minerais métalliques	2.3	98	96	97	94	91	98	98	99
Total manufacturing [3]									
Total, industries manufacturières [3]	**88.5**	**95**	**92**	**93**	**93**	**92**	**94**	**94**	**95**
Light industry									
Industrie légère	43.1	97	96	98	98	98	100	100	102
Heavy industry									
Industrie lourde	45.4	93	89	89	88	86	88	88	88
Selected manufacturing									
Industries manufacturières déterminées									
Food, beverages, tobacco									
Industries alimentaires, boissons, tabac	16.1	99	96	95	96	94	97	96	97
Textiles									
Textiles	3.8	91	89	95	92	90	90	84	80
Apparel, leather, footwear									
Articles d'habillement, cuir et chaussures	6.0	92	92	96	96	94	94	93	95

6
Index numbers of industrial employment: world and regions [*cont.*]
Indices de l'emploi industriel: monde et régions [*suite*]
1980=100

Region ʃ and industry [ISIC] Région ʃ et industrie [CITI]	Weight(%) Pond.(%)	1983	1984	1985	1986	1987	1988	1989	1990
Wood products, furniture									
Bois, meubles	6.6	99	98	103	103	102	107	109	112
Paper, printing, publishing									
Papier, imprimerie, édition	8.3	102	102	102	103	105	106	107	110
Chemicals and related products									
Produits chimiques et alliés	7.8	99	97	97	97	95	96	95	95
Non-metallic mineral products									
Produits minéraux non métalliques	3.4	93	89	91	91	90	88	90	90
Basic metals									
Métallurgie de base	6.0	91	85	84	82	80	82	78	76
Metal products									
Ouvrages en métaux	29.1	92	88	87	87	85	88	89	90
Electricity, gas, water [4]									
Electricité, gaz et eau [4]	6.7	**114**	**114**	**112**	**111**	**113**	**111**	**107**	**106**

Source:
Industrial statistics database of the Statistical Office of the
United Nations Secretariat.

Source:
Base de données de statistiques industrielles du Bureau de statistique du
Secrétariat de l'ONU.

+   For Member States of this grouping, see Annex I -
    Other groupings.

+   Les Etats membres de ce groupement, voir annexe I -
    Autres groupements.

ʃ   All series exclude Albania, China, Democratic
    People's Republic of Korea and Viet Nam.
    Series for "Developed regions" include North America
    (Canada and the United States), Western Europe, Australia,
    Israel, Japan, New Zealand and South Africa. Eastern
    Europe and the former USSR are not included in "Developed
    regions" totals but are shown separately and include
    Bulgaria, Czechoslovakia, Hungary, Poland,
    Romania and the former USSR. Series for "Developing
    regions" exclude Australia, Israel, Japan, New Zealand and
    South Africa, and include Yugoslavia.

ʃ   Aucune série ne comprend l'Albanie, la Chine,
    la République populaire démocratique de Corée et le Viet Nam.
    Les régions developpées comprennent l'Amérique du nord
    (le Canada et les Etats-Unis), l'Europe de l'ouest, l'Australie,
    Israël, le Japon, la Nouvelle-Zélande et l'Afrique du sud.
    L'Europe de l'est et l'ancienne URSS ne sont pas compris dans les
    totaux des "Régions developpées" mais sont indiqués séparément et
    comprennent la Bulgarie, la Tchécoslovaquie, la Hongrie, la Pologne,
    la Roumanie et l'ancienne URSS. Les séries pour les "Régions en voie
    de développement excluent l'Australie, Israël, le Japon,
    la Nouvelle-Zélande et l'Afrique du sud, et incluent la Yougoslavie.

1   Excluding printing and publishing.
2   Including Australia and New Zealand.

1   Non compris l'imprimerie et l'édition.
2   Y compris l'Australie et la Nouvelle-Zélande.

7

Index numbers of labour productivity in industry: world and regions
Indices de la productivité du travail dans l'industrie :
monde et régions

1980 = 100

Region ʃ and industry [ISIC] Région ʃ et industrie [CITI]	1983	1984	1985	1986	1987	1988	1989	1990
World · Monde								
Total industry [2-4] **Total, industrie [2-4]**	**109**	**117**	**119**	**125**	**129**	**137**	**141**	...
Total mining [2] **Total, industries extractives [2]**	**98**	**102**	**105**	**110**	**114**	**120**	**123**	...
Coal Houille	99	101	105	110	114	118	122	...
Petroleum, gas Pétrole, gaz	77	79	76	78	81	81	82	...
Metal Minerais métalliques	106	114	119	126	131	143	144	...
Total manufacturing [3] **Total, industries manufacturières [3]**	**110**	**118**	**120**	**126**	**130**	**138**	**143**	...
Light industry Industrie légère	111	118	118	122	123	129	132	...
Heavy industry Industrie lourde	109	118	123	130	137	146	153	...
Selected manufacturing **Industries manufacturières déterminées**								
Food, beverages, tobacco Industries alimentaires, boissons, tabac	117	129	124	124	124	134	139	...
Textiles Textiles	118	121	129	135	141	146	151	...
Apparel, leather, footwear Articles d'habillement, cuir et chaussures	107	110	108	114	114	118	121	...
Wood products, furniture Bois, meubles	106	106	105	110	111	112	112	...
Paper, printing, publishing Papier, imprimerie, édition	108	116	120	126	131	138	142	...
Chemicals and related products Produits et alliés	108	115	114	119	124	133	136	...
Non-metallic mineral products Produits minéraux non métalliques	105	112	113	118	123	130	136	...
Basic metals Métallurgie de base	108	116	122	125	133	142	147	...
Metal products Ouvrages en métaux	109	120	126	134	141	152	158	...
Electricity, gas, water [4] **Electricité, gaz et eau [4]**	**104**	**108**	**115**	**118**	**121**	**126**	**131**	**134**
A. Developed Regions · Régions developpées								
Total industry [2-4] **Total, industrie [2-4]**	**107**	**113**	**118**	**120**	**123**	**130**	**134**	**103**
Total mining [2] **Total, industries extractives [2]**	**102**	**108**	**116**	**119**	**126**	**132**	**139**	**140**
Coal Houille	108	111	124	130	136	146	155	150
Petroleum, gas Pétrole, gaz	83	87	87	86	92	87	90	92
Metal Minerais métalliques	102	112	118	118	126	132	137	145
Total manufacturing [3] **Total, industries manufacturières [3]**	**107**	**114**	**118**	**121**	**124**	**130**	**134**	**132**
Light industry Industrie légère	107	111	113	116	117	120	122	119
Heavy industry Industrie lourde	108	116	121	124	128	137	143	141

7
Index numbers of labour productivity in industry: world and regions
Indices de la productivité du travail dans l'industrie : monde et régions
1980 = 100

Region ₠ and industry [SITC] Région ₠ et industrie [CITI]	1983	1984	1985	1986	1987	1988	1989	1990
Selected manufacturing **Industries manufacturières déterminées**								
Food, beverages, tobacco Industries alimentaires, boissons, tabac	107	109	112	114	115	117	119	119
Textiles Textiles	112	114	120	122	124	125	128	122
Apparel, leather, footwear Articles d'habillement, cuir et chaussures	103	107	109	111	109	109	111	105
Wood products, furniture Bois, meubles	108	111	114	116	122	125	127	124
Paper, printing, publishing Papier, imprimerie, édition	107	114	115	120	124	130	132	132
Chemicals and related products Produits et alliés	107	113	116	119	122	128	130	125
Non-metallic mineral products Produits minéraux non métalliques	104	109	112	116	116	124	127	125
Basic metals Métallurgie de base	112	122	128	128	135	147	151	147
Metal products Ouvrages en métaux	107	116	121	124	128	138	144	143
Electricity, gas, water [4] **Electricité, gaz et eau [4]**	**101**	**105**	**109**	**111**	**114**	**118**	**122**	**124**

North America · Amérique du Nord

	1983	1984	1985	1986	1987	1988	1989	1990
Total industry [2-4] **Total, industrie [2-4]**	**109**	**119**	**123**	**128**	**130**	**136**	**140**	**138**
Total mining [2] **Total, industries extractives [2]**	**101**	**120**	**127**	**132**	**149**	**154**	**163**	**172**
Coal Houille	110	132	138	147	164	177	194	218
Petroleum, gas Pétrole, gaz	78	83	82	78	86	82	88	92
Metal Minerais métalliques	123	160	180	192	225	228	233	226
Total manufacturing [3] **Total, industries manufacturières [3]**	**110**	**119**	**124**	**129**	**130**	**136**	**140**	**138**
Light industry Industrie légère	110	117	122	128	126	128	131	127
Heavy industry Industrie lourde	109	120	125	130	134	141	146	144
Selected manufacturing **Industries manufacturières déterminées**								
Food, beverages, tobacco Industries alimentaires, boissons, tabac	110	115	119	122	122	124	125	125
Textiles Textiles	112	109	117	126	127	127	132	128
Apparel, leather, footwear Articles d'habillement, cuir et chaussures	101	114	118	128	128	131	138	130
Wood products, furniture Bois, meubles	125	133	140	146	142	147	146	139
Paper, printing, publishing Papier, imprimerie, édition	108	116	116	122	124	127	130	130
Chemicals and related products Produits et alliés	107	113	117	124	124	127	128	125
Non-metallic mineral products Produits minéraux non métalliques	112	121	126	132	124	127	125	118
Basic metals Métallurgie de base	117	126	133	133	145	154	153	149
Metal products Ouvrages en métaux	108	121	125	130	132	141	147	145
Electricity, gas, water [4] **Electricité, gaz et eau [4]**	**97**	**102**	**101**	**101**	**101**	**107**	**109**	**109**

7

Index numbers of labour productivity in industry: world and regions
Indices de la productivité du travail dans l'industrie : monde et régions
1980 = 100

Regiong and industry [SITC] Régiong et industrie [CITI]	1983	1984	1985	1986	1987	1988	1989	1990
Eastern Europe and former USSR • Europe de l'Est et l'ancienne URSS								
Total industry [2-4] **Total, industrie [2-4]**	**106**	**114**	**118**	**123**	**128**	**136**	**141**	**135**
Total mining [2] **Total, industries extractives [2]**	**87**	**87**	**87**	**89**	**90**	**92**	**90**	**82**
Coal Houille	88	88	88	90	90	92	90	81
Petroleum, gas Pétrole, gaz	84	82	80	80	84	81	79	72
Metal Minerais métalliques	82	83	83	86	91	96	97	91
Total manufacturing [3] **Total, industries manufacturières [3]**	**108**	**117**	**120**	**126**	**131**	**140**	**145**	**139**
Light industry Industrie légère	107	116	118	121	124	131	137	128
Heavy industry Industrie lourde	108	117	122	129	136	145	150	146
Selected manufacturing **Industries manufacturières déterminées**								
Food, beverages, tobacco Industries alimentaires, boissons, tabac	106	108	107	109	113	117	120	117
Textiles Textiles	122	123	128	132	136	146	154	144
Apparel, leather, footwear Articles d'habillement, cuir et chaussures	106	112	114	115	116	124	128	124
Wood products, furniture Bois, meubles	110	114	118	124	126	135	140	137
Paper, printing, publishing [1] Papier, imprimerie, édition [1]	112	117	124	132	138	149	157	144
Chemicals and related products Produits et alliés	105	108	104	108	113	120	123	116
Non-metallic mineral products Produits minéraux non métalliques	107	110	112	118	122	129	138	134
Basic metals Métallurgie de base	103	107	108	113	116	124	128	122
Metal products Ouvrages en métaux	109	120	128	135	143	153	158	154
Electricity, gas, water [4] **Electricité, gaz et eau [4]**	**104**	**108**	**109**	**110**	**113**	**115**	**116**	**114**
Western Europe • Europe de l'Ouest								
Total industry [2-4] **Total, industrie [2-4]**	**107**	**111**	**116**	**119**	**121**	**127**	**131**	**126**
Total mining [2] **Total, industries extractives [2]**	**105**	**101**	**116**	**120**	**121**	**125**	**131**	**120**
Coal Houille	108	100	119	124	126	132	139	124
Petroleum, gas Pétrole, gaz	104	104	108	117	111	106	96	92
Metal Minerais métalliques	94	105	107	102	100	104	112	114
Total manufacturing [3] **Total, industries manufacturières [3]**	**107**	**111**	**116**	**119**	**122**	**128**	**132**	**127**
Light industry Industrie légère	106	108	112	114	116	119	121	117
Heavy industry Industrie lourde	108	113	119	122	125	133	139	133
Selected manufacturing **Industries manufacturières déterminées**								
Food, beverages, tobacco Industries alimentaires, boissons, tabac	107	110	112	115	118	121	122	123

7
Index numbers of labour productivity in industry: world and regions
Indices de la productivité du travail dans l'industrie : monde et régions
1980 = 100

Region ᵍ and industry [SITC] Région ᵍ et industrie [CITI]	1983	1984	1985	1986	1987	1988	1989	1990
Textiles Textiles	108	110	118	120	122	124	125	115
Apparel, leather, footwear Articles d'habillement, cuir et chaussures	105	106	108	108	104	102	102	96
Wood products, furniture Bois, meubles	100	103	103	109	114	120	123	119
Paper, printing, publishing Papier, imprimerie, édition	106	114	114	120	124	131	133	131
Chemicals and related products Produits et alliés	109	114	118	120	122	129	131	121
Non-metallic mineral products Produits minéraux non métalliques	103	106	109	114	117	124	129	126
Basic metals Métallurgie de base	114	126	132	131	136	151	156	150
Metal products Ouvrages en métaux	107	111	118	121	124	132	138	133
Electricity, gas, water [4] **Electricité, gaz et eau [4]**	**103**	**104**	**112**	**115**	**119**	**121**	**124**	**124**

European Economic Community ⁺ · Communauté économique européenne ⁺

	1983	1984	1985	1986	1987	1988	1989	1990
Total industry [2-4] **Total, industrie [2-4]**	**108**	**112**	**118**	**121**	**124**	**131**	**135**	**129**
Total mining [2] **Total, industries extractives [2]**	**108**	**102**	**121**	**127**	**129**	**136**	**143**	**130**
Coal Houille	108	101	122	128	132	139	148	133
Petroleum, gas Pétrole, gaz	109	109	116	128	115	105	88	83
Metal Minerais métalliques	95	109	115	104	98	100	113	116
Total manufacturing [3] **Total, industries manufacturières [3]**	**108**	**112**	**118**	**121**	**124**	**131**	**135**	**129**
Light industry Industrie légère	107	110	114	117	119	123	124	119
Heavy industry Industrie lourde	109	114	120	124	127	136	142	136
Selected manufacturing **Industries manufacturières déterminées**								
Food, beverages, tobacco Industries alimentaires, boissons, tabac	109	112	115	119	121	125	126	126
Textiles Textiles	110	112	122	124	127	129	130	120
Apparel, leather, footwear Articles d'habillement, cuir et chaussures	107	108	111	112	108	106	107	100
Wood products, furniture Bois, meubles	101	103	103	110	116	123	125	120
Paper, printing, publishing Papier, imprimerie, édition	106	112	112	118	123	129	132	130
Chemicals and related products Produits et alliés	110	115	119	121	124	130	132	121
Non-metallic mineral products Produits minéraux non métalliques	103	107	110	116	118	126	131	128
Basic metals Métallurgie de base	116	128	136	134	141	157	164	158
Metal products Ouvrages en métaux	108	112	119	123	126	134	141	135
Electricity, gas, water [4] **Electricité, gaz et eau [4]**	**103**	**104**	**112**	**117**	**120**	**123**	**126**	**126**

European Free Trade Association ⁺
Association européenne de libre échange ⁺

	1983	1984	1985	1986	1987	1988	1989	1990
Total industry [2-4] **Total, industrie [2-4]**	**106**	**112**	**116**	**119**	**123**	**129**	**134**	**139**

7

Index numbers of labour productivity in industry: world and regions

Indices de la productivité du travail dans l'industrie : monde et régions

1980 = 100

Region ȝ and industry [SITC] Région ȝ et industrie [CITI]	1983	1984	1985	1986	1987	1988	1989	1990
Total mining [2] **Total, industries extractives [2]**	**118**	**128**	**141**	**149**	**170**	**177**	**224**	**244**
Coal Houille	176	182	229	217	218	183	323	359
Petroleum, gas Pétrole, gaz	86	86	82	84	96	105	130	135
Metal Minerais métalliques	109	124	132	147	178	198	222	242
Total manufacturing [3] **Total, industries manufacturières [3]**	**106**	**112**	**116**	**119**	**122**	**128**	**132**	**137**
Light industry Industrie légère	103	108	110	113	116	120	124	129
Heavy industry Industrie lourde	108	115	120	123	127	134	139	144
Selected manufacturing **Industries manufacturières déterminées**								
Food, beverages, tobacco Industries alimentaires, boissons, tabac	97	101	101	102	106	108	113	117
Textiles Textiles	110	115	119	121	125	131	135	140
Apparel, leather, footwear Articles d'habillement, cuir et chaussures	104	110	111	116	113	116	117	126
Wood products, furniture Bois, meubles	100	107	107	113	116	120	125	131
Paper, printing, publishing Papier, imprimerie, édition	113	121	125	129	135	142	142	147
Chemicals and related products Produits et alliés	110	119	122	122	126	136	145	150
Non-metallic mineral products Produits minéraux non métalliques	105	108	113	114	118	119	121	128
Basic metals Métallurgie de base	117	131	133	135	138	152	155	149
Metal products Ouvrages en métaux	106	111	117	121	124	130	136	143
Electricity, gas, water [4] **Electricité, gaz et eau [4]**	**109**	**112**	**119**	**119**	**128**	**130**	**135**	**140**

B. Developing Regions · Régions en voie de développement

	1983	1984	1985	1986	1987	1988	1989	1990
Total industry [2-4] **Total, industrie [2-4]**	**113**	**122**	**122**	**131**	**135**	**145**	**150**	...
Total mining [2] **Total, industries extractives [2]**	**104**	**111**	**113**	**124**	**128**	**137**	**142**	...
Coal Houille	110	118	120	129	138	141	153	...
Petroleum, gas Pétrole, gaz	70	72	68	72	71	76	79	...
Metal Minerais métalliques	115	125	130	144	146	163	163	...
Total manufacturing [3] **Total, industries manufacturières [3]**	**114**	**123**	**123**	**132**	**136**	**145**	**150**	...
Light industry Industrie légère	114	123	120	126	127	135	138	...
Heavy industry Industrie lourde	113	124	128	144	156	169	177	...
Selected manufacturing **Industries manufacturières déterminées**								
Food, beverages, tobacco Industries alimentaires, boissons, tabac	125	145	136	134	132	147	154	...
Textiles Textiles	120	124	133	142	151	157	161	...
Apparel, leather, footwear Articles d'habillement, cuir et chaussures	111	111	104	116	115	122	124	...

7

Index numbers of labour productivity in industry: world and regions
Indices de la productivité du travail dans l'industrie : monde et régions
1980 = 100

Regions and industry [SITC] Régions et industrie [CITI]	1983	1984	1985	1986	1987	1988	1989	1990
Wood products, furniture Bois, meubles	102	98	92	98	97	92	90	...
Paper, printing, publishing Papier, imprimerie, édition	109	120	131	141	148	156	164	...
Chemicals and related products Produits et alliés	113	122	118	127	135	149	153	...
Non-metallic mineral products Produits minéraux non métalliques	104	117	114	120	128	134	141	...
Basic metals Métallurgie de base	104	112	120	129	144	148	157	
Metal products Ouvrages en métaux	118	130	138	162	176	193	202	...
Electricity, gas, water [4] **Electricité, gaz et eau [4]**	...	...	...	...	...	...	...	...

Latin America and Caribbean • Amérique latine et Caraïbes

	1983	1984	1985	1986	1987	1988	1989	1990
Total industry [2-4] **Total, industrie [2-4]**	**100**	**100**	**104**	**112**	**118**	**120**	**126**	...
Total mining [2] **Total, industries extractives [2]**	...	...	...	...	...	...	...	
Coal Houille	...	...	...	...	...	...	...	...
Petroleum, gas Pétrole, gaz	83	83	83	88	86	90	88	...
Metal Minerais métalliques	...	...	...	...	...	...	...	
Total manufacturing [3] **Total, industries manufacturières [3]**	**100**	**99**	**104**	**111**	**117**	**118**	**125**	...
Light industry Industrie légère	101	100	105	108	113	113	119	...
Heavy industry Industrie lourde	97	97	103	114	124	126	132	...
Selected manufacturing **Industries manufacturières déterminées**								
Food, beverages, tobacco Industries alimentaires, boissons, tabac	105	106	110	113	118	119	133	
Textiles Textiles	105	107	112	122	128	129	136	
Apparel, leather, footwear Articles d'habillement, cuir et chaussures	95	83	84	82	84	83	84	
Wood products, furniture Bois, meubles	...	...	...	...	...	...	...	
Paper, printing, publishing Papier, imprimerie, édition	101	111	137	142	151	159	168	
Chemicals and related products Produits et alliés	108	107	112	122	137	144	153	
Non-metallic mineral products Produits minéraux non métalliques	103	95	99	106	117	114	120	
Basic metals Métallurgie de base	99	108	109	121	135	143	167	
Metal products Ouvrages en métaux	90	90	96	107	113	111	112	...
Electricity, gas, water [4] **Electricité, gaz et eau [4]**	...	...	...	...	...	...	...	...

Asia • Asie

	1983	1984	1985	1986	1987	1988	1989	1990
Total industry [2-4] **Total, industrie [2-4]**	**117**	**129**	**127**	**137**	**141**	**153**	**159**	...
Total mining [2] **Total, industries extractives [2]**	**92**	**97**	**99**	**106**	**115**	**119**	**132**	...
Coal Houille	107	115	119	128	140	141	157	...

7

Index numbers of labour productivity in industry: world and regions

Indices de la productivité du travail dans l'industrie : monde et régions

1980 = 100

Region ᶲ and industry [SITC] Région ᶲ et industrie [CITI]	1983	1984	1985	1986	1987	1988	1989	1990
Petroleum, gas Pétrole, gaz	64	65	59	67	67	76	83	...
Metal Minerais métalliques	83	86	88	90	100	109	118	...
Total manufacturing [3] **Total, industries manufacturières [3]**	**118**	**130**	**128**	**137**	**142**	**154**	**160**	**...**
Light industry Industrie légère	117	128	123	129	130	140	143	...
Heavy industry Industrie lourde	120	137	140	161	176	194	203	...
Selected manufacturing **Industries manufacturières déterminées**								
Food, beverages, tobacco Industries alimentaires, boissons, tabac	132	159	144	139	136	157	164	...
Textiles Textiles	121	125	133	142	153	160	164	...
Apparel, leather, footwear Articles d'habillement, cuir et chaussures	114	117	109	123	124	131	133	...
Wood products, furniture Bois, meubles	101	95	86	92	90	80	78	...
Paper, printing, publishing Papier, imprimerie, édition	117	133	139	152	159	170	179	...
Chemicals and related products Produits et alliés	116	128	121	129	134	150	152	...
Non-metallic mineral products Produits minéraux non métalliques	106	124	118	124	131	138	146	...
Basic metals Métallurgie de base	106	113	128	137	154	157	159	...
Metal products Ouvrages en métaux	132	152	162	197	219	248	260	...
Electricity, gas, water [4] **Electricité, gaz et eau [4]**	**115**	**118**	**139**	**146**	**146**	**155**	**166**	**...**

Oceania[2] · Océanie[2]

	1983	1984	1985	1986	1987	1988	1989	1990
Total industry [2-4] **Total, industrie [2-4]**	**98**	**106**	**112**	**111**	**116**	**122**	**126**	**124**
Total mining [2] **Total, industries extractives [2]**	**111**	**120**	**147**	**159**	**192**	**218**	**257**	**294**
Coal Houille	81	85	90	89	108	118	153	139
Petroleum, gas Pétrole, gaz	...	...	...	...	...	...	...	...
Metal Minerais métalliques	134	148	193	219	267	309	357	434
Total manufacturing [3] **Total, industries manufacturières [3]**	**98**	**106**	**110**	**108**	**112**	116	**118**	**114**
Light industry Industrie légère	106	111	111	111	112	114	112	109
Heavy industry Industrie lourde	89	101	109	106	112	118	123	118
Selected manufacturing **Industries manufacturières déterminées**								
Food, beverages, tobacco Industries alimentaires, boissons, tabac	102	107	111	112	116	113	116	117
Textiles Textiles	106	113	111	113	110	115	116	120
Apparel, leather, footwear Articles d'habillement, cuir et chaussures	96	100	99	92	87	87	81	70
Wood products, furniture Bois, meubles	97	104	104	100	106	107	110	104
Paper, printing, publishing Papier, imprimerie, édition	141	144	144	146	148	152	144	141

7

Index numbers of labour productivity in industry: world and regions
Indices de la productivité du travail dans l'industrie : monde et régions
1980 = 100

Region ₰ and industry [SITC] Région ₰ et industrie [CITI]	1983	1984	1985	1986	1987	1988	1989	1990
Chemicals and related products Produits et alliés	92	95	96	99	108	121	124	121
Non-metallic mineral products Produits minéraux non métalliques	96	110	115	110	116	147	142	134
Basic metals Métallurgie de base	94	107	115	116	118	126	142	134
Metal products Ouvrages en métaux	86	99	108	103	108	110	114	106
Electricity, gas, water [4] **Electricité, gaz et eau [4]**	**99**	**105**	**113**	120	**123**	**132**	142	144

Source:
Industrial statistics database of the Statistical Office of the
United Nations Secretariat.

+   For Member States of this grouping, see Annex I -
     Other groupings.

₰   All series exclude Albania, China, Democratic
     People's Republic of Korea and Viet Nam.
     Series for "Developed regions" include North America
     (Canada and the United States), Western Europe, Australia,
     Israel, Japan, New Zealand and South Africa. Eastern
     Europe and the former USSR are not included in "Developed
     regions" totals but are shown separately and include
     Bulgaria, Czechoslovakia, Hungary, Poland,
     Romania and the former USSR. Series for "Developing
     regions" exclude Australia, Israel, Japan, New Zealand
     and South Africa, and include Yugoslavia.

1   Excluding printing and publishing.
2   Including Australia and New Zealand.

Source:
Base de données de statistiques industrielles du Bureau de statistique
du Secrétariat de l'ONU.

+   Les Etats membres de ce groupement, voir annexe I -
     Autres groupements.

₰   Aucune série ne comprend l'Albanie, la Chine,
     la République populaire démocratique de Corée et le Viet Nam.
     Les régions developpées comprennent l'Amérique du nord
     (le Canada et les Etats-Unis), l'Europe de l'ouest, l'Australie,
     Israël, le Japon, la Nouvelle-Zélande et l'Afrique du sud.
     L'Europe de l'est et l'ancienne URSS ne sont pas compris dans les
     totaux des "Régions developpées" mais sont indiqués séparément et
     comprennent la Bulgarie, la Tchécoslovaquie, la Hongrie, la Pologne,
     la Roumanie et l'ancienne URSS. Les séries pour les "Régions en voie
     de développement" excluent l'Australie, Israël, le Japon,
     la Nouvelle-Zélande et l'Afrique du sud, et incluent la Yougoslavie.

1   Non compris l'imprimerie et l'édition.
2   Y compris l'Australie et la Nouvelle-Zélande.

8
Production, trade and consumption of commercial energy
Production, commerce et consommation d'énergie commerciale
Thousand metric tons of coal equivalent and kilograms per capita
Milliers de tonnes métriques d'équivalent houille et kilogrammes par habitant

| Regions | Year | Primary energy production - Production d'energie primaire | | | | | Changes in stocks Variations des stocks | Imports Importations | Exports Exportations |
		Total Totale	Solids Solides	Liquids Liquides	Gas Gaz	Electricity Electricité			
World	**1980**	**9 186 719**	**2 623 157**	**4 421 569**	**1 842 351**	**299 642**	**74 436**	**3 245 428**	**3 214 321**
	1988	**10 486 858**	**3 281 191**	**4 386 032**	**2 329 963**	**489 672**	**22 001**	**3 272 523**	**3 226 110**
	1989	**10 697 942**	**3 335 283**	**4 452 239**	**2 415 467**	**494 953**	**7 355**	**3 435 751**	**3 367 691**
	1990	**10 877 829**	**3 318 109**	**4 547 014**	**2 499 832**	**512 873**	**105 318**	**3 516 653**	**3 524 664**
Africa	1980	577 343	98 232	436 396	35 227	7 488	6 336	71 109	425 651
	1988	604 508	142 238	386 127	69 856	6 288	-891	72 716	386 609
	1989	648 113	138 860	426 328	76 525	6 401	662	74 614	431 816
	1990	702 903	139 368	462 025	94 870	6 640	5 977	73 521	461 482
America, North	1980	2 531 811	661 861	972 818	792 627	104 505	22 799	748 836	348 523
	1988	2 697 911	784 902	1 019 177	746 487	147 344	-23 874	703 196	416 581
	1989	2 690 844	804 771	972 446	761 892	151 735	-24 517	763 819	422 694
	1990	2 745 139	848 669	960 104	776 694	159 673	49 658	757 812	430 202
America, South	1980	337 312	9 032	260 510	43 759	24 011	-1 662	92 630	161 493
	1988	420 286	22 571	288 113	70 710	38 892	2 205	85 451	177 183
	1989	429 726	26 474	291 236	72 494	39 521	1 937	82 443	180 391
	1990	460 686	26 682	316 034	77 257	40 713	4 709	80 433	202 029
Asia	1980	2 433 339	619 155	1 653 480	115 292	45 412	17 489	761 442	1 447 220
	1988	2 747 435	967 380	1 454 492	244 741	80 823	33 270	913 363	1 166 817
	1989	2 939 193	1 029 632	1 558 827	268 655	82 079	18 348	975 440	1 266 814
	1990	3 065 672	1 049 594	1 647 713	280 929	87 435	25 836	1 034 593	1 364 163
Europe	1980	1 266 753	653 614	203 830	324 964	84 344	31 274	1 524 395	456 071
	1988	1 392 108	638 244	313 024	284 153	156 687	1 344	1 431 781	556 210
	1989	1 370 134	619 980	304 309	289 194	156 651	-8 924	1 479 520	566 766
	1990	1 320 626	562 053	308 802	290 924	158 847	7 901	1 517 593	585 203
Oceania	1980	119 616	70 817	31 288	13 587	3 924	-2 920	31 153	40 408
	1988	197 165	124 414	41 404	26 617	4 730	-10 010	24 503	97 398
	1989	206 959	136 748	37 957	27 220	5 034	677	28 499	92 472
	1990	226 019	145 297	42 938	32 882	4 903	5 742	28 673	101 300
former USSR	1980	1 920 545	510 446	863 246	516 895	29 958	1 120	15 776	334 956
	1988	2 427 445	601 443	883 696	887 398	54 907	19 956	41 417	425 313
	1989	2 412 973	578 817	861 136	919 488	53 532	19 172	31 321	406 738
	1990	2 356 784	546 446	809 399	946 277	54 662	5 496	23 931	380 285

Source:
Energy statistics database of the Statistical Division of the United Nations Secretariat.

Source:
Base de données pour les statistiques énergétiques de la Division de statistique du Secrétariat de l'ONU.

Bunkers - Soutes			Consumption - Consommation						
Air Avion	Sea Maritime	Unallocated Nondistribué	Per capita Par habitant	Total Totale	Solids Solides	Liquids Liquides	Gas Gaz	Electricity Electricité	Année	Régions
50 796	**163 288**	**334 049**	**1 928**	**8 595 257**	**2 623 477**	**3 836 921**	**1 835 437**	**299 422**	**1980**	*Monde*
58 648	**151 219**	**250 259**	**1 996**	**10 051 098**	**3 312 679**	**3 932 893**	**2 313 961**	**491 566**	**1988**	
63 011	**151 601**	**276 033**	**2 004**	**10 268 045**	**3 386 689**	**3 982 133**	**2 402 268**	**496 956**	**1989**	
64 328	**161 581**	**241 765**	**1 974**	**10 296 592**	**3 327 917**	**3 982 514**	**2 471 030**	**515 131**	**1990**	
3 610	9 429	12 653	400	190 774	74 221	85 070	24 019	7 464	1980	Afrique
3 346	9 438	28 387	427	250 334	103 049	106 527	34 679	6 079	1988	
3 765	10 013	22 257	420	254 215	100 570	109 991	37 539	6 114	1989	
3 720	10 085	22 683	437	272 476	100 823	112 255	53 107	6 291	1990	
11 377	57 846	50 139	7 463	2 789 964	573 699	1 315 532	796 230	104 503	1980	Amérique du Nord
12 228	36 951	19 532	7 144	2 939 690	717 658	1 329 882	744 475	147 674	1988	
12 514	37 267	16 920	7 174	2 989 786	724 506	1 343 492	769 945	151 843	1989	
12 666	38 659	8 257	7 022	2 963 510	735 191	1 313 362	755 187	159 771	1990	
1 524	3 159	18 527	1 025	246 901	15 115	163 966	43 826	23 993	1980	Amérique du Sud
1 128	3 333	22 673	1 070	299 215	23 632	164 097	70 681	40 805	1988	
1 178	3 722	25 661	1 049	299 280	24 826	159 896	72 441	42 117	1989	
1 345	3 622	23 192	1 052	306 224	23 451	162 036	77 075	43 661	1990	
11 250	39 755	115 246	603	1 563 820	699 951	705 136	113 141	45 592	1980	Asie
11 502	45 386	110 902	776	2 292 920	1 113 239	849 894	248 963	80 824	1988	
12 385	46 336	129 887	810	2 440 863	1 186 621	896 106	276 099	82 037	1989	
13 237	50 491	117 265	824	2 529 272	1 212 179	941 381	288 469	87 243	1990	
20 470	44 029	48 238	4 523	2 191 067	729 588	981 172	394 009	86 298	1980	Europe
27 831	47 448	6 229	4 416	2 184 782	732 173	874 788	416 536	161 285	1988	
30 313	46 154	14 887	4 436	2 200 501	736 195	869 295	433 957	161 055	1989	
30 787	50 897	17 227	4 316	2 145 970	677 373	853 483	452 098	163 016	1990	
2 565	2 611	2 873	4 616	105 232	40 062	47 727	13 520	3 924	1980	Océanie
2 614	1 489	-2 967	5 255	133 144	50 680	51 123	26 611	4 730	1988	
2 857	1 643	-3 009	5 476	140 817	54 817	53 748	27 219	5 034	1989	
2 573	1 646	-4 404	5 665	147 835	55 843	57 947	29 142	4 903	1990	
...	6 372	86 374	5 677	1 507 500	490 841	538 318	450 692	27 649	1980	ancienne URSS
...	7 080	65 503	6 916	1 951 009	572 248	556 579	772 015	50 168	1988	
...	6 372	69 432	6 832	1 942 579	559 154	549 602	785 068	48 756	1989	
...	6 089	57 545	6 742	1 931 301	523 058	542 045	815 951	50 247	1990	

9
Total exports and imports (SNA countries)
Exportations et importations totales (pays au SCN)

Index numbers of quantum, unit value and terms of trade (1980 = 100)
Indices du quantum, de la valeur unitaire et des termes de l'échange (1980 =100)

Regions[1] Régions[1]	Weight[2] Pondera-tion[2]	1975	1983	1984	1985	1986	1987	1988	1989	1990
A. Exports • Exportations										
Quantum indices[3]• Indices du quantum[3]										
Total	100	77	100	109	112	120	127	137	146	154
Developed economies Econ. développées	69	73	104	114	119	122	128	136	146	153
North America Amérique du Nord	16	73	88	97	98	102	111	128	137	146
Europe Europe	43	74	108	117	123	125	131	138	149	155
EEC+ CEE+	37	75	108	116	122	125	131	137	148	154
EFTA+ AELE+	6	74	109	119	126	129	134	141	151	158
Africa[4] Afrique[4]	1	61	88	96	105	115	106	109	124	121
Asia Asie	7	65	116	135	141	141	142	148	154	162
Oceania Océanie	2	77	104	122	131	132	141	143	148	157
Developing economies Econ. en dévelop.	32	89	92	98	97	116	125	139	146	154
America Amérique		83	120	129	123	137	139	138	153	159
Europe[5] Europe[5]		66	98	107	112	111	117	112	116	111
Africa Afrique		83	70	72	73	86	82	87	88	91
Asia Asie		87	90	94	94	117	131	152	157	169
Middle East Moyen-Orient		109	62	57	52	71	71	84	86	75
Other Asia Autres pays d'Asie		58	126	143	150	177	209	241	250	291

Unit value indices in US dollars[3]• Indices de la valeur unitaire en dollars E-U[3]

		1975	1983	1984	1985	1986	1987	1988	1989	1990
Total		56	88	87	85	88	98	104	105	114
Developed economies Econ. développées		63	88	86	85	97	109	116	116	127
North America Amérique du Nord		68	109	110	107	106	108	117	121	122
Europe Europe		61	81	76	77	92	107	112	111	128
EEC+ CEE+		60	81	76	77	93	107	112	111	128
EFTA+ AELE+		62	82	78	78	92	107	113	111	127
Africa[4] Afrique[4]		60	88	77	71	77	94	96	97	108
Asia Asie		66	96	96	95	113	124	138	137	136

9

Total exports and imports (SNA countries)
Index numbers of quantum, unit value and terms of trade (1980 = 100) [cont.]
Exportations et importations totales (pays au SCN)
Indices du quantum, de la valeur unitaire et des termes de l'échange (1980 =100) [suite]

Regions[1] Régions[1]	Weight[2] Pondera- tion[2]	1975	1983	1984	1985	1986	1987	1988	1989	1990
Oceania Océanie		65	91	88	80	78	87	107	112	113
Developing economies Econ. en dévelop.		**43**	**89**	**88**	**85**	**66**	**75**	**77**	**80**	**86**
America Amérique		50	83	82	80	60	66	76	75	79
Europe[5] Europe[5]		57	113	107	106	103	109	126	129	143
Africa Afrique		40	81	81	77	51	59	55	60	69
Asia Asie		**42**	**92**	**92**	**87**	**71**	**79**	**80**	**83**	**90**
Middle East Moyen-Orient		36	94	92	89	55	59	50	55	71
Other Asia Autres pays d'Asie		57	91	92	87	80	88	94	95	97

B. Imports • Importations
Quantum indices[3] • Indices du quantum[3]

Total	**100**	**74**	**100**	**110**	**114**	**123**	**131**	**144**	**153**	**162**
Developed economies Econ. développées	**75**	**74**	**100**	**111**	**117**	**126**	**135**	**145**	**155**	**161**
North America Amérique du Nord	17	75	104	128	139	152	159	167	173	177
Europe Europe	**48**	**72**	**99**	**105**	**111**	**119**	**129**	**137**	**148**	**158**
EEC+ CEE+	41	71	98	104	110	118	128	137	149	159
EFTA+ AELE+	7	78	101	108	115	124	131	138	147	152
Africa[4] Afrique[4]	1	77	80	89	69	77	74	89	89	78
Asia Asie	8	81	100	110	110	121	133	153	164	155
Oceania Océanie	1	82	103	124	130	128	131	142	170	166
Developing economies Econ. en dévelop.	**25**	**74**	**102**	**104**	**106**	**113**	**120**	**141**	**148**	**166**
Europe[5] Europe[5]		74	72	69	69	74	75	69	76	88

Unit value indices in US dollars[6] • Indices de la valeur unitaire en dollars E-U[6]

Total		58	90	88	86	87	95	100	101	110
Developed economies Econ. développées		**58**	**88**	**86**	**85**	**88**	**97**	**102**	**103**	**115**
North America Amérique du Nord		58	101	103	100	97	102	108	111	113
Europe Europe		**59**	**82**	**78**	**78**	**86**	**98**	**102**	**102**	**116**
EEC+ CEE+		59	83	79	78	86	97	102	102	116

9
Total exports and imports (SNA countries)
Index numbers of quantum, unit value and terms of trade (1980 = 100) [cont.]
Exportations et importations totales (pays au SCN)
Indices du quantum, de la valeur unitaire et des termes de l'échange (1980 = 100) [suite]

Regions[1] Régions[1]	Weight[2] Pondera-tion[3]	1975	1983	1984	1985	1986	1987	1988	1989	1990
EFTA+ AELE+		58	79	74	74	87	100	106	104	118
Africa[4] Afrique[4]		53	97	90	81	83	103	105	104	118
Asia Asie		52	91	89	84	75	82	88	92	108
Oceania Océanie		62	93	92	88	92	101	111	108	113
Developing economies[7] Econ. en dévelop.[7]		58	96	94	90	84	90	91	94	98
Europe[5] Europe[5]		57	112	115	117	106	111	127	129	142

C. Terms of trade · Termes de l'échange

	Weight	1975	1983	1984	1985	1986	1987	1988	1989	1990
Developed economies Econ. développées		109	101	100	101	111	112	113	112	111
North America Amérique du Nord		117	108	107	108	110	106	109	109	108
Europe Europe		104	98	98	99	107	109	110	109	110
EEC+ CEE+		102	97	97	98	108	110	110	109	110
EFTA+ AELE+		106	104	105	105	106	107	107	107	108
Africa[4] Afrique[4]		113	90	85	87	93	91	92	93	91
Asia Asie		127	107	109	113	150	151	156	150	126
Oceania Océanie		105	97	95	91	85	86	97	104	100
Developing economies Econ. en dévelop.		75	92	94	94	79	83	84	85	88
Europe[5] Europe[5]		100	101	93	91	97	98	99	100	101

Source:
Trade statistics database of the Statistical Division of the
United Nations Secretariat.
+   For Member States of this grouping, see
    Annex I - Other groupings.

1 For the composition of the regions see table 117 of this
   issue.
2 Based on the value of the trade, in US dollars, in 1980.

3 Quantum indices are derived from value data and unit value
   indices. They are base period weighted.

4 South African Customs Union.
5 Yugoslavia only.
6 Regional aggregates are current period weighted.

7 Indices, except those for Europe, are based on estimates
   prepared by the International Monetary Fund.

Source:
Base de données pour les statistiques du commerce extérieur
de la Division de statistique du Secrétariat de l'ONU.
+   Les Etats membres de ce groupement, voir
    annexe I - Autres groupements.

1 Pour la composition des régions, voir le tableau 117 du
   présent numéro.
2 Calculée d'après la valeur du commerce en 1980, en dollars
   E-U.
3 Les indices du quantum sont calculés à partir des chiffres
   de la valeur et des indices de la valeur unitaire. Ils sont
   à coéfficients de pondération correspondant à la période en
   base.
4 L'Union Douanière d'Afrique Méridionale.
5 Yougoslavie seulement.
6 Les totaux régionaux sont à coéfficients de pondération
   correspondant à la période en cours.
7 Le calcul des indices, sauf ceux pour l'Europe, sont basés
   sur les estimations préparées par le Fonds monétaire
   international.

10
Export price index numbers of primary commodities non-ferrous base metals
Indices des prix des exportations des matières premières et des métaux non-férreux

1980 = 100

Items Elements	1975	1982	1983	1984	1985	1986	1987	1988	1989	1990
A. Primary commodities • Matières premières										
Total **Totaux**	**48**	**99**	**91**	**90**	**85**	**63**	**66**	**70**	**72**	**79**
Developed economies[1] Econ. développées[1]	67	90	87	87	80	75	79	90	92	97
Developing economies[1] Econ. en voie de dévelop[1]	40	103	92	91	87	58	59	61	63	70
B. Food • Produits alimentaires										
Total **Totaux**	**68**	**78**	**77**	**74**	**66**	**74**	**75**	**87**	**87**	**88**
Developed economies[1] Econ. développées[1]	71	83	81	75	68	73	79	93	94	98
Developing economies[1] Econ. en voie de dévelop[1]	61	67	70	70	63	77	66	75	72	69
Meat Viande	61	84	78	73	68	82	93	101	100	110
Beef Boeuf	56	86	84	78	69	84	99	112	108	115
Mutton Mouton	48	88	71	66	63	74	79	87	86·	94
Pork Porc	73	85	73	69	68	83	86	89	97	108
Poultry Volaille	68	72	68	67	66	85	91	92	87	109
Bacon Bacon	73	86	71	68	64	73	79	83	93	105
Dairy products Produits laitiers	74	87	82	74	73	83	100	106	101	110
Milk Lait	92	95	90	81	88	85	102	109	104	114
Butter Beurre	56	88	82	71	66	81	93	100	100	103
Cheese Fromage	70	81	76	68	65	85	107	113	103	115
Eggs Oeufs	68	76	76	76	65	72	88	78	84	97
Cereals Céreales	77	86	85	80	70	63	60	83	89	84
Wheat Blé	92	89	89	87	81	71	65	86	101	82
Rice Riz	84	71	71	69	66	61	60	82	81	74
Barley Orge	104	88	89	93	69	60	58	75	82	80
Maize Mais	46	88	84	72	57	54	54	80	78	92
Oats Avoine	97	83	83	93	78	57	56	97	100	76
Vegetables Légumes	76	75	77	92	61	78	96	102	95	112

10

Export price index numbers of primary commodities non-ferrous base metals
[*cont.*]

Indices des prix des exportations des matières premières et des métaux
1980 = 100 [*suite*]

Items Elements	1975	1982	1983	1984	1985	1986	1987	1988	1989	1990
Potatoes Pommes de terre	72	71	69	187	60	79	115	112	99	129
Cassava Manioc	64	75	89	59	52	86	83	74	63	78
Beans Haricots secs	78	49	57	64	69	67	59	82	117	117
Peas Pois secs	135	92	77	75	87	97	114	154	137	153
Tomatoes Tomates	61	92	93	73	54	79	117	119	92	107
Onions Oignons	59	67	68	73	45	53	78	61	68	90
Beverage crops Stimulants	47	77	83	96	85	105	70	72	62	55
Coffee Café	42	78	77	83	81	114	64	67	56	46
Cocoa Cacao	55	65	81	95	86	78	75	67	51	49
Tea Thé	65	98	126	187	112	107	100	112	122	127
Fruits Fruits	68	88	87	81	84	94	105	112	122	131
Citrus fruits Agrumes	64	83	83	77	83	86	98	100	100	118
Oranges Oranges	62	87	85	83	84	87	102	103	103	122
Lemons Citrons	75	63	76	53	78	83	88	91	92	108
Grapefruit Pamplemousses	62	80	78	74	81	84	83	93	88	108
Other fruits Autres fruits	71	94	91	86	85	101	112	125	145	144
Bananas Bananes	64	97	112	95	98	103	105	130	181	150
Apples Pommes	78	90	70	76	71	99	119	121	108	139
Animal feed Alimentation animale	62	88	102	81	63	76	85	113	102	87
Oilseed cake Tourteaux	65	92	105	83	65	79	87	114	107	87
Fishmeal Farine de poisson	50	73	90	73	56	65	76	108	84	88
Pepper Poivre	95	75	80	118	201	254	278	208	147	102
Fish Poissons	55	88	80	77	77	89	113	119	107	117
Sugar Sucre	72	30	30	20	15	22	25	37	46	45
Wine Vin	71	90	88	75	78	103	105	113	124	155

10

Export price index numbers of primary commodities non-ferrous base metals
[*cont.*]

Indices des prix des exportations des matières premières et des métaux
1980 = 100 [*suite*]

Items Elements	1975	1982	1983	1984	1985	1986	1987	1988	1989	1990

C. Agricultural non-food • Produits agr. non-alimentaires

Items Elements	1975	1982	1983	1984	1985	1986	1987	1988	1989	1990
Total **Totaux**	**63**	**82**	**86**	**91**	**74**	**75**	**86**	**100**	**98**	**100**
Developed economies[1] Econ. développées[1]	67	86	88	93	75	77	89	105	105	106
Developing economies[1] Econ. en voie de dévelop[1]	57	76	85	90	73	70	82	93	88	89
Oilseeds, oils and fats Oléagineux	79	80	91	109	84	65	69	91	83	78
Groundnuts Arachides	92	69	84	120	107	68	59	69	91	113
Soyabeans Feves de soya	73	83	94	96	76	69	71	100	89	81
Linseeds Graines de lin	113	91	90	94	84	64	53	93	106	101
Copra Coprah	57	69	109	158	85	44	69	88	77	51
Soyabean oil Huile de soya	97	76	91	123	100	61	59	83	75	80
Cottonseed oil Huile de coton	110	77	100	137	115	94	94	108	107	123
Groundnut oil Huile d' arachide	92	69	84	120	107	68	59	69	91	113
Olive oil Huile d' olive	74	99	84	81	79	94	101	108	108	115
Palm oil Huile de palme	74	77	85	127	86	44	59	75	60	49
Coconut oil Huile de coprah	63	66	107	182	103	43	63	81	74	48
Palmkernel oil Huile de palmiste	59	67	105	155	82	42	63	79	69	49
Linseed oil Huile de lin	101	73	68	88	110	83	69	123	166	169
Sunflower oil Huile de tournesol	118	84	90	125	97	58	57	76	76	77
Animal fats Graisses animales	64	78	75	75	58	51	59	73	61	64
Forest products Produits forestiers	59	76	74	75	66	82	94	103	105	114
Lumber Bois	57	70	76	72	62	79	88	100	103	113
Woodpulp Pâte a papier	73	83	73	85	70	80	91	103	110	113
Logs Grumes	42	77	74	68	67	89	107	110	102	116
Textile fibres Fibres textiles	63	82	86	87	71	61	81	95	95	95
Wool Laine	57	90	88	85	60	59	82	123	106	95
Cotton Coton	64	77	85	86	75	61	79	77	88	93

10
Export price index numbers of primary commodities non-ferrous base metals
[*cont.*]

Indices des prix des exportations des matières premières et des métaux
1980 = 100 [*suite*]

Items Elements	1975	1982	1983	1984	1985	1986	1987	1988	1989	1990
Jute Jute	93	93	86	130	140	78	97	100	101	113
Flax Lin	85	77	91	104	99	107	108	116	102	124
Sisal Sisal	83	77	74	77	74	74	73	75	85	89
Tobacco Tabac	73	128	130	130	129	113	110	115	122	125
Hides and skins Cuirs et peaux	49	84	106	137	121	142	174	209	213	192
Natural rubber Caoutchouc naturel	39	58	75	67	53	57	69	83	66	60

D. Minerals • Minéraux

Items Elements	1975	1982	1983	1984	1985	1986	1987	1988	1989	1990
Total Totaux	39	110	97	95	93	57	59	59	62	71
Developed economies[1] Econ. développées[1]	58	113	102	105	106	74	70	70	74	85
Developing economies[1] Econ. en voie de dévelop[1]	36	109	96	94	92	55	57	57	59	69
Iron ore Minerai de fer	83	96	92	85	84	82	82	86	98	124
Chrome ore Minerai de chrôme	85	103	93	93	93	93	93	132	189	144
Manganese ore Minerai de manganese	86	101	84	83	85	84	78	101	175	243
Fuels Combustibles	37	110	97	95	94	57	58	58	60	70
Coal Charbon	92	93	83	82	82	88	87	90	95	108
Crude petroleum Petrole brut	34	109	96	93	91	54	56	56	59	68
Natural gas Gaz naturel	55	148	142	161	170	99	75	68	72	77
Crude fertilizers Engrais bruts	131	96	70	74	65	67	60	69	79	78

E. Non-ferrous base metals • Métaux non-ferreux

Items Elements	1975	1982	1983	1984	1985	1986	1987	1988	1989	1990
Total Totaux	54	77	78	76	71	70	87	130	124	112
Developed economies[1] Econ. développées[1]	54	78	79	78	71	73	90	137	127	116
Developing economies[1] Econ. en voie de dévelop[1]	52	75	77	72	69	64	78	114	114	101
Copper Cuivre	56	67	72	62	64	62	81	119	131	123
Nickel Nickel	54	94	94	94	108	123	138	284	258	191
Aluminum Aluminium	50	85	84	89	72	75	93	133	100	93
Lead Plomb	46	60	46	49	43	45	66	73	75	90

10

Export price index numbers of primary commodities non-ferrous base metals
[cont.]

Indices des prix des exportations des matières premières et des métaux
1980 = 100 [suite]

Items Elements	1975	1982	1983	1984	1985	1986	1987	1988	1989	1990
Zinc Zinc	97	96	99	113	94	88	108	161	216	206
Tin Etain	41	76	75	69	68	44	47	52	61	42

F. Pri. commodities excl. crude petroleum • Matières pre., non compris le pet. brut

	1975	1982	1983	1984	1985	1986	1987	1988	1989	1990
Total Totaux	68	85	85	85	77	77	79	90	90	94
Developed economies[1] Econ. développées[1]	71	89	87	86	78	77	82	94	96	100
Developing economies[1] Econ. en voie de dévelop[1]	62	76	79	81	72	76	73	81	79	79

G. Minerals excl. crude petroleum • Minéraux, non-compris le pet. brut

	1975	1982	1983	1984	1985	1986	1987	1988	1989	1990
Total Totaux	77	116	108	114	117	90	80	80	87	99
Developed economies[1] Econ. développées[1]	76	116	108	115	118	91	81	81	87	98
Developing economies[1] Econ. en voie de dévelop[1]	80	117	108	113	116	88	76	78	89	102

Source:
Trade statistics database of the Statistical Division of the
United Nations Secretariat.

1 This classification is intended for statistical convenience
and does not necessarily express a judgement about the stage
reached by a particular country in the development process.

Source:
Base de données pour les statistiques du commerce extérieur
de la Division de statistique du Secrétariat de l'ONU.

1 Cette classification est utilisée pour plus de commodité
dans la présentation des statistiques et n'implique pas
nécessairement un jugement au stade de développment auquel
et parvenu un pays donné.

Technical notes, tables 1-10

Table 1: These series of world aggregates on population and production have been compiled from statistical publications of the United Nations and the specialized agencies.[1, 4, 5, 6, 9, 10, 14, 21, 22] Reference should be made to these sources for details of compilation and coverage.

Table 2 presents for the world and regions estimates of population size, rates of population increase, crude birth and death rates, surface area and population density. Unless otherwise specified all figures are estimates of the order of magnitude and are subject to a substantial margin of error.

All population estimates and rates presented in this table were prepared by the Population Division of the United Nations Secretariat and published in *World Population Prospects 1990*.[25]

The average annual percentage rates of population growth were calculated by the Population Division of the United Nations Secretariat, using an exponential rate of increase formula.

Crude birth and crude death rates are expressed in terms of the average annual number of births and deaths respectively, per 1,000 mid-year population. These rates are estimated.

Surface area totals were obtained by summing the figures for individual countries or areas.

Density is the number of persons in the 1990 total population per square kilometre of total surface area.

The scheme of regionalization used for the purpose of making these estimates is presented in annex I. Although some continental totals are given, and all can be derived, the basic scheme presents eight macro regions that are so drawn as to obtain greater homogeneity in sizes of population, types of demographic circumstances and accuracy of demographic statistics.

Tables 3-4: The index numbers in table 3 refer to agricultural production, which is defined to include both crop and livestock products. Seeds and feed are excluded. The index numbers of food refer to commodities which are considered edible and contain nutrients. Coffee, tea and other inedible commodities are excluded.

The index-numbers of agricultural output and food production in table 4 are calculated by the Laspeyres formula with the base year period 1979-1981. The latter is provided in order to diminish the impact of annual fluctuations in agricultural output during base years on the indices for the period. Production quantities of each commodity are weighted by 1979-1981 average national producer prices and summed for each year. The index numbers are based on production data for a calendar year. As in the past, the series include a large number of estimates made by FAO in cases where figures are not available from official country sources.

Notes techniques, tableaux 1-10

Tableau 1 : Ces séries d'agrégats mondiaux sur la population et la production ont été établies à partir de publications statistiques des Nations Unies et d'institutions spécialisées [1, 4, 5, 6, 9, 10, 14, 21, 22]. On doit se référer à ces sources pour tous renseignements détaillés sur les méthodes de calcul et la portée des statistiques.

Le *Tableau 2* présente les estimations mondiales et régionales de la population, des taux d'accroissement de la population, des taux bruts de natalité et de mortalité, de la superficie et de la densité de population. Sauf indication contraire, tous les chiffres sont des estimations de l'ordre de grandeur et comportent une assez grande marge d'erreur.

Toutes les estimations de la population et tous les taux présentés dans ce tableau ont été établis par la Division de la population du Secrétariat des Nations Unies et publiés dans "World Population Prospects 1990" [25].

Les pourcentages annuels moyens de l'accroissement de la population ont été calculés par la Division de la population du Secrétariat des Nations Unies, sur la base d'une formule de taux d'accroissement exponentiel.

Les taux bruts de natalité et de mortalité sont exprimés, respectivement, sur la base du nombre annuel moyen de naissances et de décès par tranche de 1.000 habitants en milieu d'année. Ces taux sont estimatifs.

On a déterminé les superficies totales en additionnant les chiffres correspondant aux différents pays ou régions.

La densité est le nombre de personnes de la population totale de 1990 par kilomètre carré de la superficie totale.

Le schéma de régionalisation utilisé aux fins de l'établissement de ces estimations est présenté à l'Annexe I. Bien que les totaux de certains continents soient donnés et que tous puissent être déterminés, le schéma de base présente huit grandes régions qui sont établies de manière à obtenir une plus grande homogénéité en ce qui concerne l'ampleur des populations, les types de conditions démographiques et la précisions des statistiques démographiques.

Tableaux 3-4 : Les indices du Tableau 3 se rapportent à la production agricole, qui est définie comme comprenant à la fois les produits de l'agriculture et de l'élevage. Les semences et les aliments pour les animaux sont exclus de cette définition. Les indices de la production alimentaire se rapportent aux produits considérés comme comestibles et contenant des éléments nutritifs. Le café, le thé et les produits non comestibles sont exclus.

Les indices de la production agricole et de la production alimentaire présentés au Tableau 4 sont calculés selon la formule de Laspeyres avec les années 1979-1981 comme période de référence, cela afin de limiter l'incidence, sur les indices correspondant à la période considérée, des fluctuations annuelles de la production agricole enregistrée pendant les années de référence. Les chiffres de production de chaque produit sont pondérés par les prix nationaux moyens à la production pour la période 1979-81 et additionnés pour chaque année. Les indices sont fondés sur les données de production de l'année. Comme dans le passé, les séries comprennent un grand nombre d'estimations établies par la FAO lorsqu'elle n'avait pu obtenir de chiffres de sources officielles dans les pays eux-mêmes.

Index numbers for the world and regions are computed in a similar way to the country index numbers except that, instead of using different commodity prices for each country group, "international commodity prices" derived from the Gheary-Khamis formula are used for all country groupings. This method assigns a single "price" to each commodity.

The indexes in table 4 are calculated as a ratio between the index numbers of total agricultural and food production in table 3 described above and the corresponding index numbers of population.

For further information on the series presented in these tables, see the production yearbook published by FAO.[6].

For the purpose of FAO's analytical studies, in its statistics, "developed regions" includes Israel and South Africa. In the "developing regions", Egypt, Libyan Arab Jamahiriya and Sudan are included in the Near East. In comparison with annex I, the Near East includes western Asia, plus Afghanistan and Islamic Republic of Iran. The Far East includes the balance of Asia.

Table 5: The indices of industrial production are classified according to divisions, major groups or combinations of major groups of the International Standard Industrial Classification of All Economic Activities (ISIC) for mining, manufacturing and electricity, gas and water.[46] The indices indicate trends in value added in constant US dollars. The measure of value added used is the national accounts concept, which is defined as the gross value of output less the cost of materials, supplies, fuels and electricity consumed and services received.

Each series is compiled by use of the Laspeyres formula, that is, the indices are base-weighted arithmetic means. The weight base year is 1980 and value added, generally at factor values, is used in weighting. For countries using the System of National Accounts, the estimates of value added used for weighting purposes are derived, in most cases, from the results of industrial censuses or other inquiries around 1980. These census results are adjusted to ISIC where necessary and the estimates of value added in units of national currency are then converted to United States dollars.

Value added estimates for eastern Europe and former USSR are made using official data supplied by the statistical authorities or from data extracted from national publications (number of persons engaged in industrial activity, average wages and salaries, cost structure, industrial origin of net material product etc.). The value added estimates are computed separately for their major components, "compensation of employees, operating surplus and consumption of fixed capital", and adjusted to ISIC major groups. Efforts are made to ensure consistency with the international standards. Estimates in national currencies have been converted to United States dollars by using official exchange rates.

Les indices pour le monde et les régions sont calculés de la même façon que les indices par pays, mais au lieu d'appliquer des prix différents aux produits de base pour chaque groupe de pays, on a utilisé des "prix internationaux" établis d'après la formule de Gheary-Khamis pour tous les groupes de pays. Cette méthode attribue un seul "prix" à chaque produit de base.

Les indices du Tableau 4 sont calculés comme ratio entre les indices de la production alimentaire et de la production agricole totale du Tableau 3 décrits ci-dessus et les indices de population correspondants.

Pour tout renseignement complémentaire sur les séries présentées dans ces tableaux, voir l'Annuaire publié par la FAO [6].

Pour ces études analytiques, dans ses statistiques, la FAO inclut Israël et l'Afrique du Sud dans les "régions développées". Dans les "régions en développement", l'Egypte, la Jamahiriya arabe libyenne et le Soudan sont inclus dans le Proche-Orient. Dans les comparaisons avec l'Annexe I, le Proche-Orient comprend l'Asie de l'Ouest, plus l'Afghanistan et la République islamique d'Iran. L'Extrême-Orient est constitué du reste de l'Asie.

Tableau 5 : Les indices de la production industrielle sont classés par catégorie, classe ou groupement de classes de la Classification internationale type, par industrie, de toutes les branches d'activité économique (CITI) et portent sur les industries extractives et manufacturières et les industries de l'électricité, du gaz et de l'eau [46]. Ces indices mesurent les variations de la valeur ajoutée en dollars constants des Etats-Unis. La mesure de la valeur ajoutée utilisée est celle de la comptabilité nationale, qui se définit comme la valeur brute de la production moins le coût des matières premières et des fournitures, des combustibles et de l'électricité consommés ainsi que des services reçus.

Chaque série d'indices est calculée à l'aide de la formule de Laspeyres, c'est-à-dire sous forme de moyennes arithmétiques pondérées. L'année de base pour la pondération est 1980 et la valeur ajoutée utilisée dans la pondération est généralement calculée au coût des facteurs. Pour les pays utilisant le Système de comptabilité nationale, les estimations de la valeur ajoutée utilisées pour les pondérations sont tirées, le plus souvent, des résultats de recensements industriels ou d'autres enquêtes effectuées autour de 1980. Ces données sont alignées sur la CITI en cas de besoin et les estimations de la valeur ajoutée, exprimées en monnaie nationale, sont ensuite converties en dollars des Etats-Unis.

Les estimations de la valeur ajoutée pour l'Europe de l'Est et l'ancienne URSS sont faites sur la base des données officielles fournies par les services de statistiques ou de données extraites de publications nationales (nombre de personnes travaillant dans une branche d'activité, salaires et traitements moyens, structure des coûts, produit matériel net par branche d'activité, etc.). Les estimations de la valeur ajoutée sont calculées séparément par grande composante, "rémunération des employés, excédent d'exploitation et consommation de capital fixe", et ajustées aux grandes catégories de la CITI. On s'efforce d'assurer la conformité aux normes internationales. Les estimations en monnaie nationale ont été converties en dollars des Etats-Unis aux taux de change officiels.

The elementary series used in compiling indices for ISIC major groups are, in general, indices for the corresponding category, or its sub-divisions, compiled by national statistical authorities. Adjustments are made, when necessary, to align the national industrial classification with ISIC. In the case of SNA economies, most of the national indices used as elementary series indicate trends in value added at constant prices, in most instances, at factor values. In the case of eastern Europe and former USSR, the indices that serve as elementary series measure the movement in gross industrial output valued at constant enterprise prices, that is, excluding turnover taxes.

The indices for divisions, major groups or combinations of major groups of the ISIC are calculated in three main stages. First, the indices for both the developed and developing SNA economies are calculated. Secondly, these two sets of indices are combined to yield the indices for all market economies. Finally, these indices, in turn are combined with similarly derived indices for Eastern Europe and the former USSR to yield indices for the world as a whole.

Table 6: The indices of industrial employment shown are classified according to the same divisions, major group or combination of major groups, of the International Standard Industrial Classification of All Economic Activities (ISIC) [46] as the indices of industrial production in table 5. The indices are designed to measure trends in the average number of persons engaged. Estimates of numbers engaged, including working proprietors and active business partners, unpaid family workers and homeworkers as well as employees, are used as weighting coefficients. These data are obtained in most cases from the results of industrial censuses or other inquiries.

In order to make the coverage of the weights and the indices coincide as closely as possible, the figures are adjusted to ISIC and projected to the base year, wherever necessary. The weight base year is 1980. The elementary series used in compiling the indices for each division, major group, or combination of major groups, consist of national indices or absolute figures of employment. Available series of national indices are given preference where these series related to the average number of persons engaged during the year and cover the same field of industrial units as the national indices of industrial production used as indicators for the indices of industrial production (table 5). Substitute series are based on the number of persons engaged during one period of the year and average number of employees or operatives during the year. As in the case of the production indices, adjustments are sometimes required to bring the industrial classification of the national data into line with ISIC. The procedures used in the compilation of the world indices of employment are the same as those used for the production index (table 5).

Les séries élémentaires utilisées pour établir les indices pour les principales catégories de la CITI sont généralement les indices de la catégorie correspondante, ou de ses subdivisions, calculés par les services nationaux de statistiques. Le cas échéant, des ajustements sont effectués pour aligner la classification nationale par branche d'activité sur la CITI. Dans le cas des économies SCN (Système de comptabilité nationale), la plupart des indices nationaux utilisés comme séries élémentaires indiquent les tendances de la valeur ajoutée aux prix constants, dans la plupart des cas, aux coûts des facteurs. Dans le cas de l'Europe de l'Est et de l'ancienne URSS, les indices qui servent de séries élémentaires mesurent le mouvement de la production industrielle brute évaluée à prix constants par l'entreprise, c'est-à-dire abstraction faite de l'impôt sur le chiffre d'affaires.

Les indices correspondant aux catégories et classes ou groupements de classes de la CITI sont calculés en trois étapes. Premièrement, on calcule les indices pour les économies SCN développées et en développement. Deuxièmement, on combine ces deux séries d'indices pour déterminer les indices correspondant à l'ensemble des économies de marché. Enfin, on combine ces indices à des indices équivalents pour l'Europe de l'Est et l'ancienne URSS pour déterminer les indices pour l'ensemble du monde.

Tableau 6 : Les indices de l'emploi industriel indiqués sont classés selon les mêmes catégories et classes et groupements de classes de la Classification internationale type, par industrie, de toutes les branches d'activité économique (CITI) [46] comme les indices de la production industrielle du Tableau 5. Ces indices sont conçus pour mesurer les tendances du nombre moyen de personnes occupées. On utilise les estimations des nombres de personnes occupées, y compris les propriétaires et associés exerçant une activité dans l'entreprise, les travailleurs familiaux non rémunérés et les travailleurs à domicile, ainsi que les salariés, comme coefficients de pondération. Dans la plupart des cas, ces coefficients ont été tirés des résultats de recensements industriels ou d'autres enquêtes.

Afin de faire coïncider autant que possible le champ des coefficients de pondération et celui des indices, les chiffres obtenus sont alignés sur la CITI et projetés, le cas échéant, sur l'année de base. L'année de base pour la pondération est 1980. Les séries élémentaires utilisées pour le calcul des indices relatifs à chaque catégorie (ou groupements de classes principales) sont des indices nationaux ou des chiffres absolus de l'emploi. La préférence a été donnée aux séries d'indices nationaux chaque fois que ces séries se rapportaient au nombre moyen de personnes occupées au cours de l'année et couvraient les mêmes industries que les indices nationaux de la production industrielle utilisés comme indicateur pour l'établissement des indices de la production industrielle (Tableau 5). D'autres séries sont fondées sur le nombre de personnes occupées au cours d'une période de l'année et sur le nombre moyen d'employés ou d'ouvriers pour l'année. Parfois, comme pour les indices de la production, il a fallu aligner la classification industrielle des données nationales sur la CITI. La méthode utilisée pour le calcul des séries mondiales d'indices de l'emploi est la même que celle utilisée pour le calcul des indices de la production (Tableau 5).

For each division, major group, or combination of major groups, of mining, manufacturing and electricity, gas and water, indices for individual countries are multiplied by the appropriate weight. The products are summed into aggregates and indices compiled for each division for regions, sub-regions, and, in the case of SNA economies for countries in each of the developed and developing SNA economy groupings, are combined to yield aggregates and indices for market economies. These aggregates are combined with aggregates for eastern Europe and former USSR to yield aggregates and indices for each division, major group or combination of major groups for the world as a whole. The aggregates for the constituent divisions are summed to derive the corresponding indices for all mining, light, heavy and total manufacturing and total industrial activity.

Table 7: The indexes of labour productivity in industry are derived from the indexes of industrial production and industrial employment (tables 5 and 6), described above. The indexes of labour productivity for divisions, major groups, or combinations of major groups, of industrial activity for each geographical region are calculated by dividing the index of industrial production for each division, major groups, or combination of major-groups, for the region concerned by the corresponding index of industrial employment. The indexes for divisions, major groups or combinations of major groups, for the developed SNA economies and for the developing SNA economies are calculated by combining the corresponding indexes for division, major groups, or combinations of major groups, for the various regions included in each of these categories with weights proportional to the number of persons engaged in each region. The resulting indexes for division, major groups, or combinations of major groups, for the two categories are then combined to yield division indexes for the SNA economies, with weights proportional to the number of persons engaged in the divisions, major groups, or combinations of major groups, of each category.

The indexes for divisions, major groups, or combinations of major groups for the world as a whole are obtained by combining the series of indexes for the SNA economies and MPS economies, with weights proportional to the number of persons engaged in the two groupings. The indexes for the major divisions of mining and manufacturing are calculated for each region, for SNA economies, for MPS economies, and for the world as a whole by combining the appropriate industry indexes, with weights proportional to the number of persons engaged in each division, major group, or combination of major groups, for the region or grouping of countries concerned. The resulting major division indexes are combined with the index for electricity, gas and water to from the total index for each grouping with weights proportional to the number of person engaged in each major division. The weights used in these calculations are number of persons engaged in 1980.

Pour chaque catégorie et classe (ou groupements de classes) des industries extractives et manufacturières et des industries de l'électricité, du gaz et de l'eau, on a multiplié les indices relatifs à chaque pays par le coefficient de pondération approprié. Les produits ainsi obtenus sont groupés en agrégats et les indices relatifs à chaque catégorie, calculés pour chaque région, sous-région et, en ce qui concerne les pays à économie SCN, pour les pays de chacun des deux groupements des pays développés et en développement à économie SCN, sont combinés en agrégats et indices relatifs aux pays à économie de marché. Ces agrégats sont combinés aux agrégats pour l'Europe de l'Est et l'ancienne URSS pour donner des agrégats et des indices pour chaque catégorie, classe ou groupement de classes pour l'ensemble du monde. Pour calculer les indices correspondants des industries extractives, des industries légères, des industries lourdes et de l'ensemble de l'activité industrielle, on additionne les agrégats relatifs aux diverses catégories d'industries.

Tableau 7 : Les indices de la productivité du travail dans l'industrie sont établis à partir d'indices de la production industrielle et de l'emploi dans l'industrie (Tableaux 5 et 6), décrits ci-dessus. Les indices de la productivité du travail par catégorie, classe ou groupe de classes d'activité industrielle et par région géographique se calculent en divisant l'indice de la production industrielle de chaque catégorie, classe ou groupe de classes pour la région en question par l'indice correspondant de l'emploi dans l'industrie. Pour calculer les indices par catégorie, classe ou groupe de classes en ce qui concerne, d'une part, les pays développés à économie SCN et, d'autre part, les pays en développement à économie SCN, on affecte aux indices respectifs de catégories, classes ou groupes de classes relatifs aux diverses régions, pour chacune de ces catégories, des coefficients de pondération proportionnels au nombre de personnes occupées dans chaque région. On combine ensuite les indices ainsi obtenus pour chaque catégorie, classe ou groupe de classes pour les deux ensembles de pays pour obtenir les indices par catégorie pour les pays à économie SCN en affectant des pondérations proportionnelles au nombre de personnes occupées dans chaque catégorie, classe ou groupe de classes de chaque ensemble de pays.

On obtient les indices par catégorie, classe ou groupe de classes pour l'ensemble du monde en combinant les séries d'indices correspondant aux pays à économie SCN à ceux des pays à économie BPM, en affectant des pondérations proportionnelles au nombre de personnes occupées dans les deux ensembles de pays. Les indices pour les grandes catégories des industries extractives et manufacturières se calculent pour chaque région, pour les pays à économie SCN, pour les pays à économie BPM et pour l'ensemble du monde en combinant les indices industriels pertinents, affectés de coefficients proportionnels au nombre de personnes occupées dans chaque catégorie, classe ou groupe de classes pour la région ou le groupe de pays dont il s'agit. Les indices ainsi obtenus par grande catégorie sont combinés à l'indice pour l'électricité, le gaz et l'eau pour former un indice total pour chaque groupe de pays avec des pondérations proportionnelles au nombre de personnes occupées dans chaque grande catégorie. Les pondérations utilisées dans ces calculs sont le nombre de personnes occupées en 1980.

It should be noted that it has not been feasible, in all instances, to obtain consistency of scope and coverage in the indexes of industrial production and the indexes of industrial employment. Consequently, the series of indexes shown should not be interpreted as precise indicators of changes in labour productivity.

Table 8: For description of the series in table 8, see technical notes to table 106 of chapter XIV.

Table 9: For description of the series in table 9, see technical notes to chapter XVII. The composition of the regions is presented in table 117.

Table 10: The export price indexes for primary commodities and non-ferrous base metals are weighted averages of price indexes of individual commodities, each of which is based on specific price quotations for representative grades of the commodity in countries that are major traders of the commodity. The weights are determined by the patterns of primary commodities exports in the base year 1980. All indexes are calculated in terms of United States dollars. The indexes are computed using various categories of prices such as export prices f.o.b., import prices c.i.f., wholesale prices, commodity exchange prices, producer's prices and auction prices. For more details and comparable historical series see the United Nations publication *Methods Used in Compiling the United Nations Price Indexes for External Trade*, vol. 1.[48]

On notera qu'il n'a pas toujours été possible de réaliser la concordance de portée et de champ des indices de la production industrielle et de l'emploi dans l'industrie. En conséquence, on ne doit pas interpréter les séries d'indices présentées comme constituant des indicateurs précis des changements intervenus dans la productivité de la main-d'oeuvre.

Tableau 8 : On trouvera une description de la série de statistiques du Tableau 8 dans les Notes techniques du Tableau 106 du Chapitre XIV.

Tableau 9 : On trouvera une description de la série de statistiques du Tableau 9 dans les notes techniques du Chapitre XVII. La composition des régions est présentée au Tableau 117.

Tableau 10 : Les indices des prix à l'exportation pour les produits primaires et les métaux communs non ferreux sont les moyennes pondérées des indices des prix des différents produits, chaque indice étant fondé sur des cotations obtenues pour des qualités représentatives du produit dans les pays qui sont les principaux exportateurs du produit. Les coefficients de pondération sont déterminés par la structure des exportations de produits primaires de données de référence 1980. Tous les indices sont calculés en dollars des Etats-Unis, au moyen de diverses catégories de prix telles que prix à l'exportation f.o.b., prix à l'importation c.a.f., prix de gros, prix pratiqués par les bourses des produits de base, prix au producteur et prix des ventes aux enchères. Pour plus de détails et pour des séries historiques comparables, voir la publication des Nations Unies intitulée *"Méthodes utilisées par les Nations Unies pour établir les indices des prix des produits de base entrant dans le commerce international,"* vol. 1 [48].

Part Two
Population and Social Statistics

II
**Population and human settlements
(tables 11 and 12)**
III
Education and literacy (tables 13-18)
IV
**Health and child-bearing; nutrition
(tables 19-22)**
V
Culture (tables 23-28)

Part Two of the *Yearbook* presents statistical series on a wide range of population and social topics for all countries or areas of the world for which data are available. These include population and population growth, surface area and density, and urbanization; illiteracy, education at first, second and third levels, and expenditure on education; births and deaths and life expectancy; and books, newspapers, television and radio.

Deuxième partie
Population et statistiques sociales

II
**Population et établissements humains
(tableaux 11 et 12)**
III
**Education et alphabétisation (tableaux
13 à 18)**
IV
**Santé et maternité; nutrition (tableaux
19 à 22)**
V
Culture (tableaux 23 à 28)

La deuxième partie de l'*Annuaire* présente, pour tous les pays ou zones du monde pour lesquels des données sont disponibles, des séries statistiques intéressant une large gamme de questions démographiques et sociales : population et croissance démographique, superficie, densité et urbanisation; analphabétisme, enseignement des premier, second et troisième degrés, et dépenses relatives à l'enseignement; naissances, décès et espérance de vie; livres, journaux, télévision et radio.

11
Population by sex, rate of population increase, surface area and density
Population selon le sexe, taux d'accroissement de la population, superficie et densité

Country or area Pays ou zone	Date	Latest Census Dernier recensement Both sexes Les deux sexes	Male Masculin	Female Féminin	Midyear estimates (thousands) Estimations au milieu de l'année (milliers) 1985	1990	Type[1] 1990	Annual rate of increase Taux d'accrois- sement annuel % 1985-90	Surface area (km²) Superficie (km²) 1990	Density Densité 1990[2]
Africa · Afrique										
Algeria[3] Algérie[3]	1-IV-87	*22 971 000	...	...	21 850	x24 961	A3 c1	2.7	2 381 741	10
Angola[4] Angola[4]	15-XII-70	5 646 166	2 943 974	2 702 192	x8 754	*10 020	A20c1	2.7	1 246 700	8
Benin Bénin	20-III-79	3 331 210	1 596 939	1 734 271	4 041	* 736	A11c3	3.2	112 622	42
Botswana Botswana	12-VIII-81	941 027	443 104	497 923	1 081	* 291	A9 c1	3.6	581 730	2
British Indian Territory[5] Territoire britannique de l'océan indien[5]	([6])	([6])	([6])	([6])	x3	x3	D25d	0.0	78	32
Burkina Faso Burkina Faso	10-XII-85	7 964 705	3 833 237	4 131 468	7 886	*9 001	A5 c3	2.6	274 200	33
Burundi Burundi	16-VIII-79	4 114 135	1 988 292	2 125 843	4 718	*5 458	A11c3	2.9	27 834	196
Cameroon Cameroun	IV-87	*10 493 655	...	...	10 166	x11 834	A14c3	3.0	475 442	25
Cape Verde Cap-Vert	2-VI-80	289 027	131 266	157 761	333	x370	A10c1	2.1	4 033	92
Central African Republic République centrafricaine	8-XII-75	2 054 610	985 224	1 069 386	2 608	x3 039	A15c3	3.1	622 984	5
Chad Tchad	1963-1964	3 254 000[7]	...	...	x5 019	x5 679	B27c3	2.5	1 284 000	4
Comoros Comores	15-IX-80	385 890[8]	167 089	168 061	x463	x551	A10c3	3.5	2 235	246
Congo Congo	22-XII-84	1 843 421	...	...	x1 940	x2 271	A63	3.2	342 000	7
Côte d'Ivoire Côte d'Ivoire	30-IV-75	6 709 600	3 474 750	3 234 850	x9 934	x11 998	A15c3	3.8	322 463	37
Djibouti Djibouti	1960-1961	81 200	...	...	430	x409[9]	A30d	([10])	23 200	18
Egypt Egypte	18-IX-86	47 995 265	24 512 701	23 482 564	48 349	*53 153	A4 c1	1.9	1 001 449	53
Equatorial Guinea[11] Guinée équatoriale[11]	4-VII-83	300 000	144 268	155 732	x312	*348	A7 c3	2.2	28 051	12
Ethiopia Ethiopie	9-V-84	*42 169 203	*21 018 900	21 150 303		x49 241[9]	A6 c3	([10])	1 221 900	40
Gabon Gabon	1960-1961	448 564	211 350	237 214	x986	x1 172	A30c3	3.5	267 667	4
Gambia Gambie	15-IV-83	687 817	342 134	345 683	x745	x861	A7 c1	2.9	11 295	76
Ghana Ghana	11-III-84	12 296 081	6 063 848	6 232 233	12 717	x15 028	A6 c1	3.3	238 533	63
Guinea[12] Guinée[12]	4-II-83	*4 533 240	...	...	4 661	x5 756	A7 c3	4.2	245 857	23
Guinea-Bissau Guinée-Bissau	16-IV-79	753 313	362 589	390 724	869	x965	A11c1	2.1	36 125	27
Kenya Kenya	24-VIII-89	*21 400 000	...	...	20 333	x24 872[9]	A11c2	([10])	580 367	43
Lesotho Lesotho	12-IV-86	*1 447 000	...	...	1 528	x1 774	A14c3	3.0	30 355	58
Liberia Libéria	1-II-84	*2 101 628	...	...	2 189	*2 607	A6 c3	3.5	111 369	23
Libyan Arab Jamahiriya Jamahiriya arabe libyenne	31-VII-84	*3 637 488	*1 950 152	*1 687 336	3 363	x4 545[9]	A6 c3	([10])	1 759 540	3
Madagascar Madagascar	1-I-75	7 603 790	3 805 288	3 798 502	9 985	*11 197	A16c3	2.3	587 041	19

11
Population by sex, rate of population increase, surface area and density [*cont.*]
Population selon le sexe, taux d'accroissement de la population, superficie et densité [*suite*]

Country or area Pays ou zone	Date	Latest Census Dernier recensement Both Sexes Les deux sexes	Male Masculin	Female Féminin	Mid-year estimates (thousands) Estimations au milieu de l'année (milliers) 1985	1990	Type[1] 1990	Annual rate of increase Taux d'accrois-sement annual % 1985-90	Surface area (km²) Superficie (km²) 1990	Density Densité 1990[2]
Malawi Malawi	1-IX-87	*7 982 607	3 880 100	4 102 507	7 059	8 289*	A3 c3	3.2	118 484	70
Mali Mali	1-IV-87	7 696 348	3 760 711	3 935 637	8 206[9]	8 156*	A3 c3	([10])	1 240 192	7
Mauritania Mauritanie	1-I-77	1 338 830[13]	658 361[13]	680 469[13]	1 767x	2 025x	A13c3	2.7	1 025 520	2
Mauritius Maurice	30-VI-72	851 334	426 122	425 212	1 020	1 082*	A18b1	1.2	2 040	531
Island of Mauritius Ile Maurice	11-VII-83	966 863	481 368	485 495	985	1 037*	A7 b1	1.0	1 865	556
Rodrigues Rodrigues	11-VII-83	33 082	16 552	16 530	35	38*	A7 b1	1.5	104	364
Others[14] Autres[14]	30-VI-72	366	272	94	...	...	..	...	71	...
Morocco Maroc	3-IX-82	20 449 551	10 236 078	10 213 473	x22 026	x25 061	A8 c2	2.6	446 550	56
Mozambique[12] Mozambique[12]	1-VIII-80	11 673 725	5 670 484	6 003 241	13 810	x15 656	A10c1	2.5	801 590	20
Namibia[12] Nambie[12]	6-V-70	762 184	...	...	x1 519	x1 781	A20c3	3.2	824 292	2
Niger Niger	10-V-88	*7 249 596	...	...	x6 609	x7 732	A2 c3	3.1	1 267 000	6
Nigeria[15] Nigéria[15]	5-XI-63	55 670 055[16]	28 111 852[16]	27 558 203[16]	95 690	x108 542	A27c2	2.5	923 768	117
Reunion[3] Réunion[3]	9-III-82	515 798	252 997	262 801	546	x599	A8 b3	1.8	2 510	239
Rwanda Rwanda	16-VIII-78	4 800 433	2 331 147	2 469 286	x6 103	*7 181	A12c3	3.3	26 338	273
St. Helena ex. dep. Sainte-Hélène sans dép.	22-II-87	5 644	2 769	2 875	6	x7	A3 b3	2.2	122	...
Ascension Ascension	31-XII-78	849	608	241	...	...	..	...	88	...
Tristan da Cunha Tristan da Cunha	31-XII-88	297	...	...	...	...	..	...	104	...
Sao Tome and Principe Sao Tomé-et-Principe	15-VIII-81	96 611	48 031	48 580	108	x121	A9 c1	2.3	964	126
Senegal Sénégal	27-V-88	*6 928 405	...	...	6 547	x7 327	A2 c3	2.2	196 722	37
Seychelles Seychelles	17-VIII-87	68 598	...	...	65	*67	A13b2	0.6	280	241
Sierra Leone[12] Sierra Leone[12]	15-XII-85	*3 515 812	*1 746 055	*1 769 757	x3 665	x4 151	A5 c1	2.5	71 740	58
Somalia Somalie	1986-1987	*7 114 431	*3 741 664	*3 372 767	x6 371	*7 497[9]	A15c3	([10])	637 657	12
South Africa[12] Afrique du Sud[12]	5-III-85	23 385 645[17]	11 545 282[17]	11 840 363[17]	x31 569	x35 282	A5 c1	2.2	1 221 037	29
Sudan Soudan	1-II-83	20 594 197	10 512 884	10 081 313	x21 822	x25 204	A7 c3	2.9	2 505 813	10
Swaziland Swaziland	25-VIII-86	681 059	321 579	359 480	638	*768	A4 c1	3.7	17 364	44
Togo Togo	22-XI-81	2 703 250	...	...	x3 029	x3 531	A9 c1	3.1	56 785	62
Tunisia Tunisie	30-III-84	6 966 173	3 547 315	3 418 858	7 261	x8 180	A6 c1	2.4	163 610	50
Uganda Ouganda	18-I-80	12 636 179	6 259 837	6 376 342	x15 647	x18 795	A10c1	3.7	235 880	80
United Rep. of Tanzania Rép.-Unie de Tanzanie	26-VIII-78	17 512 611	8 587 086	8 925 525	21 733	*25 635	A12c3	3.3	945 087	27
Tanganyika Tanganyika	26-VIII-78	17 036 498	8 350 492	8 686 006	21 162	*24 972	A12c3	3.3	942 626	26

11
Population by sex, rate of population increase, surface area and density [*cont.*]
Population selon le sexe, taux d'accroissement de la population, superficie et densité [*suite*]

Country or area Pays ou zone	Date	Latest Census Dernier recensement Both Sexes Les deux sexes	Male Masculin	Female Féminin	Mid-year estimates (thousands) Estimations au milieu de l'année (milliers) 1985	1990	Type[1] 1990	Annual rate of increase Taux d'accrois- sement annual % 1985-90	Surface area (km²) Superficie (km²) 1990	Density Densité 1990[2]
Zanzibar Zanzibar	26-VIII-78	476 113	236 594	239 519	571	*663	A12c3	3.0	2 461	269
Western Sahara[18] Sahara occidental[18]	31-XII-70	76 425	43 981	32 444	x156	x179	A20c1	2.7	266 000	1
Zaire Zaïre	1-VII-84	29 441 169	14 632 261	14 808 908	30 981	*35 562	A6 c3	2.8	2 345 409	15
Zambia Zambie	20-VIII-90	*7 818 447	...	...	6 725	*8 073	A10c1	3.7	752 614	11
Zimbabwe Zimbabwe	18-VIII-82	*7 546 071	...	...	8 379	*9 369	A8 c1	2.2	390 580	24
America, North · Amérique du Nord										
Anguilla Anguilla	..	...	...	...	x7	x8	..	1.3	96	81
Antigua and Barbuda Antigua-et-Barbuda	7-IV-70	65 525	31 054	34 471	76	x77	A20b1	0.3	440	175
Aruba[3] Aruba[3]	1-II-81	60 312	29 340	30 972	61	x60	A9 b1	-0.4	193	311
Bahamas Bahamas	2-V-90	254 685	123 507	131 178	232	*253	A10b1	1.8	13 878	18
Barbados Barbade	2-V-90	257 082	...	...	253	x255	A10b1	0.2	430	594
Belize Belize	12-V-80	145 353	...	...	166	x188	A10c1	2.4	22 965	8
Bermuda Bermudes	12-V-80	67 761	33 621	34 140	56[19]	61[19]	A10b1	1.6	53	1 144
British Virgin Islands Iles Vierges britanniques	12-V-80	11 697	...	...	12	x13	A10b1	2.2	153	87
Canada[3] Canada[3]	3-VI-86	*25 309 330	*12 485 650	12 823 680	25 165	x26 522	A4 b1	1.0	9 976 139	3
Cayman Islands Iles Caïmanes	15-X-89	25 355	12 372	12 983	21	*27	A11c1	5.0	259	105
Costa Rica[3] Costa Rica[3]	10-VI-84	2 416 809	1 208 216	1 208 593	2 489	*2 994	A6 b2	3.7	51 100	59
Cuba Cuba	11-IX-81	9 723 605	4 914 873	4 808 732	10 098	x10 609	A9 b1	1.0	110 861	96
Dominica Dominique	7-IV-81	74 625	...	...	80	x83	A9 b1	0.7	751	110
Dominican Republic Rép. dominicaine	12-XII-81	5 647 977	2 832 454	2 815 523	6 416	x7 170	A9 c1	2.2	48 734	147
El Salvador El Salvador	28-VI-71	3 554 648	1 763 190	1 791 458	4 916	x5 252	A19b1	1.3	21 041	250
Greenland[3] Groenland[3]	26-X-76	49 630	26 856	22 774	53	x57	A14a1	1.2	2 175 600	-
Grenada[20] Grenade[20]	30-IV-81	89 088	...	...	91	x85[9]	A9 b1	(¹⁰)	344	248
Guadeloupe[3][21] Guadeloupe[3][21]	9-III-82	327 002	160 112	166 890	333	x344	A8 b1	0.6	1 705	201
Guatemala[12] Guatemala[12]	26-III-81	6 054 227	3 015 826	3 038 401	7 963	*9 197	A9 b2	2.9	108 889	84
Haiti[3] Haïti[3]	30-VIII-82	5 053 792	2 448 370	2 605 422	5 285	x5 963	A8 c3	2.4	27 750	215
Honduras Honduras	6-III-74	2 656 948[3]	1 317 307[3]	1 339 641[3]	4 372	*5 105	A16c1	3.1	112 088	46
Jamaica Jamaïque	8-VI-82	2 205 507	1 079 640	1 125 867	2 311	*2 420	A8 b1	0.9	10 990	220
Martinique[3] Martinique[3]	9-III-82	326 717	158 415	168 302	331	x341	A8 b1	0.6	1 102	310
Mexico[3] Mexique[3]	12-III-90	*81 140 922	*39 878 536	41 262 386	77 938	*86 154	A10c1	2.0	1 958 201	44
Montserrat Montserrat	12-V-80	11 932	...	...	12	x13	A10b1	1.2	102	124

11
Population by sex, rate of population increase, surface area and density [*cont.*]
Population selon le sexe, taux d'accroissement de la population, superficie et densité [*suite*]

Country or area Pays ou zone	Date	Latest Census Dernier recensement Both Sexes Les deux sexes	Male Masculin	Female Féminin	Mid-year estimates (thousands) Estimations au milieu de l'année (milliers) 1985	1990	Type[1] 1990	Annual rate of increase Taux d'accroissement annual % 1985-90	Surface area (km²) Superficie (km²) 1990	Density Densité 1990[2]
Netherlands Antilles[3 12 22] Antilles néerlandaises[3 12 22]	1-II-81	171 620	82 808	88 812	x182	x189	A9 c1	0.7	800	236
Nicaragua[3] Nicaragua[3]	20-IV-71	1 877 952	921 543	956 409	x3 273	*3 871	A19b3	3.4	130 000	30
Panama[12] Panama[12]	11-V-80	1 831 399	928 285	903 114	2 180	*2 418	A10c1	2.1	77 082	31
Puerto Rico[23] Porto Rico[23]	1-IV-80	3 196 520	1 556 727	1 639 793	3 283	*3 599	A10b1	1.8	8 897	405
Saint Kitts and Nevis Saint-Kitts-et-Nevis	12-V-80	44 224	...	...	44	x44	A10b1	0.1	261	170
Saint Lucia[12] Sainte-Lucie[12]	12-V-80	115 153	...	...	137	x151	A10b1	1.9	622	242
St. Pierre and Miquelon Saint-Pierre et Miquelon	9-III-82	6 037	2 981	3 056	x7	x7	A8 d	0.0	242	27
St. Vincent and Grenadines[24] St.-Vincent-et-Grenadines[24]	12-V-80	97 914	...	...	109	x116	A10b1	1.2	388	299
Trinidad and Tobago Trinité-et-Tobago	2-V-90	*1 234 388	...	...	1 178	*1 227	A10b1	0.8	5 130	239
Turks and Caicos Islands Iles Turques et Caïques	12-V-80	7 435	3 602	3 833	x9	x10	A10d	2.5	430	24
United States[25] Etats-Unis[25]	1-V-90	248 709 873*	...	...	238 466	*249 975	A0 b1	0.9	9 372 614	27
United States Virgin[23] Iles Vierges américaines[23]	1-IV-80	96 569	46 204	50 365	109	x117	A10c1	1.4	342	342
America, South · Amérique du Sud										
Argentina[12] Argentine[12]	22-X-80	27 947 446	13 755 983	14 191 463	30 331	*32 322	A10c1	1.3	2 766 889	12
Bolivia[12] Bolivie[12]	29-IX-76	4 613 486	2 276 029	2 337 457	6 429	*7 400	A14c3	2.8	1 098 581	7
Brazil[26] Brésil[26]	1-IX-80	121 148 582	60 298 897	60 849 685	135 564	*150 368	A10c1	2.1	8 511 965	18
Chile Chili	21-IV-82	11 329 736	5 553 409	5 776 327	12 122	*13 173	A8 b1	1.7	756 945	17
Colombia[27] Colombie[27]	15-X-85	27 837 932	13 777 700	14 060 232	28 624	*32 987	A5 b3	2.8	1 138 914	29
Ecuador[28] Equateur[28]	25-X-90	*9 622 608	*4 788 118	*4 834 490	9 378	*10 782	A8 b3	2.8	283 561	38
Falkland Is. (Malvinas)[29 30] Iles Falkland (Malvinas)[29 30]	16-XI-86	1 878	994	884	x2	x2	A4 d	0.0	12 173	-
French Guyana[3] Guyane française[3]	9-III-82	73 012	38 448	34 564	83	x99	A8 c1	3.5	90 000	1
Guyana Guyana	12-V-80	758 619	375 841	382 778	790	x796	A10b1	0.2	214 969	4
Paraguay Paraguay	11-VII-82	3 029 830	1 521 409	1 508 421	3 693	*4 277	A8 c2	2.9	406 752	11
Peru[12 26] Pérou[12 26]	12-VII-81	17 005 210	8 489 867	8 515 343	19 418	*22 332	A9 c2	2.8	1 285 216	17
Suriname Suriname	1-VII-80	352 041	173 083	178 958	x383	x422	A10c2	1.9	163 265	3
Uruguay[12] Uruguay[12]	23-X-85	2 955 241	1 439 021	1 516 220	3 008	*3 094	A5 b3	0.6	177 414	17
Venezuela[26] Venezuela[26]	20-X-90	*18 105 265	*9 004 717	*9 100 548	17 317	*19 735	A9 c1	2.6	912 050	22
Asia · Asie										
Afghanistan Afghanistan	23-VI-79	13 051 358[31]	6 712 377[31]	6 338 981[31]	18 136	*1 612[31]	A11c3	([10])	652 090	25
Armenia Arménie	12-I-89	3 304 776	1 619 308	1 685 468	3 339	3 324	A1b1	-0.1	29 800	112
Azerbaijan Azerbaïdjan	12-I-89	7 021 178	3 423 793	3 597 385	6 661	7 153	A1b1	1.4	86 600	83

11
Population by sex, rate of population increase, surface area and density [*cont.*]
Population selon le sexe, taux d'accroissement de la population, superficie et densité [*suite*]

Country or area Pays ou zone	Date	Latest Census Dernier recensement Both Sexes Les deux sexes	Male Masculin	Female Féminin	Mid-year estimates (thousands) Estimations au milieu de l'année (milliers) 1985	1990	Type[1] 1990	Annual rate of increase Taux d'accrois- sement annual % 1985-90	Surface area (km²) Superficie (km²) 1990	Density Densité 1990[2]
Bahrain Bahreïn	5-IV-81	350 798	204 793	146 005	425	*503	A9 c1	3.4	678	742
Bangladesh[12] Bangladesh[12]	6-III-81	87 119 965	44 919 191	42 200 774	99 434	x115 594	A9 c1	3.0	143 998	803
Bhutan Bhoutan	XI-XII-69	1 034 774	...	...	x1 363	x1 517	A21c3	2.1	47 000	32
Brunei Darussalam[12 32] Brunéi Darussalam[12 32]	26-VIII-81	192 832	102 942	89 890	222	x266	A9 c2	3.6	5 765	46
Cambodia[33] Cambodge[33]	17-IV-62	5 728 771	2 862 939	2 865 832	x7 285	x8 246	A28c3	2.5	181 035	46
China[34] Chine[34]	1-VII-90	116 0017381	...	...	x1059520	x1139060	A8 c3	1.4	9 596 961	12
Cyprus[3] Chypre[3]	30-IX-76	612 851	306 144	306 707	665	702	A14b2	1.1	9 251	76
East Timor Timor oriental	31-X-90	747 750	386 939	360 811	x659	x737	A10c1	2.2	14 874	50
Georgia Géorgie	12-I-89	5 400 841	2 562 040	2 838 801	5 218	5 464	A1b1	0.9	69 700	78
Hong Kong[35] Hong-kong[35]	11-III-86	5 395 997	2 772 464	2 623 533	5 456	*5 801	A9 b2	1.2	1 045[36]	5 551
India[37] Inde[37]	1-III-91	*843 930 861	...	...	750 859	*827 057	A9 c1	1.9	3 287 590	252
Indonesia[38] Indonésie[38]	31-X-90	179 321 641	89 448 235	89 873 406	164 630	*179 300	A10c1	1.7	1 904 569	94
Iran, Islamic Republic of Iran, Rép. islamique d'	22-IX-86	49 445 010	25 280 961	24 164 049	47 820	x54 608	A4 c1	2.7	1 648 000	33
Iraq Iraq	17-X-87	16 335 199	8 395 889	7 939 310	15 585	x18 920	A3 c1	3.9	438 317	43
Israel[3 39] Israël[3 39]	4-VI-83	4 037 620	2 011 590	2 026 030	4 233	*4 659	A7 b1	1.9	20 770	224
Japan[40] Japon[40]	1-X-85	121 048 923	59 497 316	61 551 607	120 837	*123 537	A5 b1	0.4	377 801	327
Jordan[41] Jordanie[41]	10-XI-79	2 100 019[42]	1 086 591[42]	1 013 428[42]	3 407	x4 010	A11b3	3.3	97 740	41
Kazakhstan Kazakhstan	12-I-89	16 464 464	7 974 004	8 490 460	15 935	16 742	A1b1	1.0	2 717 300	6
Korea[43] Corée[43]	1-V-44	25 120 174[3]	12 521 173[3]	12 599 001[3]	x60 694	x64 566	..	1.2	220 277	293
Korea, Dem. People's Rep. Corée, Rép. pop. dém. de	1-V-44	...	...	...	x19 888	x21 773	D27c3	1.8	120 538	181
Korea, Republic of[12 44] Corée, Rép. de[12 44]	1-XI-85	40 448 486	20 243 765	20 204 721	40 806	*42 793	A5 c1	1.0	99 016	432
Kuwait Koweït	20-IV-85	1 697 301	965 297	732 004	1 712	*2 143	A5 c1	4.5	17 818	120
Kyrgyzstan Kirghizistan	12-I-89	4 257 755	2 077 623	2 180 132	4 009	4 394	A1b1	1.8	198 500	22
Lao People's Dem. Rep. Rép. dém. populaire Lao	1-III-85	*3 584 803	*1 757 115	*1 827 688	x3 595	x4 139	A5 c3	2.8	236 800	17
Lebanon[45] Liban[45]	15-XI-70	2 126 325[46]	1 080 015[46]	1 046 310[46]	x2 668	x2 701	B20c3	0.2	10 400	260
Macau[47] Macao[47]	16-III-81	247 826	127 650	120 176	392	x479	A9 c1	4.0	16	29 962
Malaysia Malaisie	10-VI-80	13 136 109	6 588 756	6 547 353	15 681	*17 861	A10c2	2.6	329 749	54
Maldives Maldives	25-III-85	180 088	93 482	86 606	180	x215	A5 c1	3.6	298	722
Mongolia Mongolie	5-I-89	2 043 400	...	...	1 878	x2 190	A1 c1	3.1	1 566 500	1
Myanmar Myanmar	31-III-83	35 307 913[3]	17 518 255[3]	17 789 658[3]	38 541	x41 675	A7 c2	1.6	676 578	62

11
Population by sex, rate of population increase, surface area and density [*cont.*]
Population selon le sexe, taux d'accroissement de la population, superficie et densité [*suite*]

Country or area Pays ou zone	Date	Latest Census Dernier recensement Both Sexes Les deux sexes	Male Masculin	Female Féminin	Mid-year estimates (thousands) Estimations au milieu de l'année (milliers) 1985	1990	Type[1] 1990	Annual rate of increase Taux d'accrois- sement annual % 1985-90	Surface area (km[2]) Superficie (km[2]) 1990	Density Densité (km[2]) 1990[2]
Nepal[3] Népal[3]	22-VI-81	15 022 839	7 695 336	7 327 503	16 687	*18 916	A9 c1	2.5	140 797	134
Oman Oman	..	...	...	...	2 000	*1 502[9]	(..)	([10])	212 457	7
Pakistan[48] Pakistan[48]	1-III-81	84 253 644	44 232 677	40 020 967	96 180	*112 049	A9 c1	3.1	796 095	141
Palestine[49] Palestine[49]	18-XI-31	1 035 821	524 268[50]	509 028[50]	...	...	...	..	...	...
Gaza Strip[51] Zone de Gaza[51]	14-IX-67	356 261	172 511	183 750	...	...	...	...	378	...
Philippines[3] Philippines[3]	1-V-90	60 684 887	...	...	54 668	*61 480	A10c2	2.3	300 000	205
Qatar Qatar	16-III-86	*369 079	*247 852	*121 227	x299	x368	A4 c3	4.2	11 000	33
Saudi Arabia Arabie saoudite	9-IX-74	*7 012 642	...	...	x11 596	*14 870[9]	A16c3	([10])	2 149 690	7
Singapore[52] Singapour[52]	30-VI-90	*3 002 800	...	...	2 558[9]	*3 003	A0 b2	([10])	618	4 859
Sri Lanka Sri Lanka	17-III-81	14 846 750	7 568 253	7 278 497	15 842	*16 993	A9 c1	1.4	65 610	259
Syrian Arab Republic[53] République arabe syrienne[53]	7-IX-81	9 046 144	4 621 852	4 424 292	10 267	*12 116	A9 c1	3.3	185 180	65
Tajikistan Tadjikistan	12-I-89	5 092 603	2 530 245	2 562 358	4 574	5 303	A1b1	3.0	143 100	37
Thailand Thaïlande	1-IV-90	*54 532 000	*27 031 000	27 501 000	51 683	*57 196	A10c1	2.0	513 115	111
Turkey Turquie	20-X-85	50 664 458	25 671 975	24 992 483	50 231	*58 687	A5 c1	3.1	779 452	75
Turkmenistan Turkménistan	12-I-89	3 522 717	1 735 179	1 787 538	3 230	3 670	A1b1	2.6	488 100	8
United Arab Emirates[54] Emirats arabes unis[54]	15-XII-80	1 043 225	720 360	322 865	x1 350	x1 589	A10c3	3.3	83 600	19
Uzbekistan Ouzbékistan	12-I-89	19 810 077	9 784 156	10 025 921	18 231	20 531	A1b1	2.4	447 400	46
Viet Nam Viet Nam	1-IV-89	*64 411 713	*31 336 568	33 075 145	x59 902	*66 200	A1 c3	2.0	331 689	200
Yemen Yémen	...	...	...	...	...	*11 282	..	...	...	...
former Dem. Yemen ancienne Yémen dém.	29-III-88	2 345 266	1 184 359	1 160 907	2 164	*2 460	A2 c3	2.6	332 968	7
former Yemen Arab Rep.[3] anc. Yémen rép. arabe[3]	1-II-86	9 274 173	4 647 310	4 626 863	x7 621	x9 196	A15c3	3.8	195 000	47
Europe · Europe										
Albania Albanie	12-IV-89	*3 182 400	1 638 900	1 543 500	2 957	*3 250	A1 b1	1.9	28 748	113
Andorra Andorre	XI-54	5 664	...	...	45	52	A36c3	2.9	453	114
Austria[3] Autriche[3]	12-V-81	7 555 338	3 572 426	3 982 912	7 555	*7 712	A9 b1	0.4	83 853	92
Belarus Bélarus	12-I-89	10 151 806	4 749 324	5 402 482	9 999	10 278	A1b1	0.6	207 600	20
Belgium[3] Belgique[3]	1-III-81	9 848 647	4 810 349	5 038 298	9 858	x9 845	A9 b1	-0.0	30 519	323
Bulgaria Bulgarie	4-XII-85	8 948 388	4 430 061	4 518 327	8 960	x9 011	A5 b1	0.1	110 912[55]	81
Channel Islands Iles Anglo-Normandes	23-III-86	135 694	65 610	70 084	133	x139	A4 b1	0.8	195	710
Guernsey[56] Guernesey[56]	23-III-86	55 482	26 859	28 623	53	60	A4 b1	2.2	78	763
Jersey Jersey	12-III-89	82 809	40 086	42 723	79	...	..	...	116	...

11
Population by sex, rate of population increase, surface area and density [*cont.*]
Population selon le sexe, taux d'accroissement de la population, superficie et densité [*suite*]

Country or area Pays ou zone	Date	Latest Census Dernier recensement Both Sexes Les deux sexes	Male Masculin	Female Féminin	Mid-year estimates (thousands) Estimations au milieu de l'année (milliers) 1985	1990	Type[1] 1990	Annual rate of increase Taux d'accroissement annual % 1985-90	Surface area (km²) Superficie (km²) 1990	Density Densité 1990[2]
Czechoslovakia Tchécoslovaquie	1-XI-80	15 283 095	7 441 160	7 841 935	15 499	*15 662	A10b1	0.2	127 876	122
Denmark[3][57] Danemark[3][57]	1-I-81	5 123 989	2 528 225	2 595 764	5 114	5 140	A9 a1	0.1	43 077	119
Estonia Estonie	12-I-89	1 565 662	731 392	834 270	1 519	1 571	A1b1	0.7	45 100	35
Faeroe Islands[3] Iles Féroe[3]	22-IX-77	41 969	21 997	19 972	46	x48	A13b1	0.8	1 399	34
Finland[3] Finlande[3]	17-XI-85	4 910 619	2 377 978	2 532 641	4 902	*4 986	A5 b1	0.3	338 127	15
France[58][59] France[58][59]	5-III-90	*56 556 000[60]	...	...	55 170	*56 440	A8 b1	0.5	551 500	102
Germany ſ · Allemagne ſ Federal Rep. of Germany[3] Rép. fédérale d'Allemagne[3]	25-V-87	61 077 042	29 322 923	31 754 119	60 975	*63 232	A3 b1	0.7	248 577	254
former German Dem Rep.[3] ancienne R. d. allemande[3]	31-XII-81	16 705 635	7 849 112	8 856 523	16 644	*16 247	A9 b2	-0.5	108 333	150
Gibraltar[61] Gibraltar[61]	9-XI-81	29 616	14 992	14 624	29	x31	A9 b1	1.3	6	5 083
Greece Grèce	17-III-91	*10 269 074[62]	...	...	9 934[62]	x10 048[62]	A9 b2	0.2	131 990	76
Holy See Saint-Siège	30-IV-48	890	548	342	1	x1	D2 d	6.0	0[63]	...
Hungary Hongrie	1-I-80	10 709 463	5 188 709	5 520 754	10 649	*10 553	A10b1	-0.2	93 032	113
Iceland[3] Islande[3]	1-XII-70	204 930	103 621	101 309	241	*255	A20a1	1.1	103 000	2
Ireland Irlande	13-IV-86	3 540 643	1 769 690	1 770 953	3 540	*3 503	A4 b2	-0.2	70 284	50
Isle of Man Ile de Man	6-IV-86	62 096	29 696	32 400	63	x64	A4 b1	0.2	588	108
Italy Italie	25-X-81	56 556 911	27 506 354	29 050 557	57 141[3]	*57 062[3]	A9 b1	-0.0	301 268	189
Latvia Lettonie	12-I-89	2 666 567	1 238 806	1 427 761	2 594	2 683	A1b1	0.7	64 500	42
Liechtenstein Liechtenstein	2-XII-80	25 215	...	...	27	x29	A10b1	1.2	160	178
Lithuania Lithuanie	12-I-89	3 674 802	1 783 953	1 935 849	3 545	3 722	A1b1	1.0	65 200	57
Luxembourg[3] Luxembourg[3]	31-III-81	364 602	177 869	186 733	367	x373	A9 b2	0.3	2 586	144
Malta[64] Malte[64]	16-XI-85	345 418	169 832	175 586	336	*354	A5 b2	1.0	316	1 120
Monaco[3] Monaco[3]	4-III-82	27 063	12 598	14 465	x28	x29	A8 c1	0.7	1[65]	28500
Netherlands[3] Pays-Bas[3]	28-II-71	13 060 115	...	...	14 484	*14 943	A19a1	0.6	40 844	366
Norway[3] Norvège[3]	1-XI-80	4 091 132	2 027 083	2 064 049	4 153	*4 242	A10a1	0.4	323 895	13
Poland[66] Pologne[66]	6-XII-88	37 878 641	...	...	37 203	*38 180	A12b1	0.5	312 677	122
Portugal[67] Portugal[67]	16-III-81	9 833 014	4 737 715	5 095 299	10 157	*10 525	A9 b1	0.7	92 389	114
Republic of Moldova Moldova, Rép. de	12-I-89	4 335 360	2 063 192	2 272 168	4 214	4 364	A1b1	0.7	33 700	129
Romania Roumanie	5-I-77	21 559 910	10 626 055	10 933 855	22 725	*23 200	A13b2	0.4	237 500	98
Russian Federation Fédération Russe	12-I-89	147 021 869	68 713 869	78 308 000	143 585	148 263	A1b1	0.6	17075400	9
San Marino Saint-Marin	30-XI-76	19 149	9 654	9 495	22	x24	A14a2	1.0	61	385

11
Population by sex, rate of population increase, surface area and density [*cont.*]
Population selon le sexe, taux d'accroissement de la population, superficie et densité [*suite*]

Country or area Pays ou zone	Date	Latest Census Dernier recensement Both Sexes Les deux sexes	Male Masculin	Female Féminin	Mid-year estimates (thousands) Estimations au milieu de l'année (milliers) 1985	1990	Type[1] 1990	Annual rate of increase Taux d'accrois- sement annual % 1985-90	Surface area (km²) Superficie (km²) 1990	Density Densité 1990[2]
Spain[68] Espagne[68]	1-III-81	37 746 260	18 529 764	19 216 496	38 474	*38 959	A9 c1	0.3	504 782	77
Svalbard and Jan Mayen Isl.[69] Savalbard et Ile Jan-Mayen[69]	1-XI-60	3 431	2 545	886	...	...	..	...	62 422	...
Sweden[3] Suède[3]	1-XI-85	8 360 178	4 128 367	4 231 811	8 350	*8 559	A5 a1	0.5	449 964	19
Switzerland[3] Suisse[3]	2-XII-80	6 365 960	3 114 812	3 251 148	6 470	*6 712	A10b1	0.7	41 293	163
Ukraine Ukraine	12-I-89	51 704 000	23 959 000	27 745 000	50 917	51 872	A1b1	0.4	603 700	86
United Kingdom[70] Royaume-Uni[70]	5-IV-81	55 678 079	27 030 383	28 647 696	56 618	x57 237	A9 b1	0.2	244 100	234
Yugoslavia[3] Yougoslavie[3]	31-III-81	22 424 687	11 083 768	11 340 919	23 124	*23 809	A9 b1	0.6	255 804	93
Oceania · Océanie										
American Samoa[23] Samoa américaines[23]	1-IV-80	32 297	16 384	15 913	36	*39	A10b1	1.8	199	196
Australia[12] Australie[12]	30-VI-86	15 602 156	7 768 313	7 833 843	15 788	*17 086	A4 b1	1.6	7 713 364	2
Canton and Enderbury Isl.[71] Iles Canton et Enderbury[71]	1-IV-70	-[72]	-	-	...	...	..	...	70	...
Christmas Island Iles Christmas	30-VI-81	2 871	1 918	953	2	...	..	...	135	...
Cocos (Keeling) Islands Iles des Cocos (Keeling)	30-VI-81	555	298	257	1	...	..	...	14	...
Cook Islands[73] Iles Cook[73]	1-XII-86	17 614	...	...	18	x8	A4 b1	0.6	236	77
Fiji Fidji	31-VIII-86	715 375	362 568	352 807	697	x765	A4 b1	1.9	18 274	42
French Polynesia[74] Polynésie française[74]	6-IX-88	188 814	...	...	174	x206	A7 c1	3.4	4 000	52
Guam[23] Guam[23]	1-IV-80	105 979	55 321	50 658	x114	x119	A10b1	0.9	541	220
Johnston Island Ile Johnston	1-IV-70	1 007	...	...	...	...	..	...	1	...
Kiribati[75] Kiribati[75]	9-V-85	63 883	...	...	x63	x66	A5 c1	1.1	726	91
Marshall Islands Iles Marshall	..	...	...	...	x35	x40	..	2.3	...	...
Micronesia, Federated States of Micronésie, Etats fédérés de	..	...	...	...	x88	x99	..	2.3	...	...
Midway Is. Iles Midway	1-IV-70	2 220	...	...	...	...	..	...	5	...
Nauru Nauru	22-I-77	7 254	...	...	x9	x10	A13d	1.5	21	457
New Caledonia[76] Nouvelle-Calédonie[76]	4-IV-89	164 173	...	...	151	x168	A1 c1	2.0	...	...
New Zealand[77] Nouvelle-Zélande[77]	4-III-86	3 307 083	1 638 354	1 668 729	3 247	*3 346	A4 b1	0.6	270 986	12
Niue Nioué	29-X-86	2 531	...	...	3	x3	A4 d	2.9	260	12
Norfolk Island Ile Norfolk	30-VI-86	2 367	1 170	1 197	...	...	..	...	36	...
Northern Mariana Islands Iles Mariannes du Nord	..	...	...	...	20	x26	..	5.1	...	...
Pacific Islands (Palau) Iles du Pacifique (Palaos)	..	...	...	...	x16	x18	..	2.2	...	99
Papua New Guinea[78] Papouasie-Nouv.-Guinée[78]	22-IX-80	3 010 727	1 575 672	1 435 055	3 337	*3 699	A10c3	2.1	462 840	8
Pitcairn Pitcairn	31-XII-89	52	...	...	...	...	A0 b1	...	5	...

11
Population by sex, rate of population increase, surface area and density [*cont.*]
Population selon le sexe, taux d'accroissement de la population, superficie et densité [*suite*]

Country or area Pays ou zone	Date	Latest Census Dernier recensement Both Sexes Les deux sexes	Male Masculin	Female Féminin	Mid-year estimates (thousands) Estimations au milieu de l'année (milliers) 1985	1990	Type[1] 1990	Annual rate of increase Taux d'accroissement annual % 1985-90	Surface area (km^2) Superficie (km^2) 1990	Density Densité 1990[2]
Samoa Samoa	3-XI-81	156 349	81 027	75 322	x163	x164	A9 c1	0.1	2 831	58
Solomon Islands[79] Iles Salomon[79]	23-XI-86	285 176	147 972	137 204	x272	x321	A4 c1	3.3	28 896	11
Tokelau Tokélau	1-X-82	1 552	751	801	...	...	...	...	12	...
Tonga Tonga	28-XI-86	94 649	47 611	47 038	97	x95	A4 c1	-0.4	747	127
Tuvalu Tuvalu	27-V-79	7 300	...	...	x9	x10	A11c1	1.9	26	373
Vanuatu Vanuatu	16-V-89	142 630	73 590	69 040	136	*147	A1 c1	1.6	12 189	12
Wake Island Ile de Wake	1-IV-70	1 647	...	...	...	...	...	...	8	...
Wallis and Futuna Islands Iles Wallis et Futuna	15-II-83	12 408	6 266	6 142	x14	x18	A7 c1	4.4	200	89
former USSR · ancienne URSS										
former USSR ancienne URSS	12-I-89	286 730 817	135 360 790	151 370 027	277 334	x288 595	A1 b1	0.8	22402200	13
Asia Asie	15-I-70	*59 245 000	...	...	...	...	..	...	*16831000	...
Europe Europe	15-I-70	*182 503 000	...	...	...	...	..	...	*5571000	...

Source:
Demographic statistics database of the Statistical Division of
the United Nations Secretariat.

Source:
Base de données pour les statistiques démographique de la
Division de statistique du Secrétariat de l'ONU.

ʃ   All data shown which pertain to Germany prior to 3 October
1990 are indicated separately for the Federal Republic of
Germany and the former German Democratic Republic based on
their respective territories at the time indicated. Where
data for united Germany (3 October 1990 and thereafter) are
not available, available data are shown separately under the
designations Federal Republic of Germany and former German
Democratic Republic and pertain to the territorial
boundaries prior to 3 October 1990. For detailed explanatory notes
on data pertaining to Germany, see Annex I - Country Nomenclature.
*   Provisional.
x   Estimate prepared by the Population Division of the United
Nations.
1   For explanation of code, see general note to this chapter.
2   Population per squre kilometre of surface area in 1990.
Figures are merely the quotients of population divided by
surface area and are not to be considered either as
reflecting density in the urban sense or as indicating the
supporting power of a territory's land and resources.
3   De jure population.
4   Including the enclave of Cabinda.
5   Comprising Chagos Archipelago (formerly dependency of
Mauritius).
6   Census of Chagos Archipelago taken 30 June 1962 gave
total population of 747 persons.
7   Estimate for de jure African population based on results of a
sample survey covering 5 per cent of the population in 549
rural villages and 10 per cent of the population in 10 urban
communes and a complete enumeration of the population of

ʃ   Toutes les données se rapportant à l'Allemagne avant le 3
octobre 1990 figurent dans deux rubriques séparées basées sur les
territoires respectifs de la République fédérale d'Allemagne et l'ancienne
République démocratique allemande selon la période indiquée.   En l'absence
de données pour l'Allemagne unifiée (à compter du 3 octobre 1990), les
données disponibles sont fournies séparément sous les rubriques République
fédérale d'Allemagne et ancienne République démocratique allemande et se
rapportent aux limites territoriales antérieures au 3 octobre 1990.   Pour
les notes explicatives en détail sur les données concernant
l'Allemagne, voir Annexe I - Nomenclature des pays.
*   Données provisoires.
x   Estimation établie par la Division de la population de
l'Organisation des Nations Unies.
1   Pour l'explication du code, voir la remarque générale concerment
ce chapître.
2   Nombre d'habitants au kilomètre carré en 1990.   Il s'agit
simplement du quotient du chiffre de la population divisé
par celui de la superficie: il ne faut pas y voir d'indication de la densité au
sens urbain du terme ni de population que les terres et les ressources
du territoire sont capables de nourrir.
3   Population de droit.
4   Y compris l'enclave de Cabinda.
5   Comprend l'archipel de Chagos (ancienne dépendance de
Maurice).
6   Le recensement de la population de l'archipel de Chagos au 30 juin 1962
a donnée comme population total 747 personnes.
7   Estimation pour la population de droit africaine fondée sur les résultats d'une
enquête par sondage ayant porté sur 5 p.100 de la population de 549 villages
ruraux et 10 p. 100 de la population de 10 comunes urbaines et sur un

11
Population by sex, rate of population increase, surface area and density [*cont.*]
Population selon le sexe, taux d'accroissement de la population, superficie et densité [*suite*]

Fort-Archambault, Doba, Moundou, Koumra, Bongor and Abeche. Including estimates of 100,000 for Fort-Lamy enumerated in 1962, and 630,000 for other areas not covered by survey.

8 Including an estimated figure of 50,740 for the island of Mayotte, not covered by the census.

9 Estimate not in accord with the latest census and/or the latest estimate.

10 Rate not computed because of apparent lack of comparability between estimates shown for 1985 and 1990.

11 Comprising Bioko (which includes Pagalu) and Rio Muni (which includes Corisco and Elobeys).

12 Mid-year estimates have been adjusted for under-enumeration, estimated as follows:

dénombrement complet de la population de Fort-Archambault, Moundou, Koumra, Bongor et Abeche. Y compris une estimation de 100 000 pour Fort-Lamy, dénombrées en 1962, et de 630 000 pour d'autres régions sur qui l'enquête n'a pas porté.

8 Y compris un chiffre estimés à 50 740 pour l'île de Mayotte, non couverte par le recensement.

9 L'estimation ne s'accorde avec le dernier recensement, et/ou avec la dernière estimation.

10 On n'a pas calculé le taux parce que les estimations pour 1985 et 1990 ne paraissent pas comparables.

11 Comprend Bioko (qui comprend Pagalu) et Rio Muni (qui comprend Corisco et Elobeys).

12 Les estimations au milieu de l'année tiennent compte d'un ajustement destiné à compenser les lacunes du dénombrement. Les données de recensement ne tiennent pas compte de cet ajustement. En voici le détail:

	Percentage adjustment	Adjusted census total		Adjustement (en pourcentage)	Chiffre de recensement ajusté
Argentina	1.0	...	Argentine	1,0	...
Australia	1.8	...	Australie	1,8	...
Bamgladesh	3.1	*89 949 000	Bangladesh	3,1	*89 949 000
Bolivia	6.99	...	Bolivie	6,99	...
Brunei Darussalam	1.06	...	Brunéi Darussalam	1,06	...
Guatemala	13.7	...	Guatemala	13,7	...
Guinea	...	...	Guinée	...	...
Korea, Republic of	1.9	...	Corée, Rép. de	1,9	...
Mozambique	3.8	...	Mozambique	3,8	...
Namibia	25.6	...	Namibie	25,6	...
Netherlands Antilles	2.0	...	Antilles néerlandaises	2,0	...
Panama	6.6	...	Panama	6,6	...
Peru	...	...	Pérou	...	...
Saint Lucia	7.24	...	Sainte-Lucie	7,24	...
Sierra Leone	10.0	*3 002 426	Sierra Leone	10,0	*3 002 426
South Africa	...	...	Afrique du Sud	...	...
Uruguay	2.6	...	Uruguay	2,6	...

13 Including an estimate of 444,000 for nomad population.

14 Comprising the islands of Agalega and St. Brandon.

15 Estimates based on results of census of 5-8 November 1963. Data have been adjusted for estimated over-enumeration. Official estimates for 1975 and 1979 based on the unadjusted census results are: 74,870,000 and 82,643,000.

16 Census figures believed to be over-enumerated.

17 Excluding Bophuthatswana, Ciskei, Transkei and Venda.

18 Comprising the Northern Region (former Saguia el Hamra) and Southern Region (former Rio de Oro).

19 De jure population, but excluding persons residing in institutions.

20 Including Carriacou and other dependencies in the Grenadines.

21 Including dependencies: Marie-Galante, la Désirade, les Saintes, Petite-Terre, St. Barthélemy and French part of St. Martin.

22 Comprising Bonaire, Curaçao, Saba, St. Eustatius and Dutch part of St. Martin.

23 De jure population, but including armed forces in the area.

24 Including Bequia and other islands in the Grenadines.

25 De jure population, but excluding civilian citizens absent from country for extended period of time. Census figures also exclude armed forces overseas.

26 Excluding Indian jungle population.

27 Mid-year estimates for 24 October.

28 Excluding nomadic Indian tribes.

29 Excluding dependencies, of which South Georgia (area 3,755 km^2) had an estimated population of 499 in 1964 (494 males, 5 females). The other dependencies namely, the South Sandwich group (surface area 337 km^2) and a number of smaller islands, are presumed to be uninhabited.

13 Y compris une estimation de 444 000 personnes pour la population nomade.

14 Y compris les îles Agalega et Saint-Brandon.

15 Estimations fondées sur les résultats du recemse,emt di 5 au 8 novembre 1963. Les données ont été ajustées pour compenser le chiffre jugé trop élevé de la population dénombrée. Les estimations officielles pour 1975 et 1979 fondées sur résultats non ajustés du recensement sont les suivantes: 74 870 000 et 82 643 000.

16 Chiffres du recensement où l'on pense qu'il y a eu surdénombrement.

17 Non compris Bophuthatswana, Ciskei, Transkei et Venda.

18 Comprend la région septentrionale (ancien Saguia-el-Hamura) et la région méridionale (ancien Rio de Oro).

19 Population de droit, mais non compris les personnes dans les institutions.

20 Y compris Carriacou et les autres dépendances du groupe des îles Grenadines.

21 Y compris les dépendances: Marie-Galante, la Désirade, les Désirade, les Saintes, Petite-Terre, Saint-Barthélemy et la partie française de Saint-Martin.

22 Comprend Bonaire, Curaçao, Saba, Saint-Eustache et la partie néederlandaise de Saint-Martin.

23 Population de droit, mais y compris les militaires en garnison sur le territoire.

24 Y compris Bequia et des autres îles dans les Grenadines.

25 Population de droit, mais non compris les civils hors du pays pendant une période prolongée. Les chiffres de recensement ne comprennent pas également les militaires à l'étranger.

26 Non compris les Indiens de la jungle.

27 Estimations au milieu de l'années pour le 24 Octobre.

28 Non compris les tribus d'Indiens nomades.

29 Non compris les dépendances, parmi lesquelles figure la Georgie du Sud

11
Population by sex, rate of population increase, surface area and density [*cont.*]
Population selon le sexe, taux d'accroissement de la population, superficie et densité [*suite*]

30 A dispute exists between the governments of Argentina and the United Kingdom of Great Britain and Northern Ireland concerning sovereignty over the Falkland Islands (Malvinas).
31 Excluding nomad population.
32 Excluding transients afloat.
33 Excluding foreign diplomatic personnel and their dependants.
34 This total population of China, as given in the communiqué of the State Statistical Bureau releasing the major figures of the census, includes a population of 6,130,000 for Hong Kong and Macau.
35 Comprising Hong Kong island, Kowloon and the New (leased) Territories.
36 Land area only. Total including ocean area within administrative boundaries is 2,916 km^2.
37 Including data for the Indian-held part of Jammu and Kashmir, the final status of which has not yet been determined.
38 Figures provided by Indonesia including East Timor, shown separately.
39 Including data for East Jerusalem and Israeli residents in certain other territories under occupation by Israeli military forces since June 1967.
40 Comprising Hokkaido, Honshu, Shikoku, Kyushu. Excluding diplomatic personnel outside the country and foreign military and civilian personnel and their dependants stationed in the area.
41 Including military and diplomatic personnel and their families abroad, numbering 933 at 1961 census, but excluding foreign military and diplomatic personnel and their families in the country, numbering 389 at 1961 census. Also including registered Palestinian refugees number 654,092 and 722,687 at 30 June 1963 and 31 May 1967, respectively.
42 Excluding data for Jordanian territory under occupation since June 1967 by Israeli military forces.
43 Including the area of the demilitarized zone (1,262 km^2).
44 Excluding alien armed forces, civilian aliens employed by armed forces, foreign diplomatic personnel and their dependants and Korean diplomatic personnel and their dependants outside the country.
45 Excluding Palestinian refugees in camps.
46 Based on results of sample survey.
47 Comprising Macau City and islands of Taipa and Coloane.
48 Excluding data for Jammu and Kashmir, the final status of which has not yet been determined, Junagardh, Manavadar, Gilgit and Baltistan.
49 Former mandated territory administered by the United Kingdom until 1948.
50 Excluding United Kingdom armed forces, numbering 2,507.
51 Comprising that part of Palestine under Egyptian administration following the Armistice of 1949 until June 1967, when it was occupied by Israeli military forces.
52 Excluding transients afloat and non-locally domiciled military and civilian services personnel and their dependants and visitors, numbering 5,553, 5,187 and 8,895 respectively at 1980 census.
53 Including Palestinian refugees numbering 193,000 on 1 July 1977.
54 Comprising 7 sheikdoms of Abu Dhabi, Dubai, Sharjah, Ajaman, Umm al Qaiwain, Ras al Khaimah and Fujairah, and the area lying within the modified Riyadh line as announced in October 1955.
57 Excluding surface area of frontier rivers.
56 Including dependencies: Alderey, Brechou, Herm, Jethou, Lithou and Sark Island.
57 Excluding Faeroe Islands and Greenland, shown separately.
58 Excluding Overseas Departments, namely French Guiana, Guadeloupe, Martinique and Réunion, shown separately.

(3 755 km^2) avec une population estimée à 499 personnes en 1964 (494 du sexe masculin et 5 du sexe féminin). Les autres dépendances, c'est-à-dire le groupe des Sandwich de Sud (superficie: 337 km^2) et certaines petite-îles, sont présumées inhabitées.
30 La souveraineté sur les îles Falkland (Malvinas) fait l'objet d'un différend entre le Gouvernement argentin et le Gouvernement du Royaume-Uni de Grande-Bretagne et d'Irlande du Nord.
31 Non compris la population nomade.
32 Non compris les personnes de passage à bord des navires.
33 Non compris le personnel diplomatique étranger et les membres de leur famille les accompagnant.
34 Le chiffre indiqué pour la population totale de la Chine, qui figure dans le communiqué du Bureau du statistique de l'Etat publiant les principaux chiffres du recensement, comprennent la population de Hong-kong et Macao qui s'élève à 6 130 000 personnes.
35 Comprend les îles de Hong-kong, Kowloon et les Nouveaux Territoires (à bail).
36 Superficie terrestre seulement. La superficie totale, qui comprend la zone maritime se trouvant à l'intérieur des limites administratives, est de 2 916 km^2.
37 Y compris les données pour la partie du Jammu et du Cacehmire occupée par l'Inde dont le statut définitif n'a pas encore été déterminé.
38 Les chiffres fournis par l'Indonésie comprennent le Timor oriental, qui fait l'objet d'une rubrique distincte.
39 Y compris les données pour Jérusalem-Est et les résidents israéliens dans certains autres territoires occupés depuis juin 1967 par les forces armées israéliennes.
40 Comprend Hokkaido, Honshu, Shikoku, Kyushu. Non compris le personnel diplomatique hors du pays, les militaires et agents civils étrangers en poste sur le territoire et les membres de leur famille les accompagnant.
41 Y compris les militaires et le personnel diplomatique à l'étranger et les membres de leur famille les accompagnant, au nombre de 933 personnes au recensement de 1961, mais non compris les militaires et le personnel diplomatique étrangers sur le territoire et les membres de leur famille les accompagnant, au nombre de 389 personnes au recensement de 1961. Y compris également les réfugiés de Palestine immatriculés: 654 092 au 30 juin 1963 et 722 687 au 31 mai 1967.
42 Non compris les données pour le territoire jordanien occupé depuis juin 1967 par les forces armées israéliennes.
43 Y compris la zone démilitarisée (superficie: 1 262 km^2).
44 Non compris les militaires étrangers, les civils étrangers employés par les armées, le personnel diplomatique étranger et les membres de leur famille les accompagnant et le personnel diplomatique coréen hors du pays et les membres de leur familles les accompagnant.
45 Non compris les réfugiés de Palestine dans les camps.
46 D'après les résultats d'une enquête par sondage.
47 Comprend la ville de Macao et les îles de Taipa et de Colowane.
48 Non compris les données pour le Jammu et le Cachemire, dont le status définitif n'a pas encore été déterminé, le Junagardh, le Manavadar, le Gilgit et le Baltistan.
49 Ancien territoire sous mandat administré par le Royaume-Uni jusqu'à 1948.
50 Non compris les forces armées du Royaume-Uni au nombre de 2 507 personnes.
51 Comprend la partie de la Palestine administrée par l'Egypt depuis l'armistice de 1949 jusqu'en juin 1967, date laquelle elle a été occupée par les forces armées israéliennes.
52 Non compris les personnes de passage à bord de navires, les militaires et agents civils non résidents et les membres de leur famille les accompagnant, et les visiteurs, soit: 5 553, 5 187 et 8 895 personnes respectivement au recensement de 1980.
53 Y compris les réfugiés de Palestine au nombre de 193 000 au 1er juillet 1977.
54 Comprend les sept cheikhats de Abou Dhabi, Dabai, Ghârdja, Adjmân, Oumm-al-Quiwaïn, Ras al Khaïma et Foudjaïra, ainsi que la zone délimitée par la ligne de Riad modifiée comme il a été annoncé en octobre 1955.

11
Population by sex, rate of population increase, surface area and density [*cont.*]
Population selon le sexe, taux d'accroissement de la population, superficie et densité [*suite*]

59 De jure population, but excluding diplomatic personnel outside the country and including foreign diplomatic personnel not living in embassies or consulates.
60 Excluding military personnel stationed outside the country who do not have a personal residence in France.
61 Excluding armed forces.
62 Including armed forces stationed outside the country, but excluding alien armed forces stationed in the area.
63 Surface area is 0.44 km^2.
64 Including Gozo and Comino Islands and civilian nationals temporarily outside the country.
65 Surface area is 1.49 km^2.
66 Excluding civilian aliens within the country, but including civilian nationals temporarily outside the country.
67 Including the Azores and Madeira Islands.
68 Including the Balearic and Canary Islands, and Alhucemas, Ceuta, Chafarinas, Melilla and Penon de Vélez de la Gomera.
69 Inhabited only during the winter season. Census data are for total population while estimates refer to Norwegian population only. Included also in the de jure population of Norway.
70 Excluding Channel Islands and Isle of Man, shown separately.
71 Part of Phoenix Islands group (see also Kiribati and Tuvalu shown separately). Jointly administered as a condominium by United Kingdom and United States.
72 Both islands were uninhabited at time of census.
73 Excluding Niue, shown separately, which is part of Cook Islands, but because of remoteness is administered separately.
74 Comprising Austral, Gambier, Marquesas, Rapa, Society and Tuamotu Islands.
75 Including Christmas, Fanning, Ocean and Washington Islands.
76 Including the islands of Huon, Chesterfield, Loyalty, Walpole and Belep Archipelago.
77 Including Campbell and Kermadec Islands (population 20 in 1961, surface area 148 km^2) as well as Antipodes, Auckland, Bounty, Snares, Solander and Three Kings island, all of which are uninhabited. Excluding diplomatic personnel and armed forces outside the country, the latter numbering 1,936 at 1966 census; also excluding alien armed forces within the country.
78 Comprising eastern part of New Guinea, the Bismarck Archipelago, Bougainville and Buka of Solomon Islands group and about 600 smaller islands.
79 Comprising the Solomon islands group (except Bougainville and Buka which are included with Papua New Guinea shown separately), Ontong, Java, Rennel and Santa Cruz Islands.

55 Non compris la surface des cours d'eau frontières.
56 Y compris les dépendances: Aurigny, Brecqhou, Herm, Jethou, Lihou et l'île de Sercq.
57 Non compris les îles Féroé et le Groenland, qui font l'objet de rubriques distinctes.
58 Non compris les départements d'outre-mer, c'est-à-dire la Guyane française, la Guadeloupe, la Martinique et la Réunion, qui font l'objet de rubriques distinctes.
59 Population de droit, non compris le personnel diplomatique hors du pays et y compris le personnel diplomatique étranger qui ne vit pas dans les ambassades ou les consulats.
60 Non compris les militaires en garnison hors du pays et sans résidence personnelle en France.
61 Non compris les militaires.
62 Y compris les militaires en garnison hors du pays, mais non compris les militaires étrangers en garnison sur le territoire. territoire.
63 Superficie: 0,44 km^2.
64 Y compris les îles de Gozo et de Comino et les civils nationaux temporairement hors du pays.
65 Superficie: 1,49 km^2.
66 Non compris les civils étrangers dans le pays, mais y compris les civils nationaux temporairement hors du pays.
67 Y compris les Açores et Madère.
68 Y compris les Baléares et les Canaries, Al Hoceima, Ceuta, les îles Zaffarines, Melilla et Penon de Vélez de la Gomera.
69 N'est habitée pendant la saison d'hiver. Les données de recensement se rapportent à la populatio totale, mais les estimations ne concernent que la population norvégienne, comprise également dans la population de droit de la Norvège.
70 Non compris les îles Anglo-Normandes et l'île de Man, qui font l'objet de rubriques distinctes.
71 Iles du groupe des Phoenix (voir également Kiribati et Tuvalu qui font l'objet d'une rubrique distincte). Condominium exercé en commun par le Royaume-Uni et les Etats-Unis.
72 Les deux îles étaient inhabitées au moment du recensement.
73 Non compris Nioué, qui fait l'objet d'une rubrique distincte et qui fait partie des îles Cook, mais qui, en raison de son éloignement, est administrée séparément.
74 Comprend les îles Australes, Gambier, Marquises, Rapa, de la Societé et Tuamotou.
75 Y compris les îles Christmas, Fanning, Océan et Washington.
76 Y compris les îles Huon, Chesterfield, Loyauté et Walpole, et l'archipel Belep.
77 Y compris les îles Campbell et Kermadec (20 habitants en 1961, superficie: 148 km^2) ainsi que les îles Antipodes, Auckland, Bounty, Snares, Solander et Three Kings, qui sont toutes inhabitées. Non compris le personnel diplomatique et les militaires hors du pays, ces derniers au nombre de 1 936 au recensement de 1966; non compris également les militaires étrangers dans le pays.
78 Comprend l'est de la Nouvelle-Guinée, l'archipel Bismarck, Bougainville et Buka (ces deux dernières du groupe des Salomon) et environ 600 îlots.
79 Comprend les îles Salomon (à l'exception de Bougainville et de Buka dont la population est comprise dans celle de Papouasie-Nouvelle Guinée qui font l'objet d'une rubrique distincte), ainsi que les îles Ontong, Java, Rennel et Santa Cruz.

12
Population in urban and rural areas, rates of growth and largest city population
Population urbaine, population rurale, taux d'accroissement et population de la ville la plus peuplée

Country or area Pays ou zone	Year Année	Population estimates Estimations de la population				Year Année	Largest city population Ville la plus peuplée		
		R%	U%	Δ p.a.(%)[1]			Number (000s) Nombre (000s)	% u	% t
				Pop.r.	Pop.u.				
Africa · Afrique									
Algeria	1980	56.6	43.4	2.1	4.6	...	...	...	...
Algérie	1990	48.3	51.7	1.1	4.4	1992	3300	23.4	12.5
Angola	1980	79.0	21.0	1.9	6.0	...	...	...	...
Angola	1990	71.7	28.3	1.8	5.7	1992	1900	62.6	18.7
Benin	1980	68.4	31.6	0.8	7.4	...	...	...	...
Bénin	1990	61.8	38.2	1.9	4.8	1990	210	12.0	4.5
Botswana	1980	84.9	15.1	2.8	8.2	...	...	...	...
Botswana	1990	75.0	25.0	1.5	7.9	1990	130	36.2	10.0
Burkina Faso	1980	91.5	8.5	1.8	8.1	...	...	...	...
Burkina Faso	1990	84.8	15.2	1.8	8.4	1990	410	50.6	4.6
Burundi	1980	95.7	4.3	2.1	8.0	...	...	...	...
Burundi	1990	94.7	5.3	2.8	5.1	1990	250	82.5	4.6
Cameroon	1980	68.6	31.4	1.5	5.9	...	...	...	...
Cameroun	1990	59.7	40.3	1.4	5.3	1992	1100	21.9	9.2
Cape Verde	1980	76.5	23.5	0.3	2.6	...	...	...	...
Cap-Vert	1990	71.3	28.7	1.5	4.4	1990	60	56.6	16.2
Central African Rep.	1980	61.7	38.3	1.1	4.6	...	...	...	...
Rép. centrafricaine	1990	53.3	46.7	1.2	4.6	1990	720	50.8	23.7
Chad	1980	79.5	20.5	0.9	7.6	...	...	...	...
Tchad	1990	68.4	31.6	0.5	6.0	1990	730	43.6	12.9
Comoros	1980	76.8	23.2	3.4	5.7	...	...	...	...
Comores	1990	72.2	27.8	2.9	5.4	1990	20	13.1	3.6
Congo	1980	64.2	35.8	2.2	4.0	...	...	...	...
Congo	1990	59.5	40.5	2.1	4.3	1990	630	68.6	27.7
Côte d'Ivoire	1980	65.2	34.8	3.1	5.5	...	...	...	...
Côte d'Ivoire	1990	59.6	40.4	2.8	5.2	1992	2400	44.5	18.5
Djibouti	1980	26.3	73.7	0.9	6.0	...	...	...	...
Djibouti	1990	19.3	80.7	0.1	3.7	1990	330	100.0	80.7
Egypt	1980	56.2	43.8	2.2	2.6	...	...	...	...
Egypte	1990	56.1	43.9	2.4	2.4	1992	9000	37.3	16.5
Equatorial Guinea	1980	72.8	27.2	-0.8	-0.5	...	...	...	...
Guinée équatoriale	1990	71.3	28.7	2.1	3.1	...	...	...	...
Ethiopia	1980	89.5	10.5	2.2	4.4	...	...	...	...
Ethiopie	1990	87.7	12.3	2.7	4.2	1992	2000	29.2	3.7
Gabon	1980	64.3	35.7	3.2	7.8	...	...	...	...
Gabon	1990	54.4	45.6	1.6	5.5	1990	300	56.1	25.6
Gambia	1980	81.7	18.3	2.8	5.0	...	...	...	...
Gambie	1990	77.4	22.6	2.3	5.2	1990	200	100.0	23.2
Ghana	1980	68.8	31.2	1.4	2.5	...	...	...	...
Ghana	1990	66.0	34.0	2.6	4.2	1992	1500	27.1	9.5
Guinea	1980	80.9	19.1	0.8	4.6	...	...	...	...
Guinée	1990	74.2	25.8	1.9	5.8	1992	1300	76.1	20.7
Guinea-Bissau	1980	83.1	16.9	4.5	5.8	...	...	...	...
Guinée-Bissau	1990	80.2	19.8	1.6	3.9	1990	70	36.6	7.3
Kenya	1980	83.9	16.1	3.1	8.2	...	...	...	...
Kenya	1990	76.4	23.6	2.4	7.1	1992	1700	27.0	6.8
Lesotho	1980	86.8	13.2	1.9	6.5	...	...	...	...
Lesotho	1990	80.6	19.4	1.7	6.3	1990	60	16.7	3.4
Liberia	1980	65.0	35.0	1.7	5.9	...	...	...	...
Libéria	1990	54.6	45.4	1.3	5.7	1990	670	56.6	26.0
Libyan Arab Jamahiriya	1980	30.4	69.6	-0.7	7.0	...	...	...	...
Jamah. arabe libyenne	1990	17.6	82.4	-1.6	5.0	1992	2900	69.9	58.7
Madagascar	1980	81.7	18.3	2.4	5.5	...	...	...	...
Madagascar	1990	76.2	23.8	2.4	5.8	1990	675	23.6	5.6
Malawi	1980	90.9	9.1	3.0	6.8	...	...	...	...
Malawi	1990	88.2	11.8	5.0	7.9	...	...	...	...
Mali	1980	81.5	18.5	1.6	4.7	...	...	...	...
Mali	1990	76.2	23.8	2.3	5.5	1990	660	37.4	7.2
Mauritania	1980	71.0	29.0	0.1	9.6	...	...	...	...
Mauritanie	1990	53.2	46.8	-0.3	6.8	1990	760	80.3	37.5

12
Population in urban and rural areas, rates of growth and largest city population [*cont.*]
Population urbaine, population rurale, taux d'accroissement et population de la ville la plus peuplée [*suite*]

| Country or area | Year | Population estimates Estimations de la population | | | | Year | Largest city population Ville la plus peuplée | | |
Pays ou zone	Année	R%	U%	Δ p.a.(%)[1] Pop.r.	Pop.u.	Année	Number (000s) Nombre (000s)	% u	% t
Mauritius	1980	57.7	42.3	2.0	1.1	...	...	...	...
Maurice	1990	59.4	40.6	1.3	0.7	1990	160	36.5	14.8
Morocco	1980	59.0	41.0	1.2	4.0	...	...	...	...
Maroc	1990	53.9	46.1	1.8	3.6	1992	3000	24.2	11.4
Mozambique	1980	86.9	13.1	1.8	11.2	...	...	...	...
Mozambique	1990	73.2	26.8	-1.0	7.3	1992	1800	40.7	12.1
Namibia	1980	77.2	22.8	2.3	4.8	...	...	...	...
Namibie	1990	72.2	27.8	2.3	5.1	1990	150	30.3	8.4
Niger	1980	86.8	13.2	2.6	7.5	...	...	...	...
Niger	1990	80.5	19.5	2.3	6.9	1990	580	38.5	7.5
Nigeria	1980	72.9	27.1	2.4	6.3	...	...	...	...
Nigéria	1990	64.8	35.2	2.1	5.8	1992	8700	20.4	7.5
Réunion	1980	45.3	54.7	-1.4	3.1	...	...	...	...
Réunion	1990	36.1	63.9	-0.4	3.3	1990	230	60.2	38.5
Rwanda	1980	95.3	4.7	3.1	6.7	...	...	...	...
Rwanda	1990	94.4	5.6	3.2	4.9	1990	300	54.1	4.1
Sao Tome and Principe	1980	67.0	33.0	1.6	6.1	...	...	...	...
Sao Tomé-et-Principe	1990	57.9	42.1	0.8	4.7	1990	51	100.0	42.1
Senegal	1980	64.1	35.9	2.3	3.8	...	...	...	...
Sénégal	1990	60.2	39.8	2.1	3.8	1992	1800	55.9	22.8
Seychelles	1980	...	...	-2.2	6.0	...	...	...	...
Seychelles	1990	...	...	-2.4	3.7	...	...	...	...
Sierra Leone	1980	75.5	24.5	1.3	5.1	...	...	...	...
Sierra Leone	1990	67.8	32.2	1.4	5.1	1990	690	51.7	16.6
Somalia	1980	77.8	22.2	3.9	5.0	...	...	...	...
Somaliae	1990	75.8	24.2	1.7	2.8	1990	720	26.4	9.6
South Africa	1980	51.9	48.1	2.6	2.7	...	...	...	...
Afrique du Sud	1990	50.8	49.2	2.1	2.8	1992	2400	12.4	6.1
St. Helena	1980	...	...	0.9	-3.8	...	...	...	...
Sainte Hélène	1990	...	...	3.3	3.3	...	...	...	...
Sudan	1980	80.0	20.0	2.8	4.1	...	...	...	...
Soudan	1990	77.5	22.5	2.5	4.3	1992	2100	34.5	8.0
Swaziland	1980	82.1	17.9	2.3	8.0	...	...	...	...
Swaziland	1990	73.6	26.4	1.4	6.4	1990	50	19.2	6.3
Togo	1980	77.1	22.9	1.1	9.5	...	...	...	...
Togo	1990	71.5	28.5	2.5	4.5	1990	510	56.1	14.4
Tunisia	1980	49.7	50.3	1.5	3.7	...	...	...	...
Tunisie	1990	44.0	56.0	0.8	3.2	1992	1900	39.3	22.5
Uganda	1980	91.2	8.8	3.1	4.3	...	...	...	...
Ouganda	1990	88.8	11.2	2.8	5.5	1990	690	35.2	3.7
United Rep. of Tanzania	1980	85.2	14.8	2.1	10.7	...	...	...	...
Rép.-Unie de Tanzanie	1990	79.2	20.8	2.6	6.8	1992	1600	25.4	5.6
Western Sahara	1980	50.4	49.6	0.7	3.6	...	...	...	...
Sahara occidental	1990	43.3	56.7	2.8	5.7	1990	88	87.1	49.4
Zaire	1980	71.3	28.7	3.2	2.5	...	...	...	...
Zaïre	1990	71.9	28.1	3.3	3.4	1992	3700	33.1	9.4
Zambia	1980	60.2	39.8	1.8	6.1	...	...	...	...
Zambie	1990	58.0	42.0	3.0	3.9	1990	990	23.5	11.7
Zimbabwe	1980	77.7	22.3	2.3	5.6	...	...	...	...
Zimbabwe	1990	71.5	28.5	2.6	5.9	1990	850	31.8	8.8
America, North · Amérique du Nord									
Antigua and Barbuda	1980	...	...	-0.8	-0.8	...	...	...	...
Antigua-et-Barbuda	1990	...	...	0.3	1.4	...	...	...	...
Aruba	1980	...	...	-0.5	-0.5	...	...	...	...
Aruba	1990	...	...	-0.3	0.2	...	...	...	...
Bahamas	1980	39.5	60.5	1.8	2.9	...	...	...	...
Bahamas	1990	35.7	64.3	0.7	2.5	...	...	...	...
Barbados	1980	59.8	40.2	-0.2	1.1	...	...	...	...
Barbade	1990	55.3	44.7	-0.6	1.5	1990	110	96.5	43.1
Belize	1980	...	...	2.8	2.2	...	...	...	...
Belize	1990	...	...	2.2	3.0	...	...	...	...
Bermuda	1980	...	...	...	0.4	...	...	...	...
Bermudes	1990	...	...	...	1.7	...	...	...	...
Canada	1980	24.3	75.7	1.1	1.2	...	...	...	...
Canada	1990	22.9	77.1	0.5	1.3	1992	3600	16.8	13.0

12
Population in urban and rural areas, rates of growth and largest city population [*cont.*]
Population urbaine, population rurale, taux d'accroissement et population de la ville la plus peuplée [*suite*]

Country or area Pays ou zone	Year Année	Population estimates Estimations de la population				Year Année	Largest city population Ville la plus peuplée		
		R%	U%	Δ p.a.(%)[1] Pop.r.	Pop.u.		Number (000s) Nombre (000s)	% u	% t
Cayman Islands	1980	...	...	...	4.2	...	...	...	...
Iles Caïmanes	1990	...	...	...	4.7	...	...	...	...
Costa Rica	1980	56.9	43.1	2.4	3.8	...	...	...	...
Costa Rica	1990	52.9	47.1	2.0	3.7	1990	1020	71.8	33.8
Cuba	1980	31.9	68.1	-1.5	2.0	...	...	...	...
Cuba	1990	26.4	73.6	-0.8	1.7	1992	2200	26.9	20.1
Dominican Republic	1980	49.5	50.5	0.4	4.6	...	...	...	...
Rép. dominicaine	1990	39.6	60.4	...	3.9	1992	2400	50.8	31.6
El Salvador	1980	58.5	41.5	1.7	2.6	...	...	...	...
El Salvador	1990	55.6	44.4	1.1	2.5	1990	590	25.3	11.2
Greenland	1980	...	...	-0.7	0.6	...	...	...	...
Groenland	1990	...	...	-0.2	1.4	...	...	...	...
Guadeloupe	1980	56.3	43.7	-0.6	0.6	...	...	...	...
Guadeloupe	1990	51.5	48.5	0.9	3.0	1990	38	22.9	11.1
Guatemala	1980	62.6	37.4	2.6	3.1	...	...	...	...
Guatemala	1990	60.6	39.4	2.4	3.6	1990	840	23.2	9.1
Haiti	1980	76.3	23.7	1.2	3.4	...	...	...	...
Haïti	1990	71.4	28.6	1.3	3.9	1992	1100	56.0	16.7
Honduras	1980	64.0	36.0	2.4	5.6	...	...	...	...
Honduras	1990	56.3	43.7	1.8	5.1	1990	780	34.7	15.2
Jamaica	1980	53.2	46.8	0.2	2.3	...	...	...	...
Jamaïque	1990	47.7	52.3	-0.3	2.1	1990	640	49.8	26.1
Martinique	1980	34.0	66.0	-3.4	1.8	...	...	...	...
Martinique	1990	25.3	74.7	-1.8	2.2	1990	27	10.6	7.9
Mexico	1980	33.7	66.3	0.6	3.7	...	...	...	...
Mexique	1990	27.4	72.6	0.1	3.1	1992	15300	23.5	17.3
Montserrat	1980	...	...	-0.3	-0.3	...	...	...	...
Montserrat	1990	...	...	-0.8	0.9	...	...	...	...
Nicaragua	1980	46.6	53.4	1.6	4.1	...	...	...	...
Nicaragua	1990	40.2	59.8	1.1	3.7	1990	1010	43.7	26.1
Panama	1980	50.3	49.7	1.8	2.7	...	...	...	...
Panama	1990	47.1	52.9	1.4	2.7	1990	480	37.2	19.9
Puerto Rico	1980	33.0	67.0	-1.0	2.7	...	...	...	...
Porto Rico	1990	26.1	73.9	-1.4	1.8	1992	1400	53.0	39.8
Saint Kitts and Nevis	1980	...	...	-1.7	1.3	...	...	...	...
Saint-Kitts-et-Nevis	1990	...	...	-2.0	1.1	...	...	...	...
Saint Lucia	1980	58.1	41.9	1.0	1.7	...	...	...	...
Sainte-Lucie	1990	53.3	46.7	0.9	2.1	1990	70	100.0	46.7
St. Pierre and Miquelon	1980	...	...	-1.2	0.5	...	...	...	...
Saint-Pierre-et-Miquelon	1990	...	...	-0.8	1.0	...	...	...	...
St. Vincent and the Grenadines	1980	...	...	0.8	2.7	...	...	...	...
St. Vincent-et-Grenadines	1990	...	...	0.5	2.3	...	...	...	...
Trinidad and Tobago	1980	37.0	63.0	1.3	1.3	...	...	...	...
Trinité-et-Tobago	1990	35.2	64.8	0.6	1.7	1990	100	11.3	7.8
Turks and Caicos Islands	1980	...	...	2.1	4.1	...	...	...	...
Iles Turques et Caïques	1990	...	...	3.1	5.6	...	...	...	...
United States	1980	26.3	73.7	1.0	1.1	...	...	...	...
Etats-Unis	1990	24.8	75.2	0.4	1.1	1992	16200	8.4	6.3
US Virgin Islands	1980	...	...	0.7	0.7	...	...	...	...
Iles Vierges américaines	1990	...	...	-0.6	0.7	...	...	...	...
America, South · Amérique du Sud									
Argentina	1980	17.1	82.9	-0.8	2.1	...	...	...	...
Argentine	1990	13.9	86.1	-0.8	1.6	1992	11800	41.0	35.5
Bolivia	1980	55.8	44.2	1.7	3.9	...	...	...	...
Bolivie	1990	49.0	51.0	1.1	3.8	1992	1100	27.3	14.3
Brazil	1980	33.8	66.2	-0.5	3.9	...	...	...	...
Brésil	1990	24.8	75.2	-1.3	3.1	1992	19200	16.3	12.5
Chile	1980	18.8	81.2	-1.4	2.2	...	...	...	...
Chili	1990	15.4	84.6	-0.1	2.0	1992	5100	43.8	37.3
Colombia	1980	36.1	63.9	0.5	3.2	...	...	...	...
Colombie	1990	30.0	70.0	-0.1	2.7	1992	5200	21.7	15.4
Ecuador	1980	53.0	47.0	1.2	4.9	...	...	...	...
Equateur	1990	43.7	56.3	0.5	4.2	1992	1800	28.4	16.5
French Guiana	1980	29.0	71.0	2.4	4.0	...	...	...	...
Guyane française	1990	25.5	74.5	1.9	3.9	1990	66	90.4	67.3

12
Population in urban and rural areas, rates of growth and largest city population [*cont.*]
Population urbaine, population rurale, taux d'accroissement et population de la ville la plus peuplée [*suite*]

Country or area Pays ou zone	Year Année	Population estimates Estimations de la population				Year Année	Largest city population Ville la plus peuplée		
		R%	U%	Δ p.a.(%)[1]			Number (000s) Nombre (000s)	% u	% t
				Pop.r.	Pop.u.				
Guyana	1980	70.0	30.0	0.6	0.9	...	...	...	...
Guyana	1990	67.2	32.8	-0.4	1.2	1990	210	76.4	26.4
Paraguay	1980	58.3	41.7	2.3	4.5	...	...	...	...
Paraguay	1990	52.5	47.5	1.8	4.3	1990	970	47.8	22.7
Peru	1980	35.4	64.6	1.0	3.6	...	...	...	...
Pérou	1990	30.2	69.8	0.5	2.8	1992	6900	43.4	30.7
Suriname	1980	55.1	44.9	-0.7	-0.7	...	...	...	...
Suriname	1990	52.6	47.4	1.2	2.8	1990	110	55.0	26.1
Uruguay	1980	14.8	85.2	-2.2	1.1	...	...	...	...
Uruguay	1990	11.1	88.9	-2.3	1.0	1992	1300	46.5	41.6
Venezuela	1980	16.7	83.3	-2.2	4.8	...	...	...	...
Venezuela	1990	9.5	90.5	-3.3	3.1	1992	2800	15.4	14.1
Asia · Asie									
Afghanistan	1980	84.3	15.7	0.3	4.2				
Afghanistan	1990	81.8	18.2	2.3	4.2	1992	1700	48.3	9.1
Armenia	1980	...	...	...	...				
Arménie	1990	...	...	...	...	1992	1200	52.4	35.6
Azerbaijan	1980	...	...	...	...	...	...	...	...
Azerbaïdjan	1990	...	...	...	...	1992	1800	45.4	24.5
Bahrain	1980	19.6	80.4	3.6	5.2	...	...	...	...
Bahreïn	1990	17.1	82.9	1.8	3.5	1990	130	30.4	25.2
Bangladesh	1980	88.7	11.3	2.4	6.8	...	...	...	...
Bangladesh	1990	83.6	16.4	1.8	6.1	1992	7400	35.3	6.2
Bhutan	1980	96.1	3.9	1.9	4.6	...	...	...	...
Bhoutan	1990	94.7	5.3	2.0	5.5	1990	20	24.7	1.3
Brunei Darussalam	1980	39.9	60.1	4.8	3.0	...	...	...	...
Brunéi Darussalam	1990	42.0	58.0	2.6	2.6	1990	65	42.5	24.4
Cambodia	1980	89.7	10.3	-1.8	-1.8	...	...	...	...
Cambodge	1990	88.4	11.6	2.4	4.1	1990	967	100.0	11.7
China	1980	80.4	19.6	0.9	4.0	...	...	...	...
Chine	1990	73.8	26.2	0.5	4.5	1992	14100	4.3	1.2
Cyprus	1980	53.7	46.3	-0.4	2.0	...	...	...	...
Chypre	1990	47.2	52.8	-0.3	2.4	1990	160	43.2	22.8
East Timor	1980	89.2	10.8	-3.0	-2.0	...	...	...	...
Timor oriental	1990	86.9	13.1	2.5	5.0	1990	97	100.0	13.2
Gaza Strip	1980	...	...	-3.2	3.2	...	...	...	...
Zone de Gaza	1990	...	...	-0.1	4.1	...	...	...	...
Georgia	1980	...	...	...	...	...	...	...	...
Géorgie	1990	...	...	...	...	1992	1300	42.9	24.0
Hong Kong	1980	8.5	91.5	-1.1	3.1	...	...	...	...
Hong-kong	1990	5.9	94.1	-2.8	1.2	1992	5500	100.0	94.5
India	1980	76.9	23.1	1.6	3.7	...	...	...	...
Inde	1990	74.5	25.5	1.7	3.0	1992	13300	5.8	1.5
Indonesia	1980	77.8	22.2	1.4	4.9	...	...	...	...
Indonésie	1990	71.2	28.8	1.0	4.5	1992	10000	17.3	5.2
Iran, Islamic Rep. of	1980	50.4	49.6	1.8	4.8	...	...	...	...
Iran, Rép. islamique d'	1990	43.1	56.9	1.9	4.8	1992	7000	19.5	11.4
Iraq	1980	34.5	65.5	1.1	4.6	...	...	...	...
Iraq	1990	28.2	71.8	1.2	4.2	1992	4200	29.9	21.8
Israel	1980	11.4	88.6	-0.8	2.8	...	...	...	...
Israël	1990	8.4	91.6	-1.0	2.2	1992	2000	43.2	39.8
Japan	1980	23.8	76.2	0.5	1.1	...	...	...	...
Japon	1990	22.8	77.2	...	0.6	1992	25800	26.7	20.7
Jordan	1980	40.1	59.9	0.2	4.0	...	...	...	...
Jordanie	1990	32.0	68.0	0.9	4.4	1990	1020	37.4	25.4
Kazakhstan	1980	...	...	...	...	...	...	...	...
Kazakhstan	1990	...	...	...	...	1992	1200	12.4	7.0
Korea, Dem. People's Rep.	1980	43.1	56.9	1.8	2.1	...	...	...	...
Corée, Rép. pop. dém. de	1990	40.2	59.8	1.3	2.2	1992	2300	17.0	10.3
Korea, Republic of	1980	43.1	56.9	-2.2	4.9	...	...	...	...
Corée, République de	1990	27.9	72.1	-3.4	3.3	1992	11600	35.2	26.2
Kuwait	1980	9.8	90.2	-3.7	7.7	...	...	...	...
Koweït	1990	4.2	95.8	-3.6	4.8	1992	1100	58.0	55.9
Lao People's Dem. Rep.	1980	86.6	13.4	0.7	4.5	...	...	...	...
République dém. pop. lao	1990	81.4	18.6	2.5	6.3	1990	410	53.2	9.9

12

Population in urban and rural areas, rates of growth and largest city population [*cont.*]

Population urbaine, population rurale, taux d'accroissement et population de la ville la plus peuplée [*suite*]

Country or area Pays ou zone	Year Année	Population estimates Estimations de la population				Year Année	Largest city population Ville la plus peuplée		
		R%	U%	Δ p.a.(%)[1] Pop.r.	Pop.u.		Number (000s) Nombre (000s)	% u	% t
Lebanon	1980	26.6	73.4	-5.1	1.2	...	...	...	...
Liban	1990	16.2	83.8	-4.6	1.7	1992	1600	68.0	58.0
Macau	1980	...	...	-0.1	4.4	...	...	...	...
Macao	1990	...	...	0.3	3.4	...	...	...	...
Malaysia	1980	65.3	34.7	1.1	4.8	...	...	...	...
Malaisie	1990	57.0	43.0	1.2	4.7	1992	1900	22.0	9.9
Maldives	1980	77.8	22.2	1.8	7.0	...	...	...	...
Maldives	1990	70.4	29.6	2.0	5.8	1990	63	100.0	29.3
Mongolia	1980	47.9	52.1	1.4	4.1	...	...	...	...
Mongolie	1990	42.1	57.9	1.4	3.8	1990	470	41.0	21.5
Myanmar	1980	76.0	24.0	2.1	2.2	...	...	...	...
Myanmar	1990	75.2	24.8	2.0	2.8	1992	3500	31.9	8.0
Nepal	1980	93.5	6.5	2.3	8.1	...	...	...	...
Népal	1990	89.1	10.9	2.1	7.6	1990	360	19.6	1.9
Oman	1980	92.4	7.6	4.8	8.9	...	...	...	...
Oman	1990	89.0	11.0	3.4	7.4	1990	70	44.0	4.7
Pakistan	1980	71.9	28.1	2.2	3.9	...	...	...	...
Pakistan	1990	68.0	32.0	2.6	4.6	1992	8600	21.0	6.9
Philippines	1980	62.6	37.4	1.9	3.5	...	...	...	...
Philippines	1990	57.3	42.7	1.5	3.7	1992	9600	33.5	14.7
Qatar	1980	14.4	85.6	2.5	6.5	...	...	...	...
Qatar	1990	10.1	89.9	...	4.0	1990	329	100.0	89.4
Saudi Arabia	1980	33.2	66.8	0.8	7.7	...	...	...	...
Arabie saoudite	1990	22.7	77.3	0.2	4.8	1992	2200	17.9	14.0
Singapore	1980	...	100.0	...	1.3	...	...	...	...
Singapour	1990	...	100.0	...	1.2	1992	2800	100.0	100.0
Sri Lanka	1980	78.4	21.6	1.8	1.3	...	...	...	...
Sri Lanka	1990	78.6	21.4	1.3	1.6	1990	620	16.9	3.6
Syrian Arab Republic	1980	53.3	46.7	2.5	3.9	...	...	...	...
Rép. arabe syrienne	1990	49.8	50.2	2.9	4.3	1992	1900	27.8	14.2
Thailand	1980	82.9	17.1	2.0	4.9	...	...	...	...
Thaïlande	1990	77.8	22.2	0.6	4.0	1992	7600	58.0	13.6
Turkey	1980	56.2	43.8	1.3	3.1	...	...	...	...
Turquie	1990	39.1	60.9	-1.8	5.1	1992	7000	18.7	12.0
United Arab Emirates	1980	28.5	71.5	10.1	15.8	...	...	...	...
Emirats arabes unis	1990	19.1	80.9	-0.7	4.3	1990	260	21.0	16.4
Uzbekistan	1980	...	...	...	...	...	...	...	...
Ouzbékistan	1990	...	...	...	...	1992	2200	24.6	10.1
Viet Nam	1980	80.8	19.2	2.1	2.7	...	...	...	...
Viet Nam	1990	80.1	19.9	2.1	2.5	1992	3300	23.9	4.8
Yemen	1980	79.8	20.2	2.3	7.4	...	...	...	...
Yémen	1990	71.1	28.9	2.4	7.0	1990	360	15.7	3.9
Europe · Europe									
Albania	1980	66.2	33.8	1.6	2.6	...	...	...	...
Albanie	1990	64.2	35.8	1.6	2.4	1990	240	21.0	7.4
Andorra	1980	...	...	7.2	5.1	...	...	...	...
Andorre	1990	...	...	2.4	0.3	...	...	...	...
Austria	1980	45.2	54.8	-0.8	0.5	...	...	...	...
Autriche	1990	41.6	58.4	-0.5	1.1	1992	2100	46.6	27.6
Belarus	1980	...	...	...	...	...	...	...	...
Bélarus	1990	...	...	...	...	1992	1700	24.8	16.6
Belgium	1980	4.6	95.4	-2.0	0.2	...	...	...	...
Belgique	1990	3.7	96.3	-1.9	0.3	1992	1300	13.8	13.3
Bulgaria	1980	38.8	61.2	-1.5	1.6	...	...	...	...
Bulgarie	1990	32.3	67.7	-1.8	1.0	1992	1400	21.9	15.1
Channel Islands	1980	...	...	0.6	-0.7	...	...	...	...
Iles Anglo-Normandes	1990	...	...	1.4	1.4	...	...	...	...
Czechoslovakia	1980	32.5	67.5	-2.8	2.6	...	...	...	...
Tchécoslovaquie	1990	22.9	77.1	-3.4	1.4	1992	1200	9.9	7.8
Denmark	1980	16.3	83.7	-2.0	0.7	...	...	...	...
Danemark	1990	15.2	84.8	-0.6	0.2	1992	1300	30.3	25.8
Estonia	1980	30.3	69.7	-0.6	1.2	...	...	...	...
Estonie	1990	28.2	71.8	-0.1	0.9	...	...	...	...
Faeroe Islands	1980	...	...	1.2	1.3	...	...	...	...
Iles Féroé	1990	...	...	0.2	1.7	...	...	...	...

12
Population in urban and rural areas, rates of growth and largest city population [*cont.*]
Population urbaine, population rurale, taux d'accroissement et population de la ville la plus peuplée [*suite*]

Country or area Pays ou zone	Year Année	Population estimates Estimations de la population				Year Année	Largest city population Ville la plus peuplée		
		R%	U%	Δ p.a.(%)[1] Pop.r.	Pop.u.		Number (000s) Nombre (000s)	% u	% t
Finland	1980	40.2	59.8	-0.5	0.8	...	...	...	...
Finlande	1990	40.3	59.7	0.4	0.3	1992	1000	34.2	20.5
France	1980	26.7	73.3	0.2	0.5	...	...	...	...
France	1990	27.3	72.7	0.9	0.4	1992	9400	22.6	16.4
Germany	1980	17.1	82.9	-1.7	0.3	...	...	...	...
Allemagne	1990	14.8	85.2	-1.2	0.8	1992	6400	9.3	8.0
Gibraltar	1980	...	...	...	0.1	...	...	...	...
Gibraltar	1990	...	...	...	1.6	...	...	...	...
Greece	1980	42.4	57.6	0.2	2.1	...	...	...	...
Grèce	1990	37.5	62.5	-0.9	1.2	1992	3500	54.7	34.7
Holy See	1980	...	...	...	...	...	...	...	...
Saint-Siège	1990	...	...	...	...	...	...	...	...
Hungary	1980	43.0	57.0	-1.5	1.9	...	...	...	...
Hongrie	1990	35.7	64.3	-2.1	1.0	1992	2100	30.9	20.3
Iceland	1980	11.8	88.2	-1.6	1.3	...	...	...	...
Islande	1990	9.4	90.6	-1.2	1.3	1990	150	65.5	59.3
Ireland	1980	44.7	55.3	0.6	2.0	...	...	...	...
Irlande	1990	42.9	57.1	-0.7	...	1990	930	43.8	25.0
Isle of Man	1980	...	...	-7.5	6.4	...	...	...	...
Ile de Man	1990	...	...	0.6	1.9	...	...	...	...
Italy	1980	33.4	66.6	-0.2	0.7	...	...	...	...
Italie	1990	31.1	68.9	-0.7	0.6	1992	5300	13.1	9.1
Latvia	1980	31.8	68.2	-1.2	1.3	...	...	...	...
Lettonie	1990	28.8	71.2	-0.4	1.0	...	...	...	...
Liechtenstein	1980	...	...	1.4	1.2	...	...	...	...
Liechtenstein	1990	...	...	0.7	1.3	...	...	...	...
Lithuania	1980	38.9	61.1	-1.9	2.6	...	...	...	...
Lituanie	1990	31.2	68.8	-1.5	1.9	...	...	...	...
Luxembourg	1980	21.7	78.3	-3.8	1.3	...	...	...	...
Luxembourg	1990	15.5	84.5	-2.6	1.0	1990	80	25.5	21.5
Malta	1980	16.4	83.6	-1.7	1.9	...	...	...	...
Malte	1990	12.7	87.3	-1.9	0.9	1990	10	3.2	2.8
Monaco	1980	...	...	...	0.8	...	...	...	...
Monaco	1990	...	...	...	0.4	...	...	...	...
Netherlands	1980	11.6	88.4	0.7	0.7	...	...	...	...
Pays-Bas	1990	11.4	88.6	0.4	0.7	1992	1100	8.0	7.1
Norway	1980	29.5	70.5	-1.2	1.1	...	...	...	...
Norvège	1990	25.0	75.0	-1.2	1.0	1990	660	20.9	15.7
Poland	1980	41.9	58.1	-0.4	1.8	...	...	...	...
Pologne	1990	38.2	61.8	-0.4	1.1	1992	3500	14.5	9.1
Portugal	1980	70.6	29.4	0.9	2.7	...	...	...	...
Portugal	1990	66.4	33.6	-0.8	1.4	1992	1600	47.5	16.4
Romania	1980	50.9	49.1	-0.2	2.1	...	...	...	...
Roumanie	1990	46.3	53.7	-0.5	1.3	1992	2200	17.5	9.6
San Marino	1980	...	...	-5.8	4.0	...	...	...	...
Saint-Marin	1990	...	...	-8.1	1.6	...	...	...	...
Spain	1980	27.2	72.8	-1.2	2.0	...	...	...	...
Espagne	1990	21.6	78.4	-2.0	0.9	1992	5400	17.3	13.7
Sweden	1980	16.9	83.1	-0.1	0.4	...	...	...	...
Suède	1990	16.0	84.0	-0.2	0.6	1992	1700	23.5	19.8
Switzerland	1980	43.0	57.0	-0.6	0.4	...	...	...	...
Suisse	1990	38.5	61.5	-0.4	1.5	1990	290	7.3	4.4
Ukraine	1980	...	...	...	...	...	...	...	...
Ukraine	1990	...	...	...	...	1992	2700	7.8	5.2
United Kingdom	1980	11.2	88.8	-0.1	0.1	...	...	...	...
Royaume-Uni	1990	10.9	89.1	-0.1	0.3	1992	7300	14.3	12.7
Yugoslavia	1980	54.7	45.3	-1.0	3.5	...	...	...	...
Yougoslavie	1990	43.9	56.1	-1.7	2.5	1992	1700	12.0	7.0
Oceania · Océanie									
American Samoa	1980	...	...	1.4	2.3	...	...	...	...
Samoa américaines	1990	...	...	3.1	4.4	...	...	...	...
Australia	1980	14.2	85.8	1.7	1.5	...	...	...	...
Australie	1990	14.8	85.2	2.0	1.6	1992	3800	25.1	21.4
Cook Islands	1980	...	...	-0.6	-3.6	...	...	...	...
Iles Cook	1990	...	...	-0.4	-0.4	...	...	...	...

12
Population in urban and rural areas, rates of growth and largest city population [*cont.*]
Population urbaine, population rurale, taux d'accroissement et population de la ville la plus peuplée [*suite*]

Country or area Pays ou zone	Year Année	Population estimates Estimations de la population		Δ p.a.(%)[1]		Year Année	Largest city population Ville la plus peuplée Number (000s) Nombre (000s)	% u	% t
		R%	U%	Pop.r.	Pop.u.				
Fiji	1980	62.3	37.7	1.6	2.5	...	...	...	...
Fidji	1990	60.7	39.3	0.5	1.2	1990	200	66.7	26.2
French Polynesia	1980	39.7	60.3	2.1	3.7	...	...	...	...
Polynésie française	1990	35.4	64.6	1.3	3.3	1990	114	85.7	55.3
Guam	1980	59.8	40.2	-0.1	6.4	...	...	...	...
Guam	1990	47.0	53.0	-0.5	4.6	...	...	...	...
Kiribati	1980	...	...	1.0	2.4	...	...	...	...
Kiribati	1990	...	...	1.2	3.4	...	...	...	...
New Caledonia	1980	51.1	48.9	...	2.1	...	...	...	...
Nouvelle-Calédonie	1990	64.7	35.3	0.7	2.3	1990	77	...	46.1
New Zealand	1980	16.6	83.4	-0.6	0.4	...	...	...	...
Nouvelle-Zélande	1990	16.1	83.9	0.6	0.9	1990	330	11.6	9.7
Niue	1980	...	...	-3.7	-3.5	...	...	...	...
Nioué	1990	...	...	-5.7	-3.6	...	...	...	...
Papua New Guinea	1980	86.9	13.1	2.2	4.3	...	...	...	...
Papouasie-Nouvelle-Guinée	1990	84.2	15.8	1.9	4.3	1990	200	32.6	5.2
Samoa	1980	...	...	0.6	0.8	...	...	...	...
Samoa	1990	...	...	-0.1	0.8	...	...	...	...
Solomon Islands	1980	89.4	10.6	3.2	6.5	...	...	...	...
Iles Salomon	1990	85.3	14.7	2.9	6.6	...	...	...	...
Tonga	1980	...	...	-0.1	3.9	...	...	...	...
Tonga	1990	...	...	-1.3	4.3	...	...	...	...
Vanuatu	1980	...	...	2.4	5.5	...	...	...	...
Vanuatu	1990	...	...	2.4	2.8	...	...	...	...
former USSR · ancienne URSS									
former USSR	1980	37.5	62.5	-0.6	1.8	...	...	...	...
ancienne URSS	1990	33.9	66.1	-0.2	1.4	1992	9200	8.4	6.2

Source:
World Population Prospects (United Nations forthcoming publication), World Urbanization Prospects 1990, (United Nations publication, Sales No. E.91.XIII.11) and World Urbanization 1992 (wall poster, Sales No. E.93.XIII.2).

Source:
"World Population Prospects" (publication des Nations Unies), "World Urbanization Prospects 1990," (publication des Nations Unies, No de vente E.91.XIII.11) et "World Urbanization 1992" (tableau mural, No de vente E.93.XIII.2).

₲ All data shown pertain to Germany prior to 3 October 1990 are indicated separately for the Federal Republic of Germany and the former German Democratic Republic based on their respective territories at the time indicated. Where data for united Germany (3 October 1990 and thereafter) are not available, available data are shown separately under the designations Federal Republic of Germany and former German Democratic Republic and pertain to the territorial boundaries prior to 3 October 1990. For detailed explanatory notes on data pertaining to Germany, see Annex I-Country Nomenclature.

₲ Toutes les données se rapportant à l'Allemagne avant le 3 octobre 1990 figurent dans deux rubriques séparées basées sur les territoires respectifs de la République fédérale d'Allemagne et l'ancienne République démocratique allemande selon la période indiquée. En l'absence de données pour l'Allemagne unifiée (à compter du 3 octobre 1990), les données disponibles sont fournies séparément sous les rubriques République fédérale d'Allemagne et ancienne République démocratique allemande et se rapportent aux limites territoriales antérieures au 3 octobre 1990. Pour les notes explicatives en détail sur les données concernant l'Allemagne, voir Annexe I - Nomenclature des pays.

1 Annual rates of growth calculated for periods 1975-1980 and 1985-1990.

1 Ces taux d'accroissement annuel ont été calculés pour les périods 1975-1980 et 1985-1990.

Technical notes, tables 11 and 12

Table 11 is based on detailed data on population and its growth and distribution published in the United Nations *Demographic Yearbook*, which also provides a comprehensive description of methods of evaluation and limitations of data. A brief explanation of the quality code used for total population estimates in table 11 is given below. [19]

For "Type of estimate", the code indicates the method by which official estimates of population for the year 1990 were prepared, so far as could be ascertained. The letters A–D indicate the nature of the base figure; numerals to capital letters indicate the lapse of time since the establishment of the base figure; letters a–d indicate the method of time adjustment by which the base figure is brought up to date; numerals to small letters indicate the quality of time adjustment.

The details of the code classification for this column are given below:

Nature of base data (capital letter)

A. Complete census of individuals.
B. Sample survey.
C. Partial census or partial registration of individuals.
D. Conjecture.
... Nature of base data not determined.

Recency of base data (subscript numeral following capital letter)

Numeral indicates time elapsed (in years) since establishment of base figure.

Method of time adjustment (lower-case letter)

a. Adjustment by continuous population register.
b. Adjustment based on calculated balance of births, deaths and migration.
c. Adjustment of assumed rate of population increase.
d. No adjustment: base figure held constant at least two consecutive years.
... Method of time adjustment not determined.

Quality of adjustment for types a and b (numeral following letter a or b)

1. Population balance adequately accounted for.
2. Adequacy of accounting for population balance not determined but assumed to be adequate.
3. Population balance not adequately accounted for.

Quality of adjustment for type c (numeral following letter c)

1. Two or more censuses taken at decennial intervals or less.
2. Two or more censuses taken, but latest interval exceeds a decennium.
3. One or no census taken.

Unless otherwise indicated, figures refer to de facto (present-in-area) population for present territory; surface area estimates include inland waters.

Notes techniques, tableaux 11 et 12

Le *Tableau 11* : est fondé sur des données détaillées sur la population, sa croissance et sa distribution, publiées dans l'*Annuaire démographique des Nations Unies*, qui offre également une description complète des méthodes d'évaluation et une indication des limites des données. On trouvera ci-après une brève explication du code de qualité utilisé pour les estimations de la population totale présentées au Tableau 11 [19].

Pour le "Type d'estimation", le code indique, dans la mesure où elle a pu être déterminée, la méthode selon laquelle les estimations officielles de la population ont été établies pour l'année 1990. Les lettres A-D indiquent la nature du chiffre de base; les nombres placés à la droite de ces lettres indiquent le temps écoulé depuis l'établissement du chiffre de base; les lettres a-d indiquent la méthode d'actualisation du chiffre de base; les numéros qui suivent ces lettres en petits caractères indiquent la qualité de l'actualisation.

Le détail de la classification codée est donné ci-dessous :

Nature des données de base (majuscules)

A. Recensement complet de la population.
B. Enquête par échantillonnage.
C. Recensement partiel ou enregistrement partiel de la population.
D. Conjecture.
... La nature des données de base n'est pas déterminée.

Actualité relative des données (nombre en indice suivant la lettre majuscule)

Le nombre indique le temps écoulé (en années) depuis l'établissement du chiffre de base.

Méthode d'actualisation (lettre minuscule)

a. Actualisation par enregistrement continu de la population.
b. Actualisation basée sur le calcul de la balance des naissances, des décès et des migrations.
c. Actualisatin du taux présumé d'accroissement de la population.
d. Absence d'actualisation : le chiffre de base est maintenu constant pendant au moins deux années consécutives.
... Méthode d'actualisation non déterminée.

Qualité de l'actualisation pour les types a et b (chiffre suivant la lettre a ou b)

1. Balance démographique convenablement établie.
2. La qualité de l'établissement de la balance démographique n'est pas déterminée, mais on suppose qu'elle est convenable.
3. Balance démographique non convenablement établie.

In *table 12*, statistics of urban and rural population and population in the largest urban agglomeration by each country or area are from estimates by the Population Division of the United Nations Secretariat [25, 26, 27]. As noted above, because of national differences in the specific characteristics that distinguish urban from rural areas, there are no internationally agreed definitions of urban and rural. In most countries, the distinction is mainly based on size of locality. For the latest available census definition of urban areas in such country or area, reference should be made to the *Demographic Yearbook 1990* [19].

For this table, data for largest city refer to the urban agglomeration. An urban agglomeration comprises the city or town proper and also the suburban fringe or thickly settled territory lying outside, but adjacent to, its boundaries.

Annual rates of change in urban and rural population are computed as average annual percentage changes using mid-year population estimates.

Qualité de l'actualisation pour le type c (définie par le chiffre suivant la lettre c)
1. Deux recensements ou plus ont été effectués en dix ans ou moins.
2. Deux recensements ou plus ont été effectués, mais à plus de dix ans d'intervalle.
3. Un recensement effectué ou aucun.

Sauf indication contraire, les chiffres se rapportent à la population effectivement présente sur le territoire tel qu'il est actuellement défini; les estimations de superficie comprennent les étendues d'eau intérieures.

Au Tableau 12, les statistiques de la population urbaine, de la population rurale et de la population de la ville la plus peuplée de chaque pays ou zone sont tirées d'estimations de la Division de la population du Secrétariat des Nations Unies [25, 26, 27]. Comme il a été indiqué précédemment, il n'existe pas de définition reconnue à l'échelle internationale des zones urbaines et rurales parce que les caractéristiques retenues pour distinguer ces deux types de zone diffèrent d'un pays à un autre. Dans la plupart des pays, cette distinction est essentiellement fonction de la taille des agglomérations. Pour la définition la plus récente des zones urbaines utilisée dans une région ou un pays donné, se reporter à l'*Annuaire démographique 1990* [19].

Sur ce tableau, les données concernant la ville la plus peuplée se rapportent à l'agglomération urbaine. L'agglomération urbaine comprend la ville proprement dite et ses faubourgs ou banlieues, et tout territoire à forte densité de population situé à sa périphérie.

Les taux annuels de variation des populations urbaines et rurales se calculent sur la base de la variation annuelle moyenne en pourcentage déterminée à partir des estimations de la population en milieu d'année.

13
Education preceding the first level
Enseignement précédant le premier degré
Institutions, teachers and pupils
Etablissements, personnel enseignant et élèves

Country or area Pays ou zone	Years Années	Schools Ecoles	Teaching staff Personnel enseignant		Pupils enrolled Elèves inscrits	
			Total	% F	Total	% F
Africa · Afrique						
Angola[1]	1980	...	...	...	390 512	40
Angola[1]	1985	...	...	...	227 654	...
	1989	...	...	...	191 882	...
Benin	1975	-	-	-	-	-
Bénin	1980	92	174	...	3 779	...
	1985	241	526	55	11 302	45
	1988	308	652	57	13 433	45
Burkina Faso	1975	...	19	100	497	48
Burkina Faso	1980	...	...	...	732	44
	1985	...	...	...	1 075	...
	1989	95	259[2]	98[2]	7 655	51
Burundi	1975[2]	...	10	...	594	49
Burundi	1980[2]	...	15	100	1 004	50
	1985	...	32	94	1 774	50
	1987	...	...	...	2 140	48
Cameroon	1975	...	...	...	21 752	49
Cameroun	1980	436	1 512	99	40 574	49
	1985	536	2 454	100	73 486	49
	1989	745	3 444	100	92 966	50
Cape Verde	1981	...	...	...	1 823	...
Cap-Vert	1983	13	69	100	1 973	...
	1986	58	136	100	4 523	...
Central African Rep.	1975	...	213	100	10 673	*50
Rép. centrafricaine	1986	164	360	100	11 450	39
	1987	173	572	...	11 677	...
Comoros						
Comores	1980	...	600	33	17 778	48
Congo	1980	36	334	100	3 498	50
Congo	1985	51	537	100	5 595	49
	1989	53	600	...	5 983	...
Côte d'Ivoire	1975	...	108	100	4 656	49
Côte d'Ivoire	1980	...	179	100	6 291	47
	1985	...	...	...	8 570	48
Djibouti	1975	1	2	100	159	57
Djibouti	1985	1	...	...	234	50
	1989	1	4	...	220	58
Egypt	1975	...	...	...	41 948	49
Egypte	1980	433	...	...	74 921	49
	1985	602	...	...	128 272	49
	1989	936	5 094	97	177 740	48
Ethiopia	1982	573	1 125	...	52 749	...
Ethiopie	1985	708	1 532	...	69 736	...
	1988	848	1 888	...	87 355	...
Ghana	1974	1 055	917	...	56 089	...
Ghana	1980	1 160	...	...	158 395	46
	1986	3 049	11 656	97	236 665	51
	1989	4 735	15 152	...	323 406	48
Guinea-Bissau[3]	1980	6	46	98	459	49
Guinée-Bissau[3]	1984	7	38	97	751	51
	1986	17	61	87	1 246	...
	1988	5	43	86	754	49
Kenya	1982	*9 141	*11 594	...	*485 194	*49
Kenya	1985	11 780	15 676	100	582 505	50
	1987	12 466	17 473	100	660 973	51
Liberia	1975	...	...	...	53 785	40
Libéria	1980	60	...	...	80 215	42
	1984	...	...	...	70 507	...
Libyan Arab Jamahiriya	1975	62	364	100	7 727	43
Jamah. arabe libyenne	1980	45	515	100	9 008	46
	1982	56	701	100	12 493	48
	1985	78	1 051	100	15 028	48

13
Education preceding the first level
Institutions, teachers and pupils [*cont.*]
      Enseignement précédant le premier degré
            Etablissements, personnel enseignant et élèves [*suite*]

Country or area Pays ou zone	Years Années	Schools Ecoles	Teaching staff Personnel enseignant		Pupils enrolled Elèves inscrits	
			Total	% F	Total	% F
Mauritius	1975	283	358	99	9 233	49
Maurice	1980	349	453	100	11 704	49
	1982	340	543	98	10 617	49
Morocco	1975[4]	...	...	...	375 567	...
Maroc	1980	27 245	29 611[4]	5[4]	579 547	25
	1985	30 180	33 919	10	723 770	30
	1989	32 988	38 321	...	787 472	31
Mozambique[1]	1981	...	...	...	90 151	47
Mozambique[1]	1985	...	...	...	62 940	47
	1986	...	...	...	45 100	46
Niger	1975	6	19	100	874	49
Niger	1980	21	85	100	2 561	50
	1985	24	115	100	3 980	49
	1990	71	277	100	11 038	48
Réunion	1975	...	514[2]	...	26 700	...
Réunion	1980	...	...	...	33 816	...
	1985	142	1 212	97	37 694	49
	1988	...	...	...	40 257	...
Rwanda						
Rwanda	1987	150	...	...	*8 000	...
St. Helena	1976	3	10	100	85	54
Sainte-Hélène	1980	3	9	100	60	60
	1985	3	6	100	88	68
Sao Tome and Principe	1976	...	...	...	1 627	47
Sao Tomé-et-Principe	1980	...	...	...	2 430	...
	1986	...	...	...	3 240	50
	1989	13	116	93	3 446	51
Senegal	1980	98	85[2]	100[2]	8 445	51
Sénégal	1985	124	547	...	12 764	51
	1989	144	629	76	15 964	51
Seychelles	1976	41	...	...	1 787	51
Seychelles	1980	34	91	100	2 568	...
	1985	38	109	100	3 180	...
	1989	35	173	100	3 456	50
Somalia	1975	9	68	90	1 080	46
Somalie	1980	17	143	94	2 089	52
	1985	16	133	94	1 558	57
Sudan[5]	1980	3 135	3 183	...	148 879	...
Soudan[5]	1985	4 003	5 569	41	235 943	...
Swaziland	1976	39	67	100	1 605	53
Swaziland	1978	50	32	100	1 880	...
Togo	1975	94	133	89	6 723	49
Togo	1980	148	237	100	8 424	49
	1985	212	364	100	9 740	49
	1987	212	353	100	10 483	49
Tunisia	1987	520	1 376	100	39 819	...
Tunisie	1989	649	1 726	99	43 765	...
America, North · Amérique du Nord						
Antigua and Barbuda	1975	26	43	100	985	51
Antigua-et-Barbuda	1983	...	...	...	677	54
Barbados	1975	*114	86[2]	...	*2 661	*53
Barbade	1980	132	*151[2]	...	3 936	50
	1982	126	132[2]	...	3 052	49
Belize	1975	25	...	...	1 084	...
Belize	1980	55	*130	...	*2 000	...
	1986	72	130	...	*2 100	*52
Bermuda	1976	29	70	100	1 009	49
Bermudes	1980	31	...	...	1 067	51
	1984	32	37[2]	100[2]	1 287	49
British Virgin Islands	1975	6	10	100	155	43
Iles Vierges brit.	1980	8	15	80	299	45
	1984	...	...	...	226	49

13
Education preceding the first level
Institutions, teachers and pupils [cont.]
Enseignement précédant le premier degré
Etablissements, personnel enseignant et élèves [suite]

Country or area Pays ou zone	Years Années	Schools Ecoles	Teaching staff Personnel enseignant Total	% F	Pupils enrolled Elèves inscrits Total	% F
Canada	1975	...	*16 600	...	398 476	49
Canada	1980	...	...	...	397 266	49
	1985	...	...	...	422 085	49
	1989	...	14 950	69	475 000	49
Costa Rica	1975	318	500	100	15 608	50
Costa Rica	1980	370	673	...	21 857	49
	1985	536	1 302	...	36 356	50
	1989	763	1 479	...	43 641	49
Cuba	1975	...	4 358	...	126 565	49
Cuba	1980	...	5 047	99	123 741	49
	1985	...	4 847	96	108 881	...
	1988	...	7 076	90	143 705	...
Dominica	1975	52	60	97	2 300	55
Dominique	1985	54	86	100	2 500	...
	1989	49	89	100	2 192	52
Dominican Republic	1975	...	...	...	24 015	...
République dominicaine	1980	286	...	...	27 278	...
	1985	...	...	...	67 615	...
	1989	...	...	...	22 237[2]	48[2]
El Salvador	1975	320	592[2]	100[2]	42 227	52
El Salvador	1980	459	1 036	100	48 684	52
	1984	770	1 144	97	61 223	52
	1988	1 054	1 769	95	72 238	52
Grenada	1976	71	94	100	2 487	52
Grenade	1980	67	115	98	2 500[2]	...
	1986	68	146	99	3 283	50
	1987	70	150	99	3 584	51
Guadeloupe	1975	...	...	...	11 313	...
Guadeloupe	1980	...	...	...	16 875	...
	1985	96	615	...	18 866	49
	1990	119	768	...	19 983	49
Guatemala[6]	1975	342	999	93	63 869	44
Guatemala[6]	1980	564	1 700	...	48 869	50
	1985	2 864	4 407	...	133 726	48
	1987	2 992	*5 029	...	*144 312	...
Haiti	1982	150	...	...	*10 000	...
Haïti	1984	360	956	100	21 000	...
Honduras	1975	234	406	100	16 136	50
Honduras	1980	441	833	...	33 034	*42
	1985	...	...	...	*48 610	...
	1986	766	1 446	100	52 831	51
Jamaica	1975	1 705	3 163	...	126 217	54
Jamaïque	1980	...	...	...	119 508	52
	1985	1 581	...	...	125 046	...
	1989	1 673	3 942	...	136 671	50
Martinique	1975	...	...	...	21 459	...
Martinique	1980	...	...	...	17 678	...
	1986	75	598	98	18 422	48
	1989	81	...	...	19 546	50
Mexico	1975	4 156	14 073	*100	537 090	49
Mexique	1980	13 021	32 368	100	1 071 619	50
	1985	35 649	80 529	100	2 381 412	50
	1990	46 736	104 967	*100	2 734 065	...
Montserrat[2]	1976	10	23	100	444	...
Montserrat[2]	1981	9	20	95	278	45
Netherlands Antilles	1981	94	348	...	8 694	49
Antilles néerlandaises	1982	93	349	...	8 707	50
Nicaragua	1975	...	...	...	8 986	...
Nicaragua	1980	463	924	99	30 524	51
	1985	686	1 983	96	62 784	52
	1989	879	2 135	99	64 916	52

13
Education preceding the first level
Institutions, teachers and pupils [*cont.*]
Enseignement précédant le premier degré
Etablissements, personnel enseignant et élèves [*suite*]

Country or area Pays ou zone	Years Années	Schools Ecoles	Teaching staff Personnel enseignant		Pupils enrolled Elèves inscrits	
			Total	% F	Total	% F
Panama	1975	224	457	100	12 398	51
Panama	1980	365	645	100	18 136	50
	1985	637	1 121	99	27 501	50
	1989	797	1 369	100	32046	49
St. Kitts-Nevis	1984	37	52	98	1 151	50
St. Kitts-Nevis	1985	36	*46	*98	*1 501	*49
	1988	50	94	100	1 618	52
Saint Lucia	1985	105	202	...	3 711	...
Sainte-Lucie	1989	125	225	...	4 500	...
St. Pierre and Miquelon	1980	4	18	100	356	47
St. Pierre et Miquelon	1985	4	18	100	361	52
	1986	5	18	94	373	44
St. Vincent and the Grenadines	1981	30	...	...	967	48
Saint-Vincent-et-Grenadines	1983	40	...	...	*1 415	...
	1986	66	171	99	2 150	51
Trinidad and Tobago	1980[2]	39	117	100	1 739	...
Trinité-et-Tobago	1985[2]	49	147	100	2 035	...
	1989	158	360	96	5 049	61
United States	1975	...	...	...	5 141 000	...
Etats-Unis	1980	...	...	...	5 163 000	47
	1985	...	...	...	6 306 000	49
	1986	...	...	...	6 515 000	48
US Virgin Islands	1980	...	...	...	2 561	...
Iles Vierges américaine	1985	58	71[2]	...	2 656	...
America, South · Amérique du Sud						
Argentina	1975	5 694	18 991	100	369 082	50
Argentine	1980	6 622	...	...	480 216	...
	1985	8 015	36 287	100	693 259	50
	1988	9 137	44 584	99	798 235	50
Bolivia	1974	345	1 434	...	74 031	...
Bolivie	1980	...	...	...	90 031	49
	1986	2 038	2 386	...	133 677	49
	1989	...	3 522	80	112 086	49
Brazil	1975	9 158	26 393	99	566 008	50
Brésil	1980	15 320	58 788	98	1 335 317	49
	1985	38 418	115 140	...	2 650 490	...
	1989	*49 998	*149 150	...	*3 530 000	...
Chile	1975	1 761	2 512	100	124 697	...
Chili	1980	...	...	...	174 909	37
	1985	...	...	...	202 252	49
	1989	4 722	9 205	...	278 443	49
Colombia	1975	2 013	3 887	...	95 908	...
Colombie	1980	3 281	6 742	...	174 369	...
	1985	5 127	10 891	...	285 286	50
	1989	6 920	13 794	...	328 425	50
Ecuador	1975	254	778	95	23 864	51
Equateur	1980	736	1 858	95	50 819	51
	1985	...	3 846	95	93 665	50
	1987	...	4 756	91	108 348	50
French Guiana	1975	8	56[2]	...	3 359	...
Guyane française	1980	...	...	...	3 879	...
	1983	18	152	98	4 117	...
Guyana	1976	356	1 662	99	25 784	50
Guyana	1980	374	2 018	98	27 955	50
	1985	...	1 399	99	25 685	51
	1986	349	...	...	25 316	50
Paraguay	1982	32	...	...	13 590	51
Paraguay	1985	70	...	...	19 052	50
	1989	94	...	...	30 019	49
Peru	1975	2 098	4 459	98	172 051	49
Pérou	1980	3 271	6 778	99	228 168	50
	1985	5 268	11 206	99	342 779	50
	1990	...	22 271	...	*603 757	...

13
Education preceding the first level
Institutions, teachers and pupils [cont.]
Enseignement précédant le premier degré
Etablissements, personnel enseignant et élèves [suite]

Country or area Pays ou zone	Years Années	Schools Ecoles	Teaching staff Personnel enseignant		Pupils enrolled Elèves inscrits	
			Total	% F	Total	% F
Suriname	1975	309	589	100	17 581	50
Suriname	1979	279	653	...	18 191	...
	1984	281	...	...	16 935	50
	1986	255	...	...	*16 570	*50
Uruguay	1975	...	921[2]	...	40 239	...
Uruguay	1980	...	1 001[2]	...	42 444	...
	1985	968	2 012	...	55 092	50
	1988	...	1 393	...	61 187	...
Venezuela	1975	327	6 246	100	224 600	50
Venezuela	1980	...	16 487	...	421 183	...
	1985	1 407	22 102	98	561 846	50
	1988	...	24 174	99	555 933	50
Asia · Asie						
Afghanistan	1976	27	164	100	2 891	*42
Afghanistan	1980	36	369	100	4 470	45
	1985	...	873	100	17 000	44
	1988	263	...	...	19 660	49
Bahrain	1975	...	...	...	1 983	49
Bahreïn	1980	...	...	...	3 730	47
	1985	56	255	100	7 608	47
	1988[7]	...	298	100	8 994	48
Bangladesh						
Bangladesh	1988	51 495	63 054	18	2 317 181	45
Brunei Darussalam	1975	14	39	100	1 496	46
Brunéi Darussalam	1980	143	324	77	6 760	48
	1985	141	366	...	8 055	...
	1989	155	...	...	8 634	49
China	1975	171 749	236 500	...	6 200 000	...
Chine	1980	170 419	410 700	82	11 507 700	...
	1985	172 262	549 900	96	14 796 900	47
	1989	172 634	709 100	95	18 476 600	46
Cyprus[8]	1975	96	222	98	4 229	49
Chypre[8]	1980	259	418	100	10 397	48
	1985	423	676	99	16 810	48
	1989	540	917	99	22 008	48
Hong Kong	1975	839	4 168	97	160 184	48
Hong-kong	1980	761	5 177	98	197 410	48
	1985	787	6 959	98	229 089	48
	1988	814	...	...	214 700	...
India	1975	5 658	...	...	569 296	45
Inde	1980	10 802	...	...	918 238	45
	1985	11 187	...	...	1 235 750	45
	1989	13 858	...	...	1 353 023	46
Indonesia	1976	12 935	24 503	100	579 876	55
Indonésie	1980	19 868	37 100	...	1 005 226	...
	1985	26 419	58 341	...	1 258 468	...
	1988	36 190	81 426	...	1 568 450	...
Iran, Islamic Rep. of	1975	1 804	6 985	100	175 424	46
Iran, Rép. islamique d'	1980	2 791	9 356	...	172 000	...
	1985	1 732	5 795	100	106 986	47
	1989	3 210	8 405	100	217 496	48
Iraq	1975	245	1 913	100	44 413	46
Iraq	1980	387	3 235	100	76 507	48
	1985	584	4 657	100	81 431	47
	1988	614	4 654	100	85 096	48
Israel	1975	5 289	6 122	100[2]	200 710	49
Israël	1980	...	...	...	269 506	*47
	1985	...	...	...	292 000	...
	1989	...	...	...	318 160	...
Japan	1975	13 108	93 853	88	2 292 591	49
Japon	1980	14 893	110 037	88	2 407 093	49
	1985	15 220	107 606	88	2 067 951	49
	1989	15 080	109 226	89	2 037 614	49

13
Education preceding the first level
Institutions, teachers and pupils [cont.]
Enseignement précédant le premier degré
Etablissements, personnel enseignant et élèves [suite]

Country or area Pays ou zone	Years Années	Schools Ecoles	Teaching staff Personnel enseignant		Pupils enrolled Elèves inscrits	
			Total	% F	Total	% F
Jordan	1975	158	453	100	14 952	42
Jordanie	1980	207	737	99	19 598	43
	1985	358	1 336	100	27 954	45
	1989	529	1 795	100	42 223	45
Korea, Dem.People's Rep.						
Corée, Rép. pop. dém. de	1987	16 964	35 000	100	728 000	48
Korea, Republic of[9]	1975	611	2 153	78	32 032	44
Corée, République de[9]	1980	901	3 339	85	66 433	45
	1985	6 242	9 281	90	314 692	47
	1990	8 354	18 511	94	414 532	47
Kuwait	1975	101	1 299	100	24 097	47
Koweït	1980	109	1 696	100	29 965	48
	1985	161	2 539	100	42 830	48
	1988	190	2 935	100	52 105	49
Lao People's Dem. Rep.	1980	153	252	100	5 296	51
République dém. pop. lao	1985	500	1 327	100	21 625	44
	1987	604	1 366	100	27 298	49
Lebanon	1980	...	6 604	...	123 530	...
Liban	1984	1 771	...	...	116 344	49
	1986	1 776	5 257	...	129 590	...
Malaysia	1980	3 087	...	...	170 955	...
Malaisie	1985	5 757	9 056	98	293 801	...
	1989	6 959	15 571	...	331 520	49
Maldives	1981	1	...	...	1 651	...
Maldives	1986	...	...	...	2 327	...
Mongolia	1975	542	1 699	...	36 974	53
Mongolie	1980	617	1 813	100	49 800	...
	1984	657	...	...	59 100	...
	1986	696	2 349	...	61 668	...
Nepal	1976	-	-	-	-	-
Népal	1983	153	706	61	14 952	38
	1984	176	733	61	16 864	41
Oman	1975	2	4	100	160	43
Oman	1980	3	11	91	396	44
	1985	7	71	100	1 665	44
	1989	6	139	100	3 050	42
Palestine . Palestine						
Gaza Strip						
Zone de Gaza	1986	56	173	...	6 773	43
Philippines	1975	...	...	...	86 443	...
Philippines	1980	...	...	...	*123 571	...
	1985	2 334	4 636	...	189 654	52
	1989	3 827	10 988	...	321 459	...
Qatar	1975	12	38	100	1 434	44
Qatar	1980	19	...	...	2 587	47
	1985	36	247	100	4 859	46
	1989	52	358	99	6 105	46
Saudi Arabia	1975	92	444	96	15 485	42
Arabie saoudite	1980	195	1 127	97	28 045	44
	1985	492	3 001	98	51 604	46
	1988	619	4 375	100	68 426	44
Singapore	1975	52	278	99	4 883	48
Singapour	1980	122	...	...	11 142	47
	1985	108	684	100	15 658	47
	1989	104	776	100	17 858	47
Syrian Arab Republic	1975	323	1 012	97	33 477	45
Rép. arabe syrienne	1980	351	1 082	97	33 611	45
	1985	610	2 028	100	61 988	45
	1988	790	2 705	98	75 374	45
Thailand	1975	2 864	...	...	214 620	48
Thaïlande	1980	...	...	...	367 313	49
	1985	12 996	*33 119	...	672 080	...
	1989	25 705	...	...	1 224 259	...

13
Education preceding the first level
Institutions, teachers and pupils [*cont.*]
Enseignement précédant le premier degré
Etablissements, personnel enseignant et élèves [*suite*]

Country or area Pays ou zone	Years Années	Schools Ecoles	Teaching staff Personnel enseignant		Pupils enrolled Elèves inscrits	
			Total	% F	Total	% F
Turkey	1980[7]	117	262	97	4 691	45
Turquie	1985	3 551	5 903	99	117 819	48
	1989	3 601	7 244	100	112 053	47
United Arab Emirates	1975	11[2]	186[2]	100[2]	7 603	51
Emirats arabes unis	1980	20[2]	359[2]	100[2]	17 263	46
	1985	32[2]	590[2]	100[2]	35 360	47
	1989	...	2 169[2]	98[2]	47 007	47
Viet Nam	1975	...	*22 300	...	*764 400	...
Viet Nam	1980	6 121	57 605	100	1 595 724	52
	1985	6 446	65 718	100	1 701 681	52
Yemen[10]	1975	11	157	92	2 820	45
Yémen[10]	1980	26	311	99	5 541	46
	1985	29	387	...	11578	48
	1989	39	475	99	11 500	49
Europe · Europe						
Albania	1980	2 667	4 162	100	92 490	...
Albanie	1985	3 064	4 850	100	110 603	...
	1989	3 330	5 440	100	125 312	...
Austria	1975	2 882	5 578	99	154 318[11]	49[11]
Autriche	1980	3 423	7 069	99	165 611	49
	1985	3 667	8 159	99	181 582	49
	1989	3 876	9 044	100	192 946	49
Belgium	1975	5 226	17 460	100	437 838	49
Belgique	1980	4 325	17 116	100	383 955	49
	1985	4 087	19 793	99	391 848	49
	1987	4 060	...	...	371 509	49
Bulgaria	1975	7 550	24 137	100	392 625	49
Bulgarie	1980	6 185	28 996	100	420 804	48
	1985	5 054	28 864	100	360 395	48
	1989	4 562	28 312	100	317 559	49
Czechoslovakia	1975	9 226	34 755	100	475 004	...
Tchécoslovaquie	1980	11 119	47 290	100	694 720	...
	1985	11 477	51 104	100	681 515	...
	1989	11 380	50 519	100	636 622	...
Denmark	1975	...	...	...	44 859	49
Danemark	1980	...	...	...	62 936	49
	1985	...	3 592	57	56 735	49
	1988	...	...	...	51 814	49
Finland	1975	...	...	...	23 230	...
Finlande	1980	...	...	...	31 282	...
	1985	...	...	...	41 971	...
	1989	...	...	...	49 772	...
France[12]	1975	13 051	57 658[2]	100[2]	2 591 142	49
France[12]	1980	16 080	66 948[2]	...	2 383 465	49
	1985	17 776	71 705[2]	96[2]	2 563 464	49
	1989	18 676	74 504[2]	96[2]	2 535 955	49
Germany ʃ · Allemagne ʃ						
Federal Republic of Germany	1975	26 305	...	...	1 655 825	48
Rép. féd. d'Allemagne	1980	26 793	...	...	*1 535 959	*49
	1985	*26 874	...	...	*1 582 051	*48
	1988	29 089	84 764	85	1 645 959	48
former German Dem. Rep.	1975	11 648	51 998	*100	693 163	...
ancienne Rép. dém. allemande	1980	12 145	56 448	100	663 491	...
	1985	13 148	69 612	100	788 232	...
	1989	13 452	73 383	100	747 140	...
Gibraltar[2]	1975	1	1	100	13	46
Gibraltar[2]	1980	1	2	100	28	39
	1984	2	5	100	150	35

13
Education preceding the first level
Institutions, teachers and pupils [cont.]
        Enseignement précédant le premier degré
        Etablissements, personnel enseignant et élèves [suite]

Country or area Pays ou zone	Years Années	Schools Ecoles	Teaching staff Personnel enseignant Total	% F	Pupils enrolled Elèves inscrits Total	% F
Greece	1975	3 279	4 137	100	108 357	48
Grèce	1980	4 576	6 514	100	145 924	49
	1985	5 203	7 617	100	160 079	49
	1987	5 389	7 942	100	155 246	49
Hungary	1975	4 077	20 512	100	329 408	48
Hongrie	1980	4 690	29 437	100	478 100	48
	1985	4 823	33 548	100	424 678	49
	1898	4 748	33835	100	392 273	48
Iceland	1975	...	...	...	3 502	...
Islande	1980	...	...	...	4 041	...
	1985	...	...	...	4 528	49
	1989	...	...	...	4 355	...
Ireland	1975	...	4 408	73	135 783	49
Irlande	1980	...	4 782	74	137 533	48
	1985	...	5 164	76	147 908	48
	1988	...	4 597	76	135 923	48
Italy	1975	27 485	63 523	100	1 822 527	49
Italie	1980	30 295	108 261	...	1 870 477	49
	1985	28 943	108 184	...	1 660 986	49
	1989	28 038	109 565	...	1 566 364	49
Luxembourg	1975	...	394	100	8 625	49
Luxembourg	1980	...	415	99	7 621	49
	1985	...	440	...	7 779	...
	1989	...	466	...	7 965	...
Malta	1975	43	406	85	6 237	47
Malte	1980	42	428	87	7 691	47
	1985	43	324	100	7 899	48
	1988	57	532	100	11 322	47
Monaco	1976[2]	...	...	...	352	...
Monaco	1980[2]	5	...	...	330	46
	1990	...	...	64	887	47
Netherlands[13]	1975	7 568	20 565	100	518 890	49
Pays-Bas[13]	1980	8 050	22 361	100	409 576	49
	1985	-	-	-	-	-
	1988	-	-	-	-	-
Norway	1975	...	5 516	...	30 479	...
Norvège	1980	2 554	16 866	...	78 189	...
	1985	3 281	...	...	98 454	...
	1989	4 310	33 075	95	128 237	...
Poland	1975	31 176	44 542	99	1 107 648	...
Pologne	1980	31 014	57 730	99	1 349 528	...
	1985	26 344	78 092	98	1 360 044	...
	1989	26 212	90 015	* 99	1 316 699	...
Portugal	1975	679	1 903	97	44 832	48
Portugal	1980	1 916	5 047	99	100 178	48
	1985	2 547	6 408	99	128 089	49
	1989	2 822	...	...	128 877	...
Romania	1975	13 537	33 789	100	812 420	50
Roumanie	1980	13 467	38 512	100	935 711	49
	1985	12 811	33 522	100	864 332	49
	1989	12 108	31 293	100	835 890	49
San Marino	1975	16	68	100	735	48
Saint-Marin	1980	17	91	100	877	49
	1985	16	109	100	814	48
	1989	15	108	100	773	50
Spain	1975	...	24 621	99	920 336	51
Espagne	1980	...	35 588	97	1 182 425	50
	1985	...	39 573	93	1 127 348	49
	1987	15 948	39 513	94	1 054 241	49
Sweden	1975	5 913	...	...	206 726	...
Suède	1980	8 504	...	...	226 571	...
	1985	...	...	...	262 417	...
	1989	12 850	...	...	307 816	...

13
Education preceding the first level
Institutions, teachers and pupils [cont.]
   Enseignement précédant le premier degré
   Etablissements, personnel enseignant et élèves [suite]

Country or area Pays ou zone	Years Années	Schools Ecoles	Teaching staff Personnel enseignant		Pupils enrolled Elèves inscrits	
			Total	% F	Total	% F
Switzerland	1976	...	...	...	130 116	46
Suisse	1980	...	...	...	120 315	49
	1985	...	...	...	123 128	49
	1989	...	...	...	136 845	49
United Kingdom[14]	1975	1 040	...	...	571 545	49
Royaume-Uni[14]	1980	1 251	...	...	326 400	49
	1985	1 262	...	...	665 000	47
	1988	1 312	27 000	100	716 000	49
Yugoslavia	1975	2 308	17 794	95	188 526	48
Yougoslavie	1980	3 177	29 436	94	290 870	48
	1985	3 852	37 686	94	374 740	48
	1989	4 157	46 413	95	435 932	48
Oceania · Océanie						
American Samoa	1980	111	...	...	1 922	46
Samoa américaines	1985[2]	97	98	100	2 001	47
	1989[2]	95	103	99	2 874	51
Australia[15]	1975	...	...	...	186 652	49
Australie[15]	1980	...	...	...	165 742	49
	1985	...	...	...	161 974	48
	1989	...	...	...	179 344	49
Cook Islands	1982	...	...	...	446	...
Iles Cook	1985	...	...	...	412	...
	1988	24	23	96	360	...
Fiji	1975	117	256	100	3 339	48
Fidji	1980	163	196	100	4 493	49
	1985	244	308	99	4 206	31
	1986	214	325	100	5 400	30
French Polynesia	1975	9	265	100	8 117	49
Polynésie française	1981	47	350	100	9 354	49
	1984	67	462	*100	11 937	49
	1989	74	623	...	15 379	49
Guam	1979	...	...	...	2 665	...
Guam	1983	...	99	94	2 892	43
Nauru						
Nauru	1985	4	20	100	383	49
New Caledonia	1975	...	...	...	6 195	...
Nouvelle-Calédonie	1980	...	...	...	10 313	50
	1985	52	355	...	8 647	49
	1990	73	428	...	10 745	49
New Zealand	1975	1 098	1 882	100	54 757	49
Nouvelle-zélande	1980	1 208	...	...	56 858	49
	1985	1 380	1 459	*99	60 666	49
	1989	1 410	1 672	99	73 390	49
Papua New Guinea	1975	...	...	...	51	41
Papouasie-Nouv.-Guinée	1982	...	...	...	370	44
	1987	...	...	...	1 196	43
	1989	...	...	...	1 108	46
Tokelau	1979	3	3	100	75	51
Tokélaou	1981	3	3	100	56	48
	1983	3	5	60	71	35
Vanuatu	1976	26	41	100	1 076	49
Vanuatu	1980	34	49	100	1 187	47
	1981	...	...	...	1 092	48
former USSR · ancienne URSS						
former USSR	1975	99 392	730 723	100	8 403 162	...
ancienne URSS	1980	116 000	1 081 000	...	10 212 000	...
	1985	131 000	1 389 000	...	11 546 000	...
	1989	143 000	1 579 000	100	12 718 000	49
Belarus	1976	...	...	...	371 601	...
Bélarus	1980	3 488	36 400	...	345 300	...
	1985	4 476	52 000	...	407 100	...
	1989	5 036	59 600	...	465 600	...

13
Education preceding the first level
Institutions, teachers and pupils [*cont.*]
Enseignement précédant le premier degré
Etablissements, personnel enseignant et élèves [*suite*]

| Country or area | Years | Schools | Teaching staff Personnel enseignant | | Pupils enrolled Elèves inscrits | |
Pays ou zone	Années	Ecoles	Total	% F	Total	% F
Ukraine	1975	17 707	...	...	1 886 000	...
Ukraine	1980	21 200	...	...	1 779 300	...
	1985	22 600	...	...	1 902 600	...
	1989	23 500	...	...	2 000 000	...

Source:
United Nations Educational, Scientific and Cultural Organization (Paris).

Source:
Organisation des Nations Unies pour l'éducation, la science et la culture (Paris).

§   All data shown which pertain to Germany prior to 3 October 1990 are indicated separately for the Federal Republic of Germany and the former German Democratic Republic based on their respective territories at the time indicated. For detailed explanatory notes on data pertaining to Germany, see Annex I-Country Nomenclature.

1   Data refer to initiation classes where pupils learn Portuguese.
2   Data refer to public education only.
3   Except for 1986, data refer to 'Sector autonomo Bissau' only.
4   Data refer to Koranic schools only.
5   Data include Koranic schools 'khalwas' which accept pupils of all ages.
6   For 1975, data on teachers and all data for 1980 do not include classes where pupils learn Spanish (la castellanización).
7   Data refer to kindergarten only.
8   Not including Turkish schools.
9   Due to a change of school system, from 1987 data are not comparable with those of previous years.
10 Data refer to former Democratic Yemen only.
11 Including special education and exclude 'Vorschulklassen'.
12 Data from 1985 do not include pre-primary classes in primary schools.
13 From 1985, preprimary education and education at the first level constitute together basic education which has a duration of 8 years.
14 Data for 1980 refer to full-time pupils only.
15 Data refer only to preprimary classes in primary schools (pre-year 1).

§   Toutes les données se rapportant à l'Allemagne avant le 3 octobre 1990 figurent dans deux rubriques séparées basées sur les territoires respectifs de la République fédérale d'Allemagne et l'ancienne République démocratique allemande selon la période indiquée. Pour les notes explicatives en détail sur les données concernant l'Allemagne, voir Annexe I - Nomenclature des pays.

1   Les données se réfèrent aux classes d'initiation où les élèves apprennent le portugais.
2   Les données se réfèrent à l'enseignement public seulement.
3   A l'exception de 1986, les données se réfèrent au 'Sector autonomo Bissau' seulement.
4   Les données se réfèrent aux écoles coraniques seulement.
5   Les données incluent les écoles coraniques 'khalwas' qui acceptent des élèves de tous âges.
6   Pour 1975, les données relatives au personnel enseignant et toutes les données de 1980 n'incluent pas les classes où les élèves apprennent l'espagnol (la castellanización).
7   Les données se réfèrent aux jardins d'enfants seulement.
8   Non compris les écoles turques.
9   A la suite d'un changement dans le système scolaire, à partir de 1987 les données ne sont pas comparables à celles des années précédentes.
10 Les données se réfèrent à l'ancien Yémen démocratique seulement.
11 Y compris l'éducation spéciale et non compris les 'Vorschulklassen'.
12 Les données à partir de 1985 ne comprennent pas les classes préprimaires rattachées aux écoles primaires.
13 A partir de 1985, l'enseignement préprimaire et l'enseignement de premier degré font partie de l'enseignement de base dont la durée est de 8 années.
14 Les données pour 1980 se réfèrent aux élèves à plein temps seulement.
15 Les données se réfèrent seulement aux classes maternelles rattachées aux écoles primaires ('pre-year 1').

14
Education at the first level
Enseignement du premier degré
Institutions, teachers and pupils
Etablissements, personnel enseignant et élèves

Country or area Pays ou zone	Years Années	Schools Ecoles	Teaching staff Personnel enseignant		Pupils enrolled Elèves inscrits	
			Total	%F	Total	%F
Africa · Afrique						
Algeria	1975	7 798	65 043	...	2 663 248	40
Algérie	1980	9 263	88 481	37	3 118 827	42
	1985	11 360	125 034	40	3 481 288	44
	1990	13 135	151 262	39	4 189 152	45
Angola	1980	6 090	...	...	1 300 673	...
Angola	1985	...	31 161	...	970 698	...
	1989	...	...	...	1 038 126	...
Benin	1975	1 325	4 864	...	259 880	31
Bénin	1980	2 275	7 994	23	379 926	32
	1985	2 715	13 452	24	444 163	34
	1988	2 879	13 821	25	482 451	34
Botswana	1975	323	3 509	66	116 293	55
Botswana	1980	415	5 316	72	171 914	55
	1985	528	6 980	78	223 608	52
	1990	636	9 411	...	296 370	52
Burkina Faso	1975	712	2 997	18	141 177	37
Burkina Faso	1980	936	3 700	20	201 595	37
	1985	1 758	...	...	351 807	37
	1989	2 362	8 572	27	472 979	38
Burundi	1975	580	4 199	43	129 597	39
Burundi	1981	792	5 252	47	206 408	38
	1985	1 023	6 866	47	385 936	42
	1988[5]	...	...	...	560 095	...
Cameroon	1975	4 506	22 209	14	1 122 900	45
Cameroun	1980	4 971	26 763	20	1 379 205	45
	1985	5 856	33 598	27	1 705 319	46
	1989	6 549	37 804	30	1 946 301	46
Cape Verde[1]	1975	477	1 243	...	64 794	*48
Cap-Vert[1]	1980	...	1 436	...	57 587	49
	1986	545	1 791	61	60 226	49
	1989	...	...	...	67 761	...
Central African Rep.	1975	732	3 329	18	221 432	36
Rép. centrafricaine	1980	825	4 130	25	246 174	37
	1985	961	4 718	25	309 656	39
	1988	1 040	4 226	25	297 457	38
Chad	1975	...	*2 512[2]	*5[2]	*212 983	*26
Tchad	1985	1 243	4 779	4	337 616	28
	1989	1 868	7 327	6	492 231	30
Comoros	1975	161	756	...	35 818	...
Comores	1980	236	1 292	7	59 709	41
	1985	253	1 901	21	66 084	43
	1987	257	1 777	...	64 737	44
Congo	1975	1 033	5 434	17	319 101	47
Congo	1980	1 335	7 186	25	390 676	48
	1985	1 558	7 745	30	475 805	49
	1989	1 604	7 704	...	492 595	...
Côte d'Ivoire	1975	2 904	15 358	12	672 707	38
Côte d'Ivoire	1980	4 807	26 460	15	1 024 585	40
	1985	5 796	33 500	...	1 214 511	41
Djibouti	1975	...	268	35	9 764	35
Djibouti	1980	45	419	...	16 841	...
	1985	58	...	...	25 212	41
	1989	68	722	...	31 378	42
Egypt[3]	1975	10 346	118 251	48	4 181 198	38
Egypte[3]	1980	12 120	...	...	4 662 816	40
	1985	14 057	194 929	48	6 214 250	43
	1989	14 767	241 119	53	6 155 100	45

14
Education at the first level
Institutions, teachers and pupils [cont.]
Enseignement du premier degré
Etablissements, personnel enseignant et élèves [suite]

Country or area Pays ou zone	Years Années	Schools Ecoles	Teaching staff Personnel enseignant		Pupils enrolled Elèves inscrits	
			Total	%F	Total	%F
Equatorial Guinea	1975	...	...	...	*39 000	...
Guinée équatoriale	1980	...	647	...	44 499	...
	1983	...	...	...	61 532	...
Ethiopia	1975	3 706	24 469	15	1 084 307	32
Ethiopie	1980	5 822	33 322	22	2 130 716	35
	1985	7 900	50 922	26	2 448 778	39
	1988	8 584	65 993	23	2 855 846	39
Gabon	1975	746	2 664	20	128 552	49
Gabon	1980	864	3 441	27	155 081	49
	1985	946	4 008	35	183 607	49
	1987	992	4 229	37	195 049	49
Gambia	1975	103	948	...	24 617	33
Gambie	1980	148	1 808	34	43 432	35
	1986	226	2 590	*31	74 455	*39
	1989	232	2 451	32	75 177	43
Ghana[2]	1975	6 966	38 381	33	1 156 758	43
Ghana[2]	1980	7 848	47 921	42	1 377 734	44
	1985	9 004	64 795	40	1 505 819	...
	1989	9 831	62 859	37	1 703 074	45
Guinea	1975	2 115	4 977	10	198 849	34
Guinée	1980	2 555	7 165	14	257 547	33
	1985	2 285	7 605	19	276 438	32
	1989	2 442	8 113	22	310 064	31
Guinea-Bissau	1976	*633	*2 493	...	84 793	33
Guinée-Bissau	1980	...	3 257	24	74 539	32
	1986	795	3 121	23	77 004	35
	1988	...	...	...	79 035	36
Kenya	1975	8 161	86 107	29	2 881 155	46
Kenya	1980	10 268	102 489	...	3 926 629	47
	1985	12 936	138 374	34	4 702 414	48
	1988	14 288	155 694	36	5 123 581	49
Lesotho	1975	1 079	4 226	70	221 922	59
Lesotho	1980	1 074	5 097	75	244 838	59
	1985	...	5 663	77	314 003	56
	1990	1 190	6 448	80	351 632	55
Liberia	1975	...	*3 832[5]	*30[5]	104 056	34
Libéria	1980	...	...	...	147 216	35
	1984	...	...	...	132 889	*35
	1986[2]	...	...	...	80 048	...
Libyan Arab Jamahiriya	1975	2 042	24 331	29	556 169	46
Jamah. arabe libyenne	1980	2 607	36 591	47	662 843	47
	1985	...	41 515	...	788 780	...
Madagascar	1975	7 960	20 134	...	1 210 841	43
Madagascar	1980	13 594	39 474	...	1 723 779	49
	1984	13 973	42 462	...	1 625 216	48
	1989	13 555	37 932	...	1 512 322	49
Malawi	1975	2 140	10 588	...	641 709	40
Malawi	1980	2 340	12 540	...	809 862	41
	1985	2 520	15 440	33	942 539	43
	1988	...	...	...	1 202 836	45
Mali	1975	1 063	6 213	18	252 393	36
Mali	1980	1 160	6 862	20	291 159	36
	1985	1 259	8 593	22	292 395	37
	1989	1 428	8 405	24	324 369	37
Mauritania	1975	...	1 439	...	50 465	36
Mauritanie	1980	599	2 183	9	90 530	35
	1985	875	2 785	15	140 871	40
	1988	1 121	3 254	18	158 800	41
Mauritius	1975	...	5 791	...	150 573	49
Maurice	1980	267	6 379	...	128 758	49
	1985	280	6 450	43	140 714	49
	1989	289	5 705	45	137 929	49

14
Education at the first level
Institutions, teachers and pupils [cont.]
Enseignement du premier degré
Etablissements, personnel enseignant et élèves [suite]

Country or area Pays ou zone	Years Années	Schools Ecoles	Teaching staff Personnel enseignant Total	%F	Pupils enrolled Elèves inscrits Total	%F
Morocco	1975	1 928	37 226[2]	21[2]	1 547 647	36
Maroc	1980	2 332	56 908[2]	30[2]	2 172 289	37
	1985	3 570	81 867	33	2 279 887	38
	1989	3 903	83 616[2]	35[2]	2 163 185	39
Mozambique	1976[5]	5 853	15 000	...	1 276 500	41
Mozambique	1980[5]	5 730	17 030	22	1 387 192	43
	1985	...	20 286[5]	22[5]	1 248 074	44
	1987	3 637	...	...	1 287 681	44
Niger	1975	1 249	3 617	31	142 182	35
Niger	1980	1 686	5 518	30	228 855	35
	1985	1 850	7 383	33	275 902	36
	1989	2 215	8 462	32	344 848	36
Nigeria	1975	21 223	177 221	...	6 165 547	42
Nigéria	1980	36 254	369 636	33	13 760 030	43
	1985	35 433	292 821	...	12 914 870	44
	1989	34 904	344 221	45	12 712 087	45
Réunion	1975	371	3 661	...	95 810	...
Réunion	1980	...	...	...	79 143	...
	1985	349	3 811	67	73 985	48
	1988	...	...	...	73 747	...
Rwanda	1975	1 668	8 022	29	401 521	46
Rwanda	1980	1 567	11 912	38	704 924	48
	1985	1 594	14 896	44	836 877	49
	1989	1 671	18 524	46	1 058 529	50
St. Helena	1975	8	...	...	755	48
Sainte-Hélène	1980	8	39	95	717	51
	1985	8	32	97	582	55
Sao Tome and Principe	1975	...	422[5]	...	14 290	...
Sao Tomé-et-Principe	1980	...	588[5]	...	16 376	...
	1986	63	574[5]	59[5]	17 010	48
	1989	64	559	52	19 822	47
Senegal	1975	...	7 577	...	311 913	42
Sénégal	1980	1 672	9 175	24	419 748	40
	1985	2 322	12 559	27	583 890	40
	1989	2 422	11 859	26	682 925	42
Seychelles	1975	36	428	95	10 232	51
Seychelles	1980	27	658	80	14 468	51
	1985	25	652	85	14 368	49
	1989	25	781	82	14 595	49
Sierra Leone	1975	1 074	6 373	24	205 910	40
Sierra Leone	1980	1 199	9 528	...	315 145	42
	1984	1 267	12 215	22	416 024	41
	1988	1 312	12 601	31	399 018	38
Somalia	1975	730	3 481	16	197 706	35
Somalie	1980	1 408	8 122	29	271 704	36
	1985	1 224	10 338	45	196 496	34
Sudan	1975	4 719	31 695	32	1 169 279	36
Soudan	1980	6 027	43 451	31	1 464 227	40
	1985	6 775	50 089	44	1 738 341	40
	1986	7 009	50 389	...	1 766 738	...
Swaziland	1975	412	2 363	76	89 528	49
Swaziland	1980	450	3 278	79	112 019	50
	1985	466	4 107	80	139 345	50
	1989	490	4 890	78	157 208	50
Togo	1975	1 362	6 080	20	362 895	35
Togo	1980	2 205	9 193	21	506 356	38
	1985	2 336	10 049	20	462 858	38
	1988	2 429	10 426	20	569 388	39
Tunisia	1975	2 319	23 320	*24	932 787	39
Tunisie	1980	2 661	27 375	29	1 054 027	42
	1985	3 373	40 887	38	1 291 490	45
	1989	3 798	46 366	43	1 376 519	45

14
Education at the first level
Institutions, teachers and pupils [cont.]
Enseignement du premier degré
Etablissements, personnel enseignant et élèves [suite]

Country or area Pays ou zone	Years Années	Schools Ecoles	Teaching staff Personnel enseignant		Pupils enrolled Elèves inscrits	
			Total	%F	Total	%F
Uganda[6]	1975	3 472	28 681	29	973 604	40
Ouganda[6]	1980	4 276	38 422	*30	1 292 377	*43
	1985	7 025	61 424	...	2 117 000	...
	1988	7 905	75 561	...	2 632 764	...
United Republic of Tanzania[7]	1975	5 804	29 735	33	1 592 396	42
Rép.-Unie de Tanzanie[7]	1981	9 980	81 659	37	3 538 183	48
	1985	10 173	92 586	...	3 169 759	50
	1989	10 431	98 392	41	3 258 601	50
Zaire	1975	...	...	...	3 544 498	40
Zaïre	1980	10 536	...	...	4 195 699	42
	1985	10 068	...	...	4 650 756	39
	1987	10 819	...	...	4 356 516	42
Zambia	1975	2 710	18 096	37	872 392	45
Zambie	1980	2 819	21 455	40	1 041 938	47
	1985	3 128	27 302	43	1 348 318	47
	1988	3 392	32 348	45	1 426 135	48
Zimbabwe	1975	3 623	21 202[6]	34[6]	862 736	46
Zimbabwe	1980	3 157	28 118	...	1 235 036	...
	1985	4 216	56 067	43	2 214 963	48
	1989	4 507	58 362	40	2 214 939	49
America, North · Amerique du Nord						
Bahamas	1975	...	1 294[2]	...	31 707	50
Bahamas	1980	...	1 261[2]	...	32 854	...
	1985	88	1 767	...	32 848	49
	1986	90	1 409	...	29 518	49
Barbados	1975	...	...	...	*32 884	*50
Barbade	1980	134	*1 172[2]	...	31 147	50
	1984	125	1 421	...	30 161	48
Belize	1975	194	1 207[6]	72[6]	33 444	...
Belize	1980[6]	197	1 421	...	34 615	...
	1985	...	1 555	72	39 212	48
	1986	...	1 611	*74	40 729	48
Bermuda	1975	22	...	...	6 808	48
Bermudes	1980	22	312	...	5 934	49
	1984	22	309	90	5 398	50
British Virgin Islands	1975	22	113	90	2 096	50
Iles Vierges brit.	1980	18	109	84	1 974	50
	1984	19	...	...	2 069	48
Canada[9]	1975	...	...	...	2 440 016	49
Canada[9]	1980	...	...	...	2 184 919	49
	1985	15 632	299 025	58	2 254 887	48
	1989	...	149 500	69	2 345 000	48
Costa Rica	1975	2 770	12 429	...	361 303	49
Costa Rica	1980	2 936	12 596	...	348 674	49
	1985	3 091	11 526	...	362 877	48
	1989	3 234	13 073	...	422 102	48
Cuba	1975	14 886	77 472	72	1 795 752	48
Cuba	1980	12 196	84 041	75	1 468 538	48
	1985	10 187	77 111	77	1 077 213	47
	1989	9 417	71 887	78	885 576	48
Dominica	1975	57	674[10]	71[10]	17 166	48
Dominique	1980	...	...	...	14 815	50
	1985	66	808[10]	*67[10]	12 340	48
	1990	65	439	81	12 836	49
Dominican Republic	1975	5 487	17 932	...	911 142	...
République dominicaine	1980	4 606	...	...	1 105 730	...
	1985	*6 299	*27 952	...	1 219 681	50
	1989[2 4]	4 854	21 850	...	1 032 055	49
El Salvador	1975	3 103	14 256[2]	68[2]	759 460	48
El Salvador	1980	3 196	17 364	65	834 101	49
	1984	2 631	21 145	66	883 329	50
	1989	4 160	25 318	67	1 016 181	50

14
Education at the first level
Institutions, teachers and pupils [cont.]
Enseignement du premier degré
Etablissements, personnel enseignant et élèves [suite]

Country or area Pays ou zone	Years Années	Schools Ecoles	Teaching staff Personnel enseignant		Pupils enrolled Elèves inscrits	
			Total	%F	Total	%F
Grenada	1975	63	...	...	21 195	48
Grenade	1980	57	776	65	18 076	48
	1985	...	...	...	20 808	...
	1987	63	806	72	20 976	48
Guadeloupe	1975	257	2 018	...	62 813	...
Guadeloupe	1980	...	...	...	53 581	...
	1985	230	1 927	...	42 734	...
	1990	222	2 064	...	39 290	49
Guatemala	1975	6 122	18 129	62	627 126	45
Guatemala	1980	6 959	23 770	...	803 404	45
	1985	8 016	27 809	...	1 016 474	45
	1988	...	...	...	1 149 134	...
Haiti	1975	2 788	11 816	...	487 135	...
Haïti	1980	3 271	14 581	49	642 391	46
	1985	3 734	23 200	...	872 500	47
	1987	4 799	22 421	45	780 660	48
Honduras	1975	4 602	13 045	...	460 744	49
Honduras	1980	5 524	16 385	74	601 337	50
	1985	...	...	...	765 809	50
	1988	8 319	...	...	863 313	50
Jamaica	1976	915	9 889	...	367 525	50
Jamaïque	1980	894	8 676[2]	87[2]	359 488	50
	1985	...	...	...	340 059	...
	1989	873	10 076[2]	88[2]	339 023	49
Martinique	1975	238	2 273	...	58 747	...
Martinique	1980	...	...	...	47 382	...
	1986	218	...	...	33 492	48
	1989	210	...	...	32 649	48
Mexico	1975	55 618	255 939	*62	11 461 415	48
Mexique	1980	76 179	375 220	...	14 666 257	49
	1985	76 690	449 760	...	15 124 160	49
	1990	82 280	471 623	...	14 508 116	...
Montserrat	1975	16	106	86	2 635	48
Monteserrat	1980	16	...	...	1 846	48
	1981	15	86	87	1 725	50
Netherlands antilles	1980	...	...	...	32 856	49
Antilles néerlandaises	1982	122	...	...	32 380	49
Nicaragua	1975	...	...	...	341 533	*51
Nicaragua	1980	4 421	13 318	78	472 167	51
	1985	4 008	16 872	75	561 551	52
	1989	3 840	18 746	87	595 612	52
Panama	1975	2 171	12 459	81	334 607	48
Panama	1980	2 306	12 361	80	337 522	48
	1985	2 476	13 359	78	340 135	48
	1989	2 619	17 756	63	350 277	48
St. Kitts and Nevis	1975	...	...	...	8 804	48
St. Kitts-Nevis	1980	...	...	...	7 149	49
	1985	32	353	...	7 810	...
	1988	32	331	76	7 473	49
Saint Lucia	1975	76	953	...	29 859	49
Sainte-Lucie	1980	77	957	80	29 605	51
	1985	83	1 084[2]	80[2]	32 817	49
	1989	88	1 137	80	33 148	48
St. Pierre and Miquelon	1980	5	36	72	747	49
St. Pierre et Miquelon	1985	5	38	74	558	47
	1986	5	37	68	556	47
St. Vincent and the Grenadines	1975	61	1 210	43	21 854	49
Saint-Vincent-et-Grenadines	1980[8]	...	...	...	24 158	...
	1985[8]	61	1 263	63	24 561	49
	1989[8]	...	...	...	25 742	47

14
Education at the first level
Institutions, teachers and pupils [cont.]
        Enseignement du premier degré
        Etablissements, personnel enseignant et élèves [suite]

Country or area Pays ou zone	Years Années	Schools Ecoles	Teaching staff Personnel enseignant		Pupils enrolled Elèves inscrits	
			Total	%F	Total	%F
Trinidad and Tobago[6]	1975	473	6 471	62	199 033	50
Trinité-et-Tobago[6]	1980	464	7 002	66	167 039	50
	1985	468	7 627	70	168 308	50
	1989	472	6 839	73	189 623	50
Turks and Caicos Is.	1975	17	83	90	1 764	...
Iles Turques et Caïques	1980[2]	17	80	93	1 483	48
	1984[2]	17	68	99	1 429	49
United States[12]	1975	*77 619	1 352 000	...	30 446 000	49
Etats-Unis[12]	1980	...	1 365 000	...	27 449 000	49
	1985	...	1 371 000	...	26 870 000	49
	1986	...	...	...	27 117 000	48
US Virgin Islands[2]	1980	...	819[5]	...	21 738	...
Iles Vierges américaines[2]	1985	57	711	...	20 548	...
America, South · Amerique du Sud						
Argentina	1975	20 646	195 997	92	3 571 180	49
Argentine	1980	...	...	...	3 917 449	49
	1985	20 700	229 715	92	4 589 291	49
	1988	21 207	259 579	91	4 998 963	51
Bolivia	1975	9 519	38 737	...	859 413	45
Bolivie	1980	...	48 894	48	978 250	47
	1986	12 451	47 363	...	1 204 534	47
	1989	...	*48 432	*59	1 225 843	47
Brazil	1975	188 260	896 652	85	19 549 249	49
Brésil	1980	201 926	884 257	85	22 598 254	49
	1985	187 274	1 040 566	...	24 769 736	...
	1989	*202 950	*1 201 100	...	*27 640 000	...
Chile	1975	8 461	65 817	74	2 298 998	49
Chili	1980	...	...	...	2 185 459	49
	1985	8 586	...	...	2 062 344	49
	1990	...	...	...	1 991 178	49
Colombia	1975	33 202	121 957	...	3 911 244	50
Colombie	1980	33 557	136 381	79	4 168 200	50
	1985	34 004	132 940	79	4 039 533	50
	1989	39 634	140 681	...	4 205 657	50
Ecuador	1975	9 479	32 279	65	1 216 233	49
Equateur	1980	11 451	42 415	65	1 534 258	49
	1985	...	53 683	65	1 738 549	49
	1987	...	58 326	65	1 822 252	49
Falkland Islands (Malvinas)	1975	18	23	48	206	48
Iles Falkland (Malvinas)	1980	...	15	33	223	58
French Guiana	1975	46	259	...	7 594	...
Guyane française	1980	...	...	...	9 276	...
	1983	51	423	72	9 780	...
Guyana	1975	...	4 052	69	130 240	49
Guyana	1980	425	3 909	70	130 832	49
	1985	...	3 879[8]	69[8]	113 857	49
	1986	415	3 948[8]	71[8]	112 581	49
Paraguay	1975	...	15 398	...	452 249	47
Paraguay	1980	...	18 948	...	518 968	48
	1985	3 923	22 764	...	570 775	48
	1989	4 411	26 317	...	656 877	48
Peru	1975	19 701	72 641	...	2 840 625	...
Pérou	1980	20 776	84 360	60	3 161 375	48
	1985	24 327	106 600	60	3 711 592	48
	1990	...	143 025	...	4 019 483	...
Suriname	1975	309	2 552	65	80 171	48
Suriname	1980	...	2 803	...	74 538	...
	1986	256	...	...	*62 633	*49
	1988	301	2 921	83	61 570	49

14
Education at the first level
Institutions, teachers and pupils [cont.]
Enseignement du premier degré
Etablissements, personnel enseignant et élèves [suite]

Country or area Pays ou zone	Years Années	Schools Ecoles	Teaching staff Personnel enseignant		Pupils enrolled Elèves inscrits	
			Total	%F	Total	%F
Uruguay	1975	2 308	13 572	...	322 602	49
Uruguay	1980	2 294	14 768	...	331 247	49
	1985	2 360	14 193	...	356 002	49
	1988	2 381	15 188	...	351 984	49
Venezuela	1975	11 532	69 466	...	2 108 413	49
Venezuela	1980	...	92 551	...	2 530 263	...
	1985	13 184	108 125	83	2 770 520	49
	1988	...	...	...	2 967 110	49
Asia · Asie						
Afghanistan	1975	3 371	...	...	784 568	15
Afghanistan	1980	3 824	35 364	21	1 115 993	18
	1985	792	15 581	53	580 499	31
	1989	...	...	...	726 287	33
Bahrain	1975	...	2 044	46	44 857	44
Bahreïn	1980	...	2 577[2]	48[2]	48 451	46
	1985	...	2 856[2]	48[2]	57 330	49
	1988	104	2 882[2]	49[2]	60 179	49
Bangladesh	1975	39 914	164 717	5	8 349 834	34
Bangladesh	1980	43 936	153 859	8	8 240 169	37
	1985	44 180	189 900	8	8 920 293	40
	1989	45 383	186 872	18	11 285 445	44
Bhutan	1976	...	...	...	16 671	27
Bhoutan	1980	...	...	...	29 899	...
	1985	145	...	...	45 395	34
	1988	150	1 513	...	55 340	37
Brunei Darussalam	1975	139	1 582	43	30 109	48
Brunéi Darussalam	1980	137	1 671	45	30 513	48
	1985	140	2 165	...	34 815	...
	1989	146	2 842[5]	61[5]	39 862	48
China	1975	1 093 317	5 203 000	36	150 941 000	45
Chine	1980	917 316	5 499 400	37	146 270 000	45
	1985	832 309	5 376 800	40	133 701 800	45
	1989	777 244	5 543 800	43	123 731 000	46
Cyprus[13]	1975	402	2 101	44	56 602	48
Chypre[13]	1980	443	2 193	45	48 701	49
	1985	380	2 239	49	50 990	48
	1989	378	2 846	57	60 841	48
Hong Kong	1975	1 126	20 666	69	642 611	48
Hong-kong	1980	803	17 937	73	540 260	48
	1985	...	19 404	74	534 903	48
	1987	...	19 625	74	534 309	48
India	1975	453 530	1 559 137	25	65 660 022	38
Inde	1980	485 538	1 345 376[14]	26[14]	73 873 184	39
	1985	528 079	1 509 910[14]	27[14]	87 440 514	40
	1989	550 700	1 601 717[14]	28[14]	97 318 114	41
Indonesia	1975	72 760	603 327	...	17 776 617	45
Indonésie	1980	128 875	787 400	...	25 537 053	46
	1985	168 555	1 181 807	...	29 897 115	48
	1988	167 638	1 278 889	50	30 130 564	48
Iran, Islamic Rep. of	1975	36 738	152 106	52	4 468 299	38
Iran, Rép. islamique d'	1980	39 213	...	...	4 799 000	...
	1985	50 432	309 736	52	6 788 323	44
	1989	56 537	361 878	51	8 817 145	46
Iraq	1975	7 595	69 812	37	1 776 095	33
Iraq	1980	11 284	94 000	48	2 615 910	46
	1985	8 127	118 442	67	2 816 326	45
	1988	8 052	130 777	69	3 023 132	44
Israel	1975	1 503	32 657	...	535 320	49
Israël	1980	1 576	41 468	...	621 912	...
	1985	1 621	41 943	...	699 476	49
	1989	1 153	37 627	82	702 472	49

14
Education at the first level
Institutions, teachers and pupils [*cont.*]
    Enseignement du premier degré
    Etablissements, personnel enseignant et élèves [*suite*]

Country or area Pays ou zone	Years Années	Schools Ecoles	Teaching staff Personnel enseignant		Pupils enrolled Elèves inscrits	
			Total	%F	Total	%F
Japan	1975	24 650	402 553	57	10 364 846	49
Japon	1980	24 945	470 991	57	11 826 573	49
	1985	25 040	464 173	56	11 095 372	49
	1989	24 851	454 109	58	9 606 627	49
Jordan[15]	1975	1 165	11 136	51	386 012	46
Jordanie[15]	1980	1 115	14 303	59	454 391	48
	1985	1 239	16 979	66	530 906	48
	1989	...	21 073	64	290 275	48
Korea, Dem. People's Rep.	1976	4 700	...	...	2 561 674	49
Corée, Rép. pop. dém. de	1987	4 813	59 000	90	1 543 000	49
Korea, Republic of	1975	6 367	108 126	34	5 599 074	48
Corée, République de	1980	6 487	119 064	37	5 658 002	49
	1985	6 519	126 785	43	4 856 752	49
	1990	6 335	136 800	50	4 868 520	49
Kuwait	1975	177	6 360	55	111 820	46
Koweït	1980	238	8 035	56	148 983	48
	1985	270	9 623	68	172 975	49
	1988	297	10 288	69	185 464	49
Lao People's Dem. Rep.	1975	...	11 848	...	317 126	...
République dém. pop. lao	1980	6 339	16 109	30	479 291	45
	1985	8 011	21 033	32	523 347	45
	1987	8 316	20 384	35	558 852	44
Lebanon[15]	1980	...	22 646	...	405 402	...
Liban[15]	1984	2 130	...	...	329 340	47
	1986	...	...	...	399 029	...
Malaysia	1975	6 387	59 343	...	1 893 323	48
Malaisie	1980	6 414	73 664	44	2 008 973	49
	1985	6 685	91 424	50	2 199 096	49
	1989	6 838	111 204	56	2 397 816	49
Maldives	1980	...	...	...	30 621	...
Maldives	1986	...	...	...	39 775	...
Mongolia	1975	...	...	...	129 802	49
Mongolie	1980	...	...	...	145 200	49
	1986	...	5 045	...	155 740	50
	1990	638	5 917	58	166 200	50
Myanmar	1975	18 670	66 251	...	3 475 749	48
Myanmar	1980	21 999	80 343	...	4 148 342	48
	1985	...	...	...	4 710 616	48
	1987	31 329	116 950	64	5 046 471	48
Nepal	1975	8 314	18 874	...	542 524	15
Népal	1980	10 130	27 805	10	1 067 912	28
	1985	11 946	51 266	...	1 812 098	30
	1988	13 514	57 204	11	2 108 739	32
Oman	1975	181	2 055	27	54 611	27
Oman	1980	178	3 959	34	91 895	34
	1985	349	6 681	44	177 541	44
	1989	426	8 972	46	247 128	47
Pakistan[5]	1975	52 800	130 295	34	5 236 203	30
Pakistan[5]	1980	59 165	150 004	32	5 473 578	33
	1985	86 142	199 700	32	7 735 000	32
	1989	118 607	209 754	33	8 614 857	34
Palestine · Palestine Gaza Strip						
Zone de Gaza	1986	155	2 615	...	109 521	...
Philippines	1975	30 839	261 817	...	7 597 279	...
Philippines	1981	31 729	273 492	...	8 518 283	49
	1985	33 104	289 251	...	8 925 959	49
	1989	34 377	314 838	...	10 284 861	49
Qatar	1975	100	1 252	54	23 615	48
Qatar	1980	101	2 029	57	30 078	48
	1985	113	3 154	66	40 636	48
	1989	147	4 037	68	49 657	47

14
Education at the first level
Institutions, teachers and pupils [*cont.*]
Enseignement du premier degré
Etablissements, personnel enseignant et élèves [*suite*]

Country or area Pays ou zone	Years Années	Schools Ecoles	Teaching staff Personnel enseignant		Pupils enrolled Elèves inscrits	
			Total	%F	Total	%F
Saudi Arabia	1975	3 460	34 481	30	677 803	36
Arabie saoudite	1980	5 719	50 511	39	926 531	39
	1985	7 813	83 420	42	1 344 076	43
	1988	8 631	105 937	47	1 694 394	45
Singapore	1975	391	10 777	67	328 401	47
Singapour	1980	335	9 463	66	291 649	48
	1985	236	10 363	69	278 060	47
	1989	203	9 998	71	257 833	47
Sri Lanka	1975	...	99 067[10]	52[10]	1 367 860[2]	47[2]
Sri Lanka	1980	8 772	...	...	2 081 391	48
	1985	9 349	144 707[10]	...	2 242 645	48
	1989	...	...	...	2 114 800	48
Syrian Arab Republic[15]	1975	7 018	37 621	45	1 273 944	40
Rép. arabe syrienne[15]	1980	7 846	55 346	54	1 555 921	43
	1985	9 039	78 388	61	2 029 752	46
	1989	9 524	90 272	64	2 357 981	46
Thailand	1975	42 179	239 128	...	6 686 477	47
Thaïlande	1980	...	299 473[25]	49[25]	7 392 563	48
	1985	33 964	*369 822	...	7 150 489	...
	1990	31 349	338 020	...	6 676 562	49
Turkey	1975	41 981	171 032	39	5 463 684	45
Turquie	1980	45 549	212 456	41	5 656 494	45
	1985	47 631	212 717	42	6 635 858	47
	1989	51 170	224 672	42	6 848 083	47
United Arab Emirates	1975	...	3 191[2]	49[2]	52 207	45
Emirats arabes unis	1980	200	5 424[2]	54[2]	88 617	48
	1985	...	6 123[2]	54[2]	152 125	48
	1989	...	11 921	64	215 532	48
Viet Nam	1975	...	204 998	56	7 403 715	49
Viet Nam	1980	...	204 104	66	7 887 439	47
	1985	12 511	235 791	70	8 125 836	48
Yemen[16]	1975	2 139	6 604[2]	9[2]	254 651	11
Yémen[16]	1980	3 094	9 826[2]	11[2]	435 913	13
	1985	6 252	18 279	9	981 127	20
	1990	...	35 350	12	1 291 372	24
Europe · Europe						
Albania	1975	...	...	...	579 303	47
Albanie	1980	1 559	25 980	50	552 651	47
	1985	1 641	27 167	52	543 775	48
	1989	1 700	28 440	54	550 656	48
Austria	1975	...	...	...	501 843	49
Autriche	1980	3 450	27 525	76	400 397	49
	1985	3 760	32 806	79	343 823	48
	1989	3 717	33 796	81	367 006	49
Belgium	1975	7 773	48 625	57	941 941	49
Belgique	1980	4 968	46 430	59	842 117	49
	1985	4 386	44 190	63	730 288	49
	1987	4 263	71 064[5]	75[5]	728 718	49
Bulgaria	1975	3 419	48 445	71	980 318	48
Bulgarie	1980	3 247	51 581	72	994 018	49
	1985	2 973	61 153	75	1 080 979	48
	1989	2 884	61 154	78	991 626	48
Czechoslovakia	1975	9 285	95 634	77	1 881 414	49
Tchécoslovaquie	1980	6 753	90 380	79	1 904 476	49
	1985	6 332	96 414	82	2 074 403	49
	1989	6 206	98 038	83	1 961 742	49
Denmark	1975	...	...	...	490 891	48
Danemark	1980	2 346	...	...	434 635	49
	1985	2 556	34 744	57	402 707	49
	1989	...	30 000	...	355 311	49

14
Education at the first level
Institutions, teachers and pupils [cont.]
Enseignement du premier degré
Etablissements, personnel enseignant et élèves [suite]

Country or area Pays ou zone	Years Années	Schools Ecoles	Teaching staff Personnel enseignant		Pupils enrolled Elèves inscrits	
			Total	%F	Total	%F
Finland	1975	...	24 494	61	453 737	48
Finlande	1980	4 245	25 949	...	373 347	49
	1985	...	...	...	379 339	49
	1989	4 235	...	...	389 067	49
France	1975	55 886	204 311[2]	68[2]	4 601 550	51
France	1980	51 448	192 438[2]	68[2]	4 610 361	48
	1985	47 923	199 445[2 5]	68[2 5]	4 115 846	48
	1989	44 972	265 600[20]	71[20]	4 163 161	48
Germany ₰ · Allemagne ₰						
Federal Republic of Germany	1975	18 107	...	...	3 903 196	49
Rép. féd. d'Allemagne	1980	18 411	...	...	2 783 867	49
	1985	19 594	*133 471	*79	2 271 546	49
	1989	...	...	...	2 473 700	49
former German Dem. Rep.	1975	5 636	53 734	87	1 060 301	48
ancienne Rép. dém. allemande	1980	5 624	53 396	86	852 109	48
	1985	5 649	58 406	88	859 830	48
	1988	5 703	57 458	89	956 170	48
Gibraltar	1975	14	158	88	2 808	50
Gibraltar	1980	13	157	83	2 750	50
	1984	13	183	77	2 830	48
Greece	1975	9 633	30 953	47	935 730	48
Grèce	1980	9 461	37 315	48	900 641	48
	1985	8 675	37 994	49	887 735	48
	1987	8 178	39 125	49	868 335	49
Hungary	1975	4 468	66 861	...	1 051 095	49
Hongrie	1980	3 633	75 422	80	1 162 203	49
	1985	3 546	88 066	82	1 297 818	49
	1989	3 527	90 602	84	1 183 573	49
Iceland	1975	183	...	...	26 418	49
Islande	1980	...	...	...	24 736	...
	1985	...	...	...	24 603	49
	1989	...	...	...	25 525	...
Ireland	1975	3 558	13 060	65	404 818	49
Irlande	1980	3 385	14 636	74	419 998	49
	1985	3 334	15 674	76	420 236	49
	1988	3 310	15 395	76	423 662	49
Italy	1975	33 233	255 267	83	4 833 415	49
Italie	1980	30 305	273 744	87	4 422 888	49
	1985	27 748	273 800	89	3 703 108	49
	1989	25 163	257 961	90	3 140 113	49
Luxembourg	1975	...	1 521	53	29 430	49
Luxembourg	1980	...	1 765	50	24 628	49
	1985	...	1 755	...	22 003	49
	1987	...	...	...	23 375	49
Malta	1975	104	*1 421	*69	29 834	48
Malte	1980	89	1 557	64	33 063	48
	1985	107	1 665	67	36 240	47
	1988	112	1 805	78	36 726	48
Monaco	1976[2]	...	...	...	1 145	...
Monaco	1980[2]	3	...	...	1 017	46
	1990	7	107	91	1 796	50
Netherlands[17]	1975	8 568	52 700	46	1 453 467	49
Pays-Bas[17]	1980	8 727	57 536	46	1 333 342	49
	1985	8 401	87 814	64	1 468 720	49
	1988	8 426	82 567	63	1 428 577	49
Norway[18]	1975	3 424	46 901	51	390 129	49
Norvège[18]	1980	3 518	47 739	56	390 186	49
	1985	3 524	48 763	58	335 373	49
	1989	3 442	50 904	62	310 600	49

14
Education at the first level
Institutions, teachers and pupils [cont.]
Enseignement du premier degré
Etablissements, personnel enseignant et élèves [suite]

Country or area Pays ou zone	Years Années	Schools Ecoles	Teaching staff Personnel enseignant Total	%F	Pupils enrolled Elèves inscrits Total	%F
Poland	1975	14 738	208 173	81	4 309 823	48
Pologne	1980	12 593	212 050	83	4 167 313	49
	1985	16 254	307 798	82	4 801 307	48
	1989	17 537	...	...	5 141 434	49
Portugal	1975	13 111	59 485	81	1 204 567	48
Portugal	1980	12 460	68 746	...	1 240 307	48
	1985	12 741	73 343	*83	1 235 312	48
	1989	...	...	...	1 003 569	48
Romania	1975	14 695	144 978	67	2 889 208	49
Roumanie	1980	14 381	156 817	70	3 236 808	49
	1985	14 076	147 147	70	3 030 666	49
	1989	13 357	138 981	73	2 891 810	49
San Marino	1975	15	116	86	1 692	47
Saint-Marin	1980	14	145	88	1 509	49
	1985	13	158	87	1 411	49
	1989	14	199	85	1 227	48
Spain	1975	...	...	...	3 653 320	49
Espagne	1980	...	*127 679	*67	3 609 623	49
	1985	18 851	*137 807	*69	3 483 948	48
	1987	18 532	*131 389	*70	3 246 655	48
Sweden	1975	...	34 185	81	698 677	49
Suède	1980	4 928	...	...	666 679	49
	1985	...	...	...	612 704	...
	1989[18]	4 640	92 885	69	580 200	49
Switzerland	1976	...	...	...	503 153	49
Suisse	1980	...	...	...	450 942	49
	1985	...	...	...	376 512	49
	1988	...	...	...	394 061	49
United Kingdom	1975	26 981	285 786[5]	81[5]	5 725 167	49
Royaume-Uni	1980	26 504	260 283[5]	78[5]	4 910 724	49
	1985	24 756	244 000[5]	80[5]	4 296 000	49
	1988	24 482	225 000	78	4 414 966	49
Yugoslavia	1975	13 442	60 904	67	1 494 825	49
Yougoslavie	1980	12 671	59 391	70	1 431 582	48
	1985	12 148	61 288	71	1 448 562	48
	1989	11 841	62 132	72	1 406 630	48
Oceania · Océanie						
American Samoa	1975	...	*270[2]	...	6 052[2]	...
Samoa américaine	1981	27	...	...	6 744	48
	1985	31	454	63	7 704	47
	1989	26	461	64	8 574	47
Australia	1975	8 009	78 390[5]	72[5]	1 632 716	49
Australie	1980	...	91 280[5]	70[5]	1 718 352	49
	1985	...	97 070[5]	72[5]	1 542 101	49
	1989	7 940	92 820[5]	72[5]	1 555 230	49
Cook Islands	1975	...	...	...	5 339	...
Iles Cook	1983	...	231	...	2 909	...
	1985	28	162	...	2 713	...
	1988	28	137	69	2 376	...
Fiji	1975	641	4 274	53	134 971[19]	49[19]
Fidji	1980	652	4 097	57	116 139	49
	1985	668	4 396	58	127 286	49
	1990	681	4 272	...	143 552	...
French Polynesia	1975	167	1 213	78	28 533	49
Polynésie française	1981	...	1 544	69	29 012	48
	1984	198	1 337	...	27 401	48
	1989	179	1 936	...	27 854	48
Guam	1975	39	*919	...	*20 215	...
Guam	1983	39	672	91	15 676	48

14
Education at the first level
Institutions, teachers and pupils [*cont.*]
Enseignement du premier degré
Etablissements, personnel enseignant et élèves [*suite*]

Country or area Pays ou zone	Years Années	Schools Ecoles	Teaching staff Personnel enseignant		Pupils enrolled Elèves inscrits	
			Total	%F	Total	%F
Kiribati	1975	106	449	...	14 862	...
Kiribati	1980	100	435	48	13 235	49
	1985	112	460	51	13 440	49
	1990	104	500	59	14 709	50
Nauru	1974	...	...	...	1 392	47
Nauru	1985	7	71	61	1 451	47
New Caledonia	1975	235	1 431[5]	...	24 943	49
Nouvelle-Calédonie	1981	...	1 518[5]	...	26 779	48
	1985	211	1 131	...	22 517	48
	1990	205	1 147	...	22 958	48
New Zealand	1975	2 471	...	...	391 399	49
Nouvelle-zélande	1980	2 345	...	...	381 262	49
	1985	...	16 547	70	329 337	49
	1989	2 310	16 154	75	312 773	49
Niue[11]	1975	8	65	42	1 122	48
Nioué[11]	1980	...	45	...	666	...
	1985	...	29	...	503	...
	1988	8	32	63	453	49
Pacific Islands (Palau)[21]	1975	248	1 526	33	30 285	48
Iles du Pacifique (Palaos)[21]	1980	249	...	...	30 159	...
	1981	245	*1 374	*27	31 099	...
Papua New Guinea	1975	1 762	7 544	23	238 267	37
Papouasie-Nouv.-Guinée	1980	*2 130	*9 549	*27	299 823	*41
	1987	2 584	12 294	33	384 367	44
	1989	2 692	13 171	33	417 818	44
Samoa	1975	152	1 216[19]	64[19]	32 642	51
Samoa	1980	152	1 438[19]	71[19]	33 012	48
	1986	...	1 511[19]	74[19]	31 412	48
	1989	...	...	...	37 833	48
Solomon Islands	1976	338	1 060	...	27 021	39
Iles Salomon	1980	370	1 148	26	28 870	41
	1985	423	1 496	...	38 716	...
	1986	430	1 849	34	39 563	44
Tokelau	1976	...	...	...	386	52
Tokélaou	1981	...	21	76	434	49
	1983	...	48[10]	58[10]	411	48
Tonga	1975	126	688	53	19 260	48
Tonga	1980	110	781	...	19 012	...
	1985	112	744	62	17 019	48
	1990	115	689	69	16 522	48
Tuvalu	1976	...	...	...	1 540	...
Tuvalu	1980	...	...	...	1 327	...
	1986	9	58	74	1 280	49
	1987	11	64	75	1 364	46
Vanuatu	1975	...	...	...	20 095	45
Vanuatu	1980	278	986	...	23 264	46
	1983[2]	244	934	...	22 244	...
	1990[2]	261	...	...	24 471	47
former USSR · ancienne URSS						
former USSR	1975	147 083[10]	2 399 299[10]	*71[10]	21 366 200	...
ancienne URSS	1980	130 000[10]	2 321 000[10]	*71[10]	21 713 900	...
	1985	128 000[10]	2 520 000[10]	...	23 585 000	...
	1989	128 000[10]	3 004 000[10]	...	25 040 000	49
Belarus	1975	8 408[10]	106 893[10]	...	770 500	...
Bélarus	1980	6 600[10]	90 800[10]	...	750 300	...
	1985	5 900[10]	95 600[10]	...	796 600	...
	1989	5 400[10]	109 600[10]	...	868 000	...
Ukraine	1975	23 435[10]	445 400[10]	...	5 970 000	...
Ukraine	1980	21 000[10]	389 900[10]	...	3 595 200	...
	1985	20 500[10]	404 800[10]	...	3 739 900	...
	1989	20 700[10]	...	...	3 931 600	...

14
Education at the first level
Institutions, teachers and pupils [*cont.*]
Enseignement du premier degré
Etablissements, personnel enseignant et élèves [*suite*]

Source:
United Nations Educational, Scientific and Cultural
Organization (Paris).

§   All data shown which pertain to Germany prior to 3 October
1990 are indicated separately for the Federal Republic of
Germany and the former German Democratic Republic based
on their respective territories at the time indicated. For detailed
explanatory notes on data pertaining to Germany, see Annex I-
Country Nomenclature.

1   From 1975 and 1980 data on teachers refer to 'ensino básico
elementar' (grades 1-4) only.
2   Data refer to public education only.
3   Data on teaching staff and all data for 1989 do not include
Al Azhar.
4   Data include intermediate education (grades 7 and 8 of
'tradicional' education).
5   Including education preceding the first level.
6   Data refer to government-maintained and aided schools only.
7   Data refer to Tanzania mainland only.
8   Including secondary classes attached to primary schools.
9   Until 1988 data on schools and teaching staff include all
levels of education.
10   Including general education at the second level.
11   1975 data refer to grades 1 to 8; for the other years to
grades 1 to 7.
12   Data on pupils refer to grades 1 to 8, while data on teaching
staff refer to kindergarten and 'elementary' schools whose
duration is six or eight grades, depending on the State.
13   Excluding Turkish schools.
14   Excluding primary classes attached to secondary schools.
15   Including UNRWA schools.
16   Excluding former Democratic Yemen.
17   From 1985, preprimary education and education at the first
level have been replaced by basic education which has a
duration of 8 years.
18   Data on teachers include the first stage of general education
at the second level.
19   Including Forms I and II (education at the second level).
20   Include private pre-primary education.
21   Including data for Federated States of Micronesia, Marshall Is.
and Northern Marianna Is.

Source:
Organisation des Nations Unies pour l'éducation, la science et la culture (F

§   Toutes les données se rapportant à l'Allemagne avant le 3 octobre 199(
figurent dans deux rubriques séparées basées sur les territoires respect
de la République fédérale d'Allemagne et l'ancienne République
démocratique allemande selon la période indiquée. Pour les notes
explicatives en détail sur les données concernant l'Allemagne, voir
Annexe I - Nomenclature des pays.

1   De 1975 et 1980 les données sur le personnel enseignant se réfèrent à
l''básico elementar' (années d'études 1 à 4) seulement.
2   Les données se réfèrent à l'enseignment public seulement.
3   Les données relatives au personnel enseignant ainsi que toutes les
données pour 1989 ne comprennent pas Al Azhar.
4   Les données incluent l'enseignement intermédiaire (sep:ième et huitiè
(années d'études de l'enseignement 'tradicional').
et réfèrent également à l'enseignment public seulement.
5   Y compris l'enseignment précédant le premier degré.
6   Les données se réfèrent aux écoles publiques et subventionnées seulen
7   Les données se réfèrent à la Tanzanie continentale seulement.
8   Y compris les classes secondaires rattachées aux écoles primaires.
9   Jusqu' à 1988, les chiffres sur les écoles et le personnel enseignant
incluent tous les niveaux d'enseignement.
10   Y comrpis l'enseignment général du second degré.
11   En 1975, les données se rapportent aux classes allant de la première à l
huitième année d'études; pour les autres années aux classes allant de la
première à la septième année.
12   Les données relatives aux élèves se réfèrent aux classes allant de la
première à la huitième année d'études; les données relatives au personi
enseignant se réfèrent aux jardins d'enfants et aux écoles 'élémentaires
dont la durée est de six ou huit années, selon les Etats.
13   Non compris les écoles turques.
14   Non compris les classes primaires rattachées aux écoles secondaires.
15   Y compris les écoles de l'UNRWA.
16   Non compris l'ancien Yémen démocratique.
17   A partir de 1985, l'enseignement préprimaire et l'enseignement du pre:
degré ont été remplaceés par l'enseignement de base dont la durée est
8 années.
18   Les données relatives au personnel enseignant incluent le premier cycl
l'enseignement général du second degré.
19   Y compris les 'Forms' I et II (enseignement du second degré).
20   Inclut l'enseignement privé du préprimaire.
21   Y compris les données pour les Etats féférés de Micronésie, les îles
Marshall et les îles Mariannes du Nord.

15
Education at the second level
Enseignement du second degré
Teachers and students
Personnel enseignant et élèves

Country or area Pays ou zone	Years Années	Total second level Total du second degré				Vocational education Enseignement technique			
		Teachers Personnel enseignant		Students Elèves		Teachers Personnel enseignant		Students Elèves	
		Total	%F	Total	%F	Total	%F	Total	%F
Africa · Afrique									
Algeria	1975	19 764	...	509 258	34	767	...	12 801	21
Algérie	1980	41 137	...	1 028 294	39	...	...	14 493	21
	1985	82 218	36	1 823 392	42	2 163	14	66 886	30
	1989	120 722	39	2 162 469	43	5 883	15	165 182	31
Angola	1980	...	...	190 702	...	...	...	2 712	...
Angola	1985	...	...	177 960	...	...	...	4 675	...
	1989	...	...	170 280	...	...	...	8 216	...
Benin	1975	...	...	43 123	...	...	...	1 151	...
Bénin	1980	...	...	89 969	...	...	...	5 974	...
	1985	...	...	107 172	29	...	...	6 495	37
	1986	3 657	...	102 171	29	714	...	6 115	40
Botswana	1975	860	30	14 286	52	242	29	1 699	38
Botswana	1980	1 137	37	20 969	55	227	45	1 800	25
	1985	1 675	40	39 713	52	319	58	2 784	36
	1988	2 573	39	44 306	51	460	29	2 633	32
Burkina Faso	1975	818	...	16 227	32	210	19	2 669	36
Burkina Faso	1980	...	...	27 539	33	194	...	3 871	40
	1985	...	...	53 565	34	...	...	4 276	52
	1989	...	...	91 336	...	341	...	8 055	38
Burundi	1975	731	21	13 623	31	103	-	1 099	2
Burundi	1980	...	...	19 013	*32	*239	*10	3 265	*18
	1985	1 849	21	25 939	34	566	11	4 870	21
	1987	...	...	31 413	37	511	12	5 501	28
Cameroon	1975	4 805	20	143 812	33	1 364	23	36 262	36
Cameroun	1980	8 926	...	234 090	35	2 764	...	62 674	39
	1985	11 096	22	343 720	38	3 084	26	83 209	42
	1989	17 667	22	457 161	41	5 652	24	89 289	40
Cape Verde	1975	...	...	...	...	33	...	409	47
Cap-Vert	1980	184	...	3 341	...	40	...	504	39
	1985	...	...	...	...	...	...	361	48
	1989	...	...	...	...	56	...	588	...
Central African Rep.	1975	...	...	23 011	18	...	...	1 972	23
Rép. centrafricaine	1980	724	16	45 211	26	126	25	2 723	49
	1985	922	...	59 273	27	125	...	2 225	47
	1988	1 052	...	45 340	29	112	34	1 989	44
Chad	1975	...	...	16 391	...	...	...	714	...
Tchad	1985	...	...	...	...	...	...	2 896	24
	1989	...	...	58 570	18	...	...	2 802	24
Comoros	1980	449	20	13 798	34	9	11	151	26
Comores	1985	...	...	21 056	39	...	...	421	29
	1986	...	...	21 168	40	31	...	302	26
Congo	1975	2 413	11	102 110	36	337	23	7 129	40
Congo	1980	*5 117	...	187 585	41	*1 239	...	16 933	54
	1985	6 322	...	222 633	44	1 358	...	23 342	53
	1988	8 495	...	200 466	44	...	...	19 583	55
Côte d'Ivoire	1975	...	...	119 482	...	...	...	15 758	...
Côte d'Ivoire	1980	...	...	221 940	30	...	...	21 296[1]	49[1]
	1985	...	...	...	...	...	...	22 861[1]	37[1]
	1986	...	...	...	...	...	...	25 328[1]	39[1]
Djibouti	1975	148	39	1 994	27	59	25	560	14
Djibouti	1980	278	...	5 133	...	88	...	1 279	...
	1985	306	...	7 041	39	72[2]	...	1 877	54
	1988	323	...	8 902	41	69[2]	...	2 250	60
Egypt	1975	78 789	27	2 176 362	34	24 294	16	377 495	34
Egypte	1980	121 999	31	2 929 168	37	34 487	21	633 909	38
	1985	187 580	35	3 826 601	40	52 237	28	877 399	40
	1989	...	...	4 998 615	43	65 554	33	950 133	42
Equatorial Guinea	1975	165	11	4 523	17	29	-	370	-
Guinée équatoriale	1979	158	...	2 729	...	...	...	...	...
	1982	...	...	4 368	...	...	...	...	...

15
Education at the second level
Teachers and students [*cont.*]
    Enseignement du second degré
    Personnel enseignant et élèves [*suite*]

| Country or area | Years | Total second level Total du second degré | | | | Vocational education Enseignement technique | | | |
| Pays ou zone | Années | Teachers Personnel enseignant | | Students Elèves | | Teachers Personnel enseignant | | Students Elèves | |
		Total	%F	Total	%F	Total	%F	Total	%F
Ethiopia	1985	15 861	...	666 169	...	390	...	4 969	...
Ethiopie	1988	21 983	...	882 243	...	492	...	4 101	...
Gabon	1975	1 016	...	22 542	35	168	23	2 450	23
Gabon	1980	1 587	24	29 406	40	372	17	5 530	28
	1985	2 074	22	44 124	42	447	15	8 656	31
	1987	2 271	19	48 274	43	475	13	9 967	33
Gambia	1975	347	26	6 618	*27	30	20	329	*15
Gambie	1980	620	25	9 657	30	54	20	314	19
	1984	914	24	15 913	30	148	28	1 107	38
Ghana	1975	25 142	22	555 980	38	1 022	10	18 919	19
Ghana	1980	31 636	21	693 159	38	1 113[3]	21[3]	27 123	25
	1985	37 290	26	749 980	...	1 194[2 3]	9[2 3]	16 367[2]	...
	1988	45 429	...	793 388	39	1 279[2 3]	...	14 915[2]	13[2]
Guinea	1976	...	...	83 020	...	...	...	1 518	10
Guinée	1980	...	...	98 305	28	...	...	...	...
	1985	4 642	7	92 754	26	777	7	5 368	36
	1989	...	...	78 659	25	...	...	5 671	23
Guinea-Bissau	1975	213	...	2 153	31	40	13	343	22
Guinée-Bissau	1980	462	21	4 757	20	29	3	277	14
	1986	764	13	6 450	25	71	10	351	7
	1988	...	...	6 330	32	74	-	649	9
Kenya	1975	9 730	...	240 969	35	...	...	5 468	-
Kenya	1980	17 081	...	428 023	41	433	...	8 575	...
	1985	23 055	36	457 767	38	535	18	7 840	14
	1988	...	...	563 440	41	...	...	8 880	24
Lesotho	1975	698	...	16 476	56	66	59	547	70
Lesotho	1980	1 299	...	25 292	60	122	47	1 236	56
	1985	1 897	...	37 343	60	128	48	1 263	52
	1989	2 351	...	47 212	60	147	46	1 690	54
Liberia	1975	...	...	34 151	25	...	...	851	13
Libéria	1980	...	...	54 623	28	...	...	2 322	27
Libyan Arab Jamahiriya	1975	11 819	...	166 122	34	523	...	4 888	-
Jamah. arabe libyenne	1980	24 323	24	296 197	40	1 508	12	16 008	25
	1985	35 825	...	430 885	...	2 250	...	27 000	...
Madagascar	1975	...	...	...	...	858	...	7 504	...
Madagascar	1983	...	...	518 538	...	1 023	...	9 646	...
	1987	13 388	...	366 455	46	1 150	...	14 518	36
	1989	14 382	...	331 238	49	1 590	...	17 169	40
Malawi	1975	794	...	15 018	27	46	-	529	-
Malawi	1980	878	...	18 653	28	44	-	647	-
	1985	1 192	...	25 737	32	51	-	560	-
	1988	...	...	29 588	...	...	...	1 024	...
Mali	1975	...	...	55 444	26	...	...	*5 008	*31
Mali	1979	...	...	78 707	...	...	...	5 482	35
	1986	...	...	63 768	30	...	...	7 112	33
	1987	...	...	66 431	29	...	...	7 636	27
Mauritania	1980	...	...	22 102	20	...	...	1 004	7
Mauritanie	1985	...	...	35 955	...	...	...	1 854	...
	1988	1 984	...	39 154	30	80[20]	...	1 089[20]	-
Mauritius	1975	2 177	...	65 113	44	93	28	1 032	42
Maurice	1980	...	...	81 926	48	...	...	270	22
	1985	...	...	72 551	47	...	...	865	38
	1989	...	...	78 776	49	...	...	1 065	25
Morocco	1975	...	...	486 173	34	...	...	13 350	38
Maroc	1980	36 526	...	797 110	38	...	...	10 106[4]	23[4]
	1985	64 079	28	1 201 858	39	1 175	31	27 259	51
	1989[2]	72 434	...	1 366 999	40	1 080[4]	17[4]	16 537[4]	35[4]
Mozambique	1976	...	...	46 656	32	...	...	9 401	...
Mozambique	1981	3 388	22	107 849	28	907	15	13 778	17
	1985	4 688	21	151 888	31	961	17	11 643	18
	1987	...	...	116 928	33	852	19	9 318	18
Niger	1975	637	23	14 462	28	25	20	233	6
Niger	1980	1 284	21	38 861	29	40	15	521	8

15
Education at the second level
Teachers and students [cont.]
Enseignement du second degré
Personnel enseignant et élèves [suite]

Country or area Pays ou zone	Years Années	Total second level Total du second degré				Vocational education Enseignement technique			
		Teachers Personnel enseignant		Students Elèves		Teachers Personnel enseignant		Students Elèves	
		Total	%F	Total	%F	Total	%F	Total	%F
	1985	...	...	...	...	...	...	615	12
	1988	2 443	18	65 816	30	92	...	859	7
Nigeria Nigéria	1975	16 686	...	745 717	32	4 769	...	27 843	12
	1980	81 492	20	2 345 604	35	10 785	38	67 943	17
	1985	...	...	3 102 126	43	...	...	...	...
	1989	136 677	32	2 723 791	42	...	...	...	...
Reunion Réunion	1975	2 363	...	50 467	55	...	...	5 943	47
	1980	...	...	62 613	...	...	...	...	...
	1985	3 994	...	69 863	54	1 016	...	23 313	47
	1986	3 982	...	69 585	53	...	...	17 759	43
Rwanda Rwanda	1975	1 133	17	19 936	52	...	...	9 680	73
	1980	1 454	16	20 672	46	...	...	12 070	55
	1985	3 120	9	46 998	42	...	...	33 645	45
	1989	...	9	65 323	42	...	...	38 463	44
St. Helena Sainte-Hélène	1975	...	...	524	54	...	...	10	-
	1980	47	74	638	48	3	-	32	13
	1985	54	69	513	49	4	-	32	-
Sao Tome and Principe Sao Tomé-et-Principe	1975	...	...	4 010	...	...	...	234	...
	1980	...	...	3 815	...	...	...	130	...
	1986	...	...	5 255	...	...	...	108	...
	1989	...	...	...	...	18	-	101	33
Senegal Sénégal	1975	...	...	...	...	...	...	8 182	...
	1980	4 302	...	95 604	33	...	...	9 932	25
	1985	...	...	130 338	33	...	...	8 770	27
	1989	5 050	16	173 044	34	179[2]	7[2]	5 658	30
Seychelles Seychelles	1975	177	58	3 778	56	32	38	313	72
	1980	127	37	924	*47	60	28	446	*48
	1985	364	29	3 975	50	149	26	1 370	45
	1989	329	39	4 052	47	...	...	1 256	41
Sierra Leone Sierra Leone	1975	2 596	35	50 478	32	98	13	799	6
	1984	4 927	22	101 056	33	119	39	1 396	40
Somalia Somalie	1975	1 529	9	31 857	24	187	15	1 824	17
	1980	2 089	7	43 841	27	625	0	7 704	20
	1985	2 786	11	45 686	35	637	14	5 933	23
Sudan Soudan	1975	13 166	...	281 839	31	649	-	8 996	7
	1980	18 831	...	384 194	37	684	...	15 545	21
	1985	23 035	33	556 587	42	1 214	16	25 610	24
	1986	...	...	...	...	*1 280	...	*23 150	...
Swaziland Swaziland	1975	...	...	16 876	...	...	...	444	...
	1980	...	...	23 665	...	...	...	184	...
	1985	...	...	31 109	...	...	...	395	15
	1987	...	...	33 670	...	...	...	467	23
Togo Togo	1975	1 634	*16	64 404	24	251	24	5 118	30
	1981	...	...	130 366	25	...	...	7 067	31
	1985	4 351	13	97 120	24	279	14	5 511	26
	1988	...	...	109 791	24	357	13	5 956	27
Tunisia Tunisie	1975	8 769[2]	28[2]	201 845	34	...	...	55 974	31
	1980	14 328	29	293 351	37	...	...	80 190	30
	1985	25 245	31	457 630	40	...	...	86 700	36
	1989	31 587	32	546 953	43	...	...	54 573	37
Uganda Ouganda	1975	2 599	...	55 263	26	275	0	3 296	3
	1980	3 833	...	86 560	...	243	...	3 441	...
	1985	8 252	...	179 185	...	443	...	6 932	...
	1988	15 437	...	260 069	...	652	...	6 556	...
United Rep. of Tanzania Rép.-Unie de Tanzanie	1975	3 218	...	62 031	31	...	...	...	...
	1980	3 837	...	78 715	...	...	...	...	...
	1985	5 267	...	92 945	...	...	...	...	...
	1989	7 863	...	145 748	42	...	...	...	...
Zaire Zaïre	1975	...	...	511 481	26	...	...	54 905	33
	1980	...	...	861 774	27	...	...	...	...
	1986	49 153	...	959 934	30	...	...	227 007	31
	1987	...	...	1 066 351	32	...	...	291 743	35

15
Education at the second level
Teachers and students [cont.]
Enseignement du second degré
Personnel enseignant et élèves [suite]

Country or area Pays ou zone	Years Années	Total second level Total du second degré				Vocational education Enseignement technique			
		Teachers Personnel enseignant		Students Elèves		Teachers Personnel enseignant		Students Elèves	
		Total	%F	Total	%F	Total	%F	Total	%F
Zambia	1975	...	...	77 672	...	...	...	2 377	...
Zambie	1980	4 882	...	102 019	35	235	3	2 506	26
	1986	...	...	...	...	505[2]	...	4 567[2]	29[2]
	1988	...	...	170 299	37	...	...	4 181	22
Zimbabwe	1975	3 788	40	70 005	42	51	96	1 312	81
Zimbabwe	1980	3 782	...	74 746	...	46	...	734	100
	1985	...	...	482 000	...	...	...	292	...
	1988	...	...	651 772	...	473[21]	...	11 104[21]	...
America, North · Amérique du Nord									
Bahamas[22]	1975	...	...	30 610	...	92	...	1 823[10]	...
Bahamas[22]	1980	1 018	...	28 136	...	-	-	-	-
	1985	1 472	...	27 604	52	-	-	-	-
	1986	1 555	...	29 765	51	-	-	-	-
Barbados	1975	1 421	...	29 025	52	...	...	...	...
Barbade	1980	1 231[2]	...	28 818	50	...	...	...	...
	1984	1 449	45	28 695	50	...	...	...	...
Belize	1976	349	48	5 420	70	6	100	51	100
Belize	1982	...	...	6 308	*52	...	...	190	...
	1985	534	43	7 048	54	6	17	101	13
	1986	599	44	7 560	53	10	20	93	16
British Virgin Islands	1975	48	52	821	56	...	...	...	...
Iles Vierges brit.	1980	55	60	791	57	...	...	281	60
	1983	100	60	1 323	58	...	...	309	66
Canada	1975	*144 300	*42	2 589 862	49	...	...	...	...
Canada	1980	...	...	2 323 228	49	...	...	...	...
	1985	...	...	2 250 941	49	...	...	...	...
	1989	158 618	54	2 254 654	49	...	...	...	...
Cayman Islands	1975	106[2]	52[2]	1 495	53	...	...	...	...
Iles Caïmanes	1980	207	55	2 075	48	...	...	...	...
Costa Rica	1975	4 929	...	111 538	52	1 063	...	20 311	44
Costa Rica	1980	7 157	...	135 830	53	2 254	...	30 610	50
	1985	6 613	...	112 531	52	...	...	25 493	50
	1989	6 661	...	123 052	50	2 183	...	28 012	48
Cuba	1975	42 306	47	629 398	43	6 911	32	114 653	22
Cuba	1980	88 017	46	1 146 414	50	17 602	25	228 487	46
	1985	100 673	48	1 156 555	51	27 461	33	307 129	45
	1989	108 560	50	1 073 119	52	30 252	39	311 964	47
Dominica	1975	...	...	6 487	59	...	...	548	93
Dominique	1983	...	...	7 562	...	25	...	376	...
	1985	...	...	7 370	54	27	26	259	36
	1986	...	...	6 308	54	...	...	70	93
Dominican Republic	1975	...	...	260 133	...	...	...	19 320	...
République dominicaine	1980	...	...	356 091	...	...	...	22 898	...
	1985	...	...	463 511	...	...	...	21 156	...
El Salvador	1975	2 869	...	51 731	44	...	...	21 552	44
El Salvador	1980	3 080	27	73 030	48	1 210	32	45 299	48
	1984	3 590	31	85 081	53	...	...	57 448	54
	1989	...	...	95 078	50	...	...	65 682	51
Grenada	1975	...	...	10 197	...	...	...	...	...
Grenade	1980	...	...	8 626	59	...	...	...	...
	1985	...	...	6 341	...	...	...	...	...
	1987	317	...	6 497	...	...	...	...	...
Guadeloupe	1975	2 147	...	43 805	53	...	...	6 796	49
Guadeloupe	1980	...	...	49 398	...	...	...	...	...
	1985	...	...	51 634	53	...	...	13 124	49
	1990	3 237	...	49 846	53	...	...	10 638	49
Guatemala	1975	5 994	38	99 233	46	...	...	11 655	51
Guatemala	1980	9 613	...	171 903	45	...	...	29 768	39
	1985	14 629	...	204 049	...	...	...	...	...
	1987	*16 332	...	*241 053	...	...	...	...	...
Haiti	1980	4 392	...	99 894	...	235	...	2 575	...
Haïti	1985	...	...	143 758	...	...	...	3 469	...

15
Education at the second level
Teachers and students [cont.]
Enseignement du second degré
Personnel enseignant et élèves [suite]

Country or area Pays ou zone	Years Années	Total second level Total du second degré				Vocational education Enseignement technique			
		Teachers Personnel enseignant		Students Elèves		Teachers Personnel enseignant		Students Elèves	
		Total	%F	Total	%F	Total	%F	Total	%F
Honduras	1975	*3 132	...	56 705	...	...	...	17 745	...
Honduras	1980	4 489	48	127 293	50	...	...	28 331	49
	1985	...	...	184 112	...	...	...	...	...
	1986	6 945	...	179 444	...	...	...	58 514	...
Jamaica	1975	6 473	...	216 248	54	292	...	4 939	49
Jamaïque	1980	7 525	66	248 001	53	415[2]	56[2]	14 278	65
	1985	8 012	66	237 713	52	527[2]	52[2]	8 690[2]	50[2]
	1988	9 061	66	241 000	52	...	...	...	...
Martinique	1975	2 357	...	45 260	54	...	...	6 026	58
Martinique	1980	...	...	47 745	...	...	...	...	...
	1985	...	...	47 500	...	...	...	...	...
	1989	...	...	43 480	53	...	...	11 590	53
Mexico	1975	169 781	33	2 938 972	39	19 655	51	321 456	72
Mexique	1980	268 178	...	4 741 850	47	28 658	...	491 665	66
	1985	380 774	...	6 549 105	48	50 824	...	766 833	61
	1990	402 473	...	6 704 647	...	58 180	...	792 481	...
Montserrat	1975	...	...	535	...	8	38	53	45
Montserrat	1980	37	...	887	...	5	...	59	...
Netherlands Antilles	1980	...	...	...	...	...	...	10 532	42
Antilles néerlandaises	1982	...	...	...	...	747	...	10 088	42
Nicaragua	1975	2 308	...	80 202	50	625	...	12 422	56
Nicaragua	1980	4 221	...	139 743	53	...	...	16 661	56
	1985	5 204	56	128 499	67	1 520	51	23 919	75
	1989	...	...	161 212	62	847	48	14 581	58
Panama	1975	5 666	55	133 181	52	1 950	52	37 967	51
Panama	1980	8 138	53	171 273	52	2 085	47	39 793	54
	1985	9 681	53	184 536	52	2 727	47	49 154	54
	1988	9 796	54	190 166	52	2 727	49	49 238	53
Saint Lucia	1975	297	39	4 522	...	41	12	230	26
Sainte-Lucie	1980	...	...	4 485	55	...	...	179	34
	1985	...	...	6 833	61	...	...	594	100
St. Pierre and miquelon	1980	62	50	748	52	13	23	221	52
St. Pierre et Miquelon	1985	74	49	821	53	16	25	265	51
	1986	71	46	800	53	16	31	252	51
St. Vincent and the Grenadines	1975	243	43	5 084	58	12	17	108	29
Saint-Vincent-et-Grenadines	1981	...	...	8 058	59	...	...	104	32
	1985	...	...	6 782	59	...	...	128	53
	1987	446	...	7 237	60	35	37	270	62
Trinidad and Tobago	1975	...	...	66 583	49	...	...	2 544	17
Trinité-et-Tobago	1980	4 377	...	89 272	...	...	...	2 439	38
	1985	...	...	95 302	...	...	...	738	35
	1987	...	...	99 615	50	...	...	731	33
Turks and Caicos Is.	1975	35	*37	671	...	...	...	...	...
Iles Turques et Caïques	1980	47	60	691	...	...	...	...	...
	1984	51	67	707	...	...	...	...	...
United States	1975	1 099 000	...	15 683 000	49	...	...	...	...
Etats-Unis	1980	1 074 000	...	14 556 000	50	...	...	...	...
	1985	1 042 000	...	13 977 000	49	...	...	...	...
	1986	...	...	13 913 000	49	...	...	...	...
US Virgin Islands[5]	1975	592	...	*10 590	...	...	...	...	...
Iles Vierges américaines[5]	1980	617[2]	...	6 737	...	...	...	...	...
	1985	...	...	7 948	...	...	...	...	...
America, South · Amérique du Sud									
Argentina	1975	161 859	63	1 243 058	52	99 525	56	788 864	48
Argentine	1980	...	...	1 326 680	...	...	...	...	...
	1985	230 093	66	1 800 049	52	136 418	60	1 084 531	46
	1987	247 804	68	1 862 325	51	136 383[23]	63[23]	1 088 710[23]	42[23]
Bolivia	1975	7 143	...	130 029	...	...	...	...	...
Bolivie	1980	...	...	170 710	43	...	...	...	...
	1986	8 523	...	209 293	46	...	...	...	...
	1989	*9 585	*48	207 824	46	...	...	...	...
Brazil	1975	*133 070	...	1 935 903	53	...	...	782 507	28
Brésil	1980	198 087	53	2 819 182	54	...	...	...	...

15
Education at the second level
Teachers and students [cont.]
Enseignement du second degré
Personnel enseignant et élèves [suite]

Country or area Pays ou zone	Years Années	Total second level Total du second degré				Vocational education Enseignement technique			
		Teachers Personnel enseignant		Students Elèves		Teachers Personnel enseignant		Students Elèves	
		Total	%F	Total	%F	Total	%F	Total	%F
	1985	206 124	...	3 016 175	...	...	...	1 480 997	...
	1989	*238 700	...	*3 441 000	...	...	...	...	...
Chile Chili	1975	29 567	50	448 911	53	11 768	42	163 105	44
	1980	...	...	538 309	53	...	...	169 129	47
	1985	...	...	667 797	52	...	...	128 647	50
	1989	...	...	742 010	51	...	...	134 301	49
Colombia Colombie	1975	70 451	42	1 370 567	50	15 074	40	256 487	43
	1980	85 135	42	1 733 192	50	19 203	42	352 605	45
	1985	95 981	...	1 934 032	50	22 293	...	404 602	49
	1989	114 839	...	2 282 816	50	26 630	...	479 301	49
Ecuador[6] Equateur[6]	1975	23 446	35	383 624	48	4 981	36	42 940	60
	1980	34 868	...	591 969	...	7 562	...	70 476	...
	1985	47 506	41	730 226	50	13 949	41	234 471	56
	1987	53 568	41	771 928	50	16 838	41	260 850	55
Falkland Islands (Malvinas) Falkland Islands (Malvinas)	1975	...	...	126	44	...	...	...	...
	1980	11	36	90	56	...	...	...	...
French Guiana Guyane Française	1975	338	...	5 534	52			1 536	46
	1982	537[2]	51[2]	8 014	52	148[2]	37[2]	2 477	49
	1983	578	52	8 485	52	...	...	2 623	48
Guyana Guyana	1975	...	...	71 327	51	...	...	5 001	50
	1985	2 324	47	76 546	51	237	19	3 867	34
	1986	...	...	76 012	51	121	...	2 594	25
Paraguay Paraguay	1975	...	...	75 424	50	...	...	5 376	23
	1980	...	...	118 828	...	...	...	6 923	...
	1985	...	...	150 736	...	...	...	9 275	...
	1989	...	...	155 434	50	...	...	9 000	...
Peru Pérou	1975	34 136	...	813 489	...	8 103	...	186 430	...
	1980	...	...	1 203 116	45	...	...	51 368	40
	1985	68 541	...	1 427 261	47	-	-	-	-
	1990	83 469	...	1 746 182	...	-	-	-	-
Suriname Suriname	1975	1 793	50	30 603	55	267	16	2 267	19
	1980	...	...	24 027	...	...	...	9 751	...
	1984	2 330	...	34 608	...	1 114	...	13 968	...
	1986	...	...	35 878	54	...	...	9 524	39
Uruguay Uruguay	1975	...	...	182 195	...	...	...	37 698	38
	1980	...	...	148 294	53	...	...	22 856	27
	1985	...	...	213 774	...	...	...	25 598	44
	1988	...	...	243 135	...	...	...	34 299	47
Venezuela Venezuela	1975	37 232	...	669 138	53	...	...	30 265	...
	1980	48 910	...	850 470	55	...	...	43 227	...
	1985	60 112	55	1 037 950	54	...	...	54 883	55
	1988	...	...	1 138 916	54	...	...	50 204	53
Asia · Asie									
Afghanistan Afghanistan	1975	8 089	13	93 497	11	664	*6	5 960	11
	1980	7 532	...	136 898	...	1 262	...	12 410	...
Bahrain Bahreïn	1975	...	...	18 617	48	...	...	1 655	28
	1980	1 184	51	26 528	46	...	...	2 810	35
	1985	2 056	46	38 577	48	566	23	7 870	35
	1988	2 486	47	41 038	49	559	9	6 725	30
Bangladesh Bangladesh	1976	...	...	2 183 413	...	...	...	11 475	...
	1980	111 927	7	2 659 208	24	1 059	5	19 600	2
	1985	112 700	8	3 125 219	28	1 427	3	19 045	5
	1988	124 760	10	3 340 120	31	1 980	64	24 880	5
Brunei Darussalam Brunéi Darussalam	1975	782	37	14 614	48	61	-	314	2
	1980	1 413	34	17 441	50	118	3	459	15
	1985	1 893	...	20 642	...	221	...	975	...
	1989	...	...	...	...	279	16	1 241	28
China Chine	1975	2 164 601	...	45 368 318	39	47 828	...	405 030	...
	1980	3 171 564	25	56 778 008	39	114 200	25	1 214 900	34
	1985	2 966 400	28	50 926 400	40	268 800	29	3 308 600	41
	1989	3 422 900	31	50 541 400	42	384 800	34	4 315 600	44
Cyprus[7] Chypre[7]	1975	2 451	40	49 373	47	385	19	6 112	6
	1980	2 953	42	47 599	49	504	19	5 805	14

15
Education at the second level
Teachers and students [*cont.*]
Enseignement du second degré
Personnel enseignant et élèves [*suite*]

Country or area Pays ou zone	Years Années	Total second level Total du second degré				Vocational education Enseignement technique			
		Teachers Personnel enseignant		Students Elèves		Teachers Personnel enseignant		Students Elèves	
		Total	%F	Total	%F	Total	%F	Total	%F
	1985	3 138	44	46 159	49	...	...	3 874	12
	1989	3 610	45	43 219	49	465	17	3 316	16
Hong Kong	1975	15 149	41	368 655	47	...	...	21 509	29
Hong-kong	1980	15 986	49	468 975	49	...	...	31 019	32
	1985	18 773	49	450 367	50	...	...	38 979	30
	1987	20 183	49	458 444	49	2 488	28	45 943	32
India	1975	...	...	23 638 666	30	...	...	171 590	41
Inde	1980	...	...	30 531 881	33	...	...	409 875	32
	1985	...	...	...	...	...	...	677 164	30
Indonesia	1975	...	...	3 570 080	38	...	...	757 280	29
Indonésie	1980	385 186	...	5 721 815	36	...	...	610 430	*27
	1985	620 857	...	*9 479 086	...	...	...	*869 841	...
	1988	788 929	33	11 693 361	44	...	...	1 240 907	*42
Iran, Islamic Rep. of	1975	81 855	39	2 183 137	36	7 066	18	150 509	19
Iran, Rép. islamique d'	1981	...	...	2 836 144	38	12 836	10	158 510	14
	1985	...	...	3 401 274	40	20 665	13	195 352	24
	1989	263 399	38	4 668 442	40	23 297	13	212 100	20
Iraq	1975	21 454	40	525 255	29	1 654	16	23 775	23
Iraq	1980	33 514	40	1 033 418	32	4 148	24	56 924	29
	1985	42 998	53	1 190 833	35	6 745	42	125 439	26
	1988	53 937	53	1 166 859	38	9 741	43	160 278	29
Israel[8]	1975	...	...	170 168	52	...	...	73 543	46
Israël[8]	1980	31 650	57	199 859	...	...	...	82 332	46
	1985	37 735	...	251 466	51	...	...	100 470	45
	1989	43 235	63	291 754	51	...	...	110 209	46
Japan	1975	511 590	24	8 843 511	49	...	...	1 505 032	47
Japon	1980	554 078	26	9 557 563	49	...	...	1 410 718	47
	1985	619 105	28	11 058 133	49	...	...	1 423 017	46
	1989	651 728	29	11 143 930	49	...	...	1 450 704	47
Jordan[9]	1975	7 768	39	164 186	41	358	23	6 441	29
Jordanie[9]	1980	12 848	43	266 430	45	849	28	14 063	30
	1985	19 174	45	335 835	48	2 100	38	30 789	39
	1989	...	...	384 279	48	2 135	39	26 525	35
Korea, Republic of	1975	83 811	20	3 111 510	41	16 479	12	436 538	33
Corée, République de	1980	109 546	26	4 285 889	45	27 208	20	881 287[10]	44[10]
	1985	140 942	30	4 834 339	47	31 819	23	834 628	48
	1990	180 724	24	4 559 557	48	32 944	23	723 193[10]	50[10]
Kuwait	1975	9 371	49	108 219	45	363	18	1 328	15
Koweït	1980	15 342	50	181 882	46	85	-	421	-
	1985	18 795	52	239 581	47	145	12	1 161	8
	1988	21 112	52	265 001	48	133	12	863	9
Lao People's Dem. Rep.	1976	...	...	48 669	33	...	...	894	26
République dém. pop. lao	1980	4 703	...	90 435	39	289	...	2 002	28
	1985	10 146	35	113 630	41	868	20	6 799	24
	1987	11 173	...	135 583	41	895	...	6 864	31
Lebanon[9]	1980	...	...	287 310	...	...	...	31 203	...
Liban[9]	1984	...	...	...	...	3 506	8	37 036	38
	1986	...	...	...	...	...	...	30 407	...
Malaysia	1975	34 133	...	933 411	...	...	...	...	...
Malaisie	1980	47 625	45	1 083 818	48	1 462	22	18 031	29
	1985	58 630	47	1 294 990	49	1 699	24	20 720	29
	1989	72 167	50	1 480 490	50	2 692	31	27 284	25
Maldives	1980	...	...	998	...	...	...	123	...
Maldives	1983	...	...	2 756	...	...	...	412	...
Mongolia	1975	...	...	184 688	52	...	...	10 936	61
Mongolie	1980	...	...	245 600	...	...	...	...	...
	1982	11 500	...	256 700	53	900	44	18 400	63
	1986	...	...	...	...	1 916	...	25 036	47
Myanmar	1975	24 911	...	933 486	...	810	...	10 700	...
Myanmar	1980	31 248	...	1 066 300	...	806	...	*14 500	...
	1985	63 168	...	1 284 900	...	1 224	...	17 200	...
	1987	61 089	...	1 358 788	...	1 257	...	15 631	...

15
Education at the second level
Teachers and students [cont.]
Enseignement du second degré
Personnel enseignant et élèves [suite]

Country or area	Years	Total second level Total du second degré				Vocational education Enseignement technique			
		Teachers Personnel enseignant		Students Elèves		Teachers Personnel enseignant		Students Elèves	
Pays ou zone	Années	Total	%F	Total	%F	Total	%F	Total	%F
Nepal	1976	11 295	*9	262 748	17	513	...	16 815	...
Népal	1980	16 376	9	512 434	20	...	...	...	...
	1985	18 362	...	496 921	23	...	...	...	...
	1988	21 132	8	612 943	27	...	...	...	...
Oman	1975	208	16	1 379	16	20	-	84	-
Oman	1980	...	...	16 776	...	...	...	1 013	...
	1985	3 845	...	48 096	32	403	...	3 004	9
	1989	5 645	39	85 230	41	500	-	2 907	-
Pakistan	1975	106 960	29	1 935 849	23	1 887	20	25 715	22
Pakistan	1980	123 817	30	2 165 832	26	1 994	20	33 492	17
	1985	...	...	2 933 422	27	...	...	51 426	14
	1989	188 281	32	3 637 466	28	...	...	65 000	25
Palestine · Palestine									
Gaza Strip									
Zone de Gaza	1986	2 169	...	59 241	...	102	...	1 094	...
Philippines	1975	72 778	...	2 291 707	...	...	...	...	...
Philippines	1980	85 779	...	2 928 525	53	...	...	...	...
	1985	99 468	...	3 214 159	50	...	...	...	...
	1989	...	...	3 961 639	50	...	...	...	...
Qatar	1975	829	45	10 109	48	75	-	369	-
Qatar	1980	1 624	50	15 901	48	86	-	440	-
	1985	2 539	53	22 574	50	105	-	700	-
	1989	3 338	55	28 685	51	127	-	927	-
Saudi Arabia	1975	13 956	24	202 741	33	646	5	4 322	6
Arabie saoudite	1980	26 634	34	348 996	38	761	-	5 106	-
	1985	42 892	40	603 127	38	1 345[24]	-	10 677[24]	-
	1986	45 798	43	654 202	39	1 371[24]	-	12 302[24]	-
Singapore	1975	7 951	...	183 364	49	740	...	7 140	7
Singapour	1980	9 298	52	180 817	50	1 023	24	9 391	23
Sri Lanka[11]									
Sri Lanka[11]	1976	...	...	1 088 089	51	1 239	6	4 778	33
Syrian Arab Republic[9]	1975	24 895	30	488 409	31	1 677	11	22 046	17
Rép. arabe syrienne[9]	1980	...	...	604 327	37	3 280	...	26 190	29
	1985	53 250	25	870 383	40	7 338	16	55 466	25
	1989	...	...	923 532	41	...	...	66 590	37
Thailand	1975	43 830	...	1 193 741	44	*9 715	*41	191 066	45
Thaïlande	1980	...	...	1 919 967	...	...	...	297 114	...
	1985	...	...	...	...	*19 278	...	373 013	...
	1988	*123 152	...	2 071 107	...	*19 775	...	336 420	...
Turkey	1975	...	...	1 746 160	31	...	...	*321 271	*28
Turquie	1980	112 178	...	2 217 909	...	33 969	...	520 332	...
	1985	138 640	36	2 927 692	35	43 293	35	616 283	28
	1989	155 661	38	3 620 982	37	47 637	37	830 716	33
United Arab Emirates	1975	...	...	12 562	37	90	1	296	-
Emirats arabes unis	1980	...	...	32 362	45	...	...	422	-
	1985	4 237	49	62 082	48	...	...	604	-
	1989	7 614	54	95 669	50	...	...	690	-
Viet Nam									
Viet Nam	1976	127 635	57	3 200 912	49	5 911	26	66 553	35
Yemen[12]	1975	...	...	24 606	13	...	...	566	-
Yémen[12]	1980	...	...	41 155	*13	...	...	1 314	7
	1985	7 197	...	146 133	11	279	-	2 197	7
	1990	13 353	10	420 697	15	464	4	5 068	9
Europe · Europe									
Albania	1975	...	...	110 519	46	...	...	73 568	42
Albanie	1980	5 392	35	163 866	45	4 363	32	132 482	41
	1985	7 072	40	177 679	45	5 500	35	133 129	41
	1989	9 367	40	202 864	44	...	...	137 704	39
Austria	1975	...	...	755 670	50	...	...	119 535	51
Autriche	1980	63 678	51	739 702	50	15 129	41	150 981	52
	1985	69 550	54	671 198	50	17 270	46	170 080	53
	1989	71 129	56	601 140	50	17 881	48	160 918	52

15
Education at the second level
Teachers and students [*cont.*]
　　Enseignement du second degré
　　Personnel enseignant et élèves [*suite*]

Country or area Pays ou zone	Years Années	Total second level Total du second degré				Vocational education Enseignement technique			
		Teachers Personnel enseignant		Students Elèves		Teachers Personnel enseignant		Students Elèves	
		Total	%F	Total	%F	Total	%F	Total	%F
Belgium	1975	...	...	795 203	49	...	...	300 519	52
Belgique	1980	...	...	835 524	50	...	...	...	...
	1985	...	...	824 997	49	...	...	374 335	47
	1987	114 628	53	805 647	49	...	...	372 383	47
Bulgaria	1975	27 045	52	344 015	...	19 408	48	242 809	42
Bulgarie	1980	25 666	53	314 753	48	18 507	49	222 890	40
	1985	26 851	57	374 565	49	17 459	52	211 148	37
	1989	27 616	60	397 362	49	18 185	54	239 572	39
Czechoslovakia	1975	56 468	34	656 843	47	23 720	21	330 621	34
Tchécoslovaquie	1980	60 820	37	780 811	48	27 593	26	346 246	37
	1985	74 021	35	744 059	51	41 326	23	344 252	43
	1989	86 885	...	856 971	51	49 929	21	395 806	43
Denmark[13]	1975	...	...	365 561	...	...	...	37 973	...
Danemark[13]	1980	...	...	498 944	49	...	...	125 996	41
	1985	...	...	487 526	49	...	...	150 772	43
	1988	...	...	483 502	49	...	...	155 772	44
Finland	1975	...	...	419 808	52	...	...	81 530	46
Finlande	1980	...	...	444 165	52	13 948	42	102 239	47
	1985	...	...	415 825	53	15 700	...	114 123	52
	1989	...	...	410 582	53	...	...	110 388	54
France	1975	316 341	...	4 890 152	*49	58 546	...	1 009 432	...
France	1980	256 369[2]	...	5 013 666	53	...	...	1 102 612	68
	1985	...	...	5 371 593	51	...	...	1 328 347	46
	1989	434 018	57	5 390 599	50	...	...	1 220 424	46
Germany ₣ · Allemagne ₣									
Federal Republic of Germany[15]	1975	...	...	6 158 274	49	...	...	459 846	54
Rép. féd. d'Allemagne[15]	1980	...	...	6 561 297	50	60 621	35	610 400	54
	1985	...	...	7 101 250	48	...	...	2 528 245	45
	1988	443 731	41	6 219 158	48	101 467	31	2 258 980	46
former German Dem. Rep.	1975	147 786	43	2 018 845	49	49 263	26	429 354	44
ancienne Rép. dém. allemande	1980	164 382	46	1 895 579	48	55 730	29	482 366	43
	1985	167 310	51	1 519 152	48	55 234	33	378 761	43
	1988	158 331	54	1 418 272	48	53 373	36	360 468	43
Gibraltar	1975	132	40	1 629	51	18	-	42	2
Gibraltar	1980	...	...	1 811	50	...	...	41	-
	1984	144	36	1 806	49	19	5	57	7
Greece	1975	...	...	661 796	43	...	...	132 591	13
Grèce	1980	39 571	49	740 058	46	7 834	24	100 425	20
	1985	50 388	53	813 534	48	8 138	34	109 415	29
	1987	54 173	54	840 020	47	9 286	39	131 471	30
Hungary[16]	1975	22 781	...	371 898	45	...	...	267 329	37
Hongrie[16]	1980	...	...	357 334	46	...	...	262 037	39
	1985	...	...	422 323	49	...	...	311 448	42
	1989	...	...	487 208	49	...	...	365 964	43
Iceland	1975	2 387	29	*25 853	*45	819	21	*5 351	*24
Islande	1980	...	...	26 643	47	...	...	7 384	...
	1985	...	...	27 559	47	...	...	7 165	35
	1988	...	...	29 059	...	...	...	...	...
Ireland	1975	18 913	50	270 956	51	...	...	9 957	83
Irlande	1980	...	...	300 601	52	...	...	13 982	72
	1985	...	...	338 256	51	...	...	22 672	64
	1988	...	...	341 800	51	...	...	23 106	63
Italy	1975	432 867	...	4 875 179	46	119 646	...	1 333 668	36
Italie	1980	519 128	58	5 307 989	48	171 102	45	1 586 179	41
	1985	562 196	61	5 361 579	49	193 074	48	1 782 796	43
	1989	571 607	61	5 245 132	49	212 384	49	1 939 133	44
Luxembourg	1975	...	...	22 652	49	...	...	7 461	46
Luxembourg	1980	...	...	27 487	...	...	...	18 397	...
	1985	1 908	...	25 656	48	...	...	17 619	45
	1987	1 990	...	22 496	49	1 040	...	14 790	46

15
Education at the second level
Teachers and students [cont.]
Enseignement du second degré
Personnel enseignant et élèves [suite]

Country or area	Years	Total second level Total du second degré				Vocational education Enseignement technique			
Pays ou zone	Années	Teachers Personnel enseignant		Students Elèves		Teachers Personnel enseignant		Students Elèves	
		Total	%F	Total	%F	Total	%F	Total	%F
Malta	1975	2 498	37	32 409	46	409	9	4 387	17
Malte	1980	2 141	37	25 501	45	450	20	4 124	22
	1985	2 315	34	27 779	48	591	20	6 358	21
	1988	2 509	35	30 210	47	704	26	6 880	27
Monaco	1976[2]	...	...	1 993	...	...	...	598	...
Monaco	1980	...	...	2 065	...	...	...	751	51
	1982	...	...	3 132	...	...	...	1 218	...
	1990	...	...	2 909	48	145	43	700	41
Netherlands	1975	...	...	1 283 585	47	...	...	506 364	41
Pays-Bas	1980	...	...	1 391 485	48	...	...	560 565	41
	1985	...	...	1 439 275	48	...	...	633 355	42
	1988	...	...	1 289 368	48	...	...	571 027	43
Norway	1975	...	...	326 640	49	10 332	33	62 699	44
Norvège	1980	...	...	360 776	50			81 165	47
	1985	...	...	387 990	50			107 003	48
	1989	...	...	375 095	50			117 061	47
Poland	1975	169 635	46	1 946 366	50	...	...	1 457 184	43
Pologne	1980	139 412	52	1 673 869	50	...	...	1 309 952	44
	1985	147 782	53	1 567 641	51	...	...	1 207 355	44
	1989	...	...	1 829 747	50	...	...	1 389 423	43
Portugal	1975	29 714	56	466 491	49	13 985	52	127 317	37
Portugal	1980	32 028	59	398 320	48	-	-		
	1985	...	...	580 170	...	...	...	7 473	...
	1989	...	...	544 943	50	...	...	16 118	36
Romania[17]	1975	46 907	44	870 161	50	30 946	38	595 128	40
Roumanie[17]	1980	48 082	43	871 257	47	38 694	41	784 061	45
	1985	50 333	48	1 537 548	46	42 251	46	1 429 994	44
	1989	44 417	50	1 652 441	48	...	...	...	...
San Marino	1975	108	57	1 211	47	-	-	-	-
Saint-Marin	1980	112	62	1 219	49	-	-	-	-
	1985	...	...	1 248	48	-	-	-	-
	1989	...	...	1 204	49			113	19
Spain	1975			3 188 619	48	37 744	27	550 068	43
Espagne	1980	190 251	40	3 976 747	50	40 696	31	888 721	46
	1985	217 364	45	4 555 541	51	56 641	35	1 206 330	51
	1987	229 145	47	4 798 337	51	59 659	39	*1 290 290	*52
Sweden	1975	...	...	507 642	51	...	...	137 552	50
Suède	1980	...	...	606 833	51	...	...	163 405	52
	1986	...	...	616 312	50	...	...	220 334	47
	1989	...	...	597 971	50	...	...	212 627	46
Switzerland	1975[2]	...	...	371 978	49	...	...	26 279	48
Suisse	1980	...	...	459 590	51	...	...	24 259	63
	1985	...	...	415 952	51	...	...	26 489	62
	1989	...	...	371 097	51	...	...	27 212	59
United Kingdom	1975	...	...	5 154 371	49	...	...	208 601	54
Royaume-Uni	1980	...	...	5 341 849	50	...	...	254 813	57
	1985	...	...	4 877 000	50	...	...	403 000	57
	1988	...	...	4 365 912	50	...	...	429 675	54
Yugoslavia[18]	1975	...	...	2 264 101	47	18 867	51	561 219	43
Yougoslavie[18]	1980	131 348	49	2 426 164	47	...	...	587 808	45
	1985	132 912	51	2 352 985	47	...	...	587 156	45
	1989	139 465	52	2 361 660	48	...	...	785 818	44
Oceania · Océanie									
American Samoa	1976	...	...	*2 602	...	...	...	...	...
Samoa américaines	1980	...	...	3 000	...	...	...	...	...
	1985	203	36	3 342	47	...	...	...	...
	1989	222	39	3 437	48	15	27	139	8
Australia	1975	74 041	47	1 099 922	49	...	...	...	...
Australie	1980	85 340	45	1 100 468	50	...	...	...	...
	1985	105 955	48	1 278 272	49	...	...	...	...
	1989	101 321	49	1 276 969	50	...	...	...	...
Fiji	1975	...	...	30 545	49	...	...	1 938	48
Fidji	1980	2 564	41	49 963	51	235	26	2 330	59

15
Education at the second level
Teachers and students [cont.]
Enseignement du second degré
Personnel enseignant et élèves [suite]

Country or area Pays ou zone	Years Années	Total second level Total du second degré				Vocational education Enseignement technique			
		Teachers Personnel enseignant		Students Elèves		Teachers Personnel enseignant		Students Elèves	
		Total	%F	Total	%F	Total	%F	Total	%F
	1985	2 954	43	45 093	50	285	25	3 588	48
	1986	2 838	42	46 457	50	287	27	4 241	51
French Polynesia Polynésie française	1975	569	44	9 035	56	129	47	1 644	50
	1979	786	...	12 012	...	181	...	2 400	48
	1986	1 240	44	17 878	55	385	31	4 113	49
	1989	1 335	45	19 655	57	514	39	4 642	50
Guam Guam	1975	*595	...	*13 242	...	...	...	...	...
	1983	720	63	13 375	48	66	47	868	45
Kiribati Kiribati	1980	154	39	2 440	46	71	37	1 376	41
	1985	160	38	2 196	52	60	20	662	45
	1989	236	33	2 903	...	79	20	555	...
Nauru Nauru	1985	40	45	482	50	4	-	17	47
New Caledonia Nouvelle-Calédonie	1975	...	...	7 960	50	...	...	2 221	40
	1980	839	...	11 945	54	294	32	2 806	52
	1985	1 265	...	18 351	52	...	...	5 429	46
	1989	1 509	41	20 151	52	...	...	6 477	46
New Zealand Nouvelle-Zélande	1975[25]	...	...	354 107	49	...	...	2 387	78
	1980	...	...	352 427	49	...	...	3 071	82
	1985	...	...	354 080	50	...	...	2 462	82
	1989	...	...	341 249	50	...	...	4 108	70
Niue Nioué	1975	22	41	271	49	...	...	...	...
	1980	25	...	397	...	...	...	...	...
	1985	31	...	321	...	...	...	...	...
	1988	27	48	194	45	...	...	...	...
Pacific Islands (palau)[14] Iles du Pacifique (palaos)[14]	1975	528	30	7 951	44	...	...	...	...
	1980	...	...	6 885	...	...	...	...	...
	1981	*445	*32	6 872	...	...	...	...	...
Papua New Guinea Papouasie-Nouv.-Guinée	1975	2 034	34	41 391	*28	586	24	9 639	*17
	1981	2 289	32	49 334	...	499	31	7 984	...
	1987	2 922	33	63 391	*35	568	28	7 999	*20
	1989	3 057	33	67 007	36	584	30	7 494	*24
Samoa Samoa	1975	...	...	15 943	*51	*48	*6	355	*25
	1980	...	...	19 785	49	35	46	264	41
	1986	...	...	20 604	...	19	26	156	...
Solomon Islands Iles Salomon	1975	131	28	2 014	25	26	15	313	14
	1980	257	26	4 030	...	37	8	367	5
	1986	...	...	6 615	*38	...	...	1 144	37
Tokelau Tokélaou	1979	18	50	277	47	12	50	197	44
	1983	...	...	488	50	...	...	380	50
Tonga Tonga	1975	...	...	11 351	48	23	17	549	33
	1981	...	...	16 566	47	...	...	624	40
	1985	840	45	15 232	51	58	43	465	62
	1990	832	49	14 749	48	45	51	662	44
Tuvalu Tuvalu	1976	...	...	248	...	...	...	...	...
	1980	...	...	248	...	...	...	...	...
Vanuatu Vanuatu	1975	...	...	1 505	42	...	...	139	22
	1980	185	...	2 426	42	45	...	350	27
	1983	...	...	2 904	38	...	...	633	22
	1990	...	...	...	...	...	...	199	38
former USSR · ancienne URSS									
former USSR[19] ancienne URSS[19]	1975	...	...	23 171 700	...	218 428	50	2 206 800	...
	1980	...	...	20 274 500	...	...	...	2 758 400	...
	1985	...	...	20 513 200	...	...	...	3 085 000	...
	1989	...	...	21 124 000	...	...	...	2 959 000	...
Belarus[19] Bélarus[19]	1975	...	...	882 600	...	8 210	...	79 300	...
	1980	...	...	759 700	...	...	...	110 800	...
	1985	...	...	716 700	...	...	...	122 800	...
	1989	...	...	711 700	...	...	...	121 700	...
Ukraine[19] Ukraine[19]	1975	...	...	1 821 900	...	40 400	...	783 800	...
	1980	...	...	3 406 400	...	...	...	479 900	...
	1985	...	...	3 401 100	...	...	...	557 300	...
	1989	...	...	3 451 800	...	...	...	547 800	...

15
Education at the second level
Teachers and students [*cont.*]
Enseignement du second degré
Personnel enseignant et élèves [*suite*]

Source:
United Nations Educational, Scientific and Cultural Organization
(Paris).

§ All data shown which pertain to Germany prior to 3 October 1990
are indicated separately for the Federal Republic of Germany
and the former German Democratic Republic based on their
respective territories at the time indicated. For detailed
explanatory notes on data pertaining to Germany, see Annex I-
Country Nomenclature.

1   Data refer to schools attached to Ministry of Education only.
2   Data refer to public education only.
3   From 1980, commercial schools have been converted into general
secondary schools and certain vocational courses are classified
at the third level of education.
4   Data exclude professional schools.
5   From 1980, data refer to grades 9 to 12, for 1975 to grades 7 to 12.
6   Due to a change in classification, from 1984 data are not
comparable with those of previous years.
7   Not including Turkish schools. From 1985, data on commercial
schools are included with general education.
8   Data on teaching staff include intermediate classes attached
to primary schools.
9   Including UNRWA schools.
10 Including part-time education.
11 For Vocational education, data refer to technical institutes
attached to the Ministry of Education only.
12 Excluding former Democratic Yemen.
13 For Vocational education, except for 1975, data include
apprenticeship training.
14 Including data for Federated States of Micronesia, Marshall Is.
and Northern Mariana Is.
15 For 1975 and 1980, data do not include technical education
consisting of both on the job training and school education.
16 For Vocational education, data include apprenticeship training.
17 For vocational education, from 1985, data include evening
courses.
18 For vocational education, for 1980, data on pupils include a part
of enrolment in teacher training.
19 For vocational education, data on teachers include teacher
training as well as evening and correspondence courses.
20 Data exclude health-related programmes.
21 In 1988 a new vocational education system was launch in some
pilot schools.
22 Data on teaching staff do not include senior departments of
all-age school (included in first level).
23 Data exclude fine and applied art.
24 Data include evening commercial school.
25 Data on pupils include special education.

Source:
Organisation des Nations Unies pour l'éducation, la science
et la culture (Paris).

§ Toutes les données se rapportant à l'Allemagne avant le 3 octobre
1990 figurent dans deux rubriques séparées basées sur les
territoires respectifs de la République fédérale d'Allemagne et
l'ancienne République démocratique allemande selon la période
indiquée. Pour les notes explicatives en détail sur les données
concernant l'Allemagne, voir Annexe I - Nomenclature des pays.

1   Les données se réfèrent aux écoles rattachées au Ministère
de l'Education seulement.
2   Les données se réfèrent à l'enseignement public seulement.
3   A partir de 1980, les écoles commerciales ont été convertis
en écoles d'enseignement général et certain cours techniques sont
classés dans l'enseignement du trosième degré.
4   Les données ne comprennent pas les écoles professionnelles.
5   A partir de 1980, les données se réfèrent aux classes allant de la
neuvième à la douzième année d'études, pour 1975 aux classes
allant de la septième à la douzième année d'études.
6   Suite à un changement de classification, à partir de 1984, les données
ne sont pas comparables à celles des années antérieures.
7   Compte non tenu des écoles turques. A partir de 1985, les données
relatives aux écoles commerciales sont comprises
avec l'enseignement général.
8   Les données relatives au personnel enseignant incluent les classes
intermédiaires rattachées aux écoles primaires.
9   Y compris les écoles de l'UNRWA.
10 Y compris l'enseignement à temps partiel.
11 Pour l'enseignement technique, les données se réfèrent aux
instituts techniques rattachés au Ministère de l'Education seulement.
12 Non compris l'ancien Yémen démocratique.
13 Pour la enseignement technique, à l'exception de 1975, les données
comprennent la formation pratique.
14 Y compris les données pour les Etats fédérés de Micronésie, les îles
Marshall et les îles Mariannes du Nord.
15 Pour 1975 et 1980, les données n'incluent pas l'enseignement
technique dispensé à la fois dans les institutions scolaires et
auprès des entreprises.
16 Pour la enseignement technique, les données comprennent la
formation pratique.
17 Pour l'enseignement technique, à partir de 1985, les données
incluent les cours du soir.
18 Pour l'enseignement technique, pour 1980, les données sur les
élèves incluent une partie des effectifs de l'enseignement normal.
19 Pour l'enseignement technique, les données relatives au personnel
personnel enseignant comprennent l'enseignement normal ainsi que
les cours du soir et par correspondance.
20 Les données excluent la formation des services de santé.
21 En 1988, un nouveau système d'enseignement technique a été
introduit dans des écoles pilotes.
22 Les données sur le personnel enseignant n'incluent pas les classes
supérieures des 'all-age schools' (comprises dans le premier degré).
23 Les données excluent les beaux-arts et les arts appliqués.
24 Les données incluent les cours commerciaux du soir.
25 Les données relatives aux élèvès incluent l'éducation spéciale.

16
Education at the third level
Enseignement du troisième degré
Teachers and students
Professeurs et étudiants

		All institutions Total des établissements				Universities and equivalent institutions Universités et établissements équivalents			
		Teachers Professeurs		Students Etudiants		Teachers Professeurs		Students Etudiants	
Country or area Pays ou zone	Years Années	Total	%F	Total	%F	Total	%F	Total	%F
Africa · Afrique									
Algeria[1]	1975	...	...	41 847	...	...	...	41 847	...
Algérie[1]	1980	8 962	...	79 351	26	8 962	...	79 351	26
	1985	11 464	...	132 057	31[2]	11 464	...	132 057	31[2]
	1988	14 087	20	180 755	34[2]	...	...	...	...
Angola	1980	225	...	2 183	...	225	...	2 183	...
Angola	1984	...	...	4 493	...	...	...	4 493	...
Benin	1975	153	...	2 118	15	148	...	2 102	15
Bénin	1980	...	...	4 822	...	...	...	...	...
	1985	...	...	9 063	16	...	...	...	...
	1989	625	...	8 883	14	...	...	...	...
Botswana	1975	56	21	469	32[3]	56	21	469	32[3]
Botswana	1980	140	...	1 078	35[3]	140	...	1 078	35[3]
	1985	154	...	1 938	...	142	...	1 773	...
	1989	...	...	...	45[3]	250	...	2 837	44[3]
Burkina Faso	1975	166	13	1 067	20	166	13	1 067	20
Burkina Faso	1980	140	...	1 643	22	140	...	1 643	22
	1986	280	7	4 498	23	253	8	4 405	24
	1989	...	...	...	...	205	7	5 675	23
Burundi	1975	223	...	1 002	11	123	...	652	11
Burundi	1980	...	...	1 879	25	...	...	1 793	24
	1985	467	7	2 783	24	315	8	2 111	24
	1989	516	9	3 080	26	...	...	...	...
Cameroon[4]	1975	...	...	...	...	...	...	7 191	15
Cameroun[4]	1980	...	...	11 686	...	...	...	10 631	...
	1985	871	...	...	...	871	...	17 071	...
	1988	...	...	26 783	...	924	...	24 651	...
Central African Rep.	1975	...	...	669	...	85	24	555	6
Rép. centrafricaine	1980	444	15	1 719	8	379	17	1 394	9
	1985	489	8	2 651	11	...	...	2 374	11
	1988	386	14	3 075	13	330	13	2 563	14
Chad[6]	1975	...	...	547	5	...	...	547	5
Tchad[6]	1984	141	8	1 643	9	104	5	1 470	9
	1988	329	...	2 983	...	320	...	2 923	...
Comoros Comores	1989	32	31	248	15	...	...	...	...
Congo	1975	165	15	3 249	10	165	15	3 249	10
Congo	1980	292	9	7 255	15	292	9	7 255	15
	1985	...	...	10 684	16	...	...	10 684	16
	1988	...	...	10 310	15	...	...	10 310	15
Côte d'Ivoire	1975[1]	...	...	7 174	17	...	...	6 274	18
Côte d'Ivoire	1980	...	...	19 633	...	625	...	12 742	19
	1984	...	...	19 660	...	...	...	11 300	21
Egypt[5]	1975	...	...	480 016	30	...	...	451 187	30
Egypte[5]	1980	...	...	715 701	32	...	...	663 418	32
	1985	31 903	...	854 584	30	29 889	27	753 190	30
	1988	...	...	764 539	33	33 106	28	656 179	33
Ethiopia	1976	...	...	6 966	...	...	...	4 971	...
Ethiopie	1980	1 051	...	14 368	...	787	...	9 291	...
	1985	1 314	9	27 338	18	1 034	...	21 601	14
	1989	1 717	7	33 486	18	1 368	8	25 562	14
Gabon	1975	...	...	1 014	20	...	...	651	23
Gabon	1985	...	...	...	...	...	...	2 607	...
	1988	...	...	4 077	30	363	16	2 896	34
Ghana	1975	1 103	...	9 079	16	963	...	7 179	13
Ghana	1980	...	...	...	...	987	...	7 951	20
	1985	...	...	...	...	1 097	...	8 324	17
	1989	...	...	...	...	700	...	9 274	18
Guinea	1975	...	...	12 411	18	...	...	...	...
Guinée	1980	1 289	3	18 270	19	577	5	5 319	14
	1985	1 107	4	8 801	14	...	...	...	...
	1988	...	...	...	...	805	3	6 245	10

16
Education at the third level
Teachers and students [*cont.*]
Enseignement du troisième degré
Professeurs et étudiants [*suite*]

Country or area Pays ou zone	Years Années	All institutions Total des établissements				Universities and equivalent institutions Universités et établissements équivalents			
		Teachers Professeurs		Students Etudiants		Teachers Professeurs		Students Etudiants	
		Total	%F	Total	%F	Total	%F	Total	%F
Guinea-Bissau Guinée-Bissau	1988	...	...	404	6	...	...	...	...
Kenya Kenya	1975	...	...	...	...	...	...	6 327	...
	1980	...	...	12 986	...	...	...	9 155	25
	1985	...	...	21 756	26	...	...	9 148	29
	1989	...	...	31 287	...	1 015	...	22 840	31
Lesotho Lesotho	1975	83	...	543	42	73	...	502	42
	1980	192	...	1 889	...	137	...	995	...
	1984	202	...	2 339	63	141	...	1 127	48
	1988	295	...	5 577	72	244	14	4 224	70
Liberia Libéria	1975	...	...	2 404	22	...	...	2 404	22
	1979	...	...	3 789	28	...	...	3 789	28
	1987	472	18	5 095	23	444	20	4 855	24
Libyan Arab Jamahiriya Jamah. arabe libyenne	1975	951	...	13 427	18	951	...	13 427	18
	1980	...	...	20 166	25	...	...	20 166	25
	1985	...	...	30 000	...	...	...	30 000	...
Madagascar Madagascar	1975	...	...	8 385	52	...	...	8 385	52
	1980	451	...	22 632	...	451	...	22 632	...
	1985	773	...	38 310	38	773	...	38 310	38
	1989	965	26	37 046	44	960	26	37 046	44
Malawi • Malawi	1975	244	...	2 198	14	150	9	1 148	12
	1980	281	...	3 476	31	173	20	1 722	25
	1985	402	...	3 928	29	278	...	1 974	18
	1988	407	...	4 951	25	242	...	2 330	14
Mali Mali	1975	...	...	2 936	10	...	...	2 936	10
	1981	475	...	4 498	11	475	...	4 498	11
	1985	...	...	6 768	13	...	...	6 768	13
	1986	715	...	5 536	13	...	...	5 536	13
Mauritania Mauritanie	1985	...	...	4 526	...	...	...	...	...
	1986	250	...	5 378	...	237	...	5 326	...
	1988	267	...	5 808	13	...	...	...	...
Mauritius Maurice	1975	155	6	1 096	14	155	6	1 096	14
	1980	210	15	1 038	31	125	6	470	19
	1985	236	11	1 161	36	130	6	507	22
	1989	382	16	2 179	34	246	11	1 487	27
Morocco Maroc	1975	1 642	...	45 322	19	937	17	35 081	19
	1980	...	...	112 405	...	2 757	18	86 731	25
	1985	...	...	181 087	32	5 310	17	150 795	33
	1989	...	19	239 923	36	7 191	19	206 485	36
Mozambique Mozambique	1976	164	24	906	34	164	24	906	34
	1980	300	...	1 000	*21[3]	300	...	1 000	*21[3]
	1985	331	21	1 442	23	331	21	1 442	23
	1987	368	26	2 335	22	368	26	2 335	22
Niger Niger	1975	74	15	541	10	74	15	541	10
	1980	224	13	1 435	20	224	13	1 435	20
	1986	349	13	3 317	18	349	13	3 317	17
	1989	...	...	4 506	15	...	...	4 163	14
Nigeria Nigéria	1975	...	...	44 964	...	5 019	...	32 971	16
	1980	10 742	...	150 072	...	5 475	...	70 395	...
	1985	...	...	266 679	27	...	...	135 783	24
	1987[14]	14 756	14	...	...	11 521	13	160 767	26
Rwanda Rwanda	1975	175	...	1 108	...	89	...	672	...
	1980	240	9	1 243	10	126	9	920	11
	1985	331	6	1 987	14	284	7	1 669	15
	1989	646	5	3 389	19	469	6	2 489	21
St. Helena Sainte-Hélène	1980	9	44	36	25	...	...	...	...
	1981	9	56	37	14	...	...	...	...
Senegal Sénégal	1976	...	...	8 921	20	...	...	8 102	20
	1980	1 084	...	13 626	18	652	...	12 673	19
	1985	...	...	13 354	...	701	13	12 711	21
	1988	...	...	...	...	770	...	14 833	21

16
Education at the third level
Teachers and students [cont.]
Enseignement du troisième degré
Professeurs et étudiants [suite]

Country or area Pays ou zone	Years Années	All institutions Total des établissements				Universities and equivalent institutions Universités et établissements équivalents			
		Teachers Professeurs		Students Etudiants		Teachers Professeurs		Students Etudiants	
		Total	%F	Total	%F	Total	%F	Total	%F
Seychelles	1976	16	56	142	93	...	15	...	...
Seychelles	1980	28	57	144	89	...	...	...	...
	1981[7]	-	-	-	-	...	...	...	...
Sierra Leone	1975	289	13	1 642	16	289	13	1 642	16
Sierra Leone	1980	270	...	1 809	...	270	...	1 809	...
	1987	341	...	2 334	...	324	...	2 314	...
Somalia	1975	324	...	2 040	11	286	...	1 936	10
Somalie	1986	817	...	15 672	20	817	...	15 672	20
Sudan	1975	1 420	7	21 342	16	1 178	4	19 208	15
Soudan	1980	1 276	8	28 788	27	1 027	4	25 699	27
	1985	2 165	10	37 367	37	1 635	8	33 934	39
	1989	2 522	16	60 134	40	1 933	12	54 558	41
Swaziland[8]	1975	136	18	1 012	...	86	14	336	...
Swaziland[8]	1980	183	35	1 875	40	108	34	1 009	37
	1986	...	...	2 172	...	115	...	1 289	42
	1987	...	...	2 363	...	...	...	1 289	42
Togo	1975	236[7]	17[7]	2 353	14	236	17	2 167	14
Togo	1980	297[7]	12[7]	4 750	15	272	13	4 345	15
	1986	...	...	6 223	13	268	12	6 105	12
	1989	...	...	...	...	...	...	7 732	13
Tunisia	1975	1 427	15	20 505	26	1 427	15	20 505	26
Tunisie	1980	4 031	9	31 827	30	4 031	9	31 827	30
	1985	5 194	...	41 594	36	5 194	...	41 594	36
	1989	4 225[3]	19[3]	62 658	38	4 225[3]	...	62 658	38
Uganda	1975	617	9	5 474	18	444	7	3 914	16
Ouganda	1980	...	...	5 856	23	...	...	4 035	20
	1985	...	...	10 103	23	*500	...	5 390	23
	1988	1 484	...	14 194	28	743	...	6 433	23
United Rep. of Tanzania	1975	...	...	3 064	14	434	...	2 644	10
Rép.-Unie de Tanzanie	1981	...	...	...	...	893	3	3 622	21
	1985	1 239	...	4 863[2 3]	15[2 3]	1 025	...	3 414[2 3]	16[2 3]
	1989	1 206	...	5 254[2 3]	...	939	...	3 327[2 3]	...
Zaire	1975	2 010	...	24 853	...	...	...	...	...
Zaïre	1980	...	...	28 493	...	...	...	9 927	...
	1985	3 272	...	40 878	...	1 387	...	16 239	...
	1988	3 873	...	61 422	...	1 636	...	27 166	...
Zambia	1975	...	...	8 403	*14	...	...	2 354	*17
Zambie	1980	...	...	...	...	...	...	3 425	...
	1986	1 345	...	14 492	...	399	...	4 857	17
	1989	...	...	14 465	28	...	...	6 247	17
Zimbabwe	1975	...	...	8 479	...	...	...	1 355	...
Zimbabwe	1980	...	...	8 339	...	...	...	1 873	...
	1985	...	...	30 843[17]	...	...	...	4 742	...
	1990	2 308	...	49 361	...	558	...	9 300	...
America, North · Amérique du Nord									
Bahamas	1976	128	...	5 660	...	...	...	...	...
Bahamas	1980	127	...	4 093	...	...	...	...	...
	1985	...	...	4 531	...	...	...	...	...
	1987	249	48	5 305	68	...	...	...	...
Barbados	1975	...	...	...	...	...	...	1 065	...
Barbade	1980	317	28	4 033	54	140	23	1 606	48
	1984	544	...	5 227	49	242	...	1 767	55
Bermuda	1980	67	33	608	51	...	...	...	...
Bermudes	1982[9]	110	36	2 664	...	...	...	...	...
Canada	1975	...	...	818 153	...	30 732[3]	14[3]	546 769	43
Canada	1980	53 434[3]	24[3]	888 444[3]	...	33 015[3]	15[3]	627 617[3]	50[3]
	1985	57 443[3]	24[3]	1 241 976	...	35 245[3]	17[3]	752 276	52
	1989	59 741[3]	25[3]	1 322 917	56	36 500[3]	18[3]	819 044	55
Costa Rica[10]	1975	...	...	33 239	...	...	...	32 794	...
Costa Rica[10]	1980	...	...	55 593	...	4 382	...	47 340	...
	1985	...	...	63 771	...	...	...	50 047	...
	1989	...	...	77 576	...	...	...	67 539	...

16
Education at the third level
Teachers and students [cont.]
Enseignement du troisième degré
Professeurs et étudiants [suite]

Country or area Pays ou zone	Years Années	All institutions Total des établissements				Universities and equivalent institutions Universités et établissements équivalents			
		Teachers Professeurs		Students Etudiants		Teachers Professeurs		Students Etudiants	
		Total	%F	Total	%F	Total	%F	Total	%F
Cuba	1975	5 380	...	82 688	...	5 380	...	82 688	...
Cuba	1980	10 680	...	151 733	48	10 680	...	151 733	48
	1985	19 552	43	235 224	54	19 552	43	235 224	54
	1989	24 499	45	242 366	58	24 499	45	242 366	58
Dominica	1980	17	71	63	22	...	...	...	...
Dominique	1984	*17	*65	60	67	...	...	...	...
	1988	12	75	68	78	...	...	...	...
Dominican Republic	1975	...	...	...	...	1 435	...	28 628	41
République dominicaine	1985	6 539	...	123 748	...	...	...	...	...
El Salvador[11]	1975	2 137	23	28 281	33	1 880	20	26 909	33
El Salvador[11]	1980	893	23	16 838	31	445	16	12 740	31
	1985	4 197	25	70 499	44	3 109	24	57 131	42
	1989	...	23	80 818	...	...	21	70 548	...
Grenada[12]	1981	71	27	928	61	...	...	...	...
Grenade[12]	1983	53	30	535	55	...	...	...	...
Guatemala[13]	1975	1 411	...	22 881	23	1 411	...	22 881	23
Guatemala[13]	1980	4 024	...	50 890	...	4 024	...	50 890	...
	1985	...	...	48 283	...	...	...	48 283	...
	1986	...	...	51 860	...	...	...	51 860	...
Haiti	1975	408	12	2 881	24	366	13	2 467	27
Haïti	1980	690	11	4 671	30	523	13	3 441	35
	1985	654	37	6 288	26	479	17	4 471	29
Honduras	1975	817	...	11 907	34	648	18	10 635	32
Honduras	1980	1 653	...	25 825	38	1 439	...	24 021	38
	1985	2 662	34	36 620	...	2 274	34	30 623	42
	1989	3 319	...	44 849	...	2 912	29	37 706	38
Jamaica	1975	...	...	...	...	...	...	3 963	...
Jamaïque	1980	...	...	13 999	...	397	...	4 548	...
	1985	...	...	10 969	...	295	29	5 126	57
	1989	...	...	...	...	...	...	5 504	60
Mexico	1975	47 529	...	562 056	...	45 025	...	520 194	...
Mexique	1980	77 653	...	929 865	33	72 742	...	817 558	30
	1985	108 002	...	1 207 779	...	...	...	1 199 120	38
	1989	133 068	...	1 314 027	...	...	...	1 258 725	42
Nicaragua	1975	1 066	...	18 282	34	911	...	15 579	38
Nicaragua	1980	...	...	35 268	...	...	...	32 958	...
	1985	2 526	...	29 001	56	2 151	...	24 430	57
	1987	1 930	...	26 878	55	1 678	...	23 873	53
Panama	1975	1 519	...	26 289	*50	1 519	...	26 289	*50
Panama	1980	2 673	...	40 369	55	2 673	...	40 369	55
	1985	3 986	...	55 303	58	3 986	...	55 303	58
	1988	3 336	30	51 086	58	3 336	30	51 086	58
Puerto Rico	1975	...	...	97 517	53	...	...	91 254	53
Porto Rico	1980	...	...	131 184	59	...	...	...	...
St. Kitts and Nevis	1979	8	63	40	70	...	...	...	...
St. Kitts-Nevis	1985	27	33	212	42	...	...	...	...
	1987	37	35	194	40	...	...	...	...
Saint Lucia	1976	...	...	298	53	...	...	...	...
Sainte-Lucie	1980	...	...	301	52	...	...	...	...
	1986	...	...	367	55	...	...	...	...
	1987	62	34	389	54	...	...	...	...
St. Vincent and the Grenadines	1985	35[3]	37[3]	736	69	...	...	...	...
Saint-Vincent-et-Grenadines	1986	34[3]	32[3]	795	64	...	...	...	...
	1989	96	53	677	68	...	...	...	...
Trinidad and Tobago	1975	...	...	4 940	38	178	...	2 229	31
Trinité-et-Tobago	1980	...	...	5 649	43	...	...	2 923	38
	1985	...	...	6 282	40	...	...	3 663	45
	1989	...	...	...	...	235	11	4 166	49
United States	1975	670 000	...	11 184 859	45	...	...	7 223 037	45
Etats-Unis	1980	395 992[3]	26[3]	12 096 895	51	305 982[3]	24[3]	7 572 657	49
	1985	694 000	...	12 247 055	52	494 000	...	7 715 978	51
	1989	...	...	13 824 592	...	...	...	...	...

16
Education at the third level
Teachers and students [cont.]
Enseignement du troisième degré
Professeurs et étudiants [suite]

Country or area Pays ou zone	Years Années	All institutions Total des établissements				Universities and equivalent institutions Universités et établissements équivalents			
		Teachers Professeurs		Students Etudiants		Teachers Professeurs		Students Etudiants	
		Total	%F	Total	%F	Total	%F	Total	%F
US Virgin Islands	1975	...	...	2 069	63	...	...	2 069	63
Iles Vierges américaines	1980	...	...	2 148	71	...	...	2 148	71
	1985	221	43	2 602	72	221	43	2 602	72
	1986	221	47	2 545	73	221	47	2 545	73
America, South · Amérique du Sud									
Argentina	1975	45 204	39	596 736	48	33 176	29	536 959	43
Argentine	1980	46 267	43	491 473	50	30 602	30	397 828	43
	1985	70 699	46	846 145	53	44 038	32	664 200	46
	1987	75 244	50	958 542	53	41 797	35	755 206	47
Bolivia[15]	1975	...	...	*49 850	...	2 178	...	34 350	...
Bolivie[15]	1980	...	...	60 900	...	...	...	60 900	...
	1985	...	...	...	...	4 924	...	95 052	...
	1989	...	...	140 890	...	...	...	106 001	...
Brazil	1975	92 546	...	1 089 808	...	92 546	...	1 089 808	...
Brésil	1980	109 788	30	1 409 243	48[2]	109 788	30	1 409 243	48[2]
	1986	117 211	...	1 451 191	...	117 211	...	1 451 191	...
	1989	128 029	36	1 518 904[2]	52[2]	128 029	36	1 518 904[2]	52[2]
Chile	1975	11 419[3]	...	149 647	*45	11 419[3]	...	149 647	*45
Chili	1980	...	...	145 497	43	...	...	120 168	40
	1985	...	...	197 437	43	...	...	...	...
	1988	...	...	234 973	44	...	...	158 278	...
Colombia[16]	1975	21 153	14	176 098	36	19 821	13	167 503	35
Colombie[16]	1980	31 136	20	271 630	45	26 930	19	234 705	43
	1985	43 227	24	391 490	49	35 890	23	317 987	45
	1989	51 725	25	474 787	52	43 248	24	410 396	51
Ecuador	1975	...	...	170 173	...	...	...	170 173	...
Equateur	1980	...	...	269 775	...	11 326	...	264 136	37
	1984	...	...	280 594	...	11 495	...	277 799	...
	1989	12 670	...	197 614	...	12 520	...	186 456	...
Guyana	1975	...	...	2 852	35	172	22	1 749	28
Guyana	1980	442	27	2 465	44	322	21	1 681	32
	1985	527	23	2 328	48	390	17	1 598	40
	1987	509	32	3 700	44	371	29	2 245	43
Paraguay	1975	...	...	18 302	...	1 741	...	17 153	...
Paraguay	1980	...	...	26 915	...	1 893	...	25 333	...
	1985	...	...	32 090	...	...	...	29 154	...
	1989	...	...	31 117	...	...	...	28 784	46
Peru	1975	11 598	14	195 641	32	10 844	12	186 511	32
Pérou	1980	15 816	...	306 353	35	14 384	16	246 510	34
	1985	26 118	...	452 462	...	20 123	...	354 888	...
	1990	58 131	...	743 569	...	45 707	...	504 700	...
Suriname	1975	...	...	...	...	...	...	465	12
Suriname	1980	...	...	2 378	...	...	...	1 217	...
	1985	...	...	2 751	54	187	...	1 070	43
	1986	484	...	3 402	51	187	10	1 198	56
Uruguay	1975	2 332	23	32 627	44	2 332	23	32 627	44
Uruguay	1980	...	...	...	...	3 847	30	36 298	53
	1987	...	...	...	...	...	...	61 450	57
	1988	...	...	...	...	6 141	...	62 461	58
Venezuela	1975	15 792	...	213 542	...	12 849	...	185 518	...
Venezuela	1980	28 052	...	307 133	...	23 984	...	271 583	...
	1985	30 844	...	443 064	41	23 951	...	347 618	...
	1988	37 564	...	500 295	47	...	...	...	...
Asia · Asie									
Afghanistan	1975	...	...	*12 256	*14	...	...	8 681	9
Afghanistan	1982	1 724	...	19 652	...	1 212	...	13 611	...
	1986	...	...	22 306	14	...	...	...	...
	1987	1 418	...	17 509	...	...	...	...	...
Bahrain	1975	79	14	703	53	...	...	...	...
Bahreïn	1980	159	13	1 908	41	70	26	317	82
	1985	434	37	4 180	60	294	44	2 011	75
	1988	...	...	...	...	430	18	5 556	54

16
Education at the third level
Teachers and students [*cont.*]
Enseignement du troisième degré
Professeurs et étudiants [*suite*]

Country or area Pays ou zone	Years Années	All institutions Total des établissements				Universities and equivalent institutions Universités et établissements équivalents			
		Teachers Professeurs		Students Etudiants		Teachers Professeurs		Students Etudiants	
		Total	%F	Total	%F	Total	%F	Total	%F
Bangladesh Bangladesh	1976	13 503	10	158 604	13	2 103	8	27 553	17
	1980	12 428	11	240 181	14	2 421	8	36 530	18
	1985	16 187	18	457 862	19	2 705	9	41 780	17
	1989	...	...	370 900	16	2 901	12	48 780	22
Bhutan Bhoutan	1980	37	27	322	22	15	20	180	8
	1984	...	...	...	..	19	...	220	20
Brunei Darussalam Brunéi Darussalam	1980	57	32	143	50	...	...	...	...
	1986	151	16	601	50	98	17	446	53
	1987	174	14	945	51	130	18	747	55
China[3] Chine[3]	1975	155 723	24	500 993	33	69	25	602	40
	1980	246 862	25	1 161 440	23	227	25	1 940	42
	1985	344 262	27	1 778 608	30	343	31	3 134	48
	1989	397 365	29	2 173 112	33	481	42	5 852	55
Hong Kong[27] Hong-kong[27]	1975	3 043	15	44 482	25	814	17	8 264	29
	1980	3 060	...	38 153	26	1 073	19	11 689	34
	1984	5 928	24	76 844	35	1 569	18	14 436	35
India Inde	1975	235 822	17	4 615 992	23	...	...	...	...
	1983	298 977	20	4 211 176	28	...	...	...	...
	1985	302 843	21	4 470 844	30	...	...	...	...
Indonesia Indonésie	1975	...	...	278 200	...	...	...	...	...
	1981	60 584	...	565 501	31	...	...	526 110	30
	1984	75 589	18	980 162	32	...	...	852 104	32
Iran, Islamic Rep. of Iran, Rép. islamique d'	1975	13 392	14	151 905	28	6 253	14	57 264	30
	1985	15 040[7]	15[7]	195 942	27	10 229[7]	13[7]	121 459	30
	1988	20 515	17	315 657	30	14 380	19	215 898	31
Iraq Iraq	1975	3 801	20	86 111	33	2 965	13	71 456	29
	1980	6 703	17	106 709	32	4 627	16	81 782	32
	1985	8 818	24	169 665	36	...	...	...	...
	1988	11 072	25	209 818	38	...	...	...	...
Israel Israël	1976	...	...	85 081	46	...	...	52 980	44
	1980	...	...	97 097	51	10 237	*32	55 840	47
	1985	...	...	116 062	47	...	...	62 514	49
	1988	...	...	117 454	47	...	...	65 080	50
Japan[18] Japon[18]	1975	191 551	12	2 248 903	32	149 349	8	1 840 708	22
	1980	213 537	14	2 412 117	33	168 739	9	1 937 124	23
	1985	243 507	14	2 347 463	35	191 533	10	1 932 785	24
	1989	271 109	16	2 683 035	39	210 791	11	2 156 528	27
Jordan Jordanie	1975	*797	*21	11 873	33	344	9	5 307	32
	1980	...		36 549	46	...		17 103	41
	1985	2 307[3]	20[3]	53 753	45	1 295[3]	13[3]	26 711	39
	1989	...	...	...	...	1 738	14	31 757	45
Korea, Dem.People's Rep. Corée, Rép. pop. dém. de	1987	27 000	19	390 000	34	23 000	17	325 000	29
Korea, Republic of Corée, République de	1975	15 317	15	318 683	25	11 578	14	222 856	26
	1980	21 173	15	647 505	23	14 969	15	436 918	22
	1985	34 300	...	1 455 759	30	27 082	16	1 018 236	27
	1990	42 911	57	1 691 429	32	34 034	18	1 143 037	29
Kuwait Koweït	1975	596	28	8 104	57	327	5	6 246	56
	1980	1 151	23	13 630	57	608	10	9 388	58
	1985	...	...	23 678	54	858	...	16 359	55
	1988	...	...	26 080	57	1 181	...	17 988	65
Lao People's Dem. Rep. République dém. pop. lao	1980	140	18	1 408	31	140	18	1 408	31
	1985	534	25	5 382	36	446	27	3 915	43
	1987	620	...	5 322	37	478	...	4 492	36
Lebanon Liban	1980	...	...	79 073	36	...	...	79 073	36
	1984	7 460	31	70 510	39	7 460	31	70 510	39
Malaysia Malaisie	1980	5 541	26	57 650	39	3 299	23	26 287	35
	1985	8 213	22	93 249	44	4 718	24	43 295	39
	1988	8 355	...	108 091	46	4 211	...	51 460	42
Mongolia Mongolie	1975	807	...	9 861	51	625	...	7 677	49
	1981	2 400	38	38 200	63	1 300	31	18 700	67
	1986	2 712	39	39 072	60	1 488	30	17 358	59

16
Education at the third level
Teachers and students [*cont.*]
Enseignement du troisième degré
Professeurs et étudiants [*suite*]

Country or area Pays ou zone	Years Années	All institutions Total des établissements				Universities and equivalent institutions Universités et établissements équivalents			
		Teachers Professeurs		Students Etudiants		Teachers Professeurs		Students Etudiants	
		Total	%F	Total	%F	Total	%F	Total	%F
Myanmar	1975	...	...	56 083	...	...	...	...	...
Myanmar	1981	...	...	165 000	...	...	...	...	...
	1987	9 028	...	202 381	...	...	...	...	...
Nepal[28]	1975	1 516	...	23 504	...	1 516	...	23 504	...
Népal[28]	1980	2 918	16	34 094	22	2 918	16	34 094	22
	1985	...	...	54 452	...	...	...	54 452	...
	1988	...	...	95 240	...	...	...	95 240	...
Oman	1976	2	-	20	-	...	...	...	...
Oman	1980	8	...	18	-	...	...	...	...
	1985	154	...	990	38	...	...	...	...
	1988	508	41	4 285	41	262	...	1 819	39
Pakistan[19]	1975	5 327	14	127 932	24	5 327	14	127 932	24
Pakistan[19]	1979	7 042	12	156 558	27	7 042	12	156 558	27
	1985	7 805	17	267 742	26	7 744	17	233 989	26
	1987	...	...	315 793	27	...	...	272 267	27
Palestine · Palestine Gaza Strip									
Zone de Gaza	1986	258	...	5 313	...	203	...	4 483	36
Philippines	1975	31 783	...	769 749	...	...	...	...	...
Philippines	1980	43 770	53	1 276 016	53	...	...	1 143 702	54
	1985	57 000	...	1 402 000	...	50 821	...	1 167 000	...
	1988	70 012	...	1 579 938	...	56 348	...	1 308 000	...
Qatar	1975	69	19	779	57	69	19	779	57
Qatar	1980	283	31	2 269	62	283	31	2 269	62
	1985	...	...	5 344	62	...	...	5 344	62
	1989	451	25	6 469	71	451	25	6 371	71
Saudi Arabia	1975	2 133	15	26 437	20	2 133	15	26 437	20
Arabie saoudite	1980	7 448	19	62 074	28	6 598	20	56 552	29
	1985	10 923	24	113 529	39	9 297	24	104 046	39
	1988	11 231	20	123 848	...	9 767	23	111 436	44
Singapore	1975	1 448	18	22 607	40	927	18	8 539	44
Singapour	1980	2 270	19	23 256	39	1 433	17	9 078	44
	1983	3 141	21	35 192	42	1 613	20	14 179	49
Sri Lanka	1975	2 000	17	15 426	36	2 000	17	15 426	36
Sri Lanka	1980[3]	4 818	...	42 694	43	1 827	...	18 111	40
	1985	3 359	...	59 377	40	2 189	...	24 222	43
	1988	...	...	...	...	1 822	28	34 858	43
Syrian Arab Republic	1975	...	...	73 660	25	1 332	8	65 348	22
Rép. arabe syrienne	1980	...	...	140 180	29	...	...	110 832	27
	1985	...	...	179 473	35	4 504	17	135 191	32
	1989	...	...	...	...	...	...	152 422	33
Thailand[20]	1975	9 070	56	130 965	40	9 070	56	78 229	43
Thaïlande[20]	1980	*19 594	...	361 400	...	10 350	...	100 401	...
	1985	30 905	...	1 026 952	...	14 666	...	150 355	...
	1989	...	...	884 237	...	...	...	172 545	...
Turkey	1975	15 560	24	327 082	16	9 596	24	93 541	23
Turquie	1980	21 577	25	246 183	25	16 162	23	165 647	26
	1985	22 968	30	469 992	32	20 353	29	300 836	34
	1989	32 029	31	685 500	34	30 630	32	394 534	36
United Arab Emirates	1980	257	8	2 861	48	207	5	2 646	45
Emirats arabes unis	1985	445	11	7 772	58	295	5	7 248	56
	1988	478	9	7 655	66	402	6	7 383	65
Viet Nam	1975	9 642	18	80 323	39	9 642	18	80 323	39
Viet Nam	1980[18]	17 242	22	114 701	24	17 242	22	114 701	24
Yemen[21]	1975	...	...	2 408	10	...	...	2 408	10
Yémen[21]	1980	157	4	4 519	11	157	4	4 519	11
	1985	386	...	12 589	...	386	...	12 589	...
	1988	470	...	23 457	...	470	...	23 457	...
Europe · Europe									
Albania[22]	1980	1 103	20	14 568	50	1 103	20	14 568	50
Albanie[22]	1985	1 468	24	21 995	45	1 468	24	21 995	45
	1990	1 806	28	22 059	52	1 806	28	22 059	52

16
Education at the third level
Teachers and students [*cont.*]
Enseignement du troisième degré
Professeurs et étudiants [*suite*]

Country or area Pays ou zone	Years Années	All institutions Total des établissements				Universities and equivalent institutions Universités et établissements équivalents			
		Teachers Professeurs		Students Etudiants		Teachers Professeurs		Students Etudiants	
		Total	%F	Total	%F	Total	%F	Total	%F
Austria Autriche	1975	...	...	96 736	38	10 001	13	86 123	34
	1980	14 086	15	136 774	42	12 572	13	127 423	39
	1985	12 135	20	173 215	45	10 252[3]	18[3]	160 904	43
	1989	13 013	23	199 845	46	11 773[3]	20[3]	186 602	44
Belgium Belgique	1975[3]	...	...	159 660	41	...	...	83 360	33
	1980[3]	...	...	196 153	44	...	...	95 246	37
	1985	...	...	247 499	46	...	...	103 598	41
	1988	...	...	...	...	...	...	106 463	42
Bulgaria[22] Bulgarie[22]	1975	12 230	32	128 593	57	11 248	30	108 814	53
	1980	14 412	39	101 359	56	12 622	37	87 335	53
	1985	15 252	37	113 795	55	14 409	36	104 259	53
	1989	20 752	38	157 861	52	19 213	37	137 994	51
Czechoslovakia Tchécoslovaquie	1975	21 298	25	155 059	40	21 298	25	155 059	40
	1980	22 478	26	197 041	42	22 478	26	197 041	42
	1985	23 944	28	169 344	43	23 944	28	169 344	43
	1989	25 350	29	186 142	43	25 350	29	186 142	43
Denmark[23] Danemark[23]	1975	6 328	...	110 271	44	4 777	...	60 106	36
	1980	...	...	106 241	49	...	...	85 388	43
	1985	...	...	116 319	49	...	...	91 450	44
	1988	...	...	126 662	51	...	...	100 543	47
Finland[29] Finlande[29]	1975	...	...	114 272[3]	...	5 225	...	75 765[3]	49[3]
	1980	...	...	123 165[3]	48[3]	6 194	...	84 176[3]	50[3]
	1985	...	...	127 976[3]	49[3]	7 169	...	92 230[3]	51[3]
	1989	...	...	155 313[3]	52[3]	7 744	...	108 125[3]	51[3]
France[24] France[24]	1975	...	...	1 038 576	*48	40 512	...	811 258	*48
	1980	...	...	1 076 717	...	...	...	869 788	...
	1985	...	...	1 278 581	50	45 211	*26	978 519	52
	1989	...	...	1 587 202	53	46 338	27	1 124 051	54
Germany ₰ · Allemagne ₰ Federal Republic of Germany Rép. féd. d'Allemagne	1975	144 834	...	1 041 225	39	103 578	...	836 002	34
	1980	171 708	...	1 223 221	41	127 383	...	1 031 590	37
	1985	...	...	1 550 211	42	...	...	1 336 395	38
	1988	198 241	22	1 686 725	41	150 980	18	1 464 594	38
former German Dem. Rep.[22] l'ancienne Rép. dém. allemande[22]	1975	34 566	25	386 000	...	26 115	22	142 567	47
	1980	38 699	27	400 799	58	28 848	23	137 554	47
	1985	42 336	...	432 672	54	30 082	26	148 650	47
	1988	42 702	...	438 930	52	30 921	28	156 296	45
Greece Grèce	1975	...	...	117 246	37	5 956	35	95 385	37
	1980	10 542	32	121 116	41	6 924	35	85 718	42
	1985	11 878	30	181 901	49	6 934	28	110 917	48
	1987	12 760	31	189 173	49	7 435	28	117 193	51
Holy See[23] Saint-Siège[23]	1975	1 280	3	7 758	27	1 280	3	7 758	27
	1980	1 349	7	9 104	39	1 349	7	9 104	39
	1985	1 498	7	9 775	33	1 498	7	9 775	33
	1989	1 478	7	11 000	33	1 478	7	11 000	33
Hungary[22] Hongrie[22]	1975	12 135	27	107 555	48	9 494	25	67 983	49
	1980	13 890	29	101 166	50	10 616	27	61 767	47
	1985	14 850	30	99 344	54	11 460	28	61 163	49
	1989	16 319	33	100 868	51	12 101	30	65 531	49
Iceland Islande	1975	575	12	2 970	37	527	10	2 789	35
	1980	...	...	3 633	60	...	...	...	...
	1985	...	...	4 724	55	...	...	...	...
	1989	...	...	5 407	56	...	...	...	...
Ireland Irlande	1975	4 088	...	46 174	34	2 261	...	24 976	41
	1980	...	...	54 746	41	...	...	33 173	48
	1985	6 002	...	70 301	43	3 332	...	39 120	48
	1988	5 086	...	81 133	44	2 538	...	43 556	49
Italy Italie	1975	...	...	976 712	39	41 824	...	968 119	39
	1980	...	...	1 117 742	43	42 531	...	1 110 547	42
	1985	51 539	...	1 185 304	46	50 996	...	1 176 726	46
	1989	54 473	...	1 358 254	48	53 760	...	1 348 500	48

16
Education at the third level
Teachers and students [*cont.*]
Enseignement du troisième degré
Professeurs et étudiants [*suite*]

Country or area	Years	All institutions Total des établissements				Universities and equivalent institutions Universités et établissements équivalents			
		Teachers Professeurs		Students Etudiants		Teachers Professeurs		Students Etudiants	
Pays ou zone	Années	Total	%F	Total	%F	Total	%F	Total	%F
Luxembourg[25]	1975	137	8	483	42	...	...	...	...
Luxembourg[25]	1980	250	...	748	35	...	...	...	...
	1985	366	*11	759	34	...	...	...	...
Malta	1975	236	6	1 425	28	179	5	844	30
Malte	1980	129	5	947	24	129	5	947	24
	1985	156	6	1 474	33	156	6	1 474	33
	1988	128	9	1 682	38	128	9	1 682	38
Netherlands	1975	...	...	288 026	33	...	...	120 134	25
Pays-Bas	1980	...	...	360 033	40	...	...	149 524	31
	1985	...	...	404 866	41	...	...	168 858	37
	1988	...	...	415 847	44	...	...	171 196	40
Norway	1975	6 650	16	66 628	38	3 757	12	40 774	36
Norvège	1980	7 763	19	79 117	48	3 903	13	40 620	41
	1985	8 898	26	94 658	52	4 265	17	41 658	47
	1988	9 572	27	114 855	53	4 277	19	47 884	50
Poland[22]	1975	...	...	575 499	54	50 272	34	468 129	49
Pologne[22]	1980	...	...	589 134	56	57 083	35	453 652	50
	1985	...	...	454 190	56	57 280	35	359 245	50
	1989	...	...	505 727	59	65 917	...	408 064	54
Portugal[30]	1975	7 891	32	79 702	45	4 168	25	51 489	46
Portugal[30]	1980	10 695	31	92 152	48	6 906	30	67 652	48
	1985	12 476	37	103 585	54	7 614	32	70 244	52
	1989	13 248	...	156 701	...	10 629	...	113 704	...
Romania[22]	1975	14 066	29	164 567	45	14 066	29	164 567	45
Roumanie[22]	1980	14 592	30	192 769	43	14 592	30	192 769	43
	1985	12 961	29	159 798	45	12 961	29	159 798	45
	1989	11 696	...	164 507	...	11 696	...	164 507	...
Spain	1975	29 701	19	540 238	36	29 438	19	515 732	37
Espagne	1980	42 831	21	697 789	44	42 260	21	690 801	43
	1985	47 504	26	935 126	49	46 740	26	882 798	50
	1987	52 206	28	1 036 439	50	51 453	28	978 337	51
Sweden	1975	...	...	162 640	40	...	...	113 348	37
Suède	1980	...	...	171 356	...	...	...	...	...
	1985	...	...	183 697	...	...	...	...	...
	1989	...	...	184 815	53	...	...	...	...
Switzerland	1975	...	...	64 720	...	5 414	7	52 623	27
Suisse	1980	...	...	85 127	30	5 942	8	61 374	32
	1985	...	...	110 111	32	6 236	...	74 806	36
	1989	...	...	132 753	34	...	...	83 263	38
United Kingdom	1975	...	...	732 947	36	32 208	...	295 031	33
Royaume-Uni	1980	...	...	827 146	37	34 297	18	339 925	37
	1985	79 621	15	1 032 491	46	31 412	12	352 419	40
	1988	84 666	18	1 113 341	47	34 242	17	383 644	42
Yugoslavia	1975	...	...	394 992	40	...	...	271 517	40
Yougoslavie	1980	24 449	24	411 995	45	19 981	24	310 650	46
	1985	25 862	26	350 334	46	22 204	27	287 907	46
	1989	26 540	28	342 643	49	23 461	29	295 302	49
Oceania · Océanie									
American Samoa	1975	...	...	689	49	...	...	...	...
Samoa américaines	1980	...	...	976	56	...	...	...	...
	1985	...	...	758	52	...	...	...	...
	1988	...	...	909	54	...	...	...	...
Australia	1975	19 920[3]	...	274 738	41	19 920[3]	...	274 738	41
Australie	1980	22 134[3]	...	323 716	45	22 134[3]	...	323 716	45
	1985	22 659[3]	23[3]	370 048	48	22 659[3]	23[3]	370 048	48
	1989	25 032[3]	29[3]	441 076	52	25 032[3]	29[3]	441 076	52
Cook Islands									
Iles Cook	1980	41	24	360	45	32	25	303	44
Fiji	1975	166	17	1 653	29	105	16	1 229	31
Fidji	1980	...	...	1 666	...	...	...	1 391	...
	1985	249	18	2 313	38	148	17	1 932	37
	1989	...	...	...	...	265	...	2 711	...

16
Education at the third level
Teachers and students [*cont.*]
Enseignement du troisième degré
Professeurs et étudiants [*suite*]

| Country or area | Years | All institutions Total des établissements | | | | Universities and equivalent institutions Universités et établissements équivalents | | | |
| | | Teachers Professeurs | | Students Etudiants | | Teachers Professeurs | | Students Etudiants | |
Pays ou zone	Années	Total	%F	Total	%F	Total	%F	Total	%F
French Polynesia	1980	14	7	27	26	...	...	...	...
Polynésie française	1983	...	...	180		...	...	...	...
Guam	1975	...	...	3 800	43	...	...	3 800	43
Guam	1980	...	...	3 217	55	...	...	3 217	55
	1988	...	...	...	...	202	34	4 257	59
New Caledonia	1975	23	13	178	35	...	...	...	...
Nouvelle-Calédonie	1980	97	48	438	39	...	...	...	...
	1985	63		761	44	...	...	...	...
New Zealand[18]	1975	...	...	66 178	36	4 108	11	36 931	37
Nouvelle-Zélande[18]	1980	7 694	18	76 643	41	4 780	14	43 933	42
	1985	8 300	24	95 793	46	5 226	20	47 799	46
	1989	10 337	42	120 821	50	6 015	26	62 875	49
Pacific Islands (Palau)[31]									
Iles du Pacifique (Palaos)[31]	1980	...	...	2 129	24	...	...	...	...
Papua New Guinea	1975	...	...	...	...	353	...	2 869	12
Papouasie-Nouv.-Guinée	1980	*638	...	*5 040	*22	473	...	2 872	11
	1985	...	...	5 068	23	...	...	3 181	16
	1986	902	21	6 397	24	400	14	3 413	17
Samoa[26]	1975	*40	-	249	*5	...	...	...	...
Samoa[26]	1981	79		694	7	22	...	295	3
	1983	37	30	562	47	11	...	136	50
Tonga									
Tonga	1985	...	...	705	56	17	18	85	22
former USSR · ancienne URSS									
former USSR[22]	1975	317 152	37	4 853 958	50	...	...	...	...
ancienne URSS[22]	1980	365 300	...	5 235 200	...	...	...	...	...
	1985	377 300	...	5 147 200	...	...	...	...	...
	1989	404 000	...	5 273 000	50	...	...	...	...
Belarus[22]	1975	...	...	159 903	...	...	...	...	...
Bélarus[22]	1980	12 900	...	177 000	...	...	...	...	...
	1985	13 500	...	181 900	...	...	...	...	...
	1989	16 800	...	192 400	53	...	...	...	...
Ukraine[22]	1975	...	...	831 300	49	...	...	...	...
Ukraine[22]	1980	...	...	880 400	...	...	...	...	...
	1985	...	...	853 100	...	...	...	...	...
	1989	74 300	...	902 100	...	...	...	...	...

Source:
United Nations Educational, Scientific and Cultural Organization
(Paris).

Source:
Organisation des Nations Unies pour l'éducation, la science et la culture
(Paris).

§ All data shown which pertain to Germany prior to 3 October 1990 are indicated separately for the Federal Republic of Germany and the former German Democratic Republic based on their respective territories at the time indicated. For detailed explanatory notes on data pertaining to Germany, see Annex I-Country Nomenclature.

1 Data refer to institutions under the Ministry of Education.
2 Excluding post-graduate students.
3 Full time only.
4 In 1980 data exclude 'ENAM'.
5 Data on students in 1975, and teaching staff since 1985 exclude Al Azhar University. In 1985 data on female students exclude private higher institions.
6 In 1988, not including the University of Law.
7 Excluding teacher training.
8 Data on students include nationals studying abroad.
9 Data include adult education.
10 From 1980, data refer only to institutions recognized by the National Council for Higher Education.
11 In 1980, data exclude the National University which was closed.
12 In 1983, data exclude Grenada Teachers' College.

§ Toutes les données se rapportant à l'Allemagne avant le 3 octobre 1990 figurent dans deux rubriques séparées basées sur les territoires respectifs de la République fédérale d'Allemagne et l'ancienne République démocratique allemande selon la période indiquée. Pour les notes explicatives en détail sur les données concernant l'Allemagne, voir Annexe I - Nomenclature des pays.

1 Les données se rapportent aux institutions relevant du Ministère de l'Education seulement.
2 Non compris les étudiants du niveau universitaire supérieur.
3 Enseignement à plein temps seulement.
4 En 1980 les données exculent 'ENAM'.
5 Les données relatives aux étudiants en 1975 et à tous les professeurs depuis 1985 excluent l'Université Al Azhar. En 1985, les données relatives aux étudiantes excluent les établissements privés d'enseignement supérieur.
6 En 1988, non compris l'Université des Sciences juridiques.
7 Non compris l'enseignement normal.
8 Les données relatives aux étudiants incluent les étudiants à l'étranger.
9 Les données incluent l'éducation des adultes.
10 A partir de 1980, les données se réfèrent seulement aux institutions reconnues par le Conseil National pour l'Education supérieure.

16
Education at the third level
Teachers and students [*cont.*]
Enseignement du troisième degré
Professeurs et étudiants [*suite*]

13 Except for students in 1980, data refer to the University of San Carlos only.
14 In 1987, data on other third level institutions refer to teacher training only.
15 Teaching staff in 1975 excludes 'la Universidad Católica Boliviana'.
16 In 1980 and 1989, data on universities include the Open university; data on teaching staff refer to teaching posts only.
17 Data covering 'other third level institutions' refer to teacher training only.
18 Including correspondence courses.
19 For 1975 and 1979, data exclude teaching staff in arts and science colleges. From 1985 data refer to universities only.
20 In 1980, data exclude one Open university. In 1981, data on universities include open universities.
21 Excluding former Democratic Yemen.
22 Including evening and corresponding courses.
23 Data refer to teaching staff and students enrolled in higher institutions under the authority of the Holy See.
24 The total number of students (all institutions) is overestimated due to some students enrolled at institutions, considered here as non-university being enrolled also at the universities. From 1985 data on teaching staff refer to public universities only.
25 Data refer to students enrolled in institutions located in Luxembourg. At university level, the majority of the students pursue their studies in the following countries: Austria, Belgium, France, Federal Republic of Germany and Switzerland.
26 In 1983, data exclude school of agriculture.
27 In 1980, data in 'other third level institutions' refer to Hong Kong Polytechnic only.
28 From 1980 to 1985 data refer to public universities only.
29 From 1985, the data are not strictly comparable with those of the previous years due to classification changes.
30 In 1985, excluding the University of Porto.
31 Including data for Federated States of Micronesia, Marshall Is. and Northern Mariana Is.

11 En 1980, les données excluent l'Université Nationale qui était fermée.
12 En 1983, les données excluent 'Grenada Teachers' College'.
13 A l'exception des étudiants en 1980, les données se réfèrent à l'Université de San Carlos seulement.
14 En 1987, les données sur les autres établissements du troisième degré se réfèrent à l'enseignement normal seulement.
15 Le personnel enseignant en 1975 exclut 'la Universidad Católica Boliviana'.
16 En 1980 et 1989, les données relatives aux universités incluent l'enseignement à distance; celles du personnel enseignant se réfèrent aux postes d'enseignants.
17 Les données des 'autres établissements du troisième degré' se réfèrent à l'enseignement normal seulement.
18 Y compris les cours par correspondance.
19 En 1975 et 1979, non compris les données relatives au personnel enseignant des 'Arts and science colleges'. A partir de 1985 les données se réfèrent aux universités seulement.
20 En 1980, les données excluent une université d'enseignement à distance. En 1981, les données relatives aux universités incluent les universités à distance.
21 Non compris l'ancien Yémen démocratique.
22 Y compris les cours du soir et par correspondance.
23 Les données se réfèrent aux enseignants et étudiants dans les institutions du troisième degré sous l'autorité du Saint-Siège.
24 Le nombre total d'étudiants (ensemble d'établissements) est surestimé du fait que certains étudiants inscrits dans les établissements considérés ici comme non-universitaires sont également inscrits dans les universités. A partir de 1985 les données relatives au personnel enseignant se réfèrent aux universités publiques seulement.
25 Les données se réfèrent seulement aux étudiants inscrits dans les institutions du Luxembourg. La plus grande partie des étudiants luxembourgeois poursuivent leurs études universitaires dans les universités des pays suivants : Autriche, Belgique, France, République fédérale d'Allemagne et Suisse.
26 En 1983, les données excluent l'école d'agriculture.
27 En 1980, les données relatives à 'l'autre enseignement du troisième degré' se réfèrent à 'Hong Kong Polytechnic' seulement.
28 De 1980 à 1985 les données se réfèrent aux universités publiques seulement.
29 A partir de 1985, les données ne sont pas comparables avec celles des années précédentes à la suite d'un changement de classification.
30 En 1985, non compris l'Université de Porto.
31 Y compris les données pour les Etats fédérés de Micronésie, les îles Marshall et les îles Mariannes du Nord.

17
Public expenditure on education at current market prices
Dépenses publiques afférentes à l'enseignement aux prix courant du marché

Country or area Pays ou zone (Currency unit · Unité monétaire)	Years Années	Total educational expenditure Dépenses totales d'éducation			$ US per cap. $ E-U par hab.	Current educational expenditure Dépenses ordinaires d'éducation		
		Amount Montant (000)	% of GNP % du PNB	% total gov't exp. % dép. totales gouv.		Amount Montant (000)	% of GNP % du PNB	% current gov't exp. % dép. ordinaires gouv.
Africa · Afrique								
Algeria	1975	4 080 500	6.7	23.0	...	2 951 000	4.8	23.9
Algérie	1980	12 354 500	7.8	24.3	...	8 259 100	5.2	29.8
(dinar)	1985	24 248 000	8.5	24.3	...	16 814 000	5.9	30.8
	1989	32 626 000	9.4	27.0	...	25 696 000	7.4	35.7
Angola	1985	9 643 000	...	10.8	...	9 419 000	...	14.0
Angola	1987	12 854 000	...	13.8	...	10 561 000	...	15.5
(kwansa)	1989[1]	12 076 000	...	10.7	...	10 856 000	...	...
Benin								
Bénin	1975	...	...	...	...	5 627 477	4.8	38.9
(CFA franc · franc CFA)	1980	...	...	...	...	12 426 246	5.1	36.8
Botswana	1975	15 927	8.5	18.8	...	8 688	4.6	17.6
Botswana	1979	36 546	7.8	16.1	...	27 483	5.8	21.4
(pula)	1985	110 812	8.0	15.4	...	87 907	6.4	18.6
	1989[1]	359 359	8.2	16.3	...	249 063	5.6	20.9
Burkina Faso	1976	4 020 947	1.9	19.0	...	3 674 455	1.7	22.0
Burkina Faso	1980	7 994 307	2.2	19.8	3.2	7 435 813	2.1	21.1
(CFA franc · franc CFA)	1985	12 901 470	2.0	21.0	...	12 292 250	1.9	22.4
	1989	18 780 000	2.3	17.5	...	18 727 000	2.3	21.9
Burundi	1975	...	...	...	...	721 831	2.2	22.7
Burundi	1979	2 066 175	3.5	17.5	...	1 796 482	2.6	...
(franc)	1985	3 466 904	3.0	15.5	...	3 212 494	2.3	17.1
	1988	4 876 000	2.5	16.7	...	4 797 000	3.2	19.0
Cameroon	1975	21 924 800	3.9	21.3	...	18 310 000	3.2	22.7
Cameroun	1980	45 099 400	3.2	20.3	17.3	36 653 000	2.6	...
(CFA franc · franc CFA)	1985[1]	109 344 000	3.0	14.8	...	90 045 000	2.5	...
	1989[1]	112 103 000	3.3	18.7	...	101 537 000	3.0	23.9
Cape Verde								
Cap-Vert	1976	...	...	...	...	99 660	4.2	20.2
(escudo)	1985	341 039	2.8	...	...	325 013	2.6	15.2
	1987	493 000	2.9	...	...	471 668	2.8	14.2
Central African Rep.	1975	3 905 000	4.9	20.1	...	3 393 000	4.2	...
Rép. centrafricaine	1979	5 669 867	3.8	20.9	...	5 509 867	3.6	...
(CFA franc · franc CFA)	1985[1]	8 129 000	2.6	...	...	...	...	...
	1987	9 197 000	2.9	16.8	...	9 002 000	2.8	24.7
Comoros								
Comores								
(CFA franc · franc CFA)	1986	3 652 680	6.5	13.2	...	2 426 680	4.3	23.4
Congo	1975	12 752 237	8.1	18.2	...	10 537 237	6.7	24.7
Congo	1980	22 941 531	7.1	23.6	...	21 517 291	6.6	24.1
(CFA franc · franc CFA)	1984	44 441 800	5.1	9.8	...	41 033 300	4.7	14.7
Côte d'Ivoire	1975	50 879 000	6.6	19.0	...	43 076 000	5.6	33.7
Côte d'Ivoire	1980	147 478 000	7.0	22.6	...	123 196 000	5.9	36.4
(CFA franc · franc CFA)	1985	...	...	...	...	179 447 000	6.3	...
Djibouti	1974	1 624 821	...	...	...	1 479 821	...	...
Djibouti	1979	1 367 509	...	11.5	...	1 045 949	...	9.6
(franc)	1985[1]	1 689 778	...	7.5	...	1 689 778	...	...
	1989	2 595 557	...	10.7	...	2 595 557	...	11.9
Egypt[2]	1975	262 328	5.1	...	...	226 668	4.4	...
Egypte[2]	1981	918 679	5.7	9.4	...	722 108	4.5	10.1
(pound · livre)	1985	1 877 850	6.3	...	...	1 774 700	5.9	10.8
	1989	4 020 000	6.8	...	...	3 599 000	6.1	...
Equatorial Guinea								
Guinée équatoriale								
(CFA francs · francs CFA)	1988	619 860	...	3.9	...	527 060	...	3.5
Ethiopia	1975	179 833	3.3	14.5	...	147 837	2.7	16.5
Ethiopie	1980	278 521	3.3	10.4	...	221 504	2.6	12.7
(birr)	1985	419 582	4.3	9.5	...	353 823	3.6	14.3
	1987	484 780	4.4	9.4	...	405 422	3.6	13.1

17
Public expenditure on education at current market prices [cont.]
Dépenses publiques afférentes à l'enseignement aux prix courants du marché [suite]

Country or area Pays ou zone (Currency unit · Unité monétaire)	Years Années	Total educational expenditure Dépenses totales d'éducation			$US per cap. $E-U par hab.	Current educational expenditure Dépenses ordinaires d'éducation		
		Amount Montant (000)	% of GNP % du PNB	% total gov't exp. % dép. totales gouv.		Amount Montant (000)	% of GNP % du PNB	% current gov't exp. % dép. ordinaires gouv.
Gabon Gabon (CFA franc · franc CFA)	1975	8 902 838	2.1	...	...	7 569 838	1.8	...
	1980	22 203 707	2.7	...	...	16 054 707	2.0	...
	1985[1]	69 500 000	4.5	9.4	...	47 500 000	3.1	21.7
	1987	53 372 000	5.6	...	...	50 860 000	5.4	...
Gambia Gambie (dalasi)	1975	7 007	3.2	...	...	5 471	2.5	...
	1980	12 869	3.2	...	...	11 336	2.9	...
	1985	30 857	4.2	...	...	24 706	3.4	16.4
	1988	53 441	4.0	8.8	...	37 462	2.8	9.3
Ghana Ghana (cedi)	1975	309 024	5.9	21.5	...	240 682	4.6	24.1
	1980	1 318 860	3.1	17.1	...	...	...	...
	1985[1]	8 674 617	2.6	19.0	...	...	...	...
	1989	47 791 000	3.4	...	...	43 857 000	3.2	...
Guinea Guinée (franc)	1984	1 491 077	...	15.3	...	1 486 054	...	17.2
	1987	10 968 000	...	13.0	...	7 268 000	...	10.4
	1988	19 743 000	...	21.5	...	15 330 000	...	...
Guinea-Bissau Guinée-Bissau (peso)	1976	...	...	...	...	175 917	3.9	...
	1980	...	...	...	...	207 512	4.0	...
	1984	...	...	...	...	539 092	3.2	11.2
	1987	2 533 118	2.8	...	...	2 473 235	2.8	...
Kenya Kenya (shilling)	1975	1 444 760	6.3	19.4	...	1 378 580	6.0	27.7
	1980	3 526 420	6.8	18.1	28.5	3 247 000	6.2	23.6
	1985	6 171 405	6.4	...	...	5 789 282	6.0	...
	1988	9 300 000	6.5	27.0	...	8 926 000	6.2	31.5
Lesotho Lesotho (maloti)	1975	8 912	4.5	23.5	...	6 777	3.4	26.8
	1980	25 074	5.1	14.8	24.6	20 037	4.1	19.0
	1984	34 841	3.7	...	...	29 169	3.1	...
	1988	66 329	4.0	...	11.7[3]	60 682	3.6	...
Liberia Libéria (dollar)	1975	13 866	1.9	11.6	...	...	...	...
	1980	61 778	5.7	24.3	...	53 074	4.9	27.0
Libyan Arab Jamahiriya Jamah. arabe libyenne (dinar)	1975	224 908	5.9	14.5	...	122 485	3.2	...
	1980	356 131	3.4	...	301.1	224 431	2.1	...
	1985	574 638	8.2	19.8	...	457 238	6.5	38.1
	1986	636 260	10.1	20.8	...	506 260	8.1	37.1
Madagascar Madagascar (franc)	1975	12 313 747	2.5	18.5	...	11 726 747	2.4	23.0
	1980	36 895 761	4.4	...	...	31 547 842	3.7	...
	1985	52 182 368	2.8	...	...	49 805 768	2.7	...
	1988[4]	59 842 000	1.8	...	...	58 542 000	1.8	...
Malawi Malawi (kwacha)	1975	13 052	2.4	9.6	...	11 992	2.2	18.3
	1980	31 208	3.4	8.4	...	23 595	2.6	...
	1985	64 260	3.5	9.6	...	46 396	2.5	...
	1988	114 254	3.3	...	...	83 239	2.4	...
Mali Mali (CFA franc · franc CFA)	1975	...	...	...	...	5 852 452	3.5	30.6
	1980	12 903 454	3.8	30.8	...	12 752 392	3.7	32.1
	1985	17 183 900	3.7	...	...	17 047 900	3.7	...
	1987	18 693 000	3.3	17.3	...	18 285 000	3.3	18.5
Mauritania Mauritanie (ouguiya)	1976	847 400	3.8	14.3	...	836 800	3.8	15.4
	1980	...	...	...	...	1 546 000	5.0	...
	1985	...	...	...	...	3 973 200	7.7	33.2
	1988	...	...	...	...	3 188 000	4.7	22.0
Mauritius Maurice (rupee · roupie)	1975	144 000	3.6	9.6	...	125 800	3.1	11.7
	1980	453 901	5.3	11.6	40.7	407 926	4.7	15.5
	1985	597 600	3.8	9.8	...	554 500	3.5	12.4
	1988	1 046 800	4.1	10.4	33.8[3]	991 600	3.9	13.0
Morocco Maroc (dirham)	1975	1 846 709	5.1	14.3	...	1 591 988	4.4	20.4
	1980	4 358 718	6.1	18.5	...	3 528 566	4.9	23.3
	1985	8 961 170	7.4	26.7	...	7 121 470	5.8	33.5
	1987	11 103 134	7.3	25.5	...	8 098 234	5.4	34.2
Niger Niger (CFA franc · franc CFA)	1975	5 249 700	2.4	18.7	...	4 566 400	2.1	17.9
	1980	16 532 722	3.1	22.9	...	7 762 522	1.5	16.8
	1989[4]	19 873 000	3.1	9.0	...	15 545 000	2.5	13.6

17
Public expenditure on education at current market prices [cont.]
Dépenses publiques afférentes à l'enseignement aux prix courants du marché [suite]

Country or area Pays ou zone (Currency unit · Unité monétaire)	Years Années	Total educational expenditure Dépenses totales d'éducation			$ US per cap. $ E-U par hab.	Current educational expenditure Dépenses ordinaires d'éducation		
		Amount Montant (000)	% of GNP % du PNB	% total gov't exp. % dép. totales gouv.		Amount Montant (000)	% of GNP % du PNB	% current gov't exp. % dép. ordinaires gouv.
Nigeria[5]	1974	601 795	2.9	...	...	326 361	1.6	...
Nigéria[5]	1981	3 152 687	5.5	24.7	...	2 522 740	4.4	68.7
(naira)	1985	814 470	1.0	8.7	...	699 450	0.9	19.7
	1986	1 170 060	1.5	12.0	...	728 010	0.9	19.0
Réunion	1980	1 314 343	...	...	...	1 263 595	...	...
Réunion	1985	2 416 435	...	...	...	2 291 828	...	...
(franc)	1987	2 288 493	...	...	...	2 284 520	...	...
Rwanda	1975	1 198 064	2.3	25.3	...	1 197 164	2.3	28.4
Rwanda	1980	2 880 120	2.7	21.6	...	2 438 563	2.3	21.5
(franc)	1987	6 010 000	3.5	22.9	...	5 642 000	3.3	25.0
	1989	7 222 000	4.2	25.4	...	6 793 000	3.9	29.1
Sao Tome and Principe								
Sao Tomé et Principe	1981	90 639	8.0	...	...	...	...	...
(dobra)	1986	100 191	4.3	18.8	...			
Senegal	1975	...	...	...	...	15 949 600	4.0	21.4
Sénégal	1980	...	...	...	...	27 485 411	4.5	23.5
(CFA franc · franc CFA)	1985	...	...	...	...	46 118 000	4.2	...
	1988	...	...	...	...	51 906 000	3.7	...
Seychelles	1975	12 632	4.4	9.5	...	11 202	3.9	15.1
Seychelles	1980	52 485	5.8	14.4	...	50 215	5.5	14.0
(rupee · roupie)	1985	124 588	10.7	21.3	...	120 150	10.3	21.9
	1989	143 477	9.1	15.2	...	133 477	8.5	16.9
Sierra Leone	1975	19 300	3.4	...	...	17 410	3.1	16.3
Sierra Leone	1980	42 734	3.8	11.8	13.8	40 731	3.6	14.5
(leone)	1985	...	...	...	5.2	105 901	2.3	15.5
	1989	...	...	...	...	576 552	1.4	13.7
Somalia	1975	92 400	2.1	12.5	...	77 300	1.7	12.8
Somalie	1980[6]	169 384	1.0	8.7	...	154 384	0.9	...
(shilling)	1985[6]	371 256	0.5	4.1	...	273 956	0.3	...
	1986[6]	433 784	0.4	2.8	...	290 360	0.3	...
Sudan	1974	68 257	5.5	14.8	...	61 157	4.9	...
Soudan	1980	187 005	4.8	9.1	19.2	172 399	4.4	12.6
(pound · livre)	1985[6]	...	...	...	...	579 537	4.0	15.0
Swaziland	1975	7 920	3.7	...	...	6 186	2.9	...
Swaziland	1980	25 697	6.2	...	...	19 627	4.7	23.1
(lilangeni)	1986	59 954	6.0	...	...	56 905	5.7	25.9
	1988	89 887	6.2	...	...	72 215	5.0	18.3
Togo	1975	4 622 873	3.5	15.2	...	4 450 794	3.4	21.3
Togo	1980	13 049 223	5.5	19.4	...	12 574 743	5.3	21.0
(CFA franc · franc CFA)	1985	15 879 511	4.9	19.4	...	15 027 511	4.6	19.2
	1988	20 401 000	5.2	21.2	...	19 145 000	4.9	22.2
Tunisia	1975	87 533	5.2	16.4	...	78 045	4.6	22.5
Tunisie	1980	185 351	5.4	16.4	...	162 291	4.7	23.5
(dinar)	1985	388 794	5.9	14.1	...	350 609	5.3	20.8
	1988	...	...	...	...	449 012	5.5	19.2
Uganda[1]	1975	5 686	2.5	17.0	...	5 290	2.4	22.0
Ouganda[1]	1980	15 475	1.2	11.3	...	13 667	1.1	12.8
(shilling)	1984	287 970	3.1	...	...	204 630	2.2	...
	1987	6 348 700	3.4	22.5	...	6 187 800	3.4	30.6
United Rep. of Tanzania	1975	1 029 100	5.4	17.8	...	827 800	4.4	22.5
Rep. Unie de Tanzania	1980[1]	1 839 929	4.4	11.2	10.7	1 521 972	3.6	16.3
(shilling)	1985[1]	4 234 000	3.6	14.0	...	3 643 000	3.1	15.6
	1989	13 997 000	3.7	14.0	2.9[3]	13 069 000	3.5	17.0
Zaire	1975	...	...	...	...	107 794	2.1	27.0
Zaïre	1980	1 015 127	2.5	24.2	...	997 760	2.5	25.3
(zaire · zaïre)	1985	3 291 428	1.0	7.3	...	3 239 170	1.0	7.3
	1988	15 006 000	0.9	6.4	...	14 357 000	0.8	6.4
Zambia	1975	98 020	6.7	11.9	...	75 420	5.2	13.0
Zambie	1980	126 637	4.5	7.6	...	120 377	4.2	11.1
(kwacha)	1984	246 706	5.5	16.3	...	230 681	5.2	17.5

17
Public expenditure on education at current market prices [*cont.*]
Dépenses publiques afférentes à l'enseignement aux prix courants du marché [*suite*]

Country or area Pays ou zone (Currency unit · Unité monétaire)	Years Années	Total educational expenditure Dépenses totales d'éducation				Current educational expenditure Dépenses ordinaires d'éducation		
		Amount Montant (000)	% of GNP % du PNB	% total gov't exp. % dép. totales gouv.	$ US per cap. $ E-U par hab.	Amount Montant (000)	% of GNP % du PNB	% current gov't exp. % dép. ordinaires gouv.
Zimbabwe	1975	70 827	3.6	...	...	65 015	3.3	14.2
Zimbabwe	1980	223 845	6.6	13.7	21.6	218 037	6.4	14.1
(dollar)	1986	726 361	9.2	16.0	31.0	720 587	9.1	...
	1989[6]	1 035 000	8.5	...	...	1 025 000	8.4	...
America, North · Amérique du Nord								
Antigua and Barbuda	1975	6 346	4.5	14.4	...	5 681	4.0	15.9
Antigua-et-Barbuda	1980	8 621	2.9	...	...	8 529	2.8	...
(dollar)	1984	11 870	2.5	...	...	11 371	2.4	...
	1988	...	...	...	...	30 777	3.7	...
Bahamas	1975	32 495	6.2	...	...	29 938	5.7	22.2
Bahamas (dollar)	1980	...	...	...	...	53 386	4.4	22.1
Barbados	1975	48 827	6.0	20.9	...	42 256	5.2	...
Barbade	1980	108 718	6.5	20.5	...	90 042	5.4	...
(dollar)	1984	139 260	6.1	...	...	122 760	5.4	...
	1989	232 032	6.9	...	...	196 467	5.9	...
Bermuda	1974	11 192	...	18.0	...	9 735	...	16.7
Bermudes	1980	26 247	...	...	...	25 713	...	...
(dollar)	1989	51 195	...	16.7	...	43 691	...	15.9
British Virgin Islands	1975	1 002	...	...	...	967	...	...
Iles Vierges brit.	1980	...	...	...	...	1 622	...	13.3
(US dollars · dollars E-U)	1984	4 161	...	15.3	...	4 161	...	16.9
Canada	1975	12 790 683	7.6	17.8	...	11 526 821	6.8	...
Canada	1980	22 100 070	7.3	17.3	...	20 450 870	6.8	...
(dollar)	1985	32 429 019	7.0	12.7	...	30 175 973	6.5	...
	1989	44 187 000	7.0	15.3	...	41 091 000	6.5	...
Cayman Islands								
Iles Caïmanes	1975	1 625	...	12.0	...	1 384	...	14.6
(dollar)	1980	4 895	...	...	...	3 398	...	...
Costa Rica	1975	1 113 826	6.8	31.1	...	1 041 297	6.4	39.7
Costa Rica	1980	3 069 078	7.8	22.2	...	2 802 162	7.1	26.7
(colon)	1985	8 180 767	4.5	22.7	...	7 786 585	4.2	26.2
	1989	17 611 000	4.4	20.9	...	17 126 000	4.3	27.0
Cuba[7]	1975	808 500	5.7	30.1	...	...	...	...
Cuba[7]	1980	1 267 000	7.2	...	...	1 134 500	6.4	...
(peso)	1985	1 690 400	6.3	...	...	1 587 800	5.9	...
	1989	1 777 900	6.6	12.8	...	1 659 300	6.2	15.4
Dominica	1986	17 086	5.9	16.7	...	16 389	5.7	18.5
Dominique	1987	18 649	5.7	14.1	...	17 548	5.4	17.5
(dollar)	1989	22 103	...	10.6	...	20 104	...	19.9
Dominican Republic	1976	75 977	1.9	14.3	...	62 595	1.6	19.7
Rép. dominicaine	1980[1]	138 515	2.2	16.0	...	104 387	1.6	...
(peso)	1985[1]	234 297	1.8	14.0	...	203 639	1.5	...
	1986[1]	227 701	1.5	10.0	...	219 417	1.5	...
El Salvador	1975	150 893	3.4	22.2	...	138 364	3.1	28.1
El Salvador	1980	340 136	3.9	17.1	...	320 009	3.6	22.9
(colon)	1984	335 535	3.0	12.5	...	292 928	2.6	16.3
	1989	625 110	2.0	...	...	615 074	1.9	...
Grenada	1975	5 507	6.9	12.5	...	5 478	6.8	23.4
Grenade	1979	...	...	...	...	11 096	6.5	20.9
(dollar)	1985	...	...	...	...	14 037	4.6	...
Guadeloupe	1980	862 034	...	...	...	818 124	...	...
Guadeloupe	1983	1 273 075	15.0	...	...	1 225 431	14.4	...
(franc)	1987	1 419 000	...	...	...	1 324 000	...	...
Guatemala	1975	56 200	1.6	15.7	...	...	...	...
Guatemala	1979	130 586	1.9	16.6	...	116 780	1.7	23.4
(quetzal)	1984	163 367	1.8	12.4	...	159 778	1.7	23.1
Haiti	1975	...	...	...	...	34 138	1.0	15.8
Haïti	1980	107 136	1.5	14.9	...	85 781	1.2	17.2
(gourde)	1985	117 557	1.2	16.5	...	117 314	1.2	16.7
	1989	213 397	1.8	19.7	...	213 109	1.8	20.0

17
Public expenditure on education at current market prices [cont.]
Dépenses publiques afférentes à l'enseignement aux prix courants du marché [suite]

Country or area Pays ou zone (Currency unit · Unité monétaire)	Years Années	Total educational expenditure Dépenses totales d'éducation				Current educational expenditure Dépenses ordinaires d'éducation		
		Amount Montant (000)	% of GNP % du PNB	% total gov't exp. % dép. totales gouv.	$ US per cap. $ E-U par hab.	Amount Montant (000)	% of GNP % du PNB	% current gov't exp. % dép. ordinaires gouv.
Honduras	1975	80 078	3.7	20.3	...	...	...	...
Honduras	1980	155 003	3.2	14.2	20.6	141 075	3.0	19.5
(lempira)	1985	289 816	4.4	13.8	30.7	285 824	4.3	...
	1987[8]	376 097	4.9	19.5	...	369 664	4.8	38.2
Jamaica	1975[1]	154 676	5.9	16.0	...	121 854	4.7	...
Jamaïque	1980[1]	303 837	6.9	13.1	...	302 669	6.8	19.4
(dollar)	1985	549 938	5.7	12.1	...	515 352	5.3	15.8
	1989	1 225 160	6.6	12.9	...	1 019 604	5.5	17.1
Martinique								
Martinique	1980	930 613	...	...	...	886 861	...	...
(franc)	1983	1 349 457	13.5	...	...	1 324 757	13.3	...
Mexico[19]	1975	41 185	3.6	11.9	...	26 783	2.4	...
Mexique[19]	1982[1]	477 092	5.2	14.6	...	331 448[1]	3.6	...
(peso)	1985[1]	1 724 237	3.8	13.2	...	1 206 708[1]	2.7	...
	1989[1]	17 318 179	3.8	...	...	12 187 290[1]	2.7	...
Montserrat								
Montserrat								
(dollar)	1988	6 335	...	9.3	...	6 335	...	18.9
Netherlands Antilles[19]	1980	43 000	6.3	13.6	...	42 000	6.2	20.2
Antilles néerlandaises[19]	1985	57 000	7.2	14.4	...	57 000	7.2	16.3
(guilder · florin)	1989	69 000	5.2	...	...	54 000	4.1	18.8
Nicaragua	1975	258	2.4	13.1	...	206	1.9	...
Nicaragua	1980	662	3.4	10.4	...	580	3.0	...
(córdoba)	1985[1]	6 409	6.1	10.2	...	6 196	5.9	12.2
	1988[1]	10 717 000	3.9	...	...	10 559 000	3.8	...
Panama	1975	103 316	5.7	21.3	...	96 693	5.3	31.6
Panama	1980	166 140	5.2	19.0	80.2	155 716	4.8	19.8
(balboa)	1985	236 969	6.3	18.7	...	231 459	5.1	19.9
	1989	251 167	6.2	...	116.6[3]	264 005	6.0	...
St. Kitts and Nevis	1975	2 752	4.0	10.5	...	2 732	4.0	12.5
Saint-Kitts-et-Nevis	1980	6 595	5.2	9.4	...	6 563	5.2	13.6
(dollar)	1985	11 927	5.9	18.5	...	11 887	5.9	19.1
	1989[6]	11 540	...	12.0	...	11 218	3.0	12.9
Saint Lucia	1975	8 338	...	16.8	...	7 321	...	20.0
Sainte lucie	1980	...	...	...	...	18 596	6.9	...
(dollar)	1986	38 419	7.5	...	...	36 270	6.8	...
	1988	...	...	...	...	42 169	6.5	...
St. Pierre-Miquelon	1980	18 432	...	...	...	16 725	...	...
St. Pierre-Miquelon	1985	40 721	...	...	...	36 831	...	...
(franc)	1986	43 131	...	...	...	39 899	...	...
St. Vincent and the Grenadines	1975	...	...	...	...	5 196	7.3	...
Saint-Vincent-et-Grenadines	1986	19 253	5.8	11.6	...	17 983	5.4	16.8
(dollar)	1989[6]	26 126	...	10.5	...	23 091		17.1
Trinidad and Tobago	1975	157 055	3.1	14.7	...	138 295	2.7	17.7
Trinité-et-Tobago	1980	563 931	4.0	11.5	...	430 831	3.0	17.6
(dollar)	1985	1 042 291	6.2	...	291.8	911 869	5.4	...
	1989	772 343	4.9	12.1	...	701 022	4.4	13.4
Turks and Caicos Is.	1975	718	...	14.4	...	601	...	16.2
Iles Turques et Caïques	1980	1 018	...	...	...	832	...	...
(US dollars · dollars E-U)	1983	1 394	...	10.3	...	1 287	...	12.4
United States[9][19]	1975	118 706	7.4	21.3	...	...	...	...
Etats-Unis[9][19]	1980	182 849	6.7	...	539.3	...	...	...
(dollar)	1985	269 485	6.7	21.0	...	247 220	6.2	22.0
	1987	308 800	6.8	...	785.3[3]	...	...	...
US Virgin Islands	1975	40 575	...	...	...	40 138	...	...
Iles Vierges américaine	1980	57 541	...	...	...	56 774	...	...
(US dollars · dollars E-U)	1984	74 207	...	...	...	...	...	...
America, South · Amérique du Sud								
Argentina[10]	1975	3.577	2.5	9.5	...	3.356	2.4	10.9
Argentine[10]	1980	1 018	3.6	15.1	...	860	3.1	18.8
(austral)	1985[1]	740 032	2.0	8.6	...	666 854	1.8	8.9
	1989[1]	348 167 000	1.5	8.0	...	326 659 000	1.4	8.8

17
Public expenditure on education at current market prices [*cont.*]
Dépenses publiques afférentes à l'enseignement aux prix courants du marché [*suite*]

Country or area Pays ou zone (Currency unit · Unité monétaire)	Years Années	Total educational expenditure Dépenses totales d'éducation				Current educational expenditure Dépenses ordinaires d'éducation		
		Amount Montant (000)	% of GNP % du PNB	% total gov't exp. % dép. totales gouv.	$ US per cap. $ E-U par hab.	Amount Montant (000)	% of GNP % du PNB	% current gov't exp. % dép. ordinaires gouv.
Bolivia	1975	1 661	3.5	...	...	1 659	3.5	...
Bolivie	1980	5 122	4.4	25.3	...	4 919	4.2	27.0
(boliviano)	1983	33 729	2.6	21.3	...	33 702	2.6	...
	1989[8]	268 869	2.3	...	...	267 640	2.3	...
Brazil	1975	30 207	3.0	...	...	...	...	...
Brésil	1980	431 194	3.6	...	...	...	...	...
(cruzado)	1985	49 521 047	3.7	...	...	...	...	...
	1988	3 271 013 000	3.7	...	...	...	...	...
Chile	1975	1 399 626	4.1	12.0	...	1 321 482	3.9	16.0
Chili	1980	47 960 718	4.6	11.9	...	45 503 514	4.4	13.4
(peso)	1985	101 492 580	4.4	15.3	...	...	...	...
	1989[1]	181 378 000	...	...	...	181 998 000	2.9	...
Colombia[1]	1975	8 763 702	2.2	16.4	...	7 486 158	1.9	25.0
Colombie[1]	1980	29 240 258	1.9	14.3	34.9	27 286 015	1.7	19.9
(peso)	1985	136 570 000	2.8	...	...	127 908 000	2.7	...
	1989	409 215 000	2.9	21.4	...	389 489 000	2.8	38.5
Ecuador	1975[1]	3 386 500	3.2	25.9	...	...	...	...
Equateur	1980	15 579 992	5.6	33.3	64.7	14 649 139	5.3	36.0
(sucre)	1985	38 008 583	3.7	20.6	65.7	35 611 421	3.5	25.8
	1989	134 554 000	2.6	19.1	...	127 394 000	2.5	24.7
French Guiana								
Guyane français	1980	158 509	...	...	...	152 333	...	...
(franc)	1983	263 197	...	...	...	252 655	...	...
Guyana	1975	57 064	4.9	9.8	...	47 944	4.2	14.9
Guyana	1979	121 395	9.7	14.0	...	89 369	7.2	15.2
(dollar)	1985	162 080	9.8	10.4	...	134 610	8.1	13.0
	1987	231 664	8.8	8.1	...	176 593	6.7	8.0
Paraguay	1975[1]	2 963 772	1.6	14.0	...	2 292 296	1.2	...
Paraguay	1980[1]	8 793 247	1.5	16.4	...	6 267 238	1.1	...
(guarani)	1985	20 662 000	1.5	16.7	...	16 822 000	1.2	18.8
Peru	1975	21 756	3.3	16.6	...	20 970	3.2	23.2
Pérou	1980	175 915	3.1	15.2	33.3	166 040	2.9	18.5
(inti)	1985	5 042 400	2.7	15.7	...	4 854 900	2.6	17.9
	1987	24 853 000	3.5	22.9	38.4[3]	23 507 000	3.3	25.8
Suriname	1975	50 700	5.6	14.1	...	44 800	4.9	17.9
Suriname	1980	105 193	6.7	22.5	...	105 193	6.7	22.8
(guilder · florin)	1985	157 513	9.4	...	...	...	...	...
	1989	223 646	9.5	...	...	222 646	9.5	...
Uruguay	1980	2 035 103	2.2	10.0	...	1 926 769	2.1	...
Uruguay	1985	12 564 912	2.6	9.3	...	12 067 807	2.5	9.3
(peso)	1988	85 884 000	3.1	15.1	...	78 989 000	2.9	15.5
Venezuela	1975	6 225 161	4.5	...	...	5 965 747	4.3	...
Venezuela	1980	13 162 451	4.4	14.7	177.5	12 523 608	4.2	24.3
(bolivar)	1984	20 478 400	5.0	20.6	131.8	19 629 100	4.8	29.3
	1989	59 965 000	4.2	18.8	...	...	...	...
Asia · Asie								
Afghanistan	1974	1 264 216	...	...	...	1 142 630	...	15.3
Afghanistan	1980	3 205 456	...	12.7	...	2 886 301	...	14.4
(afghani)	1987	3 810 000	...	4.0	...	3 150 000	...	3.7
Bahrain	1975	10 541	...	8.8	...	9 606	...	14.2
Bahreïn	1980	32 574	2.9	10.3	...	28 166	2.5	14.7
(dinar)	1985	51 681	4.0	...	...	48 681	3.8	8.5
	1988[6]	59 555	5.4	...	...	55 631	5.1	11.4
Bangladesh	1975	1 379 862	1.1	13.6	...	929 862	0.7	18.9
Bangladesh	1980[1]	3 009 311	1.5	8.2	...	2 010 009	1.0	15.4
(taka)	1985[1]	7 790 093	1.9	...	...	7 465 509	1.8	...
	1989[1]	14 125 000	2.2	10.5	...	10 939 000	1.7	14.3
Bhutan								
Bhoutan								
(ngultrum)	1987	122 141	3.7	...	...	...	...	...

17
Public expenditure on education at current market prices [*cont.*]
Dépenses publiques afférentes à l'enseignement aux prix courants du marché [*suite*]

Country or area Pays ou zone (Currency unit · Unité monétaire)	Years Années	Total educational expenditure Dépenses totales d'éducation			$ US per cap. $ E-U par hab.	Current educational expenditure Dépenses ordinaires d'éducation		
		Amount Montant (000)	% of GNP % du PNB	% total gov't exp. % dép. totales gouv.		Amount Montant (000)	% of GNP % du PNB	% current gov't exp. % dép. ordinaires gouv.
Brunei Darussalam	1975	55 908	...	12.2	...	47 354	...	12.8
Brunéi Darussalam	1980	129 428	...	11.8	...	114 917	...	12.5
(dollar)	1986	242 833	...	...	...	217 474	...	...
	1988	296 100	...	...	...	272 188	...	...
China	1975	5 193 000	1.7	6.3	...	4 826 000	1.6	...
Chine	1980	11 319 000	2.5	9.3	...	10 263 000	2.3	...
(yuan)	1985	22 489 000	2.6	12.2	...	19 770 000	2.3	...
	1989	37 299 000	2.4	12.4	...	34 254 000	2.2	...
Cyprus[11]	1975	11 421	4.4	14.3	...	10 325	4.0	15.2
Chypre[11]	1980	27 066	3.5	12.9	106.7	25 452	3.3	16.2
(pound · livre)	1985	55 383	3.7	12.2	119.7	52 814	3.6	13.4
	1989	79 424	3.6	11.7	...	75 757	3.4	12.8
Hong Kong	1975	1 246 272	...	20.7	...	1 144 538	...	25.7
Hong-kong	1980	3 446 432	...	14.6	...	3 036 139	...	25.5
(dollar)	1984	6 990 588	...	18.7	...	6 189 691	...	23.0
	1989	13 283 000	...	15.9	...	11 705 000	...	...
India	1975	21 036 400	2.7	8.6	...	20 852 100	2.7	15.1
Inde	1980	37 924 238	2.8	10.0	2.6	37 461 638	2.8	...
(rupee · roupie)	1985	87 257 000	3.3	9.4	3.7	85 198 000	3.3	13.0
	1987	106 434 000	3.2	8.5	...	104 807 000	3.2	9.9
Indonesia[19]	1975	357 110	2.7	13.1	...	277 076	2.1	...
Indonésie[19]	1980	808 087	1.7	8.9	...	...	...	...
(rupiah)	1988[1]	1 238 315	0.9	4.3	...	1 095 493	0.8	5.5
Iran, Islamic Rep. of	1976	259 995 564	5.5	14.1	...	230 748 564	4.9	...
Iran, Rép. islamique d'	1980	498 268 000	7.5	15.7	123.3	440 298 000	6.6	20.1
(rial)	1985	575 519 000	3.5	17.2	...	510 081 000	3.1	20.6
	1989	1 045 478 000	3.1	21.9	...	915 742 000	2.7	23.9
Iraq	1976	204 493	...	6.9	...	155 763	...	10.5
Iraq	1980	418 370	...	...	...	...	...	...
(dinar)	1985	550 922	...	6.5	...	...	...	...
	1988	690 134	...	...	...	624 933	...	...
Israel	1975	551	6.7	7.6	...	464	5.7	8.3
Israël	1980	8 861	7.9	7.3	503.3	8 182	7.3	8.9
(shekel)	1985	1 876 000	6.4	8.6	425.1	1 720 000	5.9	8.3
	1989[9]	7 430 000	8.6	...	...	6 910 000	8.0	...
Japan[12 19]	1975	8 156 673	5.5	22.4	...	5 826 604	3.9	...
Japon[12 19]	1980	13 908 111	5.8	19.6	339.5	9 416 591	3.9	...
(yen)	1985	16 142 654	5.1	17.9	...	15 280 808	4.8	...
	1988	17 461 095	4.8	16.2	574.1[3]	17 278 783	4.7	...
Jordan	1975[8]	16 698	...	8.1	...	13 900	...	11.1
Jordanie	1980	63 822	...	11.3	44.9	50 543	...	15.0
(dinar)	1985	105 478	5.6	13.0	...	91 907	4.9	18.9
	1989	137 738	5.9	...	56.9[3]	129 225	5.5	...
Korea, Republic of[19]	1975	220 282	2.2	13.9	...	163 800	1.6	...
Corée, République de[19]	1980	1 374 736	3.7	23.7	35.0	1 158 967	3.2	...
(won)	1985	3 530 101	4.5	28.2	...	2 811 861	3.6	28.3
	1989	5 139 366	3.6	23.3	53.9[3]	4 348 137	3.1	24.7
Kuwait	1975	112 023	3.0	10.0	...	90 069	2.4	10.8
Koweït	1980	218 510	2.4	8.1	530.7	203 758	2.3	11.2
(dinar)	1986	360 288	5.0	*11.9	641.0	343 800	4.8	...
	1987	382 869	5.0	*12.1	...	365 767	4.8	...
Lao People's Dem. Rep.	1980	23 500	...	1.3	...	...	...	...
Rép. dém. pop. lao	1985	463 000	...	4.5	...	...	...	...
(kip)	1988	2 782 071	...	...	...	2 096 955	...	...
Lebanon[1]	1975	263 500	...	16.1	...	...	...	...
Liban[1]	1980	510 800	...	13.2	...	...	...	...
(pound · livre)	1985	1 639 467	...	16.8	...	...	...	...
Malaysia	1975	1 362 900	6.0	19.3	...	1 157 500	5.1	23.6
Malaysie	1980	3 104 258	6.0	14.7	...	2 575 488	5.0	18.4
(ringgit)	1985	4 754 057	6.6	16.3	...	4 062 257	5.6	...
	1989[1]	5 482 000	5.6	...	...	4 335 000	4.4	...

17
Public expenditure on education at current market prices [cont.]
Dépenses publiques afférentes à l'enseignement aux prix courants du marché [suite]

Country or area Pays ou zone (Currency unit · Unité monétaire)	Years Années	Total educational expenditure Dépenses totales d'éducation				Current educational expenditure Dépenses ordinaires d'éducation		
		Amount Montant (000)	% of GNP % du PNB	% total gov't exp. % dép. totales gouv.	$ US per cap. $ E-U par hab.	Amount Montant (000)	% of GNP % du PNB	% current gov't exp. % dép. ordinaires gouv.
Maldives	1975	...	...	...	...	1 115	...	11.5
Maldives	1986	23 800	...	*7.2	...	19 600	...	...
(rufiyaa)	1987	33 400	...	*8.5	...	29 500	...	...
Myanmar	1975	405 257	...	15.3	...	385 990	...	16.3
Myanmar								
(kyat)	1987	1 290 519	...	...	...	1 080 606	...	...
Nepal	1975	245 990	1.5	11.5	...	...	...	...
Népal	1981	568 896	2.0	13.6	...	...	...	...
(rupee · roupie)	1985	1 240 717	2.8	10.8	...	...	...	...
Oman	1975	9 396	1.6	1.9	...	6 605	1.1	2.0
Oman	1980	38 235	2.1	4.1	92.3	31 067	1.7	4.6
(rial)	1985	123 471	4.0	...	...	77 432	2.5	...
	1989	108 639	3.7	12.4	196.3[3]	102 180	3.5	18.8
Pakistan	1975	2 488 206	2.2	5.2	...	1 731 089	1.6	...
Pakistan	1980	4 619 071	2.0	5.0	4.0	3 378 591	1.5	5.2
(rupee · roupie)	1985	12 645 000	2.7	...	...	9 390 000	2.0	...
	1989	19 443 000	2.6	...	5.6[3]	14 805 000	2.0	...
Philippines[13]	1975	2 149 896	1.9	11.4	...	1 753 958	1.5	...
Philippines[13]	1980	4 190 750	1.6	10.3	...	4 022 850	1.5	...
(peso)	1985	7 524 490	1.3	...	...	7 026 164	1.2	...
	1989	27 602 000	2.9	...	...	23 382 000	2.4	...
Qatar	1975	177 124	...	4.0	...	123 524	...	4.1
Qatar	1980	792 200	...	7.2	...	598 000	...	7.8
(riyal)	1985	1 012 359	...	...	...	766 634	...	...
	1989	917 355	...	...	...	868 728	...	...
Saudi Arabia	1975	12 940 937	9.3	11.7	...	5 528 566	4.0	15.1
Arabie saoudite	1980	21 294 465	5.5	8.7	...	13 526 367	3.5	...
(riyal)	1985	23 540 000	6.7	*12.0	...	19 283 000	5.5	...
	1988	22 909 000	7.6	16.2	...	20 818 000	6.9	...
Singapore	1975	391 264	2.9	8.6	...	339 870	2.5	11.4
Singapour	1980	686 380	2.8	7.3	...	587 469	2.4	10.3
(dollar)	1985	1 775 580	4.4	...	...	1 388 325	3.4	...
	1988	1 718 370	3.4	...	...	1 522 624	3.0	...
Sri Lanka	1975	729 590	2.8	10.1	...	682 680	2.6	13.0
Sri Lanka	1980	2 073 012	3.1	8.8	4.8	1 736 787	2.6	15.3
(rupee · roupie)	1985	4 860 151	3.0	8.0	...	3 728 820	2.3	12.0
	1988	6 719 000	3.0	7.8	7.5[3]	5 027 000	2.3	12.0
Syrian Arab Republic[6]	1975	813 245	3.9	7.8	...	462 514[6]	2.2	...
Rép. arabe syrienne[6]	1980	2 346 984	4.6	8.1	...	1 272 481[6]	2.5	...
(pound · livre)	1985	5 060 425	6.1	11.8	...	2 798 708[6]	3.4	...
	1989	8 439 000	4.1	...	...	5 748 000[6]	...	...
Thailand	1975	10 605 251	3.5	21.0	...	7 775 347	2.6	18.9
Thaïlande	1980	22 489 450	3.4	20.6	22.1	15 867 328	2.4	19.1
(baht)	1985	39 366 950	3.9	18.5	...	33 829 610	3.4	19.2
	1988	47 283 000	3.2	16.6	27.1[3]	40 916 000	2.8	16.7
Turkey[19]	1980	117 744	2.8	10.5	...	98 593	2.3	...
Turquie[19]	1985	627 104	2.3	...	...	523 102	1.9	...
(lira · livre)	1989[6]	2 967 077	1.8	...	...	2 418 100	1.6	...
United Arab Emirates	1975	346 585	0.9	...	...	271 382	0.7	...
Emirats arabes unis	1980	1 459 688	1.3	...	...	1 153 063	1.0	...
(dirham)	1985	1 738 155	1.7	10.4	...	1 636 705	1.6	10.6
	1989	2 179 000	2.1	14.9	...	2 029 000	2.0	14.3
Yemen[14]	1981	1 108 004	6.2	15.8	...	...	...	...
Yémen[14]	1985	2 039 400	6.0	21.8	...	1 788 700	5.3	27.4
(rial)	1986	2 504 800	6.1	23.5	...	2 227 500	5.4	30.2
Europe · Europe								
Albania	1980	767 000	...	10.3	...	...	...	...
Albanie	1987	946 000	...	11.2	...	...	...	...
(lek)	1988	952 000	...	11.1	...	...	...	...
Andorra	1975	41 315	...	...	...	39 185	...	...
Andorre	1982	628 134	...	10.0	...	495 836	...	16.5
(peseta)	1986	951 310	...	15.8	...	800 365	...	24.4

17
Public expenditure on education at current market prices [cont.]
Dépenses publiques afférentes à l'enseignement aux prix courants du marché [suite]

		Total educational expenditure Dépenses totales d'éducation				Current educational expenditure Dépenses ordinaires d'éducation		
Country or area Pays ou zone (Currency unit · Unité monétaire)	Years Années	Amount Montant (000)	% of GNP % du PNB	% total gov't exp. % dép. totales gouv.	$ US per cap. $ E-U par hab.	Amount Montant (000)	% of GNP % du PNB	% current gov't exp. % dép. ordinaires gouv.
Austria	1975	37 409 300	5.7	8.5	...	29 366 500	4.5	9.8
Autriche	1980	55 016 300	5.6	8.0	394.7	46 955 400	4.8	8.4
(schilling)	1985	78 638 600	5.9	7.9	361.1	70 846 500	5.3	8.6
	1989	91 624 000	5.5	7.6	...	84 911 000	5.1	8.9
Belgium[1]	1975	143 856 800	6.2	22.2	...	131 906 200	5.7	23.0
Belgique[1]	1980	208 469 200	5.9	16.3	...	206 226 600	5.9	18.6
(franc)	1985	287 387 700	6.0	15.2	...	272 843 000	5.7	16.0
	1988	274 028 000	4.9	...	...	271 902 000	4.9	...
Bulgaria	1975[15]	787 599	5.5	8.5	...	725 259	5.1	...
Bulgarie	1980	1 145 118	5.4	...	...	1 097 806	5.4	...
(lev)	1985	1 784 008	5.5	...	...	1 597 966	4.9	...
	1989	2 111 950	5.5	...	...	1 945 867	5.1	...
Czechoslovakia[15]	1975	19 104 801	4.7	7.0	...	17 288 021	4.3	...
Tchécoslovaquie[15]	1980	23 230 783	4.8	...	...	21 801 883	4.5	...
(koruna · couronne)	1985	28 201 234	5.1	7.9	...	26 959 444	4.9	8.6
	1988	32 254 000	5.4	8.0	...	31 005 000	5.2	8.7
Denmark	1975	16 801 000	7.8	15.2	...	14 600 000	6.8	...
Danemark	1980	25 020 000	6.9	9.5	...	22 188 000	6.1	9.0
(krone · couronne)	1985	42 672 000	7.2	...	...	...	...	...
	1989	55 448 000	7.6	13.0	...	52 270 000	7.1	12.9
Finland	1975	6 497 103	6.3	13.0	...	5 596 511	5.4	14.3
Finlande	1980	10 434 855	5.5	...	526.5	9 565 205	5.1	12.5
(markka)	1985	18 966 714	5.8	12.9	...	17 604 190	5.4	13.9
	1989	28 060 000	5.8	16.1	741.8[3]	26 326 000	5.4	...
France[16 19]	1975	76 357	5.2	...	...	66 873	4.5	...
France[16 19]	1980	142 099	5.0	...	...	131 441	4.7	...
(franc)	1985	268 846	5.8	...	472.6	253 572	5.4	...
	1988	302 056	5.3	...	...	283 339	5.0	...
Germany ʃ · Allemagne ʃ								
Federal Republic of Germany	1975	52 860 900	5.1	10.7	...	42 456 700	4.1	9.4
Rép. féd. d'Allemagne	1980	70 098 600	4.7	10.1	539.2	60 557 900	4.1	9.5
(deutsche mark)	1985	83 691 000	4.5	9.2	404.2	75 566 000	4.1	9.4
	1988	89 747 000	4.2	8.8	...	81 123 000	3.8	8.9
former German Dem. Republic[15]	1975	...	...	...	...	8 276 353	...	...
ancienne Rép. dém. allemande[15]	1980	...	...	...	...	9 836 257	...	...
(DDR mark)	1985	...	...	...	...	12 404 100	...	...
	1988	15 461 500	...	...	...	14 096 800	...	...
Gibraltar	1975	997	...	...	...	909	...	...
Gibraltar	1980	5 421	...	...	...	3 305	...	...
(pound · livre)	1984	5 134	...	...	...	5 134	...	...
Greece	1975	13 559 508	2.0	8.0	...	12 336 357	1.8	9.0
Grèce	1979	31 754 307	2.2	8.4	90.6[17]	29 950 864	2.0	9.6
(drachma · drachme)	1985	133 091 000	2.9	7.5	111.4	126 749 000	2.8	8.5
	1987	166 808 000	2.7	6.1	...	156 807 000	2.5	6.5
Hungary	1975	19 325 492	4.1	4.2	...	16 719 792	3.5	5.3
Hongrie	1980	33 099 379	4.7	5.2	...	27 516 379	3.9	6.4
(forint)	1985	54 061 000	5.4	6.4	...	48 125 000	4.8	7.4
	1989	98 301 000	6.0	7.1	...	66 887 000	5.3	7.8
Iceland	1975	78 100	3.9	12.2	...	...	...	...
Islande	1980	699 000	4.6	14.0	514.3	...	...	...
(krona · couronne)	1985	5 684 000	5.0	13.8	...	...	...	...
	1988	13 357 000	5.4	14.1	566.7[3]	...	...	...
Ireland	1975	230 483	6.1	10.8	...	199 666	5.3	12.8
Irlande	1980	594 520	6.6	...	...	515 000	5.7	...
(pound · livre)	1985	1 057 530	6.7	8.9	...	962 795	6.1	10.5
	1987	1 263 225	7.0	9.5	...	1 168 959	6.5	11.2
Italy[19]	1975	5 675 784	4.1	9.4	...	5 060 156	3.7	12.6
Italy[19]	1979	13 633 022	4.4	11.1	342.3[17]	11 791 069	3.8	10.7
(lira · lire)	1985[1]	40 533 126	5.0	8.3	...	36 859 217	4.6	10.0
	1986[1]	44 480 428	5.0	8.3	470.4	40 498 628	4.6	9.7

17
Public expenditure on education at current market prices [cont.]
Dépenses publiques afférentes à l'enseignement aux prix courants du marché [suite]

Country or area Pays ou zone (Currency unit · Unité monétaire)	Years Années	Total educational expenditure Dépenses totales d'éducation				Current educational expenditure Dépenses ordinaires d'éducation		
		Amount Montant (000)	% of GNP % du PNB	% total gov't exp. % dép. totales gouv.	$ US per cap. $ E-U par hab.	Amount Montant (000)	% of GNP % du PNB	% current gov't exp. % dép. ordinaires gouv.
Luxembourg Luxembourg (franc)	1975	5 023 200	5.1	15.0	...	3 903 300	4.0	15.8
	1980	9 791 500	6.1	14.9	...	9 305 300	5.8	19.8
	1986	12 269 000	4.1	15.7	...	10 712 000	3.6	15.4
	1989	16 363 000	4.7	...	...	13 440 000	3.8	...
Malta Malte (lira · lire)	1975	7 105	3.9	7.6	...	6 881	3.7	11.9
	1980	12 558	3.0	7.8	121.7	12 475	2.9	9.7
	1985	17 423	3.4	7.7	...	17 109	3.3	9.1
	1988	22 803	3.6	8.3	152.1[3]	22 081	3.5	9.9
Monaco Monaco (franc)	1974	35 282	...	10.4	...	20 538	...	...
	1980	...	...	...	...	42 629	...	...
	1989	123 500	...	5.3	...	112 500	...	7.8
Netherlands Pays-Bas (guilder · florin)	1975	18 096 000	8.2	...	...	14 881 000	6.8	...
	1980	26 606 000	7.9	...	751.2	23 079 000	6.9	...
	1985	28 298 000	6.8	...	...	24 796 000	5.9	...
	1988	30 618 000	6.8	...	613.7[3]	27 907 000	6.2	...
Norway[19] Norvège[19] (krone · couronne)	1975	10 456	7.1	14.7	...	8 427	5.7	13.9
	1980	19 731	7.2	13.8	712.2	16 448	6.0	14.5
	1985	31 680	6.5	13.6	...	27 979	5.7	14.1
	1989	46 037	7.5	13.4	870.4[3]	40 768	6.7	13.9
Poland[19] Pologne[19] (zloty)	1975[15]	...	...	...	...	50 449	...	8.7
	1980	...	...	...	...	79 984	3.3	7.0
	1985	497 497	4.9	12.2	...	405 597	4.0	11.6
	1989	4 342 528	4.6	12.9	...	...	...	...
Portugal Portugal (escudo)	1975	15 214 700	4.0	11.2	...	14 269 981	3.8	13.9
	1980	53 233 500	4.4	...	87.9	45 442 800	3.8	...
	1985	152 886 000	4.6	...	...	135 612 000	4.1	...
	1989	342 381 000	4.9	...	...	313 506 000	4.5	...
Romania Roumanie (leu)	1975	15 194 700	...	6.4	...	12 892 700	...	...
	1980	19 930 200	...	6.7	...	17 691 200	...	...
	1985	17 940 900	...	...	...	17 344 900	...	...
	1988	...	...	...	...	19 797 397	...	...
San Marino Saint-Marin (lira · lire)	1975	2 775 902	...	13.1	...	2 499 565	...	15.0
	1980	7 251 937	...	7.5	...	6 263 207	...	9.5
	1986	14 315 001	...	...	...	13 477 313	...	...
	1988	19 140 000	...	...	...	17 946 000	...	...
Spain[19] Espagne[19] (peseta)	1975	105 582	1.8	13.6	...	...	...	...
	1979	342 376	2.6	16.4	...	295 443	2.3	16.7
	1985	917 076	3.3	14.1	...	821 118	3.0	...
	1988	1 677 825	4.3	12.1	...	1 491 083	3.8	...
Sweden Suède (krona · couronne)	1975	21 230 200	7.3	13.4	...	19 281 100	6.6	13.9
	1980	47 322 300	9.0	14.1	886.9	40 885 700	7.8	...
	1985	65 000 600	7.7	12.6	646.5	57 702 700	6.8	...
	1989	87 290 000	7.3	13.1	...	79 777 000	6.7	...
Switzerland Suisse (franc)	1975	7 392 300	5.1	19.4	...	5 978 800	4.1	19.8
	1980	8 872 700	5.0	18.8	...	7 936 800	4.5	20.0
	1985	11 696 300	4.8	18.6	...	10 638 400	4.4	19.9
	1988	13 757 300	4.9	18.8	...	12 311 000	4.4	20.0
United Kingdom[19] Royaume-Uni[19] (pound · livre)	1975	7 020	6.6	14.3	...	6 292	5.9	...
	1980	12 856	5.6	13.9	406.7	12 094	5.2	...
	1985	17 501	4.9	...	...	16 764	4.7	...
	1988	22 148	4.7	...	403.2[3]	21 275	4.6	...
Yugoslavia[19] Yugoslavie[19] (dinar)	1975	28 897	5.4	24.4	...	25 515	4.8	...
	1980	84 051	4.7	32.5	...	71 514	4.0	...
	1985	419 799	3.4	...	...	391 345	3.1	...
	1989	95 549 000	4.4	...	...	90 019 000	4.2	...
Oceania · Océanie								
American Samoa Samoa américaines (US dollars · dollars E-U)	1975	9 642	...	...	...	7 962	...	...
	1981	11 470	...	16.0	...	11 248	...	17.9
	1986	22 635	...	42.8	...	22 635	...	54.4
	1988	25 887	...	23.7	...	...	...	...

17
Public expenditure on education at current market prices [*cont.*]
Dépenses publiques afférentes à l'enseignement aux prix courants du marché [*suite*]

Country or area Pays ou zone (Currency unit · Unité monétaire)	Years Années	Total educational expenditure Dépenses totales d'éducation				Current educational expenditure Dépenses ordinaires d'éducation		
		Amount Montant (000)	% of GNP % du PNB	% total gov't exp. % dép. totales gouv.	$ US per cap. $ E-U par hab.	Amount Montant (000)	% of GNP % du PNB	% current gov't exp. % dép. ordinaires gouv.
Australia	1975	4 495 000	6.5	14.8	...	3 735 000	5.4	18.4
Australie	1980	7 592 000	5.9	14.8	487.7	6 899 000	5.3	16.8
(dollar)	1985	12 925 000	5.9	12.8	459.5	11 848 000	5.4	14.2
	1987	14 726 000	5.5	12.5	...	13 586 000	5.0	13.4
Cook Islands	1981	2 768	...	13.1	...	2 712	...	...
Iles Cook	1984	3 481	...	10.2	...	3 474	...	...
(NZ dollar)	1986	4 509	...	9.5	...	4 468	...	9.5
Fiji	1975	26 032	4.7	*19.5	...	24 039	4.3	23.2
Fidji	1980	49 641	5.1	...	91.4	47 861	4.9	21.5
(dollar)	1986	84 935	6.0	...	...	83 243	5.9	...
	1989	86 024	5.0	15.4	...	81 740	4.7	19.4
French Polynesia	1975	207 802	...	...	...	181 573	...	...
Polynésia française	1984	1 030 255	...	...	...	934 589	...	...
(CFP franc · franc CFP)	1987[1]	770 348	...	...	...	747 630	...	...
Guam	1975	52 874	...	...	...	52 723	...	...
Guam	1981	48 761	...	...	...	47 888	...	...
(US dollars · dollars E-U)	1985	60 200	...	...	...	58 820	...	...
Kiribati	1975	2 077	...	...	...	2 022	...	6.7
Kiribati	1980	2 550	...	...	...	2 550	...	17.4
(dollar)	1985	3 021	...	18.5	...	3 021	...	...
	1989	3 511	...	17.0	...	3 511	...	...
New Caledonia	1981	627 856	...	...	...	595 836	...	...
Nouvelle-Calédonie	1985	1 007 342	...	...	...	918 951	...	...
(CFP franc · franc CFP)	1987	1 140 089	...	...	...	1 072 770	...	...
New Zealand	1975	631 751	6.1	17.1	...	503 643	4.9	...
Nouvelle-zélande	1980	1 302 324	6.1	23.1	...	1 171 226	5.5	27.9
(dollar)	1985	2 028 416	4.8	18.4	...	1 849 002	4.4	25.5
	1988	3 637 863	5.9	...	...	3 302 339	5.4	...
Niue	1975	631	...	18.6	...	595	...	28.7
Nioué	1980	777	...	13.2	...	755	...	15.9
(NZ dollar)	1986	1 394	...	...	...	1 283	...	...
	1988	1 366	...	10.8	...	1 166	...	10.1
Norfolk Island	1975	220	...	13.7	...	209	...	17.7
Ile Norfolk	1980	425	...	15.1	...	425	...	17.3
(dollar)	1982	455	...	14.0	...	455	...	14.9
Pacific Islands (Palau)[18] Iles du Pacific (Palaos)[18]	1975	23 469	...	*24.3	...	21 501	...	...
(US dollars · dollars E-U)	1981	26 129	...	...	...	26 129	...	...
Solomon Islands	1975	2 069	5.4	14.7	...	1 634	4.3	17.6
Iles Salomon	1980	5 049	5.6	11.2	...	3 846	4.3	14.1
(dollar)	1984	9 965	4.7	12.4	...	4 009	1.9	8.0
Tokelau Tokélaou (NZ dollar)	1981	669	...	...	...	630	...	...
Tonga	1975	709	...	12.7	...	686	...	14.4
Tonga	1980	1 558	...	11.6	...	1 399	...	13.3
(pa'anga)	1985	3 659	4.4	16.1	...	3 659	4.4	...
	1986	4 330	4.2	...	...	4 330	4.2	...
Vanuatu Vanuatu	1987	923 594	7.5	24.6	...	923 594	7.5	25.4
(vatu)	1989	810 802	5.5	...	...	810 802	5.5	19.5
former USSR · ancienne URSS								
former USSR[15]	1975	27 747 100	7.6	12.9	...	23 415 900	6.4	...
ancienne URSS[15]	1980	33 026 200	7.3	11.2	...	28 098 600	6.2	...
(rouble)	1985	39 866 000	7.0	...	...	33 319 400	5.9	...
	1989	51 542 600	7.9	...	...	40 247 100	6.2	...
Belarus[15]	1975	1 040 600	...	...	...	859 300	...	...
Bélarus[15]	1980	1 253 300	...	...	...	1 051 300	...	...
(rouble)	1985	1 499 000	...	...	...	1 264 000	...	...
	1989	1 966 600	...	...	...	1 551 000	...	...

17
Public expenditure on education at current market prices [*cont.*]
Dépenses publiques afférentes à l'enseignement aux prix courants du marché [*suite*]

Country or area Pays ou zone (Currency unit · Unité monétaire)	Years Années	Total educational expenditure Dépenses totales d'éducation				Current educational expenditure Dépenses ordinaires d'éducation		
		Amount Montant (000)	% of GNP % du PNB	% total gov't exp. % dép. totales gouv.	$ US per cap. $ E-U par hab.	Amount Montant (000)	% of GNP % du PNB	% current gov't exp. % dép. ordinaires gouv.
Ukrain[15]	1975	5 014 000	...	27.4	...	4 312 200	...	...
Ukraine[15]	1980	5 926 700	...	24.5	...	5 114 300	...	...
(rouble)	1985	6 721 200	...	21.1	...	5 708 100	...	...
	1989	7 911 100	...	...	...	6 245 000	...	...

Source:
United Nations Educational, Scientific and Cultural Organization
(Paris).

§ All data shown which pertain to Germany prior to 3 October 1990
are indicated separately for the Federal Republic of Germany
and the former German Democratic Republic based on their
respective territories at the time indicated. For detailed
explanatory notes on data pertaining to Germany, see Annex I-
Country Nomenclature.

1 Data refer to expenditure of the Ministry of Education only.
2 From 1985 to 1989, expenditure of Al-Azhar is not included.
3 1986.
4 Data refer to expenditure of the Ministry of Primary and Secondary
Education only.
5 From 1985 and 1986, data refer to expenditure of the Federal
government only.
6 Expenditure on third level education is not included.
7 Expenditure on education is calculated as percentage of
global social product.
8 Expenditure on Universities is not included.
9 Data refer to total public and private expenditure on education.
10 For 1975, figures are in pesos (1 austral = 1000 pesos).
11 Expenditure by the Office of Greek Education only.
12 For 1975 and 1980, data on current and capital expenditure do not
include public subsidies to private education. For 1985, these
data refer to total public and private expenditure on education.
13 For 1975, data on current and capital expenditure do not include
state universities and colleges.
14 Excluding former Democratic Yemen.
15 Expenditure on education is calculated as percentage of net
material product.
16 Metropolitan France.
17 1980.
18 Including data for Federated States of Micronesia, Marshall Is.
and Northern Mariana Is.
19 In millions.

Source:
Organisation des Nations Unies pour l'éducation, la science et
la culture (Paris).

§ Toutes les données se rapportant à l'Allemagne avant le 3 octobre
1990 figurent dans deux rubriques séparées basées sur les
territoires respectifs de la République fédérale d'Allemagne et
l'ancienne République démocratique allemande selon la période
indiquée. Pour les notes explicatives en détail sur les données
concernant l'Allemagne, voir Annexe I - Nomenclature des pays.

1 Les données se réfèrent aux dépenses du Ministère de l'Education
seulement.
2 De 1985 à 1989, les dépenses relatives à Al-Azhar ne sont pas incluses.
3 1986.
4 Les données de réfèrent aux dépenses du Ministère des enseignements
primaire et secondaire seulement.
5 De 1985 et 1986, les dépenses se réfèrent aux dépenses du
gouvernement fédéral seulement.
6 Les dépenses relatives à l'enseignement du troisième degré ne sont
pas incluses.
7 Les dépenses de l'enseignement sont calculés en pourcentage du
produit social global.
8 Les dépenses relatives aux Universités ne sont pas incluses.
9 Les données se réfèrent à la totalité des dépenses publiques et
privées afférentes à l'enseignement.
10 En 1975, les chiffres sont exprimés en pesos (1 austral = 1 000 pesos).
11 Dépenses du bureau grec de l'Education seulement.
12 En 1975 et 1980, les données relatives aux dépenses ordinaires et en
capital ne comprennent pas les subventions publiques à
l'enseignement privé. En 1985, ces données se réfèrent à la
totalité des dépenses publiques et privées afférentes à l'enseignement.
13 En 1975, les données relatives aux dépenses ordinaires et en capital
ne comprennent pas les universités et collèges d'état.
14 Non compris l'ancien Yémen démocratique.
15 Les dépenses de l'enseignement sont calculés en pourcentage du
produit matériel net.
16 France métropolitaine.
17 1980.
18 Y compris les données pour les Etats fédérés de Micronésie, les
îles Marshall et les îles Mariannes du Nord.
19 En millions.

18
Illiterate population by sex
Population analphabète selon le sexe

Country or area Pays ou zone	Year Année	Age group Groupe d'âge	Illiterate population Population analphabète			Percentage of illiterates Percentage d'analphabètes		
			Total	M	F	Total	M	F
Africa · Afrique								
Algeria Algérie	1987	15+	6373688	2320756	4052932	50.4	36.6	64.2
Angola[1] Angola[1]	1985	15+	...	...	...	59.0	51.0	...
Benin Bénin	1979	15+	1418051	563351	854700	83.5	74.8	90.5
Botswana[1] Botswana[1]	1990	15+	174500	49600	124900	26.4	16.3	34.9
Burkina Faso[1] Burkina Faso[1]	1975	15+	2803440	1272593	1530847	91.2	85.3	96.7
Burundi[1] Burundi[1]	1982	10+	...	...	...	66.2	57.2	74.3
Cameroon Cameroun	1976	15+	2360088	863884	1496204	58.8	45.4	70.9
Cape Verde[1] Cap-Vert[1]	1989	15+	64510	...	...	33.5	...	...
Central African Rep.[1] République centrafricaine[1]	1975	15+	841995	336544	505451	81.8	70.4	91.6
Chad[1] Tchad[1]	1990	15+	2280300	916800	1364400	70.2	57.8	82.1
Comoros Comores	1980	15+	88780	36429	52351	52.1	44.0	60.0
Congo[1] Congo[1]	1990	15+	484700	163500	321100	43.4	30.0	56.1
Côte d'Ivoire[1] Côte d'Ivoire[1]	1990	15+	2941000	1080600	1860400	46.2	33.1	59.8
Egypt Egypte	1986	15+	15931816	6279533	9652283	55.5	43.0	68.6
Equatorial Guinea Guinée équatoriale	1983	15+	58847	16288	42559	38.0	22.6	51.5
Ethiopia Ethiopie	1983	10+	8500000	...	...	37.6	...	...
Gabon[1] Gabon[1]	1990	15+	311400	103000	208400	39.3	26.5	51.5
Gambia[1] Gambie[1]	1990	15+	349900	143600	206300	72.8	61.0	84.0
Ghana[1] Ghana[1]	1990	15+	3258200	1215100	2043100	39.7	30.0	49.0
Guinea[1] Guinée[1]	1990	15+	2947000	1237000	1710000	76.0	65.1	86.6
Guinea-Bissau Guinée-Bissau	1979	15+	342393	130922	211471	80.0	66.7	91.4
Kenya[1] Kenya[1]	1990	15+	3728300	1207400	2520900	31.0	20.2	41.5
Liberia[1] Libéria[1]	1990	15+	839000	352800	486200	60.5	50.2	71.2
Libyan Arab Jamahiriya[1] Jamahiriya arabe libyenne[1]	1990	15+	890300	324500	565700	36.2	24.6	49.6
Madagascar[1] Madagascar[1]	1990	15+	1304500	395300	909200	19.8	12.3	27.1
Mali Mali	1990	15+	3398100	1401700	1996400	68.0	59.2	76.1
Mauritania Mauritanie	1976	6+	895877	...	...	82.6	...	...
Morocco Maroc	1982	15+	8119233	3187079	4932154	69.7	56.3	82.5
Mozambique Mozambique	1980	15+	4557751	1650952	2906799	72.8	56.0	87.8
Niger[1] Niger[1]	1990	15+	2683000	1099400	1583600	71.6	59.6	83.2
Nigeria[1] Nigéria[1]	1990	15+	28722600	10758200	17964400	49.3	37.7	60.5

18
Illiterate population by sex [*cont.*]
Population analphabète selon le sexe [*suite*]

Country or area Pays ou zone	Year Année	Age group Groupe d'âge	Illiterate population Population analphabète			Percentage of illiterates Percentage d'analphabètes		
			Total	M	F	Total	M	F
Réunion Réunion	1982	15+	73220	38861	34359	21.4	23.5	19.5
Rwanda Rwanda	1978	15+	1619117	620852	998265	61.8	49.2	73.4
Sao Tome and Principe Sao Tomé-et-Principe	1981	15+	22080	6755	15325	42.6	26.8	57.6
Senegal[1] Sénégal[1]	1990	15+	2525400	966400	1559100	61.7	48.1	74.9
Seychelles Seychelles	1971	15+	12494	6465	6029	42.3	44.4	40.2
Sierra Leone[1] Sierra Leone[1]	1990	15+	1829500	776800	1052800	79.3	69.3	88.7
Somalia[1] Somalie[1]	1990	15+	3002500	1153700	1848800	75.9	63.9	86.0
South Africa[8] Afrique du Sud[8]	1980	15+	3711776	1796523	1915253	23.8	22.5	25.2
St. Helena Sainte-Hélène	1987	20+	104	65	39	2.7	3.3	2.1
Sudan[1] Soudan[1]	1990	15+	10061100	3947700	6113500	72.9	57.3	88.3
Swaziland Swaziland	1976	15+	115836	49281	66555	44.8	42.7	46.5
Togo Togo	1981	15+	927712	328497	599215	68.6	53.3	81.5
Tunisia Tunisie	1984	15+	2076900	840400	1236500	49.3	39.5	59.4
Uganda[1] Ouganda[1]	1990	15+	4908100	1762200	3145900	51.7	37.8	65.1
United Rep. Tanzania[1] République-Unie de Tanzanie[1]	1986	15+	1366721	489830	878891	9.6	...	...
Zaire[1] Zaïre[1]	1990	15+	5465900	1563300	3902700	28.2	16.4	39.3
Zambia[1] Zambie[1]	1990	15+	1170100	398000	772100	27.2	19.2	34.7
Zimbabwe[10] Zimbabwe[10]	1982	15+	852120	292790	559330	22.2	15.8	28.1
America, North · Amérique du Nord								
Barbados[2] Barbade[2]	1970	15+	1093	493	600	0.7	0.7	0.7
Belize[2] Belize[2]	1970	15+	5353	2656	2699	8.8	8.8	8.8
Bermuda[2] Bermudes[2]	1970	15+	586	391	195	1.6	2.1	1.1
British Virgin Islds[2] Iles Vierges britanniques[2]	1970	15+	100	61	39	1.7	1.9	1.5
Cayman Islands[2] Iles Caïmanes[2]	1970	15+	152	70	82	2.5	2.5	2.4
Costa Rica Costa Rica	1984	15+	112946	55431	57515	7.4	7.3	7.4
Cuba[3] Cuba[3]	1981	10+	...	...	...	3.8	3.8	3.8
Dominica[2] Dominica[2]	1970	15+	2083	944	1139	5.9	6.0	5.8
Dominican Republic République dominicaine	1981	5+	1519198	770758	748440	31.4	31.8	30.9
El Salvador El Salvador	1980	15+	818100	...	...	32.7	...	...
Grenada[2] Grenade[2]	1970	15+	1070	424	646	2.2	2.0	2.4
Guadeloupe Guadeloupe	1982	15+	22359	11231	11128	10.0	10.4	9.6
Guatemala[1] Guatemala[1]	1990	15+	2253200	931900	1321300	44.9	36.9	52.9
Haiti Haïti	1982	15+	2004791	926751	1078040	65.2	62.7	67.5

18

Illiterate population by sex [*cont.*]

Population analphabète selon le sexe [*suite*]

Country or area Pays ou zone	Year Année	Age group Groupe d'âge	Illiterate population Population analphabète			Percentage of illiterates Percentage d'analphabètes		
			Total	M	F	Total	M	F
Honduras[1] Honduras[1]	1990	15+	766000	348800	417200	26.9	24.5	29.4
Jamaica[1,2] Jamaïques[1,2]	1990	15+	26500	14700	11800	1.6	1.8	1.4
Martinique Martinique	1982	15+	16814	8824	7990	7.2	8.0	6.6
Mexico Méxique	1980	15+	6451740	2545171	3906569	17.0	13.8	20.1
Montserrat[2] Montserrat[2]	1970	15+	231	100	131	3.4	3.2	3.4
Netherlands Antilles Antilles néerlandaises	1981	15+	10236	4497	5739	6.2	5.8	6.6
Nicaragua[4] Nicaragua[4]	1971	15+	410755	193475	217277	42.5	42.0	42.9
Panama Panama	1980	15+	156531	74737	81794	14.4	13.7	15.1
Puerto Rico Porto Rico	1980	15+	239095	107372	131723	10.9	10.3	11.5
St. Kitts-and-Nevis[2] Saint-Kitts-et-Nevis[2]	1970	15+	546	247	299	2.4	2.4	2.3
Saint Lucia[2] Sainte-Lucie[2]	1970	15+	9195	4251	4944	18.3	19.2	17.6
St. Vincent-and-Grenadines[2] Saint-Vincent-et-Grenadines[2]	1970	15+	1839	779	1060	4.4	4.2	4.5
Trinidad and Tobago Trinité-et-Tobago	1980	15+	34800	11890	22910	5.1	3.5	6.6
Turks and Caicos Is.[2] Iles Turques et Caïques[2]	1970	15+	56	18	38	1.9	1.4	2.3
United States Etats-Unis	1979	14+	...	...	...	0.5	...	...
America, South · Amérique du Sud								
Argentina Argentine	1980	15+	1184964	543174	641790	6.1	5.7	6.4
Bolivia Bolivie	1976	15+	993437	315460	677977	36.8	24.2	48.6
Brazil[1] Brésil[1]	1985	15+	...	...	...	22.2	20.9	23.6
Chile Chili	1982	15+	681039	315538	365501	8.9	8.5	9.2
Colombia Colombie	1981	15+	2407458	1091407	1316051	14.8	13.6	16.1
Ecuador[5] Equateur[5]	1982	15+	758272	298018	460254	16.1	12.8	19.4
French Guiana Guyane française	1982	15+	8372	4321	4051	17.0	16.4	17.7
Guyana[1,2] Guyane[1,2]	1990	15+	24500	8800	15700	3.6	2.5	4.6
Paraguay Paraguay	1982	15+	219120	84340	134780	12.5	9.7	15.2
Peru[6] Pérou[6]	1981	15+	1799458	485486	1313972	18.1	9.9	26.1
Suriname[1] Suriname[1]	1990	15+	13400	6300	7100	5.1	4.9	5.3
Uruguay Uruguay	1990	15+	87700	39100	48700	3.8	3.4	4.1
Venezuela Venezuela	1981	15+	1331260	584023	747237	15.3	13.5	17.0
Asia · Asie								
Afghanistan[7] Afghanistan[7]	1979	15+	5832988	2583581	3249407	81.8	69.7	95.0
Bahrain Bahreïn	1981	15+	71160	34398	36762	30.2	23.5	41.4
Bangladesh Bangladesh	1981	15+	32923083	14501583	18421500	70.8	60.3	82.0
Brunei Darussalam Brunéi Darussalam	1981	15+	26224	9574	16650	22.2	14.8	31.0

18
Illiterate population by sex [*cont.*]
Population analphabète selon le sexe [*suite*]

Country or area Pays ou zone	Year Année	Age group Groupe d'âge	Illiterate population Population analphabète			Percentage of illiterates Percentage d'analphabètes		
			Total	M	F	Total	M	F
China[10] Chine[10]	1987	15+	217532300	65104800	152427500	28.6	16.9	40.6
Cyprus Chypre	1987	15+	...	...	...	6.0	2.0	9.0
Hong Kong[2] Hong-kong[2]	1971	15+	571840	126152	445688	22.7	9.9	35.9
India Inde	1981	15+	238097747	93899834	144197913	59.2	45.2	74.3
Indonesia Indonésie	1980	15+	28325026	9490915	18834111	32.7	22.5	42.3
Iran, Islamic Rep. of[7] Iran, Rép. islamique d'[7]	1986	15+	12919534	5074037	7845497	48.0	37.0	59.4
Iraq[1] Iraq[1]	1985	15-45	...	...	...	10.7	9.8	12.5
Israel[2] Israël[2]	1987	15+	...	...	...	3.6	...	...
Jordan Jordanie	1979	15+	344002	104670	239332	33.2	19.5	47.9
Korea, Republic of[1][11] Corée, République de[1][11]	1990	15+	1185300	134300	1050900	3.7	0.9	6.5
Kuwait Koweït	1985	15+	273513	141082	132431	25.5	21.8	31.2
Lao People's Dem. Rep.[1] Rép. dém. pop. lao[1]	1985	15-45	...	...	...	16.1	8.0	24.2
Lebanon[1] Liban[1]	1990	15+	382300	110900	271300	19.9	12.2	26.9
Macau Macao	1970	15+	31917	11894	20023	20.6	15.2	26.1
Malaysia Malaisie	1980	15+	2399790	791000	1608790	30.4	20.4	40.3
Maldives Maldives	1985	15+	8568	4565	4003	8.7	8.8	8.5
Myanmar Myanmar	1983	15+	4492769	1460457	3032312	21.4	14.2	28.3
Nepal Népal	1981	15+	6998148	3053083	3945065	79.4	68.3	90.8
Pakistan Pakistan	1981	15+	34713824	16051771	18662053	74.3	64.6	85.2
Philippines[12] Philippines[12]	1980	15+	4626922	2200485	2426437	16.7	16.1	17.2
Qatar Qatar	1986	15+	64891	45253	19638	24.3	23.2	27.5
Saudi Arabia Arabie saoudite	1982	15+	...	...	...	48.9	28.9	69.2
Singapore Singapour	1980	15+	300994	75422	225572	17.1	8.4	26.0
Sri Lanka Sri Lanka	1981	15+	1271984	424424	847560	13.2	8.7	18.0
Syrian Arab Republic[13] République arabe syrienne[13]	1981	15+	1982265	601390	1380875	44.4	26.4	63.0
Thailand[12] Thaïlande[12]	1980	15+	3296606	1049664	2246942	12.0	7.7	16.0
Turkey Turquie	1985	15+	7579671	1959483	5620188	24.0	12.4	35.7
United Arab Emirates Emirats arabes unis	1975	15+	186058	126586	59472	46.5	41.6	61.9
Viet Nam Viet Nam	1979	15+	4846849	1340445	3506404	16.0	9.5	21.7
Yemen[1][9] Yémen[1][9]	1990	15+	2558600	876000	1682600	61.5	46.7	73.7
Europe · Europe								
Greece Grèce	1981	15+	701056	140544	560512	9.5	3.9	14.7
Hungary Hongrie	1980	15+	95542	27756	67786	1.1	0.7	1.5

18
Illiterate population by sex [*cont.*]
Population analphabète selon le sexe [*suite*]

Country or area Pays ou zone	Year Année	Age group Groupe d'âge	Illiterate population Population analphabète			Percentage of illiterates Percentage d'analphabètes		
			Total	M	F	Total	M	F
Italy Italie	1981	15+	1572556	539781	1032775	3.5	2.5	4.5
Liechtenstein Liechtenstein	1981	10+	68	33	35	0.3	0.3	0.3
Malta Malte	1985	20+	33740	16802	16938	14.3	14.8	13.9
Poland Pologne	1978	15+	334586	92609	241977	1.2	0.7	1.7
Portugal Portugal	1981	15+	1506206	524461	981745	20.6	15.2	25.4
San Marino San Marino	1976	10+	640	260	380	3.9	3.2	4.7
Spain Espagne	1986	15+	1260789	360483	900306	4.2	2.5	5.8
Yugoslavia Yougoslavie	1981	15+	1764042	370558	1393484	10.4	4.5	16.1
Oceania · Océanie								
American Samoa Samoa américaines	1980	15+	507	240	267	2.7	2.5	2.8
Fiji Fidji	1976	15+	65957	24305	41652	21.0	16.0	26.0
Guam Guam	1980	15+	2470	1317	1153	3.6	3.6	3.5
New Caledonia Nouvelle-Calédonie	1976	15+	7133	3370	3763	8.7	7.8	9.7
Pacific Islands (Palau)[14] Iles du Pacifique (Palaos)[14]	1980	15+	5798	2454	3344	8.1	6.7	9.5
Papua New Guinea[1] Papouasie-Nouvelle-Guinée[1]	1990	15+	1119000	426500	692500	48.0	35.1	62.2
Samoa Samoa	1971	15+	1581	819	762	2.2	2.2	2.1
Tonga Tonga	1976	15+	193	81	112	0.4	0.3	0.5
Vanuatu Vanuatu	1979	15+	28647	13823	14824	47.1	42.7	52.2
former USSR · ancienne URSS								
former USSR ancienne URSS	1989	15+	4282023	644964	3637059	2.0	0.7	3.2
Belarus Bélarus	1989	15+	...	...	...	2.1	0.6	3.4
Ukraine Ukraine	1989	15+	...	...	...	1.6	0.5	2.6

Source:
United Nations Educational, Scientific and Cultural Organization
(Paris).

1 Estimates.
2 Persons with no schooling are defined as illiterates.
3 Excluding functionally and physically handicapped.
4 In 1980, after the National Literacy Campaign, the Ministry
   of Education estimated that the illiteracy rate of the
   population aged 10 years and over had been reduced to 12.96%.
5 Excluding nomadic Indian tribes.
6 Excluding Indian jungle population.
7 Excluding nomadic population.
8 Not including Bothuthatswana, Transkei and Veda.
9 Excluding former Democratic Yemen.
10 Based on a 10% sample of census returns.
11 Based on a sample survey.
12 Based on a 20% sample of census returns.
13 National population only.
14 Including data for Federated States of Micronesia, Marshall Is.
   and Northern Mariana Is.

Source:
Organisation des Nations Unies pour l'éducation, la science
et la culture (Paris).

1 Estimation.
2 Les personnes sans scolarité ont été considérées comme
   étant analphabètes.
3 Non compris les handicapés physiques et fonctionnels.
4 En 1980, à la fin de la compagne nationale d'alphabétisation,
   le Ministère de l'Education a estimé que le taux d'analphabétisation
   de la population âgée de 10 ans et plus a été réduit à 12.96%.
5 Non compris les tribus indiennes nomades.
6 Non compris les indiens de la jungle.
7 Non compris les populations nomades.
8 Non compris le Bothuthatswana, le Transkei and Veda.
9 Non compris l'ancien Yémen démocratique.
10 D'après un échantillon portant sur 10% des bulletins de recensement.
11 Basées sur une enquête par sondage.
12 D'après un échantillon portant sur 20% des bulletins de recensement.
13 Population nationale seulement.
14 Y compris les données pour les Etats fédérés de Micronésie, les îles
   Marshall et les îles Mariannes du Nord.

Technical notes, tables 13-18

Detailed data on education accompanied by explanatory notes can be found in the UNESCO *Statistical Yearbook.* [32] Brief notes pertaining to major items of statistical information on education shown in the present edition of the United Nations *Statistical Yearbook* are given below.

The tables included in this chapter cover basic data on education preceding the first level as well as education at the first, second and third levels. Data are also provided on public expenditures on education and on illiteracy. Definitions for the different levels and types of education given below are based on the "Revised recommendations concerning the international standardization of educational statistics" adopted by the General Conference of UNESCO at its twentieth session (1978). [32, 50]

Table 13: Data on education preceding the first level refer to kindergartens, nursery schools and infant classes attached to schools at higher levels. Nursery play centres are excluded whenever possible. Figures on teachers refer to both full-time and part-time teachers. Unless otherwise stated data cover both public and private schools.

Table 14: Data on education at the first level refer to education whose main function is to provide basic instruction in the tools of learning (for example at elementary and primary schools). Its length may vary from 4 to 9 years, depending on the organization of the school system in each country. Unless otherwise stated, data cover both public and private schools. Figures on teachers refer to both full-time and part-time but exclude classes organized for adults or for handicapped children.

Table 15: Data on education at the second level refer to education which is based upon at least four years of previous instruction at first level and provides general or specialized instruction, or both (for example at middle and secondary schools, high schools, teachers' training schools at this level, schools of a vocational or technical nature). Unless otherwise stated data cover both public and private schools. In most cases data include part-time teachers.

"General education" refers to secondary schools that provide general or specialized education based upon at least four years previous instruction at the first level and which do not aim at preparing the pupils directly for a given trade or occupation. "Teacher training" refers to education in schools whose purpose is the training of students for the teaching profession and which do not require as a prerequisite the completion of secondary school. "Other second level" refers to technical and vocational education provided in those secondary schools which aim at preparing the pupils directly for a trade or occupation other than teaching.

Table 16: Data on education at the third level refer to education which requires as a minimum condition of admission the successful completion of education at the second level or proof of equivalent knowledge or

Notes techniques, tableaux 13-18

On trouvera des données détaillées sur l'éducation accompagnées de notes explicatives dans l'*Annuaire statistique* de l'UNESCO [32]. Des notes sommaires concernant les principaux éléments d'information statistique sur l'éducation figurant dans la présente édition de l'*Annuaire statistique* des Nations Unies sont présentées ci-dessous.

Les tableaux de ce chapitre contiennent des données de base sur l'enseignement précédant le premier degré ainsi que sur l'enseignement aux premier, second et troisième degrés. Ils contiennent également des données sur les dépenses publiques consacrées à l'éducation et à l'analphabétisme. Les définitions des différents niveaux et types d'enseignement données ci-dessous sont fondées sur les "recommandations révisées concernant la normalisation internationale des statistiques de l'éducation" adoptées par la Conférence générale de l'UNESCO à sa vingtième session (1978) [32, 50].

Tableau 13 : Les données sur l'éducation avant le premier niveau portent sur le jardin d'enfants, les écoles maternelles ainsi que les cours préparatoires rattachés aux établissements du niveau suivant. Dans la mesure du possible, les garderies d'enfants ne sont pas incluses. Les chiffres relatifs aux enseignants portent à la fois sur les enseignants à plein temps et à temps partiel. Sauf indication contraire, les données portent sur les écoles publiques et privées.

Tableau 14 : Les données relatives à l'enseignement du premier degré portent sur l'enseignement dont la fonction principale est d'offrir les premiers éléments de l'instruction (par exemple, dans les écoles élémentaires ou primaires). Sa durée peut varier de quatre à neuf ans, selon l'organisation du système scolaire de chaque pays. Sauf indication contraire, les données portent à la fois sur les établissements d'enseignement publics et privés. Les chiffres relatifs aux enseignants désignent à la fois ceux qui enseignent à plein temps et à temps partiel mais ne comprennent pas les classes organisées à l'intention des adultes ou des enfants handicapés.

Tableau 15 : Les données relatives à l'enseignement secondaire se rapportent à un enseignement fondé sur au moins quatre années d'enseignement préalable au niveau primaire et donnant une formation générale ou spécialisée, ou les deux (par exemple, dans les écoles moyennes et secondaires, les lycées, les écoles normales de ce niveau, les écoles professionnelles ou techniques).

Les établissements d'"enseignement général" désignent les écoles secondaires qui dispensent un enseignement général ou spécialisé nécessitant au moins quatre années d'études préalables dans un établissement primaire et ne visant pas à préparer directement les élèves à l'exercice d'une profession ou d'un métier. La "formation d'enseignant" désigne l'éducation dispensée dans les écoles dont le but est de former les élèves à la profession d'enseignant et qui n'exigent pas comme condition préalable des études secondaires complètes. Par "autres types d'enseignement secondaire", on entend l'enseignement

experience. It can be given in different types of institutions such as universities, teacher-training institutes and technical institutes. The figures include, as a rule, both full-time and part-time teachers and students. Correspondence courses are not normally included. Figures on teachers include auxiliary teachers (assistants, demonstrators and the like) but exclude staff with no teaching duties (such as administrators and laboratory technicians).

Table 17: Data on total expenditure refer to public expenditure on public education plus subsidies for private education. Total expenditures cover both current and capital expenditure.

Current expenditures include expenditures on administration, emoluments of teachers and supporting teaching staff, school books and other teaching materials, scholarships, welfare services and maintenance of school buildings.

Capital expenditures include outlays on purchases of land, building construction expenditures and so forth. This item also includes loan transactions.

Data include, unless otherwise indicated, educational expenditure at every level of administration. In general, the data do not include development assistance expenditures on education. Data on gross national product (GNP) used to derive the ratio of total public expenditure on education to GNP are World Bank estimates. For certain countries with centrally planned economies, the concept of material product rather than GNP was employed.

Table 18: Except where otherwise stated, data refer to population 15 years of age and over.

Ability to both read and write a simple sentence on everyday life is used as the criterion of literacy; hence semi-literates (persons who can read but not write) are included with illiterates. Persons for whom literacy is not known are excluded from calculations; consequently the percentage of illiteracy for a given country is based on the number of reported illiterates, divided by the total number of reported literates and illiterates. The data are based on population censuses or surveys.

technique et professionnel dispensé dans les écoles secondaires visant à préparer les élèves directement à l'exercice d'un métier ou d'une profession autre que la profession d'enseignant.

Tableau 16 : Les données relatives à l'enseignement du troisième degré se rapportent à un enseignement qui exige comme condition minimum d'admission l'achèvement avec succès d'un enseignement secondaire complet ou la preuve de connaissances ou d'une expérience équivalentes. Il peut être dispensé dans différents types d'établissement tels que les universités, les instituts de formation pédagogique et les instituts techniques. En règle générale, les chiffres portent à la fois sur les enseignants et les étudiants à plein temps et à temps partiel. Normalement, les cours par correspondance ne sont pas compris. Les données relatives aux enseignants englobent les professeurs auxiliaires (assistants, maîtres de travaux pratiques, etc.), mais pas le personnel n'exerçant pas de fonctions d'enseignement (tel que les administrateurs et les techniciens de laboratoire).

Tableau 17 : Les données relatives aux dépenses totales se rapportent aux dépenses publiques consacrées à l'enseignement public et aux subventions à l'enseignement privé. Les totaux englobent à la fois les dépenses ordinaires et les dépenses d'équipement.

Les dépenses ordinaires comprennent les dépenses d'administration, les émoluments du personnel enseignant et auxiliaire, les manuels scolaires et autres matériels didactiques, les bourses d'études, les services sociaux et l'entretien des bâtiments scolaires.

Les dépenses d'équipement comprennent les dépenses consacrées à l'achat de terrains, à la construction de bâtiments, etc. Les transactions de prêt sont également incluses dans cette rubrique.

Sauf indication contraire, les données comprennent les dépenses effectuées à tous les niveaux administratifs. En règle générale, elles ne comprennent pas les dépenses d'enseignement financées au titre de l'aide au développement. Les données relatives au produit national brut (PNB), utilisées pour déterminer le ratio du volume total de dépenses publiques consacrées à l'éducation au PNB, sont des estimations de la Banque mondiale. Pour certains pays à économie planifiée, on a utilisé la notion de produit matériel à la place du PNB.

Tableau 18 : Sauf indication contraire, les données se réfèrent à la population âgée de 15 ans et plus.

On utilise l'aptitude à lire et à écrire une phrase simple sur la vie quotidienne comme critère d'alphabétisme; par conséquent, les semi-alphabètes (c'est-à-dire les personnes qui savent lire, mais non écrire) sont assimiliés aux analphabètes. Les personnes dont on ne sait pas si elles savent lire ou écrire sont exclues de ces calculs; par conséquent, le pourcentage d'analphabétisme d'un pays donné est fondé sur le nombre d'analphabètes connus divisé par le total des alphabètes et analphabètes connus. Ces données sont fondées sur les recensements ou enquêtes de population.

19
Vital statistics summary and expectation of life at birth
Aperçu des statistiques de l'état civil et espérance de vie à la naissance

Country or area Pays ou zone	Year Année	Live births Naissances vivantes Number Nombre	Live births Naissances vivantes Rate Taux	Deaths Décès Number Nombre	Deaths Décès Rate Taux (000)	Natural increase Accroisse- ment naturel	Year Année	Infant deaths Décès d'enfants de moins d'un an Number Nombre	Infant deaths Décès d'enfants de moins d'un an Rate Taux (000)
Africa · Afrique									
Algeria Algérie	1985-90	...	35.5[1]	...	8.3[1]	27.2[1]	1985-90	...	74.0
Angola[1] Angola	1985-90	...	47.2	...	20.2	27.0	1985-90	...	137.0
Benin[1] Bénin	1985-90	...	49.2	...	19.3	29.9	1985-90	...	90.0
Botswana Botswana	1985-90	...	48.5[1]	...	11.6[1]	36.9[1]	1985-90	...	67.0[1]
Burkina Faso[1] Burkina Faso[1]	1985-90	...	47.1	...	18.4	28.7	1985-90	...	138.0
Burundi[1] Burundi[1]	1985-90	...	47.6	...	17.9	29.7	1985-90	...	119.0
Cameroon[1] Cameroun[1]	1985-90	...	47.5	...	14.9	32.6	1985-90	...	94.0
Cape Verde Cap-Vert	1985	12 639	37.9	2 735	8.2	29.7	1985	863	68.3
Central African Rep.[1] Rép. centrafricaine[1]	1985-90	...	45.5	...	17.8	27.7	1985-90	...	104.0
Chad[1] Tchad[1]	1985-90	...	44.2	...	19.5	24.7	1985-90	...	132.0
Comoros[1] Comores[1]	1985-90	...	47.5	...	13.0	34.5	1985-90	...	99.0
Congo[1] Congo[1]	1985-90	...	46.1	...	14.6	31.5	1985-90	...	73.0
Côte d'Ivoire[1] Côte d'Ivoire[1]	1985-90	...	49.9	...	14.5	35.4	1985-90	...	96.0
Djibouti[1] Djibouti[1]	1985-90	...	46.5	...	17.8	28.7	1985-90	...	122.0
Egypt Egypte	1987	1 902 604	38.8	466 161	9.5	29.3	1987	94 044	49.4
Equatorial Guinea[1] Guinée équatoriale[1]	1985-90	...	43.8	...	19.6	24.2	1985-90	...	127.0
Ethiopia[1] Ethiopie[1]	1985-90	...	48.6	...	20.7	27.9	1985-90	...	137.0
Gabon[1] Gabon[1]	1985-90	...	39.4	...	16.8	22.6	1985-90	...	103.0
Gambia[1] Gambie[1]	1985-90	...	47.4	...	21.4	26.0	1985-90	...	143.0
Ghana[1] Ghana[1]	1985-90	...	44.4	...	13.1	31.3	1985-90	...	90.0
Guinea[1] Guinée[1]	1985-90	...	51.0	...	22.0	29.0	1985-90	...	145.0
Guinea-Bissau Guinée-Bissau	1985-90	...	42.9[1]	...	23.0[1]	19.9[1]	1985-90	...	151.0[1]
Kenya[1] Kenya[1]	1985-90	...	47.0	...	11.3	35.7	1985-90	...	72.0
Lesotho[1] Lesotho[1]	1985-90	...	40.8	...	12.4	28.4	1985-90	...	100.0
Liberia[1] Libéria[1]	1985-90	...	47.3	...	15.8	31.5	1985-90	...	142.0
Libyan Arab Jamahiriya Jamahiriya arabe libyenne	1985-90	...	44.0[1]	...	9.4[1]	34.6[1]	1985-90	...	82.0
Madagascar[1] Madagascar[1]	1985-90	...	45.8	...	14.0	31.8	1985-90	...	120.0
Malawi Malawi	1977[2]	267 805	48.3	138 694	25.0	23.3	1977[2]	34 808	130.0
Mali Mali	1977[2]	375 117	48.7	96 221	12.5	36.2	1987[2]	26 731	71.3
Mauritania[1] Mauritanie[1]	1985-90	...	46.2	...	19.0	27.2	1985-90	...	127.0

Year Année	Expectation of life at birth Esp. de vie à la naissance		Fertility Fécondité		Marriages Mariages			Divorces		
	Male Masculin	Female Féminin	Year Année	Rate Taux	Year Année	Number Nombre	Rate Taux (000)	Year Année	Number Nombre	Rate Taux (000)
1983	61.57	63.32	1985-90	5.430[1]	1985	123 688	5.7	...	...	...
1985-90	42.90	46.10	1985-90	6.390	...	...	...	...	...	...
1985-90	44.40	47.60	1985-90	7.100	...	...	...	...	...	...
1981	52.32	59.70	1981	7.070	1987	1 862	1.6	...	...	...
1985-90	45.60	48.90	1985-90	6.500	...	...	...	...	...	...
1985-90	45.90	49.20	1985-90	6.790	...	...	...	...	...	...
1985-90	51.00	54.00	1985-90	6.900	...	...	...	...	...	...
1979-81	58.95	61.04	1985	4.605	1975	1 604	5.4	...	...	...
1985-90	46.00	51.00	1985-90	6.190	...	...	...	...	...	...
1985-90	43.90	47.10	1985-90	5.890	...	...	...	...	...	...
1985-90	53.50	54.50	1985-90	7.030	...	...	...	...	...	...
1985-90	50.10	55.30	1985-90	6.290	...	...	...	...	...	...
1985-90	50.80	54.20	1985-90	7.410	...	...	...	...	...	...
1985-90	45.40	48.70	1985-90	6.600	...	...	...	...	...	...
1985-90	57.80[1]	60.30[1]	1986	5.075	1985	442 280	9.1	1985	79 189	1.6
1985-90	44.40	47.60	1985-90	5.890	...	...	...	...	...	...
1985-90	42.40	45.60	1985-90	6.780	...	...	...	...	...	...
1985-90	49.90	53.20	1985-90	4.990	...	...	...	...	...	...
1985-90	41.40	44.60	1985-90	6.500	...	...	...	...	...	...
1985-90	52.20	55.80	1985-90	6.390	...	...	...	...	...	...
1985-90	42.00	43.00	1985-90	7.000	...	...	...	...	...	...
1985-90	39.90[1]	43.10[1]	1985-90	5.790[1]	1981	100	0.1	...	...	...
1985-90	56.50	60.50	1985-90	7.000	...	...	...	...	...	...
1985-90	51.50	60.50	1985-90	5.790	...	...	...	...	...	...
1985-90	52.00	54.00	1985-90	6.800	...	...	...	...	...	...
1985-90	59.10[1]	62.50[1]	1985-90	6.870[1]	1988	16 989	4.5	1988	2 264	0.6
1985-90	52.00	55.00	1985-90	6.600	...	...	...	...	...	...
1977[2]	38.12	41.16	1985-90	7.600[1]	...	...	...	...	...	...
1976[2]	46.91	49.66	1985-90	7.100[1]	...	...	...	...	...	...
1985-90	44.40	47.60	1985-90	6.500	...	...	...	...	...	...

19
Vital statistics summary and expectation of life at birth [cont.]
Aperçu des statistiques de l'état civil et espérance de vie à la naissance [suite]

Country or area Pays ou zone	Year Année	Live births Naissances vivantes		Deaths Décès		Natural increase Accroisse- ment naturel	Year Année	Infant deaths Décès d'enfants de moins d'un an	
		Number Nombre	Rate Taux	Number Nombre	Rate Taux (000)			Number Nombre	Rate Taux (000)
Mauritius[1] Maurice[1]	1985-90	...	18.6	...	6.4	12.2	1985-90	...	23.0
Island of Mauritius Ile Maurice	1990	21 801	21.0	6 857	6.6	14.4	1990	428	19.6
Rodrigues Rodrigues	1990	803	21.2	177	4.7	16.6	1990	28	34.9
Morocco[1] Maroc[1]	1985-90	...	35.5	...	9.8	25.7	1985-90	...	82.0
Mozambique[1] Mozambique[1]	1985-90	...	45.0	...	18.5	26.5	1985-90	...	141.0
Namibia[1] Namibie[1]	1985-90	...	44.0	...	12.1	31.9	1985-90	...	106.0
Niger[1] Niger[1]	1985-90	...	51.7	...	20.4	31.3	1985-90	...	135.0
Nigeria[1] Nigéria[1]	1985-90	...	48.5	...	15.6	32.9	1985-90	...	105.0
Réunion Réunion	1988	13 559	22.6	3 253	5.7	17.9	1987	124	9.8
Rwanda Rwanda	1985-90	...	51.2	...	17.2	34.0	1985-90	...	122.0
St. Helena ex. dep. Sainte-Hélène sans dép.	1989	89	12.7	51	7.3	5.4	1989	3	33.7
Ascension Ascension	1981	15	14.6	2	2.0	12.7	1980	1	200.0
Tristan da Cunha Tristan da Cunha	1988	4	13.1	3	9.8	3.3	1980	2	400.0
Sao Tome and Principe Sao Tomé-et-Principe	1989	4 047	35.0	1 179	10.2	24.8	1989	291	71.9
Senegal[1] Sénégal[1]	1985-90	...	45.5	...	17.7	27.8	1985-90	...	87.0
Seychelles Seychelles	1990	1 617	24.0	543	8.1	15.9	1990	21	13.0
Sierra Leone[1] Sierra Leone[1]	1985-90	...	48.2	...	23.4	24.8	1985-90	...	154.0
Somalia[1] Somalie[1]	1985-90	...	50.1	...	20.2	29.9	1985-90	...	132.0
South Africa[1] Afrique du Sud[1]	1985-90	...	32.1	...	9.9	22.2	1985-90	...	72.0
Sudan[1] Soudan[1]	1985-90	...	44.6	...	15.8	28.8	1985-90	...	108.0
Swaziland Swaziland	1985-90	...	46.8[1]	...	12.5[1]	34.3[1]	1985-90	...	118.0[1]
Togo Togo	1985-90	...	44.7[1]	...	14.1[1]	30.6[1]	1985-90	...	94.0[1]
Tunisia Tunisie	1985-90	...	31.1[1]	...	7.3[1]	23.8[1]	1985-90	...	52.0[1]
Uganda[1] Ouganda[1]	1985-90	...	52.2	...	15.6	36.6	1985-90	...	103.0
United Rep. of Tanzania[1] Rép.-Unie de Tanzanie[1]	1985-90	...	50.5	...	14.0	36.5	1985-90	...	106.0
Zaire[1] Zaïre[1]	1985-90	...	45.6	...	14.2	31.4	1985-90	...	83.0
Zambia Zambie	1985-90	...	51.1[1]	...	13.7[1]	37.4[1]	1985-90	...	80.0[1]
Zimbabwe[1] Zimbabwe[1]	1985-90	...	41.7	...	10.3	31.4	1985-90	...	66.0
America, North · Amérique du Nord									
Anguilla Anguilla	1985	177	25.3	7.3	10.4	14.9	1985	6	33.9
Antigua and Barbuda Antigua-et-Barbuda	1987	1 094	13.1	364	4.4	8.7	1985	29	24.4
Aruba Aruba	1988	949	15.6	335	5.5	10.1		...	...

	Expectation of life at birth Esp. de vie à la naissance		Fertility Fécondité		Marriages Mariages			Divorces		
Year Année	Male Masculin	Female Féminin	Year Année	Rate Taux	Year Année	Number Nombre	Rate Taux (000)	Year Année	Number Nombre	Rate Taux (000)
1985-90	66.40	71.70	1985-90	2.000	...	...	...	...	...	...
1986-88	64.74	72.22	1989	2.169	1990	11 253	10.9	1990	692	0.7
1981-85	64.47	68.95	1989	3.099	1990	173	4.6	...	...	...
1985-90	59.10	62.50	1985-90	4.820	...	...	...	...	...	...
1985-90	44.90	48.10	1985-90	6.390	...	...	...	...	...	...
1985-90	55.00	57.50	1985-90	6.090	...	...	...	...	...	...
1985-90	42.90	46.10	1985-90	7.100	...	...	...	...	...	...
1985-90	48.80	52.20	1985-90	6.900	...	...	...	...	...	...
1985-90	67.00[1]	75.30[1]	1986	2.707	1987	3 001	5.3	1987	694	1.2
1978	45.10[2]	47.70[2]	1978	8.737	1982	14 313	2.6	...	...	...
...	...	...	...	...	1986	29	4.5	1986	6	0.9
...	...	...	...	...	1981	3	2.9	...	...	...
...	...	...	...	...	1988	1	3.3	...	...	...
...	...	...	...	...	1988	49	0.4	...	...	...
1985-90	46.30	48.30	1985-90	6.500	...	...	...	...	...	...
1981-85	65.26	74.05	1990	2.750	1989	777	11.6	1989	45	0.7
1985-90	39.40	42.60	1985-90	6.500	...	...	...	...	...	...
1985-90	43.40	46.60	1985-90	6.600	...	...	...	...	...	...
1985-90	57.50	63.50	1985-90	4.480	...	...	...	...	...	...
1985-90	48.60	51.00	1985-90	6.440	...	...	...	...	...	...
1976	42.90	49.50	1985-90	6.500[1]	1986	2 243	3.4	...	...	...
1985-90	51.30[1]	54.80[1]	1985-90	6.580[1]	1979	5 753	2.3	...	...	...
1985-90	64.90[1]	66.40[1]	1980	4.512	1989	53 280	6.7	1989	12 695	1.6
1985-90	49.40	52.70	1985-90	7.300	...	...	...	...	...	...
1985-90	51.30	54.70	1985-90	7.100	...	...	...	...	...	...
1985-90	50.30	53.70	1985-90	6.090	...	...	...	...	...	...
1980	50.36	52.46	1985-90	7.200[1]	...	...	...	...	...	...
1985-90	56.50	60.10	1985-90	5.790	...	...	...	...	...	...
...	...	...	...	...	1985	101	13.8	1985	6	0.9
...	...	...	...	...	1987	343	4.1	1985	35	0.5
1972-78	68.30	75.40	...	...	1988	390	6.4	1988	196	3.2

19
Vital statistics summary and expectation of life at birth [*cont.*]
Aperçu des statistiques de l'état civil et espérance de vie à la naissance [*suite*]

Country or area Pays ou zone	Year Année	Live births Naissances vivantes		Deaths Décès		Natural increase Accroissement naturel	Year Année	Infant deaths Décès d'enfants de moins d'un an	
		Number Nombre	Rate Taux	Number Nombre	Rate Taux (000)			Number Nombre	Rate Taux (000)
Bahamas Bahamas	1989	4 971	20.0	1 348	5.4	14.6	1989	111	22.3
Barbados Barbade	1989	4 015	15.7	2 277	8.9	6.8	1989	36	9.0
Belize Belize	1989	6 810	37.2	762	4.2	33.0	1989	132	19.4
Bermuda Bermudes	1989	912	15.2	462	7.7	7.5	1989	6	6.6
British Virgin Islands Iles Vierges britanniques	1989	244	19.5	77	6.1	13.3	1988	7	29.5
Canada Canada	1988	375 743	14.5	190 011	7.3	7.2	1988	2 705	7.2
Cayman Islands Iles Caïmanes	1989	438	16.9	122	4.7	12.2	1989	4	9.1
Costa Rica Costa Rica	1989	83 460	28.2	11 273	3.8	24.4	1989	1 160	13.9
Cuba Cuba	1989	184 891	17.6	67 212	6.4	11.2	1989	2 049	11.1
Dominica Dominique	1988	1 731	21.3	424	5.2	16.1	1990	30	18.4
Dominican Republic Rép. dominicaine	1985-90	...	31.3[1]	...	6.8[1]	24.5[1]	1985-90	...	65.0[1]
El Salvador El Salvador	1985-90	...	36.3[1]	...	8.5[1]	27.8[1]	1985-90	...	64.0[1]
Greenland Groenland	1988	1 217	22.2	442	8.1	14.1	1987	29	26.6
Grenada Grenade	1979	2 664	24.5	739	6.8	17.7	1979	41	15.4
Guadeloupe Guadeloupe	1986	6 374	19.2	2 238	6.7	12.4	1986	98	15.4
Guatemala Guatemala	1988	341 382	39.3	64 837	7.5	31.9	1988	15 892	46.6
Haiti[1] Haïti[1]	1985-90		36.2		13.2	23.0	1985-90	...	97.0
Honduras Honduras	1985-90	...	39.8[1]	...	8.1[1]	31.7[1]	1985-90	...	69.0[1]
Jamaica Jamaïque	1988	53 623	22.7	12 167	5.2	17.6	1984	758	13.2
Martinique Martinique	1988	6 386	19.0	2 092	6.2	12.8	1988	55	8.6
Mexico Mexique	1985-90	...	29.0[1]	...	5.8[1]	23.2[1]	1985-90	...	43.0[1]
Montserrat Montserrat	1986	200	16.8	123	10.3	6.5	1986	1	5.0
Netherlands Antilles Antilles néerlandaises	1989	3 480	18.5	1 193	6.4	12.2	1989	22	6.3
Nicaragua Nicaragua	1985-90	...	41.8[1]		8.0[1]	33.8[1]	1985-90	...	62.0[1]
Panama Panama	1985-90	...	26.7[1]	...	5.2[1]	21.5[1]	1985-90	...	23.0[1]
Puerto Rico Porto Rico	1989	66 692	18.7	25 987	7.3	11.4	1989	952	14.3
Saint Kitts and Nevis Saint-Kitts-et-Nevis	1988	944	21.3	465	10.5	10.8	1988	23	24.4
Saint Lucia Sainte-Lucie	1989	3 159	21.3	816	5.5	15.8	1989	56	17.7
St. Pierre and Miquelon Saint-Pierre-et-Miquelon	1983	81	13.5	25	4.2	9.3	1981	1	9.2
St. Vincent and the Grenadines Saint-Vincent-et-Grenadines	1986	2 708	24.5	655	5.9	18.5	1988	55	21.7
Trinidad and Tobago Trinité-et-Tobago	1988	27 301	22.5	8 036	6.6	15.9	1987	332	11.4
Turks and Caicos Islands Iles Turques et Caïques	1980	214	28.9	15	2.0	26.8	1982	5	24.5

Year Année	Expectation of life at birth Esp. de vie à la naissance		Fertility Fécondité		Marriages Mariages			Divorces		
	Male Masculin	Female Féminin	Year Année	Rate Taux	Year Année	Number Nombre	Rate Taux (000)	Year Année	Number Nombre	Rate Taux (000)
...	...	...	1985	2.474	1989	2 131	8.6	1989	275	1.1
1980	67.15	72.46	1987	1.598	1989	2 047	8.0	1989	416	1.6
1980	69.85	71.78	1989	5.200	1989	1 138	6.2	1989	114	0.6
1980	68.81	76.28	1989	1.790	1989	877	14.6	1989	172	2.9
...	...	...	1988	1.932	1988	176	14.2	1988	9	0.7
1985-87	73.02	79.79	1988	1.692	1988	187 860	7.3	1987	78 160	3.1
...	...	...	...	...	1989	267	10.3	1989	76	2.9
1985-90	72.40[1]	77.00[1]	1984	3.538	1989	22 984	7.8	1988	2 482	0.9
1983-84	72.66	76.10	1987	1.819	1989	85 492	8.1	1989	37 493	3.6
...	...	...	...	...	...	...	...	...	...	...
1985-90	63.90[1]	68.10[1]	1980	5.552	1985	21 301	3.3	1985	7 808	1.2
1985	50.74	63.89	1986	3.965	1986	19 630	3.9	1986	2 020	0.4
1981	60.40	66.30	1987	2.051	1988	376	6.9	1988	95	1.7
...	...	...	...	...	1979	360	3.3	1979	21	0.2
1975	66.40	72.40	1985	2.579	1986	1 692	5.1	1986	511	1.5
1979	55.11	59.43	1985	5.879	1988	46 155	5.3	1988	1 614	0.2
1985-90	53.10	56.40	1985-90	4.990	...	...	...	...	...	...
1985-90	61.90[1]	66.10[1]	1981	5.881	1983	19 875	4.9	1983	1 520	0.4
1985-90	70.40[1]	74.80[1]	1985-90	2.650[1]	1990	13 037	5.4	1989	672	0.3
1975	67.00	73.50	1985	2.134	1988	1 558	4.6	1988	374	1.1
1979	62.10	66.00	1985	4.157	1990	663 344	7.7	1990	54 012	0.6
...	...	...	1982	1.623	1986	40	3.4	...	...	...
1981	71.13	75.75	...	...	1989	1 219	6.5	1989	413	2.2
1985-90	62.00[1]	64.60[1]	1985-90	5.500[1]	1986	11 919	3.5	1990	866	0.2
1985-90	70.15	74.10	1985	3.140	1990	12 467	5.2	1989	1 872	0.8
1986	71.12	79.44	1988	2.408	1989	31 642	8.9	1989	13 838	3.9
1988	65.87	70.98	1986	2.883	1977	172	3.9	1977	8	0.2
1986	68.00	74.80	1986	3.823	1989	396	2.7	1989	44	0.3
...	...	...	...	...	1983	37	6.2	1983	11	1.8
...	...	...	1980	3.873	1986	425	3.8	1980	19	0.2
1980	66.88	71.62	1984	3.013	1988	7 327	6.0	1987	1 163	1.0
...	...	...	...	...	1980	39	5.3	1980	10	1.3

19
Vital statistics summary and expectation of life at birth [*cont.*]
Aperçu des statistiques de l'état civil et espérance de vie à la naissance [*suite*]

Country or area Pays ou zone	Year Année	Live births Naissances vivantes		Deaths Décès		Natural increase Accroisse- ment naturel	Year Année	Infant deaths Décès d'enfants de moins d'un an	
		Number Nombre	Rate Taux	Number Nombre	Rate Taux (000)			Number Nombre	Rate Taux (000)
United States Etats-Unis	1990	4 179 000	16.7	2 162 000	8.6	8.1	1990	38 100	9.1
United States Virgin Is. Iles Vierges américaines	1987	2 375	22.4	558	5.3	17.1	1987	46	19.4
America, South · Amérique du Sud									
Argentina Argentine	1988	653 576	20.7	263 655	8.4	12.4	1988	16 824	25.7
Bolivia Bolivie	1975	...	46.6[3]	...	18.0[3]	28.6[3]	1985-90	...	110.0[1]
Brazil Brésil	1985-90	...	28.6[1]	...	7.9[1]	20.7[1]	1985-90	...	63.0[1]
Chile Chile	1989	303 798	23.4	75 453	5.8	17.6	1989	5 183	17.1
Colombia Colombie	1985-90	...	27.4[1]	...	6.1[1]	21.3[1]	1985-90	...	40.0[1]
Ecuador Equateur	1985-90	...	32.9[1]	...	7.4[1]	25.5[1]	1985-90	...	63.0[1]
Falkland Is. (Malvinas) Iles Falkland (Malvinas)	1988	18	7.5	28	11.7	-4.2	1981	-	-
French Guiana Guyane française	1986	2 392	27.9	491	5.7	22.2	1986	53	22.2
Guyana Guyana	1978	23 200	28.3	6 000	7.3	21.0	1985-90	...	56.0[1]
Paraguay Paraguay	1985-90	...	34.8[1]	...	6.6[1]	28.2[1]	1985-90	...	42.0[1]
Peru Pérou	1990	734 000	32.9	186 000	8.3	24.5	1990	80 700	109.9
Suriname Suriname	1982	11 295	31.0	2 377	6.5	24.5	1985-90	...	33.0[1]
Uruguay Uruguay	1989	55 324	18.0	29 658	9.6	8.3	1989	1 209	21.9
Venezuela Venezuela	1989	529 015	27.5	84 761	4.4	23.1	1989	12 322	23.3
Asia · Asie									
Afghanistan Afghanistan	1979[2]	627 619	40.4	290 974	18.7	21.6	1979[2]	113 954	181.6
Armenia Arménie	1989	75 250	22.9	20 853	6.3	16.5	1989	1 534	20.4
Azerbaijan Azerbaïdjan	1989	181 631	25.6	44 016	6.2	19.4	1989	4 749	26.1
Bahrain Bahreïn	1985-90	...	28.4[1]	...	3.8[1]	24.6[1]	1985-90	...	16.0[1]
Bangladesh Bangladesh	1985-90	...	42.2[1]	...	15.5[1]	26.7[1]	1985-90	...	119.0[1]
Bhutan[1] Bhoutan[1]	1985-90	...	38.3	...	16.8	21.5	1985-90	...	128.0
Brunei Darussalam Brunéi Darussalam	1988	6 881	28.5	777	3.2	25.3	1988	46	6.7
Cambodia[1] Cambodge[1]	1985-90	...	41.4	...	16.6	24.8	1985-90	...	130.0
China[1] Chine[1]	1985-90	...	21.2	...	6.7	14.5	1985-90	...	32.0
Cyprus Chypre	1985-90	...	18.6[1]	...	8.2[1]	10.4[1]	1985-90	...	12.0[1]
East Timor[1] Timor oriental[1]	1985-90	...	43.8	...	21.5	22.3	1985-90	...	166.0
Georgia Géorgie	1989	91 138	16.7	47 077	8.6	8.1	1989	1 787	19.6
Hong Kong Hong-kong	1990	67 911	11.7	28 688	4.9	6.8	1990	417	6.1
India Inde	1989	...	30.5[4]	...	10.2[4]	20.3[4]	1989	...	91.0[4]

	Expectation of life at birth Esp. de vie à la naissance		Fertility Fécondité			Marriages Mariages			Divorces		
Year Année	Male Masculin	Female Féminin	Year Année	Rate Taux	Year Année	Number Nombre	Rate Taux (000)	Year Année	Number Nombre	Rate Taux (000)
1988	71.50	78.30	1988	1.923	1990	2 448 000	9.8	1990	1 175 000	4.7
...	...	...	1980	3.124	1987	1 906	18.0	1987	263	2.5
1980-81	65.48	72.70	1985	2.899	1983	185 578	6.3	...	...	...
1985-90	50.90[1]	55.40[1]	1985-90	6.060[1]	1980	26 990	4.8	...	...	...
1985-90	62.30[1]	67.60[1]	1985-90	3.400[1]	1988	951 236	6.6	1988	33 437	0.2
1985-90	68.05	75.05	1990	2.660	1989	103 710	8.0	1988	5 413	0.4
1980-85	63.39	69.23	1985-90	3.130[1]	1986	70 350	2.4	...	...	...
1985	63.39	67.59	1985	4.650	1988	66 468	6.5	1988	4 424	0.4
...	...	...	...	...	1981	11	5.9	1981	7	3.8
...	...	...	...	...	1986	332	3.9	1986	34	0.4
1985-90	60.40[1]	66.10[1]	1985-90	2.770[1]	...	...	...	...	...	...
1980-85	64.42	68.51	1985-90	4.580[1]	1987	17 741	4.5	...	...	...
1980-85	56.77	66.50	1992	3.970	1982	109 200	6.0	...	...	...
1985-90	66.40[1]	71.30[1]	1985-90	2.970[1]	...	...	...	...	...	...
1984-86	68.43	74.88	1985-90	2.430	1987	22 728	7.5	1987	4 611	1.5
1985	66.68	72.80	1989	3.300	1989	111 970	5.8	1989	21 876	1.1
1985-90	41.00[1]	42.00[1]	1979	7.603[2]	...	...	...	...	...	...
1989	69.00	74.70	...	...	1989	27 257	8.3	1989	4 134	1.3
1989	66.60	74.20	...	...	1989	71 874	10.1	1989	11 436	1.6
1986-91	66.83	69.43	1985-90	4.140[1]	1989	3 033	6.2	1989	726	1.5
1988	56.91	55.97	1985-90	5.530[1]	1988	1 183 710	11.3	...	...	...
1985-90	48.60	47.10	1985-90	5.530	...	...	...	...	...	...
1981	70.13	72.69	1988	3.183	1987	1 845	7.9	1988	190	0.8
1985-90	47.00	49.90	1985-90	4.710	...	...	...	...	...	...
1985-90	68.00	70.90	1985-90	2.450	...	...	...	...	...	...
1985-90	73.92	78.33	1989	2.339	1990	6 500	9.3	1990	350	0.5
1985-90	41.60	43.40	1985-90	5.410	...	...	...	...	...	...
1989	68.10	75.70	...	...	1989	38 288	7.0	1989	7 358	1.4
1989	74.25	80.05	1989	1.231	1990	47 168	8.1	1990	5 551	1.0
1976-80	52.50	52.10	1987	4.200[4]	...	...	...	...	...	...

19
Vital statistics summary and expectation of life at birth [*cont.*]
Aperçu des statistiques de l'état civil et espérance de vie à la naissance [*suite*]

Country or area Pays ou zone	Year Année	Live births Naissances vivantes Number Nombre	Rate Taux	Deaths Décès Number Nombre	Rate Taux (000)	Natural increase Accroisse- ment naturel	Year Année	Infant deaths Décès d'enfants de moins d'un an Number Nombre	Rate Taux (000)
Indonesia Indonésie	1985-90	...	28.6[1]	...	9.4[1]	19.2[1]	1985-90	...	75.0[1]
Iran, Islamic Republic of Iran, Rép. islamique d'	1974-75	...	42.5[5]	...	11.5[5]	31.0	1974-75	...	108.1[1]
Iraq Iraq	1985-90	...	42.6[1]	...	7.8[1]	34.8[1]	1985-90	...	69.0[1]
Israel[7] Israël[7]	1989	100 757	22.3	28 600	6.3	16.0	1990	994	9.7
Japan Japon	1990	1 228 000	9.9	826 000	6.7	3.3	1990	5 500	4.5
Jordan Jordanie	1985-90	...	38.9[1]	...	6.4[1]	32.5[1]	1985-90	...	44.0[1]
Kazakhstan Kazakhstan	1989	382 269	23.0	126 378	7.6	15.4	1989	9 949	26.0
Korea, Dem. People's Rep.[1] Corée, Rép. pop. dém. de[1]	1985-90	...	23.5	...	5.4	18.1	1985-90	...	28.0
Korea, Republic of Corée, République de	1989[8]	613 240	14.5	230 207	5.4	9.0	1985-90	...	25.0[1]
Kuwait Koweït	1987	50 198	26.8	4 113	2.2	24.6	1986	841	15.6
Kyrgyzstan Kirghizistan	1989	131 508	30.4	31 156	7.2	23.2	1989	4 258	32.4
Lao People's Dem. Rep.[1] Rép. dém. populaire Lao[1]	1985-90	...	45.1	...	16.9	28.2	1985-90	...	110.0
Lebanon[1] Liban[1]	1985-90	...	31.7	...	8.7	23.0	1985-90	...	48.0
Macau Macao	1989	7 568	16.9	1 516	3.4	13.5	1990	58	8.4
Malaysia[1] Malaisie[1]	1985-90	...	31.9	...	5.6	26.3	1985-90	...	24.0
Peninsular Malaysia Malaisie Péninsulaire	1988	407 801	29.2	68 930	4.9	24.2	1988	5 729	14.0
Sabah Sabah	1986	51 410	40.4	5 114	4.0	36.4	1986	1 089	21.2
Sarawak Sarawak	1986	41 702	27.5	5 184	3.4	24.1	1986	426	10.2
Maldives Maldives	1988	8 237	41.2	1 526	7.6	33.6	1990	290	33.6
Mongolia Mongolie	1985-90	...	36.1[1]	...	8.8[1]	27.3[1]	1985-90	...	68.0[1]
Myanmar Myanmar	1985-90	...	30.6[1]	...	9.7[1]	20.9[1]	1985-90	...	70.0[1]
Nepal Népal	1985-90	...	39.6[1]	...	14.8[1]	24.8[1]	1985-90	...	128.0[1]
Oman[1] Oman[1]	1985-90	...	45.6	...	7.8	37.8	1985-90	...	40.0
Pakistan[6] Pakistan[6]	1988	3 194 926	30.3	852 341	8.1	22.2	1988	344 058	107.7
Philippines Philippines	1985-90	...	33.2[1]	...	7.7[1]	25.5[1]	1985-90	...	45.0[1]
Qatar Qatar	1989	10 908	25.8	847	2.0	23.8	1985-90	...	31.0[1]
Saudi Arabia[1] Arabie saoudite[1]	1985-90	...	42.1	...	7.6	34.5	1985-90	...	71.0
Singapore Singapour	1989	47 735	17.8	13 946	5.2	12.6	1989	360	7.5
Sri Lanka Sri Lanka	1989	357 964	21.3	104 590	6.2	15.1	1988	6 658	19.4
Syrian Arab Republic République arabe syrienne	1985-90	...	44.6[1]	...	7.0[1]	37.6[1]	1985-90	...	48.0[1]
Tajikistan Tadjikistan	1989	200 430	38.7	33 395	6.4	32.3	1989	8 673	43.3

| Year Année | Expectation of life at birth Esp. de vie à la naissance | | Fertility Fécondité | | Marriages Mariages | | | Divorces | | |
	Male Masculin	Female Féminin	Year Année	Rate Taux	Year Année	Number Nombre	Rate Taux (000)	Year Année	Number Nombre	Rate Taux (000)
1985-90	58.50[1]	62.00[1]	1985-90	3.480[1]	1986	1 249 034	7.4	1986	131 886	0.8
1976	55.75[6]	55.04[6]	1985-90	5.220[1]	1987	346 647	6.8	1987	33 433	0.7
1985-90	63.00[1]	64.80[1]	1985-90	6.350[1]	1988	145 885	8.5	1981	1 476	0.1
1988	73.87	77.44	1989	3.031	1989	31 700	7.0	1989	5 700	1.3
1989	75.91	81.77	1989	1.540	1990	721 000	5.8	1990	157 000	1.3
1985-90	64.20[1]	67.80[1]	1985-90	6.150[1]	1976	15 773	5.7	1976	2 638	0.9
1989	63.90	73.10	...	...	1989	165 380	10.0	1989	45 772	2.8
1985-90	66.20	72.70	1985-90	2.500	...	...	...	...	...	...
1989	66.92	74.96	1989	1.502	1989	309 872	7.3	1989	32 474	0.8
1985-90	71.20[1]	75.40[1]	1986	4.034	1988	10 283	5.3	1988	2 834	1.4
1989	64.30	72.40	...	...	1989	41 790	9.7	1989	8 231	1.9
1985-90	47.00	50.00	1985-90	6.690	...	...	...	...	...	...
1985-90	63.10	67.00	1985-90	3.790	...	...	...	...	...	...
1988	75.01	80.26	1988	1.502	1989	1 728	3.9	1989	70	0.2
1985-90	67.50	71.60	1985-90	4.000	...	...	...	...	...	...
1988	68.58	72.94	1988	3.591	1988	44 904	3.2	1976	221	...
...	...	...	...	...	...	...	...	...	...	...
...	...	...	1986	3.363	...	...	...	...	...	...
1985	62.20	59.48	...	...	1981	5 428	34.4	1981	4 010	25.4
1985-90	60.00[1]	62.50[1]	1985-90	5.000[1]	1989	15 600	7.5	1989	1 000	0.5
1978	58.93	63.66	1985-90	4.020[1]	...	...	...	...	...	...
1981	50.88	48.10	1985-90	5.940[1]	...	...	...	...	...	...
1985-90	62.20	65.80	1985-90	7.170	...	...	...	...	...	...
1976-78	59.04	59.20	1988	6.486	...	...	...	...	...	...
1989	62.50	66.10	1985-90	4.330[1]	1989	302 109	5.0	...	...	...
1985-90	66.90[1]	71.80[1]	1985-90	5.640[1]	1989	1 330	3.2	1989	406	1.0
1985-90	61.70	65.20	1985-90	7.170	...	...	...	...	...	...
1980	68.70	74.00	1988	1.975	1988	24 853	9.4	1988	2 536	1.0
1981	67.78	71.66	1985	2.969	1989	141 533	8.4	1988	2 732	0.2
1981	64.42	68.05	1985-90	6.760[1]	1989	102 557	8.8	1989	8 568	0.7
1989	66.80	71.70	...	...	1989	47 616	9.2	1989	7 576	1.5

19
Vital statistics summary and expectation of life at birth [*cont.*]
Aperçu des statistiques de l'état civil et espérance de vie à la naissance [*suite*]

Country or area Pays ou zone	Year Année	Live births Naissances vivantes		Deaths Décès		Natural increase Accroisse-ment naturel	Year Année	Infant deaths Décès d'enfants de moins d'un an	
		Number Nombre	Rate Taux	Number Nombre	Rate Taux (000)			Number Nombre	Rate Taux (000)
Thailand Thaïlande	1985-90	...	22.3[1]	...	7.7[1]	15.3[1]	1985-90	...	28.0[1]
Turkey Turquie	1985-90	...	29.2[1]	...	8.4[1]	20.8[1]	1985-90	...	76.0[1]
Turkmenistan Turkménistan	1989	124 992	34.9	27 609	7.7	27.2	1989	6 847	54.8
United Arab Emirates[1] Emirats arabes unis[1]	1985-90	...	22.8	...	3.8	19.0	1985-90	...	26.0
Uzbekistan Ouzbékistan	1989	668 807	33.3	126 862	6.3	26.9	1989	25 459	38.1
Viet Nam Viet Nam	1985-90	...	31.8[1]	...	9.5[1]	22.3[1]	1985-90	...	64.0[1]
Yemen Yémen	1990	577 781	51.2	239 001	21.2	30.0	...	...	...
Europe · Europe									
Albania Albanie	1989	78 862	24.7	18 168	5.7	19.0	1989	2 432	30.8
Andorra Andorre	1990	628	12.2	74	1.4	10.7	1990	2	3.2
Austria Austriche	1990	89 649	11.6	82 442	10.7	0.9	1990	708	7.9
Belarus Bélarus	1989	153 449	15.0	103 479	10.1	4.9	1989	1 835	12.0
Belgium Belgique	1988	118 764	12.0	104 551	10.6	1.4	1990	982	7.9
Bulgaria Bulgarie	1989	112 289	12.5	106 902	11.9	0.6	1989	1 614	14.4
Channel Islands Iles Anglo-Normandes	1989	1 761	12.4	1 465	10.3	2.1	1989	8	4.5
Guernsey Guernesey	1990	754	12.7	602	10.1	2.6	1990	1	1.3
Jersey Jersey	1989	1 074	13.0	827	10.0	3.0	1989	4	3.7
Czechoslovakia Tchécoslovaquie	1990	210 527	13.4	183 710	11.7	1.7	1990	2 388	11.3
Denmark Danemark	1990	63 554	12.4	60 979	11.9	0.5	1990	479	7.5
Estonia Estonie	1990	22 308	14.2	19 530	12.4	1.8	1990	276	12.4
Faeroe Islands Iles Féroé	1988	870	18.4	410	8.7	9.7	1987	4	5.1
Finland Finlande	1990	65 639	13.2	50 125	10.1	3.1	1989	367	5.8
France France	1989	765 000	13.6	528 000	9.4	4.2	1989	5 628	7.4
Germany₰ · Allemagne₰ F. R. Germany R. f. Allemagne	1990	723 634	11.4	709 703	11.2	0.2	1989	5 074	7.4
former German D. R. anc. R. d. allemande	1989	198 922	12.0	205 709	12.4	-0.4	1989	1 504	7.6
Gibraltar Gibraltar	1989	530	17.3	219	7.1	10.1	1976	5	9.8
Greece Grèce	1989	101 149	10.1	92 717	9.2	0.8	1988	1 188	11.0
Hungary Hongrie	1990	122 251	11.6	141 791	13.4	-1.9	1990	1 815	14.8
Iceland Islande	1990	4 770	18.7	1 730	6.8	11.9	1990	19	4.0
Ireland Irlande	1990	52 825	15.1	31 765	9.1	6.0	1989	390	7.5
Isle of Man Ile de Man	1989	817	12.1	988	14.6	-2.5	1989	5	6.1

Year Année	Expectation of life at birth Esp. de vie à la naissance		Fertility Fécondité		Marriages Mariages			Divorces		
	Male Masculin	Female Féminin	Year Année	Rate Taux	Year Année	Number Nombre	Rate Taux (000)	Year Année	Number Nombre	Rate Taux (000)
1985-86	63.82	68.85	1985-90	2.600[1]	1989	406 134	7.3	1986	36 602	0.7
1985-90	62.50[1]	65.80[1]	1985-90	3.690[1]	1989	450 763	7.9	1989	25 376	0.4
1989	61.80	68.40	...	...	1989	34 890	9.8	1989	4 940	1.4
1985-90	68.60	72.90	1985-90	4.820	...	...	...	...	...	...
1989	66.00	72.10	...	...	1989	200 681	10.0	1989	29 953	1.5
1979	63.66	67.89	1985-90	4.100[1]	...	...	...	...	...	...
...	...	...	...	...	...	...	...	...	...	...
1988-89	69.60	75.50	1989	3.072	1989	27 655	8.6	1989	2 628	0.8
...	...	...	...	...	1990	153	3.0	...	...	...
1989	72.09	78.78	1989	1.446	1990	45 156	5.9	1989	15 489	2.0
1989	66.80	76.40	...	...	1989	97 929	9.6	1989	34 573	3.4
1979-82	70.04	76.79	1983	1.572	1988	59 093	6.0	1986	18 316	1.9
1987-89	68.33	74.70	1989	1.862	1989	63 263	7.0	1989	12 611	1.4
...	...	...	...	...	1987	1 099	8.1	1989	370	2.6
...	...	...	...	...	1990	403	6.8	1990	196	3.3
...	...	...	...	...	1989	651	7.9	1989	186	2.2
1988	67.76	75.29	1989	1.929	1990	131 445	8.4	1990	40 922	2.6
1987-88	71.80	77.70	1988	1.560	1990	31 293	6.1	1989	15 132	2.9
1990	64.72	74.94	...	...	1990	11 774	7.5	1990	5 785	3.7
1981-85	73.30	79.60	1987	2.326	1988	260	5.5	1988	43	0.9
1987	70.66	78.68	1987	1.591	1990	24 150	4.8	1989	14 558	2.9
1988	72.33	80.46	1989	1.808	1989	281 000	5.0	1988	106 096	1.9
1985-87	71.81	78.37	1988	1.434	1990	414 048	6.5	1989	126 628	2.0
1987-88	69.81	75.91	1988	1.689	1989	130 989	7.9	1989	50 063	3.0
...	...	...	...	...	1989	754	24.6	1981	93	3.1
1980	72.15	76.35	1984	1.821	1989	59 955	6.0	1986	8 650	0.9
1989	65.44	73.79	1989	1.777	1990	66 304	6.3	1989	24 935	2.4
1988-89	75.23	79.93	1989	2.212	1990	1 150	4.5	1990	470	1.8
1985-87	71.01	76.70	1989	2.105	1989	17 769	5.1	...	...	...
...	...	...	...	...	1989	483	7.1	1989	190	2.8

19
Vital statistics summary and expectation of life at birth [*cont.*]
Aperçu des statistiques de l'état civil et espérance de vie à la naissance [*suite*]

Country or area Pays ou zone	Year Année	Live births Naissances vivantes		Deaths Décès		Natural increase Accroisse- ment naturel	Year Année	Infant deaths Décès d'enfants de moins d'un an	
		Number Nombre	Rate Taux	Number Nombre	Rate Taux (000)			Number Nombre	Rate Taux (000)
Italy Italie	1989	555 535	9.7	525 960	9.1	0.5	1989	4 889	8.8
Latvia Lettonie	1989	38 922	14.5	32 584	12.1	2.4	1989	438	11.3
Liechtenstein Liechtenstein	1989	373	13.4	172	6.2	7.2	1989	1	2.7
Lithuania Lituanie	1991	56 219	15.0	41 013	11.0	4.1	1991	806	14.3
Luxembourg Luxembourg	1989	4 665	12.4	3 984	10.6	1.8	1990	21	4.3
Malta Malte	1990	5 378	15.2	2 733	7.7	7.5	1990	61	11.3
Monaco Monaco	1979	489	19.5	578	23.1	-3.6	1980	1	1.9
Netherlands Pays-Bas	1990	197 905	13.2	128 790	8.6	4.6	1990	1 399	7.1
Norway Norvège	1990	60 594	14.3	45 376	10.7	3.6	1989	463	7.8
Poland Pologne	1990	546 129	14.3	388 440	10.2	4.1	1990	8 737	16.0
Portugal Portugal	1988	122 121	11.9	98 236	9.5	2.3	1988	1 595	13.1
Republic of Moldova Moldova, République de	1989	82 221	18.9	40 113	9.2	9.7	1989	1 705	20.7
Romania Roumanie	1990	314 746	13.6	247 086	10.7	2.9	1990	8 471	26.9
Russian Federation Fédération Russe	1989	2 160 559	14.6	1 583 743	10.7	3.9	1989	39 030	18.1
San Marino Saint-Marin	1989	231	10.1	173	7.6	2.5	1989	5	21.6
Spain Espagne	1988	415 844	10.7	318 848	8.2	2.5	1988	3 356	8.1
Sweden Suède	1990	124 000	14.5	95 000	11.1	3.4	1990	700	5.6
Switzerland Suisse	1990	84 000	12.5	63 800	9.5	3.0	1990	600	7.1
Ukraine Ukraine	1989	690 981	13.4	600 590	11.6	1.7	1989	9 039	13.1
United Kingdom Royaume-Uni	1989	777 285	13.6	657 733	11.5	2.1	1989	6 542	8.4
Yugoslavia Yougoslavie	1990	333 746	14.0	213 841	9.0	5.0	1990	6 743	20.2
Oceania · Océanie									
American Samoa Samoa américaines	1988	1 625	43.4	197	5.3	38.1	1988	17	10.5
Australia Australie	1989	250 853	14.9	124 232	7.4	7.5	1989	2 004	8.0
Christmas Island Ile Christmas	1985	36	15.8	2	0.9	14.9	1981	1	32.3
Cocos (Keeling) Islands Iles des Cocos (Keeling)	1986	12	19.8	2	3.3	16.5	...	...	...
Cook Islands Iles Cook	1988	430	24.3	94	5.3	19.0	1988	4	9.3
Fiji Fidji	1986	19 045	26.7	3 917	5.5	21.2	1987	189	9.8
French Polynesia Polynésie française	1989	5 364	27.9	1 021	5.3	22.6	1989	88	16.4
Guam Guam	1987	3 348	26.5	486	3.8	22.6	1987	40	11.9
Nauru Nauru	1976	158	19.8	36	4.5	15.3	1976	3	19.0

Year Année	Expectation of life at birth Esp. de vie à la naissance		Fertility Fécondité			Marriages Mariages			Divorces		
	Male Masculin	Female Féminin	Year Année	Rate Taux	Year Année	Number Nombre	Rate Taux (000)	Year Année	Number Nombre	Rate Taux (000)
1985	72.01	78.61	1988	1.326	1989	311 613	5.4	1988	25 092	0.4
1989	65.30	75.20	...	...	1989	24 496	9.1	1989	11 249	4.2
1980-84	66.07	72.94	1987	1.446	1989	315	11.3	1989	29	1.0
1990	66.55	76.22	...	...	1991	34 241	9.2	1989	12 295	3.3
1985-87	70.61	77.87	1989	1.520	1989	2 184	5.8	1989	855	2.3
1989	73.79	78.04	1989	2.109	1990	2 609	7.4	...	...	...
...	...	...	...	...	1979	177	7.1	1979	54	2.2
1988-89	73.66	80.23	1989	1.559	1990	95 653	6.4	1990	28 500	1.9
1989	73.34	79.85	1989	1.889	1989	20 755	4.9	1989	9 238	2.2
1988	67.15	75.67	1988	2.126	1990	255 369	6.7	1990	42 436	1.1
1979-82	68.35	75.20	1988	1.534	1988	71 098	6.9	1988	9 022	0.9
1989	65.50	72.30	...	...	1989	39 928	9.2	1989	12 401	2.9
1987-89	66.51	72.41	1989	2.200	1990	192 662	8.3	1990	32 966	1.4
1989	64.20	74.50	...	...	1989	1 384 307	9.4	1989	582 500	3.9
1977-86	73.16	79.12	1989	3.740	1989	169	7.4	1989	22	1.0
1980-82	72.52	78.61	1986	1.544	1988	214 898	5.5	1982	21 544	0.6
1988	74.15	79.96	1988	1.961	1990	40 000	4.7	1990	19 000	2.2
1987-89	73.90	80.70	1988	1.570	1990	46 600	6.9	1990	13 000	1.9
1989	66.10	75.20	...	...	1990	482 800	9.3	1989	193 676	3.7
1986-89	72.15	77.88	1989	1.810	1989	392 042	6.8	1989	163 942	2.9
1988-90	68.64	74.48	1989	1.873	1990	149 498	6.3	1990	19 418	0.8
...	...	...	...	...	1988	342	9.1	1988	42	1.1
1989	73.30	79.55	1989	1.846	1989	117 176	7.0	1989	41 383	2.5
...	...	...	...	...	1985	32	14.0	...	...	...
...	...	...	...	...	1985	3	4.8	...	...	...
...	...	...	...	...	1988	122	6.9	1976	8	0.4
1976	60.72	63.87	1986	3.024	1986	5 997	8.4	1979	410	0.7
...	...	...	...	...	1989	1 093	5.7	...	...	...
1979-81	69.53	75.59	...	...	1987	1 512	12.0	1987	1 279	10.1
...	...	...	...	...	...	...	...	...	...	...

19
Vital statistics summary and expectation of life at birth [*cont.*]
Aperçu des statistiques de l'état civil et espérance de vie à la naissance [*suite*]

Country or area Pays ou zone	Year Année	Live births Naissances vivantes		Deaths Décès		Natural increase Accroisse- ment naturel	Year Année	Infant deaths Décès d'enfants de moins d'un an	
		Number Nombre	Rate Taux	Number Nombre	Rate Taux (000)			Number Nombre	Rate Taux (000)
New Caledonia									
Nouvelle-Calédonie	1987	3 881	24.5	864	5.5	19.0	1987	59	15.2
New Zealand									
Nouvelle-Zélande	1989	58 091	17.5	27 042	8.2	9.4	1989	592	10.2
Niue									
Nioué	1987	50	20.9	13	5.4	15.5	1986	2	41.7
Norfolk Is.									
Ile Norfolk	1981	20	10.0	14	7.0	3.0	...	...	...
Northern Mariana Islands									
Iles Mariannes du Nord	1989	989	39.5	122	4.9	34.6	1989	2	2.0
Papua New Guinea[1]									
Papouasie-Nouvelle-Guinée[1]	1985-90	...	34.2	...	11.6	22.6	1985-90	...	59.0
Pitcairn									
Pitcairn	1978	1	16.1	1	16.1	-	...	...	...
Samoa									
Samoa	1982-83	...	31.0[9]	...	7.4[9]	23.6[9]	1982-83	...	33.0[9]
Solomon Islands									
Iles Salomon	1980-84	...	42.0[2]	...	10.0[2]	32.0[2]	...	...	...
Tokelau									
Tokélau	1983	35	21.9	8	5.0	16.9	1983	1	28.6
Tonga									
Tonga	1985	2 810	28.9	343	3.5	25.4	1985	14	5.0
Wallis and Futuna Islands									
Iles Wallis et Futuna	1978	370	41.1	...	...	...	1978	15	40.5

Source:
Demographic statistics database of the Statistical Division of the
United Nations Secretariat.

ℊ All data shown pertain to Germany prior to 3 October 1990 are
indicated separately for the Federal Republic of Germany and the
former German Democratic Republic based on their respective
territories at the time indicated. Where data for united Germany
(3 October 1990 and thereafter) are not available, available data
are shown separately under the designations Federal Republic of
Germany and former German Democratic Republic and pertain to
the territorial boundaries prior to 3 October 1990. For detailed
explanatory notes on data pertaining to Germany, see Annex I-
Country Nomenclature.

1 Estimate(s) for 1985-1990 prepared by the Population Division
of the United Nations.
2 Estimate(s) based on results of the population census.
3 Based on National Sample Survey.
4 Based on a Sample Registration Scheme.
5 Based on the results of the Population Growth Survey, second
survey year.
6 Based on the results of the Population Growth Survey.
7 Including data for East Jerusalem and Israeli residents in certain
other territories under occupation by Israeli military forces since
June 1967.
8 Based on the results of the continuous Demographic Sample
survey.
9 Estimate(s) based on results of a sample survey.

Source:
Base de données pour les statistiques démographiques de la Division de
statistique du Secrétariat de l'ONU.

ℊ Toutes les données se rapportant à l'Allemagne avant le 3 octobre 1990
figurent dans deux rubriques séparées basées sur les territoires
respectifs de la République fédérale d'Allemagne et l'ancienne
République démocratique allemande selon la période indiquée. En
l'absence de données pour l'Allemagne unifiée (à compter du 3 octobre
1990), les données disponibles sont fournies séparément sous les rubriques
République fédérale d'Allemagne et ancienne République démocratique
allemande et se rapportent aux limites territoriales antérieures au 3
octobre 1990. Pour les notes explicatives en détail sur les données
concernant l'Allemagne, voir Annexe I - Nomenclature des pays.

1 Estimation(s) pour 1985-1990 établie(s) par la Division de la
population de l'Organisation des Nations Unies.
2 Estimation(s) fondée(s) sur les résultats du recensement de la
population.
3 D'après l'enquête nationale par sondage.
4 D'après le Programme d'enregistrement par sondage.
5 D'après les résultats de la "Population Growth Survey," deuxième
année de l'enquête.
6 D'après les résultats de la "Population Growth Survey."
7 Y compris les données pour Jérusalem-Est et les résidents israéliens
dans certains autres territoires occupés depuis juin 1967 par les forces
armées israéliennes.
8 D'après les résultats de l'enquête démographique par sondage continue.
9 Estimation(s) fondée(s) sur les résultats d'un enquête par sondage.

	Expectation of life at birth Esp. de vie à la naissance		Fertility Fécondité			Marriages Mariages				Divorces	
Year Année	Male Masculin	Female Féminin	Year Année	Rate Taux	Year Année	Number Nombre	Rate Taux (000)	Year Année	Number Nombre	Rate Taux (000)
...	...	...	...	...	1986	780	5.1	1986	158	1.0
1987-89	71.57	77.59	1989	2.101	1989	22 733	6.9	1989	8 555	2.6
...	...	...	...	...	1987	10	4.2	1981	3	0.9
...	...	...	...	...	1981	16	8.0	...	...	...
...	...	...	...	...	1989	713	28.5	1986	62	2.9
1985-90	53.20	54.70	1985-90	5.250	...	...	...	...	...	...
...	...	...	...	...	...	...	...	...	...	...
1976	61.00	64.30	1977	3.666	1981	656	4.2	1980	49	0.3
1980-84	59.90	61.40	1980	6.400	...	...	...	...	...	...
...	...	...	...	...	1983	4	2.5	...	...	...
...	...	...	...	...	1985	645	6.6	1985	63	0.6
...	...	...	...	...	...	...	...	...	...	...

20
Selected indicators of life expectancy, child-bearing and mortality
Choix d'indicateurs de l'espérance de vie, de la maternité et de la mortalité

| Country or area
Pays or zone | Year
Année | Life expectancy
at birth (years)
Espérance de vie
à la naissance
(en années) | | Infant
mortality
rate
Taux de
mortalité
infantile | Total
fertility
rate
Taux de
fécondité | Year
Année | Child mortality
rate
Taux
de mortalité
juvénile | | Maternal mortality
per 100000 live births
Mort. lié à la mater-
nité pour 100000
naissances vivantes | |
		M	F				M	F	Year[1] Année[1]	Rate Taux
Africa · Afrique										
Algeria	1980	56.5	58.5	112	7.2	...	...	...	1978	136
Algérie	1990	63.1	65.0	74	5.4	1982[2][3]	12.5	12.8	1980/88	130
Angola	1980	38.5	41.6	160	6.8	...	...	...	...	...
Angola	1990	42.4	45.6	137	7.2	...	...	...	...	...
Benin	1980	40.4	43.6	108	7.1	...	...	...	1975	170
Bénin	1990	43.9	47.1	91	7.1	...	...	...	...	...
Botswana	1980	51.0	56.0	83	7.0	...	...	...	1980	250[4]
Botswana	1990	55.5	61.5	67	5.5	...	...	...	...	...
Burkina Faso	1980	41.6	44.9	162	6.5	...	...	...	...	...
Burkina Faso	1990	45.3	48.8	126	6.5	...	...	...	1980/88	810
Burundi	1980	44.4	47.6	127	6.8	...	...	...	...	...
Burundi	1990	46.7	50.1	112	6.8	...	...	...	...	...
Cameroon	1980	47.0	50.0	102	6.5	...	...	...	1978	300
Cameroun	1990	52.0	55.0	74	6.1	...	...	...	...	...
Cape Verde	1980	58.5	61.5	70	6.7	...	...	...	1975	134
Cap-Vert	1990	65.0	67.0	54	4.8	1985	21.1[5]	19.3[5]	1980	107
Central African Rep.	1980	42.0	47.0	122	5.9	...	...	...	(1983)	600
Rép. centrafricaine	1990	46.0	51.0	109	6.2	...	...	...	1980/88	600
Chad	1980	39.5	42.6	154	5.9	...	...	...	...	...
Tchad	1990	43.9	47.1	132	5.9	...	...	...	1980/88	960
Comoros	1980	49.5	50.5	120	7.1	...	...	...	...	...
Comores	1990	53.5	54.5	99	7.1	...	...	...	...	...
Congo	1980	46.1	51.3	88	6.3	...	...	...	...	...
Congo	1990	49.4	54.7	84	6.3	...	...	...	1980/88	900
Côte d'Ivoire	1980	46.4	49.7	117	7.4	...	...	...	...	...
Côte d'Ivoire	1990	50.3	53.7	98	7.4	...	...	...	...	...
Djibouti	1980	41.4	44.6	143	6.6	...	...	...	...	...
Djibouti	1990	45.4	48.6	122	6.6	...	...	...	...	...
Egypt	1980	52.9	55.3	131	5.3	1976	15.6	19.1	1980	93
Egypte	1990	57.9	60.3	65	4.5	1986	7.0	7.6	1980/88	318
Equatorial Guinea	1980	40.4	43.6	149	5.7	...	...	...	...	...
Guinée équatoriale	1990	44.4	47.6	127	5.9	...	...	...	...	...
Ethiopia	1980	40.4	43.6	149	7.0	...	...	...	...	...
Ethiopie	1990	43.4	46.6	132	7.0	...	...	...	...	...
Gabon	1980	45.4	48.7	122	4.4	...	...	...	...	...
Gabon	1990	49.9	53.2	103	5.0	...	...	...	...	...
Gambia	1980	37.5	40.6	167	6.5	...	...	...	...	...
Gambie	1990	41.4	44.6	143	6.5	...	...	...	...	...
Ghana	1980	49.3	52.8	103	6.5	...	...	...	1978	484
Ghana	1990	52.3	55.8	90	6.4	...	...	...	1980/88	1000
Guinea	1980	38.3	39.3	167	7.0	...	...	...	...	...
Guinée	1990	42.0	43.0	145	7.0	...	...	...	...	...
Guinea-Bissau	1980	36.0	39.1	176	5.6	...	...	...	...	...
Guinée-Bissau	1990	39.9	43.1	151	5.8	...	...	...	...	...
Kenya	1980	51.5	55.5	88	8.1	...	...	...	...	...
Kenya	1990	56.0	59.9	72	6.8	...	...	...	1977	168
Lesotho	1980	50.5	55.5	121	5.7	...	...	...	...	...
Lesotho	1990	55.5	60.5	89	5.0	...	...	...	...	...
Liberia	1980	48.0	51.0	167	6.8	...	...	...	...	...
Libéria	1990	52.0	54.0	142	6.8	...	...	...	...	...
Libyan Arab Jamahiriya	1980	54.1	57.5	107	7.4	...	...	...	1978	80
Jamah. arabe libyenne	1990	59.1	62.5	82	6.9	...	...	...	...	...
Madagascar	1980	48.0	51.0	150	6.6	...	...	...	1979	300
Madagascar	1990	52.0	55.0	120	6.6	...	...	...	1980/88	240
Malawi	1980	42.4	43.7	177	7.6	1977	97.7	87.2	(1975)	250
Malawi	1990	44.6	46.2	149	7.6	...	...	...	1980/88	100[6]
Mali	1980	38.5	41.6	191	7.1	1976	44.8	41.9	...	...
Mali	1990	42.4	45.6	169	7.1	...	...	...	...	...

20
Selected indicators of life expectancy, child-bearing and mortality [*cont.*]
Choix d'indicateurs de l'espérance de vie, de la maternité et de la mortalité [*suite*]

Country or area Pays or zone	Year Année	Life expectancy at birth (years) Espérance de vie à la naissance (en années) M	F	Infant mortality rate Taux de mortalité infantile	Total fertility rate Taux de fécondité	Year Année	Child mortality rate Taux de mortalité juvénile M	F	Maternal mortality per 100000 live births Mort. lié à la maternité pour 100000 naissances vivantes Year[1] Année[1]	Rate Taux
Mauritania	1980	40.4	43.6	149	6.5	...	...	...	...	...
Mauritanie	1990	44.4	47.6	127	6.5	...	...	...	...	...
Mauritius	1980	62.4	67.6	38	3.1	...	...	...	1980	110
Maurice	1990	65.3	72.2	24	2.1	1989	0.9	0.9	1987	99
Morocco	1980	54.1	57.5	110	5.9	...	...	...	...	...
Maroc	1990	59.1	62.5	82	4.8	...	...	...	...	...
Mozambique	1980	41.9	45.1	160	6.5	...	...	...	...	...
Mozambique	1990	44.5	47.8	155	6.5	...	...	...	1980/88	300
Namibia	1980	50.0	52.5	102	6.0	...	...	...	...	...
Namibie	1990	55.0	57.5	80	6.0	...	...	...	...	...
Niger	1980	39.0	42.1	157	7.1	...	...	...	1980	135
Niger	1990	42.9	46.1	135	7.1	...	...	...	1982	420
Nigeria	1980	44.9	48.1	124	6.9	...	...	...	1981/83	1500
Nigéria	1990	48.8	52.2	105	6.9	...	...	...	1980/88	800
Réunion	1980	62.9	71.3	21	3.3	...	...	...	...	...
Réunion	1990	67.9	77.0	8	2.5	1987[3]	0.6	0.6	...	...
Rwanda	1980	43.4	46.6	133	8.5	...	...	...	1976	600[7]
Rwanda	1990	45.6	48.9	116	8.5	...	...	...	1980/88	210[7]
Sao Tome and Principe	1980	...	...	50	...	...	...	...	...	...
Sao Tomé-et-Principe	1990	...	...	72	...	...	...	...	1987	76.7
Senegal	1980	41.8	43.8	112	7.0	...	...	...	1975	530
Sénégal	1990	46.3	48.3	87	6.5	...	...	...	1980/88	600
Seychelles	1980	64.6	71.1	...	4.3	...	...	...	...	...
Seychelles	1990	...	...	18	2.9	...	...	...	...	...
Sierra Leone	1980	35.5	38.6	179	6.5	...	...	...	1981	450
Sierra Leone	1990	39.5	42.6	154	6.5	...	...	...	1986	450
Somalia	1980	40.4	43.6	149	6.7	...	...	...	1981	1100
Somalie	1990	43.4	46.6	132	6.7	...	...	...	...	...
South Africa	1980	53.0	59.0	72	5.1	...	...	...	...	...
Afrique du Sud	1990	57.5	63.5	58	4.4	...	...	...	1980/82	83[8]
Sudan	1980	43.9	46.4	131	6.7	...	...	...	1980/88	660
Soudan	1990	48.6	51.0	108	6.4	...	...	...	...	...
Swaziland	1980	47.6	52.3	108	6.8	...	...	...	...	...
Swaziland	1990	53.7	57.3	83	5.3	...	...	...	...	...
Togo	1980	46.4	49.7	117	6.6	...	...	...	...	...
Togo	1990	51.3	54.8	94	6.6	...	...	...	...	...
Tunisia	1980	59.6	60.6	88	5.7	1980	12.6[5]	11.2[5]	...	...
Tunisie	1990	64.9	66.4	49	3.9	...	...	...	...	...
Uganda	1980	45.4	48.6	114	6.9	...	...	...	...	...
Ouganda	1990	43.2	46.1	108	7.3	...	...	...	1984	300
United Rep. Tanzania	1980	47.3	50.7	125	6.8	...	...	...	1980/88	340[9]
Rép.-Unie de Tanzanie	1990	50.1	53.5	108	6.8	...	...	...	...	...
Zaire	1980	46.4	49.7	115	6.5	...	...	...	...	...
Zaïre	1990	49.8	53.3	102	6.7	...	...	...	...	...
Zambia	1980	47.7	51.0	94	7.2	...	...	...	1979	140
Zambie	1990	47.7	49.5	86	6.8	...	...	...	1980/88	150
Zimbabwe	1980	52.0	55.6	86	6.6	...	...	...	1979	480
Zimbabwe	1990	55.1	58.6	67	5.8	1982	5.3[5]	4.7[5]	...	...
America, North · Amérique du Nord										
Antigua and Barbuda										
Antigua-et-Barbuda	...	...	...	...	...	1975	2.1	1.3	...	...
Aruba										
Aruba	1980	68.3	75.4	...	...	...	...	...	...	...
Bahamas	1980	63.5	71.2	29	2.6	...	...	...	1981	38
Bahamas	1990	67.5	74.9	26	2.2	1985	7.4[5]	6.8[5]	1987	69
Barbados	1980	68.7	73.9	27	2.2	...	...	...	1980	24
Barbade	1990	71.9	76.9	12	1.6	1988	0.5	0.5	1988	27
Belize	1980	69.8	71.8	...	5.8	...	...	...	...	...
Belize	1990	...	...	21	5.2	1984	2.3	2.2	...	...
Bermuda	1980	68.8	76.3	...	...	...	...	...	...	...
Bermudes	1990	...	...	7	1.8	...	...	...	...	...

20
Selected indicators of life expectancy, child-bearing and mortality [*cont.*]
Choix d'indicateurs de l'espérance de vie, de la maternité et de la mortalité [*suite*]

Country or area Pays or zone	Year Année	Life expectancy at birth (years) Espérance de vie à la naissance (en années) M	F	Infant mortality rate Taux de mortalité infantile	Total fertility rate Taux de fécondité	Year Année	Child mortality rate Taux de mortalité juvénile M	F	Maternal mortality per 100000 live births Mort. lié à la maternité pour 100000 naissances vivantes Year[1] Année[1]	Rate Taux
Canada	1980	70.5	78.1	12	1.8	...	...	...	1980	8
Canada	1990	73.4	80.2	7	1.7	1988	0.5	0.4	1988	5
Cayman Islands										
Iles Caïmanes	1990	...	...	9	...	...	...	...	...	...
Costa Rica	1980	68.9	73.3	30	3.9	...	...	...	1980	24
Costa Rica	1990	73.1	77.7	16	3.4	1984	0.8	0.7	1988	18
Cuba	1980	71.5	74.8	22	2.1	...	...	...	1980	60
Cuba	1990	73.5	77.0	15	1.8	1988	0.8	0.7	1988	39
Dominica	1980	...	...	20	...	...	...	...	...	...
Dominique	1990			18	...	1984	0.5	0.3	...	...
Dominican Republic	1980	60.3	64.0	84	4.7	...	...	...	1979	84
République dominicaine	1990	63.9	68.1	65	3.8	1984	3.2	2.9	1984	66
El Salvador	1980	52.4	62.6	87	5.7	...	...	...	1980	71
El Salvador	1990	58.1	66.9	59	4.5	1984	2.9	2.7	1984	70
Greenland	1980	59.7	67.3	27	2.3	...	...	...	...	...
Groenland	1990	...	...	...	2.1					
Grenada										
Grenade	1980	...	...	15	...	1980	1.4	1.6	...	...
Guadeloupe	1980	66.4	73.4	25	3.1	...	...	...	...	...
Guadeloupe	1990	70.1	77.1	14	2.4	1985[3]	4.8[5]	4.2[5]	1978	106
Guatemala	1980	54.5	58.4	82	6.4	...	...	...	1980	91
Guatemala	1990	59.7	64.4	· 59	5.8	1985	22.2[5]	20.1[5]	1980/88	110
Haiti	1980	49.1	52.2	121	5.4	...	...	...	1974/78	367
Haïti	1990	53.1	56.4	97	5.0	...	...	...	1980/88	230
Honduras	1980	55.8	59.6	90	6.6	...	...	...	1979	82
Honduras	1990	61.9	66.1	68	5.6	1983	2.8	2.9	1980/88	50
Jamaica	1980	68.0	72.3	26	4.0	...	...	...	1977	50
Jamaïque	1990	70.4	74.8	17	2.7	1982	1.4	1.5	1981/83	106
Martinique	1980	68.5	75.0	22	2.7	...	...	...	1975	27
Martinique	1990	72.0	78.7	11	2.1	1985	0.3	0.5	1987	95
Mexico	1980	62.7	68.4	58	5.0	...	...	...	1980	94
Mexique	1990	65.7	72.1	41	3.6	1983	2.3	2.3	1986	65
Netherlands Antilles										
Antilles néerlandaises	1990	...	...	14	...	1981	6.5[5]	5.9[5]	...	...
Nicaragua	1980	55.3	60.0	97	6.4	1975	3.4	3.3	1978	65
Nicaragua	1990	59.0	66.0	71	5.5	...	...	...	1980/88	47
Panama	1980	67.6	70.9	32	4.1	...	...	...	1980	72
Panama	1990	70.2	74.1	23	3.1	1989	1.4	1.3	1987	38
Puerto Rico	1980	70.2	77.0	20	2.7	...	...	...	1980	8
Porto Rico	1990	70.9	77.7	14	2.2	1988	0.6	0.4	1989	20
St. Kitts and Nevis										
Saint-Kitts-et-Nevis	1990	65.9	71.0	24	2.9	1984	0.5	2.1	...	...
Saint Lucia	1980	...	...	29	4.7	...	...	...	...	...
Sainte-Lucie	1990	68.0	74.8	18	3.8	1981	1.6	1.6	...	...
St. Vincent and the Grenadines	1980	...	...	49[10]	3.9	...	...	...	...	...
St.-Vincent-et-Grenadines	1990	...	...	22	...	1985	1.3	0.7	...	...
Trinidad and Tobago	1980	64.7	69.7	38	3.4	...	...	...	1982	43
Trinité-et-Tobago	1990	67.7	72.7	24	3.0	1983	0.9	0.9	1983	54
United States	1980	69.4	77.2	14	1.8	...	...	...	1980	9
Etats-Unis	1990	71.6	78.5	10	1.9	1989	0.5	0.4	1988	8
US Virgin Islands	1980	...	...	23	3.1	...	...	...	...	...
Iles Vierges américaines	1990	...	...	19	...	1980	0.8	0.6	...	...
America, South · Amérique du Sud										
Argentina	1980	65.4	72.1	41	3.4	...	...	...	1980	70
Argentine	1990	67.3	74.0	32	3.0	1982	1.6	1.3	1985	59
Bolivia	1980	48.8	53.3	131	6.2	1976	8.7	7.8	1973/77	480
Bolivie	1990	56.6	61.2	98	5.0	...	...	...	...	...
Brazil	1980	59.5	64.3	79	4.2	...	...	...	1980	92
Brésil	1990	62.3	67.6	63	3.2	1988	8.3[5]	6.4[5]	1980/88	120
Chile	1980	63.9	70.6	47	2.9	...	...	...	1980	75
Chili	1990	68.1	75.1	18	2.7	1989	1.0[5]	0.7[5]	1987	48

20
Selected indicators of life expectancy, child-bearing and mortality [*cont.*]
Choix d'indicateurs de l'espérance de vie, de la maternité et de la mortalité [*suite*]

Country or area Pays or zone	Year Année	Life expectancy at birth (years) Espérance de vie à la naissance (en années) M	F	Infant mortality rate Taux de mortalité infantile	Total fertility rate Taux de fécondité	Year Année	Child mortality rate Taux de mortalité juvénile M	F	Maternal mortality per 100000 live births Mort. lié à la maternité pour 100000 naissances vivantes Year[1] Année[1]	Rate Taux
Colombia	1980	61.8	66.3	59	4.1	...	...	...	1981	126
Colombie	1990	65.5	71.1	40	2.9	1985[11]	2.3	2.1	1980/88	110
Ecuador	1980	59.7	63.2	82	5.4	...	...	...	1981[12]	186
Equateur	1990	63.4	67.6	63	4.1	1985[12]	3.5	3.3	1988[12]	156
French Guiana Guyane française	1990	...	...	22	...	1984	1.4	0.5	...	...
Guyana	1980	58.3	63.2	67	3.9	...	...	...	...	...
Guyana	1990	60.4	66.1	56	2.8	1984	2.1	1.7	1976	153
Paraguay	1980	64.1	68.1	53	5.0	...	...	...	1980	469
Paraguay	1990	64.8	69.1	49	4.6	1985	3.3	3.2	1986	380
Peru	1980	55.2	58.8	105	5.4	...	...	...	1980[13]	108
Pérou	1990	59.5	63.4	88	4.0	1983[13]	5.6	5.7	1983[13]	89
Suriname	1980	62.8	67.7	44	4.2	...	...	...	1980	82
Suriname	1990	66.4	71.3	33	3.0	1985	2.0	2.2	1985	60
Uruguay	1980	66.4	73.2	42	2.9	...	...	...	1981	59
Uruguay	1990	68.9	75.3	24	2.4	1985	7.5[5]	5.9[5]	1987	28
Venezuela	1980	64.9	70.7	43	4.4	...	...	...	1980[13]	65
Venezuela	1990	66.7	72.8	36	3.5	1987[13]	6.2[5]	5.2[5]	1987[13]	55
Asia · Asie										
Afghanistan	1980	40.0	40.0	183	7.2	1979	28.9	25.8	1975	690
Afghanistan	1990	41.0	42.0	172	6.9	...	...	...	...	...
Bahrain	1980	65.6	69.5	38	5.2	...	...	...	1981	19
Bahreïn	1990	68.6	72.9	20	4.1	1989	4.1[5]	4.1[5]	1987	8
Bangladesh	1980	47.1	46.1	137	6.7	...	...	...	(1981)	3000
Bangladesh	1990	51.1	50.4	119	5.1	1981	14.1	15.8	(1985)	600
Bhutan	1980	41.6	43.1	167	6.0	...	...	...	1984	1710
Bhoutan	1990	45.6	46.6	143	5.9	...	...	...	...	...
Brunei Darussalam	1980	68.1	71.4	23	4.4	...	...	...	...	...
Brunéi Darussalam	1990	71.5	75.3	10	3.4	...	...	...	...	...
Cambodia	1980	30.0	32.5	263	4.1	...	...	...	...	...
Cambodge	1990	47.0	49.9	130	4.6	...	...	...	...	...
China	1980	65.5	66.2	42	2.9	...	...	...	...	...
Chine	1990	68.0	70.9	32	2.4	...	...	...	1980/88	44
Cyprus	1980	72.0	75.5	20	2.2	...	...	...	...	...
Chypre	1990	73.4	78.4	12	2.4	1989	2.3[5]	1.7[5]	...	...
East Timor	1980	30.0	32.5	254	4.3	...	...	...	...	...
Timor oriental	1990	41.7	43.4	166	5.4	...	...	...	...	...
Hong Kong	1980	70.5	76.8	13	2.3	...	...	...	1980[14]	5
Hong-kong	1990	74.3	79.8	7	1.4	1989[14]	0.3	0.3	1987[14]	4
India	1980	53.3	52.4	129	4.8	...	...	...	1978	480
Inde	1990	57.8	57.9	96	4.2	...	...	...	1985	340
Indonesia	1980	51.5	54.0	105	4.7	...	...	...	1976	300
Indonésie	1990	58.5	62.0	75	3.5	...	...	...	1980/88	450
Iran, Islamic Rep. of	1980	58.2	59.0	100	6.5	...	...	...	...	...
Iran, Rép. islamique d'	1990	65.0	65.5	52	6.5	1986	3.9	1.4	1980/88	120
Iraq	1980	60.5	62.3	84	6.6	1977	2.2	1.9	...	...
Iraq	1990	63.5	66.5	64	6.2	...	...	...	...	...
Israel	1980	71.4	74.9	18	3.4	...	...	...	1980[15]	5
Israël	1990	73.8	77.4	11	3.1	1988[15]	0.5	0.5	1987[15]	3
Japan	1980	72.8	78.2	9	1.8	...	...	...	1980[17]	21
Japon	1990	75.4	81.2	5	1.7	1989[16]	0.5	0.4	1989[17]	11
Jordan	1980	59.4	63.0	65	7.4	1979[18]	2.4	2.0	1976	50
Jordanie	1990	64.2	67.8	44	6.2	...	...	...	...	...
Korea, Dem.People's Rep.	1980	62.4	68.8	35	3.5	...	...	...	1981	41
Corée, Rép. pop. dém.	1990	66.2	72.7	28	2.5	...	...	...	...	...
Korea, Republic of	1980	62.4	68.8	35	2.8	...	...	...	1983	38
Corée, République de	1990	66.2	72.5	25	1.7	1989	1.1	1.0	1989	10
Kuwait	1980	67.5	71.7	34	5.9	...	...	...	1980	8
Koweït	1990	72.6	76.3	14	3.9	1986	0.1	0.7	1987	2
Lao People's Dem. Rep.	1980	42.1	45.0	135	6.7	...	...	...	1984	2
République dém. pop. lao	1990	47.0	50.0	110	6.7	...	...	...	...	...

20
Selected indicators of life expectancy, child-bearing and mortality [*cont.*]
Choix d'indicateurs de l'espérance de vie, de la maternité et de la mortalité [*suite*]

Country or area Pays or zone	Year Année	Life expectancy at birth (years) Espérance de vie à la naissance (en années) M	F	Infant mortality rate Taux de mortalité infantile	Total fertility rate Taux de fécondité	Year Année	Child mortality rate Taux de mortalité juvénile M	F	Maternal mortality per 100000 live births Mort. lié à la maternité pour 100000 naissances vivantes Year[1] Année[1]	Rate Taux
Lebanon	1980	63.1	67.0	48	4.3	...	...	...	...	...
Liban	1990	65.1	69.0	40	3.4	...	...	...	...	...
Macau	1980	...	...	...	...		...	...	1981	48
Macao	1990	75.0	80.3	9	1.5	1988[19]	2.9[5]	2.6[5]	1982	40
Malaysia	1980	63.5	67.1	34	4.2	...	...	...	...	...
Malaisie	1990	67.5	71.6	17	4.0	1988[3 20]	1.1	1.1	1981	59
Maldives	1980	55.9	53.2	96	7.0	...	...	...	...	...
Maldives	1990	62.2	59.5	66	6.5	1985	7.8	9.3	1987	646
Mongolia	1980	55.0	57.5	88	5.5	...	...	...	1978	140
Mongolie	1990	60.0	62.5	68	5.0	...	...	...	1980/88	100
Myanmar	1980	49.7	52.9	114	5.3	...	...	...	1978	120
Myanmar	1990	53.5	56.8	94	4.5	...	...	...	1985	135
Nepal	1980	46.5	45.0	138	6.5	...	...	...	1979	850
Népal	1990	51.5	50.3	110	6.0	...	...	...	1980/88	830
Oman	1980	53.8	56.1	95	7.2	...	...	...	...	...
Oman	1990	66.2	69.8	36	7.2	...	...	...	...	...
Pakistan	1980	52.0	51.0	130	7.0	1976	12.5	15.8	(1978)	700[4]
Pakistan	1990	56.5	56.5	109	6.8	...	...	...	(1984)	500[4]
Philippines	1980	58.3	61.5	62	5.0	...	...	...	1980	125
Philippines	1990	61.6	65.4	45	4.3	1988	5.2	4.7	1984	80
Qatar	1980	63.5	67.6	46	6.0	...	...	...	...	...
Qatar	1990	66.9	71.8	31	4.8	...	...	...	...	...
Saudi Arabia	1980	57.6	59.9	75	7.3	...	...	...	...	...
Arabie saoudite	1990	66.4	69.1	37	6.8	...	...	...	...	...
Singapore	1980	68.6	73.1	13	1.9	...	...	...	1980	5
Singapour	1990	70.8	76.4	8	1.7	1988	0.4	0.5	1987	7
Sri Lanka	1980	65.0	68.5	44	3.8	...	...	...	1979	80
Sri Lanka	1990	68.3	72.5	28	2.7	1985	6.9[5]	6.0[5]	1980/88	60
Syrian Arab Republic	1980	58.3	61.9	67	7.4	...	...	...	1980	7
Rép. arabe syrienne	1990	63.2	66.9	49	6.7	1984[21]	2.8	2.9	(1984)	280
Thailand	1980	59.3	63.2	56	4.3	...	...	...	1981	81
Thaïlande	1990	64.7	70.0	32	2.6	1988	2.4[5]	1.9[5]	1980/88	50
Turkey	1980	58.0	62.5	120	4.5	...	...	...	...	...
Turquie	1990	62.8	68.0	68	3.8	...	...	...	(1986)	208
United Arab Emirates	1980	64.7	68.9	38	5.7	...	...	...	...	...
Emirats arabes unis	1990	68.6	72.9	26	4.8	...	...	...	...	...
Viet Nam	1980	53.7	58.1	82	5.6	...	...	...	1982	100
Viet Nam	1990	60.6	64.8	47	4.2	...	...	...	1980/88	140
Yemen[22]	1980	44.9	45.4	149	7.8	...	...	...	...	...
Yémen[22]	1990	49.8	50.2	120	7.7	...	...	...	...	...
Europe · Europe										
Albania	1980	66.8	71.2	50	4.2	...	...	...	...	...
Albanie	1990	69.2	75.5	31	3.1	...	...	...	...	...
Andorra Andorre	1990	...	...	3	...	...	...	...	...	...
Austria	1980	68.5	75.9	17	1.6	...	...	...	1980	8
Autriche	1990	71.6	78.3	9	1.4	1989	0.5	0.4	1989	8
Belgium	1980	69.1	75.7	14	1.7	...	...	...	1980	6
Belgique	1990	71.9	78.5	9	1.6	1984	0.6	0.6	1986	3
Bulgaria	1980	68.7	74.1	22	2.2	...	...	...	1980	21
Bulgarie	1990	68.4	74.7	15	1.9	1989	1.0	0.8	1989	19
Czechoslovakia	1980	67.0	74.1	19	2.4	...	...	...	1980	9
Tchécoslovaquie	1990	67.6	75.2	13	2.0	1989	0.5	0.4	1989	10
Denmark	1980	71.3	77.3	9	1.7	...	...	...	1980	2
Danemark	1990	71.8	77.8	8	1.5	1987	0.5	0.5	1988	3
Finland	1980	68.0	76.6	9	1.6	...	...	...	1980	2
Finlande	1990	70.7	78.8	6	1.7	1986	0.3	0.2	1988	11
France	1980	69.7	77.8	11	1.9	...	...	...	1980	13
France	1990	72.0	80.3	8	1.8	1988	0.5	0.4	1989	9
Germany ʃ · Allemagne ʃ Federal Republic of Germany	1980	69.0	75.8	15	1.4	...	...	...	1980	21
Rép. féd. d'Allemagne	1990	71.6	78.2	9	1.4	1988	0.5	0.4	1989	5

20
Selected indicators of life expectancy, child-bearing and mortality [*cont.*]
Choix d'indicateurs de l'espérance de vie, de la maternité et de la mortalité [*suite*]

Country or area Pays or zone	Year Année	Life expectancy at birth (years) Espérance de vie à la naissance (en années) M	F	Infant mortality rate Taux de mortalité infantile	Total fertility rate Taux de fécondité	Year Année	Child mortality rate Taux de mortalité juvénile M	F	Maternal mortality per 100000 live births Mort. lié à la maternité pour 100000 naissances vivantes Year[1] Année[1]	Rate Taux
former German Dem. Rep.	1980	68.8	74.6	13	1.8	...	...	...	1980	18
l'ancienne Rép. dém. allemande	1990	70.4	76.2	9	1.7	1988	0.6	0.5	1988	15
Greece	1980	71.7	75.8	25	2.3	...	...	...	1980	18
Grèce	1990	74.2	79.4	11	1.5	1984	3.2[5]	2.6[5]	1988	6
Hungary	1980	66.8	73.3	27	2.1	...	...	...	1980	21
Hongrie	1990	65.6	73.7	17	1.8	1989	0.6	0.5	1989	15
Iceland	1980	73.4	79.3	9	2.3	...	...	...	...	...
Islande	1990	75.3	80.3	6	2.1	1984	0.2	0.1	1987	24
Ireland	1980	69.6	74.6	15	3.5	...	...	...	1980[23]	7
Irlande	1990	71.6	77.3	8	2.3	1988[23]	0.6	0.3	1988[23]	2
Italy	1980	70.4	76.9	18	1.9	...	...	...	1980	13
Italie	1990	73.1	79.6	10	1.3	1987	0.5	0.5	1988	8
Liechtenstein										
Liechtenstein	1990	...	...	3	1.4	...	...	...	...	...
Luxembourg	1980	68.2	75.5	13	1.5	...	...	...	...	...
Luxembourg	1990	70.8	78.3	9	1.5	1987	0.6	0.5	1987	24
Malta	1980	70.0	74.3	15	2.0	...	...	...	1981	18
Malte	1990	72.8	77.3	9	2.0	1989	2.5[5]	2.0[5]	1985	36
Netherlands	1980	72.1	78.6	10	1.6	...	...	...	1980	9
Pays-Bas	1990	73.5	80.0	9	1.6	1989	0.4	0.4	1989	5
Norway	1980	72.2	78.6	9	1.8	...	...	...	1980	12
Norvège	1990	73.0	79.8	8	1.8	1989	0.5	0.3	1988	4
Poland	1980	67.0	75.0	23	2.3	...	...	...	1980	12
Pologne	1990	66.8	75.3	17	2.2	1988	0.6	0.5	1989	11
Portugal	1980	66.7	73.8	30	2.4	...	...	...	1980	20
Portugal	1990	70.1	77.2	14	1.6	1988	1.0	0.8	1988	7
Romania	1980	67.3	72.0	31	2.6	...	...	...	1980	132
Roumanie	1990	66.5	72.5	26	2.3	1989	2.2[5]	2.0[5]	1984	149
San Marino	1980	...	...	...	1.8					
Saint-Marin	1990	...	...	22	1.2					
Spain	1980	71.4	77.4	16	2.6	...	...	...	1980	11
Espagne	1990	73.8	80.0	9	1.5	1981	3.0[5]	2.3[5]	1986	6
Sweden	1980	72.3	78.3	8	1.6	...	...	...	1980	8
Suède	1990	74.3	80.3	6	1.9	1988	0.3	0.3	1988	9
Switzerland	1980	72.0	78.6	10	1.5	...	...	...	1980	5
Suisse	1990	73.9	80.7	7	1.5	1987	0.4	0.4	1989	4
United Kingdom	1980	69.7	76.0	14	1.7	...	...	...	1980[24]	11
Royaume-Uni	1990	72.3	77.9	9	1.8	1989	0.4	0.4	1988[24]	6
Yugoslavia	1980	67.6	72.9	35	2.2	...	...	...	1980	18
Yougoslavie	1990	68.5	74.1	25	2.0	1989	0.9	0.8	1989	16
Oceania · Océanie										
Australia	1980	70.1	77.0	13	2.1	...	...	...	1980	10
Australie	1990	73.0	79.4	9	1.9	1989	0.4	0.4	1988	5
Cook Islands	1980	63.2	67.1	...	4.4	...	...	...	...	...
Iles Cook	1990	...	...	9	...	...	...	...	...	...
Fiji	1980	65.5	69.0	37	4.0	...	...	...	1980	32
Fidji	1990	68.3	72.5	27	3.2	1986	1.2	1.3	1985	41
French Polynesia	1980	60.5	64.9	56	4.2	...	...	...	...	...
Polynésie française	1990	65.8	71.1	18	3.6	...	...	...	...	...
Guam	1980	68.2	74.8	16	3.5	...	...	...	...	...
Guam	1990	72.0	77.6	10	2.9	...	...	...	1983	31
New Caledonia	1980	...	...	30	3.9	...	...	...	...	...
Nouvelle-Calédonie	1990	...	...	15	...	1976	11.3[5]	8.5[5]	1984	160
New Zealand	1980	69.3	75.7	14	2.2	...	...	...	1980	14
Nouvelle-Zélande	1990	71.7	77.9	11	2.0	1988	0.6	0.5	1987	13
Pacific Islands (Palau)[25]										
Iles du Pacifique (Palau)[25]	1980	...	...	32[26]	5.0	...	...	...	...	...
Papua New Guinea	1980	49.5	50.0	70	5.9	...	...	...	1980	900
Papouasie-Nouv.-Guinée	1990	53.2	54.7	59	5.3	...	...	...	...	...
Samoa	1980	61.0	64.3	...	3.7	...	...	...	...	...
Samoa		...	...	...	...	...	...	...	1983	40

20
Selected indicators of life expectancy, child-bearing and mortality [*cont.*]
Choix d'indicateurs de l'espérance de vie, de la maternité et de la mortalité [*suite*]

Country or area Pays or zone	Year Année	Life expectancy at birth (years) Espérance de vie à la naissance (en années) M	F	Infant mortality rate Taux de mortalité infantile	Total fertility rate Taux de fécondité	Year Année	Child mortality rate Taux de mortalité juvénile M	F	Maternal mortality per 100000 live births Mort. lié à la maternité pour 100000 naissances vivantes Year[1] Année[1]	Rate Taux
Solomon Islands	1980	63.4	67.3	47	7.1	...	...	...	...	...
Iles Salomon	1990	67.2	71.2	32	5.8	...	...	...	1984	10
Vanuatu										
Vanuatu	...	...	...	...	...	...	...	...	1983	107
former USSR · ancienne URSS										
former USSR	1980	63.0	73.0	28	2.3	...	...	...	...	...
ancienne URSS	1990	64.7	73.7	24	2.4	1989	6.7[5]	5.2[5]	1980/88	48
Belarus										
Bélarus	1990	67.0	75.9	13	2.0	...	...	...	1987	7
Ukraine										
Ukraine	1990	66.4	74.8	13	2.0	...	...	...	1988	9

Sources:
World Population Prospects, 1992 (United Nations publication, forthcoming); Demographic Yearbook (United Nations publication various years up to 1990); World Health Organization, "Maternal mortality rates; a tabulation of available information (second edition)" (Geneva, FHE/86.3) United Nations Children's fund, "State of the world's chidren 1991" (New York and Oxford, Oxford University press, 1991) and unpublished data from the Pan American Health Organization.

Sources:
"World Population Prospects, 1992", (publication des Nations Unies, à paraître); Annuaire démographique (publication des Nations Unies, différentes années jusqu'à 1990); Organisation mondiale de la santé, "Maternal mortality rates; a tabulation of available information (second edition)" (Genève, FHE/86.3); Fonds des Nations Unies pour l'enfance, "State of the world's children 1991" (New York and Oxford, Oxford university press, 1991) et données inédites de l'Organisation panaméricaine de la santé.

ɠ   All data shown pertain to Germany prior to 3 October 1990 are indicated separately for the Federal Republic of Germany and the former German Democratic Republic based on their respective territories at the time indicated. Where data for united Germany (3 October 1990 and thereafter) are not available, available data are shown separately under the designations Federal Republic of Germany and former German Democratic Republic and pertain to the territorial boundaries prior to 3 October 1990. For detailed explanatory notes on data pertaining to Germany, see Annex I- Country Nomenclature.

1   When the reference year is not known, publication year is given instead and enclosed in ( ).
2   For Algerian population only.
3   Excluding live-born infants dying before registration of birth.
4   This figure represents the midpoint of a given range;
5   0-4 years old.
6   All health institutions.
7   All hospitals.
8   From 267 hospitals.
9   From 48 hospitals, all regions.
10 Data tabulated by year of registration rather than occurence.
11 Based on burial permits.
12 Excluding nomadic Indian tribes.
13 Excluding Indian jungle population.
14 Excluding Vietnamese refugees.
15 Including data for East Jerusalem and Israeli residents in certain other territories under occupation by Israel since June 1967.
16 For Japanese nationals in Japan only; however, rates computed on population including foreigners except foreign military and civilian personnel and their dependants stationed in the area.
17 For Japanese nationals in Japan only.
18 Excluding data for Jordanian territory under occupation since June 1967 by Israel.
19 Events registered by Health Service only.
20 Peninsular Malaysia only.
21 Excluding deaths for which cause is unknown.
22 Excluding former Democratic Yemen.

ɠ   Toutes les données se rapportant à l'Allemagne avant le 3 octobre 1990 figurent dans deux rubriques séparées basées sur les territoires respectifs de la République fédérale d'Allemagne et l'ancienne République démocratique allemande selon la période indiquée. En l'absence de données pour l'Allemagne unifiée (à compter du 3 octobre 1990), les données disponibles sont fournies séparément sous les rubriques République fédérale d'Allemagne et ancienne République démocratique allemande et se rapportent aux limites territoriales antérieures au 3 octobre 1990. Pour les notes explicatives en détail sur les données concernant l'Allemagne, voir Annexe I - Nomenclature des pays.

1   Lorsqu'on ne connaît pas l'année de référence, c'est l'année de publication qui est donnée entre parenthèses.
2   Pour la population algérienne uniquement.
3   Non compris les enfants nés vivants, décédés avant l'enregistrement de leur naissance.
4   Ce chiffre représente le milieu de l'étendue donnée (150-300).
5   De 0 à 4 ans.
6   Toutes les établissements sanitaires.
7   Tous les hôpitaux.
8   Données communiquées par 267 hôpitaux.
9   Données communiquées par 48 hôpitaux situés dans toutes les régions.
10 Données exploitées selon l'année de l'enregistrement et non la date de l'événement.
11 D'aprés les permis d'inhumer.
12 Non compris les tribus indiennes nomades.
13 Non compris la population indienne de la jungle.
14 Non compris les réfugiés vietnamiens.
15 Y compris les données concernant la partie orientale de Jérusalem et les résidents israéliens de certains autres territoires occupés par les forces militaires israéliennes depuis juin 1967.
16 Pour les nationaux japonais au Japon uniquement; néanmoins, les taux sont calculés pour la population comprenant les étrangers, à l'exception personnels civils et militaires étrangers (et des personnes à leur charge) en poste dans la région.
17 Pour les nationaux japonais au Japon uniquement.
18 Non compris les données pour le territoire jordanien occupé depuis juin 1967 par les forces armées israéliennes.
19 Evénements enregistrés par le Service sanitaire uniquement.

20
Selected indicators of life expectancy, child-bearing and mortality [*cont.*]
Choix d'indicateurs de l'espérance de vie, de la maternité et de la mortalité [*suite*]

23 Deaths registered within one year of occurrence.
24 England and Wales only.
25 Including data for Federated States of Micronesia, Marshall Is. and Northern Mariana Is.
26 Excluding infant deaths among United States military personnel, their dependants and contract employees.

20 Péninsulaire Malasie seulement.
21 Non compris les décès dont la cause est inconnue.
22 Non compris l'ancien Yémen démocratique.
23 Décès enregistrés dans l'année qui suit.
24 Angleterre et Galles uniquement.
25 Y compris les données pour les Etats fédérés de Micronésie, les îles Marshall et les îles Mariannes du Nord.
26 Non compris de la mortalité infantile pour les personnels militaires des Etats Unis, et des personnes à leur charge et des employées.

21
Estimates/projections of total AIDS cases and cumulative HIV infections and number of reported AIDS cases
Estimations/projections du nombre total de cas de SIDA et du nombre cumulé de personnes infectées par le VIH et nombre de cas de SIDA déclarés

A. Estimated cumulative HIV infection to the end of 1991,
   and projections for year 2000
   Estimations du nombre cumulé de personnes infectées par le VIH jusqu'à la
   fin 1991 et projections pour l'an 2000

Regions Régions	Total	F/M	Pattern of infection Schéma d'infection
World · Monde			
Adults	at least 9-11 million	1 to 3	
Adultes	au moins 9-11 millions	1/3	
Children	approx. 1 million		
Enfants	à peu près 1 million		
Projections, 2000	30-40 million		
Projections, 2000	30-40 millions		
Sub-Saharan Africa	more than 6.5 million	approx. 1 to 1	Extensive HIV spread probably beagn in the late 1970s or early 1980s, predominantly among heterosexuals. This pattern has remain unchanged, with continuing increase in the numbers of individuals HIV-infected.
Afrique subsaharienne	plus de 6,5 millions	à peu près 1/1	La propagation massive du virus a probablement commencé à la fin des années 70 ou au début des années 80, essentiellement parmi les hétérosexuels. Ce schéma n'a pas changé et le nombre de personnes contaminées a augmenté régulièrement.
Latin America and Caribbean	more than 1 million	more than 1 to 4	Extensive HIV spread began in the late 1970s or early 1980s, predomonantly among homosexual or bisexual men. Since the mid-1980s, however, this epidemiologic pattern has changed, with increasing transmission among heterosexuals and injecting drug users.
Amérique latine et Caraïbes	plus d'un million	plus de 1/4	La propagation massive du virus a probablement commencé à la fin des années 70 ou au début des années 80, essentiellement parmi les hommes homosexuels ou bisexuels. Depuis le milieu des années 80, ce schéma épidémiologique change et la contagion se fait de plus en plus parmi les hétérosexuels et les toxicomanes s'injectant la drogue par intraveineuse.
South-eastern and southern Asia	more than 1 million	more than 1 to 3	Additional epidemiological patterns of HIV transmission are in the process of evolution, particularly in these regions, where extensive spread of HIV infections has been documented since the late 1980s. Although widespread transmission was first noted in some countries among injecting drug users, heterosexual transmission now appears to be the predominant mode of spread.
Asie du Sud-Est et Asie du Sud	plus d'un million	plus de 1/3	Les modes de transmission du virus sont en train de changer, surtout dans les régions où la propagation massive de l'infection est attestée depuis la fin des années 80. Si c'est parmi les toxicomanes s'administrant la drogue par voie intraveineuse qu'on a d'abord observé dans certains pays une transmission rapide du virus, le mode de propagation prédominant semble maintenant être la transmission hétérosexuelle.
Northern America; western Europe; Australia and New Zealand	about 1.5 million	less than 1 to 8	Extensive HIV spread began in the late 1970s or early 1980s predominantly among homosexual or bisexual men and injecting drug users and their sexual partners. Although the annual rate of increase in HIV infection is believed to have peaked in the mid-1980s, there is increasing evidence of heterosexual transmission in urban areas where there is a high prevalence of infection in injecting drug users.
Amérique du Nord, Europe occidentale, Australie et Nouvelle-Zélande	à peu près 1,5 million	moins de 1/8	La propagation massive du virus a commencé à la fin des années 70 ou au début des années 80, essentiellement parmi les hommes homosexuels ou bisexuels et les toxicomanes s'injectant la drogue par intraveineuse et leurs partenaires sexuels. La progression annuelle du nombre de séropositifs a sans doute atteint un sommet au milieu des années 80, mais on observe de plus en plus de cas de transmission hétérosexuelle dans les zones urbaines où le pourcentage de séropositifs est élève parmi les toxicomanes qui s'administrent la drogue par voie intraveineuse.

21
Estimates/projections of total AIDS cases and cumulative HIV infections, and number of reported AIDS cases [*cont.*]
Estimations/projections du nombre total de cas de SIDA et du nombre cumulé de personnes infectées par le VIH
et nombre de cas de SIDA déclarés [*suite*]

A. Estimated cumulative HIV infection to the end of 1991, and projections for year
Estimations du nombre cumulé de personnes infectées par le VIH jusqu'à la fin 1991 et projections pour l'an 2000

Regions Régions	Total	F/M	Pattern of infection Schéma d'infection
Eastern Europe and former USSR; eastern and western Asia and northern Africa; Oceania	less than 100,000	about 1 to 7	Although data from some regions are incomplete, there is increasing evidence in some countries of HIV infection in persons with multiple sexual partners and in groups of injecting drug users. Careful monitoring of the HIV situation in these regions is required.
Europe de l'Est et l'ancienne URSS, Asie de l'Est, Asie de l'Ouest et Afrique du Nord, Océanie	moins de 100 000	à peu près 1/7	Bien que, pour certaines régions, les données dont on dispose soient incomplètes, on observe dans certains pays des cas de plus en plus fréquents de séropositivité parmi les personnes qui ont des partenaires sexuels multiples et les toxicomanes qui s'administrent la drogue par injection. La situation de l'infection par le VIH dans ces régions demande à être suivie de près.

B. Reported, estimated and projected AIDS cases to December 1990
Cas de SIDA : nombre de cas déclarés jusqu'à fin décembre 1990, estimations et projections

Year Année	Number of cases Nombre de cas
Cases reported to World Health Organization to end-1990 (cumulative) Cas déclarés à l'Organisation mondiale de la santé jusqu'à la fin de 1990 (cumulatif)	314 610
Estimated cases to end 1991: Estimation du nombre jusqu'à la fin de 1991: Adults (approx.) Adultes (à peu près)	1.5 million
Children (more than) Enfants (plus de)	0.5 million
Projected cases, 2000 Nombre de cas projeté, 2000	12 - 18 million
Adults Adultes	8 - 10 million
Children Enfants	4 - 8 million

21
Estimates/projections of total AIDS cases and cumulative HIV infections, and number of reported AIDS cases [*cont.*
Estimations/projections du nombre total de cas de SIDA et du nombre cumulé de personnes infectées pa
et nombre de cas de SIDA déclarés [*suite*]

C. Reported AIDS cases, 1991 • Cas de SIDA déclarés, 1991

Country or area Pays ou zone	Total reported to Dec.1989 Total des cas déclarés jusqu' en déc. 1989	New cases reported for 1990 Nouveaux cas déclarés en 1990	New cases reported for 1991 Nouveaux cas déclarés en 1991	Cumulative total to Dec.1991 Total cumulatif en déc. 1991
Africa • Afrique				
Algeria Algérie	45	47	0	92
Angola Angola	198	93	130	421
Benin Bénin	84	50	51	185
Botswana Botswana	87	91	72	250
Burkina Faso Burkina Faso	906	72	0	978
Burundi Burundi	2784	521	0	3305
Cameroon Cameroun	134	61	234	429
Cape Verde Cap-Vert	28	4	0	32
Central African Rep. Rép. centrafricaine	1162	702	0	1864
Chad Tchad	21	38	165	224
Comoros Comores	1	1	1	3
Congo Congo	1940	465	1077	3482
Côte d'Ivoire Côte d'Ivoire	3709	3189	3894	10792
Djibouti Djibouti	7	51	107	165
Egypt Egypte	20	7	12	39
Equatorial Guinea Guinée équatoriale	3	2	2	7
Ethiopia Ethiopie	294	448	889	1631
Gabon Gabon	51	66	98	215
Gambia Gambie	78	46	56	180
Ghana Ghana	1226	1011	903	3140
Guinea Guinée	82	138	118	338
Guinea-Bissau Guinée-Bissau	123	34	0	157
Kenya Kenya	9139	0	0	9139
Lesotho Lesotho	13	10	21	44
Liberia Libéria	5	0	19	24
Libyan Arab Jamahiriya Jamah. arabe libyenne	1	4	2	7
Madagascar Madagascar	2	0	0	2
Malawi Malawi	7848	4226	0	12074
Mali Mali	234	104	0	338
Mauritania Mauritanie	11	5	10	26

21
Estimates/projections of total AIDS cases and cumulative HIV infections, and number of reported AIDS cases [*cont.*]
Estimations/projections du nombre total de cas de SIDA et du nombre cumulé de personnes infectées par le VIH
et nombre de cas de SIDA déclarés [*suite*]

C. Reported AIDS cases, 1991 · Cas de SIDA déclarés, 1991

Country or area Pays ou zone	Total reported to Dec.1989 Total des cas déclarés jusqu' en déc. 1989	New cases reported for 1990 Nouveaux cas déclarés en 1990	New cases reported for 1991 Nouveaux cas déclarés en 1991	Cumulative total to Dec.1991 Total cumulatif en déc. 1991
Mauritius Maurice	4	1	5	10
Morocco Maroc	44	26	28	98
Mozambique Mozambique	64	98	178	340
Namibia Namibie	189	122	0	311
Niger Niger	80	213	204	497
Nigeria Nigéria	48	36	100	184
Réunion Réunion	47	2	0	49
Rwanda Rwanda	2285	2204	2089	6578
Sao Tome and Principe Sao Tomé-et-Principe	0	2	4	6
Senegal Sénégal	307	118	127	552
Seychelles Seychelles	0	0	0	0
Sierra Leone Sierra Leone	27	7	6	40
Somalia Somalie	8	5	0	13
South Africa Afrique du Sud	383	320	316	1019
Sudan Soudan	188	130	182	500
Swaziland Swaziland	10	20	41	71
Togo Togo	192	458	628	1278
Tunisia Tunisie	73	27	5	105
Uganda Ouganda	13278	8441	8471	30190
United Rep. Tanzania Rép.-Unie de Tanzanie	14107	7948	5341	27396
Zaire Zaïre	12337	2425	0	14762
Zambia Zambie	2809	1393	1601	5803
Zimbabwe Zimbabwe	1602	4362	4587	10551
America, North · Amérique du Nord				
Anguilla Anguilla	3	1	0	4
Antigua and Barbuda Antigua-et-Barbuda	3	3	0	6
Bahamas Bahamas	437	162	235	834
Barbados Barbade	111	61	78	250
Belize Belize	11	1	0	12
Bermuda Bermude	135	33	23	191
British Virgin Is. Iles Vierges britanniques	1	2	1	4

21

Estimates/projections of total AIDS cases and cumulative HIV infections, and number of reported AIDS cases [*cont.*]
Estimations/projections du nombre total de cas de SIDA et du nombre cumulé de personnes infectées par le VIH
et nombre de cas de SIDA déclarés [*suite*]

C. Reported AIDS cases, 1991 · Cas de SIDA déclarés, 1991

Country or area Pays ou zone	Total reported to Dec.1989 Total des cas déclarés jusqu' en déc. 1989	New cases reported for 1990 Nouveaux cas déclarés en 1990	New cases reported for 1991 Nouveaux cas déclarés en 1991	Cumulative total to Dec.1991 Total cumulatif en déc. 1991
Canada Canada	4238	1050	788	6076
Cayman Is. Iles Caïmanes	5	2	3	10
Costa Rica Costa Rica	148	69	83	300
Cuba Cuba	63	10	30	103
Dominica Dominique	10	2	0	12
Dominican Republic République dominicaine	1222	238	162	1622
El Salvador El Salvador	129	54	107	290
Grenada Grenade	19	5	7	31
Guadeloupe Guadeloupe	189	6	0	195
Guatemala Guatemala	64	78	94	236
Haiti Haïti	2456	630	0	3086
Honduras Honduras	556	586	453	1595
Jamaica Jamaïque	139	62	133	334
Martinique Martinique	127	45	28	200
Mexico Mexique	5720	2403	950	9073
Montserrat Montserrat	1	0	0	1
Netherlands Antilles Antilles néerlandaises	46	31	0	77
Nicaragua Nicaragua	4	7	13	24
Panama Panama	187	73	70	330
St. Kitts and Nevis Saint-Kitts-et-Nevis	24	8	1	33
Saint Lucia Sainte-Lucie	18	15	7	40
St. Vincent and the Grenadines St.-Vincent-et-Grenadines	21	4	14	39
Trinidad and Tobago Trinité-et-Tobago	563	173	235	971
Turks and Caicos Islands Iles Turques et Caiques	18	1	2	21
United States Etats-Unis	140318	39249	35696	215263
America, South · Amérique du Sud				
Argentina Argentine	542	383	373	1298
Bolivia Bolivie	18	7	16	41
Brazil Brésil	12445	6177	5914	24536
Chile Chili	236	117	147	500
Colombia Colombie	900	620	669	2189

21
Estimates/projections of total AIDS cases and cumulative HIV infections, and number of reported AIDS cases [*cont.*]
Estimations/projections du nombre total de cas de SIDA et du nombre cumulé de personnes infectées par le VIH
et nombre de cas de SIDA déclarés [*suite*]

C. Reported AIDS cases, 1991 • Cas de SIDA déclarés, 1991

Country or area Pays ou zone	Total reported to Dec.1989 Total des cas déclarés jusqu' en déc. 1989	New cases reported for 1990 Nouveaux cas déclarés en 1990	New cases reported for 1991 Nouveaux cas déclarés en 1991	Cumulative total to Dec.1991 Total cumulatif en déc. 1991
Ecuador Equateur	86	42	51	179
French Guiana Guyane française	191	41	0	232
Guyana Guyana	84	61	85	230
Paraguay Paraguay	14	12	10	36
Peru Pérou	245	141	155	541
Suriname Suriname	48	35	16	99
Uruguay Uruguay	83	76	86	245
Venezuela Venezuela	896	426	251	1573
Asia • Asie				
Afghanistan Afghanistan	0	0	0	0
Bahrain Bahreïn	0	0	0	0
Bangladesh Bangladesh	0	1	0	1
Bhutan Bhoutan	0	0	0	0
Brunei Darussalam Brunéi Darussalam	1	1	0	2
Cambodia Cambodge	0	0	0	0
China Chine	3	2	3	8
Cyprus Chypre	16	3	4	23
Hong Kong Hong-kong	32	12	15	59
India Inde	40	17	45	102
Indonesia Indonésie	6	6	9	21
Iran, Islamic Rep. of Iran, Rép. islamique d'	9	10	25	44
Iraq Iraq	0	0	7	7
Israel Israël	113	32	24	169
Japan Japon	182	189	82	453
Jordan Jordanie	10	1	6	17
Korea, Dem.People's Rep. Corée, Rép. pop. dém. de	0	0	0	0
Korea, Republic of Corée, République de	5	2	1	8
Kuwait Koweït	2	1	3	6
Lao People's Dem. Rep. Rép. dém. pop. lao	0	0	1	1
Lebanon Liban	16	8	5	29
Macau Macao	1	0	1	2

21
Estimates/projections of total AIDS cases and cumulative HIV infections, and number of reported AIDS cases [*cont.*]
Estimations/projections du nombre total de cas de SIDA et du nombre cumulé de personnes infectées par le VIH
et nombre de cas de SIDA déclarés [*suite*]

C. Reported AIDS cases, 1991 · Cas de SIDA déclarés, 1991

Country or area Pays ou zone	Total reported to Dec.1989 Total des cas déclarés jusqu' en déc. 1989	New cases reported for 1990 Nouveaux cas déclarés en 1990	New cases reported for 1991 Nouveaux cas déclarés en 1991	Cumulative total to Dec.1991 Total cumulatif en déc. 1991
Malaysia Malaisie	12	12	23	47
Maldives Maldives	0	0	0	0
Mongolia Mongolie	0	0	0	0
Myanmar Myanmar	0	0	10	10
Nepal Népal	2	2	1	5
Oman Oman	16	7	1	24
Pakistan Pakistan	13	1	4	18
Philippines Philippines	41	15	11	67
Qatar Qatar	23	7	1	31
Saudi Arabia Arabie saoudite	25	7	8	40
Singapore Singapour	15	8	12	35
Sri Lanka Sri Lanka	5	3	2	10
Syrian Arab Republic Rép. arabe syrienne	9	1	7	17
Thailand Thaïlande	43	54	82	179
Turkey Turquie	32	12	18	62
United Arab Emirates Emirats arabes unis	8	0	0	8
Viet Nam Viet Nam	0	0	0	0
Yemen Yémen	0	0	0	0
Europe · Europe				
Albania Albanie	0	0	0	0
Austria Autriche	387	152	168	707
Belgium Belgique	668	183	195	1046
Bulgaria Bulgarie	7	2	4	13
Czechoslovakia Tchécoslovaquie	19	5	2	26
Denmark Danemark	542	195	210	947
Finland Finlande	59	15	26	100
France France	10877	3722	3237	17836
Germany Allemagne	5060	1274	1199	7533
Greece Grèce	277	135	147	559
Hungary Hongrie	32	17	33	82
Iceland Islande	13	3	6	22

21
Estimates/projections of total AIDS cases and cumulative HIV infections, and number of reported AIDS cases [*cont.*]
Estimations/projections du nombre total de cas de SIDA et du nombre cumulé de personnes infectées par le VIH
et nombre de cas de SIDA déclarés [*suite*]

C. Reported AIDS cases, 1991 · Cas de SIDA déclarés, 1991

Country or area Pays ou zone	Total reported to Dec.1989 Total des cas déclarés jusqu' en déc. 1989	New cases reported for 1990 Nouveaux cas déclarés en 1990	New cases reported for 1991 Nouveaux cas déclarés en 1991	Cumulative total to Dec.1991 Total cumulatif en déc. 1991
Ireland Irlande	124	55	62	241
Italy Italie	5852	2955	2802	11609
Luxembourg Luxembourg	24	9	12	45
Malta Malte	14	1	7	22
Monaco Monaco	3	2	2	7
Netherlands Pays-Bas	1205	408	407	2020
Norway Norvège	145	53	54	252
Poland Pologne	29	21	37	87
Portugal Portugal	414	202	200	816
Romania Roumanie	222	947	535	1704
San Marino Saint-Marin	1	0	0	1
Spain Espagne	6157	2612	2786	11555
Sweden Suède	401	119	125	645
Switzerland Suisse	1458	429	341	2228
United Kingdom Royaume-Uni	3349	1068	1034	5451
former Yugoslavia ancienne Yougoslavie	109	70	75	254
Oceania · Océanie				
Australia Australie	1889	631	651	3171
Cook Islands Iles Cook	0	0	0	0
Fiji Fidji	1	2	1	4
French Polynesia Polynésie française	13	9	5	27
Guam Guam	6	2	2	10
Kiribati Kiribati	0	0	0	0
Marshall Islands Iles Marshall	2	0	0	2
Micronesia, Federated state of Micronésie, Etats fédérés de	1	1	0	2
New Caledonia Nouvelle-Calédonie	13	3	2	18
New Zealand Nouvelle-Zélande	159	73	78	310
Papua New Guinea Papouasie-Nouv.-Guinée	16	13	13	42
Samoa Samoa	0	1	0	1
Solomon Islands Iles Salomon	0	0	0	0
Tonga Tonga	2	0	0	2

21

Estimates/projections of total AIDS cases and cumulative HIV infections, and number of reported AIDS cases [*cont.*]
Estimations/projections du nombre total de cas de SIDA et du nombre cumulé de personnes infectées par le VIH
et nombre de cas de SIDA déclarés [*suite*]

C. Reported AIDS cases, 1991 · Cas de SIDA déclarés, 1991

Country or area Pays ou zone	Total reported to Dec.1989 Total des cas déclarés jusqu' en déc. 1989	New cases reported for 1990 Nouveaux cas déclarés en 1990	New cases reported for 1991 Nouveaux cas déclarés en 1991	Cumulative total to Dec.1991 Total cumulatif en déc. 1991
Tuvalu 　Tuvalu	0	0	0	0
Vanuatu 　Vanuatu	0	0	0	0
Wallis and Futana Islands 　Iles Wallis e Futna	0	0	0	0
former USSR · ancienne URSS				
former USSR 　ancienne URSS	29	23	20	72

Source:
World Health Organization (Geneva), Global Programme on AIDS, Surveillance, Forecasting and Impact Assessement Unit, 1991.

Source:
Organisation mondiale de la santé (Gèneve), Programme mondial de lutte contre le SIDA, equipe de surveillance, prévision et étude d'impact, 1991.

22
Food Supply
Disponibilités alimentaires
Calories and protein: per capita, per day
Calories et protéine : par habitant, par jour

Country or area Pays ou zone	Calories - Calories Number - Nombre			% of requirements - % des besoins			Protein (grams) Protéine (grammes)		
	1980	1986	1990	1967/70	1980	1984/86	1980	1986	1990
World *Monde*	**2 622.7**	**2 680.9**	**2 689.1**	...	...	...	**70.9**	**72.4**	**72.5**
Afghanistan Afghanistan	2 141.0	1 890.6	1 709.7	86.0	73.0	94.0	60.3	51.3	47.6
Albania Albanie	2 755.3	2 708.0	2 392.1	104.0	110.0	114.0	85.4	84.4	73.7
Algeria Algérie	2 639.0	2 790.6	2 988.6	77.0	101.0	112.0	67.4	73.0	75.7
Angola Angola	2 135.5	1 919.9	1 876.9	84.0	89.0	82.0	46.5	45.4	44.3
Antigua and Barbuda Antigua-et-Barbuda	2 105.8	2 214.2	2 381.4	86.0	88.0	...	65.8	73.0	79.5
Argentina Argentine	3 202.2	3 116.1	3 075.0	126.0	128.0	136.0	106.4	100.2	98.3
Australia Australie	3 042.0	3 296.8	3 385.0	122.0	118.0	125.0	93.7	99.7	100.7
Austria Autriche	3 423.0	3 479.2	3 459.1	129.0	135.0	130.0	94.2	97.3	97.3
Bahamas Bahamas	2 580.9	2 676.8	2 782.1	102.0	98.0	...	74.5	79.5	79.0
Bangladesh Bangladesh	1 988.1	1 962.1	2 099.6	91.0	88.0	83.0	44.1	41.6	43.6
Barbados Barbade	3 144.2	3 147.0	3 221.5	114.0	129.0	...	84.3	95.8	98.9
Belgium-Luxembourg Belgique-Luxembourg	3 529.3	3 692.3	3 921.8	...	...	...	100.2	103.0	107.6
Belize Belize	2 675.0	2 422.2	2 579.1	110.0	119.0	...	67.8	65.3	67.8
Benin Bénin	2 078.6	2 303.9	2 358.5	96.0	101.0	...	50.5	53.7	56.0
Bermuda Bermudes	3 005.4	3 071.7	2 975.3	...	...	...	106.7	105.4	105.6
Bolivia Bolivie	2 090.9	2 054.7	1 981.7	81.0	87.0	89.0	53.7	53.1	51.3
Botswana Botswana	2 129.7	2 180.4	2 272.4	85.0	101.0	96.0	68.2	66.7	69.0
Brazil Brésil	2 734.5	2 692.8	2 722.9	105.0	106.0	111.0	61.1	60.2	61.7
Brunei Darussalam Brunéi Darussalam	2 662.1	2 859.2	2 869.0	103.0	119.0	...	70.6	83.1	75.2
Bulgaria Bulgarie	3 608.4	3 654.4	3 712.2	139.0	146.0	145.0	104.7	108.5	110.7
Burkina Faso Burkina Faso	1 704.1	2 242.1	2 136.8	85.0	85.0	86.0	55.0	68.5	65.8
Burundi Burundi	2 039.1	2 065.2	1 923.3	94.0	93.0	97.0	68.5	64.8	53.4
Cambodia Cambodge	1 734.3	2 101.8	2 113.8	100.0	90.0	98.0	41.1	50.4	50.8
Cameroon Cameroun	2 351.6	2 264.3	2 201.4	91.0	105.0	88.0	59.1	55.7	55.3

22
Food Supply
Calories and protein: per capita, per day [*cont.*]
Disponibilités alimentaires
Calories et protéine : par habitant, par jour [*suite*]

Country or area	Calories - Calories			% of requirements - % des besoins			Protein (grams) Protéine (grammes)		
Pays ou zone	Number - Nombre								
	1980	1986	1990	1967/70	1980	1984/86	1980	1986	1990
Canada Canada	3 132.9	3 285.7	3 221.9	124.0	127.0	129.0	95.0	99.3	101.4
Cape Verde Cap-Vert	2 593.2	2 848.0	2 871.8	80.0	121.0	...	66.8	66.0	66.7
Central African Rep. Rép. centrafricaine	2 138.0	1 829.7	1 867.4	94.0	96.0	86.0	42.9	43.1	46.9
Chad Tchad	1 732.3	1 813.6	1 640.9	95.0	74.0	69.0	56.4	56.6	49.6
Chile Chili	2 657.4	2 444.7	2 480.7	111.0	114.0	106.0	70.3	64.2	69.6
China Chine	2 315.9	2 579.9	2 703.3	89.0	107.0	111.0	54.3	61.7	66.0
Colombia Colombie	2 424.4	2 410.3	2 492.2	90.0	109.0	110.0	52.0	53.4	56.5
Comoros Comores	1 781.0	1 698.5	1 757.5	94.0	99.0	...	38.6	37.5	38.3
Congo Congo	2 243.3	2 339.0	2 321.4	94.0	98.0	117.0	41.2	48.3	47.0
Costa Rica Costa Rica	2 563.8	2 713.8	2 711.5	105.0	120.0	124.0	63.6	62.2	64.0
Côte d'Ivoire Côte d'Ivoire	2 860.0	2 653.1	2 410.9	111.0	116.0	110.0	60.5	55.6	50.4
Cuba Cuba	2 984.2	3 148.2	3 153.2	107.0	120.0	135.0	72.7	74.7	71.6
Czechoslovakia Tchécoslovaquie	3 338.5	3 454.2	3 547.6	140.0	141.0	141.0	98.8	101.8	105.4
Denmark Danemark	3 505.2	3 558.9	3 647.5	125.0	133.0	131.0	85.1	97.3	98.6
Djibouti Djibouti	1 749.1	2 100.8	2 424.7	...	...	...	42.6	50.6	59.9
Dominica Dominique	2 398.2	2 737.4	2 916.9	89.0	90.0	...	60.3	66.3	75.3
Dominican Republic Rép. dominicaine	2 299.0	2 362.0	2 296.6	85.0	96.0	109.0	50.4	49.0	50.4
Ecuador Equateur	2 297.6	2 384.1	2 409.8	86.0	91.0	89.0	48.6	47.9	50.0
Egypt Egypte	3 081.8	3 326.3	3 318.3	101.0	118.0	132.0	75.8	84.5	84.5
El Salvador El Salvador	2 316.0	2 385.1	2 306.1	80.0	94.0	94.0	57.1	56.2	55.2
Ethiopia Ethiopie	1 850.9	1 763.0	1 693.9	87.0	75.0	71.0	60.3	55.9	52.9
Fiji Fidji	2 599.9	2 728.0	2 737.9	91.0	109.0	...	65.8	63.8	64.4
Finland Finlande	3 106.9	3 027.8	3 026.6	115.0	118.0	113.0	95.2	93.6	97.6
France France	3 431.2	3 483.9	3 618.3	134.0	134.0	130.0	108.6	110.9	112.8
French Guiana Guyane française	2 468.7	2 691.1	2 823.4	...	...	...	76.1	89.8	100.8

22
Food Supply
Calories and protein: per capita, per day [cont.]
Disponibilités alimentaires
Calories et protéine : par habitant, par jour [suite]

Country or area Pays ou zone	Calories - Calories						Protein (grams) Protéine (grammes)		
	Number - Nombre			% of requirements - % des besoins					
	1980	1986	1990	1967/70	1980	1984/86	1980	1986	1990
French Polynesia Polynésie française	2 777.8	2 842.9	2 735.0	106.0	100.0	...	74.3	80.5	75.7
Gabon Gabon	2 381.4	2 481.0	2 419.6	94.0	122.0	107.0	62.9	65.8	61.3
Gambia Gambie	2 071.0	2 389.5	2 249.4	96.0	95.0	...	49.0	58.6	56.2
Germany†• Allemagne† F. R. Germany R. f. Allemagne	3 303.6	3 458.5	3 465.4	123.0	133.0	145.0	92.9	98.3	102.9
former German D. R. anc. R. d. allemande	3 556.5	3 718.4	3 597.4	130.0	144.0	130.0	102.4	109.2	102.3
Ghana Ghana	1 948.2	2 088.7	1 974.3	97.0	87.0	76.0	44.2	44.6	43.8
Greece Grèce	3 347.0	3 532.1	3 778.3	123.0	147.0	147.0	104.5	106.6	112.4
Grenada Grenade	2 225.2	2 378.5	2 377.6	92.0	87.0	...	61.9	68.0	61.5
Guadeloupe Guadeloupe	2 442.9	2 641.7	2 816.1	95.0	115.0	...	80.2	86.1	90.6
Guatemala Guatemala	2 130.3	2 255.9	2 254.2	92.0	93.0	105.0	52.9	54.4	55.0
Guinea Guinée	2 224.1	2 259.1	2 228.7	88.0	83.0	77.0	50.1	51.8	51.1
Guinea-Bissau Guinée-Bissau	2 000.5	2 250.9	2 229.7	88.0	99.0	...	42.2	44.0	43.4
Guyana Guyana	2 503.7	2 556.0	2 393.0	102.0	110.0	...	61.8	64.6	65.2
Haiti Haïti	2 034.6	2 084.7	1 986.8	82.0	83.0	84.0	48.3	50.2	46.7
Honduras Honduras	2 131.9	2 109.5	2 258.5	93.0	96.0	92.0	52.7	48.6	55.0
Hong Kong Hong-kong	2 651.6	2 787.7	2 856.8	114.0	127.0	121.0	83.8	87.2	85.4
Hungary Hongrie	3 487.0	3 535.1	3 610.0	127.0	134.0	135.0	98.0	101.3	100.6
Iceland Islande	3 201.3	3 611.0	3 460.1	108.0	109.0	...	130.8	144.0	127.9
India Inde	2 091.5	2 239.8	2 243.4	87.0	86.0	100.0	51.7	56.1	55.7
Indonesia Indonésie	2 514.5	2 663.1	2 630.7	89.0	108.0	116.0	51.7	57.3	57.3
Iran, Islamic Rep. of Iran, Rép. islamique d'	2 906.2	3 029.4	3 037.7	90.0	124.0	138.0	75.1	79.5	81.1
Iraq Iraq	2 730.8	3 231.4	2 835.8	91.0	111.0	124.0	72.6	83.6	71.0
Ireland Irlande	3 904.7	3 891.7	3 987.3	137.0	148.0	146.0	116.1	114.1	119.7
Israel Israël	2 977.6	3 096.4	3 203.8	115.0	118.0	118.0	97.7	99.5	103.1

22
Food Supply
Calories and protein: per capita, per day [*cont.*]
Disponibilités alimentaires
Calories et protéine : par habitant, par jour [*suite*]

Country or area Pays ou zone	Calories - Calories						Protein (grams) Protéine (grammes)		
	Number - Nombre			% of requirements - % des besoins					
	1980	1986	1990	1967/70	1980	1984/86	1980	1986	1990
Italy Italie	3 588.2	3 473.8	3 483.6	135.0	148.0	139.0	107.7	106.2	107.2
Jamaica Jamaïque	2 676.7	2 590.3	2 527.2	106.0	118.0	116.0	62.7	64.2	61.6
Japan Japon	2 777.6	2 865.3	2 926.3	116.0	124.0	122.0	86.9	91.3	95.6
Jordan Jordanie	2 532.2	2 678.0	2 703.9	96.0	96.0	121.0	67.8	73.0	68.4
Kenya Kenya	2 164.4	2 099.2	2 047.5	97.0	89.0	92.0	56.4	57.3	55.7
Kiribati Kiribati	2 769.5	2 687.4	2 498.1	...	...	...	65.8	65.1	66.2
Korea, Dem. P. R. Corée, R. p. dém. de	2 690.2	2 803.7	2 859.8	103.0	90.0	135.0	75.9	79.9	82.0
Korea, Republic of Corée, République de	2 766.8	2 851.9	2 840.3	102.0	129.0	122.0	75.2	75.8	79.0
Kuwait Koweït	3 030.4	3 040.3	2 756.7	...	...	...	93.3	91.4	81.9
Lao People's Dem. Rep. Rép. dém. pop. lao	2 419.1	2 404.6	2 475.2	95.0	89.0	104.0	65.2	66.1	66.8
Lebanon Liban	2 694.9	2 994.1	3 160.4	101.0	100.0	125.0	76.1	79.0	82.2
Lesotho Lesotho	2 412.8	2 155.7	2 100.2	90.0	107.0	101.0	70.2	62.3	59.1
Liberia Libéria	2 404.2	2 406.4	2 067.2	98.0	109.0	102.0	46.8	45.1	37.1
Libyan Arab Jamah. Jamah. arabe libyenne	3 474.3	3 360.1	3 353.4	102.0	147.0	153.0	88.5	79.4	80.3
Macau Macao	2 242.6	2 258.1	2 321.5	84.0	106.0	...	64.9	62.8	61.3
Madagascar Madagascar	2 483.0	2 322.7	2 161.7	106.0	109.0	106.0	59.3	55.5	52.3
Malawi Malawi	2 250.9	2 088.5	2 041.6	94.0	92.0	102.0	65.6	60.3	58.4
Malaysia Malaisie	2 669.2	2 648.9	2 697.0	110.0	119.0	121.0	55.4	54.6	53.6
Maldives Maldives	2 208.1	2 343.5	2 415.6	81.0	83.0	...	71.1	88.8	88.0
Mali Mali	1 875.2	2 228.1	2 233.1	89.0	82.0	86.0	56.3	63.6	62.0
Malta Malte	2 950.9	3 034.4	3 147.7	121.0	124.0	...	91.8	91.9	90.0
Martinique Martinique	2 576.2	2 727.8	2 771.9	97.0	117.0	...	80.8	84.3	88.2
Mauritania Mauritanie	2 099.3	2 377.2	2 469.3	88.0	90.0	92.0	70.4	74.6	71.7
Mauritius Maurice	2 688.3	2 879.5	2 893.6	105.0	121.0	121.0	61.8	65.8	72.1
Mexico Mexique	3 010.2	3 098.0	2 985.7	116.0	120.0	135.0	78.4	79.9	81.5

22
Food Supply
Calories and protein: per capita, per day [cont.]
Disponibilités alimentaires
Calories et protéine : par habitant, par jour [suite]

Country or area Pays ou zone	Calories - Calories						Protein (grams) Protéine (grammes)		
	Number - Nombre			% of requirements - % des besoins					
	1980	1986	1990	1967/70	1980	1984/86	1980	1986	1990
Mongolia Mongolie	2 478.8	2 533.1	2 302.8	99.0	111.0	116.0	92.5	92.2	85.1
Morocco Maroc	2 688.1	2 945.7	3 051.6	99.0	109.0	118.0	70.8	78.9	81.9
Mozambique Mozambique	1 958.7	1 761.5	1 802.5	89.0	80.0	69.0	33.2	30.1	30.3
Myanmar Myanmar	2 347.6	2 513.9	2 447.6	100.0	107.0	119.0	60.0	65.5	61.5
Namibia Namibie	1 922.3	1 943.3	1 944.9	100.0	96.0	...	64.2	64.6	62.2
Nepal Népal	1 847.2	2 001.8	2 246.4	92.0	88.0	93.0	48.1	51.6	57.5
Netherlands Pays-Bas	3 097.3	3 100.3	3 023.5	128.0	131.0	121.0	93.3	93.6	89.6
Netherlands Antilles Antilles néerlandaises	2 875.9	2 769.6	2 650.6	100.0	108.0	...	90.6	83.2	81.1
New Caledonia Nouvelle-Calédonie	2 831.7	2 806.6	2 942.8	110.0	94.0	...	72.1	72.8	79.6
New Zealand Nouvelle-Zélande	3 453.7	3 375.9	3 503.6	134.0	132.0	129.0	108.4	98.7	107.1
Nicaragua Nicaragua	2 318.5	2 316.1	2 213.7	112.0	97.0	110.0	59.5	56.0	53.3
Niger Niger	2 202.2	2 223.4	2 262.8	90.0	95.0	100.0	63.5	59.9	64.4
Nigeria Nigéria	2 157.2	2 287.5	2 146.8	93.0	100.0	90.0	47.0	50.8	42.9
Norway Norvège	3 383.5	3 261.7	3 219.2	116.0	124.0	120.0	102.7	100.4	99.9
Pakistan Pakistan	2 116.5	2 140.5	2 377.0	92.0	100.0	97.0	58.0	57.6	64.6
Panama Panama	2 269.3	2 584.0	2 290.5	107.0	98.0	107.0	56.9	63.0	59.3
Papua New Guinea Papouasie-Nvl-Guinée	2 316.9	2 495.4	2 609.3	80.0	85.0	96.0	45.9	50.7	51.7
Paraguay Paraguay	2 624.1	2 643.4	2 644.3	117.0	126.0	123.0	73.4	68.7	67.9
Peru Pérou	2 078.1	2 074.8	1 890.0	95.0	93.0	93.0	54.1	56.8	50.8
Philippines Philippines	2 252.6	2 175.7	2 452.1	86.0	103.0	104.0	51.9	50.3	56.5
Poland Pologne	3 571.1	3 393.0	3 351.1	128.0	134.0	126.0	110.5	101.7	100.2
Portugal Portugal	2 929.4	3 139.5	3 420.4	124.0	128.0	128.0	77.5	91.8	97.9
Romania Roumanie	3 454.9	3 286.6	3 043.1	114.0	126.0	127.0	106.9	99.1	92.9
Rwanda Rwanda	2 045.9	1 934.5	1 961.0	86.0	95.0	81.0	51.5	50.2	48.9
Réunion Réunion	2 825.1	2 962.4	3 112.0	108.0	130.0	...	75.8	79.2	87.3

22
Food Supply
Calories and protein: per capita, per day [cont.]
Disponibilités alimentaires
Calories et protéine : par habitant, par jour [suite]

Country or area Pays ou zone	Calories - Calories						Protein (grams) Protéine (grammes)		
	Number - Nombre			% of requirements - % des besoins					
	1980	1986	1990	1967/70	1980	1984/86	1980	1986	1990
Saint Kitts and Nevis Saint-Kitts-et-Nevis	2 168.5	2 441.9	2 423.5	...	...	...	60.3	66.3	69.7
Saint Lucia Sainte-Lucie	2 116.7	2 395.2	2 429.2	88.0	99.0	...	53.1	65.8	69.7
St. Vincent-Grenadines St. Vincent-Grenadines	2 479.3	2 417.9	2 470.4	91.0	91.0	...	54.4	55.0	57.2
Samoa Samoa	2 514.6	2 501.7	* 2 695.0	79.0	86.0	...	61.3	62.8	* 66.3
Sao Tome and Principe Sao Tomé-et-Principe	2 084.6	2 073.3	2 171.0	93.0	101.0	...	47.4	42.1	45.6
Saudi Arabia Arabie saoudite	2 772.2	2 721.0	3 022.8	87.0	120.0	125.0	76.7	79.8	88.6
Senegal Sénégal	2 411.4	2 442.6	2 328.4	99.0	100.0	99.0	68.9	73.6	65.3
Seychelles Seychelles	2 271.2	2 342.2	2 343.7	...	...	...	63.2	65.2	58.7
Sierra Leone Sierra Leone	2 076.2	1 908.1	1 939.5	96.0	89.0	81.0	44.8	40.6	39.9
Singapore Singapour	2 722.8	3 079.6	3 114.3	114.0	136.0	124.0	76.1	87.2	87.1
Solomon Islands Iles Salomon	2 287.8	2 318.8	2 234.6	81.0	81.0	...	57.8	55.9	53.3
Somalia Somalie	1 910.5	1 962.8	1 830.1	95.0	92.0	90.0	62.5	60.8	59.6
South Africa Afrique du Sud	2 980.1	3 060.0	3 158.3	112.0	116.0	120.0	77.6	76.7	79.0
Spain Espagne	3 268.2	3 311.1	3 493.7	116.0	136.0	137.0	96.0	96.8	101.4
Sri Lanka Sri Lanka	2 169.6	2 300.9	2 285.6	105.0	102.0	110.0	43.1	46.7	47.6
Sudan Soudan	2 273.6	2 098.7	1 964.1	86.0	102.0	88.0	64.6	58.4	56.5
Suriname Suriname	2 545.6	2 438.9	2 431.3	106.0	108.0	...	61.6	58.9	60.0
Swaziland Swaziland	2 468.4	2 533.9	2 647.6	93.0	108.0	...	63.1	61.0	62.3
Sweden Suède	3 034.8	2 965.1	2 961.2	113.0	119.0	113.0	98.0	95.6	93.7
Switzerland Suisse	3 573.3	3 497.4	3 446.7	127.0	132.0	128.0	97.3	95.4	94.5
Syrian Arab Republic Rép. arabe syrienne	2 922.8	3 163.9	3 106.6	99.0	117.0	131.0	78.3	85.2	81.5
Thailand Thaïlande	2 303.2	2 309.7	2 270.6	101.0	104.0	105.0	47.6	49.7	47.5
Togo Togo	2 244.8	2 078.3	2 279.3	96.0	93.0	97.0	47.8	47.7	51.1
Tonga Tonga	2 822.7	2 896.2	2 977.7	93.0	120.0	...	65.2	71.1	77.5
Trinidad and Tobago Trinité-et-Tobago	2 994.8	3 016.7	2 721.2	96.0	113.0	126.0	80.8	79.2	64.6

22
Food Supply
Calories and protein: per capita, per day [cont.]
Disponibilités alimentaires
Calories et protéine : par habitant, par jour [suite]

Country or area Pays ou zone	Calories - Calories						Protein (grams) Protéine (grammes)		
	Number - Nombre			% of requirements - % des besoins					
	1980	1986	1990	1967/70	1980	1984/86	1980	1986	1990
Tunisia Tunisie	2 829.0	3 018.1	3 168.6	93.0	116.0	123.0	78.3	82.0	83.3
Turkey Turquie	3 043.8	3 087.7	3 261.6	111.0	120.0	125.0	85.7	84.3	85.5
Uganda Ouganda	2 094.2	2 007.1	2 213.2	95.0	79.0	95.0	50.2	45.5	53.0
former USSR ancienne URSS	3 379.9	3 356.2	3 391.0	130.0	132.0	133.0	102.9	105.6	108.2
United Arab Emirates Emirats arabes unis	3 347.1	3 214.1	3 330.7	...	...	...	104.2	101.2	105.8
United Kingdom Royaume-Uni	3 146.4	3 241.5	3 281.5	132.0	132.0	128.0	86.8	92.1	93.7
United Rep.Tanzania Rép. Unie de Tanzanie	2 269.1	2 279.3	2 181.1	88.0	86.0	96.0	55.2	54.6	54.2
United States Etats-Unis	3 351.5	3 557.6	3 680.1	130.0	139.0	138.0	101.0	107.1	111.1
Uruguay Uruguay	2 822.3	2 605.6	2 678.0	109.0	109.0	100.0	85.5	79.0	82.5
Vanuatu Vanuatu	2 571.5	2 699.2	2 740.9	90.0	94.0	...	66.6	64.9	63.5
Venezuela Venezuela	2 710.5	2 393.0	2 382.8	96.0	107.0	102.0	68.9	59.8	57.9
Viet Nam Viet Nam	2 050.5	2 189.5	2 215.1	98.0	91.0	105.0	45.7	49.7	51.2
Yemen Yémen	2 064.0	2 089.7	2 280.2	86.0	93.0	94.0	63.1	60.4	65.0
Yugoslavia Yougoslavie	3 592.9	3 644.3	3 530.5	131.0	140.0	139.0	101.9	100.0	98.4
Zaire Zaïre	2 141.9	2 178.1	2 094.3	99.0	96.0	98.0	34.8	36.2	33.7
Zambia Zambie	2 197.2	2 017.2	2 019.4	92.0	91.0	92.0	59.5	54.1	53.9
Zimbabwe Zimbabwe	2 181.1	2 201.9	2 247.1	87.0	80.0	89.0	55.8	52.2	53.7

Source:
Food and Agriculture Organization of the United Nations
(Rome).

† All data shown which pertain to Germany prior to 3 October
1990 are indicated separately for the Federal Republic of
Germany and the former German Democratic Republic based on
their respective territories at the time indicated. Where
data for united Germany (3 October 1990 and thereafter) are
not available, available data are shown separately under the
designations Federal Republic of Germany and former German
Democratic Republic and pertain to the territorial
boundaries prior to 3 October 1990. For detailed
explanatory notes on data pertaining to Germany, see Annex I
- Country Nomenclature.

Source:
Organisation des Nations Unies pour l'alimentation et
l'agriculture (Rome).

† Toutes les données se rapportant à l'Allemagne avant le 3
octobre 1990 figurent dans deux rubriques séparées basées
sur les territoires respectifs de la République fédérale
d'Allemagne et l'ancienne République démocratique allemande
selon la période indiquée. En l'absence de données pour
l'Allemagne unifiée (à compter du 3 octobre 1990), les
données disponibles sont fournies séparément sous les
rubriques République fédérale d'Allemagne et ancienne
République démocratique allemande et se rapportent aux
limites territoriales antérieures au 3 octobre 1990. Pour
les notes explicatives en détail sur les données concernant
l'Allemagne, voir Annexe I – Nomenclature des pays.

Technical notes, tables 19-22

Table 19: The definitions of vital events and rates shown are as follows: [43]

— "Live Birth" is the completed expulsion or extraction from its mother of a product of conception, irrespective of the duration of pregnancy, which after such separation breathes or shows any other evidence of life such as beating of the heart, pulsation of the umbilical cord, or definite movement of voluntary muscles, whether or not the umbilical cord has been cut or the placenta is attached. Each product of such a birth is considered live-born regardless of gestational age;

— "Death" is the permanent disappearance of all evidence of life at any time after live birth has taken place (post-natal cessation of vital functions without capability of resuscitation). This definition therefore excludes foetal deaths;

— "Infant deaths" are deaths of live-born infants under one year of age;

— "Rates of natural increase" are the difference between the crude birth rate and the crude death rate.

Rates calculated by the Statistical Division of the United Nations Secretariat presented in this table are not limited to those countries or areas having a minimum number of events in a given year. However, rates based on 30 or fewer live births, infant deaths, marriages or divorces are identified by the symbol (♦).

Rates of natural increase which were calculated using crude birth rates and crude death rates considered unreliable, as described above, are set in italics rather than roman type.

Countries or areas not listed may be assumed to lack vital statistics of national scope. Crude birth, death, marriage, divorce and natural increase rates are computed per 1,000 mid-year population; infant mortality rates are per 1,000 live births and total fertility rates are the sum of the age-specific fertility rates per woman.

Table 20: The definitions of series shown are as follows: [19, 25, 43]

— "Life expectation at birth" is defined as the average number of years of life for males and females if they continued to be subject to the same mortality experienced in the year(s) to which these life expectancies refer;

— "Infant mortality rate" is the annual number of deaths of infants under one year of age per 1 000 live births (as shown in table 19) in the same year;

— "Total fertility rate" is the average number of children that would be born alive to a hypothetical cohort of women if, throughout their reproductive years, the age-specific fertility rates for the specified year remained unchanged;

Notes techniques, tableaux 19-22

Tableau 19 : On trouvera ci-dessous les définitions des événements et des taux concernant la vie [43] :

— La "naissance vivante" est l'expulsion ou l'extraction complète du corps de la mère, indépendamment de la durée de gestation, d'un produit de la conception qui, après cette séparation, respire ou manifeste tout autre signe de vie, tel que battements de coeur, pulsations du cordon ombilical ou contractions effectives d'un muscle soumis à l'action de la volonté, que le cordon ombilical ait été coupé ou non et que le placenta soit ou non demeuré attaché. Tout produit d'une telle naissance est considéré comme "enfant né vivant".

— Le "décès" est la disparition permanente de tout signe de vie à un moment quelconque postérieur à la naissance vivante (cessation des fonctions vitales après la naissance sans possibilité de réanimation). Cette définition ne comprend donc pas les morts foetales.

— Les "décès d'enfants de moins d'un an" sont les décès d'enfants nés vivants survenus dans leur première année.

— Le "taux d'accroissement naturel" est égal à la différence entre le taux brut de natalité et le taux brut de mortalité.

Les taux calculés par la Division de statistique de l'ONU qui sont présentés dans ce tableau ne se rapportent pas aux seuls pays ou zones où l'on a enregistré un certain nombre minimal d'événements au cours d'une année donnée. Toutefois, les taux qui sont fondés sur un maximum de 30 naissances vivantes, décès d'enfants de moins d'un an, mariages ou divorces sont identifiés par le signe (♦).

Les taux d'accroissement naturel calculés à partir de taux bruts de natalité et de taux bruts de mortalité jugés douteux d'après les normes mentionnées plus haut sont indiqués en italique plutôt qu'en caractères romains.

On peut supposer que les pays ou zones non énumérés ne disposent pas de statistiques de portée nationale. Les taux de natalité brute, de mortalité, de mariage, de divorce et d'accroissement naturel sont calculés par tranches de 1.000 habitants en milieu d'année; les taux de mortalité infantile sont donnés pour 1.000 naissances vivantes et les indices synthétiques de fécondité sont la somme des indices de fécondité par femme d'un âge donné.

Tableau 20 : Les définitions des séries présentées sont données ci-après [19, 25, 43] :

— L'"espérance de vie à la naissance" est définie comme le nombre moyen d'années de vie pour les hommes et les femmes s'ils restent soumis aux mêmes conditions de mortalité caractérisant les années auxquelles se rapporte l'espérance de vie;

— Le "taux de mortalité infantile" est le nombre annuel de décès d'enfants de moins d'un an pour 1.000 naissances vivantes (indiqué au tableau 19) survenus la même année;

The above estimates and rates were prepared by the Population Division of the United Nations Secretariat and published in the forthcoming *World Population Prospects 1992* [25].

— "Child mortality rates" refer to the number of children aged one to five years who die per year per 1000 population in that age group. These series have been compiled by the Statistical Division of the United Nations Secretariat for the *Demographic Yearbook* [19: various years up to *1990*], by the United Nations Children's Fund [28] and by the Pan American Health Organization (unpublished), and are subject to the limitations of national reporting in this field described in those sources;

Data on maternal mortality are compiled in the *Demographic Yearbook* [19: *1980 and 1985*] and by WHO. Maternal mortality is calculated on the basis of maternal deaths and live births for a given year. These statistics are based on national civil registration and demographic survey statistics on births and deaths computed by national statistical services. The statistical concepts and definitions used are those recommended by WHO in the International Classification of Diseases, Injuries and Causes of Death. [54]

Table 21: Data on acquired immunodeficiency syndrome (AIDS) have been compiled and estimated by WHO from official national reports and special studies (unpublished).

Table 22: Data on per capita food supply are expressed in terms of nutrient elements such as calories and protein. These series represent average supply for the population as a whole and do not show what is actually consumed by individuals. The series on calorie supply in relation to requirements have been compiled from estimates prepared by the Food and Agriculture Organization of the United Nations (FAO). "Requirements" are the energy intake that is considered adequate to meet the energy needs of the average health person in a specified age/sex category. They have been derived by FAO and/or countries considering four interrelated variables: (a) physical activity; (b) body size and composition; (c) age; and (d) climate or other ecological factors. [37]

— L'"indice synthétique de fécondité" est le nombre moyen d'enfants que mettrait au monde une cohorte hypothétique de femmes si, pendant toutes leurs années d'âge reproductif, les taux de fécondité par âge de l'année en question restaient inchangés;

Les estimatifs et taux susmentionnés ont été établis par la Division de la population du Secrétariat des Nations Unies et ont été publiés dans "World Population Prospect 1992" (à paraître). [25]

— Le "taux de mortalité juvénile" désigne le nombre d'enfants âgés de un à cinq ans qui meurent par an, pour 1.000 enfants de ce groupe d'âges. Cette série a été compilée par la Division de statistique du Secrétariat de l'ONU pour l'*Annuaire démographique* [19 : diverses années jusqu'en 1990], par le Fonds des Nations Unies pour l'enfance (UNICEF) [28] et par l'Organisation panaméricaine de la santé (document non publié), et les données qui sont incluses sont présentées sous réserve de mises en garde formulées à leur sujet dans les documents d'origine nationale.

Les données sur la mortalité maternelle proviennent de l'*Annuaire démographique* [19 : 1980 et 1985] et de l'OMS. La mortalité maternelle est calculée sur la base du nombre de décès de mères et de naissances vivantes au cours d'une année donnée. Ces statistiques proviennent des registres d'état civil des pays et des informations produites par les enquêtes démographiques sur les naissances et les décès rassemblées par les services statistiques nationaux. Les concepts statistiques et les définitions utilisés sont ceux que recommande l'OMS dans la Classification internationale des maladies, traumatismes et causes de décès [54].

Tableau 21 : Les données sur le syndrome d'immuno-déficience acquise (SIDA) ont été compilées et estimées par l'OMS sur la base des rapports officiels soumis par les pays et d'études spéciales (non publiées).

Tableau 22 : Les données sur les disponibilités alimentaires par habitant sont exprimées en éléments nutritifs tels que calories et protéines. Ces séries représentent la ration alimentaire moyenne de l'ensemble de la population et n'indiquent pas la consommation effective des individus. Les séries sur l'apport calorique par référence aux besoins ont été compilées à partir d'estimations préparées par l'Organisation des Nations Unies pour l'alimentation et l'agriculture (FAO). Les "besoins" correspondent à l'apport énergétique jugé nécessaire pour répondre aux besoins en énergie d'une personne de santé moyenne appartenant à une catégorie particulière définie par l'âge et le sexe. Ils ont été établis par la FAO et/ou les pays, compte tenu de quatre variables interdépendantes : a) l'activité physique; b) la taille et la structure du corps; c) l'âge; et d) le climat et d'autres facteurs tenant à l'environnement [37].

23
Newsprint consumption
Consommation de papier journal

Country or area Pays ou zone	Total: metric tons Totale : tonnes métriques					Per 1 000 inhabitants: kilograms Par 1 000 habitants : kilogrammes				
	1970	1975	1980	1985	1989	1970	1975	1980	1985	1989
Africa · Afrique										
Algeria Algérie	6 200	8 900	12 000	15 600	4 400	451	556	643	719	181
Angola Angola	...	1 100	1 500	500	500	...	169	194	57	51
Benin Bénin	...	100	100	100	100	...	33	29	25	22
Cameroon Cameroun	100	...	500	2 100	...	15	...	58	209	...
Côte d'Ivoire Côte d'Ivoire	900	800	800	1 400	2 200	163	118	96	137	191
Egypt Egypte	32 900	40 200	67 000	35 300	84 300	995	1 108	1 614	742	1 648
Ethiopia Ethiopie	900	400	2 000	2 200	2 300	29	12	52	52	48
Ghana Ghana	3 600	7 300	1 500	4 000	1 800	418	743	140	312	124
Guinea-Bissau Guinée-Bissau	...	100	100	200	200	...	159	123	225	212
Kenya Kenya	5 000	3 900	7 900	9 500	13 400	435	284	475	467	578
Liberia Libéria	100	...	...	200	100	73	...	...	92	40
Libyan Arab Jamah. Jamah. arabe libyenne	800	100	800	2 000	2 000	403	41	263	528	457
Madagascar Madagascar	300	817	3 500	800	240	44	108	399	78	21
Malawi Malawi	200	500	600	600	500	44	95	99	84	59
Mali Mali	200	100	...	...	...	35	16	...	...	...
Mauritius Maurice	600	600	500	1 400	2 100	708	690	523	1 351	1 959
Morocco Maroc	3 100	2 800	5 400	4 800	11 800	202	162	279	217	483
Mozambique Mozambique	200	1 700	1 500	100	800	21	162	124	7	52
Niger Niger	...	150	100	500	500	...	32	19	82	67
Nigeria Nigéria	17 100	23 700	29 000	14 800	52 800	299	350	360	155	503
Réunion Réunion	400	500	1 000	3 100	2 200	905	1 033	1 969	5 678	3 735
Senegal Sénégal	500	1 800	900	1 000	1 000	125	377	159	155	140
Sierra Leone Sierra Leone	200	100	200	200	200	75	34	61	55	49
Somalia Somalie	500	300	200	100	100	136	72	37	16	14

23
Newsprint consumption
[cont.]

Consommation de papier journal
[suite]

Country or area Pays ou zone	Total: metric tons Totale : tonnes métriques					Per 1 000 inhabitants: kilograms Par 1 000 habitants : kilogrammes				
	1970	1975	1980	1985	1989	1970	1975	1980	1985	1989
South Africa Afrique du Sud	212 500	195 800	154 000	120 100	204 100	9 462	7 737	5 445	3 801	5 913
Sudan Soudan	2 700	3 400	1 500	5 000	2 300	195	212	80	229	94
Tunisia Tunisie	...	2 900	5 500	9 900	10 000	...	517	862	1 363	1 252
Uganda Ouganda	1 000	500	200	200	200	102	45	15	13	11
United Rep.Tanzania Rép. Unie de Tanzanie	1 500	4 500	3 500	2 500	11 500	111	283	186	110	437
Zaire Zaïre	500	1 000	1 000	1 000	1 000	26	45	38	33	29
Zambia Zambie	2 100	4 300	3 000	3 000	3 000	501	888	523	428	369
Zimbabwe Zimbabwe	10 000	12 000	16 000	18 400	19 300	1 897	1 950	2 242	2 216	2 052
America, North · Amérique du Nord										
Bahamas Bahamas	1 000	600	9 000	400	400	5 848	2 941	42 857	1 653	1 606
Barbados Barbade	900	500	1 500	1 400	2 400	3 766	2 041	6 024	5 534	9 412
Belize Belize	200	200	200	200	200	1 667	1 527	1 379	1 227	1 099
Canada Canada	656 700	661 200	918 200	716 300	1 493 000	30 795	29 093	38 354	28 224	56 809
Costa Rica Costa Rica	11 100	11 200	12 000	11 000	8 500	6 409	5 691	5 256	4 164	2 893
Cuba Cuba	22 800	26 700	32 000	43 200	38 600	2 660	2 862	3 288	4 343	3 670
Dominican Republic Rép. dominicaine	4 200	6 600	12 400	7 800	11 700	950	1 307	2 177	1 216	1 667
El Salvador El Salvador	13 000	10 200	15 000	12 700	11 500	3 623	2 498	3 314	2 664	2 226
Guatemala Guatemala	8 300	8 100	17 000	7 200	12 000	1 582	1 345	2 458	904	1 342
Haiti Haïti	700	800	300	400	400	156	161	55	68	63
Honduras Honduras	2 700	2 100	6 500	6 400	7 300	1 028	681	1 774	1 460	1 466
Jamaica Jamaïque	8 600	8 700	3 500	4 500	28 900	4 601	4 258	1 610	1 926	11 893
Martinique Martinique	...	...	800	1 200	3 000	...	...	2 446	3 659	8 824
Mexico Mexique	158 800	214 600	226 000	297 500	446 000	3 009	3 466	3 210	3 748	5 141
Netherlands Antilles Antilles néerlandaises	600	500	600	2 100	2 100	2 703	2 092	2 381	11 667	11 230

23
Newsprint consumption
[*cont.*]

Consommation de papier journal
[*suite*]

Country or area Pays ou zone	Total: metric tons Totale : tonnes métriques					Per 1 000 inhabitants: kilograms Par 1 000 habitants : kilogrammes				
	1970	1975	1980	1985	1989	1970	1975	1980	1985	1989
Nicaragua Nicaragua	3 700	4 000	3 000	5 000	6 300	1 801	1 661	1 082	1 528	1 683
Panama Panama	5 900	3 400	2 600	6 300	1 700	3 854	1 945	1 330	2 889	717
Trinidad and Tobago Trinité-et-Tobago	5 600	6 300	5 100	11 900	5 000	5 858	6 250	4 653	10 042	3 965
United States Etats-Unis	9 032[1]	8 503[1]	10 673[1]	12 324[1]	12 854[1]	44 048	39 372	46 861	51 502	52 026
America, South · Amérique du Sud										
Argentina Argentine	277 400	148 800	270 800	203 300	262 500	11 577	5 712	9 590	6 703	8 222
Bolivia Bolivie	4 700	4 700	6 000	3 000	8 100	1 087	961	1 077	471	1 138
Brazil Brésil	211 600	240 700	271 600	299 000	369 600	2 208	2 228	2 239	2 206	2 508
Chile Chili	46 100	41 700	68 000	56 000	49 400	4 850	4 029	6 101	4 620	3 812
Colombia Colombie	59 200	44 400	71 000	77 500	72 800	2 846	1 916	2 753	2 699	2 248
Ecuador Equateur	13 900	10 800	32 600	39 600	22 000	2 297	1 535	4 013	4 223	2 130
Guyana Guyana	1 200	1 400	1 500	500	500	1 690	1 793	1 734	525	629
Paraguay Paraguay	4 100	3 100	8 000	8 000	6 700	1 744	1 157	2 542	2 166	1 614
Peru Pérou	49 400	51 300	36 500	39 000	85 000	3 744	3 383	2 110	1 980	4 023
Suriname Suriname	500	400	600	1 000	900	1 337	1 099	1 690	2 667	2 174
Uruguay Uruguay	20 800	10 600	15 200	7 500	5 500	7 407	3 747	5 227	2 490	1 785
Venezuela Venezuela	84 300	85 600	141 000	136 000	96 000	7 951	6 758	9 386	7 854	4 992
Asia · Asie										
Afghanistan Afghanistan	...	1 200	100	...	...	...	78	6	...	...
Bangladesh Bangladesh	36 000	14 700	19 000	28 900	49 000	540	192	215	286	435
Brunei Darussalam Brunéi Darussalam	100	200	200	400	400	769	1 282	1 081	1 786	1 563
China Chine	392 600	395 400	499 800	784 200	999 900	473	426	502	740	891
Cyprus Chypre	1 700	1 300	2 600	3 200	3 100	2 764	2 131	4 134	4 812	4 473
Hong Kong Hong-kong	45 000	53 300	75 800	113 900	167 000	11 416	12 125	15 043	20 876	28 983

23
Newsprint consumption
[*cont.*]

Consommation de papier journal
[*suite*]

Country or area Pays ou zone	Total: metric tons Totale : tonnes métriques					Per 1 000 inhabitants: kilograms Par 1 000 habitants: kilogrammes				
	1970	1975	1980	1985	1989	1970	1975	1980	1985	1989
India Inde	181 500	152 800	310 000	378 000	510 000	327	246	450	491	610
Indonesia Indonésie	43 600	46 700	66 000	157 200	133 800	362	344	437	944	740
Iran, Islamic Rep. of Iran, Rép. islamique d'	10 900	32 800	14 000	18 000	10 000	384	984	360	378	188
Iraq Iraq	3 100	4 900	14 000	20 000	34 000	331	445	1 053	1 258	1 860
Israel Israël	34 000	37 700	38 100	49 700	69 000	11 429	10 912	9 825	11 741	15 269
Japan Japon	1 973[1]	2 082[1]	2 703[1]	2 841[1]	3 560[1]	18 911	18 670	23 144	23 530	28 977
Jordan Jordanie	600	700	2 600	8 400	8 900	261	269	889	2 396	2 290
Korea, Dem. P. R. Corée, R. p. dém. de	1 300	1 300	1 300	3 200	3 200	94	82	72	157	150
Korea, Republic of Corée, République de	108 300	150 600	227 000	230 800	427 500	3 392	4 269	5 954	5 622	10 070
Kuwait Koweït	...	4 800	19 000	18 000	20 000	...	4 762	13 818	10 526	10 204
Lao People's Dem. Rep. Rép. dém. pop. lao	300	200	200	...	...	111	66	62	...	...
Lebanon Liban	5 100	6 300	7 000	4 700	3 900	2 066	2 277	2 621	1 762	1 447
Macau Macao	3 700	2 800	6 200	...	...	15 102	10 769	19 195	...	...
Malaysia Malaisie	38 200	33 900	65 900	100 000	78 000	3 520	2 766	4 788	6 473	4 480
Mongolia Mongolie	2 400	2 600	2 300	3 800	3 800	1 923	1 799	1 382	1 992	1 783
Myanmar Myanmar	15 900	6 500	4 000	7 500	11 600	587	214	118	200	284
Nepal Népal	...	...	...	1 200	1 300	...	...	...	71	70
Oman Oman	...	...	...	700	600	...	...	...	564	416
Pakistan Pakistan	500	6 700	33 100	40 400	52 700	8	90	388	391	445
Philippines Philippines	75 500	68 800	100 000	58 000	103 300	2 011	1 616	2 070	1 052	1 696
Saudi Arabia Arabie saoudite	500	2 300	13 000	28 000	28 000	87	317	1 387	2 415	2 064
Singapore Singapour	23 600	28 200	55 700	56 900	78 600	11 373	12 461	23 074	22 235	29 274
Sri Lanka Sri Lanka	18 300	6 500	8 000	17 200	10 000	1 462	478	540	1 068	588
Syrian Arab Republic Rép. arabe syrienne	1 500	900	3 700	5 600	4 600	240	121	420	535	380

23
Newsprint consumption
[*cont.*]

Consommation de papier journal
[*suite*]

Country or area Pays ou zone	Total: metric tons Totale : tonnes métriques					Per 1 000 inhabitants: kilograms Par 1 000 habitants: kilogrammes				
	1970	1975	1980	1985	1989	1970	1975	1980	1985	1989
Thailand Thaïlande	36 300	63 400	91 600	137 800	136 700	1 016	1 533	1 961	2 670	2 490
Turkey Turquie	26 100	98 200	157 400	166 000	136 000	739	2 453	3 542	3 297	2 485
Viet Nam Viet Nam	21 700	2 000	2 500	10 000	11 000	508	42	47	167	168
Yemen[2] Yémen[2]	...	100	500	300	300	...	60	269	140	124
Europe · Europe										
Austria Autriche	103 400	120 800	138 400	247 300	238 600	13 883	16 062	18 329	32 965	31 503
Belgium Belgique	179 700	161 900	197 000	221 500	281 000	18 610	16 527	19 998	22 367	28 554
Bulgaria Bulgarie	35 400	49 000	46 200	42 300	37 200	4 170	5 618	5 214	4 721	4 137
Czechoslovakia Tchécoslovaquie	49 700	68 000	72 000	74 500	79 700	3 467	4 594	4 702	4 806	5 106
Denmark Danemark	149 000	125 300	152 600	187 800	218 800	30 229	24 763	29 793	36 665	42 568
Finland Finlande	118 900	215 800	137 100	167 600	219 000	25 814	45 798	28 688	34 197	44 189
France France	605 500	402 000	628 800	568 300	757 000	11 926	7 628	11 670	10 301	13 554
Germany† · Allemagne† F. R. Germany R. f. Allemagne	1 055[1]	1 077[1]	1 394[1]	1 406[1]	1 696[1]	17 397	17 414	22 642	23 040	27 704
former German D. R. anc. R. d. allemande	117 300	140 200	141 300	156 500	128 200	6 873	8 320	8 442	9 403	7 865
Greece Grèce	31 600	47 900	41 100	79 000	87 400	3 594	5 295	4 263	7 952	8 718
Hungary Hongrie	53 600	56 300	61 000	67 500	67 000	5 177	5 341	5 695	6 339	6 348
Iceland Islande	3 100	3 000	4 000	5 800	4 800	15 196	13 761	17 544	24 066	19 124
Ireland Irlande	54 200	56 700	60 600	51 000	62 000	18 348	17 686	17 818	14 358	16 802
Italy Italie	282 300	244 900	327 700	382 000	610 000	5 245	4 417	5 807	6 687	10 683
Malta Malte	300	200	700	1 500	1 200	920	615	1 928	4 360	3 419
Netherlands Pays-Bas	378 700	359 700	457 800	384 200	517 600	29 057	26 321	32 351	26 526	34 855
Norway Norvège	78 400	54 500	66 200	115 700	160 200	20 217	13 601	16 206	27 859	38 152

23
Newsprint consumption
[*cont.*]

Consommation de papier journal
[*suite*]

Country or area Pays ou zone	Total: metric tons Totale : tonnes métriques					Per 1 000 inhabitants: kilograms Par 1 000 habitants : kilogrammes				
	1970	1975	1980	1985	1989	1970	1975	1980	1985	1989
Poland Pologne	87 900	121 000	127 000	135 400	38 900	2 702	3 556	3 570	3 639	1 022
Portugal Portugal	44 000	27 400	41 000	40 300	65 000	4 865	3 013	4 198	3 968	6 327
Romania Roumanie	51 700	44 100	56 000	60 000	78 000	2 539	2 076	2 522	2 640	3 375
Spain Espagne	193 800	203 900	162 400	263 100	288 800	5 737	5 728	4 326	6 816	7 386
Sweden Suède	343 400	292 200	295 600	242 300	324 200	42 701	35 665	35 567	29 018	38 517
Switzerland Suisse	167 500	140 800	207 000	247 000	299 000	26 732	21 979	32 717	38 176	45 448
United Kingdom Royaume-Uni	1 544[1]	1 257[1]	1 381[1]	1 561[1]	2 054[1]	27 757	22 358	24 518	27 571	35 972
Yugoslavia Yougoslavie	88 500	85 100	30 000	30 600	29 700	4 344	3 986	1 345	1 324	1 257
Oceania · Océanie										
Australia Australie	448 500	520 300	556 600	651 600	650 000	35 734	38 179	37 874	41 350	39 074
Fiji Fidji	800	900	1 700	2 000	1 900	1 538	1 560	2 703	2 894	2 530
French Polynesia Polynésie française	400	300	100	1 100	1 100	3 670	2 256	676	6 748	5 556
New Caledonia Nouvelle-Calédonie	...	1 100	1 300	300	1 100	...	8 271	9 353	1 961	6 707
New Zealand Nouvelle-Zélande	95 800	114 900	78 100	132 100	139 300	33 984	37 209	25 096	40 684	41 409
Papua New Guinea Papouasie-Nvl-Guinée	600	500	...	...	...	248	185	...	...	...
former USSR · ancienne URSS										
former USSR ancienne URSS	928[1]	1 139[1]	1 064[1]	1 204[1]	1 204[1]	3 820	4 468	4 007	4 347	4 338

Source:
United Nations Educational, Scientific and Cultural
Organizations (Paris).

Source:
Organisation des Nations Unies pour l'education, la science
et la culture (Paris).

† All data shown which pertain to Germany prior to 3 October
1990 are indicated separately for the Federal Republic of
Germany and the former German Democratic Republic based on
their respective territories at the time indicated. For
detailed explanatory notes on data pertaining to Germany,
see Annex I – Country Nomenclature.

1 Data in thousand metric tons.
2 Former Democratic Yemen only.

† Toutes les données se rapportant à l'Allemagne avant le 3
octobre 1990 figurent dans deux rubriques séparées basées
sur les territoires respectifs de la République fédérale
d'Allemagne et l'ancienne République démocratique allemande
selon la période indiquée. Pour les notes explicatives en
détail sur les données concernant l'Allemagne, voir Annexe I
- Nomenclature des pays.

1 Les données sont en milliers de tonnes métriques.
2 L'ancien Yémen démocratique seulement.

24
Book production: number of titles by UDC classes
Production de livres : nombre de titres classés d'après la CDU

Country or area Pays ou zone	Year Année	Total	Gener-alities Géné-ralités	Philo-sophy Philo-sophie	Reli-gion	Social sciences Sciences sociales	Philo-logy Philo-logie	Pure sciences Sciences pures	Applied sciences Sciences appli-quées	Arts Beaux arts	Litera-ture Litté-rature	Geogr/History Géogr., histoire
Africa · Afrique												
Angola Angola	1986	14	-	-	-	-	-	-	-	-	14	-
Botswana [1] Botswana [1]	1987	289	5	-	-	144	5	22	89	4	12	8
Burundi Burundi	1986	54	-	-	10	28	-	-	16	-	-	-
Cape Verde Cap-Vert	1989	10	-	-	-	2	-	-	1	-	6	1
Egypt Egypte	1988	1451	19	47	191	305	65	73	159	65	403	124
Equatorial Guinea [1] Guinée équatoriale [1]	1988	17	-	-	-	8	3	1	3	-	-	2
Ethiopia Ethiopie	1988	560	8	-	21	284	49	49	82	13	17	37
Gambia Gambie	1988	42	1	-	4	16	-	2	18	-	-	1
Kenya Kenya	1986	933	23	3	97	229	91	102	94	47	131	116
Libyan Arab Jamahiriya [1] Jamahiriya arabe libyenne [1]	1988	121	6	7	4	41	6	20	6	-	27	4
Madagascar Madagascar	1989	146	1	2	43	41	3	5	23	12	11	5
Malawi Malawi	1989	141	11	-	31	53	2	7	22	-	11	4
Mauritius Maurice	1989	100	5	2	9	24	12	1	17	2	15	13
Nigeria Nigéria	1989	1466	86	18	240	440	66	50	246	82	146	92
Rwanda Rwanda	1987	207	18	5	42	72	9	20	31	3	4	3
Tunisia [2] Tunisie [2]	1988	293	21	8	22	31	8	6	16	22	137	22
Zimbabwe Zimbabwe	1989	337	2	-	11	142	43	18	71	6	29	15
America, North · Amérique du Nord												
Costa Rica [3] Costa Rica [3]	1989	198	2	4	22	76	10	4	29	7	30	14
Cuba Cuba	1989	2199	98	11	2	525	111	200	678	195	315	64
El Salvador [2] El Salvador [2]	1988	15	-	-	-	8	-	1	-	-	4	2
Haiti Haïti	1989	271	9	4	6	104	5	8	21	18	81	15
Mexico [2] Mexique [2]	1989	3490	525	213	184	1264	138	316	330	181	63	276
Nicaragua [4] Nicaragua [4]	1987	41	-	-	-	15	-	-	3	2	21	-
St. Kitts and Nevis [5] St. Kitts-et-Nevis [5]	1988	3	-	-	-	3	-	-	-	-	-	-
America, South · Amérique du Sud												
Argentina [1] Argentine [1]	1987	4836	-	278	368	1642	31	119	513	177	1410	298
Bolivia Bolivie	1988	447	25	7	26	171	10	24	56	25	56	47
Chile Chili	1989	2350	25	36	123	908	42	72	296	69	567	212
Colombia Colombie	1989	1486	81	44	88	431	34	43	206	58	363	138

24
Book production: number of titles by UDC classes [*cont.*]
Production de livres : nombre de titres classés d'après la CDU [*suite*]

Country or area Pays ou zone	Year Année	Total	Gener- alities Géné- ralités	Philo- sophy Philo- sophie	Reli- gion	Social sciences Sciences sociales	Philo- logy Philo- logie	Pure sciences Sciences pures	Applied sciences Sciences appli- quées	Arts Beaux arts	Litera- ture Litté- rature	Geogr/ History Géogr., histoire
Guyana Guyana	1989	46	1	-	-	29	-	-	4	3	4	5
Peru Pérou	1988	481	10	12	17	177	8	19	87	17	71	63
Uruguay Uruguay	1989	805	-	11	26	330	19	54	90	19	154	102
Venezuela Venezuela	1987	1202	34	64	83	355	30	64	281	66	195	30
Asia · Asie												
Bangladesh [1] [2] Bangladesh [1] [2]	1988	1209	63	21	60	457	27	66	107	14	300	94
Brunei Darussalam Brunéi Darussalam	1989	16	3	-	-	3	-	-	-	-	7	3
China [6] Chine [6]	1989	74973	1026	1225	...	33699	1886	2556	11841	4355	7648	2597
Cyprus Chypre	1989	561	23	4	26	116	72	26	109	14	113	58
India [2] Inde [2]	1989	11851	261	447	1065	3342	217	528	865	323	3591	1212
Indonesia [7] Indonésie [7]	1989	1396	199	51	132	345	42	63	301	33	128	102
Iran, Islamic Rep. of [7] Iran, Rép. islamique d' [7]	1989	6289	168	283	1354	484	632	387	967	310	1292	412
Japan [8] Japon [8]	1987	36346	1190	1156	630	8015	1523	1924	6875	5046	8170	1817
Korea, Republic of Corée, République de	1989	39267	1573	1068	3122	6178	2867	2568	3382	6597	9487	2425
Kuwait Koweït	1988	793	8	9	73	144	134	309	74	19	1	22
Malaysia Malaisie	1989	3348	94	16	304	734	585	454	316	116	560	169
Mongolia Mongolie	1986	889	-	-	-	467	6	16	181	41	178	-
Nepal [9] Népal [9]	1989	122	-	4	7	19	25	18	27	3	13	6
Philippines Philippines	1988	1072	101	10	72	248	70	38	170	272	58	33
Qatar Qatar	1987	461	13	8	94	67	36	85	33	15	52	58
Sri Lanka Sri Lanka	1989	2188	23	17	245	1075	79	41	145	127	344	92
Thailand Thaïlande	1989	11217	693	192	644	4740	295	837	2295	390	727	404
United Arab Emirates [9] Emirats arabes unis [9]	1989	152	17	2	26	2	32	58	3	-	-	12
Europe · Europe												
Albania [10] Albanie [10]	1986	959	5	8	-	175	...	186	279	16	254	36
Andorra [1] [2] Andorre [1] [2]	1989	45	-	-	4	7	-	-	4	12	3	15
Austria Autriche	1989	9462	225	388	371	2078	317	1422	1794	836	1256	775
Belgium [11] Belgique [11]	1989	6822	200	186	237	1178	143	244	779	459	2870	526
Bulgaria Bulgarie	1989	4543	178	67	9	1235	140	461	1146	159	829	319
Czechoslovakia Tchécoslovaquie	1989	9294	537	117	69	1809	334	920	2951	742	1454	361
Denmark Danemark	1989	10762	197	391	284	2058	284	718	2543	703	2595	989

24
Book production: number of titles by UDC classes [*cont.*]
Production de livres : nombre de titres classés d'après la CDU [*suite*]

Country or area Pays ou zone	Year Année	Total	Gener- alities Géné- ralités	Philo- sophy Philo- sophie	Reli- gion	Social sciences Sciences sociales	Philo- logy Philo- logie	Pure sciences Sciences pures	Applied sciences Sciences appli- quées	Arts Beaux arts	Litera- ture Litté- rature	Geogr/ History Géogr., histoire
Finland Finlande	1989	10097	397	165	326	2661	154	900	2677	543	1772	502
France [12] France [12]	1989	40115	725	1397	1001	10426	848	1886	4983	3029	10911	4909
Germany ℊ · Allemagne ℊ Federal Republic of Germany Rép. féd. d'Allemagne	1989	65980	5511	2890	3709	14568	2894	1958	9048	5296	11806	8300
former German Dem. Rep. [13] ancienne Rép. dém. allemande [13]	1989	6018	87	103	281	702	287	458	910	448	1399	287
Holy See Saint-Siège	1989	176	4	38	101	23	6	-	-	2	-	2
Hungary Hongrie	1989	8631	174	125	184	1667	355	636	2136	683	1996	675
Iceland Islande	1989	1250	41	23	31	189	82	95	131	77	444	137
Italy Italie	1989	22647	611	1190	1444	4994	494	715	3115	2594	5260	2230
Luxembourg Luxembourg	1989	520	22	6	16	178	9	12	28	100	68	81
Malta Malte	1989	386	10	1	60	180	13	9	12	38	32	31
Monaco Monaco	1989	48	12	1	1	5	1	4	2	12	5	5
Netherlands [2 14] Pays-Bas [2 14]	1989	15392	118	619	806	1690	280	361	2367	899	3048	1189
Norway [15] Norvège [15]	1989	5331	264	104	217	1522	127	315	705	312	1413	352
Poland Pologne	1989	10286	171	206	478	1948	389	929	2827	564	1888	886
Portugal [10] Portugal [10]	1987	7733	935	192	415	1445	...	604	939	458	2053	692
Romania Roumanie	1989	3867	84	45	49	389	152	502	1322	136	1042	146
Spain Espagne	1989	38353	995	1527	1674	5931	2166	2455	4615	3097	11859	4034
Sweden [16] Suède [16]	1989	11197	287	233	385	1864	309	688	2397	793	3236	1005
Switzerland [14] Suisse [14]	1989	13270	213	627	728	2799	213	1002	2839	1418	1667	1008
Yugoslavia Yougoslavie	1989	11339	492	239	369	3169	23	413	1712	1197	3171	554
Oceania · Océanie												
Australia Australie	1987	10963	390	76	391	4000	624	596	1766	994	1074	1052
New Caledonia Nouvelle-Calédonie	1987	14	-	-	-	4	2	-	1	3	-	4
former USSR · ancienne URSS												
former USSR [17] ancienne URSS [17]	1989	76711	2244	1596	299	19676	1933	6425	27458	2655	11692	2219
Belarus [17] Bélarus [17]	1988	2962	130	63	11	946	96	141	1071	82	376	46
Ukraine [17] Ukraine [17]	1988	8311	351	179	91	2241	180	708	2936	194	1191	171

Source:
United Nations Educational, Scientific and Cultural Organization (Paris).

ℊ   All data shown which pertain to Germany prior to 3 October 1990 are indicated separately for the Federal Republic of Germany and the former German Democratic Republic based on their respective

Source:
Organisation des Nations Unies pour l'éducation, la science et la culture (Paris).

ℊ   Toutes les données se rapportant à l'Allemagne avant le 3 octobre 1990 figurent dans deux rubriques séparées basées sur les territoires

24
Book production: number of titles by UDC classes [*cont.*]
Production de livres : nombre de titres classés d'après la CDU [*suite*]

territories at the time indicated. For detailed explanatory notes on data pertaining to Germany, see Annex I - Country Nomenclature.

1   All first editions.
2   Data do not include pamphlets.
3   University theses are included in the total but not in the class breakdown.
4   Data on pamphlets refer to the first editions of school text books and children's books.
5   Data refer only to first editions of school textbooks.
6   Data include 8140 titles for which a class breakdown is not available.
7   Pamphlets are included in the total but not in the class breakdown.
8   Data do not include pamphlets and government publications and refer only to first editions.
9   Data refer to school textbooks only.
10   Data on philology are included with those on literature.
11   Data do not include pamphlets; children's books are included in the total but not in the class breakdown.
12   Government publications are included in the total but not in the class breakdown.
13   Data include only books and pamphlets shown in series A of the German National Bibliography (publications of the book market); those of the series B (publications outside the book market) and C (books published by universities) are excluded. School textbooks and children's books are included in the total but not in the class breakdown.
14   School textbooks and children's books are included in the total but not in the class breakdown.
15   Data do not include schools textbooks.
16   Children's books are included in the total but not in the class breakdown.
17   Books for the popularization of science for children are included in the total but not in the class breakdown.

respectifs de la République fédérale d'Allemagne et l'ancienne République démocratique allemande selon la période indiquée. Pour les notes explicatives en détail sur les données concernant l'Allemagne voir Annexe I - Nomenclature des pays.

1   Tous les ouvrages recensés sont des premières éditions.
2   Les données ne comprennent pas les brochures.
3   Les thèses universitaires sont comprises dans le total mais ne sont pas réparties par catégories.
4   Les données dans les brochures se réfèrent aux premieres éditions des manuels scolaires et les livres pour enfants.
5   Les données se réfèrent seulement aux premieres éditions des manuels scolaires.
6   Les données incluent 8140 titres pour lesquels aucune répartition par catégorie n'est disponible.
7   Les brochures sont comprises dans le total mais ne sont pas réparties par catégories.
8   Les données ne comprennent ni les brochures ni publications officielles, et ne se referent qu'aux premieres éditions.
9   Les données se réfèrent aux manuels scolaires seulement.
10   Les données relatives à la philologie sont comprises avec celles de la littérature.
11   Les données ne comprennent pas les brochures; les livres pour enfants sont comprises dans le total mais non répartis par catégories.
12   Les publications officielles sont comprises dans le total mais ne sont pas réparties par catégories.
13   Les données se réfèrent aux livres et brochures de la série A de la Bibliographie nationale allemande (publications en vente dans le commerce). Les données relatives aux séries B (publications non vendues dans le commerce) et C (livres publiés par les universités) ne sont pas prises en compte. Les manuels scolaires et les livres pour enfants sont compris dans le total mais ne sont pas répartis par catégories.
14   Les manuels scolaires et les livres pour enfants sont compris dans le total mais ne sont pas répartis par catégories.
15   Les données n'incluent pas les manuels scolaires.
16   Les livres pour enfants sont compris dans le total mais ne sont pas répartis par catégories.
17   Les livres pour la vulgarisation des sciences pour les enfants sont compris dans le total mais ne sont pas répartis par catégories.

25
Book production: number of titles by language of publication
Production de livres : nombre de titres par langue de publication

Country or area / Pays ou zone	Year / Année	Total	National language / Langue nationale	Foreign languages / Languages étrangères						Two or more languages / Deux langues ou plus
				English / Anglais	French / Français	German / Allemand	Spanish / Espagnol	Russian / Russe	Others / Autres	
Africa · Afrique										
Burundi [1] / Burundi [1]	1986	54	54	-	...[2]	-	-	-	-	-
Cape Verde [3] / Cap-Vert [3]	1989	10	10	-	-	-	-	-	-	-
Egypt / Egypte	1988	1451	1344	83	23	-	-	-	1	-
Equatorial Guinea [4] / Guinée équatoriale [4]	1988	17	17							
Ethiopia [5] / Ethiopie [5]	1988	560	409	...[2]	-	-	-	-	151	-
Gambia / Gambie	1988	42	42	...[2]	-	-	-	-	-	-
Libyan Arab Jamahiriya [4] / Jamah. arabe libyenne [4]	1988	121	121	-	-	-	-	-	-	-
Madagascar [6] / Madagascar [6]	1988	146	134	-	...[2]	1	-	-	-	12
Malawi [7] / Malawi [7]	1989	141	87	...[2]	-	-	-	-	13	-
Mauritius [8] / Maurice [8]	1989	100	100	...[2]	...[2]	-	-	-	-	-
Nigeria [9] / Nigéria [9]	1989	1466	1444	...[2]	-	-	-	-	-	22
Rwanda [10] / Rwanda [10]	1987	207	191	-	...[2]	-	-	-	11	5
Zimbabwe [11] / Zimbabwe [11]	1989	337	333	...[2]	-	-	-	-	2	2
America, North · Amérique du Nord										
Costa Rica / Costa Rica	1989	198	190	-	-	-	...[2]	-	-	8
Cuba / Cuba	1989	2199	2040	104	3	2	...[2]	17	-	33
El Salvador [12] / El Salvador [12]	1988	15	15	-	-	-	...[2]	-	-	-
Haiti [13] / Haïti [13]	1989	271	258	-	...[2]	-	-	-	23	-
St. Kitts and Nevis [14] / Saint Kitts-et-Nevis [14]	1989	3	3	...[2]	-	-	-	-	-	-
America, South · Amérique du Sud										
Colombia [12] / Colombie [12]	1989	1486	1471	10	-	-	...[2]	-	5	-
Peru [15] / Pérou [15]	1988	481	472	6	1	-	...[2]	-	2	-
Uruguay / Uruguay	1989	805	801	4	-	-	...[2]	-	-	-
Asia · Asie										
Bangladesh [17] / Bangladesh [17]	1988	1209	1209	...[2]	-	-	-	-	-	-
Brunei Darussalam [17] / Brunéi Darussalam [17]	1989	16	10	-	-	-	-	-	-	-
Cyprus [4] / Chypre [4]	1989	561	431	88	15	10	-	-	17	-
Indonesia [12] / Indonésie [12]	1988	1687	1627	20	-	-	-	-	-	40
Korea, Republic of / Corée, République de	1989	39267	38845	262	35	20	-	-	90	15
Kuwait / Koweït	1988	793	793	-	-	-	-	-	-	-

25
Book production: number of titles by language of publication [*cont.*]
Production de livres : nombre de titres par langue de publication [*suite*]

Country or area Pays ou zone	Year Année	Total	National language Langue nationale	English Anglais	French Français	German Allemand	Spanish Espagnol	Russian Russe	Others Autres	Two or more languages Deux langues ou plus
Malaysia Malaysie	1989	3348	1908	893	-	2	-	-	351	194
Nepal [18] Népal [18]	1989	122	86	8	-	-	-	-	28	-
Sri Lanka [19] Sri Lanka [19]	1989	2188	1339	582	-	-	-	-	4	263
Thailand Thaïlande	1989	11217	10058	1159	-	-	-	-	-	-
United Arab Emirates [18] Emirats arabes unis [18]	1989	152	152	-	-	-	-	-	-	-
Europe · Europe										
Albania Albanie	1986	959	881	18	10	4	3	11	23	9
Andorra [20] Andorre [20]	1989	45	39	1	2	-	3	-	-	-
Austria [21] Autriche [21]	1989	9462	9108	274	31	...[2]	12	4	32	1
Belgium [22] Belgique [22]	1989	6822	6330	149	...[2]	...[2]	34	2	145	162
Bulgaria Bulgarie	1989	4543	3900	139	60	75	54	188	62	65
Czechoslovakia [23] Tchécoslovaquie [23]	1989	9294	8533	168	29	73	32	101	273	85
Denmark Danemark	1989	10762	9221	1016	37	85	-	-	135	268
Finland [24] Finlande [24]	1989	10097	8252	1233	19	27	5	13	82	466
France France	1989	40115	31819	567	...[2]	123	62	28	188	...[2]
Holy See [25] Saint-Siège [25]	1989	176	88	43	9	6	17	-	6	7
Hungary Hongrie	1989	8631	7886	280	37	150	13	41	80	144
Italy Italie	1989	22647	21257	464	146	109	41	-	115	515
Luxembourg [26] Luxembourg [26]	1989	520	383	-	...[2]	...[2]	-	-	39	98
Malta [27] Malte [27]	1989	386	368	...[2]	1	-	-	-	6	7
Netherlands [12 28] Pays-Bas [12 28]	1989	15392	13740	1335	65	179	-	-	73	-
Norway [29] Norvège [29]	1989	5331	4555	612	7	7	-	-	150	-
Poland Pologne	1989	10286	9762	310	34	63	7	70	25	15
Romania Roumanie	1989	3867	3304	122	44	81	-	-	316	-
Spain [30] Espagne [30]	1989	38353	34840	1329	698	350	...[2]	-	564	572
Sweden Suède	1988	11197	9254	1408	19	42	16	-	132	326
Switzerland [31] Suisse [31]	1989	13270	11318	1633	...[2]	...[2]	-	-	319	-
Yugoslavia [32] Yougoslavie [32]	1989	11339	9299	197	31	65	-	22	759	966
Oceania · Océanie										
Australia Australie	1987	10963	10723	...[2]	3	4	10	3	86	134

25
Book production: number of titles by language of publication [*cont.*]
Production de livres : nombre de titres par langue de publication [*suite*]

Country or area Pays ou zone	Year Année	Total	National language Langue nationale	English Anglais	French Français	German Allemand	Spanish Espagnol	Russian Russe	Others Autres	Two or more languages Deux langues ou plus
				Foreign languages Languages étrangères						
former USSR · ancienne URSS										
former USSR [33] ancienne URSS [33]	1989	76711	72981	1021	349	371	380	...[2]	1356	253
Belarus [34] Bélarus [34]	1988	2962	2844	-	-	-	-	...[2]	118	-
Ukraine [35] Ukraine [35]	1988	8311	8068	72	24	37	6	...[2]	72	32

Source:
United Nations Educational, Scientific and Cultural Organization (Paris).

1   The titles in column "National language" have been published in the following languages: Kirundi, French and Kirundi-French.
2   Data are included in column "National language".
3   The titles in column "National language" have been published in the following languages: Portuguese, Caboverdiano.
4   All first editions.
5   The titles in column "National language" have been published in the following languages: Amharic and English.
6   The titles in column "National language" have been published in the following languages: Malagasy and French.
7   The titles in column "National language" have been published in the following languages: Chichewa and English.
8   The titles in column "National language" have been published in the following languages: French, English, Creole, Hindi, Marathi, Telugu, Modern Chinese and in more than one national or official language.
9   The titles in column "National language" have been published in the following languages: English, Yoruba and Hausa.
10   The titles in column "National language" have been published in the following languages: Kinyarwanda and French.
11   The titles in column "National language" have been published in the following languages: English, Shona and Ndebale.
12   Data do not include pamphlets.
13   The titles in column "National language" have been published in the following languages: French, Creole and French-Creole.
14   Data refer only to the first editions of school textbooks.
15   The titles in column "National language" have been published in the following languages: Spanish and Quichua.
16   Data do not include pamphlets and are all first editions. The titles in column "National language" have been published in the following languages: Bengali and English.
17   Language breakdown refers only to ten titles of first editions of school textbooks, children's books and university theses.
18   Data refer only to school textbooks.
19   The titles in column "National language" have been published in the following languages: Sinhala and Tamil.
20   Data refer only to the first editions of books.
21   The titles in column "National language" have been published in the following languages: German and Slovenian.
22   Data do not include pamphlets. The titles in column "National language" have been published in the following languages: Dutch, French, German and Walloon.
23   The titles in column "National language" have been published in the following languages: Czech, Slovak, and Czech and Slovak.
24   The titles in column "National language" have been published in the following languages: Finnish and Swedish.
25   The titles in column "National language" have been published in the following languages: Italian and Latin.
26   The titles in column "National language" have been published in the following languages: French, German and Luxemburgish.

Source:
Organisation des Nations Unies pour l'éducation, la science et la culture (Paris).

1   Les titres de la colonne "langue nationale" ont été publiés dans les langues suivantes: kirundi, français et kirundi-français.
2   Les chiffres sont inclus dans la colonne "langue nationale".
3   Les titres de la colonne "langue nationale" ont été publiés dans les langues suivantes: portugais et caboverdiano.
4   Tous les ouvrages recensés sont des premières éditions.
5   Les titres de la colonne "langue nationale" ont été publiés dans les langues suivantes: amharic et anglais.
6   Les titres de la colonne "langue nationale" ont été publiés dans les langues suivantes: malgache et français.
7   Les titres de la colonne "langue nationale" ont été publiés dans les langues suivantes: chichewa et anglais.
8   Les titres de la colonne "langue nationale" ont été publiés dans les langues suivantes: français, anglais, créole, hindi, marathi, telegou chinois moderne et dans plus d'une langue nationale ou officielle.
9   Les titres de la colonne "langue nationale" ont été publiés dans les langues suivantes: anglais, yoruba et hausa.
10   Les titres de la colonne "langue nationale" ont été publiés dans les langues suivantes: kinyarwanda et français.
11   Les titres de la colonne "langue nationale" ont été publiés dans les langues suivantes: anglais, shona et ndebale.
12   Les données ne tiennent pas compte des brochures.
13   Les titres de la colonne "langue nationale" ont été publiés dans les langues suivantes: français, créole et français-créole.
14   Les données se réfèrent seulement aux premières éditions des manuels scolaires.
15   Les titres de la colonne "langue nationale" ont été publiés dans les langues suivantes: espagnol et quechua.
16   Les données ne tiennent pas compte des brochures et sont des premières éditions. Les titres de la colonne "langue nationale" ont été publiés dans les langues suivantes: bengali et anglais.
17   La répartition par langue ne se réfère qu'aux 10 titres des premières éditions des manuels scolaires, des livres pour enfants et thèses universitaires.
18   Les données se réfèrent seulement aux manuels scolaires.
19   Les titres de la colonne "langue nationale" ont été publiés dans les langues suivantes: sinhala et tamoul.
20   Les données se réfèrent seulement aux premières éditions des livres.
21   Les titres de la colonne "langue nationale" ont été publiés dans les langues suivantes: allemand et slovène.
22   Les données ne tiennent pas compte des brochures. Les titres de la colonne "langue nationale" ont été publiés dans les langues suivantes: néerlandais, français, allemand et wallon.
23   Les titres de la colonne "langue nationale" ont été publiés dans les langues suivantes: tchèque, slovaque, et tchèque et slavaque.
24   Les titres de la colonne "langue nationale" ont été publiés dans les langues suivantes: finnois et suédois.
25   Les titres de la colonne "langue nationale" ont été publiés dans les langues suivantes: italien et latin.

25
Book production: number of titles by language of publication [*cont.*]
Production de livres : nombre de titres par langue de publication [*suite*]

27 The titles in column "National language" have been published
in the following languages: Maltese and English.
28 The titles in column "National language" have been published
in the following languages: Dutch and Frisian.
29 Data do not include schools textbooks.The titles in column
"National language" have been published in the following
languages: Norwegian and New-Norwegian.
30 The titles in column "National language" have been published
in the following languages: Spanish, Galician, Catalan,
Basque and other national langues.
31 The titles in column "National language" have been published
in the following languages: German, French, Italian and Romansch.
32 The titles in column "National language" have been published
in the following languages: Serbo-Croat, Slovenian and Macedonian.
33 The titles in column "National language" have been published
in the following languages: Russian, Ukrainian, Lithuanian,
Estonian, Georgian, Latvian, Uzbek, Kazakh, Azerbaidjan,
Armenian, Moldavian, Kirghiz, Byelorussian, Turkoman, Tadzhik
and other languages of former USSR.
34 The titles in column "National language" have been published
in the following languages: Russian and Byelorussian.
35 The titles in column "National language" have been published
in the following languages: Russian, Ukrainian and other
languages of former USSR.

26 Les titres de la colonne "langue nationale" ont été publiés dans les
suivantes: français, allemand et luxembourgois.
27 Les titres de la colonne "langue nationale" ont été publiés dans les
langues suivantes: maltais et anglais.
28 Les titres de la colonne "langue nationale" ont été publiés dans les
langues suivantes: néerlandais et frison.
29 Les manuels scolaires ne sont pas inclus dans les données. Les
titres de la colonne "langue nationale" ont été publiés dans les
langues suivantes: norvégien et nouveau norvégien.
30 Les titres de la colonne "langue nationale" ont été publiés dans les
langues suivantes: espagnol, galicien, catalan, basque et autres
langues nationales.
31 Les titres de la colonne "langue nationale" ont été publiés dans les
langues suivantes: allemand, français, italien et romanche.
32 Les titres de la colonne "langue nationale" ont été publiés dans les
langues suivantes: serbo-croate, slovène et macédonien.
33 Les titres de la colonne "langue nationale" ont été publiés dans les
langues suivantes: russe, ukrainien, lituanien, estonien, géorgien,
letton, ouzbek, kazakh, azerbaidjanais, arménien, moldave, kirghiz,
biélorussien, turkmène, tadjik et autres langues de l'ancienne
URSS.
34 Les titres de la colonne "langue nationale" ont été publiés dans les
langues suivantes: russe et biélorussien.
35 Les titres de la colonne "langue nationale" ont été publiés dans les
langues suivantes: russe, ukrainien et autres langues de l'ancienne
URSS.

26
Daily newspapers
Journaux quotidiens

Country or area Pays ou zone	Number Nombre				Estimated circulation Tirage (estimation) Total (000)				Per 1000 inhabitants Pour 1000 habitants			
	1975	1979	1986	1988	1975	1979	1986	1988	1975	1979	1986	1988
Africa · Afrique												
Algeria Algérie	4	4	6	5	285	425	812	510	18	23	36	21
Angola Angola	4	5	4	4	14[1]	120	103	85[1]	...	16	11	...
Benin Bénin	1	1	1	3	1	1	1	12[1]	0.3	0.4	0.3	...
Botswana Botswana	1	1	1	3	14	17	18	31[1]	19	20	16	...
Burkina Faso Burkina Faso	1	1	1	1	2	2	3	5	0.2	0.2	0.4	1
Burundi Burundi	1	1	2	1	2	...	2[1]	20	0.4	...	...	4
Cameroon Cameroun	2	3	1	1	25	28	35	65	3	3	3	6
Central African Rep. Rép. centrafricaine	-	-	-	1	-	-	-	...	-	-	-	...
Chad Tchad	4	4	1	1	...	...	1	1	...	...	0.2	0.2
Congo Congo	3	1	5	5	1	-	8[1]	8[1]	0.4	-	...	...
Côte d'Ivoire Côte d'Ivoire	3	1	1	1	35[1]	53	90	90	...	7	8	8
Egypt Egypte	12	9	8	13	1095	2475[1]	2638	1953	30	...	54	38
Equatorial Guinea Guinée équatoriale	...	2	2	2	...	...	1[1]	1[1]	...	...	...	...
Ethiopia Ethiopie	3	5	3	3	44	52	41	42	1	1	1	1
Gabon Gabon	1	1	1	1	...	...	15	15	...	...	15	14
Gambia Gambie	-	-	6	2	-	-	4	1[1]	-	-	6	...
Ghana Ghana	4	5	5	4	500	345[1]	460[1]	460	51	...	...	33
Guinea Guinée	1	1	1	1	5	20	13	13	1	4	2	2
Guinea-Bissau Guinée-Bissau	1	1	1	1	6	6	6	6	10	8	7	6
Kenya Kenya	3	3	4	5	134	156	283	303[1]	10	10	13	...
Lesotho Lesotho	1	3	4	4	1	32	47	47	1	24	29	28
Liberia Libéria	3	3	5	7	13	11	23[1]	23[1]	8	6	...	...
Libyan Arab Jamahiriya Jamah. arabe libyenne	2	...	3	3	41	...	40[1]	40[1]	17	...	...	...
Madagascar Madagascar	...	5	7	6	...	23[1]	67	69	...	...	6	6
Malawi Malawi	...	...	1	1	...	...	15	25	...	...	2	3
Mali Mali	...	2	2	2	...	...	4[1]	40[1]	...	...	...	...
Mauritania Mauritanie	-	-	-	1	-	-	-	...	-	-	-	...
Mauritius Maurice	12	8	7	7	82	74	*75	75	94	79	*71	69
Morocco Maroc	7	9	14	12	245[1]	230[1]	310[1]	290[1]	...	...	...	...
Mozambique Mozambique	...	2	2	2	...	42	81	81	...	4	6	5

26
Daily newspapers [cont.]
Journaux quotidiens [suite]

Country or area / Pays ou zone	Number / Nombre				Estimated circulation / Tirage (estimation) Total (000)				Per 1000 inhabitants / Pour 1000 habitants			
	1975	1979	1986	1988	1975	1979	1986	1988	1975	1979	1986	1988
Namibia / Namibie	...	...	3	3	...	...	21	18	...	...	13	10
Niger / Niger	2	1	1	1	...	...	5	4	...	...	1	0.5
Nigeria / Nigéria	12	15	19	31	613[1]	...	898[1]	1108[1]	...	...	...	...
Réunion / Réunion	2	3	2	3	27	52	49	49[1]	56	103	88	...
Rwanda / Rwanda	1	...	1[2]	1	0.2	...	0.3[2]	0.6	0.0	...	0.1[2]	0.1
Senegal / Sénégal	1	1	3	3	25	25	53	58	5	5	8	8
Seychelles / Seychelles	2	2	1	1	4	4	3	3	62	64	53	48
Sierra Leone / Sierra Leone	2	1	1	1	30	10	10	10	10	3	3	3
Somalia / Somalie	...	1	2	1	...	...	...	...	...	...	...	...
South Africa / Afrique du Sud	...	...	24	22	...	...	1440	1617	...	...	45	48
Sudan / Soudan	4	...	5	5	...	...	220[1]	590[1]	...	...	...	...
Swaziland / Swaziland	1	1	2	2	-	9	8[1]	50	9	16	...	68
Togo / Togo	1	1	2	1	7	7	10[1]	10	3	3	...	3
Tunisia / Tunisie	4	5	6	6	190	271	272[1]	172[1]	34	44	...	...
Uganda / Ouganda	...	1	1	6	...	20	25	49[1]	...	2	2	...
United Rep. of Tanzania / Rép.-Unie de Tanzanie	3	2	2	2	70	189	101	180	4	10	4	7
Zaire / Zaïre	...	6	4	9	...	45[1]	45[1]	45[1]	...	...	...	...
Zambia / Zambie	2	2	2	2	106	109	100	89	22	20	14	11
Zimbabwe / Zimbabwe	2	2	3	2	116	111	204	214	19	16	24	24
America, North · Amérique du Nord												
Antigua and Barbuda / Antigua-et-Barbuda	1	1	2	1	4	6	6[1]	6	59	81	...	71
Bahamas / Bahamas	2	3	3	3	31	33	39	35	151	160	159	138
Barbados / Barbade	1	1	2	2	24	21	40	...	98	86	155	...
Belize / Belize	1	1	1	-	4	3	3	-	31	20	18	-
Bermuda / Bermudes	1	1	1	1	11	13	18	...	228	260	321	...
Canada / Canada	121	126	116	107	4872[1]	5700	5747	5993	214	241	225	231
Cayman Islands / Iles Caïmanes	-	-	1	1	-	-	...	6	-	-	...	262
Costa Rica / Costa Rica	6	4	6	4	174	155	248[1]	245	88	70	...	86
Cuba / Cuba	15	9	17	17	53[1]	891	1075	1315	...	92	107	129
Dominican Republic / République dominicaine	10	7	7	12	197[1]	220	276	267[1]	...	39	42	...
El Salvador / El Salvador	10	7	4	5	211	291	243	229	53	60	50	45
Grenada / Grenade	1	1	-	-	...	...	-	-	...	...	-	-
Guadeloupe / Guadeloupe	2	1	2	1	24	18	32[1]	20	73	56	...	58

26
Daily newspapers [*cont.*]
  Journaux quotidiens [*suite*]

Country or area	Number Nombre				Estimated circulation Tirage (estimation)							
					Total (000)				Per 1000 inhabitants Pour 1000 habitants			
Pays ou zone	1975	1979	1986	1988	1975	1979	1986	1988	1975	1979	1986	1988
Guatemala												
Guatemala	10	9	9	7	249	91	203[1]	135[1]	41	14	...	...
Haiti												
Haïti	7	4	5	5	93	32	45[1]	45[1]	19	6	...	...
Honduras												
Honduras	8	7	7	5	99[1]	223	293	199	...	63	65	41
Jamaica												
Jamaïque	3	3	4	4	131	128	95[1]	155[1]	64	59	...	...
Martinique												
Martinique	2	1	1	1	27	26	32	32	82	80	96	95
Mexico												
Mexique	216	317	308	286	5499	8322	10356	10539	91	120	127	127
Netherlands Antilles												
Antilles néerlandaises	5	5	6	5	55	54	54[1]	33[1]	228	...	...	...
Nicaragua												
Nicaragua	7	8	5	6	91[1]	170	36[1]	80[1]	...	63	...	...
Panama												
Panama	6	6	9	9	131	148	171[1]	161	75	78	...	69
Puerto Rico												
Porto Rico	...	4	5	5	...	475	599	667	...	150	171	185
Saint Lucia												
Sainte Lucie	...	-	1	1	...	-	4	8	...	-	31	60
Trinidad and Tobago												
Trinité-et-Tobago	3	4	4	4	135	196	173	134	125	181	147	108
United States												
Etats-Unis	1775	1787	1657	...	60655	62223	62502	...	281	276	259	...
US Virgin Islands												
Iles Vierges américaines	3	4	3	2	17	19	21	19	179	196	198	175
America, South · Amérique du Sud												
Argentina												
Argentine	164	133	218	194	2773[1]	451	2633[1]	2652[1]	...	...	...	...
Bolivia												
Bolivie	14	14	14	16	199	214	231[1]	353[1]	41	39	...	...
Brazil [3]												
Brésil [3]	299	328	322	366	4895	5094	6534	7944[1]	45	44	48	...
Chile												
Chili	47	37	...	39[4]	921	945	...	840[4]	89	86	...	67[4]
Colombia												
Colombie	40	38	46	45	1248[1]	1273[1]	1172[1]	1631[1]	...	...	...	...
Ecuador												
Equateur	...	38	26	26	...	400	742[1]	887[1]	...	51	...	...
French Guiana												
Guyane française	1	1	1	1	2	2	...	1	27	29	...	11
Guyana												
Guyana	...	3	2	2	...	67	78	58[1]	...	79	80	...
Paraguay												
Paraguay	8	5	6	5	73[1]	88[1]	158[1]	53[1]	...	...	...	...
Peru [5]												
Pérou [5]	49	59	70	14	1377	...	1427[1]	669	91	...	71	31
Suriname												
Suriname	7	5	5	2	33[1]	32[1]	31[1]	40	...	...	...	102
Uruguay												
Uruguay	30	28	33	33	637[1]	...	574[1]	694[1]	...	...	...	...
Venezuela												
Venezuela	49	69	55	56	1067[1]	2383	1982[1]	2225[1]	...	164	...	...
Asia · Asie												
Afghanistan												
Afghanistan	...	14	13	14	...	69[1]	71[1]	151[1]	...	...	...	...
Bahrain												
Bahreïn	-	-	2	2	-	-	19	29	-	-	43	61
Bangladesh												
Bangladesh	29	40	62	72	356[1]	274	622	1016	...	3	6	10
Cambodia												
Cambodge	16	17	10[2]	...	...	...	...	...	...	...	...	...

26
Daily newspapers [cont.]
Journaux quotidiens [suite]

Country or area Pays ou zone	Number Nombre				Estimated circulation Tirage (estimation) Total (000)				Per 1000 inhabitants Pour 1000 habitants			
	1975	1979	1986	1988	1975	1979	1986	1988	1975	1979	1986	1988
China Chine	...	...	*73	78	...	...	*37860[1]	39597[1]	...	...	...	...
Cyprus Chypre	12	12	10	11	78[1]	67[1]	83	86	...	...	124	125
Hong Kong Hong-kong	82	41	46	40	...	...	3594[1]	3146[1]	...	...	...	...
India Inde	835	1087	1978	...	9383	13033	21857	...	15	19	28	...
Indonesia Indonésie	*60	84	61	60	*2200	2281	2733	3716	*16	16	16	21
Iran, Islamic Rep. of Iran, Rép. islamique d'	19	35	16	10	...	...	...	640[1]	...	...	...	...
Iraq Iraq	7	5	6	6	192[1]	325[1]	572[1]	572[1]	...	...	...	...
Israel Israël	23	24	27	30	...	...	1017[1]	1133[1]	...	...	...	...
Japan Japon	176	178	124	158	60782	65881	68653	71228[1]	545	570	566	...
Jordan Jordanie	4	5	4	4	58	59	155	210	22	20	42	53
Korea, Dem.People's Rep. Corée, Rép. pop. dém. de	...	11	11[3]	...	...	...	...	...	...	...	...	...
Korea, Republic of Corée, République de	...	29	35	39	...	6496[1]	8654[1]	10429[1]	...	...	...	...
Kuwait Koweït	8	8	8	8	180	305	380	410	178	223	222	209
Lao People's Dem. Rep. République dém. pop. lao	8	3	3	3	5[1]	...	13[1]	13[1]	...	...	...	...
Lebanon Liban	33	25	13	14	283[1]	...	234[1]	276[1]	102	...	86	97
Macau Macao	...	6	9	8	...	59[1]	...	16[1]	...	...	...	...
Malaysia Malaisie	31	44	40	47	1038	1796[1]	1546[1]	2462[1]	85	...	...	...
Maldives Maldives	1	...	2	...	...	...	2	...	...	...	8	...
Mongolia Mongolie	1	1	2	2	112	112	177	177	78	69	90	85
Myanmar Myanmar	7	7	7	7	319	329	439[1]	329[1]	10	10	...	...
Nepal Népal	29	29	28	28	96[1]	...	122[1]	122[1]	...	...	...	...
Oman Oman	-	-	3	4	-	-	51	62	-	-	40	45
Pakistan Pakistan	102	119	121	183	358[1]	1095	3887	7314	...	13	36	64
Philippines Philippines	15	19	23	38	850	972[1]	1968[1]	3298	20	...	...	56
Qatar Qatar	1	3	4	5	...	...	60	10	...	...	195	206
Saudi Arabia Arabie saoudite	...	12	13	11	...	...	663[1]	490[1]	...	...	...	...
Singapore Singapour	10	11	10	8	449	588	924	763	198	246	357	289
Sri Lanka Sri Lanka	15	21	17	21	480	450	500	580	36	31	31	35
Syrian Arab Republic Rép. arabe syrienne	6	6	7	10	...	104	163	244[1]	...	12	15	...
Thailand Thaïlande	22	27	32	40	2540	2680	4350	4750	61	57	84	87
Turkey Turquie	...	1115[6]	336	426	...	3880[6]	3020	3094	...	...	...	55
United Arab Emirates Emirats arabes unis	2	9	13	10	10	152	290	300	20	150	215	200

26
Daily newspapers [*cont.*]
Journaux quotidiens [*suite*]

Country or area	Number Nombre				Estimated circulation Tirage (estimation)							
					Total (000)				Per 1000 inhabitants Pour 1000 habitants			
Pays ou zone	1975	1979	1986	1988	1975	1979	1986	1988	1975	1979	1986	1988
Viet Nam												
Viet Nam	...	3	4	5	...	500[1]	545[1]	545[1]	...	...	...	...
Yemen [7]												
Yémen [7]	...	...	1	2	...	...	110	120	...	...	15	16
Europe · Europe												
Albania												
Albanie	2	2	2	2	115	145	135	135	47	55	45	43
Austria												
Autriche	30	31	33	34	2405	2634	2685	2712	320	349	358	362
Belgium												
Belgique	30	26	24	23	2340	2242	2186	2174	239	228	221	219
Bulgaria												
Bulgarie	14	12	17	20	2109	2093	2834	2396	234	237	316	267
Czechoslovakia												
Tchécoslovaquie	29	30	30	30	4436	4641	5139	5364	300	306	332	345
Denmark												
Danemark	49	49	47	46	1723	1876	1880	1842	341	367	367	359
Finland												
Finlande	60	58	66	66	2100	2289	2665	2707	446	480	543	547
France												
France	92	90	92	96	10615	10619	10670	9328[1]	238	201	193	...
Germany ʃ · Allemagne ʃ												
Federal Republic of Germany												
Rép. féd. d'Allemagne	350	329	318	319	*20200	20611	20987	20247	*327	335	344	331
former German Dem. Rep.												
l'ancienne Rép. dém. allemande	40	39	39	39	7946	8658	9467	9706	472	517	570	585
Gibraltar												
Gibraltar	1	1	1	2	3	2	3	3[1]	111	69	103	...
Greece												
Grèce	108	116	142	117	921	...	...	609[1]	102	...	...	...
Holy See [8]												
Saint-Siège [8]	1	1	1	1	70	70	70	70	70000	70000	70000	70000
Hungary												
Hongrie	27	27	29	28	2455	2585	2778	2888	233	242	262	273
Iceland												
Islande	5	6	6	5	93	127	113	125	427	562	471	500
Ireland												
Irlande	7	7	7	8	693	771	648	724	216	229	181	205
Italy												
Italie	78	75	72	73	*6497	5308	5636	6005	117	94	99	105
Liechtenstein												
Liechtenstein	2	2	2	2	12	12	14	16	483	477	504	557
Luxembourg												
Luxembourg	7	5	6	5	130	130	140	145	358	357	*389	395
Malta												
Malte	6	5	4	3	...	...	54[1]	54	...	...	...	155
Monaco												
Monaco	2	2	2	2	...	11	...	...	...	423	...	...
Netherlands												
Pays-Bas	84	83	90	86	4194	4553	4527	4592	307	324	311	311
Norway												
Norvège	80	83	84	83	1657	1859	2209	2309	414	457	530	551
Poland												
Pologne	44	44	45	45	8429	8433	7480	6939	248	239	200	184
Portugal												
Portugal	30	30	27	28	*612	*493	418	859[1]	*67	*51	41	...
Romania												
Roumanie	20	35	36	34	*3015	*3998	3637	3648	*142	*182	159	159
San Marino												
Saint-Marin	4	3	16	16	1	1	2	2	50	50	75	120
Spain												
Espagne	115	105	105	102	3491	...	2910	3200	98	...	75	82
Sweden												
Suède	112	114	114	107	4413	4378	4462	4387	539	529	534	526

26
Daily newspapers [*cont.*]
Journaux quotidiens [*suite*]

Country or area Pays ou zone	Number Nombre				Estimated circulation Tirage (estimation)							
					Total (000)				Per 1000 inhabitants Pour 1000 habitants			
	1975	1979	1986	1988	1975	1979	1986	1988	1975	1979	1986	1988
Switzerland Suisse	95	88	118	98	2574	2501	3241	3280	402	395	500	504
United Kingdom Royaume-Uni	109	110	104	104	805	24087	22519	22494	444	431	397	394
Yugoslavia Yougoslavie	26	27	28	28	1896	2282	2508	2354	89	103	108	100
Oceania · Océanie												
American Samoa Samoa américaines	1	2	3	...	4	10	9[1]	...	120	313	...	...
Australia Australie	70	63	62	68	*5336	4851	4213	4121[1]	*392	335	264	252
Cook Islands Iles Cook	1	1	1	1	1	2	2	2	44	111	100	100
Fiji Fidji	1	2	3	6	20	54	68	76	35	87	96	109
French Polynesia Polynésie française	4	2	3	2	11	13	23	6[1]	83	...	139	...
Guam Guam	1	1	1	1	18	25	18	18	191	...	157	151
New Caledonia Nouvelle-Calédonie	...	1	1	1	...	15	19	19	...	109	122	119
New Zealand Nouvelle-Zélande	40	37	33	35	...	1068	1075	1090	...	343	328	327
Papua New Guinea Papouasie-Nouv.-Guinée	1	1	2	2	...	19	45	65	...	6	12	17
Tonga Tonga	-	-	-	1	-	-	-	7	-	-	-	60
former USSR · ancienne URSS												
former USSR[5] ancienneURSS[5]	691	...	723	723	100928	...	122982	133979	396	...	442	474
Belarus Bélarus	25	27	29	28	2214	2397	2587	2738	236	251	259	266
Ukraine[5] Ukraine[5]	2029	1755	1789[3]	1794	24344	23837	24000[3]	24000	497	479	470[3]	466

Source:
United Nations Educational, Scientific and Cultural Organization (Paris).

Source:
Organisation des Nations Unies pour l'éducation, la science et la culture (Paris).

§ All data shown which pertain to Germany prior to 3 October 1990 are indicated separately for the Federal Republic of Germany and the former German Democratic Republic based on their respective territories at the time indicated. For detailed explanatory notes on data pertaining to Germany, see Annex I- Country Nomenclature.

1 The circulation figures relate to only a part of the total number of daily newspapers.
2 Data refer to 1984.
3 Data shown for 1975, 1979 and 1986 refer to 1976, 1978 and 1985 respectively.
4 Data refer to 1987.
5 Data shown for 1988 refer to Lima only.
6 Data include non-daily newspapers.
7 Excluding former Democratic Yemen.
8 The figures refer to the State of the Vatican City.

§ Toutes les données se rapportant à l'Allemagne avant le 3 octobı 1990 figurent dans deux rubriques séparées basées sur les territoires respectifs de la République fédérale d'Allemagne et l'ancienne République démocratique allemande selon la période indiquée. Pour les notes explicatives en détail sur les données concernant l'Allemagne, voir Annexe I - Nomenclature des pays.

1 Les chiffres relatifs au tirage se rapportent seulement à une partie du nombre total des journaux quotidiens.
2 Les données se réfèrent à 1984.
3 Les données pour 1975, 1979 et 1986 se réfèrent à 1976, 1978 et 1985 respectivement.
4 Les données se réfèrent à 1987.
5 Les données pour 1988 se réfèrent à Lima seulement.
6 Les données incluent les journaux non quotidiens.
7 Non compris l'ancien Yémen démocratique.
8 Les chiffres se réfèrent à l'Etat de la Cité du Vatican.

27
Non-daily newspapers and periodicals
Journaux non quotidiens et périodiques

Country or area Pays ou zone	Year Année	Non-daily newspapers Journaux non quotidiens			Periodicals Périodiques		
			Estimated circulation Tirage (estimation)			Estimated circulation Tirage (estimation)	
		Number Nombre	Total (000)	Per 1000 inhabitants Pour 1000 habitants	Number Nombre	Total (000)	Per 1000 inhabitants Pour 1000 habitants
Africa · Afrique							
Algeria [1] Algérie [1]	1987	20	800	34	26	...	...
Botswana Botswana	1984	...	...	...	20	153	147
Burkina Faso [1] Burkina Faso [1]	1986	14	30	4	...	...	...
Cameroon Cameroun	1988	25	315	30	58	127	12
Cape Verde Cap-Vert	1988	3	7	19	...	...	...
Chad Tchad	1988	1	1	0.2	10	...	...
Congo [2] Congo [2]	1986	2	15	9	21	64	36
Egypt [3] Egypte [3]	1988	33	2031	39	241	1696	34
Ethiopia Ethiopie	1988	4	40	1	...	...	...
Gabon Gabon	1988	1	20	18	...	...	...
Gambia Gambie	1986	6	...	...	...	...	...
Madagascar Madagascar	1988	32	220	20	121	261	23
Malawi Malawi	1986	5	121	16	14	124	17
Mauritius Maurice	1988	18	45	42	58	...	...
Mozambique Mozambique	1988	2	...	...	5	2263	151
Nigeria Nigéria	1988	...	...	...	92	495	5
Rwanda Rwanda	1988	23	171	25	9	91	13
St. Helena Saint-Hélène	1984	1	2	300	2	2	320
Senegal Sénégal	1988	9	60	9	123	381	55
Seychelles Seychelles	1984	1	3	46	2	2	23
Sudan Soudan	1984	9	121	6	10	136	6
Tunisia Tunisie	1986	6	...	...	...	...	...
Zaire Zaïre	1984	9	115	4	106	225	8
Zambia Zambie	1986	12	133	18	...	...	...
Zimbabwe Zimbabwe	1988	16	428	48	...	...	...
America, North · Amérique du Nord							
Barbados Barbade	1988	3	35	136	52	...	...
Belize Belize	1988	6	25	139	...	...	...
Bermuda Bermudes	1984	1	14	284	...	...	...
British Virgin Islands Iles Vierges brit.	1984	2	4	300	20	*23	*1950

27
Non-daily newspapers and periodicals [*cont.*]
Journaux non quotidiens et périodiques [*suite*]

Country or area Pays ou zone	Year Année	Non-daily newspapers Journaux non quotidiens			Periodicals Périodiques		
			Estimated circulation Tirage (estimation)			Estimated circulation Tirage (estimation)	
		Number Nombre	Total (000)	Per 1000 inhabitants Pour 1000 habitants	Number Nombre	Total (000)	Per 1000 inhabitants Pour 1000 habitants
Canada [4] Canada [4]	1988	1425	19719	760	1444	38540	1493
Cuba [5] Cuba [5]	1988	1	216	21	108	2894	278
Dominica Dominique	1988	2	...	...	...	...	...
Dominican Republic Rép. dominicaine	1986	39	...	...	277	...	...
Mexico Mexique	1988	30	889	11	203	27203	320
Panama [3] Panama [3]	1988	3	...	...	18	...	...
St. Kitts and Nevis Saint Kitts-et-Nevis	1986	6	22	468	3	6	110
St. Lucia Sainte Lucie	1984	1	5	35	...	...	...
St. Vincent and the Grenadines Saint Vincent-et-Grenadines	1988	2	11	100			
Trinidad and Tobago Trinité-et-Tobago	1988	6	302	244	...	...	...
United States Etats-Unis	1987	7260	...	...	11593	...	...
America, South · Amérique du Sud							
Brazil Brésil	1985	1307	4766	35	3782	980	7
Chile Chili	1987	37	118	9	255	...	...
Falkland Islands (Malvinas) Iles Falkland (Malvinas)	1984	...	...	...	3	2	800
Suriname Suriname	1983	...	...	...	22	44	119
Uruguay Uruguay	1986	93	...	...	465	...	...
Asia · Asie							
Afghanistan Afghanistan	1988	...	...	...	71	...	...
Bangladesh Bangladesh	1986	155	350	3	41	163	2
Brunei Darussalam [1] Brunéi Darussalam [1]	1989	3	83	346	4	...	...
Cyprus Chypre	1988	40	124	181	59	237	345
India Inde	1986	21638	42464	54	...	...	...
Indonesia Indonésie	1988	89	3445	20	1456	3622[6]	...
Iran, Islamic Rep. of Iran, Rép. islamique d'	1987	...	...	...	314	...	...
Iraq Iraq	1986	22	...	...	807	...	...
Israel Israël	1985	83	...	...	807	...	...
Japan Japon	1984	...	...	...	2138	36293	303
Jordan Jordanie	1988	4	90	23	29	36	9
Kuwait Koweït	1988	59	420	...	73	257	146
Malaysia Malaisie	1984	20	4292	284	1631	1689	112
Maldives [8] Maldives [8]	1986	23	7	37	64	70	...

27
Non-daily newspapers and periodicals [cont.]
Journaux non quotidiens et périodiques [suite]

Country or area Pays ou zone	Year Année	Non-daily newspapers Journaux non quotidiens			Periodicals Périodiques		
		Number Nombre	Estimated circulation Tirage (estimation)		Number Nombre	Estimated circulation Tirage (estimation)	
			Total (000)	Per 1000 inhabitants Pour 1000 habitants		Total (000)	Per 1000 inhabitants Pour 1000 habitants
Mongolia Mongolie	1984	35	718	387	38	663	357
Oman Oman	1986	2	10	8	19	117	91
Pakistan Pakistan	1988	264	4138	36	282	7674	67
Philippines Philippines	1988	292	581	10	...	...	...
Qatar Qatar	1986	1	6	20	13	320	1042
Saudi Arabia Arabie saoudite	1984	7	...	...	58	...	...
Singapore Singapour	1984	7	425	168	1786	...	...
Sri Lanka Sri Lanka	1986	12	1041	64	170	1770	108
Thailand Thaïlande	1988	245	7275	134	976	...	...
Turkey Turquie	1989	771	...	...	2670	...	...
Yemen [9] Yémen [9]	1984	40	...	...	40	...	...
Europe · Europe							
Andorra [10] [11] Andorre [10] [11]	1988	2	6	128	15	15	374
Austria Autriche	1988	143	...	...	2549	...	...
Belgium Belgique	1988	3	36	4	12222	...	...
Bulgaria [1] [12] Bulgarie [1] [12]	1989	284	4008	446	994	6091	679
Czechoslovakia Tchécoslovaquie	1988	105	1139	73	...	...	...
Denmark Danemark	1988	13	1493	291	209	7283	1421
Faeroe Islands Iles Féroé	1983	8	36	878	...	...	...
Finland Finlande	1988	360	...	...	6941	...	...
Germany ʤ · Allemagne ʤ Federal Republic of Germany Rép. féd. d'Allemagne	1988	37	4279	70	...	...	...
former German Dem. Rep. l'ancienne Rép. dém. allemande	1988	30	9431	568	1209	23872	1439
Gibraltar Gibraltar	1984	5	*6	*190	15	4	127
Greece Grèce	1988	1051	...	...	309	...	...
Holy See Saint-Siège	1988	-	-	-	...	...	...
Hungary Hongrie	1986	95	5845	551	1559	13390	1261
Iceland Islande	1989	82	...	...	...	...	...
Ireland [11] Irlande [11]	1984	59	1785	506	257	2975	844
Italy Italie	1988	203	2027	35	9158	3958	69
Liechtenstein [7] Liechtenstein [7]	1990	1	2	67	79	220	8148
Luxembourg Luxembourg	1988	1	3	8	430	...	...

27
Non-daily newspapers and periodicals [cont.]
Journaux non quotidiens et périodiques [suite]

Country or area Pays ou zone	Year Année	Non-daily newspapers Journaux non quotidiens			Periodicals Périodiques		
		Number Nombre	Estimated circulation Tirage (estimation)		Number Nombre	Estimated circulation Tirage (estimation)	
			Total (000)	Per 1000 inhabitants Pour 1000 habitants		Total (000)	Per 1000 inhabitants Pour 1000 habitants
Malta Malte	1988	8	...	...	336	...	...
Netherlands Pays-Bas	1984	*110	*800	*55	...	...	...
Norway Norvège	1988	70	375	90	...	...	...
Poland Pologne	1988	52	3106	82	3031	44599	1180
Portugal Portugal	1986	349	6942	681	805	...	...
Romania [13] Roumanie [13]	1986	24	756	33	422	*17090	*752
San Marino Saint-Marin	1988	9	10	500	16	...	...
Spain [7] Espagne [7]	1988	85	...	...	1998	...	...
Sweden [14] Suède [14]	1988	71	452	54	46	4947	593
Switzerland Suisse	1988	380	7303	1123	*3079	...	...
United Kingdom [11] Royaume-Uni [11]	1988	818	29047	509	6408	...	...
Yugoslavia [1] Yugoslavie [1]	1989	2976	23505	998	1719	4485	191
Oceania · Océanie							
Australia Australie	1984	465	15208	978	...	...	...
New Zealand [11] Nouvelle-Zéalande [11]	1988	139	...	...	5788	...	...
Tokelau Tokélaou	1983	1	1	500	...	...	...
Vanuatu Vanuatu	1988	1	2	12	...	...	...
former USSR · ancienne URSS							
former USSR [1] ancienne URSS [1]	1989	8092	92710	325	5413	4260612	15079
Belarus Bélarus	1986	186	2567	...	110	2567	

Source:
United Nations Educational, Scientific and Cultural Organization
(Paris).

Source:
Organisation des Nations Unies pour l'éducation, la science et la
culture (Paris).

§ All data shown which pertain to Germany prior to 3 October 1990 are indicated separately for the Federal Republic of Germany and the former German Democratic Republic based on their respective territories at the time indicated. For detailed explanatory notes on data pertaining to Germany, see Annex I-Country Nomenclature.

1 Data on periodicals refer to 1988.
2 Data on non-daily newspapers refer to 1984.
3 Data on periodicals refer to 1987.
4 Data on non-daily newspapers refer to 1987.
5 Data on non-daily newspapers refer to 1986.
6 The circulation figure for periodicals refer only to 114 periodicals for the general public.
7 Data on periodicals refer to 1986.
8 The circulation figure for periodicals refer only to 64 periodicals for the general public.
9 Excluding former Democratic Yemen.
10 Data on non-daily newspapers refer to Catalan newspapers only.

§ Toutes les données se rapportant à l'Allemagne avant le 3 octobre 1990 figurent dans deux rubriques séparées basées sur les territoires respectifs de la République fédérale d'Allemagne et l'ancienne République démocratique allemande selon la période indiquée. Pour les notes explicatives en détail sur les données concernant l'Allemagne voir Annexe I - Nomenclature des pays.

1 Les données relatives aux périodiques se réfèrent à 1988.
2 Les données relatives aux non quotidiens se réfèrent à 1984.
3 Les données relatives aux périodiques se réfèrent à 1987.
4 Les données relatives aux non quotidiens se réfèrent à 1987.
5 Les données relatives aux non quotidiens se réfèrent à 1986.
6 Le chiffre relatif au tirage pour les périodique se réfère seulement à 114 périodiques destinés au grand public.
7 Les données relatives aux périodiques se réfèrent à 1986.
8 Le chiffre relatif au tirage pour les périodiques se réfèrent à 64 périodiques destinés au grand public.
9 Non compris l'ancien Yémen démocratique.
10 Les données relatives aux non quotidiens se réfèrent aux journaux

27
Non-daily newspapers and periodicals [*cont.*]
Journaux non quotidiens et périodiques [*suite*]

11 Data on periodicals refer to 1984.
12 Data on non-daily newspapers include regional editions.
13 Data on periodicals refer to 1985.
14 Data for periodicals refer only to periodicals for the general public.

catalans seulement.

11 Les données relatives aux périodiques se réfèrent à 1984.
12 Les données relatives aux non quotidiens comprennent les éditions régionales.
13 Les données relatives aux périodiques se réfèrent à 1985.
14 Les données relatives aux périodiques se réfèrent seulement aux périodiques destinés au grand public.

28
Television and radio receivers
Postes récepteurs de télévision et de radio

Country or area Pays ou zone	Code §	Number (000) Nombre (000)				Per 1000 inhabitants Par 1000 habitants			
		1975	1980	1985	1989	1975	1980	1985	1989
Africa · Afrique									
Algeria	T	500	975	1500	1770	31	52	69	73
Algérie	R	3000	3700	4800	5645	187	197	220	232
Angola	T	-	30	37	55	-	4	4	6
Angola	R	118	145	230	520	18	19	26	53
Benin	T	-	5	15	20	-	1	4	5
Bénin	R	150	230	300	400	49	67	75	89
Botswana	T	-	-	-	15	-	-	-	12
Botswana	R	57	75	115	140	76	83	106	111
Burkina Faso	T	9	20	37	45	2	3	5	5
Burkina Faso	R	100	125	150	225	16	18	19	26
Burundi	T	-	-	0.3	3	-	-	0	1
Burundi	R	100	150	250	300	27	36	53	57
Cameroon	T	-	-	-	250	-	-	-	22
Cameroun	R	232	760	1200	1500	31	88	119	131
Cape Verde									
Cap-Vert	R	31	41	50	57	112	142	154	158
Central African Rep.	T	-	1	5	10	-	0	2	3
Rép. centrafricaine	R	70	120	150	180	34	52	57	61
Chad	T	-	-	-	6	-	-	-	1
Tchad	R	500	750	1150	1310	124	168	229	237
Comoros	T	-	-	-	0.2	-	-	-	0
Comores	R	36	44	53	61	113	113	115	115
Congo	T	3	4	5	10	2	2	3	5
Congo	R	81	100	138	240	56	60	71	109
Côte d'Ivoire	T	110	300	500	675	16	37	50	59
Côte d'Ivoire	R	300	1000	1300	1600	44	122	131	139
Djibouti	T	3	5	12	22	13	17	34	55
Djibouti	R	13	21	30	35	54	69	85	88
Egypt	T	620	1400	3860	5000	17	34	83	98
Egypte	R	4900	6000	12000	16450	135	147	258	322
Equatorial Guinea	T	...	1	2	3	2	5	7	9
Guinée équatoriale	R	78	80	116	128	347	372	373	374
Ethiopia	T	20	30	70	100	1	1	2	2
Ethiopie	R	1000	3000	8000	9000	29	77	186	188
Gabon	T	7	9	22	40	11	11	22	36
Gabon	R	80	105	132	155	125	130	134	138
Gambia									
Gambie	R	59	73	105	140	108	114	141	168
Ghana	T	33	57	150	211	3	5	12	15
Ghana	R	1060	1700	2500	4300	108	158	195	295
Guinea	T	-	6	8	30	-	1	2	5
Guinée	R	110	135	180	230	27	30	36	41
Guinea-Bissau									
Guinée-Bissau	R	10	25	30	37	16	31	34	39
Kenya	T	38	62	100	200	3	4	5	9
Kenya	R	400	650	1600	2200	29	32	80	95
Lesotho	T	-	-	1	5	-	-	0	3
Lesotho	R	22	33	70	118	19	25	46	68
Liberia	T	9	21	35	45	6	11	16	18
Libéria	R	264	335	475	560	164	179	216	225
Libyan Arab Jamahiriya	T	85	168	235	400	35	55	62	91
Jamah. arabe libyenne	R	110	200	800	980	45	66	211	224
Madagascar	T	8	45	100	230	1	5	10	20
Madagascar	R	720	1600	1950	2300	95	182	190	198
Malawi									
Malawi	R	170	275	*1500	2000	32	44	*204	237
Mali	T	-	-	1	4	-	-	0	0
Mali	R	81	105	250	350	13	15	32	39
Mauritania	T	-	-	1	45	-	-	0	23
Mauritanie	R	82	150	250	282	60	97	142	143

28
Television and radio receivers [*cont.*]
Postes récepteurs de télévision et de radio [*suite*]

Country or area Pays ou zone	Code §	Number (000) Nombre (000)				Per 1000 inhabitants Par 1000 habitants			
		1975	1980	1985	1989	1975	1980	1985	1989
Mauritius	T	50	92	140	230	56	95	137	215
Maurice	R	162	260	325	380	182	269	319	354
Morocco	T	535	890	1370	1700	31	46	62	70
Maroc	R	1400	3000	3850	5100	81	155	175	209
Mozambique	T	1	2	7	35	0	0	1	2
Mozambique	R	200	254	450	620	19	21	33	41
Namibia	T	...	5	16	27	...	4	11	16
Namibie	R	...	...	150	230	...	...	99	133
Niger	T	...	5	12	30	0	1	2	4
Niger	R	120	250	300	440	25	45	45	59
Nigeria	T	100	550	1000	3000	2	7.0	11	29
Nigéria	R	4500	7000	14500	18000	68	89	158	171
Réunion	T	41	81	88	96	84	159	161	163
Réunion	R	91	100	122	140	188	197	223	238
Rwanda Rwanda	R	65	175	335	415	15	34	55	59
St. Helena Sainte-Hélène	R	1	1	2	2	180	280	342	375
Sao Tome and Principe Sao Tomé-et-Principe	R	17	22	27	31	210	234	248	256
Senegal	T	2	8	200	250	0	1	31	35
Sénégal	R	290	360	700	802	60	65	110	113
Seychelles	T	-	-	2	5	-	-	31	74
Seychelles	R	16	21	25	31	271	333	385	449
Sierra Leone	T	8	20	30	40	3	6	8	10
Sierra Leone	R	280	450	800	890	96	138	218	220
Somalia	T	-	-	1	100	-	-	0	14
Somalie	R	68	112	200	300	16	21	31	41
South Africa	T	100	2010	3000	3500	4	71	95	101
Afrique du Sud	R	2400	8000	10000	11200	95	283	317	324
Sudan	T	100	800	1100	1500	6	43	50	61
Soudan	R	2000	3500	5060	5755	125	187	232	235
Swaziland	T	-	1	8	13	-	2	11	16
Swaziland	R	55	81	101	117	114	144	151	154
Togo	T	1	10	15	20	0	4	5	6
Togo	R	350	530	630	719	153	203	208	210
Tunisia	T	191	300	400	600	34	47	55	75
Tunisie	R	775	1000	1185	1500	138	157	163	188
Uganda	T	59	72	90	150	5	6	6	8
Ouganda	R	250	400	1250	1800	22	30	80	99
United Rep. of Tanzania	T	4	7	8	25	0	0	0	1
Rép.-Unie de Tanzanie	R	232	290	365	565	15	15	16	21
Western Sahara	T	2	2	3	4	17	18	19	23
Sahara occidental	R	16	28	34	40	137	207	219	230
Zaire	T	7	10	13	30	0	0	0	1
Zaïre	R	1000	1500	2800	3480	44	57	92	101
Zambia	T	23	60	90	200	5	11	13	25
Zambie	R	100	135	500	603	21	24	71	74
Zimbabwe	T	60	73	178	250	10	10	21	27
Zimbabwe	R	180	240	500	801	29	34	60	85
America, North · Amérique du Nord									
Anguilla Anguilla	R	...	...	...	2	...	...	...	300
Antigua and Barbuda	T	15	16	19	22	211	213	250	290
Antigua-et-Barbuda	R	15	17	21	...	211	227	276	...
Bahamas	T	-	31	51	56	-	149	220	225
Bahamas	R	95	102	120	134	503	486	517	538
Barbados	T	46	50	60	67	188	201	237	188
Barbade	R	94	135	200	224	384	542	791	878
Belize	T	-	-	-	30	-	-	-	165
Belize	R	59	71	88	106	457	486	530	580
Bermuda	T	20	30	45	55	377	556	804	948
Bermudes	R	50	60	76	80	943	1111	1357	1379

28
Television and radio receivers [*cont.*]
Postes récepteurs de télévision et de radio [*suite*]

Country or area Pays ou zone	Code §	Number (000) Nombre (000)				Per 1000 inhabitants Par 1000 habitants			
		1975	1980	1985	1989	1975	1980	1985	1989
British Virgin Islands	T	1	2	3	3	64	200	223	231
Iles Vierges brit.	R	5	6	7	8	432	527	567	577
Canada	T	9200	10617	14028	16459	405	442	553	626
Canada	R	16300	17734	23237	26878	717	738	916	1023
Cayman Islands	T	-	2	4	5	-	106	191	208
Iles Caïmanes	R	4	12	19	22	286	706	905	923
Costa Rica	T	128	155	200	400	65	68	76	136
Costa Rica	R	151	190	650	760	77	83	246	259
Cuba	T	595	1273	1940	2140	64	132	193	203
Cuba	R	1805	2914	3282	3608	194	301	326	343
Dominica	T	-	-	-	4	-	-	-	49
Dominica	R	6	30	38	41	82	405	475	507
Dominican Republic	T	180	400	500	575	36	70	78	82
République dominicaine	R	770	900	1020	1180	153	158	159	168
El Salvador	T	135	300	350	450	33	66	73	87
El Salvador	R	1100	1550	1900	2080	269	342	398	403
Greenland	T	3	4	7	10	56	70	132	182
Groënland	R	13	15	19	22	260	300	358	391
Grenada									
Grenade	R	22	35	45	53	234	389	517	624
Guadeloupe	T	18	37	78	100	54	113	234	292
Guadeloupe	R	35	50	80	85	106	153	240	249
Guatemala	T	110	175	207	400	18	25	26	45
Guatemala	R	300	350	450	570	50	51	57	64
Haiti	T	13	16	20	29	3	3	3	5
Haïti	R	93	105	140	270	19	20	24	42
Honduras	T	34	65	280	350	11	18	64	70
Honduras	R	142	500	1600	1910	46	136	365	384
Jamaica	T	110	170	215	300	55	80	93	124
Jamaïque	R	550	800	920	995	273	375	398	409
Martinique	T	22	38	43	47	67	116	130	137
Martinique	R	57	62	67	71	172	190	202	207
Mexico	T	2700	3820	8500	11000	44	54	107	127
Mexique	R	6900	9000	15000	21000	111	128	189	242
Montserrat									
Montserrat	R	5	6	6	6	413	458	500	521
Netherlands Antilles	T	35	43	58	62	208	247	320	332
Antilles néerlandaises	R	132	175	190	206	786	1006	1050	1099
Nicaragua	T	83	160	190	230	35	58	58	61
Nicaragua	R	400	670	802	925	166	242	245	247
Panama	T	185	225	350	390	106	115	161	165
Panama	R	260	300	400	527	149	153	183	222
Puerto Rico	T	630	725	850	915	210	226	259	266
Porto Rico	R	1760	2000	2300	2480	588	624	701	720
St. Kitts and Nevis	T	4	4	5	8	78	96	114	182
Saint-Kitts-et-Nevis	R	...	...	21	26	...	...	477	580
Saint Lucia	T	2	2	2	3	15	15	16	18
Sainte-Lucie	R	70	81	92	98	625	653	672	667
St. Pierre and Miquelon	T	2	3	4	4	283	533	600	667
Saint-Pierre-et-Miquelon	R	2	4	4	4	350	583	633	692
St. Vincent and the Grenadines	T	4	5	6	9	44	51	55	79
Saint-Vincent-et-Grenadines	R	32	42	55	73	327	408	505	636
Trinidad and Tobago	T	105	210	320	380	104	194	272	301
Trinité-et-Tobago	R	225	300	500	580	222	277	424	460
Turks and Caicos Is.									
Iles Turques et Caïques	R	3	4	5	6	500	529	533	613
United States	T	121000	155800	190000	201000	560	684	794	814
Etats-Unis	R	401000	453000	500000	524200	1857	1989	2090	2122
US Virgin Islands	T	30	50	59	64	316	510	547	561
Iles Vierges américaines	R	75	85	93	100	789	867	867	873
America, South · Amérique du Sud									
Argentina	T	4000	5140	6500	7000	154	182	214	219
Argentine	R	9890	12000	18000	21500	380	425	593	673

28
Television and radio receivers [*cont.*]
Postes récepteurs de télévision et de radio [*suite*]

Country or area Pays ou zone	Code §	Number (000) Nombre (000)				Per 1000 inhabitants Par 1000 habitants			
		1975	1980	1985	1989	1975	1980	1985	1989
Bolivia	T	45	300	420	700	9	54	66	98
Bolivie	R	1150	2800	3675	4250	235	503	577	597
Brazil	T	8400	15000	25000	30000	78	124	184	204
Brésil	R	16980	35000	49000	55000	157	289	361	373
Chile	T	700	1225	1750	2600	68	110	144	201
Chili	R	1700	3250	4000	4400	164	292	330	340
Colombia	T	1600	2250	2750	3500	67	84	92	108
Colombie	R	2808	3300	4000	5400	117	123	134	167
Ecuador	T	252	500	600	850	36	62	64	82
Equateur	R	2000	2425	2850	3240	284	298	306	314
French Guiana	T	6	13	17	20	107	188	204	211
Guyane française	R	14	30	59	71	241	435	705	751
Guyana	T	-	-	-	25	-	-	-	31
Guyana	R	266	303	355	386	362	399	449	486
Paraguay	T	54	68	85	200	20	21	23	48
Paraguay	R	185	350	600	700	69	111	162	169
Peru	T	610	895	1500	2000	40	52	77	95
Pérou	R	2050	2750	4000	5300	135	159	206	251
Suriname	T	34	40	45	55	93	113	118	133
Suriname	R	110	189	230	262	302	535	601	633
Uruguay	T	351	364	500	700	124	125	166	227
Uruguay	R	1500	1630	1760	1850	530	559	585	600
Venezuela	T	1284	1710	2250	3000	101	114	130	156
Venezuela	R	4775	5900	7000	8300	377	393	404	432
Asia · Asie									
Afghanistan	T	-	45	100	130	-	3	7	8
Afghanistan	R	...	1200	1450	1670	...	75	100	104
Bahrain	T	30	90	170	198	110	260	396	402
Bahreïn	R	85	125	210	260	313	361	490	527
Bangladesh	T	25	80	261	500	0	1	3	4
Bangladesh	R	500[1]	730[1]	4000	4650	7[1]	8[1]	40	41
Bhutan									
Bhoutan	R	4	7	18	23	3	6	13	15
Brunei Darussalam	T	14	26	45	58	90	142	201	225
Brunéi Darussalam	R	24	41	50	60	154	222	223	234
Cambodia	T	30	35	52	65	4	6	7	8
Cambodge	R	110	600	780	860	15	94	107	107
China	T	1185	4000	10000	30000	1	4	9	27
Chine	R	15000	55000	120000	206000	16	55	113	184
Cyprus	T	59	85	92	98	97	135	138	141
Chypre	R	180	163	190	200	295	259	286	289
Hong Kong	T	837	1114	1275	1500	190	221	234	260
Hong-kong	R	2200	2550	3250	3700	500	506	596	642
India	T	515	3000	10000	22539	1	4	13	27
Inde	R	17228[1]	26000	50000	65000	28[1]	38	65	78
Indonesia	T	300	3000	6438	10000	2	20	39	55
Indonésie	R	5010	15000	19454	26000	37	99	116	144
Iran, Islamic Rep. of	T	1500	2000	2500	3500	45	51	53	66
Iran, Rép. islamique d'	R	2050	6400	10000	13000	61	165	210	245
Iraq	T	415	650	900	1250	38	49	57	68
Iraq	R	1252	2100	3000	3700	114	158	189	202
Israel	T	680	900	1100	1200	197	232	260	266
Israël	R	595	950	1800	2115	172	245	425	468
Japan	T	40000	62976	70000	75000	359	539	579	610
Japon	R	58026	79200	95000	110000	520	678	786	895
Jordan	T	120	171	240	300	46	59	70	77
Jordanie	R	450	536	791	980	173	183	232	252
Korea, Dem. People's Rep.	T	100	130	200	300	6	7	10	14
Corée, Rép. pop. dém. de	R	13509[1]	20000[1]	38605	2500	...	...	...	117
Korea, Republic of	T	2500	6300	7721	8800	71	165	189	207
Corée, République de	R	13509	20000	38605	42570	383	525	946	1003
Kuwait	T	150	353	450	550	149	257	262	281
Koweït	R	203	360	497	660	201	262	289	337

28
Television and radio receivers [*cont.*]
Postes récepteurs de télévision et de radio [*suite*]

Country or area Pays ou zone	Code §	Number (000) Nombre (000)				Per 1000 inhabitants Par 1000 habitants			
		1975	1980	1985	1989	1975	1980	1985	1989
Lao People's Dem. Rep.	T	-	-	...	20	-	-	...	5
République dém. pop. lao	R	150	350	430	500	50	109	120	124
Lebanon	T	410	750	800	880	148	281	300	327
Liban	R	1321	2000	2050	2247	477	749	768	834
Macau	T	-	-	-	30	-	-	-	65
Macao	R	61	76	93	113	235	235	237	246
Malaysia	T	452	1200	1800	2500	37	87	115	144
Malaisie	R	1420	5760	6600	7460	116	419	421	428
Maldives	T	-	1	3	5	-	7	18	24
Maldives	R	3	7	19	24	21	45	105	115
Mongolia	T	4	5	45	80	2	3	24	38
Mongolie	R	114	166	200	280	79	100	105	131
Myanmar	T	-	1	20	70	-	0	1	2
Myanmar	R	662	774	2500	3300	22	23	67	81
Nepal	T	-	-	20	30	-	-	1	2
Népal	R	113	300	450	625	9	20	27	33
Oman	T	3	35	900	1100	3	36	725	762
Oman	R	15	300	800	932	20	305	644	645
Pakistan	T	380	938	1304	1875	5	11	13	16
Pakistan	R	4000	5500	9000	10200	54	64	87	86
Philippines	T	756	1050	1500	2500	18	22	27	41
Philippines	R	1800	2100	5000	8300	42	43	91	136
Qatar	T	20	80	120	180	117	349	401	514
Qatar	R	50	100	150	179	292	437	502	510
Saudi Arabia	T	800	2100	3100	3750	110	224	267	277
Arabie saoudite	R	950	2500	3220	3800	131	267	278	280
Singapore	T	421	750	850	1000	186	311	332	372
Singapour	R [1]	345	459	720	822	152	190	281	306
Sri Lanka	T	-	35	450	550	-	2	28	32
Sri Lanka	R	700	1454	2551	3300	51	98	158	194
Syrian Arab Republic	T	224	385	600	710	30	44	57	59
Rép. arabe syrienne	R	1400	1700	2200	3000	188	193	210	248
Thailand	T	500	1000	4122	6000	12	21	80	109
Thaïlande	R	4000	5910	8000	10000	97	127	155	182
Turkey	T	1050	3500	8000	9500	26	79	159	174
Turquie	R	4154[1]	5000	7000	8800	104[1]	113	139	161
United Arab Emirates	T	25	93	130	165	50	92	96	109
Emirats arabes unis	R	52	240	350	490	103	236	259	322
Viet Nam	T	...	...	2000	2500	...	...	33	38
Viet Nam	R	...	...	6000	7000	...	...	100	107
Yemen [2]	T	-	5	28	150	-	1	4	17
Yémen [2]	R	87	110	150	290	16	17	20	33
Europe · Europe									
Albania	T	5	96	232	265	2	36	78	83
Albanie	R	175	400	493	550	72	150	166	172
Andorra	T	2	4	6	7	74	114	133	149
Andorre	R	5	7	9	10	185	194	196	219
Austria	T	2550	2950	3260	3600	337	391	431	475
Autriche	R	3600	3830	4203	4710	475	507	556	622
Belgium	T	3315	3815	3950	4400	338	387	401	447
Belgique	R	6200	7200	7450	7640	633	731	756	776
Bulgaria	T	1960	2150	2220	2240	225	243	248	249
Bulgarie	R	3150	3500	3750	3920	361	395	419	436
Czechoslovakia	T	4987	5960	6080	6400	337	389	392	410
Tchécoslovaquie	R	7425	7800	8050	9100	502	509	519	583
Denmark	T	2110	2550	2675	2715	417	498	522	528
Danemark	R	4625	4750	4875	5200	914	927	952	1012
Faeroe Islands	T	-	4	9	11	-	93	196	236
Iles Féroé	R	15	18	22	24	366	407	478	511
Finland	T	1658	1980	2300	2420	353	414	469	488
Finlande	R	2300	4000	4830	4944	488[1]	837	986	998
France	T	15000	19000	21500	22350	285	353	390	400
France	R	30000	39900	48000	50000	569	741	870	895

28
Television and radio receivers [*cont.*]
Postes récepteurs de télévision et de radio [*suite*]

Country or area Pays ou zone	Code §	Number (000) Nombre (000)				Per 1000 inhabitants Par 1000 habitants			
		1975	1980	1985	1989	1975	1980	1985	1989
Germany *§* · Allemagne *§*									
Federal Republic of Germany	T	25000	27000	29500	31000	404	439	483	506
Rép. féd. d'Allemagne	R	52500	550009	57500	58080	849	893	942	949
former German Dem. Rep.	T	7835	8600	12300	12700	465	514	739	779
ancienne Rép. dém. allemande	R	8650	8970	10250	11200	513	536	616	687
Gibraltar	T	6	7	8	10	200	232	272	328
Gibraltar	R	30	33	34	35	1000	1087	1179	1182
Greece	T	1155	1650	1896	1950	128	171	191	195
Grèce	R	2500	3310	4000	4200	276	343	403	419
Hungary	T	2850	3320	4250	4320	271	310	399	409
Hongrie	R	4100	5340	6144	6250	389	499	577	592
Iceland	T	56	65	75	80	257	258	311	319
Islande	R	140	162	185	197	642	711	768	785
Ireland	T	600	785	910	1000	189	231	256	271
Irlande	R	907	1275	2050	2150	285	375	577	583
Italy	T	15000	22000	23600	24150	271	390	413	423
Italie	R	32500	34000	37000	45350	586	602	648	794
Liechtenstein	T	5	7	9	10	188	269	322	343
Liechtenstein	R	7	13	18	20	308	500	656	718
Luxembourg	T	87	90	92	94	238	247	250	252
Luxembourg	R	179	200	228	231	493	548	621	623
Malta	T	176	202	235	260	542	557	683	741
Malte	R	141	175	179	184	432	482	520	524
Monaco	T	16	17	19	22	648	654	704	786
Monaco	R	20	26	29	30	800	1004	1056	1084
Netherlands	T	4230	5650	6700	7200	310	399	463	485
Pays-Bas	R	7710	9200	12000	13400	565	650	829	902
Norway	T	1252	1430	1640	1775	313	350	395	423
Norvège	R	2575	2700	3230	3342	643	661	778	796
Poland	T	7120	8750	10400	11100	209	246	280	292
Pologne	R	9700	10615	12250	16300	285	298	329	428
Portugal	T	860	1540	1765	1810	95	158	174	176
Portugal	R	1530	1660	2000	2220	168	170	197	216
Romania	T	2950	4085	4375	4480	139	184	193	194
Roumanie	R	3725	3940	4100	4500	175	177	180	195
San Marino	T	4	6	7	8	200	300	314	326
Saint-Marin	R	6	10	12	13	300	476	527	583
Spain	T	6640	9525	10400	15200	187	254	269	389
Espagne	R	9050	9700	11300	11900	254	258	293	304
Sweden	T	3760	3830	3880	3961	459	461	465	471
Suède	R	4200	7000	7250	7450	513	842	868	885
Switzerland	T	2050	2300	2550	2670	323	364	394	406
Suisse	R	4500	5140	5420	5600	710	813	838	851
United Kingdom	T	20200	22600	24500	24800	359	401	433	434
Royaume-Uni	R	39000	53500	57000	65400	694	950	1007	1145
Yugoslavia	T	3385	4245	4500	4650	159	190	195	197
Yougoslavie	R	4415	5000	5580	5800	207	224	241	245
Oceania · Océanie									
American Samoa	T	5	6	7	9	167	180	194	240
Samoa américaines	R	...	32	38	47	...	991	1056	1245
Australia	T	4549	5600	7000	8050	334	381	444	484
Australie	R	13900	16000	19500	21000	1020	1089	1237	1262
Cook Islands									
Iles Cook	R	7	8	10	11	368	417	578	583
Fiji	T	-	-	-	11	-	-	-	14
Fidji	R	...	300	370	430	...	473	529	573
French Polynesia	T	14	18	27	32	106	124	156	159
Polynésie française	R	65	78	93	105	492	527	529	532
Guam	T	50	75	81	86	521	708	717	735
Guam	R	86	100	150	185	896	943	1327	1581
Kiribati									
Kiribati	R	8	12	13	15	152	195	210	225

28
Television and radio receivers [*cont.*]
Postes récepteurs de télévision et de radio [*suite*]

Country or area Pays ou zone	Code §	Number (000) Nombre (000)				Per 1000 inhabitants Par 1000 habitants			
		1975	1980	1985	1989	1975	1980	1985	1989
Nauru									
Nauru	R	...	4	5	6	...	500	600	633
New Caledonia	T	15	25	40	44	113	180	263	268
Nouvelle-Calédonie	R	60	76	80	92	451	547	526	561
New Zealand	T	958	1035	1165	1250	311	333	359	372
Nouvelle-zélande	R	2670	2755	2950	3100	866	885	909	922
Pacific Islands (Palau) [3]	T	3	6	8	9	25	44	51	52
Iles du Pacifique (Palaos) [3]	R	72	...	...	101	595	...	...	598
Papua New Guinea	T	-	-	-	8	-	-	-	2
Papouasie-Nouv.-Guinée	R	110	180	220	260	40	58	64	69
Samoa	T	-	3	5	6	-	16	31	36
Samoa	R	23	32	68	73	154	206	420	437
Solomon Islands									
Iles Salomon	R	10	20	25	36	52	89	92	117
Tokelau									
Tokélaou	R	...	...	1	1	...	...	500	525
Tonga									
Tonga	R	...	70	30	52	...	722	313	547
Tuvalu									
Tuvalu	R	...	...	2	3	...	...	250	278
Vanuatu	T	-	-	-	1	-	-	-	9
Vanuatu	R	13	23	36	41	135	197	261	267
former USSR · ancienne URSS									
former USSR	T	55200	76500	82400	92354	217	288	297	323
ancienne URSS	R	122477	130000	182800	195800	481	490	659	685
Belarus	T	1600	2100	2500	2700	...	218	255	...
Bélarus	R	...	2145	2900	3100	...	222	290	303
Ukraine	T	11035	12722	15190	16752	225	254	292	...
Ukraine	R	23539	28941	36434	40414	480	578	714	781

Source:
United Nations Educational, Scientific and Cultural Organization
(Paris).

Source:
Organisation des Nations Unies pour l'éducation, la science et la culture
(Paris).

ᶴ All data shown which pertain to Germany prior to 3 October 1990
are indicated separately for the Federal Republic of Germany
and the former German Democratic Republic based on their
respective territories at the time indicated. For detailed
explanatory notes on data pertaining to Germany, see Annex I-
Country Nomenclature.

§ T: Estimated number of television receivers in use.
R: Estimated number of radio receivers in use.

1   Number of licenses issued or sets declared.
2   Excluding former Democratic Yemen.
3   Including data for Federated States of Micronesia,
Marshall Islands and Northern Mariana Islands.

ᶴ Toutes les données se rapportant à l'Allemagne avant le 3 octobre 1990
figurent dans deux rubriques séparées basées sur les territoires
respectifs de la République fédérale d'Allemagne et l'ancienne
République démocratique allemande selon la période indiquée. Pour
les notes explicatives en détail sur les données concernant l'Allemagne,
voir Annexe I - Nomenclature des pays.

§ T: Estimation du nombre de récepteurs de télévision en service.
R: Estimation du nombre de récepteurs de radio en service.

1   Nombre de licences délivrées ou de postes déclarées.
2   Non compris l'ancien Yémen démocratique.
3   Y compris les données pour les Etats fédérés de Micronésie,
les îles Marshall et les îles Mariannes du Nord.

Technical notes, tables 23-28

These tables are compiled from the UNESCO *Statistical Yearbook*. [32] The technical notes in the UNESCO *Yearbook* concerning these tables are summarized below.

Table 23: The data represent apparent consumption (that is, production plus imports minus exports). The term "newsprint" designates the bleached, unsized or slack-sized printing paper, without coating, of the type usually used for newspapers.

Table 24: Data on books by subject groups cover printed books and pamphlets and, unless otherwise stated, refer to first editions and re-editions of originals and translations. The grouping by subject follows the Universal Decimal Classification (UDC).

Table 25: Unless otherwise stated, the data refer to the total number of titles (first editions and re-editions of original works and translations) by language of publication.

Table 26: For the purposes of this table, a daily general interest newspaper is defined as a publication devoted primarily to recording general news. It is considered to be "daily" if it appears at least four times a week.

It is known or believed that no daily general-interest newspapers are published in the following 32 countries and territories: *Africa*: Cape Verde, Comoros, Djibouti, St. Helena, Sao Tome and Principe, Western Sahara; *Latin America and Caribbean*: Belize, British Virgin Islands, Dominica, Falkland Islands (Malvinas), Greenland, Grenada, Montserrat, Panama-former Canal Zone, St. Kitts and Nevis, St. Pierre and Miquelon, St. Vincent and the Grenadines, Turks and Caicos Islands. *Asia*: Bhutan, Brunei Darussalam, East Timor; *Europe*: Faeroe Islands; *Oceania*: Kiribati, Nauru, Niue, Norfolk Island, Pacific Islands (Palau), Samoa, Solomon Islands, Tokelau, Tuvalu, Vanuatu.

Table 27: For the purposes of this table a non-daily general interest newspaper is defined as a publication which is devoted primarily to recording general news and which is published three times a week or less. Under the category of periodicals are included publications of periodical issue, other than newspapers, containing information of a general or of a specialized nature.

Table 28: Data show either the estimated number of receivers in use (indicated by R) or, when estimations are not available, the number of licenses issued (L). The figures refer to 31 December of the year stated. In these tables the term "receivers" relates to all types of receivers for broadcasts to the general public, including receivers connected to a redistribution system (wired receivers).

Notes techniques, tableaux 23-28

Ces tableaux ont été établis à partir de l'*Annuaire statistique* de l'UNESCO [32]. Les notes techniques de l'*Annuaire* de l'UNESCO concernant ces tableaux sont résumées ci-dessous.

Tableau 23 : Les données correspondent à la consommation apparente (c'est-à-dire la production plus les importations moins les exportations). L'expression "papier journal" désigne le papier d'impression blanchi, non collé ou peu encollé, non couché, du type ordinaire utilisé pour les journaux.

Tableau 24 : Les données concernant la production de livres par groupes de sujets se rapportent aux livres et brochures imprimés, sauf indication contraire, aux premières éditions et aux rééditions d'originaux et aux traductions. Les sujets sont groupés selon la Classification décimale universelle (CDU).

Tableau 25 : Sauf indication contraire, les données se rapportent au nombre total de titres (premières éditions et rééditions d'ouvrages originaux et de traductions) par langue de publication.

Tableau 26 : Dans ce tableau, par "journal quotidien d'information générale", on entend une publication qui a essentiellement pour objet de rendre compte des événements courants. Il est considéré comme "quotidien" s'il paraît au moins quatre fois par semaine.

On sait ou l'on croit savoir qu'il ne paraît aucun journal quotidien d'information générale dans les 32 pays ou territoires suivants : *Afrique* : Cap-Vert, Comores, Djibouti, Sahara occidental, Sainte-Hélène, Sao Tomé-et-Principe; *Amérique latine et Caraïbes* : Bélize, Dominique, Grenade, Groenland, Iles Falkland (Malvinas), Iles turques et caïques, Iles vierges britanniques, Montserrat, Panama-ancienne zone du Canal, Saint Kitts-et-Nevis, Saint-Pierre-et-Miquelon, Saint-Vincent-et-Grenadines; *Asie* : Bhoutan, Brunëi Darussalam, Timor oriental; *Europe* : Iles Faeroe; *Océanie* : Ile Norfolk, Iles du Pacifique (Palaos), Iles Salomon, Kirabiti, Nauru, Niue, Samoa, Tokelau, Tuvalu, Vanuatu.

Tableau 27 : Aux fins de ce tableau, par "journal non quotidien d'information générale", on entend une publication qui a essentiellement pour objet de rendre compte des événements courants et qui est publié trois fois par semaine ou moins. La catégorie périodique comprend les publications périodiques autres que les journaux, contenant des informations de caractère général ou spécialisé.

Tableau 28 : Les données de ces tableaux indiquent le nombre estimatif de récepteurs en usage (indiqués par un R), ou, lorsqu'on ne dispose pas de telles estimations, le nombre de licences délivrées (L). Les chiffres se rapportent au 31 décembre de l'année indiquée. Dans ces tableaux, le terme "récepteurs" désigne tous les types de récepteur permettant de capter les émissions destinées au grand public, y compris les récepteurs reliés (par fil) à un réseau de redistribution.

Part Three
Economic Activity

VI
National accounts (tables 29 and 30)
VII
Government finance (tables 31-35)
VIII
Labour force (tables 36 and 37)
IX
Wages and prices (tables 38-40)
X
Agriculture, hunting, forestry and fishing (tables 41-60)
XI
Mining and quarrying (tables 61-67)
XII
Manufacturing (tables 68-99)
XIII
Transport and communications (tables 100-105)
XIV
Energy (tables 106 and 107)
XV
Environment and land use (tables 108-110)
XVI
Science and technology; intellectual property (tables 111-115)

Part Three of the *Yearbook* presents statistical series on economic production and consumption for a wide range of economic activities, and other basic series on major economic topics, for all countries or areas of the world for which data are available. Included are basic tables on national accounts, government finance, labour force, wages and prices, a wide range of agricultural, mined and manufactured commodities, transport and communications, energy, environment, land use and science and technology. In most cases, tables present statistics on production; in a few cases, data are presented on stocks and consumption.

International economic topics such as external trade are covered in Part Four.

Troisième partie
Activité économique

VI
Comptabilités nationales (tableaux 29 et 30)
VII
Finances publiques (tableaux 31 à 35)
VIII
Main-d'oeuvre (tableaux 36 et 37)
IX
Salaires et prix (tableaux 38 à 40)
X
Agriculture, chasse, forêts et pêche (tableaux 41 à 60)
XI
Industries extractives (tableaux 61 à 67)
XII
Industries manufacturières (tableaux 68 à 99)
XIII
Transports et communications (tableaux 100 à 105)
XIV
Energie (tableaux 106 et 107)
XV
Environnement et utilisation des terres (tableaux 108 à 110)
XVI
Science et technologie; propriété intellectuelle (tableaux 111 à 115)

La troisième partie de l'*Annuaire* présente, pour une large gamme d'activités économiques, des séries statistiques sur la production économique et la consommation, et, pour tous les pays ou zones du monde pour lesquels des données sont disponibles, d'autres séries fondamentales ayant trait à des questions économiques importantes. Y figurent des tableaux de base consacrés à la comptabilité nationale, aux finances publiques, à la main-d'oeuvre, aux salaires et aux prix, à un large éventail de produits agricoles, miniers et manufacturés, aux transports et communications, à l'énergie, à l'environnement, à l'utilisation des terres et à la science et à la technologie. On y trouve, dans la plupart des cas, des statistiques sur la production et parfois des données relatives aux stocks et à la consommation.

Les questions économiques internationales comme le commerce extérieur sont traitées dans la quatrième partie.

29
Gross domestic product and net material product
Produit intérieur brut et produit matériel net

A. Million national currency units • Million d'unités monetaires nationales

Country or area Pays ou zone	1982	1983	1984	1985	1986	1987	1988	1989	1990
Albania: lek Albanie : lek	16 537	16 718	16 503	16 856	17 383	17 246	17 001	18 674	16 812
Algeria: dinar Algérie : dinar	207 552	233 752	259 900	287 400	286 500	307 900	320 000	...	...
Angola: kwanza Angola : kwanza	...	126 357	141 570	144 747	...	...	...	...	...
Anguilla: EC dollar Anguilla : EC dollar	...	...	39	48	63	77	97	114	...
Antigua and Barbuda: EC dollar Antigua-et-Barbuda: EC dollar	372	414	468	541	642	...	...	...	...
Argentina: peso Argentine : peso	148	683	5 281	39 593	74 309	173 109	784 793	25 580[1]	515 644[1]
Australia: dollar[2] Australie : dollar[2]	171 706	194 777	216 150	240 319	264 564	298 335	340 440	370 805	377 114
Austria: schilling Autriche : schilling	1 133 535	1 201 217	1 276 775	1 348 425	1 422 497	1 481 388	1 561 700	1 663 891	1 789 386
Bahamas: dollar Bahamas : dollar	1 516	1 493	1 615	1 868	2 098	2 324	2 579	2 661	2 811
Bahrain: dinar Bahreïn : dinar	1 371	1 404	1 468	1 393	1 198	1 192	1 263	1 347	1 467
Bangladesh: taka[2] Bangladesh : taka[2]	292 208	354 642	406 933	466 227	539 201	597 136	659 598	737 571	807 372
Barbados: dollar Barbade : dollar	1 990	2 113	2 303	2 410	2 646	2 914	3 099	3 427	3 440
Belarus: rouble[3] Bélarus : rouble[3]	17 429	18 864	20 814	19 699	21 279	21 941	22 122	25 225	28 529
Belgium: franc Belgique : franc	3 888 986	4 122 259	4 429 036	4 738 039	4 985 952	5 205 543	5 542 673	6 015 975	6 429 346
Belize: dollar Belize : dollar	335	348	390	388	422	510	577	673	733
Benin: CFA franc Bénin : franc CFA	#416 638	417 438	459 340	469 778	462 535	469 554	482 434	487 525	...
Bermuda: dollar[4] Bermudes : dollar[4]	905	1 003	1 073	1 190	1 290	1 401	1 500	1 572	...
Bhutan: ngultrum Bhoutan : ngultrum	1 522	1 789	2 106	2 391	2 802	3 607	3 934	4 355	...
Bolivia: boliviano Bolivie : boliviano	0	1	23	2 867	8 924	10 180	12 303	14 743	...
Botswana: pula[5] Botswana : pula[5]	900	1 153	1 391	1 829	2 421	2 810	3 398	4 988	...
Brazil: cruzeiro Brésil : cruzeiro	51	117	386	1 383	3 662	11 537	86 197	1 266 348	...
British Virgin Islands: dollar Iles Vierges britanniques: dollar	70	79	86	90	98	117	131	...	...
Brunei Darussalam: dollar Brunéi Darussalam : dollar	9 125	8 163	8 068	...	...	...	...	...	...
Bulgaria: lev Bulgarie : lev	29 013	29 815	31 671	32 595	34 424	36 531	38 345	39 579	45 390

29
Gross domestic product and net material product [*cont.*]
Produit intérieur brut et produit matériel net [*suite*]

A. Million national currency units • Million d'unités monetaires nationales

Country or area Pays ou zone	1982	1983	1984	1985	1986	1987	1988	1989	1990
Burkina Faso: CFA franc Burkina Faso : franc CFA	359 604	381 013	390 565	469 313	503 500	...	...	...	...
Burundi: franc Burundi : franc	94 094	102 892	120 451	141 347	140 842	143 590	152 907	173 295	189 142
Cameroon: CFA franc[5] Cameroun : franc CFA[5]	2 172 800	2 618 100	3 195 000	3 838 900	4 135 100	4 004 900	3 769 900	3 560 000	3 638 300
Canada: dollar Canada : dollar	371 820	402 229	441 312	474 339	501 426	546 777	599 970	643 406	665 236
Cape Verde: escudo Cap-Vert : escudo	8 438	10 140	11 548	13 081	15 558	17 984	20 640	...	...
Cayman Islands: dollar Iles Caïmanes : dollar	...	221	244	264	290	346	429	498	596
Central African Rep.: CFA franc Rép. centrafricaine : franc CFA	245 966	251 055	275 730	311 430	334 470	313 210	326 420	...	...
Chile: peso Chili : peso	1 202 808	1 535 676	1 902 702	# 2 577[1]	3 246[1]	4 160[1]	5 411[1]	6 778[1]	8 478[1]
China: yuan renminbi[3] Chine : yuan renminbi[3]	425 800	473 600	565 200	702 000	785 900	931 300	1 173 800	1 317 600	1 442 900
Colombia: peso Colombie : peso	2 497 000	3 054 000	3 857 000	4 966 000	6 788 000	8 824 000	11 731[1]	15 190[1]	20 654[1]
Congo: CFA franc Congo : franc CFA	710 020	799 245	958 509	970 850	640 407	690 523	658 964	773 524	...
Costa Rica: colon Costa Rica : colon	97 505	129 314	163 011	197 920	246 579	284 533	346 559	421 781	520 741
Côte d'Ivoire: CFA franc Côte d'Ivoire : franc CFA	2 486 544	2 605 913	2 989 400	3 134 800	3 244 419	...	...	...	...
Cuba: peso[3] Cuba : peso[3]	12 175	12 926	13 695	13 952	12 857	12 284	12 764	12 791	...
Cyprus: pound Chypre : livre	1 025	1 137	1 337	1 481	1 599	1 779	1 994	2 253	2 519
Czechoslovakia: koruna[3] Tchécoslovaquie : couronne[3]	491 389	503 317	534 014	548 723	562 164	574 657	598 609	609 059	664 762
Denmark: krone Danemark : couronne	464 467	512 540	565 284	615 072	666 496	699 908	732 054	769 801	800 014
Dominica: EC dollar Dominique : EC dollar	195	216	243	266	303	339	393	423	462
Dominican Republic: peso Rép. dominicaine : peso	7 964	8 623	10 355	13 972	15 780	19 536	28 353	42 393	60 555
Ecuador: sucre Equateur : sucre	415 715	560 271	812 629	1 109 940	1 383 232	1 794 501	3 019 724	5 325 155	8 349 688
Egypt: £[5] Egypte : £[5]	20 221	25 772	31 246	36 618	40 820	44 052	...	...	...
El Salvador: colon El Salvador : colon	8 966	10 152	11 657	14 331	19 763	23 141	27 366	32 230	41 057
Equatorial Guinea: CFA franc Guinée équatoriale : franc CFA	...	25 649	29 532	# 38 067	37 205	39 322	42 749	42 257	44 350
Ethiopia: birr[4] Ethiopie : birr[4]	9 167	10 031	10 001	9 890	10 832	11 196	11 767	12 491	...

29
Gross domestic product and net material product [cont.]
Produit intérieur brut et produit matériel net [suite]

A. Million national currency units • Million d'unités monetaires nationales

Country or area Pays ou zone	1982	1983	1984	1985	1986	1987	1988	1989	1990
Fiji: dollar Fidji : dollar	1 113	1 142	1 275	1 316	1 462	1 465	1 588	1 861	...
Finland: markka Finlande : markka	245 716	274 647	308 357	334 986	357 566	391 597	441 539	496 935	524 781
France: franc France : franc	3 626 021	4 006 498	4 361 913	4 700 143	5 069 296	5 336 652	5 723 206	6 136 070	6 484 109
French Polynesia: CFP franc Polynésie française : franc CFP	141 574	170 489	201 807	226 772	266 935	255 966	268 076	281 667	297 754
Gabon: CFA franc Gabon : franc CFA	1 188 900	1 292 600	1 556 000	1 645 800	1 201 100	1 020 600	1 013 600	1 168 066	...
Gambia: dalasi[2] Gambie : dalasi[2]	606	618	782	1 085	1 486	1 636	1 942	2 345	2 689
Germany†• Allemagne† F. R. Germany: deutsche mark R. f. Allemagne : deutsche mark	1 588 090	1 668 540	1 750 890	1 823 180	1 925 290	1 990 480	2 095 980	2 220 880	2 404 540
former German D. R.: mark anc. R. d. allemande : mark	263 858	276 505	293 271	311 755	321 935	332 810	346 129	...	...
Ghana: cedi Ghana : cedi	86 451	184 038	270 561	343 048	511 400	746 000	1 051 200	1 417 200	2 031 700
Greece: drachma Grèce : drachme	2 574 652	3 079 178	3 805 686	4 617 816	5 514 766	6 258 478	7 526 471	8 777 535	10 455[1]
Grenada: EC dollar Grenade : EC dollar	239	253	275	311	350	406	449	491	541
Guadeloupe: franc Guadeloupe : franc	7 520	8 413	9 047	9 802	10 958	...	...	...	...
Guatemala: quetzal Guatemala : quetzal	8 717	9 050	9 470	11 180	15 838	17 711	20 545	23 685	34 289
Guinea-Bissau: peso Guinée-Bissau : peso	...	...	...	...	46 973	92 375	171 949	...	...
Guyana: dollar Guyana : dollar	1 446	1 455	1 700	1 964	2 219	3 424	4 138	6 957	10 131
Haiti: gourde[6] Haïti : gourde[6]	7 425	8 148	9 082	10 047	11 177	9 946	9 795	10 812	11 404
Honduras: lempira Honduras : lempira	5 762	6 035	6 462	7 008	7 596	8 128	8 937	9 782	...
Hong Kong: dollar Hong-kong : dollar	185 728	206 217	247 933	261 070	298 515	367 603	433 657	496 385	553 243
Hungary: forint Hongrie : forint	847 871	896 367	978 456	1 033 658	1 088 800	1 226 370	# 1 435[1]	1 711[1]	2 080[1]
Iceland: krona Islande : couronne	38 132	65 837	87 508	119 091	158 597	208 099	254 639	300 595	341 790
India: rupee[7] Inde : roupie[7]	1 781 320	2 075 890	2 313 870	2 619 200	2 919 740	3 326 160	3 949 920	4 427 690	...
Indonesia: rupiah[1] Indonésie : rupiah[1]	62 476	77 623	89 885	96 997	102 683	124 817	142 105	167 495	197 721
Iran, Islamic Rep. of: rial[18] Iran, Rép. islamique d' : rial[18]	10 756	13 750	15 030	16 556	18 126	21 270	23 588	28 288	...
Iraq: dinar Iraq : dinar	12 777	13 256	14 922	15 494	15 063	17 901	20 032	20 010	...

29
Gross domestic product and net material product *[cont.]*
Produit intérieur brut et produit matériel net *[suite]*

A. Million national currency units • Million d'unités monetaires nationales

Country or area Pays ou zone	1982	1983	1984	1985	1986	1987	1988	1989	1990
Ireland: pound Irlande : livre	13 382	14 779	16 407	17 790	18 877	20 263	21 815	24 308	25 693
Israel: new shekel Israël : nouveau shekel	635	1 639	8 047	30 402	47 813	61 267	74 385	89 165	108 810
Italy: lira[1] Italie : lire[1]	545 124	633 436	725 760	810 580	899 903	983 803	1 091 837	1 192 725	1 306 833
Jamaica: dollar Jamaïque : dollar	5 867	6 993	9 358	11 203	13 388	16 002	18 748	22 224	28 691
Japan: yen[1] Japon : yen[1]	270 602	281 767	300 542	320 419	334 609	348 425	371 428	396 197	425 735
Jordan: dinar Jordanie : dinar	1 321	# 1 728	1 868	1 898	2 040	2 088	2 201	2 541	2 567
Kenya: shilling Kenya : shilling	3 515	3 983	4 466	5 043	5 874	6 558	7 551	8 495	10 033
Korea, Republic of: won[1] Corée, République de : won[1]	54 443	63 833	72 644	80 847	93 426	108 428	127 963	143 001	169 701
Kuwait: dinar Koweït : dinar	6 214	6 083	6 425	6 450	5 141	6 154	5 586	6 779	...
Lebanon: pound Liban : livre	12 599	...	...	...	...	...	...	...	...
Lesotho: loti Lesotho : maloti	373	391	455	551	631	753	1 026	1 287	1 505
Liberia: dollar Libéria : dollar	1 093	1 086	1 070	1 055	1 037	1 089	1 158	1 194	...
Libyan Arab Jamah.: dinar Jamah. arabe libyenne : dinar	9 096	8 805	8 013	8 277	...	...	...	...	...
Luxembourg: franc Luxembourg : franc	158 786	174 683	193 666	205 255	223 304	227 846	248 387	279 033	291 505
Madagascar: franc Madagascar : franc	996 100	1 221 100	1 369 100	1 553 400	1 806 900	...	...	...	...
Malawi: kwacha Malawi : kwacha	1 245	1 437	1 707	1 950	2 194	2 731	3 574	4 082	5 023
Malaysia: ringgit Malaisie : ringgit	62 599	69 941	79 550	77 470	71 594	79 625	90 861	101 463	114 616
Maldives: rufiyaa Maldives : rufiyaa	331	382	543	613	697	...	...	...	...
Mali: CFA franc Mali : franc CFA	403 600	411 300	463 500	520 200	585 100	600 700	615 800	661 400	683 300
Malta: lira Malte : lire	462	458	461	476	512	549	606	670	735
Martinique: franc Martinique : franc	8 848	10 266	10 920	12 484	13 824	...	...	...	...
Mauritania: ouguiya Mauritanie : ouguiya	42 669	44 899	46 127	52 665	59 715	67 216	72 635	83 520	...
Mauritius: rupee Maurice : roupie	11 725	12 763	14 360	16 618	19 700	23 576	27 803	32 145	37 700
Mexico: peso[1] Mexique : peso[1]	9 798	17 879	29 472	47 392	79 536	192 802	389 259	503 668	678 923

29
Gross domestic product and net material product [cont.]
Produit intérieur brut et produit matériel net [suite]

A. Million national currency units • Million d'unités monetaires nationales

Country or area Pays ou zone	1982	1983	1984	1985	1986	1987	1988	1989	1990
Mongolia: tugrik[3] Mongolie : tughrik[3]	6 826	7 325	7 378	7 636	7 248	7 479	7 900	8 646	8 327
Montserrat: EC dollar Montserrat : EC dollar	81	86	94	100	114	...	...	...	...
Morocco: dirham Maroc : dirham	92 900	99 140	112 340	129 510	154 290	156 340	181 250	191 270	207 490
Mozambique: metical Mozambique : metical	92 000	91 000	109 000	147 000	167 000	426 000	659 000	...	...
Myanmar: kyat Myanmar : kyat	46 811	49 823	53 597	55 989	59 028	68 698	76 243	119 281	138 136
Nepal: rupee[9] Népal : roupie[9]	30 988	33 761	39 390	44 417	50 428	59 246	68 858	78 259	88 711
Netherlands: guilder Pays-Bas : florin	368 860	381 020	400 250	418 180	428 610	430 170	449 820	475 300	508 310
Netherlands Antilles: guilder[10] Antilles néerlandaises : florin[10]	1 919	1 910	1 929	1 966	...	...	...	...	...
New Caledonia: CFP franc Nouvelle-Calédonie : franc CFP	108 093	114 161	126 489	139 650	151 281	162 627	224 498	...	...
New Zealand: dollar[7] Nouvelle-Zélande : dollar[7]	31 537	34 896	39 529	45 435	55 088	61 850	66 605	71 549	73 746
Nicaragua: córdoba Nicaragua : córdoba	28 350	32 920	45 030	115 404	435 742	2 389 500	...	...	...
Niger: CFA franc Niger : franc CFA	663 022	687 142	638 406	647 100	685 200	662 600	672 900	676 900	...
Nigeria: naira[7] Nigéria : naira[7]	51 709	57 142	63 608	72 355	73 062	108 885	145 243	224 797	260 637
Norway: krone Norvège : couronne	362 269	402 199	452 512	500 199	513 718	561 480	583 277	621 383	661 670
Oman: rial Omani Oman : rial omani	2 614	2 740	3 047	3 454	2 800	3 003	2 926	3 231	4 084
Pakistan: rupee[2] Pakistan : roupie[2]	364 387	419 802	472 157	514 532	572 479	675 389	769 745	862 452	1 016 728
Panama: balboa Panama : balboa	4 279	4 374	4 565	4 901	5 145	5 310	4 551	4 582	4 949
Papua New Guinea: kina Papouasie-Nvl-Guinée : kina	1 749	# 2 145	2 282	2 403	2 572	2 854	3 170	3 046	3 059
Paraguay: guarani Paraguay : guaraní	737 040	818 114	1 070 444	1 393 890	1 833 800	2 493 601	3 319 124	4 608 400	6 474 400
Peru: new sol Pérou : nouveau sol	17 909	32 448	72 410	197 903	373 976	739 439	4 942 317	115 340[1]	7 672 972[1]
Philippines: peso Philippines : peso	317 177	369 077	524 481	571 883	608 887	685 068	802 519	922 569	1 066 224
Poland: zloty[13] Pologne : zloty[13]	4 753	5 924	7 182	# 8 658	10 697	14 013	24 995	104 952	506 253
Portugal: escudo Portugal : escudo	1 850 407	2 301 713	2 815 728	3 523 945	4 420 400	5 174 732	6 002 751	7 130 260	8 507 434
Puerto Rico: dollar Porto Rico : dollar	17 277	19 163	20 289	21 969	23 878	26 178	28 267	30 536	32 469
Qatar: riyal Qatar : riyal	27 705	23 605	25 008	22 398	18 393	19 825	# 21 979	23 616	26 865

29
Gross domestic product and net material product [*cont.*]
Produit intérieur brut et produit matériel net [*suite*]

A. Million national currency units • Million d'unités monetaires nationales

Country or area Pays ou zone	1982	1983	1984	1985	1986	1987	1988	1989	1990
Réunion: franc Réunion : franc	12 286	13 722	15 785	17 315	19 520	21 881	23 719	...	...
Romania: leu Roumanie : leu	727 400	768 700	816 100	817 400	838 600	845 100	857 000	798 000	844 000
Rwanda: franc Rwanda : franc	130 960	142 190	159 110	173 700	168 990	171 430	177 920	190 220	...
Saint Kitts and Nevis: EC dollar Saint-Kitts-et-Nevis : EC dollar	158	154	167	...	...	...	...	...	...
Saint Lucia: EC dollar Sainte-Lucie : EC dollar	364	380	408	...	...	...	...	...	...
St. Vincent-Grenadines:EC dollar St. Vincent-Grenadines : EC dollar	227	255	277	305	344	384	434	469	...
Samoa: tala Samoa : tala	...	...	212	237	243	...	...	...	...
Sao Tome and Principe: dobra Sao Tomé-et-Principe : dobra	1 549	1 523	1 609	...	2 478	3 003	4 221	...	...
Saudi Arabia: riyal[5] Arabie saoudite : riyal[5]	561 136	458 124	373 882	351 397	313 941	271 091	275 452	285 150	310 820
Senegal: CFA franc Sénégal : franc CFA	848 800	944 700	1 021 300	1 158 100	1 302 900	1 382 300	1 483 300	1 475 400	...
Seychelles: rupee Seychelles : roupie	965	989	1 068	1 205	1 284	1 385	1 524	1 703	...
Sierra Leone: leone[5] Sierra Leone : leone[5]	1 604	1 876	2 729	4 785	7 481	19 700	29 080	43 947	70 133
Singapore: dollar Singapour : dollar	32 670	36 733	40 048	38 923	38 653	42 662	49 694	56 235	62 711
Solomon Islands: dollar Iles Salomon : dollar	158	141	222	237	252	293	367	...	...
Somalia: shilling Somalie : shilling	29 111	35 129	62 253	87 290	118 781	169 608	...	...	...
South Africa: rand[11] Afrique du Sud : rand[11]	80 531	91 457	107 221	123 126	142 135	164 524	197 910	233 134	263 812
Spain: peseta[1] Espagne : peseta[1]	19 567	22 235	25 111	# 28 201	32 324	36 144	40 164	45 025	50 074
Sri Lanka: rupee Sri Lanka : roupie	97 527	119 201	147 344	157 763	172 440	188 822	218 774	248 230	317 904
Sudan: pound[5] Soudan : livre[5]	6 720	9 186	11 329	13 913	22 009	31 157	...	...	...
Suriname: guilder Suriname : florin	1 830	1 767	1 728	1 746	1 803	1 974	2 321	2 707	3 078
Swaziland: lilangeni[5] Swaziland : lilangeni[5]	584	619	729	802	1 051	1 118	1 325	...	...
Sweden: krona Suède : couronne	633 682	709 852	794 298	865 788	945 583	1 019 547	1 110 471	1 226 313	1 350 138
Switzerland: franc Suisse : franc	195 980	203 865	213 230	227 950	243 350	254 685	268 410	290 360	312 355
Syrian Arab Republic: pound Rép. arabe syrienne : livre	68 788	73 291	75 342	83 225	99 933	127 712	186 047	208 741	278 038

29
Gross domestic product and net material product [cont.]
Produit intérieur brut et produit matériel net [suite]

A. Million national currency units • Million d'unités monetaires nationales

Country or area Pays ou zone	1982	1983	1984	1985	1986	1987	1988	1989	1990
Thailand: baht Thaïlande : baht	820 002	910 054	973 412	1 014 399	1 095 368	1 253 147	1 506 977	1 775 978	2 051 208
Togo: CFA franc Togo : franc CFA	269 700	281 300	304 800	332 500	363 600	...	...	...	...
Tonga: pa'anga[5] Tonga : pa'anga[5]	64	73	74	80	...	...	...	...	...
Trinidad and Tobago: dollar Trinité-et-Tobago : dollar	19 176	18 719	18 615	17 801	17 260	17 301	17 333	18 440	21 650
Tunisia: dinar Tunisie : dinar	4 804	5 497	6 240	6 910	7 004	7 959	8 605	9 497	...
Turkey: lira[1] Turquie : livre[1]	8 620	11 532	18 212	27 552	39 288	58 299	100 825	167 770	282 800
Uganda: shilling Ouganda : shilling	3 350	5 178	8 300	21 502	50 644	174 443	491 432	995 579	...
Ukraine: rouble[3] Ukraine : rouble[3]	87 500	92 300	96 200	94 000	95 000	98 700	102 500	108 900	117 200
former USSR: rouble[3] ancienne URSS : rouble[3]	523 900	548 300	570 500	578 500	587 400	599 600	630 800	673 700	730 400
United Arab Emirates: dirham Emirats arabes unis : dirham	112 433	102 909	101 843	99 416	79 566	87 366	87 106	100 976	124 008
United Kingdom: pound Royaume-Uni : livre	278 087	303 377	324 081	354 911	381 731	419 983	466 477	510 020	549 181
United Rep.Tanzania: shilling[12] Rép. Unie de Tanzanie : shilling[12]	58 226	70 509	88 892	120 621	159 721	227 456	331 217	406 542	495 863
United States: dollar Etats-Unis : dollar	3 118 560	3 349 390	3 717 770	3 962 220	4 176 100	4 452 880	4 809 080	5 132 000	5 392 200
Uruguay: new peso Uruguay : nouveau peso	128 696	# 175 417	271 025	478 641	890 451	1 661 477	2 725 507	4 734 131	9 623 666
Vanuatu: vatu Vanuatu : vatu	...	10 150	12 339	12 534	12 179	13 404	15 006	16 367	17 899
Venezuela: bolivar Venezuela : bolívar	291 268	290 492	# 420 072	464 741	489 172	696 421	873 283	1 510 361	2 264 039
Viet Nam: dong[1] Viet Nam : dong[1]	...	...	...	...	...	...	13 174	24 307	38 365
Yemen: rial[13] Yémen : rial[13]	15 869	17 170	17 950	30 939	37 472	...	...	# 61 406	77 159
Yugoslavia: new dinar Yougoslavie : nouveau dinar	316	428	665	1 195	2 340	5 234	15 833	235 395	1 147 787
Zaire: zaire Zaïre : zaïre	31 110	59 134	99 723	147 263	203 416	326 946	622 822	1 146 811	...
Zambia: kwacha Zambie : kwacha	3 595	4 181	4 931	7 072	12 963	19 778	30 021	60 024	113 341
Zimbabwe: dollar Zimbabwe : dollar	5 197	6 306	6 404	7 297	8 376	9 273	11 005	13 067	15 174

29
Gross domestic product and net material product [*cont.*]
Produit intérieur brut et produit matériel net [*suite*]

B. Million US dollars • Millions de dollars E-U

Country or area Pays ou zone	1982	1983	1984	1985	1986	1987	1988	1989	1990
Albania Albanie	2 362	2 388	2 358	2 408	2 483	2 464	2 549	3 046	2 169
Algeria Algérie	45 199	48 810	52 157	57 160	60 932	63 485	54 100	...	...
Angola Angola	...	4 223	4 732	4 838	...	...	...	...	...
Anguilla Anguilla	...	...	14	18	23	28	36	42	...
Antigua and Barbuda Antigua-et-Barbuda	138	153	173	200	238	...	...	...	...
Argentina Argentine	56 949	64 829	78 064	65 769	78 801	80 741	89 660	60 473	105 751
Australia[2] Australie[2]	173 440	175 792	190 126	168 416	177 496	209 103	266 973	293 863	294 639
Austria Autriche	66 448	66 872	63 810	65 173	93 175	117 171	126 474	125 757	157 378
Bahamas Bahamas	1 516	1 493	1 615	1 868	2 098	2 324	2 579	2 661	2 811
Bahrain Bahreïn	3 646	3 735	3 906	3 705	3 187	3 170	3 359	3 584	3 903
Bangladesh[2] Bangladesh[2]	13 211	14 405	16 050	16 654	17 733	19 294	20 786	22 856	23 355
Barbados Barbade	990	1 051	1 145	1 198	1 316	1 449	1 541	1 704	1 711
Belarus[3] Bélarus[3]	23 875	25 561	25 570	23 175	30 013	34 444	36 445	40 231	42 836
Belgium Belgique	85 115	80 620	76 648	79 795	111 612	139 432	150 747	152 674	192 392
Belize Belize	167	174	195	194	211	255	289	336	367
Benin Bénin	# 1 268	1 095	1 051	1 046	1 336	1 562	1 620	1 528	...
Bermuda[4] Bermudes[4]	905	1 003	1 073	1 190	1 290	1 401	1 500	1 572	...
Bhutan Bhoutan	161	170	185	193	222	278	283	268	...
Bolivia Bolivie	6 271	6 317	7 406	6 487	4 648	4 954	5 235	5 477	...
Botswana[5] Botswana[5]	874	1 052	1 083	968	1 296	1 675	1 871	2 478	...
Brazil Brésil	283 333	202 773	208 874	223 065	268 082	294 311	328 495	447 473	...
British Virgin Islands Iles Vierges britanniques	70	79	86	90	98	117	131	...	...
Brunei Darussalam Brunéi Darussalam	3 908	3 517	3 442	...	...	...	...	...	...
Bulgaria Bulgarie	19 872	20 421	31 991	32 273	31 581	28 765	26 085	21 747	20 726

29
Gross domestic product and net material product [cont.]
Produit intérieur brut et produit matériel net [suite]

B. Million US dollars • Millions de dollars E-U

Country or area Pays ou zone	1982	1983	1984	1985	1986	1987	1988	1989	1990
Burkina Faso Burkina Faso	1 094	1 000	894	1 045	1 454	...	...	...	...
Burundi Burundi	1 045	1 107	1 006	1 171	1 234	1 162	1 089	1 092	1 104
Cameroon[5] Cameroun[5]	6 612	6 871	7 312	8 545	11 941	13 326	12 657	11 160	13 363
Canada Canada	301 313	326 485	340 755	347 374	360 868	412 351	487 503	543 417	570 137
Cape Verde Cap-Vert	150	141	136	143	194	248	286	...	...
Central African Rep. Rép. centrafricaine	749	659	631	693	966	1 042	1 096	...	...
Chile Chili	23 626	19 478	19 285	# 15 996	16 817	18 948	22 081	25 372	27 791
China[3] Chine[3]	224 934	239 676	243 621	239 044	227 612	250 208	315 360	349 951	301 660
Colombia Colombie	38 961	38 732	38 256	34 896	34 943	36 371	39 212	39 705	41 122
Congo Congo	2 161	2 097	2 194	2 161	1 849	2 298	2 212	2 425	...
Costa Rica Costa Rica	2 607	3 147	3 660	3 923	4 404	4 531	4 572	5 175	5 686
Côte d'Ivoire Côte d'Ivoire	7 567	6 839	6 841	6 978	9 369	...	...	...	...
Cuba[3] Cuba[3]	14 324	15 030	15 562	15 165	15 491	15 629	16 839	16 399	...
Cyprus Chypre	2 135	2 161	2 279	2 430	3 094	3 701	4 276	4 568	5 511
Czechoslovakia[3] Tchécoslovaquie[3]	35 842	35 545	32 150	32 014	37 503	41 976	41 686	40 469	37 034
Denmark Danemark	55 745	56 046	54 580	58 048	82 375	102 326	108 742	105 308	129 264
Dominica Dominique	72	80	90	99	112	126	146	157	171
Dominican Republic Rép. dominicaine	7 964	8 623	10 355	4 488	5 434	5 081	4 638	6 687	7 103
Ecuador Equateur	13 845	12 700	12 995	15 958	11 266	10 527	10 012	10 117	10 876
Egypt[5] Egypte[5]	28 888	36 818	44 638	52 311	58 314	62 932	...	...	...
El Salvador El Salvador	3 586	4 061	4 663	5 732	4 075	4 628	5 473	6 446	5 113
Equatorial Guinea Guinée équatoriale	...	67	68	# 85	107	131	144	132	163
Ethiopia[4] Ethiopie[4]	4 429	4 846	4 831	4 778	5 233	5 409	5 685	6 034	...
Fiji Fidji	1 194	1 124	1 178	1 141	1 290	1 178	1 110	1 255	...

29
Gross domestic product and net material product [cont.]
Produit intérieur brut et produit matériel net [suite]

B. Million US dollars • Millions de dollars E-U

Country or area Pays ou zone	1982	1983	1984	1985	1986	1987	1988	1989	1990
Finland Finlande	50 978	49 308	51 307	54 048	70 533	89 088	105 561	115 803	137 251
France France	551 738	525 718	499 126	523 098	731 912	887 859	960 769	961 751	1 190 772
French Polynesia Polynésie française	1 185	1 230	1 270	1 388	2 120	2 346	2 475	2 428	2 567
Gabon Gabon	3 618	3 392	3 561	3 663	3 468	3 396	3 403	3 662	...
Gambia[2] Gambie[2]	265	234	218	279	214	231	290	309	341
Germany† • Allemagne† F. R. Germany R. f. Allemagne	654 343	653 561	615 211	619 287	886 618	1 107 422	1 193 475	1 181 319	1 488 234
former German D. R. anc. R. d. allemande	105 543	109 724	103 997	106 039	147 002	183 873	198 925	...	...
Ghana Ghana	5 510	6 135	7 730	6 346	5 733	4 853	5 195	5 249	6 226
Greece Grèce	38 541	34 967	33 762	33 433	39 397	46 212	53 056	54 042	65 958
Grenada Grenade	89	94	102	115	130	150	166	182	200
Guadeloupe Guadeloupe	1 144	1 104	1 035	1 091	1 582	...	...	...	...
Guatemala Guatemala	8 717	9 050	9 470	11 180	8 470	7 084	7 843	8 410	7 644
Guinea-Bissau Guinée-Bissau	...	...	...	...	230	165	155	...	...
Guyana Guyana	482	485	444	462	520	349	414	256	256
Haiti[6] Haïti[6]	1 485	1 630	1 816	2 009	2 235	1 989	1 959	2 162	2 281
Honduras Honduras	2 881	3 017	3 231	3 504	3 798	4 064	4 468	4 891	...
Hong Kong Hong-kong	30 598	28 365	31 705	33 513	38 255	47 139	55 554	63 640	70 048
Hungary Hongrie	23 147	21 007	20 367	20 626	23 756	26 109	# 28 469	28 963	32 901
Iceland Islande	3 087	2 650	2 761	2 869	3 858	5 382	5 920	5 270	5 864
India[7] Inde[7]	188 300	205 554	203 632	211 755	231 523	256 609	283 820	272 876	...
Indonesia Indonésie	94 457	85 369	87 612	87 339	80 061	75 932	84 300	94 625	107 294
Iran, Islamic Rep. of[8] Iran, Rép. islamique d'[8]	128 661	159 215	166 940	181 828	230 136	297 655	343 431	392 807	...
Iraq Iraq	42 876	42 623	47 982	49 819	48 434	57 558	64 413	64 340	...
Ireland Irlande	19 036	18 451	17 833	18 726	25 510	28 947	31 165	34 726	42 612

29
Gross domestic product and net material product [cont.]
Produit intérieur brut et produit matériel net [suite]

B. Million US dollars • Millions de dollars E-U

Country or area Pays ou zone	1982	1983	1984	1985	1986	1987	1988	1989	1990
Israel Israël	26 167	29 159	27 445	25 791	32 137	38 422	46 523	45 390	53 968
Italy Italie	403 046	417 050	413 077	424 512	603 634	759 049	838 842	869 270	1 090 755
Jamaica Jamaïque	3 294	3 620	2 373	2 015	2 444	2 916	3 416	3 868	3 994
Japan Japon	1 086 406	1 186 337	1 265 333	1 343 251	1 985 574	2 408 912	2 898 385	2 871 825	2 940 362
Jordan Jordanie	3 753	# 4 761	4 864	4 818	5 830	6 166	5 925	4 454	3 869
Kenya Kenya	6 438	5 980	6 196	6 137	7 240	7 972	8 509	8 259	8 756
Korea, Republic of Corée, République de	74 469	82 286	90 131	92 925	105 991	131 816	174 940	212 970	239 772
Kuwait Koweït	21 576	20 904	21 706	21 429	17 728	20 513	18 620	22 842	...
Lebanon Liban	2 656	...	...	...	...	...	...	...	...
Lesotho Lesotho	342	351	316	252	278	370	454	492	582
Liberia Libéria	1 093	1 086	1 070	1 055	1 037	1 089	1 158	1 194	...
Libyan Arab Jamah. Jamah. arabe libyenne	30 730	29 747	27 072	27 963	...	...	...	...	...
Luxembourg Luxembourg	3 475	3 416	3 352	3 457	4 999	6 103	6 756	7 081	8 723
Madagascar Madagascar	2 848	2 837	2 374	2 345	2 672	...	...	...	...
Malawi Malawi	1 179	1 223	1 208	1 134	1 179	1 237	1 395	1 479	1 841
Malaysia Malaisie	26 752	30 134	33 938	31 200	27 735	31 602	34 696	37 457	42 373
Maldives Maldives	46	54	77	86	98	...	...	...	...
Mali Mali	1 228	1 079	1 061	1 158	1 690	1 999	2 067	2 073	2 510
Malta Malte	1 121	1 059	1 001	1 018	1 304	1 592	1 835	1 924	2 316
Martinique Martinique	1 346	1 347	1 250	1 389	1 996	...	...	...	...
Mauritania Mauritanie	824	819	723	683	803	910	965	1 005	...
Mauritius Maurice	1 078	1 090	1 041	1 076	1 463	1 831	2 069	2 108	2 537
Mexico Mexique	173 720	148 878	175 604	184 498	130 010	139 894	171 246	204 618	241 386
Mongolia[3] Mongolie[3]	2 087	2 247	2 120	2 031	2 251	2 570	2 724	2 882	1 933

29
Gross domestic product and net material product [*cont.*]
Produit intérieur brut et produit matériel net [*suite*]

B. Million US dollars • Millions de dollars E-U

Country or area Pays ou zone	1982	1983	1984	1985	1986	1987	1988	1989	1990
Montserrat Montserrat	30	32	35	37	42	...	...	...	...
Morocco Maroc	15 483	13 963	12 750	12 871	16 947	18 703	22 079	22 534	25 175
Mozambique Mozambique	2 436	2 265	2 568	3 639	4 131	1 465	1 256	...	...
Myanmar Myanmar	6 008	6 200	6 392	6 608	8 052	10 325	11 923	17 790	21 793
Nepal[9] Népal[9]	2 340	2 321	2 393	2 434	2 375	2 715	2 957	2 878	3 021
Netherlands Pays-Bas	138 150	133 504	124 727	125 920	174 943	212 356	227 573	224 124	279 153
Netherlands Antilles[10] Antilles néerlandaises[10]	1 066	1 061	1 072	1 092	...	...	...	...	...
New Caledonia Nouvelle-Calédonie	905	824	796	855	1 201	1 488	2 073	...	...
New Zealand[7] Nouvelle-Zélande[7]	23 712	23 342	22 862	22 650	28 861	36 628	43 693	42 822	44 026
Nicaragua Nicaragua	1 929	2 239	3 063	4 274	6 504	34 136	...	...	...
Niger Niger	2 018	1 803	1 461	1 440	1 979	2 205	2 259	2 122	...
Nigeria[7] Nigéria[7]	76 834	79 035	82 931	80 934	41 631	27 113	32 013	30 522	32 426
Norway Norvège	56 131	55 126	55 445	58 182	69 471	83 337	89 501	89 997	105 703
Oman Oman	7 467	7 828	8 705	8 983	7 283	7 809	7 610	8 402	10 622
Pakistan[2] Pakistan[2]	30 758	32 004	33 615	32 304	34 387	38 818	42 756	41 987	46 839
Panama Panama	4 279	4 374	4 565	4 901	5 145	5 310	4 551	4 582	4 949
Papua New Guinea Papouasie-Nvl-Guinée	2 370	# 2 573	2 553	2 403	2 649	3 144	3 657	3 559	3 201
Paraguay Paraguay	5 850	6 493	5 326	4 545	5 407	4 534	6 035	4 363	5 265
Peru Pérou	25 584	19 907	20 867	18 040	26 713	44 014	38 372	43 263	40 835
Philippines Philippines	37 140	33 211	31 408	30 735	29 868	33 307	38 043	42 442	43 858
Poland[3] Pologne[3]	56 036	64 672	63 445	# 58 858	61 021	52 859	58 060	72 924	53 290
Portugal Portugal	23 283	20 777	19 234	20 682	29 550	36 731	41 700	45 283	59 680
Puerto Rico Porto Rico	17 277	19 163	20 289	21 969	23 878	26 178	28 267	30 536	32 469
Qatar Qatar	7 611	6 485	6 870	6 153	5 053	5 446	# 6 038	6 488	7 380
Réunion Réunion	1 869	1 801	1 806	1 927	2 818	3 647	3 982	...	...

29
Gross domestic product and net material product [*cont.*]
Produit intérieur brut et produit matériel net [*suite*]

B. Million US dollars • Millions de dollars E-U

Country or area Pays ou zone	1982	1983	1984	1985	1986	1987	1988	1989	1990
Romania Roumanie	48 493	44 746	38 351	47 687	51 916	58 055	60 027	53 478	37 625
Rwanda Rwanda	1 411	1 507	1 588	1 715	1 928	2 152	2 327	2 378	...
Saint Kitts and Nevis Saint-Kitts-et-Nevis	59	57	62	...	...	...	...	...	...
Saint Lucia Sainte-Lucie	135	141	151	...	...	...	...	...	...
St. Vincent-Grenadines St. Vincent-Grenadines	84	94	103	113	127	142	161	174	...
Samoa Samoa	...	...	115	105	109	...	...	...	...
Sao Tome and Principe Sao Tomé-et-Principe	38	36	36	...	64	55	49	...	...
Saudi Arabia[5] Arabie saoudite[5]	163 740	132 597	106 096	97 017	84 780	72 387	73 552	76 142	82 996
Senegal Sénégal	2 583	2 479	2 337	2 578	3 762	4 599	4 980	4 625	...
Seychelles Seychelles	147	146	151	169	208	247	283	302	...
Sierra Leone[5] Sierra Leone[5]	1 296	1 118	1 087	939	465	579	894	735	463
Singapore Singapour	15 266	17 384	18 775	17 692	17 752	20 258	24 694	28 834	34 599
Solomon Islands Iles Salomon	163	123	174	160	145	146	176	...	...
Somalia Somalie	2 708	2 225	3 110	2 211	1 650	1 612	...	...	...
South Africa[11] Afrique du Sud[11]	74 428	82 246	74 563	56 196	62 656	80 849	87 532	89 101	102 004
Spain Espagne	178 111	155 021	156 204	# 165 849	230 803	292 664	344 784	380 342	491 260
Sri Lanka Sri Lanka	4 686	5 066	5 792	5 808	6 154	6 413	6 878	6 886	7 935
Sudan[5] Soudan[5]	7 165	7 066	8 715	6 081	8 804	10 386	...	...	...
Suriname Suriname	1 025	990	968	978	1 010	1 106	1 300	1 516	1 725
Swaziland Swaziland	540	556	507	367	463	559	576	...	...
Sweden[5] Suède[5]	100 857	92 585	96 022	100 626	132 732	160 802	181 236	190 217	228 110
Switzerland Suisse	96 542	97 125	90 736	92 772	135 277	170 792	183 428	177 493	224 845
Syrian Arab Republic Rép. arabe syrienne	17 526	18 673	19 195	21 204	25 461	32 538	16 574	18 596	24 770
Thailand Thaïlande	35 652	39 568	41 178	37 350	41 651	48 717	59 578	69 099	80 172

29
Gross domestic product and net material product [*cont.*]
Produit intérieur brut et produit matériel net [*suite*]

B. Million US dollars • Millions de dollars E-U

Country or area Pays ou zone	1982	1983	1984	1985	1986	1987	1988	1989	1990
Togo Togo	821	738	698	740	1 050	...	...	...	...
Tonga[5] Tonga[5]	65	66	65	56	...	...	...	...	...
Trinidad and Tobago Trinité-et-Tobago	7 990	7 800	7 756	7 266	4 794	4 806	4 509	4 339	5 094
Tunisia Tunisie	8 129	8 096	8 031	8 275	8 821	9 604	10 031	10 004	...
Turkey Turquie	53 032	51 148	49 668	52 783	58 247	68 011	70 889	79 073	108 411
Uganda Ouganda	3 564	3 362	2 306	3 200	3 617	4 072	4 630	4 463	...
Ukraine[3] Ukraine[3]	119 863	125 068	118 182	110 588	133 992	154 945	168 863	173 684	175 976
former USSR[3] ancienne URSS[3]	717 671	742 954	700 860	680 588	828 491	941 287	1 039 209	1 074 482	1 096 697
United Arab Emirates Emirats arabes unis	30 627	28 033	27 743	27 081	21 674	23 799	23 728	27 506	33 780
United Kingdom Royaume-Uni	487 018	460 360	433 263	460 326	559 723	688 305	710 952	836 277	980 124
United Rep.Tanzania[12] Rép. Unie de Tanzanie[12]	6 272	6 328	5 813	6 904	4 884	3 540	3 336	2 835	2 542
United States Etats-Unis	3 118 560	3 349 390	3 717 770	3 962 220	4 176 100	4 452 880	4 809 080	5 132 000	5 392 200
Uruguay Uruguay	9 253	# 5 079	4 829	4 719	5 859	7 329	7 583	7 819	8 218
Vanuatu Vanuatu	0	102	124	118	115	122	144	141	154
Venezuela Venezuela	67 847	67 588	# 59 856	61 965	60 519	48 029	60 226	43 551	48 274
Viet Nam Viet Nam	...	...	...	...	...	...	7 528	8 041	8 526
Yemen[13] Yémen[13]	3 478	3 750	3 353	4 181	3 903	...	...	# 6 266	7 873
Yugoslavia Yougoslavie	62 827	46 133	43 538	44 237	61 705	71 018	62 764	81 848	101 413
Zaire Zaïre	5 410	4 588	2 760	2 953	3 412	2 909	3 329	3 007	...
Zambia Zambie	3 874	3 342	2 749	2 619	1 776	2 222	3 661	4 653	3 910
Zimbabwe Zimbabwe	6 865	6 237	5 148	4 527	5 031	5 455	6 114	6 184	6 199

29

Gross domestic product and net material product [*cont.*]
Produit intérieur brut et produit matériel net [*suite*]

B. Million US dollars • Millions de dollars E-U

Source:
National accounts database of the Statistical Division of
the United Nations Secretariat.

† All data shown which pertain to Germany prior to 3 October
1990 are indicated separately for the Federal Republic of
Germany and the former German Democratic Republic based on
their respective territories at the time indicated. Where
data for united Germany (3 October 1990 and thereafter) are
not available, available data are shown separately under the
designations Federal Republic of Germany and former German
Democratic Republic and pertain to the territorial
boundaires prior to 3 October 1990. For detailed
explanatory notes on data pertaining to Germany, see Annex I
- Country Nomenclature.

1 Billion national currency unit.
2 Year beginning 1 July.
3 NMP.
4 Year ending 7 July.
5 Year ending 30 June.
6 Year ending 30 September.
7 Year beginning 1 April.
8 Year beginning 21 March.
9 Year ending 15 July.
10 Estimates exclude Aruba.
11 Estimates exclude Namibia.
12 Estimates refer to former Tanganyika only.

13 For 1982-1986, data exclude former Democratic Yemen.

Source:
Base de données sur les comptes nationaux de la Division de
statistique du Secrétariat de l'ONU.

† Toutes les données se rapportant à l'Allemagne avant le 3
octobre 1990 figurent dans deux rubriques séparées basées
sur les territoires respectifs de la République fédérale
d'Allemagne et l'ancienne République démocratique allemande
selon la période indiquée. En l'absence de données pour
l'Allemagne unifiée (à compter du 3 octobre 1990, les
données disponibles sont fournies séparément sous les
rubriques République fédérale d'Allemagne et ancienne
République démocratique allemande et se rapportent aux
limites territoriales antérieures au 3 octobre 1990. Pour
les notes explicatives en détail sur les données concernant
l'Allemagne, voir Annexe I – Nomenclature des pays.

1 Milliards d'unités monétaires nationales.
2 L'année commençant le 1er juillet.
3 PMN.
4 L'année finissant le 7 juillet.
5 L'année finissant le 30 juin.
6 L'année finissant le 30 septembre.
7 L'année commençant le 1er avril.
8 L'année commençant le 21 mars.
9 L'année finissant le 15 juillet.
10 Ces montants estimatifs ne comprennent pas l'Aruba.
11 Ces montants estimatifs ne comprennent pas le Namibie.
12 Ces montants estimatifs concernent seulement l'ancien
    Tanganyika.
13 Pour les années 1982-1986, non compris les données relatives
    à l'ancien Yémen démocratique.

30
Index numbers of industrial production
Indices de la production industrielle
1980=100

Country or area and industry [SITC] Pays ou zone et industrie [CITI]	1982	1983	1984	1985	1986	1987	1988	1989	1990	1991[1]
Africa · Afrique										
Algeria Algérie										
Total industry [2-4]										
Total, industrie [2-4]	**125**	**140**	**152**	**157**	**164**	**161**	**165**	162	166	...
Total mining [2]										
Total, industries extractives [2]	**108**	**96**	**117**	**124**	**132**	**128**	**122**	124	127	...
Total manufacturing [3]										
Total, industries manufacturières [3]	**138**	**150**	**174**	**179**	**189**	**186**	**182**	172	174	...
Food, beverages, tobacco										
Aliments, boissons, tabac	110	118	141	...	...	...	...	...	...	...
Textiles										
Textiles	137	140	165	...	...	...	...	...	...	...
Petroleum products										
Produits pétroliers	187	171	192	...	...	...	...	...	...	...
Non-metallic mineral products										
Produits minéraux non métalliques	105	112	128	...	...	...	...	...	...	...
Electricity [4]										
Electricité [4]	**131**	**143**	**155**	**174**	**186**	**199**	**216**	226	239	...
Côte d'Ivoire Côte d'Ivoire										
Total industry [2-4]										
Total, industrie [2-4]	**108**	**112**	**117**	...	...	...	...	...	...	...
Total mining [2]										
Total, industries extractives [2]	**107**	**112**	**115**	...	...	...	...	...	...	...
Total manufacturing [3]										
Total, industries manufacturières [3]	**108**	**112**	**118**	...	...	...	...	...	...	...
Food, beverages, tobacco										
Aliments, boissons, tabac	94	90	88	114	129	133	84	130	121	97
Textiles and clothing										
Textiles, habillement	99	108	96	122	120	126	140	145	118	101
Chemicals and petroleum										
Produits chimiques et pétrole	132	92	99	109	112	117	129	117	119	103
Metal products										
Produits métalliques	71	49	70	92	124	94	87	89	71	78
Electricity, gas and water										
Electricité, gaz et eau	101	110	103	110	120	124	136	...	...	...
Egypt Egypte										
Total industry [2-4]										
Total, industrie [2-4]	**117**	**125**	**139**	**152**	**139**	**144**	**146**	142	...	...
Total mining [2]										
Total, industries extractives [2]	**111**	**115**	**130**	**146**	**142**	**142**	**145**	141	143	...
Total manufacturing [3]										
Total, industries manufacturières [3]	**119**	**129**	**142**	**152**	**136**	**142**	**143**	139	130	...
Food, beverages, tobacco										
Aliments, boissons, tabac	130	164	174	210	237	203	197	233	185	...
Textiles										
Textiles	131	128	138	145	149	130	108	112	106	...
Chemicals, coal, petroleum products										
Produits chimiques, houillers, pétroliers	125	134	154	168	143	184	185	169	171	...
Basic metals										
Métaux de base	90	95	119	134	148	169	169	182	179	...
Metal products										
Produits métalliques	136	160	172	206	179	185	190	182	162	...
Electricity [4]										
Electricité [4]	**121**	**136**	**153**	**185**	**167**	**178**	**198**	189	...	...
Ethiopia[18] Ethiopie[18]										
Total industry [2-4]										
Total, industrie [2-4]	**114**	**113**	**136**	**143**	**152**	...	...	...	...	...
Food, beverages, tobacco										
Aliments, boissons, tabac	118	112	153	165	167	170	166	153	...	...

30
Index numbers of industrial production [*cont.*]
Indices de la production industrielle [*suite*]
1980=100

Country or area and industry [SITC] Pays ou zone et industrie [CITI]	1982	1983	1984	1985	1986	1987	1988	1989	1990	1991[1]
Textiles Textiles	107	108	108	116	127	135	115	124	...	...
Chemicals Produits chimiques	146	156	184	156	238	188	152	122	...	...
Metal products Produits métalliques	114	121	137	130	139	144	108	92	...	...
Ghana Ghana										
Total industry [2-4][2] **Total, industrie [2-4]**[2]	**75**	**54**	**60**	**74**	**83**	**87**	**94**	**...**	**...**	**...**
Total mining [2] **Total, industries extractives [2]**	**81**	**68**	**78**	**88**	**82**	**88**	**94**	**...**	**...**	**...**
Total manufacturing [3] **Total, industries manufacturières [3]**	**73**	**51**	**57**	**71**	**79**	**82**	**90**	**...**	**...**	**...**
Food, beverages, tobacco Aliments, boissons, tabac	62	67	65	74	82	94	101	94	101	...
Chemical and rubber products Produits chimiques ou en caoutchouc	49	54	116	92	110	150	194	177	166	...
Non-metallic mineral products Produits minéraux non métalliques	89	96	81	122	91	95	141	192	225	...
Electricity [4] **Electricité [4]**	**100**	**91**	**85**	**103**	**153**	**160**	**161**	**...**	**...**	**...**
Kenya Kenya										
Total mining [2] **Total, industries extractives [2]**	**153**	**181**	**171**	**93**	**101**	**106**	**102**	**104**	**107**	**...**
Total manufacturing [3][3] **Total, industries manufacturières [3]**[3]	**106**	**111**	**115**	**109**	**116**	**125**	**132**	**138**	**143**	**...**
Food, beverages, tobacco Aliments, boissons, tabac	102	106	113	119	130	143	151	155	156	...
Textiles Textiles	183	91	103	108	116	119	122	125	141	...
Chemicals, coal, petroleum products Produits chimiques, houillers, pétroliers	96	99	117	121	130	140	156	178	203	...
Metal products Produits métalliques	91	99	93	92	92	94	106	113	118	...
Malawi Malawi										
Total industry [2-4] **Total, industrie [2-4]**	**90**	**100**	**97**	**100**	**102**	**97**	**104**	**113**	**127**	**...**
Total manufacturing [3] **Total, industries manufacturières [3]**	**88**	**98**	**94**	**97**	**99**	**92**	**98**	**107**	**121**	**...**
Food, beverages, tobacco Aliments, boissons, tabac	126	146	138	139	149	151	148	176	189	85
Textiles[4] Textiles[4]	121	125	117	122	115	104	108	112	129	112
Electricity and water [4] **Electricité et eau [4]**	**106**	**115**	**120**	**123**	**132**	**144**	**146**	**159**	**175**	**...**
Morocco Maroc										
Total industry [2-4] **Total, industrie [2-4]**	**103**	**111**	**113**	**117**	**121**	**124**	**136**	**123**	**133**	**...**
Total mining [2][5] **Total, industries extractives [2]**[5]	**102**	**111**	**121**	**120**	**118**	**117**	**136**	**104**	**111**	**...**
Total manufacturing [3][6] **Total, industries manufacturières [3]**[6]	**103**	**110**	**110**	**116**	**121**	**124**	**135**	**125**	**138**	**...**
Food, beverages, tobacco Aliments, boissons, tabac	99	100	107	105	110	95	96	95	97	112
Textiles Textiles	111	113	102	117	128	143	147	132	109	92
Chemicals and petroleum products Produits chimiques et pétroliers	109	120	118	122	122	109	105	104	90	100
Basic metals Métaux de base	84	91	95	81	67	124	259	319	113	104

30
Index numbers of industrial production [*cont.*]
Indices de la production industrielle [*suite*]
1980=100

Country or area and industry [SITC] Pays ou zone et industrie [CITI]	1982	1983	1984	1985	1986	1987	1988	1989	1990	1991 [1]
Metal products Produits métalliques	93	98	77	80	77	97	124	283	143	103
Electricity [4] [7] **Electricité [4] [7]**	**107**	**115**	**119**	**127**	**136**	**142**	**148**	**160**	**148**	...
Senegal Sénégal										
Total industry [2-4] **Total, industrie [2-4]**	**113**	**118**	**115**	**118**	**106**	**118**	**124**	**112**	**118**	...
Total mining [2] **Total, industries extractives [2]**	**76**	**94**	**120**	**125**	**130**	**115**	**149**	**147**	**130**	...
Total manufacturing [3] [8] [2] **Total, manufactures [3] [8] [2]**	**120**	**134**	**113**	**117**	**101**	**117**	**109**	**105**	**115**	...
Food, beverages, tobacco Aliments, boissons, tabac	129	137	104	88	73	114	145	130	135	...
Textiles Textiles	159	153	125	150	91	124	140	73	89	...
Chemicals, coal, petroleum products [9] Produits chimiques, houillers, pétroliers [9]	87	80	79	85	83	98	95	80	94	...
Electricity and water [4] **Electricité et eau [4]**	**109**	**115**	**118**	**120**	**128**	**148**	**144**	**140**	**140**	...
South Africa Afrique du Sud										
Total mining [2] **Total, industries extractives [2]**	**99**	**100**	**104**	**105**	**102**	**97**	**100**	**98**	**97**	...
Total manufacturing [3] **Total, industries manufacturières [3]**	**113**	**98**	**103**	**101**	**99**	**102**	**110**	**110**	**109**	...
Food, beverages, tobacco Aliments, boissons, tabac	107	104	109	112	113	118	124	125	127	100
Textiles Textiles	105	88	84	82	81	86	90	92	78	92
Chemicals Produits chimiques	102	92	99	100	98	94	101	105	102	100
Basic metals Métaux de base	92	90	97	99	107	106	107	109	106	99
Metal products Produits métalliques	107	96	95	83	77	79	86	88	86	98
Tunisia Tunisie										
Total industry [2-4] **Total, industrie [2-4]**	**103**	**112**	**109**	**109**	**110**	**110**	**112**	**114**	**114**	...
Total mining [2] **Total, industries extractives [2]**	**94**	**103**	**100**	**97**	**97**	**94**	**91**	**93**	**86**	...
Total manufacturing [3] **Total, industries manufacturières [3]**	**109**	**116**	**119**	**123**	**123**	**126**	**134**	**137**	**144**	...
Food, beverages, tobacco Aliments, boissons, tabac	114	117	118	125	132	133	143	144	147	101
Textiles Textiles	100	105	112	118	112	127	134	134	128	89
Chemicals and petroleum products Produits chimiques et pétroliers	104	113	114	114	127	134	150	157	163	101
Basic metals Métaux de base	97	98	100	96	112	107	107	105	108	99
Metal products Produits métalliques	133	127	133	135	103	94	101	109	125	100
Electricity and water [4] **Electricité et eau [4]**	**113**	**124**	**135**	**140**	**138**	**130**	**130**	**131**	**141**	...
Zambia Zambie										
Total industry [2-4] **Total, industrie [2-4]**	**100**	**97**	**83**	**92**	**88**	**95**	**97**	**96**	**97**	**95**
Total mining [2] [10] **Total, industries extractives [2] [10]**	**96**	**95**	**86**	**80**	**76**	**84**	**81**	**83**	**79**	**100**
Total manufacturing [3] [11] **Total, manufactures [3] [11]**	**106**	**100**	**100**	**109**	**113**	**112**	**120**	**120**	**125**	**90**
Food, beverages, tobacco Aliments, boissons, tabac	101	105	102	102	100	104	114	110	122	84

30
Index numbers of industrial production [*cont.*]
Indices de la production industrielle [*suite*]
1980=100

Country or area and industry [SITC] Pays ou zone et industrie [CITI]	1982	1983	1984	1985	1986	1987	1988	1989	1990	1991 [1]
Textiles and clothing										
Textiles, habillement	120	112	120	152	133	114	142	148	162	60
Basic metals										
Métaux de base	80	81	82	96	90	92	90	72	53	99
Electricity and water [4]										
Electricité et eau [4]	115	110	105	109	104	91	91	73	84	111
Zimbabwe Zimbabwe										
Total industry [2-4]										
Total, industrie [2-4]	**105**	**102**	**100**	**108**	**113**	**118**	**122**	**129**	**134**	**...**
Total mining [2] [10]										
Total, industries extractives [2] [10]	**96**	**93**	**97**	**97**	**99**	**103**	**102**	**107**	**108**	**...**
Total manufacturing [3] [11]										
Total, manufactures [3] [11]	**109**	**106**	**101**	**112**	**115**	**118**	**124**	**131**	**138**	**...**
Food, beverages, tobacco										
Aliments, boissons, tabac	110	111	106	106	112	121	124	124	138	102
Textiles										
Textiles	118	109	124	175	190	196	203	208	217	102
Chemicals and petroleum products										
Produits chimiques et pétroliers	118	121	113	122	122	119	131	146	159	104
Basic metals and metal products										
Métaux de base et produits métalliques	110	103	93	100	98	93	101	113	117	107
Electricity and water [4]										
Electricité et eau [4]	**91**	**95**	**97**	**111**	**132**	**171**	**177**	**208**	**202**	**...**
America, North · Amérique du Nord										
Barbados Barbade										
Total industry [2-4]										
Total, industrie [2-4]	**93**	**97**	**100**	**97**	**102**	**97**	**102**	**107**	**109**	**101**
Total mining [2]										
Total, industries extractives [2]	**95**	**116**	**154**	**165**	**155**	**143**	**134**	**126**	**132**	**104**
Total manufacturing [3]										
Total, industries manufacturières [3]	**89**	**92**	**92**	**87**	**92**	**87**	**91**	**99**	**99**	**99**
Food, beverages, tobacco										
Aliments, boissons, tabac	89	88	84	85	89	95	93	94	103	105
Wearing apparel										
Habillement	103	109	114	96	77	90	93	72	48	64
Chemicals and petroleum products										
Produits chimiques et pétroliers	96	93	89	88	99	93	105	116	119	103
Electricity and gas										
Electricité et gaz	**102**	**118**	**124**	**132**	**139**	**145**	**157**	**162**	**164**	**107**
Canada Canada										
Total industry [2-4]										
Total, industrie [2-4]	**91**	**97**	**109**	**115**	**115**	**121**	**127**	**128**	**122**	**94**
Total mining [2]										
Total, industries extractives [2]	**92**	**99**	**112**	**117**	**113**	**123**	**134**	**135**	**135**	**102**
Total manufacturing [3]										
Total, industries manufacturières [3]	**88**	**94**	**106**	**112**	**113**	**118**	**125**	**125**	**118**	**91**
Food, beverages, tobacco										
Aliments, boissons, tabac	100	97	101	106	104	105	106	105	105	98
Textiles										
Textiles	80	102	81	104	115	115	118	116	104	86
Paper and paper products										
Papier, produits en papier	85	95	98	97	102	108	111	111	108	98
Chemicals, coal, petroleum products										
Produits chimiques, houillers, pétroliers	92	105	114	114	116	125	126	128	127	93
Basic metals										
Métaux de base	77	85	104	111	107	119	129	128	118	96
Metal products										
Produits métalliques	90	91	110	119	120	125	138	140	133	91
Electricity, gas and water [4]										
Electricité, gaz, eau [4]	**100**	**107**	**112**	**120**	**123**	**127**	**131**	**131**	**127**	**104**

30
Index numbers of industrial production [cont.]
Indices de la production industrielle [suite]
1980=100

Country or area and industry [SITC] Pays ou zone et industrie [CITI]	1982	1983	1984	1985	1986	1987	1988	1989	1990	1991 [1]
Costa Rica [12] **Costa Rica** [12]										
Total industry [2-4]										
Total, industrie [2-4]	90	93	102	104	113	119	122	127	...	...
Total mining [2]										
Total, industries extractives [2]	77	78	87	90	88	84	87	98	...	...
Total manufacturing [3]										
Total, industries manufacturières [3]	88	90	99	101	108	114	117	122	125	...
Electricity, gas and water [4]										
Electricité, gaz et eau [4]	112	135	139	129	137	147	151	159	167	...
El Salvador [14] **El Salvador** [14]										
Total industry [2-4]										
Total, industrie [2-4]	84	86	87	91	...	...	...	...	...	...
Total mining [2]										
Total, industries extractives [2]	97	103	103	103	...					
Total manufacturing [3]										
Total, industries manufacturières [3]	82	84	85	88	...	...	...	...	...	...
Food, beverages, tobacco										
Aliments, boissons, tabac	88	90	95	102	103	105	109	...	...	...
Textiles										
Textiles	54	58	55	47	56	59	53	...	...	...
Chemical products and petroleum										
Produits chimiques et pétroliers	89	98	100	91	94	96	96	...	...	...
Basic metals										
Métaux de base	57	66	91	103	120	122	124	...	...	...
Metal products										
Produits métalliques	52	71	63	67	68	71	72	...	...	...
Electricity, gas and water [4]										
Electricité, gaz et eau [4]	94	99	102	107	...	...	...	...	...	...
Honduras Honduras										
Total industry [2-4]										
Total, industrie [2-4]	96	101	110	109	...	...	...	...	...	...
Total mining [2]										
Total, industries extractives [2]	109	118	132	135	...	...	...	...	...	...
Total manufacturing [3]										
Total, industries manufacturières [3]	94	99	107	105	110	116	121	127	129	...
Food, beverages, tobacco										
Aliments, boissons, tabac	102	108	114	114	121	120	130	135	142	...
Textiles										
Textiles	69	73	80	80	70	86	105	118	121	...
Chemicals										
Produits chimiques	116	112	138	124	138	156	150	172	166	...
Electricity [4]										
Electricité [4]	102	102	106	112	...	...	...	...	...	...
Mexico Méxique										
Total industry [2-4] [16]										
Total, industrie [2-4] [16]	106	96	102	108	103	107	109	114	120	104
Total mining [2]										
Total, industries extractives [2]	125	124	128	129	126	131	131	131	136	101
Total manufacturing [3] [17]										
Total, manufactures [3] [17]	104	96	101	108	105	109	112	119	125	104
Food, beverages, tobacco										
Aliments, boissons, tabac	108	107	109	115	117	118	117	124	128	104
Textiles [4]										
Textiles [4]	97	91	93	98	92	92	96	97	96	95
Chemicals, coal, petroleum products [9]										
Produits chimiques, houillers, pétroliers [9]	109	108	113	119	116	121	123	132	140	105
Basic metals										
Métaux de base	94	87	100	97	89	102	108	113	120	104
Metal products										
Produits métalliques	101	81	88	101	94	103	113	125	136	114
Electricity [4]										
Electricité [4]	116	118	126	135	140	148	156	169	178	105

30
Index numbers of industrial production [*cont.*]
Indices de la production industrielle [*suite*]
1980=100

Country or area and industry [SITC] Pays ou zone et industrie [CITI]	1982	1983	1984	1985	1986	1987	1988	1989	1990	1991 [1]
Panama Panama										
Total mining [2]										
Total, industries extractives [2]	**136**	**105**	**83**	**79**	**81**	**87**	**51**	**48**	**64**	...
Total manufacturing [3]										
Total, industries manufacturières [3]	**110**	**109**	**105**	**110**	**113**	**121**	**92**	**97**	**109**	...
Food, beverages, tobacco										
Aliments, boissons, tabac	111	113	105	105	105	112	97	104	111	101
Textiles										
Textiles	86	76	74	93	97	96	68	95	110	106
Non-metallic mineral products										
Produits minéraux non-métalliques	119	109	88	89	108	116	46	45	62	194
Basic metals										
Métaux de base	83	62	58	62	70	107	36	34	58	142
Metal products										
Produits métalliques	100	110	104	120	128	145	76	75	104	103
Trinidad and Tobago Trinité-et-Tobago										
Total mining [2]										
Total, industries extractives [2]	**71**	**39**	**42**	**42**	**41**	**45**	**41**	**38**	**38**	...
Total manufacturing [3] [13]										
Total, industries manufacturières [3][13]	**104**	**109**	**103**	**99**	**121**	**125**	**120**	**121**	**124**	...
Food, beverages, tobacco										
Aliments, boissons, tabac	96	98	98	85	88	89	87	89	90	101
Textiles										
Textiles	66	57	55	43	53	38	29	33	42	101
Chemicals and petroleum products										
Produits chimiques et pétroliers	77	52	56	64	67	70	72	70	71	102
Metal products										
Produits métalliques	94	82	75	56	64	54	50	46	45	113
United States Etats-Unis										
Total industry [2-4]										
Total, industrie [2-4]	**97**	**101**	**110**	**112**	**113**	**119**	**125**	**128**	**130**	**97**
Total mining [2]										
Total, industries extractives [2]	**99**	**95**	**102**	**99**	**92**	**91**	**92**	**91**	**93**	**100**
Total manufacturing [3]										
Total, industries manufacturières [3]	**97**	**103**	**113**	**116**	**120**	**127**	**134**	**138**	**139**	**97**
Food, beverages, tobacco										
Aliments, boissons, tabac	101	104	108	111	113	116	120	122	125	100
Textiles										
Textiles	90	102	100	99	105	112	112	114	112	95
Paper and paper products										
Papier, produits en papier	100	109	115	116	124	131	135	135	138	98
Chemicals, coal, petroleum products										
Produits chimiques, houillers, pétroliers	94	100	106	109	115	121	127	130	133	99
Basic metals										
Métaux de base	72	82	91	89	83	90	99	98	97	91
Metal products										
Produits métalliques	92	97	114	117	118	122	132	138	139	97
Electricity [4]										
Electricité [4]	**96**	**98**	**101**	**104**	**100**	**104**	**109**	**112**	**113**	**101**
America, South · Amérique du Sud										
Argentina Argentine										
Total industry [2-4] [2]										
Total, industrie [2-4] [2]	**82**	**90**	**94**	**86**	**96**	**95**	**85**	**79**	...	...
Total mining [2]										
Total, industries extractives [2]	**101**	**101**	**101**	**98**	**94**	**94**	**104**	**107**	...	...
Total manufacturing [3]										
Total, industries manufacturières [3]	**81**	**89**	**93**	**84**	**95**	**88**	**82**	**76**	...	...
Food, beverages, tobacco										
Aliments, boissons, tabac	90	93	98	99	106	102	95	97	...	...
Textiles										
Textiles	82	97	101	76	96	91	89	88	...	...
Chemicals, coal, petroleum products										
Produits chimiques, houillers, pétroliers	97	104	111	108	119	117	115	106	...	...

30
Index numbers of industrial production [*cont.*]
Indices de la production industrielle [*suite*]
1980=100

Country or area and industry [SITC] Pays ou zone et industrie [CITI]	1982	1983	1984	1985	1986	1987	1988	1989	1990	1991 [1]
Basic metals Métaux de base	99	105	104	99	111	125	130	141	...	...
Metal products Produits métalliques	65	76	81	66	77	79	68	53	...	...
Electricity and gas [4] Electricité et gaz [4]	102	110	116	117	127	136	139	135	...	...
Bolivia [19] Bolivie [19]										
Total industry [2-4] **Total, industrie [2-4]**	**88**	**85**	**74**	**67**	...	...	...	...	...	...
Total mining [2] **Total, industries extractives [2]**	**95**	**92**	**72**	**59**	**42**	**43**	**62**	**77**	...	...
Total manufacturing [3] **Total, industries manufacturières [3]**	**87**	**84**	**71**	**64**	**67**	**69**	**72**	**74**	**76**	...
Electricity, gas and water [4] Electricité, gaz et eau [4]	129	131	141	133	134	126	132	137	...	...
Brazil Brésil										
Total industry [2-4] **Total, industrie [2-4]**	**90**	**85**	**91**	**99**	**110**	**111**	**107**	**110**	**100**	**97**
Total mining [2] **Total, industries extractives [2]**	**104**	**120**	**157**	**175**	**182**	**180**	**181**	**186**	**193**	**103**
Total manufacturing [3] [20] **Total, manufactures [3] [20]**	**89**	**84**	**89**	**97**	**108**	**109**	**105**	**108**	**98**	**97**
Food, beverages, tobacco Aliments, boissons, tabac	103	105	104	106	109	120	113	116	118	104
Textiles Textiles	90	81	78	89	101	100	94	94	85	101
Chemicals and petroleum products Produits chimiques et pétroliers	107	105	116	124	125	132	127	127	116	103
Basic metals [21] Métaux de base [21]	78	76	86	93	104	104	101	106	93	96
Chile Chili										
Total mining [2] [22] [23] **Total, industries extractives [2] [22] [23]**	**119**	**119**	**125**	**129**	**130**	**131**	**137**	**148**	**150**	**114**
Total manufacturing [3] [24] **Total, manufactures [3] [24]**	**85**	**89**	**98**	**98**	**106**	**110**	**119**	**129**	**128**	**103**
Food, beverages, tobacco Aliments, boissons, tabac	103	104	119	116	125	121	128	140	134	100
Textiles Textiles	69	78	98	105	124	130	124	126	118	101
Chemicals and petroleum products Produits chimiques et pétroliers	86	90	95	93	103	108	120	134	133	103
Basic metals Métaux de base	100	103	110	107	111	114	122	129	131	105
Metal products Produits métalliques	63	59	64	68	75	86	92	104	105	99
Colombia Colombie										
Total industry [2-4] **Total, industrie [2-4]**	**100**	**101**	**108**	**110**	**127**	**139**	**141**	...	...	...
Total mining [2] **Total, industries extractives [2]**	**115**	**130**	**161**	**153**	**246**	**287**	**283**	**299**	**291**	...
Total manufacturing [3] **Total, industries manufacturières [3]**	**93**	**93**	**102**	**104**	**112**	**120**	**122**	**125**	**132**	
Food, beverages, tobacco Aliments, boissons, tabac	92	96	106	114	116	119	118	128	130	...
Textiles Textiles	96	88	100	104	115	125	118	112	115	...
Chemicals, coal, petroleum products Produits chimiques, houillers, pétroliers	106	118	125	134	152	164	163	165	168	...
Basic metals [11] Métaux de base [11]	108	116	106	106	116	134	143	141	145	...
Metal products Produits métalliques	93	84	95	87	96	104	117	109	115	...

30
Index numbers of industrial production [*cont.*]
Indices de la production industrielle [*suite*]
1980=100

Country or area and industry [SITC] Pays ou zone et industrie [CITI]	1982	1983	1984	1985	1986	1987	1988	1989	1990	1991 [1]
Electricity, gas and water [4] **Electricité, gaz et eau [4]**	**108**	**113**	**119**	**117**	**118**	**126**	**137**	**139**	...	...
Ecuador Equateur										
Total manufacturing [3] **Total, industries manufacturières [3]**	**105**	**101**	**107**	**113**	**117**	**120**	**126**	**131**	...	...
Food, beverages, tobacco Aliments, boissons, tabac	92	87	92	96	99	102	102	105	110	...
Textiles Textiles	107	105	94	98	95	91	94	96	100	...
Chemicals, coal, petroleum products Produits chimiques, houillers, pétroliers	104	101	114	123	128	128	132	128	133	...
Basic metals Métaux de base	149	159	187	150	142	129	146	129	132	...
Metal products Produits métalliques	107	101	106	122	136	136	149	163	174	...
Paraguay Paraguay										
Total industry [2-4] [25] **Total, industrie [2-4] [25]**	**99**	**98**	**99**	**105**	**110**	**119**	**139**	...	...	...
Total manufacturing [3] **Total, industries manufacturières [3]**	**100**	**97**	**98**	**102**	**101**	**106**	**114**	**148**	**135**	...
Food, beverages, tobacco Aliments, boissons, tabac	114	113	109	111	116	123	111	148	131	...
Textiles Textiles	104	87	114	169	115	95	196	215	224	...
Chemicals, coal, petroleum products Produits chimiques, houillers, pétroliers	68	65	66	74	67	76	87	78	81	...
Basic metals Métaux de base	214	211	251	271	254	560	622	615	509	...
Metal products Produits métalliques	140	136	184	212	192	190	181	167	145	...
Electricity [4] **Electricité [4]**	**92**	**122**	**127**	**181**	**342**	**453**	**830**	...	...	...
Peru Pérou										
Total manufacturing [3] **Total, industries manufacturières [3]**	**96**	**78**	**84**	**89**	**105**	**120**	**104**	**84**	**81**	...
Food, beverages, tobacco Aliments, boissons, tabac	98	87	92	94	114	132	118	94	94	96
Textiles Textiles	108	92	98	113	121	134	126	116	100	93
Chemicals, coal, petroleum products Produits chimiques, houillers, pétroliers	92	77	79	83	103	124	111	78	78	95
Basic metals Métaux de base	94	84	94	99	96	100	78	86	75	103
Metal products Produits métalliques	84	45	45	54	81	101	70	42	45	86
Uruguay Uruguay										
Total industry [2-4] **Total, industrie [2-4]**	**81**	**76**	**78**	**77**	**86**	**96**	**94**	**91**	**109**	...
Total manufacturing [3] [25] **Total, manufactures [3] [25]**	**79**	**70**	**76**	**75**	**84**	**93**	**91**	**89**	**98**	...
Food, beverages, tobacco Aliments, boissons, tabac	103	95	88	95	96	97	101	103	103	105
Textiles [4] Textiles [4]	53	71	86	85	99	106	101	96	100	108
Chemicals, coal, petroleum products Produits chimiques, houillers, pétroliers	86	74	77	75	83	97	100	99	102	100
Metal products Produits métalliques	49	41	46	48	62	85	75	69	69	94
Electricity, gas and water [4] **Electricité, gaz et eau [4]**	**107**	**109**	**108**	**112**	**116**	**135**	**145**	**127**	**118**	...
Venezuela Venezuela										
Total industry [2-4] **Total, industrie [2-4]**	**122**	**123**	**156**	**181**	**217**	**310**	**392**	...	...	...

30
Index numbers of industrial production [*cont.*]
Indices de la production industrielle [*suite*]
1980=100

Country or area and industry [SITC] Pays ou zone et industrie [CITI]	1982	1983	1984	1985	1986	1987	1988	1989	1990	1991 [1]
Total mining [2]										
Total, industries extractives [2]	**91**	**82**	**91**	**118**	**177**	**319**	**454**	...	...	...
Total manufacturing [3] [26]										
Total, manufactures [3] [26]	**119**	**119**	**152**	**176**	**218**	**311**	**393**	...	...	...
Food, beverages, tobacco										
Aliments, boissons, tabac	131	138	161	192	219	282	350	624	...	...
Textiles										
Textiles	103	109	166	185	216	331	451	613	...	...
Chemicals, coal, petroleum products										
Produits chimiques, houillers, pétroliers	119	114	117	150	203	299	406	493	...	...
Basic metals [11]										
Métaux de base [11]	132	107	124	119	173	272	386	608	...	...
Metal products										
Produits métalliques	114	102	121	135	170	255	311	332	...	...
Asia · Asie										
Bangladesh [27] **Bangladesh** [27]										
Total industry [2-4]										
Total, industrie [2-4]	**98**	**109**	**124**	**127**	**146**	**148**	**153**	**171**	**170**	**101**
Total mining [2]										
Total, industries extractives [2]	**134**	**156**	**178**	**199**	**236**	**277**	**295**	**316**	**332**	**109**
Total manufacturing [3]										
Total, industries manufacturières [3]	**97**	**107**	**122**	**126**	**144**	**145**	**148**	**166**	**166**	**147**
Food, beverages, tobacco										
Aliments, boissons, tabac	109	105	104	108	124	127	129	156	158	76
Textiles										
Textiles	93	90	88	77	89	88	87	91	80	79
Basic metals [11]										
Métaux de base [11]	51	53	70	55	50	45	51	57	44	69
Electricity [4]										
Electricité [4]	**130**	**149**	**172**	**180**	**210**	**246**	**285**	**298**	**310**	**116**
Cyprus Chypre										
Total industry [2-4]										
Total, industrie [2-4]	**109**	**112**	**118**	**116**	**119**	**131**	**142**	**147**	**154**	...
Total mining [2]										
Total, industries extractives [2]	**79**	**75**	**65**	**71**	**64**	**68**	**66**	**55**	**55**	...
Total manufacturing [3] [28]										
Total, manufactures [3] [28]	**109**	**112**	**118**	**115**	**118**	**130**	**138**	**144**	**151**	**97**
Food, beverages, tobacco										
Aliments, boissons, tabac	117	122	126	126	125	137	146	157	168	108
Textiles										
Textiles	101	97	101	95	94	102	124	113	125	92
Chemicals										
Produits chimiques	127	143	125	140	156	166	221	234	212	97
Metal products										
Produits métalliques	134	123	128	134	123	141	157	143	141	100
Electricity [4]										
Total, industries manufacturières [3]	**108**	**114**	**118**	**123**	**134**	**150**	**167**	**181**	**198**	...
India [27] **Inde** [27]										
Total industry [2-4]										
Total, industrie [2-4]	**112**	**117**	**128**	**140**	**149**	**165**	**178**	**188**	**210**	...
Total mining [2]										
Total, industries extractives [2]	**127**	**144**	**157**	**164**	**176**	**184**	**194**	**210**	**216**	...
Total manufacturing [3]										
Total, industries manufacturières [3]	**109**	**113**	**122**	**135**	**143**	**160**	**173**	**181**	**205**	...
Food, beverages, tobacco										
Aliments, boissons, tabac	133	127	120	125	129	129	136	137	153	106
Textiles										
Textiles	89	99	99	109	107	114	107	108	123	106
Chemicals and petroleum products										
Produits chimiques et du pétroliers	119	123	138	149	172	197	227	238	254	104
Basic metals										
Métaux de base	118	109	122	133	140	156	169	167	180	112

30
Index numbers of industrial production [*cont.*]
Indices de la production industrielle [*suite*]
1980=100

Country or area and industry [SITC] Pays ou zone et industrie [CITI]	1982	1983	1984	1985	1986	1987	1988	1989	1990	1991 [1]
Metal products Produits métalliques	107	116	128	146	157	194	209	227	279	92
Electricity [4] Electricité [4]	115	121	138	149	164	178	192	216	233	...
Indonesia Indonésie										
Total mining [2] **Total, industries extractives [2]**	**85**	**83**	**87**	**82**	**86**	**77**	**82**	**88**	...	...
Total manufacturing [3] **Total, industries manufacturières [3]**	**110**	**113**	**121**	**130**	**145**	**162**	**185**	**208**	...	...
Food, beverages, tobacco Aliments, boissons, tabac	110	117	114	116	132	147	168	177	204	...
Textiles Textiles	102	96	100	97	127	134	163	184	221	...
Chemicals Produits chimiques	130	126	145	169	175	163	144	174	187	...
Iran, Islamic Rep. of [29] Iran, Rép. islamique d' [29]										
Total manufacturing [3] [30] [31] **Total, manufactures [3] [30] [31]**	**120**	**146**	**157**	**152**	**121**	**113**	**104**	**109**	...	...
Food, beverages, tobacco Aliments, boissons, tabac	103	119	120	125	110	104	98	97	...	...
Textiles Textiles	126	154	146	148	135	118	105	96	...	...
Chemicals Produits chimiques	98	131	140	141	140	131	123	137	...	...
Non-metallic mineral products Produits minéraux non-métalliques	125	137	141	145	138	144	136	142	...	...
Israel Israël										
Total industry [2-4] **Total, industrie [2-4]**	**107**	**111**	**116**	**120**	**124**	**130**	**126**	**124**	**132**	...
Total mining [2] [22] **Total, industries extractives [2] [22]**	**107**	**98**	**103**	**108**	**118**	**126**	**115**	**118**	**125**	...
Total manufacturing [3] **Total, industries manufacturières [3]**	**107**	**111**	**116**	**120**	**124**	**130**	**126**	**124**	**132**	...
Food and beverages Aliments et boissons	113	123	128	125	144	161	161	157	159	100
Textiles Textiles	106	103	103	99	98	99	92	92	97	100
Chemicals and petroleum products Produits chimiques et du pétroliers	106	118	128	128	131	146	148	156	163	103
Basic metals Métaux de base	88	87	99	85	83	86	90	85	106	112
Metal products Produits métalliques	106	111	118	123	111	122	118	114	120	108
Japan Japon										
Total industry [2-4] **Total, industrie [2-4]**	**101**	**104**	**114**	**118**	**118**	**122**	**134**	**142**	**149**	**104**
Total mining [2] **Total, industries extractives [2]**	**96**	**94**	**94**	**94**	**94**	**85**	**81**	**77**	**74**	**103**
Total manufacturing [3] [33] **Total, manufactures [3] [33]**	**101**	**104**	**114**	**119**	**118**	**122**	**134**	**142**	**149**	**104**
Food and beverages Aliments et boissons	101	102	101	101	102	103	105	106	106	101
Textiles Textiles	97	97	99	98	94	92	92	91	89	98
Chemicals and petroleum products Produits chimiques et du pétroliers	95	98	107	107	107	111	119	127	133	104
Basic metals Métaux de base	92	91	99	101	96	100	108	111	114	105
Metal products Produits métalliques	108	114	134	142	143	149	170	183	194	106
Electricity and gas [4] **Electricité et gaz [4]**	**103**	**108**	**116**	**120**	**120**	**126**	**133**	**139**	**150**	**107**

30
Index numbers of industrial production [*cont.*]
Indices de la production industrielle [*suite*]
1980=100

Country or area and industry [SITC] Pays ou zone et industrie [CITI]	1982	1983	1984	1985	1986	1987	1988	1989	1990	1991 [1]
Jordan Jordanie										
Total industry [2-4]										
Total, industrie [2-4]	120	126	152	155	157	172	158	71	84	102
Food and beverages										
Aliments et boissons	104	94	112	108	80	84	95	64	93	94
Textiles										
Textiles	73	86	112	149	132	149	156	72	89	75
Chemicals and petroleum products										
Produits chimiques et du pétroliers	130	127	140	135	139	145	148	75	89	105
Basic metals										
Métaux de base	148	162	130	155	162	172	152	75	94	86
Electricity and gas [4]										
Electricité et gaz [4]	139	170	197	216	266	313	290	63	82	108
Korea, Rep. of Corée, Rép. de										
Total industry [2-4]										
Total, industrie [2-4]	**118**	**137**	**158**	**164**	**198**	**236**	**268**	**91**	**102**	**103**
Total mining [2] [34]										
Total, industries extractives [2] [34]	**97**	**98**	**105**	**115**	**123**	**125**	**124**	**101**	**101**	**90**
Total manufacturing [3]										
Total, industries manufacturières [3]	**120**	**139**	**161**	**167**	**204**	**244**	**278**	**91**	**102**	**103**
Food, beverages, tobacco										
Aliments, boissons, tabac	116	132	142	153	169	188	207	86	104	108
Textiles										
Textiles	114	118	123	123	148	164	170	81	90	97
Chemicals and petroleum products										
Produits chimiques et du pétroliers	108	123	135	141	154	175	211	87	100	105
Basic metals										
Métaux de base	140	160	178	184	204	233	252	71	92	102
Metal products										
Produits métalliques	162	209	274	297	419	557	674	129	142	98
Electricity [4]										
Electricité [4]	**116**	**131**	**144**	**156**	**174**	**199**	**229**	**85**	**96**	**110**
Malaysia Malaisie										
Total industry [2-4]										
Total, industrie [2-4]	**109**	**122**	**142**	**138**	**152**	**164**	**186**	**205**	**231**	**112**
Total mining [2]										
Total, industries extractives [2]	**104**	**130**	**163**	**163**	**182**	**184**	**197**	**214**	**223**	**104**
Total manufacturing [3] [35]										
Total, manufactures [3] [35]	**109**	**116**	**130**	**122**	**133**	**150**	**177**	**197**	**232**	**116**
Food, beverages, tobacco										
Aliments, boissons, tabac	105	104	115	123	129	136	145	164	175	102
Textiles										
Textiles	106	105	132	128	144	170	189	218	228	105
Chemicals										
Produits chimiques	90	92	105	111	126	143	148	154	157	115
Basic metals										
Métaux de base	105	124	165	131	102	125	166	198	226	108
Metal products										
Produits métalliques	120	134	181	149	183	218	269	317	418	121
Electricity [4]										
Electricité [4]	110	123	133	143	156	168	186	210	238	111
Mongolia Mongolie										
Total industry [2-4]										
Total, industrie [2-4]	**122**	**133**	**145**	**156**	**168**	**176**	**182**	**187**	**183**	...
Food, beverages, tobacco										
Aliments, boissons, tabac	121	128	140	137	142	144	148	152	139	...
Textiles										
Textiles	149	175	201	227	240	246	244	276	248	...
Chemicals										
Produits chimiques	112	120	126	130	150	162	192	198	187	...
Metal products										
Produits métalliques	117	129	140	159	165	184	186	186	158	...

30
Index numbers of industrial production [*cont.*]
Indices de la production industrielle [*suite*]
1980=100

Country or area and industry [SITC] Pays ou zone et industrie [CITI]	1982	1983	1984	1985	1986	1987	1988	1989	1990	1991 [1]
Electricity [4] Électricité [4]	97	113	144	181	202	214	217	224	236	...
Pakistan [27] Pakistan [27]										
Total industry [2-4] Total, industrie [2-4]	122	131	144	160	170	186	191	204	...	...
Total mining [2] Total, industries extractives [2]	118	122	147	183	187	212	218	249	...	...
Total manufacturing [3] Total, industries manufacturières [3]	123	133	143	154	165	179	183	192	203	106
Philippines Philippines										
Total industry [2-4] Total, industrie [2-4]	128	142	193	231	258	296	358	395	451	114
Total mining [2] Total, industries extractives [2]	91	84	75	85	71	87	92	116	157	173
Total manufacturing [3] [3] Total, industries manufacturières [3] [3]	131	153	207	241	291	333	280	472	506	112
Food, beverages, tobacco Aliments, boissons, tabac	140	154	233	272	264	288	343	379	411	115
Textiles Textiles	122	140	178	166	221	256	274	288	289	101
Non-metallic mineral products Produits minéraux non-métalliques	128	147	172	173	182	213	256	310	355	134
Metal products Produits métalliques	128	145	231	357	598	669	826	953	1072	109
Electricity [4] Électricité [4]	117	126	126	126	130	141	148	152	155	103
Singapore Singapour										
Total manufacturing [3] Total, industries manufacturières [3]	104	106	116	107	116	136	162	178	196	109
Food, beverages, tobacco Aliments, boissons, tabac	97	90	87	90	93	96	112	132	132	108
Textiles Textiles	66	50	42	30	30	35	37	39	36	96
Chemicals, coal, petroleum products Produits chimiques, houillers, pétroliers	118	122	137	136	145	143	149	169	198	116
Basic metals Métaux de base	134	127	125	123	124	129	136	139	151	105
Metal products Produits métalliques	103	108	119	103	108	132	158	182	181	107
Sri Lanka Sri Lanka										
Total manufacturing [3] [36] Total, manufactures [3] [36]	110	94	100	95	102	98	99	110	120	...
Food, beverages, tobacco Aliments, boissons, tabac	94	83	59	55	52	46	46	52	56	...
Textiles and clothing Textiles, habillement	162	151	168	221	248	238	235	234	309	...
Chemicals and rubber products Produits chimiques et en caoutchouc	106	84	100	88	92	92	101	91	84	...
Basic metals Métaux de base	54	55	37	20	32	36	32	28	34	...
Metal products Produits métalliques	84	67	76	88	119	86	83	89	102	...
Syrian Arab Rep. Rép. arabe syrienne										
Total industry [2-4] Total, industrie [2-4]	119	139	143	136	141	146	150	169	186	...
Total mining [2] Total, industries extractives [2]	108	102	98	101	125	143	170	218	270	...
Total manufacturing [3] Total, industries manufacturières [3]	127	167	178	163	161	160	152	158	163	...
Food, beverages, tobacco Aliments, boissons, tabac	154	180	175	147	150	131	121	...	...	...
Textiles Textiles	110	134	141	118	133	118	92	...	...	...

30
Index numbers of industrial production [*cont.*]
Indices de la production industrielle [*suite*]
1980=100

Country or area and industry [SITC] Pays ou zone et industrie [CITI]	1982	1983	1984	1985	1986	1987	1988	1989	1990	1991 [1]
Non-metallic mineral products										
Produits minéraux non-métalliques	137	165	191	191	191	174	145	...	...	...
Electricity and water [4]										
Electricité et eau [4]	**136**	**156**	**170**	**183**	**172**	**179**	**214**	**232**	**258**	**...**
Turkey Turquie										
Total industry [2-4]										
Total, industrie [2-4]	**117**	**127**	**141**	**149**	**166**	**184**	**187**	**94**	**212**	**99**
Total mining [2]										
Total, industries extractives [2]	**123**	**120**	**124**	**138**	**153**	**161**	**152**	**71**	**182**	**113**
Total manufacturing [3]										
Total, industries manufacturières [3]	**119**	**131**	**145**	**152**	**168**	**186**	**188**	**191**	**210**	**98**
Food, beverages, tobacco										
Aliments, boissons, tabac	142	132	158	154	151	165	170	179	188	105
Textiles										
Textiles	134	148	162	173	190	209	216	226	231	91
Chemicals, coal, petroleum products										
Produits chimiques, houillers, pétroliers	140	161	170	176	211	249	244	251	261	99
Basic metals										
Métaux de base	121	137	168	188	220	244	241	247	288	93
Metal products										
Produits métalliques	142	168	195	206	225	245	233	228	298	100
Electricity, gas and water [4]										
Electricité, gaz et eau [4]	**112**	**131**	**147**	**162**	**170**	**190**	**206**	**222**	**246**	**102**
Europe · Europe										
Austria Autriche										
Total industry [2-4]										
Total, industrie [2-4]	**98**	**99**	**104**	**109**	**110**	**112**	**116**	**123**	**132**	**102**
Total mining [2] [37]										
Total, industries extractives [2] [37]	**109**	**107**	**113**	**110**	**108**	**112**	**101**	**100**	**104**	**94**
Total manufacturing [3]										
Total, industries manufacturières [3]	**97**	**99**	**104**	**109**	**110**	**109**	**116**	**124**	**133**	**102**
Food, beverages, tobacco										
Aliments, boissons, tabac	107	107	109	112	115	116	118	124	135	110
Textiles										
Textiles	91	86	89	91	91	86	90	94	98	101
Chemicals and petroleum products										
Produits chimiques et pétroliers	98	105	121	116	113	118	131	136	138	103
Basic metals										
Métaux de base	91	97	106	107	102	102	112	118	115	96
Metal products										
Produits métalliques	99	100	106	118	123	119	127	140	157	104
Electricity [4]										
Electricité [4]	**103**	**103**	**106**	**110**	**110**	**123**	**119**	**123**	**125**	**102**
Belgium Belgique										
Total industry [2-4]										
Total, industrie [2-4]	**98**	**99**	**102**	**104**	**105**	**107**	**114**	**118**	**122**	**98**
Total mining [2] [39]										
Total, industries extractives [2] [39]	**94**	**89**	**90**	**84**	**75**	**65**	**56**	**45**	**32**	**62**
Total manufacturing [3] [33]										
Total, manufactures [3] [33]	**77**	**100**	**102**	**104**	**105**	**108**	**115**	**120**	**125**	**99**
Food, beverages, tobacco										
Aliments, boissons, tabac	110	112	115	118	120	124	127	118	118	96
Textiles										
Textiles	96	103	104	104	105	105	106	110	115	92
Chemicals and petroleum products										
Produits chimiques et pétroliers	101	103	109	112	117	119	126	118	118	101
Basic metals										
Métaux de base	83	85	93	92	86	88	99	100	103	96
Metal products										
Produits métalliques	98	100	101	105	106	106	113	121	125	100
Electricity [4]										
Electricité [4]	**94**	**98**	**102**	**106**	**109**	**117**	**121**	**125**	**132**	**104**

30
Index numbers of industrial production [*cont.*]
Indices de la production industrielle [*suite*]
1980=100

Country or area and industry [SITC] Pays ou zone et industrie [CITI]	1982	1983	1984	1985	1986	1987	1988	1989	1990	1991 [1]
Bulgaria Bulgarie										
Total industry [2-4]										
Total, industrie [2-4]	**110**	**115**	**120**	**124**	**129**	**134**	**141**	**139**	...	**77**
Total mining [2] [10] [40]										
Total, industries extractives [2] [10] [40]	**102**	**103**	**106**	**105**	**109**	**109**	**108**	**108**	...	...
Total manufacturing [3] [11] [41]										
Total, manufactures [3] [11] [41]	**110**	**115**	**120**	**124**	**130**	**135**	**142**	**140**	...	...
Food, beverages, tobacco										
Aliments, boissons, tabac	108	105	104	107	107	107	110	111	100	...
Textiles										
Textiles	110	113	115	112	114	121	125	132	130	...
Chemicals										
Produits chimiques	131	139	149	153	164	162	167	168	142	...
Basic metals										
Métaux de base	106	113	118	117	120	120	123	119	70	...
Metal products										
Produits métalliques	115	126	137	153	167	183	200	196	149	...
Electricity and steam [4]										
Electricité et vapeur [4]	**119**	**126**	**132**	**124**	**127**	**133**	**140**	**137**	...	...
Czechoslovakia Tchécoslovaquie										
Total industry [2-4]										
Total, industrie [2-4]	**104**	**107**	**111**	**115**	**119**	**121**	**124**	**125**	**120**	**83**
Total mining [2]										
Total, industries extractives [2]	**99**	**101**	**102**	**101**	**101**	**102**	**103**	**100**	**91**	**89**
Total manufacturing [3] [41]										
Total, manufactures [3] [41]	**104**	**107**	**111**	**116**	**119**	**122**	**125**	**126**	**121**	**83**
Food, beverages, tobacco										
Aliments, boissons, tabac	101	104	107	107	109	110	110	113	111	78
Textiles										
Textiles	104	106	108	111	114	115	118	120	120	76
Chemicals, coal, petroleum products										
Produits chimiques, houillers, pétroliers	101	103	107	112	117	120	123	123	128	71
Basic metals										
Métaux de base	100	101	102	104	105	107	108	108	107	86
Metal products										
Produits métalliques	117	123	131	139	146	152	156	156	150	81
Electricity, gas and water [4]										
Electricité, gaz et eau [4]	**103**	**105**	**108**	**112**	**118**	**120**	**122**	**124**	**123**	**99**
Denmark Danemark										
Total industry [2-4]										
Total, industrie [2-4]	**103**	**106**	**116**	**121**	**129**	**125**	**127**	**130**	**131**	...
Total mining [2]										
Total, industries extractives [2]	**84**	**89**	**100**	**110**	**138**	**122**	**124**	**132**	**116**	...
Total manufacturing [3] [42]										
Total, manufactures [3] [42]	**102**	**106**	**117**	**122**	**131**	**127**	**129**	**133**	**133**	...
Food, beverages, tobacco										
Aliments, boissons, tabac	107	109	113	117	122	120	123	122	126	107
Textiles										
Textiles	103	107	117	116	120	116	111	113	106	99
Chemicals, coal, petroleum products										
Produits chimiques, houillers, pétroliers	109	115	121	127	134	134	137	137	138	97
Basic metals										
Métaux de base	84	80	101	97	93	87	91	103	96	96
Metal products										
Produits métalliques	101	103	116	125	136	129	131	138	139	101
Electricity [4]										
Electricité [4]	**88**	**81**	**83**	**107**	**113**	**108**	**103**	**84**	**122**	...
Finland Finlande										
Total industry [2-4]										
Total, industrie [2-4]	**104**	**107**	**112**	**116**	**118**	**123**	**128**	**132**	**130**	**91**
Total mining [2]										
Total, industries extractives [2]	**115**	**115**	**119**	**123**	**127**	**125**	**147**	**142**	**141**	**79**

30
Index numbers of industrial production [*cont.*]
Indices de la production industrielle [*suite*]
1980=100

Country or area and industry [SITC] Pays ou zone et industrie [CITI]	1982	1983	1984	1985	1986	1987	1988	1989	1990	1991 [1]
Total manufacturing [3]										
Total, industries manufacturières [3]	**104**	**107**	**112**	**116**	**118**	**123**	**128**	**133**	**130**	**90**
Food, beverages, tobacco										
Aliments, boissons, tabac	105	108	109	110	114	116	120	122	122	113
Textiles										
Textiles	89	85	84	79	75	78	73	70	64	83
Paper and paper products										
Papier, produits en papier	95	104	114	113	116	121	129	130	132	97
Chemicals, coal, petroleum products										
Produits chimiques, houillers, pétroliers	96	103	107	110	109	118	125	129	132	90
Basic metals										
Métaux de base	102	108	116	120	121	125	132	136	139	93
Metal products										
Produits métalliques	116	116	121	132	135	146	152	163	161	91
Electricity, gas and water [4]										
Electricité, gaz et eau [4]	**100**	**104**	**111**	**121**	**121**	**130**	**128**	**130**	**135**	**106**
France France										
Total industry [2-4]										
Total, industrie [2-4]	**98**	**98**	**100**	**101**	**101**	**103**	**108**	**112**	**114**	**100**
Total mining [2]										
Total, industries extractives [2]	**93**	**91**	**90**	**86**	**81**	**81**	**77**	**75**	**72**	**98**
Total manufacturing [3] [45]										
Total, manufactures [3] [45]	**97**	**96**	**97**	**97**	**98**	**99**	**104**	**108**	**110**	**97**
Food, beverages, tobacco										
Aliments, boissons, tabac	105	103	106	106	105	108	111	113	118	94
Textiles										
Textiles	91	89	89	86	84	81	80	80	77	93
Chemicals, coal, petroleum products										
Produits chimiques, houillers, pétroliers	100	104	109	110	110	114	120	126	127	101
Basic metals										
Métaux de base	87	82	84	82	80	80	86	88	88	98
Metal products										
Produits métalliques	98	96	93	92	92	94	101	107	110	98
Electricity and gas [4]										
Electricité et gaz [4]	106	113	122	130	136	142	145	150	154	113
Germany ş Allemagne ş										
Germany, Fed. Rep. of Allemagne, Rép. féd. d'										
Total industry [2-4]										
Total, industrie [2-4]	**95**	**96**	**99**	**104**	**107**	**107**	**111**	**117**	**123**	**105**
Total mining [2] [49]										
Total, industries extractives [2] [49]	**97**	**92**	**92**	**93**	**91**	**87**	**85**	**84**	**84**	**98**
Total manufacturing [3] [49]										
Total, manufactures [3] [49]	**95**	**96**	**99**	**104**	**107**	**107**	**112**	**118**	**124**	**105**
Food, beverages, tobacco										
Aliments, boissons, tabac	100	100	102	105	108	109	113	116	129	118
Textiles										
Textiles	88	89	91	94	95	94	92	93	95	102
Chemicals, coal, petroleum products										
Produits chimiques, houillers, pétroliers	94	99	103	104	103	104	111	112	115	103
Basic metals										
Métaux de base	87	86	93	97	92	90	99	102	98	100
Metal products										
Produits métalliques	98	98	101	111	116	117	120	130	137	105
Electricity and gas [4]										
Electricité et gaz [4]	**99**	**102**	**107**	**112**	**111**	**115**	**118**	**121**	**125**	**104**
former German Dem. Rep. ancienne Rép. dém. allemande										
Total industry [2-4]										
Total, industrie [2-4]	**105**	**110**	**114**	**119**	**123**	**126**	**130**	**133**	**94**	...
Total mining [2] [46]										
Total, industries extractives [2] [46]	**106**	**109**	**115**	**118**	**120**	**119**	**120**	**116**	**82**	...
Total manufacturing [3] [46] [47]										
Total, manufactures [3] [46] [47]	**105**	**109**	**114**	**119**	**123**	**127**	**131**	**135**	**95**	...

30
Index numbers of industrial production [*cont.*]
Indices de la production industrielle [*suite*]
1980=100

Country or area and industry [SITC] Pays ou zone et industrie [CITI]	1982	1983	1984	1985	1986	1987	1988	1989	1990	1991 [1]
Food, beverages, tobacco Aliments, boissons, tabac	101	104	107	110	113	113	114	115	66	...
Textiles Textiles	104	105	107	109	111	114	117	119	74	...
Chemicals, coal, petroleum products Produits chimiques, houillers, pétroliers	100	103	108	112	113	114	119	122	65	...
Basic metals Métaux de base	107	108	108	112	113	117	117	116	69	...
Metal products Produits métalliques	111	117	125	134	141	149	158	165	75	...
Electricity and gas [4] Electricité et gaz [4]	105	111	114	118	121	122	123	123	93	...
Greece Grèce										
Total industry [2-4] **Total, industrie [2-4]**	**102**	**101**	**104**	**107**	**107**	**106**	**111**	**113**	**110**	**97**
Total mining [2] [50] **Total, industries extractives [2]** [50]	**148**	**161**	**178**	**183**	**185**	**182**	**189**	**180**	**74**	**100**
Total manufacturing [3] [51] **Total, manufactures [3]** [51]	**99**	**97**	**98**	**101**	**100**	**98**	**103**	**106**	**103**	**97**
Food, beverages, tobacco Aliments, boissons, tabac	101	103	110	116	111	102	112	120	114	108
Textiles Textiles	94	92	92	95	102	104	101	99	95	91
Chemicals, coal, petroleum products Produits chimiques, houillers, pétroliers	98	104	110	115	115	117	125	132	133	94
Basic metals Métaux de base	83	85	90	88	86	85	95	96	93	108
Metal products Produits métalliques	101	94	88	84	87	79	83	82	83	94
Electricity and gas [4] **Electricité et gaz [4]**	**102**	**105**	**109**	**123**	**126**	**136**	**144**	**151**	**155**	**95**
Hungary Hongrie										
Total industry [2-4] **Total, industrie [2-4]**	**105**	**106**	**109**	**110**	**112**	**116**	**116**	**111**	**102**	**84**
Total mining [2] [46] **Total, industries extractives [2]** [46]	**100**	**97**	**97**	**99**	**99**	**99**	**95**	**88**	**78**	**98**
Total manufacturing [3] [46] [52] **Total, manufactures [3]** [46] [52]	**106**	**107**	**110**	**110**	**112**	**117**	**117**	**111**	**101**	**81**
Food, beverages, tobacco Aliments, boissons, tabac	106	108	111	109	111	114	111	112	109	90
Textiles Textiles	98	98	100	101	102	104	107	103	85	73
Chemicals, coal, petroleum products Produits chimiques, houillers, pétroliers	109	113	119	125	128	134	142	135	120	90
Basic metals Métaux de base	104	104	106	107	110	111	116	117	83	82
Metal products Produits métalliques	111	112	115	119	123	129	130	132	110	67
Electricity, gas and water [4] **Electricité, gaz et eau [4]**	**109**	**112**	**117**	**120**	**123**	**130**	**129**	**130**	**132**	**108**
Ireland Irlande										
Total industry [2-4] [53] **Total, industrie [2-4]** [53]	**105**	**113**	**124**	**128**	**131**	**143**	**158**	**176**	**184**	**103**
Total mining [2] **Total, industries extractives [2]**	**88**	**92**	**108**	**72**	**79**	**85**	**69**	**89**	**84**	**82**
Total manufacturing [3] [30] **Total, manufactures [3]** [30]	**105**	**114**	**125**	**131**	**135**	**149**	**167**	**187**	**96**	**103**
Food, beverages, tobacco Aliments, boissons, tabac	103	106	110	117	121	132	139	145	149	104
Textiles Textiles	101	92	96	93	92	95	99	103	111	100
Chemicals Produits chimiques	110	132	145	166	165	172	200	240	247	127

30
Index numbers of industrial production [*cont.*]
Indices de la production industrielle [*suite*]
1980=100

Country or area and industry [SITC] Pays ou zone et industrie [CITI]	1982	1983	1984	1985	1986	1987	1988	1989	1990	1991 [1]
Basic metals										
Métaux de base	102	113	121	130	120	120	136	150	158	89
Metal products										
Produits métalliques	114	132	157	158	166	201	244	282	300	93
Electricity, gas and water [4]										
Electricité, gaz et eau [4]	**113**	**118**	**125**	**133**	**127**	**124**	**128**	**135**	**144**	**109**
Italy Italie										
Total industry [2-4]										
Total, industrie [2-4]	**95**	**92**	**95**	**97**	**100**	**104**	**110**	**114**	**114**	**96**
Total mining [2]										
Total, industries extractives [2]	**98**	**97**	**99**	**100**	**105**	**118**	**130**	**128**	**130**	**92**
Total manufacturing [3] [54]										
Total, manufactures [3] [54]	**95**	**92**	**95**	**96**	**100**	**103**	**109**	**113**	**112**	**97**
Food, beverages, tobacco										
Aliments, boissons, tabac	101	102	101	105	109	115	119	120	121	98
Textiles										
Textiles	98	89	96	97	101	105	106	112	110	101
Chemicals, coal, petroleum products										
Produits chimiques, houillers, pétroliers	85	85	90	90	93	96	98	98	99	96
Basic metals										
Métaux de base	90	85	92	93	93	96	104	108	104	101
Metal products										
Produits métalliques	95	95	97	101	110	113	122	128	127	93
Electricity and gas [4]										
Electricité et gaz [4]	**99**	**98**	**100**	**103**	**107**	**113**	**116**	**121**	**125**	**103**
Luxembourg Luxembourg										
Total industry [2-4]										
Total, industrie [2-4]	**95**	**99**	**113**	**121**	**124**	**122**	**133**	**144**	**143**	**101**
Total mining [2]										
Total, industries extractives [2]	**38**	**37**	**32**	**29**	**29**	**33**	**36**	**140**	**144**	**103**
Total manufacturing [3] [55]										
Total, manufactures [3] [55]	**95**	**99**	**114**	**122**	**124**	**123**	**134**	**144**	**143**	**100**
Food, beverages, tobacco										
Aliments, boissons, tabac	110	127	123	129	134	130	128	133	138	108
Chemicals										
Produits chimiques	92	101	118	143	204	269	268	308	280	108
Basic metals [11]										
Métaux de base [11]	82	80	94	102	97	90	102	104	100	97
Metal products										
Produits métalliques	94	102	139	151	155	147	161	176	183	103
Electricity and gas [4]										
Electricité et gaz [4]	**111**	**114**	**121**	**126**	**132**	**138**	**148**	**154**	**158**	**105**
Malta Malte										
Total industry [2-4] [2]										
Total, industrie [2-4] [2]	**107**	**110**	**118**	**120**	**127**	...	...	...	...	...
Total mining [2]										
Total, industries extractives [2]	**113**	**104**	**86**	**64**	**84**	...	...	...	...	...
Total manufacturing [3]										
Total, industries manufacturières [3]	**103**	**104**	**113**	**118**	**128**	...	...	...	...	...
Food, beverages, tobacco										
Aliments, boissons, tabac	131	136	143	146	152	...	...	...	...	...
Metal products [56]										
Produits métalliques [56]	108	112	125	135	156	...	...	...	...	...
Electricity, gas and water [4]										
Electricité, gaz et eau [4]	**114**	**125**	**125**	**132**	**143**	...	...	...	...	...
Netherlands Pays-Bas										
Total industry [2-4]										
Total, industrie [2-4]	**94**	**97**	**102**	**106**	**106**	**107**	**107**	**112**	**116**	**106**
Total mining [2]										
Total, industries extractives [2]	**78**	**84**	**86**	**93**	**87**	**90**	**79**	**84**	**86**	**118**
Total manufacturing [3]										
Total, industries manufacturières [3]	**99**	**101**	**106**	**109**	**112**	**112**	**118**	**123**	**129**	**101**

30
Index numbers of industrial production [*cont.*]
Indices de la production industrielle [*suite*]
1980=100

Country or area and industry [SITC] Pays ou zone et industrie [CITI]	1982	1983	1984	1985	1986	1987	1988	1989	1990	1991 [1]
Food, beverages, tobacco Aliments, boissons, tabac	105	104	108	105	112	112	116	119	124	104
Textiles Textiles	92	90	98	100	94	92	97	101	104	100
Chemicals, coal, petroleum products [57] Produits chimiques, houillers, pétroliers [57]	104	120	131	142	145	152	160	166	169	107
Basic metals Métaux de base	97	95	108	110	108	112	123	130	132	96
Metal products Produits métalliques	100	99	105	110	110	108	111	118	125	101
Electricity, gas and water [4] **Electricité, gaz et eau [4]**	**103**	**104**	**105**	**108**	**112**	**108**	**110**	**114**	**118**	**107**
Norway Norvège										
Total industry [2-4] **Total, industrie [2-4]**	**99**	**108**	**118**	**121**	**126**	**135**	**142**	**165**	**771**	**...**
Total mining [2] **Total, industries extractives [2]**	**97**	**114**	**131**	**138**	**147**	**164**	**181**	**230**	**243**	**...**
Total manufacturing [3] **Total, industries manufacturières [3]**	**99**	**98**	**104**	**106**	**108**	**110**	**109**	**109**	**110**	**...**
Food, beverages, tobacco Aliments, boissons, tabac	81	76	72	66	66	65	60	60	55	105
Textiles Textiles	75	72	72	76	77	72	63	57	59	99
Paper and paper products Papier, produits en papier	101	107	127	132	130	129	130	139	138	101
Chemicals, coal, petroleum products Produits chimiques, houillers, pétroliers	124	136	166	166	163	175	173	183	208	101
Basic metals Métaux de base	98	123	128	126	124	131	142	144	144	97
Metal products Produits métalliques	101	94	99	103	106	107	104	105	105	101
Electricity and gas [4] **Electricité et gaz [4]**	**112**	**127**	**121**	**118**	**111**	**119**	**125**	**136**	**138**	**...**
Poland Pologne										
Total industry [2-4] **Total, industrie [2-4]**	**85**	**90**	**95**	**99**	**103**	**106**	**111**	**110**	**81**	**87**
Total mining [2] **Total, industries extractives [2]**	**95**	**99**	**102**	**104**	**104**	**106**	**105**	**106**	**78**	**92**
Total manufacturing [3] [58] **Total, manufactures [3] [58]**	**84**	**89**	**94**	**97**	**102**	**105**	**111**	**108**	**80**	**86**
Food, beverages, tobacco [59] Aliments, boissons, tabac [59]	83	87	90	94	98	100	101	94	70	110
Textiles Textiles	77	78	83	88	89	90	98	102	61	88
Chemicals, coal, petroleum products Produits chimiques, houillers, pétroliers	87	92	95	98	102	106	112	107	79	90
Basic metals Métaux de base	80	86	89	89	91	90	93	89	69	85
Metal products Produits métalliques	86	93	100	106	114	122	132	131	100	85
Electricity [4] **Electricité [4]**	**105**	**113**	**121**	**128**	**136**	**135**	**135**	**134**	**122**	**108**
Portugal [60] Portugal [60]										
Total industry [2-4] **Total, industrie [2-4]**	**110**	**114**	**117**	**119**	**128**	**133**	**138**	**144**	**161**	**102**
Total mining [2] **Total, industries extractives [2]**	**94**	**89**	**100**	**109**	**98**	**86**	**95**	**237**	**422**	**126**
Total manufacturing [3] [33] **Total, manufactures [3] [33]**	**111**	**114**	**117**	**118**	**118**	**135**	**138**	**142**	**150**	**100**
Food, beverages, tobacco Aliments, boissons, tabac	101	106	98	98	101	107	116	122	131	...
Chemicals, coal, petroleum products Produits chimiques, houillers, pétroliers	109	105	110	119	130	136	145	151	162	...

30
Index numbers of industrial production [*cont.*]
Indices de la production industrielle [*suite*]
1980=100

Country or area and industry [SITC] Pays ou zone et industrie [CITI]	1982	1983	1984	1985	1986	1987	1988	1989	1990	1991 [1]
Basic metals Métaux de base	112	102	107	104	105	113	121	119	120	...
Metal products Produits métalliques	110	99	87	85	85	87	92	92	100	...
Electricity [4] Electricité [4]	103	121	127	127	130	132	145	168	184	104
Romania Roumanie										
Total industry [2-4] **Total, industrie [2-4]**	**104**	**109**	**116**	**120**	**129**	**132**	**136**	**134**	**108**	...
Total mining [2] [61] **Total, industries extractives [2]** [61]	**104**	**109**	**112**	**114**	**118**	**123**	**126**	**128**	**100**	...
Total manufacturing [3] [62] **Total, manufactures [3]** [62]	**104**	**109**	**116**	**121**	**130**	**133**	**137**	**134**	**109**	...
Food, beverages, tobacco [63] Aliments, boissons, tabac [63]	97	102	107	108	112	121	121	121	114	91
Textiles Textiles	109	111	110	114	121	123	125	124	99	91
Chemicals, coal, petroleum products Produits chimiques, houillers, pétroliers	103	109	117	118	127	124	128	125	95	64
Basic metals [11] Métaux de base [11]	104	106	119	117	127	121	124	121	84	...
Metal products Produits métalliques	105	110	119	128	136	141	145	134	108	...
Electricity and steam [4] **Electricité et vapeur [4]**	**102**	**106**	**112**	**114**	**124**	**126**	**134**	**133**	**137**	...
Spain Espagne										
Total industry [2-4] **Total, industrie [2-4]**	**98**	**101**	**101**	**103**	**107**	**112**	**115**	**120**	**120**	**98**
Total mining [2] **Total, industries extractives [2]**	**122**	**129**	**131**	**130**	**124**	**107**	**102**	**110**	**104**	**94**
Total manufacturing [3] **Total, industries manufacturières [3]**	**96**	**99**	**99**	**101**	**105**	**111**	**114**	**119**	**119**	**97**
Food, beverages, tobacco Aliments, boissons, tabac	104	111	112	117	116	125	129	127	133	104
Textiles Textiles	92	94	92	96	104	108	100	105	102	91
Chemicals, coal, petroleum products Produits chimiques, houillers, pétroliers	96	97	102	104	107	108	109	115	116	98
Basic metals Métaux de base	99	101	105	107	99	98	101	108	104	99
Metal products Produits métalliques	94	93	89	91	99	110	120	130	130	97
Electricity and gas [4] **Electricité et gaz [4]**	**103**	**106**	**109**	**115**	**117**	**122**	**127**	**132**	**136**	**105**
Sweden Suède										
Total industry [2-4] [53] **Total, industrie [2-4]** [53]	**97**	**101**	**107**	**110**	**110**	**113**	**116**	**118**	**114**	**90**
Total mining [2] [64] **Total, industries extractives [2]** [64]	**74**	**87**	**98**	**103**	**102**	**103**	**100**	**100**	**101**	**91**
Total manufacturing [3] **Total, industries manufacturières [3]**	**97**	**102**	**107**	**110**	**110**	**113**	**116**	**118**	**114**	**90**
Food, beverages, tobacco Aliments, boissons, tabac	101	102	103	104	104	104	106	108	105	96
Textiles Textiles	90	91	94	96	94	94	92	88	83	91
Paper and paper products Papier, produits en papier	93	102	108	108	111	118	120	120	117	102
Chemicals, coal, petroleum products Produits chimiques, houillers, pétroliers	102	111	113	113	111	123	130	130	134	96
Basic metals Métaux de base	96	100	108	109	103	105	113	114	109	98
Metal products Produits métalliques	100	104	111	119	118	120	124	131	124	96

30
Index numbers of industrial production [*cont.*]
Indices de la production industrielle [*suite*]
1980=100

Country or area and industry [SITC] Pays ou zone et industrie [CITI]	1982	1983	1984	1985	1986	1987	1988	1989	1990	1991 [1]
Electricity and gas [4]										
Electricité et gaz [4]	110	126	141	154	156	169	...	...	...	...
Switzerland Suisse										
Total industry [2-4]										
Total, industrie [2-4]	**96**	**95**	**97**	**103**	**107**	**108**	**117**	**119**	**122**	**99**
Total manufacturing [3]										
Total, industries manufacturières [3]	**94**	**94**	**97**	**103**	**106**	**107**	**117**	**120**	**123**	**99**
Food, beverages, tobacco										
Aliments, boissons, tabac	101	100	101	101	102	104	106	109	111	101
Textiles										
Textiles	96	96	102	105	107	105	106	104	99	95
Chemicals, coal, petroleum products										
Produits chimiques, houillers, pétroliers	103	109	115	123	125	129	146	169	171	100
Basic metals										
Métaux de base	91	90	98	101	103	105	115	116	117	96
Metal products										
Produits métalliques	92	88	88	94	99	99	106	106	114	105
Electricity and gas [4]										
Electricité et gaz [4]	108	108	102	114	116	121	123	110	112	103
United Kingdom Royaume Uni										
Total industry [2-4]										
Total, industrie [2-4]	**98**	**102**	**102**	**108**	**111**	**114**	**118**	**119**	**118**	**96**
Total mining [2] [65]										
Total, industries extractives [2] [65]	**117**	**124**	**118**	**129**	**133**	**130**	**120**	**102**	**100**	**94**
Total manufacturing [3] [65]										
Total, manufactures [3] [65]	**94**	**97**	**101**	**104**	**105**	**111**	**118**	**124**	**123**	**94**
Food, beverages, tobacco										
Aliments, boissons, tabac	100	101	102	102	102	105	106	107	108	**101**
Textiles										
Textiles	90	92	94	98	98	103	102	98	94	92
Chemicals, coal, petroleum products										
Produits chimiques, houillers, pétroliers	98	104	109	113	116	121	127	132	131	99
Basic metals										
Métaux de base	100	98	99	101	100	108	120	122	116	90
Metal products										
Produits métalliques	93	95	99	104	104	107	117	124	125	93
Electricity, gas and water [4]										
Electricité, gaz et eau [4]	**98**	**107**	**91**	**104**	**120**	**124**	**120**	**121**	**123**	**111**
Yugoslavia Yougoslavie										
Total industry [2-4]										
Total, industrie [2-4]	**104**	**105**	**111**	**114**	**119**	**120**	**119**	**120**	**107**	...
Total mining [2]										
Total, industries extractives [2]	**105**	**107**	**109**	**115**	**116**	**117**	**116**	**113**	**108**	...
Total manufacturing [3]										
Total, industries manufacturières [3]	**104**	**105**	**112**	**114**	**117**	**119**	**117**	**120**	**106**	...
Food, beverages, tobacco										
Aliments, boissons, tabac	106	106	110	108	110	113	109	83	108	99
Textiles										
Textiles	100	100	107	109	116	117	115	112	92	76
Chemicals, coal, petroleum products										
Produits chimiques, houillers, pétroliers	101	109	117	118	126	128	134	131	118	82
Basic metals										
Métaux de base	107	114	122	127	129	124	128	131	113	79
Metal products										
Produits métalliques	108	112	117	122	131	131	129	132	110	70
Electricity and gas [4]										
Electricité et gaz [4]	**103**	**109**	**116**	**119**	**124**	**128**	**132**	**131**	**131**	...
Oceania · Océanie										
Australia [27] Australie [27]										
Total industry [2-4]										
Total, industrie [2-4]	**97**	**101**	**108**	**115**	**114**	**121**	**129**	...	...	...
Total manufacturing [2]										
Total, industries manufacturières [2]	**94**	**96**	**100**	**104**	**104**	**110**	**117**	...	...	...

30
Index numbers of industrial production [*cont.*]
Indices de la production industrielle [*suite*]
1980=100

Country or area and industry [SITC] Pays ou zone et industrie [CITI]	1982	1983	1984	1985	1986	1987	1988	1989	1990	1991 [1]
Food, beverages, tobacco Aliments, boissons, tabac	99	100	105	105	107	108	109	111	113	102
Textiles Textiles	92	97	101	105	104	99	103	98	97	100
Chemicals, coal, petroleum products Produits chimiques, houillers, pétroliers	99	101	103	105	106	116	126	132	129	99
Basic metals Métaux de base	79	85	91	97	95	94	103	111	112	98
Metal products Produits métalliques	86	79	86	94	89	91	96	102	96	95
Electricity and gas [4] **Electricité et gaz [4]**	**110**	**114**	**121**	**126**	**130**	**137**	**143**	...	...	...
Fiji Fidji										
Total industry [2-4] **Total, industrie [2-4]**	**109**	**97**	**116**	**103**	**123**	**110**	**117**	**128**	**138**	...
Total mining [2] **Total, industries extractives [2]**	**184**	**161**	**195**	**241**	**369**	**370**	**552**	**545**	**532**	...
Total manufacturing [3] **Total, industries manufacturières [3]**	**108**	**95**	**113**	**98**	**118**	**102**	**103**	**114**	**126**	...
Food, beverages, tobacco Aliments, boissons, tabac	119	88	120	95	126	117	112	128	124	111
Electricity [4] **Electricité [4]**	**108**	**114**	**124**	**129**	**138**	**136**	**147**	**157**	**166**	...
				former USSR · ancienne URSS						
former USSR ancienne URSS										
Total industry [2-4] [66] **Total, industrie [2-4] [66]**	**106**	**111**	**115**	**119**	**125**	**129**	**134**	**136**	**135**	**94**
Total mining [2] [67] **Total, industries extractives [2] [67]**	**102**	**104**	**105**	**107**	**111**	**113**	**116**	**115**	**111**	**92**
Total manufacturing [3] [47] **Total, manufactures [3] [47]**	**107**	**112**	**116**	**120**	**126**	**131**	**136**	**139**	**138**	**94**
Food, beverages, tobacco [68] Aliments, boissons, tabac [68]	106	111	115	112	111	115	119	123	124	94
Textiles Textiles	102	104	103	106	108	110	114	117	115	93
Chemicals, coal, petroleum products Produits chimiques, houillers, pétroliers	106	110	112	115	120	124	129	131	130	94
Basic metals Métaux de base	103	107	110	113	118	121	125	125	122	94
Metal products Produits métalliques	111	118	126	135	144	152	160	164	165	100
Electricity and steam [4] **Electricité et vapeur [4]**	**106**	**109**	**116**	**120**	**123**	**129**	**132**	**133**	**135**	**100**
Ukraine Ukraine										
Total industry [2-4] **Total, industrie [2-4]**	**106**	**110**	**115**	**119**	**124**	...	...	...	...	...

Source:
Industrial statistics database of the Statistical Office of the
United Nations Secretariat.

Source:
Base de données pour les statistiques du Bureau de statistique
du Secrétariat de l'ONU.

¶ All data shown which pertain to Germany prior to 3 October 1990
are indicated separately for the Federal Republic of Germany and
the former German Democratic Republic based on their respective
territories at the time indicated. Where data for united Germany
(3 October 1990 and thereafter) are not available, available data
are shown separately under the designations Federal Republic
of Germany and former German Democratic Republic and pertain
to the territorial boundaries prior to 3 October 1990. For detailed
explanatory notes on data pertaining to Germany, see Annex I -
Country Nomenclature.

¶ Toutes les données se rapportant à l'Allemagne avant
le 3 octobre 1990 figurent dans deux rubriques séparées basées sur les
territoires respectifs de la République fédérale d'Allemagne et
l'ancienne République démocratique allemande selon la période
indiquée. En l'absence de données pour l'Allemagne unifiée
(à compter du 3 octobre 1990), les données disponibles sont fournies
séparément sous les rubriques République fédérale d'Allemagne et
ancienne République démocratique allemande et se rapportent aux
limites territoriales antérieures au 3 octobre 1990. Pour les notes
explicatives en détail sur les données concernant l'Allemagne, voir
Annexe I - Nomenclature des pays.

30
Index numbers of industrial production [*cont.*]
Indices de la production industrielle [*suite*]
1980=100

1 Percentage ratio of the index for January-June 1991 to that for January-June 1990.
2 Calculated by the Statistical Office of the United Nations.
3 Excluding basic metals.
4 Including clothing and footwear.
5 Excluding coal, stone quarrying, clay and sand pits.
6 Excluding wearing apparel except footwear, furniture, printing and publishing, petroleum refineries and some miscellaneous manufacturing products.
7 Including coal mining and petroleum refineries.
8 Including food, beverages, tobacco, textiles, shoes, chemicals, non-metallic mineral products, beds, mattresses and some metal products only.
9 Including rubber and plastic products.
10 Including non-ferrous metal basic industries.
11 Excluding non-ferrous metal basic industries.
12 Index numbers of gross domestic product (at constant prices of 1966).
13 Including oil and sugar.
14 Index numbers of gross domestic product (at constant prices of 1962).
15 Index numbers of gross domestic product (at constant prices of 1958).
16 Including construction.
17 Excluding furniture.
18 Index numbers of gross domestic product (at constant prices of 1978/79).
19 Index numbers of gross domestic product (at constant prices of 1970).
20 Excluding leather and leather products, wood products, furniture, printing and publishing and miscellaneous manufacturing industries.
21 Manufacture of fabricated metal products, except machinery and equipment, is included in basic metals.
22 Excluding the extraction of natural gas.
23 Including the smelting and refining of copper.
24 Excluding slaughtering, sawmills and the smelting and refining of copper.
25 Including mining.
26 Excluding footwear: including cold storage and production of motion pictures.
27 Figures relate to 12 months beginning 1 July of year stated.
28 Excluding coal, petroleum products and basic metals.
29 Figures relate to 12 months beginning 21 March of year stated.
30 Excluding petroleum refineries.
31 Excluding tobacco industry.
32 Extraction of non-metallic minerals only.
33 Excluding printing and publishing.
34 Excluding saltern.
35 Excluding made-up textiles and other wearing apparel, leather, fur and leather products, footwear, furniture and fixtures, printing and publishing, non-electrical machinery and miscellaneous industries.
36 Excluding machinery and transport equipment.
37 Inlcuding magnesite products.
38 Including stone quarrying, clay and sand pits, excluding sawmills, printing and publishing, and coal products.
39 Excluding metal mining.
40 Including logging and fishing.
41 Excluding publishing.
42 Excluding shipbuilding and repairing.
43 Excluding iron foundries.
44 Including iron foundries.
45 Excluding clothing.
46 Stone quarrying, clay and sand pits are included in manufacturing. Coal briquetting is included in mining.
47 Including commercial fishing. Excluding publishing.
48 Including fishing.
49 Stone quarrying, clay and sand pits are included in manufacturing.
50 Including magnesite roasting.

1 Rapport en pourcentage entre l'indice du premier semestre de 1991 et celui du premier semestre de 1990.
2 Calculé par le Bureau de Statistique de l'Organisation des Nations Unies.
3 Non compris les métaux de base.
4 Y compris l'industrie d'habillement et des chaussures.
5 Non compris l'extraction du charbon, l'extraction de la pierre à bâtir, de l'argile et du sable.
6 Non compris l'industrie d'articles d'habillement, à l'exclusion des chaussures, l'industrie de meuble, l'imprimerie et l'édition, raffineries de pétrole et quelques produits de l'industrie manufacturière diverse.
7 Y compris l'extraction du charbon et les raffineries de pétrole.
8 Y compris aliments, boissons, tabac, textiles, chaussures, produits chimiques, produits minéraux non-métalliques, lits, sommiers et quelques produits métalliques seulement.
9 Y compris l'industrie du caoutchouc et articles en matière plastique.
10 Y compris la métallurgie de base de métaux non-ferreux.
11 Non compris la métallurgie de base de métaux non-ferreux.
12 Indices du produit intérieur brut (aux prix constants de 1966).
13 Y compris le pétrole et le sucre.
14 Indices du produit intérieur brut (aux prix constants de 1962).
15 Indices du produit intérieur brut (aux prix constants de 1958).
16 Y compris la construction.
17 Non compris l'industrie du meuble.
18 Indices du produit intérieur brut (aux prix constants de 1978/79).
19 Indices du produit intérieur brut (aux prix constants de 1970).
20 Non compris le cuir et les articles en cuir, les produits de l'industrie du bois, les meubles, l'imprimerie et l'édition et industries manufacturières diverses.
21 Fabrication d'ouvrages en métaux, à l'exclusion des machines et du matériel est comprise dans les industries des métaux de base.
22 Non compris l'extraction du gaz naturel.
23 Y compris la fonderie et l'affinage du cuivre.
24 Non compris l'abattage du bétail, les scieries, la fonderie et l'affinage du cuivre.
25 Y compris les industries extractives.
26 Non compris les chaussures, y compris les entrepôts frigorifiques et la production cinématographique.
27 Les chiffres se rapportent à 12 mois commençant le 1er juillet de l'année indiquée.
28 Non compris les produits dérivés du charbon, des produits pétroliers et des métaux de base.
29 Les chiffres se rapportent à 12 mois commençant le 21 mars de l'année indiquée.
30 Non compris les raffineries de pétrole.
31 Non compris l'industrie du tabac.
32 Extraction des minéraux non-métalliques seulement.
33 Non compris l'imprimerie et l'édition.
34 Non compris les salines.
35 Non compris la confection d'ouvrages en tissue et autres industries d'habillement, l'industrie du cuir, de la fourrure et des articles en cuir, la fabrication des chaussures, l'industrie des meubles, l'imprimerie, l'édition, la construction de machines non électriques et les industries diverses.
36 Non compris la construction de machines et du matériel de transport.
37 Y compris les articles de magnésite.
38 Y compris l'extraction de la pierre à bâtir, de l'argile et du sable, non compris les scieries, l'imprimerie, l'édition et les produits du charbon.
39 Non compris l'extraction des minerais métalliques.
40 Y compris l'exploitation forestière et la pêche.
41 Non compris l'édition.
42 Non compris la construction navale et la réparation des navires.
43 Non compris les fonderies de fer.
44 Y compris les fonderies de fer.
45 Non compris les articles d'habillement.
46 L'extraction de la pierre à bâtir, de l'argile et du sable est comprise dans les industries manufacturières. La fabrication des briquettes de charbon est comprise dans les industries extractives.

30
Index numbers of industrial production [*cont.*]
Indices de la production industrielle [*suite*]
1980=100

51 Including car repairs.
52 Including waterworks and gasworks, excluding publishing.
53 Excluding electricity and gas.
54 Excluding some groups of miscellaneous manufacturing.
55 Excluding paper and paper products and miscellaneous manufacturing industries.
56 Including basic metals.
57 Including plastic products.
58 Including also fishing, film laboratories, repair services and gas production.
59 Including fishing, extraction of salt, Excluding animal fodder.
60 Annual figures are monthly averages.
61 Excluding stone quarrying, clay and sand pits, extraction of salt and chemical minerals.
62 Including fishing, stone quarrying, clay and sand pits, gas works, the extraction of salt and chemical minerals. Excluding publishing.
63 Including fishing and the extraction of salt.
64 Excluding clay and sand pits.
65 These figures do not include an adjustment to correct mixed sales and production data to true production index numbers; consequently, they are not strictly comparable with the United Kingdom's output of production industries' series.
66 Including logging, motion picture production, cleaning and dyeing.
67 Excluding oil and natural gas drilling, prospecting and preparing sites for the extraction of minerals.
68 Including commercial fishing and the processing and cold storage of fish and fish products by factory-type vessels.

47 Y compris la prise de poissons. Non compris l'édition.
48 Y compris la pêche.
49 L'extraction de la pierre à bâtir, de l'argile et du sable est comprise dans les industries manufacturières.
50 Y compris le rôtissage du magnésite.
51 Y compris la réparation des véhicules automobiles.
52 Y compris les usines des eaux et les usines à gaz, non compris l'édition.
53 Non compirs l'électricité et le gaz.
54 Non compris quelques groupes d'industries manufacturières diverses.
55 Non compris le papier et les ouvrages en papier et les industries manufacturières diverses.
56 Y compris les métaux de base.
57 Y compris articles en matière plastique.
58 Y compris également la pêche, les laboratoires de films, les services de réparation et les usines à gaz.
59 Y compris la pêche, l'extraction du sel. Non compris la préparation des fourrages.
60 Les chiffres annuels sont des moyennes mensuelles.
61 Non compris l'extraction de la pierre à bâtir, de l'argile et du sable, l'extraction du sel et l'extraction des minéraux destinés à l'industrie chimique.
62 Y compris la pêche, l'extraction de la pierre à bâtir, de l'argile et du sable, les usines à gaz, l'extraction du sel et des minéraux chimiques. Non compris l'édition.
63 Y compris la pêche et l'extraction du sel.
64 Non compris l'extraction de l'argile et du sable.
65 Ces chiffres ne comprennent pas un ajustement corrigeant les ventes mixtes et les données de production aux indices réels de production. En conséquence, ils ne sont pas strictement comprables avec ceux du Royaume Uni relatifs au rendement de la production des industries.
66 Y compris l'exploitation forestière, la production cinématographique, le nettoyage et la teinture.
67 L'exploitation des puits de pétrole et des puits de gaz naturel, la prospection et la préparation du terrain avant l'extraction des minéraux ne sont pas comprises.
68 Y compris la pêche commerciale, le traitement des poissons et des produits poissonniers et les entrepôts frigorifiques dans les usines flottantes.

Technical Notes, tables 29 and 30

Detailed internationally comparable data on national accounts are compiled and published annually by the Statistical Division, Department of Economic and Social Development, of the United Nations Secretariat. Data on the major national accounts aggregates for countries using the United Nations System of National Accounts are based on the concepts and definitions contained in *A System of National Accounts*. [53] For the countries using the Material Product System (MPS) in the period covered by the present publication, the figures are shown in accordance with the methodology for MPS published by the United Nations. [40] These countries comprise Albania, eastern Europe and former USSR in the developed regions, and Afghanistan, China, Cuba, Korea, Democratic People's Republic, Mongolia and Viet Nam in the developing regions. Brief general notes with regard to the scope of major aggregates of national accounts and balances are given below, supplemented by more specific explanatory notes pertaining to the individual tables.

Definitions of major aggregates

"Gross domestic product in purchasers' values" is the total of the gross expenditures on the final uses of domestic supply of goods and services, valued at purchasers' values. This consists of private final consumption, general government consumption, gross capital formation and exports, less imports of goods and services valued c.i.f.; or the sum of compensation of employees, consumption of fixed assets, operating surplus and indirect taxes (net), of resident producers, and import duties.

The gross domestic product can also be computed as the difference between the producers' value of gross output of resident producers' less the purchasers' value of the intermediate consumption plus import duties.

The major components of disposition of gross domestic product are defined below.

"Final consumption expenditures" consists of (a) private final consumption which includes final consumption of households and of private non-profit institutions serving households and (b) final consumption of general government.

"Private final consumption" covers the outlays of resident households on new durable and non-durable goods and services less their net sales of second-hand goods, scraps and wastes and the value of goods and services produced by private non-profit organizations for own use on current account.

"Final consumption expenditures of general government" covers value of goods and services produced by general government services for own use on current account.

Notes techniques, tableaux 29 et 30

La Division de statistique du Département du développement économique et social du Secrétariat des Nations Unies établit et publie chaque année des données détaillées sur les comptes nationaux qui permettent des comparaisons entre pays. Les données sur les principaux agrégats des comptes nationaux pour les pays utilisant le Système de comptes nationaux des Nations Unies sont fondées sur les concepts et définitions figurant dans *"Le système de comptabilité nationale"* [53]. Pour les pays qui utilisaient le Système du produit matériel (BPM) pendant la période couverte par les présentes publications, les chiffres sont indiqués selon la méthodologie de la BPM publiée par les Nations Unies [40]. Ces pays sont l'Albanie, l'Europe de l'Est et l'ancienne URSS pour les régions développées, et l'Afghanistan, la Chine, Cuba, la Mongolie, la République populaire démocratique de Corée et le Vietnam pour les régions en développement. De brèves notes de caractère général concernant la portée des principaux agrégats des comptes nationaux et des balances sont données ci-dessous et sont complétées par des notes explicatives plus précises sur les différents tableaux.

Définitions des principaux agrégats

Le "produit intérieur brut aux prix d'acquisition" est égal au total des dépenses brutes afférentes aux utilisations finales de l'offre intérieure de biens et de services, à leurs valeurs d'acquisition. Cela comprend la consommation finale privée, la consommation générale des administrations publiques, la formation brute de capital et les exportations, moins les importations de biens et de services à leur valeur c.a.f.; cet agrégat peut également se définir comme la somme de la rémunération des employés, de la consommation de capital fixe, de l'excédent d'exploitation et des impôts indirects (nets) des producteurs résidents, et des droits d'importation.

Le produit intérieur brut peut également se calculer en soustrayant la valeur au prix d'acquisition de la consommation intermédiaire majorée des droits d'importation de la valeur brute au producteur de la production du producteur résident.

Les principaux éléments de l'emploi du produit intérieur brut sont définis ci-après.

Les "dépenses de consommation finale" comprennent : (a) la consommation finale privée des ménages et des institutions privées sans but lucratif au service des ménages, et (b) la consommation finale des administrations publiques.

La "consommation finale du secteur privé" désigne les dépenses consacrées par les ménages résidents à l'achat de biens neufs, durables et non durables, et de services, diminuées des ventes nettes de biens d'occasion, de rebuts et de déchets, et de la valeur des biens et services produits par des institutions privées sans but lucratif pour leur propre usage courant.

"Gross capital formation" is defined to include gross fixed capital formation and increase in stocks.

"Gross fixed capital formation" includes outlays of industries, producers of government services and producers of private non-profit services to households on additions to their stocks of fixed assets, less net sales of similar second-hand and scrapped goods.

"Increase in stocks" is defined as the difference between levels of stocks of materials, supplies, work in progress, finished products, livestock raised for slaughter etc. at the beginning and end of the period, both valued at the approximate average prices ruling over the given period.

"Net exports" is the excess of exports over imports of goods and services.

"Imports of goods and services" include broadly the equivalent of general imports of merchandise as defined in external trade statistics, plus imports of services and direct purchases abroad made by resident households and by the government on current account. Transfer of migrants' household and personal effects and gifs between households are also included. The valuation of imports is c.i.f.

"Exports of goods and services" are defined to be parallel to the definition of imports given above. Exports are, however, valued f.o.b., whereas imports are valued c.i.f.

"Net material product" (NMP) is defined to include the sum of the final use of goods and material services, that is final consumption, net capital formation, losses and the excess of exports over imports; or it can be computed as the sum of the primary incomes of the population and the primary incomes of the enterprises engaged in the material production sphere. The latter includes agriculture, industry, construction, transportation, communication, trade, other activities of material production. Non-material services such as general government, finance, science, education, medical services, housing and personal services are excluded. Net material product can also be calculated as the difference between gross output produced by enterprises of the material sphere and their intermediate material input including consumption of fixed assets at purchasers' prices.

Major classifications

In tables where gross domestic product is classified by kind of economic activity, the categories of the classification correspond to the major divisions of the *International Standard Industrial Classification of All Economic Activities* [46]. In general, the category agriculture includes agriculture, hunting, forestry and fishing; industrial activity includes mining and quarrying, manufacturing, electricity, gas and water; "other" comprises financing, insurance, real estate and business services.

Par "dépenses de consommation finale des administrations publiques", on entend la valeur des biens et services que produisent les administrations publiques pour leur propre usage courant.

La "formation brute de capital" se définit comme la formation brute de capital fixe et l'accroissement des stocks.

La "formation brute de capital fixe" compend l'ensemble des dépenses des branches d'activités marchandes, des branches non marchandes des administrations publiques et des branches non marchandes des institutions privées sans but lucratif au service des ménages, ayant pour but d'ajouter des biens durables neufs à leur capital fixe, déduction faite des ventes de ces branches, nettes de leurs achats de biens analogues, d'occasion ou de rebuts.

La "variation des stocks" se définit comme la différence entre les niveaux des stocks (matières premières, produits semi-ouvrés, travaux en cours, produits finis, bétail élevé pour la boucherie, etc.) au début et à la fin de la période, évalués aux prix moyens approximatifs de la période.

Les "exportations nettes" représentent l'excédent des exportations sur les importations de biens et de services.

Les "importations de biens et de services" désignent en gros l'équivalent des importations générales de marchandises définies dans les statistiques du commerce extérieur, plus les importations de services et les achats directs à l'étranger des ménages résidents et des administrations publiques pour leur propre usage courant. Les importations comprennent également la valeur des services de transport et d'assurances des biens familiaux et personnels importés. Leur valeur est établie en prix c.a.f.

La définition des "exportations de biens et de services" fait pendant à la définition des importations donnée ci-dessus. Toutefois, les valeurs des exportations s'entendent f.o.b., alors que les importations sont évaluées en prix c.a.f.

Le "produit matériel net" (PMN) est égal à la somme des utilisations finales des biens et services matériels, c'est-à-dire de la consommation finale, de la formation nette de capital, des pertes et de l'excédent des exportations sur les importations, ou encore à la somme des revenus primaires de la population et des revenus primaires des entreprises du secteur de la production matérielle. Celui-ci comprend l'agriculture, l'industrie, la construction, les transports, les communications, le commerce et les autres branches de la production matérielle. Les services non matériels tels que ceux fournis par les administrations publiques, les services financiers, scientifiques, éducatifs, médicaux, les services de logement et les services personnels sont exclus. Le produit matériel net peut également se calculer comme la différence entre le produit brut des entreprises du secteur de la production matérielle et leur consommation intermédiaire matérielle, y compris l'usure des immobilisations à leur prix d'acquisition.

Description of tables

Table 29 consists of two parts, Part A contains figures on gross domestic product (GDP) for countries using the SNA and net material product (NMP) for countries using MPS. The figures are shown in national currencies, in final purchasers' values. The data on GDP are shown in terms of concepts and definitions of the SNA, whereas data on NMP are computed on the basis of MPS methodology and for this reason are not directly comparable. There are also some other deviations from international standards. For example, not all countries using MPS allocate transportation of passengers to the material sphere.

Part B contains data on GDP expressed in United States dollars using official exchange rates. In converting data into United States dollars, the annual average of market rates as reported by the International Monetary Fund has been used. For countries with multiple exchange rates, the conversion rate was normally obtained by comparing the implicit price deflators of the country and those of the United States. In certain cases the rate employed for conversion was arrived at after a close examination of the pattern of international trade, production and the level of internal trade.

It should be emphasized that figures in part B of the table must to be treated with caution due to possible deviation of the official exchange rates from the actual purchasing power parities.

Only official data are shown in this table.

Table 30: Detailed descriptions of national practices in the compilation of production index numbers are given in the United Nations *1977 Supplement to the Statistical Yearbook and Monthly Bulletin of Statistics.* [49] Some differences in national practice in compilation, as well as major deviations from ISIC in the scope of the indexes, are indicated in the footnotes to this table.

Principales classifications

Dans les tableaux où le produit intérieur brut est classé par types d'activité économique, les rubriques correspondent aux grandes subdivisions de la *Classification internationale type, par industrie, de toutes les branches d'activité économique* [46]. En général, la rubrique "agriculture" comprend l'agriculture, la chasse, la foresterie et la pêche; la rubrique "activité industrielle" recouvre les industries extractives (mines et carrières), les industries manufacturières, l'électricité, le gaz et l'eau; la rubrique "divers" comprend les banques, les assurances, les affaires immobilières et les services fournis aux entreprises.

Description des tableaux

Le *Tableau 29* est divisé en deux parties. La Partie A donne les chiffres du produit intérieur brut (PIB) des pays utilisant le Système des comptes nationaux et le produit matériel net (PMN) des pays utilisant le Système des balances du produit matériel (BPM). Les chiffre sont indiqués en monnaie nationale, au prix final d'acquisition. Les données relatives au PIB sont indiquées conformément aux concepts et aux définitions du SCN, tandis que les données relatives au PMN sont calculées selon les méthodes du Système BPM, ce qui explique qu'elles ne soient pas directement comparables. Il se produit également certains autres écarts par rapport aux normes internationales. C'est ainsi que les pays qui appliquent le Système BPM n'imputent pas tous le transport de passagers au secteur matériel.

La Partie B contient des données sur le PIB exprimées en dollars des Etats-Unis, en utilisant pour la conversion les taux de change officiels. Pour convertir les données en dollars des Etats-Unis, on s'est servi de la moyenne annuelle des taux du marché donnée par le Fonds monétaire international. Pour les pays à taux de change multiples, on a généralement déterminé le taux de conversion en comparant les déflateurs implicites des prix du pays à ceux des Etats-Unis. Dans certains cas, le taux de conversion a été déterminé après un examen minutieux des structures du commerce international, de la production et du volume du commerce intérieur.

Il convient de noter que les chiffres de la Partie B doivent être interprétés avec prudence, vu l'écart possible entre les taux de change officiels et les parités de pouvoir d'achat réelles.

Ce tableau ne présente que des données officielles.

Tableau 30 : Des descriptions détaillées des pratiques nationales employées pour la compilation des indices de production sont données dans le *Supplément 1977 à l'Annuaire statistique et au Bulletin mensuel de statistique* des Nations Unies [49]. Certaines différences dans la méthode nationale de compilation, ainsi que les écarts importants par rapport à la CITI dans la portée des indices, sont indiqués dans les notes figurant au bas de ce tableau.

31
Money supply
Disponibilités monétaires
Million national currency units, end of period
Millions d'unités monétaires nationales, fin de la période

Country or area and unit Pays ou zone et unité	1982	1983	1984	1985	1986	1987	1988	1989	1990	1991
Afghanistan: afghani Afghanistan : afghani										
Domestic Credit										
Crédit intérieur	32 964	...	66 732	83 082	103 956	155 278	210 073	301 072	447 411	...
Money										
Monnaie	52 973	...	68 638	76 359	85 113	131 419	179 414	251 062	351 025	...
Quasi-Money										
Quasi-monnaie	11 439	...	13 476	14 282	16 678	21 866	26 732	31 604	46 305	...
Algeria: dinar Algérie : dinar										
Domestic Credit										
Crédit intérieur	148 970	185 660	223 772	251 247	278 009	303 797	339 239	366 592	414 021	...
Money										
Monnaie	125 300	152 756	180 433	202 230	204 817	223 906	252 207	250 013	270 082	...
Quasi-Money										
Quasi-monnaie	12 590	13 170	14 284	21 630	22 199	33 990	40 758	58 134	* 72 923	...
Antigua and Barbuda: EC dollar Antigua-et-Barbuda : EC dollar										
Domestic Credit[1]										
Crédit intérieur[1]	197 817	230 437	287 373	341 638	391 755	460 372	524 264	593 821	632 431	...
Money[1]										
Monnaie[1]	41 698	51 095	58 402	68 445	87 367	110 623	128 191	141 092	156 739	...
Quasi-Money[1]										
Quasi-monnaie[1]	149 554	185 709	227 680	251 472	290 758	338 821	371 645	425 031	436 336	...
Argentina: peso Argentine : peso										
Money										
Monnaie	0	0	0	0	1	1	6	231	2 706	...
Quasi-Money										
Quasi-monnaie	0	0	0	1	1	4	21	365	4 490	...
Aruba: florin Aruba : florin										
Domestic Credit										
Crédit intérieur	...	...	...	...	235	348	465	547	615	683
Money										
Monnaie	...	...	...	...	100	158	190	243	264	314
Quasi-Money										
Quasi-monnaie	...	...	...	...	216	286	336	393	481	570
Australia: dollar Australie : dollar										
Domestic Credit[2]										
Crédit intérieur[2]	66 874	76 620	86 837	112 132	133 981	157 040	185 404	260 509	285 368	293 334
Money[2]										
Monnaie[2]	18 032	20 796	22 492	23 298	25 947	31 218	40 470	43 514	46 699	49 983
Quasi-Money[2]										
Quasi-monnaie[2]	48 194	54 196	61 300	75 505	82 305	94 385	111 923	151 346	172 678	170 963
Austria: schilling Autriche : schilling										
Domestic Credit[3]										
Crédit intérieur[3]	1 123	1 216	1 331	1 435	1 561	1 720	1 865	2 031	2 232	2 383
Money[3]										
Monnaie[3]	153	170	176	182	194	213	232	235	247	266
Quasi-Money[3]										
Quasi-monnaie[3]	783	814	871	928	1 012	1 083	1 140	1 234	1 364	1 465
Bahamas: dollar Bahamas : dollar										
Domestic Credit										
Crédit intérieur	633	692	682	714	780	964	1 096	1 240	1 496	1 579
Money										
Monnaie	160	177	188	208	249	278	297	298	328	354
Quasi-Money										
Quasi-monnaie	393	456	483	496	537	628	681	765	891	959
Bahrain: dinar Bahreïn : dinar										
Domestic Credit										
Crédit intérieur	155	209	147	58	86	258	22	−85	−17	...

31
Money supply
Million national currency units, end of period [cont.]
Disponibilités monétaires
Millions d'unités monétaires nationales, fin de la période [suite]

Country or area and unit Pays ou zone et unité	1982	1983	1984	1985	1986	1987	1988	1989	1990	1991
Money 　Monnaie	267	250	239	243	236	247	239	235	259	...
Quasi-Money 　Quasi-monnaie	513	593	588	660	650	722	769	818	672	...
Bangladesh: taka　Bangladesh : taka										
Domestic Credit 　Crédit intérieur	71 000	87 385	112 894	133 536	148 884	160 215	* 179 186	220 400	* 239 631	* 251 581
Money 　Monnaie	23 336	31 636	42 269	45 955	49 996	51 000	* 53 165	60 004	* 65 735	* 70 804
Quasi-Money 　Quasi-monnaie	29 442	42 262	58 314	68 322	82 794	106 694	125 955	152 535	168 474	194 874
Barbados: dollar　Barbade : dollar										
Domestic Credit 　Crédit intérieur	884	965	1 019	1 029	1 106	1 244	1 394	1 511	1 676	1 811
Money 　Monnaie	251	313	306	353	395	467	525	459	526	495
Quasi-Money 　Quasi-monnaie	589	618	696	740	800	899	992	1 075	1 219	1 232
Belgium: franc　Belgique : franc										
Domestic Credit[3] 　Crédit intérieur[3]	2 456	2 807	2 992	3 276	3 580	3 930	4 177	4 764	5 005	...
Money[3] 　Monnaie[3]	856	930	933	962	1 037	1 086	1 146	1 207	1 217	...
Quasi-Money[3] 　Quasi-monnaie[3]	942	1 017	1 102	1 180	1 329	1 495	1 573	1 787	1 900	...
Belize: dollar　Belize : dollar										
Domestic Credit 　Crédit intérieur	160 934	187 386	205 034	206 812	210 112	230 339	208 367	209 185	221 528	306 486
Money 　Monnaie	39 395	42 408	50 010	58 607	71 032	84 303	81 784	99 866	105 991	118 296
Quasi-Money 　Quasi-monnaie	85 674	105 751	104 620	109 125	128 064	157 347	170 328	187 403	224 804	245 600
Benin: CFA franc　Bénin : franc CFA										
Domestic Credit[3] 　Crédit intérieur[3]	116	132	131	151	# 141	137	144	101	112	...
Money[3] 　Monnaie[3]	83	82	89	87	79	62	72	84	105	...
Quasi-Money[3] 　Quasi-monnaie[3]	15	15	21	25	30	35	26	20	29	...
Bhutan: ngultrum　Bhoutan : ngultrum										
Domestic Credit 　Crédit intérieur	...	...	...	118	71	−185	−227	−27	−54	...
Money 　Monnaie	...	...	...	262	277	317	412	546	540	...
Quasi-Money 　Quasi-monnaie	...	...	...	202	222	236	312	430	539	...
Bolivia: boliviano　Bolivie : boliviano										
Domestic Credit 　Crédit intérieur	164[1]	445[1]	4 881[1]	# 136	327	# 1 280	1 937	3 202	4 749	6 267
Money 　Monnaie	58[1]	178[1]	3 370[1]	# 198	369	# 516	698	715	997	1 447
Quasi-Money 　Quasi-monnaie	39[1]	88[1]	684[1]	# 89	448	# 1 060	1 329	1 762	2 787	4 249
Botswana: pula　Botswana : pula										
Domestic Credit 　Crédit intérieur	43	−8	−130	−419	−997	−1 532	−1 981	−2 533	−3 250	...
Money 　Monnaie	127	137	151	188	243	312	407	507	586	614
Quasi-Money 　Quasi-monnaie	118	179	218	369	363	701	822	1 291	960	1 575

31
Money supply
Million national currency units, end of period [*cont.*]
Disponibilités monétaires
Millions d'unités monétaires nationales, fin de la période [*suite*]

Country or area and unit Pays ou zone et unité	1982	1983	1984	1985	1986	1987	1988	1989	1990	1991
Brazil: cruzeiro Brésil : cruzeiro										
Domestic Credit										
Crédit intérieur	14	39	108	432	1 675	7 299	...	...	...	...
Money										
Monnaie	4	8	24	106	457	1 440	...	...	...	...
Quasi-Money										
Quasi-monnaie	1	5	22	97	275	854	...	...	...	...
Burkina Faso: CFA franc Burkina Faso : franc CFA										
Domestic Credit[3]										
Crédit intérieur[3]	78	73	72	78	78	72	81	102	* 103	...
Money[3]										
Monnaie[3]	54	60	67	70	85	91	101	105	* 104	...
Quasi-Money[3]										
Quasi-monnaie[3]	17	20	26	24	29	36	48	49	50	...
Burundi: franc Burundi : franc										
Domestic Credit										
Crédit intérieur	19 907	23 915	25 544	28 655	27 860	29 297	34 891	34 457	37 954	...
Money										
Monnaie	10 827	13 845	14 762	18 453	19 909	19 786	20 507	21 112	23 233	...
Quasi-Money										
Quasi-monnaie	4 478	5 593	5 444	5 730	4 438	5 067	7 734	10 856	12 148	...
Cameroon: CFA franc Cameroun : franc CFA										
Domestic Credit[3]										
Crédit intérieur[3]	563	701	732	836	997	939	# 975	1 062	1 045	...
Money[3]										
Monnaie[3]	298	377	411	427	448	387	# 422	452	419	...
Quasi-Money[3]										
Quasi-monnaie[3]	185	235	325	438	383	291	# 303	317	338	...
Canada: dollar Canada : dollar										
Domestic Credit[3]										
Crédit intérieur[3]	214	213	230	244	258	275	302	337	367	...
Money[3]										
Monnaie[3]	40	44	52	69	80	85	90	95	96	...
Quasi-Money[3]										
Quasi-monnaie[3]	146	140	143	137	142	156	177	208	230	242
Cape Verde: escudo Cap-Vert : escudo										
Domestic Credit										
Crédit intérieur	2 194	2 768	3 182	3 874	4 726	5 381	5 858	7 112	8 182	...
Money										
Monnaie	3 228	3 956	4 202	4 843	5 621	5 710	6 038	6 411	6 926	...
Quasi-Money										
Quasi-monnaie	616	628	915	1 297	1 966	2 527	3 592	4 659	5 754	...
Central African Rep.: CFA franc Rép. centrafricaine : franc CFA										
Domestic Credit[3]										
Crédit intérieur[3]	45	49	47	52	54	48	# 49	54	52	...
Money[3]										
Monnaie[3]	41	45	48	52	52	54	# 50	57	55	...
Quasi-Money[3]										
Quasi-monnaie[3]	3	3	4	5	6	6	# 7	8	7	...
Chad: CFA franc Tchad : franc CFA										
Domestic Credit[3]										
Crédit intérieur[3]	42	43	54	67	82	78	71	61	48	...
Money[3]										
Monnaie[3]	33	41	65	68	69	71	62	66	66	...
Quasi-Money[3]										
Quasi-monnaie[3]	1	2	3	4	4	5	4	4	3	...

31
Money supply
Million national currency units, end of period [*cont.*]
Disponibilités monétaires
Millions d'unités monétaires nationales, fin de la période [*suite*]

Country or area and unit Pays ou zone et unité	1982	1983	1984	1985	1986	1987	1988	1989	1990	1991
Chile: peso Chili : peso										
Domestic Credit[3]										
Crédit intérieur[3]	1 089	1 361	2 057	3 063	3 713	4 454	5 003	5 574	6 708	...
Money[3]										
Monnaie[3]	92	113	130	148	215	229	386	442	518	...
Quasi-Money[3]										
Quasi-monnaie[3]	392	476	590	913	1 114	1 565	1 895	2 549	3 178	...
China: yuan renminbi Chine : yuan renminbi										
Domestic Credit[3]										
Crédit intérieur[3]	305	344	451	593	795	973	1 156	1 353	1 675	...
Money[3]										
Monnaie[3]	149	175	245	302	386	457	549	583	701	...
Quasi-Money[3]										
Quasi-monnaie[3]	78	96	115	186	249	338	411	556	767	...
Colombia: peso Colombie : peso										
Domestic Credit[3]										
Crédit intérieur[3]	499	686	997	1 218	...	1 891	2 695	...	...	...
Money[3]										
Monnaie[3]	321	397	492	545	...	1 020	1 282	...	...	...
Quasi-Money[3]										
Quasi-monnaie[3]	204	262	324	437	...	796	916	...	...	...
Comoros: franc Comores : franc										
Domestic Credit										
Crédit intérieur	4 993	6 120	8 399	5 820	5 984	6 516	8 163	10 067	11 918	12 873
Money										
Monnaie	5 202	7 292	6 254	7 177	7 232	7 812	8 879	10 256	10 888	10 080
Quasi-Money										
Quasi-monnaie	1 168	1 229	959	1 015	1 432	3 035	3 559	4 608	4 574	5 829
Congo: CFA franc Congo : franc CFA										
Domestic Credit[3]										
Crédit intérieur[3]	151	170	218	231	237	217	# 204	231	222	...
Money[3]										
Monnaie[3]	98	91	101	112	98	102	# 98	97	121	...
Quasi-Money[3]										
Quasi-monnaie[3]	18	23	21	36	33	36	# 39	45	48	...
Costa Rica: colon Costa Rica : colon										
Domestic Credit										
Crédit intérieur	31 180	56 421	65 944	70 977	87 888	104 720	120 939	130 084	157 008	...
Money										
Monnaie	18 448	25 619	30 132	32 439	42 487	42 611	65 267	63 975	66 484	...
Quasi-Money										
Quasi-monnaie	22 545	30 524	35 632	43 563	49 671	64 535	84 959	110 859	156 464	...
Côte d'Ivoire: CFA franc Côte d'Ivoire : franc CFA										
Domestic Credit[3]										
Crédit intérieur[3]	1 120	1 336	1 352	1 320	1 360	1 384	1 457	1 374	1 307	...
Money[3]										
Monnaie[3]	460	488	575	620	636	599	578	511	526	...
Quasi-Money[3]										
Quasi-monnaie[3]	200	204	251	319	327	331	366	357	319	...
Cyprus: pound Chypre : livre										
Domestic Credit										
Crédit intérieur	637	724	826	958	1 039	1 234	# 1 497	1 559	1 969	2 281
Money										
Monnaie	219	249	259	285	283	314	359	385	429	462
Quasi-Money										
Quasi-monnaie	491	543	640	707	813	924	# 1 103	1 301	1 540	1 782

31
Money supply
Million national currency units, end of period [cont.]
          Disponibilités monétaires
          Millions d'unités monétaires nationales, fin de la période [suite]

Country or area and unit Pays ou zone et unité	1982	1983	1984	1985	1986	1987	1988	1989	1990	1991
Czechoslovakia: koruna Tchécoslovaquie : couronne										
Domestic Credit[3]										
Crédit intérieur[3]	406	421	438	449	468	489	521	586	640	750
Money[3]										
Monnaie[3]	237	246	258	260	261	271	309	318	291	371
Quasi-Money[3]										
Quasi-monnaie[3]	127	141	156	171	186	204	220	237	260	326
Denmark: krone Danemark : couronne										
Domestic Credit[3]										
Crédit intérieur[3]	230	281	351	399	464	# 443	469	519	520	566
Money[3]										
Monnaie[3]	92	113	128	156	168	# 188	225	226	244	# 258
Quasi-Money[3]										
Quasi-monnaie[3]	115	151	182	210	233	# 239	226	* 231	242	# 248
Djibouti: franc Djibouti : franc										
Domestic Credit										
Crédit intérieur	...	...	26 565	29 975	34 061	31 603	32 485	33 466	32 641	33 633
Money										
Monnaie	...	...	19 934	20 533	23 290	25 331	27 066	25 184	27 362	31 649
Quasi-Money										
Quasi-monnaie	...	...	18 264	24 694	27 050	26 346	28 391	31 619	31 462	29 554
Dominica: EC dollar Dominique : EC dollar										
Domestic Credit[1]										
Crédit intérieur[1]	121 545	139 762	147 624	161 223	160 340	152 647	132 363	187 207	267 838	...
Money[1]										
Monnaie[1]	25 405	25 404	32 402	31 222	37 028	58 666	57 192	56 081	69 918	...
Quasi-Money[1]										
Quasi-monnaie[1]	70 855	80 305	90 187	96 977	110 239	132 179	127 063	150 047	183 649	...
Dominican Republic: peso Rép. dominicaine : peso										
Domestic Credit										
Crédit intérieur	2 773	3 234	3 502	3 844	4 873	5 881	7 077	9 704	11 883	...
Money										
Monnaie	731	781	1 160	1 355	1 989	2 609	4 168	5 285	7 341	9 250
Quasi-Money										
Quasi-monnaie	804	895	1 010	1 235	2 297	2 324	3 338	4 613	6 290	9 636
Ecuador: sucre Equateur : sucre										
Domestic Credit[3]										
Crédit intérieur[3]	107	170	217	358	419	545	956	1 100	1 224	...
Money[3]										
Monnaie[3]	73	95	129	153	183	246	373	540	853	...
Quasi-Money[3]										
Quasi-monnaie[3]	20	23	40	64	85	138	249	343	504	...
Egypt: £ Egypte : £										
Domestic Credit										
Crédit intérieur	22 008	26 104	31 553	37 674	44 747	52 476	63 052	78 799	99 501	107 101
Money										
Monnaie	9 552	10 933	12 443	14 696	15 973	18 241	20 579	22 471	26 205	28 338
Quasi-Money										
Quasi-monnaie	8 240	10 885	13 486	15 980	21 129	26 637	33 970	41 623	56 303	70 127
El Salvador: colon El Salvador : colon										
Domestic Credit										
Crédit intérieur	5 344	5 341	5 818	6 966	7 457	8 290	9 110	10 859	11 667	14 065
Money										
Monnaie	1 633	1 596	1 830	2 306	2 758	2 762	2 996	3 370	4 153	4 872
Quasi-Money										
Quasi-monnaie	1 600	1 979	2 405	3 068	4 147	4 609	5 223	5 760	7 867	9 684

31
Money supply
Million national currency units, end of period [*cont.*]
Disponibilités monétaires
Millions d'unités monétaires nationales, fin de la période [*suite*]

Country or area and unit Pays ou zone et unité	1982	1983	1984	1985	1986	1987	1988	1989	1990	1991
Equatorial Guinea: CFA franc Guinée équatoriale : franc CFA										
Domestic Credit[3]										
Crédit intérieur [3]	...	...	...	11	13	11	# 12	12	11	...
Money[3]										
Monnaie[3]	...	...	...	8	10	9	# 5	7	3	...
Quasi-Money [3]										
Quasi-monnaie[3]	...	...	...	1	1	0	0	1	1	...
Ethiopia: birr Ethiopie : birr										
Domestic Credit										
Crédit intérieur	3 565	4 118	4 514	4 900	5 379	5 981	6 629	7 510	8 960	...
Money										
Monnaie	1 892	2 142	2 309	2 702	3 273	3 341	3 722	4 322	5 273	...
Quasi-Money										
Quasi-monnaie	797	1 056	1 140	1 292	1 195	1 413	1 570	1 727	1 894	...
Fiji: dollar Fidji : dollar										
Domestic Credit										
Crédit intérieur	346	394	413	442	472	541	501	662	781	...
Money										
Monnaie	124	135	136	138	172	168	273	264	265	...
Quasi-Money										
Quasi-monnaie	258	298	344	352	403	430	448	526	723	...
Finland: markka Finlande : markka										
Domestic Credit										
Crédit intérieur	123 365	147 606	170 100	199 670	222 610	261 679	337 909	388 564	434 574	...
Money										
Monnaie	19 917	21 427	24 945	27 694	27 838	30 342	35 921	41 444	44 428	...
Quasi-Money										
Quasi-monnaie	87 140	99 912	115 300	137 895	150 895	169 911	210 458	228 315	238 901	151 653
France: franc France : franc										
Domestic Credit[3]										
Crédit intérieur[3]	3 186	3 597	3 991	4 329	4 787	5 305	5 877	6 349	6 971	...
Money[3]										
Monnaie[3]	985	1 108	1 219	1 312	1 407	1 471	1 532	1 633	1 702	...
Quasi-Money [3]										
Quasi-monnaie[3]	1 593	1 753	1 888	2 006	2 164	2 333	2 484	2 479	2 509	...
Gabon: CFA franc Gabon : franc CFA										
Domestic Credit[3]										
Crédit intérieur [3]	150	207	229	308	362	391	363	378	325	...
Money[3]										
Monnaie[3]	126	143	168	177	152	133	163	171	181	...
Quasi-Money [3]										
Quasi-monnaie[3]	74	93	105	129	123	105	100	108	107	...
Gambia: dalasi Gambie : dalasi										
Domestic Credit										
Crédit intérieur	291	369	432	547	388	206	242	216	85	−0
Money										
Monnaie	87	100	100	162	166	198	214	260	296	394
Quasi-Money										
Quasi-monnaie	40	61	70	95	109	146	181	216	220	255
Germany† · Allemagne†										
F. R. Germany: deutsche mark R. f. Allemagne : deutsche mark										
Domestic Credit[3]										
Crédit intérieur[3]	1 703	1 817	1 927	2 046	2 121	2 205	2 341	2 473	2 858	3 141
Money[3]										
Monnaie[3]	257	278	295	315	340	366	408	432	551	577
Quasi-Money[3]										
Quasi-monnaie[3]	634	664	700	760	804	847	875	916	1 047	1 125

31
Money supply
Million national currency units, end of period [*cont.*]
Disponibilités monétaires
Millions d'unités monétaires nationales, fin de la période [*suite*]

Country or area and unit Pays ou zone et unité	1982	1983	1984	1985	1986	1987	1988	1989	1990	1991
Ghana: cedi Ghana : cedi										
Domestic Credit[3]										
Crédit intérieur[3]	19	32	49	78	119	205	196	243	253	...
Money[3]										
Monnaie[3]	11	17	27	38	55	84	122	186	206	...
Quasi-Money[3]										
Quasi-monnaie[3]	4	4	5	8	14	22	33	53	65	...
Greece: drachma Grèce : drachme										
Domestic Credit[3]										
Crédit intérieur[3]	1 878	2 225	2 672	3 343	3 963	5 715	6 704	8 250	9 582	...
Money[3]										
Monnaie[3]	459	526	631	743	896	1 058	1 206	1 486	1 848	...
Quasi-Money[3]										
Quasi-monnaie[3]	1 118	1 383	1 758	2 211	2 641	3 102	3 841	4 705	5 229	...
Grenada: EC dollar Grenade : EC dollar										
Domestic Credit[1]										
Crédit intérieur[1]	151 111	145 420	157 127	184 628	224 811	257 912	284 955	344 687	347 484	...
Money[1]										
Monnaie[1]	59 574	58 100	50 708	54 120	69 336	74 307	83 002	87 993	91 092	...
Quasi-Money[1]										
Quasi-monnaie[1]	87 033	88 863	97 709	123 093	157 611	179 238	216 631	234 260	263 324	...
Guatemala: quetzal Guatemala : quetzal										
Domestic Credit										
Crédit intérieur	2 629	3 004	3 520	3 829	3 560	3 760	4 060	4 552	5 462	...
Money										
Monnaie	787	834	869	1 347	1 608	1 766	2 019	2 438	3 242	...
Quasi-Money										
Quasi-monnaie	1 404	1 321	1 530	1 846	2 267	2 402	2 976	3 363	4 055	...
Guyana: dollar Guyana : dollar										
Domestic Credit										
Crédit intérieur	2 467	3 169	4 007	5 351	6 182	8 168	13 125	11 995	13 657	...
Money										
Monnaie	436	509	619	740	881	1 333	2 095	2 923	4 262	...
Quasi-Money										
Quasi-monnaie	818	1 009	1 176	1 400	1 608	2 488	3 312	5 107	8 030	...
Haiti: gourde Haïti : gourde										
Domestic Credit										
Crédit intérieur	3 153	3 699	3 725	2 633	3 720	4 003	...	4 974		...
Money										
Monnaie	1 164	1 175	1 388	1 076	1 790	2 098	1 635	2 634	...	...
Quasi-Money										
Quasi-monnaie	1 083	1 173	1 218	...	1 487	1 596	1 678	1 925	...	...
Honduras: lempira Honduras : lempira										
Domestic Credit										
Crédit intérieur	1 943	2 346	2 653	2 890	3 131	3 720	4 208	4 740	5 293	5 393
Money										
Monnaie	717	815	846	856	955	1 119	1 253	1 524	1 970	2 277
Quasi-Money										
Quasi-monnaie	761	919	1 065	1 015	1 095	1 377	1 617	1 778	2 046	2 324
Hungary: forint Hongrie : forint										
Domestic Credit[3]										
Crédit intérieur[3]	861	892	934	993	1 113	1 252	1 330	1 538	1 704	...
Money[3]										
Monnaie[3]	188	192	201	240	266	307	302	355	449	...
Quasi-Money[3]										
Quasi-monnaie[3]	230	235	248	259	276	288	311	351	464	...

31
Money supply
Million national currency units, end of period [cont.]
Disponibilités monétaires
Millions d'unités monétaires nationales, fin de la période [suite]

Country or area and unit Pays ou zone et unité	1982	1983	1984	1985	1986	1987	1988	1989	1990	1991
Iceland: krona Islande : couronne										
Domestic Credit Crédit intérieur	14 340	26 959	39 476	53 660	61 742	89 534	124 328	# 165 017	179 184	...
Money Monnaie	2 076	3 698	5 282	6 722	9 674	12 883	15 177	# 19 673	24 816	...
Quasi-Money Quasi-monnaie	8 780	15 750	20 749	31 731	42 328	57 540	72 127	91 528	103 040	
India: rupee Inde : roupie										
Domestic Credit[3] Crédit intérieur[3]	820	953	1 150	1 351	1 603	1 844	2 175	2 531	2 836	3 238
Money[3] Monnaie[3]	274	309	366	412	479	543	633	747	836	1 022
Quasi-Money[3] Quasi-monnaie[3]	424	507	597	712	848	1 000	1 192	1 365	1 537	1 776
Indonesia: rupiah Indonésie : rupiah										
Domestic Credit[3] Crédit intérieur[3]	8 434	11 069	11 978	14 799	20 329	28 739	40 835	60 564	95 896	114 052
Money[3] Monnaie[3]	7 120	7 576	8 581	10 124	11 631	12 705	14 392	20 559	23 819	26 693
Quasi-Money[3] Quasi-monnaie[3]	3 954	7 093	9 356	13 054	15 984	21 200	27 681	37 967	60 811	72 717
Iran, Islamic Rep. of: rial Iran, Rép. islamique d' : rial										
Domestic Credit[3] Crédit intérieur[3]	7 134	7 144	...	...	10 849	13 414	15 807	...	22 751	...
Money[3] Monnaie[3]	3 293	3 922	...	...	5 509	6 462	7 118	...	9 729	...
Quasi-Money[3] Quasi-monnaie[3]	2 591	3 061	...	...	4 339	5 283	6 988	...	10 663	...
Ireland: pound Irlande : livre										
Domestic Credit Crédit intérieur	5 923	6 589	7 420	7 700	8 696	8 707	9 741	10 642	11 448	11 641
Money Monnaie	1 838	2 048	2 245	2 288	2 382	2 640	2 826	3 112	3 346	3 390
Quasi-Money Quasi-monnaie	3 696	3 855	4 188	4 464	5 459	5 892	6 258	7 245	7 934	9 181
Israel: new shekel Israël : nouveau shekel										
Domestic Credit Crédit intérieur	947	2 826	17 722	49 825	57 822	69 727	85 188	97 589	112 422	129 411
Money Monnaie	28	68	305	1 051	2 238	3 346	3 723	5 374	7 023	7 981
Quasi-Money Quasi-monnaie	515	1 603	9 885	26 310	30 801	38 691	47 667	56 842	67 280	79 557
Italy: lira Italie : lire										
Domestic Credit[4] Crédit intérieur[4]	450	504	562	619	671	704	746	824	891	...
Money[4] Monnaie[4]	212	239	270	298	331	357	386	433	467	514
Quasi-Money[4] Quasi-monnaie[4]	186	208	232	260	280	306	335	372	418	466
Jamaica: dollar Jamaïque : dollar										
Domestic Credit Crédit intérieur	4 605	6 012	6 736	6 482	7 597	6 808	7 006	7 616	7 227	7 285
Money Monnaie	876	1 066	1 319	1 520	2 140	2 252	3 445	6 153	4 016	7 818
Quasi-Money Quasi-monnaie	1 897	2 461	2 878	3 718	4 548	5 275	6 513	7 460	8 875	11 698

31
Money supply
Million national currency units, end of period [*cont.*]
Disponibilités monétaires
Millions d'unités monétaires nationales, fin de la période [*suite*]

Country or area and unit Pays ou zone et unité	1982	1983	1984	1985	1986	1987	1988	1989	1990	1991
Japan: yen Japon : yen										
Domestic Credit[4]										
Crédit intérieur[4]	298	320	348	379	413	452	500	557	609	626
Money[4]										
Monnaie[4]	81	81	86	89	98	103	112	114	120	131
Quasi-Money[4]										
Quasi-monnaie[4]	166	183	195	218	237	270	298	343	375	376
Jordan: dinar Jordanie : dinar										
Domestic Credit										
Crédit intérieur	1 162	1 360	1 588	1 743	1 834	2 191	2 637	2 781	2 935	...
Money										
Monnaie	788	869	878	848	897	980	1 167	1 302	1 425	...
Quasi-Money										
Quasi-monnaie	619	748	884	1 031	1 182	1 424	1 611	1 935	2 080	...
Kenya: shilling Kenya : shilling										
Domestic Credit										
Crédit intérieur	26 058	25 944	28 628	32 174	41 025	49 139	52 529	54 686	66 589	77 992
Money										
Monnaie	10 635	11 473	13 095	12 923	17 522	18 916	19 160	21 647	27 529	31 667
Quasi-Money										
Quasi-monnaie	10 735	10 953	12 198	13 976	18 172	20 750	23 696	26 746	30 571	37 804
Korea, Republic of: won Corée, République de : won										
Domestic Credit[3]										
Crédit intérieur[3]	29 005	33 642	38 042	44 789	51 318	59 693	66 612	81 807	102 108	124 974
Money[3]										
Monnaie[3]	5 799	6 783	6 821	7 558	8 809	10 107	12 152	14 328	15 905	21 752
Quasi-Money[3]										
Quasi-monnaie[3]	14 105	16 155	17 885	21 008	25 024	30 172	36 787	44 309	52 802	61 994
Kuwait: dinar Koweït : dinar										
Money										
Monnaie	1 248	1 180	968	944	979	1 036	958	939	...	1 230
Quasi-Money										
Quasi-monnaie	2 935	3 188	3 508	3 491	3 568	3 726	4 110	4 338	...	4 088
Lebanon: pound Liban : livre										
Domestic Credit[3]										
Crédit intérieur[3]	35	50	67	92	180	602	948	1 478	2 617	3 253
Money[3]										
Monnaie[3]	11	13	14	20	30	69	183	287	450	689
Quasi-Money[3]										
Quasi-monnaie[3]	38	49	63	99	294	1 402	1 992	2 178	3 372	4 810
Lesotho: loti Lesotho : maloti										
Domestic Credit										
Crédit intérieur	129	137	143	187	241	300	418	502	437	414
Money										
Monnaie	77	85	103	133	155	157	220	245	264	313
Quasi-Money										
Quasi-monnaie	108	132	147	178	197	230	270	311	338	345
Liberia: dollar Libéria : dollar										
Domestic Credit[1]										
Crédit intérieur[1]	385 352	452 025	469 497	539 345	584 350	677 005	844 066	974 360	...	...
Quasi-Money[1]										
Quasi-monnaie[1]	65 040	69 131	54 961	57 041	53 937	57 484	67 469	75 427	...	...
Libyan Arab Jamah.: dinar Jamah. arabe libyenne : dinar										
Domestic Credit										
Crédit intérieur	2 707	2 848	3 345	3 430	3 138	3 325	3 688	6 745	7 230	7 580
Money										
Monnaie	3 232	2 885	2 711	3 492	3 041	3 439	3 012	3 521	4 452	4 297
Quasi-Money										
Quasi-monnaie	1 073	1 242	1 464	1 561	1 770	1 635	1 743	1 635	1 749	1 691

31
Money supply
Million national currency units, end of period [cont.]
Disponibilités monétaires
Millions d'unités monétaires nationales, fin de la période [suite]

Country or area and unit Pays ou zone et unité	1982	1983	1984	1985	1986	1987	1988	1989	1990	1991
Madagascar: franc Madagascar : franc										
Domestic Credit[3] Crédit intérieur[3]	498	607	727	812	927	1 143	1 175	1 144	1 245	...
Money[3] Monnaie[3]	208	193	240	239	290	372	455	598	574	...
Quasi-Money[3] Quasi-monnaie[3]	21	15	18	54	78	63	78	114	169	...
Malawi: kwacha Malawi : kwacha										
Domestic Credit Crédit intérieur	506	593	606	685	835	847	713	851	877	1 095
Money Monnaie	131	128	154	167	221	298	436	453	482	634
Quasi-Money Quasi-monnaie	146	165	234	218	268	370	377	409	475	567
Malaysia: ringgit Malaisie : ringgit										
Domestic Credit Crédit intérieur	36 726	42 623	50 501	53 444	58 213	60 497	63 694	76 375	90 103	...
Money Monnaie	12 477	13 432	13 357	14 132	14 523	16 375	18 730	21 978	25 405	...
Quasi-Money Quasi-monnaie	25 141	27 731	32 502	34 788	39 779	39 977	41 405	47 315	51 256	...
Maldives: rufiyaa Maldives : rufiyaa										
Domestic Credit Crédit intérieur	275	328	394	457	454	440	362	485	607	686
Money Monnaie	93	89	114	172	193	204	230	270	318	404
Quasi-Money Quasi-monnaie	48	49	69	94	135	162	193	231	265	302
Mali: CFA franc Mali : franc CFA										
Domestic Credit[3] Crédit intérieur[3]	171	193	164	186	202	193	152	153	127	...
Money[3] Monnaie[3]	68	81	107	114	118	109	114	110	98	...
Quasi-Money[3] Quasi-monnaie[3]	4	6	10	14	18	21	26	32	37	...
Malta: lira Malte : lire										
Domestic Credit Crédit intérieur	119	134	171	175	231	290	336	431	557	668
Money Monnaie	306	325	326	321	323	355	364	369	385	407
Quasi-Money Quasi-monnaie	239	263	304	341	378	416	474	555	643	723
Mauritania: ouguiya Mauritanie : ouguiya										
Domestic Credit Crédit intérieur	16 441	18 085	19 703	20 212	21 428	23 574	25 022	32 224	38 651	...
Money Monnaie	7 135	8 090	9 641	12 173	11 392	13 397	14 032	15 062	17 034	...
Quasi-Money Quasi-monnaie	2 437	1 995	1 682	1 674	3 471	4 244	3 971	3 562	5 769	...
Mauritius: rupee Maurice : roupie										
Domestic Credit Crédit intérieur	6 897	8 051	9 270	10 082	10 947	11 620	13 620	15 279	17 709	22 171
Money Monnaie	1 741	1 804	2 050	2 039	2 432	3 304	3 820	4 511	5 578	6 677
Quasi-Money Quasi-monnaie	3 185	3 622	4 145	6 109	8 085	10 351	13 748	15 765	18 990	23 278

31
Money supply
Million national currency units, end of period [*cont.*]
Disponibilités monétaires
Millions d'unités monétaires nationales, fin de la période [*suite*]

Country or area and unit Pays ou zone et unité	1982	1983	1984	1985	1986	1987	1988	1989	1990	1991
Mexico: peso Mexique : peso										
Domestic Credit[3]										
Crédit intérieur[3]	4 180	6 528	9 879	16 767	33 815	69 118	115 328	167 930	237 896	...
Money[3]										
Monnaie[3]	1 031	1 447	2 315	3 462	5 790	12 627	21 191	29 087	47 439	...
Quasi-Money[3]										
Quasi-monnaie[3]	2 024	3 498	6 017	8 474	15 509	40 029	22 257	64 731	117 513	...
Morocco: dirham Maroc : dirham										
Domestic Credit										
Crédit intérieur	43 719	52 324	56 759	64 946	72 114	77 243	84 026	92 644	94 787	111 557
Money										
Monnaie	30 064	34 197	36 779	42 820	50 029	54 489	62 032	68 395	80 914	91 946
Quasi-Money										
Quasi-monnaie	8 691	11 412	13 491	16 592	18 819	21 056	24 627	28 458	33 597	41 938
Myanmar: kyat Myanmar : kyat										
Domestic Credit										
Crédit intérieur	19 737	23 986	27 326	31 141	41 484	41 819	46 829	37 693	49 699	...
Money										
Monnaie	9 803	11 067	12 777	11 551	16 337	9 474	15 668	21 317	30 587	...
Quasi-Money										
Quasi-monnaie	3 752	4 603	5 659	6 558	7 447	8 446	7 615	9 467	11 789	...
Nepal: rupee Népal : roupie										
Domestic Credit										
Crédit intérieur	6 693	8 799	10 461	13 629	16 220	18 075	21 884	27 005	30 550	...
Money										
Monnaie	3 705	4 366	4 942	5 616	6 951	8 682	9 826	11 720	14 205	...
Quasi-Money										
Quasi-monnaie	4 283	5 228	5 899	7 398	8 592	10 343	13 394	16 386	19 098	...
Netherlands: guilder Pays-Bas : florin										
Domestic Credit[3]										
Crédit intérieur[3]	302	317	334	351	378	394	502	528	* 554	584
Money[3]										
Monnaie[3]	72	80	85	91	97	104	111	119	* 124	130
Quasi-Money[3]										
Quasi-monnaie[3]	196	202	215	229	240	240	257	290	311	325
Netherlands Antilles: guilder Antilles néerlandaises : florin										
Domestic Credit										
Crédit intérieur	1 181	1 303	1 325	1 244	939	1 148	1 296	1 463	1 677	...
Money										
Monnaie	475	493	467	528	514	569	578	569	643	...
Quasi-Money										
Quasi-monnaie	884	943	920	893	820	883	1 064	1 071	1 163	...
New Zealand: dollar Nouvelle-Zélande : dollar										
Domestic Credit										
Crédit intérieur	8 035	8 522	8 081	13 349	17 457	21 996	# 23 930	36 238	56 120	...
Money										
Monnaie	3 030	3 426	3 761	4 104	4 668	6 667	# 14 668	21 408	23 694	...
Quasi-Money										
Quasi-monnaie	5 419	5 581	7 079	10 267	13 286	14 448	# 3 722	6 020	22 627	...
Nicaragua: córdoba Nicaragua : córdoba										
Domestic Credit										
Crédit intérieur	20	36	48	95	240	1 351	...	...	...	...
Money										
Monnaie	7	11	21	55	193	1 422	...	...	...	...
Quasi-Money										
Quasi-monnaie	3	6	10	20	57	200	...	...	...	...
Niger: CFA franc Niger : franc CFA										
Domestic Credit[3]										
Crédit intérieur[3]	131	137	131	128	134	120	118	114	109	...

31
Money supply
Million national currency units, end of period [*cont.*]
Disponibilités monétaires
Millions d'unités monétaires nationales, fin de la période [*suite*]

Country or area and unit Pays ou zone et unité	1982	1983	1984	1985	1986	1987	1988	1989	1990	1991
Money[3] Monnaie[3]	71	67	78	81	83	73	82	88	78	...
Quasi-Money[3] Quasi-monnaie[3]	12	16	23	28	38	41	50	51	56	...
Nigeria: naira Nigéria : naira										
Domestic Credit Crédit intérieur	21 286	27 067	30 056	31 623	35 935	39 762	50 603	43 508	61 709	...
Money Monnaie	10 049	11 283	12 204	13 227	12 663	14 906	21 446	25 813	34 540	...
Quasi-Money Quasi-monnaie	6 645	7 752	9 039	9 926	10 942	13 989	16 960	16 707	23 014	...
Norway: krone Norvège : couronne										
Domestic Credit[3] Crédit intérieur[3]	204	226	263	296	342	434	446	476	487	479
Money[3] Monnaie[3]	59	66	82	99	102	153	187	218	238	256
Quasi-Money[3] Quasi-monnaie[3]	140	154	181	203	207	207	190	192	195	190
Oman: rial Omani Oman : rial omani										
Domestic Credit Crédit intérieur	188	210	219	459	615	568	711	675	657	608
Money Monnaie	249	299	284	329	313	335	315	345	391	406
Quasi-Money Quasi-monnaie	322	406	490	608	561	584	659	720	781	830
Pakistan: rupee Pakistan : roupie										
Domestic Credit Crédit intérieur	157 089	179 914	206 648	242 769	282 908	328 818	357 706	394 776	435 401	520 320
Money Monnaie	87 341	100 566	105 780	123 060	145 251	173 016	189 834	217 026	254 620	305 978
Quasi-Money Quasi-monnaie	44 892	59 316	61 532	68 925	77 592	86 380	89 543	83 028	80 372	92 475
Panama: balboa Panama : balboa										
Domestic Credit Crédit intérieur	3 057	3 195	3 422	3 514	3 856	3 813	3 221	3 198	2 799	...
Money Monnaie	379	373	381	409	450	442	304	307	433	...
Quasi-Money Quasi-monnaie	1 370	1 376	1 482	1 543	1 910	1 836	1 345	1 293	1 752	
Papua New Guinea: kina Papouasie-Nvl-Guinée : kina										
Domestic Credit Crédit intérieur	450	456	500	595	719	740	885	1 013	1 102	1 357
Money Monnaie	189	206	249	244	257	281	322	345	344	417
Quasi-Money Quasi-monnaie	379	444	499	576	672	662	663	693	738	855
Paraguay: guaraní Paraguay : guaraní										
Domestic Credit Crédit intérieur	116 001	139 773	169 859	192 381	257 417	333 851	414 613	...	690 062	...
Money Monnaie	60 200	75 587	97 807	125 202	158 674	243 667	328 488	...	840 091	
Quasi-Money Quasi-monnaie	105 487	117 838	128 131	147 735	188 976	225 673	234 737	...	726 403	...
Peru: new sol Pérou : nouveau sol										
Domestic Credit Crédit intérieur	3 291	8 717	16 845	30 973	54 721	134 821	859[1]	# 17 040[1]	...	...
Money Monnaie	1 509	2 965	6 060	23 098	43 424	96 390	593[1]	# 10 403[1]	...	...
Quasi-Money Quasi-monnaie	2 191	4 560	11 038	20 765	24 117	43 091	413[1]	# 9 883[1]	...	...

31
Money supply
Million national currency units, end of period [*cont.*]
Disponibilités monétaires
Millions d'unités monétaires nationales, fin de la période [*suite*]

Country or area and unit Pays ou zone et unité	1982	1983	1984	1985	1986	1987	1988	1989	1990	1991
Philippines: peso Philippines : peso										
Domestic Credit										
Crédit intérieur	150 977	198 826	206 247	195 108	166 151	148 599	157 548	191 415	250 274	243 862
Money										
Monnaie	23 525	33 546	34 442	36 756	43 167	53 804	61 196	81 276	92 936	107 693
Quasi-Money										
Quasi-monnaie	55 999	90 873	111 490	123 267	119 678	130 774	168 856	218 091	273 912	322 670
Poland: zloty Pologne : zloty										
Domestic Credit										
Crédit intérieur	...	...	...	6 079	7 116	8 411	12 348	39 627	116 462	...
Money										
Monnaie	...	...	...	2 449	2 998	3 786	5 748	19 975	101 670	...
Quasi-Money										
Quasi-monnaie	...	...	...	1 917	2 534	3 631	6 362	54 458	96 334	...
Portugal: escudo Portugal : escudo										
Domestic Credit[3]										
Crédit intérieur[3]	2 012	2 483	3 068	3 602	4 155	4 541	# 5 613	5 691	7 063	8 534
Money[3]										
Monnaie[3]	628	679	784	1 000	1 359	1 537	# 1 804	1 903	2 454	2 880
Quasi-Money[3]										
Quasi-monnaie[3]	1 086	1 316	1 706	2 084	2 331	2 678	# 2 947	3 363	3 804	4 928
Qatar: riyal Qatar : riyal										
Money										
Monnaie	3 795	3 625	4 135	4 017	4 487	4 778	3 399	3 403	4 055	...
Quasi-Money										
Quasi-monnaie	4 772	4 863	6 220	7 289	8 082	8 872	9 109	10 847	9 544	...
Romania: leu Roumanie : leu										
Domestic Credit[3]										
Crédit intérieur[3]	504	544	589	666	697	774	813	810	# 684	1 370
Money[3]										
Monnaie[3]	167	156	162	166	170	171	199	204	# 233	569
Quasi-Money[3]										
Quasi-monnaie[3]	122	131	143	157	171	182	190	206	# 280	449
Rwanda: franc Rwanda : franc										
Domestic Credit										
Crédit intérieur	8 247	12 038	13 597	16 683	17 611	21 626	26 698	31 653	36 569	31 208
Money										
Monnaie	11 460	12 317	13 343	14 699	17 334	17 791	18 332	16 052	16 846	17 991
Quasi-Money										
Quasi-monnaie	4 712	5 759	6 549	8 731	9 274	11 548	13 159	14 132	15 020	15 484
Saint Kitts-Nevis: EC dollar Saint-Kitts-et-Nevis : EC dollar										
Domestic Credit										
Crédit intérieur	110	133	149	184	206	169	202	252	283	...
Money										
Monnaie	25	28	29	33	50	55	48	61	60	...
Quasi-Money										
Quasi-monnaie	105	123	142	171	183	157	182	219	241	...
Saint Lucia: EC dollar Sainte-Lucie : EC dollar										
Domestic Credit[1]										
Crédit intérieur[1]	232	248	300	320	351	387	395	507	543	...
Money[1]										
Monnaie[1]	58	58	63	71	98	123	140	158	167	...
Quasi-Money[1]										
Quasi-monnaie[1]	157	187	211	258	315	357	375	444	509	...
St. Vincent-Grenadines: EC dollar St. Vincent-Grenadines : EC dollar										
Domestic Credit[1]										
Crédit intérieur[1]	139	151	176	176	168	191	156	194	194	...

31
Money supply
Million national currency units, end of period [cont.]
Disponibilités monétaires
Millions d'unités monétaires nationales, fin de la période [suite]

Country or area and unit Pays ou zone et unité	1982	1983	1984	1985	1986	1987	1988	1989	1990	1991
Money[1] Monnaie[1]	35	41	48	53	63	53	63	70	73	...
Quasi-Money[1] Quasi-monnaie[1]	100	111	124	144	164	191	184	210	245	...
Samoa: tala Samoa : tala										
Domestic Credit Crédit intérieur	56	51	53	58	48	33	15	−2	0	−11
Money Monnaie	17	17	19	20	22	29	30	33	47	43
Quasi-Money Quasi-monnaie	27	22	24	32	42	52	57	68	74	76
Saudi Arabia: riyal Arabie saoudite : riyal										
Money[3] Monnaie[3]	84	85	83	82	86	91	93	91	102	...
Quasi-Money[3] Quasi-monnaie[3]	40	52	62	65	74	77	84	88	86	95
Senegal: CFA franc Sénégal : franc CFA										
Domestic Credit[3] Crédit intérieur[3]	457	490	510	554	553	558	595	575	524	...
Money[3] Monnaie[3]	189	189	192	193	227	214	215	231	204	...
Quasi-Money[3] Quasi-monnaie[3]	73	84	95	107	107	118	120	138	147	...
Seychelles: rupee Seychelles : roupie										
Domestic Credit Crédit intérieur	267	288	314	374	422	445	558	686	764	839
Money Monnaie	143	131	135	155	155	155	188	220	217	266
Quasi-Money Quasi-monnaie	140	158	188	208	250	286	352	413	507	541
Sierra Leone: leone Sierra Leone : leone										
Domestic Credit Crédit intérieur	738	919	1 172	1 748	2 957	4 446	6 852	10 923	15 011	23 197
Money Monnaie	253	359	486	900	1 851	2 888	4 637	8 675	14 253	24 785
Quasi-Money Quasi-monnaie	166	193	223	313	433	859	1 241	1 565	3 565	6 306
Singapore: dollar Singapour : dollar										
Domestic Credit Crédit intérieur	20 189	28 789	33 029	30 134	30 471	33 530	34 973	36 782	41 317	47 085
Money Monnaie	8 157	8 607	8 866	8 785	9 822	11 031	11 958	13 745	15 261	16 436
Quasi-Money Quasi-monnaie	14 647	16 918	18 254	19 363	21 134	26 059	30 130	37 801	46 584	53 112
Solomon Islands: dollar Iles Salomon : dollar										
Domestic Credit Crédit intérieur	28	28	33	61	63	71	91	128	155	204
Money Monnaie	16	18	28	28	30	37	49	51	64	80
Quasi-Money Quasi-monnaie	24	29	36	38	42	60	80	77	76	94
Somalia: shilling Somalie : shilling										
Domestic Credit Crédit intérieur	5 024	5 261	9 616	11 519	13 902	34 492	51 067	98 930	...	...
Money Monnaie	4 033	4 309	5 334	9 773	12 143	30 046	45 436	139 827	...	...
Quasi-Money Quasi-monnaie	1 083	1 192	1 600	2 785	4 691	8 188	14 678	18 159	...	...

31
Money supply
Million national currency units, end of period [cont.]
Disponibilités monétaires
Millions d'unités monétaires nationales, fin de la période [suite]

Country or area and unit Pays ou zone et unité	1982	1983	1984	1985	1986	1987	1988	1989	1990	1991
South Africa: rand Afrique du Sud : rand										
Domestic Credit										
Crédit intérieur	28 161	32 390	40 942	49 639	51 850	60 000	79 872	106 051	126 296	...
Money										
Monnaie	13 124	16 586	23 413	21 332	23 207	32 026	39 934	43 343	49 858	...
Quasi-Money										
Quasi-monnaie	15 649	15 611	17 229	24 757	24 267	26 807	39 881	65 551	71 049	...
Spain: peseta Espagne : peseta										
Domestic Credit[3]										
Crédit intérieur[3]	19 797	20 851	23 455	26 878	30 098	33 887	38 362	42 739	47 631	52 699
Money[3]										
Monnaie[3]	4 850	5 277	5 746	6 589	7 580	8 899	10 573	12 177	14 674	16 573
Quasi-Money[3]										
Quasi-monnaie[3]	11 409	10 806	11 137	12 279	14 448	15 151	17 116	18 255	19 992	24 101
Sri Lanka: rupee Sri Lanka : roupie										
Domestic Credit										
Crédit intérieur	47 695	55 604	54 748	65 707	72 280	83 743	107 368	117 201	136 660	...
Money										
Monnaie	11 672	14 589	16 647	18 662	21 051	24 901	32 155	35 087	39 596	...
Quasi-Money										
Quasi-monnaie	20 342	24 123	28 265	31 994	31 750	36 052	37 857	44 207	51 905	...
Sudan: pound Soudan : livre										
Domestic Credit										
Crédit intérieur	2 761	3 574	4 272	5 830	8 111	11 274	15 024	23 689	...	...
Money										
Monnaie	2 091	2 336	2 764	4 145	5 849	7 768	11 218	18 899	...	...
Quasi-Money										
Quasi-monnaie	443	774	955	1 963	1 964	2 895	2 946	2 817	...	...
Suriname: guilder Suriname : florin										
Domestic Credit										
Crédit intérieur	792	1 121	1 431	1 840	2 335	2 868	3 526	4 120	4 266	...
Money										
Monnaie	421	455	577	880	1 229	1 562	1 945	2 165	2 251	...
Quasi-Money										
Quasi-monnaie	455	516	591	664	705	913	1 115	1 453	1 521	...
Swaziland: lilangeni Swaziland : lilangeni										
Domestic Credit										
Crédit intérieur	115	136	151	168	196	184	178	124	174	128
Money										
Monnaie	58	61	68	77	115	125	143	169	198	212
Quasi-Money										
Quasi-monnaie	103	137	171	216	211	246	367	472	452	446
Sweden: krona Suède : couronne										
Domestic Credit[3]										
Crédit intérieur[3]	473	521	601	623	716	796	933	1 106	1 177	1 176
Switzerland: franc Suisse : franc										
Domestic Credit[3]										
Crédit intérieur[3]	286	309	337	364	384	419	459	521	568	590
Money[3]										
Monnaie[3]	69	76	76	74	75	86	88	86	84	83
Quasi-Money										
Quasi-monnaie	161	177	198	211	219	240	256	280	284	294
Syrian Arab Republic: pound Rép. arabe syrienne : livre										
Domestic Credit										
Crédit intérieur	50 795	60 900	72 942	88 828	97 101	106 670	126 763	134 518	...	...
Money										
Monnaie	29 518	36 978	45 606	54 976	61 214	67 163	79 814	95 030	...	...
Quasi-Money										
Quasi-monnaie	3 993	4 309	7 167	8 516	9 907	11 528	17 043	21 340	28 054	...

31
Money supply
Million national currency units, end of period [*cont.*]
Disponibilités monétaires
Millions d'unités monétaires nationales, fin de la période [*suite*]

Country or area and unit Pays ou zone et unité	1982	1983	1984	1985	1986	1987	1988	1989	1990	1991
Thailand: baht Thaïlande : baht										
Domestic Credit[3] Crédit intérieur[3]	433	547	644	698	740	871	1 007	1 207	1 531	...
Money[3] Monnaie[3]	78	82	89	86	102	132	148	175	195	...
Quasi-Money[3] Quasi-monnaie[3]	285	366	449	508	569	676	808	1 032	1 334	...
Togo: CFA franc Togo : franc CFA										
Domestic Credit[3] Crédit intérieur[3]	85	80	74	68	101	102	107	83	95	...
Money[3] Monnaie[3]	90	83	91	83	90	91	64	63	75	...
Quasi-Money[3] Quasi-monnaie[3]	27	35	45	60	76	73	80	83	85	...
Tonga: pa'anga Tonga : pa'anga										
Domestic Credit[1] Crédit intérieur[1]	8 257	7 011	5 487	11 664	12 247	15 501	21 399	29 441	33 131	39 530
Money[1] Monnaie[1]	7 417	9 619	10 054	10 929	13 291	15 198	15 685	17 171	22 296	25 674
Quasi-Money[1] Quasi-monnaie[1]	8 928	10 035	11 870	16 632	19 489	23 959	22 523	23 725	27 017	29 557
Trinidad and Tobago: dollar Trinité-et-Tobago : dollar										
Domestic Credit Crédit intérieur	1 773	4 699	6 112	6 057	7 547	9 183	9 509	9 902	10 113	...
Money Monnaie	2 547	2 435	2 294	2 261	2 073	2 170	1 893	2 144	2 576	...
Quasi-Money Quasi-monnaie	4 848	5 499	6 048	6 183	6 125	6 323	6 674	6 952	7 085	...
Tunisia: dinar Tunisie : dinar										
Domestic Credit Crédit intérieur	2 561	3 101	3 592	4 175	4 521	4 917	5 131	6 367	6 765	7 318
Money Monnaie	1 455	1 699	1 814	2 059	2 112	2 126	2 494	2 524	2 678	2 696
Quasi-Money Quasi-monnaie	672	776	950	1 103	1 206	1 687	1 986	2 650	2 892	3 196
Turkey: lira Turquie : livre										
Domestic Credit[3] Crédit intérieur[3]	3 648	5 071	8 965	14 456	22 347	35 948	54 668	80 441	118 733	...
Money[3] Monnaie[3]	1 413	2 090	2 487	3 468	5 432	8 960	11 956	20 302	32 419	...
Quasi-Money[3] Quasi-monnaie[3]	1 275	1 397	3 047	5 125	6 925	9 032	15 925	27 656	40 199	...
Uganda: shilling Ouganda : shilling										
Domestic Credit Crédit intérieur	582	803	1 152	2 876	5 299	...	...	...	...	...
Money Monnaie	298	435	987	2 256	6 062	...	...	...	...	...
Quasi-Money Quasi-monnaie	87	109	176	366	1 056	2 014	3 227	8 480	19 400	30 837
United Arab Emirates: dirham Emirats arabes unis : dirham										
Domestic Credit Crédit intérieur	28 831	34 330	34 777	40 134	40 614	38 839	41 617	44 872	43 466	...
Money Monnaie	9 739	9 124	8 892	9 505	9 201	10 096	10 753	11 056	10 762	...
Quasi-Money Quasi-monnaie	23 907	27 218	37 978	40 382	42 875	44 844	47 403	52 132	47 247	...

31
Money supply
Million national currency units, end of period [*cont.*]
Disponibilités monétaires
Millions d'unités monétaires nationales, fin de la période [*suite*]

Country or area and unit Pays ou zone et unité	1982	1983	1984	1985	1986	1987	1988	1989	1990	1991
United Kingdom: pound Royaume-Uni : livre										
Domestic Credit										
Crédit intérieur	126 554	143 181	168 633	192 842	225 224	402 388	489 237	607 942	674 440	689 470
Money										
Monnaie	40 657	42 462	48 049	56 674	69 265	154 118	170 673	195 310	214 943	229 231
Quasi-Money										
Quasi-monnaie	65 741	74 853	82 765	88 850	109 094	189 087	231 942	286 061	318 659	314 472
United Rep.Tanzania: shilling Rép. Unie de Tanzanie : shilling										
Domestic Credit										
Crédit intérieur	27 494	31 650	38 512	51 399	56 658	87 512	120 973	...	...	...
Money										
Monnaie	18 323	20 564	20 608	25 468	35 809	47 131	64 125	...	...	...
Quasi-Money										
Quasi-monnaie	6 406	8 563	9 591	13 876	14 499	19 309	23 655	...	...	...
United States: dollar Etats-Unis : dollar										
Domestic Credit[3]										
Crédit intérieur[3]	2 409	2 698	3 018	3 319	3 630	3 902	4 192	4 300	4 296	4 258
Money[3]										
Monnaie[3]	491	538	570	641	747	767	812	824	854	924
Quasi-Money[3]										
Quasi-monnaie[3]	1 296	1 540	1 696	1 842	1 960	2 047	2 173	2 278	2 345	2 375
Uruguay: new peso Uruguay : nouveau peso										
Domestic Credit[3]										
Crédit intérieur[3]	125	168	267	443	677	996	1 774	3 090	5 946	...
Money[3]										
Monnaie[3]	14	15	22	46	84	135	222	383	770	...
Quasi-Money[3]										
Quasi-monnaie[3]	59	67	111	214	394	568	1 148	2 399	5 243	...
Vanuatu: vatu Vanuatu : vatu										
Domestic Credit										
Crédit intérieur	3 082	2 895	2 746	2 265	3 062	1 986	2 331	3 289	4 537	...
Money										
Monnaie	1 746	2 274	3 027	2 643	2 811	4 219	3 516	4 369	3 894	...
Quasi-Money										
Quasi-monnaie	6 192	6 353	7 926	9 749	12 828	10 386	11 953	18 460	20 203	...
Venezuela: bolivar Venezuela : bolívar										
Domestic Credit										
Crédit intérieur	96 669	104 487	120 083	126 164	165 909	216 052	323 975	402 299	480 923	610 568
Money										
Monnaie	60 807	76 252	96 675	105 224	109 850	122 562	145 452	169 952	269 263	347 573
Quasi-Money										
Quasi-monnaie	52 759	64 054	70 837	82 509	101 586	137 620	153 272	266 455	473 941	644 500
Yemen: rial Yémen : rial										
Domestic Credit[5]										
Crédit intérieur[5]	12 230	15 952	20 895	25 992	29 778	35 477	41 509	46 162	...	...
Money[5]										
Monnaie[5]	10 669	13 148	16 395	18 823	23 683	26 641	27 342	29 086	...	...
Quasi-Money[5]										
Quasi-monnaie[5]	2 246	2 820	3 970	5 608	6 969	7 102	8 490	8 364	...	...
Yugoslavia: new dinar Yougoslavie : nouveau dinar										
Domestic Credit										
Crédit intérieur	239	319	470	706	1 224	2 772	9 416	219 761	326 559	...
Money										
Monnaie	73	87	125	182	383	764	2 407	51 030	126 283	...

31
Money supply
Million national currency units, end of period [*cont.*]
Disponibilités monétaires
Millions d'unités monétaires nationales, fin de la période [*suite*]

Country or area and unit Pays ou zone et unité	1982	1983	1984	1985	1986	1987	1988	1989	1990	1991
Quasi-Money Quasi-monnaie	142	211	311	520	896	2 196	7 715	196 653	218 685	...
Zaire: zaire Zaïre : zaïre										
Domestic Credit[3] Crédit intérieur[3]	10	14	34	52	86	183	385	374	1 741	...
Money[3] Monnaie[3]	7	13	17	22	36	68	150	263	746	...
Quasi-Money[3] Quasi-monnaie[3]	1	1	1	2	3	7	25	30	140	...
Zambia: kwacha Zambie : kwacha										
Domestic Credit Crédit intérieur	2 910	3 180	3 554	4 223	5 834	7 047	# 9 619	15 328	19 059	...
Money Monnaie	690	795	870	1 232	2 304	3 225	# 5 244	7 947	12 543	...
Quasi-Money Quasi-monnaie	620	659	834	870	1 758	3 041	# 4 882	8 781	11 847	...
Zimbabwe: dollar Zimbabwe : dollar										
Domestic Credit Crédit intérieur	1 889	2 123	2 073	2 319	2 474	3 093	3 600	4 478	5 715	...
Money Monnaie	827	751	866	1 005	1 103	1 225	1 603	1 905	2 418	...
Quasi-Money Quasi-monnaie	921	939	971	1 152	1 138	1 620	1 860	2 387	2 506	...

Source:
International Monetary Fund (Washington, DC).

† All data shown which pertain to Germany prior to 3 October 1990 are indicated separately for the Federal Republic of Germany and the former German Democratic Republic based on their respective territories at the time indicated. Where data for united Germany (3 October 1990 and thereafter) are not available, available data are shown separately under the designations Federal Republic of Germany and former German Democratic Republic and pertain to the territorial boundaires prior to 3 October 1990. For detailed explanatory notes on data pertaining to Germany, see Annex I - Country Nomenclature.

1 Thousand national currency units.
2 Average of figures for last month of period (Australia: weekly figures).
3 Billion national currency units.
4 Trillion national currency units.
5 Excluding former Democratic Yemen.

Source:
Fonds monétaire international (Washington, DC).

† Toutes les données se rapportant à l'Allemagne avant le 3 octobre 1990 figurent dans deux rubriques séparées basées sur les territoires respectifs de la République fédérale d'Allemagne et l'ancienne République démocratique allemande selon la période indiquée. En l'absence de données pour l'Allemagne unifiée (à compter du 3 octobre 1990, les données disponibles sont fournies séparément sous les rubriques République fédérale d'Allemagne et ancienne République démocratique allemande et se rapportent aux limites territoriales antérieures au 3 octobre 1990. Pour les notes explicatives en détail sur les données concernant l'Allemagne, voir Annexe I – Nomenclature des pays.

1 Milliers d'unités monétaires nationales.
2 Moyenne des chiffres du dernier mois de la période (Australie: chiffres hebdomadaires).
3 Milliards d'unités monétaires nationales.
4 Trillions d'unités monétaires.
5 Non compris l'ancien Yémen démocratique.

32
Rates of discount of central banks
Taux d'escompte des banques centrales
Per cent per annum, end of period
Pour cent par année, fin de la période

Country or area Pays ou zone	1982	1983	1984	1985	1986	1987	1988	1989	1990	1991
Aruba Aruba	...	...	...	...	9.50	9.50	9.50	9.50	9.50	9.50
Australia Australie	15.76	12.14	12.03	15.98	16.92	14.95	13.20	17.17	15.24	10.99
Austria Autriche	4.75	3.75	4.50	4.00	4.00	3.00	4.00	6.50	6.50	8.00
Bahamas Bahamas	10.00	9.00	9.50	8.50	7.50	7.50	9.00	9.00	9.00	9.00
Bangladesh Bangladesh	10.50	10.50	10.50	11.25	10.75	10.75	10.75	10.75	9.75	9.25
Barbados Barbade	20.00	16.00	16.00	13.00	8.00	8.00	8.00	13.50	13.50	16.00
Belgium Belgique	11.50	10.00	11.00	9.75	8.00	7.00	7.75	10.25	10.50	8.50
Belize Belize	13.50	11.50	12.00	20.00	12.00	12.00	10.00	12.00	12.00	12.00
Benin Bénin	12.50	10.50	10.50	10.50	8.50	8.50	9.50	11.00	11.00	...
Botswana Botswana	12.00	10.50	9.00	9.00	9.00	8.50	6.50	6.50	8.50	12.00
Brazil Brésil	172.40	197.20	266.80	385.30	89.50	401.40	2 282.00	38 341.10	1 082.80	2 494.30
Burkina Faso Burkina Faso	12.50	10.50	10.50	10.50	8.50	8.50	9.50	11.00	11.00	...
Burundi Burundi	7.00	7.00	7.00	7.00	5.00	7.00	7.00	...	...	...
Cameroon Cameroun	8.50	8.50	8.50	9.00	8.00	8.00	9.50	9.50	...	...
Canada Canada	10.26	10.04	10.16	9.49	8.49	8.66	11.17	12.47	11.78	8.00
Central African Rep. Rép. centrafricaine	8.50	8.50	8.50	9.00	8.00	8.00	9.50	9.50	...	...
Chad Tchad	8.50	9.00	9.00	9.00	8.00	8.00	9.50	9.50	...	...
Colombia Colombie	27.00	27.00	27.00	27.00	...	30.00	30.00	30.00	...	...
Comoros Comores	...	10.00	10.00	10.00	10.00	8.50	8.50	...	...	...
Congo Congo	8.50	8.50	8.50	9.00	8.00	8.00	9.50	9.50	...	...
Costa Rica Costa Rica	30.00	30.00	28.00	28.00	27.50	31.38	31.50	31.61	37.80	42.50
Côte d'Ivoire Côte d'Ivoire	12.50	10.50	10.50	10.50	8.50	8.50	9.50	11.00	11.00	...
Cyprus Chypre	6.00	6.00	6.00	6.00	6.00	6.00	6.00	6.50	6.50	6.50
Czechoslovakia Tchécoslovaquie	...	...	...	...	...	...	...	...	8.50	9.50
Denmark Danemark	10.00	7.00	7.00	7.00	7.00	7.00	7.00	7.00	8.50	9.50

32
Rates of discount of central banks
Per cent per annum, end of period [cont.]
Taux d'escompte des banques centrales
Pour cent par année, fin de la période [suite]

Country or area Pays ou zone	1982	1983	1984	1985	1986	1987	1988	1989	1990	1991
Ecuador Equateur	15.00	19.00	23.00	23.00	23.00	23.00	23.00	32.00	35.00	49.00
Egypt Egypte	13.00	13.00	13.00	13.00	13.00	13.00	13.00	14.00	14.00	...
Equatorial Guinea Guinée équatoriale	...	...	...	9.00	8.00	8.00	9.50	9.50	...	...
Ethiopia Ethiopie	...	...	...	6.00	6.00	3.00	3.00	3.00	3.00	...
Fiji Fidji	9.50	10.17	11.00	11.00	8.00	11.00	11.00	8.00	8.00	...
Finland Finlande	8.50	9.50	15.07	9.00	7.00	7.00	8.00	8.50	8.50	8.50
France France	9.50	9.50	9.50	9.50	9.50	9.50	9.50	9.50	9.50	9.50
Gabon Gabon	8.50	8.50	8.50	9.00	8.00	8.00	9.50	9.50	...	...
Gambia Gambie	9.50	9.50	9.50	15.00	20.00	21.00	19.00	15.00	16.50	15.50
Germany†·Allemagne† F. R. Germany R. f. Allemagne	5.00	4.00	4.50	4.00	3.50	2.50	3.50	6.00	6.00	8.00
Ghana Ghana	10.50	14.50	18.00	18.50	20.50	23.50	26.00	26.00	33.00	...
Greece Grèce	20.50	20.50	20.50	20.50	20.50	20.50	19.00	19.00	19.00	19.00
Guatemala Guatemala	9.00	9.00	9.00	9.00	9.00	9.00	9.00	13.00	18.50	...
Guyana Guyana	14.00	14.00	14.00	14.00	14.00	14.00	14.00	...	30.00	...
Honduras Honduras	24.00	24.00	24.00	24.00	24.00	24.00	24.00	24.00	...	...
Hungary Hongrie	...	...	...	10.50	9.50	10.00	10.50	14.00	20.00	...
Iceland Islande	28.00	22.00	16.50	30.00	21.00	49.20	24.10	38.40	21.00	21.00
India Inde	10.00	10.00	10.00	10.00	10.00	10.00	10.00	10.00	10.00	12.00
Ireland Irlande	14.00	12.25	14.00	10.25	13.25	9.25	8.00	12.00	11.25	10.75
Israel Israël	108.20	311.00	690.30	79.60	31.40	26.80	30.90	15.00	13.00	14.20
Italy Italie	18.00	17.00	16.50	15.00	12.00	12.00	12.50	13.50	12.50	12.00
Jamaica Jamaïque	11.00	11.00	16.00	21.00	21.00	21.00	21.00	21.00	21.00	...
Japan Japon	5.50	5.00	5.00	5.00	3.00	2.50	2.50	4.25	6.00	4.50
Jordan Jordanie	6.50	6.25	6.25	6.25	6.25	6.25	* 6.25	8.00	* 8.50	8.50

32
Rates of discount of central banks
Per cent per annum, end of period [*cont.*]
Taux d'escompte des banques centrales
Pour cent par année, fin de la période [*suite*]

Country or area Pays ou zone	1982	1983	1984	1985	1986	1987	1988	1989	1990	1991
Kenya Kenya	15.00	15.00	12.50	12.50	12.50	12.50	16.02	16.50	19.43	20.27
Korea, Republic of Corée, République de	5.00	5.00	5.00	5.00	7.00	7.00	8.00	7.00	7.00	7.00
Kuwait Koweït	6.00	6.00	6.00	6.00	6.00	6.00	7.50	7.50	...	7.50
Lebanon Liban	12.00	12.00	12.00	19.70	21.85	21.85	21.85	21.85	21.85	18.04
Lesotho Lesotho	12.00	12.00	15.00	12.00	9.50	9.00	15.50	17.00	15.75	18.00
Libyan Arab Jamah. Jamah. arabe libyenne	5.00	5.00	5.00	5.00	5.00	5.00	5.00	...	...	...
Madagascar Madagascar	12.50	13.00	13.00	11.50	11.50	11.50	11.50	...	...	...
Malawi Malawi	10.00	10.00	10.00	11.00	11.00	14.00	11.00	11.00	14.00	13.00
Malaysia Malaisie	5.12	5.20	5.06	4.13	3.89	3.20	3.33	4.44	6.79	...
Mali Mali	12.50	10.50	10.50	10.50	8.50	8.50	9.50	11.00	11.00	...
Malta Malte	6.50	6.50	6.50	6.00	6.00	5.50	5.50	5.50	5.50	5.50
Mauritania Mauritanie	6.00	6.00	6.00	6.50	6.50	6.50	6.50	...	...	...
Mauritius Maurice	12.00	11.00	11.00	11.00	11.00	10.00	10.00	12.00	12.00	11.30
Morocco Maroc	7.00	7.00	7.00	8.50	8.50	8.50	8.50	...	...	...
Nepal Népal	15.00	15.00	15.00	15.00	11.00	11.00	11.00	11.00	11.00	...
Netherlands Pays-Bas	5.00	5.00	5.00	5.00	4.50	3.75	4.50	7.00	7.25	8.50
Netherlands Antilles Antilles néerlandaises	9.00	8.00	8.00	8.00	8.00	6.00	6.00	6.00	6.00	...
New Zealand Nouvelle-Zélande	13.00	7.50	13.50	19.80	24.60	18.55	15.10	15.00	13.25	8.30
Niger Niger	12.50	10.50	10.50	10.50	8.50	8.50	8.50	11.00	11.00	...
Nigeria Nigéria	8.00	8.00	10.00	10.00	10.00	...	...	...	...	...
Norway Norvège	9.20	10.00	10.20	10.70	14.80	13.80	12.00	11.00	10.50	10.00
Pakistan Pakistan	10.00	10.00	10.00	10.00	10.00	10.00	10.00	10.00	10.00	10.00
Papua New Guinea Papouasie-Nvl-Guinée	...	8.75	8.75	9.75	11.40	8.80	10.80	9.55	9.30	9.30
Peru Pérou	44.50	60.00	60.00	42.60	36.10	29.80	748.00	865.60	289.60	...
Philippines Philippines	6.30	8.05	12.11	11.50	9.63	9.08	8.94	9.64	10.60	10.75

32
Rates of discount of central banks
Per cent per annum, end of period [cont.]
Taux d'escompte des banques centrales
Pour cent par année, fin de la période [suite]

Country or area Pays ou zone	1982	1983	1984	1985	1986	1987	1988	1989	1990	1991
Portugal Portugal	18.75	23.17	25.00	23.50	17.00	14.96	13.71	14.33	14.50	14.50
Rwanda Rwanda	9.00	9.00	9.00	9.00	9.00	9.00	9.00	9.00	14.00	14.00
Senegal Sénégal	12.50	10.50	10.50	10.50	8.50	8.50	9.50	11.00	11.00	...
Seychelles Seychelles	6.00	6.00	6.00	6.00	6.00	6.00	6.00	6.00	9.50	9.50
Sierra Leone Sierra Leone	12.00	14.00	14.00	14.00	16.00	16.00	16.00	16.00	55.00	55.00
Somalia Somalie	8.00	8.00	8.00	12.00	12.00	12.00	45.00	45.00	...	...
South Africa Afrique du Sud	14.35	17.75	20.75	13.00	9.50	9.50	14.50	18.00	18.00	...
Spain Espagne	18.40	21.40	12.50	10.50	11.84	13.50	12.40	14.52	14.71	12.50
Sri Lanka Sri Lanka	14.00	13.00	13.00	11.00	11.00	10.00	10.00	14.00	15.00	...
Swaziland Swaziland	16.00	13.50	19.00	12.50	9.50	9.00	11.00	12.00	12.00	13.00
Sweden Suède	10.00	8.50	9.50	10.50	7.50	7.50	8.50	10.50	11.50	8.00
Switzerland Suisse	4.50	4.00	4.00	4.00	4.00	2.50	3.50	6.00	6.00	7.00
Syrian Arab Republic Rép. arabe syrienne	5.00	5.00	5.00	5.00	5.00	5.00	5.00	...	...	...
Thailand Thaïlande	12.50	13.00	12.00	11.00	8.00	8.00	8.00	8.00	12.00	11.00
Togo Togo	12.50	10.50	10.50	10.50	8.50	8.50	9.50	11.00	11.00	...
Trinidad and Tobago Trinité-et-Tobago	6.00	7.50	7.50	7.50	5.97	7.50	9.50	9.50	9.50	
Tunisia Tunisie	7.00	7.00	7.00	9.25	9.25	9.25	9.25	...	...	...
Turkey Turquie	31.50	48.50	52.00	52.00	48.00	45.00	54.00	54.00	45.00	...
Uganda Ouganda	11.00	15.50	24.00	24.00	36.00	31.00	45.00	55.00	50.00	46.00
United Rep.Tanzania Rép. Unie de Tanzanie	4.00	4.00	4.00	4.25	6.50	11.31	12.67	15.17	...	...
United States Etats-Unis	8.50	8.50	8.00	7.50	5.50	6.00	6.50	7.00	6.50	3.50
Uruguay Uruguay	83.70	112.70	133.20	145.10	138.40	143.40	154.50	219.60	251.60	219.00
Venezuela Venezuela	13.00	11.00	11.00	8.00	8.00	8.00	8.00	45.00	43.00	43.00
Yugoslavia Yougoslavie	14.00	30.00	47.00	61.00	56.00	131.00	372.00	8 187.00	...	...
Zaire Zaïre	15.00	20.00	20.00	26.00	26.00	26.00	26.00	50.00	...	...

32
Rates of discount of central banks
Per cent per annum, end of period [cont.]
Taux d'escompte des banques centrales
Pour cent par année, fin de la période [suite]

Country or area Pays ou zone	1982	1983	1984	1985	1986	1987	1988	1989	1990	1991
Zambia Zambie	7.50	10.00	14.50	25.00	30.00	15.00	15.00	...	...	...
Zimbabwe Zimbabwe	9.00	9.00	9.00	9.00	9.00	9.00	9.00	9.00	10.25	20.00

Source:
International Monetary Fund (Washington, DC).

† All data shown which pertain to Germany prior to 3 October 1990 are indicated separately for the Federal Republic of Germany and the former German Democratic Republic based on their respective territories at the time indicated. Where data for united Germany (3 October 1990 and thereafter) are not available, available data are shown separately under the designations Federal Republic of Germany and former German Democratic Republic and pertain to the territorial boundaries prior to 3 October 1990. For detailed explanatory notes on data pertaining to Germany, see Annex I - Country Nomenclature.

Source:
Fonds monétaire international (Washington, DC).

† Toutes les données se rapportant à l'Allemagne avant le 3 octobre 1990 figurent dans deux rubriques séparées basées sur les territoires respectifs de la République fédérale d'Allemagne et l'ancienne République démocratique allemande selon la période indiquée. En l'absence de données pour l'Allemagne unifiée (à compter du 3 octobre 1990), les données disponibles sont fournies séparément sous les rubriques République fédérale d'Allemagne et ancienne République démocratique allemande et se rapportent aux limites territoriales antérieures au 3 octobre 1990. Pour les notes explicatives en détail sur les données concernant l'Allemagne, voir Annexe I – Nomenclature des pays.

33
Government bonds
Obligations d'Etat
Yields of long-term bonds: per cent per annum
Taux de rendement des obligations à long terme : pour cent par année

Country or area Pays ou zone	1982	1983	1984	1985	1986	1987	1988	1989	1990	1991
Australia Australie	15.35	14.33	13.83	14.10	13.56	13.47	12.32	13.62	13.18	10.68
Austria Autriche	9.92	8.17	8.02	7.77	7.33	6.91	6.67	7.14	8.74	8.62
Belgium Belgique	13.56	11.86	11.98	10.61	7.93	7.82	7.85	8.64	10.06	9.28
Cameroon Cameroun	...	...	...	4.25	6.25	7.25	7.25	7.25	...	...
Canada Canada	14.26	11.79	12.75	11.04	9.52	9.95	10.22	9.92	10.85	9.76
Congo Congo	...	...	...	6.50	6.50	6.50	6.50	6.50	...	...
Denmark Danemark	20.38	14.46	13.96	11.31	9.91	11.06	9.78	9.75	10.74	9.59
Ethiopia Ethiopie	...	...	...	...	6.00	5.00	5.00	5.00	5.00	...
France France	15.69	13.63	12.54	10.94	8.62	9.43	9.06	8.79	9.96	9.05
Gabon Gabon	...	...	...	7.50	7.50	7.50	7.50	7.50	...	...
Germany†·Allemagne† F. R. Germany R. f. Allemagne	8.95	7.89	7.77	6.87	5.92	5.84	6.10	7.09	8.88	8.63
Honduras Honduras	...	9.40	10.30	10.40	10.40	10.40	10.40	10.40	10.40	10.40
Ireland Irlande	17.06	13.90	14.61	12.64	11.07	11.27	9.49	8.94	10.08	9.17
Italy Italie	20.89	18.02	14.95	13.00	10.52	9.68	10.16	10.72	11.51	10.09
Jamaica Jamaïque	13.67	15.16	17.14	22.47	22.62	20.82	20.40	20.17	25.46	...
Japan Japon	8.06	7.42	6.80	6.34	4.94	4.21	4.27	5.05	7.36	6.53
Korea, Republic of Corée, République de	17.40	13.10	14.30	13.60	11.60	12.40	13.00	14.70	15.00	16.50
Luxembourg Luxembourg	10.50	9.83	10.22	9.53	8.67	7.96	7.13	7.68	8.51	8.15
Malawi Malawi	9.73	10.27	10.58	11.50	11.50	11.50	11.50	11.50	11.50	11.50
Nepal Népal	10.50	10.50	10.50	10.50	10.50	10.50	10.50	10.50	10.50	...
Netherlands Pays-Bas	10.10	8.61	8.33	7.34	6.35	6.38	6.29	7.21	8.93	8.74
Netherlands Antilles Antilles néerlandaises	...	10.65	10.63	9.46	9.29	10.36	10.74	10.63	10.74	...
New Zealand Nouvelle-Zélande	12.91	12.18	12.57	17.71	16.52	15.69	13.11	12.78	12.45	9.83
Norway Norvège	13.20	12.86	12.16	12.58	13.47	13.56	12.97	10.83	10.72	9.87

33
Government bonds
Yields of long-term bonds: per cent per annum [*cont.*]
Obligations d'Etat
Taux de rendement des obligations à long terme : pour cent par année [*suite*]

Country or area Pays ou zone	1982	1983	1984	1985	1986	1987	1988	1989	1990	1991
Pakistan Pakistan	9.36	9.31	9.25	9.19	8.77	8.26	8.32	8.18	8.05	7.88
Portugal Portugal	16.79	19.22	21.50	20.75	15.54	15.02	13.87	14.74	15.17	14.26
Samoa Samoa	12.30	14.90	17.50	15.00	14.20	13.50	13.50	13.50	13.50	13.50
Solomon Islands Iles Salomon	11.00	11.00	11.00	12.00	13.00	12.33	12.00	12.44	12.75	12.92
South Africa Afrique du Sud	13.51	12.67	15.23	16.79	16.37	15.30	16.37	16.90	16.15	16.34
Spain Espagne	15.99	16.91	16.52	13.37	11.36	12.81	11.74	13.70	14.68	12.43
Sri Lanka Sri Lanka	...	...	14.67	15.33	12.00	12.00	11.49	11.71	12.20	...
Sweden Suède	13.04	12.30	12.28	13.09	10.26	11.68	11.35	11.18	13.08	10.69
Switzerland Suisse	4.83	4.51	4.70	4.78	4.29	4.12	4.15	5.20	6.68	6.35
Thailand Thaïlande	13.85	11.13	12.41	12.11	9.11	7.48	7.50	8.08	10.60	10.75
Trinidad and Tobago Trinité-et-Tobago	9.84	9.88	9.89	9.89	9.62	9.54	...	10.77	10.73	...
Uganda Ouganda	12.00	12.63	18.00	24.00	38.33	40.00	38.50	45.33	44.50	42.00
United Kingdom Royaume-Uni	12.88	10.80	10.69	10.62	9.87	9.47	9.36	9.58	11.08	9.92
United States Etats-Unis	13.00	11.10	12.52	10.62	7.68	8.38	8.85	8.50	8.55	7.86
Vanuatu Vanuatu	...	...	...	9.50	9.50	9.50	9.50	# 8.00	8.00	...
Venezuela Venezuela	...	...	13.15	12.55	12.07	13.49	14.86	17.32	17.03	...
Zimbabwe Zimbabwe	13.00	13.08	13.29	13.26	13.20	13.87	14.00	14.00	15.24	...

Source:
International Monetary Fund (Washington, DC).

† All data shown which pertain to Germany prior to 3 October
1990 are indicated separately for the Federal Republic of
Germany and the former German Democratic Republic based on
their respective territories at the time indicated. Where
data for united Germany (3 October 1990 and thereafter) are
not available, available data are shown separately under the
designations Federal Republic of Germany and former Germany
Democratic Republic and pertain to the territorial
boundaries prior to 3 October 1990. For detailed
explanatory notes on data pertaining to Germany, see Annex I
- Country Nomenclature.

Source:
Fonds monétaire international (Washington, DC).

† Toutes les données se rapportant à l'Allemagne avant le 3
octobre 1990 figurent dans deux rubriques séparées basées
sur les territoires respectifs de la République fédérale
d'Allemagne et l'ancienne République démocratique allemande
selon la période indiquée. En l'absence de données pour
l'Allemagne unifiée (à compter du 3 october 1990), les
données disponibles sont fournies séparément sous les
rubriques République fédérale d'Allemagne et ancienne
République démocratique allemande et se rapportent aux
limites territoriales antérieures au 3 oxtober 1990. Pour
les notes explicatives en détail sur les données concernant
l'Allemagne, voir Annexe I – Nomenclature des pays.

34
Market prices of industrial shares
Cours des actions industrielles
Index numbers: 1985 = 100
Indices : 1985 = 100

Country or area Pays ou zone	1981	1982	1983	1984	1986	1987	1988	1989	1990	1991
Australia Australie	73	55	70	82	135	193	165	177	166	168
Austria Autriche	53	47	51	52	125	107	101	175	284	242
Belgium Belgique	47	56	70	90	144	170	168	194	179	174
Canada Canada	80	61	87	86	111	132	122	140	126	128
Chile Chili	90	70	61	81	220	401	595	958	1 480	2 676
Colombia Colombie	120	154	117	95	...	...	368	405	...	...
Denmark Danemark	40	51	89	96	101	84	95	132	146	157
Finland Finlande	41	54	83	116	153	241	289	328	254	...
France France	48	47	63	86	153	178	162	235	208	221
Germany† · Allemagne† F. R. Germany R. f. Allemagne	50	50	67	75	135	125	104	133	153	136
India Inde	96	94	99	67	122	112	115	174	241	325
Ireland Irlande	...	...	...	...	157	227	223	...	486	447
Israel Israël	1	4	8	21	154	...	...	...	...	...
Italy Italie	53	43	53	60	233	225	185	214	194	153
Japan Japon	55	55	65	82	133	196	214	258	219	185
Korea, Republic of Corée, République de	91	88	88	95	164	301	499	661	538	473
Luxembourg Luxembourg	57	51	58	67	151	138	138	166	153	146
Mexico Mexique	15	10	38	56	358	2 490	2 902	5 060	8 766	16 665
Netherlands[1] Pays-Bas[1]	42	42	61	77	129	129	120	151	148	152
New Zealand Nouvelle-Zélande	46	45	59	85	170	198	125	139	120	...
Norway Norvège	37	34	53	76	105	129	113	181	224	209
Pakistan[2] Pakistan[2]	55	61	78	103	104	129	150	161	171	259
Peru Pérou	112	101	68	64	...	...	...	...	...	...
Philippines[3] Philippines[3]	156	128	127	113	471	1 086	1 134	1 622	1 422	1 490

34
Market prices of industrial shares
Index numbers: 1985 = 100 [cont.]
Cours des actions industrielles
Indices: 1985 = 100 [suite]

Country or area Pays ou zone	1981	1982	1983	1984	1986	1987	1988	1989	1990	1991
Portugal[4] Portugal[4]	...	...	...	...	...	...	100	104	91	63
South Africa[3] Afrique du Sud[3]	86	76	96	94	128	188	148	207	217	289
Spain Espagne	62	57	58	79	204	293	325	353	303	311
Sweden Suède	41	51	98	107	163	195	218	301	278	262
Switzerland Suisse	53	52	68	77	122	126	# 119	145	138	139
United Kingdom Royaume-Uni	47	54	68	81	124	164	147	177	173	190
United States[5] Etats-Unis[5]	69	64	87	87	126	159	148	178	188	215
Venezuela Venezuela	93	69	69	90	158	350	304	161	169	...
Zimbabwe Zimbabwe	100	73	60	51	120	146	...	...	...	...

Source:
International Monetary Fund (Washington, DC).

† All data shown which pertain to Germany prior to 3 October
1990 are indicated separately for the Federal Republic of
Germany and the former German Democratic Republic based on
their respective territories at the time indicated. Where
data for united Germany (3 October 1990 and thereafter) are
not available, available data are shown separately under the
designations Federal Republic of Germany and former German
Democratic Republic and pertain to the territorial
boundaries prior to 3 october 1990. For detailed
explanatory notes on data pertaining to Germany, see Annex I
- Country Nomenclature.

1 General index.
2 Last Friday of the month quotations.
3 Commercial and industrial quotations.
4 Base: January 5, 1988=100.
5 Source: Standard and Poor's Corp.

Source:
Fonds monétaire international (Washington, DC).

† Toutes les données se rapportant à l'Allemagne avant le 3
octobre 1990 figurent dans deux rubriques séparées basées
sur les territoires respectifs de la République fédérale
d'Allemagne et l'ancienne République démocratique allemande
selon la période indiquée. En l'absence de données pour
l'Allemagne unifiée (à compter du 3 october 1990), les
données disponibles sont fournies séparément sous les
rubriques République fédérale d'Allemagne et ancienne
République démocratique allemande et se rapportent aux
limites territoriales antérieures au 3 octobre 1990. Pour
les notes explicatives en détail sur les données concernant
l'Allemagne, voir Annexe I – Nomenclature des pays.
1 Indice général.
2 Cotations du dernier vendredi du mois.
3 Cotations commerciales et industrielles.
4 Base: 5 janvier, 1988=100.
5 Source: "Standard and Poor's Corp."

35
Money market rates
Taux de l'argent hors banque
Treasury bill and call money rates: per cent per annum
Taux d'intérêt des bons du Trésor et de l'argent au jour le jour : pour cent par année

Country or area Pays ou zone	1982	1983	1984	1985	1986	1987	1988	1989	1990	1991
Antigua and Barbuda Antigua-et-Barbuda										
Treasury bill										
Bons du Trésor	7.00	7.00	7.00	7.00	7.00	7.00	7.00	7.00	7.00	...
Australia Australie										
Treasury bill										
Bons du Trésor	14.64	11.06	10.98	15.34	15.39	12.80	12.14	16.80	14.15	9.96
Call money										
Argent au jour le jour	13.90	9.50	10.83	14.70	15.75	13.06	11.90	16.75	14.76	10.46
Austria Autriche										
Call money										
Argent au jour le jour	8.00	5.36	6.57	6.11	5.19	4.35	4.59	7.46	8.53	9.10
Bahamas Bahamas										
Treasury bill										
Bons du Trésor	8.76	9.11	6.88	5.90	3.47	2.40	4.46	5.21	5.85	6.49
Bahrain Bahrein										
Treasury bill										
Bons du Trésor	...	...	...	...	...	...	7.40	9.10	...	...
Call money										
Argent au jour le jour	...	...	...	...	7.20	7.10	7.90	9.20	8.50	6.30
Barbados Barbade										
Treasury bill										
Bons du Trésor	13.25	7.45	6.92	0.55	4.42	4.84	4.75	4.90	7.07	9.34
Belgium Belgique										
Treasury bill										
Bons du Trésor	13.96	10.38	11.60	9.44	8.09	7.00	6.61	8.45	9.62	...
Call money										
Argent au jour le jour	11.44	8.18	9.47	8.27	6.64	5.66	5.04	7.00	8.29	...
Belize Belize										
Treasury bill										
Bons du Trésor	11.06	10.51	9.55	12.76	10.80	8.80	8.32	7.36	7.37	6.71
Benin Bénin										
Call money										
Argent au jour le jour	13.50	12.23	11.84	10.66	8.58	8.37	8.72	10.07	10.98	...
Brazil Brésil										
Treasury bill										
Bons du Trésor	93.10	127.40	190.20	215.10	150.60	195.40	482.60	381.80	...	...
Burkina Faso Burkina Faso										
Call money										
Argent au jour le jour	13.50	12.23	11.84	10.67	8.58	8.37	8.72	10.07	10.98	...
Canada Canada										
Treasury bill										
Bons du Trésor	13.66	9.31	11.06	9.43	8.97	8.15	9.48	12.05	12.81	8.73
Call money										
Argent au jour le jour	10.38	9.07	10.05	9.84	8.16	8.50	10.35	12.06	11.62	7.40
Côte d'Ivoire Côte d'Ivoire										
Call money										
Argent au jour le jour	13.50	12.23	11.84	10.66	8.58	8.37	8.72	10.07	10.98	...
Denmark Danemark										
Treasury bill										
Bons du Trésor	19.65	14.77	13.03	10.78	9.59	10.99	9.19	...	...	...
Call money										
Argent au jour le jour	16.92	12.81	11.77	10.33	9.22	10.20	8.52	9.66	10.97	9.78
Dominica Dominique										
Treasury bill										
Bons du Trésor	6.50	6.50	6.50	6.50	6.50	6.50	6.50	6.50	6.50	...

35
Money market rates
Treasury bill and call money rates: per cent per annum [cont.]
Taux de l'argent hors banque
Taux d'intérêt des bons du Trésor et de l'argent au jour le jour : pour cent par année [suite]

Country or area Pays ou zone	1982	1983	1984	1985	1986	1987	1988	1989	1990	1991
Ethiopia Ethiopie										
Treasury bill Bons du Trésor	3.00	3.00	3.00	3.00	3.00	3.00	3.00	3.00	3.00	...
Fiji Fidji										
Treasury bill Bons du Trésor	5.96	6.17	7.08	7.03	6.36	9.76	1.78	2.74	4.40	...
Call money Argent au jour le jour	5.07	6.19	8.73	6.61	6.55	9.02	1.49	2.33	2.92	...
Finland Finlande										
Call money Argent au jour le jour	11.66	14.67	16.50	13.46	11.90	11.19	11.49	13.84	14.00	13.08
France France										
Treasury bill Bons du Trésor	...	...	...	...	...	8.22	7.82	9.38	10.16	9.70
Call money Argent au jour le jour	14.87	12.53	11.74	9.93	7.74	7.98	7.52	9.07	9.85	9.49
Germany† · Allemagne† **F. R. Germany R. f. Allemagne**										
Treasury bill Bons du Trésor	8.31	5.62	5.66	5.04	3.86	3.28	3.62	6.28	8.10	8.33
Call money Argent au jour le jour	8.67	5.36	5.54	5.19	4.57	3.72	4.01	6.59	7.92	8.84
Ghana Ghana										
Treasury bill Bons du Trésor	13.00	13.00	14.20	17.10	18.50	21.70	19.80	19.80	21.80	...
Greece Grèce										
Treasury bill Bons du Trésor	...	...	...	...	17.00	17.30	16.30	16.50	18.50	18.80
Grenada Grenade										
Treasury bill Bons du Trésor	6.50	6.50	6.50	6.50	6.50	6.50	6.50	6.50	6.50	...
Guyana Guyana										
Treasury bill Bons du Trésor	12.28	12.75	12.75	12.75	12.75	11.33	11.03	...	30.00	...
Iceland Islande										
Call money Argent au jour le jour	...	...	36.90	0.00	0.00	31.52	34.49	21.57	12.73	15.10
India Inde										
Call money Argent au jour le jour	7.27	8.30	9.95	10.00	9.97	9.83	...	...	15.57	19.35
Indonesia Indonésie										
Call money Argent au jour le jour	17.24	13.17	18.63	10.33	13.00	14.51	15.00	12.57	14.37	15.12
Ireland Irlande										
Treasury bill Bons du Trésor	16.33	13.26	13.13	11.78	11.85	10.70	7.81	9.70	10.90	10.12
Call money Argent au jour le jour	17.65	14.45	12.93	11.87	12.28	10.84	7.84	9.55	11.10	10.45
Israel Israël										
Treasury bill Bons du Trésor	...	...	217.30	210.10	19.90	20.00	16.00	12.90	...	14.50
Italy Italie										
Treasury bill Bons du Trésor	19.44	17.89	15.37	13.71	11.40	10.73	11.13	12.58	12.38	12.54

35

Money market rates
Treasury bill and call money rates: per cent per annum [cont.]
Taux de l'argent hors banque
Taux d'intérêt des bons du Trésor et de l'argent au jour le jour : pour cent par année [suite]

Country or area Pays ou zone	1982	1983	1984	1985	1986	1987	1988	1989	1990	1991
Call money Argent au jour le jour	20.16	18.44	17.27	15.25	13.41	11.51	11.29	12.69	12.38	12.18
Jamaica Jamaïque Treasury bill Bons du Trésor	8.61	12.38	13.29	19.03	20.88	18.16	18.50	19.10	26.21	...
Japan Japon Call money Argent au jour le jour	6.94	6.39	6.10	6.46	4.79	3.51	3.62	4.87	7.24	7.46
Kenya Kenya Treasury bill Bons du Trésor	12.58	14.15	13.24	13.90	13.22	12.86	13.48	13.86	14.78	16.59
Korea, Republic of Corée, République de Call money Argent au jour le jour	14.20	13.00	11.40	9.30	9.70	8.90	9.60	13.30	14.00	17.00
Kuwait Koweït Treasury bill Bons du Trésor	6.42	5.69	5.69	5.69	5.69	5.61	...	...	...	...
Call money Argent au jour le jour	10.22	6.78	8.90	...	...	6.08	6.12	8.70	...	...
Lebanon Liban Treasury bill Bons du Trésor	14.02	9.52	13.08	14.96	17.83	25.17	23.58	18.00	18.00	16.85
Lesotho Lesotho Treasury bill Bons du Trésor	10.67	18.00	18.42	17.60	11.21	10.75	11.42	15.75	16.33	...
Libyan Arab Jamah. Jamah. arabe libyenne Call money Argent au jour le jour	4.00	4.00	4.00	4.00	4.00	4.00	4.00	...	...	...
Malawi Malawi Treasury bill Bons du Trésor	9.00	11.00	11.00	12.31	12.75	14.25	15.75	15.75	12.92	11.50
Malaysia Malaisie Call money Argent au jour le jour	7.90	8.97	8.96	7.57	8.02	2.85	3.22	4.72	6.81	...
Maldives Maldives Call money Argent au jour le jour	...	...	11.00	9.00	9.00	8.67	8.50	7.33	7.00	...
Mali Mali Call money Argent au jour le jour	13.50	12.23	11.84	10.66	8.58	8.37	8.72	10.07	10.98	...
Malta Malte Treasury bill Bons du Trésor	...	...	...	...	...	4.50	4.24	4.24	4.25	4.46
Mauritius Maurice Call money Argent au jour le jour	9.90	10.80	11.00	11.20	11.10	10.30	...	...	13.30	12.20
Mexico Mexique Treasury bill Bons du Trésor	45.75	59.07	49.32	63.20	...	103.07	# 69.14	44.99	34.76	19.28
Call money Argent au jour le jour	45.86	57.50	49.93	62.43	88.01	95.59	69.01	# 47.43	37.36	23.58
Morocco Maroc Treasury bill Bons du Trésor	8.30	8.50	8.50	10.00	10.50	10.50	10.50	...	...	...

35
Money market rates
Treasury bill and call money rates: per cent per annum [cont.]
Taux de l'argent hors banque
Taux d'intérêt des bons du Trésor et de l'argent au jour le jour : pour cent par année [suite]

Country or area Pays ou zone	1982	1983	1984	1985	1986	1987	1988	1989	1990	1991
Call money Argent au jour le jour	10.92	...	...	9.41	9.44	...	...	...	...	...
Nepal Népal										
Treasury bill Bons du Trésor	5.00	5.00	5.00	5.00	5.00	5.00	5.00	5.62	7.93	...
Netherlands Pays-Bas										
Treasury bill Bons du Trésor	8.12	5.48	5.97	6.23	5.49	5.18	4.34	6.80	...	...
Call money Argent au jour le jour	8.06	5.28	5.78	6.30	5.83	5.16	4.48	6.99	8.29	9.01
Netherlands Antilles Antilles néerlandaises										
Treasury bill Bons du Trésor	7.50	7.25	7.35	7.21	7.34	6.36	5.79	5.96	6.10	...
New Zealand Nouvelle-Zélande										
Treasury bill Bons du Trésor	11.25	10.12	...	...	19.97	20.49	...	13.51	13.78	9.73
Call money Argent au jour le jour	...	12.90	15.00	20.30	24.10	17.40	14.18	14.10	12.61	7.59
Niger Niger										
Call money Argent au jour le jour	13.50	12.23	11.84	10.66	8.58	8.37	8.72	10.07	10.98	...
Norway Norvège										
Treasury bill Bons du Trésor	11.30	11.30	...	...	...	...	...	...	...	...
Call money Argent au jour le jour	13.91	12.27	12.67	12.29	14.15	14.66	13.29	11.31	11.45	10.58
Pakistan Pakistan										
Call money Argent au jour le jour	9.51	8.15	8.97	8.13	6.59	6.25	6.31	6.30	7.29	7.64
Papua New Guinea Papouasie-Nvl-Guinée										
Treasury bill Bons du Trésor	13.80	10.92	9.28	10.40	12.32	10.44	10.12	10.50	11.40	10.33
Philippines Philippines										
Treasury bill Bons du Trésor	13.78	14.23	28.53	26.73	16.08	11.51	14.67	18.65	23.67	21.48
Portugal Portugal										
Treasury bill Bons du Trésor	14.37	18.14	21.15	20.90	15.56	13.89	12.96	...	13.52	14.19
Call money Argent au jour le jour	12.42	18.24	21.27	20.17	14.52	13.69	12.34	12.84	13.73	15.81
Saint Kitts-Nevis Saint-Kitts-et-Nevis										
Treasury bill Bons du Trésor	6.50	6.50	6.50	6.50	6.50	6.50	6.50	6.50	6.50	...
Call money Argent au jour le jour	...	...	...	...	...	...	...	...	...	...
Saint Lucia Sainte-Lucie										
Treasury bill Bons du Trésor	6.50	6.50	6.50	7.00	7.00	7.00	7.00	7.00	7.00	...
St. Vincent-Grenadines St. Vincent-Grenadines										
Treasury bill Bons du Trésor	6.50	6.50	6.50	6.50	6.50	6.50	6.50	6.50	6.50	...
Senegal Sénégal										
Call money Argent au jour le jour	13.50	12.23	11.84	10.66	8.58	8.37	8.72	10.07	10.98	...

35

Money market rates
Treasury bill and call money rates: per cent per annum [cont.]
Taux de l'argent hors banque
Taux d'intérêt des bons du Trésor et de l'argent au jour le jour : pour cent par année [suite]

Country or area Pays ou zone	1982	1983	1984	1985	1986	1987	1988	1989	1990	1991
Seychelles Seychelles										
Treasury bill										
Bons du Trésor	10.10	12.07	12.61	12.44	12.91	15.15	13.90	13.41	* 13.00	13.00
Sierra Leone Sierra Leone										
Treasury bill										
Bons du Trésor	10.00	11.00	12.00	12.00	14.50	16.50	18.00	22.00	47.50	50.00
Singapore Singapour										
Call money										
Argent au jour le jour	7.92	7.11	7.67	5.38	4.27	3.89	4.30	5.34	6.61	4.76
Solomon Islands Iles Salomon										
Treasury bill										
Bons du Trésor	7.50	8.92	9.00	9.58	12.00	11.33	11.00	11.00	11.00	14.75
South Africa Afrique du Sud										
Treasury bill										
Bons du Trésor	15.59	13.44	19.33	17.56	10.43	8.71	12.03	16.84	17.80	...
Call money										
Argent au jour le jour	16.90	13.98	20.31	18.21	10.92	9.50	13.90	18.77	19.46	...
Spain Espagne										
Treasury bill										
Bons du Trésor	15.70	19.80	13.43	10.90	8.63	8.03	# 10.79	13.57	14.17	12.45
Call money										
Argent au jour le jour	17.21	19.40	12.59	11.60	11.50	16.07	11.30	14.39	14.76	...
Sri Lanka Sri Lanka										
Treasury bill										
Bons du Trésor	12.28	12.38	13.08	13.39	10.47	7.30	13.59	14.81	14.08	...
Call money										
Argent au jour le jour	16.88	23.88	21.42	14.56	12.95	13.14	18.75	19.28	...	...
Swaziland Swaziland										
Treasury bill										
Bons du Trésor	14.60	13.04	17.74	16.47	9.76	5.96	7.28	10.16	11.14	12.67
Call money										
Argent au jour le jour	...	...	...	...	...	...	...	8.39	10.50	10.61
Sweden Suède										
Treasury bill										
Bons du Trésor	13.22	12.34	11.93	14.17	9.83	9.39	10.08	11.50	13.66	11.59
Call money										
Argent au jour le jour	13.29	10.85	11.77	13.85	10.15	9.16	10.08	11.52	13.45	11.81
Switzerland Suisse										
Treasury bill										
Bons du Trésor	3.87	3.04	3.58	4.15	3.54	3.18	3.01	6.60	8.32	7.74
Call money										
Argent au jour le jour	1.32	1.84	3.34	3.75	3.17	2.51	2.22	6.50	8.33	7.73
Thailand Thaïlande										
Treasury bill										
Bons du Trésor	11.64	9.35	10.00	11.02	6.76	3.63	5.08	...	...	...
Call money										
Argent au jour le jour	14.95	12.15	13.57	13.48	8.07	5.91	8.66	9.82	12.73	10.58
Togo Togo										
Call money										
Argent au jour le jour	13.50	12.23	11.84	10.66	8.58	8.37	8.72	10.07	10.98	...
Trinidad and Tobago Trinité-et-Tobago										
Treasury bill										
Bons du Trésor	3.05	3.08	3.39	3.47	3.99	4.63	...	7.13	7.50	...

35

Money market rates
Treasury bill and call money rates: per cent per annum [cont.]
Taux de l'argent hors banque
Taux d'intérêt des bons du Trésor et de l'argent au jour le jour : pour cent par année [suite]

Country or area Pays ou zone	1982	1983	1984	1985	1986	1987	1988	1989	1990	1991
Tunisia Tunisie										
Call money Argent au jour le jour	8.25	8.38	8.89	10.28	9.95	10.00	9.15	9.40	11.53	11.79
Turkey Turquie										
Treasury bill Bons du Trésor	...	...	...	...	...	41.92	54.56	48.01	43.46	67.01
Uganda Ouganda										
Treasury bill Bons du Trésor	9.50	11.17	18.00	22.00	30.67	30.50	33.00	42.17	41.00	34.17
United Kingdom Royaume-Uni										
Treasury bill Bons du Trésor	11.47	9.59	9.30	11.55	10.36	9.25	9.78	13.05	14.08	10.96
Call money Argent au jour le jour	11.36	9.09	7.62	10.78	10.68	9.66	10.31	13.88	14.68	11.74
United States Etats-Unis										
Treasury bill Bons du Trésor	10.72	8.62	9.57	7.49	5.97	5.83	6.67	8.11	7.51	5.41
Call money [1] Argent au jour le jour [1]	12.26	9.09	10.22	8.10	6.80	6.66	7.61	9.22	8.10	5.70
Vanuatu Vanuatu										
Call money Argent au jour le jour	...	...	...	7.00	6.96	6.50	7.50	7.08	7.00	...
Yugoslavia Yougoslavie										
Call money Argent au jour le jour	...	...	...	...	64.00	93.30	423.30	4 150.80	...	...
Zambia Zambie										
Treasury bill Bons du Trésor	6.00	7.50	7.67	13.21	24.25	16.50	15.17	...	...	...
Zimbabwe Zimbabwe										
Treasury bill Bons du Trésor	8.50	8.52	8.49	8.48	8.71	8.73	8.38	...	8.39	...
Call money Argent au jour le jour	9.50	9.09	8.90	8.80	9.10	9.30	9.07	8.72	8.68	...

Source:
International Monetary Fund (Washington, DC).

† All data shown which pertain to Germany prior to 3 October 1990 are indicated separately for the Federal Republic of Germany and the former German Democratic Republic based on their respective territories at the time indicated. Where data for united Germany (3 October 1990 and thereafter) are not available, available data are shown separately under the designations Federal Republic of Germany and former German Democratic Republic and pertain to the territorial boundaries prior to 3 October 1990. For detailed explanatory notes on data pertaining to Germany, see Annex I - Country Nomenclature.

1 Federal funds rate.

Source:
Fonds monétaire international (Washington, DC).

† Toutes les données se rapportant à l'Allemagne avant le 3 octobre 1990 figurent dans deux rubriques séparées basées sur les territoires respectifs de la République fédérale d'Allemagne et l'ancienne République démocratique allemande selon la période indiquée. En l'absence de données pour l'Allemagne unifiée (à compter du 3 octobre 1990), les données disponibles sont fournies séparément sous les rubriques République fédérale d'Allemagne et ancienne République démocratique allemande et se rapportent aux limites territoriales antérieures au 3 octobre 1990. Pour les notes explicatives en détail sur les données concernant l'Allemagne, voir Annexe I – Nomenclature des pays.

1 Taux des fonds du système fédéral.

Technical notes, tables 31-35

Detailed information and current figures relating to tables 31-35 are contained in *International Financial Statistics*, published monthly by the International Monetary Fund [9] and in the United Nations *Monthly Bulletin of Statistics*. [23]

Table 31: "Money" relates to the liabilities of the monetary system in currency and demand deposits to the domestic private sector. "Quase-Money" comprises time, savings and foreign currency deposits of resident sectors other than central government.

Table 32: Rates shown represent those rates at which the central bank either discounts or makes advances against eligible commercial paper and/or government securities for commercial banks or brokers. For countries with more than one rate applicable to such discounts or advances, the rate shown is the one at which the largest proportion of central bank credit operations is understood to be transacted.

Table 33: These series measure the trend of long-term interest rates. They usually comprise the yield to maturity (YM), yield in perpetuity (YP) or the current yield (CY) of a single issue of government or quasi-government bonds whose ownership is widely distributed and with a final maturity of not less than 10 years.

Table 34: Index numbers of market prices of industrial shares are based primarily on common shares traded on the leading exchange or exchanges of each country and are intended to cover a representative sample of industrial companies. In some cases where an industrial index is not available, a general index, including shares of companies in the utilities, transportation, distribution, and finance fields, is reported.

The indexes are intended to show the increase in value occurring to an investor who bought in the base period a list of shares corresponding in selection and amount to those contained in the index, who subsequently made only such changes in holdings as the substitution of new issues for retired issues, and who retained, in the form of shares, the proceeds of all rights, warrants and share dividends made available since the base period.

Table 35: Treasury bill rates represent the average tender rates per annum on new issues of bills (ordinarily 3-month issues) offered by the treasury during the period. Call money rates are the average rates per annum on loans which were available on demand in the open market during the period.

Notes techniques, tableaux 31-35

Les informations détaillées et les chiffres courants concernant les tableaux 31-35 figurent dans les *Statistiques financières internationales* publiées chaque mois par le Fonds monétaire international [9] et dans le *Bulletin mensuel de statistique* des Nations Unies [23].

Tableau 31 : Par "monnaie", on entend le passif du système monétaire sous forme de monnaie fiduciaire et de dépôts à vue à l'égard du secteur privé national. La Quasi-monnaie comprenant les dépôts à terme, d'épargne et en devises des secteurs résidents autres que l'administration centrale.

Tableau 32 : Les taux indiqués représentent les taux pratiqués par la banque centrale à l'escompte, ou pour avance de fonds, dans toute transaction portant sur des effets de commerce ou des obligations de l'Etat détenus par les banques commerciales ou des courtiers. Pour les pays où il existe plus d'un taux applicable à de telles transactions, le tableau indique le taux que la Banque centrale semble pratiquer pour la plupart de ses opérations de crédit.

Tableau 33 : Ces séries décrivent l'évolution des taux d'intérêt des obligations à long terme. Elles représentent habituellement le taux de capitalisation (YM), le rendement actuel des obligations perpétuelles (YP) ou le rendement courant (CY) d'une seule émission d'obligations de l'Etat ou d'organismes parapublics qui ont été largement diffusées et dont l'échéance finale n'est pas à moins de dix ans de la date d'émission.

Tableau 34 : Les indices des cours des actions industrielles se rapportent principalement aux actions ordinaires qui font l'objet de transactions à la Bourse principale ou aux Bourses principales de chaque pays et qui ont été réunies en un échantillon aussi représentatif que possible de sociétés industrielles. Dans certains cas, à défaut d'un indice industriel, il est donné un indice général, comprenant les actions de sociétés d'utilité publique, de transport, de distribution et de services financiers.

Les indices visent à mesurer l'accroissement de la valeur du portefeuille d'un investisseur qui a acheté pendant l'année choisie comme base un ensemble d'actions correspondant par sa composition et son montant à celles utilisées dans l'indice, dont il n'a changé la composition qu'en substituant de nouvelles actions en remplacement de celles qui sont retirées du marché et qui a retenu sous forme d'actions le montant des droits, warrants et dividendes distribués depuis l'année choisie comme base.

Tableau 35 : Les taux d'intérêt des bons du Trésor représentent les taux moyens annuels d'adjudication des nouvelles émissions de bons (généralement à trois mois) offertes par le Trésor au cours de la période considérée. Les taux d'intérêt de l'argent au jour le jour représentent les taux moyens annuels des prêts pouvant être obtenus sans préavis sur l'"open market" pendant la période considérée.

Table 36 follows overleaf
Le tableau 36 est présenté au verso

36
Employment by industry
Emploi par industrie

		Persons employed, by branch of economic activity (000s) Personnes employées, par branches d'activité économique (000s)								
Country or area Pays ou zone	Year Année	Total employment (000s) Emploi total (000s)		Agriculture, hunting, forestry and fishing Agriculture, chasse sylviculture, pêche		Mining and quarrying Industries extractives		Manufacturing industries manufacturières		Electricity, gas, water Electricité, gaz, eau	
		M	F	M	F	M	F	M	F	M	F
Albania[1][2]	1980	397.2	262.4	75.0	62.6	...	...	126.7[3]	102.0[3]	...	...
Albanie[1][2]	1991	494.1	356.4	107.2	88.0	...	...	151.2[3]	141.1[3]	...	...
Angola[4]	1983	355.5	...	84.1	...	...	...	57.8	...	...	...
Angola[4]	1986	367.6	...	75.5	...	...	...	66.1	...	...	...
Australia[5][6]	1980	3997.8	2286.5	313.0	94.3	76.8	6.2	937.3	309.3	119.5	10.1
Australie[5][6]	1991	4489.3	3223.8	295.9	125.5	84.7	8.7	811.6	316.0	91.9	11.5
Austria[1][6]	1983	1947.0	1211.4	162.2	151.0	13.5	1.7	657.6	236.7	37.7	3.7
Autriche[1][6]	1990	2024.8*	1396.3*	137.8	130.8	10.9	0.7	676.5	245.3	35.2	5.0
Bahrain[7]	1988	80.5	5.9	2.2	0.1	0.9	0.0	17.0	0.6	2.7	0.0
Bahreïn[7]	1991	88.1	7.5	0.7	0.3	0.1	0.4	18.2	1.3	3.1	0.4
Barbados[5][6]	1981	57.4	42.8	5.6	3.8	...	...	6.4	7.7	1.1	0.1
Barbade[5][6]	1991	56.2	45.2	3.6	2.0	...	...	5.2	5.0	1.4	0.3
Belgium[1][6][9]	1980	2409.7	1290.5	90.2	25.4	27.8	0.5	702.2	219.1	30.3	2.4
Belgique[1][6][9]	1990	2268.2	1495.8	73.8	26.2	7.7	0.4	596.8	185.4	26.7	3.3
Benin[4]	1980	66.2	...	5.0	...	0.1	...	6.5	...	0.7	...
Bénin[4]	1985	80.8	...	6.1	...	0.4	...	8.3	...	1.1	...
Bermuda[9][10]	1980	16.9	12.8	0.2	0.0	0.1	0.0	0.7	0.4	0.3	0.1
Bermudes[9][10]	1989	19.0	17.3	0.4	0.0	0.1	0.0	0.7	0.4	0.4	0.1
Bolivia[5][6]	1980	1316.7	402.9	686.1	113.5	70.7	5.0	107.7	69.4	6.3	0.5
Bolivie[5][6]	1990	1375.2	468.2	743.8	129.6	43.1	4.4	79.5	50.8	8.2	0.8
Botswana[5][9]	1980	64.0	19.4	3.7	0.5	6.9	0.3	4.6	0.9	1.4	0.0
Botswana[5][9]	1991	146.7	76.1	4.8	1.9	7.3	0.5	16.4	9.6	2.3	0.2
Brazil[5][6]	1981	31266.0	14199.0	10495.0	2805.0	...	...	5791.0[3]	1771.0[3]	...	...
Brésil[5][6]	1988	38222.0	20507.0	11224.0	3010.0	...	...	6582.0[8]	2404.0[8]	889.0	107.0
British Virgin Islands[4]	1983	4.6	...	0.1	...	0.0	...	0.2	...	0.1	...
Iles Vierges brit.[4]	1987	6.6	...	0.1	...	0.0	...	0.4	...	0.2	...
Brunei Darussalam[4][9][12]	1980	27.1	...	0.5	...	4.8	...	2.2	...	...	...
Brunéi Darussalam[4][9][12]	1986	30.0	...	0.7	...	4.8	...	2.7	...	...	...
Bulgaria[4][13]	1980	4363.9	...	1062.7	...	108.6	...	1401.0	...	25.4	...
Bulgarie[4][13]	1991	3466.4	...	659.5	...	...	...	1216.9[3]	...	...	...
Burundi[4][9][14]	1984	44.1	...	7.8	...	0.5	...	4.9	...	0.8	...
Burundi[4][9][14]	1990	47.3	...	6.9	...	0.4	...	7.0	...	1.3	...
Canada[5][6]	1980	6459.0	4249.0	450.0	134.0	172.0	23.0	1543.0[15]	568.0[15]	105.0[16]	19.0[16]
Canada[5][6]	1991	6751.0	5589.0	395.0	159.0	150.0	23.0	1331.0[15]	534.0[15]	105.0[16]	31.0[16]
Central African Rep.[4]	1980	14.5	...	4.7	...	...	...	2.1	...	0.6	...
Rép. centrafricaine[4]	1986	16.1	...	2.9	...	...	...	5.3	...	0.9	...
Chad[10]	1986	8.4	0.5	1.6	0.1	...	...	0.5	0.0	0.5	0.0
Tchad[10]	1989	10.7	0.6	0.9	0.0	0.3	0.0	2.0	0.0	0.1	0.0
Chile[5][6][17]	1980	2297.9	959.3	503.0	26.9	69.8	2.0	374.8	149.1	22.9	1.5
Chili[5][6][17]	1991	3154.3	1386.1	783.2	83.4	92.6	5.0	562.9	190.1	19.2	2.3
China[9][18]	1985	60795.0	29100.0	5055.0	2776.0	...	...	25221.0[3]	12924.0[3]	...	...
Chine[9][18]	1989	66867.0	34220.0	5076.0	2867.0	6218.0	1589.0	20292.0	13144.0	1089.0	395.0
Colombia[6][9]	1985	1916.5	1183.4	33.5	8.6	9.3	1.4	443.5	247.4	19.9	3.2
Colombie[6][9]	1991	2716.5	1894.1	47.2	14.5	15.7	3.0	622.2	455.5	32.2	6.3
Costa Rica[5][6][9]	1980	548.3	176.4	188.5	10.4	...	...	83.0[8]	34.9[8]	...	...
Costa Rica[5][6][9]	1991	711.4	295.2	236.6	19.8	1.3	0.2	113.8	74.9	9.6	1.8
Côte d'Ivoire[4]	1982	446.8	...	72.4	...	...	...	71.5[3]	...	...	...
Côte d'Ivoire[4]	1990	385.0	...	53.3	...	...	...	60.0[3]	...	...	...
Cuba[2]	1980	1784.6	822.0	437.3[21]	74.1[21]	...	...	420.4[3]	145.3[3]	11.1	2.6
Cuba[2]	1988	2135.6	1309.8	519.4	142.7	29.9	5.3	452.2	230.9	29.2	9.7
Cyprus[5]	1980	122.8	69.7	18.3	18.6	1.6	0.1	21.5	18.4	1.4	0.1
Chypre[5]	1991	157.8*	101.3*	18.2	16.2	0.7	...	25.3	23.2	1.3	0.1
Czechoslovakia	1980[1]	4389.0	3772.0	592.0	443.0	171.0	35.0	1459.0	1274.0	76.0	31.0
Tchécoslovaquie	1990[13]	4317.0	3932.0	576.0	367.0	219.0	41.0	1349.0	1247.0	89.0	35.0
Denmark[6][22]	1981	1307.8	1060.9	137.6	36.7	2.1	0.2	357.5	147.6	14.1	2.0
Danemark[6][22]	1990	1448.3	1221.7	112.6	34.4	2.2	...	359.5	172.2	17.3	2.5
Ecuador[4][6][9]	1987	1147.5	...	28.2[8]	...	...	...	243.4	...	...	...
Ecuador[4][6][9]	1989	2192.3	...	151.8[8]	...	...	...	380.8	...	...	...

Construction Construction		Trade, restaurants and hotels Commerce, restaurants, hôtels		Transport, storage, communications Transports, entrepôts, communications		Finance, insurance, real est.,bus. services Services financierès, immob.,et apparentées		Community, social and personal services Services à collectivité services soc. et pers.	
M	F	M	F	M	F	M	F	M	F
65.7	6.5	23.5	24.5	23.4	4.4	10.6	4.8	72.3	57.6
69.7	6.4	27.7	29.6	32.1	6.4	12.8	13.9	93.4	71.0
19.8	...	30.0	...	24.1	...	1.1	...	132.7	...
21.6	...	31.9	...	27.4	...	1.6	...	135.2	...
439.1	47.8	813.6	680.7	378.2	82.1	286.5	231.2	628.3	822.3
469.6	68.7	1030.1	908.7	415.0	119.4	463.1	427.2	820.5	1237.5
251.8	22.5	231.4	308.3	170.2	35.6	94.8	78.8	326.6	372.2
261.2	25.1	263.0	370.8	175.4	42.5	111.1	109.6	346.8	457.8
29.3	0.2	13.1	0.8	4.1	0.4	4.4	1.5	7.4	2.2
29.2	0.3	15.4	1.2	7.4	1.6	4.3	1.5	8.4	1.6
6.1[8]	0.2[8]	12.6	10.8	3.6	1.2	1.7	2.2	20.3	16.8
8.0[8]	0.5[8]	8.0	8.0	3.5	1.3	1.5	2.4	20.7	21.1
274.9	12.3	300.6	282.2	238.9	36.1	141.7	84.3	603.1	628.2
221.1	14.6	319.5	314.9	213.2	43.9	190.6	137.0	618.8	770.2
7.4	...	8.0	...	8.8	...	2.2	...	27.6	...
5.5	...	9.1	...	9.0	...	2.5	...	40.0	...
1.7	0.1	6.1	5.3	1.6	0.6	1.4	2.0	4.6	4.3
2.7	0.2	5.9	6.0	1.7	0.8	1.7	3.3	5.3	6.5
93.8	0.8	58.5	68.7	89.9	3.0	11.2	2.6	192.5	139.4
49.3	0.5	68.5	81.7	130.4	5.1	12.5	3.2	239.9	192.0
13.1	0.3	6.2	4.3	3.2	0.3	3.3	1.1	21.6	11.6
30.1	3.7	20.3	20.8	7.2	1.9	11.0	5.2	45.2	31.0
3593.0	71.0	3301.0	1387.0	1619.0	149.0	825.0	398.0	5643.0[11]	7617.0[11]
3629.0	97.0	4384.0	2405.0	2020.0	189.0	2349.0	1147.0	7145.0[11]	11148.0[11]
0.5	...	0.4	...	0.4	...	0.2	...	2.6	...
0.5	...	0.6	...	0.4	...	0.3	...	4.1	...
9.6	...	4.5	...	1.7	...	1.3	...	2.7	...
9.2	...	6.4	...	1.4	...	2.0	...	2.8	...
328.7	...	370.1	...	295.3	...	54.4	...	717.7	...
231.0	...	324.7	...	259.8	...	51.1	...	723.4	...
6.3	...	3.7	...	3.3	...	1.5	...	13.1	...
5.8	...	3.7	...	2.9	...	2.8	...	16.5	...
570.0	55.0	1274.0	1137.0	613.0	170.0	492.0	526.0	1241.0	1618.0
615.0	80.0	1502.0	1433.0	571.0	210.0	701.0	770.0	1381.0	2350.0
0.8	...	4.8	...	1.2	...	0.0	...	0.2	...
0.8	...	3.6	...	0.8	...	0.4	...	...	...
1.1	0.0	1.2	0.0	1.2	0.1	0.1	0.1	1.9	0.1
1.7	0.0	3.3	0.1	0.2	0.0	0.3	0.1	1.9	0.3
148.4	3.4	365.3	224.6	194.3	16.6	74.1	26.8	539.8	507.0
314.6	7.2	451.4	323.2	278.9	29.5	151.4	76.5	499.2	668.4
4602.0	1192.0	5028.0	3407.0	4623.0	1225.0	599.0	331.0	14837.0[19]	7021.0[19]
4550.0	1233.0	5639.0	4167.0	4947.0	1453.0	864.0	492.0	17396.0[19]	8637.0[19]
184.0	10.9	481.2	299.9	179.6	18.1	135.6	74.9	438.8	518.2
235.1	13.7	683.4	513.5	250.9	27.2	223.8	129.5	601.3	727.6
55.7	0.5	90.8[19]	40.5[19]	44.3[20]	3.2[20]	...	...	84.5	86.5
62.7	0.7	97.9	58.9	40.4	3.2	28.1	8.7	113.1	125.6
28.4	...	34.7	...	63.8	...	...	...	176.0[19]	...
17.4	...	25.9	...	53.5	...	...	...	174.9[19]	...
243.4	24.2	190.9	135.5	163.4	32.9	19.1	15.0	299.0	392.4
280.1	42.5	223.3	213.2	187.4	54.8	33.5	29.6	380.6	581.1
20.3	1.7	20.7	13.0	7.9	1.6	5.2	3.0	22.2	12.4
21.8	1.4	34.3	26.3	11.6	3.8	9.1	7.9	32.5	21.6
583.0	96.0	250.0	647.0	344.0	170.0	128.0	126.0	777.0	938.0
548.0	99.0	228.0	700.0	359.0	192.0	140.0	147.0	795.0	1091.0
155.9	16.0	164.7	171.9	125.2	38.0	83.7	73.7	256.7	567.4
152.7	19.4	205.5	185.4	136.0	52.8	132.8	113.5	319.3	633.0
91.8	...	282.6	...	67.5[20]	...	71.5	...	361.8	...
148.9	...	577.5	...	143.4[20]	...	111.4	...	678.4	...

36
Employment by industry [*cont.*]
Emploi par industrie [*suite*]

		Total employment (000s) Emploi total (000s)		Agriculture, hunting, forestry and fishing Agriculture, chasse sylviculture, pêche		Mining and quarrying Industries extractives		Manufacturing industries industries manufacturières		Electricity, gas, water Electricité, gaz, eau	
Country or area Pays ou zone	Year Année	M	F	M	F	M	F	M	F	M	F
Egypt	1980	9106.3	692.8	4086.2	65.7	17.8	2.1	1364.2	74.8	78.5	4.7
Egypte	1984	9732.9	2085.7	3826.1	971.2	31.3	0.8	1439.5	203.2	84.0	7.5
El Salvador[4 7 23]	1980	240.4	...	...	...	...	...	54.3	...	3.6	...
El Salvador[4 7 23]	1989	233.3	...	...	...	...	...	47.0	...	5.0	...
Estonia[4 13]	1980	788.0	...	109.7	...	...	...	267.5[3]	...	...	...
Estonie[4 13]	1991	790.0	...	100.9	...	...	...	257.0[3]	...	...	...
Fiji[4 9]	1980	80.5	...	2.6	...	1.1	...	15.4	...	2.3	...
Fidji[4 9]	1991	91.5	...	2.6	...	1.1	...	23.4	...	2.7	...
Finland[16]	1980	1271.0	1088.0	186.0	127.0	...	...	395.0[3]	233.0[3]	...	...
Finlande[16]	1991	1232.0	1134.0	126.0	72.0	4.0	...	309.0	161.0	22.0	6.0
France[1]	1980	13208.0	8429.6	1221.0	632.8	...	...	5797.9[3]	1865.5[3]	...	...
France[1]	1991	12702.7	9364.9	853.7	402.8	...	...	4842.5[3]	1581.8[3]	...	...
French Polynesia[4]	1982	34.0	...	0.6	...	0.2	...	1.6	...	0.3	...
Polynésie française[4]	1989	41.7	...	0.9	...	0.1	...	2.2	...	0.4	...
Gambia[7 9]	1983	25.0	4.3	2.2	0.3	...	...	3.1	0.9	1.4	0.0
Gambie[7 9]	1987	21.1	5.0	1.4	0.6	...	...	1.9	0.5	0.8	0.1
Germany **ℊ** · Allemagne **ℊ**											
F. R. Germany[6 9]	1980	16782.0	10092.0	732.0	706.0	335.0	26.0	6386.0	2747.0	218.0	36.0
R. f. Allemagne[6 9]	1990	17585.0	11750.0	601.0	469.0	230.0	14.0	6535.0	2735.0	225.0	46.0
former German D. R.	1980	3847.7	3909.9	201.8	135.6	...	...	1988.5[3]	1507.7[3]	...	...
anc. R. d. allemande	1988	3998.5	3951.3	197.5	117.1	...	...	2092.3[3]	1442.9[3]	...	...
Ghana[4 9]	1980	337.2	...	54.9	...	23.8	...	35.1	...	6.6	...
Ghana[4 9]	1988	306.8	...	27.1	...	21.9	...	43.6	...	6.0	...
Gibraltar[9 23]	1980	8.9	2.9	...	...	...	...	2.6	0.5	0.2	0.0
Gibraltar[9 23]	1990	9.3	4.9	...	...	...	...	0.8	0.1	0.3	0.0
Greece[16 17]	1981	2423.5	1107.5	622.4	461.1	18.8	0.4	486.0[15]	194.8[15]	26.1	4.2
Grèce[16 17]	1990	2409.3	1309.7	492.7	396.5	20.7	1.9	501.8[15]	218.1[15]	32.0	4.6
Guam[9 13]	1982	18.2	12.5	0.1	...	...	...	0.9	0.3	...	...
Guam[9 13]	1987	25.8	18.3	0.2	...	...	...	1.0	0.7	...	...
Guatemala[4 13]	1980	755.5	...	373.5	...	3.7	...	83.1	...	13.5[16]	...
Guatemala[4 13]	1991	786.9	...	237.5	...	2.8	...	118.8	...	14.8[16]	...
Haiti[5 9]	1980	999.2	954.4	812.4	507.2	1.0	0.2	57.6	74.5	1.5	0.1
Haïti[5 9]	1988	1123.3	665.1	854.9	329.9	9.0	8.6	66.3	49.2	2.2	0.7
Honduras[6]	1985	162.1	118.6	2.5	0.3	0.6	0.4	32.2	22.4	2.6	0.4
Honduras[6]	1990	903.6	309.4	589.1	18.0	2.0	0.2	69.0	68.6	7.3	0.4
Hong Kong[5 6]	1980	1458.0	779.8	22.5	9.5	0.6	...	516.0	427.1	10.7	0.9
Hong-kong[5 6]	1991	1719.0	1029.5	15.7	6.5	0.4	...	424.0	291.7	17.0	1.1
Hungary[25]	1980	2782.4	2261.7	679.1	433.4	...	...	927.3[3]	748.3[3]	...	...
Hongrie[25]	1990	2569.9	2408.1	545.9	360.8	...	...	834.4[3]	661.3[3]	...	...
Iceland[4 6]	1980	105.9	...	14.0	...	...	...	25.9[8]	...	0.9	...
Islande[4 6]	1990	124.6	...	13.1	...	...	...	23.3[8]	...	1.1	...
India[9 27 28]	1980	19603.0	2702.0	829.0	461.0	836.0	86.0	5309.0	563.0	679.0	16.0
Inde[9 27 28]	1989	22442.0	3544.0	960.0	475.0	976.0	78.0	5669.0	568.0	879.0	28.0
Indonesia[16]	1982	37064.6	20738.2	20443.2	11150.1	323.9	66.8	3138.9	2883.1	54.6	1.1
Indonésie[16]	1990	46427.9	29422.7	25827.4	16550.9	442.5	85.7	4210.1	3483.2	123.9	10.9
Ireland[16 9]	1983	777.6	346.4	167.6	21.8	9.1	0.5	160.4	57.6	13.8	1.3
Irlande[16 9]	1989	737.7	352.2	151.2	12.1	6.6	0.5	150.9	64.5	12.0	1.5
Israel[5 6 29]	1980	801.9	452.2	60.7	18.6	...	...	229.9[8]	64.3[8]	11.2	1.7
Israël[5 6 29]	1991	953.0	630.3	43.7	11.8	...	...	254.5[8]	85.2[8]	14.5	2.4
Italy[16]	1980	14135.0	6540.0	1851.0	1048.0	201.0[20]	19.0[20]	3715.0	1724.0	...	...
Italie[16]	1990	14028.0	7427.0	1216.0	679.0	207.0[20]	23.0[20]	3198.0	1559.0	...	...
Jamaica[5 6]	1980	424.7	273.8	198.3	62.3	7.0	1.2	53.6	19.3	...	...
Jamaïque[5 6]	1990	514.2	379.3	174.4	58.4	6.4	0.8	96.3	39.8	...	...
Japan[16]	1980	33940.0	21420.0	2940.0	2830.0	100.0	10.0	8400.0	5270.0	260.0	40.0
Japon[16]	1991	37760.0	25920.0	2260.0	2010.0	50.0	10.0	9410.0	6090.0	280.0	40.0
Jordan[7 9 23]	1980	96.2	20.4	...	...	5.9	0.0	11.9	1.3	1.9	0.0
Jordanie[7 9 23]	1989	172.3	52.7	...	...	6.6	0.1	30.6	4.0	4.5	0.2
Kenya[9 30]	1980	829.0	176.8	186.4	44.9	2.2	0.1	128.2	13.1	9.6	0.5
Kenya[9 30]	1990	1100.5	308.5	205.3	64.4	3.3	0.9	166.5	21.2	19.3	3.1

Construction Construction		Trade, restaurants and hotels Commerce, restaurants, hôtels		Transport, storage, communications Transports, entrepôts, communications		Finance, insurance, real est.,bus. services Services financierès, immob.,et apparentées		Community, social and personal services Services à collectivité services soc. et pers.	
M	F	M	F	M	F	M	F	M	F
419.5	6.1	836.4	47.9	475.7	27.6	106.5	20.3	1600.3	381.5
595.8	14.2	848.8	160.5	530.1	26.2	130.6	32.3	1988.8	561.9
26.2	...	17.2	...	12.0	...	7.8	...	119.2	...
21.6	...	16.8	...	12.2	...	7.8	...	123.0	...
71.5	...	72.1	...	75.2	...	4.0	...	176.8	...
78.0	...	69.0	...	67.5	...	4.0	...	198.5	...
9.0	...	13.4	...	8.1	...	4.4	...	24.1	...
7.0	...	14.5	...	9.0	...	5.9	...	25.4	...
160.0	16.0	137.0	191.0	135.0	48.0	44.0	83.0	207.0	386.0
162.0	17.0	156.0	203.0	128.0	47.0	89.0	110.0	234.0	517.0
...	...	6189.1[19]	5931.3[19]	...	...	...	...	...	...
...	...	7006.5[19]	7380.3[19]	...	...	...	...	...	...
4.7	...	7.4	...	2.3	...	1.5	...	15.4	...
5.3	...	9.5	...	9.7	...	2.8	...	15.3	...
1.9	0.0	2.5	0.4	5.5	0.3	0.4	0.1	8.0	2.2
2.9	0.0	3.7	0.8	2.7	0.4	0.8	0.3	7.0	2.3
1972.0	188.0	1651.0	2127.0	1201.0	326.0	814.0	735.0	3474.0	3202.0
1720.0	217.0	1820.0	2487.0	1254.0	436.0	1282.0	1113.0	3918.0	4234.0
484.3	92.7	219.5[19]	632.5[19]	407.9	230.8	...	...	545.8[11][24]	1310.6[11][24]
447.2	91.0	234.7[19]	641.9[19]	427.8	227.9	...	...	599.0[11][24]	1430.6[11][24]
22.4	...	22.1	...	17.0	...	11.0	...	144.3	...
16.6	...	15.0	...	15.6	...	15.7	...	145.3	...
2.3	0.1	1.4	0.8	0.6	0.1	0.2	0.2	1.3	1.0
2.1	0.1	1.8	1.4	0.5	0.1	0.6	0.9	2.6	1.8
290.5	2.1	359.7	168.9	247.4	26.4	78.1	38.6	292.9	210.9
248.4	3.9	405.2	248.8	221.3	28.1	110.6	73.5	375.9	383.6
1.6	0.1	2.9	2.6	1.3[20]	0.3[20]	0.5	0.8	2.3[11]	1.8[11]
4.2	0.2	4.0	4.0	1.5[20]	0.7[20]	0.8	1.2	4.0[11]	3.7[11]
27.7	...	58.9[19]	...	21.6	...	...	...	173.5	...
14.0	...	99.5[19]	...	23.2	...	...	...	276.3	...
21.9	0.3	28.0	293.2	14.0	1.4	3.2	0.5	59.6	77.0
19.1	3.3	65.2	195.6	14.6	2.0	2.2	1.3	62.9	52.5
13.9	0.5	38.1	30.6	11.3	1.8	7.8	4.6	52.8	57.9
52.5	1.1	68.8	102.3	25.3	3.4	10.7	5.5	76.6	109.7
157.0	10.1	319.1	129.4	147.5	15.7	64.9	38.0	219.3	149.0
213.6	10.9	431.0	299.7	232.7	40.4	127.0	101.6	257.6	277.5
327.0	71.2	173.8	314.2	303.1	100.5	...	...	372.1[19][26]	594.1[19][26]
264.4	72.0	188.5	389.7	289.7	137.3	...	...	447.0[19][26]	787.0[19][26]
10.7	...	14.2	...	7.7	...	5.7	...	26.8	...
12.4	...	18.1	...	8.4	...	10.0	...	38.2	...
1077.0	64.0	364.0	20.0	2646.0	76.0	825.0	72.0	7048.0	1344.0
1186.0	59.0	403.0	28.0	2946.0	131.0	1206.0	153.0	8217.0	2025.0
2103.0	43.2	4328.8	4225.2	1778.6	17.5	92.4	20.5	4800.9	2324.5
2000.1	59.4	5290.2	5777.1	2259.4	53.1	353.7	124.7	5863.0	3207.3
84.2	3.3	110.3	75.8	56.5	13.2	44.5	33.3	128.9	137.6
68.2	2.1	110.0	59.7	53.6	12.1	27.7	24.4	144.4	168.6
74.1	5.1	94.9	50.8	70.5	15.5	52.5	50.4	202.2	242.5
90.9	5.3	143.7	80.4	75.3	21.3	85.8	75.0	237.4	345.8
1972.0	69.0	2533.0	1265.0	1005.0	129.0	369.0	153.0	2489.0	2133.0
1760.0	99.0	2856.0	1681.0	983.0	163.0	528.0	368.0	3281.0	2855.0
24.3[15]	0.7[15]	...	...	24.7[16][20]	7.5[16][20]	29.3[11]	58.3[11]	84.7	123.0
57.2[15]	1.8[15]	...	...	32.3[16][20]	9.3[16][20]	49.2[11]	95.4[11]	95.4	171.6
4720.0	770.0	6720.0	5760.0	3090.0	410.0	1830.0	1350.0	5810.0[11]	4940.0[11]
5030.0	1010.0	7280.0	7050.0	3200.0	580.0	2950.0	2410.0	7110.0[11]	6610.0[11]
3.4	0.1	7.1[19]	0.3[19]	5.9	0.8	4.4	0.8	55.8	17.1
4.8	0.1	10.1[19]	0.7[19]	10.9	1.0	8.2	2.9	96.7	43.6
60.7	2.4	63.2	7.3	48.9	6.3	33.4	6.3	296.3	95.8
67.4	4.0	95.5	18.5	63.8	10.4	50.8	14.4	428.6	171.6

36
Employment by industry [cont.]
Emploi par industrie [suite]

Country or area Pays ou zone	Year Année	Total employment (000s) Emploi total (000s)		Agriculture, hunting, forestry and fishing Agriculture, chasse sylviculture, pêche		Mining and quarrying Industries extractives		Manufacturing industries industries manufacturières		Electricity, gas, water Electricité, gaz, eau	
		M	F	M	F	M	F	M	F	M	F
Korea, Republic of[56] Corée, Rép. de[56]	1980 1991	8462.0 11068.0	5222.0 7508.0	2620.0 1706.0	2034.0 1396.0	113.0 62.0	11.0 6.0	1800.0 2878.0	1155.0 2058.0	39.0 58.0	5.0 9.0
Luxembourg[1] Luxembourg[1]	1983 1990	106.0 125.0	51.9 64.6	4.9 4.3	2.6 1.9	0.2 0.2	... ...	34.1 33.1	5.2 4.1	1.3 1.2	0.1 0.2
Malawi[31] Malawi[31]	1980 1989	328.2 379.0	42.2 62.6	157.2 177.7	25.1 39.4	0.6 0.2	... 0.0	38.0 50.3	2.0 5.5	3.8 4.5	0.3 0.4
Malaysia[56] Malaisie[56]	1980 1990	3185.8 4310.7	1601.6 2374.3	1079.5 1137.8	701.1 599.8	40.1 32.7	6.3 4.2	457.6 697.3	294.1 635.5	62.1 42.2	3.6 4.4
Malta[9] Malte[9]	1981 1989	85.8 93.9	29.4 32.2	5.7 2.8	0.9 0.4	1.2 0.7	... ...	23.4[15] 24.2[15]	13.5[15] 11.2[15]	1.2[16] 1.7[16]	... 0.1[16]
Mauritius[9] Maurice[9]	1980 1991	145.5 180.8	52.1 101.7	42.7[32] 34.3[32]	15.4[32] 12.5[32]	0.1 0.1	0.1 0.1	16.0 43.7	20.4 64.8	4.3 3.4	0.1 0.1
Montserrat[45] Montserrat[45]	1980 1987	4.6 5.2	... ...	0.5 0.5	... ...	0.0 0.0	... ...	0.5 0.6	... ...	0.1 0.1	... ...
Mozambique[4 25] Mozambique[4 25]	1987 1988	192.7 201.6	... ...	15.1 16.9	... ...	5.0 4.9	... ...	111.6 117.0	... ...	3.8 2.9	... ...
Myanmar[45] Myanmar[45]	1980 1990	13208.0 15221.0	... ...	8864.0 10614.0	... ...	68.0 78.0	... ...	1009.0 1137.0	... ...	16.0 17.0	... ...
Netherlands[16] Pays-Bas[16]	1981 1991	3518.0 4004.0	1590.0 2517.0	229.0 211.0	42.0 82.0	8.0 13.0	1.0 ...	865.0 932.0	177.0 237.0	43.0 38.0	4.3 6.0
Netherlands Antilles[45] Antilles néerlandaises[45]	1980 1989[33]	76.0 43.8	... ...	0.3 ...	... ...	0.4 0.4[34]	... ...	10.1 4.2	... ...	1.2 0.8	... ...
New Caledonia[4] Nouvelle-Calédonie[4]	1985 1990	33.4 44.4	... ...	1.6 2.1	... ...	1.0 1.1	... ...	4.6[20] 4.6	... ...	... 0.6	... ...
New Zealand[56] Nouvelle-Zélande[56]	1986 1991	903.0 815.0	641.0 636.0	116.0 107.0	49.0 49.0	6.0 3.0	0.0 1.0	217.0 173.0	101.0 74.0	14.0 11.0	2.0 2.0
Nicaragua[4] Nicaragua[4]	1980 1991	146.4 229.0	... ...	4.5 30.5	... ...	3.4 1.5	... ...	30.1 36.8	... ...	3.4 5.6	... ...
Niger Niger	1980 1989	25.1 25.5	0.9 2.6	1.6 1.7	... 0.1	5.8 3.9	0.0 0.1	2.6 2.8	0.0 0.1	1.2 3.3	0.0 0.2
Norway[16] Norvège[16]	1980 1991	1133.0 1095.0	775.0 914.0	112.0 86.0	47.0 30.0	10.0 17.0	1.0 4.0	291.0 219.0	92.0 75.0	17.0 17.0	2.0 5.0
Pakistan[56] Pakistan[56]	1985 1991	24360.0 26070.0	2601.0 3758.0	11677.0 11671.0	1954.0 2481.0	46.0 45.0	... ...	3390.0 3147.0	296.0 501.0	183.0 245.0	3.0 3.0
Panama[5 69] Panama[5 69]	1982 1991	399.1 497.4	162.0 224.7	148.6 185.9	8.9 6.2	0.3 0.8	... 0.3	38.1 46.8	16.2 21.4	6.6 7.4	1.0 2.2
Paraguay[56] Paraguay[56]	1982 1991	156.6 292.9	104.5 202.5	2.3 5.0	0.4 0.8	0.3 1.1	0.1 ...	25.4 51.1	13.1 27.0	3.8 4.9	0.4 0.8
Peru[16] Pérou[16]	1987 1991	1233.3 1459.0	827.8 918.0	10.5 16.8	5.9 3.7	13.0 8.6	0.6 1.6	314.9 312.6	138.5 140.4	5.5 7.7	0.6 0.8
Philippines[5 69] Philippines[5 69]	1980 1990	11083.0 14347.0	6070.0 8185.0	6629.0 7620.0	2265.0 2565.0	117.0 121.0	13.0 12.0	979.0 1187.0	871.0 1001.0	47.0 75.0	8.0 16.0
Poland[45] Pologne[45]	1981 1990	18005.7 17552.1	... ...	5389.7 4597.0	... ...	539.0 607.9	... ...	5089.3 4335.9	... ...	154.4 148.7	... ...
Portugal[16 35] Portugal[16 35]	1980[17] 1990	2419.0 2723.7	1542.0 1993.8	540.0 425.4	542.0 420.3	22.0 33.3	1.0 2.7	651.0 664.5	378.0 497.3	17.0 37.9	2.0 4.4
Puerto Rico[56] Porto Rico[56]	1980 1991	487.0 567.0	273.0 359.0	39.0 32.0	1.0 1.0	... ...	... ...	76.0 88.0	65.0 68.0	12.0 13.0	1.0 3.0
Romania[5] Roumanie[5]	1980 1990	5803.3 5838.2	4546.8 5801.3	1298.6 1388.1	1789.0 1708.8	226.8 250.6	31.9 55.4	1969.4 1968.5	1406.3 1661.3	36.7 64.1	7.6 15.2
San Marino[69] Saint-Marin[69]	1980 1991	6.2 7.8	3.4 5.1	0.3 0.2	0.2 0.1	... ...	... ...	2.3 3.0	1.5 1.5	0.0 ...	0.0 ...
Seychelles[4] Seychelles[4]	1980 1989	17.9 22.3	... ...	1.9 2.2	... ...	... ...	... ...	1.1 2.5	... ...	0.7 ...	... ...
Sierra Leone[47 9 27] Sierra Leone[47 9 27]	1980 1987	69.9 67.3	... ...	6.4 7.1	... ...	6.9 6.9	... ...	7.4 7.5	... ...	1.8 2.2	... ...
Singapore[69] Singapour[69]	1980 1991	695.0 917.8	373.9 606.5	10.6 3.6	3.5 0.7	1.3 0.4	0.3 0.1	168.8 240.2	143.9 189.4	8.6 5.9	1.1 1.2

Construction Construction		Trade, restaurants and hotels Commerce, restaurants, hôtels		Transport, storage, communications Transports, entrepôts, communications		Finance, insurance, real est.,bus. services Services financierès, immob.,et apparentées		Community, social and personal services Services à collectivité services soc. et pers.	
M	F	M	F	M	F	M	F	M	F
770.0	72.0	1350.0	1275.0	564.0	56.0	229.0	103.0	978.0	511.0
1388.0	155.0	1921.0	2161.0	891.0	93.0	618.0	399.0	1545.0	1229.0
14.0	0.8	17.1	16.8	8.9	1.6	8.7	6.6	16.8[19]	18.2[19]
17.7	1.1	20.9	19.3	10.6	2.3	9.1	7.7	27.9[19]	28.0[19]
32.6	0.2	23.9	3.2	16.6	0.9	11.2	1.0	44.5	9.5
34.4	0.3	23.6	2.5	19.3	1.2	13.4	1.5	55.6	11.7
257.8	15.6	499.5	192.9	193.0	16.6	...	...	578.4	371.2
404.2	19.7	761.1	456.7	267.2	34.6	163.9	94.6	804.1	524.7
4.9	...	10.3[19]	4.2[19]	7.0	0.9	...	...	32.2[11]	9.9[11]
5.7	0.1	9.2	3.4	8.0	1.0	2.7	1.8	38.8[11]	13.9[11]
7.3	0.1	7.2	2.0	7.7	0.4	3.3	1.2	50.5	12.4
10.0	0.2	14.2	4.1	11.9	1.5	6.9	2.7	52.1	15.6
0.7	...	0.6	...	0.2	...	0.2	...	1.8	...
0.9	...	0.7	...	0.3	...	0.2	...	2.0	...
19.6	...	5.5	...	29.8	...	...	...	2.1	...
21.5	...	6.4	...	29.3	...	...	...	2.7	...
195.0	...	1262.0	...	443.0	...	772.0	...	579.0	...
174.0	...	1409.0	...	385.0	...	956.0	...	455.0	...
447.0	25.0	564.0	347.0	279.0	43.0	257.0	145.0	791.0	792.0
387.0	31.0	628.0	510.0	319.0	85.0	425.0	258.0	1022.0	1291.0
6.1	...	23.8	...	3.9	...	4.1	...	25.7	...
3.9	...	10.7	...	3.8	...	4.3	...	15.5	...
3.1	...	5.6	...	1.6	...	2.8	...	9.2	...
5.5	...	7.4	...	1.9	...	3.3	...	11.6	...
92.0	11.0	141.0	156.0	78.0	31.0	64.0	70.0	172.0	218.0
68.0	8.0	152.0	148.0	63.0	31.0	77.0	74.0	158.0	248.0
6.8	...	16.5	...	8.1	...	8.5	...	61.7	...
9.0	...	20.9	...	10.0	...	14.3	...	94.8	...
7.7	0.2	2.7	0.1	1.6	0.1	1.2	0.1	0.8	0.3
3.5	0.1	2.6	0.4	2.1	0.2	0.9	0.3	4.8	1.0
134.0	8.0	147.0	173.0	136.0	38.0	56.0	45.0	227.0	366.0
120.0	10.0	162.0	192.0	114.0	47.0	86.0	67.0	271.0	482.0
1502.0	8.0	3074.0	37.0	1394.0	8.0	226.0	11.0	2701.0	284.0
1927.0	48.0	3832.0	116.0	1545.0	21.0	260.0	9.0	3380.0	579.0
37.0	1.5	45.5	28.4	28.4	6.1	15.8	8.6	65.8	88.6
22.4	0.3	92.5	52.2	42.5	6.2	18.3	12.7	78.1	121.9
25.6	0.3	32.5	28.7	12.9	2.0	10.9	4.6	42.6	54.7
45.4	...	68.6	63.4	24.9	3.2	20.0	13.5	72.0	93.5
112.8	7.1	309.4	318.7	118.3	14.2	73.4	24.1	275.4	318.3
110.9	...	388.5	389.5	141.3	12.2	106.0	30.1	366.6	339.6
590.0	10.0	601.0	1197.0	695.0	31.0	204.0	104.0	1221.0[11]	1570.0[11]
957.0	18.0	1142.0	2003.0	1092.0	45.0	269.0	175.0	1874.0[11]	2346.0[11]
1414.6	...	1581.5	...	1416.8	...	371.5	...	2434.3	...
1320.9	...	1652.9	...	1127.2	...	337.8	...	3084.3	...
363.0	9.0	282.0	184.0	135.0	25.0	59.0	21.0	345.0	379.0
372.5	12.2	401.1	325.2	172.2	40.3	134.2	74.1	481.7	616.6
45.0	...	101.0	39.0	30.0	5.0	13.0	9.0	170.0[11]	151.0[11]
50.0	1.0	125.0	59.0	31.0	10.0	15.0	16.0	214.0[11]	200.0[11]
761.7	95.9	263.6	356.3	607.5	102.2	11.6	18.3	627.4	739.3
562.0	91.1	246.6	431.9	617.1	135.2	14.4	29.4	726.8	873.0
0.9	0.0	0.7	0.7	0.1	0.0	0.1	0.0	1.7	1.0
1.0	0.0	1.1	1.2	0.2	0.0	0.2	0.1	2.2	2.1
2.7	...	3.4[8]	...	2.2	...	0.8	...	2.7	...
1.7	...	4.4[8]	...	3.1	...	0.7	...	4.6	...
7.8	...	5.6	...	7.9	...	1.9	...	24.2	...
7.3	...	5.0	...	7.5	...	1.8	...	22.2	...
52.9	5.3	158.6	86.1	101.7	20.8	41.9	37.1	147.1	75.7
90.4	8.6	207.9	137.4	118.2	34.8	83.1	80.2	167.8	154.2

36
Employment by industry [cont.]
Emploi par industrie [suite]

Country or area Pays ou zone	Year Année	Total employment (000s) Emploi total (000s)		Persons employed, by branch of economic activity (000s) Personnes employées, par branches d'activité économique (000s)							
				Agriculture, hunting, forestry and fishing Agriculture, chasse sylviculture, pêche		Mining and quarrying Industries extractives		Manufacturing industries manufacturières		Electricity, gas, water Electricité, gaz, eau	
		M	F	M	F	M	F	M	F	M	F
Solomon Islands[4 9 25]	1980	19.8	...	6.2	...	...	...	2.2[8]	...	0.3[36]	...
Iles Salomon[4 9 25]	1990	26.1	...	7.5	...	...	...	2.3[8]	...	0.3[36]	...
South Africa[49]	1980	4812.9				709.0	...	1421.4		45.3	
Afrique du Sud[49]	1991	5128.4	...	...	...	653.1	...	1430.8		47.9	...
Spain[16]	1980	8258.3	3298.8	1618.9	607.8	89.6	2.2	2265.8	675.4	77.8	3.7
Espagne[16]	1991	8530.8	4078.6	982.8	362.3	72.3	3.1	2103.4	624.9	79.1	7.4
Sri lanka[7]	1980	715.3	363.1	273.1	263.9	4.8	0.4	124.3	55.8	5.0	0.1
Sri lanka[7]	1990	425.1	337.8	155.2	164.9	3.9	0.5	110.5	126.2	10.5	0.7
Suriname[14]	1983	68.5	...	...	...	5.1	...	5.5	...	1.3	...
Suriname[14]	1990	67.4	...	...	...	2.7	...	4.9	...	1.3	...
Swaziland[9]	1980	56.1	19.0	22.7	7.3	2.6	0.0	6.9	2.4	1.1	0.1
Swaziland[9]	1986	55.0	21.4	18.1	5.0	2.3	0.1	7.5	3.4	1.3	0.1
Sweden[16]	1980[22]	2327.0	1906.0	178.0	59.0	14.0	1.0	750.0	276.0	31.0	6.0
Suède[16]	1991	2299.0	2132.0	105.0	38.0	10.0	1.0	644.0	246.0	31.0	6.0
Switzerland[5]	1980	2021.0	1145.0	160.0	58.0	...	...	943.0[3]	264.0[3]	...	...
Suisse[5]	1991	2198.0	1363.0	139.0	58.0	...	...	647.0[3]	247.0[3]	...	...
Thailand[5 6]	1980[22]	11866.3	10657.5	8048.7	7893.9	27.7	8.8	1036.0[15]	752.7[15]	54.2[16]	5.5[16]
Thaïlande[5 6]	1988[17]	15718.9	13745.1	10296.5	9279.6	36.4	6.5	1346.2[15]	1114.3[15]	105.4[16]	14.2[16]
Togo[4]	1980	46.0	...	1.2	...	3.4	...	6.7	...	1.7	...
Togo[4]	1987	63.9	...	5.7	...	2.6	...	5.1	...	2.4	...
Trinidad and Tobago[6]	1980	270.2	117.7	29.5	10.1	...		44.5[8]	17.7[8]	...	...
Trinité-et-Tobago[6]	1989	246.9	118.8	41.5	9.1	15.9	1.2	26.4	10.0	6.7	0.6
Turkey[5 6 9]	1982	4681.0	647.0	173.0	55.0	54.0	2.0	1208.0	192.0	14.0	1.0
Turquie[5 6 9]	1991	13373.0	6119.0	4780.0	4745.0	133.0	5.0	2319.0	578.0	15.0	...
former USSR[4]	1980	125626.0	...	25236.0	...	...	...	36891.0[3]	...	...	...
ancienne URSS[4]	1990	124971.0	...	22761.0	...	...	...	35400.0[3]	...	...	...
United Kingdom[16 9 10]	1980	15248.0	10078.0	523.0	131.0	408.0	32.0	4961.0	2055.0	282.0	71.0
Royaume-Uni[16 9 10]	1989	15062.0	11622.0	456.0	110.0	200.0	23.0	3840.0	1605.0	221.0	62.0
United Rep. of Tanzania[9]	1980	398.3	73.9	72.3	5.9	4.3	0.4	70.3	9.6	12.6	1.2
Rép.-Unie de Tanzanie[9]	1984	448.7	86.2	67.6	6.8	3.7	0.5	82.0	9.1	18.6	1.8
United States[5 6]	1980	57186.0	42117.0	2842.0	687.0	845.0	134.0	15036.0	6906.0	1153.0[16]	252.0[16]
Etats-Unis[5 6]	1991	63593.0	53284.0	2683.0	708.0	604.0	129.0	13752.0	6682.0	1256.0[16]	328.0[16]
US Virgin Islands[4]	1980	37.5	...	0.2	...	...	...	3.2	...	...	...
Iles Vierges amé.[4]	1991	43.2	...	0.2	...	0.0	...	2.7	...	...	...
Uruguay[16]	1984[17]	569.7	362.9	...	...	35.6[20 34]	5.5[20 34]	117.8	72.1	...	...
Uruguay[16]	1990	686.9	449.3	...	...	50.1[20 34]	7.7[20 34]	153.0	89.0	...	...
Venezuela[5 6 17]	1980	3061.1	1183.8	609.1	28.3	57.0	5.8	483.8	190.9	42.5	7.5
Venezuela[5 6 17]	1990	4324.2	1951.3	777.4	42.2	58.4	7.4	697.1	259.7	51.1	13.1
Yugoslavia[22 37]	1980	3663.0	2019.0	223.0	55.0	115.0	11.0	1322.0	746.0	92.0	16.0
Yugoslavie[22 37]	1989	4012.0	2685.0	253.0	85.0	129.0	15.0	1570.0	1025.0	125.0	23.0
Zambia[4 17 31]	1980	379.3	...	32.6	...	63.1	...	47.8	...	8.0	...
Zambie[4 17 31]	1989	359.6	...	37.2	...	54.2	...	50.9	...	8.7	...
Zimbabwe[1 30]	1980	838.5	171.4	242.0	85.0	65.0	1.2	147.7	11.7	6.5	0.2
Zimbabwe[1 30]	1989	957.0	210.1	209.4	75.2	54.1	1.6	181.2	14.1	8.4	0.4

Source:
International Labour Office (Geneva).

Source:
Bureau international du travail (Genève).

§ All data shown pertain to Germany prior to 3 October 1990 are indicated separately for the Federal Republic of Germany and the former German Democratic Republic based on their respective territories at the time indicated. Where data for united Germany (3 October 1990 and thereafter) are not available, available data are shown separately under the designations Federal Republic of Germany and former German Democratic Republic and pertain to the territorial boundaries prior to 3 October 1990. For detailed explanatory notes on data pertaining to Germany, see Annex I- Country Nomenclature.

§ Toutes les données se rapportant à l'Allemagne avant le 3 octobre 1990 figurent dans deux rubriques séparées basées sur les territoires respectifs de la République fédérale d'Allemagne et l'ancienne République démocratique allemande selon la période indiquée. En l'absence de données pour l'Allemagne unifiée (à compter du 3 octobre 1990), les données disponibles sont fournies séparément sous les rubriques République fédérale d'Allemagne et ancienne République démocratique allemande et se rapportent aux limites territoriales antérieures au 3 octobre 1990. Pour les notes explicatives en détail sur les données concernant l'Allemagne, voir Annexe I - Nomenclature des pays.

1   Including armed forces.

1   Y compris les forces armées.

Construction Construction		Trade, restaurants and hotels Commerce, restaurants, hôtels		Transport, storage, communications Transports, entrepôts, communications		Finance, insurance, real est.,bus. services Services financierès, immob.,et apparentées		Community, social and personal services Services à collectivité services soc. et pers.	
M	F	M	F	M	F	M	F	M	F
1.6	...	2.0	...	1.5	...	0.3	...	5.8	...
1.4	...	2.6	...	1.3	...	0.9	...	9.8	...
364.1	...	756.3	...	452.5	...	122.0	...	942.3[31]	...
391.0	...	781.3	...	343.3	...	186.5	...	1300.8[31]	...
1017.7	20.5	1185.5	824.8	607.6	57.5	315.8	81.4	1073.0	1023.6
1229.1	44.4	1479.0	1112.5	641.2	85.8	504.8	228.9	1439.1	1609.4
88.9	9.6	80.3	18.0	88.4	3.2	29.9	7.3	20.9	4.7
19.7	4.3	52.9	18.3	21.9	0.7	31.3	14.0	18.8	8.2
3.0	...	6.7	...	3.0	...	1.5	...	42.3	...
2.4	...	3.4	...	3.4	...	1.8	...	47.5	...
5.9	0.2	3.5	2.4	2.9	0.4	1.6	0.8	9.0	5.5
5.1	0.2	4.3	3.1	4.9	0.7	2.2	1.2	9.2	7.6
260.0	26.0	281.0	300.0	214.0	81.0	155.0	128.0	444.0	1028.0
286.0	26.0	308.0	316.0	218.0	101.0	219.0	178.0	475.0	1217.0
...	...	919.0[19]	823.0[19]	...	...	...	...	...	...
309.0	23.0	336.0	394.0	166.0	54.0	229.0	154.0	372.0	433.0
374.2	61.6	881.3[19]	1034.5[19]	425.7	30.2	...	...	1017.6[11]	869.2[11]
585.3	116.8	1344.0[19]	1552.6[19]	581.9	58.5	...	...	1417.5[11]	1598.8[11]
6.2	...	...	...	2.1	...	7.1[11]	...	17.5	...
5.3	...	8.2[19]	...	4.2	...	...	...	30.5	...
73.6[20]	7.4[20]	39.5	39.7	29.2	3.2	...	...	53.0[19]	39.6[19]
34.8	3.2	31.6	29.7	22.6	4.5	12.0	11.1	55.2	49.2
397.0	3.0	1105.0	48.0	362.0	20.0	169.0	59.0	1185.0	264.0
864.0	11.0	2030.0	144.0	746.0	34.0	303.0	111.0	2184.0	491.0
1240.0	...	9694.0	...	11950.0	...	649.0	...	28522.0	...
2550.0	...	9800.0	...	10225.0	...	720.0	...	31425.0	...
1500.0	117.0	2282.0	2508.0	1289.0	291.0	1033.0	831.0	2662.0	4027.0
1648.0	157.0	2513.0	2853.0	1194.0	330.0	1563.0	1436.0	3136.0	5029.0
23.0	0.9	27.7	4.0	49.4	2.8	10.1	3.0	128.7	46.2
19.7	1.4	35.4	2.6	52.6	2.8	11.3	4.7	157.7	56.5
5717.0	498.0	10838.0	9353.0	3725.0	1395.0	3800.0	4551.0	13228.0[11]	18341.0[11]
6485.0	602.0	12766.0	11289.0	4586.0	2034.0	6026.0	7146.0	15436.0[11]	24366.0[11]
3.5[8]	...	10.5	...	2.1[20]	...	2.2	...	2.5	...
2.6[8]	...	13.4	...	2.6[20]	...	3.1	...	5.0	...
50.0	0.7	109.8	61.0	56.7	8.0	32.8	15.2	167.0	200.3
74.2	1.2	124.5	78.9	59.8	9.5	34.4	19.6	190.8	243.5
373.9	12.8	548.4	254.5	275.9	26.5	115.3	71.6	554.1	585.2
440.6	18.2	860.9	438.6	357.6	32.0	226.0	150.3	850.9	987.8
580.0	52.0	384.0	400.0	369.0	71.0	80.0	90.0	498.0	579.0
480.0	60.0	408.0	514.0	416.0	105.0	91.0	119.0	540.0	739.0
43.8	...	31.4	...	23.9	...	22.7	...	106.1	...
20.8	...	26.6	...	26.1	...	24.7	...	110.2	...
41.6	0.6	58.2	12.1	42.6	3.0	7.5	5.0	227.4[19]	52.6[19]
65.1	1.4	78.0	13.9	48.2	3.5	11.8	5.3	300.8[19]	94.7[19]

2   State sector.
3   Including data for mining and quarrying, electricity, gas and water (Cuba: and fishing, but electricity is not included; France and Switzerland (1980): and construction).
4   Both sexes.
5   Civilian labour force employed.
6   Persons aged 15 years and over (Bolivia, Brazil, Honduras, Indonesia, Pakistan: 10 years and over; Colombia, Costa Rica, Ecuador, Paraguay, Portugal, Turkey: 12 years and over; Denmark, Finland: 15 to 74 years; Greece, Italy, Jamaica, Peru, San Marino Uruguay: 14 years and over; Iceland, Puerto Rico, Spain, United Kingdom, United States: 16 years and over; Malaysia, Netherlands: 15 to 64 years; Norway: 16 to 74 years; Sweden: 16 to 64 years; Thailand: 11 years and over).
7   Establishments with 5 or more persons employed (Bahrain: 10 or more persons; Sierra Leone: 6 or more persons).

2   Secteur d'état.
3   Y compris les données concernant les industries extractives, l'électricité, le gaz at l'eau (Cuba: et la pêche, mais l'électricité n'est pas incluse; France et Suisse (1980): et la construction).
4   Les deux sexes.
5   Main-d'oeuvre civile occupée.
6   Personnes âgée 15 ans et plus (Bolivie, Brésil, Honduras, Indonésie, Pakistan: 10 ans et plus; Colombie, Costa Rica, Ecuador, Paraguay, Portugal, Turquie: 12 ans et plus; Danemark, Finlande: de 15 à 74 ans; Grèce, Italie, Jamaïque, Pérou, Saint-Marin Uruguay: 14 ans et plus; Islande, Porto Rico, Espagne, Royaume-Uni, Etats-Unis: 16 ans et plus; Malaisie, Pays-Bas: de 15 à 64 ans; Norvège: de 16 à 74 ans; Suéde: de 16 à 64 ans; Thaïlande: 11 ans et plus).
7   Etablissements occupant cinq personnes et plus (Bahreïn: 10 et plus personnes; Sierra Leone: 6 et plus personnes).

36
Employment by industry [*cont.*]
Emploi par industrie [*suite*]

8   Including mining and quarrying (Barbados: quarrying only).
9   One month of each year.
10   Excluding unpaid family workers and employees in private domestic services (Bermuda, Chad: unpaid family workers only).
11   Including data for restaurants and hotels (Guam, Japan, Puerto Rico, United States: hotels only; Malta: and business services; Jamaica, Togo: and trade).
12   Excluding government and personal services.
13   Excluding armed forces.
14   Bujumbara.
15   Including repair and installation services (Greece: repairs only).
16   Including sanitary services.
17   One quarter of each year.
18   State owned enterprises.
19   Including data for finance, insurance, real estate and business services (former German Democratic Republic, Jordan, Malta, Thailand: finance, insurance and real estate only; Guatemala: finance and insurance only; France, Switzerland: and transport, storage, communication, community, social and personal services; Zimbabwe: business services only; China, Luxembourg: real estate and business services only).
20   Including data for electricity, gas and water.
21   Excluding hunting and fishing.
22   Average of less than twelve months.
23   Non-agricultural activities.
24   Excluding repair and installation services and sanitary services.
25   Employees only.
26   Non-material activities.
27   Employees including working proprietors.
28   Public sector and establishments of non-agricultural private sector with ten or more persons employed.
29   Including the residents of East Jerusalem.
30   Excluding small establishments in rural areas.
31   Excluding domestic services.
32   Including data for sugar and tea factories.
33   Curaçao.
34   Including data for agriculture, hunting, forestry and fishing.
35   Including Azores and Madiera.
36   Excluding gas.
37   Socialized sector.

8   Y compris les industries extractives (Barbade: les carrières seulement).
9   Un mois de chaque année.
10   Non compris les travailleurs familiaux non rémunérés et les personnes occupées à des services domestiques privés (Bermudes, Tchad: les travailleurs familiaux non rémunérés seulement).
11   Y compris les données concernant les restaurants et les hôtels (Guam, Japon, Porto Rico, Etats-Unis: les hôtels seulement; Malte: et les services entreprises; Jamaïque, Togo: et le commerce).
12   Non compris les services gouvernementaux et personnels.
13   Non compris les forces armées.
14   Bujumbara.
15   Y compris les services de réparation et d'installation (Grèce: les services de réparation seulement).
16   Y compris les services sanitaires.
17   Une trimstre de chaque année.
18   Entreprises d'Etat.
19   Y compris les données concernant les banques, les assurances, les affaires immobilières et les services aux entreprises (l'anc. R.d. allemande, Jordanie, Malte, Thaïlande: les banques, les assurances, les affaires immobilières seulement; Guatemala: les banques et les assurances seulement; France, Suisse: et les transports, les entrepôts, les communications, services à collectivité et les services social et personnel; Zimbabwe: les services aux entreprises seulement; Chine, Luxembourg: les affaires immobilières et les services aux entreprises seulement).
20   Y compris les données concernant l'électricité, le gaz et l'eau.
21   Non compris la chasse et la pêche.
22   Moyenne de moins de douze mois.
23   Activités non agricoles.
24   Non compris les services de réparation et d'installation et les services sanitaires.
25   Salariés seulement.
26   Activités non matérielles.
27   Salariés y compris les propriétaires exploitants.
28   Secteur public et établissements du secteur privé non agricole occupant dix personnes et plus.
29   Y compris les résidents de Jerusalem Est.
30   Non compris les petites entreprises des zones rurales.
31   Non compris les services domestiques.
32   Y compris les données concernant les fabriques de sucre et de thé.
33   Curaçao.
34   Y compris les données concernant l'agriculture, la chasse, le sylviculture et la pêche.
35   Y compris Açores et Madère.
36   Non compris le gaz.
37   Secteur socialisé.

37
Unemployment
Chômage
Number (thousands) and percentage of unemployed
Nombre (milliers) et pourcentage des chômeurs

Country or area Pays ou zones	1982	1983	1984	1985	1986	1987	1988	1989	1990	1991
Albania Albanie										
MF [II]	45.9	58.6	77.5	94.0	92.0	89.4	105.8	113.4	150.7	139.8
% MF [II]	3.6	4.4	5.7	6.7	6.4	6.1	7.0	7.3	9.5	9.1
American Samoa Samoa américaines										
MF [II][1]	1.3	1.4	1.4	1.5	1.6	...	...	...	...	...
% MF [II][1]	12.8	13.0	13.1	12.5	13.4	...	...	...	...	...
Angola Angola										
MF [I][2 3]	...	...	52.0	56.1	69.4	...	...	...	...	...
M [I][2 3]	...	...	34.8	34.9	43.6	...	...	...	...	...
F [I][2 3]	...	...	17.3	21.2	25.8	...	...	...	...	...
% MF [I][2 3]	...	...	14.2	14.9	18.9	...	...	...	...	...
% M [I][2 3]	...	...	11.8	11.5	14.8	...	...	...	...	...
% F [I][2 3]	...	...	24.7	28.5	35.1	...	...	...	...	...
Argentina Argentine										
MF [IV][1 4]	183.6	159.4	152.1	216.2	177.8[5]	230.5	251.2	322.6	...	...
M [IV][1 4]	112.3	98.9	92.8	143.0	107.7[5]	126.8	137.2	195.2	...	...
F [IV][1 4]	71.3	60.5	59.3	73.3	70.1[5]	103.7	114.0	127.4	...	...
% MF [IV][1 4]	4.8	4.2	3.8	5.3	4.4[5]	5.3	5.9	7.3	...	...
% M [IV][1 4]	...	...	...	...	...	4.5	5.2	7.0	...	...
% F [IV][1 4]	...	...	...	...	...	6.6	7.2	7.7	...	...
Australia Australie										
MF [IV][1]	494.9	697.0	# 641.2	602.9	# 613.1[6]	628.9	576.2	509.1	587.1	821.1
M [IV][1]	275.2	424.8	# 384.6	355.9	# 351.8[6]	361.0	321.1	277.4	334.1	494.8
F [IV][1]	219.7	272.3	# 256.7	246.9	# 261.3[6]	267.9	255.1	231.7	253.0	326.2
% MF [IV][1]	7.2	10.0	# 9.0	8.3	# 8.1[6]	8.1	7.2	6.2	6.9	9.6
% M [IV][1]	6.4	9.7	# 8.7	7.9	# 7.7[6]	7.8	6.8	5.7	6.7	9.9
% F [IV][1]	8.5	10.4	# 9.5	8.8	# 8.7[6]	8.6	7.9	6.9	7.2	9.2
Austria Autriche										
MF [I][1 7]	105.4	127.4	130.5	139.5	152.0	164.5	158.6	149.2	165.8	185.0
M [I][1 7]	65.1	79.8	80.6	84.2	88.9	95.0	89.8	81.0	89.0	99.0
F [I][1 7]	40.2	47.6	49.9	55.3	63.1	69.5	68.8	68.2	76.8	86.0
% MF [I][1 7]	3.7	4.5	4.5	4.8	5.2	5.6	5.3	5.0	5.4	5.8
% M [I][1 7]	3.8	4.7	4.7	4.9	5.1	5.5	5.1	4.6	4.9	5.3
% F [I][1 7]	3.5	4.1	4.3	4.7	5.2	5.7	5.6	5.5	6.0	6.5
Bahamas Bahamas										
MF [IV][1]	...	...	...	...	# 13.5[3]	...	13.7	14.9	...	...
M [IV][1]	...	...	...	...	# 5.7[3]	...	5.4	7.4	...	...
F [IV][1]	...	...	...	...	# 7.8[3]	...	8.3	7.5	...	...

37
Unemployment
Number (thousands) and percentage of unemployed [cont.]
Chômage
Nombre (milliers) et pourcentage des chômeurs [suite]

Country or area Pays ou zones	1982	1983	1984	1985	1986	1987	1988	1989	1990	1991
% MF [IV][1]	...	...	...	...	# 12.2[3]	...	11.0	11.7	...	...
% M [IV][1]	...	...	...	...	# 9.7[3]	...	8.2	11.0	...	...
% F [IV][1]	...	...	...	...	# 15.0[3]	...	14.2	12.5	...	...
Bahrain Bahreïn										
MF [I][2][8]	2.6	3.5	4.1	6.3	6.7	4.0	4.5	3.4	3.0	3.3
M [I][2][8]	...	...	...	...	...	3.1	3.6	2.5	2.1	2.4
F [I][2][8]	...	...	...	...	...	0.9	0.9	0.9	0.8	0.9
Barbados Barbade										
MF [IV][1]	15.5	16.9	19.2	21.2	20.7	21.4	21.2[9]	17.1[9]	18.6[9]	20.9
M [IV][1]	6.2	6.7	7.9	8.6	8.0	8.4	7.7[9]	5.9[9]	6.6[9]	8.6
F [IV][1]	9.3	10.2	11.3	12.6	12.7	13.0	13.5[9]	11.2[9]	12.0[9]	12.3
% MF [IV][1]	13.7	15.0	17.1	18.7	17.7	17.9	17.4[9]	13.7[9]	15.0[9]	17.1
% M [IV][1]	10.1	10.9	13.0	14.1	13.0	13.3	12.3[9]	9.1[9]	10.3[9]	13.3
% F [IV][1]	18.0	19.8	22.1	24.0	23.0	23.1	22.9[9]	18.7[9]	20.2[9]	21.4
Belgium Belgique										
MF [I][7]	535.0	589.5	595.8	# 558.3	516.8	500.8	459.4	419.3	402.8	429.5
M [I][7]	243.0	273.8	273.7	# 246.0	217.6	208.9	187.8	167.5	161.3	178.0
F [I][7]	292.1	315.6	322.1	# 312.3	299.1	292.0	271.6	251.8	241.5	251.5
% MF [I][7]	13.0	14.2	14.4	# 13.6	12.6	12.2	11.1	10.1	9.6	10.3
% M [I][7]	9.5	10.8	10.9	# 9.9	8.9	8.6	7.7	6.9	6.6	7.3
% F [I][7]	18.6	19.7	19.9	# 19.1	18.1	17.4	16.0	14.7	14.1	14.7
Bermuda Bermudes										
MF [I][7]	0.0	0.0	0.0	0.0	0.0	0.1	...	...	...	...
Bolivia Bolivie										
MF [IV][10]	47.6	76.1	85.3	58.2	46.2	78.0	...	96.1	79.2	67.2
M [IV][10]	...	...	...	...	...	...	...	51.1	48.2	37.2
F [IV][10]	...	...	...	...	...	...	...	45.0	31.0	30.0
% MF [IV][10]	...	...	...	...	...	...	...	...	9.4	7.3
% M [IV][10]	...	...	...	...	...	...	...	...	9.9	6.9
% F [IV][10]	...	...	...	...	...	...	...	...	8.8	7.8
Brazil Brésil										
MF [IV][11][12]	2 533.0	2 474.0	2 234.0	1 875.0[5]	1 380.0	2 133.0	2 319.0	...	...	...
M [IV][11][12]	1 678.0	1 668.0	1 437.0	1 172.0[5]	854.0	1 315.0	1 410.0	...	...	...
F [IV][11][12]	855.0	806.0	797.0	704.0[5]	526.0	818.0	909.0	...	...	...
% MF [IV][11][12]	...	4.9	4.3	3.4[5]	2.4	3.7	3.9	...	...	...
% M [IV][11][12]	...	4.9	4.1	3.3[5]	2.3	3.5	3.7	...	...	...
% F [IV][11][12]	...	4.8	4.6	3.9[5]	2.8	4.1	4.4	...	...	...

37
Unemployment
Number (thousands) and percentage of unemployed [*cont.*]
Chômage
Nombre (milliers) et pourcentage des chômeurs [*suite*]

Country or area Pays ou zones	1982	1983	1984	1985	1986	1987	1988	1989	1990	1991
Bulgaria Bulgarie										
MF [I]	...	...	...	...	...	...	...	...	65.1	419.1
M [I]	...	...	...	...	...	...	...	...	22.7	190.7
F [I]	...	...	...	...	...	...	...	...	42.4	228.4
% MF [I]	...	...	...	...	...	...	...	...	1.7	12.0
Burundi Burundi										
MF [I][2 13]	6.7	3.4	2.6	1.9	6.8	8.2	9.3	11.1	14.5	...
M [I][2 13]	...	3.0	2.2	1.7	6.0	7.4	8.4	9.8	...	...
F [I][2 13]	...	0.4	0.4	0.3	0.8	0.8	0.9	1.3	...	...
Cameroon Cameroun										
MF [I][17]	38.6	44.8	37.8	14.3	25.5	19.2	...	...		...
Canada Canada										
MF [IV][1]	1 308.0	1 434.0	1 384.0	1 311.0	1 215.0	1 150.0	1 031.0	1 018.0	1 109.0	1 417.0
M [IV][1]	773.0	849.0	792.0	739.0	677.0	623.0	546.0	548.0	613.0	818.0
F [IV][1]	534.0	585.0	592.0	572.0	539.0	527.0	485.0	470.0	496.0	599.0
% MF [IV][1]	11.0	11.8	11.2	10.5	9.5	8.8	7.8	7.5	8.1	10.3
% M [IV][1]	11.0	12.0	11.2	10.3	9.3	8.5	7.4	7.3	8.1	10.8
% F [IV][1]	10.9	11.6	11.3	10.7	9.8	9.3	8.3	7.9	8.1	9.7
Central African Rep. Rép. centrafricaine										
MF [I][2 14]	8.3	11.0	9.1	8.2	9.1	8.7	8.2	...	...	...
M [I][2 14]	7.2	...	...	7.5	...	8.1	7.5	...	...	...
F [I][2 14]	1.0	...	...	0.7	...	0.7	0.8	...	...	...
Chad Tchad										
MF [I]	...	4.5	...	...	4.5	3.7	5.0	10.7	...	...
M [I]	...	4.5	...	...	4.5	3.7	4.9	10.6	...	...
F [I]	...	0.0	...	...	0.0	0.0	0.0	0.1		...
Chile Chili										
MF [IV][19]	# 717.6	551.9	541.3	# 516.5	374.2	343.5	286.1	249.8	268.9	253.6
M [IV][19]	# 517.7	381.7	351.1	# 346.7	250.0	222.5	177.4	162.4	184.8	168.6
F [IV][19]	# 199.9	170.2	190.1	# 169.8	124.0	120.9	108.6	87.4	84.0	85.0
% MF [IV][19]	# 19.6	14.6	13.9	# 12.1	8.8	7.9	6.3	5.3	5.6	5.3
% M [IV][19]	# 20.2	14.6	...	# 11.7	8.4	7.3	5.6	5.0	5.7	5.1
% F [IV][19]	# 18.3	14.7	...	# 13.4	9.7	9.3	7.8	6.1	5.7	5.8
China Chine										
MF [II][5 10 15]	3 794.0	2 714.0	2 357.0	2 385.0	2 644.0	2 766.0	2 960.0	3 779.0	3 832.0	3 522.0
M [II][5 10 15]	...	883.0	752.0	783.0	805.0	953.0	1 001.0	1 942.0	1 313.0	1 207.0
F [II][5 10 15]	...	1 337.0	1 207.0	1 186.0	1 288.0	1 398.0	1 452.0	1 837.0	1 814.0	1 677.0
% MF [II][5 10 15]	3.2	2.3	1.9	1.8	2.0	2.0	2.0	2.6	2.5	2.3

37
Unemployment
Number (thousands) and percentage of unemployed [cont.]
Chômage
Nombre (milliers) et pourcentage des chômeurs [suite]

Country or area Pays ou zones	1982	1983	1984	1985	1986	1987	1988	1989	1990	1991
% M [II][5 10 15]	...	0.7	0.6	0.6	0.6	0.7	0.7	1.3	0.9	0.8
% F [II][10 15]	...	1.1	1.0	0.9	1.0	1.0	1.0	1.3	1.2	1.1
Colombia Colombie										
MF [IV][5 16 17]	375.8	377.0	463.2	499.9	482.8	429.0	403.0	356.5	491.6	501.6
M [IV][5 16 17]	...	186.1	230.5	228.7	221.0	190.7	175.5	159.4	232.8	216.8
F [IV][5 16 17]	...	190.9	232.7	271.2	261.8	238.3	227.5	197.1	258.8	254.8
Costa Rica Costa Rica										
MF [IV][5 17]	78.6	76.2	44.4	60.8	56.7	# 54.5	54.9	38.7	49.5	59.1
M [IV][5 17]	53.5	55.5	32.0	42.5	40.3	# 33.1	31.9	23.4	31.7	35.5
F [IV][5 17]	25.1	20.7	12.4	18.3	16.5	# 21.4	23.1	15.3	17.8	23.5
% MF [IV][5 17]	9.4	9.0	5.0	6.8	6.2	# 5.6	5.5	3.8	4.6	5.5
% M [IV][5 17]	8.6	8.8	5.0	6.5	6.0	# 4.7	4.4	3.2	4.2	4.8
% F [IV][5 17]	11.4	9.6	5.0	7.9	6.9	# 7.9	8.0	5.3	5.9	7.4
Côte d'Ivoire Côte d'Ivoire										
MF [I][5 7 18]	59.7	59.1	73.5	86.4	92.0	107.8	120.6	128.5	140.3	...
M [I][5 7 18]	...	...	55.8	62.5	...	73.2	81.9	89.4	99.0	...
F [I][5 7 18]	...	...	17.7	23.9	...	34.6	38.7	39.1	41.3	...
Cyprus Chypre										
MF [I][3 7]	6.4	7.8	8.0	8.3	9.2	8.7	7.4	6.2	5.1	8.3
M [I][3 7]	3.6	4.4	4.5	4.6	4.8	4.4	3.7	2.9	2.5	3.8
F [I][3 7]	2.9	3.4	3.5	3.7	4.4	4.3	3.7	3.3	2.6	4.5
% MF [I][3 7]	2.8	3.3	3.3	3.3	3.7	3.4	2.8	2.3	1.8	3.0
% M [I][3 7]	2.4	2.9	2.9	2.9	2.9	2.7	2.2	1.7	1.4	2.2
% F [I][3 7]	4.5	4.1	4.1	4.2	5.0	4.7	3.8	3.3	2.5	4.4
Czechoslovakia Tchécoslovaquie										
MF [I]	...	...	...	...	...	...	...	1.0	77.0	524.0
M [I]	...	...	...	...	...	...	...	...	38.0	239.0
F [I]	...	...	...	...	...	...	...	...	39.0	285.0
% MF [I]	...	...	...	...	...	...	...	...	1.0	6.6
% M [I]	...	...	...	...	...	...	...	...	0.9	5.9
% F [I]	...	...	...	...	...	...	...	...	1.0	7.3
Denmark Danemark										
MF [I][7 19]	262.8	283.0	276.0	251.8	220.4	221.9	243.9	264.9	271.7	296.1
M [I][7 19]	141.3	143.7	129.7	110.9	90.8	96.1	108.8	120.0	124.0	137.2
F [I][7 19]	121.5	139.3	146.3	140.9	129.6	125.7	135.1	145.0	147.7	158.9
% MF [I][7 19]	10.0	10.5	10.1	9.1	7.9	7.9	8.7	9.5	9.7	10.6
% M [I][7 19]	9.7	9.8	8.8	7.5	6.1	6.4	7.3	8.1	8.4	9.3
% F [I][7 19]	10.0	11.3	11.7	11.0	10.0	9.6	10.3	11.1	11.3	12.1

37
Unemployment
Number (thousands) and percentage of unemployed [*cont.*]
Chômage
Nombre (milliers) et pourcentage des chômeurs [*suite*]

Country or area Pays ou zones	1982	1983	1984	1985	1986	1987	1988	1989	1990	1991
Ecuador Equateur										
MF [IV][5 10 17]	...	...	...	...	...	89.5	155.4	187.0	...	...
M [IV][5 10 17]	...	...	...	...	...	39.8	73.0	88.1	...	...
F [IV][5 10 17]	...	...	...	...	...	49.7	82.3	98.9	...	...
% MF [IV][5 10 17]	...	...	...	...	...	7.2	7.0	7.9	...	...
% M [IV][5 10 17]	...	...	...	...	...	5.2	5.1	5.9	...	...
% F [IV][5 10 17]	...	...	...	...	...	10.4	10.3	11.1	...	...
El Salvador El Salvador										
MF [IV][11]	...	...	...	280.2	# 28.3[10]	...	# 74.1[10]	72.0	97.9	72.5
M [IV]	...	...	...	128.0	# 14.5[11 20]	...	# 50.9[10 11]	46.2[11]	54.8[11]	43.9[11]
F [IV][11]	...	...	...	152.3	# 13.8[20]	...	# 23.2[10]	25.8	43.1	28.6
% MF [IV][11]	...	...	...	16.9	# 7.9[20]	...	# 9.4[10]	8.3	10.0	7.5
% M [IV][11]	...	...	...	12.4	# 7.5[20]	...	# 11.0[10]	9.9	10.1	8.3
% F [IV][11]	...	...	...	24.3	# 8.5[20]	...	# 7.1[10]	6.8	9.8	6.6
Ethiopia Ethiopie										
MF [I][7 21 22]	57.7	57.2	54.7	56.4	52.6	58.2	55.3	51.3	44.2	44.3
M [I][7 21 22]	35.2	36.1	33.2	33.3	30.7	37.9	34.1	28.6	25.8	24.9
F [I][7 21 22]	22.5	21.2	21.5	23.1	21.9	20.3	21.2	22.7	18.4	19.4
Fiji Fidji										
MF [II]	13.3	14.9	16.8	18.6	18.2	23.0	23.0	15.0	16.0	15.0
% MF [II]	6.5	7.5	7.5	8.1	7.5	9.3	9.4	6.1	6.4	5.9
Finland Finlande										
MF [IV][23]	135.0	138.0	133.0	129.0	138.0	130.0	116.0	89.0	88.0	193.0
M [IV][23]	73.0	76.0	72.0	73.0	82.0	78.0	67.0	48.0	54.0	124.0
F [IV][23]	62.0	62.0	61.0	56.0	56.0	53.0	48.0	41.0	34.0	69.0
% MF [IV][23]	5.4	5.5	5.2	5.0	5.4	5.1	4.5	3.5	3.4	7.5
% M [IV][23]	5.5	5.7	5.4	5.5	6.1	5.8	5.1	3.6	4.0	9.2
% F [IV][23]	5.2	5.2	5.0	4.6	4.6	4.3	4.0	3.3	2.8	5.7
France France										
MF [II]	1 922.8	1 973.8	2 323.1	2 442.2	2 489.5	2 531.7	2 410.2	2 284.9	2 180.7	2 297.1
M [II]	844.4	888.8	1 082.5	1 154.7	1 168.9	1 145.6	1 063.4	972.3	929.7	1 002.8
F [II]	1 078.4	1 085.0	1 240.6	1 287.5	1 320.6	1 386.1	1 346.8	1 312.6	1 251.0	1 294.3
% MF [II]	8.1	8.3	9.7	10.2	10.4	10.5	10.0	9.4	8.9	9.3
% M [II]	6.0	6.3	7.7	8.3	8.4	8.3	7.7	7.0	6.7	7.2
% F [II]	11.2	11.2	12.6	12.9	13.1	13.6	13.1	12.6	11.9	12.1
French Guiana Guyane française										
MF [I][7 24]	2.3	2.4	3.7	4.2[5]	3.7	3.4	3.3	3.8	4.4	4.7
M [I][7 24]	1.0	1.2	1.8	2.1[5]	1.8	1.6	1.6	1.9	2.3	2.5

37
Unemployment
Number (thousands) and percentage of unemployed [*cont.*]
Chômage
Nombre (milliers) et pourcentage des chômeurs [*suite*]

Country or area Pays ou zones	1982	1983	1984	1985	1986	1987	1988	1989	1990	1991
F [I][7 24]	1.3	1.2	1.9	2.1[5]	1.9	1.8	1.7	2.0	2.1	2.2
% MF [I][7 24]	15.3	7.7	12.2	13.5[5]	12.0	11.0	10.6	12.2	13.9	9.6
% M [I][7 24]	...	6.3	6.0	10.5[5]	9.2	8.4	...	9.4	11.7	8.2
% F [I][7 24]	...	10.1	16.3	18.5[5]	16.7	15.4	...	16.8	17.6	11.6
French Polynesia Polynésie française										
MF [I][2 3]	0.8	0.9	0.9	0.9	1.0	0.5	0.7	0.6	0.6	...
Germany† · Allemagne†										
F. R. Germany R. f. Allemagne										
MF [I][7 23]	1 833.2	2 258.2	2 265.6	2 304.0	2 228.0	2 228.8	2 241.6	2 037.8	1 883.1	1 689.4
M [I][7 23]	1 021.1	1 273.1	1 276.7	1 289.1	1 200.0	1 207.4	1 198.8	1 069.8	967.7	897.7
F [I][7 23]	812.2	985.1	988.9	1 015.0	1 028.0	1 021.4	1 042.8	968.0	915.4	791.7
% MF [I][7 23]	7.5	9.1	9.1	9.3	9.0	8.9	8.7	7.9	7.2	6.3
% M [I][7 23]	6.8	8.4	8.5	8.6	8.0	8.0	7.8	6.9	6.3	5.8
% F [I][7 23]	8.6	10.1	10.2	10.4	10.5	10.2	10.0	9.4	8.4	7.0
former German D. R. anc. R. d. allemande										
MF [I][7]	...	...	...	...	...	...	...	...	642.0	913.0
M [I][7]	...	...	...	...	...	...	...	...	290.0	383.0
F [I][7]	...	...	...	...	...	...	...	...	352.0	530.0
% MF [I][7]	...	...	...	...	...	...	...	...	7.3	10.3
Ghana Ghana										
MF [I][7 25]	23.5	23.7	21.2	24.2	25.8	...	28.7	27.4	...	...
M [I][7 25]	16.3	17.1	15.4	19.2	22.5	...	26.1	24.7	...	...
F [I][7 25]	7.2	6.6	5.8	5.1	3.3	...	2.6	2.7	...	...
% MF [I][7 25]	0.7	0.7	0.4	0.4	0.5	...	...	...	...	...
% M [I][7 25]	0.9	0.9	0.6	0.7	0.8	...	...	...	...	...
% F [I][7 25]	0.5	0.4	0.2	0.2	0.1	...	...	...	...	...
Gibraltar Gibraltar										
MF [I][7 26]	0.4	0.5	0.5	0.5	0.5	0.3	0.5	0.6	...	...
M [I][7 26]	0.3	0.3	0.3	0.3	0.4	0.2	0.5	0.5	...	...
F [I][7 26]	0.1	0.2	0.2	0.1	0.1	0.1	0.1	0.1	...	...
Greece Grèce										
MF [I][17]	50.6	61.7	71.2	89.0	110.5	117.9	115.3	133.9	* 140.2	* 173.2
M [I][17]	30.5	37.2	42.0	51.2	61.0	64.1	58.7	64.6	* 68.1	* 83.9
F [I][17]	20.1	24.5	29.2	37.8	49.4	53.9	56.6	69.3	* 72.1	* 89.3
% MF [I][17]	3.2	3.8	4.2	5.1	6.1	6.4	6.0	6.5	* 6.4	* 7.3
Greenland Groënland										
MF [I][7]	1.3	1.8	2.3	2.4	...	...	...	...	...	...
M [I][7]	0.6	0.9	1.2	1.2	...	...	...	...	...	...

37
Unemployment
Number (thousands) and percentage of unemployed [*cont.*]
Chômage
Nombre (milliers) et pourcentage des chômeurs [*suite*]

Country or area Pays ou zones	1982	1983	1984	1985	1986	1987	1988	1989	1990	1991
F [I][7]	0.6	0.9	1.1	1.2	...	...	...	...	...	...
Guadeloupe Guadeloupe										
MF [I][7 24]	21.0	18.6	21.5	24.2	27.2	27.8	29.5	30.8	29.4	34.3
M [I][7 24]	9.4	...	...	11.4	12.2	11.7	11.5	13.5	11.7	13.9
F [I][7 24]	11.6	...	...	12.8	15.0	16.1	18.0	21.3	17.6	20.4
% MF [I][7 24]	19.4	17.5	20.3	22.0	27.0	23.0	24.0	24.0	17.0	19.9
% M [I][7 24]	14.5	...	...	19.0	22.0	16.0	16.0	16.0	...	...
% F [I][7 24]	27.2	...	...	27.0	33.0	31.0	33.0	34.0	...	...
Guam Guam										
MF [IV][24]	3.5	3.3	2.9	2.7	2.2	1.6	1.6	1.1	1.3	...
M [IV][24]	2.0	2.0	1.7	1.4	1.2	0.9	...	...	...	...
F [IV][24]	1.5	1.4	1.1	1.3	1.0	0.6	...	...	...	...
% MF [IV][24]	9.6	9.6	8.4	7.8	6.1	4.4	4.2	2.8	2.8	...
% M [IV][24]	9.4	9.4	8.5	5.6	7.1	4.3	...	...	...	...
% F [IV][24]	10.8	9.9	8.3	8.8	6.9	4.5	...	...	...	...
Guatemala Guatemala										
MF [I][7 11 27]	3.0	3.6	4.8	2.7	2.7	2.2	1.9	1.7	1.8	...
M [I][7 11 27]	2.3	2.4	3.4	1.7	1.8	1.4	1.3	1.2	1.3	...
F [I][7 11 27]	0.7	1.2	1.4	1.0	0.9	0.8	0.6	0.5	0.5	...
Guyana Guyana										
MF [I][7 28]	27.0	7.2	11.7	13.3	10.6	9.2	...	...	...	...
M [I][7 28]	15.6	4.6	7.2	5.8	5.2	4.8	...	...	...	...
F [I][7 28]	11.4	2.6	4.4	7.5	5.4	4.4	...	...	...	...
Haïti Haïti										
MF [II][11]	260.2	276.2	...	...	...	...	561.8	...	339.7	...
M [II][11]	141.4	144.7	...	...	...	...	265.0	...	191.3	...
F [II][11]	118.9	131.5	...	...	...	...	296.7	...	148.3	...
Honduras Honduras										
MF [IV][10 11]	...	...	...	...	56.0	50.3	...	...	61.7	72.1
M [IV][10 11]	...	...	...	...	34.6	31.1	...	...	41.4	46.0
F [IV][10 11]	...	...	...	...	21.4	19.3	...	...	20.3	26.1
Hong Kong Hong-kong										
MF [IV][1]	91.1	113.8	101.0	# 83.6	76.1	47.5	38.0	30.0	37.0	50.3
M [IV][1]	63.5	81.6	68.8	# 58.6	51.1	29.8	24.1	19.3	23.5	33.7
F [IV][1]	27.6	32.3	32.2	# 25.1	24.9	17.7	13.8	10.7	13.5	16.6
% MF [IV][1]	3.6	4.5	3.9	# 3.2	2.8	1.7	1.4	1.1	1.3	1.8
% M [IV][1]	4.0	5.0	4.2	# 3.5	3.0	1.7	1.4	1.1	1.3	1.9
% F [IV][1]	3.1	3.5	3.4	# 2.6	2.5	1.8	1.4	1.1	1.3	1.6

37
Unemployment
Number (thousands) and percentage of unemployed [*cont.*]
    Chômage
    Nombre (milliers) et pourcentage des chômeurs [*suite*]

Country or area Pays ou zone§	1982	1983	1984	1985	1986	1987	1988	1989	1990	1991
Hungary Hongrie										
MF [I][57]	...	...	...	...	...	...	...	...	79.5	406.1
M [I][57]	...	...	...	...	...	...	...	...	45.8	239.0
F [I][57]	...	...	...	...	...	...	...	...	30.3	167.1
% MF [I][57]	...	...	...	...	...	...	...	...	1.7	8.5
% M [I][57]	...	...	...	...	...	...	...	...	1.8	9.2
% F [I][57]	...	...	...	...	...	...	...	...	1.4	7.6
Iceland Islande										
MF [I][7 24]	# 0.8	1.2	1.5	1.1	0.8	0.6	0.8	2.1	2.3	1.9
M [I][7 24]	# 0.3	0.6	0.7	0.5	0.4	0.3	0.3	0.9	1.1	1.0
F [I][7 24]	# 0.4	0.6	0.8	0.7	0.4	0.3	0.5	1.2	1.2	0.9
% MF [I][7 24]	# 0.8	1.0	1.2	0.9	0.7	0.4	0.6	1.7	1.8	1.5
% M [I][7 24]	# 0.5	0.9	0.9	0.6	0.5	0.3	0.4	1.2	1.4	1.3
% F [I][7 24]	# 1.0	1.3	1.8	1.3	0.8	0.6	0.9	2.2	2.2	1.7
India Inde										
MF [I][23]	18 646.0	20 802.0	23 034.0	24 861.0	28 261.0	30 542.0	30 050.0	32 776.0	34 632.0	36 300.0
M [I][23]	15 722.0	17 454.0	19 186.0	20 628.0	23 476.0	25 251.0	24 590.0	26 668.0	27 932.0	28 992.0
F [I][23]	2 924.0	3 348.0	3 848.0	4 233.0	4 785.0	5 291.0	5 461.0	6 109.0	6 700.0	7 308.0
Indonesia Indonésie										
MF [I][2 29]	367.6	353.5	577.0	785.2	855.0	1 017.2	1 352.4	1 518.5	1 238.7	...
M [I][2 29]	287.0	279.9	430.7	552.3	586.3	672.8	900.8	991.7	735.9	...
F [I][2 29]	80.6	73.6	146.3	233.0	268.7	344.4	451.6	526.8	502.9	...
MF [IV][11]	1 796.0	...	...	1 368.0	1 855.0	1 843.0	2 106.0	2 083.0	1 952.0	...
M [IV][11]	1 023.0	...	...	898.0	1 127.0	1 148.0	1 269.0	1 251.0	1 155.0	...
F [IV][11]	773.0	...	...	470.0	728.0	695.0	837.0	832.0	796.0	...
Ireland Irlande										
MF [IV][15]	...	183.0	203.5	226.0	227.4	231.6	218.5	202.2	178.9	208.4
M [IV][15]	...	140.0	156.4	172.8	173.7	176.3	169.6	157.3	138.1	156.3
F [IV][15]	...	43.0	47.1	53.2	53.8	55.2	49.0	44.9	40.8	52.1
% MF [IV][15]	...	14.0	15.6	17.4	17.4	17.6	16.7	15.6	13.7	15.6
% M [IV][15]	...	15.3	17.0	18.8	19.0	19.4	18.6	17.6	15.5	17.3
% F [IV][15]	...	11.0	12.2	13.8	13.7	13.6	12.3	11.3	9.9	12.1
Isle of Man Ile de Man										
MF [I][7]	# 1.6	1.9	2.0	2.2	2.1	1.6	0.9	0.6	0.6	1.0
M [I][7]	# 1.1	1.4	1.4	1.5	1.4	1.1	0.6	0.4	0.4	0.7
F [I][7]	# 0.6	0.6	0.6	0.7	0.7	0.6	0.3	0.2	0.1	0.3
% MF [I][7]	# 5.9	7.1	7.4	8.0	7.7	5.9	3.4	2.0	2.1	3.0

37
Unemployment
Number (thousands) and percentage of unemployed [cont.]
Chômage
Nombre (milliers) et pourcentage des chômeurs [suite]

Country or area Pays ou zones	1982	1983	1984	1985	1986	1987	1988	1989	1990	1991
% M [I][7]	# 6.3	8.0	8.3	8.8	8.4	6.3	3.7	2.3	2.6	3.9
% F [I][7]	# 5.3	5.6	6.0	6.7	6.6	5.3	2.9	1.6	1.3	1.9
Israel Israël										
MF [IV][1 30 31]	68.4	63.2	85.1	# 97.0[3]	104.2	90.1	100.0	142.5	158.0	187.2
M [IV][1 30 31]	37.8	35.2	46.5	# 56.5[3]	59.1	47.6	53.0	75.7	82.2	89.8
F [IV][1 30 31]	30.4	27.7	38.6	# 40.1[3]	45.2	42.5	47.0	66.8	75.8	97.4
% MF [IV][1 30 31]	5.0	4.5	5.9	# 6.7[3]	7.1	6.1	6.4	8.9	9.6	10.6
% M [IV][1 30 31]	4.4	4.0	5.2	# 6.3[3]	6.5	5.2	5.7	7.9	8.4	8.6
% F [IV][1 30 31]	6.0	5.3	7.0	# 7.2[3]	7.9	7.3	7.6	10.3	11.3	13.4
Italy Italie										
MF [IV][3]	2 052.0	2 264.0	2 303.0	2 382.0	2 611.0	2 832.0	2 885.0	2 865.0	2 621.0	...
M [IV][3]	909.0	992.0	986.0	1 024.0	1 115.0	1 228.0	1 240.0	1 220.0	1 102.0	...
F [IV][3]	1 143.0	1 272.0	1 317.0	1 358.0	1 496.0	1 604.0	1 645.0	1 646.0	1 519.0	...
% MF [IV][3]	9.1	9.9	10.0	10.3	11.1	11.9	12.0	12.0	11.0	...
% M [IV][3]	6.1	6.6	6.6	6.8	7.4	8.1	8.1	8.1	7.3	...
% F [IV][3]	14.9	16.2	16.5	16.7	17.8	18.5	15.4	18.7	17.1	...
Jamaica Jamaïque										
MF [IV][3]	278.5	266.0	266.8	260.8	250.4	224.3	203.3	177.4	166.6	...
M [IV][3]	87.0	88.2	87.8	88.4	85.3	77.0	68.0	54.1	52.8	...
F [IV][3]	191.5	177.8	179.0	172.3	165.2	147.3	135.3	123.3	113.8	...
% MF [IV][3]	27.6	26.4	25.5	25.0	23.6	21.0	18.9	16.8	15.7	...
% M [IV][3]	16.1	16.1	15.8	15.7	14.9	13.2	11.9	9.5	9.3	...
% F [IV][3]	40.6	38.3	36.5	35.9	33.8	30.4	27.0	25.2	23.1	...
Japan Japon										
MF [IV][1]	1 360.0	1 560.0	1 610.0	1 560.0	1 670.0	1 730.0	1 550.0	1 420.0	1 340.0	1 360.0
M [IV][1]	840.0	950.0	960.0	930.0	990.0	1 040.0	910.0	830.0	770.0	780.0
F [IV][1]	520.0	610.0	650.0	630.0	670.0	690.0	640.0	590.0	570.0	590.0
% MF [IV][1]	2.4	2.6	2.7	2.6	2.8	2.8	2.5	2.3	2.1	2.1
% M [IV][1]	2.4	2.7	2.7	2.6	2.7	2.8	2.5	2.2	2.0	2.0
% F [IV][1]	2.3	2.6	2.8	2.7	2.8	2.8	2.6	2.3	2.2	2.2
Korea, Republic of Corée, République de										
MF [IV][1]	654.0	613.0	568.0	622.0	611.0	519.0	435.0	459.0	451.0	436.0
M [IV][1]	509.0	486.0	444.0	480.0	480.0	397.0	315.0	326.0	318.0	287.0
F [IV][1]	145.0	128.0	124.0	141.0	131.0	122.0	120.0	134.0	133.0	150.0
% MF [IV][1]	4.4	4.1	3.8	4.0	3.8	3.1	2.5	2.6	2.4	2.3
% M [IV][1]	5.5	5.2	4.8	5.0	4.9	3.9	3.0	3.0	2.9	2.5
% F [IV][1]	2.5	2.2	2.2	2.4	2.1	1.8	1.7	1.8	1.8	2.0

37
Unemployment
Number (thousands) and percentage of unemployed [cont.]
  Chômage
  Nombre (milliers) et pourcentage des chômeurs [suite]

Country or area Pays ou zone	1982	1983	1984	1985	1986	1987	1988	1989	1990	1991
Luxembourg Luxembourg										
MF [I][7 32]	2.0	2.5	2.7	2.6	2.3	2.7	2.5	2.3	2.1	...
M [I][7 32]	1.1	1.3	1.4	1.3	1.2	1.5	1.5	1.4	1.2	...
F [I][7 32]	0.9	1.1	1.3	1.2	1.1	1.1	1.0	0.9	0.8	...
% MF [I][7 32]	1.3	1.6	1.8	1.7	1.5	1.7	1.6	1.4	1.3	...
% M [I][7 32]	1.1	1.3	1.4	1.3	1.2	1.5	1.5	...	...	...
% F [I][7 32]	1.8	2.1	2.5	2.3	1.8	2.0	1.7	...	...	...
Madagascar Madagascar										
MF [I][2 5]	25.9	29.5	29.0	28.8	24.2	18.4	16.1	15.7	...	...
Malaysia Malaisie										
MF [I][1 2 5]	80.5	67.5	72.1	80.7	86.9	78.6	...	...	...	...
M [I][1 2 5]	52.3	43.0	44.6	49.3	...	49.6	...	...	...	...
F [I][1 2 5]	28.2	24.5	27.5	31.4	...	29.6	...	...	...	...
Malta Malte										
MF [I][5 7 33]	10.4	10.3	10.4	9.9	8.5	5.6	5.2	4.8	...	...
M [I][5 7 33]	7.8	7.5	8.1	7.8	6.6	4.6	4.2	4.1	...	...
F [I][5 7 33]	2.5	2.8	2.3	2.2	1.9	1.0	1.0	0.8	...	...
% MF [I][5 33]	...	8.5[7]	...	8.1[7]	6.8[7]	4.4[7]	4.0[3 7]	3.7[3]	...	...
% M [I][5 33]	...	8.2[7]	...	8.4[7]	6.9[7]	4.8[7]	4.3[3 7]	4.1[3]	...	...
% F [I][5 33]	...	9.3[7]	...	7.2[7]	6.2[7]	3.1[7]	3.1[3 7]	2.3[3]	...	...
Martinique Martinique										
MF [I][5 7 24]	23.1	24.6	25.6	29.5	33.9	30.7	...	...	...	...
M [I][5 7 24]	10.5	11.2	13.6	13.6	...	13.1	...	...	...	...
F [I][5 7 24]	15.4	14.2	15.6	17.2	...	19.4	...	...	...	...
Mauritius Maurice										
MF [I][17 34]	73.5	73.0	70.2	64.8	54.6	46.8	27.7	18.1	12.8	10.6
M [I][17 34]	52.5	53.1	52.5	48.8	42.2	36.3	19.9	11.8	7.3	5.2
F [I][17 34]	21.1	19.9	17.8	16.0	12.4	10.5	7.8	6.3	5.5	5.4
Mexico Mexique										
% MF [IV][15]	4.2	6.8	6.0	4.4	4.3	3.9	3.6	3.0	2.8	2.6
% M [IV][15]	3.9	6.0	5.3	3.5	3.7	3.4	3.0	2.6	2.6	2.5
% F [IV][15]	4.9	8.5	7.6	5.7	5.3	4.8	4.7	3.8	3.1	3.0
Montserrat Montserrat										
MF [II]	0.3	0.4	0.3	0.3	0.2	0.1	...	...	...	...
% MF [II]	5.6	7.0	5.8	5.3	4.2	2.0	...	...	...	...
Morocco Maroc										
MF [I][7 35]	20.8	23.3	40.6	30.2	...	...	...	...	...	...

37
Unemployment
Number (thousands) and percentage of unemployed [cont.]
Chômage
Nombre (milliers) et pourcentage des chômeurs [suite]

Country or area Pays ou zone§	1982	1983	1984	1985	1986	1987	1988	1989	1990	1991
Myanmar Myanmar										
MF [I][5 7 36]	656.4	430.4	441.0	338.0	354.4	331.4	312.7	485.8	555.3	...
Netherlands Pays-Bas										
MF [I][7 23]	541.7	# 800.6	822.4	761.0	710.7	685.5	# 433.0	390.0	346.0	319.0
M [I][7 23]	376.2	# 549.9	555.2	498.0	453.5	428.7	# 278.0	241.0	209.0	187.0
F [I][7 23]	165.5	# 250.7	267.2	263.0	257.2	256.8	# 155.0	149.0	137.0	132.0
% MF [I][7 23]	9.7	# 13.9	14.1	12.9	12.0	11.5	# 6.5	5.8	5.0	4.5
% M [I][7 23]	10.1	14.5	14.6	13.0	11.8	11.1	6.8	5.8	5.0	4.4
% F [I][7 23]	8.9	# 12.7	13.3	12.8	12.3	12.1	# 6.1	5.8	5.1	4.7
Netherlands Antilles Antilles néerlandaises										
MF [II][37]	...	20.1	# 15.8	17.2	17.9	17.6	13.3[38]	11.7[38]	11.2[38]	9.4[38]
M [II][37]	...	10.0	# 6.7	7.3	7.8	...	...	...	5.7[38]	4.3[38]
F [II][37]	...	10.1	# 9.1	9.9	10.1	...	...	...	5.5[38]	5.1[38]
% MF [II][37]	...	19.4	# 20.8	22.2	...	24.6[38]	24.4[38]	21.0	19.8[38]	16.4[38]
% M [II][37]	...	15.8	# 15.8	15.9	...	...	...	...	17.2[38]	13.1[38]
% F [II][37]	...	24.9	# 29.4	31.3	...	...	...	...	23.6[38]	20.9[38]
New Caledonia Nouvelle-Calédonie										
MF [I][7 24]	1.3	1.5	1.1	1.5	1.6	# 4.5	5.0	5.2	5.7	...
M [I][7 24]	0.7	1.2	0.8	0.9	...	# 2.4	...	...	...	...
F [I][7 24]	0.6	0.3	0.3	0.4	...	# 2.1	...	...	...	...
New Zealand Nouvelle-Zélande										
MF [I][7 25]	52.1	76.5	66.5	53.2	67.2	88.1	120.9	153.6	163.8	196.0
M [I][7 25]	31.1	48.3	41.4	33.5	45.0	60.8	84.6	106.6	112.1	135.9
F [I][7 25]	21.0	28.2	25.2	19.7	22.2	27.3	36.4	47.0	51.7	60.1
Nicaragua Nicaragua										
MF [II][11]	...	...	...	34.6	51.8	66.6	71.8	107.2	145.6	194.2
M [II][11]	...	...	...	18.6	27.9	35.9	38.9	57.8	78.5	104.8
F [II][11]	...	...	...	16.0	23.9	30.7	32.9	49.4	67.1	89.4
% MF [II][11]	...	...	...	3.2	4.7	5.8	6.0	8.4	11.1	14.0
% M [II][11]	...	...	...	2.6	3.8	4.7	4.9	6.9	9.0	11.3
% F [II][11]	...	...	...	4.6	6.5	8.0	8.3	12.0	15.4	19.4
Niger Niger										
MF [I][7]	16.0	19.5	27.4	29.0	27.7	27.2	26.1	24.6	20.9	...
M [I]	15.6[7]	19.0[7]	26.8[7]	28.2[7]	26.7[7]	25.7[7]	24.8[7]	23.4	19.9	...
F [I][7]	0.4	0.5	0.5	0.8	1.0	1.5	1.3	1.3	1.1	...
% MF [I][7]	38.4	49.0	56.1	56.3	51.5	49.3	50.1	46.8	45.4	...
% M [I]	38.8[7]	52.0[7]	57.5[7]	57.4[7]	52.2[7]	49.9[7]	50.8[7]	47.8	46.8	...
% F [I][7]	25.7	15.2	26.2	33.2	39.1	40.5	39.3	32.9	28.5	...

37
Unemployment
Number (thousands) and percentage of unemployed [*cont.*]
Chômage
Nombre (milliers) et pourcentage des chômeurs [*suite*]

Country or area Pays ou zones	1982	1983	1984	1985	1986	1987	1988	1989	1990	1991
Nigeria Nigéria										
MF [I][17]	16.3	31.0	34.3	28.3	32.5	57.3	60.5	57.6	57.1	...
Norway Norvège										
MF [I][7,24]	41.4	63.5	66.6	51.4	36.2	32.4	49.3	82.9	92.7	100.7
M [I][7,24]	25.3	40.9	41.1	29.8	20.2	18.4	30.0	51.6	57.1	62.8
F [I][7,24]	16.1	22.7	25.5	21.7	16.0	14.0	19.3	31.3	35.6	37.9
% MF [I][7,24]	2.0	3.1	3.2	2.5	1.8	1.5	2.3	3.8	4.3	4.7
Pakistan Pakistan										
MF [IV][11]	952.0	1 049.0	1 082.0	1 042.0	1 018.0	903.0	937.0	966.0	996.0	1 999.0
M [IV][11]	705.0	990.0	1 021.0	1 005.0	970.0	862.0	907.0	935.0	964.0	1 238.0
F [IV][11]	247.0	59.0	61.0	37.0	48.0	41.0	30.0	31.0	32.0	761.0
% MF [IV][11]	3.6	3.9	3.9	3.7	3.6	3.1	3.1	3.1	3.1	6.3
% M [IV][11]	3.0	4.2	4.2	4.0	3.9	3.3	3.4	3.4	3.4	4.5
% F [IV][11]	7.5	1.9	1.9	1.5	1.7	1.1	0.9	0.9	0.9	16.8
Panama Panama										
MF [IV][15]	51.5	64.2	68.8	88.3	75.8	91.1	127.8	133.7	...	134.0
M [IV][15]	26.7	35.6	38.9	46.3	42.9	48.5	74.9	75.0	...	72.6
F [IV][15]	24.8	28.6	29.9	42.0	32.9	42.7	52.9	58.7	...	61.4
% MF [IV][15]	8.4	9.7	10.1	12.3	10.5	11.8	16.3	16.3	...	15.7
% M [IV][15]	6.3	7.7	8.2	9.5	8.7	9.4	14.0	13.7	...	12.7
% F [IV][15]	13.3	14.5	14.2	18.5	14.5	16.8	21.4	21.6	...	21.5
Paraguay Paraguay										
MF [IV][17,20]	15.5	30.1	28.5	22.0	26.7	24.8	22.2	32.0	34.1	26.6
M [IV][17,20]	8.9	21.6	20.8	12.4	15.1	16.1	12.8	19.3	20.3	16.6
F [IV][17,20]	6.6	8.4	7.7	9.6	11.6	8.7	9.4	12.7	13.8	10.0
% MF [IV][5,20]	5.6	8.3	7.3	5.1	6.1	5.5	4.7	6.1	6.6	5.1
% M [IV][5,20]	5.4	9.8	9.1	5.1	6.3	6.5	4.8	6.6	6.6	5.4
% F [IV][5,20]	5.9	5.9	4.8	5.1	5.8	4.3	4.6	5.6	6.5	4.7
Peru Pérou										
MF [IV][3,39]	92.1[40]	137.8[40]	153.1[40]	...	111.7	103.5	...	186.7	...	146.3
M [IV][3,39]	44.4[40]	...	76.1[40]	...	40.9	48.6	...	84.8	...	73.7
F [IV][3,39]	47.7[40]	...	76.9[40]	...	70.8	54.9	...	101.9	...	72.6
% MF [IV][3,39]	...	...	...	...	5.3	4.8	...	7.9	...	5.8
% M [IV][3,39]	...	...	...	...	3.4	3.8	...	6.0	...	4.8
% F [IV][3,39]	...	...	...	...	8.0	6.2	...	10.7	...	7.3
Philippines Philippines										
MF [IV][15]	1 083.0	1 003.0	1 465.0	1 316.0	1 438.0	# 2 085.0[9]	1 954.0	2 009.0	1 993.0	2 267.0
M [IV][15]	434.0	458.0	675.0	644.0	686.0	# 1 163.0[9]	1 131.0	1 101.0	1 099.0	1 290.0

37
Unemployment
Number (thousands) and percentage of unemployed [*cont.*]
Chômage
Nombre (milliers) et pourcentage des chômeurs [*suite*]

Country or area Pays ou zones	1982	1983	1984	1985	1986	1987	1988	1989	1990	1991
F [IV][15]	649.0	545.0	790.0	672.0	752.0	# 922.0[9]	823.0	908.0	893.0	977.0
% MF [IV][15]	5.5	4.9	7.0	6.1	6.4	# 9.1[9]	8.3	8.4	8.1	9.0
% M [IV][15]	3.6	3.7	5.2	4.8	4.9	# 8.1[9]	7.6	7.3	7.1	8.1
% F [IV][15]	8.6	6.7	10.0	8.2	8.9	# 10.9[9]	9.5	10.3	9.8	10.5
Poland Pologne										
MF [I]	...	...	...	...	...	...	...	...	1 126.1[2]	2 155.6[5]
M [I]	...	...	...	...	...	...	...	...	552.4[2]	1 021.5[5]
F [I]	...	...	...	...	...	...	...	...	573.7[2]	1 134.1[5]
% MF [I]	...	...	...	...	...	...	...	...	6.1[2]	11.5[5]
% M [I]	...	...	...	...	...	...	...	...	5.5[2]	12.1[5]
% F [I]	...	...	...	...	...	...	...	...	6.8[2]	12.8[5]
Portugal Portugal										
MF [IV][17 41]	315.5	# 365.7[11]	393.9	397.0	393.4	329.0	...	243.4	231.1	...
M [IV][17 41]	92.5	# 127.5[11]	161.2	171.0	176.2	143.4	...	95.1	90.0	...
F [IV][17 41]	223.0	# 238.2[11]	232.7	225.9	217.2	185.6	...	148.3	141.1	...
% MF [IV][17 41]	7.4	# 7.3[11]	8.5	8.5	8.3	7.0	...	5.0	4.7	...
% M [IV][17 41]	3.7	# 4.6[11]	5.9	6.3	6.4	5.2	...	3.4	3.2	...
% F [IV][17 41]	12.3	# 12.2[11]	12.2	11.6	10.9	9.3	...	7.2	6.6	...
Puerto Rico Porto Rico										
MF [IV][24 42]	208.0	220.0	198.0	211.0	188.0	171.0	158.0	155.0	152.0	176.0
M [IV][24 42]	159.0	165.0	148.0	156.0	140.0	125.0	115.0	112.0	109.0	124.0
F [IV][24 42]	49.0	55.0	50.0	55.0	48.0	46.0	43.0	43.0	43.0	52.0
% MF [IV][24 42]	22.8	23.4	20.7	21.8	18.9	16.8	15.0	14.6	14.2	16.0
% M [IV][24 42]	26.3	26.7	23.7	24.7	21.9	19.4	17.5	16.9	16.2	17.9
% F [IV][24 42]	16.0	17.0	15.0	16.2	13.4	12.4	10.8	10.8	10.7	12.7
Réunion Réunion										
MF [I][7 24]	31.7	33.8	36.4	45.0	51.6	52.8	56.7	59.5	56.5	...
M [I][7 24]	19.9	20.4	24.3	28.2	26.8	25.8	29.2	30.2	28.4	...
F [I][24]	13.5[7]	13.5[7]	15.9[7]	19.6[7]	24.5[7]	26.3	29.2	27.3[7]	25.4[7]	...
Saint Pierre and Miquelon Saint-Pierre-et-Miquelon										
MF [I][7 43]	0.2	0.2	0.3	0.3	0.3	0.3	0.4	0.3	...	...
M [I][7 43]	0.1	0.1	0.1	0.2	0.2	0.2	0.2	0.2	...	...
F [I][7 43]	0.1	0.1	0.1	0.1	0.1	0.2	0.2	0.2	...	...
% MF [I][7 43]	11.0	10.3	10.3	10.9	11.0	11.7	13.2	11.3	...	...
San Marino Saint-Marin										
MF [II][3 5 7]	0.5	0.7	0.7	0.7	0.7	0.7	0.7	0.6	0.6	0.5
M [II][3 5 7]	0.1	0.2	0.2	0.2	0.2	0.2	0.2	0.1	0.2	0.1

37
Unemployment
Number (thousands) and percentage of unemployed [cont.]
Chôma e
Nombr (milliers) et pourcentage des chômeurs [suite]

Country or area Pays ou zones	1982	1983	1984	1985	1986	1987	1988	1989	1990	1991
F [II][357]	0.3	0.5	0.5	0.5	0.5	0.5	0.5	0.5	0.5	0.3
% MF [II][357]	4.9	6.8	7.2	6.9	6.9	6.2	6.0	5.3	5.5	4.3
% M [II][357]	2.2	3.5	3.9	3.0	3.2	2.8	2.7	2.1	2.4	2.3
% F [II][357]	9.0	11.7	12.0	12.5	12.0	10.9	10.5	9.5	9.7	6.9
Senegal Sénégal										
MF [I][2 18 44]	10.6	11.8	12.8	10.8	10.2	8.1	17.3	8.3	10.4	14.4
M [I][2 18 44]	...	10.5	11.1	9.0	8.4	6.4	15.0	7.1	8.3	13.1
F [I][2 18 44]	...	1.3	1.7	1.8	1.9	1.8	2.4	1.2	2.1	1.3
Seychelles Seychelles										
MF [I][7 45]	3.0	4.0	5.6[5]	5.7	...	...	...	...	...	...
M [I][7 45]	1.2	2.0	3.0[5]	2.4	...	...	...	...	...	...
F [I][7 45]	1.8	2.0	2.6[5]	3.3	...	...	...	...	...	...
% MF [I][7 45]	12.2	15.4	20.8[5]	22.5	...	...	...	...	...	...
% M [I][7 45]	9.7	14.3	19.9[5]	16.2	...	...	...	...	...	...
% F [I][7 45]	16.2	16.6	21.0[5]	31.6	...	...	...	...	...	...
Sierra Leone Sierra Leone										
MF [I][2 17 46]	2.9	2.5	2.5	2.8	2.7	2.4	...	...	...	...
Singapore Singapour										
MF [I][37]	7.0	7.4	6.2	9.2	9.5	8.1	4.5	2.7	1.7	1.2
M [I][37]	3.5	4.2	3.7	5.6	6.5	5.6	3.2	1.9	1.2	0.8
F [I][37]	3.5	3.1	2.5	3.6	2.9	2.5	1.4	0.8	0.5	0.4
MF [IV][15]	30.0	38.8	32.5	49.8	79.5	58.8	42.9	28.1	* 22.7	30.0
M [IV][15]	18.0	25.1	20.2	31.9	54.4	39.7	30.1	18.5	* 14.3	18.7
F [IV][15]	11.9	13.6	12.3	17.9	25.1	19.1	12.8	9.6	* 8.3	11.3
% MF [IV][15]	2.6	3.2	2.7	4.1	6.5	4.7	3.3	2.2	* 1.7	1.9
% M [IV][15]	2.4	3.2	2.6	4.2	7.0	5.1	3.8	2.3	* 1.7	2.0
% F [IV][15]	2.9	3.2	2.8	4.1	5.5	4.0	2.6	1.9	* 1.6	1.8
South Africa Afrique du Sud										
MF [IV][5 47 48]	427.0	483.0	492.0	495.0	519.0	# 1 018.0	874.0	755.0	99.0	...
M [IV][5 47 48]	195.0	231.0	222.0	226.0	253.0	# 475.0	438.0	360.0	59.0	...
F [IV][5 47 48]	232.0	252.0	270.0	269.0	266.0	# 543.0	436.0	394.0	...	...
% MF [IV][5 47 48]	5.0	7.6	6.2	8.1	10.7	14.1	10.4	7.9	7.9	8.6
% M [IV][5 47 48]	4.5	7.0	5.2	7.9	9.2	13.0	9.7	7.3	7.9	8.0
% F [IV][5 47 48]	6.0	8.5	7.8	8.5	13.0	15.8	11.5	9.0	7.8	9.3
Spain Espagne										
MF [I][7 32]	1 872.6	2 198.9	2 475.2	2 642.0	2 758.7	2 924.1	2 858.3	2 550.3	2 350.0	2 289.0
M [I][7 32]	1 240.4	1 400.1	1 529.0	1 599.3	1 572.2	1 525.4	1 359.7	1 087.5	942.5	910.7

37
Unemployment
Number (thousands) and percentage of unemployed [cont.]
Chômage
Nombre (milliers) et pourcentage des chômeurs [suite]

Country or area Pays ou zones	1982	1983	1984	1985	1986	1987	1988	1989	1990	1991
F [I][7 32]	632.2	798.8	946.2	1 042.7	1 186.4	1 398.8	1 498.6	1 462.8	1 407.5	1 378.3
% MF [I][7 32]	14.2	16.5	18.4	19.5	20.0	20.4	19.5	17.2	15.7	15.2
% M [I][7 32]	13.4	15.0	16.3	16.9	16.5	15.9	14.1	11.2	9.7	9.4
% F [I][7 32]	16.1	20.2	23.5	25.5	28.0	29.7	30.0	28.6	26.7	25.8
Sri Lanka Sri Lanka										
MF [IV]	...	...	...	840.3	...	...	...	...	1 005.1[9]	...
M [IV]	...	...	...	433.2	...	...	...	...	395.8[9]	...
F [IV]	...	...	...	407.0	...	...	...	...	609.2[9]	...
% MF [IV][9]	...	...	...	...	...	...	...	...	14.4	...
% M [IV][9]	...	...	...	...	...	...	...	...	9.1	...
% F [IV][9]	...	...	...	...	...	...	...	...	23.5	...
Sudan Soudan										
MF [I][2]	48.8	38.8	63.4	48.8	63.1	...	...	25.5	70.1	19.9[49]
M [I][2]	35.7	25.7	48.9	36.6	48.6	...	...	15.5	44.4	10.2[49]
F [I][2]	13.1	13.1	14.4	12.3	14.5	...	...	9.9	25.7	9.7[49]
Suriname Suriname										
MF [I][7]	# 7.4[50]	10.7	12.9	17.0	13.4	# 2.8	3.0	2.4	...	...
M [I][7]	# 3.7[50]	5.3	6.5	8.6	7.3	# 1.3	1.4	1.2	...	...
F [I][7]	# 3.7[50]	5.4	6.5	8.3	6.1	# 1.5	1.6	1.2	...	...
Sweden Suède										
MF [I][7 32]	80.4	91.7	91.9	84.9	84.2	78.0	61.1	56.3	66.4	114.8
M [I][7 32]	43.1	50.3	48.1	42.3	42.4	38.9	31.2	29.2	35.6	46.5
F [I][7 32]	37.3	41.3	43.8	42.6	41.9	39.2	29.9	27.1	30.9	48.3
% MF [I][7 32]	2.5	2.8	2.8	2.5	2.5	2.3	1.7	1.6	1.9	3.3
% M [I][7 32]	2.5	3.0	2.8	2.6	2.4	2.2	1.8	1.6	2.0	3.9
% F [I][7 32]	2.7	2.7	2.7	2.6	2.5	2.2	1.7	1.6	1.7	2.7
Switzerland Suisse										
MF [I][17]	13.2	26.3	# 35.2	30.3	25.7	24.7	22.2	17.5	18.1	39.2
M [I][17]	7.5	15.8	# 19.7	16.4	13.4	12.5	11.4	9.1	9.8	22.7
F [I][17]	5.7	10.5	# 15.5	13.9	12.3	12.1	10.8	8.4	8.3	16.5
% MF [I][17]	0.4	0.9	# 1.1	1.0	0.8	0.8	0.7	0.6	0.6	1.3
% M [I][17]	0.4	0.8	# 1.0	0.8	0.7	0.6	0.6	0.5	0.5	1.2
% F [I][17]	0.6	0.9	# 1.4	1.2	1.1	1.1	1.0	0.8	0.7	1.5
Thailand Thaïlande										
MF [IV][51]	653.3	1 148.3	1 279.8	1 337.0	968.7[9]	1 721.6[9]	929.2[9]	433.1	...	...
M [IV][51]	300.9	515.4	541.1	595.8	463.9[9]	672.7[9]	419.3[9]	...	...	...
F [IV][51]	352.3	632.9	738.7	741.1	504.8[9]	1 048.9[9]	509.8[9]	...	...	...

37
Unemployment
Number (thousands) and percentage of unemployed [*cont.*]
Chômage
Nombre (milliers) et pourcentage des chômeurs [*suite*]

Country or area Pays ou zone §	1982	1983	1984	1985	1986	1987	1988	1989	1990	1991
% MF [IV][51]	2.8	2.9	2.9	2.6	3.5[9]	5.9[9]	3.1[9]	...	...	...
% M [IV][51]	2.4	2.6	2.5	2.3	3.1[9]	4.3[9]	2.6[9]	...	...	...
% F [IV][51]	3.1	3.3	3.4	2.9	3.9[9]	7.6[9]	3.6[9]	...	...	...
Togo Togo										
MF [I][7 18]	4.3	4.8	4.7	4.1	...	...	...	...	...	...
M [I][7 18]	3.8	4.2	4.2	3.6	...	...	...	...	...	...
F [I][7 18]	0.5	0.6	0.5	0.5	...	...	...	...	...	...
Trinidad and Tobago Trinité-et-Tobago										
MF [IV][1]	44.1	50.0	63.2	72.8	81.2	106.6	104.7	103.4	93.6	91.2
M [IV][1]	24.2	27.6	37.6	46.4	51.5	65.7	66.5	64.8	55.1	49.6
F [IV][1]	19.9	22.4	25.6	26.4	29.7	40.9	38.2	38.6	38.5	41.5
% MF [IV][1]	10.0	11.0	13.5	15.5	17.2	22.3	22.0	22.0	20.0	18.5
% M [IV][1]	8.3	9.0	12.0	15.0	16.4	20.7	21.1	20.8	17.8	15.7
% F [IV][1]	13.8	15.0	16.0	17.0	18.9	25.3	23.6	24.5	24.2	23.4
Tunisia Tunisie										
MF [I][2 36]	78.0	72.1	74.3	84.0	80.2	83.7	91.5	105.9	...	...
M [I][2 36]	64.2	57.1	59.7	69.1	65.8	67.3	72.8	80.5	...	...
F [I][2 36]	13.7	14.9	14.6	14.9	14.3	16.4	18.7	25.5	...	...
Turkey Turquie										
MF [I][3 7]	426.0	549.0	761.0	935.0	1 053.0	1 124.0	1 155.0	1 076.0	980.0	859.0
M [I][3 7]	346.0	455.0	639.0	782.0	877.0	929.0	952.0	888.0	809.0	707.0
F [I][3 7]	79.0	94.0	122.0	153.0	177.0	195.0	204.0	189.0	171.0	152.0
United Kingdom Royaume-Uni										
MF [III][42 52]	2 916.9	# 3 104.7	3 159.8	3 271.2	# 3 292.9	2 953.4	# 2 370.4	# 1 798.7	1 664.5	2 291.9
M [III][42 52]	2 133.2	# 2 218.6	2 197.4	2 251.7	# 2 254.7	2 045.8	# 1 650.5	# 1 290.8	1 232.3	1 737.1
F [III][42 52]	783.6	886.0	962.5	1 019.5	# 1 036.6	907.6	# 719.9	507.9	432.2	554.9
% MF [III][42 52]	10.9	# 11.7	11.6	11.8	# 11.8	10.6	# 8.4	# 6.3	5.9	8.1
% M [III][42 52]	13.1	# 13.8	13.5	13.7	# 13.8	12.5	# 10.1	# 7.9	7.6	10.7
% F [III][42 52]	7.5	8.4	8.8	9.1	# 9.0	7.8	# 6.1	4.2	3.5	4.6
United States Etats-Unis										
MF [IV][24]	10 678.0	10 717.0	8 539.0	8 312.0	8 237.0	7 425.0	6 701.0	6 528.0	6 874.0	8 426.0
M [IV][24]	6 179.0	6 260.0	4 744.0	4 521.0	4 530.0	4 101.0	3 655.0	3 525.0	3 799.0	4 817.0
F [IV][24]	4 499.0	4 457.0	3 794.0	3 791.0	3 707.0	3 324.0	3 046.0	3 003.0	3 075.0	3 609.0
% MF [IV][24]	9.5	9.5	7.4	7.1	6.9	6.1	5.4	5.2	5.4	6.6
% M [IV][24]	9.7	9.7	7.3	6.9	6.8	6.1	5.3	5.1	5.4	6.9
% F [IV][24]	9.4	9.2	7.6	7.4	7.1	6.2	5.5	5.3	5.4	6.3

37
Unemployment
Number (thousands) and percentage of unemployed [*cont.*]
Chômage
Nombre (milliers) et pourcentage des chômeurs [*suite*]

Country or area Pays ou zone§	1982	1983	1984	1985	1986	1987	1988	1989	1990	1991
United States Virgin Is. **Iles Vierges américaines**										
MF [I][7 53]	3.4	3.6	3.3	2.6	2.1	1.4	1.5	1.7	1.3	1.3
% MF [I][7 53]	7.9	8.2	7.6	6.0	4.7	3.0	3.3	3.7	2.8	2.9
Uruguay Uruguay										
MF [IV][3 10]	...	...	145.3	...	122.0[9]	108.7	104.1	98.4	105.7	...
M [IV][3 10]	...	...	68.0	...	58.1[9]	48.6	46.3	44.9	50.6	...
F [IV][3 10]	...	...	77.3	...	63.9[9]	60.1	57.8	53.5	55.1	...
% MF [IV][3 10]	...	...	...	...	10.7[9]	9.1	8.6	8.0	8.5	...
% M [IV][3 10]	...	...	...	...	8.5[9]	6.7	6.3	6.1	6.9	...
% F [IV][3 10]	...	...	...	...	13.9[9]	12.6	11.9	10.8	10.9	...
Venezuela Venezuela										
MF [IV][1]	377.3	552.0	734.9	767.1	668.1	573.5	478.2	621.1	741.7[9]	...
M [IV][1]	...	...	...	577.5	502.2	448.9	366.2	471.2	535.8[9]	...
F [IV][1]	...	...	...	189.6	165.8	124.6	110.6	151.9	205.8[9]	...
% MF [IV][1]	7.1	10.1	13.0	13.1	11.0	9.2	7.3	9.2	10.4[9]	...
% M [IV][1]	...	...	...	13.5	11.4	9.9	7.8	9.8	10.9[9]	...
% F [IV][1]	...	...	...	11.6	9.9	7.2	6.1	7.6	9.3[9]	...
Yugoslavia Yougoslavie										
MF [I][2]	862.5	910.3	974.8	1 039.6	1 086.7	1 080.6	1 131.8	1 201.2	1 308.5	...
M [I][2 26]	369.7	392.4	430.0	460.4	482.9	477.4	511.0	550.4	610.2	...
F [I][2 25]	492.8	518.0	544.7	579.3	603.8	603.1	620.7	650.8	698.3	...
% MF [I][2]	12.4	12.8	13.3	13.8	14.1	13.6	14.1	14.9	16.4	...
% M [I][2 26]	8.7	9.1	9.8	10.2	10.4	10.2	10.9	11.7	...	...
% F [I][2 25]	18.1	18.4	18.6	19.1	19.0	18.5	18.7	19.3	...	...
Zambia Zambie										
MF [I]	46.5	39.9	38.5	31.3	31.8	24.4	18.1	15.9	...	...

Source:
International Labour Office (Geneva).

§ I - Employment office statistics.
   II - Official estimates.
   III - Social insurance statistics.
   IV - Labour force sample surveys and general
   household sample surveys.

Source:
Bureau international du Travail (Genève).

§ I - Statistiques des bureaux de placement.
   II - Evaluations officielles.
   III - Statistiques d'assurances sociales.
   IV - Enquêtes par sondage sur la main-d'oeuvre
   et enquêtes générales par sondage auprès
   des ménages.

† All data shown which pertain to Germany prior to 3 October
1990 are indicated separately for the Federal Republic of
Germany and the former German Democratic Republic based on
their respective territories at the time indicated. Where
data for united Germany (3 October 1990 and thereafter) are
not available, available data are shown separately under the
designations Federal Republic of Germany and former German
Democratic Republic and pertain to the territorial
boundaries prior to 3 October 1990. For detailed
explanatory notes on data pertaining to Germany, see Annex I
- Country Nomenclature.

† Toutes les données se rapportant à l'Allemagne avant le 3
octobre 1990 figurent dans deux rubriques séparées basées
sur les territoires respectifs de la République fédérale
d'Allemagne et l'ancienne République démocratique allemande
selon la période indiquée. En l'absence de données pour
l'Allemagne unifiée (à compter du 3 octobre 1990), les
données disponibles sont fournies séparément sous les
rubriques République fédérale d'Allemagne et ancienne
République démocratique allemande et se rapportent aux
limites territoriales antérieures au 3 octobre 1990. Pour
les notes explicatives en détail sur les données concernant
l'Allemagne, voir Annexe I - Nomenclature des pays.

37

Unemployment
Number (thousands) and percentage of unemployed [cont.]

Chômage
Nombre (milliers) et pourcentage des chômeurs [suite]

1 Persons aged 15 years and over (Netherlands: prior to 1987).
2 Applicants for work (incl. persons in employment).
3 Persons aged 14 years and over (Israel: prior to 1985).
4 Gran Buenos Aires.
5 One month of each year.
6 Including unpaid family workers who worked for less than 15 hours.
7 Registered unemployed.
8 Private sector.
9 Average of less than twelve months.
10 Urban areas.
11 Persons aged 10 years and over.
12 Excl. the rural population of the northern region.
13 Bujumbura.
14 Bangui.
15 Young people aged 16 to 25 years.
16 7 main cities of the country.
17 Persons aged 12 years and over.
18 Persons aged 14 to 55 years.
19 Persons aged 16 to 66 years.
20 Metropolitan area.
21 Year ending in June of the year indicated.
22 Persons aged 18 to 55 years.
23 Persons aged 15 to 64 years.
24 Persons aged 16 years and over.
25 Persons aged 15 to 60 years.
26 Persons aged 15 to 65 years.
27 Guatemala city only.
28 Persons aged 14 to 60 years.
29 Persons aged 10 to 56 years.
30 Incl. the residents of East Jerusalem.
31 Incl. persons who did not work in the country during the previous 12 months.
32 Persons aged 16 to 64 years.
33 Persons aged 16 to 61 years.
34 Excluding Rodrigues.
35 Persons aged 18 to 60 years.
36 Persons aged 18 years and over.
37 Excl. Aruba. Prior to 1984, including Aruba.
38 Curaçao.
39 Lima.
40 Excl. domestic services.
41 Including the Azores and Madeira.
42 Excl. persons temporarily laid off.
43 Persons aged 16 to 60 years.
44 Dakar.
45 Persons aged 15 to 63 years.
46 Excluding persons registered at the Maritime Pool.

47 Blacks.
48 Excluding Transkei, Bophuthatswana, Venda, Ciskei; elsewhere persons enumerated at de facto dwelling place.
49 Khartoum province.
50 Paramaribo only.
51 Persons aged 11 years and over.
52 Claimants at unemployment benefits offices.
53 Persons aged 16 to 65 years.

1 Personnes âgées de 15 ans et plus (Pays-Bas: avant 1987).
2 Demandeurs d'emploi (y compris les personnes d'un emploi).
3 Personnes âgées de 14 ans et plus (Israël: avant 1985).
4 Gran Buenos Aires.
5 Un mois de chaque année.
6 Y compris les travailleurs familiaux non rémunérés qui ont travaillé moins de 15 heures.
7 Les chômeurs enregistrés.
8 Secteur privé.
9 Moyenne de moins de douze mois.
10 Les régions urbaines.
11 Personnes âgées de 10 ans et plus.
12 Non compris la population rurale de la région du Nord.
13 Bujumbura.
14 Bangui.
15 Jeunes gens de 16 à 25 ans.
16 7 villes principales du pays.
17 Personnes âgées de 12 ans et plus.
18 Personnes âgées de 14 à 55 ans.
19 Personnes âgées de 16 à 66 ans.
20 Région métropolitaine.
21 Année se terminant en juin de l'année indiquée.
22 Personnes âgées de 18 à 55 ans.
23 Personnes âgées de 15 à 64 ans.
24 Personnes âgées de 16 ans et plus.
25 Personnes âgées de 15 à 60 ans.
26 Personnes âgées de 15 à 65 ans.
27 Ville de Guatemala seulement.
28 Personnes âgées de 14 à 60 ans.
29 Personnes âgées de 10 à 56 ans.
30 Y compris les résidents de Jérusalem-Est.
31 Y compris les personnes qui n'ont pas travaillé dans le pays pendant les 12 mois précédents.
32 Personnes âgées de 16 à 64 ans.
33 Personnes âgées de 16 à 61 ans.
34 Non compris Rodrigues.
35 Personnes âgées de 18 à 60 ans.
36 Personnes âgées de 18 ans et plus.
37 Non compris Aruba. Avant 1984, y compris Aruba.
38 Curaçao.
39 Lima.
40 Non compris les services domestiques.
41 Y compris le Açores et Madère.
42 Non compris les personnes temporairement mises à pied.
43 Personnes âgées de 16 à 60 ans.
44 Dakar.
45 Personnes âgées de 15 à 63 ans.
46 Non compris les demandeurs d'emploi enregistrés au bureau de placement maritime.
47 Blacks.
48 Non compris Transkei, Bophuthatswana, Venda, Ciskei; ailleurs personnes énumérées aux logis de facto.
49 Province de Khartoum.
50 Paramaribo seulement.
51 Personnes âgées de 11 ans et plus.
52 Demandeurs auprès des bureaux des prestations de chômage.
53 Personnes âgées de 16 à 65 ans.

Technical notes, tables 36 and 37

Detailed data on labour force and related topics are published in the ILO *Year Book of Labour Statistics*.[7] The series shown in the *Statistical Yearbook* give an overall picture of the availability and disposition of labour resources and, in conjunction with other macro-economic indicators, can be useful for an overall assessment of economic performance. The ILO *Year Book of Labour Statistics* provides a comprehensive description of the methodology underlying the labour series. Brief definitions of the major categories of labour statistics are given below.

"Employment" is defined to include persons above a specified age who, during a specified period of time, were in one of the following categories:

(a) "Paid employment", comprising persons who perform some work for pay or profit during the reference period or persons with a job but not at work due to temporary absence, such as vacation, strike, education leave;

(b) "Self-employment", comprising employers, own-account workers, members of producers' cooperatives, persons engaged in production of goods and services for own consumption and unpaid family workers;

(c) Members of the armed forces. Students, homemakers and others mainly engaged in non-economic activities during the reference period who, at the same time, were in paid employment or self-employment are considered as employed on the same basis as other categories.

"Unemployment" is defined to include persons above a certain age and who during a specified period of time were:

(a) "Without work", i.e. were not in paid employment or self-employment;

(b) "Currently available for work", i.e. were available for paid employment or self-employment during the reference period; and

(c) "Seeking work", i.e. had taken specific steps in a specified period to find paid employment or self-employment.

The following categories of persons are not considered to be unemployed:

(a) Persons intending to establish their own business or farm, but who had not yet arranged to do so and who were not seeking work for pay or profit;

(b) Former unpaid family workers not at work and not seeking work for pay or profit.

For various reasons, national definitions of employment and unemployment often differ from the recommended international standard definitions and thereby limit international comparability. Intercountry comparisons are also complicated by a variety of types of data collection systems used to obtain information on employed and unemployed persons.

Notes techniques, tableaux 36 et 37

Des données détaillées sur la main-d'oeuvre et des sujets connexes sont publiés dans l'*Annuaire des Statistiques du Travail* de l'OIT [7]. Les séries indiqués dans l'*Annuaire des Statistiques* donnent un tableau d'ensemble des disponibilités de main-d'oeuvre et de l'emploi de ces ressources et, combinées à d'autres indicateurs économiques, elles peuvent être utiles pour une évaluatation générale de la performance économique. L'*Annuaire des statistiques du Travail* de l'OIT donne une description complète de la méthodologie employée pour établir les séries sur la main-d'oeuvre. On trouvera ci-dessous quelques brèves définitions des grandes catégories de statistiques du travail.

Le terme "Emploi" désigne les personnes dépassant un âge déterminé qui, au cours d'une période donnée, se trouvaient dans l'une des catégories suivantes :

(a) La catégorie "emploi rémunéré", composée des personnes faisant un certain travail en échange d'une rémunération ou d'un profit pendant la période de référence, ou les personnes ayant un emploi, mais qui ne travaillaient pas en raison d'une absence temporaire (vacances, grève, congé d'études);

(b) La catégorie "emploi indépendant" regroupe les employeurs, les travailleurs indépendants, les membres de coopératives de producteurs et les personnes s'adonnant à la production de biens et de services pour leur propre consommation et la main-d'oeuvre familiale non rémunérée;

(c) Les membres des forces armées, les étudiants, les aides familiales et autres personnes qui s'adonnaient essentiellement à des activités non économiques pendant la période de référence et qui, en même temps, avaient un emploi rémunéré ou indépendant, sont considérés comme employés au même titre que les personnes des autres catégories.

Par "chômeurs", on entend les personnes dépassant un âge déterminé et qui, pendant une période donnée, appartenaient :

(a) "sans emploi", c'est-à-dire sans emploi rémunéré ou indépendant;

(b) "disponible", c'est-à-dire qui pouvaient être engagées pour un emploi rémunéré ou pouvaient s'adonner à un emploi indépendant au cours de la période de référence; et

(c) "à la recherche d'un emploi", c'est-à-dire qui avaient pris des mesures précises à un certain moment pour trouver un emploi rémunéré ou un emploi indépendant.

Ne sont pas considérés comme chômeurs :

(a) Les personnes qui, pendant la période de référence, avaient l'intention de créer leur propre entreprise ou exploitation agricole, mais n'avaient pas encore pris les dispositions nécessaires à cet effet et qui n'étaient pas à la recherche d'un emploi en vue d'une rémunération ou d'un profit;

For various reasons, national definitions of employment and unemployment often differ from the recommended international standard definitions and thereby limit international comparability. Intercountry comparisons are also complicated by a variety of types of data collection systems used to obtain information on employed and unemployed persons.

Table 36 presents absolute figures on the distribution of employed persons by major divisions of economic activity. Data are arranged as far as possible according to the major divisions of economic activity of the *International Standard Industrial Classification of All Economic Activities*. [46] The column for total employment includes economic activities not adequately defined.

Table 37: Figures are presented in absolute numbers and in percentages. Data are normally annual averages of monthly, quarterly or semi-annual data.

The series generally represent the total number of persons wholly unemployed or temporarily laid off. Percentage figures, where given, are calculated by comparing the number of unemployed to the total members of that group of the labour force on which the unemployment data are based.

(b)   Les anciens travailleurs familiaux non rémunérés qui n'avaient pas d'emploi et n'étaient pas à la recherche d'un emploi en vue d'une rémunération ou d'un profit.

Pour diverses raisons, les définitions nationales de l'emploi et du chômage diffèrent souvent des définitions internationales types recommandées, limitant ainsi les possibilités de comparaison entre pays. Ces comparaisons se trouvent en outre compliquées par la diversité des systèmes de collecte de données utilisés pour recueillir des informations sur les personnes employées et les chômeurs.

Le *Tableau 36* présente des chiffres en valeur absolue sur la répartition des personnes employées par branche d'activité économique. Les données sont présentées autant que possible selon les principales divisions de l'activité économique de la *Classification internationale type, par Industrie, de toutes les branches d'activité économique* [46]. La colone indiquant l'emploi total englobe les activités économiques non convenablement définies.

Tableau 37 : Les chiffres sont présentés en valeur absolue et en pourcentage. Les données sont normalement des moyennes annuelles des données mensuelles, trimestrielles ou semestrielles.

Les séries représentent généralement le nombre total des chômeurs complets ou des personnes temporairement mises à pied. Les données en pourcentage, lorsqu'elles figurent au tableau, sont calculées par comparaison du nombre de chômeurs au nombre total des personnes du groupe de main-d'oeuvre sur lequel sont basées les données relatives au chômage.

38
Earnings in manufacturing
Gains dans les industries manufacturières
By hour, day, week or month
Par heure, jour, semaine ou mois

Country or area and unit Pays ou zone et unité	1982	1983	1984	1985	1986	1987	1988	1989	1990	1991
Albania: lek Albanie : lek										
MF - month mois[1 2 3]	522.0	520.0	513.0	521.0	528.0	533.0	533.0	542.0	554.0	* 665.0
American Samoa: dollar Samoa américaines : dollar										
MF - hour heure[1 4]	2.3	2.6	2.7	2.8	2.8	...	...	...	...	...
Antigua and Barbuda: EC dollar Antigua-et-Barbuda : EC dollar										
MF - week semaine	...	...	...	...	...	456.3	502.0	932.7	...	...
Argentina: peso Argentine : peso										
MF - hour heure[5 6]	2.6	14.6	131.2	# 0.9	1.6	...	...	...	9 557.3	...
Australia: dollar Australie : dollar										
MF - hour heure * [7 8]	8.0	8.4	8.9	9.4	9.9	10.5	11.5	12.5	...	...
M - hour heure * [7 8]	8.4	8.9	9.3	9.9	10.4	11.1	12.1	13.0	...	...
F - hour heure * [7 8]	6.6	6.8	7.4	7.9	8.3	8.8	9.6	10.5	...	...
Austria: schilling Autriche : schilling										
MF - month mois[2 5]	14 069.0	14 715.0	15 453.0	16 395.0	17 116.0	17 646.0	18 318.0	19 130.0	20 496.0	21 547.0
Barbados: dollar Barbade : dollar										
MF - week semaine	...	...	...	...	...	...	238.5[7]	235.3[7]	251.5[7]	255.7
Belarus: rouble Bélarus : rouble										
MF - month mois[1 2 3]	174.9	177.9	184.2	191.0	196.8	204.3	225.6	246.3	279.4	564.0
Belgium: franc Belgique : franc										
MF - hour heure[5 7]	256.8	270.3	281.8	293.8	295.3	302.2	309.8	327.1	342.4	...
M - hour heure[5 7]	273.7	287.5	299.7	312.5	314.2	321.2	329.3	348.2	...	...
F - hour heure[5 7]	201.3	213.8	223.3	232.3	232.7	239.9	245.3	257.6	...	...
Botswana: pula Botswana : pula										
MF - month mois[1 7 9]	158.6	166.7	211.9	249.2	282.4	285.3	330.9	345.0	383.0	427.0
Brazil: cruzeiro Brésil : cruzeiro										
MF - month mois[1 10]	...	...	423.8	1 607.2	3 507.0	10 949.0	75 579.0	...	...	...
Bulgaria: lev Bulgarie : lev										
MF - month mois[1 3]	205.6	210.5	217.1	227.0	239.9	243.8	262.8	281.8	335.8	...
Burundi: franc Burundi : franc										
MF - month mois[1 11 12]	13 599.0	12 439.0	17 229.0	17 421.0	17 532.0	17 600.0	...	...	...	...
Canada: dollar Canada : dollar										
MF - hour heure[1 13]	10.3[14]	# 10.6[15]	11.2	11.6	12.0	# 12.2	12.8	13.5	14.3	15.1
Chile: peso Chili : peso										
MF - month mois[5 7 8]	13 005.0	17 461.0	20 379.0	24 409.0	29 157.0	34 137.0	41 497.0	50 432.0	64 447.0	83 908.0
China: yuan renminbi Chine : yuan renminbi										
MF - month mois[1 2 16]	66.8	68.3	82.3	# 92.6[17]	106.4	118.1	148.5	166.8	174.2	191.3
Colombia: peso Colombie : peso										
MF - month mois[1 18]	167.8	212.3	262.5	317.9	393.6	483.1	614.1	787.9	...	...
Costa Rica: colon Costa Rica : colon										
MF - month mois[17]	3 427.0	5 520.0	7 110.0	8 673.0	9 588.0	13 211.0	14 658.0	16 784.0	20 037.0	27 229.0
M - month mois[17]	3 767.0	5 948.0	7 699.0	9 422.0	10 370.0	14 207.0	16 573.0	18 534.0	21 887.0	30 151.0
F - month mois[17]	2 627.0	4 440.0	5 645.0	6 932.0	7 815.0	11 021.0	11 102.0	13 505.0	16 262.0	21 733.0
Cuba: peso Cuba : peso										
MF - month mois[1 3]	183.0[2]	188.0[2]	192.0[2]	193.0[2]	# 189.0[2]	183.0[2]	# 183.0	...	...	...

38
Earnings in manufacturing
By hour, day, week or month [cont.]
Gains dans les industries manufacturières
Par heure, jour, semaine ou mois [suite]

Country or area and unit Pays ou zone et unité	1982	1983	1984	1985	1986	1987	1988	1989	1990	1991
Cyprus: pound Chypre : livre										
MF - week semaine[7 8 12 19]	43.8	47.0	48.9	52.4	54.0	58.1	61.9	68.0	74.2	* 80.9
M - week semaine[7 8 12 19]	56.5	61.9	65.5	70.1	72.5	76.7	80.6	89.4	98.8	* 107.7
F - week semaine[7 8 12 19]	31.8	33.7	36.4	39.1	40.7	44.3	47.7	52.1	56.8	* 62.1
Czechoslovakia: koruna Tchécoslovaquie : couronne										
MF - month mois[3]	2 666.0	2 728.0	2 795.0	2 854.0	2 903.0	2 958.0	3 018.0	3 085.0	3 184.0	3 716.0
M - month mois[3]	3 049.0	3 120.0	3 197.0	3 264.0	3 320.0	3 383.0	3 452.0	3 529.0	3 642.0	4 250.0
F - month mois[3]	2 070.0	2 118.0	2 170.0	2 216.0	2 254.0	2 297.0	2 343.0	2 395.0	2 472.0	2 885.0
Denmark: krone Danemark : couronne										
MF - hour heure[15 19 20]	65.0	69.3	72.3	75.8	78.7	86.8	92.1	96.0	99.9	104.6
M - hour heure[15 19 20]	67.1	71.6	74.7	78.3	81.5	90.2	95.6	99.6	103.8	108.5
F - hour heure[15 19 20]	57.1	61.1	64.1	67.1	69.2	75.9	80.7	84.3	87.8	92.1
Dominican Republic: peso Rép. dominicaine : peso										
MF - month mois[1]	185.9	190.8	213.8	252.6	...	...	...	...	...	...
Egypt: £ Egypte : £										
MF - week semaine[7 14]	19.5	...	...	27.0	...	38.0	...	...	...	...
M - week semaine[7 14]	20.2	...	...	28.0	...	38.0	...	...	...	...
F - week semaine[7 14]	12.6	...	...	19.0	...	37.0	...	...	...	...
El Salvador: colon El Salvador : colon										
MF - hour heure[21]	2.6	2.8	3.3	3.3	# 3.2	3.2	3.2	3.3	3.3	...
M - hour heure[21]	2.8	3.3	3.3	3.6	# 3.5	3.3	3.3	3.4	3.4	...
F - hour heure[21]	2.5	2.5	2.9	2.9	# 3.0	3.0	3.0	3.1	3.2	...
Estonia: rouble Estonie : rouble										
MF - month mois[1]	214.0	216.6	224.6	230.0	237.2	246.2	268.0	292.6	359.8	850.7
Fiji: dollar Fidji : dollar										
MF - day jour[7]	11.2	11.9	12.0	12.2	11.8	12.3	12.5	11.4	...	...
Finland: markka Finlande : markka										
MF - hour heure[2]	24.6	26.9	29.7	32.1	34.0	36.5	39.7	43.5	47.7	50.7
M - hour heure[2]	26.6	29.8	32.0	34.6	36.6	39.2	42.6	46.7	51.1	53.9
F - hour heure[2]	20.5	22.5	24.7	26.6	28.3	30.3	32.9	35.8	39.5	42.1
France: franc France : franc										
MF - hour heure[7]	29.8	33.6	35.7	# 37.8	39.3	41.0	# 41.8	43.4	45.5	...
M - hour heure[7]	31.8	35.8	38.1	# 40.2	41.6	43.4	# 44.4	46.2	48.4	...
F - hour heure[7]	24.7	28.1	29.9	# 31.8	33.0	34.4	# 35.1	36.3	38.2	...
French Polynesia: CFP franc Polynésie française : franc CFP										
MF - month mois[4]	41 771.0	47 000.0	62 530.0	70 000.0	77 700.0	82 000.0	84 000.0	86 500.0	...	...
Gambia: dalasi Gambie : dalasi										
MF - day jour[1 7 14]	...	7.4	7.0	# 12.3	11.5	11.1	...	...	...	...
Germany† · Allemagne†										
F. R. Germany: deutsche mark R. f. Allemagne : deutsche mark										
MF - hour heure[12]	14.6	15.1	15.5	16.2	16.8	17.5	18.3	19.1	20.1	21.3
M - hour heure[12]	15.6	16.1	16.5	17.2	17.9	18.6	19.5	20.3	21.3	22.6

38
Earnings in manufacturing
By hour, day, week or month [cont.]
Gains dans les industries manufacturières
Par heure, jour, semaine ou mois [suite]

Country or area and unit Pays ou zone et unité	1982	1983	1984	1985	1986	1987	1988	1989	1990	1991
F - hour heure[12]	11.4	11.7	12.0	12.5	13.0	13.6	14.2	14.7	15.5	16.5
Ghana: cedi Ghana : cedi										
MF - month mois[17]	787.0	1 312.0	3 441.0	5 059.0	8 787.0	15 216.0	*21 411.0	...	...	...
Gibraltar: pound Gibraltar : livre										
MF - week semaine[7]	121.1	130.7	138.8	140.9	150.2	152.7	190.6	190.1	238.5	...
M - week semaine[7]	128.4	140.3	146.8	149.0	157.9	158.7	203.3	194.3	247.7	...
F - week semaine[7]	83.7	89.5	93.3	97.5	108.5	118.6	122.0	137.7	151.7	...
Greece: drachma Grèce : drachme										
MF - hour heure[5 14]	173.9	# 207.6	262.2	# 314.2	354.1	388.2	459.7	554.0	661.3	...
M - hour heure[5 14]	195.3	# 233.1	292.6	# 347.2	393.2	430.2	509.2	614.5	733.8	...
F - hour heure[5 14]	142.8	# 174.0	222.9	# 269.5	302.3	333.7	397.4	481.0	575.3	...
Guam: dollar Guam : dollar										
MF - hour heure[17]	6.4	6.3	5.5	6.2	6.3	6.0	6.3	7.3	8.1	9.1
Guatemala: quetzal Guatemala : quetzal										
MF - hour heure	1.1	1.2	0.9	0.9	...	...	...	...	...	...
Honduras: lempira Honduras : lempira										
MF - week semaine[1 14]	174.4	272.3	97.0	107.1	...	...	...	115.4	...	...
Hong Kong: dollar Hong-kong : dollar										
MF - day jour[15]	72.3	79.5	90.1	98.3	106.4	119.4	136.9	157.0	179.5	200.7
M - day jour[15]	85.7	92.9	104.2	115.1	125.9	143.3	166.1	191.7	224.5	249.9
F - day jour[15]	66.6	73.7	84.4	91.2	98.1	108.2	123.7	140.3	155.8	173.6
Hungary: forint Hongrie : forint										
MF - month mois[8 16 22]	4 131.0	4 321.0	4 913.0	5 366.0	5 716.0	6 214.0	# 7 761.0[23]	9 121.0	11 167.0	*13 983.0
India: rupee Inde : roupie										
MF - month mois[24]	623.6	703.3	799.5	740.2	825.0	839.3	...	...	...	...
Ireland: pound Irlande : livre										
MF - hour heure[7 25]	3.2	3.5	3.9	# 4.2	4.5	4.7	4.9	5.1	5.4	* 5.7
M - hour heure[7 19]	3.6	3.9	4.4	# 4.8	5.1	5.3	5.5	5.7	6.0	...
F - hour heure[7 19]	2.4	2.7	3.0	# 3.2	3.5	3.6	3.8	4.0	4.2	...
Israel: new shekel Israël : nouveau shekel										
MF - day jour[1 26 27]	670.5	1 710.0	8 470.0	# 29.9	47.5	64.9	78.8	95.2	...	...
Italy: lira Italie : lire										
MF - hour heure[8]	6 693.0	7 712.0	8 539.0	9 451.0	...	...	...	...	...	...
Jamaica: dollar Jamaïque : dollar										
MF - week semaine[1]	...	...	...	...	271.9	309.4	345.9	391.8	450.6[15]	...
Japan: yen Japon : yen										
MF - month mois[1 12 28]	269 583.0	279 106.0	292 255.0	299 531.0	305 414.0	313 170.0	318 663.0	336 648.0	352 020.0	368 011.0
M - month mois[1 12 28]	327 977.0	340 047.0	356 561.0	367 182.0	373 324.0	381 138.0	393 804.0	414 981.0	436 135.0	450 336.0
F - month mois[1 12 28]	141 352.0	146 903.0	152 519.0	154 571.0	158 550.0	163 944.0	164 673.0	173 097.0	180 253.0	193 112.0
Jordan: dinar Jordanie : dinar										
MF - day jour[1 7 19]	3.5	3.7	3.7	4.0	4.2	4.5	4.2	4.4	...	...

38
Earnings in manufacturing
By hour, day, week or month [*cont.*]
Gains dans les industries manufacturières
Par heure, jour, semaine ou mois [*suite*]

Country or area and unit Pays ou zone et unité	1982	1983	1984	1985	1986	1987	1988	1989	1990	1991
Kenya: shilling Kenya : shilling										
MF - month mois[1 7 8]	1 545.7	1 673.8	1 836.2	1 928.7	2 078.1	2 293.8	2 469.7	2 797.5	3 064.6	...
M - month mois[1 7 8]	1 579.5	1 709.0	1 879.3	2 025.2	2 137.8	2 376.7	2 576.0	3 160.9	3 474.6	...
F - month mois[1 7 8]	1 197.8	1 332.7	1 439.7	1 531.0	1 554.3	1 551.1	1 751.7	2 212.0	2 565.9	
Korea, Republic of: won Corée, République de : won										
MF - month mois[1 8 12]	202 117.0	226 790.0	245 261.0	269 652.0	294 485.0	328 696.0	393 056.0	491 632.0	590 760.0	690 310.0
M - month mois[1 8 12]	267 332.0	295 982.0	317 273.0	346 852.0	374 786.0	413 348.0	490 542.0	608 929.0	726 476.0	842 828.0
F - month mois[1 8 12]	120 522.0	136 810.0	149 718.0	162 705.0	181 795.0	207 906.0	249 739.0	307 452.0	364 259.0	428 063.0
Luxembourg: franc Luxembourg : franc										
MF - hour heure[5 7]	277.8	301.4	310.7	320.0	332.2	339.2	357.0	374.0	391.0	...
M - hour heure[5 7]	287.4	311.8	321.9	331.1	344.3	352.4	375.0	393.0	411.0	...
F - hour heure[5 7]	172.8	191.7	197.7	207.3	209.1	217.2	218.0	234.0	243.0	...
Macau: pataca Macao : pataca										
MF - month mois[1]	...	...	...	...	...	...	2 198.6	2 479.3	2 727.9	...
Malawi: kwacha Malawi : kwacha										
MF - month mois[1]	99.2	79.6	72.4	90.2	101.3	125.8	140.5	153.8	176.8	...
Mauritius: rupee Maurice : roupie										
MF - day jour[5 15 29 30]	27.0	30.0	32.0	35.0	36.0	# 41.0[31]	47.0	55.0	70.0	86.0
Mexico: peso Mexique : peso										
MF - month mois[28]	18 792.0	27 853.0	44 277.0	63 446.0	106 230.0	254 437.0	514 452.0	632 292.0	752 924.0	...
Myanmar: kyat Myanmar : kyat										
M - month mois[1 15]	247.6	263.1	255.7	260.1	282.2	325.7	533.1	729.5	...	...
F - month mois[1 15]	224.1	242.4	246.7	256.8	243.6	261.1	496.9	705.6	...	...
Netherlands: guilder Pays-Bas : florin										
MF - hour heure[5 7 25]	15.6	15.9	16.2	16.6	16.9	17.3	17.7	18.1	18.9	...
M - hour heure[5 7 19]	16.5	16.7	17.0	17.4	17.7	18.1	18.5	19.0	19.8	...
F - hour heure[5 7 19]	13.0	13.2	13.4	13.5	14.0	14.1	14.5	14.7	15.3	...
Netherlands Antilles: guilder Antilles néerlandaises : florin										
MF - month mois[1]	1 967.0	2 128.0	# 2 287.0[32]	2 243.0	2 158.0	2 122.0	2 407.0[33]	...	...	...
M - month mois[1]	2 018.0	2 203.0	# 2 337.0[32]	2 388.0	2 260.0	...	2 510.0[33]	...	...	...
F - month mois[1]	1 321.0	1 420.0	# 1 561.0[32]	1 628.0	1 456.0	...	1 627.0[33]	...	...	...
New Zealand: dollar Nouvelle-Zélande : dollar										
MF - hour heure[1 34]	7.3	7.5	7.7	8.4	10.1	11.0	12.0	12.6	13.4	13.9
M - hour heure[1 34]	7.9	8.1	8.4	9.1	10.9	11.9	12.9	13.5	14.4	14.8
F - hour heure[1 34]	5.6	5.7	5.9	6.4	7.8	8.6	9.6	10.2	10.8	11.2
Norway: krone Norvège : couronne										
MF - hour heure[8 19]	48.4	52.6	57.1	61.5	67.7	78.6	83.0	87.3	92.5	97.3
M - hour heure[8 19]	49.8	54.0	58.6	63.3	69.7	81.0	85.4	89.5	94.6	99.5
F - hour heure[8 19]	41.4	45.4	49.2	52.9	58.4	67.8	72.0	76.5	81.8	88.7
Pakistan: rupee Pakistan : roupie										
MF - month mois[1]	702.2	751.9	713.4	881.6	1 026.4	1 115.4	1 130.9	...	...	...

38
Earnings in manufacturing
By hour, day, week or month [cont.]
Gains dans les industries manufacturières
Par heure, jour, semaine ou mois [suite]

Country or area and unit Pays ou zone et unité	1982	1983	1984	1985	1986	1987	1988	1989	1990	1991
Panama: balboa Panama : balboa										
MF - hour heure	1.5	1.7	1.8	1.8	...	...	...	...	...	...
Paraguay: guarani Paraguay : guaraní										
MF - month mois[1]	...	30 150.0	35 958.0	53 100.0	67 466.0	...	115 897.0	177 464.0	220 548.0	273 537.0
M - month mois[1]	...	31 319.0	37 359.0	54 216.0	68 493.0	...	118 473.0	170 222.0	234 234.0	292 787.0
F - month mois[1]	...	24 681.0	29 364.0	47 127.0	62 324.0	...	105 252.0	212 743.0	155 744.0	196 738.0
Peru: new sol Pérou : nouveau sol										
MF - day jour[1 7 35]	4.1	7.1	12.6	26.7	# 80.6	161.0	758.2	18 636.6		...
Philippines: peso Philippines : peso										
MF - month mois[1 14 36]	1 091.0	# 1 248.0	1 642.0	1 951.0	2 183.0	2 537.0		...	...	...
Poland: zloty Pologne : zloty										
MF - month mois[1 8 16]	10 845.0	14 123.0	16 717.0	19 901.0	24 076.0	29 382.0	54 708.0	212 170.0	995 541.0	...
Portugal: escudo Portugal : escudo										
MF - hour heure	116.9	136.6	163.1	199.6	235.9	270.9	* 258.2	...	...	...
Puerto Rico: dollar Porto Rico : dollar										
MF - hour heure	4.6	4.8	5.0	5.2	5.3	5.4	5.6	5.8	6.0	* 6.3
Romania: leu Roumanie : leu										
MF - month mois[16 22]	2 433.0	2 504.0	2 697.0	2 718.0	2 728.0	2 719.0	2 835.0	2 920.0	3 139.0	...
Saint Lucia: EC dollar Sainte-Lucie : EC dollar										
MF - week semaine	66.5	72.0	90.0	99.0	...	...	...	...	...	...
San Marino: lira Saint-Marin : lire										
MF - day jour[1]	47 498.0	...	60 134.0	65 742.0	72 911.0	76 133.0	...	84 782.0	84 835.0	102 579.0
Seychelles: rupee Seychelles : roupie										
MF - month mois[1 24]	1 779.0	1 783.0	1 856.0	1 956.0	2 003.0	2 075.0	# 1 863.0[37]	1 975.0[37]	2 187.0[37]	...
Sierra Leone: leone Sierra Leone : leone										
MF - week semaine[15 19]	23.5	23.9	31.9	45.7	79.7	80.5	...	...	...	...
Singapore: dollar Singapour : dollar										
MF - month mois[1]	...	...	...	...	975.8	1 008.8	1 115.9	1 242.9	1 395.0	1 551.8
M - month mois[1]	...	...	...	...	...	...	...	1 623.0	1 797.5	1 970.1
F - month mois[1]	...	...	...	...	...	...	...	876.2	983.3	1 096.8
South Africa: rand Afrique du Sud : rand										
MF - month mois[1]	551.0	632.0	713.0	804.0	916.0	1 057.0	1 238.0	1 448.0	1 654.0	1 863.0
Spain: peseta Espagne : peseta										
MF - hour heure[1]	399.0	457.0	514.0	564.0	620.0	678.0	735.0	780.0	902.0	994.0
Sri Lanka: rupee Sri Lanka : roupie										
MF - hour heure[15]	3.4	3.9	4.4	5.1	5.5	6.0	6.6	7.5	9.5	...
M - hour heure[15]	3.5	4.1	4.9	5.9	5.8	6.4	6.9	8.2	9.8	...
F - hour heure[15]	2.9	2.9	3.4	4.3	4.5	4.6	4.8	5.4	8.9	...
Swaziland: lilangeni Swaziland : lilangeni										
M - month mois[7 38]	766.0	807.0	806.0	881.0	946.0	1 032.0	...	...	...	...
F - month mois[7 38]	318.0	290.0	...	265.0	572.0	577.0	...	...	...	...
Sweden: krona Suède : couronne										
MF - hour heure[8 15 19]	45.7	48.5	54.0	58.6	62.7	67.0	72.2	79.3	87.3	91.7
M - hour heure[8 15 19]	46.6	49.4	55.1	59.8	63.9	68.5	73.8	81.1	89.5	93.8

38
Earnings in manufacturing
By hour, day, week or month [cont.]
Gains dans les industries manufacturières
Par heure, jour, semaine ou mois [suite]

Country or area and unit Pays ou zone et unité	1982	1983	1984	1985	1986	1987	1988	1989	1990	1991
F - hour heure[8 15 19]	42.1	44.6	49.6	53.7	57.8	61.7	66.4	72.6	79.5	83.7
Switzerland: franc Suisse : franc										
M - hour heure[7 12 19]	17.3	18.0	18.5	19.0	19.9	20.5	21.3	22.1	23.4	...
F - hour heure[7 12 19]	11.6	12.0	12.3	12.8	13.4	13.8	14.3	15.0	15.9	...
Thailand: baht Thaïlande : baht										
MF - month mois[1]	1 696.0	1 706.0	2 717.0	2 826.0	2 631.0	...	...	2 950.0	3 383.0	
Tonga: pa'anga Tonga : pa'anga										
MF - week semaine[1 7]	...	...	24.7	26.4	28.6	30.8	33.8	43.0	42.6	51.1
Turkey: lira Turquie : livre										
MF - day jour[1 7 39]	722.9	989.5	1 336.3	...	...	...	8 461.0	17 820.5	30 582.2	
M - day jour[1 7 39]	724.3	989.4	1 372.5	...	...	...	8 642.9	18 679.1	31 231.1	
F - day jour[1 7 39]	713.3	989.8	1 280.4	...	...	...	6 877.3	13 830.0	25 299.5	
Ukraine: rouble Ukraine : rouble										
MF - month mois[1 16]	188.2[2]	191.8[2]	196.5[2]	201.5[2]	# 193.5	198.2	215.8	...	...	...
former USSR: rouble ancienne URSS : rouble										
MF - month mois[1 2 16]	188.9	192.3	198.3	203.3	208.1	213.6	232.2	253.9	...	...
United Kingdom: pound Royaume-Uni : livre										
MF - hour heure[7 40 41]	3.0	# 3.2	3.4	3.7	4.0	4.2	4.5	4.9	5.4	...
M - hour heure[7 40 41]	3.2	# 3.4	3.7	4.0	4.3	4.6	4.9	5.3	5.8	
F - hour heure[7 40 41]	2.2	# 2.4	2.5	2.7	2.9	3.1	3.3	3.6	3.9	
United States: dollar Etats-Unis : dollar										
MF - hour heure	8.5	8.8	9.2	9.5	9.7	9.9	# 10.2[42]	10.5	10.8	11.2
United States Virgin Is.: dollar Iles Vierges américaines : dollar										
MF - hour heure[1]	9.8	10.0	9.5	9.4	9.6	9.4	9.9	10.9	11.9	12.5
Uruguay: new peso Uruguay : nouveau peso										
MF - month mois[1 15 43]	...	...	100.0	169.8	322.2	572.5	949.3	1 751.6	3 453.2	...
Venezuela: bolívar Venezuela : bolívar										
MF - month mois	...	...	...	...	5 084.0	7 440.0	...	...	...	...
Yugoslavia: new dinar Yougoslavie : nouveau dinar										
MF - month mois[1 16]	1.2	1.4	2.2	3.9	8.0	16.6	44.6	773.0	...	...
Zimbabwe: dollar Zimbabwe : dollar										
MF - month mois[8 44]	300.0	341.0	382.2	435.4	471.2	527.9	590.3	665.5	...	...

Source:
International Labour Office (Geneva).

† All data shown which pertain to Germany prior to 3 October
1990 are indicated separately for the Federal Republic of
Germany and the former German Democratic Republic based on
their respective territories at the time indicated. Where
data for united Germany (3 October 1990 and thereafter) are
not available, available data are shown separately under the
designations Federal Republic of Germany and former German
Democratic Republic and pertain to the territorial
boundaries prior to 3 October 1990. For detailed
explanatory notes on data pertaining to Germany, see Annex I
- Country Nomenclature.

Source:
Bureau international du travail (Genève).

† Toutes les données se rapportant à l'Allemagne avant le 3
octobre 1990 figurent dans deux rubriques séparées basées
sur les territoires respectifs de la République fédérale
d'Allemagne et l'ancienne République démocratique allemande
selon la période indiquée. En l'absence de données pour
l'Allemagne unifiée (à compter du 3 octobre 1990), les
données disponibles sont fournies séparément sous les
rubriques République fédérale d'Allemagne et ancienne
République démocratique allemande et se rapportent aux
limites territoriales antérieures au 3 octobre 1990. Pour
les notes explicatives en détail sur les données concernant
l'Allemagne, voir Annexe I – Nomenclature des pays.

38

Earnings in manufacturing
By hour, day, week or month [cont.]

Gains dans les industries manufacturères
Par heure, jour, semaine ou mois [suite]

1 Including salaried employees.	1 Y compris les employés.
2 Including mining and quarrying (Finland: also electricity; China: also electricity, gas and water, geological and prospecting activities; Cuba: also fishing and gas).	2 Y compris les industries extractives (Finlande: aussi l'électricité; Chine: aussi l'électricité, le gaz et l'eau, les recherches géologiques et de prospection; Cuba: aussi la pêche, l'eau et le gaz).
3 State sector (Bulgaria: and cooperative sector).	3 Secteur d'etat (Bulgarie: et secteur de coopératif).
4 Minimum rates.	4 Taux minima.
5 Wage earners.	5 Ouvriers.
6 Prior to 1985, pesos argentinos. One austral is equivalent to 1,000 pesos argentinos.	6 Avant 1985, pesos argentinos, un austral équivaut à 1,000 pesos argentinos.
7 One month of each year.	7 Un mois de chaque année.
8 Including the value of payments in kind (Sweden: including holidays and sick-leave payments).	8 Y compris la valeur des paiements en nature (Suède: y compris les versements pour les vacances et congés de maladie).
9 Citizens only.	9 Nationaux seulement.
10 Figures in thousands.	10 Données in milliers.
11 Bujumbura only.	11 Bujumbura seulement.
12 Including family allowances (F.R. Germany: paid directly by the employers).	12 Y compris les allocations familiales (Allemagne, Rép. féd. d': payées directement par les employeurs).
13 Paid by the hour.	13 Rémunérés à l'heure.
14 Establishments with ten or more persons employed only (Canada: 20 or more persons; Gambia: 5 or more persons; Singapore: 25 or more persons).	14 Etablissements occupant dix personnes et plus seulement (Canada: 20 personnes et plus; Gambie: 5 personnes et plus; Singapour: 25 personnes et plus).
15 Average of less than twelve months.	15 Moyenne de moins de douze mois.
16 Socialized sector.	16 Secteur socialisé.
17 Prior to 1985, excluding geological and prospecting activities.	17 Avant 1985, non compris les recherches géologiques et de prospection.
18 Index numbers; Base: 1980 equals 100.	18 Indices; Base: 1980 égal à 100.
19 Adults only.	19 Adultes seulement.
20 Excluding vacation pay.	20 Non compris les versements pour congés payés.
21 San Salvador only, prior to 1986.	21 San Salvador seulement, avant 1986.
22 Net earnings after deduction of income taxes.	22 Gains nets après déduction de l'impôt sur le revenu.
23 Prior to 1988: gross earnings before income tax deduction.	23 Avant 1988: gains bruts avant déduction de l'impôt sur le revenu.
24 Including electricity, gas, water and services (Seychelles: electricity and water only).	24 Y compris l'électricite, le gaz, l'eau et les services (Seychelles: l'électricite et l'eau seulement).
25 Including juveniles.	25 Y compris les jeunes gens.
26 Beginning 1985, new shekels. One new shekel is equivalent to 1,000 shekels.	26 A partir de 1985, nouveaux shekels. Un nouveaux shekel équivant à 1000 shekels.
27 Including payments subject to income tax.	27 Y compris les versements soumis à l'impôt sur le revenu.
28 Break in series: Japan at 1985; Mexico at 1986.	28 Discontinuité dans la série: Japon 1985; Mexique 1986.
29 Excluding sugar and tea factories.	29 Non compris les fabriques de sucre et de thé.
30 Daily rates of pay.	30 Rémunérés sur la base de taux de salaire journaliers.
31 Prior to Sept. 1987: including workers on piece rates of pay.	31 Avant sept. 1987: y compris les travailleurs aux pièces.
32 Prior to 1984: including Aruba, and oil storage and transhipment.	32 Avant 1984: y compris Aruba, et l'entreposage et le transbordement du pétrole.
33 Curaçao.	33 Curaçao.
34 Establishments with the equivalent of more than 2 full-time paid employees.	34 Etablissements occupant plus de l'équivalent de 2 salariés à plein temps.
35 Lima only.	35 Lima seulement.
36 Computed on the basis of annual wages.	36 Calculés sur la base de salaires annuels.
37 Earnings are exempted from income tax.	37 Les gains sont exempts de l'impôt sur le revenu.
38 Skilled workers only.	38 Ouvriers qualifiés seulement.
39 Insurance statistics.	39 Statistiques d'assurances.
40 Beginning 1983, including quarrying.	40 A partir de 1983, y compris les carrières.
41 Full-time workers on adult rates of pay.	41 Travailleurs à plein temps rémunéres sur la base de taux de salaire pour adultes.
42 New industrial classification.	42 Nouvelle classification industrielle.
43 Index of average monthly earnings (Oct.-Dec. 1984=100).	43 Indices des gains mensuels moyens (oct.-déc. 1984=100).
44 All persons engaged.	44 Ensemble de l'effectif occupé.

39
Producers prices and wholesale prices
Prix à la production et des prix de gros
Index numbers: 1980 = 100
Indices: 1980 = 100

Country or area and groups	1985	1986	1987	1988	1989	1990	1991	Pays ou zone et groupes
Argentina								**Argentine**
Domestic supply[123]	100	163	365	1 872	66 123	1 128 670	2 375 572	Offre intérieure[123]
Domestic production[3]	100	164	365	1 859	65 021	1 150 802	2 443 959	Production intérieure[3]
Agricultural products[23]	100	212	457	2 187	78 419	1 248 401	2 435 898	Produits agricoles[23]
Industrial products[234]	100	158	352	1 815	63 215	1 137 647	2 445 044	Produits industriels[234]
Import products[34]	100	160	370	1 991	76 632	917 612	1 723 428	Produits importés[34]
Australia								**Australie**
Industrial products[245]	143	151	162	174	186	204	208	Produits industriels[245]
Austria								**Autriche**
Domestic supply[267]	119	113	111	111	113	116	117	Offre intérieure[267]
Agricultural products	115	105	109	107	107	115	117	Produits agricoles
Producers' material[267]	122	113	108	108	112	113	112	Matériaux de production[267]
Consumers' goods[26]	120	113	115	113	115	121	123	Biens de consommation[26]
Capital goods[26]	109	111	108	107	105	107	108	Biens d'équipement[26]
Bangladesh								**Bangladesh**
Domestic supply[267]	174	183	197	209	226	245	255	Offre intérieure[267]
Agricultural products[268]	183	191	213	226	244	265	269	Produits agricoles[268]
Industrial products[2468]	157	166	168	177	190	206	227	Produits industriels[2468]
Raw materials[6]	140	143	146	150	171	190	196	Matières premières[6]
Finished goods[6]	138	147	155	161	169	188	196	Produits finis[6]
Belgium								**Belgique**
Domestic supply	145	132	126	127	135	136	135	Offre intérieure
Agricultural products[2]	137	137	131	133	140	137	138	Produits agricoles[2]
Industrial products	142	129	125	127	136	136	134	Produits industriels
Import products	160	138	131	133	...	...	...	Produits importés
Intermediate products	150	127	119	120	128	129	124	Produits intermédiaires
Consumers' goods	138	138	135	137	143	143	145	Biens de consommation
Capital goods	135	138	139	141	146	150	155	Biens d'équipement
Brazil								**Brésil**
Domestic supply[39]	100	239	748	5 956	83 111	2 378 647	12 004 206	Offre intérieure[39]
Agricultural products[39]	100	277	744	5 824	72 408	2 356 680	13 874 940	Produits agricoles[39]
Industrial products[39]	100	224	713	5 764	82 271	2 188 515	10 303 545	Produits industriels[39]
Raw materials[39]	100	212	608	4 543	55 609	1 067 362	5 689 838	Matières premières[39]
Producers' material[39]	100	214	678	5 376	73 031	2 033 744	10 125 359	Matériaux de production[39]
Consumers' goods[39]	100	259	797	6 394	93 316	2 689 839	13 854 613	Biens de consommation[39]
Capital goods[3]	100	216	773	7 265	99 321	2 883 393	12 071 429	Biens d'équipement[3]
Canada								**Canada**
Agricultural products[2]	113	112	...	...	112	115	109	Produits agricoles[2]
Industrial products[24]	131	132	136	142	144	145	143	Produits industriels[24]
Raw materials[10]	139	115	123	119	123	128	120	Matières premières[10]
Intermediate products[11]	115	115	119	127	129	128	125	Produits intermédiaires[11]
Finished goods[11]	124	127	129	130	133	137	139	Produits finis[11]
Chile								**Chili**
Domestic supply	303	363	433	459	528	643	...	Offre intérieure
Domestic production	288	355	431	454	528	650	808	Production intérieure

39
Producers prices and wholesale prices
Index numbers: 1980 = 100 [cont.]
Prix à la production et des prix de gros
Indices : 1980 = 100 [suite]

Country or area and groups	1985	1986	1987	1988	1989	1990	1991	Pays ou zone et groupes
Agricultural products	252	343	424	397	484	600	732	Produits agricoles
Industrial products[7]	303	364	440	497	562	702	...	Produits industriels [7]
Import products	374	401	444	483	531	613	...	Produits importés
Colombia								**Colombie**
Domestic supply[2 6]	255	304	385	407	511	663	817	Offre intérieure [2 6]
Domestic production[6]	260	315	395	495	509	647	809	Production intérieure[6]
Agricultural products[6]	277	338	417	562	681	848	...	Produits agricoles [6]
Industrial products[6]	235	285	351	466	580	760	...	Produits industriels [6]
Import products[6]	285	361	449	529	673	848	962	Produits importés [6]
Exported goods[6]	274	424	407	525	602	626	607	Produits exportés [6]
Raw materials[6]	222	280	350	430	540	656	...	Matières premières [6]
Intermediate products[6]	270	330	413	547	711	899	...	Produits intermédiaires [6]
Finished goods[6]	268	339	423	541	702	752	...	Produits finis[6]
Costa Rica								**Costa Rica**
Domestic supply[7 12]	516	560	863	1 053	1 207	1 387	1 237	Offre intérieure [7 12]
Cyprus								**Chypre**
Industrial products	132	114	109	134	142	149	155	Produits industriels
Denmark								**Danemark**
Domestic supply[2 9]	148	138	138	143	152	153	155	Offre intérieure [2 9]
Domestic production[2 9]	147	142	144	149	158	160	162	Production intérieure [2 9]
Import products[9]	150	131	128	133	142	143	143	Produits importés [9]
Producers' material	150	134	133	139	148	150	151	Matériaux de production
Consumers' goods	145	145	146	150	157	158	161	Biens de consommation
Capital goods	151	156	162	169	177	184	190	Biens d'équipement
Dominican Republic								**Rép. dominicaine**
Domestic supply[2 6 7 13]	202	230	284	413	666	1 263	1 577	Offre intérieure [2 6 7 13]
Industrial products[3]	100	101	119	221	335	...	...	Produits industriels [3]
Consumers' goods[3]	100	116	144	202	416	335	405	Biens de consommation[3]
Ecuador								**Equateur**
Agricultural products	328	414	473	733	1 264	1 499	3 650	Produits agricoles
Industrial products	233	300	395	694	1 221	1 754	2 676	Produits industriels
Egypt								**Egypte**
Domestic supply[6 7]	164	192	218	275	351	409	483	Offre intérieure [6 7]
Raw materials[6]	164	195	203	270	365	475	569	Matières premières [6]
Intermediate products[6]	150	162	175	216	282	380	466	Produits intermédiaires [6]
Finished goods[6]	153	174	212	288	350	432	525	Produits finis [6]
Capital goods	126	153	164	211	261	305	418	Biens d'équipement
El Salvador								**El Salvador**
Domestic supply[6 7 14]	154	203	206	216	237	282	...	Offre intérieure [6 7 14]
Finland								**Finlande**
Domestic supply	140	136	138	143	151	156	156	Offre intérieure
Domestic production	143	140	142	150	159	166	166	Production intérieure
Import products	133	113	111	118	123	124	125	Produits importés
Raw materials	138	122	121	130	138	140	137	Matières premières

39
Producers prices and wholesale prices
Index numbers: 1980 = 100 [*cont.*]
Prix à la production et des prix de gros
Indices : 1980 = 100 [*suite*]

Country or area and groups	1985	1986	1987	1988	1989	1990	1991	Pays ou zone et groupes
Finished goods	140	134	136	144	152	155	156	Produits finis
Consumers' goods	146	146	148	152	157	162	166	Biens de consommation
Capital goods	142	148	154	162	173	184	189	Biens d'équipement
France								**France**
Agricultural products	143	143	140	141	151	152	153	Produits agricoles
Germany†								**Allemagne†**
F. R. Germany								**R. f. Allemagne**
Domestic production	119	114	111	114	119	118	118	Production intérieure
Agricultural products[15]	103	98	95	95	103	99	98	Produits agricoles [15]
Industrial products	122	118	116	117	121	123	126	Produits industriels
Import products	125	101	99	101	105	103	103	Produits importés
Exported goods	119	117	116	119	122	122	122	Produits exportés
Raw materials	123	95	89	92	99	95	91	Matières premières
Intermediate products	122	113	110	113	117	116	116	Produits intermédiaires
Finished goods[3]	100	99	99	101	103	106	109	Produits finis [3]
Producers' material	124	121	117	119	123	125	128	Matériaux de production
Consumers' goods	117	113	112	113	117	119	123	Biens de consommation
Capital goods	120	123	126	128	132	136	140	Biens d'équipement
Greece								**Grèce**
Domestic supply[16 17 18]	255	298	327	360	409	474	553	Offre intérieure [16 17 18]
Domestic production[17 18]	245	293	320	352	398	474	563	Production intérieure [17 18]
Agricultural products[17 19]	259	288	321	353	409	503	619	Produits agricoles [17 19]
Industrial products[17 18]	248	293	320	352	395	468	551	Produits industriels [17 18]
Import products[16 17 18]	281	349	392	428	481	542	621	Produits importés [16 17 18]
Exported goods[17 18 20]	270	275	301	333	387	413	458	Produits exportés [17 18 20]
Honduras								**Honduras**
Domestic supply	132	135	136	...	...	...	...	Offre intérieure
Domestic production	123	126	127	...	...	...	...	Production intérieure
Agricultural products	106	111	113	...	...	...	...	Produits agricoles
Import products	152	154	155	...	...	...	...	Produits importés
India								**Inde**
Domestic supply	144	152	163	173	183	199	222	Offre intérieure
Agricultural products[21]	158	169	183	200	208	220	254	Produits agricoles [21]
Industrial products[4]	137	144	154	163	177	193	213	Produits industriels [4]
Raw materials	144	150	162	176	184	202	239	Matières premières
Indonesia								**Indonésie**
Domestic supply[6 18]	163	163	200	210	228	250	259	Offre intérieure [6 18]
Agricultural products[6]	172	187	212	237	258	278	301	Produits agricoles [6]
Industrial products[4 6]	165	178	205	223	238	252	276	Produits industriels [4 6]
Import products[6 18]	164	178	218	226	246	263	278	Produits importés [6 18]
Exported goods[6]	155	116	162	171	179	218	204	Produits exportés [6]
Ireland								**Irlande**
Domestic supply[2 16 22]	154	150	151	158	166	162	164	Offre intérieure [2 16 22]
Agricultural products[2 22]	136	135	140	155	163	144	140	Produits agricoles [2 22]
Industrial products[2 4 22]	155	153	156	162	170	167	169	Produits industriels [2 4 22]
Capital goods[8]	149	153	158	164	173	178	184	Biens d'équipement [8]

39
Producers prices and wholesale prices
Index numbers: 1980 = 100 [cont.]
Prix à la production et des prix de gros
Indices : 1980 = 100 [suite]

Country or area and groups	1985	1986	1987	1988	1989	1990	1991	Pays ou zone et groupes
Israel								**Israël**
Industrial products[3]	100	144	170	200	247	268	313	Produits industriels[3]
Italy								**Italie**
Domestic supply[2 18]	172	171	175	184	196	210	221	Offre intérieure [2 18]
Agricultural products[2 8]	161	165	164	169	186	196	214	Produits agricoles [2 8]
Industrial products[2 18]	174	172	177	186	197	212	222	Produits industriels [2 18]
Producers' material	169	160	162	169	180	195	202	Matériaux de production
Consumers' goods	175	180	186	195	208	220	235	Biens de consommation
Capital goods	183	194	206	217	231	244	255	Biens d'équipement
Japan								**Japon**
Domestic supply[2 6]	100	92	88	88	90	92	92	Offre intérieure [2 6]
Domestic production[6]	101	96	93	92	94	95	97	Production intérieure [6]
Agricultural products[6 8]	105	104	99	98	99	100	101	Produits agricoles [6 8]
Industrial products[6 8]	100	95	93	92	94	96	97	Produits industriels [6 8]
Import products[6 16]	92	77	81	86	89	91	90	Produits importés [6 16]
Exported goods[20]	98	102	106	113	114	110	112	Produits exportés [20]
Raw materials[6]	97	65	61	58	63	67	64	Matières premières [6]
Intermediate products[6]	97	90	86	86	88	90	91	Produits intermédiaires [6]
Finished goods[6]	104	102	100	98	99	101	102	Produits finis [6]
Producers' material[6]	97	86	82	82	84	87	87	Matériaux de production[6]
Consumers' goods[6]	104	102	100	98	98	99	101	Biens de consommation[6]
Capital goods[6]	104	102	99	99	101	103	104	Biens d'équipement[6]
Jordan								**Jordanie**
Domestic supply[2 6]	121	120	122	133	178	203	215	Offre intérieure[2 6]
Korea, Republic of								**Corée, République de**
Domestic supply	128	126	127	130	132	138	145	Offre intérieure
Agricultural products[8 23]	139	135	132	155	158	179	198	Produits agricoles [8 23]
Raw materials	144	149	156	164	168	181	198	Matières premières
Intermediate products	124	120	121	121	120	122	126	Produits intermédiaires
Finished goods	132	131	132	139	143	152	162	Produits finis
Producers' material	127	124	124	124	124	126	131	Matériaux de production
Consumers' goods	132	131	131	139	142	153	165	Biens de consommation
Capital goods	123	123	125	131	139	142	146	Biens d'équipement
Kuwait								**Koweït**
Domestic supply[6 9]	107	108	111	116	126	...	...	Offre intérieure [6 9]
Agricultural products[6 9]	117	116	107	107	109	...	...	Produits agricoles [6 9]
Raw materials[6]	97	93	88	85	96	...	...	Matières premières [6]
Intermediate products[6]	102	102	102	114	114	...	...	Produits intermédiaires [6]
Finished goods[6]	109	111	116	119	127	...	...	Produits finis [6]
Producers' material[6]	100	96	95	101	110	...	...	Matériaux de production[6]
Consumers' goods[6]	110	112	114	121	130	...	...	Biens de consommation[6]
Capital goods[6]	109	114	124	125	139	...	...	Biens d'équipement[6]
Luxembourg								**Luxembourg**
Industrial products	150	146	137	140	151	148	144	Produits industriels
Import products	143	148	146	149	155	162	161	Produits importés

39
Producers prices and wholesale prices
Index numbers: 1980 = 100 [cont.]
Prix à la production et des prix de gros
Indices : 1980 = 100 [suite]

Country or area and groups	1985	1986	1987	1988	1989	1990	1991	Pays ou zone et groupes
Exported goods	151	146	135	139	150	146	141	Produits exportés
Intermediate products	152	146	133	136	148	143	136	Produits intermédiaires
Consumers' goods	143	148	151	156	161	161	170	Biens de consommation
Capital goods	141	148	149	151	159	163	170	Biens d'équipement
Mexico								**Mexique**
Domestic supply[7 24]	1 001	1 797	4 407	8 784	9 905	12 158	14 478	Offre intérieure[7 24]
Agricultural products	1 043	1 675	3 290	6 476	8 259	17 805	21 865	Produits agricoles
Exported goods	936	1 395	3 721	5 938	7 213	9 275	9 210	Produits exportés
Raw materials	1 093	2 013	4 865	10 343	11 878	13 907	16 255	Matières premières
Consumers' goods[8 24]	1 052	1 967	4 738	9 637	10 917	13 526	16 628	Biens de consommation[8 24]
Capital goods[8 24 25]	923	1 669	4 085	8 714	9 350	10 909	13 058	Biens d'équipement[8 24 25]
Netherlands								**Pays-Bas**
Agricultural products[15 26]	116	107	105	106	114	111	110	Produits agricoles[15 26]
Industrial products	121	115	113	115	119	118	120	Produits industriels
Import products	131	100	95	95	103	101	99	Produits importés
Exported goods[20]	123	106	102	104	110	109	108	Produits exportés[20]
Raw materials	125	109	104	106	112	108	107	Matières premières
Intermediate products	121	112	109	112	116	114	115	Produits intermédiaires
Producers' material	129	104	99	99	107	104	102	Matériaux de production
Consumers' goods	122	118	116	117	121	122	125	Biens de consommation
Capital goods	119	121	121	122	126	128	130	Biens d'équipement
New Zealand								**Nouvelle-Zélande**
Domestic supply[2 27]	180	195	213	225	242	251	251	Offre intérieure[2 27]
Agricultural products[28]	165	150	161	164	197	199	176	Produits agricoles[28]
Industrial products[28]	167	175	186	193	206	215	217	Produits industriels[28]
Norway								**Norvège**
Domestic supply	139	144	152	160	169	174	180	Offre intérieure
Raw materials	132	132	138	156	173	172	171	Matières premières
Finished goods	131	139	150	159	166	170	170	Produits finis
Consumers' goods	146	155	166	175	182	190	...	Biens de consommation
Pakistan								**Pakistan**
Domestic supply[2 6 7]	137	144	158	166	180	194	219	Offre intérieure[2 6 7]
Agricultural products[6]	140	147	160	169	182	194	215	Produits agricoles[6]
Industrial products[6]	132	141	152	159	176	198	230	Produits industriels[6]
Raw materials[6]	117	129	156	163	177	185	207	Matières premières[6]
Panama								**Panama**
Domestic supply	128	125	126	118	121	132	133	Offre intérieure
Peru								**Pérou**
Domestic supply[3]	100	160	243	1 718	59 467	3 333 333	12 094 700	Offre intérieure[3]
Domestic production[3]	100	163	255	1 873	52 900	3 333 333	13 584 333	Production intérieure[3]
Agricultural products[3 29]	100	222	393	1 688	43 425	2 500 000	11 251 500	Produits agricoles[3 29]
Industrial products[3 4 8]	100	147	216	2 018	48 567	3 333 333	13 157 667	Produits industriels[3 4 8]
Import products[3]	100	151	194	1 353	42 267	3 333 333	13 099 000	Produits importés[3]
Philippines								**Philippines**
Domestic supply[2 6 7 30]	290	291	315	354	391	431	...	Offre intérieure[2 6 7 30]
Singapore								**Singapour**
Domestic supply[7]	93	79	85	83	86	87	82	Offre intérieure[7]
Domestic production[2 4 31]	88	75	77	76	77	80	76	Production intérieure[2 4 31]

39
Producers prices and wholesale prices
Index numbers: 1980 = 100 [cont.]
Prix à la production et des prix de gros
Indices : 1980 = 100 [suite]

Country or area and groups	1985	1986	1987	1988	1989	1990	1991	Pays ou zone et groupes
Agricultural products[2][31]	101	95	96	99	99	103	...	Produits agricoles[2][31]
Import products[16][31]	91	81	87	88	88	87	82	Produits importés[16][31]
Exported goods	90	78	80	79	79	79	75	Produits exportés
South Africa								**Afrique du Sud**
Domestic supply[32]	181	217	247	279	322	361	402	Offre intérieure[32]
Domestic production[32]	180	213	245	279	320	361	405	Production intérieure[32]
Spain								**Espagne**
Domestic supply[18]	180	182	183	189	196	201	204	Offre intérieure[18]
Producers' material	189	183	179	183	190	191	190	Matériaux de production
Consumers' goods	169	176	182	188	196	203	210	Biens de consommation
Capital goods	175	187	197	207	216	225	234	Biens d'équipement
Sri Lanka								**Sri Lanka**
Domestic supply	164	159	181	213	232	284	310	Offre intérieure
Domestic production	172	185	192	217	263	334	373	Production intérieure
Import products	150	149	153	170	193	244	259	Produits importés
Exported goods	168	138	191	245	229	259	279	Produits exportés
Producers' material	173	174	180	199	236	286	313	Matériaux de production
Consumers' goods	162	156	183	220	261	284	310	Biens de consommation
Capital goods	155	152	150	164	190	265	300	Biens d'équipement
Sweden								**Suède**
Domestic supply[2][18][33]	158	154	158	166	179	188	191	Offre intérieure[2][18][33]
Domestic production[2][18][33]	161	161	165	175	188	196	200	Production intérieure[2][18][33]
Import products[18][33]	156	140	143	148	157	162	163	Produits importés[18][33]
Exported goods[18][33]	153	154	160	168	180	186	187	Produits exportés[18][33]
Producers' material	157	160	166	175	190	198	201	Matériaux de production
Switzerland								**Suisse**
Domestic supply[27]	115	111	108	111	116	117	118	Offre intérieure[27]
Domestic production[27]	116	115	114	116	120	123	125	Production intérieure[27]
Agricultural products[2]	118	115	114	117	116	118	119	Produits agricoles[2]
Import products[7]	112	101	95	98	105	104	102	Produits importés[7]
Raw materials	114	110	106	110	116	116	116	Matières premières
Consumers' goods	118	118	119	120	123	127	129	Biens de consommation
Syrian Arab Republic								**Rép. arabe syrienne**
Domestic supply	166	229	336	491	563	684	787	Offre intérieure
Raw materials	203	212	370	694	801	819	1 147	Matières premières
Intermediate products	130	210	262	321	516	776	861	Produits intermédiaires
Consumers' goods	154	158	298	327	557	702	716	Biens de consommation
Thailand								**Thaïlande**
Domestic supply[2][9]	109	109	115	125	131	136	144	Offre intérieure[2][9]
Agricultural products	95	98	110	122	129	131	147	Produits agricoles
Industrial products[4]	114	110	114	120	124	135	141	Produits industriels[4]
Exported goods[9]	84	86	90	100	101	100	102	Produits exportés[9]
Raw materials	93	93	103	121	127	122	134	Matières premières
Intermediate products	107	102	107	115	119	125	135	Produits intermédiaires
Finished goods	118	120	126	133	139	145	155	Produits finis

39
Producers prices and wholesale prices
Index numbers: 1980 = 100 [cont.]
Prix à la production et des prix de gros
Indices : 1980 = 100 [suite]

Country or area and groups	1985	1986	1987	1988	1989	1990	1991	Pays ou zone et groupes
Consumers' goods	117	118	124	129	135	141	152	Biens de consommation
Trinidad and Tobago								**Trinité-et-Tobago**
Domestic supply[28]	110	117	122	129	141	143	...	Offre intérieure[28]
Tunisia								**Tunisie**
Domestic supply[2 6 7]	161	170	177	...	...	...	...	Offre intérieure[2 6 7]
Domestic production[2 6 7]	166	177	183	...	...	...	...	Production intérieure[2 6 7]
Agricultural products[28]	117	125	132	136	146	160	169	Produits agricoles[28]
Industrial products[28]	116	126	143	164	178	186	191	Produits industriels[28]
Import products[2 6 25]	153	159	166	...	...	...	...	Produits importés[2 6 25]
Turkey								**Turquie**
Domestic supply[2 11 18 34]	357	462	610	1 026	1 740	2 596	4 033	Offre intérieure[2 11 18 34]
Agricultural products[11]	354	443	575	868	1 574	2 425	3 657	Produits agricoles[11]
Industrial products[11]	345	458	611	1 104	1 815	2 683	...	Produits industriels[11]
United Kingdom								**Royaume-Uni**
Agricultural products[5]	124	123	123	125	133	135	138	Produits agricoles[5]
Industrial products[4 5]	136	142	148	155	163	173	182	Produits industriels[4 5]
Raw materials	138	127	131	136	143	143	141	Matières premières
Finished goods	139	145	151	158	166	176	186	Produits finis
United States								**Etats-Unis**
Domestic supply[2]	115	112	114	119	125	130	130	Offre intérieure[2]
Agricultural products[2]	92	90	93	102	108	109	103	Produits agricoles[2]
Industrial products[2 35]	118	114	117	121	127	132	132	Produits industriels[2 35]
Raw materials	101	92	98	101	108	114	106	Matières premières
Intermediate products	114	110	112	119	124	127	127	Produits intermédiaires
Finished goods	119	117	120	123	129	135	138	Produits finis
Consumers' goods	117	114	117	120	127	133	136	Biens de consommation
Capital goods	125	128	130	133	138	143	148	Biens d'équipement
Uruguay								**Uruguay**
Domestic supply[2 25 36]	757	1 266	2 066	3 253	5 634	11 700	...	Offre intérieure[2 25 36]
Agricultural products[36]	654	1 202	2 016	3 001	5 223	8 663	12 998	Produits agricoles[36]
Industrial products[25 36]	792	1 287	2 083	3 337	5 771	12 116	23 599	Produits industriels[25 36]
Venezuela								**Venezuela**
Domestic supply[6 7]	178	208	303	362	714	909	...	Offre intérieure[6 7]
Domestic production[6 25]	182	212	291	351	706	920	...	Production intérieure[6 25]
Agricultural products[6 7]	196	250	374	463	654	1 028	...	Produits agricoles[6 7]
Industrial products[6 7]	173	201	292	347	703	882	...	Produits industriels[6 7]
Import products[6 7]	172	204	339	396	750	901	...	Produits importés[6 7]
Yemen								**Yémen**
Domestic supply[37]	113	97	115	...	...	...	...	Offre intérieure[37]
Agricultural products[37]	114	115	118	...	...	...	...	Produits agricoles[37]
Industrial products[37]	89	114	114	...	...	...	...	Produits industriels[37]
Raw materials[37]	99	100	100	...	...	...	...	Matières premières[37]
Yugoslavia								**Yougoslavie**
Domestic supply[28]	0	17	33	100	1 483	7 800	16 667	Offre intérieure[28]
Agricultural products[28]	7	14	36	100	1 286	7 143	...	Produits agricoles[28]
Producers' material[28]	0	17	32	100	1 318	6 923	15 385	Matériaux de production[28]
Consumers' goods[28]	0	17	33	100	1 500	8 167	16 667	Biens de consommation[28]

39
Producers prices and wholesale prices
Index numbers: 1980 = 100 [cont.]
Prix à la production et des prix de gros
Indices : 1980 = 100 [suite]

Country or area and groups	1985	1986	1987	1988	1989	1990	1991	Pays ou zone et groupes
Capital goods[20]	0	17	50	100	1 517	8 450	16 667	Biens d'équipement[20]
Zambia								**Zambie**
Agricultural products	286	494	658	765	1 642	3 170	...	Produits agricoles
Industrial products	267	567	1 194	851	1 567	3 378	...	Produits industriels
Exported goods	246	587	631	1 138	1 802	4 046	...	Produits exportés
Producers' material	238	411	1 038	1 442	3 043	5 728	...	Matériaux de production
Consumers' goods	252	380	764	713	1 607	3 245	...	Biens de consommation
Capital goods	321	909	1 929	1 560	2 616	6 302	...	Biens d'équipement

Source:
Price statistics database of the Statistical Division of the
United Nations Secretariat.

Source:
Base de données pour les statistiques des prix de la
Division de statistique du Secrétariat de l'ONU.

† All data shown which pertain to Germany prior to 3 October
1990 are indicated separately for the Federal Republic of
Germany and the former German Democratic Republic based on
their respective territories at the time indicated. Where
data for united Germany (3 October 1990 and thereafter) are
not available, available data are shown separately under the
designations Federal Republic of Germany and former German
Democratic Republic and pertain to the territorial
boundaries prior to 3 October 1990. For detailed
explanatory notes on data pertaining to Germany, see Annex I
- Country Nomenclature.

1 Domestic agricultural products only.
2 Including exported products.
3 Base: 1985 = 100.
4 Manufacturing industry only.
5 Prices relate only to products for sale or transfer to other
sectors or for use as capital equipment.

6 Prices are collected from wholesalers.
7 Exclusive of products of mining and quarrying.
8 Including imported products.
9 Agricultural products and products of manufacturing
industry.
10 Valued at purchasers' values.
11 Base: 1981 = 100.
12 San José.
13 St.Domingo.
14 San Salvador.
15 Excluding forestry, fishing and hunting.
16 Imports are valued c.i.f.
17 Finished products only.
18 Excluding electricity, gas and water.
19 Including mining and quarrying.
20 Exports are valued f.o.b.
21 Primary articles include food articles, non food articles
and minerals.
22 Excluding value added tax.
23 Including marine foods.
24 Mexico city.
25 Excluding mining and quarrying.
26 Crop growing production only, excluding live stock
production.

† Toutes les données se rapportant à l'Allemagne avant le 3
octobre 1990 figurent dans deux rubriques séparées basées
sur les territoires respectifs de la République fédérale
d'Allemagne et l'ancienne République démocratique allemande
selon la période indiquée. En l'absence de données pour
l'Allemagne unifiée (à compter du 3 octobre 1990), les
données disponibles sont fournies séparément sous les
rubriques République fédérale d'Allemagne et ancienne
République démocratique allemande et se rapportent aux
limites territoriales antérieures au 3 octobre 1990. Pour
les notes explicatives en détail sur les données concernant
l'Allemagne, voir Annexe I – Nomenclature des pays.

1 Produits agricoles intérieurs seulement.
2 Y compris les produits exportés.
3 Base: 1985 = 100.
4 Industries manufacturières seulement.
5 Uniquement les prix des produits destinés à être vendus ou
transférés à d'autres secteurs ou à être utilisés comme
biens d'équipement.
6 Prix recueillis auprès des grossistes.
7 Non compris les produits des industries extractives.
8 Y compris les produits importés.
9 Produits agricoles et produits des industries
manufacturières.
10 A la valeur d'acquisition.
11 Base: 1981 = 100.
12 San José.
13 Saint-Domingue.
14 San Salvador.
15 Non compris sylviculture, pêche et chasse.
16 Les importations sont évaluées à leur valeur c.a.f.
17 Produits finis uniquement.
18 Non compris l'électricité, le gaz et l'eau.
19 Y compris les industries extractives.
20 Les exportations sont évaluées f.o.b.
21 Les articles premiéres comprennent des articles des produits
alimentaires, non alimentaires et des minereaux.
22 Non compris taxe sur la valeur ajoutée.
23 Y compris l'alimentation marine.
24 Mexico.
25 Non compris les industries extractives.
26 Cultures uniquement, non compris les produits de l'élevage.

39
Producers prices and wholesale prices
Index numbers: 1980 = 100 [*cont.*]

Prix à la production et des prix de gros
Indices : 1980 = 100 [*suite*]

27 Output price index.
28 Base:1983 = 100.
29 Excluding fishing.
30 Manila.
31 Not a sub-division of the domestic supply index.
32 Excluding gold mining.
33 Excluding agricultural products.
34 Exclusive of industrial finished goods.
35 Excluding foods and feeds production.
36 Montevideo.
37 Excluding former Democratic Yemen.

27 Indice des prix de production.
28 Base: 1983 = 100.
29 Non compris la pêche.
30 Manille.
31 Pas un élément de l'indice de l'offre intérieure.
32 Non compris l'extraction de l'or.
33 Non compris les produits agricoles.
34 Non compris les produits finis industriels.
35 Non compris les produits alimentaires et d'affouragement.
36 Montevideo.
37 Non compris l'ancien Yémen démocratique.

40
Consumer price index numbers
Indices des prix à la consommation
All items and food; 1980 = 100
Ensemble et aliments; 1980 = 100

Country or area Pays ou zone	1982	1983	1984	1985	1986	1987	1988	1989	1990	1991
Afghanistan[1][2] **Afghanistan**[1][2]	**145**	**167**	**202**	**219**	**215**	**254**	**324**	**555**	**818**	...
Food Aliments[1]	155	181	214	237	233	289	354	596	1 063	...
Albania[3][4] **Albanie**[3][4]	...	...	...	...	...	...	...	**98**	**100**	**204**
Food Aliments	...	...	...	...	...	...	...	97	100	211
Algeria[5] **Algérie**[5]	**122**	**130**[6]	**140**	**155**	**174**	**187**	**198**	**216**	**252**	**310**
Food[5] Aliments[5]	127	102[6]	110	123	144	155	161	176	208	249
American Samoa[2][7] **Samoa américaines**[2][7]	**100**	**101**	**103**	**104**	**105**	**110**	**115**	**120**	**130**	...
Food[7] Aliments[7]	100	100	102	101	103	105	111	116	124	...
Angola[3][8] **Angola**[3][8]	...	...	...	...	...	...	...	...	...	**184**
Food[3][8] Aliments[3][8]	...	...	...	...	...	...	...	...	...	194
Anguilla[9] **Anguilla**[9]	...	...	...	**100**	**102**	**105**	**110**	**115**	**121**	...
Food[9] Aliments[9]	...	...	...	100	103	104	110	121	128	...
Antigua and Barbuda[10] **Antigua-et-Barbuda**[10]	**161**	**165**	**171**	**173**	**173**	**180**	**191**	**199**	**213**	**225**
Food[10] Aliments[10]	158	161	170	174	171	182	192	207	220	234
Argentina[11][12] **Argentine**[11][12]	...	**0**	**1**	**5**	**10**	**23**	**100**[6]	**3 295**	**79 531**	...
Food[11][12] Aliments[11][12]	...	0	1	5	10	23	100[6]	3 187	62 808	...
Aruba[13] **Aruba**[13]	...	...	**100**	**104**	**105**	**109**	**113**	**117**	**124**	**131**
Food[13] Aliments[13]	...	...	100	100	103	110	118	128	139	145
Australia **Australie**	**122**	**134**	**140**	**149**	**162**	**176**	**189**[6]	**203**	**218**	**225**
Food Aliments	118	130	137	145	158	167	180[6]	196	204	211
Austria **Autriche**	**113**	**116**	**123**	**127**	**129**	**131**[6]	**133**	**137**	**141**	**146**
Food Aliments	111	113	120	123	125	126[6]	127	129	133	138
Bahamas **Bahamas**	**118**	**123**	**128**	**134**	**141**	**149**	**155**	**164**	**171**	**184**
Food Aliments	123	124	126	133	150	157	168	177	190	207
Bahrain **Bahreïn**	**119**	**119**	**120**	**# 102**[14][15]	**100**	**98**	**98**	**100**	**101**	**102**
Food Aliments	102	100	104	# 100[14][15]	100	96	95	96	96	98
Bangladesh[16][17][18] **Bangladesh**[16][17][18]	**131**	**143**	**158**	**175**	**194**	**213**	**233**	**256**	**277**	**296**
Food[16][17][18] Aliments[16][17][18]	126	138	155	170	194	214	227	248	262	278

40
Consumer price index numbers
All items and food; 1980 = 100 [cont.]
Indices des prix à la consommation
Ensemble et aliments; 1980 = 100 [suite]

Country or area Pays ou zone	1982	1983	1984	1985	1986	1987	1988	1989	1990	1991
Barbados **Barbade**	**126**	**133**	**139**	**145**	**147**	**152**	**159**	**169**	**174**	**185**
Food Aliments	123	127	132	138	142	148	158	172	179	188
Belarus **Bélarus**	**104**	**105**	**105**	**105**	**107**	**109**	**110**	**111**	**116**	**227**
Food Aliments	105	107	107	109	115	120	121	122	126	237
Belgium **Belgique**	**117**	**126**	**134**	**141**	**142**	**145**	**146**	**151**	**156**	**161**[6]
Food Aliments	116	126	136[6]	141	143	143	143	147	153	156[6]
Belize **Belize**	**119**	**125**	**129**	**134**	**135**	**138**	**142**	**145**	**150**	**158**
Food Aliments	114	119	122	123	123	125	130	135	138	146
Bermuda **Bermudes**	**121**	**128**	**134**	**139**[6]	**145**	**151**	**159**	**168**	**178**	**185**
Food Aliments	118	123	128	129[6]	132	140	146	157	166	171
Bolivia[19 20] **Bolivie**[19 20]	**3**	**11**	**153**	**18 158**	**68 337**	**78 299**	**90 829**	**104 611**	**122 520**	**148 793**
Food[19 20] Aliments[19 20]	3	12	173	19 688	74 160	81 762	91 117	104 193	123 728	150 457
Botswana[21] **Botswana**[21]	**111**	**123**	**133**	**144**	**159**[6]	**174**	**189**	**211**	**234**	**262**
Food[21] Aliments[21]	113	128	139	152	167[6]	184	200	219	244	273
Brazil[19 22] **Brésil**[19 22]	**0**	**1**	**2**	**7**	**17**	**53**	**364**	**4 733**	**142 023**	**723 218**
Food[19 22] Aliments[19 22]	0	1	3	8	20	58	429	5 997	163 608	809 085
Brunei Darussalam[16] **Brunéi Darussalam**[16]	**116**	**117**	**121**	**124**	**126**	**128**	**129**	**131**	...	...
Food[16] Aliments[16]	123	124	128	129	130	131	133	135	...	...
Bulgaria **Bulgarie**	**101**	**102**	**103**	**105**	**108**	**108**	**110**	**117**	**# 100**[3 23]	**439**
Food Aliments	101	104	105	108	112	112	114	116	# 100[3]	477
Burkina Faso[14 24 25] **Burkina Faso**[14 24 25]	**95**	**100**	**105**	**112**	**109**	**106**	**110**	**110**	**109**	...
Food[14 24 25] Aliments[14 24 25]	96	100	110	113	105	93	103	99	98	...
Burundi[26] **Burundi**[26]	**119**	**128**	**147**	**152**	**155**	**166**	**174**	**194**	**207**	**226**
Food[26] Aliments[26]	118	128	145	157	145	144	155	184	198	211
Cameroon[14 27] **Cameroun**[14 27]	**125**	**146**	**162**	**181**	**195**	**220**	**224**	**219**	...	...
Food[14 27] Aliments[14 27]	135	154	162	160	162	172	173	160	...	...
Canada **Canada**	**125**	**132**	**138**	**143**	**149**	**155**	**162**	**170**	**178**	**188**
Food Aliments	120	124	131	135	141	148	151	157	163	171

40
Consumer price index numbers
All items and food; 1980 = 100 [*cont.*]
Indices des prix à la consommation
Ensemble et aliments; 1980 = 100 [*suite*]

Country or area Pays ou zone	1982	1983	1984	1985	1986	1987	1988	1989	1990	1991
Cape Verde[2][24][28] **Cap-Vert**[2][24][28]	...	100	112	118	130	136	140	148	164	179
Food[24][28] Aliments[24][28]	...	100	112	111	125	131	135	147	166	185
Cayman Islands **Iles Caïmanes**	120	126	130	133	137	142	150	159	171	185
Food[9] Aliments[9]	...	...	...	100	103	104	106	111	121	125
Central African Rep.[2][21][29] **Rép. centrafricaine**[2][21][29]	104	130	133	147	150	140	134	135	135	131
Food[21][29] Aliments[21][29]	108	133	133	147	147	133	127	128	129	124
Chad[24][30] **Tchad**[24][30]	...	100	125	129	107	101[14]	# 100[31]	95	96	...
Food[24][30] Aliments[24][30]	...	100	135	136	100	86[14]	# 100[31]	92	91	...
Chile[32] **Chili**[32]	132	168	201	262	313	376	431	504	636	774
Food[32] Aliments[32]	118	149	180	231	287	356	403	486	613	771
China **Chine**	105	107	110	123	131	143	172	200	203	213
Food Aliments	102	103	112	126	136	148	178	209	211	216
Colombia[33][34] **Colombie**[33][34]	160	192	223	279	330	404	526	...	...	...
Food[33][34] Aliments[33][34]	156	187	216	288	343	431	579	...	...	...
Congo[2][35] **Congo**[2][35]	130	142	160	170	174	177	184	192	197	200
Food[35] Aliments[35]	137	153	169	174	179	178	184	191	197	198
Cook Islands[36] **Iles Cook**[36]	136	146	164	184	202	224	243	# 100[37]	105	111
Food[36] Aliments[36]	136	145	162	179	198	219	237	# 100[37]	104	106
Costa Rica[11][38] **Costa Rica**[11][38]	261	346	387	443	498	582	703	819	975	1 255
Food[11][38] Aliments[11][38]	292	386	420	472	527	605	727	858	1 014	1 277
Côte d'Ivoire[33][39] **Côte d'Ivoire**[33][39]	117	123	129	131	144[6]	154	165	167[14]	165	168
Food[33][39] Aliments[33][39]	110	114	120	122	140[6]	156	175	177[14]	175	...
Cyprus **Chypre**	118[6]	124	131	138	140	144[6]	148	154	161	169
Food Aliments	121[6]	126	136	143	147	151[6]	158	165	173	184
Czechoslovakia **Tchécoslovaquie**	106	107	108	110	111	111[6]	111	113	124	196[6]
Food Aliments	109	110	111	114	114	114[6]	114	114	127	184[6]
Denmark **Danemark**	123	132	140	146	152	158	165	173	177	182
Food Aliments	124	130	142	148	151	152	158	164	165	166

40
Consumer price index numbers
All items and food; 1980 = 100 [cont.]
Indices des prix à la consommation
Ensemble et aliments; 1980 = 100 [suite]

Country or area Pays ou zone	1982	1983	1984	1985	1986	1987	1988	1989	1990	1991
Dominica **Dominique**	**118**	**123**	**126**	**128**	**132**	**139**	**142**	**151**	**156**	...
Food Aliments	118	117	119	120	122	133	137	148	145	...
Dominican Republic[23] **Rép. dominicaine**[23]	**116**	**124**	**154**	**212**	**233**	**269**	**389**	**566**	...	...
Food Aliments	108	115	141	202	226	273	418	620	...	...
Ecuador[21] **Equateur**[21]	**116**	**173**	**227**	**290**	**357**	**462**	**731**	**1 284**	**1 906**	**2 835**
Food[21] Aliments[21]	117	208	286	373	460	592	973	1 828	2 697	4 007
Egypt Egypte Food Aliments	131	155	181	205	255	313	373	473	548	639
El Salvador[40] **El Salvador**[40]	**128**	**145**	**162**	**198**	**262**	**327**	**391**	**460**	**571**	**653**
Food[40] Aliments[40]	130	148	169	200	264	331	428	544	684	807
Estonia[3] **Estonie**[3]	...	...	...	...	...	...	...	...	**100**	**302**
Food[3] Aliments[3]	...	...	...	...	...	...	...	...	100	333
Ethiopia[2 41] **Ethiopie**[2 41]	**111**	**112**	**121**	**144**	**130**	**127**	**136**	**146**	**154**	**209**
Food[41] Aliments[41]	111	112	124	155	132	124	133	142	150	211
Faeroe Islands **Iles Féroé**	**128**	**138**	**149**	**156**	**156**	**156**	**161**	**171**	**180**	**187**
Food Aliments	129	141	159	162	169	176	181	195	206	217
Falkland Is. (Malvinas)[2 42] **Iles Falkland (Malvinas)**[2 42]	**121**	**129**	**138**	**151**	**161**	**165**	**177**	...	**# 100**[3]	**105**
Food[2 42] Aliments[2 42]	116	121	127	135	138	140	158	161	# 100[3]	104
Fiji **Fidji**	**119**	**127**	**134**	**140**	**102**[9]	**108**	**120**	**128**	**138**	**147**
Food Aliments	123	131	137	148	98[9]	104	123	136	147	149
Finland **Finlande**	**122**	**133**	**142**[6]	**151**	**156**	**162**	**170**[6]	**181**	**192**	**200**
Food Aliments	127	136	146[6]	157	163	166	170[6]	176	183	187
France **France**	**127**	**139**	**149**	**158**	**162**	**167**	**172**	**178**	**184**	**190**
Food Aliments	128	140	151	159	164	167	170	177	184	190
French Guiana[43] **Guyane française**[43]	**131**	**145**[6]	**159**	**170**	**174**	**181**	**185**	**193**	**200**	**204**
Food[43] Aliments[43]	129	143[6]	157	168	170	178	180	188	195	197
French Polynesia **Polynésie française**	**130**	**148**	**166**	**184**	**186**	**188**	**192**[6]	**197**	**200**	**202**
Food Aliments	115	132	150	163	158	154	157[6]	164	167	166

40
Consumer price index numbers
All items and food; 1980 = 100 [cont.]
Indices des prix à la consommation
Ensemble et aliments; 1980 = 100 [suite]

Country or area Pays ou zone	1982	1983	1984	1985	1986	1987	1988	1989	1990	1991
Gabon[44] **Gabon**[44]	**127**[14]	**140**	**148**	**159**	**169**	**168**	**153**	**164**[14]	**176**[14]	**176**
Food[44] Aliments[44]	...	...	...	157	166	165	143	199	208[14]	220
Gambia[45] **Gambie**[45]	**118**	**130**	**159**	**188**	**294**	**364**	**406**	**440**	**493**	...
Food[45] Aliments[45]	115	130	158	186	297	367	413	442	504	...
Germany† · Allemagne† **F. R. Germany** **R. f. Allemagne**	**112**	**116**	**118**	**121**	**121**	**121**	**123**	**126**	**129**	**134**
Food Aliments	111	114	116	117	118	117	117	120	124	127
former German D. R. **anc. R. d. allemande**	**100**	**100**	**100**	**100**	**100**	**100**	**100**	**100**	...	...
Food Aliments	100	100	100	100	100	100	100	100	...	...
Ghana **Ghana**	**265**	**590**	**824**	**909**	**1 132**	**1 583**	**2 080**	**2 605**	**3 575**	**4 219**
Food Aliments	287	702	779	692	833	1 153	1 547	1 935	2 711	2 955
Gibraltar[21] **Gibraltar**[21]	**109**	**115**	**123**	**130**	**135**	**141**	**146**	**153**	**162**	**174**
Food[21] Aliments[21]	109	113	119	124	128	134	139	145	...	...
Greece **Grèce**	**151**	**181**	**215**	**256**	**315**	**366**	**416**	**473**	**570**	**677**
Food Aliments	158	186	220	263	316	356	395	467	567	666
Greenland[21] **Groënland**[21]	**113**	**126**	**137**	**147**	**155**	**160**	**172**	**181**	**190**	**198**
Food[21] Aliments[21]	116	128	140	150	150	154	159	169	180	190
Grenada **Grenade**	...	...	...	**147**	**148**	**147**	**# 104**[46]	**110**	**113**	**116**
Food Aliments	...	...	...	142	148	145	# 105[46]	114	118	121
Guadeloupe[40] **Guadeloupe**[40]	**128**	**139**	**150**	**161**	**165**	**170**	**174**	**178**	**183**	**188**
Food[40] Aliments[40]	127	138	150	160	165	170	172	174	179	182
Guam **Guam**	**127**	**131**	**143**	**148**	**152**	**159**	**167**	**185**	**201**	**222**
Food Aliments	136	142	157	165	177	196	217	239	286	339
Guatemala[40] **Guatemala**[40]	**112**	**# 97**[13][14]	**100**	**119**	**163**	**183**	**202**	**225**	**318**	**424**
Food[40] Aliments[40]	108	# 98[13][14]	100	121	168	194	221	248	369	483
Guinea[46][47] **Guinée**[46][47]	...	...	...	...	...	**100**	**127**	**163**	**194**	**233**
Food[46][47] Aliments[46][47]	...	...	...	...	...	100	133	169	199	231
Guinea-Bissau **Guinée-Bissau** Food[46][48] Aliments[46][48]	...	...	...	...	52	100[14]	160	290	386	608

40
Consumer price index numbers
All items and food; 1980 = 100 [cont.]
Indices des prix à la consommation
Ensemble et aliments; 1980 = 100 [suite]

Country or area Pays ou zone	1982	1983	1984	1985	1986	1987	1988	1989	1990	1991
Guyana[40]										
Guyana[40]	148	170	209[14]	245	264	339	...	...	...	...
Food[40]										
Aliments[40]	165	200	259[14]	320	348	452	...	...	...	...
Haiti[11]										
Haïti[11]	119	131	140	155	160	141	147	158	191	...
Food[11]										
Aliments[11]	114	127	135	150	155	126	134	144	177	...
Honduras										
Honduras	120	130	137	139	144	148	156	172	212	283
Food										
Aliments	115	121	121	123	126	127	138	155	196	282
Hong Kong										
Hong-kong	126	139	150	155	159[6]	168	180	199	218	243[6]
Food										
Aliments	130	141	151	151	153[6]	160	176	198	217	242[6]
Hungary										
Hongrie	112	120	130	139	147	159	184	215	277	374
Food										
Aliments	108	114	128	136	138	151	175	206	279	340
Iceland[49]										
Islande[49]	227	419	540	715	867	1 030	1 291[6]	1 564	1 796	1 919
Food[49]										
Aliments[49]	228	440	577	794	975	1 138	1 524[6]	1 816	2 047	2 099
India[50]										
Inde[50]	122	136	148	156	170	184	199[14]	216[6]	235	268
Food[50]										
Aliments[50]	123	139	149	154	169	185	199[14]	215[6]	234	272
Indonesia										
Indonésie	123	137	152	159	168	184	199	211	113[51]	123
Food										
Aliments	122	133	146	149	162	180	203	220	110[51]	118
Iran, Islamic Rep. of										
Iran, Rép. islamique d'	...	...	...	84	100	129	165	202	218	255
Food										
Aliments	...	...	...	79	100	124	141	169	174	213
Iraq[52]										
Iraq[52]	158	177	191	199	202	230	279	296	# 161[31]	462[14]
Food[52]										
Aliments[52]	164	187	197	202	203	250	309	331	# 176[31]	640[14]
Ireland										
Irlande	141	156[6]	169	178	185	191	195	203	210[6]	217
Food										
Aliments	130	141[6]	154	160	167	172	177	185	462[6]	468
Isle of Man										
Ile de Man	123	130	137	145	149	155	163	174	189	203
Food										
Aliments	118	124	134	139	145	154	164	177	194	210
Israel										
Israël	478	1 174	5 560	22 500	33 323[6]	39 938	46 447[6]	55 833	65 418	77 838
Food										
Aliments	430	1 104	5 196	21 640	33 499[6]	38 909	46 029[6]	55 484	60 231	68 518
Italy										
Italie	137	157	174	190	202	211[6]	222	236	251	267[6]
Food										
Aliments	135	152	166	180	190	198[6]	206	219	233	248[6]

40
Consumer price index numbers
All items and food; 1980 = 100 [cont.]
Indices des prix à la consommation
Ensemble et aliments; 1980 = 100 [suite]

Country or area Pays ou zone	1982	1983	1984	1985	1986	1987	1988	1989	1990	1991
Jamaica										
Jamaïque	**120**	**133**	**170**	**214**	**246**	**262**	**284**	**115**[31]	**139**	**211**
Food										
Aliments	117	131	168	211	249	267	292	120[31]	147	227
Japan										
Japon	**108**	**110**	**112**	**114**	**115**	**115**	**116**	**119**	**122**	**126**
Food										
Aliments	107	109	113	115	115	114	115	117	122	128
Jordan										
Jordanie	**116**	**122**	**126**	**130**	**130**[6]	**130**	**138**	**174**	**202**	**219**
Food										
Aliments	113	116	118	121	123[6]	121	128	154	186	206
Kenya[33][53]										
Kenya[33][53]	**135**	**150**	**166**	**187**	**194**	**204**	**221**	**243**	**# 100**[3]	**119**
Food[33][53]										
Aliments[33][53]	135	148	164	195	202	209	225	246	# 100[3]	124
Kiribati[54]										
Kiribati[54]	...	**119**	**125**	**131**	**140**	**149**	**153**	**158**	**166**	...
Food[54]										
Aliments[54]	...	115	120	125	134	142	148	152	157	...
Korea, Republic of										
Corée, République de	**130**	**135**	**138**	**141**	**145**	**149**	**160**	**169**	**184**	**201**
Food										
Aliments	131	133	134	140	143	147	162	174	191	211
Kuwait										
Koweït	**116**	**121**	**123**	**124**	**126**	**126**	**128**	**133**	...	...
Food										
Aliments	110	112	112	112	112	110	111	115	...	...
Lesotho[33]										
Lesotho[33]	**128**	**151**	**166**	**193**	**229**	**257**	**287**	**# 104**[14][55]	**114**	...
Food[33]										
Aliments[33]	126	150	166	185	219	256	286	# 103[14][55]	114	...
Liberia[56]										
Libéria[56]	**115**	**118**	**120**	**119**	**123**	**129**	**142**	**150**	**162**[14]	...
Food[56]										
Aliments[56]	109	112	113	109	108	108	129	...	161[14]	...
Luxembourg[2]										
Luxembourg[2]	**118**	**128**	**136**	**140**	**140**	**140**[57]	**142**	**147**	**152**	**# 103**[3]
Food[2]										
Aliments[2]	120	129	138	104	107	105[57]	106	110	114	# 103[3]
Macau[2][13]										
Macao[2][13]	...	...	**100**	**102**	**104**	**109**	**117**	**128**	**138**	**151**
Food[13]										
Aliments[13]	...	...	100	100	102	106	115	127	138	150
Madagascar[2][58]										
Madagascar[2][58]	**172**	**205**	**226**	**249**	**286**	**330**	**417**	**454**	**508**	**551**
Food[58]										
Aliments[58]	173	201	223	251	297	321	385	420	479	523
Malawi[2][59][60]										
Malawi[2][59][60]	**120**	**134**	**143**	**170**	**195**	**252**	**327**	**384**	**415**	**458**
Food[59][60]										
Aliments[59][60]	128	142	151	183	220	279	355	410	449	506
Malawi[2][33][60]										
Malawi[2][33][60]	**123**	**139**	**167**	**185**	**211**	**264**	**354**	**398**	**444**	**492**
Food[33][60]										
Aliments[33][60]	115	132	153	167	193	245	324	375	423	478

40
Consumer price index numbers
All items and food; 1980 = 100 [*cont.*]
Indices des prix à la consommation
Ensemble et aliments; 1980 = 100 [*suite*]

Country or area Pays ou zone	1982	1983	1984	1985	1986	1987	1988	1989	1990	1991
Malaysia **Malaisie**	**116**	**120**	**125**	**125**	**126**	**127**	**130**	**134**	**138**[6]	**144**
Food Aliments	120	121	125	122	122	122	126	131	137[6]	143
Mali[31 61] **Mali**[31 61]	...	...	...	...	...	...	**100**	**100**	**100**	**102**
Food[31 61] Aliments[31 61]	...	...	...	...	...	...	100	97	98	102
Malta **Malte**	**118**	**117**	**# 100**[24]	**99**	**101**	**102**	**103**	**104**	**107**	...
Food Aliments	122	119	# 99[24]	98	100	101	101	100	...	...
Martinique **Martinique**	**130**	**143**	**155**	**167**	**173**	**179**	**183**	**189**	**196**	**202**
Food Aliments	127	141	153	164	169	177	178	181	188	194
Mauritius **Maurice**	**100**[7]	**106**	**113**	**121**	**123**	**# 100**[46]	**109**	**123**	**140**	**149**
Food Aliments	100[7]	107	117	125	128	# 100[46]	112	127	141	147
Mexico **Mexique**	**203**	**411**	**679**	**1 072**	**1 995**	**4 626**	**9 907**	**11 889**	**15 058**	**18 470**
Food Aliments	194	370	648	1 035	1 921	4 441	9 305	11 192	14 034	16 855
Montserrat **Montserrat**	**118**	**123**[6]	**130**	**134**	**138**	**143**	...	...	...	...
Food Aliments	111	111[6]	119	118	125	131	...	...	...	...
Morocco **Maroc**	**124**	**132**	**149**	**160**	**174**	**179**	**183**	**189**	**201**	**218**
Food Aliments	130	136	154	166	181	182	183	187	200	217
Mozambique[62] **Mozambique**[62]	**120**	**155**	**202**	**261**	**305**	...	...	...	...	...
Food[62] Aliments[62]	123	155	194	290	326	...	...	...	...	...
Myanmar[63] **Myanmar**[63]	**105**	**110**	**116**	**124**	**135**	**167**	**196**	**249**	**292**	**387**
Food[63] Aliments[63]	99	108	110	121	128	162	196	252	300	413
Nepal **Népal**	**124**	**140**	**144**	**155**	**185**	**204**	**223**	**242**[6]	**264**	**306**
Food Aliments	125	142	143	152	187	210	...	250[6]	268	317
Netherlands **Pays-Bas**	**113**	**117**	**120**	**123**	**123**	**122**	**123**	**125**	**128**	**133**
Food Aliments	112	112	117	118	117	115	115	116	119	123
Netherlands Antilles[64 65] **Antilles néerlandaises**[64 65]	**119**	**122**[6]	**125**	**126**	**127**	**132**	**135**	**141**	**146**	**# 102**[3]
Food[64] Aliments[64]	120	122[6]	125	123	125	132	142	151	...	# 104[3]
New Caledonia[66] **Nouvelle-Calédonie**[66]	**131**	**147**	**160**	**169**	**172**	**173**	**178**	**185**	**189**	**196**
Food[66] Aliments[66]	136	154	167	176	174	173	178	186	187	193

40
Consumer price index numbers
All items and food; 1980 = 100 [*cont.*]
Indices des prix à la consommation
Ensemble et aliments; 1980 = 100 [*suite*]

Country or area Pays ou zone	1982	1983	1984	1985	1986	1987	1988	1989	1990	1991
New Zealand **Nouvelle-Zélande**	**134**	**144**	**153**	**176**	**200**	**231**	**246**	**260**	**276**	**283**
Food Aliments	131	137	145	166	185	209	222	242	259	261
Nicaragua **Nicaragua**	**155**	**203**	**274**	**877**	**6 853**	**69 256**	**# 1**[19 31]	**49**	**3 694**	**104 998**
Food Aliments	167	236	334	1 156	11 413	124 481	# 1[19 31]	41	2 994	85 364
Niger[2 67] **Niger**[2 67]	**137**	**134**	**145**	**144**	**139**	**130**	**128**	**124**	**# 123**[31]	**114**
Food[67] Aliments[67]	142	130	142	139	131	116	113	106	# 105[31]	93
Nigeria[68] **Nigéria**[68]	**130**	**160**	**224**	**236**	**249**	**274**	**379**	**534**	**162**	**183**
Food[68] Aliments[68]	137	169	242	251	252	273	414	578	158	177
Niue **Nioué**	**132**[6]	**149**	**163**	**181**	**194**	**208**	**219**	**240**	**255**	**268**
Food Aliments	132[6]	145	158	176	191	202	210	222	235	244
Mariana Islands **Iles Mariannes**	**140**	**150**	**155**	**159**	**167**	**...**	**...**	**181**	**...**	**...**
Food Aliments	127	130	133	139	143	...	...	146	...	...
Norway **Norvège**	**127**	**137**	**146**	**154**	**165**	**180**	**192**	**200**	**209**	**216**
Food Aliments	134	145	154	164	179	193	205	211	218	221
Oman Oman										
Food Aliments	...	...	...	96	104	105	106	108	107	...
Pakistan[7] **Pakistan**[7]	**100**	**106**	**113**	**119**	**123**	**129**	**141**	**152**	**165**	**185**
Food[7] Aliments[7]	100	106	113	118	121	129	144	156	170	188
Panama[69] **Panama**[69]	**112**	**114**	**116**	**117**	**117**	**118**	**119**	**119**	**119**	**121**
Food[69] Aliments[69]	116	118	120	120	121	124	124	123	123	127
Papua New Guinea **Papouasie-Nvl-Guinée**	**114**	**123**	**132**	**137**	**145**	**149**	**158**	**165**	**176**	**188**
Food Aliments	114	120	128	133	137	140	145	151	165	178
Paraguay[70] **Paraguay**[70]	**122**	**138**	**166**	**208**	**274**	**334**	**411**	**517**	**715**	**888**
Food[70] Aliments[70]	114	134	173	220	316	391	484	583	853	1 026
Peru[11 19 71] **Pérou**[11 19 71]	**3**	**6**	**13**	**34**	**60**	**112**	**855**	**# 1**[3]	**100**	**509**
Food[11 19 71] Aliments[11 19 71]	3	6	13	31	58	96	611	# 1[3]	100	448
Philippines **Philippines**	**125**	**137**	**206**	**254**	**256**	**265**	**289**	**319**	**360**	**423**
Food Aliments	122	133	204	250	248	258	286	323	357	406

40
Consumer price index numbers
All items and food; 1980 = 100 [*cont.*]
Indices des prix à la consommation
Ensemble et aliments; 1980 = 100 [*suite*]

Country or area Pays ou zone	1982	1983	1984	1985	1986	1987	1988	1989	1990	1991
Poland **Pologne**	**243**	**297**	**342**	**393**	**463**	**580**	**929**	**3 261**	**22 362**	**38 082**
Food Aliments	294	353	401	448	517	638	986	3 946	24 985	37 727
Portugal[2] **Portugal**[2]	**147**	**184**	**238**	**284**	**318**	**347**	**381**[6]	**429**	**486**	**541**
Food Aliments	148	186	243	286	312	339	371[6]	425	483	530
Puerto Rico **Porto Rico**	**114**	**115**	**117**	**117**	**117**	**120**	**124**	**128**	**136**	**140**
Food Aliments	112	113	116	117	118	121	125	131	143	149
Qatar[21 72] **Qatar**[21 72]	**106**	**109**	**110**	**112**	**114**	**118**	**124**	**...**	**...**	**...**
Food[21 72] Aliments[21 72]	106	110	114	116	117	118	121	...	...	...
Réunion[40] **Réunion**[40]	**126**	**137**	**148**	**158**	**162**	**167**	**169**	**176**	**183**	**190**
Food[40] Aliments[40]	125	135	143	152	156	158	154	160	163	170
Romania **Roumanie**	**122**	**127**	**128**	**127**	**129**	**130**	**134**	**134**	**139**	**# 274**[73]
Food Aliments	136	141	143	152	152	153	154	155	161	# 289[73]
Rwanda[21 74] **Rwanda**[21 74]	**112**	**120**	**127**	**129**	**127**	**133**	**137**	**138**	**144**	**172**
Food[21 74] Aliments[21 74]	98	147	160	161	146	158	171	176	185	210
Saint Helena[21] **Sainte-Hélène**[21]	**116**	**126**	**136**	**146**	**140**	**145**	**150**	**# 105**[31]	**108**	**112**
Food[21] Aliments[21]	116	127	139	146	144	151	156	# 105[31]	109	111
Saint Kitts and Nevis[75] **Saint-Kitts-et-Nevis**[75]	**117**	**120**	**123**	**126**	**126**	**127**	**128**	**134**	**140**	**146**
Food[75] Aliments[75]	115	118	121	121	121	122	122	128	131	138
Saint Lucia **Sainte-Lucie**	**120**	**122**	**# 100**[9 14]	**100**	**102**	**109**	**111**	**116**	**120**	**127**
Food Aliments	123	123	# 101[9 14]	100	103	115	114	119	124	134
Saint Pierre and Miquelon[21] **Saint-Pierre-et-Miquelon**[21]	**116**	**136**	**155**	**168**	**164**	**166**	**172**	**183**	**191**	**...**
Food[21] Aliments[21]	117	136	159	172	167	168	175	191	194	...
St. Vincent-Grenadines[15 76] **St. Vincent-Grenadines**[15 76]	**...**	**...**	**...**	**...**	**100**	**103**	**103**	**106**	**114**	**120**
Food[15 76] Aliments[15 76]	...	...	...	...	100	104	104	104	112	123
Samoa[2] **Samoa**[2]	**143**	**166**	**186**	**203**	**213**	**224**	**243**	**259**	**299**	**294**
Food Aliments	146	168	182	200	214	219	239	247	296	...
San Marino **Saint-Marin**	**141**	**160**	**175**	**190**	**202**[6]	**212**	**222**	**236**	**252**	**270**
Food Aliments	138	154	167	184	196[6]	203	211	224	237	250

40
Consumer price index numbers
All items and food; 1980 = 100 [cont.]
Indices des prix à la consommation
Ensemble et aliments; 1980 = 100 [suite]

Country or area Pays ou zone	1982	1983	1984	1985	1986	1987	1988	1989	1990	1991
Saudi Arabia **Arabie saoudite**	**101**	**101**	**100**	**97**	**94**	**92**	**93**	**94**	**96**	**100**
Food Aliments	102	102	101	99	97	96	96	98	100	107
Senegal[77] **Sénégal[77]**	**124**	**139**	**155**	**175**	**186**	**178**	**175**	**176**	**177**	**173**
Food[77] Aliments[77]	122	135	151	165	175	162	161	162	163	158
Seychelles **Seychelles**	**110**	**116**	**121**	**122**	**122**[6]	**126**	**128**	**130**	**135**	**138**
Food Aliments	107	115	119	120	121[6]	125	126	128	135	139
Sierra Leone[78] **Sierra Leone[78]**	**157**	**264**	**439**	**776**	**1 404**	**3 952**	**5 189**	**8 446**	**17 817**	**...**
Food[78] Aliments[78]	162	268	409	736	1 307	3 625	4 859	8 069	17 231	...
Singapore **Singapour**	**113**	**114**	**117**	**117**	**116**	**116**	**118**	**121**	**125**	**129**
Food Aliments	115	115	117	116	114	114	115	117	118	120
Solomon Islands[79] **Iles Salomon[79]**	**132**	**141**	**156**	**171**	**194**	**216**	**252**	**289**	**318**	**362**
Food[79] Aliments[79]	135	144	165	176	196	212	255	300	326	380
Somalia[80] **Somalie[80]**	**177**	**241**	**464**	**640**	**868**	**1 112**	**...**	**...**	**...**	**...**
Food[80] Aliments[80]	151	210	451	511	631	824	...	...	...	...
South Africa **Afrique du Sud**	**132**	**148**	**166**	**193**	**228**	**265**	**299**	**343**	**392**	**452**
Food Aliments	136	152	169	188	227	278	322	358	415	496
Spain **Espagne**	**131**	**147**	**164**	**178**	**194**[6]	**204**	**214**	**228**	**244**	**258**
Food Aliments	131	145	163	178	197[6]	207	215	231	246	255
Sri Lanka[81] **Sri Lanka[81]**	**131**	**149**	**174**	**176**	**190**	**205**	**234**	**261**	**317**	**356**
Food[81] Aliments[81]	133	149	176	176	189	205	236	260	321	359
Sudan[33] **Soudan[33]**	**156**	**204**	**274**	**398**	**516**	**650**	**965**	**1 604**	**385**	**790**
Food[33] Aliments[33]	157	212	287	421	554	679	1 107	1 602	392	804
Suriname[82] **Suriname[82]**	**117**	**122**	**126**	**140**	**166**	**255**	**274**	**276**	**336**	**423**
Food[82] Aliments[82]	112	117	118	130	163	293	307	311	404	480
Swaziland[2 33 83] **Swaziland[2 33 83]**	**133**	**149**	**168**	**199**	**225**	**253**	**283**	**...**	**...**	**...**
Food[33 83] Aliments[33 83]	136	152	174	197	220	254	286	...	...	...
Sweden **Suède**	**122**	**133**	**143**	**154**	**160**	**167**	**177**	**188**	**208**	**227**
Food Aliments	129	144	161	173	185	191	202	213	229	239

40
Consumer price index numbers
All items and food; 1980 = 100 [*cont.*]
Indices des prix à la consommation
Ensemble et aliments; 1980 = 100 [*suite*]

Country or area Pays ou zone	1982	1983	1984	1985	1986	1987	1988	1989	1990	1991
Switzerland **Suisse**	**113**	**116**[6]	**119**	**123**	**124**	**126**	**128**	**133**	**140**	**148**
Food Aliments	118	121[6]	125	129	131	132	135	138	145	152
Syrian Arab Republic [84] **Rép. arabe syrienne** [84]	**135**	**144**	**157**	**184**	**251**	**400**	**538**	**599**	**715**	**770**
Food [84] Aliments [84]	137	143	152	181	256	419	583	633	796	824
Thailand [85] **Thaïlande** [85]	**120**	**124**	**124**	**129**	**131**	**134**	**139**	**148**	**158**	**166**
Food [85] Aliments [85]	115	120	119	117	117	119	125	137	151	160
Togo [86] **Togo** [86]	**133**	**146**	**140**	**138**	**144**	**144**	**144**	**142**	**144**	**144**
Food [86] Aliments [86]	143	159	143	131	136	135	155	127	130	124
Tonga [2] **Tonga** [2]	**127**	**# 98**[13]	**100**	**117**	**142**	**149**	**164**	**170**	**187**	**206**
Food Aliments	139	# 106[13]	100	119	149	154	170	168	178	192
Trinidad and Tobago **Trinité-et-Tobago**	**127**	**149**[6]	**169**	**181**	**195**	**216**	**233**	**260**	**289**	**300**
Food Aliments	132	163[6]	180	195	215	257	290	355	416	441
Tunisia **Tunisie**	**124**	**135**	**147**[6]	**158**	**167**	**179**	**191**	**205**	**218**	**237**
Food Aliments	123	133	146[6]	160	169	180	194	211	225	245
Turkey [87] **Turquie** [87]	**100**	**131**	**195**	**283**	**381**	**100**	**# 174**[46]	**284**	**455**	**754**
Food [87] Aliments [87]	100	126	198	279	363	100	# 183[46]	310	510	852
Tuvalu [88] **Tuvalu** [88]	**121**	**132**	**137**[6]	**142**	**154**	**...**	**181**	**186**	**193**	**...**
Food [88] Aliments [88]	116	125	130[6]	137	149	...	177	175	179	...
Uganda [2 7 33 89] **Ouganda** [2 7 33 89]	**100**	**134**	**189**	**499**	**1 225**	**3 736**	**11 066**	**...**	**...**	**...**
Food [7 33 89] Aliments [7 33 89]	100	139	196	540	1 233	3 660	10 875	...	...	...
Ukraine **Ukraine**	**104**	**105**	**104**	**104**	**105**	**107**	**107**	**109**	**...**	**...**
Food Aliments	104	105	105	107	112	117	117	118	...	...
former USSR [90] **ancienne URSS** [90]	**105**	**105**	**104**	**105**	**107**	**108**	**110**	**112**	**118**	**...**
Food [90] Aliments [90]	106	107	107	109	114	119	121	122	125	...
United Kingdom **Royaume-Uni**	**122**	**127**	**133**	**142**	**146**	**152**[6]	**160**	**172**	**189**	**200**
Food Aliments	117	121	127	131	136	140[6]	145	153	165	174
United Rep.Tanzania [91] **Rép. Unie de Tanzanie** [91]	**162**	**206**	**280**	**374**	**495**	**643**	**843**	**1 061**	**1 271**	**1 554**
Food [91] Aliments [91]	164	209	288	372	501	654	881	1 095	1 209	1 482

40
Consumer price index numbers
All items and food; 1980 = 100 [*cont.*]
Indices des prix à la consommation
Ensemble et aliments; 1980 = 100 [*suite*]

Country or area Pays ou zone	1982	1983	1984	1985	1986	1987	1988	1989	1990	1991
United States **Etats-Unis**	**117**	**121**	**126**	**131**	**133**	**138**	**144**	**151**	**159**	**165**
Food Aliments	112	115	119	122	126	131	136	144	152	158
United States Virgin Is. **Iles Vierges américaines**	**123**	**126**	**127**	**128**	**131**	**132**	**138**	**145**	**152**	**162**
Food Aliments	119	121	123	127	129	128	135	141	146	156
Uruguay[92] **Uruguay**[92]	**160**	**238**	**370**	**637**	**1 123**[6]	**1 836**	**2 978**	**5 374**	**11 422**	**23 069**
Food[92] Aliments[92]	140	216	365	595	1 141[6]	1 821	2 877	5 194	11 456	21 245
Vanuatu[21 33 40] **Vanuatu**[21 33 40]	**107**	**108**	**114**	**115**	**121**	**140**	**150**	**163**	**171**	**...**
Food[21 33 40] Aliments[21 33 40]	102	99	100	98	104	125	134	148	155	...
Venezuela[11 93] **Venezuela**[11 93]	**127**	**135**	**152**	**# 111**[13]	**124**	**159**	**206**	**380**	**535**	**...**
Food[11 93] Aliments[11 93]	130	140	164	# 122[13]	146	206	289	652	961	...
Yemen[94] **Yémen**[94]	**114**	**126**	**128**	**134**	**135**	**139**	**139**	**...**	**...**	**...**
Food[94] Aliments[94]	115	119	120	120	121	126	125	...	...	...
Yugoslavia **Yougoslavie**	**185**	**261**	**400**	**694**	**1 312**	**2 891**	**8 517**	**115 142**	**783 401**	**1 399 241**[14]
Food Aliments	198	287	423	720	1 370	2 890	8 738	118 413	748 223	1 263 166[14]
Zambia[33] **Zambie**[33]	**128**	**153**	**184**	**253**	**390**	**568**	**878**	**2 009**	**4 239**	**8 164**
Food[33] Aliments[33]	131	158	188	256	383	554	877	2 027	4 262	8 139
Zimbabwe[33] **Zimbabwe**[33]	**125**	**154**	**185**	**201**	**230**	**258**	**278**	**313**	**368**	**457**
Food[33] Aliments[33]	124	160	199	212	240	276	300	343	406	509

Source:
International Labour Office (Geneva).

† All data shown which pertain to Germany prior to 3 October 1990 are indicated separately for the Federal Republic of Germany and the former German Democratic Republic based on their respective territories at the time indicated. Where data for united Germany (3 October 1990 and thereafter) are not available, available data are shown separately under the designations Federal Republic of Germany and former German Democratic Republic and pertain to the territorial boundaries prior to 3 October 1990. For detailed explanatory notes on data pertaining to Germany, see Annex I - Country Nomenclature.

1 Kabul.
2 Excluding "Rent".
3 Index base: 1990 = 100.
4 One month of each year (Peru: beginning 1989).
5 Algiers.
6 Series linked to former series.

Source:
Bureau international du travail (Genève).

† Toutes les données se rapportant à l'Allemagne avant le 3 octobre 1990 figurent dans deux rubriques séparées basées sur les territoires respectifs de la République fédérale d'Allemagne et l'ancienne République démocratique allemande selon la période indiquée. En l'absence de données pour l'Allemagne unifiée (à compter du 3 octobre 1990), les données disponibles sont fournies séparément sous les rubriques République fédérale d'Allemagne et ancienne République démocratique allemande et se rapportent aux limites territoriales antérieures au 3 octobre 1990. Pour les notes explicatives en détail sur les données concernant l'Allemagne, voir Annexe I - Nomenclature des pays.

1 Kaboul.
2 Non compris le groupe "Loyer".
3 Indices base: 1990 = 100.
4 Un mois de chaque année (Pérou: à partir de 1989).
5 Algers.
6 Série enchaînée à la précédente.

40
Consumer price index numbers
All items and food; 1980 = 100 [*cont.*]

Indices des prix à la consommation
Ensemble et aliments; 1980 = 100 [*suite*]

7 Index base: 1982 = 100	7 Indices base: 1982 = 100.
8 Luanda.	8 Luanda.
9 Index base: 1985 = 100 (Saint-Lucia: beginning 1984).	9 Indices base: 1985 = 100 (Sainte-Lucie: à partir de 1984).
10 Index base: 1978 = 100.	10 Indices base: 1978 = 100.
11 Metropolitan area.	11 Région métropolitaine.
12 Buenos Aires.	12 Buenos Aires.
13 Index base: 1984 = 100 (Guatemala, Tonga: beginning 1983; Venezuela: beginning 1985).	13 Indices base: 1984 = 100 (Guatemala, Tonga: à partir de 1983; Venezuela: à partir de 1985).
14 Average of less than twelve months.	14 Moyenne de moins de douze mois.
15 Index base: 1986 = 100 (Bahrain: beginning 1985).	15 Indices base: 1986 = 100 (Bahreïn: à partir de 1985).
16 Government officials.	16 Fonctionnaires.
17 Middle income group.	17 Familles à revenu moyen.
18 Dhaka.	18 Dhaka.
19 Due to lack of space, multiply each figure by 100 (Brazil: by 1000).	19 En raison du manque de place, multiplier chaque chiffre par 100 (Brésil: par 1000).
20 La Paz.	20 La Paz.
21 Index base: 1981 = 100.	21 Indices base: 1981 = 100.
22 Sao Paulo.	22 Sao Paulo.
23 Including direct taxes.	23 Y compris les impôts directs.
24 Index base: 1983 = 100.	24 Indices base: 1983 = 100.
25 Ouagadougou.	25 Ouagadougou.
26 Bujumbura.	26 Bujumbura.
27 Yaounde, Africans.	27 Yaoundé, Africains.
28 Praia.	28 Praya.
29 Bangui.	29 Bangui.
30 N'Djamena.	30 N'Djamena.
31 Index base: 1988 = 100 (Chad: beginning 1988; Iraq, Niger: beginning 1990; Saint Helena: beginning 1989).	31 Indices base: 1988 = 100 (Tchad: à partir de 1988; Iraq, Niger: à partir de 1990; Sainte-Hélène: à partir de 1989).
32 Santiago.	32 Santiago.
33 Low income group.	33 Familles à revenu modique.
34 Bogota.	34 Bogotá.
35 Brazzaville, Europeans.	35 Brazzaville, Européens.
36 Rarotonga.	36 Rarotonga.
37 Index base: 1989 = 100.	37 Indices base: 1989 = 100.
38 San Jose.	38 San José.
39 Abidjan, Africans.	39 Abidjan, Africains.
40 Urban areas.	40 Régions urbaines.
41 Addis Ababa.	41 Addis Abéba.
42 Stanley.	42 Stanley.
43 Cayenne.	43 Cayenne.
44 Libreville, Africans.	44 Libreville, Africains.
45 Banjul, Kombo St. Mary.	45 Banjul, Kombo St. Mary.
46 Index base: 1987 = 100 (Grenada, Turkey: beginning 1988).	46 Indices base: 1987 = 100 (Grenade, Turquie: à partir de 1988).
47 Conakry.	47 Conakry.
48 Bissau.	48 Bissau.
49 Reykjavik.	49 Reykjavik.
50 Industrial workers.	50 Travailleurs de l'industrie.
51 Base: April 1988 - March 1989 = 100 (Indonesia: beginning 1990).	51 Base: avril 1988 - mars 1989 = 100 (Indonésie: à partir de 1990).
52 Index base: 1979 = 100.	52 Indices base: 1979 = 100.
53 Nairobi.	53 Nairobi.
54 Tarawa.	54 Tarawa.
55 Base: April 1989 = 100.	55 Base: avril 1989 = 100.
56 Monrovia.	56 Monrovia.
57 Beginning July 1987: including "Rent".	57 A partir de juillet 1987: y compris le groupe "Loyers".
58 Antananarivo.	58 Antananarivo.
59 High income group.	59 Familles à revenu élévé.
60 Blantyre.	60 Blantyre.
61 Bamako.	61 Bamako.
62 Maputo.	62 Maputo.

40
Consumer price index numbers
All items and food; 1980 = 100 [cont.]

Indices des prix à la consommation
Ensemble et aliments; 1980 = 100 [suite]

63 Yangon.	63 Yangon.
64 Curaçao, Aruba and Bonaire (beginning 1983: Curaçao and Bonaire).	64 Curaçao, Aruba et Bonaire (à partir de 1983: Curaçao et Bonaire).
65 Excluding compulsory social security.	65 Non compris la sécurité sociale obligatoire.
66 Nouméa.	66 Nouméa.
67 Niamey, Africans.	67 Niamey, Africains.
68 Rural and urban areas.	68 Régions rurales et urbaines.
69 Panama City.	69 Panamá.
70 Asuncion.	70 Asunción.
71 Lima.	71 Lima.
72 Doha.	72 Doha.
73 Base: October 1990 = 100.	73 Base: octobre 1990 = 100.
74 Kigali.	74 Kigali.
75 Saint Christopher.	75 Saint Christophe.
76 St. Vincent.	76 St. Vincent.
77 Dakar.	77 Dakar.
78 Freetown.	78 Freetown.
79 Honiara.	79 Honiara.
80 Mogadishu.	80 Mogadishu.
81 Colombo.	81 Colombo.
82 Paramaribo.	82 Paramaribo.
83 Mbabane-Manzini.	83 Mbabane-Manzini.
84 Damascus.	84 Damas.
85 Bangkok Metrolopis.	85 Bangkok.
86 Lome.	86 Lomé.
87 Ankara.	87 Ankara.
88 Funafuti.	88 Funafuti.
89 Kampala.	89 Kampala.
90 Including Belarus and Ukraine shown separately in this table.	90 Y compris le Bélarus et l'Ukraine, figurant séparément dans ce tableau.
91 Tanganyika.	91 Tanganyika.
92 Montevideo.	92 Montevideo.
93 Caracas.	93 Caracas.
94 Aden.	94 Aden.

Technical notes, tables 38-40

In *Table 38*, the series generally relate to the average earnings of wage earners in manufacturing industries. Earnings generally include bonuses, cost of living allowances, taxes, social insurance contributions payable by the employed person and, in some cases, payments in kind, and normally exclude social insurance contributions payable by the employers, family allowances and other social security benefits. The time of year to which the figures refer is not the same for all countries. Unless otherwise stated, the series relate to wage earners of both sexes, irrespective of age.

Some of the series do not conform to the above for one or more of the following reasons: inclusion of salaried employees, inclusion of non-manufacturing industries and use of wage rates instead of earnings.

In the case of countries with widely fluctuating exchange rates or with multiple exchange systems it is advisable to consult table 126 on exchange rates.

For international definitions, further details and current figures, see the International Labour Office *Year Book of Labour Statistics* and the United Nations *Monthly Bulletin of Statistics*.[7, 23]

In *table 39*, producer prices are prices at which producers sell their output on the domestic market or for export. Wholesale prices, in the strict sense, are prices at which wholesalers sell on the domestic market or for export. In practice, many national wholesale price indexes are a mixture of producer and wholesale prices for domestic goods representing prices for purchases in large quantities from either source. In addition, these indexes may cover the prices of goods imported in quantity for the domestic market either by producers or by retail or wholesale distribution.

Producer or wholesale price indexes normally cover the prices of the characteristic products of agriculture, forestry and fishing, mining and quarrying, manufacturing, and electricity, gas and water supply. Prices are normally measured in terms of transaction prices, including non-deductible indirect taxes less subsidies, in the case of domestically-produced goods and import duties and other non-deductible indirect taxes less subsidies in the case of imported goods.

The Laspeyres index number formula is generally used and, for the purpose of the presentation, the national index numbers have been recalculated, where necessary, on the reference base 1980=100.

The price index numbers for each country are arranged according to the following scheme:

(a) Components of supply
    000 Domestic supply
    010 Domestic production for domestic market
    011 Agricultural products
    012 Industrial products
    020 Import products
    110 Exported goods

Notes techniques, tableaux 38-40

Au *Tableau 38*, les séries se rapportent généralement aux gains moyens des salariés des industries manufacturières. Ces gains, en général, comprennent normalement les primes, les indemnités de vie chère, les impôts, les cotisations des travailleurs à une caisse d'assurance sociale et, dans certains cas, des paiements en nature; ils excluent normalement les contributions de l'employeur à la caisse d'assurance sociale, les allocations familiales et autres prestations de la sécurité sociale. La période de l'année à laquelle se rapportent les chiffres n'est pas la même pour tous les pays. Sauf indication contraire, les séries se rapportent aux salariés des deux sexes et ne tiennent pas compte de l'âge.

Certaines des séries s'écartent des normes indiquées ci-dessus pour une ou plusieurs des raisons suivantes : l'inclusion des employés salariés, l'inclusion des industries non manufacturières et l'utilisation des taux de rémunération au lieu des gains.

Dans le cas des pays à larges fluctuations des cours des changes ou à système de changes multiples, il est recommandé de consulter le tableau 126 sur les cours de changes.

Pour les définitions internationales, plus de détails et pour les chiffres courants, voir l'*Annuaire des Statistiques du Travail* du Bureau international du travail et le *Bulletin mensuel de statistique* des Nations Unies [7, 23].

Au *tableau 39*, les prix à la production sont les prix auxquels les producteurs vendent leur production sur le marché intérieur ou à l'exportation. Les prix de gros, au sens strict du terme, sont les prix auxquels les grossistes vendent sur le marché intérieur ou à l'exportation. En pratique, les indices nationaux des prix de gros combinent souvent les prix à la production et les prix de gros de biens nationaux représentant les prix d'achat par grandes quantités au producteur ou au grossiste. En outre, ces indices peuvent s'appliquer aux prix de biens importés en quantités pour être vendus sur le marché intérieur par les producteurs, les détaillants ou les grossistes.

Les indices de prix de gros ou de prix à la production comprennent aussi en général les prix des produits provenant de l'agriculture, de la sylviculture et de la pêche, des industries extractives (mines et carrières), de l'industrie manufacturière ainsi que les prix de l'électricité, de gaz et de l'eau. Les prix sont normalement ceux auxquels s'effectue la transaction, y compris les impôts indirects non déductibles, mais non compris les subventions dans le cas des biens produits dans le pays et y compris les taxes à l'importation et autres impôts indirects non déductibles, mais non compris les subventions dans le cas des biens importés.

On utilise généralement la formule de Laspeyres et, pour la présentation, on a recalculé les indices nationaux, le cas échéant, en prenant comme base de référce 1980=100.

(b) Stage of processing
210 Raw materials
220 Intermediate products
230 Finished goods
(c) End-use
310 Producers' material
320 Consumers' goods
330 Capital goods

The leading number in each line of the tables identifies the position in the classification scheme of the price index numbers concerned.

Description of the general methods used in compiling the related national indexes is given in the United Nations *1977 Supplement to the Statistical Yearbook and the Monthly Bulletin of Statistics*.[49]

In *table 40*, unless otherwise stated, the index covers all the main classes of expenditure on all items and on food. Monthly data for many of these series and descriptions of them may be found in the United Nations *Monthly Bulletin of Statistics* and the United Nations *1977 Supplement to the Statistical Yearbook and the Monthly Bulletin of Statistics*.[49]

Les indices des prix pour chaque pays sont présentés suivant la classification ci-après :
(a) Eléments de l'offre
000 Offre intérieure
010 Production nationale pour le marché intérieur
011 Produits agricoles
012 Produits industriels
020 Produits importés
110 Produits exportés
(b) Stade de la transformation
210 Matières premières
220 Produits intermédiaires
230 Produits finis
(c) Utilisation finale
310 Biens de production
320 Biens de consommation
330 Biens d'équipement

Le premier chiffre de chacune des entrées du tableau dénote la position de l'indice en question dans la classification.

Les méthodes générales utilisées pour calculer les indices nationaux correspondants sont exposées dans : *1977 Supplément à l'Annuaire statistique et au Bulletin mensuel de statistique* des Nations Unies [49].

Au *tableau 40*, sauf indication contraire, les indices donnés englobent tous les groupes principaux de dépenses pour l'ensemble et les aliments. Les données mensuelles pour plusieurs de ces séries et définitions figurent dans le *Bulletin mensuel de statistique* (ONU) et dans le *1977 Supplément à l'Annuaire statistique et au Bulletin mensuel de statistique* des Nations Unies [49].

41
Agricultural production
Production agricole
Index numbers: 1979-81 = 100
Indices : 1979-81 = 100

Country or area Pays ou zone	All commodities Ensemble des produits					Food Produit alimentaires				
	1987	1988	1989	1990	1991	1987	1988	1989	1990	1991
Africa · Afrique										
Algeria Algérie	137	133	127	126	148	136	132	126	125	147
Angola Angola	100	100	99	100	102	103	103	102	104	106
Benin Bénin	131	156	163	165	173	128	150	158	158	164
Botswana Botswana	87	114	118	109	102	87	114	119	109	102
Burkina Faso Burkina Faso	134	151	148	138	162	132	148	146	134	158
Burundi Burundi	127	127	121	120	125	127	128	122	121	125
Cameroon Cameroun	110	115	108	112	106	112	111	110	112	110
Cape Verde Cap-Vert	168	156	152	152	142	170	157	153	154	144
Central African Rep. Rép. centrafricaine	111	118	119	124	126	111	117	119	124	127
Chad Tchad	116	132	127	125	136	113	130	124	121	133
Comoros Comores	118	121	124	127	130	118	121	123	126	129
Congo Congo	119	122	118	126	128	119	122	118	126	129
Côte d'Ivoire Côte d'Ivoire	125	134	135	137	137	129	143	139	141	141
Egypt Egypte	137	137	139	143	138	147	148	151	155	150
Ethiopia Ethiopie	103	105	108	112	110	103	106	109	113	113
Gabon Gabon	108	112	115	122	124	108	112	114	122	123
Gambia Gambie	124	127	135	113	125	125	125	134	112	124
Ghana Ghana	136	149	149	132	166	139	151	151	133	168
Guinea Guinée	108	110	109	112	117	111	111	107	114	120
Guinea-Bissau Guinée-Bissau	125	122	123	131	132	125	123	123	132	132
Kenya Kenya	133	146	149	154	150	134	146	150	158	154
Lesotho Lesotho	100	116	99	103	97	96	115	98	102	95
Liberia Libéria	118	121	117	80	72	121	124	120	94	94
Libyan Arab Jamah. Jamah. arabe libyenne	109	114	122	117	124	109	114	122	118	124
Madagascar Madagascar	115	117	120	122	120	115	118	121	124	121

41
Agricultural production
Index numbers: 1979-81 = 100 [cont.]
Production agricole
Indices : 1979-81 = 100 [suite]

Country or area Pays ou zone	All commodities Ensemble des produits					Food Produit alimentaires				
	1987	1988	1989	1990	1991	1987	1988	1989	1990	1991
Malawi Malawi	107	113	111	110	122	105	109	104	101	110
Mali Mali	115	127	129	128	138	113	126	126	125	133
Mauritania Mauritanie	105	111	114	108	107	105	111	114	108	107
Mauritius Maurice	120	112	113	117	117	120	112	113	118	118
Morocco Maroc	136	168	170	161	187	135	167	169	160	186
Mozambique Mozambique	103	105	106	109	101	103	106	107	110	102
Namibia Namibie	96	94	99	99	136	97	96	100	100	139
Niger Niger	81	107	96	91	111	81	107	96	91	111
Nigeria Nigéria	122	143	148	165	176	122	143	148	165	176
Réunion Réunion	76	93	80	81	94	75	93	80	81	94
Rwanda Rwanda	111	113	120	123	126	107	108	116	118	122
Sao Tome and Principe Sao Tomé-et-Principe	83	95	89	81	83	83	95	88	81	83
Senegal Sénégal	143	126	142	135	134	143	125	143	134	134
Sierra Leone Sierra Leone	120	123	127	127	115	116	118	122	123	110
Somalia Somalie	119	124	127	124	113	119	124	127	124	113
South Africa Afrique du Sud	101	105	111	104	105	102	106	112	104	104
Sudan Soudan	93	117	93	85	110	91	117	92	85	111
Swaziland Swaziland	122	125	120	120	122	123	126	121	120	124
Togo Togo	116	125	136	137	138	111	120	132	134	133
Tunisia Tunisie	142	106	121	142	148	142	105	121	142	148
Uganda Ouganda	120	127	136	143	147	120	127	136	143	146
United Rep.Tanzania Rép. Unie de Tanzanie	114	115	125	121	116	115	115	126	124	117
Zaire Zaïre	122	125	128	129	132	122	125	127	129	132
Zambia Zambie	118	147	147	128	151	118	145	145	127	146
Zimbabwe Zimbabwe	109	139	131	129	125	92	130	121	125	109

41
Agricultural production
Index numbers: 1979-81 = 100 [cont.]
Production agricole
Indices : 1979-81 = 100 [suite]

Country or area Pays ou zone	All commodities Ensemble des produits					Food Produit alimentaires				
	1987	1988	1989	1990	1991	1987	1988	1989	1990	1991
America, North · Amérique du Nord										
Antigua and Barbuda Antigua-et-Barbuda	125	125	92	100	107	124	124	92	100	107
Bahamas Bahamas	112	111	101	104	105	112	111	101	104	105
Barbados Barbade	76	77	80	79	83	76	77	80	79	83
Belize Belize	112	108	111	119	120	112	108	111	119	120
Canada Canada	116	103	114	126	125	117	104	115	126	126
Costa Rica Costa Rica	114	117	126	132	129	110	112	120	130	124
Cuba Cuba	106	110	111	111	107	106	110	110	110	107
Dominica Dominique	160	175	150	156	157	160	175	150	156	157
Dominican Republic Rép. dominicaine	111	114	120	120	115	112	115	123	125	122
El Salvador El Salvador	85	82	81	93	92	101	114	111	116	119
Grenada Grenade	88	81	80	80	81	88	81	80	80	81
Guadeloupe Guadeloupe	117	130	115	92	104	117	129	115	92	104
Guatemala Guatemala	104	107	107	113	110	118	120	116	124	121
Haiti Haïti	111	111	113	107	103	115	113	115	110	105
Honduras Honduras	111	120	125	133	139	114	120	124	126	132
Jamaica Jamaïque	110	104	105	115	113	110	103	105	115	112
Martinique Martinique	120	121	123	132	133	121	122	124	133	134
Mexico Mexique	112	117	116	123	122	113	117	118	123	124
Nicaragua Nicaragua	74	73	74	75	71	78	78	83	85	88
Panama Panama	112	102	109	112	112	111	101	108	111	110
Puerto Rico Porto Rico	103	101	103	97	102	100	100	101	96	101
Saint Lucia Sainte-Lucie	140	186	169	180	180	140	186	169	180	180
St. Vincent-Grenadines St. Vincent-Grenadines	125	171	150	158	159	125	171	151	158	159
Trinidad and Tobago Trinité-et-Tobago	85	88	85	99	101	86	90	87	99	103

41
Agricultural production
Index numbers: 1979-81 = 100 [*cont.*]
Production agricole
Indices : 1979-81 = 100 [*suite*]

Country or area	All commodities Ensemble des produits					Food Produit alimentaires				
Pays ou zone	1987	1988	1989	1990	1991	1987	1988	1989	1990	1991
United States Etats-Unis	100	94	102	105	104	100	94	103	106	103
America, South · Amérique du Sud										
Argentina Argentine	106	112	104	111	111	107	111	103	111	110
Bolivia Bolivie	124	129	131	142	165	126	131	133	144	167
Brazil Brésil	166	162	162	163	164	167	165	165	166	167
Chile Chili	115	123	131	136	140	116	124	132	136	141
Colombia Colombie	111	118	125	133	135	115	123	133	137	139
Ecuador Equateur	120	132	137	147	152	120	130	137	147	154
Guyana Guyana	80	73	66	62	68	80	73	66	62	68
Paraguay Paraguay	135	158	167	171	174	140	152	159	163	159
Peru Pérou	114	124	123	110	114	116	126	123	113	117
Suriname Suriname	104	99	103	91	93	104	99	103	91	93
Uruguay Uruguay	107	116	129	128	120	104	116	128	127	116
Venezuela Venezuela	118	121	130	134	135	118	121	129	134	137
Asia · Asie										
Afghanistan Afghanistan	80	77	76	76	77	81	78	76	77	78
Bangladesh Bangladesh	113	113	124	124	127	112	114	126	125	129
Bhutan Bhoutan	122	97	99	105	107	122	97	99	105	107
Brunei Darussalam Brunéi Darussalam	157	158	160	158	158	157	158	160	158	158
Cambodia Cambodge	157	182	189	194	191	152	177	185	187	185
China Chine	141	143	146	159	164	139	140	144	157	159
Cyprus Chypre	92	109	109	102	113	92	109	109	102	113
Hong Kong Hong-kong	57	111	113	120	124	57	111	113	120	124
India Inde	122	137	146	147	149	123	139	147	149	151
Indonesia Indonésie	137	146	151	159	165	139	148	155	162	168

41
Agricultural production
Index numbers: 1979-81 = 100 [cont.]
Production agricole
Indices : 1979-81 = 100 [suite]

Country or area Pays ou zone	All commodities Ensemble des produits					Food Produit alimentaires				
	1987	1988	1989	1990	1991	1987	1988	1989	1990	1991
Iran, Islamic Rep. of Iran, Rép. islamique d'	147	142	144	159	165	148	142	145	160	166
Iraq Iraq	125	122	131	145	99	126	123	132	146	100
Israel Israël	119	115	113	130	109	129	123	125	144	126
Japan Japon	101	98	99	100	96	104	100	102	103	99
Jordan Jordanie	155	167	123	153	134	157	168	123	155	134
Korea, Dem. P. R. Corée, R. p. dém. de	123	127	125	127	127	122	127	125	126	126
Korea, Republic of Corée, République de	106	111	109	106	107	107	113	111	108	108
Lao People's Dem. Rep. Rép. dém. pop. lao	127	119	140	148	147	127	119	140	149	147
Lebanon Liban	138	129	140	132	136	140	131	144	136	140
Macau Macao	99	99	96	96	96	99	99	96	96	96
Malaysia Malaisie	152	163	172	174	174	172	186	204	210	211
Maldives Maldives	121	123	125	128	129	121	123	125	128	129
Mongolia Mongolie	109	107	111	113	105	111	109	113	116	105
Myanmar Myanmar	140	130	117	119	124	142	132	119	121	126
Nepal Népal	126	148	152	154	164	128	150	154	157	167
Pakistan Pakistan	136	141	150	155	167	130	136	146	150	157
Philippines Philippines	106	108	114	117	116	105	108	114	117	117
Saudi Arabia Arabie saoudite	304	423	468	507	532	308	428	474	515	541
Singapore Singapour	107	94	117	119	104	107	94	117	120	104
Sri Lanka Sri Lanka	97	101	97	106	104	97	102	100	109	106
Syrian Arab Republic Rép. arabe syrienne	105	137	94	117	118	105	137	91	116	114
Thailand Thaïlande	115	125	129	123	133	113	124	127	118	128
Turkey Turquie	114	123	116	124	126	115	123	116	125	127
Viet Nam Viet Nam	137	141	149	156	161	135	139	147	153	157
Yemen Yémen	99	112	111	107	100	99	112	111	106	98

41
Agricultural production
Index numbers: 1979-81 = 100 [*cont.*]
Production agricole
Indices : 1979-81 = 100 [*suite*]

Country or area Pays ou zone	All commodities Ensemble des produits					Food Produit alimentaires				
	1987	1988	1989	1990	1991	1987	1988	1989	1990	1991
Europe · Europe										
Albania Albanie	113	109	114	91	76	112	107	114	92	76
Austria Autriche	109	115	111	108	105	109	115	111	108	105
Belgium-Luxembourg Belgique-Luxembourg	109	112	112	113	118	109	112	112	113	118
Bulgaria Bulgarie	99	98	99	94	88	100	100	106	101	94
Czechoslovakia Tchécoslovaquie	121	125	129	126	124	121	125	129	126	126
Denmark Danemark	114	122	127	136	134	114	122	127	136	134
Finland Finlande	99	103	114	120	105	99	103	114	120	105
France France	110	106	103	106	107	110	106	104	106	107
Germany† · Allemagne† F. R. Germany R. f. Allemagne	111	117	116	119	123	111	117	116	119	123
former German D. R. anc. R. d. allemande	118	113	115	108	79	117	113	114	107	78
Greece Grèce	105	111	115	97	111	101	107	111	94	107
Hungary Hongrie	109	115	112	105	112	109	116	113	106	113
Iceland Islande	100	89	88	88	89	100	89	87	88	89
Ireland Irlande	117	114	110	121	123	117	114	110	121	122
Italy Italie	104	100	104	96	105	104	100	104	95	104
Malta Malte	113	105	116	117	117	113	105	116	117	117
Netherlands Pays-Bas	114	111	120	114	113	114	111	120	114	113
Norway Norvège	107	107	107	119	116	107	107	107	120	116
Poland Pologne	111	114	116	119	109	111	114	118	121	110
Portugal Portugal	115	94	123	129	123	115	94	124	130	123
Romania Roumanie	97	104	97	83	78	97	104	96	82	77
Spain Espagne	123	120	119	123	118	123	119	119	123	118
Sweden Suède	93	92	100	107	99	93	92	100	107	99

41
Agricultural production
Index numbers: 1979-81 = 100 [cont.]
Production agricole
Indices : 1979-81 = 100 [suite]

Country or area Pays ou zone	All commodities Ensemble des produits					Food Produit alimentaires				
	1987	1988	1989	1990	1991	1987	1988	1989	1990	1991
Switzerland Suisse	106	106	115	110	113	106	106	115	110	113
United Kingdom Royaume-Uni	110	106	110	110	112	110	106	110	109	112
Yugoslavia Yougoslavie	105	99	104	94	92	105	100	104	95	93
Oceania · Océanie										
Australia Australie	109	116	116	124	119	104	110	109	114	106
Fiji Fidji	88	93	110	115	106	88	93	111	115	107
French Polynesia Polynésie française	98	96	97	104	106	100	97	98	106	108
New Caledonia Nouvelle-Calédonie	104	99	101	103	105	105	103	104	106	108
New Zealand Nouvelle-Zélande	108	110	108	101	106	111	114	112	106	112
Papua New Guinea Papouasie-Nvl-Guinée	116	122	129	128	128	116	122	129	129	129
Samoa Samoa	95	97	100	92	98	95	97	100	92	98
Solomon Islands Iles Salomon	107	115	127	130	135	107	115	127	131	135
Tonga Tonga	90	87	89	89	88	90	87	89	89	88
Vanuatu Vanuatu	114	105	97	125	116	114	106	97	126	116
former USSR · ancienne URSS										
former USSR ancienne URSS	117	117	119	119	105	119	119	122	121	107

Source:
Food and Agriculture Organization of the United Nations
(Rome).

Source:
Organisation des Nations Unies pour l'alimentation et
l'agriculture (Rome).

† All data shown which pertain to Germany prior to 3 October
1990 are indicated separately for the Federal Republic of
Germany and the former German Democratic Republic based on
their respective territories at the time indicated. Where
data for united Germany (3 October 1990 and thereafter) are
not available, available data are shown separately under the
designations Federal Republic of Germany and former German
Democratic Republic and pertain to the territorial
boundaries prior to 3 October 1990. For detailed
explanatory notes on data pertaining to Germany, see Annex I
- Country Nomenclature.

† Toutes les données se rapportant à l'Allemagne avant le 3
octobre 1990 figurent dans deux rubriques séparées basées
sur les territoires respectifs de la République fédérale
d'Allemagne et l'ancienne République démocratique allemande
selon la période indiquée. En l'absence de données pour
l'Allemagne unifiée (à compter du 3 octobre 1990), les
données disponibles sont fournies séparément sous les
rubriques République fédérale d'Allemagne et ancienne
République démocratique allemande et se rapportent aux
limites territoriales antérieures au 3 octobre 1990. Pour
les notes explicatives en détail sur les données concernant
l'Allemagne, voir Annexe I – Nomenclature des pays.

42
Cereals
Céréales

Production: thousand metric tons
Production : milliers de tonnes métriques

Country or area Pays ou zone	1982	1983	1984	1985	1986	1987	1988	1989	1990	1991
World *Monde*	**1 707 449**	**1 640 773**	**1 801 020**	**1 839 905**	**1 854 155**	**1 788 419**	**1 744 329**	**1 884 166**	**1 971 460**	**1 883 889**
Africa **Afrique**	**72 668**	**63 965**	**64 505**	**83 564**	**87 895**	**80 028**	**92 882**	**95 934**	**88 828**	**99 397**
Algeria Algérie	1 524	1 291	1 461	2 919	2 404	2 066	1 038	1 704	1 884	3 623
Angola Angola	325[1]	351[1]	335[1]	325[1]	356	382	352	289	264	381
Benin Bénin	352	351	483	535	497	401	540	563	553	* 534
Botswana Botswana	18	15	8	19	22	23	106	76	54	45
Burkina Faso Burkina Faso	1 210	1 119	1 089	1 583	1 890	1 637	2 101	1 952	* 1 519	* 2 234
Burundi Burundi	221	227	224	256	268	287	301	265	293	342
Cameroon Cameroun	961	889	657	771	1 094	715	900	841	865[1]	1 003
Cape Verde Cap-Vert	4	3	3	1	12	21	17	10	11	* 4
Central African Rep. Rép. centrafricaine	120	103	102	111	178	126	143	124	166	165[1]
Chad Tchad	393	450	354	704	687	563	782	582	602	891
Comoros Comores	19[1]	20[1]	19	19	19	18	19	19	19[1]	19[1]
Congo Congo	8	12	14	16	22	20	21	26	26[1]	26[1]
Côte d'Ivoire Côte d'Ivoire	935	818	1 105	1 089	1 049	1 086	1 144	1 193	1 247	1 292
Egypt Egypte	8 522	8 700	8 455	8 587	8 754	9 318	9 764	11 113	12 993	13 670
Ethiopia Ethiopie	6 718	5 527	4 240	4 820	6 504	6 195	6 384	6 355	6 457[1]	6 420[1]
Gabon Gabon	13[1]	13[1]	14	14	21	25	25	21	21[1]	21[1]
Gambia Gambie	101	68	87	116	102	92	100	96	93	108
Ghana Ghana	544	308	1 066	909	867	1 057	1 146	1 217	844	1 436
Guinea Guinée	722	639	700	739	814	814	810	740	809[1]	888
Guinea-Bissau Guinée-Bissau	160	141	159	158[1]	162	153[1]	152	144	167	* 165
Kenya Kenya	2 943	2 705	1 728	2 900	3 371	2 863	3 280	3 364	3 105	2 747
Lesotho Lesotho	125	123	132	167	132	146	224	147	161	125
Liberia Libéria	284	290	298	289	288	298	298	* 280	100[1]	110[1]

42
Cereals
Production: thousand metric tons [cont.]
Céréales
Production : milliers de tonnes métriques [suite]

Country or area Pays ou zone	1982	1983	1984	1985	1986	1987	1988	1989	1990	1991
Libyan Arab Jamah. Jamah. arabe libyenne	293	418	275	235	285	276	283	322	273	298[1]
Madagascar Madagascar	2 085	2 281	2 274	2 320	2 385	2 338	2 307	2 542	2 577	2 352[1]
Malawi Malawi	1 483	1 405	1 457	1 423	1 364	1 255	1 489	1 588	1 415	* 1 679
Mali Mali	1 324	1 509	1 113	1 719	1 728	1 639	2 197	2 157	1 772	2 232
Mauritania Mauritanie	45	48	52	108	127	152	174	184	105	117
Mauritius Maurice	1	1	4	5	8	4	4	2	2	2[1]
Morocco Maroc	4 917	3 581	3 759	5 312	7 825	4 337	7 944	7 428	6 313	8 648
Mozambique Mozambique	633[1]	623[1]	665[1]	746[1]	829	569	562	607	734	546
Namibia Namibie	92[1]	96[1]	101[1]	104[1]	110[1]	124	133	137	135	114
Niger Niger	1 702	1 716	1 082	1 849	1 834	1 439	2 389	1 843	1 490	2 415
Nigeria Nigéria	8 322	8 842	10 531	11 911	12 744	12 626	14 560	15 122	13 773	14 200
Réunion Réunion	14	12	11	12	13	11	13	13	10	13[1]
Rwanda Rwanda	316	310	296	337	297	292	282	277	282	330
Sao Tome and Principe[1] Sao Tomé-et-Principe[1]	1	1	1	1	1	1	1	1	1	1
Senegal Sénégal	782	523	709	1 249	887	1 054	867	1 067	950	* 914
Sierra Leone Sierra Leone	568	514	561	488	577	517	546	573	562	442
Somalia Somalie	405	361	497	514	586	543	601	654	578[1]	251
South Africa Afrique du Sud	11 441	6 548	7 802	10 870	11 349	11 392	11 464	14 911	11 205	10 945
Sudan Soudan	2 286	2 264	1 440	4 036	3 799	1 699	5 105	1 948	1 676	3 993
Swaziland Swaziland	56	33	154	178	164	103	157	119	89	159
Togo Togo	308	291	440	371	366	368	504	566	484	447
Tunisia Tunisie	1 285	951	1 053	2 098	624	1 917	295	640	1 637	2 554
Uganda Ouganda	1 093	1 399	944	1 171	1 058	1 220	1 398	1 619	1 578	1 652[1]
United Rep.Tanzania Rép. Unie de Tanzanie	3 008	2 875	3 143	3 622	3 864	4 038	3 685	4 793	3 842	3 826
Zaire Zaïre	981	1 010	1 058	1 093	1 141	1 181	1 227	1 272	1 303	1 359

42
Cereals
Production: thousand metric tons [cont.]
Céréales
Production : milliers de tonnes métriques [suite]

Country or area Pays ou zone	1982	1983	1984	1985	1986	1987	1988	1989	1990	1991
Zambia Zambie	798	983	925	1 193	1 319	1 158	2 055	1 967	1 207	1 603
Zimbabwe Zimbabwe	2 205	1 206	1 424	3 551	3 096	1 465	2 993	2 459	2 580	2 054
America, North **Amérique du Nord**	**411 782**	**283 230**	**386 916**	**428 433**	**401 471**	**361 443**	**269 396**	**359 746**	**402 338**	**363 994**
Bahamas[1] Bahamas[1]	1	1	1	1	1	1	1	1	1	1
Barbados Barbade	* 2	2[1]	2[1]	2[1]	2[1]	2[1]	2[1]	2[1]	2[1]	2[1]
Belize Belize	29	24	22	25	23	28	29	24	22[1]	24[1]
Canada Canada	53 336	47 450	42 777	48 239	57 031	51 682	35 796	48 206	58 590	55 969
Costa Rica Costa Rica	256	372	363	381	342	299	315	246	293	256
Cuba Cuba	616	614	651	620	667	562	585	632	543	526
Dominican Republic Rép. dominicaine	496	585	634	608	560	611	556	596	465	363
El Salvador El Salvador	574	610	731	697	631	647	807	802	825	729
Guatemala Guatemala	1 269	1 229	1 106	1 298	1 268	1 390	1 573	1 431	1 457	1 301
Haiti Haïti	391	428	432	445	476	462	469	453	361	335
Honduras Honduras	462	552	626	416	426	533	552	644	686	734
Jamaica Jamaïque	5	7	9	8	7	6	4	3	2	3
Mexico Mexique	20 193	22 606	23 707	27 403	23 553	23 636	21 073	21 506	25 545	23 056
Nicaragua Nicaragua	451	437	451	506	478	473	512	488	407	455
Panama Panama	256	291	263	301	293	304	298	328	333	299
Puerto Rico Porto Rico	6	6	5	5	6	2	1	0	0	0
St. Vincent-Grenadines St. Vincent-Grenadines	1	1	1	1	1	1	1	1	1	1[1]
Trinidad and Tobago Trinité-et-Tobago	6	6	7	7	7	10	10	15	16	17[1]
United States Etats-Unis	333 433	208 010	315 129	347 470	315 699	280 797	206 811	284 366	312 787	279 923
America, South **Amérique du Sud**	**80 313**	**72 043**	**77 823**	**77 818**	**77 931**	**82 913**	**82 135**	**78 532**	**68 530**	**72 908**
Argentina Argentine	34 276	31 613	31 801	28 096	26 158	22 868	22 301	17 599	19 581	20 351
Bolivia Bolivie	700	491	863	973	826	855	809	845	788	1 010
Brazil Brésil	33 838	29 198	32 711	36 011	37 319	44 112	42 921	43 943	32 468	35 991

42

Cereals
Production: thousand metric tons [*cont.*]
Céréales
Production : milliers de tonnes métriques [*suite*]

Country or area Pays ou zone	1982	1983	1984	1985	1986	1987	1988	1989	1990	1991
Chile Chili	1 448	1 437	2 116	2 360	2 675	2 819	2 800	3 148	2 981	2 866
Colombia Colombie	3 618	3 379	3 258	3 144	3 067	3 206	3 555	3 790	4 314	3 948
Ecuador Equateur	786	561	819	832	1 127	1 297	1 495	1 384	1 419	1 583
French Guiana Guyane française	0	2	6	8	9	13	14	17	22	29
Guyana Guyana	303	247	301	281	243	229	229	240	146	253
Paraguay Paraguay	725	842	1 008	1 152	1 143	1 447	1 595	1 550	1 628	1 508
Peru Pérou	1 754	1 618	2 130	1 812	1 887	2 357	2 374	2 439	1 801	1 757
Suriname Suriname	301	268	302	299	300	272	265	260	196	190[1]
Uruguay Uruguay	1 075	1 084	1 082	1 028	925	1 021	1 291	1 488	1 406	1 098
Venezuela Venezuela	1 487	1 302	1 428	1 822	2 250	2 418	2 485	1 830	1 780	2 324
Asia Asie	679 181	749 190	773 391	757 912	776 606	766 429	804 838	833 013	875 603	865 794
Afghanistan Afghanistan	* 3 724	* 3 591	* 3 418	* 3 242	3 084	3 394	2 997	2 834	2 655	2 724
Bangladesh Bangladesh	22 340	22 990	23 256	24 135	24 266	24 304	24 449	28 429	27 747	29 655
Bhutan Bhoutan	166	172[1]	179	167	164	140	94	94	105[1]	105[1]
Brunei Darussalam Brunéi Darussalam	2	3	3	1	1	1	1	1[1]	1[1]	1[1]
Cambodia Cambodge	2 000	2 082	1 308	1 863	2 131	1 862	2 447	2 605	2 555	2 450
China Chine	315 470	345 661	365 984	339 924	352 103	359 319	351 746	367 560	403 940	392 919
Cyprus Chypre	92	71	93	112	68	126	158	148	111	65
Gaza Strip (Palestine) Zone de Gaza (Palestine)	2	6	2	1	1	1	1	1	1	1
India Inde	136 101	166 782	164 478	165 682	164 955	156 114	183 867	199 413	194 701	195 109
Indonesia Indonésie	36 822	40 393	43 427	43 364	45 649	45 236	48 330	50 920	51 915	50 732
Iran, Islamic Rep. of Iran, Rép. islamique d'	10 222	9 321	10 052	10 728	11 899	12 197	12 103	10 736	13 876	14 626
Iraq Iraq	2 066	1 822	1 100	2 932	2 281	1 728	2 591	1 498	3 470	1 248
Israel Israël	197	427	171	163	203	326	223	208	303	167
Japan Japon	14 000	14 060	16 012	15 856	15 805	14 527	13 867	14 318	14 451	13 165

42
Cereals
Production: thousand metric tons [*cont.*]
Céréales
Production : milliers de tonnes métriques [*suite*]

Country or area Pays ou zone	1982	1983	1984	1985	1986	1987	1988	1989	1990	1991
Jordan Jordanie	75	161	63	85	43	117	131	82	128	68
Korea, Dem. P. R. Corée, R. p. dém. de	9 628	9 849[1]	10 100	10 101[1]	10 202[1]	10 300[1]	10 600[1]	10 345[1]	10 205[1]	10 080[1]
Korea, Republic of Corée, République de	8 274	8 653	8 943	8 586	8 462	8 264	8 943	8 748	8 280	7 905
Kuwait Koweït	1	1	1	3	3	3[1]	3[1]	3[1]	3[1]	...
Lao People's Dem. Rep. Rép. dém. pop. lao	1 127	1 132	1 355	1 431	1 491	1 243	1 054	1 448	1 558	1 460[1]
Lebanon Liban	30	27	26	26	39	70	73	79	77	71[1]
Malaysia Malaisie	1 893	1 754	1 594	1 873	1 773	1 730	1 815	1 778	1 690	1 586
Mongolia Mongolie	551	813	597	884	869	689	814	641	521	591
Myanmar Myanmar	14 839	15 024	14 960	15 067	14 861	14 239	13 648	14 256	14 418	13 667
Nepal Népal	3 221	4 311	4 310	4 374	4 000	4 760	5 307	5 673	5 847	5 919
Oman Oman	2	2	2	1	4	4	4	5	5	5[1]
Pakistan Pakistan	18 076	19 101	17 536	17 699	20 866	18 454	19 240	21 018	20 963	21 198
Philippines Philippines	10 867	11 193	11 640	13 019	13 049	12 818	13 399	13 981	14 173	14 325
Qatar Qatar	1	1	1	2	2	* 3	3[1]	3[1]	3[1]	3[1]
Saudi Arabia Arabie saoudite	489	875	1 444	2 188	2 461	2 865	3 632	3 628	4 046	4 476
Sri Lanka Sri Lanka	2 194	2 528	2 464	2 703	2 640	2 178	2 525	2 102	2 580	2 440
Syrian Arab Republic Rép. arabe syrienne	2 283	2 696	1 444	2 543	3 167	2 300	5 001	1 404	3 100	3 275
Thailand Thaïlande	20 126	23 438	24 516	25 613	23 399	21 412	26 169	24 818	21 277	24 279
Turkey Turquie	26 558	24 492	26 314	26 493	29 358	29 282	30 894	23 499	30 179	31 051
United Arab Emirates Emirats arabes unis	5	6	6	5	6[1]	6	5	6[1]	6[1]	7[1]
Viet Nam Viet Nam	14 832	15 204	16 065	16 466	16 578	15 667	17 820	19 840	19 901	20 084
Yemen Yémen	862	471	480	536	700	718	843	864	767	298
Europe **Europe**	**268 583**	**257 245**	**305 405**	**283 562**	**282 177**	**274 977**	**285 925**	**292 584**	**284 697**	**298 636**
Albania Albanie	970	1 067	1 040[1]	991[1]	1 001[1]	1 017	989	1 036	741[1]	559
Austria Autriche	5 026	5 076	5 353	5 551	5 108	4 965	5 359	5 009	5 290	4 921

42

Cereals
Production: thousand metric tons [*cont.*]
Céréales
Production : milliers de tonnes métriques [*suite*]

Country or area Pays ou zone	1982	1983	1984	1985	1986	1987	1988	1989	1990	1991
Belgium-Luxembourg Belgique-Luxembourg	2 222	1 962	2 533	2 227	2 403	2 062	2 351	2 368	2 217	2 467
Bulgaria Bulgarie	9 967	7 932	9 251	5 384	8 492	7 278	7 820	9 527	8 115	8 918
Czechoslovakia Tchécoslovaquie	10 277	11 058	11 984	11 775	10 805	11 777	11 907	12 047	12 567	11 853
Denmark Danemark	7 993	6 380	9 300	7 956	7 969	7 184	8 068	8 795	9 606	9 115
Finland Finlande	3 418	3 877	3 647	3 642	3 520	2 183	2 826	3 809	4 296	3 399
France France	48 676	46 409	58 329	55 996	50 390	52 964	56 305	57 596	55 023	60 442
Germany† · Allemagne† F. R. Germany R. f. Allemagne	24 625	23 011	26 489	25 915	25 593	23 843	27 112	26 113	25 883	27 611
former German D. R. anc. R. d. allemande	10 021	10 067	11 349	11 640	11 664	11 218	9 820	10 768	11 697	11 659
Greece Grèce	5 591	4 605	5 513	4 492	5 285	5 183	5 585	5 746	4 509	5 167
Hungary Hongrie	14 919	13 765	15 731	14 809	14 301	14 168	14 966	15 417	12 561	15 505
Ireland Irlande	2 178	1 991	2 514	2 096	1 955	2 108	2 194	2 050	2 109	2 084
Italy Italie	18 337	18 081	19 928	18 029	18 698	18 400	17 400	17 133	17 411	19 043
Malta Malte	12	12	9	9	9[1]	9[1]	9	9	9[1]	9[1]
Netherlands Pays-Bas	1 379	1 309	1 407	1 132	1 266	1 107	1 222	1 368	1 350	1 205
Norway Norvège	1 201	1 072	1 407	1 302	1 109	1 285	1 066	1 180	1 568	1 467
Poland Pologne	21 166	22 099	24 392	23 741	25 036	26 060	24 504	26 958	28 014	27 861
Portugal Portugal	1 291	1 146	1 469	1 380	1 632	1 696	1 434	1 843	1 379	1 403
Romania Roumanie	19 954	17 666	20 045	19 503	19 725	16 889	19 286	18 379	17 189	19 270
Spain Espagne	13 148	13 759	21 032	20 972	16 520	20 697	23 657	19 697	18 812	19 195
Sweden Suède	5 924	5 407	6 897	5 629	5 811	5 170	4 743	5 493	6 380	5 092
Switzerland Suisse	929	900	1 118	1 054	960	939	1 244	1 411	1 268	1 281
United Kingdom Royaume-Uni	21 919	21 306	26 614	22 486	24 509	21 698	21 063	22 723	22 539	22 649
Yugoslavia Yougoslavie	17 439	17 292	18 051	15 850	18 416	15 076	14 996	16 110	14 165	16 461
Oceania **Océanie**	**15 035**	**32 076**	**29 808**	**26 393**	**25 905**	**21 140**	**22 971**	**23 037**	**24 250**	**18 157**
Australia Australie	14 113	31 169	28 653	25 168	24 683	20 110	22 163	22 328	23 348	17 212

42

Cereals
Production: thousand metric tons [*cont.*]
Céréales
Production : milliers de tonnes métriques [*suite*]

Country or area Pays ou zone	1982	1983	1984	1985	1986	1987	1988	1989	1990	1991
Fiji Fidji	21	18	24	29	27	24	34	33	28	35
New Caledonia Nouvelle-Calédonie	3	3	1	2	2	1	1	1	1[1]	1[1]
New Zealand Nouvelle-Zélande	882	872	1 119	1 184	1 189	1 001	770	672	870	906
Papua New Guinea Papouasie-Nvl-Guinée	3	4	2	2	2	3[1]	3[1]	3[1]	3[1]	3[1]
Solomon Islands Iles Salomon	11	9	7	6	2	...	...	...	...	...
Vanuatu[1] Vanuatu[1]	1	1	1	1	1	1	1	1	1	1
former USSR ancienne URSS	179 887	183 023	163 172	182 223	202 170	201 490	186 183	201 320	227 214	165 003

Source:
Food and Agriculture Organization of the United Nations
(Rome).

† All data shown which pertain to Germany prior to 3 October
1990 are indicated separately for the Federal Republic of
Germany and the former German Democratic Republic based on
their respective territories at the time indicated. Where
data for united Germany (3 October 1990 and thereafter) are
not available, available data are shown separately under the
designations Federal Republic of Germany and former German
Democratic Republic and pertain to the territorial
boundaries prior to 3 October 1990. For detailed
explanatory notes on data pertaining to Germany, see Annex I
- Country Nomenclature.

1 FAO estimate.

Source:
Organisation des Nations Unies pour l'alimentation et
l'agriculture (Rome).

† Toutes les données se rapportant à l'Allemagne avant le 3
octobre 1990 figurent dans deux rubriques séparées basées
sur les territoires respectifs de la République fédérale
d'Allemagne et l'ancienne République démocratique allemande
selon la période indiquée. En l'absence de données pour
l'Allemagne unifiée (à compter du 3 octobre 1990), les
données disponibles sont fournies séparément sous les
rubriques République fédérale d'Allemagne et ancienne
République démocratique allemande et se rapportent aux
limites territoriales antérieures au 3 octobre 1990. Pour
les notes explicatives en détail sur les données concernant
l'Allemagne, voir Annexe I – Nomenclature des pays.

1 Estimation de la FAO.

43
Wheat
Froment

Production: thousand metric tons
Production : milliers de tonnes métriques

Country or area Pays ou zone	1982	1983	1984	1985	1986	1987	1988	1989	1990	1991
World *Monde*	**481 475**	**493 058**	**516 024**	**504 611**	**534 986**	**510 973**	**506 835**	**542 722**	**601 723**	**550 993**
Africa *Afrique*	**10 381**	**8 740**	**9 190**	**10 395**	**11 699**	**12 604**	**13 166**	**12 503**	**13 939**	**17 689**
Algeria Algérie	977	790	886	1 478	1 229	1 175	614	850	1 005	1 741
Angola Angola	5[1]	4[1]	3[1]	3[1]	2	2	2	2	3	3
Botswana Botswana	1[1]	1[1]	1[1]	1[1]	1[1]	1[1]	1[1]	1	1	* 2
Burundi Burundi	6	6	8	8	8	8	9	8	9	* 14
Cameroon Cameroun	1	0	0	0	0	0[1]	0[1]	0[1]	0[1]	0[1]
Chad Tchad	5[1]	10	1	* 5	* 4	3[1]	* 2	* 1	* 2	4
Egypt Egypte	2 017	1 996	1 815	1 872	1 928	2 721	2 838	3 182	4 268	4 483
Ethiopia Ethiopie	917	666	676	774	856	813	836	845	867[1]	890[1]
Kenya Kenya	248	253	145	225	252	207	234	243	* 185	* 210
Lesotho Lesotho	14	15	17	18	11	19	21	23	* 20	10
Libyan Arab Jamah. Jamah. arabe libyenne	188	210	184	* 149	* 190	172	161	185	129	150[1]
Madagascar Madagascar	0[1]	0[1]	0[1]	0[1]	1	0	0[1]	1[1]	1[1]	1[1]
Malawi Malawi	1[1]	1[1]	2	1	1	2	2	1	2	* 2
Mali Mali	2[1]	2[1]	2[1]	1	2	1	2[1]	2	* 2	* 2
Mauritania Mauritanie	0	0	1	1[1]	1[1]	1[1]	1[1]	1[1]	1[1]	1[1]
Morocco Maroc	2 183	1 970	1 989	2 358	3 809	2 427	4 019	3 927	3 614	4 939
Mozambique[1] Mozambique[1]	8	6	6	5	5	3	5	5	5	3
Namibia Namibie	1[1]	1[1]	2[1]	2[1]	3[1]	5	5	4	4	* 6
Niger Niger	2[1]	3	8	7	* 8	6[1]	2	2	13	12
Nigeria Nigéria	* 30	21	* 45	30[1]	30[1]	* 30	* 50	* 60	* 90	* 85
Rwanda Rwanda	2	3	3	5	6	7	8	10	* 10	* 10
Somalia Somalie	1[1]	1[1]	1[1]	1[1]	1[1]	1	1[1]	2[1]	1[1]	1[1]

43

Wheat
Production: thousand metric tons [cont.]
Froment
Production : milliers de tonnes métriques [suite]

Country or area Pays ou zone	1982	1983	1984	1985	1986	1987	1988	1989	1990	1991
South Africa Afrique du Sud	2 434	1 774	2 335	1 684	2 322	3 146	3 539	2 026	1 702	* 2 245
Sudan Soudan	142	176	157	79	199	157	181	247	408	680
Swaziland[1] Swaziland[1]	1	1	1	1	1	1	1	1	1	0
Tunisia Tunisie	916	618	711	1 380	474	1 360	220	420	1 122	1 786
Uganda Ouganda	10	12	7	8	8	10	13	11	8	8[1]
United Rep.Tanzania Rép. Unie de Tanzanie	59	74	83	72	72	76	97	106	84	75[1]
Zaire Zaïre	3	3	3	3	8	7	7	7	7[1]	7
Zambia Zambie	14	11	13	17	18	28	37	47	* 52	* 68
Zimbabwe Zimbabwe	192	111	84	205	248	215	257	284	325	253
America, North **Amérique du Nord**	**106 472**	**95 839**	**96 363**	**95 511**	**93 099**	**87 824**	**69 025**	**84 435**	**111 148**	**90 881**
Canada Canada	26 715	26 465	21 188	24 252	31 378	25 991	15 996	24 578	32 709	32 822
Guatemala Guatemala	43	55	51	69	54	55	43	54	32	28
Honduras Honduras	1	1[1]	1[1]	1[1]	1[1]	1[1]	1[1]	1[1]	1[1]	1[1]
Mexico Mexique	4 462	3 460	4 506	5 214	4 770	4 415	3 665	4 374	3 931	4 115
United States Etats-Unis	75 251	65 858	70 618	65 975	56 897	57 362	49 320	55 428	74 475	53 915
America, South **Amérique du Sud**	**18 217**	**16 610**	**17 354**	**14 932**	**16 849**	**17 848**	**17 082**	**18 919**	**17 029**	**14 555**
Argentina Argentine	15 000	13 000	13 600	8 700	8 700	9 000	8 360	10 300	* 11 014	* 9 000
Bolivia Bolivie	66	40	75	74	81	77	63	61	54	103
Brazil Brésil	1 827	2 237	1 983	4 320	5 690	6 035	5 738	5 553	3 094	* 3 077
Chile Chili	650	586	988	1 165	1 626	1 874	1 734	1 766	1 718	1 589
Colombia Colombie	71	78	59	76	82	74	63	80	105	94
Ecuador Equateur	39	27	25	18	33	31	34	26	30	28
Paraguay Paraguay	99	139	187	240	284	318	524	432	* 375	* 400
Peru Pérou	103	84	87	92	121	131	153	159	99	128

43

Wheat
Production: thousand metric tons [cont.]
Froment
Production : milliers de tonnes métriques [suite]

Country or area Pays ou zone	1982	1983	1984	1985	1986	1987	1988	1989	1990	1991
Uruguay Uruguay	363	419	349	246	232	308	414	542	539	136
Asia **Asie**	**150 105**	**168 667**	**176 530**	**176 891**	**188 768**	**180 301**	**184 554**	**192 295**	**203 283**	**205 652**
Afghanistan Afghanistan	* 2 391	* 2 306	* 2 194	* 2 081	1 925	* 2 300	1 900[1]	1 800[1]	1 650	1 726
Bangladesh Bangladesh	967	1 095	1 211	1 464	1 042	1 091	1 048	1 022	890	1 004
Bhutan Bhoutan	10[1]	11[1]	12	* 11	* 11	8[1]	4	4	5[1]	5[1]
China Chine	68 472	81 392	87 817	85 807	90 044	85 905	85 433	90 810	98 232	95 003[1]
Cyprus Chypre	10	9	9	9	7	14	13	8	10	* 5
Gaza Strip (Palestine) Zone de Gaza (Palestine)	1	3	1	0	1[1]	1[1]	1[1]	1[1]	1[1]	0[1]
India Inde	37 452	42 794	45 476	44 069	47 052	44 323	46 169	54 110	49 850	54 522
Iran, Islamic Rep. of Iran, Rép. islamique d'	6 660	5 956	6 207	6 631	7 556	7 600	7 265	6 010	8 218	* 8 900
Iraq Iraq	965	841	471	1 406	1 036	722	929	491	1 196	* 525
Israel Israël	147	335	130	128	169	298	211	202	291	* 160
Japan Japon	742	695	741	874	876	864	1 021	985	952	* 860
Jordan Jordanie	52	131	50	63	31	80	79	55	83	* 40
Korea, Dem. P. R.[1] Corée, R. p. dém. de[1]	130	156	200	200	200	205	205	210	220	195
Korea, Republic of Corée, République de	66	112	17	11	5	4	2	1	1	* 1
Lebanon Liban	23	20	18	* 23	30[1]	49	51	56	52[1]	50[1]
Mongolia Mongolie	440	648	460	689	664	543	672	526	467	536
Myanmar Myanmar	124	130	214	206	190	192	157	130	124	123
Nepal Népal	526	657	634	534	598	701	745	830	855	836
Oman Oman	1	1	1	0	0	0	1	1	1	1[1]
Pakistan Pakistan	11 304	12 414	10 882	11 703	13 923	12 016	12 675	14 419	14 316	14 505
Saudi Arabia Arabie saoudite	417	817	1 402	2 135	2 290	2 653	3 200	3 192	* 3 600	* 4 000
Syrian Arab Republic Rép. arabe syrienne	1 556	1 612	1 068	1 714	1 969	1 656	2 067	1 020	2 070	* 2 135

43

Wheat
Production: thousand metric tons [cont.]
Froment
Production : milliers de tonnes métriques [suite]

Country or area Pays ou zone	1982	1983	1984	1985	1986	1987	1988	1989	1990	1991
Turkey Turquie	17 542	16 437	17 235	17 032	19 032	18 932	20 523	16 221	20 000	20 400
United Arab Emirates Emirats arabes unis	2	1	2	1	1[1]	1	1	1[1]	1[1]	1[1]
Yemen Yémen	82	50	52	81	95	113	142	163	155	77
Europe **Europe**	**102 831**	**103 367**	**128 974**	**112 326**	**115 108**	**116 378**	**124 297**	**127 913**	**131 133**	**132 407**
Albania Albanie	524	583	580[1]	550[1]	560[1]	589	633	611	450[1]	* 300
Austria Autriche	1 236	1 417	1 501	1 563	1 415	1 451	1 560	1 363	1 404	* 1 341
Belgium-Luxembourg Belgique-Luxembourg	1 063	1 062	1 330	1 215	1 324	1 115	1 320	1 478	* 1 409	* 1 620
Bulgaria Bulgarie	4 913	3 608	4 836	3 068	4 327	4 149	4 743	5 425	5 292	4 503
Czechoslovakia Tchécoslovaquie	4 606	5 820	6 170	6 023	5 305	6 154	6 547	6 356	6 707	6 205
Denmark Danemark	1 207	1 548	2 446	1 972	2 177	2 285	2 080	3 224	3 953	3 629
Finland Finlande	435	550	478	472	529	281	285	507	627	431
France France	25 358	24 745	32 977	28 784	26 475	27 415	29 677	31 813	33 312	34 483
Germany† · Allemagne† F. R. Germany R. f. Allemagne	8 632	8 998	10 223	9 866	10 406	9 932	11 922	11 032	11 053	11 948
former German D. R. anc. R. d. allemande	2 739	3 550	3 903	3 936	4 195	4 040	3 699	3 477	4 189	4 721
Greece Grèce	3 039	2 059	2 316	1 807	2 389	2 213	2 498	2 592	1 965	2 750
Hungary Hongrie	5 762	5 985	7 392	6 578	5 793	5 748	7 026	6 540	6 198	5 954
Ireland Irlande	400	387	602	495	424	402	475	477	625	703
Italy Italie	8 968	8 717	10 057	8 461	9 102	9 381	7 952	7 413	8 109	9 289
Malta Malte	8	8	* 5	* 5	5[1]	5[1]	5	5	5[1]	5[1]
Netherlands Pays-Bas	967	1 043	1 131	851	940	769	827	1 047	1 076	916
Norway Norvège	76	97	170	170	159	249	148	140	224	254
Poland Pologne	4 476	5 165	6 010	6 461	7 502	7 942	7 582	8 462	9 026	9 269
Portugal Portugal	428	329	468	397	502	534	396	616	296	323

43

Wheat
Production: thousand metric tons [*cont.*]
Froment
Production : milliers de tonnes métriques [*suite*]

Country or area Pays ou zone	1982	1983	1984	1985	1986	1987	1988	1989	1990	1991
Romania Roumanie	6 122	4 935	7 388	5 532	6 278	6 632	8 528	7 840	7 290	5 442
Spain Espagne	4 410	4 268	6 052	5 329	4 392	5 791	6 514	5 468	4 760	5 392
Sweden Suède	1 490	1 721	1 776	1 338	1 731	1 558	1 296	1 751	2 243	1 524
Switzerland Suisse	434	446	597	549	492	462	565	649	563	574
United Kingdom Royaume-Uni	10 320	10 800	14 970	12 046	13 911	11 940	11 720	14 030	14 000	14 300
Yugoslavia Yougoslavie	5 218	5 525	5 595	4 859	4 776	5 345	6 300	5 599	6 359	6 530
Oceania **Océanie**	**9 168**	**22 318**	**18 981**	**16 477**	**17 158**	**12 706**	**14 266**	**14 350**	**15 590**	**9 809**
Australia Australie	8 876	22 016	18 666	16 167	16 778	12 369	14 060	14 214	15 402	* 9 633
New Caledonia Nouvelle-Calédonie	1	1	0	0	1	0	0	0	0[1]	0[1]
New Zealand Nouvelle-Zélande	292	301	315	310	380	337	206	135	188	176
former USSR **ancienne URSS**	**84 300**	**77 519**	**68 633**	**78 078**	**92 306**	**83 312**	**84 445**	**92 307**	**109 600**	**80 000**[1]

Source:
Food and Agriculture Organization of the United Nations
(Rome).

† All data shown which pertain to Germany prior to 3 October
  1990 are indicated separately for the Federal Republic of
  Germany and the former German Democratic Republic based on
  their respective territories at the time indicated. Where
  data for united Germany (3 October 1990 and thereafter) are
  not available, available data are shown separately under the
  designations Federal Republic of Germany and former German
  Democratic Republic and pertain to the territorial
  boundaries prior to 3 October 1990. For detailed
  explanatory notes on data pertaining to Germany, see Annex I
  - Country Nomenclature.

1 FAO estimate.

Source:
Organisation des Nations Unies pour l'alimentation et
l'agriculture (Rome).

† Toutes les données se rapportant à l'Allemagne avant le 3
  octobre 1990 figurent dans deux rubriques séparées basées
  sur les territoires respectifs de la République fédérale
  d'Allemagne et l'ancienne République démocratique allemande
  selon la période indiquée. En l'absence de données pour
  l'Allemagne unifiée (à compter du 3 octobre 1990), les
  données disponibles sont fournies séparément sous les
  rubriques République fédérale d'Allemagne et ancienne
  République démocratique allemande et se rapportent aux
  limites territoriales antérieures au 3 octobre 1990. Pour
  les notes explicatives en détail sur les données concernant
  l'Allemagne, voir Annexe I – Nomenclature des pays.

1 Estimation de la FAO.

44
Rice (rough or paddy)
Riz (brut ou paddy)
Production: thousand metric tons
Production : milliers de tonnes métriques

Country or area Pays ou zone	1982	1983	1984	1985	1986	1987	1988	1989	1990	1991
World Monde	424 362	451 600	468 590	471 601	471 365	464 373	491 045	517 272	521 703	519 869
Africa Afrique	8 931	8 961	8 956	9 419	10 016	10 250	10 547	12 686	12 396	13 066
Algeria[1] Algérie[1]	1	1	1	1	1	1	1	1	2	2
Angola[1] Angola[1]	20	22	22	22	20	20	20	20	18	18
Benin Bénin	9	5	8	7	9	8	10	9	11	* 8
Burkina Faso Burkina Faso	43	40	41	50	38	22	39	42	* 48	* 50
Burundi Burundi	9	9	18	20	23	28	35	37	40	39[1]
Cameroon Cameroun	77	89	* 61	* 63	90	65	54	62	70[1]	* 90
Central African Rep. Rép. centrafricaine	15	12[1]	17	13	9	12	14	15	15	15[1]
Chad Tchad	23	18	2	8	* 31	* 42	* 74	* 57	* 66	86
Comoros[1] Comores[1]	14	15	15	16	15	15	15	15	15	15
Congo Congo	2	2	1	1	1	1	1	1	1[1]	1[1]
Côte d'Ivoire Côte d'Ivoire	450	360	514	540	560	580	610	635	687	690[1]
Egypt Egypte	2 441	2 442	2 236	2 311	2 445	2 279	2 132	2 679	3 167	* 3 152
Gabon Gabon	1[1]	1[1]	1	1	1	0	0	1	1[1]	1[1]
Gambia Gambie	34	26	27	23	24	20	29	21	21	* 21
Ghana Ghana	36	40	65	68	70	81	95	74	81	151
Guinea Guinée	490	396	403	437	510	515	521	* 426	500[1]	* 628
Guinea-Bissau Guinée-Bissau	103	85	105	105[1]	110[1]	105[1]	100	106	123	* 118
Kenya Kenya	38	37	39	50	38	41	45	58	59[1]	60[1]
Liberia Libéria	284	290	298	289	288	298	298	* 280	100[1]	110
Madagascar Madagascar	1 970	2 147	2 131	2 178	2 230	2 178	2 149	2 380	2 420	2 200[1]
Malawi Malawi	37	26	35	34	37	28	32	46	45	* 61
Mali Mali	153	216	109	214	225	237	* 288	338	282	* 445
Mauritania Mauritanie	20	22	20	25	33	53	51	55	52	52[1]

44
Rice (rough or paddy)
Production: thousand metric tons [*cont.*]
Riz (brut ou paddy)
Production : milliers de tonnes métriques [*suite*]

Country or area Pays ou zone	1982	1983	1984	1985	1986	1987	1988	1989	1990	1991
Morocco Maroc	4	4	5	2	20	49	33	4	33	33[1]
Mozambique Mozambique	80[1]	82[1]	84[1]	86[1]	* 93	90[1]	93[1]	95[1]	96	56
Niger Niger	41	45	49	57	75	61	53	79	* 73	71
Nigeria Nigéria	1 250	1 280	* 1 300	1 430	1 416	1 780	2 082	3 303	2 500	* 3 185
Rwanda Rwanda	6	7	8	4	8	6	7	8	8[1]	8[1]
Senegal Sénégal	119	109	136	147	143	136	146	168	156	* 160
Sierra Leone Sierra Leone	524	460	504	430	525	466	493	518	504	* 386
Somalia Somalie	19	4	4	11	12	12	12	20	12[1]	5[1]
South Africa[1] Afrique du Sud[1]	3	3	3	3	3	3	3	3	3	3
Sudan Soudan	4	3	2	2[1]	2[1]	* 1	* 1	* 1	1	1
Swaziland Swaziland	2	2	3	3[1]	3[1]	3[1]	3[1]	3[1]	3[1]	3[1]
Togo Togo	16	10	18	15	20	23	29	28	25	* 34
Uganda Ouganda	19	22	20	19	21	20	23	27	62	64[1]
United Rep.Tanzania Rép. Unie de Tanzanie	320	350	356	427	547	644	615	720	740	664[1]
Zaire Zaïre	251	271	286	297	307	319	329	341	345[1]	365
Zambia Zambie	5	10	9	11	11	8	10	12	9	14
Zimbabwe Zimbabwe	0	0	1	0	0	1	1	1[1]	1[1]	1[1]
America, North **Amérique du Nord**	**9 200**	**6 853**	**8 678**	**8 803**	**8 390**	**8 223**	**9 469**	**9 446**	**9 178**	**8 903**
Belize Belize	8	6	6	6	4	5	6	5	4[1]	4[1]
Costa Rica Costa Rica	146	247	223	224	185	154	205	157	219	193
Cuba Cuba	520	518	555	524	571	466	489	536	447	430
Dominican Republic Rép. dominicaine	447	501	507	494	468	515	460	462	405	303
El Salvador El Salvador	35	43	63	69	47	42	57	64	62	61
Guatemala Guatemala	49	46	44	38	48	59	69	45	45	* 42
Haiti Haïti	113	124	124	129	135	121	121	124	130	120

44
Rice (rough or paddy)
Production: thousand metric tons [cont.]
Riz (brut ou paddy)
Production : milliers de tonnes métriques [suite]

Country or area Pays ou zone	1982	1983	1984	1985	1986	1987	1988	1989	1990	1991
Honduras Honduras	38	46	49	46	34	53	48	71	45	56
Jamaica Jamaïque	2	3	5	4	2	2	2	1	0	0[1]
Mexico Mexique	511	416	484	808	545	591	456	637	394	354
Nicaragua Nicaragua	176	171	138	144	112	149	* 111	* 118	* 117	* 140
Panama Panama	176	199	175	186	180	180	183	207	216	180[1]
Puerto Rico Porto Rico	6[1]	6	5	5	6	2	1	0	...	...
Trinidad and Tobago Trinité-et-Tobago	3	3	4	4	4	7	7	12	13	14[1]
United States Etats-Unis	6 969	4 523	6 296	6 122	6 049	5 879	7 253	7 007	7 080	7 006
America, South Amérique du Sud	**15 317**	**12 444**	**14 565**	**14 348**	**15 228**	**15 848**	**17 770**	**17 220**	**13 522**	**15 334**
Argentina Argentine	437	277	480	400	439	371	415	469	467	347
Bolivia Bolivie	86	62	166	173	137	164	171	227	211	257
Brazil Brésil	9 735	7 742	9 027	9 025	10 374	10 419	11 809	11 044	7 419	9 503
Chile Chili	131	116	165	157	127	147	162	185	136	117
Colombia Colombie	2 023	1 814	1 715	1 742	1 521	1 473	1 777	1 884	2 117	1 739
Ecuador Equateur	384	274	437	397	576	781	955	867	840	841
French Guiana Guyane française	0	2	6	8	9	13	14	17	22	29
Guyana Guyana	303	246	300	280	242	226	227	237	144	250
Paraguay Paraguay	63	73	80	97	62	105	81	87	86	99
Peru Pérou	826	799	1 140	878	726	1 169	1 129	1 091	966	814
Suriname Suriname	301	268	302	299	300	272	265	260	196	190[1]
Uruguay Uruguay	419	323	340	421	394	335	381	537	517	540
Venezuela Venezuela	609	449	408	472	322	373	383	313	401	* 608
Asia Asie	**385 670**	**418 461**	**431 047**	**433 384**	**432 071**	**424 582**	**447 414**	**472 457**	**480 769**	**477 267**
Afghanistan Afghanistan	* 364	* 350	* 334	* 317	336	324	343	320	333	335
Bangladesh Bangladesh	21 325	21 761	21 933	22 556	23 110	23 120	23 316	27 324	26 778	* 28 575
Bhutan Bhoutan	59[1]	60[1]	66	* 62	* 63	65[1]	43	43	43[1]	43[1]

44
Rice (rough or paddy)
Production: thousand metric tons [*cont.*]
Riz (brut ou paddy)
Production : milliers de tonnes métriques [*suite*]

Country or area Pays ou zone	1982	1983	1984	1985	1986	1987	1988	1989	1990	1991
Brunei Darussalam Brunéi Darussalam	2	3	3	1	1	1	1	1[1]	1[1]	1[1]
Cambodia Cambodge	1 949	2 039	1 260	1 812	2 093	1 815	2 400	* 2 555	* 2 500	* 2 400
China Chine	164 848	172 096	181 193	171 416	174 790	176 798	171 416	182 461	191 197	187 450[1]
India Inde	70 772	90 048	87 553	95 818	90 779	85 339	106 369	110 311	111 953	* 110 945
Indonesia Indonésie	33 584	35 303	38 136	39 033	39 727	40 078	41 676	44 726	45 179	44 321
Iran, Islamic Rep. of Iran, Rép. islamique d'	1 605	1 216	1 484	1 776	1 784	1 803	1 419	1 854	2 273	2 100
Iraq Iraq	163	111	109	149	141	196	141	232	229	* 125
Japan Japon	12 838	12 958	14 848	14 578	14 559	13 284	12 419	12 934	13 124	* 12 005
Korea, Dem. P. R. Corée, R. p. dém. de	* 5 000	5 100[1]	* 5 000	5 100[1]	5 200[1]	5 300[1]	5 600[1]	5 400[1]	5 300[1]	5 100[1]
Korea, Republic of Corée, République de	7 308	7 608	7 970	7 855	7 871	7 596	8 260	8 100	7 732	* 7 478
Lao People's Dem. Rep. Rép. dém. pop. lao	1 092	1 101	1 321	1 395	1 449	1 207	1 003	1 404	1 491	1 400[1]
Malaysia Malaisie	1 884	1 734	1 572	1 849	1 747	1 700	1 783	1 744	1 655	* 1 550
Myanmar Myanmar	14 373	14 288	14 255	14 317	14 126	13 640	13 168	13 803	13 969	13 201
Nepal Népal	1 833	2 757	2 709	2 804	2 372	2 982	3 283	3 390	3 502	3 600[1]
Pakistan Pakistan	5 167	5 009	4 973	4 378	5 230	4 861	4 800	4 830	4 897	4 903
Philippines Philippines	7 731	7 841	8 200	9 097	8 958	8 540	8 971	9 459	9 319	9 670
Sri Lanka Sri Lanka	2 156	2 484	2 414	2 661	2 588	2 128	2 477	2 063	2 538	* 2 397
Thailand Thaïlande	16 879	19 549	19 905	20 264	18 868	18 428	21 263	20 177	* 17 300	* 20 040
Turkey Turquie	350	315	280	270	275	275	263	330	230	200
Viet Nam Viet Nam	14 390	14 732	15 528	15 875	16 003	15 103	17 000	18 996	19 225	* 19 428
Europe **Europe**	**1 886**	**1 736**	**1 967**	**2 174**	**2 284**	**2 151**	**2 194**	**2 121**	**2 415**	**2 340**
Albania Albanie	12	13	13[1]	12[1]	11[1]	11	9	8	6[1]	5[1]
Bulgaria Bulgarie	75	74	61	55	62	53	46	43	25	27
France France	27	38	36	62	60	54	65	106	121	109
Greece Grèce	83	86	95	104	121	137	114	103	110	* 127

44
Rice (rough or paddy)
Production: thousand metric tons [*cont.*]
Riz (brut ou paddy)
Production : milliers de tonnes métriques [*suite*]

Country or area Pays ou zone	1982	1983	1984	1985	1986	1987	1988	1989	1990	1991
Hungary Hongrie	48	47	33	38	47	40	47	28	39	38[1]
Italy Italie	1 008	1 021	1 009	1 123	1 137	1 064	1 093	1 246	1 291	1 236
Portugal Portugal	143	109	134	147	149	144	146	147	159	153
Romania Roumanie	46	83	110	137	153	116	132	70	67	31
Spain Espagne	402	224	440	462	496	483	507	342	569	582
Yugoslavia Yougoslavie	42	40	36	36	48	49	36	27	28	* 32
Oceania **Océanie**	**888**	**545**	**662**	**900**	**744**	**637**	**784**	**782**	**951**	**760**
Australia Australie	857	519	632	866	716	613	751	749	924	726
Fiji Fidji	20	16	22	28	25	23	32	32	26	33
Papua New Guinea Papouasie-Nvl-Guinée	1[1]	1	1[1]	1[1]	1[1]	1[1]	1[1]	1[1]	1[1]	1[1]
Solomon Islands Iles Salomon	11	9	7	6	2	...	...	...	...	...
former USSR **ancienne URSS**	**2 470**	**2 600**	**2 715**	**2 572**	**2 633**	**2 683**	**2 866**	**2 560**	**2 473**	***2 200**

Source:
Food and Agriculture Organization of the United Nations
(Rome).

1 FAO estimate.

Source:
Organisation des Nations Unies pour l'alimentation et
l'agriculture (Rome).

1 Estimation de la FAO.

45
Maize (corn)
Maïs
Production: thousand metric tons
Production : milliers de tonnes métriques

Country or area Pays ou zone	1982	1983	1984	1985	1986	1987	1988	1989	1990	1991
World *Monde*	**450 245**	**348 473**	**451 973**	**487 000**	**478 010**	**450 917**	**400 152**	**474 095**	**479 340**	**478 775**
Africa **Afrique**	**27 085**	**22 339**	**23 058**	**30 589**	**31 753**	**28 114**	**32 615**	**38 740**	**33 266**	**33 038**
Algeria Algérie	1	3	5	1	1	2	* 2	* 2	* 2	2[1]
Angola Angola	250[1]	275[1]	260[1]	250[1]	280	300	270	204	180	299
Benin Bénin	273	282	379	435	378	277	407	424	418	* 390
Botswana Botswana	12	8	0	1	4	3	8	20	8	4
Burkina Faso Burkina Faso	111	71	77	142	155	131	227	257	* 258	* 296
Burundi Burundi	144	148	139	157	164	174	180	138	168	* 190
Cameroon Cameroun	503	489	375	310	389	387	367	371	380[1]	* 450
Cape Verde Cap-Vert	4	3	3	1	12	21	17	10	11	* 4
Central African Rep. Rép. centrafricaine	48	40	42	46	94	66	70	62	99	100[1]
Chad Tchad	30	29	18	35	35	29	43	19	29	44
Comoros Comores	5[1]	5[1]	3	4	4	3	4	4	4[1]	4[1]
Congo Congo	* 6	11	13	15	21	19	20	25[1]	25[1]	25[1]
Côte d'Ivoire Côte d'Ivoire	430	410	520	480	420	435	460	480	484	* 510
Egypt Egypte	3 347	3 509	3 698	3 699	3 608	3 619	4 088	4 529	4 799	* 5 270
Ethiopia Ethiopie	1 603	1 533	1 088	1 037	2 003	2 006	1 948	2 147	1 636[1]	1 590[1]
Gabon Gabon	12[1]	12[1]	13[1]	13	20	25	25	20	20[1]	20[1]
Gambia Gambie	17	9	13	26	17	15	16	14	14	* 25
Ghana Ghana	346	172	696	584	559	598	751	749	553	932
Guinea Guinée	90[1]	90[1]	* 100	* 100	* 100	90[1]	80[1]	108	100[1]	79[1]
Guinea-Bissau Guinée-Bissau	10	10	10	10[1]	10[1]	10[1]	8	10	14	* 13
Kenya Kenya	2 502	2 300	1 422	2 430	2 898	2 416	2 761	2 836	* 2 630	* 2 250
Lesotho Lesotho	83	76	79	92	86	95	146	98	111	* 95
Libyan Arab Jamah. Jamah. arabe libyenne	1	1	1	1[1]	1[1]	1[1]	1[1]	1[1]	0	1[1]

45
Maize (corn)
Production: thousand metric tons [cont.]
Maïs
Production : milliers de tonnes métriques [suite]

Country or area Pays ou zone	1982	1983	1984	1985	1986	1987	1988	1989	1990	1991
Madagascar Madagascar	113	132	141	140	153	158	156	160	155	150[1]
Malawi Malawi	1 415	1 369	1 398	1 355	1 295	1 202	1 424	1 510	1 343	* 1 590
Mali Mali	89	144	102	140	213	179	215	225	197	* 226
Mauritania Mauritanie	4	3	3	1	3	1	7	3	3	1
Mauritius Maurice	1	1	3	5	8	4	4	2	2	2[1]
Morocco Maroc	247	258	264	321	307	240	358	403	436	335
Mozambique Mozambique	350[1]	330[1]	350[1]	400[1]	* 459	* 271	* 322	* 330	453	327
Namibia Namibie	48[1]	48[1]	50[1]	52[1]	55[1]	62[1]	65[1]	65[1]	65[1]	50
Niger Niger	7	7	7	4	6	8	5	3	4	6
Nigeria Nigéria	626	1 027	1 196	1 826	1 735	1 430	2 080	2 132	1 832	1 900[1]
Réunion Réunion	14	12	11	12	13	11	13	13	10	13[1]
Rwanda Rwanda	92	110	102	99	90	91	88	94	* 110	* 104
Sao Tome and Principe[1] Sao Tomé-et-Principe[1]	1	1	1	1	1	1	1	1	1	1
Senegal Sénégal	76	61	98	147	108	114	123	131	133	* 106
Sierra Leone Sierra Leone	15[1]	15[1]	* 14	* 14	8	* 12	11	11	13	11
Somalia Somalie	150	236	270	280	336	286	353	299	315[1]	* 100
South Africa Afrique du Sud	8 503	4 318	4 714	8 295	8 318	7 372	7 253	12 061	* 8 900	* 8 200
Sudan Soudan	29	26	24	18	36	25	3	3	2	63
Swaziland Swaziland	52	30	148	172	156	* 95	149	112	81	* 153
Togo Togo	151	145	222	182	127	172	296	287	285	* 236
Uganda Ouganda	393	413	338	354	322	357	440	624	584	600[1]
United Rep.Tanzania Rép. Unie de Tanzanie	1 654	1 651	1 939	2 093	2 210	2 359	2 339	3 125	2 445	2 332
Zaire Zaïre	666	673	704	726	756	786	816	846	870[1]	906
Zambia Zambie	750	935	872	1 122	1 231	1 063	1 943	1 845	1 093	1 448

45
Maize (corn)
Production: thousand metric tons [*cont.*]
Maïs
Production : milliers de tonnes métriques [*suite*]

Country or area Pays ou zone	1982	1983	1984	1985	1986	1987	1988	1989	1990	1991
Zimbabwe Zimbabwe	1 808	910	1 133	2 960	2 546	1 094	2 253	1 927	1 994	1 586
America, North **Amérique du Nord**	**228 274**	**127 664**	**217 381**	**249 251**	**229 232**	**202 822**	**144 406**	**211 696**	**226 458**	**213 664**
Bahamas[1] Bahamas[1]	1	1	1	1	1	1	1	1	1	1
Barbados Barbade	* 2	2[1]	2[1]	2[1]	2[1]	2[1]	2[1]	2[1]	2[1]	2[1]
Belize Belize	21	18	16	20	18	23	23	19	18[1]	20[1]
Canada Canada	6 522	5 931	6 778	6 970	5 912	7 015	5 369	6 379	7 157	7 319
Costa Rica Costa Rica	82	94	103	119	120	127	98	83	72	60
Cuba Cuba	* 95	* 95	* 95	* 95	* 95	* 95	* 95	95[1]	95[1]	95[1]
Dominican Republic Rép. dominicaine	31	42	84	67	47	47	56	84	38	43
El Salvador El Salvador	414	443	527	495	437	578	596	589	603	504
Guatemala Guatemala	1 100	1 046	922	1 088	1 077	1 217	1 324	1 247	1 293	1 150[1]
Haiti Haïti	171	186	186	196	206	205	202	196	* 163	145[1]
Honduras Honduras	366	458	524	337	358	443	449	511	559	584
Jamaica Jamaïque	3	4	4	4	5	4	2	3	2	* 3
Mexico Mexique	10 030	13 061	12 932	14 103	11 721	11 607	10 600	10 945	14 635	13 527
Nicaragua Nicaragua	190	163	205	208	192	213	298	293	214	245
Panama Panama	62	74	71	96	93	98	92	90	94	* 94
St. Vincent-Grenadines St. Vincent-Grenadines	1[1]	1[1]	1[1]	1[1]	1[1]	1[1]	1[1]	1	1	1[1]
Trinidad and Tobago Trinité-et-Tobago	3	3	3[1]	3[1]	3[1]	3[1]	3[1]	3[1]	3[1]	3[1]
United States Etats-Unis	209 180	106 041	194 928	225 445	208 943	181 142	125 194	191 156	201 508	189 867
America, South **Amérique du Sud**	**35 350**	**31 511**	**35 202**	**38 850**	**37 996**	**41 726**	**39 677**	**36 637**	**32 221**	**36 533**
Argentina Argentine	9 600	9 000	9 500	11 900	12 100	9 250	9 200	4 260	5 049	* 7 768
Bolivia Bolivie	450	337	497	554	457	481	446	400	407	510
Brazil Brésil	21 842	18 731	21 164	22 018	20 531	26 803	24 748	26 573	21 339	22 604
Chile Chili	425	512	721	772	721	617	661	938	823	836
Colombia Colombie	899	864	864	763	788	860	908	1 044	1 213	1 274

45
Maize (corn)
Production: thousand metric tons [cont.]
        Maïs
        Production : milliers de tonnes métriques [suite]

Country or area Pays ou zone	1982	1983	1984	1985	1986	1987	1988	1989	1990	1991
Ecuador Equateur	324	229	326	363	467	431	443	428	501	665
Guyana Guyana	1	1	1	1	1	3	3[1]	3[1]	3[1]	3[1]
Paraguay Paraguay	553	619	730	801	* 778	1 001	961	1 000	1 139	* 980
Peru Pérou	659	626	740	702	876	909	908	1 010	632	669
Uruguay Uruguay	97	104	112	108	103	104	118	60	112	124
Venezuela Venezuela	501	488	547	868	1 173	1 267	1 281	921	1 002	* 1 100
Asia **Asie**	**85 739**	**97 759**	**104 963**	**94 191**	**103 824**	**108 272**	**112 491**	**114 957**	**132 549**	**127 514**
Afghanistan Afghanistan	* 663	* 639	* 608	* 577	567	* 514	* 494	* 458	430	420
Bangladesh Bangladesh	1	1	1	3	3	3	3	3	3	3[1]
Bhutan Bhoutan	81	83[1]	80	76[1]	74[1]	50[1]	31	31	40[1]	40[1]
Cambodia Cambodge	51	43	48	51	38	47[1]	47	50[1]	55[1]	50[1]
China Chine	60 678	68 353	73 600	64 052	71 128	79 547	77 672	79 259	97 158	93 350[1]
India Inde	6 549	7 922	8 442	6 644	7 593	5 721	8 229	9 651	9 073	8 200
Indonesia Indonésie	3 235	5 087	5 288	4 330	5 920	5 156	6 652	6 193	6 734	6 409
Iran, Islamic Rep. of Iran, Rép. islamique d'	* 28	* 90	* 50	6	33	40	6	7[1]	7[1]	7[1]
Iraq Iraq	28	28	31	41	53	61	77	104	* 185	* 74
Israel Israël	22	28	26	* 20	21	7	2	3	2	3[1]
Japan Japon	2	1	2	2	1	1	1[1]	1[1]	1[1]	1[1]
Jordan Jordanie	0	0	0	0	2	0	4	6	0[1]	0[1]
Korea, Dem. P. R. Corée, R. p. dém. de	4 200[1]	4 300[1]	* 4 600	4 500[1]	4 500[1]	4 500[1]	4 500[1]	4 450[1]	4 400[1]	4 500[1]
Korea, Republic of Corée, République de	117	101	133	132	113	127	106	121	120	75
Kuwait Koweït	0[1]	0[1]	0	2	2	2[1]	2[1]	2[1]	2[1]	0[1]
Lao People's Dem. Rep. Rép. dém. pop. lao	35	32	34	36	42	36	51	44	67	60[1]
Lebanon Liban	0	0[1]	1	1[1]	2[1]	2	3	3	3[1]	3[1]
Malaysia Malaisie	9	20	22	* 24	* 26	* 30	* 32	* 34	* 35	* 36

45

Maize (corn)
Production: thousand metric tons [*cont.*]

Maïs
Production : milliers de tonnes métriques [*suite*]

Country or area Pays ou zone	1982	1983	1984	1985	1986	1987	1988	1989	1990	1991
Myanmar Myanmar	239	310	303	299	285	224	193	194	187	190 [1]
Nepal Népal	718	761	820	874	868	902	1 072	1 201	1 231	1 235 [1]
Pakistan Pakistan	1 005	1 014	1 028	1 009	1 111	1 127	1 204	1 179	1 185	* 1 190
Philippines Philippines	3 126	3 346	3 439	3 922	4 091	4 278	4 428	4 522	4 854	4 655
Saudi Arabia Arabie saoudite	1	1	1	1	1	2	3 [1]	* 2	* 3	* 4
Sri Lanka Sri Lanka	24	29	38	30	41	42	39	31	33	34 [1]
Syrian Arab Republic Rép. arabe syrienne	50	27	60	79	74	57	90	109	180	185
Thailand Thaïlande	3 002	3 552	4 226	4 934	4 309	2 781	4 675	4 393	3 722	3 990
Turkey Turquie	1 360	1 480	1 500	1 900	2 300	2 400	2 000	2 000	2 100	2 100
United Arab Emirates Emirats arabes unis	* 2	* 4	* 3	3 [1]	3 [1]	3 [1]	3 [1]	4 [1]	4 [1]	4 [1]
Viet Nam Viet Nam	438	467	532	587	570	559	815	838	671	* 652
Yemen Yémen	74	40	50	56	52	54	58	68	66	46
Europe **Europe**	**58 783**	**55 521**	**57 346**	**59 255**	**62 328**	**54 787**	**54 576**	**56 399**	**44 578**	**59 168**
Albania Albanie	342	366	340 [1]	320 [1]	320 [1]	306	233	302	200 [1]	180 [1]
Austria Autriche	1 551	1 454	1 542	1 727	1 740	1 685	1 700	1 491	1 620	* 1 524
Belgium-Luxembourg Belgique-Luxembourg	52	39	53	50	57	40	54	62	60	62 [1]
Bulgaria Bulgarie	3 418	3 115	2 994	1 350	2 848	1 858	1 557	2 265	1 221	2 718
Czechoslovakia Tchécoslovaquie	941	722	940	1 114	992	1 160	996	1 000	468	862
France France	10 400	10 525	10 493	12 409	11 641	12 470	14 120	13 335	9 291	12 787
Germany† · Allemagne† F. R. Germany R. f. Allemagne	1 054	934	1 026	1 204	1 302	1 217	1 535	1 573	1 545	1 809
former German D. R. anc. R. d. allemande	1	0	0	1	2	0	2	0	7	119
Greece Grèce	1 550	1 758	2 162	1 908	1 994	2 156	2 251	2 327	1 992	1 700
Hungary Hongrie	7 959	6 426	6 686	6 818	7 261	7 234	6 256	6 996	4 500	7 509
Italy Italie	6 793	6 699	6 672	6 357	6 401	5 764	6 289	6 360	5 864	6 208

45
Maize (corn)
Production: thousand metric tons [*cont.*]
Maïs
Production : milliers de tonnes métriques [*suite*]

Country or area Pays ou zone	1982	1983	1984	1985	1986	1987	1988	1989	1990	1991
Netherlands Pays-Bas	1	2	1	5	* 5	* 5	5[1]	* 5	* 3	* 5
Poland Pologne	68	64	57	69	113	146	204	244	290	340
Portugal Portugal	464	461	521	550	628	655	658	679	666	677
Romania Roumanie	10 550	10 296	9 891	11 903	10 901	7 527	7 182	6 762	6 810	10 493
Spain Espagne	2 330	1 803	2 529	3 414	3 424	3 557	3 577	3 328	3 086	3 151
Switzerland Suisse	173	139	126	157	174	144	256	256	233	225
Yugoslavia Yougoslavie	11 137	10 719	11 312	9 901	12 526	8 863	7 700	9 415	6 724	8 800[1]
Oceania **Océanie**	**314**	**386**	**450**	**457**	**399**	**388**	**358**	**361**	**366**	**358**
Australia Australie	139	238	291	278	206	208	217	219	200	159
Fiji Fidji	1[1]	2	2	2	2	1	1	1	2	2
New Caledonia Nouvelle-Calédonie	2	2	1	2	1	0	1	1	1[1]	1[1]
New Zealand Nouvelle-Zélande	170	143	154	175	188	176	137	139	162	* 195
Papua New Guinea Papouasie-Nvl-Guinée	1	1	1	1	1	1[1]	1[1]	1[1]	1[1]	2[1]
Vanuatu[1] Vanuatu[1]	1	1	1	1	1	1	1	1	1	1
former USSR **ancienne URSS**	**14 700**	**13 293**	**13 573**	**14 406**	**12 479**	**14 808**	**16 030**	**15 305**	**9 900**	**8 500**[1]

Source:
Food and Agriculture Organization of the United Nations
(Rome).

Source:
Organisation des Nations Unies pour l'alimentation et
l'agriculture (Rome).

† All data shown which pertain to Germany prior to 3 October
1990 are indicated separately for the Federal Republic of
Germany and the former German Democratic Republic based on
their respective territories at the time indicated. Where
data for united Germany (3 October 1990 and thereafter) are
not available, available data are shown separately under the
designations Federal Republic of Germany and former German
Democratic Republic and pertain to the territorial
boundaries prior to 3 October 1990. For detailed
explanatory notes on data pertaining to Germany, see Annex I
- Country Nomenclature.

1 FAO estimate.

† Toutes les données se rapportant à l'Allemagne avant le 3
octobre 1990 figurent dans deux rubriques séparées basées
sur les territoires respectifs de la République fédérale
d'Allemagne et l'ancienne République démocratique allemande
selon la période indiquée. En l'absence de données pour
l'Allemagne unifiée (à compter du 3 octobre 1990), les
données disponibles sont fournies séparément sous les
rubriques République fédérale d'Allemagne et ancienne
République démocratique allemande et se rapportent aux
limites territoriales antérieures au 3 octobre 1990. Pour
les notes explicatives en détail sur les données concernant
l'Allemagne, voir Annexe I – Nomenclature des pays.

1 Estimation de la FAO.

46
Soybeans
Soya
Production: thousand metric tons
Production : milliers de tonnes métriques

Country or area Pays ou zone	1982	1983	1984	1985	1986	1987	1988	1989	1990	1991
World *Monde*	92 108	79 453	90 747	101 139	94 448	100 025	93 410	107 003	108 134	103 065
Africa **Afrique**	397	349	351	372	370	393	421	436	488	577
Côte d'Ivoire Côte d'Ivoire	1	2	2	0[1]	0[1]	0[1]	0[1]	0[1]	0[1]	0[1]
Egypt Egypte	178	162	143	140	133	134	129	91	107	* 135
Ethiopia Ethiopie	12	6	0	5	3	6	8	25	25[1]	26[1]
Gabon Gabon	...	...	...	2	3	3	3	3	3	3[1]
Liberia[1] Libéria[1]	2	2	2	2	2	2	2	2	2	2
Morocco Maroc	0	1	1[1]	1[1]	1[1]	1[1]	1[1]	1[1]	1[1]	1[1]
Nigeria Nigéria	* 60	42	43	60	68	* 70	* 40	* 55	* 60	* 65
Rwanda Rwanda	8	5	5	8	6	7	7	9	9[1]	9[1]
South Africa Afrique du Sud	21	29	39	40	38	35	63	80	* 94	* 126
Uganda Ouganda	6	7	8	8	10	8	14	16	37	38[1]
United Rep.Tanzania Rép. Unie de Tanzanie	0	0	0	1	0	1	2	1	1[1]	1[1]
Zaire Zaïre	9	5	5	4	7	8	10	11	12[1]	12[1]
Zambia Zambie	7	8	13	15	16	13	22	21	27	61
Zimbabwe Zimbabwe	92	81	90	87	83	104	120	121	110	97
America, North **Amérique du Nord**	61 136	45 950	52 261	59 083	54 565	54 874	43 578	54 618	54 334	56 216
Canada Canada	848	735	917	1 012	960	1 270	1 153	1 219	1 292	1 406
Costa Rica Costa Rica	...	2	2	1	1	1	* 0	* 0	* 0	0[1]
El Salvador El Salvador	1[1]	1[1]	1[1]	1[1]	1	1	1	* 2	* 2	* 2
Guatemala * Guatemala *	1	2	3	5	6	26	28	32	37	37
Honduras Honduras	...	...	...	...	...	...	...	2	...	...
Mexico Mexique	672	686	685	929	709	828	226	992	575	718
Nicaragua Nicaragua	3[1]	5[1]	6[1]	8[1]	9	12	17	16	12	14
United States Etats-Unis	59 611	44 518	50 648	57 128	52 880	52 737	42 152	52 354	52 416	54 039
America, South **Amérique du Sud**	18 002	19 635	23 748	26 225	21 674	25 445	29 817	32 833	33 072	28 124

46

Soybeans
Production: thousand metric tons [*cont.*]
Soya
Production : milliers de tonnes métriques [*suite*]

Country or area Pays ou zone	1982	1983	1984	1985	1986	1987	1988	1989	1990	1991
Argentina Argentine	4 150	4 000	7 000	6 500	7 100	6 700	9 900	6 500	10 700	* 11 250
Bolivia Bolivie	86	52	78	83	150	122	151	260	233	384
Brazil Brésil	12 836	14 582	15 541	18 279	13 330	16 969	18 016	24 071	19 888	14 771
Chile * Chili *	1	...	...	...	...	...	...	...	...	...
Colombia Colombie	99	122	94	104	167	128	115	177	232	194
Ecuador Equateur	37	14	47	63	76	146	131	153	167	169
Paraguay Paraguay	757	850	975	1 172	* 810	* 1 310	1 407	1 615	1 795	* 1 304
Peru Pérou	8	2	2	2	4	6	6	3	3	1
Uruguay Uruguay	28	12	11	21	35	63	78	45	52	* 47
Venezuela Venezuela	...	...	...	...	1	1	11	8	4	4[1]
Asia **Asie**	**11 256**	**12 214**	**12 908**	**13 970**	**15 415**	**16 005**	**16 356**	**15 381**	**16 830**	**15 385**
Bhutan Bhoutan	1[1]	2[1]	3	2[1]	2[1]	2[1]	1[1]	1[1]	1[1]	1[1]
Cambodia Cambodge	1	5[1]	8	6[1]	5	8[1]	12	12[1]	15[1]	16[1]
China Chine	9 042	9 769	9 705	10 512	11 629	12 202	11 660	10 239	11 008	* 9 807
India Inde	491	614	955	1 024	891	898	1 547	1 806	2 419	* 2 100
Indonesia Indonésie	521	536	769	870	1 227	1 161	1 270	1 315	1 487	1 549
Iran, Islamic Rep. of Iran, Rép. islamique d'	104	130	* 110	* 105	* 95	58	95	* 96	105	105[1]
Iraq[1] Iraq[1]	2	2	2	2	2	2	2	2	2	2
Japan Japon	226	217	238	228	245	287	277	272	220	* 260
Korea, Dem. P. R. Corée, R. p. dém. de	360[1]	380[1]	* 440	425[1]	432[1]	445[1]	448[1]	420[1]	455[1]	460[1]
Korea, Republic of Corée, République de	233	226	254	234	199	203	239	252	233	183
Lao People's Dem. Rep. Rép. dém. pop. lao	4	4	4	5	3	4	4	5	4	4[1]
Myanmar Myanmar	19	21	22	23	23	27	27	27	26	29
Nepal Népal	...	...	6	7	8	10	12	13	13[1]	13
Pakistan Pakistan	2	1	2	2	3	4	2	1	1	1[1]
Philippines Philippines	11	8	8	8	7	7	6	4	7	8[1]

46

Soybeans
Production: thousand metric tons [*cont.*]
Soya
Production : milliers de tonnes métriques [*suite*]

Country or area Pays ou zone	1982	1983	1984	1985	1986	1987	1988	1989	1990	1991
Sri Lanka Sri Lanka	11	11	8	3	4	4	2	2	* 8	* 8
Thailand Thaïlande	113	179	246	309	356	338	517	672	578	605
Turkey Turquie	36	46	60	125	200	250	150	161	162	150
Viet Nam Viet Nam	77	64	69	79	85	96	85	82	86	85[1]
Europe **Europe**	**704**	**693**	**916**	**913**	**1 616**	**2 507**	**2 287**	**2 649**	**2 453**	**1 933**
Austria Autriche	...	...	...	...	...	...	9	10	18	* 24
Bulgaria Bulgarie	116	82	72	37	54	33	17	22	15	18
France France	21	29	46	56	96	210	255	306	245	150
Germany†・Allemagne† F. R. Germany R. f. Allemagne	...	...	...	...	...	...	...	5	5	3
Greece Grèce	...	...	...	...	...	4	5	23	23	* 16
Hungary Hongrie	54	52	50	46	52	69	105	118	54	57
Italy Italie	9	59	110	286	806	1 589	1 408	1 624	1 751	1 325
Romania Roumanie	301	259	407	308	380	361	295	304	141	179
Spain Espagne	5	2	5	5	3	4	10	27	42	10
Switzerland Suisse	...	...	...	...	...	...	2	2	3	4
Yugoslavia Yougoslavie	198	210	228	174	225	237	180	209	157	* 147
Oceania **Océanie**	**77**	**53**	**89**	**110**	**105**	**90**	**69**	**130**	**77**	**70**
Australia Australie	77	53	89	110	105	90	69	130	77	70
former USSR **ancienne URSS**	**536**	**560**	**473**	**465**	**703**	**712**	**884**	**956**	*** 880**	*** 760**

Source:
Food and Agriculture Organization of the United Nations
(Rome).

† All data shown which pertain to Germany prior to 3 October
1990 are indicated separately for the Federal Republic of
Germany and the former German Democratic Republic based on
their respective territories at the time indicated. Where
data for united Germany (3 October 1990 and thereafter) are
not available, available data are shown separately under the
designations Federal Republic of Germany and former German
Democratic Republic and pertain to the territorial
boundaries prior to 3 October 1990. For detailed
explanatory notes on data pertaining to Germany, see Annex I
- Country Nomenclature.

1 FAO estimate.

Source:
Organisation des Nations Unies pour l'alimentation et
l'agriculture (Rome).

† Toutes les données se rapportant à l'Allemagne avant le 3
octobre 1990 figurent dans deux rubriques séparées basées
sur les territoires respectifs de la République fédérale
d'Allemagne et l'ancienne République démocratique allemande
selon la période indiquée. En l'absence de données pour
l'Allemagne unifiée (à compter du 3 octobre 1990), les
données disponibles sont fournies séparément sous les
rubriques République fédérale d'Allemagne et ancienne
République démocratique allemande et se rapportent aux
limites territoriales antérieures au 3 octobre 1990. Pour
les notes explicatives en détail sur les données concernant
l'Allemagne, voir Annexe I – Nomenclature des pays.

1 Estimation de la FAO.

47
Coffee
Café

Production: thousand metric tons
Production : milliers de tonnes métriques

Country or area Pays ou zone	1982	1983	1984	1985	1986	1987	1988	1989	1990	1991
World *Monde*	**5 018.3**	**5 618.3**	**5 219.4**	**5 866.4**	**5 298.2**	**6 469.9**	**5 801.2**	**6 113.2**	**6 281.8**	**6 087.8**
Africa *Afrique*	**1 203.2**	**1 122.2**	**1 018.4**	**1 183.2**	**1 256.7**	**1 243.3**	**1 220.6**	**1 269.4**	**1 258.6**	**1 161.8**
Angola Angola	* 17.4	* 13.0	* 15.0	* 12.0	14.8	8.7	8.2	* 5.0	5.0[1]	5.5[1]
Benin Bénin	1.6[1]	1.0[1]	* 3.0	* 0.7	* 1.1	* 1.9	* 2.7	* 0.9	* 0.8	1.0[1]
Burundi Burundi	20.3	36.0	27.0	32.5	31.8	37.3	35.3	32.5	34.0	38.0
Cameroon Cameroun	128.2	63.7	137.9	96.1	132.0	96.0	138.0	* 86.4	101.9	* 58.0
Cape Verde Cap-Vert	...	...	...	0.0	0.1	0.0	0.1	0.0	0.0[1]	...
Central African Rep. Rép. centrafricaine	17.0	15.4[1]	18.4	13.3	20.1	21.3	24.5	20.8	* 16.0	17.0[1]
Comoros[1] Comores[1]	0.1	0.1	0.1	0.1	0.1	0.1	0.1	0.1	0.1	0.1
Congo Congo	* 1.9	* 3.4	* 2.0	* 3.5	1.4	1.0	1.3	1.3	* 1.0	* 1.0
Côte d'Ivoire Côte d'Ivoire	247.7	270.6	85.2	277.1	265.2	270.1	186.7	239.3	* 284.0	* 240.0
Equatorial Guinea[1] Guinée équatoriale[1]	6.7	6.7	6.8	7.0	7.0	7.0	7.0	7.0	7.0	7.0
Ethiopia Ethiopie	202.0	157.8	145.3	155.2	186.0	186.0	190.0	200.0	* 206.3	* 168.0
Gabon Gabon	1.9	1.4	1.2[1]	1.3	1.3	1.7	1.8	1.8[1]	1.8[1]	1.8[1]
Ghana Ghana	* 1.5	* 1.3	* 0.7	* 0.5	0.5	0.8	0.4	* 0.7	* 1.0	* 1.3
Guinea Guinée	14.5	14.7	14.8	14.9	* 6.5	* 6.5	* 10.0	24.2	* 7.5	7.5[1]
Kenya Kenya	87.4	86.1	118.5	93.6	113.9	104.7	128.7	116.9	* 104.7	* 90.0
Liberia Libéria	11.7	* 7.5	* 11.5	* 9.0	* 9.0	* 3.9	* 4.6	4.2[1]	1.5[1]	1.5[1]
Madagascar Madagascar	81.2	80.9	81.4	78.5	82.3	80.8	83.5	88.2	80.0[1]	80.0[1]
Malawi Malawi	* 1.0	* 1.1	* 1.9	* 3.5	3.7	5.0	4.3	* 6.9	* 6.2	* 7.7
Mozambique[1] Mozambique[1]	1.0	1.0	1.0	1.0	1.0	1.0	1.0	1.0	1.0	1.0
Nigeria Nigéria	* 3.0	* 2.5	* 1.3	* 0.8	* 1.2	* 1.5	* 1.5	* 1.5	* 1.2	1.2[1]
Rwanda Rwanda	19.8	33.5	26.2[1]	* 43.0	41.3	41.8	42.7	* 39.0	* 45.0	43.0[1]
Sao Tome and Principe Sao Tomé-et-Principe	0.1[1]	0.1[1]	0.1	0.1[1]	0.1[1]	0.1[1]	0.1[1]	0.1[1]	0.1[1]	0.1[1]
Sierra Leone Sierra Leone	8.6	16.5	18.0	26.0	23.1	24.2	25.3	25.6	25.8	26.2

47

Coffee
Production: thousand metric tons [cont.]
Café
Production : milliers de tonnes métriques [suite]

Country or area Pays ou zone	1982	1983	1984	1985	1986	1987	1988	1989	1990	1991
Togo Togo	9.2	5.9	2.7	10.0	8.2	13.6	14.5	12.1	9.2	12.0[1]
Uganda Ouganda	166.6	157.4	138.7	155.0	143.3	159.4	156.4	174.0	129.4	180.0
United Rep.Tanzania Rép. Unie de Tanzanie	53.1	52.1	56.9	45.8	54.8	57.7	40.9	57.6	52.0	55.9
Zaire Zaïre	93.4	84.2	92.7	91.6	95.0	97.2	99.0	106.9	* 120.0	* 101.7
Zambia Zambie	0.1	0.1	0.2[1]	0.4	0.6	* 0.5	* 0.5	* 1.5	* 1.4	* 1.8
Zimbabwe Zimbabwe	6.1	8.2	10.0	10.7	11.4	13.5	11.6	13.9	14.6	13.6
America, North **Amérique du Nord**	**1 086.5**	**1 054.8**	**1 017.0**	**984.7**	**1 115.0**	**1 077.6**	**1 167.1**	**1 119.0**	**1 263.3**	**1 088.2**
Costa Rica Costa Rica	115.1	124.0	136.9	124.0	128.2	138.0	144.9	157.0	151.1	* 158.0
Cuba Cuba	28.7	18.4	22.1	23.8	24.5	26.2	28.8	28.9	* 27.1	* 26.4
Dominica Dominique	0.2	0.4	0.4	0.4	0.1	0.4	0.4[1]	0.4[1]	0.4[1]	0.4[1]
Dominican Republic Rép. dominicaine	63.5	68.0	72.1	72.0	68.5	67.1	67.9	64.6	59.4	* 45.6
El Salvador El Salvador	174.6	154.6	163.9	148.8	138.2	147.9	120.3	121.9	156.4	149.5
Guatemala Guatemala	189.3	183.0	193.9	180.0	196.6	182.3	179.5	193.2	202.4	* 195.0
Haiti Haïti	32.3	36.0	37.3	36.9	37.8	30.1	37.7	38.4	37.2	37.0[1]
Honduras Honduras	72.4	79.5	72.5	75.1	76.3	79.9	94.0	99.0	118.4	121.5
Jamaica Jamaïque	1.5	1.6	1.7	1.3	1.7	1.7	2.2	* 1.3	* 1.4	* 2.2
Martinique[1] Martinique[1]	0.6	0.5	0.5	0.5	0.5	0.5	0.5	0.5	0.5	0.5
Mexico Mexique	312.7	312.6	239.9	260.2	374.8	* 336.2	423.0	343.4	440.0	299.0
Nicaragua Nicaragua	72.1	49.2	51.3	35.4	43.3	38.6	42.9	43.4	42.9	27.6
Panama Panama	8.1	8.9	10.8	9.4	10.6	10.2	10.3	9.8	* 10.1	* 10.8
Puerto Rico Porto Rico	13.0	15.6	12.2	14.1	11.3	16.0	13.2	14.5	12.9	12.7
St. Vincent-Grenadines[1] St. Vincent-Grenadines[1]	0.1	0.1	0.1	0.1	0.1	0.1	0.1	0.1	0.1	0.2
Trinidad and Tobago Trinité-et-Tobago	1.8	1.4	0.9	2.1	1.3	1.8	0.6	1.2	* 1.9	* 0.8

47
Coffee
Production: thousand metric tons [cont.]
Café
Production : milliers de tonnes métriques [suite]

Country or area Pays ou zone	1982	1983	1984	1985	1986	1987	1988	1989	1990	1991
United States Etats-Unis	0.4	1.0	0.6	0.7	1.1	0.7	0.7	1.2	1.0	1.0[1]
America, South Amérique du Sud	1 989.3	2 703.0	2 510.7	2 871.4	2 078.2	3 178.2	2 437.1	2 546.0	2 647.0	2 679.1
Bolivia Bolivie	21.2	21.1	21.4	23.4	23.6	25.2	26.2	26.7	28.6	30.0
Brazil Brésil	957.9	1 671.6	1 420.3	1 910.6	1 041.4	2 202.7	1 368.8	1 529.8	1 463.1	1 497.2
Colombia Colombie	773.6	768.6	807.8	643.1	713.5	651.6	708.7	664.0	845.0	870.0
Ecuador Equateur	83.9	81.1	97.3	120.9	118.0	* 111.7	144.4	129.3	135.0	* 114.0
Guyana Guyana	* 1.3	* 1.0	0.4	0.4	0.4	* 0.2	* 0.3	* 0.3	* 0.3	* 0.3
Paraguay Paraguay	13.9	15.0	18.4	18.1	19.1	18.3	18.3	17.6	17.6	19.0[1]
Peru Pérou	79.4	85.6	83.4	90.6	95.9	98.2	99.3	105.6	81.0	82.5
Suriname Suriname	0.0	0.0	0.0	0.0	0.0	0.0	0.0	0.0	0.0	0.1[1]
Venezuela Venezuela	58.0	58.8	61.9	64.4	66.2	70.2	71.0	72.6	76.4	* 66.0
Asia Asie	697.8	681.7	629.1	770.2	803.4	907.7	914.0	1 107.8	1 044.7	1 095.9
Cambodia[1] Cambodge[1]	0.1	0.1	0.1	0.1	0.1	0.1	0.1	0.2	0.2	0.2
China[1] Chine[1]	14.1	15.0	17.0	20.0	23.0	26.0	27.0	30.0	33.0	35.0
India Inde	152.1	130.0	105.0	195.1	122.4	192.3	123.0	215.0	* 118.0	173.0
Indonesia Indonésie	281.3	306.0	315.0	311.0	357.5	388.7	391.1	401.0	411.3	* 408.0
Lao People's Dem. Rep. Rép. dém. pop. lao	5.2	5.3	5.8	6.0	4.7	5.3	7.8	5.4	5.2	5.3[1]
Malaysia Malaisie	11.3	11.9	11.9	10.9	10.3	11.5	9.5	6.7	5.2	5.2[1]
Myanmar Myanmar	1.2	1.2	1.2	1.3	1.4	1.4	1.5	1.6	1.4	1.4
Philippines Philippines	171.4	146.9	116.8	135.4	145.3	140.1	141.9	155.9	134.1	113.0[1]
Saudi Arabia[1] Arabie saoudite[1]	0.2	0.2	0.2	0.2	0.2	0.2	0.2	0.2	0.2	0.2
Sri Lanka Sri Lanka	13.2	16.6	10.5	9.3	5.6	6.0	5.6	6.8	7.6	6.8[1]
Thailand Thaïlande	18.5	18.0	18.1	23.5	26.0	20.2	30.7	59.4	61.3	55.0
Viet Nam Viet Nam	25.0	26.6	22.4	52.6	101.8	110.7	169.2	218.9	260.0	285.0[1]

47
Coffee
Production: thousand metric tons [*cont.*]
Café
Production : milliers de tonnes métriques [*suite*]

Country or area Pays ou zone	1982	1983	1984	1985	1986	1987	1988	1989	1990	1991
Yemen Yémen	4.3	3.9	5.2	4.9	5.0	5.1	6.5	6.8	7.4	7.8
Oceania **Océanie**	**41.6**	**56.5**	**44.2**	**56.9**	**44.9**	**63.1**	**62.2**	**70.9**	**68.1**	**62.8**
Fiji Fidji	* 0.0	* 0.0	0.0	0.0[1]	0.0	0.0	0.1	0.1	0.7	0.6[1]
French Polynesia Polynésie française	0.1	0.1	0.1	0.1	0.0	0.0	0.0	0.0	0.0	0.0[1]
New Caledonia Nouvelle-Calédonie	0.3	0.4	0.3	0.6	0.5	0.4	0.1	0.3	0.3	0.3[1]
Papua New Guinea Papouasie-Nvl-Guinée	41.1	55.9	43.7	56.2	44.3	* 62.5	* 62.0	* 70.5	* 67.1	* 61.8
Vanuatu Vanuatu	0.0	0.0	0.1	0.1	0.1	0.1	0.0	0.0	0.0[1]	0.0[1]

Source:
Food and Agriculture Organization of the United Nations
(Rome).

1 FAO estimate.

Source:
Organisation des Nations Unies pour l'alimentation et
l'agriculture (Rome).

1 Estimation de la FAO.

48
Coffee
Café
Apparent consumption
Consommation apparente

Country or area Pays ou zone	Total: thousand metric tons Totale : milliers de tonnes métriques					Per capita: kilograms Par habitant : kilogrammes				
	1980	1988	1989	1990	1991	1980	1988	1989	1990	1991
Austria Autriche	52.3	60.9	80.2	80.4	78.1	6.93	8.01	10.53	10.32	9.99
Belgium-Luxembourg Belgique-Luxembourg	72.8	72.6	62.2	23.5	38.5	7.13	7.08	6.08	2.30	3.76
Cyprus Chypre	1.7	2.9	2.3	2.0	2.8	2.76	4.26	3.30	2.91	4.03
Denmark Danemark	56.5	52.3	55.1	51.9	54.5	11.03	10.20	10.74	10.10	10.61
Fiji Fidji	0.1	0.1	0.1	0.1	0.1	0.19	0.08	0.08	0.08	0.08
Finland Finlande	63.7	57.8	63.5	64.2	57.9	13.32	11.68	12.80	12.87	11.58
France France	316.4	323.0	317.4	312.2	333.4	5.87	5.78	5.65	5.53	5.89
Germany Allemagne	463.5	580.6	592.8	544.7	628.6	5.92	7.44	7.54	6.82	7.86
Greece Grèce	25.6	31.7	34.9	37.9	23.2	2.66	3.17	3.48	3.74	2.29
Ireland Irlande	3.5	6.5	5.8	7.1	6.8	1.02	1.85	1.66	2.04	1.94
Italy Italie	220.7	253.0	258.8	291.5	253.7	3.91	4.40	4.50	5.06	4.39
Japan Japon	201.8	305.2	306.0	314.2	362.3	1.73	2.49	2.49	2.54	2.92
Netherlands Pays-Bas	108.9	146.8	134.6	153.2	149.2	7.70	9.95	9.08	10.25	9.90
Norway Norvège	39.8	38.6	42.7	43.6	44.0	9.74	9.16	10.09	10.29	10.35
Portugal Portugal	9.2	25.3	26.9	31.3	31.3	0.94	2.46	2.57	2.97	2.96
Spain Espagne	85.7	140.5	155.5	162.8	161.1	2.28	3.62	4.00	4.18	4.08
Sweden Suède	95.0	93.5	92.9	101.4	96.2	11.43	11.08	10.95	11.85	11.18
Switzerland Suisse	40.3	51.3	56.1	54.5	56.9	6.38	7.78	8.44	8.12	8.41
United Kingdom Royaume-Uni	121.3	139.9	130.6	140.9	140.5	2.15	2.45	2.28	2.45	2.44
United States Etats-Unis	1 044.8	1 073.3	1 112.6	1 138.4	1 133.4	4.63	4.40	4.51	4.59	4.52

Source:
International Coffee Organization (London).

Source:
Organisation internationale du café (Londres).

49

Tea

Thé

Production: thousand metric tons
Production : milliers de tonnes métriques

Country or area Pays ou zone	1982	1983	1984	1985	1986	1987	1988	1989	1990	1991
World *Monde*	1 949	2 047	2 187	2 291	2 288	2 391	2 461	2 428	2 533	2 576
Africa **Afrique**	215	226	233	268	261	267	282	304	325	335
Burundi Burundi	2	2	3	4	4	4	4	4	4	4[1]
Cameroon Cameroun	2	2	2	3	3	3	* 3	3[1]	3[1]	3[1]
Ethiopia Ethiopie	0[1]	0[1]	0[1]	0[1]	0[1]	0[1]	0	0[1]	1[1]	1[1]
Kenya Kenya	96	119	116	147	143	156	164	181	197	* 204
Malawi Malawi	38	32	38	40	39	32	40	39	39	* 41
Mauritius Maurice	5	6	8	8	8	7	7	6	6	6[1]
Mozambique Mozambique	* 21	* 15	* 11	* 7	* 5	* 3	* 2	2[1]	2[1]	2[1]
Rwanda Rwanda	7	7	6	8	10	12	12	13	13[1]	13[1]
South Africa Afrique du Sud	7	7	8	9	* 10	* 10	* 10	13	13[1]	14[1]
Uganda Ouganda	3	3	5	6	3	4	4	5	7	8
United Rep.Tanzania Rép. Unie de Tanzanie	18	15	17	16	14	17	16	17	20	21[1]
Zaire Zaïre	4	5	5	5	5	3	3	3	3[1]	3[1]
Zambia Zambie	0	0	0	0	0	0	1	* 1	1[1]	1[1]
Zimbabwe Zimbabwe	11	11	12	14	* 16	* 15	* 17	18	17	* 16
America, North **Amérique du Nord**	0	0	0	0	0	0	1	1	1	1
America, South **Amérique du Sud**	49	54	56	63	59	63	47	54	60	65
Argentina Argentine	33	41	41	47	41	45	32	38	* 43	48[1]
Bolivia Bolivie	2[1]	2[1]	1	1	1	2	2	2	2	3
Brazil Brésil	11	9	11	11	12	12	10	10	* 10	* 10
Ecuador Equateur	2	1	1	1	1	1	1	1	1	* 1
Peru Pérou	2	2	2	* 3	* 3	* 3	* 3	3[1]	3[1]	3[1]
Asia **Asie**	1 536	1 611	1 736	1 799	1 813	1 896	2 000	1 930	2 004	2 049
Bangladesh Bangladesh	41	42	38	43	38	38	41	44	39	38[1]

49

Tea
Production: thousand metric tons [cont.]
Thé
Production : milliers de tonnes métriques [suite]

Country or area Pays ou zone	1982	1983	1984	1985	1986	1987	1988	1989	1990	1991
China Chine	421	425	438	456	484	535	569	557	562	566
India Inde	561	581	640	656	621	665	701	684	715	* 730
Indonesia Indonésie	93	110	126	127	136	126	134	141	149	158
Iran, Islamic Rep. of Iran, Rép. islamique d'	35	36	43	33	41	43	43	28	* 44	* 45
Japan Japon	99	103	93	96	94	96	90	91	90	90[1]
Korea, Republic of Corée, République de	0[1]	0[1]	0[1]	1[1]	1[1]	0	0	0	0	0[1]
Lao People's Dem. Rep. Rép. dém. pop. lao	0	0	0	0	1	1	1	1	2	2[1]
Malaysia Malaisie	3	3	5	4[1]	5	5	5	5	* 5	5[1]
Nepal Népal	1	1	1	1	1	1	1[1]	1[1]	1[1]	1[1]
Sri Lanka Sri Lanka	188	179	208	214	211	213	227	207	238	* 241
Thailand Thaïlande	* 2	* 2	2[1]	3	* 3	3[1]	5	5	5[1]	5[1]
Turkey Turquie	68	101	114	137	148	141	153	137	123	136
Viet Nam Viet Nam	25	27	27	28	30	29	30	30	31	32[1]
Oceania **Océanie**	**8**	**10**	**10**	**8**	**8**	**8**	**8**	**8**	**8**	**9**
Papua New Guinea Papouasie-Nvl-Guinée	8	10	10	8	8	8	8	8[1]	8[1]	9[1]
former USSR **ancienne URSS**	**140**	**146**	**151**	**152**	**146**	**156**	**123**	**131**	**136**	**118[1]**

Source:
Food and Agriculture Organization of the United Nations
(Rome).

1 FAO estimate.

Source:
Organisation des Nations Unies pour l'alimentation et
l'agriculture (Rome).

1 Estimation de la FAO.

50
Tobacco
Tabac
Production: thousand metric tons
Production : milliers de tonnes métriques

Country or area Pays ou zone	1982	1983	1984	1985	1986	1987	1988	1989	1990	1991
World *Monde*	6 934.2	5 979.7	6 518.0	7 047.5	6 068.6	6 170.6	6 864.6	7 061.6	7 076.2	7 662.3
Africa Afrique	269.7	292.8	333.2	314.5	303.7	320.2	323.4	357.4	377.9	445.6
Algeria Algérie	4.5	3.3	6.3	4.0	3.9	4.6	3.5	3.5	3.6	* 5.0
Angola Angola	3.0[1]	3.0[1]	3.0[1]	* 3.0	* 3.0	* 4.0	* 5.0	* 4.0	* 4.0	* 4.0
Benin Bénin	0.2	0.2	0.3	0.4	0.4	0.3	0.4	0.3	0.3[1]	0.3[1]
Burkina Faso[1] Burkina Faso[1]	1.2	1.2	1.2	1.2	1.2	1.2	1.2	1.2	1.2	1.2
Burundi Burundi	3.4	3.5	3.0	3.6	3.7	4.0	4.0[1]	4.0[1]	4.0[1]	4.0[1]
Cameroon Cameroun	1.8	1.8	2.2	1.6	3.3	1.7	* 2.0	* 2.0	3.0[1]	4.0[1]
Central African Rep. Rép. centrafricaine	1.2	0.7	0.8	1.1	1.1	0.6	0.6	0.4	0.5[1]	0.5[1]
Chad[1] Tchad[1]	0.2	0.2	0.2	0.2	0.2	0.2	0.2	0.2	0.2	0.2
Congo Congo	0.4	0.2	0.3	0.1	0.1	0.1	0.1[1]	* 2.0	* 2.0	* 2.0
Côte d'Ivoire Côte d'Ivoire	1.8	1.8[1]	1.9[1]	2.0[1]	2.0[1]	2.0[1]	2.0[1]	* 2.0	* 2.0	* 2.0
Ethiopia Ethiopie	2.9[1]	2.9[1]	3.0[1]	3.1[1]	3.3[1]	3.4[1]	3.5[1]	* 4.0	* 4.0	* 4.0
Ghana Ghana	1.2[1]	1.8[1]	2.4[1]	2.8	* 2.0	* 2.0	* 2.0	* 2.0	* 2.0	* 2.0
Guinea[1] Guinée[1]	1.6	1.7	1.7	1.7	1.8	1.8	1.8	1.8	1.8	1.8
Kenya Kenya	5.2	6.6	8.6	5.1	5.1	* 7.0	* 8.0	* 10.6	* 8.8	* 10.0
Libyan Arab Jamah.[1] Jamah. arabe libyenne[1]	0.9	1.0	1.0	1.0	1.0	1.1	1.2	1.2	1.3	1.3
Madagascar Madagascar	2.6	2.1	3.7	3.8	5.5	4.5	2.5	3.9	4.0[1]	4.2[1]
Malawi Malawi	58.6	72.4	73.3	73.4	63.5	72.5	75.0	86.3	101.0	* 125.4
Mali[1] Mali[1]	0.5	0.5	0.5	0.5	0.6	0.6	0.6	0.6	0.6	0.6
Mauritius Maurice	0.6	0.8	0.9	0.8	0.9	0.9	1.0	1.0	0.8	0.8[1]
Morocco Maroc	* 9.6	9.5[1]	9.4[1]	9.3[1]	* 4.0	* 5.0	* 6.0	6.6	7.3	* 7.0
Mozambique Mozambique	3.0[1]	3.0[1]	3.0[1]	3.0[1]	3.0[1]	3.0[1]	3.0[1]	3.0[1]	* 3.0	* 3.0
Niger Niger	0.8[1]	0.8[1]	0.9[1]	0.8	0.9[1]	0.9[1]	0.9[1]	1.0[1]	1.0[1]	1.0[1]
Nigeria Nigéria	* 4.1	* 2.5	* 9.5	* 10.5	* 9.0	* 8.0	* 10.0	9.9	* 9.0	* 9.0
Réunion Réunion	0.2	0.2	0.2	0.2	0.2	0.2	0.2	0.2[1]	0.2[1]	0.2[1]

50

Tobacco
Production: thousand metric tons [cont.]
Tabac
Production : milliers de tonnes métriques [suite]

Country or area Pays ou zone	1982	1983	1984	1985	1986	1987	1988	1989	1990	1991
Rwanda[1] Rwanda[1]	2.4	2.5	2.6	3.0	3.0	3.1	3.2	3.3	3.4	3.5
Sierra Leone[1] Sierra Leone[1]	0.5	0.5	0.5	0.5	0.6	0.7	0.7	0.7	0.7	0.7
Somalia[1] Somalie[1]	0.1	0.1	0.1	0.1	0.1	0.1	0.1	0.1	0.1	0.1
South Africa Afrique du Sud	34.2	38.7	35.9	37.3	32.2	27.2	29.8	34.2	33.9	* 35.0
Swaziland Swaziland	0.2	0.2	0.1	0.1	0.1	0.1	0.1[1]	0.1[1]	0.1[1]	0.1[1]
Togo Togo	2.0[1]	2.0[1]	2.0[1]	2.0[1]	2.0[1]	2.0[1]	2.0[1]	* 2.0	* 2.0	* 2.0
Tunisia Tunisie	5.5	4.4	3.4	4.0	4.5	5.4	5.0	5.5	6.8	5.8[1]
Uganda Ouganda	0.6	1.7	2.0	1.6	0.9	1.2	2.6	3.5	3.3	* 4.0
United Rep.Tanzania Rép. Unie de Tanzanie	13.6	11.0	13.4	12.6	16.5	12.9	14.5	14.7	* 14.0	* 14.0
Zaire Zaïre	7.8	8.1	8.2	8.2	3.2	2.9	2.8	2.9	* 4.0	4.0
Zambia Zambie	2.6	2.9	3.1	2.7	3.9	3.6	4.4	3.6	4.4	5.0
Zimbabwe Zimbabwe	90.6	99.0	124.9	109.1	116.9	131.5	123.7	135.2	139.8	178.0
America, North **Amérique du Nord**	**1 153.3**	**906.0**	**1 016.3**	**930.7**	**753.4**	**739.9**	**840.7**	**837.6**	**947.8**	**932.6**
Canada Canada	70.2	111.7	89.4	96.7	67.5	61.4	58.3	75.6	63.1	* 67.7
Costa Rica Costa Rica	1.3	1.9	2.4	2.0	1.6	1.7	1.8	* 1.4	* 1.6	* 1.2
Cuba Cuba	44.9	30.2	44.6	44.6	45.6	38.8	39.4	41.6	* 44.0	* 44.0
Dominican Republic Rép. dominicaine	34.2	33.6	27.9	31.4	26.0	28.7	28.0	29.7	19.3	25.3
El Salvador El Salvador	4.4	4.7	1.4	1.3	1.1	1.4	* 1.0	* 0.7	* 0.7	* 0.7
Guatemala Guatemala	* 9.7	9.0	10.3	* 8.7	* 7.4	10.1	11.5	8.0	* 6.8	* 6.5
Haiti Haïti	* 0.6	* 0.7	* 0.7	* 0.7	* 0.7	* 0.7	* 0.7	* 0.7	0.7[1]	0.7[1]
Honduras Honduras	7.4	7.4	5.4	5.3	4.7	4.2	3.6	4.8	5.5	6.8
Jamaica * Jamaïque *	1.8	1.5	2.4	1.4	2.3	2.3	2.3	2.3	2.3	2.3
Mexico Mexique	69.5	53.3	41.4	47.8	64.0	47.0	69.0	49.0	62.0	20.0
Nicaragua Nicaragua	2.9	2.4	4.7	3.4	2.7	2.9	2.1	1.7	2.1	2.5
Panama[2] Panama[2]	1.4	1.1	1.3	1.4	1.5	1.2	1.5	1.8	1.8[1]	1.8[1]
Puerto Rico Porto Rico	0.2	0.2	0.4	0.3	0.2	0.1	0.1	0.1	0.0	0.0

50
Tobacco
Production: thousand metric tons [*cont.*]
Tabac
Production : milliers de tonnes métriques [*suite*]

Country or area Pays ou zone	1982	1983	1984	1985	1986	1987	1988	1989	1990	1991
St. Vincent-Grenadines St. Vincent-Grenadines	0.1	0.1	0.1	0.1	0.1	0.1	0.1	0.0	0.0[1]	0.0[1]
Trinidad and Tobago Trinité-et-Tobago	0.1	0.0	0.1	0.0	0.1	0.1	0.1	0.1	0.1	0.1[1]
United States Etats-Unis	904.7	648.2	783.8	685.7	528.0	539.3	621.2	620.2	737.7	753.0
America, South Amérique du Sud	**582.4**	**560.9**	**579.3**	**555.1**	**532.8**	**551.6**	**580.8**	**597.4**	**589.1**	**591.2**
Argentina Argentine	68.7	74.4	74.8	60.5	66.4	71.0	72.2	* 80.5	* 68.0	* 94.0
Bolivia Bolivie	1.1	1.2	1.1	1.0	0.8	0.9	1.0	1.1	* 1.0	* 1.0
Brazil Brésil	420.3	392.6	413.6	410.5	386.8	397.5	431.0	446.0	444.4	413.8
Chile Chili	5.7	5.8	8.1	7.8	7.8	8.3	10.4	10.5	14.4	15.4
Colombia Colombie	48.7	47.8	34.6	27.2	28.6	34.9	35.7	33.7	33.0	40.5
Ecuador Equateur	3.3	1.8	4.4	3.1	3.5	3.7	3.5	3.2	3.6	* 4.0
Guyana Guyana	* 0.1	0.1	0.2	0.3	0.1	* 0.1	* 0.1	* 0.1	* 0.1	0.1[1]
Paraguay Paraguay	14.5	18.6	22.1	24.9	17.5	14.5	6.8	2.4	5.1	5.0[1]
Peru Pérou	2.4	2.4	4.0	* 3.1	* 3.1	* 3.1	* 3.1	* 3.1	* 3.1	* 3.1
Uruguay Uruguay	* 1.4	1.3	* 1.4	* 1.4	* 1.4	* 1.4	1.8	1.9	1.9	2.0
Venezuela Venezuela	16.3	15.0	14.8	15.5	16.7	16.2	15.3	14.9	14.6	* 12.4
Asia Asie	**3 817.2**	**3 103.2**	**3 391.6**	**4 041.1**	**3 262.1**	**3 434.5**	**4 098.0**	**4 326.4**	**4 235.5**	**4 735.3**
Bangladesh Bangladesh	51.0	50.4	47.8	49.3	46.5	40.0	41.5	39.3	37.8	36.0[1]
Bhutan Bhoutan	0.1[1]	0.1[1]	0.1	0.1	0.1[1]	0.1[1]	...	...	...	...
Cambodia[1] Cambodge[1]	5.0	7.0	7.5	9.0	9.5	10.0	11.0	12.0	13.0	14.0
China Chine	2 204.6	1 403.3	1 815.5	2 450.2	1 731.5	1 967.0	2 754.2	2 848.8	* 2 616.5	* 3 121.0
Cyprus Chypre	0.2	0.2	0.3	0.3	0.3	0.4	0.2	0.3	0.3	0.3[1]
India Inde	520.1	581.6	492.5	485.9	441.2	461.8	367.4	492.8	550.0	560.0
Indonesia Indonésie	109.0	110.0	108.0	161.0	164.5	112.7	116.9	80.9	149.6	* 159.0
Iran, Islamic Rep. of Iran, Rép. islamique d'	25.0	21.1	22.0	28.2	27.7	25.3	20.7	15.4	* 25.0	* 25.0
Iraq Iraq	12.3	14.3	13.6	17.0	12.8	9.1	1.8	3.2	4.4	* 2.0
Israel Israël	0.5	0.6	0.6	0.6	0.6	0.5	0.3	0.3	0.2	0.2[1]

50
Tobacco
Production: thousand metric tons [*cont.*]
Tabac
Production : milliers de tonnes métriques [*suite*]

Country or area Pays ou zone	1982	1983	1984	1985	1986	1987	1988	1989	1990	1991
Japan Japon	139.4	136.7	135.5	116.2	116.8	104.4	85.8	74.4	77.5	* 71.0
Jordan Jordanie	3.5	5.9	2.5	2.8	1.7	2.9	3.7	2.9	2.9	2.9[1]
Korea, Dem. P. R.[1] Corée, R. p. dém. de[1]	48.0	50.0	52.0	56.0	58.0	60.0	62.0	64.0	65.0	66.0
Korea, Republic of Corée, République de	115.2	100.7	94.2	75.7	83.0	78.0	73.0	78.4	70.1	* 73.4
Lao People's Dem. Rep. Rép. dém. pop. lao	3.1	2.2	2.3	3.1	2.5[1]	3.6[1]	3.8[1]	4.2[1]	5.0[1]	5.5[1]
Lebanon Liban	4.1	4.1	* 4.0	* 3.5	3.5[1]	3.3	3.1	2.2	1.7	1.8[1]
Malaysia Malaisie	8.6	8.7	6.5	9.4	14.3	11.4	7.3	13.9	11.0	* 11.0
Myanmar Myanmar	51.8	58.0	64.8	69.4	74.1	65.0	48.8	50.3	40.0	40.3
Nepal Népal	4.8	6.6	6.9	6.4	4.7	4.9	4.5	5.4	6.6	7.0
Oman Oman	1.0[1]	1.0[1]	1.9	1.8	2.0	2.1	2.1	0.7	2.0	2.1[1]
Pakistan Pakistan	69.2	64.7	79.6	87.2	78.2	69.2	69.5	73.9	68.1	* 75.9
Philippines Philippines	86.1	86.3	98.1	74.3	74.2	82.8	76.4	79.9	81.7	* 78.8
Sri Lanka Sri Lanka	16.7	14.5	15.3	15.3	15.1	12.0	12.9	11.0	* 10.3	10.5[1]
Syrian Arab Republic Rép. arabe syrienne	13.7	14.4	13.3	13.9	16.6	16.8	15.4	13.0	13.1	16.4
Thailand Thaïlande	86.0	93.0	* 90.0	* 90.0	* 85.0	* 67.0	* 54.2	* 60.2	* 70.9	* 70.5
Turkey Turquie	207.7	233.8	177.5	170.5	158.5	184.7	219.1	269.9	287.5	247.4
United Arab Emirates Emirats arabes unis	1.4	2.3[1]	1.0	0.7	0.6[1]	0.5	0.5	0.6[1]	0.7[1]	* 2.0
Viet Nam Viet Nam	21.2	24.9	33.0	38.2	33.4	33.4	35.6	23.9	17.6	* 28.0
Yemen Yémen	7.5	6.4	5.0	5.1	5.1	5.3	6.4	4.4	6.8	7.0
Europe **Europe**	**789.2**	**716.2**	**803.0**	**810.0**	**830.2**	**806.9**	**761.7**	**696.8**	**652.3**	**702.3**
Albania Albanie	17.5	17.5	* 19.0	* 20.0	* 20.0	* 20.0	* 26.2	* 20.0	14.8[1]	11.5[1]
Austria Autriche	0.4	0.5	0.5	0.5	0.5	0.4	0.5	0.4	0.4	* 0.4
Belgium-Luxembourg Belgique-Luxembourg	1.7	1.6	1.8	1.9	2.0	1.1	1.5	1.5	1.5	1.5[1]
Bulgaria Bulgarie	149.0	112.0	140.8	126.0	126.1	133.1	116.1	81.1	76.5	74.1
Czechoslovakia Tchécoslovaquie	6.2	5.8	5.3	6.1	5.1	6.1	5.4	5.5	5.0	5.1
France France	43.0	36.2	35.5	35.8	37.5	37.6	32.6	28.6	27.6	* 27.0

50

Tobacco
Production: thousand metric tons [cont.]
Tabac
Production : milliers de tonnes métriques [suite]

Country or area Pays ou zone	1982	1983	1984	1985	1986	1987	1988	1989	1990	1991
Germany† · Allemagne† F. R. Germany R. f. Allemagne	8.1	6.8	7.2	8.1	7.8	6.2	7.1	7.0	6.3	5.8[1]
former German D. R. anc. R. d. allemande	3.9	4.5	5.3	5.7	6.0	5.5	5.7	5.4	4.1	0.7[1]
Greece Grèce	138.4	115.8	145.3	150.6	161.0	155.0	149.0	133.0	130.0	* 177.7
Hungary Hongrie	24.7	20.5	20.2	21.0	21.3	19.7	16.0	14.7	13.6	* 19.0
Italy Italie	145.0	156.0	161.3	166.5	145.7	162.1	184.4	197.3	194.0	192.0[1]
Poland Pologne	95.9	99.8	98.3	111.5	125.2	113.9	90.4	56.1	59.0	57.4
Portugal Portugal	1.8	2.6	3.5	3.9	4.3	3.8	3.9	4.8	4.9	* 6.0
Romania Roumanie	32.9	24.7	37.1	26.1	32.1	33.1	35.8	27.5	32.0[1]	34.0[1]
Spain Espagne	42.2	43.2	43.2	42.2	37.7	31.9	33.7	55.1	* 35.7	* 43.2
Switzerland Suisse	1.6	1.7	1.8	1.5	1.7	1.3	1.5	1.6	1.1	1.3
Yugoslavia Yougoslavie	77.0	67.0	77.0	82.8	96.3	76.0	52.0	57.3	45.7	45.5[1]
Oceania **Océanie**	**15.4**	**15.7**	**16.7**	**15.1**	**12.4**	**14.7**	**14.5**	**12.7**	**13.6**	**15.3**
Australia Australie	13.3	13.4	14.4	12.5	10.7	12.2	12.7	11.4	12.3	14.0
Fiji Fidji	0.4	0.3	0.2	0.4	0.3	0.2	0.1	0.2	0.2	0.2
New Zealand Nouvelle-Zélande	1.5	1.7	1.8	2.0	1.2	2.0	1.5	0.9	0.8	0.8[1]
Samoa[1] Samoa[1]	0.2	0.2	0.2	0.2	0.2	0.2	0.2	0.2	0.2	0.2
Solomon Islands[1] Iles Salomon[1]	0.1	0.1	0.1	0.1	0.1	0.1	0.1	0.1	0.1	0.1
former USSR **ancienne URSS**	**307.0**	**385.0**	**378.0**	**381.0**	**374.0**	**303.0**	**245.4**	**233.2**	***260.0**	**240.0**[1]

Source:
Food and Agriculture Organization of the United Nations
(Rome).

† All data shown which pertain to Germany prior to 3 October
1990 are indicated separately for the Federal Republic of
Germany and the former German Democratic Republic based on
their respective territories at the time indicated. Where
data for united Germany (3 October 1990 and thereafter) are
not available, available data are shown separately under the
designations Federal Republic of Germany and former German
Democratic Republic and pertain to the territorial
boundaries prior to 3 October 1990. For detailed
explanatory notes on data pertaining to Germany, see Annex I
- Country Nomenclature.

1 FAO estimate.
2 Excluding former Canal Zone.

Source:
• Organisation des Nations Unies pour l'alimentation et
l'agriculture (Rome).

† Toutes les données se rapportant à l'Allemagne avant le 3
octobre 1990 figurent dans deux rubriques séparées basées
sur les territoires respectifs de la République fédérale
d'Allemagne et l'ancienne République démocratique allemande
selon la période indiquée. En l'absence de données pour
l'Allemagne unifiée (à compter du 3 octobre 1990), les
données disponibles sont fournies séparément sous les
rubriques République fédérale d'Allemagne et ancienne
République démocratique allemande et se rapportent aux
limites territoriales antérieures au 3 octobre 1990. Pour
les notes explicatives en détail sur les données concernant
l'Allemagne, voir Annexe I – Nomenclature des pays.
1 Estimation de la FAO.
2 Non compris l'ancienne Zone du Canal.

51
Cotton (lint)
Coton (fibre)
Production: thousand metric tons
Production : milliers de tonnes métriques

Country or area Pays ou zone	1982	1983	1984	1985	1986	1987	1988	1989	1990	1991
World *Monde*	**14 888**	**14 264**	**18 229**	**17 360**	**15 229**	**16 586**	**18 325**	**17 073**	**18 447**	**20 641**
Africa **Afrique**	**1 136**	**1 176**	**1 243**	**1 346**	**1 368**	**1 354**	**1 423**	**1 351**	**1 274**	**1 375**
Angola[1] Angola[1]	11	11	11	11	11	11	11	11	11	11
Benin * Bénin *	12	17	33	34	48	27	44	43	59	67
Botswana[1] Botswana[1]	1	1	1	1	1	1	1	1	1	1
Burkina Faso Burkina Faso	29	30	34	46	66	59	59	55	* 77	* 77
Burundi Burundi	2	2	2	3	3	3	3	3	2	* 3
Cameroon Cameroun	29	37	* 38	* 46	* 48	* 45	* 75	* 42	* 44	35[1]
Central African Rep. Rép. centrafricaine	* 13	* 12	* 17	13	10	8	11	11	13	* 11
Chad Tchad	38	60	36	39	34	48	53	58	* 60	* 60
Côte d'Ivoire Côte d'Ivoire	56	66	58	88	82	93	114	128	116	* 133
Egypt Egypte	460	400	399	447	419	365	322	296	303	* 294
Ethiopia * Ethiopie *	27	20	20	22	20	20	21	18	19	19
Gambia Gambie	1	0[1]	1	1	1[1]	0[1]	1[1]	1[1]	1[1]	1[1]
Ghana Ghana	* 2	* 2	* 3	1[1]	3	3	2	3	5	8
Guinea Guinée	...	0[1]	0[1]	0[1]	0[1]	0	1	* 1	* 1	* 2
Guinea-Bissau Guinée-Bissau	2	1	2	2[1]	2[1]	2[1]	2[1]	1	* 1	1[1]
Kenya Kenya	8	9	8	* 13	8	8	3	* 4	* 9	* 10
Madagascar Madagascar	9	10	13	* 16	* 19	* 10	* 12	* 15	* 13	* 13[1]
Malawi Malawi	6	3	8	12	9	5	* 8	* 9	* 9	* 11
Mali Mali	38	50	54	55	67	79	75	97	99	* 115
Morocco * Maroc *	7	7	4	8	8	13	11	11	12	12
Mozambique Mozambique	* 15	* 15	* 5	* 11	* 30	* 32	* 29	* 27	* 28	25[1]
Niger Niger	1	2	1	2	3	* 3	* 3	2	1	* 2
Nigeria Nigéria	21	20	13	11	36	54	* 48	35	* 36	* 45

51
Cotton (lint)
Production: thousand metric tons [cont.]
Coton (fibre)
Production : milliers de tonnes métriques [suite]

Country or area Pays ou zone	1982	1983	1984	1985	1986	1987	1988	1989	1990	1991
Senegal * Sénégal *	19	12	19	11	11	15	15	12	18	18
Somalia Somalie	2	1	1	1	1	2	* 2	* 1	* 1	* 1
South Africa Afrique du Sud	33	24	30	40	41	56	69	54	* 53	* 61
Sudan Soudan	154	201	219	203	145	163	136	* 142	* 83	* 91
Swaziland[1] Swaziland[1]	11	11	11	11	11	11	11	11	11	10
Togo Togo	* 8	* 12	* 11	23	26	* 32	* 31	* 33	* 34	* 42
Uganda Ouganda	5	10	12	3	5	3	2	3	4	* 8
United Rep.Tanzania Rép. Unie de Tanzanie	43	47	46	35	73	63	* 85	* 84	* 47	* 64
Zaire Zaïre	23	26	26[1]	26[1]	26[1]	26[1]	26[1]	26[1]	26[1]	26[1]
Zambia Zambie	5	7	16	11	* 12	* 7	* 21	* 17	* 10	* 25
Zimbabwe Zimbabwe	46	50	* 91	* 103	* 89	* 87	* 116	94	* 67	73[1]
America, North **Amérique du Nord**	**2 965**	**2 098**	**3 305**	**3 317**	**2 392**	**3 572**	**3 768**	**2 897**	**3 652**	**4 099**
Costa Rica Costa Rica	2	1	1	1	1	0	1	* 1	* 1	* 1
Dominican Republic * Rép. dominicaine *	2	2	2	2	2	2	2	2	2	2
El Salvador El Salvador	40	41	30	30	18	11	10	7	6	4
Guatemala Guatemala	78	47	61	63	54	28	48	42	41	38
Haiti[1] Haïti [1]	2	2	2	2	1	1	1	1	1	1
Honduras * Honduras *	7	4	6	5	3	3	3	1	2	2
Mexico Mexique	166	229	289	220	144	263	312	162	201	202
Nicaragua Nicaragua	64	81	87	69	51	50	36	26	23	30
United States Etats-Unis	2 605	1 692	2 827	2 924	2 119	3 214	3 355	2 655	3 375	3 819
America, South **Amérique du Sud**	**1 025**	**864**	**1 213**	**1 532**	**1 246**	**963**	**1 532**	**1 295**	**1 404**	**1 500**
Argentina Argentine	152	112	180	171	120	100	282	* 195	* 270	* 290
Bolivia Bolivie	4	3	2	5	4	2	4	1	2	9
Brazil * Brésil *	636	527	713	943	764	552	837	625	660	700

51
Cotton (lint)
Production: thousand metric tons [cont.]
Coton (fibre)
Production : milliers de tonnes métriques [suite]

Country or area Pays ou zone	1982	1983	1984	1985	1986	1987	1988	1989	1990	1991
Colombia Colombie	* 43	* 77	* 112	118	107	* 134	* 97	* 105	* 140	* 142
Ecuador Equateur	* 9	* 1	* 3	* 7	* 13	6	8	12	13	* 12
Paraguay Paraguay	91	* 96	* 121	166	107	84	187	220	* 215	* 259
Peru * Pérou *	85	35	67	96	100	67	93	103	73	65
Venezuela Venezuela	6	12	* 15	* 27	* 30	* 18	* 24	* 34	* 31	* 23
Asia Asie	**6 671**	**7 252**	**9 778**	**7 885**	**7 013**	**7 735**	**8 192**	**8 232**	**8 884**	**10 531**
Afghanistan Afghanistan	15	17	23	* 19	* 18	* 14	* 16	12	9	12[1]
Bangladesh Bangladesh	10	8	6	5	4	7	7	* 11	* 12	* 12
China Chine	3 598	4 637	6 258	4 147	3 540	4 245	4 149	3 788	4 508	5 663
India Inde	1 280	1 086	1 446	1 484	1 174	1 085	1 486	1 940	1 659	1 700
Indonesia [1] Indonésie [1]	6	7	4	15	18	8	7	13	13	6
Iran, Islamic Rep. of * Iran, Rép. islamique d' *	94	91	112	106	111	106	116	114	138	146
Iraq[1] Iraq[1]	5	4	2	2	7	5	4	5	5	4
Israel Israël	88	93	88	99	69	59	63	46	51	* 22
Korea, Dem. P. R.[1] Corée, R. p. dém. de[1]	4	4	4	5	6	7	8	8	9	10
Korea, Republic of Corée, République de	1	* 1	* 1	* 1	* 0	* 0	* 0	* 0	* 0	* 0
Lao People's Dem. Rep. Rép. dém. pop. lao	5	5	5	5	3	4	4	4	5	5[1]
Myanmar Myanmar	32	33	35	42	33	27	24	20	21	21
Pakistan Pakistan	824	495	1 009	1 217	1 320	1 468	1 426	1 456	1 637	* 2 112
Philippines Philippines	9	6	8	6	3	5	4	3	3	* 7
Sri Lanka Sri Lanka	2	1	0	0	0	0[1]	0[1]	0[1]	...	...
Syrian Arab Republic Rép. arabe syrienne	158	194	* 160	* 170	159	123	180	155	159	200
Thailand Thaïlande	41	40	26	34	19	25	35	29	32	35
Turkey Turquie	489	522	580	518	518	537	650	617	611	565

51
Cotton (lint)
Production: thousand metric tons [*cont.*]
Coton (fibre)
Production : milliers de tonnes métriques [*suite*]

Country or area Pays ou zone	1982	1983	1984	1985	1986	1987	1988	1989	1990	1991
Viet Nam Viet Nam	3	4	5	4	5	4	4	3	4	5[1]
Yemen Yémen	7	7	6	6	7	7	7	7	7	7
Europe **Europe**	**170**	**187**	**206**	**249**	**304**	**290**	**364**	**327**	**294**	**283**
Albania[1] Albanie[1]	7	5	5	6	6	8	5	7	6	5
Bulgaria Bulgarie	7	6	5	5	6	7	4	4	3	5
Greece Grèce	101	134	142	168	* 205	* 194	* 235	* 255	* 209	* 190
Romania[1] Roumanie[1]	0	0	0	0	0	1	1	1	1	1
Spain Espagne	55	41	53	70	86	80	119	60	* 75	* 83
Oceania **Océanie**	**134**	**101**	**141**	**248**	**259**	**213**	**284**	**286**	**305**	**433**
Australia Australie	134	101	141	248	259	213	284	286	305	433
former USSR **ancienne URSS**	**2 786**	**2 586**	**2 343**	**2 782**	**2 647**	**2 460**	**2 762**	**2 686**	*** 2 634**	*** 2 420**

Source:
Food and Agriculture Organization of the United Nations
(Rome).

Source:
Organisation des Nations Unies pour l'alimentation et
l'agriculture (Rome).

† All data shown which pertain to Germany prior to 3 October
1990 are indicated separately for the Federal Republic of
Germany and the former German Democratic Republic based on
their respective territories at the time indicated. Where
data for united Germany (3 October 1990 and thereafter) are
not available, available data are shown separately under the
designations Federal Republic of Germany and former German
Democratic Republic and pertain to the territorial
boundaries prior to 3 October 1990. For detailed
explanatory notes on data pertaining to Germany, see Annex I
- Country Nomenclature.

1 FAO estimate.

† Toutes les données se rapportant à l'Allemagne avant le 3
octobre 1990 figurent dans deux rubriques séparées basées
sur les territoires respectifs de la République fédérale
d'Allemagne et l'ancienne République démocratique allemande
selon la période indiquée. En l'absence de données pour
l'Allemagne unifiée (à compter du 3 octobre 1990), les
données disponibles sont fournies séparément sous les
rubriques République fédérale d'Allemagne et ancienne
République démocratique allemande et se rapportent aux
limites territoriales antérieures au 3 octobre 1990. Pour
les notes explicatives en détail sur les données concernant
l'Allemagne, voir Annexe I – Nomenclature des pays.

1 Estimation de la FAO.

52
Livestock
Cheptel
Thousand head
Milliers de têtes

Country or area	1984	1985	1986	1987	1988	1989	1990	1991	Pays ou zone
World									***Monde***
Cattle	1252528	1260248	1268044	1268194	1263598	1277345	1293641	1294605	**Bovine**
Sheep	1126121	1126004	1131810	1153042	1169357	1199038	1215633	1202920	**Ovine**
Pigs	786043	791332	819960	837344	830259	846976	855870	857099	**Porcine**
Horses	59684	60337	60507	60630	60803	61095	61164	61620	**Chevaline**
Asses	40031	40523	41269	42030	42729	43408	43862	44143	**Asine**
Mules	13791	13953	14119	14359	14521	14682	14775	14925	**Mulassière**
Africa									**Afrique**
Cattle	172148	174705	178461	176370	178837	183721	189655	191471	**Bovine**
Sheep	181766	186356	190526	195329	197049	202346	209376	216510	**Ovine**
Pigs	11092	12579	13241	13943	14298	15463	16232	17268	**Porcine**
Horses	4233	4467	4543	4636	4742	4793	4890	4943	**Chevaline**
Asses	11712	11673	11884	12231	12704	13003	13263	13505	**Asine**
Mules	1196	1224	1260	1306	1300	1314	1344	1378	**Mulassière**
Algeria									**Algérie**
Cattle	1404	1416	1557	1523	1520[1]	* 1410	1427[1]	1443[1]	Bovine
Sheep	14725	15000	14795	14300	* 14325	* 12500	* 13350	13350[1]	Ovine
Pigs[1]	5	5	5	5	5	5	5	5	Porcine[1]
Horses	92	166	170	185	187[1]	190[1]	195[1]	202[1]	Chevaline
Asses	403	351	340	354	327	322	330[1]	340[1]	Asine
Mules	155	135	131	127	113	101	105[1]	107[1]	Mulassière
Angola									**Angola**
Cattle	3350[1]	3360[1]	3400	3300	3200	3100	3100[1]	3100[1]	Bovine
Sheep[1]	245	250	255	260	265	270	275	280	Ovine[1]
Pigs[1]	460	465	470	475	480	485	490	495	Porcine[1]
Horses[1]	1	1	1	1	1	1	1	1	Chevaline[1]
Asses[1]	5	5	5	5	5	5	5	5	Asine[1]
Benin									**Bénin**
Cattle	892	912	895	896	925	932	951	* 955	Bovine
Sheep	1085	1122	830	831	821	890[1]	921[1]	970[1]	Ovine
Pigs	541	570	600[1]	617	648	680	714	730[1]	Porcine
Horses[1]	6	6	6	6	6	6	6	6	Chevaline[1]
Asses[1]	1	1	1	1	1	1	1	1	Asine[1]
Botswana									**Botswana**
Cattle	2685	2459	2332	2263	2408	2543	2696	2500[1]	Bovine
Sheep	167	200	229	240	259	286	317	320[1]	Ovine
Pigs	8	9	11	11	13	15	16	16[1]	Porcine
Horses	23	24	24	24	29	32	34	34[1]	Chevaline
Asses	138	143	143[1]	144[1]	148	151	152[1]	153[1]	Asine
Mules	4[1]	4[1]	3[1]	3[1]	2	2	3[1]	3[1]	Mulassière
Burkina Faso									**Burkina Faso**
Cattle	2986	3045	3106	2754	2809	2850[1]	2900[1]	2900[1]	Bovine
Sheep	2086	2148	2800	2885	2972	3050[1]	3150[1]	3339[1]	Ovine
Pigs	400[1]	450[1]	500	500	500	496	496[1]	520[1]	Porcine
Horses	70	70	70	70[1]	70[1]	70[1]	70[1]	70[1]	Chevaline
Asses	200	200	200	270[1]	330[1]	403	450[1]	476[1]	Asine
Burundi									**Burundi**
Cattle	415	378	479	422	429	423	432	435[1]	Bovine
Sheep	369	316	329	313	350	327	361	365[1]	Ovine
Pigs	76[1]	60	77	80	115	91	103	103[1]	Porcine
Cameroon									**Cameroun**
Cattle	3561	4151	4255	4362	4471	4582	4697	4700[1]	Bovine
Sheep	2100[1]	2248	2473	2597	2897	3170	* 3500	3550[1]	Ovine
Pigs	800	1724	1451	1178	1237	1299	1364	1414[1]	Porcine
Horses	21	20[1]	18[1]	15	13	14[1]	14[1]	15[1]	Chevaline
Asses	42	37[1]	34[1]	25	33	35[1]	35[1]	36[1]	Asine
Cape Verde									**Cap-Vert**
Cattle	9	10	12	12[1]	13	19	19[1]	19[1]	Bovine
Sheep	2	2	2	2[1]	4	6	6[1]	6[1]	Ovine
Pigs	53	58	67	69	67	86	86[1]	86[1]	Porcine
Horses	1[1]	0[1]	0	0[1]	1	1	1[1]	1[1]	Chevaline
Asses	6[1]	6[1]	6[1]	6	7	11	6[1]	11[1]	Asine
Mules	2[1]	1[1]	1	1[1]	1	2	2[1]	2[1]	Mulassière

52
Livestock
Thousand head [*cont.*]
Cheptel
Milliers de têtes [*suite*]

Country or area	1984	1985	1986	1987	1988	1989	1990	1991	Pays ou zone
Central African Rep.									**Rép. centrafricaine**
Cattle	2043	2128	2216	2306	2398	2495	2595	2677	Bovine
Sheep	97	108	113	117	122	128	134	135[1]	Ovine
Pigs	338	344	360	371	382	397	413	426[1]	Porcine
Chad									**Tchad**
Cattle	3705	3794	3886	4002	4098	4197	4297	4400	Bovine
Sheep	1750	1770	1800	1800	1815	1870	1964	1983	Ovine
Pigs	10	11	11	12	13	13	14	15	Porcine
Horses	175	179	182	186	188	192	195	182	Chevaline
Asses	221	223	227	222	234	259	264	269	Asine
Comoros									**Comores**
Cattle	52[1]	50[1]	40[1]	44	45	47	47[1]	47[1]	Bovine
Sheep	9[1]	9[1]	10[1]	12	12	13	13[1]	14[1]	Ovine
Asses[1]	4	4	4	4	4	4	4	5	Asine[1]
Congo									**Congo**
Cattle	66	67	70	70	69	62	68	68[1]	Bovine
Sheep	89	92	94	97	101	104	105[1]	108[1]	Ovine
Pigs	29	35	44	47	50	44	50[1]	52[1]	Porcine
Côte d'Ivoire									**Côte d'Ivoire**
Cattle	820	843	* 885	917	992	1028	1046	1064	Bovine
Sheep	1040[1]	1040[1]	1050[1]	1051	1090	1102	* 1133	* 1150	Ovine
Pigs	328[1]	330[1]	333[1]	335	342	351	360	369	Porcine
Horses[1]	1	1	1	1	1	1	1	1	Chevaline[1]
Asses[1]	1	1	1	1	1	1	1	1	Asine[1]
Djibouti[1]									**Djibouti[1]**
Cattle	94	122	138	143	155	206	170	170	Bovine
Sheep	451	477	410	412	414	400	416	420	Ovine
Asses	7	7	8	8	8	8	8	8	Asine
Egypt									**Egypte**
Cattle	3033	3103	3174	3245	3317	3389	3463	3500[1]	Bovine
Sheep	3472	3576	3683	3793	3908	4026	4806	4900[1]	Ovine
Pigs	63	69	75	81	88	95	102	110[1]	Porcine
Horses	9	9	9	9[1]	10[1]	10[1]	10[1]	10[1]	Chevaline
Asses	1844	1879	1879	1900[1]	1950[1]	1960[1]	1980[1]	2000[1]	Asine
Mules	1	1	1	1[1]	1[1]	1[1]	1[1]	1[1]	Mulassière
Equatorial Guinea[1]									**Guinée équatoriale[1]**
Cattle	4	4	4	5	5	5	5	5	Bovine
Sheep	34	35	35	35	35	35	35	36	Ovine
Pigs	5	5	5	5	5	5	5	5	Porcine
Ethiopia									**Ethiopie**
Cattle	26000[1]	28000[1]	* 30000	* 27000	* 27000	* 28900	30000[1]	30000[1]	Bovine
Sheep	23450[1]	23500[1]	* 23000	* 24000	* 24000	* 24000	* 22960	23000[1]	Ovine
Pigs	19[1]	19[1]	19[1]	19[1]	19[1]	20[1]	20[1]	20[1]	Porcine
Horses	2300[1]	2400[1]	2450[1]	2500[1]	2550[1]	2600[1]	2650[1]	2700[1]	Chevaline
Asses	4400[1]	4500[1]	4600[1]	4700[1]	4800[1]	4900[1]	5000[1]	5100[1]	Asine
Mules	470[1]	490[1]	510[1]	530[1]	550[1]	570[1]	590[1]	610[1]	Mulassière
Gabon									**Gabon**
Cattle	6	11	19	21	24	27	27[1]	28[1]	Bovine
Sheep	130	138	146	149	153	157	160[1]	165[1]	Ovine
Pigs	114	125	136	141	147	159	160[1]	162[1]	Porcine
Gambia									**Gambie**
Cattle	280[1]	300[1]	320[1]	350[1]	387	390[1]	400[1]	410[1]	Bovine
Sheep	148[1]	151[1]	154[1]	157[1]	163	166[1]	170[1]	175[1]	Ovine
Pigs	11[1]	11[1]	11[1]	11[1]	11	11[1]	11[1]	11[1]	Porcine
Ghana									**Ghana**
Cattle	1078	1132	1135	1170	1145	1136	1145	1300[1]	Bovine
Sheep	1900	2000	1814	1989	2046	2212	2224	2500[1]	Ovine
Pigs	407	413	469	399	478	559	474	620[1]	Porcine
Horses	3[1]	3[1]	2	2	2	2	1	1[1]	Chevaline
Asses	15[1]	14[1]	13	14	10	11	10	10[1]	Asine
Guinea									**Guinée**
Cattle[1]	1850	1800	1800	1800	1800	1800	1800	1800	Bovine[1]
Sheep	455[1]	480[1]	460[1]	490[1]	500[1]	506	510[1]	518[1]	Ovine
Pigs	45[1]	45[1]	40[1]	38[1]	35[1]	33	33[1]	33[1]	Porcine
Horses	1[1]	1[1]	1[1]	2[1]	2[1]	2	2[1]	2[1]	Chevaline
Asses	2[1]	2[1]	2[1]	2[1]	1[1]	1	1[1]	1[1]	Asine

52

Livestock
Thousand head [*cont.*]
Cheptel
Milliers de têtes [*suite*]

Country or area	1984	1985	1986	1987	1988	1989	1990	1991	Pays ou zone
Guinea-Bissau									**Guinée-Bissau**
Cattle	318[1]	325[1]	333	350[1]	370[1]	390[1]	410	410[1]	Bovine
Sheep	193[1]	195[1]	200	210[1]	220[1]	230[1]	242	245[1]	Ovine
Pigs	277[1]	282[1]	286	290[1]	290[1]	290[1]	290[1]	293[1]	Porcine
Horses	1[1]	1[1]	1	1[1]	1[1]	1[1]	1[1]	1[1]	Chevaline
Asses	3[1]	3[1]	3	3[1]	3[1]	3[1]	3[1]	3[1]	Asine
Kenya									**Kenya**
Cattle	11000[1]	11500[1]	12000[1]	12645	13050	13457	13793	13700[1]	Bovine
Sheep	6419	* 7000	6500[1]	* 6040	* 6317	* 6325	* 6516	6550[1]	Ovine
Pigs	95	95[1]	92[1]	94	97	100	105	105[1]	Porcine
Horses[1]	2	2	2	2	2	2	2	2	Chevaline[1]
Lesotho									**Lesotho**
Cattle	522	525	525[1]	527[1]	530[1]	530[1]	535[1]	540[1]	Bovine
Sheep	1412	1392	1400[1]	1430	1440[1]	1450[1]	1460[1]	1470[1]	Ovine
Pigs	53	54	60[1]	70[1]	72[1]	73[1]	74[1]	75[1]	Porcine
Horses	115[1]	116[1]	117[1]	118[1]	119[1]	120[1]	121[1]	122[1]	Chevaline
Asses	122[1]	123[1]	124[1]	125[1]	126[1]	127[1]	128[1]	129[1]	Asine
Mules[1]	1	1	1	1	1	1	1	1	Mulassière[1]
Liberia[1]									**Libéria[1]**
Cattle	42	42	42	42	42	40	38	38	Bovine
Sheep	235	238	239	240	240	230	220	220	Ovine
Pigs	120	127	130	140	140	130	120	120	Porcine
Libyan Arab Jamah.									**Jamah. arabe libyenne**
Cattle	190	200	140[1]	90	95	102	120	150[1]	Bovine
Sheep	5000	5500	5000[1]	4500	4500	5000	5200	5500[1]	Ovine
Horses	40	41	30[1]	20	25	20	20	25[1]	Chevaline
Asses[1]	60	60	61	61	61	62	62	62	Asine[1]
Madagascar									**Madagascar**
Cattle	10363	10194	10207	10220	10332	10243	10254	10265[1]	Bovine
Sheep	455	478	540	664	684	700[1]	737[1]	753[1]	Ovine
Pigs	1379	1339	1412	1532	1427	1400	1431	1461[1]	Porcine
Horses	1	0	1	0	0	0	0[1]	0[1]	Chevaline
Malawi									**Malawi**
Cattle	949	1020	1011	1055	860	1000[1]	1100[1]	1150[1]	Bovine
Sheep	150	185	165	180[1]	180[1]	200[1]	220[1]	230[1]	Ovine
Pigs	186	185	282	313	250	260[1]	270[1]	280[1]	Porcine
Asses	0	1	2	1	1	2[1]	2[1]	2[1]	Asine
Mali									**Mali**
Cattle	4899	4344	4475	4589	4738	4880	5000	5000[1]	Bovine
Sheep	* 5200	* 5000	* 5340	* 5329	* 5500	* 5650	* 5850	5850[1]	Ovine
Pigs	52	55	54	56	58	60[1]	60[1]	61[1]	Porcine
Horses	61	55	67	54	56	58[1]	62[1]	62[1]	Chevaline
Asses	549	436	383	348	510	530[1]	550[1]	550[1]	Asine
Mauritania									**Mauritanie**
Cattle	950	1050	1200	1220	1260	1300	1350[1]	1360[1]	Bovine
Sheep	* 3800	* 3900	3950[1]	4000[1]	4100[1]	4200[1]	4200[1]	4200[1]	Ovine
Horses[1]	16	16	16	16	17	17	18	18	Chevaline[1]
Asses	145[1]	147[1]	149[1]	149[1]	149[1]	150[1]	151[1]	153[1]	Asine
Mauritius									**Maurice**
Cattle	27	30	32	32	33	34	33	34[1]	Bovine
Sheep	7	7	7	7	7	7	7	7[1]	Ovine
Pigs	13[1]	10[1]	10	13	10	10	10	10[1]	Porcine
Morocco									**Maroc**
Cattle	2363	2501	2851	3178	3137	3324	* 3400	* 3500	Bovine
Sheep	11493	12862	14545	16136	12733	13761	14000[1]	14000[1]	Ovine
Pigs[1]	8	8	8	9	9	9	9	9	Porcine[1]
Horses	153	173	172	198	197	184	190[1]	190[1]	Chevaline
Asses	707	741	785	854	904	918	920[1]	942[1]	Asine
Mules	449	475	494	524	511	515	520[1]	531[1]	Mulassière
Mozambique[1]									**Mozambique[1]**
Cattle	1320	1330	1340	1350	1360	1370	1380	1370	Bovine
Sheep	114	115	116	117	119	120	121	118	Ovine
Pigs	140	145	150	155	160	165	170	165	Porcine
Asses	20	20	20	20	20	20	20	20	Asine

52
Livestock
Thousand head [cont.]
Cheptel
Milliers de têtes [suite]

Country or area	1984	1985	1986	1987	1988	1989	1990	1991	Pays ou zone
Namibia									**Namibie**
Cattle	* 1880	* 1850	* 1980	* 1840	* 1830	* 2070	* 2087	2131	Bovine
Sheep	2600	2550	2750	* 2800	2800	3200	3328	6700[1]	Ovine
Pigs[1]	44	45	46	47	48	49	50	51	Porcine[1]
Horses	48[1]	49[1]	49[1]	49[1]	50[1]	50[1]	52	53[1]	Chevaline
Asses[1]	67	68	68	68	68	68	68	68	Asine[1]
Mules[1]	6	6	6	6	6	6	6	6	Mulassière[1]
Niger									**Niger**
Cattle	2114	1832	1978	2097	2195	2296	2200[1]	2200[1]	Bovine
Sheep	2369	1538	2549	2676	2810	2951	2970[1]	2970[1]	Ovine
Pigs[1]	35	36	36	37	37	37	37	38	Porcine[1]
Horses	285[1]	290[1]	292[1]	294[1]	296[1]	298[1]	302[1]	305[1]	Chevaline
Asses	370	383	367	382	397	413	415	415[1]	Asine
Nigeria									**Nigéria**
Cattle	11583	11872	12169	12600[1]	13000[1]	13400[1]	13947	14500[1]	Bovine
Sheep	10000[1]	12000[1]	14000[1]	16000[1]	18000[1]	20000[1]	22104	24000[1]	Ovine
Pigs	1300[1]	1600[1]	2000[1]	2400[1]	2600[1]	3000[1]	3410	4000[1]	Porcine
Horses	220[1]	218[1]	216[1]	214[1]	212[1]	210[1]	208	206[1]	Chevaline
Asses	730[1]	760[1]	800[1]	830[1]	860[1]	900[1]	936	960[1]	Asine
Réunion									**Réunion**
Cattle	19	19	19	19	19	19	19	19[1]	Bovine
Sheep	3	3	3	3	2	2	2[1]	2[1]	Ovine
Pigs	71	70	70	99	97	84	86	87[1]	Porcine
Rwanda									**Rwanda**
Cattle	628	650	614	583	579	630	630[1]	630[1]	Bovine
Sheep	338	357	349	363	364	392	393[1]	394[1]	Ovine
Pigs	82	82	89	105	126	135	137[1]	139[1]	Porcine
Saint Helena									**Sainte-Hélène**
Cattle	1[1]	1[1]	1[1]	1	1	1	1	1[1]	Bovine
Sheep	1[1]	1[1]	1[1]	1	2	2	2	2[1]	Ovine
Pigs	1[1]	1[1]	1[1]	1	1	1	1	1[1]	Porcine
Sao Tome and Principe[1]									**Sao Tomé-et-Principe[1]**
Cattle	3	3	3	3	3	4	4	4	Bovine
Sheep	2	2	2	2	2	2	2	2	Ovine
Pigs	3	3	3	3	3	3	3	3	Porcine
Senegal									**Sénégal**
Cattle	2200	2200	2484	2543	2608	2673	2740	2813[1]	Bovine
Sheep	2700[1]	3100[1]	3509	3700	3792	3886	3920	4000[1]	Ovine
Pigs	260[1]	300[1]	423	450[1]	470[1]	490[1]	500[1]	547[1]	Porcine
Horses	260[1]	300[1]	314	338	360	380[1]	400[1]	400[1]	Chevaline
Asses	208	209	255	278	290	300[1]	310[1]	325[1]	Asine
Seychelles[1]									**Seychelles[1]**
Cattle	1	2	2	2	2	2	2	2	Bovine
Pigs	14	15	16	16	17	17	18	18	Porcine
Sierra Leone									**Sierra Leone**
Cattle	330[1]	330[1]	330[1]	330[1]	330[1]	330[1]	330[1]	330[1]	Bovine
Sheep	310[1]	320[1]	325[1]	330[1]	330[1]	330[1]	330[1]	330[1]	Ovine
Pigs[1]	40	44	47	50	50	50	50	52	Porcine[1]
Somalia									**Somalie**
Cattle	4296	4494	4571	4770	4983	5000[1]	5100[1]	4900[1]	Bovine
Sheep	11800	11800	12274	13195	14304	14400[1]	14000[1]	13800[1]	Ovine
Pigs[1]	10	10	10	10	10	10	10	10	Porcine[1]
Horses[1]	1	1	1	1	1	1	1	1	Chevaline[1]
Asses[1]	24	24	25	25	25	25	25	25	Asine[1]
Mules[1]	23	23	23	23	23	24	24	24	Mulassière[1]
South Africa									**Afrique du Sud**
Cattle	12895	* 12000	* 11750	* 11799	11820[1]	11850[1]	* 13398	* 13512	Bovine
Sheep	31265	* 30256	* 29481	* 29753	* 29640	* 30935	* 32665	* 32580	Ovine
Pigs	1412	* 1426	* 1445	* 1455	1460[1]	1470[1]	1480[1]	1490[1]	Porcine
Horses[1]	230	230	230	230	230	230	230	230	Chevaline[1]
Asses[1]	210	210	210	210	210	210	210	210	Asine[1]
Mules[1]	14	14	14	14	14	14	14	14	Mulassière[1]
Sudan									**Soudan**
Cattle	21033	20942	19632	19738	19858	20167	20583	21028	Bovine
Sheep	19970	19709	19691	18807	19207	19668	20168	20700	Ovine
Horses	20	18[1]	20[1]	21[1]	20[1]	21[1]	22[1]	22[1]	Chevaline

52
Livestock
Thousand head [*cont.*]
Cheptel
Milliers de têtes [*suite*]

Country or area	1984	1985	1986	1987	1988	1989	1990	1991	Pays ou zone
Asses	* 700	600[1]	650[1]	660[1]	650[1]	670[1]	675[1]	680[1]	Asine
Mules	* 1	0[1]	1[1]	1[1]	1[1]	1[1]	1[1]	1[1]	Mulassière
Swaziland									**Swaziland**
Cattle	614	648	653	641	640	679	716	750	Bovine
Sheep	35	30	28	28	30[1]	33[1]	35[1]	35[1]	Ovine
Pigs	19	16	18	21	22[1]	19	23	24	Porcine
Horses	2	2	2	1	2[1]	1	2[1]	2[1]	Chevaline
Asses	13	12	13	13	13[1]	12	14[1]	14[1]	Asine
Togo									**Togo**
Cattle	247	228	232	235	238	246	250	250[1]	Bovine
Sheep	621	716	1048	1094	1048	1147	1200	1220[1]	Ovine
Pigs	222	318	236	245	233	433	500	500[1]	Porcine
Horses	1	1	1	2	2[1]	2[1]	2[1]	2[1]	Chevaline
Asses	2	2	3	2	3	4	3	3[1]	Asine
Tunisia									**Tunisie**
Cattle	613	620	624	666	634	626	622	631	Bovine
Sheep	5561	5846	5409	5707	5581	5548	5966	6290	Ovine
Pigs	4[1]	4[1]	4[1]	4[1]	3[1]	5[1]	7	6[1]	Porcine
Horses[1]	55	55	55	55	55	55	55	56	Chevaline[1]
Asses[1]	210	215	217	218	220	224	226	229	Asine[1]
Mules[1]	72	74	75	75	76	77	78	79	Mulassière[1]
Uganda									**Ouganda**
Cattle	4993	5000	5200	3905	4260	4417	4913	5000[1]	Bovine
Sheep	1602	1674	1680	1500[1]	1690[1]	1710[1]	1920[1]	1950[1]	Ovine
Pigs	227	238	250	470	452	716	824	850[1]	Porcine
Asses[1]	16	16	17	17	17	17	17	17	Asine[1]
United Rep.Tanzania									**Rép. Unie de Tanzanie**
Cattle	12493	12600	12688	12777	12866	12956	13047	13138	Bovine
Sheep	3080	3487	3499	3500[1]	3526	3541	3557	3556	Ovine
Pigs	275	276[1]	277[1]	278[1]	279[1]	280[1]	281[1]	282[1]	Porcine
Asses[1]	168	169	170	171	172	173	174	175	Asine[1]
Zaïre									**Zaïre**
Cattle	1250	1300	1350	1400	1450	1500	1550[1]	1600[1]	Bovine
Sheep	737	800	824	849	880[1]	890[1]	900[1]	910[1]	Ovine
Pigs	697	700	737	776	800[1]	810[1]	820[1]	830[1]	Porcine
Zambia									**Zambie**
Cattle	2215	2470	2520	2601	2684	2800[1]	2920[1]	3045[1]	Bovine
Sheep	* 36	* 41	* 44	* 47	* 51	55[1]	60[1]	65[1]	Ovine
Pigs	166	178	187	196	207	210[1]	220[1]	230[1]	Porcine
Asses[1]	1	1	1	1	2	2	2	2	Asine[1]
Zimbabwe									**Zimbabwe**
Cattle	5465	5499	5783	5918	5820	5846	5900[1]	5950[1]	Bovine
Sheep	431	569	550[1]	567	671	539	545[1]	550[1]	Ovine
Pigs	178	171	180[1]	216	238	304	320[1]	340[1]	Porcine
Horses	21[1]	21[1]	* 22	22[1]	23[1]	23[1]	24[1]	24[1]	Chevaline
Asses[1]	97	98	99	100	101	102	103	104	Asine[1]
Mules[1]	1	1	1	1	1	1	1	1	Mulassière[1]
America, North									**Amérique du Nord**
Cattle	**177754**	**174223**	**169017**	**165522**	**163133**	**161618**	**162673**	**161040**	**Bovine**
Sheep	**20022**	**19375**	**18112**	**18768**	**18980**	**19021**	**19572**	**19582**	**Ovine**
Pigs	**92452**	**87684**	**87082**	**86860**	**87854**	**89223**	**86642**	**88203**	**Porcine**
Horses	**14061**	**14142**	**14142**	**14135**	**14440**	**14490**	**14299**	**14559**	**Chevaline**
Asses	**3639**	**3642**	**3650**	**3652**	**3658**	**3689**	**3691**	**3691**	**Asine**
Mules	**3596**	**3606**	**3620**	**3632**	**3646**	**3655**	**3667**	**3674**	**Mulassière**
Antigua and Barbuda									**Antigua-et-Barbuda**
Cattle	16[1]	18[1]	18[1]	18[1]	18[1]	18[1]	16[1]	16[1]	Bovine
Sheep	12[1]	13[1]	13[1]	13[1]	13[1]	13[1]	12[1]	13[1]	Ovine
Pigs[1]	4	4	4	4	4	4	4	4	Porcine[1]
Horses[1]	1	1	1	1	1	1	1	1	Chevaline[1]
Asses[1]	2	2	2	2	2	2	1	2	Asine[1]
Bahamas[1]									**Bahamas[1]**
Cattle	4	4	5	5	5	5	5	5	Bovine
Sheep	38	39	39	40	40	40	40	40	Ovine
Pigs	19	19	19	20	20	20	20	20	Porcine

52
Livestock
Thousand head [*cont.*]
Cheptel
Milliers de têtes [*suite*]

Country or area	1984	1985	1986	1987	1988	1989	1990	1991	Pays ou zone
Barbados[1]									**Barbade**[1]
Cattle	21	21	21	22	21	21	21	21	Bovine
Sheep	54	54	55	56	56	56	56	56	Ovine
Pigs	45	42	44	43	44	44	45	45	Porcine
Horses	1	1	1	1	1	1	1	1	Chevaline
Asses	2	2	2	2	2	2	2	2	Asine
Mules	2	2	2	2	2	2	2	2	Mulassière
Belize									**Belize**
Cattle	50[1]	48	49	* 50	* 50	* 50	* 51	* 51	Bovine
Sheep	3[1]	3	4	4[1]	4[1]	4[1]	4[1]	4[1]	Ovine
Pigs	22[1]	24	25	25[1]	26[1]	26[1]	26[1]	26[1]	Porcine
Horses[1]	5	5	5	5	5	5	5	5	Chevaline[1]
Mules[1]	4	4	4	4	4	4	4	4	Mulassière[1]
Bermuda									**Bermudes**
Cattle	1[1]	1[1]	1[1]	1	1[1]	1	1[1]	1[1]	Bovine
Pigs[1]	2	2	2	2	2	2	2	2	Porcine[1]
Horses	1[1]	1[1]	0[1]	0	0[1]	1	1[1]	1[1]	Chevaline
British Virgin Islands									**Iles Vierges britanniques**
Cattle	2[1]	2[1]	2[1]	2[1]	2[1]	2[1]	2[1]	2[1]	Bovine
Sheep	8[1]	7[1]	6[1]	6	6[1]	6[1]	6[1]	6[1]	Ovine
Pigs	3[1]	3[1]	2[1]	2	2[1]	2[1]	2[1]	2[1]	Porcine
Canada									**Canada**
Cattle	12582	12160	11788	11750	12041	12162	12249	12369	Bovine
Sheep	769	720	694	701	697	729	759	780	Ovine
Pigs	10272	10154	9885	10493	10890	10763	10370	10516	Porcine
Horses[1]	380	390	390	400	405	415	415	415	Chevaline[1]
Mules[1]	4	4	4	4	4	4	4	4	Mulassière[1]
Cayman Islands									**Iles Caïmanes**
Cattle	2	2	2	2	2	2	2	2[1]	Bovine
Costa Rica									**Costa Rica**
Cattle	* 2146	* 1850	* 1846	* 1836	* 1753	* 1735	* 1762	* 1741	Bovine
Sheep	3	3	3	3	3	3[1]	3[1]	3[1]	Ovine
Pigs	* 223	* 220	* 222	* 238	* 223	223[1]	224[1]	224[1]	Porcine
Horses[1]	113	113	114	114	114	114	114	114	Chevaline[1]
Asses[1]	7	7	7	7	7	7	7	7	Asine[1]
Mules[1]	5	5	5	5	5	5	5	5	Mulassière[1]
Cuba									**Cuba**
Cattle	5101	5115	5020	5007	4984	4927	4920	4920[1]	Bovine
Sheep[1]	375	378	380	382	382	385	385	385	Ovine[1]
Pigs[1]	1500	1650	1650	1750	1750	1850	1850	1900	Porcine[1]
Horses	781	759	740	718	703	630	629	629[1]	Chevaline
Asses	4	4	4	4	4	5	5	5[1]	Asine
Mules	28	29	30	31	32	31	31	32[1]	Mulassière
Dominica									**Dominique**
Cattle	8	9	9[1]	9[1]	9[1]	9[1]	9[1]	9[1]	Bovine
Sheep	7	7[1]	7[1]	7[1]	8[1]	7[1]	8[1]	8[1]	Ovine
Pigs	5	5	5[1]	5[1]	5[1]	5[1]	5[1]	5[1]	Porcine
Dominican Republic									**Rép. dominicaine**
Cattle	2020	2019	2055	2092	2129	2245	* 2240	2250[1]	Bovine
Sheep	80[1]	86	88[1]	* 95	* 100	* 110	115[1]	120[1]	Ovine
Pigs	505	344	368	389	409	429	431[1]	435[1]	Porcine
Horses	250[1]	300[1]	300[1]	300[1]	310[1]	310[1]	315[1]	320[1]	Chevaline
Asses	132[1]	136	140[1]	140[1]	142[1]	142[1]	143[1]	143[1]	Asine
Mules	120[1]	128	130[1]	130[1]	132[1]	132[1]	133[1]	133[1]	Mulassière
El Salvador									**El Salvador**
Cattle	929	980	1050	1088	1144	1176	1220	1243	Bovine
Sheep[1]	4	4	5	5	5	5	5	5	Ovine[1]
Pigs	375	397	411	418	377	289	317	320[1]	Porcine
Horses[1]	91	92	92	93	93	93	93	93	Chevaline[1]
Asses[1]	2	2	2	2	2	2	2	2	Asine[1]
Mules[1]	23	23	23	23	23	23	23	23	Mulassière[1]
Greenland									**Groënland**
Sheep	21	21[1]	21[1]	21[1]	21[1]	21[1]	22[1]	22[1]	Ovine

52
Livestock
Thousand head [cont.]
    Cheptel
    Milliers de têtes [suite]

Country or area	1984	1985	1986	1987	1988	1989	1990	1991	Pays ou zone
Grenada									**Grenade**
Cattle	5[1]	4[1]	4[1]	4[1]	4[1]	4[1]	4[1]	4[1]	Bovine
Sheep[1]	14	14	13	12	12	11	11	11	Ovine[1]
Pigs[1]	8	8	8	8	8	7	7	7	Porcine[1]
Asses[1]	1	1	1	1	1	1	1	1	Asine[1]
Guadeloupe									**Guadeloupe**
Cattle	94	94	82	76	74	66	65	65[1]	Bovine
Sheep	4	4	5[1]	5[1]	6[1]	5[1]	5[1]	5[1]	Ovine
Pigs	46	46	44	41	43	29	32[1]	36[1]	Porcine
Horses[1]	1	1	1	1	1	1	1	1	Chevaline[1]
Asses[1]	1	1	1	1	1	1	1	1	Asine[1]
Guatemala									**Guatemala**
Cattle	2100[1]	2029	2022	2004	* 1966	2023	* 1900	* 1695	Bovine
Sheep	660[1]	660[1]	665[1]	666[1]	667[1]	660	670[1]	675[1]	Ovine
Pigs	806	834	862	850[1]	820[1]	800	* 1100	* 1110	Porcine
Horses[1]	100	100	100	110	112	112	113	114	Chevaline[1]
Asses[1]	9	9	9	9	9	9	9	9	Asine[1]
Mules	37[1]	37[1]	38[1]	38[1]	38[1]	38[1]	38[1]	38[1]	Mulassière
Haiti									**Haïti**
Cattle	1350[1]	1350[1]	1400[1]	1450[1]	1500[1]	1550[1]	1450[1]	1400[1]	Bovine
Sheep[1]	92	92	92	93	94	94	93	92	Ovine[1]
Pigs	500[1]	500[1]	700[1]	750[1]	900[1]	950[1]	950[1]	930[1]	Porcine
Horses[1]	425	425	425	430	430	432	435	435	Chevaline[1]
Asses[1]	212	212	215	216	216	217	218	218	Asine[1]
Mules[1]	83	83	84	84	85	86	86	86	Mulassière[1]
Honduras									**Honduras**
Cattle	* 2659	* 2618	* 2574	* 2532	* 2489	2424	* 2424	* 2388	Bovine
Sheep	6[1]	* 6	* 7	* 7	7[1]	7[1]	8[1]	8[1]	Ovine
Pigs	491	* 558	* 563	* 567	* 600	706	734	740[1]	Porcine
Horses[1]	168	169	170	170	170	170	170	170	Chevaline[1]
Asses[1]	22	22	22	22	22	22	22	22	Asine[1]
Mules[1]	68	68	68	69	69	69	69	69	Mulassière[1]
Jamaica[1]									**Jamaïque**[1]
Cattle	290	280	290	290	290	290	310	300	Bovine
Sheep	4	4	3	3	2	2	2	2	Ovine
Pigs	240	245	215	210	220	240	250	250	Porcine
Horses	4	4	4	4	4	4	4	4	Chevaline
Asses	24	23	23	23	23	23	23	23	Asine
Mules	10	10	10	10	10	10	10	10	Mulassière
Martinique									**Martinique**
Cattle	41	41	38	37	37	35	35	35[1]	Bovine
Sheep[1]	110	94	93	73	73	63	75	63	Ovine[1]
Pigs	47[1]	54[1]	47[1]	47[1]	48[1]	48[1]	49[1]	49[1]	Porcine
Horses	2[1]	2[1]	2[1]	2[1]	2[1]	2[1]	2[1]	2[1]	Chevaline
Mexico									**Mexique**
Cattle	30479	31489	* 31123	* 31156	* 31200	* 30900	32054	* 29847	Bovine
Sheep	6120	6373	5699	5926	5761	5863	5846	6003	Ovine
Pigs	19393	17233	18397	18722	15884	16157	15203	15902	Porcine
Horses	6134	6135[1]	6140[1]	6150[1]	6160[1]	6170[1]	6170[1]	6175[1]	Chevaline
Asses	3182	3183[1]	3183[1]	3184[1]	3185[1]	3186[1]	3187[1]	3188[1]	Asine
Mules	3130	3130[1]	3140[1]	3150[1]	3160[1]	3170[1]	3180[1]	3186[1]	Mulassière
Montserrat									**Montserrat**
Cattle[1]	9	9	9	9	9	9	10	10	Bovine[1]
Sheep	4	4[1]	4[1]	4[1]	4[1]	4[1]	5[1]	5[1]	Ovine
Pigs	1	1[1]	1[1]	1[1]	1[1]	1[1]	1[1]	1[1]	Porcine
Netherlands Antilles[1]									**Antilles néerlandaises**[1]
Cattle	1	1	1	1	1	1	1	1	Bovine
Sheep	9	5	4	5	4	4	4	4	Ovine
Pigs	6	4	4	3	3	3	3	3	Porcine
Asses	3	3	3	3	3	3	3	3	Asine
Nicaragua									**Nicaragua**
Cattle	2344	2369	* 2110	* 1885	* 1700	1800[1]	1680[1]	1680[1]	Bovine
Sheep[1]	3	3	3	3	4	4	4	4	Ovine[1]
Pigs	744	745	* 750	* 749	* 700	680[1]	690[1]	709[1]	Porcine
Horses[1]	265	263	260	255	250	250	250	250	Chevaline[1]

52
Livestock
Thousand head [*cont.*]
Cheptel
Milliers de têtes [*suite*]

Country or area	1984	1985	1986	1987	1988	1989	1990	1991	Pays ou zone
Asses[1]	8	8	8	8	8	8	8	8	Asine[1]
Mules[1]	45	45	45	45	45	45	45	45	Mulassière[1]
Panama									**Panama**
Cattle	1452	1447	1430	1410	1423	1417	1388	1399	Bovine
Pigs	195	208	250	229	211	202	226	256	Porcine
Horses	156[1]	156[1]	155[1]	156[1]	156[1]	156[1]	156[1]	156	Chevaline
Mules	4[1]	4[1]	4[1]	4[1]	4[1]	4[1]	4[1]	4	Mulassière
Puerto Rico									**Porto Rico**
Cattle	593	590	600	580	559	586	601	599	Bovine
Sheep[1]	6	6	7	7	7	7	7	7	Ovine[1]
Pigs	203	211	206	199	195	199	203	209	Porcine
Horses[1]	21	21	21	22	22	22	22	23	Chevaline[1]
Asses[1]	2	2	2	2	2	2	2	2	Asine[1]
Mules[1]	2	2	2	2	2	3	3	3	Mulassière[1]
Saint Kitts-Nevis[1]									**Saint-Kitts-et-Nevis[1]**
Cattle	5	5	5	5	5	5	5	5	Bovine
Sheep	14	14	14	15	15	15	15	15	Ovine
Pigs	2	2	2	2	2	2	2	2	Porcine
Saint Lucia									**Sainte-Lucie**
Cattle	12	12[1]	12[1]	12[1]	12[1]	12[1]	12[1]	12[1]	Bovine
Sheep	15	15[1]	15[1]	15[1]	15[1]	16[1]	16[1]	16[1]	Ovine
Pigs	12	12[1]	12[1]	12[1]	12[1]	12[1]	12[1]	12[1]	Porcine
Horses[1]	1	1	1	1	1	1	1	1	Chevaline[1]
Asses[1]	1	1	1	1	1	1	1	1	Asine[1]
Mules[1]	1	1	1	1	1	1	1	1	Mulassière[1]
Saint Pierre and Miquelon									**Saint-Pierre-et-Miquelon**
Pigs	...	1	1	1	1	0	0	0[1]	Porcine
St. Vincent-Grenadines									**St. Vincent-Grenadines**
Cattle	8	8	7	6	7[1]	7[1]	7[1]	8[1]	Bovine
Sheep	16	16	16	15	15[1]	15[1]	15[1]	15[1]	Ovine
Pigs	8	9	9	9[1]	9[1]	9[1]	9[1]	9[1]	Porcine
Asses[1]	1	1	1	1	1	1	1	1	Asine[1]
Trinidad and Tobago									**Trinité-et-Tobago**
Cattle[1]	60	55	57	60	70	65	60	60	Bovine[1]
Sheep[1]	9	11	11	12	13	13	14	14	Ovine[1]
Pigs[1]	76	73	60	68	60	50	50	50	Porcine[1]
Horses[1]	1	1	1	1	1	1	1	1	Chevaline[1]
Asses[1]	2	2	2	2	2	2	2	2	Asine[1]
Mules[1]	2	2	2	2	2	2	2	2	Mulassière[1]
United States									**Etats-Unis**
Cattle	113360	109582	105378	102118	99622	98065	98162	98896	Bovine
Sheep	11559	10716	10145	10572	10945	10858	11364	11200	Ovine
Pigs	56694	54073	52313	51001	54384	55469	53821	54427	Porcine
Horses	5160[1]	5203	5203[1]	5203[1]	5500[1]	5600[1]	5400[1]	5650[1]	Chevaline
Asses	24[1]	24	24[1]	24[1]	26[1]	55[1]	53[1]	52[1]	Asine
Mules	28[1]	28	28[1]	28[1]	28[1]	28[1]	28[1]	28[1]	Mulassière
United States Virgin Is.									**Iles Vierges américaines**
Cattle	9[1]	10[1]	11	5	8[1]	8[1]	8[1]	8[1]	Bovine
Sheep	4[1]	3[1]	3	3	3[1]	3[1]	3[1]	3[1]	Ovine
Pigs	5[1]	4[1]	2	3	3[1]	3[1]	3[1]	3[1]	Porcine
America, South									**Amérique du Sud**
Cattle	249779	251221	253677	258170	262957	266405	271128	274554	**Bovine**
Sheep	101352	102075	106209	109410	110918	111929	110994	109890	**Ovine**
Pigs	51362	50312	50637	51461	52268	53555	54790	54935	**Porcine**
Horses	13264	13527	13709	13876	13988	14410	14666	14897	**Chevaline**
Asses	3800	3793	3800	3884	3919	3976	3997	4011	**Asine**
Mules	3214	3219	3190	3237	3279	3312	3351	3376	**Mulassière**
Argentina									**Argentine**
Cattle	54594	54636	52537	51683	* 50782	* 49500	* 50582	* 50080	Bovine
Sheep	28000	28750	29167	28750	29167	* 29345	* 28571	* 27552	Ovine
Pigs	* 3800	* 3800	* 4000	4100[1]	4100[1]	4200[1]	4400[1]	* 3400	Porcine
Horses	* 2970	* 3000	* 3000	* 3000	* 2900	* 3200	* 3400	* 3400	Chevaline
Asses[1]	90	90	90	90	90	90	90	90	Asine[1]
Mules[1]	165	165	165	167	168	170	172	172	Mulassière[1]

52
Livestock
Thousand head [*cont.*]
Cheptel
Milliers de têtes [*suite*]

Country or area	1984	1985	1986	1987	1988	1989	1990	1991	Pays ou zone
Bolivia									**Bolivie**
Cattle	5215	5515	5055	5239	5402	5476	5538	* 5600	Bovine
Sheep	9287	9413	* 9700	11600[1]	12000[1]	12260[1]	12220[1]	12300[1]	Ovine
Pigs	1694	1725	1788	1902	2019	2127	2199	* 2340	Porcine
Horses	293	311	310[1]	315[1]	315[1]	320[1]	320[1]	320[1]	Chevaline
Asses[1]	600	600	600	610	620	630	630	630	Asine[1]
Mules[1]	80	80	80	80	80	80	80	80	Mulassière[1]
Brazil									**Brésil**
Cattle	127655	128423	132222	135726	139599	144154	* 148000	* 152000	Bovine
Sheep	18447	18659	19660	19860	20085	20041	* 20100	* 20300	Ovine
Pigs	32327	32248	32539	32480	32121	33015	* 34000	* 35000	Porcine
Horses	5442	5550	5735	5855	5971	* 6000	* 6000	6200	Chevaline
Asses	1246	1274	1286	1295	1304	1320[1]	1330[1]	1340[1]	Asine
Mules	1946	1943	1921	1952	1984	2000[1]	2030[1]	2050[1]	Mulassière
Chile									**Chili**
Cattle	3650	3191	3220	3371	3466	3330	3404	3300	Bovine
Sheep	6000	5800	5806	6470	6429	6600	6650	6650[1]	Ovine
Pigs	1070	1100	1150	1300	1360	1400	1251	1300[1]	Porcine
Horses	480[1]	490[1]	490[1]	490[1]	490[1]	500[1]	520[1]	520[1]	Chevaline
Asses[1]	28	28	28	28	28	28	28	28	Asine[1]
Mules[1]	10	10	10	10	10	10	10	10	Mulassière[1]
Colombia									**Colombie**
Cattle	23349	23271	23593	23971	24245	24598	24550	* 24875	Bovine
Sheep	2466	2500	2538	2576	2614	2650	2690	* 2745	Ovine
Pigs	2312	2381	2440	2511	2580	2600	2640	2700	Porcine
Horses	* 1815	* 1906	1860	1898	1936	1974	1975[1]	1980[1]	Chevaline
Asses	650[1]	650[1]	672	682	693	703	703[1]	705[1]	Asine
Mules	600[1]	600[1]	600	606	612	618	618[1]	620[1]	Mulassière
Ecuador									**Équateur**
Cattle	3575	3730	3765	3884	3997	4177	4359	* 4200	Bovine
Sheep	1020	1080	1195	1293	1226	1329	1420	* 1250	Ovine
Pigs	3792	2464	1443	1620	1922	2092	2220	* 2125	Porcine
Horses	331	337	372	404	427	460	492	512[1]	Chevaline
Asses	218	189	162	215	221	242	253	255[1]	Asine
Mules	104	112	105	113	116	124	131	134[1]	Mulassière
Falkland Is. (Malvinas)									**Iles Falkland (Malvinas)**
Cattle	7	8	7	6	6	6	5	5	Bovine
Sheep	679	692	699	692	705	745	740	729	Ovine
Horses	2	2	2	2	2	2	2	2	Chevaline
French Guiana									**Guyane française**
Cattle	14	14	17	17	15	16	16[1]	19[1]	Bovine
Sheep	3[1]	3[1]	3[1]	3[1]	4[1]	4	4[1]	* 4	Ovine
Pigs	13[1]	14[1]	12[1]	10[1]	11[1]	9	11[1]	11[1]	Porcine
Guyana									**Guyana**
Cattle	180[1]	199	200[1]	* 210	220[1]	230[1]	230[1]	230[1]	Bovine
Sheep[1]	122	125	130	130	130	130	130	130	Ovine[1]
Pigs[1]	75	80	80	80	85	80	75	80	Porcine[1]
Horses[1]	2	2	2	2	2	2	2	2	Chevaline[1]
Asses[1]	1	1	1	1	1	1	1	1	Asine[1]
Paraguay									**Paraguay**
Cattle	6795	6956	7151	7374	7780	8074	8254	8260[1]	Bovine
Sheep	372	378	388	411	432	451	457	460[1]	Ovine
Pigs	1109	1278	1508	1809	2108	2305	2444	2450[1]	Porcine
Horses	313	314	317	323	328	334	335	335[1]	Chevaline
Asses[1]	30	30	30	31	31	31	31	31	Asine[1]
Mules[1]	13	13	13	14	14	14	14	14	Mulassière[1]
Peru									**Pérou**
Cattle	4051	4002	4012	4027	4174	4234	4102	* 3630	Bovine
Sheep	13950[1]	13052	13117	13126	12922	12970	12257	* 11250	Ovine
Pigs	2214	2066	2453	2222	2376	2434	2400	* 2250	Porcine
Horses	655[1]	655[1]	655[1]	655[1]	655[1]	660[1]	660[1]	660[1]	Chevaline
Asses	490[1]	490[1]	490[1]	490[1]	490[1]	490[1]	490[1]	490[1]	Asine
Mules	220[1]	220[1]	220[1]	220[1]	220[1]	220[1]	220[1]	220[1]	Mulassière

52
Livestock
Thousand head [*cont.*]
Cheptel
Milliers de têtes [*suite*]

Country or area	1984	1985	1986	1987	1988	1989	1990	1991	Pays ou zone
Suriname									**Suriname**
Cattle	58	60	67	76	84	89	92	98	Bovine
Sheep	4	3	4	6	8	8	10	10[1]	Ovine
Pigs	22	21	19	19	21	25	32	30	Porcine
Uruguay									**Uruguay**
Cattle	9062	9371	9300	9945	10331	9446	8723	8889	Bovine
Sheep	20637	21196	23337	24006	24689	24872	25220	25986	Ovine
Pigs	235	200	195	220	215	* 215	215[1]	215[1]	Porcine
Horses	469	464	469	437	466	462	465[1]	470[1]	Chevaline
Asses[1]	1	1	1	1	1	1	1	1	Asine[1]
Mules[1]	4	4	4	4	4	4	4	4	Mulassière[1]
Venezuela									**Venezuela**
Cattle	11575	11844	12531	12641	12856	13076	13272	* 13368	Bovine
Sheep	* 365	* 422	467	488	508	523	525[1]	525[1]	Ovine
Pigs	2699	2935	3011	3187	3349	3053	2904	* 1971	Porcine
Horses	* 491	* 495	495[1]	495[1]	495[1]	495[1]	495[1]	495[1]	Chevaline
Asses[1]	445	440	440	440	440	440	440	440	Asine[1]
Mules[1]	73	72	72	72	72	72	72	72	Mulassière[1]
Asia									**Asie**
Cattle	369689	375191	382821	386922	384304	389987	395911	399274	**Bovine**
Sheep	332593	320153	325211	333269	336687	355482	360443	357561	**Ovine**
Pigs	369121	378783	404230	411721	405093	420266	433545	435708	**Porcine**
Horses	16899	17013	17182	17105	16852	16719	16637	16555	**Chevaline**
Asses	19387	19981	20536	20894	21102	21410	21589	21619	**Asine**
Mules	5319	5463	5637	5788	5912	6036	6063	6157	**Mulassière**
Afghanistan									**Afghanistan**
Cattle	2700[1]	2200[1]	1532	1500[1]	1550[1]	1600[1]	1650[1]	1650[1]	Bovine
Sheep	14900[1]	12700[1]	10500	10500[1]	11500[1]	12500[1]	13500[1]	13500[1]	Ovine
Horses	410[1]	410[1]	* 410	400[1]	400[1]	400[1]	400[1]	400[1]	Chevaline
Asses	1318[1]	1321[1]	* 1325	1300	1300[1]	1300[1]	1300[1]	1300[1]	Asine
Mules	30[1]	30[1]	* 30	30[1]	30[1]	30[1]	30[1]	30[1]	Mulassière
Bahrain									**Bahreïn**
Cattle	6	8[1]	8[1]	9[1]	13[1]	15[1]	15[1]	15[1]	Bovine
Sheep	7	7[1]	7[1]	8[1]	8[1]	8[1]	8[1]	9[1]	Ovine
Bangladesh									**Bangladesh**
Cattle	21920	22132	22348	22567	22789	23015	23244	23500[1]	Bovine
Sheep	667	709	739	770	803	837	873	900[1]	Ovine
Bhutan									**Bhoutan**
Cattle	358	379	376	387	393	409[1]	406[1]	413[1]	Bovine
Sheep	* 19	* 23	* 16	36	47	50[1]	54[1]	59[1]	Ovine
Pigs	59	60	89	70	66	70[1]	72[1]	73[1]	Porcine
Horses	19[1]	21[1]	24	26	26	25[1]	25[1]	27[1]	Chevaline
Asses[1]	18	18	18	18	18	18	18	18	Asine[1]
Mules[1]	8	8	9	9	9	9	9	9	Mulassière[1]
Brunei Darussalam									**Brunéi Darussalam**
Cattle[1]	1	1	1	1	1	1	1	1	Bovine[1]
Pigs[1]	22	20	18	15	17	15	14	14	Porcine[1]
Cambodia									**Cambodge**
Cattle	1436	1560	1645	1837	1950[1]	2000[1]	2100[1]	2150[1]	Bovine
Pigs	1008	1205	1228	1434	1500[1]	1550[1]	1585[1]	1610[1]	Porcine
Horses[1]	11	12	13	14	15	16	17	18	Chevaline[1]
China									**Chine**
Cattle	59020	62711	66991	70964	73963	77025	79493	81407	Bovine
Sheep	98920	95194	94210	99009	102655	110571	113508	112820	Ovine
Pigs	304424	313361	338070	344248	334862	349172	360594	363975	Porcine
Horses	10806	10978	11081	10988	10691	10540	10294	10174	Chevaline
Asses	9449	9962	10415	10689	10846	11052	11136	11198	Asine
Mules	4593	4790	4972	5113	5248	5366	5391	5494	Mulassière
Cyprus									**Chypre**
Cattle	34	38	42	44	45	46	49	52[1]	Bovine
Sheep	325	334	325	325	310	300	325	325[1]	Ovine
Pigs	243	235	201	225	266	284	281	289[1]	Porcine
Horses	1	1	1	1	1	1	1[1]	1[1]	Chevaline
Asses	6	6	6	6	6	5	5[1]	5[1]	Asine
Mules	2	2	2	2	2	2	2[1]	2[1]	Mulassière

52
Livestock
Thousand head [*cont.*]
Cheptel
Milliers de têtes [*suite*]

Country or area	1984	1985	1986	1987	1988	1989	1990	1991	Pays ou zone
Gaza Strip (Palestine)									**Zone de Gaza (Palestine)**
Cattle	4	3	4[1]	4[1]	4[1]	4[1]	4[1]	4[1]	Bovine
Sheep	21	17	11[1]	9[1]	10[1]	10[1]	10[1]	10[1]	Ovine
Hong Kong									**Hong-kong**
Cattle	3	3	3	2	1	1	2	2[1]	Bovine
Pigs	507	499	372	353	358	350	304	304[1]	Porcine
Horses	1	1	1	1	1	1	1[1]	1[1]	Chevaline
India									**Inde**
Cattle	* 195610	* 197950	* 200330	* 199300	* 193000	* 195500	* 197300	* 198400	Bovine
Sheep	* 51130	* 52770	* 54460	* 55482	* 51684	* 53486	* 54588	* 55700	Ovine
Pigs	10120[1]	10150[1]	10200[1]	10200[1]	10300[1]	10300[1]	10400[1]	10450[1]	Porcine
Horses	900[1]	910[1]	920[1]	950[1]	953[1]	955[1]	960[1]	965[1]	Chevaline
Asses	1070[1]	1100[1]	1200[1]	1300[1]	1328[1]	1400[1]	1450[1]	1500[1]	Asine
Mules	130[1]	132[1]	132[1]	134[1]	135[1]	138[1]	139[1]	140[1]	Mulassière
Indonesia									**Indonésie**
Cattle	9439	9526	9635	9742	9776	10094	10550	10350[1]	Bovine
Sheep	4789	4890	5280	5363	5825	5910	5900	5750[1]	Ovine
Pigs	5290	5560	6220	6340	6484	6936	7650	6800[1]	Porcine
Horses	659	668	715	658	675	683	740[1]	750[1]	Chevaline
Iran, Islamic Rep. of									**Iran, Rép. islamique d'**
Cattle	6300[1]	6500[1]	6691	6500[1]	6368	* 6500	* 6650	* 6800	Bovine
Sheep	38000[1]	40000[1]	42695	41000	40665	45000	* 45000	* 45000	Ovine
Horses	285[1]	282[1]	279[1]	276[1]	274[1]	271	270[1]	270[1]	Chevaline
Asses	2150[1]	2100[1]	2050[1]	2000[1]	1970[1]	* 1943	1940[1]	1937[1]	Asine
Mules	142[1]	141[1]	139[1]	138[1]	137[1]	* 136	135[1]	134[1]	Mulassière
Iraq									**Iraq**
Cattle	1698	1600[1]	1578	1580[1]	* 1600	1650[1]	1675[1]	1400[1]	Bovine
Sheep	* 9723	9200[1]	8981	9000	* 9000	9500[1]	9600[1]	7800[1]	Ovine
Horses	53	50[1]	53[1]	55[1]	55[1]	58[1]	60[1]	48[1]	Chevaline
Asses	* 400	400[1]	400[1]	400[1]	410[1]	415[1]	416[1]	350[1]	Asine
Mules	* 25	25[1]	25[1]	25[1]	26[1]	26[1]	27[1]	21[1]	Mulassière
Israel									**Israël**
Cattle	309	312	316	325	345	348	342	331	Bovine
Sheep	273	267	262	306	372	394	380	375	Ovine
Pigs	110	120	130	130[1]	130[1]	130[1]	115[1]	100[1]	Porcine
Horses[1]	4	4	4	4	4	4	4	4	Chevaline[1]
Asses[1]	5	5	5	5	5	5	5	5	Asine[1]
Mules[1]	2	2	2	2	2	2	2	2	Mulassière[1]
Japan									**Japon**
Cattle	4682	4698	4742	4694	4667	4682	4760	* 4863	Bovine
Sheep	22	24	26	27	29	30	31	32[1]	Ovine
Pigs	10423	10718	11061	11354	11725	11866	11816	11335	Porcine
Horses	24	23	23	22	22	22	23	24[1]	Chevaline
Jordan									**Jordanie**
Cattle	34	35	31	29	30	29	29[1]	29[1]	Bovine
Sheep	960	1121	930	1219	1279	1523	1400[1]	1400[1]	Ovine
Horses	12	3	3	3	3	3	3[1]	3[1]	Chevaline
Asses	19[1]	19[1]	19[1]	19	19	19	19[1]	19[1]	Asine
Mules	3	3	3[1]	3	3	3	3[1]	3[1]	Mulassière
Korea, Dem. P. R.									**Corée, R. p. dém. de**
Cattle[1]	1025	1100	1150	1200	1250	1280	1300	1350	Bovine[1]
Sheep[1]	340	350	360	368	372	380	385	390	Ovine[1]
Pigs[1]	2700	2800	2900	3050	3100	3145	3200	3300	Porcine[1]
Horses[1]	40	40	41	42	43	44	44	45	Chevaline[1]
Asses[1]	3	3	3	3	3	3	3	3	Asine[1]
Mules[1]	1	2	2	2	2	2	2	2	Mulassière[1]
Korea, Republic of									**Corée, République de**
Cattle	2215	2652	2944	2807	2386	2039	2051	2126	Bovine
Sheep	6	5	5	4	3	3	3	5	Ovine
Pigs	3649	2958	2853	3347	4281	4852	4801	4528	Porcine
Horses	3	3	3	3	3	4	5	5	Chevaline
Kuwait									**Koweït**
Cattle	20	23	23	28	28[1]	28[1]	25[1]	0[1]	Bovine
Sheep	326	393	227	231	250[1]	260[1]	200[1]	0[1]	Ovine
Horses	2	3	3[1]	3[1]	3[1]	4[1]	3[1]	0[1]	Chevaline

52
Livestock
Thousand head [*cont.*]
Cheptel
Milliers de têtes [*suite*]

Country or area	1984	1985	1986	1987	1988	1989	1990	1991	Pays ou zone
Lao People's Dem. Rep.									**Rép. dém. pop. lao**
Cattle	544	576	646	703	764	817	842	865[1]	Bovine
Pigs	1360	1434	1280	1420	1268	1350	1372	1390[1]	Porcine
Horses[1]	37	40	42	42	42	43	44	45	Chevaline[1]
Lebanon									**Liban**
Cattle	45	48[1]	55[1]	61	59	59	60[1]	57[1]	Bovine
Sheep	160[1]	185[1]	200[1]	208	204	210	210[1]	205[1]	Ovine
Pigs	29[1]	35[1]	40[1]	49	47	49	49[1]	45[1]	Porcine
Horses	2	2[1]	2[1]	2[1]	2[1]	2[1]	2[1]	2[1]	Chevaline
Asses	* 10	10[1]	10[1]	10[1]	11[1]	11[1]	12[1]	12[1]	Asine
Mules	* 4	4[1]	4[1]	4[1]	4[1]	5[1]	5[1]	5[1]	Mulassière
Macau									**Macao**
Pigs[1]	1	1	1	1	1	1	1	1	Porcine[1]
Malaysia									**Malaisie**
Cattle	618	619	607	663	637	661	686	658[1]	Bovine
Sheep	69	79	91	130	150	183	220	200[1]	Ovine
Pigs	2030	2111	2176	2261	2113	2345	2548	2400[1]	Porcine
Horses[1]	5	5	5	5	5	5	5	5	Chevaline[1]
Mongolia									**Mongolie**
Cattle	2374	2374	2408	2480	2526	2541	2693	2849	Bovine
Sheep	14110	13391	13249	13194	13234	13451	14265	15083	Ovine
Pigs	45	48	56	80	120	171	192	185[1]	Porcine
Horses	1960	1961	1971	2018	2047	2103	2200	2255	Chevaline
Myanmar									**Myanmar**
Cattle	9502	9718	9758	9919	10091	9126	9298	9310[1]	Bovine
Sheep	316	332	317	304	313	269	274	280[1]	Ovine
Pigs	2706	2953	2986	3059	3199	2449	2243	2250[1]	Porcine
Horses	126[1]	130[1]	133	136	138	119	120[1]	121[1]	Chevaline
Mules[1]	9	9	9	9	9	9	9	9	Mulassière[1]
Nepal									**Népal**
Cattle	6550[1]	6357	6372	6363	6343	6285	6281	6350[1]	Bovine
Sheep	770[1]	785	808	837	873	910	892	925[1]	Ovine
Pigs	420[1]	442	456	476	516	548	574	575[1]	Porcine
Oman									**Oman**
Cattle	126	130[1]	130[1]	135[1]	136[1]	136[1]	137	138[1]	Bovine
Sheep	136	200[1]	210[1]	215[1]	218[1]	220[1]	250	280[1]	Ovine
Asses[1]	23	23	24	24	25	25	25	26	Asine[1]
Pakistan									**Pakistan**
Cattle	16352	16549	16749	16951	17156	17363	17573	17785	Bovine
Sheep	24272	25037	25826	26640	27479	28345	29239	30160	Ovine
Horses	450	451	452	454	455	457	460[1]	461[1]	Chevaline
Asses	2701	2778	2857	2938	3022	3108	3200[1]	3279[1]	Asine
Mules	64	64	65	65	65	66	66[1]	66[1]	Mulassière
Philippines									**Philippines**
Cattle	1849	1786	1814	1747	1700	1682	1629	1677	Bovine
Sheep[1]	30	30	30	30	30	30	30	30	Ovine[1]
Pigs	7613	7304	7275	7038	7581	7909	8124	8007	Porcine
Horses	189	186	195	195[1]	200[1]	200[1]	200[1]	200[1]	Chevaline
Qatar									**Qatar**
Cattle	6	6	8	9	10	10[1]	10[1]	10[1]	Bovine
Sheep	50	* 68	119	120[1]	125[1]	128[1]	130[1]	132[1]	Ovine
Horses	1[1]	1	1	1[1]	1[1]	1[1]	1[1]	1[1]	Chevaline
Saudi Arabia									**Arabie saoudite**
Cattle	271	290	303	216	216[1]	* 217	* 191	* 176	Bovine
Sheep	6940	6684	7117	7405	* 7466	* 7084	* 6457	* 5692	Ovine
Horses[1]	3	3	3	3	3	3	3	3	Chevaline[1]
Asses[1]	110	109	108	107	106	105	104	103	Asine[1]
Mules[1]	6	6	6	6	6	6	6	6	Mulassière[1]
Singapore									**Singapour**
Cattle	1	0	0	0[1]	0[1]	0[1]	0[1]	0[1]	Bovine
Pigs	800	689	516	459	321	350[1]	380[1]	380[1]	Porcine
Sri Lanka									**Sri Lanka**
Cattle	1738	1783	1783	1808	1788	1820	1800[1]	1814[1]	Bovine
Sheep	29	27	29	28	28	30	31[1]	31[1]	Ovine
Pigs	85	84	86	97	95	94	100[1]	102[1]	Porcine
Horses[1]	2	2	2	2	2	2	2	2	Chevaline[1]

52
Livestock
Thousand head [*cont.*]
Cheptel
Milliers de têtes [*suite*]

Country or area	1984	1985	1986	1987	1988	1989	1990	1991	Pays ou zone
Syrian Arab Republic									**Rép. arabe syrienne**
Cattle	736	742	706	710	763	800	787	* 786	Bovine
Sheep	12693	10993	11669	12669	13691	14011	14509	15321	Ovine
Pigs	0[1]	1[1]	1[1]	1[1]	1[1]	1[1]	1[1]	1[1]	Porcine
Horses	46	45	42	41	40	43	41	45[1]	Chevaline
Asses	192	196	199	184	177	178	168	175[1]	Asine
Mules	28	30	29	28	27	27	26	30[1]	Mulassière
Thailand									**Thaïlande**
Cattle	4789	4829	4879	4969	5072	5285	5669	6052	Bovine
Sheep	45	58	73	95	131	156	162	178	Ovine
Pigs	4263	4224	4201	4209	4685	4679	* 4900	* 5000	Porcine
Horses	20	21	19	19	19	18	18[1]	18[1]	Chevaline
Turkey									**Turquie**
Cattle	14099	12410	12466	12713	12713	12562	12173	11377	Bovine
Sheep	48707	40391	42500	43758	43796	45384	43647	40553	Ovine
Pigs	11	12	8	8	7	9	8	12	Porcine
Horses	703	623	604	600	590	557	545	513	Chevaline
Asses	1208	1226	1192	1188	1153	1119	1084	985	Asine
Mules	269	213	206	216	205	208	210	202	Mulassière
United Arab Emirates									**Emirats arabes unis**
Cattle	33	35	41	44	46	48[1]	50[1]	53[1]	Bovine
Sheep	169	181	194	207	222	250[1]	260[1]	270[1]	Ovine
Viet Nam									**Viet Nam**
Cattle	2174	2418	2598	2784	2979	3126	3199	3282	Bovine
Pigs	11202	11760	11808	11796	12051	11643	12221	12583	Porcine
Horses	122	130	133	137	136	133	143	145[1]	Chevaline
Yemen									**Yémen**
Cattle	1056	1081	1100	1120	1137	1170	1175	1180[1]	Bovine
Sheep	3394[1]	3443[1]	3483[1]	3488	3602	3720	3756	3800[1]	Ovine
Horses	3[1]	3[1]	3[1]	3[1]	3[1]	3[1]	3[1]	3[1]	Chevaline
Asses	...	...	...	...	...	690[1]	690[1]	690[1]	Asine
Europe									**Europe**
Cattle	**133094**	**132620**	**130883**	**128618**	**125577**	**125181**	**124043**	**120453**	**Bovine**
Sheep	**136138**	**137563**	**136650**	**140649**	**147890**	**147381**	**148694**	**145596**	**Ovine**
Pigs	**178974**	**179690**	**182592**	**189379**	**188769**	**185873**	**181288**	**181016**	**Porcine**
Horses	**4926**	**4802**	**4630**	**4498**	**4391**	**4296**	**4275**	**4287**	**Chevaline**
Asses	**1152**	**1123**	**1088**	**1059**	**1038**	**1020**	**1013**	**1008**	**Asine**
Mules	**463**	**440**	**411**	**396**	**383**	**364**	**349**	**339**	**Mulassière**
Albania									**Albanie**
Cattle	620[1]	630[1]	650[1]	672	696	700	680[1]	650[1]	Bovine
Sheep	1350[1]	1370[1]	1400[1]	1432	1525	1598	1600[1]	1600[1]	Ovine
Pigs	200[1]	208[1]	210[1]	214	197	182	180[1]	170[1]	Porcine
Horses	65[1]	70[1]	80[1]	* 94	* 100	* 101	102[1]	100[1]	Chevaline
Asses[1]	52	52	52	53	53	53	53	53	Asine[1]
Mules[1]	22	22	22	23	23	23	23	23	Mulassière[1]
Austria									**Autriche**
Cattle	2633	2669	2651	2637	2590	2541	2562	2584	Bovine
Sheep	216	220	243	256	261	256	287	309	Ovine
Pigs	3881	4027	3926	3801	3947	3874	3773	3688	Porcine
Horses	42	41	45	44	45	44	48	49	Chevaline
Belgium-Luxembourg									**Belgique-Luxembourg**
Cattle	3184	3210	3163	3190	3159	3174	3069[1]	3000[1]	Bovine
Sheep	122	150	166	161	166	164	165[1]	166[1]	Ovine
Pigs	5253	5340	* 5484	5838	5958	6310	6440	* 6421	Porcine
Horses	29	28	26	24	23	22	21	20[1]	Chevaline
Mules	1	1	1[1]	1[1]	1[1]	1[1]	1[1]	1[1]	Mulassière
Bulgaria									**Bulgarie**
Cattle	1778	1751	1706	1678	1649	1613	1575	1457	Bovine
Sheep	10978	10501	9724	9563	8886	8609	8130	7938	Ovine
Pigs	3769	3734	3912	4050	4034	4119	4352	4187	Porcine
Horses	119	118	120	121	123	122	115	113	Chevaline
Asses	351	349	345	341	333	329	329	329	Asine
Mules	29	27	26	26	25	24	19	17	Mulassière
Czechoslovakia									**Tchécoslovaquie**
Cattle	5190	5150	5065	5073	5044	5075	5129	4923	Bovine
Sheep	1041	1068	1087	1104	1075	1047	1051	1030	Ovine

52
Livestock
Thousand head [*cont.*]
Cheptel
Milliers de têtes [*suite*]

Country or area	1984	1985	1986	1987	1988	1989	1990	1991	Pays ou zone
Pigs	7070	6743	6651	6833	7235	7384	7498	7090	Porcine
Horses	45	46	46	46	45	44	42	39	Chevaline
Denmark									**Danemark**
Cattle	2750	2623	2623	2490	2323	2232	2241	* 2220	Bovine
Sheep	55	52	52	70	73	86	100	* 111	Ovine
Pigs	8717	9104	9104	9422	9048	9120	9282	9489	Porcine
Horses	33	32	30	29	34	35	38	38[1]	Chevaline
Faeroe Islands									**Iles Féroé**
Cattle	2	2[1]	2[1]	2[1]	2[1]	2[1]	2[1]	2[1]	Bovine
Sheep	65	65	65	66	67[1]	67[1]	67[1]	67[1]	Ovine
Finland									**Finlande**
Cattle	1588	1592	1567	1485	1434	1379	1363	1315	Bovine
Sheep	60	68	70	66	63	59	61	57	Ovine
Pigs	1404	1256	1211	1309	1291	1327	1348	1290	Porcine
Horses	34	38	37	39	36	40	44	45	Chevaline
France									**France**
Cattle	* 23519	23481	23290	22803	22189	21780	21419	21446	Bovine
Sheep	13035	12676	12432	12044	12105	11943	11790	11490	Ovine
Pigs	11251	10975	11842	12419	12643	12410	12366	12239	Porcine
Horses	298	294	271	268	266	269	319	322	Chevaline
Asses[1]	23	23	24	24	25	25	25	25	Asine[1]
Mules	12	12[1]	12[1]	12[1]	12[1]	12[1]	12[1]	12[1]	Mulassière
Germany ʃ									**Allemagne ʃ**
F. R. Germany									**R. f. Allemagne**
Cattle	15552	15688	15627	15305	14887	14659	14563	14541	Bovine
Sheep	1218	1300	1296	1383	1414	1464	1533	1784	Ovine
Pigs	23449	23617	24282	24502	23670	22589	22165	22036	Porcine
Horses	354	370	370	367	365	375	391[1]	406	Chevaline
former German D. R.									**anc. R. d. allemande**
Cattle	5768	5848	5827	5804	5721	5710	5724	4947	Bovine
Sheep	2359	2528	2587	2647	2656	2634	2603	1456	Ovine
Pigs	13058	13191	12946	12840	12503	12464	12013	8783	Porcine
Horses	88	101	105	105	104	102	93[1]	85	Chevaline
Greece									**Grèce**
Cattle	755	725	723	703	741	731	* 687	* 634	Bovine
Sheep	8252	8258	8342	8617	10816	10376	* 10150	* 9759	Ovine
Pigs	1065	1061	1009	1000	1139	1226	1160	1143	Porcine
Horses	80	74	67	62	57	50[1]	50[1]	50[1]	Chevaline
Asses	199	188	177	167	157	150[1]	150[1]	145[1]	Asine
Mules	94	89	84	80	75	70[1]	65[1]	60[1]	Mulassière
Hungary									**Hongrie**
Cattle	1907	1901	1766	1725	1664	1690	1598	1571	Bovine
Sheep	2977	2832	2465	2337	2336	2215	2069	1865	Ovine
Pigs	9844	9237	8280	8687	8216	8327	7660	8000	Porcine
Horses	111	102	98	95	88	76	77	79	Chevaline
Asses	4	4[1]	4[1]	5	5[1]	4[1]	4[1]	4[1]	Asine
Iceland									**Islande**
Cattle	73	73	71	69	71	73	73[1]	73[1]	Bovine
Sheep	715	709	676	624	587	561	700[1]	700[1]	Ovine
Pigs[1]	12	14	15	18	18	17	19	19	Porcine[1]
Horses	52	54	56	59	64	69	70[1]	69[1]	Chevaline
Ireland									**Irlande**
Cattle	5812	5835	5779	5626	5580	5637	5899	6029	Bovine
Sheep	2813	3082	3304	3672	4301	4991	5782	6001	Ovine
Pigs	1053	1020	994	980	960	961	999	1069	Porcine
Horses	58	58	56	55	53	52	54	53[1]	Chevaline
Asses[1]	18	18	17	17	16	16	15	15	Asine[1]
Mules[1]	2	1	1	1	1	1	1	1	Mulassière[1]
Italy									**Italie**
Cattle	9113	9106	8908	8819	8794	8737	8746	8647[1]	Bovine
Sheep	10745	11098	11293	11451	11457	11623	11569	11575[1]	Ovine
Pigs	9187	9041	9169	9278	9383	9359	9254	* 9520	Porcine
Horses	253	246	248	253	250	256	269	280[1]	Chevaline
Asses	101	98	94	91	86	81	76	75[1]	Asine
Mules	59	57	56	52	50	47	43	40[1]	Mulassière

52
Livestock
Thousand head [*cont.*]
Cheptel
Milliers de têtes [*suite*]

Country or area	1984	1985	1986	1987	1988	1989	1990	1991	Pays ou zone
Liechtenstein									**Liechtenstein**
Cattle	6	6	6	6	6	6	6	6[1]	Bovine
Sheep	2	3	2	2	2	2	3	3[1]	Ovine
Pigs	4	3	3	3	3	3	3	3[1]	Porcine
Malta									**Malte**
Cattle	* 14	* 14	16[1]	18[1]	19[1]	21	21[1]	22[1]	Bovine
Sheep	* 5	* 5	5[1]	6[1]	6[1]	6	6[1]	6[1]	Ovine
Pigs	80[1]	95[1]	95[1]	95[1]	98[1]	101	101[1]	102[1]	Porcine
Horses	1[1]	1[1]	1[1]	1[1]	1[1]	1	1[1]	1[1]	Chevaline
Asses	1[1]	1[1]	1[1]	1[1]	1[1]	1	1[1]	1[1]	Asine
Netherlands									**Pays-Bas**
Cattle	5516	5248	5123	4895	4546	4606	* 4731	* 4830	Bovine
Sheep	766	814	868	985	* 1169	* 1405	* 1702	* 1800	Ovine
Pigs	11146	12383	13481	14349	* 14226	* 13820	* 13634	* 13788	Porcine
Horses	64	62	63	64	64[1]	64[1]	65[1]	65[1]	Chevaline
Norway									**Norvège**
Cattle	976	970	968	945	932	949	953	953[1]	Bovine
Sheep	2351	2415	2340	2248	2210	2183	2211	2211[1]	Ovine
Pigs	720	692	738	779	745	697	708	715[1]	Porcine
Horses	16	16	16	16	17	17	19	19[1]	Chevaline
Poland									**Pologne**
Cattle	11197	11055	10919	10523	10322	10733	10049	8844	Bovine
Sheep	4534	4837	4991	4739	4377	4409	4158	3234	Ovine
Pigs	16657	17614	18949	18546	19605	18835	19464	21868	Porcine
Horses	1537	1404	1272	1141	1051	973	941	939	Chevaline
Portugal									**Portugal**
Cattle	* 1283	* 1297	* 1310	* 1332	* 1356	* 1331	1335	1375	Bovine
Sheep	4900[1]	4950[1]	5000[1]	* 5100	* 5298	* 5354	* 5567	* 5673	Ovine
Pigs	* 3000	* 3127	* 3092	* 2454	2455	2331	2531	2664	Porcine
Horses	28[1]	28[1]	26[1]	26[1]	26[1]	26[1]	26[1]	26[1]	Chevaline
Asses[1]	178	175	175	170	170	170	170	170	Asine[1]
Mules[1]	90	85	85	80	80	80	80	80	Mulassière[1]
Romania									**Roumanie**
Cattle	* 6537	* 6791	* 6692	6703	6559	* 6416	* 6291	5381	Bovine
Sheep	18451	18637	17342	17219	16839	16210	15435	14062	Ovine
Pigs	14347	14777	13651	14095	14328	14351	11671	12003	Porcine
Horses	635[1]	660[1]	672	686	693	702	663	670	Chevaline
Asses[1]	35	35	35	36	36	36	35	35	Asine[1]
Spain									**Espagne**
Cattle	5004	5007	5086	5095	5061	5187	5126	* 5126	Bovine
Sheep	17554	17520	17894	20310	23798	22739	24037	* 24500	Ovine
Pigs	12001	11390	13387	17303	16614	16911	16002	* 16100	Porcine
Horses	252	253	248	249[1]	250[1]	241[1]	241[1]	241[1]	Chevaline
Asses	165	155	140	130[1]	131[1]	130[1]	130[1]	130[1]	Asine
Mules	148	139	117	115[1]	110[1]	100[1]	100[1]	100[1]	Mulassière
Sweden									**Suède**
Cattle	1878	1837	1716	1656	1662	1688	1718	* 1675	Bovine
Sheep	438	426	407	397	395	401	406	406[1]	Ovine
Pigs	2685	2589	2439	2234	2274	2264	2264	* 2170	Porcine
Horses[1]	57	57	57	57	58	58	58	58	Chevaline[1]
Switzerland									**Suisse**
Cattle	1943	1926	1902	1858	1837	1850	1855	1829	Bovine
Sheep	361	357	365	355	367	371	395	409	Ovine
Pigs	2004	1988	1973	1917	1941	1869	1787	1723	Porcine
Horses	48	46	48	48	49	48	45	49	Chevaline
Asses	2	2	2	2	2	2[1]	2[1]	2[1]	Asine
United Kingdom									**Royaume-Uni**
Cattle	13157	12985	12695	12476	11855	11902	11922	11846	Bovine
Sheep	23317	23946	24540	25976	27820	29045	29521	29954	Ovine
Pigs	7782	7793	7930	7955	7915	7626	7383	7379	Porcine
Horses	165[1]	165[1]	163[1]	165[1]	168[1]	168[1]	169[1]	170[1]	Chevaline
Asses[1]	10	10	10	10	10	10	10	10	Asine[1]
Yugoslavia									**Yougoslavie**
Cattle	5341	5199	5034	5030	4881	4759	4705	4527	Bovine
Sheep	7458	7679	7693	7819	7824	7564	7596	7431	Ovine
Pigs	9336	8673	7821	8459	8323	7396	7231	7358	Porcine

52

Livestock
Thousand head [*cont.*]
Cheptel
Milliers de têtes [*suite*]

Country or area	1984	1985	1986	1987	1988	1989	1990	1991	Pays ou zone
Horses	463	438	409	384	362	340	314	* 300	Chevaline
Asses	* 13	* 14	* 13	* 13	* 14	* 13	* 14	14[1]	Asine
Mules	6[1]	6[1]	6[1]	5[1]	5[1]	5[1]	4[1]	4[1]	Mulassière
Oceania									**Océanie**
Cattle	30505	31234	32298	30489	30489	30834	31831	32213	Bovine
Sheep	208986	217607	214251	213407	217050	222179	228156	219781	Ovine
Pigs	4320	4371	4405	4479	4575	4497	4474	4369	Porcine
Horses	590	586	517	478	503	482	478	480	Chevaline
Asses	12	12	12	9	8	9	9	9	Asine
American Samoa									**Samoa américaines**
Pigs[1]	10	10	11	11	11	11	11	11	Porcine[1]
Australia									**Australie**
Cattle	22161	22738	23436	21915	21851	22434	23191	23430	Bovine
Sheep	139242	149747	146776	149157	152443	161603	170297	162774	Ovine
Pigs	2527	2512	2553	2611	2706	2671	2648	2530	Porcine
Horses	429	416	346	313	335	317	310	310[1]	Chevaline
Asses	5[1]	5[1]	5[1]	2	1	2	2	2[1]	Asine
Cook Islands									**Iles Cook**
Pigs	17	17	17	17[1]	16	16	17	17	Porcine
Horses	2[1]	1[1]	0	0[1]	0	0	0	0	Chevaline
Fiji									**Fidji**
Cattle[1]	157	158	159	159	159	160	160	158	Bovine[1]
Pigs[1]	16	14	14	13	11	11	12	15	Porcine[1]
Horses[1]	40	41	41	41	42	42	43	43	Chevaline[1]
French Polynesia									**Polynésie française**
Cattle	7	7[1]	10	8	8	9	7	7[1]	Bovine
Sheep	1[1]	1[1]	0	0[1]	0[1]	0[1]	0	0[1]	Ovine
Pigs	34	* 45	32	32	32	32	32	32[1]	Porcine
Horses	2[1]	2[1]	2	2[1]	2[1]	2[1]	2	2[1]	Chevaline
Guam									**Guam**
Cattle	1	0	0	0	0	0	0[1]	0[1]	Bovine
Pigs	14	14[1]	14[1]	14[1]	14[1]	15[1]	15[1]	15[1]	Porcine
Kiribati									**Kiribati**
Pigs	10[1]	10[1]	* 10	10[1]	10[1]	9[1]	9[1]	9[1]	Porcine
Nauru									**Nauru**
Pigs[1]	2	2	2	2	2	2	3	3	Porcine[1]
New Caledonia									**Nouvelle-Calédonie**
Cattle	121	122[1]	125[1]	130[1]	136	121	120[1]	122[1]	Bovine
Sheep	3	3[1]	3[1]	2	3[1]	3[1]	3[1]	3[1]	Ovine
Pigs	35	35[1]	35[1]	35	36[1]	36[1]	37[1]	38[1]	Porcine
Horses	9	9[1]	9[1]	9[1]	9[1]	9[1]	9[1]	9[1]	Chevaline
New Zealand									**Nouvelle-Zélande**
Cattle	7777	7921	8279	7999	8058	7828	8065	8200	Bovine
Sheep	69739	67854	67470	64244	64600	60569	57852	57000	Ovine
Pigs	436	454	435	426	414	411	395	400	Porcine
Horses	95	99	100	94	97	98[1]	99[1]	100[1]	Chevaline
Niue									**Nioué**
Pigs	1[1]	1[1]	1[1]	1[1]	2[1]	2	2	2[1]	Porcine
Pacific Islands (Palau) [2]									**Iles du Pacifique (Palaos)**[2]
Cattle[1]	10	10	11	12	12	12	13	13	Bovine[1]
Pigs[1]	27	28	28	29	30	30	30	31	Porcine[1]
Papua New Guinea									**Papouasie-Nvl-Guinée**
Cattle	* 120	* 117	* 116	105[1]	* 101	101[1]	103[1]	105[1]	Bovine
Sheep	2	4	3	3[1]	3[1]	3[1]	4[1]	4[1]	Ovine
Pigs[1]	920	940	950	960	975	990	1000	1000	Porcine[1]
Horses	1	1	1	1[1]	1[1]	1[1]	1[1]	2[1]	Chevaline
Samoa									**Samoa**
Cattle	20	22[1]	25[1]	* 27	28[1]	29[1]	30[1]	31[1]	Bovine
Pigs[1]	50	50	52	53	54	55	57	57	Porcine[1]
Horses	3[1]	3[1]	3[1]	3[1]	3[1]	3[1]	3[1]	3[1]	Chevaline
Asses	7[1]	7[1]	7[1]	7[1]	7[1]	7[1]	7[1]	7[1]	Asine
Solomon Islands									**Iles Salomon**
Cattle	23	23	20	14	* 13	13[1]	13[1]	13[1]	Bovine
Pigs[1]	48	49	50	51	52	52	53	53	Porcine[1]

52
Livestock
Thousand head [*cont.*]
Cheptel
Milliers de têtes [*suite*]

Country or area	1984	1985	1986	1987	1988	1989	1990	1991	Pays ou zone
Tokelau									**Tokélaou**
Pigs[1]	1	1	1	1	1	1	1	1	Porcine[1]
Tonga									**Tonga**
Cattle	8	9	9[1]	10[1]	10	10	10[1]	10[1]	Bovine
Pigs	66	81	90[1]	105[1]	120	60	60[1]	60[1]	Porcine
Horses	9	11	11[1]	11[1]	11	6	6[1]	6[1]	Chevaline
Tuvalu									**Tuvalu**
Pigs	8	9	9	8	8	11	11[1]	12[1]	Porcine
Vanuatu									**Vanuatu**
Cattle	* 100	105	107	110	113	116	120[1]	123[1]	Bovine
Pigs	74[1]	75[1]	76[1]	77	58	58	60	60	Porcine
Horses	1	2[1]	3[1]	3	3[1]	4[1]	4[1]	4[1]	Chevaline
Wallis and Futuna Islands									**Îles Wallis et Futuna**
Pigs	24[1]	24[1]	24[1]	24[1]	24[1]	24[1]	24[1]	25[1]	Porcine
former USSR									**ancienne URSS**
Cattle	119558	121055	120888	122103	118300	119600	118400	115600	Bovine
Sheep	145265	142876	140850	142210	140783	140700	138400	* 134000	Ovine
Pigs	78722	77914	77772	79501	77403	78100	78900	75600	Porcine
Horses	5711	5800	5800	5900	5885	5904	5920	5900[1]	Chevaline
Asses	330	300	300	300	300	300[1]	300[1]	300[1]	Asine
Mules	2	1	1	1	1[1]	1[1]	1[1]	1[1]	Mulassière

Source:
Food and Agriculture Organization of the United Nations
(Rome).

Source:
Organisation des Nations Unies pour l'alimentation et l'agriculture
(Rome).

ℊ All data shown which pertain to Germany prior to 3 October 1990 are indicated separately for the Federal Republic of Germany and the former German Democratic Republic based on their respective territories at the time indicated. Where data for united Germany (3 October 1990 and thereafter) are not available, available data are shown separately under the designations Federal Republic of Germany and former German Democratic Republic and pertain to the territorial boundaries prior to 3 October 1990. For detailed explanatory notes on data pertaining to Germany, see Annex I - Country Nomenclature.

1 FAO estimate.
2 Including data for Federated States of Micronesia, Marshall Is. and Northern Mariana Is.

ℊ Toutes les données se rapportant à l'Allemagne avant le 3 octobre 1990 figurent dans deux rubriques séparées basées sur les territoires respectifs de la République fédérale d'Allemagne et l'ancienne République démocratique allemande selon la période indiquée. En l'absence de données pour l'Allemagne unifiée (à compter du 3 octobre 1990), les données disponibles sont fournies séparément sous les rubriques République fédérale d'Allemagne et ancienne République démocratique allemande et se rapportent aux limites territoriales antérieures au 3 octobre 1990. Pour les notes explicatives en détail sur les données concernant l'Allemagne, voir Annexe I-Nomenclature des pays.

1 Estimation de la FAO.
2 Y compris les données pour les Etats fédérés de Micronésie, les îles Marshall et les îles Mariannes du Nord.

53

Wool
Laine

Production: thousand metric tons
Production : milliers de tonnes métriques

Country or area Pays ou zone	1981/82	1982/83	1983/84	1984/85	1985/86	1986/87	1987/88	1988/89	1989/90	1990/91
World *Monde*	2 896	2 910	2 930	3 010	3 002	3 069	3 128	3 219	3 352	3 358
Africa Afrique	209	214	212	227	225	217	225	231	238	242
Lesotho Lesotho	3	3	3	3	3	3	3	3	3	3
Morocco Maroc	21	26	24	27	31	32	34	34	35	35
South Africa Afrique du Sud	110	112	108	104	98	90	92	96	99	102
Not specified[1] Non-dénommée[1]	75	73	77	93	93	92	96	98	101	102
America, North Amérique du Nord	53	51	49	46	41	40	40	42	43	42
Canada Canada	2	2	2	2	1	1	1	1	2	2
United States Etats-Unis	51	49	47	44	40	39	39	41	41	40
America, South Amérique du Sud	330	329	324	305	322	319	327	330	326	322
Argentina Argentine	168	162	162	150	152	150	157	164	149	146
Brazil Brésil	30	28	25	30	29	26	26	26	26	26
Chile Chili	22	22	21	21	20	19	19	20	20	20
Peru Pérou	12	12	12	11	11	11	11	12	12	12
Uruguay Uruguay	76	82	82	71	87	90	89	83	94	93
Not specified[2] Non-dénommée[2]	22	23	22	22	23	23	25	25	25	25
Asia Asie	483	501	496	477	470	472	496	516	534	545
China Chine	189	202	194	183	178	185	209	222	237	240
India Inde	35	35	35	35	35	30	27	30	32	32
Iran, Islamic Rep. of Iran, Rép. islamique d'	34	34	33	32	32	32	32	32	32	32
Iraq Iraq	18	18	17	17	17	17	17	17	17	17
Pakistan Pakistan	40	41	45	48	49	50	53	55	57	61
Turkey Turquie	84	84	85	85	85	85	85	85	82	85

53

Wool
Production: thousand metric tons [cont.]
Laine
Production : milliers de tonnes métriques [suite]

Country or area Pays ou zone	1981/82	1982/83	1983/84	1984/85	1985/86	1986/87	1987/88	1988/89	1989/90	1990/91
Not specified[3] Non-dénommée[3]	83	87	87	77	74	73	73	75	77	78
Europe **Europe**	**281**	**290**	**295**	**303**	**309**	**312**	**317**	**322**	**323**	**322**
Bulgaria Bulgarie	35	35	36	36	34	33	32	31	28	26
France France	25	25	25	24	24	24	24	23	23	23
Germany† · Allemagne† F. R. Germany R. f. Allemagne	5	5	5	5	5	6	6	6	7	7
former German D. R. anc. R. d. allemande	12	13	12	15	15	16	17	17	17	14
Greece Grèce	10	10	9	9	9	10	10	10	9	9
Hungary Hongrie	12	13	13	12	11	10	10	10	10	9
Ireland Irlande	7	7	7	8	9	10	11	13	15	17
Italy Italie	13	13	13	13	13	13	14	14	14	14
Poland Pologne	11	12	13	15	17	18	17	16	14	11
Portugal Portugal	9	9	9	9	9	9	9	9	9	9
Romania Roumanie	36	39	39	42	44	40	39	38	35	33
Spain Espagne	28	30	31	31	32	32	33	38	39	42
United Kingdom Royaume-Uni	51	50	54	56	58	59	62	65	70	75
Yugoslavia Yougoslavie	10	10	10	10	10	10	10	10	10	10
Not specified[4] Non-dénommée[4]	17	19	19	18	19	22	23	22	23	23
Oceania **Océanie**	**1 080**	**1 073**	**1 092**	**1 187**	**1 188**	**1 240**	**1 262**	**1 300**	**1 409**	**1 414**
Australia Australie	717	702	728	814	830	890	916	959	1 100	1 110
New Zealand Nouvelle-Zélande	363	371	364	373	358	350	346	341	309	304
former USSR **ancienne URSS**	**460**	**452**	**462**	**465**	**447**	**469**	**461**	**478**	**479**	**471**
Ukraine Ukraine	(29)	(28)	(30)	(32)	(31)	(32)	(32)	(32)	(32)	(29)

53

Wool
Production: thousand metric tons [*cont.*]

Laine
Production : milliers de tonnes métriques [*suite*]

Source:
Commonwealth Secretariat (London).

† All data shown which pertain to Germany prior to 3 October
1990 are indicated separately for the Federal Republic of
Germany and the former German Democratic Republic based on
their respective territories at the time indicated. Where
data for united Germany (3 October 1990 and thereafter) are
not available, available data are shown separately under the
designations Federal Republic of Germany and former German
Democratic Republic and pertain to the territorial
boundaries prior to 3 October 1990. For detailed
explanatory notes on data pertaining to Germany, see Annex I
- Country Nomenclature.

1 Principally Algeria, Ethiopia, Tunisia and Namibia.
2 Principally Bolivia and Falkland Islands (Malvinas).
3 Principally Afghanistan, Mongolia, Nepal, Saudi Arabia and
Syria.
4 Principally small Western European producers.

Source:
"Commonwealth Secretariat" (Londres).

† Toutes les données se rapportant à l'Allemagne avant le 3
octobre 1990 figurent dans deux rubriques séparées basées
sur les territoires respectifs de la République fédérale
d'Allemagne et l'ancienne République démocratique allemande
selon la période indiquée. En l'absence de données pour
l'Allemagne unifiée (à compter du 3 octobre 1990), les
données disponibles sont fournies séparément sous les
rubriques République fédérale d'Allemagne et ancienne
République démocratique allemande et se rapportent aux
limites territoriales antérieures au 3 octobre 1990. Pour
les notes explicatives en détail sur les données concernant
l'Allemagne, voir Annexe I – Nomenclature des pays.

1 Principalement Algérie, Ethiopie, Namibie et Tunisie.
2 Principalement la Bolivie et les Isles Falkland (Malvinas).
3 Principalement Afghanistan, Jordanie, Mongolie, Népal,
Arabie saoudite et Syrie.
4 Principalement les producteurs peu importants de l'Europe
occidentale.

54
Roundwood
Bois rond
Production (solid volume of roundwood without bark): million cubic metres
Production (volume solide de bois rond sans écorce) : millions de mètres cubes

Country or area Pays ou zone	1981	1982	1983	1984	1985	1986	1987	1988	1989	1990
World *Monde*	2 937.5	2 933.7	3 048.0	3 163.3	3 199.7	3 304.9	3 371.9	3 425.9	3 455.6	3 450.4
Africa **Afrique**	397.6	408.6	421.9	432.9	447.3	459.3	472.6	485.7	498.1	511.2
Algeria [1] Algérie [1]	1.7	1.7	1.8	1.8	1.9	1.9	2.0	2.1	2.1	2.2
Angola Angola	5.2	5.2	5.4	5.6	5.7	5.8	6.0	6.1	6.3	6.4
Benin Bénin	3.9 [1]	4.0 [1]	4.1 [1]	4.2	4.3	4.5 [1]	4.6	4.7 [1]	4.9 [1]	5.0 [1]
Botswana [1] Botswana [1]	1.0	1.0	1.1	1.1	1.2	1.2	1.2	1.3	1.3	1.4
Burkina Faso Burkina Faso	6.9 [1]	7.1 [1]	7.3 [1]	7.5	7.7	7.9 [1]	8.1 [1]	8.3 [1]	8.5 [1]	8.7
Burundi Burundi	3.3	3.3	3.4	3.5	3.6	3.8	3.9 [1]	4.0	4.1	4.2
Cameroon Cameroun	10.5	10.8	11.2	11.6	12.1	12.4	12.7	13.0	13.5	14.2
Central African Rep. Rép. centrafricaine	3.1	3.1	3.2	3.3	3.4	3.4	3.5	3.5 [1]	3.5	3.5
Chad [1] Tchad [1]	3.3	3.3	3.4	3.5	3.6	3.7	3.7	3.8	3.9	4.0
Congo Congo	2.3	2.4	2.4	2.5	2.6	2.8	3.1	3.5	3.6	3.6 [1]
Côte d'Ivoire Côte d'Ivoire	11.5	11.9	12.1	12.3	12.0	12.1	12.0	11.8	12.2	12.7
Egypt [1] Egypte [1]	1.8	1.8	1.9	1.9	2.0	2.0	2.1	2.1	2.2	2.2
Equatorial Guinea Guinée équatoriale	0.5	0.5 [1]	0.6	0.5	0.6	0.6 [1]	0.6 [1]	0.6 [1]	0.6 [1]	0.6 [1]
Ethiopia Ethiopie	34.3 [1]	35.2	36.0	36.7 [1]	37.5 [1]	38.3 [1]	39.3 [1]	40.3	41.4	42.5 [1]
Gabon Gabon	3.0	3.1	3.3	3.6	3.5	3.5	3.5	3.6	3.7	3.8
Gambia Gambie	0.9 [1]	0.7	0.8	0.9	0.9	0.9 [1]	0.9 [1]	0.9 [1]	0.9 [1]	0.9 [1]
Ghana Ghana	13.4	14.3	16.2	16.4	16.5	16.6	16.8	17.0	17.2	17.2 [1]
Guinea Guinée	3.2 [1]	3.1	3.4 [1]	3.4 [1]	3.5 [1]	3.6 [1]	3.7 [1]	3.8 [1]	3.9 [1]	4.0 [1]
Guinea-Bissau [1] Guinée-Bissau [1]	0.6	0.6	0.6	0.6	0.6	0.6	0.6	0.6	0.6	0.6
Kenya Kenya	25.6	26.6	27.6	28.8	29.8 [1]	30.9 [1]	32.0 [1]	33.2	34.4	35.6
Lesotho [1] Lesotho [1]	0.5	0.5	0.5	0.5	0.5	0.5	0.6	0.6	0.6	0.6
Liberia Libéria	4.3	4.4	4.1	4.2	4.5	5.3	5.6	5.9	6.0 [1]	6.1 [1]
Libyan Arab Jamah. [1] Jamah. arabe libyenne [1]	0.6	0.6	0.6	0.6	0.6	0.6	0.6	0.6	0.6	0.6
Madagascar Madagascar	6.3	6.5	6.6	6.8	7.0	7.2	7.4 [1]	7.6	7.9	8.1

54
Roundwood
Production (solid volume of roundwood without bark): million cubic metres [*cont.*]
Bois rond
Production (volume solide de bois rond sans écorce) : millions de mètres cubes [*suite*]

Country or area Pays ou zone	1981	1982	1983	1984	1985	1986	1987	1988	1989	1990
Malawi Malawi	6.1	6.2	6.4	6.6	6.9	7.1	7.4[1]	7.6	7.9	8.2
Mali Mali	4.3	4.4	4.5	4.7	4.8	4.9	5.1	5.3	5.4[1]	5.6[1]
Morocco Maroc	1.5	1.6	1.8	2.0[1]	2.2	2.0	2.0[1]	1.9	2.1	2.1[1]
Mozambique Mozambique	13.7	14.4	14.8	15.0	15.2	15.6	16.0	16.0	16.0[1]	16.0[1]
Niger[1] Niger[1]	3.7	3.8	4.0	4.1	4.2	4.4	4.5	4.7	4.8	5.0
Nigeria Nigéria	83.0[1]	84.8[1]	86.7[1]	89.5[1]	92.5	95.3	98.3[1]	101.3[1]	104.5[1]	107.7[1]
Rwanda Rwanda	5.0	5.2	5.3	5.6	5.8	5.8[1]	5.8[1]	5.8[1]	5.8[1]	5.8[1]
Senegal Sénégal	3.6	3.6	3.9	4.0	4.0	4.1[1]	4.2[1]	4.3[1]	4.4[1]	4.5[1]
Sierra Leone Sierra Leone	2.5	2.6	2.6	2.7[1]	2.7	2.8	2.9[1]	2.9[1]	3.0[1]	3.1[1]
Somalia[1] Somalie[1]	5.3	5.5	5.7	5.9	6.1	6.3	6.5	6.7	6.9	7.1
South Africa[2] Afrique du Sud[2]	19.5	19.5	20.6	19.0	19.0[1]	18.6	19.0	19.4	19.4[1]	19.4[1]
Sudan Soudan	17.5	18.0	18.6	19.2	19.8	20.4	21.0[1]	21.6	22.2	22.8
Swaziland Swaziland	2.2	2.2[1]	2.2[1]	2.2[1]	2.2[1]	2.2[1]	2.2[1]	2.2[1]	2.2[1]	2.2[1]
Togo[1] Togo[1]	0.7	0.7	0.7	0.8	0.8	0.8	0.8	0.9	0.9	0.9
Tunisia Tunisie	2.6	2.6	2.7	2.8	2.9	2.9	3.0	3.1	3.2	3.2
Uganda Ouganda	11.0	11.4	11.8	12.2	12.6	13.1	13.6[1]	14.1	14.6	15.1
United Rep.Tanzania Rép. Unie de Tanzanie	24.3	25.2	26.2	27.2	28.2	29.5	30.8	31.9[1]	33.1[1]	34.3[1]
Zaire Zaïre	27.4	28.3	29.1	30.0	33.7	34.7	35.6[1]	37.1[1]	37.7[1]	38.9[1]
Zambia Zambie	9.2[1]	9.6[1]	10.0[1]	10.4[1]	10.8[1]	11.3[1]	11.7[1]	12.2[1]	12.2	12.2
Zimbabwe Zimbabwe	6.8	7.1	6.8	7.1	7.2	7.3	7.6	7.8	7.9[1]	7.9[1]
America, North **Amérique du Nord**	**615.9**	**579.7**	**648.2**	**697.3**	**701.5**	**745.9**	**766.2**	**769.9**	**763.0**	**717.9**
Bahamas[1] Bahamas[1]	0.1	0.1	0.1	0.1	0.1	0.1	0.1	0.1	0.1	0.1
Belize Belize	0.1	0.1	0.1	0.2	0.2	0.2	0.2	0.2[1]	0.2[1]	0.2[1]
Canada Canada	144.6	129.7	157.0	167.5	168.7	177.1	177.1	180.1	169.1	155.5
Costa Rica Costa Rica	3.5	3.3	3.2	3.4	3.5	3.6	3.9	4.0	4.0	4.1
Cuba Cuba	3.3	3.4	3.3	3.3	3.3	3.4	3.2	3.3	3.1	3.1

54
Roundwood
Production (solid volume of roundwood without bark): million cubic metres [cont.]
Bois rond
Production (volume solide de bois rond sans écorce) : millions de mètres cubes [suite]

Country or area Pays ou zone	1981	1982	1983	1984	1985	1986	1987	1988	1989	1990
Dominican Republic Rép. dominicaine	0.9	0.9	1.0	1.0	1.0	1.0	1.0[1]	1.0	1.0	1.0
El Salvador[1] El Salvador[1]	4.0	4.0	4.0	4.0	4.1	4.2	4.2	4.4	4.5	4.6
Guatemala Guatemala	6.2	6.3	6.5	6.6	6.9	7.0	7.2	7.4	7.6	7.8
Haiti[1] Haïti[1]	4.9	5.0	5.1	5.2	5.3	5.4	5.5	5.6	5.7	5.8
Honduras Honduras	5.0	5.1	4.8	5.1[1]	5.4	5.5	5.8	6.0	6.1	6.2
Jamaica Jamaïque	0.1	0.0	0.1	0.1	0.1[1]	0.2	0.2	0.2	0.2	0.2
Mexico Mexique	18.9	19.5	19.8	20.8	21.4	21.3[1]	21.9[1]	22.1[1]	22.3	22.2
Nicaragua[1] Nicaragua[1]	3.2	3.3	3.4	3.5	3.6	3.7	3.8	3.9	4.0	4.1
Panama[3] Panama[3]	1.9[1]	1.9[1]	1.9[1]	1.9[1]	1.9[1]	1.8[1]	1.8[1]	1.8[1]	1.9	1.9
Trinidad and Tobago Trinité-et-Tobago	0.1[1]	0.1	0.1	0.1	0.1	0.1	0.1[1]	0.1	0.1	0.1
United States Etats-Unis	419.0	396.9	437.8	474.6	476.1	511.5	530.1	529.8	533.2	501.0
America, South Amérique du Sud	**283.1**	**287.4**	**296.6**	**305.1**	**311.4**	**318.7**	**325.2**	**331.6**	**336.3**	**341.9**
Argentina Argentine	8.9	9.3	11.5	11.3	11.1	10.8	10.8[1]	10.8	10.8	10.8
Bolivia Bolivie	1.4	1.3	1.3	1.3	1.4	1.4	1.4[1]	1.5	1.6	1.6
Brazil Brésil	215.1[1]	220.1[1]	225.4[1]	231.5	236.3	241.1	246.3	251.7	255.5	259.2[1]
Chile Chili	13.9	12.8	13.9	15.0	15.7	16.4	16.6	16.8	17.4	18.7
Colombia Colombie	16.9	16.9	17.2[1]	17.5[1]	17.8[1]	18.1[1]	18.4[1]	18.7[1]	19.1[1]	19.4[1]
Ecuador Equateur	7.6	7.6	7.9	8.2	8.6	8.9	9.3	9.5	9.7	10.2
French Guiana[1] Guyane française[1]	0.3	0.3	0.3	0.3	0.3	0.3	0.3	0.3	0.3	0.3
Guyana Guyana	0.2[1]	0.2[1]	0.2[1]	0.2	0.2	0.2	0.2	0.2[1]	0.2[1]	0.2[1]
Paraguay Paraguay	6.8[1]	6.8[1]	6.8[1]	7.5	7.7	8.2	8.5	8.4	8.4[1]	8.4[1]
Peru Pérou	7.8	7.8	7.8[1]	7.7	7.7	8.5	8.7	8.8	8.6	8.1
Suriname Suriname	0.3	0.3	0.2	0.2	0.2	0.2	0.2	0.2	0.2	0.1
Uruguay Uruguay	2.8	2.9	2.9	3.1	3.3	3.3	3.3[1]	3.3	3.3	3.3
Venezuela Venezuela	1.2[1]	1.2[1]	1.3[1]	1.3[1]	1.3	1.3	1.3[1]	1.5	1.5[1]	1.5[1]
Asia Asie	**913.4**	**932.6**	**953.1**	**979.0**	**987.9**	**1 011.7**	**1 031.7**	**1 050.7**	**1 067.6**	**1 074.7**

54

Roundwood
Production (solid volume of roundwood without bark): million cubic metres [*cont.*]
Bois rond
Production (volume solide de bois rond sans écorce) : millions de mètres cubes [*suite*]

Country or area Pays ou zone	1981	1982	1983	1984	1985	1986	1987	1988	1989	1990
Afghanistan[1] Afghanistan[1]	6.2	6.1	5.9	5.8	5.8	5.8	5.9	6.0	6.2	6.5
Bangladesh Bangladesh	24.5[1]	25.3	25.8	26.5	27.1	27.9	28.6[1]	29.4[1]	30.1[1]	30.9[1]
Bhutan[1] Bhoutan[1]	3.2	3.2	3.2	3.2	3.2	3.2	3.2	3.2	3.2	3.2
Brunei Darussalam Brunéi Darussalam	0.3	0.3	0.3	0.3	0.3	0.3	0.3[1]	0.3	0.3	0.3
Cambodia[1] Cambodge[1]	4.8	4.9	5.0	5.2	5.3	5.4	5.6	5.7	5.8	5.9
China[1] Chine[1]	233.5	239.0	245.4	258.4	265.3	271.4	276.5	279.6	281.8	277.0
Cyprus Chypre	0.1	0.1	0.1	0.1	0.1	0.1	0.1	0.1	0.1	0.1
Hong Kong[1] Hong-kong[1]	0.2	0.2	0.2	0.2	0.2	0.2	0.2	0.2	0.2	0.2
India[1] Inde[1]	226.8	232.3	237.9	243.6	249.4	254.3	259.2	264.2	269.3	274.5
Indonesia Indonésie	144.5	146.0	151.7	155.6	154.7	161.2	161.8	165.6	169.8	171.5
Iran, Islamic Rep. of Iran, Rép. islamique d'	6.7[1]	6.7[1]	6.7[1]	6.7[1]	6.7[1]	6.7[1]	6.7[1]	6.7	6.7	6.7[1]
Iraq[1] Iraq[1]	0.1	0.1	0.1	0.1	0.1	0.1	0.1	0.1	0.1	0.2
Israel Israël	0.1	0.1	0.1[1]	0.1[1]	0.1[1]	0.1[1]	0.1[1]	0.1[1]	0.1	0.1
Japan Japon	32.0	32.8	31.6	33.1	33.5	32.1	31.3	31.1	31.0	29.8
Korea, Dem. P. R.[1] Corée, R. p. dém. de[1]	4.3	4.4	4.4	4.5	4.5	4.5	4.6	4.6	4.6	4.7
Korea, Republic of Corée, République de	8.6	8.4	8.4	7.7	6.7	6.7[1]	6.9[1]	6.8[1]	6.8	6.6
Lao People's Dem. Rep. Rép. dém. pop. lao	3.2	3.3	3.4	3.5	3.7	3.7	3.8[1]	3.9[1]	4.0[1]	4.1[1]
Lebanon Liban	0.5	0.5	0.5	0.5[1]	0.5[1]	0.5[1]	0.5	0.5	0.5	0.5
Malaysia Malaisie	38.7	41.0	41.2	39.7	37.5	39.0	44.4	48.5	50.8	51.0[1]
Mongolia[1] Mongolie[1]	2.4	2.4	2.4	2.4	2.4	2.4	2.4	2.4	2.4	2.4
Myanmar Myanmar	18.5	18.9	18.9	19.6	19.8	20.0	20.2	20.6	21.8	23.2
Nepal[1] Népal[1]	14.6	15.0	15.4	15.8	16.2	16.6	17.0	17.4	17.8	18.2
Pakistan Pakistan	18.0	18.8	19.6	20.6	21.5	22.2	22.8	23.6	24.8	26.6
Philippines Philippines	34.7	34.5	35.4	36.1	35.5	36.1	37.6	38.0	38.2	38.6
Sri Lanka Sri Lanka	8.1	8.3	8.5	8.6	8.5	8.6	8.8[1]	8.9[1]	9.0	9.1
Syrian Arab Republic Rép. arabe syrienne	0.0[1]	0.0	0.1	0.0[1]	0.0[1]	0.0[1]	0.0[1]	0.0[1]	0.0	0.1

54

Roundwood
Production (solid volume of roundwood without bark): million cubic metres [*cont.*]

Bois rond
Production (volume solide de bois rond sans écorce) : millions de mètres cubes [*suite*]

Country or area Pays ou zone	1981	1982	1983	1984	1985	1986	1987	1988	1989	1990
Thailand Thaïlande	33.7	34.3	35.0	35.9	36.3	37.1	37.8	38.2	37.6	37.7
Turkey Turquie	21.9	21.9	20.8	19.7	16.3	17.9	16.8	16.4	15.5	15.5[1]
Viet Nam[1] Viet Nam[1]	23.0	23.8	24.7	25.5	26.4	27.2	28.0	28.2	28.6	29.2
Yemen[14] Yémen[14]	0.3	0.3	0.3	0.3	0.3	0.3	0.3	0.3	0.3	0.3
Europe Europe	**333.4**	**333.8**	**338.0**	**345.7**	**345.4**	**347.3**	**347.5**	**355.0**	**368.4**	**398.6**
Albania[1] Albanie[1]	2.3	2.3	2.3	2.3	2.3	2.3	2.3	2.3	2.3	2.3
Austria Autriche	14.3	13.4	13.6	14.2	14.2	13.7	13.6	15.0	16.3	17.3
Belgium-Luxembourg Belgique-Luxembourg	2.7	2.9	3.0	3.1	3.3	3.3	3.7	4.0	4.8	4.7
Bulgaria Bulgarie	5.0	4.9	4.8	4.8	4.8	4.5	4.4	4.5	4.3	4.1
Czechoslovakia Tchécoslovaquie	18.8	18.9	18.8	18.9	19.0	18.9	18.7	18.1	18.2	17.9
Denmark Danemark	2.2	3.7	3.0	2.6	2.3	2.3	2.1	2.2	2.1	2.1[1]
Finland Finlande	43.9	37.7	38.4	40.5	41.7	40.7	41.9	44.9	46.3	41.6
France France	37.9	37.4	38.7	38.8	38.9	39.9	41.1	43.0	44.7	44.7[1]
Germany† · Allemagne† F. R. Germany R. f. Allemagne	31.3	31.0	28.2	30.7	30.7[1]	30.3	31.0	32.6	36.9	73.5
former German D. R. anc. R. d. allemande	10.1	10.3	10.4	10.6	10.9	10.8	10.6	10.9	11.3	11.3[1]
Greece Grèce	2.7	2.4	2.9	2.7	3.0	3.2	2.9	2.8	2.4	2.0
Hungary Hongrie	6.5	6.5	6.4	6.3	6.8	7.0	6.8	6.6	6.6	6.6[1]
Ireland Irlande	0.4	1.1	1.0	1.3	1.3	1.2	1.2[1]	1.4	1.5	1.5[1]
Italy Italie	9.1	8.7	8.4	9.2	9.4	9.6	9.1	9.1	8.8	8.0
Netherlands Pays-Bas	0.9	0.9	0.9	0.9	1.1	1.1	1.2	1.3	1.4	1.4
Norway Norvège	10.4	9.5	9.6	10.0	9.5	9.9	10.4	11.0	11.5	11.8
Poland Pologne	20.7	22.1	24.7	24.0	23.3	24.3	23.3	22.8	21.4	19.6
Portugal Portugal	7.6	7.8	8.6	8.5	9.4	9.9	9.4	9.4	10.3	10.4
Romania Roumanie	20.4	23.0	22.9	24.1	23.0	20.7	19.9	18.2	17.3	17.3[1]
Spain Espagne	13.3	14.5	15.0	14.6	14.0	15.0	15.5	15.2	17.4	17.8
Sweden Suède	49.9	50.9	53.0	53.0	51.5	52.4	53.1	53.9	55.9	55.9

54
Roundwood
Production (solid volume of roundwood without bark): million cubic metres [*cont.*]
Bois rond
Production (volume solide de bois rond sans écorce) : millions de mètres cubes [*suite*]

Country or area Pays ou zone	1981	1982	1983	1984	1985	1986	1987	1988	1989	1990
Switzerland Suisse	4.6	4.2	4.0	5.0	4.2	4.7	4.6	4.6	4.6	6.8
United Kingdom Royaume-Uni	4.3	4.1	4.0	3.9	4.8	5.2	5.5	6.1	6.5	6.5
Yugoslavia Yougoslavie	14.1	15.3	15.4	15.9	15.9	16.1	15.2	15.2	15.6	13.6
Oceania **Océanie**	**36.0**	**35.7**	**34.5**	**35.3**	**38.1**	**39.2**	**38.9**	**39.4**	**40.0**	**41.5**
Australia Australie	17.7	16.9	16.3	18.0	19.7	20.0	20.0[1]	20.3	20.0	20.3
Fiji Fidji	0.3	0.2	0.2	0.2	0.2	0.2	0.3	0.3	0.3	0.3[1]
New Zealand Nouvelle-Zélande	10.3	10.0	9.7	8.8	9.8	10.3	9.7	9.8	10.7	12.0
Papua New Guinea Papouasie-Nvl-Guinée	7.1	7.8	7.6	7.6	7.6	7.9[1]	8.2[1]	8.2[1]	8.2[1]	8.2
Samoa[1] Samoa[1]	0.1	0.1	0.1	0.1	0.1	0.1	0.1	0.1	0.1	0.1
Solomon Islands Iles Salomon	0.5	0.5	0.5	0.5	0.5	0.6	0.4	0.4	0.4	0.4
Vanuatu Vanuatu	0.0	0.0	0.0	0.0[1]	0.0[1]	0.0[1]	0.1[1]	0.1[1]	0.1[1]	0.1[1]
former USSR **ancienne URSS**	358.2	355.9	355.7	367.9	368.0	382.8[1]	389.8[1]	393.7	382.1	364.6

Source:
Food and Agriculture Organization of the United Nations (Rome).

Source:
Organisation des Nations Unies pour l'alimentation et l'agriculture (Rome).

† All data shown which pertain to Germany prior to 3 October 1990 are indicated separately for the Federal Republic of Germany and the former German Democratic Republic based on their respective territories at the time indicated. Where data for united Germany (3 October 1990 and thereafter) are not available, available data are shown separately under the designations Federal Republic of Germany and former German Democratic Republic and pertain to the territorial boundaries prior to 3 October 1990. For detailed explanatory notes on data pertaining to Germany, see Annex I - Country Nomenclature.

1 FAO estimate.
2 Including data for Namibia.
3 Excluding former Canal Zone.
4 Former Democratic Yemen only.

† Toutes les données se rapportant à l'Allemagne avant le 3 octobre 1990 figurent dans deux rubriques séparées basées sur les territoires respectifs de la République fédérale d'Allemagne et l'ancienne République démocratique allemande selon la période indiquée. En l'absence de données pour l'Allemagne unifiée (à compter du 3 octobre 1990), les données disponibles sont fournies séparément sous les rubriques République fédérale d'Allemagne et ancienne République démocratique allemande et se rapportent aux limites territoriales antérieures au 3 octobre 1990. Pour les notes explicatives en détail sur les données concernant l'Allemagne, voir Annexe I – Nomenclature des pays.

1 Estimation de la FAO.
2 Y compris les données pour la Namibie.
3 Non compris l'ancienne Zone du Canal.
4 L'ancien Yémen démocratique seulement.

55
Natural rubber
Caoutchouc naturel
Production: thousand metric tons
Production : milliers de tonnes métriques

Country or area Pays ou zone	1982	1983	1984	1985	1986	1987	1988	1989	1990	1991
World *Monde*	3 750.0	4 030.0	4 255.0	4 400.0	4 490.0	4 850.0	5 130.0	5 240.0	5 210.0	5 340.0
Brazil Brésil	32.8	35.2	36.0	40.4	32.6	26.6	32.9	33.0	* 33.0	* 35.0
Cambodia Cambodge	* 9.0	* 12.0	13.4	22.0	* 24.0	* 25.0	* 26.5	* 28.0	* 38.0	* 31.5
Cameroon Cameroun	17.4	15.6	17.9	17.9	19.9	26.3	33.2	35.9	38.3	43.4
China Chine	152.6	172.4	188.8	187.9	209.7	237.6	239.8	242.8	264.2	* 269.0
Côte d'Ivoire Côte d'Ivoire	26.7	30.5	35.5	40.6	49.2	54.6	61.0	67.0	69.2	* 76.0
Ghana Ghana	* 2.0	* 1.5	* 1.0	* 1.0	* 1.0	* 1.3	1.5	2.9	* 3.6	* 3.9
Guatemala * Guatemala *	10.0	11.0	11.0	12.0	12.0	14.0	15.0	18.0	21.0	21.0
India Inde	165.9	168.0	183.9	198.3	219.0	227.4	254.8	288.6	323.5	360.2
Indonesia Indonésie	* 880.0	997.2	1 115.6	1 130.6	1 049.2	1 203.3	1 235.0	1 256.0	* 1 262.0	* 1 346.0
Liberia Libéria	68.0	76.5	86.4	84.4	88.7	* 93.0	108.4	* 106.0	* 40.0	* 19.0
Malaysia[1] Malaisie[1]	1 494.2	1 563.9	1 530.6	1 469.4	1 538.6	1 578.7	1 661.6	1 415.3	1 291.5	1 252.9
Myanmar Myanmar	15.3	15.8	15.2	* 17.0	* 17.0	14.9	* 16.0	14.4	14.9	* 15.0
Nigeria Nigéria	* 47.0	* 50.0	* 51.0	51.6	36.8	55.3	80.5	118.4	152.0	* 161.0
Papua New Guinea[1] Papouasie-Nvl-Guinée[1]	2.3	2.7	3.0	5.1	4.9	3.8	4.5	4.1	* 3.5	* 3.7
Philippines[2,3] Philippines[2,3]	101.0	123.0	135.0	150.0	146.0	147.2	156.4	171.9	* 168.0	* 156.0
Sri Lanka Sri Lanka	126.2	140.0	141.9	137.5	137.8	121.8	122.4	110.7	113.1	103.9
Thailand[4] Thaïlande[4]	552.2	587.2	628.6	724.0	786.0	925.6	978.9	1 178.9	1 271.1	* 1 351.0
Viet Nam Viet Nam	44.9	48.0	* 47.0	52.5	* 54.0	* 60.0	* 66.0	* 71.0	* 60.0	* 64.5
Zaire Zaïre	16.3	21.6	16.0	* 16.0	* 17.0	* 18.0	* 19.0	* 24.0	* 10.0	* 13.5

Source:
International Rubber Study Group (London).

1 Net exports (Malaysia: Sabah and Sarawak only).
2 Crop year ending June 30th.
3 Beginning 1983 revised data are provided by the Ministry of Agriculture.
4 Exports plus consumption adjusted for variation in stocks.

Source:
Groupe international d'études du caoutchouc (Londres).

1 Exportations nettes (Malaysie: Sabah et Sarawak seulement).
2 Production par campgne se terminant au 30 juin.
3 A partir de 1983, les données révisées sont fournies par le Ministère de l'Agriculture.
4 Exportations plus consommation ajustée pour tenir compte de la variation des stocks

56
Fish catches
Quantités pêchées
All fishing areas: thousand metric tons
Toutes les zones de pêche : milliers de tonnes métriques

Country or area Pays ou zone	1981	1982	1983	1984	1985	1986	1987	1988	1989	1990
World Monde	74 627.3	76 817.4	77 525.9	83 924.5	86 378.1	92 828.9	94 398.6	99 062.2	100 332.8	97 245.7
Afghanistan Afghanistan	1.5	1.5	1.5	1.5	1.5	1.5	1.5	1.5	1.5	1.5
Albania Albanie	9.4	9.5	9.0	8.0	11.4	11.9	13.1	14.6	12.0	12.0[1]
Algeria Algérie	56.0[1]	64.5	65.0	65.6	66.1	65.5	94.4	106.7	99.7	91.1
Angola Angola	131.5	112.0	110.9	72.7	74.5	57.4	82.9	101.8	111.1	106.9
Antigua and Barbuda Antigua-et-Barbuda	1.2	1.0	1.1	1.5	2.4	2.4[1]	2.4[1]	2.4[1]	2.4[1]	2.2[1]
Argentina Argentine	361.5	475.0	416.4	315.2	406.8	420.7	559.8	493.4	486.6	555.6
Aruba Aruba	0.8[1]	0.8[1]	0.8[1]	0.8[1]	0.8	0.8[1]	0.8[1]	0.8[1]	0.8[1]	0.8[1]
Australia[2] Australie[2]	146.5	166.4	169.2	169.4	161.0[1]	180.3[1]	204.2	213.3	175.9	210.4
Austria Autriche	4.4	4.5	4.7	4.4	4.5	4.6	4.6	5.1	5.0	4.8
Bahamas Bahamas	4.4	4.7	5.2	5.3	7.6	5.9	7.1	7.2	8.1	7.5
Bahrain Bahreïn	5.7	5.6	4.8	5.6	7.8	8.1	7.8	6.7	9.2	8.3
Bangladesh Bangladesh	651.3	689.5	726.6	756.0	775.6	796.9	817.0	829.9	843.6	847.8
Barbados Barbade	3.4	3.5	6.5	5.8	3.9	4.2	3.7	9.1	2.5	3.0
Belgium Belgique	49.3	47.8	48.6	47.9	44.6	39.5	40.4	41.8	39.9	41.6
Belize Belize	1.4	1.4	1.5	1.3	1.4	1.5	1.5	1.5	1.8	1.5
Benin Bénin	37.8[1]	37.5	34.6	35.3[1]	36.4[1]	38.8[1]	41.9	37.3	41.9	41.7[1]
Bermuda Bermudes	0.4	0.4	0.5	0.5	0.7	0.8	0.8	0.8	0.8	0.5
Bhutan Bhoutan	1.0	1.0	1.0	1.0	1.0	1.0	1.0	1.0	1.0	1.0
Bolivia Bolivie	5.6	4.1[1]	4.1[1]	4.1[1]	4.2	3.9	4.3	4.4	6.0	7.4
Botswana Botswana	1.5	1.4	1.3	1.5	1.5	1.7	1.9	1.9	1.9	1.9[1]
Brazil Brésil	808.9	827.4	875.9	954.3	966.8	940.9	933.8	829.5	850.0	800.0[1]
British Virgin Islands Iles Vierges britanniques	0.7	0.8	0.9	1.0	1.2	1.2	1.2	1.3	1.4	1.4
Brunei Darussalam Brunéi Darussalam	2.4	2.4	3.1	2.2	3.0	2.8	2.7	2.2	2.3	2.3[1]
Bulgaria Bulgarie	106.7	115.6	121.1	113.0	100.2	109.3	110.7	117.1	102.0	56.1

56
Fish catches
All fishing areas: thousand metric tons [cont.]
Quantités pêchées
Toutes les zones de pêche : milliers de tonnes métriques [suite]

Country or area Pays ou zone	1981	1982	1983	1984	1985	1986	1987	1988	1989	1990
Burkina Faso Burkina Faso	7.5	7.0	7.1[1]	7.4[1]	7.4[1]	7.6[1]	7.8[1]	7.9[1]	8.0	7.0
Burundi Burundi	11.9	12.1	11.4	11.4	11.4	11.8	12.0	11.7	11.7[1]	17.4
Cambodia Cambodge	51.6	68.7	68.2	62.3	67.6	71.4	79.6	82.2	76.6	105.0
Cameroon Cameroun	91.8	95.0	89.3	87.3	86.0	84.0	82.5	82.5[1]	77.6	77.6[1]
Canada Canada	1 416.7	1 403.3	1 349.4	1 284.1	1 453.3	1 512.8	1 565.2	1 609.9	1 572.7	1 624.3
Cape Verde Cap-Vert	14.7	12.5	11.9	10.7	10.2	6.4	6.9	6.1	8.5	7.0
Cayman Islands Iles Caïmanes	4.0	0.9	0.7	0.4	0.4	0.5	1.1	0.4	0.4[1]	0.4[1]
Central African Rep. Rép. centrafricaine	13.0	13.0	13.0	13.0	13.0	13.0	13.0	13.0	13.0	13.0
Chad Tchad	115.0	115.0	110.0	110.0	115.0	110.0	110.0	110.0	110.0	115.0
Channel Islands Iles Anglo-Normandes	2.8	2.2	3.5	3.7	3.0	2.3	2.4	2.8[1]	2.9	3.0[1]
Chile Chili	3 393.5	3 673.3	3 978.2	4 499.3	4 804.4	5 571.6	4 814.6	5 209.9	6 454.1	5 195.4
China Chine	4 378.0	4 926.7	5 213.3	5 926.8	6 778.8	8 000.1	9 346.2	10 358.7	11 220.0	12 095.4
Colombia Colombie	94.7	71.4	57.5	79.1	71.5	83.4	85.5	88.3	98.3	101.1
Comoros Comores	4.4[1]	4.6[1]	4.8[1]	5.0[1]	5.2[1]	5.3[1]	5.3[1]	5.5[1]	6.8	8.0
Congo Congo	29.7	30.8	35.3	32.8	29.9	31.6	37.9	42.0	45.8	48.2
Cook Islands[1] Iles Cook[1]	0.9	0.9	1.0	1.0	1.0	1.1	1.1	1.1	1.2	1.2
Costa Rica Costa Rica	17.4	13.2	10.6	16.6	20.5	21.0	20.3[1]	20.4[1]	20.4[1]	21.1[1]
Côte d'Ivoire Côte d'Ivoire	92.0	91.2	94.6	88.6	111.2	104.9	102.3	91.0	100.7	108.9
Cuba Cuba	164.8	194.6	198.3	199.6	219.8	244.7	215.0	231.2	192.1	188.2
Cyprus Chypre	1.5	1.6	2.0	2.3	2.4	2.6	2.6	2.5	2.6	2.7
Czechoslovakia Tchécoslovaquie	16.5	18.0	19.5	19.7	20.0	21.3	20.7	21.2	21.6	22.4
Denmark Danemark	1 852.4	1 926.6	1 862.1	1 847.8	1 764.8	1 848.7	1 706.4	1 971.8	1 927.5	1 517.2
Djibouti Djibouti	0.4	0.4	0.4	0.4	0.4	0.4	0.4	0.5	0.4	0.4[1]
Dominica Dominique	1.5	1.5	0.8[1]	0.7[1]	0.6	0.6	0.7[1]	0.7[1]	0.7[1]	0.7[1]
Dominican Republic Rép. dominicaine	12.0	13.2	15.3	14.6	18.3	17.2	20.3	12.8	21.8	20.0[1]

56
Fish catches
All fishing areas: thousand metric tons [*cont.*]
Quantités pêchées
Toutes les zones de pêche : milliers de tonnes métriques [*suite*]

Country or area Pays ou zone	1981	1982	1983	1984	1985	1986	1987	1988	1989	1990
Ecuador Equateur	538.8	607.4	371.9	882.8	1 087.0	1 003.4	680.1	876.0	739.9	391.1
Egypt Egypte	141.7	155.7	157.3	163.8	215.9	229.1	231.0	284.2	293.6	313.0
El Salvador El Salvador	20.3	13.5	7.6	12.2	16.1	20.5	21.5	11.7	13.6	13.2[1]
Equatorial Guinea Guinée équatoriale	2.5	2.0[1]	2.3[1]	4.0	3.6	4.4	4.0[1]	4.0[1]	4.0[1]	4.0[1]
Ethiopia Ethiopie	3.8[1]	3.8[1]	3.9[1]	4.3[1]	4.0[1]	4.1[1]	4.0[1]	4.1	4.3	5.0[1]
Faeroe Islands Iles Féroé	242.0	249.1	329.9	346.8	374.0	350.7	386.0	358.6	299.6	282.9
Falkland Is. (Malvinas) Iles Falkland (Malvinas)	0.0	0.0	0.0	0.0	0.0	0.0	0.0	2.6	4.6	5.9
Fiji Fidji	24.2	28.0	27.8	27.9	27.6	27.0	35.3	32.4	32.8	35.0
Finland Finlande	109.8	118.8	130.3	134.5	135.2	131.1	106.3	120.9	110.5	97.4
France France	775.3	756.4	782.0	770.0	837.7	879.3	851.5	893.0[1]	913.5[1]	896.8[1]
French Guiana Guyane française	1.4	2.0	2.1	2.2	2.5	3.3	5.4	4.9	5.7	5.7
French Polynesia Polynésie française	2.2	2.3	2.2	2.8	2.3	2.3	2.7	3.1	3.3	3.0
Gabon Gabon	20.0[1]	20.6[1]	19.4[1]	21.0	21.0	21.0	22.3[1]	22.6[1]	22.9[1]	22.0[1]
Gambia Gambie	14.1	9.2	11.7	11.9	10.7	13.3	14.4	11.0	17.6	16.8
Gaza Strip (Palestine) Zone de Gaza (Palestine)	1.1	1.2	0.8	1.6	0.4	0.5[1]	0.5[1]	0.5[1]	0.5[1]	0.5[1]
Germany† · Allemagne† F. R. Germany R. f. Allemagne	330.7	313.5	305.5	327.2	225.3	202.4	201.8	209.5	234.0	250.2
former German D. R. anc. R. d. allemande	244.6	236.3	237.4	226.0	200.5	211.9	196.5	182.9	177.1	140.6
Ghana Ghana	239.9	242.7	250.9	269.1	276.0	319.5	382.0	362.0	361.7	391.8
Greece Grèce	101.7	104.8	99.9	108.1	115.1	123.4	133.0	125.6	138.8	140.0[1]
Greenland Groënland	107.3	105.8	107.7	86.3	93.5	101.0	100.0	120.3	162.5	138.2
Grenada Grenade	0.8	1.0	1.5	1.6	1.7	2.7	2.2	2.0	1.7	1.8[1]
Guadeloupe Guadeloupe	8.3[1]	8.7	8.8	9.0	8.4	8.5	8.6	8.2	8.3	8.5
Guam Guam	0.3	0.3	0.3	0.4	0.5	0.6	0.5	0.5	0.6	0.6[1]
Guatemala Guatemala	4.3	4.3	2.4	3.0	2.7	2.1	2.4	2.8	3.3	6.9

56
Fish catches
All fishing areas: thousand metric tons [cont.]
Quantités pêchées
Toutes les zones de pêche : milliers de tonnes métriques [suite]

Country or area Pays ou zone	1981	1982	1983	1984	1985	1986	1987	1988	1989	1990
Guinea[1] Guinée[1]	22.0	24.0	26.0	28.0	30.0	32.0	34.0	34.0	34.0	32.0
Guinea-Bissau Guinée-Bissau	2.7	4.0	2.7	2.8	3.7[1]	3.7[1]	4.1[1]	4.7[1]	5.4[1]	5.4[1]
Guyana Guyana	32.8	32.8	35.4	37.2	37.6	37.4	36.8	36.5	35.3	36.9
Haiti[1] Haïti[1]	5.5	6.0	6.5	7.0	7.5	8.0	8.1	8.1	8.1	7.5
Honduras Honduras	6.3	8.0	11.4	8.4	9.6	20.6	23.1	19.9	17.1	15.4
Hong Kong Hong-kong	182.3	181.0	189.3	199.7	198.2	213.6	228.1	238.2	242.5	234.5
Hungary Hongrie	39.3	42.0	43.9	39.0	36.9	36.1	36.8	38.3	35.5	33.9
Iceland Islande	1 441.7	788.7	839.2	1 535.0	1 680.4	1 658.6	1 632.7	1 759.5	1 504.8	1 507.6
India Inde	2 448.3	2 369.3	2 508.6	2 864.5	2 826.1	2 923.2	2 907.8	3 126.6	3 641.3	3 790.6
Indonesia Indonésie	1 907.3	1 990.1	2 204.9	2 251.9	2 332.7	2 457.0	2 583.9	2 789.1	2 948.4	3 080.5
Iran, Islamic Rep. of Iran, Rép. islamique d'	44.6	95.7	111.9	115.8	118.5	152.1	211.4	235.4	260.5	250.0[1]
Iraq[1] Iraq[1]	26.2	24.0	22.5	21.0	21.5	20.6	20.5	18.0	18.2	14.0
Ireland Irlande	190.5	212.2	203.4	210.0	230.7	231.5	250.7	255.7	200.2	230.5[1]
Isle of Man Ile de Man	8.6	6.3	8.3	7.7	7.0	5.8	5.6	5.6	5.7	4.0
Israel Israël	26.3	24.9	22.2	22.9	26.9	25.2	28.4	28.2[1]	26.5	26.1[1]
Italy Italie	514.7	547.4	551.8	578.4	589.3	568.1	560.5	576.7	551.0	525.2
Jamaica Jamaïque	7.8	7.9	8.7	9.7	10.4	10.8	10.7	9.7	10.6	10.4[1]
Japan Japon	10 740.1	10 826.5	11 254.7	12 021.0	11 408.9	11 976.3	11 848.6	11 966.2	11 173.4	10 353.6
Kenya Kenya	57.7	81.5	98.1	91.0	106.0	119.8	131.2	138.1	146.4	142.4
Kiribati Kiribati	11.4	11.0	19.3	14.6	20.6	26.3	36.9	23.8	33.1	30.0[1]
Korea, Dem. P. R.[1] Corée, R. p. dém. de[1]	1 500.0	1 550.0	1 600.0	1 650.0	1 700.0	1 700.2	1 700.3	1 700.0	1 700.1	1 750.1
Korea, Republic of Corée, République de	2 366.0	2 280.8	2 400.3	2 476.8	2 649.9	3 103.4	2 876.5	2 727.1	2 833.0	2 750.0[1]
Kuwait Koweït	3.7	6.6	8.7	9.6	10.1	7.6	7.7	10.8	7.7	4.5[1]
Lao People's Dem. Rep. Rép. dém. pop. lao	20.0	20.0	20.0	20.0	20.0	20.0	20.0	20.0	20.0	20.0
Lebanon Liban	1.6[1]	1.5[1]	1.4[1]	1.3[1]	1.5[1]	1.6	1.8	1.8	1.8	1.5

56
Fish catches
All fishing areas: thousand metric tons [*cont.*]
Quantités pêchées
Toutes les zones de pêche : milliers de tonnes métriques [*suite*]

Country or area Pays ou zone	1981	1982	1983	1984	1985	1986	1987	1988	1989	1990
Liberia Libéria	13.0	13.6	15.3	14.7	11.5	16.1	18.7	16.1	17.0[1]	16.0[1]
Libyan Arab Jamah. Jamah. arabe libyenne	13.1	14.7[1]	7.7[1]	7.8[1]	9.7[1]	9.7[1]	8.6[1]	9.7[1]	7.8[1]	7.8[1]
Macau Macao	7.7[1]	7.1[1]	8.2[1]	11.8[1]	12.4[1]	8.0	3.5	2.5	3.5	2.6
Madagascar Madagascar	55.2	59.3	63.4	67.5	69.2	84.3	92.4	101.1	99.6	106.7
Malawi Malawi	51.4	58.4	67.0	65.1	62.1	72.9	88.6	87.9[1]	85.0[1]	80.0[1]
Malaysia Malaisie	803.6	682.1	740.7	669.6	639.7	621.9	619.3[1]	612.4[1]	607.3[1]	602.5[1]
Maldives Maldives	34.3	30.5	38.6	55.1	61.9	59.3	57.0	71.5	71.2	78.3
Mali Mali	75.6	73.5	61.3	54.7	54.2	61.0	55.7	55.9	71.8	65.0[1]
Malta Malte	0.9	1.2	1.0	1.2	2.5	1.1	1.0	0.8	0.9	0.7
Martinique Martinique	4.7[1]	5.5	5.1	5.2	4.6	4.1	3.2	3.1	3.3[1]	3.4[1]
Mauritania Mauritanie	60.8	56.3	81.6	93.8	103.3	98.2	99.4	97.6	92.6[1]	91.0[1]
Mauritius Maurice	7.1	9.6	9.8	10.6	12.4	12.9	18.2	17.2	17.2	14.7
Mexico Mexique	1 540.8	1 314.7	1 064.6	1 105.0	1 226.5	1 315.7	1 419.2	1 372.1	1 469.8	1 401.0
Morocco Maroc	390.5	363.6	453.9	467.5	473.2	595.4	494.0	551.5	520.4	565.5
Mozambique Mozambique	42.1	39.7	42.5	35.8	36.3	31.9	36.1	33.5	33.3[1]	35.0[1]
Myanmar Myanmar	594.5	584.4	587.6	613.7	648.8	686.5	685.9	704.5	733.8	743.8
Namibia Namibie	10.9[1]	11.7[1]	12.1[1]	12.6[1]	13.1[1]	14.1[1]	31.6	32.6	20.3	289.8
Nepal[2] Népal[2]	3.9	4.4	4.7	4.9	9.1	9.4	10.7	12.1	12.5	14.5
Netherlands Pays-Bas	434.4	505.5	506.0	432.4	504.2	454.8	446.1	398.8	451.1[1]	438.3[1]
Netherlands Antilles Antilles néerlandaises	1.0[1]	1.0[1]	1.0[1]	1.0[1]	1.0	1.1[1]	1.1[1]	1.2[1]	1.2[1]	1.2[1]
New Caledonia Nouvelle-Calédonie	1.7	2.9	1.3	3.3	2.7	4.0	4.8	3.7	3.3	4.8
New Zealand Nouvelle-Zélande	235.8	250.1	282.1	322.5	304.8	345.2	418.5	551.0	565.4	565.4
Nicaragua Nicaragua	5.9	5.0	4.5	4.3	4.2	2.5	5.0	4.7	4.6	2.8
Niger Niger	8.2	6.8	3.3	3.0	2.0	2.4	2.3	2.5	4.8	3.4
Nigeria Nigéria	260.1[1]	269.7[1]	272.3[1]	263.2[1]	244.5	271.5	260.9	279.4[1]	299.7[1]	316.3

56
Fish catches
All fishing areas: thousand metric tons [cont.]
    Quantités pêchées
    Toutes les zones de pêche : milliers de tonnes métriques [suite]

Country or area Pays ou zone	1981	1982	1983	1984	1985	1986	1987	1988	1989	1990
Norway Norvège	2 552.0	2 500.6	2 835.8	2 466.0	2 119.0	1 913.9	1 949.5	1 839.9	1 908.8	1 747.1
Oman[2] Oman[2]	83.7	89.4	108.8	105.0	101.2	96.4	136.1	165.6	117.7	120.2
Pacific Islands (Palau)[13] Iles du Pacifique (Palaos)[13]	3.1	3.1	3.4	3.5	5.2	5.4	5.4	5.5	5.6	5.6
Pakistan Pakistan	317.8	337.3	343.4	372.3	408.4	415.7	427.7	445.4	446.2	479.0
Panama Panama	149.5	109.6	166.7	131.6	286.4	128.0	168.4	122.6	188.6	161.7
Papua New Guinea[1] Papouasie-Nvl-Guinée[1]	44.9	17.7	15.5	17.8	25.5	25.0	25.1	25.0	25.2	25.0
Paraguay Paraguay	3.4[1]	3.4[1]	3.5	5.0	7.5	13.0	10.0	10.0	11.0	12.5
Peru Pérou	2 717.2	3 513.2	1 569.5	3 319.9	4 138.1	5 616.2	4 587.4	6 641.7	6 853.8	6 875.1
Philippines Philippines	1 686.1	1 786.8	1 976.1	1 933.7	1 865.0	1 916.3	1 988.7	2 010.4	2 098.8	2 208.8
Poland Pologne	630.1	608.1	735.1	719.2	683.5	645.2	670.9	654.9	564.8	473.0
Portugal Portugal	260.5	254.9	247.6	291.7	305.8	402.0	389.6	346.7	331.8	321.9
Puerto Rico Porto Rico	1.8	2.2	2.7	2.4	1.5	1.3	1.2	1.7	2.0	2.1
Qatar Qatar	2.6	2.3	2.1	3.2	2.5	2.0	2.7	3.1	4.4	5.7
Réunion Réunion	3.6	5.3	2.6	2.6	2.2	1.7	1.8	2.1	2.0	2.0
Romania Roumanie	192.0	235.7	242.6	232.2	237.6	271.1	264.4	267.6	224.8	127.7
Rwanda Rwanda	1.0	1.2	1.2	0.8	0.9	1.5	1.7	1.3	1.5	2.5
Saint Helena Sainte-Hélène	0.7	0.9	0.6	0.7	0.6	0.6	0.7	0.8	1.0	0.8
Saint Kitts and Nevis Saint-Kitts-et-Nevis	1.9	1.9	1.2	1.4[1]	1.6	1.6[1]	1.7[1]	1.7[1]	1.7[1]	1.7[1]
Saint Lucia Sainte-Lucie	0.9	0.9	0.9	0.9	1.1	0.8	0.7	0.8	0.8	1.0
Saint Pierre and Miquelon Saint-Pierre-et-Miquelon	11.1	10.5	10.1	12.3	12.6	23.8	23.7	14.0	18.5	7.4
St. Vincent-Grenadines St. Vincent-Grenadines	0.5[1]	0.5[1]	0.5[1]	0.5[1]	0.5	0.6	0.7	4.6	5.9	8.4
Samoa Samoa	3.1	4.0	3.8	3.7[1]	3.6	3.7[1]	3.4[1]	3.5[1]	3.5[1]	3.6[1]
Sao Tome and Principe Sao Tomé-et-Principe	2.2	2.7	4.0	4.4	4.0	2.8	2.8	2.9	3.1	3.6
Saudi Arabia Arabie saoudite	29.0[1]	33.0[1]	36.0[1]	40.0[1]	43.7	45.5	47.9	47.1	53.4[1]	46.4
Senegal Sénégal	225.5	229.3	263.5	247.3	246.0	255.6	255.0	260.7	288.2	299.7

56
Fish catches
All fishing areas: thousand metric tons [*cont.*]
     Quantités pêchées
         Toutes les zones de pêche : milliers de tonnes métriques [*suite*]

Country or area Pays ou zone	1981	1982	1983	1984	1985	1986	1987	1988	1989	1990
Seychelles Seychelles	5.2	4.0	3.9	3.8	4.1	4.5	4.0	4.3	4.4	5.4
Sierra Leone Sierra Leone	51.0[1]	53.0	51.1[1]	52.7[1]	53.5[1]	53.3[1]	53.2[1]	53.3[1]	53.2[1]	50.0[1]
Singapore Singapour	16.1	19.3	19.5	26.2	23.9	21.4	16.7	15.2	12.6	13.3
Solomon Islands Iles Salomon	37.5	32.8	47.3	48.9	44.0	55.5	44.6	55.1	57.0	55.8
Somalia Somalie	9.8	9.0	11.5	19.9	16.8	16.9[1]	17.5[1]	18.2[1]	18.2[1]	17.5
South Africa Afrique du Sud	863.0	823.6	935.3	734.5	775.4	819.5	1 424.5	1 299.7	878.6	536.4
Spain Espagne	1 356.4	1 474.4	1 412.8	1 440.6	1 482.8	1 488.5	1 524.9	1 592.7	1 559.8[1]	1 458.1[1]
Sri Lanka Sri Lanka	206.8	216.3	218.8	169.2	179.2	181.5	185.7	197.5	205.3	165.4
Sudan Soudan	28.5	29.7	29.5	29.8	26.3	23.9	26.2[1]	29.2[1]	31.2[1]	38.8
Suriname Suriname	3.4	2.9	3.6	4.1	4.1	3.7	5.2	3.7	3.7[1]	4.0[1]
Sweden Suède	265.3	263.8	269.5	282.2	239.6	215.1	214.5	251.0	257.8	260.1
Switzerland Suisse	3.7	3.4	3.5	4.1	4.5	4.6	4.5	4.2	4.4	4.2
Syrian Arab Republic Rép. arabe syrienne	3.8	4.1	4.4	5.3	5.8	5.3	5.4	5.5	5.1	5.8
Thailand Thaïlande	1 989.0	2 120.1	2 260.0	2 134.8	2 225.1	2 536.3	2 779.1	2 642.1	2 781.8	2 650.0[1]
Togo Togo	10.3	14.5	14.6	14.5	15.5	14.8	15.2	15.5	16.5	15.8
Tonga Tonga	2.1	2.1	2.2	2.3	2.5	2.7	2.7	2.7	2.7	2.6[1]
Trinidad and Tobago Trinité-et-Tobago	3.8	4.6	4.2	3.6	2.9	3.0	3.2[1]	3.5[1]	3.0[1]	3.3[1]
Tunisia Tunisie	57.4	62.7	67.1	74.9	88.9	92.6	99.2	102.7	95.1	92.1
Turkey Turquie	470.2	503.5	557.3	566.9	578.1	582.9	627.9	676.0	457.1	382.2
Turks and Caicos Islands Iles Turques et Caiques	1.0	1.1	1.2	1.2	1.3	1.5	1.3	1.3	1.3	1.0
Tuvalu Tuvalu	0.2	0.4	0.8	0.7[1]	0.7[1]	0.6[1]	0.9	1.4	1.4[1]	1.5[1]
Uganda Ouganda	166.6	170.0	172.0	212.3	160.8	197.6	200.0	214.3	212.2	245.2
former USSR ancienne URSS	9 566.3	9 990.6	9 816.7	10 592.9	10 522.8	11 260.0	11 159.6	11 332.1	11 310.1	10 389.0
United Arab Emirates Emirats arabes unis	67.8	69.8	72.7	72.7	72.3	79.3	85.2	89.5	91.2	95.1
United Kingdom Royaume-Uni	871.2	903.4	840.0	836.2	891.4	850.4	945.4	937.1	823.4	803.5

56
Fish catches
All fishing areas: thousand metric tons [cont.]
Quantités pêchées
Toutes les zones de pêche : milliers de tonnes métriques [suite]

Country or area Pays ou zone	1981	1982	1983	1984	1985	1986	1987	1988	1989	1990
United Rep.Tanzania Rép. Unie de Tanzanie	230.7	227.8	239.2	277.3	300.6	309.9	342.3	393.0	377.1	377.0
United States Etats-Unis	3 793.6	4 032.5	4 319.2	4 990.9	4 949.3	5 166.6	5 986.2	5 937.5	5 763.3	5 856.0
United States Virgin Is.[2] Iles Vierges américaines[2]	0.6	0.9	0.6	0.7	0.6	0.9	0.9	0.7	0.8	0.8[1]
Uruguay Uruguay	147.0	119.0	143.4	133.0	138.2	140.7	137.8	107.3	121.6	90.8
Vanuatu Vanuatu	2.7	2.7	2.5	2.9	3.6	3.2	3.3	3.4[1]	3.3[1]	3.4[1]
Venezuela Venezuela	179.7	221.3	230.4	259.4	263.6	284.2	297.6	285.5	329.3	332.3
Viet Nam Viet Nam	596.9	661.2	757.1	776.3	808.0	824.7	871.4	874.0[1]	868.0[1]	850.0[1]
Wallis and Futuna Islands[1] Iles Wallis et Futuna[1]	0.8	0.9	0.9	0.9	1.0	1.0	1.0	1.0	1.0	1.0
Yemen Yémen	68.2[1]	58.1[1]	63.7	65.7	71.3	72.7	72.4	73.2[1]	72.9[1]	89.1
Yugoslavia Yougoslavie	71.6	66.6	79.7	73.4	75.0	77.5	81.3	71.8	71.7	65.5
Zaire Zaïre	102.6[1]	100.7[1]	102.0[1]	148.3	148.5	156.5	162.0[1]	162.0[1]	166.0[1]	162.0[1]
Zambia Zambie	38.8	55.8	67.2	64.6	67.7	68.2	63.6	60.6	66.7	64.5
Zimbabwe Zimbabwe	16.4	17.5	13.6	16.4	17.4	18.9	19.2	22.2	24.0	25.0[1]

Source:
Food and Agriculture Organization of the United Nations
(Rome).

† All data shown which pertain to Germany prior to 3 October
  1990 are indicated separately for the Federal Republic of
  Germany and the former German Democratic Republic based on
  their respective territories at the time indicated. Where
  data for united Germany (3 October 1990 and thereafter) are
  not available, available data are shown separately under the
  designations Federal Republic of Germany and former German
  Democratic Republic and pertain to the territorial
  boundaries prior to 3 October 1990. For detailed
  explanatory notes on data pertaining to Germany, see Annex I
  - Country Nomenclature.

1 FAO estimate.
2 Data refer to a split year period.
3 Including data for Federated States of Micronesia, Marshall
  Is. and Northern Mariana Is.

Source:
Organisation des Nations Unies pour l'alimentation et
l'agriculture (Rome).

† Toutes les données se rapportant à l'Allemagne avant le 3
  octobre 1990 figurent dans deux rubriques séparées basées
  sur les territoires respectifs de la République fédérale
  d'Allemagne et l'ancienne République démocratique allemande
  selon la période indiquée. En l'absence de données pour
  l'Allemagne unifiée (à compter du 3 octobre 1990), les
  données disponibles sont fournies séparément sous les
  rubriques République fédérale d'Allemagne et ancienne
  République démocratique allemande et se rapportent aux
  limites territoriales antérieures au 3 octobre 1990. Pour
  les notes explicatives en détail sur les données concernant
  l'Allemagne, voir Annexe I – Nomenclature des pays.
1 Estimation de la FAO.
2 Les données se réfèrent à une année fractionnée.
3 Y compris les données pour les Etats fédérés de Micronésie,
  les îles Marshall et les îles Mariannes du Nord.

57
Tractors in use
Tracteurs en service
Number

Nombre

Country or area Pays ou zone	1981	1982	1983	1984	1985	1986	1987	1988	1989	1990
World[1] *Monde*[1]	22 503	23 087	23 616	24 149	24 846	25 283	25 642	26 064	26 405	26 544
Africa **Afrique**	462 156	471 997	480 793	497 029	516 620	526 831	535 316	545 329	552 050	563 845
Algeria Algérie	48 000[2]	49 200[2]	50 279	61 319	75 310	82 808	89 271	94 000[2]	96 000[2]	100 000[2]
Angola[2] Angola[2]	10 250	10 250	10 250	10 250	10 250	10 270	10 270	10 270	10 270	10 290
Benin[2] Bénin[2]	108	110	112	114	116	118	120	123	125	127
Botswana Botswana	2 290[2]	2 460[2]	2 570[2]	2 710[2]	2 870[2]	3 010[2]	3 150	4 500	5 000	5 900
Burkina Faso Burkina Faso	120[2]	120[2]	120[2]	120[2]	120[2]	121	123[2]	125[2]	127[2]	130[2]
Burundi Burundi	95[2]	100[2]	105[2]	110[2]	115[2]	120[2]	125[2]	130[2]	133	133
Cameroon Cameroun	675	750[2]	780[2]	820[2]	880[2]	920[2]	960[2]	1 000[2]	1 050[2]	1 080[2]
Cape Verde Cap-Vert	15[2]	15[2]	16	16[2]	16[2]	16[2]	16[2]	16[2]	16[2]	16[2]
Central African Rep.[2] Rép. centrafricaine[2]	160	165	170	175	180	185	190	195	200	205
Chad[2] Tchad[2]	160	160	160	160	160	160	165	165	165	165
Congo[2] Congo[2]	673	676	678	680	685	687	690	695	700	700
Côte d'Ivoire[2] Côte d'Ivoire[2]	3 100	3 150	3 200	3 250	3 300	3 350	3 400	3 450	3 500	3 550
Djibouti Djibouti	6[2]	6[2]	6	6[2]	6[2]	6[2]	8	8[2]	8[2]	8[2]
Egypt Egypte	38 639	41 900[2]	45 200[2]	48 500[2]	51 856	52 000[2]	52 290	52 500[2]	52 700[2]	52 900[2]
Equatorial Guinea[2] Guinée équatoriale[2]	98	98	98	98	98	98	100	100	100	100
Ethiopia[2] Ethiopie[2]	3 950	3 900	3 900	3 900	3 900	3 900	3 900	3 900	3 900	3 900
Gabon[2] Gabon[2]	1 270	1 290	1 310	1 330	1 350	1 370	1 400	1 430	1 450	1 460
Gambia Gambie	44[2]	44[2]	43	43[2]	43[2]	43[2]	43[2]	43[2]	43[2]	43[2]
Ghana[2] Ghana[2]	3 550	3 600	3 650	3 700	3 750	3 800	3 850	3 900	3 950	4 000
Guinea[2] Guinée[2]	170	190	200	210	220	230	240	250	260	270
Guinea-Bissau[2] Guinée-Bissau[2]	46	47	47	47	48	48	48	48	48	48
Kenya Kenya	6 399	6 650	8 568	8 394	8 668	9 224	9 414	9 500[2]	9 700[2]	9 900[2]
Lesotho Lesotho	1 450	1 500	1 500[2]	1 550[2]	1 600[2]	1 650[2]	1 700[2]	1 750[2]	1 800[2]	1 830[2]

57
Tractors in use
Number [cont.]

Tracteurs en service
Nombre [suite]

Country or area Pays ou zone	1981	1982	1983	1984	1985	1986	1987	1988	1989	1990
Liberia [2] Libéria [2]	300	305	308	310	315	318	320	325	328	330
Libyan Arab Jamah. Jamah. arabe libyenne	24 800 [2]	26 000 [2]	26 900 [2]	27 000 [2]	27 100 [2]	27 200 [2]	27 320	29 216	29 696	33 272
Madagascar [2] Madagascar [2]	2 680	2 700	2 730	2 750	2 780	2 800	2 820	2 850	2 870	2 890
Malawi [2] Malawi [2]	1 250	1 250	1 280	1 300	1 330	1 350	1 370	1 380	1 390	1 400
Mali [2] Mali [2]	830	830	830	830	830	835	835	835	835	840
Mauritania Mauritanie	312	312 [2]	312 [2]	312 [2]	320 [2]	330	330 [2]	330 [2]	335 [2]	335 [2]
Mauritius [2] Maurice [2]	332	335	337	339	342	344	346	350	353	355
Morocco Maroc	29 712	31 084	31 586	31 299	32 000 [2]	32 800 [2]	33 500 [2]	34 278	36 700 [2]	39 155
Mozambique [2] Mozambique [2]	5 750	5 750	5 750	5 750	5 750	5 750	5 750	5 750	5 750	5 750
Namibia [2] Namibie [2]	2 600	2 650	2 700	2 750	2 800	2 850	2 900	2 950	3 000	3 050
Niger Niger	105	121	128	140	150	158	158	162	174	176 [2]
Nigeria [2] Nigéria [2]	8 800	9 000	9 500	10 000	10 300	10 500	10 800	11 000	11 300	11 500
Réunion Réunion	1 309	1 489	1 569	1 703	1 838	1 952	2 087	2 264	2 270 [2]	2 280 [2]
Rwanda [2] Rwanda [2]	84	84	84	84	85	85	86	88	90	90
Saint Helena Sainte-Hélène	5	5 [2]	5 [2]	5	6 [2]	8	8	9	10	10
Sao Tome and Principe [2] Sao Tomé-et-Principe [2]	123	123	123	124	124	124	124	124	124	124
Senegal [2] Sénégal [2]	460	460	460	460	460	470	470	480	480	490
Seychelles [2] Seychelles [2]	37	37	38	38	38	38	38	39	39	40
Sierra Leone Sierra Leone	345 [2]	380 [2]	410	440 [2]	470 [2]	480 [2]	500 [2]	510 [2]	520 [2]	530 [2]
Somalia Somalie	1 720 [2]	* 1 800	* 1 820	* 1 915	1 990 [2]	2 060 [2]	2 080 [2]	2 100 [2]	2 120 [2]	2 130 [2]
South Africa [2] Afrique du Sud [2]	181 000	181 400	181 800	182 200	182 500	182 800	183 000	183 200	183 500	183 700
Sudan Soudan	9 700 [2]	9 900 [2]	9 900 [2]	10 000 [2]	10 100 [2]	10 300 [2]	9 800 [2]	9 700 [2]	9 500 [2]	9 182
Swaziland Swaziland	2 873	3 180 [2]	3 480 [2]	3 787	3 500 [2]	3 354	3 300 [2]	3 300 [2]	3 280 [2]	3 280 [2]
Togo [2] Togo [2]	220	240	260	280	300	320	350	350	360	370
Tunisia Tunisie	26 200 [2]	26 600	26 000	26 000 [2]	26 100 [2]	26 100 [2]	26 000 [2]	26 000 [2]	25 900 [2]	25 800 [2]

57
Tractors in use
Number [*cont.*]

Tracteurs en service
Nombre [*suite*]

Country or area Pays ou zone	1981	1982	1983	1984	1985	1986	1987	1988	1989	1990
Uganda[2] Ouganda[2]	2 750	3 000	3 300	3 500	3 650	3 750	4 000	4 200	4 400	4 500
United Rep.Tanzania Rép. Unie de Tanzanie	9 500[2]	9 300[2]	8 700[2]	8 500[2]	8 000[2]	7 500[2]	7 262	7 050[2]	6 900[2]	6 800[2]
Zaire[2] Zaïre[2]	2 000	2 050	2 150	2 200	2 250	2 280	2 300	2 330	2 370	2 400
Zambia Zambie	4 780[2]	4 920[2]	5 060[2]	5 200[2]	5 340[2]	5 480[2]	5 628	5 700[2]	5 800[2]	5 900[2]
Zimbabwe[2] Zimbabwe[2]	20 300	20 300	20 300	20 300	20 300	20 350	20 350	20 350	20 400	20 400
America, North[1] Amérique du Nord[1]	5 603	5 593	5 615	5 637	5 649	5 712	5 769	5 786	5 806	5 819
Antigua and Barbuda[2] Antigua-et-Barbuda[2]	235	235	235	235	236	236	236	236	238	238
Bahamas[2] Bahamas[2]	74	75	75	75	75	75	77	77	77	78
Barbados[2] Barbade[2]	560	570	575	580	585	590	600	610	615	620
Belize Belize	825[2]	845[2]	870[2]	893[2]	940	980[2]	1 020[2]	1 050[2]	1 080[2]	1 090[2]
Bermuda Bermudes	40[2]	40[2]	40[2]	40[2]	40	44[2]	48	48[2]	47	47[2]
British Virgin Islands Iles Vierges britanniques	4[2]	4	4[2]	4[2]	3[2]	3[2]	3	3[2]	3[2]	3[2]
Canada Canada	657 606	672 000[2]	686 000[2]	700 000[2]	714 000[2]	728 074	742 200	756 300	770 400	780 000[2]
Costa Rica[2] Costa Rica[2]	6 000	6 050	6 100	6 150	6 200	6 250	6 300	6 350	6 400	6 450
Cuba Cuba	65 943	66 509	66 262	65 900	68 585	73 739	74 403	74 919	76 783	77 800[2]
Dominica[2] Dominique[2]	89	89	90	90	90	90	90	90	90	90
Dominican Republic Rép. dominicaine	2 170[2]	2 190[2]	2 210[2]	2 230[2]	2 250[2]	2 270	2 290[2]	2 310[2]	2 320[2]	2 330[2]
El Salvador[2] El Salvador[2]	3 320	3 340	3 360	3 380	3 390	3 400	3 400	3 410	3 410	3 420
Greenland Groënland	78[2]	* 80	82[2]	84[2]	85[2]	86[2]	86[2]	88[2]	88[2]	88[2]
Grenada[2] Grenade[2]	26	26	27	27	27	27	28	28	28	28
Guadeloupe Guadeloupe	921	950	950	1 040	1 390	1 450	1 550	1 036	1 020[2]	1 000[2]
Guatemala[2] Guatemala[2]	4 020	4 040	4 060	4 080	4 100	4 120	4 140	4 160	4 180	4 200
Haiti Haïti	180[2]	185[2]	190[2]	195[2]	200[2]	205[2]	210[2]	215[2]	220	220[2]
Honduras[2] Honduras[2]	3 280	3 300	3 310	3 330	3 350	3 370	3 390	3 420	3 450	3 470
Jamaica[2] Jamaïque[2]	2 870	2 900	2 930	2 950	2 970	3 000	3 020	3 040	3 050	3 060

57
Tractors in use
Number [*cont.*]

Tracteurs en service
Nombre [*suite*]

Country or area Pays ou zone	1981	1982	1983	1984	1985	1986	1987	1988	1989	1990
Martinique Martinique	865	900[2]	945	990	980[2]	958	958	930[2]	901	900[2]
Mexico Mexique	143 078	146 083	152 319	155 000[2]	157 000[2]	160 000[2]	163 000[2]	165 000[2]	168 000[2]	170 000[2]
Montserrat Montserrat	12[2]	12[2]	12[2]	12[2]	12[2]	12[2]	12	12[2]	12[2]	12[2]
Netherlands Antilles[2] Antilles néerlandaises[2]	20	20	20	20	20	20	20	20	20	20
Nicaragua[2] Nicaragua[2]	2 250	2 300	2 350	2 400	2 430	2 450	2 480	2 500	2 550	2 600
Panama[2 3] Panama[2 3]	5 420	5 380	5 350	5 310	5 270	5 240	5 200	5 160	5 120	5 090
Puerto Rico Porto Rico	3 556	3 427	3 112	2 688	2 234	1 879	2 042	3 027	3 773	3 853
Saint Kitts-Nevis[2] Saint-Kitts-et-Nevis[2]	215	215	216	216	216	216	216	216	216	216
Saint Lucia[2] Sainte-Lucie[2]	85	85	85	85	85	85	85	87	87	87
St. Vincent-Grenadines St. Vincent-Grenadines	* 75	75[2]	76[2]	76[2]	76[2]	77[2]	77[2]	77[2]	78[2]	78[2]
Trinidad and Tobago[2] Trinité-et-Tobago[2]	2 420	2 470	2 500	2 540	2 580	2 610	2 620	2 620	2 620	2 620
United States[1] Etats-Unis[1]	4 697	4 669	4 671	4 676	4 670	4 710[2]	* 4 749	4 749[2]	4 749[2]	4 749[2]
United States Virgin Is. Iles Vierges américaines	70[2]	71[2]	72[2]	73[2]	74[2]	75[2]	76	77[2]	77[2]	77[2]
America, South[1] **Amérique du Sud[1]**	**957**	**976**	**1 001**	**1 030**	**1 057**	**1 077**	**1 095**	**1 102**	**1 113**	**1 123**
Argentina Argentine	213 000	203 700	201 800	203 700	204 000[2]	203 000[2]	205 000[2]	204 000[2]	204 000[2]	203 000[2]
Bolivia Bolivie	4 200[2]	4 400[2]	4 500[2]	4 600[2]	4 750[2]	4 900	5 000	5 100	5 100	5 200
Brazil Brésil	569 000[2]	593 000[2]	617 000[2]	641 000[2]	666 309	680 000[2]	690 000[2]	700 000[2]	710 000[2]	720 000[2]
Chile Chili	34 370[2]	34 365[2]	34 360[2]	34 350[2]	34 340	37 920	41 270	37 450	36 620	35 750
Colombia Colombie	29 500[2]	31 000[2]	32 000[2]	33 066	33 450	33 757	34 232	34 700[2]	35 200[2]	35 600[2]
Ecuador Equateur	6 844	7 186	7 400[2]	7 600[2]	7 800[2]	8 000[2]	8 200[2]	8 400[2]	8 600[2]	8 700[2]
Falkland Is. (Malvinas) Iles Falkland (Malvinas)	117[2]	117[2]	117[2]	* 117	* 117	* 117	150	150	150	150
French Guiana Guyane française	106	125[2]	150	190	209	230[2]	250[2]	260[2]	280[2]	304
Guyana[2] Guyana[2]	3 480	3 500	3 520	3 540	3 550	3 560	3 570	3 580	3 590	3 600
Paraguay[2] Paraguay[2]	7 400	8 000	8 500	9 000	9 500	9 900	10 000	10 000	10 050	10 100
Peru[2] Pérou[2]	15 600	16 500	15 400	14 800	14 200	15 100	15 500	15 000	15 500	16 000

57
Tractors in use
Number [*cont.*]

Tracteurs en service
Nombre [*suite*]

Country or area Pays ou zone	1981	1982	1983	1984	1985	1986	1987	1988	1989	1990
Suriname Suriname	1 120[2]	1 140[2]	1 160[2]	1 180[2]	1 202	1 220[2]	1 230[2]	1 250[2]	1 270[2]	1 290[2]
Uruguay Uruguay	33 160[2]	33 450[2]	33 750[2]	34 000[2]	34 300[2]	34 599	34 900[2]	35 200[2]	35 500[2]	35 800[2]
Venezuela[2] Venezuela[2]	39 000	40 000	41 500	42 500	43 500	44 500	46 000	47 000	47 500	48 000
Asia[1] Asie[1]	3 657	3 921	4 112	4 326	4 652	4 763	4 955	5 140	5 418	5 587
Afghanistan[2] Afghanistan[2]	780	780	800	800	820	820	820	840	840	850
Bangladesh[2] Bangladesh[2]	4 500	4 600	4 700	4 800	4 900	4 950	5 000	5 100	5 150	5 200
Brunei Darussalam Brunéi Darussalam	72	72	72	72	72	72	72	72	72	72
Cambodia[2] Cambodge[2]	1 350	1 350	1 350	1 355	1 355	1 360	1 360	1 365	1 365	1 365
China Chine	796 860	817 403	848 133	862 071	861 357	876 463	891 952	882 187	861 220	827 512
Cyprus Chypre	10 900[2]	11 000[2]	11 100[2]	12 840	13 316	13 400[2]	13 500[2]	13 600[2]	13 650[2]	13 700[2]
East Timor[2] Timor oriental[2]	114	115	115	115	115	115	115	115	115	115
Gaza Strip (Palestine) Zone de Gaza (Palestine)	703	540	610	615[2]	620[2]	627	680	726	734	745[2]
Hong Kong Hong-kong	7	7	7	7	7	7	7	7	7	7[2]
India Inde	417 769	461 567	502 581	553 555	607 773	648 932	697 568	750 935	925 365	988 070
Indonesia Indonésie	9 268	9 835	9 895	11 363	12 033	15 394	15 700[2]	16 000[2]	16 600[2]	17 000[2]
Iran, Islamic Rep. of Iran, Rép. islamique d'	85 000[2]	93 051	98 000[2]	105 000[2]	108 000[2]	109 000[2]	111 000[2]	113 000[2]	114 000[2]	115 000[2]
Iraq Iraq	26 600[2]	29 956	* 35 552	36 005	38 000[2]	40 915	40 000[2]	39 000[2]	38 186	37 000[2]
Israel Israël	26 500	26 800	27 475	26 870	26 254	28 248	28 662	28 444	28 400	27 400
Japan[1] Japon[1]	1 413	1 526	1 584	1 650	1 854	1 834	1 904	1 985	2 049	2 120[2]
Jordan Jordanie	4 799	* 4 748	* 4 778	* 4 795	4 914	5 553	5 673	5 673	5 700[2]	5 750[2]
Korea, Dem. P. R. Corée, R. p. dém. de	50 100[2]	55 900[2]	61 700[2]	* 67 500	68 000[2]	69 000[2]	70 000[2]	71 000[2]	72 000[2]	73 000[2]
Korea, Republic of Corée, République de	3 862	5 575	7 469	9 684	12 389	16 167	19 863	24 616	31 328	41 203
Kuwait Koweït	24	24[2]	23[2]	23	60	102	110[2]	120[2]	120[2]	120[2]
Lao People's Dem. Rep. Rép. dém. pop. lao	664	680[2]	720[2]	750[2]	780[2]	800[2]	820[2]	840[2]	860[2]	870[2]
Lebanon[2] Liban[2]	3 000	3 000	3 000	3 000	3 000	3 000	3 000	3 000	3 000	3 000

57
Tractors in use
Number [*cont.*]
Tracteurs en service
Nombre [*suite*]

Country or area Pays ou zone	1981	1982	1983	1984	1985	1986	1987	1988	1989	1990
Malaysia Malaisie	8 009	7 974	8 909	10 211	11 400[2]	11 600[2]	11 700[2]	11 800[2]	12 000[2]	12 100[2]
Mongolia Mongolie	10 000	10 400	10 800	10 800	11 100	11 129	11 697	11 869	11 477	11 500[2]
Myanmar Myanmar	8 280	8 528	8 745	9 859	10 026	10 204	10 617	12 000	10 000	7 000
Nepal Népal	2 420[2]	2 591	2 690[2]	2 770[2]	* 2 783	3 200[2]	3 600[2]	* 4 018	4 200[2]	4 400[2]
Oman Oman	109	107	102	103	125	125	130	137	140	144
Pakistan Pakistan	112 000[2]	122 004	133 000[2]	141 000[2]	156 633	178 400[2]	200 000[2]	222 000[2]	244 000[2]	265 728
Philippines Philippines	10 208	9 832	9 374	8 670	8 050[2]	7 600[2]	7 250[2]	7 730[2]	9 250[2]	10 700[2]
Qatar Qatar	68	68	75[2]	* 82	84[2]	86[2]	87[2]	90[2]	92[2]	92[2]
Saudi Arabia Arabie saoudite	1 300[2]	1 400[2]	1 500[2]	1 600[2]	1 700	1 750[2]	1 800[2]	1 850[2]	1 900[2]	1 950[2]
Singapore [2] Singapour [2]	46	48	50	52	55	56	58	59	60	62
Sri Lanka Sri Lanka	24 985	25 555	26 075	26 710	27 374	28 200[2]	29 000[2]	29 800[2]	30 700[2]	* 31 510
Syrian Arab Republic Rép. arabe syrienne	31 387	35 533	37 216	37 920	43 959	47 573	52 400	54 900	58 919	70 100
Thailand Thaïlande	89 202	107 528	113 116	120 918	125 000[2]	130 000[2]	136 000[2]	142 000[2]	150 000[2]	158 000[2]
Turkey Turquie	457 425	489 813	512 282	555 227	582 291	611 052	635 526	654 636	672 845	690 000[2]
United Arab Emirates Emirats arabes unis	200[2]	205[2]	210[2]	212[2]	213	213	215	225[2]	242	245[2]
Viet Nam Viet Nam	38 000	38 600	37 800	39 600	45 000	43 700	35 800	35 800	35 000[2]	35 000[2]
Yemen Yémen	4 630	4 750	4 920	5 030	5 170	5 190	5 240	5 320	5 350	5 410
Europe [1] Europe [1]	**8 707**	**8 965**	**9 225**	**9 444**	**9 719**	**9 941**	**10 109**	**10 293**	**10 410**	**10 427**
Albania Albanie	10 200[2]	10 300[2]	* 10 420	10 500[2]	10 550[2]	10 580[2]	* 10 630	* 11 600	* 12 100	12 300[2]
Austria Autriche	322 300	326 060	326 060	326 060	326 060	326 060	326 060	351 444	351 444	351 444
Belgium-Luxembourg Belgique-Luxembourg	114 088	116 227	120 138	123 509	123 191	123 507	120 352	123 060	122 277	121 700[2]
Bulgaria Bulgarie	60 522	59 584	58 251	56 700	55 161	54 180	53 640	53 679	53 210	53 100[2]
Czechoslovakia Tchécoslovaquie	134 100	132 286	134 366	135 563	137 054	138 731	140 090	141 191	140 202	138 634
Denmark Danemark	181 349	183 135	176 000[2]	169 711	166 314	168 574	164 949	167 773	165 908	162 555
Finland Finlande	218 000	227 000	234 000	238 000	240 000	240 000	240 000	244 000	244 000	* 244 000

57
Tractors in use
Number [cont.]

Tracteurs en service
Nombre [suite]

Country or area Pays ou zone	1981	1982	1983	1984	1985	1986	1987	1988	1989	1990
France[1] France[1]	1 485	1 494	1 495	1 491	1 491	1 485	1 481	1 475	1 470[2]	1 465[2]
Germany†·Allemagne† F. R. Germany[1] R. f. Allemagne[1]	1 468	1 472	1 482	1 483	1 484	1 483	1 470	1 449	1 424	1 395
former German D. R. anc. R. d. allemande	147 384	149 523	153 412	156 076	158 025	161 515	164 512	167 529	170 967	173 000[2]
Greece Grèce	152 254	159 516	168 715	178 857	183 410	193 257	195 000[2]	200 000[2]	205 000[2]	210 000[2]
Hungary Hongrie	54 912	55 335	55 611	55 414	55 317	53 947	53 507	52 765	50 388	49 400
Iceland Islande	13 500	13 800	14 000	14 000	13 200	13 200	13 000	12 000	11 400	11 000[2]
Ireland[2] Irlande[2]	148 000	150 000	152 000	155 000	158 000	160 000	162 000	164 000	165 000	166 000
Italy[1] Italie[1]	1 106	1 139	1 170	1 198	1 227	1 269	1 315	1 363	1 403	1 440[2]
Liechtenstein Liechtenstein	444[2]	442[2]	440[2]	436[2]	432	430[2]	428[2]	426[2]	425[2]	424[2]
Malta Malte	433	437[2]	440	443[2]	445[2]	447[2]	448[2]	448[2]	448[2]	448[2]
Netherlands Pays-Bas	178 000[2]	180 000[2]	183 000[2]	186 000[2]	188 399	190 000[2]	192 000	194 000[2]	196 000[2]	197 000[2]
Norway Norvège	135 900	139 206	144 439	145 848	150 000[2]	154 209	150 605	153 093	154 696	155 000[2]
Poland[1] Pologne[1]	670	710	757	806	925	990	1 044	1 101	1 153	1 185
Portugal Portugal	90 300[2]	95 600[2]	100 800[2]	106 100[2]	111 400[2]	116 700[2]	122 000[2]	127 200[2]	132 548	137 000[2]
Romania Roumanie	155 993	169 000	167 691	174 005	184 408	185 208	183 850	165 100	151 700	127 100
Spain Espagne	548 080	571 174	592 010	611 433	633 210	657 826	678 680	701 928	722 000	740 000[2]
Sweden Suède	189 654	188 500[2]	187 300[2]	186 200[2]	185 000[2]	183 828	183 000[2]	183 000	183 000	183 000[2]
Switzerland Suisse	98 000	102 472	106 300	103 413	105 314	106 000	108 000	109 000	110 000	112 700
United Kingdom Royaume-Uni	515 455	520 808	529 438	523 740	525 549	520 495	519 495	515 000[2]	509 780	505 000[2]
Yugoslavia[1] Yougoslavie[1]	510[2]	600[2]	706	809	882	955	1 017	1 066	1 108	1 092
Oceania Océanie	435 357	439 716	425 214	423 754	422 281	420 839	419 326	417 862	416 492	415 511
American Samoa[2] Samoa américaines[2]	12	12	12	12	12	12	12	14	14	14
Australia[2] Australie[2]	332 000	332 000	332 000	332 000	332 000	332 000	332 000	332 000	332 000	332 000
Cook Islands Iles Cook	143[2]	147[2]	151[2]	155[2]	159[2]	163[2]	167[2]	171	171	180

57
Tractors in use
Number [*cont.*]

Tracteurs en service
Nombre [*suite*]

Country or area Pays ou zone	1981	1982	1983	1984	1985	1986	1987	1988	1989	1990
Fiji[2] Fidji[2]	4 200	4 200	4 200	4 220	4 240	4 260	4 280	4 290	4 300	4 310
French Polynesia Polynésie française	150	150[2]	152[2]	152[2]	152[2]	154[2]	155[2]	155[2]	155[2]	155[2]
Guam Guam	80[2]	80	80[2]	80[2]	80[2]	80[2]	80[2]	80[2]	80[2]	80[2]
Kiribati[2] Kiribati[2]	17	17	17	17	17	18	18	18	18	18
New Caledonia Nouvelle-Calédonie	1 050[2]	1 110[2]	1 175[2]	1 239	1 260[2]	1 270[2]	1 280[2]	1 300[2]	1 320[2]	1 330[2]
New Zealand Nouvelle-Zélande	96 000[2]	100 434	85 926	84 400[2]	82 900[2]	81 441	79 900[2]	78 400[2]	77 000[2]	76 000[2]
Niue Nioué	* 10	* 10	10	10	10[2]	10[2]	10[2]	10[2]	10[2]	10[2]
Norfolk Island[2] Ile Norfolk[2]	11	11	12	12	12	12	12	12	12	12
Pacific Islands (Palau)[2,4] Iles du Pacifique (Palaos)[2,4]	52	52	52	52	53	53	53	53	53	53
Papua New Guinea Papouasie-Nvl-Guinée	1 429	1 290	1 224	1 200[2]	1 180[2]	1 160[2]	1 150[2]	1 150[2]	1 150[2]	1 140[2]
Samoa[2] Samoa[2]	30	30	30	32	32	32	35	35	35	35
Tonga Tonga	115[2]	115[2]	115[2]	115[2]	115	115[2]	115[2]	115[2]	115[2]	115[2]
Tuvalu Tuvalu	1[2]	1[2]	1	1	1	1	1	1[2]	1[2]	1[2]
Vanuatu[2] Vanuatu[2]	57	57	57	57	58	58	58	58	58	58
former USSR[1] ancienne URSS[1]	2 682[2]	2 720[2]	2 756[2]	2 792[2]	2 830	2 844	2 759	2 780	2 689	2 609

Source:
Food and Agriculture Organization of the United Nations
(Rome).

† All data shown which pertain to Germany prior to 3 October
1990 are indicated separately for the Federal Republic of
Germany and the former German Democratic Republic based on
their respective territories at the time indicated. Where
data for united Germany (3 October 1990 and thereafter) are
not available, available data are shown separately under the
designations Federal Republic of Germany and former German
Democratic Republic and pertain to the territorial
boundaries prior to 3 October 1990. For detailed
explanatory notes on data pertaining to Germany, see Annex I
- Country Nomenclature.

1 Number in thousands.
2 FAO estimate.
3 Excluding former Canal Zone.
4 Including data for Federated States of Micronesia, Marshall
Is. and Northern Mariana Is.

Source:
Organisation des Nations Unies pour l'alimentation et
l'agriculture (Rome).

† Toutes les données se rapportant à l'Allemagne avant le 3
octobre 1990 figurent dans deux rubriques séparées basées
sur les territoires respectifs de la République fédérale
d'Allemagne et l'ancienne République démocratique allemande
selon la période indiquée. En l'absence de données pour
l'Allemagne unifiée (à compter du 3 octobre 1990), les
données disponibles sont fournies séparément sous les
rubriques République fédérale d'Allemagne et ancienne
République démocratique allemande et se rapportent aux
limites territoriales antérieures au 3 octobre 1990. Pour
les notes explicatives en détail sur les données concernant
l'Allemagne, voir Annexe I – Nomenclature des pays.

1 En milliers d'unités.
2 Estimation de la FAO.
3 Non compris l'ancienne Zone du Canal.
4 Y compris les données pour les Etats fédérés de Micronésie,
les îles Marshall et les îles Mariannes du Nord.

58
Nitrogenous fertilizers
Engrais azotés
Consumption in terms of nitrogen: thousand metric tons
Consommation exprimée en azote : milliers de tonnes métriques

Country or area Pays ou zone	1981/82	1982/83	1983/84	1984/85	1985/86	1986/87	1987/88	1988/89	1989/90	1990/91
World *Monde*	60 451.6	61 191.6	67 665.3	70 647.1	69 829.6	71 492.4	75 595.7	79 602.7	79 185.9	77 051.3
Africa **Afrique**	1 870.8	1 878.1	1 820.9	1 858.1	1 957.1	1 930.1	1 887.3	2 090.8	2 040.9	2 063.4
Algeria Algérie	* 57.7	48.7	59.3	87.7	98.2	112.7	80.0	76.0	* 46.3	* 63.2
Angola Angola	* 4.6	* 2.8	* 3.8	3.7	12.8	* 5.0	* 4.0	5.2	9.2	* 2.5
Benin Bénin	0.5	* 1.1	* 3.0	3.6	* 3.8	4.3	3.4	2.1	* 1.1	* 2.5
Botswana Botswana	* 0.6	* 0.6	* 0.5	* 0.4	0.1	0.2	* 0.3	* 0.3	* 0.3	* 0.3
Burkina Faso Burkina Faso	3.1	3.3	4.1	4.2	4.7	6.8	8.3	8.0	8.2	* 6.0
Burundi Burundi	* 0.3	* 0.5	1.2	0.6	1.6	* 1.1	* 1.0	* 1.0	* 2.0	* 1.0
Cameroon Cameroun	16.6	22.3	24.9	22.0	* 26.9	* 20.7	* 24.6	22.6	* 18.2	* 9.0
Cape Verde Cap-Vert	...	...	...	0.0	0.1	0.1	0.0	...	...	...
Central African Rep. Rép. centrafricaine	* 1.2	* 0.7	0.3	* 1.1	* 2.5	0.6	0.8	0.5	0.7	* 0.8
Chad Tchad	* 1.6	* 1.7	* 2.5	* 3.0	* 3.2	* 2.6	* 2.0	2.2	2.0	* 2.0
Congo Congo	0.2	* 0.9	* 0.8	* 0.9	* 1.3	...	* 0.1	* 0.5	0.2	* 0.5
Côte d'Ivoire * Côte d'Ivoire *	11.3	7.4	10.0	11.0	8.2	7.0	4.5	8.2	10.0	13.5
Djibouti Djibouti	0.0	...	...	...	...	* 0.1	* 0.1	...	...	...
Egypt Egypte	585.0	* 667.8	722.2	649.2	640.3	655.5	* 677.0	* 799.1	754.1	745.1
Ethiopia Ethiopie	* 16.0	* 15.0	* 18.0	* 13.0	25.8	24.2	22.4	* 30.2	* 38.0	* 42.1
Gabon Gabon	0.8	* 0.2	* 0.6	* 0.5	* 0.5	1.1	0.8	0.3	* 0.4	0.1
Gambia Gambie	0.3	0.8	* 1.1	* 1.0	* 3.0	* 3.0	* 0.8	0.3	0.7	* 0.2
Ghana Ghana	* 13.0	* 12.5	* 11.0	* 5.0	* 5.0	* 4.1	5.1	7.0	5.8	* 8.0
Guinea Guinée	0.4	0.6	...	0.1	0.2	0.3	* 0.3	0.3	0.4	0.3
Guinea-Bissau Guinée-Bissau	* 1.2	* 0.2	* 0.3	...	...	...	...	...	0.4	0.2
Kenya Kenya	* 37.0	* 34.2	* 31.0	36.0	57.5	63.7	45.9	67.2	* 45.0	* 57.0
Lesotho * Lesotho *	0.5	0.5	0.5	0.6	0.1	0.1	0.4	0.4	0.5	0.5
Liberia Libéria	* 2.0	* 0.7	* 0.9	* 0.4	1.0	0.8	* 1.6	* 1.6	* 0.1	* 0.1

58
Nitrogenous fertilizers
Consumption in terms of nitrogen: thousand metric tons [*cont.*]
Engrais azotés
Consommation exprimée en azote : milliers de tonnes métriques [*suite*]

Country or area Pays ou zone	1981/82	1982/83	1983/84	1984/85	1985/86	1986/87	1987/88	1988/89	1989/90	1990/91
Libyan Arab Jamah. Jamah. arabe libyenne	27.2	35.2	33.9	56.4	20.9	24.8	* 25.1	* 32.5	* 30.0	* 35.0
Madagascar Madagascar	* 3.5	* 6.6	* 7.4	* 2.0	* 5.6	6.2	3.6	3.8	2.5	3.3
Malawi Malawi	25.7	* 24.2	23.1	23.6	15.9	22.7	29.8	27.2	* 30.3	* 30.0
Mali Mali	3.9	5.6	3.3	20.3	11.2	8.0	9.0	* 9.0	* 8.7	* 9.2
Mauritania Mauritanie	...	...	0.4	0.4	1.9	* 0.9	* 0.9	* 2.0	* 1.7	* 1.5
Mauritius Maurice	9.5	9.9	10.7	10.6	11.0	9.9	12.6	11.0	* 11.5	11.2
Morocco Maroc	96.6	108.1	103.1	115.3	136.6	141.4	146.5	139.6	148.3	144.2
Mozambique Mozambique	* 20.1	* 20.0	* 7.5	1.7	1.7	* 2.0	* 3.0	* 1.1	* 1.7	* 2.2
Niger Niger	2.4	1.3	* 1.4	1.5	2.1	1.2	2.0	1.1	1.8	0.5
Nigeria Nigéria	* 96.7	* 91.8	* 123.5	* 131.0	* 135.0	* 140.0	* 148.0	* 160.0	197.4	210.0
Réunion Réunion	5.3	3.5	6.3	4.7	5.1	5.5	4.8	5.2	5.7	* 3.6
Rwanda Rwanda	* 0.1	* 0.4	* 0.5	* 1.0	* 1.0	0.4	1.3	0.3	0.3	1.2
Senegal Sénégal	* 6.3	* 3.4	* 8.0	* 8.0	* 7.0	* 7.5	* 7.0	* 8.0	6.2	6.0
Seychelles Seychelles	...	0.9	0.3	...	...	...	...	...	...	...
Sierra Leone Sierra Leone	1.3	0.2	0.4	* 0.5	* 1.6	* 1.8	* 0.2	0.2	0.9	* 0.6
Somalia Somalie	* 1.3	* 0.8	* 1.8	* 2.6	3.2	0.9	* 3.0	* 1.3	* 1.4	* 1.7
South Africa Afrique du Sud	528.6	474.5	368.1	406.7	379.7	373.2	328.6	380.9	375.2	366.6
Sudan Soudan	74.5	79.0	36.8	41.4	92.0	* 46.0	* 49.0	45.6	46.9	73.7
Swaziland Swaziland	6.4	6.6	* 10.0	* 4.5	5.3	6.0	* 3.1	* 4.0	* 4.0	* 4.0
Togo Togo	0.8	1.0	1.2	2.6	3.7	* 4.0	4.3	5.2	* 6.5	* 5.0
Tunisia Tunisie	30.5	36.7	* 32.8	* 41.5	44.9	47.1	* 50.6	* 46.4	* 49.5	* 36.7
Uganda Ouganda	* 0.5	...	...	* 0.4	* 0.2	0.3	0.8	0.1	0.3	0.1
United Rep.Tanzania Rép. Unie de Tanzanie	* 17.5	15.8	15.5	* 23.5	24.7	* 29.5	32.1	27.0	* 28.7	* 31.0
Zaire Zaïre	* 2.5	3.1	4.1	* 4.5	* 3.6	1.0	3.1	1.9	* 4.0	* 4.0
Zambia Zambie	57.6	54.1	43.3	37.8	53.4	54.0	61.9	56.2	* 51.8	* 38.0

58
Nitrogenous fertilizers
Consumption in terms of nitrogen: thousand metric tons [cont.]
Engrais azotés
Consommation exprimée en azote : milliers de tonnes métriques [suite]

Country or area Pays ou zone	1981/82	1982/83	1983/84	1984/85	1985/86	1986/87	1987/88	1988/89	1989/90	1990/91
Zimbabwe Zimbabwe	98.0	73.1	* 81.3	* 71.6	93.0	81.9	73.4	88.2	82.2	89.1
America, North **Amérique du Nord**	**12 645.7**	**11 062.0**	**12 847.0**	**13 449.8**	**12 545.0**	**12 395.7**	**12 743.9**	**12 715.1**	**13 252.9**	**13 106.4**
Bahamas * Bahamas *	0.5	0.4	0.3	0.3	0.2	0.2	0.2	0.2	0.2	...
Barbados Barbade	* 2.0	* 2.0	1.6	1.6	1.3	* 1.4	* 1.5	* 1.5	* 1.5	* 1.5
Belize Belize	0.5	0.4	0.7	0.8	1.1	1.4	1.5	1.8	* 1.9	* 1.5
Bermuda Bermudes	0.1	0.1	0.1	0.1	0.0	* 0.1	* 0.1	* 0.1	* 0.1	...
Canada Canada	965.9	1 001.9	1 157.2	1 255.1	1 225.0	1 144.6	1 160.2	1 160.2	1 197.2	1 157.8
Costa Rica * Costa Rica *	45.2	49.0	54.0	50.6	49.0	52.0	55.0	59.0	63.0	55.6
Cuba Cuba	315.0	276.7	248.7	294.4	293.6	328.4	306.4	304.7	* 366.7	* 283.0
Dominica Dominique	* 1.0	1.3	0.8	* 0.8	* 0.9	* 1.0	* 1.0	* 1.0	* 2.0	* 2.2
Dominican Republic Rép. dominicaine	* 31.4	* 30.7	* 25.0	* 31.3	* 33.0	* 37.0	* 44.0	25.1	32.8	47.0
El Salvador El Salvador	66.5	44.2	59.4	39.8	61.2	49.1	64.5	67.8	56.2	53.2
Guadeloupe Guadeloupe	3.3	2.3	3.1	2.9	3.5	3.6	* 2.3	4.1	* 3.9	* 3.1
Guatemala Guatemala	47.0	* 58.7	* 41.7	* 55.5	* 60.0	* 80.0	* 83.8	* 82.2	* 90.5	* 88.0
Haiti * Haïti *	3.0	2.6	1.6	2.0	1.6	1.0	1.3	1.4	2.3	0.9
Honduras Honduras	* 14.1	12.8	19.0	* 23.0	14.1	11.6	17.3	21.5	20.3	* 29.0
Jamaica Jamaïque	10.2	3.4	6.7	* 8.8	* 5.9	* 6.6	* 12.0	* 16.0	* 16.0	* 7.4
Martinique Martinique	4.3	4.4	5.4	* 3.0	* 5.0	* 6.0	* 7.0	6.4	* 6.4	* 8.1
Mexico Mexique	1 111.7	1 254.6	* 1 087.8	* 1 193.1	1 262.6	1 324.9	1 378.1	* 1 269.6	* 1 292.8	* 1 163.0
Nicaragua Nicaragua	* 39.0	* 14.2	51.4	38.0	44.9	55.1	* 44.1	* 55.0	21.9	31.8
Panama[1] Panama[1]	* 13.9	* 13.9	* 11.0	* 12.2	* 14.0	* 20.6	* 20.3	* 21.1	18.4	21.7
Saint Kitts-Nevis Saint-Kitts-et-Nevis	* 0.4	* 0.4	* 0.4	* 0.5	* 0.5	* 0.5	* 0.5	* 0.5	0.4	0.4
Saint Lucia Sainte-Lucie	1.0	1.4	1.2	1.5	1.7	* 1.5	* 1.5	* 1.5	2.9	2.4
St. Vincent-Grenadines* St. Vincent-Grenadines*	2.4	2.4	2.4	2.5	2.5	2.6	2.6	2.6	1.5	0.5
Trinidad and Tobago Trinité-et-Tobago	2.8	* 3.4	* 3.9	* 5.0	* 5.5	* 3.5	1.8	1.3	* 5.0	* 6.0
United States[2] Etats-Unis[2]	9 963.6	8 279.8	10 062.6	10 425.8	9 456.9	9 261.9	9 535.9	9 609.4	10 047.9	10 141.4

58
Nitrogenous fertilizers
Consumption in terms of nitrogen: thousand metric tons [cont.]
Engrais azotés
Consommation exprimée en azote : milliers de tonnes métriques [suite]

Country or area Pays ou zone	1981/82	1982/83	1983/84	1984/85	1985/86	1986/87	1987/88	1988/89	1989/90	1990/91
United States Virgin Is. * Iles Vierges américaines *	0.9	0.9	1.0	1.0	1.0	1.0	1.0	1.0	1.0	1.0
America, South **Amérique du Sud**	**1 155.1**	**1 104.9**	**1 066.8**	**1 433.9**	**1 566.2**	**1 773.6**	**1 876.0**	**1 860.7**	**1 791.7**	**1 735.9**
Argentina Argentine	51.2	* 53.2	* 64.6	* 90.0	* 104.6	* 93.0	* 94.5	* 99.1	* 99.4	* 101.4
Bolivia Bolivie	2.7	* 1.3	3.1	2.5	* 2.7	* 2.8	4.7	3.1	4.6	4.8
Brazil Brésil	667.8	642.3	568.0	812.9	824.5	* 895.2	* 880.8	815.0	823.3	779.3
Chile Chili	47.7	48.1	64.9	86.3	104.5	136.0	* 150.0	* 160.0	158.3	152.9
Colombia Colombie	* 143.0	* 149.8	* 157.4	* 180.9	184.9	211.5	* 247.2	236.0	* 269.0	* 312.0
Ecuador Equateur	* 34.8	* 33.2	* 47.0	* 43.0	43.2	24.0	* 26.0	* 40.0	41.1	36.7
French Guiana Guyane française	0.2	0.2	0.4	* 0.4	0.4	0.5	* 0.5	0.5	* 0.3	* 0.7
Guyana Guyana	9.3	9.2	9.6	9.7	7.3	12.4	* 8.0	* 10.6	* 12.0	* 8.0
Paraguay Paraguay	2.3	2.0	* 1.7	* 1.3	3.5	* 3.8	* 1.9	2.2	3.5	2.7
Peru Pérou	100.6	71.2	52.9	53.8	50.6	121.5	159.5	156.1	113.2	* 104.1
Suriname Suriname	* 5.4	* 7.1	7.1	10.5	7.5	7.8	7.4	1.6	1.3	* 0.6
Uruguay Uruguay	* 20.4	* 16.5	* 16.7	* 19.5	* 17.1	* 15.0	* 15.5	30.9	24.5	27.7
Venezuela Venezuela	69.6	* 70.8	* 73.4	* 123.1	215.5	250.0	* 279.9	305.6	* 241.1	* 205.0
Asia **Asie**	**21 690.4**	**23 043.6**	**26 166.6**	**28 004.7**	**27 069.0**	**27 747.7**	**31 275.2**	**34 939.7**	**35 909.7**	**37 324.9**
Afghanistan Afghanistan	28.7	38.8	36.8	52.8	53.2	50.5	* 52.2	49.8	50.0	* 44.1
Bahrain Bahreïn	* 0.3	0.0	0.2	0.3	0.2	0.9	0.2	0.2	0.2	0.3
Bangladesh Bangladesh	251.6	306.0	356.4	386.4	367.8	423.0	474.6	524.0	629.8	608.6
Bhutan Bhoutan	* 0.1	* 0.1	0.1	0.1	* 0.1	* 0.1	* 0.1	* 0.1	* 0.1	* 0.1
Brunei Darussalam Brunéi Darussalam	0.1	0.2	0.1	0.2	0.1	0.2	* 0.3	* 0.3	* 0.1	* 1.4
Cambodia * Cambodge *	10.5	7.3	1.6	0.5	...	...	...	...	...	1.6
China Chine	* 11 528.3	* 12 210.7	* 13 678.9	* 15 075.0	* 13 650.2	* 13 576.1	16 803.9	18 514.1	* 18 855.3	* 19 449.8
Cyprus[3] Chypre[3]	9.0	11.6	10.6	10.1	10.1	11.3	12.0	12.1	12.5	12.4
India Inde	4 068.7	4 224.2	5 204.4	5 486.0	5 660.8	5 716.0	5 716.8	7 251.0	7 385.9	8 021.3

58
Nitrogenous fertilizers
Consumption in terms of nitrogen: thousand metric tons [cont.]
Engrais azotés
Consommation exprimée en azote : milliers de tonnes métriques [suite]

Country or area Pays ou zone	1981/82	1982/83	1983/84	1984/85	1985/86	1986/87	1987/88	1988/89	1989/90	1990/91
Indonesia Indonésie	997.1	1 082.5	1 049.1	* 1 285.4	* 1 299.0	* 1 359.0	* 1 460.3	* 1 495.0	* 1 474.0	* 1 610.0
Iran, Islamic Rep. of Iran, Rép. islamique d'	402.1	493.9	596.6	489.5	465.3	529.2	524.1	548.4	* 668.3	* 558.1
Iraq Iraq	* 63.6	62.1	61.3	* 73.0	120.0	131.5	* 148.5	* 146.5	138.4	* 130.0
Israel Israël	35.8	43.2	* 47.1	55.5	52.8	49.8	* 52.0	51.9	53.8	49.6
Japan Japon	643.0	683.0	701.0	697.0	680.0	693.0	669.0	640.0	641.0	612.0
Jordan Jordanie	2.9	7.1	8.8	7.0	10.7	10.8	9.2	14.3	* 15.0	* 10.0
Korea, Dem. P. R. * Corée, R. p. dém. de *	564.1	564.1	582.8	597.0	623.0	605.5	602.0	634.0	644.0	655.4
Korea, Republic of Corée, République de	* 431.8	* 308.6	* 367.3	* 401.8	414.0	423.0	428.0	439.0	455.0	465.2
Kuwait Koweït	* 0.3	0.6	0.8	0.5	0.7	0.4	0.3	0.8	* 0.8	* 1.0
Lao People's Dem. Rep. Rép. dém. pop. lao	* 4.0	0.2	0.5	...	* 1.4	...	* 0.3	* 0.1	* 0.1	* 1.1
Lebanon Liban	* 16.9	24.6	18.0	17.4	11.9	* 10.6	* 10.0	* 11.3	* 14.0	* 11.0
Malaysia Malaisie	* 127.9	* 138.0	* 235.0	* 216.0	* 243.0	* 247.0	* 262.0	* 273.0	288.0	* 313.0
Mongolia Mongolie	9.5	9.7	11.7	11.8	13.1	12.8	14.3	11.3	10.0	9.1
Myanmar Myanmar	* 92.9	* 114.7	* 115.0	* 127.4	132.5	115.0	92.0	86.6	63.7	* 53.0
Nepal Népal	18.0	22.9	28.1	31.7	31.7	32.9	38.1	39.8	49.2	52.0
Oman Oman	0.5	0.8	0.6	0.4	1.1	0.8	0.8	2.4	2.4	4.8
Pakistan Pakistan	832.6	952.6	914.3	934.8	1 128.1	1 332.6	1 281.7	1 324.9	1 467.6	1 471.6
Philippines Philippines	210.7	231.4	240.2	177.9	205.4	* 298.4	371.8	371.0	376.0	400.6
Qatar Qatar	0.8	0.7	0.6	0.6	0.5	0.7	* 0.7	* 0.6	1.2	1.4
Saudi Arabia Arabie saoudite	41.5	59.0	113.3	* 138.6	169.0	181.8	* 219.0	* 256.2	* 265.0	* 273.0
Singapore * Singapour *	2.0	2.0	2.0	2.0	2.2	2.2	2.5	2.6	2.6	2.6
Sri Lanka Sri Lanka	* 78.5	* 79.4	83.0	100.3	101.7	99.3	103.9	* 109.0	108.5	92.4
Syrian Arab Republic Rép. arabe syrienne	83.1	95.9	109.5	126.7	137.0	143.6	* 116.6	160.6	153.6	184.8
Thailand Thaïlande	151.1	174.8	233.4	227.7	252.9	308.5	342.8	439.7	494.9	576.5
Turkey Turquie	798.9	863.4	1 021.4	954.8	* 916.1	* 951.7	1 141.7	1 081.6	1 140.4	1 199.7
United Arab Emirates Emirats arabes unis	2.4	* 2.9	* 2.5	3.0	2.8	1.0	* 2.0	* 2.0	7.8	9.2

58
Nitrogenous fertilizers
Consumption in terms of nitrogen: thousand metric tons [*cont.*]
Engrais azotés
Consommation exprimée en azote : milliers de tonnes métriques [*suite*]

Country or area Pays ou zone	1981/82	1982/83	1983/84	1984/85	1985/86	1986/87	1987/88	1988/89	1989/90	1990/91
Viet Nam Viet Nam	169.1	214.7	318.3	299.1	293.4	413.9	313.3	* 428.9	424.0	* 419.0
Yemen Yémen	11.9	11.9	15.1	16.6	17.3	14.7	8.2	16.6	16.5	19.2
Europe Europe	14 419.0	14 749.8	15 149.5	15 237.0	15 356.4	15 764.6	15 598.7	15 958.4	15 775.1	13 578.7
Albania Albanie	* 72.5	70.8	* 73.0	* 75.0	* 75.0	* 75.0	66.1	69.3	79.9	* 73.4
Austria Autriche	161.5	146.3	152.5	* 161.1	165.1	137.8	146.3	140.9	135.6	* 135.0
Belgium-Luxembourg Belgique-Luxembourg	195.4	* 197.0	* 198.8	* 199.0	* 195.0	* 199.3	* 199.0	* 196.0	* 190.5	* 186.0
Bulgaria Bulgarie	518.0	536.0	550.0	479.0	474.0	440.0	418.0	548.0	* 495.0	* 478.0
Czechoslovakia Tchécoslovaquie	615.0	646.0	682.0	691.7	670.8	* 646.2	589.0	642.0	704.8	* 588.3
Denmark Danemark	376.0	391.4	419.0	398.1	382.1	381.3	* 367.0	* 377.0	* 400.4	* 394.9
Finland Finlande	183.6	216.0	204.8	196.1	202.2	218.0	214.6	199.4	231.5	206.8
France France	* 2 193.0	* 2 196.4	* 2 320.0	* 2 336.8	* 2 408.0	* 2 568.4	* 2 557.1	2 605.0	2 660.0	2 492.0
Germany† · Allemagne† F. R. Germany⁴ R. f. Allemagne⁴	1 323.0	1 464.5	1 377.9	1 451.7	1 515.7	1 578.3	1 601.4	1 539.9	1 487.2	# * 1 788.0
former German D. R. anc. R. d. allemande	749.8	607.4	693.9	697.0	770.3	708.9	773.9	873.2	* 813.0	...
Greece Grèce	373.3	* 383.8	* 417.9	* 428.3	450.0	432.1	383.6	409.2	425.7	426.6
Hungary Hongrie	562.9	646.5	625.4	625.9	558.4	593.4	613.8	646.3	583.0	358.0
Iceland Islande	15.2	* 14.0	14.8	13.5	12.9	12.7	11.4	11.5	10.4	11.8
Ireland Irlande	* 275.2	* 296.0	* 331.4	* 329.7	313.7	* 343.0	* 340.0	* 349.0	* 378.5	* 370.0
Italy Italie	988.0	967.8	996.1	1 026.3	1 054.5	1 010.6	1 046.6	924.0	827.3	845.8
Malta Malte	0.2	0.2	0.5	0.5	0.5	0.5	0.6	0.5	0.4	* 0.7
Netherlands Pays-Bas	477.3	456.7	478.3	505.3	499.7	504.0	* 458.2	455.7	412.4	* 392.0
Norway Norvège	106.7	114.1	110.3	113.3	107.3	109.7	113.6	* 110.1	* 110.4	* 110.0
Poland Pologne	1 213.0	1 244.9	1 321.5	1 238.5	1 336.6	1 388.1	1 335.4	1 520.6	* 1 478.6	* 671.0
Portugal Portugal	145.0	138.6	125.7	* 123.2	* 137.2	* 149.8	* 153.0	* 156.5	* 145.4	* 150.1
Romania Roumanie	882.0	884.0	* 878.0	* 857.0	703.0	* 716.0	* 720.0	739.0	778.2	* 656.1
Spain Espagne	818.3	798.9	808.5	916.8	* 961.6	* 1 062.5	* 1 147.8	1 168.4	1 109.4	* 1 063.5

58
Nitrogenous fertilizers
Consumption in terms of nitrogen: thousand metric tons [cont.]
Engrais azotés
Consommation exprimée en azote : milliers de tonnes métriques [suite]

Country or area Pays ou zone	1981/82	1982/83	1983/84	1984/85	1985/86	1986/87	1987/88	1988/89	1989/90	1990/91
Sweden Suède	248.1	249.3	257.9	253.2	246.0	240.9	241.5	240.4	221.5	211.7
Switzerland Suisse	* 62.9	* 60.2	67.4	70.7	71.9	71.1	72.6	71.6	70.3	63.4
United Kingdom Royaume-Uni	1 386.0	1 560.0	1 601.0	1 580.0	1 568.0	1 671.0	* 1 525.0	* 1 462.0	* 1 582.0	* 1 525.0
Yugoslavia Yougoslavie	477.0	463.0	443.0	469.0	477.0	506.0	503.1	503.0	443.7	380.7
Oceania **Océanie**	**287.7**	**315.1**	**312.5**	**384.6**	**385.8**	**405.6**	**427.6**	**451.0**	**497.6**	**503.9**
Australia Australie	245.0	272.6	* 269.0	* 330.0	* 340.0	* 360.0	* 372.1	* 393.9	* 440.3	* 439.4
Fiji Fidji	* 12.8	* 10.3	* 7.5	* 10.1	* 7.8	11.3	10.6	10.1	* 13.2	* 12.0
French Polynesia Polynésie française	0.3	0.3	0.4	* 0.8	* 0.4	* 0.3	* 0.3	* 0.3	* 0.3	* 0.3
New Caledonia Nouvelle-Calédonie	0.5	* 0.5	0.6	* 0.2	* 0.2	* 0.2	* 0.5	* 0.5	* 0.5	* 1.0
New Zealand Nouvelle-Zélande	* 21.7	28.2	31.4	* 40.0	* 32.0	* 27.0	* 36.8	* 40.0	* 37.0	* 45.8
Papua New Guinea Papouasie-Nvl-Guinée	* 7.3	* 3.2	* 3.5	3.4	* 5.3	* 6.8	* 7.3	* 6.2	* 6.3	* 5.4
Samoa Samoa	0.2	...	...	...	...	...	...	...	...	...
Tonga Tonga	...	0.1	0.1	* 0.1	* 0.1	...	...	...	...	...
former USSR **ancienne URSS**	**8 383.0**	**9 038.0**	**10 302.0**	**10 279.0**	**10 950.0**	**11 475.0**	**11 787.0**	**11 587.0**	**9 918.0**	**8 738.0**

Source:
Food and Agriculture Organization of the United Nations
(Rome).

† All data shown which pertain to Germany prior to 3 October 1990 are indicated separately for the Federal Republic of Germany and the former German Democratic Republic based on their respective territories at the time indicated. Where data for united Germany (3 October 1990 and thereafter) are not available, available data are shown separately under the designations Federal Republic of Germany and former German Democratic Republic and pertain to the territorial boundaries prior to 3 October 1990. For detailed explanatory notes on data pertaining to Germany, see Annex I - Country Nomenclature.

1 Excluding former Canal Zone.
2 Data include Puerto Rico.
3 Data do not cover the whole country.
4 Data are for united Germany.

Source:
Organisation des Nations Unies pour l'alimentation et l'agriculture (Rome).

† Toutes les données se rapportant à l'Allemagne avant le 3 octobre 1990 figurent dans deux rubriques séparées basées sur les territoires respectifs de la République fédérale d'Allemagne et l'ancienne République démocratique allemande selon la période indiquée. En l'absence de données pour l'Allemagne unifiée (à compter du 3 octobre 1990), les données disponibles sont fournies séparément sous les rubriques République fédérale d'Allemagne et ancienne République démocratique allemande et se rapportent aux limites territoriales antérieures au 3 octobre 1990. Pour les notes explicatives en détail sur les données concernant l'Allemagne, voir Annexe I – Nomenclature des pays.

1 Non compris l'ancienne Zone du Canal.
2 Les chiffres comprennent Porto Rico.
3 Les données ne couvrent plus la totalité du pays.
4 Les données sont pour l'Allemagne unifiée.

59
Phosphate fertilizers
Engrais phosphatés
Consumption in terms of phosphoric acid: thousand metric tons
Consommation exprimée en acide phosphorique: milliers de tonnes métriques

Country or area Pays ou zone	1981/82	1982/83	1983/84	1984/85	1985/86	1986/87	1987/88	1988/89	1989/90	1990/91
World *Monde*	30 946.3	30 863.9	33 083.8	34 022.3	33 224.4	34 658.1	36 683.6	37 987.1	37 392.6	36 024.8
Africa **Afrique**	1 227.9	1 187.7	1 140.8	1 196.8	1 256.0	1 178.3	1 137.9	1 172.9	1 077.0	1 066.4
Algeria Algérie	* 75.0	* 62.0	* 71.0	91.9	135.6	105.8	108.7	60.2	* 47.7	* 35.5
Angola Angola	* 4.8	* 2.0	* 4.0	2.7	7.2	* 3.5	* 2.8	6.7	* 8.0	* 5.0
Benin Bénin	0.4	* 1.3	* 1.5	2.2	5.7	4.2	4.0	2.9	* 1.1	* 2.0
Botswana Botswana	* 0.8	* 0.8	* 0.4	* 0.4	0.2	0.3	* 0.4	* 0.4	* 0.4	* 0.4
Burkina Faso Burkina Faso	4.4	4.6	6.0	4.2	4.5	6.0	4.8	3.9	7.9	* 4.5
Burundi Burundi	* 0.4	* 0.5	0.5	0.6	0.5	* 1.7	* 1.4	* 0.9	* 2.4	* 1.0
Cameroon Cameroun	7.7	8.2	9.0	8.1	* 9.6	* 11.7	* 14.3	6.2	* 4.4	* 3.0
Central African Rep. Rép. centrafricaine	0.0	...	0.1	...	* 0.2	0.0	0.0	0.0	0.0	...
Chad Tchad	* 1.4	* 1.4	* 1.5	* 2.0	* 1.5	* 0.8	* 1.3	1.0	1.1	* 1.9
Congo Congo	0.2	...	...	...	...	...	...	...	0.2	* 0.5
Côte d'Ivoire * Côte d'Ivoire *	8.8	7.3	6.5	8.5	7.3	7.0	6.1	8.0	11.0	8.2
Egypt Egypte	* 110.0	* 149.6	159.6	181.0	198.2	* 181.5	* 185.4	* 203.7	164.9	184.1
Ethiopia Ethiopie	* 30.0	* 20.6	* 30.0	* 20.0	40.2	38.3	31.9	* 48.3	* 59.0	* 69.7
Gabon Gabon	0.1	* 0.1	* 0.3	* 0.7	* 0.8	0.3	0.2	0.0	* 0.1	0.5
Gambia Gambie	0.9	1.6	* 1.0	* 1.0	* 0.9	* 0.4	* 1.3	0.3	0.7	* 0.2
Ghana Ghana	* 8.0	* 5.6	* 7.0	* 1.4	* 4.3	* 2.0	2.7	2.8	1.2	* 3.0
Guinea Guinée	0.1	0.1	...	0.1	0.1	0.2	* 0.3	0.1	0.1	0.1
Guinea-Bissau Guinée-Bissau	* 0.3	* 0.5	* 0.2	...	...	...	...	...	0.3	0.2
Kenya Kenya	* 40.0	* 30.8	* 49.1	38.0	43.7	45.4	50.3	51.1	* 62.0	* 51.0
Lesotho * Lesotho *	4.0	4.0	4.0	4.0	3.1	3.5	3.5	4.0	4.0	4.0
Liberia Libéria	* 2.4	* 0.4	* 1.2	* 0.1	0.2	0.4	* 0.3	* 0.1	* 0.2	* 0.1
Libyan Arab Jamah. Jamah. arabe libyenne	43.9	49.3	53.5	44.3	39.1	46.3	62.9	54.2	* 47.0	* 41.0
Madagascar Madagascar	* 1.4	* 3.0	* 3.5	* 1.0	* 1.4	2.2	1.5	3.5	1.0	2.2
Malawi Malawi	4.9	* 5.2	12.6	* 16.0	* 13.0	8.2	10.7	17.3	* 18.0	* 13.0

59
Phosphate fertilizers
Consumption in terms of phosphoric acid: thousand metric tons [cont.]
Engrais phosphatés
Consommation exprimée en acide phosphorique : milliers de tonnes métriques [suite]

Country or area Pays ou zone	1981/82	1982/83	1983/84	1984/85	1985/86	1986/87	1987/88	1988/89	1989/90	1990/91
Mali Mali	2.8	3.0	3.7	15.0	5.0	3.7	3.8	* 3.4	* 9.1	* 6.0
Mauritania * Mauritanie *	...	0.1	...	0.1	0.1	0.1	0.1	0.6	0.6	0.4
Mauritius Maurice	2.9	3.1	3.2	3.4	3.5	3.7	4.5	4.1	* 4.7	3.8
Morocco Maroc	80.0	104.6	96.9	103.8	113.5	118.4	117.5	119.3	116.6	109.3
Mozambique Mozambique	* 15.0	* 12.8	* 4.3	1.7	1.5	* 2.5	* 2.5	* 0.3	* 0.3	* 0.2
Niger Niger	2.4	1.0	* 0.2	0.7	1.1	0.5	0.7	0.4	0.7	0.3
Nigeria Nigéria	* 79.5	* 67.1	* 92.6	* 96.0	* 105.0	* 92.0	* 95.0	* 100.0	93.5	96.1
Réunion Réunion	5.0	2.4	5.5	4.1	4.6	4.0	3.2	3.5	3.7	* 2.7
Rwanda Rwanda	* 0.1	* 0.3	* 0.1	* 0.5	* 0.2	0.3	0.4	0.2	0.1	1.0
Senegal Sénégal	* 8.0	* 9.5	* 10.0	* 5.3	* 7.5	* 7.5	* 8.0	* 10.0	3.2	3.5
Sierra Leone Sierra Leone	1.3	0.1	0.4	* 0.4	* 1.4	* 1.3	* 0.2	0.2	0.4	* 0.4
Somalia * Somalie *	...	0.1	0.3	0.8	0.3	0.4	0.3	0.5	0.8	0.5
South Africa Afrique du Sud	547.1	503.7	388.1	416.2	363.3	337.1	274.7	317.7	273.0	277.0
Sudan Soudan	0.6	0.5	* 0.7	1.9	1.4	* 0.2	* 0.9	1.8	2.4	7.2
Swaziland Swaziland	6.6	3.7	* 5.0	* 2.0	1.9	2.6	* 2.0	* 2.0	* 2.0	* 2.0
Togo Togo	0.7	0.9	1.0	2.3	3.4	* 3.9	3.6	* 4.0	* 4.5	* 5.0
Tunisia Tunisie	49.1	40.7	* 36.8	* 41.7	47.2	57.8	* 48.2	* 49.9	* 45.9	* 44.2
Uganda Ouganda	* 0.1	...	...	* 0.1	...	0.1	0.4	0.0	0.1	* 0.1
United Rep.Tanzania Rép. Unie de Tanzanie	* 8.5	4.8	* 5.2	* 9.0	11.0	* 12.1	12.5	11.1	* 16.3	* 12.4
Zaire Zaïre	* 2.5	1.8	2.7	* 5.0	* 2.2	1.0	2.0	0.5	* 1.4	* 0.5
Zambia Zambie	20.9	23.7	17.2	14.1	18.7	16.5	23.6	20.4	* 19.4	* 15.9
Zimbabwe Zimbabwe	45.0	45.0	* 44.5	* 44.5	45.9	42.8	39.1	47.5	35.7	46.7
America, North Amérique du Nord	5 575.2	5 048.3	5 656.2	5 514.8	5 052.4	4 866.6	4 986.7	4 958.7	5 093.5	4 872.8
Bahamas * Bahamas *	0.4	0.6	0.7	0.5	0.2	0.2	0.2	0.2	0.2	...
Barbados Barbade	* 1.0	* 1.0	1.4	1.4	1.0	* 1.2	* 0.3	* 0.3	* 0.1	* 0.2
Belize Belize	0.6	0.5	0.6	0.6	0.6	1.2	1.2	1.0	* 1.2	* 1.7

59
Phosphate fertilizers
Consumption in terms of phosphoric acid: thousand metric tons [cont.]
Engrais phosphatés
Consommation exprimée en acide phosphorique : milliers de tonnes métriques [suite]

Country or area Pays ou zone	1981/82	1982/83	1983/84	1984/85	1985/86	1986/87	1987/88	1988/89	1989/90	1990/91
Bermuda Bermudes	0.1	0.1	0.1	0.1	0.0	...	0.0	...	...	...
Canada Canada	636.3	651.7	713.0	727.7	703.4	626.1	634.5	614.4	609.2	578.2
Costa Rica * Costa Rica *	10.2	12.8	14.0	17.0	11.6	14.0	10.0	15.9	14.0	15.0
Cuba Cuba	82.9	79.5	77.7	84.1	75.9	81.8	81.8	80.6	83.0	* 82.0
Dominica Dominique	* 0.7	0.3	0.8	* 0.9	* 1.0	* 1.0	* 1.0	* 1.0	* 1.2	* 0.6
Dominican Republic Rép. dominicaine	* 11.0	* 6.7	* 7.0	* 13.8	* 15.0	* 17.0	* 20.0	* 18.1	* 20.0	18.8
El Salvador El Salvador	20.1	12.5	21.1	11.2	20.8	14.8	26.6	24.9	17.9	18.9
Guadeloupe Guadeloupe	3.2	2.2	2.7	2.7	3.0	2.9	* 2.1	3.6	* 2.5	* 1.4
Guatemala Guatemala	30.3	18.1	* 15.8	* 21.9	* 22.0	* 28.0	* 26.6	* 30.0	* 18.0	* 27.0
Haiti * Haïti *	1.4	0.9	0.4	0.6	0.7	0.7	0.5	0.3	0.7	...
Honduras Honduras	* 4.2	* 2.6	3.4	4.8	2.7	5.6	6.6	8.2	6.0	* 10.6
Jamaica Jamaïque	2.9	3.1	2.0	* 5.3	* 2.0	* 2.0	* 4.8	4.9	* 4.1	* 3.6
Martinique Martinique	2.7	0.6	0.1	* 1.8	* 2.1	* 3.0	* 4.0	4.2	* 4.5	* 5.2
Mexico Mexique	384.2	486.7	* 330.5	* 382.0	382.8	410.0	404.0	* 394.9	* 354.3	* 330.0
Nicaragua Nicaragua	* 7.5	* 6.4	* 10.0	5.1	8.6	12.2	* 7.9	* 12.0	7.6	6.1
Panama[1] Panama[1]	* 6.3	* 6.4	* 4.0	* 5.7	* 5.8	* 6.0	* 5.6	* 7.1	4.0	4.9
Saint Kitts-Nevis Saint-Kitts-et-Nevis	* 0.7	* 0.7	* 0.7	* 0.8	* 0.8	* 0.9	* 0.9	* 1.0	0.2	0.2
Saint Lucia Sainte-Lucie	* 0.1	* 0.1	* 0.1	* 0.1	0.1	* 0.1	* 0.1	* 0.1	2.1	1.7
St. Vincent-Grenadines* St. Vincent-Grenadines *	0.5	0.5	0.5	0.6	0.6	0.7	0.7	0.7	0.9	0.9
Trinidad and Tobago Trinité-et-Tobago	0.6	* 0.8	* 3.3	* 0.6	1.3	* 0.7	1.6	0.2	* 0.2	* 0.2
United States[2] Etats-Unis[2]	4 367.1	3 753.5	4 446.2	4 225.3	3 790.1	3 636.2	3 745.3	3 734.7	3 941.4	3 765.3
United States Virgin Is. * Iles Vierges américaines *	0.2	0.2	0.2	0.3	0.3	0.3	0.3	0.3	0.3	0.3
America, South **Amérique du Sud**	**1 616.9**	**1 518.5**	**1 302.8**	**1 854.7**	**1 766.1**	**2 149.2**	**2 193.5**	**2 134.2**	**1 861.3**	**1 693.1**
Argentina Argentine	* 40.0	* 48.4	* 56.1	* 46.5	* 50.0	51.0	51.5	* 51.2	* 42.7	* 51.5
Bolivia Bolivie	* 3.5	* 1.1	4.8	2.3	* 2.8	* 3.0	1.0	* 1.3	5.3	8.4
Brazil Brésil	* 1 318.3	* 1 210.4	991.8	* 1 476.4	1 309.1	* 1 610.8	* 1 576.1	1 507.4	1 296.2	1 185.8

59
Phosphate fertilizers
Consumption in terms of phosphoric acid: thousand metric tons [cont.]
Engrais phosphatés
Consommation exprimée en acide phosphorique: milliers de tonnes métriques [suite]

Country or area Pays ou zone	1981/82	1982/83	1983/84	1984/85	1985/86	1986/87	1987/88	1988/89	1989/90	1990/91
Chile Chili	54.6	47.3	60.9	80.0	86.0	108.8	131.3	* 139.2	125.9	113.8
Colombia Colombie	* 70.0	* 97.5	* 82.8	* 90.5	87.3	94.1	* 115.3	101.3	* 121.8	* 126.7
Ecuador Equateur	* 17.5	* 18.4	* 13.1	18.9	20.8	* 20.4	16.4	* 19.0	21.9	6.6
French Guiana Guyane française	0.2	0.2	0.3	...	0.3	0.2	* 0.2	0.2	* 0.2	* 0.2
Guyana Guyana	3.9	* 1.1	* 0.8	4.0	3.8	7.7	* 3.4	* 1.6	* 1.8	* 2.2
Paraguay Paraguay	4.9	3.8	* 5.4	* 7.3	4.8	* 5.0	* 8.2	3.8	12.8	8.9
Peru Pérou	18.3	12.6	16.2	14.5	14.2	29.1	43.4	37.6	18.8	12.8
Suriname Suriname	* 0.3	* 0.3	1.3	0.4	2.0	2.0	1.8	0.2	0.2	* 0.2
Uruguay Uruguay	* 39.7	* 30.0	* 23.0	* 30.0	* 40.5	* 38.0	* 42.0	35.7	42.3	41.2
Venezuela Venezuela	45.8	* 47.4	* 46.3	* 84.0	144.7	179.1	* 203.0	235.7	171.3	* 135.0
Asia Asie	7 180.7	7 958.5	9 068.0	9 509.9	8 545.9	9 123.6	10 883.2	12 254.9	12 763.6	13 784.3
Afghanistan Afghanistan	16.9	* 14.1	19.8	19.8	19.8	16.6	26.2	* 5.9	* 5.6	* 0.4
Bahrain Bahreïn	...	0.1	0.1	0.1	0.2	0.3	0.1	0.1	0.1	0.2
Bangladesh Bangladesh	120.2	130.4	163.0	160.9	136.9	154.0	179.5	191.4	221.1	236.3
Bhutan Bhoutan	...	...	0.1	0.0	...	...	...	...	...	...
Brunei Darussalam Brunéi Darussalam	0.1	0.2	0.0	0.2	0.2	0.1	* 0.1	* 0.2	* 0.1	* 1.4
Cambodia * Cambodge *	7.5	3.4	3.2	0.6	...	...	...	...	3.5	6.2
China Chine	* 2 931.3	* 3 208.8	* 3 728.9	* 3 869.0	* 2 764.5	* 3 025.7	4 489.0	5 162.2	* 5 272.1	* 5 878.5
Cyprus[3] Chypre[3]	5.5	6.5	7.3	6.6	6.9	7.3	7.8	8.2	8.3	8.3
India Inde	1 339.7	1 454.3	1 750.7	1 910.5	2 035.4	2 113.5	2 224.8	2 757.8	3 052.9	3 210.0
Indonesia Indonésie	320.4	367.7	* 358.7	* 437.7	* 494.8	* 556.6	569.2	* 610.4	* 598.0	* 585.0
Iran, Islamic Rep. of Iran, Rép. islamique d'	282.0	394.2	427.3	432.5	434.6	373.7	412.3	521.5	* 512.4	* 587.0
Iraq Iraq	* 13.1	12.1	26.2	40.3	51.0	* 56.0	* 64.3	* 62.0	75.1	* 75.0
Israel Israël	16.6	14.7	* 18.1	19.5	18.3	17.3	* 18.0	22.3	21.4	20.7
Japan Japon	701.0	721.0	765.0	777.0	741.0	753.0	766.0	726.0	728.0	690.0
Jordan Jordanie	3.3	6.3	6.5	6.5	3.2	1.2	3.9	9.2	* 5.0	* 5.0

59
Phosphate fertilizers
Consumption in terms of phosphoric acid: thousand metric tons [cont.]
Engrais phosphatés
Consommation exprimée en acide phosphorique : milliers de tonnes métriques [suite]

Country or area Pays ou zone	1981/82	1982/83	1983/84	1984/85	1985/86	1986/87	1987/88	1988/89	1989/90	1990/91
Korea, Dem. P. R. * Corée, R. p. dém. de *	130.0	129.4	130.0	132.0	135.0	137.0	137.0	147.6	163.2	158.9
Korea, Republic of Corée, République de	137.8	* 149.0	* 171.1	* 181.2	186.0	192.0	193.0	194.0	235.1	228.0
Kuwait Koweït	* 0.1	0.1	...	...	...	...	...	...	...	...
Lao People's Dem. Rep. Rép. dém. pop. lao	...	0.1	0.0	...	* 0.6	...	* 0.1	* 0.1	* 0.1	* 0.3
Lebanon Liban	8.0	6.3	6.5	22.7	11.9	* 3.5	* 6.7	* 7.0	* 10.0	* 12.0
Malaysia Malaisie	* 98.6	* 110.8	* 123.0	* 111.0	* 118.4	* 154.8	* 149.0	* 150.0	150.1	* 163.5
Mongolia Mongolie	3.6	3.9	3.5	3.3	3.1	3.4	3.5	3.2	3.5	3.7
Myanmar Myanmar	* 29.9	* 43.1	* 32.7	* 46.9	47.6	49.2	24.6	18.7	17.3	14.8
Nepal Népal	5.1	7.5	8.5	10.6	11.1	11.9	15.2	15.3	16.7	19.1
Oman Oman	0.2	0.4	0.4	0.3	0.9	0.6	0.6	1.4	1.1	1.5
Pakistan Pakistan	225.5	265.3	* 261.4	293.8	349.8	409.8	393.4	390.4	382.5	388.5
Philippines Philippines	47.8	51.0	54.6	45.4	* 42.8	* 45.1	65.2	77.5	84.1	105.0
Qatar Qatar	...	0.1	0.1	...	0.1	...	...	...	...	...
Saudi Arabia Arabie saoudite	23.6	33.1	87.7	109.6	142.3	196.7	* 180.0	* 220.4	175.0	* 193.0
Singapore * Singapour *	0.5	0.5	0.5	0.5	0.5	0.5	0.5	0.5	0.5	0.5
Sri Lanka Sri Lanka	* 27.6	31.2	34.9	37.5	38.6	38.0	42.6	* 39.1	40.7	29.2
Syrian Arab Republic Rép. arabe syrienne	50.1	53.5	63.7	86.6	85.6	95.5	* 101.2	109.3	91.6	111.9
Thailand Thaïlande	116.3	134.2	154.0	142.6	125.0	132.5	148.3	200.8	188.8	318.3
Turkey Turquie	489.3	565.4	617.6	554.8	* 476.9	* 520.2	585.4	490.2	599.7	624.8
United Arab Emirates Emirats arabes unis	1.0	1.0	0.8	0.5	0.8	0.4	* 1.0	* 1.6	* 1.6	* 1.2
Viet Nam Viet Nam	28.2	* 38.4	* 41.6	47.5	61.1	56.0	73.6	* 109.6	* 97.7	103.3
Yemen Yémen	0.1	0.3	0.5	1.9	1.2	* 1.4	* 1.0	* 1.1	1.0	2.8
Europe **Europe**	8 380.2	7 972.4	8 299.5	8 219.9	7 999.0	7 972.8	7 817.9	7 882.0	7 372.0	5 987.3
Albania Albanie	* 6.0	22.8	* 25.0	* 18.0	* 18.0	* 18.0	26.7	25.0	24.3	* 25.5
Austria Autriche	93.3	83.4	94.6	* 95.1	90.5	75.8	80.1	78.3	74.4	* 71.5
Belgium-Luxembourg * Belgique-Luxembourg*	92.7	94.1	98.0	93.6	91.0	89.0	87.0	87.0	86.5	85.5

59
Phosphate fertilizers
Consumption in terms of phosphoric acid: thousand metric tons [cont.]
Engrais phosphatés
Consommation exprimée en acide phosphorique: milliers de tonnes métriques [suite]

Country or area Pays ou zone	1981/82	1982/83	1983/84	1984/85	1985/86	1986/87	1987/88	1988/89	1989/90	1990/91
Bulgaria Bulgarie	409.0	361.0	345.0	355.0	287.0	247.0	233.0	258.2	220.3	* 150.0
Czechoslovakia Tchécoslovaquie	494.0	479.0	529.0	524.4	536.6	* 523.3	452.0	480.4	* 437.0	* 359.0
Denmark Danemark	105.3	113.3	118.7	111.3	105.6	107.1	* 95.5	* 92.0	* 94.8	* 88.6
Finland Finlande	142.4	159.9	162.8	158.4	154.9	159.2	156.2	140.6	143.1	117.2
France France	* 1 677.1	* 1 630.7	* 1 679.4	* 1 580.0	* 1 466.0	* 1 425.2	* 1 405.1	1 460.0	1 494.0	1 349.0
Germany†·Allemagne† F. R. Germany R. f. Allemagne	752.7	739.8	744.7	732.3	736.8	683.4	679.4	643.5	594.4	# * 609.0 [4]
former German D. R. anc. R. d. allemande	374.3	304.1	332.9	319.0	318.4	352.6	323.7	348.8	356.3	...
Greece Grèce	166.1	* 164.1	* 174.5	180.5	180.0	181.5	169.7	176.4	188.8	187.3
Hungary Hongrie	398.9	391.9	410.4	431.1	335.9	355.0	332.4	322.1	266.0	127.0
Iceland Islande	8.3	* 9.0	8.2	7.5	7.0	6.8	6.5	5.9	5.2	5.9
Ireland * Irlande *	141.6	145.2	151.7	151.2	133.0	149.6	142.0	148.0	145.5	136.8
Italy Italie	705.8	656.6	682.3	694.0	692.2	667.4	785.7	715.5	607.9	600.1
Malta Malte	0.1	0.1	0.2	0.3	0.1	0.1	0.0	0.0	0.0	0.0
Netherlands Pays-Bas	80.6	77.5	86.6	89.2	81.3	87.6	* 79.6	86.3	76.1	* 74.0
Norway Norvège	62.3	65.5	59.9	56.2	53.9	52.8	45.3	* 39.8	* 36.6	* 33.5
Poland Pologne	817.3	826.1	900.3	885.1	908.2	961.2	836.6	943.7	751.8	* 332.6
Portugal Portugal	84.7	80.6	* 67.2	* 61.4	* 69.6	* 78.9	* 83.9	* 89.1	* 81.1	* 80.3
Romania Roumanie	551.0	288.0	360.0	371.0	444.0	* 440.0	* 440.0	395.6	360.0	* 313.1
Spain Espagne	340.3	422.4	398.4	436.1	461.7	* 493.5	* 537.3	542.2	559.4	* 534.2
Sweden Suède	121.7	114.8	111.9	101.3	85.5	78.8	76.4	69.3	69.7	57.8
Switzerland Suisse	* 41.7	* 43.7	42.8	43.9	41.8	39.0	40.4	40.0	39.4	38.3
United Kingdom Royaume-Uni	445.0	464.0	474.0	469.0	439.0	446.0	* 435.0	* 433.0	* 428.0	* 415.0
Yugoslavia Yougoslavie	268.0	235.0	241.0	255.0	261.0	254.0	268.3	261.4	231.2	196.1
Oceania Océanie	1 086.5	1 063.6	1 151.5	1 101.1	989.9	1 013.7	1 100.4	1 028.4	1 050.2	806.0
Australia Australie	754.1	729.7	* 772.6	* 764.0	* 685.0	* 758.0	* 830.8	* 836.0	* 790.5	* 578.9

59
Phosphate fertilizers
Consumption in terms of phosphoric acid: thousand metric tons [cont.]
Engrais phosphatés
Consommation exprimée en acide phosphorique: milliers de tonnes métriques [suite]

Country or area Pays ou zone	1981/82	1982/83	1983/84	1984/85	1985/86	1986/87	1987/88	1988/89	1989/90	1990/91
Fiji Fidji	* 3.3	* 2.2	* 1.9	* 2.7	2.8	7.2	8.2	10.4	2.7	* 3.0
French Polynesia Polynésie française	0.1	0.2	0.3	* 0.5	* 0.3	* 0.3	* 0.3	* 0.3	* 0.3	* 0.3
New Caledonia Nouvelle-Calédonie	0.0	...	0.2	* 0.2	* 0.1	* 0.2	* 0.5	* 0.5	* 0.5	* 0.5
New Zealand Nouvelle-Zélande	* 326.5	330.2	374.6	* 332.0	* 300.0	* 246.0	* 258.1	* 178.5	253.2	220.3
Papua New Guinea Papouasie-Nvl-Guinée	2.3	* 1.3	* 1.9	* 1.7	* 1.7	* 2.0	* 2.5	* 2.7	* 3.0	* 3.0
Samoa Samoa	0.2	...	...	...	...	...	...	...	...	...
former USSR ancienne URSS	5 879.0	6 115.0	6 465.0	6 625.0	7 615.0	8 354.0	8 564.0	8 556.0	8 175.0	7 815.0

Source:
Food and Agriculture Organization of the United Nations
(Rome).

† All data shown which pertain to Germany prior to 3 October
1990 are indicated separately for the Federal Republic of
Germany and the former German Democratic Republic based on
their respective territories at the time indicated. Where
data for united Germany (3 October 1990 and thereafter) are
not available, available data are shown separately under the
designations Federal Republic of Germany and former German
Democratic Republic and pertain to the territorial
boundaries prior to 3 October 1990. For detailed
explanatory notes on data pertaining to Germany, see Annex I
- Country Nomenclature.

1 Excluding former Canal Zone.
2 Data include Puerto Rico.
3 Data do not cover the whole country.
4 Data are for united Germany.

Source:
Organisation des Nations Unies pour l'alimentation et
l'agriculture (Rome).

† Toutes les données se rapportant à l'Allemagne avant le 3
octobre 1990 figurent dans deux rubriques séparées basées
sur les territoires respectifs de la République fédérale
d'Allemagne et l'ancienne République démocratique allemande
selon la période indiquée. En l'absence de données pour
l'Allemagne unifiée (à compter du 3 octobre 1990), les
données disponibles sont fournies séparément sous les
rubriques République fédérale d'Allemagne et ancienne
République démocratique allemande et se rapportent aux
limites territoriales antérieures au 3 octobre 1990. Pour
les notes explicatives en détail sur les données concernant
l'Allemagne, voir Annexe I – Nomenclature des pays.

1 Non compris l'ancienne Zone du Canal.
2 Les chiffres comprennent Porto Rico.
3 Les données ne couvrent plus la totalité du pays.
4 Les données sont pour l'Allemagne unifiée.

60
Potash fertilizers
Engrais potassiques
Consumption: thousand metric tons
Consommation : milliers de tonnes métriques

Country or area Pays ou zone	1981/82	1982/83	1983/84	1984/85	1985/86	1986/87	1987/88	1988/89	1989/90	1990/91
World *Monde*	23 749.2	22 985.7	25 542.7	25 895.6	25 559.2	26 103.9	27 302.5	28 040.6	26 928.8	24 443.6
Africa Afrique	436.3	413.2	411.0	432.6	473.4	452.9	458.3	455.5	487.2	487.9
Algeria Algérie	33.4	15.5	29.3	24.3	49.0	53.2	52.5	33.8	* 23.0	* 29.0
Angola Angola	* 2.7	* 0.2	* 0.8	0.6	0.4	* 3.5	* 3.5	4.0	6.4	* 2.0
Benin Bénin	1.9	* 0.7	* 0.9	1.5	2.0	2.3	2.0	1.9	* 1.1	* 2.6
Botswana Botswana	...	* 0.4	* 0.4	* 0.4	0.1	0.2	* 0.2	* 0.2	* 0.2	* 0.2
Burkina Faso Burkina Faso	2.4	2.5	3.2	3.1	2.9	3.6	4.8	3.3	4.5	* 3.5
Burundi Burundi	* 0.4	* 0.3	* 1.0	0.6	0.2	* 0.3	* 0.2	* 0.7	* 0.3	* 0.1
Cameroon Cameroun	10.7	11.7	12.2	13.2	* 20.0	* 20.0	* 10.9	5.6	* 6.0	* 9.8
Central African Rep. Rép. centrafricaine	0.0	...	0.1	* 0.1	* 0.2	0.0	0.0	0.0	0.0	* 0.1
Chad Tchad	* 1.6	* 2.3	* 1.8	* 2.0	* 2.4	* 0.8	* 2.1	1.7	1.8	* 1.9
Congo Congo	0.2	0.8	* 0.8	* 2.0	* 3.4	* 2.5	* 1.6	* 0.2	0.2	* 1.0
Côte d'Ivoire * Côte d'Ivoire *	29.0	19.1	21.2	22.0	26.0	15.0	22.0	25.0	18.0	14.0
Egypt Egypte	12.9	9.2	* 17.5	* 21.0	* 25.0	* 30.1	* 35.0	* 30.7	* 46.1	* 41.0
Ethiopia * Ethiopie *	...	0.1	0.1	...	...	...	...	...	...	...
Gabon Gabon	0.6	* 0.1	* 1.4	* 1.5	* 1.5	0.8	1.1	0.6	* 0.7	* 0.5
Gambia Gambie	0.2	0.2	* 0.3	* 0.1	...	* 0.4	* 0.5	0.3	0.7	* 0.2
Ghana Ghana	* 8.0	* 9.0	* 3.4	* 2.0	* 3.2	* 1.5	3.1	2.8	1.5	* 2.0
Guinea Guinée	0.3	0.1	...	0.1	0.1	0.1	* 0.3	0.3	0.3	0.1
Guinea-Bissau Guinée-Bissau	...	* 0.2	...	...	...	...	...	...	0.2	0.1
Kenya Kenya	* 5.8	* 4.0	* 6.7	5.0	7.8	13.7	5.8	6.3	* 9.8	* 8.0
Lesotho * Lesotho *	...	...	...	...	0.3	0.3	0.1	0.1	0.1	0.1
Liberia Libéria	* 0.7	* 0.2	* 0.7	* 0.6	0.3	0.5	* 1.6	* 1.7	* 2.4	* 0.1
Libyan Arab Jamah. Jamah. arabe libyenne	3.2	3.1	3.2	0.5	...	1.4	1.2	1.2	* 1.8	* 1.6
Madagascar Madagascar	* 2.1	* 6.0	* 3.1	* 3.6	* 2.7	2.5	1.2	3.4	2.7	2.6
Malawi Malawi	3.6	* 2.9	7.3	5.2	5.1	6.5	7.8	6.7	* 6.5	* 5.0

60
Potash fertilizers
Consumption: thousand metric tons [cont.]
Engrais potassiques
Consommation : milliers de tonnes métriques [suite]

Country or area Pays ou zone	1981/82	1982/83	1983/84	1984/85	1985/86	1986/87	1987/88	1988/89	1989/90	1990/91
Mali Mali	2.0	2.6	3.1	14.9	3.5	3.0	2.8	* 1.0	...	...
Mauritania * Mauritanie *	...	...	...	...	...	...	0.1	0.6	...	...
Mauritius Maurice	10.0	12.7	13.2	13.0	13.5	11.7	15.8	13.1	* 16.0	12.8
Morocco Maroc	35.5	42.0	43.8	47.6	54.1	55.0	57.1	56.4	55.9	56.3
Mozambique Mozambique	* 6.3	* 7.1	* 2.9	0.5	0.6	* 1.4	* 1.0	* 0.2	* 0.4	* 0.2
Niger Niger	0.8	0.5	* 0.2	* 0.1	0.3	0.3	0.2	0.2	0.3	0.2
Nigeria Nigéria	* 37.0	* 42.9	* 47.7	* 50.0	* 52.0	* 30.0	* 50.0	* 52.0	87.1	94.2
Réunion Réunion	4.8	3.1	5.5	4.1	4.6	5.8	6.3	4.5	5.3	* 5.8
Rwanda Rwanda	* 0.1	* 0.3	* 0.1	* 0.1	* 0.2	0.3	0.3	0.1	0.1	0.7
Senegal Sénégal	* 10.0	* 5.5	* 8.6	* 5.0	* 6.0	* 6.0	* 6.0	* 8.0	3.2	2.3
Sierra Leone Sierra Leone	0.7	0.1	0.4	* 0.4	* 0.6	* 0.9	* 0.1	0.2	0.3	* 0.3
Somalia * Somalie *	...	0.1	0.3	0.5	0.3	0.4	0.4	0.4	0.5	0.5
South Africa Afrique du Sud	157.2	153.5	122.6	140.9	135.7	127.4	109.7	137.2	127.8	137.0
Sudan Soudan	0.2	...	0.3	0.5	...	...	...	...	...	...
Swaziland Swaziland	4.4	3.8	* 2.3	* 2.0	1.7	3.3	* 1.4	* 1.5	* 1.5	* 1.5
Togo Togo	0.6	0.9	0.8	2.0	2.7	* 3.2	3.0	3.0	* 1.0	* 1.5
Tunisia Tunisie	* 4.3	3.9	3.9	3.5	* 1.4	* 1.0	* 5.0	* 5.0	* 8.3	* 2.0
Uganda Ouganda	...	...	...	...	...	...	0.0	0.0	0.1	...
United Rep.Tanzania Rép. Unie de Tanzanie	* 3.1	3.2	2.3	* 2.2	3.3	* 3.8	3.4	3.0	* 4.0	* 5.0
Zaire Zaïre	* 2.6	2.6	2.6	* 2.0	* 1.0	0.5	1.1	0.4	* 2.5	* 1.7
Zambia Zambie	6.7	7.0	5.4	4.5	8.1	6.7	9.9	8.1	* 8.0	* 5.7
Zimbabwe Zimbabwe	30.4	32.8	* 29.5	* 29.6	31.2	33.2	27.3	30.1	30.6	34.7
America, North Amérique du Nord	**5 835.0**	**5 085.4**	**5 981.0**	**5 829.7**	**5 355.8**	**5 212.6**	**5 378.9**	**5 155.1**	**5 509.7**	**5 277.2**
Bahamas * Bahamas *	0.3	0.1	0.2	0.3	0.1	0.1	0.1	0.1	0.1	0.1
Barbados Barbade	* 3.0	* 3.0	1.4	1.7	1.1	* 1.2	* 1.3	* 1.3	* 1.4	* 1.0
Belize Belize	0.4	0.3	0.4	0.6	0.7	1.2	1.3	1.6	* 1.8	* 1.9

60
Potash fertilizers
Consumption: thousand metric tons [cont.]
Engrais potassiques
Consommation : milliers de tonnes métriques [suite]

Country or area Pays ou zone	1981/82	1982/83	1983/84	1984/85	1985/86	1986/87	1987/88	1988/89	1989/90	1990/91
Bermuda Bermudes	0.1	0.1	0.1	0.1	0.0	...	0.0	...	...	...
Canada Canada	343.6	342.1	370.3	396.5	396.3	389.2	405.6	356.1	359.8	337.9
Costa Rica * Costa Rica *	17.0	10.2	20.3	28.7	20.0	19.0	30.0	26.0	30.0	40.0
Cuba Cuba	209.5	198.0	201.7	201.4	215.8	253.0	265.9	213.4	211.7	* 215.0
Dominica Dominique	* 1.4	0.3	0.8	* 0.8	* 0.8	* 0.9	* 1.0	* 1.0	* 1.2	* 0.6
Dominican Republic Rép. dominicaine	* 15.8	* 13.8	* 10.0	* 14.3	* 13.0	* 10.0	* 18.0	17.3	* 20.0	* 23.0
El Salvador El Salvador	1.8	3.4	1.6	3.8	2.7	2.4	1.4	4.5	4.0	3.2
Guadeloupe Guadeloupe	3.2	2.3	2.9	2.7	3.0	3.0	* 1.6	3.7	* 2.5	* 1.4
Guatemala Guatemala	12.0	12.1	10.6	* 12.8	13.1	* 15.0	* 12.0	* 15.7	* 16.0	* 17.0
Haiti * Haïti *	1.4	1.1	1.2	1.2	0.9	0.4	0.5	0.5	0.7	0.1
Honduras Honduras	* 10.1	* 8.8	5.7	8.7	5.9	7.9	* 10.0	10.1	7.2	* 10.0
Jamaica Jamaïque	5.9	5.5	4.0	* 9.0	* 3.9	* 8.7	* 7.8	8.4	* 11.0	* 8.1
Martinique Martinique	6.9	5.9	* 3.4	* 5.8	* 8.0	* 10.0	* 12.0	9.8	* 8.0	* 11.8
Mexico Mexique	65.1	83.2	* 67.4	* 85.5	68.7	90.7	80.4	78.5	* 92.8	* 66.0
Nicaragua Nicaragua	* 13.8	* 2.8	* 9.2	5.5	9.4	0.6	* 2.9	* 5.0	5.8	2.1
Panama [1] Panama [1]	* 10.0	* 7.0	* 8.1	* 10.3	* 5.9	* 8.7	* 11.9	* 10.3	10.9	11.9
Saint Kitts-Nevis Saint-Kitts-et-Nevis	* 1.3	* 1.3	* 1.3	* 1.4	* 1.5	* 1.5	* 1.5	* 1.6	0.3	0.3
Saint Lucia Sainte-Lucie	0.0	* 0.1	* 0.1	* 0.1	0.1	* 0.1	* 0.1	* 0.1	2.2	1.7
St. Vincent-Grenadines* St. Vincent-Grenadines*	1.0	1.0	1.0	1.0	1.0	0.5	0.5	0.5	0.9	0.9
Trinidad and Tobago Trinité-et-Tobago	3.1	* 0.4	* 0.6	* 0.5	0.3	* 0.9	2.0	0.6	* 1.6	* 1.6
United States [2] Etats-Unis [2]	5 108.3	4 382.6	5 258.8	5 037.1	4 583.6	4 387.6	4 511.1	4 389.0	4 719.9	4 521.7
America, South Amérique du Sud	920.0	1 045.6	890.5	1 278.1	1 322.0	1 571.5	1 667.9	1 778.0	1 623.8	1 527.9
Argentina Argentine	5.3	11.0	* 10.4	* 11.4	* 7.1	* 10.4	* 15.4	* 17.0	* 11.0	* 12.6
Bolivia Bolivie	0.7	* 0.3	* 0.5	* 0.3	* 0.3	* 0.9	0.9	0.2	0.3	0.2
Brazil Brésil	766.6	876.4	727.1	1 073.5	1 063.6	* 1 276.0	* 1 302.3	1 406.3	1 263.7	1 183.2
Chile Chili	12.0	11.5	11.3	13.8	14.2	16.8	* 22.0	* 25.0	27.7	28.8

60
Potash fertilizers
Consumption: thousand metric tons [cont.]
Engrais potassiques
Consommation : milliers de tonnes métriques [suite]

Country or area Pays ou zone	1981/82	1982/83	1983/84	1984/85	1985/86	1986/87	1987/88	1988/89	1989/90	1990/91
Colombia Colombie	* 67.0	* 77.2	* 77.5	* 91.5	92.1	* 115.0	* 139.9	129.1	153.5	* 163.8
Ecuador Equateur	* 18.1	20.5	* 13.7	11.2	8.3	7.8	* 19.1	* 25.8	14.4	19.9
French Guiana Guyane française	0.2	0.2	0.3	* 0.3	0.3	0.2	* 0.3	0.2	* 0.2	* 0.2
Guyana Guyana	1.0	0.0	0.0	1.2	1.2	1.3	* 1.9	* 2.1	* 2.5	* 1.8
Paraguay Paraguay	2.1	1.8	* 1.9	* 0.6	3.1	* 3.6	* 4.9	2.2	3.4	2.8
Peru Pérou	12.8	9.5	9.8	9.7	9.4	20.9	28.8	21.6	19.8	8.3
Suriname Suriname	* 0.3	* 0.3	0.9	0.6	2.2	2.0	1.8	0.2	0.2	* 0.2
Uruguay Uruguay	* 3.5	* 2.0	* 2.5	* 3.5	* 2.6	2.3	* 3.1	2.4	3.8	3.0
Venezuela Venezuela	30.3	* 35.0	* 34.6	* 60.5	117.5	114.3	* 127.6	146.0	123.4	* 103.0
Asia Asie	2 785.8	2 720.5	3 183.8	3 290.4	2 964.0	3 353.4	4 121.7	4 697.3	4 516.2	5 294.1
Afghanistan * Afghanistan *	0.1	...	...	...	...	...	...	...	...	...
Bahrain Bahreïn	...	0.1	0.1	0.1	0.1	0.2	0.1	0.1	0.1	0.2
Bangladesh Bangladesh	28.2	31.4	38.5	43.3	36.0	57.8	51.7	56.5	56.1	88.2
Bhutan Bhoutan	...	...	0.1	0.0	...	...	...	...	...	...
Brunei Darussalam Brunéi Darussalam	0.0	0.2	0.1	0.5	0.7	0.9	...	* 0.2	* 0.2	* 1.4
Cambodia * Cambodge *	1.0	0.2	...	0.6	...	...	0.6	0.1	0.3	...
China Chine	* 692.9	* 616.0	* 810.3	* 831.0	* 436.9	* 721.4	1 394.8	1 645.7	* 1 300.8	* 1 748.4
Cyprus[3] Chypre[3]	1.4	1.1	* 2.1	1.0	1.2	1.4	1.4	1.3	1.7	2.0
India Inde	676.2	726.5	775.4	838.5	808.1	850.0	880.5	1 068.4	1 168.0	1 330.0
Indonesia Indonésie	136.2	80.5	107.8	* 151.2	* 178.0	* 163.0	* 236.1	* 286.8	* 274.0	* 315.0
Iran, Islamic Rep. of Iran, Rép. islamique d'	1.0	4.0	14.1	3.7	2.7	6.8	4.6	5.8	* 0.2	* 15.9
Iraq Iraq	* 0.1	0.5	2.6	4.0	5.9	3.6	* 3.6	* 0.7	1.9	* 4.2
Israel Israël	23.0	19.0	18.5	21.0	21.0	27.0	28.0	33.5	33.9	32.1
Japan Japon	535.0	581.0	632.0	631.0	613.0	607.0	602.0	577.0	569.0	537.0
Jordan Jordanie	1.0	0.9	1.1	1.6	1.6	0.5	1.9	3.7	* 4.0	* 4.0
Korea, Dem. P. R. * Corée, R. p. dém. de *	92.0	46.4	77.8	45.0	86.4	13.3	6.6	30.4	7.5	18.1

60
Potash fertilizers
Consumption: thousand metric tons [cont.]
Engrais potassiques
Consommation : milliers de tonnes métriques [suite]

Country or area Pays ou zone	1981/82	1982/83	1983/84	1984/85	1985/86	1986/87	1987/88	1988/89	1989/90	1990/91
Korea, Republic of Corée, République de	* 199.1	* 156.5	* 179.1	* 195.1	207.0	215.0	219.0	* 246.0	275.8	277.1
Kuwait Koweït	* 0.1	0.1	...	...	...	...	...	...	...	...
Lao People's Dem. Rep. Rép. dém. pop. lao	...	0.0	0.0	...	...	...	* 0.1	* 0.1	* 0.1	* 0.1
Lebanon Liban	* 10.3	13.6	11.0	10.7	11.9	* 3.2	* 3.5	* 4.3	* 3.6	* 2.4
Malaysia Malaisie	* 173.6	* 193.9	* 208.0	* 229.0	* 250.0	* 304.6	* 288.0	* 312.0	* 391.0	475.0
Mongolia Mongolie	...	...	0.8	1.8	2.4	3.4	6.8	5.3	3.6	1.8
Myanmar Myanmar	* 2.8	* 10.6	* 12.0	* 13.9	14.0	14.5	8.8	1.4	1.8	2.9
Nepal Népal	0.8	0.9	0.8	0.6	0.7	0.2	0.9	1.8	1.3	1.6
Oman Oman	0.2	0.4	0.4	0.3	0.9	0.6	0.6	1.5	1.8	3.1
Pakistan Pakistan	21.8	25.7	28.5	24.7	33.2	42.9	45.1	24.5	40.1	32.8
Philippines Philippines	61.1	57.7	64.6	38.6	35.0	* 46.3	48.7	54.9	77.1	82.5
Qatar Qatar	...	* 0.1	0.1	...	...	...	...	...	...	...
Saudi Arabia Arabie saoudite	2.3	2.2	14.2	13.2	32.5	34.0	* 35.0	* 30.0	* 35.0	* 23.0
Singapore * Singapour *	2.2	2.2	2.2	2.5	2.5	2.5	2.5	2.5	2.5	2.5
Sri Lanka Sri Lanka	* 41.3	* 45.0	49.8	53.1	55.2	54.3	* 60.0	* 61.0	61.6	49.6
Syrian Arab Republic Rép. arabe syrienne	5.1	5.8	5.8	5.6	6.2	7.1	9.4	10.5	4.6	6.5
Thailand Thaïlande	45.8	57.6	83.7	67.9	55.7	70.3	96.2	137.5	117.8	148.9
Turkey Turquie	8.7	30.9	24.9	31.4	* 33.9	* 47.3	50.7	41.8	58.0	63.0
United Arab Emirates Emirats arabes unis	0.3	* 0.4	* 1.2	0.2	0.2	0.1	* 0.1	* 1.8	* 2.7	* 2.1
Viet Nam Viet Nam	22.1	9.0	16.4	28.3	31.1	54.0	34.3	* 50.0	* 20.0	22.2
Yemen Yémen	0.0	0.2	0.0	1.0	0.2	* 0.1	* 0.1	* 0.1	* 0.1	0.5
Europe **Europe**	**8 601.7**	**8 483.2**	**8 607.5**	**8 612.6**	**8 391.9**	**8 614.5**	**8 394.7**	**8 646.1**	**8 148.8**	**6 521.3**
Albania Albanie	* 2.9	6.2	* 4.7	* 1.0	* 1.1	* 3.3	2.2	2.2	* 2.3	* 3.1
Austria Autriche	140.0	123.2	135.1	* 134.2	132.9	102.7	107.9	103.0	97.8	* 94.0
Belgium-Luxembourg Belgique-Luxembourg	127.9	* 132.4	* 148.3	* 124.5	* 135.0	* 137.0	* 131.0	* 132.0	* 131.0	* 130.0
Bulgaria Bulgarie	117.0	141.0	114.0	103.0	103.0	121.0	94.1	112.0	91.6	* 90.0

60
Potash fertilizers
Consumption: thousand metric tons [cont.]
Engrais potassiques
Consommation : milliers de tonnes métriques [suite]

Country or area Pays ou zone	1981/82	1982/83	1983/84	1984/85	1985/86	1986/87	1987/88	1988/89	1989/90	1990/91
Czechoslovakia Tchécoslovaquie	611.0	617.0	565.0	533.1	526.6	533.0	515.0	484.0	460.6	* 355.9
Denmark Danemark	136.4	147.9	156.5	149.9	145.9	153.7	* 143.3	* 145.0	* 155.4	* 149.7
Finland Finlande	132.6	152.6	155.8	153.2	149.9	151.9	151.0	133.3	140.9	119.1
France France	* 1 699.5	* 1 744.3	* 1 833.6	* 1 863.1	* 1 820.7	* 1 878.0	* 1 856.2	1 935.0	1 949.0	1 842.0
Germany† • Allemagne† F. R. Germany R. f. Allemagne	1 054.9	1 041.8	1 013.9	988.1	932.0	931.7	864.8	887.1	791.6	# * 875.0 [4]
former German D. R. anc. R. d. allemande	601.5	496.5	424.1	550.0	550.2	580.1	566.7	583.3	595.5	...
Greece Grèce	41.0	* 42.8	* 47.8	52.3	55.0	61.4	54.1	62.4	73.2	71.3
Hungary Hongrie	523.6	489.9	550.6	467.0	443.7	434.9	426.4	449.1	372.0	186.0
Iceland Islande	6.6	* 6.9	6.4	5.6	5.3	5.4	5.4	4.5	4.6	4.8
Ireland * Irlande *	175.5	175.0	194.7	196.6	173.6	198.6	187.9	193.6	182.3	183.8
Italy Italie	362.9	379.2	386.1	370.4	355.5	383.4	470.9	452.5	377.7	339.2
Malta Malte	0.1	0.1	0.2	0.3	0.1	0.1	0.1	0.1	0.0	0.0
Netherlands Pays-Bas	105.9	102.0	117.4	125.4	119.8	104.9	* 97.8	104.8	98.3	* 94.6
Norway Norvège	83.0	88.3	83.1	82.3	77.0	69.4	72.6	* 67.9	* 65.5	* 60.0
Poland Pologne	1 315.3	1 110.9	1 202.1	1 156.5	1 167.8	1 233.7	1 104.7	1 160.6	1 003.1	* 537.8
Portugal Portugal	45.1	40.2	32.0	* 32.0	* 34.0	* 40.7	* 45.7	* 48.9	* 47.5	* 48.0
Romania Roumanie	185.0	237.0	230.0	278.0	227.0	* 230.0	* 230.0	291.8	240.0	* 133.9
Spain Espagne	218.7	263.6	249.6	285.9	310.8	298.7	335.6	383.2	383.3	* 378.0
Sweden Suède	115.5	115.2	111.5	101.9	88.0	85.8	83.1	71.9	71.4	58.7
Switzerland Suisse	* 64.8	* 66.2	66.8	65.3	66.0	63.1	64.4	66.0	67.4	66.2
United Kingdom Royaume-Uni	470.0	521.0	542.0	541.0	517.0	549.0	* 524.0	* 521.0	* 525.0	* 510.0
Yugoslavia Yougoslavie	265.0	242.0	236.0	252.0	254.0	263.0	259.6	250.9	221.8	190.1
Oceania **Océanie**	**265.5**	**246.8**	**267.8**	**285.2**	**230.1**	**222.0**	**229.1**	**264.6**	**262.1**	**254.2**
Australia Australie	145.1	124.1	* 136.1	* 136.0	* 130.0	* 136.0	* 146.0	* 157.3	* 162.6	* 145.4
Fiji Fidji	* 2.7	* 1.3	* 1.5	* 2.9	3.0	3.2	2.7	4.7	7.2	* 8.0
French Polynesia Polynésie française	0.3	0.4	0.3	* 0.5	* 0.4	* 0.3	* 0.3	* 0.3	* 0.3	* 0.3

60
Potash fertilizers
Consumption: thousand metric tons [cont.]
Engrais potassiques
Consommation : milliers de tonnes métriques [suite]

Country or area Pays ou zone	1981/82	1982/83	1983/84	1984/85	1985/86	1986/87	1987/88	1988/89	1989/90	1990/91
New Caledonia Nouvelle-Calédonie	0.1	* 0.2	0.1	* 0.1	* 0.1	* 0.2	* 0.2	* 0.2	* 0.2	* 0.3
New Zealand Nouvelle-Zélande	* 114.7	119.7	128.4	* 144.0	* 95.0	* 79.0	* 75.0	* 97.0	85.7	96.3
Papua New Guinea Papouasie-Nvl-Guinée	2.4	* 1.1	* 1.4	* 1.7	* 1.6	* 3.3	* 4.9	* 5.1	* 6.2	* 3.9
Samoa Samoa	0.2	...	...	...	...	...	...	...	...	...
former USSR ancienne URSS	4 905.0	4 991.0	6 201.0	6 167.0	6 822.0	6 677.0	7 052.0	7 044.0	6 381.0	5 081.0

Source:
Food and Agriculture Organization of the United Nations (Rome).

† All data shown which pertain to Germany prior to 3 October 1990 are indicated separately for the Federal Republic of Germany and the former German Democratic Republic based on their respective territories at the time indicated. Where data for united Germany (3 October 1990 and thereafter) are not available, available data are shown separately under the designations Federal Republic of Germany and former German Democratic Republic and pertain to the territorial boundaries prior to 3 October 1990. For detailed explanatory notes on data pertaining to Germany, see Annex I - Country Nomenclature.

1 Excluding former Canal Zone.
2 Data include Puerto Rico.
3 Data do not cover the whole country.
4 Beginning 1990/91, data are for United Germany.

Source:
Organisation des Nations Unies pour l'alimentation et l'agriculture (Rome).

† Toutes les données se rapportant à l'Allemagne avant le 3 octobre 1990 figurent dans deux rubriques séparées basées sur les territoires respectifs de la République fédérale d'Allemagne et l'ancienne République démocratique allemande selon la période indiquée. En l'absence de données pour l'Allemagne unifiée (à compter du 3 octobre 1990), les données disponibles sont fournies séparément sous les rubriques République fédérale d'Allemagne et ancienne République démocratique allemande et se rapportent aux limites territoriales antérieures au 3 octobre 1990. Pour les notes explicatives en détail sur les données concernant l'Allemagne, voir Annexe I – Nomenclature des pays.
1 Non compris l'ancienne Zone du Canal.
2 Les chiffres comprennent Porto Rico.
3 Les données ne couvrent plus la totalité du pays.
4 A partir de 1990/91, les données sont pour l'Allemagne unifiée.

Technical notes, tables 41-60

The series shown on agriculture and fishing have been furnished by the Food and Agriculture Organization of the United Nations (FAO). They refer to the following three topics:

(a) Long-term trends in growth of agricultural output and food supply;

(b) Output of principal agricultural commodities and in a few cases consumption;

(c) Basic means of production.

Agricultural output is defined to include all crop and livestock products except those used for seed and fodder and other intermediate uses in agriculture; for example deductions are made for eggs used for hatching. Intermediate input of seeds and fodder and similar items refer to both domestically produced and imported commodities.

Detailed data and technical notes are published by FAO in its yearbooks.[4, 6]

The index-numbers of agricultural output and food production are calculated by the Laspeyres formula with the base year period 1979-1981. The latter is provided in order to diminish the impact of annual fluctuations in agricultural output during base years on the indices for the period. Production quantities of each commodity are weighted by 1979-1981 average national producer prices and summed for each year. The index numbers are based on production data for a calendar year. These may differ in some instances from those actually produced and published by the individual countries themselves due to variations in concepts, coverage, weights and methods of calculation. Efforts have been made to estimate these methodological differences to achieve a better international comparability of data. The series include a large amount of estimates made by FAO in cases where no official or semi-official figures are available from the countries.

In *table 41*, The "All commodities index" relates to the production of all crops and livestock products. The "Food Index" includes those commodities which are considered edible and contain nutrients.

In *table 42*, Cereals, production data relate to crops harvested for grain only. Cereals harvested for hay, green feed or used for grazing are excluded.

In *table 43*, Wheat, data on production of wheat include spelt, except for the former USSR.

In *table 47*, Coffee production, figures relate to green beans. Data for a few countries reported in terms of cherries or as parchment coffee have been converted into clean coffee by using appropriate coefficient factors.

In *table 48*, Coffee consumption, data are shown for importing members of the International Coffee Organization only. These account for about 90 per cent of world consumption of coffee. The figures represent residuals covering all forms of coffee, measured by deducting re-exports from imports and adjusting the resultant figures for changes in the levels of known inventories.

Notes techniques, tableaux 41-60

Les séries présentées sur l'agriculture et la pêche ont été fournies par l'Organisation des Nations Unies pour l'alimentation et l'agriculture (FAO). Elles portent sur les trois aspects suivants :

(a) Les tendances à long terme de la croissance de la production agricole et des approvisionnements alimentaires;

(b) La production des principales denrées agricoles et, dans certains cas, la consommation;

(c) Les moyens essentiels de production.

La production agricole se définit comme comprenant l'ensemble des produits agricoles et des produits de l'élevage à l'exception de ceux utilisés comme semences et comme aliments pour les animaux, et pour les autres utilisations intermédiaires en agriculture; par exemple, on déduit les oeufs utilisés pour la reproduction. L'apport intermédiaire de semences et d'aliments pour les animaux et d'autres éléments similaires se rapporte à la fois à des produits locaux et importés.

Des données détaillées et des notes techniques sont publiées par la FAO dans ses annuaires [4, 6].

Les indices de la production agricole et de la production alimentaire sont calculés selon la formule de Laspeyres avec les années 1979-1981 pour période de base. Le choix d'une période de plusieurs années permet de diminuer l'incidence des fluctuations annuelles de la production agricole pendant les années de base sur les indices pour cette période. Les quantités produites de chaque denrée sont pondérées par les prix nationaux moyens à la production de 1979-1981, et additionnées pour chaque année. Les indices sont fondés sur les données de production d'une année civile. Ils peuvent différer dans certains cas des indices effectivement établis et publiés par les pays eux-mêmes par suite de différences dans les concepts, la couverture, les pondérations et les méthodes de calcul. On s'est efforcé d'estimer ces différences méthodologiques afin de rendre les données plus facilement comparables à l'échelle internationale. Les séries comprennent une grande quantité d'estimations faites par la FAO dans les cas où les pays n'avaient pas fourni de chiffres officiels ou semi-officiels.

Au *Tableau 41*, l'"Indice relatif à l'ensemble des produits" se rapporte à la production de tous les produits de l'agriculture et de l'élevage. L'"Indice des produits alimentaires" comprend les produits considérés comme comestibles et qui contiennent des éléments nutritifs.

Au *Tableau 42*, Céréales : les données sur la production se rapportent uniquement aux céréales récoltées pour le grain; celles cultivées pour le foin, le fourrage vert ou le pâturage en sont exclues.

Au *Tableau 43*, Froment : les données sur l'épeautre sont comprises dans celles sur le froment, sauf pour l'ancienne URSS.

Au *Tableau 47*, Café : les chiffres de la production se rapportent aux grains verts. Pour quelques pays fournissant des données exprimées en cerises ou en café en parche, les chiffres ont été convertis en café nettoyé, à l'aide de coefficients de conversion appropriés.

In *table 50*, Tobacco, production figures refer to farm sales weight. Data on a dry weight basis have been converted into farm sales weight at about 90 parts to 100.

In *table 51*, Cotton, for most countries production figures are those officially reported as lint and do not include cotton linters. In exceptional cases, when production was reported in terms of unginned cotton and when no special conversion factor for lint was available, the lint equivalent was taken to be equal to one third.

In *table 52*, Livestock, data refer to livestock numbers grouped into twelve-month periods ending 30 September of the year stated and cover all animals irrespective of their age and place or purpose of their breeding.

In *table 53*, Wool, data refer to the production of raw wool (greasy basis) from sheep and cover both shorn and pulled wool together with the wool element of "woolled sheepskins". Many of the figures for countries with limited production are estimates based on sheep population.

In *table 55*, Natural rubber, unless otherwise stated the figures relate to production of natural rubber, including latex but excluding reclaimed rubber. For production of synthetic and reclaimed rubber, see table 82.

In *table 56*, the data cover as far as possible both sea and inland fisheries and are expressed in terms of live weight. They generally include crustaceans and molluscs but exclude seaweed and aquatic mammals such as whales and dolphins. Data include landings by domestic craft in foreign ports and exclude the landings by foreign craft in domestic ports. The flag of the vessel is considered as the paramount indication of the nationality of the catch.

In *table 57*, Tractors in use, data generally refer to total wheel and crawl tractors (excluding garden tractors) used in agriculture. Data refer as far as possible to the position at the end of the year stated.

In *tables 58-60*, data generally refer to the fertilizer year 1 July-30 June.

In *table 58*, Nitrogenous fertilizers, data refer to the nitrogen content of commercial inorganic fertilizers.

In *table 59*, Phosphate fertilizers, data refer to commercial phosphoric acid (P_2O_5) and cover the P_2O_5 of superphosphates, ammonium phosphate and basic slag.

In *table 60*, Potash fertilizers, data refer to K_2O content of commercial potash, muriate, nitrate and sulphate of potash, manure salts, kainit and nitrate of soda potash.

Data on production of fertilizers are shown in tables 85, 86 and 87 in chapter XII below.

Au *Tableau 48*, Café : les chiffres de consommation ne sont indiqués que pour les membres importateurs de l'Organisation internationale du café. Ces membres représentent environ 90 % de la consommation mondiale de café. Ces chiffres couvrent toutes les formes de café, et se mesurent en déduisant les réexportations des importations et en ajustant les chiffres ainsi obtenus en fonction des variations des niveaux des stocks connus.

Au *Tableau 50*, Tabac : les chiffres de production se rapportent au poids des ventes à la ferme. Les données sur le poids sec ont été converties en poids des ventes à la ferme à raison de 90 % environ.

Au *Tableau 51*, Coton : pour la plupart des pays, les chiffres de production sont ceux qui ont été communiqués officiellement comme représentant la production de coton égrené et ne comprennent pas les linters. Dans certains cas exceptionnels, lorsque les renseignements reçus concernaient le coton non égrené et que l'on ne disposait pas d'un coefficient spécifique du rendement en coton égrené, on a considéré que ce rendement était égal à un tiers.

Au *Tableau 52*, Elevage : les statistiques sur les effectifs du cheptel sont groupées en périodes de 12 mois se terminant le 30 septembre de l'année indiquée et s'entendent de tous les animaux, quels que soient leur âge et l'emplacement ou le but de leur élevage.

Au *Tableau 53*, Laine : les données se rapportent à la production de laine brute de mouton (en suint) et englobent la laine provenant de la tonte et du délainage, ainsi que la laine sur peau. Souvent, les données des pays de faible production sont des estimations basées sur l'effectif du troupeau.

Au *Tableau 55*, Caoutchouc naturel : sauf avis contraire, les données se rapportent à la production de caoutchouc naturel, y compris le latex, mais à l'exclusion du caoutchouc régénéré. Pour la production de caoutchouc synthétique et de caoutchouc régénéré, voir tableau 82.

Au *Tableau 56*, les données englobent autant que possible la pêche maritime et intérieure, et sont exprimées en poids vif. Elles comprennent, en général, crustacés et mollusques, mais excluent les plantes marines et les mammifères aquatiques (baleines, dauphins, etc.). Les données comprennent les quantités débarquées par des bateaux nationaux dans des ports étrangers et excluent les quantités débarquées par des bateaux étrangers dans des ports nationaux. Le pavillon du navire est considéré comme la principale indication de la nationalité de la prise.

Au *Tableau 57*, Tracteurs en service : les données portent généralement sur tous les tracteurs à pneus ou à chenilles, utilisés dans l'agriculture, à l'exclusion des motoculteurs. Les données se réfèrent autant que possible à la situation en fin d'année.

Aux *Tableaux 58-60*, Engrais : les données se rapportent en général à une période d'un an comptée du 1er juillet au 30 juin.

Au *Tableau 58*, Engrais azotés : les données se rapportent à la teneur en azote des engrais commerciaux inorganiques.

Au *Tableau 59*, Engrais phosphatés : les données se rapportent à l'acide phosphorique (P_2O_5) et englobent la teneur en (P_2O_5) des superphosphates, du phosphate d'ammonium et des scories de déphosphoration.

Au *Tableau 60*, Engrais potassiques: les données se rapportent à la teneur en K_2O des produits potassiques commerciaux, muriate, nitrate et sulfate de potasse, sels d'engrais, kainite et nitrate de soude potassique.

Les données relatives à la production d'engrais sont présentées aux tableaux 85, 86 et 87 du chapitre XII ci-dessous.

61
Diamonds
Diamants
Production: thousand carats
Production : milliers de carats

Country or area Pays ou zone	Industrial Industriels					Gem Précieux				
	1980	1987	1988	1989	1990	1980	1987	1988	1989	1990
Total	**30 058**	**32 373**	**32 396**	**32 714**	**33 423**	**12 067**	**26 542**	**27 474**	**27 518**	**28 694**
Angola[1] Angola[1]	* 370	* 10	* 50	* 80	* 80	* 1 110	* 180	* 950	1 165	1 200
Botswana Botswana	4 336[1]	3 844	4 186	4 576	5 206	765[1]	9 482	11 043	10 676	12 146
Brazil Brésil	414[1]	200[1]	180[1]	150[1]	150[1]	158	300[1]	353[1]	* 350[1]	350[1]
Central African Rep. Rép. centrafricaine	115	109	59	81	80	227	304	* 284	334	300
Côte d'Ivoire[1] Côte d'Ivoire[1]	...	6	3	3	3	0	15	* 8	9	9
Ghana[1] Ghana[1]	1 132	400	465	370	386	126	65	155	124	129
Guinea[1] Guinée[1]	26	12	10	10	5	* 12	* 163	* 136	* 138	130
Guyana Guyana	...	...	...	...	...	10	7	4	* 2	7
India Inde	3	3	2	3	12	12	13	12	13	3
Indonesia[1] Indonésie[1]	* 12	* 22	* 22	* 25	* 23	* 3	6	6	6	6
Lesotho[1] Lesotho[1]	4	...	...	...	...	50	...	...	...	...
Liberia Libéria	175	231	263	93	60	123[1]	42[1]	67[1]	62[1]	70[1]
Namibia[1] Namibie[1]	* 78	* 50	* 37	* 17	* 13	1 482	971	901	910	735
Sierra Leone[1] Sierra Leone[1]	* 275	* 75	* 6	* 36	* 50	317	150	12	39	50
South Africa[1] Afrique du Sud[1]	5 708	4 990	4 687	5 106	4 868	2 813	4 063	3 817	* 4 010	3 826
former USSR[1] ancienne URSS[1]	* 8 600	* 6 713	* 6 804	* 6 804	* 6 804	* 2 250	6 713	6 804	6 804	6 804
United Rep.Tanzania[1] Rép. Unie de Tanzanie[1]	137	45	45	* 45	45	137	105	* 105	* 105	105
Venezuela Venezuela	666	113	128	213	328	238	34	49	64	80
Zaire Zaïre	8 001	15 540[1]	15 439[1]	15 092[1]	15 300[1]	2 234	3 885[1]	2 724[1]	2 663[1]	2 700[1]

Source:
Industrial statistics database of the Statistical Division of the United Nations Secretariat.

1 Source: US Bureau of Mines, (Washington, DC).

Source:
Base de données pour les statistiques industrielles de la Division de statistique du Secrétariat de l'ONU.

1 Source: "US Bureau of Mines," (Washington, DC).

62
Uranium
Uranium
Reserves and production: metric tons
Réserves et production : tonnes métriques

Country or area Pays ou zone	Reserves Réserves 1/1991	Production 1983	1984	1985	1986	1987	1988	1989	1990	1991[1]
Total	1 973 600	36 993	38 846	34 794	37 270	36 803	41 005	38 608	31 904	29 124
Algeria Algérie	26 000	-	-	-	-	-	-	-	-	-
Argentina Argentine	8 740	172	129	126	173	95	142	52	9	60
Australia Australie	469 000	3 211	4 324	3 206	4 154	3 780	3 532	3 655	3 530	3 530
Belgium Belgique	-	45	40	37	41	44	43	43	38	40
Brazil Brésil	162 000	189	117	115	115	-	18	35	5	-
Canada Canada	146 000	7 140	11 170	10 880	11 720	12 440	12 393	11 323	8 729	8 940
Central African Rep. Rép. centrafricaine	8 000	-	-	-	-	-	-	-	-	-
China[2] Chine[2]	51 000	...	...	...	...	...	...	...	...	...
France France	23 800	3 271	3 168	3 189	3 248	3 376	3 394	3 241	2 841	2 508
Gabon[3] Gabon[3]	11 000	1 006	918	940	900	800	930	870	700	700
Germany† · Allemagne† F. R. Germany R. f. Allemagne	600	47	32	30	26	53	38	36	2 972	1 190
former German D. R. anc. R. d. allemande	-	...	...	...	...	...	3 927	3 871	-	-
Greece Grèce	300	-	-	-	-	-	-	-	-	-
Hungary Hongrie	1 620	...	...	...	...	...	576	530	524	440
India[3] Inde[3]	50 610[4]	200	200	200	200	200	200	200	200	230
Indonesia Indonésie	4 320	-	-	-	-	-	-	-	-	-
Italy Italie	4 800	-	-	-	-	-	-	-	-	-
Japan Japon	-	4	4	7	6	8	-	-	-	-
Namibia Namibie	84 750	3 719	3 700	3 400	3 300	3 500	3 510	3 077	3 211	3 000
Niger Niger	166 070	3 426	3 276	3 181	3 110	2 970	2 965	2 962	2 831	2 960
Pakistan[3] Pakistan[3]	...	30	30	30	30	30	30	-	30	30

62

Uranium
Reserves and production: metric tons [cont.]
Uranium
Réserves et production : tonnes métriques [suite]

Country or area Pays ou zone	Reserves Réserves 1/1991	Production								
		1983	1984	1985	1986	1987	1988	1989	1990	1991[1]
Peru Pérou	1 790	-	-	-	-	-	-	-	-	-
Portugal Portugal	7 300	104	115	119	110	141	159	137	111	32
Romania[4] Roumanie[4]	18 000	...	...	...	...	...	...	...	...	...
South Africa Afrique du Sud	247 600	6 060	5 732	4 880	4 602	3 963	3 800	2 943	2 487	1 750
Spain Espagne	17 850	170	196	201	215	223	228	227	213	212
Sweden Suède	2 000	-	-	-	-	-	-	-	-	-
former USSR[5] ancienne URSS[5]	465 000	...	...	...	...	...	...	...	...	...
United States Etats-Unis	101 900	8 200	5 700	4 300	5 200	5 000	5 040	5 320	3 420	3 500
Yugoslavia Yougoslavie	-	-	-	30	59	72	80	86	53	2
Zaire Zaïre	1 800	-	-	-	-	-	-	-	-	-
Zimbabwe Zimbabwe	1 800	-	-	-	-	-	-	-	-	-

Source:
International Atomic Energy Agency (Vienna).

† All data shown which pertain to Germany prior to 3 October 1990 are indicated separately for the Federal Republic of Germany and the former German Democratic Republic based on their respective territories at the time indicated. Where data for united Germany (3 October 1990 and thereafter) are not available, available data are shown separately under the designations Federal Republic of Germany and former German Democratic Republic and pertain to the territorial boundaries prior to 3 October 1990. For detailed explanatory notes on data pertaining to Germany, see Annex I - Country Nomenclature.

1 Preliminary data.
2 As "defined reserves" unassigned to any cost category and without information on their recoverability.
3 Includes estimated data.
4 Unassigned to any cost category and without information on their recoverability.
5 As "known resources" corresponding to the resource categories B + C1 + C2 as in-situ resources; a portion of these resources are exhausted.

Source:
Agence internationale de l'énergie atomique (Vienne).

† Toutes les données se rapportant à l'Allemagne avant le 3 octobre 1990 figurent dans deux rubriques séparées basées sur les territoires respectifs de la République fédérale d'Allemagne et l'ancienne République démocratique allemande selon la période indiquée. En l'absence de données pour l'Allemagne unifiée (à compter du 3 octobre 1990), les données disponibles sont fournies séparément sous les rubriques République fédérale d'Allemagne et ancienne République démocratique allemande et se rapportent aux limites territoriales antérieures au 3 octobre 1990. Pour les notes explicatives en détail sur les données concernant l'Allemagne, voir Annexe I – Nomenclature des pays.
1 Données préliminaires.
2 Considéré comme réserves prouvées, sans affectation à une catégorie de coût ni information sur leur recupérabilité.
3 Y compris les données estimatifs.
4 Sans affectation à une quelconque catégorie de coût ni information sur leur recupérabilité.
5 Considérées comme "ressources B+C1+C2"; une partie de ces ressources est épuisée.

63
Iron-bearing ores
Minerais ferrifères

Production: thousand metric tons
Production : milliers de tonnes métriques

Country or area Pays ou zone	1981	1982	1983	1984	1985	1986	1987	1988	1989	1990
Total	534 919	481 116	461 183	532 863	561 407	568 038	547 585	565 585	569 968	587 027
Algeria Algérie	1 850	2 103	1 989	1 979	1 823	1 814	1 827	1 685	1 485	1 589
Argentina Argentine	249	389	390	346	389	514	360	379	414	...
Australia[1] Australie[1]	59 064	54 886	45 302	60 288	66 264	63 930	61 481	65 080	62 142	70 328
Austria Autriche	948	1 045	1 107	1 138	1 019	976	1 006	767	804	762
Bolivia Bolivie	4	5	* 7	0	0	0	8	* 21	...	...
Brazil[2] Brésil[2]	83 442	81 559	77 649	97 813	113 718	119 493	91 458	99 285	104 516	...
Bulgaria Bulgarie	537	474	554	622	607	661	559	528	483	321
Canada[3] Canada[3]	30 654	20 501	20 105	24 357	24 096	22 062	23 060	24 302	24 061	23 724
Chile Chili	5 190	3 874	3 602	4 250	3 945	4 311	4 078	4 801	5 313	5 035
China Chine	52 295	53 660	56 695	63 350	69 095	73 755	80 715	77 190	* 81 078	86 080
Colombia Colombie	189	205	201	204	202	234	279	283	261	...
Czechoslovakia Tchécoslovaquie	520	499	507	494	489	472	483	474	476	490
Denmark Danemark	* 3	* 3	0	0	0	0	0	0	0	0
Egypt Egypte	972	1 070	1 112	950	975	1 068	1 024	1 054	1 290	1 202
Finland Finlande	554	562	# 715	742	595	431	345	405	28	78
France France	6 800	6 200	5 174	4 679	4 544	3 861	3 483	3 225	2 905	2 655
Germany†•Allemagnet† F. R. Germany[4] R. f. Allemagne[4]	477	387	280	293	309	212	...	...	...	...
former German D. R. anc. R. d. allemande	* 9	* 9	8	7	0	0	0	0	0	...
Greece Grèce	550	215	559	624	736	559	* 440	* 460	...	...
Hong Kong Hong-kong	17	17	17	0	0	0	0	0	0	0
Hungary Hongrie	88	101	96	82	68	0	0	0	0	...
India Inde	26 219	26 887	23 693	26 750	27 918	32 392	32 550	32 085	33 986	34 950
Indonesia[5] Indonésie[5]	49	78	76	47	75	87	111	118	* 82	...

63
Iron-bearing ores
Production: thousand metric tons [*cont.*]
Minerais ferrifères
Production : milliers de tonnes métriques [*suite*]

Country or area Pays ou zone	1981	1982	1983	1984	1985	1986	1987	1988	1989	1990
Iran, Islamic Rep. of Iran, Rép. islamique d'	453	602	918	1 430	1 175	1 062	889	1 046	...	...
Italy Italie	50	2	0	0	0	0	0	0	0	...
Japan[6] Japon[6]	275	227	185	202	212	182	167	61	25	21
Korea, Dem. P. R. *[7] Corée, R. p. dém. de *[7]	3 250	3 250	3 200	3 200	3 200	3 200	3 200	3 600	...	...
Korea, Republic of Corée, République de	311	310	331	310	350	315	316	372	379	304
Liberia Libéria	12 259	11 177	9 671	9 211	8 909	8 825	8 565	8 011	...	...
Luxembourg Luxembourg	146	0	0	0	0	0	0	0	0	...
Malaysia Malaisie	298	191	64	109	102	117	90	113	108	192
Mauritania[7] Mauritanie[7]	5 243	* 4 750	* 4 800	5 754	6 066	5 804	5 851	6 500	...	...
Mexico Mexique	5 749	5 382	5 306	5 489	5 161	4 817	4 965	5 564	3 393	3 902
Morocco Maroc	43	132	102	96	106	115	164	92	93	87
Norway Norvège	2 670	1 648	2 307	2 500	2 321	2 294	2 022	1 681	1 548	1 349
Peru Pérou	4 073	3 904	2 949	2 784	3 421	3 473	3 358	2 839	2 954	2 181
Philippines Philippines	4	4	2	0	0	15	0	0	0	6
Poland Pologne	32	15	3	3	3	3	2	2	2	0
Portugal Portugal	* 13	9	12	14	29	22	9	8	...	...
Romania Roumanie	579	535	490	460	497	530	504	496	503	386
Sierra Leone Sierra Leone	0	6	192	224	37	0	0	0	...	...
South Africa Afrique du Sud	17 728	15 371	10 395	15 429	15 283	15 326	13 777	15 805	18 754	18 962
Spain Espagne	3 816	3 690	3 552	3 557	2 903	2 761	2 109	1 925	2 128	1 506
Sweden Suède	14 791	9 421	8 442	11 378	13 115	13 118	12 185	* 12 670	* 13 455	* 12 382
Thailand Thaïlande	34	16	23	35	54	21	56	57	103	75
Tunisia Tunisie	212	147	169	166	165	166	157	173	* 146	* 153
Turkey Turquie	1 587	1 646	1 978	2 199	2 130	2 839	2 991	2 983	1 947	2 788

63
Iron-bearing ores
Production: thousand metric tons [cont.]
Minerais ferrifères
Production : milliers de tonnes métriques [suite]

Country or area Pays ou zone	1981	1982	1983	1984	1985	1986	1987	1988	1989	1990
Ukraine Ukraine	68 714	68 793	68 395	67 701	66 208	66 128	65 156	64 371	61 665	58 988
former USSR ancienne URSS	131 071	132 055	133 563	134 809	135 633	137 252	138 216	138 217	134 789	133 578
United Kingdom Royaume-Uni	158	103	85	82	60	61	58	* 60	2	2
United States[1] Etats-Unis[1]	47 286	23 005	24 554	33 640	31 797	25 293	30 525	36 468	* 37 188	* 34 942
Uruguay Uruguay	3	4	0	0	...	4	...	...	...	...
Venezuela Venezuela	9 935	6 605	5 949	8 355	9 442	10 722	11 380	12 116	11 770	13 034
Yugoslavia Yougoslavie	1 510	1 680	1 529	1 837	1 685	2 181	1 764	1 844	1 683	1 444
Zimbabwe[7] Zimbabwe[7]	660	500	574	575	682	688	824	610	686	756

Source:
Industrial statistics database of the Statistical Division
of the United Nations Secretariat.

† All data shown which pertain to Germany prior to 3 October
1990 are indicated separately for the Federal Republic of
Germany and the former German Democratic Republic based on
their respective territories at the time indicated. Where
data for united Germany (3 October 1990 and thereafter) are
not available, available data are shown separately under the
designations Federal Republic of Germany and former German
Democratic Republic and pertain to the territorial
boundaries prior to 3 October 1990. For detailed
explanatory notes on data pertaining to Germany, see Annex I
- Country Nomenclature.

1 Twelve months ending 30 June of year stated.

2 Content of ores.
3 Shipments.
4 Beginning 1987, data are confidential.
5 Content of iron sand concentrate.
6 Including iron content of iron sand and pyrites.

7 Source: US Bureau of Mines, (Washington, DC).
8 Including metal content of by-product ore.

Source:
Base de données pour les statistiques industrielles de la
Division de statistique du Secrétariat de l'ONU.

† Toutes les données se rapportant à l'Allemagne avant le 3
octobre 1990 figurent dans deux rubriques séparées basées
sur les territoires respectifs de la République fédérale
d'Allemagne et l'ancienne République démocratique allemande
selon la période indiquée. En l'absence de données pour
l'Allemagne unifiée (à compter du 3 octobre 1990), les
données disponibles sont fournies séparément sous les
rubriques République fédérale d'Allemagne et ancienne
République démocratique allemande et se rapportent aux
limites territoriales antérieures au 3 octobre 1990. Pour
les notes explicatives en détail sur les données concernant
l'Allemagne, voir Annexe I – Nomenclature des pays.

1 Période de douze mois finissant le 30 juin de l'année
indiquée.
2 Teneur des minerais.
3 Expéditions.
4 A partir de 1987, les données sont confidentielles.
5 La teneur en fer des concentrés de sables ferrugineux.
6 Y compris la teneur en fer des sables ferrugineux et des
pyrites.
7 Source: "US Bureau of Mines," (Washington, DC).
8 Y compris la teneur en métal des minerais de récupération.

64
Bauxite
Bauxite

Production: thousand metric tons
Production : milliers de tonnes métriques

Country or area Pays ou zone	1981	1982	1983	1984	1985	1986	1987	1988	1989	1990
Total	**88 248**	**78 525**	**78 332**	**93 378**	**89 679**	**88 204**	**94 165**	**98 393**	**102 845**	**109 115**
Australia[1] Australie[1]	25 450	24 690	24 373	31 537	31 839	31 864	33 168	35 142	37 355	39 914
Brazil Brésil	6 969	6 290	7 199	10 355	9 963	6 463	8 750	8 083	8 665	9 876
China[3] Chine[3]	1 500[2]	1 600[2]	1 600[2]	1 600[2]	1 650[2]	1 650[2]	2 400[2]	3 200[2]	2 055	3 655
Colombia Colombie	...	...	1	0	0	0	0	0	0	0
Denmark Danemark	25	23	14	...	0	0	0	0	0	0
Dominican Republic[4] Rép. dominicaine[4]	405	139	0	0	0	0	211	185	164	85
France France	1 824	1 610	1 595	1 527	1 452	1 230	1 271	878	550	490
Ghana Ghana	181	64	70	49	170	204	195	287	347	381
Greece Grèce	3 218	2 845	2 455	2 286	2 341	2 315	2 477	2 542	2 522	2 455
Guinea[5] Guinée[5]	12 822	11 827	12 986	14 738	13 956	14 961	16 413	17 859	17 547	17 524
Guyana Guyana	1 511	1 172	1 088	1 349	1 601	1 467	1 362	1 339	1 340[5]	1 424[5]
Haiti Haïti	488	431	0	0	0	0	0	0	0	0
Hungary Hongrie	2 914	2 627	2 917	2 994	2 815	3 022	3 101	2 906	2 644	2 559
India Inde	1 955	1 998	1 976	2 903	2 341	2 662	2 814	4 013	4 492	4 853
Indonesia[4] Indonésie[4]	1 203	700	778	1 003	830	649	635	513	862	1 164
Italy Italie	19	24	3	0	0	2	15	19	15	0
Jamaica[4] Jamaïque[4]	11 683	8 158	7 725	8 605	6 219	6 953	7 702	7 315	9 487	10 937
Malaysia[6] Malaisie[6]	701	589	502	680	492	566	482	361	355	398
Mozambique[7] Mozambique[7]	0	0	0	0	5	4	5	7	6	7
Pakistan[1] Pakistan[1]	2	3	3	4	2	2	3	3	2	2
Romania Roumanie	346	355	356	355	314	343	419	435	345	247
Sierra Leone Sierra Leone	606	632	785	1 042	1 185	1 242	1 390	1 380	1 562	1 445
Spain Espagne	9	7	5	7	7	1	1	1	1	0

64

Bauxite
Production: thousand metric tons [*cont.*]
Bauxite
Production : milliers de tonnes métriques [*suite*]

Country or area Pays ou zone	1981	1982	1983	1984	1985	1986	1987	1988	1989	1990
Suriname Suriname	4 125	3 060	2 793	3 375	3 738	3 731	2 522	3 434	3 457	3 267 [5]
Turkey Turquie	590	508	306	131	214	280	259	269	550	784
former USSR * [7] ancienne URSS * [7]	4 600	4 600	4 600	4 600	4 600	4 600	4 600	4 600	4 600	4 200
United States Etats-Unis	1 847	896	679	856	674	510	576	588	670 [5]	495 [5]
Yugoslavia Yougoslavie	3 249	3 668	3 500	3 347	3 250	3 459	3 394	3 034	3 252	2 953
Zimbabwe Zimbabwe	5	8	23	23	21	* 24	0	0	0	0

Source:
Industrial statistics database of the Statistical Division
of the United Nations Secretariat.

1 Twelve months ending 30 June of year stated.

2 Diasporic bauxite for production of aluminium only.
3 Including an estimated 160000 metric tons annually of
   production for refractory applications.
4 Dried equivalent of crude ore.
5 Source: World Metal Statistics (London).
6 Data refer to Peninsular Malaysia only.
7 Source: US Bureau of Mines, (Washington, DC).

Source:
Base de données pour les statistiques industrielles de la
Division de statistique du Secrétariat de l'ONU.

1 Période de douze mois finissant le 30 juin de l'année
   indiquée.
2 Bauxite (diaspore) pour la production d'aluminium seulement.
3 Une quantité estimative annuelle de 160000 tonnes réservée à
   la production d'éléments réfractaires.
4 Equivalent desséché de minerai brut.
5 Source: "World Metal Statistics," (Londres).
6 Données se rapportant à la Malaisie péninsulaire seulement.
7 Source: "US Bureau of Mines," (Washington, DC).

65
Phosphate rock
Phosphates naturels
Production: thousand metric tons
Production : milliers de tonnes métriques

Country or area Pays ou zone	1981	1982	1983	1984	1985	1986	1987	1988	1989	1990
Total	144 732	127 162	139 049	149 734	149 498	142 644	145 541	156 675	160 656	157 226
Algeria Algérie	915	965	876	1 000	1 221	1 203	1 073	1 332	1 223	1 102
Australia Australie	35[1]	150[1]	127[1]	11[1]	33[1]	34[1]	11[1]	6[1]	* 8	* 14
Brazil Brésil	2 658	2 767	3 208	3 798	4 148	4 509	4 777	4 672	3 655	2 968
Chile Chili	...	...	...	...	...	7	10	9	14	12
China Chine	* 11 500[2]	* 11 720[2]	* 12 500[2]	* 11 800[2]	* 6 970[2]	* 6 700[2]	* 9 000[2]	18 237	19 827	21 552
Christmas Is.(Aust)[2][3] Ile Christmas (Aust)[2][3]	1 423	1 328	1 094	1 259	1 187	880	* 842	...	...	...
Colombia[2] Colombie[2]	17	20	17	11	24	29	* 34	* 30	* 31	37
Denmark * Danemark *	...	...	...	...	...	...	...	2	...	...
Egypt Egypte	737	691	783	946	1 038	1 162	1 310	1 330	1 347	1 505
Finland[2] Finlande[2]	201	233	381	477	512	527	553	583	564	546
France France	12	11	13	...	...	...	...	...	...	...
India Inde	565	633	790	889	949	667	692	725	708	659
Indonesia[2] Indonésie[2]	8	5	6	2	1	1	3	1	11	2
Israel Israël	1 919	2 148	1 966	2 065	2 195	2 518	2 731	2 648	2 762	2 428
Jordan Jordanie	4 244	4 390	4 746	6 263	6 067	6 249	6 841	5 628	6 642	5 925
Korea, Dem. P. R.[2] Corée, R. p. dém. de[2]	* 550	* 500	* 500	* 500	* 500	* 500	* 500	* 500	* 500	500
Mexico Mexique	331	512	498	518	645	660	633	667	655	604
Morocco Maroc	20 150	17 835	19 842	21 351	23 789	24 088	23 163	20 078	18 687	21 396
Nauru[2] Nauru[2]	1 480	1 359	1 684	1 358	1 508	1 494	1 376	* 1 540	1 181	926
Netherlands Antilles[2] Antilles néerlandaises[2]	0	0	3	19	* 20	* 20	16	...	...	...
Pakistan Pakistan	171	211	238	...	317	308	323	334	331	333
Peru[2] Pérou[2]	12	29	3	13	12	25	61	* 60	* 21	5
Philippines Philippines	8	6	4	7	6	2	6	8	4	13

65
Phosphate rock
Production: thousand metric tons [cont.]
Phosphates naturels
Production : milliers de tonnes métriques [suite]

Country or area Pays ou zone	1981	1982	1983	1984	1985	1986	1987	1988	1989	1990
Senegal Sénégal	1 792	1 179	1 397	1 893	2 198	* 2 920	2 077	2 322	2 273	2 147
South Africa[2] Afrique du Sud[2]	2 718	3 161	2 887	2 585	2 433	2 920	2 623	2 850	2 963	3 165
Sweden[2] Suède[2]	124	131	107	133	187	192	* 221	142	71	7
Syrian Arab Republic Rép. arabe syrienne	1 319	1 462	1 231	1 515	1 224	1 600	1 985	2 186	2 250	1 633
Thailand Thaïlande	3	4	5	3	4	5	5	8	7	10
Togo Togo	2 244	2 007	2 081	2 696	2 456	2 314	2 644	3 464	3 355	2 314
Tunisia Tunisie	4 924	4 729	5 924	5 346	4 530	5 951	6 390	6 027	* 6 610	6 259
Turkey Turquie	43	26	105	169	59	85	19	74	85	87
former USSR[2] ancienne URSS[2]	* 30 700	* 31 300	* 33 100	* 33 300	* 33 750	* 33 900	* 34 100	34 400	34 400	33 500
United States Etats-Unis	53 624	37 414	42 573	49 197	50 835	40 320	40 954	45 389	48 866	46 343
Venezuela Venezuela	0[4]	0[4]	...	3[2]	9[2]	174[2]	99[2]	...	* 237[2]	165[2]
Viet Nam[2] Viet Nam[2]	* 181	* 110	* 200	* 200	* 516	* 530	* 300	330	* 500	274
Zimbabwe Zimbabwe	122	122	133	134	135	136	155	124	134	148

Source:
Industrial statistics database of the Statistical Division
of the United Nations Secretariat.

1 Twelve months ending 30 June of year stated.

2 Source: US Bureau of Mines, (Washington, DC).
3 Exports.
4 Source: International Superphosphate Manufacturers'
Association, (London).

Source:
Base de données pour les statistiques industrielles de la
Division de statistique du Secrétariat de l'ONU.

1 Période de douze mois finissant le 30 juin de l'année
indiquée.
2 Source: "US Bureau of Mines," (Washington, DC).
3 Exportations.
4 Source: Association internationale des fabricants de
superphosphates et d'engrais composés, (Londres).

66
Silver
Argent

Production: metric tons
Production : tonnes métriques

Country or area Pays ou zone	1981	1982	1983	1984	1985	1986	1987	1988	1989	1990
Total	11 267	11 527	12 368	12 664	12 970	12 904	13 485	14 123	14 391	14 692
Algeria[1] Algérie[1]	* 3	* 3	* 4	* 4	* 4	* 4	* 4	* 4	* 4	4
Argentina Argentine	78	84	78	62	68	66	60	79	83	80
Australia[2] Australie[2]	759	888	1 033[1]	972[1]	1 044	1 074	1 037	1 135	* 1 075	1 273
Bolivia Bolivie	205	170	187	142	111	100	140	* 232	* 267	280
Brazil Brésil	19	23	15	26	49	60	61	90	60	60
Bulgaria[3] Bulgarie[3]	* 23	88	89	89	83	82	80	71	59	54
Canada Canada	1 203	1 291	1 197	1 327	1 197	1 088	1 375	1 371	1 312	1 381
Chile Chili	361	382	468	490	517	500	500	507	545	655
China[1] Chine[1]	* 78	* 78	* 78	* 78	* 78	* 90	* 100	* 110	* 125	125
Colombia Colombie	4	4	3	4	5	* 5	* 5	7	* 7	7
Czechoslovakia[3] Tchécoslovaquie[3]	* 35	* 33	* 30	* 32	* 32	* 33	* 34	* 25	* 20	15
Dominican Republic Rép. dominicaine	64	66	42	38	49	41	36	38	* 23	22
Ecuador[1] Equateur[1]	1	...	...	...	...	...	...	...	...	...
El Salvador[1] El Salvador[1]	4	3	1	* 1	0	...	...	...	...	...
Fiji Fidji	0	1	0	1	1	1	1	1	1	1
Finland[4,5] Finlande[4,5]	38	37	23	26	24	29	33	28	35	29
France France	3	6	6	6	5	5	4	5	7	8
Germany† · Allemagne† F. R. Germany[3] R. f. Allemagne[3]	39	40	36	38	34	28	31	* 20	* 9	8
former German D. R.[1] anc. R. d. allemande[1]	* 45	* 45	* 43	* 40	* 41	* 41	41	40	39	35
Ghana[1] Ghana[1]	* 1	1	1	* 1	* 1	0	* 1	* 1	1	1
Greece[3] Grèce[3]	60	49	56	57	52	54	52	61	* 61	63
Honduras Honduras	56	53	83	89	83	84	* 23	* 58	* 50	36
Hungary * [3] Hongrie * [3]	1	1	1	0	0	0	0	0	...	...

66

Silver
Production: metric tons [*cont.*]
Argent
Production : tonnes métriques [*suite*]

Country or area Pays ou zone	1981	1982	1983	1984	1985	1986	1987	1988	1989	1990
India Inde	17	14	16	25	26	35	38	41	35	38
Indonesia [4] Indonésie [4]	2	3	# 35	39	38	46	50	* 58	* 74	67
Ireland Irlande	19	11	10	9	9	8	7	6	7	7
Italy Italie	55	56	73	48	72	56	82	92	96	98
Japan Japon	280	306	307	324	339	351	281	252	156	150
Korea, Dem. P. R. [1] Corée, R. p. dém. de [1]	* 50	* 50	* 50	* 50	* 50	* 50	* 50	* 50	50	50
Korea, Republic of Corée, République de	100	101	105	117	124	157	209	227	239	238
Malaysia [3] Malaisie [3]	15	16	15	15	16	14	16	* 10	* 13	13
Mexico Mexique	1 655	1 550	1 911	1 987	2 153	2 303	2 415	2 359	* 2 306	2 346
Morocco Maroc	41	53	88	120	85	* 49	* 44	131	...	190
Myanmar [6] Myanmar [6]	14 [3]	16 [3]	17	16	16	11	8	9	6	7
Namibia [1] Namibie [1]	107	87	110	101	106	108	103	117	108	90
Nicaragua [1] Nicaragua [1]	4	2	2	2	1	1	1	* 1	1	1
Papua New Guinea Papouasie-Nvl-Guinée	42	43	47	45	46	54	66	* 70	92	130
Peru Pérou	1 470	1 663	1 739	1 651	1 899	1 853	1 905	1 568	1 748	1 725
Philippines Philippines	63	62	57	49	52	52	51	55	51	47
Poland Pologne	640	657	678	744	831	829	831	1 063	1 003	832
Portugal Portugal	1	1	1	5	3	3	5	4	19	1
Romania [1] Roumanie [1]	* 27	* 27	28	25	* 25	* 23	* 23	* 23	* 25	25
South Africa Afrique du Sud	235	215	170	215	* 220	222	208	* 200	180	161
Spain Espagne	191	115	177	218	156	173	215	220	251	270
Sweden Suède	166	185	203	235	250	262	254	208	228	220
Tunisia Tunisie	3	4	3	2	1	2	2	* 2	* 2	2
Turkey [3] Turquie [3]	* 9	* 9	7	7	7	9	9	53	35	40

66

Silver
Production: metric tons [cont.]
Argent
Production : tonnes métriques [suite]

Country or area Pays ou zone	1981	1982	1983	1984	1985	1986	1987	1988	1989	1990
former USSR[1] ancienne URSS[1]	* 1 446	* 1 459	* 1 468	* 1 474	* 1 490	* 1 499	* 1 499	* 1 493	* 1 500	1 400
United Kingdom Royaume-Uni	3	3	3	2	2	2	2	2	2	2
United States Etats-Unis	1 265	1 252	1 351	1 387	1 227	1 074	1 241	1 661	2 007	2 168
Yugoslavia[5] Yougoslavie[5]	138	104	124	128	156	177	151	139	133	105
Zaire[3] Zaïre[3]	80	59	40	47	47	40	45	74	60	84
Zambia[7] Zambie[7]	22	28	29	25	19	27	30	29	* 25	25
Zimbabwe Zimbabwe	27	29	29	28	25	27	25	22	22	21

Source:
Industrial statistics database of the Statistical Division
of the United Nations Secretariat.

† All data shown which pertain to Germany prior to 3 October
1990 are indicated separately for the Federal Republic of
Germany and the former German Democratic Republic based on
their respective territories at the time indicated. Where
data for united Germany (3 October 1990 and thereafter) are
not available, available data are shown separately under the
designations Federal Republic of Germany and former German
Democratic Republic and pertain to the territorial
boundaries prior to 3 October 1990. For detailed
explanatory notes on data pertaining to Germany, see Annex I
- Country Nomenclature.

1 Source: US Bureau of Mines, (Washington, DC).
2 Twelve months ending 30 June of year stated.

3 Source: "Metallgesellschaft Aktiengesellschaft",
(Frankfurt).
4 Silver content of concentrates only.
5 Silver refined from copper, lead and zinc ore.

6 Government production only.
7 Silver recovered.

Source:
Base de données pour les statistiques industrielles de la
Division de statistique du Secrétariat de l'ONU.

† Toutes les données se rapportant à l'Allemagne avant le 3
octobre 1990 figurent dans deux rubriques séparées basées
sur les territoires respectifs de la République fédérale
d'Allemagne et l'ancienne République démocratique allemande
selon la période indiquée. En l'absence de données pour
l'Allemagne unifiée (à compter du 3 octobre 1990), les
données disponibles sont fournies séparément sous les
rubriques République fédérale d'Allemagne et ancienne
République démocratique allemande et se rapportent aux
limites territoriales antérieures au 3 octobre 1990. Pour
les notes explicatives en détail sur les données concernant
l'Allemagne, voir Annexe I – Nomenclature des pays.
1 Source: "US Bureau of Mines," (Washington, DC).
2 Période de douze mois finissant le 30 juin de l'année
indiquée.
3 Source: "Metallgesellschaft Aktiengesellschaft",
(Francfort).
4 Teneur en argent des concentrés seulement.
5 Argent affiné provenant de minerais de cuivre, plomb et
zinc.
6 Production de l'Etat seulement.
7 Argent récupéré.

67
Gold
Or

Production: kilograms
Production : kilogrammes

Country or area Pays ou zone	1981	1982	1983	1984	1985	1986	1987	1988	1989	1990
Total	1 260 286	1 311 864	1 401 055	1 416 687	1 479 750	1 539 221	1 605 280	1 771 353	1 916 350	1 987 317
Argentina Argentine	459	632	727	688	882	944	990	938	1 150	1 200
Australia[1] Australie[1]	15 991	22 328	25 825	33 881	48 853	64 780	* 110 696	* 152 003	* 203 563	242 299
Bolivia Bolivie	2 064	1 249	1 531	1 270	561	800	2 755	* 4 889	3 595	6 000
Botswana Botswana	10	15	11	18	13	25	32	21	67	46
Brazil[2] Brésil[2]	17 276	25 517	53 684	37 218	29 673	23 361	35 780	56 447	...	...
Burundi Burundi	3[3]	3[3]	9[3]	35[3]	26[3]	* 30[3]	* 31[3]	* 31[3]	19	9
Cameroon Cameroun	2[4]	4[3]	8[3]	* 8[3]	* 7[3]	* 8[3]	* 8[3]	8[3]	8[3]	8[3]
Canada Canada	49 500	64 733	73 513	83 445	87 561	102 900	115 818	130 487	159 494	167 373
Central African Rep. Rép. centrafricaine	56	47	95	236	240	* 186	224	* 382	* 328	300
Chile Chili	12 456	16 907	17 759	16 829	17 240	17 947	17 035	20 614	22 559	27 503
China[3] Chine[3]	# 52 876	* 55 986	* 57 541	* 59 097	* 59 097	* 66 000	* 72 000	* 78 000	90 000	100 000
Colombia Colombie	16 460[4]	14 702[4]	13 642	24 879	35 546	40 000	26 546	29 020[3]	* 27 090[3]	28 000[3]
Congo Congo	2	3	8	3	13	9	6	5	* 16	16
Costa Rica[3] Costa Rica[3]	* 622	* 840	* 933	* 1 089	* 498	* 700	* 600	* 600	* 700	800
Dominican Republic Rép. dominicaine	12 846	11 776	11 240	10 448	10 204	8 838	7 663	5 833	* 5 238	4 312
Ecuador Equateur	73[4]	50[4]	# 2 109[3]	8 709[3]	9 331[3]	9 870[3]	9 000[3]	8 050[3]	13 000[3]	10 000[3]
El Salvador[3] El Salvador[3]	121	103	20	* 31	...	...	...	...	...	...
Ethiopia Ethiopie	386[4]	373[4]	435[13]	467[13]	467[13]	923[13]	642[13]	728[13]	745[13]	800[13]
Fiji Fidji	960	1 423	1 248	1 509	1 865	2 856	2 864	4 274	4 221	4 116
Finland Finlande	992	1 144	# 442	523	498	501	678	594	1 785	1 489
France France	1 131	1 651	2 744	2 416	2 380	1 879	* 2 164	2 653	3 303	4 236
French Guiana Guyane française	167	162	249	315	407	326	* 342	* 342	* 550	550
Gabon[3] Gabon[3]	* 17	* 17	17	41	50	62	* 78	138	81	80
Germany†・Allemagne† F. R. Germany[56] R. f. Allemagne[56]	* 9 490	* 9 356	* 10 945	* 47	* 37	* 37	* 26	* 16	* 16	18

67

Gold
Production: kilograms [cont.]
Or
Production : kilogrammes [suite]

Country or area Pays ou zone	1981	1982	1983	1984	1985	1986	1987	1988	1989	1990
Ghana Ghana	10 596	10 280	8 601	8 923	9 311	8 950	10 228	11 631	13 265	16 840
Guyana Guyana	599	268	143	328	323	436	666	585	...	...
Honduras Honduras	11[4]	53[3]	67[3]	87[3]	156[3]	63[3]	131[3]	* 123[3]	1 244[3]	1 300[3]
Hungary Hongrie	1 900[7]	1 600[7]	* 933[3]	* 622[3]	* 622[3]	* 600[3]	* 600[3]	* 600[3]	500[3]	500[3]
India Inde	2 495	2 244	2 154	1 989	1 853	1 931	1 863	1 944	1 827	1 850
Indonesia Indonésie	183	223	# 2 391	2 447	2 619	3 304	3 752	4 738	6 155	11 158
Japan Japon	3 087	3 239	3 139	3 220	5 309	10 280	8 590	7 308	6 097	7 303
Kenya[3] Kenya[3]	1	1	3	19	14	73	* 278	* 17	* 15	25
Korea, Dem. P. R. *[3] Corée, R. p. dém. de *[3]	5 000	5 000	5 000	5 000	5 000	5 000	5 000	5 000	5 000	...
Korea, Republic of Corée, République de	1 342	1 734	2 242	2 462	2 403	4 648	7 600	11 121	14 270	20 760
Liberia Libéria	463	381	449	328	146	626	593	677	700	700
Madagascar Madagascar	1	3	3	3	4	4	* 40	* 90	61	40
Malaysia[8] Malaisie[8]	180	180	180	234	357	528	799	...	...	...
Mali[3] Mali[3]	* 498	* 404	* 404	* 500	* 500	* 725	* 950	* 2 650	* 3 000	5 200
Mexico Mexique	6 319	6 104	6 930	7 058	7 524	7 795	7 988	9 098	8 613	8 338
Namibia[39] Namibie[39]	* 187	* 230	* 232	* 196	* 194	* 184	* 172	* 158	* 336	1 700
New Zealand[3] Nouvelle-Zélande[3]	189	242	301	672	1 400	1 265	1 148	2 404	4 963	5 000
Nicaragua[3] Nicaragua[3]	1 926	1 691	1 444	787	762	892	984	* 878	* 1 232	1 200
Papua New Guinea Papouasie-Nvl-Guinée	16 830	17 540	18 067	17 169	39 235	34 105	36 413	* 38 130	* 27 538	31 035
Peru Pérou	6 084	4 188	5 242	5 829	6 621	9 312	8 933	9 720	9 898	6 850
Philippines Philippines	23 435	25 954	25 397	24 475	33 063	35 430	32 600	30 490	30 040	24 590
Portugal Portugal	244	211	199	193	229	303	320	* 280	* 295	350
Rwanda Rwanda	46	15	39	8	1	6	9	15	732[3]	700
Solomon Islands Iles Salomon	42[3]	37[3]	34[3]	80	729	117	74	50	36	35
South Africa Afrique du Sud	655 728	662 628	679 700	683 300	677 491	642 071	601 775	* 617 749	605 466	602 997
Spain Espagne	3 408	3 700	4 105	4 595	3 888	4 092	5 505	5 571	6 710	...

67

Gold
Production: kilograms [cont.]
Or
Production : kilogrammes [suite]

Country or area Pays ou zone	1981	1982	1983	1984	1985	1986	1987	1988	1989	1990
Sudan[3] Soudan[3]	* 9	* 12	* 16	* 47	* 47	* 50	* 85	* 500	* 50	75
Suriname Suriname	25	17	12	8	4	3	22	* 30[3]	* 31[3]	30
Sweden Suède	2 039	2 447	3 369	4 408	4 710	4 332	* 4 670	* 4 200	* 5 120	5 000
former USSR[3] ancienne URSS[3]	* 262 050	* 265 930	* 267 460	* 269 045	* 270 600	* 275 266	* 275 266	* 280 000	* 285 000	250 000
United Rep.Tanzania[3] Rép. Unie de Tanzanie[3]	* 12	* 19	* 25	* 83	* 55	* 85	* 201	* 52	* 116	100
United States Etats-Unis	42 892	45 588	62 300	64 839	75 495	116 296	153 870	200 914	* 265 541	290 202
Venezuela Venezuela	877	902	1 084	1 740	2 210	2 493	3 329	* 4 670	* 3 860	...
Yugoslavia[10] Yougoslavie[10]	3 582	4 211	4 238	3 892	4 460	4 498	5 348	4 620	4 349	4 500
Zaire Zaïre	2 027[4]	1 889[3]	6 000[3]	3 643[3]	1 960[3]	* 5 220[3]	* 4 372[3]	* 3 422[3]	2 032[3]	225[3]
Zambia[11] Zambie[11]	328	311	316	379	247	258	350	262	* 225	...
Zimbabwe Zimbabwe	11 691	13 367	14 101	14 877	14 691	14 930	14 710	* 14 774	16 003	16 900

Source:
Industrial statistics database of the Statistical Division
of the United Nations Secretariat.

† All data shown which pertain to Germany prior to 3 October
1990 are indicated separately for the Federal Republic of
Germany and the former German Democratic Republic based on
their respective territories at the time indicated. Where
data for united Germany (3 October 1990 and thereafter) are
not available, available data are shown separately under the
designations Federal Republic of Germany and former German
Democratic Republic and pertain to the territorial
boundaries prior to 3 October 1990. For detailed
explanatory notes on data pertaining to Germany, see Annex I
- Country Nomenclature.

1 Twelve months ending 30 June of year stated.

2 Gold refined from domestic ores only.
3 Source: US Bureau of Mines, (Washington, DC).
4 Source: International Monetary Fund (Washington, DC).
5 Gold refined, including gold recovered as a by-product.
6 Official data are confidential. Data are estimates.

7 Source: World Metal Statistics (London).
8 Data refer to Peninsular Malaysia and Sarawak only.

9 Smelter production only.
10 Gold refined from copper, lead and zinc ores only.

11 Recovered from refinery sludges.

Source:
Base de données pour les statistiques industrielles de la
Division de statistique du Secrétariat de l'ONU.

† Toutes les données se rapportant à l'Allemagne avant le 3
octobre 1990 figurent dans deux rubriques séparées basées
sur les territoires respectifs de la République fédérale
d'Allemagne et de l'ancienne République démocratique allemande
selon la période indiquée. En l'absence de données pour
l'Allemagne unifiée (à compter du 3 octobre 1990), les
données disponibles sont fournies séparément sous les
rubriques République fédérale d'Allemagne et ancienne
République démocratique allemande et se rapportent aux
limites territoriales antérieures au 3 octobre 1990. Pour
les notes explicatives en détail sur les données concernant
l'Allemagne, voir Annexe I – Nomenclature des pays.

1 Période de douze mois finissant le 30 juin de l'année
indiquée.

2 Or affiné provenant de minerais indigène seulement.
3 Source: "US Bureau of Mines," (Washington, DC).
4 Source: Fonds monétaire international, (Washington, DC).
5 Or affiné, y compris l'or récupéré comme sous produit.
6 Les chiffres sont confidentiels. Ces chiffres sont
estimatifs.
7 Source: "World Metal Statistics," (Londres).
8 Données se rapportant à la Malaisie péninsulaire et au
Sarawak seulement.
9 Production des fonderies seulement.
10 Or affiné provenant de minerais de cuivre, plomb et zinc
seulement.
11 Récupéré dans les boues des raffineries.

Technical notes, tables 61-67

Table 61: The data relate to mine and alluvial production of uncut diamonds and cover both gem and industrial stones. Industrial diamonds are small and impure diamonds, bort, carbonado, etc., suitable only for industrial use as abrasives, in cutting tools and the like .

Table 62: Reserves (defined in terms of "reasonably assured resources") refer in general to recoverable resources, measured as tonnes of uranium contained in ores recoverable at costs of less than $80/kg U. Total reserves include an adjustment to account for estimated mining and milling losses not incorporated in certain reserve estimates.

Production refers to tonnes of uranium contained in concentrates.

Table 63: Data refer to Fe content of iron ores and concentrates and all other iron-bearing ores and concentrates intended for treatment for iron recovery.

Table 64 refers to gross weight of bauxite crude ore mined.

Table 65 refers to the production of natural phosphate rock and apatite. Guano is excluded.

Table 66 refers to Ag content of silver ores and concentrates and all other silver-bearing ores and concentrates intended for treatment for silver recovery.

Table 67 refers to Au content of gold ores and concentrates and all other gold-bearing ores and concentrates intended for treatment for gold recovery.

Notes techniques, tableaux 61-67

Tableau 61 : Les données se rapportent à la production minière et alluviale de diamants non taillés et couvrent à la fois les diamants naturels et industriels. Les diamants industriels sont des diamants petits et impurs, bort, carbonado (diamant noir), etc. convenant seulement pour une utilisation industrielle comme abrasifs, dans les outils de coupe et autres.

Tableau 62 : Les réserves (définies comme "ressources raisonnablement assurées") se rapportent en général aux ressources récupérables, mesurées en tonnes d'uranium contenues dans les minerais récupérables à un coût inférieur à 80 dollars le kilo d'uranium. Les réserves totales comportent un ajustement pour tenir compte des pertes estimatives à l'extraction et au broyage non incorporées dans certaines estimations des réserves.

La production se réfère aux tonnes d'uranium contenues dans les concentrés.

Tableau 63 : Teneur en fer des minerais de fer et concentrés et de tous autres minerais ferrifères et concentrés destinés à être traités en vue de l'extraction du fer.

Tableau 64, Bauxite : poids brut du minerai brut naturel.

Tableau 65, Phosphates naturels : production de phosphates naturels et d'apatite. Le guano n'est pas compris.

Tableau 66, Argent : teneur en argent des minerais d'argent et concentrés et des autres minerais et concentrés argentifères destinés à être traités pour la récupération de l'argent.

Tableau 67, Or : teneur en or des minerais d'or et concentrés et de tous les autres minerais et concentrés aurifères destinés à être traités pour la récupération de l'or.

68
Sugar
Sucre

Production: thousand metric tons
Production : milliers de tonnes métriques

Country or area Pays ou zone	1981	1982	1983	1984	1985	1986	1987	1988	1989	1990
World *Monde*	92 764	100 542	95 595	98 392	97 116	999 051	103 378	103 813	106 323	109 051
Afghanistan Afghanistan	3	* 2	* 2	* 3	* 3	0	0	0	0	* 0
Albania Albanie	* 25	* 20	* 38	* 40	* 33	* 35	* 40	* 45	25	17
Algeria * Algérie *	4	4	5	7	0	0	0	0	0	0
Angola * Angola *	50	35	60	50	50	50	30	30	25	25
Argentina Argentine	1 624	1 623	1 624	1 545	1 188	1 120	1 063	1 283	1 017	1 351
Australia Australie	3 509	3 652	3 256	3 627	3 439	3 439	3 511	3 759	3 887	3 612
Austria Autriche	449	536	515	464	468	307	390	357	390	432
Bangladesh Bangladesh	172	218	186	157	94	* 180	* 200	* 200	* 130	* 190
Barbados Barbade	97	88	86	98	101	113	84	81	67	70
Belgium[1] Belgique[1]	1 038	1 246	847	910	961	915	915	910	1 312	1 105
Belize Belize	104	114	120	109	110	105	88	89	94	108
Benin * Bénin *	0	0	0	3	5	5	5	5	7	5
Bolivia Bolivie	260	228	197	198	* 175	* 180	161	162	* 170	* 225
Brazil Brésil	8 726	8 941	9 555	9 259	8 455	7 999	9 266	7 874	* 7 326	8 007
Bulgaria * Bulgarie *	145	170	110	110	115	115	110	105	80	40
Burkina Faso * Burkina Faso *	30	28	28	27	10	10	25	25	20	30
Burundi Burundi	...	...	...	...	...	...	...	3	9	8
Cameroon Cameroun	66	* 70	60	* 59	* 50	* 40	* 28	67	* 35	* 75
Canada Canada	99	129	132	110	* 60	* 106	* 147	* 110	* 117	* 140
Chad * Tchad *	20	25	26	15	8	10	20	20	25	25
Chile Chili	235	132	229	360	351	481	437	443	445	373
China * Chine *	3 450	3 700	3 900	4 300	4 800	5 700	5 450	4 875	5 400	6 200
Colombia Colombie	1 212	1 318	1 340	1 177	1 367	1 272	1 293	1 364	1 523	1 593
Congo Congo	* 16	* 29	21	31	* 25	* 32	* 35	* 40	* 35	* 40
Costa Rica Costa Rica	* 190	194	206	* 245	* 230	220	* 230	237	* 220	* 246

68

Sugar
Production: thousand metric tons [cont.]
Sucre
Production : milliers de tonnes métriques [suite]

Country or area Pays ou zone	1981	1982	1983	1984	1985	1986	1987	1988	1989	1990
Côte d'Ivoire Côte d'Ivoire	147	170	181	121	* 125	* 120	* 165	* 165	* 160	* 160
Cuba Cuba	7 926	8 039	7 460	7 783	7 889	7 467	7 232	8 119	7 579	8 445
Czechoslovakia * Tchécoslovaquie *	850	700	790	833	840	850	775	660	756	717
Denmark[2] Danemark[2]	* 522	* 584	376	595	574	547	422	* 549	* 530	525
Dominican Republic Rép. dominicaine	1 108	1 285	1 209	1 133	921	895	816	777	693	590
Ecuador Equateur	330	246	164	329	* 300	286	341	292	300	334
Egypt Egypte	658	745	723	* 780	* 900	* 950	* 1 000	* 1 035	* 947	* 955
El Salvador El Salvador	182	199	259	242	279	292	262	178	196	* 220
Ethiopia Ethiopie	170	177	207	200	191	193	* 195	169	183	184
Fiji Fidji	488	495	300	484	367	508	426	377	466	378
Finland Finlande	99	116	155	129	103	133	70	147	173	169
France[13] France[13]	5 130	4 446	3 560	3 956	3 953	3 734	* 3 973	4 372	4 198	4 595
Gabon Gabon	15	15	17	* 11	* 12	18	* 19	* 20	* 15	* 20
Germany† · Allemagne† F. R. Germany[2] R. f. Allemagne[2]	3 702	3 586	2 725	3 151	3 454	3 479	2 963	3 004	3 337	3 396
former German D. R. anc. R. d. allemande	681	822	750	750	798	805	* 750	658	610	...
Ghana * Ghana *	7	4	3	0	0	0	0	0	0	0
Greece[2] Grèce[2]	351	322	326	237	345	* 312	* 197	* 235	* 421	* 312
Guadeloupe Guadeloupe	59[14]	72[14]	57[14]	41[14]	53[14]	65[14]	63[14]	76[2]	78[2]	30[2]
Guatemala Guatemala	474	580	614	555	583	651	639	720	735	939
Guinea * Guinée *	22	22	14	10	5	5	10	10	20	20
Guyana Guyana	320	305	265	256	258	261	234	178	170	134
Haiti Haïti	50	* 55	* 43	* 41	* 50	* 40	35	* 30	* 30	* 35
Honduras Honduras	196	217	221	207	235	227	* 190	* 180	* 180	* 205
Hungary Hongrie	584	580	584	493	579	510	538	513	630	580
India Inde	5 596	9 190	8 452	6 635	7 016	7 594	9 215	10 207	* 9 912	12 068
Indonesia Indonésie	* 1 200	* 1 585	1 507	1 759	1 705	2 150	* 2 200	* 2 205	2 171	2 346

68

Sugar
Production: thousand metric tons [cont.]
Sucre
Production : milliers de tonnes métriques [suite]

Country or area Pays ou zone	1981	1982	1983	1984	1985	1986	1987	1988	1989	1990
Iran, Islamic Rep. of * Iran, Rép. islamique d' *	400	400	500	600	700	600	600	725	600	620
Ireland[2] Irlande[2]	* 183	* 242	* 215	* 239	* 189	* 202	* 242	* 212	233	245
Italy [2] Italie[2]	* 2 207	* 1 282	* 1 352	* 1 385	* 1 352	* 1 868	* 1 867	* 1 607	* 1 880	1 584
Jamaica Jamaïque	204	198	202	188	209	199	189	222	205	209
Japan Japon	812	822	868	876	928	953	960	944	998	982
Kenya Kenya	* 370	* 320	* 345	371	* 260	* 200	* 365	* 430	* 440	* 440
Liberia[2] Libéria [2]	0	* 5	* 5	* 5	* 5	* 5	* 5	* 5	* 5	5
Madagascar Madagascar	112	87	103	79	99	98	107	122	120	118
Malawi Malawi	177	183	187	160	154	168	181	187	173	204
Malaysia * Malaisie *	55	60	70	75	70	70	90	90	100	114
Mali Mali	9	5	* 12	10	21	21	17	21	22	* 25
Martinique [1] Martinique [1]	3	2	4	5	9	* 8	* 8	* 8	* 7	* 8
Mauritius Maurice	610	729	640	610	684	748	733	672	602	661
Mexico Mexique	2 642	2 739	3 076	3 308	3 492	4 068	4 061	3 909	3 570	3 384
Morocco Maroc	384	* 380	* 428	441	433	352	* 450	590	469	520
Mozambique Mozambique	178	126	74	39	* 60	* 40	* 25	* 40	* 25	32
Myanmar * Myanmar *	5	65	75	65	55	60	40	30	35	30
Nepal[1] Népal[1]	15	22	17	11	15	25	30	27	45	30
Netherlands Pays-Bas	* 1 139[1]	1 195[1]	* 807[1]	* 1 000[1]	* 1 000[1]	1 325[2]	1 064[2]	1 075[2]	1 240[2]	1 337[2]
Nicaragua Nicaragua	214	247	249	267	240	256	199	209	160	* 212
Nigeria * Nigéria *	25	20	60	60	50	45	40	31	55	55
Pakistan Pakistan	970	* 1 530	* 1 200	* 1 355	* 1 450	1 151	* 1 425	1 943	2 052	1 989
Panama Panama	186	239	* 215	176	160	139	* 100	* 90	* 110	* 90
Papua New Guinea Papouasie-Nvl-Guinée	0	* 12	* 33	34	30	* 10	24	51	30	28
Paraguay Paraguay	77	85	92	* 92	* 80	* 80	* 112	* 112	* 118	89
Peru Pérou	492	623	452	605	* 710	585	* 560	571	* 625	* 590

68

Sugar
Production: thousand metric tons [cont.]
Sucre
Production : milliers de tonnes métriques [suite]

Country or area Pays ou zone	1981	1982	1983	1984	1985	1986	1987	1988	1989	1990
Philippines Philippines	2 376	2 709	2 112	2 578	1 665	1 514	1 304	1 495	1 878	1 686
Poland Pologne	1 824	1 932	2 141	1 933	1 841	* 1 881	* 1 820	1 815	1 939	1 865
Portugal[1] Portugal[1]	1	1	* 8	* 8	* 9	* 4	* 2	* 1 [2]	1 [2]	2 [2]
Réunion[1] Réunion[1]	251	258	224	246	228	244	226	252	183	213
Romania Roumanie	610	* 550	* 450	* 600	* 540	* 500	* 500	* 430	* 520	* 380
Rwanda[1] Rwanda[1]	3	2	2	2	4	* 3	* 4	5	5	5
Saint Kitts and Nevis Saint-Kitts-et-Nevis	33	37	28	31	27	28	25	26	25	* 25
Samoa Samoa	3	3	3	1	3	2	2	2	2	2
Senegal Sénégal	37	* 40	50	47	* 65	73	* 71	* 72	* 70	* 60
Sierra Leone Sierra Leone	* 1	* 3	* 3	6	* 5	* 5	* 6	* 5	* 5	* 5
Somalia * Somalie *	50	55	60	45	54	30	35	40	47	35
South Africa Afrique du Sud	1 987	2 371	1 584 [5]	2 276 [5]	2 540 [5]	2 248	2 235	2 470	2 293	2 226
Spain [2] Espagne[2]	1 111	1 243	1 356	1 166	976	1 111	1 093	* 1 306	* 1 038	1 042
Sri Lanka Sri Lanka	25	24	22	20	* 17	35	* 30	24	* 29	* 26
Sudan * Soudan *	230	270	400	360	450	550	525	500	385	425
Suriname * Suriname *	8	11	11	10	7	10	10	10	10	10
Swaziland Swaziland	368	402	403	429	396	537	461	464	504	527
Sweden Suède	370	399	308	399	350	391	274	385	424	445
Switzerland Suisse	135	120	124	131	139	129	123	150	152	160
Syrian Arab Republic Rép. arabe syrienne	* 35	81	* 110	* 110	50	* 50	* 40	* 40	* 30	* 16
Thailand Thaïlande	1 702	3 017	2 113	2 550	2 393	2 718	2 532	2 638	4 338	3 542
Togo Togo	0	0	0	0	0	0	* 4	* 5	* 5	* 5
Trinidad and Tobago Trinité-et-Tobago	93	79	79	* 67	* 80	* 95	* 100	* 100	* 100	* 122
Tunisia Tunisie	* 15	* 10	* 6	16	17	21	27	26	22	25
Turkey Turquie	1 211	1 642	1 844	1 654	1 398	1 414	1 784	1 414	1 565	1 565
Uganda * Ouganda *	10	15	15	20	20	10	20	40	40	25

68
Sugar
Production: thousand metric tons [cont.]
Sucre
Production : milliers de tonnes métriques [suite]

Country or area Pays ou zone	1981	1982	1983	1984	1985	1986	1987	1988	1989	1990
former USSR ancienne URSS	6 413	7 391	8 696	8 587	8 261	8 696	9 565	8 913	9 533	9 159
United Kingdom[1] Royaume-Uni[1]	1 187	1 541	1 154	1 428	1 315	1 433	1 333	1 417	1 377	1 358
United Rep.Tanzania Rép. Unie de Tanzanie	122	* 105	116	* 129	* 105	* 100	* 95	* 80	* 100	* 115
United States[6] Etats-Unis[6]	5 789	5 418	5 215	5 342	5 415	5 676	6 631	6 415	6 193	5 743
Uruguay Uruguay	97	103	91	* 100	* 90	98	103	63	88	* 84
Venezuela Venezuela	303	397	377	402	495	588	634	521	569	542
Viet Nam * Viet Nam *	110	370	380	410	470	440	440	460	465	465
Yugoslavia Yougoslavie	860[1]	696[1]	710[1]	* 930	933	801	* 920	691	930	* 945
Zaire Zaïre	* 60	* 58	43	59	* 65	* 55	* 75	* 75	* 90	* 85
Zambia Zambie	102	117	132	141	143	119	130	136	132	* 145
Zimbabwe Zimbabwe	391	401	437	463	456	507	459	453	502	493

Source:
International Sugar Organization (London).

† All data shown which pertain to Germany prior to 3 October 1990 are indicated separately for the Federal Republic of Germany and the former German Democratic Republic based on their respective territories at the time indicated. Where data for united Germany (3 October 1990 and thereafter) are not available, available data are shown separately under the designations Federal Republic of Germany and former German Democratic Republic and pertain to the territorial boundaries prior to 3 October 1990. For detailed explanatory notes on data pertaining to Germany, see Annex I - Country Nomenclature.

1 Official figures communicated directly to the Statistical Division of the United Nations.
2 Source: Food and Agriculture Organization of the United Nations.
3 Crop year ending 30 September of year stated.

4 Crop year production.
5 Excluding high-test molasses.
6 Including data for Puerto Rico.

Source:
L'organisation internationale du sucre (Londres).

† Toutes les données se rapportant à l'Allemagne avant le 3 octobre 1990 figurent dans deux rubriques séparées basées sur les territoires respectifs de la République fédérale d'Allemagne et l'ancienne République démocratique allemande selon la période indiquée. En l'absence de données pour l'Allemagne unifiée (à compter du 3 octobre 1990), les données disponibles sont fournies séparément sous les rubriques République fédérale d'Allemagne et ancienne République démocratique allemande et se rapportent aux limites territoriales antérieures au 3 octobre 1990. Pour les notes explicatives en détail sur les données concernant l'Allemagne, voir Annexe I – Nomenclature des pays.

1 Données officielles fournies directement à la Division de statistique des Nations Unies.
2 Source: Organisation des Nations Unies pour l'alimentation et l'agriculture.
3 Production par campagne se terminant au 30 septembre de l'année indiquée.
4 Production par campagne.
5 Non les présences de mélasse.
6 Y compris les données de Porto Rico.

69
Sugar
Sucre

Consumption of centrifugal sugar
Consommation de sucre centrifugé

Country or area Pays ou zone	Total: thousand metric tons Totale : milliers de tonnes métriques					Per capita: kilograms Par habitant : kilogrammes				
	1980	1987	1988	1989	1990	1980	1987	1988	1989	1990
World Monde	88 646	106 360	106 250	107 857	108 873	20.2	21.3	20.9	20.8	20.6
Afghanistan * Afghanistan *	75	80	80	55	39	4.7	3.9	5.2	3.5	2.4
Albania Albanie	* 42	* 60	* 65	68	70	* 15.4	* 19.5	* 20.7	21.3	* 21.6
Algeria * Algérie *	550	600	650	700	800	27.9	25.9	27.3	28.6	31.7
Angola * Angola *	110	77	75	90	95	15.5	8.4	7.9	9.3	9.5
Argentina Argentine	1 037	1 104	895	914	1 070	36.8	33.5	27.1	27.7	33.1
Australia Australie	783	817	844	855	864	53.6	51.0	51.1	50.9	50.6
Austria Autriche	384	370	375	373	396	46.9	39.3	40.3	39.7	40.9
Bahamas Bahamas	8	12	9	14	11	31.3	47.9	37.5	56.0	44.0
Bangladesh * Bangladesh *	140	340	300	275	250	1.4	3.3	2.9	2.6	2.3
Barbados Barbade	16	13	14	13	12	58.4	52.9	55.7	51.5	48.2
Belize Belize	7	7	7	7	9	49.4	39.4	40.2	40.4	45.4
Benin Bénin	* 7	20	20	* 15	* 15	* 2.0	* 4.7	4.5	* 3.3	* 3.2
Bermuda Bermudes	2	2	4	2	2	38.3	33.3	66.7	30.0	33.3
Bolivia Bolivie	171	134	183	* 170	* 185	* 29.9	27.9	26.2	* 23.6	* 25.0
Botswana Botswana	37	40	45	45	50	45.2	34.2	37.2	35.7	38.3
Brazil Brésil	6 264	6 572	6 241	* 7 401	6 615	50.9	46.5	43.2	* 50.2	44.0
Brunei Darussalam Brunéi Darussalam	6	7	8	8	10	31.6	30.4	33.3	32.0	38.8
Bulgaria * Bulgarie *	425	475	475	450	340	48.0	53.0	53.0	50.0	37.7
Burkina Faso Burkina Faso	27	* 30	* 30	* 40	* 35	3.9	* 4.4	* 3.5	* 4.5	* 3.9
Burundi Burundi	5	15	14	15	17	1.0	3.0	3.1	3.0	3.1
Cambodia Cambodge	6	5	5	5	5	0.7	0.6	0.6	0.6	0.6
Cameroon Cameroun	56	* 40	63	* 40	* 75	6.6	* 3.7	* 6.3	* 3.5	* 6.3
Canada Canada	1 014	1 120	* 1 100	* 1 050	* 1 050	42.4	43.5	* 42.4	* 40.0	* 39.7
Cape Verde * Cap-Vert *	5	15	13	12	11	15.6	44.1	36.1	32.4	28.9

69
Sugar
Consumption of centrifugal sugar [cont.]
Sucre
Consommation de sucre centrifugé [suite]

Country or area Pays ou zone	Total: thousand metric tons Totale : milliers de tonnes métriques					Per capita: kilograms Par habitant : kilogrammes				
	1980	1987	1988	1989	1990	1980	1987	1988	1989	1990
Central African Rep. * Rép. centrafricaine *	2	5	4	3	3	0.7	1.9	1.4	1.0	1.0
Chad * Tchad *	14	20	20	30	50	3.1	3.8	3.7	5.4	8.8
Chile Chili	* 420	467	466	498	* 490	* 38.7	36.8	36.4	38.3	* 37.2
China * Chine *	3 600	7 500	7 700	7 200	7 100	3.8	6.9	7.0	6.4	6.2
Colombia Colombie	992	1 208	1 143	1 163	1 195	35.8	36.8	37.8	34.4	35.7
Comoros Comores	3	3	3	3	3	8.8	6.0	6.3	5.9	5.7
Congo * Congo *	12	20	20	25	20	7.8	10.9	10.6	12.9	10.0
Costa Rica Costa Rica	139	* 166	* 170	* 168	178	62.1	* 59.7	* 59.3	* 58.6	58.8
Côte d'Ivoire Côte d'Ivoire	80	* 130	* 155	* 160	* 160	9.9	* 12.4	* 13.4	* 13.3	* 12.8
Cuba Cuba	530	773	746	882	937	53.9	68.7	67.1	68.0	88.1
Cyprus Chypre	* 19	21	17	* 20	* 25	* 30.6	30.2	24.5	* 29.0	* 35.7
Czechoslovakia * Tchécoslovaquie *	705	750	750	800	741	46.0	48.2	48.0	51.2	49.0
Djibouti Djibouti	8	9	10	10	10	38.1	18.4	25.6	25.0	25.3
Dominican Republic Rép. dominicaine	209	351	223	244	201	38.4	52.3	31.8	35.4	28.5
Ecuador Equateur	300	383	313	327	360	36.0	38.6	30.5	31.1	34.4
Egypt Egypte	1 121	* 1 650	* 1 775	* 1 650	* 1 725	25.7	* 32.5	* 34.2	* 31.1	* 32.4
El Salvador El Salvador	155	161	173	163	* 175	32.2	34.0	31.4	31.2	* 32.9
Ethiopia Ethiopie	161	* 160	139	162	* 160	5.4	3.5	2.9	3.3	* 3.2
EEC+ CEE+	10 972	11 922	12 240	12 752	13 067	40.3	37.8	35.9	39.0	38.1
Fiji Fidji	33	36	38	37	40	52.9	48.9	53.1	50.6	54.0
Finland Finlande	225	207	223	217	189	43.2	41.9	44.9	43.6	38.0
Gabon Gabon	7	* 20	* 20	* 18	* 20	16.4	* 16.8	* 19.8	* 15.9	* 17.1
Gambia * Gambie *	18	40	60	50	45	30.0	59.7	74.1	60.2	52.7
Germany†·Allemagne† former German D. R. anc. R. d. allemande	756	* 740	787	* 782	...	45.2	* 44.5	47.2	* 46.3	...

69
Sugar
Consumption of centrifugal sugar [cont.]
Sucre
Consommation de sucre centrifugé [suite]

Country or area Pays ou zone	Total: thousand metric tons Totale : milliers de tonnes métriques					Per capita: kilograms Par habitant : kilogrammes				
	1980	1987	1988	1989	1990	1980	1987	1988	1989	1990
Ghana * Ghana *	50	65	80	80	90	4.4	4.5	5.8	5.5	6.2
Gibraltar Gibraltar	1	1	1	2	3	30.0	33.3	33.3	66.7	100.0
Guatemala Guatemala	256	320	332	353	360	35.2	38.0	38.3	39.5	39.1
Guinea * Guinée *	25	55	40	50	50	5.0	8.6	7.9	9.6	9.3
Guinea-Bissau Guinée-Bissau	2	4	4	4	3	3.5	3.2	4.3	4.2	3.1
Guyana Guyana	34	45	38	34	29	39.8	55.7	47.1	41.2	34.8
Haiti * Haïti *	62	55	55	70	85	12.4	11.0	10.0	12.5	14.9
Honduras Honduras	111	* 120	* 150	* 160	* 175	38.4	* 25.8	* 31.3	* 32.3	* 34.2
Hong Kong * Hong-kong *	90	140	150	150	160	17.8	24.9	26.4	26.0	27.6
Hungary Hongrie	541	555	488	572	657	50.5	52.2	46.0	54.1	61.0
Iceland Islande	10	* 13	* 14	* 15	* 14	45.3	* 52.0	* 56.0	* 59.6	* 56.0
India Inde	5 042	9 732	10 175	* 10 677	11 121	7.5	12.3	13.0	* 13.2	13.3
Indonesia * Indonésie *	1 550	2 350	2 545	2 600	2 650	10.4	13.5	14.7	14.5	14.8
Iran, Islamic Rep. of * Iran, Rép. islamique d' *	1 150	1 300	1 200	1 000	1 200	30.7	25.1	21.7	18.5	21.5
Iraq * Iraq *	550	600	575	600	525	42.1	35.3	33.3	29.5	28.5
Israel * Israël *	195	250	250	250	250	50.4	57.2	56.4	55.4	54.0
Jamaica Jamaïque	117	108	113	125	114	55.8	46.9	48.0	53.0	48.0
Japan Japon	2 982	2 690	2 905	2 801	2 833	25.5	22.1	23.7	22.8	23.0
Jordan Jordanie	89	* 150	* 160	* 160	* 170	28.0	* 39.5	* 38.0	* 39.2	* 40.1
Kenya * Kenya *	325	450	460	475	500	19.8	19.6	19.3	19.1	19.1
Korea, Dem. P. R. Corée, R. p. dém. de	* 100	120	120	120	120	* 5.6	5.7	5.5	5.4	5.2
Korea, Republic of Corée, République de	447	668	673	745	727	11.7	15.9	16.0	17.5	17.0
Kuwait * Koweït *	60	65	60	60	60	44.1	34.8	30.6	29.3	28.0
Lao People's Dem. Rep. * Rép. dém. pop. lao *	6	6	6	6	6	1.6	1.4	1.6	1.5	1.5
Lebanon * Liban *	85	95	110	110	110	26.9	34.5	39.0	38.2	37.4

69
Sugar
Consumption of centrifugal sugar [cont.]
Sucre
Consommation de sucre centrifugé [suite]

Country or area Pays ou zone	Total: thousand metric tons Totale : milliers de tonnes métriques					Per capita: kilograms Par habitant : kilogrammes				
	1980	1987	1988	1989	1990	1980	1987	1988	1989	1990
Liberia Libéria	7	15	17	15	11	3.7	6.4	6.8	6.0	4.2
Libyan Arab Jamah. * Jamah. arabe libyenne *	120	175	160	150	190	40.2	45.7	37.8	34.2	41.8
Macau Macao	3	3	3	3	3	10.7	7.0	6.8	6.5	6.2
Madagascar Madagascar	100	82	79	76	86	11.4	8.1	6.9	6.5	7.3
Malawi Malawi	43	89	108	107	124	6.9	11.8	12.7	12.6	12.6
Malaysia Malaisie	489	* 575	* 625	* 650	* 675	36.4	* 34.7	* 37.7	* 37.4	* 37.9
Maldives Maldives	5	9	9	7	9	33.3	47.4	45.0	33.3	41.0
Mali Mali	35	* 55	* 50	* 70	* 75	5.4	* 7.0	* 6.2	* 7.6	* 9.2
Malta Malte	14	15	15	17	18	40.0	44.7	43.4	49.4	51.0
Mauritania * Mauritanie *	28	70	65	70	55	17.2	34.8	34.0	35.5	27.2
Mauritius Maurice	39	40	41	40	41	41.4	38.1	37.0	37.0	38.1
Mexico Mexique	3 152	3 657	4 070	4 023	4 424	44.7	45.0	49.2	47.6	54.5
Mongolia Mongolie	39	45	50	50	50	23.4	22.6	23.9	23.1	22.4
Morocco Maroc	664	* 700	756	740	775	32.5	* 30.4	32.3	30.9	31.8
Mozambique Mozambique	136	* 60	* 45	* 55	* 40	11.3	* 4.1	* 3.0	* 3.6	* 2.5
Myanmar * Myanmar *	65	40	35	30	30	1.9	2.4	0.6	0.7	0.6
Nepal Népal	18	35	40	35	40	1.2	2.0	2.2	1.9	2.1
Netherlands Antilles * Antilles néerlandaises *	7	8	8	9	14	25.9	42.1	42.1	30.0	45.4
New Zealand Nouvelle-Zélande	159	* 170	* 170	* 165	* 170	50.7	* 51.8	* 51.7	* 49.8	* 50.7
Nicaragua Nicaragua	126	151	136	* 150	* 150	46.5	43.2	37.6	* 40.1	* 38.8
Niger Niger	7	30	25	20	* 25	1.3	4.4	3.7	2.9	* 2.5
Nigeria * Nigéria *	750	625	425	335	400	9.8	6.1	4.0	3.1	3.6
Norway Norvège	174	173	158	168	171	42.6	41.4	35.0	39.7	40.3
Pakistan * Pakistan *	781	2 005	1 978	2 089	2 290	9.6	19.0	18.5	19.2	20.4
Panama * Panama *	70	70	80	100	120	35.9	30.8	34.5	42.2	49.6

69

Sugar
Consumption of centrifugal sugar [cont.]
    Sucre
    Consommation de sucre centrifugé [suite]

Country or area Pays ou zone	Total: thousand metric tons Totale : milliers de tonnes métriques					Per capita: kilograms Par habitant : kilogrammes				
	1980	1987	1988	1989	1990	1980	1987	1988	1989	1990
Papua New Guinea Papouasie-Nvl-Guinée	* 28	30	29	29	27	* 9.1	8.6	8.3	8.2	7.3
Paraguay Paraguay	70	* 100	* 100	* 110	* 79	22.8	* 20.4	* 24.8	* 26.4	* 18.5
Peru Pérou	585	* 860	* 750	* 750	* 620	31.4	* 41.5	* 37.5	* 34.4	* 27.8
Philippines Philippines	1 209	1 438	1 225	1 471	1 582	25.2	25.1	20.9	24.5	24.5
Poland Pologne	1 534	* 1 850	* 1 850	1 793	1 404	42.7	* 49.1	* 48.9	47.4	37.0
Romania * Roumanie *	600	600	500	600	700	27.0	26.2	21.7	25.9	30.2
Rwanda Rwanda	3	12	12	10	10	0.6	1.9	2.0	1.4	1.4
Saint Kitts and Nevis Saint-Kitts-et-Nevis	2	2	2	2	* 2	34.3	43.1	40.0	37.6	* 38.1
Samoa Samoa	3	3	3	3	3	22.0	18.8	18.8	17.6	17.1
Saudi Arabia * Arabie saoudite *	325	380	400	400	400	34.7	27.9	28.5	27.7	26.9
Senegal * Sénégal *	75	80	80	90	110	13.2	11.9	11.6	12.3	14.6
Sierra Leone * Sierra Leone *	20	17	20	16	18	5.8	4.5	5.1	4.0	4.3
Singapore Singapour	120	* 175	* 200	* 190	* 190	50.4	* 67.0	* 75.5	* 70.9	* 70.1
Somalia * Somalie *	55	80	60	50	40	15.1	16.4	8.5	6.8	5.3
South Africa Afrique du Sud	1 291	1 433	1 417	1 390	1 433	42.1	34.0	36.3	30.9	34.0
Sri Lanka Sri Lanka	205	* 300	* 220	* 285	* 240	13.9	* 18.3	* 13.7	* 17.0	* 14.1
Sudan * Soudan *	385	600	550	440	450	20.6	26.3	23.1	18.0	17.8
Suriname * Suriname *	12	14	15	15	22	30.8	36.8	27.8	37.5	54.0
Swaziland Swaziland	21	35	41	49	47	37.9	53.0	57.4	64.7	62.0
Sweden Suède	364	372	383	377	379	43.8	44.3	45.4	44.4	44.3
Switzerland Suisse	270	287	282	291	305	41.8	42.8	44.1	43.4	45.6
Syrian Arab Republic * Rép. arabe syrienne *	345	385	375	360	375	38.4	30.8	33.1	30.7	30.9
Thailand Thaïlande	632	883	886	981	* 1 105	13.6	16.4	16.4	17.7	* 19.6
Togo * Togo *	22	40	45	45	35	8.7	11.1	13.9	16.4	10.1
Trinidad and Tobago Trinité-et-Tobago	39	* 60	* 65	* 65	62	32.9	* 49.2	* 51.6	* 51.2	48.2

69
Sugar
Consumption of centrifugal sugar [*cont.*]
   Sucre
   Consommation de sucre centrifugé [*suite*]

Country or area Pays ou zone	Total: thousand metric tons Totale : milliers de tonnes métriques					Per capita: kilograms Par habitant : kilogrammes				
	1980	1987	1988	1989	1990	1980	1987	1988	1989	1990
Tunisia Tunisie	* 150	212	202	192	212	* 24.4	27.6	27.0	24.6	25.0
Turkey Turquie	1 097	1 658	1 534	1 641	1 775	24.3	32.3	29.3	30.7	30.2
Uganda * Ouganda *	16	70	70	70	35	1.1	4.8	4.1	3.9	1.9
former USSR * ancienne URSS *	12 760	14 950	14 350	13 150	13 400	46.3	52.9	50.2	45.7	46.2
United Rep.Tanzania Rép. Unie de Tanzanie	120	* 100	* 85	* 100	* 95	6.9	* 5.2	* 3.5	* 4.0	* 3.7
United States Etats-Unis	9 330	7 409	7 428	7 538	7 859	41.0	30.2	30.2	30.3	31.4
Uruguay Uruguay	102	83	78	68	* 75	36.4	29.6	25.9	22.0	* 24.3
Venezuela Venezuela	704	777	853	706	732	50.6	42.5	45.5	36.7	37.1
Viet Nam * Viet Nam *	200	500	500	500	520	3.8	4.8	7.8	7.6	7.9
Yemen * Yémen *	127	285	305	240	250	35.6	54.2	57.8	44.0	47.0
Yugoslavia Yougoslavie	858	* 950	* 900	907	* 950	38.4	* 40.6	* 38.2	* 38.3	39.9
Zaire * Zaïre *	67	120	120	110	95	2.4	3.7	3.6	3.2	2.7
Zambia Zambie	111	107	115	105	* 125	19.0	15.0	16.0	13.5	* 15.5
Zimbabwe Zimbabwe	154	253	270	283	297	20.6	28.7	29.7	30.2	30.7

Source:
International Sugar Organization (London).

+   For Member States of this grouping, see
   Annex I - Other groupings.

†   All data shown which pertain to Germany prior to 3 October
   1990 are indicated separately for the Federal Republic of
   Germany and the former German Democratic Republic based on
   their respective territories at the time indicated. Where
   data for united Germany (3 October 1990 and thereafter) are
   not available, available data are shown separately under the
   designations Federal Republic of Germany and former German
   Democratic Republic and pertain to the territorial
   boundaries prior to 3 October 1990. For detailed
   explanatory notes on data pertaining to Germany, see Annex I
   - Country Nomenclature.

Source:
Organisation internationale du sucre (Londres).

+   Les Etats membres de ce groupement, voir
   annexe I: Autres groupements.

†   Toutes les données se rapportant à l'Allemagne avant le 3
   octobre 1990 figurent dans deux rubriques séparées basées
   sur les territoires respectifs de la République fédérale
   d'Allemagne et l'ancienne République démocratique allemande
   selon la période indiquée. En l'absence de données pour
   l'Allemagne unifiée (à compter du 3 octobre 1990), les
   données disponibles sont fournies séparément sous les
   rubriques République fédérale d'Allemagne et ancienne
   République démocratique allemande et se rapportent aux
   limites territoriales antérieures au 3 octobre 1990. Pour
   les notes explicatives en détail sur les données concernant
   l'Allemagne, voir Annexe I – Nomenclature des pays.

70
Meat
Viande

Production: thousand metric tons
Production : milliers de tonnes métriques

Country or area Pays ou zone	1982	1983	1984	1985	1986	1987	1988	1989	1990	1991
World *Monde*	**107 496**	**111 099**	**114 132**	**117 767**	**120 574**	**123 490**	**128 416**	**130 250**	**133 154**	**134 198**
Beaf and veal *Boeuf et veau*	**46 536**	**47 577**	**48 507**	**49 656**	**50 770**	**51 448**	**52 446**	**53 047**	**53 679**	**53 612**
Pork *Porc*	**53 233**	**55 528**	**57 540**	**59 811**	**61 429**	**63 361**	**66 977**	**67 916**	**69 883**	**70 852**
Mutton and lamb *Mouton et agneau*	**7 727**	**7 994**	**8 085**	**8 300**	**8 375**	**8 681**	**8 992**	**9 288**	**9 592**	**9 734**
Africa *Afrique*	**4 728**	**4 883**	**5 042**	**5 230**	**5 164**	**5 398**	**5 596**	**5 676**	**5 902**	**6 084**
Beaf and veal *Boeuf et veau*	**3 098**	**3 198**	**3 328**	**3 458**	**3 369**	**3 456**	**3 594**	**3 619**	**3 771**	**3 839**
Pork *Porc*	**372**	**400**	**420**	**449**	**479**	**513**	**525**	**562**	**590**	**623**
Mutton and lamb *Mouton et agneau*	**1 258**	**1 286**	**1 295**	**1 322**	**1 315**	**1 429**	**1 477**	**1 495**	**1 542**	**1 622**
Algeria *Algérie*	**133**	**147**	**151**	**158**	**160**	**195**	**204**	**145**	**147**	**147**
Beaf and veal *Boeuf et veau*	46[1]	54[1]	58[1]	67	66	74	78	60[1]	60[1]	60[1]
Mutton and lamb[1] *Mouton et agneau*[1]	87	93	93	91	94	121	127	85	86	86
Angola *Angola*	**69**	**70**	**72**	**73**	**74**	**75**	**76**	**77**	**78**	**79**
Beaf and veal[1] *Boeuf et veau*[1]	52	53	53	54	55	55	56	56	56	57
Pork[1] *Porc*[1]	14	15	15	16	16	16	17	17	18	18
Mutton and lamb[1] *Mouton et agneau*[1]	3	3	3	4	4	4	4	4	4	4
Benin *Bénin*	**24**	**25**	**25**	**26**	**25**	**25**	**26**	**27**	**28**	**28**
Beaf and veal[1] *Boeuf et veau*[1]	12	13	13	13	13	13	13	14	14	14
Pork[1] *Porc*[1]	6	6	6	6	7	7	7	8	8	8
Mutton and lamb[1] *Mouton et agneau*[1]	6	6	6	7	5	5	5	6	6	6
Botswana *Botswana*	**50**	**46**	**45**	**40**	**38**	**30**	**43**	**51**	**45**	**38**
Beaf and veal[1] *Boeuf et veau*[1]	47	42	41	36	33	25	38	45	38	32
Mutton and lamb[1] *Mouton et agneau*[1]	3	3	4	4	4	5	5	6	6	6
Burkina Faso *Burkina Faso*	**42**	**48**	**47**	**49**	**56**	**54**	**56**	**58**	**60**	**62**
Beaf and veal[1] *Boeuf et veau*[1]	25	28	27	28	28	25	26	28	29	29
Pork[1] *Porc*[1]	6	8	9	10	11	11	11	11	11	12
Mutton and lamb[1] *Mouton et agneau*[1]	11	11	11	11	17	18	18	19	20	22
Burundi *Burundi*	**13**	**14**	**15**	**13**	**16**	**15**	**17**	**16**	**17**	**17**
Beaf and veal[1] *Boeuf et veau*[1]	6	6	6	6	7	7	7	7	7	7
Pork[1] *Porc*[1]	3	5	5	4	5	5	7	5	6	6

70

Meat
Production: thousand metric tons [*cont.*]
Viande
Production : milliers de tonnes métriques [*suite*]

Country or area Pays ou zone	1982	1983	1984	1985	1986	1987	1988	1989	1990	1991
Mutton and lamb[1] Mouton et agneau[1]	3	3	4	4	4	4	4	4	4	4
Cameroon Cameroun	**71**	**71**	**83**	**91**	**90**	**96**	**106**	**110**	**113**	**115**
Beaf and veal Boeuf et veau	46	49	60	56	53	62	68	70[1]	70[1]	70[1]
Pork[1] Porc[1]	11	9	10	21	17	14	15	16	16	17
Mutton and lamb[1] Mouton et agneau[1]	13	13	14	15	19	20	22	25	27	28
Cape Verde Cap-Vert	**2**	**2**	**2**	**2**	**2**	**3**	**2**	**6**	**6**	**6**
Beaf and veal Boeuf et veau	0	0	0	0[1]	0[1]	0[1]	0[1]	3[1]	3[1]	3[1]
Pork[1] Porc[1]	1	2	2	2	2	2	2	3	3	3
Central African Rep. Rép. centrafricaine	**45**	**51**	**49**	**50**	**52**	**57**	**60**	**62**	**64**	**66**
Beaf and veal[1] Boeuf et veau[1]	30	35	34	34	35	40	42	43	44	46
Pork[1] Porc[1]	11	12	12	12	13	13	14	14	15	15
Mutton and lamb[1] Mouton et agneau[1]	3	3	4	4	4	4	4	5	5	5
Chad Tchad	**52**	**49**	**41**	**44**	**68**	**68**	**80**	**75**	**81**	**79**
Beaf and veal[1] Boeuf et veau[1]	31	31	26	29	52	51	62	57	63	65
Mutton and lamb[1] Mouton et agneau[1]	21	18	14	16	16	17	17	18	18	14
Comoros Comores	**2**	**2**	**1**	**1**	**1**	**1**	**1**	**1**	**1**	**2**
Beaf and veal[1] Boeuf et veau[1]	1	1	1	1	1	1	1	1	1	1
Congo Congo	**4**	**4**	**5**	**5**	**5**	**5**	**5**	**5**	**5**	**5**
Beaf and veal[1] Boeuf et veau[1]	2	2	3	2	2	2	2	2	2	2
Pork[1] Porc[1]	1	1	1	2	2	2	2	2	2	2
Mutton and lamb[1] Mouton et agneau[1]	1	1	1	1	1	1	1	1	1	1
Côte d'Ivoire Côte d'Ivoire	**57**	**64**	**65**	**66**	**63**	**59**	**59**	**57**	**66**	**69**
Beaf and veal[1] Boeuf et veau[1]	35	42	43	44	41	37	37	34	42	45
Pork[1] Porc[1]	13	13	13	13	13	13	14	14	14	15
Mutton and lamb[1] Mouton et agneau[1]	9	9	9	9	9	8	8	9	9	9
Djibouti Djibouti	**5**	**5**	**5**	**5**	**6**	**6**	**6**	**6**	**6**	**6**
Beaf and veal[1] Boeuf et veau[1]	2	1	1	1	2	2	2	2	2	2
Mutton and lamb[1] Mouton et agneau[1]	3	4	4	4	4	4	4	4	4	4
Egypt Egypte	323	417	467	528	539	550	560	535	509	524

70

Meat
Production: thousand metric tons [cont.]
Viande
Production : milliers de tonnes métriques [suite]

Country or area Pays ou zone	1982	1983	1984	1985	1986	1987	1988	1989	1990	1991
Beaf and veal Boeuf et veau	262[1]	345[1]	390[1]	448	455	463	469	440[1]	408[1]	420[1]
Pork Porc	* 3	3[1]	3[1]	3	3	2	2	2[1]	2[1]	3[1]
Mutton and lamb Mouton et agneau	* 59[1]	69[1]	75[1]	77	81	85	89	93[1]	99[1]	101[1]
Ethiopia Ethiopie	**363**	**370**	**365**	**383**	**394**	**353**	**353**	**387**	**394**	**395**
Beaf and veal Boeuf et veau	215[1]	220[1]	214[1]	230[1]	246[1]	* 206	* 206	* 237	* 245	* 245
Pork[1] Porc[1]	1	1	1	1	1	1	1	1	1	1
Mutton and lamb[1] Mouton et agneau[1]	147	149	151	152	147	* 146	* 146	* 149	* 148	149
Gabon Gabon	**2**	**3**	**3**	**3**	**3**	**3**	**4**	**4**	**4**	**4**
Beaf and veal[1] Boeuf et veau[1]	0	0	0	0	1	1	1	1	1	1
Pork[1] Porc[1]	2	2	2	2	2	2	2	2	2	2
Mutton and lamb[1] Mouton et agneau[1]	1	1	1	1	1	1	1	1	1	1
Gambia Gambie	**6**	**6**	**5**	**6**	**6**	**7**	**7**	**7**	**7**	**7**
Beaf and veal[1] Boeuf et veau[1]	4	4	4	4	5	5	6	6	6	6
Mutton and lamb[1] Mouton et agneau[1]	1	1	1	1	1	1	1	1	1	1
Ghana Ghana	**32**	**33**	**35**	**37**	**39**	**39**	**41**	**44**	**41**	**50**
Beaf and veal[1] Boeuf et veau[1]	13	14	16	17	18	20	20	20	20	22
Pork[1] Porc[1]	8	8	9	9	11	9	11	12	11	14
Mutton and lamb[1] Mouton et agneau[1]	11	10	10	10	10	10	10	12	11	14
Guinea Guinée	**22**	**22**	**22**	**21**	**21**	**21**	**21**	**21**	**21**	**21**
Beaf and veal[1] Boeuf et veau[1]	19	19	19	18	18	18	18	18	18	18
Pork[1] Porc[1]	1	1	1	1	1	1	1	1	1	1
Mutton and lamb[1] Mouton et agneau[1]	2	2	2	2	2	2	2	2	2	2
Guinea-Bissau Guinée-Bissau	**12**	**12**	**12**	**12**	**12**	**13**	**13**	**13**	**13**	**13**
Beaf and veal[1] Boeuf et veau[1]	2	2	3	3	3	3	3	3	3	3
Pork[1] Porc[1]	8	8	8	9	9	9	9	9	9	9
Mutton and lamb[1] Mouton et agneau[1]	1	1	1	1	1	1	1	1	1	1
Kenya Kenya	**228**	**193**	**228**	**244**	**250**	**270**	**294**	**286**	**391**	**388**
Beaf and veal Boeuf et veau	186[1]	156[1]	182[1]	190[1]	199	219	238	228	329	326[1]
Pork[1] Porc[1]	4	4	4	5	5	5	5	5	5	5

70

Meat
Production: thousand metric tons [*cont.*]
Viande
Production : milliers de tonnes métriques [*suite*]

Country or area Pays ou zone	1982	1983	1984	1985	1986	1987	1988	1989	1990	1991
Mutton and lamb[1] Mouton et agneau[1]	38	33	42	49	46	46	50	54	56	57
Lesotho **Lesotho**	**20**	**21**	**21**	**21**	**22**	**22**	**22**	**23**	**23**	**23**
Beaf and veal[1] Boeuf et veau[1]	12	12	12	12	12	12	13	13	13	13
Pork[1] Porc[1]	3	2	2	3	3	3	3	3	3	3
Mutton and lamb[1] Mouton et agneau[1]	6	6	6	6	7	7	7	7	7	7
Liberia **Libéria**	**7**	**8**	**7**	**7**	**7**	**7**	**7**	**7**	**6**	**6**
Beaf and veal[1] Boeuf et veau[1]	2	3	2	2	1	1	1	1	1	1
Pork[1] Porc[1]	4	4	4	4	4	4	4	4	4	4
Mutton and lamb[1] Mouton et agneau [1]	1	1	1	1	1	1	1	1	1	1
Libyan Arab Jamah. **Jamah. arabe libyenne**	**91**	**104**	**101**	**74**	**67**	**69**	**75**	**81**	**78**	**91**
Beaf and veal[1] Boeuf et veau[1]	37	44	41	28	26	20	19	18	19	23
Mutton and lamb[1] Mouton et agneau[1]	54	60	60	46	41	49	56	63	59	68
Madagascar **Madagascar**	**176**	**176**	**180**	**176**	**179**	**187**	**187**	**188**	**189**	**190**
Beaf and veal[1] Boeuf et veau[1]	135	136	136	134	134	139	141	142	142	142
Pork[1] Porc[1]	31	32	37	36	38	41	39	38	39	39
Mutton and lamb[1] Mouton et agneau[1]	11	8	7	6	7	7	8	8	9	9
Malawi **Malawi**	**29**	**26**	**26**	**21**	**31**	**32**	**30**	**33**	**35**	**37**
Beaf and veal[1] Boeuf et veau[1]	18	15	15	10	16	16	16	18	20	21
Pork[1] Porc[1]	8	8	7	7	11	13	10	10	11	11
Mutton and lamb[1] Mouton et agneau[1]	3	3	3	4	3	4	4	4	4	5
Mali **Mali**	**97**	**92**	**94**	**88**	**96**	**96**	**104**	**111**	**115**	**119**
Beaf and veal[1] Boeuf et veau[1]	46	45	57	54	62	61	65	69	72	75
Pork[1] Porc[1]	1	2	2	2	2	2	2	2	2	2
Mutton and lamb[1] Mouton et agneau[1]	50	46	35	32	33	33	37	40	42	42
Mauritania **Mauritanie**	**28**	**27**	**24**	**24**	**26**	**27**	**28**	**28**	**28**	**26**
Beaf and veal[1] Boeuf et veau[1]	16	15	13	13	14	16	17	17	17	14
Mutton and lamb[1] Mouton et agneau[1]	12	12	11	12	12	12	12	12	12	11
Mauritius **Maurice**	**2**	**2**	**2**	**2**	**2**	**2**	**3**	**3**	**3**	**3**
Beaf and veal Boeuf et veau	1	1	1	1	1	1	2	2	2	2[1]

70

Meat
Production: thousand metric tons [*cont.*]
Viande
Production : milliers de tonnes métriques [*suite*]

Country or area Pays ou zone	1982	1983	1984	1985	1986	1987	1988	1989	1990	1991
Pork Porc	1	1	1	1	1	1	1	1	1[1]	1[1]
Morocco **Maroc**	**210**	**183**	**177**	**199**	**201**	**220**	**204**	**210**	**213**	**217**
Beaf and veal Boeuf et veau	150[1]	* 124	* 115	* 132	134[1]	135[1]	137[1]	139[1]	141[1]	145[1]
Pork Porc	1	1	1	0	0	1	1	1	1	1
Mutton and lamb Mouton et agneau	59[1]	59	61	67	67[1]	84	67[1]	71[1]	71[1]	72[1]
Mozambique **Mozambique**	**48**	**47**	**48**	**49**	**51**	**52**	**53**	**54**	**55**	**55**
Beaf and veal[1] Boeuf et veau[1]	36	35	36	37	38	38	39	40	41	41
Pork[1] Porc[1]	9	9	9	10	10	10	11	11	11	11
Mutton and lamb[1] Mouton et agneau[1]	3	3	3	3	3	3	3	3	3	3
Namibia **Namibie**	**49**	**49**	**51**	**52**	**54**	**67**	**69**	**73**	**82**	**126**
Beaf and veal Boeuf et veau	34[1]	35[1]	36[1]	37[1]	38[1]	50[1]	52[1]	55[1]	63	66
Pork[1] Porc[1]	3	3	3	3	3	4	4	4	4	4
Mutton and lamb[1] Mouton et agneau[1]	12	10	12	12	13	13	14	15	15	55
Niger **Niger**	**79**	**80**	**52**	**38**	**47**	**50**	**52**	**54**	**54**	**54**
Beaf and veal Boeuf et veau	35	35	20[1]	17[1]	19[1]	20[1]	20[1]	21[1]	21[1]	21[1]
Pork[1] Porc[1]	1	1	1	1	1	1	1	1	1	1
Mutton and lamb[1] Mouton et agneau[1]	42	43	31	20	27	29	30	32	32	32
Nigeria **Nigéria**	**438**	**425**	**433**	**470**	**513**	**545**	**581**	**619**	**661**	**703**
Beaf and veal[1] Boeuf et veau[1]	256	243	237	245	255	256	269	275	282	288
Pork[1] Porc[1]	37	37	48	59	75	90	97	112	128	150
Mutton and lamb[1] Mouton et agneau[1]	144	144	148	166	184	199	215	232	251	266
Réunion **Réunion**	**6**	**6**	**7**	**7**	**7**	**8**	**8**	**9**	**9**	**9**
Beaf and veal Boeuf et veau	1	1	1	1	1	1	1	1	1	1[1]
Pork Porc	5	5	6	5	6	6	7	7	8[1]	7[1]
Rwanda **Rwanda**	**20**	**20**	**19**	**21**	**20**	**19**	**21**	**22**	**22**	**22**
Beaf and veal[1] Boeuf et veau[1]	13	13	14	15	14	12	14	14	14	14
Pork[1] Porc[1]	3	3	2	2	2	2	3	3	3	3
Mutton and lamb[1] Mouton et agneau[1]	4	4	4	4	4	4	4	5	5	5
Senegal **Sénégal**	**62**	**64**	**64**	**68**	**77**	**77**	**79**	**82**	**85**	**88**

70
Meat
Production: thousand metric tons [cont.]
Viande
Production : milliers de tonnes métriques [suite]

Country or area Pays ou zone	1982	1983	1984	1985	1986	1987	1988	1989	1990	1991
Beaf and veal[1] Boeuf et veau[1]	42	43	41	41	44	43	44	46	47	48
Pork[1] Porc[1]	8	9	10	11	16	17	18	18	19	20
Mutton and lamb[1] Mouton et agneau[1]	12	13	13	16	17	17	18	19	20	20
Seychelles **Seychelles**	**1**	**1**	**1**	**1**	**1**	**1**	**1**	**1**	**1**	**1**
Pork[1] Porc[1]	1	1	1	1	1	1	1	1	1	1
Sierra Leone **Sierra Leone**	**9**	**9**	**8**	**9**	**9**	**9**	**9**	**9**	**9**	**9**
Beaf and veal[1] Boeuf et veau[1]	6	6	5	5	5	5	5	5	5	5
Pork[1] Porc[1]	2	2	2	2	2	2	2	2	2	2
Mutton and lamb[1] Mouton et agneau[1]	2	2	2	2	2	1	2	2	2	2
Somalia **Somalie**	**105**	**112**	**113**	**117**	**105**	**130**	**137**	**143**	**151**	**152**
Beaf and veal[1] Boeuf et veau[1]	41	42	40	42	35	44	50	53	57	57
Mutton and lamb[1] Mouton et agneau[1]	64	70	73	75	69	86	87	90	94	95
South Africa **Afrique du Sud**	**845**	**872**	**902**	**924**	**878**	**910**	**936**	**946**	**953**	**972**
Beaf and veal Boeuf et veau	* 582	588	620	* 638	* 616	* 628	650[1]	655[1]	* 661	* 678
Pork Porc	* 100	* 120	* 118	120[1]	120[1]	122[1]	123[1]	124[1]	125[1]	126[1]
Mutton and lamb[1] Mouton et agneau[1]	163	164	164	166	142	160	163	167	167	168
Sudan **Soudan**	**286**	**326**	**362**	**371**	**239**	**289**	**304**	**312**	**322**	**338**
Beaf and veal Boeuf et veau	193	231	261	267	150	198	205	211	218	231
Mutton and lamb Mouton et agneau	93	95	101	104	89	92	99	101	104	107
Swaziland **Swaziland**	**19**	**20**	**16**	**18**	**19**	**19**	**15**	**16**	**18**	**18**
Beaf and veal[1] Boeuf et veau[1]	15	16	13	14	15	15	12	12	14	15
Pork[1] Porc[1]	1	1	1	1	1	1	1	1	1	1
Mutton and lamb[1] Mouton et agneau[1]	3	3	3	3	3	3	3	3	3	3
Togo **Togo**	**11**	**8**	**11**	**13**	**13**	**14**	**13**	**17**	**19**	**19**
Beaf and veal[r] Boeuf et veau[1]	5	4	5	5	5	5	4	5	5	6
Pork[1] Porc[1]	3	2	3	4	3	3	3	6	7	7
Mutton and lamb[1] Mouton et agneau[1]	3	2	3	4	5	5	5	6	6	7
Tunisia **Tunisie**	**57**	**69**	**72**	**80**	**81**	**83**	**89**	**83**	**84**	**88**
Beaf and veal Boeuf et veau	28	31	34	36	39	38	39	38	39	40

70

Meat
Production: thousand metric tons [*cont.*]
Viande
Production : milliers de tonnes métriques [*suite*]

Country or area Pays ou zone	1982	1983	1984	1985	1986	1987	1988	1989	1990	1991
Mutton and lamb Mouton et agneau	29	38	39	44	42	45	50	44	45	* 48[1]
Uganda **Ouganda**	**114**	**116**	**119**	**123**	**97**	**101**	**103**	**123**	**139**	**142**
Beaf and veal Boeuf et veau	86[1]	86	88	90	60	59	64[1]	70[1]	77[1]	79[1]
Pork Porc	9	10	* 10	* 11	11[1]	21	20[1]	32[1]	37[1]	38[1]
Mutton and lamb[1] Mouton et agneau[1]	* 19	* 20	* 21	* 22	26	21	18	21	24	25
United Rep.Tanzania **Rép. Unie de Tanzanie**	**169**	**178**	**189**	**200**	**208**	**209**	**224**	**232**	**235**	**237**
Beaf and veal Boeuf et veau	140	146	154[1]	163[1]	170[1]	171[1]	186[1]	194[1]	195[1]	197[1]
Pork[1] Porc[1]	2	5	8	8	8	8	8	8	8	8
Mutton and lamb[1] Mouton et agneau[1]	27	27	27	29	29	30	30	31	31	32
Zaire **Zaïre**	**58**	**59**	**60**	**61**	**62**	**64**	**65**	**66**	**68**	**69**
Beaf and veal Boeuf et veau	21	22	22	23	24	25	26	26	27[1]	28[1]
Pork[1] Porc[1]	27	28	28	28	29	29	29	29	30	30
Mutton and lamb[1] Mouton et agneau[1]	10	10	10	10	10	10	10	10	11	11
Zambia **Zambie**	**42**	**38**	**35**	**39**	**40**	**42**	**43**	**45**	**47**	**49**
Beaf and veal[1] Boeuf et veau[1]	33	30	28	32	32	33	34	36	37	39
Pork[1] Porc[1]	8	6	5	6	6	6	7	7	7	8
Mutton and lamb[1] Mouton et agneau[1]	1	2	1	2	2	2	2	2	2	2
Zimbabwe **Zimbabwe**	**91**	**93**	**103**	**99**	**90**	**103**	**99**	**93**	**99**	**97**
Beaf and veal[1] Boeuf et veau[1]	77	77	88	84	73	84	79	72	78	77
Pork[1] Porc[1]	10	12	9	8	8	10	11	11	11	11
Mutton and lamb[1] Mouton et agneau[1]	4	5	6	6	8	9	9	9	9	9
America, North **Amérique du Nord**	**22 124**	**23 124**	**23 083**	**23 105**	**23 047**	**22 751**	**23 842**	**23 933**	**23 152**	**23 263**
Beaf and veal **Boeuf et veau**	**12 865**	**13 276**	**13 413**	**13 542**	**14 134**	**13 719**	**14 192**	**14 350**	**13 783**	**13 563**
Pork **Porc**	**9 011**	**9 596**	**9 418**	**9 316**	**8 671**	**8 806**	**9 410**	**9 337**	**9 115**	**9 442**
Mutton and lamb **Mouton et agneau**	**248**	**251**	**251**	**248**	**242**	**227**	**241**	**247**	**253**	**258**
Antigua and Barbuda **Antigua-et-Barbuda**	**1**	**1**	**1**	**1**	**1**	**1**	**1**	**1**	**1**	**1**
Beaf and veal Boeuf et veau	0	0	0[1]	1[1]	1[1]	1[1]	1[1]	0[1]	1[1]	1[1]
Bahamas **Bahamas**	**1**	**1**	**1**	**1**	**1**	**1**	**1**	**1**	**1**	**1**
Barbados **Barbade**	**6**	**7**	**6**	**5**	**6**	**5**	**5**	**5**	**6**	**6**

70
Meat
Production: thousand metric tons [cont.]
Viande
Production : milliers de tonnes métriques [suite]

Country or area Pays ou zone	1982	1983	1984	1985	1986	1987	1988	1989	1990	1991
Pork[1] Porc[1]	6	6	5	5	5	5	5	5	5	
Belize Belize	2	2	2	2	2	3	2	2	2	2
Beaf and veal Boeuf et veau	1	1	1	1	1	1	1	1	1[1]	1[1]
Pork[1] Porc[1]	1	1	1	1	1	1	1	1	1	1
Canada Canada	2 039	2 071	2 043	2 125	2 141	2 116	2 171	2 172	2 065	1 998
Beaf and veal Boeuf et veau	1 025	1 032	991	1 029	1 036	977	973	979	924	* 879
Pork Porc	1 006	1 030	1 044	1 088	1 097	1 131	1 190	1 184	1 133	* 1 110
Mutton and lamb Mouton et agneau	8	8	9	8	8	8	8	8	9	9
Costa Rica Costa Rica	80	75	88	106	102	107	102	101	100	106
Beaf and veal Boeuf et veau	66	61	77	94	* 92	97	* 86	86	* 85	* 91
Pork Porc	* 14	14	11	12	10	10	16	* 15	* 15	* 15
Cuba Cuba	199	204	212	220	230	227	227	229	232	232
Beaf and veal Boeuf et veau	141	142	142	142	148	140	141	138	141[1]	140[1]
Pork Porc	56	61	68	76	80	85	84	89[1]	89[1]	90[1]
Mutton and lamb[1] Mouton et agneau[1]	2	2	2	2	2	2	2	2	2	2
Dominica Dominique	1	1	1	1	1	1	1	1	1	1
Dominican Republic Rép. dominicaine	56	60	62	78	81	85	89	93	91	90
Beaf and veal Boeuf et veau	54	56	54	63	64	67	69	71[1]	68[1]	67[1]
Pork Porc	* 0	* 2	7	13[1]	14[1]	17[1]	18[1]	20[1]	21[1]	21[1]
Mutton and lamb[1] Mouton et agneau[1]	2	2	2	2	2	2	2	2	2	2
El Salvador El Salvador	35	35	35	36	38	36	39	43	43	46
Beaf and veal Boeuf et veau	22	22	22	21	22	19	22	28	27	29
Pork[1] Porc[1]	13	13	13	15	16	17	16	14	16	16
Guadeloupe Guadeloupe	7	8	8	7	7	7	6	5	6	6
Beaf and veal Boeuf et veau	4	4	4	4	4	3	3	3	3[1]	3[1]
Pork Porc	2	3	3	3	3	3	3	2	2[1]	2[1]
Guatemala Guatemala	67	71	70	69	53	62	75	72	78	72
Beaf and veal Boeuf et veau	47	51	50	49	34	44	57	53	* 59	* 53
Pork Porc	17	16	16	16	16	14	* 14	15	* 15	15[1]

70

Meat
Production: thousand metric tons [cont.]
Viande
Production : milliers de tonnes métriques [suite]

Country or area Pays ou zone	1982	1983	1984	1985	1986	1987	1988	1989	1990	1991
Mutton and lamb[1] Mouton et agneau[1]	4	4	3	4	4	4	4	4	4	4
Haiti **Haïti**	**43**	**45**	**47**	**48**	**52**	**54**	**56**	**58**	**57**	**54**
Beaf and veal[1] Boeuf et veau[1]	30	32	33	33	34	34	34	35	34	32
Pork[1] Porc[1]	8	7	9	10	13	15	16	18	18	17
Mutton and lamb[1] Mouton et agneau[1]	5	5	5	5	5	5	5	5	5	5
Honduras **Honduras**	**57**	**44**	**46**	**46**	**38**	**56**	**58**	**59**	**59**	**59**
Beaf and veal Boeuf et veau	46	35	34	35	27	* 45	* 46	* 46	* 46	45[1]
Pork Porc	11	9	11	10	11	* 11	* 12	* 13	13	13[1]
Jamaica **Jamaïque**	**21**	**22**	**23**	**22**	**22**	**22**	**24**	**23**	**24**	**22**
Beaf and veal Boeuf et veau	12	14	15	14	15	14	14	13	15	13
Pork Porc	7	7	7	7	6	6	8	8	7	7[1]
Mutton and lamb[1] Mouton et agneau[1]	2	2	2	2	2	2	2	2	2	2
Martinique **Martinique**	**5**	**5**	**6**	**5**	**5**	**5**	**5**	**5**	**5**	**5**
Beaf and veal Boeuf et veau	3	3	4	3	3	3	3	3[1]	3[1]	3[1]
Pork[1] Porc[1]	2	2	2	2	2	2	2	2	2	2
Mutton and lamb[1] Mouton et agneau[1]	1	1	1	1	1	0	0	0	0	0
Mexico **Mexique**	**2 285**	**2 485**	**2 433**	**2 279**	**2 269**	**2 245**	**2 677**	**2 928**	**2 608**	**2 427**
Beaf and veal Boeuf et veau	862	944	925	927	1 248	1 273	* 1 754	* 2 140	* 1 790	* 1 550
Pork Porc	1 365	1 486	1 455	1 293	959	915	861	727	757	812
Mutton and lamb Mouton et agneau	57	55	52	59	62	58	62	62	61	66
Montserrat **Montserrat**	**1**	**1**	**1**	**1**	**1**	**1**	**1**	**1**	**1**	**1**
Beaf and veal[1] Boeuf et veau[1]	1	1	1	1	1	1	1	1	1	1
Netherlands Antilles **Antilles néerlandaises**	**1**	**1**	**1**	**1**	**0**	**0**	**0**	**0**	**1**	**1**
Nicaragua **Nicaragua**	**66**	**67**	**66**	**58**	**43**	**45**	**46**	**59**	**51**	**51**
Beaf and veal Boeuf et veau	52	53	51	45	29	30	33	46	38	38
Pork Porc	* 14	14[1]	14[1]	13[1]	14[1]	15[1]	13[1]	13[1]	13[1]	13[1]
Panama **Panama**	**61**	**61**	**64**	**67**	**69**	**65**	**58**	**60**	**68**	**70**
Beaf and veal Boeuf et veau	52	52	53	55	56	54	48	50	58	60
Pork Porc	9	9	10	12	13	10	10	10	10	11

70
Meat
Production: thousand metric tons [cont.]
Viande
Production : milliers de tonnes métriques [suite]

Country or area Pays ou zone	1982	1983	1984	1985	1986	1987	1988	1989	1990	1991
Puerto Rico Porto Rico	**37**	**38**	**40**	**43**	**51**	**52**	**45**	**45**	**49**	**52**
Beaf and veal Boeuf et veau	17	21	24	24	25	28	22	20	21	22
Pork Porc	20	17	15	18	25	24	22	25	28[1]	29[1]
Saint Lucia Sainte-Lucie	**1**	**1**	**1**	**1**	**1**	**1**	**1**	**1**	**1**	**1**
Beaf and veal[1] Boeuf et veau[1]	1	1	1	1	1	1	1	1	1	1
Pork[1] Porc[1]	0	1	1	1	1	1	1	1	1	1
St. Vincent-Grenadines St. Vincent-Grenadines	**1**	**1**	**1**	**1**	**1**	**1**	**1**	**1**	**1**	**1**
Pork[1] Porc[1]	0	0	0	1	1	1	1	1	1	1
Trinidad and Tobago Trinité-et-Tobago	**5**	**5**	**5**	**5**	**5**	**5**	**5**	**4**	**4**	**4**
Beaf and veal Boeuf et veau	2	1	1	1	1	1	2	1	1[1]	1[1]
Pork Porc	3	4	4	3	3	3	3	2	3[1]	3[1]
Mutton and lamb[1] Mouton et agneau[1]	1	1	1	1	1	0	0	0	0	0
United States Etats-Unis	**17 045**	**17 812**	**17 818**	**17 873**	**17 824**	**17 547**	**18 145**	**17 963**	**17 595**	**17 954**
Beaf and veal Boeuf et veau	10 425	10 746	10 927	10 996	11 292	10 884	10 879	10 633	10 465	10 531
Pork Porc	6 454	6 896	6 719	6 715	6 379	6 520	7 114	7 173	6 965	7 258
Mutton and lamb Mouton et agneau	166	171	172	162	153	143	152	157	165	165
United States Virgin Is. Iles Vierges américaines	**1**	**1**	**1**	**1**	**1**	**0**	**1**	**1**	**1**	**1**
Beaf and veal[1] Boeuf et veau[1]	1	1	1	1	1	0	0	0	0	0
America, South Amérique du Sud	**8 810**	**8 685**	**8 337**	**8 841**	**8 614**	**8 916**	**9 556**	**9 920**	**10 057**	**10 234**
Beaf and veal Boeuf et veau	6 844	6 705	6 463	6 944	6 654	6 737	7 209	7 606	7 653	7 762
Pork Porc	1 645	1 665	1 566	1 591	1 654	1 863	2 022	1 970	2 053	2 123
Mutton and lamb Mouton et agneau	321	315	308	306	306	316	325	345	350	349
Argentina Argentine	**2 898**	**2 707**	**2 889**	**3 145**	**3 087**	**2 996**	**2 943**	**2 925**	**2 921**	**2 948**
Beaf and veal Boeuf et veau	2 551	2 384	2 553	2 848	2 813	2 700	2 650	2 623	2 610	* 2 640
Pork Porc	230	207	220[1]	193	170	200	200	200[1]	216[1]	216[1]
Mutton and lamb[1] Mouton et agneau[1]	117	117	* 115	104	104	96	* 93	* 102	* 95	* 92
Bolivia Bolivie	**173**	**187**	**195**	**207**	**196**	**210**	**224**	**232**	**247**	**257**
Beaf and veal Boeuf et veau	100[1]	113[1]	116	125	115	121	131	135	* 146	* 152
Pork[1] Porc[1]	48	49	53	55	54	57	61	64	67	70

70

Meat
Production: thousand metric tons [cont.]
Viande
Production : milliers de tonnes métriques [suite]

Country or area Pays ou zone	1982	1983	1984	1985	1986	1987	1988	1989	1990	1991
Mutton and lamb[1] Mouton et agneau[1]	25	25	26	27	27	32	33	34	34	35
Brazil **Brésil**	**3 318**	**3 301**	**2 905**	**3 060**	**2 835**	**3 307**	**3 743**	**3 870**	**4 000**	**4 038**
Beaf and veal Boeuf et veau	2 397	2 365	2 096	2 223	1 958	2 262	2 581	2 748	2 775	* 2 800
Pork Porc	870[1]	885[1]	757[1]	780[1]	825[1]	986[1]	* 1 100	* 1 050	* 1 150	* 1 160
Mutton and lamb[1] Mouton et agneau[1]	52	52	52	58	52	59	63	72	76	78
Chile **Chili**	**272**	**285**	**273**	**259**	**270**	**282**	**316**	**353**	**385**	**402**
Beaf and veal Boeuf et veau	195	208	197	175	177	175	197	221	242	240
Pork Porc	58	59	59	66	75	88	100	113	123	142
Mutton and lamb[1] Mouton et agneau[1]	20	18	17	18	18	19	19	19	20	20
Colombia **Colombie**	**709**	**688**	**725**	**740**	**788**	**755**	**814**	**942**	**921**	**952**
Beaf and veal Boeuf et veau	596	564	599	609	652	621	669	805	* 795	* 823
Pork Porc	* 101	* 111	114	118	123	* 122	* 131	124	112	115
Mutton and lamb Mouton et agneau	* 12	* 13	12	13	* 12	* 13	* 13	14	14	14[1]
Ecuador **Equateur**	**141**	**146**	**150**	**159**	**163**	**166**	**169**	**167**	**171**	**183**
Beaf and veal Boeuf et veau	78	81	83	89	90	92	95	98	100	* 110
Pork[1] Porc[1]	57	59	62	65	68	68	65	64	65	67
Mutton and lamb[1] Mouton et agneau[1]	6	6	5	6	5	6	8	5	6	6
Falkland Is. (Malvinas) **Iles Falkland (Malvinas)**	**1**	**1**	**1**	**1**	**1**	**1**	**1**	**1**	**1**	**1**
Mutton and lamb[1] Mouton et agneau[1]	1	1	1	1	1	1	1	1	1	1
French Guiana **Guyane française**	**1**	**1**	**1**	**1**	**1**	**1**	**1**	**1**	**1**	**1**
Beaf and veal Boeuf et veau	0	0	0	0	0	0	1	0	0	0[1]
Pork Porc	1	1	1	1	1	1	1	1	1	1[1]
Guyana **Guyana**	**4**	**4**	**3**	**3**	**4**	**4**	**4**	**4**	**4**	**4**
Beaf and veal Boeuf et veau	2	2	2	2	2	2	2	2	2[1]	2[1]
Pork Porc	1	1	1	1	1	1	1	1	1	1[1]
Mutton and lamb[1] Mouton et agneau[1]	1	1	1	1	1	1	1	1	1	1
Paraguay **Paraguay**	**201**	**210**	**219**	**225**	**228**	**213**	**244**	**253**	**270**	**304**
Beaf and veal Boeuf et veau	100	106	112	115	118	101	131	* 135	148[1]	160[1]
Pork[1] Porc[1]	98	101	104	107	107	108	109	114	118	141

70

Meat
Production: thousand metric tons [*cont.*]
Viande
Production : milliers de tonnes métriques [*suite*]

Country or area Pays ou zone	1982	1983	1984	1985	1986	1987	1988	1989	1990	1991
Mutton and lamb[1] Mouton et agneau[1]	3	3	3	3	3	3	3	3	3	3
Peru **Pérou**	**197**	**216**	**201**	**195**	**191**	**216**	**234**	**230**	**232**	**229**
Beaf and veal Boeuf et veau	91	111	103	101	90	107	117	112	117	112
Pork * Porc *	76	74	72	70	77	83	91	91	84	91
Mutton and lamb Mouton et agneau	30[1]	31[1]	* 26	* 24	* 24	26[1]	* 27	* 28	* 31	* 26
Suriname **Suriname**	**2**	**3**	**3**	**3**	**3**	**3**	**3**	**4**	**4**	**4**
Beaf and veal Boeuf et veau	1	1	1	1	1	1	1	2	2	2
Pork Porc	1	1	1	2	1	1	1	2	2	2
Uruguay **Uruguay**	**463**	**493**	**362**	**394**	**391**	**346**	**396**	**458**	**409**	**430**
Beaf and veal Boeuf et veau	397	432	301	332	324	279	326	386	333	350
Pork Porc	17	18	18	16	16	15	13	14	15	16[1]
Mutton and lamb Mouton et agneau	48	43	43	45	52	53	56	58	61	* 64
Venezuela **Venezuela**	**430**	**445**	**410**	**448**	**457**	**417**	**464**	**478**	**490**	**480**
Beaf and veal Boeuf et veau	336	337	298	324	313	276	307	338	* 382	* 370
Pork Porc	87	100	105	117	136	133	148	132	* 99	* 101
Mutton and lamb Mouton et agneau	8	7	7	7	7	8	8	8	9[1]	9[1]
Asia **Asie**	**24 803**	**25 663**	**27 693**	**30 330**	**32 144**	**33 277**	**35 441**	**37 219**	**39 568**	**41 709**
Beaf and veal Boeuf et veau	**4 777**	**4 868**	**5 117**	**5 372**	**5 645**	**5 876**	**6 029**	**6 267**	**6 557**	**6 972**
Pork Porc	**17 362**	**18 031**	**19 704**	**22 017**	**23 454**	**24 188**	**26 090**	**27 393**	**29 244**	**30 883**
Mutton and lamb Mouton et agneau	**2 663**	**2 764**	**2 872**	**2 942**	**3 044**	**3 213**	**3 322**	**3 559**	**3 767**	**3 853**
Afghanistan **Afghanistan**	**229**	**218**	**207**	**195**	**183**	**185**	**191**	**197**	**203**	**203**
Beaf and veal Boeuf et veau	* 69	68[1]	68[1]	68[1]	68	65[1]	65[1]	65[1]	65[1]	65[1]
Mutton and lamb Mouton et agneau	* 160[1]	150[1]	139[1]	127[1]	* 115	120[1]	126[1]	132[1]	138[1]	138[1]
Bahrain **Bahreïn**	**5**	**6**	**6**	**6**	**6**	**7**	**7**	**7**	**7**	**7**
Beaf and veal[1] Boeuf et veau[1]	1	1	1	1	1	1	1	1	1	1
Mutton and lamb[1] Mouton et agneau[1]	4	5	5	5	5	6	6	6	6	6
Bangladesh **Bangladesh**	**156**	**160**	**165**	**188**	**194**	**199**	**205**	**190**	**200**	**188**
Beaf and veal Boeuf et veau	* 126	* 127	125	135	137	138	140	* 141	* 143	145[1]
Mutton and lamb Mouton et agneau	30[1]	33[1]	40	53	57	61	65	* 49[1]	* 58[1]	44[1]

70

Meat
Production: thousand metric tons [*cont.*]
Viande
Production : milliers de tonnes métriques [*suite*]

Country or area Pays ou zone	1982	1983	1984	1985	1986	1987	1988	1989	1990	1991
Bhutan **Bhoutan**	**5**	**6**	**6**	**6**	**6**	**7**	**7**	**7**	**7**	**7**
Beaf and veal[1] Boeuf et veau[1]	4	5	5	5	5	5	5	5	5	6
Pork[1] Porc[1]	1	1	1	1	1	1	1	1	1	1
Brunei Darussalam **Brunéi Darussalam**	**1**	**2**	**1**	**2**	**2**	**2**	**2**	**2**	**2**	**2**
Beaf and veal[1] Boeuf et veau[1]	1	1	0	2	1	1	2	1	1	1
Pork[1] Porc[1]	0	0	1	1	1	0	1	0	0	0
Cambodia **Cambodge**	**31**	**39**	**45**	**52**	**52**	**57**	**60**	**62**	**64**	**66**
Beaf and veal[1] Boeuf et veau[1]	20	21	22	25	25	26	26	28	29	30
Pork[1] Porc[1]	12	18	23	27	27	32	34	34	35	36
China **Chine**	**14 171**	**14 697**	**16 150**	**18 442**	**20 044**	**20 802**	**22 802**	**24 188**	**26 079**	**28 087**
Beaf and veal Boeuf et veau	314[1]	* 335	* 385	* 471	* 593	* 796	* 913	* 1 078	* 1 255	* 1 505
Pork Porc	13 332	13 817	15 179	17 378	18 829	19 287	21 087	22 148	23 756	* 25 460
Mutton and lamb Mouton et agneau	525[1]	* 545	* 586	* 593	* 622	* 719	* 802	* 962	* 1 068	* 1 122[1]
Cyprus **Chypre**	**27**	**29**	**32**	**34**	**35**	**38**	**40**	**42**	**44**	**46**
Beaf and veal Boeuf et veau	2	2	2	3	4	4	4	4	4	4[1]
Pork Porc	20	21	23	24	23	25	27	29	32	33[1]
Mutton and lamb Mouton et agneau	6	6	6	7	8	9	9	8	8	9[1]
Gaza Strip (Palestine) **Zone de Gaza (Palestine)**	**3**	**3**	**3**	**2**	**2**	**2**	**2**	**2**	**3**	**3**
Beaf and veal Boeuf et veau	* 1	* 1	* 1	* 1	* 1	* 1	* 1	* 1	* 1	1[1]
Mutton and lamb Mouton et agneau	* 1	* 1	* 1	* 1	* 1	* 1	* 1	* 1	* 2	2[1]
Hong Kong **Hong-kong**	**224**	**221**	**217**	**221**	**237**	**238**	**232**	**231**	**228**	**230**
Beaf and veal Boeuf et veau	37	37	37	38	40	40	40	37	39	40[1]
Pork Porc	187	184	180	182	197	197	192[1]	194[1]	188[1]	190[1]
India **Inde**	**2 489**	**2 519**	**2 575**	**2 633**	**2 688**	**2 701**	**2 631**	**2 684**	**2 737**	**2 785**
Beaf and veal[1] Boeuf et veau[1]	1 692	1 703	1 740	1 779	1 817	1 804	1 748	1 780	1 805	1 835
Pork[1] Porc[1]	350	352	354	354	354	355	357	359	361	364
Mutton and lamb[1] Mouton et agneau[1]	447	464	481	499	517	542	527	545	572	587
Indonesia **Indonésie**	**406**	**460**	**495**	**518**	**559**	**548**	**524**	**588**	**608**	**616**
Beaf and veal Boeuf et veau	193	204	216	227	239	* 216[1]	* 192	* 227	* 223	243[1]

70
Meat
Production: thousand metric tons [cont.]
Viande
Production : milliers de tonnes métriques [suite]

Country or area Pays ou zone	1982	1983	1984	1985	1986	1987	1988	1989	1990	1991
Pork[1] Porc[1]	143	165	204	215	242	253	256	281	303	275
Mutton and lamb Mouton et agneau	70[1]	92	75[1]	76[1]	78[1]	* 79[1]	* 76[1]	* 81[1]	* 83[1]	98[1]
Iran, Islamic Rep. of **Iran, Rép. islamique d'**	**423**	**431**	**444**	**461**	**483**	**483**	**493**	**526**	**570**	**600**
Beaf and veal[1] Boeuf et veau[1]	178	174	174	177	180	183	190	* 211	* 240	* 260
Mutton and lamb Mouton et agneau	244[1]	257[1]	270[1]	284[1]	303[1]	300[1]	304[1]	* 315	* 330	* 340
Iraq **Iraq**	**81**	**74**	**87**	**81**	**74**	**77**	**77**	**82**	**84**	**65**
Beaf and veal[1] Boeuf et veau[1]	42	36	52	49	46	48	50	51	52	40
Mutton and lamb[1] Mouton et agneau[1]	39	38	35	32	28	29	28	31	32	25
Israel **Israël**	**34**	**31**	**46**	**45**	**45**	**47**	**51**	**50**	**51**	**52**
Beaf and veal Boeuf et veau	20	18	34	32	32	33	35	36	36	* 38
Pork Porc	10	9	8	9	8	9	9	9	9	8[1]
Mutton and lamb * Mouton et agneau *	4	4	4	4	5	5	6	6	6	6
Japan **Japon**	**1 909**	**1 924**	**1 960**	**2 087**	**2 111**	**2 147**	**2 149**	**2 142**	**2 105**	**2 063**
Beaf and veal Boeuf et veau	481	495	536	555	559	565	570	548	549	* 573
Pork Porc	1 428	1 429	1 424	1 532	1 552	1 582	1 579	1 594	1 555	* 1 490
Jordan **Jordanie**	**8**	**10**	**10**	**10**	**6**	**8**	**8**	**9**	**9**	**9**
Beaf and veal Boeuf et veau	1	2	2	1	1	1	1	1	* 1	1[1]
Mutton and lamb Mouton et agneau	7	8	8	9	5	7	7	8	* 8	8[1]
Korea, Dem. P. R. **Corée, R. p. dém. de**	**153**	**167**	**180**	**189**	**195**	**206**	**209**	**211**	**213**	**217**
Beaf and veal[1] Boeuf et veau[1]	32	35	36	38	39	41	42	44	45	53
Pork[1] Porc[1]	118	130	141	148	153	161	164	164	164	161
Mutton and lamb[1] Mouton et agneau[1]	3	3	3	3	3	3	3	4	4	4
Korea, Republic of **Corée, République de**	**398**	**464**	**555**	**603**	**533**	**587**	**604**	**606**	**681**	**661**
Beaf and veal Boeuf et veau	96	94	123	166	208	208	177	125	130	130
Pork Porc	300	369	429	434	322	377	425	480	550	530
Mutton and lamb[1] Mouton et agneau[1]	2	1	3	3	3	3	2	1	1	1
Kuwait **Koweït**	**36**	**35**	**45**	**35**	**34**	**35**	**37**	**37**	**30**	**0**
Beaf and veal Boeuf et veau	10[1]	8[1]	5[1]	2	1	2	3[1]	3[1]	3[1]	0[1]
Mutton and lamb[1] Mouton et agneau[1]	26	* 27	40	* 33	* 33	* 33	34	34	27	0

70
Meat
Production: thousand metric tons [cont.]
Viande
Production : milliers de tonnes métriques [suite]

Country or area Pays ou zone	1982	1983	1984	1985	1986	1987	1988	1989	1990	1991
Lao People's Dem. Rep. **Rép. dém. pop. lao**	**74**	**79**	**82**	**88**	**88**	**90**	**83**	**83**	**85**	**88**
Beaf and veal[1] Boeuf et veau[1]	30	31	33	36	37	36	33	32	31	33
Pork[1] Porc[1]	44	47	49	52	50	54	50	51	53	54
Mutton and lamb[1] Mouton et agneau[1]	0	0	0	0	0	0	0	0	0	1
Lebanon **Liban**	**28**	**29**	**29**	**29**	**30**	**31**	**30**	**31**	**31**	**31**
Beaf and veal[1] Boeuf et veau[1]	14	15	14	14	15	15	14	15	15	15
Pork[1] Porc[1]	1	1	1	1	1	1	1	1	1	1
Mutton and lamb[1] Mouton et agneau[1]	13	13	14	14	15	15	15	16	16	16
Macau **Macao**	**9**	**9**	**9**	**8**	**8**	**8**	**9**	**9**	**9**	**9**
Beaf and veal Boeuf et veau	1	1	1	1[1]	1[1]	1[1]	2[1]	2[1]	2[1]	2[1]
Pork Porc	8	8	8	7	7	6	7	7	8[1]	8[1]
Malaysia **Malaisie**	**155**	**162**	**160**	**170**	**175**	**191**	**180**	**202**	**221**	**212**
Beaf and veal[1] Boeuf et veau[1]	19	17	17	16	15	15	17	15	16	17
Pork[1] Porc[1]	134	144	143	153	159	175	163	186	205	195
Mutton and lamb[1] Mouton et agneau[1]	1	1	1	1	1	1	1	1	1	1
Mongolia **Mongolie**	**199**	**193**	**196**	**187**	**201**	**196**	**190**	**200**	**206**	**194**
Beaf and veal Boeuf et veau	66	68	68	68	75	73	73	73	66	63
Pork Porc	1	2	2	2	2	3	4	6	8	7[1]
Mutton and lamb Mouton et agneau	132	124	126	116	123	121	112	122	* 132	* 125
Myanmar **Myanmar**	**183**	**186**	**188**	**190**	**193**	**196**	**199**	**189**	**193**	**198**
Beaf and veal[1] Boeuf et veau[1]	96	97	98	99	101	103	105	105	106	109
Pork[1] Porc[1]	81	82	83	83	85	86	87	77	80	83
Mutton and lamb[1] Mouton et agneau[1]	6	7	7	8	7	7	7	7	7	7
Nepal **Népal**	**110**	**116**	**121**	**125**	**128**	**130**	**135**	**139**	**140**	**147**
Beaf and veal[1] Boeuf et veau[1]	77	83	86	90	92	93	96	98	98	99
Pork Porc	7	7	7	7	7	8	9	9	10	10[1]
Mutton and lamb Mouton et agneau	27[1]	27[1]	28[1]	28	29	29	31	31	32	38[1]
Oman **Oman**	**9**	**10**	**9**	**10**	**11**	**12**	**11**	**11**	**12**	**13**
Beaf and veal[1] Boeuf et veau[1]	4	4	3	3	3	3	3	3	3	3

70
Meat
Production: thousand metric tons [cont.]
Viande
Production : milliers de tonnes métriques [suite]

Country or area Pays ou zone	1982	1983	1984	1985	1986	1987	1988	1989	1990	1991
Mutton and lamb[1] Mouton et agneau[1]	6	7	6	7	7	9	9	8	9	10
Pakistan **Pakistan**	**711**	**740**	**784**	**831**	**881**	**934**	**1 013**	**1 078**	**1 144**	**1 216**
Beaf and veal Boeuf et veau	407	421	443	466	490	516	568	598	629	662
Mutton and lamb Mouton et agneau	304	319	341	365	391	418	445	480	515	554
Philippines **Philippines**	**495**	**555**	**671**	**503**	**576**	**628**	**705**	**798**	**863**	**866**
Beaf and veal Boeuf et veau	* 87	* 86	* 85	87	96	108	113	* 132	* 125	* 129
Pork Porc	* 392	* 452	569	397	460	500	566	639	712	* 710
Mutton and lamb[1] Mouton et agneau[1]	16	17	17	19	19	20	25	27	27	27
Qatar **Qatar**	**4**	**4**	**4**	**4**	**5**	**5**	**5**	**6**	**6**	**6**
Beaf and veal[1] Boeuf et veau[1]	0	0	0	0	0	0	1	1	1	1
Mutton and lamb[1] Mouton et agneau[1]	3	3	3	4	4	4	5	5	5	5
Saudi Arabia **Arabie saoudite**	**62**	**76**	**84**	**104**	**106**	**102**	**112**	**117**	**124**	**127**
Beaf and veal * Boeuf et veau *	15	21	26	25	25	18	24	25	28	30
Mutton and lamb Mouton et agneau	* 47	* 55	* 59	* 79	* 81[1]	84[1]	88[1]	92[1]	* 96	* 97
Singapore **Singapour**	**63**	**65**	**70**	**74**	**74**	**79**	**75**	**76**	**77**	**79**
Pork Porc	61	63	68	72	72	77	74	75	* 76	78[1]
Mutton and lamb Mouton et agneau	2	2[1]	2	1	1	1	1	1	1[1]	1[1]
Sri Lanka **Sri Lanka**	**35**	**35**	**37**	**36**	**35**	**34**	**30**	**29**	**28**	**28**
Beaf and veal[1] Boeuf et veau[1]	* 30	* 31	* 32	* 32	32	31	27	26	25	25
Pork Porc	* 2	* 1	* 1	1	1	1	1	1	2[1]	2[1]
Mutton and lamb[1] Mouton et agneau[1]	* 3	* 3	* 3	* 3	2	2	2	2	2	2
Syrian Arab Republic **Rép. arabe syrienne**	**128**	**128**	**128**	**118**	**125**	**124**	**126**	**126**	**126**	**126**
Beaf and veal[1] Boeuf et veau[1]	36	34	30	30	30	29	30	29	* 27	28
Mutton and lamb Mouton et agneau	92[1]	94[1]	98[1]	88[1]	95	95[1]	96[1]	97[1]	98	99[1]
Thailand **Thaïlande**	**499**	**438**	**500**	**602**	**503**	**554**	**563**	**563**	**574**	**584**
Beaf and veal[1] Boeuf et veau[1]	218	219	223	224	226	227	228	227	235	242
Pork[1] Porc[1]	280	219	277	378	276	326	334	335	338	340
Mutton and lamb[1] Mouton et agneau[1]	0	1	1	1	1	1	1	1	1	1
Turkey **Turquie**	**576**	**602**	**614**	**622**	**633**	**639**	**646**	**658**	**668**	**676**

70

Meat
Production: thousand metric tons [*cont.*]
Viande
Production : milliers de tonnes métriques [*suite*]

Country or area Pays ou zone	1982	1983	1984	1985	1986	1987	1988	1989	1990	1991
Beaf and veal Boeuf et veau	219	231	238	* 242	* 247	* 257	* 266	* 283	* 298	* 311 [1]
Pork[1] Porc[1]	1	1	1	1	0	0	0	0	0	0
Mutton and lamb * Mouton et agneau *	356	370	375	380	385	382	380	375	370	365
United Arab Emirates **Emirats arabes unis**	**19**	**19**	**25**	**24**	**23**	**30**	**30**	**32**	**29**	**31**
Beaf and veal[1] Boeuf et veau[1]	3	3	3	3	3	3	3	3	3	3
Mutton and lamb[1] Mouton et agneau[1]	16	16	22	22	20	27	27	29	26	27
Viet Nam **Viet Nam**	**568**	**632**	**659**	**697**	**762**	**816**	**861**	**901**	**994**	**1 053**
Beaf and veal[1] Boeuf et veau[1]	116	120	127	135	134	142	196	184	191	199
Pork Porc	451	511	531	560	625	671	662	714	800	850 [1]
Mutton and lamb[1] Mouton et agneau [1]	2	2	2	2	3	3	3	3	3	3
Yemen **Yémen**	**75**	**77**	**79**	**81**	**85**	**87**	**89**	**92**	**95**	**99**
Beaf and veal Boeuf et veau	16	17	18	20	22	24	25	27	28	29
Mutton and lamb Mouton et agneau	59	60	61	62	63	63	64	66	67	70
Europe **Europe**	**30 781**	**31 541**	**32 680**	**32 807**	**33 399**	**34 012**	**34 006**	**33 379**	**34 333**	**33 669**
Beaf and veal Boeuf et veau	**10 230**	**10 452**	**11 151**	**11 161**	**11 262**	**11 291**	**10 634**	**10 343**	**10 947**	**10 969**
Pork Porc	**19 267**	**19 761**	**20 174**	**20 240**	**20 767**	**21 328**	**21 956**	**21 563**	**21 881**	**21 185**
Mutton and lamb Mouton et agneau	**1 283**	**1 327**	**1 356**	**1 405**	**1 370**	**1 393**	**1 416**	**1 473**	**1 505**	**1 514**
Albania **Albanie**	**61**	**62**	**61**	**63**	**63**	**63**	**66**	**65**	**59**	**51**
Beaf and veal Boeuf et veau	26 [1]	* 27	27 [1]	27 [1]	27 [1]	27 [1]	29 [1]	27 [1]	26 [1]	23 [1]
Pork[1] Porc[1]	9	9	9	9	9	9	9	8	8	7
Mutton and lamb[1] Mouton et agneau[1]	26	26	26	26	26	27	28	30	25	21
Austria **Autriche**	**666**	**663**	**695**	**701**	**698**	**705**	**745**	**735**	**669**	**654**
Beaf and veal Boeuf et veau	200	204	224	221	231	230	219	212	224	* 220
Pork Porc	* 463	* 456	* 467	* 476	* 462	* 470	* 522	* 518	* 439	429 [1]
Mutton and lamb Mouton et agneau	* 3 [1]	* 3 [1]	* 4 [1]	* 3 [1]	* 4 [1]	* 5 [1]	5	5	6	5 [1]
Belgium-Luxembourg **Belgique-Luxembourg**	**969**	**1 005**	**1 062**	**1 059**	**1 081**	**1 121**	**1 137**	**1 143**	**1 156**	**1 237**
Beaf and veal Boeuf et veau	282	291	319	326	326	326	317	305	* 323	* 329
Pork Porc	680	707	734	725	747	788	813	830	* 825	900 [1]
Mutton and lamb Mouton et agneau	6	8	8	8	8	7	7	7	8 [1]	8 [1]

70

Meat
Production: thousand metric tons [cont.]
Viande
Production : milliers de tonnes métriques [suite]

Country or area Pays ou zone	1982	1983	1984	1985	1986	1987	1988	1989	1990	1991
Bulgaria **Bulgarie**	**527**	**541**	**549**	**563**	**591**	**581**	**586**	**608**	**590**	**542**
Beaf and veal Boeuf et veau	126	129	130	134	130	129	122	125	121	114
Pork Porc	324	334	335	334	372	372	394	412	406	366
Mutton and lamb Mouton et agneau	77	78	84	94	* 89	80	71	72	63	62
Czechoslovakia **Tchécoslovaquie**	**1 123**	**1 189**	**1 231**	**1 247**	**1 254**	**1 281**	**1 331**	**1 352**	**1 327**	**1 072**
Beaf and veal Boeuf et veau	373	384	406	413	412	412	405	407	404	294
Pork Porc	743	796	814	823	832	858	915	934	913	771
Mutton and lamb Mouton et agneau	8	10	11	11	10	11	11	11	10	7
Denmark **Danemark**	**1 216**	**1 283**	**1 282**	**1 320**	**1 388**	**1 384**	**1 385**	**1 369**	**1 412**	**1 469**
Beaf and veal Boeuf et veau	230	239	247	236	243	235	217	205	202	* 212
Pork Porc	986	1 043	1 035	1 083	1 144	1 149	1 167	1 163	1 208	* 1 255
Mutton and lamb Mouton et agneau	0	0	1	1	1	1	1	1	1	* 2
Faeroe Islands **Iles Féroé**	**1**	**1**	**1**	**1**	**1**	**1**	**1**	**1**	**1**	**1**
Mutton and lamb[1] Mouton et agneau[1]	1	0	0	0	0	1	1	1	1	1
Finland **Finlande**	**302**	**305**	**296**	**300**	**300**	**310**	**290**	**291**	**306**	**300**
Beaf and veal Boeuf et veau	117	118	124	126	125	127	115	110	118	122
Pork Porc	184	185	170	172	174	181	174	180	187	177
Mutton and lamb Mouton et agneau	1	1	1	2	1	1	1	1	1	1
France **France**	**3 615**	**3 675**	**3 860**	**3 753**	**3 770**	**3 874**	**3 871**	**3 702**	**3 799**	**3 931**
Beaf and veal Boeuf et veau	1 745	1 812	1 991	1 893	1 911	1 963	1 828	1 673	1 750	1 934
Pork Porc	1 675	1 676	1 684	1 662	1 677	1 729	1 852	1 844	1 871	1 820
Mutton and lamb Mouton et agneau	195	187	* 185	198	182	182	191	185	178	177[1]
Germany† · Allemagne† **F. R. Germany** **R. f. Allemagne**	**4 645**	**4 735**	**4 864**	**4 845**	**5 058**	**5 075**	**4 980**	**4 769**	**5 187**	**5 388**
Beaf and veal Boeuf et veau	1 477	1 494	1 613	1 576	1 696	1 681	1 609	1 576	1 793	2 024
Pork Porc	3 140	3 211	3 222	3 243	3 336	3 365	3 342	3 161	3 357	3 320
Mutton and lamb Mouton et agneau	27[1]	29[1]	28[1]	27	26	29	29	31	37	44
former German D. R. **anc. R. d. allemande**	**1 540**	**1 573**	**1 665**	**1 772**	**1 794**	**1 802**	**1 712**	**1 705**	**1 432**	**751**
Beaf and veal Boeuf et veau	* 378	* 360	* 369	* 417	* 413	* 418	379	377	319	159

70

Meat
Production: thousand metric tons [cont.]
Viande
Production : milliers de tonnes métriques [suite]

Country or area Pays ou zone	1982	1983	1984	1985	1986	1987	1988	1989	1990	1991
Pork Porc	* 1 148	* 1 201	* 1 281	* 1 338	* 1 364	* 1 365	1 321	1 317	1 100	589
Mutton and lamb Mouton et agneau	* 14	* 12	* 16	* 16	* 17	* 18	12	11	13	3
Greece Grèce	**367**	**366**	**366**	**349**	**352**	**366**	**367**	**364**	**360**	**357**
Beaf and veal Boeuf et veau	99	95	91	84	85[1]	85[1]	82[1]	81[1]	82[1]	* 80[1]
Pork Porc	146	147	148	138	142	150	160	151	148	* 151
Mutton and lamb Mouton et agneau	122	124	127	126	125	132	* 125	* 132	* 130	* 126
Hungary Hongrie	**1 094**	**1 184**	**1 263**	**1 165**	**1 118**	**1 171**	**1 137**	**1 135**	**1 145**	**1 126**
Beaf and veal Boeuf et veau	141	127	132	146	124	129	110	114	* 101	* 92
Pork Porc	947	1 049	1 124	1 011	986	1 037	1 022	1 014	1 040[1]	1 030[1]
Mutton and lamb Mouton et agneau	7	7	8	9	8	5	5	7	4[1]	4[1]
Iceland Islande	**18**	**18**	**17**	**18**	**19**	**18**	**16**	**15**	**16**	**16**
Beaf and veal Boeuf et veau	2	3	3	3	3	3	3	3	3[1]	3[1]
Pork Porc	1	1	1	2	2	2	2	3	3[1]	3[1]
Mutton and lamb Mouton et agneau	14	14	13	13	13	13	11	10	* 10	10[1]
Ireland Irlande	**539**	**554**	**587**	**632**	**693**	**672**	**656**	**639**	**734**	**808**
Beaf and veal Boeuf et veau	345	353	402	448	509	484	459	432	491	* 550
Pork Porc	153	161	144	135	137	141	148	144	* 159	* 168
Mutton and lamb Mouton et agneau	41	40	41	48	46	48	50	63	84	* 90
Italy Italie	**2 283**	**2 382**	**2 471**	**2 462**	**2 420**	**2 476**	**2 507**	**2 520**	**2 583**	**2 577**
Beaf and veal Boeuf et veau	1 107	1 149	1 182	1 205	1 180	1 175	1 165	1 146	1 165	* 1 164[1]
Pork Porc	1 108	1 166	1 218	1 187	1 172	1 231	1 269	1 295	1 333	* 1 330
Mutton and lamb Mouton et agneau	68	67	71	70	67	70	73	79	85	* 83[1]
Malta Malte	**3**	**5**	**7**	**8**	**9**	**9**	**10**	**10**	**10**	**10**
Beaf and veal Boeuf et veau	2	2	2	1	1[1]	2[1]	2	2	2[1]	2[1]
Pork Porc	1	3	5	7	7[1]	8[1]	8	8	8[1]	8[1]
Netherlands Pays-Bas	**1 644**	**1 710**	**1 831**	**1 933**	**1 994**	**2 062**	**2 151**	**2 106**	**2 216**	**2 150**
Beaf and veal Boeuf et veau	420	450	515	511	539	540	507	486	* 540	* 495
Pork Porc	1 211	1 248	1 306	1 412	1 443	1 511	1 631	1 606	1 661	* 1 639
Mutton and lamb Mouton et agneau	13	11	10	11	11	* 11[1]	13[1]	* 14[1]	* 15[1]	* 16[1]

70

Meat
Production: thousand metric tons [*cont.*]
Viande
Production : milliers de tonnes métriques [*suite*]

Country or area Pays ou zone	1982	1983	1984	1985	1986	1987	1988	1989	1990	1991
Norway **Norvège**	**187**	**181**	**180**	**183**	**186**	**196**	**191**	**184**	**190**	**192**
Beaf and veal Boeuf et veau	81	76	71	74	75	78	77	76	83	84
Pork Porc	81	81	84	84	85	92	90	84	83	83
Mutton and lamb Mouton et agneau	24	24	25	25	26	26	25	24	25	26
Poland **Pologne**	**2 154**	**2 056**	**2 012**	**2 236**	**2 511**	**2 495**	**2 544**	**2 513**	**2 609**	**2 597**
Beaf and veal Boeuf et veau	674	642	687	724	752	735	689	637	725	708[1]
Pork Porc	1 462	1 396	1 304	1 486	1 728	1 729	1 828	1 854	1 855	* 1 869
Mutton and lamb Mouton et agneau	18	18	21	26	31	30	26[1]	22[1]	29[1]	* 20[1]
Portugal **Portugal**	**331**	**312**	**308**	**299**	**309**	**339**	**334**	**365**	**386**	**372**
Beaf and veal Boeuf et veau	127	110	99	100	105	105	115	131	115	* 108
Pork Porc	179	176	183	174	178	208	191	211	242	* 236
Mutton and lamb Mouton et agneau	25	26	25	24	26	26	28	23	* 28	* 28
Romania **Roumanie**	**1 205**	**1 302**	**1 396**	**1 244**	**1 299**	**1 296**	**1 209**	**1 024**	**1 173**	**1 090**
Beaf and veal Boeuf et veau	* 229	* 269	* 297	* 255	* 235	* 218	* 208	* 176	* 181[1]	165[1]
Pork Porc	* 908	* 949	* 1 010	* 909	* 985	* 1 002	* 927	* 773	* 920	850[1]
Mutton and lamb * Mouton et agneau *	68	84	89	80	79	76	74	75	72[1]	75[1]
Spain **Espagne**	**1 949**	**1 966**	**2 034**	**1 998**	**2 050**	**2 164**	**2 404**	**2 384**	**2 537**	**2 587**
Beaf and veal Boeuf et veau	420	422	398	401	440	450	450	459	514	* 490
Pork Porc	1 336	1 342	1 429	1 388	1 399	1 489	1 722	1 703	1 789	* 1 850
Mutton and lamb Mouton et agneau	193	202	207	210	211	225	231	222	234	247[1]
Sweden **Suède**	**491**	**484**	**484**	**495**	**462**	**427**	**432**	**450**	**444**	**426**
Beaf and veal Boeuf et veau	161	161	155	158	147	135	127	139	146	* 146
Pork Porc	325	318	324	332	309	288	299	307	293	* 275
Mutton and lamb Mouton et agneau	5	5	5	5	5	5	5	5	5	5[1]
Switzerland **Suisse**	**454**	**447**	**446**	**459**	**461**	**454**	**438**	**441**	**439**	**444**
Beaf and veal Boeuf et veau	160	152	165	170	171	172	154	157	164	172
Pork Porc	289	291	276	285	286	278	279	280	270	266
Mutton and lamb Mouton et agneau	4	4	4	5	5	5	5	5	5	6
United Kingdom **Royaume-Uni**	**2 184**	**2 332**	**2 387**	**2 454**	**2 335**	**2 421**	**2 285**	**2 283**	**2 318**	**2 381**
Beaf and veal Boeuf et veau	965	1 038	1 152	1 179	1 062	1 118	946	978	1 001	1 019

70

Meat
Production: thousand metric tons [*cont.*]
Viande
Production : milliers de tonnes métriques [*suite*]

Country or area Pays ou zone	1982	1983	1984	1985	1986	1987	1988	1989	1990	1991
Pork Porc	955	1 007	947	971	983	1 007	1 017	939	946	980
Mutton and lamb Mouton et agneau	264	287	288	304	290	296	322	366	370	382
Yugoslavia Yougoslavie	**1 213**	**1 213**	**1 327**	**1 248**	**1 187**	**1 252**	**1 224**	**1 202**	**1 237**	**1 141**
Beaf and veal Boeuf et veau	343	345	350	333	317	317	301	309	352	* 260
Pork Porc	* 811	* 808	* 919	* 853	* 807	* 870	* 853	* 824	* 817	815[1]
Mutton and lamb Mouton et agneau	59	60	58	62	63	65	70	69	67	66[1]
Oceania Océanie	**3 545**	**3 595**	**3 260**	**3 404**	**3 425**	**3 656**	**3 764**	**3 623**	**3 743**	**3 939**
Beaf and veal Boeuf et veau	**2 104**	**2 067**	**1 791**	**1 810**	**1 866**	**2 088**	**2 172**	**2 061**	**2 168**	**2 307**
Pork Porc	303	314	332	344	356	365	380	392	400	395
Mutton and lamb Mouton et agneau	**1 137**	**1 214**	**1 137**	**1 250**	**1 203**	**1 203**	**1 211**	**1 170**	**1 175**	**1 237**
Australia Australie	**2 318**	**2 314**	**2 066**	**2 090**	**2 241**	**2 395**	**2 479**	**2 351**	**2 631**	**2 755**
Beaf and veal Boeuf et veau	1 576	1 543	1 345	1 310	1 385	1 521	1 588	1 491	1 677	1 760
Pork Porc	228	239	253	260	271	283	297	308	317	312
Mutton and lamb[1] Mouton et agneau[1]	513	532	468	520	585	591	594	551	637	684
Cook Islands Iles Cook	**0**	**0**	**0**	**1**	**1**	**1**	**0**	**0**	**1**	**1**
Fiji Fidji	5	5	5	5	5	5	5	4	4	4
Beaf and veal Boeuf et veau	3	3	3	3	4	4	4	3	3	2[1]
Pork Porc	1	1	1	1	1	1	1	1	1	1[1]
Mutton and lamb Mouton et agneau	1	1	1	1	1	1	1	1	1	1[1]
French Polynesia Polynésie française	**1**	**1**	**1**	**1**	**1**	**1**	**1**	**1**	**1**	**2**
Pork Porc	1	1	1	1	1	1	1	1	1	1[1]
Guam Guam	**1**	**1**	**1**	**0**	**0**	**0**	**0**	**0**	**0**	**0**
Pork Porc	0	1	1	0	0	0	0	0[1]	0[1]	0[1]
New Caledonia Nouvelle-Calédonie	**3**	**3**	**3**	**4**	**4**	**3**	**3**	**3**	**3**	**3**
Beaf and veal Boeuf et veau	3	2	3	3	3	2	2	2	2	2[1]
Pork Porc	1	1	1	1	1	1	1	1	1	1[1]
New Zealand Nouvelle-Zélande	**1 179**	**1 233**	**1 145**	**1 263**	**1 133**	**1 210**	**1 233**	**1 220**	**1 059**	**1 130**
Beaf and veal Boeuf et veau	516	512	433	487	468	555	572	557	479	535
Pork Porc	40	40	43	48	48	44	46	45	43	43
Mutton and lamb Mouton et agneau	623[1]	681[1]	668[1]	729[1]	617[1]	611	616	617	537	552[1]

70

Meat
Production: thousand metric tons [*cont.*]
Viande
Production : milliers de tonnes métriques [*suite*]

Country or area Pays ou zone	1982	1983	1984	1985	1986	1987	1988	1989	1990	1991
Pacific Islands (Palau)[2] **Iles du Pacifique (Palaos)**[2]	**1**	**1**	**1**	**1**	**1**	**1**	**1**	**1**	**1**	**1**
Pork[1] Porc[1]	1	1	1	1	1	1	1	1	1	1
Papua New Guinea **Papouasie-Nvl-Guinée**	**26**	**26**	**27**	**27**	**28**	**28**	**28**	**29**	**29**	**29**
Beaf and veal[1] Boeuf et veau[1]	2	2	2	2	2	2	2	2	2	2
Pork[1] Porc[1]	24	24	25	25	26	26	26	27	27	27
Samoa **Samoa**	**2**	**2**	**2**	**2**	**2**	**2**	**2**	**2**	**2**	**2**
Beaf and veal Boeuf et veau	1	1	* 1	* 1	1[1]	1[1]	1[1]	1[1]	1[1]	1[1]
Pork Porc	* 1	* 1	* 1	1[1]	1[1]	1[1]	1[1]	1[1]	1[1]	1[1]
Solomon Islands **Iles Salomon**	**2**	**2**	**2**	**2**	**2**	**2**	**2**	**2**	**2**	**3**
Beaf and veal[1] Boeuf et veau[1]	1	1	1	1	1	1	1	1	1	1
Pork[1] Porc[1]	1	1	1	1	2	2	2	2	2	2
Tonga **Tonga**	**2**	**2**	**2**	**2**	**2**	**2**	**2**	**2**	**2**	**2**
Pork[1] Porc[1]	1	1	1	1	1	1	1	2	2	2
Vanuatu **Vanuatu**	**4**	**5**	**5**	**5**	**5**	**5**	**5**	**5**	**5**	**6**
Beaf and veal Boeuf et veau	2	2	2	2	2	3	3	3	3	3
Pork[1] Porc[1]	2	2	2	3	3	2	2	2	2	2
former USSR **ancienne URSS**	**12 707**	**13 608**	**14 037**	**14 050**	**14 781**	**15 480**	**16 211**	**16 500**	**16 400**	**15 300**
Beaf and veal Boeuf et veau	6 618	7 011	7 244	7 370	7 840	8 281	8 616	8 800	8 800	8 200[1]
Pork Porc	5 273	5 760	5 927	5 853	6 047	6 299	6 595	6 700	6 600	6 200[1]
Mutton and lamb Mouton et agneau	* 816	* 837	* 866	* 827	* 894	* 900	* 1 000	* 1 000	* 1 000	900[1]

Source:
Food and Agriculture Organization of the United Nations
(Rome).

Source:
Organisation des Nations Unies pour l'alimentation et
l'agriculture (Rome).

† All data shown which pertain to Germany prior to 3 October
  1990 are indicated separately for the Federal Republic of
  Germany and the former German Democratic Republic based on
  their respective territories at the time indicated. Where
  data for united Germany (3 October 1990 and thereafter) are
  not available, available data are shown separately under the
  designations Federal Republic of Germany and former German
  Democratic Republic and pertain to the territorial
  boundaries prior to 3 October 1990. For detailed
  explanatory notes on data pertaining to Germany, see Annex I
  - Country Nomenclature.

1 FAO estimate.
2 Including data for Federated States of Micronesia, Marshall
  Is. and Northern Mariana Is.

† Toutes les données se rapportant à l'Allemagne avant le 3
  octobre 1990 figurent dans deux rubriques séparées basées
  sur les territoires respectifs de la République fédérale
  d'Allemagne et l'ancienne République démocratique allemande
  selon la période indiquée. En l'absence de données pour
  l'Allemagne unifiée (à compter du 3 octobre 1990), les
  données disponibles sont fournies séparément sous les
  rubriques République fédérale d'Allemagne et ancienne
  République démocratique allemande et se rapportent aux
  limites territoriales antérieures au 3 octobre 1990. Pour
  les notes explicatives en détail sur les données concernant
  l'Allemagne, voir Annexe I – Nomenclature des pays.
1 Estimation de la FAO.
2 Y compris les données pour les Etats fédérés de Micronésie,
  les iles Marshall et les iles Mariannes du Nord.

71

Butter
Beurre

Production: thousand metric tons
Production : milliers de tonnes métriques

Country or area Pays ou zone	1982	1983	1984	1985	1986	1987	1988	1989	1990	1991
World *Monde*	7 039	7 582	7 472	7 625	7 801	7 435	7 406	7 620	7 775	7 450
Africa **Afrique**	153	160	159	162	168	168	166	173	180	178
Algeria[1] Algérie[1]	1	1	1	1	1	1	1	1	1	1
Angola[1] Angola[1]	1	1	1	1	1	1	1	1	1	1
Botswana[1] Botswana[1]	1	1	1	1	1	1	1	1	1	1
Burkina Faso[1] Burkina Faso[1]	1	1	1	1	1	1	1	1	1	1
Egypt[1] Egypte[1]	68	70	73	75	75	77	77	78	79	80
Ethiopia[1] Ethiopie[1]	9	9	9	9	9	9	9	9	10	10
Kenya Kenya	* 3	* 4	* 2	* 3	4	5	* 4	* 4	4[1]	4[1]
Mauritania[1] Mauritanie[1]	1	1	0	1	1	1	1	1	1	1
Morocco[1] Maroc[1]	12	13	12	13	14	16	14	15	15	16
Niger[1] Niger[1]	3	4	3	2	3	3	3	4	3	3
Nigeria[1] Nigéria[1]	8	8	8	8	8	8	8	8	8	8
Rwanda[1] Rwanda[1]	1	1	1	1	1	1	1	1	1	1
Somalia[1] Somalie[1]	10	9	9	9	10	10	10	11	11	10
South Africa Afrique du Sud	15	19	19	16	18	13	12	16	* 21	* 18
Sudan[1] Soudan[1]	13	13	13	13	13	13	13	13	13	14
United Rep.Tanzania[1] Rép. Unie de Tanzanie[1]	3	3	4	4	4	4	4	4	4	4
Zimbabwe[1] Zimbabwe[1]	1	1	2	2	3	3	3	4	4	3
America, North **Amérique du Nord**	740	742	659	712	698	647	705	748	763	789
Canada Canada	126	108	112	100	103	98	109	102	104	101
Costa Rica[1] Costa Rica[1]	4	4	4	4	4	4	4	4	4	4
Cuba Cuba	10	10	11	10	10	9	9	9	* 9	9[1]
Dominican Republic Rép. dominicaine	1	1	1	1	1	2[1]	2[1]	2[1]	2[1]	2[1]

71

Butter
Production: thousand metric tons [cont.]
Beurre
Production : milliers de tonnes métriques [suite]

Country or area Pays ou zone	1982	1983	1984	1985	1986	1987	1988	1989	1990	1991
Guatemala Guatemala	0	0	0	0	0[1]	1[1]	1[1]	1[1]	1[1]	1[1]
Honduras[1] Honduras[1]	4	4	4	4	4	4	4	4	4	4
Mexico Mexique	* 23	* 24	* 25	* 25	29	29	29	29	* 31	* 34
Nicaragua[1] Nicaragua[1]	1	1	1	1	1	1	1	1	1	1
United States Etats-Unis	570	589	500	566	545	501	548	598	608	634
America, South **Amérique du Sud**	**147**	**148**	**141**	**148**	**148**	**154**	**157**	**169**	**160**	**153**
Argentina Argentine	37	34	28	32	32	34	34	* 45	* 40	* 38
Bolivia Bolivie	* 0	1[1]	1[1]	1[1]	1[1]	1[1]	1[1]	1[1]	1[1]	1[1]
Brazil * Brésil *	70	70	73	74	72	77	79	80	75	70
Chile Chili	3	4	4	4	6	5	5	5	6	7[1]
Colombia[1] Colombie[1]	13	13	14	14	14	14	14	15	15	15
Ecuador[1] Equateur[1]	4	4	4	4	4	5	5	5	5	5
Peru Pérou	* 4	* 3	* 2	* 2	* 2	2[1]	2[1]	2[1]	2[1]	2[1]
Uruguay Uruguay	9	13	10	12	12	12	13	13	14	14[1]
Venezuela Venezuela	6	7	5	5	5	5	4	3	* 3	* 3
Asia **Asie**	**1 194**	**1 240**	**1 281**	**1 321**	**1 368**	**1 408**	**1 489**	**1 553**	**1 703**	**1 791**
Afghanistan[1] Afghanistan[1]	12	13	11	11	9	9	10	10	11	11
Bangladesh Bangladesh	14	15	14	13	13	13	13	13[1]	13[1]	13[1]
China[1] Chine[1]	40	41	48	49	53	57	60	63	64	67
India * Inde *	650	670	690	700	720	750	800	840	970	1 040
Iran, Islamic Rep. of[1] Iran, Rép. islamique d'[1]	54	56	59	61	64	63	62	65	66	67
Iraq[1] Iraq[1]	8	8	8	8	8	8	8	8	8	6
Israel Israël	4	4	5	3	5	5	4	6	7	7[1]
Japan Japon	64	74	78	89	88	69	68	78	76	* 70

71

Butter
Production: thousand metric tons [cont.]
Beurre
Production : milliers de tonnes métriques [suite]

Country or area Pays ou zone	1982	1983	1984	1985	1986	1987	1988	1989	1990	1991
Korea, Republic of[1] Corée, République de[1]	14	16	18	21	29	35	43	38	44	48
Mongolia Mongolie	4	4	5	4	5	5	5	5	4	4[1]
Myanmar[1] Myanmar[1]	8	10	12	13	13	13	13	10	10	10
Nepal[1] Népal[1]	11	11	12	12	12	12	14	15	16	16
Pakistan[1] Pakistan[1]	180	184	195	208	221	234	248	264	281	298
Saudi Arabia[1] Arabie saoudite[1]	1	1	1	1	1	1	1	1	1	1
Syrian Arab Republic Rép. arabe syrienne	15	17	11	13	11	14	17[1]	13	12	13[1]
Turkey Turquie	* 110	* 111	* 109	* 111	* 112	* 115	* 118	117	113	114[1]
Yemen Yémen	4	4	4	4	4	4	4	4	5	5
Europe Europe	3 086	3 391	3 251	3 285	3 321	2 956	2 722	2 846	2 803	2 572
Albania[1] Albanie[1]	4	4	4	4	4	4	4	4	3	2
Austria Autriche	43	47	45	44	45	41	42	40	41	* 42
Belgium-Luxembourg Belgique-Luxembourg	101	113	110	106	108	93	79	89	* 87	* 85
Bulgaria Bulgarie	23	24	25	25	24	26	24	22	22[1]	21[1]
Czechoslovakia Tchécoslovaquie	138	149	152	152	156	149	148	157	158	133
Denmark Danemark	121	131	104	110	112	96	94	92	93	* 82
Finland Finlande	70	84	80	72	72	68	61	62	62	59
France France	619	622	597	606	640	571	516	539	* 527	* 500
Germany† · Allemagne† F. R. Germany R. f. Allemagne	556	627	572	515	566	464	392	398	393	420[1]
former German D. R. anc. R. d. allemande	266	291	309	316	322	310	306	313	272	233[1]
Greece Grèce	6	5	5	5	5	5	5	5	5	* 7
Hungary Hongrie	32	33	33	31	33	33	35	38	39	* 29
Iceland Islande	1	1	1	1	1	1	* 2	* 1	* 2	* 2

71
Butter
Production: thousand metric tons [*cont.*]
Beurre
Production : milliers de tonnes métriques [*suite*]

Country or area Pays ou zone	1982	1983	1984	1985	1986	1987	1988	1989	1990	1991
Ireland Irlande	140	164	171	167	153	134	124	139	148	* 151
Italy Italie	76	78	81	80	80	84	82	81	79	76[1]
Netherlands Pays-Bas	216	271	241	229	265	199	170	180	178	165[1]
Norway Norvège	24	27	25	25	25	26	23	23	20	* 21
Poland Pologne	263	294	289	398	* 289	* 293	* 293	* 325	* 315	225[1]
Portugal Portugal	7	5	8	10	11	8	10	12	15	* 17
Romania Roumanie	38	47	49	47	43	38	40	46	* 42	* 43
Spain Espagne	17	13	16	17	29	28	24	31	46	37[1]
Sweden Suède	69	73	78	75	68	64	61	67	72	* 65
Switzerland Suisse	33	35	38	37	37	34	36	39	38	* 37
United Kingdom Royaume-Uni	216	242	206	202	222	176	140	130	138	114
Yugoslavia Yougoslavie	8	11	12	13	11	10	11	12	* 8	* 7
Oceania **Océanie**	**316**	**338**	**393**	**393**	**398**	**346**	**362**	**323**	**365**	**397**
Australia Australie	* 76	* 88	* 111	* 114	* 105	* 104	* 94	* 96	* 106	104
Fiji Fidji	1	0	0	1	1	1	1	1	1	1[1]
New Zealand Nouvelle-Zélande	239	250	282	278	292	241	267	226	258	292
former USSR **ancienne URSS**	**1 403**	**1 563**	**1 588**	**1 605**	**1 700**	**1 755**	**1 805**	**1 808**	**1 802**	**1 570**[1]

Source:
Food and Agriculture Organization of the United Nations
(Rome).

† All data shown which pertain to Germany prior to 3 October
1990 are indicated separately for the Federal Republic of
Germany and the former German Democratic Republic based on
their respective territories at the time indicated. Where
data for united Germany (3 October 1990 and thereafter) are
not available, available data are shown separately under the
designations Federal Republic of Germany and former German
Democratic Republic and pertain to the territorial
boundaries prior to 3 October 1990. For detailed
explanatory notes on data pertaining to Germany, see Annex I
- Country Nomenclature.

1 FAO estimate.

Source:
Organisation des Nations Unies pour l'alimentation et
l'agriculture (Rome).

† Toutes les données se rapportant à l'Allemagne avant le 3
octobre 1990 figurent dans deux rubriques séparées basées
sur les territoires respectifs de la République fédérale
d'Allemagne et l'ancienne République démocratique allemande
selon la période indiquée. En l'absence de données pour
l'Allemagne unifiée (à compter du 3 octobre 1990), les
données disponibles sont fournies séparément sous les
rubriques République fédérale d'Allemagne et ancienne
République démocratique allemande et se rapportent aux
limites territoriales antérieures au 3 octobre 1990. Pour
les notes explicatives en détail sur les données concernant
l'Allemagne, voir Annexe I - Nomenclature des pays.

1 Estimation de la FAO.

72
Cheese
Fromage
Production: thousand metric tons
Production : milliers de tonnes métriques

Country or area Pays ou zone	1982	1983	1984	1985	1986	1987	1988	1989	1990	1991
World *Monde*	12 052	12 391	12 665	13 157	13 395	13 842	14 296	14 440	14 539	14 163
Africa **Afrique**	384	400	414	414	428	440	446	459	469	479
Algeria[1] Algérie[1]	1	1	1	1	1	1	1	1	1	1
Angola[1] Angola[1]	3	3	3	3	3	3	3	3	3	3
Botswana[1] Botswana[1]	1	1	1	1	1	1	1	1	1	1
Egypt[1] Egypte[1]	246	259	275	284	290	296	301	306	311	319
Ethiopia[1] Ethiopie[1]	4	4	4	4	4	4	4	4	4	5
Mauritania[1] Mauritanie[1]	2	2	1	1	2	2	2	2	2	2
Morocco[1] Maroc[1]	6	6	5	6	7	8	7	7	7	8
Niger[1] Niger[1]	16	16	12	8	9	10	10	11	11	11
Nigeria[1] Nigéria[1]	7	6	6	6	6	6	6	6	6	6
South Africa Afrique du Sud	35	36	36	36	34	35	40	40	42	40[1]
Sudan[1] Soudan[1]	58	60	61	53	59	61	57	63	65	69
Tunisia Tunisie	3	2	3	4	4	4	4	5	6	6[1]
United Rep.Tanzania[1] Rép. Unie de Tanzanie[1]	1	1	1	1	1	1	1	1	1	1
Zambia[1] Zambie[1]	1	1	1	1	1	1	1	1	1	1
Zimbabwe[1] Zimbabwe[1]	3	3	4	5	6	7	7	7	7	7
America, North **Amérique du Nord**	2 846	2 985	2 931	3 135	3 210	3 301	3 406	3 393	3 583	3 551
Canada Canada	203	216	223	247	258	278	284	280	286	291
Costa Rica[1] Costa Rica[1]	6	6	6	6	6	6	6	6	6	6
Cuba Cuba	10	11	14	14	15	16	16	16	16[1]	16[1]
Dominican Republic Rép. dominicaine	3	3	3	2	2	3[1]	3[1]	3[1]	3[1]	3[1]
El Salvador[1] El Salvador[1]	2	2	2	2	2	2	2	2	2	2
Guatemala Guatemala	9	10	9	10	* 10	* 11	11[1]	11[1]	11[1]	11[1]

72

Cheese
Production: thousand metric tons [*cont.*]
Fromage
Production : milliers de tonnes métriques [*suite*]

Country or area Pays ou zone	1982	1983	1984	1985	1986	1987	1988	1989	1990	1991
Honduras[1] Honduras[1]	8	8	8	8	8	8	8	8	8	8
Mexico Mexique	101[1]	100[1]	100[1]	97[1]	95	116	113	115	115[1]	115[1]
Nicaragua[1] Nicaragua[1]	5	5	5	5	6	6	6	6	6	6
Panama[2] Panama[2]	...	...	...	...	...	23	20	24	23	24
United States Etats-Unis	2 498	2 621	2 558	2 740	2 803	2 853	2 953	2 943	3 127	3 090
America, South **Amérique du Sud**	**490**	**477**	**474**	**482**	**502**	**536**	**543**	**529**	**563**	**568**
Argentina Argentine	239	248	244	246	258	277	265	* 260	* 270	* 280
Bolivia[1] Bolivie[1]	7	7	7	7	7	7	7	7	7	7
Brazil[1] Brésil[1]	59	59	59	59	59	59	60	60	60	60
Chile Chili	21	18	17	18	25	24	27	29	32	31[1]
Colombia[1] Colombie[1]	47	47	48	50	50	51	51	51	51	51
Ecuador[1] Equateur[1]	12	12	12	12	13	13	13	13	13	13
Peru Pérou	33	25	15	15	16	17[1]	17[1]	17[1]	17[1]	17[1]
Uruguay Uruguay	* 12	* 10	* 11	* 16	* 17	* 16	16	19	16	17[1]
Venezuela Venezuela	60	50	61	59	59	72	88	73	* 96	* 92
Asia **Asie**	**680**	**703**	**703**	**732**	**740**	**765**	**793**	**810**	**813**	**814**
Afghanistan[1] Afghanistan[1]	19	20	18	16	13	14	15	16	18	18
Bangladesh Bangladesh	1	1	1	1	1	1	1	1[1]	1[1]	1[1]
Bhutan[1] Bhoutan[1]	2	2	2	2	2	2	2	2	2	2
China[1] Chine[1]	83	87	95	107	113	123	131	137	145	143
Cyprus Chypre	7	7	7	7	7	8	8	7	8[1]	8[1]
Iran, Islamic Rep. of[1] Iran, Rép. islamique d'[1]	150	159	166	172	179	174	172	181	183	184
Iraq[1] Iraq[1]	33	33	33	33	33	33	33	34	34	27
Israel Israël	60	63	63	65	67	71	74	74	76	76[1]

72

Cheese
Production: thousand metric tons [*cont.*]
Fromage
Production : milliers de tonnes métriques [*suite*]

Country or area Pays ou zone	1982	1983	1984	1985	1986	1987	1988	1989	1990	1991
Japan Japon	67	68	69	68	73	77	82	88	81	85[I]
Jordan[I] Jordanie[I]	2	2	2	2	2	2	2	3	3	3
Lebanon[I] Liban[I]	11	10	10	9	8	7	7	7	7	7
Mongolia[I] Mongolie[I]	1	1	1	1	1	1	1	1	1	1
Myanmar[I] Myanmar[I]	23	28	35	36	36	36	37	27	28	29
Syrian Arab Republic Rép. arabe syrienne	61	61	47	55	46	57	67	69	65	67[I]
Turkey Turquie	139	140	132	135	136	137	139	140	139[I]	141[I]
Yemen Yémen	22	22	22	23	23	23	23	23	23	23
Europe **Europe**	**5 831**	**5 932**	**6 207**	**6 309**	**6 383**	**6 588**	**6 825**	**6 884**	**6 746**	**6 601**
Albania Albanie	12[I]	12[I]	13	13	14[I]	13	14	14	12[I]	11[I]
Austria Autriche	102	101	105	103	98	98	105	114	113	113
Belgium-Luxembourg Belgique-Luxembourg	* 53	* 45	* 47	* 55	* 54	* 59	64[I]	67[I]	69[I]	72[I]
Bulgaria Bulgarie	174	187	187	181	193	187	197	195	192[I]	181[I]
Czechoslovakia Tchécoslovaquie	187	188	193	195	202	216	228	220	205	181
Denmark Danemark	245	251	295	256	254	272	260	277	295	* 290
Finland Finlande	74	72	76	79	83	85	87	90	93	85
France France	1 179	1 233	1 248	1 296	1 297	1 365	1 393	1 427	1 363	1 425
Germany† · Allemagne† F. R. Germany R. f. Allemagne	840	848	879	914	924	956	1 022	1 064	1 134	1 135[I]
former German D. R. anc. R. d. allemande	223	225	237	246	264	264	272	275	208[I]	58[I]
Greece Grèce	196	198	197	198	203	200	208	212	199	207[I]
Hungary Hongrie	75	78	91	91	89	92	91	80	80	81
Iceland Islande	4	4	5	5	5	5	4	5	5	5
Ireland Irlande	59	54	58	83	63	69	77	76	72	75
Italy Italie	637	637	661	667	670	694	714	702	700	692

72

Cheese
Production: thousand metric tons [*cont.*]
Fromage
Production : milliers de tonnes métriques [*suite*]

Country or area Pays ou zone	1982	1983	1984	1985	1986	1987	1988	1989	1990	1991
Netherlands Pays-Bas	484	489	519	518	528	545	553	559	584	614
Norway Norvège	74	64	68	69	73	75	75	83	84	82
Poland Pologne	297	328	366	389	413	451	471	451	333	293
Portugal Portugal	44	45	47	49	53	51	50	54	55	54 [1]
Romania Roumanie	122	101	107	99	97	91	86	94	99	102 [1]
Spain Espagne	127	132	153	155	160	156	163	161	156	155
Sweden Suède	114	115	116	115	113	114	123	117	116	113
Switzerland Suisse	125	130	130	126	131	128	130	133	132	132 [1]
United Kingdom Royaume-Uni	244	245	247	260	263	268	301	280	312	310
Yugoslavia Yougoslavie	137	149	161	147	140	135	138	136	135 [1]	135 [1]
Oceania **Océanie**	**265**	**273**	**270**	**287**	**284**	**316**	**304**	**310**	**300**	**305**
Australia Australie	153	158	161	160	170	177	176	190	175	176
New Zealand Nouvelle-Zélande	112	114	109	127	113	138	128	120	125	* 129
former USSR ancienne URSS	**1 555**	**1 621**	**1 665**	**1 798**	**1 848**	**1 896**	**1 978**	**2 055**	**2 064**	**1 845** [1]

Source:
Food and Agriculture Organization of the United Nations
(Rome).

Source:
Organisation des Nations Unies pour l'alimentation et
l'agriculture (Rome).

† All data shown which pertain to Germany prior to 3 October
1990 are indicated separately for the Federal Republic of
Germany and the former German Democratic Republic based on
their respective territories at the time indicated. Where
data for united Germany (3 October 1990 and thereafter) are
not available, available data are shown separately under the
designations Federal Republic of Germany and former German
Democratic Republic and pertain to the territorial
boundaries prior to 3 October 1990. For detailed
explanatory notes on data pertaining to Germany, see Annex I
- Country Nomenclature.

1 FAO estimate.
2 Excluding former Canal Zone.

† Toutes les données se rapportant à l'Allemagne avant le 3
octobre 1990 figurent dans deux rubriques séparées basées
sur les territoires respectifs de la République fédérale
d'Allemagne et l'ancienne République démocratique allemande
selon la période indiquée. En l'absence de données pour
l'Allemagne unifiée (à compter du 3 octobre 1990), les
données disponibles sont fournies séparément sous les
rubriques République fédérale d'Allemagne et ancienne
République démocratique allemande et se rapportent aux
limites territoriales antérieures au 3 octobre 1990. Pour
les notes explicatives en détail sur les données concernant
l'Allemagne, voir Annexe I – Nomenclature des pays.

1 Estimation de la FAO.
2 Non compris l'ancienne Zone du Canal.

73
Wheat flour and flour of other cereals
Farine de froment et farine d'autres céréales
Production: thousand metric tons
Production : milliers de tonnes métriques

Country or area Pays ou zone	Wheat flour Farine de froment					Cereal flour Farine de céréale				
	1980	1987	1988	1989	1990	1980	1987	1988	1989	1990
Total	**187 238**	**200 861**	**205 358**	**203 234**	**203 411**	**5 826**	**5 289**	**5 277**	**5 242**	**5 157**
Afghanistan Afghanistan	113	203	166	...	...	...	...	...	...	...
Algeria Algérie	643[1]	...	...	...	...	1	...	...	...	...
Angola Angola	* 35[2]	32	...	...	...	...	24	...	...	...
Argentina Argentine	2 438	2 796	2 751	2 882	* 2 818	...	...	...	...	...
Australia Australie	1 045[3]	1 217[3]	1 266[3]	* 1 308[3]	...	160	123[4]	...	...	...
Austria Autriche	209	209	212	216	259	102	91	88	86	72
Bangladesh Bangladesh	...	45	103	52	...	...	...	...	...	...
Barbados Barbade	...	12	12	12	14	...	...	...	...	...
Belarus[1] Bélarus[1]	...	2 943	3 059	2 927	3 091	...	...	...	...	...
Belgium[5] Belgique[5]	859	932	896	1 010	...	...	...	...	...	...
Belize Belize	7	8	9	10	10	...	...	...	...	...
Bolivia[2] Bolivie[2]	225	...	...	...	...	...	...	...	...	...
Brazil Brésil	5 154	5 110	...	...	...	1 188[4]	...	...	...	...
Bulgaria Bulgarie	1 356[1]	1 274[1]	1 254[1]	...	1 293[1]	23	102	26	...	73
Burkina Faso Burkina Faso	5	21	24	13	23	2	...	...	...	...
Burundi Burundi	...	6	6	1	...	...	...	...	...	...
Cameroon Cameroun	50	22	49	...	...	...	...	...	...	...
Canada Canada	1 868	* 1 801	1 861	1 784	...	14[6]	...	...	...	...
Central African Rep. Rép. centrafricaine	1	...	...	...	...	...	...	...	...	...
Chad Tchad	4	...	...	...	...	...	...	...	...	...
Chile Chili	874	1 012	1 024	1 022	...	...	...	...	...	...
China[2] Chine[2]	50 000	...	...	...	...	...	...	...	...	...

73
Wheat flour and flour of other cereals
Production: thousand metric tons [*cont.*]
Farine de froment et farine d'autres céréales
Production : milliers de tonnes métriques [*suite*]

Country or area Pays ou zone	Wheat flour Farine de froment					Cereal flour Farine de céréale				
	1980	1987	1988	1989	1990	1980	1987	1988	1989	1990
Colombia Colombie	329	438	462	...	...	...	113	109	...	...
Congo Congo	4	16	...	...	...	...	...	...	...	...
Costa Rica[2] Costa Rica[2]	80	...	...	...	...	...	...	...	...	...
Côte d'Ivoire Côte d'Ivoire	115	...	...	...	...	...	...	...	...	...
Cuba Cuba	271	454	442	398	...	33[7]	3[7]	2[7]	3[7]	...
Cyprus Chypre	26	* 67	* 72	67	* 72	23	* 33	* 33	12	* 13
Czechoslovakia Tchécoslovaquie	1 236	1 354	427	1 405	1 418	293	287	285	305	289
Denmark Danemark	223	260	* 250	...	...	114	85	* 84	...	...
Dominican Republic Rép. dominicaine	115	150	...	...	...	10	13	...	...	...
Ecuador Equateur	164 [13]	323	329	...	...	9	1	3	...	...
Egypt Egypte	3 138	3 706	3 675	3 629	3 532	...	...	...	...	...
El Salvador El Salvador	82	...	...	...	...	...	...	...	...	...
Ethiopia[8] Ethiopie[8]	175	181	195	...	...	...	...	...	...	...
Fiji Fidji	19	26	26	28	25	...	...	...	...	...
Finland Finlande	189 [9 10]	230 [9 10]	226 [9 10]	233 [9 10]	213 [9 10]	15	108	72	72	84
France[6] France[6]	3 489	* 5 021	* 5 203	* 5 227	...	...	...	...	...	...
Gabon Gabon	17	...	...	...	...	...	...	...	...	...
Germany† · Allemagne† F. R. Germany R. f. Allemagne	2 713	2 339	2 443	2 429	2 617	378	465	436	414	452
former German D. R. anc. R. d. allemande	910	937	941	931	...	...	455	466	442	...
Ghana Ghana	90	72	95	...	...	...	...	...	...	...
Greece Grèce	1 016	...	...	...	...	...	...	...	...	...
Grenada Grenade	...	...	...	8	8	...	...	...	...	...

73
Wheat flour and flour of other cereals
Production: thousand metric tons [cont.]
Farine de froment et farine d'autres céréales
Production : milliers de tonnes métriques [suite]

Country or area Pays ou zone	Wheat flour Farine de froment					Cereal flour Farine de céréale				
	1980	1987	1988	1989	1990	1980	1987	1988	1989	1990
Guadeloupe Guadeloupe	36	38	...	...	...	...	...	...	...	...
Guatemala Guatemala	90	11	77	...	...	...	...	...	...	...
Guyana Guyana	36	31	32	...	...	...	...	...	...	...
Haiti Haïti	117	* 92	...	...	...	...	...	...	...	...
Honduras Honduras	40	...	...	...	...	...	...	...	...	...
Hong Kong Hong-kong	93	87	91	80	...	...	26	17	0	...
Hungary Hongrie	1 197	1 169	1 147	1 185	1 191	28	30	31	45	41
India[11] Inde[11]	2 409	3 427	4 360	4 701	...	...	...	...	...	...
Indonesia Indonésie	998	1 182	...	* 1 032	...	69	...	...	...	...
Iran, Islamic Rep. of[2] Iran, Rép. islamique d'[2]	4 674	...	...	...	...	...	...	...	...	...
Iraq Iraq	1 716[2]	...	5 765	...	...	...	...	...	...	...
Ireland Irlande	221	* 159	* 136	138	...	...	...	...	...	...
Israel Israël	455	529	* 528	521	...	...	...	...	...	...
Italy[2] Italie[2]	4 921	...	...	...	...	...	...	...	...	...
Jamaica Jamaïque	49	137	130	129	...	25	51	31	56	...
Japan Japon	4 184[12]	4 499[12]	4 558[12]	4 582[12]	4 652[12]	98	97	103	104	106
Jordan Jordanie	183	...	...	...	...	...	...	...	...	...
Kenya Kenya	199	* 260	267	189	172	369	198	261	220	241
Korea, Republic of Corée, République de	1 472	1 613	1 692	1 614	1 616	...	...	...	...	...
Kuwait Koweït	156	142	147	153	...	...	...	...	...	...
Lebanon[2] Liban[2]	180	...	...	...	...	...	...	...	...	...
Libyan Arab Jamah. Jamah. arabe libyenne	102	...	...	...	...	...	...	...	...	...
Luxembourg Luxembourg	34	41	...	...	...	0	1	...	...	...

73
Wheat flour and flour of other cereals
Production: thousand metric tons [cont.]
Farine de froment et farine d'autres céréales
Production : milliers de tonnes métriques [suite]

Country or area Pays ou zone	Wheat flour Farine de froment					Cereal flour Farine de céréale				
	1980	1987	1988	1989	1990	1980	1987	1988	1989	1990
Malaysia[13] Malaisie[13]	308	441	500	515	539	...	...	...	...	...
Malta Malte	23	...	...	...	...	...	...	...	...	...
Mexico Mexique	2 147	2 257	2 383	2 487	2 511	434	# 815	885	939	932
Mongolia Mongolie	83	194	196	200	...	...	...	...	...	...
Morocco[12] Maroc[12]	1 503	...	...	...	...	...	...	...	...	...
Mozambique Mozambique	98	89	...	...	...	...	63	56	...	...
Myanmar[14] Myanmar[14]	50	...	...	...	...	...	...	...	...	...
Netherlands[1] Pays-Bas[1]	990[2]	1 046	1 048	1 110	1 135	...	...	...	...	...
New Zealand Nouvelle-Zélande	223	...	...	...	...	...	...	...	...	...
Nicaragua Nicaragua	40	...	...	...	...	...	...	...	...	...
Nigeria Nigéria	1 092[2]	79	...	...	...	...	...	...	...	...
Norway Norvège	237	...	...	...	...	214[4]	...	...	...	...
Panama Panama	45[15]	69[15]	68[15]	...	...	2	3	2	...	...
Paraguay Paraguay	82	115	105	108	127	...	...	...	...	...
Peru Pérou	733	891	824	598	481	...	...	...	...	...
Philippines Philippines	570	789	746	746	...	...	...	...	...	...
Poland Pologne	2 642	2 726	2 898	2 893	1 807	876	780	858	706	576
Portugal Portugal	522[1]	609[1]	564[1]	* 448[1]	...	42	43	14	...	...
Qatar Qatar	28	33[16]	37[16]	38[16]	112[16]	...	...	...	...	...
Romania Roumanie	2 500	2 677	2 861	2 642	2 543	...	...	...	...	...
Senegal Sénégal	76	87	...	...	...	...	...	...	...	...
Sierra Leone[2] Sierra Leone[2]	20	...	...	...	...	...	...	...	...	...
South Africa[17] Afrique du Sud[17]	638	830	959	1 122	1 229	...	...	...	...	...

73
Wheat flour and flour of other cereals
Production: thousand metric tons [*cont.*]
Farine de froment et farine d'autres céréales
Production : milliers de tonnes métriques [*suite*]

Country or area Pays ou zone	Wheat flour Farine de froment					Cereal flour Farine de céréale				
	1980	1987	1988	1989	1990	1980	1987	1988	1989	1990
Spain Espagne	3 127	2 317	2 277	2 455	...	298	585	584	612	...
Sri Lanka Sri Lanka	58	461	478	538	526	28 [18]	...	...	...	...
Sudan Soudan	244	414	437	...	...	...	...	...	...	...
Sweden Suède	459 [9 10]	466 [9 10]	* 461 [9 10]	* 401 [9 10]	...	75	63	...	...	...
Switzerland Suisse	412	427	424	431	451	...	...	...	...	...
Syrian Arab Republic Rép. arabe syrienne	447	1 016	1 221	1 161	1 171	...	...	...	...	...
Thailand Thaïlande	111	...	...	...	...	...	...	...	...	...
Togo Togo	20	58	...	...	...	...	...	...	...	...
Tunisia Tunisie	340	513	* 533	* 560	...	...	...	...	...	...
Turkey Turquie	1 437	1 528	1 616	1 587	...	...	...	...	...	...
Uganda Ouganda	8	9	12	14	13	...	...	...	...	...
Ukraine [1] Ukraine [1]	7 845	7 784	7 534	7 614	7 671	...	...	...	...	...
former USSR [1 2] ancienne URSS [1 2]	43 000	...	...	...	...	...	...	...	...	...
United Kingdom Royaume-Uni	3 702	3 874	3 998	3 930	...	1 [19]	3 [19]	3 [19]	2 [19]	...
United Rep.Tanzania Rép. Unie de Tanzanie	32	22	30	21	...	...	...	...	...	...
United States Etats-Unis	12 821	15 493	15 610	15 547	16 312	73 [20]	71 [20]	...	...	...
Uruguay Uruguay	240	...	164	...	...	...	...	...	...	...
Venezuela Venezuela	437	306	319	302	310	...	...	...	...	...
Yemen [21] Yémen [21]	32	31	...	...	...	...	...	...	...	...
Yugoslavia Yougoslavie	2 293	2 277	2 202	2 321	...	21	16	12	11	...
Zaire Zaïre	...	...	...	...	...	181	...	...	...	...
Zambia Zambie	81	...	* 48	...	...	...	...	...	...	...
Zimbabwe [2] Zimbabwe [2]	120	...	...	...	...	...	...	...	...	...

73

Wheat flour and flour of other cereals
Production: thousand metric tons [*cont.*]

Farine de froment et farine d'autres céréales
Production : milliers de tonnes métriques [*suite*]

Source:
Industrial statistics database of the Statistical Division
of the United Nations Secretariat.

† All data shown which pertain to Germany prior to 3 October
1990 are indicated separately for the Federal Republic of
Germany and the former German Democratic Republic based on
their respective territories at the time indicated. Where
data for united Germany (3 October 1990 and thereafter) are
not available, available data are shown separately under the
designations Federal Republic of Germany and former German
Democratic Republic and pertain to the territorial
boundaries prior to 3 October 1990. For detailed
explanatory notes on data pertaining to Germany, see Annex I
- Country Nomenclature.

1 Including flour produced from other grains.
2 Source: World Wheat Statistics, International Wheat Council,
(London).
3 Twelve months ending 30 June of year stated.

4 Including meal and groats of all cereals.
5 Industrial production only.
6 Shipments.
7 Including corn flour.
8 Twelve months ending 7 July of the year stated.

9 Production for sale only.
10 Including wheat flour mixed with rye flour.
11 Production by large and medium scale establishments only.
12 Twelve months beginning 1 April of year stated.

13 Beginning 1987, scope of series revised.
14 Twelve months ending 30 September of year stated.

15 Including groats.
16 Including bran.
17 Twelve months ending 31 October of year stated.

18 State sector only.
19 Oat flour only.
20 Rye flour only.
21 Former Democratic Yemen only.

Source:
Base de données pour les statistiques industrielles de la
Division de statistique du Secrétariat de l'ONU.

† Toutes les données se rapportant à l'Allemagne avant le 3
octobre 1990 figurent dans deux rubriques séparées basées
sur les territoires respectifs de la République fédérale
d'Allemagne et l'ancienne République démocratique allemande
selon la période indiquée. En l'absence de données pour
l'Allemagne unifiée (à compter du 3 octobre 1990), les
données disponibles sont fournies séparément sous les
rubriques République fédérale d'Allemagne et ancienne
République démocratique allemande et se rapportent aux
limites territoriales antérieures au 3 octobre 1990. Pour
les notes explicatives en détail sur les données concernant
l'Allemagne, voir Annexe I – Nomenclature des pays.
1 Y compris les farines d'autres grains.
2 Source: "World Wheat Statistics, International Wheat
Council," (Londres).
3 Période de douze mois finissant le 30 juin de l'année
indiquée.
4 Y compris les farines et gruaux de toutes céréales.
5 Production industrielle seulement.
6 Expéditions.
7 Y compris la farine de maïs.
8 Période de douze mois finissant le 7 juillet de l'année
indiquée.
9 Production destinée à la vente seulement.
10 Y compris la farine de froment mélangée avec du seigle.
11 Production des grandes et moyennes entreprises seulement.
12 Période de douze mois commençant le 1er avril de l'année
indiquée.
13 A partir de 1987, portée de la série révisée.
14 Période de douze mois finissant le 30 septembre de l'année
indiquée.
15 Y compris le gruau.
16 Y compris le son.
17 Période de douze mois finissant le 31 octobre de l'année
indiquée.
18 Secteur public seulement.
19 Farine d'avoine seulement.
20 Farine de seigle seulement.
21 L'ancien Yémen démocratique seulement.

74

Wine
Vin
Production: thousand metric tons
Production : milliers de tonnes métriques

Country or area Pays ou zone	1982	1983	1984	1985	1986	1987	1988	1989	1990	1991
World *Monde*	37 405	34 842	32 109	29 383	31 980	31 768	27 103	28 381	29 010	27 767
Africa **Afrique**	**1 143**	**1 216**	**1 163**	**1 031**	**955**	**1 071**	**1 051**	**1 081**	**1 070**	**1 097**
Algeria Algérie	151	188	139	94	91	92	62	50	48	50[1]
Egypt Egypte	2	2	2	2[1]	2[1]	2[1]	2[1]	2[1]	2[1]	2[1]
Ethiopia[1] Ethiopie[1]	1	1	1	1	1	1	1	1	1	1
Madagascar[1] Madagascar[1]	7	7	8	8	8	8	8	9	9	9
Morocco Maroc	35	42	35	37	45	45	40	50	30	30[1]
South Africa Afrique du Sud	895	917	908	831	767	880	916	944	* 952	* 963
Tunisia Tunisie	51	58	68	57	40	42	20	23	27	41
Zimbabwe[1] Zimbabwe[1]	2	1	2	1	1	1	1	1	2	2
America, North **Amérique du Nord**	**2 221**	**1 760**	**1 933**	**1 990**	**2 176**	**1 872**	**2 074**	**1 752**	**1 794**	**1 685**
Canada[1] Canada[1]	46	53	61	50	57	55	54	27	46	50
Mexico[1] Mexique[1]	225	231	206	219	192	197	197	174	163	145
United States Etats-Unis	1 950	1 476	1 666	1 720	1 927	1 620	1 823	1 551	1 585	1 490[1]
America, South **Amérique du Sud**	**3 540**	**3 244**	**2 595**	**2 466**	**2 505**	**3 054**	**2 958**	**2 798**	**2 726**	**2 263**
Argentina Argentine	2 498	2 472	1 881	1 574	1 857	2 444	* 2 063	* 2 032	1 910[1]	1 465[1]
Bolivia[1] Bolivie[1]	2	2	2	2	2	2	2	2	2	2
Brazil Brésil	330	188	243	366	225	218	349	274	311	* 311
Chile Chili	610	490	385	438	335	300	* 423	* 390	* 398	390[1]
Paraguay[1] Paraguay[1]	8	8	8	5	5	6	6	6	6	6
Peru Pérou	* 9	* 9	10[1]	10[1]	10[1]	10[1]	10[1]	10[1]	10[1]	10[1]
Uruguay Uruguay	84	75	67	* 71	* 71	* 74	106	85	90	80
Asia **Asie**	**203**	**204**	**206**	**208**	**239**	**254**	**255**	**263**	**254**	**260**
China Chine	47[1]	50[1]	51[1]	62[1]	* 65	75[1]	80[1]	85[1]	90[1]	95[1]
Cyprus Chypre	* 48	* 45	* 44	* 39	* 67	* 72	* 63	* 72	* 60	60[1]

74

Wine
Production: thousand metric tons [cont.]
  Vin
  Production : milliers de tonnes métriques [suite]

Country or area Pays ou zone	1982	1983	1984	1985	1986	1987	1988	1989	1990	1991
Israel Israël	16	17	18	18	17	17	16	15	13	13[1]
Japan Japon	* 67	* 70	* 67	* 62	* 60	* 53	* 61	55[1]	55[1]	55[1]
Lebanon Liban	5	* 5	6[1]	7[1]	8[1]	9	10	11	10[1]	11[1]
Macau[1] Macao[1]	2	2	2	2	2	2	2	2	2	2
Turkey Turquie	18	14	18	17	19	25	23	22	24	24[1]
Europe **Europe**	**26 357**	**24 510**	**22 363**	**20 527**	**24 258**	**23 609**	**18 538**	**20 013**	**21 103**	**20 213**
Albania Albanie	21[1]	24[1]	23[1]	22[1]	19[1]	22[1]	* 25	* 26	22[1]	18[1]
Austria Autriche	491	370	252	113	223	218	350	258	317	300[1]
Belgium-Luxembourg Belgique-Luxembourg	26	19	16	11	17	18[1]	* 14	23	15	15[1]
Bulgaria Bulgarie	574	455	516	386	390	359	340	257	* 293	293[1]
Czechoslovakia Tchécoslovaquie	130	138	156	* 18	140	* 74	142	139	142	134[1]
France France	7 995	6 880	6 415	7 029	7 422	6 944	5 753	6 100	6 553	6 200[1]
Germany† • Allemagne† F. R. Germany R. f. Allemagne	1 613	1 339	888	610	1 092	971	998	1 449	949	1 015
Greece Grèce	462	501	501	480	433	447	435	453	353	450[1]
Hungary Hongrie	677	626	507	289	442	326	471	371	547	547[1]
Italy Italie	7 265	8 328	7 090	6 234	7 709	7 587	6 101	6 033	5 487	5 915
Malta[1] Malte[1]	2	2	2	2	2	2	3	2	3	3
Portugal Portugal	1 021	848	866	989	802	1 085	370	750	1 097	991
Romania Roumanie	1 307	949	1 004	535	1 185	806	642	391	598	600[1]
Spain Espagne	3 743	3 091	3 396	3 238	3 508	3 998	2 213	3 113	4 090	* 3 107
Switzerland Suisse	172	151	103	109	118	110	105	160	* 120	124[1]
United Kingdom Royaume-Uni	1	1	1	1	1	1	1	2[1]	1[1]	2[1]
Yugoslavia Yougoslavie	858	788	629	461	756	642	576	486	517	500[1]
Oceania **Océanie**	**450**	**398**	**438**	**512**	**438**	**439**	**447**	**545**	**494**	**449**
Australia Australie	403	340	396	451	389	401	408	500	445	400

74

Wine
Production: thousand metric tons [*cont.*]
Vin
Production : milliers de tonnes métriques [*suite*]

Country or area Pays ou zone	1982	1983	1984	1985	1986	1987	1988	1989	1990	1991
New Zealand Nouvelle-Zélande	47	58	42	60	49	38	39	45	49	49
former USSR ancienne URSS	3 490	3 510	3 410	2 650	1 408	1 469	1 780	1 930	1 570	1 800[1]

Source:
Food and Agriculture Organization of the United Nations (Rome).

† All data shown which pertain to Germany prior to 3 October 1990 are indicated separately for the Federal Republic of Germany and the former German Democratic Republic based on their respective territories at the time indicated. Where data for united Germany (3 October 1990 and thereafter) are not available, available data are shown separately under the designations Federal Republic of Germany and former German Democratic Republic and pertain to the territorial boundaries prior to 3 October 1990. For detailed explanatory notes on data pertaining to Germany, see Annex I - Country Nomenclature.

1 FAO estimate.

Source:
Organisation des Nations Unies pour l'alimentation et l'agriculture (Rome).

† Toutes les données se rapportant à l'Allemagne avant le 3 octobre 1990 figurent dans deux rubriques séparées basées sur les territoires respectifs de la République fédérale d'Allemagne et l'ancienne République démocratique allemande selon la période indiquée. En l'absence de données pour l'Allemagne unifiée (à compter du 3 octobre 1990), les données disponibles sont fournies séparément sous les rubriques République fédérale d'Allemagne et ancienne République démocratique allemande et se rapportent aux limites territoriales antérieures au 3 octobre 1990. Pour les notes explicatives en détail sur les données concernant l'Allemagne, voir Annexe I – Nomenclature des pays.

1 Estimation de la FAO.

75

Beer

Bière

Production: thousand hectolitres
Production : milliers d'hectolitres

Country or area Pays ou zone	1981	1982	1983	1984	1985	1986	1987	1988	1989	1990
Total	935 092	951 413	961 066	945 252	954 069	973 418	990 171	1 028 784	1 049 185	1 088 742
Algeria Algérie	497	470	468	520	526	503	491	475	365	325
Angola Angola	674	563	690	* 650	653	583	466	...	...	...
Argentina Argentine	2 102	2 225	3 157	4 079	3 827	5 548	5 861	5 232	5 110	3 949
Australia [12] Australie [12]	19 868	19 677	19 724	18 729	18 553	18 627	18 589	18 912	19 508	19 390
Austria Autriche	8 181	8 313	8 403	8 440	8 836	9 017	8 638	8 938	9 174	9 799
Barbados Barbade	77	73	74	58	55	63	73	76	80	91
Belarus Bélarus	3 095	3 206	3 241	3 296	3 360	2 746	3 044	3 093	3 222	3 283
Belgium Belgique	13 811	14 629	14 224	14 311	13 931	13 715	13 988	13 792	13 164	14 141
Belize Belize	34	31	32	24	23	21	22	24	27	32
Bolivia Bolivie	1 141	901	717	701	606	803	...	...	...	...
Botswana Botswana	471	591	648	667	...	...	830	1 002	1 076	1 214
Brazil Brésil	28 928	...	25 861	25 980	27 092	34 005	33 893	36 445	42 343	...
Bulgaria Bulgarie	5 128	5 477	5 505	5 764	5 838	6 023	6 212	6 332	6 720	6 507
Burkina Faso Burkina Faso	748	693	701	701	525	422	389	401	398	350
Burundi Burundi	687	684	644	798	817	897	937	953	919	1 107
Cameroon Cameroun	...	2 859	3 354	3 976	4 904	5 308	5 857	5 105	...	...
Canada Canada	23 085	22 658	23 332	23 558	23 237	23 547	...	...	...	...
Central African Rep. Rép. centrafricaine	203	187	211	218	276	280	299	...	...	...
Chad Tchad	...	125	114	135	152	118	107	109	115	116
Chile Chili	1 904	1 797	1 760	1 781	1 892	2 050	2 548	2 650	2 765	2 653
China [3] Chine [3]	9 100	11 730	16 300	22 400	31 000	41 301	54 043	65 600	64 339	69 221
Colombia Colombie	13 536	13 383	14 494	...	15 509	16 915	15 355	11 973	...	...
Congo Congo	646	691	788	906	882	897	762	744	...	...
Côte d'Ivoire Côte d'Ivoire	1 348	1 285	1 400	1 245	...	...	...	...	...	...
Cuba Cuba	2 243	2 421	2 582	2 607	2 736	2 931	3 288	3 324	3 333	...

75

Beer
Production: thousand hectolitres [cont.]
Bière
Production : milliers d'hectolitres [suite]

Country or area Pays ou zone	1981	1982	1983	1984	1985	1986	1987	1988	1989	1990
Cyprus Chypre	198	210	228	232	247	257	271	296	318	342
Czechoslovakia Tchécoslovaquie	23 934	24 912	24 957	23 768	22 354	22 789	22 228	22 670	23 333	23 527
Denmark Danemark	9 747	10 122	11 038	8 671	8 286	9 064	8 754	* 9 160	...	...
Dominican Republic Rép. dominicaine	1 101	1 119	999	945	1 038	1 099	1 269	...	...	...
Ecuador Equateur	728	769	626	...	...	...	...	...	...	...
Egypt Egypte	459	506	314	360	423	460	470	510	490	500
El Salvador El Salvador	427	420	344	...	...	...	...	...	...	...
Ethiopia[4] Ethiopie[4]	604	616	655	613	818	797	797	877	...	...
Fiji Fidji	182	195	191	185	178	160	148	157	175	194
Finland Finlande	2 818	2 810	2 840	2 968	3 110	3 238	3 509	3 752	3 947	4 151
France France	21 852	22 409	22 086	20 280	19 300	19 000	...	...	...	...
French Polynesia Polynésie française	82	91	96	94	95	118	120	125	121	...
Gabon Gabon	628	696	733	770	850	888	...	...	...	...
Germany† · Allemagne† F. R. Germany R. f. Allemagne	90 857	90 916	91 382	87 309	87 895	88 476	86 962	87 044	87 761	99 150
former German D. R. anc. R. d. allemande	24 091	25 404	25 313	24 500	24 288	24 316	24 128	24 521	24 843	...
Ghana Ghana	527	339	310	452	429	546	589	614	...	...
Greece[3] Grèce[3]	3 100	2 930	2 850	2 970	3 262	3 103	...	...	...	...
Grenada Grenade	12	13	12	12	14	16	17	20	21	25
Guatemala Guatemala	* 789	718	645	672	715	613	795	869	...	...
Guyana Guyana	92	74	45	77	80	80	95	134	...	...
Honduras Honduras	471	397	469	511	* 500	* 548	...	...	...	...
Hungary Hongrie	7 987	7 882	7 830	7 962	8 740	8 963	9 045	9 425	9 722	9 918
Iceland Islande	33	34	30	36	34	31	38	51	...	...
India[5] Inde[5]	1 575	1 924	1 673	1 923	2 030	2 036	1 492	1 635	1 972	...
Indonesia Indonésie	778	901	942	847	797	718	833	...	...	...

75

Beer
Production: thousand hectolitres [cont.]
Bière
Production : milliers d'hectolitres [suite]

Country or area Pays ou zone	1981	1982	1983	1984	1985	1986	1987	1988	1989	1990
Ireland[6] Irlande[6]	* 5 000	* 4 000	4 223	4 214	4 352	4 479	4 360	4 624	5 094	...
Israel Israël	439	402	408	415	511	523	536	556	511	567
Italy Italie	9 016	10 543	10 320	9 201	10 381	11 372	11 503	11 589	10 615	...
Jamaica Jamaïque	563	581	604	517	568	632	700	809	852	...
Japan[7] Japon[7]	45 416	48 389	50 534	45 978	48 522	50 754	54 922	58 572	62 869	65 636
Jordan Jordanie	74	77	63	50	46	45	46	45	46	...
Kenya Kenya	2 283	2 337	2 200	2 280	2 630	3 010	3 070	3 140	3 150	3 311
Korea, Republic of Corée, République de	5 992	6 268	7 096	7 626	7 919	8 040	8 789	10 312	12 108	13 045
Lebanon * Liban *	...	...	...	...	...	...	...	...	...	130
Liberia Libéria	156	123	117	111	97	106	136	158	...	...
Luxembourg Luxembourg	779	751	651	634	738	732	662	635	618	600
Madagascar Madagascar	210	190	236	229	241	255	240	201	232	298
Malawi Malawi	729	657	889	622	658	627	675	659	757	752
Malaysia Malaisie	1 176	1 181	954	998	1 026	* 1 013	988	1 115	1 263	1 401
Mali Mali	10	9	9	...	...	...	...	...	...	...
Mauritius Maurice	132	137	150	166	172	188	238	261	254	...
Mexico Mexique	28 635	28 028	24 139	25 616	27 215	27 353	31 482	33 261	37 355	38 734
Mongolia Mongolie	101	94	91	97	89	65	50	49	67	...
Morocco[8] Maroc[8]	402	376	383	393	411	...	...	...	...	...
Mozambique Mozambique	511	450	445	375	228	230	214	297	...	...
Myanmar[9] Myanmar[9]	36	33	34	40	57	50	17	10	21	...
Nepal[10] Népal[10]	13	20	31	23	60	37	52	63	73	...
Netherlands[11] Pays-Bas[11]	16 639	16 180	17 330	17 050	17 530	17 990	17 550	17 530	18 810	20 160
New Caledonia Nouvelle-Calédonie	42	47	...	...	...	...	...	...	...	...
New Zealand[12] Nouvelle-Zélande[12]	3 826	3 755	3 679	3 839	3 863	4 062	4 087	3 925	3 916	3 890
Nicaragua Nicaragua	480	470	503	522	...	...	...	...	...	...

75

Beer
Production: thousand hectolitres [cont.]
Bière
Production : milliers d'hectolitres [suite]

Country or area Pays ou zone	1981	1982	1983	1984	1985	1986	1987	1988	1989	1990
Niger Niger	86	90	100	...	...	...	...	...	...	...
Nigeria Nigéria	8 185	8 376	7 516	7 355	8 052	10 160	6 695	...	...	...
Norway Norvège	2 131	2 013	1 911	1 944	1 991	...	2 198	2 238	2 229	...
Panama Panama	694	713	708	734	797	925	1 014	871	...	...
Papua New Guinea Papouasie-Nvl-Guinée	469	522	461	* 474	* 550	* 555	...	...	...	...
Paraguay Paraguay	632	696	717	753	768	887	918	903	106	107
Peru Pérou	5 141	5 556	5 230	5 388	5 724	7 477	8 559	7 018	5 548	5 732
Philippines Philippines	5 219	...	9 260	19 030	11 395	8 780	1 175	...	...	...
Poland Pologne	10 514	10 617	10 306	9 867	11 078	11 307	11 897	12 238	12 082	11 294
Portugal Portugal	4 035	3 608	4 265	3 682	3 795	4 128	4 977	5 462	...	...
Puerto Rico [1] Porto Rico [1]	692	546	377	347	260	261	266	350	525	625
Romania Roumanie	9 730	9 911	9 928	9 845	9 847	10 603	10 364	10 655	11 513	10 527
Rwanda Rwanda	599	629	555	576	631	578	596	711	...	...
Saint Kitts and Nevis Saint-Kitts-et-Nevis	12	10	9	9	10	11	14	15	16	...
Sao Tome and Principe * Sao Tomé-et-Principe *	...	...	23	17	31	30	29	...	...	...
Senegal Sénégal	178	176	189	164	168	197	157	170	...	...
Seychelles Seychelles	46	44	39	38	41	42	47	50	52	53
Sierra Leone Sierra Leone	104	62	65	102	60	31	50	40	...	...
South Africa Afrique du Sud	9 203	11 440	11 586	12 400	12 284	13 751	16 710	18 330	18 610	17 750
Spain Espagne	19 296	19 122	20 823	21 464	22 475	23 510	24 788	26 141	27 546	...
Sri Lanka Sri Lanka	52	51	55	61	74	85	92	* 92	60	...
Sudan Soudan	11	...	...	...	...	...	...	...	...	...
Suriname Suriname	142	147	151	150	142	112	124	99	117	...
Sweden Suède	3 684	3 694	3 705	3 616	3 784	3 979	4 106	...	...	...
Switzerland [12] Suisse [12]	4 201	4 205	4 171	4 078	4 076	4 087	4 045	4 049	4 121	4 143

75

Beer
Production: thousand hectolitres [cont.]
Bière
Production : milliers d'hectolitres [suite]

Country or area Pays ou zone	1981	1982	1983	1984	1985	1986	1987	1988	1989	1990
Syrian Arab Republic Rép. arabe syrienne	89	83	81	89	94	101	91	83	95	99
Thailand Thaïlande	1 054	1 217	1 456	1 639	1 052	863	973	1 303	1 801	2 635
Togo Togo	493	482	392	359	423	464	452	...	...	...
Trinidad and Tobago Trinité-et-Tobago	239	367	356	347	274	246	371	349	...	...
Tunisia Tunisie	323	357	423	396	373	355	347	405	...	...
Turkey Turquie	2 932	3 036	3 226	2 622	1 965	1 890	2 440	2 671	3 004	...
Uganda Ouganda	7	98	142	148	81	66	169	215	195	194
Ukraine Ukraine	13 804	14 256	14 627	14 460	14 756	10 926	11 850	12 951	13 749	13 778
former USSR ancienne URSS	62 976	64 673	66 081	65 385	65 721	48 907	50 711	55 811	60 182	62 507
United Kingdom Royaume-Uni	61 716	59 784	60 324	60 105	59 655	59 439	59 895	60 156	54 950	70 800
United Rep.Tanzania Rép. Unie de Tanzanie	643	642	650	670	758	652	588	530	537	...
United States[6] Etats-Unis[6]	227 303	228 007	228 945	228 955	228 960	230 588	229 321	231 985	...	...
Uruguay Uruguay	...	...	...	...	468	678	709	603	...	...
Viet Nam Viet Nam	590	558	527	845	866	872	840	976	892	1 000
Yemen[13] Yémen[13]	3	55	61	66	70	59	50	...	...	...
Yugoslavia Yougoslavie	12 163	13 469	12 398	10 925	10 656	11 643	12 054	11 970	11 286	...
Zaire Zaïre	2 874	2 954	3 069	3 699	4 222	4 284	...	...	...	...
Zambia Zambie	22 006	22 760	21 861	* 4 169	* 717[14]	* 782[14]	714[14]	825[14]	...	...
Zimbabwe Zimbabwe	963	1 074	1 235	920	1 100	...	...	...	...	...

Source:
Industrial statistics database of the Statistical Division
of the United Nations Secretariat.

† All data shown which pertain to Germany prior to 3 October
1990 are indicated separately for the Federal Republic of
Germany and the former German Democratic Republic based on
their respective territories at the time indicated. Where
data for united Germany (3 October 1990 and thereafter) are
not available, available data are shown separately under the
designations Federal Republic of Germany and former German
Democratic Republic and pertain to the territorial

Source:
Base de données pour les statistiques industrielles de la
Division de statistique du Secrétariat de l'ONU.

† Toutes les données se rapportant à l'Allemagne avant le 3
octobre 1990 figurent dans deux rubriques séparées basées
sur les territoires respectifs de la République fédérale
d'Allemagne et l'ancienne République démocratique allemande
selon la période indiquée. En l'absence de données pour
l'Allemagne unifiée (à compter du 3 octobre 1990), les
données disponibles sont fournies séparément sous les
rubriques République fédérale d'Allemagne et ancienne

75

Beer
Production: thousand hectolitres [*cont.*]

Bière
Production : milliers d'hectolitres [*suite*]

boundaries prior to 3 October 1990. For detailed explanatory notes on data pertaining to Germany, see Annex I - Country Nomenclature.	République démocratique allemande et se rapportent aux limites territoriales antérieures au 3 octobre 1990. Pour les notes explicatives en détail sur les données concernant l'Allemagne, voir Annexe I – Nomenclature des pays.
1 Twelve months ending 30 June of year stated.	1 Période de douze mois finissant le 30 juin de l'année indiquée.
2 Excluding light beer containing less than 1.15% by volume of alcohol.	2 Non compris la bière légère contenant moins de 1.15 p. 100 en poids d'alcohol.
3 Original data in metric tons.	3 Données d'origine exprimées en tonnes.
4 Twelve months ending 7 July of the year stated.	4 Période de douze mois finissant le 7 juillet de l'année indiquée.
5 Production by large and medium scale establishments only.	5 Production des grandes et moyennes entreprises seulement.
6 Twelve months ending 30 September of year stated.	6 Période de douze mois finissant le 30 septembre de l'année indiquée.
7 Twelve months beginning 1 April of year stated.	7 Période de douze mois commençant le 1er avril de l'année indiquée.
8 Including some carbonated drinks.	8 Y compris certaines boissons gazeuses.
9 Government production only.	9 Production de l'Etat seulement.
10 Twelve months beginning 16 July of year stated.	10 Période de douze mois commençant le 16 juillet de l'année indiquée.
11 Industrial sales.	11 Ventes industrielles.
12 Sales by breweries.	12 Ventes des brasseries.
13 Former Democratic Yemen only.	13 L'ancien Yémen démocratique seulement.
14 Incomplete coverage.	14 Couverture incomplète.

76
Cigarettes
Cigarettes
Production: millions
Production : millions

Country or area Pays ou zone	1981	1982	1983	1984	1985	1986	1987	1988	1989	1990
Total	4 401 863	4 428 266	4 431 500	4 582 358	4 725 932	4 847 503	4 972 776	5 063 208	5 120 643	5 100 906
Albania Albanie	* 6 200[1]	* 6 200[1]	* 6 100[1]	* 6 200[1]	5 348	* 6 300[1]	* 6 300[1]	* 6 500[1]	* 6 700[1]	6 800[1]
Algeria[2] Algérie[2]	* 16 178	17 542	17 500	18 000	18 500	20 223	17 699	17 016	15 950	18 779
Angola[1] Angola[1]	* 2 400	* 2 400	* 2 400	* 2 400	* 2 400	* 2 400	* 2 400	* 2 400	* 2 400	2 400
Argentina[2] Argentine[2]	29 085	26 930	28 241	30 843	31 092	32 112	29 745	32 000	33 722	33 612
Austria[2] Autriche[2]	14 671	15 538	15 625	14 907	16 051	15 354	15 067	14 324	14 402	14 961
Bangladesh[3] Bangladesh[3]	14 906	15 778	14 031	14 843	14 393	14 365	14 762	14 031	14 088	14 000
Barbados[2] Barbade[2]	260	271	241	238	200	199	162	149	143	135
Belarus Bélarus	19 218	19 254	19 285	19 525	19 843	19 902	19 835	15 959	15 953	16 399
Belgium[4] Belgique[4]	28 714	30 461	29 738	29 422	30 184	28 619	28 953	28 683	27 489	25 600
Belize Belize	68	56	57	65	74	76	99	94	97	100
Bolivia[1] Bolivie[1]	1 122	624	628	368	533	899	900	1 200	1 200	1 200
Brazil Brésil	137 997	132 300[1]	129 200[1]	127 800[1]	146 300[1]	168 000[1]	161 400[1]	157 900[1]	162 700[1]	174 000[1]
Bulgaria[5] Bulgarie[5]	88 593	88 132	91 296	92 055	93 975	89 918	90 298	89 215	85 800	75 812
Burkina Faso Burkina Faso	565	708	700	660	671	510	430	533	609	822
Burundi Burundi	260	235	294	334	293	257	284	333	286	384
Cambodia[1] Cambodge[1]	* 4 100	* 4 100	* 4 100	* 4 100	* 4 150	* 4 175	* 4 175	* 4 200	* 4 200	4 200
Cameroon Cameroun	* 1 500	* 1 550	1 954	2 319	2 128	* 3 800	* 4 280	* 4 300[1]	* 4 300[1]	4 300[1]
Canada Canada	68 611	68 232	63 900	61 600	63 486	55 632	54 030	53 858[1]	48 792[1]	46 000[1]
Central African Rep. Rép. centrafricaine	435	431	403	413	539	407	460	...	...	...
Chad Tchad	...	...	...	14 902	15 456	11 268	9 921	10 202	9 350	12 378
Chile Chili	9 201	7 669	7 680	8 107	8 053	8 296	8 183	9 061	9 686	10 011
China[12] Chine[12]	866 000	942 000	968 800	1 066 000	1 180 000	1 296 500	1 440 500	1 545 000	1 597 800	1 648 765
Colombia[1] Colombie[1]	19 255	17 907	19 141	23 400	24 050	24 181	21 987	18 253	18 300	17 500
Congo Congo	830	801	895	903	1 027	888	787	770	1 000	1 000[2]

76
Cigarettes
Production: millions [cont.]
Cigarettes
Production : millions [suite]

Country or area Pays ou zone	1981	1982	1983	1984	1985	1986	1987	1988	1989	1990
Costa Rica[1] Costa Rica[1]	2 320	1 976	2 200	2 200	2 200	2 000	2 135	2 135	2 050	2 030
Côte d'Ivoire Côte d'Ivoire	3 700	3 500	3 640	3 210	4 000	* 4 200	* 4 400	* 4 500	* 4 500	4 500
Cuba Cuba	15 409	17 044	16 802	18 697	17 961	16 841	15 398	16 885	16 520	...
Cyprus Chypre	3 320	3 353	2 878	2 586	2 366	1 914	3 717	4 412	3 935	4 601
Czechoslovakia Tchécoslovaquie	23 734	24 006	25 016	24 603	23 840	24 998	25 365	25 502	25 428	26 708
Denmark Danemark	9 660	10 353	9 846	10 583	10 966	11 246	11 162	* 1 144	* 1 170	11 170
Dominica Dominique	30	...	27	27	29	29	...	...	...	...
Dominican Republic Rép. dominicaine	3 439	3 612	3 604	3 696	3 826	4 164	4 473	4 802[1]	4 473[1]	4 692[1]
Ecuador Equateur	4 227	5 149	4 980	5 000	4 800	3 967	* 4 600	* 4 600	* 4 600	4 600
Egypt Egypte	33 249	36 008	35 571	44 599	47 520	44 394	47 500	43 846	43 208	39 837
El Salvador El Salvador	2 328	2 291	2 128	* 2 500	* 2 300	* 2 100	* 2 100	1 967[1]	1 970[1]	1 961[1]
Ethiopia[6] Ethiopie[6]	1 575	1 741	1 946	2 036	2 229	2 619	3 040	2 972	* 2 300[1]	2 300
Fiji Fidji	566	536	533	513	513	478	466	466	505	531
Finland Finlande	12 710	8 491	8 375	8 345	8 185	8 540	9 061	9 619	8 931	8 974
France France	62 454	62 510	62 147	60 729	67 376	59 122	54 120	* 53 307	* 54 225	53 000
Gabon Gabon	262	288	328	317	355	268	...	...	...	...
Germany† • Allemagne† F. R. Germany R. f. Allemagne	163 979	146 713	155 883	160 680	163 267	166 665	157 586	159 499	159 477	177 905
former German D. R. anc. R. d. allemande	26 004	25 605	27 387	28 018	26 909	27 364	27 625	28 576	28 625	28 300
Ghana Ghana	1 611	1 209	1 074	2 008	1 942	1 826	1 734	1 831	2 100[2]	2 100
Greece Grèce	22 861[2]	17 660	24 286	25 699	27 635	27 118	28 900	* 28 780	* 28 553	28 450
Grenada Grenade	20	21	26	25	24	23	22	23	24	22
Guatemala Guatemala	2 162	2 273	2 156	1 968	1 988	1 926	2 270	1 858[1]	1 997[1]	1 997[1]
Guyana Guyana	602	430	408	373	467	481	477	470	600	600
Haiti Haïti	1 061	921	909	886	806	829	903	870[1]	870[1]	870[1]
Honduras Honduras	2 160	2 279	2 022	2 145	2 311	* 2 300	* 2 200	2 319[1]	2 582[1]	2 788[1]

76
Cigarettes
Production: millions [cont.]
Cigarettes
Production : millions [suite]

Country or area Pays ou zone	1981	1982	1983	1984	1985	1986	1987	1988	1989	1990
Hong Kong[4] Hong-kong[4]	5 250	5 116	6 577	8 681	12 496	11 841	15 309	24 000	20 901	21 700
Hungary Hongrie	27 102	26 135	25 676	26 679	26 430	26 351	26 541	26 171	26 540	28 212
India[7] Inde[7]	88 283	97 391	86 900	86 217	80 600	74 991	64 782	54 013	58 066	85 000
Indonesia Indonésie	85 274	85 068	91 463	95 634	102 300	114 312	124 432	136 271	148 000	155 000
Iran, Islamic Rep. of[8] Iran, Rép. islamique d'[8]	12 549	13 546	15 104	16 154	16 168	15 239	15 068	13 790	9 923	15 000
Iraq[1] Iraq[1]	7 725	7 000	7 900	7 000	7 000	10 000	13 000	20 250	27 000	27 000
Ireland Irlande	8 823	8 136	7 531	7 391	8 037	6 825	6 659	6 420	6 161	7 850
Israel Israël	5 985	6 043	6 373	6 714	6 709	6 723	6 888	5 882	5 245	5 440
Italy[2] Italie[2]	72 168	80 551	83 700	80 452	78 774	75 541	70 447	67 394	67 942	65 000
Jamaica Jamaïque	1 246	1 414	1 359	1 270	1 314	* 1 140	1 273	1 303	1 383	1 273
Japan[9] Japon[9]	306 624	309 063	306 320	306 867	303 000	309 200	273 700	267 000	268 400	268 100
Jordan Jordanie	4 711	4 614	4 535	5 027	3 905	3 732	4 378	3 678	2 926	4 100
Kenya Kenya	4 972	4 904	5 584	5 391	5 409	5 821	6 372	6 641	6 661	6 648
Korea, Republic of Corée, République de	72 139	71 528	75 275	77 984	75 532	78 694	81 816	86 014	86 259	91 923
Lao People's Dem. Rep.[1] Rép. dém. pop. lao[1]	1 100	1 100	1 100	1 100	1 125	1 200	1 200	1 200	1 200	1 200
Lebanon[1] Liban[1]	280	200	200	200	1 800	* 1 800	* 1 800	* 1 800	* 1 800	1 800
Liberia Libéria	18	20	59	123	82	92	22[1]	22[1]	22[1]	22[1]
Libyan Arab Jamah. Jamah. arabe libyenne	2 306	2 924[1]	3 400[1]	3 500[1]	3 500[1]	3 500[1]	3 500[1]	3 500[1]	3 500[1]	3 500[1]
Madagascar[2] Madagascar[2]	1 867	2 065	1 780	2 137	2 368	2 188	2 669	1 817	2 341	1 955
Malawi Malawi	646	743	778	824	857	874	908	956	1 080[1]	1 061
Malaysia[2] Malaisie[2]	13 744	14 054[10]	13 502[10]	14 671[10]	13 839[10]	13 706[10]	13 729[10]	15 904[10]	16 169[10]	17 331[10]
Malta Malte	1 209	1 307	1 128	1 129	1 300	* 1 300	* 1 400	* 1 425	* 1 450	1 475
Mauritius[2] Maurice[2]	1 117	1 010	982	1 253	1 418	1 266	1 392	1 304	1 318	1 400
Mexico Mexique	54 660	54 979	49 337	51 666	54 332	49 898	54 644	50 507	53 920	55 380

76
Cigarettes
Production: millions [cont.]
Cigarettes
Production : millions [suite]

Country or area Pays ou zone	1981	1982	1983	1984	1985	1986	1987	1988	1989	1990
Morocco Maroc	12 441	11 493	# 586	548	593	663	671	684	615	640
Mozambique Mozambique	1 300	1 081	889	792	710	1 053	899	670	1 060[1]	...
Myanmar[11] Myanmar[11]	2 792	3 117	2 949	2 515	3 506	1 574	553	952	505	...
Nepal Népal	2 835	3 209	3 741	4 252	4 741	5 600	6 046	5 664	6 706	...
Netherlands Pays-Bas	39 189	43 772	46 337	47 761	51 321	53 339	59 801	61 693	68 824	78 052
New Zealand Nouvelle-Zélande	6 191	6 334	6 196	6 274	5 767	5 471	5 400	* 6 250	* 6 250	6 250
Nicaragua Nicaragua	2 117[1]	2 200[1]	1 919	2 318	* 2 400[1]	* 2 400[1]	* 2 400[1]	* 2 400[1]	* 2 400[1]	2 400[1]
Nigeria[1] Nigéria[1]	* 8 640	* 9 000	* 9 800	* 9 000	* 9 100	* 9 600	* 10 000	* 10 000	* 10 000	10 000
Norway[1] Norvège[1]	740	* 740	* 792	* 790	* 1 082	* 1 400	* 1 400	* 1 655	* 1 700	1 740
Pakistan[3] Pakistan[3]	35 891	38 132	38 199	40 096	38 921	39 593	39 929	40 697	31 567	32 279
Panama Panama	1 050	1 001	981	911	873	873	826	671	1 150	1 150
Paraguay Paraguay	757	758	931	878	834	750	1 134	2 730[1]	2 730[1]	2 730[1]
Peru Pérou	3 965	3 996	3 582	3 489	3 102	3 741	3 420	2 671	2 439	2 672
Philippines[3] Philippines[3]	55 050	70 565	57 812	58 600	62 300	60 700	61 072	66 850[1]	69 700[1]	71 500[1]
Poland Pologne	83 010	87 460	82 823	86 363	90 021	94 212	98 666	89 681	81 342	91 497
Portugal[2] Portugal[2]	13 605	15 016	15 531	14 798	14 900	15 166	15 481	15 592	* 15 170	15 220
Romania[5] Roumanie[5]	35 000[1]	36 000[1]	36 000[1]	36 000[1]	32 477	31 013	32 918	33 349	22 121	18 090
Rwanda Rwanda	...	...	812	696	698	649	698	458	...	...
Senegal[2] Sénégal[2]	2 400	2 176	2 516	2 501	2 979	3 300	1 872	1 348	3 350	3 350
Seychelles Seychelles	45	52	52	65	56	60	68	61	58	67
Sierra Leone Sierra Leone	1 774	1 156	966	1 341	1 189	940	1 100	933	1 200[1]	1 200
Singapore[2] Singapour[2]	3 203	3 792	3 165	* 3 000	* 2 400	* 2 000	* 2 340	* 4 700	* 5 982	6 880
South Africa Afrique du Sud	30 025	32 463	31 101	32 501	31 704	30 680	17 773	28 002	36 665	40 792
Spain Espagne	67 135	65 269	77 396	83 884	80 495	81 705	87 362	85 612	75 817	79 500
Sri Lanka Sri Lanka	* 5 200	5 802	5 858	5 200	6 168	6 111	5 894	5 028	5 136	5 621

76
Cigarettes
Production: millions [cont.]
Cigarettes
Production : millions [suite]

Country or area Pays ou zone	1981	1982	1983	1984	1985	1986	1987	1988	1989	1990
Sudan Soudan	1 100 [2]	759 [2]	1 354 [2]	* 1 884 [2]	2 700	2 200	2 200	1 800	750 [1]	750
Suriname Suriname	402	444	460	521	514	557	492	501	526	380
Sweden Suède	10 269	11 092	10 423	10 420	10 090	10 197	10 103	* 10 389	* 10 170	9 970
Switzerland Suisse	27 569	26 497	25 681	25 449	23 150	23 555	24 863	27 075	28 059	31 771
Syrian Arab Republic[2] Rép. arabe syrienne[2]	9 939	10 880	13 211	11 274	12 127	12 112	9 143	7 056	6 345	6 855
Thailand Thaïlande	32 803	26 867	28 941	29 170	29 192	29 530	31 407	33 992	37 365	38 316
Trinidad and Tobago[2] Trinité-et-Tobago[2]	850	1 087	1 148	1 064	987	920	583	766	...	1 250
Tunisia Tunisie	3 965	6 016	5 900	5 527	6 334	8 097	6 937	7 125	7 339 [1]	7 530
Turkey Turquie	71 697	62 254	61 497	62 085	62 000	59 740	58 617	60 153	60 891	62 400
Uganda Ouganda	233	745	645	966	1 416	1 420	1 435	1 638	1 586	1 290
Ukraine Ukraine	78 602	77 858	79 716	81 084	82 020	82 104	82 389	81 659	78 446	69 397
former USSR ancienne URSS	365 218	359 440	369 193	373 793	381 274	383 878	378 475	358 218	343 288	313 082
United Kingdom[2 12] Royaume-Uni[2 12]	108 300	100 600	93 400	93 400	88 600	83 300	89 900	87 900	104 155	112 000
United Rep.Tanzania Rép. Unie de Tanzanie	3 865	4 693	3 800	4 000	2 666	2 748	2 635	2 785	2 845	...
United States[3] Etats-Unis[3]	736 500	694 200	667 000	668 800	655 300	652 000	677 971	676 331	677 200	701 272
Uruguay Uruguay	* 4 242 [1]	* 3 700 [1]	* 3 750 [1]	* 3 800 [1]	3 098	3 583	3 383	3 446	3 900	3 900
Venezuela[1] Venezuela[1]	19 800	19 487	20 156	20 643	19 760	* 18 400	* 18 100	18 824	18 035	17 675
Viet Nam Viet Nam	10 906	13 886	18 476	21 236	21 012	22 364	19 630	17 754	23 288	24 990
Yemen[13] Yémen[13]	1 150	1 142	1 204	1 202	1 467	1 147	1 193	...	...	...
Yugoslavia Yougoslavie	63 096	58 941	58 493	55 965	57 652	55 787	56 128	60 169	51 287	58 200
Zaire[14] Zaïre[14]	3 130	3 282	3 472	3 500	3 525	3 600	3 600	* 5 200	* 5 200	5 200
Zambia Zambie	1 239	* 1 350	* 1 400	* 1 400	* 1 400	* 1 450	* 1 500	* 1 500	* 1 500	1 500
Zimbabwe Zimbabwe	3 750 [1]	3 050 [1]	2 400 [1]	2 300 [1]	2 318 [1]	* 2 300	* 2 400	* 2 500	* 2 623	2 500

76

Cigarettes
Production: millions [*cont.*]

Cigarettes
Production : millions [*suite*]

Source:
Industrial statistics database of the Statistical Division
of the United Nations Secretariat.

† All data shown which pertain to Germany prior to 3 October
 1990 are indicated separately for the Federal Republic of
 Germany and the former German Democratic Republic based on
 their respective territories at the time indicated. Where
 data for united Germany (3 October 1990 and thereafter) are
 not available, available data are shown separately under the
 designations Federal Republic of Germany and former German
 Democratic Republic and pertain to the territorial
 boundaries prior to 3 October 1990. For detailed
 explanatory notes on data pertaining to Germany, see Annex I
 - Country Nomenclature.

1 Source: US Department of Agriculture, (Washington, DC).
2 Original data in units of weight. Computed on the basis of
 one million cigarettes per ton.
3 Twelve months ending 30 June of year stated.

4 Including cigarillos.
5 Including cigars.
6 Twelve months ending 7 July of the year stated.

7 Production by large and medium scale establishments only.
8 Production by establishments employing 50 or more persons.
9 Twelve months beginning 1 April of year stated.

10 Beginning 1982, scope of series revised.
11 Government production only.
12 Sales by manufacturers employing 25 or more persons.
13 Former Democratic Yemen only.
14 Including cut tobacco.

Source:
Base de données pour les statistiques industrielles de la
Division de statistique du Secrétariat de l'ONU.

† Toutes les données se rapportant à l'Allemagne avant le 3
 octobre 1990 figurent dans deux rubriques séparées basées
 sur les territoires respectifs de la République fédérale
 d'Allemagne et l'ancienne République démocratique allemande
 selon la période indiquée. En l'absence de données pour
 l'Allemagne unifiée (à compter du 3 octobre 1990), les
 données disponibles sont fournies séparément sous les
 rubriques République fédérale d'Allemagne et ancienne
 République démocratique allemande et se rapportent aux
 limites territoriales antérieures au 3 octobre 1990. Pour
 les notes explicatives en détail sur les données concernant
 l'Allemagne, voir Annexe I – Nomenclature des pays.

1 Source: "US Department of Agriculture," (Washington, DC).
2 Données d'origine exprimées en poids. Calcul sur la base
 d'un million cigarettes par tonne.
3 Période de douze mois finissant le 30 juin de l'année
 indiquée.
4 Y compris les cigarillos.
5 Y compris les cigares.
6 Période de douze mois finissant le 7 juillet de l'année
 indiquée.
7 Production des grandes et moyennes entreprises seulement.
8 Production des établissements occupant 50 personnes ou plus.
9 Période de douze mois commençant le 1er avril de l'année
 indiquée.
10 A partir de 1982, portée de la série révisée.
11 Production de l'Etat seulement.
12 Ventes des fabricants employant 25 personnes ou plus.
13 L'ancien Yémen démocratique seulement.
14 Y compris le tabac haché.

77
Cotton
Coton

Industrial consumption: thousand metric tons
Consommation industrielle : milliers de tonnes métriques

Country or area Pays ou zone	1982/83	1983/84	1984/85	1985/86	1986/87	1987/88	1988/89	1989/90	1990/91	1991/92
World *Monde*	**14 482**	**14 665**	**15 112**	**16 568**	**18 252**	**18 178**	**18 529**	**18 798**	**18 634**	**18 688**
Afghanistan Afghanistan	11	11	11	8	11	11	11	11	17	17
Albania Albanie	9	9	...	...	...	...	...	...	...	...
Algeria Algérie	29	33	26	22	24	26	30	26	33	34
Angola Angola	2	2	2	2	3	3	3	3	3	3
Argentina Argentine	102	112	101	117	131	125	133	133	137	147
Armenia Arménie	...	...	...	...	26	25	26	6	6	5
Australia Australie	17	21	20	21	22	22	22	22	25	28
Austria Autriche	21	21	20	20	23	25	23	24	25	32
Azerbaijan Azerbaijan	...	...	...	...	33	31	32	30	30	30
Bangladesh Bangladesh	49	50	50	50	51	52	87	93	98	104
Belarus Bélarus	...	...	...	...	35	33	34	34	33	32
Belgium Belgique	40	35	40	42	54	46	43	46	40	39
Benin Bénin	1	1	1	1	1	1	1	1	1	1
Bolivia Bolivie	6	7	11	11	11	11	11	11	10	10
Brazil Brésil	567	556	599	692	759	811	822	764	723	719
Bulgaria Bulgarie	65	65	61	57	61	66	65	65	39	26
Burkina Faso Burkina Faso	1	1	1	1	1	1	1	1	1	1
Burundi Burundi	0	1	1	2	2	2	3	3	2	3
Cameroon Cameroun	7	7	8	7	6	6	5	5	5	5
Canada Canada	55	55	48	54	43	44	41	45	41	42
Central African Rep. Rép. centrafricaine	0	0	0	1	1	1	1	1	1	1
Chad Tchad	2	2	2	2	2	5	3	3	3	3
Chile Chili	11	11	16	22	24	25	26	24	24	25
China Chine	3 663	3 400	3 484	4 117	4 567	4 369	4 376	4 150	4 225	4 351

77

Cotton
Industrial consumption: thousand metric tons [*cont.*]
Coton
Consommation industrielle : milliers de tonnes métriques [*suite*]

Country or area Pays ou zone	1982/83	1983/84	1984/85	1985/86	1986/87	1987/88	1988/89	1989/90	1990/91	1991/92
Colombia Colombie	23	52	72	64	78	92	90	93	103	110
Costa Rica Costa Rica	2	2	2	2	2	2	3	3	3	3
Côte d'Ivoire Côte d'Ivoire	18	18	18	20	21	14	14	15	15	15
Cuba Cuba	40	44	44	44	49	45	46	40	39	33
Czechoslovakia Tchécoslovaquie	120	125	122	121	122	121	123	123	97	63
Denmark Danemark	2	2	2	2	2	2	2	2	2	2
Dominican Republic Rép. dominicaine	1	1	1	1	1	2	1	2	1	1
Ecuador Equateur	11	11	13	11	13	13	17	17	18	19
Egypt Egypte	266	275	297	316	302	306	288	307	315	327
El Salvador El Salvador	14	14	13	14	14	15	15	18	18	19
Estonia Estonie	...	...	...	...	49	47	48	46	44	30
Ethiopia Ethiopie	22	22	22	23	22	20	21	22	21	17
Finland Finlande	10	10	9	8	4	7	4	4	2	2
France France	167	162	159	153	157	152	141	124	113	100
Germany† · Allemagne† F. R. Germany R. f. Allemagne	202	215	218	213	243	230	202	206	223	178
former German D. R. anc. R. d. allemande	105	100	98	87	85	89	89	72	...	...
Georgia Géorgie	...	...	...	...	25	24	24	21	19	15
Ghana Ghana	7	5	4	7	9	11	9	9	9	9
Greece Grèce	148	144	160	165	178	170	165	170	180	180
Guatemala Guatemala	7	10	10	11	13	15	16	17	21	27
Guinea Guinée	2	2	...	...	...	...	...	...	...	...
Guinea-Bissau Guinée-Bissau	0	0	...	...	...	...	...	...	...	...
Haiti Haïti	2	2	...	...	...	...	...	...	...	...
Honduras Honduras	2	2	2	2	2	2	2	2	2	2
Hong Kong Hong-kong	164	164	156	183	234	232	244	226	191	200

77
Cotton
Industrial consumption: thousand metric tons [cont.]
Coton
Consommation industrielle : milliers de tonnes métriques [suite]

Country or area Pays ou zone	1982/83	1983/84	1984/85	1985/86	1986/87	1987/88	1988/89	1989/90	1990/91	1991/92
Hungary Hongrie	61	62	62	64	67	68	65	58	41	30
India Inde	1 368	1 433	1 550	1 564	1 702	1 708	1 762	1 876	1 954	1 879
Indonesia Indonésie	121	133	136	163	180	198	250	290	336	361
Iran, Islamic Rep. of Iran, Rép. islamique d'	94	90	80	91	115	120	108	119	120	120
Iraq Iraq	27	25	24	22	27	29	44	46	16	11
Ireland Irlande	17	21	21	20	22	22	24	25	26	26
Israel Israël	20	17	21	20	21	14	16	21	21	21
Italy Italie	229	248	264	258	309	329	310	315	333	320
Japan Japon	717	713	693	674	747	747	759	698	650	603
Kazakhstan Kazakhstan	...	...	...	...	37	35	36	38	39	39
Kenya Kenya	10	9	11	11	12	12	12	13	14	14
Korea, Dem. P. R. Corée, R. p. dém. de	39	40	38	37	36	36	36	36	36	36
Korea, Republic of Corée, République de	338	335	356	394	408	434	455	452	441	435
Kyrgyzstan Kirghizistan	...	...	...	...	23	22	22	26	26	26
Latvia Lettonie	...	...	...	...	14	14	14	14	13	12
Lebanon Liban	2	2	...	...	...	...	...	...	...	...
Lithuania Lituanie	...	...	...	...	29	27	28	24	22	20
Luxembourg Luxembourg	0	0	...	...	...	...	...	...	...	...
Madagascar Madagascar	15	15	15	16	16	11	11	11	11	10
Malawi Malawi	4	5	...	...	...	...	...	...	...	...
Malaysia Malaisie	27	29	28	26	31	34	38	41	42	40
Mali Mali	3	3	3	3	3	2	3	3	3	3
Malta Malte	1	1	...	...	...	...	...	...	...	...
Mexico Mexique	137	115	121	138	120	149	169	164	167	178
Moldova, Rep. of Moldova, Rép. de	...	...	...	...	39	37	38	38	36	32

77
Cotton
Industrial consumption: thousand metric tons [*cont.*]
Coton
Consommation industrielle : milliers de tonnes métriques [*suite*]

Country or area Pays ou zone	1982/83	1983/84	1984/85	1985/86	1986/87	1987/88	1988/89	1989/90	1990/91	1991/92
Morocco Maroc	20	17	17	21	29	28	30	34	35	34
Mozambique Mozambique	8	8	1	2	4	2	5	5	6	6
Myanmar Myanmar	17	13	12	15	12	11	28	29	31	26
Nepal Népal	1	1	...	...	...	...	...	...	...	...
Netherlands Pays-Bas	17	11	10	12	11	9	7	8	5	4
Nicaragua Nicaragua	4	5	4	3	4	3	4	2	2	1
Niger Niger	1	1	1	1	1	1	1	1	1	1
Nigeria Nigéria	45	58	60	55	45	54	58	53	74	68
Norway Norvège	2	2	3	3	3	3	3	3	3	3
Pakistan Pakistan	528	503	545	533	700	767	874	1 100	1 210	1 334
Paraguay Paraguay	7	6	8	10	10	10	10	10	12	15
Peru Pérou	47	47	48	59	85	69	68	65	53	65
Philippines Philippines	23	23	22	29	44	54	64	67	53	50
Poland Pologne	133	135	135	144	155	163	166	137	88	72
Portugal Portugal	140	144	154	166	178	187	191	190	154	150
Romania Roumanie	103	82	96	93	87	82	81	81	78	65
Russian Federation Fédération de Russie	...	...	...	...	1 296	1 235	1 262	1 275	1 239	1 034
Senegal Sénégal	3	3	3	3	3	2	2	2	2	2
Singapore Singapour	16	19	20	20	21	22	22	23	24	25
Somalia Somalie	1	1	...	...	...	...	...	...	...	...
South Africa Afrique du Sud	65	66	68	77	78	76	75	80	74	63
Spain Espagne	108	108	106	138	159	151	135	154	145	144
Sri Lanka Sri Lanka	9	9	9	10	11	10	10	9	9	9
Sudan Soudan	14	19	22	23	16	18	15	12	14	13

77
Cotton
Industrial consumption: thousand metric tons [*cont.*]
Coton
Consommation industrielle : milliers de tonnes métriques [*suite*]

Country or area Pays ou zone	1982/83	1983/84	1984/85	1985/86	1986/87	1987/88	1988/89	1989/90	1990/91	1991/92
Swaziland Swaziland	0	0	...	...	...	...	...	...	...	...
Sweden Suède	5	6	7	6	5	5	5	5	5	5
Switzerland Suisse	52	56	60	67	74	64	64	61	62	61
Syrian Arab Republic Rép. arabe syrienne	50	43	47	61	62	66	57	57	55	65
Tajikistan Tadjikistan	...	...	...	...	31	29	30	32	32	32
Thailand Thaïlande	114	198	159	205	247	252	280	295	328	390
Togo Togo	2	2	2	2	3	3	3	3	3	2
Tunisia Tunisie	11	13	14	17	17	19	19	27	26	27
Turkey Turquie	339	386	414	422	521	551	552	600	557	575
Turkmenistan Turkménistan	...	...	...	...	6	6	6	6	7	15
Uganda Ouganda	4	5	2	2	2	2	3	4	3	3
former USSR ancienne URSS	...	...	1 879	2 003	2 047	1 960	2 003	1 995	1 960	1 700
Ukraine Ukraine	...	...	...	...	212	205	210	213	208	171
United Kingdom Royaume-Uni	45	45	44	47	50	48	41	33	23	14
United Rep.Tanzania Rép. Unie de Tanzanie	14	12	16	13	11	12	10	17	14	16
United States Etats-Unis	1 200	1 290	1 206	1 393	1 622	1 658	1 694	1 907	1 885	2 068
Uruguay Uruguay	5	5	5	5	7	9	9	9	9	7
Uzbekistan Ouzbékistan	...	...	...	...	193	189	193	200	205	207
Venezuela Venezuela	22	30	33	35	37	44	44	47	45	47
Viet Nam Viet Nam	40	44	47	52	57	59	57	57	41	44
Yemen Yémen	2	2	...	...	...	...	...	...	...	...
Yugoslavia Yougoslavie	102	120	127	128	114	118	112	98	77	70
Zaire Zaïre	11	9	9	10	10	10	10	10	9	8
Zambia Zambie	7	7	7	7	7	7	9	10	10	10
Zimbabwe Zimbabwe	12	23	25	25	30	31	38	39	35	49

77

Cotton
Industrial consumption: thousand metric tons [cont.]
Coton
Consommation industrielle : milliers de tonnes métriques [suite]

Source:
International Cotton Advisory Committee (Washington, DC).

† All data shown which pertain to Germany prior to 3 October 1990 are indicated separately for the Federal Republic of Germany and the former German Democratic Republic based on their respective territories at the time indicated. Where data for united Germany (3 October 1990 and thereafter) are not available, available data are shown separately under the designations Federal Republic of Germany and former German Democratic Republic and pertain to the territorial boundaries prior to 3 October 1990. For detailed explanatory notes on data pertaining to Germany, see Annex I - Country Nomenclature.

Source:
Le Comité consultatif international du coton (Washington, DC).

† Toutes les données se rapportant à l'Allemagne avant le 3 octobre 1990 figurent dans deux rubriques séparées basées sur les territoires respectifs de la République fédérale d'Allemagne et l'ancienne République démocratique allemande selon la période indiquée. En l'absence de données pour l'Allemagne unifiée (à compter du 3 octobre 1990), les données disponibles sont fournies séparément sous les rubriques République fédérale d'Allemagne et ancienne République démocratique allemande et se rapportent aux limites territoriales antérieures au 3 octobre 1990. Pour les notes explicatives en détail sur les données concernant l'Allemagne, voir Annexe I – Nomenclature des pays.

78
Wool
Laine

Industrial consumption: thousand metric tons
Consommation industrielle : milliers de tonnes métriques

Country or area Pays ou zone	1981	1982	1983	1984	1985	1986	1987	1988	1989	1990
World *Monde*	1 668.0	1 628.0	1 682.0	1 672.0	1 610.0	1 680.0	1 725.0	1 743.0	1 703.0	1 501.0
Afghanistan Afghanistan	3.8	3.8	4.3	7.8	7.3	11.0	11.0	8.1	11.7	11.5
Algeria Algérie	16.4	17.4	19.2	17.1	20.8	19.9	19.6	21.3	21.3	21.0
Argentina[1] Argentine[1]	13.3	20.0	25.4	26.7	24.6	31.8	30.7	28.4	27.4	24.4
Australia Australie	17.6	18.5	16.2	17.7	20.6	20.7	20.3	20.3	17.9	17.0
Austria Autriche	9.8	7.9	7.2	7.8	3.1	3.3	4.9	4.5	4.1	4.0
Belgium Belgique	30.0	34.7	40.8	40.0	38.1	36.6	31.0	34.0	39.0	35.4
Bolivia Bolivie	4.8	4.8	4.8	4.8	4.9	4.9	4.9	4.9	4.9	4.9
Brazil Brésil	6.5	4.4	3.8	4.2	8.3	7.2	6.3	5.5	12.0	11.5
Bulgaria Bulgarie	17.0	16.9	17.4	17.5	16.0	15.5	14.6	14.6	13.0	12.9
Canada Canada	8.1	6.5	6.8	6.4	6.9	7.3	6.3	6.4	6.6	4.4
Chile Chili	8.0	6.2	7.9	8.6	6.6	5.3	5.3	6.3	8.3	7.7
China Chine	133.8	156.2	165.7	176.2	188.6	243.1	247.5	293.0	239.4	173.6
Colombia Colombie	3.8	5.5	4.0	3.3	3.6	3.7	3.7	2.5	2.7	1.9
Czechoslovakia Tchécoslovaquie	20.6	22.2	22.6	20.0	21.4	21.9	21.8	24.6	22.4	20.1
Denmark Danemark	4.9	4.5	5.4	5.6	4.7	4.6	4.0	3.2	3.7	2.8
Ecuador Equateur	0.8	1.6	1.6	1.6	1.2	1.2	1.0	1.0	1.0	1.0
Egypt Egypte	7.7	9.5	8.4	9.7	7.9	4.8	3.4	5.3	3.6	3.2
Finland Finlande	2.8	2.6	2.0	2.3	1.7	1.5	1.5	1.1	1.0	0.6
France France	52.7	46.5	43.6	44.6	42.7	37.8	35.8	32.7	28.2	21.6
Germany†· Allemagne† F. R. Germany R. f. Allemagne	65.9	57.1	59.1	62.8	62.9	59.8	62.3	58.0	60.1	52.2
former German D. R. anc. R. d. allemande	21.6	19.0	19.4	20.5	19.4	20.2	19.8	21.5	22.9	18.4
Greece Grèce	18.0	18.5	15.9	15.2	13.6	15.6	12.4	13.7	13.1	14.3
Hong Kong Hong-kong	5.2	3.7	4.0	4.5	5.2	6.6	7.2	5.7	5.2	4.7

78

Wool
Industrial consumption: thousand metric tons [cont.]
Laine
Consommation industrielle : milliers de tonnes métriques [suite]

Country or area Pays ou zone	1981	1982	1983	1984	1985	1986	1987	1988	1989	1990
Hungary Hongrie	9.4	8.5	8.3	7.5	7.5	8.4	6.7	5.9	4.6	4.5
Iceland Islande	1.8	1.6	2.1	1.7	1.9	0.9	0.7	0.7	0.6	0.4
India Inde	35.7	35.3	36.5	36.5	35.0	34.5	33.0	33.7	39.3	38.3
Iran, Islamic Rep. of Iran, Rép. islamique d'	51.5	27.9	51.7	31.4	18.8	21.9	21.3	19.8	24.4	21.3
Iraq Iraq	7.6	5.2	6.1	6.0	6.3	6.1	5.9	4.9	4.8	5.2
Ireland Irlande	7.9	6.8	6.0	6.8	8.0	8.2	6.6	7.8	8.8	9.8
Israel Israël	2.3	2.0	1.8	1.8	1.3	1.2	0.8	0.6	0.8	0.8
Italy Italie	133.0	116.2	111.4	127.1	130.1	125.0	140.2	139.9	133.6	137.6
Japan Japon	119.6	135.7	105.6	118.1	123.2	113.6	126.8	126.3	121.8	110.2
Jordan Jordanie	1.2	1.1	1.3	1.2	1.4	1.4	1.4	1.1	1.1	1.1
Korea, Republic of Corée, République de	30.9	31.1	33.6	31.2	27.6	39.5	46.3	54.2	54.1	48.5
Libyan Arab Jamah. Jamah. arabe libyenne	2.1	2.1	2.2	2.2	2.2	2.1	2.0	1.5	2.1	2.1
Malaysia Malaisie	0.1	0.1	1.0	0.7	0.7	1.0	0.8	0.9	1.4	1.9
Maldives Maldives	2.4	2.2	2.1	2.6	3.0	3.0	2.9	2.1	2.6	2.7
Mexico Mexique	8.1	6.1	4.6	5.0	9.2	6.9	6.1	6.3	6.0	5.8
Mongolia Mongolie	3.0	4.2	2.9	3.4	3.6	1.9	7.2	7.9	9.1	8.8
Morocco Maroc	11.4	11.4	14.2	11.5	15.1	17.6	19.4	19.8	19.1	19.0
Nepal Népal	0.7	0.7	1.0	1.7	1.7	4.0	6.0	3.8	7.3	5.8
Netherlands Pays-Bas	6.4	6.4	5.6	5.2	5.8	6.1	6.5	4.7	4.4	4.3
New Zealand Nouvelle-Zélande	17.4	18.3	18.0	23.3	25.4	22.4	21.5	19.8	20.3	17.9
Norway Norvège	3.8	3.8	2.6	2.8	3.1	3.0	2.8	2.0	2.2	2.1
Pakistan Pakistan	18.3	15.4	19.7	15.3	13.6	16.2	19.5	25.4	24.4	20.8
Peru Pérou	4.6	4.1	4.4	4.4	5.3	6.1	6.0	5.0	4.9	4.5
Poland Pologne	23.6	25.5	25.6	27.9	26.3	23.2	25.0	24.8	21.5	13.6
Portugal Portugal	11.1	10.2	10.4	9.6	11.5	13.2	12.7	10.9	12.7	10.8

78

Wool
Industrial consumption: thousand metric tons [cont.]
Laine
Consommation industrielle : milliers de tonnes métriques [suite]

Country or area Pays ou zone	1981	1982	1983	1984	1985	1986	1987	1988	1989	1990
Romania Roumanie	22.6	21.6	21.1	24.1	22.7	21.3	17.4	17.9	17.5	15.4
Saudi Arabia Arabie saoudite	1.4	1.5	1.0	0.5	0.9	2.1	2.4	1.8	1.7	1.6
South Africa Afrique du Sud	12.5	12.4	12.1	10.1	10.7	11.2	13.5	9.9	12.5	11.1
Spain Espagne	28.5	27.6	30.0	30.4	21.5	21.4	14.7	12.0	14.0	12.8
Sweden Suède	1.5	1.8	1.8	1.7	1.7	1.3	0.8	0.6	0.5	0.4
Switzerland Suisse	11.5	10.4	9.9	11.6	12.9	12.5	10.1	7.4	6.7	7.1
Syrian Arab Republic Rép. arabe syrienne	12.5	11.5	12.5	11.7	12.8	12.3	11.5	11.3	11.3	11.3
Tunisia Tunisie	5.6	4.9	4.9	5.3	5.9	5.7	5.8	6.0	5.9	5.9
Turkey Turquie	47.0	47.1	51.3	57.8	40.4	53.4	58.2	57.1	51.8	63.4
former USSR ancienne URSS	341.8	333.1	360.6	302.9	308.3	298.5	314.1	310.6	324.7	257.4
United Kingdom Royaume-Uni	80.6	81.0	82.0	83.9	77.0	78.0	87.0	87.6	80.7	79.8
United States Etats-Unis	70.0	58.3	72.3	77.0	51.8	62.1	62.5	53.0	53.3	50.4
Uruguay Uruguay	5.7	7.1	8.0	6.6	5.4	7.4	7.0	6.3	6.3	5.5
Yugoslavia Yougoslavie	26.4	23.7	13.0	19.7	20.1	21.3	16.8	15.5	14.9	12.3

Source:
Commonwealth Secretariat (London).

† All data shown which pertain to Germany prior to 3 October 1990 are indicated separately for the Federal Republic of Germany and the former German Democratic Republic based on their respective territories at the time indicated. Where data for united Germany (3 October 1990 and thereafter) are not available, available data are shown separately under the designations Federal Republic of Germany and former German Democratic Republic and pertain to the territorial boundaries prior to 3 October 1990. For detailed explanatory notes on data pertaining to Germany, see Annex I - Country Nomenclature.

1 Twelve months ending 30 September of the year stated.

Source:
"Commonwealth Secretariat" (Londres).

† Toutes les données se rapportant à l'Allemagne avant le 3 octobre 1990 figurent dans deux rubriques séparées basées sur les territoires respectifs de la République fédérale d'Allemagne et l'ancienne République démocratique allemande selon la période indiquée. En l'absence de données pour l'Allemagne unifiée (à compter du 3 octobre 1990), les données disponibles sont fournies séparément sous les rubriques République fédérale d'Allemagne et ancienne République démocratique allemande et se rapportent aux limites territoriales antérieures au 3 octobre 1990. Pour les notes explicatives en détail sur les données concernant l'Allemagne, voir Annexe I - Nomenclature des pays.

1 Douze mois finissant le 30 septembre de l'année indiquée.

79
Sawnwood
Sciages

Production (sawn): thousand cubic metres
Production (sciés) : milliers de mètres cubes

Country or area Pays ou zone	1981	1982	1983	1984	1985	1986	1987	1988	1989	1990
World *Monde*	430 161	422 882	441 619	460 954	467 697	483 085	504 728	506 145	501 995	485 946
Africa **Afrique**	7 994	7 543	7 321	7 495	7 815	8 175	8 484	8 638	8 726	8 716
Algeria[1] Algérie[1]	13	13	13	13	13	13	13	13	13	13
Angola Angola	10	8	6	2	5	5	5	5	5	5
Benin Bénin	9[1]	9[1]	9[1]	5	8	11	11[1]	11[1]	11[1]	11[1]
Burkina Faso Burkina Faso	2[1]	2[1]	2[1]	2	1	1[1]	1[1]	1[1]	1[1]	1[1]
Burundi Burundi	1	1	1	4	3	3	3[1]	3	3	3
Cameroon Cameroun	468	468[1]	508	539	637	665	577	574	574[1]	574[1]
Central African Rep. Rép. centrafricaine	70	63	61	58	56	54	52	52[1]	57	63
Chad[1] Tchad[1]	1	1	1	1	1	1	1	1	1	1
Congo Congo	73	66	66	60	50	77	60	57	46	46[1]
Côte d'Ivoire Côte d'Ivoire	611	748	718	679	753	765	775	784	777	753
Equatorial Guinea Guinée équatoriale	13[1]	18[1]	23[1]	24[1]	39	51[1]	51[1]	51[1]	56[1]	52[1]
Ethiopia Ethiopie	65[1]	45	45	45[1]	45[1]	45[1]	45[1]	39	34	34[1]
Gabon Gabon	108[1]	108[1]	108[1]	117	126	126[1]	126[1]	126[1]	126[1]	126[1]
Gambia[1] Gambie[1]	1	1	1	1	1	1	1	1	1	1
Ghana Ghana	225	215	275	285	345	355	455	455[1]	537	537[1]
Guinea[1] Guinée[1]	90	90	90	90	90	90	90	90	90	90
Guinea-Bissau[1] Guinée-Bissau[1]	16	16	16	16	16	16	16	16	16	16
Kenya Kenya	149	149	142	183	192	173	195	188	185	185[1]
Liberia Libéria	199	172	161	153	169	191	411	411[1]	411[1]	411[1]
Libyan Arab Jamah.[1] Jamah. arabe libyenne[1]	31	31	31	31	31	31	31	31	31	31
Madagascar Madagascar	234	234	234	234	234	234	234[1]	234	234	234
Malawi Malawi	43	38[1]	23[1]	16	19	23	30	31	31[1]	48
Mali Mali	6[1]	6	6[1]	6	4	6	11	13	13[1]	13[1]

79
Sawnwood
Production (sawn): thousand cubic metres [cont.]
Sciages
Production (sciés) : milliers de mètres cubes [suite]

Country or area Pays ou zone	1981	1982	1983	1984	1985	1986	1987	1988	1989	1990
Mauritius Maurice	3	4	3	0	1	1	4	4	5	4
Morocco Maroc	114	149	130	120	100	90	80[1]	53	83	83
Mozambique Mozambique	65	42	38	37	35	39	42	36	30	35
Nigeria Nigéria	2 962	2 667	2 402	2 512	2 712	2 712	2 712[1]	2 712	2 712	2 712
Réunion Réunion	1[1]	1[1]	1[1]	1[1]	2[1]	2	2	1	1	1[1]
Rwanda Rwanda	5[1]	8[1]	10[1]	12	13	13[1]	13[1]	13[1]	11	8
Sao Tome and Principe Sao Tomé-et-Principe	3	1	3	3[1]	3[1]	3[1]	4[1]	5	5[1]	5[1]
Senegal[1] Sénégal[1]	11	11	11	11	11	11	11	11	11	11
Sierra Leone Sierra Leone	22[1]	20[1]	19	19[1]	14	12	12[1]	12[1]	12[1]	12[1]
Somalia[1] Somalie[1]	14	14	14	14	14	14	14	14	14	14
South Africa Afrique du Sud	1 772	1 563	1 620	1 635	1 510	1 734	1 734	1 873	1 873[1]	1 873[1]
Sudan Soudan	6	6	13	13	13	13	13[1]	13	7	6
Swaziland Swaziland	106	103	103	103	103	103	103[1]	103	103	103
Togo[1] Togo[1]	5	5	5	5	5	5	5	5	5	5
Tunisia Tunisie	3	3	3	4	6	9	11[1]	12	20	16
Uganda Ouganda	23	23	23[1]	23[1]	23[1]	23[1]	23[1]	28	28[1]	28[1]
United Rep.Tanzania Rép. Unie de Tanzanie	95	86	91	102	109	154	156	156	156	156
Zaire Zaïre	* 105	* 95	112	120	118	120	127	135	131	131[1]
Zambia Zambie	42	42	50	50	50	67	51	76	76	76
Zimbabwe Zimbabwe	200	198	131	149	138	114	175	190	190[1]	190[1]
America, North **Amérique du Nord**	**119 846**	**110 838**	**127 191**	**141 747**	**146 633**	**158 011**	**173 415**	**172 347**	**166 270**	**160 312**
Bahamas[1] Bahamas[1]	1	1	1	1	1	1	1	1	1	1
Belize Belize	19[1]	21[1]	16[1]	19[1]	22	14	14[1]	14[1]	14[1]	14[1]
Canada Canada	39 675	37 168	42 070	49 869	54 586	54 853	61 775	60 737	59 225	52 600
Costa Rica Costa Rica	534	378	306[1]	412	412	412[1]	503	515	439	412

79
Sawnwood
Production (sawn): thousand cubic metres [cont.]
Sciages
Production (sciés) : milliers de mètres cubes [suite]

Country or area Pays ou zone	1981	1982	1983	1984	1985	1986	1987	1988	1989	1990
Cuba Cuba	108	107	104	108	104	108	114	118	130	130[1]
El Salvador El Salvador	47	45	39	46[1]	43[1]	44[1]	47[1]	54[1]	70[1]	70[1]
Guadeloupe[1] Guadeloupe[1]	1	1	1	1	1	1	1	1	1	1
Guatemala Guatemala	136	130	104	103	131	83	83	83	83	83
Haiti[1] Haïti[1]	14	14	14	14	14	14	14	14	14	14
Honduras Honduras	560	489	468	427	436	405	464	447	412	343
Jamaica Jamaïque	25	25	23	31	31	26	30	44	40	40
Martinique Martinique	0[1]	0[1]	0[1]	0[1]	0[1]	0[1]	1	2	1	1[1]
Mexico Mexique	1 928	1 669	1 827	1 975	2 205	2 143	2 410	2 528[1]	2 447[1]	2 366
Nicaragua[1] Nicaragua[1]	402	402	222	222	222	222	222	222	222	222
Panama Panama	53[1]	53[1]	53[1]	46	45	30[1]	25[1]	18	50[1]	48
Trinidad and Tobago Trinité-et-Tobago	33[1]	28	22	21	18	22	19	21	80	53
United States Etats-Unis	76 310	70 307	81 920	88 451	88 361	99 632	107 691	107 527	103 040	103 913
America, South **Amérique du Sud**	**22 708**	**22 303**	**23 612**	**24 647**	**24 890**	**25 522**	**26 273**	**26 412**	**26 788**	**27 747**
Argentina Argentine	1 047	1 192	1 237	1 117	1 067	1 446	1 446[1]	1 446[1]	1 446[1]	1 446[1]
Bolivia Bolivie	172	117	97	97	97	90	95	95	95	95
Brazil Brésil	15 852	16 470	17 199	17 781	17 781	18 063	18 063[1]	18 179	18 179	18 179
Chile Chili	1 735	1 176	1 610	2 001	2 194	2 028	2 680	2 713[1]	2 684	3 331
Colombia Colombie	1 006	721	721[1]	721[1]	721[1]	721[1]	721[1]	721[1]	721[1]	721[1]
Ecuador Equateur	986	980	1 142	1 212	1 215	1 258	1 265	1 280	1 492	1 641
French Guiana[1] Guyane française[1]	19	19	19	19	19	19	19	19	19	19
Guyana Guyana	70	70	70	60	65	60	57	57	57	57
Paraguay Paraguay	655[1]	655[1]	655[1]	834	758	766	862	906	906[1]	906[1]
Peru Pérou	653	577	577[1]	479	535	617	630	542	764	926
Suriname Suriname	64	60	59	57	73	61	42[1]	73[1]	44	45

79
Sawnwood
Production (sawn): thousand cubic metres [cont.]
Sciages
Production (sciés) : milliers de mètres cubes [suite]

Country or area Pays ou zone	1981	1982	1983	1984	1985	1986	1987	1988	1989	1990
Uruguay Uruguay	100	47	16	59	57	57[1]	57[1]	57[1]	57[1]	57[1]
Venezuela Venezuela	349[1]	220	210	210[1]	308	336	336[1]	* 325	325[1]	325[1]
Asia Asie	**91 023**	**95 358**	**95 309**	**96 654**	**99 168**	**100 102**	**105 044**	**105 533**	**107 586**	**103 831**
Afghanistan[1] Afghanistan[1]	400	400	400	400	400	400	400	400	400	400
Bangladesh Bangladesh	176[1]	203[1]	162[1]	154[1]	99	79	79[1]	79[1]	79[1]	79[1]
Bhutan Bhoutan	5	6[1]	6[1]	6[1]	6[1]	6[1]	6[1]	10[1]	20[1]	40[1]
Brunei Darussalam Brunéi Darussalam	95	99	90	90	90	90	90[1]	90	90	90
Cambodia Cambodge	43	43	43	43	43	43	43[1]	43	43	43
China[1] Chine[1]	22 007	23 064	23 905	25 761	27 087	26 391	26 344	26 282	24 958	23 037
Cyprus Chypre	64	73	55	59	63	59	57	55	60	22
Hong Kong Hong-kong	273[1]	319[1]	271[1]	248[1]	248[1]	248[1]	248[1]	248[1]	587	587[1]
India[1] Inde[1]	12 040	13 209	14 495	15 907	17 460	17 460	17 460	17 460	17 460	17 460
Indonesia Indonésie	5 269	6 825	6 315	6 620	7 118	7 549	9 887	10 290	10 499	9 145
Iran, Islamic Rep. of Iran, Rép. islamique d'	163[1]	163[1]	163[1]	172[1]	186[1]	202[1]	219[1]	239	262	262[1]
Iraq[1] Iraq[1]	8	8	8	8	8	8	8	8	8	8
Japan Japon	32 519	32 537	29 670	28 667	28 472	29 105	30 159	* 30 138	* 30 542	29 842[1]
Korea, Dem. P. R.[1] Corée, R. p. dém. de[1]	280	280	280	280	280	280	280	280	280	280
Korea, Republic of Corée, République de	2 978	3 010	3 518	2 974	3 018	3 563	4 145	4 014	4 194	3 963
Lao People's Dem. Rep. Rép. dém. pop. lao	16	22	25	26	16[1]	16	16[1]	16[1]	16[1]	16[1]
Lebanon Liban	38	29	33	33[1]	33[1]	33[1]	29	27	21	18
Malaysia Malaisie	5 703	6 398	7 282	5 933	5 494	5 525	6 285	6 662	8 275	8 275[1]
Mongolia[1] Mongolie[1]	470	470	470	470	470	470	470	470	470	470
Myanmar Myanmar	708	722	674	664	615	568	392	283	375	466
Nepal[1] Népal[1]	220	220	220	220	220	220	220	220	220	220
Pakistan Pakistan	65	75	85	95	153	165	227	456	751	1 153

79
Sawnwood
Production (sawn): thousand cubic metres [*cont.*]
    Sciages
    Production (sciés) : milliers de mètres cubes [*suite*]

Country or area Pays ou zone	1981	1982	1983	1984	1985	1986	1987	1988	1989	1990
Philippines Philippines	1 219	1 200	1 222	1 234	1 062	978	* 1 233	1 033	975	841
Singapore Singapour	242	229[1]	232[1]	211[1]	191[1]	206[1]	206[1]	206[1]	206[1]	206[1]
Sri Lanka Sri Lanka	31	21	29	23[1]	23[1]	20[1]	20[1]	20[1]	20[1]	20[1]
Syrian Arab Republic Rép. arabe syrienne	9	9	9	9	9	9	9[1]	9	9	9
Thailand Thaïlande	921	911	950	1 036	958	1 027	1 095	1 044	1 279	1 356
Turkey Turquie	4 633	4 424	4 343	4 923	4 923	4 923[1]	4 923[1]	4 923[1]	4 923[1]	4 923[1]
Viet Nam[1] Viet Nam[1]	429	390	354	389	424	459	494	529	564	600
Europe **Europe**	**84 464**	**83 443**	**85 796**	**87 749**	**85 199**	**84 401**	**84 253**	**84 641**	**86 739**	**87 567**
Albania Albanie	200	200	200[1]	200[1]	200[1]	200[1]	200[1]	200[1]	200[1]	200[1]
Austria Autriche	6 426	5 958	6 269	6 315	6 001	5 818	5 944	6 478	6 920	7 332
Belgium-Luxembourg Belgique-Luxembourg	653	676	753	746	795	824	939	1 009	1 144	1 184
Bulgaria Bulgarie	1 473	1 511	1 504	1 483	1 515	1 338	1 490	1 459	1 426	985
Czechoslovakia Tchécoslovaquie	4 956	5 093	5 143	5 227	5 219	5 251	5 186	5 128	4 860	4 702
Denmark Danemark	829	829	829	829[1]	879	879[1]	861	861[1]	861[1]	861[1]
Finland Finlande	8 280	7 322	8 023	8 265	7 333	7 143	7 563	7 823	7 763	7 503
France France	9 016	9 049	9 005	9 038	9 087	9 318	9 612	10 248	10 655	10 655[1]
Germany† · Allemagne† F. R. Germany R. f. Allemagne	9 416	8 714	9 413	9 825	9 541	9 805	9 754	10 395	11 388	12 203
former German D. R. anc. R. d. allemande	2 394	2 422	2 449	2 491	2 491	2 431	2 465	2 489	2 521	2 521[1]
Greece Grèce	350	363	323	321	305	454	428	410	417	355
Hungary Hongrie	1 327	1 127	1 204	1 322	1 284	1 277	1 225	1 237	1 257	1 257[1]
Ireland Irlande	215	238	258	290	300	300[1]	300[1]	300[1]	356	393
Italy Italie	2 600	2 330	2 025	2 234	2 599	1 919	1 905	2 095	1 998	1 950
Netherlands Pays-Bas	300	204	314	335	412	425	387	435	465	455
Norway Norvège	2 462	2 262	2 312	2 364	2 230	2 260	2 362	2 387	2 492	2 413
Poland Pologne	6 751	6 368	6 762	6 765	6 639	6 645	6 442	5 577	4 878	4 630

79

Sawnwood
Production (sawn): thousand cubic metres [*cont.*]
Sciages
Production (sciés) : milliers de mètres cubes [*suite*]

Country or area Pays ou zone	1981	1982	1983	1984	1985	1986	1987	1988	1989	1990
Portugal Portugal	1 870	2 229	2 360	2 606	1 860	2 070	2 095	1 460	1 690	1 790
Romania Roumanie	4 500	4 568	4 878	4 868	4 425	3 538	2 858	2 758	2 851	2 851[1]
Spain Espagne	2 387	2 798	2 218	2 150	2 383	2 613	2 643	2 427	2 724	2 826
Sweden Suède	10 502	11 213	11 762	12 382	11 531	11 641	11 524	11 267	11 487	11 830
Switzerland Suisse	1 785	1 850	1 760	1 515	1 689	1 719	1 650	1 693	1 700	1 985
United Kingdom Royaume-Uni	1 543	1 687	1 619	1 520	1 717	1 807	1 823	1 919	2 191	2 191[1]
Yugoslavia Yougoslavie	4 229	4 432	4 413	4 658	4 764	4 726	4 597	4 587	4 496	4 496[1]
Oceania **Océanie**	**6 027**	**5 897**	**5 391**	**5 362**	**5 791**	**5 874**	**5 259**	**5 573**	**5 886**	**5 772**
Australia Australie	3 553	3 364	2 991	3 003	3 216	3 220	3 131	3 426	3 490	3 313
Fiji Fidji	100	76[1]	82	80[1]	91[1]	78	88	103	94	94[1]
New Caledonia Nouvelle-Calédonie	4	4	4	4	4	4	4[1]	5	5	5
New Zealand Nouvelle-Zélande	2 203	2 288	2 150	2 109	2 320	2 412	1 876	1 881	2 135	2 198
Papua New Guinea Papouasie-Nvl-Guinée	124	124[1]	124[1]	124[1]	* 117	117[1]	117[1]	117[1]	117[1]	117[1]
Samoa Samoa	21	21	21	21	21	21	21[1]	21	21	21
Solomon Islands Iles Salomon	20	15	15	17	17	15	13	12	16	16[1]
Tonga Tonga	...	3	1	1[1]	1	2	2[1]	2[1]	1[1]	1[1]
Vanuatu Vanuatu	2	2	3	3	4	6	7	7	7	7
former USSR **ancienne URSS**	**98 100**	**97 500**	**97 000**	**97 300**	**98 200**	**101 000**	**102 000**	**103 000**	**100 000**	**92 000**

Source:
Food and Agriculture Organization of the United Nations
(Rome).

Source:
Organisation des Nations Unies pour l'alimentation et
l'agriculture (Rome).

† All data shown which pertain to Germany prior to 3 October
1990 are indicated separately for the Federal Republic of
Germany and the former German Democratic Republic based on
their respective territories at the time indicated. Where
data for united Germany (3 October 1990 and thereafter) are
not available, available data are shown separately under the
designations Federal Republic of Germany and former German
Democratic Republic and pertain to the territorial
boundaries prior to 3 October 1990. For detailed
explanatory notes on data pertaining to Germany, see Annex I
- Country Nomenclature.

1 FAO estimate.

† Toutes les données se rapportant à l'Allemagne avant le 3
octobre 1990 figurent dans deux rubriques séparées basées
sur les territoires respectifs de la République fédérale
d'Allemagne et l'ancienne République démocratique allemande
selon la période indiquée. En l'absence de données pour
l'Allemagne unifiée (à compter du 3 octobre 1990), les
données disponibles sont fournies séparément sous les
rubriques République fédérale d'Allemagne et ancienne
République démocratique allemande et se rapportent aux
limites territoriales antérieures au 3 octobre 1990. Pour
les notes explicatives en détail sur les données concernant
l'Allemagne, voir Annexe I – Nomenclature des pays.

1 Estimation de la FAO.

80
Paper and paperboard
Papiers et cartons
Production: thousand metric tons
Production : milliers de tonnes métriques

Country or area Pays ou zone	1981	1982	1983	1984	1985	1986	1987	1988	1989	1990
World *Monde*	170 954	167 264	177 237	189 973	192 643	201 943	212 699	225 638	231 682	238 238
Africa **Afrique**	**1 849**	**1 860**	**2 062**	**2 030**	**2 139**	**2 331**	**2 371**	**2 602**	**2 754**	**2 766**
Algeria Algérie	104	102	107	135	110	120	120[1]	120[1]	120[1]	120[1]
Angola[1] Angola[1]	13	...	...	...	...	...	...	...	...	...
Cameroon Cameroun	5	5[1]	5[1]	5[1]	5[1]	5[1]	5[1]	5[1]	5[1]	5[1]
Egypt Egypte	146[1]	110	110[1]	* 145	* 145	145[1]	* 160	160[1]	* 216	* 223
Ethiopia Ethiopie	8	8	9	10	10	* 10	10[1]	9[1]	9[1]	8
Kenya Kenya	63	66	69	75	75	85	89	100	108	108[1]
Libyan Arab Jamah. Jamah. arabe libyenne	5[1]	5[1]	5[1]	5[1]	* 6	* 6	6[1]	6[1]	6[1]	6[1]
Madagascar Madagascar	4	1	10	10	10	15	8	6[1]	6[1]	6[1]
Morocco Maroc	96	97	96	101	* 107	* 109	105[1]	103[1]	102[1]	105
Mozambique Mozambique	2	2	2	1[1]	* 2	* 2	2[1]	2[1]	2[1]	2[1]
Nigeria Nigéria	18	4	18	15	41	76	76[1]	95	73	73[1]
South Africa[2] Afrique du Sud[2]	1 290	1 350	1 520	1 422	1 489	* 1 611	1 600	1 800	1 899	* 1 904
Sudan Soudan	9[1]	9[1]	9[1]	9[1]	9[1]	9[1]	* 10	10[1]	10[1]	10[1]
Tunisia Tunisie	24	26	30	28	48	53	62	70	82	78
Uganda Ouganda	...	...	...	...	* 2	* 2	2[1]	2[1]	2[1]	2[1]
United Rep.Tanzania Rép. Unie de Tanzanie	...	...	...	...	...	...	29	28	28[1]	25
Zaire Zaïre	2[1]	2	3	2	2[1]	3	3	2[1]	1	1[1]
Zambia Zambie	...	8	5	3	2	5	3	2	4	4
Zimbabwe Zimbabwe	62	67	65	64	76	* 75	81	82	82[1]	86
America, North **Amérique du Nord**	**73 590**	**69 458**	**74 400**	**79 067**	**78 007**	**82 415**	**86 379**	**89 823**	**89 695**	**91 556**
Canada Canada	13 835	12 408	13 353	14 222	14 448	15 259	16 044	16 639	16 555	16 466
Costa Rica Costa Rica	15	18	13	13	13	13[1]	13[1]	17	18	19
Cuba Cuba	73[1]	112	109	122	132	150	148	141	168	168[1]

80
Paper and paperboard
Production: thousand metric tons [*cont.*]
Papiers et cartons
Production : milliers de tonnes métriques [*suite*]

Country or area Pays ou zone	1981	1982	1983	1984	1985	1986	1987	1988	1989	1990
Dominican Republic Rép. dominicaine	9[1]	9[1]	9[1]	10	10[1]	10[1]	10[1]	10[1]	10[1]	10[1]
El Salvador El Salvador	16	16[1]	16[1]	16[1]	16[1]	16[1]	17[1]	17[1]	17[1]	17[1]
Guatemala Guatemala	29	12	16	18	14	17	14	14	14	14
Jamaica Jamaïque	10	18	18	18	15[1]	11[1]	2	3	4	4
Mexico Mexique	1 893	1 924	2 019	2 239	2 376	2 469	2 573	3 375	3 375[1]	2 873
Panama Panama	43	43[1]	43[1]	43[1]	24	26	26[1]	20	20	20
United States Etats-Unis	* 57 667	* 54 899	58 804	62 366	60 959	64 444	67 532	69 587	69 514	71 965
America, South **Amérique du Sud**	**5 364**	**5 572**	**5 728**	**6 292**	**6 504**	**7 234**	**7 651**	**7 702**	**7 621**	**7 680**
Argentina Argentine	669	730	879	942	864	998	1 027	974[1]	917	891
Bolivia Bolivie	1[1]	1[1]	1[1]	1[1]	1[1]	1[1]	2	2[1]	2[1]	5
Brazil Brésil	3 102	3 329	3 426	3 768	4 022	4 525	4 712	4 685	4 806	4 844
Chile Chili	318	306	333	381	385	394	444	454	442	462
Colombia Colombie	407	366	366	413	* 446	457	488	501	501[1]	534
Ecuador Equateur	31	34	34[1]	34[1]	34	34	35	36	36	38
Paraguay Paraguay	13	13[1]	13[1]	13[1]	8	8	10	11	11[1]	11
Peru Pérou	272	272[1]	146	138	* 150	152	209	260	311	311
Uruguay Uruguay	48	39	43	45	44	* 54	70	70[1]	70[1]	60
Venezuela Venezuela	503	482	487	* 557	551	612	654	708	524	524[1]
Asia **Asie**	**29 030**	**30 092**	**32 382**	**35 013**	**37 936**	**40 046**	**43 556**	**47 521**	**51 805**	**54 708**
Bangladesh Bangladesh	79	94	148	149	* 104	* 113	* 114	96	97	92
China Chine	6 900[1]	7 446[1]	8 333	9 489	11 130	11 874	12 996	14 144	15 427	16 058[1]
Hong Kong Hong-kong	17[1]	17[1]	32	36	* 40	* 40	40[1]	40[1]	40[1]	40[1]
India Inde	1 266	1 288	1 491	1 557	* 1 590	1 810	1 910	1 940	2 135[1]	2 200[1]
Indonesia Indonésie	258	329	374	403	500	* 611	813	974	1 158	* 1 438
Iran, Islamic Rep. of Iran, Rép. islamique d'	78[1]	78[1]	78[1]	78[1]	78[1]	80[1]	90[1]	100	100	100[1]

80
Paper and paperboard
Production: thousand metric tons [cont.]
Papiers et cartons
Production : milliers de tonnes métriques [suite]

Country or area Pays ou zone	1981	1982	1983	1984	1985	1986	1987	1988	1989	1990
Iraq[1] Iraq[1]	28	28	28	28	28	28	28	28	28	28
Israel Israël	107	107	157	148	131	151	160	170	179	194
Japan Japon	16 980	17 453	18 442	19 345	20 469	21 062	22 537	24 625	26 809	28 088
Jordan Jordanie	6[1]	6[1]	5	7	13	* 14	12	10	10[1]	12
Korea, Dem. P. R.[1] Corée, R. p. dém. de[1]	80	80	80	80	80	80	80	80	80	80
Korea, Republic of Corée, République de	1 783	1 737	1 982	2 207	2 312	2 773	3 163	3 659	4 018	4 524
Lebanon Liban	45	45[1]	45[1]	45[1]	* 45	42	42[1]	37	37[1]	37[1]
Malaysia Malaisie	* 62	* 68	* 45	50	* 53	* 73	97	120	251	251
Myanmar Myanmar	10[1]	10[1]	10[1]	10[1]	15[1]	23	9	8	10	11
Nepal Népal	2[1]	2[1]	2[1]	2[1]	...	* 2	2[1]	2[1]	2[1]	2[1]
Pakistan Pakistan	68	74	76	73	82	80	96	147	151	229
Philippines Philippines	290	242	280	290	268	218	358	314	239	175
Singapore Singapour	* 50	* 50	* 10	* 10	* 10	10[1]	10[1]	10[1]	10[1]	10[1]
Sri Lanka Sri Lanka	20	19	22[1]	23[1]	23	25	25[1]	28[1]	17	18
Syrian Arab Republic Rép. arabe syrienne	3[1]	3[1]	3[1]	3[1]	* 5	* 5	10[1]	19	19[1]	19[1]
Thailand Thaïlande	349	329	300[1]	437	466	432[1]	465	514	520	576
Turkey Turquie	512	537	391	488	434	435	440	400	413	436
Viet Nam Viet Nam	37	50[1]	48[1]	55[1]	60[1]	65[1]	59	56	56[1]	90
Europe **Europe**	**50 017**	**49 116**	**51 008**	**55 523**	**55 711**	**57 563**	**60 406**	**64 677**	**66 548**	**68 057**
Albania Albanie	8	8	8	8	8	8	8	22	19	21
Austria Autriche	1 671	1 708	1 789	1 922	2 127	2 183	2 396	2 650	2 754	2 872
Belgium-Luxembourg Belgique-Luxembourg	886	828	840	847	843	850	1 031	1 133	1 133	1 195
Bulgaria Bulgarie	427	435	442	445	454	458	456	477	438	322
Czechoslovakia Tchécoslovaquie	1 201	1 228	1 231	1 237	1 259	1 255	1 273	1 266	1 305	1 307
Denmark Danemark	227	301	322	332	302	293	326	326	326	326

80
Paper and paperboard
Production: thousand metric tons [*cont.*]
Papiers et cartons
Production : milliers de tonnes métriques [*suite*]

Country or area Pays ou zone	1981	1982	1983	1984	1985	1986	1987	1988	1989	1990
Finland Finlande	6 135	5 895	6 388	7 318	7 447	7 549	8 011	8 652	8 579	8 781
France France	5 148	5 067	5 263	5 566	5 150	5 583	5 581	6 313	6 754	7 006
Germany† · Allemagne† F. R. Germany R. f. Allemagne	7 828	7 784	8 273	9 145	9 178	9 409	9 938	10 576	11 259	11 873
former German D. R. anc. R. d. allemande	1 258	1 261	1 244	1 293	1 297	1 320	1 340	1 362	1 351	1 351[1]
Greece Grèce	287	266	277	294	282	283	280	282	282[1]	282[1]
Hungary Hongrie	457	462	479	506	494	517	522	535	504	443
Ireland Irlande	52	20	20	20	22	37	29	33	34	34
Italy Italie	4 855	4 503	4 259	4 722	4 587	4 631	4 882	5 512	5 640	5 582
Netherlands Pays-Bas	1 590	1 671	1 746	1 885	1 905	2 041	2 168	2 460	2 572	2 770
Norway Norvège	1 373	1 305	1 368	1 562	1 604	1 573	1 590	1 670	1 789	1 819
Poland Pologne	1 125	1 182	1 231	1 257	1 292	1 327	1 380	1 448	1 406	1 065
Portugal Portugal	485	527	592	671	706	590	628	681	740	740
Romania Roumanie	834	802	802[1]	806	801	811	816	819	819[1]	819[1]
Spain Espagne	2 589	2 685	2 754	2 952	2 913	3 152	3 251	3 408	3 446	3 446
Sweden Suède	6 131	5 919	6 349	6 870	7 001	7 364	7 811	8 161	8 362	8 426
Switzerland Suisse	920	887	918	986	1 014	1 087	1 147	1 216	1 259	1 295
United Kingdom Royaume-Uni	3 379	3 227	3 208	3 591	3 712	3 941	4 184	4 295	4 475	4 980
Yugoslavia Yougoslavie	1 151	1 145	1 205	1 288	1 313	1 301	1 358	1 381	1 302	1 302[1]
Oceania Océanie	**2 151**	**2 188**	**2 101**	**2 214**	**2 316**	**2 267**	**2 170**	**2 492**	**2 605**	**2 813**
Australia Australie	1 427	1 465	1 430	1 520	1 546	1 596	1 526	1 792	1 870	2 056
New Zealand Nouvelle-Zélande	724	723	671	694	* 770	671	644	700	735	757
former USSR ancienne URSS	**8 954**	**8 978**	**9 556**	**9 835**	**10 031**	**10 087**	**10 166**	**10 821**	**10 654**	**10 657**

80

Paper and paperboard
Production: thousand metric tons [*cont.*]

Papiers et cartons
Production : milliers de tonnes métriques [*suite*]

Source:
Food and Agriculture Organization of the United Nations
(Rome).

† All data shown which pertain to Germany prior to 3 October
1990 are indicated separately for the Federal Republic of
Germany and the former German Democratic Republic based on
their respective territories at the time indicated. Where
data for united Germany (3 October 1990 and thereafter) are
not available, available data are shown separately under the
designations Federal Republic of Germany and former German
Democratic Republic and pertain to the territorial
boundaries prior to 3 October 1990. For detailed
explanatory notes on data pertaining to Germany, see Annex I
- Country Nomenclature.

1 FAO estimate.
2 Including data for Namibia.

Source:
Organisation des Nations Unies pour l'alimentation et
l'agriculture (Rome).

† Toutes les données se rapportant à l'Allemagne avant le 3
octobre 1990 figurent dans deux rubriques séparées basées
sur les territoires respectifs de la République fédérale
d'Allemagne et l'ancienne République démocratique allemande
selon la période indiquée. En l'absence de données pour
l'Allemagne unifiée (à compter du 3 octobre 1990), les
données disponibles sont fournies séparément sous les
rubriques République fédérale d'Allemagne et ancienne
République démocratique allemande et se rapportent aux
limites territoriales antérieures au 3 octobre 1990. Pour
les notes explicatives en détail sur les données concernant
l'Allemagne, voir Annexe I – Nomenclature des pays.

1 Estimation de la FAO.
2 Y compris les données pour la Namibie.

81
Wood-based panels
Panneaux à base de bois
Production: thousand metric tons
Production : milliers de tonnes métriques

Country or area Pays ou zone	1981	1982	1983	1984	1985	1986	1987	1988	1989	1990
World *Monde*	100 338	96 283	105 620	108 888	112 284	117 847	121 809	127 530	129 586	124 939
Africa **Afrique**	1 533	1 615	1 680	1 692	1 832	1 831	1 836	1 860	1 891	1 935
Algeria[1] Algérie[1]	50	50	50	50	50	50	50	50	50	50
Angola Angola	28[1]	33[1]	43[1]	47[1]	51	41	21	11	11	11[1]
Cameroon Cameroun	68[1]	70[1]	80[1]	95[1]	99	102	89	71	80	80[1]
Central African Rep. Rép. centrafricaine	8	12	11	7	6	5	4	4[1]	3	2
Congo Congo	71	67	79	69	66	58	54	59	54	54[1]
Côte d'Ivoire Côte d'Ivoire	153	150	189	177	220	204	211	232	241	248
Egypt Egypte	39	32	37	45	51	49	72	71	90	114
Equatorial Guinea Guinée équatoriale	1[1]	1[1]	2[1]	6[1]	10	10[1]	10[1]	10[1]	10[1]	10[1]
Ethiopia Ethiopie	14[1]	15	15	15[1]	15	14	15	16	15	15[1]
Gabon Gabon	167	201	201[1]	207	228	228[1]	228[1]	228[1]	228[1]	228[1]
Ghana Ghana	74	68	50	58	60	64	64	64	53	53[1]
Guinea[1] Guinée[1]	2	2	2	2	...	...	...	...	...	...
Kenya Kenya	24	22	21	24	33	40	35	49	52	52[1]
Liberia Libéria	4	4	5	5[1]	5[1]	5[1]	5[1]	5[1]	5[1]	5[1]
Madagascar Madagascar	1	1	2	3	5	1	7	5	5	5
Malawi Malawi	10	8	6[1]	4	4	5	6	6	14[1]	14[1]
Morocco Maroc	95	105	105[1]	105[1]	135	135	135	145	147	160
Mozambique Mozambique	4	3	2	6	7	8	9	5	5	5
Nigeria Nigéria	209	209	209	209	203	233	233[1]	233[1]	233[1]	233[1]
Rwanda Rwanda	2	2	2[1]	2[1]	2[1]	2[1]	2[1]	2[1]	2[1]	2[1]
Somalia[1] Somalie[1]	2	2	2	2	2	2	2	2	...	...

81
Wood-based panels
Production: thousand metric tons [cont.]
Panneaux à base de bois
Production : milliers de tonnes métriques [suite]

Country or area Pays ou zone	1981	1982	1983	1984	1985	1986	1987	1988	1989	1990
South Africa Afrique du Sud	382 [1]	398	398	398 [1]	398 [1]	398 [1]	398 [1]	398 [1]	398 [1]	398 [1]
Sudan [1] Soudan [1]	2	2	2	2	2	2	2	2	2	2
Swaziland Swaziland	5	8	8	8	8	8	8 [1]	8	8	8
Tunisia Tunisie	47	67	76	64	85	86	87	96	97	97 [1]
Uganda Ouganda	1 [1]	2	2	2	2 [1]	3	4	3	3 [1]	3 [1]
United Rep.Tanzania Rép. Unie de Tanzanie	7	6	5	9	8	11	10	14	15	15 [1]
Zaire Zaïre	30	28	33	39	40	41	38	39	37	37 [1]
Zambia Zambie	4	3	8	6	8	7	6	9	8	8 [1]
Zimbabwe Zimbabwe	32	46	39	29	32	23	34	25	26	26 [1]
America, North **Amérique du Nord**	**32 806**	**29 290**	**35 715**	**37 326**	**39 219**	**41 798**	**41 487**	**41 328**	**40 992**	**39 153**
Canada Canada	4 711	4 134	5 324	5 474	6 063	5 921	6 500	6 224	6 917	6 266
Costa Rica Costa Rica	57	52	39	46	46	46 [1]	49	58	58 [1]	70
Cuba Cuba	6	90	98	135	123	136	137	126	149	149 [1]
Guatemala Guatemala	6	6	6	5	8	8	6	6	6	6
Honduras Honduras	* 11	5	6	8	7	8	9	9	9	9
Jamaica [1] Jamaïque [1]	4	4	4	4	4	3	2	1	...	...
Mexico Mexique	683	761	696	732	764	756	768	753	645	550
Nicaragua Nicaragua	14 [1]	22	14	10 [1]	5	6	5	5	3	9
Panama Panama	13	13	12	9	5	6	5	4	5	8
United States Etats-Unis	27 300	24 204	29 518	30 904	32 194	34 908	34 006	34 142	33 200	32 086
America, South **Amérique du Sud**	**3 627**	**3 370**	**3 527**	**3 570**	**3 496**	**3 655**	**3 955**	**4 089**	**4 181**	**4 232**
Argentina Argentine	357	347	390	360	331	389	391	387	386	386
Bolivia Bolivie	11	7	5	1	3	5	5	4	4	4 [1]

81
Wood-based panels
Production: thousand metric tons [cont.]
Panneaux à base de bois
Production : milliers de tonnes métriques [suite]

Country or area Pays ou zone	1981	1982	1983	1984	1985	1986	1987	1988	1989	1990
Brazil Brésil	2 576	2 398	2 523	2 512	2 490	2 523	2 758	2 847	2 892	2 892[1]
Chile Chili	142	120	140	183	202	218	251	261	298	349
Colombia Colombie	106	123	113	111	112	113	112	113	113[1]	113[1]
Ecuador Equateur	95	97[1]	97[1]	97[1]	99	106	128	137	145	145
Paraguay Paraguay	69[1]	83[1]	60[1]	97	88	89	100	107	107[1]	107[1]
Peru Pérou	78	58	37	42	34	45	48	46	57	60
Suriname Suriname	25	26	* 21	18	17	* 10	8	10	9	7
Uruguay Uruguay	17	10	12	16	10	9	10	11	10	10
Venezuela Venezuela	151	101	129	133[1]	110	· 148	144	167	160	160[1]
Asia Asie	18 569	19 184	21 151	21 354	22 186	23 224	25 754	27 599	27 797	27 732
Afghanistan[1] Afghanistan[1]	1	1	1	1	1	1	1	1	1	1
Bangladesh Bangladesh	13	11	10	9[1]	8	8	8	8	8	8[1]
Bhutan Bhoutan	...	...	...	...	...	...	...	...	9	13
Cambodia[1] Cambodge[1]	2	2	2	2	2	2	2	2	2	2
China Chine	2 448	2 494	2 654	2 478	2 553	2 835	3 397	3 831	3 650	3 396
Cyprus Chypre	...	21	22	23	24	23	24	24[1]	22	22
Hong Kong[1] Hong-kong[1]	12	12	12	12	12	12	12	12	12	12
India Inde	357	377	384	442	442[1]	442[1]	442[1]	442[1]	442[1]	442[1]
Indonesia Indonésie	1 553	2 488	3 278	3 876	4 919	6 077	6 715	8 103	9 115	9 617
Iran, Islamic Rep. of Iran, Rép. islamique d'	66	83	103	132[1]	155	228	203	184	185	202
Iraq Iraq	2[1]	2[1]	2[1]	2[1]	3	3	3	3	3	3[1]
Israel Israël	138	162	157	145	142	148	148[1]	148	170	177
Japan Japon	9 082	8 627	9 196	8 932	8 964	8 660	9 533	9 554	8 949	8 616

81
Wood-based panels
Production: thousand metric tons [cont.]
Panneaux à base de bois
Production : milliers de tonnes métriques [suite]

Country or area Pays ou zone	1981	1982	1983	1984	1985	1986	1987	1988	1989	1990
Korea, Republic of Corée, République de	1 676	1 487	1 568	1 403	1 351	1 241	1 360	1 520	1 369	1 343
Lao People's Dem. Rep. Rép. dém. pop. lao	3	5	5[1]	5[1]	6	6	8	9	10	10[1]
Lebanon[1] Liban[1]	46	46	46	46	46	46	46	46	46	46
Malaysia Malaisie	1 124	1 351	1 556	1 445[1]	1 383	1 266	* 1 479	* 1 455	* 1 630	1 630[1]
Mongolia[1] Mongolie[1]	4	4	4	4	4	4	4	4	4	4
Myanmar[1] Myanmar[1]	15	15	15	15	15	15	15	15	15	15
Pakistan Pakistan	48	54	54	54	67	72	83	85	86	70
Philippines Philippines	672	649	676	737	492	564	661	542	451	488
Singapore Singapour	617	550	572	501[1]	527[1]	489[1]	489[1]	489[1]	489[1]	489[1]
Sri Lanka Sri Lanka	15	13	16	17	16	10	10[1]	10[1]	10[1]	10[1]
Syrian Arab Republic[1] Rép. arabe syrienne[1]	27	27	27	27	27	27	27	27	27	27
Thailand Thaïlande	182	238	264	315	206	223	263	265	271	270
Turkey Turquie	446	444	479	691	781	781	781	781	781	781
Viet Nam Viet Nam	21	23	49	40	40[1]	40[1]	40[1]	41[1]	41[1]	39
Europe **Europe**	**31 572**	**30 545**	**30 963**	**31 577**	**31 678**	**32 434**	**33 305**	**36 518**	**38 305**	**37 983**
Albania[1] Albanie[1]	12	12	12	12	12	12	12	12	12	12
Austria Autriche	1 286	1 165	1 234	1 263	1 309	1 353	1 434	1 563	1 538	1 753
Belgium-Luxembourg Belgique-Luxembourg	1 741	1 755	1 840	1 864	1 966	2 036	2 090	2 148	2 262	2 358
Bulgaria Bulgarie	558	638	597	593	595	584	536	528	493	401
Czechoslovakia Tchécoslovaquie	1 215	1 250	1 338	1 418	1 393	1 420	1 427	1 445	1 434	1 354
Denmark Danemark	342	273	276	276	329	334	328	326	326	326[1]
Finland Finlande	1 551	1 491	1 446	1 373	1 315	1 333	1 394	1 447	1 516	1 341
France France	3 000	2 725	2 658	2 489	2 473	2 537	2 558	2 841	3 018	3 118

81
Wood-based panels
Production: thousand metric tons [*cont.*]
Panneaux à base de bois
Production : milliers de tonnes métriques [*suite*]

Country or area Pays ou zone	1981	1982	1983	1984	1985	1986	1987	1988	1989	1990
Germany† · Allemagne† F. R. Germany R. f. Allemagne	6 589	6 295	6 745	6 988	6 815	6 832	6 990	7 738	8 577	8 454
former German D. R. anc. R. d. allemande	1 096	1 198	1 263	1 282	1 219	1 243	1 207	1 210	1 179	1 181
Greece Grèce	398	363	371	397	388	413	378	422	398	425
Hungary Hongrie	363	390	404	375	396	383	381	340	449	503
Ireland Irlande	36	42	30	120	174	211	236	236	240	235
Italy Italie	2 670	2 392	2 208	2 205	2 223	2 310	2 920	4 295	4 277	4 292
Netherlands Pays-Bas	145	104	102	86	77	95	94	84	96	96
Norway Norvège	574	614	629	651	668	681	680	644	611	632
Poland Pologne	1 814	1 890	1 981	2 054	2 105	2 208	2 026	2 130	1 834	1 503
Portugal Portugal	493	471	523	496	657	776	827	851	1 035	1 101
Romania Roumanie	1 554	1 590	1 467	1 618	1 413	1 277	1 277	1 277	1 542	1 495
Spain Espagne	1 840	1 910	1 805	1 896	1 924	2 037	2 010	1 980	2 330	2 330
Sweden Suède	1 672	1 469	1 501	1 466	1 380	1 296	1 346	1 318	1 306	1 303
Switzerland Suisse	703	604	617	640	638	666	667	709	918	856
United Kingdom Royaume-Uni	627	627	646	682	971	1 090	1 229	1 690	1 676	1 676[1]
Yugoslavia Yougoslavie	1 293	1 277	1 270	1 333	1 238	1 307	1 258	1 284	1 238	1 238
Oceania **Océanie**	**1 246**	**1 258**	**1 083**	**1 242**	**1 325**	**1 370**	**1 538**	**1 665**	**1 784**	**1 772**
Australia Australie	865	871	733	858	899	923	1 004	988	1 081	1 023
Fiji Fidji	11	11	11	13	14	16	16	16	16	16[1]
New Zealand Nouvelle-Zélande	350	357	320	352	393	407	494	632	658	687
Papua New Guinea Papouasie-Nvl-Guinée	19	19[1]	19[1]	19[1]	19[1]	24[1]	24[1]	29[1]	29[1]	46
Solomon Islands Iles Salomon	1	...	...	...	...	...	...	...	...	...
former USSR ancienne URSS	**10 986**	**11 020**	**11 501**	**12 128**	**12 549**	**13 535**	**13 935**	**14 471**	**14 635**	**12 131**

81

Wood-based panels
Production: thousand metric tons [*cont.*]

Panneaux à base de bois
Production : milliers de tonnes métriques [*suite*]

Source:
Food and Agriculture Organization of the United Nations
(Rome).

† All data shown which pertain to Germany prior to 3 October
1990 are indicated separately for the Federal Republic of
Germany and the former German Democratic Republic based on
their respective territories at the time indicated. Where
data for united Germany (3 October 1990 and thereafter) are
not available, available data are shown separately under the
designations Federal Republic of Germany and former German
Democratic Republic and pertain to the territorial
boundaries prior to 3 October 1990. For detailed
explanatory notes on data pertaining to Germany, see Annex I
- Country Nomenclature.

1 FAO estimate.

Source:
Organisation des Nations Unies pour l'alimentation et
l'agriculture (Rome).

† Toutes les données se rapportant à l'Allemagne avant le 3
octobre 1990 figurent dans deux rubriques séparées basées
sur les territoires respectifs de la République fédérale
d'Allemagne et l'ancienne République démocratique allemande
selon la période indiquée. En l'absence de données pour
l'Allemagne unifiée (à compter du 3 octobre 1990), les
données disponibles sont fournies séparément sous les
rubriques République fédérale d'Allemagne et ancienne
République démocratique allemande et se rapportent aux
limites territoriales antérieures au 3 octobre 1990. Pour
les notes explicatives en détail sur les données concernant
l'Allemagne, voir Annexe I – Nomenclature des pays.

1 Estimation de la FAO.

82
Rubber: synthetic and reclaimed
Caoutchouc : synthétique et régénéré
Production: thousand metric tons
Production : milliers de tonnes métriques

Country or area Pays ou zone	1981	1982	1983	1984	1985	1986	1987	1988	1989	1990
A. Synthetic • Synthétique										
Total	8 479.7	7 677.4	8 238.4	9 010.4	8 900.6	9 096.5	9 353.6	1 060.9	9 916.7	9 784.8
Argentina[1] Argentine[1]	28.0	45.7	47.5	47.0	51.3	52.9	46.4	52.0	48.0	57.6
Australia[1] Australie[1]	43.0	43.0	32.7	22.8	31.2	35.7	34.0	45.4	42.2	40.8
Austria *[1] Autriche *[1]	5.0	5.0	5.0	5.0	5.0	5.0	5.0	5.0	5.0	5.0
Belgium *[1] Belgique *[1]	108.0	100.0	111.0	102.0	92.3	94.4	102.9	126.0	120.0	125.0
Brazil Brésil	222.9	228.1	220.9	258.4	265.9	270.7	250.0	283.8	260.3	* 252.0
Bulgaria Bulgarie	* 30.0[1]	* 25.0[1]	* 25.0[1]	* 26.0[1]	* 26.0[1]	* 25.0[1]	* 24.0[1]	28.1	25.0	15.5
Canada Canada	263.3	181.7	183.4	217.4	209.2	* 187.1	* 179.8	* 197.0	* 188.0	213.0
China Chine	124.9	136.0	168.9	174.1	181.1	188.4	218.7	257.6	292.2	317.6
Czechoslovakia Tchécoslovaquie	63.2	65.2	67.2	70.9	69.6	76.0	76.5	77.1	75.7	69.4
Finland *[1] Finlande *[1]	9.0	9.0	9.0	8.0	8.5	8.5	8.5	10.0	17.0	29.0
France France	486.3	389.6	# 513.5	554.8	597.9	541.8	538.2	569.0	587.4	514.6
Germany† • Allemagne† F. R. Germany R. f. Allemagne	415.4	403.8	432.4	448.9	457.8	466.6	470.6	500.3	508.7	523.3
former German D. R. anc. R. d. allemande	156.6	148.2	161.4	162.7	157.8	119.8	98.7	149.2	146.0	...
Greece Grèce	...	...	...	1.0	0.9	...	...	...	...	...
India Inde	28.7	29.0	29.0	37.8	36.6	29.6	29.6	44.7	50.7	52.2
Italy Italie	243.2	215.9	229.8	234.0[1]	234.7[1]	235.0[1]	245.0[1]	260.0[1]	295.0[1]	300.0[1]
Japan Japon	1 010.3	930.7	1 002.5	1 160.5	1 158.0	1 153.4	1 191.9	1 298.9	1 352.7	1 425.8
Korea, Republic of Corée, République de	82.1	64.0	91.2	99.8	99.6	111.5	136.4	147.9	139.7	183.8
Mexico Mexique	104.3	106.7	131.9	125.0	* 140.0	* 130.0	* 144.0	* 146.0	134.2	119.5
Netherlands Pays-Bas	211.0	201.0	196.0	207.0	235.0	209.4	204.8	188.6	212.3	236.6
Poland Pologne	111.1	99.9	119.7	119.4	126.1	116.1	116.7	127.4	125.1	103.1
Romania[1] Roumanie[1]	145.7	137.5	147.2	159.3	155.9	173.0	152.0	161.0	149.0	102.0

82
Rubber: synthetic and reclaimed
Production: thousand metric tons [cont.]
Caoutchouc: synthétique et régénéré
Production : milliers de tonnes métriques [suite]

Country or area Pays ou zone	1981	1982	1983	1984	1985	1986	1987	1988	1989	1990
South Africa [1] Afrique du Sud[1]	36.5	32.4	27.5	42.4	42.8	44.9	51.6	61.8	62.4	45.9
Spain Espagne	55.1	55.3	54.5	55.4	59.1	56.0	63.9	67.7	64.0	* 66.0
Sweden Suède	* 17.1[1]	* 18.8[1]	* 29.5[1]	36.1	46.2	47.3	47.8	* 31.0	32.0	* 48.0
Turkey Turquie	29.0	30.3	24.0	27.3	28.0	22.2	31.2	39.3	36.3	39.5
former USSR * [1] ancienne URSS * [1]	2 000.0	1 950.0	1 970.0	2 085.0	2 125.0	2 320.0	2 366.0	2 435.0	2 280.0	2 365.0
United Kingdom Royaume-Uni	224.0	207.6	219.5	237.2	232.7	249.2	257.0	312.8	311.4	292.1
United States Etats-Unis	2 225.1	1 817.2	1 987.4	2 285.1	2 026.2	2 126.4	2 261.4	2 437.5	2 355.5	2 114.5

B. Reclaimed • Régénéré

	1981	1982	1983	1984	1985	1986	1987	1988	1989	1990
Total	367.3	340.0	345.1	340.2	327.3	297.2	273.1	267.4	260.8	230.3
Australia Australie	0.2	0.1	...	...	...	...	...	...	...	...
Belarus Bélarus	...	...	...	...	...	23.6	22.9	23.7	21.7	16.1
Brazil Brésil	26.0	23.5	23.0	25.2						
Canada Canada	3.0	4.2	...	...	...	...	...	...	...	...
Colombia Colombie	0.4	0.4	0.4	...	...	...	...	...	...	...
Czechoslovakia Tchécoslovaquie	20.3	20.1	20.1	19.1	18.4	18.3	16.4	15.9	14.6	12.8
France France	15.5	15.1	...	...	...	...	...	...	...	...
Germany† • Allemagne† F. R. Germany R. f. Allemagne	6.3	5.6	5.0	4.3	5.4	4.4	...	...	...	...
Hungary Hongrie	0.0	0.0	0.1	0.1	0.1	0.2	1.3	1.4	3.7	3.7
India Inde	23.0	22.0	24.4	26.4	...	...	...	...	...	...
Japan Japon	56.6	54.5	58.0	60.7	63.5	54.1	51.1	47.4	46.9	42.1
Poland Pologne	...	...	...	...	...	5.8	5.4	6.1	5.7	3.9
Romania Roumanie	...	...	...	...	7.0	6.6	5.9	6.1	5.8	3.6
Spain Espagne	11.9	9.9	11.9	8.5	9.5	9.9	10.0	10.0	12.6	...

82
Rubber: synthetic and reclaimed
Production: thousand metric tons [cont.]
Caoutchouc: synthétique et régénéré
Production : milliers de tonnes métriques [suite]

Country or area Pays ou zone	1981	1982	1983	1984	1985	1986	1987	1988	1989	1990
Sri Lanka Sri Lanka	6.2	...	...	...	...	...	...	...	...	...
Switzerland Suisse	...	...	...	...	...	10.3	10.1	10.5	11.5	11.2
Ukraine Ukraine	...	...	...	...	...	34.6	31.2	31.4	29.1	26.7
former USSR ancienne URSS	...	...	...	...	...	58.2	54.0	55.0	50.8	42.7
United States Etats-Unis	88.8	73.6	75.3	72.7	62.4	47.4	34.1	31.7	25.5	...
Yugoslavia Yougoslavie	7.8	7.1	8.0	7.9	8.4	7.7	6.4	5.5	6.1	...

Source:
Industrial statistics database of the Statistical Division
of the United Nations Secretariat.

† All data shown which pertain to Germany prior to 3 October
1990 are indicated separately for the Federal Republic of
Germany and the former German Democratic Republic based on
their respective territories at the time indicated. Where
data for united Germany (3 October 1990 and thereafter) are
not available, available data are shown separately under the
designations Federal Republic of Germany and former German
Democratic Republic and pertain to the territorial
boundaries prior to 3 October 1990. For detailed
explanatory notes on data pertaining to Germany, see Annex I
- Country Nomenclature.

1 Source: International Rubber Study Group, (London).

Source:
Base de données pour les statistiques industrielles de la
Division de statistique du Secrétariat de l'ONU.

† Toutes les données se rapportant à l'Allemagne avant le 3
octobre 1990 figurent dans deux rubriques séparées basées
sur les territoires respectifs de la République fédérale
d'Allemagne et l'ancienne République démocratique allemande
selon la période indiquée. En l'absence de données pour
l'Allemagne unifiée (à compter du 3 octobre 1990), les
données disponibles sont fournies séparément sous les
rubriques République fédérale d'Allemagne et ancienne
République démocratique allemande et se rapportent aux
limites territoriales antérieures au 3 octobre 1990. Pour
les notes explicatives en détail sur les données concernant
l'Allemagne, voir Annexe I – Nomenclature des pays.

1 Source: "International Rubber Study Group", (Londres).

83
Rubber
Caoutchouc
Total and per capita consumption
Consommation totale et par habitant

Country or area Pays ou zone	Total: thousand metric tons Totale : milliers de tonnes métriques					Per capita: kilograms Par habitant : kilogrammes				
	1980	1987	1988	1989	1990	1980	1987	1988	1989	1990
World Monde										
natural										
naturel	3 780.0	4 790.0	5 160.0	5 290.0	5 230.0	0.8	1.0	1.0	1.0	1.0
synthetic										
synthtique	8 760.0	9 610.0	10 000.0	10 060.0	9 870.0	2.0	1.9	1.9	1.9	1.8
Argentina Argentine										
natural [1 2]										
naturel [1 2]	* 22.0	25.7	27.6	21.9	...	* 0.8	0.8	0.9	0.7	...
synthetic * [1]										
synthtique * [1]	16.0	16.2	15.2	13.3	...	0.6	...	0.5	0.4	...
Australia Australie										
natural										
naturel	42.0	* 35.5	* 40.5	* 43.5	* 34.0	2.9	* 2.2	* 2.5	* 2.6	* 2.0
synthetic										
synthtique	58.7	* 49.0	* 57.0	* 58.0	* 55.0	4.0	* 3.0	* 3.5	* 3.5	* 3.3
Belgium-Luxembourg Belgique-Luxembourg										
natural * [1 2]										
naturel * [1 2]	27.0	43.0	47.0	48.0	44.0	2.6	4.2	4.6	4.7	4.3
synthetic * [1]										
synthtique * [1]	90.0	80.0	85.0	90.0	91.0	8.8	7.8	8.3	8.8	8.9
Brazil Brésil										
natural										
naturel	81.1	115.4	125.3	124.2	* 114.0	0.7	0.8	0.9	0.8	* 0.8
synthetic										
synthtique	243.8	273.6	282.9	284.5	* 288.0	2.0	1.9	2.0	1.9	* 1.9
Canada Canada										
natural										
naturel	80.0	98.0	* 82.5	84.9	* 84.0	3.3	3.8	* 3.2	3.2	* 3.2
synthetic										
synthtique	200.0	210.0	202.0	191.1	185.4	8.3	8.1	7.7	7.3	7.0
China Chine										
natural *										
naturel *	340.0	555.0	660.0	675.0	600.0	0.3	0.5	0.6	0.6	0.5
synthetic										
synthtique	155.0	* 265.0	* 290.0	* 330.0	* 361.0	0.2	* 0.2	* 0.3	* 0.3	* 0.3
Czechoslovakia Tchécoslovaquie										
natural *										
naturel *	58.0	40.0	40.0	39.0	42.0	3.8	2.6	2.6	2.5	2.7
synthetic *										
synthtique *	118.0	130.0	138.0	139.0	131.0	7.7	8.4	8.8	8.9	8.4
France France										
natural										
naturel	187.7	170.0	181.0	184.0	179.0	3.5	3.1	3.2	3.3	3.2
synthetic										
synthtique	341.9	316.4	315.0	358.0	351.0	6.3	5.7	5.6	6.4	6.3
Germany† · Allemagne†										
F. R. Germany R. f. Allemagne										
natural										
naturel	179.7	198.5	203.6	221.1	208.7	2.9	3.2	3.3	3.6	3.4
synthetic										
synthtique	421.3	453.4	471.0	476.0	511.0	6.8	7.4	7.7	7.8	8.3
former German D. R. anc. R. d. allemande										
natural										
naturel	* 37.0	32.0	33.0	41.0	14.5	* 2.2	1.9	2.0	2.5	0.9
synthetic										
synthtique	* 123.0	128.0	131.0	125.0	51.0	* 7.3	7.8	8.0	7.7	3.1

83
Rubber
Total and per capita consumption [cont.]
Caoutchouc
Consommation totale et par habitant [suite]

Country or area	Total: thousand metric tons Totale : milliers de tonnes métriques					Per capita: kilograms Par habitant : kilogrammes				
Pays ou zone	1980	1987	1988	1989	1990	1980	1987	1988	1989	1990
India Inde										
natural										
naturel	170.8	277.6	311.1	333.2	358.3	0.2	0.3	0.4	0.4	0.4
synthetic										
synthtique	46.2	75.8	82.9	90.0	97.0	0.1	0.1	0.1	0.1	0.1
Indonesia Indonésie										
natural *										
naturel *	46.0	98.0	103.0	105.0	108.0	0.3	0.6	0.6	0.6	0.6
synthetic *										
synthtique *	16.0	35.0	42.0	48.0	49.0	0.1	0.2	0.2	0.3	0.3
Italy Italie										
natural										
naturel	132.0	136.0	140.0	143.0	130.0	2.3	2.4	2.5	2.5	2.3
synthetic										
synthtique	288.0	299.0	312.0	325.0	309.5	5.1	5.2	5.5	5.7	5.4
Japan Japon										
natural										
naturel	427.0	568.0	623.0	657.0	677.0	3.7	4.7	5.1	5.3	5.5
synthetic										
synthtique	885.0	946.0	1 042.0	1 103.0	1 133.0	7.6	7.8	8.5	9.0	9.2
Korea, Republic of Corée, République de										
natural *										
naturel *	118.0	200.0	235.0	230.0	252.0	3.1	4.8	5.6	5.4	5.9
synthetic *										
synthtique *	141.0	215.0	255.0	235.0	279.0	3.7	5.2	6.1	5.5	6.5
Malaysia Malaisie										
natural										
naturel	45.0	82.4	103.4	121.6	182.3	3.3	5.0	6.1	7.0	10.2
synthetic										
synthtique	* 6.0	7.4	8.9	11.6	14.6	* 0.4	0.4	0.5	0.7	0.8
Mexico Mexique										
natural										
naturel	* 53.0	74.0	78.0	83.0	* 60.0	* 0.8	0.9	0.9	1.0	* 0.7
synthetic										
synthtique	* 112.0	106.0	105.0	107.0	114.0	* 1.6	1.3	1.2	1.2	1.3
Netherlands Pays-Bas										
natural										
naturel	20.1	11.5	13.5	* 11.5	...	1.4	0.8	0.9	* 0.8	...
synthetic										
synthtique	69.9	62.1	64.6	* 64.0	...	4.9	4.2	4.4	* 4.3	...
Poland Pologne										
natural *										
naturel *	50.0	43.0	35.0	36.0	16.0	1.4	1.1	0.9	0.9	0.4
synthetic *										
synthtique *	146.0	140.0	145.0	130.0	72.0	4.1	3.7	3.8	3.4	1.9
Romania Roumanie										
natural *										
naturel *	44.0	22.0	21.0	12.0	24.0	2.0	1.0	0.9	0.5	1.0
synthetic *										
synthtique *	117.0	140.0	150.0	143.0	96.0	5.3	6.1	6.5	6.2	4.1
South Africa Afrique du Sud										
natural * [1 2]										
naturel * [1 2]	52.4	36.0	38.0	33.0	44.0	1.9	1.1	1.1	1.0	1.2
synthetic * [1]										
synthtique * [1]	63.0	60.0	69.0	68.0	68.0	2.2	1.8	2.0	2.0	1.9

83
Rubber
Total and per capita consumption [cont.]
Caoutchouc
Consommation totale et par habitant [suite]

Country or area Pays ou zone	Total: thousand metric tons Totale : milliers de tonnes métriques					Per capita: kilograms Par habitant : kilogrammes				
	1980	1987	1988	1989	1990	1980	1987	1988	1989	1990
Spain Espagne										
natural[12] naturel[12]	* 91.9	* 112.0	* 117.0	* 122.0	118.7	* 2.4	* 2.9	* 3.0	* 3.1	3.0
synthetic[1] synthtique[1]	* 141.0	* 166.0	* 165.0	* 157.0	160.4	* 3.8	* 4.3	* 4.2	* 4.0	4.1
Sweden Suède										
natural[12] naturel[12]	12.1	12.0	10.1	10.0	...	1.5	1.4	1.2	1.2	...
synthetic *[1] synthtique *[1]	43.5	40.9	41.9	40.2	...	5.2	4.9	5.0	4.8	...
Thailand Thaïlande										
natural naturel	28.0	47.1	57.3	77.6	99.1	0.6	0.9	1.1	1.4	1.8
synthetic * synthtique *	9.0	16.0	21.0	27.0	32.0	0.2	0.3	0.4	0.5	0.6
Turkey Turquie										
natural * naturel *	23.0	55.0	57.0	49.0	46.0	0.5	1.0	1.1	0.9	0.8
synthetic * synthtique *	26.0	61.0	60.0	55.0	52.0	0.6	1.2	1.1	1.0	0.9
former USSR ancienne URSS										
natural naturel	215.0	165.0	100.0	140.0	150.0	0.8	0.6	0.4	0.5	0.5
synthetic synthtique	1 955.0	2 300.0	2 375.0	2 280.0	2 340.0	7.4	8.2	8.4	8.0	8.1
United Kingdom Royaume-Uni										
natural naturel	130.8	134.0	140.0	* 132.5	* 136.0	2.3	2.3	2.4	* 2.3	* 2.4
synthetic synthtique	248.2	216.0	226.5	240.0	* 251.0	4.4	3.8	4.0	4.2	* 4.4
United States Etats-Unis										
natural naturel	585.0	789.0	858.3	866.9	807.5	2.6	3.2	3.5	3.5	3.2
synthetic synthtique	1 980.0	2 017.3	2 016.8	2 051.0	1 820.8	8.7	8.3	8.2	8.3	7.3
Yugoslavia Yougoslavie										
natural * naturel *	36.0	37.0	34.0	36.0	34.0	1.6	1.6	1.4	1.5	1.4
synthetic * synthtique *	67.0	73.0	64.0	75.0	60.0	3.0	3.1	2.7	3.2	2.5

Source:
International Rubber Study Group (London).

Source:
Groupe international d'études du caoutchouc (Londres).

† All data shown which pertain to Germany prior to 3 October 1990 are indicated separately for the Federal Republic of Germany and the former German Democratic Republic based on their respective territories at the time indicated. Where data for united Germany (3 October 1990 and thereafter) are not available, available data are shown separately under the designations Federal Republic of Germany and former German Democratic Republic and pertain to the territorial boundaries prior to 3 October 1990. For detailed explanatory notes on data pertaining to Germany, see Annex I - Country Nomenclature.

1 For synthetic rubber production, see table 82.

2 Net imports.

† Toutes les données se rapportant à l'Allemagne avant le 3 octobre 1990 figurent dans deux rubriques séparées basées sur les territoires respectifs de la République fédérale d'Allemagne et l'ancienne République démocratique allemande selon la période indiquée. En l'absence de données pour l'Allemagne unifiée (à compter du 3 octobre 1990), les données disponibles sont fournies séparément sous les rubriques République fédérale d'Allemagne et ancienne République démocratique allemande et se rapportent aux limies territoriales antérieures au 3 octobre 1990. Pour les notes explicatives en détail sur les données concernant l'Allemagne, voir Annexe I - Nomenclature des pays.

1 Pour la production du caoutchouc synthétique, voir tableau 82.

2 Importation nettes.

84

Tires

Pneumatiques : enveloppes

Production: thousands
Production : milliers

Country or area Pays ou zone	1981	1982	1983	1984	1985	1986	1987	1988	1989	1990
Total	662 404	655 234	693 882	752 158	753 291	771 916	814 201	863 631	889 885	886 584
Algeria Algérie	6	* 5	* 6	...	...	...	...	...	...	...
Argentina Argentine	3 883	3 799	4 463	5 024	3 741	4 594	5 007	4 238	4 527	4 554
Australia[12] Australie[12]	7 245	6 943	5 821	6 165	5 846	...	5 576	...	...	...
Belarus Bélarus	...	...	...	...	...	4 130	4 166	4 405	4 360	4 018
Brazil Brésil	16 585	17 438	18 230	21 020	22 827	24 376	28 224	30 012	29 172	29 160
Bulgaria Bulgarie	1 567	1 577	1 613	1 666	1 659	1 668	1 857	1 858	* 1 800	1 279
Canada[3] Canada[3]	21 574	22 965	23 533	25 519	26 655	28 765	27 553	24 571	...	...
Chile Chili	870	482	595	913	858	862	1 221	1 347	1 562	1 632
China[4] Chine[4]	7 290	8 640	12 710	15 690	19 260	19 243	23 332	29 910	32 262	32 091
Colombia Colombie	1 703	1 750	1 625	1 950	1 856	1 851	1 955	1 915	...	...
Cuba Cuba	252	184	255	328	345	319	232	320	230	...
Cyprus Chypre	...	81	113	119	105	91	* 91	* 92	89	* 99
Czechoslovakia Tchécoslovaquie	5 097	4 506	4 465	4 621	4 547	4 737	4 858	5 058	5 263	5 315
Côte d'Ivoire Côte d'Ivoire	25	16	12	...	...	...	...	...	...	...
Ecuador Equateur	601	643	12	12	15	17	97	32	...	...
Egypt Egypte	601	661	730	775	845	1 011	1 038	1 006	1 126	1 171
Finland[5] Finlande[5]	912	1 341	1 352	1 427	1 339	...	...	...	...	...
France France	43 196	41 478	45 606	47 817	47 316	51 138	49 344	54 043	61 678	54 536
Germany† · Allemagne† F. R. Germany R. f. Allemagne	36 843	36 335	37 878	39 254	40 475	42 826	47 083	48 644	49 467	48 247
former German D. R. anc. R. d. allemande	7 026	7 051	7 379	7 784	8 362	8 582	8 760	8 845	8 878	...
Ghana Ghana	...	112	...	...	...	...	...	...	...	...
Hungary Hongrie	615	549	537	552	578	635	674	686	736	605
India Inde	4 744	5 323	5 529	5 952	5 042	6 383	5 616	7 537	8 066	8 076

84

Tires
Production: thousands [*cont.*]
Pneumatiques: enveloppes
Production : milliers [*suite*]

Country or area Pays ou zone	1981	1982	1983	1984	1985	1986	1987	1988	1989	1990
Indonesia Indonésie	4 131	3 098	3 332	3 406	3 420	4 740	4 791	* 6 564	* 8 028	10 080
Iran, Islamic Rep. of[6] Iran, Rép. islamique d'[6]	2 943	29	35	48	36	4 291	4 587	4 638	4 814	...
Israel Israël	1 106	1 227	1 045	1 009	930	744	920	573	751	778
Italy Italie	27 513	27 146	26 598	27 460	27 955	30 860	32 005	32 053	33 625	38 292
Jamaica Jamaïque	232	216	196	209	210	217	265	243	290	268
Japan Japon	110 924	109 968	123 094	131 199	135 745	135 323	139 086	150 562	155 038	153 226
Kenya[7] Kenya[7]	...	...	...	...	...	725	617	631	797	813
Korea, Republic of Corée, République de	9 101	7 855	12 033	15 126	15 207	18 214	20 060	24 250	24 535	27 907
Luxembourg[8] Luxembourg[8]	2 118	2 088	...	...	...	...	...	...	...	...
Malaysia[4 9] Malaisie[4 9]	5 375	4 010	4 262	3 700	3 622	3 846	5 173	6 222	6 156	6 764
Mexico Mexique	8 328	9 522	8 722	9 779	10 472	9 330	10 164	10 474	11 038	11 855
Morocco Maroc	681	745	816	834	779	821	943	...	...	...
Mozambique Mozambique	155	128	111	51	80	60	...	...	...	...
New Zealand[10] Nouvelle-Zélande[10]	1 583	1 737	1 236	1 282	1 222	1 389	...	...	...	...
Nigeria Nigéria	2 246	2 046	2 205	1 604	2 154	* 2 941	...	...	...	...
Pakistan[2] Pakistan[2]	227	193	217	238	307	412	382	679	907	915
Panama Panama	65	57	49	49	37	...	22	23	...	...
Peru Pérou	968	834	676	782	784	914	946	868	710	637
Philippines Philippines	1 387	1 632	1 737	1 450	1 300	934	1 159	* 1 968	* 2 016	2 208
Poland Pologne	4 804	4 043	5 084	5 902	6 254	6 217	6 020	6 276	6 025	4 704
Portugal Portugal	1 878	1 852	2 011	2 154	2 286	2 385	2 754	3 338	3 120	2 976
Romania[10] Roumanie[10]	5 178	5 392	5 346	5 882	5 642	5 789	5 247	5 552	4 804	3 600
South Africa Afrique du Sud	5 761	5 923	5 666	5 825	5 041	5 432	6 066	6 813	6 817	7 478
Spain Espagne	17 470	15 479	15 670	20 610	19 386	20 034	23 204	24 323	24 696	...

84

Tires
Production: thousands [*cont.*]
      Pneumatiques: enveloppes
      Production: milliers [*suite*]

Country or area Pays ou zone	1981	1982	1983	1984	1985	1986	1987	1988	1989	1990
Sri Lanka Sri Lanka	193	189	197	164	180	200	340	343	339	382
Sweden Suède	2 030	2 028	2 616	2 741	2 996	3 001	2 852	...	...	...
Thailand Thaïlande	1 968	1 583	2 202	2 177	2 126	2 264	3 063	3 980	4 320	4 183
Tunisia Tunisie	106	132	172	205	261	358	374	471	504	576
Turkey Turquie	2 433	2 905	3 777	4 690	4 922	4 871	7 251	7 008	6 636	4 596
Ukraine Ukraine	...	...	...	...	...	7 474	8 228	8 783	8 730	8 539
former USSR[11] ancienne URSS[11]	60 476	61 675	62 025	63 737	65 171	66 023	67 802	69 125	69 705	...
United Kingdom Royaume-Uni	23 724	24 376	23 976	24 120	24 216	25 644	27 624	30 204	31 080	29 376
United Rep.Tanzania Rép. Unie de Tanzanie	...	...	...	...	113	139	197	188	213	...
United States Etats-Unis	181 762	178 500	186 924	209 375	195 972	190 296	202 980	211 356	212 868	210 660
Venezuela Venezuela	3 681	3 590	3 358	2 985	4 492	4 989	5 447	5 203	4 177	3 951
Yugoslavia[1] Yougoslavie[1]	10 512	9 388	9 420	9 858	11 194	11 632	11 718	12 548	13 201	12 744

Source:
Industrial statistics database of the Statistical Division
of the United Nations Secretariat.

† All data shown which pertain to Germany prior to 3 October
1990 are indicated separately for the Federal Republic of
Germany and the former German Democratic Republic based on
their respective territories at the time indicated. Where
data for united Germany (3 October 1990 and thereafter) are
not available, available data are shown separately under the
designations Federal Republic of Germany and former German
Democratic Republic and pertain to the territorial
boundaries prior to 3 October 1990. For detailed
explanatory notes on data pertaining to Germany, see Annex I
- Country Nomenclature.

1 Including motorcycle tires.
2 Twelve months ending 30 June of year stated.

3 Source: International Rubber Study Group, (London).
4 Tires of all types.
5 Beginning 1986, data are confidential.
6 Production by establishments employing 50 or more persons.
7 Including retreaded tires.
8 Beginning 1983, data for hard, antimonial lead are
confidential.
9 Beginning 1982, scope of series revised.
10 Including tires for vehicles operating off-the-road.
11 Including tires for agricultural vehicles, motorcycles and
scooter.

Source:
Base de données pour les statistiques industrielles de la
Division de statistique du Secrétariat de l'ONU.

† Toutes les données se rapportant à l'Allemagne avant le 3
octobre 1990 figurent dans deux rubriques séparées basées
sur les territoires respectifs de la République fédérale
d'Allemagne et l'ancienne République démocratique allemande
selon la période indiquée. En l'absence de données pour
l'Allemagne unifiée (à compter du 3 octobre 1990), les
données disponibles sont fournies séparément sous les
rubriques République fédérale d'Allemagne et ancienne
République démocratique allemande et se rapportent aux
limites territoriales antérieures au 3 octobre 1990. Pour
les notes explicatives en détail sur les données concernant
l'Allemagne, voir Annexe I – Nomenclature des pays.

1 Y compris les pneumatiques pour motocyclettes.
2 Période de douze mois finissant le 30 juin de l'année
indiquée.
3 Source: "International Rubber Study Group", (Londres).
4 Pneumatiques de tous genres.
5 A partir de 1986, les données sont confidentielles.
6 Production des établissements occupant 50 personnes ou plus.
7 Y compris les pneumatiques rechapés.
8 A partir de 1983, les données pour le plomb antimonieux
affiné sont confidentielles.
9 A partir de 1982, portée de la série révisée.
10 Y compris les pneumatiques pour véhicules tous terrains.
11 Y compris les pneumatiques pour véhicules agricoles,
motocyclettes et scooters.

85
Nitrogenous fertilizers
Engrais azotés
Production in terms of nitrogen: thousand metric tons
Production exprimée en azote : milliers de tonnes métriques

Country or area Pays ou zone	1981/82	1982/83	1983/84	1984/85	1985/86	1986/87	1987/88	1988/89	1989/90	1990/91
World *Monde*	**62 295.1**	**63 419.2**	**67 815.7**	**74 507.6**	**73 100.8**	**77 426.8**	**82 264.7**	**85 718.2**	**84 640.2**	**82 270.1**
Africa **Afrique**	**1 160.9**	**1 415.0**	**1 477.2**	**1 804.9**	**1 666.7**	**1 653.4**	**1 901.6**	**2 144.2**	**2 143.0**	**2 223.3**
Algeria[1] Algérie[1]	23.7	51.0	47.0	101.8	105.8	112.5	117.0	105.5	* 88.0	* 80.0
Cameroon * [1] Cameroun * [1]	8.5	...	...	...	...	...	...	...	...	...
Côte d'Ivoire * [1] Côte d'Ivoire * [1]	2.0	3.0	3.9	1.0	2.6	1.0	1.6	...	...	...
Egypt Egypte	* 482.0	622.8	639.2	625.9	575.5	601.9	* 657.0	* 676.7	* 678.0	676.1
Libyan Arab Jamah.[1] Jamah. arabe libyenne[1]	...	106.6	151.9	331.4	278.3	239.9	* 302.3	* 144.3	* 124.0	* 89.0
Mauritius Maurice	8.3	8.0	9.0	7.8	9.2	10.1	11.4	11.4	* 9.0	12.0
Morocco[1] Maroc[1]	* 26.2	39.2	55.1	42.7	63.5	* 61.8	* 82.6	* 232.9	* 273.1	* 344.1
Mozambique * Mozambique *	3.1	2.0	0.5	...	...	...	...	...	...	...
Nigeria * [1] Nigéria * [1]	...	...	...	...	...	...	73.0	243.4	272.4	284.0
Senegal Sénégal	* 5.2	* 2.9	* 3.5	* 10.0	* 12.0	* 5.6	* 16.0	* 12.0	14.4	15.7
Somalia * [1] Somalie * [1]	...	...	...	2.0	3.2	...	...	...	...	...
South Africa * [1] Afrique du Sud * [1]	447.0	409.0	397.0	482.0	420.0	368.0	380.0	440.0	400.0	430.0
Swaziland * Swaziland *	3.3	4.5	5.0	...	...	...	...	...	...	...
Tunisia[1] Tunisie[1]	68.7	73.1	83.4	* 124.2	120.1	160.9	172.8	197.5	206.3	202.0
United Rep.Tanzania[1] Rép. Unie de Tanzanie[1]	* 8.3	1.9	5.0	* 2.5	4.3	5.7	2.6	0.7	* 4.1	* 1.1
Zambia[1] Zambie[1]	4.7	7.0	5.7	5.0	2.4	12.1	14.8	* 5.7	* 1.9	* 6.0
Zimbabwe[1] Zimbabwe[1]	69.9	84.0	* 71.0	* 68.6	69.7	73.9	70.5	74.1	71.7	83.3
America, North **Amérique du Nord**	**13 441.7**	**12 148.5**	**12 974.7**	**15 386.4**	**13 956.6**	**15 326.7**	**16 678.1**	**17 509.1**	**17 185.7**	**18 021.6**
Canada Canada	* 1 825.0	* 1 897.0	2 371.0	2 543.0	2 825.3	2 466.5	2 762.0	* 3 000.0	2 706.0	2 683.4
Costa Rica * [1] Costa Rica * [1]	42.0	46.0	36.0	40.5	27.0	32.0	30.0	27.7	24.0	30.2
Cuba[1] Cuba[1]	142.2	96.1	91.6	169.1	175.5	167.2	156.9	142.1	145.9	* 140.0
El Salvador[1] El Salvador[1]	...	...	...	...	...	...	3.8	7.7	5.5	11.2
Guatemala Guatemala	* 8.9	* 12.4	* 8.0	* 9.0	* 10.0	* 10.0	6.2	* 7.0	* 7.0	* 7.0

85
Nitrogenous fertilizers
Production in terms of nitrogen: thousand metric tons [cont.]
Engrais azotés
Production exprimée en azote : milliers de tonnes métriques [suite]

Country or area Pays ou zone	1981/82	1982/83	1983/84	1984/85	1985/86	1986/87	1987/88	1988/89	1989/90	1990/91
Mexico[1] Mexique[1]	877.2	1 067.1	1 044.5	1 188.0	1 304.8	1 249.5	1 339.3	1 348.6	* 1 497.9	* 1 366.0
Trinidad and Tobago[1] Trinité-et-Tobago[1]	23.4	29.9	22.6	66.9	156.9	225.5	217.7	* 248.0	* 223.4	* 231.8
United States[2 3] Etats-Unis[2 3]	10 523.0	9 000.0	9 401.0	11 370.0	9 457.0	11 176.0	12 162.0	12 728.0	12 576.0	13 552.0
America, South **Amérique du Sud**	**769.0**	**849.6**	**952.4**	**1 143.8**	**1 171.9**	**1 364.1**	**1 451.5**	**1 415.6**	**1 430.3**	**1 372.8**
Argentina[1] Argentine[1]	25.1	* 29.0	* 32.2	27.9	29.8	45.0	43.4	* 51.0	* 50.0	* 40.5
Brazil[1] Brésil[1]	349.6	396.8	533.3	669.2	696.2	714.2	746.1	705.1	748.5	737.2
Chile[1] Chili[1]	98.2	90.7	98.1	112.8	* 113.0	122.3	* 125.0	* 130.0	* 130.0	* 126.0
Colombia[1] Colombie[1]	* 42.2	* 48.3	53.9	60.4	65.1	75.6	91.4	105.8	95.7	* 99.7
Ecuador *[1] Equateur *[1]	2.5	3.0	...	...	...	...	...	...	...	...
Peru[1] Pérou[1]	86.1	73.5	* 28.0	* 18.7	* 13.0	86.3	65.9	79.7	36.2	32.4
Venezuela[1] Venezuela[1]	* 165.3	* 208.3	* 206.9	* 254.7	254.9	320.8	379.8	* 344.0	* 370.0	* 337.0
Asia **Asie**	**19 352.2**	**20 252.2**	**21 571.6**	**24 117.9**	**23 964.5**	**25 891.2**	**28 256.3**	**30 212.5**	**31 391.9**	**32 076.3**
Afghanistan Afghanistan	48.3	48.3	48.3	48.8	55.2	* 56.1	* 56.7	55.2	55.2	* 53.0
Bangladesh Bangladesh	193.4	236.6	337.1	343.1	386.2	391.3	592.7	667.4	677.8	653.6
China[1] Chine[1]	10 154.0	10 456.0	11 289.0	12 419.0	11 649.6	11 858.5	13 685.1	13 954.0	14 515.0	14 914.5
Cyprus[1] Chypre[1]	...	2.6	8.9	...	...	...	5.3	16.5	11.2	...
India[4] Inde[4]	3 143.3	3 429.7	3 491.5	3 917.3	4 322.9	5 412.2	5 465.6	6 712.4	6 747.4	6 993.1
Indonesia[1] Indonésie[1]	970.6	940.3	1 077.2	* 1 402.4	* 1 749.1	* 1 971.1	* 1 978.9	2 032.7	2 369.0	2 462.3
Iran, Islamic Rep. of Iran, Rép. islamique d'	* 21.4	22.2	23.6	13.3	* 20.0	70.6	98.9	115.9	* 308.1	* 376.2
Iraq[1] Iraq[1]	...	...	14.0	* 11.0	38.8	64.7	* 76.9	* 333.4	* 450.0	* 409.0
Israel Israël	62.4	74.1	81.7	80.5	79.0	* 75.0	* 76.0	* 73.3	* 75.0	* 75.0
Japan Japon	1 253.0	1 126.0	1 077.0	1 211.0	1 054.0	988.0	986.0	977.0	946.0	957.0
Jordan[1] Jordanie[1]	...	21.0	54.3	97.4	91.9	99.2	108.9	110.7	108.5	107.2
Korea, Dem. P. R. *[1] Corée, R. p. dém. de *[1]	600.0	588.0	608.0	620.0	630.0	640.0	650.0	660.0	660.0	660.0
Korea, Republic of[1] Corée, République de[1]	* 666.0	* 615.5	* 580.3	* 659.9	* 635.0	657.0	* 593.7	* 678.8	583.4	554.4

85
Nitrogenous fertilizers
Production in terms of nitrogen: thousand metric tons [cont.]
Engrais azotés
Production exprimée en azote : milliers de tonnes métriques [suite]

Country or area Pays ou zone	1981/82	1982/83	1983/84	1984/85	1985/86	1986/87	1987/88	1988/89	1989/90	1990/91
Kuwait[1] Koweït[1]	* 183.0	234.1	* 223.5	* 280.8	* 262.6	343.9	399.8	384.5	* 386.4	* 204.0
Malaysia * Malaisie *	20.0	19.8	44.0	47.0	60.0	180.0	282.0	247.0	249.0	211.0
Myanmar * Myanmar *	60.0	51.2	60.0	86.5	115.0	140.3	150.1	111.9	88.3	60.0
Pakistan Pakistan	715.0	999.4	1 014.6	1 027.9	1 033.1	1 117.7	1 097.1	1 112.5	1 156.3	1 120.3
Philippines[1] Philippines[1]	41.1	18.2	23.4	15.4	88.9	* 120.0	119.9	127.8	127.0	* 120.8
Qatar[1] Qatar[1]	264.5	304.7	315.2	337.6	342.2	343.6	320.2	* 358.6	350.1	342.7
Saudi Arabia[1] Arabie saoudite[1]	158.7	159.4	266.3	396.7	421.3	435.6	449.1	417.5	* 428.4	* 584.0
Sri Lanka[1] Sri Lanka[1]	38.4	97.0	57.6	64.3	* 5.5	...	...	...	...	...
Syrian Arab Republic Rép. arabe syrienne	36.9	80.0	74.9	117.0	119.4	104.1	* 80.6	81.3	108.1	31.8
Turkey[1] Turquie[1]	717.7	723.0	783.1	759.5	* 631.3	* 603.1	* 730.0	* 725.2	700.0	947.1
United Arab Emirates[1] Emirats arabes unis[1]	...	...	* 9.0	* 157.9	166.5	211.6	* 243.8	* 243.8	266.4	228.3
Viet Nam[1] Viet Nam[1]	4.4	5.1	9.3	3.6	6.9	7.7	9.0	15.2	25.3	* 11.0
Europe **Europe**	**16 784.3**	**17 073.3**	**17 744.4**	**18 642.6**	**18 065.4**	**17 942.3**	**18 150.7**	**18 522.3**	**17 927.6**	**15 194.6**
Albania[1] Albanie[1]	* 73.0	70.8	* 73.0	* 75.0	* 75.0	* 75.0	66.1	69.3	79.9	* 73.4
Austria * Autriche *	238.1	254.8	220.0	232.5	240.0	194.0	249.6	221.5	230.0	227.0
Belgium-Luxembourg * Belgique-Luxembourg *	739.0	755.0	706.0	760.0	750.0	795.0	760.0	666.0	678.0	725.0
Bulgaria[1] Bulgarie[1]	752.9	758.0	813.0	836.1	837.7	817.9	* 808.0	956.3	926.0	* 914.5
Czechoslovakia Tchécoslovaquie	* 674.0	* 656.0	* 685.7	* 687.5	* 689.6	* 700.0	625.4	585.2	* 675.0	* 604.0
Denmark Danemark	130.7	161.3	202.5	186.7	169.3	158.0	* 176.0	* 191.0	* 200.4	* 184.9
Finland Finlande	* 235.7	262.6	293.5	302.7	282.2	294.0	* 281.2	* 293.4	* 277.6	* 268.0
France France	* 1 553.0	* 1 530.0	* 1 600.0	* 1 695.0	* 1 690.0	* 1 530.0	* 1 435.0	1 675.0	1 572.0	1 524.0
Germany† • Allemagne† F. R. Germany[5] R. f. Allemagne[5]	1 108.3	985.0	1 111.2	1 195.1	1 117.2	1 019.2	1 032.5	918.6	787.8	# * 1 165.0
former German D. R.[1] anc. R. d. allemande[1]	966.6	948.4	968.3	959.1	1 078.2	1 252.1	1 317.5	1 382.2	* 1 133.0	...
Greece[1] Grèce[1]	306.8	* 364.4	* 402.4	434.2	420.0	355.9	415.6	* 388.0	* 407.0	396.2

85
Nitrogenous fertilizers
Production in terms of nitrogen: thousand metric tons [cont.]
Engrais azotés
Production exprimée en azote : milliers de tonnes métriques [suite]

Country or area Pays ou zone	1981/82	1982/83	1983/84	1984/85	1985/86	1986/87	1987/88	1988/89	1989/90	1990/91
Hungary [1] Hongrie [1]	690.7	670.8	699.8	687.8	700.5	672.2	692.4	591.4	591.2	469.8
Iceland [1] Islande [1]	8.8	* 8.0	6.1	12.0	11.6	11.6	10.0	9.6	10.3	10.1
Ireland * Irlande *	205.0	235.0	265.0	260.0	238.0	255.4	285.9	252.0	296.7	279.0
Italy Italie	1 149.9	1 132.5	1 096.0	1 332.8	1 242.4	1 144.9	1 279.0	1 297.1	1 168.6	762.3
Netherlands Pays-Bas	1 463.4	1 504.7	1 699.6	1 751.3	1 570.7	1 742.4	* 1 757.4	1 836.9	1 847.9	* 1 875.0
Norway Norvège	424.0	458.0	494.1	478.6	408.3	384.1	425.9	* 450.7	* 494.0	* 509.0
Poland Pologne	1 273.5	1 297.7	1 341.5	1 369.3	1 253.7	1 445.2	1 543.1	1 622.2	1 642.6	1 303.0
Portugal Portugal	181.0	141.7	108.8	* 130.0	* 164.0	* 166.0	* 180.0	* 158.9	* 133.0	* 126.0
Romania [1] Roumanie [1]	1 822.0	2 008.0	2 091.0	2 212.0	2 197.0	* 1 900.0	* 1 916.0	2 130.2	2 035.4	* 1 249.1
Spain Espagne	895.6	851.9	858.0	980.2	982.2	* 987.8	* 975.3	935.5	966.7	* 885.7
Sweden [6] Suède [6]	170.9	188.2	193.5	200.6	196.2	164.3	166.0	127.7	154.4	176.0
Switzerland Suisse	* 32.5	* 29.5	33.5	34.1	39.6	36.2	32.8	33.2	30.2	26.7
United Kingdom Royaume-Uni	1 270.0	1 399.0	1 372.0	1 360.0	1 209.0	1 318.0	* 1 105.0	* 1 100.0	* 1 070.0	* 980.0
Yugoslavia [1] Yougoslavie [1]	419.0	402.0	* 410.0	* 470.0	* 503.0	* 523.0	* 615.0	630.4	519.8	460.8
Oceania Océanie	**206.0**	**199.6**	**221.3**	**269.0**	**275.7**	**253.1**	**288.0**	**300.2**	**289.8**	**287.5**
Australia * Australie *	206.0	194.6	173.1	214.0	215.7	202.1	220.0	230.2	230.0	224.5
New Zealand * Nouvelle-Zélande *	...	5.0	48.2	55.0	60.0	51.0	68.0	70.0	59.8	63.0
former USSR [1] ancienne URSS [1]	**10 581.0**	**11 481.0**	**12 874.0**	**13 143.0**	**14 000.0**	**14 996.0**	**15 538.5**	**15 614.3**	**14 272.0**	**13 094.0**

Source:
Food and Agriculture Organization of the United Nations
(Rome).

Source:
Organisation des Nations Unies pour l'alimentation et
l'agriculture (Rome).

† All data shown which pertain to Germany prior to 3 October
1990 are indicated separately for the Federal Republic of
Germany and the former German Democratic Republic based on
their respective territories at the time indicated. Where
data for united Germany (3 October 1990 and thereafter) are
not available, available data are shown separately under the
designations Federal Republic of Germany and former German

† Toutes les données se rapportant à l'Allemagne avant le 3
octobre 1990 figurent dans deux rubriques séparées basées
sur les territoires respectifs de la République fédérale
d'Allemagne et l'ancienne République démocratique allemande
selon la période indiquée. En l'absence de données pour
l'Allemagne unifiée (à compter du 3 octobre 1990), les
données disponibles sont fournies séparément sous les

85
Nitrogenous fertilizers
Production in terms of nitrogen: thousand metric tons [*cont.*]

Engrais azotés
Production exprimée en azote : milliers de tonnes métriques [*suite*]

Democratic Republic and pertain to the territorial boundaries prior to 3 October 1990. For detailed explanatory notes on data pertaining to Germany, see Annex I - Country Nomenclature.

1 Calendar year referring to the first part of the split year.

2 Excluding sodium nitrate.
3 Including data for Puerto Rico.
4 Fertilizer year - April to March.
5 Beginning 1990/91, data are for United Germany.

6 Fertilizer year - June to May.

rubriques République fédérale d'Allemagne et ancienne République démocratique allemande et se rapportent aux limites territoriales antérieures au 3 octobre 1990. Pour les notes explicatives en détail sur les données concernant l'Allemagne, voir Annexe I – Nomenclature des pays.

1 L'année civile se rapporte à la première partie de l'année fractionnée.
2 Non compris le nitrate de sodium.
3 Y compris les données de Porto Rico.
4 Périodes de douze mois - avril à mars.
5 A partir de 1990/91, les données sont pour l'Allemagne unifiée.
6 Périodes de douze mois - juin à mai.

86
Phosphate fertilizers
Engrais phosphatés

Production in terms of phosphoric acid: thousand metric tons
Production exprimée en acide phosphorique: milliers de tonnes métriques

Country or area Pays ou zone	1981/82	1982/83	1983/84	1984/85	1985/86	1986/87	1987/88	1988/89	1989/90	1990/91
World *Monde*	31 686.4	32 117.3	35 097.0	36 491.0	34 640.0	37 370.3	39 138.9	41 378.8	39 732.6	38 906.3
Africa **Afrique**	**1 412.3**	**1 551.1**	**1 701.3**	**1 582.8**	**1 798.9**	**1 869.3**	**1 923.6**	**2 506.8**	**2 442.6**	**2 594.8**
Algeria [1] Algérie [1]	22.7	48.0	53.0	77.5	71.1	53.8	55.1	70.7	* 45.3	* 50.0
Côte d'Ivoire * [1] Côte d'Ivoire * [1]	2.7	3.3	2.0	3.1	2.6	2.7	0.9	1.5	2.5	2.5
Egypt [2] Egypte [2]	* 116.6	145.8	122.8	112.0	* 149.1	* 188.0	* 189.3	* 202.0	* 217.4	* 182.2
Morocco [1] Maroc [1]	* 199.0	* 295.9	493.4	335.7	* 468.5	466.5	* 536.6	* 969.8	* 935.8	* 1 179.9
Mozambique * Mozambique *	2.0	0.6	0.2	...	...	...	...	...	...	...
Nigeria [1] Nigéria [1]	* 9.5	* 7.0	* 4.2	* 5.0	* 5.0	* 5.0	* 5.0	* 27.4	44.1	50.2
Senegal Sénégal	* 25.0	* 16.1	* 30.0	* 30.0	* 37.5	* 30.0	* 35.0	* 35.0	27.5	* 50.0
South Africa * [1] Afrique du Sud * [1]	510.0	500.0	420.0	500.0	445.0	370.0	330.0	404.0	375.0	380.0
Tunisia [1,2] Tunisie [1,2]	* 473.8	495.5	527.1	475.8	567.4	701.5	726.0	746.9	753.5	657.1
United Rep.Tanzania [1] Rép. Unie de Tanzanie [1]	* 11.0	1.4	4.1	* 4.0	4.3	4.7	2.1	0.7	* 3.0	* 1.5
Zimbabwe [1] Zimbabwe [1]	40.0	37.5	* 44.5	* 39.7	48.3	47.1	43.6	48.9	38.4	41.4
America, North **Amérique du Nord**	**7 999.6**	**8 144.2**	**9 643.7**	**10 579.3**	**8 711.1**	**9 713.3**	**10 105.0**	**10 440.5**	**10 499.5**	**10 954.5**
Canada Canada	* 658.0	* 581.0	655.0	644.8	630.3	465.2	436.4	* 463.9	440.0	347.9
Cuba [1] Cuba [1]	5.6	4.1	2.1	16.2	19.1	14.1	17.5	14.6	14.9	* 7.0
El Salvador [1] El Salvador [1]	...	...	...	...	...	0.4	0.1	0.1	0.1	0.1
Guatemala * Guatemala *	7.5	10.6	9.0	10.0	10.0	10.0	6.2	7.0	7.0	7.0
Mexico [1] Mexique [1]	236.4	252.5	258.6	232.2	267.7	240.0	368.4	375.8	* 447.5	* 497.5
United States [3] Etats-Unis [3]	7 092.0	7 296.0	8 719.0	9 676.0	7 784.0	8 984.0	9 276.0	9 579.0	9 590.0	10 095.0
America, South **Amérique du Sud**	**1 281.7**	**1 214.5**	**1 135.2**	**1 528.6**	**1 395.5**	**1 631.9**	**1 605.3**	**1 450.0**	**1 215.0**	**1 175.9**
Argentina [1] Argentine [1]	0.1	* 0.1	...	* 0.1	* 0.1	...	...	...	...	...
Brazil [1,2] Brésil [1,2]	* 1 186.6	* 1 115.2	1 050.8	* 1 419.5	* 1 281.7	1 501.5	1 471.6	1 356.9	1 109.4	1 057.0
Chile [1] Chili [1]	...	...	...	* 3.7	* 3.7	5.4	4.1	* 4.0	* 4.0	* 5.0

86
Phosphate fertilizers
Production in terms of phosphoric acid: thousand metric tons [cont.]
Engrais phosphatés
Production exprimée en acide phosphorique: milliers de tonnes métriques [suite]

Country or area Pays ou zone	1981/82	1982/83	1983/84	1984/85	1985/86	1986/87	1987/88	1988/89	1989/90	1990/91
Colombia[1] Colombie[1]	* 49.5	* 30.5	36.3	* 40.0	15.6	30.5	* 30.5	* 20.7	* 39.0	* 40.0
Ecuador[1] Equateur[1]	* 7.5	* 9.5	* 0.3	0.3	...	...	...	...	...	...
Peru[1] Pérou[1]	4.0	3.3	2.2	1.3	1.4	7.5	* 4.1	2.9	4.0	4.7
Uruguay[1 2] Uruguay[1 2]	* 20.2	* 17.0	* 6.0	* 10.0	* 14.0	20.5	* 25.0	15.5	8.6	* 12.0
Venezuela[1] Venezuela[1]	13.8	* 38.9	* 39.5	* 53.7	79.0	66.4	70.0	* 50.0	* 50.0	* 57.2
Asia Asie	5 618.5	5 937.9	6 626.1	6 879.5	6 577.0	7 619.6	8 555.5	9 682.6	9 041.4	9 656.2
Bangladesh Bangladesh	27.0	32.0	37.4	26.5	46.3	57.8	53.8	65.6	67.3	50.9
Cambodia *[1 2] Cambodge *[1 2]	...	...	...	...	...	...	...	...	3.5	3.7
China[1] Chine[1]	2 593.6	2 589.1	2 722.2	2 420.9	1 830.3	2 410.6	3 341.4	3 766.2	3 807.7	4 196.0
Cyprus[1] Chypre[1]	...	3.2	11.4	...	...	...	8.2	38.2	24.1	...
India[2 4] Inde[2 4]	967.1	1 002.3	1 084.5	1 341.9	1 460.3	1 696.5	1 702.5	2 289.6	1 834.0	2 088.9
Indonesia[1] Indonésie[1]	257.2	270.6	367.9	* 461.0	* 463.3	513.8	554.4	551.5	551.3	588.8
Iran, Islamic Rep. of Iran, Rép. islamique d'	* 2.6	7.0	3.9	...	...	...	6.4	1.8	* 49.7	* 80.5
Iraq[1] Iraq[1]	...	* 5.5	* 86.0	185.7	198.2	* 290.0	* 220.0	* 396.0	* 415.0	* 207.0
Israel Israël	38.4	* 60.0	* 97.0	119.0	127.8	* 120.0	* 160.0	* 145.0	* 180.0	* 200.0
Japan Japon	580.0	625.0	647.0	641.0	623.0	577.0	531.0	490.0	445.0	429.0
Jordan[1] Jordanie[1]	14.7	53.5	138.7	248.9	234.8	253.5	278.3	282.9	277.2	274.1
Korea, Dem. P. R. *[1] Corée, R. p. dém. de *[1]	130.0	130.0	130.0	132.0	135.0	137.0	137.0	137.0	137.0	137.0
Korea, Republic of[1] Corée, République de[1]	* 323.7	* 447.1	* 457.5	* 487.5	493.0	480.0	515.0	488.0	* 393.8	414.4
Lebanon[1] Liban[1]	* 51.0	9.4	4.7	13.9	...	...	...	* 20.0	* 15.0	* 34.0
Pakistan Pakistan	66.9	73.6	91.8	90.0	93.0	93.8	96.2	100.6	105.1	104.7
Philippines[1] Philippines[1]	38.2	18.9	23.8	14.2	109.7	* 256.1	192.5	198.5	191.0	* 198.9
Saudi Arabia *[1] Arabie saoudite *[1]	...	...	...	...	...	...	...	...	...	27.0
Sri Lanka[1 2] Sri Lanka[1 2]	4.8	4.4	4.9	4.4	4.6	6.2	8.1	* 6.5	6.8	9.1

86
Phosphate fertilizers
Production in terms of phosphoric acid: thousand metric tons [cont.]
Engrais phosphatés
Production exprimée en acide phosphorique: milliers de tonnes métriques [suite]

Country or area Pays ou zone	1981/82	1982/83	1983/84	1984/85	1985/86	1986/87	1987/88	1988/89	1989/90	1990/91
Syrian Arab Republic Rép. arabe syrienne	26.3	53.8	57.5	87.9	86.5	79.0	* 78.2	30.0	14.9	21.3
Turkey[1] Turquie[1]	468.9	518.0	621.5	580.1	* 604.3	* 587.1	* 610.2	615.4	468.0	524.8
Viet Nam[1 2] Viet Nam[1 2]	28.2	34.5	38.4	24.6	67.0	61.3	62.3	* 59.7	55.0	66.0
Europe **Europe**	**7 882.3**	**7 762.1**	**8 055.0**	**7 925.2**	**7 479.9**	**7 231.0**	**7 177.3**	**7 381.3**	**6 802.4**	**5 126.7**
Albania[1] Albanie[1]	* 6.0	22.8	* 25.0	* 18.0	* 18.0	* 18.0	26.7	25.0	24.3	* 25.5
Austria * Autriche *	95.5	95.4	110.1	99.1	75.3	109.3	105.0	110.0	85.0	70.0
Belgium-Luxembourg * Belgique-Luxembourg*	530.0	460.0	515.0	364.0	361.0	389.0	368.0	325.0	332.5	353.0
Bulgaria[1] Bulgarie[1]	257.6	238.9	204.1	211.1	171.7	132.4	128.0	178.6	168.8	* 46.6
Czechoslovakia Tchécoslovaquie	340.5	335.4	326.2	343.7	359.6	* 335.0	306.9	313.0	* 320.0	* 290.0
Denmark Danemark	107.8	130.4	146.6	170.2	151.2	121.3	* 150.0	* 130.0	* 136.2	* 81.1
Finland Finlande	184.6	216.9	* 238.7	250.4	243.3	225.8	* 230.6	* 216.3	* 183.3	* 171.9
France[2 5] France[2 5]	* 1 300.0	* 1 200.0	* 1 230.0	* 1 168.0	* 1 023.0	* 1 000.0	* 955.0	1 116.0	1 025.0	921.0
Germany† · Allemagne† F. R. Germany R. f. Allemagne	559.4	563.9	552.2	492.5	446.3	379.8	362.0	339.2	307.2	# * 294.0[6]
former German D. R.[1] anc. R. d. allemande[1]	359.7	286.2	315.0	308.4	299.2	309.2	291.2	299.7	287.5	...
Greece[1] Grèce[1]	143.4	* 166.7	* 193.7	186.4	185.0	191.2	203.7	* 206.3	* 198.0	* 198.6
Hungary[1] Hongrie[1]	* 233.6	* 223.5	* 233.3	* 220.4	* 192.4	* 207.0	* 244.0	* 230.4	220.0	* 109.0
Ireland * Irlande *	37.3	12.6	15.3	8.0	3.0	3.0	3.0	...	...	...
Italy Italie	395.1	475.8	452.5	427.2	380.9	351.2	450.2	412.1	266.0	221.7
Netherlands Pays-Bas	324.6	344.6	* 387.0	* 400.0	* 360.0	* 325.0	* 334.0	* 378.7	* 378.4	* 357.0
Norway Norvège	147.5	162.6	163.3	164.0	165.3	151.5	168.0	195.9	229.9	* 245.0
Poland Pologne	866.1	868.1	871.6	868.7	888.8	947.9	942.5	962.1	946.0	467.0
Portugal Portugal	93.9	90.4	* 79.0	* 75.6	* 81.0	* 69.8	* 79.6	* 60.8	* 76.7	* 61.0
Romania[1] Roumanie[1]	717.0	* 584.0	733.0	765.0	788.0	* 790.0	* 690.0	725.3	647.7	* 387.3

86

Phosphate fertilizers

Production in terms of phosphoric acid: thousand metric tons [cont.]

Engrais phosphatés

Production exprimée en acide phosphorique: milliers de tonnes métriques [suite]

Country or area Pays ou zone	1981/82	1982/83	1983/84	1984/85	1985/86	1986/87	1987/88	1988/89	1989/90	1990/91
Spain Espagne	371.6	449.2	409.8	463.9	480.7	429.1	* 451.1	419.0	368.2	* 326.1
Sweden[7] Suède[7]	119.3	129.9	140.1	139.9	112.7	99.0	120.2	122.1	108.0	108.0
Switzerland Suisse	* 2.8	* 2.7	3.7	3.7	3.5	2.7	2.6	2.7	2.4	2.5
United Kingdom[2 7] Royaume-Uni[2 7]	346.0	324.0	327.0	347.0	286.0	264.0	* 200.0	* 248.0	* 172.0	* 128.0
Yugoslavia[1] Yougoslavie[1]	343.0	378.0	* 382.7	* 430.0	* 404.0	* 380.0	* 365.0	365.2	319.3	262.3
Oceania **Océanie**	**1 149.1**	**947.4**	**1 054.8**	**899.6**	**846.6**	**828.1**	**932.7**	**973.3**	**716.6**	**437.2**
Australia Australie	785.0	* 590.7	* 677.7	* 568.0	545.4	609.2	702.7	809.8	* 511.0	* 269.6
Christmas Is.(Aust) Ile Christmas (Aust)	27.7	* 26.7	19.4	29.6	25.3	* 18.9	* 20.0	* 10.0	...	...
New Zealand Nouvelle-Zélande	* 336.4	* 330.0	357.7	* 302.0	* 276.0	* 200.0	* 210.0	* 153.5	* 205.6	* 167.6
former USSR[1 2] **ancienne URSS[1 2]**	**6 343.0**	**6 560.0**	**6 881.0**	**7 096.0**	**7 831.0**	**8 477.0**	**8 839.6**	**8 944.3**	**9 015.0**	**8 961.0**

Source:
Food and Agriculture Organization of the United Nations
(Rome).

Source:
Organisation des Nations Unies pour l'alimentation et
l'agriculture (Rome).

† All data shown which pertain to Germany prior to 3 October 1990 are indicated separately for the Federal Republic of Germany and the former German Democratic Republic based on their respective territories at the time indicated. Where data for united Germany (3 October 1990 and thereafter) are not available, available data are shown separately under the designations Federal Republic of Germany and former German Democratic Republic and pertain to the territorial boundaries prior to 3 October 1990. For detailed explanatory notes on data pertaining to Germany, see Annex I - Country Nomenclature.

1 Calendar year referring to the first part of the split year.

2 Including ground rock phosphate.
3 Including data for Puerto Rico.
4 Fertilizer year - April to March.
5 Fertilizer year - May to April.
6 Beginning 1990/91, data are for United Germany.

7 Fertilizer year - June to May.

† Toutes les données se rapportant à l'Allemagne avant le 3 octobre 1990 figurent dans deux rubriques séparées basées sur les territoires respectifs de la République fédérale d'Allemagne et l'ancienne République démocratique allemande selon la période indiquée. En l'absence de données pour l'Allemagne unifiée (à compter du 3 octobre 1990), les données disponibles sont fournies séparément sous les rubriques République fédérale d'Allemagne et ancienne République démocratique allemande et se rapportent aux limites territoriales antérieures au 3 octobre 1990. Pour les notes explicatives en détail sur les données concernant l'Allemagne, voir Annexe I - Nomenclature des pays.

1 L'année civile se rapporte à la première partie de l'année fractionnée.
2 Y compris les phosphates naturels broyés.
3 Y compris les données de Porto Rico.
4 Périodes de douze mois - avril à mars.
5 Périodes de douze mois - mai à avril.
6 A partir de 1990/91, les données sont pour l'Allemagne unifiée.
7 Périodes de douze mois - juin à mai.

87
Potash fertilizers
Engrais potassiques
Production: thousand metric tons
Production : milliers de tonnes métriques

Country or area Pays ou zone	1981/82	1982/83	1983/84	1984/85	1985/86	1986/87	1987/88	1988/89	1989/90	1990/91
World *Monde*	**25 654.0**	**24 417.2**	**27 899.0**	**28 681.5**	**28 285.7**	**28 759.2**	**30 818.8**	**31 157.9**	**28 327.8**	**26 720.4**
America, North **Amérique du Nord**	**7 678.0**	**7 064.8**	**8 654.0**	**8 567.0**	**7 475.5**	**8 249.5**	**9 132.3**	**9 201.9**	**7 780.5**	**8 528.3**
Canada Canada	6 043.0	5 377.8	7 154.0	7 285.0	6 521.5	7 042.5	7 841.3	8 088.9	* 6 773.5	* 7 520.3
United States[1] Etats-Unis[1]	1 635.0	1 687.0	1 500.0	1 282.0	954.0	1 207.0	1 291.0	1 113.0	1 007.0	1 008.0
America, South **Amérique du Sud**	**23.6**	**21.8**	**21.3**	**23.1**	**2.0**	**10.5**	**37.3**	**55.7**	**109.4**	**68.1**
Brazil[2] Brésil[2]	...	...	...	...	2.0	10.5	37.3	55.7	109.4	68.1
Chile Chili	23.6	21.8	21.3	23.1	...	...	...	...	...	...
Asia **Asie**	**909.3**	**974.0**	**1 154.5**	**1 362.6**	**1 706.5**	**1 926.4**	**2 011.7**	**2 052.2**	**2 167.4**	**2 191.1**
China[2] Chine[2]	24.4	25.0	29.0	32.0	23.0	24.0	39.7	53.8	56.0	47.0
Israel Israël	* 884.9	940.0	955.8	1 039.0	1 138.6	1 240.0	1 252.0	* 1 218.4	* 1 301.0	* 1 295.0
Jordan[2] Jordanie[2]	...	9.0	169.7	291.6	544.9	662.4	720.0	* 780.0	* 810.4	* 849.1
Europe **Europe**	**8 594.2**	**8 277.6**	**8 775.3**	**8 952.8**	**8 734.7**	**8 344.8**	**8 749.3**	**8 547.8**	**8 037.9**	**6 895.7**
France *[3] France *[3]	1 726.0	1 601.4	1 685.0	1 729.0	1 719.0	1 549.0	1 564.0	1 411.0	1 199.4	1 292.4
Germany† · Allemagne† F. R. Germany[4] R. f. Allemagne[4]	2 286.3	2 226.3	2 565.5	2 627.0	2 380.2	2 061.0	2 319.4	2 267.9	2 291.4	# * 4 462.0
former German D. R.[2] anc. R. d. allemande[2]	3 460.0	3 434.0	3 431.0	3 465.0	3 465.0	3 485.0	3 510.1	3 510.4	3 199.8	...
Italy Italie	119.9	134.8	125.3	140.6	139.9	112.2	128.1	159.4	130.5	6.0
Spain Espagne	720.9	620.1	* 653.5	662.2	686.5	728.6	754.1	754.2	728.7	* 642.3
United Kingdom[5] Royaume-Uni[5]	281.0	261.0	315.0	329.0	344.0	409.0	* 473.6	* 444.9	* 488.0	* 493.0
former USSR[2] **ancienne URSS[2]**	**8 449.0**	**8 079.0**	**9 294.0**	**9 776.0**	**10 367.0**	**10 228.0**	**10 888.2**	**11 300.3**	**10 232.6**	**9 037.2**

Source:
Food and Agriculture Organization of the United Nations
(Rome).

Source:
Organisation des Nations Unies pour l'alimentation et
l'agriculture (Rome).

† All data shown which pertain to Germany prior to 3 October
1990 are indicated separately for the Federal Republic of
Germany and the former German Democratic Republic based on
their respective territories at the time indicated. Where
data for united Germany (3 October 1990 and thereafter) are
not available, available data are shown separately under the
designations Federal Republic of Germany and former German

† Toutes les données se rapportant à l'Allemagne avant le 3
octobre 1990 figurent dans deux rubriques séparées basées
sur les territoires respectifs de la République fédérale
d'Allemagne et l'ancienne République démocratique allemande
selon la période indiquée. En l'absence de données pour
l'Allemagne unifiée (à compter du 3 octobre 1990), les
données disponibles sont fournies séparément sous les

87

Potash fertilizers
Production: thousand metric tons [*cont.*]

Engrais potassiques
Production : milliers de tonnes métriques [*suite*]

Democratic Republic and pertain to the territorial
boundaries prior to 3 October 1990. For detailed
explanatory notes on data pertaining to Germany, see Annex I
- Country Nomenclature.

1 Including data for Puerto Rico.
2 Calendar year referring to the first part of the split year.

3 Fertilizer year - May to April.
4 Beginning 1990/91, data are for united Germany.

5 Fertilizer year - June to May.

rubriques République fédérale d'Allemagne et ancienne
République démocratique allemande et se rapportent aux
limites territoriales antérieures au 3 octobre 1990. Pour
les notes explicatives en détail sur les données concernant
l'Allemagne, voir Annexe I – Nomenclature des pays.
1 Y compris les données de Porto Rico.
2 L'année civile se rapporte à la première partie de l'année
   fractionnée.
3 Périodes de douze mois - mai à avril.
4 A partir de 1990/91, les données sont pour l'Allemagne
   unifiée.
5 Périodes de douze mois - juin à mai.

88
Cement
Ciment

Production: thousand metric tons
Production : milliers de tonnes métriques

Country or area Pays ou zone	1981	1982	1983	1984	1985	1986	1987	1988	1989	1990
Total	**878 718**	**880 290**	**906 038**	**933 052**	**949 269**	**995 721**	**1 042 738**	**1 108 022**	**1 143 780**	**1 152 493**
Afghanistan[1] Afghanistan[1]	77	87	130	112	128	103	104	70	* 100	100
Albania Albanie	* 790[2]	* 830[2]	* 840[2]	* 860[2]	642	848	* 860[2]	746	754	800
Algeria Algérie	4 457	3 743	4 776	5 538	6 096	6 448	7 541	7 195	6 819	6 337
Angola Angola	244	176	125	126	205	* 354	* 354	* 1 000	* 1 000	1 000
Argentina Argentine	6 913	5 818	5 882	5 220	4 795	5 558	6 302	6 024	4 439	4 199
Australia Australie	5 656	6 136	5 351	4 655	5 680	6 106	5 920	6 158	6 901	7 075
Austria Autriche	5 288	5 012	4 907	4 899	4 560	4 569	4 518	4 763	4 749	4 903
Bahamas[2] Bahamas[2]	# 29	64	26	...	...	...	...	...	...	...
Bangladesh[3] Bangladesh[3]	345	326	307	273	240	292	310	310	344	336
Barbados Barbade	...	...	...	64	148	199	147	185	225	155
Belarus Bélarus	...	...	...	2 090	2 119	2 141	2 204	2 266	2 283	2 258
Belgium Belgique	6 691	6 320	5 719	5 708	5 537	5 760	5 689	6 451	6 720	6 924
Benin[2] Bénin[2]	297	315	300	300	300	* 300	* 300	* 500	* 500	500
Bolivia Bolivie	374	326	310	220	241	343	386	448	499	560
Brazil Brésil	26 051	25 644	20 870	19 497	20 635	25 252	25 470	25 332	25 917	25 850
Bulgaria Bulgarie	5 433	5 614	5 644	5 717	5 296	5 702	5 494	5 535	5 036	4 710
Burundi Burundi	1	0	...	...	...	...	...	...	...	...
Cameroon Cameroun	* 516[2]	530	598	694	785	779	707	586	...	...
Canada Canada	10 152	8 136	7 089	8 609	9 559	10 379	11 841	11 927	12 591	10 953
Chile Chili	1 863	1 131	1 255	1 390	1 430	1 441	1 500	1 885	2 010	2 115
China Chine	82 897	95 200	108 250	123 020	145 950	166 060	186 249	210 140	210 295	209 711
Colombia Colombie	4 610	4 707	4 787	5 280	5 412	5 916	5 892	6 312	6 648	6 360
Congo Congo	52	39	28	* 45	62	* 58	38	* 58	* 58	58

88

Cement
Production: thousand metric tons [cont.]
Ciment
Production : milliers de tonnes métriques [suite]

Country or area Pays ou zone	1981	1982	1983	1984	1985	1986	1987	1988	1989	1990
Costa Rica[4] Costa Rica[4]	460	424	386	350	350	306	285	309	315	...
Côte d'Ivoire Côte d'Ivoire	1 317	988	636	552	535	775	652	* 700	* 700	500
Cuba Cuba	3 292	3 163	3 231	3 347	3 182	3 305	3 535	3 566	3 759	3 696
Cyprus Chypre	1 036	1 068	943	853	659	857	854	868	1 042	1 133
Czechoslovakia Tchécoslovaquie	10 646	10 325	10 498	10 530	10 265	10 298	10 369	10 974	10 888	10 215
Denmark Danemark	1 602	1 770	1 654	1 668	1 983	2 029	1 886	1 681	2 004	1 656
Dominican Republic Rép. dominicaine	902	955	1 057	1 109	1 001	952	1 209	1 235	* 1 269	1 189
Ecuador Equateur	1 380	1 800	1 494	1 755	2 008	2 118	2 875	2 126	* 1 548	...
Egypt Egypte	3 452	3 631	3 794	4 600	5 275	7 612	8 762	9 794	12 480	14 111
El Salvador[4] El Salvador[4]	459	429	435	399	450	460	480	455	447	444
Ethiopia[5] Ethiopie[5]	142	146	120	165	228	270	350	* 406	* 400	400
Fiji Fidji	92	88	110	98	93	92	59	44	58	78
Finland Finlande	1 863	1 906	1 979	1 656	1 695	1 495	1 579	1 619	1 693	1 649
France France	28 229	26 141	24 503	22 724	22 224	21 584	23 544	* 25 374	25 994	26 497
Gabon Gabon	150	178	194	207	244	210	141	* 132[2]	* 115[2]	115
Germany†·Allemagne† F. R. Germany R. f. Allemagne	31 498	30 079	30 466	28 909	25 758	26 580	25 268	26 215	28 499	30 456
former German D. R. anc. R. d. allemande	12 204	11 721	11 782	11 555	11 608	11 988	12 430	12 510	12 229	10 000
Ghana Ghana	396	252	278	235	356	219	294	412	* 565	600
Greece Grèce	13 366	13 297	14 196	13 521	12 855	12 494	11 760	13 128	12 528	13 944
Guadeloupe Guadeloupe	160	* 160	211	188	194	197	222	* 200	200	200
Guatemala Guatemala	514	613	491	455	574	1 392	1 260	880	* 591	611
Haiti Haïti	229	213	219	248	263	248	254	245	* 234	226
Honduras Honduras	307	278	485	368	348	360	375	* 268	* 321	326

88
Cement
Production: thousand metric tons [cont.]
Ciment
Production : milliers de tonnes métriques [suite]

Country or area Pays ou zone	1981	1982	1983	1984	1985	1986	1987	1988	1989	1990
Hong Kong Hong-kong	1 517	1 436	1 717	1 847	1 835	2 236	2 226	2 189	2 141	1 808
Hungary Hongrie	4 635	4 369	4 243	4 145	3 678	3 846	4 153	3 873	3 857	3 933
Iceland Islande	115	128	120	114	117	115	127	* 134	* 116	120
India Inde	20 908	22 650	25 261	29 541	31 971	34 983	37 135	41 136	44 197	45 720
Indonesia Indonésie	5 719	5 832	8 095	8 893	9 940	11 323	11 814	12 096	15 660	15 972
Iran, Islamic Rep. of[6] Iran, Rép. islamique d'[6]	9 319	10 103	10 655	12 064	11 954	12 148	12 852	11 926	12 587	12 520
Iraq Iraq	5 600[2]	5 600[2]	5 600[2]	8 000[2]	8 000[2]	7 992	9 780	9 162	12 500	13 000
Ireland Irlande	1 944	1 619	1 486	1 377	1 457	1 250	1 446	* 1 332	* 1 600	1 600
Israel Israël	2 361	2 189	2 058	1 889	1 596	1 624	2 226	2 326	2 289	2 868
Italy Italie	42 094	40 245	39 763	38 307	37 155	36 393	37 788	38 220	39 692	40 788
Jamaica Jamaïque	159	237	278	259	241	247	261	339	360	...
Japan Japon	84 828	80 686	80 891	78 860	72 847	71 264	71 551	77 554	79 717	84 445
Jordan Jordanie	965	788	1 269	2 026	2 022	1 837	2 472	1 780	1 930	1 780
Kenya Kenya	1 280	1 238	1 180	1 134	1 115	1 174	1 243	1 201	1 316	1 512
Korea, Dem. P. R.[2] Corée, R. p. dém. de[2]	* 8 000	* 8 000	* 8 000	* 8 000	* 8 000	* 8 000	* 9 000	11 800	16 300	16 300
Korea, Republic of Corée, République de	15 617	17 887	21 282	20 413	20 509	23 530	25 946	29 611	30 821	33 914
Kuwait Koweït	1 549	1 553	1 124	1 184	1 066	1 014	888	984	1 108	800
Lebanon[2] Liban[2]	2 391	1 700	1 500	1 250	* 1 000	* 900	* 900	* 900	* 900	900
Liberia Libéria	100	66	88	86	104	97	95	130	* 85	50
Libyan Arab Jamah. Jamah. arabe libyenne	2 722	3 139	3 093	6 000[2]	6 500[2]	2 077[2]	2 700[2]	2 700[2]	2 700[2]	2 700[2]
Luxembourg Luxembourg	342	344	353	340	295	389	509	563	590	636
Madagascar Madagascar	36	36	36	37	28	32	44	33	24	20
Malawi Malawi	78	53	70	70	62	69	75	62	84	101
Malaysia[7] Malaisie[7]	2 833	3 123	3 231	3 469	3 128	3 569	3 316	3 861	4 794	5 881

88
Cement
Production: thousand metric tons [*cont.*]
Ciment
Production : milliers de tonnes métriques [*suite*]

Country or area Pays ou zone	1981	1982	1983	1984	1985	1986	1987	1988	1989	1990
Mali Mali	35	27^2	20^2	25^2	19^2	20^2	22^2	25^2	20^2	20^2
Martinique Martinique	180^2	208	207	189	191	209	221	247	* 200	200
Mexico Mexique	18 173	19 343	17 363	18 702	20 255	19 825	23 482	23 606	24 210	24 683
Mongolia Mongolie	106	179	165	141	151	425	541	502	513	510
Morocco Maroc	3 606	3 739	3 848	3 573	3 704	3 709	3 879	4 260	4 641	5 381
Mozambique Mozambique	261	270	188	105	77	73	73	64	* 80	100
Myanmar Myanmar	317	344	334	310	429	444	389	349	454	420
Nepal Népal	30	37	39	32	96	152	215	217	114	107
Netherlands Pays-Bas	3 316	3 103	3 108	3 176	2 911	3 099	2 929	3 418	3 541	3 729
New Caledonia Nouvelle-Calédonie	46	49	41	* 41	31	40	* 58	63	67	64
New Zealand Nouvelle-Zélande	759	781	760	823	863	895	880	793	729	681
Nicaragua[4] Nicaragua[4]	255	180	298	280	245	284	265	256	225	...
Niger Niger	41^2	36^2	30^2	* 38^2	* 38^2	26	29	40	* 27	27
Nigeria Nigéria	$2 568^8$	$3 012^8$	2 760	2 184	3 348	3 624	3 085	* 3 400	* 3 500	3 500
Norway Norvège	1 837	1 704	1 666	1 547	1 343	1 752	1 703	1 667	1 380	1 260
Pakistan[3] Pakistan[3]	3 538	3 657	3 938	4 503	4 732	5 773	6 508	7 072	7 125	7 488
Panama[2] Panama[2]	520	350	327	304	305	336	350	200	* 169	...
Paraguay Paraguay	156	111	153	109	46	179	269	256	336	468
Peru Pérou	2 584	2 473	1 967	1 947	1 757	2 207	2 584	2 514	2 105	2 184
Philippines Philippines	4 008	4 393	4 560	3 660	3 072	3 288	3 984	4 092	3 624	6 360
Poland Pologne	14 224	16 035	16 163	16 649	14 990	15 831	16 090	16 984	17 125	12 518
Portugal Portugal	6 033	5 969	6 060	5 514	5 279	5 425	5 853	6 423	* 6 000	6 000
Puerto Rico[3] Porto Rico[3]	1 225	1 044	855	896	865	896	1 094	1 163	1 257	1 302
Qatar Qatar	188	229	162	313	385	324	293	291	295	267

88
Cement
Production: thousand metric tons [cont.]
Ciment
Production : milliers de tonnes métriques [suite]

Country or area Pays ou zone	1981	1982	1983	1984	1985	1986	1987	1988	1989	1990
Réunion Réunion	162	...	...	...	...	...	...	...	322	336
Romania Roumanie	13 762	13 931	13 027	12 991	11 189	13 054	12 435	13 124	12 225	9 468
Rwanda Rwanda	...	...	...	8	32	47	57	51	...	...
Saudi Arabia Arabie saoudite	4 753	7 090	8 265	7 504	9 232	9 232	* 8 595	* 9 525	* 9 500	10 000
Senegal Sénégal	396	375	388	414	408	372	362	393	* 380	380
Sierra Leone Sierra Leone	...	...	42	0	17	34	24	...	9	...
Singapore Singapour	2 253	2 695	3 126	2 511	1 897	1 875	1 550	1 684	1 704	1 848
South Africa Afrique du Sud	8 095	8 080	7 897	8 084	6 880	6 246	5 999	6 760	7 261	6 563
Spain Espagne	28 752	29 604	30 637	25 435	21 876	22 007	23 012	24 372	27 375	28 092
Sri Lanka Sri Lanka	706	538	480	401	380	558	619	633	596	566
Sudan Soudan	150	170	232	200	148	175	178	110	* 150	150
Suriname Suriname	71	74	70	50	79	61	40	34	* 49	50
Sweden Suède	2 318	2 304	2 240	2 393	2 101	2 044	2 238	* 2 200	* 2 200	2 360
Switzerland Suisse	4 348	4 100	4 138	4 181	4 254	4 393	4 617	4 965	5 461	5 206
Syrian Arab Republic Rép. arabe syrienne	2 310	2 850	3 719	4 279	4 357	4 316	3 870	3 330	3 976	3 500
Thailand Thaïlande	6 323	6 656	7 301	8 271	7 951	8 005	9 870	11 519	15 042	18 040
Togo Togo	279	272	232	243	284	* 338	370	378	* 389	400
Trinidad and Tobago Trinité-et-Tobago	139	189	393	406	329	338	326	360	384	438
Tunisia Tunisie	2 023	1 834	2 532	2 777	3 033	2 962	3 215	3 600	3 780	4 140
Turkey Turquie	15 043	15 778	13 595	15 738	17 581	20 004	21 980	22 675	23 800	24 636
Uganda Ouganda	6	18	31	25	12	16	16	15	17	27
Ukraine Ukraine	...	20 747	21 800	22 672	22 444	23 069	23 193	23 533	23 416	22 729
former USSR ancienne URSS	127 169	123 681	128 156	129 866	130 772	135 119	137 404	139 499	140 436	137 321
United Arab Emirates Emirats arabes unis	2 220	2 068	2 068	4 005 [2]	4 205 [2]	2 748 [2]	3 106 [2]	2 980 [2]	3 112 [2]	3 110 [2]

88

Cement
Production: thousand metric tons [cont.]
Ciment
Production : milliers de tonnes métriques [suite]

Country or area Pays ou zone	1981	1982	1983	1984	1985	1986	1987	1988	1989	1990
United Kingdom Royaume-Uni	12 729	12 962	13 396	13 481	13 339	13 413	14 311	16 506	16 849	14 736
United Rep.Tanzania Rép. Unie de Tanzanie	396	336	247	369	376	435	498	592	595	300
United States Etats-Unis	65 054	57 475	63 927	70 452	70 284	71 112	67 380	71 544	71 308	70 944
Uruguay Uruguay	604	550	401	374	317	329	420	435	* 465	469
Venezuela Venezuela	4 904	5 432	4 430	4 783	5 121	5 875	5 975	6 199	5 259	5 996
Viet Nam Viet Nam	538	710	907	1 296	1 503	1 526	1 665	1 954	2 087	2 534
Yemen Yémen	85	243	623	709	698	708	* 760	* 646	* 700	700
Yugoslavia Yougoslavie	9 780	9 720	9 588	9 315	9 028	9 128	8 963	8 840	8 560	7 956
Zaire Zaïre	494	541	513	534	444	* 445[2]	* 492[2]	* 495[2]	* 460[2]	...
Zambia Zambie	302	319	392	220	316	334	375	405	385	432
Zimbabwe Zimbabwe	588	576	501	650	614	659	811	780	910	996

Source:
Industrial statistics database of the Statistical Division
of the United Nations Secretariat.

† All data shown which pertain to Germany prior to 3 October
1990 are indicated separately for the Federal Republic of
Germany and the former German Democratic Republic based on
their respective territories at the time indicated. Where
data for united Germany (3 October 1990 and thereafter) are
not available, available data are shown separately under the
designations Federal Republic of Germany and former German
Democratic Republic and pertain to the territorial
boundaries prior to 3 October 1990. For detailed
explanatory notes on data pertaining to Germany, see Annex I
- Country Nomenclature.

1 Twelve months beginning 21 March of year stated.

2 Source: US Bureau of Mines, (Washington, DC).
3 Twelve months ending 30 June of year stated.

4 Source: United Nations Economic Commission for Latin America
(ECLA), (Santiago).
5 Twelve months ending 7 July of the year stated.

6 Production by establishments employing 50 or more persons.
7 Beginning 1982, scope of series revised.
8 Incomplete coverage.

Source:
Base de données pour les statistiques industrielles de la
Division de statistique du Secrétariat de l'ONU.

† Toutes les données se rapportant à l'Allemagne avant le 3
octobre 1990 figurent dans deux rubriques séparées basées
sur les territoires respectifs de la République fédérale
d'Allemagne et l'ancienne République démocratique allemande
selon la période indiquée. En l'absence de données pour
l'Allemagne unifiée (à compter du 3 octobre 1990), les
données disponibles sont fournies séparément sous les
rubriques République fédérale d'Allemagne et ancienne
République démocratique allemande et se rapportent aux
limites territoriales antérieures au 3 octobre 1990. Pour
les notes explicatives en détail sur les données concernant
l'Allemagne, voir Annexe I – Nomenclature des pays.
1 Période de douze mois commençant le 21 mars de l'année
indiquée.
2 Source: "US Bureau of Mines," (Washington, DC).
3 Période de douze mois finissant le 30 juin de l'année
indiquée.
4 Source: Commission économique des Nations Unies pour
l'Amérique Latine (CEPAL), (Santiago).
5 Période de douze mois finissant le 7 juillet de l'année
indiquée.
6 Production des établissements occupant 50 personnes ou plus.
7 A partir de 1982, portée de la série révisée.
8 Couverture incomplète.

89
Pig iron and crude steel
Fonte et acier brut
Production: thousand metric tons
Production : milliers de tonnes métriques

Country or area Pays ou zone	Pig-iron and ferro-alloys Fonte et ferro-alliages					Crude steel Acier brut				
	1980	1987	1988	1989	1990	1980	1987	1988	1989	1990
Total	**522 036**	**519 106**	**555 390**	**673 025**	**543 435**	**718 208**	**729 301**	**770 595**	**777 287**	**762 015**
Algeria Algérie	669	1 493	* 1 215	* 1 115	...	370[1]	1 380	* 1 303	* 945	* 769
Angola * [23] Angola * [23]	...	...	...	...	...	10	10	10	10	10
Argentina Argentine	1 080[45]	1 777[245]	1 641[245]	* 2 154[245]	* 1 950[245]	2 556	3 463	3 527	* 3 874	* 3 624
Australia[7] Australie[7]	7 407[268]	* 5 897[268]	* 5 583[268]	* 6 009[268]	* 6 346[268]	7 895[1]	6 188[1]	6 093[1]	6 651[1]	6 681[1]
Austria Autriche	* 3 495[29]	* 3 458[29]	* 3 685[29]	3 838[29]	* 3 464[29]	4 624	4 301[9]	4 376[9]	4 901[9]	4 291
Bangladesh[3] Bangladesh[3]	...	...	...	...	...	137[7]	82[7]	70[7]	86[7]	90[10]
Belarus Bélarus	...	...	...	...	...	...	1 090	1 109	1 104	1 112
Belgium Belgique	9 845	8 242	9 147	8 863	9 416	12 344	9 844	11 306	11 053	11 546
Brazil Brésil	13 116[46]	21 485[46]	24 427[246]	25 395[246]	23 800[246]	15 337	22 228	24 657	25 055	* 20 504
Bulgaria Bulgarie	1 583[11]	1 706[11]	1 484[11]	1 487	1 160[11]	2 565[1]	3 045[1]	2 875[1]	2 899[1]	2 184[1]
Canada Canada	* 11 182[4]	* 10 057[246]	* 9 808[246]	* 10 579[246]	* 7 719[246]	15 901	14 737	14 778	* 15 005	* 12 184
Chile Chili	* 708[412]	* 628[4]	* 794[4]	* 694[4]	* 738[24]	704[12]	726[12]	899[12]	800[12]	800[12]
China Chine	39 604[46]	* 57 388[246]	* 59 619[246]	* 61 132[246]	* 65 372[246]	38 996	58 928	62 147	64 209	68 858
Colombia Colombie	280[256]	372[26]	351[26]	338[26]	391[26]	420[1]	691[1]	777[1]	711[1]	733[1]
Cuba[3] Cuba[3]	...	...	...	...	...	292	402	321	314	270
Czechoslovakia Tchécoslovaquie	9 992	9 948	9 868	10 077	9 836	15 225[1]	15 416[1]	15 379[1]	15 465[1]	14 877[1]
Denmark[9] Danemark[9]	...	...	...	...	...	734	606	650	624	610
Ecuador[3] Equateur[3]	...	...	...	...	...	16	25	24	23	20
Egypt Egypte	* 254[213]	155[13]	140[13]	119[13]	116[13]	800[2]	1 603[2]	* 1 703[2]	* 1 403[2]	* 1 403[2]
Finland Finlande	2 072	2 207	2 328	2 420	2 438	2 509[9]	2 669[9]	2 798[9]	2 921[9]	2 880[9]
France France	19 595	13 658	15 018	15 365	14 718	23 176	17 689	19 108	19 535	19 304
Germany†·Allemagne† F. R. Germany R. f. Allemagne	34 055	28 615[14]	32 558[14]	32 890[14]	30 206[14]	43 838	36 248	41 023	41 078	38 433
former German D. R. anc. R. d. allemande	2 571[9]	2 879[9]	2 913[9]	2 851[9]	* 2 000	7 308	8 243	8 131	7 829	...

89
Pig iron and crude steel
Production: thousand metric tons [cont.]
Fonte et acier brut
Production : milliers de tonnes métriques [suite]

Country or area Pays ou zone	Pig-iron and ferro-alloys Fonte et ferro-alliages					Crude steel Acier brut				
	1980	1987	1988	1989	1990	1980	1987	1988	1989	1990
Greece Grèce	* 387[15]	* 233[2 13 15]	* 248[2 13 15]	* 259[2 13 15]	* 255[2 13 15]	1 067[1]	908[1]	959[1]	960[1]	999[1]
Hungary Hongrie	2 227	2 124	2 107	1 973	1 715	3 766	3 495	3 480	3 263	2 924
India Inde	8 718[2 4 6]	11 246[4 6]	12 015[2 4 6]	12 355[2 4 6]	* 12 460[2 4 6]	9 427	12 262	13 019	12 881	* 13 000
Indonesia[3] Indonésie[3]	...	...	...	...	...	360	* 1 453	* 2 050	* 2 000	* 2 100
Iran, Islamic Rep. of Iran, Rép. islamique d'	1	1	1	* 1	* 1	565	783	883	* 1 000	* 1 000
Ireland[13] Irlande[13]	...	...	...	...	...	2	220	200	324	325
Israel[1 2 3] Israël[1 2 3]	...	...	...	...	...	* 115	116	* 120	* 118	* 144
Italy Italie	12 411	11 573	11 592	12 010	* 12 024	26 501	22 858	23 943	25 567	* 25 513
Japan Japon	89 130[16]	74 512[16]	80 553[16]	81 542[16]	81 511[16]	111 395	98 513	105 680	107 907	110 339
Korea, Dem. P. R. *[2] Corée, R. p. dém. de *[2]	5 470[4 6 17]	5 870[4 6 17]	6 570[4 6 17]	6 570[4 6 17]	6 570[4 6 17]	5 800[1]	6 500[1]	7 980[1]	8 000[1]	8 000[1]
Korea, Republic of Corée, République de	5 686[6]	11 007	12 740	15 130	14 519	14 447	19 481	21 318	23 705	24 459
Luxembourg Luxembourg	3 568	2 305	2 520	2 684	2 645	4 624	3 302	3 661	3 721	3 560
Mexico Mexique	5 330	5 377[4]	* 3 913[4]	* 3 523[4]	* 4 313[4]	7 003	7 211	7 312	7 329	8 221
Morocco *[2] Maroc *[2]	11	15	15	15	15	6	6	7	7	7
Netherlands Pays-Bas	4 328	4 575	4 994	5 163	4 960	5 272	5 083	5 518	5 681	5 412
Nigeria[3] Nigéria[3]	...	...	...	...	...	15	200	192	213	220
Norway Norvège	1 428	1 325	1 458	1 341	930	853	853	903	677	376
Peru Pérou	* 263[2 6]	* 187[2 6]	* 204[2 6]	* 229[2 6]	* 226[2 6]	447[1]	503[1]	496[1]	364[1]	284[1]
Philippines Philippines	* 30	* 0[2]	* 0[2]	* 28[2]	* 32[2]	430	* 250[1]	300[1]	* 300[1]	* 300[1]
Poland Pologne	11 682	10 310	10 117	9 340	8 564	18 648	16 263	* 15 943	12 466	11 501
Portugal Portugal	578[4]	* 461[2 4]	* 455[2 4]	* 385[2 4]	* 392[2 4]	382	399	430	* 420	* 417
Romania Roumanie	9 116[18]	8 853[18]	9 121[18]	9 251[18]	6 497[18]	10 735	9 913	9 805	14 411	9 761
Saudi Arabia[3] Arabie saoudite[3]	...	...	...	...	...	* 50	1 365[2]	1 614[2]	* 1 810[2]	1 833[2]
South Africa Afrique du Sud	* 8 847[15]	9 068[15]	8 874[2 6 15]	9 452[2 6 15]	8 995[2 6 15]	* 9 064	* 9 032	8 837[1 2]	9 567[1 2]	8 738[1 2]

89
Pig iron and crude steel
Production: thousand metric tons [cont.]
Fonte et acier brut
Production : milliers de tonnes métriques [suite]

Country or area Pays ou zone	Pig-iron and ferro-alloys Fonte et ferro-alliages					Crude steel Acier brut				
	1980	1987	1988	1989	1990	1980	1987	1988	1989	1990
Spain Espagne	6 817[4]	* 5 049[4]	* 4 803[24]	* 5 639[24]	* 5 910[4]	12 553	* 11 629	* 11 679	* 12 564	* 12 818
Sweden Suède	2 596	2 503	* 2 645	* 2 772	* 2 906	* 4 231	4 683[9]	4 779[9]	4 692[9]	4 435
Switzerland Suisse	* 38[29 13]	* 145[2 13]	* 139[2 13]	* 146[2 13]	* 134[2 13]	929[9]	866[9]	988[9]	916[9]	970[9]
Thailand Thaïlande	17[6]	0[6]	0[6]	0[6]	0[6]	450[12]	534[12]	552[12]	689[12]	685[12]
Tunisia Tunisie	152	168	133	156	* 150	178	188	159	192	200
Turkey Turquie	* 2 072[2]	* 4 495[2]	* 12 959[2]	* 18 933[2]	* 5 565[2]	1 700[1]	7 044[1]	8 009[1]	7 853[1]	9 408[1]
Ukraine Ukraine	...	48 043[5]	48 058[5]	47 092[5]	45 521[5]	51 473[1]	56 263	56 438	54 786	52 622
former USSR ancienne URSS	107 283[19]	114 489[19]	115 180[19]	114 555[19]	110 762[19]	147 941[1]	161 887[1]	163 037[1]	160 096[1]	154 436[1]
United Kingdom Royaume-Uni	6 316[20]	12 008[20]	13 050[20]	12 694[20]	12 463[9 20]	11 277	17 414	18 950	18 740	17 841
United States Etats-Unis	63 748[21]	44 692[6 17]	51 517[6]	51 617[6]	50 543[6]	101 456[1 22]	80 876[1 22]	91 765[22]	* 88 852[22]	* 88 596[22]
Uruguay[3] Uruguay[3]	...	...	...	...	...	18	30	29	37	34
Venezuela Venezuela	559[246]	551[246]	588[246]	* 540[246]	* 399[246]	1 784[1]	3 721[1]	3 650[1]	3 404[1]	3 140[1]
Viet Nam[3] Viet Nam[3]	...	...	...	...	...	...	70	74	84	101
Yugoslavia Yougoslavie	2 668	3 158	3 278	3 246	2 631[9]	3 636	4 368	4 488	4 542	3 576
Zimbabwe[2] Zimbabwe[2]	* 612[4]	* 790[4]	* 855[4]	718[4]	716[4]	804	515	602	592	580

Source:
Industrial statistics database of the Statistical Division
of the United Nations Secretariat.

† All data shown which pertain to Germany prior to 3 October
1990 are indicated separately for the Federal Republic of
Germany and the former German Democratic Republic based on
their respective territories at the time indicated. Where
data for united Germany (3 October 1990 and thereafter) are
not available, available data are shown separately under the
designations Federal Republic of Germany and former German
Democratic Republic and pertain to the territorial
boundaries prior to 3 October 1990. For detailed
explanatory notes on data pertaining to Germany, see Annex I
- Country Nomenclature.

1 Including crude steel for casting.
2 Source: US Bureau of Mines, (Washington, DC).
3 Ingots only.
4 Ferro-manganese only.
5 Ferro-silicon only.

Source:
Base de données pour les statistiques industrielles de la
Division de statistique du Secrétariat de l'ONU.

† Toutes les données se rapportant à l'Allemagne avant le 3
octobre 1990 figurent dans deux rubriques séparées basées
sur les territoires respectifs de la République fédérale
d'Allemagne et l'ancienne République démocratique allemande
selon la période indiquée. En l'absence de données pour
l'Allemagne unifiée (à compter du 3 octobre 1990), les
données disponibles sont fournies séparément sous les
rubriques République fédérale d'Allemagne et ancienne
République démocratique allemande et se rapportent aux
limites territoriales antérieures au 3 octobre 1990. Pour
les notes explicatives en détail sur les données concernant
l'Allemagne, voir Annexe I – Nomenclature des pays.

1 Y compris l'acier brut pour moulages.
2 Source: "US Bureau of Mines," (Washington, DC).
3 Les lingots seulement.
4 Ferro-manganèse seulement.
5 Ferro-silicium seulement.

89

Pig iron and crude steel
Production: thousand metric tons [*cont.*]

Fonte et acier brut
Production : milliers de tonnes métriques [*suite*]

6 Including foundry pig iron.	6 Y compris la fonte de moulage.
7 Twelve months ending 30 June of year stated.	7 Période de douze mois finissant le 30 juin de l'année indiquée.
8 Twelve months ending 30 November of the year stated.	8 Période de douze mois finissant le 30 novembre de l'année indiquée.
9 Source: Annual Bulletin of Steel Statistics for Europe, United Nations Economic Commission of Europe (Geneva).	9 Source: Bulletin annuel de statistiques de l'acier pour l'Europe, Commission économique des Nations Unies pour l'Europe (Genève).
10 Mild steel products.	10 Produits en acier doux.
11 Including other ferro-alloys.	11 Y compris les autres ferro-alliages.
12 Source: "Instituto Latino Americano del Fierro y el Acero", (Santiago).	12 Source: "Instituto Latino Americano del Fierro y el Acero", (Santiago).
13 Including pig-iron for steel making.	13 Y compris la fonte d'affinage.
14 Including other raw iron, ferrosilicon and iron sponge.	14 Y compris les fontes brutes, le ferro-silicium et le fer spongieux.
15 Including ferro-manganese.	15 Y compris le ferro-manganèse.
16 Including silico-chromium.	16 Y compris le silico-chrome.
17 Including silico-manganese.	17 Y compris le silico-manganèse.
18 Ferro-alloys produced in electric furnaces.	18 Ferro-alliages produits dans les fourneaux électriques.
19 Including foundry pig iron, spiegeleisen, ferro-manganese and other ferro-alloys.	19 Y compris fonte de moulage, fonte spiegel, ferro-manganèse et autres ferro-alliages.
20 Including refined pig iron.	20 Y compris la fonte affinée.
21 Gross weight, including fused salt electrolytic low carbon ferro-manganese.	21 Poids brut. Y compris le ferro-manganèse à faible teneur en carbone obtenu par électrolyse en présence de sels fondus.
22 Excluding steel for castings made in foundries operated by companies not producing ingots.	22 Non compris l'acier pour les moulages fabriqués dans des fonderies exploitées par des entreprises ne produisant pas de lingots.

90
Aluminium
Aluminium
Production: thousand metric tons
Production : milliers de tonnes métriques

Country or area Pays ou zone	1981	1982	1983	1984	1985	1986	1987	1988	1989	1990
Total	19 187.5	17 342.6	18 377.5	20 320.3	20 062.5	20 134.7	21 658.2	22 933.9	23 589.6	23 690.6
Argentina[1] **Argentine**[1]	133.9[2]	140.5	136.4	137.8	139.9	150.6	155.1	157.4	164.2[2]	165.6[2]
Australia[3] **Australie**[3]	388.8	422.0	441.6	681.8	926.4	* 925.0	* 960.0	* 1 120.5	* 1 289.2	* 1 268.0
Primary[3] 1re fusion[3]	344.8	377.6	403.9	617.9	822.3	870.0	921.0	1 074.0	1 240.8	1 235.1
Secondary[3] 2ème fusion[3]	44.0	44.4	37.7	41.0	45.0	55.0	39.0[4]	46.5[4]	48.4[4]	32.9
Austria **Autriche**	358.9	357.9	482.6	380.7	285.0	283.7	322.3	391.1	334.1	246.3
Primary 1re fusion	94.2	93.9	94.2	151.9	152.7	148.8	156.1	165.9	169.0	159.1
Secondary[5] 2ème fusion[5]	264.7	264.0	388.4	228.8	132.4	134.9	166.1	225.2	165.2	87.2
Bahrain[12] **Bahreïn**[12]	130.0	171.0	171.7	177.3	174.8	178.2	180.3	182.8	186.9	212.5
Belgium[26] **Belgique**[26]	0.0	0.0	1.0	2.0	2.0	2.0	3.2	3.0	3.0	3.0
Brazil **Brésil**	292.5	345.3	443.8	503.9	594.2	805.3	893.8[4]	938.4[4]	952.8[4]	982.7[4]
Primary 1re fusion	256.4	299.1	400.7	455.0	549.4	757.4	843.5[4]	873.5[4]	887.9[4]	930.6[4]
Secondary 2ème fusion	36.0	46.2	43.0	48.9	44.8	48.0	50.3[4]	64.9[4]	64.9[4]	52.1[4]
Cameroon[1] **Cameroun**[1]	* 65.4	64.1	53.6	125.4	81.8	51.0	79.0	80.0[2]	87.3[2]	87.5[2]
Canada[7] **Canada**[7]	1 175.0	1 126.8	1 154.2	1 286.0	1 347.3	1 420.2	1 582.2	1 590.6	1 615.2	1 635.1
Primary[7] 1re fusion[7]	1 115.7	1 064.8	1 091.2	1 222.0	1 282.3	1 355.2	1 540.4	1 534.5	1 554.8	1 567.4
Secondary[7] 2ème fusion[7]	59.3	62.0	63.0	64.0	65.0	65.0	41.7	56.1	60.4	67.7
China[1] **Chine**[1]	* 350.0[7]	* 400.0[7]	* 425.0[7]	* 400.0[7]	* 410.0[7]	* 410.0[7]	* 615.0[7]	718.4	758.4	854.3
Czechoslovakia **Tchécoslovaquie**	66.8	62.6	66.3	66.0	66.0	66.1	66.4	67.4	69.3	69.8
Primary 1re fusion	32.7	33.8	36.2	31.6	31.7	33.1	32.4	31.4	32.6	30.1
Egypt[1] **Egypte**[1]	134.0	140.0	140.2	128.4[8]	135.5[8]	114.0[8]	148.7[8]	142.6[8]	146.2[8]	141.1[8]
Finland **Finlande**	11.0	11.6	14.9	# 1.6	1.0	0.2	0.2	* 2.0	* 4.1	* 3.8
France[9] **France**[9]	591.3	554.8	522.9	504.0	457.3	* 495.2	568.2	541.0	554.8	533.4
Primary 1re fusion	435.6	400.0	360.8	341.0	293.1	321.8	381.8	328.0	329.3	325.2
Secondary[9] 2ème fusion[9]	155.7	154.8	162.1	163.0	164.3	173.4	186.3	213.0	225.5	208.3
Germany† · Allemagne† **F. R. Germany**[2] **R. f. Allemagne**[2]	1 126.4	1 129.0	1 168.8	1 219.4	1 202.6	1 246.2	1 238.9	1 274.8	1 279.4	1 259.2
Primary 1re fusion	728.9	723.7	743.4	777.2	745.5	765.1	737.7	744.1	742.2	720.3[2]

90
Aluminium
Production: thousand metric tons [*cont.*]
Aluminium
Production : milliers de tonnes métriques [*suite*]

Country or area Pays ou zone	1981	1982	1983	1984	1985	1986	1987	1988	1989	1990
Secondary[2] 2ème fusion[2]	397.5	406.2	425.5	442.2	457.3	482.5	501.2	530.7	537.4	538.9
former German D. R. anc. R. d. allemande	60.0[1,2]	58.0[1,2]	59.0[1,2]	125.9	127.1	125.0	122.3	115.8	107.7	...
Primary 1re fusion	60.0[2]	58.0[2]	59.0[2]	67.8	65.7	66.0	67.9	61.2	53.9	...
Secondary 2ème fusion	...	...	...	58.1	61.4	59.0	54.5	54.6	53.8	...
Ghana[1] Ghana[1]	173.3	188.5	42.5	0.0	48.6	124.6	150.0	161.0	169.0	174.0
Greece Grèce	185.6	177.4	157.9	189.0	175.5	175.9	132.0	* 154.4	* 154.3	151.2
Hungary Hongrie	83.8	84.6	87.2	89.0	86.9	86.4	85.2	83.8	85.1	51.1
Primary 1re fusion	74.3	74.2	74.0	74.2	73.9	73.9	73.5	74.7	75.2	75.2
Iceland[1] Islande[1]	73.6	75.2	76.1	80.4	73.4	75.9	83.5	82.0	88.7	87.6
India[1] Inde[1]	212.0	216.5	203.9	268.3	259.9	234.9	245.3	289.7	425.3	428.4
Indonesia[1] Indonésie[1]	0.0	0.0	138.6	206.9	216.8	220.0	* 219.9	180.0	196.9[2]	192.1
Iran, Islamic Rep. of[1] Iran, Rép. islamique d'[1]	12.5	45.0	39.2	42.0	43.0	37.4	33.6	38.2	* 45.0	20.0
Italy Italie	523.8	474.9	473.7	513.2	506.1	543.6	567.6	604.1	609.5	594.4
Primary 1re fusion	273.8	232.9	195.7	230.2	224.1	242.6	232.6	226.3	219.5	244.8
Secondary 2ème fusion	250.0	242.0	278.0	283.0	282.0	301.0	335.0	377.8	390.0	349.6
Japan[9] Japon[9]	1 616.4	1 153.2	1 096.3	1 131.4	1 098.5	1 013.6	1 085.1	1 053.3	1 081.5	1 140.6
Primary 1re fusion	776.8	355.1	258.6	291.1	231.3	148.3	52.8	49.0	50.7	50.5
Secondary[9] 2ème fusion[9]	839.6	798.1	837.7	840.3	867.2	865.3	1 032.3	1 004.3	1 030.8	1 090.1
Korea, Republic of[1] Corée, République de[1]	17.5	15.2	12.6	18.3	17.1	18.3	16.8	16.1	15.7	13.3
Mexico[1] Mexique[1]	43.2	42.3	39.8	43.6	42.8	46.2	79.3	75.4	67.6	56.8
Secondary[2] 2ème fusion[2]	23.5	25.8	15.1	19.6	22.3	16.4	8.8	4.5	13.2	...
Netherlands Pays-Bas	312.2	300.8	293.6	309.1	312.9	362.6	377.3	394.1	409.4	392.1
Primary 1re fusion	262.0	250.9	235.4	249.2	250.6	265.8	275.9	278.2	279.2	257.9
Secondary 2ème fusion	50.2	49.8	58.2	59.9	62.3	96.8	101.4	115.9	130.2	134.2
New Zealand[1] Nouvelle-Zélande[1]	154.0	163.4	218.6	242.9	240.8	236.3	252.0	264.0	260.0	259.7
Norway Norvège	648.1	645.6	717.6	770.7	748.7	733.0	805.0	833.8	866.2	878.1
Primary 1re fusion	633.6	636.9	713.0	765.1	742.7	725.8	797.8	826.6	867.3	871.1
Secondary 2ème fusion	14.5	8.7	4.6	5.6	6.0	7.2	7.2	7.2	7.2	7.0

90
Aluminium
Production: thousand metric tons [cont.]
Aluminium
Production : milliers de tonnes métriques [suite]

Country or area Pays ou zone	1981	1982	1983	1984	1985	1986	1987	1988	1989	1990
Poland[1] Pologne[1]	66.0	42.7	44.4	46.0	47.0	47.5	47.5	47.7	47.8	46.0
Portugal[6] Portugal[6]	1.9	1.9	2.4	2.4	2.5	2.1	2.3	4.7	* 2.0	* 2.0
Romania[9][10] Roumanie[9][10]	261.0	228.0	244.0	264.0	265.0	269.0	275.0	279.0	282.0	178.0
Primary[10] 1re fusion[10]	242.0	208.0	223.0	244.0	247.0	253.0	260.0	266.0	269.0	168.0
Secondary 2ème fusion	19.0	20.0	21.0	20.0	18.0	16.0	15.0	13.0	11.0	9.8
South Africa[17] Afrique du Sud[17]	83.7	105.5	161.3	167.4	165.0	* 169.6	* 170.6	* 170.4	* 165.9	* 166.0
Spain[4] Espagne[4]	431.1	402.2	394.9	421.4	412.6	402.9	411.0	378.9	430.0	434.5
Primary[4] 1re fusion[4]	396.6	366.5	357.6	380.8	370.1	354.7	341.0	293.9	352.4	355.3
Secondary[4] 2ème fusion[4]	34.5	35.7	37.4	40.6	42.5	48.2	70.0	85.0	77.6	79.2
Suriname[1] Suriname[1]	40.5	42.5	33.6	23.0	28.8	28.7	1.9	9.8	28.4	31.3
Sweden Suède	107.1	* 104.5	109.7	106.5	108.7	* 109.9	* 115.4	* 130.6	* 130.0	* 124.8
Primary 1re fusion	82.3	81.3	82.0	82.8	83.7	* 77.1	* 81.5	* 98.6	* 97.0	* 92.3
Secondary[2] 2ème fusion[2]	24.7	26.9	27.7	30.8	30.4	32.8	33.9	32.0	* 33.0	28.5
Switzerland[1] Suisse[1]	82.2	75.3	76.0	79.2	72.7	80.3	73.2	71.8	71.3	71.6
Turkey[1] Turquie[1]	40.4	36.3	30.4	37.9	54.1	60.0	41.7	56.7	57.2	* 60.0
former USSR *[7] ancienne URSS *[7]	1 980.0	2 040.0	2 400.0	2 550.0	2 700.0	2 850.0	3 000.0	3 000.0	3 000.0	...
Primary *[7] 1re fusion *[7]	1 800.0	1 850.0	2 000.0	2 100.0	2 200.0	2 300.0	2 400.0	2 400.0	2 400.0	2 200.0
Secondary *[7] 2ème fusion *[7]	180.0	190.0	400.0	450.0	500.0	550.0	600.0	600.0	600.0	...
United Kingdom[10] Royaume-Uni[10]	487.2	355.4	380.8	431.8	403.0	382.3	411.1	405.9	366.9	362.8
Primary[10] 1re fusion[10]	339.2	240.8	252.5	287.9	275.4	275.9	294.4	300.2	297.3	289.8
Secondary 2ème fusion	148.0	114.6	128.3	143.9	127.6	116.4	116.7	105.8	69.5	73.0
United States[9] Etats-Unis[9]	6 156.1	4 822.3	5 001.6	5 859.0	5 262.0	4 810.0	5 329.0	6 066.0	6 084.0	5 962.8
Primary 1re fusion	4 488.7	3 273.7	3 353.2	4 099.0	3 500.0	3 037.0	3 343.0	3 944.0	4 030.0	4 047.6
Secondary[9] 2ème fusion[9]	1 667.4	1 548.6	1 648.4	1 760.0	1 762.0	1 773.0	1 986.0	2 122.0	2 054.0	1 915.2
Venezuela Venezuela	313.5	273.6	335.3	385.2	402.8	421.4	430.2	443.3	565.6	598.8
Yugoslavia Yougoslavie	196.8	246.4	283.6	301.6	316.1	319.7	281.1	313.3	331.7	291.0
Primary 1re fusion	172.7	220.1	258.2	292.3	306.3	309.2	280.6	312.7	331.0	* 290.0
Secondary 2ème fusion	24.1	26.3	25.4	9.3	9.7	10.5	0.5	0.7	0.6	* 1.0

90

Aluminium
Production: thousand metric tons [cont.]

Aluminium
Production : milliers de tonnes métriques [suite]

Source:
Industrial statistics database of the Statistical Division
of the United Nations Secretariat.

† All data shown which pertain to Germany prior to 3 October
  1990 are indicated separately for the Federal Republic of
  Germany and the former German Democratic Republic based on
  their respective territories at the time indicated. Where
  data for united Germany (3 October 1990 and thereafter) are
  not available, available data are shown separately under the
  designations Federal Republic of Germany and former German
  Democratic Republic and pertain to the territorial
  boundaries prior to 3 October 1990. For detailed
  explanatory notes on data pertaining to Germany, see Annex I
  - Country Nomenclature.

1 Primary metal production only.
2 Source: "Metallgesellschaft Aktiengesellschaft",
  (Frankfurt).
3 Twelve months ending 30 June of year stated.

4 Source: World Metal Statistics (London).
5 Secondary aluminium produced from old scrap only.

6 Secondary metal production only.
7 Source: US Bureau of Mines, (Washington, DC).
8 Including aluminium plates, shapes and bars.
9 Including alloys.
10 Including pure content of virgin alloys.

Source:
Base de données pour les statistiques industrielles de la
Division de statistique du Secrétariat de l'ONU.

† Toutes les données se rapportant à l'Allemagne avant le 3
  octobre 1990 figurent dans deux rubriques séparées basées
  sur les territoires respectifs de la République fédérale
  d'Allemagne et l'ancienne République démocratique allemande
  selon la période indiquée. En l'absence de données pour
  l'Allemagne unifiée (à compter du 3 octobre 1990), les
  données disponibles sont fournies séparément sous les
  rubriques République fédérale d'Allemagne et ancienne
  République démocratique allemande et se rapportent aux
  limites territoriales antérieures au 3 octobre 1990. Pour
  les notes explicatives en détail sur les données concernant
  l'Allemagne, voir Annexe I – Nomenclature des pays.
1 Production du métal de première fusion seulement.
2 Source: "Metallgesellschaft Aktiengesellschaft",
  (Francfort).
3 Période de douze mois finissant le 30 juin de l'année
  indiquée.
4 Source: "World Metal Statistics," (Londres).
5 Aluminium de deuxième fusion obtenu à partir de vieux
  déchets seulement.
6 Production du métal de deuxième fusion seulement.
7 Source: "US Bureau of Mines," (Washington, DC).
8 Y compris les tôles, les profilés et les barres d'aluminium.
9 Y compris les alliages.
10 Y compris la teneur pure des alliages de première fusion.

91
Copper
Cuivre

Production: thousand metric tons
Production : milliers de tonnes métriques

Country or area Pays ou zone	Smelter production Production de fonderie					Refined production Production affinée				
	1980	1987	1988	1989	1990	1980	1987	1988	1989	1990
Total	**8 360.4**	**8 918.2**	**8 904.2**	**9 498.5**	**9 034.0**	**9 395.8**	**10 031.6**	**10 374.9**	**10 981.1**	**10 711.7**
Albania Albanie	...	11.0	15.0	15.0[1]	12.0[1]	...	...	...	...	...
Australia[2] Australie[2]	163.6	174.0	180.1	191.0	199.9	158.4	199.8	212.7	255.0	244.9
Austria Autriche	6.5[3]	2.3[3]	2.3[3]	1.2[3]	1.2[3]	36.8	40.8	38.4	46.3	49.5
Belgium[4] Belgique[4]	...	...	...	...	...	526.3	475.9	504.6	563.3	542.5
Brazil Brésil	0.0[5]	134.0[5]	132.0[5]	146.5[5]	* 147.6[5]	63.0	201.7	185.9	195.6[6]	* 200.0[6]
Bulgaria[1] Bulgarie[1]	59.0	49.8	57.4	59.2	30.3	63.0	52.8	* 50.0	55.8	24.1
Canada Canada	492.7	504.6	542.9	500.2	525.2	505.2	491.1	528.7	515.2	515.8
Chile Chili	953.1[7]	911.0[7]	837.8[7]	1 226.6[67]	1 110.0[7]	810.7[8]	795.0[8]	852.9[8]	* 1 071.0[8]	990.8[8]
China *[1] Chine *[1]	175.0	417.0	425.0	430.0	425.0	295.0	450.0	460.0	470.0	490.0
Czechoslovakia Tchécoslovaquie	31.0	34.4	34.8	36.8	33.0	25.6	27.2	27.1	26.9	24.6
France France	...	...	...	...	...	47.1	39.3	43.0	43.0	43.8
Finland Finlande	47.2	102.0	95.7	99.1	2.5	40.5	160.1	153.7	173.0	177.9
Germany† • Allemagne† F. R. Germany R. f. Allemagne	153.9	168.9[1]	162.5[1]	176.9[1]	183.6[1]	373.8	399.8	426.4	475.2	476.3
former German D. R. anc. R. d. allemande	29.0[1]	44.9	46.5	46.1	* 14.3[1]	51.0[1]	48.4	48.1	44.4	* 56.7[1]
Hungary Hongrie	2.1	3.5	3.6	4.3	0.3	27.8	23.3	15.6	13.1	12.8
India Inde	24.6	32.1	38.9	41.0	38.8	25.6[6]	30.8[6]	40.1[6]	41.8[6]	36.5[6]
Iran, Islamic Rep. of Iran, Rép. islamique d'	* 0.8[69]	* 40.0[69]	* 52.0[69]	* 76.0[69]	* 80.3[69]	0.0	47.1	58.0	* 40.0	* 47.8
Italy[10] Italie[10]	...	...	...	...	...	12.2	65.0	75.4	83.3	* 86.0
Japan Japon	1 155.5	1 215.9	1 270.3	1 323.9	1 351.4	1 014.3[11]	980.3[11]	955.1[11]	989.6[11]	1 008.0[11]
Korea, Dem. P. R. *[6] Corée, R. p. dém. de *[6]	18.0	18.0	18.0	18.0	18.0	22.0	22.0	22.0	22.0	22.0
Korea, Republic of Corée, République de	72.9[3]	154.6[3]	168.3[3]	178.7[3]	192.2[3]	72.9	154.6	168.3	178.7	192.2
Mexico Mexique	87.9	126.3	150.3	174.3	210.0	85.6	120.4	125.9	133.6	125.9
Namibia[6] Namibie[6]	40.0	35.5	40.0	38.0	37.0	...	...	...	...	...
Norway Norvège	33.7	30.1	31.7	35.0	36.5	33.2	29.4[8]	31.7[8]	32.3[8]	36.0[8]

91

Copper
Production: thousand metric tons [cont.]
Cuivre
Production : milliers de tonnes métriques [suite]

Country or area Pays ou zone	Smelter production Production de fonderie					Refined production Production affinée				
	1980	1987	1988	1989	1990	1980	1987	1988	1989	1990
Peru[5] Pérou[5]	348.6[7]	311.0[7]	239.2[7]	309.3[7]	261.4[7]	230.6	224.8	174.7	224.3	181.8
Philippines Philippines	304.1	124.7[6]	* 159.2[6]	* 105.0[6]	* 110.0[6]	0.0[1]	132.1[1]	132.2[1]	132.2[1]	125.9[1]
Poland Pologne	373.3	403.0	406.6	396.0	351.4	357.3	390.2	400.6	390.3	346.1
Portugal Portugal	3.7	2.3	3.1	2.7[6]	3.0[6]	4.5	5.0	5.4	* 6.0	* 6.0
Romania Roumanie	40.7	38.0	38.7	39.4	28.3	36.2	32.2	32.0	33.2	24.7
South Africa[5] Afrique du Sud[5]	185.8[7]	189.8[7]	180.6[7]	184.8[7]	176.0[7]	147.9	146.5	136.7	144.2	133.0
Spain[5] Espagne[5]	103.1	138.7	145.6	152.3	150.4	153.7	151.4	158.8	165.7	170.6
Sweden Suède	56.4[5]	105.6[5]	115.9[5]	94.6[5]	108.0[5]	55.7	91.9	90.3	* 94.7	* 97.3
Switzerland Suisse	...	57.2	55.7	54.0	50.4	...	...	...	...	...
Turkey Turquie	16.3[7]	19.4[7]	12.9[7]	22.6[7]	20.9[7]	18.8	* 75.6	* 68.4	* 86.4	* 84.2
former USSR *[1] ancienne URSS *[1]	1 130.0	1 140.0	1 110.0	1 075.0	990.0	1 300.0	1 410.0	1 380.0	1 355.0	1 260.0
United States Etats-Unis	1 008.4[7]	972.1[7]	1 043.0[7]	1 120.4[7]	1 158.5[7]	1 730.3	1 541.6	1 852.0	1 953.8	2 017.4
United Kingdom Royaume-Uni	...	...	...	...	...	161.3	120.3	122.3	118.6	121.5
Yugoslavia Yougoslavie	172.4	165.8	172.0	173.2	151.4	131.3	138.9	145.4	151.0	151.2
Zaire[6] Zaïre[6]	447.8[7]	487.4[7]	* 466.8[7]	* 442.8[7]	339.4[7]	144.0[8]	210.1[8]	202.6[8]	181.9[8]	140.0[8]
Zambia Zambie	609.3[1 12]	523.2[1 12]	422.3[1 12]	485.2[1 12]	461.4[1 12]	607.2	483.0	422.3	450.8	424.8
Zimbabwe Zimbabwe	26.3[5]	30.0[5]	27.5[5]	23.1[5]	24.4[5]	27.0[8]	* 23.0[6]	* 27.5[6]	* 24.0[6]	* 23.2[6]

Source:
Industrial statistics database of the Statistical Division
of the United Nations Secretariat.

† All data shown which pertain to Germany prior to 3 October
1990 are indicated separately for the Federal Republic of
Germany and the former German Democratic Republic based on
their respective territories at the time indicated. Where
data for united Germany (3 October 1990 and thereafter) are
not available, available data are shown separately under the
designations Federal Republic of Germany and former German
Democratic Republic and pertain to the territorial
boundaries prior to 3 October 1990. For detailed
explanatory notes on data pertaining to Germany, see Annex I
- Country Nomenclature.

1 Source: "Metallgesellschaft Aktiengesellschaft",

Source:
Base de données pour les statistiques industrielles de la
Division de statistique du Secrétariat de l'ONU.

† Toutes les données se rapportant à l'Allemagne avant le 3
octobre 1990 figurent dans deux rubriques séparées basées
sur les territoires respectifs de la République fédérale
d'Allemagne et l'ancienne République démocratique allemande
selon la période indiquée. En l'absence de données pour
l'Allemagne unifiée (à compter du 3 octobre 1990), les
données disponibles sont fournies séparément sous les
rubriques République fédérale d'Allemagne et ancienne
République démocratique allemande et se rapportent aux
limites territoriales antérieures au 3 octobre 1990. Pour
les notes explicatives en détail sur les données concernant
l'Allemagne, voir Annexe I – Nomenclature des pays.

1 Source: "Metallgesellschaft Aktiengesellschaft",

91

Copper
Production: thousand metric tons [*cont.*]

Cuivre
Production : milliers de tonnes métriques [*suite*]

(Frankfurt).

2 Twelve months ending 30 June of year stated.

3 Production at the refined stage.
4 Including alloys and processing of refined copper imported from Zaire.
5 Source: World Metal Statistics (London).
6 Source: US Bureau of Mines, (Washington, DC).
7 Including some production at the refined stage.
8 Primary metal production only.
9 Twelve months beginning 21 March of year stated.

10 Secondary metal production only.
11 Electrolytic cathode copper.
12 Including copper obtained from ores by leaching in electrowinning plants.

(Francfort).

2 Période de douze mois finissant le 30 juin de l'année indiquée.

3 Données se rapportant au cuivre affiné.
4 Y compris les alliages et le traitement ultérieur du cuivre raffiné importé du Zaïre.
5 Source: "World Metal Statistics," (Londres).
6 Source: "US Bureau of Mines," (Washington, DC).
7 Y compris certaines productions au stade affiné.
8 Production du métal de première fusion seulement.
9 Période de douze mois commençant le 21 mars de l'année indiquée.

10 Production du métal de deuxième fusion seulement.
11 Cuivre électrolytique à cathode.
12 Y compris le cuivre obtenu à partir des minerais après lixiviation dans les fabriques de récupération éléctrolytique.

92

Lead

Plomb

Production: thousand metric tons
Production : milliers de tonnes métriques

Country or area Pays ou zone	1981	1982	1983	1984	1985	1986	1987	1988	1989	1990
Total	4 961.8	4 962.5	5 030.2	5 139.4	5 304.5	5 280.3	5 314.6	5 447.2	5 488.9	5 480.4
Argentina Argentine	34.6	30.6	31.3	28.0	28.7	30.7	32.2	* 29.0	* 26.0	* 27.0
Primary 1re fusion	19.0	15.9	15.2	16.3	15.1	15.7	16.2	14.0	13.0	14.0
Secondary 2ème fusion	15.6	14.6	16.1	11.7	13.6	15.0	16.0	15.0	13.0	13.0
Australia[1] Australie[1]	239.2	252.2	238.9	220.6	226.1	222.7	157.0	199.2	* 199.4	* 214.2
Primary[1] 1re fusion[1]	207.7	218.8	212.2	190.1	200.1	206.0	142.0	182.0	184.0	197.2
Secondary[1,2] 2ème fusion[1,2]	31.5	33.4	26.8	19.6	26.0	16.7	15.0	17.2	* 15.4	* 17.0
Austria Autriche	16.7	18.4	17.8	19.6	18.8	18.3	21.8	24.3	23.4	23.2
Primary 1re fusion	5.3	7.8	10.9	10.6	10.2	9.4	12.9	15.2	14.0	12.6
Secondary 2ème fusion	11.4	10.8	11.0	13.0	12.4	12.2	15.0	16.0	14.1	14.8
Remelted Refondu	0.2	0.3	0.5	0.5	0.9	0.5	1.1	0.8	0.6	0.3
Belgium[3] Belgique[3]	109.9	99.7	134.2	127.7	114.3	98.3	108.0	126.6	109.4	106.8
Bolivia Bolivie	0.3	0.2[4]	0.3[4]	0.2[4]	0.2[4]	0.2[4]	0.2[4]	0.4[4]	0.3[4]	0.2[4]
Brazil Brésil	70.8	53.4	63.1	63.8	81.6	85.4	88.2	98.2	89.6	* 88.0
Primary 1re fusion	34.7	21.9	20.6	26.0	29.8	32.7	29.8	29.5	34.2	* 33.0
Secondary 2ème fusion	31.0	27.4	28.9	45.7	51.8	52.7	58.4	68.7	55.4	* 55.0
Bulgaria *[4,5] Bulgarie *[4,5]	123.0	118.0	116.0	128.2	98.0	96.0	100.6	101.0	101.5	66.6
Canada Canada	238.1	241.9	242.0	252.0	241.6	257.7	230.7	268.1	242.8	283.6
Primary 1re fusion	168.5	174.3	178.0	175.0	173.2	169.9	139.5	179.5	157.3	187.2
Secondary 2ème fusion	69.7[6]	67.6[6]	63.9[6]	79.0[6]	68.4[6]	87.7[6]	91.2[6]	88.6	85.5	96.5
China Chine	* 175.0[6]	* 175.0[6]	* 195.0[6]	* 195.0[6]	* 210.0[6]	* 240.0[6]	* 245.0[6]	241.4	304.2	296.5[6]
Primary *[6,7] 1re fusion *[6,7]	150.0	155.0	165.0	165.0	170.0	200.0	200.0	200.0	200.0	235.0
Secondary *[6,7] 2ème fusion *[6,7]	25.0	20.0	30.0	30.0	40.0	40.0	45.0	45.0	50.0	55.0
Colombia[8] Colombie[8]	3.3	3.1	2.2	...	...	...	...	...	...	...
Czechoslovakia Tchécoslovaquie	20.7	21.1	21.0	21.1	21.4	23.6	26.0	26.0	26.0	23.7
Denmark[9] Danemark[9]	12.9	15.9	10.1	13.1	4.5	0.5	0.4	0.0	0.0	0.0
Recovered in alloys Récupéré en alliages	11.2	...	...	...	...	...	...	...	...	...
Finland Finlande	3.5	2.9	4.0	4.8	# 0.2	0.1	0.0	0.0	0.0	0.0
Recovered in alloys[10] Récupéré en alliages[10]	1.5	0.9	1.8	1.7	# 0.2	0.2	0.0	...	...	...

92

Lead
Production: thousand metric tons [cont.]
Plomb
Production : milliers de tonnes métriques [suite]

Country or area / Pays ou zone	1981	1982	1983	1984	1985	1986	1987	1988	1989	1990
France / **France**	**228.0**	**220.9**	**204.2**	**203.0**	**205.8**	**224.9**	**178.0**	**194.0**	**198.7**	**197.0**
Primary / 1re fusion	128.6	144.2	114.9	118.0	133.0	131.9	138.7	147.0	149.3	138.9
Secondary / 2ème fusion	99.4	76.7	89.3	85.0	72.8	92.9	39.3	47.0	49.4	58.1
Germany† · Allemagne† F. R. Germany[11] / R. f. Allemagne[11]	**304.2**	**365.9**	**368.7**	**374.1**	**372.5**	**388.6**[12]	**362.1**[12]	**352.4**[12]	**367.3**[12]	***372.7**[12]
Primary / 1re fusion	107.5	110.7	116.2	102.3	109.7	111.1	113.6	116.4	116.1	* 106.0
Secondary[11] / 2ème fusion[11]	196.7	255.2	252.5	271.9	262.8	277.5	248.5	236.0	251.3	* 266.8
Remelted[13] / Refondu[13]	56.8	...	...	...	...	...	...	...	...	...
Recovered in alloys / Récupéré en alliages	5.0	4.0	3.4	2.9	2.4	1.9	1.7	1.0	1.2	...
former German D. R. *[4][14] / anc. R. d. allemande *[4][14]	43.6	41.5	41.3	42.9	41.5	66.5	62.1	39.6	40.1	45.5
Greece / **Grèce**	**20.2**	**4.2**	**0.0**	**11.9**	**13.9**	**18.7**	***2.7**[15]	***15.1**[15]	***7.0**[15]	***0.0**[15]
Primary / 1re fusion	20.2	4.2	0.0	9.1	10.2	16.2	0.7[15]	13.1[15]	5.6[15]	0.0[15]
Secondary / 2ème fusion	...	...	...	2.8	3.7	2.4	2.0[15]	2.0[15]	1.4[15]	0.0[15]
Recovered in alloys / Récupéré en alliages	7.6	0.7	...	3.0	...	...	...	...	...	...
Guatemala[69] / **Guatemala**[69]	**0.0**	**0.0**	**0.1**	**0.1**	**0.1**	**0.1**	**0.1**	**0.1**	**0.2**	**0.2**
Hong Kong Hong-kong Recovered in alloys / Récupéré en alliages	...	...	...	...	...	...	...	...	2.4	...
Hungary / **Hongrie**	**0.6**	**0.8**	**0.0**	**0.0**	**0.0**	**0.0**	**0.0**	**0.0**	**0.0**	**0.0**
India[16] / **Inde**[16]	**14.3**	**14.4**	**15.0**	**15.7**	**15.6**	**18.7**	**20.4**	**18.8**	***19.0**	***22.0**
Italy / **Italie**	**133.0**	**133.7**	**130.0**	**140.5**	**135.0**	**126.4**	**168.3**	**168.4**	**180.8**	**166.7**
Primary[7] / 1re fusion[7]	35.6	36.4	40.0	37.6	38.3	28.4	56.9	73.0	76.5	64.5
Secondary[7] / 2ème fusion[7]	97.4	97.3	90.0	102.9	96.7	98.0	111.4	95.4	104.3	102.2
Recovered in alloys / Récupéré en alliages	1.4	1.7	2.8	2.6	3.4	3.6	3.6	4.5	6.3	...
Japan / **Japon**	**226.2**	**223.9**	**241.3**	**278.5**	**285.4**	**283.1**	**268.5**	**267.4**	**260.0**	**261.0**
Primary / 1re fusion	211.1	216.5	237.1	273.9	279.4	276.9	261.1	258.8	250.5	253.1
Secondary / 2ème fusion	15.1	7.4	4.2	4.6	6.0	6.3	7.4	8.6	9.5	7.9
Remelted / Refondu	80.0	71.5	75.7	81.1	79.7	75.0	67.6	70.0	71.7	64.6
Korea, Dem. P. R. *[6] / **Corée, R. p. dém. de *[6]**	**65.0**	**60.0**	**60.0**	**95.0**	**95.0**	**95.0**	**95.0**	**95.0**	**95.0**	**95.0**
Primary *[6] / 1re fusion *[6]	65.0	60.0	60.0	95.0	95.0	95.0	95.0	95.0	95.0	95.0
Korea, Republic of[16] / **Corée, République de**[16]	**14.7**	**16.1**	**17.8**	**20.3**	**23.6**	**52.2**	**89.8**	**92.4**	**66.0**	**62.8**

92

Lead
Production: thousand metric tons [cont.]
Plomb
Production : milliers de tonnes métriques [suite]

Country or area Pays ou zone	1981	1982	1983	1984	1985	1986	1987	1988	1989	1990
Mexico[16] Mexique[16]	138.3	128.8	152.6	153.8	185.2	173.3	180.3	180.6	177.7	172.7
Morocco[15 17] Maroc[15 17]	52.3	58.6	57.6	48.3	63.5	55.3	64.9	71.0	66.2	66.9
Myanmar[6] Myanmar[6]	5.3	8.1	7.9	7.2	9.9	5.7	4.3	4.4	3.4	2.8
Namibia[6 16] Namibie[6 16]	41.7	40.6	35.4	28.9	38.5	40.0	40.6	44.4	44.2	44.0
Netherlands Pays-Bas	19.7	23.5	25.6[15]	33.6[15]	37.3[15]	35.7[15]	40.3[15]	39.5[15]	41.5[15]	44.1[15]
Primary 1re fusion	19.7[2 18]	23.5[2 18]	2.0	0.0	0.0	0.0	0.0	0.0	0.0	...
Secondary[15] 2ème fusion[15]	...	...	23.6	33.6	37.3	35.7	40.3	39.5	41.5	44.1
Nigeria[6 9] Nigéria[6 9]	2.0	2.0	0.4	0.6	0.8	1.0	0.3	* 0.5	* 0.5	* 0.3
Norway[19] Norvège[19]	0.1	0.0	0.0	0.0	0.0	0.0	0.0	0.0	...	...
Secondary[20] 2ème fusion[20]	0.1	0.0	0.0	0.0	0.0	0.0	0.0	0.0	0.0	...
Peru[4] Pérou[4]	84.2	82.0	65.1	71.2	83.5	67.4	71.3	53.6	74.4	69.3
Poland[14] Pologne[14]	69.0	78.8	81.0	83.4	87.3	88.3	89.8	90.7	78.2	64.8
Portugal *[4] Portugal *[4]	3.0	4.0	6.0	6.0	7.0	6.0	7.0	7.0	7.0	5.0
Primary 1re fusion	0.1	0.1	0.0	0.0	0.0	0.1	0.0	0.0	0.0	0.0
Secondary * 2ème fusion *	2.9	3.9	6.0	6.0	7.0	5.9	6.9	6.9	6.9	5.0
Romania[2] Roumanie[2]	40.7	42.9	31.9	29.4	30.1[7]	51.5[7]	43.2[7]	42.1[7]	41.1[7]	20.7[7]
South Africa[8] Afrique du Sud[8]	32.7	31.7	27.8	30.8	32.8	40.4	38.3	36.7	30.8	37.2
Spain[6] Espagne[6]	117.1	131.6	143.8	160.0	156.1	130.0	122.7	* 120.8	* 113.3	* 120.0
Primary[6] 1re fusion[6]	83.0	99.5	107.8	110.1	112.8	88.0	71.4	* 68.8	68.3	70.0
Secondary[6] 2ème fusion[6]	34.1	32.1	36.0	49.9	43.3	42.0	51.3	* 52.0	45.0	50.0
Recovered in alloys[10] Récupéré en alliages[10]	4.7	2.9	2.5	4.1	5.2	7.5	8.0	11.1	11.6	...
Sweden[16] Suède[16]	21.8	34.1	30.3	15.9	43.2	49.1	61.1	60.0	57.6	47.5
Remelted Refondu	21.5	20.4	17.8	23.8	27.2	28.8	30.1	...	...	...
Tunisia Tunisie	17.8	15.3	10.5	8.4	2.0	2.3	1.6	0.0	0.0	0.0
Turkey[4 17] Turquie[4 17]	13.0	12.7	13.0	11.7	12.6	12.5	12.0	11.0	9.0	9.0
former USSR *[6] ancienne URSS *[6]	715.0	730.0	745.0	755.0	750.0	755.0	750.0	727.0	745.0	700.0
Primary *[6] 1re fusion *[6]	480.0	485.0	490.0	495.0	485.0	485.0	475.0	447.0	465.0	420.0
Secondary *[6] 2ème fusion *[6]	235.0	245.0	255.0	260.0	265.0	270.0	275.0	280.0	280.0	280.0

92

Lead
Production: thousand metric tons [cont.]
Plomb
Production : milliers de tonnes métriques [suite]

Country or area Pays ou zone	1981	1982	1983	1984	1985	1986	1987	1988	1989	1990
United Kingdom[21] Royaume-Uni[21]	333.4	306.2	322.2	338.4	327.2	328.6	347.0	373.8	350.0	329.4
Primary 1re fusion	135.4	131.0	136.9	147.1	148.1	156.1	145.8	172.2	156.5	155.9
Secondary[7] 2ème fusion[7]	198.0	175.2	185.3	191.3	179.1	172.5	201.1	201.6	193.5	173.5
United States Etats-Unis	1 084.9	1 041.6	1 022.7	1 022.8	1 109.7	995.1	1 083.7	1 129.1	1 205.0	1 326.6
Primary 1re fusion	498.3[22]	521.4[22]	519.2	389.4	494.0	370.3	373.6	392.1	396.5	403.7
Secondary 2ème fusion	586.5	524.8	461.2	575.7	559.6	561.5	668.4[20]	698.4[20]	* 808.6[20]	* 922.9[20]
Remelted Refondu	17.3	9.4	11.8	15.5	13.4	20.0	...	...	...	...
Recovered in alloys Récupéré en alliages	40.1	30.7	27.2	...	...	...	...	...	...	...
Venezuela *[69] Venezuela *[69]	12.0	15.0	15.0	17.0	18.0	16.0	17.0	18.5	17.0	17.0
Yugoslavia Yougoslavie	86.4	81.2	97.5	82.8	100.0	113.2	112.4	109.9	97.2	82.8
Zambia Zambie	9.9	14.6	14.6	8.8[16]	8.8[16]	6.6[16]	8.0[16]	6.4[16]	3.8[16]	4.8[16]

Source:
Industrial statistics database of the Statistical Division
of the United Nations Secretariat.

† All data shown which pertain to Germany prior to 3 October
1990 are indicated separately for the Federal Republic of
Germany and the former German Democratic Republic based on
their respective territories at the time indicated. Where
data for united Germany (3 October 1990 and thereafter) are
not available, available data are shown separately under the
designations Federal Republic of Germany and former German
Democratic Republic and pertain to the territorial
boundaries prior to 3 October 1990. For detailed
explanatory notes on data pertaining to Germany, see Annex I
- Country Nomenclature.

1 Twelve months ending 30 June of year stated.

2 Including secondary refined soft lead.
3 Including lead remelted and recovered in alloys.

4 Source: "Metallgesellschaft Aktiengesellschaft",
(Frankfurt).
5 Including secondary antimonial lead.
6 Source: US Bureau of Mines, (Washington, DC).
7 Including antimonial lead.
8 Recovered in alloys.
9 Secondary metal production only.
10 Including other lead alloys.
11 Including production from imported bullion.

12 Excluding antimonial lead.
13 Beginning 1982, series discontinued.
14 Including secondary metal production.
15 Source: World Metal Statistics (London).
16 Primary metal production only.

Source:
Base de données pour les statistiques industrielles de la
Division de statistique du Secrétariat de l'ONU.

† Toutes les données se rapportant à l'Allemagne avant le 3
octobre 1990 figurent dans deux rubriques séparées basées
sur les territoires respectifs de la République fédérale
d'Allemagne et l'ancienne République démocratique allemande
selon la période indiquée. En l'absence de données pour
l'Allemagne unifiée (à compter du 3 octobre 1990), les
données disponibles sont fournies séparément sous les
rubriques République fédérale d'Allemagne et ancienne
République démocratique allemande et se rapportent aux
limites territoriales antérieures au 3 octobre 1990. Pour
les notes explicatives en détail sur les données concernant
l'Allemagne, voir Annexe I – Nomenclature des pays.

1 Période de douze mois finissant le 30 juin de l'année
indiquée.
2 Y compris le plomb raffiné de deuxième fusion doux.
3 Y compris le plomb refondu et le plomb récupéré à partir des
alliages.
4 Source: "Metallgesellschaft Aktiengesellschaft",
(Francfort).
5 Y compris le plomb antimonié de deuxième fusion.
6 Source: "US Bureau of Mines," (Washington, DC).
7 Y compris le plomb antimonié.
8 Récupéré en alliages.
9 Production du métal de deuxième fusion seulement.
10 Y compris autres alliages de plomb.
11 Y compris la production obtenue à partir de gueuses
importées.
12 Non compris les plomb antimonieux.
13 A partir de 1982, la série a été abandonnée.
14 Y compris la production de métal de deuxième fusion.
15 Source: "World Metal Statistics," (Londres).
16 Production du métal de première fusion seulement.

92

Lead
Production: thousand metric tons [*cont.*]

Plomb
Production : milliers de tonnes métriques [*suite*]

17 Including primary and secondary antimonial lead and
   secondary refined soft lead.
18 Including secondary refined hard (antimonial) lead.
19 Secondary lead, including remelted lead.
20 Including remelted lead.
21 Excluding hard lead.
22 Beginning 1983, data for hard, antimonial lead are
   confidential.

17 Y compris le plomb antimonieux de première et deuxième
   fusion et le plomb doux de deuxième fusion.
18 Y compris le plomb raffiné dur antimonié de deuxième fusion.
19 Plomb de deuxième fusion, y compris le plomb refondu.
20 Y compris le plomb refondu.
21 Non compris le plomb antimonié.
22 A partir de 1983, les données pour le plomb antimonieux
   affiné sont confidentielles.

93
Magnesium
Magnésium
Production: metric tons
Production : tonnes métriques

Country or area Pays ou zone	1981	1982	1983	1984	1985	1986	1987	1988	1989	1990
Total	**389 694**	**324 588**	**319 322**	**389 331**	**392 803**	**387 087**	**392 195**	**409 066**	**433 085**	**437 708**
Brazil[1] Brésil[1]	0	300	500	1 200	2 600	4 400	5 500	5 900	6 200	8 700
Canada[2] Canada[2]	8 775	7 900	6 000	8 000	* 7 200	* 5 100	* 8 800	* 7 600	* 7 200	* 25 300
China[4] Chine[4]	7 000[3]	7 000[3]	7 000[3]	7 000[3]	7 000[3]	7 000[3]	7 000[3]	3 200	8 500	5 900
Denmark Danemark	0	2	0	0	0	0	0	0	0	0
Finland Finlande	...	...	...	...	189	145	114	113	19	0
France France	7 263	9 610	11 075	12 953	13 873	13 361	13 605	* 13 970	14 609	14 640
India *[1] Inde *[1]	100	100	100	100	100	100	300	300	300	1 000
Italy Italie	10 800	9 943	9 799	8 148	7 862	7 245	7 626	5 589	5 768	* 5 900
Japan Japon	34 104	27 225	19 038	22 758	29 349	22 531	18 304	24 111	28 651	36 151
Norway[5] Norvège[5]	47 455	35 000	* 29 844	49 301	54 704	56 864	56 853	50 317	52 872	48 222
Poland[1] Pologne[1]	500	500	0	0	0	0	0	0	0	0
former USSR *[3] ancienne URSS *[3]	86 180[4]	88 900[4]	82 600[4]	85 300[4]	87 100[4]	# 97 000	98 000	99 000	99 000	95 500
United Kingdom[6] Royaume-Uni[6]	1 902	1 758	1 724	1 000	900	998	600	* 600[1]	* 600[1]	* 1 000[1]
United States Etats-Unis	181 472	131 931	146 746	188 299	177 026	167 446	169 561	192 190	203 266	189 595
Yugoslavia[1] Yougoslavie[1]	3 900	4 200	4 700	5 100	4 900	4 897	5 932	6 176	6 100	5 800

Source:
Industrial statistics database of the Statistical Division of the United Nations Secretariat.

1 Source: "Metallgesellschaft Aktiengesellschaft", (Frankfurt).
2 Shipments.
3 Source: US Bureau of Mines, (Washington, DC).
4 Primary metal production only.
5 Including magnesium-based alloys.
6 Secondary metal production only.

Source:
Base de données pour les statistiques industrielles de la Division de statistique du Secrétariat de l'ONU.

1 Source: "Metallgesellschaft Aktiengesellschaft", (Francfort).
2 Expéditions.
3 Source: "US Bureau of Mines," (Washington, DC).
4 Production du métal de première fusion seulement.
5 Y compris les alliages à base de magnésium.
6 Production du métal de deuxième fusion seulement.

94

Tin
Etain

Production: metric tons
Production : tonnes métriques

Country or area Pays ou zone	1981	1982	1983	1984	1985	1986	1987	1988	1989	1990
Total	231 638	213 690	196 710	197 937	198 120	193 422	196 586	215 425	228 244	225 161
Argentina[1] Argentine[1]	457	323	282	290	366	393	296	318[2]	311[2]	310[2]
Australia Australie	4 669	3 617	2 898	2 937	2 683	2 208	784	501	372	381
Belgium[1] Belgique[1]	2 508	2 208	2 214	2 408	2 298	2 712	3 904	4 972	5 978	6 063
Bolivia Bolivie	20 104	17 149	13 542	15 842	14 205	10 500	2 667[2]	5 481[2]	9 718[2]	13 364[2]
Brazil Brésil	7 789	9 298	12 950	18 877	24 701	25 104	29 046	41 857	44 240	35 111
Canada[2] Canada[2]	...	...	...	...	...	...	...	41	27	* 25
China Chine	* 16 500	* 16 500	* 16 500	* 17 000	* 19 000	* 20 000	* 25 000	29 500	31 300	35 800
Czechoslovakia[3] Tchécoslovaquie[3]	140	170	192	325	420	240	545	515	562	613
Finland[1] Finlande[1]	52	58	66	55	# 108	109	152	2	0	0
Germany† · Allemagne† F. R. Germany *[2] R. f. Allemagne *[2]	562	...	...	...	...	...	...	...	...	...
former German D. R. anc. R. d. allemande	2 000[4]	2 200[4]	2 600[4]	2 310	2 310	2 410	2 449	2 537	2 509	...
Indonesia[2] Indonésie[2]	32 519	29 755	28 390	22 467	20 418	22 080	24 200	28 365	29 916	* 31 000
Japan Japon	1 314	1 296	1 260	1 354	1 391	1 280	895	846	808	816
Malaysia Malaisie	70 326	62 836	53 338	46 300	45 500	43 788	44 363	47 403	51 934	48 864
Mexico[2] Mexique[2]	866	944	1 216	1 531	1 533	1 483	1 723	1 514	3 000	* 2 900
Myanmar *[2] Myanmar *[2]	...	...	...	...	...	...	300	100	500	500
Netherlands Pays-Bas	3 500	2 757	5 398	6 517	6 033	5 104	3 834	3 478	4 529[2]	3 400[2]
Nigeria Nigéria	2 483	1 677	* 1 244[2]	* 1 334[2]	* 1 085[2]	91[2]	* 560[2]	245[2]	257[2]	320[2]
Portugal Portugal	424	503	421	409	460	137	137	11	50[2]	* 30[2]
Singapore *[2] Singapour *[2]	...	...	...	...	400	500	1 100	900	600	600
South Africa[2] Afrique du Sud[2]	2 174	2 197	2 200	2 200	2 056	1 816	1 608	2 330	2 500	* 2 300
Spain[2] Espagne[2]	* 3 070	* 2 750	2 812	3 426	3 291	1 725	1 431	246	1 409	1 349
Thailand Thaïlande	32 636	25 479	18 467	19 364	17 996	19 672	15 344	14 674	14 571	15 512

94

Tin
Production: metric tons [*cont.*]
Etain
Production : tonnes métriques [*suite*]

Country or area Pays ou zone	1981	1982	1983	1984	1985	1986	1987	1988	1989	1990
former USSR * [4] ancienne URSS * [4]	16 000	17 000	18 500	18 500	18 500	17 000	18 000	17 000	16 500	14 000
United Kingdom Royaume-Uni	6 863	8 164	6 467	7 105	7 548	9 227	12 135	9 014	3 584 [2]	6 122 [2]
United States Etats-Unis	2 000	3 500	2 500	4 000	3 000	3 213	3 905	1 467	1 000 [2]	0 [2]
Viet Nam Viet Nam	...	...	...	...	...	503	526	573	659	1 774
Zaire [2] Zaïre [2]	450	352	134	170	85	56	90	118	0	0
Zimbabwe Zimbabwe	1 157	1 197	1 235	1 209	1 207	1 128	999 [2]	855 [2]	848	839

Source:
Industrial statistics database of the Statistical Division
of the United Nations Secretariat.

† All data shown which pertain to Germany prior to 3 October
1990 are indicated separately for the Federal Republic of
Germany and the former German Democratic Republic based on
their respective territories at the time indicated. Where
data for united Germany (3 October 1990 and thereafter) are
not available, available data are shown separately under the
designations Federal Republic of Germany and former German
Democratic Republic and pertain to the territorial
boundaries prior to 3 October 1990. For detailed
explanatory notes on data pertaining to Germany, see Annex I
- Country Nomenclature.

1 Including secondary metal production.
2 Source: International Tin Statistics, (London).
3 Secondary metal production only.
4 Source: "Metallgesellschaft Aktiengesellschaft",
   (Frankfurt).

Source:
Base de données pour les statistiques industrielles de la
Division de statistique du Secrétariat de l'ONU.

† Toutes les données se rapportant à l'Allemagne avant le 3
octobre 1990 figurent dans deux rubriques séparées basées
sur les territoires respectifs de la République fédérale
d'Allemagne et l'ancienne République démocratique allemande
selon la période indiquée. En l'absence de données pour
l'Allemagne unifiée (à compter du 3 octobre 1990), les
données disponibles sont fournies séparément sous les
rubriques République fédérale d'Allemagne et ancienne
République démocratique allemande et se rapportent aux
limites territoriales antérieures au 3 octobre 1990. Pour
les notes explicatives en détail sur les données concernant
l'Allemagne, voir Annexe I – Nomenclature des pays.

1 Y compris la production de métal de deuxième fusion.
2 Source: "International Tin Statistics", (Londres).
3 Production du métal de deuxième fusion seulement.
4 Source: "Metallgesellschaft Aktiengesellschaft",
   (Francfort).

95

Tin
Etain

Industrial consumption: metric tons
Consommation industrielle : tonnes métriques

Country or area Pays ou zone	1982	1983	1984	1985	1986	1987	1988	1989	1990	1991
World[1] *Monde*[1]	195 000	196 100	207 700	206 800	213 300	217 000	226 200	232 422	228 902	217 139
Algeria Algérie	...	...	...	...	...	...	500	500	450	400
Argentina Argentine	1 256	1 179	* 1 207	* 800	700	* 1 000	863	943	950	850
Australia Australie	* 2 700	* 2 500	* 2 800	* 2 700	* 2 700	* 2 600	2 380	2 160	2 160	1 750
Austria[2] Autriche[2]	434	388	453	493	506	486	454	467	548	500
Belgium-Luxembourg[2] Belgique-Luxembourg[2]	1 889	1 815	1 697	920	1 141	* 1 220	1 900	3 300	2 100	2 100
Bolivia Bolivie	1 500	2 400	1 800	1 900	1 100	300	117	115	204	170
Brazil Brésil	5 061	4 010	4 270	4 655	6 200	7 900	9 047	9 000	9 000	7 000
Bulgaria * Bulgarie *	800	800	1 000	1 000	800	900	500	500	450	350
Canada[2] Canada[2]	3 528	3 381	4 086	4 000	3 300	3 800	3 489	3 567	3 600	* 4 030
Chile * Chili *	720	720	720	720	1 000	400	800	800	800	800
China Chine	10 000	11 000	11 000	11 500	11 000	12 500	13 300	15 000	17 000	18 000
Colombia * Colombie *	270	270	270	270	270	270	...	...	...	...
Cuba * Cuba *	100	100	100	100	100	100	...	...	...	...
Czechoslovakia Tchécoslovaquie	* 3 500	* 3 200	* 3 300	* 3 100	* 3 700	* 3 300	3 100	2 900	2 500	2 500
Denmark Danemark	70	20	6	46	45	* 45	70	24	40	4
Egypt * Egypte *	400	400	400	400	400	400	400	400	400	400
Finland Finlande	130	108	110	101	203	109	300	92	64	60
France France	8 187	7 564	7 799	6 900	7 461	7 389	7 800	8 100	8 300	8 100
Germany†・Allemagne† F. R. Germany R. f. Allemagne	13 163	13 792	15 641	15 668	17 400	17 600	19 142	18 333	18 740	18 790
former German D. R. anc. R. d. allemande	3 200	2 800	3 000	3 300	3 300	3 400	4 100	4 000	4 000	-
Greece Grèce	* 300	* 400	* 300	* 400	600	400	500	500	500	500
Hong Kong * Hong-kong *	400	600	1 100	1 500	1 800	2 500	3 000	2 500	2 000	2 000
Hungary Hongrie	1 800	2 000	* 1 700	* 1 281	1 353	1 300	1 100	700	500	400

95

Tin
Industrial consumption: metric tons [*cont.*]
Etain
Consommation industrielle : tonnes métriques [*suite*]

Country or area Pays ou zone	1982	1983	1984	1985	1986	1987	1988	1989	1990	1991
India Inde	2 076	2 300	* 2 400	* 2 300	* 2 900	2 700	* 2 940	* 2 940	* 2 940	3 000
Indonesia Indonésie	556	550	850	1 012	1 114	882	1 338	1 600	1 200	1 200
Iran, Islamic Rep. of * Iran, Rép. islamique d' *	500	500	500	500	500	500	500	500	500	500
Ireland Irlande	...	...	100	100	100	100	100	100	100	100
Israel * Israël *	140	100	100	100	100	100	100	100	100	100
Italy Italie	4 200	4 500	4 500	5 000	5 600	* 6 000	6 000	5 900	6 900	6 000
Japan[2] Japon[2]	28 705	30 394	33 278	31 598	31 521	32 600	* 31 317	* 33 834	* 34 814	* 34 785
Korea, Republic of Corée, République de	2 093	2 628	3 632	* 2 600	4 335	* 4 000	6 684	11 585	7 399	6 000
Malaysia Malaisie	434	777	1 520	1 557	1 944	1 993	2 420	2 359	2 554	3 300
Mexico * Mexique *	1 400	1 600	1 600	1 000	1 200	1 200	1 200	1 300	1 200	1 200
Morocco * Maroc *	240	100	400	100	160	100	116	131	120	100
Netherlands[2] Pays-Bas[2]	5 142	4 672	4 842	* 4 253	4 300	4 900	4 660	5 300	5 200	5 200
New Zealand * Nouvelle-Zélande *	170	170	100	100	100	200	100	100	100	100
Nigeria * Nigéria *	80	80	75	100	200	200	29	63	24	20
Norway Norvège	379	392	396	425	398	392	406	426	460	410
Pakistan * Pakistan *	200	200	70	300	500	70	500	400	400	350
Peru Pérou	600	400	300	600	500	500	400	* 400	400	350
Philippines * Philippines *	1 000	700	700	400	500	700	700	700	700	650
Poland Pologne	4 575	4 500	3 613	3 029	3 624	2 700	3 500	3 000	1 000	800
Portugal * Portugal *	800	600	600	700	900	800	800	650	750	650
Romania Roumanie	* 2 500	* 2 400	* 2 000	* 2 000	* 2 000	* 1 500	1 500	1 270	1 084	700
Singapore * Singapour *	500	500	500	500	500	500	500	500	500	500
South Africa Afrique du Sud	* 1 870	1 563	1 745	* 1 900	* 1 900	* 1 900	2 350	2 200	2 100	1 800
Spain Espagne	* 4 700	* 3 700	* 3 900	* 3 100	* 2 600	* 2 600	2 800	2 900	3 200	3 000

95

Tin
Industrial consumption: metric tons [cont.]
Etain
Consommation industrielle : tonnes métriques [suite]

Country or area Pays ou zone	1982	1983	1984	1985	1986	1987	1988	1989	1990	1991
Sweden Suède	200	242	* 300	* 440	530	* 400	541	341	390	370
Switzerland[2] Suisse[2]	647	693	598	780	1 208	912	844	866	978	840
Syrian Arab Republic * Rép. arabe syrienne *	60	60	100	100	100	100	100	100	100	100
Thailand Thaïlande	700	700	631	518	1 481	1 860	1 955	2 587	2 743	* 4 200
Turkey * Turquie *	700	1 000	900	900	1 100	1 100	1 000	800	1 000	1 000
former USSR ancienne URSS	27 000	27 000	27 000	31 500	32 000	31 000	26 000	27 000	26 000	20 000
United Kingdom[2] Royaume-Uni[2]	6 800	5 943	5 838	* 6 000	* 6 000	* 6 200	6 400	6 300	6 500	6 400
United States[2] Etats-Unis[2]	33 019	34 301	37 819	37 187	32 548	35 597	37 529	36 518	36 616	37 100
Uruguay * Uruguay *	80	80	80	80	80	80	...	...	...	...
Venezuela * Venezuela *	200	300	600	1 000	800	1 000	1 776	841	764	550
Yugoslavia * Yougoslavie *	1 200	1 100	1 100	1 400	1 500	1 100	1 400	1 000	1 050	1 050
Zaire Zaïre	65	3	6	6	7	8	7	10	10	10
Zimbabwe * Zimbabwe *	600	600	600	600	600	600	...	...	...	...

Source:
Former International Tin Statistics and United Nations
Conference on Trade and Development (Geneva).

† All data shown which pertain to Germany prior to 3 October
1990 are indicated separately for the Federal Republic of
Germany and the former German Democratic Republic based on
their respective territories at the time indicated. Where
data for united Germany (3 October 1990 and thereafter) are
not available, available data are shown separately under the
designations Federal Republic of Germany and former German
Democratic Republic and pertain to the territorial
boundaries prior to 3 October 1990. For detailed
explanatory notes on data pertaining to Germany, see Annex I
- Country Nomenclature.

1 Excluding Albania, Democratic People's Republic of Korea,
Mongolia and Vietnam but including tonnages for minor
consumers not listed above.

2 Reported actual consumption.

Source:
"Former International Tin Statistics" et Conférence des
Nations Unies sur le commerce et le développement (Genève).

† Toutes les données se rapportant à l'Allemagne avant le 3
octobre 1990 figurent dans deux rubriques séparées basées
sur les territoires respectifs de la République fédérale
d'Allemagne et l'ancienne République démocratique allemande
selon la période indiquée. En l'absence de données pour
l'Allemagne unifiée (à compter du 3 octobre 1990), les
données disponibles sont fournies séparément sous les
rubriques République fédérale d'Allemagne et ancienne
République démocratique allemande et se rapportent aux
limites territoriales antérieures au 3 octobre 1990. Pour
les notes explicatives en détail sur les données concernant
l'Allemagne, voir Annexe I – Nomenclature des pays.

1 Non compris l'Albanie, la République populaire démocratique
de Corée, la Mongolie et le Viet-Nam, mais y compris les
tonnages de l'étain pour les petits consommateurs non
mentionnés ci-dessus.

2 Consommation effective rapportée par le pays.

96

Zinc
Zinc

Production: thousand metric tons
Production : milliers de tonnes métriques

Country or area Pays ou zone	1981	1982	1983	1984	1985	1986	1987	1988	1989	1990
Total	5 987.6	5 838.4	6 128.3	6 407.6	6 612.1	6 601.9	6 796.6	6 967.4	6 952.9	6 876.7
Algeria[1] **Algérie**[1]	32.3	28.8	31.2	35.0	35.8	30.3	21.0	26.5	16.8	15.0
Argentina **Argentine**	29.3	31.4	32.0	29.9	32.9	31.5	34.3	* 31.4	* 31.5	* 31.5
Australia[2] **Australie**[2]	279.0	305.5	292.8	302.6	292.0	301.5	304.5	309.7	* 308.5	* 299.3
Primary 1re fusion	274.0	301.0	288.3	299.7	288.7	297.0	300.0	306.0	303.0	294.7
Secondary 2ème fusion	* 5.0	* 4.5	* 4.5	* 2.8	3.3	4.5[3]	4.5[3]	3.7[3]	5.5[3]	4.5[3]
Austria **Autriche**	22.7	20.1	19.7	21.0	20.5	19.5	19.8	20.7	22.1	23.2
Primary 1re fusion	21.9	19.3	18.8	19.4	19.0	18.4	18.3	19.0	20.8	22.2
Secondary 2ème fusion	0.8	0.8	0.8	1.1	1.2	1.1	1.1	0.7	0.2	0.2
Remelted Refondu	0.7	0.5	0.2	0.5	0.3	0.0	0.4	1.1	1.1	0.9
Recovered in alloys Récupéré en alliages	3.1	3.5	4.6	4.5	5.1	4.1	5.2	5.6	5.3	5.1
Belgium[4] **Belgique**[4]	250.4	240.9	275.8	285.3	290.5	288.8	308.6	323.8	306.0	356.5
Brazil **Brésil**	110.8	110.4	111.0	114.4	120.7	135.6	148.1	144.0	162.3	* 155.5
Primary 1re fusion	91.8	96.0	99.9	106.9	116.1	129.7	138.7	139.7	155.9	* 150.0
Secondary 2ème fusion	19.0	14.4	11.0	7.5	4.6	5.9	9.4	4.3	6.4	* 5.5
Bulgaria[5] **Bulgarie**[5]	90.0	92.0	93.0	91.0	92.1	92.7	89.9	82.0	89.6	73.5
Canada[1] **Canada**[1]	618.7	511.9	617.0	683.1	692.4	571.0	610.5	703.2	669.7	591.8
China **Chine**	160.0[6]	160.0[6]	175.0[6]	185.0[6]	275.0[6]	336.0[6]	383.0[6]	425.2	446.5	551.8
Czechoslovakia[1] **Tchécoslovaquie**[1]	2.0	2.4	2.1	1.2	1.2	1.3	1.1	1.4	1.3	1.3
Denmark[7] **Danemark**[7]	0.0	0.0	0.0	0.0	0.4	0.8	0.7	* 1.0	0.0	0.0
Finland **Finlande**	140.0	144.0	128.5	141.7	144.6	138.8	158.8	149.8	154.2	163.1
Remelted Refondu	0.9	0.7	...	...	...	...	...	...	...	...
Recovered in alloys[8] Récupéré en alliages[8]	0.7	2.0	2.5	3.0	3.8	5.6	6.2	5.2	6.2	5.8
France **France**	270.7	283.2	273.2	290.0	278.1	288.2	290.8	304.0	308.5	276.8
Primary 1re fusion	257.1	243.8	237.6	257.6	247.3	257.4	249.3	264.0	263.0	264.7
Secondary 2ème fusion	13.5	# 39.5	35.6	32.4	30.8	30.8	41.5	40.0	45.5	12.1
Germany† · Allemagne† **F. R. Germany** **R. f. Allemagne**	366.6	335.1	349.7	356.4	367.8	371.9	385.3	378.2	360.9	360.0

96

Zinc
Production: thousand metric tons [cont.]
Zinc
Production : milliers de tonnes métriques [suite]

Country or area Pays ou zone	1981	1982	1983	1984	1985	1986	1987	1988	1989	1990
Primary 1re fusion	227.0	235.3	246.8	260.8	269.3	268.3	270.8	275.6	272.7	259.2
Secondary 2ème fusion	296.5	307.8	314.1	324.1	324.1	321.9	329.1	337.9	331.5	334.1
Remelted Refondu	35.4	31.6	27.8	30.8	27.9	26.6	26.0	42.4	45.4	38.1
former German D. R. anc. R. d. allemande	21.2[5]	20.4[5]	19.8[5]	24.5	25.0	24.9	25.3	24.2	21.9	12.7[5]
Greece Grèce	0.4	0.4	0.4	0.4	0.3	0.3	0.0	0.0	0.0	0.0
Hungary[9] Hongrie[9]	4.5	4.3	4.1	4.1	3.3	2.3	1.6	1.5	1.4	1.1
Remelted Refondu	0.0	0.0	0.2	0.7	0.4	0.9	1.6	1.4	1.5	1.4
India[1] Inde[1]	57.4	52.6	53.3	55.4	70.8	73.8	68.9	68.9	71.4	79.1
Italy Italie	180.9	158.2	155.6	166.8	210.1	230.4	247.0	242.1	246.0	259.4
Recovered in alloys Récupéré en alliages	55.0	53.9	52.1	48.0	48.3	50.6	50.0	55.9	59.6	...
Japan Japon	670.2	662.4	701.3	754.4	739.6	708.0	665.6	678.2	664.5	687.5
Primary 1re fusion	626.1	606.2	637.1	685.4	675.5	664.5	642.2	656.6	640.1	664.2
Secondary 2ème fusion	44.1	56.1	64.2	69.0	64.2	43.5	23.4	21.6	24.4	23.3
Remelted Refondu	50.3	46.0	47.9	51.1	49.8	45.2	42.2	45.2	47.5	42.3
Korea, Dem. P. R. *[16] Corée, R. p. dém. de *[16]	120.0	120.0	120.0	120.0	180.0	180.0	210.0	210.0	210.0	210.0
Korea, Republic of[1] Corée, République de[1]	83.9	99.2	107.9	108.5	111.0	127.2	186.0	226.0	241.7	257.0
Mexico[1] Mexique[1]	126.5	127.0	174.1	166.5	175.0	174.3	96.8	108.2	102.8	108.2
Netherlands[1] Pays-Bas[1]	177.4	186.0	187.5	209.7	201.7	196.2	207.1	211.0	203.0	207.1
Norway[1] Norvège[1]	80.3	78.7	90.7	94.2	92.8	90.5	116.6	128.3	120.5	125.0
Recovered in alloys[1] Récupéré en alliages[1]	8.7	9.8	8.6	7.1	7.7	6.3	9.5	6.1	5.0	...
Peru[16] Pérou[16]	126.2	160.2	154.0	148.4	162.7	155.8	144.2	123.1	* 137.8	120.6
Recovered in alloys Récupéré en alliages	...	...	...	22.2	...	...	...	...	...	...
Poland Pologne	167.1	165.4	170.3	176.1	180.1	179.2	176.5	173.6	163.7	132.1
Romania Roumanie	45.2	39.9	42.0	38.2	35.4	46.4	46.0	39.6	29.8	11.5
Spain Espagne	179.5[6]	181.8[6]	195.5	205.0	206.2	193.0	222.6	245.4	250.1	252.7
Primary 1re fusion	156.8	173.4	189.1	199.4	202.3	184.6	213.0	232.8	* 237.5	* 239.7
Secondary 2ème fusion	22.7	8.4	6.4	5.5	3.9	8.4	9.6	12.6	12.9	* 13.0
Recovered in alloys Récupéré en alliages	0.3	19.6	21.3	26.8	21.1	23.3	23.6	25.2	25.6	...

96
Zinc
Production: thousand metric tons [cont.]
Zinc
Production : milliers de tonnes métriques [suite]

Country or area Pays ou zone	1981	1982	1983	1984	1985	1986	1987	1988	1989	1990
Switzerland **Suisse**	...	...	...	...	...	9.6	9.6	9.7	10.4	10.3
former USSR * [6] **ancienne URSS * [6]**	905.0	940.0	965.0	995.0	990.0	985.0	1 000.0	963.0	977.0	890.0
Primary * [6] 1re fusion * [6]	870.0	850.0	870.0	900.0	890.0	880.0	890.0	848.0	862.0	780.0
Secondary * [6] 2ème fusion * [6]	85.0	90.0	95.0	95.0	100.0	105.0	110.0	115.0	115.0	110.0
United Kingdom **Royaume-Uni**	81.7	79.3	87.7	85.6	74.3	85.9	81.4	76.7	79.8	93.3
Remelted Refondu	6.7	6.8	6.5	6.5	7.8	6.9	6.6	6.9	...	...
Recovered in alloys Récupéré en alliages	27.2	31.9	29.5	32.5	30.8	29.8	...	...	...	...
United States **Etats-Unis**	396.8	302.5	305.1	331.2	333.8	316.3	343.9	329.8	358.2	358.4
Primary 1re fusion	346.6	228.2	235.7	253.1	261.2	253.4	261.3	241.3	263.1	262.7
Secondary 2ème fusion	50.2	74.3	69.4	78.1	72.6	62.9	82.6	88.5	95.1	95.7
Remelted Refondu	2.8	2.6	2.9	...	...	...	...	...	...	...
Recovered in alloys Récupéré en alliages	185.7	121.0	174.3	...	...	...	...	...	...	...
Yugoslavia [10] **Yougoslavie [10]**	96.4	86.8	88.0	86.9	83.4	133.3	118.1	127.5	119.4	113.8
Zaire * [1] **Zaïre * [1]**	57.6	64.4	62.5	66.1	64.0	61.4	53.6	61.1	54.0	38.2
Zambia [1] **Zambie [1]**	33.3	39.2	37.9	29.2	22.8	22.5	21.0	20.2	12.8	10.4

Source:
Industrial statistics database of the Statistical Division
of the United Nations Secretariat.

† All data shown which pertain to Germany prior to 3 October
1990 are indicated separately for the Federal Republic of
Germany and the former German Democratic Republic based on
their respective territories at the time indicated. Where
data for united Germany (3 October 1990 and thereafter) are
not available, available data are shown separately under the
designations Federal Republic of Germany and former German
Democratic Republic and pertain to the territorial
boundaries prior to 3 October 1990. For detailed
explanatory notes on data pertaining to Germany, see Annex I
- Country Nomenclature.

1 Primary metal production only.
2 Twelve months ending 30 June of year stated.

3 Source: World Metal Statistics (London).
4 Including zinc remelted and recovered in alloys.
5 Source: "Metallgesellschaft Aktiengesellschaft",
(Frankfurt).
6 Source: US Bureau of Mines, (Washington, DC).
7 Secondary metal production only.
8 Including other zinc alloys.
9 Recovered in alloys only.
10 Including zinc dust.

Source:
Base de données pour les statistiques industrielles de la
Division de statistique du Secrétariat de l'ONU.

† Toutes les données se rapportant à l'Allemagne avant le 3
octobre 1990 figurent dans deux rubriques séparées basées
sur les territoires respectifs de la République fédérale
d'Allemagne et l'ancienne République démocratique allemande
selon la période indiquée. En l'absence de données pour
l'Allemagne unifiée (à compter du 3 octobre 1990), les
données disponibles sont fournies séparément sous les
rubriques République fédérale d'Allemagne et ancienne
République démocratique allemande et se rapportent aux
limites territoriales antérieures au 3 octobre 1990. Pour
les notes explicatives en détail sur les données concernant
l'Allemagne, voir Annexe I – Nomenclature des pays.

1 Production du métal de première fusion seulement.
2 Période de douze mois finissant le 30 juin de l'année
indiquée.
3 Source: "World Metal Statistics," (Londres).
4 Y compris le zinc refondu et récupéré à partir des alliages.
5 Source: "Metallgesellschaft Aktiengesellschaft",
(Francfort).
6 Source: "US Bureau of Mines," (Washington, DC).
7 Production du métal de deuxième fusion seulement.
8 Y compris les autres alliages de zinc.
9 Récupére en alliages seulement.
10 Y compris la poudre de zinc.

97

Radio and television receivers
Radiodiffusion et télévision : postes récepteurs
Production: thousands
Production : milliers

Country or area Pays ou zone	Radio receivers Radiodiffusion : postes récepteurs					Television receivers Télévision : postes récepteurs				
	1980	1987	1988	1989	1990	1980	1987	1988	1989	1990
Total	185 538	140 345	135 552	141 709	149 263	72 170	115 464	117 291	121 829	126 951
Albania Albanie	...	...	25	30	...	...	...	17	23	...
Algeria Algérie	112	160	156	120	213	94	318	319	219	283
Argentina Argentine	...	...	...	...	...	454	695	650	674	281
Australia[1] Australie[1]	199	...	...	...	...	332[2]	211[2]	177[2]	162[2]	158[2]
Austria[3] Autriche[3]	...	...	...	...	...	...	754	...	...	...
Bangladesh Bangladesh	228	144	121	102	...	8	77	60	79	...
Belarus Bélarus	499	726	798	882	...	552	1 064	1 040	1 102	1 302
Belgium[4][5] Belgique[4][5]	2 246	1 165	836	859	986	746	917	946	979	1 084
Brazil Brésil	6 769	8 676	7 632	7 210	5 151	3 254	2 902	2 722	2 920	3 196
Bulgaria Bulgarie	51	56	50	52	43	91	199	181	185	219
Cameroon Cameroun	76	...	...	...	...	...	...	...	...	...
Canada[5] Canada[5]	...	...	...	...	...	444	...	...	...	...
Chad Tchad	...	...	6	9	13	...	...	...	...	...
Chile Chili	...	...	...	...	...	107	35	42	24	0
China Chine	30 038[6]	17 638[6]	15 489[6]	18 347[6]	21 000[6]	2 492	19 344	25 051	27 665	26 847
Colombia Colombie	3	...	...	...	...	103	112	135	* 79	...
Cuba Cuba	200	227	153	173	...	40	56	65	71	...
Cyprus Chypre	...	...	...	...	...	...	* 2	* 2	0	0
Czechoslovakia Tchécoslovaquie	250[7]	200[7]	184[7]	173[7]	194[7]	389	507	482	524	505
Denmark Danemark	88	47	...	...	...	77	110	* 112	...	...
Ecuador Equateur	...	10	14	...	...	36	44	45	...	...
Egypt Egypte	171	199	161	43	59	308	333	319	194	333

97
Radio and television receivers
Production: thousands [cont.]
Radiodiffusion et télévision : postes récepteurs
Production : milliers [suite]

Country or area Pays ou zone	Radio receivers Radiodiffusion : postes récepteurs					Television receivers Télévision : postes récepteurs				
	1980	1987	1988	1989	1990	1980	1987	1988	1989	1990
El Salvador[8] El Salvador[8]	...	...	...	...	...	47	...	...	...	...
Finland Finlande	257	...	# 30	...	...	341	627	535	413	412
France France	2 141	2 080	1 983	2 039	2 059	1 928	2 184	2 081	2 447	2 838
Gabon Gabon	7	...	...	...	...	1	...	...	...	...
Germany† · Allemagne† F. R. Germany R. f. Allemagne	3 707	5 141	4 758	4 975	5 955	4 425	3 537	3 737	3 236	3 595
former German D. R. anc. R. d. allemande	915	1 240	1 217	1 151	...	578	723	774	775	...
Greece Grèce	...	...	...	...	...	137	...	...	...	...
Hong Kong Hong-kong	67 478	...	...	...	...	108	...	1 908	683	...
Hungary Hongrie	271	100	134	145	83	417	446	433	502	492
India Inde	1 918[9]	1 178[9]	1 068[9]	672[9]	...	88	972	1 269	1 237	...
Indonesia Indonésie	1 530[10]	997[10]	...	...	...	607	575	...	...	...
Iran, Islamic Rep. of[11] Iran, Rép. islamique d'[11]	57[12 13]	254[12 13]	344[12 13]	221[12 13]	...	246	400	303	334	...
Iraq Iraq	...	...	...	...	...	...	...	58	...	...
Israel[14] Israël[14]	...	...	...	...	...	24	35	21	...	...
Italy Italie	...	...	...	...	...	1 984	2 233	2 337	2 391	...
Japan Japon	15 343	10 496	10 769	10 690	10 955	15 205	14 777	13 299	...	...
Kenya Kenya	...	...	...	...	...	...	10 714	5 978	2 077	4 186
Korea, Republic of Corée, République de	3 972	1 425	1 414	1 112	1 462	6 819	14 922	14 820	15 469	15 847
Malaysia Malaisie	5 000[15]	15 879[15]	21 070[15]	28 450[15]	37 019[15]	157	1 240	1 221	1 585	2 169
Mexico Mexique	1 029	207	...	...	...	964	609	697	536	633
Morocco Maroc	47[16]	...	...	...	...	100[17]	...	...	...	...
Mozambique Mozambique	164	113	91	...	...	...	...	...	...	...

97
Radio and television receivers
Production: thousands [cont.]
Radiodiffusion et télévision : postes récepteurs
Production : milliers [suite]

Country or area Pays ou zone	Radio receivers Radiodiffusion : postes récepteurs					Television receivers Télévision : postes récepteurs				
	1980	1987	1988	1989	1990	1980	1987	1988	1989	1990
Myanmar Myanmar	1[18]	...	...	...	...	...	1	1	0	...
New Zealand Nouvelle-Zélande	...	...	...	...	...	126	90	...	...	...
Nicaragua * Nicaragua *	1	...	...	...	...	1	...	...	...	...
Nigeria Nigéria	...	...	...	...	...	208	34	...	...	...
Pakistan Pakistan	...	...	...	...	...	74	199	246	211	200
Peru Pérou	...	...	...	...	...	75	197	72	55	...
Philippines Philippines	...	...	...	...	...	206	300	...	...	...
Poland Pologne	2 695	2 833	2 684	2 523	1 433	900	647	749	772	748
Portugal Portugal	593	1 188	1 058	...	...	467	485	401	...	...
Romania Roumanie	863	618	623	526	384	541	484	511	511	401
Rwanda Rwanda	0	12	10	...	...	...	...	...	...	...
Singapore Singapour	17 070[19]	19 868[19]	19 618[19]	...	...	1 889	2 123	2 807	3 040	...
South Africa Afrique du Sud	861	776	1 092	941	775	338	231	351	344	373
Spain Espagne	188	16	17	15	...	763	1 362	1 604	2 100	...
Sri Lanka Sri Lanka	72	...	...	...	...	...	...	...	...	...
Sweden Suède	26	...	...	...	...	308	354	...	...	...
Syrian Arab Republic Rép. arabe syrienne	...	...	...	...	...	72	...	...	90	18
Thailand Thaïlande	1 113	...	341	933	1 105	248	544	702	1 476	2 351
Trinidad and Tobago Trinité-et-Tobago	8	10	10	6	...	13	10	6	2	...
Tunisia Tunisie	60	28	41	...	...	88	58	93	...	...
Turkey Turquie	52[20]	441[20]	294[20]	219[20]	...	327	694	738	999	...
Ukraine Ukraine	315	395	432	574	...	2 526	3 110	3 434	3 572	3 773
former USSR ancienne URSS	8 478	8 143	8 025	8 561	9 168	7 528	9 081	9 637	9 938	10 540

97
Radio and television receivers
Production: thousands [*cont.*]
Radiodiffusion et télévision : postes récepteurs
Production : milliers [*suite*]

Country or area Pays ou zone	Radio receivers Radiodiffusion : postes récepteurs					Television receivers Télévision : postes récepteurs				
	1980	1987	1988	1989	1990	1980	1987	1988	1989	1990
United Kingdom Royaume-Uni	439 [21]	...	...	34 [21]	...	2 364	3 022	3 024	...	...
United Rep.Tanzania Rép. Unie de Tanzanie	223	72	...	...	...	...	...	...	...	...
United States[5] Etats-Unis[5]	7 672	4 315	3 061	3 408	2 989	10 320	12 871	12 938	15 478	15 168
Yugoslavia Yougoslavie	125	198	129	86	...	543	591	549	503	...
Zambia Zambie	78	...	...	...	...	...	...	...	...	...

Source:
Industrial statistics database of the Statistical Division
of the United Nations Secretariat.

† All data shown which pertain to Germany prior to 3 October
1990 are indicated separately for the Federal Republic of
Germany and the former German Democratic Republic based on
their respective territories at the time indicated. Where
data for united Germany (3 October 1990 and thereafter) are
not available, available data are shown separately under the
designations Federal Republic of Germany and former German
Democratic Republic and pertain to the territorial
boundaries prior to 3 October 1990. For detailed
explanatory notes on data pertaining to Germany, see Annex I
- Country Nomenclature.

1 Twelve months ending 30 June of year stated.

2 Colour television receivers only.
3 Source: United Nations Economic Commission for Europe (ECE),
  (Geneva).
4 Production by establishments employing 5 or more persons.
5 Shipments.
6 Portable battery sets only.
7 Including record players.
8 Including radio receivers.
9 Production by large and medium scale establishments only.
10 Including radio with tape recording unit.

11 Production by establishments employing 50 or more persons.
12 Including sound reproducers.
13 Including tape recorders.
14 Marketed local production.
15 Data refer to Peninsular Malaysia only.
16 Data refer to assembly.
17 Including assembly.
18 Government production only.
19 Including radio with tape recording unit and clock.

20 Excluding radio-phonograph combinations.
21 Including car tape-players.

Source:
Base de données pour les statistiques industrielles de la
Division de statistique du Secrétariat de l'ONU.

† Toutes les données se rapportant à l'Allemagne avant le 3
octobre 1990 figurent dans deux rubriques séparées basées
sur les territoires respectifs de la République fédérale
d'Allemagne et l'ancienne République démocratique allemande
selon la période indiquée. En l'absence de données pour
l'Allemagne unifiée (à compter du 3 octobre 1990), les
données disponibles sont fournies séparément sous les
rubriques République fédérale d'Allemagne et ancienne
République démocratique allemande et se rapportent aux
limites territoriales antérieures au 3 octobre 1990. Pour
les notes explicatives en détail sur les données concernant
l'Allemagne, voir Annexe I – Nomenclature des pays.

1 Période de douze mois finissant le 30 juin de l'année
  indiquée.
2 Récepteurs de télévision en couleur seulement.
3 Source: Commission économique des Nations Unies pour
  l'Europe (CEE), (Genève).
4 Production des établissements occupant 5 personnes ou plus.
5 Expéditions.
6 Appareils portatifs à piles seulement.
7 Y compris les tourne-disques.
8 Y compris les récepteurs de radio.
9 Production des grandes et moyennes entreprises seulement.
10 Y compris les récepteurs de radio avec appareil enregistreur
   à bande magnétique incorporés.
11 Production des établissements occupant 50 personnes ou plus.
12 Y compris les lecteurs de son.
13 Y compris les magnétophones.
14 Production locale commercialisée.
15 Données se rapportant à la Malaisie péninsulaire seulement.
16 Données se rapportant à l'assemblage.
17 Y compris le montage.
18 Production de l'Etat seulement.
19 Y compris les récepteurs de radio avec appareil enregistreur
   à bande magnétique et pendule incorporés.
20 Non compris les radios-phonos.
21 Y compris les lecteurs de cassettes pour automobiles.

98
Merchant vessels
Navires marchands
Tonnage launched: thousand gross registered tons
Tonnage lancé : milliers de tonneaux de jauge brute

Country or area Pays ou zone	1982	1983	1984	1985	1986	1987	1988	1989	1990	1991
World [1] • **Monde** [1]	**17290**	**14888**	**17732**	**17247**	**14914**	**9770**	**11997**	**13041**	**14894**	**16860**
Steam Vapeur	552	445	102	-	-	20	338	-	-	-
Motor Moteur	16737	14444	17630	17247	14914	9750	11659			
Oil tankers Pétroliers	3412	1854	1925	2928	3564	3094	4594	5368	4686	7648
Bulk carriers Transp. de vracs	9000	7106	9715	8036	6013	2471	2782	3681	5161	3152
General cargo Marchandises diverses	1830	2471	2240	2076	1797	2267	1310	1094	1635	1611
Container ships Navires porte-conteneurs	902	1350	1524	1846	1785	659	1660	1138	1548	2367
Argentina **Argentine**	**108**	**18**	**71**	**96**	**35**	**37**	**60**	**37**	**10**	**1**
Oil tankers Pétroliers	86	-	4	-	-	3	-	-	9	-
Bulk carriers Transp. de vracs	-	-	66	68	35	32	35	35	-	-
General cargo Marchandises diverses	21	18	-	4	-	-	-	-	0	-
Australia **Australie**	**8**	**8**	**11**	**11**	**15**	**10**	**10**	**14**	**26**	**25**
Belgium **Belgique**	**260**	**118**	**194**	**77**	**102**	**53**	**28**	**45**	**56**	**47**
Oil tankers Pétroliers	-	-	-	52	-	52	26	-	-	-
Bulk carriers Transp. de vracs	248	53	123	-	86	-	-	-	-	-
General cargo Marchandises diverses	10	12	-	1	10	-	1	-	0	16
Brazil **Brésil**	**455**	**359**	**460**	**405**	**317**	**61**	**269**	**260**	**105**	**277**
Oil tankers Pétroliers	40	-	122	81	18	53	88	201	90	202
Bulk carriers Transp. de vracs	347	329	326	300	278	-	178	41	-	-
General cargo Marchandises diverses	64	27	10	-	17	-	-	15	15	74
Container ships Navires porte-conteneurs	-	-	-	23	-	-	-	-	-	-
Bulgaria **Bulgarie**	**154**	**114**	**159**	**141**	**125**	**97**	**52**	**71**	**81**	**64**
Oil tankers Pétroliers	-	-	-	4	8	27	4	26	14	26
Bulk carriers Transp.de vracs	146	85	117	104	96	41	23	22	25	14
General cargo Marchandises diverses	-	-	24	24	12	-	16	3	42	23
Container ships Navires porte-conteneurs	8	28	17	9	9	28	9	20	-	-
Canada **Canada**	**108**	**46**	**45**	**43**	**7**	**13**	**3**	**32**	**5**	**4**
Oil tankers Pétroliers	8	-	-	-	-	-	-	-	-	-
Bulk carriers Transp.de vracs	67	23	25	20	-	-	-	-	-	-

98
Merchant vessels
Tonnage launched: thousand gross registered tons [*cont.*]
Navires marchands
Tonnage lancé : milliers de tonneaux de jauge brute [*suite*]

Country or area Pays ou zone	1982	1983	1984	1985	1986	1987	1988	1989	1990	1991
China **Chine**	**211**	**196**	**291**	**140**	**343**	**231**	**287**	**225**	**513**	**333**
Oil tankers Pétroliers	...	...	37	3	115	85	114	0	197	106
Bulk carriers Transp.de vracs	158	114	137	73	133	52	61	89	210	135
General cargo Marchandises diverses	46	33	37	38	77	69	111	101	90	70
Container ships Navires porte-conteneurs	8	37	45	15	18	10	-	33	10	5
Denmark **Danemark**	**434**	**525**	**393**	**430**	**313**	**270**	**315**	**314**	**421**	**416**
Oil tankers Pétroliers	50	115	28	165	236	191	86	129	174	-
Bulk carriers Transp.de vracs	271	215	142	36	-	-	-	-	-	165
General cargo Marchandises diverses	33	62	86	158	54	72	31	14	58	52
Container ships Navires porte-conteneurs	37	77	87	43	-	-	157	157	183	180
Egypt **Egypte**	**15**	**12**	**7**	**45**	**17**	**2**	**7**	**7**	**20**	**1**
General cargo Marchandises diverses	15	10	6	-	16	2	7	7	19	-
Finland **Finlande**	**199**	**268**	**317**	**252**	**213**	**128**	**274**	**208**	**175**	**97**
Oil tankers Pétroliers	19	27	33	39	108	3	9	12	6	-
General cargo Marchandises diverses	130	186	112	19	60	20	47	9	63	5
France **France**	**308**	**197**	**229**	**186**	**114**	**196**	**100**	**77**	**108**	**81**
Oil tankers Pétroliers	41	16	-	0	39	-	-	-	-	-
Bulk carriers Transp.de vracs	-	-	-	23	53	91	-	-	-	-
General cargo Marchandises diverses	15	4	78	-	-	-	-	-	-	-
Container ships Navires porte-conteneurs	41	31	92	76	-	-	-	-	-	-
Germany ʂ · Allemagne ʂ **Fed. Rep. Germany** **Rép. féd. allemagne**	**722**	**651**	**528**	**627**	**496**	**222**	**526**	**546**	**867**	**870**
Oil tankers Pétroliers	0	1	-	4	10	-	-	10	15	14
Bulk carriers Transp.de vracs	159	159	38	19	-	-	9	-	-	-
General cargo Marchandises diverses	224	253	274	238	134	52	51	95	278	242
Container ships Navires porte-conteneurs	226	158	142	240	209	94	369	256	392	436
former German Dem.Rep. [2] l'anc. Rép. dém. allemande [2]	**328**	**343**	**362**	**407**	**334**	**302**	**283**	**219**	-	-
Bulk carriers Transp.de vracs	68	76	62	34	50	17	-	-	-	-
General cargo Marchandises diverses	115	139	197	218	203	222	170	129	-	-
Container ships Navires porte-conteneurs	53	54	36	77	18	18	37	-	-	-

98
Merchant vessels
Tonnage launched: thousand gross registered tons [*cont.*]
Navires marchands
Tonnage lancé : milliers de tonneaux de jauge brute [*suite*]

Country or area Pays ou zone	1982	1983	1984	1985	1986	1987	1988	1989	1990	1991
Greece **Grèce**	**38**	**37**	**41**	**20**	**18**	**9**	**11**	**2**	**4**	**4**
General cargo Marchandises diverses	-	-	3	-	-	9	9	-	-	3
India **Inde**	**17**	**95**	**35**	**78**	**51**	**20**	**19**	**54**	**1**	**5**
Bulk carriers Transp.de vracs	-	93	32	57	42	15	15	50	-	
General cargo Marchandises diverses	12	-	-	1	4	-	-	-	-	-
Indonesia **Indonésie**	**2**	**2**	**9**	**0**	**5**	**5**	**8**	**5**	**12**	**4**
Oil tankers Pétroliers	-	-	7	0	2	2	-	1	1	1
Italy **Italie**	**277**	**165**	**241**	**38**	**89**	**281**	**334**	**212**	**344**	**534**
Oil tankers Pétroliers	69	32	126	18	18	67	4	11	27	27
Bulk carriers Transp.de vracs	168	106	30	-	52	87	150	-	203	336
General cargo Marchandises diverses	10	7	53	12	9	115	43	31	26	33
Container ships Navires porte-conteneurs	-	5	5	-	-	-	94	76	-	34
Japan **Japon**	**8247**	**7071**	**9408**	**9299**	**7739**	**4170**	**4546**	**6023**	**6564**	**7718**
Oil tankers Pétroliers	1538	810	698	1428	1891	1426	1265	2566	2405	4263
Bulk carriers Transp.de vracs	5067	3569	6436	5083	3268	1193	1568	2047	2245	1300
General cargo Marchandises diverses	505	1071	697	798	682	795	525	368	554	547
Container ships Navires porte-conteneurs	221	465	305	834	995	288	540	327	507	723
Korea,Republic of **Corée, République de**	**1530**	**1201**	**2515**	**2777**	**2517**	**2298**	**3406**	**2679**	**3318**	**3674**
Oil tankers Pétroliers	220	156	489	702	475	919	2623	1588	871	1972
Bulk carriers Transp.de vracs	1081	531	1238	1483	1516	620	351	888	1898	825
General cargo Marchandises diverses	69	154	137	70	94	638	149	65	63	41
Container ships Navires porte-conteneurs	100	326	509	370	311	81	226	67	405	713
Malaysia **Malaisie**	**11**	**11**	**1**	**10**	**1**	**3**	**0**	**2**	**38**	**9**
General cargo Marchandises diverses	10	10	-	5	-	3	-	2	-	-
Mexico **Mexique**	**19**	**-**	**-**	**38**	**-**	**23**	**-**	**-**	**-**	**4**
Oil tankers Pétroliers	-	-	-	23	-	23	-	-	-	-
Netherlands **Pays-Bas**	**217**	**190**	**161**	**190**	**107**	**50**	**68**	**108**	**171**	**182**
Oil tankers Pétroliers	104	31	1	25	2	3	4	1	3	8
General cargo Marchandises diverses	37	53	71	77	37	21	19	71	110	129
Container ships Navires porte-conteneurs	-	48	24	-	17	-	-	9	-	2

98
Merchant vessels
Tonnage launched: thousand gross registered tons [*cont.*]
Navires marchands
Tonnage lancé : milliers de tonneaux de jauge brute [*suite*]

Country or area Pays ou zone	1982	1983	1984	1985	1986	1987	1988	1989	1990	1991
Norway Norvège	**307**	**159**	**97**	**101**	**68**	**59**	**32**	**31**	**87**	**145**
Oil tankers Pétroliers	138	78	12	1	-	-	3	6	19	19
Bulk carriers Transp.de vracs	5	-	4	26	-	-	-	-	-	-
General cargo Marchandises diverses	36	12	19	5	20	5	-	10	31	21
Peru Pérou	**20**	**-**	**1**	**17**	**2**	**2**	**1**	**1**	**2**	**10**
Oil tankers Pétroliers	17	0	-	17	-	-	-	-	-	-
Poland Pologne	**304**	**420**	**320**	**293**	**241**	**204**	**179**	**122**	**155**	**225**
Oil tankers Pétroliers	181	21	75	75	-	-	17	17	62	51
Bulk carriers Transp.de vracs	40	229	81	107	113	44	66	13	-	26
General cargo Marchandises diverses	67	106	76	50	91	94	31	21	49	53
Container ships Navires porte-conteneurs	-	-	-	-	-	19	25	-	-	49
Portugal Portugal	**22**	**3**	**36**	**69**	**13**	**5**	**19**	**67**	**25**	**11**
Oil tankers Pétroliers	12	-	-	-	-	-	-	50	-	-
Bulk carriers Transp.de vracs	-	-	23	46	-	-	-	-	-	-
General cargo Marchandises diverses	9	0	12	-	12	3	17	14	23	10
Romania [2] Roumanie [2]	**102**	**133**	**263**	**201**	**227**	**80**	**4**	**60**	**5**	**196**
Oil tankers Pétroliers	-	-	6	6	138	24	-	58	-	51
Bulk carriers Transp.de vracs	73	116	230	111	39	39	-	-	-	131
General cargo Marchandises diverses	30	17	23	84	49	16	4	-	-	10
Singapore Singapour	**56**	**35**	**39**	**18**	**4**	**17**	**14**	**33**	**44**	**127**
Oil tankers Pétroliers	0	-	9	4	-	1	7	14	29	50
General cargo Marchandises diverses	8	5	-	1	1	-	3	16	2	3
Spain Espagne	**722**	**671**	**150**	**104**	**181**	**127**	**141**	**276**	**384**	**461**
Oil tankers Pétroliers	131	2	2	-	2	0	-	159	271	321
Bulk carriers Transp.de vracs	424	558	89	-	83	17	45	-	-	-
General cargo Marchandises diverses	107	48	27	64	57	24	18	63	69	21
Container ships Navires porte-conteneurs	7	-	13	26	-	-	-	-	-	-
Sweden Suède	**271**	**292**	**177**	**210**	**65**	**45**	**20**	**4**	**27**	**59**
Oil tankers Pétroliers	7	86	-	99	15	15	-	-	-	-
Bulk carriers Transp.de vracs	174	100	98	-	-	-	-	-	-	-
General cargo Marchandises diverses	28	103	52	51	-	14	-	-	19	19

98
Merchant vessels
Tonnage launched: thousand gross registered tons [*cont.*]
Navires marchands
Tonnage lancé : milliers de tonneaux de jauge brute [*suite*]

Country or area Pays ou zone	1982	1983	1984	1985	1986	1987	1988	1989	1990	1991
Turkey **Turquie**	**42**	**34**	**74**	**22**	**36**	**40**	**4**	**36**	**8**	**99**
Oil tankers Pétroliers	10	-	2	-	-	-	1	-	1	-
Bulk carriers Transp.de vracs	-	-	-	-	-	15	-	15	-	17
General cargo Marchandises diverses	31	31	67	14	28	7	-	18	7	79
former USSR [2] **ancienne URSS** [2]	**36**	**36**	**37**	...	...	**1**	**37**	**89**	**95**	**108**
Oil tankers Pétroliers	...	...	37	...	...	...	37	57	95	76
General cargo Marchandises diverses	36	31	...	...	...	-	-	-	-	-
United Kingdom **Royaume-Uni**	**528**	**527**	**191**	**145**	**238**	**36**	**91**	**100**	**79**	**151**
Oil tankers Pétroliers	142	132	-	-	-	-	32	2	4	82
Bulk carriers Transp.de vracs	263	321	87	93	91	-	-	-	-	-
General cargo Marchandises diverses	57	52	87	39	61	-	3	11	9	13
United States **Etats- Unis**	**276**	**177**	**118**	**178**	**328**	**6**	**7**	**5**	**7**	**5**
Oil tankers Pétroliers	53	116	68	59	240	-	-	-	-	-
Bulk carriers Transp.de vracs	24	22	-	-	-	-	-	-	-	-
General cargo Marchandises diverses	44	-	35	104	35	-	-	-	-	-
Container ships Navires porte-conteneurs	81	-	-	-	47	-	-	-	-	-
Yugoslavia **Yougoslavie**	**389**	**243**	**214**	**215**	**301**	**343**	**330**	**479**	**468**	**411**
Oil tankers Pétroliers	271	146	131	103	169	198	213	314	349	247
Bulk carriers Transp.de vracs	76	57	-	48	78	35	35	109	28	5
General cargo Marchandises diverses	17	17	43	-	25	68	28	23	66	73

Source:
Lloyd's Register of Shipping (London).

§ All data shown which pertain to Germany prior to 3 October 1990
are indicated separately for the Federal Republic of Germany and
the former German Democratic Republic based on their respective
territories at the time indicated. Where data for united Germany (3
October 1990 and thereafter) are not available, available data are shown
separately under the designations Federal Republic of Germany and
former German Democratic Republic and pertain to the territorial
boundaries prior to 3 October 1990. For detailed explanatory notes
on data pertaining to Germany, see Annex I - Country Nomenclature.

1 Excluding former USSR in 1985 and 1986.
2 Incomplete information (former German Democratic Republic;
beginning 1988).

Source:
"Lloyd's Register of Shipping" (Londres).

§ Toutes les données se rapportant à l'Allemagne avant le 3
octobre 1990 figurent dans deux rubriques séparées basées
sur les territoires respectifs de la République fédérale
d'Allemagne et l'ancienne République démocratique
allemande selon la période indiquée. En l'absence de
données pour l'Allemagne unifiée (à compter du 3 octobre
1990), les données disponibles sont fournies séparément
sous les rubriques République fédérale d'Allemagne et
ancienne République démocratique allemande et se
rapportent aux limites territoriales antérieures au 3 octobre
1990. Pour les notes explicatives en détail sur les données
concernant l'Allemagne, voir Annexe I - Nomenclature des
pays.

1 Non compris l'ancienne URSS en 1985 et 1986.
2 Information incomplète (l'ancienne République
démocratique allemande; à partir de 1988).

99
Passenger cars
Voitures de tourisme
Production: thousands
Production : milliers

Country or area Pays ou zone	1981	1982	1983	1984	1985	1986	1987	1988	1989	1990
Total	27 797	26 982	29 920	30 366	31 948	32 396	32 775	33 945	35 261	34 919
Argentina[1] Argentine[1]	138	110	132	143	120	147	158	125	107	* 293
Australia[12] Australie[12]	317	375	363	341	376	365	302	315	333	386
Austria Autriche	7	7	6	7	7	7	7	7	7	15
Brazil Brésil	586	673	748	679	759	815	683	782	731	663
Canada Canada	803	808	971	1 022	1 075	1 061	810	1 008	984	94
China Chine	3	5	6	6	9	10	12	11	...	...
Czechoslovakia Tchécoslovaquie	181	174	178	180	184	185	172	164	189	191
Finland Finlande	23	30	33	32	39	43	46	45	37	30
France France	2 953	3 086	3 228	2 909	2 631	2 773	3 052	3 228	3 415	3 293
Germany† · Allemagne† F. R. Germany R. f. Allemagne	3 590	3 771	3 875	3 783	4 165	4 269	4 348	4 312	4 536	4 634
former German D. R. anc. R. d. allemande	180	183	188	202	210	218	217	218	217	...
India Inde	42	43	45	64	89	101	123	157	178	221
Italy Italie	1 254	1 296	1 395	1 439	1 384	1 653	1 712	1 883	1 971	1 875
Japan Japon	6 974	6 882	7 152	7 073	7 647	7 810	7 891	8 198	9 052	9 948
Korea, Republic of Corée, République de	72	99	128	167	262	457	778	868	846	956
Mexico Mexique	369	324	214	247	285	198	278	345	488	611
Netherlands Pays-Bas	78	91	106	109	108	119	125	120	133	123
Poland Pologne	240	228	269	278	283	290	293	293	285	266
Romania Roumanie	92	104	90	125	134	124	129	141	144	100
Spain Espagne	877	944	1 110	1 137	1 220	1 290	1 444	1 498	1 651	1 736
Sweden Suède	306	323	374	378	402	415	416	407	384	336

99
Passenger cars
Production: thousands [cont.]
    Voitures de tourisme
    Production : milliers [suite]

Country or area Pays ou zone	1981	1982	1983	1984	1985	1986	1987	1988	1989	1990
Ukraine Ukraine	162	160	162	166	168	168	167	155	155	156
former USSR ancienne URSS	1 324	1 307	1 315	1 327	1 332	1 326	1 332	1 262	1 217	1 259
United Kingdom Royaume-Uni	955	888	1 045	909	1 048	1 019	1 143	1 227	1 308	1 302
United States Etats-Unis	6 253	5 073	6 781	7 622	8 002	7 516	7 085	7 105	6 808	6 052
Yugoslavia Yougoslavie	180	158	168	187	177	185	219	226	227	289

Source:
Industrial statistics database of the Statistical Division
of the United Nations Secretariat.

† All data shown which pertain to Germany prior to 3 October
1990 are indicated separately for the Federal Republic of
Germany and the former German Democratic Republic based on
their respective territories at the time indicated. Where
data for united Germany (3 October 1990 and thereafter) are
not available, available data are shown separately under the
designations Federal Republic of Germany and former German
Democratic Republic and pertain to the territorial
boundaries prior to 3 October 1990. For detailed
explanatory notes on data pertaining to Germany, see Annex I
- Country Nomenclature.

1 Including assembly.
2 Twelve months ending 30 June of year stated.

Source:
Base de données pour les statistiques industrielles de la
Division de statistique du Secrétariat de l'ONU.

† Toutes les données se rapportant à l'Allemagne avant le 3
octobre 1990 figurent dans deux rubriques séparées basées
sur les territoires respectifs de la République fédérale
d'Allemagne et l'ancienne République démocratique allemande
selon la période indiquée. En l'absence de données pour
l'Allemagne unifiée (à compter du 3 octobre 1990), les
données disponibles sont fournies séparément sous les
rubriques République fédérale d'Allemagne et ancienne
République démocratique allemande et se rapportent aux
limites territoriales antérieures au 3 octobre 1990. Pour
les notes explicatives en détail sur les données concernant
l'Allemagne, voir Annexe I – Nomenclature des pays.

1 Y compris le montage.
2 Période de douze mois finissant le 30 juin de l'année
indiquée.

Technical notes, tables 68-99

Industrial activity comprises mining and quarrying, manufacturing and the production of electricity, gas and water. These activities correspond to the major divisions 2, 3 and 4 respectively of the *International Standard Industrial Classification of All Economic Activities*.[46]

Many of the tables are based on data compiled for the United Nations *Industrial Statistics Yearbook*, vol. I, *General Industrial Statistics*, and vol. II, *Commodity Production Statistics*.[21] Exceptions are indicated in notes at the end of the tables.

The methods used by countries for the computation of industrial output are, as a rule, consistent with the recommendations on this subject by the United Nations and provide a satisfactory basis for comparative analysis.[45] In some cases, however, the definitions and procedures underlying computations of output differ from approved guidelines. The differences, where known, are indicated in the footnotes to each table.

A. *Food, beverages and tobacco*

Table 68 covers the production of centrifugal sugar from both beet and cane and the figures are expressed as far as possible in terms of raw sugar. However, where exact information about polarisation is lacking, data are expressed in terms of sugar "tel quel" and are footnoted accordingly. Unless otherwise stated, the data refer to calendar years.

Table 69: The data relate to the apparent consumption of centrifugal sugar in the country concerned, including sugar used for the manufacture of sugar-containing products whether exported or not and sugar used for purposes other than human consumption as food. Unless otherwise specified the statistics are expressed in terms of raw value (i.e. sugar polarizing at 96 degrees). However, where exact information is lacking, data are expressed in terms of sugar "tel quel" and are footnoted accordingly. The world and regional totals also include data for countries whose sugar consumption was less than 10 thousand metric tons.

Table 70 refers to meat from animals slaughtered within the national boundaries irrespective of the origin of the animals. Production figures of beef and veal (including buffalo meat), pork (including bacon and ham) and mutton and lamb (including goat meat), are in terms of carcass weight, excluding edible offals, tallow and lard. All data refer to total meat production, i.e from both commercial and farm slaughter.

Table 73: Wheat flour. Sifted (bolted) flour made from wheat or meslin. Cereal flour. Sifted (bolted) flour of cereals other than wheat or meslin.

Table 75 refers to beer made from malt, including ale, stout, porter.

Table 76 refers to cigarettes only.

Notes techniques, tableaux 68-99

L'activité industrielle comprend les industries extractives (mines et carrières), les industries manufacturières et la production d'électricité, de gaz et d'eau. Ces activités correspondent aux grandes divisions 2, 3 et 4, respectivement, de la *Classification internationale type par industrie de toutes les branches d'activité économique* [46].

Un grand nombre de ces tableaux sont établis sur la base de données compilées pour l'*Annuaire de statistiques industrielles* des Nations Unies, vol. I, *Statistiques industrielles générales*, et vol. II, *Statistiques sur la production de matières premières* [21]. Les exceptions sont indiquées dans des notes au bas des tableaux.

En règle générale, les méthodes employées par les pays pour le calcul de leur production industrielle sont conformes aux recommandations des Nations Unies à ce sujet et offrent une base satisfaisante pour une analyse comparative [45]. Toutefois, dans certains cas, les définitions des méthodes sur lesquelles reposent les calculs de la production diffèrent des directives approuvées. Lorsqu'elles sont connues, les différences sont indiquées dans les notes au bas des tableaux.

A. *Alimentation, boisson et tabacs*

Le *Tableau 68* porte sur la production de sucre centrifugé à partir de la betterave et de la canne à sucre, et les chiffres sont exprimés autant que possible en sucre brut. Toutefois, en l'absence d'informations exactes sur la polarisation, les données sont exprimées en sucre tel quel, accompagnées d'une note au bas du tableau. Sauf indication contraire, les chiffres se rapportent à des années civiles.

Tableau 69 : Les données se rapportent à la consommation apparente de sucre centrifugé dans le pays en question, y compris le sucre utilisé pour la fabrication de produits à base de sucre, exportés ou non, et le sucre utilisé à d'autres fins que pour la consommation alimentaire humaine. Sauf indication contraire, les statistiques sont exprimées en valeur brute (sucre polarisant à 96°). Toutefois, en l'absence d'informations exactes, les données sont exprimées en sucre tel quel, accompagnées d'une note au bas du tableau. Les totaux mondiaux et régionaux comprennent également les données relatives aux pays où la consommation de sucre est inférieure à 10.000 tonnes métriques.

Le *Tableau 70* indique la production de viande provenant des animaux abattus à l'intérieur des frontières nationales, quelle que soit leur origine. Les chiffres de production de viande de boeuf et de veau (y compris la viande de buffle), de porc (y compris le bacon et le jambon) et de mouton et d'agneau (y compris la viande de chèvre) se rapportent à la production en poids de carcasses et ne comprennent pas le saindoux, le suif et les abats comestibles. Toutes les données se rapportent à la production totale de viande, c'est-à-dire à la fois aux animaux abattus à des fins commerciales et des animaux sacrifiés à la ferme

B. *Textile, wearing apparel and leather industries*

Table 77 refers to 12-month periods 1 August to 31 July and to cotton consumed in spinning mills and other factories, plus estimates of non-commercial and household consumption. Data do not necessarily represent the final domestic consumption of cotton in the country specified since a proportion of the finished cotton goods may be exported or alternatively foreign-made finished cotton goods may be imported.

Table 78: In conformity with international agreement to switch from the carding to the spinning stage as the point for monitoring raw wool consumption, the statistics in the table have been re-calculated. It is considered that the spinning stage represents the best point for comparison with man-made fibres and best indicates the size of the wool textile industry in a particular country. The data thus now refer to spinning-stage consumption of virgin wool by the textile industries of the countries specified and are expressed in terms of clean wool. They do not necessarily represent the final domestic consumption of wool in the country specified since a proportion of the finished woollen goods may be exported or foreign-made woollen goods may be imported for home consumption. In many cases the data are estimates based on available supplies of raw wool, tops, noils and soft waste, i.e. domestic production of wool plus or minus their net import or export balance in raw wool, tops, noils and soft waste. For the most part, changes in stocks or strategic reserves are not taken into account.

Production of raw wool in *table 53* above is shown on a "greasy" basis. Conversion factors from a "greasy" to a "clean" basis vary from country to country, for example, some approximate conversion factors are Argentina, 0.60; Australia, 0.66; New Zealand, 0.75; South Africa, 0.59; United States of America, 0.51; Uruguay, 0.68; and for world production, about 0.59.

C. *Wood and wood products; paper and paper products*

Table 79 refers to aggregate of sawnwood-coniferous, non-coniferous and sleepers. Data cover wood planed, unplaned, grooved, tongued and the like, sawn lengthwise or produced by a profile-chipping process, and planed wood which may also be finger-jointed, tongued or grooved, chamfered, rabbeted, V-jointed, beaded and so on. Wood flooring is excluded. Sleepers may be sawn or hewn.

Table 80 refers to the production of all paper and paper board. Data cover newsprint, printing and writing paper, construction paper and paperboard, household and sanitary paper, special thin paper, wrapping and packaging paper and paperboard.

Table 81 refers to the second major group of industrial wood products. Data cover veneer sheets, plywood, particle board and fibre board compressed or non-compressed.

Tableau 73 : Farine de froment. La farine tamisée (blutée) de froment ou de méteil. Farine de céréales : farine tamisée (blutée) de céréales autres que le froment ou le méteil.

Tableau 75 : Bière produite à partir du malte, y compris ale, stout et porter (bière anglaise, blonde et brune).

Le *Tableau 76* se rapporte seulement aux cigarettes.

B. *Textile, habillement et cuir*

Le *Tableau 77* porte sur des périodes de 12 mois comptés du 1er août au 31 juillet; les données ont trait au coton consommé dans les filatures et autres usines, et aux estimations de la consommation non commerciale et domestique. Elles ne représentent pas nécessairement la consommation domestique finale du coton dans le pays indiqué car une proportion des produits finis à base de coton peut être exportée ou, inversement, des produits finis à base de coton peuvent être importés de l'étranger.

Tableau 78 : Conformément à une décision prise au niveau internatinal de ne plus considérer la consommation de laine brute au stade du cardage mais au stade du filage, les statistiques de ce tableau ont été recalculées. On estime que le stade du filage représente le meilleur point de comparaison avec les fibres artificielles et fournit la meilleure indication de l'ampleur de l'industrie textile lainière dans un pays particulier. Les données se rapportent donc à la consommation de laine vierge au stade du filage par les industries textiles des pays indiqués et sont exprimées en laine propre. Elles ne représentent pas nécessairement la consommation domestique finale de laine du pays spécifié car une proportion des produits finis à base de laine peut être exportée ou, inversement, des produits à base de laine peuvent être importés de l'étranger pour la consommation nationale. Dans beaucoup de cas, les données sont des estimations fondées sur l'offre de laine brute, ruban, blousse et déchets, c'est-à-dire la production intérieure de laine plus ou moins le solde des importations ou exportations nettes de laine brute, ruban, blousse et déchets. Le plus souvent, il n'est pas tenu compte des variations des stocks ou des réserves stratégiques.

La production de laine brute du *tableau 53* ci-dessus se rapporte à la laine brute "en suint". Les coefficients de conversion de la laine "en suint" en laine "propre" varient d'un pays à un autre. Par exemple, ces coefficients de conversion sont de 0,60 pour l'Argentine; 0,66 pour l'Australie; 0,75 pour la Nouvelle-Zélande; 0,59 pour l'Afrique du Sud; 0,51 pour les Etats-Unis d'Amérique; 0,68 pour l'Uruguay; et environ 0,59 pour la production mondiale.

C. *Bois et produits dérivés; papier et produits dérivés*

Les données du *Tableau 79* sont un agrégat des sciages de bois de cônifères et de non-cônifères et des traverses de chemins de fer. Elles comprennent les bois rabotés, non rabotés, rainés, languetés, etc. sciés en long ou obtenus à l'aide d'un procédé de profilage par enlèvement de copeaux et les bois rabotés qui peuvent être également à joints digitiformes

D. *Chemicals and related products*

Table 82: Series on synthetic rubber refer to copolymers of butadiene with styrene and acrylonitrile, and neoprene and butyl-type rubber. Lattices are included (dry weight). Series on reclaimed rubber include both natural and synthetic reclaimed rubber.

Table 83 refers to the quantities of raw rubber employed in the countries specified for making manufactured goods.

Table 84 refers to the production of rubber tires for passenger cars and commercial vehicles. Unless otherwise stated, data do not cover tires for vehicles operating off the road, motorcycles, bicycles and animal-drawn road vehicles, or the production of inner tubes.

Table 85: Data refer to the estimated nitrogen content of nitrogenous fertilizers only and do not include nitrogen for industrial purposes. Years relate to 12-month periods, 1 July-30 June.

Table 86: Data refer to the production of commercial fertilizers in terms of plant nutrient content and include a small amount of products other than processed phosphate. Unless otherwise stated, ground rock phosphate is excluded. Years relate to 12-month periods, 1 July-30 June.

Table 87: Data refer to the production of commercial fertilizers in terms of plant nutrient content and include a small amount of technical potash. Years relate to 12-month periods, 1 July-30 June.

Data on consumption of fertilizers are shown in tables 58, 59 and 60 in chapter X of the present *Yearbook*.

Table 88 refers to all hydraulic cements used for construction (portland, metallurgic, aluminous, natural, and so on).

E. *Basic metal industries*

Table 89 includes foundry and steel making pig-iron, spiegeleisen and ferro manganese and other ferro-alloys. Figures on crude steel include both ingots and steel for castings. In selected cases data are obtained from the United States of America Bureau of Mines (Washington, DC), Instituto Latino-Americano del Ferro y el Acero (Santiago) and the United Nations Economic Commission for Europe. Detailed references to sources of data are given in the United Nations *Industrial Statistics Yearbook*.[21]

Table 90 refers to aluminium obtained by electrolytic reduction of alumina (primary) and remelting metal waste or scrap (secondary).

Table 91: Smelter production refers to unwrought product obtained by smelting (black or blister copper), by precipitation or concentration (cement copper or copper precipitate) of ores and concentrates. Refined production refers to refined copper obtained by fire-refining or electrolysis of primary crude metal (blister, black or cement copper), recovered from secondary materials (scrap) and electrolytic copper obtained directly from ores by leaching.

languetés ou rainés, chanfreinés, à feuillures, à joints en V, à rebords, etc. Cette rubrique ne comprend pas les éléments de parquet en bois. Les traverses de chemin de fer comprennent les traverses sciées ou équaries à la hache.

Le *Tableau 80* se rapporte à la production de tout papier et carton. Les données comprennent le papier journal, les papiers d'impression et d'écriture, les papiers et cartons de construction, les papiers de ménage et les papiers hygiéniques, les papiers minces spéciaux, les papiers d'empaquetage et d'emballage et carton.

Les données du *Tableau 81* se rapportent au deuxième grand groupe des produits de bois industriel. Les données comprennent les feuilles de placage, les contre-plaqués, les panneaux de particules et de fibres, comprimés ou non.

D. *Produits chimiques et apparentés*

Tableau 82 : Les données relatives au caoutchouc synthétique se rapportent aux copolymères du butadiène avec styrène et acrylonitrile, au néoprène et au caoutchouc de type butyl. Elles comprennent le latex (poids à sec). Les statistiques relatives au caoutchouc régénéré portent à la fois sur les caoutchoucs naturel et synthétique régénérés.

Le *Tableau 83* se rapporte aux quantités de caoutchouc brut consommées dans les pays indiqués pour la fabrication de produits manufacturés.

Le *Tableau 84* se rapporte à la production de pneus en caoutchouc pour voitures particulières et véhicules utilitaires. Sauf indication contraire, elles ne couvrent pas les pneus pour véhicules non routiers, motocyclettes, bicyclettes et véhicules routiers à traction animale, ni la production de chambres à air.

Tableau 85 : Les données se rapportent seulement à la teneur estimative en azote des engrais azotés et n'englobent pas l'azote consacré à des usages industriels. Les années portent sur des périodes de 12 mois comptées du 1er juillet au 30 juin.

Tableau 86 : Les données se rapportent à la production d'engrais commerciaux exprimée en principes fertilisants et comprennent une petite quantité de produits autres que le phosphate transformé. Sauf indication contraire, les phosphates naturels broyés sont exclus. Années : 1er juillet - 30 juin.

Tableau 87 : Les données se rapportent à la production d'engrais commerciaux exprimée en principes fertilisants et comprennent une petite quantité de potasse utilisée à des fins techniques. Années : 1er juillet - 30 juin.

Les données relatives à la consommation d'engrais sont indiquées aux tableaux 58, 59 et 60, au chapitre X du présent *Annuaire*.

Le *Tableau 88* se rapporte à tous les ciments hydrauliques utilisés dans la construction (portland, métallurgique, alumineux, naturel, etc.).

Table 92 refers to soft lead recovered directly from lead ores and concentrates (primary) or from scrap (secondary). Unless otherwise stated, remelted lead (soft and hard lead produced by simple remelting without further processing), lead bullions produced for export and lead recovered in the form of alloys are excluded.

Table 93 refers to magnesium recovered from both domestic and imported ores (primary) and metal derived from scrap and waste (secondary).

Table 94 refers to production of primary (virgin) metal recovered from the ores and concentrates, unless otherwise stated.

Table 95 refers to the consumption of tin metal by the manufacturing industries of the country specified. The figures usually represent apparent or estimated consumption (including estimates supplied by the country concerned) and thus do not necessarily represent the final domestic consumption of tin, since a proportion of the goods thus manufactured may be exported for final consumption elsewhere or tin-containing goods of foreign manufacture may be imported for home consumption.

Table 96 refers to zinc recovered from zinc ores and concentrates, distilled zinc and zinc produced directly from ores by electrolysis (primary) and zinc recovered from scrap (secondary). Where available, separate series are shown for remelted zinc (metal produced by simple remelting without further processing) and zinc recovered in the form of alloys and are not included in the total.

F. Non-metallic mineral products and fabricated metal products, machinery and equipment

Table 97 refers to total production of radio and television receivers of all kinds.

In *table 98* figures are given in gross registered tons (100 cubic feet or 2.83 cubic metres) and represent the total volume of all the enclosed spaces of the vessels launched. All ships built of wood and non-self propelled vessels are excluded but sailing vessels fitted with auxiliary power are included.

In *table 99*, passenger cars include three-and four-wheeled road motor vehicles other than motor-cycle combinations intended for the transport of passengers and seating not more than nine persons (including the driver), which are manufactured wholly or mainly from domestically-produced parts and passenger cars shipped in "knocked-down" form for assembly abroad.

E. Industries métallurgiques de base

Les données du *tableau 89* se rapportent à la production de fonte et d'acier, de spiegeleisen, de ferro manganèse et autres ferro-alliages. Les données sur l'acier brut comprennent les lingots et l'acier pour moulage. Dans certains cas, les données proviennent du United States of America Bureau of Mines (Washington, DC), de l'Instituto Latino-Americano del Ferro y el Acero (Santiago) et de la United States Economic Commission for Europe. Pour plus de détails sur les sources de données, se reporter à l'*Annuaire des statistiques industrielles* des Nations Unies [21].

Le *Tableau 90* se rapporte à la production d'aluminium obtenue par réduction électrolytique de l'alumine (production primaire) et par refusion de déchets métalliques (production secondaire).

Tableau 91 : La production de fonderie se rapporte au métal brut obtenu par fusion (cuivre noir ou cuivre ampoulé) par précipitation ou concentration (cuivre de cémentation ou précipité de cuivre) des minerais et concentrés. La production affinée se rapporte au cuivre raffiné obtenu par raffinage au feu ou électrolyse du métal brut primaire (cuivre ampoulé, cuivre noir ou cuivre de cémentation), au cuivre récupéré de matériaux secondaire (déchets) et au cuivre électrolytique obtenu directement à partir des minerais par lessivage.

Le *Tableau 92* se rapporte au plomb doux obtenu directement à partir des minerais et des concentrés de plomb (formes primaires) ou des déchets de plomb (formes secondaires). Sauf indication contraire, le plomb refondu (plomb doux et antimonieux obtenu par simple refonte sans processus ultérieur), le plomb d'oeuvre pour l'exportation et les quantités de plomb obtenu sous forme d'alliages sont exclus.

Le *Tableau 93* se rapporte au magnésium obtenu à partir des minerais nationaux et importés (formes primaires) et du métal récupéré de déchets (formes secondaires).

Tableau 94 : Sauf indication contraire, production de métal primaire (vierge) extrait des minerais et concentrés.

Tableau 95 : Consommation d'étain par les industries manufacturières du pays indiqué. Les chiffres représentent généralement la consommation apparente ou estimée (y compris les estimations fournies par le pays en question) et n'indiquent pas nécessairement la consommation nationale définitive, car une partie des biens ainsi manufacturés peut être exportée pour la consommation ou, inversement, des biens contenant de l'étain et manufacturés à l'étranger peuvent être importés pour la consommation nationale.

Tableau 96 : Zinc extrait des minerais et des concentrés de zinc, zinc distillé et zinc extrait directement des minerais par électrolyse (formes primaires) ou obtenu par récupération à partir de déchets (formes secondaires). Dans certains cas, des données distinctes sont fournies pour les zinc refondu (métal obtenu par simple refonte sans traitement ultérieur) et pour le zinc récupéré sous forme d'alliage; ces données ne sont pas incluses dans le total.

F. *Produits minéraux non métalliques et fabrications métallurgiques, machines et équipements*

Tableau 97 : Production totale de postes récepteurs de radiodiffusion et de télévision de toutes sortes.

Tableau 98 : Les chiffres sont exprimés en tonneaux de jauge brute (100 pieds cubes ou 2,83 mètres cubes) et représentent le volume total de tous les espaces clos des navires lancés. Tous les navires en bois et les navires sans auto-propulsion sont exclus, mais les voiliers à moteur auxiliaire sont inclus.

Tableau 99 : Les voitures de tourime comprennent les véhicules automobiles routiers à trois ou quatre roues, autres que les motocycles, destinés au transport de passagers, dont le nombre de places assises (y compris celle du conducteur) n'est pas supérieur à neuf et qui sont construits entièrement ou principalement avec des pièces fabriqués dans le pays, et les voitures destinées au transport de passagers exportées en pièces détachées pour être montées à l'étranger.

100
Railways: traffic
Chemins de fer : trafic
Passenger and net ton-kilometres: millions
Voyageurs et tonnes-kilomètres : millions

Country or area Pays ou zone	1981	1982	1983	1984	1985	1986	1987	1988	1989	1990
Albania Albanie										
Passenger-kilometres										
Voyageurs-kilomètres	369	449	533	581	564	619	662	703	753	779
Net ton-kilometres										
Tonnes-kilomètres	517	590	619	636	605	622	629	626	674	584
Algeria Algérie										
Passenger-kilometres										
Voyageurs-kilomètres	2 156	1 774	1 804	1 909	1 938	2 035	1 972	2 439	2 724	2 991
Net ton-kilometres										
Tonnes-kilomètres	2 687	2 750	2 652	2 631	3 048	2 934	2 937	2 814	2 698	2 690
Angola Angola										
Passenger-kilometres										
Voyageurs-kilomètres	376	361	320	300	331	326	...	...	...	...
Net ton-kilometres										
Tonnes-kilomètres	2 000	2 000	1 890	1 700	1 615	1 720	...	...	...	...
Argentina Argentine										
Passenger-kilometres										
Voyageurs-kilomètres	11 261	10 153	10 260	10 469	10 743	12 459	12 475	10 271	10 533	* 10 507
Net ton-kilometres										
Tonnes-kilomètres	9 238	11 472	13 368	11 208	9 501	8 761	7 952	8 983	8 237	...
Austria Autriche										
Passenger-kilometres										
Voyageurs-kilomètres	7 246	7 372	7 212	7 210	7 499	7 542	7 568	7 994	8 663	9 017
Net ton-kilometres										
Tonnes-kilomètres	10 479	10 245	10 361	11 398	12 066	11 436	11 263	11 331	11 962	12 796
Bangladesh Bangladesh										
Passenger-kilometres[1]										
Voyageurs-kilomètres[1]	5 197	5 366	6 427	6 284	6 031	6 005	6 027	5 052	4 338	5 070
Net ton-kilometres[1]										
Tonnes-kilomètres[1]	774	844	814	779	813	612	503	678	666	643
Belarus Bélarus										
Passenger-kilometres[2]										
Voyageurs-kilomètres[2]	...	11 874	12 878	13 154	13 731	14 199	14 965	15 989	16 525	16 852
Net ton-kilometres										
Tonnes-kilomètres	...	67 126	69 205	71 603	73 213	77 943	79 862	82 231	81 734	75 430
Belgium Belgique										
Passenger-kilometres										
Voyageurs-kilomètres	7 078	6 879	6 631	6 444	6 572	6 069	6 270	6 348	6 400	6 539
Net ton-kilometres										
Tonnes-kilomètres	7 528	6 788	6 870	7 905	8 254	7 423	7 266	7 694	8 049	8 354
Benin Bénin										
Passenger-kilometres										
Voyageurs-kilomètres	188	197	157	162	150	167	120	...	...	...
Net ton-kilometres										
Tonnes-kilomètres	176	157	145	177	179	186	191	...	...	...
Bolivia Bolivie										
Passenger-kilometres										
Voyageurs-kilomètres	482	555	771	684	748	657	500	369	353	388
Net ton-kilometres										
Tonnes-kilomètres	620	484	577	548	494	464	505	424	510	541
Botswana Botswana										
Net ton-kilometres[3]										
Tonnes-kilomètres[3]	1 115	1 279	1 432	1 337	1 297	1 328	1 401	...	...	...
Brazil Brésil										
Passenger-kilometres										
Voyageurs-kilomètres	13 133	13 266	13 797	15 578	16 362	15 782	15 273	13 891[4]	14 159[4]	13 568[4]
Net ton-kilometres[5]										
Tonnes-kilomètres[5]	79 269	77 865	74 792	92 167	99 881	103 877	109 433	119 754	124 733	120 108

100
Railways: traffic
Passenger and net ton-kilometres: millions [*cont.*]
Chemins de fer : trafic
Voyageurs et tonnes-kilomètres : millions [*suite*]

Country or area Pays ou zone	1981	1982	1983	1984	1985	1986	1987	1988	1989	1990
Bulgaria Bulgarie										
Passenger-kilometres										
Voyageurs-kilomètres	6 962	7 092	7 255	7 538	7 785	8 004	8 075	8 143	7 601	7 793
Net ton-kilometres[5]										
Tonnes-kilomètres[5]	17 982	18 208	17 978	18 055	18 078	18 199	17 735	17 462	16 894	14 018
Cameroon Cameroun										
Passenger-kilometres										
Voyageurs-kilomètres	279	366	427	492	440	432	433	466	458	442
Net ton-kilometres										
Tonnes-kilomètres	702	817	850	868	999	756	758	622	743	684
Canada Canada										
Passenger-kilometres										
Voyageurs-kilomètres	3 276	2 640	2 932	2 915	3 040	2 831	2 709	2 989	3 178	2 004
Net ton-kilometres[5]										
Tonnes-kilomètres[5]	234 374	224 165	231 871	257 561	245 284	246 722	272 122	274 571	252 075	250 117
Chile Chili										
Passenger-kilometres										
Voyageurs-kilomètres	1 582	1 506	1 575	1 424	1 522	1 274	1 176	1 013	1 058	1 102
Net ton-kilometres										
Tonnes-kilomètres	1 765	1 773	2 248	2 315	2 577	2 555	2 657	2 809	2 946	2 787
China Chine										
Passenger-kilometres[6]										
Voyageurs-kilomètres[6]	146 987	157 500	177 600	204 600	241 600	258 696	284 304	326 000	303 700	261 263
Net ton-kilometres[6]										
Tonnes-kilomètres[6]	570 134	612 000	664 600	724 800	812 600	876 504	947 196	987 740	1 039 423	1 062 238
Colombia Colombie										
Passenger-kilometres										
Voyageurs-kilomètres	230	158	175	189	228	178	171	148	152	141
Net ton-kilometres[5]										
Tonnes-kilomètres[5]	641	553	641	733	777	691	562	464	361	391
Congo Congo										
Passenger-kilometres										
Voyageurs-kilomètres	357	390	381	408	437	456	400	419	434	410
Net ton-kilometres										
Tonnes-kilomètres	539	815	437	477	518	536	449	477	467	421
Côte d'Ivoire Côte d'Ivoire										
Passenger-kilometres[7]										
Voyageurs-kilomètres[7]	996	888	972	996	1 008	1 015	* 1 021	...	...	...
Net ton-kilometres[5 7]										
Tonnes-kilomètres[5 7]	624	612	469	530	544	562	578	...	...	...
Cuba Cuba										
Passenger-kilometres										
Voyageurs-kilomètres	1 916	2 073	2 144	2 360	2 257	2 200	2 189	2 627	2 891	...
Net ton-kilometres										
Tonnes-kilomètres	2 307	2 281	2 279	2 304	2 409	2 155	2 105	2 087	2 048	...
Czechoslovakia Tchécoslovaquie										
Passenger-kilometres										
Voyageurs-kilomètres	17 909	19 043	18 884	19 323	19 839	19 935	20 029	19 408	19 669	19 335
Net ton-kilometres										
Tonnes-kilomètres	72 258	71 585	73 069	74 015	73 598	75 152	73 525	75 294	71 985	64 326
Denmark Danemark										
Passenger-kilometres[8]										
Voyageurs-kilomètres[8]	4 207	4 528	4 456	4 618	4 910	4 876	4 860	4 850	4 733	4 851
Net ton-kilometres[9]										
Tonnes-kilomètres[9]	1 551	1 612	1 621	1 670	1 768	1 800	1 644	1 671	1 723	1 787
Ecuador Equateur										
Passenger-kilometres										
Voyageurs-kilomètres	58	50	43	49	52	55	63	77	83	82

100
Railways: traffic
Passenger and net ton-kilometres: millions [*cont.*]
Chemins de fer : trafic
Voyageurs et tonnes-kilomètres : millions [*suite*]

Country or area Pays ou zone	1981	1982	1983	1984	1985	1986	1987	1988	1989	1990
Net ton-kilometres Tonnes-kilomètres	15	14	8	6	9	7	8	8	6	5
Egypt Egypte										
Passenger-kilometres[3] Voyageurs-kilomètres[3]	11 000	12 479	14 468	15 627	16 853	18 485	23 796	28 743	27 083	28 684
Net ton-kilometres[3] Tonnes-kilomètres[3]	2 334	2 307	2 303	2 597	2 756	2 908	3 021	3 029	2 853	3 045
El Salvador El Salvador										
Passenger-kilometres Voyageurs-kilomètres	14	6	4	5	5	...	...	...	...	...
Net ton-kilometres Tonnes-kilomètres	31	31	31	24	25	24	...	...	...	...
Estonia Estonie										
Passenger-kilometres Voyageurs-kilomètres	1 593	1 580	1 649	1 672	1 649	1 721	1 764	1 722	1 562	1 510
Net ton-kilometres Tonnes-kilomètres	6 192	6 274	6 410	6 515	6 446	6 736	7 134	7 989	9 609	6 977
Ethiopia Ethiopie[10 11]										
Passenger-kilometres[10 11] Voyageurs-kilomètres[10 11]	310	307	360	392	383	395	398	342	298	277
Net ton-kilometres[5 10 11] Tonnes-kilomètres[5 10 11]	131	108	122	117	128	131	136	141	129	126
Finland Finlande										
Passenger-kilometres Voyageurs-kilomètres[12]	3 274	3 326	3 339	3 276	3 224	2 676	3 106	3 147	3 208	3 331
Net ton-kilometres[12] Tonnes-kilomètres[12]	8 391	8 004	8 094	7 979	8 066	6 951	7 402	7 815	7 958	8 357
France France										
Passenger-kilometres Voyageurs-kilomètres	55 667	56 854	58 430	60 390	62 070	59 860	59 970	63 290	64 490	63 960
Net ton-kilometres[13] Tonnes-kilomètres[13]	63 730	60 554	59 376	57 470	55 780	51 690	51 330	52 290	53 270	51 530
Georgia Géorgie										
Passenger-kilometres Voyageurs-kilomètres	3 646	3 694	3 731	3 678	3 724	3 684	3 614	3 442	2 858	2 497
Net ton-kilometres Tonnes-kilomètres	13 898	12 849	13 492	13 316	13 487	13 132	12 500	13 020	12 671	12 355
Germany† · Allemagne†										
F. R. Germany R. f. Allemagne										
Passenger-kilometres Voyageurs-kilomètres	40 268	40 840	39 097	39 575	43 451	42 129	39 965	41 760	42 023	44 588
Net ton-kilometres Tonnes-kilomètres	61 854	57 261	55 840	59 835	63 873	57 916	58 947	59 922	61 981	61 729
former German D. R. anc. R. d. allemande										
Passenger-kilometres Voyageurs-kilomètres	21 644	22 705	22 603	22 919	22 451	22 402	22 563	22 775	23 588	17 397
Net ton-kilometres Tonnes-kilomètres	...	...	53 675	55 422	57 582	57 916	58 096	59 374	58 027	39 112
Ghana Ghana										
Passenger-kilometres Voyageurs-kilomètres	314	290	380	157	244	294	318	...	...	...
Net ton-kilometres Tonnes-kilomètres	92	74	61	44	80	101	114	...	...	...
Greece Grèce										
Passenger-kilometres Voyageurs-kilomètres	1 515	1 501	1 546	1 652	1 732	1 950	1 973	1 963	2 011	...
Net ton-kilometres[14] Tonnes-kilomètres[14]	693	586	670	770	733	702	599	604	657	...

100
Railways: traffic
Passenger and net ton-kilometres: millions [cont.]
   Chemins de fer : trafic
   Voyageurs et tonnes-kilomètres : millions [suite]

Country or area Pays ou zone	1981	1982	1983	1984	1985	1986	1987	1988	1989	1990
Guatemala Guatemala										
Passenger-kilometres Voyageurs-kilomètres	313	427	981	1 040	380	318	293	329	...	...
Net ton-kilometres Tonnes-kilomètres	596	588	495	397	590	582	426	456	...	...
Hong Kong Hong-kong										
Passenger-kilometres Voyageurs-kilomètres	399	407	867	1 480	1 781	1 933	2 136	2 360	2 469	2 533
Net ton-kilometres Tonnes-kilomètres	36	36	45	47	52	69	72	70	69	70
Hungary Hongrie										
Passenger-kilometres Voyageurs-kilomètres	12 373	11 871	10 394	10 511	10 463	10 452	10 486	10 758	10 414	11 403 [15 16]
Net ton-kilometres Tonnes-kilomètres	23 773	22 725	22 554	22 308	21 814	22 095	21 253	20 573	19 364	16 781 [16]
India Inde										
Passenger-kilometres [17] Voyageurs-kilomètres [17]	220 787	226 930	222 935	226 582	240 614	256 535	269 389	263 731	280 848	295 644
Net ton-kilometres [17] Tonnes-kilomètres [17]	164 253	167 781	168 849	172 632	196 600	214 096	222 528	222 374	229 602	235 785
Indonesia Indonésie										
Passenger-kilometres Voyageurs-kilomètres	6 166	6 105	6 096	6 379	6 774	7 327	7 516	7 863	8 427	...
Net ton-kilometres Tonnes-kilomètres	970	885	936	1 173	1 333	1 465	1 759	2 359	2 921	...
Iran, Islamic Rep. of Iran, Rép. islamique d'										
Passenger-kilometres Voyageurs-kilomètres	2 728	4 735	5 784	6 130	5 585	4 638	3 674	4 661	4 752	4 573
Net ton-kilometres Tonnes-kilomètres	3 811	5 566	6 762	7 566	6 888	7 316	8 625	8 047	7 963	11 354
Iraq Iraq										
Passenger-kilometres Voyageurs-kilomètres	1 215	1 481	1 375	1 227	1 118	1 005	1 150	1 570	...	...
Net ton-kilometres [18] Tonnes-kilomètres [18]	2 081	1 136	1 065	1 245	1 245	1 294	1 534	2 023	...	...
Ireland Irlande										
Passenger-kilometres Voyageurs-kilomètres	995	887	846	903	1 023	1 075	1 196	1 180	1 220	1 226
Net ton-kilometres Tonnes-kilomètres	678	654	582	601	601	574	563	545	556	589
Israel Israël										
Passenger-kilometres Voyageurs-kilomètres	246	221	212	215	209	179	173	166	151	...
Net ton-kilometres Tonnes-kilomètres	803	827	883	970	942	952	1 062	1 032	1 016	...
Italy Italie										
Passenger-kilometres Voyageurs-kilomètres	40 090	39 542	38 840	39 045	39 194	40 500	41 395	43 343	44 452	45 513
Net ton-kilometres [18] Tonnes-kilomètres [18]	17 115	16 904	16 746	17 870	18 024	17 410	18 427	19 567	20 587	21 217
Jamaica Jamaïque										
Passenger-kilometres Voyageurs-kilomètres	...	20	20	28	25	42	29	22	24	...
Net ton-kilometres Tonnes-kilomètres	...	48	59	...	62	117	123	72	18	...
Japan Japon										
Passenger-kilometres Voyageurs-kilomètres	306 148	314 939	320 177	324 991	328 450	333 425	341 439	356 468	369 642	383 735

100
Railways: traffic
Passenger and net ton-kilometres: millions [cont.]
Chemins de fer : trafic
Voyageurs et tonnes-kilomètres : millions [suite]

Country or area Pays ou zone	1981	1982	1983	1984	1985	1986	1987	1988	1989	1990
Net ton-kilometres Tonnes-kilomètres	34 211	31 323	29 133	23 191	22 099	20 917	20 307	22 911	24 767	26 656
Jordan Jordanie										
Passenger-kilometres Voyageurs-kilomètres	5	4	3	1	1	1	1	1	1	2
Net ton-kilometres Tonnes-kilomètres	444	600	718	870	692	771	720	625	610	711
Kazakhstan Kazakhstan										
Passenger-kilometres Voyageurs-kilomètres	14 893	15 460	15 245	15 019	15 749	16 922	17 888	18 637	18 921	19 743
Net ton-kilometres Tonnes-kilomètres	364 950	352 957	367 591	373 074	382 507	397 907	404 583	416 875	409 573	406 963
Kenya Kenya										
Passenger-kilometres Voyageurs-kilomètres	2 245	2 307	2 388	2 409	2 500	2 534	2 608	...	...	...
Net ton-kilometres Tonnes-kilomètres	2 241	2 097	2 091	2 034	1 858	1 973	2 080	...	...	...
Korea, Republic of Corée, République de										
Passenger-kilometres Voyageurs-kilomètres	21 529	21 034	21 688	21 884	22 595	23 563	24 457	25 978	27 390	29 894
Net ton-kilometres Tonnes-kilomètres	10 815	10 892	11 629	12 033	12 296	12 813	13 061	13 784	13 605	13 663
Latvia Lettonie										
Passenger-kilometres[19] Voyageurs-kilomètres[19]	4 880	4 978	5 149	5 310	5 214	5 573	5 679	5 761	5 449	5 366
Net ton-kilometres[9] Tonnes-kilomètres[9]	17 694	19 112	20 300	20 799	19 933	20 691	21 380	21 689	21 132	18 538
Lithuania Lituanie										
Passenger-kilometres Voyageurs-kilomètres	3 314	3 324	3 331	3 377	3 417	3 494	3 565	3 665	3 470	3 640
Net ton-kilometres Tonnes-kilomètres	17 902	18 669	18 952	19 999	20 927	21 076	21 205	22 595	21 749	19 258
Luxembourg Luxembourg										
Passenger-kilometres Voyageurs-kilomètres	311	311	297	286	283	278	269	277	280	261
Net ton-kilometres Tonnes-kilomètres	587	552	505	584	645	604	593	639	704	709
Madagascar Madagascar										
Passenger-kilometres Voyageurs-kilomètres	245	184	224	202	178	208	209	242	204	198
Net ton-kilometres[5] Tonnes-kilomètres[5]	177	145	186	226	208	188	174	174	207	209
Malawi Malawi										
Passenger-kilometres Voyageurs-kilomètres	78	97	104[17]	109[17]	123[17]	112[17]	113[17]	113[17]	112[17]	...
Net ton-kilometres Tonnes-kilomètres	225	182	115[17]	121[17]	94[17]	125[17]	107[17]	77[17]	65[17]	...
Malaysia Malaisie										
Passenger-kilometres[20] Voyageurs-kilomètres[20]	1 640	1 615	1 499	1 512	1 409	1 369	1 425	1 518	1 701	1 839
Net ton-kilometres[20] Tonnes-kilomètres[20]	1 123	1 091	1 072	1 077	1 018	1 042	1 119	1 326	1 361	1 407
Mali Mali										
Passenger-kilometres Voyageurs-kilomètres	...	147	157	149	173	177	196	177	184	184
Net ton-kilometres Tonnes-kilomètres	...	...	148	239	241	225	199	227	279	273

100
Railways: traffic
Passenger and net ton-kilometres: millions [cont.]
Chemins de fer : trafic
Voyageurs et tonnes-kilomètres : millions [suite]

Country or area Pays ou zone	1981	1982	1983	1984	1985	1986	1987	1988	1989	1990
Mauritania Mauritanie										
Net ton-kilometres Tonnes-kilomètres	12 831	11 319	11 016	14 217	13 929	13 396	13 434	14 931	16 623	...
Mexico Mexique										
Passenger-kilometres Voyageurs-kilomètres	5 303	5 261	5 630	5 951	6 015	5 874	5 828	5 619	5 383	5 336
Net ton-kilometres Tonnes-kilomètres	43 513	38 800	42 377	44 592	45 306	40 608	40 475	41 177	38 570	36 408
Moldova, Rep. of Moldova, Rép. de										
Passenger-kilometres Voyageurs-kilomètres	1 302	1 293	1 513[2]	1 587[2]	1 648[2]	1 685[2]	1 766[2]	1 616	1 514	1 464
Net ton-kilometres Tonnes-kilomètres	15 043	15 008	15 778	16 838	16 614	16 890	15 820	15 989	15 632	15 007
Mongolia Mongolie										
Passenger-kilometres Voyageurs-kilomètres	327	350	386	420	436	467	486	531	579	...
Net ton-kilometres Tonnes-kilomètres	3 609	3 914	4 498	5 121	5 960	6 333	6 180	6 241	5 956	...
Morocco Maroc										
Passenger-kilometres[21] Voyageurs-kilomètres[21]	1 140	1 375	1 407	1 620	1 933	1 958	2 069	2 092	2 168	2 237
Net ton-kilometres[21] Tonnes-kilomètres[21]	3 970	3 851	4 180	4 517	4 562	4 953	4 880	5 706	4 519	5 107
Mozambique Mozambique										
Passenger-kilometres Voyageurs-kilomètres	628	458	507	284	161	183	105	75	74	...
Net ton-kilometres Tonnes-kilomètres	1 583	1 261	767	538	290	301	353	306	403	...
Myanmar Myanmar										
Passenger-kilometres Voyageurs-kilomètres	3 503	3 720	4 005	3 859	3 834	3 554	4 486	3 830	3 920	4 370
Net ton-kilometres[5] Tonnes-kilomètres[5]	692	650	655	640	600	507	545	356	458	525
Netherlands Pays-Bas										
Passenger-kilometres Voyageurs-kilomètres	9 230	9 376	9 052	8 790	9 007	8 919	9 396	9 664	10 162	* 11 064
Net ton-kilometres Tonnes-kilomètres	3 254	2 887	2 835	3 103	3 274	3 050	3 010	3 194	3 094	* 2 556
New Zealand Nouvelle-Zélande										
Passenger-kilometres[22] Voyageurs-kilomètres[22]	396	396	...	...	...	...	...	...	...	...
Net ton-kilometres Tonnes-kilomètres	3 139[22]	3 252[22]	3 164[22]	3 165[22]	3 192[22]	3 051[22]	2 912[22]	2 924[22]	2 682[22]	2 744[3]
Nicaragua Nicaragua										
Passenger-kilometres Voyageurs-kilomètres	20	25	45	60	* 66	...	...	...	...	...
Net ton-kilometres Tonnes-kilomètres	14	7	2	5	* 4	...	...	...	...	...
Nigeria Nigéria										
Net ton-kilometres Tonnes-kilomètres	1 698	1 630	1 600	1 644	1 709	1 725	1 743	...	...	...
Norway Norvège										
Passenger-kilometres Voyageurs-kilomètres	2 425	2 242	2 175	2 198	2 241	2 225	2 187	2 110	2 136	2 136
Net ton-kilometres Tonnes-kilomètres	2 825	2 545	2 459	2 650	2 932	3 015	2 822	2 617	2 780	2 354
Pakistan Pakistan										
Passenger-kilometres[1] Voyageurs-kilomètres[1]	16 387	18 031	18 287	17 806	16 859	16 919	18 544	19 732	20 373	19 963

100
Railways: traffic
Passenger and net ton-kilometres: millions [*cont.*]
Chemins de fer : trafic
Voyageurs et tonnes-kilomètres : millions [*suite*]

Country or area Pays ou zone	1981	1982	1983	1984	1985	1986	1987	1988	1989	1990
Net ton-kilometres[1] Tonnes-kilomètres[1]	7 918	7 323	7 385	7 203	8 270	7 819	8 033	8 364	7 226	5 704
Paraguay Paraguay										
Passenger-kilometres Voyageurs-kilomètres	1	2	2	6	2	2	2	2	2	2
Net ton-kilometres Tonnes-kilomètres	23	34	30	17	13	17	17	18	14	4
Peru Pérou										
Passenger-kilometres Voyageurs-kilomètres	494	458	454	483	475	490	594	596	668	...
Net ton-kilometres Tonnes-kilomètres	1 039	974	897	1 044	1 036	1 022	1 134	967	929	...
Philippines Philippines										
Passenger-kilometres Voyageurs-kilomètres	359	284	246	230	146	173	219	230	230	...
Net ton-kilometres Tonnes-kilomètres	32	21	17	17	13	15	16	15	13	...
Poland Pologne										
Passenger-kilometres Voyageurs-kilomètres	48 238	49 266	50 153	53 179	51 978	48 526	48 285	52 134	55 888	50 373
Net ton-kilometres Tonnes-kilomètres	109 835	112 689	118 034	123 503	120 642	121 775	121 381	122 204	111 140	83 530
Portugal Portugal										
Passenger-kilometres Voyageurs-kilomètres	5 856	5 414	5 195	5 456	5 725	5 803	5 907	6 036	5 908	5 664
Net ton-kilometres Tonnes-kilomètres	1 003	1 060	1 044	1 239	1 306	1 448	1 615	1 708	1 719	1 588
Romania Roumanie										
Passenger-kilometres[23] Voyageurs-kilomètres[23]	24 379	25 578	27 676	28 785	31 082	32 304	33 520	34 643	35 456	30 582
Net ton-kilometres[5] Tonnes-kilomètres[5]	75 251	71 110	72 316	75 159	74 215	79 092	78 070	80 607	81 131	57 253
Saudi Arabia Arabie saoudite										
Passenger-kilometres Voyageurs-kilomètres	96	105	87	79	72	71	81	92	...	...
Net ton-kilometres Tonnes-kilomètres	272	393	199	525	415	321	491	470	...	...
Senegal Sénégal										
Passenger-kilometres Voyageurs-kilomètres	152	130	136	140	139	143	155	...	...	...
Net ton-kilometres Tonnes-kilomètres	255	278	384	400	468	492	524	...	...	...
South Africa Afrique du Sud										
Passenger-kilometres[22][24] Voyageurs-kilomètres[22][24]	...	45 461	39 873	27 416	24 009	23 032	21 413	...	...	...
Net ton-kilometres[22][24] Tonnes-kilomètres[22][24]	101 295	101 295	82 026	83 329	90 162	93 331	92 178	...	...	...
Spain Espagne										
Passenger-kilometres Voyageurs-kilomètres	14 261	14 703	15 092	15 574	15 979	15 693	15 394	15 716	14 715	15 476
Net ton-kilometres[5] Tonnes-kilomètres[5]	10 603	10 504	10 599	11 645	11 654	11 292	11 475	11 716	11 619	11 206
Sri Lanka Sri Lanka										
Passenger-kilometres[25] Voyageurs-kilomètres[25]	2 985	3 121	2 447	2 280	2 111	1 972	1 881	1 859	1 734	2 803
Net ton-kilometres[25] Tonnes-kilomètres[25]	219	217	224	263	247	203	195	197	162	164

100
Railways: traffic
Passenger and net ton-kilometres: millions [*cont.*]
Chemins de fer : trafic
Voyageurs et tonnes-kilomètres : millions [*suite*]

Country or area Pays ou zone	1981	1982	1983	1984	1985	1986	1987	1988	1989	1990
Sudan Soudan										
Passenger-kilometres Voyageurs-kilomètres	1 175	900	629	836	849	...	...	...	...	...
Net ton-kilometres Tonnes-kilomètres	1 530	1 700	1 750	1 770	1 860	...	...	...	...	...
Swaziland Swaziland										
Passenger-kilometres Voyageurs-kilomètres	980	990	1 009	1 053	1 157	...	...	...	...	...
Net ton-kilometres Tonnes-kilomètres	2 573	2 520	2 633	2 598	2 624	1 082	3 987	6 113	8 002	7 287
Sweden Suède										
Passenger-kilometres Voyageurs-kilomètres	7 062	6 381	6 460	6 483	6 586	6 152	6 013	6 081	6 060	5 981
Net ton-kilometres Tonnes-kilomètres	15 290	13 745	14 951	16 944	17 587	17 755	17 630	18 094	18 539	18 941
Switzerland Suisse										
Passenger-kilometres Voyageurs-kilomètres	10 449	10 310	10 340	10 394	10 801	10 737	12 203	12 391	12 673	12 710
Net ton-kilometres Tonnes-kilomètres	7 473	6 825	6 708	7 224	7 379	7 279	7 124	7 875	8 560	8 794
Syrian Arab Republic Rép. arabe syrienne										
Passenger-kilometres Voyageurs-kilomètres	443	413	486	757	944	904	1 029	1 133	1 113	1 140
Net ton-kilometres Tonnes-kilomètres	704	710	742	966	1 251	1 418	1 508	1 569	1 350	1 265
Thailand Thaïlande										
Passenger-kilometres[25] Voyageurs-kilomètres[25]	9 483	9 231	9 699	9 643	9 140	9 274	9 583	10 301	10 935	11 612
Net ton-kilometres[25] Tonnes-kilomètres[25]	2 601	2 421	2 413	2 618	2 718	2 583	2 729	2 867	3 065	3 291
Togo Togo										
Passenger-kilometres Voyageurs-kilomètres	86	95	99	100	102	109	117	...	...	...
Net ton-kilometres[9] Tonnes-kilomètres[9]	13	17	12	10	10	11	12	...	...	...
Tunisia Tunisie										
Passenger-kilometres[14] Voyageurs-kilomètres[14]	1 011	944	802	742	744	750	798	1 014	1 039	1 019
Net ton-kilometres[5 26] Tonnes-kilomètres[5 26]	1 720	1 588	1 810	1 695	1 710	1 877	1 986	2 156	2 064	1 834
Turkey Turquie										
Passenger-kilometres Voyageurs-kilomètres	6 105	5 440	5 722	6 277	6 489	6 052	6 174	6 708	6 845	6 410
Net ton-kilometres Tonnes-kilomètres	5 936	6 212	6 301	7 679	7 959	7 396	7 403	8 149	7 707	8 031
Uganda Ouganda										
Passenger-kilometres Voyageurs-kilomètres	227	342	338	246	234	195	212	118	69	69
Net ton-kilometres Tonnes-kilomètres	51	69	73	79	60	71	77	83	90	90
Ukraine Ukraine										
Passenger-kilometres Voyageurs-kilomètres	60 214	61 295	65 013	64 799	66 954	68 580	71 425	72 859	73 218	76 038
Net ton-kilometres Tonnes-kilomètres	476 297	475 006	488 796	499 320	497 916	506 123	496 001	504 689	497 333	473 953

100
Railways: traffic
Passenger and net ton-kilometres: millions [cont.]
Chemins de fer : trafic
Voyageurs et tonnes-kilomètres : millions [suite]

Country or area Pays ou zone	1981	1982	1983	1984	1985	1986	1987	1988	1989	1990
former USSR ancienne URSS										
Passenger-kilometres										
Voyageurs-kilomètres	...	...	...	...	91	100	108	113	113	114
Net ton-kilometres										
Tonnes-kilomètres	900	948	1 036	1 126	1 288	1 281	1 251	1 316	1 336	1 276
United Kingdom Royaume-Uni[27]										
Passenger-kilometres[27]										
Voyageurs-kilomètres[27]	29 724	27 231	29 536	29 752	29 684	30 984	32 318	34 412	33 232	34 063
Net ton-kilometres[27]										
Tonnes-kilomètres[27]	17 505	15 880	17 144	12 720	15 370	16 473	17 289	18 203	17 295	15 830
United Rep.Tanzania Rép. Unie de Tanzanie										
Passenger-kilometres										
Voyageurs-kilomètres	604	976	1 240	1 192	1 194	1 024	647	855	832	809
Net ton-kilometres										
Tonnes-kilomètres	876	765	501	565	660	814	789	936	990	956
United States Etats-Unis										
Passenger-kilometres[28]										
Voyageurs-kilomètres[28]	18 371	16 966	17 606	16 564	17 649	8 069	8 639	* 9 156	* 9 396	* 9 864
Net ton-kilometres[29]										
Tonnes-kilomètres[29]	1 359 582	1 191 844	1 237 428	1 345 488	1 280 394	1 283 736	1 388 388	* 1 467 672	* 1 482 024	* 1 513 776
Uruguay Uruguay										
Passenger-kilometres[30]										
Voyageurs-kilomètres[30]	339	274	312	331	241	196	140	0	0	...
Net ton-kilometres										
Tonnes-kilomètres	218	186	220	273	185	210	210	213	243	204
Venezuela Venezuela										
Passenger-kilometres										
Voyageurs-kilomètres	10	19	22	12	8	17	22	29	38	64
Net ton-kilometres										
Tonnes-kilomètres	13	29	22	11	14	12	18	40	39	35
Yugoslavia Yougoslavie										
Passenger-kilometres										
Voyageurs-kilomètres	10 510	11 265	11 643	11 734	11 999	12 399	11 827	11 449	11 653	11 325
Net ton-kilometres[5]										
Tonnes-kilomètres[5]	25 720	26 166	27 860	28 731	28 719	27 573	26 070	25 413	25 921	23 149
Zaire Zaïre										
Passenger-kilometres										
Voyageurs-kilomètres	432	480	420	450	504	511	522	...	...	...
Net ton-kilometres										
Tonnes-kilomètres	1 520	1 652	1 423	1 509	1 599	1 624	1 681	...	...	...
Zambia Zambie										
Passenger-kilometres										
Voyageurs-kilomètres	68	371	434	475	488	496	510	512	...	...
Net ton-kilometres										
Tonnes-kilomètres	37	1 409	1 403	1 417	1 401	1 407	1 420	1 431	...	...
Zimbabwe Zimbabwe										
Net ton-kilometres[3][31]										
Tonnes-kilomètres[3][31]	6 612	6 264	6 288	6 408	6 204	6 576	6 600	5 551	5 287	5 290

Source:
Transport statistics database of the Statistical Division of
the United Nations Secretariat.

Source:
Base de données pour les statistiques des transports de la
Division de statistique du Secrétariat de l'ONU.

100
Railways: traffic
Passenger and net ton-kilometres: millions [*cont.*]

Chemins de fer : trafic
Voyageurs et tonnes-kilomètres : millions [*suite*]

† All data shown which pertain to Germany prior to 3 October 1990 are indicated separately for the Federal Republic of Germany and the former German Democratic Republic based on their respective territories at the time indicated. Where data for united Germany (3 October 1990 and thereafter) are not available, available data are shown separately under the designations Federal Republic of Germany and former German Democratic Republic and pertain to the territorial boundaries prior to 3 October 1990. For detailed explanatory notes on data pertaining to Germany, see Annex I - Country Nomenclature.

1 Twelve months beginning 1 July of year stated.

2 Including passengers carried without revenues.
3 Twelve months ending 30 June of year stated.
4 Including urban railways traffic.

5 Including service traffic.
6 May include service traffic.
7 Abidjan-Ouagadougou line, which lies in Burkina Faso.

8 Including ferry traffic.
9 Including passengers' baggage and parcel post (Denmark and Latvia: also mail).
10 Including traffic of Djibouti portion of Djibouti-Addis Ababa line.
11 Twelve months beginning 8 July of year stated.
12 Beginning 1984, excluding local transport.
13 Including passengers' baggage.
14 Including military traffic (Greece: also government traffic).
15 Including military, government and railway personnel.

16 Excluding suburban railways.
17 Twelve months beginning 1 April of year stated.
18 Excluding livestock.
19 Including railway personnel.
20 Peninsular Malaysia only.
21 Principal railways.
22 Twelve months ending 31 March of year stated.
23 Including military and government personnel.
24 Including Namibia.
25 Twelve months ending 30 September of year stated.
26 Ordinary goods only.
27 Excluding Northern Ireland.
28 Excluding commuter railroads beginning 1986.

29 Class I railways only.
30 Beginning 1988, passenger transport suspended.
31 Including traffic in Botswana.

† Toutes les données se rapportant à l'Allemagne avant le 3 octobre 1990 figurent dans deux rubriques séparées basées sur les territoires respectifs de la République fédérale d'Allemagne et l'ancienne République démocratique allemande selon la période indiquée. En l'absence de données pour l'Allemagne unifiée (à compter du 3 octobre 1990), les données disponibles sont fournies séparément sous les rubriques République fédérale d'Allemagne et ancienne République démocratique allemande et se rapportent aux limites territoriales antérieures au 3 octobre 1990. Pour les notes explicatives en détail sur les données concernant l'Allemagne, voir Annexe I – Nomenclature des pays.

1 Douze mois commençant le premier juillet de l'année indiquée.
2 Y compris passagers transportés gratuitement.
3 Douze mois finissant le 30 juin de l'année indiquée.
4 Y compris les lignes situées a l'intérieur d'une agglomération urbaine.
5 Y compris le trafic de service.
6 Le trafic de service peut être compris.
7 Ligne Abidjan-Ouagadougou dont un tronçon passe en Burkina Faso.
8 Y compris le trafic par ferry.
9 Y compris les bagages des voyageurs et les colis postaux (Danemark et Lettonie : courrier aussi).
10 Y compris le trafic de la ligne Djibouti-Addis Abéba en Djibouti.
11 Douze mois commençant le 8 juillet de l'année indiquée.
12 A compter de 1984 non compris le transport local.
13 Y compris les bagages des voyageurs.
14 Y compris le trafic militaire (Grèce: et de l'Etat aussi).
15 Y compris les militaires, les fonctionnaires et le personnel de chemin de fer.
16 Non compris les lignes de banlieues.
17 Douze mois commençant le premier avril de l'année indiquée.
18 Non compris le bétail.
19 Y compris le personnel de chemins de fer.
20 Malasie péninsulaire seulement.
21 Chemins de fer principaux.
22 Douze mois finissant le 31 mars de l'année indiquée.
23 Y compris les militaires et les fonctionnaires.
24 Y compris Namibie.
25 Douze mois finissant le 30 septembre de l'année indiquée.
26 Petite vitesse seulement.
27 Non compris l'Irlande du Nord.
28 A compter de 1986 non compris les chemins de fer de banlieue.
29 Réseaux de catégorie 1 seulement.
30 Transport passager interrompu à partir de 1988.
31 Y compris le trafic en Botswana.

101
Motor vehicles in use
Véhicules automobiles en circulation
Passenger cars and commercial vehicles: thousand units
Voitures de tourisme et véhicules utilitaires : milliers de véhicules

Country or area Pays or zone	1981	1982	1983	1984	1985	1986	1987	1988	1989	1990
World Monde										
Passenger cars[1,2]										
Voitures de tourisme[1,2]	335 101.7	342 574.0	351 353.7	364 042.2	373 667.2	393 351.7	395 129.4	407 959.3	422 240.3	438 525.0
Commercial vehicles[1,2]										
Véhicules utilitaires[1,2]	96 976.9	100 360.1	104 991.1	109 891.4	115 164.7	120 859.7	124 036.9	129 592.7	133 831.8	139 856.6
Afghanistan Afghanistan										
Passenger cars[1]										
Voitures de tourisme[1]	34.1	33.3	33.3	32.8	32.0	31.1	31.0	31.0	31.0	31.0
Commercial vehicles[1]										
Véhicules utilitaires[1]	27.9	27.0	27.0	26.2	25.5	24.8	24.8	25.0	25.0	25.0
Albania Albanie										
Commercial vehicles										
Véhicules utilitaires	...	...	...	2.8	2.5	2.5	2.5	2.8	2.8	* 2.8
Algeria Algérie										
Passenger cars										
Voitures de tourisme	574.0	574.0	580.0	582.0	611.0	639.0	667.0	725.0[1]	725.0[1]	725.0[1]
Commercial vehicles										
Véhicules utilitaires	266.0	266.0	270.0	275.0	300.0	317.0	324.0	480.0	480.0[1]	480.0[1]
American Samoa Samoa américaines										
Passenger cars										
Voitures de tourisme	3.1	2.8	2.9	3.4	4.0	...	...	...	...	...
Commercial vehicles										
Véhicules utilitaires	0.4	0.3	0.4	0.4	0.5	...	...	...	...	...
Angola Angola										
Passenger cars[1]										
Voitures de tourisme[1]	140.2	136.5	132.6	129.0	125.9	122.4	122.0	122.0	122.0	122.0
Commercial vehicles[1]										
Véhicules utilitaires[1]	43.7	44.5	43.6	42.8	42.3	41.5	41.0	41.0	41.0	41.0
Antigua and Barbuda Antigua-et-Barbuda										
Passenger cars										
Voitures de tourisme	...	...	8.4	9.6	11.0	12.4	14.2	...	...	...
Commercial vehicles										
Véhicules utilitaires	...	...	1.5	1.7	2.0	2.4	2.7	...	...	...
Argentina Argentine										
Passenger cars										
Voitures de tourisme	3 319.1	3 516.7	3 620.2	3 759.3	3 878.2	4 037.9	4 137.2	4 079.8	4 235.1	4 283.7
Commercial vehicles										
Véhicules utilitaires	1 292.6	1 357.2	1 376.6	1 409.4	1 432.0	1 459.2	1 483.5	1 470.6	1 483.8	1 500.8
Australia Australie										
Passenger cars[3]										
Voitures de tourisme[3]	5 800.6	6 021.1	6 308.1	6 469.6	6 636.2	6 842.5	6 985.4	7 072.8	7 243.6	7 442.2
Commercial vehicles[3]										
Véhicules utilitaires[3]	1 425.1	1 505.2	1 617.7	1 673.2	1 751.0	1 838.2	1 881.1	1 897.1	1 925.1	1 995.1
Austria Autriche										
Passenger cars[4]										
Voitures de tourisme[4]	2 312.9	2 361.1	2 414.5	2 468.5	2 530.8	2 609.4	2 684.8	2 784.8	2 902.9	2 991.3
Commercial vehicles[4,5]										
Véhicules utilitaires[4,5]	538.9	547.5	558.2	570.9	580.2	591.6	605.8	624.9	643.1	648.3
Bahamas Bahamas										
Passenger cars										
Voitures de tourisme	50.0	58.0	52.6	48.1	54.1	54.5	59.3	69.0[1]	69.0[1]	69.0[1]
Commercial vehicles										
Véhicules utilitaires	8.9	10.6	8.6	7.1	8.5	9.5	12.6	14.0[1]	14.0[1]	14.0[1]
Bahrain Bahreïn										
Passenger cars										
Voitures de tourisme	51.1	60.1	66.6	73.2	79.4	82.8	84.9	90.3	94.9	98.6
Commercial vehicles										
Véhicules utilitaires	17.6[6]	20.0[6]	18.6	20.4	21.8	22.3	23.0	21.3	22.7	24.2

101
Motor vehicles in use
Passenger cars and commercial vehicles: thousand units [cont.]
Véhicules automobiles en circulation
Voitures de tourisme et véhicules utilitaires : milliers de véhicules [suite]

Country or area Pays or zone	1981	1982	1983	1984	1985	1986	1987	1988	1989	1990
Bangladesh Bangladesh										
Passenger cars Voitures de tourisme	12.4	12.9	26.7	28.1	29.4	30.9	32.4	34.2	39.1	41.4
Commercial vehicles Véhicules utilitaires	20.7	22.2	39.2	40.7	42.4	44.2	45.9	47.7	53.7	55.7
Barbados Barbade										
Passenger cars[4] Voitures de tourisme[4]	...	...	...	31.0	32.8	34.8	37.0	38.7	43.1	...
Commercial vehicles[4,7] Véhicules utilitaires[4,7]	...	...	...	4.9	4.8	4.7	5.1	4.3	4.8	...
Belgium Belgique										
Passenger cars Voitures de tourisme	3 162.6	3 188.7	3 221.4	3 258.4	3 300.5	3 336.2	3 453.1	3 567.4	3 688.1	3 874.6
Commercial vehicles Véhicules utilitaires	303.6	303.8	307.3	310.7	318.9	326.9	344.1	358.7	377.7	396.0
Belize Belize										
Passenger cars[1] Voitures de tourisme[1]	7.9	8.0	7.8	3.1	3.1	3.5	3.0	3.0	1.6	1.7
Commercial vehicles[1] Véhicules utilitaires[1]	3.7	4.0	3.9	3.4	3.3	3.6	4.0	4.0	2.7	2.8
Benin Bénin										
Passenger cars Voitures de tourisme	21.1	23.0	24.0	25.0	25.0	25.0	26.0	22.0[1]	22.0[1]	22.0[1]
Commercial vehicles Véhicules utilitaires	10.0	10.0	11.0	11.0	12.0	12.0	13.0	12.0[1]	12.0[1]	12.0[1]
Bermuda Bermudes										
Passenger cars Voitures de tourisme	15.0	15.7	16.4	17.3	17.8	17.7	18.2	18.9	19.5	19.7
Commercial vehicles Véhicules utilitaires	2.6	2.7	2.9	3.1	3.2	3.2	3.3	3.4	3.6	3.8
Bolivia Bolivie										
Passenger cars[1] Voitures de tourisme[1]	34.4	33.6	32.8	32.3	33.2	33.0	74.7	75.0	250.0	261.1
Commercial vehicles[1] Véhicules utilitaires[1]	49.1	48.0	46.6	46.0	46.9	46.6	135.8	136.0	54.0	55.9
Botswana Botswana										
Passenger cars Voitures de tourisme	8.9	10.1	11.0	12.8	14.3	16.6	17.1	17.8	19.0	21.0[1]
Commercial vehicles Véhicules utilitaires	16.1	17.7	19.8	22.0	23.1	24.9	26.5	30.2	33.4	26.0[1]
Brazil Brésil										
Passenger cars Voitures de tourisme	8 251.1	8 543.2	9 007.9	9 198.4	9 527.3	* 9 885.2	* 10 035.2	* 10 274.4	10 475.3	10 597.5
Commercial vehicles Véhicules utilitaires	2 062.8	2 149.2	2 237.9	2 285.2	2 410.0	* 2 350.0	* 2 373.6	* 2 417.8	2 451.6	2 472.5
British Virgin Islands Iles Vierges britanniques										
Passenger cars[8] Voitures de tourisme[8]	2.7	2.8	2.9	3.3	3.6	4.0	4.3	4.8	5.5	6.2
Brunei Darussalam Brunéi Darussalam										
Passenger cars Voitures de tourisme	48.7	53.9	62.5	70.6	78.6	83.4	87.8	92.1	98.8	106.6
Commercial vehicles Véhicules utilitaires	8.3	8.4	8.9	9.3	9.7	10.0	10.2	10.4	10.8	11.4
Bulgaria Bulgarie										
Passenger cars Voitures de tourisme	...	...	...	...	...	...	...	...	...	1 309.7
Commercial vehicles Véhicules utilitaires	...	...	...	...	...	...	...	...	...	145.1

101
Motor vehicles in use
Passenger cars and commercial vehicles: thousand units [cont.]
Véhicules automobiles en circulation
Voitures de tourisme et véhicules utilitaires : milliers de véhicules [suite]

Country or area Pays or zone	1981	1982	1983	1984	1985	1986	1987	1988	1989	1990
Burkina Faso Burkina Faso										
Passenger cars[1]										
Voitures de tourisme[1]	12.6	11.0	11.0	11.0	10.9	10.6	11.0	11.0	11.0	11.0
Commercial vehicles[1]										
Véhicules utilitaires[1]	13.4	13.0	13.0	13.0	13.2	12.8	13.0	13.0	13.0	13.0
Burundi Burundi										
Passenger cars										
Voitures de tourisme	6.5	7.2	7.9	8.5	9.2	10.1	11.1	11.7	12.4	13.5
Commercial vehicles										
Véhicules utilitaires	3.4	4.2	4.6	5.1	5.6	6.1	6.8	7.2	7.7	8.3
Cameroon Cameroun										
Passenger cars										
Voitures de tourisme	55.4	60.4	66.9	72.4	77.1	80.8	78.3	73.7	68.7	63.4
Commercial vehicles										
Véhicules utilitaires	34.4	37.1	40.1	41.3	43.5	44.9	43.9	41.0	37.7	34.3
Canada Canada										
Passenger cars[4]										
Voitures de tourisme[4]	10 199.4	10 530.4	10 731.5	10 780.7	11 118.1	11 477.0	11 772.5	12 086.0	12 811.3	12 622.0
Commercial vehicles[4]										
Véhicules utilitaires[4]	3 192.2	3 293.3	3 365.2	3 099.1	3 148.5	3 212.0	3 567.8	3 765.9	3 458.4	3 931.3
Cape Verde Cap-Vert										
Passenger cars										
Voitures de tourisme	1.0	1.0	2.0	2.0	2.0	2.0	2.0	...	...	...
Commercial vehicles										
Véhicules utilitaires	1.0	1.0	1.0	1.0	1.0	1.0	1.0	...	...	...
Cayman Islands Iles Caïmanes										
Passenger cars										
Voitures de tourisme	...	...	...	...	7 913.0	8 021.0	9 769.0	9 055.0	9 681.0	10 743.0
Commercial vehicles										
Véhicules utilitaires	...	...	...	...	1 646.0	1 653.0	2 064.0	1 856.0	2 088.0	2 351.0
Central African Rep. Rép. centrafricaine										
Passenger cars										
Voitures de tourisme	23.8	38.9	41.3	44.0	45.0	45.0	46.0	11.7	11.1	11.6
Commercial vehicles										
Véhicules utilitaires	3.1	3.3	3.8	4.0	5.0	5.0	5.0	3.7	3.4	3.6
Chad Tchad										
Passenger cars										
Voitures de tourisme	10.0	10.0	10.0	10.0	11.0	11.0	11.0	...	...	...
Commercial vehicles										
Véhicules utilitaires	2.0	2.0	2.0	3.0	3.0	3.0	3.0	...	...	...
Chile Chili										
Passenger cars										
Voitures de tourisme	573.8	605.6	618.7	630.4	624.9	590.7	618.5	669.1	660.2	710.6
Commercial vehicles[6]										
Véhicules utilitaires[6]	259.2	256.5	248.8	235.4	257.9	242.1	267.9	297.0	330.0	357.8
Colombia Colombie										
Passenger cars[9]										
Voitures de tourisme[9]	599.3	669.9	723.4	767.8	805.5	842.0	...	...	...	...
Commercial vehicles[9]										
Véhicules utilitaires[9]	328.2	353.9	368.3	381.1	390.9	400.6	...	...	...	...
Comoros Comores										
Passenger cars										
Voitures de tourisme	1.0	1.0	1.0	1.0	1.0	1.0	1.0	...	...	...
Commercial vehicles										
Véhicules utilitaires	3.0	3.0	3.0	3.0	4.0	4.0	4.0	...	...	...
Congo Congo										
Passenger cars[1]										
Voitures de tourisme[1]	29.6	32.8	35.7	37.8	25.8	26.1	26.0	26.0	26.0	26.0

101
Motor vehicles in use
Passenger cars and commercial vehicles: thousand units [cont.]
Véhicules automobiles en circulation
Voitures de tourisme et véhicules utilitaires : milliers de véhicules [suite]

Country or area Pays or zone	1981	1982	1983	1984	1985	1986	1987	1988	1989	1990
Commercial vehicles[1] Véhicules utilitaires[1]	18.1	20.1	21.2	22.1	19.6	20.0	20.0	20.0	20.0	20.0
Costa Rica Costa Rica										
Passenger cars[4] Voitures de tourisme[4]	89.1	91.4	101.3	106.2	111.5	119.1	127.2	135.0	143.9	168.8
Commercial vehicles[4] Véhicules utilitaires[4]	65.6	65.9	65.7	66.7	69.5	76.3	84.2	89.6	94.6	95.1
Côte d'Ivoire Côte d'Ivoire										
Passenger cars Voitures de tourisme	157.0	167.0	171.0	176.0	176.0	178.0	178.0	168.0[1]	155.0[1]	155.0[1]
Commercial vehicles Véhicules utilitaires	85.0	88.0	93.0	97.0	99.0	90.0	90.0	91.0[1]	90.0[1]	90.0[1]
Cuba Cuba										
Passenger cars Voitures de tourisme	171.4	182.2	190.4	200.1	206.3	217.2	229.5	241.3	...	...
Commercial vehicles Véhicules utilitaires	142.8	152.4	158.9	164.5	172.8	184.2	194.9	208.4	...	...
Cyprus Chypre										
Passenger cars[4] Voitures de tourisme[4]	96.1	104.5	110.6	118.1	125.7	131.6	142.6	152.7	165.4	178.2
Commercial vehicles[4] Véhicules utilitaires[4]	27.8	31.9	38.1	43.0	47.3	50.7	56.5	61.8	68.8	76.6
Czechoslovakia Tchécoslovaquie										
Passenger cars Voitures de tourisme	2 373.0	2 441.5	2 511.3	2 639.6	2 726.3	2 812.4	2 904.0	3 000.0	3 122.3	3 242.3
Commercial vehicles Véhicules utilitaires	354.6	357.9	362.5	378.1	388.2	400.8	413.0	423.0	446.7	461.6
Denmark Danemark										
Passenger cars[4 10] Voitures de tourisme[4 10]	1 375.4	1 367.1	1 401.3	1 451.7	1 513.3	1 571.1	1 601.6	1 610.3	1 611.7	1 603.8
Commercial vehicles[4 10] Véhicules utilitaires[4 10]	258.0	251.8	246.8	255.3	270.3	286.2	298.0	305.0	306.5	305.9
Djibouti Djibouti										
Passenger cars Voitures de tourisme	6.0	6.0	6.0	6.0	6.0	7.0	7.0	...	...	...
Commercial vehicles Véhicules utilitaires	1.0	1.0	1.0	1.0	1.0	1.0	1.0	...	...	...
Dominican Republic Rép. dominicaine										
Passenger cars[11] Voitures de tourisme[11]	105.0	97.2	92.4	108.7	101.5	132.9	151.6	...	...	...
Commercial vehicles Véhicules utilitaires	57.6	60.6	58.0	60.4	52.0	76.6	84.6	...	...	...
Ecuador Equateur										
Passenger cars Voitures de tourisme	95.1	99.0	102.5	120.2	121.3	136.5	140.7	146.5	176.2	165.6
Commercial vehicles Véhicules utilitaires	161.6	159.5	177.1	179.6	176.0	185.1	188.5	190.2	221.6	207.3
Egypt Egypte										
Passenger cars Voitures de tourisme	519.0	633.0	733.0	847.0	900.0	933.0	965.0	980.0	1 019.0	1 054.0
Commercial vehicles Véhicules utilitaires	146.0	177.0	207.0	238.0	265.0	287.0	302.0	318.0	353.0	380.0
El Salvador El Salvador										
Passenger cars Voitures de tourisme	82.6	82.7	88.0	85.3	52.9[1]	52.1[1]	52.0[1]	52.0[1]	52.0[1]	52.0[1]
Commercial vehicles Véhicules utilitaires	64.1	63.9	60.8	65.2	64.8[1]	64.0[1]	65.0[1]	65.0[1]	65.0[1]	65.0[1]

101
Motor vehicles in use
Passenger cars and commercial vehicles: thousand units [cont.]
Véhicules automobiles en circulation
Voitures de tourisme et véhicules utilitaires : milliers de véhicules [suite]

Country or area Pays or zone	1981	1982	1983	1984	1985	1986	1987	1988	1989	1990
Estonia Estonie										
Passenger cars										
Voitures de tourisme	133.0	143.0	153.0	164.0	173.0	182.0	194.0	206.0	222.0	242.0
Ethiopia Ethiopie										
Passenger cars										
Voitures de tourisme	41.2[12]	42.3[12]	46.2[12]	47.5[12]	41.1[12]	41.2[12]	48.3[12]	43.5[12]	42.8[12]	38.5
Commercial vehicles										
Véhicules utilitaires	11.8[12]	11.6[12]	11.3[12]	11.1[12]	18.2[12]	18.7[12]	10.9[12]	21.8[12]	21.6[12]	20.3
Fiji Fidji										
Passenger cars[4]										
Voitures de tourisme[4]	25.6	27.4	29.3	31.0	32.5	33.6	34.4	34.9	37.5	40.3
Commercial vehicles[4,13]										
Véhicules utilitaires[4,13]	17.5	18.8	24.4	25.8	27.2	28.2	28.8	29.4	30.5	32.4
Finland Finlande										
Passenger cars										
Voitures de tourisme	1 279.2	1 352.1	1 410.4	1 474.0	1 546.1	1 619.8	1 698.7	1 795.9	1 908.9	1 939.9
Commercial vehicles										
Véhicules utilitaires	164.1	170.7	176.9	182.9	188.7	196.6	207.4	222.9	254.0	273.5
France France										
Passenger cars										
Voitures de tourisme	18 800.0	19 300.0	20 600.0	20 800.0	21 090.0	21 500.0	21 970.0	22 520.0	23 010.0	23 550.0
Commercial vehicles										
Véhicules utilitaires	2 835.0	2 951.0	3 008.0	3 072.0	3 209.0	3 298.0	3 419.0	3 547.0	3 674.0	3 810.0
French Guiana Guyane française										
Passenger cars[1]										
Voitures de tourisme[1]	19.0	19.9	20.8	21.6	22.4	23.4	24.0	24.0	24.0	25.0
Commercial vehicles[1]										
Véhicules utilitaires[1]	5.2	5.6	6.0	6.4	6.7	6.9	7.0	7.0	8.0	9.0
Gabon Gabon										
Passenger cars										
Voitures de tourisme	16.0	16.0	15.0	16.0	16.0	16.0	17.0	...	...	...
Commercial vehicles										
Véhicules utilitaires	10.0	11.0	11.0	11.0	11.0	11.0	11.0	...	...	...
Gambia Gambie										
Passenger cars										
Voitures de tourisme	5.7	5.7	4.6	5.3	4.2	3.8	3.9	4.5	5.3	...
Commercial vehicles										
Véhicules utilitaires	1.4	1.6	1.2	1.3	1.6	1.4	1.4	1.5	2.0	...
Georgia Géorgie										
Passenger cars										
Voitures de tourisme	...	...	...	...	...	...	425.7	440.8	448.7	481.9
Commercial vehicles										
Véhicules utilitaires	...	...	...	...	...	...	99.9	101.7	101.7	110.0
Germany† · Allemagne† **F. R. Germany R. f. Allemagne**										
Passenger cars										
Voitures de tourisme	23 730.6	24 104.5	24 580.5	25 217.8	25 844.5	26 917.4	27 908.2	28 878.2	29 755.4	30 684.8
Commercial vehicles										
Véhicules utilitaires	1 611.6	1 605.5	1 603.2	1 614.6	1 629.0	1 664.3	1 704.4	1 753.0	1 811.9	1 895.4
former German D. R. anc. R. d. allemande										
Passenger cars										
Voitures de tourisme	2 812.0	2 921.6	3 019.9	3 157.1	3 306.2	3 462.2	3 600.4	* 3 743.6	* 3 898.9	4 817.0
Commercial vehicles[5]										
Véhicules utilitaires[5]	641.0	642.8	640.1	647.0	656.8	668.5	685.3	* 706.0	* 732.3	774.8
Ghana Ghana										
Passenger cars[1]										
Voitures de tourisme[1]	64.6	63.1	61.6	60.2	59.1	58.3	58.0	58.0	66.0	82.2

101
Motor vehicles in use
Passenger cars and commercial vehicles: thousand units [cont.]
Véhicules automobiles en circulation
Voitures de tourisme et véhicules utilitaires : milliers de véhicules [suite]

Country or area Pays or zone	1981	1982	1983	1984	1985	1986	1987	1988	1989	1990
Commercial vehicles[1] Véhicules utilitaires[1]	46.9	46.1	45.6	45.0	45.5	45.7	46.0	46.0	40.0	42.1
Gibraltar Gibraltar										
Passenger cars Voitures de tourisme	8.7	8.6	8.7	8.5	10.6	13.0	13.1	15.5	17.7	19.8
Commercial vehicles Véhicules utilitaires	0.9	0.9	0.9	0.8	1.0	1.0	1.6	2.0	2.3	2.5
Greece Grèce										
Passenger cars Voitures de tourisme	912.4	996.3	1 069.3	1 154.9	1 263.4	1 359.2	1 428.5	1 503.9	1 605.1	1 737.5
Commercial vehicles Véhicules utilitaires	458.4	514.3	555.1	588.9	619.2	645.7	663.8	697.1	732.2	784.2
Greenland Groënland										
Passenger cars[4] Voitures de tourisme[4]	1.4	1.4	1.6	1.6	1.7	2.1	2.0	2.1	2.0	1.9
Commercial vehicles[4] Véhicules utilitaires[4]	1.2	1.2	1.4	1.4	1.4	1.3	1.6	1.5	1.6	1.5
Guadeloupe Guadeloupe										
Passenger cars Voitures de tourisme	80.4	87.8	75.4	82.7	89.4	71.3[1]	79.0[1]	79.0[1]	83.0[1]	86.0[1]
Commercial vehicles Véhicules utilitaires	30.7	33.4	23.7	26.3	28.3	28.1[1]	31.0[1]	31.0[1]	33.0[1]	34.0[1]
Guam Guam										
Passenger cars Voitures de tourisme	39.6	39.3	44.0	57.6	60.6	54.9	55.9	56.2	62.0	...
Commercial vehicles Véhicules utilitaires	16.3	13.4	14.7	15.3	16.4	15.1	19.5	16.7	22.8	...
Guatemala Guatemala										
Passenger cars[1] Voitures de tourisme[1]	96.2	95.1	95.5	96.0	96.7	95.1	95.0	95.0	95.0	95.0
Commercial vehicles[1] Véhicules utilitaires[1]	95.4	94.0	93.3	94.0	93.7	92.7	93.0	93.0	93.0	93.0
Guinea Guinée										
Passenger cars Voitures de tourisme	9.0	9.0	10.0	10.0	11.0	11.0	11.0	...	...	...
Commercial vehicles Véhicules utilitaires	10.0	10.0	10.0	10.0	11.0	12.0	12.0	...	...	...
Guinea-Bissau Guinée-Bissau										
Passenger cars Voitures de tourisme	4.0	4.0	4.0	3.0	4.0	4.0	4.0	...	...	...
Commercial vehicles Véhicules utilitaires	3.0	3.0	3.0	3.0	3.0	3.0	3.0	...	...	...
Guyana Guyana										
Passenger cars[1] Voitures de tourisme[1]	31.6	30.8	30.1	30.0	29.4	28.9	20.3	22.0	24.0	24.0
Commercial vehicles[1] Véhicules utilitaires[1]	12.6	12.3	12.0	12.0	11.7	11.7	8.5	9.0	9.0	9.0
Haiti Haïti										
Passenger cars Voitures de tourisme	15.1	15.6	18.1	24.6	26.1	19.5	19.7	20.6	27.7	* 25.8
Commercial vehicles Véhicules utilitaires	6.8	7.8	8.1	10.6	10.7	13.0	22.6	22.8	22.8	* 9.6
Honduras Honduras										
Passenger cars Voitures de tourisme	25.5[4]	27.0[4]	26.8[4]	28.5[4]	34.2[4]	26.8[1]	27.0[1]	27.0[1]	27.0[1]	27.0[1]
Commercial vehicles Véhicules utilitaires	47.1[4 14]	49.9[4 14]	49.0[4 14]	...	...	50.2[1]	51.0[1]	52.0[1]	52.0[1]	52.0[1]

101
Motor vehicles in use
Passenger cars and commercial vehicles: thousand units [*cont.*]
Véhicules automobiles en circulation
Voitures de tourisme et véhicules utilitaires : milliers de véhicules [*suite*]

Country or area Pays or zone	1981	1982	1983	1984	1985	1986	1987	1988	1989	1990
Hong Kong Hong-kong										
Passenger cars										
Voitures de tourisme	201.4	197.3	173.8	163.4	160.9	155.6	162.3	177.4	197.2	214.9
Commercial vehicles										
Véhicules utilitaires	65.4	70.7	71.0	73.9	81.4	91.7	107.3	118.4	126.1	131.8
Hungary Hongrie										
Passenger cars										
Voitures de tourisme	1 105.4	1 181.7	1 258.5	1 344.1	1 435.9	1 538.9	1 660.3	1 789.6	1 732.4	1 944.6
Commercial vehicles										
Véhicules utilitaires	175.9	184.4	197.5	213.4	223.4	238.3	254.2	259.6	267.8	288.5
Iceland Islande										
Passenger cars										
Voitures de tourisme	90.3	94.7	96.0	100.2	103.0	112.3	120.1	125.2	124.3	119.7
Commercial vehicles										
Véhicules utilitaires	10.7	11.7	12.3	13.0	14.2	13.1	12.9	13.2	13.5	14.5
India Inde										
Passenger cars										
Voitures de tourisme	1 116.7	1 207.2	1 346.8	1 430.0	1 579.2	1 747.8	1 978.7	2 266.0	2 471.1	2 789.9
Commercial vehicles[15]										
Véhicules utilitaires[15]	1 288.7	1 471.6	1 625.7	1 822.2	2 080.5	2 065.7	2 508.2	2 818.0	3 058.2	3 455.5
Indonesia Indonésie										
Passenger cars										
Voitures de tourisme	719.3[6]	791.0[6]	862.4[6]	927.0[6]	990.7[6]	1 064.0[6]	1 170.1[6]	1 073.1[6]	1 182.2[6]	1 293.8[1]
Commercial vehicles										
Véhicules utilitaires	702.9	791.5	878.1	982.5	1 072.6	1 138.9	1 257.1	1 278.3	1 391.3	1 478.0[1]
Iran, Islamic Rep. of Iran, Rép. islamique d'										
Passenger cars[9 16]										
Voitures de tourisme[9 16]	1 514.0	1 591.0	1 655.0	1 745.0	1 851.0	1 919.0	1 958.0	1 981.0	2 000.0	...
Commercial vehicles[9 16]										
Véhicules utilitaires[9 16]	597.0	615.0	836.0	695.0	760.0	819.0	853.0	862.0	870.0	...
Iraq Iraq										
Passenger cars[1]										
Voitures de tourisme[1]	213.5	255.7	250.3	250.7	258.4	251.3	251.0	630.3	672.2	672.0
Commercial vehicles[1]										
Véhicules utilitaires[1]	238.0	277.1	271.3	265.3	273.4	269.2	269.0	344.8	368.0	368.0
Ireland Irlande										
Passenger cars[3]										
Voitures de tourisme[3]	778.3	713.8	723.8	716.8	715.3	717.2	742.8	755.7	779.8	802.7
Commercial vehicles[3]										
Véhicules utilitaires[3]	72.2	73.4	75.1	89.3	98.7	106.9	116.6	124.4	135.9	149.5
Israel Israël										
Passenger cars										
Voitures de tourisme	453.6	512.7	571.5	599.3	613.9	648.8	696.7	753.5	777.9	813.0[1]
Commercial vehicles										
Véhicules utilitaires	96.1	104.1	110.8	113.1	114.7	120.8	132.0	143.8	149.2	163.0[1]
Italy Italie										
Passenger cars										
Voitures de tourisme	18 603.4	19 616.1	20 388.6	20 888.2	22 494.6	23 495.5	24 320.2	25 290.3	26 267.4	27 300.0
Commercial vehicles										
Véhicules utilitaires	1 810.4	1 912.5	2 098.6	2 182.1	2 308.8	2 428.2	2 002.7	2 120.5	2 350.0	2 504.0
Jamaica Jamaïque										
Passenger cars										
Voitures de tourisme	41.2	40.3	35.0	42.0	42.9	44.5	52.9	63.1	64.8	68.5
Commercial vehicles										
Véhicules utilitaires	17.4	20.2	17.4	23.2	26.1	20.7	23.0	26.9	24.5	28.2
Japan Japon										
Passenger cars[17 18]										
Voitures de tourisme[17 18]	24 613.0	25 539.0	26 386.0	27 144.0	27 844.0	28 654.0	29 478.0	30 776.0	32 621.0	34 924.0

101
Motor vehicles in use
Passenger cars and commercial vehicles: thousand units [*cont.*]
Véhicules automobiles en circulation
Voitures de tourisme et véhicules utilitaires : milliers de véhicules [*suite*]

Country or area Pays or zone	1981	1982	1983	1984	1985	1986	1987	1988	1989	1990
Commercial vehicles[17] Véhicules utilitaires[17]	14 197.0	14 956.0	15 674.0	16 477.0	17 377.0	18 346.0	19 401.0	20 592.0	21 330.0	21 571.0
Jordan Jordanie										
Passenger cars[4] Voitures de tourisme[4]	102.8	117.6	130.7	139.5	143.4	149.8	153.6	158.9	159.9	172.0
Commercial vehicles[4] Véhicules utilitaires[4]	37.7	39.2	52.2	56.2	55.8	57.8	60.9	63.2	63.4	68.3
Kenya Kenya										
Passenger cars Voitures de tourisme	114.2[49 19]	115.3[49 19]	116.9[49 19]	122.0[49 19]	129.0[49]	128.0[49]	129.0[49]	133.0[1]	136.0[1]	136.0[1]
Commercial vehicles Véhicules utilitaires	87.4[49 14]	88.7[49 14]	88.9[49 14]	95.0[49 14]	104.0[49]	106.0[49]	106.0[49]	149.0[1]	154.0[1]	158.0[1]
Korea, Republic of Corée, République de										
Passenger cars Voitures de tourisme	267.6	305.8	381.0	465.1	556.7	664.2	844.4	1 118.0	1 558.7	2 074.9
Commercial vehicles Véhicules utilitaires	294.4	330.3	391.4	468.4	541.0	627.2	746.9	895.0	1 092.3	1 308.4
Kuwait Koweït										
Passenger cars Voitures de tourisme	435.3	478.8	519.5	384.1	417.4	410.0	429.2	458.1	504.4	...
Commercial vehicles Véhicules utilitaires	155.2	229.9	178.7	134.3	140.4	133.6	135.2	110.5	117.9	...
Latvia Lettonie										
Passenger cars Voitures de tourisme	177.0	184.0	190.0	201.0	215.0	226.0	236.0	251.0	264.0	283.0
Commercial vehicles Véhicules utilitaires	66.0	68.0	69.0	70.0	72.0	73.0	74.0	75.0	77.0	79.0
Lesotho Lesotho										
Passenger cars Voitures de tourisme	6.3[4]	5.8[4]	5.0[4]	5.1[4]	6.1[4]	6.7[4]	5.0	...	...	...
Commercial vehicles Véhicules utilitaires	14.7[4]	15.2[4]	14.4[4]	16.1[4]	14.6[4]	16.3[4]	* 13.0	...	...	...
Liberia Libéria										
Passenger cars Voitures de tourisme	13.0	14.0	15.0	16.0	16.0	17.0	18.0	...	...	...
Commercial vehicles Véhicules utilitaires	12.0	12.0	13.0	14.0	14.0	15.0	15.0	...	...	...
Libyan Arab Jamah. Jamah. arabe libyenne										
Passenger cars Voitures de tourisme	485.3	403.0	410.0	408.0	420.0	428.0	433.0	448.0[1]	448.0[1]	448.0[1]
Commercial vehicles Véhicules utilitaires	275.8	203.0	206.0	207.0	210.0	216.0	223.0	322.0[1]	322.0[1]	322.0[1]
Lithuania Lituanie										
Passenger cars Voitures de tourisme	...	...	...	...	...	...	...	...	452.0	493.0
Commercial vehicles Véhicules utilitaires	...	...	...	...	...	...	...	...	90.0	95.0
Luxembourg Luxembourg										
Passenger cars Voitures de tourisme	133.3	137.9	141.1	145.8	151.6	156.0	162.5	168.5	177.0	183.4
Commercial vehicles[5] Véhicules utilitaires[5]	24.9	25.5	30.8	26.3	26.7	27.4	28.0	28.7	30.0	31.6
Macau Macao										
Passenger cars[6] Voitures de tourisme[6]	14.3	15.6	16.9	17.5	17.9	19.5	19.6	21.0	22.4	24.7
Commercial vehicles[6] Véhicules utilitaires[6]	3.7	4.0	4.3	4.6	4.4	4.8	4.9	5.3	6.3	6.4

101

Motor vehicles in use
Passenger cars and commercial vehicles: thousand units [cont.]
Véhicules automobiles en circulation
Voitures de tourisme et véhicules utilitaires : milliers de véhicules [suite]

Country or area Pays or zone	1981	1982	1983	1984	1985	1986	1987	1988	1989	1990
Madagascar Madagascar										
Passenger cars										
Voitures de tourisme	48.0	46.0	44.0	45.0	46.0	46.0	46.0	...	...	...
Commercial vehicles										
Véhicules utilitaires	49.0	45.0	43.0	44.0	45.0	48.0	47.0	...	...	...
Malawi Malawi										
Passenger cars[4]										
Voitures de tourisme[4]	14.1	14.0	13.7	13.7	13.6	...	14.9	15.6	16.1	...
Commercial vehicles[4]										
Véhicules utilitaires[4]	17.2	17.1	15.7	15.1	15.4	...	16.7	12.9	19.3	...
Malaysia Malaisie										
Passenger cars										
Voitures de tourisme	957.2	1 022.9	1 153.9	1 292.0	1 384.0	1 453.6	1 504.2	1 578.9	1 689.4	1 845.6
Commercial vehicles[20]										
Véhicules utilitaires[20]	318.8	353.5	262.9	286.3	311.1	330.1	339.0	351.9	374.6	407.1
Mali Mali										
Passenger cars										
Voitures de tourisme	16.0	17.0	17.0	18.0	19.0	19.0	19.0	20.0	...	...
Commercial vehicles										
Véhicules utilitaires	11.0	11.0	12.0	12.0	13.0	13.0	13.0	13.0	...	...
Malta Malte										
Passenger cars										
Voitures de tourisme	74.8	79.8	76.2	80.3	82.3	85.6	89.5	97.6	110.6	109.2
Commercial vehicles										
Véhicules utilitaires	17.9	18.3	16.6	17.4	18.2	17.8	17.2	19.3	19.6	20.5
Martinique Martinique										
Passenger cars[1]										
Voitures de tourisme[1]	64.1	67.2	70.5	72.7	74.6	77.0	80.0	84.0	80.0	92.0
Commercial vehicles[1]										
Véhicules utilitaires[1]	22.1	22.8	23.7	24.5	25.3	26.0	27.0	28.0	29.0	30.0
Mauritania Mauritanie										
Passenger cars										
Voitures de tourisme	8.0	9.0	10.0	10.0	11.0	11.0	12.0	...	...	...
Commercial vehicles										
Véhicules utilitaires	3.0	4.0	4.0	5.0	5.0	6.0	6.0	...	...	...
Mauritius Maurice										
Passenger cars										
Voitures de tourisme	31.1	31.6	32.4	32.7	33.1	34.2	36.5	39.4	42.1	45.5
Commercial vehicles										
Véhicules utilitaires	10.1	10.4	10.6	10.6	10.8	11.2	12.2	13.7	15.1	16.9
Mexico Mexique										
Passenger cars										
Voitures de tourisme	4 591.3	4 797.4	4 563.3	4 802.1	5 102.4	5 202.9	5 336.2	5 806.9	6 219.1	6 893.3
Commercial vehicles										
Véhicules utilitaires	1 650.0	1 754.2	1 819.0	1 933.1	2 033.4	2 213.0	2 292.0	2 435.9	2 704.1	2 982.0
Moldova, Rep. of Moldova, Rép. de										
Passenger cars										
Voitures de tourisme	109.2	113.6	124.9	129.2	148.1	163.1	171.3	184.9	195.7	210.1
Commercial vehicles										
Véhicules utilitaires	16.2	16.2	16.2	16.3	16.1	16.3	16.1	15.6	15.2	15.0
Morocco Maroc										
Passenger cars[6]										
Voitures de tourisme[6]	445.0	462.6	477.4	491.1	508.3	527.4	554.0	588.9	634.4	669.6
Commercial vehicles[6]										
Véhicules utilitaires[6]	207.4	217.5	226.6	232.7	239.9	247.7	256.1	263.4	272.4	282.9
Mozambique Mozambique										
Passenger cars										
Voitures de tourisme	58.0	56.0	50.0	49.0	46.0	45.0	45.0	...	...	...

101
Motor vehicles in use
Passenger cars and commercial vehicles: thousand units [cont.]
Véhicules automobiles en circulation
Voitures de tourisme et véhicules utilitaires : milliers de véhicules [suite]

Country or area Pays or zone	1981	1982	1983	1984	1985	1986	1987	1988	1989	1990
Commercial vehicles Véhicules utilitaires	21.0	21.0	20.0	19.0	18.0	18.0	18.0	...	...	...
Myanmar Myanmar										
Passenger cars[4] Voitures de tourisme[4]	45.4	47.8	50.6	54.8	59.8	60.9	60.4	64.9	71.3	78.1
Commercial vehicles[4] Véhicules utilitaires[4]	46.2	47.2	49.2	50.2	51.9	53.5	50.7	51.6	53.7	54.9
Netherlands Pays-Bas										
Passenger cars[4 21 22] Voitures de tourisme[4 21 22]	4 594.0	4 630.0	4 728.0	4 772.4	4 901.1	4 949.9	5 117.7	5 250.6	5 371.4	...
Commercial vehicles[4 21 22] Véhicules utilitaires[4 21 22]	349.0	354.0	390.2	381.0	428.2	463.8	506.5	538.2	556.8	...
New Caledonia Nouvelle-Calédonie										
Passenger cars[1] Voitures de tourisme[1]	39.2	40.9	42.0	43.9	43.8	46.4	48.0	50.0	53.0	54.0
Commercial vehicles[1] Véhicules utilitaires[1]	15.1	15.9	16.2	16.5	16.3	16.7	17.0	18.0	19.0	19.0
New Zealand Nouvelle-Zélande										
Passenger cars Voitures de tourisme	1 341.8	1 376.5	1 411.5	1 466.2	1 495.1	1 531.4	...	1 382.3[23]	1 438.7[23]	1 497.7[23]
Commercial vehicles Véhicules utilitaires	274.3	280.7	288.2	294.2	302.1	373.8	...	300.1[23]	300.4[23]	306.6[23]
Nicaragua Nicaragua										
Passenger cars[1] Voitures de tourisme[1]	35.6	34.6	33.6	32.7	31.9	31.1	31.0	30.0	31.1	31.1
Commercial vehicles[1] Véhicules utilitaires[1]	30.4	29.5	28.9	28.5	28.3	27.9	28.0	42.0	43.0	43.0
Niger Niger										
Passenger cars Voitures de tourisme	29.0	32.0	34.0	34.0	35.0	35.0	35.0	...	...	...
Commercial vehicles Véhicules utilitaires	8.0	8.0	9.0	9.0	9.0	9.0	9.0	...	...	...
Nigeria Nigéria										
Passenger cars Voitures de tourisme	350.0	310.0	340.0	360.0	350.0	377.0	391.0	...	...	...
Commercial vehicles Véhicules utilitaires	30.0	32.0	30.0	33.0	34.0	36.0	40.0	...	...	...
Norway Norvège										
Passenger cars[4 24] Voitures de tourisme[4 24]	127.8	1 336.9	1 382.3	1 428.7	1 514.0	1 592.2	1 623.1	1 622.0	1 612.7	1 612.0
Commercial vehicles[4 24] Véhicules utilitaires[4 24]	171.3	180.1	195.7	215.1	249.7	282.8	303.2	313.9	320.4	330.6
Pakistan Pakistan										
Passenger cars[4] Voitures de tourisme[4]	302.1	325.2	362.0	404.4	452.1	500.2	540.4	590.5	660.0	715.0
Commercial vehicles[4] Véhicules utilitaires[4]	123.8	128.7	138.3	144.2	153.2	172.8	175.0	194.4	259.6	265.4
Panama Panama										
Passenger cars Voitures de tourisme	104.3	110.2	115.9	121.0	129.0	134.3	129.4	129.5	120.8	132.9
Commercial vehicles Véhicules utilitaires	34.8	36.4	37.6	41.2	40.3	41.9	44.6	41.2	40.0	42.2
Papua New Guinea Papouasie-Nvl-Guinée										
Passenger cars[4] Voitures de tourisme[4]	18.9	17.8	16.7	17.4	16.0	16.6	17.1	...	...	...
Commercial vehicles[4 13] Véhicules utilitaires[4 13]	28.2	26.9	26.7	29.0	26.6	27.0	26.1	...	...	...

101
Motor vehicles in use
Passenger cars and commercial vehicles: thousand units [cont.]
Véhicules automobiles en circulation
Voitures de tourisme et véhicules utilitaires : milliers de véhicules [suite]

Country or area Pays or zone	1981	1982	1983	1984	1985	1986	1987	1988	1989	1990
Paraguay Paraguay										
Passenger cars										
Voitures de tourisme	57.5	62.9	78.7	85.0	104.9	111.9	122.5	80.2	149.2	165.2
Commercial vehicles										
Véhicules utilitaires	15.5	12.6	23.6	24.5	22.0	23.2	27.2	18.9	25.2	25.7
Peru Pérou										
Passenger cars										
Voitures de tourisme	330.9	359.7	371.7	374.0	376.0	377.2	377.4	376.8	372.8	...
Commercial vehicles										
Véhicules utilitaires	191.0	204.6	212.4	217.0	220.3	226.5	233.4	239.8	239.5	...
Philippines Philippines										
Passenger cars										
Voitures de tourisme	318.1	342.0	367.0	360.7	347.9	356.7	358.8	376.6	413.0	454.6 [1]
Commercial vehicles										
Véhicules utilitaires	485.7	509.5	556.0	534.1	514.5	526.7	554.7	598.2	671.7	764.9 [1]
Poland Pologne										
Passenger cars										
Voitures de tourisme	2 634.3	2 881.7	3 178.9	3 425.8	3 671.4	3 964.0	4 231.7	4 519.1	4 846.4	5 260.6
Commercial vehicles [25]										
Véhicules utilitaires [25]	709.9	692.1	733.6	813.4	864.3	914.4	955.0	1 010.5	1 069.6	1 138.1
Portugal Portugal [3 26]										
Passenger cars [3 26]										
Voitures de tourisme [3 26]	1 346.0	1 428.8	1 517.6	1 600.7	1 701.7	1 813.0	1 947.3	2 152.5	2 343.4	2 552.3
Commercial vehicles [3 26]										
Véhicules utilitaires [3 26]	527.4 [20]	586.0 [20]	618.5 [20]	642.5 [20]	669.1 [20]	705.9 [20]	607.9	669.6	739.2	812.8
Puerto Rico Porto Rico										
Passenger cars										
Voitures de tourisme	1 022.1	1 031.0	1 066.9	896.6	1 142.6	1 225.1	1 304.3	1 337.7	1 321.9	1 332.1
Commercial vehicles										
Véhicules utilitaires	142.8	155.7	161.0	121.5	168.0	178.8	190.6	196.1	192.1	206.7
Qatar Qatar										
Passenger cars										
Voitures de tourisme	74.5	91.9	...	74.8	80.0	85.4	91.0	96.7	102.0	105.8
Commercial vehicles										
Véhicules utilitaires	43.8	45.9	...	32.5	33.4	36.8	39.3	43.9	45.7	46.8
Réunion Réunion										
Passenger cars										
Voitures de tourisme	94.2	108.7	118.9	128.2	138.1	147.4	159.2	...	...	...
Commercial vehicles										
Véhicules utilitaires	33.3	36.1	39.0	41.8	45.0	48.5	53.3	...	...	...
Rwanda Rwanda										
Passenger cars										
Voitures de tourisme	5.7	5.7	5.8	5.6	6.1	6.7	7.1	7.2	7.9	7.2
Commercial vehicles										
Véhicules utilitaires	7.9	6.9	6.9	5.6	6.3	7.0	7.3	7.9	7.0	7.0
Saint Kitts and Nevis Saint-Kitts-et-Nevis										
Passenger cars										
Voitures de tourisme	2.7	2.7	2.9	3.0	3.1	3.3	3.3	3.4	3.9	4.1
Commercial vehicles										
Véhicules utilitaires	0.7	0.8	0.9	0.9	0.9	1.0	1.0	1.2	1.5	1.6
Saint Lucia Sainte-Lucie										
Passenger cars										
Voitures de tourisme	4.9	5.8	6.1	5.5	5.4	5.5	6.0	6.5	7.2	8.1
Commercial vehicles										
Véhicules utilitaires	2.7	3.1	3.2	3.5	3.7	4.2	5.1	5.5	5.9	6.3
St. Vincent-Grenadines St. Vincent-Grenadines										
Passenger cars										
Voitures de tourisme	4.5	4.8	5.1	5.0	4.9	5.1	4.9	5.2	5.3	5.3

101
Motor vehicles in use
Passenger cars and commercial vehicles: thousand units [cont.]
Véhicules automobiles en circulation
Voitures de tourisme et véhicules utilitaires : milliers de véhicules [suite]

Country or area Pays or zone	1981	1982	1983	1984	1985	1986	1987	1988	1989	1990
Commercial vehicles Véhicules utilitaires	1.3	1.5	1.8	1.9	2.0	2.3	2.4	2.6	2.7	2.8
Samoa Samoa										
Passenger cars Voitures de tourisme	1.8	1.4	1.5	1.8	1.8	1.9	1.8	1.9	2.5	...
Commercial vehicles Véhicules utilitaires	2.7	2.5	2.5	2.7	2.8	2.2	2.7	2.6	3.2	...
Sao Tome and Principe Sao Tomé-et-Principe										
Passenger cars Voitures de tourisme	1.8	1.9	1.9	2.0	2.2	2.4	2.6	...	...	...
Commercial vehicles Véhicules utilitaires	0.3	0.3	0.3	0.3	0.3	0.3	0.3	...	...	...
Saudi Arabia Arabie saoudite										
Passenger cars[1] Voitures de tourisme[1]	885.5	1 067.6	1 202.0	1 150.0	1 300.0	1 310.1	1 337.0	1 378.0	1 420.0	1 468.0
Commercial vehicles[1] Véhicules utilitaires[1]	1 109.1	1 338.6	1 534.2	1 200.0	1 450.0	1 452.6	1 477.0	1 477.0	1 499.0	1 536.0
Senegal Sénégal										
Passenger cars Voitures de tourisme	59.0	61.0	62.0	62.0	63.0	63.0	63.0	...	...	...
Commercial vehicles Véhicules utilitaires	33.0	34.0	35.0	36.0	36.0	36.0	36.0	...	...	...
Seychelles Seychelles										
Passenger cars Voitures de tourisme	3.2	3.5	3.3	3.4	3.5	3.5	3.4	3.6	4.0	4.3
Commercial vehicles Véhicules utilitaires	0.8	1.1	1.1	1.2	1.3	1.0	1.0	1.0	1.0	1.2
Sierra Leone Sierra Leone										
Passenger cars Voitures de tourisme	28.0	27.0	25.4	23.5	31.0	33.0	33.0	...	...	...
Commercial vehicles Véhicules utilitaires	11.7	11.5	10.7	10.0	14.0	15.0	15.0	...	...	...
Singapore Singapour										
Passenger cars Voitures de tourisme	175.1	194.4	216.9	232.3	236.2	234.6	236.1	251.4	271.2	286.8
Commercial vehicles Véhicules utilitaires	94.7	104.5	113.1	119.5	118.3	114.3	113.7	117.4	122.8	126.9
Somalia Somalie										
Passenger cars Voitures de tourisme	6.0	6.0	6.0	6.0	6.0	5.0	5.0	...	...	...
Commercial vehicles Véhicules utilitaires	8.0	8.0	8.0	7.0	7.0	8.0	8.0	...	...	...
South Africa Afrique du Sud										
Passenger cars[27] Voitures de tourisme[27]	2 529.4	2 721.9	2 827.5	3 018.7	3 096.6	3 237.2	3 286.8	3 222.4	3 498.2	3 599.8
Commercial vehicles[28] Véhicules utilitaires[28]	1 269.4	1 335.5	1 366.2	1 430.4	1 461.6	1 433.9	1 454.9	1 456.7	1 462.5	1 486.9
Spain Espagne										
Passenger cars Voitures de tourisme	7 943.3	8 354.1	8 714.1	8 874.4	9 273.7	9 645.4	10 218.5	10 787.5	11 467.7	11 995.6
Commercial vehicles Véhicules utilitaires	1 440.1	1 504.9	1 572.9	1 486.0	1 570.9	1 715.4	1 864.8	2 019.8	2 207.6	2 378.7
Sri Lanka Sri Lanka										
Passenger cars Voitures de tourisme	126.3[4]	131.7[4]	136.9[4]	141.7[4]	148.6[4]	155.2[4]	147.8[4]	155.2[4]	163.8[4]	173.5
Commercial vehicles Véhicules utilitaires	88.2[4]	97.6[4]	109.6[4]	121.8[4]	132.4[4]	141.3[4]	135.4[4]	139.2[4]	142.1[4]	146.0

101
Motor vehicles in use
Passenger cars and commercial vehicles: thousand units [*cont.*]
Véhicules automobiles en circulation
Voitures de tourisme et véhicules utilitaires : milliers de véhicules [*suite*]

Country or area Pays or zone	1981	1982	1983	1984	1985	1986	1987	1988	1989	1990
Sudan Soudan										
Passenger cars										
Voitures de tourisme	143.0	150.0	155.0	160.0	171.0	177.0	185.0	...	...	...
Commercial vehicles										
Véhicules utilitaires	18.0	22.0	23.0	23.0	23.0	23.0	24.0	...	...	...
Suriname Suriname										
Passenger cars[4]										
Voitures de tourisme[4]	30.0[29]	31.6[29]	31.2	28.8	31.6	32.1	32.1	35.1	36.6	...
Commercial vehicles										
Véhicules utilitaires	...	...	12.8	11.6	12.8	13.2	13.0	13.4	14.0	...
Swaziland Swaziland										
Passenger cars										
Voitures de tourisme	14.5	15.5	17.2	18.5	19.6	20.9	22.6	23.5	25.3	26.9
Commercial vehicles										
Véhicules utilitaires	14.3	15.6	16.7	17.4	20.1	20.6	21.9	23.6	24.4	26.3
Sweden Suède										
Passenger cars[4]										
Voitures de tourisme[4]	2 893.2	2 936.0	3 006.8	3 081.0	3 151.2	3 253.6	3 366.6	3 482.7	3 578.0	3 600.5
Commercial vehicles										
Véhicules utilitaires	502.7	514.2	526.6	540.5	552.1	567.1	584.9	611.6	643.8	658.0
Switzerland Suisse										
Passenger cars[3 4 17]										
Voitures de tourisme[3 4 17]	2 394.5	2 473.3	2 520.6	2 552.1	2 617.2	2 678.9	2 732.7	2 745.4	2 899.6	2 993.5
Commercial vehicles[3 4 17]										
Véhicules utilitaires[3 4 17]	179.0	189.7	201.2	203.6	211.3	217.8	228.7	252.1	277.6	303.7
Syrian Arab Republic Rép. arabe syrienne										
Passenger cars										
Voitures de tourisme	85.1	89.9	127.3	126.4	120.1	123.9	125.3	124.3	124.7	126.0
Commercial vehicles										
Véhicules utilitaires	93.3	102.7	108.7	111.3	109.2	116.8	117.3	126.4	118.0	118.5
Thailand Thaïlande										
Passenger cars[3]										
Voitures de tourisme[3]	450.5	512.3	558.9	689.8	732.6	770.4	...	1 147.0	1 000.0	1 222.0
Commercial vehicles[3]										
Véhicules utilitaires[3]	293.8[14]	353.1[14]	635.9	674.8	684.7	689.9	...	789.0	835.0	976.0
Togo Togo										
Passenger cars										
Voitures de tourisme	2.7	2.6	2.3	2.1	2.1	2.2	2.6	...	...	...
Commercial vehicles										
Véhicules utilitaires	0.9	0.9	0.9	2.1	0.8	0.8	0.9	...	...	...
Tonga Tonga										
Passenger cars										
Voitures de tourisme	0.6	0.6	1.4	0.8	0.9	1.1	1.4	1.6	1.4	2.0
Commercial vehicles										
Véhicules utilitaires	1.2	2.0	2.2	1.6	1.7	2.1	2.4	3.5	2.2	2.6
Trinidad and Tobago Trinité-et-Tobago										
Passenger cars										
Voitures de tourisme	170.9	188.2	210.4	229.4	241.6[1]	244.1[1]	244.0[1]	244.0[1]	244.0[1]	244.0[1]
Commercial vehicles										
Véhicules utilitaires	58.0	65.5	70.1	60.6[1]	79.2[1]	79.0[1]	79.0[1]	79.0[1]	79.0[1]	79.0[1]
Tunisia Tunisie										
Passenger cars										
Voitures de tourisme	132.0	141.0	148.0	145.0	166.0	171.0	179.0	...	...	...
Commercial vehicles										
Véhicules utilitaires	132.0	148.0	152.0	150.0	160.0	181.0	188.0	...	...	...
Turkey Turquie										
Passenger cars										
Voitures de tourisme	776.4	811.5	856.4	919.6	983.4	1 087.2	1 193.0	1 310.3	1 434.8	1 649.9

101
Motor vehicles in use
Passenger cars and commercial vehicles: thousand units [*cont.*]
Véhicules automobiles en circulation
Voitures de tourisme et véhicules utilitaires : milliers de véhicules [*suite*]

Country or area Pays or zone	1981	1982	1983	1984	1985	1986	1987	1988	1989	1990
Commercial vehicles[14] Véhicules utilitaires[14]	313.7	326.4	342.5	361.6	381.1	401.8	414.9	427.3	438.0	452.9
Uganda Ouganda										
Passenger cars Voitures de tourisme	11.0	10.0	10.0	10.0	10.0	12.0	12.0	13.0	13.0	...
Commercial vehicles Véhicules utilitaires	8.0	8.0	9.0	10.0	10.0	11.0	12.0	13.0	14.0	...
Ukraine Ukraine										
Passenger cars Voitures de tourisme	1 780.0	1 949.0	2 119.0	2 269.0	2 447.0	2 619.0	2 816.0	3 012.0	3 195.0	3 362.0
former USSR ancienne URSS										
Passenger cars Voitures de tourisme	9 687.0	10 448.0	11 225.0	12 085.0	12 979.0	13 846.0	14 744.0	15 632.0	16 562.0	...
United Kingdom Royaume-Uni										
Passenger cars Voitures de tourisme	15 822.0	16 282.4	16 611.8	17 213.3	17 737.1	18 355.1	18 859.1	19 940.0	20 925.0	20 807.0
Commercial vehicles Véhicules utilitaires	1 897.9	1 866.9	1 957.4	1 974.5	1 983.8	2 011.1	2 053.4	2 193.7	2 684.7	2 907.8
United Rep.Tanzania Rép. Unie de Tanzanie										
Passenger cars Voitures de tourisme	43.0	49.0	48.0	49.0	49.0	49.0	49.0	49.0	49.0	49.0
Commercial vehicles Véhicules utilitaires	29.0	32.0	33.0	32.0	33.0	33.0	33.0	33.0	33.0	33.0
United States Etats-Unis										
Passenger cars Voitures de tourisme	123 461.0[30]	123 698.0[30]	126 444.0	128 158.0	131 864.0	135 431.0	137 323.0	141 251.7[31]	143 081.4[31]	143 549.6[31]
Commercial vehicles Véhicules utilitaires	34 995.0[30]	35 852.0[30]	36 723.0	37 507.0	39 196.0	40 166.0	41 119.0	43 145.0[31]	44 179.1[31]	45 105.8[31]
Uruguay Uruguay										
Passenger cars Voitures de tourisme	281.3	298.4	291.8	292.4	306.3	318.4	332.7	350.2	360.3	379.6
Commercial vehicles Véhicules utilitaires	47.8	52.5	45.4	45.1	46.5	47.0	46.6	49.8	50.2	49.9
Vanuatu Vanuatu										
Passenger cars Voitures de tourisme	2.3[4,32]	2.3[4,32]	2.3[4,32]	2.3[4,32]	2.3[4,32]	3.8[1]	4.0[1]	4.0[1]	4.0[1]	4.0[1]
Commercial vehicles Véhicules utilitaires	0.9[4,32]	1.0[4,32]	1.1[4,32]	1.1[4,32]	1.2[4,32]	2.4[1]	2.0[1]	2.0[1]	2.0[1]	2.0[1]
Venezuela Venezuela										
Passenger cars Voitures de tourisme	1 635.0	1 846.0	1 955.4	1 559.0	1 598.0	1 656.0	1 718.0	1 740.0	1 615.0	1 582.0
Commercial vehicles Véhicules utilitaires	795.9	853.0	952.0	405.0	418.0	430.0	448.0	421.0	459.0	464.0
Yemen Yémen										
Passenger cars[33] Voitures de tourisme[33]	17.8	20.4	22.0	24.7	25.6	26.5	27.6	...	...	...
Commercial vehicles[33] Véhicules utilitaires[33]	19.4	22.2	24.6	27.2	29.2	31.2	32.4	...	...	...
Yugoslavia Yougoslavie										
Passenger cars[34] Voitures de tourisme[34]	2 568.0	2 702.6	2 770.7	2 874.0	2 824.3	2 957.1	3 023.7	3 089.7	3 323.9	3 510.8
Commercial vehicles[34] Véhicules utilitaires[34]	257.2	268.1	277.8	290.2	286.7	308.4	317.0	315.2	331.5	338.9
Zaire Zaïre										
Passenger cars Voitures de tourisme	91.0	92.0	92.0	92.0	92.0	92.0	92.0	...	...	...
Commercial vehicles Véhicules utilitaires	80.0	80.0	80.0	80.0	80.0	80.0	80.0	...	...	...

101
Motor vehicles in use
Passenger cars and commercial vehicles: thousand units [cont.]
Véhicules automobiles en circulation
Voitures de tourisme et véhicules utilitaires : milliers de véhicules [suite]

Country or area Pays or zone	1981	1982	1983	1984	1985	1986	1987	1988	1989	1990
Zambia Zambie										
Passenger cars[9]										
Voitures de tourisme[9]	67.5	68.0	68.1	...	71.8	74.0	75.2	...	...	...
Commercial vehicles[9]										
Véhicules utilitaires[9]	31.1	32.5	31.4	33.7	35.2	37.7	39.1	...	...	...
Zimbabwe Zimbabwe										
Passenger cars										
Voitures de tourisme	176.0	224.0	220.0	200.0	241.0	266.0	274.0	* 281.0	* 282.0	283.0
Commercial vehicles										
Véhicules utilitaires	76.0	75.0	77.0	75.0	78.0	80.0	81.0	* 81.0	* 82.0	83.0

Source:
Transport statistics database of the Statistical Division of
the United Nations Secretariat.

† All data shown which pertain to Germany prior to 3 October
1990 are indicated separately for the Federal Republic of
Germany and the former German Democratic Republic based on
their respective territories at the time indicated. Where
data for united Germany (3 October 1990 and thereafter) are
not available, available data are shown separately under the
designations Federal Republic of Germany and former German
Democratic Republic and pertain to the territorial
boundaries prior to 3 October 1990. For detailed
explanatory notes on data pertaining to Germany, see Annex I
- Country Nomenclature.

1 Source: World Automotive Market, Automobile International
(New York).
2 Excluding Democratic People's Republic of Korea.
3 Twelve months ending 30 September of year indicated.
4 Including vehicles operated by police or other governmental
security organizations.
5 Including farm tractors.
6 Including special-purpose vehicles.
7 Including jeeps.
8 Including commercial vehicles.
9 Including vehicles no longer in circulation.
10 Including Faeroe Islands.
11 Excluding jeeps.
12 Twelve months ending 7 July.
13 Including ambulances (Fiji: fire engines also).

14 Excluding tractors and semi-trailer combinations.
15 Including 3-wheeled passengers and goods vehicles.

16 Twelve months ending 20 March of year indicated.
17 Excluding small vehicles.
18 Including cars with a seating capacity of up to 10 persons.
19 Including light commercial vehicles.
20 Excluding tractors.
21 Excluding diplomatic corps vehicles.
22 Twelve months ending 31 July.
23 Twelve months ending 31 March.
24 Including hearses (Norway: registered before 1981).

Source:
Base de données pour les statistiuqes des transports de la
Division de statistique du Secrétariat de l'ONU.

† Toutes les données se rapportant à l'Allemagne avant le 3
octobre 1990 figurent dans deux rubriques séparées basées
sur les territoires respectifs de la République fédérale
d'Allemagne et l'ancienne République démocratique allemande
selon la période indiquée. En l'absence de données pour
l'Allemagne unifiée (à compter du 3 octobre 1990), les
données disponibles sont fournies séparément sous les
rubriques République fédérale d'Allemagne et ancienne
République démocratique allemande et se rapportent aux
limites territoriales antérieures au 3 octobre 1990. Pour
les notes explicatives en détail sur les données concernant
l'Allemagne, voir Annexe I – Nomenclature des pays.

1 Source : "World Automotive Market, Automobile International"
(New York).
2 Non compris la République populaire démocratique de Corée.
3 Douze mois finissant le 30 septembre de l'année indiquée.
4 Y compris véhicules de la police ou d'autres services
gouvernementaux d'ordre public.
5 Y compris tracteurs agricoles.
6 Y compris véhicules à usages spéciaux.
7 Y compris jeeps.
8 Y compris véhicules utilitaires.
9 Y compris véhicules retirés de la circulation.
10 Y compris les Iles Féroé.
11 Non compris jeeps.
12 Douze mois finissant le 7 juillet de l'année indiquée.
13 Y compris ambulances (Fidji : aussi les voitures de
pompiers).
14 Non compris ensembles tracteur-remorque et semi-remorque.
15 Y compris véhicules à trois roues (passagers et
marchandises).
16 Douze mois finissant le 20 mars de l'année indiquée.
17 Non compris véhicules petites.
18 Y compris véhicules comptant jusqu'à 10 places.
19 Y compris véhicules utilitaires légers.
20 Non compris tracteurs.
21 Non compris véhicules des diplomates.
22 Douze mois finissant le 31 juillet.
23 Douze mois finissant le 31 mars.
24 Y compris corbillards (Norvège : enregistrés avant de 1981).

101
Motor vehicles in use
Passenger cars and commercial vehicles: thousand units [*cont.*]

Véhicules automobiles en circulation
Voitures de tourisme et véhicules utilitaires : milliers de véhicules [*suite*]

25 Excluding buses and tractors, but including special lorries.

26 Excluding Madeira and Azores.

27 Passenger cars include mini-buses equipped for transport of nine to fifteen passengers.

28 Commercial vehicles include hearses, ambulances, fire-engines and jeeps specifically registered as commercial vehicles.

29 Including vehicles seating 9 to 17 persons.

30 Source: Road and Motor Vehicles Statistics, International Road Federation (Washington, DC).

31 Source: MVMA Motor Vehicle Facts and Figures, Motor Vehicle Manufacturers Association of the United States, Inc. (Michigan).

32 Limited coverage.

33 Former Democratic Yemen only.

34 Beginning 1985, data are not strictly comparable to data of previous years.

25 Non compris autobus et tracteurs, mais y compris camions spéciaux.

26 Non compris Madère et Azores.

27 Voitures de tourisme comprennent mini-buses ayant une capacité de neuf à quinze passagers.

28 Véhicules utilitaires comprennent corbillards, ambulances, voitures de pompiers et jeeps spécifiquement immatriculés comme véhicules utilitaires.

29 Y compris véhicules comptant 9 à 17 places.

30 Source: Statistiques des routes et des véhicules automobiles, Fédération Internationale Routières (Washington, DC).

31 Source : "MVMA Motor Vehicle Facts and Figures", Motor Vehicle Manufacturers Association of the United States, Inc. (Michigan).

32 Portée limitée.

33 L'ancien Yémen démocratique seulement.

34 A compter de 1985, les données ne sont plus exactement comparables à celles des années précédants.

102
Merchant shipping: fleets
Transports maritimes : flotte marchande
Total, Oil tankers and Ore and bulk carrier fleets: thousand gross registered tons
Total, Pétroliers et Minéraliers et transporteurs de vracs : milliers de tonneaux de jauge brute

Flag Pavillon	1982	1983	1984	1985	1986	1987	1988	1989	1990	1991
A. Total • Totale										
World *Monde*	424 742	422 590	418 682	416 269	404 910	403 498	403 406	410 481	423 627	436 027
Steam **Vapeur**	120 184	108 956	99 114	89 857	76 896	72 228	67 989	65 070	63 974	62 662
Motor **Moteur**	304 558	313 634	319 569	326 412	328 014	331 270	335 417	345 410	359 653	373 364
Africa • Afrique										
Algeria Algérie	1 365	1 369	1 372	1 347	882	893	897	848	906	921
Angola Angola	90	92	97	91	92	92	91	93	93	93
Cameroon Cameroun	38	42	74	76	77	58	57	33	33	34
Côte d'Ivoire Côte d'Ivoire	152	154	143	142	121	119	119	83	82	82
Egypt Egypte	636	663	779	953	1 063	1 074	1 227	1 230	1 257	1 257
Gabon Gabon	78	78	97	98	98	24	25	25	24	25
Ghana Ghana	257	254	186	163	166	142	125	126	126	135
Liberia Libéria	70 718	67 564	62 025	58 180	52 649	51 412	49 734	47 893	54 700	52 427
Libyan Arab Jamah. Jamah. arabe libyenne	912	904	855	854	825	817	830	831	835	840
Madagascar Madagascar	77	77	78	74	74	64	92	70	74	73
Mauritius Maurice	31	31	42	38	152	163	157	130	99	82
Morocco Maroc	394	396	434	461	416	418	437	454	488	483
Mozambique Mozambique	41	43	46	41	43	36	36	38	40	37
Nigeria Nigéria	463	464	442	443	564	594	587	500	496	493
Senegal Sénégal	40	44	44	51	50	46	49	51	52	55
South Africa Afrique du Sud	776	765	712	632	600	533	486	397	352	340
Sudan Soudan	93	94	96	96	96	97	97	97	58	45
Tunisia Tunisie	136	270	277	284	286	285	281	282	278	276
United Rep.Tanzania Rép. Unie de Tanzanie	59	59	59	51	51	32	32	32	32	39
Zaire Zaïre	92	92	85	85	66	56	56	56	56	56
Other Autres	143	180	174	233	212	191	194	187	211	359

102
Merchant shipping: fleets
Total, Oil tankers and Ore and bulk carrier fleets: thousand gross registered tons [cont.]
Transports maritimes : flotte marchande
Total, Pétroliers et Minéraliers et transporteurs de vracs : milliers de tonneaux de jauge brute [suite]

Flag Pavillon	1982	1983	1984	1985	1986	1987	1988	1989	1990	1991
America, North · Amérique du Nord										
Bahamas Bahamas	433	861	3 192	3 907	5 985	9 105	8 963	11 579	13 626	17 541
Barbados Barbade	45	84	8	8	8	8	8	8	8	8
Bermuda Bermudes	474	819	822	981	1 208	1 925	3 774	4 076	4 258	3 037
Canada Canada	3 213	3 385	3 449	3 344	3 160	2 971	2 902	2 825	2 744	2 685
Cayman Islands Iles Caïmanes	311	330	348	414	1 390	706	477	411	415	395
Cuba Cuba	949	961	959	965	959	966	912	900	836	770
Dominican Republic Rép. dominicaine	35	36	37	47	42	44	48	44	36	12
Honduras Honduras	234	222	277	357	555	506	582	691	712	816
Mexico Mexique	1 252	1 475	1 489	1 467	1 520	1 532	1 448	1 388	1 320	1 196
Panama Panama	32 600	34 666	37 244	40 674	41 305	43 255	44 604	47 365	39 298	44 949
United States Etats-Unis	19 111	19 358	19 292	19 518	19 900	20 178	20 832	20 588	21 328	20 291
Other Autres	152	199	212	345	327	309	301	292	401	458
America, South · Amérique du Sud										
Argentina Argentine	2 256	2 470	2 422	2 457	2 117	1 901	1 877	1 833	1 890	1 709
Brazil Brésil	5 678	5 808	5 722	6 057	6 212	6 324	6 123	6 078	6 016	5 883
Chile Chili	495	488	473	454	567	547	604	590	616	619
Colombia Colombie	314	359	374	366	380	424	412	379	372	313
Ecuador Equateur	354	388	412	444	438	421	428	402	385	384
Paraguay Paraguay	32	34	38	43	43	42	39	39	37	35
Peru Pérou	836	781	788	818	754	788	675	638	617	605
Uruguay Uruguay	202	217	190	173	150	144	170	100	104	105
Venezuela Venezuela	911	973	1 003	985	998	999	982	1 087	935	970
Other Autres	59	58	61	60	49	51	43	49	114	182
Asia · Asie										
Bangladesh Bangladesh	411	380	367	358	379	411	432	439	464	456
China Chine	8 057	8 675	9 300	14 896	15 840	12 341	12 920	13 514	13 899	14 299
Cyprus Chypre	2 150	3 450	6 728	8 196	10 617	15 650	18 390	18 134	18 336	20 298

102
Merchant shipping: fleets
Total, Oil tankers and Ore and bulk carrier fleets: thousand gross registered tons [*cont.*]
Transports maritimes : flotte marchande
Total, Pétroliers et Minéraliers et transporteurs de vracs : milliers de tonneaux de jauge brute [*suite*]

Flag Pavillon	1982	1983	1984	1985	1986	1987	1988	1989	1990	1991
Hong Kong Hong-kong	3 499	4 384	5 784	6 858	8 180	8 035	7 329	6 151	6 565	5 876
India Inde	6 214	6 227	6 415	6 605	6 540	6 726	6 161	6 315	6 476	6 517
Indonesia Indonésie	1 847	1 950	1 857	1 936	2 086	2 121	2 126	2 035	2 179	2 337
Iran, Islamic Rep. of Iran, Rép. islamique d'	1 313	1 795	2 106	2 380	2 911	3 977	4 337	4 733	4 738	4 583
Iraq Iraq	1 521	1 561	1 074	1 012	1 016	1 002	953	1 056	1 044	931
Israel Israël	676	690	563	550	557	515	546	505	530	604
Japan Japon	41 594	40 752	40 358	39 940	38 488	35 932	32 074	28 030	27 078	26 407
Korea, Dem. P. R. Corée, R. p. dém. de	279	439	460	513	407	407	406	396	442	511
Korea, Republic of Corée, République de	5 529	6 386	6 771	7 169	7 184	7 214	7 334	7 832	7 783	7 821
Kuwait Koweït	2 014	2 548	2 551	2 350	2 581	2 088	735	1 865	1 855	1 373
Lebanon Liban	368	459	458	505	485	461	405	384	307	274
Malaysia Malaisie	1 195	1 475	1 664	1 773	1 744	1 689	1 608	1 668	1 717	1 755
Maldives Maldives	218	183	137	133	85	100	104	94	78	42
Myanmar Myanmar	88	107	108	116	126	239	273	582	827	1 046
Pakistan Pakistan	580	555	507	451	434	394	366	366	354	358
Philippines Philippines	2 774	2 964	3 441	4 594	6 922	8 681	9 312	9 385	8 515	8 626
Qatar Qatar	234	315	333	353	307	306	309	306	359	485
Saudi Arabia Arabie saoudite	4 302	5 297	3 863	3 137	2 978	2 692	2 269	2 119	1 683	1 321
Singapore Singapour	7 183	7 009	6 512	6 505	6 268	7 098	7 209	7 273	7 928	8 488
Sri Lanka Sri Lanka	125	380	746	635	622	594	410	287	350	333
Syrian Arab Republic Rép. arabe syrienne	43	48	56	58	63	63	64	74	80	109
Thailand Thaïlande	442	567	517	586	533	511	515	539	615	725
Turkey Turquie	2 128	2 524	3 125	3 684	3 424	3 336	3 281	3 240	3 719	4 107
United Arab Emirates Emirats arabes unis	231	301	766	869	654	732	825	839	750	889
Viet Nam Viet Nam	262	269	279	299	339	360	338	358	470	574
Other Autres	2 296	2 979	4 087	4 460	4 334	3 987	3 950	4 003	4 877	5 206

102
Merchant shipping: fleets
Total, Oil tankers and Ore and bulk carrier fleets: thousand gross registered tons [*cont.*]
Transports maritimes : flotte marchande
Total, Pétroliers et Minéraliers et transporteurs de vracs : milliers de tonneaux de jauge brute [*suite*]

Flag Pavillon	1982	1983	1984	1985	1986	1987	1988	1989	1990	1991
	Europe · Europe									
Albania Albanie	56	56	56	56	56	56	56	56	56	59
Austria Autriche	101	114	129	134	125	194	201	204	139	139
Belgium Belgique	2 271	2 274	2 407	2 400	2 420	2 268	2 118	2 044	1 954	314
Bulgaria Bulgarie	1 248	1 293	1 283	1 322	1 385	1 551	1 392	1 375	1 360	1 367
Czechoslovakia Tchécoslovaquie	185	184	184	184	198	157	158	191	326	361
Denmark Danemark	5 214	5 115	5 211	4 942	4 651	4 754	4 322	4 842	5 064	5 757
Faeroe Islands Iles Féroé	68	78	91	103	115	119	130	121	124	114
Finland Finlande	2 377	2 358	2 168	1 974	1 470	1 122	838	944	1 069	1 053
France France	10 771	9 868	8 945	8 237	5 936	5 371	4 506	4 413	3 832	3 988
Germany† · Allemagne† F. R. Germany R. f. Allemagne	7 707	6 897	6 242	6 177	5 565	4 318	3 917	3 967	4 301	5 971
former German D. R. anc. R. d. allemande	1 439	1 421	1 422	1 434	1 519	1 494	1 443	1 500	1 437	...
Greece Grèce	40 035	37 478	35 059	31 032	28 391	23 560	21 979	21 324	20 522	22 753
Hungary Hongrie	82	82	80	77	86	77	76	76	98	104
Iceland Islande	181	179	179	180	176	174	175	183	177	168
Ireland Irlande	239	223	221	194	149	154	173	167	181	195
Italy Italie	10 375	10 015	9 158	8 343	7 897	7 817	7 794	7 602	7 991	8 122
Malta Malte	426	907	1 366	1 856	1 710	1 726	2 686	3 329	4 519	6 916
Netherlands Pays-Bas	5 393	4 940	4 586	4 301	3 898	3 908	3 726	3 655	3 785	3 872
Norway Norvège	21 862	19 230	17 663	15 339	8 387	6 359	9 350	15 597	23 429	23 586
Poland Pologne	3 651	3 686	3 267	3 315	3 437	3 470	3 489	3 416	3 369	3 348
Portugal Portugal	1 402	1 358	1 571	1 437	947	1 048	989	726	854	891
Romania Roumanie	2 203	2 391	2 667	3 024	3 234	3 264	3 561	3 783	4 005	3 828
Spain Espagne	8 131	7 505	7 005	6 256	5 422	4 949	4 415	3 962	3 807	3 617
Sweden Suède	3 788	3 433	3 520	3 162	2 517	2 270	2 116	2 167	2 775	3 174
Switzerland Suisse	315	321	319	342	346	355	259	220	287	286

102
Merchant shipping: fleets
Total, Oil tankers and Ore and bulk carrier fleets: thousand gross registered tons [*cont.*]
Transports maritimes : flotte marchande
Total, Pétroliers et Minéraliers et transporteurs de vracs : milliers de tonneaux de jauge brute [*suite*]

Flag Pavillon	1982	1983	1984	1985	1986	1987	1988	1989	1990	1991
United Kingdom Royaume-Uni	22 505	19 121	15 874	14 344	11 567	8 505	8 260	7 646	6 716	6 611
Yugoslavia Yougoslavie	2 532	2 547	2 682	2 699	2 873	3 165	3 476	3 681	3 816	3 293
Other Autres	19	247	253	587	578	529	498	457	531	689
Oceania · Océanie										
Australia Australie	1 875	2 022	2 173	2 088	2 368	2 405	2 366	2 494	2 512	1 709
Nauru Nauru	62	70	67	67	67	66	60	41	32	15
New Zealand Nouvelle-Zélande	250	251	285	296	314	334	337	257	260	275
Other Autres	81	103	195	249	247	228	207	189	259	317
former USSR · ancienne URSS										
former USSR ancienne URSS	23 789	24 549	24 492	24 745	24 961	25 232	25 784	25 854	26 737	26 405

B. Oil tankers · Pétroliers

World *Monde*	166 828	157 279	144 380	138 448	128 426	127 600	127 843	129 578	134 836	138 897
Africa · Afrique										
Algeria Algérie	595	595	594	594	119	134	119	40	39	39
Egypt Egypte	106	102	103	97	99	95	255	244	263	262
Gabon Gabon	74	74	74	74	74	0	0	0	0	1
Liberia Libéria	41 223	38 605	33 420	31 585	28 675	28 249	27 961	26 667	28 763	26 700
Libyan Arab Jamah. Jamah. arabe libyenne	797	797	745	745	708	708	708	581	707	581
Morocco Maroc	113	62	62	62	10	10	10	10	10	10
Nigeria Nigéria	147	148	152	154	223	227	227	225	225	225
South Africa Afrique du Sud	38	38	38	38	39	21	21	20	1	1
Other Autres	55	161	160	157	161	150	148	161	207	255
America, North · Amérique du Nord										
Bahamas Bahamas	187	424	2 588	3 010	4 201	5 357	4 556	6 110	6 780	8 738
Bermuda Bermudes	173	173	220	220	301	895	2 836	3 273	3 285	1 987
Canada Canada	304	310	284	281	279	272	259	260	242	233
Cayman Islands Iles Caïmanes	59	60	47	46	737	218	75	43	79	72
Cuba Cuba	68	68	68	68	68	68	68	80	80	78

102
Merchant shipping: fleets
Total, Oil tankers and Ore and bulk carrier fleets: thousand gross registered tons [*cont.*]
Transports maritimes : flotte marchande
Total, Pétroliers et Minéraliers et transporteurs de vracs : milliers de tonneaux de jauge brute [*suite*]

Flag Pavillon	1982	1983	1984	1985	1986	1987	1988	1989	1990	1991
Honduras Honduras	33	38	48	51	67	58	51	87	112	149
Mexico Mexique	565	744	658	540	606	587	522	533	507	507
Panama Panama	8 723	8 433	7 920	8 414	9 192	9 966	10 659	11 418	10 080	13 976
United States Etats-Unis	8 220	8 156	7 504	7 472	7 296	7 428	7 949	7 956	8 532	8 069
Other Autres	13	13	14	2	4	5	11	9	31	47
America, South · Amérique du Sud										
Argentina Argentine	748	893	860	842	654	585	586	543	568	542
Brazil Brésil	1 773	1 759	1 750	1 834	1 938	1 943	1 849	1 838	1 897	1 947
Chile Chili	36	29	13	15	15	15	19	28	26	26
Colombia Colombie	31	31	31	32	36	12	14	14	14	11
Ecuador Equateur	164	163	160	160	157	159	159	120	120	116
Peru Pérou	147	134	166	183	147	197	197	197	190	177
Uruguay Uruguay	95	95	95	96	76	77	118	47	47	47
Venezuela Venezuela	458	517	551	498	470	470	463	463	463	478
Other Autres	5	6	4	7	5	7	9	11	50	43
Asia · Asie										
Bangladesh Bangladesh	48	50	51	38	40	50	60	49	50	51
China Chine	1 179	1 226	1 329	1 476	1 701	1 751	1 826	1 790	1 810	1 840
Cyprus Chypre	560	1 268	3 162	3 327	4 481	4 984	5 618	5 639	5 390	5 996
Hong Kong Hong-kong	540	522	528	440	884	1 095	945	827	1 000	756
India Inde	1 257	1 286	1 305	1 717	1 814	1 828	1 772	1 718	1 734	1 805
Indonesia Indonésie	318	367	411	491	617	644	660	580	582	595
Iran, Islamic Rep. of Iran, Rép. islamique d'	631	916	918	918	1 242	2 347	2 717	3 102	3 101	2 945
Iraq Iraq	1 141	1 141	797	747	777	770	729	831	829	725
Japan Japon	17 296	16 317	14 974	14 089	12 365	10 798	9 628	7 879	7 584	7 204
Korea, Dem. P. R. Corée, R. p. dém. de	59	171	171	171	59	59	13	13	13	13
Korea, Republic of Corée, République de	1 104	961	1 013	1 002	977	963	951	808	593	536

102
Merchant shipping: fleets
Total, Oil tankers and Ore and bulk carrier fleets: thousand gross registered tons [cont.]
Transports maritimes : flotte marchande
Total, Pétroliers et Minéraliers et transporteurs de vracs : milliers de tonneaux de jauge brute [suite]

Flag Pavillon	1982	1983	1984	1985	1986	1987	1988	1989	1990	1991
Kuwait Koweït	1 188	1 684	1 395	1 290	1 629	1 261	133	1 092	1 101	1 044
Malaysia Malaisie	38	154	216	224	238	241	180	163	179	249
Pakistan Pakistan	43	43	43	44	43	43	43	43	43	43
Philippines Philippines	577	565	555	562	656	755	480	403	372	376
Qatar Qatar	73	77	77	112	112	112	112	108	160	160
Saudi Arabia Arabie saoudite	2 893	3 234	1 933	1 578	1 604	1 636	1 300	1 214	928	561
Singapore Singapour	2 583	2 429	2 036	2 049	1 653	2 305	2 443	2 553	3 165	3 543
Thailand Thaïlande	140	141	137	147	62	69	77	73	85	100
Turkey Turquie	709	842	1 158	1 582	1 037	846	828	793	777	772
United Arab Emirates Emirats arabes unis	73	87	513	629	372	396	450	441	332	334
Viet Nam Viet Nam	35	35	37	38	40	41	15	18	18	91
Other Autres	519	598	844	791	810	802	794	801	792	933
Europe · Europe										
Belgium Belgique	274	266	141	232	266	224	252	273	272	12
Bulgaria Bulgarie	338	338	312	312	317	441	292	285	288	293
Denmark Danemark	2 557	2 423	2 311	2 199	2 044	2 166	2 053	2 041	2 024	2 061
Finland Finlande	1 135	1 227	948	926	597	457	211	155	245	209
France France	6 557	5 443	4 785	4 346	2 603	2 465	1 953	1 944	1 717	1 674
Germany† · Allemagne† F. R. Germany R. f. Allemagne	2 652	2 025	1 490	1 394	750	317	265	283	228	249
former German D. R. anc. R. d. allemande	56	56	31	36	36	36	36	36	6	...
Greece Grèce	13 175	12 056	10 768	9 366	10 259	9 247	8 492	8 229	7 856	9 095
Italy Italie	4 128	3 872	3 475	3 601	2 561	2 631	2 670	2 461	2 560	2 685
Netherlands Pays-Bas	2 181	1 600	950	745	931	774	615	575	590	618
Norway Norvège	11 698	9 447	8 665	7 263	3 304	2 785	4 397	7 074	10 794	10 904
Poland Pologne	547	548	209	318	318	317	235	154	154	136

102
Merchant shipping: fleets
Total, Oil tankers and Ore and bulk carrier fleets: thousand gross registered tons [*cont.*]
Transports maritimes : flotte marchande
Total, Pétroliers et Minéraliers et transporteurs de vracs : milliers de tonneaux de jauge brute [*suite*]

Flag Pavillon	1982	1983	1984	1985	1986	1987	1988	1989	1990	1991
Portugal Portugal	769	781	1 023	860	533	533	486	323	393	460
Romania Roumanie	384	295	295	384	384	384	523	596	645	679
Spain Espagne	4 919	4 234	3 620	2 906	2 372	2 104	1 616	1 486	1 472	1 488
Sweden Suède	1 616	1 369	1 243	894	504	370	191	211	534	862
United Kingdom Royaume-Uni	10 371	8 285	6 428	5 937	4 394	2 822	2 835	2 604	2 367	2 316
Yugoslavia Yougoslavie	231	231	231	217	308	317	312	312	306	264
Other Autres	29	151	207	511	524	504	489	461	577	704
Oceania · Océanie										
Australia Australie	392	473	572	588	662	702	702	678	677	708
New Zealand Nouvelle-Zélande	52	52	73	73	73	73	80	80	80	80
Other Autres	7	7	7	7	8	8	7	10	68	85
former USSR · ancienne URSS										
former USSR ancienne URSS	4 805	4 858	4 662	4 591	4 087	4 207	4 368	4 128	4 167	4 068

C. Ore and bulk carrier fleets · Minéraliers et transporteurs de vracs

World Monde	1982	1983	1984	1985	1986	1987	1988	1989	1990	1991
World Monde	119 298	124 397	128 334	133 983	132 908	131 028	129 635	129 482	133 190	135 884
Africa · Afrique										
Algeria Algérie	81	81	81	57	57	57	77	95	153	172
Liberia Libéria	22 922	22 766	21 651	20 773	18 028	16 788	15 647	14 374	16 099	15 629
Morocco Maroc	59	80	103	125	125	92	92	92	92	92
South Africa Afrique du Sud	185	206	161	129	125	88	88	...	...	...
Other Autres	43	60	175	327	309	287	316	340	410	395
America, North · Amérique du Nord										
Bahamas Bahamas	51	163	233	405	902	2 105	2 368	2 822	3 698	4 872
Bermuda Bermudes	39	444	326	258	319	432	294	157	161	199
Canada Canada	1 889	1 997	2 056	2 000	1 851	1 654	1 590	1 550	1 462	1 414
Cayman Islands Iles Caïmanes	46	41	51	117	283	184	103	111	89	54
Cuba Cuba	62	62	62	62	62	62	62	62	62	62

102
Merchant shipping: fleets
Total, Oil tankers and Ore and bulk carrier fleets: thousand gross registered tons [*cont.*]
Transports maritimes : flotte marchande
Total, Pétroliers et Minéraliers et transporteurs de vracs : milliers de tonneaux de jauge brute [*suite*]

Flag Pavillon	1982	1983	1984	1985	1986	1987	1988	1989	1990	1991
Mexico Mexique	129	129	213	310	311	328	271	226	178	48
Panama Panama	10 809	12 315	14 558	17 431	17 400	17 733	17 924	17 513	13 293	13 669
United States Etats-Unis	1 914	2 047	2 130	2 075	2 020	1 989	1 935	1 884	2 140	2 168
Other Autres	16	68	88	195	208	226	219	251	273	291
America, South · Amérique du Sud										
Argentina Argentine	373	447	451	505	515	457	465	459	502	365
Brazil Brésil	2 433	2 511	2 405	2 748	2 802	2 894	2 859	2 943	2 971	2 855
Chile Chili	213	241	224	240	314	278	320	295	296	296
Peru Pérou	231	184	185	192	190	160	134	129	129	129
Venezuela Venezuela	36	36	54	75	86	86	109	157	147	147
Other Autres	42	61	72	58	61	43	79	71	91	83
Asia · Asie										
Bangladesh Bangladesh	56	0	0	8	8	0	0	...	...	...
China Chine	2 293	2 589	2 961	3 531	3 871	4 143	4 391	4 726	4 907	5 206
Cyprus Chypre	297	764	1 711	2 741	3 739	6 788	8 789	8 821	9 226	10 234
Hong Kong Hong-kong	2 346	3 238	4 367	5 428	6 090	5 748	5 355	4 363	4 396	3 925
India Inde	2 792	2 855	3 053	2 954	2 943	3 152	2 729	3 025	3 182	3 134
Indonesia Indonésie	219	206	149	128	129	129	129	145	138	149
Iran, Islamic Rep. of Iran, Rép. islamique d'	124	266	527	777	1 109	1 080	1 068	1 059	1 059	1 059
Israel Israël	232	193	74	64	74	42	42	32	32	22
Japan Japon	13 714	13 534	13 227	13 900	13 895	12 611	10 847	9 234	8 788	8 652
Korea, Republic of Corée, République de	2 675	3 577	3 856	4 258	4 269	4 140	4 227	4 492	4 708	4 463
Malaysia Malaisie	396	426	428	468	456	378	378	347	347	319
Maldives Maldives	34	34	52	54	45	43	43	43	32	21
Philippines Philippines	961	1 191	1 595	2 756	4 805	6 388	6 906	6 987	6 344	6 262
Saudi Arabia Arabie saoudite	312	491	455	387	373	193	170	170	26	...

102
Merchant shipping: fleets
Total, Oil tankers and Ore and bulk carrier fleets: thousand gross registered tons [*cont.*]
Transports maritimes : flotte marchande
Total, Pétroliers et Minéraliers et transporteurs de vracs : milliers de tonneaux de jauge brute [*suite*]

Flag Pavillon	1982	1983	1984	1985	1986	1987	1988	1989	1990	1991
Singapore Singapour	1 866	1 937	2 044	2 258	2 478	2 465	2 287	2 083	2 190	2 132
Turkey Turquie	758	895	1 069	1 129	1 353	1 419	1 360	1 413	1 932	2 331
Other Autres	869	1 738	2 629	2 611	2 597	2 583	2 498	2 604	2 194	2 812
Europe · Europe										
Austria Autriche	63	59	63	63	63	128	128	136	71	71
Belgium Belgique	1 228	1 320	1 424	1 365	1 416	1 350	1 164	1 091	979	...
Bulgaria Bulgarie	477	508	508	525	572	620	612	612	611	601
Czechoslovakia Tchécoslovaquie	103	103	103	103	116	75	75	96	241	276
Denmark Danemark	481	485	446	423	290	251	163	326	353	525
Finland Finlande	423	341	373	280	121	70	70	78	120	89
France France	1 705	1 815	1 625	1 400	957	858	698	641	357	406
Germany† · Allemagne† F. R. Germany R. f. Allemagne	1 438	1 138	842	768	559	402	344	318	397	695
former German D. R. anc. R. d. allemande	296	296	321	323	353	362	338	324	324	...
Greece Grèce	16 841	16 785	16 439	15 324	13 202	10 557	10 060	9 987	9 783	10 802
Ireland Irlande	101	90	91	57	0	0	0	0	9	9
Italy Italie	3 886	3 775	3 389	3 041	3 059	2 792	2 561	2 351	2 346	2 228
Malta Malte	141	336	623	806	952	882	1 019	1 172	1 774	2 772
Netherlands Pays-Bas	639	783	782	699	524	318	295	328	328	359
Norway Norvège	6 241	5 391	4 965	3 930	2 479	1 085	1 684	4 170	7 283	7 091
Poland Pologne	1 262	1 274	1 244	1 337	1 499	1 551	1 604	1 610	1 603	1 636
Portugal Portugal	133	133	133	158	231	286	267	178	223	197
Romania Roumanie	872	1 024	1 250	1 439	1 590	1 590	1 667	1 758	1 891	1 698
Spain Espagne	1 206	1 199	1 278	1 321	1 175	1 060	1 105	955	850	732
Sweden Suède	418	359	442	394	271	134	127	176	383	415
Switzerland Suisse	181	181	202	247	264	284	210	183	252	252
United Kingdom Royaume-Uni	5 557	4 617	3 374	3 014	2 150	1 492	1 286	1 249	749	743

102
Merchant shipping: fleets
Total, Oil tankers and Ore and bulk carrier fleets: thousand gross registered tons [cont.]
Transports maritimes : flotte marchande
Total, Pétroliers et Minéraliers et transporteurs de vracs : milliers de tonneaux de jauge brute [suite]

Flag Pavillon	1982	1983	1984	1985	1986	1987	1988	1989	1990	1991
Yugoslavia Yougoslavie	908	923	1 045	1 115	1 182	1 369	2 212	1 915	2 018	1 707
Other Autres	0	13	59	282	197	205	201	188	247	302
Oceania · Océanie										
Australia Australie	980	1 010	984	932	1 185	1 173	1 178	1 106	1 112	1 004
Nauru Nauru	37	37	37	37	37	37	37	17	17	...
Other Autres	0	19	57	118	120	98	84	0	124	207
former USSR · ancienne URSS										
former USSR ancienne URSS	2 121	2 429	2 744	2 974	3 433	3 535	3 805	4 115	4 183	3 902

Source:
Lloyd's Register of Shipping (London).

† All data shown which pertain to Germany prior to 3 October
1990 are indicated separately for the Federal Republic of
Germany and the former German Democratic Republic based on
their respective territories at the time indicated. Where
data for united Germany (3 October 1990 and thereafter) are
not available, available data are shown separately under the
designations Federal Republic of Germany and former German
Democratic Republic and pertain to the territorial
boundaries prior to 3 October 1990. For detailed
explanatory notes on data pertaining to Germany, see Annex I
- Country Nomenclature.

Source:
"Lloyd's Register of Shipping" (Londres).

† Toutes les données se rapportant à l'Allemagne avant le 3
octobre 1990 figurent dans deux rubriques séparées basées
sur les territoires respectifs de la République fédérale
d'Allemagne et l'ancienne République démocratique allemande
selon la période indiquée. En l'absence de données pour
l'Allemagne unifiée (à compter du 3 octobre 1990), les
données disponibles sont fournies séparément sous les
rubriques République fédérale d'Allemagne et ancienne
République démocratique allemande et se rapportent aux
limites territoriales antérieures au 3 octobre 1990. Pour
les notes explicatives en détail sur les données concernant
l'Allemagne, voir Annexe I – Nomenclature des pays.

103
International maritime transport
Transports maritimes internationaux
Vessels entered and cleared: thousand net registered tons
Navires entrés et sortis : milliers de tonneaux de jauge nette

Country or area Pays ou zone	1981	1982	1983	1984	1985	1986	1987	1988	1989	1990
Algeria Algérie										
Vessels entered Navires entrés	38 398	45 233	51 529	53 154	51 012	49 408	48 657	49 213	50 721	44 659
Vessels cleared Navires sortis	38 202	45 080	51 837	53 133	50 962	49 161	48 516	48 992	50 646	44 611
American Samoa Samoa américaines										
Vessels entered[1] Navires entrés[1]	836	925	913	944	851	...	...	...	...	...
Argentina Argentine										
Vessels entered[2] Navires entrés[2]	32 753	28 504	36 935	36 336	37 850	34 044	29 294	32 535	31 253	38 283
Bangladesh Bangladesh										
Vessels entered[3] Navires entrés[3]	4 321	4 846	5 320	4 803	5 456	5 372	5 646	5 184	5 655	5 512
Vessels cleared[3] Navires sortis[3]	4 978	5 299	5 433	4 437	4 795	3 833	3 254	3 323	3 796	3 208
Belgium Belgique										
Vessels entered Navires entrés	79 155	83 706	82 191	86 817	88 098	87 458	90 822	96 548	190 757	204 857
Vessels cleared Navires sortis	68 386	69 491	68 173	74 965	77 086	72 235	75 037	76 919	155 017	162 236
Benin Bénin										
Vessels entered Navires entrés	1 903	2 268	2 377	2 411	...	...	...	...	...	...
Brazil Brésil										
Vessels entered Navires entrés	...	...	...	...	143 333	134 303	145 425	159 468	...	...
Vessels cleared Navires sortis	...	...	...	...	52 206	63 499	65 519	61 992	...	...
Brunei Darussalam Brunéi Darussalam										
Vessels entered Navires entrés	11 824	9 533	13 837	23 929	24 542	...	...	...	...	...
Vessels cleared Navires sortis	12 254	10 841	12 141	13 548	10 524	...	...	...	...	...
Cameroon Cameroun										
Vessels entered[2][4] Navires entrés[2][4]	6 152	6 369	6 061	5 865	6 362	6 427	6 059	5 562	5 333	5 521
Canada Canada										
Vessels entered[5] Navires entrés[5]	68 667	56 444	63 007	67 728	67 643	62 523	61 975	67 688	68 940	64 163
Vessels cleared[5] Navires sortis[5]	115 202	104 611	115 337	118 980	116 796	109 616	114 546	120 442	116 769	114 272
Cape Verde Cap-Vert										
Vessels entered Navires entrés	4 785	3 660	4 560	4 154	3 539	4 054	4 493	...	...	...
Christmas Is.(Aust) Ile Christmas (Aust)										
Vessels entered[1][2] Navires entrés[1][2]	...	...	...	560	...	...	...	...	...	...
Colombia Colombie										
Vessels entered[2] Navires entrés[2]	12 663	15 883	15 112	14 990	16 092	17 497	19 691	18 781	20 410	22 713
Vessels cleared Navires sortis	...	...	15 228	14 791	15 799	17 218	19 331	18 826	20 437	22 448
Congo Congo										
Vessels entered Navires entrés	5 936	6 671	6 686	7 592	7 564	7 504	7 181	7 793	7 386	7 148

103
International maritime transport
Vessels entered and cleared: thousand net registered tons [cont.]
Transports maritimes internationaux
Navires entrés et sortis : milliers de tonneaux de jauge nette [suite]

Country or area Pays ou zone	1981	1982	1983	1984	1985	1986	1987	1988	1989	1990
Costa Rica Costa Rica										
Vessels entered										
Navires entrés	1 421	1 366	1 326	...	...	...	...	...	...	...
Cuba Cuba										
Vessels entered										
Navires entrés	11 000	10 700	11 300	11 000	12 400	12 900	15 400	12 100	...	...
Vessels cleared										
Navires sortis	11 000	10 400	10 900	10 900	12 100	12 700	15 300	12 000	...	...
Cyprus Chypre										
Vessels entered										
Navires entrés	6 489	8 623	10 634	12 296	11 436	12 429	12 839	13 231	14 793	14 964
Djibouti Djibouti										
Vessels entered[2]										
Navires entrés[2]	5 547	6 376	5 258	4 273	4 197	...	...	...	...	...
Dominican Republic Rép. dominicaine										
Vessels entered[2]										
Navires entrés[2]	8 405	7 426	7 878	7 112	...	...	...	...	...	...
Vessels cleared										
Navires sortis	8 678	7 455	7 942	7 328	...	...	...	...	...	...
Ecuador Equateur										
Vessels entered										
Navires entrés	12 143	12 391	11 753	11 885	12 653	12 694	12 524	16 303	17 008	1 958
Egypt Egypte										
Vessels entered										
Navires entrés	22 529	29 671	25 191	27 037	26 025	30 199	23 807	19 683	23 868	23 813
Vessels cleared										
Navires sortis	10 759	10 798	10 340	12 382	10 737	9 724	10 560	9 514	11 309	10 591
El Salvador El Salvador										
Vessels entered										
Navires entrés	2 774	2 304	2 381	2 500	2 341	...	...	...	...	...
Vessels cleared										
Navires sortis	...	1 306	1 466	1 397	909	...	...	...	...	...
Fiji Fidji										
Vessels entered										
Navires entrés	2 613	3 522	2 573	3 552	699	1 590	2 025	1 739	2 603	3 012
Finland Finlande										
Vessels entered										
Navires entrés	49 452	49 750	51 108	50 722	55 740	65 098	68 203	70 420	85 265	102 500
Vessels cleared										
Navires sortis	49 844	49 479	51 214	50 717	55 713	65 509	68 807	70 877	84 438	102 995
France France										
Vessels entered										
Navires entrés	...	...	1 151 185	1 474 968	1 475 319	1 597 747	1 619 267	1 554 890	1 564 075	1 681 491
Gambia Gambie										
Vessels entered[1]										
Navires entrés[1]	649	670	806	749	659	954	892	982	969	...
Germany† · Allemagne†										
F. R. Germany R. f. Allemagne										
Vessels entered										
Navires entrés	140 153	139 486	135 040	136 819	138 149	147 513	155 938	162 784	169 542	171 906
Vessels cleared										
Navires sortis	111 008	119 300	115 983	119 243	119 678	123 942	135 987	141 305	150 792	150 514
former German D. R. anc. R. d. allemande										
Vessels entered										
Navires entrés	5 814	6 501	8 718	9 643	10 343	11 023	10 324	10 469	10 820	8 713
Vessels cleared										
Navires sortis	13 688	13 724	14 407	15 052	14 778	14 486	14 480	15 076	14 303	8 327

103
International maritime transport
Vessels entered and cleared: thousand net registered tons [cont.]
Transports maritimes internationaux
Navires entrés et sortis : milliers de tonneaux de jauge nette [suite]

Country or area Pays ou zone	1981	1982	1983	1984	1985	1986	1987	1988	1989	1990
Gibraltar Gibraltar										
Vessels entered Navires entrés	195	174	184	152	176	251	403	304	231	209
Greece Grèce										
Vessels entered Navires entrés	23 729	23 644	23 071	25 176	27 800	28 700	31 365	31 638	31 608	35 152
Vessels cleared Navires sortis	21 552	22 218	19 672	20 524	26 276	24 777	23 917	24 448	25 365	22 262
Guadeloupe Guadeloupe										
Vessels entered[2] Navires entrés[2]	2 806	2 379	2 966	3 558	...	...	...	...	...	...
Guatemala Guatemala										
Vessels entered Navires entrés	1 466	1 259	1 184	1 255	1 345	1 544	1 826	1 877	2 018	...
Vessels cleared Navires sortis	1 920	2 119	1 871	1 302	1 424	1 562	1 557	1 579	1 705	...
Haiti Haïti										
Vessels entered Navires entrés	735	767	714	714	728	725	707	734	709	1 975
Hong Kong Hong-kong										
Vessels entered[6] Navires entrés[6]	70 339	74 787	80 927	82 457	91 608	98 137	106 396	115 737	126 307	131 801
Vessels cleared[6] Navires sortis[6]	70 239	74 551	80 777	82 533	91 631	98 078	106 451	115 641	126 562	131 636
India Inde										
Vessels entered[7][8] Navires entrés[7][8]	25 913	27 188	24 786	31 036	35 566	33 563	30 171	27 894	28 593	23 703
Vessels cleared[7][8] Navires sortis[7][8]	24 709	26 614	23 002	26 766	26 924	28 386	26 253	31 031	34 529	27 594
Indonesia Indonésie										
Vessels entered Navires entrés	...	...	...	66 801	54 814	63 589	65 245	...	...	...
Vessels cleared Navires sortis	...	...	...	12 427	15 886	20 302	21 449	...	...	...
Ireland Irlande										
Vessels entered[2] Navires entrés[2]	18 373	18 880	19 894	20 804	20 455	20 915	22 557	25 630	25 940	31 769
Israel Israël										
Vessels entered[2][9] Navires entrés[2][9]	10 009	10 915	12 746	13 747	13 230	14 408	15 857	15 962	...	...
Italy Italie										
Vessels entered Navires entrés	152 684	149 360	143 864	149 118	152 426	155 299	161 403	160 251	161 145	173 360
Vessels cleared Navires sortis	59 826	58 710	57 657	61 807	63 165	65 509	65 863	67 927	67 170	74 000
Japan Japon										
Vessels entered[2] Navires entrés[2]	337 657	334 267	332 147	347 907	352 589	345 284	347 605	361 530	378 291	385 110
Jordan Jordanie										
Vessels entered Navires entrés	1 522	2 311	2 012	1 900	6 370	7 153	8 743	9 193	8 694	6 164
Vessels cleared Navires sortis	222	288	442	429	8 177	9 699	11 271	10 952	9 985	8 871
Kenya Kenya										
Vessels entered[2][4] Navires entrés[2][4]	6 513	5 959	5 792	5 756	...	...	...	...	...	...

103
International maritime transport
Vessels entered and cleared: thousand net registered tons [cont.]
Transports maritimes internationaux
Navires entrés et sortis : milliers de tonneaux de jauge nette [suite]

Country or area Pays ou zone	1981	1982	1983	1984	1985	1986	1987	1988	1989	1990
Korea, Republic of Corée, République de										
Vessels entered										
Navires entrés	120 408	138 352	156 218	166 217	179 760	201 857	238 186	252 390	262 791	279 004
Vessels cleared										
Navires sortis	121 247	137 601	152 158	163 723	179 286	204 090	240 505	254 627	264 114	278 423
Macau Macao										
Vessels entered										
Navires entrés	9 774	11 011	9 142	9 598	11 482	...	...	...	...	...
Madagascar Madagascar										
Vessels entered[2]										
Navires entrés[2]	1 385	1 484	1 409	1 202	1 302	1 383	1 400	1 494	1 654	2 161
Vessels cleared[2]										
Navires sortis[2]	1 366	1 490	1 406	1 173	1 301	1 355	1 430	1 500	1 639	2 160
Malaysia Malaisie										
Vessels entered[10]										
Navires entrés[10]	57 574	67 509	70 809	69 803	69 032	55 197[11]	59 900[11]	79 521	86 033	95 724
Vessels cleared[10]										
Navires sortis[10]	57 715	63 566	70 470	70 596	69 012	54 925[11]	59 709[11]	79 735	85 301	94 713
Malta Malte										
Vessels entered										
Navires entrés	2 285	2 292	2 106	1 885	2 016	2 370	2 327	2 726	2 681	4 068
Vessels cleared										
Navires sortis	753	727	966	747	763	888	880	942	897	1 670
Mauritius Maurice										
Vessels entered[2]										
Navires entrés[2]	2 559	2 906	2 464	2 515	2 715	2 988	3 733	4 064	4 195	4 364
Vessels cleared										
Navires sortis	2 609	2 879	2 780	2 538	2 696	3 051	3 464	3 677	3 978	4 357
Mexico Mexique										
Vessels entered										
Navires entrés	15 656	12 820	11 147	11 008	10 619	9 009	10 880	14 262	17 318	19 274
Vessels cleared										
Navires sortis	62 556	73 600	76 503	72 387	84 139	82 796	88 950	90 432	84 575	89 111
Myanmar Myanmar										
Vessels entered										
Navires entrés	1 192	1 255	895	875	821	866	636	597	248	610
Vessels cleared										
Navires sortis	1 622	1 495	1 305	1 109	960	1 243	1 311	750	683	577
Netherlands Pays-Bas										
Vessels entered										
Navires entrés	215 061	215 198	205 127	210 353	213 668	215 340	210 547	220 323	232 153	...
Vessels cleared										
Navires sortis	211 328	209 774	201 668	205 105	209 119	212 823	180 611	190 524	200 373	...
New Caledonia Nouvelle-Calédonie										
Vessels entered										
Navires entrés	2 383	2 416	2 337	2 908	2 526	...	...	...	...	...
Vessels cleared										
Navires sortis	2 370	2 386	2 324	2 885	2 524	...	...	...	...	...
New Zealand Nouvelle-Zélande										
Vessels entered										
Navires entrés	12 738	13 031	13 040	14 001	14 607	13 388	14 113	27 844	30 890	32 592
Vessels cleared										
Navires sortis	12 867	13 010	13 053	13 934	14 613	13 365	14 107	27 247	29 753	31 967
Nicaragua Nicaragua										
Vessels entered										
Navires entrés	...	1 053	1 353	1 300	1 326	...	...	...	...	...
Vessels cleared										
Navires sortis	...	365	454	430	442	...	...	...	...	...

103
International maritime transport
Vessels entered and cleared: thousand net registered tons [*cont.*]
Transports maritimes internationaux
Navires entrés et sortis : milliers de tonneaux de jauge nette [*suite*]

Country or area Pays ou zone	1981	1982	1983	1984	1985	1986	1987	1988	1989	1990
Nigeria Nigéria										
Vessels entered Navires entrés	59 475	53 213	49 933	48 299	...	...	...	...	...	...
Vessels cleared Navires sortis	...	...	51 190	48 830	...	...	...	...	...	...
Norway Norvège										
Vessels entered Navires entrés	...	...	42 289	...	48 734	...	...	...	...	...
Papua New Guinea Papouasie-Nvl-Guinée										
Vessels entered[2] Navires entrés[2]	2 711	...	...	...	...	...	...	...	...	...
Pakistan Pakistan										
Vessels entered[3] Navires entrés[3]	11 795	12 655	12 478	13 194	14 493	14 736	15 168	15 190	15 331	15 057
Vessels cleared[3] Navires sortis[3]	5 476	6 341	6 376	6 046	7 155	6 899	7 141	6 728	7 265	6 743
Panama Panama										
Vessels entered Navires entrés	1 108	1 043	1 041	1 049	1 204	1 322	1 362	1 019	1 141	1 337
Vessels cleared Navires sortis	1 017	1 042	1 169	1 111	1 247	1 134	1 228	1 159	1 210	1 377
Peru Pérou										
Vessels entered Navires entrés	...	...	...	3 303	3 038	4 241	5 547	5 960	4 075	...
Vessels cleared Navires sortis	...	...	...	8 692	10 386	11 501	10 478	8 854	9 722	...
Philippines Philippines										
Vessels entered Navires entrés	20 963	21 641	21 674	18 577	18 395	19 380	21 787	24 130	25 729	...
Vessels cleared Navires sortis	20 695	21 152	21 946	19 949	18 927	20 282	20 593	20 830	20 441	...
Poland Pologne										
Vessels entered Navires entrés	15 008	10 680	10 777	11 550	11 279	11 467	12 989	14 394	15 129	12 959
Vessels cleared Navires sortis	12 732	13 589	17 417	21 692	18 320	17 939	18 298	19 844	19 477	21 131
Portugal Portugal										
Vessels entered[2 12] Navires entrés[2 12]	27 041	21 380	18 987	24 324	25 860	30 333	29 202	32 404	32 836	35 757
Réunion Réunion										
Vessels entered[2] Navires entrés[2]	1 651	1 689	1 895	1 849	1 892	1 995	2 189	2 391	2 547	...
Saint Helena Sainte-Hélène										
Vessels entered Navires entrés	42	28	62	55	91	276	174	79	49	78
Saint Lucia Sainte-Lucie										
Vessels entered Navires entrés	2 708	1 792	1 782	2 009	2 257	2 255	2 408	2 329	2 276	2 063
Vessels cleared Navires sortis	...	1 793	1 782	2 004	2 252	2 256	3 126	...	...	...
St. Vincent-Grenadines St. Vincent-Grenadines										
Vessels entered Navires entrés	863	1 074	1 212	1 532	1 179	1 224	1 309	1 345	1 388	...
Vessels cleared Navires sortis	...	1 046	1 171	1 501	1 150	1 182	1 184	1 208	1 221	...

103
International maritime transport
Vessels entered and cleared: thousand net registered tons [*cont.*]
Transports maritimes internationaux
Navires entrés et sortis : milliers de tonneaux de jauge nette [*suite*]

Country or area Pays ou zone	1981	1982	1983	1984	1985	1986	1987	1988	1989	1990
Samoa Samoa										
Vessels entered										
Navires entrés	558	584	574	575	453	434	659	...	...	...
Senegal Sénégal										
Vessels entered										
Navires entrés	10 722	10 183	10 144	10 229	9 750	5 258	9 872	...	...	...
Vessels cleared										
Navires sortis	10 647	...	...	10 275	9 824	5 237	9 858	...	...	...
Seychelles Seychelles										
Vessels entered										
Navires entrés	339	336	407	515	578	796	909	916	959	953
Singapore Singapour										
Vessels entered										
Navires entrés	62 880	69 845	74 666	73 607	75 520	77 529	81 712	...	...	...
Vessels cleared										
Navires sortis	54 924	59 726	61 644	60 142	62 239	65 501	68 990	...	...	...
South Africa Afrique du Sud										
Vessels entered[13]										
Navires entrés[13]	14 391	11 813	11 854	15 206	11 026	10 277	11 467	...	...	...
Vessels cleared[13]										
Navires sortis[13]	63 582	62 279	57 560	70 263	82 206	81 819	77 715	...	...	...
Spain Espagne										
Vessels entered										
Navires entrés	95 228	94 648	93 589	88 696	93 553	100 658	104 011	108 637	119 595	125 881
Vessels cleared										
Navires sortis	39 902	40 342	46 446	46 720	48 992	47 549	43 967	45 349	43 301	43 208
Sri Lanka Sri Lanka										
Vessels entered										
Navires entrés	11 249	12 139	12 732	12 790	13 608	15 422	13 870	14 161	17 084	20 148
Vessels cleared										
Navires sortis	12 093	...	...	...	...	...	...	...	...	...
Syrian Arab Republic Rép. arabe syrienne										
Vessels entered[2 4]										
Navires entrés[2 4]	16 166	11 357	11 778	11 180	10 893	...	...	...	...	...
Vessels cleared[2]										
Navires sortis[2]	16 418	10 691	11 564	10 922	10 809	...	...	...	...	...
Thailand Thaïlande										
Vessels entered										
Navires entrés	12 135	11 862	14 175	13 304	12 643	12 826	15 259	17 706	21 511	24 486
Vessels cleared										
Navires sortis	10 690	12 534	11 663	12 066	12 531	13 682	12 937	14 677	17 624	16 210
Togo Togo										
Vessels entered[2]										
Navires entrés[2]	996	1 163	1 023	...	...	...	...	...	...	...
Tonga Tonga										
Vessels entered										
Navires entrés	1 218	1 236	1 542	1 662	1 765	1 765	1 778	1 611	1 874	1 816
Turkey Turquie										
Vessels entered										
Navires entrés	30 076	42 296	43 822	54 653	50 004	52 736	63 457	64 447	83 749	62 689
Vessels cleared										
Navires sortis	30 127	40 576	41 529	49 546	44 814	51 152	60 477	63 695	81 515	60 651
United States Etats-Unis										
Vessels entered[5 14]										
Navires entrés[5 14]	307 355	264 732	273 720	304 170	294 110	321 761	344 710	...	...	...
Vessels cleared[5 14]										
Navires sortis[5 14]	274 475	278 530	267 045	273 505	273 444	265 475	289 767	...	...	...

103
International maritime transport
Vessels entered and cleared: thousand net registered tons [cont.]
Transports maritimes internationaux
Navires entrés et sortis : milliers de tonneaux de jauge nette [suite]

Country or area Pays ou zone	1981	1982	1983	1984	1985	1986	1987	1988	1989	1990
Venezuela Venezuela										
Vessels entered Navires entrés	22 091	21 316	15 212	18 852	21 337	20 579	22 663	26 101	17 529	15 774
Vessels cleared Navires sortis	3 852	4 260	5 508	6 966	7 044	13 320	8 826	8 223	6 641	18 294
Yemen Yémen										
Vessels entered[2][15] Navires entrés[2][15]	10 487	12 532	10 098	10 157	9 315	8 344	8 788	...	...	...
Vessels cleared[2][15] Navires sortis[2][15]	...	2 360	2 135	2 088	1 921	1 737	1 892	...	...	...
Yugoslavia Yougoslavie										
Vessels entered Navires entrés	15 150	14 111	14 562	15 135	16 605	15 580	15 452	18 498	20 887	24 422
Vessels cleared Navires sortis	8 505	8 638	9 144	9 668	9 778	9 431	10 438	14 906	15 478	14 929

Source:
Transport statistics data base of the Statistical Division
of the United Nations Secretariat.

† All data shown which pertain to Germany prior to 3 October
1990 are indicated separately for the Federal Republic of
Germany and the former German Democratic Republic based on
their respective territories at the time indicated. Where
data for united Germany (3 October 1990 and thereafter) are
not available, available data are shown separately under the
designations Federal Republic of Germany and former German
Democratic Republic and pertain to the territorial
boundaries prior to 3 October 1990. For detailed
explanatory notes on data pertaining to Germany, see Annex I
- Country Nomenclature.

1 Twelve months ending 30 June of year stated.
2 Including vessels in ballast.
3 Twelve months beginning 1 July of year stated.

4 All entrances counted.
5 Including Great Lakes international traffic (Canada: also
St. Lawrence).
6 Including ocean-going vessels and river vessels as well as
vessels in ballast.
7 Twelve months beginning 1 April of year stated.
8 Excluding minor and intermediate ports.
9 Excluding tankers.
10 Data for Sarawak include vessels in ballast and all
entrances counted.
11 Peninsular Malaysia and Sarawak only.
12 Including traffic with Portuguese overseas provinces.

13 Including transshipments.
14 Excluding traffic with United States Virgin Islands.
15 Former Democratic Yemen only.

Source:
Base de données pour les statistiques des transports de la
Division de statistique du Secrétariat de l'ONU.

† Toutes les données se rapportant à l'Allemagne avant le 3
octobre 1990 figurent dans deux rubriques séparées basées
sur les territoires respectifs de la République fédérale
d'Allemagne et l'ancienne République démocratique allemande
selon la période indiquée. En l'absence de données pour
l'Allemagne unifiée (à compter du 3 octobre 1990), les
données disponibles sont fournies séparément sous les
rubriques République fédérale d'Allemagne et ancienne
République démocratique allemande et se rapportent aux
limites territoriales antérieures au 3 octobre 1990. Pour
les notes explicatives en détail sur les données concernant
l'Allemagne, voir Annexe I – Nomenclature des pays.

1 Douze mois finissant le 30 juin de l'année indiquée.
2 Y compris navires sur lest.
3 Douze mois commençant le premier juillet de l'année
indiquée.
4 Toutes entrées comprises.
5 Y compris trafic international des Grands Lacs (Canada: et
du St. Laurent).
6 Y compris les grandes navigations et les navigations
fluviales et aussi les navires sur lest.
7 Douze mois commençant le premier avril de l'année indiquée.
8 Non compris les ports petits et moyens.
9 Non compris les bateaux citernes.
10 Les données pour Sarawak comprennent navires sur lest et
toutes entrées comprises.
11 Malaisie péninsulaire et Sarawak seulement.
12 Y compris le trafic avec les provinces portugaises
d'outre-mer.
13 Y compris les transbordements.
14 Non compris le trafic avec les Iles Vierges américaines.
15 L'ancien Yémen démocratique seulement.

104
Civil Aviation
Aviation civile

Scheduled airline traffic: Passengers (000); Kilometres (million)
Vols réguliers: Passagers (000); Kilomètres (millions)

Country or area and traffic	Total Totale 1980	1989	1990	1991	International Internationaux 1980	1989	1990	1991	Pays ou zone et trafic
World[1]									**Monde[1]**
Kilometres flown	9362	13564	14339	14167	3613	5317	5779	5832	Kilomètres parcourus
Passengers carried	645201	986873	1027320	992571	160698	257697	275897	262387	Passagers transportés
Passengers km	929004	1555679	1652297	1615137	457512	808933	875625	844143	Passagers-km
Freight ton-km	26866	54616	56278	56200	20003	44599	45956	46074	Fret tonnes-km
Mail ton-km	3107	4497	4779	4710	1440	2043	2162	2194	Courrier tonnes-km
Total ton-km	113486	200167	210379	207355	63140	121338	128729	126393	Total tonnes-km
Africa[2]									**Afrique[2]**
Kilometres flown	351	400	403	373	236	281	287	266	Kilomètres parcourus
Passengers carried	21235	26465	26694	24742	9003	12324	12652	12128	Passagers transportés
Passengers km	29724	40789	42337	38321	22434	31765	33465	31150	Passagers-km
Freight ton-km	797	1188	1126	1104	721	1080	1035	1025	Fret tonnes-km
Mail ton-km	48	57	56	52	40	41	36	36	Courrier tonnes-km
Total ton-km	3544	4939	4957	4643	2811	4023	4069	3854	Total tonnes-km
Algeria									**Algérie**
Kilometres flown	28	...	32	27	19	...	17	15	Kilomètres parcourus
Passengers carried	2950	...	3748	3385	1500	...	1851	1673	Passagers transportés
Passengers km	2300	...	3463	3092	1600	...	2351	2071	Passagers-km
Freight ton-km	13	...	15	24	11	...	12	20	Fret tonnes-km
Mail ton-km	2	...	0	0	1	...	0	0	Courrier tonnes-km
Total ton-km	220	...	330	303	156	...	224	206	Total tonnes-km
Angola									**Angola**
Kilometres flown	...	14	10	10	...	12	8	8	Kilomètres parcourus
Passengers carried	...	510	452	456	...	126	106	110	Passagers transportés
Passengers km	...	822	1189	1241	...	614	1008	1061	Passagers-km
Freight ton-km	...	52	40	42	...	51	39	40	Fret tonnes-km
Mail ton-km	...	1	0	0	...	1	0	0	Courrier tonnes-km
Total ton-km	...	124	145	151	...	107	129	136	Total tonnes-km
Benin[3]									**Bénin[3]**
Kilometres flown	2	2	2	2	2	2	2	2	Kilomètres parcourus
Passengers carried	61	74	76	64	61	74	76	64	Passagers transportés
Passengers km	178	224	232	203	178	224	232	203	Passagers-km
Freight ton-km	18	18	18	16	18	18	18	16	Fret tonnes-km
Mail ton-km	1	1	1	1	1	1	1	1	Courrier tonnes-km
Total ton-km	35	39	39	35	35	39	39	35	Total tonnes-km
Botswana									**Botswana**
Kilometres flown	1	3	3	4	0	2	2	2	Kilomètres parcourus
Passengers carried	39	82	101	102	20	55	67	68	Passagers transportés
Passengers km	15	50	63	73	7	37	45	55	Passagers-km
Freight ton-km	0	0	3	0	0	0	3	0	Fret tonnes-km
Total ton-km	1	5	9	7	1	4	7	6	Total tonnes-km
Burkina Faso[3]									**Burkina Faso[3]**
Kilometres flown	2	3	3	3	2	2	3	2	Kilomètres parcourus
Passengers carried	73	132	137	124	65	108	112	100	Passagers transportés
Passengers km	183	254	264	235	181	246	256	227	Passagers-km
Freight ton-km	18	18	18	16	18	18	18	16	Fret tonnes-km
Mail ton-km	1	1	1	1	1	1	1	1	Courrier tonnes-km
Total ton-km	36	41	42	38	35	41	42	37	Total tonnes-km
Burundi									**Burundi**
Passengers carried	...	...	8	...	...	...	7	...	Passagers transportés
Passengers km	...	...	2	...	...	...	2	...	Passagers-km
Cameroon									**Cameroun**
Kilometres flown	7	...	6	5	5	...	4	4	Kilomètres parcourus
Passengers carried	480	...	378	357	126	...	167	175	Passagers transportés
Passengers km	477	...	533	301	358	...	440	226	Passagers-km
Freight ton-km	28	...	9	10	26	...	8	9	Fret tonnes-km
Mail ton-km	2	...	0	0	2	...	0	0	Courrier tonnes-km
Total ton-km	72	...	57	37	60	...	48	30	Total tonnes-km
Cape Verde									**Cap-Vert**
Kilometres flown	...	...	2	...	...	...	1	...	Kilomètres parcourus
Passengers carried	...	...	177	...	...	...	46	...	Passagers transportés
Passengers km	...	...	161	...	...	...	137	...	Passagers-km
Freight ton-km	...	...	1	...	...	...	0	...	Fret tonnes-km
Total ton-km	...	...	15	...	...	...	13	...	Total tonnes-km

104
Civil Aviation
Scheduled airline traffic: Passengers (000); Kilometres (million) [*cont.*]
Aviation civile
Vols réguliers: Passagers (000); Kilomètres (millions) [*suite*]

Country or area and traffic	Total Totale				International Internationaux				Pays ou zone et trafic
	1980	1989	1990	1991	1980	1989	1990	1991	
Central African Rep.[3]									**Rép. centrafricaine[3]**
Kilometres flown	3	3	3	3	2	2	2	2	Kilomètres parcourus
Passengers carried	136	128	130	118	61	74	76	64	Passagers transportés
Passengers km	190	236	245	216	178	224	232	203	Passagers-km
Freight ton-km	18	18	18	16	18	18	18	16	Fret tonnes-km
Mail ton-km	1	1	1	1	1	1	1	1	Courrier tonnes-km
Total ton-km	36	40	41	36	35	39	39	35	Total tonnes-km
Chad[3]									**Tchad[3]**
Kilometres flown	3	2	3	2	2	2	2	2	Kilomètres parcourus
Passengers carried	106	91	93	81	61	76	79	66	Passagers transportés
Passengers km	210	231	240	211	178	225	234	205	Passagers-km
Freight ton-km	19	18	18	16	18	18	18	16	Fret tonnes-km
Mail ton-km	1	1	1	1	1	1	1	1	Courrier tonnes-km
Total ton-km	39	39	40	36	35	39	40	35	Total tonnes-km
Comoros									**Comores**
Passengers carried	...	...	26	...	...	...	5	...	Passagers transportés
Passengers km	...	...	3	...	...	...	1	...	Passagers-km
Congo[3]									**Congo[3]**
Kilometres flown	3	3	3	3	2	2	2	2	Kilomètres parcourus
Passengers carried	117	234	239	227	67	79	81	69	Passagers transportés
Passengers km	197	272	282	253	182	226	234	205	Passagers-km
Freight ton-km	18	19	18	17	18	18	18	16	Fret tonnes-km
Mail ton-km	1	1	1	1	1	1	1	1	Courrier tonnes-km
Total ton-km	37	44	45	40	36	39	40	35	Total tonnes-km
Côte d'Ivoire[3]									**Côte d'Ivoire[3]**
Kilometres flown	3	4	4	4	2	3	3	3	Kilomètres parcourus
Passengers carried	151	218	200	175	61	131	129	122	Passagers transportés
Passengers km	215	319	317	286	178	276	285	261	Passagers-km
Freight ton-km	18	18	18	16	18	18	18	16	Fret tonnes-km
Mail ton-km	1	1	1	1	1	1	1	1	Courrier tonnes-km
Total ton-km	39	47	47	43	35	44	44	40	Total tonnes-km
Djibouti									**Djibouti**
Kilometres flown	...	1	1	...	...	1	1	...	Kilomètres parcourus
Passengers carried	...	130	131	...	...	105	106	...	Passagers transportés
Passengers km	...	66	67	...	...	65	65	...	Passagers-km
Total ton-km	...	6	6	...	...	6	6	...	Total tonnes-km
Egypt									**Egypte**
Kilometres flown	31	42	43	35	27	37	37	31	Kilomètres parcourus
Passengers carried	2028	3419	3239	2602	1270	2137	2033	1755	Passagers transportés
Passengers km	2870	6186	5998	5230	2536	5616	5446	4840	Passagers-km
Freight ton-km	29	138	144	132	29	138	144	131	Fret tonnes-km
Mail ton-km	2	4	4	4	2	4	4	4	Courrier tonnes-km
Total ton-km	299	706	694	611	268	654	644	576	Total tonnes-km
Equatorial Guinea									**Guinée équatoriale**
Passengers carried	...	...	14	...	...	...	14	...	Passagers transportés
Passengers km	...	...	7	...	...	...	7	...	Passagers-km
Total ton-km	...	...	1	...	...	...	1	...	Total tonnes-km
Ethiopia									**Ethiopie**
Kilometres flown	11	23	21	21	10	17	18	18	Kilomètres parcourus
Passengers carried	243	651	620	636	186	434	341	352	Passagers transportés
Passengers km	647	1606	1529	1568	597	1504	1391	1431	Passagers-km
Freight ton-km	25	93	67	79	22	75	61	67	Fret tonnes-km
Mail ton-km	1	3	2	2	1	3	2	2	Courrier tonnes-km
Total ton-km	85	251	220	234	78	223	201	209	Total tonnes-km
Gabon									**Gabon**
Kilometres flown	6	6	6	...	5	4	4	...	Kilomètres parcourus
Passengers carried	331	407	398	...	95	120	120	...	Passagers transportés
Passengers km	374	447	445	...	313	370	370	...	Passagers-km
Freight ton-km	27	29	26	...	25	28	25	...	Fret tonnes-km
Mail ton-km	1	0	0	...	1	0	0	...	Courrier tonnes-km
Total ton-km	61	70	67	...	54	61	59	...	Total tonnes-km

104
Civil Aviation
Scheduled airline traffic: Passengers (000); Kilometres (million) [*cont.*]
Aviation civile
Vols réguliers: Passagers (000); Kilomètres (millions) [*suite*]

Country or area and traffic	Total Totale				International Internationaux				Pays ou zone et trafic
	1980	1989	1990	1991	1980	1989	1990	1991	
Ghana									**Ghana**
Kilometres flown	4	4	4	5	3	4	4	5	Kilomètres parcourus
Passengers carried	279	232	188	192	179	213	182	192	Passagers transportés
Passengers km	330	382	366	331	304	375	364	331	Passagers-km
Freight ton-km	4	14	16	19	4	14	16	19	Fret tonnes-km
Mail ton-km	1	0	0	1	1	0	0	1	Courrier tonnes-km
Total ton-km	34	48	49	58	32	48	49	58	Total tonnes-km
Guinea									**Guinée**
Kilometres flown	...	2	1	...	...	1	1	...	Kilomètres parcourus
Passengers carried	...	40	41	...	...	22	24	...	Passagers transportés
Passengers km	...	27	29	...	...	20	22	...	Passagers-km
Total ton-km	...	3	3	...	...	2	2	...	Total tonnes-km
Guinea-Bissau									**Guinée-Bissau**
Kilometres flown	...	...	1	...	...	...	0	...	Kilomètres parcourus
Passengers carried	...	...	21	...	...	...	8	...	Passagers transportés
Passengers km	...	...	10	...	...	...	6	...	Passagers-km
Total ton-km	...	...	1	...	...	...	1	...	Total tonnes-km
Kenya									**Kenya**
Kilometres flown	12	14	16	13	10	12	13	10	Kilomètres parcourus
Passengers carried	393	760	794	760	204	411	440	389	Passagers transportés
Passengers km	863	1399	1652	1479	782	1256	1509	1343	Passagers-km
Freight ton-km	18	48	52	39	18	47	52	39	Fret tonnes-km
Mail ton-km	2	3	2	2	2	3	2	2	Courrier tonnes-km
Total ton-km	98	177	204	175	90	164	191	162	Total tonnes-km
Lesotho									**Lesotho**
Kilometres flown	1	1	1	1	0	1	1	1	Kilomètres parcourus
Passengers carried	52	63	53	56	10	32	27	28	Passagers transportés
Passengers km	11	33	13	14	4	33	10	10	Passagers-km
Freight ton-km	0	1	0	0	0	1	0	0	Fret tonnes-km
Total ton-km	1	4	1	1	0	4	1	1	Total tonnes-km
Liberia									**Libéria**
Passengers carried	...	32	32	...		0	0	...	Passagers transportés
Passengers km	...	7	7	...	...	0	0	...	Passagers-km
Total ton-km	...	1	1	...	...	0	0	...	Total tonnes-km
Libyan Arab Jamah.									**Jamah. arabe libyenne**
Kilometres flown	12	17	19	19	7	9	11	11	Kilomètres parcourus
Passengers carried	1169	1617	1803	1884	468	547	632	685	Passagers transportés
Passengers km	1101	1680	1968	1251	632	946	1190	1251	Passagers-km
Freight ton-km	11	5	10	9	8	4	9	8	Fret tonnes-km
Total ton-km	116	146	178	183	70	88	116	120	Total tonnes-km
Madagascar									**Madagascar**
Kilometres flown	7	6	6	6	2	2	2	3	Kilomètres parcourus
Passengers carried	448	342	424	315	54	72	85	68	Passagers transportés
Passengers km	379	442	513	387	226	323	373	290	Passagers-km
Freight ton-km	20	29	30	26	17	28	29	25	Fret tonnes-km
Mail ton-km	1	1	2	1	1	1	1	1	Courrier tonnes-km
Total ton-km	54	69	77	61	38	58	64	52	Total tonnes-km
Malawi									**Malawi**
Kilometres flown	2	2	2	3	1	1	1	1	Kilomètres parcourus
Passengers carried	94	121	120	118	52	60	58	57	Passagers transportés
Passengers km	68	82	80	87	56	66	65	58	Passagers-km
Freight ton-km	1	1	1	1	1	1	1	1	Fret tonnes-km
Total ton-km	7	8	8	9	6	7	6	6	Total tonnes-km
Mauritania[3]									**Mauritanie**[3]
Kilometres flown	3	3	4	3	2	2	3	2	Kilomètres parcourus
Passengers carried	141	220	223	210	76	96	99	86	Passagers transportés
Passengers km	218	298	307	278	188	247	255	226	Passagers-km
Freight ton-km	19	18	18	16	18	18	18	16	Fret tonnes-km
Mail ton-km	1	1	1	1	1	1	1	1	Courrier tonnes-km
Total ton-km	39	46	46	42	36	41	42	37	Total tonnes-km

104
Civil Aviation
Scheduled airline traffic: Passengers (000); Kilometres (million) [*cont.*]
Aviation civile
Vols réguliers: Passagers (000); Kilomètres (millions) [*suite*]

Country or area and traffic	Total Totale				International Internationaux				Pays ou zone et trafic
	1980	1989	1990	1991	1980	1989	1990	1991	
Mauritius									**Maurice**
Kilometres flown	3	12	15	16	3	12	14	16	Kilomètres parcourus
Passengers carried	84	424	520	535	77	406	500	513	Passagers transportés
Passengers km	185	1828	2279	2464	181	1818	2268	2451	Passagers-km
Freight ton-km	2	55	64	81	2	55	64	81	Fret tonnes-km
Mail ton-km	0	1	2	2	0	1	2	2	Courrier tonnes-km
Total ton-km	20	230	282	305	20	229	281	304	Total tonnes-km
Morocco									**Maroc**
Kilometres flown	21	27	30	27	20	27	28	26	Kilomètres parcourus
Passengers carried	947	1361	1580	1430	906	1251	1349	1227	Passagers transportés
Passengers km	1868	2695	2889	2533	1855	2663	2797	2449	Passagers-km
Freight ton-km	26	34	32	56	26	34	31	55	Fret tonnes-km
Mail ton-km	1	1	1	0	1	1	1	0	Courrier tonnes-km
Total ton-km	209	293	255	257	208	290	246	249	Total tonnes-km
Mozambique									**Mozambique**
Kilometres flown	6	5	5	5	3	3	3	3	Kilomètres parcourus
Passengers carried	282	243	280	283	52	70	83	95	Passagers transportés
Passengers km	467	494	502	465	260	335	323	284	Passagers-km
Freight ton-km	9	10	9	9	5	8	7	7	Fret tonnes-km
Mail ton-km	1	1	1	1	0	1	1	0	Courrier tonnes-km
Total ton-km	51	56	56	52	29	39	37	33	Total tonnes-km
Namibia									**Namibie**
Kilometres flown	...	...	...	3	...	...	...	3	Kilomètres parcourus
Passengers carried	...	...	...	455	...	...	...	400	Passagers transportés
Passengers km	...	...	...	423	...	...	...	395	Passagers-km
Freight ton-km	...	...	...	2	...	...	...	2	Fret tonnes-km
Total ton-km	...	...	...	40	...	...	...	38	Total tonnes-km
Niger[3]									**Niger[3]**
Kilometres flown	3	2	2	2	2	2	2	2	Kilomètres parcourus
Passengers carried	111	74	76	64	61	74	76	64	Passagers transportés
Passengers km	199	224	232	203	178	224	232	203	Passagers-km
Freight ton-km	18	18	18	16	18	18	18	16	Fret tonnes-km
Mail ton-km	1	1	1	1	1	1	1	1	Courrier tonnes-km
Total ton-km	37	39	39	35	35	39	39	35	Total tonnes-km
Nigeria									**Nigéria**
Kilometres flown	24	14	17	14	11	7	10	9	Kilomètres parcourus
Passengers carried	1939	849	965	930	301	211	344	374	Passagers transportés
Passengers km	1877	1007	1287	1391	996	592	866	1012	Passagers-km
Freight ton-km	9	17	24	28	9	15	22	26	Fret tonnes-km
Mail ton-km	1	4	2	2	1	3	1	2	Courrier tonnes-km
Total ton-km	179	112	143	159	99	72	103	122	Total tonnes-km
Rwanda									**Rwanda**
Passengers carried	...	10	8	...	...	3	3	...	Passagers transportés
Passengers km	...	2	2	...	...	1	1	...	Passagers-km
Sao Tome and Principe									**Sao Tomé-et-Principe**
Passengers carried	...	22	22	...	...	13	13	...	Passagers transportés
Passengers km	...	8	8	...	...	4	4	...	Passagers-km
Total ton-km	...	1	1	...	...	0	0	...	Total tonnes-km
Senegal[3]									**Sénégal[3]**
Kilometres flown	3	3	3	3	2	2	3	2	Kilomètres parcourus
Passengers carried	113	140	148	136	71	114	120	108	Passagers transportés
Passengers km	196	243	253	224	182	235	245	216	Passagers-km
Freight ton-km	18	18	18	16	18	18	18	16	Fret tonnes-km
Mail ton-km	1	1	1	1	1	1	1	1	Courrier tonnes-km
Total ton-km	37	40	41	37	36	40	41	36	Total tonnes-km
Seychelles									**Seychelles**
Kilometres flown	...	...	4	3	...	...	3	3	Kilomètres parcourus
Passengers carried	...	...	242	243	...	...	64	51	Passagers transportés
Passengers km	...	...	440	351	...	...	431	342	Passagers-km
Freight ton-km	...	...	10	9	...	...	10	9	Fret tonnes-km
Mail ton-km	...	...	0	1	...	...	0	1	Courrier tonnes-km
Total ton-km	...	...	53	42	...	...	53	41	Total tonnes-km

104
Civil Aviation
Scheduled airline traffic: Passengers (000); Kilometres (million) [*cont.*]
Aviation civile
Vols réguliers: Passagers (000); Kilomètres (millions) [*suite*]

Country or area and traffic	Total Totale				International Internationaux				Pays ou zone et trafic
	1980	1989	1990	1991	1980	1989	1990	1991	
Somalia									**Somalie**
Kilometres flown	4	3	3	1	2	2	2	1	Kilomètres parcourus
Passengers carried	90	89	88	46	70	81	80	40	Passagers transportés
Passengers km	140	248	255	131	120	240	247	125	Passagers-km
Freight ton-km	1	8	9	5	0	8	9	5	Fret tonnes-km
Total ton-km	13	30	32	17	11	30	31	16	Total tonnes-km
South Africa									**Afrique du Sud**
Kilometres flown	67	70	67	67	37	28	29	30	Kilomètres parcourus
Passengers carried	4116	5641	5365	4819	801	707	804	842	Passagers transportés
Passengers km	8920	9201	9049	8413	6088	4867	5043	4887	Passagers-km
Freight ton-km	251	206	179	191	214	164	136	157	Fret tonnes-km
Mail ton-km	22	25	27	23	17	12	9	9	Courrier tonnes-km
Total ton-km	1074	1037	997	950	778	603	587	593	Total tonnes-km
Sudan									**Soudan**
Kilometres flown	11	8	10	6	7	6	6	4	Kilomètres parcourus
Passengers carried	519	363	454	363	311	201	239	165	Passagers transportés
Passengers km	710	494	589	426	530	405	453	302	Passagers-km
Freight ton-km	12	17	13	11	10	6	5	5	Fret tonnes-km
Mail ton-km	0	1	1	0	0	0	0	0	Courrier tonnes-km
Total ton-km	76	66	67	50	58	45	46	32	Total tonnes-km
Swaziland									**Swaziland**
Kilometres flown	1	1	1	1	1	1	1	1	Kilomètres parcourus
Passengers carried	31	51	55	59	31	51	55	59	Passagers transportés
Passengers km	30	40	45	45	30	40	45	45	Passagers-km
Total ton-km	3	4	4	4	3	4	4	4	Total tonnes-km
Togo[3]									**Togo[3]**
Kilometres flown	2	2	2	2	2	2	2	2	Kilomètres parcourus
Passengers carried	65	74	76	64	65	74	76	64	Passagers transportés
Passengers km	179	224	232	203	179	224	232	203	Passagers-km
Freight ton-km	18	18	18	16	18	18	18	16	Fret tonnes-km
Mail ton-km	1	1	1	1	1	1	1	1	Courrier tonnes-km
Total ton-km	35	39	39	35	35	39	39	35	Total tonnes-km
Tunisia									**Tunisie**
Kilometres flown	14	15	15	14	13	15	14	13	Kilomètres parcourus
Passengers carried	978	1315	1313	1201	898	1174	1149	1059	Passagers transportés
Passengers km	1241	1571	1502	1407	1213	1524	1449	1359	Passagers-km
Freight ton-km	12	19	19	18	12	19	19	18	Fret tonnes-km
Mail ton-km	1	1	1	1	1	1	1	1	Courrier tonnes-km
Total ton-km	123	158	154	144	120	154	149	140	Total tonnes-km
Uganda									**Ouganda**
Kilometres flown	3	...	4	0	2	...	3	0	Kilomètres parcourus
Passengers carried	83	...	116	26	43	...	100	25	Passagers transportés
Passengers km	68	...	278	13	58	...	270	13	Passagers-km
Freight ton-km	9	...	22	0	9	...	22	0	Fret tonnes-km
Total ton-km	15	...	47	1	14	...	46	1	Total tonnes-km
United Rep.Tanzania									**Rép. Unie de Tanzanie**
Kilometres flown	8	4	4	5	3	2	2	3	Kilomètres parcourus
Passengers carried	388	267	292	290	94	54	63	97	Passagers transportés
Passengers km	284	184	215	284	180	95	117	195	Passagers-km
Freight ton-km	2	2	1	4	1	1	1	3	Fret tonnes-km
Total ton-km	28	18	21	30	18	9	11	21	Total tonnes-km
Zaire									**Zaïre**
Kilometres flown	10	7	7	5	4	4	4	3	Kilomètres parcourus
Passengers carried	439	203	207	150	92	74	74	73	Passagers transportés
Passengers km	834	506	500	384	464	336	316	269	Passagers-km
Freight ton-km	35	60	57	33	22	41	40	26	Fret tonnes-km
Mail ton-km	1	0	1	1	0	0	0	0	Courrier tonnes-km
Total ton-km	110	105	102	68	64	71	69	50	Total tonnes-km
Zambia									**Zambie**
Kilometres flown	11	9	10	7	9	8	8	5	Kilomètres parcourus
Passengers carried	257	365	407	293	114	202	249	157	Passagers transportés
Passengers km	467	983	985	655	425	915	942	611	Passagers-km
Freight ton-km	47	25	30	22	47	25	29	22	Fret tonnes-km

104
Civil Aviation
Scheduled airline traffic: Passengers (000); Kilometres (million) [*cont.*]
Aviation civile
Vols réguliers: Passagers (000); Kilomètres (millions) [*suite*]

Country or area and traffic	Total Totale				International Internationaux				Pays ou zone et trafic
	1980	1989	1990	1991	1980	1989	1990	1991	
Mail ton-km	1	1	1	1	1	1	1	1	Courrier tonnes-km
Total ton-km	89	114	122	82	85	108	118	78	Total tonnes-km
Zimbabwe									**Zimbabwe**
Kilometres flown	6	11	12	12	4	9	10	10	Kilomètres parcourus
Passengers carried	412	583	601	606	180	233	240	245	Passagers transportés
Passengers km	362	760	797	817	286	623	655	675	Passagers-km
Freight ton-km	3	65	65	64	3	64	64	63	Fret tonnes-km
Mail ton-km	1	1	1	1	1	1	1	1	Courrier tonnes-km
Total ton-km	33	131	134	135	27	119	123	123	Total tonnes-km
America, North[3]									**Amérique du Nord**[3]
Kilometres flown	**5074**	**7384**	**7946**	**7708**	**812**	**1370**	**1523**	**1549**	**Kilomètres parcourus**
Passengers carried	**336904**	**495944**	**508282**	**491883**	**38700**	**60607**	**66544**	**61817**	**Passagers transportés**
Passengers km	**467244**	**771958**	**812497**	**789185**	**112724**	**215593**	**241337**	**232107**	**Passagers-km**
Freight ton-km	**9546**	**16252**	**16511**	**16077**	**3969**	**8724**	**8809**	**8601**	**Fret tonnes-km**
Mail ton-km	**2094**	**2886**	**3040**	**2861**	**576**	**750**	**782**	**751**	**Courrier tonnes-km**
Total ton-km	**53926**	**89167**	**93276**	**90555**	**14691**	**29113**	**31573**	**30512**	**Total tonnes-km**
Antigua and Barbuda									**Antigua-et-Barbuda**
Kilometres flown	...	...	8	...	...	...	8	...	Kilomètres parcourus
Passengers carried	...	...	755	...	...	...	755	...	Passagers transportés
Passengers km	...	...	195	...	...	...	195	...	Passagers-km
Total ton-km	...	...	18	...	...	...	18	...	Total tonnes-km
Bahamas									**Bahamas**
Kilometres flown	10	...	6	...	4	...	4	...	Kilomètres parcourus
Passengers carried	345	...	1090	...	160	...	635	...	Passagers transportés
Passengers km	539	...	346	...	459	...	290	...	Passagers-km
Freight ton-km	3	...	0	...	3	...	0	...	Fret tonnes-km
Total ton-km	52	...	31	...	44	...	26	...	Total tonnes-km
Barbados									**Barbade**
Kilometres flown	1	...	...	...	1	...	...	...	Kilomètres parcourus
Passengers carried	46	...	...	...	46	...	...	...	Passagers transportés
Passengers km	330	...	...	...	330	...	...	...	Passagers-km
Freight ton-km	1	...	...	...	1	...	...	...	Fret tonnes-km
Total ton-km	30	...	...	...	30	...	...	...	Total tonnes-km
Canada									**Canada**
Kilometres flown	338	433	433	372	114	167	177	155	Kilomètres parcourus
Passengers carried	22453	21274	20601	16586	5530	6581	6806	5739	Passagers transportés
Passengers km	36234	50372	47115	39082	16293	27310	26701	22042	Passagers-km
Freight ton-km	689	1349	1385	1241	414	984	1034	924	Fret tonnes-km
Mail ton-km	130	125	138	121	52	32	40	32	Courrier tonnes-km
Total ton-km	4095	6043	5797	4908	1944	3493	3497	2956	Total tonnes-km
Costa Rica									**Costa Rica**
Kilometres flown	8	10	12	12	8	10	11	11	Kilomètres parcourus
Passengers carried	431	441	467	504	332	391	417	454	Passagers transportés
Passengers km	495	914	983	1050	485	908	976	1044	Passagers-km
Freight ton-km	22	33	39	36	22	33	39	36	Fret tonnes-km
Mail ton-km	0	1	1	1	0	1	1	1	Courrier tonnes-km
Total ton-km	68	134	148	151	67	134	147	150	Total tonnes-km
Cuba									**Cuba**
Kilometres flown	15	22	21	17	7	14	13	11	Kilomètres parcourus
Passengers carried	676	990	1137	831	120	237	462	195	Passagers transportés
Passengers km	932	2017	1832	1598	666	1616	1419	1226	Passagers-km
Freight ton-km	10	22	19	18	8	21	17	16	Fret tonnes-km
Mail ton-km	2	5	4	3	2	5	3	3	Courrier tonnes-km
Total ton-km	95	219	190	172	72	187	156	141	Total tonnes-km
Dominican Republic									**Rép. dominicaine**
Kilometres flown	6	5	13	13	6	5	13	13	Kilomètres parcourus
Passengers carried	466	479	718	648	466	479	718	648	Passagers transportés
Passengers km	550	383	1375	1350	550	383	1375	1350	Passagers-km
Freight ton-km	11	0	70	70	11	0	70	70	Fret tonnes-km
Mail ton-km	0	0	1	1	0	0	1	1	Courrier tonnes-km
Total ton-km	60	38	204	201	60	38	204	201	Total tonnes-km

104
Civil Aviation
Scheduled airline traffic: Passengers (000); Kilometres (million) [cont.]
Aviation civile
Vols réguliers: Passagers (000); Kilomètres (millions) [suite]

Country or area and traffic	Total Totale				International Internationaux				Pays ou zone et trafic
	1980	1989	1990	1991	1980	1989	1990	1991	
El Salvador									**El Salvador**
Kilometres flown	6	9	11	12	6	9	11	12	Kilomètres parcourus
Passengers carried	265	618	525	590	265	618	525	590	Passagers transportés
Passengers km	289	939	1066	1257	289	939	1066	1257	Passagers-km
Freight ton-km	17	3	5	5	17	3	5	5	Fret tonnes-km
Total ton-km	41	98	113	134	41	98	113	134	Total tonnes-km
Guatemala									**Guatemala**
Kilometres flown	4	3	4	4	4	3	4	4	Kilomètres parcourus
Passengers carried	119	110	156	165	119	107	156	165	Passagers transportés
Passengers km	159	164	213	230	159	163	213	230	Passagers-km
Freight ton-km	6	23	9	9	6	23	9	9	Fret tonnes-km
Mail ton-km	0	0	0	0	0	0	0	0	Courrier tonnes-km
Total ton-km	21	38	27	29	21	38	27	29	Total tonnes-km
Haiti									**Haïti**
Kilometres flown	1	...	1	...	1	...	1	...	Kilomètres parcourus
Freight ton-km	2	...	4	...	2	...	4	...	Fret tonnes-km
Total ton-km	2	...	4	...	2	...	4	...	Total tonnes-km
Honduras									**Honduras**
Kilometres flown	7	7	7	5	7	6	6	5	Kilomètres parcourus
Passengers carried	508	574	610	447	369	473	352	253	Passagers transportés
Passengers km	387	519	468	336	369	504	426	303	Passagers-km
Freight ton-km	4	3	4	3	4	3	4	3	Fret tonnes-km
Mail ton-km	1	2	1	1	1	2	1	1	Courrier tonnes-km
Total ton-km	43	75	63	45	41	73	59	42	Total tonnes-km
Jamaica									**Jamaïque**
Kilometres flown	15	16	14	12	13	14	11	10	Kilomètres parcourus
Passengers carried	723	1211	1004	894	660	1112	897	804	Passagers transportés
Passengers km	1207	1982	1463	1311	1200	1969	1444	1294	Passagers-km
Freight ton-km	9	16	19	20	9	16	19	20	Fret tonnes-km
Total ton-km	115	196	152	139	115	195	150	137	Total tonnes-km
Mexico									**Mexique**
Kilometres flown	157	149	180	191	63	72	88	87	Kilomètres parcourus
Passengers carried	12890	12689	14341	14901	2777	3362	3939	3717	Passagers transportés
Passengers km	13870	16059	18290	18267	6594	7702	9211	8709	Passagers-km
Freight ton-km	132	114	143	163	69	65	84	95	Fret tonnes-km
Mail ton-km	4	4	8	8	3	2	2	3	Courrier tonnes-km
Total ton-km	1383	1467	1709	1726	665	737	897	872	Total tonnes-km
Nicaragua									**Nicaragua**
Kilometres flown	...	...	2	...	...	...	2	...	Kilomètres parcourus
Passengers carried	...	...	130	...	...	...	84	...	Passagers transportés
Passengers km	...	...	111	...	...	...	97	...	Passagers-km
Freight ton-km	...	...	4	...	...	...	3	...	Fret tonnes-km
Total ton-km	...	...	14	...	...	...	12	...	Total tonnes-km
Panama									**Panama**
Kilometres flown	7	...	4	4	6	...	4	4	Kilomètres parcourus
Passengers carried	355	...	205	398	307	...	205	398	Passagers transportés
Passengers km	409	...	196	272	395	...	196	272	Passagers-km
Freight ton-km	3	...	2	4	3	...	2	4	Fret tonnes-km
Total ton-km	40	...	27	35	38	...	27	35	Total tonnes-km
Trinidad and Tobago									**Trinité-et-Tobago**
Kilometres flown	16	22	23	25	16	22	23	25	Kilomètres parcourus
Passengers carried	877	1310	1285	1345	633	951	942	957	Passagers transportés
Passengers km	1505	2691	2726	3129	1485	2661	2697	3104	Passagers-km
Freight ton-km	18	14	15	15	18	14	15	15	Fret tonnes-km
Mail ton-km	1	2	2	2	1	2	2	2	Courrier tonnes-km
Total ton-km	160	304	308	351	158	301	305	349	Total tonnes-km
United States									**Etats-Unis**
Kilometres flown	4469	6675	7197	7013	543	1020	1137	1189	Kilomètres parcourus
Passengers carried	295257	452547	463821	451247	25723	43461	48626	45483	Passagers transportés
Passengers km	409480	693940	735189	719770	82639	169577	194155	189867	Passagers-km
Freight ton-km	8615	14650	14788	14480	3379	7539	7498	7392	Fret tonnes-km
Mail ton-km	1954	2745	2884	2722	517	705	731	707	Courrier tonnes-km
Total ton-km	47640	80349	84383	82514	11316	23626	25847	25327	Total tonnes-km

104
Civil Aviation
Scheduled airline traffic: Passengers (000); Kilometres (million) [*cont.*]
    Aviation civile
    Vols réguliers: Passagers (000); Kilomètres (millions) [*suite*]

Country or area and traffic	Total Totale				International Internationaux				Pays ou zone et trafic
	1980	1989	1990	1991	1980	1989	1990	1991	
America, South[2]									**Amérique du Sud[2]**
Kilometres flown	498	623	615	665	199	223	239	263	Kilomètres parcourus
Passengers carried	34008	45817	41457	41433	6049	7019	7547	7949	Passagers transportés
Passengers km	38628	56887	56853	57331	19547	30206	31109	31698	Passagers-km
Freight ton-km	1349	2373	2398	2424	1042	1828	1906	1929	Fret tonnes-km
Mail ton-km	57	67	70	115	36	43	44	52	Courrier tonnes-km
Total ton-km	4790	7698	7660	7814	2907	4749	4932	5037	Total tonnes-km
Argentina									**Argentine**
Kilometres flown	94	79	73	68	39	27	30	33	Kilomètres parcourus
Passengers carried	5589	4748	4353	4532	1300	1384	1560	1739	Passagers transportés
Passengers km	8031	9253	9431	9207	4413	5814	6482	6547	Passagers-km
Freight ton-km	195	201	200	185	166	173	173	161	Fret tonnes-km
Mail ton-km	21	12	10	10	9	7	7	7	Courrier tonnes-km
Total ton-km	939	1019	1037	1023	599	704	764	758	Total tonnes-km
Bolivia									**Bolivie**
Kilometres flown	14	11	11	12	9	7	7	7	Kilomètres parcourus
Passengers carried	1342	1273	1238	1200	268	376	397	409	Passagers transportés
Passengers km	944	1032	1068	1022	570	706	755	723	Passagers-km
Freight ton-km	38	8	8	5	35	7	7	4	Fret tonnes-km
Mail ton-km	0	1	1	1	0	1	1	0	Courrier tonnes-km
Total ton-km	121	99	103	101	89	71	76	75	Total tonnes-km
Brazil									**Brésil**
Kilometres flown	203	291	298	342	58	71	78	85	Kilomètres parcourus
Passengers carried	13008	19411	18941	19015	1330	1938	2096	2128	Passagers transportés
Passengers km	15572	27854	28500	28537	6008	11648	12769	12561	Passagers-km
Freight ton-km	588	1148	1082	1029	371	708	689	664	Fret tonnes-km
Mail ton-km	23	35	42	82	15	19	21	24	Courrier tonnes-km
Total ton-km	1956	3642	3691	3689	945	1852	1950	1919	Total tonnes-km
Chile									**Chili**
Kilometres flown	25	44	50	50	19	29	34	36	Kilomètres parcourus
Passengers carried	669	1264	1364	1406	299	452	492	497	Passagers transportés
Passengers km	1875	2807	2807	3039	1362	1970	2102	2116	Passagers-km
Freight ton-km	145	330	419	429	134	314	403	413	Fret tonnes-km
Mail ton-km	4	3	3	3	4	2	2	3	Courrier tonnes-km
Total ton-km	324	586	691	706	267	492	594	606	Total tonnes-km
Colombia									**Colombie**
Kilometres flown	45	70	72	71	22	34	35	34	Kilomètres parcourus
Passengers carried	4808	5565	5267	5540	752	744	736	685	Passagers transportés
Passengers km	4198	4537	4384	4465	2203	2321	2283	2212	Passagers-km
Freight ton-km	147	395	464	526	125	359	427	459	Fret tonnes-km
Mail ton-km	5	8	8	9	4	6	6	7	Courrier tonnes-km
Total ton-km	532	842	896	976	339	590	655	690	Total tonnes-km
Ecuador									**Equateur**
Kilometres flown	21	19	20	19	11	9	9	9	Kilomètres parcourus
Passengers carried	701	742	763	752	255	257	263	252	Passagers transportés
Passengers km	975	1209	1243	1201	851	979	1006	964	Passagers-km
Freight ton-km	40	68	63	68	35	60	55	60	Fret tonnes-km
Mail ton-km	1	1	1	1	1	1	1	1	Courrier tonnes-km
Total ton-km	131	187	185	186	115	158	156	156	Total tonnes-km
Guyana									**Guyana**
Kilometres flown	...	...	3	2	...	...	2	2	Kilomètres parcourus
Passengers carried	...	...	146	112	...	...	85	51	Passagers transportés
Passengers km	...	...	216	214	...	...	200	198	Passagers-km
Freight ton-km	...	...	3	3	...	...	2	2	Fret tonnes-km
Total ton-km	...	...	23	22	...	...	20	20	Total tonnes-km
Paraguay									**Paraguay**
Kilometres flown	...	6	6	9	...	5	4	7	Kilomètres parcourus
Passengers carried	...	272	273	309	...	239	239	275	Passagers transportés
Passengers km	...	591	591	1073	...	571	571	1054	Passagers-km
Freight ton-km	...	4	5	5	...	3	4	4	Fret tonnes-km
Total ton-km	...	57	58	102	...	55	55	99	Total tonnes-km

104
Civil Aviation
Scheduled airline traffic: Passengers (000); Kilometres (million) [*cont.*]
Aviation civile
Vols réguliers: Passagers (000); Kilomètres (millions) [*suite*]

Country or area and traffic	Total Totale				International Internationaux				Pays ou zone et trafic
	1980	1989	1990	1991	1980	1989	1990	1991	
Peru									**Pérou**
Kilometres flown	25	22	20	19	9	11	9	8	Kilomètres parcourus
Passengers carried	1980	1854	1816	1491	221	270	255	224	Passagers transportés
Passengers km	1974	2048	2025	1759	822	996	954	865	Passagers-km
Freight ton-km	40	30	26	25	24	19	19	18	Fret tonnes-km
Mail ton-km	1	1	1	1	0	1	1	1	Courrier tonnes-km
Total ton-km	217	212	216	184	98	106	110	97	Total tonnes-km
Suriname									**Suriname**
Kilometres flown	3	...	3	...	2	...	3	...	Kilomètres parcourus
Passengers carried	144	...	133	...	120	...	117	...	Passagers transportés
Passengers km	245	...	404	...	240	...	400	...	Passagers-km
Freight ton-km	4	...	15	...	4	...	15	...	Fret tonnes-km
Mail ton-km	0	...	1	...	0	...	1	...	Courrier tonnes-km
Total ton-km	26	...	54	...	25	...	53	...	Total tonnes-km
Uruguay									**Uruguay**
Kilometres flown	4	...	5	...	3	...	5	...	Kilomètres parcourus
Passengers carried	478	...	318	...	433	...	318	...	Passagers transportés
Passengers km	178	...	471	...	160	...	471	...	Passagers-km
Freight ton-km	1	...	3	...	1	...	3	...	Fret tonnes-km
Total ton-km	17	...	44	...	15	...	44	...	Total tonnes-km
Venezuela									**Venezuela**
Kilometres flown	61	69	55	64	26	21	23	33	Kilomètres parcourus
Passengers carried	5133	10099	6847	6626	960	845	989	1253	Passagers transportés
Passengers km	4367	6446	5534	5939	2671	4108	3115	3587	Passagers-km
Freight ton-km	148	161	113	132	146	160	111	128	Fret tonnes-km
Mail ton-km	3	5	3	7	3	5	3	7	Courrier tonnes-km
Total ton-km	501	931	664	727	392	603	455	520	Total tonnes-km
Asia[2]									**Asie**[2]
Kilometres flown	1270	2151	2091	2125	777	1257	1294	1317	**Kilomètres parcourus**
Passengers carried	106119	190752	209139	204948	34301	65720	67643	67679	**Passagers transportés**
Passengers km	158192	322782	344494	351943	108019	231709	241758	240572	**Passagers-km**
Freight ton-km	6249	16834	17344	18033	5748	15540	15877	16504	**Fret tonnes-km**
Mail ton-km	250	612	665	699	204	435	473	496	**Courrier tonnes-km**
Total ton-km	20458	46088	48355	49955	15977	37425	38735	39436	**Total tonnes-km**
Afghanistan									**Afghanistan**
Kilometres flown	3	...	4	4	3	...	2	2	Kilomètres parcourus
Passengers carried	76	...	241	212	51	...	91	92	Passagers transportés
Passengers km	163	...	230	205	155	...	162	151	Passagers-km
Freight ton-km	21	...	9	8	21	...	9	8	Fret tonnes-km
Total ton-km	36	...	30	27	35	...	24	22	Total tonnes-km
Bahrain[4]									**Bahreïn**[4]
Kilometres flown	7	11	13	11	7	11	13	11	Kilomètres parcourus
Passengers carried	522	753	771	876	522	753	771	876	Passagers transportés
Passengers km	714	1492	1549	1676	714	1492	1549	1676	Passagers-km
Freight ton-km	26	40	44	51	26	40	44	51	Fret tonnes-km
Mail ton-km	0	3	3	3	0	3	3	3	Courrier tonnes-km
Total ton-km	91	183	193	211	91	183	193	211	Total tonnes-km
Bangladesh									**Bangladesh**
Kilometres flown	12	12	13	...	10	11	11	...	Kilomètres parcourus
Passengers carried	614	997	1080	...	270	603	630	...	Passagers transportés
Passengers km	1179	1966	2190	...	1113	1887	2100	...	Passagers-km
Freight ton-km	20	84	92	...	19	84	91	...	Fret tonnes-km
Mail ton-km	0	1	1	...	0	1	1	...	Courrier tonnes-km
Total ton-km	126	268	296	...	119	260	287	...	Total tonnes-km
Bhutan									**Bhoutan**
Passengers carried	...	...	8	...	...	...	8	...	Passagers transportés
Passengers km	...	...	4	...	...	...	4	...	Passagers-km
Brunei Darussalam									**Brunéi Darussalam**
Kilometres flown	...	...	7	...	...	...	7	...	Kilomètres parcourus
Passengers carried	...	...	307	...	...	...	307	...	Passagers transportés
Passengers km	...	...	487	...	...	...	487	...	Passagers-km
Freight ton-km	...	...	10	...	...	...	10	...	Fret tonnes-km
Total ton-km	...	...	53	...	...	...	53	...	Total tonnes-km

104

Civil Aviation

Scheduled airline traffic: Passengers (000); Kilometres (million) [*cont.*]

Aviation civile

Vols réguliers: Passagers (000); Kilomètres (millions) [*suite*]

Country or area and traffic	Total Totale				International Internationaux				Pays ou zone et trafic
	1980	1989	1990	1991	1980	1989	1990	1991	
China									**Chine**
Kilometres flown	47	166	199	194	10	47	39	44	Kilomètres parcourus
Passengers carried	2568	11080	16596	19520	360	2080	1136	1363	Passagers transportés
Passengers km	3578	17914	23048	30132	913	5909	5169	6446	Passagers-km
Freight ton-km	121	683	818	1010	52	413	438	554	Fret tonnes-km
Total ton-km	443	1991	2500	3207	134	856	826	1038	Total tonnes-km
Cyprus									**Chypre**
Kilometres flown	10	12	14	14	10	12	14	14	Kilomètres parcourus
Passengers carried	441	748	814	820	441	748	814	820	Passagers transportés
Passengers km	798	1664	1861	1803	798	1664	1861	1803	Passagers-km
Freight ton-km	19	33	34	34	19	33	34	34	Fret tonnes-km
Mail ton-km	1	2	2	2	1	2	2	2	Courrier tonnes-km
Total ton-km	92	184	203	198	92	184	203	198	Total tonnes-km
India									**Inde**
Kilometres flown	85	128	113	108	43	50	49	43	Kilomètres parcourus
Passengers carried	6603	12740	10862	10859	1668	2253	2217	2071	Passagers transportés
Passengers km	10765	17988	16722	15667	6765	9002	8955	7859	Passagers-km
Freight ton-km	366	681	663	494	322	560	559	399	Fret tonnes-km
Mail ton-km	35	26	27	29	20	8	10	12	Courrier tonnes-km
Total ton-km	1353	2274	2156	1890	966	1397	1393	1134	Total tonnes-km
Indonesia									**Indonésie**
Kilometres flown	88	282	153	167	24	47	46	53	Kilomètres parcourus
Passengers carried	5059	8983	9223	10386	922	1966	1522	2344	Passagers transportés
Passengers km	5907	13912	14581	15965	2774	8261	8477	9589	Passagers-km
Freight ton-km	122	438	459	475	85	367	357	370	Fret tonnes-km
Mail ton-km	6	14	15	13	2	6	5	4	Courrier tonnes-km
Total ton-km	625	1679	1759	1914	337	1152	1165	1262	Total tonnes-km
Iran, Islamic Rep. of									**Iran, Rép. islamique d'**
Kilometres flown	16	29	33	33	8	9	11	11	Kilomètres parcourus
Passengers carried	1998	4883	5633	5353	396	798	932	757	Passagers transportés
Passengers km	2071	4691	5755	5551	995	1851	2462	2211	Passagers-km
Freight ton-km	20	92	114	86	17	74	93	72	Fret tonnes-km
Mail ton-km	3	12	13	13	3	11	12	12	Courrier tonnes-km
Total ton-km	210	575	644	604	109	252	326	284	Total tonnes-km
Iraq									**Iraq**
Kilometres flown	...	20	12	0	...	19	12	0	Kilomètres parcourus
Passengers carried	...	1161	702	28	...	891	540	6	Passagers transportés
Passengers km	...	2269	1370	17	...	2152	1300	7	Passagers-km
Freight ton-km	...	73	43	0	...	72	43	0	Fret tonnes-km
Mail ton-km	...	1	1	0	...	1	1	0	Courrier tonnes-km
Total ton-km	...	278	167	1	...	267	160	1	Total tonnes-km
Israel									**Israël**
Kilometres flown	30	46	46	50	28	43	43	47	Kilomètres parcourus
Passengers carried	1483	2118	2004	2047	1043	1666	1539	1624	Passagers transportés
Passengers km	4727	7595	7127	7527	4590	7482	7010	7421	Passagers-km
Freight ton-km	295	817	832	863	295	816	831	862	Fret tonnes-km
Mail ton-km	4	6	5	7	4	6	5	7	Courrier tonnes-km
Total ton-km	724	1506	1478	1548	712	1495	1467	1537	Total tonnes-km
Japan									**Japon**
Kilometres flown	365	497	491	520	150	249	252	266	Kilomètres parcourus
Passengers carried	45145	69322	76224	68347	4499	10411	11134	10631	Passagers transportés
Passengers km	51217	93235	100501	100431	22254	49471	52491	49601	Passagers-km
Freight ton-km	1871	5129	5084	5234	1609	4539	4470	4608	Fret tonnes-km
Mail ton-km	108	278	304	328	90	140	155	168	Courrier tonnes-km
Total ton-km	6184	13308	13830	14135	3778	9294	9523	9411	Total tonnes-km
Jordan									**Jordanie**
Kilometres flown	21	32	28	23	21	32	28	23	Kilomètres parcourus
Passengers carried	1113	1204	964	797	1070	1148	931	782	Passagers transportés
Passengers km	2607	3679	2782	2439	2595	3665	2773	2435	Passagers-km
Freight ton-km	80	206	223	162	80	206	223	162	Fret tonnes-km
Mail ton-km	1	3	3	2	1	3	3	2	Courrier tonnes-km
Total ton-km	316	542	479	386	315	541	478	386	Total tonnes-km

104
Civil Aviation
Scheduled airline traffic: Passengers (000); Kilometres (million) [*cont.*]
Aviation civile
Vols réguliers: Passagers (000); Kilomètres (millions) [*suite*]

Country or area and traffic	Total Totale				International Internationaux				Pays ou zone et trafic
	1980	1989	1990	1991	1980	1989	1990	1991	
Korea, Dem. P. R.									**Corée, R. p. dém. de**
Kilometres flown	...	...	3	...	...	...	1	...	Kilomètres parcourus
Passengers carried	...	...	223	...	...	...	35	...	Passagers transportés
Passengers km	...	...	182	...	...	...	101	...	Passagers-km
Freight ton-km	...	...	3	...	...	...	1	...	Fret tonnes-km
Total ton-km	...	...	20	...	...	...	10	...	Total tonnes-km
Korea, Republic of									**Corée, République de**
Kilometres flown	61	118	128	137	55	92	99	108	Kilomètres parcourus
Passengers carried	3567	13161	15684	16908	2105	4356	4828	4992	Passagers transportés
Passengers km	10833	18163	20051	20716	10240	15223	16455	16765	Passagers-km
Freight ton-km	836	2365	2459	2597	832	2320	2405	2542	Fret tonnes-km
Mail ton-km	14	48	50	39	14	47	49	39	Courrier tonnes-km
Total ton-km	1850	4041	4303	4623	1797	3737	3935	4251	Total tonnes-km
Kuwait									**Koweït**
Kilometres flown	20	28	18	14	20	28	18	14	Kilomètres parcourus
Passengers carried	1076	1641	966	840	1076	1641	966	840	Passagers transportés
Passengers km	2114	3893	2300	1908	2114	3893	2300	1908	Passagers-km
Freight ton-km	72	233	145	116	72	233	145	116	Fret tonnes-km
Mail ton-km	4	6	3	1	4	6	3	1	Courrier tonnes-km
Total ton-km	265	597	360	289	265	597	360	289	Total tonnes-km
Lao People's Dem. Rep.									**Rép. dém. pop. lao**
Kilometres flown	...	...	1	...	...	...	1	...	Kilomètres parcourus
Passengers carried	...	...	115	...	...	...	28	...	Passagers transportés
Passengers km	...	...	44	...	...	...	18	...	Passagers-km
Freight ton-km	...	...	1	...	...	...	0	...	Fret tonnes-km
Total ton-km	...	...	4	...	...	...	2	...	Total tonnes-km
Lebanon									**Liban**
Kilometres flown	43	15	16	18	43	15	16	18	Kilomètres parcourus
Passengers carried	930	183	572	536	930	183	491	536	Passagers transportés
Passengers km	1571	324	941	1150	1571	324	933	1150	Passagers-km
Freight ton-km	531	284	167	170	531	284	167	170	Fret tonnes-km
Mail ton-km	5	1	1	2	5	1	1	2	Courrier tonnes-km
Total ton-km	680	314	254	276	680	314	253	276	Total tonnes-km
Malaysia									**Malaisie**
Kilometres flown	41	71	83	98	22	39	50	60	Kilomètres parcourus
Passengers carried	4516	8695	10242	11838	1822	3834	4181	4662	Passagers transportés
Passengers km	4076	10056	11862	14226	2916	8077	9349	11284	Passagers-km
Freight ton-km	110	404	574	713	95	384	549	679	Fret tonnes-km
Mail ton-km	7	14	13	18	3	10	7	10	Courrier tonnes-km
Total ton-km	501	1326	1622	1941	384	1131	1383	1658	Total tonnes-km
Maldives									**Maldives**
Passengers carried	...	...	9	...	...	...	0	...	Passagers transportés
Passengers km	...	...	3	...	...	...	0	...	Passagers-km
Mongolia									**Mongolie**
Kilometres flown	...	...	5	5	...	...	2	2	Kilomètres parcourus
Passengers carried	...	...	552	554	...	...	92	94	Passagers transportés
Passengers km	...	...	411	415	...	...	171	175	Passagers-km
Freight ton-km	...	...	2	1	...	...	1	0	Fret tonnes-km
Mail ton-km	...	...	1	1	...	...	1	1	Courrier tonnes-km
Total ton-km	...	...	38	38	...	...	17	17	Total tonnes-km
Myanmar									**Myanmar**
Kilometres flown	...	...	4	...	...	...	0	...	Kilomètres parcourus
Passengers carried	...	...	319	...	...	...	19	...	Passagers transportés
Passengers km	...	...	140	...	...	...	14	...	Passagers-km
Freight ton-km	...	...	1	...	...	...	0	...	Fret tonnes-km
Mail ton-km	...	...	1	...	...	...	0	...	Courrier tonnes-km
Total ton-km	...	...	14	...	...	...	2	...	Total tonnes-km
Nepal									**Népal**
Kilometres flown	6	11	11	9	3	8	8	5	Kilomètres parcourus
Passengers carried	380	679	679	672	164	364	364	357	Passagers transportés
Passengers km	234	688	716	706	196	637	665	655	Passagers-km
Freight ton-km	3	11	11	11	2	10	11	11	Fret tonnes-km
Total ton-km	22	67	70	69	19	63	66	65	Total tonnes-km

104
Civil Aviation
Scheduled airline traffic: Passengers (000); Kilometres (million) [*cont.*]
Aviation civile
Vols réguliers: Passagers (000); Kilomètres (millions) [*suite*]

Country or area and traffic	Total Totale				International Internationaux				Pays ou zone et trafic
	1980	1989	1990	1991	1980	1989	1990	1991	
Oman[4]									**Oman**[4]
Kilometres flown	7	12	13	12	7	11	13	11	Kilomètres parcourus
Passengers carried	522	835	853	958	522	753	771	876	Passagers transportés
Passengers km	714	1545	1602	1729	714	1492	1549	1676	Passagers-km
Freight ton-km	26	40	44	51	26	40	44	51	Fret tonnes-km
Mail ton-km	0	3	3	3	0	3	3	3	Courrier tonnes-km
Total ton-km	91	188	198	216	91	183	193	211	Total tonnes-km
Pakistan									**Pakistan**
Kilometres flown	50	61	63	60	37	40	41	37	Kilomètres parcourus
Passengers carried	3029	5009	5180	5198	1501	2008	2124	1973	Passagers transportés
Passengers km	5696	9129	9386	9062	4522	6862	7152	6711	Passagers-km
Freight ton-km	235	419	421	373	220	388	388	341	Fret tonnes-km
Mail ton-km	6	8	8	9	5	7	7	8	Courrier tonnes-km
Total ton-km	763	1259	1284	1210	643	1021	1047	963	Total tonnes-km
Philippines									**Philippines**
Kilometres flown	42	59	57	60	26	36	35	39	Kilomètres parcourus
Passengers carried	3246	5715	5639	5438	997	1776	1673	1833	Passagers transportés
Passengers km	5959	10592	10390	11028	4880	8671	8528	9307	Passagers-km
Freight ton-km	150	327	316	308	134	290	273	267	Fret tonnes-km
Mail ton-km	5	13	15	20	5	13	14	19	Courrier tonnes-km
Total ton-km	709	1366	1351	1433	612	1185	1166	1260	Total tonnes-km
Qatar[4]									**Qatar**[4]
Kilometres flown	7	11	13	11	7	11	13	11	Kilomètres parcourus
Passengers carried	522	753	771	876	522	753	771	876	Passagers transportés
Passengers km	714	1492	1549	1676	714	1492	1549	1676	Passagers-km
Freight ton-km	26	40	44	51	26	40	44	51	Fret tonnes-km
Mail ton-km	0	3	3	3	0	3	3	3	Courrier tonnes-km
Total ton-km	91	183	193	211	91	183	193	211	Total tonnes-km
Saudi Arabia									**Arabie saoudite**
Kilometres flown	91	101	103	87	47	58	58	47	Kilomètres parcourus
Passengers carried	9241	9988	10311	9409	2348	3120	3453	2989	Passagers transportés
Passengers km	9938	15695	16068	14881	4958	10647	10830	10083	Passagers-km
Freight ton-km	166	605	610	487	138	536	543	438	Fret tonnes-km
Mail ton-km	9	21	33	22	7	16	28	18	Courrier tonnes-km
Total ton-km	1069	2038	2089	1848	591	1510	1545	1363	Total tonnes-km
Singapore									**Singapour**
Kilometres flown	69	114	120	133	69	114	120	133	Kilomètres parcourus
Passengers carried	3827	6660	7046	7745	3827	6660	7046	7745	Passagers transportés
Passengers km	14719	30466	31600	33452	14719	30466	31600	33452	Passagers-km
Freight ton-km	544	1640	1653	1741	544	1640	1653	1741	Fret tonnes-km
Mail ton-km	17	49	42	52	17	49	42	52	Courrier tonnes-km
Total ton-km	1959	4596	4702	4992	1959	4596	4702	4992	Total tonnes-km
Sri Lanka									**Sri Lanka**
Kilometres flown	8	16	18	20	8	16	18	20	Kilomètres parcourus
Passengers carried	235	744	892	893	235	744	892	893	Passagers transportés
Passengers km	691	2670	3424	3449	691	2670	3424	3449	Passagers-km
Freight ton-km	10	75	93	101	10	75	93	101	Fret tonnes-km
Mail ton-km	1	2	2	3	1	2	2	3	Courrier tonnes-km
Total ton-km	72	316	414	418	72	316	414	418	Total tonnes-km
Syrian Arab Republic									**Rép. arabe syrienne**
Kilometres flown	11	9	10	11	10	8	8	10	Kilomètres parcourus
Passengers carried	465	456	613	661	383	368	516	543	Passagers transportés
Passengers km	948	797	1105	1136	908	777	1068	1091	Passagers-km
Freight ton-km	16	10	18	14	16	10	18	14	Fret tonnes-km
Total ton-km	101	82	117	117	98	80	114	113	Total tonnes-km
Thailand									**Thaïlande**
Kilometres flown	42	93	101	107	37	81	85	91	Kilomètres parcourus
Passengers carried	2459	7365	8201	7709	1924	4729	4995	4742	Passagers transportés
Passengers km	6276	18833	19757	18246	5988	17353	17953	16554	Passagers-km
Freight ton-km	239	613	661	866	238	606	653	856	Fret tonnes-km
Mail ton-km	10	39	39	45	10	39	39	44	Courrier tonnes-km
Total ton-km	812	2354	2485	2561	789	2214	2314	2397	Total tonnes-km

104
Civil Aviation
Scheduled airline traffic: Passengers (000); Kilometres (million) [*cont.*]
Aviation civile
Vols réguliers: Passagers (000); Kilomètres (millions) [*suite*]

Country or area and traffic	Total Totale				International Internationaux				Pays ou zone et trafic
	1980	1989	1990	1991	1980	1989	1990	1991	
Turkey									**Turquie**
Kilometres flown	15	40	47	35	8	29	34	25	Kilomètres parcourus
Passengers carried	1254	3792	4337	2872	377	1460	1662	1097	Passagers transportés
Passengers km	1103	4254	5091	3351	689	3174	3785	2512	Passagers-km
Freight ton-km	10	82	101	74	8	74	92	68	Fret tonnes-km
Mail ton-km	2	4	4	4	2	3	3	3	Courrier tonnes-km
Total ton-km	113	455	543	377	72	359	429	302	Total tonnes-km
United Arab Emirates[4]									**Emirats arabes unis[4]**
Kilometres flown	7	26	31	34	7	26	31	34	Kilomètres parcourus
Passengers carried	522	1482	1686	2042	522	1482	1686	2042	Passagers transportés
Passengers km	714	3351	3876	4861	714	3351	3876	4861	Passagers-km
Freight ton-km	26	118	145	184	26	118	145	184	Fret tonnes-km
Mail ton-km	0	4	6	7	0	4	6	7	Courrier tonnes-km
Total ton-km	91	452	533	672	91	452	533	672	Total tonnes-km
Viet Nam									**Viet Nam**
Kilometres flown	...	...	2	...	...	...	0	...	Kilomètres parcourus
Passengers carried	...	...	89	...	...	...	18	...	Passagers transportés
Passengers km	...	...	87	...	...	...	15	...	Passagers-km
Freight ton-km	...	...	1	...	...	...	0	...	Fret tonnes-km
Total ton-km	...	...	8	...	...	...	1	...	Total tonnes-km
Yemen									**Yémen**
Kilometres flown	6	9	13	7	5	8	12	5	Kilomètres parcourus
Passengers carried	310	426	671	413	255	321	429	214	Passagers transportés
Passengers km	291	665	929	421	280	654	853	349	Passagers-km
Freight ton-km	1	8	12	6	1	8	11	5	Fret tonnes-km
Mail ton-km	0	1	1	0	0	1	1	0	Courrier tonnes-km
Total ton-km	27	70	99	50	26	69	92	44	Total tonnes-km
Europe[2]									**Europe[2]**
Kilometres flown	**1892**	**...**	**2865**	**2818**	**1488**	**...**	**2246**	**2223**	**Kilomètres parcourus**
Passengers carried	**128802**	**...**	**216464**	**201114**	**69391**	**...**	**115079**	**106361**	**Passagers transportés**
Passengers km	**203021**	**...**	**342866**	**321210**	**173908**	**...**	**290285**	**271462**	**Passagers-km**
Freight ton-km	**8210**	**...**	**17327**	**16918**	**7946**	**...**	**16896**	**16530**	**Fret tonnes-km**
Mail ton-km	**596**	**...**	**842**	**879**	**538**	**...**	**741**	**778**	**Courrier tonnes-km**
Total ton-km	**27042**	**...**	**49535**	**47364**	**24200**	**...**	**44381**	**42506**	**Total tonnes-km**
Austria									**Autriche**
Kilometres flown	22	37	42	48	22	36	41	47	Kilomètres parcourus
Passengers carried	1284	2284	2532	2606	1267	2169	2409	2485	Passagers transportés
Passengers km	1120	3020	3828	3605	1115	2978	3780	3559	Passagers-km
Freight ton-km	12	42	54	59	12	42	54	59	Fret tonnes-km
Mail ton-km	3	5	6	7	3	5	6	7	Courrier tonnes-km
Total ton-km	118	326	421	405	118	322	416	401	Total tonnes-km
Belgium									**Belgique**
Kilometres flown	55	69	78	73	55	69	78	73	Kilomètres parcourus
Passengers carried	1974	2812	3133	3018	1974	2812	3133	3018	Passagers transportés
Passengers km	4852	6761	7642	6223	4852	6761	7642	6223	Passagers-km
Freight ton-km	395	661	656	486	395	661	656	486	Fret tonnes-km
Mail ton-km	11	26	27	25	11	26	27	25	Courrier tonnes-km
Total ton-km	842	1295	1371	1071	842	1295	1371	1071	Total tonnes-km
Bulgaria									**Bulgarie**
Kilometres flown	...	32	31	18	...	24	23	16	Kilomètres parcourus
Passengers carried	...	1998	1907	646	...	1033	976	516	Passagers transportés
Passengers km	...	2334	2313	1171	...	1948	1937	1116	Passagers-km
Freight ton-km	...	8	8	6	...	8	8	6	Fret tonnes-km
Mail ton-km	...	3	3	1	...	2	2	1	Courrier tonnes-km
Total ton-km	...	221	219	112	...	185	184	107	Total tonnes-km
Czechoslovakia									**Tchécoslovaquie**
Kilometres flown	25	25	24	22	16	21	21	19	Kilomètres parcourus
Passengers carried	1461	1169	1096	837	585	875	812	653	Passagers transportés
Passengers km	1539	2195	2030	1736	1190	2060	1899	1655	Passagers-km
Freight ton-km	12	17	15	22	12	17	15	22	Fret tonnes-km
Mail ton-km	3	4	3	1	2	4	3	1	Courrier tonnes-km
Total ton-km	154	219	201	180	124	206	189	172	Total tonnes-km

104
Civil Aviation
Scheduled airline traffic: Passengers (000); Kilometres (million) [*cont.*]
Aviation civile
Vols réguliers: Passagers (000); Kilomètres (millions) [*suite*]

Country or area and traffic	Total Totale				International Internationaux				Pays ou zone et trafic
	1980	1989	1990	1991	1980	1989	1990	1991	
Denmark[5]									**Danemark**[5]
Kilometres flown	33	54	58	58	27	40	44	45	Kilomètres parcourus
Passengers carried	3330	4721	4840	4582	1354	2316	2426	2395	Passagers transportés
Passengers km	3296	4309	4657	4440	2637	3498	3854	3685	Passagers-km
Freight ton-km	116	122	124	119	112	117	118	112	Fret tonnes-km
Mail ton-km	14	17	17	22	11	14	14	20	Courrier tonnes-km
Total ton-km	423	529	562	544	359	448	482	467	Total tonnes-km
Finland									**Finlande**
Kilometres flown	36	55	61	62	24	38	42	42	Kilomètres parcourus
Passengers carried	2512	4299	4450	3999	969	1887	1967	1790	Passagers transportés
Passengers km	2139	4625	4859	4719	1603	3700	3874	3804	Passagers-km
Freight ton-km	48	129	135	128	46	127	133	126	Fret tonnes-km
Mail ton-km	5	8	8	8	5	8	8	8	Courrier tonnes-km
Total ton-km	243	548	575	556	194	467	489	476	Total tonnes-km
France									**France**
Kilometres flown	276	386	419	376	213	267	296	277	Kilomètres parcourus
Passengers carried	19521	33975	35964	31354	9952	13712	14059	12324	Passagers transportés
Passengers km	34130	51533	52912	48611	25938	34424	35579	32927	Passagers-km
Freight ton-km	1986	3819	3996	3747	1892	3648	3811	3600	Fret tonnes-km
Mail ton-km	107	126	176	184	92	99	135	149	Courrier tonnes-km
Total ton-km	5131	8549	8889	8281	4317	6845	7148	6722	Total tonnes-km
Germany 5									**Allemagne 5**
F. R. Germany									**R. f. Allemagne**
Kilometres flown	196	339	397	434	168	292	348	365	Kilomètres parcourus
Passengers carried	13046	19057	22147	24830	7458	11752	13619	13677	Passagers transportés
Passengers km	21056	36316	42387	43270	18932	33426	39123	38773	Passagers-km
Freight ton-km	1506	3840	3994	4109	1483	3815	3968	4081	Fret tonnes-km
Mail ton-km	79	146	151	171	68	127	130	139	Courrier tonnes-km
Total ton-km	3524	7487	8246	8445	3299	7179	7886	7973	Total tonnes-km
Greece									**Grèce**
Kilometres flown	40	54	56	49	30	37	41	36	Kilomètres parcourus
Passengers carried	4891	6632	6135	4937	1656	2274	2199	1733	Passagers transportés
Passengers km	5062	8015	7764	6193	4030	6708	6589	5231	Passagers-km
Freight ton-km	61	103	113	114	56	94	104	106	Fret tonnes-km
Mail ton-km	7	11	10	11	6	11	10	10	Courrier tonnes-km
Total ton-km	521	838	824	684	427	710	708	589	Total tonnes-km
Hungary									**Hongrie**
Kilometres flown	16	20	22	19	16	20	22	19	Kilomètres parcourus
Passengers carried	874	1316	1363	911	874	1316	1363	911	Passagers transportés
Passengers km	1020	1379	1505	1017	1020	1379	1505	1017	Passagers-km
Freight ton-km[6]	9	6	6	5	9	6	6	5	Fret tonnes-km[6]
Mail ton-km	0	2	2	2	0	2	2	2	Courrier tonnes-km
Total ton-km	111	132	144	99	111	132	144	99	Total tonnes-km
Iceland									**Islande**
Kilometres flown	11	17	19	19	9	14	16	17	Kilomètres parcourus
Passengers carried	542	736	759	773	300	484	504	526	Passagers transportés
Passengers km	1295	1631	1714	1789	1235	1568	1650	1728	Passagers-km
Freight ton-km	23	33	36	35	22	33	35	34	Fret tonnes-km
Mail ton-km	3	5	5	6	3	4	5	5	Courrier tonnes-km
Total ton-km	142	185	195	202	136	179	189	196	Total tonnes-km
Ireland									**Irlande**
Kilometres flown	22	42	45	44	21	40	43	42	Kilomètres parcourus
Passengers carried	1830	4376	4812	4765	1636	4036	4402	4407	Passagers transportés
Passengers km	2049	4299	4561	4163	2009	4242	4491	4101	Passagers-km
Freight ton-km	89	118	128	115	88	117	128	115	Fret tonnes-km
Mail ton-km	3	3	3	3	3	3	3	3	Courrier tonnes-km
Total ton-km	271	495	530	480	266	490	524	475	Total tonnes-km
Italy									**Italie**
Kilometres flown	139	175	195	201	96	112	128	133	Kilomètres parcourus
Passengers carried	9956	17438	19750	18910	4191	6336	7212	6887	Passagers transportés
Passengers km	14076	21493	23599	22653	11209	15752	17183	16429	Passagers-km
Freight ton-km	523	1119	1171	1215	504	1086	1139	1184	Fret tonnes-km

104
Civil Aviation
Scheduled airline traffic: Passengers (000); Kilometres (million) [*cont.*]
Aviation civile
Vols réguliers: Passagers (000); Kilomètres (millions) [*suite*]

Country or area and traffic	Total Totale				International Internationaux				Pays ou zone et trafic
	1980	1989	1990	1991	1980	1989	1990	1991	
Mail ton-km	24	28	30	32	22	27	28	30	Courrier tonnes-km
Total ton-km	1813	3082	3326	3286	1535	2530	2714	2692	Total tonnes-km
Luxembourg									**Luxembourg**
Kilometres flown	3	4	6	6	3	4	6	6	Kilomètres parcourus
Passengers carried	162	296	409	406	162	296	409	406	Passagers transportés
Passengers km	55	138	253	258	55	138	253	258	Passagers-km
Freight ton-km	0	1	1	1	0	1	1	1	Fret tonnes-km
Total ton-km	5	13	23	24	5	13	23	24	Total tonnes-km
Malta									**Malte**
Kilometres flown	6	8	9	10	6	8	9	10	Kilomètres parcourus
Passengers carried	401	507	598	649	401	507	598	649	Passagers transportés
Passengers km	602	737	903	1014	602	737	903	1014	Passagers-km
Freight ton-km	4	6	5	6	4	6	5	6	Fret tonnes-km
Mail ton-km	0	1	1	1	0	1	1	1	Courrier tonnes-km
Total ton-km	59	70	84	95	59	70	84	95	Total tonnes-km
Monaco									**Monaco**
Passengers carried	...	43	43	...	...	43	43	...	Passagers transportés
Passengers km	...	1	1	...	...	1	1	...	Passagers-km
Netherlands[7]									**Pays-Bas**[7]
Kilometres flown	109	158	165	178	107	156	163	177	Kilomètres parcourus
Passengers carried	4984	8253	8559	8893	4633	7812	8128	8500	Passagers transportés
Passengers km	14643	25896	29036	28197	14596	25837	28979	28147	Passagers-km
Freight ton-km	947	2003	2129	2224	947	2003	2129	2224	Fret tonnes-km
Mail ton-km	50	86	83	83	50	86	83	83	Courrier tonnes-km
Total ton-km	2347	4476	4709	4891	2342	4471	4704	4886	Total tonnes-km
Norway[5]									**Norvège**[5]
Kilometres flown	58	82	91	91	27	38	44	43	Kilomètres parcourus
Passengers carried	4804	8300	8929	8857	1354	2296	2437	2382	Passagers transportés
Passengers km	4068	5912	6502	6291	2637	3471	3840	3630	Passagers-km
Freight ton-km	120	128	130	123	112	117	118	112	Fret tonnes-km
Mail ton-km	18	21	22	27	11	14	14	20	Courrier tonnes-km
Total ton-km	493	675	729	710	359	446	481	462	Total tonnes-km
Poland									**Pologne**
Kilometres flown	35	38	39	30	26	33	36	29	Kilomètres parcourus
Passengers carried	1711	1907	1501	1051	931	1536	1315	963	Passagers transportés
Passengers km	2232	3734	3479	2878	1934	3583	3401	2845	Passagers-km
Freight ton-km	14	29	49	38	12	28	49	38	Fret tonnes-km
Mail ton-km	5	4	3	2	4	4	3	2	Courrier tonnes-km
Total ton-km	207	355	351	288	182	343	345	285	Total tonnes-km
Portugal									**Portugal**
Kilometres flown	39	52	56	60	30	41	45	48	Kilomètres parcourus
Passengers carried	1978	3187	3505	3572	971	1888	2086	2230	Passagers transportés
Passengers km	3459	6272	6881	7072	2793	5291	5845	6048	Passagers-km
Freight ton-km	106	160	167	163	93	144	150	146	Fret tonnes-km
Mail ton-km	6	13	12	13	5	10	9	9	Courrier tonnes-km
Total ton-km	424	738	799	812	349	630	684	700	Total tonnes-km
Romania									**Roumanie**
Kilometres flown	20	23	23	24	13	15	16	19	Kilomètres parcourus
Passengers carried	1112	1272	1322	1149	383	483	578	585	Passagers transportés
Passengers km	1209	1646	1834	2048	916	1281	1521	1798	Passagers-km
Freight ton-km	12	15	13	13	10	13	12	13	Fret tonnes-km
Total ton-km	109	146	159	178	83	115	133	157	Total tonnes-km
Spain									**Espagne**
Kilometres flown	164	188	200	220	96	106	113	127	Kilomètres parcourus
Passengers carried	15089	20270	21652	20945	5137	5538	5913	5443	Passagers transportés
Passengers km	15517	22848	24157	23200	10290	14762	15566	14800	Passagers-km
Freight ton-km	390	733	760	623	325	628	652	521	Fret tonnes-km
Mail ton-km	28	32	35	33	17	21	23	23	Courrier tonnes-km
Total ton-km	1808	2810	2957	2726	1267	1977	2076	1876	Total tonnes-km
Sweden[5]									**Suède**[5]
Kilometres flown	66	114	121	109	41	57	64	62	Kilomètres parcourus
Passengers carried	5209	10855	11403	9827	2031	3254	3570	3410	Passagers transportés
Passengers km	5342	8497	9118	8163	3955	5189	5722	5350	Passagers-km

104
Civil Aviation
Scheduled airline traffic: Passengers (000); Kilometres (million) [*cont.*]
Aviation civile
Vols réguliers: Passagers (000); Kilomètres (millions) [*suite*]

Country or area and traffic	Total Totale				International Internationaux				Pays ou zone et trafic
	1980	1989	1990	1991	1980	1989	1990	1991	
Freight ton-km	175	185	191	187	168	175	177	168	Fret tonnes-km
Mail ton-km	21	22	22	17	17	21	21	17	Courrier tonnes-km
Total ton-km	666	964	1025	935	538	667	718	672	Total tonnes-km
Switzerland									**Suisse**
Kilometres flown	98	134	143	146	97	128	137	141	Kilomètres parcourus
Passengers carried	5930	8229	8603	7974	5221	7242	7624	7008	Passagers transportés
Passengers km	10831	15536	16016	15327	10773	15278	15752	15087	Passagers-km
Freight ton-km	422	889	929	945	421	881	921	938	Fret tonnes-km
Mail ton-km	32	52	51	46	32	51	50	46	Courrier tonnes-km
Total ton-km	1419	2450	2529	2475	1413	2418	2497	2445	Total tonnes-km
United Kingdom[8]									**Royaume-Uni[8]**
Kilometres flown	426	578	635	614	370	495	548	527	Kilomètres parcourus
Passengers carried	25551	46354	47114	42861	18489	30279	34376	31187	Passagers transportés
Passengers km	56750	92283	104999	99856	54026	87676	99950	95191	Passagers-km
Freight ton-km	1428	3447	3825	4023	1420	3439	3816	4017	Fret tonnes-km
Mail ton-km	184	215	229	248	182	209	221	241	Courrier tonnes-km
Total ton-km	6742	12375	13930	13680	6503	11982	13500	13284	Total tonnes-km
Yugoslavia									**Yougoslavie**
Kilometres flown	35	47	48	32	21	35	36	26	Kilomètres parcourus
Passengers carried	3087	3509	3668	1888	969	1647	1941	1103	Passagers transportés
Passengers km	2984	5123	5678	3078	2094	4441	5047	2802	Passagers-km
Freight ton-km	38	136	135	59	32	130	131	57	Fret tonnes-km
Mail ton-km	2	4	4	3	2	4	4	3	Courrier tonnes-km
Total ton-km	305	638	686	367	228	576	628	342	Total tonnes-km
Oceania[2]									**Océanie[2]**
Kilometres flown	277	381	419	478	100	161	189	215	**Kilomètres parcourus**
Passengers carried	18133	22881	25286	28451	3254	6338	6432	6454	**Passagers transportés**
Passengers km	32195	49709	53250	57147	20881	36363	37672	37155	**Passagers-km**
Freight ton-km	714	1482	1571	1643	577	1354	1432	1485	**Fret tonnes-km**
Mail ton-km	63	97	107	105	46	79	86	83	**Courrier tonnes-km**
Total ton-km	3727	6156	6596	7023	2554	4814	5039	5047	**Total tonnes-km**
Australia									**Australie**
Kilometres flown	200	257	287	335	59	102	123	130	Kilomètres parcourus
Passengers carried	13649	15114	17553	21244	1961	4048	4208	4180	Passagers transportés
Passengers km	25555	37028	40797	44276	15769	26180	27686	26503	Passagers-km
Freight ton-km	517	1125	1222	1222	409	1031	1117	1093	Fret tonnes-km
Mail ton-km	54	73	89	94	39	57	69	73	Courrier tonnes-km
Total ton-km	2900	4598	5063	5384	1897	3512	3758	3635	Total tonnes-km
Cook Islands									**Iles Cook**
Kilometres flown	...	...	1	...	...	...	1	...	Kilomètres parcourus
Passengers carried	...	...	10	...	...	...	10	...	Passagers transportés
Passengers km	...	...	27	...	...	...	27	...	Passagers-km
Freight ton-km	...	...	1	...	...	...	1	...	Fret tonnes-km
Total ton-km	...	...	3	...	...	...	3	...	Total tonnes-km
Fiji									**Fidji**
Kilometres flown	7	10	11	11	3	7	8	8	Kilomètres parcourus
Passengers carried	322	459	433	414	90	278	271	255	Passagers transportés
Passengers km	250	820	210	594	141	791	191	573	Passagers-km
Freight ton-km	3	20	5	23	2	20	5	23	Fret tonnes-km
Mail ton-km	0	4	0	0	0	4	0	0	Courrier tonnes-km
Total ton-km	26	95	25	77	14	92	23	75	Total tonnes-km
Kiribati									**Kiribati**
Kilometres flown	...	1	1	...	...	0	0	...	Kilomètres parcourus
Passengers carried	...	25	25	...	...	2	2	...	Passagers transportés
Passengers km	...	9	9	...	...	5	5	...	Passagers-km
Freight ton-km	...	1	1	...	...	1	1	...	Fret tonnes-km
Total ton-km	...	2	2	...	...	1	1	...	Total tonnes-km
Marshall Islands									**Iles Marshall**
Kilometres flown	...	...	1	...	...	...	1	...	Kilomètres parcourus
Passengers carried	...	...	66	...	...	...	40	...	Passagers transportés
Passengers km	...	...	52	...	...	...	47	...	Passagers-km
Freight ton-km	...	...	3	...	...	...	3	...	Fret tonnes-km

104
Civil Aviation
Scheduled airline traffic: Passengers (000); Kilometres (million) [*cont.*]
Aviation civile
Vols réguliers: Passagers (000); Kilomètres (millions) [*suite*]

Country or area and traffic	Total Totale				International Internationaux				Pays ou zone et trafic
	1980	1989	1990	1991	1980	1989	1990	1991	
Mail ton-km	...	...	8		...	...	8	...	Courrier tonnes-km
Total ton-km	...	...	15	...	...	...	15	...	Total tonnes-km
Nauru									**Nauru**
Kilometres flown	...	...	2	2	...	...	2	2	Kilomètres parcourus
Passengers carried	...	...	55	59	...	...	55	59	Passagers transportés
Passengers km	...	...	103	110	...	...	103	110	Passagers-km
Freight ton-km	...	...	2	1	...	...	2	1	Fret tonnes-km
Total ton-km	...	...	11	11	...	...	11	11	Total tonnes-km
New Zealand									**Nouvelle-Zélande**
Kilometres flown	52	83	95	107	29	41	51	69	Kilomètres parcourus
Passengers carried	3497	5724	5866	5371	1041	1631	1722	1780	Passagers transportés
Passengers km	5725	10613	11279	11299	4563	8608	9268	9536	Passagers-km
Freight ton-km	186	317	325	380	160	290	296	354	Fret tonnes-km
Mail ton-km	8	18	17	9	6	17	16	9	Courrier tonnes-km
Total ton-km	731	1331	1404	1454	599	1126	1195	1272	Total tonnes-km
Papua New Guinea									**Papouasie-Nvl-Guinée**
Kilometres flown	11	19	17	16	4	4	4	4	Kilomètres parcourus
Passengers carried	559	963	931	907	99	132	138	141	Passagers transportés
Passengers km	520	712	681	682	301	352	357	365	Passagers-km
Freight ton-km	9	15	15	15	6	11	12	12	Fret tonnes-km
Mail ton-km	1	1	1	1	1	1	1	1	Courrier tonnes-km
Total ton-km	57	80	77	77	33	43	44	45	Total tonnes-km
Solomon Islands									**Iles Salomon**
Kilometres flown	1	...	2	...	0	...	1	...	Kilomètres parcourus
Passengers carried	42	...	69	...	0	...	8	...	Passagers transportés
Passengers km	39	...	13	...	0	...	4	...	Passagers-km
Total ton-km	3	...	1	...	0	...	0	...	Total tonnes-km
Tonga									**Tonga**
Kilometres flown	...	...	1	...	...	...	0	...	Kilomètres parcourus
Passengers carried	...	...	35	...	...	...	0	...	Passagers transportés
Passengers km	...	...	7	...	...	...	0	...	Passagers-km
Total ton-km	...	...	1	...	...	...	0	...	Total tonnes-km
Vanuatu									**Vanuatu**
Kilometres flown	...	...	1	...	...	...	1		Kilomètres parcourus
Passengers carried	...	...	19	...	...	...	19	...	Passagers transportés
Passengers km	...	...	33	...	...	...	33	...	Passagers-km
Total ton-km	...	...	3	...	...	...	3	...	Total tonnes-km
former USSR[9]									**ancienne URSS[9]**
Kilometres flown	...	147	161	143	...	147	161	143	Kilomètres parcourus
Passengers carried	103754	131420	136647	128275	2503	4251	4422	3761	Passagers transportés
Passengers km	160299	226734	240802	224947	8982	16542	17838	15869	Passagers-km
Freight ton-km	2511	2645	2545	2352	259	365	392	338	Fret tonnes-km
Mail ton-km	573	555	536	384	68	34	29	13	Courrier tonnes-km
Total ton-km	17510	23693	24754	22953	1135	1887	2031	1752	Total tonnes-km

Source:
International Civil Aviation Organization (Montreal).

Source:
Organisation de l'aviation civile internationale (Montréal).

§   All data shown which pertain to Germany prior to 3 October 1990 are indicated separately for the Federal Republic of Germany and the former German Democratic Republic based on their respective territories at the time indicated. Where data for united Germany (3 October 1990 and thereafter) are not available, available data are shown separately under the designations Federal Republic of Germany and former German Democratic Republic and pertain to the territorial boundaries prior to 3 October 1990. For detailed explanatory notes on data pertaining to Germany, see Annex I - Country Nomenclature.

§   Toutes les données se rapportant à l'Allemagne avant le 3 octobre 1990 figurent dans deux rubriques séparées basées sur les territoires respectifs de la République fédérale d'Allemagne et l'ancienne République démocratique allemande selon la période indiquée. En l'absence de données pour l'Allemagne unifiée (à compter du 3 octobre 1990), les données disponibles sont fournies séparément sous les rubriques République fédérale d'Allemagne et ancienne République démocratique allemande et se rapportent aux limites territoriales antérieures au 3 octobre 1990. Pour les notes explicatives en détail sur les données concernant l'Allemagne, voir Annexe I - Nomenclature des pays.

1   Excludes former USSR because only certain categories of data are available.
2   Regional totals add to world totals. However, individual country

1   Non compris l'ancienne URSS, la disponibilité de données étant limitée a certaines catégories seulement.
2   La somme des totaux par régions est égale dans chaque cas au total

104
Civil Aviation
Scheduled airline traffic: Passengers (000); Kilometres (million) [*cont.*]
Aviation civile
Vols réguliers: Passagers (000); Kilomètres (millions) [*suite*]

statistics do not add to regional totals because (i) not all countries are shown, and (ii) part of data shown for three countries in Europe (France, Netherlands and United Kingdom) are not included in the totals for Europe but in totals of other regions because certain of their airlines operate exclusively in other regions.

3 Includes apportionment (1/10) of the traffic of Air Afrique, a multinational airline with headquarters in Côte d'Ivoire and operated by 10 African States.

4 Includes apportionment (1/4) of the traffic of Gulf Air, a multinational airline with headquarters in Bahrain.

5 Includes an apportionment of international operations performed by Scandinavian Airlines System (SAS), Denmark (2/7), Norway (2/7), Sweden (3/7).

6 Data for 1980 include mail ton-kilometers.

7 Including data for airlines based in the Netherlands territories and dependencies.

8 Including data for airlines based in the United Kingdom territories and dependencies.

9 Figures for the former USSR are not included in the World Total.

mondial correspondant. En revanche, la somme des données statistiques par pays pour chaque région n'est pas égale ou total régional correspondant, pour deux raisons: (i) tous les pays ne figurent pas dans les statistiques, (ii) une partie des données indiquées pour trois pays européens (France, Pays-Bas et Royaume-Uni) figurent non pas dans les totaux de l'Europe, mais dans ceux d'autres régions parce que certaines de leurs compagnies aériennes fonctionnent exclusivement dans les régions en question.

3 Ces chiffres comprennent une partie du trafic (1/10) assurée par Air Afrique, compagnie aérienne multinationale dont le siège est situé en Côte d'Ivoire et est exploitée conjointement par 10 Etats Africains.

4 Ces chiffres comprennent une partie du trafic (1/4) assuré par Air Gulfe, compagnie aérienne mutlinationale dont le siège est situé en Bahreïn.

5 Y compris une partie des vols internationaux effectués par le SAS, Danemark (2/7), Norvège (2/7) et Suède (3/7).

6 Les données pour 1980 comprennent courrier tonnes-km.

7 Y compris les données relatives aux compagnies aériennes ayant des bases d'opérations dans les territoires et dépendance des Pays-Bas.

8 Y compris les données relatives aux compagnies aériennes ayant des bases d'opération dans les territoires et dépendances du Royaume-Uni.

9 Les chiffres relatifs à l'ancienne URSS ne sont pas compris dans le total mondial.

105
Telephones
Téléphones
Number in use and per 100 inhabitants
Nombre en service et par 100 habitants

Country or area Pays ou zone	Number (000) Nombre (000)					Per 100 inhabitants Par 100 habitants				
	1980	1987	1988	1989	1990	1980	1987	1988	1989	1990
Africa · Afrique										
Algeria Algérie	485	889	959	1 051	1 103	2.5	3.8	4.0	4.2	4.3
Benin Bénin	...	14[1]	16	17	16	...	0.3[1]	0.4	0.8	0.3
Botswana Botswana	...	27	34	41	48	...	2.3	2.9	3.3	3.9
Burkina Faso Burkina Faso	...	12[1]	18	...	...	...	0.1[1]	0.2	...	...
Cameroon Cameroun	...	62	61	...	...	...	0.6	0.6	...	...
Cape Verde[2] Cap-Vert[2]	2	6[1]	6[1]	...	...	0.6	1.6[1]	1.7[1]	...	...
Central African Rep.[3] Rép. centrafricaine[3]	5	4[1]	7	...	...	0.2	0.2[1]	0.3	...	...
Chad Tchad	...	5[1]	10	...	8	...	0.1[1]	0.2	...	0.1
Congo Congo	17[3]	19	23	26	16[1]	1.1[3]	1.0	1.2	1.3	0.7[1]
Djibouti Djibouti	5	9	8	8	15	1.7	2.3	1.9	1.9	3.5
Egypt Egypte	534[2]	1 455	1 455	...	2 233	1.2[2]	2.8	2.9	...	4.2
Ethiopia Ethiopie	86	137	138	153	159	0.3	0.3	0.3	0.3	0.3
Gambia[5] Gambie[5]	3[4]	4[1]	7	8	11[1]	0.6[4]	0.5[1]	1.0	1.2	1.2[1]
Ghana Ghana	67[4]	75	75	77	79	0.7[4]	0.6	0.6	0.6	0.5
Kenya Kenya	192	316	337	357	383	1.2	1.4	1.5	1.5	1.5
Lesotho Lesotho	...	18	19	...	...	...	1.1	1.2	...	...
Madagascar Madagascar	38	25[1]	44	...	...	0.4	0.2[1]	0.4	...	...
Malawi Malawi	31	47	50	50	50	0.5	0.6	0.6	0.6	0.6
Mauritius Maurice	37	67	72	74	75	3.9	6.6	7.1	7.2	7.2
Morocco Maroc	231	343	362	410	476	1.2	1.5	1.6	1.7	1.9
Mozambique Mozambique	...	63	63	64	66	...	0.4	0.4	0.4	0.4
Niger Niger	9	9[1]	12	...	...	0.2	0.1[1]	0.2	...	...
Nigeria Nigéria	...	227[1]	722	...	...	...	0.2[1]	0.7	...	...
Réunion Réunion	52	133	155	172	168	10.7	24.7	28.7	31.2	27.6

105
Telephones
Number in use and per 100 inhabitants [cont.]
Téléphones
Nombre en service et par 100 habitants [suite]

Country or area Pays ou zone	Number (000) Nombre (000)					Per 100 inhabitants Par 100 habitants				
	1980	1987	1988	1989	1990	1980	1987	1988	1989	1990
Rwanda Rwanda	5	10	11	12	14	0.1	0.2	0.2	0.2	0.2
Saint Helena Sainte-Hélène	...	1	...	...	1	...	12.4	...	...	13.2
Sao Tome and Principe Sao Tomé-et-Principe	...	3	3	...	...	...	2.3	2.4	...	...
Seychelles Seychelles	7[5]	13	14	...	...	11.3	20.1	20.9	...	...
South Africa[5] Afrique du Sud[5]	2 662	4 236	4 462	4 764	5 077	11.2	14.3	14.6	14.9	15.2
Sudan Soudan	63	63[1]	78	...	...	0.3	0.3[1]	0.3	...	...
Swaziland Swaziland	12	21[5]	22	24	25	2.2	2.9[5]	2.9[5]	2.9[5]	3.3
Togo Togo	10	9[1]	9[1]	10[1]	11[1]	0.4	0.3[1]	0.3[1]	0.3[1]	0.3[1]
Tunisia Tunisie	189	312	333	370	410	3.0	4.1	4.3	4.7	5.1
Uganda Ouganda	47	59	60	61	57	0.4	0.4	0.4	0.4	0.3
United Rep.Tanzania Rép. Unie de Tanzanie	93	123	131	137	140	0.5	0.5	0.6	0.6	0.6
Zaire Zaïre	27	29[1]	32	...	...	0.1	0.1[1]	0.1	...	...
Zambia[5] Zambie[5]	61	81[6]	92	95	100[1]	1.1[6]	1.1[6]	1.2	1.2	1.3
Zimbabwe Zimbabwe	214	278	287	...	301	2.9	3.2	3.2	...	3.2
America, North · Amérique du Nord										
Aruba Aruba	...	23	...	...	36	...	38.6	...	...	55.3
Bahamas Bahamas	72	119	126	133	140	34.4	49.6	52.5	63.6	54.9
Barbados Barbade	67[2]	94	102	111	118	26.6[2]	37.2	40.8	42.5	45.7
British Virgin Islands[5] Iles Vierges britanniques[5]	3	...	...	...	...	26.0	...	...	...	...
Canada Canada	16 708	20 126	13 290[1]	13 820[1]	15 296[1]	69.4	78.0	51.3[1]	53.1[1]	57.0[1]
Cayman Islands Iles Caïmanes	7	16	17	18	19	42.9	73.5	60.9	69.7	69.1
Costa Rica Costa Rica	236	409	410	...	450	10.7	14.7	14.3	...	14.9
Cuba Cuba	406[3]	564	537	553	610	4.2[3]	5.4	5.2	5.3	5.8
Dominica Dominique	1	...	...	...	15	1.4	...	...	...	17.9
El Salvador El Salvador	86	136	...	...	250	1.9	2.7	...	...	4.8

105
Telephones
Number in use and per 100 inhabitants [cont.]
Téléphones
Nombre en service et par 100 habitants [suite]

Country or area Pays ou zone	Number (000) Nombre (000)					Per 100 inhabitants Par 100 habitants				
	1980	1987	1988	1989	1990	1980	1987	1988	1989	1990
Grenada[2] Grenade[2]	3	...	...	...	27	5.4	...	...	...	33.8
Guadeloupe Guadeloupe	45	100	103	110	121	13.7	30.4	31.2	33.0	31.0
Guatemala[1] Guatemala[1]	...	133	138	159	192	...	1.6	1.6	1.8	2.1
Haiti Haïti	...	...	50	...	...	...	...	0.9	...	...
Honduras Honduras	...	56	69	...	92	...	1.2	1.4	...	1.9
Martinique Martinique	52	107	117	125	159	16.8	32.3	35.2	37.4	43.6
Mexico Mexique	4 992	8 016	8 422	9 359	10 103	7.2	9.7	10.0	11.1	11.8
Montserrat Montserrat	2[5]	...	4	...	...	18.3[5]	...	32.4	...	...
Panama Panama	173	240	242	252	256	9.0	10.7	10.4	10.6	11.1
Saint Lucia Sainte-Lucie	...	...	18	...	...	...	...	12.1	...	...
Saint Pierre and Miquelon Saint-Pierre-et-Miquelon	3	4	4	5	5	50.0	67.7	72.7	73.8	77.8
St. Vincent-Grenadines St. Vincent-Grenadines	6[5]	...	...	...	17	4.6[5]	...	...	...	14.0
Trinidad and Tobago Trinité-et-Tobago	...	211	212	...	226	...	17.6	17.4	...	18.4
Turks and Caicos Islands Iles Turques et Caiques	...	...	...	3	...	...	...	...	44.1	...
United States Etats-Unis	180	119 097[1]	122 275[1]	125 836[1]	127 178[1]	78.8	48.9[1]	49.6[1]	50.6[1]	50.9[1]
United States Virgin Is. Iles Vierges américaines	...	55	58	59	59	...	52.3	54.7	55.0	59.0
America, South · Amérique du Sud										
Argentina Argentine	2 588	3 655	3 694	3 922	4 622	9.3	11.6	11.5	12.9	14.5
Bolivia Bolivie	...	...	183	196	...	...	...	2.8	2.9	...
Brazil Brésil	7 323	13 162	13 905	14 060	14 125	6.2	9.3	9.6	5.5	9.4
Chile Chili	551	815	867	895	1 096	5.0	6.5	6.8	6.9	8.3
Colombia Colombie	1 718	2 438	2 499	2 775	2 909	6.4	8.0	8.1	8.6	8.8
Ecuador Equateur	272	355	...	691	540	3.3	3.6	...	5.5	5.0
Falkland Is. (Malvinas) Iles Falkland (Malvinas)	...	1	1	...	...	...	14.4	33.3	...	...
French Guiana Guyane française	16	28	33	34	31	24.9	32.7	38.8	38.8	25.4

105
Telephones
Number in use and per 100 inhabitants [cont.]
Téléphones
Nombre en service et par 100 habitants [suite]

Country or area Pays ou zone	Number (000) Nombre (000)					Per 100 inhabitants Par 100 habitants				
	1980	1987	1988	1989	1990	1980	1987	1988	1989	1990
Guyana[1] Guyana[1]	...	...	...	...	16	...	...	...	...	2.0
Paraguay Paraguay	59	100	112	122	128	1.8	2.5	2.7	2.9	3.0
Peru Pérou	475	454[1]	489[1]	736	769	2.7	2.2[1]	2.3[1]	3.4	3.5
Suriname Suriname	...	39	41	44	48	...	9.7	10.1	10.5	11.5
Uruguay Uruguay	287	437	482	529	579	9.9	14.9	16.3	17.9	19.6
Venezuela Venezuela	...	1 677	1 749	1 758	1 794	...	9.2	9.3	9.1	9.1
Asia · Asie										
Bahrain Bahreïn	72	122	134	141	153	19.8	29.2	28.2	28.9	29.6
Brunei Darussalam Brunéi Darussalam	18	40	45	48	...	8.0	16.8	18.2	19.4	...
China Chine	4 186	8 057	9 418	10 893	12 735	0.4	0.7	0.9	1.0	1.1
Cyprus Chypre	113[2]	272	304	223[1]	370	17.9[2]	39.8	43.9	33.2[1]	52.3
Hong Kong Hong-kong	1 676	2 662	2 891	3 105	3 280	32.6	47.0	50.4	53.4	56.0
India[5] Inde[5]	...	4 420	4 756	4 167[1]	5 486	...	0.6	0.6	0.5[1]	0.7
Indonesia Indonésie	513	890	938	1 015	...	0.4	0.5	0.5	0.6	...
Iran, Islamic Rep. of[7] Iran, Rép. islamique d'[7]	1 384	1 944	2 079	2 104	2 270	3.7	3.9	4.1	4.0	3.9
Israel Israël	1 140[5]	2 065	2 190	2 285	2 425	29.6[5]	46.9	48.9	50.1	50.3
Japan[5] Japon[5]	53 634	46 325[1]	48 014[1]	49 946[1]	52 034[1]	52.0	38.1[1]	39.2[1]	40.2[1]	42.1[1]
Korea, Republic of Corée, République de	3 387	10 732	12 415	14 195	15 736	9.0	25.8	29.6	33.5	36.7
Kuwait Koweït	215	330	362	...	...	15.9	18.1	18.9	18.9	...
Lao People's Dem. Rep. Rép. dém. pop. lao	...	...	6[1]	...	7	...	...	0.2[1]	...	0.2
Macau Macao	...	69	80	96	111	...	15.9	18.2	21.2	24.1
Malaysia Malaisie	598	1 501	1 646	...	2 023	4.5	9.1	9.7	...	11.3
Maldives[4] Maldives[4]	1	...	...	...	6[1]	0.6	...	...	...	2.3[1]
Oman Oman	23	80	88[1]	93[1]	104[1]	2.4	5.3	4.4[1]	4.6[1]	5.0[1]
Pakistan Pakistan	330	679	740	...	...	0.4	0.7	0.7	...	...

105
Telephones
Number in use and per 100 inhabitants [*cont.*]
Téléphones
Nombre en service et par 100 habitants [*suite*]

Country or area Pays ou zone	Number (000) Nombre (000)					Per 100 inhabitants Par 100 habitants				
	1980	1987	1988	1989	1990	1980	1987	1988	1989	1990
Philippines Philippines	702	...	994	986	1 047	1.5	...	1.7	1.7	1.7
Qatar Qatar	53	121	129	135	139	23.2	32.8	34.9	36.3	28.6
Saudi Arabia Arabie saoudite	443[2]	1 149[1]	1 120[1]	1 238[1]	1 384[1]	5.3[2]	10.6[1]	8.3[1]	9.2[1]	9.9[1]
Singapore Singapour	702[5]	1 164	1 220	982[1]	1 040[1]	29.1[5]	44.2	45.6	36.6[1]	38.7[1]
Sri Lanka Sri Lanka	85	132	185	167	166	0.6	0.8	1.1	1.0	1.0
Syrian Arab Republic Rép. arabe syrienne	335	653	669	685	695	3.7	5.9	5.8	5.8	5.6
Thailand Thaïlande	497	902[1]	1 006[1]	1 158[1]	1 325[1]	1.1	1.7[1]	1.8[1]	2.1[1]	2.3[1]
Turkey Turquie	1 902	4 827	6 388	7 467	8 517	4.2	9.1	11.7	13.6	15.1
United Arab Emirates Emirats arabes unis	209	404	463	544	655	20.1	27.8	30.9	35.1	41.2
Yemen[8] Yémen[8]	...	...	35	...	...	...	...	1.5	...	...
Europe · Europe										
Austria Autriche	3 010	3 979	4 128	4 310	4 541	40.1	52.4	54.3	56.6	58.9
Belgium Belgique	3 636	4 719	4 925	5 138	5 429	36.9	47.8	49.9	51.8	54.6
Channel Islands Iles Anglo-Normandes	90	135	84[9]	90[9]	84[1]	66.2	100.1	102.0[9]	108.7[9]	60.6[1]
Czechoslovakia Tchécoslovaquie	3 150	3 838	3 980	4 131	4 278	20.6	24.6	25.5	26.4	27.3
Denmark[10] Danemark[10]	3 283	4 434	4 509	4 398	5 000	64.1	86.4	88.2	85.7	97.2
Finland Finlande	2 374	2 365[1]	2 470[1]	2 582[1]	2 678[1]	49.6	47.9[1]	49.9[1]	52.0[1]	53.0[1]
France France	24 859	24 803[1]	25 827[1]	26 942[1]	28 085[1]	45.2	43.2[1]	45.2[1]	42.0[1]	48.2[1]
Germany† · Allemagne† F. R. Germany R. f. Allemagne	28 554	40 288	41 735	28 848[1]	41 735	46.4	65.0	68.2	46.5[1]	67.1
former German D. R. anc. R. d. allemande	3 156	3 875	3 977	1 826[1]	1 906[1]	18.9	23.3	23.9	11.0[1]	11.5[1]
Gibraltar Gibraltar	...	...	13	15	17	...	...	44.8	48.0	54.3
Greece Grèce	2 796	4 126	4 303	4 523	4 699	23.4	41.3	43.1	45.2	45.8
Hungary Hongrie	1 261	1 609	1 674	1 770	1 872	11.8	15.2	15.8	16.9	17.8
Iceland Islande	109	113[1]	117[1]	121[1]	126[1]	47.7	45.8[1]	46.6[1]	47.9[1]	49.6[1]

105
Telephones
Number in use and per 100 inhabitants [*cont.*]
Téléphones
Nombre en service et par 100 habitants [*suite*]

Country or area Pays ou zone	Number (000) Nombre (000)					Per 100 inhabitants Par 100 habitants				
	1980	1987	1988	1989	1990	1980	1987	1988	1989	1990
Ireland Irlande	650	796[1]	842[1]	903[1]	983[1]	18.7	22.5[1]	23.8[1]	25.5[1]	27.9[1]
Italy[11] Italie[11]	19 277	28 052	29 300	30 716	32 037	33.7	48.8	50.9	53.3	55.5
Malta Malte	79[5]	155	169	172	179	25.3[5]	44.7	47.1	49.1	50.2
Monaco Monaco	30	...	47	...	53	107.1	...	166.4	...	187.5
Netherlands Pays-Bas	7 357	* 9 410	* 9 750	6 691[1]	6 940[1]	51.8	64.0	65.9	44.9[1]	46.2[1]
Norway Norvège	1 881	1 948[1]	2 016[1]	2 070[1]	2 132[1]	46.0	46.4[1]	47.8[1]	48.9[1]	50.2[1]
Poland Pologne	3 387	4 618	4 830	...	5 232	9.5	12.2	12.8	...	13.7
Portugal Portugal	1 372	2 073	2 258	...	2 769	13.9	20.2	21.9	...	26.3
San Marino Saint-Marin	8[2]	14	16	...	...	36.6[2]	62.5	68.0	...	...
Spain Espagne	11 845	15 477	10 522[1]	11 292[1]	12 603[1]	31.7	39.6	28.0[1]	30.4[1]	32.3[1]
Sweden Suède	6 621	5 480[1]	5 601[1]	5 716[1]	5 849[1]	79.6	65.1[1]	66.2[1]	67.0[1]	68.1[1]
Switzerland Suisse	4 612	5 783	5 879	6 050	6 153	72.2	87.4	88.2	90.0	90.5
United Kingdom[5 12] Royaume-Uni[5 12]	26 651	22 137[1]	22 770[1]	23 848[1]	24 913[1]	47.7	39.0[1]	45.5[1]	41.7[1]	43.4[1]
Yugoslavia Yougoslavie	2 139	3 909	4 243	4 550	3 842[1]	9.5	16.6	18.0	19.2	16.1[1]
Oceania · Océanie										
Australia Australie	7 153[13]	6 965[1]	7 268[1]	7 602[1]	...	48.9[13]	42.8[1]	44.0[1]	44.8[1]	...
Cook Islands Iles Cook	...	3	3	4	4	...	15.4	16.9	20.0	23.8
Fiji Fidji	43	60	62	67	73	6.8	8.4	8.7	9.4	10.1
French Polynesia Polynésie française	23	44	...	...	...	15.5	24.5	...	...	...
Kiribati Kiribati	...	1	1	...	...	...	2.2	2.0	...	...
Nauru Nauru	2	...	2	...	...	21.8	...	19.3	...	...
New Zealand[5] Nouvelle-Zélande[5]	1 730	2 315	2 403	1 452[1]	1 473[1]	55.0	69.7	71.7	43.2[1]	43.0[1]
Papua New Guinea Papouasie-Nvl-Guinée	49	70	72	73	63	1.6	2.0	2.0	2.0	1.7
Solomon Islands Iles Salomon	...	5	...	7	...	...	1.9	...	2.2	...
Tuvalu Tuvalu	...	0	...	...	...	...	1.9	...	1.3[1]	1.9

105
Telephones
Number in use and per 100 inhabitants [*cont.*]
Téléphones
Nombre en service et par 100 habitants [*suite*]

Country or area Pays ou zone	Number (000) Nombre (000)					Per 100 inhabitants Par 100 habitants				
	1980	1987	1988	1989	1990	1980	1987	1988	1989	1990
Wallis and Futuna Islands Iles Wallis et Futuna	...	0	0	0	1	...	2.5	2.5	3.1	3.7
former USSR · ancienne URSS										
former USSR ancienne URSS	23 707	...	37 532	...	...	8.9	...	...	...	...

Source:
International Telecommunications Union (Geneva).

† All data shown which pertain to Germany prior to 3 October 1990 are indicated separately for the Federal Republic of Germany and the former German Democratic Republic based on their respective territories at the time indicated. Where data for united Germany (3 October 1990 and thereafter) are not available, available data are shown separately under the designations Federal Republic of Germany and former German Democratic Republic and pertain to the territorial boundaries prior to 3 October 1990. For detailed explanatory notes on data pertaining to Germany, see Annex I - Country Nomenclature.

1 Main lines.
2 Source: American Telephone and Telegraph Company.
3 At 31 December 1981.
4 At 31 December 1979.
5 At 31 March of year stated.
6 At 31 March of following year.
7 At 21 March of year stated.
8 Former Democratic Yemen only.
9 Jersey only.
10 Excluding Faeroe Islands and Greenland.
11 Including San Marino.
12 Excluding Channel Islands.
13 At 30 June of year stated.

Source:
Union internationale des télécommunications (Genève).

† Toutes les données se rapportant à l'Allemagne avant le 3 octobre 1990 figurent dans deux rubriques séparées basées sur les territoires respectifs de la République fédérale d'Allemagne et l'ancienne République démocratique allemande selon la période indiquée. En l'absence de données pour l'Allemagne unifiée (à compter du 3 octobre 1990), les données disponibles sont fournies séparément sous les rubriques République fédérale d'Allemagne et ancienne République démocratique allemande et se rapportent aux limites territoriales antérieures au 3 octobre 1990. Pour les notes explicatives en détail sur les données concernant l'Allemagne, voir Annexe I – Nomenclature des pays.

1 Lignes principales.
2 Source: "American Telephone and Telegraph Company".
3 Au 31 décembre 1981.
4 Au 31 décembre 1979.
5 Au 31 mars de l'année indiquée.
6 Au 31 mars de l'année suivante.
7 Au 21 mars de l'année indiquée.
8 L'ancien Yémen démocratique seulement.
9 Jersey seulement.
10 Non compris les Iles Féroé et le Groënland.
11 Y compris Saint-Marin.
12 Non compris les Iles Anglo-Normandes.
13 Au 30 juin de l'année indiquée.

Technical notes, tables 100-105

Table 100: Data refer to domestic and international traffic on all railway lines within each country shown, except railways entirely within an urban unit, and plantation, industrial mining, funicular and cable railways. The figures relating to passenger-kilometres include all passengers except military, government and railway personnel when carried without revenue. Those relating to ton-kilometres are freight net ton-kilometres and include both fast and ordinary goods services but exclude service traffic, mail, baggage and non-revenue governmental stores.

Table 101: For years in which a census or registration took place the census or registration figure is shown; for other years, unless otherwise indicated, the officially estimated number of vehicles in use is shown. The time of year to which the figures refer is variable. Special purpose vehicles such as two- or three-wheeled cycles and motorcycles, trams, trolley-buses, ambulances, hearses, military vehicles operated by police or other governmental security organizations are excluded. Passenger cars includes vehicles seating not more than nine persons (including the driver), such as taxis, jeeps and station wagons. Commercial vehicles includes: vans, lorries (trucks), buses, tractor and semi-trailer combinations but excluding trailers and farm tractors.

Table 102: Data refer to merchant fleets registered in each country on 30 June of the year stated. They are given in gross registered tons (100 cubic feet or 2.83 cubic metres) and represent the total volume of all the permanently enclosed spaces of the vessels to which the figures refer. Vessels without mechanical means of propulsion are excluded, but sailing vessels with auxiliary power are included.

Part A of the table refers to the total of merchant fleets registered. Part B shows data for oil tanker fleets and part C data for ore-oil and bulk carrier fleets.

Table 103: The figures for vessels entered and cleared, unless otherwise stated, represent the sum of the net registered tonnage of sea-going foreign and domestic merchant vessels (power and sailing) entered with cargo from or cleared with cargo to a foreign port and refer to only one entrance or clearance for each foreign voyage. The data where possible exclude vessels "in ballast", i.e. entering without unloading or clearing without loading goods.

Table 104: Data for total services cover both domestic and international scheduled services operated by airlines registered in each country. Scheduled services include supplementary services occasioned by overflow traffic on regularly scheduled trips and preparatory flights for newly scheduled services. Freight means all goods, except mail and excess baggage, carried for remuneration.

Notes techniques, tableaux 100-105

Tableau 100 : Les données se rapportent au trafic intérieur et international de toutes les lignes de chemins de fer du pays indiqué, à l'exception des lignes situées entièrement à l'intérieur d'une agglomération urbaine ou desservant une plantation ou un complexe industriel minier, des funiculaires et des téléfériques. Les chiffres relatifs aux voyageurs-kilomètres se rapportent à tous les voyageurs sauf les militaires, les fonctionnaires et le personnel des chemins de fer, qui sont transportés gratuitement. Les chiffres relatifs aux tonnes-kilomètres se rapportent aux tonnes-kilomètres nettes de fret et comprennent les services rapides et ordinaires de transport de marchandises, à l'exception des transports pour les besoins du service, du courrier, des bagages et des marchandises transportées gratuitement pour les besoins de l'Etat.

Tableau 101 : Pour les années où a eu lieu un recensement ou un enregistrement des véhicules, le chiffre indiqué est le résultat de cette opération; pour les autres années, sauf indication contraire, le chiffre indiqué correspond à l'estimation officielle du nombre de véhicules en circulation. L'époque de l'année à laquelle se rapportent les chiffres varie. Les véhicules à usage spécial, tels que les cycles à deux ou trois roues et motocyclettes, les tramways, les trolley-bus, les ambulances, les corbillards, les véhicules militaires utilisés par la police ou par d'autres services publics de sécurité ne sont pas compris dans ces chiffres. Les voitures de tourisme comprennent les véhicules automobiles dont le nombre de places assises (y compris celle du conducteur) n'est pas supérieur à neuf, tels que les taxis, jeeps et breaks. Les véhicules utilitaires comprennent les fourgons, camions, autobus et autocars, les ensembles tracteurs-remorques et semi-remorques, mais ne comprennent pas les remorques et les tracteurs agricoles.

Tableau 102 : Les données se rapportent à la flotte marchande enregistrée dans chaque pays au 30 juin de l'année indiquée. Elles sont exprimées en tonneaux de jauge brute (100 pieds cubes ou 2,83 mètres cubes) et représentent le volume total de tous les espaces clos en permanence dans les navires auxquels elle s'appliquent. Elles excluent les navires sans moteur, mais pas les voiliers avec moteurs auxiliaires.

Les données de la Partie A du tableau se rapportent au total de la flotte marchande enregistrée. Celles de la Partie B se rapportent à la flotte des pétroliers, et celles de la Partie C à la flotte des minéraliers et des transporteurs de vrac et d'huile.

Tableau 103 : sauf indication contraire, les données relatives aux navires entrés et sortis représentent la jauge nette totale des navires marchands de haute mer (à moteur ou à voile) nationaux ou étrangers, qui entrent ou sortent chargés, en provenance ou à destination d'un port étranger. On ne compte qu'une seule entrée et une seule sortie pour chaque voyage international. Dans la mesure du possible, le tableau exclut les navires sur lest (c'est-à-dire les navires entrant sans décharger ou sortant sans avoir chargé).

Table 105: The figures unless otherwise stated, relate to telephone stations (sets) of all kinds connected to the public network. Unless otherwise stated, the data refer to 31 December of the year stated. Where the American Telephone and Telegraph Company is cited as the source, the figures refer to 1 January of the following year.

Tableau 104 : Les données relatives au total des services se rapportent aux services réguliers, intérieurs ou internationaux, des compagnies de transport aérien enregistrées dans chaque pays. Les services réguliers comprennent aussi les vols supplémentaires nécessités par un surcroît d'activité des services réguliers et les vols préparatoires en vue de nouveaux services réguliers. Par fret, on entend toutes les marchandises transportées contre paiement, mais non le courrier et les excédents de bagage.

Tableau 105 : Sauf indication contraire, les données se rapportent aux postes téléphoniques de toutes natures raccordés au réseau public. Sauf indication contraire, les données se rapportent au 31 décembre de l'année indiquée. Dans les cas où l'American Telephone and Telegraph Company est indiquée comme source, les données se rapportent au 1er janvier de l'année suivante.

106
Production, trade and consumption of commercial energy
Production, commerce et consommation d'énergie commerciale
Thousand metric tons of coal equivalent and kilograms per capita
Milliers de tonnes métriques d'équivalent houille et kilogrammes par habitant

| Country or area | Year | Primary energy production - Production d'energie primaire | | | | | Changes in stocks Variations des stocks | Imports Importations | Exports Exportations |
		Total Totale	Solids Solides	Liquids Liquides	Gas Gaz	Electricity Electricité			
Africa	1980	100 916	7	74 976	25 903	31	3 575	2 593	69 193
Algeria	1988	124 565	10	78 017	46 515	22	-191	1 225	93 990
	1989	132 018	15	81 151	50 819	33	436	1 225	101 944
	1990	156 705	10	89 291	67 387	17	1 086	1 340	112 540
Angola	1980	10 849	...	10 611	102	135	0	230	9 048
	1988	32 259	...	31 886	208	165	0	21	29 817
	1989	32 734	...	32 346	222	166	0	21	30 345
	1990	34 036	...	33 647	222	167	0	21	31 626
Benin	1980	14	...	14	...	...	15	232	38
	1988	393	...	393	...	...	4	251	402
	1989	407	...	407	...	...	4	259	416
	1990	416	...	416	...	...	4	259	424
Burkina Faso	1980	...	...	...	...	...	1	207	2
	1988	...	...	...	...	...	4	252	2
	1989	...	...	...	...	...	4	263	2
	1990	...	...	...	...	...	4	262	2
Burundi	1980	2	2	...	...	13	0	58	...
	1988	19	6	...	...	13	6	81	...
	1989	20	7	...	...	13	1	93	...
	1990	18	6	...	...	13	4	99	...
Cameroon	1980	4 025	0	3 857	...	168	549	881	2 303
	1988	12 427	1	12 117	...	309	-112	17	9 448
	1989	12 660	1	12 336	...	323	-648	17	10 152
	1990	12 439	1	12 114	...	324	-998	26	10 295
Cape Verde	1980	...	...	...	...	...	0	63	0
	1988	...	...	...	...	...	0	41	0
	1989	...	...	...	...	...	0	44	0
	1990	...	...	...	...	8	0	47	0
Central African Rep.	1980	8	...	...	...	9	1	60	...
	1988	9	...	...	...	9	-13	112	...
	1989	9	...	...	...	9	-38	109	...
	1990	9	...	...	...	9	1	115	...
Chad	1980	...	...	...	...	...	0	134	...
	1988	...	...	...	...	...	1	144	...
	1989	...	...	...	...	...	-1	144	...
	1990	...	...	...	...	...	1	152	...
Comoros	1980	0	...	...	...	...	0	19	...
	1988	0	...	...	...	0	0	25	...
	1989	0	...	...	...	0	0	25	...
	1990	0	...	...	...	0	0	32	...
Congo	1980	4 721	...	4 707	2	12	136	148	4 577
	1988	10 093	0	10 054	3	35	978	27	8 245
	1989	11 426	0	11 374	3	49	950	19	9 608
	1990	11 589	0	11 537	3	49	64	19	10 641
Côte d'Ivoire	1980	338	...	171	...	167	265	2 813	740
	1988	974	...	821	...	152	13	3 583	383
	1989	466	...	271	...	195	13	4 156	368
	1990	645	...	450	...	195	16	3 931	349
Djibouti	1980	...	...	...	...	...	0	708	...
	1988	...	...	...	...	...	0	735	...
	1989	...	...	...	...	...	0	743	...
	1990	...	...	...	...	...	0	750	...
Egypt	1980	45 956	...	42 618	2 134	1 204	714	1 918	22 922
	1988	72 312	...	63 349	7 992	970	857	2 155	34 662
	1989	73 295	...	63 774	8 539	983	714	2 472	35 244
	1990	75 197	...	65 174	9 028	995	714	2 232	35 092

Bunkers - Soutes			Consumption - Consommation							
Air Avion	Sea Maritime	Unallocated Nondistribué	Per capita Par habitant	Total Totale	Solids Solides	Liquids Liquides	Gas Gaz	Electricity Electricité	Année	Pays ou zone
										Afrique
376	606	4 985	1 322	24 773	607	6 462	17 672	32	1980	Algérie
206	287	8 914	981	22 584	1 122	9 522	11 917	24	1988	
206	287	7 638	961	22 731	1 125	9 765	11 816	25	1989	
236	250	6 943	1 523	36 989	1 220	10 069	25 696	5	1990	
96	456	437	135	1 042	0	805	102	135	1980	Angola
221	978	416	92	848	0	475	208	165	1988	
228	978	334	92	870	0	482	222	166	1989	
231	974	328	92	898	0	509	222	167	1990	
15	...	0	52	179	...	164	...	14	1980	Bénin
22	...	0	51	216	...	191	...	25	1988	
22	...	0	51	223	...	200	...	23	1989	
22	...	0	50	224	...	200	...	24	1990	
...	...	...	29	204	0	204	...	0	1980	Burkina Faso
...	...	...	30	246	0	246	...	0	1988	
...	...	...	30	257	0	257	...	0	1989	
...	...	...	29	256	0	256	...	0	1990	
1	...	...	14	59	2	52	...	5	1980	Burundi
3	...	...	18	91	6	70	...	14	1988	
3	...	...	21	108	7	87	...	14	1989	
3	...	...	21	109	6	90	...	14	1990	
100	...	92	215	1 861	0	1 694	...	168	1980	Cameroun
118	...	232	258	2 759	1	2 449	...	309	1988	
72	...	207	261	2 893	1	2 569	...	323	1989	
72	...	210	252	2 886	1	2 561	...	324	1990	
0	7	...	194	56	0	56	...	...	1980	Cap-Vert
0	7	...	100	34	0	34	...	...	1988	
0	7	...	106	37	0	37	...	...	1989	
0	7	...	111	40	0	40	...	...	1990	
7	...	...	26	60	...	52	...	8	1980	Rép. centrafricaine
15	...	...	43	119	...	110	...	9	1988	
18	...	...	48	139	...	129	...	9	1989	
18	...	...	36	105	...	96	...	9	1990	
33	...	...	23	101	0	101	...	...	1980	Tchad
30	...	...	21	113	0	113	...	...	1988	
30	...	...	21	116	0	116	...	...	1989	
31	...	...	21	119	0	119	...	...	1990	
0	...	...	48	19	...	19	...	0	1980	Comores
0	...	...	51	25	...	25	...	0	1988	
0	...	...	49	25	...	25	...	0	1989	
0	...	...	62	33	...	32	...	0	1990	
6	13	0	83	138	...	123	2	13	1980	Congo
1	8	83	390	804	0	747	3	54	1988	
1	8	47	390	830	0	771	3	56	1989	
1	7	49	384	845	0	786	3	57	1990	
76	3	169	232	1 898	...	1 731	...	167	1980	Côte d'Ivoire
91	12	1 275	260	2 783	...	2 631	...	152	1988	
94	12	1 787	211	2 349	...	2 154	...	195	1989	
96	13	1 563	220	2 540	...	2 345	...	195	1990	
112	451	...	474	144	...	144	...	...	1980	Djibouti
94	466	...	467	175	...	175	...	...	1988	
94	472	...	459	177	...	177	...	...	1989	
96	476	...	448	178	...	178	...	...	1990	
236	1 465	1 607	512	20 930	685	16 907	2 134	1 204	1980	Egypte
133	1 607	2 847	703	34 360	1 149	24 248	7 992	970	1988	
486	1 993	2 748	691	34 582	1 224	23 837	8 539	983	1989	
441	2 149	3 009	703	36 024	1 030	24 971	9 028	995	1990	

106
Production, trade and consumption of commercial energy
Thousand metric tons of coal equivalent and kilograms per capita [*cont.*]
Production, commerce et consommation d'énergie commerciale
Milliers de tonnes métriques d'équivalent houille et kilogrammes par habitant [*suite*]

Country or area	Year	Primary energy production - Production d'energie primaire					Changes in stocks Variations des stocks	Imports Importations	Exports Exportations
		Total Totale	Solids Solides	Liquids Liquides	Gas Gaz	Electricity Electricité			
Equatorial Guinea	1980	0	...	...	...	0	1	28	...
	1988	0	...	...	...	0	1	50	...
	1989	0	...	...	...	0	1	54	...
	1990	0	...	...	...	0	1	55	...
Ethiopia	1980	61	...	...	...	61	57	1 209	276
	1988	95	...	...	...	95	40	1 570	300
	1989	96	...	...	...	96	29	1 585	301
	1990	97	...	...	...	97	40	1 607	202
Gabon	1980	12 759	...	12 707	20	32	-141	84	11 964
	1988	11 703	...	11 383	233	87	13	129	10 045
	1989	14 817	...	14 610	121	86	775	50	12 474
	1990	19495	...	19 276	133	87	17	53	17 950
Gambia	1980	...	...	...	...	...	0	80	4
	1988	...	...	...	...	...	0	91	3
	1989	...	...	...	...	...	0	90	3
	1990	...	...	...	...	...	0	93	3
Ghana	1980	774	...	126	...	648	13	1 568	518
	1988	597	...	0	...	597	63	1 755	134
	1989	643	...	0	...	643	53	1 779	126
	1990	662	...	0	...	662	13	1 779	130
Guinea	1980	18	...	...	...	18	0	452	...
	1988	21	...	...	...	21	6	489	...
	1989	21	...	...	...	21	6	493	...
	1990	21	...	...	...	21	6	493	...
Guinea-Bissau	1980	...	...	...	...	...	0	71	0
	1988	...	...	...	...	...	3	92	0
	1989	...	...	...	...	...	3	96	0
	1990	...	...	...	...	...	3	106	0
Kenya	1980	130	...	...	...	130	-119	4 850	1 639
	1988	325	...	...	...	325	0	3 176	999
	1989	343	...	...	...	343	0	3 259	801
	1990	353	...	...	...	353	0	3 483	774
Liberia	1980	41	...	...	...	41	14	1 026	21
	1988	40	...	...	...	40	0	450	26
	1989	39	...	...	...	39	0	397	26
	1990	30	...	...	...	30	0	249	1
Libyan Arab Jamah.	1980	132 830	...	128 245	4 585	...	547	921	122 476
	1988	81 394	...	72 079	9 315	...	-6 107	5	67 607
	1989	89 815	...	79 235	10 581	...	-1 804	7	69 149
	1990	110 523	...	97 858	12 666	...	4 853	9	80 055
Madagascar	1980	18	...	...	...	18	-399	520	160
	1988	37	...	...	...	37	82	773	77
	1989	39	...	...	...	39	-51	479	85
	1990	39	...	...	...	39	-14	506	45
Malawi	1980	48	...	...	...	48	0	315	0
	1988	70	...	...	...	70	1	229	0
	1989	70	...	...	...	70	1	244	0
	1990	70	...	...	...	70	1	256	0
Maldives	1980	...	...	...	...	...	...	20	...
	1988	...	...	...	...	...	...	41	...
	1989	...	...	...	...	...	...	41	...
	1990	...	...	...	...	...	...	45	...
Mali	1980	10	...	...	...	10	...	204	...
	1988	21	...	...	...	21	...	215	...
	1989	21	...	...	...	21	...	219	...
	1990	21	...	...	...	21	...	218	...
Mauritania	1980	3	...	...	...	3	0	314	...
	1988	3	...	...	...	3	-58	1 549	...
	1989	3	...	...	...	3	1	1 432	...
	1990	3	...	...	...	3	1	1 346	...

Bunkers - Soutes			Consumption - Consommation							
Air Avion	Sea Maritime	Unallocated Nondistribué	Per capita Par habitant	Total Totale	Solids Solides	Liquids Liquides	Gas Gaz	Electricity Electricité	Année	Pays ou zone
...	...	...	124	27	...	26	...	0	1980	Guinée équatoriale
...	...	...	144	48	...	48	...	0	1988	
...	...	...	156	53	...	53	...	0	1989	
...	...	...	157	54	...	54	...	0	1990	
33	4	106	21	795	0	733	...	61	1980	Ethiopie
78	6	-6	28	1 247	0	1 152	...	95	1988	
81	6	27	27	1 237	0	1 141	...	96	1989	
80	6	104	27	1 272	0	1 175	...	97	1990	
30	90	75	1 026	827	...	775	20	32	1980	Gabon
34	101	723	864	914	...	594	233	87	1988	
36	104	682	728	797	...	590	121	86	1989	
39	100	624	722	818	...	599	133	87	1990	
...	...	...	117	75	...	75	...	...	1980	Gambie
...	...	...	111	88	...	88	...	...	1988	
...	...	...	107	87	...	87	...	...	1989	
...	...	...	108	90	...	90	...	...	1990	
56	43	43	155	1 669	2	1 073	...	594	1980	Ghana
35	28	458	119	1 634	3	1 069	...	562	1988	
32	26	503	119	1 681	3	1 070	...	608	1989	
31	30	713	105	1 524	3	894	...	627	1990	
9	...	...	103	461	...	443	...	18	1980	Guinée
16	...	...	93	488	...	467	...	21	1988	
16	...	...	91	492	...	471	...	21	1989	
18	...	...	88	491	...	470	...	21	1990	
7	...	...	81	64	...	64	...	...	1980	Guinée-Bissau
9	...	...	88	80	...	80	...	...	1988	
7	...	...	93	86	...	86	...	...	1989	
7	...	...	102	96	...	96	...	...	1990	
374	273	535	137	2 278	18	2 091	...	169	1980	Kenya
103	151	102	99	2 147	114	1 695	...	339	1988	
88	225	37	110	2 450	132	1 962	...	357	1989	
88	233	75	115	2 665	152	2 140	...	374	1990	
12	47	35	500	938	...	897	...	41	1980	Libéria
10	60	0	168	394	...	354	...	40	1988	
10	75	0	135	325	...	286	...	39	1989	
6	33	0	96	239	...	209	...	30	1990	
251	71	3 102	2 400	7 304	1	5 695	1 608	...	1980	Jamah.arabe libyenne
295	71	4 057	3 790	15 476	2	7 695	7 780	...	1988	
295	99	3 290	4 441	18 794	4	9 847	8 943	...	1989	
310	113	3 744	4 894	21 458	5	10 438	11 016	...	1990	
1	26	-3	86	752	21	713	...	18	1980	Madagascar
1	20	44	54	587	17	532	...	37	1988	
1	19	16	40	448	13	396	...	39	1989	
1	16	16	41	481	15	427	...	39	1990	
15	...	...	56	348	65	235	...	48	1980	Malawi
15	...	...	36	283	16	197	...	70	1988	
15	...	...	37	299	18	210	...	70	1989	
16	...	...	36	308	19	219	...	70	1990	
...	...	...	129	20	...	20	...	...	1980	Maldives
...	...	...	210	41	...	41	...	...	1988	
...	...	...	203	41	...	41	...	...	1989	
...	...	...	216	45	...	45	...	...	1990	
22	...	...	28	192	...	182	...	10	1980	Mali
21	...	...	26	215	...	194	...	21	1988	
21	...	...	25	220	...	199	...	21	1989	
21	...	...	24	218	...	197	...	21	1990	
15	10	0	188	292	6	283	...	3	1980	Mauritanie
18	10	186	749	1 396	6	1 387	...	3	1988	
16	10	133	665	1 274	6	1 265	...	3	1989	
16	10	152	594	1 169	6	1 160	...	3	1990	

106
Production, trade and consumption of commercial energy
Thousand metric tons of coal equivalent and kilograms per capita [*cont.*]
Production, commerce et consommation d'énergie commerciale
Milliers de tonnes métriques d'équivalent houille et kilogrammes par habitant [*suite*]

| Country or area | Year | Primary energy production - Production d'energie primaire | | | | | Changes in stocks | | |
		Total Totale	Solids Solides	Liquids Liquides	Gas Gaz	Electricity Electricité	Variations des stocks	Imports Importations	Exports Exportations
Mauritius	1980	10	...	...	...	10	-4	446	0
	1988	12	...	...	...	12	-23	714	0
	1989	18	...	...	...	18	34	838	0
	1990	10	...	...	...	10	40	884	0
Morocco	1980	974	680	20	90	184	-351	6 026	141
	1988	879	637	29	99	115	318	8 597	9
	1989	749	504	19	84	142	598	9 627	2
	1990	765	526	21	68	150	164	9 946	2
Mozambique	1980	1 871	207	...	...	1 664	0	1 437	1 296
	1988	51	45	...	...	6	0	546	0
	1989	48	42	...	...	6	0	553	0
	1990	46	40	...	...	6	0	532	0
Niger	1980	20	20	...	...	...	0	259	...
	1988	155	155	...	...	...	1	312	...
	1989	157	157	...	...	...	1	315	...
	1990	158	158	...	...	...	1	321	...
Nigeria	1980	151 121	176	148 843	1 764	338	1 786	3 228	139 328
	1988	102 137	82	96 764	5 019	271	2 857	5 775	79 646
	1989	127 694	85	121 679	5 659	271	0	4 223	109 064
	1990	128 927	90	123 626	4 940	272	0	4 031	110 720
Reunion	1980	37	...	...	...	37	21	355	0
	1988	52	...	...	...	52	3	493	1
	1989	61	...	...	...	61	3	517	1
	1990	62	...	...	...	62	3	548	1
Rwanda	1980	15	...	...	1	14	0	133	0
	1988	21	...	...	0	21	0	197	0
	1989	21	...	...	0	21	0	197	0
	1990	21	...	...	0	21	0	204	0
Sao Tome - Principe	1980	1	...	...	...	1	...	19	...
	1988	1	...	...	...	1	...	31	...
	1989	1	...	...	...	1	...	31	...
	1990	1	...	...	...	1	...	32	...
Senegal	1980	...	...	...	...	...	3	1 985	222
	1988	...	...	...	...	...	44	1 385	48
	1989	...	...	...	...	...	12	1 637	15
	1990	...	...	...	...	...	7	1 524	51
Seychelles	1980	...	...	...	...	...	1	99	...
	1988	...	...	...	...	...	9	244	...
	1989	...	...	...	...	...	9	278	...
	1990	...	...	...	...	...	-4	208	...
Sierra Leone	1980	...	...	...	...	...	0	437	1
	1988	...	...	...	...	...	4	472	3
	1989	...	...	...	...	...	4	475	3
	1990	...	...	...	...	...	4	479	3
Somalia	1980	...	...	...	...	...	77	580	135
	1988	...	...	...	...	...	4	579	58
	1989	...	...	...	...	...	3	554	58
	1990	...	...	...	...	...	1	510	53
South Africa Customs Un.	1980	93 519	93 379	...	...	140	-775	24 080	29 358
	1988	136 161	135 581	...	...	579	0	23 040	43 307
	1989	133 058	132 475	...	...	583	0	23 327	42 869
	1990	133 708	133 131	...	...	577	0	22 474	42 981
St.Helena and Depend.	1980	0	0	...	...	...	...	1	...
	1988	0	0	...	...	...	...	3	...
	1989	0	0	...	...	...	...	3	...
	1990	0	0	...	...	...	...	3	...
Sudan	1980	83	...	...	...	83	0	1 662	21
	1988	114	...	...	...	114	14	2 000	54
	1989	115	...	...	...	115	21	1 808	54
	1990	115	...	...	...	115	14	1 863	46

Bunkers - Soutes			Consumption - Consommation							
Air Avion	Sea Maritime	Unallocated Nondistribué	Per capita Par habitant	Total Totale	Solids Solides	Liquids Liquides	Gas Gaz	Electricity Electricité	Année	Pays ou zone
78	92	...	300	290	1	279	...	10	1980	Maurice
177	140	...	415	433	38	383	...	12	1988	
178	147	...	470	497	75	404	...	18	1989	
205	114	...	500	535	75	450	...	10	1990	
170	56	474	336	6 511	642	5 595	90	184	1980	Maroc
85	4	618	364	8 442	1 698	6 530	99	115	1988	
91	0	769	374	8 916	1 801	6 888	84	142	1989	
96	0	1 200	379	9 250	1 762	7 270	68	150	1990	
37	122	35	150	1 818	288	1 046	...	485	1980	Mozambique
62	41	0	34	494	65	383	...	46	1988	
59	36	0	34	506	64	396	...	46	1989	
56	36	0	32	486	58	382	...	46	1990	
12	...	...	48	268	20	237	...	11	1980	Niger
21	...	...	63	445	155	268	...	22	1988	
21	...	...	62	450	157	271	...	23	1989	
21	...	...	61	457	158	276	...	23	1990	
562	468	245	153	11 961	180	9 690	1 764	327	1980	Nigéria
531	484	4 776	200	19 619	60	14 280	5 019	259	1988	
540	506	121	213	21 687	58	15 711	5 659	259	1989	
523	512	803	194	20 400	62	15 138	4 940	260	1990	
1	1	...	726	369	...	331	...	37	1980	Réunion
15	1	...	924	524	...	472	...	52	1988	
15	1	...	965	557	...	496	...	61	1989	
15	1	...	1 002	589	...	527	...	62	1990	
6	...	...	28	143	...	122	1	20	1980	Rwanda
13	...	...	31	204	...	181	0	23	1988	
13	...	...	30	205	...	181	0	23	1989	
13	...	...	30	212	...	188	0	23	1990	
0	...	...	213	20	...	19	...	1	1980	Sao Tomé-et-Principe
0	...	...	286	32	...	31	...	1	1988	
0	...	...	278	32	...	31	...	1	1989	
0	...	...	280	33	...	32	...	1	1990	
171	358	16	219	1 215	...	1 215	...	...	1980	Sénégal
144	171	-5	146	983	...	983	...	...	1988	
144	188	26	181	1 254	...	1 254	...	...	1989	
144	190	99	145	1 031	...	1 031	...	...	1990	
32	20	...	714	45	...	45	...	...	1980	Seychelles
0	110	...	1 881	126	...	126	...	...	1988	
28	130	...	1 642	110	...	110	...	...	1989	
0	133	...	1 162	79	...	79	...	...	1990	
30	111	33	80	261	0	261	...	...	1980	Sierra Leone
24	104	36	78	301	0	301	...	...	1988	
24	104	32	78	308	0	308	...	...	1989	
24	108	32	76	308	0	308	...	...	1990	
31	43	103	36	190	...	190	...	...	1980	Somalie
24	39	34	62	421	...	421	...	...	1988	
24	39	11	60	420	...	420	...	...	1989	
16	39	13	53	387	...	387	...	...	1990	
103	4 483	107	2 604	84 323	67 861	15 151	...	1 310	1980	Un.douan.d'Afr.mérid
147	4 445	2 962	2 841	108 338	92 581	15 448	...	309	1988	
162	4 453	3 198	2 707	105 704	89 975	15 477	...	252	1989	
155	4 438	2 361	2 658	106 247	90 531	15 480	...	236	1990	
1	...	...	0	0	0	0	...	...	1980	Ste-Hélène et dépend
1	...	...	167	1	0	1	...	...	1988	
1	...	...	167	1	0	1	...	...	1989	
1	...	...	143	1	0	1	...	...	1990	
65	7	132	81	1 520	0	1 437	...	83	1980	Soudan
41	7	215	77	1 782	0	1 668	...	114	1988	
41	7	228	66	1 570	0	1 455	...	115	1989	
46	10	241	66	1 621	0	1 506	...	115	1990	

106
Production, trade and consumption of commercial energy
Thousand metric tons of coal equivalent and kilograms per capita [*cont.*]
Production, commerce et consommation d'énergie commerciale
Milliers de tonnes métriques d'équivalent houille et kilogrammes par habitant [*suite*]

Country or area	Year	Primary energy production - Production d'énergie primaire					Changes in stocks Variations des stocks	Imports Importations	Exports Exportations
		Total Totale	Solids Solides	Liquids Liquides	Gas Gaz	Electricity Electricité			
Togo	1980	0	0	...	...	0	0	724	495
	1988	0	0	...	...	0	0	283	16
	1989	1	0	...	...	1	0	283	16
	1990	1	0	...	...	1	0	283	13
Tunisia	1980	8 668	...	8 039	626	3	411	2 836	6 594
	1988	7 626	...	7 149	471	6	277	3 430	4 993
	1989	7 688	...	7 186	497	4	-519	4 840	6 600
	1990	7 030	...	6 601	423	5	-97	5 130	5 325
Uganda	1980	80	...	...	...	80	0	308	38
	1988	72	...	...	...	72	0	408	17
	1989	84	...	...	...	84	0	390	17
	1990	73	...	...	...	73	0	396	17
United Rep. Tanzania	1980	66	1	...	...	65	0	960	84
	1988	78	3	...	...	75	4	1 013	60
	1989	79	3	...	...	76	4	1 013	60
	1990	79	4	...	...	75	3	1 035	47
Western Sahara	1980	...	...	...	...	...	...	81	...
	1988	...	...	...	...	...	...	90	...
	1989	...	...	...	...	...	...	94	...
	1990	...	...	...	...	...	...	98	...
Zaire	1980	2 121	138	1 461	...	521	2	1 309	1 267
	1988	2 853	123	2 086	...	645	3	1 446	1 855
	1989	2 801	125	1 940	...	736	41	1 484	1 642
	1990	2877	126	2 014	...	737	18	1 516	1 691
Zambia	1980	1 606	488	...	...	1 118	-65	1 109	454
	1988	1 548	529	...	...	1 018	0	881	253
	1989	1 158	335	...	...	823	0	864	247
	1990	1 267	318	...	...	950	0	793	245
Zimbabwe	1980	3 626	3 134	...	...	492	0	1 329	334
	1988	5 392	5 065	...	...	327	0	1 536	153
	1989	5 438	5 111	...	...	327	0	1 489	125
	1990	5 391	4 958	...	...	433	0	1 174	158
America, North Antigua and Barbuda	1980	...	...	...	...	...	0	172	24
	1988	...	...	...	...	...	0	197	10
	1989	...	...	...	...	...	0	197	10
	1990	...	...	...	...	...	0	204	10
Aruba	1980	...	...	...	...	...	...	...	...
	1988	...	...	...	...	...	20	260	0
	1989	...	...	...	...	...	14	248	0
	1990	...	...	...	...	...	4	239	0
Bahamas	1980	...	...	...	...	...	363	16 517	11 430
	1988	...	...	...	...	...	103	4 021	2 736
	1989	...	...	...	...	...	177	4 294	2 923
	1990	...	...	...	...	...	102	4 296	3 014
Barbados	1980	78	...	60	18	...	-29	511	0
	1988	121	...	86	35	...	0	467	4
	1989	113	...	76	38	...	0	490	4
	1990	127	...	89	39	...	0	509	7
Belize	1980	...	...	...	...	...	0	106	0
	1988	...	...	...	...	...	0	140	0
	1989	...	...	...	...	...	0	168	0
	1990	...	...	...	...	...	42	180	0
Bermuda	1980	...	...	...	...	...	3	270	0
	1988	...	...	...	...	...	0	357	0
	1989	...	...	...	...	...	0	412	0
	1990	...	...	...	...	...	0	305	0
British Virgin Islds	1980	...	...	...	...	...	...	12	...
	1988	...	...	...	...	...	...	22	...
	1989	...	...	...	...	...	...	23	...
	1990	...	...	...	...	...	...	23	...

	Bunkers - Soutes		Consumption - Consommation							
Air Avion	Sea Maritime	Unallocated Nondistribué	Per capita Par habitant	Total Totale	Solids Solides	Liquids Liquides	Gas Gaz	Electricity Electricité	Année	Pays ou zone
...	...	46	70	183	0	161	...	22	1980	Togo
...	...	0	83	267	0	234	...	32	1988	
...	...	0	81	268	0	232	...	36	1989	
...	...	0	79	270	0	234	...	36	1990	
112	10	268	643	4 108	87	3 392	626	3	1980	Tunisie
134	3	156	720	5 493	116	3 947	1 428	2	1988	
208	4	152	779	6 082	112	3 818	2 153	-1	1989	
214	6	165	819	6 548	119	4 424	2 002	2	1990	
0	...	...	27	349	...	305	...	44	1980	Ouganda
0	...	...	28	463	...	405	...	59	1988	
0	...	...	26	458	...	387	...	70	1989	
0	...	...	25	452	...	393	...	60	1990	
12	57	9	46	866	2	799	...	65	1980	Rép.Unie de Tanzanie
7	34	41	39	944	4	865	...	75	1988	
7	34	39	37	947	4	867	...	76	1989	
7	35	34	37	987	5	907	...	75	1990	
4	0	...	570	77	...	77	...	...	1980	Sahara occidental
6	0	...	512	84	...	84	...	...	1988	
6	0	...	521	88	...	88	...	...	1989	
6	0	...	538	93	...	93	...	...	1990	
167	35	0	75	1 958	316	1 133	...	509	1980	Zaïre
154	44	107	66	2 136	321	1 184	...	631	1988	
146	44	121	69	2 291	330	1 238	...	723	1989	
151	44	144	68	2 346	335	1 287	...	724	1990	
52	...	0	396	2 274	510	982	...	782	1980	Zambie
56	...	115	265	2 005	521	648	...	837	1988	
55	...	111	205	1 610	330	638	...	641	1989	
52	...	61	209	1 702	314	620	...	768	1990	
81	...	...	637	4 540	2 907	805	...	828	1980	Zimbabwe
133	...	...	752	6 643	5 052	1 098	...	492	1988	
125	...	...	733	6 676	5 130	1 053	...	493	1989	
96	...	...	671	6 311	4 946	826	...	540	1990	
										Amérique du Nord
81	0	0	893	67	...	67	...	...	1980	Antigua-et-Barbuda
50	0	0	1 789	136	...	136	...	...	1988	
50	0	0	1 789	136	...	136	...	...	1989	
50	0	0	1 895	144	...	144	...	...	1990	
...	...	...	...	...	...	...	...	...	1980	Aruba
...	...	0	4 000	240	...	240	...	...	1988	
...	...	0	3 917	235	...	235	...	...	1989	
...	...	0	3 917	235	...	235	...	...	1990	
72	1 048	2 448	5 510	1 157	0	1 157	...	...	1980	Bahamas
46	504	0	2 627	633	0	633	...	...	1988	
46	504	0	2 633	645	0	645	...	...	1989	
46	504	0	2 530	630	0	630	...	...	1990	
140	147	16	1 261	314	0	297	18	...	1980	Barbade
155	14	35	1 492	379	0	344	35	...	1988	
153	14	21	1 614	410	0	373	38	...	1989	
156	24	25	1 665	423	0	384	39	...	1990	
15	0	...	623	91	...	91	...	0	1980	Belize
21	0	...	680	119	...	119	...	0	1988	
24	0	...	804	144	...	144	...	0	1989	
16	0	...	667	122	...	122	...	0	1990	
55	10	...	3 759	203	0	203	...	...	1980	Bermudes
52	3	...	5 298	302	0	302	...	...	1988	
44	0	...	6 456	368	0	368	...	...	1989	
25	0	...	4 828	280	0	280	...	...	1990	
...	...	...	1 091	12	...	12	...	...	1980	Iles Vierges brit.
...	...	...	1 833	22	...	22	...	...	1988	
...	...	...	1 917	23	...	23	...	...	1989	
...	...	...	1 769	23	...	23	...	...	1990	

106

Production, trade and consumption of commercial energy

Thousand metric tons of coal equivalent and kilograms per capita [*cont.*]

Production, commerce et consommation d'énergie commerciale

Milliers de tonnes métriques d'équivalent houille et kilogrammes par habitant [*suite*]

Country or area	Year	Primary energy production - Production d'energie primaire					Changes in stocks Variations des stocks	Imports Importations	Exports Exportations
		Total Totale	Solids Solides	Liquids Liquides	Gas Gaz	Electricity Electricité			
Canada	1980	279 050	30 403	117 619	95 213	35 815	1 588	56 117	78 053
	1988	363 361	53 873	133 129	128 400	47 960	-478	57 664	149 939
	1989	365 329	54 288	129 522	135 906	45 614	-3 277	60 920	147 513
	1990	364 575	52 771	130 154	136 222	45 428	6 815	62 075	151 335
Cayman Islands	1980	...	...	...	...	...	...	88	...
	1988	...	...	...	...	...	...	125	...
	1989	...	...	...	...	...	...	136	...
	1990	...	...	...	...	...	...	129	...
Costa Rica	1980	259	...	...	...	259	42	1 149	58
	1988	375	...	...	...	375	-11	1 362	103
	1989	409	...	...	...	409	-45	1 452	120
	1990	432	...	...	...	432	17	1 459	45
Cuba	1980	426	...	391	22	12	-144	14 987	801
	1988	1 061	...	1 024	28	9	29	19 360	3 391
	1989	1 078	...	1 026	42	10	-47	19 913	3 397
	1990	1 266	...	1 214	41	10	0	18 580	2 913
Dominica	1980	1	...	...	...	1	...	18	...
	1988	2	...	...	...	2	...	27	...
	1989	2	...	...	...	2	...	28	...
	1990	2	...	...	...	2	...	28	...
Dominican Republic	1980	71	...	...	...	71	151	2 879	...
	1988	103	...	...	...	103	0	2 834	...
	1989	104	...	...	...	104	0	2 812	...
	1990	104	...	...	...	104	0	2 683	...
El Salvador	1980	177	...	...	...	177	14	995	0
	1988	215	...	...	...	215	-35	1 066	70
	1989	232	...	...	...	232	-37	1 076	36
	1990	257	...	...	...	257	-26	1 146	75
Greenland	1980	0	0	...	...	...	0	263	...
	1988	0	0	...	...	...	0	269	13
	1989	0	0	...	...	...	0	236	13
	1990	0	0	...	...	...	0	273	12
Grenada	1980	...	...	...	...	...	...	24	...
	1988	...	...	...	...	...	...	50	...
	1989	...	...	...	...	...	...	52	...
	1990	...	...	...	...	...	...	59	...
Guadeloupe	1980	...	...	...	...	...	-15	281	0
	1988	...	...	...	...	...	0	606	0
	1989	...	...	...	...	...	0	621	0
	1990	...	...	...	...	...	0	627	0
Guatemala	1980	324	...	290	...	34	13	2 067	168
	1988	481	...	254	...	227	0	1 656	143
	1989	514	...	257	...	257	0	1 628	143
	1990	507	...	250	...	257	0	1 569	86
Haiti	1980	27	...	...	...	27	4	316	...
	1988	39	...	...	...	39	0	316	...
	1989	39	...	...	...	39	0	319	...
	1990	40	...	...	...	40	0	322	...
Honduras	1980	96	...	...	...	96	6	915	30
	1988	108	...	...	...	108	0	808	0
	1989	108	...	...	...	108	0	814	0
	1990	109	...	...	...	109	0	789	0
Jamaica	1980	15	...	...	...	15	-47	3 975	63
	1988	12	...	...	...	12	-143	1 972	102
	1989	16	...	...	...	16	0	2 412	101
	1990	16	...	...	...	16	0	2 246	86
Martinique	1980	...	...	...	...	...	-109	525	290
	1988	...	...	...	...	...	9	1 005	294
	1989	...	...	...	...	...	11	906	298
	1990	...	...	...	...	...	7	906	296

Bunkers - Soutes			Consumption - Consommation							
Air Avion	Sea Maritime	Unallocated Nondistribué	Per capita Par habitant	Total Totale	Solids Solides	Liquids Liquides	Gas Gaz	Electricity Electricité	Année	Pays ou zone
1 016	4 874	-1 779	10 457	251 416	29 019	122 895	67 034	32 468	1980	Canada
1 073	3 211	-6 935	10 605	274 216	38 174	108 804	82 699	44 540	1988	
1 281	3 459	-8 217	10 945	285 490	38 033	113 964	89 029	44 464	1989	
1 290	3 477	-8 176	10 337	271 909	33 577	110 417	82 543	45 372	1990	
10	...	...	4 588	78	...	78	...	...	1980	Iles Caïmanes
18	...	...	4 652	107	...	107	...	...	1988	
13	...	...	5 348	123	...	123	...	...	1989	
10	...	...	4 958	119	...	119	...	...	1990	
...	0	19	564	1 289	...	1 030	...	259	1980	Costa Rica
...	0	56	569	1 589	...	1 191	...	398	1988	
...	0	164	566	1 622	...	1 196	...	426	1989	
...	0	58	602	1 771	...	1 319	...	452	1990	
...	...	1 090	1 412	13 665	143	13 488	22	12	1980	Cuba
134	173	1 258	1 501	15 436	173	15 226	28	9	1988	
142	154	1 258	1 548	16 087	224	15 811	42	10	1989	
90	149	1 347	1 461	15 346	258	15 037	41	10	1990	
0	...	...	257	19	...	18	...	1	1980	Dominique
0	...	...	358	29	...	27	...	2	1988	
0	...	...	370	30	...	28	...	2	1989	
0	...	...	366	30	...	28	...	2	1990	
1	...	9	489	2 788	0	2 717	...	71	1980	Rép. dominicaine
0	...	283	395	2 655	0	2 551	...	103	1988	
0	...	298	381	2 618	0	2 514	...	104	1989	
0	...	164	374	2 622	0	2 518	...	104	1990	
3	82	117	211	956	...	779	...	177	1980	El Salvador
0	0	35	246	1 211	...	991	...	219	1988	
0	0	88	243	1 221	...	988	...	233	1989	
0	0	58	252	1 296	...	1 039	...	257	1990	
...	...	...	5 260	263	2	261	...	...	1980	Groënland
...	...	...	4 741	256	0	256	...	...	1988	
...	...	...	4 036	222	0	222	...	...	1989	
...	...	...	4 661	261	0	261	...	...	1990	
0	...	...	267	24	...	24	...	...	1980	Grenade
1	...	...	570	49	...	49	...	...	1988	
1	...	...	581	50	...	50	...	...	1989	
1	...	...	682	58	...	58	...	...	1990	
41	...	...	777	254	...	254	...	...	1980	Guadeloupe
130	...	...	1 415	477	...	477	...	...	1988	
134	...	...	1 437	487	...	487	...	...	1989	
136	...	...	1 440	491	...	491	...	...	1990	
15	184	39	285	1 973	...	1 938	...	34	1980	Guatemala
21	181	112	199	1 681	...	1 454	...	227	1988	
22	174	85	198	1 718	...	1 461	...	257	1989	
22	181	101	189	1 685	...	1 428	...	257	1990	
13	...	...	61	325	0	298	...	27	1980	Haïti
27	...	...	54	329	0	290	...	39	1988	
27	...	...	53	332	0	292	...	39	1989	
27	...	...	52	335	0	295	...	40	1990	
...	23	49	247	904	0	809	...	96	1980	Honduras
...	0	14	193	902	0	775	...	128	1988	
...	0	31	184	891	0	763	...	128	1989	
...	0	22	176	876	0	747	...	129	1990	
29	21	37	1 822	3 887	1	3 872	...	15	1980	Jamaïque
22	42	88	789	1 872	0	1 860	...	12	1988	
25	42	188	863	2 071	0	2 056	...	16	1989	
24	35	75	841	2 042	0	2 026	...	16	1990	
...	28	-13	1 006	328	...	328	...	...	1980	Martinique
...	14	-4	2 066	692	...	692	...	...	1988	
...	14	-7	1 748	589	...	589	...	...	1989	
...	14	-14	1 779	603	...	603	...	...	1990	

106
Production, trade and consumption of commercial energy
Thousand metric tons of coal equivalent and kilograms per capita [*cont.*]
Production, commerce et consommation d'énergie commerciale
Milliers de tonnes métriques d'équivalent houille et kilogrammes par habitant [*suite*]

Country or area	Year	Primary energy production - Production d'énergie primaire					Changes in stocks Variations des stocks	Imports Importations	Exports Exportations
		Total Totale	Solids Solides	Liquids Liquides	Gas Gaz	Electricity Electricité			
Mexico	1980	195 556	5 007	153 020	35 339	2 190	1 066	1 314	67 770
	1988	249 850	7 561	205 573	33 567	3 149	863	6 020	105 711
	1989	252 595	8 214	206 710	34 055	3 616	2 642	9 394	100 477
	1990	256 764	7 146	211 247	34 498	3 874	310	8 236	101 391
Montserrat	1980	...	...	...	...	...	...	9	...
	1988	...	...	...	...	...	...	18	...
	1989	...	...	...	...	...	...	18	...
	1990	...	...	...	...	...	...	18	...
Netherland Antilles	1980	...	...	...	...	...	...	...	...
	1988	...	...	...	...	...	4	15 974	11 169
	1989	...	...	...	...	...	225	16 260	11 163
	1990	...	...	...	...	...	68	17 170	11 799
Nicaragua	1980	63	...	...	...	63	-9	931	2
	1988	70	...	...	...	70	0	1 069	1
	1989	70	...	...	...	70	0	1 031	1
	1990	68	...	...	...	68	0	972	1
Panama	1980	118	...	...	...	118	-3	6 130	417
	1988	270	...	...	...	270	0	5 056	496
	1989	268	...	...	...	268	0	4 936	473
	1990	272	...	...	...	272	0	4 894	457
Puerto Rico	1980	17	...	...	...	17	-750	13 447	5 194
	1988	34	...	...	...	34	843	12 951	624
	1989	34	...	...	...	34	316	12 782	1 230
	1990	34	...	...	...	34	209	13 712	1 470
Saint Lucia	1980	...	...	...	...	...	0	53	...
	1988	...	...	...	...	...	0	78	...
	1989	...	...	...	...	...	0	78	...
	1990	...	...	...	...	...	0	78	...
St.Kitts-Nevis	1980	...	...	...	...	...	...	...	...
	1988	...	...	...	...	...	0	31	...
	1989	...	...	...	...	...	0	31	...
	1990	...	...	...	...	...	0	31	...
St.Pierre-Miquelon	1980	...	...	...	...	...	...	49	...
	1988	...	...	...	...	...	...	100	...
	1989	...	...	...	...	...	...	128	...
	1990	...	...	...	...	...	...	120	...
S.Vincent-Grenadines	1980	2	...	...	...	2	...	18	...
	1988	4	...	...	...	4	...	33	...
	1989	4	...	...	...	4	...	37	...
	1990	4	...	...	...	4	...	38	...
Trinidad and Tobago	1980	20 700	...	15 704	4 996	...	-53	10 497	22 346
	1988	16 208	...	11 237	4 972	...	-815	1 429	10 871
	1989	16 709	...	11 031	5 678	...	66	1 097	9 687
	1990	18 598	...	11 248	7 349	...	-174	1 508	10 793
United States	1980	2 034 831	626 451	685 734	657 039	65 607	14 934	527 380	101 133
	1988	2 065 596	723 468	667 875	579 486	94 767	-25 078	544 902	117 598
	1989	2 053 219	742 269	623 825	586 173	100 952	-24 056	593 313	126 979
	1990	2 101 965	788 752	605 902	598 546	108 765	41 462	588 076	131 052
U.S. Virgin Islands	1980	...	...	...	...	...	2 916	39 454	26 659
	1988	...	...	...	...	...	816	20 953	13 306
	1989	...	...	...	...	...	-507	25 556	18 124
	1990	...	...	...	...	...	820	24 309	15 361
America, South Antarctic Fisheries	1980	...	...	...	...	...	...	87	...
	1988	...	...	...	...	...	...	97	...
	1989	...	...	...	...	...	...	96	...
	1990	...	...	...	...	...	...	97	...
Argentina	1980	50 464	329	36 833	11 155	2 148	245	7 235	2 059
	1988	64 193	431	34 651	26 459	2 652	487	6 208	2 431
	1989	64 748	316	35 415	26 761	2 256	615	4 736	3 533
	1990	67 547	228	37 133	27 064	3 122	-236	4 033	6 501

	Bunkers - Soutes		Consumption - Consommation							
Air Avion	Sea Maritime	Unallocated Nondistribué	Per capita Par habitant	Total Totale	Solids Solides	Liquids Liquides	Gas Gaz	Electricity Electricité	Année	Pays ou zone
75	50	10 833	1 663	117 076	5 829	77 376	31 617	2 254	1980	Mexique
91	57	8 013	1 700	141 134	7 604	96 955	33 652	2 924	1988	
91	61	8 105	1 774	150 614	8 335	104 159	34 666	3 454	1989	
91	61	8 021	1 788	155 125	7 516	108 860	35 043	3 706	1990	
...	1		583	7	...	7	...	...	1980	Montserrat
...	1	...	1 333	16	...	16	...	...	1988	
...	1	...	1 333	16	...	16	...	...	1989	
...	1	...	1 333	16	...	16	...	...	1990	
...	...	...	...	...	...	...	...	...	1980	Antilles néerland.
52	2 271	649	9 946	1 830	...	1 830	...	...	1988	
52	2 271	719	9 892	1 830	...	1 830	...	...	1989	
52	2 342	1 081	9 786	1 830	...	1 830	...	...	1990	
0	...	22	353	978	...	916	...	62	1980	Nicaragua
0	...	74	304	1 063	...	970	...	93	1988	
0	...	51	290	1 049	...	955	...	94	1989	
0	...	34	268	1 005	...	912	...	92	1990	
90	4 038	74	834	1 632	2	1 512	0	118	1980	Panama
38	3 155	205	630	1 432	15	1 071	74	273	1988	
37	3 134	139	612	1 421	17	1 057	70	277	1989	
38	3 105	84	625	1 481	35	1 079	81	286	1990	
75	279	-1 290	3 106	9 957	0	9 940	...	17	1980	Porto Rico
37	214	895	3 095	10 372	200	10 138	...	34	1988	
37	214	777	3 017	10 240	200	10 007	...	34	1989	
37	214	1 357	3 042	10 460	200	10 225	...	34	1990	
0	...	...	427	53	...	53	...	...	1980	Sainte-Lucie
0	...	...	549	78	...	78	...	...	1988	
0	...	...	538	78	...	78	...	...	1989	
0	...	...	527	78	...	78	...	...	1990	
...	...	...	...	...	...	...	...	...	1980	St-Kitts-Nevis
...	...	...	705	31	...	31	...	...	1988	
...	...	...	705	31	...	31	...	...	1989	
...	...	...	705	31	...	31	...	...	1990	
...	32	...	2 833	17	0	17	...	...	1980	St.Pierre-Miquelon
...	70	...	5 167	31	0	31	...	...	1988	
...	78	...	8 167	49	0	49	...	...	1989	
...	77	...	7 333	44	0	44	...	...	1990	
0	...	...	194	20	...	18	...	2	1980	S.Vincent-Grenadines
0	...	...	330	37	...	33	...	4	1988	
0	...	...	363	41	...	37	...	4	1989	
0	...	...	377	43	...	38	...	4	1990	
41	717	679	6 901	7 467	0	2 471	4 996	...	1980	Trinité-et-Tobago
1	116	441	5 757	7 023	0	2 051	4 972	...	1988	
1	97	544	5 971	7 410	0	1 733	5 678	...	1989	
0	40	395	7 178	9 052	0	1 703	7 349	...	1990	
9 454	41 565	30 736	10 381	2 364 389	538 704	1 064 253	692 544	68 889	1980	Etats-Unis
10 244	26 730	9 018	10 154	2 471 986	671 491	1 078 808	623 017	98 669	1988	
10 318	26 855	6 303	10 188	2500 134	677 698	1 079 714	640 423	102 300	1989	
10 539	28 240	-2 938	10 034	2 481 686	693 605	1 048 983	630 089	109 008	1990	
74	1 405	4 236	42 500	4 165	...	4 165	...	...	1980	Iles Vierges amér.
15	194	5 295	11 964	1 328	...	1 328	...	...	1988	
15	194	6 372	12 018	1 358	...	1 358	...	...	1989	
15	194	6 562	11 809	1358	...	1 358	...	...	1990	
										Amérique du Sud
...	87	...	0	0	...	0	...	...	1980	Pêcheries antarctiq.
...	94	...	3 000	3	...	3	...	...	1988	
...	93	...	3 000	3	...	3	...	...	1989	
...	93	...	4 000	4	...	4	...	...	1990	
6	633	5 661	1 739	49 095	1 080	32 293	13 572	2 151	1980	Argentine
0	815	4 463	1 998	62 205	1 517	28 706	29 293	2 689	1988	
0	1 052	5 047	1 878	59237	1 463	25 755	29 697	2 323	1989	
0	1 051	5 282	1 847	58 982	1 157	24 627	29 974	3 224	1990	

106

Production, trade and consumption of commercial energy

Thousand metric tons of coal equivalent and kilograms per capita [*cont.*]

Production, commerce et consommation d'énergie commerciale

Milliers de tonnes métriques d'équivalent houille et kilogrammes par habitant [*suite*]

Country or area	Year	Primary energy production - Production d'energie primaire					Changes in stocks Variations des stocks	Imports Importations	Exports Exportations
		Total Totale	Solids Solides	Liquids Liquides	Gas Gaz	Electricity Electricité			
Bolivia	1980	4 790	...	1 777	2 880	133	-38	1	2 543
	1988	5 078	...	1 494	3 429	155	-158	0	2 867
	1989	5 581	...	1 527	3 898	156	-141	1	2 971
	1990	5 724	...	1 683	3 883	158	18	1	2 962
Brazil	1980	34 058	3 655	13 263	1 306	15 834	-1 633	69 231	2 773
	1988	75 195	4 965	40 965	4 735	24 530	1 923	62 582	9 451
	1989	78 548	4 389	43 724	5 067	25 368	-244	59 965	8 200
	1990	80 028	2 939	46 337	5 022	25 730	3 246	57 881	5 708
Chile	1980	5 723	978	2 919	924	902	-254	6 112	130
	1988	7 009	1 920	2 153	1 527	1 409	99	6 849	232
	1989	7 349	1 943	1 970	2 256	1 180	150	8 794	260
	1990	7512	2 175	1 858	2 364	1 115	479	9 948	301
Colombia	1980	20 159	3 982	9 596	4 792	1 790	-9	3 005	2 155
	1988	49 745	14 022	27 272	5 442	3 010	-480	1 915	25 235
	1989	55 195	17 552	29 293	5 076	3 274	74	1 722	29 261
	1990	59 736	19 006	31 884	5 466	3 380	249	1 816	32 543
Ecuador	1980	15 057	...	14 897	51	109	113	836	9 705
	1988	23 369	...	22 648	132	589	0	382	15 836
	1989	21 602	...	20 865	132	604	0	384	14 660
	1990	22 287	...	21 511	165	611	0	381	14 744
Falkland I(Malvinas)	1980	4	4	...	...	...	...	3	...
	1988	5	5	...	...	...	...	14	...
	1989	5	5	...	...	...	...	13	...
	1990	5	5	...	...	...	...	12	...
French Guiana	1980	...	...	...	...	...	...	174	...
	1988	...	...	...	...	...	...	244	...
	1989	...	...	...	...	...	...	270	...
	1990	...	...	...	...	...	...	276	...
Guyana	1980	1	...	...	...	1	...	845	...
	1988	1	...	...	...	1	...	329	...
	1989	1	...	...	...	1	...	321	...
	1990	1	...	...	...	1	...	309	...
Paraguay	1980	83	...	...	...	83	-29	649	19
	1988	257	...	...	...	257	13	778	72
	1989	270	...	...	...	270	73	830	73
	1990	298	...	...	...	298	33	780	74
Peru	1980	15 561	41	13 946	713	861	-94	258	4 494
	1988	13 132	157	10 878	817	1 281	121	2 954	3 949
	1989	12 121	142	10 011	691	1 277	137	2 277	3 265
	1990	11 993	142	9 899	665	1 286	246	2 285	3 055
Suriname	1980	110	...	0	...	110	11	1 107	...
	1988	361	...	277	...	84	16	659	60
	1989	457	...	324	...	133	0	630	99
	1990	535	...	400	...	135	31	633	107
Uruguay	1980	279	...	...	...	279	106	2 943	3
	1988	666	...	...	...	666	-27	2 385	257
	1989	479	...	...	...	479	-93	2 341	146
	1990	783	...	...	...	783	-55	1 960	163
Venezuela	1980	191 023	44	167 280	21 938	1 761	-80	233	137 612
	1988	181 276	1 071	147 775	28 170	4 260	212	150	116 794
	1989	183 370	2 128	148 107	28 611	4 524	1 366	158	117 924
	1990	204 238	2 188	165 328	32 626	4 096	698	118	135 869
Asia Afghanistan	1980	3 710	119	12	3 491	88	36	487	3 361
	1988	4 124	138	5	3 889	92	52	869	1 176
	1989	4 183	145	6	3 938	93	50	887	1 365
	1990	4 149	143	6	3 906	94	42	904	1 365
Bahrain	1980	7 674	...	3 794	3 880	...	396	13 640	13 632
	1988	10 164	...	3 507	6 657	...	-113	14 237	12 901
	1989	9 965	...	3 301	6 664	...	2	14613	13 009
	1990	10 399	...	3 541	6 858	...	-184	14 507	13 035

Bunkers - Soutes			Consumption - Consommation							
Air Avion	Sea Maritime	Unallocated Nondistribué	Per capita Par habitant	Total Totale	Solids Solides	Liquids Liquides	Gas Gaz	Electricity Electricité	Année	Pays ou zone
9	...	-17	412	2 293	1	1 698	462	133	1980	Bolivie
0	...	119	334	2 251	0	1 500	596	155	1988	
0	...	119	381	2 634	0	1 515	962	157	1989	
0	...	119	369	2 626	0	1 512	955	159	1990	
847	867	8 199	760	92 236	7 444	67 678	1 306	15 808	1980	Brésil
559	827	11 822	800	113 196	14 363	67 353	4 745	26 734	1988	
587	919	12 411	808	116 639	14 448	69 041	5 067	28 083	1989	
669	781	12 475	780	115 030	12 725	68 625	4 903	28 777	1990	
10	82	483	1 021	11 384	2 204	7 286	992	902	1980	Chili
0	0	235	1 060	13 292	2 549	7 847	1 487	1 409	1988	
0	0	107	1 226	15 626	3 428	8 814	2 204	1 180	1989	
0	0	225	1 270	16 455	3 725	9 295	2 320	1 115	1990	
22	300	448	753	20 248	3 882	9 780	4 792	1 793	1980	Colombie
10	173	1 802	801	24 920	4 730	11 739	5 442	3 010	1988	
10	150	2 613	782	24 809	4 887	11 544	5 076	3 301	1989	
10	185	2 505	806	26 060	5 233	11 947	5 466	3 414	1990	
52	129	214	699	5 680	...	5 518	51	111	1980	Equateur
68	155	449	738	7 243	...	6 522	132	589	1988	
56	248	188	679	6 835	...	6 098	132	604	1989	
63	227	-54	744	7 687	...	6 911	165	611	1990	
...	1	...	2 500	5	4	1	...	...	1980	I Falkland (Malvinas)
...	1	...	9 000	18	5	13	...	...	1988	
...	0	...	9 000	18	5	13	...	...	1989	
...	0	...	8 500	17	5	12	...	...	1990	
...	...	...	2 522	174	...	174	...	...	1980	Guyane française
18	...	...	2 539	226	...	226	...	...	1988	
22	...	...	2 696	248	...	248	...	...	1989	
22	...	...	2 674	254	...	254	...	...	1990	
6	7	...	1 096	832	0	832	...	1	1980	Guyana
6	4	...	404	320	0	319	...	1	1988	
6	4	...	392	311	0	311	...	1	1989	
6	4	...	378	300	0	299	...	1	1990	
4	...	10	231	727	...	644	...	83	1980	Paraguay
3	...	-6	243	954	...	768	...	186	1988	
3	...	1	235	951	...	754	...	198	1989	
3	...	0	233	968	...	743	...	225	1990	
0	115	280	637	11 024	195	9 255	713	861	1980	Pérou
0	86	287	575	11 644	217	9 329	817	1 281	1988	
0	86	253	515	10 656	208	8 480	691	1 277	1989	
0	86	325	500	10 566	208	8 406	665	1 286	1990	
1	9	0	3 395	1 195	26	1 059	...	110	1980	Suriname
0	7	154	1 967	783	0	699	...	84	1988	
0	7	213	1 892	768	0	636	...	133	1989	
0	7	257	1 848	765	0	630	...	135	1990	
50	142	164	946	2 758	4	2 470	...	283	1980	Uruguay
40	141	173	811	2 467	1	2 057	...	409	1988	
44	167	148	787	2 408	0	2 070	...	338	1989	
44	144	139	750	2 307	0	1 688	...	619	1990	
516	874	3 085	3 278	49 249	276	25 277	21 938	1 757	1980	Venezuela
425	1 124	3 176	3 268	59 696	250	27 016	28 170	4 260	1988	
450	1 088	4 562	3 100	58139	387	24 616	28 611	4 524	1989	
526	1 137	1 918	3 337	64 208	397	27 088	32 626	4 096	1990	
										Asie
22	...	0	48	778	119	475	96	88	1980	Afghanistan
12	...	0	255	3 753	138	844	2 679	92	1988	
12	...	0	241	3 642	145	865	2 539	93	1989	
12	...	0	231	3 635	143	884	2 514	94	1990	
601	267	2 028	12 651	4 390	0	510	3 880	...	1980	Bahreïn
609	301	2 992	16 619	7 711	0	1 054	6 657	...	1988	
604	301	2 877	16 185	7 785	0	1 121	6 664	...	1989	
590	315	3 103	16 159	8 047	0	1 189	6 858	...	1990	

106
Production, trade and consumption of commercial energy
Thousand metric tons of coal equivalent and kilograms per capita [*cont.*]
Production, commerce et consommation d'énergie commerciale
Milliers de tonnes métriques d'équivalent houille et kilogrammes par habitant [*suite*]

Country or area	Year	Primary energy production - Production d'energie primaire					Changes in stocks Variations des stocks	Imports Importations	Exports Exportations
		Total Totale	Solids Solides	Liquids Liquides	Gas Gaz	Electricity Electricité			
Bangladesh	1980	1 617	...	9	1 537	72	-20	2 654	25
	1988	5 115	...	177	4 854	83	-13	2 980	0
	1989	5 437	...	180	5 144	113	47	3 127	0
	1990	5 544	...	180	5 255	109	-212	3 398	0
Bhutan	1980	1	...	...	...	1	...	11	...
	1988	190	0	...	...	190	...	48	171
	1989	192	2	...	...	190	...	30	173
	1990	193	2	...	...	191	...	57	173
Brunei Darussalam	1980	29 781	...	17 772	12 010	...	219	98	27 277
	1988	22 582	...	10 111	12 471	...	206	48	19 420
	1989	23 506	...	10 066	13 440	...	224	48	19 862
	1990	25 264	...	11333	13 932	...	357	48	21 475
Cambodia	1980	6	...	...	...	6	...	135	...
	1988	4	...	...	...	4	...	214	...
	1989	4	...	...	...	4	...	214	...
	1990	4	...	...	...	4	...	214	...
China	1980	612 408	434 959	151 351	18 948	7 149	...	2 782	29 418
	1988	926 882	699 212	195 293	18 970	13 407	2 439	7 508	62 475
	1989	981 900	752 104	196 357	19 988	13 450	43	7 659	59 738
	1990	1 001784	770 657	197 286	20 268	13 573	0	7 696	61 044
Cyprus	1980	...	...	...	...	...	31	1 387	0
	1988	...	...	...	...	...	-84	2 039	0
	1989	...	...	...	...	...	38	2 290	0
	1990	...	...	...	...	...	-53	2 309	0
Hong Kong	1980	...	...	...	...	...	-300	9 307	295
	1988	...	...	...	...	...	-282	15 811	1 663
	1989	...	...	...	...	...	310	19 087	3 500
	1990	...	...	...	...	...	-100	17 712	3 175
India	1980	113 322	92 081	13 427	1 726	6 087	-2 420	32 833	95
	1988	221 783	159 437	45 643	8 878	7 825	5 201	36 178	507
	1989	235 560	167 870	48 720	10 776	8 194	1 129	40 332	344
	1990	240167	170 354	47 709	13 239	8 866	1 177	44 516	619
Indonesia	1980	132 274	304	110 882	20 718	370	5 024	10 837	99 126
	1988	140 004	2 741	94 596	41 616	1 051	3 012	9 939	93 937
	1989	154 577	4 553	99 042	49 900	1 082	1 330	12 009	103 269
	1990	162 878	7 327	100 565	53 861	1 125	1 096	13 524	112 440
Iran, Islamic Rep. of	1980	116 063	900	104 966	9 506	690	-2 241	237	62 376
	1988	194 571	1 260	165 888	26 631	792	-843	9 597	122 003
	1989	235 273	1 200	203 708	29 542	823	-514	11 246	153 470
	1990	261 607	1 300	227 641	31 855	811	4 286	10 583	172 961
Iraq	1980	188 268	...	186 476	1 707	85	27	7	175 624
	1988	194 139	...	187 919	6 133	87	7 016	8	161 200
	1989	204 483	...	197 900	6 510	74	2 404	10	174 163
	1990	150 283	...	145 943	4 265	75	5 714	9	124 577
Israel	1980	220	...	29	191	...	1 762	12 802	788
	1988	72	...	26	47	...	1 546	17 461	1 660
	1989	69	...	23	47	...	-53	16 624	1 462
	1990	59	...	19	40	...	-867	16 311	1 464
Japan	1980	41 197	16 059	642	2 905	21 591	2 505	436 372	2 165
	1988	47 462	9 868	870	2 833	33 890	-852	457 564	7 234
	1989	47 114	8 958	794	2 713	34 648	3 091	489 050	10 022
	1990	47647	7 268	787	2 761	36 832	3 467	507 059	11 488
Jordan	1980	...	...	...	...	...	98	2 506	0
	1988	13	...	10	...	3	-9	4 364	0
	1989	18	...	16	...	2	6	4 328	0
	1990	25	...	23	...	2	145	4 951	0
Korea, Republic of	1980	12 909	12 239	...	...	671	2 571	45 135	0
	1988	20 986	15 622	...	...	5 364	3 152	83 929	4 365
	1989	19 743	13 365	...	...	6 378	2 151	92 517	5 425
	1990	18 348	11 071	...	...	7 278	-219	102 469	4 858

Bunkers - Soutes			Consumption - Consommation							
Air Avion	Sea Maritime	Unallocated Nondistribué	Per capita Par habitant	Total Totale	Solids Solides	Liquids Liquides	Gas Gaz	Electricity Electricité	Année	Pays ou zone
0	86	230	45	3 950	168	2 173	1 537	72	1980	Bangladesh
7	14	738	69	7 348	171	2 240	4 854	83	1988	
7	14	739	71	7 757	179	2 322	5 144	113	1989	
9	7	689	75	8 448	399	2 686	5 255	109	1990	
...	...	...	9	11	1	9	...	1	1980	Bhoutan
...	...	...	47	66	18	29	...	19	1988	
...	...	...	33	48	0	29	...	19	1989	
...	...	...	52	77	18	38	...	20	1990	
...	...	-552	15 870	2 936	2	922	2 012	...	1980	Brunéi Darussalam
...	...	-551	14 751	3 555	0	738	2 818	...	1988	
...	...	-567	16 205	4 035	0	763	3 272	...	1989	
...	...	-493	15 399	3 973	0	790	3 183	...	1990	
...	...	0	22	141	0	135	...	6	1980	Cambodge
...	...	0	28	218	0	214	...	4	1988	
...	...	0	28	218	0	214	...	4	1989	
...	...	0	27	218	0	214	...	4	1990	
...	...	40 274	557	545 499	432 174	87 189	18 948	7 187	1980	Chine
...	...	28 102	787	841 375	686 876	121 941	18 970	13 588	1988	
...	...	27 310	832	902 468	741 708	127 136	19 988	13 636	1989	
...	...	26 252	837	922 183	759 381	128 781	20 268	13 753	1990	
111	24	13	1 922	1 209	0	1 209	...	...	1980	Chypre
282	88	41	2 518	1 712	91	1 621	...	...	1988	
308	123	44	2 587	1 777	75	1 702	...	...	1989	
348	83	35	2 732	1 896	108	1 788	...	...	1990	
1 044	978	...	1 447	7 291	8	7 321	...	-38	1980	Hong-kong
1 710	1 080	...	2 071	11 641	7 084	4 733	...	-177	1988	
2 087	1 649	...	2 024	11 541	7 589	4 170	...	-217	1989	
2 444	1 831	...	1 792	10 361	6 828	3 754	...	-221	1990	
814	335	8 059	202	139 272	95 165	36 298	1 726	6 083	1980	Inde
1 041	499	14 565	294	236 149	160 295	59 003	8 878	7 973	1988	
988	335	16 806	313	256 290	172 049	65 168	10 776	8 297	1989	
929	331	16 708	317	264 920	177 550	65 155	13 239	8 977	1990	
133	362	2 364	239	36 101	267	30 067	5 398	370	1980	Indonésie
221	271	4 678	275	47 824	2 444	33 830	10 499	1 051	1988	
221	271	5 302	317	56 194	3 352	35 228	16 531	1 082	1989	
236	271	4 720	319	57 639	3 539	35 905	17 070	1 125	1990	
7	5 806	4 678	1 174	45 674	960	34 810	9 213	690	1980	Iran, Rép.islamique d'
7	639	2 964	1 566	79 398	1 460	50 515	26 631	792	1988	
9	512	4 974	1 691	88 069	1 554	56 150	29 542	823	1989	
9	512	4 083	1 693	90 339	1 754	57 673	30 101	811	1990	
6	170	1 793	802	10 654	1	8 862	1 707	85	1980	Iraq
7	283	8 365	1 013	17 276	1	15 180	2 007	87	1988	
9	283	8 278	1 096	19 356	1	17 562	1 719	74	1989	
6	212	3 238	905	16 545	2	14 933	1 535	75	1990	
348	...	1 295	2 277	8 829	6	8 652	191	-20	1980	Israël
569	...	-378	3 229	14 136	3 228	10 908	47	-46	1988	
599	...	922	3 091	13 763	3 472	10 291	47	-47	1989	
628	...	465	3 243	14 680	3709	10 978	40	-48	1990	
2 177	12 011	25 335	3 710	433 377	85 750	292 789	33 246	21 591	1980	Japon
688	6 825	11 044	3 935	480 086	113 237	273 254	59 705	33 890	1988	
647	7 106	23 970	4 011	491 328	113 268	279 223	64 189	34 648	1989	
817	7 476	19 347	4 164	512 110	114 765	292 089	68 424	36 832	1990	
283	...	108	690	2 016	...	2 016	...	...	1980	Jordanie
273	...	181	1 082	3 932	...	3 929	...	3	1988	
348	...	117	1 033	3 875	...	3 873	...	2	1989	
336	...	158	1 119	4 338	...	4 335	...	2	1990	
142	138	3 768	1 349	51 426	18 301	32 454	...	671	1980	Corée, République de
411	2 289	7 345	2 097	87 353	36 571	41 530	3 888	5 364	1988	
567	2 668	7 978	2 223	93 470	35 858	47 477	3 757	6 378	1989	
772	2 339	7 217	2 495	105 850	35 372	58 882	4 318	7 278	1990	

106

Production, trade and consumption of commercial energy
Thousand metric tons of coal equivalent and kilograms per capita [*cont.*]
Production, commerce et consommation d'énergie commerciale
Milliers de tonnes métriques d'équivalent houille et kilogrammes par habitant [*suite*]

Country or area	Year	Primary energy production - Production d'energie primaire					Changes in stocks Variations des stocks	Imports Importations	Exports Exportations
		Total Totale	Solids Solides	Liquids Liquides	Gas Gaz	Electricity Electricité			
Korea,Dem.Ppl's.Rep.	1980	44 764	42 000	...	...	2 764	...	4 077	100
	1988	51 369	47 500	...	...	3 869	...	7 814	50
	1989	52 400	48 500	...	...	3 900	...	7 814	50
	1990	52 400	48 500	...	...	3 900	...	7 671	50
Kuwait,part Ntl.Zone	1980	134 732	...	125 058	9 674	...	1 579	1	113 984
	1988	111 687	...	105 898	5 789	...	2 068	4 127	88 669
	1989	118 383	...	111 138	7 245	...	-230	4 792	96 123
	1990	96 672	...	89 713	6 960	...	-1 214	2 731	79 475
Lao Peoples Dem.Rep.	1980	114	0	...	...	114	...	90	94
	1988	60	0	...	...	60	...	101	46
	1989	82	0	...	...	82	...	111	58
	1990	101	0	...	...	101	...	111	64
Lebanon	1980	104	...	...	...	104	81	3 124	145
	1988	74	...	...	...	74	0	3 939	0
	1989	61	...	...	...	61	0	4 037	0
	1990	68	...	...	...	68	0	4 229	0
Macau	1980	...	...	...	...	...	...	234	...
	1988	...	...	...	...	...	0	455	0
	1989	...	...	...	...	...	0	487	0
	1990	...	...	...	...	...	8	501	0
Malaysia	1980	19 477	...	19 197	109	171	629	9 220	16 441
	1988	55 074	23	38 201	16 155	696	-5	8 939	43 467
	1989	58 734	114	41 113	16 864	643	6	11 602	47 505
	1990	61 176	105	43 440	17 141	490	138	13 498	48 821
Mongolia	1980	1 707	1 707	...	...	...	...	965	...
	1988	3 302	3 302	...	...	...	...	1 221	149
	1989	3 117	3 117	...	...	...	...	1 134	256
	1990	3 066	3 066	...	...	...	...	1 063	214
Myanmar	1980	2 801	21	2 229	454	98	216	37	76
	1988	2 775	43	1 090	1 527	115	...	49	0
	1989	2 914	52	1 272	1 438	152	...	49	0
	1990	2 926	55	1 287	1 431	153	...	49	0
Nepal	1980	22	...	...	...	22	...	235	1
	1988	69	...	...	...	69	-34	412	40
	1989	67	...	...	...	67	-31	378	36
	1990	87	...	...	...	87	-56	276	38
Oman	1980	21 161	...	20 043	1 118	...	269	1 227	19 769
	1988	47 761	...	44 482	3 279	...	-712	211	43 100
	1989	48 987	...	45 895	3 091	...	-252	113	43 805
	1990	52 416	...	48 791	3 625	...	-570	141	46 218
Pakistan	1980	10 816	1 057	688	8 000	1 071	71	8 328	1 164
	1988	19 872	1 853	3 136	12 801	2 081	397	11 723	31
	1989	20 482	1 781	3 282	13 330	2 088	249	12 273	196
	1990	22 259	1 854	3 722	14 565	2 118	-374	12 697	469
Philippines	1980	1 674	222	764	...	688	96	15 863	126
	1988	2 711	903	441	...	1 366	-111	17 054	436
	1989	2 757	919	381	...	1 456	-463	17 678	381
	1990	2 559	791	350	...	1 418	464	19 634	592
Qatar	1980	38 751	...	32 760	5 991	...	-157	88	32 255
	1988	31 134	...	24 531	6 604	...	-287	0	23 640
	1989	35 982	...	29 413	6 569	...	1 840	0	26 125
	1990	36 735	...	29 498	7 236	...	-733	0	28 561
Saudi Arabia, pt.Ntrl.Zn.	1980	728 545	...	726 885	1 660	...	7 921	2 839	682 244
	1988	417 549	...	382 845	34 704	...	5 435	536	298 698
	1989	417 103	...	381 542	35 561	...	5 859	571	295 473
	1990	521 803	...	484 226	37 577	...	4 602	586	391 482
Singapore	1980	...	...	...	...	...	-3 887	51 880	30 171
	1988	...	...	...	...	...	842	72 435	42 695
	1989	...	...	...	...	...	2 028	76 837	43 244
	1990	...	...	...	...	...	4 699	85 778	47 101

Bunkers - Soutes			Consumption - Consommation							
Air Avion	Sea Maritime	Unallocated Nondistribué	Per capita Par habitant	Total Totale	Solids Solides	Liquids Liquides	Gas Gaz	Electricity Electricité	Année	Pays ou zone
...	...	200	2 658	48 541	42 620	3 157	...	2 764	1980	Corée,Rép.pop.dém.de
...	...	271	2 857	58 862	50 220	4 773	...	3 869	1988	
...	...	38	2 866	60 126	51 220	5 006	...	3 900	1989	
...	...	-47	2 811	60 068	51 220	4 948	...	3 900	1990	
498	2 801	3 860	8 735	12 011	...	2 337	9 674	...	1980	Koweït et prt.Zne.N.
678	1 184	7 327	8 588	15 888	...	5 974	9 914	...	1988	
737	1 332	6 717	9 668	18 495	...	6 460	12 036	...	1989	
442	1 049	4 874	7 478	14 776	...	5 087	9 689	...	1990	
...	...	...	34	110	0	89	...	21	1980	Rép. dém. pop. lao
...	...	...	31	116	0	99	...	16	1988	
...	...	...	35	136	0	109	...	26	1989	
...	...	...	37	149	0	109	...	39	1990	
355	14	95	951	2 538	5	2 420	...	113	1980	Liban
133	35	121	1 398	3 724	0	3 646	...	78	1988	
133	14	69	1 458	3 882	0	3 816	...	66	1989	
103	14	65	1 539	4 114	0	4 042	...	72	1990	
...	...	...	724	234	4	230	...	...	1980	Macao
...	...	...	1 071	455	0	450	...	5	1988	
...	...	...	1 099	487	0	481	...	6	1989	
...	...	...	1 067	492	0	481	...	11	1990	
0	82	239	821	11 306	77	10 951	97	181	1980	Malaisie
0	89	-155	1 246	20 617	435	15 023	4 471	687	1988	
0	98	-459	1 364	23 185	1 807	15 981	4 781	616	1989	
0	127	-1 071	1 528	26 659	2 040	19 395	4 741	483	1990	
...	...	...	1 607	2 672	1 780	856	...	36	1980	Mongolie
...	...	...	2 169	4 374	3 158	1 207	...	9	1988	
...	...	...	1 927	3 994	2 867	1 108	...	19	1989	
...	...	...	1 837	3 915	2 857	1 038	...	20	1990	
0	0	525	60	2 021	58	1 412	454	98	1980	Myanmar
0	0	308	64	2 517	92	783	1 527	115	1988	
0	0	434	63	2 529	101	837	1 438	152	1989	
0	0	463	62	2 512	104	824	1 431	153	1990	
...	...	...	17	256	78	152	...	26	1980	Népal
...	...	...	27	475	89	310	...	76	1988	
...	...	...	24	440	81	280	...	79	1989	
...	...	...	20	381	35	255	...	91	1990	
171	337	0	1 872	1 842	...	724	1 118	...	1980	Oman
195	198	-11	3 870	5 201	...	1 923	3279	...	1988	
206	324	49	3 561	4 967	...	1 876	3 091	...	1989	
413	425	132	4105	5 940	...	2 315	3 625	...	1990	
304	226	755	195	16 624	1 313	6 240	8 000	1 071	1980	Pakistan
164	21	846	271	30 135	2 679	12 574	12 801	2 081	1988	
156	31	635	274	31 488	2 717	13 353	13 330	2 088	1989	
177	33	612	287	34 040	2 680	14 678	14 565	2 118	1990	
137	243	994	330	15 941	506	14 747	...	688	1980	Philippines
368	113	2 321	287	16 637	2 253	13 018	...	1 366	1988	
562	138	1 833	302	17 984	1 899	14 629	...	1 456	1989	
544	96	1 734	308	18 764	2 109	15 237	...	1 418	1990	
96	...	85	28 646	6 560	...	569	5 991	...	1980	Qatar
103	...	17	23 500	7 661	...	1 058	6 604	...	1988	
118	...	90	22 968	7 809	...	1 240	6 569	...	1989	
121	...	76	24 605	8 710	...	1 473	7 236	...	1990	
2 027	7 597	6 260	2 703	25 335	...	23 674	1 660	...	1980	Arab.saoud,p.Zne.neut
1 327	21 164	7 713	6 666	83 748	...	49 044	34 704	...	1988	
1 474	20 577	7 804	6 617	86 487	...	50 926	35 561	...	1989	
1 621	23 923	10 399	6 647	90 361	...	52 784	37 577	...	1990	
652	5 971	7 048	4 937	11 924	4	11 932	...	-11	1980	Singapour
811	8 095	7 214	4 872	12 779	18	12 761	...	0	1988	
921	8 803	8 441	5 043	13 399	17	13 382	...	0	1989	
995	9 440	8 459	5 607	15 084	30	15 055	...	0	1990	

106

Production, trade and consumption of commercial energy

Thousand metric tons of coal equivalent and kilograms per capita [*cont.*]

Production, commerce et consommation d'énergie commerciale

Milliers de tonnes métriques d'équivalent houille et kilogrammes par habitant [*suite*]

Country or area	Year	Primary energy production - Production d'energie primaire					Changes in stocks Variations des stocks	Imports Importations	Exports Exportations
		Total Totale	Solids Solides	Liquids Liquides	Gas Gaz	Electricity Electricité			
Sri Lanka	1980	182	...	...	...	182	193	2 810	228
	1988	319	...	...	...	319	164	2 748	201
	1989	344	...	...	...	344	19	2 357	125
	1990	386	...	...	...	386	44	2 712	143
Syrian Arab Republic	1980	13 518	...	13 140	64	315	38	7 534	10 317
	1988	21 098	...	20 234	273	590	206	3 677	9 223
	1989	27 070	...	26 184	307	578	246	1 427	13 742
	1990	33 735	...	32 773	375	587	196	1 077	19 375
Thailand	1980	741	573	11	0	156	-642	16 792	20
	1988	13 276	2 707	3 500	6 605	464	-682	17 388	917
	1989	14 209	3 319	3 329	6 877	684	280	22 604	964
	1990	16 261	4 631	4 029	6 990	611	120	26 882	1 060
Turkey	1980	14 086	9 334	3 329	29	1 394	896	20 148	246
	1988	22 710	15 353	3 663	129	3 564	1 717	38 657	5 565
	1989	25 563	19 038	4 109	205	2 211	37	36 360	2 984
	1990	24 291	17 052	4 143	242	2 853	-196	40 976	3 185
United Arab Emirates	1980	128 809	...	119 668	9 140	...	214	4 725	122 120
	1988	132 961	...	110 862	22 100	...	691	3	104 461
	1989	161 537	...	134 831	26 707	...	500	3	128 657
	1990	180 260	...	153 366	26 894	...	640	4	146 572
Viet Nam	1980	5 461	5 300	...	...	161	...	1 783	500
	1988	7 584	6 332	970	39	243	545	3 518	1 227
	1989	6 514	3 900	2 124	24	466	-1 250	3 449	2 602
	1990	9 513	5 000	3 850	4	660	-86	4 332	4 543
Yemen	1980	...	...	...	...	...	...	4 577	2 031
	1988	10 386	...	10 386	...	...	36	6 836	12 356
	1989	13 903	...	13 903	...	...	29	5 613	14 526
	1990	13 240	...	13 240	...	...	14	5 926	14 278
Europe Albania	1980	6 584	710	5 000	512	362	...	169	61
	1988	6 474	1 200	4 286	546	442	...	251	80
	1989	6 314	1 200	4 143	529	442	...	251	80
	1990	6 314	1 200	4 143	529	442	...	251	80
Austria	1980	9 615	1 281	2 140	2 663	3 532	1 054	20 720	906
	1988	8 499	806	1 743	1 553	4 396	22	22 254	1 391
	1989	8 505	782	1 716	1 567	4 439	-126	23 041	1 206
	1990	8 229	951	1 705	1 582	3 991	1 207	25 031	1 369
Belgium	1980	7 416	5 784	...	52	1 580	582	86 884	25 461
	1988	8 513	3 145	...	20	5 348	-1 284	77 742	25 171
	1989	8 446	3 322	...	16	5 108	281	83 471	27 358
	1990	7 452	2 156	...	15	5 281	-527	85 387	26 939
Bulgaria	1980	17 209	15 240	393	363	1 213	282	32 894	162
	1988	19 581	17 171	110	12	2 288	501	33 220	3 557
	1989	19 480	17 245	104	11	2 120	-216	33 824	3 778
	1990	17 810	15 896	86	10	1 818	1 451	26 998	1 195
Czechoslovakia	1980	66 432	64 431	137	722	1 141	619	44 242	8 479
	1988	66 924	62 301	209	1 013	3 401	1 075	43 292	7 848
	1989	64 278	59 573	210	951	3 544	131	43 726	8 361
	1990	59 325	54 822	180	792	3 531	0	41 053	5 911
Denmark	1980	431	0	426	...	5	97	30 567	2 513
	1988	9 838	0	6 764	3 032	41	-903	23 296	7 571
	1989	11 539	0	7 901	3 582	57	1 289	23 450	9 291
	1990	12 967	0	8 563	4 326	78	-529	21 833	9 474
Faeroe Islands	1980	6	...	...	...	6	...	199	...
	1988	8	...	...	...	8	...	253	...
	1989	9	...	...	...	9	...	276	...
	1990	10	...	...	...	10	...	291	...
Finland	1980	2 423	366	...	...	2 057	-496	29 230	2 829
	1988	5 350	1 306	...	...	4 043	-2 131	25 497	3 381
	1989	5 859	1 914	...	...	3 945	-186	26 414	1 562
	1990	5 630	1 927	...	...	3 703	-89	27 215	2 124

Air Avion	Sea Maritime	Unallocated Nondistribué	Per capita Par habitant	Total Totale	Solids Solides	Liquids Liquides	Gas Gaz	Electricity Electricité	Année	Pays ou zone
137	507	289	110	1 636	2	1 453	...	182	1980	Sri Lanka
83	558	28	123	2 033	0	1 714	...	319	1988	
74	443	111	115	1 930	1	1 585	...	344	1989	
83	554	45	131	2 230	1	1 843	...	386	1990	
351	...	1 712	981	8 635	4	8 261	64	306	1980	Rép. arabe syrienne
473	...	1 744	1 168	13 128	1	12 263	273	590	1988	
697	...	1 996	1 014	11 815	0	10 930	307	578	1989	
703	...	1 202	1 104	13 336	0	12 374	375	587	1990	
3	207	632	371	17 313	625	16 439	0	249	1980	Thailande
0	0	265	566	30 164	3 007	20 038	6 605	515	1988	
0	0	578	646	34 991	3 604	23 750	6 877	760	1989	
0	0	881	748	41 082	4 866	28 538	6 990	688	1990	
0	85	1 949	699	31 057	9 585	19 884	29	1 559	1980	Turquie
430	555	4 236	929	48 864	18 735	24 921	1 600	3 609	1988	
0	244	3 192	1 033	55 466	21 971	27 418	3 797	2 280	1989	
0	258	3 990	1 060	58 029	22 312	28 873	4 081	2 763	1990	
771	526	-3	9 759	9 905	...	4 240	5 665	...	1980	Emirats arabes unis
811	669	-589	18 527	26 920	...	9 162	17 758	...	1988	
811	698	-624	20 985	31 499	...	9 159	22 339	...	1989	
811	726	-556	20 745	32 071	...	9 451	22 621	...	1990	
...	...	...	126	6 744	4 807	1 775	...	161	1980	Viet Nam
...	...	2	149	9 328	5 271	3 775	39	243	1988	
...	...	2	135	8 609	4 619	3 500	24	466	1989	
...	...	2	144	9 387	4 393	4 330	4	660	1990	
62	980	-32	187	1 536	...	1 536	...	...	1980	Yémen
90	414	775	339	3 552	...	3 552	...	...	1988	
90	370	586	360	3 915	...	3 915	...	...	1989	
90	469	353	351	3 961	...	3 961	...	...	1990	
										Europe
...	...	3 406	1 230	3 286	879	1 594	512	301	1980	Albanie
...	...	2 532	1 337	4 113	1 451	1 754	546	362	1988	
...	...	2 411	1 300	4 074	1 451	1 732	529	362	1989	
...	...	2 411	1 277	4 074	1 451	1 732	529	362	1990	
99	...	-2 965	4 139	31 242	5 182	16 395	6 620	3 044	1980	Autriche
199	...	-703	3 944	29 845	5 206	14 248	6 327	4 063	1988	
252	...	-160	4 011	30 374	5 052	14 392	6 778	4 152	1989	
254	...	-377	4 065	30 808	5 491	14 426	6 956	3 935	1990	
594	3 462	1 939	6 320	62 261	15 687	31 174	14 144	1 257	1980	Belgique
957	5 382	-277	5 715	56 306	13 851	27 034	10 335	5 086	1988	
1 066	5 646	581	5 785	56 986	15 014	25 693	11 485	4 795	1989	
1 033	5 636	1 523	5 914	58 235	16 133	24 129	13 150	4 824	1990	
...	...	2 156	5 360	47 503	21 033	19 604	5 181	1 684	1980	Bulgarie
84	435	4 068	4 915	44 157	22 729	11 647	6 984	2 797	1988	
83	443	3 724	5 057	45 491	22 556	12 125	8 110	2 700	1989	
161	465	2 757	4 307	38 779	21 269	7 164	7 988	2 358	1990	
...	...	4 129	6 365	97 447	63 513	22 411	10 157	1 367	1980	Tchécoslovaquie
...	...	4 961	6 190	96 332	61 926	17 375	13 255	3 776	1988	
...	...	5 874	6 004	93 639	59 441	15 927	14 401	3 869	1989	
...	...	4 717	5 741	89 751	54 938	14 245	16 632	3 936	1990	
772	607	-651	5 399	27 659	9 437	18 369	...	-147	1980	Danemark
884	1 299	-480	4 829	24 762	9 695	12 509	2 000	558	1988	
889	1 327	-701	4 461	22 896	7 924	11 615	2 138	1 218	1989	
1 021	1 397	-538	4 665	23 975	8 780	11 435	2 817	944	1990	
...	...	...	4 767	205	0	199	...	6	1980	Iles Féroé
...	...	...	5 674	261	0	253	...	8	1988	
...	...	...	6 064	285	0	276	...	9	1989	
...	...	...	6 404	301	0	291	...	10	1990	
221	851	2 511	5 384	25 737	6 680	15 626	1 225	2 206	1980	Finlande
348	714	-232	5 826	28 765	7 331	14 260	2 224	4 950	1988	
407	763	744	5 854	28 984	7 869	13 126	2 955	5 035	1989	
472	824	2	5 946	29 512	8 327	12 573	3 588	5 024	1990	

106
Production, trade and consumption of commercial energy
Thousand metric tons of coal equivalent and kilograms per capita [*cont.*]
Production, commerce et consommation d'énergie commerciale
Milliers de tonnes métriques d'équivalent houille et kilogrammes par habitant [*suite*]

Country or area	Year	Primary energy production - Production d'energie primaire					Changes in stocks Variations des stocks	Imports Importations	Exports Exportations
		Total Totale	Solids Solides	Liquids Liquides	Gas Gaz	Electricity Electricité			
France incl. Monaco	1980	50 161	21 729	3002	9 854	15 576	5 913	232 905	19 917
	1988	67 123	14 097	5 368	4 137	43 521	5 988	176 659	23 769
	1989	66 509	13 597	5 225	4 138	43 549	449	184 150	22 682
	1990	66 846	12 351	4 864	4 006	45 625	-2 639	191 791	25 413
Germany ꞩ F. R. Germany	1980	163 957	126 239	7431	22 780	7 507	7 409	257 456	32 956
	1988	150 200	104 923	8 067	17 128	20 082	-451	218 855	20 641
	1989	148 445	103 523	7 710	16 820	20 391	-2 741	210 877	23 302
	1990	145 496	101 839	5 134	17 825	20 699	-1 519	230 912	26 085
former German D.R.	1980	82 463	78 277	77	2 445	1 664	99	48 932	6 575
	1988	97 923	93 360	57	2 851	1 655	1 270	45 232	9 438
	1989	94 613	90 575	67	2 269	1 702	-454	47 178	11 446
	1990	87 867	84 240	56	1 877	1 694	199	46 148	11 147
Gibraltar	1980	...	...	...	...	...	5	235	...
	1988	...	...	...	...	...	129	808	0
	1989	...	...	...	...	...	43	689	0
	1990	...	...	...	...	...	49	716	0
Greece	1980	4 760	4 341	0	0	418	756	34 045	13 920
	1988	11 070	8 981	1 605	193	291	33	28 392	8 070
	1989	11 332	9 640	1 316	113	264	-1 100	29 121	8 127
	1990	11 218	9 645	1 192	135	245	-801	30 653	9 158
Hungary	1980	21 898	9 826	3 978	8 080	14	626	23 310	1 894
	1988	21 287	7 920	3 540	8 154	1 672	-404	23 241	3 830
	1989	20 768	7 589	3 543	7 911	1 726	242	22 591	6 228
	1990	17 202	6 636	3 412	5 445	1 709	1 549	22 411	4 698
Iceland	1980	382	...	...	...	382	-39	836	...
	1988	549	...	...	...	549	-42	917	...
	1989	557	...	...	...	557	27	1 018	...
	1990	566	...	...	...	566	0	1 017	...
Ireland	1980	2 723	1 451	...	1 169	103	72	9 252	300
	1988	4 438	1 708	...	2 582	148	-603	9 324	758
	1989	4 743	1 709	...	2 913	122	-479	9 189	662
	1990	4 877	1 606	...	3 149	121	-49	10 643	926
Italy and San Marino	1980	25 749	696	2 610	16 287	6 156	73	189 097	16 238
	1988	31 935	393	6 919	19 248	5 375	-3 645	190 038	21 340
	1989	32 013	423	6 724	20 279	4 587	105	201 749	19 889
	1990	32 205	393	6 672	20 435	4 705	2 556	207 658	26 842
Luxembourg	1980	12	...	...	...	12	30	5 065	96
	1988	101	...	...	...	101	-27	4 472	111
	1989	101	...	...	...	101	-13	4 811	123
	1990	101	...	...	...	101	7	5 008	111
Malta	1980	...	...	...	...	...	0	615	0
	1988	...	...	...	...	...	4	791	0
	1989	...	...	...	...	...	0	836	0
	1990	...	...	...	...	...	0	834	0
Netherlands	1980	118 749	0	2 273	115 960	516	1 624	110 368	117 406
	1988	77 452	0	6 142	70 856	454	676	128 381	96 889
	1989	83 240	0	5 497	77 242	501	186	130 731	105 031
	1990	92 854	0	5 715	86 688	451	-1 097	133 516	109 741
Norway, Svlbd, J.Myn I	1980	81 944	288	35 073	36 331	10 252	347	15 637	70 853
	1988	135 435	253	80 768	41 042	13 371	14	6 177	112 788
	1989	166 285	345	107 326	44 086	14 528	190	5 951	140 099
	1990	171 258	298	117 499	38 581	14 880	2 532	6 565	144 977
Poland	1980	173 830	166 329	509	6 589	403	-3 263	34 943	29 972
	1988	180 443	174 359	289	5 279	516	2 215	35 962	30 026
	1989	167 634	161 999	286	4 887	461	-1 167	36 562	28 458
	1990	140 471	136 436	290	3 339	406	1 448	35 030	27 531
Portugal	1980	1 169	177	...	...	992	848	14 442	600
	1988	1 638	135	...	...	1 503	-258	17 037	1 267
	1989	866	151	...	...	715	-290	22 459	2 654
	1990	1 308	165	...	...	1 143	1 115	24 045	3 362

| Bunkers - Soutes | | | Consumption - Consommation | | | | | | | |
Air Avion	Sea Maritime	Unallocated Nondistribué	Per capita Par habitant	Total Totale	Solids Solides	Liquids Liquides	Gas Gaz	Electricity Electricité	Année	Pays ou zone
2 745	5 707	5 851	4 507	242 931	50 120	142 802	34 054	15 956	1980	France y comp.Monaco
4 717	3 249	-1 778	3 737	207 838	21 552	109 586	37 689	39 012	1988	
5 259	2 082	1 032	3 928	219 154	28 212	113 719	38 817	38 406	1989	
5 505	3 654	3 963	3 979	222 741	27 480	109 494	45 723	40 043	1990	
										Allemange ↯
3 740	4 103	1 598	6 036	371 608	116 690	175 768	70 935	8 214	1980	R. f. Allemagne
5 351	3 256	609	5 559	339 649	106 458	149 172	63 890	20 128	1988	
5 833	2 722	-128	5 398	330 333	105 666	137 361	66 986	20 320	1989	
5 601	2 939	1 610	5 576	341 693	103 301	141 230	76 426	20 736	1990	
...	...	4 078	7 208	120 643	86 072	23 583	9 145	1 844	1980	anc. R.d. allemande
...	...	5 293	7 711	127 156	96 180	18 348	10 768	1 860	1988	
...	...	6 320	7 592	124 479	93 949	18 396	10 323	1 811	1989	
...	...	7 061	7 088	115 607	86 321	18 024	9 446	1 817	1990	
4	196	...	1 000	30	0	30	...	...	1980	Gibraltar
10	650	...	621	18	0	18	...	0	1988	
10	618	...	621	18	0	18	...	-0	1989	
10	629	...	933	28	0	28	...	-0	1990	
1 080	1 218	1 602	2 098	20 229	4 767	14 968	0	494	1980	Grèce
1 007	3 018	-940	2 830	28 275	10 570	17 183	193	329	1988	
920	3 075	-1 795	3 119	31 227	10 869	19 934	113	312	1989	
1 134	3 710	-1 896	3 047	30 566	10 937	19 161	135	333	1990	
...	...	1 820	3 816	40 868	12 732	15 018	12 197	921	1980	Hongrie
...	...	-562	3 928	41 663	11 247	13 063	14 293	3 059	1988	
...	...	-2 186	3 691	39 075	10 437	11 000	14 550	3 087	1989	
...	...	-1 293	3 280	34 658	8 080	10 801	12 929	2 848	1990	
29	...	...	5 382	1 227	25	820	...	382	1980	Islande
0	...	...	6 130	1 508	89	869	...	549	1988	
121	...	...	5 731	1 427	88	782	...	557	1989	
0	...	...	6 307	1 583	88	930	...	566	1990	
292	109	-231	3 362	11 434	2 454	7 708	1 169	103	1980	Irlande
495	33	75	3 596	13 003	5 248	5 024	2 582	148	1988	
501	27	-164	3 667	13 385	5 251	5 099	2 913	122	1989	
500	27	491	3 697	13 625	4 933	5 423	3 149	121	1990	
2 323	6 011	1 325	3 346	188 876	15 998	129 899	36 076	6 903	1980	Italie y comp. S.Mrn
1 916	4 498	-11 931	3 669	209 794	19 888	131 899	48 793	9 214	1988	
2 432	4 186	-10 540	3 809	217 691	19 580	136 093	53 288	8 730	1989	
2 687	4 072	-6 146	3 674	209 851	19 289	128 937	52 664	8 962	1990	
93	...	...	13 349	4 859	2 410	1 414	674	361	1980	Luxembourg
153	...	...	11 753	4 337	1 462	1 746	563	566	1988	
156	...	...	12 557	4 646	1 501	1 931	644	571	1989	
159	...	...	13 022	4 831	1 506	2 135	614	576	1990	
112	40	...	1 272	463	1	462	...	...	1980	Malte
74	43	...	1 942	670	209	461	...	...	1988	
74	43	...	2 072	719	259	461	...	...	1989	
74	43	...	2 049	717	256	461	...	...	1990	
1 319	13 550	-3 298	6 965	98 514	4 902	38 255	54 880	478	1980	Pays-Bas
1 971	15 362	-8 799	6 804	99 733	11 580	43 468	43 513	1 172	1988	
2 012	14 954	-6 834	6 684	98 622	11 636	41 417	44 464	1 106	1989	
2 087	15 791	-7 944	7 257	107 790	15 266	42 024	48 918	1 582	1990	
0	397	-475	6 469	26 459	1 441	13 444	1 379	10 195	1980	Norvège,Svalbd,J.May
112	589	101	6 681	27 934	1 441	11 110	2 703	12 680	1988	
106	482	1 261	7 196	30 166	1 419	11 511	4 536	12 699	1989	
108	656	350	6 932	29 141	1 388	11 445	3 388	12 921	1990	
...	78	5 231	4969	176 755	143 178	20 816	12 386	374	1980	Pologne
...	0	4 198	4 768	179 966	144 688	20 416	13 796	1 066	1988	
...	0	4 179	4 547	172 726	137 840	20 659	13 545	681	1989	
...	0	3 649	3 739	142 873	112 624	17 255	12 716	278	1990	
427	615	608	1 281	12 513	664	10 633	...	1 216	1980	Portugal
671	676	471	1 550	15 848	2 975	11 077	...	1 796	1988	
703	820	850	1 814	18 587	3 668	14 062	...	858	1989	
414	884	370	1 871	19 208	3 963	14 097	...	1 148	1990	

106

Production, trade and consumption of commercial energy
Thousand metric tons of coal equivalent and kilograms per capita [*cont.*]
Production, commerce et consommation d'énergie commerciale
Milliers de tonnes métriques d'équivalent houille et kilogrammes par habitant [*suite*]

| Country or area | Year | Primary energy production - Production d'energie primaire | | | | | Changes in stocks | Imports | Exports |
		Total Totale	Solids Solides	Liquids Liquides	Gas Gaz	Electricity Electricité	Variations des stocks	Imports Importations	Exports Exportations
Romania	1980	87 292	17 004	17 200	51 536	1 552	...	31 423	12 407
	1988	82 798	24 931	14 030	42 164	1 673	-1 746	42 809	19 799
	1989	78 672	25 518	13 806	37 797	1 551	-1 023	47 890	20 186
	1990	63 542	17 700	11 765	32 728	1 349	344	39 683	12 937
Spain	1980	21 042	14 489	2 289	1	4 264	1 246	82 144	2 991
	1988	26 894	14 908	2 114	1 187	8 684	603	87 851	14 978
	1989	29 130	16 283	1 483	2 071	9 293	445	93 555	13 894
	1990	28940	16 211	1 136	1 713	9 881	1 850	97 827	14 412
Sweden	1980	10 537	17	36	...	10 485	3 563	45 135	6 833
	1988	17 216	28	4	' ...	17 184	-1 968	36 488	11 032
	1989	16 955	0	4	...	16 951	-3 624	37 829	12 971
	1990	17 360	0	4	...	17 356	267	38 478	13 272
Switzerland,Liechtenstein	1980	5 663	...	...	...	5 663	567	21 560	2 348
	1988	7 206	...	...	10	7 196	-313	21 766	3 034
	1989	6 468	...	...	6	6 462	70	22016	2 563
	1990	6 720	...	...	5	6 715	15	23 843	3 055
former USSR	1980	1 920 545	510 446	863 246	516 895	29 958	1 120	15 776	334 956
	1988	2 427 445	601 443	883 696	887 398	54 907	19 956	41 417	425 313
	1989	2 412 973	578 817	861 136	919 488	53 532	19 172	31 321	406 738
	1990	2 356 784	546 446	809 399	946 277	54 662	5 496	23 931	380 285
United Kingdom	1980	275 763	108 556	115 129	47 048	5 031	9 259	99 300	79 089
	1988	316 664	83 774	164 080	60 159	8 650	2 248	102 417	127 873
	1989	281 004	80 889	131 654	58 841	9 620	-1 165	107 534	95 488
	1990	278 679	73 496	131 289	64 950	8 944	556	115 481	103 225
Yugoslavia	1980	28 544	16 382	6 129	2 573	3 459	0	22 788	1 265
	1988	36 549	22 544	6 927	2 986	4 092	339	28 361	1 570
	1989	36 370	23 702	5 594	3 165	3 908	202	28 330	1 328
	1990	35 382	24 086	5 097	2 794	3 404	6	27 273	1 220
Oceania									
American Samoa	1980	...	...	...	...	...	...	201	...
	1988	...	...	...	...	...	...	245	...
	1989	...	...	...	...	...	...	257	...
	1990	...	...	...	...	...	...	267	...
Australia	1980	113 775	69 009	30 815	12 257	1 693	-3 250	19 139	39 348
	1988	184 072	122 582	39 056	20 590	1 845	-9 895	15 070	95 846
	1989	192 886	134 629	35 347	21 048	1 862	692	19 054	91 118
	1990	212 212	143 367	40 280	26 749	1 816	5 742	19 155	100 046
Cook Islands	1980	...	...	...	...	...	...	15	...
	1988	...	...	...	...	...	...	25	...
	1989	...	...	...	...	...	...	22	...
	1990	...	...	...	...	...	...	22	...
Fiji	1980	...	...	...	...	...	-18	580	101
	1988	41	...	...	...	41	0	574	132
	1989	41	...	...	...	41	-16	602	123
	1990	41	...	...	...	41	0	633	116
French Polynesia	1980	...	...	...	...	...	0	248	...
	1988	7	...	...	...	7	1	376	...
	1989	8	...	...	...	8	0	385	...
	1990	9	...	...	...	9	0	397	...
Guam	1980	...	...	...	...	...	...	2 586	864
	1988	...	...	...	...	...	...	832	0
	1989	...	...	...	...	...	...	839	0
	1990	...	...	...	...	...	...	861	0
Kiribati	1980	...	...	...	...	...	...	13	...
	1988	...	...	...	...	...	...	10	...
	1989	...	...	...	...	...	...	10	...
	1990	...	...	...	...	...	...	10	...
Nauru	1980	...	...	...	...	...	...	66	...
	1988	...	...	...	...	...	...	66	...
	1989	...	...	...	...	...	...	66	...
	1990	...	...	...	...	...	...	69	...

Bunkers - Soutes			Consumption - Consommation							
Air Avion	Sea Maritime	Unallocated Nondistribué	Per capita Par habitant	Total Totale	Solids Solides	Liquids Liquides	Gas Gaz	Electricity Electricité	Année	Pays ou zone
...	...	3 346	4 638	102 961	23 520	24 542	53 289	1 610	1980	Roumanie
...	...	2 526	4 579	105 028	32 411	23 380	46 679	2 558	1988	
...	...	3 971	4 488	103 428	32 194	22 545	46 179	2 511	1989	
...	...	3 117	3 749	86 826	24 526	18 735	41 052	2 513	1990	
1 247	2 347	8 218	2 321	87 137	18 243	62 034	2 766	4 094	1980	Espagne
1 474	4 626	4 350	2 283	88 714	21 937	53 046	5 209	8 522	1988	
1 623	4 638	4 980	2 492	97 106	27 433	53 662	6 942	9 069	1989	
1 608	5 586	4 337	2 533	98 974	25 609	55 967	7 570	9 829	1990	
237	1 232	-393	5319	44 199	2 285	31 364	...	10 550	1980	Suède
402	961	2 273	4 891	41 004	3 521	20 127	492	16 864	1988	
467	982	1 268	5 083	42 720	5 044	20 125	658	16 893	1989	
442	926	937	4 746	39 994	3 370	18 834	665	17 124	1990	
977	0	-1 000	3 835	24 331	499	18 012	1 375	4 445	1980	Suisse, Liechtenstein
1 334	32	-24	3 797	24 908	537	16 132	2 225	6 014	1988	
1 414	29	184	3 676	24 224	554	15 098	2 420	6 153	1989	
1 453	26	55	3 924	25 958	402	16 510	2 590	6 456	1990	
...	6 372	86 374	5 677	1 507 500	490 841	538 318	450 692	27 649	1980	ancienne URSS
...	7 080	65 503	6 916	1 951 009	572 248	556 579	772 015	50 168	1988	
...	6 372	69 432	6 832	1 942 579	559 154	549 602	785 068	48 756	1989	
...	6 089	57 545	6 742	1 931 301	523 058	542 045	815 951	50 247	1990	
4 157	3 505	7 218	4 809	271 837	101 594	103 879	61 332	5 031	1980	Royaume-Uni
5 672	2 624	-4 746	5 004	285 439	91 684	110 101	73 428	10 226	1988	
5 986	3 318	-3 813	5 050	288 699	93 644	111 255	72 629	11 172	1989	
6 064	3 631	-6 015	5 000	286 525	84 648	116 875	74 591	10 411	1990	
0	0	2 213	2 146	47 855	19 583	20 350	4 513	3 409	1980	Yougoslavie
0	0	5 245	2 467	57 756	26 306	19 481	8 049	3 919	1988	
0	0	3 829	2 520	59 340	27 645	19 281	8 556	3 858	1989	
0	0	4 085	2 422	57 344	26 998	19 123	7 863	3 360	1990	
										Océanie
...	...	...	6 281	201	...	201	...	...	1980	Samoa américaines
...	125	...	3 270	121	...	121	...	...	1988	
...	135	...	3 297	122	...	122	...	...	1989	
...	130	...	3 703	137	...	137	...	...	1990	
1 132	1 614	2 638	6 222	91 433	38 312	39 171	12 257	1 693	1980	Australie
1 586	812	-2 595	6 998	113 389	48 962	41 992	20 590	1 845	1988	
1 832	921	-2 658	7 306	120035	52 879	44 246	21 048	1 862	1989	
1 586	939	-3 998	7 629	127 052	54 065	48 162	23 009	1 816	1990	
...	...	...	833	15	...	15	...	...	1980	Iles Cook
15	...	...	556	10	...	10	...	...	1988	
12	...	...	556	10	...	10	...	...	1989	
12	...	...	556	10	...	10	...	...	1990	
106	45	...	546	346	21	325	...	...	1980	Fidji
133	22	...	452	328	17	270	...	41	1988	
134	39	...	490	362	13	308	...	41	1989	
133	40	...	513	385	14	330	...	41	1990	
...	112	...	919	136	...	136	...	...	1980	Polynésie française
60	48	...	1 465	274	...	267	...	7	1988	
59	46	...	1 492	288	...	280	...	8	1989	
59	46	...	1 508	300	...	291	...	9	1990	
488	302	9	8 698	922	...	922	...	...	1980	Guam
12	144	0	5 878	676	...	676	...	...	1988	
12	144	0	5 897	684	...	684	...	...	1989	
12	144	0	6 026	705	...	705	...	...	1990	
...	...	...	220	13	...	13	...	...	1980	Kiribati
...	...	...	156	10	...	10	...	...	1988	
...	...	...	156	10	...	10	...	...	1989	
...	...	...	154	10	...	10	...	...	1990	
7	...	...	7 375	59	...	59	...	...	1980	Nauru
7	...	...	6 556	59	...	59	...	...	1988	
7	...	...	6 556	59	...	59	...	...	1989	
7	...	...	6 778	61	...	61	...	...	1990	

106
Production, trade and consumption of commercial energy
Thousand metric tons of coal equivalent and kilograms per capita [*cont.*]
Production, commerce et consommation d'énergie commerciale
Milliers de tonnes métriques d'équivalent houille et kilogrammes par habitant [*suite*]

| Country or area | Year | Primary energy production - Production d'energie primaire | | | | | Changes in stocks | | |
		Total Totale	Solids Solides	Liquids Liquides	Gas Gaz	Electricity Electricité	Variations des stocks	Imports Importations	Exports Exportations
New Caledonia	1980	34	...	...	...	34	...	904	...
	1988	65	...	...	...	65	0	722	28
	1989	64	...	...	...	64	0	766	7
	1990	57	...	...	...	57	0	718	12
New Zealand	1980	5 768	1 808	473	1 330	2 157	348	5 614	95
	1988	12 919	1 832	2 348	6 028	2 711	-116	4 569	1 384
	1989	13 898	2 119	2 609	6 172	2 997	1	4 425	1 209
	1990	13 638	1 930	2 657	6 133	2 918	0	4 450	1 109
Niue	1980	...	...	...	...	...	...	1	...
	1988	...	...	...	...	...	...	1	...
	1989	...	...	...	...	...	...	1	...
	1990	...	...	...	...	...	...	1	...
Pacific Islands (Palau)[1]	1980	...	...	...	...	...	...	81	...
	1988	4	...	...	...	4	...	101	...
	1989	4	...	...	...	4	...	129	...
	1990	4	...	...	...	4	...	133	...
Papua New Guinea	1980	39	...	...	...	39	...	892	...
	1988	55	...	...	...	55	0	1 086	7
	1989	56	...	...	...	56	0	1 118	14
	1990	57	...	...	...	57	0	1 137	17
Samoa	1980	1	...	...	...	1	...	47	...
	1988	2	...	...	...	2	...	55	...
	1989	2	...	...	...	2	...	57	...
	1990	2	...	...	...	2	...	60	...
Solomon Islands	1980	...	...	...	...	...	...	53	...
	1988	...	...	...	...	...	...	76	...
	1989	...	...	...	...	...	...	79	...
	1990	...	...	...	...	...	...	80	...
Tonga	1980	...	...	...	...	...	...	19	...
	1988	...	...	...	...	...	...	38	...
	1989	...	...	...	...	...	...	41	...
	1990	...	...	...	...	...	...	41	...
Vanuatu	1980	...	...	...	...	...	...	56	...
	1988	...	...	...	...	...	...	63	...
	1989	...	...	...	...	...	...	56	...
	1990	...	...	...	...	...	...	59	...
Wake Island	1980	...	...	...	...	...	...	592	...
	1988	...	...	...	...	...	...	592	...
	1989	...	...	...	...	...	...	592	...
	1990	...	...	...	...	...	...	577	...

Source:
Energy statistics database of the Statistical Division of the United Nations Secretariat.

Source:
Base de données pour les statistiques énergétiques de la Division de statistique du Secrétariat de l'ONU.

§ All data shown which pertain to Germany prior to 3 October 1990 are indicated separately for the federal Republic of Germany and the former Germany Democratic Republic based on their respective territories at the time indicated. Where data for united Germany (3 October 1990 and thereafter) are not available, available data are shown separately under the designations Federal Republic of Germany and former German Democratic Republic and pertain to the territorial boundaries prior to 3 October 1990. For detailed explanatory notes on data pertaining to Germany, see Annex I - Country Nomenclature.

1 Including data for Federated States of Micronesia, Marshall Is. and Northern Mariana Is.

§ Toutes les données se rapportant à l'Allemagne avant le 3 octobre 1990 figurent dans deux rubriques séparées basées sur les territoires respectifs de la République fédérale d'Allemagne et l'ancienne République démocratique allemande selon la période indiquée. En l'absence de données pour l'Allemagne unifiée (à compter du 3 octobre 1990), les données disponibles sont fournies séparément sous les rubriques République fédérale d'Allemagne et ancienne République démocratique allemande et se rapportent aux limites territoriales antérieures au 3 octobre 1990. Pour les notes explicatives en détail sur les données concernant Allemagne, voir Annexe I - Nomenclature des pays.

1 Y compris les données pour les Etats fédérés de Micronésie, les îles Marshall et les îles Mariannes du Nord.

Bunkers - Soutes			Consumption - Consommation							
Air Avion	Sea Maritime	Unallocated Nondistribué	Per capita Par habitant	Total Totale	Solids Solides	Liquids Liquides	Gas Gaz	Electricity Electricité	Année	Pays ou zone
...	21	...	6 597	917	120	763	...	34	1980	Nouvelle-Calédonie
29	0	...	4 620	730	170	495	...	65	1988	
29	7	...	4 882	786	207	515	...	64	1989	
22	7	...	4 476	734	164	513	...	57	1990	
245	484	226	3 208	9 986	1 609	4 957	1 263	2 157	1980	Nouvelle-Zélande
158	307	-372	4 878	16 128	1 530	5 866	6 021	2 711	1988	
147	322	-352	5 097	16 994	1 717	6 110	6 171	2 997	1989	
133	308	-406	5 039	16 945	1 599	6 296	6 133	2 918	1990	
0	...	...	333	1	...	1	...	...	1980	Nioué
0	...	...	333	1	...	1	...	...	1988	
0	...	...	333	1	...	1	...	...	1989	
0	...	...	333	1	...	1	...	...	1990	
7	...	...	537	73	...	73	...	...	1980	I.Pacifique (Palaos) [1]
15	...	...	545	90	...	86	...	4	1988	
22	...	...	661	111	...	107	...	4	1989	
22	...	...	669	115	...	111	...	4	1990	
22	7	...	292	902	0	863	...	39	1980	Papouasie-Nvl-Guinée
38	4	...	301	1 091	1	1 035	...	55	1988	
40	4	...	301	1 115	1	1 058	...	56	1989	
40	4	...	299	1 132	1	1 074	...	57	1990	
...	...	...	310	48	0	47	...	1	1980	Samoa
...	...	...	345	57	0	55	...	2	1988	
...	...	...	361	60	0	57	...	2	1989	
...	...	...	377	63	0	60	...	2	1990	
4	...	...	213	48	...	48	...	...	1980	Iles Salomon
3	...	...	251	73	...	73	...	...	1988	
3	...	...	253	76	...	76	...	...	1989	
3	...	...	248	77	...	77	...	...	1990	
...	...	...	196	19	0	19	...	...	1980	Tonga
4	...	...	358	34	0	34	...	...	1988	
6	...	...	368	35	0	35	...	...	1989	
6	...	...	368	35	0	35	...	...	1990	
...	27	...	248	29	0	29	...	...	1980	Vanuatu
...	28	...	241	35	0	35	...	...	1988	
...	25	...	208	31	0	31	...	...	1989	
...	27	...	208	32	0	32	...	...	1990	
553	...	...	19 000	38	...	38	...	...	1980	Ile de Wake
553	...	...	19 000	38	...	38	...	...	1988	
553	...	...	19 000	38	...	38	...	...	1989	
539	...	...	19 000	38	...	38	...	...	1990	

107
Production of selected energy commodities
Production des principaux biens de l'énergie

Thousand metric tons of coal equivalent
Milliers de tonnes métriques d'équivalent houille

Region, country or area	Year	Hard coal Houille	Lignite brown coal Lignite	Fuelwood Bois de chauffage	Coke	Crude petroleum Pétrole brut	Condensate Condensat	Natural gasoline Gazoline naturelle	Natural gas Gaz naturel	Other gases Autres gaz
World	1980	2228571	373571	498201	350196	4260609	11716	39380	1842351	312572
	1988	2804208	458957	573139	349947	4157483	40121	42392	2329963	310303
	1989	2857144	459745	580659	348895	4224234	38500	44874	2415470	310595
	1990	2859937	441894	568795	339412	4308141	42732	45070	2499832	307817
Africa	1980	98230	...	115009	2504	426389	7037	0	35227	4537
	1988	142232	...	145531	2950	348164	28427	0	69856	4858
	1989	138853	...	148863	3257	389694	26614	0	76525	4903
	1990	139362	...	152427	3439	421154	30325	0	94870	4918
Algeria	1980	7	...	474	...	67654	6530	...	25903	0
	1988	10	...	601	...	43644	27065	...	46515	0
	1989	15	...	617	...	48663	25178	...	50819	0
	1990	10	...	634	...	52887	28841	...	67387	0
Angola	1980	...	...	1147	...	10611	...	...	102	...
	1988	...	...	1746	...	31886	...	...	208	...
	1989	...	...	1796	...	32346	...	...	222	...
	1990	...	...	1846	...	33647	...	...	222	...
Benin	1980	...	...	1201	...	14	...	...	...	...
	1988	...	...	1498	...	393	...	...	...	...
	1989	...	...	1544	...	407	...	...	...	...
	1990	...	...	1592	...	416	...	...	...	...
Botswana	1980	371	...	305	...	...	...	...	...	...
	1988	613	...	404	...	...	...	...	...	...
	1989	633	...	419	...	...	...	...	...	...
	1990	635	...	434	...	...	...	...	...	...
Burkina Faso	1980	...	...	2152	...	...	...	...	...	...
	1988	...	...	2636	...	...	...	...	...	...
	1989	...	...	2709	...	...	...	...	...	...
	1990	...	...	2783	...	...	...	...	...	...
Burundi	1980	...	...	1040	...	...	...	...	...	...
	1988	...	...	1308	...	...	...	...	...	...
	1989	...	...	1348	...	...	...	...	...	...
	1990	...	...	1387	...	...	...	...	...	...
Cameroon	1980	0	...	2657	...	3857	...	...	...	...
	1988	1	...	3413	...	12117	...	...	...	...
	1989	1	...	3526	...	12336	...	...	...	...
	1990	1	...	3646	...	12114	...	...	...	...
Cape Verde	1980	...	...	...	...	...	...	...	...	...
	1988	...	...	...	...	...	...	...	...	...
	1989	...	...	...	...	...	...	...	...	...
	1990	...	...	...	...	...	...	...	...	...
Central Afrian Rep.	1980	...	...	828	...	...	...	...	...	...
	1988	...	...	1157	...	...	...	...	...	...
	1989	...	...	1150	...	...	...	...	...	...
	1990	...	...	1152	...	...	...	...	...	...
Chad	1980	...	...	910	...	...	...	...	...	...
	1988	...	...	1097	...	...	...	...	...	...
	1989	...	...	1127	...	...	...	...	...	...
	1990	...	...	1154	...	...	...	...	...	...
Comoros	1980	...	...	...	...	...	...	...	...	...
	1988	...	...	...	...	...	...	...	...	...
	1989	...	...	...	...	...	...	...	...	...
	1990	...	...	...	...	...	...	...	...	...
Congo	1980	...	...	464	...	4707	...	...	2	...
	1988	...	...	650	...	10054	...	...	3	...
	1989	...	...	670	...	11374	...	...	3	...
	1990	...	...	692	...	11537	...	...	3	...
Côte d'Ivoire	1980	...	...	2256	...	171	...	...	...	...
	1988	...	...	3014	...	821	...	...	...	...
	1989	...	...	3129	...	271	...	...	...	...
	1990	...	...	3250	...	450	...	...	...	...

LPG GPL	Aviation gasoline Essence aviation	Motor gasoline Essence auto	Kerosene Kérosène	Jet fuels Carbu-réacteurs	Gas diesel oil Gas oil fluides	Residual fuel Fuel oil fluides	Lubricants Huiles lubrifiantes	Electricity Electricité	Année	Région, pays ou zone
186196	7621	976149	163735	159540	1043992	1317323	55727	1011780	1980	**Monde**
238488	4513	1075066	163583	210219	1154018	1061593	58325	1358445	1988	
244432	3963	1091946	163490	218994	1174149	1067457	59231	1406185	1989	
251840	3921	1105149	163885	230626	1187702	1061258	59074	1440193	1990	
2777	19	17214	6242	3860	27908	36303	830	23152	1980	**Afrique**
10232	231	24304	8646	4870	40030	47052	1163	36754	1988	
10674	241	26253	8807	5998	41596	48856	1254	38023	1989	
11518	247	26697	8705	6385	42502	49949	1268	38732	1990	
1092	...	1725	87	546	5403	3827	68	875	1980	Algérie
7478	...	2775	155	766	11020	8921	158	1839	1988	
7655	...	2790	162	769	10889	8977	173	1887	1989	
8443	...	2850	177	781	11020	9062	173	1978	1990	
19	...	90	31	88	305	779	...	184	1980	Angola
28	...	150	71	221	457	913	...	222	1988	
25	...	158	74	229	464	920	...	224	1989	
28	...	162	77	231	468	925	...	226	1990	
...	...	...	...	...	...	...	...	1	1980	Bénin
...	...	...	...	...	...	...	...	1	1988	
...	...	...	...	...	...	...	...	1	1989	
...	...	...	...	...	...	...	...	1	1990	
...	...	...	...	...	...	...	...	58	1980	Botswana
...	...	...	...	...	...	...	...	130	1988	
...	...	...	...	...	...	...	...	222	1989	
...	...	...	...	...	...	...	...	223	1990	
...	...	...	...	...	...	...	...	13	1980	Burkina Faso
...	...	...	...	...	...	...	...	17	1988	
...	...	...	...	...	...	...	...	19	1989	
...	...	...	...	...	...	...	...	19	1990	
...	...	...	...	...	...	...	...	0	1980	Burundi
...	...	...	...	...	...	...	...	13	1988	
...	...	...	...	...	...	...	...	13	1989	
...	...	...	...	...	...	...	...	13	1990	
16	12	225	177	15	239	425	72	178	1980	Cameroun
28	17	593	438	137	624	807	137	317	1988	
30	18	596	439	139	626	810	140	332	1989	
28	15	570	444	140	629	814	141	332	1990	
...	...	...	...	...	...	...	...	2	1980	Cap-Vert
...	...	...	...	...	...	...	...	4	1988	
...	...	...	...	...	...	...	...	4	1989	
...	...	...	...	...	...	...	...	4	1990	
...	...	...	...	...	...	...	...	8	1980	Rép. centrafricaine
...	...	...	...	...	...	...	...	11	1988	
...	...	...	...	...	...	...	...	11	1989	
...	...	...	...	...	...	...	...	12	1990	
...	...	...	...	...	...	...	...	6	1980	Tchad
...	...	...	...	...	...	...	...	10	1988	
...	...	...	...	...	...	...	...	10	1989	
...	...	...	...	...	...	...	...	10	1990	
...	...	...	...	...	...	...	...	1	1980	Comores
...	...	...	...	...	...	...	...	2	1988	
...	...	...	...	...	...	...	...	2	1989	
...	...	...	...	...	...	...	...	2	1990	
0	...	0	0	0	0	0	...	19	1980	Congo
5	68	77	0	18	155	474	7	36	1988	
5	74	77	0	18	148	484	4	49	1989	
5	80	77	0	18	149	489	4	49	1990	
16	...	300	211	140	247	969	3	215	1980	Côte d'Ivoire
20	...	467	221	149	713	845	7	284	1988	
23	...	323	221	69	703	562	7	290	1989	
22	...	353	221	69	703	701	9	291	1990	

107
Production of selected energy commodities
Thousand metric tons of coal equivalent [*cont.*]
Production des principaux biens de l'énergie
Milliers de tonnes métriques d'équivalent houille [*suite*]

Region, country or area	Year	Hard coal Houille	Lignite brown coal Lignite	Fuelwood Bois de chauffage	Coke	Crude petroleum Pétrole brut	Condensate Condensat	Natural gasoline Gazoline naturelle	Natural gas Gaz naturel	Other gases Autres gaz
Djibouti	1980	...	...	...	...	...	...	...	...	...
	1988	...	...	...	...	...	...	...	...	...
	1989	...	...	...	...	...	...	...	...	...
	1990	...	...	...	...	...	...	...	...	...
Egypt	1980	...	...	566	677	42000	507	0	2134	6
	1988	...	...	682	859	61207	1362	0	7992	7
	1989	...	...	698	1116	61427	1436	0	8539	7
	1990	...	...	715	1053	62789	1483	0	9028	10
Equatorial Guinea	1980	...	...	140	...	...	...	...	...	...
	1988	...	...	149	...	...	...	...	...	...
	1989	...	...	149	...	...	...	...	...	...
	1990	...	...	149	...	...	...	...	...	...
Ethiopia	1980	...	...	10699	...	...	...	...	...	0
	1988	...	...	12856	...	...	...	...	...	1
	1989	...	...	13217	...	...	...	...	...	1
	1990	...	...	13596	...	...	...	...	...	1
Gabon	1980	...	...	589	...	12707	...	...	20	...
	1988	...	...	800	...	11383	...	...	233	...
	1989	...	...	828	...	14610	...	...	121	...
	1990	...	...	856	...	19276	...	...	133	...
Gambia	1980	...	...	298	...	...	...	...	...	...
	1988	...	...	296	...	...	...	...	...	...
	1989	...	...	300	...	...	...	...	...	...
	1990	...	...	302	...	...	...	...	...	...
Ghana	1980	...	...	3983	...	126	...	...	...	...
	1988	...	...	5307	...	0	...	...	...	...
	1989	...	...	5355	...	0	...	...	...	...
	1990	...	...	5355	...	0	...	...	...	...
Guinea	1980	...	...	1081	...	...	...	...	...	...
	1988	...	...	1084	...	...	...	...	...	...
	1989	...	...	1117	...	...	...	...	...	...
	1990	...	...	1151	...	...	...	...	...	...
Guinea-Bissau	1980	...	...	141	...	...	...	...	...	...
	1988	...	...	141	...	...	...	...	...	...
	1989	...	...	141	...	...	...	...	...	...
	1990	...	...	141	...	...	...	...	...	...
Kenya	1980	...	...	8370	...	...	...	...	...	...
	1988	...	...	10415	...	...	...	...	...	...
	1989	...	...	10432	...	...	...	...	...	...
	1990	...	...	10432	...	...	...	...	...	...
Lesotho	1980	...	...	154	...	...	...	...	...	...
	1988	...	...	193	...	...	...	...	...	...
	1989	...	...	199	...	...	...	...	...	...
	1990	...	...	204	...	...	...	...	...	...
Liberia	1980	...	...	1326	...	...	...	...	...	...
	1988	...	...	1584	...	...	...	...	...	...
	1989	...	...	1606	...	...	...	...	...	...
	1990	...	...	1630	...	...	...	...	...	...
Libyan Arab Jamah.	1980	...	...	179	...	126177	...	...	4585	1
	1988	...	...	179	...	70769	...	...	9315	1
	1989	...	...	179	...	77600	...	...	10581	1
	1990	...	...	179	...	95946	...	...	12666	1
Madagascar	1980	...	...	1776	0	...	...	...	...	...
	1988	...	...	2279	0	...	...	...	...	...
	1989	...	...	2353	0	...	...	...	...	...
	1990	...	...	2429	0	...	...	...	...	...
Malawi	1980	...	...	3781	...	...	...	...	...	...
	1988	...	...	4345	...	...	...	...	...	...
	1989	...	...	4350	...	...	...	...	...	...
	1990	...	...	4351	...	...	...	...	...	...
Mali	1980	...	...	1330	...	...	...	...	...	...
	1988	...	...	1640	...	...	...	...	...	...
	1989	...	...	1691	...	...	...	...	...	...
	1990	...	...	1744	...	...	...	...	...	...

LPG GPL	Aviation gasoline Essence aviation	Motor gasoline Essence auto	Kerosene Kérosène	Jet fuels Carbu-réacteurs	Gas diesel oil Gas oil fluides	Residual fuel oil Fuel oil fluides	Lubricants Huiles lubrifiantes	Electricity Electricité	Année	Région, pays ou zone
...	...	...	...	...	...	...	...	14	1980	Djibouti
...	...	...	...	...	...	...	...	21	1988	
...	...	...	...	...	...	...	...	22	1989	
...	...	...	...	...	...	...	...	22	1990	
317	...	2923	2473	0	3655	9487	105	2326	1980	Egypte
1193	...	5070	3607	251	5239	14698	267	4704	1988	
1343	...	5271	3203	696	5419	15170	326	4827	1989	
1402	...	5571	3231	737	5704	16018	332	4858	1990	
...	...	...	...	...	...	...	...	2	1980	Guinée équatoriale
...	...	...	...	...	...	...	...	2	1988	
...	...	...	...	...	...	...	...	2	1989	
...	...	...	...	...	...	...	...	2	1990	
6	...	121	0	57	223	344	...	85	1980	Ethiopie
8	...	151	18	84	294	476	...	108	1988	
8	...	165	18	74	254	490	...	108	1989	
8	...	165	18	71	246	418	...	101	1990	
8	...	150	150	111	494	881	...	65	1980	Gabon
8	...	91	105	96	254	252	...	112	1988	
8	...	108	103	99	302	273	...	111	1989	
8	...	112	106	100	297	276	...	112	1990	
...	...	...	...	...	...	...	...	6	1980	Gambie
...	...	...	...	...	...	...	...	8	1988	
...	...	...	...	...	...	...	...	7	1989	
...	...	...	...	...	...	...	...	8	1990	
11	...	342	177	56	447	396	27	653	1980	Ghana
8	...	213	162	37	420	95	23	604	1988	
8	...	217	211	37	333	99	26	648	1989	
9	...	142	140	37	304	102	30	669	1990	
...	...	...	...	...	...	...	...	47	1980	Guinée
...	...	...	...	...	...	...	...	63	1988	
...	...	...	...	...	...	...	...	63	1989	
...	...	...	...	...	...	...	...	64	1990	
...	...	...	...	...	...	...	...	2	1980	Guinée-Bissau
...	...	...	...	...	...	...	...	3	1988	
...	...	...	...	...	...	...	...	4	1989	
...	...	...	...	...	...	...	...	5	1990	
37	...	643	99	457	737	1886	...	183	1980	Kenya
44	...	492	211	394	796	878	...	349	1988	
44	...	543	268	401	815	893	...	356	1989	
44	...	502	218	507	816	949	...	374	1990	
...	...	...	...	...	...	...	...	0	1980	Lesotho
...	...	...	...	...	...	...	...	0	1988	
...	...	...	...	...	...	...	...	0	1989	
...	...	...	...	...	...	...	...	0	1990	
2	...	105	12	81	189	496	...	109	1980	Libéria
0	...	0	0	0	0	0	...	102	1988	
0	...	0	0	0	0	0	...	100	1989	
0	...	0	0	0	0	0	...	69	1990	
625	...	787	159	889	1857	3421	...	594	1980	Jamah. arabe libyenne
544	...	2467	295	1327	4788	5511	...	1965	1988	
618	...	2841	295	2064	5520	6487	...	2211	1989	
653	...	3150	354	2270	5945	6797	...	2334	1990	
8	...	132	72	1	186	340	0	54	1980	Madagascar
3	...	79	56	1	129	208	10	65	1988	
2	...	43	28	1	67	122	10	69	1989	
2	...	37	25	1	62	92	6	70	1990	
...	...	...	...	...	...	...	...	51	1980	Malawi
...	...	...	...	...	...	...	...	72	1988	
...	...	...	...	...	...	...	...	72	1989	
...	...	...	...	...	...	...	...	72	1990	
...	...	...	...	...	...	...	...	13	1980	Mali
...	...	...	...	...	...	...	...	26	1988	
...	...	...	...	...	...	...	...	26	1989	
...	...	...	...	...	...	...	...	26	1990	

107
Production of selected energy commodities
Thousand metric tons of coal equivalent [cont.]
Production des principaux biens de l'énergie
Milliers de tonnes métriques d'équivalent houille [suite]

Region, country or area	Year	Hard coal Houille	Lignite brown coal Lignite	Fuelwood Bois de chauffage	Coke	Crude petroleum Pétrole brut	Condensate Condensat	Natural gasoline Gazoline naturelle	Natural gas Gaz naturel	Other gases Autres gaz
Mauritania	1980	...	...	2	...	...	...	...	...	...
	1988	...	...	2	...	...	...	...	...	...
	1989	...	...	2	...	...	...	...	...	...
	1990	...	...	2	...	...	...	...	...	...
Mauritius	1980	...	...	8	...	...	...	...	...	...
	1988	...	...	13	...	...	...	...	...	...
	1989	...	...	13	...	...	...	...	...	...
	1990	...	...	13	...	...	...	...	...	...
Morocco	1980	680	...	372	...	20	...	...	90	...
	1988	637	...	446	...	29	...	...	99	...
	1989	504	...	454	...	19	...	...	84	...
	1990	526	...	461	...	21	...	...	68	...
Mozambique	1980	207	...	4102	...	...	...	...	...	...
	1988	45	...	5007	...	...	...	...	...	...
	1989	42	...	5007	...	...	...	...	...	...
	1990	40	...	5007	...	...	...	...	...	...
Niger	1980	20	...	1065	...	...	...	...	...	...
	1988	155	...	1456	...	...	...	...	...	...
	1989	157	...	1502	...	...	...	...	...	...
	1990	158	...	1550	...	...	...	...	...	...
Nigeria	1980	176	...	24703	...	148843	...	...	1764	...
	1988	82	...	31153	...	96764	...	...	5019	...
	1989	85	...	32205	...	121679	...	...	5659	...
	1990	90	...	33285	...	123626	...	...	4940	...
Réunion	1980	...	...	10	...	...	...	...	...	...
	1988	...	...	10	...	...	...	...	...	...
	1989	...	...	10	...	...	...	...	...	...
	1990	...	...	10	...	...	...	...	...	...
Rwanda	1980	...	...	1507	...	...	...	...	1	...
	1988	...	...	1867	...	...	...	...	0	...
	1989	...	...	1867	...	...	...	...	0	...
	1990	...	...	1867	...	...	...	...	0	...
Saint Helena	1980	0	...	0	...	...	...	...	...	...
	1988	0	...	0	...	...	...	...	...	...
	1989	0	...	0	...	...	...	...	...	...
	1990	0	...	0	...	...	...	...	...	...
Sao Tome and Principe	1980	...	...	...	...	...	...	...	...	...
	1988	...	...	...	...	...	...	...	...	...
	1989	...	...	...	...	...	...	...	...	...
	1990	...	...	...	...	...	...	...	...	...
Senegal	1980	...	...	1064	...	...	...	...	...	...
	1988	...	...	1227	...	...	...	...	...	...
	1989	...	...	1257	...	...	...	...	...	...
	1990	...	...	1287	...	...	...	...	...	...
Seychelles	1980	...	...	...	...	...	...	...	...	...
	1988	...	...	...	...	...	...	...	...	...
	1989	...	...	...	...	...	...	...	...	...
	1990	...	...	...	...	...	...	...	...	...
Sierra Leone	1980	...	...	772	...	...	...	...	...	...
	1988	...	...	933	...	...	...	...	...	...
	1989	...	...	958	...	...	...	...	...	...
	1990	...	...	983	...	...	...	...	...	...
Somalia	1980	...	...	1676	...	...	...	...	...	...
	1988	...	...	2207	...	...	...	...	...	...
	1989	...	...	2280	...	...	...	...	...	...
	1990	...	...	2347	...	...	...	...	...	...
South Africa	1980	92832	...	2395	1578	...	...	...	...	4446
	1988	134803	...	2359	1825	...	...	...	...	4803
	1989	131677	...	2359	1839	...	...	...	...	4849
	1990	132336	...	2359	1843	...	...	...	...	4859
Sudan	1980	...	...	5108	...	...	...	...	...	...
	1988	...	...	6511	...	...	...	...	...	...
	1989	...	...	6700	...	...	...	...	...	...
	1990	...	...	6893	...	...	...	...	...	...

LPG GPL	Aviation gasoline Essence aviation	Motor gasoline Essence auto	Kerosene Kérosène	Jet fuels Carbu-réacteurs	Gas diesel oil Gas oil fluides	Residual fuel Fuel oil fluides	Lubricants Huiles lubrifiantes	Electricity Electricité	Année	Région, pays ou zone
...	...	...	0	...	0	0	...	11	1980	Mauritanie
42	...	418	81	...	307	446	...	15	1988	
44	...	315	77	...	283	449	...	15	1989	
44	...	304	77	...	196	442	...	17	1990	
...	...	...	...	...	...	...	...	57	1980	Maurice
...	...	...	...	...	...	...	...	78	1988	
...	...	...	...	...	...	...	...	84	1989	
...	...	...	...	...	...	...	...	95	1990	
305	...	543	105	364	1743	2488	...	605	1980	Maroc
402	...	547	68	318	2239	2693	...	1085	1988	
410	...	571	78	327	2371	2975	...	1112	1989	
362	...	588	66	368	2590	2897	...	1183	1990	
39	...	139	91	38	238	455	...	1720	1980	Mozambique
0	...	0	0	0	0	0	...	58	1988	
0	...	0	0	0	0	0	...	60	1989	
0	...	0	0	0	0	0	...	60	1990	
...	...	...	...	...	...	...	...	16	1980	Niger
...	...	...	...	...	...	...	...	20	1988	
...	...	...	...	...	...	...	...	20	1989	
...	...	...	...	...	...	...	...	20	1990	
95	...	2250	1209	59	2498	2472	...	882	1980	Nigéria
93	...	2962	1837	74	2635	2958	...	1219	1988	
93	...	4450	2266	59	3470	3255	...	1220	1989	
85	...	4350	2211	56	3335	3115	...	1222	1990	
...	...	...	...	...	...	...	...	42	1980	Réunion
...	...	...	...	...	...	...	...	94	1988	
...	...	...	...	...	...	...	...	99	1989	
...	...	...	...	...	...	...	...	100	1990	
...	...	...	...	...	...	...	...	14	1980	Rwanda
...	...	...	...	...	...	...	...	21	1988	
...	...	...	...	...	...	...	...	21	1989	
...	...	...	...	...	...	...	...	22	1990	
...	...	...	...	...	...	...	...	0	1980	Sainte-Héléne
...	...	...	...	...	...	...	...	0	1988	
...	...	...	...	...	...	...	...	0	1989	
...	...	...	...	...	...	...	...	0	1990	
...	...	...	...	...	...	...	...	1	1980	Sao Tome-et-Principe
...	...	...	...	...	...	...	...	2	1988	
...	...	...	...	...	...	...	...	2	1989	
...	...	...	...	...	...	...	...	2	1990	
12	0	186	31	174	268	382	22	78	1980	Sénégal
3	132	165	19	130	283	319	4	74	1988	
3	135	172	21	134	312	297	4	77	1989	
3	138	175	21	136	239	300	4	84	1990	
...	...	...	...	...	...	...	...	6	1980	Seychelles
...	...	...	...	...	...	...	...	10	1988	
...	...	...	...	...	...	...	...	10	1989	
...	...	...	...	...	...	...	...	10	1990	
...	...	30	7	28	104	47	14	25	1980	Sierra Leone
...	...	51	28	21	144	123	17	21	1988	
...	...	52	29	21	146	125	17	27	1989	
...	...	52	27	22	149	125	17	28	1990	
0	4	46	21	41	72	6	6	15	1980	Somalie
0	9	121	15	29	116	21	3	32	1988	
0	9	123	15	29	119	21	3	32	1989	
0	10	120	12	22	120	23	3	28	1990	
109	...	5550	663	516	6670	4956	503	11011	1980	Afrique du Sud
106	...	6532	631	597	7612	4560	503	19253	1988	
109	...	6540	634	604	7627	4574	518	19939	1989	
112	...	6525	626	590	7634	4581	521	19941	1990	
8	3	217	38	37	722	378	10	108	1980	Soudan
8	6	165	25	108	500	421	26	161	1988	
8	6	168	27	113	457	439	26	162	1989	
11	4	142	22	111	480	439	29	163	1990	

107
Production of selected energy commodities
Thousand metric tons of coal equivalent [*cont.*]
Production des principaux biens de l'énergie
Milliers de tonnes métriques d'équivalent houille [*suite*]

Region, country or area	Year	Hard coal Houille	Lignite brown coal Lignite	Fuelwood Bois de chauffage	Coke	Crude petroleum Pétrole brut	Condensate Condensat	Natural gasoline Gazoline naturelle	Natural gas Gaz naturel	Other gases Autres gaz
Swaziland	1980	176	...	181	...	...	...	...	...	...
	1988	165	...	187	...	...	...	...	...	...
	1989	165	...	187	...	...	...	...	...	...
	1990	160	...	187	...	...	...	...	...	...
Togo	1980	0	...	173	...	...	...	...	...	...
	1988	0	...	225	...	...	...	...	...	...
	1989	0	...	232	...	...	...	...	...	...
	1990	0	...	239	...	...	...	...	...	...
Tunisia	1980	...	...	801	...	8039	...	...	626	16
	1988	...	...	982	...	7011	...	...	471	4
	1989	...	...	1005	...	7023	...	...	497	1
	1990	...	...	1028	...	6431	...	...	423	2
Uganda	1980	...	...	3067	...	...	...	...	...	...
	1988	...	...	4082	...	...	...	...	...	...
	1989	...	...	4234	...	...	...	...	...	...
	1990	...	...	4395	...	...	...	...	...	...
United Rep of Tanzania	1980	1	...	7426	0	...	...	...	...	...
	1988	3	...	9986	1	...	...	...	...	...
	1989	3	...	10358	1	...	...	...	...	...
	1990	4	...	10746	1	...	...	...	...	...
Western Sahara	1980	...	...	...	...	...	...	...	...	...
	1988	...	...	...	...	...	...	...	...	...
	1989	...	...	...	...	...	...	...	...	...
	1990	...	...	...	...	...	...	...	...	...
Zaire	1980	138	...	8214	...	1461	...	...	...	...
	1988	123	...	11464	...	2086	...	...	...	...
	1989	125	...	11641	...	1940	...	...	...	...
	1990	126	...	12020	...	2014	...	...	...	...
Zambia	1980	488	...	2801	45	...	...	...	...	...
	1988	529	...	3855	31	...	...	...	...	...
	1989	335	...	3855	32	...	...	...	...	...
	1990	318	...	3855	32	...	...	...	...	...
Zimbabwe	1980	3134	...	1718	204	...	...	...	...	68
	1988	5065	...	2089	234	...	...	...	...	43
	1989	5111	...	2089	270	...	...	...	...	44
	1990	4958	...	2089	509	...	...	...	...	45
America, North	**1980**	**631738**	**30123**	**45388**	**48952**	**865773**	**3198**	**27823**	**792627**	**77926**
	1988	**718055**	**66847**	**56656**	**33611**	**899213**	**7303**	**27018**	**746487**	**75546**
	1989	**738919**	**65852**	**56845**	**34352**	**855066**	**7431**	**28031**	**761892**	**75215**
	1990	**780861**	**67808**	**46956**	**29985**	**840093**	**7725**	**28255**	**776694**	**74414**
Antigua and Barbuda	1980	...	...	...	...	...	...	...	...	...
	1988	...	...	...	...	...	...	...	...	...
	1989	...	...	...	...	...	...	...	...	...
	1990	...	...	...	...	...	...	...	...	...
Aruba	1988	...	...	...	...	...	...	...	...	...
	1989	...	...	...	...	...	...	...	...	...
	1990	...	...	...	...	...	...	...	...	...
Bahamas	1980	...	...	...	...	...	...	...	...	...
	1988	...	...	...	...	...	...	...	...	...
	1989	...	...	...	...	...	...	...	...	...
	1990	...	...	...	...	...	...	...	...	...
Barbados	1980	...	...	...	...	60	...	...	18	...
	1988	...	...	...	...	86	...	...	35	...
	1989	...	...	...	...	76	...	...	38	...
	1990	...	...	...	...	89	...	...	39	...
Belize	1980	...	...	26	...	...	...	...	...	...
	1988	...	...	42	...	...	...	...	...	...
	1989	...	...	42	...	...	...	...	...	...
	1990	...	...	42	...	...	...	...	...	...
Bermuda	1980	...	...	...	...	...	...	...	...	...
	1988	...	...	...	...	...	...	...	...	...
	1989	...	...	...	...	...	...	...	...	...
	1990	...	...	...	...	...	...	...	...	...

LPG GPL	Aviation gasoline Essence aviation	Motor gasoline Essence auto	Kerosene Kérosène	Jet fuels Carbu-réacteurs	Gas diesel oil Gas oil fluides	Residual fuel Fuel oil fluides	Lubricants Huiles lubrifiantes	Electricity Electricité	Année	Région, pays ou zone
...	...	...	...	...	...	...	...	35	1980	Swaziland
...	...	...	...	...	...	...	...	49	1988	
...	...	...	...	...	...	...	...	53	1989	
...	...	...	...	...	...	...	...	45	1990	
...	...	102	106	...	90	156	...	2	1980	Togo
...	...	0	0	...	0	0	...	6	1988	
...	...	0	0	...	0	0	...	5	1989	
...	...	0	0	...	0	0	...	5	1990	
44	...	225	170	0	637	956	...	344	1980	Tunisie
186	...	358	212	0	681	864	...	604	1988	
216	...	376	224	0	631	864	...	619	1989	
224	...	388	209	0	786	820	...	674	1990	
...	...	...	...	...	...	...	...	80	1980	Ouganda
...	...	...	...	...	...	...	...	73	1988	
...	...	...	...	...	...	...	...	85	1989	
...	...	...	...	...	...	...	...	74	1990	
6	...	37	57	44	217	354	...	94	1980	Rép.unie de Tanzanie
9	...	130	55	37	216	297	...	108	1988	
9	...	120	56	38	219	304	...	109	1989	
11	...	127	55	41	217	307	...	109	1990	
...	...	...	...	...	...	...	...	9	1980	Sahara occidental
...	...	...	...	...	...	...	...	10	1988	
...	...	...	...	...	...	...	...	10	1989	
...	...	...	...	...	...	...	...	10	1990	
2	...	72	59	59	159	149	...	546	1980	Zaïre
2	...	48	40	18	94	135	...	663	1988	
2	...	52	52	19	103	130	...	754	1989	
2	...	58	55	22	109	132	...	756	1990	
3	0	270	37	59	507	255	...	1129	1980	Zambie
14	0	178	41	59	312	137	...	1023	1988	
14	0	180	41	57	304	135	...	828	1989	
16	0	172	43	55	297	127	...	955	1990	
...	...	...	...	...	...	...	...	558	1980	Zimbabwe
...	...	...	...	...	...	...	...	986	1988	
...	...	...	...	...	...	...	...	988	1989	
...	...	...	...	...	...	...	...	1174	1990	
95481	2920	503235	19956	83930	276413	244777	16952	349307	1980	**Amérique du Nord**
115565	1941	521908	12180	108088	269981	144795	15888	436466	1988	
114796	1878	522601	11964	111157	276853	150851	15967	449973	1989	
114221	1737	526545	9846	116962	281053	149360	15895	454303	1990	
0	0	0	0	...	0	0	...	7	1980	Antigua-et-Barbuda
0	0	0	0	...	0	0	...	11	1988	
0	0	0	0	...	0	0	...	12	1989	
0	0	0	0	...	0	0	...	12	1990	
...	...	...	...	...	...	...	...	40	1988	Aruba
...	...	...	...	...	...	...	...	44	1989	
...	...	...	...	...	...	...	...	46	1990	
39	...	64	0	1598	2349	7788	...	105	1980	Bahamas
0	...	0	0	0	0	0	...	112	1988	
0	...	0	0	0	0	0	...	114	1989	
0	...	0	0	0	0	0	...	117	1990	
...	...	61	9	...	64	130	...	41	1980	Barbade
...	...	79	6	...	58	193	...	55	1988	
...	...	85	4	...	61	211	...	54	1989	
...	...	88	4	...	77	215	...	57	1990	
...	...	...	...	...	...	...	...	7	1980	Belize
...	...	...	...	...	...	...	...	11	1988	
...	...	...	...	...	...	...	...	12	1989	
...	...	...	...	...	...	...	...	13	1990	
...	...	...	...	...	...	...	...	41	1980	Bermudes
...	...	...	...	...	...	...	...	56	1988	
...	...	...	...	...	...	...	...	57	1989	
...	...	...	...	...	...	...	...	60	1990	

107
Production of selected energy commodities
Thousand metric tons of coal equivalent [*cont.*]
Production des principaux biens de l'énergie
Milliers de tonnes métriques d'équivalent houille [*suite*]

Region, country or area	Year	Hard coal Houille	Lignite brown coal Lignite	Fuelwood Bois de chauffage	Coke	Crude petroleum Pétrole brut	Condensate Condensat	Natural gasoline Gazoline naturelle	Natural gas Gaz naturel	Other gases Autres gaz
British Virign Islands	1980	...	...	...	...	...	...	...	...	...
	1988	...	...	...	...	...	...	...	...	...
	1989	...	...	...	...	...	...	...	...	...
	1990	...								
Canada	1980	20164	10239	1539	5164	100579	200	5835	95213	7296
	1988	36071	17801	2278	4358	113320	192	6306	128400	7367
	1989	36622	17666	2278	4125	109341	201	6468	135906	7276
	1990	35589	17182	2278	3465	109073	165	6369	136222	7108
Cayman Islands	1980	...	...	...	...	...	...	...	...	...
	1988	...	...	...	...	...	...	...	...	...
	1989	...	...	...	...	...	...	...	...	...
	1990	...	...	...	...	...	...	...	...	...
Costa Rica	1980	...	...	747	...	...	...	...	...	...
	1988	...	...	975	...	...	...	...	...	...
	1989	...	...	933	...	...	...	...	...	...
	1990	...	...	965	...	...	...	...	...	...
Cuba	1980	...	...	861	...	391	...	...	22	183
	1988	...	...	782	...	1024	...	...	28	214
	1989	...	...	715	...	1026	...	...	42	216
	1990	...	...	717	...	1214	...	...	41	213
Dominica	1980	...	...	...	...	...	...	...	...	...
	1988	...	...	...	...	...	...	...	...	...
	1989	...	...	...	...	...	...	...	...	...
	1990	...	...	...	...	...	...	...	...	...
Dominican Republic	1980	...	...	299	...	...	...	...	...	...
	1988	...	...	325	...	...	...	...	...	...
	1989	...	...	325	...	...	...	...	...	...
	1990	...	...	325	...	...	...	...	...	...
El Salvador	1980	...	...	1268	...	...	...	...	...	...
	1988	...	...	1197	...	...	...	...	...	...
	1989	...	...	1185	...	...	...	...	...	...
	1990	...	...	1167	...	...	...	...	...	...
Grenada	1980	...	...	...	...	...	...	...	...	...
	1988	...	...	...	...	...	...	...	...	...
	1989	...	...	...	...	...	...	...	...	...
	1990	...	...	...	...	...	...	...	...	...
Guadeloupe	1980	...	...	5	...	...	...	...	...	...
	1988	...	...	5	...	...	...	...	...	...
	1989	...	...	5	...	...	...	...	...	...
	1990	...	...	5	...	...	...	...	...	...
Guatemala	1980	...	...	1932	...	290	...	...	...	...
	1988	...	...	2425	...	254	...	...	...	...
	1989	...	...	2496	...	257	...	...	...	...
	1990	...	...	2569	...	250	...	...	...	...
Haiti	1980	...	...	1551	...	...	...	...	...	...
	1988	...	...	1793	...	...	...	...	...	...
	1989	...	...	1828	...	...	...	...	...	...
	1990	...	...	1867	...	...	...	...	...	...
Honduras	1980	...	...	1267	...	...	...	...	...	...
	1988	...	...	1671	...	...	...	...	...	...
	1989	...	...	1724	...	...	...	...	...	...
	1990	...	...	1778	...	...	...	...	...	25
Jamaica	1980	...	...	2	...	...	...	...	...	25
	1988	...	...	4	...	...	...	...	...	20
	1989	...	...	4	...	...	...	...	...	21
	1990	...	...	4	...	...	...	...	...	22
Martinique	1980	...	...	3	...	...	...	...	...	...
	1988	...	...	3	...	...	...	...	...	...
	1989	...	...	3	...	...	...	...	...	...
	1990	...	...	3	...	...	...	...	...	...
Mexico	1980	5007	...	4113	2768	142769		5337	35339	1634
	1988	7561	...	4958	1699	186679	7111	3741	33567	1056
	1989	8214	...	5067	2154	186664	7230	4273	34055	1182
	1990	7146	...	5084	1905	190121	7560	4596	34498	1215

LPG GPL	Aviation gasoline Essence aviation	Motor gasoline Essence auto	Kerosene Kérosène	Jet fuels Carbu- réacteurs	Gas diesel oil Gas oil fluides	Residual fuel Fuel oil fluides	Lubricants Huiles lubrifiantes	Electricity Electricité	Année	Région, pays ou zone
...	...	...	...	...	...	...	...	3	1980	Iles Vierges britanniques
...	...	...	...	...	...	...	...	5	1988	
...	...	...	...	...	...	...	...	5	1989	
								6	1990	
13259	256	42729	4593	5799	36617	24221	1251	46372	1980	Canada
16017	225	38805	2190	6077	31762	10764	1221	62150	1988	
16255	180	39180	2870	6514	32650	11210	1461	61361	1989	
17288	145	39535	2765	6468	33254	11607	1347	59181	1990	
...	...	...	...	...	...	...	...	8	1980	Iles Caïmanes
...	...	...	...	...	...	...	...	22	1988	
...	...	...	...	...	...	...	...	24	1989	
...	...	...	...	...	...	...	...	28	1990	
12	...	132	27	28	213	306	...	270	1980	Costa Rica
8	...	141	15	52	255	351	...	392	1988	
11	...	144	12	50	245	280	...	419	1989	
6	...	90	16	35	159	231	...	443	1990	
166	...	1210	646	57	1592	4285	193	1227	1980	Cuba
.191	...	1518	824	175	1770	5539	184	1786	1988	
202	...	1539	943	162	1707	5881	185	1872	1989	
194	...	1417	855	103	1435	5522	187	1995	1990	
...	...	...	...	...	...	...	...	1	1980	Dominique
...	...	...	...	...	...	...	...	4	1988	
...	...	...	...	...	...	...	...	4	1989	
...	...	...	...	...	...	...	...	4	1990	
79	...	433	102	62	570	797	...	407	1980	Rép. dominicaine
33	...	520	199	32	507	568	...	644	1988	
51	...	544	199	35	445	569	...	651	1989	
47	...	547	192	34	449	566	...	654	1990	
40	...	192	40	25	281	270	...	190	1980	El Salvador
39	...	210	21	47	286	329	...	251	1988	
39	...	186	25	40	293	283	...	258	1989	
40	...	168	22	43	310	317	...	282	1990	
...	...	...	...	...	...	...	...	3	1980	Grenade
...	...	...	...	...	...	...	...	3	1988	
...	...	...	...	...	...	...	...	6	1989	
...	...	...	...	...	...	...	...	6	1990	
...	...	...	...	...	...	...	...	38	1980	Guadeloupe
...	...	...	...	...	...	...	...	84	1988	
...	...	...	...	...	...	...	...	84	1989	
...	...	...	...	...	...	...	...	84	1990	
3	...	163	55	47	351	443	0	205	1980	Guatemala
11	...	195	22	53	339	235	0	265	1988	
12	...	202	24	59	333	241	0	285	1989	
14	...	198	24	56	341	238	0	286	1990	
...	...	...	...	...	...	...	...	39	1980	Haïti
...	...	...	...	...	...	...	...	58	1988	
...	...	...	...	...	...	...	...	58	1989	
...	...	...	...	...	...	...	...	58	1990	
5	...	121	59	29	293	200	...	114	1980	Honduras
8	...	85	16	15	120	120	...	134	1988	
6	...	82	15	15	116	120	...	135	1989	
8	...	85	18	13	119	113	...	136	1990	
30	...	255	118	78	294	542	7	270	1980	Jamaïque
11	...	120	71	55	209	554	22	307	1988	
14	...	186	63	47	241	668	19	335	1989	
16	...	180	66	44	232	666	17	335	1990	
17	...	202	177	...	58	170	...	34	1980	Martinique
23	...	172	186	...	196	248	...	72	1988	
25	...	175	184	...	200	248	...	75	1989	
23	...	177	186	...	206	251	...	76	1990	
5846	108	20974	2879	1915	17928	24147	581	8224	1980	Mexique
10099	70	23460	1946	3038	15137	32936	565	13495	1988	
10611	67	23791	1791	2927	17152	33328	555	14505	1989	
11033	67	26250	1769	2948	18905	33903	539	15041	1990	

107
Production of selected energy commodities
Thousand metric tons of coal equivalent [cont.]
Production des principaux biens de l'énergie
Milliers de tonnes métriques d'équivalent houille [suite]

Region, country or area	Year	Hard coal Houille	Lignite brown coal Lignite	Fuelwood Bois de chauffage	Coke	Crude petroleum Pétrole brut	Condensate Condensat	Natural gasoline Gazoline naturelle	Natural gas Gaz naturel	Other gases Autres gaz
Montserrat	1980	...	...	...	...	...	...	...	...	...
	1988	...	...	...	...	...	...	...	...	...
	1989	...	...	...	...	...	...	...	...	...
	1990	...	...	...	...	...	...	...	...	...
Netherlands Antilles	1988	...	...	...	...	...	...	...	...	...
	1989	...	...	...	...	...	...	...	...	...
	1990	...	...	...	...	...	...	...	...	...
Nicaragua	1980	...	...	763	...	...	...	...	...	28
	1988	...	...	997	...	...	...	...	...	19
	1989	...	...	1031	...	...	...	...	...	19
	1990	...	...	1066	...	...	...	...	...	18
Panama	1980	...	...	557	...	...	...	...	...	0
	1988	...	...	440	...	...	...	...	...	1
	1989	...	...	448	...	...	...	...	...	1
	1990	...	...	448	...	...	...	...	...	1
Puerto Rico	1980	...	...	...	...	...	...	...	...	...
	1988	...	...	...	...	...	...	...	...	...
	1989	...	...	...	...	...	...	...	...	...
	1990	...	...	...	...	...	...	...	...	...
Saint Kitts and Nevis	1988	...	...	...	...	...	...	...	...	...
	1989	...	...	...	...	...	...	...	...	...
	1990	...	...	...	...	...	...	...	...	...
Saint Lucia	1980	...	...	...	...	...	...	...	...	...
	1988	...	...	...	...	...	...	...	...	...
	1989	...	...	...	...	...	...	...	...	...
	1990	...	...	...	...	...	...	...	...	...
Saint Pierre - Miquelon	1980	...	...	...	...	...	...	...	...	...
	1988	...	...	...	...	...	...	...	...	...
	1989	...	...	...	...	...	...	...	...	...
	1990	...	...	...	...	...	...	...	...	...
Saint Vincent - Grenadines	1980	...	...	...	...	...	...	...	...	...
	1988	...	...	...	...	...	...	...	...	...
	1989	...	...	...	...	...	...	...	...	...
	1990	...	...	...	...	...	...	...	...	...
Trinidad and Tobago	1980	...	...	5	...	15690	...	14	4996	597
	1988	...	...	7	...	11226	...	11	4972	174
	1989	...	...	7	...	11020	...	11	5678	174
	1990	...	...	7	...	11233	...	11	7349	174
Turks and Caicos Islands	1980	...	...	...	...	...	...	...	...	...
	1988	...	...	...	...	...	...	...	...	...
	1989	...	...	...	...	...	...	...	...	...
	1990	...	...	...	...	...	...	...	...	...
United States	1980	606566	19884	30450	41019	605994	2998	16636	657039	68162
	1988	674422	49045	38752	27553	586624	...	16961	579486	66695
	1989	694083	48186	38752	28073	546681	...	17279	586173	66325
	1990	738126	50626	28630	24614	528113	...	17279	598546	65664
United States Virgin Is.	1980	...	...	...	...	...	...	...	...	...
	1988	...	...	...	...	...	...	...	...	...
	1989	...	...	...	...	...	...	...	...	...
	1990	...	...	...	...	...	...	...	...	...
America, South	**1980**	**9004**	**23**	**70513**	**4857**	**254921**	**314**	**1512**	**43759**	**6340**
	1988	**22545**	**22**	**64772**	**8641**	**279299**	**0**	**1958**	**70710**	**10052**
	1989	**26447**	**22**	**66640**	**8587**	**282117**	**0**	**1955**	**72496**	**9540**
	1990	**26731**	**23**	**60538**	**8331**	**306409**	**0**	**2015**	**77257**	**9440**
Argentina	1980	329	...	2101	468	36116	...	176	11155	855
	1988	431	...	1444	760	33027	...	411	26459	1229
	1989	316	...	1444	765	33773	...	406	26761	1331
	1990	228	...	1426	857	35406	...	414	27064	1433
Bolivia	1980	...	...	340	...	1581	...	118	2880	2
	1988	...	...	423	...	1203	...	93	3429	2
	1989	...	...	434	...	1231	...	97	3898	2
	1990	...	...	447	...	1387	...	97	3883	2

LPG GPL	Aviation gasoline Essence aviation	Motor gasoline Essence auto	Kerosene Kérosène	Jet fuels Carbu-réacteurs	Gas diesel oil Gas oil fluides	Residual fuel Fuel oil fluides	Lubricants Huiles lubrifiantes	Electricity Electricité	Année	Région, pays ou zone
...	...	...	...	...	...	...	...	2	1980	Montserrat
...	...	...	...	...	...	...	...	2	1988	
...	...	...	...	...	...	...	...	2	1989	
...	...	...	...	...	...	...	...	2	1990	
101	22	2175	29	1105	3335	6797	503	83	1988	Antilles néerlandaises
101	22	2175	29	1105	3552	7009	503	88	1989	
117	22	2325	37	1179	3770	7505	575	90	1990	
26	...	205	22	27	252	221	...	131	1980	Nicaragua
30	...	111	35	22	178	316	...	131	1988	
28	...	105	32	22	174	312	...	132	1989	
26	...	97	29	19	171	304	...	128	1990	
34	...	360	15	184	670	1417	...	223	1980	Panama
37	...	262	18	87	394	889	...	343	1988	
34	...	240	15	81	392	885	...	344	1989	
31	...	232	16	74	384	878	...	356	1990	
587	0	4351	279	209	3220	4790	411	1622	1980	Porto Rico
47	0	1747	133	553	957	1947	575	1769	1988	
54	0	2400	133	590	1015	2103	611	1758	1989	
70	0	2850	133	737	1269	2549	647	1883	1990	
...	...	...	...	...	...	...	...	5	1988	St. Kitts-et-Nevis
...	...	...	...	...	...	...	...	5	1989	
...	...	...	...	...	...	...	...	5	1990	
...	...	...	...	...	...	...	...	7	1980	Sainte-Lucie
...	...	...	...	...	...	...	...	13	1988	
...	...	...	...	...	...	...	...	13	1989	
...	...	...	...	...	...	...	...	13	1990	
...	...	...	...	...	...	...	...	2	1980	Saint-Pierre-et-Miquelon
...	...	...	...	...	...	...	...	5	1988	
...	...	...	...	...	...	...	...	6	1989	
...	...	...	...	...	...	...	...	6	1990	
...	...	...	...	...	...	...	...	3	1980	St. Vincent-Grenadines
...	...	...	...	...	...	...	...	6	1988	
...	...	...	...	...	...	...	...	6	1989	
...	...	...	...	...	...	...	...	6	1990	
117	68	2658	603	609	2777	9133	208	253	1980	Trinité-et-Tobago
132	4	1210	174	504	961	4067	3	426	1988	
124	6	798	136	392	867	4122	3	427	1989	
104	15	1081	299	271	944	3575	0	427	1990	
...	...	...	...	...	...	...	...	1	1980	Iles Turques et Caiques
...	...	...	...	...	...	...	...	1	1988	
...	...	...	...	...	...	...	...	1	1989	
...	...	...	...	...	...	...	...	1	1990	
74948	2233	420291	9503	69397	195377	123708	13398	289201	1980	Etats-Unis
88747	1618	449286	5461	95161	209845	72445	12816	353581	1988	
87196	1602	448516	5113	97201	212190	74459	12630	366684	1989	
85172	1486	448221	3091	103117	214097	74122	12583	372320	1990	
70	0	4779	694	1363	8001	17428	...	98	1980	Iles Vierges américaines
31	0	1809	834	1111	3670	6497	...	119	1988	
31	0	2250	376	1916	5220	8921	...	117	1989	
31	0	3000	324	1820	4930	6797	...	120	1990	
9029	**184**	**39823**	**5441**	**8427**	**56727**	**85369**	**2000**	**32995**	**1980**	**Amérique du Sud**
13453	**231**	**52011**	**4185**	**10806**	**71153**	**62130**	**2210**	**49934**	**1988**	
13846	**234**	**53455**	**3947**	**10868**	**69219**	**60889**	**2073**	**50719**	**1989**	
14516	**213**	**54474**	**3495**	**11342**	**70666**	**59842**	**2013**	**50943**	**1990**	
1165	31	7735	871	1153	11050	11518	398	4874	1980	Argentine
1854	28	6826	550	1035	11021	6647	414	6518	1988	
2020	18	7170	693	1054	10969	6372	367	6245	1989	
2255	21	7900	708	1026	11613	5861	313	6253	1990	
163	21	609	190	136	378	214	32	192	1980	Bolivie
253	13	544	60	113	394	38	17	231	1988	
256	15	547	63	115	396	40	19	235	1989	
256	15	547	66	115	399	40	19	240	1990	

107
Production of selected energy commodities
Thousand metric tons of coal equivalent [*cont.*]
Production des principaux biens de l'énergie
Milliers de tonnes métriques d'équivalent houille [*suite*]

Region, country or area	Year	Hard coal Houille	Lignite brown coal Lignite	Fuelwood Bois de chauffage	Coke	Crude petroleum Pétrole brut	Condensate Condensat	Natural gasoline Gazoline naturelle	Natural gas Gaz naturel	Other gases Autres gaz
Brazil	1980	3655	...	55801	3838	12976	...	...	1306	3450
	1988	4965	...	47755	7303	39790	...	...	4735	6293
	1989	4389	...	48830	7237	42636	...	...	5067	6246
	1990	3016	...	42784	6871	45160	...	...	5022	5758
Chile	1980	954	23	1873	274	2291	91	143	924	561
	1988	1898	22	2376	284	1510	0	159	1527	536
	1989	1921	22	2390	284	1351	0	155	2256	512
	1990	2152	23	2566	320	1203	0	159	2364	522
Colombia	1980	3982	...	4359	249	9287	...	121	4792	455
	1988	14022	...	5142	273	27056	...	116	5442	789
	1989	17552	...	5850	280	29117	...	116	5079	317
	1990	19006	...	5797	261	31650	...	116	5466	306
Ecuador	1980	...	...	1849	...	14881	...	12	51	...
	1988	...	...	2180	...	22580	...	34	132	...
	1989	...	...	2202	...	20794	...	28	132	...
	1990	...	...	2222	...	21337	...	43	165	...
Falkland Is. (Malvinas)	1980	...	...	...	...	...	...	...	...	...
	1988	...	...	...	...	...	...	...	...	...
	1989	...	...	...	...	...	...	...	...	...
	1990	...	...	...	...	...	...	...	...	...
French Guiana	1980	...	...	20	...	...	...	...	...	...
	1988	...	...	22	...	...	...	...	...	...
	1989	...	...	22	...	...	...	...	...	...
	1990	...	...	22	...	...	...	...	...	...
Guyana	1980	...	...	6	...	...	...	...	...	...
	1988	...	...	5	...	...	...	...	...	...
	1989	...	...	5	...	...	...	...	...	...
	1990	...	...	5	...	...	...	...	...	...
Paraguay	1980	...	...	1438	...	...	...	...	...	...
	1988	...	...	1750	...	...	...	...	...	...
	1989	...	...	1762	...	...	...	...	...	...
	1990	...	...	1774	...	...	...	...	...	...
Peru	1980	41	...	1786	23	13806	...	57	713	175
	1988	157	...	2546	21	10827	...	28	817	88
	1989	142	...	2564	21	9964	...	23	691	85
	1990	142	...	2353	21	9853	...	23	665	80
Suriname	1980	...	...	11	...	0	...	...	...	0
	1988	...	...	7	...	277	...	...	...	0
	1989	...	...	6	...	324	...	...	...	0
	1990	...	...	6	...	400	...	...	...	0
Uruguay	1980	...	...	737	5	...	...	...	...	54
	1988	...	...	882	1	...	...	...	...	43
	1989	...	...	884	0	...	...	...	...	41
	1990	...	...	883	0	...	...	...	...	41
Venezuela	1980	44	...	193	0	163983	224	885	21938	788
	1988	1071	...	241	0	143029	0	1117	28170	1072
	1989	2128	...	247	0	142986	0	1131	28611	1005
	1990	2188	...	253	0	160013	0	1163	32626	1298
Asia	1980	594183	24972	220821	98385	1625461	700	8270	115292	53060
	1988	939348	28032	255014	130117	1412114	3780	11852	244741	61474
	1989	996570	33063	259658	130588	1512566	3754	13417	268655	63189
	1990	1017124	32470	262563	129018	1596693	3972	13733	280929	64789
Afghanistan	1980	119	...	1593	...	...	12	...	3491	...
	1988	138	...	1511	...	...	5	...	3889	...
	1989	145	...	1570	...	...	6	...	3938	...
	1990	143	...	1642	...	...	6	...	3906	...
Bahrain	1980	...	...	...	...	3449	...	146	3880	...
	1988	...	...	...	...	3061	...	201	6657	...
	1989	...	...	...	...	2781	...	271	6664	...
	1990	...	...	...	...	3001	...	276	6858	...
Bangladesh	1980	...	...	1436	...	9	...	...	1537	...
	1988	...	...	1900	...	39	...	...	4854	...
	1989	...	...	1933	...	39	...	...	5144	...
	1990	...	...	1966	...	27	...	...	5255	...

LPG GPL	Aviation gasoline Essence aviation	Motor gasoline Essence auto	Kerosene Kérosène	Jet fuels Carbu- réacteurs	Gas diesel oil Gas oil fluides	Residual fuel Fuel oil fluides	Lubricants Huiles lubrifiantes	Electricity Electricité	Année	Région, pays ou zone
3639	...	12504	781	3199	24630	23385	799	17134	1980	Brésil
5439	97	13768	612	3684	31111	16808	971	26286	1988	
5218	99	13155	455	3875	30930	17452	909	27237	1989	
5470	76	12825	292	3636	30325	17040	972	27293	1990	
744	6	1485	335	233	1756	2307	...	1443	1980	Chili
692	12	1698	273	251	2340	1706	106	2078	1988	
730	22	2176	312	360	2675	2077	102	2188	1989	
671	21	2158	259	366	2913	2212	108	2257	1990	
404	75	3207	519	668	1721	3633	10	2533	1980	Colombie
609	45	4827	417	699	2830	5000	3	4078	1988	
650	48	4978	439	713	2971	4918	3	3975	1989	
718	48	5053	398	750	3016	5314	3	4138	1990	
117	...	1452	430	205	1138	3200	60	412	1980	Equateur
230	...	2010	366	261	1788	3571	66	688	1988	
222	...	1893	358	248	1618	3292	58	705	1989	
261	...	2029	388	256	1930	3830	49	777	1990	
...	...	...	...	...	...	...	...	0	1980	Iles Falkland (Malvinas)
...	...	...	...	...	...	...	...	1	1988	
...	...	...	...	...	...	...	...	1	1989	
...	...	...	...	...	...	...	...	1	1990	
...	...	...	...	...	...	...	...	14	1980	Guyane française
...	...	...	...	...	...	...	...	35	1988	
...	...	...	...	...	...	...	...	38	1989	
...	...	...	...	...	...	...	...	41	1990	
...	...	...	...	...	...	...	...	51	1980	Guyana
...	...	...	...	...	...	...	...	27	1988	
...	...	...	...	...	...	...	...	27	1989	
...	...	...	...	...	...	...	...	27	1990	
8	...	96	27	13	173	67	...	89	1980	Paraguay
11	...	123	7	44	206	86	...	258	1988	
14	...	111	7	56	228	91	...	271	1989	
14	...	112	7	56	225	85	...	299	1990	
166	0	2266	1279	572	2421	3524	16	1233	1980	Pérou
194	0	1950	1253	295	2030	5664	12	1664	1988	
194	0	1927	1142	295	1957	5239	12	1641	1989	
218	0	1950	921	295	1994	5239	12	1697	1990	
...	...	...	...	...	...	...	...	194	1980	Suriname
...	...	...	...	...	...	...	...	114	1988	
...	...	...	...	...	...	...	...	163	1989	
...	...	...	...	...	...	...	...	166	1990	
62	1	310	183	41	686	1197	10	412	1980	Uruguay
81	1	291	77	21	500	578	12	860	1988	
89	1	289	72	43	478	572	12	706	1989	
85	1	292	74	44	478	573	12	906	1990	
2559	49	10158	825	2208	12773	36325	674	4414	1980	Venezuela
4090	33	19972	570	4404	18931	22032	610	7097	1988	
4452	30	21207	404	4110	16995	20836	594	7288	1989	
4567	30	21604	382	4798	17773	19647	526	6848	1990	
37946	517	96858	66656	24508	176811	333175	9031	164764	1980	Asie
49782	328	136693	72785	36890	276864	308700	11103	273621	1988	
55517	322	148270	74326	37953	294286	318071	11703	293598	1989	
61506	402	156580	78666	41016	305261	328498	12167	315748	1990	
...	...	...	...	...	...	...	...	119	1980	Afghanistan
...	...	...	...	...	...	...	...	136	1988	
...	...	...	...	...	...	...	...	137	1989	
...	...	...	...	...	...	...	...	139	1990	
278	0	1564	105	2628	4802	5801	0	204	1980	Bahreïn
275	0	961	626	2479	5520	4762	418	428	1988	
280	0	1050	634	2491	5524	4758	431	429	1989	
311	0	1065	649	2420	5539	4772	417	429	1990	
5	0	87	458	6	225	758	45	326	1980	Bangladesh
14	0	78	360	13	284	422	42	843	1988	
14	0	90	420	10	332	389	37	914	1989	
14	0	106	380	19	252	357	46	990	1990	

107
Production of selected energy commodities
Thousand metric tons of coal equivalent [*cont.*]
Production des principaux biens de l'énergie
Milliers de tonnes métriques d'équivalent houille [*suite*]

Region, country or area	Year	Hard coal Houille	Lignite brown coal Lignite	Fuelwood Bois de chauffage	Coke	Crude petroleum Pétrole brut	Condensate Condensat	Natural gasoline Gazoline naturelle	Natural gas Gaz naturel	Other gases Autres gaz
Bhutan	1980	...	...	247	...	...	...	...	...	...
	1988	0	...	331	...	...	...	...	...	...
	1989	2	...	338	...	...	...	...	...	...
	1990	2	...	390	...	...	...	...	...	...
Brunei Darussalam	1980	...	...	26	...	16973	637	118	12010	...
	1988	...	...	26	...	9511	455	92	12471	...
	1989	...	...	26	...	9459	458	95	13440	...
	1990	...	...	26	...	10714	461	100	13932	...
Cambodia	1980	...	...	1387	...	...	...	...	...	...
	1988	...	...	1700	...	...	...	...	...	...
	1989	...	...	1741	...	...	...	...	...	...
	1990	...	...	1787	...	...	...	...	...	...
China	1980	425600	9359	51545	33026	151351	...	...	18948	...
	1988	699212	...	60381	57337	195293	...	...	18970	...
	1989	752104	...	61588	58200	196357	...	...	19988	...
	1990	770657	...	61818	59170	197286	...	...	20268	...
Cyprus	1980	...	...	5	...	...	...	...	...	16
	1988	...	...	6	...	...	...	...	...	33
	1989	...	...	7	...	...	...	...	...	31
	1990	...	...	7	...	...	...	...	...	26
Hong Kong	1980	...	...	55	...	...	...	...	...	120
	1988	...	...	62	...	...	...	...	...	418
	1989	...	...	62	...	...	...	...	...	467
	1990	...	...	64	...	...	...	...	...	514
India	1980	90581	1501	67294	10261	13427	...	0	1726	1093
	1988	156617	2820	79985	9795	45114	...	525	8878	931
	1989	164603	3267	81641	9668	48121	...	594	10776	918
	1990	167230	3124	83338	8761	47587	...	121	13239	831
Indonesia	1980	304	...	38499	0	110871	...	9	20718	314
	1988	2741	...	45251	0	94589	...	8	41616	320
	1989	4553	...	46120	0	99033	...	9	49900	329
	1990	7327	...	46998	0	100554	...	11	53861	360
Iran, Islamic Rep. of	1980	900	...	774	360	103810	...	306	9506	...
	1988	1260	...	810	360	164346	...	1072	26631	...
	1989	1200	...	814	315	200609	...	2488	29542	...
	1990	1300	...	818	360	224403	...	2604	31855	...
Iraq	1980	...	...	25	...	185890	...	184	1707	...
	1988	...	...	33	...	185914	...	766	6133	...
	1989	...	...	33	...	195147	...	1007	6510	...
	1990	...	...	35	...	143769	...	781	4265	...
Israel	1980	...	...	4	...	29	...	...	191	...
	1988	...	...	4	...	26	...	...	47	...
	1989	...	...	2	...	23	...	...	47	...
	1990	...	...	4	...	19	...	...	40	...
Japan	1980	16044	15	32	46311	611	...	6	2905	49811
	1988	9860	8	50	45945	846	...	11	2833	54857
	1989	8950	8	55	45188	779	...	0	2713	56118
	1990	7259	9	55	43014	773	...	0	2761	57095
Jordan	1980	...	...	1	...	...	...	...	...	0
	1988	...	...	1	...	10	...	...	...	65
	1989	...	...	2	...	16	...	...	...	73
	1990	...	...	2	...	23	...	...	...	73
Korea, Dem. P. Rep.	1980	36000	6000	1230	2700	...	...	...	...	...
	1988	40000	7500	1350	3150	...	...	...	...	...
	1989	40700	7800	1358	3240	...	...	...	...	...
	1990	40700	7800	1358	3240	...	...	...	...	...
Korea, Rep. of	1980	12239	...	3598	2668	...	...	...	...	21
	1988	15622	...	1659	6246	...	...	...	...	486
	1989	13365	...	1472	7133	...	...	...	...	887
	1990	11071	...	948	7569	...	...	...	...	1452
Kuwait, part Ntl. Zone	1980	...	...	...	...	120166	...	1184	9674	0
	1988	...	...	...	...	100894	...	1118	5789	471
	1989	...	...	...	...	105787	...	1164	7245	498
	1990	...	...	...	...	85071	...	919	6960	464

LPG GPL	Aviation gasoline Essence aviation	Motor gasoline Essence auto	Kerosene Kérosène	Jet fuels Carbu-réacteurs	Gas diesel oil Gas oil fluides	Residual fuel Fuel oil fluides	Lubricants Huiles lubrifiantes	Electricity Electricité	Année	Région, pays ou zone
...	...	...	...	...	...	...	...	2	1980	Bhoutan
...	...	...	...	...	...	...	...	190	1988	
...	...	...	...	...	...	...	...	190	1989	
								192	1990	
62	...	31	...	0	87	1	...	58	1980	Brunéi Darussalam
33	...	207	16	55	123	1	...	137	1988	
36	...	213	18	53	136	1	...	142	1989	
39	...	222	21	56	139	3	...	149	1990	
...	...	0	0	0	0	0	...	12	1980	Cambodge
...	...	0	0	0	0	0	...	9	1988	
...	...	0	0	0	0	0	...	9	1989	
...	...	0	0	0	0	0	...	9	1990	
...	154	15570	5867	...	26506	44491	2833	36927	1980	Chine
3652	105	28477	5641	...	35365	45170	2444	66971	1988	
3730	112	30879	5827	...	37259	45312	2516	71490	1989	
3885	115	31500	5896	...	37700	45737	2588	75912	1990	
33	...	151	15	68	222	327	...	127	1980	Chypre
47	...	216	15	44	302	333	...	205	1988	
40	...	210	15	29	283	282	...	227	1989	
39	...	183	18	28	333	246	...	243	1990	
...	...	...	...	...	...	...	...	1554	1980	Hong-kong
...	...	...	...	...	...	...	...	3133	1988	
...	...	...	...	...	...	...	...	3361	1989	
...	...	...	...	...	...	...	...	3555	1990	
570	0	2188	3454	1489	11791	8656	631	14649	1980	Inde
2667	0	4050	7541	2588	23779	12077	830	29641	1988	
2990	0	4863	8117	2438	26070	12583	799	32699	1989	
3225	0	5304	8381	2431	24825	13183	788	35135	1990	
883	22	4218	6480	355	6860	15954	102	1748	1980	Indonésie
1871	10	5250	8478	368	10150	14868	368	4885	1988	
4294	10	6270	8829	258	10875	15045	388	5136	1989	
4382	10	6714	8955	287	11107	15137	403	5437	1990	
1725	225	5440	6536	910	11368	21120	255	2749	1980	Iran, Rép. islamique d'
1399	150	6300	4569	457	13340	15576	661	4762	1988	
1703	150	6900	4717	531	14645	16992	676	5197	1989	
1896	225	7050	4761	678	15950	17134	863	6879	1990	
404	...	1977	1032	747	3450	4135	86	1404	1980	Iraq
1570	...	3754	1179	604	7714	10195	216	3367	1988	
2219	...	3900	1165	737	8265	11611	288	3550	1989	
1725	...	3750	884	604	7395	10195	259	3582	1990	
278	...	1347	339	781	1959	3449	58	1539	1980	Israël
306	...	1887	354	685	2591	3827	14	2360	1988	
312	...	1566	383	693	2711	3983	14	2493	1989	
353	...	2194	476	722	2928	4405	14	2546	1990	
12451	21	37876	28264	5482	53386	116901	2981	70940	1980	Japon
6477	15	39283	25244	4907	59256	55544	2926	92584	1988	
6754	13	42426	24523	4842	64452	53914	3025	98115	1989	
6929	13	46600	27739	5128	71044	58089	3209	105303	1990	
64	30	399	249	295	681	704	7	131	1980	Jordanie
141	21	507	295	280	1057	902	17	327	1988	
149	22	538	248	342	1003	981	17	422	1989	
159	22	600	302	368	1080	1093	17	453	1990	
...	...	900	177	...	870	425	...	4299	1980	Corée, Rép. pop. dém. de
...	...	1350	295	...	1377	850	...	6510	1988	
...	...	1425	310	...	1450	920	...	6572	1989	
...	...	1417	302	...	1421	906	...	6572	1990	
510	...	1251	1661	933	7653	19338	289	4923	1980	Corée, République de
1465	...	2403	2024	2820	14796	19190	683	11333	1988	
1458	...	3256	2406	3342	17036	22173	733	12640	1989	
1601	...	4224	2568	3109	18963	22019	891	14585	1990	
3730	...	1402	2102	793	5369	11358	...	1157	1980	Koweït et prt. zne. N.
3908	...	3645	2579	2152	13311	19062	...	2456	1988	
4211	...	3765	2786	2226	14355	19966	...	2519	1989	
3745	...	2820	2506	1636	10585	16992	...	2532	1990	

107
Production of selected energy commodities
Thousand metric tons of coal equivalent [*cont.*]
Production des principaux biens de l'énergie
Milliers de tonnes métriques d'équivalent houille [*suite*]

Region, country or area	Year	Hard coal Houille	Lignite brown coal Lignite	Fuelwood Bois de chauffage	Coke	Crude petroleum Pétrole brut	Condensate Condensat	Natural gasoline Gazoline naturelle	Natural gas Gaz naturel	Other gases Autres gaz
Lao People's Dem. Rep.	1980	0	...	987	...	...	...	...	...	...
	1988	0	...	1203	...	...	...	...	...	...
	1989	0	...	1238	...	...	...	...	...	...
	1990	0	...	1274	...	...	...	...	...	...
Lebanon	1980	...	...	149	...	...	...	...	...	...
	1988	...	...	149	...	...	...	...	...	...
	1989	...	...	150	...	...	...	...	...	...
	1990	...	...	151	...	...	...	...	...	...
Macau	1980	...	...	...	...	...	...	...	...	...
	1988	...	...	...	...	...	...	...	...	...
	1989	...	...	...	...	...	...	...	...	...
	1990	...	...	...	...	...	...	...	...	...
Malaysia	1980	...	...	2236	...	19197	...	...	109	91
	1988	23	...	2759	...	37707	...	...	16155	216
	1989	114	...	2832	...	40570	...	...	16864	180
	1990	105	...	2906	...	42897	...	...	17141	226
Maldives	1980			...	...	...	...	...	...	...
	1988			...	...	...	...	...	...	...
	1989	...		...	...	...	...	...	...	...
	1990	...		...	...	...	...	...	...	...
Mongolia	1980	392	1315	450	...	...	...	...	...	...
	1988	690	2612	450	...	...	...	...	...	...
	1989	692	2425	450	...	...	...	...	...	...
	1990	690	2376	450	...	...	...	...	...	...
Myanmar	1980	11	10	4810	...	2227	...	2	454	...
	1988	30	13	5684	...	1089	...	2	1527	...
	1989	38	14	5804	...	1270	...	2	1438	...
	1990	40	15	5927	...	1286	...	2	1431	...
Nepal	1980	...	...	4518	...	...	...	...	...	...
	1988	...	...	5472	...	...	...	...	...	...
	1989	...	...	5575	...	...	...	...	...	...
	1990	...	...	5503	...	...	...	...	...	...
Oman	1980	...	...	...	...	20043	...	...	1118	...
	1988	...	...	...	...	44004	...	15	3279	...
	1989	...	...	...	...	45433	...	0	3091	...
	1990	...	...	...	...	48597	...	0	3625	...
Pakistan	1980	1057	...	5555	...	680	8	...	8000	...
	1988	1853	...	7475	578	3127	9	...	12801	...
	1989	1781	...	7729	605	3271	11	...	13330	...
	1990	1854	...	7986	601	3711	11	...	14565	...
Philippines	1980	219	3	8624	...	764	...	...	...	62
	1988	901	2	10617	...	441	...	...	...	7
	1989	917	2	10872	...	381	...	...	...	7
	1990	789	2	11187	...	350	...	...	...	7
Qatar	1980	...	...	...	...	32526	...	71	5991	...
	1988	...	...	...	...	23339	...	276	6604	...
	1989	...	...	...	...	27234	...	250	6569	...
	1990	...	...	...	...	27231	...	260	7236	...
Saudi Arabia, part Neutral Zone	1980	...	...	...	...	708426	...	5990	1660	...
	1988	...	...	...	...	362171	...	6143	34704	...
	1989	...	...	...	...	360619	...	5943	35561	...
	1990	...	...	...	...	457679	...	6920	37577	...
Singapore	1980	...	...	...	...	...	...	...	...	75
	1988	...	...	...	...	...	...	...	...	84
	1989	...	...	...	...	...	...	...	...	89
	1990	...	...	...	...	...	...	...	...	99
Sri Lanka	1980	...	...	2373	...	...	...	...	...	5
	1988	...	...	2762	...	...	...	...	...	48
	1989	...	...	2986	...	...	...	...	...	37
	1990	...	...	3007	...	...	...	...	...	71
Syrian Arab Republic	1980	...	...	9	...	13140	...	...	64	...
	1988	...	...	5	...	20234	...	...	273	...
	1989	...	...	5	...	26184	...	...	307	...
	1990	...	...	4	...	32773	...	...	375	...

LPG GPL	Aviation gasoline Essence aviation	Motor gasoline Essence auto	Kerosene Kérosène	Jet fuels Carburéacteurs	Gas diesel oil Gas oil fluides	Residual fuel Fuel oil fluides	Lubricants Huiles lubrifiantes	Electricity Electricité	Année	Région, pays ou zone
...	...	...	...	...	...	...	...	120	1980	Rép. dém. pop. lao
...	...	...	...	...	...	...	...	65	1988	
...	...	...	...	...	...	...	...	87	1989	
...	...	...	...	...	...	...	...	107	1990	
93	...	750	37	206	500	1161	...	338	1980	Liban
31	...	435	44	118	435	581	...	553	1988	
31	...	450	44	133	449	595	...	563	1989	
34	...	465	59	133	449	595	...	582	1990	
...	...	...	...	...	...	...	...	32	1980	Macao
...	...	...	...	...	...	...	...	83	1988	
...	...	...	...	...	...	...	...	91	1989	
...	...	...	...	...	...	...	...	97	1990	
118	...	1332	332	307	2485	3196	111	1240	1980	Malaisie
718	...	1977	876	420	3870	3076	115	2372	1988	
757	...	1938	846	511	4353	3464	145	2638	1989	
912	...	2008	733	538	4969	4589	114	2823	1990	
...	...	...	...	...	...	...	...	0	1980	Maldives
...	...	...	...	...	...	...	...	3	1988	
...	...	...	...	...	...	...	...	3	1989	
...	...	...	...	...	...	...	...	4	1990	
...	...	...	...	...	...	...	...	192	1980	Mongolie
...	...	...	...	...	...	...	...	405	1988	
...	...	...	...	...	...	...	...	438	1989	
...	...	...	...	...	...	...	...	442	1990	
5	1	369	131	57	535	312	32	183	1980	Myanmar
6	0	201	10	29	371	163	22	274	1988	
6	0	189	19	32	419	170	23	304	1989	
6	0	186	18	32	413	167	22	319	1990	
...	...	...	...	...	...	...	...	26	1980	Népal
...	...	...	...	...	...	...	...	72	1988	
...	...	...	...	...	...	...	...	70	1989	
...	...	...	...	...	...	...	...	91	1990	
...	...	...	...	...	...	...	0	118	1980	Oman
508	...	606	15	165	895	1882	0	541	1988	
511	...	634	15	198	883	1961	0	578	1989	
252	...	724	16	402	884	2389	0	657	1990	
57	0	702	252	802	1670	2100	138	1839	1980	Pakistan
202	0	1240	476	685	2526	2416	275	4744	1988	
204	0	1245	557	706	2274	2291	278	4948	1989	
213	0	1447	722	775	2477	2690	292	5392	1990	
246	0	2079	687	430	3454	5845	171	2212	1980	Philippines
319	0	2062	537	945	4180	4434	183	3014	1988	
303	0	2223	649	839	4173	4739	122	3185	1989	
361	0	2517	740	650	4759	5653	0	3234	1990	
177	...	196	6	93	220	0	...	299	1980	Qatar
991	...	514	7	236	908	967	...	553	1988	
2014	...	637	12	321	908	1304	...	554	1989	
2113	...	682	15	566	1069	1420	...	568	1990	
12974	0	5274	2220	65	8926	19035	0	2323	1980	Arabie saoudite et
15691	0	13192	6349	2727	31990	35440	446	5184	1988	prt. zone Ntl.
16177	0	14025	6486	2801	32146	35542	590	5687	1989	
21117	0	14400	6574	2874	34945	37807	575	5822	1990	
250	0	3378	3912	5467	6698	17390	879	852	1980	Singapour
544	0	3000	3316	8107	13775	12744	863	1599	1988	
497	0	3045	3464	7886	14065	14160	992	1724	1989	
715	0	4599	3685	9389	15950	15576	1006	1918	1990	
11	...	163	264	91	696	1056	33	205	1980	Sri Lanka
31	...	253	234	125	747	886	26	344	1988	
23	...	178	180	72	438	646	30	351	1989	
30	...	268	243	149	771	905	32	387	1990	
82	...	252	472	215	2195	4071	...	471	1980	Rép. arabe syrienne
224	...	1533	249	631	4231	8271	...	1181	1988	
235	...	1957	271	713	4599	7370	...	1269	1989	
238	...	1959	283	722	4792	7632	...	1302	1990	

107
Production of selected energy commodities
Thousand metric tons of coal equivalent [*cont.*]
Production des principaux biens de l'énergie
Milliers de tonnes métriques d'équivalent houille [*suite*]

Region, country or area	Year	Hard coal Houille	Lignite brown coal Lignite	Fuelwood Bois de chauffage	Coke	Crude petroleum Pétrole brut	Condensate Condensat	Natural gasoline Gazoline naturelle	Natural gas Gaz naturel	Other gases Autres gaz
Thailand	1980	1	572	11318	...	11	0	...	0	...
	1988	0	2707	12054	...	1554	1102	133	6605	...
	1989	0	3319	11986	...	1630	1132	92	6877	...
	1990	0	4631	11469	...	1829	1213	146	6990	...
Turkey	1980	3137	6197	5450	2000	3329	...	...	29	1126
	1988	2982	12371	3491	3453	3663	...	...	129	2519
	1989	2810	16228	3265	2999	4109	...	...	205	2540
	1990	2539	14513	3265	3207	4143	...	...	242	2590
United Arab Emirates	1980	...	...	...	...	118273	...	213	9140	...
	1988	...	...	...	...	103614	2192	1471	22100	...
	1989	...	...	...	...	127521	2132	1486	26707	...
	1990	...	...	...	...	145656	2268	1578	26894	...
Viet Nam	1980	5300	...	6481	...	...	...	...	...	...
	1988	6332	...	7710	...	970	...	...	39	...
	1989	3900	...	7877	...	2124	...	...	24	...
	1990	5000	...	8049	...	3850	...	...	4	...
Yemen	1980	...	...	84	...	...	...	...	...	...
	1988	...	...	104	...	10386	...	...	...	...
	1989	...	...	108	...	13903	...	...	...	...
	1990	...	...	108	...	13240	...	...	...	...
Europe	**1980**	**423988**	**227716**	**17817**	**113787**	**197669**	**467**	**1766**	**324964**	**97385**
	1988	**371133**	**263946**	**21231**	**97411**	**298740**	**611**	**1564**	**284153**	**87554**
	1989	**353083**	**263124**	**20894**	**95954**	**290821**	**700**	**1471**	**289194**	**89415**
	1990	**311822**	**246540**	**20880**	**92751**	**297024**	**711**	**1068**	**290924**	**86432**
Albania	1980	...	710	536	14	5000	...	...	512	...
	1988	...	1200	536	21	4286	...	...	546	...
	1989	...	1200	536	21	4143	...	...	529	...
	1990	...	1200	536	21	4143	...	...	529	...
Austria	1980	0	1281	471	1681	2107	32	...	2663	1642
	1988	0	806	911	1684	1679	64	...	1553	1577
	1989	0	782	893	1710	1654	60	...	1567	1543
	1990	0	951	893	1641	1641	60	...	1582	1588
Belgium	1980	5784	...	111	5882	...	...	...	52	4321
	1988	3145	...	191	5395	...	...	...	20	3931
	1989	3322	...	191	5308	...	...	...	16	3822
	1990	2156	...	191	5271	...	...	...	15	3738
Bulgaria	1980	267	14973	281	1213	393	...	...	363	592
	1988	196	16976	603	1135	110	...	...	12	819
	1989	193	17053	515	1215	104	...	...	11	850
	1990	133	15763	506	1080	86	...	...	10	795
Czechoslovakia	1980	24016	40382	588	9811	133	...	5	722	7450
	1988	21347	40922	511	10085	204	...	6	1013	5749
	1989	20991	38549	459	9668	206	...	6	951	5597
	1990	19078	35711	594	10003	176	...	6	792	5357
Denmark	1980	...	0	89	59	426	...	...	...	463
	1988	...	0	160	0	6764	...	...	3032	444
	1989	...	0	142	0	7901	...	...	3582	465
	1990	...	0	142	0	8563	...	...	4326	452
Faeroe Islands	1980	...	...	...	...	...	...	...	...	...
	1988	...	...	...	...	...	...	...	...	...
	1989	...	...	...	...	...	...	...	...	...
	1990	...	...	...	...	...	...	...	...	...
Finland	1980	...	...	1366	0	...	...	...	...	483
	1988	...	...	995	423	...	...	...	...	689
	1989	...	...	995	390	...	...	...	...	746
	1990	...	...	995	438	...	...	...	...	809
France (incl. Monaco)	1980	20194	1535	3470	10006	1709	...	...	9854	11063
	1988	12893	1204	3478	6685	4793	...	...	4137	6759
	1989	12296	1301	3478	6590	4634	...	...	4138	7513
	1990	11000	1351	3480	6570	4320	...	...	4006	7527

LPG GPL	Aviation gasoline Essence aviation	Motor gasoline Essence auto	Kerosene Kérosène	Jet fuels Carbu- réacteurs	Gas diesel oil Gas oil fluides	Residual fuel Fuel oil fluides	Lubricants Huiles lubrifiantes	Electricity Electricité	Année	Région, pays ou zone
194	...	2029	349	927	3525	3383	0	1856	1980	Thaïlande
951	...	2785	155	1554	4417	3403	0	4172	1988	
747	...	2997	155	1969	5313	4691	0	4804	1989	
1152	...	3372	242	3125	4431	4797	0	5672	1990	
612	4	2856	700	215	4504	7454	227	2859	1980	Turquie
1200	0	3606	743	830	9318	13320	355	5902	1988	
1066	0	4206	511	806	8951	11783	381	6393	1989	
1119	0	4014	604	1047	9135	12886	403	7068	1990	
1214	...	243	52	...	313	188	...	773	1980	Emirats arabes unis
3761	...	1783	74	1592	3754	3809	...	1615	1988	
3978	...	1950	88	1607	3639	3710	...	1630	1989	
4211	...	1875	118	1621	3915	3965	...	1669	1990	
...	...	...	...	...	...	...	...	517	1980	Viet Nam
...	...	9	4	1	25	16	1	832	1988	
...	...	10	4	1	25	14	1	956	1989	
...	...	10	4	1	25	14	1	1071	1990	
16	...	579	295	236	1363	1458	...	62	1980	Yémen
70	...	450	251	383	1818	2110	...	205	1988	
71	...	465	243	383	1834	2131	...	206	1989	
78	...	495	251	383	1856	2741	...	214	1990	
24569	579	185590	11568	35092	337120	363816	13753	267546	1980	Europe
30303	358	219111	11382	44982	314554	219450	14519	331189	1988	
30388	334	220911	10050	48263	310588	216165	14629	339971	1989	
30202	333	224898	10465	50203	311100	207781	14128	345158	1990	
...	...	262	118	...	435	779	29	456	1980	Albanie
...	...	322	111	...	471	850	36	504	1988	
...	...	315	103	...	464	850	36	504	1989	
...	...	315	103	...	464	850	36	504	1990	
235	...	2667	6	195	3687	6056	171	5113	1980	Autriche
73	...	3568	13	324	3555	2557	0	5930	1988	
62	...	3517	19	423	3696	2242	12	6162	1989	
64	...	3946	19	450	4016	2445	99	6193	1990	
802	0	8190	34	2516	15851	15432	95	6526	1980	Belgique
802	24	7384	119	2176	13298	9449	4	7936	1988	
786	19	7999	85	2450	14132	8143	7	8214	1989	
688	12	7894	116	2120	14832	7981	4	8625	1990	
123	...	2775	339	...	5075	8496	216	4279	1980	Bulgarie
155	...	3130	324	486	4897	5236	276	5532	1988	
160	...	3249	324	548	5076	5413	257	5445	1989	
120	...	1945	324	308	2739	3509	200	5073	1990	
227	...	2277	777	...	6010	13305	529	8934	1980	Tchécoslovaquie
196	...	2529	613	...	6480	8475	575	10733	1988	
196	...	2160	622	...	6524	8134	590	10752	1989	
187	...	2106	464	...	5110	6259	590	10975	1990	
166	...	1659	38	15	4044	2996	0	3332	1980	Danemark
235	...	1948	68	340	4707	3335	1	3435	1988	
218	...	2109	78	367	4929	3448	0	2795	1989	
222	...	1977	112	367	4770	3144	0	3160	1990	
...	...	...	...	...	...	...	...	20	1980	Iles Féroé
...	...	...	...	...	...	...	...	26	1988	
...	...	...	...	...	...	...	...	27	1989	
...	...	...	...	...	...	...	...	28	1990	
180	...	2912	9	357	6207	6017	...	4755	1980	Finlande
423	...	3997	6	709	5437	2492	...	6618	1988	
305	...	3652	6	722	4559	2033	...	6618	1989	
208	...	4456	1	727	5311	2280	...	6695	1990	
5335	45	28894	201	6785	60221	48140	2345	30182	1980	France (comp. Monaco)
3946	135	28347	63	6626	40987	17580	2490	48142	1988	
4104	123	26775	71	7167	39818	17060	2597	49912	1989	
4177	145	26535	88	7366	40440	17506	2375	51540	1990	

107
Production of selected energy commodities
Thousand metric tons of coal equivalent [*cont.*]
Production des principaux biens de l'énergie
Milliers de tonnes métriques d'équivalent houille [*suite*]

Region, country or area	Year	Hard coal Houille	Lignite brown coal Lignite	Fuelwood Bois de chauffage	Coke	Crude petroleum Pétrole brut	Condensate Condensat	Natural gasoline Gazoline naturelle	Natural gas Gaz naturel	Other gases Autres gaz
Germany ₰										
F. R. Germany	1980	88256	37864	1183	28820	7431	...	...	22780	25095
	1988	74333	30430	1219	18543	8067	...	...	17128	16817
	1989	72583	30781	1219	18346	7710	...	...	16820	16591
	1990	71555	30124	1219	17596	5134	...	...	17825	16563
former German D. R.	1980	0	78277	246	5481	77	...	...	2445	4357
	1988	0	93360	209	5676	57	...	...	2851	5360
	1989	0	90575	237	5459	67	...	...	2269	5250
	1990	0	84240	237	5266	56	...	...	1877	4623
Gibraltar	1980	...	...	...	...	...	...	...	...	0
	1988	...	...	...	...	...	...	...	...	0
	1989	...	...	...	...	...	...	...	...	0
	1990	...	...	...	...	...	...	...	...	0
Greece	1980	...	4341	646	259	0	...	...	0	430
	1988	...	8981	719	0	1500	...	...	193	515
	1989	...	9640	573	0	1216	...	...	113	549
	1990	...	9645	449	0	1104	...	...	135	579
Greenland	1980	0	...	...	...	...	...	...	...	...
	1988	0	...	...	...	...	...	...	...	...
	1989	0	...	...	...	...	...	...	...	...
	1990	0	...	...	...	...	...	...	...	...
Hungary	1980	1674	8152	854	905	2901	...	734	8080	1386
	1988	1222	6698	983	947	2781	...	754	8154	1009
	1989	1149	6440	981	709	2809	...	729	7911	902
	1990	937	5699	981	461	2820	...	588	5445	858
Iceland	1980	...	...	...	...	...	...	...	...	...
	1988	...	...	...	...	...	...	...	...	...
	1989	...	...	...	...	...	...	...	...	...
	1990	...	...	...	...	...	...	...	...	...
Ireland	1980	60	...	13	0	...	...	...	1169	186
	1988	42	...	17	0	...	...	...	2582	94
	1989	43	...	17	0	...	...	...	2913	101
	1990	35	...	17	0	...	...	...	3149	96
Italy (incl. San Marino)	1980	0	696	1315	7438	2571	...	...	16287	7794
	1988	36	357	1474	6723	6873	...	...	19248	7147
	1989	66	357	1392	6743	6687	...	...	20279	7805
	1990	52	341	1234	6356	6630	...	...	20435	8060
Luxembourg	1980	...	...	7	0	...	...	...	...	762
	1988	...	...	7	0	...	...	...	...	525
	1989	...	...	7	0	...	...	...	...	551
	1990	...	...	7	0	...	...	...	...	541
Malta	1980	...	...	...	0	...	...	...	...	0
	1988	...	...	...	0	...	...	...	...	0
	1989	...	...	...	0	...	...	...	...	0
	1990	...	...	...	0	...	...	...	...	0
Netherlands	1980	0	0	30	2387	1829	435	...	115960	4879
	1988	0	0	39	2839	5584	547	...	70856	6290
	1989	0	0	50	2827	4844	640	...	77242	6431
	1990	0	0	50	2660	5047	650	...	86688	6590
Norway, Svalbard and Jan Mayen Is.	1980	288	...	198	339	35073	...	...	36331	595
	1988	253	...	312	170	77146	...	...	41042	556
	1989	345	...	306	0	103727	...	...	44086	585
	1990	298	...	306	519	113783	...	...	38581	921
Poland	1980	156428	9901	635	18434	470	...	31	6589	7898
	1988	154412	19947	1041	16233	227	...	38	5279	7349
	1989	142106	19893	810	15923	224	...	38	4887	7119
	1990	117600	18836	925	16031	229	...	38	3339	7021
Portugal	1980	177	0	178	194	...	...	...	...	256
	1988	135	0	199	258	...	...	...	...	499
	1989	151	0	199	250	...	...	...	...	590
	1990	165	0	199	220	...	...	...	...	632

LPG GPL	Aviation gasoline Essence aviation	Motor gasoline Essence auto	Kerosene Kérosène	Jet fuels Carbu-réacteurs	Gas diesel oil Gas oil fluides	Residual fuel Fuel oil fluides	Lubricants Huiles lubrifiantes	Electricity Electricité	Année	Région, pays ou zone
										Allemagne ¶
3835	...	33708	113	1812	60880	34940	2191	45146	1980	R. f. Allemagne
3601	...	29493	80	2765	53158	14440	2160	52679	1988	
3456	...	30102	62	3036	49517	11901	2194	53679	1989	
3531	...	31686	59	3393	50817	12384	2244	55854	1990	
373	...	4999	25	...	8873	12602	595	12137	1980	anc. R. d. allemande
448	...	7146	18	...	9119	6450	728	14535	1988	
438	...	7344	18	...	9302	5883	720	14615	1989	
404	...	7050	16	...	8990	5664	705	14408	1990	
...	...	...	...	...	...	...	...	7	1980	Gibraltar
...	...	...	...	...	...	...	...	9	1988	
...	...	...	...	...	...	...	...	10	1989	
...	...	...	...	...	...	...	...	10	1990	
283	...	1708	60	2022	5291	9190	148	2782	1980	Grèce
535	...	4524	31	2734	4958	7859	260	4074	1988	
589	...	4702	19	2721	5413	7918	236	4232	1989	
614	...	5068	32	2653	5630	7924	272	4299	1990	
...	...	...	...	...	...	...	...	20	1980	Groënland
...	...	...	...	...	...	...	...	25	1988	
...	...	...	...	...	...	...	...	26	1989	
...	...	...	...	...	...	...	...	27	1990	
463	...	1704	0	...	5324	4456	247	2933	1980	Hongrie
536	...	1995	466	...	4704	2475	272	3591	1988	
524	...	1947	463	...	4415	2569	267	3634	1989	
514	...	2170	453	...	3967	2182	237	3490	1990	
...	...	...	...	...	...	...	...	388	1980	Islande
...	...	...	...	...	...	...	...	550	1988	
...	...	...	...	...	...	...	...	557	1989	
...	...	...	...	...	...	...	...	566	1990	
47	...	733	0	28	782	1272	...	1298	1980	Irlande
50	...	441	0	0	729	613	...	1625	1988	
48	...	508	0	0	771	719	...	1701	1989	
50	...	510	0	0	783	715	...	1783	1990	
3239	288	21426	3572	3558	38882	55887	1478	22536	1980	Italie
3520	55	25168	3403	2914	38102	35638	1966	24652	1988	(compris San Marino)
3801	52	24589	3057	3439	38673	34774	2022	25482	1989	
3312	54	27732	3623	3984	42243	33493	1839	26642	1990	
...	...	...	...	...	...	...	...	113	1980	Luxembourg
...	...	...	...	...	...	...	...	163	1988	
...	...	...	...	...	...	...	...	169	1989	
...	...	...	...	...	...	...	...	169	1990	
...	...	...	...	...	...	...	...	65	1980	Malte
...	...	...	...	...	...	...	...	127	1988	
...	...	...	...	...	...	...	...	135	1989	
...	...	...	...	...	...	...	...	135	1990	
2535	130	13710	643	4866	25391	24137	814	7960	1980	Pays-Bas
3560	118	18541	898	6960	26723	20298	1173	8551	1988	
4002	114	18256	413	7780	25031	18956	871	8973	1989	
4103	96	19263	332	7282	22510	17945	858	8828	1990	
264	0	1794	783	177	4991	2591	35	10269	1980	Norvège, Svalbard et
1709	0	2056	351	625	6189	1508	3	13430	1988	I. Jan Mayen
1604	0	2488	240	1005	6332	2138	0	14590	1989	
1807	0	4753	231	1340	8575	1590	0	14937	1990	
317	...	4999	332	...	7385	5566	824	14970	1980	Pologne
387	...	4236	6	...	7502	4605	755	17734	1988	
435	...	4431	4	...	7025	5219	746	17868	1989	
222	...	3288	4	...	5746	5142	611	16747	1990	
342	0	1462	108	784	2706	4918	105	1875	1980	Portugal
458	0	2121	37	948	3117	3951	180	2754	1988	
552	0	2299	44	1038	4141	5153	183	3138	1989	
555	0	2556	38	1313	4327	5403	184	3504	1990	

107

Production of selected energy commodities
Thousand metric tons of coal equivalent [*cont.*]
Production des principaux biens de l'énergie
Milliers de tonnes métriques d'équivalent houille [*suite*]

Region, country or area	Year	Hard coal Houille	Lignite brown coal Lignite	Fuelwood Bois de chauffage	Coke	Crude petroleum Pétrole brut	Condensate Condensat	Natural gasoline Gazoline naturelle	Natural gas Gaz naturel	Other gases Autres gaz
Romania	1980	8060	8944	1365	3045	16444	...	751	51536	4521
	1988	8831	16100	1280	5609	13413	...	766	42164	6595
	1989	8289	17229	1280	5719	13104	...	697	37797	6545
	1990	4500	13200	1282	4385	11327	...	435	32728	5429
Spain	1980	9905	4584	505	4744	2289	...	...	1	3794
	1988	10199	4709	683	3088	2114	...	...	1187	4014
	1989	10429	5854	863	3206	1483	...	...	2071	4649
	1990	10585	5626	863	3301	1136	...	...	1713	4062
Sweden	1980	17	...	2345	1137	36	...	...	...	865
	1988	28	...	4033	868	4	...	...	...	719
	1989	0	...	4050	942	4	...	...	...	761
	1990	0	...	4050	1037	4	...	...	...	782
Switzerland, Liechtenstien	1980	...	...	312	...	...	...	...		289
	1988	...	...	283	...	...	...	...	10	198
	1989	...	...	271	...	...	...	...	6	140
	1990	...	...	292	...	...	...	...	5	149
United Kingdom	1980	108556	...	47	9310	112739	0	247	47048	6676
	1988	83774	...	63	8020	156364	0	0	60159	7920
	1989	80889	...	83	7889	124863	0	0	58841	8227
	1990	73496	...	85	7782	125729	0	0	64950	7871
Yugoslavia	1980	307	16075	1027	2627	6041	...	...	2573	1589
	1988	287	22257	1287	3009	6773	...	...	2986	1979
	1989	232	23470	1348	3039	5440	...	...	3165	2080
	1990	232	23854	1348	2113	5097	...	...	2794	1388
Oceania	**1980**	**60051**	**10766**	**2356**	**4311**	**28671**	**...**	**9**	**13587**	**6042**
	1988	**110303**	**14111**	**2911**	**3507**	**37953**	**...**	**0**	**26617**	**4936**
	1989	**121064**	**15684**	**2912**	**3796**	**34624**	**...**	**0**	**27220**	**5105**
	1990	**130245**	**15053**	**2914**	**3888**	**38911**	**...**	**0**	**32882**	**5102**
American Samoa	1980	...	...	...	...	...	...	...	...	...
	1988	...	...	...	...	...	...	...	...	...
	1989	...	...	...	...	...	...	...	...	...
	1990	...	...	...	...	...	...	...	...	...
Australia	1980	58380	10629	469	4279	28221	...	...	12257	5982
	1988	108558	14024	962	3499	35794	...	...	20590	4706
	1989	119024	15604	962	3789	32200	...	...	21048	4876
	1990	128422	14945	964	3881	36431	...	...	26749	4874
Cook Islands	1980	...	...	...	...	...	...	...	...	...
	1988	...	...	...	...	...	...	...	...	...
	1989	...	...	...	...	...	...	...	...	...
	1990	...	...	...	...	...	...	...	...	...
Fiji	1980	...	...	8	...	...	...	...	...	...
	1988	...	...	12	...	...	...	...	...	...
	1989	...	...	12	...	...	...	...	...	...
	1990	...	...	12	...	...	...	...	...	...
French Polynesia	1980	...	...	...	...	...	...	...	...	...
	1988	...	...	...	...	...	...	...	...	...
	1989	...	...	...	...	...	...	...	...	...
	1990	...	...	...	...	...	...	...	...	29
Guam	1980	...	...	...	...	...	...	...	...	0
	1988	...	...	...	...	...	...	...	...	0
	1989	...	...	...	...	...	...	...	...	0
	1990	...	...	...	...	...	...	...	...	...
Kiribati	1980	...	...	...	...	...	...	...	...	...
	1988	...	...	...	...	...	...	...	...	...
	1989	...	...	...	...	...	...	...	...	...
	1990	...	...	...	...	...	...	...	...	...
Nauru	1980	...	...	...	...	...	...	...	...	...
	1988	...	...	...	...	...	...	...	...	...
	1989	...	...	...	...	...	...	...	...	...
	1990	...	...	...	...	...	...	...	...	...

LPG GPL	Aviation gasoline Essence aviation	Motor gasoline Essence auto	Kerosene Kérosène	Jet fuels Carbu- réacteurs	Gas diesel oil Gas oil fluides	Residual fuel Fuel oil fluides	Lubricants Huiles lubrifiantes	Electricity Electricité	Année	Région, pays ou zone
345	15	7147	1279	...	10839	14487	955	8290	1980	Roumanie
328	22	10029	921	...	12283	14095	782	9252	1988	
298	22	9111	755	...	12231	14404	742	9317	1989	
337	22	6999	668	...	9279	11499	719	7899	1990	
1733	0	8016	62	3234	15645	32460	459	13414	1980	Espagne
3148	0	12156	416	5651	18759	21652	496	15179	1988	
2858	0	13849	162	5492	19566	22172	555	18006	1989	
2709	0	13845	351	6235	21563	21212	539	18502	1990	
171	...	4117	6	290	8707	10985	59	11831	1980	Suède
266	...	5148	60	960	7586	5770	119	17962	1988	
249	...	5677	46	761	8800	6960	250	17677	1989	
342	...	5611	22	768	9009	7179	381	18000	1990	
146	...	1644	7	327	2926	1136	...	5781	1980	Suisse, Liechtenstein
264	...	1464	4	367	2311	988	...	7331	1988	
214	...	1071	3	374	1756	656	...	6604	1989	
216	...	1128	3	340	1692	714	...	6860	1990	
2962	99	24913	2998	7662	32149	38552	1797	34854	1980	Royaume-Uni
5040	0	39613	3374	9913	34696	20465	1395	37861	1988	
4876	0	40855	3455	10454	33776	21278	1510	38550	1989	
5159	0	40086	3403	11115	33936	22408	1400	39182	1990	
448	1	3870	59	464	4818	9416	661	7301	1980	Yugoslavie
625	3	3750	0	483	4785	8669	848	10276	1988	
614	3	3900	0	486	4640	8142	834	10602	1989	
662	3	3975	0	442	4350	8354	834	10552	1990	
3153	100	18228	808	3723	11832	7215	795	15084	1980	Océanie
4079	223	19938	310	4583	13526	3346	788	21047	1988	
3981	202	19954	299	4755	14132	3584	808	22380	1989	
4648	238	20554	380	4718	14574	3867	805	23295	1990	
...	...	...	...	...	...	...	...	9	1980	Samoa américaines
...	...	...	...	...	...	...	...	12	1988	
...	...	...	...	...	...	...	...	12	1989	
...	...	...	...	...	...	...	...	11	1990	
3133	100	16308	808	2559	10557	5423	795	11809	1980	Australie
3888	223	17589	293	3630	11797	2950	788	17076	1988	
3795	202	17476	284	3871	12154	3104	808	18154	1989	
4469	238	17955	366	3849	12615	3315	805	18987	1990	
...	...	...	...	...	...	...	...	1	1980	Iles Cook
...	...	...	...	...	...	...	...	2	1988	
...	...	...	...	...	...	...	...	2	1989	
...	...	...	...	...	...	...	...	2	1990	
...	...	...	...	...	...	...	...	38	1980	Fidji
...	...	...	...	...	...	...	...	49	1988	
...	...	...	...	...	...	...	...	53	1989	
...	...	...	...	...	...	...	...	53	1990	
...	...	...	...	...	...	...	...	31	1980	Polynésie française
...	...	...	...	...	...	...	...	31	1988	
...	...	...	...	...	...	...	...	33	1989	
...	...	...	...	...	...	...	...	34	1990	
6	...	...	...	1164	326	680	...	141	1980	Guam
0	...	...	...	0	0	0	...	98	1988	
0	...	...	...	0	0	0	...	98	1989	
0	...	...	...	0	0	0	...	98	1990	
...	...	...	...	...	...	...	...	1	1980	Kiribati
...	...	...	...	...	...	...	...	1	1988	
...	...	...	...	...	...	...	...	1	1989	
...	...	...	...	...	...	...	...	1	1990	
...	...	...	...	...	...	...	...	3	1980	Nauru
...	...	...	...	...	...	...	...	4	1988	
...	...	...	...	...	...	...	...	4	1989	
...	...	...	...	...	...	...	...	4	1990	

107
Production of selected energy commodities
Thousand metric tons of coal equivalent [*cont.*]
Production des principaux biens de l'énergie
Milliers de tonnes métriques d'équivalent houille [*suite*]

Region, country or area	Year	Hard coal Houille	Lignite brown coal Lignite	Fuelwood Bois de chauffage	Coke	Crude petroleum Pétrole brut	Condensate Condensat	Natural gasoline Gazoline naturelle	Natural gas Gaz naturel	Other gases Autres gaz
New Caledonia	1980	...	...	0	...	...	...	...	...	...
	1988	...	...	0	...	...	...	...	...	...
	1989	...	...	...	...	...	...	...	...	...
	1990	...	...	...	...	...	...	...	...	...
New Zealand	1980	1671	137	17	32	450	...	9	1330	31
	1988	1745	87	17	8	2159	...	0	6028	229
	1989	2040	80	17	7	2424	...	0	6172	229
	1990	1823	108	17	7	2480	...	0	6133	229
Niue	1980	...	...	...	...	...	...	...	...	...
	1988	...	...	...	...	...	...	...	...	...
	1989	...	...	...	...	...	...	...	...	...
	1990	...	...	...	...	...	...	...	...	...
Pacific Islands (Palau)[1]	1980	...	...	...	...	...	...	...	...	...
	1988	...	...	...	...	...	...	...	...	...
	1989	...	...	...	...	...	...	...	...	...
	1990	...	...	...	...	...	...	...	...	...
Papua New Guinea	1980	...	...	1796	...	...	...	...	...	...
	1988	...	...	1844	...	...	...	...	...	...
	1989	...	...	1844	...	...	...	...	...	...
	1990	...	...	1844	...	...	...	...	...	...
Samoa	1980	...	...	23	...	...	...	...	...	...
	1988	...	...	23	...	...	...	...	...	...
	1989	...	...	23	...	...	...	...	...	...
	1990	...	...	23	...	...	...	...	...	...
Solomon Islands	1980	...	...	34	...	...	...	...	...	...
	1988	...	...	45	...	...	...	...	...	...
	1989	...	...	46	...	...	...	...	...	...
	1990	...	...	46	...	...	...	...	...	...
Tonga	1980	...	...	...	...	...	...	...	...	...
	1988	...	...	...	...	...	...	...	...	...
	1989	...	...	...	...	...	...	...	...	...
	1990	...	...	...	...	...	...	...	...	...
Vanuatu	1980	...	...	8	...	...	...	...	...	...
	1988	...	...	8	...	...	...	...	...	...
	1989	...	...	8	...	...	...	...	...	...
	1990	...	...	8	...	...	...	...	...	...
former USSR	1980	411377	79970	26297	77400	861724	...	...	516895	67283
	1988	500593	86000	27023	73710	882000	...	...	887398	65884
	1989	482207	82000	24847	72360	859286	...	...	919489	63229
	1990	453793	80000	22516	72000	807857	...	...	946277	62721

Source:
Energy statistics database of the Statistical Division of the United Nations Secretariat.

§ All data shown pertain to Germany prior to 3 October 1990 are indicated separately for the Federal Republic of Germany and the former German Democratic Republic based on their respective territories at the time indicated. Where data for united Germany (3 October 1990 and thereafter) are not available, available data are shown separately under the designations Federal Republic of Germany and former German Democratic Republic and pertain to the territorial boundaries prior to 3 October 1990. For detailed explanatory notes on data pertaining to Germany, see Annex I-Country Nomenclature.

1 Including data for Federated States of Micronesia, Marshall Is. and Northern Mariana Is.

Source:
Base de données pour les statistiques énergétiques de la Division de statistique du Secrétariat de l'ONU.

§ Toutes les données se rapportant à l'Allemagne avant le 3 octobre 1990 figurent dans deux rubriques séparées basées sur les territoires respectifs de la République fédérale d'Allemagne et l'ancienne République démocratique allemande selon la période indiquée. En l'absence de données pour l'Allemagne unifiée (à compter du 3 octobre, 1990), les données disponibles sont fournies séparément sous les rubriques République fédérale d'Allemagne et ancienne République démocratique allemande et se rapportent aux limites territoriales antérieures au 3 octobre 1990. Pour les notes explicatives en détail sur les données concernant l'Allemagne, voir Annexe I - Nomenclature des pays.

1 Y compris les données pour les Etats fédérés de Micronésie, les îles Marshall et les îles Mariannes du Nord.

LPG GPL	Aviation gasoline Essence aviation	Motor gasoline Essence auto	Kerosene Kérosène	Jet fuels Carbu- réacteurs	Gas diesel oil Gas oil fluides	Residual fuel Fuel oil fluides	Lubricants Huiles lubrifiantes	Electricity Electricité	Année	Région, pays ou zone
...	...	...	...	...	...	...	...	164	1980	Nouvelle-Calédonie
...	...	...	...	...	...	...	...	145	1988	
...	...	...	...	...	...	...	...	146	1989	
...	...	...	...	...	...	...	...	141	1990	
14	...	1920	...	...	948	1112	...	2700	1980	Nouvelle-Zélande
191	...	2349	16	952	1728	396	...	3378	1988	
186	...	2478	15	884	1978	480	...	3620	1989	
179	...	2599	15	870	1959	552	...	3704	1990	
...	...	...	...	...	...	...	...	0	1980	Nioué
...	...	...	...	...	...	...	...	0	1988	
...	...	...	...	...	...	...	...	0	1989	
...	...	...	...	...	...	...	...	0	1990	
...	...	...	...	...	...	...	...	17	1980	Iles du Pacifique (Palaos)[1]
...	...	...	...	...	...	...	...	22	1988	
...	...	...	...	...	...	...	...	24	1989	
...	...	...	...	...	...	...	...	25	1990	
...	...	...	...	...	...	...	...	154	1980	Papouasie-Nvl-Guinée
...	...	...	...	...	...	...	...	214	1988	
...	...	...	...	...	...	...	...	218	1989	
...	...	...	...	...	...	...	...	220	1990	
...	...	...	...	...	...	...	...	5	1980	Samoa
...	...	...	...	...	...	...	...	6	1988	
...	...	...	...	...	...	...	...	6	1989	
...	...	...	...	...	...	...	...	6	1990	
...	...	...	...	...	...	...	...	3	1980	Iles Salomon
...	...	...	...	...	...	...	...	4	1988	
...	...	...	...	...	...	...	...	4	1989	
...	...	...	...	...	...	...	...	4	1990	
...	...	...	...	...	...	...	...	1	1980	Tonga
...	...	...	...	...	...	...	...	2	1988	
...	...	...	...	...	...	...	...	3	1989	
...	...	...	...	...	...	...	...	3	1990	
...	...	...	...	...	...	...	...	2	1980	Vanuatu
...	...	...	...	...	...	...	...	3	1988	
...	...	...	...	...	...	...	...	3	1989	
...	...	...	...	...	...	...	...	3	1990	
13242	3300	115200	53064	...	157180	246667	12366	158934	1980	ancienne URSS
15074	1200	101100	54096	...	167910	276120	12653	209434	1988	
15229	750	100500	54096	...	167475	269040	12797	211522	1989	
15229	750	95400	52327	...	162545	261960	12797	212013	1990	

Technical notes, tables 106 and 107

Tables 106 and 107: Data are presented in metric tons of coal equivalent (TCE), to which the individual energy commodities are converted in the interests of international uniformity and comparability.

The procedure to convert from original units to TCE is as follows:

Data in original unit x special factors = TCE
(metric tons, TJ, kWh,m^3)

For special factors used to convert the commodities into coal equivalent and detailed description of methods, see the United Nations *Energy Statistics Yearbook* and related methodological publications. [20, 41, 42]

Table 106: The data on production refer to the first stage of production. Thus, for hard coal the data refer to mine production; for briquettes to the output of briquetting plants; for crude petroleum and natural gas to production at oil and gas wells; for natural gas liquids to production at wells and processing plants; for refined petroleum products to gross refinery output; for cokes and coke-oven gas to the output of ovens; for other manufactured gas to production at gas works, blast furnaces or refineries; and for electricity to the gross production of generating plants.

International trade of energy commodities is based on the "general trade" system, that is, all goods entering and leaving the national boundary of a country are recorded as imports and exports.

Bunkers refers to fuels supplied to ships and aircraft engaged in international transportation, irrespective of the carrier's flag.

In general, data on stocks refer to changes in stocks of producers, importers and/or industrial consumers at the beginning and end of each year.

Data on consumption refer to "apparent consumption" and are derived from the formula "production + imports – exports – bunkers +/– stock changes". Accordingly, the series on apparent consumption may in some cases represent only an indication of the magnitude of actual gross inland availability.

Table 107: Definitions of the energy commodities are as follows:

— Hard coal: Coal with a high degree of coalification, and with a gross calorific value above 24 MJ/kg (5,700 kcal/kg) on an ash-free but moist basis, and with a reflectance index of vitrinite of 0.5 and above;

— Lignite: Coal with a low degree of coalification which has retained the anatomical structure of the vegetable matter from which it was formed. Its gross calorific value is less than 24 MJ/kg (5,700 kcal/kg) on an ash-free but moist basis, and its reflectance index of vitrinite is less than 0.5;

Notes techniques, tableaux 106 et 107

Tableaux 106 et 107 : Les données relatives aux divers produits énergétiques ont été converties en tonnes métriques d'équivalent houille (TEC), dans un souci d'uniformité et pour permettre les comparaisons entre la production de différents pays.

La méthode de conversion utilisée pour passer des unités de mesure d'origine à l'unité commune est la suivante :

Données en unités d'origine (tonnes métriques, TJ, kWh, m^3) x facteurs de conversion = TEC

Pour les facteurs spéciaux utilisés pour convertir les produits énergétiques en équivalent houille et pour des descriptions détaillées des méthodes appliquées, se reporter à l'*Annuaire des statistiques de l'énergie* des Nations Unies et aux publications méthodologiques connexes [20, 41, 42].

Tableau 106 : Les données relatives à la production se rapportent au premier stade de production. Ainsi, pour la houille, les données se rapportent à la production minière; pour les briquettes, à la production des briquetteries; pour le pétrole brut et le gaz naturel, à la production des gisements de pétrole et de gaz; pour les condensats de gaz naturel, à la production au puits et aux installations de traitement; pour les produits pétroliers raffinés, à la production brute des raffineries; pour les cokes et le gaz des fours à coke, à la production des fours; pour les autres gaz manufacturés, à la production des usines à gaz, des hauts fourneaux ou des raffineries; et pour l'électricité, à la production brute des centrales.

Le commerce international des produits énergétiques est fondé sur le système du "commerce général", c'est-à-dire que tous les biens entrant sur le territoire national d'un pays ou en sortant sont respectivement enregistrés comme importations et exportations.

Les soutages se rapportent aux carburants fournis aux navires et aux avions assurant des transports internationaux, quel que soit leur pavillon.

En général, les variations des stocks se rapportent aux différences entre les stocks des producteurs, des importateurs ou des consommateurs industriels au début et à la fin de chaque année.

Les données sur la consommation se rapportent à la "consommation apparente" et sont obtenues par la formule "production + importations - exportations - soutage +/- variations des stocks". En conséquence, les séries relatives à la consommation apparente peuvent occasionnellement ne donner qu'une indication de l'ordre de grandeur des disponibilités intérieures brutes réelles.

Tableau 107 : Les définitions des produits énergétiques sont données ci-après :

— Houille : Charbon à haut degré de houillification et de pouvoir calorifique brut supérieur à 24 MJ/kg (5.700 kcal/kg) mesuré sans cendre, mais sur base humide, pour lequel l'indice de réflectance du vitrain est égal ou supérieur à 0,5;

— Fuelwood: All wood in the rough used for fuel purposes. Production data include the portion used for charcoal production, using a factor of 6 to convert from a weight basis to the volumetric equivalent (metric tons to cubic metres) of charcoal;

— Coke: The solid residue obtained from the distillation of hard coal or lignite in the total absence of air (carbonization);

— Crude petroleum: Mineral oil consisting of a mixture of hydrocarbons of natural origin, yellow to black in colour, of variable specific gravity and viscosity, including crude mineral oils extracted from bituminous minerals (shale, bituminous sand, etc.). Data for crude petroleum include lease (field) condensate (separator liquids) which is recovered from gaseous hydrocarbons in lease separation facilities;

— Plant condensate: Liquid hydrocarbons condensed from wet natural gas in natural gas processing plants used as a petroleum refinery input;

— Natural gasolene: Light spirit extracted from wet natural gas, often in association with crude petroleum. It is used as petroleum refinery and petrochemical plant input and is also used directly for blending with motor spirit without further processing;

— Natural gas: A mixture of hydrocarbon compounds and small quantities of non-hydrocarbons existing in the gaseous phase, or in solution with oil in natural underground reservoirs at reservoir conditions;

— Other gases includes gasworks gas — gas produced by carbonization or total gasification with or without enrichment with petroleum products; coke-oven gas — by-product of the carbonization process in the production of coke at coke ovens; and blast furnace gas — by-product in blast furnaces recovered on leaving the furnace;

— Liquefied petroleum gas (LPG): Hydrocarbons which are gaseous under conditions of normal temperature and pressure but are liquefied by compression or cooling to facilitate storage, handling and transportation;

— Aviation gasolene: Motor spirit prepared especially for aviation piston engines, with an octane number varying from 80 to 145 RON and a freezing point of –60°C;

— Motor gasolene: Light hydrocarbon oil used in positive engines other than aircraft, distilling between 35 and 200°C, and treated to reach a sufficiently high octane number of generally between 80 and 100 RON. Treatment may be by reforming, blending with an aromatic fraction, or the addition of benzole or other additives such as tetraethyl lead;

— Kerosene: Medium oil distilling between 150 and 300°C; at least 65 per cent in volume distils at 250°C. Its specific gravity is around 0.80 and the flash point above 38°C. It is used as an illuminant and as a fuel in certain types of spark-ignition engines, such as those used for agricultural tractors and stationary engines. Other names for this product are burning oil, vaporizing oil, power kerosene and illuminating oil;

— Lignite : Charbon d'un faible degré de houillification qui a gardé la structure anatomique des végétaux dont il est issu. Son pouvoir calorifique brut est inférieur à 24 MK/kg (5.700 kcal/kg) mesuré sans cendre, mais sur base humide, et son indice de réflectance du vitrain est égal ou inférieur à 0,5;

— Bois de chauffage : Tous les types de bois à l'état brut non dégrossis utilisés comme combustibles. Les données de production englobent les quantités utilisées pour la production de charbon de bois, utilisant un facteur de 6 pour convertir en volume le poids de charbon de bois (tonnes en mètres cubes);

— Coke : Résidu solide obtenu lors de la distillation de houille ou de lignite en l'absence totale d'air (carbonisation).

— Pétrole brut : Huile minérale constituée d'un mélange d'hydrocarbures d'origine naturelle, de couleur variant du jaune au noir, d'une densité et d'une viscosité variables. Figurent également dans cette rubrique les huiles minérales brutes extraites de minéraux bitumeux (schiste, sable, etc.). Les données relatives au pétrole brut comprennent les condensats directement récupérés sur les sites d'exploitation des hydrocarbures gazeux (dans les installations prévues pour la séparation des phases liquide et gazeuse).

— Condensats d'usine : Hydrocarbures liquides résultant de la condensation du gaz naturel humide dans les usines de traitement de gaz naturel. Ils sont utilisés comme charge d'alimentation dans les raffineries de pétrole.

— Essence naturelle : Essence légère du gaz naturel humide, souvent en association avec le pétrole brut. Elle est utilisée comme charge dans les raffineries de pétrole et les usines pétrochimiques et est aussi employée directement en mélange avec le carburant auto sans traitement spplémentaire.

— Gaz naturel : Mélanges de composés d'hydrocarbures et de petites quantités de composants autres que des hydrocarbures existant en phase gazeuse ou en solution huileuse dans des roches réservoirs souterraines naturelles, dans les conditions du réservoir.

— Les autres gaz comprennent le gaz d'usine à gaz — gaz obtenu par carbonisation ou par gazéification totale, avec ou sans enrichissement au moyen de produits pétroliers; gaz de cokerie — sous-produit du processus de carbonisation dans la production du coke dans les fours à coke; et gaz de haut-fourneau — sous-produit du fonctionnement des hauts-fourneaux, récupéré à la sortie du gueulard.

— Gaz de pétrole liquéfié (GPL) : Hydrocarbures qui sont à l'état gazeux dans des conditions de température et de pression normales mais sont liquéfiés par compression ou refroidissement pour en faciliter l'entreposage, la manipulation et le transport.

— Essence aviation : carburant fabriqué spécialement pour les moteurs d'avion à pistons, avec un indice variant de 80 à 145 IOR et dont le point de congélation est de -60°C.

— Jet fuel: Gasolene-type fuel and kerosene-type jet fuel;

— Gasolene-type jet fuel: All light hydrocarbon oils for use in aviation gas-turbine engines distilling between 100 and 250°C — at least 20 per cent in volume distils at 143°C;

— Kerosene-type jet fuel: Medium oil with the same distillation characteristics and flash point as kerosene, with a maximum aromatic content of 20 per cent in volume, and treated to give a kinematic viscosity of less than 15 cSt at −34°C and a freezing point below −50°C used in aviation gas-turbine engines;

— Gas-diesel oil (distillate fuel oil): Heavy oils distilling between 200 and 380°C, but distilling less than 65 per cent in volume at 250°C, including losses, and 85 per cent or more at 350°C. Their flash point is always above 50° and their specific gravity higher than 0.82. Heavy oils obtained by blending are grouped together with gas oils on the condition that their kinematic viscosity does not exceed 27.5 cSt at 38°C. Other names for this product are diesel fuel, diesel oil, gas oil and solar oil;

— Residual fuel oil: Heavy oils that make up the distillation residue. Comprises all fuels (including those obtained by blending) with a viscosity above 27.5 cSt at 38°C. Their flash point is always above 50° and their specific gravity higher than 0.90. Commonly used by ships and industrial large-scale heating installations as a fuel in furnaces or boilers. Also known as mazout;

— Lubricants: Viscous, liquid hydrocarbons rich in paraffin waxes, distilling between 380 and 500°C and obtained by vacuum distillation of oil residues from atmospheric distillation. Additives may be included to alter their characteristics. The main characteristics are as follows: flash point greater than 125°C; pour point between −25 and +5°C depending on the grade; strong acid number normally 0.5 mg/g; ash content less than or equal to 0.3 per cent; water content less than or equal to 0.2 per cent. Included are cutting oils, white oils, insulating oils, spindle oils and lubricating greases;

— Electricity production: Refers to gross production, which includes the consumption by station auxiliaries and any losses in the transformers that are considered integral parts of the station. Excluded is electricity produced from pumped storage.

— Essence auto : Hydrocarbure léger utilisé dans les moteurs à allumage par étincelle autres que les moteurs d'avion dont les températures de distillation se situent entre 35 et 200°C et qui est traité de façon à atteindre un indice d'octane suffisamment élevé, généralement entre 80 et 100 IOR. Le traitement peut consister en reformage, mélange avec une fraction aromatique ou adjonction de benzole ou d'autres additifs (tels que du plomb tétraéthyle).

— Pétrole lampant : Huile moyennement visqueuse dont les températures de distillation se situent entre 150 et 300°C et qui donne au moins 65 % en volume de distillat à 250°C. Sa densité se situe aux alentours de 0,80 et son point d'éclair est supérieur à 38°C. Il sert à l'éclairage et aussi de carburant dans certains moteurs à allumage par étincelle, tels que ceux utilisés dans les tracteurs agricoles et les installations stationnaires. Les données concernent les produits couramment appelés kérosène, pétrole carburant ou "power kerosene" , et huile d'éclairage.

— Carburéacteurs : Comprennent les carburéacteurs du type essence et du type kérosène.

— Carburéacteurs du type essence : Ils comprennent tous les hycrocarbures légers utilisés dans les turboréacteurs d'aviation, dont les températures de distillation se situent entre 100 et 250°C et qui donnent au moins 20 % en volume de distillat à 143°C.

— Les carburéacteurs du type kérosène : Sont des huiles moyennement visqueuses ayant les mêmes caractéristiques de distillation et le même point d'éclair que le pétrole lampant et une teneur en composés aromatiques ne dépassant pas 20 % en volume, et qui sont traitées de façon à atteindre une viscosité cinématique de moins de 15 cSt à -34°C et un point de congélation inférieur à -50°C. Ce type de carburéacteurs est également utilisé dans les turboréacteurs d'aviation.

— Gazole/carburant diesel (mazout distillé) : Huiles lourdes dont les températures de distillation se situent entre 200 et 380°C mais qui donnent moins de 65 % en volume de distillat à 250°C (y compris les pertes) et 85 % ou davantage à 350°C. Leur point d'éclair est toujours supérieur à 50°C et leur densité supérieure à 0,82. Les huiles lourdes obtenues par mélange sont classées dans la même catégorie que les gazoles à condition que leur viscosité cinématique ne dépasse pas 27,5 cSt à 38°C. Ce produit est aussi connu sous les appellations de gazole, ou gasoil, carburant ou combustible diesel et huile solaire.

— Mazout résiduel : Huiles lourdes constituant le résidu de la distillation. La rubrique comprend tous les combustibles (y compris ceux obtenus par mélange) d'une viscosité supérieure à 27,5 cSt à 38°C. Leur point d'éclair est toujours supérieur à 50°C et leur densité supérieure à 0,90. Ces produits sont couramment utilisés comme combustible dans les chaudièes des navires et des grandes installations de chauffage industriel. Ils sont également connus sous le nom de fuel oil.

— Lubrifiants : Hydrocarbures liquides et visqueux, riches en paraffines, dont les températures de distillation se situent entre 380 et 500°C et qui sont obtenus par distillation sous vide des résidus de la distillation atmosphérique du pétrole. Des additifs peuvent y être incorporés pour modifier leurs caractéristiques. Leurs principales caractéristiques sont les suivantes : point d'éclair supérieur à 125°C; point d'écoulement compris entre -25 et +5°C selon la qualité; indice d'acide fort normalement égal à 0,5mg/g; teneur en cendres inférieure ou égale à 0,3 %; et teneur en eau inférieure ou égale à 0,2 %. Figurent dans cette rubrique les huiles de coupe, les huiles blanches, les huiles isolantes, les huiles à broches et les graisses lubrifiantes.

— Production d'électricité : Se rapporte à la production brute qui comprend la consommation des équipements auxiliaires des centrales et les pertes au niveau des transformateurs considérés comme faisant partie intégrante de ces centrales. Elle ne comprend pas l'électricité produite à partir d'une accumulation par pompage.

108
Selected indicators of natural resources
Indicateurs concernant certaines ressources naturelles
Changes in land use and forests; water withdrawal and threatened species
Evolution de l'utilisation des sols et de la superficie des forêts; prélèvements d'eau et espèces menacées

Country or area Pays ou zone	Changes in land distribution 1966/68 à 1986/88 (≥ 3 percentage points) Evolution de la répartition des sols 1966/68 à 1986/88 (≥ 3 points de pourcentage)				Forests Forêts Δ p.a., 1980s		Reforest-ation Reboise-ment	Water withdrawal Prélève-ments d'eau	Number of threatened species Nombre d'espèces menacées
	Cropland Terres agricoles	Pasture Pacages	Forest Forêts	Other Autres	Deforestation Déboisement 10³ha	%	10³ha	%, 1980s	1990
World									
Monde	...	...	...	...	**15 517**	**0.4**	**14 713**	**8**	...
Africa									
Afrique	...	...	...	...	4 040	0.6	297	3	4 546
Algeria									
Algerie	...	...	...	...	40	2.3	52	16	172
Angola									
Angola	...	...	...	...	94	0.2	3	-	47
Benin									
Bénin	...	...	-10.1	7.5	67	1.7	-	-	17
Botswana									
Botswana					20	0.1	...	1	20
Burkina Faso									
Burkina Faso	4.9	...	-4.4	...	80	1.7	2	1	13
Burundi									
Burundi	6.6	10.2	...	-17.4	1	2.7	3	3	10
Cameroon									
Cameroun	...	...	-4.7	3.3	190[2]	0.8[2]	1	-	121
Cape Verde									
Cap-Vert	...	...	...	...	...	...	1	20	5
Central African Rep.									
Rép. centrafricaine	...	...	...	...	55	0.2	...	-	16
Chad									
Tchad	...	...	...	...	80	0.6	-	-	38
Comoros									
Comores	4.3	...	...	-4.3	1	3.1		1	11
Congo									
Congo	...	...	...	...	22	0.1	-	-	21
Cote d'Ivoire									
Cote d'Ivoire	...	...	-15.7	12.8	510	5.2	6	1	99
Djibouti									
Djibouti	...	...	...	...	...	...	...	2	12
Egypt									
Egypte	...	...	...	...	...	...	2	97	118
Equatorial Guinea									
Guinée équatoriale	...	...	...	...	3	0.2	...	-	29
Ethiopia									
Ethiopie	...	...	...	...	88	0.3	10	2	84
Gabon									
Gabon	...	...	...	...	15	0.1	1	-	103
Gambia									
Gambie	4.8	...	-11.8	7.0	5	2.4	-	-	10
Ghana									
Ghana	...	...	-6.1	6.9	72	0.8	2	1	57
Guinea									
Guinée	...	...	-4.9	4.7	86	0.8	-	-	61
Guinea-Bissau									
Guinée-Bissau	...	...	...	...	57	2.7	-	-	9
Kenya									
Kenya	...	...	...	...	39	1.7	10	7	181
Lesotho									
Lesotho	...	-3.6	...	5.1	...	...	1	1	16
Liberia									
Libéria	...	...	-5.2	5.1	46	2.3	2	-	31

108
Selected indicators of natural resources
Changes in land use and forests; water withdrawal and threatened species [*cont.*]
Indicateurs concernant certaines ressources naturelles
Evolution de l'utilisation des sols et de la superficie des forêts; prélèvements d'eau et espèces menacées [*suite*]

| Country or area
Pays ou zone | Changes in land distribution 1966/68 à 1986/88
(≥ 3 percentage points)
Evolution de la répartition des sols
1966/68 à 1986/88
(≥ 3 points de pourcentage) | | | | Forests
Forêts
Δ p.a., 1980s | | Reforest-ation
Reboise-ment | Water withdrawal
Prélève-ments d'eau | Number of threatened species
Nombre d'espèces menacées |
	Cropland Terres agricoles	Pasture Pacages	Forest Forêts	Other Autres	Deforestation Déboisement 10^3ha	%	10^3ha	%, 1980s	1990
Libyan Arab Jamahiriya Jam. arabe libyenne	...	...	...	...	...	...	31	374	80
Madagascar Madagascar	...	...	-5.2	3.9	156	1.2	12	41	282
Malawi Malawi	3.7	...	-12.9	9.2	150	3.5	1	2	79
Mali Mali	...	...	...	...	36	0.5	1	2	37
Mauritania Mauritanie	...	...	...	...	13	2.4	-	10	23
Mauritius Maurice	...	...	...	...	-	3.3	-	16	288
Morocco Maroc	3.6	6.5	...	-12.8	13	0.4	13	37	217
Mozambique Mozambique	...	...	-3.0		120	0.8	4	1	106
Namibia Namibia	...	...	...	...		...	...	2	37
Niger Niger	...	...	...	...	67	2.6	2	1	18
Nigeria Nigéria	...	...	-6.6	4.0	400	2.7	26	1	46
Réunion Réunion	...	...	-4.0	3.3	...	...	...	...	97
Rwanda Rwanda	18.6	-14.2	...	...	5	2.3	3	2	20
Sao Tome & Principe Sao Tomé-et-Principe	...	...	...	...	...	...	...	...	9[3]
Senegal Sénégal	...	...	-4.3	4.3	50	0.5	3	4	50
Seychelles Seychelles	3.7	...	...	-3.7	...	...	...	...	90
Sierra Leone Sierra Leone	...	...	...	...	6	0.3	...	-	34
Somalia Somalie	...	...	...	...	14	0.1	1	7	77
South Africa Afrique du Sud	...	...	...	...	...	...	63	18	1 058
St. Helena Sainte-Hélène	...	...	3.2	-3.2	...	...	...	...	1[4]
Sudan Soudan	...	4.8	...	...	504	1.1	13	14	35
Swaziland Swaziland	...	-9.4	...	8.2	...	...	5	4	31
Togo Togo	...	...	-3.7	...	12	0.7		1	12
Tunisia Tunisie	...	...	...	...	5	1.7	3	53	47
Uganda Ouganda	8.8	...	-3.5	-5.2	50	0.8	2	-	40
United Rep. of Tanzania Rép.-Unie de Tanzanie	...	...	...	...	130	0.3	9	1	217
Western Sahara Sahara Occidental	...	...	...	...	...	...	...	-	10
Zaire Zaïre	...	...	...	...	588[5]	0.3[5]	...	-	63

108

Selected indicators of natural resources
Changes in land use and forests; water withdrawal and threatened species [*cont.*]

Indicateurs concernant certaines ressources naturelles
Evolution de l'utilisation des sols et de la superficie des forêts; prélèvements d'eau et espèces menacées [*suite*]

Country or area Pays ou zone	Changes in land distribution 1966/68 à 1986/88 (≥ 3 percentage points) Evolution de la répartition des sols 1966/68 à 1986/88 (≥ 3 points de pourcentage) Cropland Terres agricoles	Pasture Pacages	Forest Forêts	Other Autres	Forests Forêts Δ p.a., 1980s Deforestation Déboisement 10³ha	%	Reforest-ation Reboise-ment 10³ha	Water withdrawal Prélève-ments d'eau %, 1980s	Number of threatened species Nombre d'espèces menacées 1990
Zambia Zambie	...	...	...	...	70	0.2	2	-	23
Zimbabwe Zimbabwe	...	...	-4.1	...	80	0.4	4	5	112
America, North Amerique du Nord	...	...	...	...	1 182	0.1	2 541	10	6 226
Antigua and Barbuda Antigua-et-Barbuda	...	...	-3.8	...	...	...	...	...	3
Bahamas Bahamas	...	...	...	...	...	...	...	...	33
Barbados Barbade	...	...	...	...	...	...	...	51	3
Belize Belize	...	...	...	...	...	...	...	-	51
Bermuda Bermudes	...	...	...	...	...	...	...	...	13
British Virgin Islands Iles Vierges brittanique	6.7	...	...	-6.7	...	...	...	...	5
Canada Canada	...	...	3.8	-4.7	...	...	720	2	23
Cayman Islands Iles Caïmanes	...	...	...	...	...	...	...	...	4
Costa Rica Costa Rica	...	20.0	-19.7	...	55[6]	3.1[6]	...	1	445
Cuba Cuba	8.8	4.7	4.3	-17.8	2	0.1	11	23	890
Dominica Dominique	...	...	-5.3	5.3	...	...	...	...	65
Dominican Republic Rép. dominicaine	6.7	...	...	-5.9	4	0.6	1	15	60
El Salvador El Salvador	5.3	...	-3.7	...	5	3.2	...	5	35[7]
Greenland Groenland	...	...	...	...	...	...	...	...	3
Grenada Grenade	-8.8	...	...	11.8	...	...	...	...	6
Guadeloupe Guadeloupe	-8.7	4.9	3.7	...	...	...	...	...	15
Guatemala Guatemala	...	...	-12.0	7.2	90	2.0	8	1	306
Haiti Haïti	3.9	-4.8	...	...	2	3.8	...	-	22
Honduras Honduras	...	3.4	-14.5	8.7	90	2.3	...	1	64
Jamaica Jamaïque	...	-4.3	...	4.1	2	3.0	1	4	20
Martinique Martinique	-5.0	6.0	...	...	...	...	...	...	15
Mexico Mexique	...	...	-5.9	5.1	615	1.3	22	15	963
Montserrat Montserrat	...	...	30.0	-30.0	...	...	...	...	2
Neth. Antilles Antilles néerladaise	...	...	...	...	...	...	...	...	5
Nicaragua Nicaragua	...	8.8	-18.9	9.5	121	2.7	1	1	85

108
Selected indicators of natural resources
Changes in land use and forests; water withdrawal and threatened species [*cont.*]
Indicateurs concernant certaines ressources naturelles
Evolution de l'utilisation des sols et de la superficie des forêts; prélèvements d'eau et espèces menacées [*suite*]

Country or area Pays ou zone	Changes in land distribution 1966/68 à 1986/88 (≥ 3 percentage points) Evolution de la répartition des sols 1966/68 à 1986/88 (≥ 3 points de pourcentage)				Forests Forêts Δ p.a., 1980s			Water withdrawal Prélève- ments d'eau %, 1980s	Number of threatened species Nombre d'espèces menacées 1990
	Cropland Terres agricoles	Pasture Pacages	Forest Forêts	Other Autres	Deforestation Déboisement 10³ha	%	Reforest- ation Reboise- ment 10³ha		
Panama Panama	...	5.4	-15.2	8.6	36	0.9	...	1	578[78]
Puerto Rico Porto Rico	-11.4	...	5.1	5.4	...	...	...	...	96
Saint Kitts and Nevis Saint-Kitts-et Nevis	3.3	...	-6.6	3.3	...	...	...	...	1
Saint Lucia Sainte Lucie	...	-5.6	...	10.2	...	...	...	...	8
St. Pierre and Miquelon Saint-Pierre-et-Miquelon	...	...	...	...	...	...	...	...	...
St. Vincent & Grenadines St. Vincent-et-Grenadines	...	...	...	-5.1	...	...	...	...	3[9]
Trinidad and Tobago Trinité-et-Tobago	4.1	...	-3.9	...	1	0.4	1	3	9
Turks and Caicos Islands Iles Turques et Caïques	...	...	...	...	...	...	...	...	2
United States Etats-Unis					159[9]	0.1[9]	1 775	19	2 379
US Virgin Is. Iles Vierges américaines	...	...	...	...	...	...	...	...	14
America, South **Amerique du Sud**	...	...	-4.8	...	5 810	0.7	608	1	2 895
Argentina Argentine	...	...	...	...	...	...	40	3	240
Bolivia Bolivie	...	...	...	...	117	0.2	1	-	88
Brazil Bresil	...	3.7	-5.7	...	3 650[10]	0.7[10]	449	1	392
Chile Chili	...	3.5	...	-4.3	50	0.7	74	4	311
Colombia Colombie	...	4.9	-5.8	...	890	1.7	8	-	431
Ecuador Equateur	...	10.2	-19.0	8.1	340	2.3	4	2	349
French Guiana Guyane Française	...	...	-9.1	8.9	...	...	...	-	64
Guyana Guyana	...	...	-9.3	7.5	3	-	-	2	92
Paraguay Paraguay	3.3	16.1	-17.1	...	212	1.1	1	-	67[7]
Peru Pérou	...	...	-3.9	3.1	300[5]	0.4[5]	6	15	471
Suriname Suriname	...	...	...	...	3	-	-	-	86
Uruguay Uruguay	...	...	...	...	...	...	5	1	32
Venezuela Venezuela	...	...	-6.6	4.6	245	0.7	19	-	162
Asia **Asie**	...	4.2	...	...	4 460	0.9	5 582	15	8 178
Afghanistan Afghanistan	...	...	...	...	...	...	...	52	32
Bahrain Bahrëin	...	...	...	...	...	...	...	...	5
Bangladesh Bangladesh	...	...	...	...	8	0.9	17	1	89

108
Selected indicators of natural resources
Changes in land use and forests; water withdrawal and threatened species [*cont.*]
Indicateurs concernant certaines ressources naturelles
Evolution de l'utilisation des sols et de la superficie des forêts; prélèvements d'eau et espèces menacées [*suite*]

| Country or area Pays ou zone | Changes in land distribution 1966/68 à 1986/88 (≥ 3 percentage points) Evolution de la répartition des sols 1966/68 à 1986/88 (≥ 3 points de pourcentage) | | | | Forests Forêts Δ p.a., 1980s | | | Water withdrawal Prélève-ments d'eau | Number of threatened species Nombre d'espèces menacées |
	Cropland Terres agricoles	Pasture Pacages	Forest Forêts	Other Autres	Deforestation Déboisement 10³ha	%	Reforest-ation Reboise-ment 10³ha	%, 1980s	1990
Bhutan Bhoutan	...	...	3.5	-4.4	1	0.1	1	-	41
Brunei Darussalam Brunéi Darussalam	...	...	-42.7	44.8	...	...	...	-	62
Cambodia Cambodie	...	...	...	...	30	0.2	-	-	51
China Chine	...	13.9	...	-10.9	...	...	4 552	16	481
Cyprus Chypre	...	...	...	...	...	...	...	60	62
East Timor Timor oriental	...	...	6.7	-6.7	...	...	...	...	...
Gaza Strip (Palestine) Gaza Strip (Palestine)	7.9	...	...	-7.9	...	...	...	...	...
Hong Kong Hong-kong	-5.3	...	...	...	...	...	...	...	17
India Inde	...	...	...	...	1 500[11]	2.3[11]	138	18	1 467
Indonesia Indonésie	...	...	-5.3	3.5	920[12]	0.8[12]	131	1	267
Iran, Islamic Rep. of Iran, Rép. islamique d'	...	...	...	...	20	0.5	...	39	340
Iraq Iraq	...	...	...	...	...	...	...	43	27
Israel Israël	...	...	...	-3.0	...	...	2	88	28
Japan Japon	...	...	...	...	...	...	240	16	78
Jordan Jordanie	...	...	...	...	...	...	3	41	768
Korea, Dem. People's Rep. Corée, rép. pop. dém. de	...	...	...	...	...	...	200	21	30
Korea, Rep. of Corée, Rép. de	...	...	...	...	...	...	67	17	61
Kuwait Koweït	...	...	...	...	...	...	...	...	13
Lao People's Dem. Rep. Rép. dém. pop. Lao	...	...	-8.7	8.4	130	1.0	1	-	49
Lebanon Liban	...	...	...	3.4	...	...	...	16	25
Malaysia Malaisie	...	...	-12.5	11.0	310[5]	1.5[5]	20	...	592
Maldives Maldives	...	...	...	...	...	...	...	2	2
Mongolia Mongolie	...	-10.1	...	10.4	...	...	...	2	22
Myanmar Myanmar	...	...	...	...	677[13]	2.1[13]	...	-	75[47]
Nepal Népal	5.4	...	...	-7.6	84	4.0	4	2	84
Oman Oman	...	...	...	...	...	...	...	22	16
Pakistan Pakistan	...	...	...	-3.6	9	0.4	7	33	60
Philippines Philippines	3.2	...	-19.0	14.5	143[14]	1.5[14]	50	9	216

108

Selected indicators of natural resources
Changes in land use and forests; water withdrawal and threatened species [*cont.*]
Indicateurs concernant certaines ressources naturelles
Evolution de l'utilisation des sols et de la superficie des forêts; prélèvements d'eau et espèces menacées [*suite*]

Country or area Pays ou zone	Changes in land distribution 1966/68 à 1986/88 (≥ 3 percentage points) Evolution de la répartition des sols 1966/68 à 1986/88 (≥ 3 points de pourcentage) Cropland Terres agricoles	Pasture Pacages	Forest Forêts	Other Autres	Forests Forêts Δ p.a., 1980s Deforestation Déboisement 10³ha	%	Reforestation Reboisement 10³ha	Water withdrawal Prélèvements d'eau %, 1980s	Number of threatened species Nombre d'espèces menacées 1990
Qatar Qatar	...	...	...	...	...	...	...	174	3
Saudi Arabia Arabie saoudité	...	...	...	...	...	...	...	106	23
Singapore Singapour	-17.5	...	...	17.5	...	...	...	32	29
Sri Lanka Sri Lanka	...	...	3.4	-3.5	58	3.5	13	15	238
Syrian Arab Republic Rép. arabe syrienne	...	...	...	...	...	...	...	9	31
Thailand Thaïlande	16.8	...	-17.2	...	397[15]	2.5[15]	24	18	137
Turkey Turquie	...	-3.0	...	...	...	...	82	18	1 973
United Arab Emirates Emirats arabes unis	...	...	...	...	...	...	...	140	11
Viet Nam Viet Nam	...	...	-12.4	10.5	173[16]	1.7[16]	29	1	409
Yemen [18] Yémen [18]	...	...	...	...	...	...	...	147	149
Europe Europe	...	...	...	...	...	...	1 031	15	3 170
Albania Albanie	4.5	-9.0	-6.8	11.3	...	...	...	1	93
Andorra Andorre	...	...	...	...	...	...	...	1	1
Austria Autriche	...	...	...	4.7	...	...	21	2	40
Belgium[17] Belgique[17]	-3.5	-4.1	...	7.2	...	...	19	72	24
Bulgaria Bulgarie	-3.6	5.3	...	-3.5	...	...	50	7	107
Czechoslovakia Tchécoslovaquie	...	...	...	...	...	...	37	6	49
Denmark Danemark	-3.0	...	...	4.5	...	...	...	9	24
Faeroe Islands Iles Féroe	...	...	...	...	...	...	...	...	2
Finland Finlande	...	...	3.3	...	...	...	158	3	26
France France	...	-3.9	...	...	...	...	51	24	173
Germany ₰ · Allemagne ₰ Germany, Fed. Rep. of Allemagne, Rép. fédérale d'	...	-3.5	...	3.6	...	...	62	28	...
former German Dem Rep anc. Rép de allemande	...	...	...	...	...	...	...	27	19[4]
Greece Grèce	...	...	...	...	...	...	...	12	552
Hungary Hongrie	-3.4	...	...	...	...	...	19	5	39
Iceland Islande	...	...	...	...	...	...	...	-	5
Ireland Irlande	-6.3	5.8	...	...	...	...	9	2	14

108
Selected indicators of natural resources
Changes in land use and forests; water withdrawal and threatened species [*cont.*]
Indicateurs concernant certaines ressources naturelles
Evolution de l'utilisation des sols et de la superficie des forêts; prélèvements d'eau et espèces menacées [*suite*]

Country or area Pays ou zone	Changes in land distribution 1966/68 à 1986/88 (≥ 3 percentage points) Evolution de la répartition des sols 1966/68 à 1986/88 (≥ 3 points de pourcentage)				Forests Forêts Δ p.a., 1980s			Water withdrawal Prélève- ments d'eau	Number of threatened species Nombre d'espèces menacées
	Cropland Terres agricoles	Pasture Pacages	Forest Forêts	Other Autres	Deforestation Déboisement		Reforest- ation Reboise- ment		
					10^3ha	%	10^3ha	%, 1980s	1990
Italy Italie	-10.3	...	...	9.4	...	...	15	30	241
Liechtenstein Liechtenstein	...	6.3	...	-6.3	...	...	...	...	3
Luxembourg Luxembourg	...	...	...	...	...	...	...	1	10
Malta Malte	-3.1	...	...	3.1	...	...	...	92	17
Netherlands Pays-Bas	...	-6.5	...	5.8	...	...	2	16	22
Norway Norvège	...	...	...	...	...	...	79	-	24
Poland Pologne	...	...	...	...	...	...	106	30	36
Portugal Portugal	...	...	...	...	...	...	4	16	265
Romania Roumanie	...	...	...	...	...	...	...	12	88
Spain Espagne	...	...	3.3	...	...	...	92	41	973
Sweden Suède	...	...	...	...	...	...	207	2	25
Switzerland Suisse	...	-4.3	...	...	...	...	7	2	36
United Kingdom Royaume-Uni	...	-3.2	...	4.0	...	...	40	12	49
Yugoslavia Yougoslavie	...	...	...	...	...	...	53	3	213
Oceania **Océanie**	...	...	...	...	26	-	115	1	2 929
American Samoa Samoa américaines	...	...	...	...	...	...	...	...	2
Australia Australie	...	...	...	...	...	...	62	5	2 113
Cook Islands Iles Cook	...	...	...	...	...	...	...	...	1
Fiji Fidji	...	...	...	...	2	0.2	7	...	36
French Polynesia Polynésie française	...	...	...	...	...	...	...	...	85
Guam Guam	...	...	...	...	...	...	...	...	18
Kiribati Kiribati	...	...	...	...	...	...	...	...	2
Marshall Islands Iles Marshall	...	...	...	...	...	...	...	...	1
Micronesia Micronésie	...	...	...	...	...	...	...	...	9
Nauru Nauru	...	...	...	...	...	...	...	...	2
New Caledonia Nouvelle-Calédonie	...	...	...	...	...	...	...	...	174
New Zealand Nouvelle-Zélande	...	...	...	...	...	...	43	-	263

108
Selected indicators of natural resources
Changes in land use and forests; water withdrawal and threatened species [*cont.*]
Indicateurs concernant certaines ressources naturelles
Evolution de l'utilisation des sols et de la superficie des forêts; prélèvements d'eau et espèces menacées [*suite*]

Country or area Pays ou zone	Changes in land distribution 1966/68 à 1986/88 (≥ 3 percentage points) Evolution de la répartition des sols 1966/68 à 1986/88 (≥ 3 points de pourcentage)				Forests Forêts Δ p.a., 1980s		Reforest-ation Reboise-ment 10³ha	Water withdrawal Prélève-ments d'eau %, 1980s	Number of threatened species Nombre d'espèces menacées 1990
	Cropland Terres agricoles	Pasture Pacages	Forest Forêts	Other Autres	Deforestation Déboisement 10³ha	%			
Norfolk Islands Iles Norfolk	...	...	-3.8	4.9	...	...	...	...	...
Northern Marianas Is Iles Marianes du nord	...	...	...	...	...	...	...	...	11
Papua New Guinea Papouasie-Nouv.-Guinée	...	55.0	-4.1	5.5	23	0.1	2	...	119
Samoa Samoa	...	...	...	...	...	...	...	...	15
Solomon Islands Iles Solomon	...	...	...	...	1	-	-	...	53
Tonga Tonga	...	...	...	...	...	...	...	...	3
Tuvalu Tuvalu	...	...	...	...	...	...	...	...	1
Vanuatu Vanuatu	...	...	...	...	...	...	...	...	13
former USSR **ancienne URSS**	...	...	...	...	...	...	4 540	8	592

Source:
United Nations Environment Programme/GEMS Monitoring and Assessment Research Centre, *Environment Data Report 1991/92* (Nairobi and London).

Source:
Programme des Nations Unies pour l'environnement/«GEMS Monitoring and Assessment Research Centre, *Environment Data Report 1991/1992*» (Nairobi et Londres).

§   All data shown which pertain to Germany prior to 3 October 1990 are indicated separately for the federal Republic of Germany and the former German Democratic Republic based on their respective territories at the time indicated. For detailed explanatory notes on data pertaining to Germany, see Annex I - Country Nomenclature.

§   Toutes les données se rapportant à l'Allemagne avant le 3 octobre 1990 figurent dans deux rubriques séparées basées sur les territoires respectifs de la République fédérale d'Allemagne et l'ancienne République démocratique allemande selon la période indiquée. Pour les notes explicatives en détail sur les données concernant Allemagne, voir Annexe I - Nomenclature des pays.

1   Estimates of a single year in 1980s.
2   Mean annual rate of deforestation during 1976-1986.
3   Excludes threatened plants in Principe.
4   Excludes numbers of threatened plant taxa.
5   Deforestation in 1989.
6   Deforestation in 1986.
7   Includes marine species.
8   Includes non-breeding species of birds.
9   Mean annual rate of deforestation during 1977-1987.
10   Mean annual rate of deforestation during 1988-1989.
11   Mean annual rate of deforestation during 1975-1982.
12   Mean annual rate of deforestation during 1979-1984.
13   Mean annual rate of deforestation during 1975-1981.
14   Mean annual rate of deforestation during 1981-1988.
15   Mean annual rate of deforestation during 1978-1985.
16   Mean annual rate of deforestation during 1976-1981.
17   Belgium includes Luxembourg.
18   Excluding former Democratic Yemen.

1   Estimations pour une seule année de la décennie 1980-1990.
2   Taux de déboisement annuel moyen sur la période 1976-1986.
3   Non compris les espèces de plantes menacées à Principe.
4   Non compris les groupes taxonomiques végétaux menacés.
5   Déboisement en 1989.
6   Déboisement en 1986.
7   Y compris les espèces marines.
8   Y compris les espèces d'oiseaux qui ne se reproduisent pas.
9   Taux de déboisement annuel moyen sur la période 1977-1987.
10   Taux de déboisement annuel moyen sur la période 1988-1989.
11   Taux de déboisement annuel moyen sur la période 1975-1982.
12   Taux de déboisement annuel moyen sur la période 1979-1984.
13   Taux de déboisement annuel moyen sur la période 1975-1981.
14   Taux de déboisement annuel moyen sur la période 1981-1988.
15   Taux de déboisement annuel moyen sur la période 1978-1985.
16   Taux de déboisement annuel moyen sur la période 1976-1981.
17   Y compris le Luxembourg.
18   Non compris l'ancien Yémen démocratique.

109
Indicators of environmental pollution and management
Indicateurs de la pollution et de la gestion de l'environnement

A. CO$_2$, CFCs and halons; drinking water and sanitation; paper consumption and recovery; and protected area
Oxyde de carbone (CO$_2$), chlorofluorocarbones (CFC) et halons; eau potable et assainissement; consommation et récupération de papier; aires protégées

Country or area Pays ou zone	CO$_2$ emissions Emissions de CO$_2$ From fossil fuel and cement Dues à la combustion de combustibles fossiles et à la production de ciment (tC/cap.)[1] 1990	Net from deforestation Dues à l'évolution de l'utilisation des sols (10^{3}tC)[1] 1989	Consumption of CFCs and halons Consommation de CFC et halons CFCs and halons CFC et halons (t) 1986	% of world consumption Pourcentage de la consommation mondiale 1986	Access to safe drinking water and sanitation Accès à l'eau salubre et à des services d'assainissement Safe drinking water Eau salubre (% pop.) 1988	Sanitation services Services d'assainissement (% pop.) 1988	Paper consumption and recovery Consommation et récupération de papier Apparent consumption Consommation apparente (10^{3}t) 1988	Recovery as % of consumption Récupération en pourcentage de la consommation 1984	1988	Protected area Aires protégées % of total land area Pourcentage de la superficie totale 1992
World *World*	...	**6 400 000**	**1 549 933**	**100.00**	...	...	...	...	...	**5.2**
Africa Afrique	...	1 500 000	26 880	1.73	60	44	...	...	...	4.5
Algeria Algérie	0.74	...	2 200[2]	0.14	...	...	...	21	...	5.3
Angola Angola	0.14	33 000	...	...	47	22	...	...	...	2.1
Benin Bénin	0.04	9 500	...	...	56	36	...	...	...	7.5
Botswana Botswana	0.36	2 600	...	...	...	59	...	...	...	17.4
Burkina Faso Burkina Faso	0.02	17 000	...	...	58	20	...	...	...	9.6
Burundi Burundi	0.01	530	...	...	67	42	...	...	...	3.1
Cameroon Cameroun	0.13	60 000	...	...	98	...	...	...	...	4.3
Cape Verde Cap-Vert	0.06	...	...	...	76	...	...	...	...	0.0
Central African Rep. République centrafricaine	0.02	13 000	...	...	12	...	...	...	...	9.4
Chad Tchad	0.01	15 000	...	...	...	...	...	...	...	0.3
Comoros Comores	0.03	...	...	...	...	...	...	...	...	0.0
Congo Congo	0.24	12 000	...	...	47	...	...	...	...	3.9
Côte d'Ivoire Côte d'Ivoire	0.19	350 000	980[2]	0.06	87	44	...	...	...	6.2
Djibouti Djibouti	0.25	...	...	...	...	...	...	...	...	0.4
Egypt[3] Egypte[3]	0.42	...	2 803	0.18	89	67	530	7	29	0.8
Equatorial Guinea Guinée équatoriale	0.09	1 800	...	...	...	...	...	...	...	0.0
Ethiopia Ethiopie	0.02	30 000	...	...	40	52	...	...	...	2.5
Gabon Gabon	...	9 300	115[2]	...	70	...	...	...	...	3.9
Gambia Gambie	0.06	1 900	...	...	82	...	...	...	...	1.7
Ghana[3] Ghana[3]	0.07	31 000	...	...	66	39	...	...	...	4.5
Guinea Guinée	0.05	37 000	...	...	39	...	...	...	...	0.7

109

Indicators of environmental pollution and management [*cont.*]
Indicateurs de la pollution et de la gestion de l'environnement [*suite*]

A. CO$_2$, CFCs and halons; drinking water and sanitation; paper consumption and recovery; and protected area
Oxyde de carbone (CO$_2$), chlorofluorocarbones (CFC) et halons; eau potable et assainissement; consommation et récupération de papier; aires protégées

Country or area Pays ou zone	CO$_2$ emissions Emissions de CO$_2$ From fossil fuel and cement Dues à la combustion de combustibles fossiles et à la production de ciment (tC/cap.)[1] 1990	Net from deforesta-tion Dues à l'évolution de l'utilisation des sols (10³tC)[1] 1989	Consumption of CFCs and halons Consommation de CFC et halons CFCs and halons CFC et halons (t) 1986	% of world consump-tion Pourcen-tage de la conso-mmation mondiale 1986	Access to safe drinking water and sanitation Accès à l'eau salubre et à des services d'assainissement Safe drinking water Eau salubre (% pop.) 1988	Sanitation services Services d'assaini-ssement (% pop.) 1988	Paper consumption and recovery Consommation et récupération de papier Apparent consump-tion Conso-mmation apparente (10³t) 1988	Recovery as % of consumption Récupération en pourcentage de la consommation 1984	1988	Protected area Aires protégées % of total land area Pourcen-tage de la superficie totale 1992
Guinea-Bissau Guinée-Bissau	0.06	18 000	...	...	22	24	...	...	...	0.0
Kenya [3] Kenya [3]	0.07	13 000	136	0.01	...	...	...	54	...	6.0
Lesotho Lesotho	...	...	...	...	52	18	...	...	...	0.2
Liberia Libéria	0.05	39 000	...	...	57	...	...	...	...	1.2
Libyan Arab Jamahiriya Jamahiriya arabe libyenne	2.57	...	...	...	90	92	...	...	...	0.1
Madagascar Madagascar	0.02	120 000	49	0.00	36	...	8	...	29	1.9
Malawi Malawi	0.02	58 000	...	...	...	...	...	...	...	11.2
Mali Mali	0.01	7 700	...	...	68	49	...	...	...	3.2
Mauritania Mauritanie	0.35	...	...	...	66	...	...	...	...	1.7
Mauritius Maurice	0.29	...	...	...	96	94	...	...	...	2.2
Morocco Maroc	0.25	...	2 200[2]	0.14	62	59	215	28	25	0.8
Mozambique Mozambique	0.02	30 000	...	...	30	36	...	...	...	0.0
Namibia Namibie	...	...	...	...	...	...	...	...	...	12.6
Niger Niger	0.04	7 400	...	...	76	21	...	...	...	8.2
Nigeria Nigéria	0.21	270 000	...	...	60	...	...	...	...	3.1
Réunion Réunion	0.42	...	...	...	...	...	...	...	...	2.4
Rwanda Rwanda	0.02	2 100	...	...	55	53	...	...	...	12.4
St. Helena Sainte-Hélène	0.12	...	...	...	...	...	...	...	...	7.3
Sao Tome and Principe Sao Tomé-et-Principe	0.15	...	...	...	32	10	...	...	...	0.0
Senegal Sénégal	0.10	11 000	600[2]	0.04	...	...	...	...	...	11.1
Seychelles Seychelles	0.66	...	...	...	99	58	...	...	...	95.5
Sierra Leone Sierra Leone	0.04	4 600	...	...	...	...	...	...	...	1.1
Somalia Somalie	0.03	5 200	...	...	39	23	...	...	...	0.3
South Africa [3] Afrique du Sud [3]	2.15	...	16 080	1.04	...	...	...	27	...	6.2
Sudan Soudan	0.04	98 000	...	...	...	...	...	...	...	3.7

109
Indicators of environmental pollution and management [*cont.*]
Indicateurs de la pollution et de la gestion de l'environnement [*suite*]

A. CO_2, CFCs and halons; drinking water and sanitation; paper consumption and recovery; and protected area
Oxyde de carbone (CO_2), chlorofluorocarbones (CFC) et halons; eau potable et assainissement; consommation et récupération de papier; aires protégées

Country or area Pays ou zone	CO₂ emissions Emissions de CO₂		Consumption of CFCs and halons Consommation de CFC et halons		Access to safe drinking water and sanitation Accès à l'eau salubre et à des services d'assainissement		Paper consumption and recovery Consommation et récupération de papier			Protected area Aires protégées
	From fossil fuel and cement Dues à la combustion de combustibles fossiles et à la production de ciment (tC/cap.)[1] 1990	Net from deforestation Dues à l'évolution de l'utilisation des sols (10³tC)[1] 1989	CFCs and halons CFC et halons (t) 1986	% of world consumption Pourcentage de la consommation mondiale 1986	Safe drinking water Eau salubre (% pop.) 1988	Sanitation services Services d'assainissement (% pop.) 1988	Apparent consumption Consommation apparente (10³t) 1988	Recovery as % of consumption Récupération en pourcentage de la consommation 1984	1988	% of total land area Pourcentage de la superficie totale 1992
Swaziland Swaziland	1.80	...	...	...	...	...	...	...	...	2.6
Togo Togo	0.05	2 900	300	0.02	80	29	...	...	...	11.4
Tunisia[3] Tunisie[3]	0.46	...	587	0.04	65	43	122	4	8	0.3
Uganda Ouganda	0.01	10 000	...	...	...	...	...	...	...	7.9
United Rep. of Tanzania Rép.-Unie de Tanzanie	0.02	21 000	...	...	60	76	...	...	...	13.8
Western Sahara Sahara Occidental	0.30	...	...	...	...	...	...	...	...	0.0
Zaire Zaire	0.03	130 000	...	...	38	14	...	...	...	3.7
Zambia Zambia	0.08	27 000	...	...	59	55	...	...	...	8.5
Zimbabwe Zimbabwe	0.71	16 000	830[2]	0.05	55	58	86	31	30	7.9
America, North Amerique du Nord	...	420 000	401 345	25.89	67	71	...	...	...	10.8
Anguilla Anguilla	...	...	...	...	...	...	...	...	...	0.0
Antigua and Barbuda Antigua-et-Barbuda	1.08	...	...	...	...	...	...	...	...	9.3
Bahamas Bahamas	1.44	...	...	...	...	...	...	...	...	8.9
Barbados Barbade	1.08	...	...	...	...	...	...	...	...	0.0
Belize Belize	0.37	...	16[2]	0.00	69	61	...	...	...	5.1
Bermuda Bermudes	2.77	...	...	...	...	...	...	...	...	...[4]
British Virgin Islands Iles Vierges britanniques	1.03	...	...	...	...	...	...	...	...	4.4
Canada[3] Canada[3]	4.35	...	23 176	1.50	...	...	5 950	20	21	5.0
Cayman Is Iles Caïmanes	2.71	...	...	...	...	...	...	...	...	19.5[5]
Costa Rica Costa Rica	0.30	26 000	...	...	92	96	...	...	...	12.2
Cuba Cuba	0.90	890	620	0.04	...	...	...	...	...	6.0
Dominica Dominique	0.19	...	2	0.00	...	...	...	...	...	9.2
Dominican Republic Rép. dominicaine	0.24	1 300	620[2]	0.04	57	56	...	...	...	19.9
El Salvador El Salvador	0.13	1 600	480[2]	0.03	43	62	...	...	...	1.2
Greenland Groenland	2.69	...	...	...	...	...	...	...	...	45.2

109
Indicators of environmental pollution and management [cont.]
Indicateurs de la pollution et de la gestion de l'environnement [suite]

A. CO_2, CFCs and halons; drinking water and sanitation; paper consumption and recovery; and protected area
Oxyde de carbone (CO_2), chlorofluorocarbones (CFC) et halons; eau potable et assainissement; consommation et récupération de papier; aires protégées

Country or area Pays ou zone	CO₂ emissions Emissions de CO₂		Consumption of CFCs and halons Consommation de CFC et halons		Access to safe drinking water and sanitation Accès à l'eau salubre et à des services d'assainissement		Paper consumption and recovery Consommation et récupération de papier			Protected area Aires protégées
	From fossil fuel and cement Dues à la combustion de combustibles fossiles et à la production de ciment (tC/cap.)[1] 1990	Net from deforestation Dues à l'évolution de l'utilisation des sols (10³tC)[1] 1989	CFCs and halons CFC et halons (t) 1986	% of world consumption Pourcentage de la consommation mondiale 1986	Safe drinking water Eau salubre (% pop.) 1988	Sanitation services Services d'assainissement (% pop.) 1988	Apparent consumption Consommation apparente (10³t) 1988	Recovery as % of consumption Récupération en pourcentage de la consommation 1984	1988	% of total land area Pourcentage de la superficie totale 1992
Grenada Grenade	0.38	...	...	...	...	...	...	...	...	0.0
Guadeloupe Guadeloupe	...	...	...	...	...	...	...	...	...	11.8
Guatemala Guatemala	0.12	41 000	1 880	0.12	66	60	...	...	...	7.7
Haiti Haïti	0.03	860	...	...	45	...	...	...	...	0.4
Honduras Honduras	0.10	42 000	160[2]	0.01	74	66	...	...	...	6.4
Jamaica Jamaïque	0.52	810	1 098	0.07	70	...	...	...	...	3.3
Martinique Martinique	1.10	...	...	...	...	...	...	...	...	65.0[5]
Mexico[3] Mexique[3]	1.01	200 000	8 922	0.58	64	56	...	41	...	5.1
Montserrat Montserrat	0.77	...	...	...	...	...	...	...	...	0.0
Netherlands Antilles Antilles néerlandaises	2.29	...	...	...	...	...	...	...	...	9.7
Nicaragua Nicaragua	0.15	59 000	300[2]	0.02	48	...	...	...	...	2.5
Panama[3] Panama[3]	0.30	19 000	304	0.02	83	84	...	43	...	16.9[5]
Puerto Rico Porto Rico	1.57	...	...	...	...	...	...	...	...	3.2
Saint Kitts and Nevis Saint-Kitts-et-Nevis	0.40	...	...	...	...	...	...	...	...	0.0
Saint Lucia Sainte-Lucie	0.30	...	...	...	...	...	...	...	...	2.4
St. Vincent and Grenadines Saint-Vincent-et-Grenadines	0.19	...	...	...	...	...	...	...	...	21.3[5]
Trinidad and Tobago Trinité-et-Tobago	3.19	330	...	...	93	98	...	...	...	3.0
Turks and Caicos Islands Iles Turques et Caïques	0.00	...	...	...	...	...	...	...	...	...[4]
United States Etats-Unis	5.26	22 000	363 767	23.47	...	...	77 057	27	30	10.5
US Virgin Islands Iles Vierges américaines	12.41	...	...	...	...	...	...	...	...	40.1
America, South Amerique du Sud	...	1 800 000	27 476	1.77	63	59	...	...	...	6.0
Argentina Argentine	0.93	...	6 091	0.39	45	64	971	36	43	3.4
Bolivia Bolivie	0.18	37 000	650[2]	0.04	46	34	26	...	17	9.0
Brazil[3] Bresil[3]	0.36	950 000	10 974	0.71	93	65	3 710	35	39	2.5
Chile Chili	0.71	...	791	0.05	60	53	370	...	54	18.3

109

Indicators of environmental pollution and management [*cont.*]
Indicateurs de la pollution et de la gestion de l'environnement [*suite*]

A. CO₂, CFCs and halons; drinking water and sanitation; paper consumption and recovery; and protected area
 Oxyde de carbone (CO₂), chlorofluorocarbones (CFC) et halons; eau potable et assainissement; consommation et récupération de papier; aires protégées

Country or area Pays ou zone	CO₂ emissions Emissions de CO₂ — From fossil fuel and cement Dues à la combustion de combustibles fossiles et à la production de ciment (tC/cap.)[1] 1990	Net from deforestation Dues à l'évolution de l'utilisation des sols (10³tC)[1] 1989	Consumption of CFCs and halons Consommation de CFC et halons — CFCs and halons CFC et halons (t) 1986	% of world consumption Pourcentage de la consommation mondiale 1986	Access to safe drinking water and sanitation Accès à l'eau salubre et à des services d'assainissement — Safe drinking water Eau salubre (% pop.) 1988	Sanitation services Services d'assainissement (% pop.) 1988	Paper consumption and recovery Consommation et récupération de papier — Apparent consumption Consommation apparente (10³t) 1988	Recovery as % of consumption Récupération en pourcentage de la consommation 1984	1988	Protected area Aires protégées % of total land area Pourcentage de la superficie totale 1992
Colombia Colombie	0.44	420 000	2 900[2]	0.19	87	51	584	...	37	8.0
Ecuador[3] Equateur[3]	0.44	160 000	385	0.02	56	54	...	...	...	6.0
Falkland Is. (Malvinas) Iles Falkland (Malvinas)	5.13	...	...	...	...	...	...	...	...	0.0
French Guiana Guyane Française	1.50	...	...	...	...	...	...	...	...	0.0
Guyana Guyana	0.22	1 100	...	...	84	85	...	...	...	0.1
Paraguay Paraguay	0.11	67 000	320[2]	0.02	36	57	...	...	...	3.0
Peru Pérou	0.27	140 000	832	0.05	50	44	...	...	...	2.1
Suriname Suriname	·1.24	1 100	...	...	69	50	...	...	...	4.5
Uruguay Uruguay	0.35	...	330	0.02	45	62	...	34	...	0.2
Venezuela Venezuela	1.40	59 000	4 203	0.27	89	83	...	26	...	31.0
Asia **Asie**	...	2 600 000	221 584	14.30	70	68	...	...	...	5.2
Afghanistan Afghanistan	0.10	...	...	...	28	...	...	...	...	0.3
Bahrain[3] Bahreïn[3]	6.93	...	108	0.01	78	97	...	...	...	0.0
Bangladesh Bangladesh	0.04	8 700	...	...	63	20	...	...	...	0.7
Bhutan Bhoutan	0.02	860	...	...	97	99	...	...	...	19.4
Brunei Darussalam Brunéi Darussalam	5.34	...	...	...	...	...	...	...	...	13.5
Cambodia Cambodge	0.01	11 000	...	...	...	...	...	...	...	0.0
China Chine	0.61	...	18 000[2]	1.16	76	97	13 200	...	22	3.0
Cyprus Cypre	1.70	...	500[2]	0.03	100	100	...	...	...	0.2
Hong Kong Hong-kong	1.26	...	...	...	85	75	...	67	...	35.6
India Inde	0.22	120 000	5 300	0.34	76	21	...	30	...	4.4
Indonesia Indonésie	0.21	870 000	2 494	0.16	50	42	...	13	...	10.0
Iran, Islamic Republic of[3] Iran, Rép. islamique d'[3]	0.90	...	4 400[2]	0.28	87	67	...	16	...	4.6
Iraq Iraq	0.75	...	1 590[2]	0.10	86	55	...	14	...	0.0
Israel Israël	2.08	...	5 000	0.32	...	...	470	19	21	10.0

109
Indicators of environmental pollution and management [*cont.*]
Indicateurs de la pollution et de la gestion de l'environnement [*suite*]

A. CO$_2$, CFCs and halons; drinking water and sanitation; paper consumption and recovery; and protected area
Oxyde de carbone (CO$_2$), chlorofluorocarbones (CFC) et halons; eau potable et assainissement; consommation et récupération de papier; aires protégées

| | CO$_2$ emissions Emissions de CO$_2$ | | Consumption of CFCs and halons Consommation de CFC et halons | | Access to safe drinking water and sanitation Accès à l'eau salubre et à des services d'assainissement | | Paper consumption and recovery Consommation et récupération de papier | | | Protected area Aires protégées |
Country or area Pays ou zone	From fossil fuel and cement Dues à la combustion de combustibles fossiles et à la production de ciment (tC/cap.)[1] 1990	Net from deforestation Dues à l' évolution de l'utilisation des sols (10³tC)[1] 1989	CFCs and halons CFC et halons (t) 1986	% of world consumption Pourcentage de la consommation mondiale 1986	Safe drinking water Eau salubre (% pop.) 1988	Sanitation services Services d'assainissement (% pop.) 1988	Apparent consumption Consommation apparente (10³t) 1988	Recovery as % of consumption Récupération en pourcentage de la consommation 1984	1988	% of total land area Pourcentage de la superficie totale 1992
Japan[3] Japon[3]	2.34	...	135 089	8.72	...	...	24 940	49	48	12.7
Jordan[3] Jordanie[3]	0.69	...	350	0.02	99	100	59	11	11	1.0
Korea, Dem. People's Rep. Corée, rép. populaire dém.	1.96		...	...	...	...	...	...	...	0.5
Korea, Republic of Corée, Rép. de	1.54	...	9 394	0.61	70	99	3 300	33	39	7.7
Kuwait Koweït	3.45		1 065	0.07	...	...	...	...	...	1.2
Lao People's Dem. Rep. Rép. dém. populaire Lao	0.01	240 000	...	...	39		...	...	...	0.0
Lebanon Liban	0.93									0.3
Malaysia[3] Malaisie[3]	0.90	280 000	3 840	0.25	80	...	...	3	...	4.5
Mongolia Mongolie	1.26	...	...	...	64	71	...	...	...	4.0
Myanmar Myanmar	0.03	380 000	...	...	33	31	20	...	14	0.3
Nepal Népal	0.01	32 000	...	...	49	...	...	...	...	8.0
Oman Oman	2.24	...	...	...	64	67	...	...	...	0.2
Pakistan Pakistan	0.14	4 000	10 000[2]	0.65	67	24	...	23	...	4.6
Philippines Philippines	0.19	190 000	4 359	0.28	87	91	...	10	...	1.9
Qatar Qatar	10.47	...	...	...	74	92	...	...	...	0.0
Saudi Arabia Arabie saoudite	3.64		5 181	0.33	87	65	...	...	...	8.8
Singapore[3] Singapour[3]	3.77	...	6 491	0.42	...	...	...	13	...	4.4
Sri Lanka[3] Sri Lanka[3]	0.06	22 000	258	0.02	63	59	75	9	22	11.9
Syria Arab Republic[3] République arabe syrinne[3]	0.66	...	1 409	0.09	...	...	...	...	...	0.0
Thailand[3] Thaïlande[3]	0.46	290 000	2 360	0.15	71	...	811	24	35	10.8
Turkey Turquie	0.69	...	2 766	0.18	...	...	979	26	24	0.3
United Arab Emirates[3] Emirats arabe unis[3]	9.05	...	1 630	0.11	...	...	...	...	...	0.0
Viet Nam Viet Nam	0.10	150 000	...	...	46	51	...	...	...	2.7
Yemen[6] Yémen[6]	0.11	...	...	...	74		...	...	...	0.0
Europe **Europe**	...	...	524 568	33.84	...	...	...	...	...	8.0

109
Indicators of environmental pollution and management [*cont.*]
Indicateurs de la pollution et de la gestion de l'environnement [*suite*]

A. CO_2, CFCs and halons; drinking water and sanitation; paper consumption and recovery; and protected area
Oxyde de carbone (CO_2), chlorofluorocarbones (CFC) et halons; eau potable et assainissement; consommation et récupération de papier; aires protégées

Country or area Pays ou zone	CO₂ emissions Emissions de CO₂ — From fossil fuel and cement Dues à la combustion de combustibles fossiles et à la production de ciment (tC/cap.)[1] 1990	Net from deforestation Dues à l'évolution de l'utilisation des sols (10³tC)[1] 1989	Consumption of CFCs and halons Consommation de CFC et halons — CFCs and halons CFC et halons (t) 1986	% of world consumption Pourcentage de la consommation mondiale 1986	Access to safe drinking water and sanitation Accès à l'eau salubre et à des services d'assainissement — Safe drinking water Eau salubre (% pop.) 1988	Sanitation services Services d'assainissement (% pop.) 1988	Paper consumption and recovery Consommation et récupération de papier — Apparent consumption Consommation apparente (10³t) 1988	Recovery as % of consumption Récupération en pourcentage de la consommation 1984	1988	Protected area Aires protégées % of total land area Pourcentage de la superficie totale 1992
EEC +[3] EEC +[3]	...	...	343 470	22.16	...	...	...	...	...	0.0
Albania Albanie	0.82	...	...	...	...	...	...	...	...	1.6
Austria[3] Autriche[3]	1.95	...	9 410	0.61	...	...	1 083	39	49	25.0
Belgium Belgique	2.87	...	...	...	...	...	1 943	32	36	2.4
Bulgaria Bulgarie	2.76	...	2 000[2]	0.13	...	...	...	...	...	2.4
Czechoslovakia[3] Tchécoslovaquie[3]	3.62	...	6 788	0.44	...	...	...	...	49	16.1
Denmark[3] Danemark[3]	2.71	...	6 953	0.45	...	...	1 085	29	31	9.5
Faeroe Islands Iles Féroe	3.56	...	...	...	...	...	...	...	...	0.0
Finland[3] Finlande[3]	2.82	...	3 899	0.25	...	...	1 115	45	34	2.4
France[3] France[3]	1.74	...	105 483	6.81	...	...	7 934	28	33	9.9
Germany ƒ · Allemagne ƒ Germany[3] Allemagne[3]	...	...	150 794	9.73	...	...	12 346	...	...	16.4
Germany, Fed. Rep. of Allemagne, Rép. fédérale d'	2.94	...	...	...	...	...	...	38	41	0.0
former German Dem. Rep. ancienne Rép. de allemande	5.05	...	...	...	...	...	...	...	...	0.0
Greece Grèce	1.88	...	...	...	...	...	...	16	...	0.6
Hungary[3] Hongrie[3]	1.49	...	7 351	0.47	...	...	680	...	29	6.2
Iceland[3] Islande[3]	2.43	...	276	0.02	...	...	5	...	...	8.9
Ireland Irlande	2.27	...	...	...	...	...	...	...	...	0.4
Italy Italie	1.82	...	...	...	...	...	...	26	27	6.7
Liechtenstein Liechtenstein	...	...	...	...	...	...	...	...	...	38.0
Luxembourg[3] Luxembourg[3]	7.10	...	174	0.01	...	...	...	...	...	0.0
Malta[3] Malte[3]	1.29	...	499	0.03	...	...	...	...	...	0.0
Monaco Monaco	...	...	...	...	...	...	...	...	...	0.0
Netherlands[3] Pays-Bas[3]	2.54	...	42 331	2.73	...	...	2 800	45	54	8.6
Norway[3] Norvège[3]	2.48	...	2 724	0.18	...	...	...	21	...	12.9

109
Indicators of environmental pollution and management [*cont.*]
Indicateurs de la pollution et de la gestion de l'environnement [*suite*]

A. CO_2, CFCs and halons; drinking water and sanitation; paper consumption and recovery; and protected area
Oxyde de carbone (CO_2), chlorofluorocarbones (CFC) et halons; eau potable et assainissement; consommation et récupération de papier; aires protégées

Country or area Pays ou zone	CO₂ emissions Emissions de CO₂ From fossil fuel and cement Dues à la combustion de combustibles fossiles et à la production de ciment (tC/cap.)[1] 1990	Net from deforestation Dues à l'évolution de l'utilisation des sols (10³tC)[1] 1989	Consumption of CFCs and halons Consommation de CFC et halons CFCs and halons CFC et halons (t) 1986	% of world consumption Pourcentage de la consommation mondiale 1986	Access to safe drinking water and sanitation Accès à l'eau salubre et à des services d'assainissement Safe drinking water Eau salubre (% pop.) 1988	Sanitation services Services d'assainissement (% pop.) 1988	Paper consumption and recovery Consommation et récupération de papier Apparent consumption Conso-mmation apparente (10³t) 1988	Recovery as % of consumption Récupération en pourcentage de la consommation 1984	Recovery 1988	Protected area Aires protégées % of total land area Pourcen-tage de la superficie totale 1992
Poland[3] Pologne[3]	2.60	...	10 556	0.68	...	...	...	...	...	7.2
Portugal Portugal	1.09	...	...	...	...	...	662	45	41	6.1
Romania Roumanie	2.12	...	4 000[2]	0.26	...	...	...	...	...	4.6
Spain[3] Espagne[3]	1.41	...	20 432	1.32	...	...	3 897	41	37	6.9
Sweden[3] Suède[3]	1.60	...	6 793	0.44	...	...	2 050	38	48	6.5
Switzerland[3] Suisse[3]	1.72	...	9 010	0.58	...	...	1 005	44	61	18.2
United Kingdom[3] Royaume-Uni[3]	2.65	...	118 514	7.65	...	...	9 286	29	30	19.0
Yugoslavia Yougoslavie	1.50	...	16 580	1.07	...	...	...	...	...	3.1
Oceania **Oceanie**	...	12 000	21 208	1.37	83	73				9.9
American Samoa Samoa américaines	2.05	...	...	...	...	...	...	...	...	18.9[4]
Australia[3] Australie[3]	4.32	...	18 560	1.20	...	...	2 570	22	30	6.1
Cook Islands Iles Cook	0.33	...	...	...	96	99	...	...	...	0.7
Fiji[3] Fidji[3]	0.27	...	70[2]	0.00	...	77	...	...	...	0.3
French Polynesia Polynésie française	0.81	...	...	...	...	...	...	...	...	3.2
Guam Guam	3.46	...	...	...	100	95	...	...	...	0.0
Kiribati Kiribati	0.09	...	...	...	71	63	...	...	...	38.9[4]
Nauru Nauru	4.00	...	...	...	...	...	...	...	...	0.0
New Caledonia Nouvelle-Calédonie	2.56	...	...	...	...	...	...	...	...	3.2
New Zealand[3] Nouvelle-Zélande[3]	2.07	...	2 578	0.17	...	...	561	20	18	11.0
Niue Nioué	0.28	...	...	...	...	...	...	...	...	0.0
Northern Marianas Islands Iles Mariannes du nord	...	...	...	...	...	...	...	...	...	2.4
Pacific Islands (Palau)[7] Iles du Pacifique (Palaos)[7]	0.36	...	...	...	...	...	...	...	...	3.3
Papua New Guinea Papouasie-Nouvelle-Guinée	0.16	12 000	...	...	58	55	...	...	...	0.1
Samoa Samoa	...	...	...	...	76	93	...	...	...	0.0
Solomon Islands Iles Solomon	0.14	...	...	...	75	30	...	...	...	0.0

109

Indicators of environmental pollution and management [*cont.*]
Indicateurs de la pollution et de la gestion de l'environnement [*suite*]

A. CO_2, CFCs and halons; drinking water and sanitation; paper consumption and recovery; and protected area
Oxyde de carbone (CO_2), chlorofluorocarbones (CFC) et halons; eau potable et assainissement; consommation et récupération de papier; aires protégées

Country or area Pays ou zone	CO_2 emissions Emissions de CO_2		Consumption of CFCs and halons Consommation de CFC et halons		Access to safe drinking water and sanitation Accès à l'eau salubre et à des services d'assainissement		Paper consumption and recovery Consommation et récupération de papier			Protected area Aires protégées
	From fossil fuel and cement Dues à la combustion de combusti- bles fossiles et à la production de ciment (tC/cap.)[1] 1990	Net from deforesta- tion Dues à l' évolution de l'utilisation des sols (10^3tC)[1] 1989	CFCs and halons CFC et halons (t) 1986	% of world consump- tion Pourcen- tage de la conso- mmation mondiale 1986	Safe drinking water Eau salubre (% pop.) 1988	Sanitation services Services d'assaini- ssement (% pop.) 1988	Apparent consump- tion Conso- mmation apparente (10^3t) 1988	Recovery as % of consumption Récupération en pourcentage de la consommation 1984		% of total land area Pourcen- tage de la superficie totale 1992
									1988	
Tonga Tonga	0.21	...	...	...	94	82	...	...	...	0.0
Tuvalu Tuvalu	0.12	...	...	...	99	78	...	...	...	0.0
Vanuatu Vanuatu	...	...	...	...	82	57	...	...	...	0.0
former USSR [3] **ancienne URSS [3]**	3.66	...	139 406	8.99	...	...	...	...	...	1.1

Source:
United Nations Environment Programme/GEMS Monitoring and Assessment Research Centre, *Environment Data Report 1991/92* (Nairobi and London).

Source:
Programme des Nations Unies pour l'environnement/«GEMS Monitoring and Assessment Research Centre, *Environment Data Report 1991/1992*» (Nairobi et Londres).

+   For Member States of this grouping, see annex - I other groupings.

+   Les Etats membres de ce groupement, voir annexe I - Autres groupements.

ℊ   All data shown which pertain to Germany prior to 3 October 1990 are indicated separately for the federal Republic of Germany and the former German Democratic Republic based on their respective territories at the time indicated. Where data for united Germany (3 October 1990 and thereafter) are not available, available data are shown separately under the designaions Federal Republic of Germany and former German Democratic Republic and pertain to the territorial boundaries prior to 3 October 1990. For detailed explanatory notes on data pertaining to Germany, see Annex I - Country Nomenclature.

ℊ   Toutes les données se rapportant à l'Allemagne avant le 3 octobre 1990 figurent dans deux rubriques séparées basées sur les territoires respectifs de la République fédérale d'Allemagne et l'ancienne République démocratique allemande selon la période indiquée. En l'absence de données pour l'Allemagne unifiée (à compter du 3 octobre 1990), les données disponibles sont fournies séparément sous les rubriques République fédérale d'Allemagne et ancienne République démocratique allemande et se rapportent aux limites territoriales antérieures au 3 octobre 1990. Pour les notes explicatives en détail sur les données concernant allemagne, voir Annexe I - Nomenclature des pays.

1   Unit tC is ton(s) of carbon.
2   Party to the Montreal Protocol.
3   UNEP estimate.
4   Nationally protected areas are predominantly marine locations.
5   The % value is inflated as the protected areas include marine areas while the national land area used for the percentage calculation excludes marine Exclusion Zones.
6   Excluding former Democratic Yemen.
7   Including data for Federated States of Micronesia, Marshall Is. and Northern Mariana Is.

1   Le symbole tC signifie tonne de carbone.
2   Partie au Protocole de Montréal.
3   Estimation du PNUE.
4   Les aires protégées à l'échelon national sont principalement des sites marins.
5   La valeur en pourcentage indiquée est plus élevée qu'elle ne l'est en réalité car les aires protégées recouvrent les sites marins tandis que les aires terrestres nationales utilisées pour calculer les pourcentages ne couvrent pas les zones d'exclusion marines.
6   Non compris l'ancien Yémen démocratique.
7   Y compris les données pour les Etats fédérés de Micronésie, les îles Marshall et les îles Mariannes du Nord.

109
Indicators of environmental pollution and management
Indicateurs de la pollution et de la gestion de l'environnement

B. Sulphur dioxide at selected sites
Teneur d'anhydride sulfureux observée en divers lieux

Country or area, city and site § Pays ou zone, ville et site §		Average mean annual values, μg m^{-3} (number of observations), 1974-1989 [1] Moyennes annuelles, microgrammes par mètre cube d'air (nombre d'observations), 1974-1989 [1]							
		1974/75	1976/77	1978/79	1980/81	1982/83	1984/85	1986/87	1988/89
America, North · Amérique du Nord									
Canada Canada									
Hamilton	CCC	... ...	58.0 (726)	43.0 (725)	30.5 (727)	36.5 (716)	31.5 (702)	19.5 (690)	24.0 (360)[2]
	SR	... ...	50.5 (696)	31.5 (705)	35.5 (696)	28.0 (704)	42.0 (259)[2]	... ...	... ...
Montréal	CCC	... ...	56.5 (646)	46.5 (638)	43.5 (614)	24.0 (640)	25.0 (360)	... ...	... ...
	SR	... ...	23.0 (720)	24.5 (638)	27.5 (557)	23.0 (629)	23.0 (597)	14.0 (299)[2]	... ...
Toronto	SI	... ...	24.5 (652)	19.5 (689)	16.0 (645)	7.0 (662)	12.0 (351)[2]	... ...	... ...
	SR	36.0 (347)	33.5 (695)	19.5 (674)	20.0 (695)	4.5 (346)	3.5 (339)	8.0 (704)	8.0 (684)
Vancouver	CCR	17.0 (548)	19.0 (330)	20.5 (502)	17.0 (614)	18.0 (335)[2]	... ...		
United States Etats-Unis									
New York	CCI	... ...	... ...	65.5 (60)	50.5 (726)	47.0 (723)	45.0 (729)	40.0 (350)[2]	... ...
	CCR	... ...	... ...	90.5 (170)	70.5 (725)	64.5 (730)	61.0 (724)	52.0 (364)[2]	... ...
	SR	... ...	... ...	42.5 (156)	40.0 (728)	35.0 (716)	31.0 (725)	27.0 (361)[2]	... ...
America, South · Amérique du Sud									
Brazil Brésil									
Sao Paolo	CCM	... ...	121.0 (731)	137.5 (730)	127.0 (723)	99.0 (723)	60.0 (725)	73.0 (120)	50.0 (61)[2]
	CCR	... ...	116.5 (729)	142.0 (728)	120.0 (729)	97.0 (728)	62.0 (727)	54.5 (120)	55.0 (61)[2]
Chile Chili									
Santiago	CCC	... ...	67.0 (665)	74.0 (574)	67.5 (659)	89.5 (628)	79.5 (691)	... ...	... ...
	CCR	... ...	45.5 (619)	53.5 (591)	37.0 (587)	43.5 (679)	46.0 (360)[2]	... ...	... ...
Venezuela Venezuela									
Caracas	CCC	... ...	20.5 (528)	27.5 (525)	29.5 (397)	34.0 (500)	27.0 (471)	21.0 (458)	21.0 (270)[2]
Asia · Asie									
China Chine									
Beijing	CCC	... ...	... ...	... ...	66.0 (113)	101.5 (374)	130.0 (86)	112.5 (326)	103.0 (351)
	CCR	... ...	... ...	... ...	98.0 (110)	147.0 (368)	161.0 (78)	125.0 (293)	114.5 (330)
	SI	... ...	... ...	... ...	38.0	60.5 (365)	79.0 (76)	61.5 (297)	53.5 (336)
	SR	... ...	... ...	... ...	6.0 (93)	15.0 (349)	27.0 (80)	28.5 (324)	29.5 (326)
Guangzhou	CCC	... ...	... ...	... ...	117.0 (76)	93.5 (349)	67.5 (336)	59.5 (345)	58.5 (353)
	CCI	... ...	... ...	... ...	12.0 (55)	14.5 (333)	113.0 (317)	122.0 (333)	135.5 (354)
	CCR	... ...	... ...	... ...	66.0 (97)	56.0 (345)	123.5 (337)	104.5 (338)	99.0 (346)
	SR	... ...	... ...	... ...	140.0 (93)	101.5 (326)	40.0 (324)	11.0 (345)	20.5 (350)
Shanghai	CCC	... ...	... ...	... ...	65.0 (85)	57.5 (355)	51.0 (358)	77.5 (358)	69.0 (343)
	CCI	... ...	... ...	... ...	23.0 (88)	52.5 (341)	49.0 (86)	40.5 (357)	55.5 (338)
	CCR	... ...	... ...	... ...	52.0 (87)	69.0 (354)	84.5 (353)	98.0 (351)	104.0 (294)
Shenyang	CCC	... ...	... ...	... ...	72.0 (72)	130.5 (288)	58.0 (72)	134.0 (288)	101.5 (288)
	CCI	... ...	... ...	... ...	136.0 (72)	238.5 (288)	320.0 (288)	223.0 (288)	207.5 (288)
	CCR	... ...	... ...	... ...	29.0 (73)	130.5 (288)	135.0 (288)	115.5 (290)	70.0 (288)
	SR	... ...	... ...	... ...	27.0 (72)	46.0 (287)	17.0 (72)	49.0 (288)	29.0 (276)
Xian	CCC	... ...	... ...	... ...	160.0 (120)	104.5 (397)	120.5 (323)	92.5 (288)	95.5 (286)
	CCR	... ...	... ...	... ...	108.0 (119)	118.0 (396)	100.5 (324)	118.0 (289)	97.5 (288)
	SI	... ...	... ...	... ...	46.0 (117)	66.5 (394)	52.0 (321)	66.5 (287)	79.5 (288)
	SR	... ...	... ...	... ...	22.0 (98)	30.0 (383)	29.5 (313)	31.0 (276)	37.0 (252)
Hong Kong Hong-kong									
Hong Kong	CCC	14.0 (357)	7.0 (720)	27.5 (722)	40.0 (731)	46.0 (716)	32.0 (208)	... ...	... ...
	SI	100.0 (364)	33.0 (730)	60.5 (730)	85.5 (731)	31.5 (714)	15.0 (208)	... ...	... ...
Iran,Islamic Rep. of Iran, Rép. islamique d'									
Teheran	CCC	... ...	74.0 (124)	47.0 (147)	169.5 (172)	145.5 (122)	97.5 (115)	90.0 (49)	211.0 (42)
	SI	... ...	82.5 (67)	41.5 (130)	137.5 (154)	161.5 (135)	87.5 (94)	120.5 (31)	149.0 (28)
	SR	... ...	85.5 (65)	46.0 (133)	140.0 (160)	111.0 (125)	47.5 (98)	45.5 (33)	81.5 (30)
Israel Israël									
Tel Aviv	CCC	29.5 (664)	25.5 (419)	22.0 (324)	4.0 (433)	9.5 (164)	53.0 (179)	... ...	... ...
Japan Japon									
Osaka	CCC	73.0 (273)	72.0 (707)	49.0 (715)	34.5 (724)	31.0 (724)	27.0 (724)	27.5 (720)	27.0 (366)
	CCI	74.0 (365)	57.5 (725)	39.5 (718)	35.5 (719)	31.0 (725)	29.5 (731)	28.5 (730)	25.0 (366)
	CCI	62.0 (362)	55.0 (343)	46.0 (705)	41.0 (731)	33.0 (729)	32.5 (730)	34.0 (728)	34.0 (366)
	SR	63.0 (330)	57.5 (706)	45.5 (729)	32.0 (730)	27.5 (729)	25.5 (731)	24.5 (729)	24.0 (363)
Tokyo	CCC	66.5 (726)	67.5 (724)	55.5 (727)	42.0 (719)	25.5 (719)	24.0 (731)	20.5 (730)	19.0 (360)
	CCI	78.0 (722)	79.5 (731)	64.5 (717)	48.5 (724)	31.0 (724)	31.0 (731)	29.5 (730)	26.0 (366)
	SR	47.0 (707)	50.0 (730)	46.5 (730)	40.0 (722)	37.0 (726)	31.0 (727)	21.5 (729)	18.0 (366)
Thailand Thaïlande									
Bangkok	SR	... ...	... ...	... ...	14.0 (155)	18.0 (310)	14.0 (166)	14.5 (121)	13.0 (190)

109
Indicators of environmental pollution and management [*cont.*]
Indicateurs de la pollution et de la gestion de l'environnement [*suite*]

B. Sulphur dioxide at selected sites
Teneur d'anhydride sulfureux observée en divers lieux

Country or area, city and site § Pays ou zone, ville et site §		Average mean annual values, μg m^{-3} (number of observations), 1974-1989 [1] Moyennes annuelles, microgrammes par mètre cube d'air (nombre d'observations), 1974-1989 [1]							
		1974/75	1976/77	1978/79	1980/81	1982/83	1984/85	1986/87	1988/89
Europe · Europe									
Belgium Belgique									
Bruxelles	CCC	97.0 (679)	101.5 (583)	93.5 (626)	67.5 (706)	70.5 (655)	... ...	... ...	... ...
	CCC	... ...	104.0 (422)	105.5 (681)	72.5 (727)	45.5 (722)	46.0 (686)	38.0 (355)	... ...
	SI	101.0 (652)	86.5 (612)	72.0 (647)	51.5 (715)	40.0 (615)	37.5 (686)	43.0 (323)	... ...
	SR	86.0 (560)	77.0 (661)	72.0 (598)	59.5 (689)	43.0 (695)	36.5 (644)	32.0 (323)	... ...
Finland Finlande									
Helsinki	CCC	... ...	... ...	25.0 (709)	20.5 (690)	28.0 (56)	... ...	29.0 (354)	30.0 (683)
Germany Allemagne									
Frankfurt am Main	CCC	104.5 (641)	85.5 (613)	79.5 (630)	70.5 (625)	58.5 (582)	59.5 (642)	54.5 (644)	26.0 (421)
	SI	81.5 (592)	66.5 (326)	49.0 (424)	35.5 (378)	40.0 (245)	... ...	... ...	
Greece Grèce									
Athinai	CCC	... ...	... ...	48.0 (540)	44.5 (539)	58.0 (397)	33.0 (614)	35.0 (122)	... ...
	SI	... ...	... ...	51.5 (354)	49.5 (262)	30.5 (569)	26.0 (362)	27.0 (130)	... ...
Ireland Irlande									
Dublin	CCI	... ...	49.0 (364)	38.5 (727)	41.5 (727)	42.5 (712)	22.5 (543)	45.0 (656)	25.0 (362)
	CCR	... ...	49.0 (364)	49.5 (729)	65.5 (722)	47.0 (725)	40.0 (726)	49.0 (671)	33.0 (351)
	SR	... ...	26.0 (347)	26.0 (726)	37.0 (725)	39.0 (704)	28.0 (696)	36.5 (620)	13.0 (341)
Netherlands Pays-Bas									
Amsterdam	CCC	38.0 (684)	34.0 (659)	39.0 (646)	28.0 (636)	26.5 (595)	21.5 (644)	30.0 (87)	... ...
	SI	34.5 (697)	32.0 (683)	35.5 (645)	27.0 (600)	26.5 (613)	22.0 (596)	33.0 (85)	... ...
Poland Pologne									
Warszawa	CCC	... ...	... ...	41.5 (371)	46.0 (367)	34.5 (422)	42.5 (538)	14.0 (516)	23.5 (574)
	CCI	... ...	... ...	40.5 (358)	39.0 (367)	46.5 (388)	45.5 (544)	16.0 (445)	29.5 (527)
	CCR	... ...	... ...	31.0 (321)	31.5 (305)	29.0 (332)	37.5 (397)	11.5 (342)	19.5 (409)
Wroclaw	CCC	... ...	... ...	45.5 (588)	39.0 (577)	40.5 (583)	62.5 (565)	35.0 (488)	53.0 (64)
	CCI	... ...	... ...	32.0 (575)	27.5 (583)	34.5 (563)	48.5 (547)	26.0 (445)	34.5 (542)
	CCR	... ...	... ...	29.0 (555)	31.5 (558)	31.0 (527)	44.0 (522)	32.0 (446)	40.0 (506)
Spain Espagne									
Madrid	CCC	149.5 (588)	68.0 (139)	80.0 (369)	111.5 (578)	80.5 (566)	47.5 (499)	50.5 (667)	31.5 (663)
	CCI	155.5 (578)	91.5 (501)	81.5 (395)	113.5 (310)	91.0 (208)	... ...	... ...	... ...
	SR	61.5 (635)	48.5 (645)	45.0 (482)	46.0 (653)	33.5 (654)	27.5 (582)	25.5 (680)	17.5 (658)
United Kingdom Royaume-Uni									
Glasgow	CCC	... ...	94.0 (365)	84.0 (359)	75.5 (730)	57.0 (730)	49.5 (731)	... ...	... ...
	CCI	... ...	86.0 (357)	76.5 (729)	64.5 (731)	40.5 (730)	46.0 (687)	... ...	... ...
London	CCC	133.0 (606)	114.0 (727)	93.0 (730)	55.0 (730)	55.0 (724)	42.5 (673)	... ...	... ...
	SI	91.0 (339)	78.0 (279)	66.0 (661)	56.5 (678)	35.0 (719)	33.5 (727)	... ...	... ...
Yugoslavia Yougoslavie									
Zagreb	CCC	151.0 (708)	128.5 (721)	96.0 (668)	87.5 (671)	85.0 (716)	104.5 (727)	110.0 (703)	82.5 (694)
	CCI	69.0 (72)	64.5 (721)	47.0 (730)	41.5 (731)	53.0 (705)	75.0 (721)	73.0 (679)	102.0 (698)
	SR	53.5 (711)	48.5 (727)	37.5 (724)	41.5 (730)	42.5 (712)	64.5 (694)	63.0 (725)	44.0 (637)
Oceania · Oceanie									
Australia Australie									
Melbourne	CCC	16.0 (292)	16.0 (651)	7.5 (562)	7.5 (585)	5.0 (432)	10.0 (157)	... ...	... ...
Sydney	SI	31.0 (340)	28.5 (595)	24.0 (658)	35.0 (692)	21.5 (725)	13.0 (697)	... ...	... ...
New Zealand Nouvelle-Zélande									
Auckland	CCC	16.0 (280)	12.0 (700)	20.0 (633)	11.0 (710)	3.5 (702)	... ...	... ...	... ...
	SI	20.0 (78)	20.0 (498)	23.5 (659)	20.5 (698)	7.5 (716)	3.0 (583)	3.0 (674)	3.0 (682)
Christchurch	SC	... ...	18.0 (72)	28.5 (672)	28.0 (646)	12.5 (665)	10.0 (629)	16.0 (193)	... ...
	SI	... ...	... ...	27.5 (563)	35.5 (674)	45.0 (472)	43.0 (347)	43.0 (97)	... ...
	SR	... ...	30.0 (319)	18.5 (653)	21.0 (646)	18.5 (660)	14.5 (679)	20.5 (442)	... ...

Source:
United Nations Environment Programme/GEMS Monitoring and Assessment Research Centre, *Environment Data Report 1991/92* (Nairobi and London).

§ CCC = City centre commercial; CCI = City centre industrial; CCM = City centre mobile; CCR = City centre residential; SC = Suburban centre; SI = Suburban industrial; SR = Suburban residental.

1 Unit μg m^{-3} is micro gram per m^3.
2 Based on one year of measurements only.

Source:
Programme des Nations Unies pour l'environnement/«GEMS Monitoring and Assessment Research Centre, *Environment Data Report 1991/1992*» (Nairobi et Londres).

§ CCC = Centre commercial urbain; CCI = Centre industriel urbain; CCM = Centre mobile urbain; SC = Zone centrale suburbaine; SI = Zone industrielle suburbaine; SR = Zone résidentielle suburbaine.

1 Le symbole μg m^{-3} signifie microgrammes par mètre cube d'air.
2 Chiffres basés sur les mesures effectuées au cours d'une seule année.

109
Indicators of environmental pollution and management
Indicateurs de la pollution et de la gestion de l'environnement

C. Global water quality in selected rivers - pollution indicators, [1]
1984-1988 (mean values)
Qualité générale de l'eau de certains cours d'eau - indicateurs de pollution, [1]
1984-1988 (moyennes)

Country or area Pays ou zone	Site	BOD DBO (mg l⁻¹ O₂)	COD DCO (mg l⁻¹ O₂)	Total Cd Cd total (mg l⁻¹)	Total Hg Hg total (mg l⁻¹)	Total Pb Pb total (mg l⁻¹)	Faecal coliforms Colibacilles fécaux (no per 100 ml) (no. par 100 ml)
America, North · Amerique du Nord							
Canada	Churchill River	...	...	...	0.020	0.001	...
Canada	Fraser River	...	...	...	0.022	0.002	...
	Mackenzie River	...	...	...	...	0.002	...
	Saskatchewan River	...	...	...	0.020	0.001	...
Mexico	Rio Atoyac	45.04	112.54	...	...	...	567 448
Mexique	Rio Balsas	5.10	130.25	...	...	...	107 192
	Rio Blanco	11.58	68.15	...	...	...	95 471
	Rio Bravo	2.61	35.05	...	...	...	1 315
	Rio Coatzacoalcos	20.26	436.93	...	...	...	83 125
	Rio Colorado	2.10	14.01	...	...	...	224
	Rio Conchos	5.33	14.96	...	...	...	...
	Rio Grijalva	5.06	22.20	...	...	...	44 681
	Rio Lerma	41.18	356.99	...	...	...	222 458
	Rio Panuco	2.81	21.05	0.050	0.001	...	1 018
	Rio Usumacinta	1.80	20.62	...	...	...	10 091
America, South · Amérique du Sud							
Argentina	Rio De La Plata Buenos Aires	1.28	...	...	...	...	1 959
Argentine	Rio Parana Corrientes	0.50	...	...	...	...	157
Brazil	Rib. Serra Asul (Faz Sobradinho)	2.27	...	0.001	1.744	...	1 447
Brésil	Rio Capibaribe	21.73	...	0.030	0.305	...	9 015
	Rio Guandu(Tomada D'agua)	1.36	12.55	0.002	...	...	1 252
	Rio Jacuľja	2.33	...	0.002	0.029	...	1 535
	Rio Paraiba Do Sul (Aparecida)	2.23	15.86	0.005	0.100	0.100	24 239
	Rio Paraiba Do Sul (Barra Mansa)	1.65	13.14	0.004	...	...	13 024
	Rio Sao Francisco (Petrolandia)	2.21	...	0.010	0.346	...	890
	Rio Velhas (Honorio Bicalho)	2.69	...	0.001	1.148	...	2 113
Chile	Rio Maipo En El Manzano	1.05	4.72	...	...	...	722
Chili	Rio Mapocho En Los Almendros	0.89	4.19	...	...	...	12
Asia · Asie							
China	Changjiang (Yangtze River)	0.88	...	...	...	...	759
Chine	Huanghe (Yellow River)	1.91	...	...	...	...	4 793
	Zhujiang (Pearl River)	0.58	...	...	0.100	...	653
India	Bhima River (Takali)	6.52	27.30	...	...	...	...
Inde	Cauveri River D/s K.r.s. Res.	1.08	28.92	...	...	...	762
	Cauveri River (Musiri)	3.89	38.03	...	...	...	194
	Cauveri River (Satyagalam)	1.32	28.21	...	...	...	1 010
	Chaliyar River (Kalpalli)	2.28	23.40	...	...	...	505
	Chaliyar River (Koolimadu)	1.52	13.76	...	...	...	534
	Godavari River (Dhalegaon)	5.68	30.29	...	...	...	...
	Godavari River (Mancherial)	3.71	31.44	...	...	...	7
	Godavari River (Polavaram)	3.11	22.71	...	...	...	2
	Kallada River (Panamthottam Kad)	2.40	17.62	...	...	...	762
	Krishna River (Gadwal)	3.50	26.17	...	...	...	16
	Krishna River (Honnali Town)	1.14	22.97	...	...	...	1 307
	Krishna River (Karad)	5.57	26.66	...	...	...	...
	Krishna River (Vijayawada)	3.35	29.93	...	...	...	8
	Mahi River (Vasad,baroda)	1.90	8.56	...	...	...	522 527
	Mahi River (Sevalia)	2.21	8.49	...	...	...	776 645
	Narmada River (Garudeshwar)	2.01	12.35	...	...	...	630 430
	Narmadh River (Sethani Ghat)	2.95	20.25	...	...	...	...
	Periyar River (Alwaye)	2.12	20.69	...	...	...	480
	Periyar River (Kaladi)	1.43	12.58	...	...	...	436
	Sabarmati River (Ahmedabad)	62.47	180.94	...	...	...	1 724 783
	Sabarmati River (Dharoi Dam)	1.89	7.02	...	...	...	148
	Subarnarekha River D/s Ranchi	6.55	42.89	...	...	...	22 403
	Subernarekha River (Mango Bridge)	3.67	49.32	...	...	...	25 519
	Subernarekha River (Jamshedpur)	1.90	35.33	...	...	...	7 103

109

Indicators of environmental pollution and management [*cont.*]
Indicateurs de la pollution et de la gestion de l'environnement [*suite*]

C. Global water quality in selected rivers - pollution indicators,[1] 1984-88 (mean values)
Qualité générale de l'eau de certains cours d'eau - indicateurs de pollution,[1] 1984-1988 (moyennes)

Country or area / Pays ou zone	Site	BOD DBO (mg l^{-1} O_2)	COD DCO (mg l^{-1} O_2)	Total Cd Cd total (mg l^{-1})	Total Hg Hg total (mg l^{-1})	Total Pb Pb total (mg l^{-1})	Faecal coliforms Colibacilles fécaux (no per 100 ml) (no. par 100 ml)
	Tapti River (Kathore)	3.24	14.70	...	...	...	19 787
	Tapti River (Burhanpur)	2.06	22.85	...	...	...	173
	Tapti River (Nepanagar)	1.33	11.47	...	...	...	93
	Tungabhadra River (Ullanuru)	1.22	34.40	...	...	...	1 393
	Wainganga River (Ashti)	6.16	33.15	...	...	...	...
Indonesia Indonésie	Barito River (South Kalimantan)	1.34	9.10	...	...	...	5 130
	River Banjir Kanal	10.40	23.49	0.002	0.200	...	884 999
	River Citarum	6.88	28.03	0.004	...	...	637 567
	River Sunter	9.35	23.23	0.002	0.200	...	865 789
	River Surabaya	14.79	21.98	0.039	...	...	...
Japan Japon	Kiso River (Asahi)	0.82	...	0.002	0.781	...	...
	Kiso River (Inuyama)	1.00	...	0.001	0.461	...	457
	Kiso River (Shimo-ochiai)	0.69	...	0.001	0.450	...	2 729
	Ohta River (Hesaka)	0.68	...	0.001	0.500	...	832
	Sagami River (Samukawa)	1.95	...	0.001	0.100	0.005	1 458
	Shinano River (Zuiun Bridge)	1.76	7.63	0.004	0.500	0.009	2 512
	Tone River (Tone-ozeki)	1.40	...	0.005	0.750	0.010	3 994
	Toyohira River (Shiraikawa)	1.35	...	...	0.500	0.010	451
	Yodo River (Hirakata Bridge)	3.69	...	0.008	0.255	...	165 196
Korea, Rep. of Corée, Rép. de	Han River	1.45	8.85	0.002	0.500	...	147
Malaysia Malaisie	Kelantan River	1.55	17.39	...	...	...	...
	Klang River	6.79	50.76	0.001	...	...	810 000
	Muda River	0.92	16.43	...	...	...	...
Pakistan Pakistan	Indus River (Kotri)	4.76	15.20	...	...	...	4 537
	Lower Chenab River / Gujra Branch	2.57	19.38	...	...	...	110
	Ravi River (Upstream Lahore)	2.33	13.77	...	...	...	396
	Ravi River (Downstream Lahore)	5.88	24.73	...	...	...	407
Philippines Philippines	Cagayan River	0.73	16.58	...	...	...	1 291
Turkey Turquie	Cark Suyu (Beskopruler)	1.35	23.80	...	...	...	145
	Porsuk River (Agackoy)	1.53	16.80	...	...	...	502
	Sakarya River (Adatepe)	2.64	31.42	...	0.590	...	39 960
Europe · Europe							
Belgium Belgique	Scheldt River (Doel)	3.19	57.59	0.001	0.004	...	...
	Sambre River (Erquelinnes)	4.93	27.28	0.001	0.002	...	...
	Escaut River (Bleharies)	3.73	45.45	0.001	0.015	...	...
	Meuse River (Heer/Agimont)	4.13	16.74	...	0.007	...	...
	Meuse River (Lanaye/Ternaaien)	3.85	22.33	0.001	0.013	...	...
	Sure River (Martelange)	3.73	15.29	0.001	0.003	...	...
	Lys River (Warneton)	5.04	77.60	0.002	0.008	...	...
	Zelzate River (Ghent/Terneuzen)	5.36	59.98	0.001	0.018	...	...
Hungary Hongrie	Danube River (Budapest)	5.99	18.75	0.001	...	...	...
	Tisza River (Szolnok)	2.58	11.42	...	...	...	254
Netherlands Pays-Bas	Maas River	3.34	22.00	0.001	0.110	...	9 050
	Rhine River	3.52	21.00	...	0.072	...	52 877
Norway Norvège	River Glama (Askim)	...	...	...	...	...	64
	River Glama (Haslemoen)	...	...	...	...	...	110
Portugal Portugal	River Tejo (Santarem)	1.61	...	...	...	...	...
United Kingdom Royaume-Uni	River Avon	...	...	0.168	...	...	...
	River Carron	...	8.19	0.192	0.129	...	...
	River Dee	...	...	1.150	...	...	...
	River Exe	1.43	...	0.576	...	...	...
	River Leven	...	...	0.527	...	...	...
	River Mersey	...	45.00	0.187	...	...	...
	River Thames	...	21.84	0.691	0.161	...	...
	River Trent	...	...	0.387	0.426	...	...
	River Tweed (Galafoot)	...	...	...	...	...	...

109
Indicators of environmental pollution and management [*cont.*]
Indicateurs de la pollution et de la gestion de l'environnement [*suite*]

C. Global water quality in selected rivers - pollution indicators,[1] 1984-88 (mean values)
Qualité générale de l'eau de certains cours d'eau - indicateurs de pollution,[1] 1984-1988 (moyennes)

Country or area Pays ou zone	Site	BOD DBO (mg l^{-1} O$_2$)	COD DCO (mg l^{-1} O$_2$)	Total Cd Cd total (mg l^{-1})	Total Hg Hg total (mg l^{-1})	Total Pb Pb total (mg l^{-1})	Faecal coliforms Colibacilles fécaux (no per 100 ml) (no. par 100 ml)
	Oceania · Océanie						
Australia Australie	River Murray (Mannum)	...	...	0.001	...	...	...
Fiji Fidji	Waimanu River	0.87	...	...	...	...	5 109
New Zealand Nouvelle-Zélande	Waikato River (Mercer Bridge)	1.39	11.37	...	...	...	459

Source:
United Nations Environment Programme/GEMS Monitoring and Assessment Research Centre, *Environment Data Report 1991/92* (Nairobi and London).

Source:
Programme des Nations Unies pour l'environnement/«GEMS Monitoring and Assessment Research Centre, *Environment Data Report 1991/1992*» (Nairobi et Londres).

1 BOD (biological oxygen demand) is the dissolved oxygen, as milligram per litre (mgl^{-1}), required by organisms for the aerobic decomposition of organic matter present in water. COD (chemical oxygen demand) is the mass concentration of oxygen, as milligram per litre (mgl^{-1}), consumed by the chemical breakdown of organic and inorganic matter. Cd is Cadmium. Hg is hydrargyrum (mercury). Pb is plumbum (lead).

1 La demande biochimique en oxygène (DBO) est la quantité d'oxygène dissous, exprimée en milligrammes par litre (mgl^{-1}) utilisée au cours des processus biologiques de décomposition de la matière organique dans l'eau. La demande chimique en oxygène (DCO) est la teneur massique d'oxygène, exprimée en milligrammes par litre (mgl^{-1}) cédée par voie chimique dans le cadre du processus de décomposition chimique de la matière organique et inorganique. Cd est le symbole du cadmium. Hg est le symbole de l'hydragyre (mercure). Pb est le symbole du plumbum (plomb).

110
Surface and land area and land use
Superficie totale, superficie des terres et utilisation des terres
Thousand hectares
Milliers d'hectares

Country or area Pays ou zone	Year Année	Surface Superficie 000Hs	Land Terres 000Hs	%	Agricultural Agricoles 000Hs	%	Forest Boisés 000Hs	%	Other Autres 000Hs
Africa · Afrique									
Algeria	1985	238174	238174	100.0	39051	16.4	4384[1]	1.8	194739
Algérie	1990	238174	238174	100.0	38778[1]	16.3	4699[1]	2.0	194697
Angola	1985	124670	124670	100.0	32400[1]	26.0	53310[1]	42.8	38960
Angola	1990	124670	124670	100.0	32000	26.0	52000	41.7	40270
Benin	1985	11262	11062	98.2	2280[1]	20.2	3720[1]	33.0	5062
Bénin	1990	11262	11062	98.2	2302[1]	20.4	3470[1]	30.8	5290
Botswana	1985	58173	56673	97.4	34360[1]	59.1	10960[1]	18.8	11353
Botswana	1990	58173	56673	97.4	34380[1]	59.1	10910[1]	18.8	11383
British Indian Ocean Ter.	1985	8	8	100.0	...	...	...	...	8
Ter. brit. de l'océan indien	1990	8	8	100.0	...	...	...	...	8
Burkina Faso	1985	27420	27380	99.9	13035[1]	47.5	6900[1]	25.2	7445
Burkina Faso	1990	27420	27380	99.9	13563[1]	49.5	6600[1]	24.1	7217[1]
Burundi	1985	2783	2565	92.2	2235[1]	80.3	64[1]	2.3	266
Burundi	1990	2783	2565	92.2	2252[1]	80.9	66[1]	2.4	247
Cameroon	1985	47544	46540	97.9	15265[1]	32.1	25090[1]	52.8	6185
Cameroun	1990	47544	46540	97.9	15308[1]	32.2	24540[1]	51.6	6692
Cape Verde	1985	403	403	100.0	65[1]	16.1	1[1]	0.2	337
Cap-Vert	1990	403	403	100.0	64[1]	15.9	1[1]	0.2	338
Central African Rep.	1985	62298	62298	100.0	4983[1]	8.0	35850[1]	57.5	21465
Rép. centrafricaine	1990	62298	62298	100.0	5006[1]	8.0	35800[1]	57.5	21492
Chad	1985	128400	125920	98.1	48155[1]	37.5	13130[1]	10.2	64635
Tchad	1990	128400	125920	98.1	48205[1]	37.5	12730[1]	9.9	64985
Comoros	1985	223	223	100.0	112[1]	50.2	35[1]	15.7	76
Comores	1990	223	223	100.0	115[1]	51.6	35[1]	15.7	73
Congo	1985	34200	34150	99.9	10162[1]	29.7	21260[1]	62.2	2728
Congo	1990	34200	34150	99.9	10168[1]	29.7	21160[1]	61.9	2822
Côte d'Ivoire	1985	32246	31800	98.6	16580[1]	51.4	8630[1]	26.8	6590
Côte d'Ivoire	1990	32246	31800	98.6	16690[1]	51.8	7380[1]	22.9	7730
Djibouti	1985	2320	2318	99.9	200[1]	8.6	6[1]	0.3	2112
Djibouti	1990	2320	2318	99.9	200[1]	8.6	6[1]	0.3	2112
Egypt	1985	100145	99545	99.4	2497[1 2]	2.5	31[1]	...	97017
Egypte	1990	100145	99545	99.4	2607[1 2]	2.6	31[1]	...	96907
Equatorial Guinea	1985	2805	2805	100.0	334[1]	11.9	1295[1]	46.2	1176
Guinée équatoriale	1990	2805	2805	100.0	334[1]	11.9	1295[1]	46.2	1176
Ethiopia	1985	122190	110100	90.1	59080[1]	48.4	27600[1]	22.6	23420
Ethiopie	1990	122190	110100	90.1	58830[1]	48.1	27100[1]	22.2	24170
Gabon	1985	26767	25767	96.3	5152[1]	19.2	20000[1]	74.7	615
Gabon	1990	26767	25767	96.3	5157[1]	19.3	20000[1]	74.7	610
Gambia	1985	1130	1000	88.5	255[1]	22.6	186[1]	16.5	559
Gambie	1990	1130	1000	88.5	268[1]	23.7	156[1]	13.8	576
Ghana	1985	23854	23002	96.4	7690[1]	32.2	8420[1]	35.3	6892
Ghana	1990	23854	23002	96.4	7720[1]	32.4	8070[1]	33.8	7212
Guinea	1985	24586	24586	100.0	6875[1]	28.0	14880[1]	60.5	2831
Guinée	1990	24586	24586	100.0	6878[1]	28.0	14580[1]	59.3	3128
Guinea-Bissau	1985	3612	2812	77.9	1400[1]	38.8	1070[1]	29.6	342
Guinée-Bissau	1990	3612	2812	77.9	1415[1]	39.2	1070[1]	29.6	327
Kenya	1985	58037	56969	98.2	40470[1]	69.7	2440[1]	4.2	14059
Kenya	1990	58037	56969	98.2	40530[1]	69.8	2340[1]	4.0	14099
Lesotho	1985	3035	3035	100.0	2300[1]	75.8	...	...	735
Lesotho	1990	3035	3035	100.0	2320[1]	76.4	...	...	715
Liberia	1985	9775	9675	99.0	6071[1]	62.1	1840[1]	18.8	1764
Libéria	1990	9775	9675	99.0	6073[1]	62.1	1740[1]	17.8	1862
Libyan Arab Jamahiriya	1985	175954	175954	100.0	15427	8.8	650	0.4	159877
Jamah. arabe libyenne	1990	175954	175954	100.0	15455[1]	8.8	690[1]	0.4	159809
Madagascar	1985	58704	58154	99.1	37040[1]	63.1	16280[1]	27.7	4834
Madagascar	1990	58704	58154	99.1	37102[1]	63.2	15530[1]	26.5	5522
Malawi	1985	11848	9408	79.4	4216[1]	35.6	4180[1]	35.3	1012
Malawi	1990	11848	9408	79.4	4259[1]	35.9	3630[1]	30.6	1519
Mali	1985	124019	122019	98.4	32073[1]	25.9	7100[1]	5.7	82846
Mali	1990	124019	122019	98.4	32093[1]	25.9	6950[1]	5.6	82976
Mauritania	1985	102552	102522	100.0	39451[1]	38.5	4480[1]	4.4	58591
Mauritanie	1990	102552	102522	100.0	39455[1]	38.5	4430[1]	4.3	58637

110
Surface and land area and land use
Thousand hectares [*cont.*]
Superficie totale, superficie des terres et utilisation des terres
Milliers d'hectares [*suite*]

Country or area Pays ou zone	Year Année	Surface Superficie 000Hs	Land Terres 000Hs	% 	Agricultural Agricoles 000Hs	% 	Forest Boisés 000Hs	% 	Other Autres 000Hs
Mauritius	1985	186[3]	185	99.5	114	61.3	58	31.2	13
Maurice	1990	186[3]	185	99.5	113[1]	60.8	57[1]	30.6	15
Morocco	1985	44655	44630	99.9	29304	65.6	7865[1]	17.6	7461
Maroc	1990	44655	44630	99.9	30227[1]	67.7	9000[1]	20.2	5403
Mozambique	1985	80159	78409	97.8	47090[1]	58.7	14860[1]	18.5	16459
Mozambique	1990	80159	78409	97.8	47130[1]	58.8	14260[1]	17.8	17019
Namibia	1985	82429	82329	99.9	38662[1]	46.9	18270[1]	22.2	25397
Namibie	1990	82429	82329	99.9	38662[1]	46.9	18120[1]	22.0	25547
Niger	1985	126700	126670	100.0	12570	9.9	2300[1]	1.8	111800
Niger	1990	126700	126670	100.0	12510[1]	9.9	2000[1]	1.6	112160
Nigeria	1985	92377	91077	98.6	71130[1]	77.0	13400[1]	14.5	6547
Nigéria	1990	92377	91077	98.6	72300[1]	78.3	11900[1]	12.9	6877
Réunion	1985	251	250	99.6	65	25.9	88	35.1	97
Réunion	1990	251	250	99.6	64	25.5	88	35.1	98
Rwanda	1985	2634	2467	93.7	1626[1]	61.7	569[1]	21.6	272
Rwanda	1990	2634	2467	93.7	1617[1]	61.4	554[1]	21.0	296
St. Helena	1985	31	31	100.0	4[1]	12.9	1[1]	3.2	26
Sainte Hélène	1990	31	31	100.0	4[1]	12.9	2[1]	6.5	25
Sao Tome and Principe	1985	96	96	100.0	38[1]	39.6	...	...	58
Sao Tomé-et-Principe	1990	96	96	100.0	38[1]	39.6	...	...	58
Senegal	1985	19672	19253	97.9	5450[1]	27.7	10800[1]	54.9	3003
Sénégal	1990	19672	19253	97.9	5450[1]	27.7	10550[1]	53.6	3253
Seychelles	1985	28	27	96.4	6[1]	21.4	5[1]	17.9	16
Seychelles	1990	28	27	96.4	6[1]	21.4	5[1]	17.9	16
Sierra Leone	1985	7174	7162	99.8	2819[1]	39.3	2090[1]	29.1	2253
Sierra Leone	1990	7174	7162	99.8	2854[1]	39.8	2060[1]	28.7	2248
Somalia	1985	63766	62734	98.4	44025[1]	69.0	9110[1]	14.3	9599
Somalie	1990	63766	62734	98.4	44039[1]	69.1	9060[1]	14.2	9635
South Africa	1985	122104	122104	100.0	94547	77.4	4515[1]	3.7	23042
Africque du Sud	1990	122104	122104	100.0	94552[1]	77.4	4515[1]	3.7	23037
Sudan	1985	250581	237600	94.8	112658[1]	45.0	46340[1]	18.5	78602
Soudan	1990	250581	237600	94.8	122900	49.0	44840[1]	17.9	69860
Swaziland	1985	1736	1720	99.1	1284[1]	74.0	104[1]	6.0	332
Swaziland	1990	1736	1720	99.1	1389[1]	80.0	104[1]	6.0	227
Togo	1985	5679	5439	95.8	2434[1]	42.9	1650[1]	29.1	1355
Togo	1990	5679	5439	95.8	2459[1]	43.3	1600[1]	28.2	1380
Tunisia	1985	16361	15536	95.0	8030[1]	49.1	590[1]	3.6	6916
Tunisie	1990	16361	15536	95.0	7936[1]	48.5	655	4.0	6945
Uganda	1985	23588	19955	84.6	8400[1]	35.6	5810[1]	24.6	5745
Ouganda	1990	23588	19955	84.6	8510[1]	36.1	5560[1]	23.6	5885
United Rep. Tanzania	1985	94509	88604	93.8	38345[1]	40.6	41540[1]	44.0	8719
Rép.-Unie de Tanzanie	1990	94509	88604	93.8	38367[1]	40.6	40940[1]	43.3	9297
Zaire	1985	234541	226760	96.7	22800[1]	9.7	175960[1]	75.0	28000
Zaïre	1990	234541	226760	96.7	22860[1]	9.7	174310[1]	74.3	29590
Zambia	1985	75261	74339	98.8	35188[1]	46.8	29200[1]	38.8	9951
Zambie	1990	75261	74339	98.8	35268[1]	46.9	28850[1]	38.3	10221
Zimbabwe	1985	39058	38667	99.0	7590[1]	19.4	19530[1]	50.0	11547
Zimbabwe	1990	39058	38667	99.0	7668[1]	19.6	19130[1]	49.0	11869
America, North · Amérique du Nord									
Antigua and Barbuda	1985	44	44	100.0	12[1]	27.3	5[1]	11.4	27
Antigua-et-Barbuda	1990	44	44	100.0	12[1]	27.3	5[1]	11.4	27
Bahamas	1985	1388	1001	72.1	12[1]	0.9	324[1]	23.3	665
Bahamas	1990	1388	1001	72.1	12[1]	0.9	324[1]	23.3	665
Barbados	1985	43	43	100.0	37[1]	86.0	...	...	6
Barbade	1990	43	43	100.0	37[1]	86.0	...	...	6
Belize	1985	2296	2280	99.3	101	4.4	1012[1]	44.1	1167
Belize	1990	2296	2280	99.3	104[1]	4.5	1012[1]	44.1	1164
Bermuda	1985	5	5	100.0	...	...	1	20.0	4
Bermudes	1990	5	5	100.0	...	...	1	20.0	4
British Virgin Islands	1985	15	15	100.0	9[1]	60.0	1[1]	6.7	5
Iles Vierges britanniques	1990	15	15	100.0	9[1]	60.0	1[1]	6.7	5
Canada	1985	997614	922097	92.4	74630[1]	7.5	350000[1]	35.1	497467
Canada	1990	997614	922097	92.4	74050[1]	7.4	359000[1]	36.0	489047
Cayman Islands	1985	26	26	100.0	2[1]	7.7	6[1]	23.1	18
Iles Caïmanes	1990	26	26	100.0	2[1]	7.7	6[1]	23.1	18

110
Surface and land area and land use
Thousand hectares [*cont.*]

Superficie totale, superficie des terres et utilisation des terres
Milliers d'hectares [*suite*]

Country or area Pays ou zone	Year Année	Surface Superficie 000Hs	Land Terres 000Hs	%	Agricultural Agricoles 000Hs	%	Forest Boisés 000Hs	%	Other Autres 000Hs
Costa Rica	1985	5110	5106	99.9	2803[1]	54.9	1640[1]	32.1	663
Costa Rica	1990	5110	5106	99.9	2859[1]	55.9	1640[1]	32.1	607
Cuba	1985	11086	10982	99.1	5932[14]	53.5	2731	24.6	2319
Cuba	1990	11086	10982	99.1	6300[14]	56.8	2760[1]	24.9	1922
Dominica	1985	75	75	100.0	19[1]	25.3	31[1]	41.3	25
Dominique	1990	75	75	100.0	19[1]	25.3	31[1]	41.3	25
Dominican Republic	1985	4873	4838	99.3	3522[1]	72.3	625[1]	12.8	691
Rép. dominicaine	1990	4873	4838	99.3	3538[1]	72.6	615[1]	12.6	685
El Salvador	1985	2104	2072	98.5	1342[1]	63.8	110[1]	5.2	620
El Salvador	1990	2104	2072	98.5	1343	63.8	104	4.9	625
Greenland	1985	34170[5]	34170	100.0	235[1]	0.7	10[1]	...	33925
Groënland	1990	34170[5]	34170	100.0	235[1]	0.7	10[1]	...	33925
Grenada	1985	34	34	100.0	15	44.1	3	8.8	16
Grenade	1990	34	34	100.0	14[1]	41.2	3[1]	8.8	17
Guadeloupe	1985	171	169	98.8	58	33.9	71	41.5	40
Guadeloupe	1990	171	169	98.8	54	31.6	66	38.6	49
Guatemala	1985	10889	10843	99.6	3185[1]	29.2	4150[1]	38.1	3508
Guatemala	1990	10889	10843	99.6	3285[1]	30.2	3750[1]	34.4	3808
Haiti	1985	2775	2756	99.3	1407[1]	50.7	48[1]	1.7	1301
Haïti	1990	2775	2756	99.3	1402[1]	50.5	38[1]	1.4	1316
Honduras	1985	11209	11189	99.8	4288[1]	38.3	3660[1]	32.7	3241
Honduras	1990	11209	11189	99.8	4380[1]	39.1	3260[1]	29.1	3549
Jamaica	1985	1099	1083	98.5	464[1]	42.2	190[1]	17.3	429
Jamaïque	1990	1099	1083	98.5	459[1]	41.8	185[1]	16.8	439
Martinique	1985	110	106	96.4	42	38.2	38	34.5	26
Martinique	1990	110	106	96.4	39	35.5	47	42.7	20
Mexico	1985	195820	190869	97.5	99199[1]	50.7	45160[1]	23.1	46510
Mexique	1990	195820	190869	97.5	99209[1]	50.7	42460[1]	21.7	49200
Montserrat	1985	10	10	100.0	3[1]	30.0	4[1]	40.0	3
Montserrat	1990	10	10	100.0	3[1]	30.0	4[1]	40.0	3
Netherlands Antilles	1985	80	80	100.0	8[1]	10.0	...	...	72
Antilles néerlandaise	1990	80	80	100.0	8[1]	10.0	...	...	72
Nicaragua	1985	13000	11875	91.3	6418[1]	49.4	3930[1]	30.2	1527
Nicaragua	1990	13000	11875	91.3	6673[1]	51.3	3380[1]	26.0	1822
Panama	1985	7708	7599	98.6	2075[1]	26.9	3680[1]	47.7	1844
Panama	1990	7708	7599	98.6	2214[1]	28.7	3300[1]	42.8	2085
Puerto Rico	1985	890	886	99.6	461[1]	51.8	179	20.1	246
Porto Rico	1990	890	886	99.6	461[1]	51.8	177[1]	19.9	248
Saint Kitts and Nevis	1985	36	36	100.0	15[1]	41.7	6[1]	16.7	15
Saint-Kitts-et-Nevis	1990	36	36	100.0	15[1]	41.7	6[1]	16.7	15
Saint Lucia	1985	62	61	98.4	20[1]	32.3	8[1]	12.9	33
Sainte Lucie	1990	62	61	98.4	21[1]	33.9	8[1]	12.9	32
Saint Pierre and Miquelon	1985	24	23	95.8	3[1]	12.5	1[1]	4.2	19
Saint-Pierre-et-Miquelon	1990	24	23	95.8	3[1]	12.5	1[1]	4.2	19
St. Vincent-Grenadines	1985	39	39	100.0	12[1]	30.8	14[1]	35.9	13
St. Vincent-Grenadines	1990	39	39	100.0	13[1]	33.3	14[1]	35.9	12
Trinidad and Tobago	1985	513	513	100.0	129[1]	25.1	225[1]	43.9	159
Trinité-et-Tobago	1990	513	513	100.0	131[1]	25.5	220[1]	42.9	162
Turks and Caicos Islands	1985	43	43	100.0	1[1]	2.3	...	...	42
Iles Turques et Caïques	1990	43	43	100.0	1[1]	2.3	...	...	42
United States	1985	937261	916660	97.8	431382	46.0	295200[1]	31.5	190078
Etats-Unis	1990	937261	916660	97.8	431382[1]	46.0	293600[1]	31.3	191678
US Virgin Islands	1985	34	34	100.0	16[1]	47.1	2[1]	5.9	16
Iles Vierges amér.	1990	34	34	100.0	16[1]	47.1	2[1]	5.9	16
America, South · Amérique du Sud									
Argentina	1985	276689	273669	98.9	169900[1]	61.4	59700[1]	21.6	44069
Argentine	1990	276689	273669	98.9	* 169400[1]	61.2	59200[1]	21.4	45069
Bolivia	1985	109858	108438	98.7	29097[1]	26.5	55890[1]	50.9	23451
Bolivie	1990	109858	108438	98.7	28908[1]	26.3	55590[1]	50.6	23940
Brazil	1985	851197	845651	99.3	230081[1]	27.0	505680[1]	59.4	109890
Brésil	1990	851197	845651	99.3	244200[1]	28.7	493030[1]	57.9	108421
Chile	1985	75695	74880	98.9	17594[1]	23.2	8800[1]	11.6	48486
Chile	1990	75695	74880	98.9	18026[1]	23.8	8800[1]	11.6	48054
Colombia	1985	113891	103870	91.2	44810[1]	39.3	51800[1]	45.5	7260
Colombie	1990	113891	103870	91.2	45820[1]	40.2	50300[1]	44.2	7750

110
Surface and land area and land use
Thousand hectares [*cont.*]
Superficie totale, superficie des terres et utilisation des terres
Milliers d'hectares [*suite*]

Country or area Pays ou zone	Year Année	Surface Superficie 000Hs	Land Terres 000Hs	%	Agricultural Agricoles 000Hs	%	Forest Boisés 000Hs	%	Other Autres 000Hs
Ecuador	1985	28356	27684	97.6	7240[1]	25.5	12400[1]	43.7	8044
Equateur	1990	28356	27684	97.6	7875[1]	27.8	10900[1]	38.4	8909
Falkland Islands (Malvinas)	1985	1217	1217	100.0	1200	98.6	...	...	17
Iles Falkland (Malvinas)	1990	1217	1217	100.0	1190	97.8	...	...	27
French Guiana	1985	9000	8815	97.9	12	0.1	7300	81.1	1503
Guyane française	1990	9000	8815	97.9	21	0.2	7300	81.1	1494
Guyana	1985	21497	19685	91.6	1725[1]	8.0	16369[1]	76.1	1591
Guyana	1990	21497	19685	91.6	1725[1]	8.0	16369[1]	76.1	1591
Paraguay	1985	40675	39730	97.7	20171[1]	49.6	17839	43.9	1720
Paraguay	1990	40675	39730	97.7	23316[1]	57.3	13800[1]	33.9	2614
Peru	1985	128522	128000	99.6	30816[1]	24.0	69650[1]	54.2	27534
Pérou	1990	128522	128000	99.6	30850[1]	24.0	68400[1]	53.2	28750
Suriname	1985	16327	15600	95.5	81[1]	0.5	14865[1]	91.0	654
Suriname	1990	16327	15600	95.5	88[1]	0.5	14855[1]	91.0	657
Uruguay	1985	17741	17481	98.5	* 14891[1]	83.9	650[1]	3.7	1940
Uruguay	1990	17741	17481	98.5	14819[1]	83.5	669[1]	3.8	1993
Venezuela	1985	91205	88205	96.7	21220[1]	23.3	31625[1]	34.7	35360
Venezuela	1990	91205	88205	96.7	21595[1]	23.7	30175[1]	33.1	36435
Asia · Asie									
Afghanistan	1985	65209	65209	100.0	38054[1]	58.4	1900	2.9	25255
Afghanistan	1990	65209	65209	100.0	38054[1]	58.4	1900[1]	2.9	25255
Bahrain	1985	68	68	100.0	6[1]	8.8	...	...	62
Bahreïn	1990	68	68	100.0	6[1]	8.8	...	...	62
Bangladesh	1985	14400	13017	90.4	9735[1]	67.6	2143	14.9	1139
Bangladesh	1990	14400	13017	90.4	9726[1]	67.5	1860	12.9	1431
Bhutan	1985	4700	4700	100.0	396[1]	8.4	2585[1]	55.0	1719
Bhoutan	1990	4700	4700	100.0	403[1]	8.6	2610[1]	55.5	1687
Brunei Darussalam	1985	577	527	91.3	13[1]	2.3	275[1]	47.7	239
Brunéi Darussalam	1990	577	527	91.3	13[1]	2.3	225[1]	39.0	289
Camobodia	1985	18104	17652	97.5	3636[1]	20.1	13372[1]	73.9	644
Cambodge	1990	18104	17652	97.5	3636[1]	20.1	13372[1]	73.9	644
China	1985	959696	932641	97.2	468268[1]	48.8	129765[1]	13.5	334608
Chine	1990	959696	932641	97.2	496563[1]	51.7	126515	13.2	309563
Cyprus	1985	925	924	99.9	163	17.6	123	13.3	638
Chypre	1990	925	924	99.9	160	17.3	123[1]	13.3	641
East Timor	1985	1487	1487	100.0	230[1]	15.5	1100[1]	74.0	157
Timor oriental	1990	1487	1487	100.0	230[1]	15.5	1100[1]	74.0	157
Gaza strip (Palestine)	1985	38	38	100.0	24[1]	63.2	4[1]	10.5	10
Zone de Gaza (Palestine)	1990	38	38	100.0	24[1]	63.2	4[1]	10.5	10
Hong Kong	1985	104	99	95.2	9	8.7	12	11.5	78
Hong-kong	1990	104	99	95.2	8	7.7	12[1]	11.5	79
India[6]	1985	328759	297319	90.4	180949	55.0	67157	20.4	49213
Inde[6]	1990	328759	297319	90.4	181130[1]	55.1	66700[1]	20.3	49489
Indonesia	1985	190457	181157	95.1	32950[1]	17.3	113433	59.6	34774
Indonesie	1990	190457	181157	95.1	33800[1]	17.7	113433[1]	59.6	33924
Iran, Islamic Rep. of	1985	164800	163600	99.3	58920[1]	35.8	18020[1]	10.9	86660
Iran, Rép. islamique d'	1990	164800	163600	99.3	59050[1]	35.8	18020[1]	10.9	86530
Iraq	1985	43832	43737	99.8	9450[1]	21.6	1900[1]	4.3	32387
Iraq	1990	43832	43737	99.8	9450[1]	21.6	1890[1]	4.3	32397
Israel	1985	2077	2033	97.9	575	27.7	110	5.3	1348
Israël	1990	2077	2033	97.9	583[1]	28.1	112[1]	5.4	1338
Japan	1985	37780	37652	99.7	5379	14.2	25105	66.5	7168
Japon	1990	37780	37652	99.7	5243	13.9	25105	66.5	7304
Jordan	1985	8921	8893	99.7	1145[1]	12.8	69	0.8	7679
Jordanie	1990	8921	8893	99.7	1191[1]	13.4	70[1]	0.8	7632
Korea, Dem. People's Rep.	1985	12054	12041	99.9	2005[1]	16.6	8970[1]	74.4	1066
Corée, Rép. p. dém. de	1990	12054	12041	99.9	2050[1]	17.0	8970[1]	74.4	1021
Korea, Republic of	1985	9902	9873	99.7	2224[1]	22.5	6531	66.0	1118
Corée, République de	1990	9902	9873	99.7	2189[1]	22.1	6476	65.4	1208
Kuwait	1985	1782	1782	100.0	137	7.7	2	0.1	1643
Koweït	1990	1782	1782	100.0	138[1]	7.7	2[1]	0.1	1642
Lao People's Dem. Rep.	1985	23680	23080	97.5	1700[1]	7.2	13200[1]	55.7	8180
Rép. dém. pop. lao	1990	23680	23080	97.5	1711[1]	7.2	12700[1]	53.6	8669
Lebanon	1985	1040	1023	98.4	310[1]	29.8	80[1]	7.7	633
Liban	1990	1040	1023	98.4	311[1]	29.9	80[1]	7.7	632

110
Surface and land area and land use
Thousand hectares [cont.]
Superficie totale, superficie des terres et utilisation des terres
Milliers d'hectares [suite]

Country or area Pays ou zone	Year Année	Surface Superficie 000Hs	Land Terres 000Hs	%￼	Agricultural Agricoles 000Hs	%￼	Forest Boisés 000Hs	%￼	Other Autres 000Hs
Macau	1985	2	2	100.0	...	...	...	...	2
Macao	1990	2	2	100.0	...	...	...	...	2
Malaysia	1985	32975	32855	99.6	4907[1]	14.9	20610[1]	62.5	7338
Malaisie	1990	32975	32855	99.6	4907[1]	14.9	19330	58.6	8618
Maldives	1985	30	30	100.0	4[1]	13.3	1[1]	3.3	25
Maldives	1990	30	30	100.0	4[1]	13.3	1[1]	3.3	25
Mongolia	1985	156650	156650	100.0	124588	79.5	15178	9.7	16884
Mongolie	1990	156650	156650	100.0	125786[1]	80.3	13915[1]	8.9	16949
Myanmar	1985	67655	65737	97.2	10429	15.4	32204	47.6	23104
Myanmar	1990	67655	65754	97.2	10428	15.4	32399	47.9	22927
Nepal	1985	14080	13680	97.2	4458[1]	31.7	2480[1]	17.6	6742
Népal	1990	14080	13680	97.2	4653[1]	33.0	2480[1]	17.6	6547
Oman	1985	21246	21246	100.0	1047[1]	4.9	...	...	20196
Oman	1990	21246	21246	100.0	1061[1]	5.0	...	...	20185
Pakistan[7]	1985	79610	77088	96.8	25610[1]	32.2	3160	4.0	48318
Pakistan[7]	1990	79610	77088	96.8	25750[1]	32.3	3550[1]	4.5	47788
Philippines	1985	30000	29817	99.4	9060[1]	30.2	11350[1]	37.8	9407
Philippines	1990	30000	29817	99.4	9230[1]	30.8	10350[1]	34.5	10237
Qatar	1985	1100	1100	100.0	54[1]	4.9	...	...	1046
Qatar	1990	1100	1100	100.0	55[1]	5.0	...	...	1045
Saudi Arabia	1985	214969	214969	100.0	87185[1]	40.6	1200[1]	0.6	126584
Arabie saoudite	1990	214969	214969	100.0	87365[1]	40.6	1200[1]	0.6	126404
Singapore	1985	62	61	98.4	5[1]	8.1	3	4.8	53
Singapour	1990	62	61	98.4	1[1]	1.6	3[1]	4.8	57
Sri Lanka	1985	6561	6463	98.5	2315[1]	35.3	1747	26.6	2401
Sri Lanka	1990	6561	6463	98.5	2339[1]	35.7	2082	31.7	2042
Syrian Arab Republic	1985	18518	18406	99.4	13951	75.3	516	2.8	3939
Rép. arabe syrienne	1990	18518	18392	99.3	13495	72.9	723	3.9	4174
Thailand	1985	51312	51089	99.6	20577	40.1	14905	29.0	15607
Thaïlande	1990	51312	51089	99.6	22920[1]	44.7	14100[1]	27.5	14069
Turkey	1985	77945	76963	98.7	36430[1]	46.7	20199	25.9	20334
Turquie	1990	77945	76963	98.7	36410[1]	46.7	20199	25.9	20354
United Arab Emirates	1985	8360	8360	100.0	235[1]	2.8	3[1]	...	8122
Emirats arabes unis	1990	8360	8360	100.0	239[1]	2.9	3[1]	...	8118
Viet Nam	1985	33169	32549	98.1	6812[1]	20.5	8490[1]	25.6	17247
Viet Nam	1990	33169	32549	98.1	6940[1]	20.9	9850[1]	29.7	15759
Yemen	1985	52797	52797	100.0	17594[1]	33.3	4060[1]	7.7	31143
Yémen	1990	52797	52797	100.0	17674[1]	33.5	4060[1]	7.7	31063
Europe · Europe									
Albania	1985	2875	2740	95.3	1110[1]	38.6	1040[1]	36.2	590
Albanie	1990	2875	2740	95.3	1110[1]	38.6	1046[1]	36.4	584
Andorra	1985	45	45	100.0	26[1]	57.8	10[1]	22.2	9
Andorre	1990	45	45	100.0	26[1]	57.8	10[1]	22.2	9
Austria	1985	8385	8273	98.7	3511	41.9	3221	38.4	1541
Autriche	1990	8385	8273	98.7	3500	41.7	3227	38.5	1546
Belgium-Luxembourg	1985	3310	3282	99.2	1524	46.0	695	21.0	1063
Belgique-Luxembourg	1990	3310	3282	99.2	1490[1]	45.0	699[1]	21.1	1093
Bulgaria	1985	11091	11055	99.7	6169	55.6	3867	34.9	1019
Bulgarie	1990	11091	11055	99.7	6159	55.5	3871	34.9	1025
Czechoslovakia	1985	12787	12540	98.1	6794	53.1	4586	35.9	1160
Tchécoslovaquie	1990	12787	12536	98.0	6736	52.7	4619	36.1	1181
Denmark	1985	4308	4238	98.4	2834	65.8	493	11.4	911
Danemark	1990	4309	4239	98.4	2788	64.7	493	11.4	958
Faeroe Islands	1985	140	140	100.0	3[1]	2.1	...	...	137
Iles Féroe	1990	140	140	100.0	3[1]	2.1	...	...	137
Finland	1985	33813	30461	90.1	2542	7.5	23222	68.7	4697
Finlande	1990	33813	30461	90.1	2558	7.6	23222	68.7	4681
Germany ʃ · Allemagne ʃ									
Fed. Rep. of Germany	1985	24862	24412	98.2	12019[8]	48.3	7360[8]	29.6	5033
Rép.féd.d'Allemagne	1990	24862	24412	98.2	11867[8]	47.7	7410[18]	29.8	5135
former German Dem. Rep.	1985	10833	10572	97.6	6225	57.5	2978	27.5	1369
ancienne R.d.allemande	1990	10833	10519	97.1	6165	56.9	2983	27.5	1371

110
Surface and land area and land use
Thousand hectares [*cont.*]
Superficie totale, superficie des terres et utilisation des terres
Milliers d'hectares [*suite*]

Country or area / Pays ou zone	Year / Année	Surface / Superficie 000Hs	Land / Terres 000Hs	%	Agricultural / Agricoles 000Hs	%	Forest / Boisés 000Hs	%	Other / Autres 000Hs
France	1985	55150	55010	99.7	31442	57.0	14617	26.5	8951
France	1990	55150	55010	99.7	30628	55.5	14811	26.9	9571
Gibraltar	1985	1	1	100.0	...	...	...	...	1
Gibraltar	1990	1	1	100.0	...	...	...	...	1
Greece	1985	13199	12890	97.7	9195	69.7	2620[1]	19.8	1075
Grèce	1990	13199	12890	97.7	9189	69.6	2620[1]	19.8	1081
Hungary	1985	9303	9234	99.3	6539	70.3	1648	17.7	1047
Hongrie	1990	9303	9234	99.3	6474	69.6	1695	18.2	1065
Iceland	1985	10300	10025	97.3	2282[1]	22.2	120[1]	1.2	7623
Islande	1990	10300	10025	97.3	2282[1]	22.2	120[1]	1.2	7623
Ireland	1985	7028	6889	98.0	5705	81.2	330	4.7	854
Irlande	1990	7028	6889	98.0	5635[1]	80.2	343[1]	4.9	911
Italy	1985	30127	29406	97.6	17095	56.7	6414	21.3	5897
Italie	1990	30127	29406	97.6	16938[1]	56.2	6737[1]	22.4	5731
Liechtenstein	1985	16	16	100.0	10[1]	62.5	3[1]	18.8	3
Liechtenstein	1990	16	16	100.0	10[1]	62.5	3[1]	18.8	3
Malta	1985	32	32	100.0	13[1]	40.6	...	...	19
Malte	1990	32	32	100.0	13[1]	40.6	...	...	19
Netherlands	1985	3733	3392	90.9	2056	55.1	300	8.0	1036
Pays-Bas	1990	3733	3392	90.9	2026[1]	54.3	300[1]	8.0	1066
Norway	1985	32390	30683	94.7	957	3.0	8330[1]	25.7	21399
Norvège	1990	32390	30683	94.7	976	3.0	8330[1]	25.7	21377
Poland	1985	31268	30449	97.4	18914	60.5	8728	27.9	2807
Pologne	1990	31268	30442	97.4	18793	60.1	8754	28.0	2895
Portugal	1985	9239	9195	99.5	4005[1 9]	43.3	2968[1]	32.1	2222
Portugal	1990	9239	9195	99.5	4022[1 9]	43.5	2968[1]	32.1	2205
Romania	1985	23750	23034	97.0	15020	63.2	6339	26.7	1675
Roumanie	1990	23750	23034	97.0	14768	62.2	6400[1]	26.9	1866
San Marino	1985	6	6	100.0	1[1]	16.7	...	...	5
Saint-Marin	1990	6	6	100.0	1[1]	16.7	...	...	5
Spain	1985	50478	49944	98.9	30712	60.8	15614	30.9	3618
Espagne	1990	50478	49944	98.9	30525[1]	60.5	15645[1]	31.0	3774
Sweden	1985	44996	41162	91.5	3494	7.8	28005	62.2	9663
Suède	1990	44996	41162	91.5	3382[1]	7.5	28020[1]	62.3	9760
Switzerland	1985	4129	3977	96.3	2021	48.9	1052	25.5	904
Suisse	1990	4129	3977	96.3	2021	48.9	1052	25.5	904
United Kingdom	1985	24488	24161	98.7	18168	74.2	2273	9.3	3720
Royaume-Uni	1990	24488	24160	98.7	17837	72.8	2400	9.8	3923
Yugoslavia	1985	25580	25540	99.8	14135	55.3	9354	36.6	2051
Yugoslavie	1990	25580	25540	99.8	14084	55.1	9379	36.7	2077
Oceania · Océanie									
American Samoa	1985	20	20	100.0	4[1]	20.0	14[1]	70.0	2
Samoa américaines	1990	20	20	100.0	4[1]	20.0	14[1]	70.0	2
Australia	1985	771336	764444	99.1	472960[1 10]	61.3	106000	13.7	185484
Australie	1990	771336	764444	99.1	466561[10]	60.5	106000	13.7	191883
Christmas Island	1985	13	13	100.0	...	...	...	...	13
Ile Christmas	1990	13	13	100.0	...	...	...	...	13
Cocos Islands	1985	1	1	100.0	...	...	...	...	1
Iles de Cocos	1990	1	1	100.0	...	...	...	...	1
Cook Islands	1985	23	23	100.0	6[1]	26.1	...	...	17
Iles Cook	1990	23	23	100.0	6	26.1	...	...	17
Fiji	1985	1827	1827	100.0	300[1]	16.4	1185[1]	64.9	342
Fidgi	1990	1827	1827	100.0	300[1]	16.4	1185[1]	64.9	342
French Polynesia	1985	400	366	91.5	47[1]	11.8	115[1]	28.8	204
Polynésie française	1990	400	366	91.5	47[1]	11.8	115[1]	28.8	204
Guam	1985	55	55	100.0	20[1]	36.4	10[1]	18.2	25
Guam	1990	55	55	100.0	20[1]	36.4	10[1]	18.2	25
Kiribati	1985	71	71	100.0	37[1]	52.1	2[1]	2.8	32
Kiribati	1990	71	71	100.0	37[1]	52.1	2[1]	2.8	32
Nauru	1985	2	2	100.0	...	...	...	...	2
Nauru	1990	2	2	100.0	...	...	...	...	2
New Caledonia	1985	1858	1828	98.4	295[1]	15.9	708[1]	38.1	825
Nouvelle-Calédonie	1990	1858	1828	98.4	302[1]	16.3	708[1]	38.1	818
New Zealand	1985	27099	26799	98.9	14392[1]	53.1	7180[1]	26.5	5227
Nouvelle-Zélande	1990	27099	26799	98.9	13902[1]	51.3	7350[1]	27.1	5547

110
Surface and land area and land use
Thousand hectares [*cont.*]
Superficie totale, superficie des terres et utilisation des terres
Milliers d'hectares [*suite*]

Country or area Pays ou zone	Year Année	Surface Superficie 000Hs	Land Terres 000Hs	%	Agricultural Agricoles 000Hs	%	Forest Boisés 000Hs	%	Other Autres 000Hs
Niue	1985	26	26	100.0	8[1]	30.8	5	19.2	13
Nioué	1990	26	26	100.0	8[1]	30.8	5[1]	19.2	13
Norfolk Islands	1985	4	4	100.0	1[1]	25.0	...	...	3
Iles Norfolk	1990	4	4	100.0	1[1]	25.0	...	...	3
Pacific Islands (Palau)[11]	1985	178	178	100.0	83[1]	46.6	40[1]	22.5	55
Iles du Pacifique (Palaos)[11]	1990	178	178	100.0	83[1]	46.6	40[1]	22.5	55
Papua New Guinea	1985	46284	45286	97.8	471[1]	1.0	38290[1]	82.7	6525
Papouasie-Nouvelle-Guinée	1990	46284	45286	97.8	478[1]	1.0	38230[1]	82.6	6578
Samoa	1985	284	283	99.6	123[1]	43.3	134[1]	47.2	26
Samoa	1990	284	283	99.6	123[1]	43.3	134[1]	47.2	26
Solomon Islands	1985	2890	2799	96.9	94[1]	3.3	2560[1]	88.6	145
Iles Salomon	1990	2890	2799	96.9	96[1]	3.3	2560[1]	88.6	143
Tokelau	1985	1	1	100.0	...	...	...	...	1
Tokélaou	1990	1	1	100.0	...	...	...	...	1
Tonga	1985	75	72	96.0	52[1]	69.3	8[1]	10.7	12
Tonga	1990	75	72	96.0	52[1]	69.3	8[1]	10.7	12
Tuvalu	1985	3	3	100.0	...	...	...	...	3
Tuvalu	1990	3	3	100.0	...	...	...	...	3
Vanuatu	1985	1219	1219	100.0	170[1]	13.9	914	75.0	135
Vanuatu	1990	1219	1219	100.0	169[1]	13.9	914[1]	75.0	136
Wallis and Futana Islands	1985	20	20	100.0	5[1]	25.0	...	...	15
Iles Wallis et Futana	1990	20	20	100.0	5[1]	25.0	...	...	15
former USSR · ancienne URSS									
former USSR	1985	2240300[12]	2227280	99.4	607055[13]	27.1	941000[1]	42.0	679225
ancienne URSS	1990	2240300[12]	2227280	99.4	598820[1][13]	26.7	947000[1]	42.3	681460

Source :
Food and Agriculture Organization of the United Nations (Rome).

Source:
Organisation des Nations Unies pour l'alimentation et l'agriculture (Rome).

§   All data shown pertain to Germany prior to 3 October 1990 are indicated separately for the Federal Republic of Germany and the former German Democratic Republic based on their respective territories at the time indicated. Where data for united Germany (3 October 1990 and thereafter) are not available, available data are shown separately under the designations Federal Republic of Germany and former German Democratic Republic and pertain to the territorial boundaries prior to 3 October 1990. For detailed explanatory notes on data pertaining to Germany, see Annex I- Country Nomenclature.

§   Toutes les données se rapportant à l'Allemagne avant le 3 octobre 1990 figurent dans deux rubriques séparées basées sur les territoires respectifs de la République fédérale d'Allemagne et l'ancienne République démocratique allemande selon la période indiquée. En l'absence de données pour l'Allemagne unifiée (à compter du 3 octobre 1990), les données disponibles sont fournies séparément sous les rubriques République fédérale d'Allemagne et ancienne République démocratique allemande et se rapportent aux limites territoriales antérieures au 3 octobre 1990. Pour les notes explicatives en détail sur les données concernant l'Allemagne, voir Annexe I - Nomenclature des pays.

1   FAO estimate.
2   Rough grazing land is included under Other land.
3   Data exclude dependencies.
4   Data refer to the State sector and to entailed land only.
5   Data refer to area free from ice.
6   Including data for the Indian-held part of Jammu and Kashmir, the final status of which has not yet been determined.
7   Excluding data for Jammu and Kashmir, the final status of which has not yet been determined.
8   Data include arable land and land under permanent crops on holdings of 1 hectare and above, and on holdings of less than 1 hectare whose production market value exceeds a fixed minimum.
9   Data include about 800000 hectares of temporary crops grown in association with permanent crops and forests.
10 Data on arable land include about 27000000 hectares of cultivated grassland. Data refer to balance of area of rural holdings.
11 Including data for Federated States of Micronesia, Marshall Is. and Northern Mariana Is.
12 Data include the White Sea (9000000 hectares) and the Azov Sea (3730000 hectares).
13 Data exclude pastures for reindeer.

1   Estimation de la FAO.
2   Les pâturages sauvages sont inclus dans 'autres-terres'.
3   Les dennées ne comprennent pas les dépendances.
4   Les données se rapportent uniquement au secteur d'Etat et aux terres substituées.
5   Les données se rapportent aux superficies non couvertes de glace.
6   Y compris les données pour la partie du Jammu et du Cachemire occupée par l'Inde dont le statut définitif n'a pas encore été déterminé.
7   Non compris les données pour le Jammu et le Cachemire, dont le statut définitif n'a pas encore été déterminé.
8   Les données se rapportent aux terres arables et cultures permanentes des exploitations d'au moins 1 hectare, et des exploitations de moins de 1 hectare dont la production a une valeur marchande qui dépasse un minimum donné.
9   Les données comprennent environ 800 000 hectares de cultures temporaires associées à des cultures permanentes ou à des forêts.
10 Les données sur les terres arables comprennent environ 27 000 000 d'hectares d'herbages cultivés.
11 Y compris les données pour les Etats fédérés de Micronésie, les îles Marshall et les îles Mariannes du Nord.
12 Les données comprennent la mer Blanche (9000000 d'hectares) et la mer d'Azov (3730000 d'hectares).
13 Les données se comprennent pas les pâturages pour les rennes.

Technical notes, tables 108-110

Table 108: Indicators on changes in land distribution give the percentage change in mean annual land use between the two three-year periods, 1966/68 and 1986/88. Land use definitions vary considerably among countries; however, in general, "cropland" refers to arable land and land under temporary or permanent cultivation; "pasture" includes pastures and lands used for five years or more for natural or cultivated forage crops respectively; "forest" refers to natural or planted stands of trees, and includes cleared forest land that is to be reforested; "other" includes barren or waste land, parks, built-on areas and roads. Regional aggregates may include estimates for countries and areas not listed. For more detailed notes on definitions and countries, refer to the most recent edition of the *FAO Yearbook: Production*.[6] Estimates of annual rates of deforestation (tropical deforestation only) are largely derived from FAO compilations. However, data from other sources have been used in place of FAO estimates for the following countries; Brazil, Cameroon, Costa Rica, India, Indonesia, Malaysia, Myanmar, Peru, The Philippines, Thailand, Viet Nam, and Zaire.

The term deforestation as used here refers to the permanent clearing of forest lands for shifting cultivation, permanent agriculture or settlements; it does not include other alterations such as selective logging. Deforestation rates for total forest areas for some countries may not always be consistent because of missing information in some forest classes. The term reforestation as used here refers to the establishment of plantations for industrial and non-industrial uses; it does not, in general, include regeneration of old tree crops, although some countries may report regeneration as reforestation. Regional and world totals include listed countries only.

Water withdrawal is shown as a proportion of estimated renewable water resources. Renewable water resources include the average annual flow of rivers, and aquifers generated from endogenous precipitation and river flows from other countries. Water withdrawal refers to water taken from both renewable and non-renewable sources; consequently water withdrawal expressed as a percentage of renewable water resources may exceed 100 per cent in countries where non-renewable resources are being tapped. Estimates of water withdrawal for the most part refer to a single year in the 1980s; however, in some cases older data are presented.

Estimates of the number of threatened species given in the table are based on the total number of mammals, birds, reptiles and amphibians plus plant taxa listed as threatened according to the 1990 IUCN (International Union for Conservation of Nature and Natural Resources) Red List(for animals) and the list of plants of the WCMC (World Conservation Monitoring Center) Threatened Plant Unit. A species is considered to be threatened when it belongs to any one of the following IUCN threat categories:

Notes techniques, tableaux 108-110

Tableau 108 : Les indicateurs de l'évolution de l'utilisation des sols donnent la variation en pourcentage des superficies moyennes annuelles affectées à différents usages entre deux périodes de trois ans, à savoir 1966–1968 et 1986–1988. Bien que les définitions des différents usages des sols diffèrent considérablement d'un pays à un autre, en règle générale, l'expression "terrains agricoles" se rapporte aux terres arables et aux terrains cultivés de manière temporaire ou en permanence; les "pacages" sont les pâturages et les terrains qui sont utilisés pendant au moins cinq années pour fournir, respectivement, des fourrages naturels et des fourrages cultivés et le terme "forêt" se rapporte aux formations naturelles ou plantées d'arbres et recouvre les aires forestières qui ont été déboisées mais qu'il est prévu de reboiser; la rubrique "Autres" recouvre les terres incultes, les parcs, les terrains bâtis et les routes. Les valeurs globales indiquées pour une région peuvent inclure des estimations pour des pays et régions qui n'apparaissent pas dans le tableau. Des notes plus détaillées concernant les définitions et les pays figurent dans la dernière édition de l'*Annuaire de la FAO* [6] : *Production*. Les estimations des taux annuels de déboisement (déboisement tropical seulement) ont, pour l'essentiel, été établies à partir des calculs de la FAO. Des données provenant d'autres sources ont néanmoins été utilisées à la place des estimations de la FAO pour le Brésil, le Cameroun, le Costa Rica, l'Inde, l'Indonésie, la Malaisie, le Myanmar, le Pérou, les Philippines, la Thaïlande, le Viet Nam et le Zaïre.

Le terme déboisement est employé ici pour indiquer le déboisement permanent de zones forestières à des fins de cultures itinérantes, de cultures permanentes ou d'établissements ruraux; il ne couvre pas les autres transformations qui peuvent prendre place dans le cadre, par exemple, d'une exploitation forestière sélective. Les taux de déboisement pour la superficie totale des zones forestières de certains pays peuvent manquer de cohérence du fait de l'absence d'informations sur certains types de forêts. Le terme "reboisement" désigne ici l'établissement de plantations à des fins industrielles ou non; il ne couvre pas, en général, la régénération d'anciennes arboricultures mais il arrive que certains pays déclarent les activités de régénération en tant qu'opérations de reboisement. Les totaux indiqués pour les régions et le monde entier ne se rapportent qu'aux pays inclus dans le tableau.

Les prélèvements d'eau sont exprimés en pourcentage des ressources renouvelables totales. Les ressources renouvelables en eau sont constituées par le débit annuel moyen des cours d'eau et des eaux souterraines provenant des eaux de pluies reçues dans le pays et des eaux des cours d'eau ayant leur source dans d'autres pays. Les prélèvements sont effectués sur les eaux provenant de sources renouvelables et non renouvelables, en conséquence de quoi les prélèvements exprimés en pourcentage des ressources renouvelables peuvent dépasser 100 % dans les pays qui exploitent des ressources en eau non renouvelables. Les estimations des prélèvements d'eau se rapportent, le plus souvent, à une seule année de la décennie 1980-1990, bien que des données plus anciennes aient été incluses dans certains cas.

— Extinct: Species not definitely located in the wild during the past 50 years;

— Endangered: Taxa in danger of extinction and whose survival is unlikely if the cause of the threat continues;

— Vulnerable: Taxa which could fall into the Endangered category in the future if the threat continues;

— Rare: Taxa with small global populations which are at risk;

— Indeterminate: Taxa known to be Endangered, Vulnerable or Rare but cannot be categorized due to a lack of information;

— Insufficient Known: Taxa which are suspected to be under threat but are not definitely known to belong to any particular group because of insufficient information;

Vertebrate data do not include extinct taxa and only full species are accounted for. Mammal data exclude marine mammals unless otherwise stated. Rare and threatened plants include plants classified as endangered, vulnerable or indeterminate.

Although the IUCN Red List is a comprehensive global compendium of animal species known to be threatened, many more species than those listed are threatened. These include species that are yet undescribed and species which have been described but whose status has not been reviewed. Only 50 per cent of mammal species, and probably less than 20 per cent of reptiles, 10 per cent of amphibians and 5 per cent of fish are estimated to have been reviewed.

For more information see WCMC *Global Biodiversity: Status of the Earth's Living Resources*, [57] and World Resources Institute, *World Resources 1992-93*. [58]

Table 109, part A: Man-made carbon dioxide(CO_2) is predominantly emitted from the combustion of fossil fuels (solid fuels, oil, natural gas, vegetal fuels) and cement production. These combustion emissions contribute to the increase of CO_2 atmospheric concentrations and thus to global air pollution problems. CO_2 is the largest contributor to the greenhouse effect. (See OECD *Environmental Indicators 1991*, pp. 16-19 [56]) National CO_2 emissions are based on estimates of net consumption of gas, liquid and solid fuels, cement and appropriate emission factors. Per capita emissions use population data contained in the United Nations *Demographic Yearbook 1990*. [19] Net CO_2 emissions from land use changes (mainly deforestation) are derived from model computations of the annual change in the carbon content of terrestrial ecosystems following land-use disturbances. For more information on full details of the model calculations see the latest edition of UNEP's *Environmental Data Report 1991–1992* [31]. In this table, emissions are in units of 10^3 t a^{-1} and kg a^{-1} as carbon; for emissions in 10^3 t a^{-1} and kg a^{-1} as carbon dioxide, multiply by 3.664.

Chlorofluorocarbons (CFCs) and halons are the main ozone-depleting chemicals. Data on consumption of CFCs and halons are based on official reports submitted to the Ozone Secretariat, UNEP, under the terms of the Montreal Protocol.

Les estimations du nombre d'espèces menacées portées dans le tableau sont basées sur le nombre total de mammifères, d'oiseaux, de reptiles et d'amphibiens ainsi que des groupes taxonomiques inscrits comme espèces menacées sur la liste rouge de 1990 de l'UICN (Union internationale pour la conservation de la nature et de ses ressources) pour les animaux, et la liste des plantes du Service des plantes menacées du Centre mondial de surveillance de la conservation. Une espèce est considérée être menacée lorsqu'elle est visée par une des catégories de menaces établies par l'UICN indiquées ci-après :

— Espèces éteintes : Espèces qui n'ont pu être observées en liberté dans la nature au cours des 50 années précédentes;

— Espèces en voie de disparition : Groupes taxonomiques menacés de disparition et qui ne pourront vraisemblablement pas survivre si la cause de la menace à laquelle ils sont exposés ne disparaît pas;

— Espèces vulnérables : Groupes taxonomiques qui pourraient, à une date future, être classés dans la catégorie des espèces en voie de disparition si la cause de la menace à laquelle ils sont exposés ne disparaît pas;

— Espèces rares : Groupes taxonomiques dont la population mondiale est très restreinte et qui sont en danger;

— Espèces dont la situation est indéterminée : Groupes taxonomiques que l'on sait être en voie de disparition, vulnérables ou rares mais qu'il n'est pas possible d'affecter à une catégorie spécifique parce que l'on ne dispose pas d'informations suffisantes;

— Espèces dont la situation n'est pas suffisamment connue : Groupes taxonomiques que l'on pense être menacés d'extinction, mais qu'il n'est pas possible d'affecter à une catégorie particulière parce que l'on ne dispose pas d'informations suffisantes;

Les données sur les espèces vertébrées ne portent pas sur les groupes taxonomiques éteints et ne se rapportent qu'à des espèces complètes. Les données sur les mammifères ne recouvrent pas les mammifères marins sauf indication contraire. Les plantes rares et en voie de disparition recouvrent les plantes classées parmi les espèces en voie de disparition, vulnérables ou dont la situation est indéterminée.

Bien que la liste rouge de l'UICN soit un registre mondial exhaustif des espèces animales que l'on sait être menacées d'extinction, de nombreuses espèces autres que celles qui sont portées sur cette liste sont également menacées. Il s'agit d'espèces qui n'ont pas encore fait l'objet d'une description ainsi que d'espèces qui ont fait l'objet d'une description mais dont la situation n'a pas encore été examinée. Il a été estimé que la situation des espèces n'a été examinée que pour seulement 50 % des mammifères, et probablement moins de 20 % des reptiles, 10 % des amphibiens et 5 % des poissons.

Pour obtenir de plus amples informations, il est possible de se reporter à l'ouvrage du Centre mondial de surveillance de la conservation, *La diversité biologique dans le monde : situation des ressources vivantes de la Planète* [57], et à la publication de "World Resources Institute : *World Resources 1992–93*". [58]

Safe drinking water includes treated surface water and untreated water from protected springs, boreholes and wells. Populations with access to sanitation services are defined as those served by connections to public sewers or household disposal systems such as pit privies, pour flush latrines, septic tanks, communal toilets, etc. See UNEP, *Environmental Data report, 1991/92*, p.227. [31]

Apparent consumption of paper and paper recovery are derived from estimates of national production of paper and paperboard, plus imports minus exports. See UNEP *Environmental Data Report 1991/92*, p.355. [31]

A nationally protected area is defined as an area over 1 000 ha that falls within IUCN (International Union for Conservation of Nature and Natural Resources) Management Categories I to V , inclusive. Areas within these categories include Scientific Reserves (strict nature reserves), National Parks, Natural Monuments (natural landmarks), Managed Nature Reserves (wildlife sanctuaries) and Protected Landscapes and Seascapes. See UNEP, *Environmental Data Report 1991/92*, p.202. [31]

Table 109, part B: Annual mean concentrations of SO_2 (sulphur dioxide) are based on data for two years, except where otherwise indicated. The total number of observations made during each two-year period is given in parentheses in order to provide an indication of the level of completeness of the data. Although many sites make observations on a daily basis, at some locations measurements are made less frequently, e.g., once every six days. Thus the use of the number of observations to assess data completeness must be viewed with some caution. See UNEP, *Environmental Data Report 1991/92* p.32. [31]

Table 109, part C: Global Environment Monitoring System water quality data presented here are extracted from the (GEMS) Water data base which is maintained by the Canada Center for Inland Waters on behalf of the GEMS/Water programme. Values presented are arithmetic means of measurements made during the five-year period 1984-1988. Only those stations which have analyzed at least 30 samples during this period are listed. In this table data for BOD (biological oxygen demand) and COD (chemical oxygen demand) are quoted to two decimal places; concentration of Cd (cadminm), Pb (plumbum) and Hg (hydrargyrum) are quoted to three decimal places. However, reporting accuracy in the GEMS/Water data base may differ. For further details see the GEMS/Water data summaries for 1982-1984 and 1985-1987. [59] Information regarding the recommended sampling frequencies and analytical methods for water quality parameters are given in the GEMS/Water Operational Guide [60].

Table 110 presents total surface area, land area, area of agricultural land and area of forest land for countries or areas.

Total surface area refers to the total area of the country, including area under inland water bodies.

Land area refers to total area, excluding area under inland water bodies. The definition of inland water bodies generally includes major rivers and lakes.

Tableau 109, partie A : Les émissions d'oxyde de carbone dues à l'homme proviennent essentiellement de la combustion de combustibles fossiles (combustibles solides, pétrole, gaz naturel, combustibles végétaux) et de la production de ciment. Ces émissions provoquées par la combustion contribuent à l'accroissement de la teneur de CO_2 dans l'atmosphère et, partant, aux problèmes que la pollution atmosphérique pose dans le monde entier. L'oxyde de carbone est le principal facteur de l'effet de serre. Se reporter à l'étude de l'OCDE de 1991 : *Indicateurs d'environnement*, pages 16 à 19 [56]. Les valeurs établies à l'échelon national pour les émissions de CO_2 sont basées sur des estimations de la consommation nette de combustibles gazeux, liquides et solides, de ciment et de facteurs d'émissions pertinents. Les valeurs par habitant sont basées sur les données relatives à la population incluses dans l'*Annuaire démographique de 1990* [19] de l'Organisation des Nations Unies. Les émissions nettes de CO_2 dues à l'évolution de l'utilisation des sols (essentiellement le déboisement) ont été calculées à partir d'un modèle de la variation annuelle de la teneur en carbone des écosystèmes terrestres à la suite de changements intervenus dans l'utilisation des terres. De plus amples informations ainsi que le détail des calculs effectués à partir du modèle figurent dans la dernière édition du rapport consacré par le PNUE aux données sur l'environnement "Environmental Data Report 1991-1992" [31]. Dans ce tableau, les emissions sout exprimées en unité de 10^3 t a^{-1} et kg a^{-1} de carbone; pour les emissions en unité de 10^3 t a^{-1} et kg a^{-1} d'oxyde de carbone, multipliez par 3,664.

Les chlorofluorocarbones (CFC) et les halons sont les principaux produits chimiques menaçant l'ozone. Les données sur la consommation de CFC et de halons sont basées sur les rapports officiels soumis au Secrétariat de l'ozone (PNUE) en application du Protocole de Montréal.

L'expression "eau salubre" englobe les eaux de surface traitées et les eaux non traitées tirées de sources, de trous de sonde et de puits protégés. Par populations qui ont accès à des services d'assainissement, on entend les populations raccordées aux égouts publics ou à des systèmes d'évacuation des déchets particuliers du type latrines à fosse, cabinets d'aisance à chasse d'eau, fosses septiques, toilettes communautaires, etc. Voir PNUE, "*Environmental Data Report 1991-1992*", page 227 [31].

Les données sur la consommation apparente et la récupération de papier proviennent d'estimations de la production nationale de papier et de carton plus les importations et moins les exportations des différents pays. Voir PNUE, "*Environmental Data Report 1991-1992*", page 355 [31].

Par aire protégée à l'échelon national, il faut entendre une zone d'au moins 1 000 hectares relevant de l'une des catégories I à V de l'UICN (Union internationale pour la conservation de la nature et de ses ressources). Ces aires couvrent les réserves scientifiques (réserves naturelles intégrales), les parcs nationaux, les monuments naturels (paysages naturels), les réserves naturelles aménagées (sanctuaires de protection d'espèces sauvages) et les paysages terrestres ou marins protégés. Voir PNUE, "*Environmental Data Report 1991-1992*", page 202 [31].

Agricultural land is comprised of the arable land, land under permanent crops and permanent meadows and pastures.

Forest land refers to land under natural or planted stands of trees, whether or not productive.

Other area includes unused but potentially productive land, built-on areas, wasteland, parks, ornamental gardens, roads, lanes, barren land, minor water bodies, and any other land area not specifically listed above.

Tableau 109, partie B : Les données relatives à la teneur annuelle moyenne d'anhydride sulfureux (SO₂) sont fondées sur les valeurs relatives à deux années, sauf indication contraire. Le nombre total d'observations effectuées pendant chaque période de deux ans est indiqué entre parenthèses de manière à fournir une indication du degré d'exhaustivité des données. Si des observations sont effectuées chaque jour dans de nombreux sites, elles sont moins fréquentes dans d'autres, par exemple une fois tous les six jours. Il importe donc de faire preuve de circonspection en considérant le nombre d'observations effectuées pour évaluer le degré d'exhaustivité des données. Voir PNUE, *"Environmental Data Report 1991-1992"*, page 32 [31].

Tableau 109, partie C : Les données sur la qualité de l'eau du Système mondial de surveillance continue de l'environnement (GEMS) présentées ici sont tirées de la base des données sur l'eau du GEMS qui est tenue par le Centre canadien des eaux intérieures pour le compte du programme GEMS/Eau. Les teneurs portées dans le tableau sont les moyennes arithmétiques de mesures effectuées pendant la période de cinq ans 1984-1988. Ne sont indiquées ici que les stations qui ont analysé au moins 30 échantillons pendant cette période. Les chiffres portés dans le tableau pour la demande biochimique en oxygène (DBO) et la demande chimique en oxygène (DCO) ont deux décimales; les teneurs en cadmium (Cd), plomb (Pb) et mercure (Hg) ont trois décimales. Le degré d'exactitude des données incluses dans la base des données du GEMS sur l'eau peut toutefois varier. Il est possible d'obtenir de plus amples détails sur la question en se reportant aux récapitulations des données sur l'eau du GEMS pour 1982–1984 et 1985–1987 [59]. Les informations concernant les fréquences d'échantillonnage et les méthodes d'analyse recommandées pour les paramètres relatifs à la qualité de l'eau figurent dans le guide des opérations du programme de l'eau du GEMS [60].

Le *Tableau 110* indique la superficie totale, la superficie des terres émergées, la superficie des terres agricoles et la superficie des terres boisées des différents pays ou sous-régions.

Par superficie totale, on entend la superficie totale du pays, y compris la superficie couverte par les eaux intérieures.

La superficie des terres émergées est égale à la superficie totale à l'exclusion de la superficie couverte par les eaux intérieures. Les eaux intérieures sont généralement définies de manière à inclure les rivières et les lacs importants.

Les terres agricoles couvrent les terres arables, les terres en culture permanente, les terres en prés et les pâturages permanents.

Les terres boisées se composent de toutes les terres ayant un couvert forestier naturel ou planté, qu'elles soient ou non productives.

Les autres superficies englobent les terres non exploitées mais potentiellement productives, les superficies construites, les terrains vagues, les parcs, les jardins d'agrément, les routes, les sentiers, les terres sans couvert végétal, les voies d'eau d'envergure limitée et tout autre terrain non mentionné ci-dessus.

111

Number of scientists, engineers and technicians engaged in research and experimental development
Nombre de scientifiques, d'ingénieurs et de techniciens employés à des travaux de recherche et de développement expérimental

Country or area Pays ou zone	Year Année	Total	Scientists and engineers Scientifiques et ingénieurs		Technicians Techniciens	
			Total	% F	Total	% F
Africa · Afrique						
Benin [1] Bénin [1]	1989	1036	794	12.6	242	26.4
Burundi [2][3] Burundi [2][3]	1989	338	170	10.0	168	...
Central African Rep. [2][3] Rép. centrafricaine [2][3]	1984	579	196	...	383	...
Congo [3][4] Congo [3][4]	1984	2335	862	...	1473	...
Egypt [4] Egypte [4]	1986	28425	20893	...	7532	...
Gabon [3] Gabon [3]	1987	...	199[2]	...	18[5]	...
Guinea Guinée	1984	1893	1282	...	611	...
Libyan Arab Jamahiriya Jamah. arabe libyenne	1980	2600	1100	...	1500	...
Madagascar [3][6] Madagascar [3][6]	1989	1225	269	31.2	956	...
Mauritius Maurice	1989	365	193	17.1	172	26.7
Nigeria [7] Nigéria [7]	1987	7380	1338	...	6042	...
Rwanda Rwanda	1985	*138	71	...	*67	...
Senegal Sénégal	1981	4610	1948	...	2662	...
Seychelles [4] Seychelles [4]	1983	24	18	...	6	...
America, North · Amérique du Nord						
British Virgin Islands Iles Vierges britanniques	1984	-	-	-	-	-
Canada [8] Canada [8]	1988	88210	61130	...	27080	...
Costa Rica [3] Costa Rica [3]	1988	...[6]	1528	...	...	...
Cuba [4] Cuba [4]	1989	20882	12052	...	8830	...
El Salvador [6][9] El Salvador [6][9]	1989	1775	142	...	1633	...
Guatemala [10] Guatemala [10]	1988	1783	858	...	925	...
Jamaica [11] Jamaïque [11]	1986	33	18	55.6	15	20.0
Mexico Mexique	1984	46146	16679	25.9	29467	25.9
Nicaragua [3] Nicaragua [3]	1987	1027	725	...	302	...
Saint Lucia Sainte-Lucie	1984	139	53	...	86	...
Trinidad and Tobago Trinité-et-Tobago	1984	529	275	21.1	254	17.3
Turks and Caicos Is. Iles Turques et Caïques	1984	-	-	-	-	-
United States [12] Etats-Unis [12]	1988	...	*949200	...	...	...

111
Number of scientists, engineers and technicians engaged in research and experimental development [cont.]
Nombre de scientifiques, d'ingénieurs et de techniciens employés à des travaux de recherche et de dévelopment expérimental [suite]

Country or area Pays ou zone	Year Année	Total	Scientists and engineers Scientifiques et ingénieurs Total	% F	Technicians Techniciens Total	% F
America, South · Amérique du Sud						
Argentina Argentine	1988	17329	11088	...	6241	...
Brazil [3] [13] Brésil [3] [13]	1985	...	52863	...	...	...
Chile [3] [4] Chili [3] [4]	1988	7570	4630	...	2940	...
Colombia [2] Colombie [2]	1982	2107	1083	...	1024	...
Guyana [4] [14] Guyana [4] [14]	1982	267	89	...	178	...
Paraguay [3] Paraguay [3]	1981	...	807	...	...	...
Peru Pérou	1981	...	4858	...	...	...
Uruguay [3] [4] Uruguay [3] [4]	1987	...	2093	34.4	...	...
Venezuela [3] [4] Venezuela [3] [4]	1983	7260	4568	32.4	2692	...
Asia · Asie						
Brunei Darussalam [3] [15] Brunéi Darussalam [3] [15]	1984	136	20	...	116	...
Cyprus [2] Chypre [2]	1984	131[16]	51	...	80[16]	...
India [17] Inde [17]	1988	199983	*119027	4.7	80956	5.2
Indonesia [3] Indonésie [3]	1988	32038	32038	...	...	...
Iran, Islamic Rep. of Iran, Rép. islamique d'	1985	5048	3194	...	1854	...
Israel Israel	1984	24400	20100	51.7 [3]	4300	32.6
Japan [8] Japon [8]	1989	742247	638817	7.1	105430	19.3
Jordan [4] Jordanie [4]	1986	447	418	12.9	29	20.7
Korea, Republic of [3] [4] [18] Corée, République de [3] [4] [18]	1988	92265	56545	6.5	35720	...
Kuwait [3] [19] Koweït [3] [19]	1984	2072	1511	22.1	561	20.1
Lebanon [20] Liban [20]	1980	186	180	...	6	...
Malaysia [18] Malaisie [18]	1988	6707	5537	...	1170	...
Maldives Maldives	1986	-	-	-	-	-
Nepal [18] [19] Népal [18] [19]	1980	409	334	...	75	...
Pakistan [21] Pakistan [21]	1988	15927	6641	6.9	9286	...
Philippines [3] Philippines [3]	1984	6685	4830	48.0	1855	...
Qatar [3] [22] Qatar [3] [22]	1986	290	229	25.3	61	3.3
Singapore [3] [23] Singapour [3] [23]	1987	4887	3361	19.3	1526	...
Sri Lanka Sri Lanka	1985	3483	2790	23.9	693	27.1
Thailand Thaïlande	1987	8324	5539	...	2785	...
Turkey Turquie	1985	18643	11276	...	7367	...
Viet Nam [23] Viet Nam [23]	1985	...	20000	...	...	...

111
Number of scientists, engineers and technicians engaged in research and experimental development [cont.]
Nombre de scientifiques, d'ingénieurs et de techniciens employés à des travaux de recherche et de dévelopment
expérimental [suite]

Country or area Pays ou zone	Year Année	Total	Scientists and engineers Scientifiques et ingénieurs		Technicians Techniciens	
			Total	% F	Total	% F
Europe · Europe						
Austria Autriche	1984	14426	7609	...	6817	...
Belgium Belgique	1988	36770[16]	16646	...	20124[16]	...
Bulgaria Bulgarie	1987	62247[6]	50585	44.0	11662[6]	58.1[6]
Czechoslovakia [24] Tchécoslovaquie [24]	1989	108351	65475	...	42876	...
Denmark Danemark	1989	25448[16]	10662	...	14786[16]	...
Finland Finlande	1989	21195[16]	11317	...	9878[16]	...
France France	1988	283099[16]	115163	...	167936[16]	...
Germany ʂ · Allemagne ʂ Federal Republic of Germany [25] Rép. féd. d'Allemagne [25]	1987	288072	165614	...	122458	...
former German Dem. Rep. [26] ancienne Rép. dém. allemande [26]	1989	195073[16]	127449	...	67624[16]	...
Greece [27] Grèce [27]	1986	1022	534	...	488	...
Hungary [24] Hongrie [24]	1989	34544[28]	20431	...	14113[28]	...
Ireland Irlande	1988	7642	6351	...	1291	...
Italy Italie	1988	113120	74833	...	38287	...
Malta [29] Malte [29]	1988	39	34	...	5	...
Netherlands [8] Pays-Bas [8]	1988	64420[16]	37520	...	26900[16]	...
Norway Norvège	1989	*20700[16]	*12100	...	*8600[16]	...
Poland [4][30] Pologne [4][30]	1989	...	32500	...	...	...
Portugal Portugal	1988	8575	5004	...	3571	...
Romania Roumanie	1989	102601	59670	41.2	42931	65.7
San Marino Saint-Marin	1986	-	-	-	-	-
Spain [31] Espagne [31]	1987	29086[6]	20890	21.4	8196[6]	12.4
Sweden [32] Suède [32]	1987	51811[16]	22725	...	29086[16]	...
Switzerland Suisse	1986	25620	14910	...	10710	...
Yugoslavia [4] Yougoslavie [4]	1989	53550	34770	...	18780	...
Oceania · Océanie						
Australia Australie	1988	55103	38568	...	16535	...
Fiji [3][33] Fidji [3][33]	1986	126	36	11.1	90	11.1
French Polynesia [33] Polynésie française [33]	1983	33	17	...	16	...
Guam [29] Guam [29]	1989	*32	*21	*19.0	*11	*36.4
Kiribati Kiribati	1980	3	2	-	1	-
New Caledonia [34] Nouvelle-Calédonie [34]	1985	148	77	9.1	71	15.5

111
Number of scientists, engineers and technicians engaged in research and experimental development [*cont.*]
Nombre de scientifiques, d'ingénieurs et de techniciens employés à des travaux de recherche et de développement expérimental [*suite*]

Country or area Pays ou zone	Year Année	Total	Scientists and engineers Scientifiques et ingénieurs		Technicians Techniciens	
			Total	% F	Total	% F
Tonga [33] Tonga [33]	1981	15	11	-	4	25.0
former USSR · ancienne URSS						
former USSR [35] ancienne URSS [35]	1990	...	1694400	...	...	...
Belarus Bélarus	1988	...	44100	42.2	...	...
Ukraine Ukraine	1989	...	348600	...	...	...

Source:
United Nations Educational, Scientific and Cultural Organization (Paris).

¶ All data shown which pertain to Germany prior to 3 October 1990 are indicated separately for the Federal Republic of Germany and the former German Democratic Republic based on their respective territories at the time indicated. For detailed explanatory notes on data pertaining to Germany, see Annex I - Country Nomenclature.

1   Not including data for the productive sector (non-integrated R&D).
2   Not including data for the productive sector.
3   Data relate to the number of full-time plus part-time scientists and engineers.
4   Not including military and defence R&D.
5   Data relate only to those in the general service sector.
6   Not including data for the higher education sector.
7   Data relate only to 23 out of 26 national research institutes under the Federal Ministry of Science and Technology.
8   Not including social sciences and humanities in the productive sector (integrated R&D).
9   Data refer to scientists and engineers and technicians.
10   Data relate to the productive sector (integrated R&D) and the higher education sector only.
11   Data refer to the Scientific Research Council only.
12   Not including data for law, humanities and education.
13   Not including data either for scientists and engineers engaged in private productive enterprises or for military and defense R&D.
14   Data for the general service sector and for medical sciences in the higher education sector are excluded.
15   Data relate to 2 research institutes only.
16   Including auxiliary personnel.
17   Data for scientists and engineers include personnel in the higher education sector. Data for women scientists and engineers and for technicians in the higher education sector are not included.
18   Not including data for social sciences and humanities.
19   Data refer to scientific and technological activities.
20   Data refer to the Faculty of Science at the University of Lebanon only.
21   Data relate to R&D activities concentrated mainly in government-financed research establishments only; social sciences and humanities in the higher education and general service sectors are excluded.
22   Not including social sciences and humanities in the higher education sector.
23   Not including data for the general service sector.
24   Not including scientists and engineers engaged in the administration of R & D; of military R&D, only that part carried out in civil establishments is included.

Source:
Organisation des Nations Unies pour l'éducation, la science et la culture (Paris).

¶ Toutes les données se rapportant à l'Allemagne avant le 3 octobre 1990 figurent dans deux rubriques séparées basées sur les territoires respectifs de la République fédérale d'Allemagne et l'ancienne République démocratique allemande selon la période indiquée. Pour les notes explicatives en détail sur les données concernant l'Allemagne, voir Annexe I - Nomenclature des pays.

1   Non compris les données relatives au secteur de la production (activités de R-D non intégrées).
2   Non compris les données relatives au secteur de la production.
3   Les données se réfèrent au nombre de scientifiques et ingénieurs et techniciens à plein temps et à temps partiel.
4   Non compris les activités de R-D de caractère militaire ou relevant de la défense nationale.
5   Les données ne se réfèrent qu'au secteur de service général.
6   Non compris les données relatives au secteur de l'enseignement supérieur.
7   Les données ne concernent que 23 des 26 instituts de recherche nationaux sous tutelle du Ministère fédéral de la Science et de la Technologie.
8   Non compris les sciences sociales et humaines dans le secteur de la production (activités de R-D intégrées).
9   Les données se réfèrent aux scientifiques et ingénieurs et techniciens employés dans les entreprises publiques.
10   Les données ne concernent que le secteur de la production (activités de R-D intégrées) et le secteur de l'enseignement supérieur.
11   Les données se réfèrent au "Scientific Research Council" seulement.
12   Non compris les données pour le droit, les sciences humaines et l'éducation.
13   Non compris les données relatives aux scientifiques et ingénieurs employés dans les entreprises privées ni les activités de R-D de caractère militaire ou relevant de la défense national.
14   Les données relatives au secteur de service général et les sciences médicales du secteur de l'enseignement supérieur sont aussi exclues.
15   Les données ne concernent que 2 instituts de recherche.
16   Y compris le personnel auxiliare.
17   Les données relatives aux scientifiques et ingénieurs comprennent le personnel dans le secteur de l'enseignement supérieur. Les données pour les femmes scientifiques et ingénieurs et pour les techniciens dans l'enseignement supérieur ne sont pas comprises.
18   Compte non tenu des sciences sociales et humaines.
19   Les données se réfèrent aux activités scientifiques et techniques.
20   Les données ne se réfèrent qu'à la Faculté des Sciences de l'Université de Liban.
21   Les données se réfèrent aux activités de R-D se trouvant pour la plupart dans les établissements de recherche financés par le gouvernement; les sciences sociales et humaines dans les secteurs de l'enseignement supérieur et de service général sont exclues.
22   Compte non tenue des sciences sociales et humaines dans le secteur de l'enseignement supérieur.

111

Number of scientists, engineers and technicians engaged in research and experimental development [*cont.*]

Nombre de scientifiques, d'ingénieurs et de techniciens employés à des travaux de recherche et de dévelopment expérimental [*suite*]

25 Not including social sciences and humanities in the productive sector.	23 Non compris les données pour le secteur de service général.
26 With the exception of economics and computer sciences, R & D in social sciences and humanities is excluded.	24 Non compris les scientifiques et ingénieurs employés dans les services administratifs de R-D; pour la R-D de caractère militaire, seule la partie effectuée dans les établissements civils a été considérée.
27 Data relate only to the productive sector.	25 Non compris les sciences sociales et humaines dans le secteur de la production.
28 Including skilled workers.	
29 Data relate to the higher education sector only.	26 A l'exception des sciences économiques et de l'informatique, la R-D dans les sciences sociales et humaines est exclue.
30 Not including data for the productive sector (integrated R&D) nor military and defence R&D.	27 Les données ne concernent que le secteur de la production.
31 Not including private non-profit organization.	28 Y compris les ouvriers qualifiés.
32 Not including social sciences and humanities in the productive and general service sectors.	29 Les données ne concernent que le secteur de l'enseignement supérieur.
33 Data relate to one research institute only.	30 Non compris le secteur de la production (données de R-D intégrées) ni les activités de R-D de caractère militaire ou relevant de la défense nationale.
34 Data refer only to 6 out of 11 institutes.	31 Non compris les organisations privées à but non lucratif.
35 Data refer to scientific workers, i.e. all persons with a higher scientific degree or scientific title, regardless of the nature of their work, persons undertaking research work in scientific establishments and scientific teaching staff in institutions of higher education; they also include persons undertaking scientific work in industrial enterprises.	32 Compte non tenu des sciences sociales et humaines dans les secteurs de la production et du service général.
	33 Les données ne concernent qu'un institut de recherche.
	34 Les données ne se rapportent qu'à 6 instituts de recherche sur 11.
	35 Les données se réfèrent aux travailleurs scientifiques, c.à.d., à toutes les personnes ayant un diplôme scientifique supérieur ou un titre scientifique, sans considération de la nature de leur travail, aux personnes qui effectuent un travail de recherche dans des insitutions scientifiques et au personnel scientifique enseignant dans des établissements d'enseignement supérieur; sont incluses aussi les personnes qui effectuent des travaux scientifiques dans les entreprises industrielles.

112
Expenditure for research and experimental development
Dépenses consacrées à la recherche
et au développement expérimental

National currency
Monnaie nationale

Country or area (Monetary unit) Pays ou zone (Unité monétaire)	Years Années	Total expenditure Dépenses totales (000)	Capital expenditure Dépenses de capital (000)	Current expenditure Dépenses courantes (000)	Current expenditure Dépenses courantes %
Africa · Afrique					
Benin (franc) [1] Bénin (franc) [1]	1989	...	...	3347695	...
Burundi (franc) [2] Burundi (franc) [2]	1989	536187	...	...	...
Central African Rep. (CFA franc) [3] Rép. centrafricaine (franc CFA) [3]	1984	680791	...	...	...
Congo (CFA franc) [4] Congo (franc CFA) [4]	1984	25530	14263	11267	44.1
Egypt (pound) [4] Egypte (livre) [4]	1982	40378	6136	34242	84.8
Gabon (CFA franc) [5] Gabon (franc CFA) [5]	1986	380000	130000	250000	65.8
Libyan Arab Jamahiriya (dinar) Jamah. arabe libyenne (dinar)	1980	22875	...	...	...
Madagascar (franc) Madagascar (franc)	1988	14371515	13378000	993515	6.9
Mauritius (rupee) Maurice (roupie)	1989	104300	50000	54300	52.1
Nigeria (naira) [6] Nigéria (naira) [6]	1987	86270	16655	69615	80.7
Rwanda (franc) Rwanda (franc)	1985	918560	819280	99280	10.8
Seychelles (rupee) [4] Seychelles (roupie) [4]	1983	12854	6771	6083	47.3
America, North · Amérique du Nord					
British Virgin Islands (US dollar) Iles Vierges brit. (E-U dollar)	1984	-	-	-	-
Canada (dollar) [7] Canada (dollar) [7]	1989	*8658000	...	...	...
Costa Rica (colon) Costa Rica (colon)	1986	612000	...	...	...
Cuba (peso) [8] Cuba (peso) [8]	1989	222000	63309	158691	71.5
El Salvador (colon) [9] El Salvador (colon) [9]	1989	290881	94917	195964	67.4
Guatemala (quetzal) [10] Guatemala (quetzal) [10]	1988	31859	...	...	...
Jamaica (dollar) [11] Jamaïque (dollar) [11]	1986	4016	130	3886	96.8
Mexico (peso) [12] Mexique (peso) [12]	1989	1050283	...	...	...
Nicaragua (córdoba) [4] Nicaragua (córdoba) [4]	1987	...	...	988970	...
Panama (balboa) [13] Panama (balboa) [13]	1986	173	-	173	100.0
Saint Lucia (dollar) Sainte-Lucie (dollar)	1984	12150	...	...	...
Trinidad and Tobago (dollar) Trinité-et-Tobago (dollar)	1984	143257	33336	109921	76.7
Turks and Caicos Is. (US dollar) Iles Turques et Caïques (E-U dollar)	1984	-	-	-	-
United States (dollar) [14] Etats-Unis (dollar) [14]	1988	139255000	4024000	135231000	97.1

112

Expenditure for research and experimental development
National currency [*cont.*]

Dépenses consacrées à la recherche et au développement expérimental
Monnaie nationale [*suite*]

Country or area (Monetary unit) Pays ou zone (Unité monétaire)	Years Années	Total expenditure Dépenses totales (000)	Capital expenditure Dépenses de capital (000)	Current expenditure Dépenses courantes (000)	Current expenditure Dépenses courantes %
America, South · Amérique du Sud					
Argentina (austral) Argentine (austral)	1988	3466700	741900	2724800	78.6
Brazil (cruzado) [12] [15] Brésil (cruzado) [12] [15]	1985	5390540	...	...	...
Chile (peso) [4] Chili (peso) [4]	1988	23161300	...	...	...
Colombia (peso) [1] Colombie (peso) [1]	1982	2754273	...	...	...
Guyana (dollar) [4] [16] Guyana (dollar) [4] [16]	1982	2800	...	...	...
Peru (sol) [17] Pérou (sol) [17]	1984	159024000	...	...	...
Venezuela (bolivar) [4] [18] Venezuela (bolivar) [4] [18]	1985	1411720	...	...	...
Asia · Asie					
Brunei Darussalam (dollar) [19] Brunéi Darussalam (dollar) [19]	1984	10880	2660	8220	75.6
Cyprus (pound) [5] Chypre (livre) [5]	1984	1173	...	...	...
India (rupee) Inde (roupie)	1988	34718100	5399930	29318170	84.4
Indonesia (rupiah) [20] Indonésie (rupiah) [20]	1988	259283000	64645000	194638000	75.1
Iran, Islamic Rep. of (rial) [18] Iran, Rép. Islamique d' (rial) [18]	1985	22010713	9464315	12546398	57.0
Israel (shekel) [21] Israël (shekel) [21]	1985	911100	...	...	...
Japan (yen) [7] [12] Japon (yen) [7] [12]	1988	10627572	1684476	8943096	84.2
Jordan (dinar) [4] Jordanie (dinar) [4]	1986	5587	1287	4300	77.0
Korea, Republic of (won) [12] [22] Corée, République de (won) [12] [22]	1988	2347000	925000	1422000	60.6
Kuwait (dinar) [17] Koweït (dinar) [17]	1984	71163	8147	63016	88.6
Lebanon (pound) [23] Liban (livre) [23]	1980	22000	...	...	...
Malaysia (ringgit) [18] Malaisie (ringgit) [18]	1989	97200	...	...	...
Maldives (rufiyaa) Maldives (rufiyaa)	1986	-	-	-	-
Pakistan (rupee) [4] [24] Pakistan (roupie) [4] [24]	1987	5582081	...	...	...
Philippines (peso) Philippines (peso)	1984	613410	98610	514800	83.9
Qatar (riyal) Qatar (riyal)	1986	6650	-	6650	100.0
Singapore (dollar) [25] Singapour (dollar) [25]	1987	374744	151021	223723	59.7
Sri Lanka (rupee) Sri Lanka (roupie)	1984	256799	82464	174335	67.9
Thailand (baht) Thaïlande (baht)	1987	2664380	543680	2120700	79.6
Turkey (lira) Turquie (livre)	1985	192465000	49150000	143315000	74.5
Viet Nam (dong) [18] Viet Nam (dong) [18]	1985	498000	...	...	...

112
Expenditure for research and experimental development
National currency [*cont.*]
Dépenses consacrées à la recherche et au développement expérimental
Monnaie nationale [*suite*]

Country or area (Monetary unit) Pays ou zone (Unité monétaire)	Years Années	Total expenditure Dépenses totales (000)	Capital expenditure Dépenses de capital (000)	Current expenditure Dépenses courantes (000)	Current expenditure Dépenses courantes %
Europe · Europe					
Austria (schilling) Autriche (schilling)	1985	17182272	3284207	13898065	80.9
Belgium (franc) [26] Belgique (franc) [26]	1988	91265100	...	...	...
Bulgaria (leva) Bulgarie (leva)	1989	1042400	109500	932900	89.5
Czechoslovakia (koruny) [27] Tchécoslovaquie (couron) [27]	1989	24721000	2621000	22100000	89.4
Denmark (kroner) Danemark (couron)	1989	11892000	1375000	10517000	88.4
Finland (markkaa) [28] Finlande (markkaa) [28]	1989	*8887800	1093400	6114400	84.8
France (franc) France (franc)	1988	130631000	13902000	116729000	89.4
Germany ℊ · Allemagne ℊ Federal Republic of Germany (D.mark) [29] Rép. féd. d'Allemagne (D.mark) [29]	1987	57240000	7018000	49578000	87.6
former German Dem. Rep. (DDR mark) [17] ancienne Rép. dém. allemande (DDR mark) [17]	1989	...	...	11880000	...
Greece (drachma) Grèce (drachme)	1986	18331000	...	...	...
Hungary (forint) [27] Hongrie (forint) [27]	1989	33441000	4031000	29410000	87.9
Ireland (pound) Irlande (livre)	1988	185800	...	...	...
Italy (lire) [12] Italie (lire) [12]	1988	13281284	1857879	11423405	86.0
Malta (lire) [30] Malte (lire) [30]	1988	10	1	9	90.0
Netherlands (guilder) [22] Pays-Bas (florin) [22]	1988	10163000	1431000	8732000	85.9
Norway (kroner) Norvège (couron)	1989	*12000000	*1300000	*10700000	89.2
Poland (zloty) [4] Pologne (zloty) [4]	1989	...	...	1226962	...
Portugal (escudo) Portugal (escudo)	1988	29910800	6390500	23520300	78.6
Romania (leu) Roumanie (leu)	1989	20866000	2417000	18449000	88.4
San Marino (lire) Saint-Marin (lire)	1986	-	-	-	-
Spain (peseta) [31] Espagne (peseta) [31]	1987	221154466	50148290	169059260	77.1
Sweden (krona) [32] Suède (couron) [32]	1987	30554000	2668000	27886000	91.3
Switzerland (franc) [33] Suisse (franc) [33]	1986	7015000	...	...	...
United Kingdom (pound) [34] Royaume-Uni (livre) [34]	1986	8777900	...	...	...
Yugoslavia (dinar) [4][12] Yougoslavie (dinar) [4][12]	1989	2152032	815082	1336950	62.1
Oceania · Océanie					
Australia (dollar) Australie (dollar)	1988	4187100	590200	3596900	85.9
Fiji (dollar) [35] Fidji (dollar) [35]	1986	3800	800	3000	78.9

112
Expenditure for research and experimental development
National currency [cont.]
Dépenses consacrées à la recherche et au développement expérimental
Monnaie nationale [suite]

Country or area (Monetary unit) Pays ou zone (Unité monétaire)	Years Années	Total expenditure Dépenses totales (000)	Capital expenditure Dépenses de capital (000)	Current expenditure Dépenses courantes (000)	Current expenditure Dépenses courantes %
French Polynesia (CFP franc) [35] Polynésie française (franc CFP) [35]	1983	324720	16280	308440	95.0
Guam (US dollar) [30] Guam (E-U dollar) [30]	1989	...	-	*1926	...
New Caledonia (CFP franc) [36] Nouvelle-Calédonie (franc CFP) [36]	1985	800820	93273	707547	88.4
Tonga (pa'anga) [35] Tonga (pa'anga) [35]	1980	426	147	279	65.5
former USSR · ancienne URSS					
former USSR (rouble) [37] ancienne URSS (rouble) [37]	1988	37800000	...	...	...

Source:
United Nations Educational, Scientific and Cultural Organization (Paris).

Source:
Organisation des Nations Unies pour l'éducation, la science et la culture (Paris).

ʃ All data shown which pertain to Germany prior to 3 October 1990 are indicated separately for the Federal Republic of Germany and the former German Democratic Republic based on their respective territories at the time indicated. For detailed explanatory notes on data pertaining to Germany, see Annex I- Country Nomenclature.

1 Not including data for the productive sector (non-integrated R&D).
2 Not including data for the productive sector nor labour costs at the Ministry of Public Health.
3 Not including data for the general service sector.
4 Not including military and defence R&D.
5 Not including data for the productive sector.
6 Data relate only to 23 out of 26 national research institutes under the Federal ministry of Science and Technology.
7 Not including social sciences and humanities in the productive sector (integrated R&D).
8 Data relate to government funds only and do not include military and defence R&D.
9 Data refer to the R&D activities performed in public enterprises. Not including data for the higher education sector and foreign funds.
10 Data refer to the productive sector (integrated R&D) and the higher education sector only.
11 Data relate to the Scientific Research Council only.
12 Figures in millions.
13 Data refer to the central government only.
14 Not including data for law, humanities and education. Data do not include capital expenditure in the productive sector.
15 Not including either military and defence R&D nor private productive enterprises.
16 Data for the general service sector and for medical sciences in the higher education sector are also excluded.
17 Data refer to scientific and technological activities.
18 Data relate to government expenditure only.
19 Data relate to 2 research institutes only.
20 Data relate to the general service sector only.
21 Data refer to the civilian sector only.
22 Not including military and defence R&D nor social sciences and humanities.
23 Data refer to the Faculty of Science at the University of Lebanon only.

ʃ Toutes les données se rapportant à l'Allemagne avant le 3 octobre 1990 figurent dans deux rubriques séparées basées sur les territoires respectifs de la République fédérale d'Allemagne et l'ancienne République démocratique allemande selon la période indiquée. Pour les notes explicatives en détail sur les données concernant l'Allemagne, voir Annexe I - Nomenclature des pays.

1 Non compris les données relatives au secteur de la production (activitiés de R-D non intégrées).
2 Non compris les données relatives au secteur de la production ni les dépenses de personnel au Ministre de la Santé Publique.
3 Non compris les données pour le secteur de service général.
4 Non compris les données de R-D de caractère militaire ou relevant de la défense nationale.
5 Non compris les données pour le secteur de la production.
6 Les données ne concernent que 23 des 26 instituts de recherche nationaux sous tutelle du Ministère fédéral de la Science et de la Technologie.
7 Non compris les données pour les sciences sociales et humaines du secteur de la production (activités de R-D intégrées).
8 Les données ne concernent que les fonds publics et ne comprennent pas les activités de R-D de caractère militaire ou relevant de la défense nationale.
9 Les données se réfèrent aux activités de R-D exercées dans les entreprises publiques. Non compris les données relatives au secteur secteur de l'enseignement supérieur et les fonds étrangers.
10 Les données ne concernent que le secteur de la production (activités de R-D intégrées) et le secteur de l'enseignement supérieur.
11 Les données se réfèrent au "Scientific Research Council" seulement.
12 Chiffres en millions.
13 Les données ne concernent que le gouvernement central.
14 Non compris les données pour le droit, les sciences humaines et l'éducation. Les données ne comprennent pas les dépenses en capital dans les secteurs de la production.
15 Non compris les activités de R-D de caractère militaire ou relevant de la défense nationale ni les entreprises privées de production.
16 Non compris les données relatives au secteur de service général et les sciences médicales du secteur de l'enseignement supérieur.
17 Non compris les données pour les sciences sociales et humaines.
18 Les données ne concernent que les dépenses du gouvernement.
19 Les données se réfèrent à 2 instituts de recherche seulement.
20 Les données ne concernent que le secteur de service général.
21 Les données ne concernent que le secteur civil.
22 Non compris les activités de R-D de caractère militaire ou relevant de la

112
Expenditure for research and experimental development
National currency [*cont.*]
Dépenses consacrées à la recherche et au développement expérimental
Monnaie nationale [*suite*]

24 Data relate to R&D activities concentrated mainly in government-financed research establishments only; social sciences and humanities in the higher education and general service sectors are excluded. Not including military and defence R&D.

25 Not including foreign funds nor social sciences and humanities.

26 Not including data from communities and regions.

27 Of military R&D, only that part carried out in civil establishments is included.

28 Data in total expenditure column include 1680 million markkas in the higher education sector for which a distribution by type of expenditure is not available; this figure has been excluded from the percentage calculation in the last column.

29 Data in Total column include 644 million Deutsche marks for which a distribution by type of expenditure is not available; this figure has been excluded from the percentage calculation in the last column. Due to methodological changes, data are not strictly comparable with the previous years. Not including data for social sciences and humanities in the productive sector.

30 Data relate to the higher education sector only.

31 Data in Total column include 1946916 thousand pesetas (disbursed by private non-profit organizations) for which a distribution by type of expenditure is not available; this figure has been excluded from the percentage calculation in the last column.

32 Not including data for social sciences and humanities in the productive and general service sectors.

33 Not including foreign and other funds.

34 Not including social sciences and humanities nor funds for R&D performed abroad.

35 Data relate to one research institute only.

36 Data refer only to 6 out of 11 research institutes.

37 Expenditure on Science from the national budget and other sources.

défense nationale ni les données pour les sciences sociales et humaines.

23 Les données ne se réfèrent qu'à la Faculté des Sciences de l'Université de Liban.

24 Les données se réfèrent aux activités de R-D se trouvant pour la plupart dans les établissements de recherche financés par le gouvernement; les sciences sociales et humaines dans les secteurs de l'enseignement supérieur et de service général sont exclues. Non compris les activités de R-D de caractère militaire ou relevant de la défense nationale.

25 Non compris les fonds étrangers ni les sciences sociales et humaines.

26 Non compris les données provenant des communautés et des régions.

27 Pour la R-D de caractère militaire, seule la partie effectuée dans les établissements civils a été considérées.

28 Les données comprennent 1680 millions de markkas du secteur de l'enseignement supérieur dont la répartition par type de dépenses n'est pas disponible; on n'a pas tenu compte de ce chiffre pour calculer le pourcentage de la dernière colonne.

29 Les données comprennent 644 millions de deutsche marks dont la répartition par type de dépenses n'est pas disponible; on n'a pas tenu compte de ce chiffre pour calculer le pourcentage de la dernière colonne. Suite à des changements de méthodologie, les données ne sont pas strictement comparables avec celles des années précédentes. Non compris les sciences sociales et humaines dans le secteur de la production.

30 Les données ne concernent que le secteur de l'enseignement supérieur.

31 Les données comprennent 1946916 milliers de pesetas (dépensées par les organisations privées à but non lucratif) dont la répartition par type de dépenses n'est pas disponible; on n'a pas tenu compte de ce chiffre pour calculer le pourcentage de la dernière colonne.

32 Non compris les données pour les sciences sociales et humaines dans les secteurs de la production et de service général.

33 Non compris les fonds étrangers et les fonds divers.

34 Non compris les sciences sociales et humaines ni les fonds de R-D exécutés à l'étranger.

35 Les données ne concernent qu'un institut de recherche.

36 Les données ne se rapportent qu'à 6 instituts de recherche sur 11.

37 Montant total des sommes dépensées pour la science d'après le budget national et autres sources.

113
Patents
Brevets
Applications, grants, patents in force: number
Demandes, délivrances, brevets en vigueur: nombre

Country or area Pays ou zone	Applications for patents Demandes de brevets			Grants of patents Brevets délivrés			Patents in force Brevets en vigueur		
	1988	1989	1990	1988	1989	1990	1988	1989	1990
African Intellectual Prop. Org [1] Org. africaine de la prop. intel. [1]	1 522	2 030	3 315	274	275	60	1 923	...	1 860
Algeria Algérie	206	204	185	...	...	592	...	...	...
Australia Australie	22 544	24 079	26 507	11 037	11 530	12 682	55 757	59 228	64 070
Austria Autriche	32 903	37 400	43 571	10 551	11 784	12 584	29 181	27 214	25 738
Bahamas Bahamas	36	...	27	36	...	...	...	...	...
Bangladesh Bangladesh	133	108	108	75	91	94	838	704	580
Barbados Barbade	1 272	1 772	3 138	...	...	...	...	...	...
Belgium Belgique	33 963	38 186	43 725	10 438	12 587	13 170	84 881	95 465	85 935
Belize Belize	6	7	...	1	...	...	1	...	...
Botswana Botswana	29	34	45	18	25	29	...	...	...
Brazil Brésil	10 192	11 035	12 434	3 040	3 510	3 355	...	...	...
Bulgaria Bulgarie	5 701	6 056	6 396	2 102	1 946	1 468	30 915	25 138	26 614
Burundi Burundi	...	2	3	...	2	2	...	9	129
Canada Canada	31 641	35 091	37 917	16 813	16 299	14 187	358 239	346 243	339 184
Chile Chili	742	833	833	445	610	607	8 077	7 985	7 878
China Chine	9 652	9 659	...	1 025	2 303	...	...	...	...
Colombia Colombie	505	606	...	538	662	...	...	...	...
Costa Rica Costa Rica	48	41	66	5	6	12	456	326	338
Cuba Cuba	342	304	260	73	50	57	...	549	340
Cyprus Chypre	50	...	43	48	...	43	585	...	517
Czechoslovakia Tchécoslovaquie	9 610	7 598	7 071	6 805	5 990	5 089	118 829	120 942	124 611
Denmark Danemark	11 214	10 900	39 447	2 815	2 616	2 376	9 674	10 964	11 759
Ecuador Equateur	121	120	...	84	70	...	645	623	...
Egypt Egypte	664	648	789	122	214	306	...	...	...
El Salvador El Salvador	...	53	54	...	6	12	...	322	...

113
Patents
Applications, grants, patents in force: number [*cont.*]
Brevets
Demandes, délivrances, brevets en vigueur : nombre [*suite*]

Country or area Pays ou zone	Applications for patents Demandes de brevets			Grants of patents Brevets délivrés			Patents in force Brevets en vigueur		
	1988	1989	1990	1988	1989	1990	1988	1989	1990
European Patent Office[2] Office européen de brevets[2]	52 312	57 765	...	19 750	22 558	...	...	...	...
Finland Finlande	9 543	10 847	12 610	2 690	2 512	2 467	14 212	14 958	16 576
France France	68 384	74 942	81 884	31 956	32 879	35 149	267 197	234 131	250 051
Gambia Gambie	24	30	40	14	22	25	...	...	...
Germany† · Allemagne† F. R. Germany R. f. Allemagne	95 998	102 427	110 349	38 890	42 233	42 860	227 106	219 168	232 791
former German D. R. anc. R. d. allemande	12 660	12 047	7 525	10 466	11 325	8 372	85 168	90 987	118 099
Ghana Ghana	39	36	62	31	33	45	...	...	176
Greece Grèce	13 758	15 538	18 908	385	503	1 975	...	...	...
Guatemala Guatemala	101	84	81	93	11	72	1 610	1 458	1 416
Guyana Guyana	14	10	6	11	7	3	226	240	218
Haiti Haïti	5	...	...	12	...	...	130	...	...
Honduras Honduras	27	...	23	32	...	19	620	...	509
Hong Kong Hong-kong	1 068	901	1 081	1 070	1 030	1 095	...	...	...
Hungary Hongrie	6 889	7 067	9 133	2 767	2 405	2 619	19 451	19 872	...
Iceland Islande	126	105	130	24	80	37	250	290	324
India Inde	3 537	3 648	3 820	3 454	1 986	1 611	13 102	13 794	10 550
Indonesia Indonésie	689	774	...	...	...	...	...	...	...
Iran, Islamic Rep. of Iran, Rép. islamique d'	340	311	355	76	179	299	...	...	...
Iraq Iraq	324	322	...	132	101	...	911	913	...
Ireland Irlande	3 901	4 230	4 735	737	925	852	6 960	6 794	6 505
Israel Israël	3 854	4 090	3 908	1 667	1 656	1 641	10 253	12 068	10 202
Italy Italie	53 675	49 091	55 569	25 195	15 832	17 794	...	...	...
Jamaica Jamaïque	...	...	81	...	...	21	...	...	491
Japan Japon	345 418	357 464	376 792	55 300	63 301	59 401	531 236	542 522	589 750

113
Patents
Applications, grants, patents in force: number [cont.]
     Brevets
     Demandes, délivrances, brevets en vigueur : nombre [suite]

Country or area Pays ou zone	Applications for patents Demandes de brevets			Grants of patents Brevets délivrés			Patents in force Brevets en vigueur		
	1988	1989	1990	1988	1989	1990	1988	1989	1990
Kenya Kenya	90	107	68	76	97	38	2 200	2 265	...
Korea, Dem. P. R. Corée, R. p. dém. de	6 492	8 526	10 033	2 682	2 659	2 677	26 642	29 295	31 972
Korea, Republic of Corée, République de	22 790	26 656	31 387	2 174	3 972	7 762	15 447	16 918	24 803
Lesotho Lesotho	17	44	35	...	21	23	...	350	...
Libyan Arab Jamah. Jamah. arabe libyenne	...	27	57	...	...	...	...	...	...
Luxembourg Luxembourg	23 826	28 355	34 777	5 187	6 169	6 749	17 706	19 444	21 899
Madagascar Madagascar	1 299	1 793	3 144	...	...	...	...	...	...
Malawi Malawi	1 343	1 836	3 204	99	74	62	294	326	288
Malaysia Malaisie	1 619	1 847	2 305	...	110	512			
Malta Malte	17	15	19	11	8	10	77	617	622
Mauritius Maurice	16	11	4	2	6	13	203	209	212
Mexico Mexique	4 549	4 741	5 289	3 411	2 268	1 752	32 521	26 398	23 219
Monaco Monaco	1 451	1 960	3 347	81	147	84	408	419	416
Mongolia Mongolie	75	92	73	37	47	38	409	460	497
Morocco Maroc	321	264	329	299	276	311	6 013	6 674	6 623
Namibia Namibie	...	...	163	...	...	139	...	...	1 082
Netherlands Pays-Bas	42 037	47 323	53 514	13 909	15 312	16 841	62 990	69 331	76 127
New Zealand Nouvelle-Zélande	4 425	4 467	4 671	2 705	2 841	3 481	...	...	...
Nigeria Nigéria	1 522	...	258	...	...	170	...	...	...
Norway Norvège	9 459	10 063	11 885	2 447	2 918	2 666	13 645	14 523	15 309
Pakistan Pakistan	...	...	547	...	...	380	...	...	...
Panama Panama	88	50	68	54	5	35	1 403	1 438	1 473
Paraguay Paraguay	44	40	52	96	44	35	3 249	836	780
Patent Cooperation Treaty[3] Traité de Coopération de brevets[3]	12 496	14 874	...	...	...	...	...	...	...
Peru Pérou	247	...	268	154	...	165	2 873	...	1 118

113

Patents
Applications, grants, patents in force: number [cont.]
Brevets
Demandes, délivrances, brevets en vigueur : nombre [suite]

Country or area Pays ou zone	Applications for patents Demandes de brevets			Grants of patents Brevets délivrés			Patents in force Brevets en vigueur		
	1988	1989	1990	1988	1989	1990	1988	1989	1990
Philippines Philippines	...	1 843	1 969	...	1 101	1 092	...	...	17 698
Poland Pologne	7 043	6 137	5 421	3 254	3 000	3 647	22 289	23 846	...
Portugal Portugal	2 464	3 397	3 642	929	1 236	563	9 295	9 823	10 385
Romania Roumanie	7 392	8 206	6 742	2 923	2 913	1 428	26 755	28 030	26 254
Rwanda Rwanda	6	1	4	6	1	4	122	123	81
Saint Lucia Sainte-Lucie	...	...	2	...	...	2	...	...	45
Saudi Arabia Arabie saoudite	...	70	455	...	...	...	...	...	...
Seychelles Seychelles	5	...	...	5	...	...	50	50	...
Singapore Singapour	...	838	1 028	...	1 064	1 238	...	...	...
Solomon Islands Iles Salomon	2	2	...	2	1	...	109	101	...
South Africa Afrique du Sud	9 734	9 976	10 469	5 571	5 586	5 465	17 728	...	...
Spain Espagne	26 229	30 685	49 026	5 129	7 134	7 597	...	...	...
Sri Lanka Sri Lanka	1 442	1 948	3 279	127	110	134	675	743	666
Sudan Soudan	1 316	1 816	3 202	18	26	36	...	...	...
Swaziland Swaziland	15	56	33	...	5	21	...	...	...
Sweden Suède	39 230	44 029	49 596	15 367	15 905	15 949	75 992	37 511	84 929
Switzerland Suisse	38 961	43 558	49 402	14 990	16 057	16 152	92 159	95 916	99 287
Thailand Thaïlande	1 119	1 424	1 940	86	164	141	237	393	...
Tunisia Tunisie	...	144	160	...	381	522	...	4 411	4 461
Turkey Turquie	900	1 048	1 228	372	481	486	6 022	6 257	6 281
Uganda Ouganda	47	46	59	39	36	41	1 606	1 645	1 675
former USSR ancienne URSS	176 913	151 808	123 797	83 983	84 577	84 658	9 608	9 437	9 369
United Kingdom Royaume-Uni	84 175	90 234	97 891	29 564	30 987	32 179	...	...	...
United Rep.Tanzania Rép. Unie de Tanzanie	37	23	...	37	23	...	2 074	...	...
United States Etats-Unis	147 344	161 660	176 100	77 924	95 539	90 366	1 141 062	1 151 178	1 154 204

113
Patents
Applications, grants, patents in force: number [cont.]
Brevets
Demandes, délivrances, brevets en vigueur: nombre [suite]

Country or area Pays ou zone	Applications for patents Demandes de brevets			Grants of patents Brevets délivrés			Patents in force Brevets en vigueur		
	1988	1989	1990	1988	1989	1990	1988	1989	1990
Uruguay Uruguay	172	134	...	89	61	...	2 316	2 040	...
Venezuela Venezuela	...	1 367	1 352	...	204	787	...	...	...
Viet Nam Viet Nam	72	70	79	14	32	14	49	82	95
Yugoslavia Yougoslavie	2 406	2 509	2 481	1 067	1 259	546	5 488	4 532	7 498
Zaire Zaïre	74	...	...	74	...	...	2 639	...	...
Zambia Zambie	109	92	105	88	140	93	745	857	903
Zimbabwe Zimbabwe	216	219	260	214	187	174	2 423	794	...

Source:
World Intellectual Property Organization (Geneva).

† All data shown which pertain to Germany prior to 3 October 1990 are indicated separately for the Federal Republic of Germany and the former German Democratic Republic based on their respective territories at the time indicated. Where data for united Germany (3 October 1990 and thereafter) are not available, available data are shown separately under the designations Federal Republic of Germany and former German Democratic Republic and pertain to the territorial boundaries prior to 3 October 1990. For detailed explanatory notes on data pertaining to Germany, see Annex I - Country Nomenclature.

1 Members of the African Intellectual Property Organization (AIPO), which includes Benin, Burkina Faso, Cameroon, Central African Republic, Chad, Congo, Côte d'Ivoire, Gabon, Mali, Mauritania, Niger, Senegal, Togo.
2 In 1989, the European Patent Office (EPO) was constituted by the following member countries: Austria, Belgium, France, Germany (Fed. Rep.), Greece, Italy, Liechtenstein, Luxembourg, Netherlands, Spain, Sweden, Switzerland, United Kingdom.
3 In 1989, the member countries of the Patent Cooperation Treaty (PCT) were as follows: Australia, Austria, Barbados, Belgium, Benin, Brazil, Bulgaria, Burkina Faso, Cameroon, Canada, Central African Republic, Chad, Congo, Democratic People's Republic of Korea, Denmark, Finland, France, Gabon, Germany (Fed. Rep.), Hungary, Italy, Japan, Liechtenstein, Luxembourg, Madagascar, Malawi, Mali, Mauritania, Monaco, Netherlands, Norway, Republic of Korea, Romania, Senegal, Spain, Sri Lanka, Sudan, Sweden, Switzerland, Togo, former USSR, United Kingdom, United States.

Source:
Organisation mondiale de la propriété intellectuelle (Genève).

† Toutes les données se rapportant à l'Allemagne avant le 3 octobre 1990 figurent dans deux rubriques séparées basées sur les territoires respectifs de la République fédérale d'Allemagne et l'ancienne République démocratique allemande selon la période indiquée. En l'absence de données pour l'Allemagne unifiée (à compter du 3 octobre 1990), les données disponibles sont fournies séparément sous les rubriques République fédérale d'Allemagne et ancienne République démocratique allemande et se rapportent aux limites territoriales antérieures au 3 octobre 1990. Pour les notes explicatives en détail sur les données concernant l'Allemagne, voir Annexe I – Nomenclature des pays.
1 Les membres de l'Organisation africaine de la propriété intellectuelle (OAPI): Bénin, Burkina Faso, Cameroun, Congo, Côte d'Ivoire, Gabon, Mali, Mauritanie, Niger, République centrafricaine, Sénégal, Tchad, Togo.
2 En 1989, l'Office européen de brevets (OEB) comprenait les pays membres suivants: Allemagne (Rép. féd.), Autriche, Belgique, Espagne, France, Grèce, Italie, Liechtenstein, Luxembourg, Pays-Bas, Royaume-Uni, Suède, Suisse.
3 En 1989, les pays membres du Traité de Coopération en matière de brevets (PCT) étaient les suivants: Allemagne (Rép. féd.), Australie, Autriche, Barbade, Belgique, Bénin, Brésil, Bulgarie, Burkina Faso, Cameroun, Canada, Congo, Danemark, Espagne, Etats-Unis, Finlande, France, Gabon, Hongrie, Italie, Japon, Liechtenstein, Luxembourg, Madagascar, Malawi, Mali, Mauritanie, Monaco, Norvège, Pays-Bas, République centrafricaine, République de Corée, République démocratique populaire de Corée, Roumanie, Royaume-Uni, Sénégal, Sri Lanka, Soudan, Suède, Suisse, Tchad, Togo, l'ancienne URSS.

114
Industrial designs
Dessins et modèles industriels
Applications, registrations, registrations in force: number
Demandes, enregistrements, enregistrements en vigueur : nombre

Country or area Pays ou zone	Applications for registrations Demandes d'enregistrements			Registrations granted Enregistrements accordés			Registrations in force Enregistrements en vigueur		
	1988	1989	1990	1988	1989	1990	1988	1989	1990
African Intellectual Prop. Org[1] Org. africaine de la prop. intel.[1]	70	50	29	37	...	...	...	...	...
Algeria Algérie	72	110	77	40	25	111	579	570	630
Argentina Argentine	1 521	...	...	1 674	...	...	10 589	...	...
Australia Australie	4 398	4 232	4 191	3 243	3 979	3 541	18 484	21 663	31 414
Austria Autriche	5 614	6 610	6 814	5 614	6 610	6 814	16 907	18 100	19 661
Bahamas Bahamas	8	...	51	8	...	...	...	...	...
Bangladesh Bangladesh	179	242	193	46	85	80	529	546	570
Benelux[2] Benelux[2]	2 998	3 079	2 966	2 907	2 892	2 710	19 039	19 506	19 743
Brazil Brésil	1 686	1 861	83	42	837	18	...	...	...
Bulgaria Bulgarie	162	298	334	73	83	187	566	555	616
Burundi Burundi	...	...	1	...	...	1	...	...	10
Canada Canada	2 873	3 083	2 681	2 326	2 788	2 700	17 285	19 510	21 692
Chile Chili	194	158	220	98	102	83	587	634	644
China Chine	1 959	2 519	...	731	1 318	...	1 488	2 648	...
Colombia Colombie	111	140	...	12	71	...	...	...	...
Cuba Cuba	52	44	32	4	31	7	47	75	79
Czechoslovakia Tchécoslovaquie	1 726	1 208	1 432	1 108	828	840	13 264	13 969	14 694
Denmark Danemark	1 551	1 557	1 466	393	1 445	1 914	6 094	6 681	7 695
Ecuador Equateur	72	62	...	34	23	...	202	204	...
Finland Finlande	1 094	1 135	1 030	1 061	978	969	6 125	6 301	6 518
France France	8 274	8 207	8 303	24 813	25 296	25 295	235 000	250 000	256 000
Germany† • Allemagne† F. R. Germany R. f. Allemagne	4 856	8 133	7 756	66 552	15 028	55 832	7 187	22 215	61 891
former German D. R. anc. R. d. allemande	1 043	905	438	761	1 001	536	...	...	1 755
Guatemala Guatemala	22	27	32	13	5	11	...	...	...
Guyana Guyana	...	...	2	...	...	...	...	...	...

114
Industrial designs
Applications, registrations, registrations in force: number [*cont.*]
Dessins et modèles industriels
Demandes, enregistrements, enregistrements en vigueur : nombre [*suite*]

Country or area Pays ou zone	Applications for registrations Demandes d'enregistrements			Registrations granted Enregistrements accordés			Registrations in force Enregistrements en vigueur		
	1988	1989	1990	1988	1989	1990	1988	1989	1990
Honduras Honduras	10	...	...	4	...	1	16	...	18
Hungary Hongrie	735	734	946	689	790	961	2 096	2 408	2 869
India Inde	1 386	1 151	1 037	1 173	960	976	14 632	15 703	15 903
Iraq Iraq	19	42	...	18	7	...	212	260	...
Ireland Irlande	439	638	546	421	326	560	2 717	2 968	3 159
Israel Israël	1 191	1 346	1 388	744	757	1 183	4 877	4 524	5 547
Italy Italie	4 438	...	...	2 851	...	...	...	...	...
Jamaica Jamaïque	6	13	14	9	3	5	38	51	56
Japan Japon	51 936	48 596	44 290	35 441	32 250	33 773	289 693	309 391	260 283
Korea, Dem. P. R. Corée, R. p. dém. de	2 544	1 023	9 093	110	865	2 480	414	...	...
Korea, Republic of Corée, République de	18 162	18 196	18 769	10 506	12 561	13 927	45 705	50 460	62 996
Liechtenstein Liechtenstein	3	0	8	3	...	8	32	...	34
Malawi Malawi	6	44	8	6	30	7	46	71	75
Malta Malte	71	32	39	48	34	34	851	872	894
Mexico Mexique	678	850	898	386	299	339	1 150	1 261	1 382
Monaco Monaco	20	11	28	15	11	12	199	204	220
Mongolia Mongolie	199	31	23	97	26	20	146	172	192
Morocco Maroc	403	322	298	403	322	298	3 492	3 761	3 970
Namibia Namibie	...	...	30	...	...	26	...	...	110
Netherlands Antilles Antilles néerlandaises	34	...	...	185	...	...	...	...	...
New Zealand Nouvelle-Zélande	673	656	637	795	691	566	...	...	...
Nigeria Nigéria	...	...	314	...	...	...	...	...	...
Norway Norvège	1 023	1 140	1 014	838	970	953	5 117	5 103	5 547
Paraguay Paraguay	91	86	...	84	67	...	255	288	...

114
Industrial designs
Applications, registrations, registrations in force: number [*cont.*]
Dessins et modèles industriels
Demandes, enregistrements, enregistrements en vigueur : nombre [*suite*]

Country or area Pays ou zone	Applications for registrations Demandes d'enregistrements			Registrations granted Enregistrements accordés			Registrations in force Enregistrements en vigueur		
	1988	1989	1990	1988	1989	1990	1988	1989	1990
Peru Pérou	89	...	89	35	...	112	198	...	360
Philippines Philippines	...	464	491	...	297	352	2 950	3 290	3 491
Poland Pologne	349	...	...	106	...	...	992	...	
Portugal Portugal	838	1 265	1 092	604	288	633	45 705	...	
Saint Lucia Sainte-Lucie	...	...	4	...	...	3	...	...	...
South Africa Afrique du Sud	2 179	1 295	1 161	...	...	...	...	...	...
Spain Espagne	3 500	3 703	3 521	3 425	3 444	3 785	...	...	...
Sri Lanka Sri Lanka	55	24	25	18	43	28	149	200	238
Swaziland Swaziland	...	15	...	...	...	...	4	13	...
Sweden Suède	3 000	2 992	2 751	2 358	1 967	2 086	16 628	15 548	15 577
Switzerland Suisse	816	952	935	711	774	779	5 569	5 794	5 978
Thailand Thaïlande	429	667	615	208	303	333	1 013	1 423	1 850
Tunisia Tunisie	...	94	83	...	94	94	705	777	830
former USSR ancienne URSS	4 355	5 340	4 744	2 738	3 138	3 661	26 867	30 005	33 582
United Kingdom Royaume-Uni	8 748	9 317	8 566	8 049	8 945	9 181	50 388	54 178	58 001
United States Etats-Unis	11 289	12 615	11 288	5 679	6 092	8 024	65 660	67 745	72 779
Uruguay Uruguay	49	61	...	18	11	...	311	289	...
Venezuela Venezuela	...	542	531	...	81	382	1 220	1 246	...
Viet Nam Viet Nam	6	66	...	...	14	...	...	14	...
Yugoslavia Yougoslavie	1 010	903	902	328	631	745	1 943	2 527	1 034
Zaire Zaïre	33	...	...	33	...	...	417	...	...
Zambia Zambie	5	13	15	2	2	13	32	34	36
Zimbabwe Zimbabwe	17	28	22	14	14	10	276	272	280

Source:
World Intellectual Property Organization (Geneva).

Source:
Organisation mondiale de la propriété intellectuelle
(Genève).

114
Industrial designs
Applications, registrations, registrations in force: number [*cont.*]
Dessins et modèles industriels
Demandes, enregistrements, enregistrements en vigueur : nombre [*suite*]

† All data shown which pertain to Germany prior to 3 October 1990 are indicated separately for the Federal Republic of Germany and the former German Democratic Republic based on their respective territories at the time indicated. Where data for united Germany (3 October 1990 and thereafter) are not available, available data are shown separately under the designations Federal Republic of Germany and former German Democratic Republic and pertain to the territorial boundaries prior to 3 October 1990. For detailed explanatory notes on data pertaining to Germany, see Annex I - Country Nomenclature.

1 Members of the African Intellectual Property Organization (AIPO), which includes Benin, Burkina Faso, Cameroon, Central African Republic, Chad, Congo, Côte d'Ivoire, Gabon, Mali, Mauritania, Niger, Senegal, Togo.
2 The Benelux Trademark Office acts as Trademark Office for Belgium, Luxembourg and the Netherlands.

† Toutes les données se rapportant à l'Allemagne avant le 3 octobre 1990 figurent dans deux rubriques séparées basées sur les territoires respectifs de la République fédérale d'Allemagne et l'ancienne République démocratique allemande selon la période indiquée. En l'absence de données pour l'Allemagne unifiée (à compter du 3 octobre 1990), les données disponibles sont fournies séparément sous les rubriques République fédérale d'Allemagne et ancienne République démocratique allemande et se rapportent aux limites territoriales antérieures au 3 octobre 1990. Pour les notes explicatives en détail sur les données concernant l'Allemagne, voir Annexe I – Nomenclature des pays.

1 Les membres de l'Organisation africaine de la propriété intellectuelle (OAPI): Bénin, Burkina Faso, Cameroun, Congo, Côte d'Ivoire, Gabon, Mali, Mauritanie, Niger, République centrafricaine, Sénégal, Tchad, Togo.
2 Le Bureau Benelux des marques sert de Bureau des marques de la Belgique, du Luxembourg et des Pays-Bas.

115
Trademarks and service marks
Marques de produits et de services
Applications, registrations, registrations in force: number
Demandes, enregistrements, enregistrements en vigueur : nombre

Country or area Pays ou zone	Applications for registrations Demandes d'enregistrements			Registrations granted Enregistrements accordés			Registrations in force Enregistrements en vigueur		
	1988	1989	1990	1988	1989	1990	1988	1989	1990
African Intellectual Prop. Org[1] Org. africaine de la prop. intel.[1]	777	893	900	...	...	...	...	...	...
Algeria Algérie	958	4 357	4 306	3 904	4 280	4 294	66 461	68 941	72 274
Aruba Aruba	...	1 895	...	...	1 878	1 597	...	...	...
Australia Australie	23 217	23 821	22 015	9 328	11 781	11 217	125 406	131 633	139 554
Austria Autriche	5 966	6 253	6 599	4 767	5 632	5 210	66 601	69 307	71 891
Bahamas Bahamas	470	...	588	470	...	264	...	...	...
Bangladesh Bangladesh	2 052	2 305	2 474	878	572	880	16 273	16 845	17 725
Benelux[2] Benelux[2]	17 585	19 106	21 044	18 178	18 068	17 101	170 784	180 358	188 100
Belize Belize	37	48	...	...	48	...	45	88	...
Brazil Brésil	81 229	69 080	63 880	33 198	24 732	34 314	275 674	274 982	307 247
Brunei Darussalam Brunéi Darussalam	779	989	880	382	819	886	19 302	19 416	21 032
Bulgaria Bulgarie	2 709	3 302	6 269	2 311	2 580	4 609	18 603	21 292	25 968
Burundi Burundi	...	65	101	...	65	101	2 367	2 451	2 552
Canada Canada	24 834	26 284	25 681	13 678	14 362	13 758	224 634	238 005	251 480
Chile Chili	...	20 107	21 220	...	11 717	13 049	...	136 296	147 724
China Chine	47 549	48 411	57 272	29 052	36 435	31 271	214 059	249 439	279 397
Colombia Colombie	13 910	16 956	...	3 435	2 537	...	...	...	...
Costa Rica Costa Rica	3 185	2 977	4 371	826	1 814	2 508	...	...	...
Cuba Cuba	620	775	1 697	508	750	1 206	12 431	13 209	14 397
Cyprus Chypre	1 359	...	1 721	622	...	773	15 127	15 257	15 732
Czechoslovakia Tchécoslovaquie	5 573	6 432	12 919	4 702	5 166	7 883	14 364	14 724	15 867
Denmark Danemark	9 166	9 684	10 034	4 621	8 294	8 513	95 623	100 837	105 942
Ecuador Equateur	2 268	3 952	...	3 256	1 628	...	124 500	129 878	...

115
Trademarks and service marks
Applications, registrations, registrations in force: number [cont.]
Marques de produits et de services
Demandes, enregistrements, enregistrements en vigueur : nombre [suite]

Country or area Pays ou zone	Applications for registrations Demandes d'enregistrements			Registrations granted Enregistrements accordés			Registrations in force Enregistrements en vigueur		
	1988	1989	1990	1988	1989	1990	1988	1989	1990
El Salvador El Salvador	...	1 409	2 973	...	1 325	247	...	...	...
Ethiopia Ethiopie	316	506	466	316	501	466	460	1 010	1 535
Finland Finlande	5 829	7 740	6 666	2 692	3 461	4 052	57 437	59 106	60 592
France France	79 948	90 310	95 091	70 581	74 719	79 052	410 547	436 297	459 797
Gambia Gambie	92	...	...	135	...	...	5 462	...	...
Germany† · Allemagne† F. R. Germany R. f. Allemagne	38 200	41 645	42 878	20 188	23 897	22 976	301 113	309 203	316 408
former German D. R. anc. R. d. allemande	526	530	8 446	416	459	605	19 251	18 950	19 101
Ghana Ghana	213	254	289	641	174	239	24 546	26 582	28 181
Guatemala Guatemala	6 007	5 459	4 795	2 354	2 628	3 081	...	...	...
Guyana Guyana	163	207	142	124	131	84	...	10 952	11 034
Honduras Honduras	1 920	...	2 041	1 590	...	1 472	52 745	52 701	53 298
Hong Kong Hong-kong	8 956	10 255	10 530	4 360	4 060	4 020	51 493	54 255	56 942
Hungary Hongrie	1 115	1 566	3 331	847	1 210	1 253	13 410	14 112	15 216
Iceland Islande	991	1 039	1 073	655	1 034	1 036	15 089	16 479	11 613
India Inde	19 582	18 846	20 681	4 366	5 335	6 429	84 132	83 337	80 789
Indonesia Indonésie	19 791	18 842	19 138	14 983	12 766	8 296	161 596	174 145	182 358
Iran, Islamic Rep. of Iran, Rép. islamique d'	1 757	2 082	2 593	1 046	1 270	1 383	...	...	...
Ireland Irlande	5 804	6 873	7 516	4 661	2 637	2 290	57 610	58 217	...
Israel Israël	3 147	3 699	3 750	1 426	1 260	1 550	25 208	26 160	26 802
Italy Italie	30 172	...	...	15 504	...	...	...	...	...
Jamaica Jamaïque	762	946	693	218	154	182	20 016	20 697	21 585
Japan Japon	172 813	172 780	171 726	119 287	119 598	117 185	1 050 324	1 094 230	1 140 933

115
Trademarks and service marks
Applications, registrations, registrations in force: number [*cont.*]
Marques de produits et de services
Demandes, enregistrements, enregistrements en vigueur : nombre [*suite*]

Country or area Pays ou zone	Applications for registrations Demandes d'enregistrements			Registrations granted Enregistrements accordés			Registrations in force Enregistrements en vigueur		
	1988	1989	1990	1988	1989	1990	1988	1989	1990
Kenya Kenya	865	932	...	611	627	...	26 480	27 070	...
Korea, Dem. P. R. Corée, R. p. dém. de	5 156	3 790	19 110	2 482	3 027	8 822	20 071	21 820	...
Korea, Republic of Corée, République de	34 681	39 832	46 826	17 272	22 263	23 789	131 023	148 116	163 775
Liechtenstein Liechtenstein	279	...	284	219	...	284	6 877	7 214	7 596
Malawi Malawi	218	422	419	197	212	302	14 475	15 190	15 841
Malaysia Malaisie	7 272	8 048	8 603	5 277	4 088	3 225	...	...	...
Malta Malte	589	585	683	246	105	134	15 421	16 312	16 776
Mauritius Maurice	341	...	515	278	...	231	...	5 714	5 945
Mexico Mexique	19 042	23 830	25 442	17 331	13 792	17 538	114 933	135 554	...
Monaco Monaco	682	523	577	396	408	542	6 058	5 298	4 622
Mongolia Mongolie	867	1 348	1 525	864	1 345	1 499	3 521	4 862	6 375
Morocco Maroc	1 986	1 805	1 909	1 986	1 805	1 909	20 230	21 645	22 741
Namibia Namibie	...	...	2 103	...	...	376	...	...	...
Netherlands Antilles Antilles néerlandaises	616	...	...	508	...	...	...	...	...
New Zealand Nouvelle-Zélande	12 325	8 329	10 072	2 085	4 524	5 525	2 085	4 587	10 322
Nigeria Nigéria	...	...	2 006	...	...	...	...	...	...
Norway Norvège	5 986	6 380	6 827	3 442	5 030	4 029	66 321	69 630	71 180
Panama Panama	2 522	2 837	3 398	2 379	2 303	1 708	48 972	50 366	51 578
Paraguay Paraguay	4 406	7 903	...	8 446	5 763	...	47 406	48 224	...
Peru Pérou	9 897	...	8 556	6 673	...	5 923	45 276	50 304	49 463
Philippines Philippines	...	3 826	3 928	...	4 687	2 511	30 820	30 685	31 756
Poland Pologne	1 732	2 257	6 101	844	844	1 803	20 519	19 409	19 883
Portugal Portugal	7 573	8 660	19 885	8 973	13 124	17 118	...	...	...

115
Trademarks and service marks
Applications, registrations, registrations in force: number [*cont.*]
Marques de produits et de services
Demandes, enregistrements, enregistrements en vigueur : nombre [*suite*]

Country or area Pays ou zone	Applications for registrations Demandes d'enregistrements			Registrations granted Enregistrements accordés			Registrations in force Enregistrements en vigueur		
	1988	1989	1990	1988	1989	1990	1988	1989	1990
Qatar Qatar	707	721	...	521	588	...	2 993	3 573	...
Romania Roumanie	2 701	2 695	4 987	2 161	2 658	3 127	147 902	151 526	158 640
Rwanda Rwanda	144	81	85	132	81	77	2 344	2 425	2 501
Saint Lucia Sainte-Lucie	...	...	190	...	...	190	...	...	...
Seychelles Seychelles	98	125	198	110	93	155	2 024	2 155	2 389
Singapore Singapour	...	8 816	8 791	...	8 803	7 901	...	...	...
Solomon Islands Iles Salomon	51	38	...	51	32	...	965	947	...
South Africa Afrique du Sud	11 762	12 213	11 268	3 304	5 093	5 985	...	...	...
Spain Espagne	66 339	76 789	78 315	22 114	23 663	36 020	...	...	...
Sri Lanka Sri Lanka	1 984	1 746	2 153	1 050	865	790	16 581	18 253	19 214
Swaziland Swaziland	...	565	...	...	565	...	98	898	...
Sweden Suède	11 363	12 225	11 920	4 058	2 614	4 913	94 941	93 873	95 344
Switzerland Suisse	6 709	7 030	6 749	15 004	16 255	15 478	282 633	284 384	285 114
Thailand Thaïlande	12 505	12 713	12 863	8 449	12 588	10 173	8 857	93 653	104 392
Tonga Tonga	...	45	47	...	45	39	32	...	...
Tunisia Tunisie	...	1 374	1 431	...	1 374	1 431	10 596	11 646	12 676
Turkey Turquie	5 798	6 658	...	4 873	5 084	...	92 482	99 247	...
Tuvalu Tuvalu	...	...	54	...	...	46	...	524	573
Uganda Ouganda	158	98	158	102	61	71	14 175	14 510	14 621
former USSR ancienne URSS	5 796	8 863	16 548	4 656	5 789	9 786	46 263	49 053	54 408
United Kingdom Royaume-Uni	38 006	40 052	39 632	14 381	22 374	28 389	277 534	287 406	307 012
United Rep.Tanzania Rép. Unie de Tanzanie	332	377	...	208	377	...	...	...	...
United States Etats-Unis	78 345	94 401	127 346	53 229	62 483	61 343	714 663	750 372	791 139

115
Trademarks and service marks
Applications, registrations, registrations in force: number [cont.]
Marques de produits et de services
Demandes, enregistrements, enregistrements en vigueur : nombre [suite]

Country or area Pays ou zone	Applications for registrations Demandes d'enregistrements			Registrations granted Enregistrements accordés			Registrations in force Enregistrements en vigueur		
	1988	1989	1990	1988	1989	1990	1988	1989	1990
Uruguay Uruguay	3 684	3 482	...	2 728	3 365	...	53 289	56 669	...
Venezuela Venezuela	...	19 465	21 770	...	4 222	...	...	...	...
Viet Nam Viet Nam	1 988	2 512	3 613	1 920	2 262	3 045	37 422	39 285	41 631
Yugoslavia Yougoslavie	4 244	5 038	8 435	965	4 573	7 394	13 876	1 285	4 298
Zaire Zaïre	340	...	...	339	...	...	5 241	...	...
Zambia Zambie	264	325	325	229	193	266	19 078	19 165	19 707
Zimbabwe Zimbabwe	832	1 065	1 107	590	799	666	22 763	22 624	22 282

Source:
World Intellectual Property Organization (Geneva).

† All data shown which pertain to Germany prior to 3 October 1990 are indicated separately for the Federal Republic of Germany and the former German Democratic Republic based on their respective territories at the time indicated. Where data for united Germany (3 October 1990 and thereafter) are not available, available data are shown separately under the designations Federal Republic of Germany and former German Democratic Republic and pertain to the territorial boundaries prior to 3 October 1990. For detailed explanatory notes on data pertaining to Germany, see Annex I - Country Nomenclature.

1 Members of the African Intellectual Property Organization (AIPO), which includes Benin, Burkina Faso, Cameroon, Central African Republic, Chad, Congo, Côte d'Ivoire, Gabon, Mali, Mauritania, Niger, Senegal, Togo.
2 The Benelux Trademark Office acts as Trademark Office for Belgium, Luxembourg and the Netherlands.

Source:
Organisation mondiale de la propriété intellectuelle (Genève).

† Toutes les données se rapportant à l'Allemagne avant le 3 octobre 1990 figurent dans deux rubriques séparées basées sur les territoires respectifs de la République fédérale d'Allemagne et l'ancienne République démocratique allemande selon la période indiquée. En l'absence de données pour l'Allemagne unifiée (à compter du 3 octobre 1990), les données disponibles sont fournies séparément sous les rubriques République fédérale d'Allemagne et ancienne République démocratique allemande et se rapportent aux limites territoriales antérieures au 3 octobre 1990. Pour les notes explicatives en détail sur les données concernant l'Allemagne, voir Annexe I – Nomenclature des pays.

1 Les membres de l'Organisation africaine de la propriété intellectuelle (OAPI): Bénin, Burkina Faso, Cameroun, Congo, Côte d'Ivoire, Gabon, Mali, Mauritanie, Niger, République centrafricaine, Sénégal, Tchad, Togo.
2 Le Bureau Benelux des marques sert de Bureau des marques de la Belgique, du Luxembourg et des Pays-Bas.

Technical notes, tables 111-115

Tables 111 and 112 present selected results of international data compilation by UNESCO in the field of science and technology, mostly obtained in response to the statistical surveys of scientific and technological activities. More comprehensive data can be found in the UNESCO *Statistical Yearbook.* [32] In utilizing these results the reader should keep in mind the factors which have an obvious bearing on the comparability and degree of accuracy of the data.

The absolute figures for R&D (research and experimental development) expenditure should not be compared country by country. Such comparisons would require the conversion of national currencies into a common currency by means of special R&D exchange rates. Official exchange rates do not always reflect the real costs of R&D activities and comparisons based on such rates can result in misleading conclusions, although they can be used to indicate a gross order of magnitude.

Abridged definitions suggested for use in the UNESCO surveys are given below, from the UNESCO *Yearbook.*

Type of personnel

Scientists and engineers are persons with scientific or technical training (usually completions of third-level education) in any field of science, including natural sciences, engineering and technology, the medical and agricultural sciences and the social sciences and humanities, who are engaged in professional work on R&D activities, administrators and other high-level personnel who direct the execution of R&D activities.

Technicians are persons who have received vocational or technical training in any branch of knowledge or technology of a specified standard (i.e. at least three years after the first stage of second-level education).

Full-time equivalent (FTE) is a measurement unit representing one person working full-time for a given period. This unit is used to convert figures relating to the number of part-time workers into the equivalent number of full-time workers. Data concerning personnel are normally calculated in FTE, especially in the case of scientists and engineers.

Research and experimental development (R&D)

In general R&D is defined as any creative systematic activity undertaken in order to increase the stock of knowledge, including knowledge of man, culture and society, and the use of this knowledge to devise new applications. It includes fundamental research, applied research in such fields as agriculture, medicine, industrial chemistry, and experimental development work leading to new devices, products or processes.

Notes techniques, tableaux 111-115

Les *Tableaux 111 et 112* présentent certains résultats d'une compilation internationale de données effectuée par l'UNESCO dans les domaines de la science et de la technologie, obtenues pour la plupart en réponse à des enquêtes statistiques sur des activités scientifiques et technologiques. On trouvera des données plus complètes dans l'*Annuaire statistique* de l'UNESCO [32]. Pour interpréter les résultats ainsi obtenus, le lecteur doit tenir compte des divers facteurs qui influent manifestement sur la comparabilité et l'exactitude des données.

Il faut éviter de comparer les chiffres absolus concernant les dépenses de R-D d'un pays à l'autre. On ne pourrait procéder à des comparaisons détaillées qu'en convertissant en une même monnaie les sommes libellées en monnaie nationale au moyen de taux de change spécialement applicables aux activités de R-D. Les taux de change officiels ne reflètent pas toujours le coût réel des activités de R-D, et les comparaisons établies sur la base de ces taux peuvent conduire à des conclusions trompeuses; toutefois, elles peuvent être utilisées pour donner une idée de l'ordre de grandeur.

On trouvera ci-dessous des définitions abrégées, tirées de l'*Annuaire* de l'UNESCO, dont l'utilisation est suggérée pour les enquêtes de cette organisation.

Classification du personnel

Scientifiques et ingénieurs : Personnes ayant une formation scientifique ou technique (ayant généralement terminé des études supérieures) dans n'importe quel domaine de la science, y compris les sciences naturelles, l'ingénierie et la technologie, les sciences médicales et agricoles et les sciences sociales et humaines, qui s'adonnent à des activités professionnelles de R-D; administrateurs et autre personnel de haut niveau qui dirigent l'exécution des activités de R-D.

Techniciens : Personnes qui ont reçu une formation professionnelle ou technique d'un niveau spécifié (à savoir au moins trois ans après achèvement du premier cycle de l'enseignement du second degré) dans n'importe quelle branche du savoir ou de la technologie.

Equivalent plein temps (EPT). Unité d'évaluation qui correspond à une personne travaillant à plein temps pendant une période donnée. Cette unité est utilisée pour convertir le nombre de travailleurs à temps partiel en un nombre de travailleurs à plein temps. En principe, les données concernant le personnel sont calculée en EPT, surtout dans le cas des scientifiques et des ingénieurs.

Recherche et développement expérimental (R-D)

En général, la recherche scientifique et le développement expérimental englobent tous les travaux systématiques et créateurs entrepris afin d'accroître la connaissance, y compris la connaissance de l'homme, de la culture et de la société, et l'utilisation de cette connaissance pour en tirer de nouvelles applications. Elle comprend la recherche fondamentale, la recherche appliquée dans des domaines tels que l'agriculture, la médecine, la chimie industrielle et le développement expérimental conduisant à la mise au point de nouveaux dispositifs, produits ou procédés.

Expenditure for research and experimental development

Total is defined as all expenditure (both current and capital) made for this purpose in the course of a reference year in institutions and installations established in the national territory, as well as installations physically situated abroad, land or experimental facilities rented or owned abroad and ships, vehicles, aircraft and satellites used by national institutions. Amounts spent on R&D activities carried out by international organizations established in the country in question are excluded from this total.

Current expenditure includes all payments made for the performance of R&D activities within units, institutions or sectors of performance, whatever the source or origin of funds, covering the cost of labor, minor equipment and expendable supplies and other current expenses including share of overheads.

Capital expenditure includes all payments made for the performance of R&D activities and relating to expenditure on major equipment and other capital expenditure. Depreciation on major instruments, equipment and buildings etc. should be excluded.

Tables 113-115: Data on patents include patent applications filed directly with the office concerned and grants made on the basis of such applications; inventors' certificates; patents of importation, including patents of introduction, revalidation patents and "patentes précautionales"; petty patents; patents applied and granted under the Patent Cooperation Treaty (PCT), the European Patent Convention, the Havana Agreement, the Harare Protocol of the African Regional Industrial Property Organization (ARIPO) and the African Intellectual Property Organization (OAPI).

Data on trademarks include trademark applications filed directly with the office concerned and registrations made on the basis of such applications; trademarks applied and registered under the Madrid Agreement.

Data on service marks include service mark applications filed directly with the office concerned and registrations made on the basis of such applications; service mark applied and registered under the Madrid Agreement.

Data on industrial design rights include applications filed directly with the office concerned and registration made on the basis of such applications; applications and registrations under the Hague Agreement and the Harare Protocol of the African Regional Industrial Property Organization (ARIPO).

Dépenses consacrées à la recherche et au développement

Le total se définit comme l'ensemble des dépenses (tant courantes qu'en capital) effectuées à ce titre, au cours d'une année de référence, dans les institutions et installations situées sur le territoire national, y compris dans les installations qui sont géographiquement situées à l'étranger : terrains ou installations d'essai acquis ou loués à l'étranger, ainsi que navires, véhicules, aéronefs et satellites utilisés par les institutions nationales. Sont exclues de ce total les dépenses pour les activités de R-D effectuées par les organisations internationales installées dans le pays considéré.

Les dépenses courantes comprennent tous les paiements effectués pour l'exécution d'activités de R-D à l'intérieur des unités, institutions ou secteurs d'exécution, quelle que soit la source ou l'origine des fonds, pour couvrir les dépenses de personnel, de petit matériel et de fournitures fongibles et autres dépenses courantes, y compris une part des frais généraux.

Les dépenses en capital comprennent tous les paiements effectués pour l'exécution d'activités de R-D qui ont trait aux dépenses de gros équipement et autres dépenses en capital. L'amortissement du gros appareillage, de l'équipement et des bâtiments doit en être exclu.

Tableaux 113-115 : Les données relatives aux brevets comprennent les demandes de brevet déposées directement au bureau intéressé et les délivrances effectuées sur la base de ces demandes; les brevets d'invention; les brevets d'importation; y compris les brevets d'introduction, les brevets de revalidation et les "patentes précautionales"; les petits brevets, les brevets demandés et délivrés en vertu du traité de coopération sur les brevets, de la Convention européenne relative aux brevets, de l'Accord de la Havane, du Protocole d'Hararé de l'Organisation régionale africaine de la propriété industrielle (ARIPO) et de l'Organisation africaine de la propriété intellectuelle (OAPI).

Les données relatives aux marques de fabrique comprennent les demandes de marque de fabrique déposées directement au bureau intéressé et les enregistrements effectués sur la base de ces demandes; et les marques de fabrique demandées et enregistrées aux termes de l'Accord de Madrid.

Les données relatives aux marques de service comprennent les demandes de marque de service déposées directement auprès du bureau intéressé et les enregistrements effectués sur la base de ces demandes; et les marques de service demandées et enregistrées aux termes de l'Accord de Madrid.

Les données relatives aux droits de conception industriels comprennent les demandes déposées directement auprès du bureau intéressé et l'enregistrement effectué sur la base de ces demandes; les demandes et enregistrements effectués aux termes de l'Accord de la Haye et du Protocole d'Hararé de l'Organisation régionale africaine de la propriété industrielle (ARIPO).

Part Four
International Economic Relations

XVII
International trade (tables 116-121)
XVIII
International tourism (tables 122-124)
XIX
Balance of payments (table 125)
XX
International finance (tables 126-129)
XXI
Development assistance (tables 130 and 131)

Part Four of the *Yearbook* presents statistics on
international economic relations in areas of international
trade, international tourism, balance of payments and
assistance to developing countries. The series cover all
countries or areas of the world for which data are
available.

Quatrième partie
Relations économiques internationales

XVII
Commerce International (tableaux 116 à 121)
XVIII
Tourisme International (tableaux 122 à 124)
XIX
Balance des paiements (tableau 125)
XX
Finances Internationales (tableaux 126 à 129)
XXI
Aide au développement (tableaux 130 et 131)

La quatrième partie de l'*Annuaire* présente des
statistiques sur les relations économiques internationales dans
les domaines du commerce international, du tourisme
international, de la balance des paiements et de l'assistance
aux pays en développement. Les séries couvrent tous les
pays ou les zones du monde pour lesquels des données sont
disponibles.

116
External trade conversion factors
Facteurs de conversion pour le commerce extérieur
US dollars per national currency
Monnaie nationale en dollars E-U

Country or area and unit Pays ou zone et monnaie	1982	1983	1984	1985	1986	1987	1988	1989	1990
A. Imports • Importations									
Algeria: dinar Algérie : dinar	0.21760	0.20890	0.20070	0.19890	0.21280	0.20580	0.17040	...	...
Angola: kwanza[1] Angola : kwanza[1]	0.03310	0.03310	0.03310	0.03310	0.03310	0.03310	...	...	...
Antigua and Barbuda: EC dollar[1][2] Antigua-et-Barbuda: EC dollar[1][2]	0.37040	0.37040	0.37040	0.37040	0.37040	0.37040	0.37040	...	...
Aruba: florin Aruba : florin	0.55560	0.55560	0.55560	0.55560	0.55870	0.55870	0.55870	...	...
Australia: dollar Australie : dollar	1.01730	0.90040	0.87700	0.69730	0.66620	0.69920	0.78400	0.77360	0.77890
Austria: schilling Autriche : schilling	0.05880	0.05560	0.05010	0.04860	0.06580	0.07940	0.08110	0.07550	0.08910
Bahamas: dollar Bahamas : dollar	1.00000	1.00000	1.00000	1.00000	1.00000	1.00000	1.00000	1.00000	1.00000
Bahrain: dinar Bahreïn : dinar	2.65960	2.65960	2.65950	2.65960	2.65960	2.65960	2.65960	2.65960	2.65960
Bangladesh: taka Bangladesh : taka	0.04530	0.04060	0.03940	0.03630	0.03320	0.03230	0.03160	0.03100	0.02890
Barbados: dollar Barbade : dollar	0.49720	0.49720	0.49720	0.49720	0.49720	0.49720	0.49720	0.49720	0.49720
Belgium-Luxembourg: franc Belgique-Luxembourg: franc	0.02200	0.01960	0.01730	0.01690	0.02240	0.02690	0.02730	0.02570	0.03000
Belize: dollar Belize : dollar	0.50000	0.50000	0.50000	0.50000	0.50000	0.50000	0.50000	0.50000	0.50000
Benin: CFA franc[1] Bénin : franc CFA[1]	0.00300	0.00260	0.00230	...	...	...	...	...	...
Bermuda: dollar[3] Bermudes : dollar[3]	1.00000	1.00000		.		.	.		
Brunei Darussalam: dollar[1] Brunéi Darussalam : dollar[1]	0.46700	0.47220	0.47020	0.45570	0.45040	0.47480	0.49690	0.51270	...
Bulgaria: lev Bulgarie : lev	1.05020	1.02650	0.99000	0.97080	1.06240	1.15240	1.19990	1.18560	1.26890
Burkina Faso: CFA franc Burkina Faso : franc CFA	0.00300	0.00260	0.00230	0.00230	0.00290	0.00330	0.00340	0.00310	...
Burundi: franc Burundi : franc	0.01110	0.01070	0.00840	0.00830	0.00880	0.00810	0.00710	0.00630	0.00590
Cameroon: CFA franc Cameroun : franc CFA	0.00310	0.00260	0.00230	0.00230	0.00290	0.00330	0.00340	0.00313	0.00367
Canada: dollar Canada : dollar	0.81110	0.81130	0.77210	0.73220	0.71970	0.75450	0.81900	0.84510	0.85680
Cape Verde: escudo Cap-Vert : escudo	0.01730	0.01400	0.01180	0.01090	0.01250	0.01380	0.01390	0.01290	...
Central African Rep.: CFA franc[1] Rép. centrafricaine: franc CFA[1]	0.00300	0.00260	0.00230	0.00220	0.00290	0.00330	0.00340	0.00310	...
Chad: CFA franc[1] Tchad : franc CFA[1]	0.00300	0.00260	0.00230	0.00220	0.00290	0.00330	0.00340	...	...
China: yuan renminbi Chine : yuan renminbi	0.53960	0.50750	0.44170	0.33590	0.28640	0.26770	0.26890	0.26880	0.20720

Country or area and unit Pays ou zone et monnaie	1982	1983	1984	1985	1986	1987	1988	1989	1990

B. Exports • Exportations

Country or area and unit Pays ou zone et monnaie	1982	1983	1984	1985	1986	1987	1988	1989	1990
Algeria: dinar Algérie : dinar	0.21660	0.20910	0.20110	0.19890	0.21260	0.20520	0.16970	...	...
Angola: kwanza[1] Angola : kwanza[1]	0.03310	0.03310	0.03310	0.03310	0.03310	0.03310	...	...	...
Antigua and Barbuda: EC dollar[1,2] Antigua-et-Barbuda: EC dollar[1,2]	0.37040	0.37040	0.37040	0.37040	0.37040	0.37040	0.37040	...	...
Aruba: florin Aruba : florin	0.55560	0.55560	0.55560	0.55560	0.55870	0.55870	0.55870	...	...
Australia: dollar Australie : dollar	1.01470	0.90230	0.87670	0.69630	0.66750	0.69950	0.78260	0.78160	0.77880
Austria: schilling Autriche : schilling	0.05880	0.05570	0.05010	0.04870	0.06580	0.07940	0.08110	0.07550	0.08910
Bahamas: dollar Bahamas : dollar	1.00000	1.00000	1.00000	1.00000	1.00000	1.00000	1.00000	1.00000	...
Bahrain: dinar Bahreïn : dinar	2.65960	2.65960	2.65960	2.65960	2.65960	2.65960	2.65960	2.65960	2.65960
Bangladesh: taka Bangladesh : taka	0.04490	0.04060	0.03940	0.03580	0.03330	0.03230	0.03160	0.03100	0.02890
Barbados: dollar Barbade : dollar	0.49720	0.49720	0.49720	0.49720	0.49720	0.49720	0.49720	0.49720	0.49720
Belgium-Luxembourg: franc Belgique-Luxembourg: franc	0.02190	0.01960	0.01730	0.01700	0.02250	0.02690	0.02740	0.02570	0.03000
Belize: dollar Belize : dollar	0.50000	0.50000	0.50000	0.50000	0.50000	0.50000	0.50000	0.50000	0.50000
Benin: CFA franc[1] Bénin : franc CFA[1]	0.00300	0.00260	0.00230	...	...	...	...	...	...
Bermuda: dollar[3] Bermudes : dollar[3]	1.00000	1.00000	.	.	.	.	.	.	.
Brunei Darussalam: dollar[1] Brunéi Darussalam : dollar[1]	0.46700	0.47220	0.47020	0.45490	0.45040	0.47480	0.49690	0.51270	...
Bulgaria: lev Bulgarie : lev	1.05040	1.02630	0.98940	0.97160	1.06300	1.15240	1.19990	1.18560	1.27160
Burkina Faso: CFA franc Burkina Faso : franc CFA	0.00310	0.00260	0.00230	0.00230	0.00290	0.00330	0.00340	0.00310	...
Burundi: franc Burundi : franc	0.01110	0.01070	0.00840	0.00820	0.00870	0.00810	0.00710	0.00630	0.00590
Cameroon: CFA franc Cameroun : franc CFA	0.00310	0.00260	0.00230	0.00220	0.00290	0.00330	0.00340	0.00313	0.00367
Canada: dollar Canada : dollar	0.81030	0.81130	0.77170	0.73220	0.71970	0.75470	0.81970	0.84520	0.85710
Cape Verde: escudo Cap-Vert : escudo	0.01730	0.01400	0.01180	0.01090	0.01250	0.01380	0.01360	0.01300	...
Central African Rep.: CFA franc[1] Rép. centrafricaine: franc CFA[1]	0.00300	0.00260	0.00230	0.00220	0.00290	0.00330	0.00340	0.00310	...
Chad: CFA franc[1] Tchad : franc CFA[1]	0.00300	0.00260	0.00230	0.00220	0.00290	0.00330	0.00340	...	...
China: yuan renminbi Chine : yuan renminbi	0.53940	0.50720	0.45030	0.33810	0.28590	0.26830	0.26900	0.26860	0.20800

116
External trade conversion factors
US dollars per national currency [*cont.*]
Facteurs de conversion pour le commerce extérieur
Monnaie nationale en dollars E-U [*suite*]

Country or area and unit Pays ou zone et monnaie	1982	1983	1984	1985	1986	1987	1988	1989	1990
Congo: CFA franc[1] Congo : franc CFA[1]	0.00300	0.00260	0.00230	0.00220	0.00290	0.00330	0.00340	0.00310	0.00370
Côte d'Ivoire: CFA franc Côte d'Ivoire : franc CFA	0.00290	0.00260	0.00230	0.00230	0.00290	0.00330	0.00340	0.00320	...
Cuba: peso Cuba : peso	1.20010	1.00000	1.00000	1.00000	1.21190	1.00000	1.00000	...	...
Cyprus: pound Chypre : livre	2.10420	1.89900	1.71200	1.63580	1.94070	2.08570	2.14250	2.01760	2.18350
Czechoslovakia: koruna Tchécoslovaquie : couronne	0.16350	0.15850	0.15020	0.05830	0.06670	0.07310	0.06960	0.06650	0.05500
Denmark: krone Danemark : couronne	0.12000	0.10930	0.09650	0.09430	0.12350	0.14610	0.14870	0.13680	0.16230
Djibouti: franc Djibouti : franc	0.00560	0.00560	0.00560	0.00560	0.00560	0.00560	0.00560	0.00560	0.00560
Dominica: EC dollar[2] Dominique : EC dollar[2]	0.37040	0.37040	0.37040	0.37040	0.37040	0.37040	0.37040	0.37040	0.37040
Egypt: £ Egypte : £	1.42860	1.42860	1.42860	0.70700	1.07810	0.66880	0.53080	0.44720	0.37070
El Salvador: colon El Salvador : colon	0.40000	0.40000	0.40000	0.40000	0.20000	0.20000	0.20000	0.20000	0.15020
Ethiopia: birr Ethiopie : birr	0.48350	0.48310	0.48310	0.48310	0.48310	0.48310	0.48310	0.48310	0.48310
Faeroe Islands: krone Iles Féroé : couronne	0.12000	0.10930	0.09650	0.09430	0.12620	0.14610	0.14850	0.13680	0.16230
Fiji: dollar Fidji : dollar	1.08240	0.98190	0.92370	0.86920	0.88170	0.81500	0.70040	0.67500	0.67780
Finland: markka Finlande : markka	0.20680	0.17970	0.16660	0.16230	0.19760	0.22850	0.23710	0.23330	0.26210
France: franc France : franc	0.15270	0.13180	0.11470	0.11200	0.14450	0.16710	0.16820	0.15690	0.18410
French Guiana: franc Guyane française : franc	0.15220	0.13040	0.11520	0.11250	0.14320	0.16620	0.16690	0.15700	0.18350
French Polynesia: CFP franc Polynésie française : franc CFP	0.00830	0.00720	0.00630	0.00620	0.00790	0.00910	0.00920	0.00860	0.01010
Gabon: CFA franc[1] Gabon : franc CFA[1]	0.00300	0.00260	0.00230	0.00220	0.00290	0.00330	...	0.00310	...
Gambia: dalasi Gambie : dalasi	0.43810	0.37890	0.27810	0.25750	0.14170	0.14140	0.14910	0.13100	0.12730
Germany† · Allemagne† F. R. Germany: deutsche mark R. f. Allemagne : deutsche mark	0.41400	0.39190	0.34830	0.34180	0.46130	0.55710	0.56970	0.53520	0.62220
former German D. R.: mark anc. R. d. allemande : mark	0.28900	0.28250	0.27470	0.27030	0.30300	0.33220	0.34010	0.12290	...
Ghana: cedi Ghana : cedi	0.36360	0.04920	0.02780	0.01840	0.01120	0.00620	0.00490	0.00370	
Gibraltar: pound[1] Gibraltar : livre[1]	1.75050	1.51700	1.37710	1.29640	1.46700	1.63890	...	...	...
Greece: drachma Grèce : drachme	0.01510	0.01120	0.00870	0.00720	0.00720	0.00740	0.00680	0.00610	0.00630

Country or area and unit Pays ou zone et monnaie	1982	1983	1984	1985	1986	1987	1988	1989	1990
Congo: CFA franc[1] Congo : franc CFA[1]	0.00300	0.00260	0.00230	0.00220	0.00290	0.00330	0.00340	0.00310	0.00370
Côte d'Ivoire: CFA franc Côte d'Ivoire : franc CFA	0.00300	0.00260	0.00230	0.00220	0.00290	0.00340	0.00340	0.00330	...
Cuba: peso Cuba : peso	1.20010	1.00000	1.00000	1.00000	1.18270	1.00000	1.00000	...	...
Cyprus: pound Chypre : livre	2.10280	1.89640	1.70620	1.63920	1.94330	2.08460	2.14360	2.01770	2.17920
Czechoslovakia: koruna Tchécoslovaquie : couronne	0.16360	0.15870	0.15020	0.05830	0.06670	0.07310	0.06960	0.06640	0.05520
Denmark: krone Danemark : couronne	0.12000	0.10930	0.09650	0.09430	0.12350	0.14610	0.14880	0.13690	0.16230
Djibouti: franc Djibouti : franc	0.00560	0.00560	0.00560	0.00560	0.00560	0.00560	0.00560	0.00560	0.00560
Dominica: EC dollar[2] Dominique : EC dollar[2]	0.37040	0.37040	0.37040	0.37040	0.37040	0.37040	0.37040	0.37040	0.37040
Egypt: £ Egypte : £	1.42860	1.42860	1.42860	0.70700	1.07810	0.66880	0.53080	0.44720	0.37130
El Salvador: colon El Salvador : colon	0.40000	0.40000	0.40000	0.40000	0.20000	0.20000	0.20000	0.20000	0.14470
Ethiopia: birr Ethiopie : birr	0.48360	0.48310	0.48310	0.48310	0.48310	0.48310	0.48310	0.48310	0.48310
Faeroe Islands: krone Iles Féroé : couronne	0.12000	0.10930	0.09650	0.09430	0.12690	0.14610	0.14850	0.13680	0.16230
Fiji: dollar Fidji : dollar	1.06870	0.97850	0.91450	0.87120	0.87570	0.78940	0.70020	0.67370	0.67800
Finland: markka Finlande : markka	0.20840	0.17960	0.16690	0.16210	0.19780	0.22870	0.23840	0.23320	0.26300
France: franc France : franc	0.15280	0.13130	0.11470	0.11210	0.14460	0.16720	0.16820	0.15690	0.18400
French Guiana: franc Guyane française : franc	0.15490	0.12920	0.11450	0.11140	0.14310	0.16630	0.16740	0.15660	0.18320
French Polynesia: CFP franc Polynésie française : franc CFP	0.00830	0.00720	0.00620	0.00620	0.00790	0.00910	0.00920	0.00860	0.01010
Gabon: CFA franc[1] Gabon : franc CFA[1]	0.00300	0.00260	0.00230	0.00220	0.00290	0.00330	...	...	...
Gambia: dalasi Gambie : dalasi	0.44710	0.38210	0.28820	0.24980	0.14740	0.14130	...	...	0.12640
Germany† · Allemagne† F. R. Germany: deutsche mark R. f. Allemagne : deutsche mark	0.41250	0.39190	0.34780	0.34260	0.46220	0.55760	0.56950	0.53510	0.61990
former German D. R.: mark anc. R. d. allemande : mark	0.28900	0.28250	0.27470	0.27030	0.30300	0.33220	0.34010	0.12290	...
Ghana: cedi Ghana : cedi	0.36360	0.04920	0.02790	0.01860	0.01120	0.00620	0.00490	0.00370	...
Gibraltar: pound[1] Gibraltar : livre[1]	1.75050	1.51700	1.37710	1.29640	1.46700	1.63890	...	...	...
Greece: drachma Grèce : drachme	0.01500	0.01120	0.00890	0.00720	0.00720	0.00740	0.00680	0.00610	0.00630

116
External trade conversion factors
US dollars per national currency [cont.]
Facteurs de conversion pour le commerce extérieur
Monnaie nationale en dollars E-U [suite]

Country or area and unit Pays ou zone et monnaie	1982	1983	1984	1985	1986	1987	1988	1989	1990
Greenland: krone Groënland : couronne	0.12000	0.10930	0.09650	0.09430	0.12350	0.14610	0.14850	0.13680	0.16230
Grenada: EC dollar[2] Grenade : EC dollar[2]	0.37040	0.37040	0.37040	0.37040	0.37040	0.37040	0.37040	...	...
Guadeloupe: franc Guadeloupe : franc	0.15160	0.13090	0.11540	0.10800	0.13960	0.16670	0.17340	0.15860	0.18460
Guyana: dollar Guyana : dollar	0.33330	0.33330	0.26690	0.23540	0.23260	...	...	0.03680	0.02530
Haiti: gourde Haïti : gourde	0.20000	0.20000	0.20000	0.20000	0.20000	0.20000	0.20000	0.20000	0.20000
Honduras: lempira Honduras : lempira	0.50000	0.50000	0.50000	0.50000	0.50000	0.50000	0.50000	0.50000	...
Hong Kong: dollar Hong-kong : dollar	0.16470	0.13750	0.12790	0.12840	0.12820	0.12820	0.12810	0.12820	0.12840
Hungary: forint Hongrie : forint	0.02710	0.02330	0.02070	0.02010	0.02190	0.02130	0.01980	0.01680	0.01610
Iceland: krona Islande : couronne	0.08090	0.03960	0.03130	0.02400	0.02430	0.02600	0.02310	0.01740	0.01720
India: rupee Inde : roupie	0.10590	0.09900	0.08860	0.08170	0.07920	0.07720	0.07100	0.06150	0.05630
Iran, Islamic Rep. of: rial Iran, Rép. islamique d' : rial	0.01200	0.01160	...	...	...	...	...	...	...
Iraq: dinar Iraq : dinar	3.34390	3.21690	3.21690	3.21690	3.21690	3.21690	...	...	3.21690
Ireland: pound Irlande : livre	1.41190	1.24740	1.08410	1.06270	1.35460	1.48790	1.52430	1.41840	1.66090
Italy: lira Italie : lire	0.00070	0.00070	0.00060	0.00050	0.00070	0.00080	0.00080	0.00070	0.00080
Jamaica: dollar Jamaïque : dollar	0.56140	0.53870	0.25420	0.18060	0.18260	0.18230	0.18220	0.17390	0.13880
Japan: yen Japon : yen	0.00400	0.00420	0.00420	0.00420	0.00590	0.00690	0.00780	0.00730	0.00690
Jordan: dinar Jordanie : dinar	2.83690	2.74630	2.59890	2.54290	2.86040	2.95220	2.72710	1.73370	1.50810
Kenya: shilling Kenya : shilling	0.08900	0.07500	0.06960	0.06090	0.06170	0.06070	0.05630	0.04850	0.04360
Kuwait: dinar Koweït : dinar	3.47370	3.43170	3.38000	3.42610	3.42330	3.59080	3.52690	3.40770	...
Liberia: dollar Libéria : dollar	1.00000	1.00000	1.00000	1.00000	1.00000	1.00000	1.00000	...	...
Libyan Arab Jamah.: dinar Jamah. arabe libyenne : dinar	3.37780	3.37780	3.37780	3.37780	3.18600	3.69510	3.50510	...	...
Macau: pataca Macao : pataca	0.15810	0.13540	0.12470	0.12570	0.12150	0.12420	0.12420	0.12450	0.12420
Madagascar: franc[1] Madagascar : franc[1]	0.00290	0.00230	0.00170	0.00150	0.00150	0.00090	0.00070	0.00060	0.00070
Malawi: kwacha Malawi : kwacha	0.96620	0.85550	0.70530	0.58340	0.53900	0.45460	0.37870	0.36300	0.36540
Malaysia: ringgit Malaisie : ringgit	0.42800	0.43080	0.42570	0.41400	0.38710	0.39710	0.38210	0.37040	0.36980

Country or area and unit Pays ou zone et monnaie	1982	1983	1984	1985	1986	1987	1988	1989	1990
Greenland: krone Groënland : couronne	0.12000	0.10930	0.09650	0.09430	0.12350	0.14610	0.14850	0.13680	0.16230
Grenada: EC dollar[2] Grenade : EC dollar[2]	0.37040	0.37040	0.37040	0.37040	0.37040	0.37040	...	...	...
Guadeloupe: franc Guadeloupe : franc	0.15150	0.13270	0.11550	0.10760	0.13960	0.16540	0.17390	0.15850	0.18220
Guyana: dollar Guyana : dollar	0.33330	0.33330	0.26010	0.23630	0.23330	0.10280	0.10000	0.03710	0.02500
Haiti: gourde Haïti : gourde	0.20000	0.20000	0.20000	0.20000	0.20000	0.20000	0.20000	0.20000	0.20000
Honduras: lempira Honduras : lempira	0.50000	0.50000	0.50000	0.50000	0.50000	0.50000	0.50000	0.50000	...
Hong Kong: dollar Hong-kong : dollar	0.16470	0.13750	0.12790	0.12840	0.12820	0.12820	0.12810	0.12820	0.12840
Hungary: forint Hongrie : forint	0.02700	0.02320	0.02070	0.02010	0.02190	0.02130	0.01980	0.01680	0.01610
Iceland: krona Islande : couronne	0.08080	0.04020	0.03140	0.02410	0.02440	0.02590	0.02310	0.01750	0.01720
India: rupee Inde : roupie	0.10600	0.09890	0.08810	0.08150	0.07930	0.07720	0.07120	0.06160	0.05640
Iran, Islamic Rep. of: rial Iran, Rép. islamique d' : rial	0.01190	0.01160	0.01100	0.01100	...	...	...	...	...
Iraq: dinar Iraq : dinar	3.34790	3.21690	...	...	...	...	...	...	3.21690
Ireland: pound Irlande : livre	1.40260	1.24140	1.08220	1.06350	1.35550	1.48950	1.52340	1.41770	1.65850
Italy: lira Italie : lire	0.00070	0.00070	0.00060	0.00050	0.00070	0.00080	0.00080	0.00070	0.00080
Jamaica: dollar Jamaïque : dollar	0.56140	0.52600	0.25790	0.18040	0.18260	0.18230	0.18220	0.17610	0.13870
Japan: yen Japon : yen	0.00400	0.00420	0.00420	0.00420	0.00590	0.00690	0.00780	0.00730	0.00690
Jordan: dinar Jordanie : dinar	2.84570	2.74840	2.58710	2.54160	2.85950	2.95220	2.71640	1.73420	1.50560
Kenya: shilling Kenya : shilling	0.08590	0.07510	0.06970	0.06090	0.06160	0.06080	0.05630	0.04870	0.04360
Kuwait: dinar Koweït : dinar	3.47180	3.43080	3.37920	3.32470	3.43140	3.58660	3.58440	3.39740	...
Liberia: dollar Libéria : dollar	1.00000	1.00000	1.00000	1.00000	1.00000	1.00000	1.00000	...	...
Libyan Arab Jamah.: dinar Jamah. arabe libyenne : dinar	3.37780	3.37780	3.37780	3.37780	3.18600	3.69510	3.50510	...	...
Macau: pataca Macao : pataca	0.15880	0.13520	0.12470	0.12570	0.12090	0.12420	0.12420	0.12450	0.12420
Madagascar: franc[1] Madagascar : franc[1]	0.00290	0.00230	0.00170	0.00150	0.00150	0.00090	0.00070	0.00060	0.00070
Malawi: kwacha Malawi : kwacha	0.95740	0.84480	0.70150	0.58940	0.53000	0.45220	0.37730	0.35990	0.37170
Malaysia: ringgit Malaisie : ringgit	0.42790	0.43050	0.42570	0.41470	0.38750	0.39700	0.38230	0.37020	0.36980

116
External trade conversion factors
US dollars per national currency [cont.]
Facteurs de conversion pour le commerce extérieur
Monnaie nationale en dollars E-U [suite]

Country or area and unit Pays ou zone et monnaie	1982	1983	1984	1985	1986	1987	1988	1989	1990
Mali: CFA franc[1] Mali : franc CFA[1]	0.00300	0.00260	0.00230	0.00220	0.00290	0.00330	0.00340	0.00310	...
Malta: lira Malte : lire	2.42650	2.31370	2.16900	2.14170	2.54960	2.89580	3.02330	2.91700	3.14730
Martinique: franc Martinique : franc	0.15210	0.13170	0.11460	0.11280	0.14500	0.16690	0.16770	0.15700	0.18370
Mauritania: ouguiya Mauritanie : ouguiya	0.01920	0.01820	0.01620	0.01310	0.01350	0.01350	0.01330	0.01200	...
Mauritius: rupee Maurice : roupie	0.09200	0.08410	0.07280	0.06510	0.07430	0.07770	0.07460	0.06560	0.06740
Morocco: dirham Maroc : dirham	0.16610	0.14060	0.11360	0.09950	0.10960	0.11990	0.12200	0.11790	0.12130
Mozambique: metical[1] Mozambique : metical[1]	0.02650	0.02490	0.02360	0.02320	0.02470	0.00350	0.00190	...	...
Myanmar: kyat Myanmar : kyat	0.12830	0.12440	0.11850	0.11770	0.13560	0.15070	0.15860	0.14930	0.15820
Nepal: rupee Népal : roupie	0.07550	0.06880	0.06090	0.05480	0.04700	0.04500	0.04290	0.03690	0.03410
Netherlands: guilder Pays-Bas : florin	0.37450	0.35130	0.31240	0.30270	0.40900	0.49500	0.50830	0.47180	0.55060
Netherlands Antilles: guilder Antilles néerlandaises : florin	0.55560	0.55560	0.55550	0.55560	0.55560	0.55560	0.55560	0.55870	0.55870
New Caledonia: CFP franc Nouvelle-Calédonie : franc CFP	0.00840	0.00720	0.00630	0.00620	0.00730	0.00910	0.00920	0.00860	0.01020
New Zealand: dollar Nouvelle-Zélande : dollar	0.74680	0.66730	0.54970	0.49670	0.52780	0.59270	0.65130	0.59660	0.59670
Niger: CFA franc[1] Niger : franc CFA[1]	0.00300	0.00260	0.00230	0.00220	...	...	...	...	...
Nigeria: naira Nigéria : naira	1.48540	1.38230	1.30850	1.12060	0.67420	0.24960	0.22040	0.13580	0.12440
Norway: krone Norvège : couronne	0.15520	0.13450	0.12280	0.11740	0.13530	0.14890	0.15370	0.14490	0.16020
Oman: rial Omani Oman : rial omani	2.89520	2.89520	2.89520	2.89520	2.60080	2.60080	2.60080	2.60080	2.60080
Pakistan: rupee Pakistan : roupie	0.08440	0.07630	0.07160	0.06280	0.06020	0.05750	0.05550	0.04860	0.04590
Panama: balboa Panama : balboa	1.00000	1.00000	1.00000	1.00000	1.00000	1.00000	1.00000	1.00000	1.00000
Papua New Guinea: kina Papouasie-Nvl-Guinée : kina	1.35300	1.19520	1.11730	0.99830	1.03270	1.10200	1.15530	1.16810	1.04590
Paraguay: guarani Paraguay : guaraní	0.00710	0.00730	0.00470	0.00420	0.00300	0.00180	0.00180	0.00095	...
Portugal: escudo Portugal : escudo	0.01270	0.00920	0.00690	0.00590	0.00670	0.00710	0.00700	0.00640	0.00710
Qatar: riyal Qatar : riyal	0.27450	0.27470	0.27470	0.27470	0.27470	0.27470	0.27470	0.27470	...
Réunion: franc Réunion : franc	0.15150	0.13080	0.11470	0.11270	0.14470	0.16740	0.16520	0.15660	0.18410
Romania: leu Roumanie : leu	0.06670	0.05870	0.04800	0.05710	0.05950	0.06250	0.06250	0.06250	0.04340

Country or area and unit Pays ou zone et monnaie	1982	1983	1984	1985	1986	1987	1988	1989	1990
Mali: CFA franc[1] Mali : franc CFA[1]	0.00300	0.00260	0.00230	0.00220	0.00290	0.00330	0.00340	0.00310	...
Malta: lira Malte : lire	2.43070	2.31370	2.17410	2.13490	2.55240	2.89500	3.02470	2.91480	3.14540
Martinique: franc Martinique : franc	0.15150	0.13210	0.11370	0.11150	0.14460	0.16620	0.16780	0.15650	0.18410
Mauritania: ouguiya Mauritanie : ouguiya	0.01920	0.01820	0.01590	0.01300	0.01350	0.01350	0.01330	0.01200	...
Mauritius: rupee Maurice : roupie	0.09200	0.08380	0.07200	0.06550	0.07450	0.07760	0.07410	0.06560	0.06750
Morocco: dirham Maroc : dirham	0.16550	0.14010	0.11370	0.09960	0.10980	0.12000	0.12190	0.11800	0.12130
Mozambique: metical[1] Mozambique : metical[1]	0.02650	0.02490	0.02360	0.02320	0.02470	0.00350	0.00190	0.00130	...
Myanmar: kyat Myanmar : kyat	0.12820	0.12430	0.11970	0.11760	0.13600	0.15090	0.15770	0.14910	0.15950
Nepal: rupee Népal : roupie	0.07560	0.06860	0.06040	0.05490	0.04700	0.04530	0.04320	0.03700	0.03400
Netherlands: guilder Pays-Bas : florin	0.37500	0.35170	0.31270	0.30270	0.40810	0.49520	0.50830	0.47200	0.55100
Netherlands Antilles: guilder Antilles néerlandaises : florin	0.55560	0.55560	0.55550	0.55560	0.55560	0.55560	0.55560	0.55870	0.55870
New Caledonia: CFP franc Nouvelle-Calédonie : franc CFP	0.00830	0.00720	0.00620	0.00620	0.00730	0.00920	0.00910	0.00870	0.01020
New Zealand: dollar Nouvelle-Zélande : dollar	0.74770	0.66820	0.55910	0.49400	0.52950	0.59170	0.65290	0.59840	0.59500
Niger: CFA franc[1] Niger : franc CFA[1]	0.00300	0.00260	0.00230	0.00220	...	...	...	...	...
Nigeria: naira Nigéria : naira	1.48590	1.37550	1.31520	1.11870	0.65200	0.24960	0.22040	0.13580	0.12400
Norway: krone Norvège : couronne	0.15540	0.13420	0.12270	0.11710	0.13520	0.14870	0.15350	0.14490	0.16090
Oman: rial Omani Oman : rial omani	2.89520	2.89520	2.89520	2.89520	2.60080	2.60080	2.60080	2.60080	2.60080
Pakistan: rupee Pakistan : roupie	0.08480	0.07630	0.07130	0.06280	0.06010	0.05750	0.05560	0.04860	0.04590
Panama: balboa Panama : balboa	1.00000	1.00000	1.00000	1.00000	1.00000	1.00000	1.00000	1.00000	1.00000
Papua New Guinea: kina Papouasie-Nvl-Guinée : kina	1.35100	1.19110	1.11580	0.99770	1.03450	1.10440	1.15520	1.16850	1.04790
Paraguay: guarani Paraguay : guaraní	0.00700	0.00660	0.00510	0.00420	0.00300	0.00180	0.00180	0.00090	...
Portugal: escudo Portugal : escudo	0.01260	0.00900	0.00690	0.00590	0.00670	0.00710	0.00700	0.00640	0.00700
Qatar: riyal Qatar : riyal	0.27470	0.27470	0.27470	0.27460	...	...	...	...	...
Réunion: franc Réunion : franc	0.15200	0.13040	0.11470	0.11190	0.14480	0.16720	0.16750	0.15620	0.18670
Romania: leu Roumanie : leu	0.06670	0.05870	0.04800	0.05710	0.05950	0.06250	0.06250	0.06250	0.04340

116
External trade conversion factors
US dollars per national currency [*cont.*]
Facteurs de conversion pour le commerce extérieur
Monnaie nationale en dollars E-U [*suite*]

Country or area and unit Pays ou zone et monnaie	1982	1983	1984	1985	1986	1987	1988	1989	1990
Rwanda: franc Rwanda : franc	0.01080	0.01060	0.01040	0.00980	0.01140	0.01260	0.01310	...	...
Saint Lucia: EC dollar[2] Sainte-Lucie : EC dollar[2]	0.37040	0.37040	0.37040	0.37040	0.37040	0.37040	0.37040	...	...
Samoa: tala Samoa : tala	0.82800	0.69950	0.54060	0.44430	0.45750	0.47490	0.48100	0.44030	0.43340
Saudi Arabia: riyal Arabie saoudite : riyal	0.29180	0.28950	0.28380	0.27610	0.27000	0.26700	0.26700	0.26700	0.26700
Senegal: CFA franc Sénégal : franc CFA	0.00300	0.00260	0.00230	0.00220	0.00290	0.00330	...	...	...
Seychelles: rupee Seychelles : roupie	0.15280	0.14800	0.14160	0.14090	0.16090	0.17930	0.18590	0.17700	0.18800
Sierra Leone: leone Sierra Leone : leone	0.80870	0.59580	0.39840	0.19700	0.06540	0.03100	0.03020	0.01670	0.00690
Singapore: dollar Singapour : dollar	0.46760	0.47320	0.46970	0.45460	0.45930	0.47590	0.49730	0.51280	0.55360
Solomon Islands: dollar Iles Salomon : dollar	1.03100	0.86730	0.78150	0.67190	0.58290	0.50260	0.48050	0.43600	0.39570
Somalia: shilling Somalie : shilling	0.09300	0.06340	0.04800	0.02530	0.01390	0.00950	...	...	...
South Africa: rand Afrique du Sud : rand	0.92280	0.89520	0.68960	0.45450	0.43740	0.49180	0.43910	0.38070	0.38650
Spain: peseta Espagne : peseta	0.00910	0.00700	0.00620	0.00610	0.00720	0.00810	0.00860	0.00840	0.00980
Sri Lanka: rupee Sri Lanka : roupie	0.04800	0.04250	0.03930	0.03630	0.03500	0.03390	0.03140	0.02770	0.02500
Sudan: pound Soudan : livre	1.05860	0.76920	0.76920	0.43500	0.40000	0.35560	0.22220	...	...
Suriname: guilder Suriname : florin	0.56020	0.56020	0.56020	0.56020	0.56020	0.56020	...	...	...
Sweden: krona Suède : couronne	0.15870	0.13040	0.12090	0.11680	0.14050	0.15920	0.16350	0.15520	0.16910
Switzerland: franc Suisse : franc	0.49380	0.47690	0.42690	0.41100	0.56150	0.67730	0.68440	0.61150	0.72130
Syrian Arab Republic: pound Rép. arabe syrienne : livre	0.25480	0.25480	0.25480	0.25480	0.25480	0.08890	0.08910	0.08910	0.08910
Thailand: baht Thaïlande : baht	0.04350	0.04350	0.04240	0.03680	0.03790	0.03890	0.03950	0.03890	0.03910
Togo: CFA franc[1] Togo : franc CFA[1]	0.00300	0.00260	0.00230	0.00220	0.00290	0.00330	0.00340	...	...
Tonga: pa'anga Tonga : pa'anga	1.00900	0.90140	0.87960	0.70000	0.66760	0.70290	0.78500	0.79030	0.78220
Trinidad and Tobago: dollar Trinité-et-Tobago : dollar	0.41670	0.41670	0.41670	0.41010	0.27780	0.27780	0.26140	0.23530	0.23530
Tunisia: dinar Tunisie : dinar	1.69960	1.47280	1.28240	1.13560	1.25910	1.21400	1.16470	1.05470	1.13460
Turkey: lira Turquie : livre	0.00610	0.00430	0.00270	0.00190	0.00150	0.00120	0.00070	0.00050	0.00040

Country or area and unit Pays ou zone et monnaie	1982	1983	1984	1985	1986	1987	1988	1989	1990
Rwanda: franc Rwanda : franc	0.01080	0.01060	0.00990	0.00990	0.01140	0.01260	0.01310	...	...
Saint Lucia: EC dollar[2] Sainte-Lucie: EC dollar[2]	0.37040	0.37040	0.37040	0.37040	0.37040	0.37040	0.37040		...
Samoa: tala Samoa : tala	0.82020	0.68960	0.53020	0.44340	0.45670	0.47610	0.48100	0.44070	0.43260
Saudi Arabia: riyal Arabie saoudite : riyal	0.29190	0.28930	0.28380	0.27610	0.27010	0.26700	0.26700	0.26700	0.26700
Senegal: CFA franc Sénégal : franc CFA	0.00300	0.00260	0.00230	0.00220	0.00290	0.00330	...	...	...
Seychelles: rupee Seychelles : roupie	0.15170	0.14810	0.14130	0.13970	0.15970	0.17860	0.18580	0.17690	0.18900
Sierra Leone: leone Sierra Leone : leone	0.81360	0.60340	0.39840	0.19340	0.07200	0.02830	0.03220	0.01670	0.00680
Singapore: dollar Singapour : dollar	0.46740	0.47300	0.46960	0.45460	0.45920	0.47600	0.49720	0.51290	0.55390
Solomon Islands: dollar Iles Salomon : dollar	1.02600	0.86120	0.78400	0.67150	0.58150	0.50200	0.47780	0.43340	0.39500
Somalia: shilling Somalie : shilling	0.09300	0.06480	0.05150	0.02530	0.01390	0.00950	...	...	...
South Africa: rand Afrique du Sud : rand	0.92280	0.89730	0.68270	0.45220	0.43880	0.49160	0.43690	0.37980	0.38690
Spain: peseta Espagne : peseta	0.00910	0.00700	0.00620	0.00610	0.00720	0.00810	0.00860	0.00850	0.00990
Sri Lanka: rupee Sri Lanka : roupie	0.04800	0.04230	0.03930	0.03640	0.03500	0.03380	0.03140	0.02750	0.02500
Sudan: pound Soudan : livre	1.03350	0.76920	0.76920	0.43460	0.40000	0.33660	0.22220	0.22220	...
Suriname: guilder Suriname : florin	0.56020	0.56020	0.56020	0.56020	0.56020	0.56020	...	...	...
Sweden: krona Suède : couronne	0.15950	0.13050	0.12100	0.11710	0.14040	0.15930	0.16370	0.15520	0.16890
Switzerland: franc Suisse : franc	0.49420	0.47640	0.42640	0.41200	0.56230	0.67790	0.68500	0.61170	0.72280
Syrian Arab Republic: pound Rép. arabe syrienne : livre	0.25480	0.25480	0.25480	0.25480	0.25480	0.08890	0.08910	0.08910	0.08910
Thailand: baht Thaïlande : baht	0.04350	0.04350	0.04230	0.03680	0.03790	0.03890	0.03950	0.03890	0.03910
Togo: CFA franc[1] Togo : franc CFA[1]	0.00300	0.00260	0.00230	0.00220	0.00290	0.00330	0.00340	...	...
Tonga: pa'anga Tonga : pa'anga	1.01140	0.89890	0.87960	0.70140	0.66470	0.69820	0.79200	0.78960	0.78230
Trinidad and Tobago: dollar Trinité-et-Tobago : dollar	0.41670	0.41670	0.41670	0.40770	0.27780	0.27780	0.26020	0.23530	0.23530
Tunisia: dinar Tunisie : dinar	1.69820	1.46190	1.28410	1.12780	1.25360	1.21230	1.16510	1.05410	1.13310
Turkey: lira Turquie : livre	0.00610	0.00440	0.00270	0.00190	0.00150	0.00120	0.00070	0.00050	0.00040

116
External trade conversion factors[cont.]
US dollars per national currency
Facteurs de conversion pour le commerce extérieur[suite]
Monnaie nationale en dollars E-U

Country or area and unit Pays ou zone et monnaie	1982	1983	1984	1985	1986	1987	1988	1989	1990
Uganda: shilling Ouganda : shilling	1.06380	0.64980	0.27780	0.14880	0.07140	0.02330	0.00940	...	...
former USSR: rouble ancienne URSS : rouble	1.37830	1.34940	1.23420	1.19750	1.42000	1.58150	1.64870	1.58820	1.70590
United Arab Emirates: dirham Emirats arabes unis : dirham	0.27240	0.27240	0.27240	0.27240	0.27240	0.27240	0.27240	0.27240	0.27240
United Kingdom: pound Royaume-Uni : livre	1.74990	1.51640	1.34180	1.28870	1.46640	1.64220	1.78050	1.63700	1.78290
United Rep.Tanzania: shilling Rép. Unie de Tanzanie : shilling	0.10720	0.08880	0.06450	0.05760	0.02920	0.01520	0.01000	...	...
Vanuatu: vatu Vanuatu : vatu	0.01050	0.01010	0.01010	0.00950	0.00940	0.00910	0.00960	0.00860	0.00860
Venezuela: bolivar Venezuela : bolívar	0.23300	0.23270	0.14450	0.13330	0.12760	0.07570	0.05390	0.02930	0.02120
Yemen: rial[4] Yémen : rial[4]	0.21920	0.21840	0.18680	0.13580	0.10370	0.09670	...	...	...
Zaire: zaire Zaïre : zaïre	0.17440	0.07300	0.02730	0.02000	0.01680	0.00890	0.00540	0.00260	0.00140
Zambia: kwacha Zambie : kwacha	1.07590	0.78750	0.54860	0.33460	0.13550	0.11160	0.12170	0.07140	0.03350
Zimbabwe: dollar Zimbabwe : dollar	1.32180	0.99060	0.79910	0.62040	0.60060	0.59870	...		0.40850

Source:
Trade statistics database of the Statistical Division of the
United Nations Secretariat.

† All data shown which pertain to Germany prior to 3 October
1990 are indicated separately for the Federal Republic of
Germany and the former German Democratic Republic based on
their respective territories at the time indicated. Where
data for united Germany (3 October 1990 and thereafter) are
not available, available data are shown separately under the
designations Federal Republic of Germany and former German
Democratic Republic and pertain to the territorial
boundaries prior to 3 October 1990. For detailed
explanatory notes on data pertaining to Germany, see Annex I
- Country Nomenclature.

1 The conversion factors are not trade weighted (for
Mauritania 1987 exports only; for Madagascar 1987 imports
only).
2 East caribbean dollar.
3 External trade data are reported in US dollars (prior to
1984, Bermudian dollar).
4 Excluding former Democratic Yemen.

Source:
Base de données pour les statistiques du commerce extérieur
de la Division de statistique du Secrétariat de l'ONU.

† Toutes les données se rapportant à l'Allemagne avant le 3
octobre 1990 figurent dans deux rubriques séparées basées
sur les territoires respectifs de la République fédérale
d'Allemagne et l'ancienne République démocratique allemande
selon la période indiquée. En l'absence de données pour
l'Allemagne unifiée (à compter du 3 octobre 1990), les
données disponisbles sont fournies séparément sous les
rubriques République fédérale d'Allemagne et ancienne
République démocratique allemande et se rapportent aux
limites territoriales antérieures au 3 octobre 1990. Pour
les notes explicatives en détail sur les données concernant
l'Allemagne, voir Annexe I – Nomenclature des pays.
1 Les facteurs de conversion ne sont pas ponderés (seulement
pour exportations 1987 de la Mauritanie; seulement pour
importations 1987 de la Madagascar).
2 Dollar des carraïbes orientales.
3 Les données du commerce extérieur sont reçues exprimées en
dollars E-U (avant 1984, dollar bermudien).
4 Non compris l'ancien Yémen démocratique.

Country or area and unit Pays ou zone et monnaie	1982	1983	1984	1985	1986	1987	1988	1989	1990
Uganda: shilling Ouganda : shilling	1.06380	0.64980	0.27780	0.14880	0.07140	0.02330	0.00940	0.00450	0.00240
former USSR: rouble ancienne URSS : rouble	1.37600	1.34540	1.23210	1.20120	1.42410	1.58310	1.64730	1.58820	1.71470
United Arab Emirates: dirham Emirats arabes unis : dirham	0.27240	0.27240	0.27240	0.27240	0.27240	...	...	...	...
United Kingdom: pound Royaume-Uni : livre	1.74730	1.51500	1.34080	1.29360	1.46570	1.64320	1.78150	1.63480	1.78980
United Rep.Tanzania: shilling Rép. Unie de Tanzanie : shilling	0.10730	0.08830	0.06460	0.05750	0.02920	0.01520	0.01000	...	...
Vanuatu: vatu Vanuatu : vatu	0.01050	0.01000	0.01010	0.00940	0.00940	0.00910	0.00960	0.00850	0.00860
Venezuela: bolivar Venezuela : bolívar	0.23300	0.22300	0.16460	0.15350	0.11130	0.08580	0.06540	0.02870	0.02110
Yemen: rial[4] Yémen : rial[4]	0.21920	0.21840	0.18680	0.13580	0.10370	0.09670	...	...	...
Zaire: zaire Zaïre : zaïre	0.17430	0.08150	0.02760	0.02010	0.01670	0.00890	0.00530	0.00260	0.00140
Zambia: kwacha Zambie : kwacha	1.07550	0.78730	0.55120	0.36290	0.13110	0.10830	0.12040	0.07290	0.03610
Zimbabwe: dollar Zimbabwe : dollar	1.31440	0.98090	0.79020	0.61780	0.59970	0.59860	0.54980	...	0.40730

117
Total imports and exports
Importations et exportations totales
Value in million US dollars
Valeur en millions de dollars E-U

Region, country or area Region, pays ou zone	1981	1982	1983	1984	1985	1986	1987	1988	1989	1990
A. Imports c.i.f. • Importations c.i.f.										
World[1] *Monde*[1]	**2 011 932**	**1 907 338**	**1 866 323**	**1 975 844**	**1 992 956**	**2 198 109**	**2 544 434**	**2 908 674**	**3 132 595**	**3 546 248**
Developed economies[1,2,3] Econ. développées[1,2,3]	1 337 398	1 254 973	1 231 236	1 337 512	1 387 259	1 557 233	1 845 647	2 079 212	2 247 479	2 577 197
Developing economies[3] Econ. en dévelop.[3]	511 789	482 615	460 361	460 836	439 366	458 912	505 153	622 615	686 903	773 251
OPEC+ OPEP+	146 830	148 471	134 324	116 954	98 420	91 399	88 758	99 758	101 936	108 627
LDC+ PMA+	20 698	19 346	17 853	18 186	17 822	18 286	19 050	21 083	20 888	22 461
America • Amérique										
Developed economies[4] **Economies développées[4]**	**338 618**	**307 446**	**329 593**	**412 264**	**436 326**	**456 835**	**508 189**	**559 252**	**599 949**	**619 662**
Canada[5] Canada[5]	66 303	55 034	61 271	73 704	76 413	80 978	87 700	107 422	114 247	116 720
United States[4] Etats-Unis[4]	273 352	254 884	269 878	341 177	361 626	382 295	424 442	459 565	493 195	516 987
Developing economies Econ. en dévelop.	**101 862**	**87 834**	**73 209**	**69 218**	**63 928**	**81 023**	**74 480**	**96 722**	**103 305**	**115 148**
LAIA+ ALAI+	**67 850**	**56 169**	**44 074**	**40 586**	**37 797**	**54 815**	**47 584**	**69 507**	**72 943**	**82 886**
Argentina[6] Argentine[6]	9 430	5 337	4 504	4 585	3 814	4 724	5 818	5 322	4 203	4 076
Bolivia[6] Bolivie[6]	917	554	577	489	691	674	766	591	620	716
Brazil[5,6] Brésil[5,6]	22 085	19 396	15 428	13 937	12 189	14 045	15 051	14 605	18 263	20 661
Chile[6] Chili[6]	7 318	4 094	3 171	3 481	3 007	3 157	4 023	4 924	6 734	7 272
Colombia[6] Colombie[6]	5 199	5 478	4 968	4 492	4 131	3 852	4 228	5 005	5 010	5 590
Ecuador Equateur	2 246	2 169	1 487	1 616	1 767	1 810	2 158	1 714	1 855	1 862
Mexico[7] Mexique[7]	25 054	15 057	9 006	11 788	13 762	11 918	12 761	18 954	23 633	29 993
Paraguay[5,6] Paraguay[5,6]	506	561	506	563	719	733	590	671	695	...
Peru[5,6] Pérou[5,6]	3 802	3 722	2 722	2 140	1 806	2 596	3 182	2 790	2 291	2 885
Uruguay[6] Uruguay[6]	1 641	1 110	788	776	708	870	1 142	1 177	1 203	1 343
Venezuela[5,1] Venezuela[5,1]	11 813	11 670	7 851	6 874	7 418	8 600	8 711	11 476	7 030	6 365
CACM+ MCAC+	**5 811**	**4 606**	**4 625**	**5 060**	**5 087**	**4 756**	**5 643**	**5 821**	**6 464**	**6 470**
Costa Rica[6] Costa Rica[6]	1 209	893	988	1 085	1 098	1 130	1 380	1 409	1 743	2 044
El Salvador[6] El Salvador[6]	985	857	893	977	961	935	994	1 007	1 161	902

Region, country or area Region, pays ou zone	1981	1982	1983	1984	1985	1986	1987	1988	1989	1990
B. Exports f.o.b. • Exportations f.o.b.										
World[1] *Monde*[1]	**1 957 686**	**1 836 688**	**1 795 823**	**1 897 266**	**1 909 538**	**2 120 550**	**2 464 651**	**2 814 472**	**3 012 955**	**3 392 230**
Developed economies[1][2][3] Econ. développées[1][2][3]	1 233 123	1 166 912	1 152 471	1 226 334	1 281 230	1 485 064	1 739 240	1 979 133	2 123 349	2 441 269
Developing economies[3] Econ. en dévelop.[3]	565 719	493 653	458 767	483 258	460 726	451 978	524 630	626 454	698 968	778 509
OPEC+ OPEP+	284 025	224 765	181 356	170 371	152 832	112 512	123 554	121 590	141 044	165 374
LDC+ PMA+	9 991	9 185	9 800	10 268	9 408	10 042	11 215	11 516	12 351	11 736
America • Amérique										
Developed economies[4] Economies développées[4]	**305 074**	**280 787**	**275 543**	**306 252**	**302 013**	**304 558**	**341 407**	**423 669**	**469 186**	**502 642**
Canada[5] Canada[5]	70 018	68 496	73 514	86 729	87 478	86 850	94 402	113 521	117 235	127 419
United States[4] Etats-Unis[4]	238 686	216 442	205 639	223 999	218 828	227 158	254 122	321 813	363 812	393 592
Developing economies Econ. en dévelop.	**93 885**	**81 692**	**80 335**	**86 424**	**81 044**	**86 348**	**76 189**	**110 224**	**121 090**	**132 006**
LAIA+ ALAI+	**68 326**	**59 194**	**58 791**	**66 184**	**63 679**	**69 764**	**60 006**	**93 301**	**102 998**	**112 971**
Argentina[6] Argentine[6]	9 143	7 625	7 836	8 107	8 396	6 852	6 360	9 135	9 579	12 353
Bolivia[6] Bolivie[6]	912	828	755	725	623	564	566	601	822	900
Brazil[5][6] Brésil[5][6]	23 293	20 175	21 899	27 005	25 639	22 349	26 224	33 782	34 383	31 414
Chile[6] Chili[6]	3 837	3 706	3 831	3 657	3 823	4 222	5 102	7 049	8 193	8 580
Colombia[6] Colombie[6]	2 956	3 095	3 081	3 483	3 552	5 018	5 024	5 026	5 739	6 745
Ecuador Equateur	2 527	2 327	2 348	2 622	2 905	2 186	1 928	2 193	2 354	2 714
Mexico[7] Mexique[7]	19 420	21 230	22 312	24 196	21 664	16 031	20 656	20 658	22 819	26 524
Paraguay[5][6] Paraguay[5][6]	296	330	284	386	403	275	379	657	1 163	...
Peru[5][6] Pérou[5][6]	3 249	3 293	3 015	3 147	2 978	2 531	2 661	2 691	3 488	3 276
Uruguay[6] Uruguay[6]	1 215	1 023	1 045	925	854	1 088	1 182	1 405	1 599	1 693
Venezuela[5][8] Venezuela[5][8]	20 125	16 443	14 529	15 997	14 438	8 660	10 577	10 113	12 867	17 586
CACM+ MCAC+	**4 260**	**3 763**	**3 892**	**3 928**	**3 771**	**3 944**	**3 835**	**4 001**	**4 202**	**4 229**
Costa Rica[6] Costa Rica[6]	1 008	870	873	978	989	1 026	1 148	1 184	1 362	1 394
El Salvador[6] El Salvador[6]	797	699	758	717	695	755	591	609	496	412

117
Total imports and exports
Value in million US dollars [cont.]
Importations et exportations totales
Valeur en millions de dollars E-U [suite]

Region, country or area Region, pays ou zone	1981	1982	1983	1984	1985	1986	1987	1988	1989	1990
Guatemala[6] Guatemala [6]	1 673	1 388	1 135	1 278	1 175	959	1 447	1 557	1 654	1 626
Honduras[6] Honduras[6]	945	692	803	893	888	875	899	933	981	...
Nicaragua Nicaragua	999	776	807	826	964	857	923	...	...	...
Other America Autres pays d'Amérique	28 201	27 060	24 510	23 572	21 044	21 452	21 253	21 393	23 898	25 792
Antigua and Barbuda Antigua-et-Barbuda	138	139	109	132	112	207	222	225	...	...
Aruba[6] Aruba[6]	3 279	2 660	2 393	2 126	184	192	236	336	...	...
Bahamas Bahamas	7 284	6 349	4 616	4 098	3 078	3 289	3 041	2 263	3 001	2 920
Barbados Barbade	580	550	625	659	607	587	515	579	673	700
Belize Belize	162	128	112	130	128	122	143	181	216	211
Bermuda[5] Bermudes[5]	323	351	378	414	402	...	420	488	535	...
British Virgin Islands Iles Vierges britanniques	50	59	...	...	...	...	...	...	...	...
Cuba[6] Cuba [6]	6 545	6 637	6 218	7 207	7 983	9 173	7 612	7 579	...	...
Dominica Dominique	50	47	45	58	55	56	66	88	107	118
Dominican Republic[5] Rép. dominicaine[5]	1 450	1 256	1 279	1 257	1 286	1 352	1 592	1 608	1 964	1 788
French Guiana[6] Guyane française[6]	250	250	279	248	257	295	394	507	568	785
Greenland Groënland	295	278	265	274	296	360	507	519	399	447
Grenada[6] Grenade[6]	54	56	57	56	69	83	89	92	...	...
Guadeloupe[6] Guadeloupe[6]	652	624	660	604	620	762	1 038	1 243	1 251	1 650
Guyana[6] Guyana[6]	426	282	246	219	248	241	...	...	426	512
Haiti[9] Haïti[9]	448	387	440	472	442	360	399	344	291	295
Jamaica Jamaïque	1 473	1 381	1 530	1 146	1 110	972	1 238	1 454	1 809	1 864
Martinique[6] Martinique[6]	776	736	747	686	683	879	1 119	1 290	1 317	1 779
Montserrat Montserrat	19	20	20	18	18	...	25	...	...	...
Netherlands Antilles[6 10] Antilles néerlandaises[6 10]	5 862	5 087	4 527	4 032	2 256	1 112	1 502	1 404	...	...
Panama[6 11] Panama[6 11]	1 562	1 568	1 413	1 430	1 392	1 284	1 306	815	965	1 489
Free zone of Colon Zone libre de Colon	1 992	1 703	1 139	1 338	1 586	1 931	2 005	1 844	...	...

Region, country or area Region, pays ou zone	1981	1982	1983	1984	1985	1986	1987	1988	1989	1990
Guatemala[6] Guatemala[6]	1 226	1 120	1 159	1 122	1 021	1 062	987	1 022	1 108	1 196
Honduras[6] Honduras[6]	729	668	672	725	765	854	808	869	912	...
Nicaragua Nicaragua	500	405	431	385	302	247	300	...	...	...
Other America **Autres pays d'Amérique**	**21 299**	**18 735**	**17 651**	**16 312**	**13 594**	**12 640**	**12 348**	**12 922**	**13 890**	**14 806**
Antigua and Barbuda Antigua-et-Barbuda	40	21	20	18	13	20	17	22	...	...
Aruba[6] Aruba[6]	3 256	2 711	2 375	2 089	290	24	26	31	...	...
Bahamas Bahamas	6 189	4 534	3 970	3 393	2 629	2 702	2 728	2 164	2 786	...
Barbados Barbade	223	258	357	391	352	275	156	173	186	209
Belize Belize	119	92	78	93	90	93	103	116	125	129
Bermuda[5] Bermudes[5]	29	17	23	41	23	...	29	31	50	...
British Virgin Islands Iles Vierges britanniques	2	1	...	...	...	...	...	...	...	...
Cuba[6] Cuba[6]	5 405	5 920	5 523	5 462	5 983	6 298	5 401	5 518	...	...
Dominica Dominique	19	24	27	26	28	43	48	54	45	55
Dominican Republic[5] Rép. dominicaine[5]	1 188	768	785	868	739	722	711	890	924	734
French Guiana[6] Guyane française[6]	35	33	38	37	38	32	54	51	57	93
Greenland Groënland	203	172	181	169	174	257	346	391	418	454
Grenada[6] Grenade[6]	19	19	19	18	22	29	32	...	...	...
Guadeloupe[6] Guadeloupe[6]	92	83	83	87	72	104	93	300	112	118
Guyana[6] Guyana[6]	346	241	189	210	167	222	267	230	227	255
Haiti[9] Haïti[9]	151	163	154	179	168	184	214	179	144	158
Jamaica Jamaïque	974	767	732	705	564	589	706	880	1 029	1 116
Martinique[6] Martinique[6]	185	154	174	154	162	216	194	196	200	278
Montserrat Montserrat	2	3	5	3	3	...	3	...	...	...
Netherlands Antilles[6 10] Antilles néerlandaises[6 10]	5 417	4 891	4 409	3 733	1 679	924	1 308	1 133	...	...
Panama[6 11] Panama[6 11]	319	310	304	258	301	327	336	292	297	321
Free zone of Colon Zone libre de Colon	2 328	2 145	1 472	1 533	1 788	2 183	2 278	2 119	...	...

117
Total imports and exports
Value in million US dollars [cont.]
Importations et exportations totales
Valeur en millions de dollars E-U [suite]

Region, country or area Region, pays ou zone	1981	1982	1983	1984	1985	1986	1987	1988	1989	1990
Saint Kitts and Nevis Saint-Kitts-et-Nevis	48	44	...	...	54	...	...	...	...	...
Saint Lucia[6] Sainte-Lucie[6]	129	118	107	119	125	155	179	221	...	...
St. Vincent-Grenadines[6] St. Vincent-Grenadines[6]	58	65	70	77	79	87	...	...	127	...
Suriname[6] Suriname[6]	566	525	440	392	338	327	294	...	...	...
Trinidad and Tobago[6] Trinité-et-Tobago[6]	3 106	3 696	2 575	1 916	1 530	1 370	1 217	1 119	1 221	1 222

Europe · Europe

Region, country or area Region, pays ou zone	1981	1982	1983	1984	1985	1986	1987	1988	1989	1990
Developed economies[1] Economies développées[1]	796 209	760 273	727 746	736 411	773 282	923 004	1 129 067	1 263 748	1 359 926	1 644 786
EEC+[12][13] CEE+[12][13]	680 954	650 839	622 508	629 995	659 609	780 938	955 794	1 072 064	1 158 375	1 407 159
Belgium-Luxembourg[6][14] Belgique-Luxembourg[6][14]	62 464	58 239	55 314	55 303	56 210	68 743	83 542	92 772	99 791	120 354
Denmark[6][15] Danemark[6][15]	17 735	16 958	16 276	16 585	18 072	22 822	25 434	26 548	26 722	31 771
France[6][16] France[6][16]	120 872	115 645	105 340	103 907	107 810	128 222	157 920	177 269	190 898	233 234
Germany† · Allemagne† F. R. Germany[6][12] R. f. Allemagne[6][12]	163 934	155 856	152 899	151 246	158 549	190 852	228 202	250 443	271 071	342 622
Greece[6] Grèce[6]	8 798	10 023	9 518	9 434	10 139	11 314	13 861	12 008	16 126	19 777
Ireland Irlande	10 608	9 624	9 189	9 662	10 019	11 679	13 622	15 571	17 423	20 727
Italy[6] Italie[6]	91 022	86 213	80 367	84 207	90 994	106 940	132 417	138 624	152 913	185 495
Netherlands[6] Pays-Bas[6]	65 921	62 669	60 965	62 102	65 382	75 690	91 495	99 522	104 439	126 195
Portugal[6] Portugal[6]	9 951	9 599	8 241	7 978	7 792	9 664	13 967	17 889	19 069	25 409
Spain[6][16] Espagne[6][16]	32 218	31 550	29 201	28 867	30 995	34 949	48 945	60 576	71 429	87 694
United Kingdom[17] Royaume-Uni[17]	102 725	99 708	100 235	105 961	109 269	126 368	154 409	189 753	197 728	224 938
EFTA+ AELE+	111 423	105 631	101 513	102 687	109 940	137 553	167 731	185 794	195 671	230 062
Austria[6] Autriche[6]	21 048	19 557	19 364	19 631	20 937	26 843	32 679	36 608	38 873	49 543
Finland Finlande	14 202	13 387	12 856	12 443	13 233	15 333	18 922	21 843	24 613	27 005
Iceland[6] Islande[6]	980	942	816	839	904	1 115	1 589	1 587	1 395	1 661
Norway Norvège	15 652	15 479	13 238	13 889	15 560	20 306	22 640	23 221	23 677	27 228
Sweden Suède	28 845	27 596	26 120	26 416	28 583	32 678	40 986	45 895	48 890	54 756

Region, country or area Région, pays ou zone	1981	1982	1983	1984	1985	1986	1987	1988	1989	1990
Saint Kitts and Nevis Saint-Kitts-et-Nevis	24	19	...	...	22	...	...	...	...	...
Saint Lucia[6] Sainte-Lucie[6]	41	42	48	48	52	83	80	119	...	...
St. Vincent-Grenadines[6] St. Vincent-Grenadines[6]	24	32	41	54	63	64	...	...	75	...
Suriname[6] Suriname[6]	457	412	349	364	329	335	306	...	...	...
Trinidad and Tobago[6] Trinité-et-Tobago[6]	3 760	3 082	2 343	2 169	2 134	1 383	1 460	1 408	1 575	2 049

Europe · Europe

Region, country or area Région, pays ou zone	1981	1982	1983	1984	1985	1986	1987	1988	1989	1990
Developed economies[1] Economies développées[1]	734 981	708 469	693 137	709 302	762 736	928 835	1 118 468	1 232 399	1 313 914	1 581 981
EEC+[12][13] CEE+[12][13]	627 252	605 349	590 475	602 402	649 825	791 647	953 175	1 049 296	1 121 030	1 341 965
Belgium-Luxembourg[6][14] Belgique-Luxembourg[6][14]	55 705	52 364	51 939	51 779	53 760	68 961	83 299	92 821	101 342	118 340
Denmark[6][15] Danemark[6][15]	16 236	15 595	16 047	15 959	16 941	21 223	25 615	27 925	28 138	35 133
France[6][16] France[6][16]	101 371	92 629	91 220	93 215	97 640	119 317	143 391	162 095	173 014	210 168
Germany† · Allemagne† F. R. Germany[6][12] R. f. Allemagne[6][12]	176 043	176 428	169 425	169 784	184 009	243 303	294 045	323 277	343 033	398 441
Greece[6] Grèce[6]	4 247	4 297	4 412	4 811	4 542	5 650	7 084	5 307	7 543	8 019
Ireland Irlande	7 677	7 983	8 620	9 629	10 362	12 707	15 973	18 745	20 694	23 788
Italy[6] Italie[6]	75 187	73 479	72 681	73 303	78 957	104 880	124 005	127 927	138 503	169 263
Netherlands[6] Pays-Bas[6]	68 732	66 288	64 836	65 872	68 418	80 512	93 096	103 586	108 285	131 839
Portugal[6] Portugal[6]	4 180	4 171	4 599	5 208	5 711	7 245	9 321	10 990	12 798	16 416
Spain[6][16] Espagne[6][16]	20 351	20 283	19 794	23 587	25 112	27 174	34 160	40 067	44 467	55 640
United Kingdom[17] Royaume-Uni[17]	102 820	97 075	91 939	94 513	109 994	106 981	131 206	145 469	152 447	185 976
EFTA+ AELE+	104 645	99 939	99 386	104 120	109 588	132 954	160 173	177 893	187 300	225 094
Austria[6] Autriche[6]	15 845	15 685	15 428	15 741	17 226	22 522	27 171	31 088	32 429	41 512
Finland Finlande	14 015	13 132	12 519	13 505	13 617	16 336	19 560	22 151	23 270	26 650
Iceland[6] Islande[6]	859	685	749	740	814	1 096	1 376	1 424	1 401	1 590
Norway Norvège	18 220	17 595	17 628	18 892	19 991	18 097	21 491	22 437	27 108	34 043
Sweden Suède	28 664	26 817	27 466	29 378	30 490	37 230	44 834	49 933	51 542	57 415

117
Total imports and exports
Value in million US dollars [*cont.*]
Importations et exportations totales
Valeur en millions de dollars E-U [*suite*]

Region, country or area Région, pays ou zone	1981	1982	1983	1984	1985	1986	1987	1988	1989	1990
Switzerland[6] Suisse[6]	30 696	28 670	29 119	29 469	30 722	41 278	50 915	56 640	58 222	69 869
Other Europe **Autres pays d'Europe**	**1 146**	**1 065**	**1 032**	**1 034**	**1 115**	**1 351**	**1 830**	**2 024**	**2 026**	**2 437**
Faeroe Islands Iles Féroé	208	202	240	262	248	336	513	478	346	338
Gibraltar Gibraltar	133	120	94	91	147	164	231	...	...	...
Malta Malte	855	789	733	717	758	887	1 138	1 353	1 505	1 953
Yugoslavia[6] Yougoslavie[6]	15 757	13 334	12 154	11 956	12 164	11 750	12 603	13 154	14 802	18 890
East Europe and the former USSR **Eur. de l'est/l'anc. URSS**	**162 745**	**169 752**	**174 728**	**177 497**	**166 333**	**181 966**	**193 634**	**206 849**	**198 215**	**195 802**
Bulgaria[5] Bulgarie[5]	10 801	11 527	12 283	12 714	13 656	15 249	16 211	16 582	14 881	12 893
Czechoslovakia[5 18] Tchécoslovaquie[5 18]	14 658	25 492	26 374	27 091	11 152	13 358	14 883	14 593	14 277	13 106
Germany† · Allemagne† former German D. R.[5 19] anc. R. d. allemande[5 19]	20 181	20 196	21 524	22 940	23 433	27 414	28 786	29 647	17 778	...
Hungary Hongrie	8 718	8 142	7 632	7 326	7 919	9 292	9 450	9 136	8 803	8 764
Poland[5] Pologne[5]	16 764	10 244	10 590	10 638	10 799	11 208	10 844	12 240	10 085	9 528
Romania[5] Roumanie[5]	10 978	8 323	7 648	7 731	8 401	8 082	8 311	7 641	8 436	9 115
former USSR[5] ancienne URSS[5]	72 960	77 752	80 412	80 680	83 140	88 871	96 061	107 229	114 567	120 651

Africa · Afrique

Region, country or area	1981	1982	1983	1984	1985	1986	1987	1988	1989	1990
South Africa[5 20 21] Afrique du Sud[5 20 21]	20 995	16 971	14 506	14 921	10 319	11 750	14 100	17 336	17 034	17 074
Developing economies **Econ. en dévelop.**	**86 111**	**75 304**	**63 044**	**59 934**	**52 299**	**55 733**	**54 631**	**59 904**	**59 492**	**67 757**
Northern Africa **Afrique du Nord**	**38 006**	**35 788**	**34 565**	**35 373**	**26 472**	**29 606**	**27 394**	**31 305**	**31 705**	**36 301**
Algeria[6] Algérie[6]	11 329	10 747	10 397	10 289	9 841	9 234	7 028	7 399	...	...
Egypt[6 22] Egypte[6 22]	8 839	9 078	10 275	10 766	5 495	8 680	7 596	8 657	7 434	9 202
Libyan Arab Jamah. Jamah. arabe libyenne	8 382	7 176	6 029	6 221	4 102	4 192	4 723	5 879	...	...
Morocco[6] Maroc[6]	4 356	4 316	3 599	3 907	3 849	3 794	4 229	4 772	5 493	6 919
Sudan[23] Soudan[23]	1 578	1 285	1 354	1 147	740	961	929	1 060	...	...
Tunisia Tunisie	3 791	3 402	3 102	3 218	2 597	2 901	3 047	3 688	4 378	5 550
Other Africa **Autres Pays d'Afrique**	**48 105**	**39 515**	**28 478**	**24 561**	**25 827**	**26 126**	**27 237**	**28 599**	**27 787**	**31 456**

Region, country or area Region, pays ou zone	1981	1982	1983	1984	1985	1986	1987	1988	1989	1990
Switzerland[6] Suisse[6]	27 042	26 024	25 595	25 863	27 451	37 674	45 742	50 861	51 549	63 884
Other Europe **Autres pays d'Europe**	**609**	**556**	**536**	**551**	**603**	**774**	**982**	**1 098**	**1 235**	**1 545**
Faeroe Islands Iles Féroé	160	149	171	158	179	249	345	348	346	418
Gibraltar Gibraltar	52	42	37	35	62	65	85	...	...	...
Malta Malte	448	411	363	394	399	497	604	714	858	1 126
Yugoslavia[6] Yougoslavie[6]	10 929	10 241	9 913	10 254	10 642	10 298	11 425	12 597	13 363	14 312
East Europe and the former USSR **Eur.de l'est/l'anc. URSS**	**158 845**	**176 123**	**184 586**	**187 674**	**167 582**	**183 510**	**200 781**	**208 886**	**190 640**	**172 453**
Bulgaria[5] Bulgarie[5]	10 689	11 428	12 129	12 850	13 348	14 192	15 905	17 223	16 013	13 347
Czechoslovakia[5,18] Tchécoslovaquie[5,18]	14 876	26 445	27 697	28 543	11 510	13 227	14 723	14 894	14 440	11 882
Germany†·Allemagne† former German D. R.[5,19] anc. R. d. allemande[5,19]	19 858	21 743	23 793	24 836	25 268	27 729	29 871	30 672	17 334	...
Hungary Hongrie	8 281	8 149	7 866	7 840	8 254	8 875	9 204	9 739	9 584	9 707
Poland[5] Pologne[5]	14 508	11 208	11 572	11 750	11 488	12 074	12 205	13 956	13 155	14 322
Romania[5] Roumanie[5]	11 180	9 848	9 847	9 899	10 173	9 761	10 491	11 391	10 486	5 870
former USSR[5] ancienne URSS[5]	79 003	86 912	91 343	91 652	87 281	97 247	107 966	110 559	109 173	104 177

Africa · Afrique

Region, country or area	1981	1982	1983	1984	1985	1986	1987	1988	1989	1990
South Africa[5,20,21] Afrique du Sud[5,20,21]	11 208	9 635	9 672	9 334	9 326	11 097	12 484	13 152	15 104	16 340
Developing economies **Econ. en dévelop.**	**76 184**	**65 463**	**60 815**	**62 403**	**59 796**	**46 663**	**51 329**	**50 143**	**55 763**	**65 674**
Northern Africa **Afrique du Nord**	**38 608**	**33 746**	**32 489**	**31 528**	**28 304**	**22 153**	**24 665**	**22 884**	**26 515**	**31 241**
Algeria[6] Algérie[6]	14 586	13 099	12 696	12 821	10 149	7 831	8 564	7 707	...	...
Egypt[6,22] Egypte[6,22]	3 233	3 120	3 215	3 140	1 838	2 214	2 037	2 120	2 565	2 582
Libyan Arab Jamah. Jamah. arabe libyenne	15 576	13 203	12 216	11 148	12 314	7 747	8 766	6 683	...	...
Morocco[6] Maroc[6]	2 320	2 059	2 062	2 172	2 165	2 428	2 807	3 625	3 337	4 229
Sudan[23] Soudan[23]	658	499	624	629	367	333	504	509	672	...
Tunisia Tunisie	2 503	1 986	1 872	1 797	1 627	1 760	2 147	2 395	2 933	3 595
Other Africa **Autres Pays d'Afrique**	**37 575**	**31 717**	**28 326**	**30 874**	**31 493**	**24 510**	**26 664**	**27 258**	**29 248**	**34 434**

117
Total imports and exports
Value in million US dollars [cont.]
Importations et exportations totales
Valeur en millions de dollars E-U [suite]

Region, country or area Region, pays ou zone	1981	1982	1983	1984	1985	1986	1987	1988	1989	1990
CEUCA+ **UDEAC+**	**2 800**	**2 744**	**2 825**	**2 512**	**2 700**	**3 318**	**3 199**	**3 001**	**2 709**	**3 198**
Cameroon[6 24] Cameroun[6 24]	1 428	1 211	1 217	1 106	1 151	1 705	1 749	1 271	...	...
Central African Rep.[6 24] Rép. centrafricaine[6 24]	94	126	68	87	113	167	204	141	150	...
Congo[6 24] Congo[6 24]	436	685	688	595	581	579	514	547	523	600
Gabon[6 24] Gabon[6 24]	841	723	853	724	855	866	732	791	767	...
ECOWAS+ **CEDEAO+**	**28 702**	**22 174**	**15 297**	**11 673**	**12 482**	**11 205**	**11 344**	**11 303**	**11 158**	**12 047**
Benin[6] Bénin[6]	543	464	294	288	...	...	...	...	...	...
Burkina Faso[6] Burkina Faso[6]	338	346	288	253	333	405	434	489	322	...
Cape Verde[6] Cap-Vert[6]	...	88	87	83	84	107	118	120	112	...
Côte d'Ivoire Côte d'Ivoire	2 384	2 090	1 808	1 511	1 742	2 054	2 241	2 100	2 185	...
Gambia Gambie	122	97	115	98	93	100	127	137	159	200
Ghana Ghana	1 106	705	542	608	866	1 046	894	907	1 275	...
Guinea-Bissau[6] Guinée-Bissau[6]	50	50	...	...	...	...	...	...	...	...
Liberia[6] Libéria[6]	447	428	412	363	284	259	308	272	...	...
Mali[6] Mali[6]	283	201	264	279	300	444	374	513	500	...
Mauritania[6] Mauritanie[6]	265	273	227	213	234	221	235	240	222	...
Niger[6] Niger[6]	510	466	324	285	345	...	...	...	...	...
Nigeria Nigéria	20 453	15 003	9 062	5 868	6 205	4 029	3 917	3 889	3 419	4 318
Senegal[6] Sénégal[6]	1 076	992	1 039	1 010	812	961	1 023	...	...	...
Sierra Leone Sierra Leone	312	298	171	167	154	132	137	157	182	164
Togo[6] Togo[6]	433	391	284	271	288	312	424	487	...	...
Rest of Africa **Afrique N.D.A.**	**16 603**	**14 597**	**10 356**	**10 376**	**10 644**	**11 604**	**12 694**	**14 294**	**13 920**	**16 211**
Angola[6] Angola[6]	1 646	878	679	713	671	619	443	...	...	...
Burundi[6] Burundi[6]	161	214	183	187	188	207	206	206	188	236
Chad[6] Tchad[6]	108	109	157	182	240	288	366	419	...	...
Comoros[6] Comores[6]	32	33	34	43	37	39	52	...	...	...

Region, country or area Région, pays ou zone	1981	1982	1983	1984	1985	1986	1987	1988	1989	1990
CEUCA+ UDEAC+	**4 212**	**4 260**	**4 055**	**4 161**	**3 853**	**2 895**	**2 764**	**3 171**	**3 688**	**4 033**
Cameroon[6][24] Cameroun[6][24]	1 122	1 000	939	882	722	781	829	924	...	...
Central African Rep.[6][24] Rép. centrafricaine[6][24]	78	108	75	85	92	66	130	66	134	...
Congo[6][24] Congo[6][24]	811	992	1 066	1 183	1 087	777	517	752	911	976
Gabon[6][24] Gabon[6][24]	2 200	2 160	1 975	2 012	1 952	1 271	1 288	1 195	1 599	2 474
ECOWAS+ CEDEAO+	**24 380**	**19 264**	**15 729**	**17 987**	**19 252**	**12 887**	**14 361**	**13 568**	**15 105**	**20 005**
Benin[6] Bénin[6]	34	24	67	167	...	...	...	...	...	...
Burkina Faso[6] Burkina Faso[6]	75	56	57	79	70	83	155	142	95	...
Cape Verde[6] Cap-Vert[6]	5	3	3	3	6	4	8	3	7	...
Côte d'Ivoire Côte d'Ivoire	2 535	2 235	2 067	2 698	2 939	3 354	3 110	2 792	2 931	...
Gambia Gambie	27	44	48	47	43	35	40	...	...	41
Ghana Ghana	1 063	873	503	540	623	876	909	1 014	1 024	...
Guinea-Bissau[6] Guinée-Bissau[6]	14	12	...	...	...	...	...	...	...	...
Liberia[6] Libéria[6]	529	477	428	452	436	408	382	...	...	...
Mali[6] Mali[6]	105	117	93	133	124	212	179	249	271	...
Mauritania[6] Mauritanie[6]	261	232	290	297	374	349	428	354	437	...
Niger[6] Niger[6]	455	332	299	274	209	...	...	...	...	...
Nigeria Nigéria	18 087	13 705	10 662	12 020	13 113	5 899	7 383	6 875	8 138	12 912
Senegal[6] Sénégal[6]	500	548	543	534	554	620	606	...	...	...
Sierra Leone Sierra Leone	137	111	122	131	129	141	133	107	138	143
Togo[6] Togo[6]	211	177	162	191	190	204	244	242	...	...
Rest of Africa Afrique N.D.A.	**8 983**	**8 192**	**8 543**	**8 726**	**8 388**	**8 727**	**9 538**	**10 519**	**10 455**	**10 396**
Angola[6] Angola[6]	1 838	1 627	1 808	2 018	2 224	1 319	2 147	...	...	...
Burundi[6] Burundi[6]	71	88	80	99	111	169	90	133	78	75
Chad[6] Tchad[6]	83	58	104	131	88	99	111	141	...	...
Comoros[6] Comores[6]	16	20	19	7	16	20	12	...	18	...

117
Total imports and exports
Value in million US dollars [*cont.*]
Importations et exportations totales
Valeur en millions de dollars E-U [*suite*]

Region, country or area Region, pays ou zone	1981	1982	1983	1984	1985	1986	1987	1988	1989	1990
Djibouti Djibouti	224	226	221	222	201	184	205	201	196	215
Ethiopia Ethiopie	737	785	876	943	989	1 101	1 101	1 085	953	1 076
Kenya Kenya	2 069	1 613	1 358	1 526	1 464	1 647	1 739	1 611	2 147	2 226
Madagascar[6] Madagascar[6]	540	425	387	366	402	353	302	362	340	568
Malawi Malawi	350	311	311	269	295	258	298	409	508	573
Mauritius Maurice	554	464	435	472	529	684	1 013	1 302	1 326	1 619
Mozambique[6] Mozambique[6]	801	836	636	540	424	543	625	715	...	...
Réunion[6] Réunion[6]	786	804	838	791	841	1 137	1 464	1 662	1 732	2 160
Rwanda Rwanda	256	269	267	283	294	352	357	369	...	...
Seychelles Seychelles	93	98	88	88	99	105	114	159	165	187
Somalia[6] Somalie[6]	512	330	349	104	112	284	132	...	...	...
Uganda Ouganda	345	377	377	344	298	307	555	544	...	...
United Rep.Tanzania Rép. Unie de Tanzanie	1 172	1 128	799	862	978	872	947	1 495	...	...
Zaire[6] Zaïre[6]	672	480	498	675	791	875	756	771	849	886
Zambia[5] Zambie[5]	1 060	1 001	546	596	714	603	739	839	779	1 243
Zimbabwe[5] Zimbabwe[5]	1 472	1 430	1 052	959	897	985	1 043	...	...	1 850

Asia • Asie

Region, country or area Region, pays ou zone	1981	1982	1983	1984	1985	1986	1987	1988	1989	1990
Developed economies **Economies développées**	149 841	138 575	133 190	142 998	135 790	133 726	158 795	196 278	220 052	245 519
Israel[6 25 26] Israël[6 25 26]	7 815	7 915	8 382	8 072	8 021	9 285	11 451	12 287	13 027	15 104
Japan Japon	143 288	131 932	126 392	136 522	129 480	126 408	149 515	187 348	210 840	234 800
Developing economies **Econ. en dévelop.**	304 520	302 927	308 810	316 409	307 660	306 704	359 299	448 486	504 411	566 244
Asia Middle East **Moyen-Orient d'Asie**	107 780	111 839	111 086	101 456	86 945	77 151	76 437	82 978	89 860	98 568
Non petroleum exports Petrole non compris	...	...	...	...	...	...	...	...	...	...
Bahrain Bahreïn	3 954	3 520	3 262	3 479	3 107	2 405	2 714	2 593	3 134	3 711
Cyprus[26] Chypre[26]	1 165	1 215	1 219	1 364	1 247	1 279	1 484	1 857	2 281	2 565
Iran, Islamic Rep. of[6 27] Iran, Rép. islamique d'[6 27]	12 537	11 989	18 296	...	...	...	...	...	...	...

Region, country or area Region, pays ou zone	1981	1982	1983	1984	1985	1986	1987	1988	1989	1990
Djibouti Djibouti	9	13	11	13	14	20	28	23	25	25
Ethiopia Ethiopie	378	404	402	417	334	464	370	421	452	294
Kenya Kenya	1 188	977	983	1 083	978	1 210	960	1 068	969	1 052
Madagascar[6] Madagascar[6]	316	310	296	333	274	304	333	274	311	308
Malawi Malawi	270	242	229	309	249	245	278	284	267	418
Mauritius Maurice	324	367	361	373	435	675	892	998	987	1 193
Mozambique[6] Mozambique[6]	281	229	132	96	77	79	97	103	101	...
Réunion[6] Réunion[6]	107	105	86	80	97	135	169	158	161	190
Rwanda Rwanda	88	93	79	156	98	118	130	101	...	...
Seychelles Seychelles	17	15	20	26	28	18	22	32	34	57
Somalia[6] Somalie[6]	152	199	111	56	91	89	104	...	...	...
Uganda Ouganda	243	349	385	385	387	436	319	274	250	152
United Rep.Tanzania Rép. Unie de Tanzanie	576	454	382	396	355	323	300	337	...	...
Zaire[6] Zaïre[6]	557	400	1 134	1 003	954	1 097	970	1 108	1 249	999
Zambia[5] Zambie[5]	1 120	1 022	825	661	547	704	873	1 178	1 347	899
Zimbabwe[5] Zimbabwe[5]	1 406	1 273	1 128	1 148	1 109	1 301	1 420	1 631	...	1 723

Asia • Asie

Region, country or area Region, pays ou zone	1981	1982	1983	1984	1985	1986	1987	1988	1989	1990
Developed economies **Economies développées**	**156 084**	**142 630**	**149 974**	**174 133**	**180 056**	**214 119**	**235 276**	**271 003**	**282 027**	**294 141**
Israel[6 25 26] Israël[6 25 26]	5 329	4 991	4 890	5 622	6 084	6 933	8 222	9 445	10 669	11 576
Japan Japon	152 016	138 911	146 668	170 107	175 683	209 153	229 224	264 915	275 173	286 949
Developing economies **Econ. en dévelop.**	**382 783**	**334 537**	**306 074**	**322 300**	**307 341**	**306 662**	**383 347**	**450 611**	**505 798**	**563 786**
Asia Middle East **Moyen-Orient d'Asie**	**203 661**	**159 373**	**122 570**	**110 378**	**98 590**	**79 606**	**87 675**	**88 126**	**100 622**	**112 557**
Non petroleum exports[35] Petrole non compris[35]	23 439	28 393	20 908	16 715	19 682	30 959	27 225	30 989	17 966	42 329
Bahrain Bahreïn	4 177	3 695	3 119	3 204	2 897	2 199	2 430	2 411	2 831	3 758
Cyprus[26] Chypre[26]	556	555	494	575	476	506	621	709	793	949
Iran, Islamic Rep. of[6 27] Iran, Rép. islamique d'[6 27]	12 587	19 414	20 247	13 128	13 403	...	...	...	...	...

117
Total imports and exports
Value in million US dollars [cont.]
Importations et exportations totales
Valeur en millions de dollars E-U [suite]

Region, country or area Region, pays ou zone	1981	1982	1983	1984	1985	1986	1987	1988	1989	1990
Iraq[6] Iraq[6]	11 344	10 596	6 636	6 693	7 619	6 360	3 854	...	...	4 834
Jordan Jordanie	3 149	3 241	3 030	2 784	2 732	2 432	2 703	2 786	2 133	2 603
Kuwait[6] Koweït[6]	6 969	8 283	7 375	6 901	6 115	5 729	5 300	6 046	6 303	...
Lebanon[6] Liban[6]	3 615	...	...	...	...	...	...	...	...	...
Oman Oman	2 288	2 682	2 492	2 748	3 153	2 384	1 823	2 202	2 255	2 681
Qatar Qatar	1 518	1 945	1 456	1 162	1 139	1 099	1 161	1 267	1 326	...
Saudi Arabia[6] Arabie saoudite[6]	35 244	40 654	39 206	33 696	23 623	19 109	20 110	21 784	21 153	24 069
Syrian Arab Republic[6] Rép. arabe syrienne[6]	5 040	4 015	4 542	4 116	3 967	2 728	2 481	2 231	2 097	2 526
Turkey[6] Turquie[6]	8 932	8 923	9 236	10 757	11 342	11 107	14 157	14 320	15 799	22 231
United Arab Emirates Emirats arabes unis	9 646	9 440	8 294	6 936	6 548	6 422	7 226	8 522	10 010	11 199
Yemen[28] Yémen[28]	2 461	2 278	2 349	2 378	1 998	1 640	1 378	...	...	...
Other Asia[29] Autres pays d'Asie[29]	196 739	191 088	197 724	214 953	220 715	229 552	282 862	365 508	414 551	467 677
Afghanistan[30] Afghanistan[30]	886	962	1 064	1 390	1 194	1 404	996	900	798	937
Bangladesh Bangladesh	1 813	1 737	1 587	2 042	2 223	2 018	2 419	2 757	3 343	3 194
Brunei Darussalam[6] Brunéi Darussalam[6]	599	734	728	626	615	656	642	721	883	...
China Chine	22 022	19 292	21 406	27 409	42 252	42 909	43 210	55 264	59 140	53 345
Hong Kong Hong-kong	24 741	23 533	24 122	28 571	29 703	35 365	48 467	63 899	72 153	82 496
India Inde	15 654	14 365	13 434	15 539	15 585	15 093	16 754	18 958	20 264	23 296
Indonesia[6] Indonésie[6]	13 272	16 859	16 352	13 882	10 259	10 718	12 370	13 249	15 922	21 931
Korea, Republic of[6] Corée, République de[6]	26 131	24 251	26 192	30 631	31 136	31 584	41 020	51 811	61 465	69 844
Lao People's Dem. Rep.[6] Rép. dém. pop. lao[6]	125	...	92	48	131	131	146	162	...	...
Macau Macao	723	708	734	798	779	890	1 121	1 290	1 484	1 534
Malaysia[31] Malaisie[31]	11 546	12 423	13 265	14 017	12 602	10 808	12 679	16 542	22 541	29 261
Maldives Maldives	31	43	57	53	53	45	81	90	105	129
Myanmar Myanmar	826	408	268	239	283	304	269	244	191	261

Region, country or area Région, pays ou zone	1981	1982	1983	1984	1985	1986	1987	1988	1989	1990
Iraq[6] Iraq[6]	10 530	10 230	9 785	...	...	...	...	...	...	392
Jordan Jordanie	732	753	579	752	790	732	932	1 036	1 098	1 063
Kuwait[6] Koweït[6]	16 380	10 959	11 574	12 275	10 479	7 221	8 357	7 765	11 476	...
Lebanon[6] Liban[6]	886	...	...	...	...	...	...	...	...	...
Oman Oman	4 696	4 421	4 248	4 422	4 972	2 842	3 776	3 268	3 933	5 215
Qatar Qatar	5 844	4 507	3 297	4 513	3 541	...	...	...	...	...
Saudi Arabia[6] Arabie saoudite[6]	120 240	79 124	45 835	36 834	27 480	20 087	22 590	23 737	28 369	44 417
Syrian Arab Republic[6] Rép. arabe syrienne[6]	2 103	2 026	1 923	1 853	1 637	1 325	1 350	1 345	3 006	4 062
Turkey[6] Turquie[6]	4 703	5 685	5 728	7 134	7 957	7 458	10 189	11 618	11 626	12 910
United Arab Emirates Emirats arabes unis	20 240	17 328	15 085	13 827	13 124	15 837	...	...	...	...
Yemen[28] Yémen[28]	69	64	56	43	54	37	101	...	...	...
Other Asia[29] Autres pays d'Asie[29]	179 122	175 164	183 505	211 923	208 751	227 056	295 672	362 485	405 176	451 229
Afghanistan[30] Afghanistan[30]	694	708	729	633	557	552	512	433	238	235
Bangladesh Bangladesh	662	667	690	934	966	961	1 194	1 344	1 417	1 631
Brunei Darussalam[6] Brunéi Darussalam[6]	4 066	3 808	3 386	3 204	2 972	1 797	1 906	1 721	1 894	...
China Chine	22 011	22 322	22 231	26 143	27 349	30 940	39 439	47 521	52 538	62 091
Hong Kong Hong-kong	21 842	20 979	22 095	28 324	30 183	35 439	48 478	63 165	73 140	82 160
India Inde	8 373	8 807	8 713	9 874	8 750	9 187	11 596	13 182	15 839	17 721
Indonesia[6] Indonésie[6]	25 164	22 328	21 146	21 888	18 587	14 805	17 136	19 218	22 026	25 675
Korea, Republic of[6] Corée, République de[6]	21 254	21 853	24 445	29 245	30 283	34 714	47 281	60 696	62 377	65 016
Lao People's Dem. Rep.[6] Rép. dém. pop. lao[6]	33	...	26	12	54	60	62	81	...	...
Macau Macao	686	718	767	913	906	1 046	1 397	1 492	1 643	1 694
Malaysia[31] Malaisie[31]	11 766	12 027	14 107	16 452	15 764	13 687	17 955	21 125	25 106	29 455
Maldives Maldives	9	10	13	18	23	25	31	40	45	52
Myanmar Myanmar	476	393	378	380	330	299	219	138	215	325

117
Total imports and exports
Value in million US dollars [cont.]
Importations et exportations totales
Valeur en millions de dollars E-U [suite]

Region, country or area Region, pays ou zone	1981	1982	1983	1984	1985	1986	1987	1988	1989	1990
Nepal Népal	38	395	464	417	453	459	560	680	581	686
Pakistan Pakistan	5 549	5 463	5 341	5 873	5 891	5 377	5 829	6 620	7 119	7 356
Philippines Philippines	8 104	8 255	7 980	6 241	5 445	5 134	6 811	8 731	10 732	13 042
Singapore[32] Singapour[32]	28 988	29 576	29 566	30 147	27 599	26 787	34 187	46 065	52 159	63 826
Sri Lanka Sri Lanka	1 849	1 771	1 787	1 845	1 783	1 795	2 027	2 211	2 088	2 634
Thailand Thaïlande	9 951	8 548	10 287	10 398	9 244	9 139	12 994	20 285	25 768	33 379

Oceania · Océanie

	1981	1982	1983	1984	1985	1986	1987	1988	1989	1990
Developed economies **Economies développées**	**30 774**	**30 920**	**25 778**	**30 441**	**31 196**	**31 297**	**34 945**	**42 249**	**49 846**	**49 730**
Australia[5] Australie[5]	23 768	24 187	19 393	23 424	23 450	24 109	26 980	33 245	40 022	38 800
New Zealand Nouvelle-Zélande	5 684	5 825	5 333	6 010	6 196	6 052	6 993	7 559	7 452	9 410
Developing economies **Econ. en dévelop.**	**3 540**	**3 216**	**3 144**	**3 319**	**3 315**	**3 702**	**4 141**	**4 349**	**4 893**	**5 210**
American Samoa[6,33] Samoa américaines[6,33]	234	198	227	284	296	313	346	339	378	...
Fiji Fidji	632	515	484	450	442	435	379	454	633	744
French Polynesia[6] Polynésie française[6]	546	520	533	539	549	736	827	808	791	929
Kiribati[34] Kiribati[34]	26	25	19	18	15	14	18	22	23	27
Guam[6] Guam[6]	355	...	...	...	...	...	...	...	...	...
New Caledonia[6] Nouvelle-Calédonie[6]	408	367	304	311	348	458	626	604	766	883
Papua New Guinea[5] Papouasie-Nvl-Guinée[5]	1 096	1 017	982	951	865	932	1 060	1 171	1 335	1 287
Samoa Samoa	56	50	56	50	51	48	62	76	77	75
Solomon Islands[6] Iles Salomon[6]	76	59	61	66	69	63	79	98	114	92
Tonga Tonga	40	42	38	41	41	40	48	55	54	62
Vanuatu Vanuatu	58	59	63	69	70	57	70	71	70	97

Source:
Trade statistics database of the Statistical Division of the
United Nations Secretariat.
+   For Member States of this grouping, see
Annex I - Other groupings.

†   All data shown which pertain to Germany prior to 3 October
1990 are indicated separately for the Federal Republic of
Germany and the former German Democratic Republic based on
their respective territories at the time indicated. Where

Source:
Base de données pour les statistiques du commerce extérieur
de la Division de statistique du Secrétariat de l'ONU.
+   Les Etats membres de ce groupement, voir
annexe I - Autres groupements.

†   Toutes les données se rapportant à l'Allemagne avant le 3
octobre 1990 figurent dans deux rubriques séparées basées
sur les territoires respectifs de la République fédérale
d'Allemagne et l'ancienne République démocratique allemande

Region, country or area Région, pays ou zone	1981	1982	1983	1984	1985	1986	1987	1988	1989	1990
Nepal Népal	143	88	93	127	160	141	149	192	159	209
Pakistan Pakistan	2 883	2 395	3 075	2 614	2 740	3 379	4 178	4 527	4 779	5 522
Philippines Philippines	5 720	5 021	5 005	5 391	4 629	4 842	5 720	7 035	7 747	8 186
Singapore[32] Singapour[32]	20 967	20 788	21 833	24 108	22 813	22 495	28 686	39 303	44 678	52 729
Sri Lanka Sri Lanka	1 044	1 015	1 066	1 454	1 275	1 194	1 348	1 481	1 529	1 912
Thailand Thaïlande	7 038	6 945	6 368	7 413	7 122	8 836	11 659	15 952	20 059	23 068

Oceania · Océanie

Region, country or area Région, pays ou zone	1981	1982	1983	1984	1985	1986	1987	1988	1989	1990
Developed economies **Economies développées**	**26 204**	**25 992**	**24 901**	**27 875**	**27 631**	**27 070**	**32 283**	**39 970**	**43 947**	**46 903**
Australia[5] Australie[5]	21 767	22 038	20 687	23 998	22 883	22 496	26 455	33 127	36 657	39 628
New Zealand Nouvelle-Zélande	5 563	5 539	5 284	5 358	5 590	5 597	7 163	8 130	8 919	9 061
Developing economies **Econ. en dévelop.**	**1 939**	**1 720**	**1 630**	**1 877**	**1 903**	**2 008**	**2 341**	**2 879**	**2 953**	**2 731**
American Samoa[6,33] Samoa américaines[6,33]	199	187	177	212	202	254	288	368	308	...
Fiji Fidji	311	286	240	256	236	274	323	343	386	533
French Polynesia[6] Polynésie française[6]	29	28	34	32	41	41	83	75	89	111
Kiribati[34] Kiribati[34]	4	3	4	11	4	2	2	5	5	3
Guam[6] Guam[6]	77	...	...	...	...	...	...	...	...	...
New Caledonia[6] Nouvelle-Calédonie[6]	343	265	154	207	271	191	225	464	675	449
Papua New Guinea[5] Papouasie-Nvl-Guinée[5]	863	771	813	899	915	1 052	1 211	1 399	1 281	1 141
Samoa Samoa	11	13	19	20	16	11	12	15	13	9
Solomon Islands[6] Iles Salomon[6]	66	58	61	93	70	67	64	81	75	70
Tonga Tonga	9	4	6	9	5	6	7	8	9	13
Vanuatu Vanuatu	32	23	29	44	31	17	18	20	22	19

data for united Germany (3 October 1990 and thereafter) are not available, available data are shown separately under the designations Federal Republic of Germany and former German Democratic Republic and pertain to the territorial boundaries prior to 3 October 1990. For detailed explanatory notes on data pertaining to Germany, see Annex I - Country Nomenclature.

1 Including trade conducted in accordance with the supplementary protocol to the treaty on the basis of relations between the Federal Republic of Germany and the

selon la période indiquée. En l'absence de données pour l'Allemagne unifiée (à compter du 3 octobre 1990), les données disponisbles sont fournies séparément sous les rubriques République fédérale d'Allemagne et ancienne République démocratique allemande et se rapportent aux limites territoriales antérieures au 3 octobre 1990. Pour les notes explicatives en détail sur les données concernant l'Allemagne, voir Annexe I – Nomenclature des pays.

1 Y compris le commerce effectué en accord avec le protocole additionnel au traite définissant la base des relations entre la République Fédérale d'Allemagne et l'ancienne

117

Total imports and exports
Value in million US dollars [cont.]

Importations et exportations totales
Valeur en millions de dollars E-U [suite]

former German Democratic Republic.

2 United States, Canada, Developed market economies of Europe, Israel, Japan, Australia, New Zealand and South African customs union.

3 This classification is intended for statistical convenience and does not necessarily express a judgement about the stage reached by a particular country in the development process.

4 Including the trade of the US Virgin Islands and Puerto Rico but excluding shipments of merchandise between the United States and its other possessions (Guam, American Samoa, etc.,).

5 Imports F.O.B.

6 Country or area using special trade system. See technical notes for explanation of trade systems.

7 Exports include revaluation, but exclude goods from customs bonded warehouses and free zones. Data exclude trade in silver.

8 Exports of petroleum and petroleum products are at realized prices.

9 Prior to 1985, year ending 30 September of the year stated.

10 Prior to 1986, includes Aruba.

11 Excluding trade of the free zone of Colon.

12 Excluding trade conducted in accordance with the supplementary protocol to the treaty on the basis of relations between the Federal Republic of Germany (FRG) and the former German Democratic Republic (GDR). Data reported by FRG are as follows: Value in million United States dollars. Exports: FRG to GDR: 1981-2475; 1982-2626; 1983-2471; 1984-2229; 1985-2721; 1986-3460; 1987-4140; 1988-4114; 1989-4351; 1990-13379; Imports: FRG from GDR: 1981-2686; 1982-2738; 1983-2694; 1984-2694; 1985-2618; 1986-3163; 1987-3712; 1988-3868; 1989-3854; 1990-5129.

13 The EEC totals have been recalculated for the whole period covered by this table to include the data for Greece which joined the community on 1 January 1981 and for Portugal and Spain which joined the community on 1 January 1986.

14 Economic union of Belgium and Luxembourg. Intertrade between the two countries is excluded.

15 Prior to 1983, General trade.

16 Including Balearic Island, Canary Island, Ceuta and Mellila.

17 Including monetary gold.

18 Beginning 1985 data are not comparable to those shown for prior periods due to revisions of the Koruna to US dollar exchange rate.

19 Beginning 1989 data are not comparable to those shown for prior periods due to revisions of the Mark to US dollar exchange rate.

20 Excluding exports of gold bullion and gold coin.

21 The South African customs union comprising Botswana, Lesotho, Namibia, South Africa and Swaziland. Trade between the component countries is excluded.

22 Import figures exclude petroleum imported without stated value.

23 Excluding exports of camels to Egypt.

24 Inter-trade among the members of Customs and Economic Union

République Démocratique Allemande.

2 Etats-Unis, Canada, Pays développés d'Europe à economie de marche, Israel, Japon, Australie, Nouvelle-Zéalande et l'union douaniere de l'Afrique du Sud.

3 Cette classification est utilisée pour plus de commodite dans la presentation des statistiques et n'implique pas necessairèment un jugement quant au stage de developpement auquel est parvenu un pays donne.

4 Y compris la commerce des Isles Virges américaines et de Porto Rico mais non compris les echanges des marchandises entre les Etats-Unis et leurs autres possessions (Guam, Samoa américaines, etc.,).

5 Importations F.O.B.

6 Pays ou zone utilisant un système spécial du commerce. Pour l'explication du système de commerce, voir les notes techniques.

7 Exportations comprennent la revaluation mais ne comprennent pas les marchandises provenant des entrepôts en douane et des zones franches. Les données ne comprennent pas le commerce en argent.

8 Les exportations de pétrole et les produits pétroliers sont exprimés en valeur realisée.

9 Avant 1985, année finissant le 30 septembre de l'année indiquée.

10 Avant 1986, comprend Aruba.

11 Non compris le commerce de la zone libre de Colon.

12 Non compris le commerce effectué en accord avec le protocole additionnel au traite définissant la base des relations entre la République fédérale d'allemagne (RFA) et l'ancienne République démocratique allemande (RDA). Les données par RFA sont les suivantes. Valeur en millions de dollars des Etats-Unis: exportations: de RFA vers RDA: 1981-2475; 1982-2626; 1983-2471; 1984-2229; 1985-2721; 1986-3460; 1987-4140; 1988-4114; 1989-4351; 1990-13379; importations: de RFA en prov. de RDA: 1981-2686; 1982-2738; 1983-2694; 1984-2694; 1985-2618; 1986-3163; 1987-3712; 1988-3868; 1989-3854; 1990-5129.

13 On a récalculé pour toute la période considerée les totaux relatifs à la CEE à fin d'y inclure les données rélatives à la Grèce qui est devenue membre da la communauté le 1 er janvier 1981 et l'Espagne et le Portugal qui sont devenus membres de la communauté le 1 er janvier 1986.

14 L'Union Economique Belgo-Luxembourgeoise. Non compris le commerce entre ces pays.

15 Avant 1983, commerce général.

16 Y compris les Iles Baléares, les îles Canaïres, Ceuta et Mellila.

17 Y compris l'or monétaire.

18 A partir de l'année 1985 les chiffres ne sont pas comparables aux chiffres indiqués pour les années antérieurs à cause des révisions de taux de change de Koruna au dollar de E-U.

19 A partir de l'année 1989 les chiffres ne sont pas comparables aux chiffres indiqués pour les années antérieurs à cause des révisions de taux de change de Mark au dollar de E-U.

20 Non compris les exportations de Lingots et pièces d'or.

21 L'Union Douaniere de l'Afrique Meridionale comprend Botswana, Lesotho, Namibie, Afrique du Sud et Swaziland. Non compris le commerce entre ces pays.

22 Non compris le petrole brut dont la valeur à la importation n'est pas stipulée.

23 Non compris les exportations des chameaux en Egypte.

24 Non compris le commerce entre les pays membres du UDEAC.

of Central Africa is excluded.

25 Imports and exports net of returned goods. The figures also exclude Judea and Samaria and the Gaza area.

26 Excluding military imports.
27 Year ending 20 December of the year stated.
28 Comprises trade of the former Democratic Yemen and former Yemen Arab Republic including any intertrade between them.

29 Includes imports and exports of China, People's Democratic Republic of Korea, Mongolia and Viet Nam which comprised former centrally planned economies of Asia.

30 Year beginning 21 March of the year stated.
31 Excluding military imports and offshore installations of petroleum industry.
32 Including transshipments to and from Peninsular Malaysia.

33 Year ending 30 September of the year stated.
34 Including Tuvalu (formerly Ellice Islands).
35 Data refer to total exports less petroleum exports of Asia Middle East countries where petroleum, in this case, is the sum of SITC groups 333, 334 and 335.

25 Importations et exportations nets. Ne comprennent pas les marchandises retournées. Sont également exclués les données de la Judee et de Samaria et ainsi que la zone de Gaza.
26 Non compris les importations des economats militaires.
27 Année finissant le 20 décembre de l'année indiquée.
28 Y compris le commerce de l'ancienne République populaire Démocratique de Yémen, le commerce de l'ancienne République Arabe de Yémen et le commerce entre eux.
29 Y compris les importations et les exportations de la Chine, de la République Démocratique de Corée, de la Mongolie et du Viet Nam auxquelles comprenait l'ancienne economie planifiée de l'Asie.
30 Année commençant le 21 mars de l'année indiquée.
31 Non compris les importations militaires et l'installation près des côtes de l'industrie pétrolière.
32 Y compris les transbordements vers et en provenance de la Malasie péninsulaire.
33 Année finissant le 30 septembre de l'année indiquée.
34 Y compris Tuvalu (anciennement îles Ellice).
35 Les données se rapportent aux exportations totales moins les exportations pétroliers. Dans ce cas, le pétrole est la somme des groupes CTCI 333, 334 et 335.

118
World exports by commodity classes and by regions
Exportations mondiales par classes de marchandises et par régions

In million US dollars f.o.b.

Total trade (SITC, Rev. 2 and Rev. 3, 0-9) /6

Exports from	Year	World Monde /1,2,3	Developed economies Econ. développées /2,3,4	Developing economies Total	Developing OPEC+ OPEP+	Eastern Europe Total /2,3	Former USSR an URSS	Europe Total /2	EEC+ CEE+ /2	EFTA+ AELE+	South Africa Afrique du Sud
World /1,2,3	1980	2000946	1335993	504031	127666	143964	61924	875878	744481	129567	13676
	1987	2477240	1727879	545810	90604	173741	71622	1071604	905029	163752	11227
	1988	2819131	1963717	641169	99025	184449	79032	1212483	1029477	179567	13715
	1989	3024779	2128572	697853	100076	172321	74856	1311633	1118356	188746	12555
	1990	3391752	2429644	772508	111868	142939	65498	1572515	1348965	218434	12661
Developed economies /2,3,4	1980	1258934	891452	316149	99886	42235	21562	650013	542052	106360	12566
	1987	1735497	1347354	334313	62926	37847	20494	908578	763481	142803	9782
	1988	1984399	1528463	391090	68352	43275	24417	1028835	869381	156807	12060
	1989	2129314	1638545	423200	68885	49004	27961	1105547	938117	164670	11442
	1990	2445204	1894297	477688	80179	49905	26382	1325794	1129538	192626	11525
Developing economies /1,3,4	1980	586897	401273	155476	22844	23015	13426	186119	172563	13342	1109
	1987	542210	338038	168159	23173	23859	14397	124897	114029	11793	1444
	1988	628671	389489	205605	26115	26061	15829	143096	130745	13409	1655
	1989	701110	436164	230196	26883	28579	18678	157958	144343	15121	1112
	1990	774605	469536	255316	28731	28054	19051	186852	171517		1134
OPEC+	1980	306770	231096	70072	3966	3716	856	113998	106075	7869	75
	1987	120232	72681	36704	4300	2243	532	33247	31215	2000	3
	1988	122697	78642	38584	4417	2615	654	33162	31395	1751	4
	1989	141018	96055	40915	4824	2554	678	39089	36550	2526	5
	1990	164512	105115	41696	4513	3053	860	47251	44586	2651	6
Eastern Europe and the former USSR /2,3	1980	155115	43269	32406	4936	78714	26937	39747	29866	9865	1
	1987	199532	42487	43339	4505	112035	36731	38130	27520	10351	1
	1988	206062	45765	44474	4558	115113	38786	40553	29352	10967	0
	1989	194355	53863	44458	4308	94739	28218	48129	35897	10667	1
	1990	171944	65812	39504	2959	64980	20064	59869	47910	10687	2
former USSR	1980	76449	24431	19740	1740	32278	.	22672	17260	5413	0
	1987	107873	22457	30977	1129	54439	.	20370	15587	4768	0
	1988	110559	24158	32249	1192	54151	.	21595	16699	4881	0
	1989	109227	26070	32626	1067	50531	.	22947	17624	5300	0
	1990	104177	37457	31128	945	35592	.	33898	28397	5467	0
Developed economies-Europe /2	1980	801865	615412	149325	60121	31447	14681	540357	446087	92771	7888
	1987	1112937	926766	150269	38335	30728	15284	778321	648607	127582	6378
	1988	1233324	1026201	159742	40146	33464	16850	866075	726435	137227	7926
	1989	1318944	1094321	171148	41216	37827	19041	926822	780983	143260	7617
	1990	1578155	1313211	204720	48755	40466	19279	1124857	952286	169216	7911
European Economic Community+ /2	1980	689597	527375	132681	54303	23888	10829	462282	384556	76305	7168
	1987	951919	793770	131096	34154	22184	10617	665686	559713	104046	5888
	1988	1054763	878057	139423	35473	24027	11740	740958	626534	112184	7469
	1989	1130391	936872	150609	37138	27979	13637	794137	674408	117330	7144
	1990	1351043	1122944	179195	43351	30039	13640	962609	820585	138967	7360
European Free Trade Association+	1980	111610	87481	16588	5777	7541	3852	77552	61041	16434	719
	1987	160058	132197	19077	4118	8521	4651	111934	88227	23501	490
	1988	177407	147215	20185	4602	9426	5106	124333	99178	24983	456
	1989	187321	156446	20423	4020	9822	5385	131783	105715	25889	473
	1990	225512	188908	25370	5330	10402	5625	160973	130477	30199	551
Other developed economies	1980	457070	276042	166825	39765	10788	6881	109658	95967	13589	4678
	1987	622562	420602	184046	24592	7119	5211	130257	114874	15221	3404
	1988	751076	502275	231361	28207	9811	7567	162760	142946	19579	4135
	1989	810371	544237	252068	27669	11177	8920	178726	157134	21411	3826
	1990	867050	581099	272985	31425	9440	7103	200938	177254	23410	3613
Canada	1980	64935	55286	7601	1869	1776	1317	9447	8565	880	175
	1987	94402	84953	8699	1211	751	605	7760	6960	799	87
	1988	113145	101717	10322	1417	1106	942	10533	9037	1494	120
	1989	116003	105585	9657	1661	761	581	11501	9824	1674	122
	1990	126897	115641	10162	1630	1081	964	12209	10269	1927	143

En millions de dollars E.-U. f.o.b.

| Economies développées /2,3,4 | | | | Developing economies / Economies en voie de développement /1,3,4 | | | | | | ← Exportations vers | |
Canada	USA E-U	Japan Japon	Australia New Zealand Australie Nouvelle-Zélande	Africa Afrique	America Amerique Total	LAIA+ ALADI+	Mid.East Moyen Orient	Other Autres	Oceania Océanie	Année	Exportations en provenance de ↓
Commerce total (CTCI, Rev. 2 et Rev. 3, 0-9) /6											
50740	240320	124484	24458	84367	126105	83358	96519	173558	2994	1980	Monde /1,2,3
83831	389754	129462	31889	63160	104816	65653	83787	268136	3394	1987	
98192	427556	162584	38381	68675	114195	75354	87803	342619	4093	1988	
107764	456019	183881	45542	70459	124310	81609	91460	381437	4639	1989	
110986	476465	199741	44400	81937	132108	87589	101501	426178	5036	1990	
43176	122805	40279	17304	65004	76243	60160	67086	97449	2118	1980	Economies dévelopées /2,3,4
75841	258152	61207	24633	45325	70424	49628	52992	155021	2569	1987	
88914	279551	79949	29408	48428	78104	57669	55558	197103	3160	1988	
97010	288124	91609	34808	50602	84473	62111	58919	215922	3440	1989	
100709	308200	102235	34222	58966	95229	70099	68028	238219	3716	1990	
7325	116139	82453	7051	15007	44698	22367	23604	68810	876	1980	Economies en voie de développement /1,3,4
7673	129716	66260	7142	13629	26154	14973	25771	100315	825	1987	
9003	145845	80067	8819	16319	27609	16824	26965	131887	933	1988	
10420	165666	89399	10540	15937	31733	18593	27375	152197	1199	1989	
9992	165841	94520	10041	19439	28648	16820	29144	174820	1318	1990	
4483	56536	53076	2925	4243	25893	10275	9282	29068	10	1980	OPEP+
585	16126	21800	907	2908	9122	4431	7672	16014	49	1987	
833	20519	23181	921	3211	8750	4797	7732	17655	52	1988	
1305	27473	26822	1344	3296	9141	4867	8646	18880	56	1989	
1413	23211	31861	1348	5823	4615	1350	9260	21218	69	1990	
239	1376	1752	103	4356	5164	832	5829	7299	0	1980	Europe de l'Est et l'ancienne URSS /2,3
317	1887	1996	113	4207	8238	1052	5025	12800	0	1987	
275	2161	2568	154	3929	8483	861	5279	13629	0	1988	
334	2230	2873	195	3920	8105	905	5167	13317	0	1989	
285	2423	2986	136	3532	8231	670	4328	13139	1	1990	
46	233	1464	14	1380	3679	130	2210	4574	0	1980	l'ancienne URSS
75	442	1540	30	1686	6498	248	1669	9709	0	1987	
27	546	1951	39	1611	6664	105	1734	10855	0	1988	
61	842	2134	53	1494	6659	249	1624	10784	0	1989	
94	953	2460	52	1514	7406	305	2554	10910	0	1990	
5776	44041	8181	6129	49626	25480	19374	41359	25147	519	1980	Economies développées-Europe /2
12206	95170	19069	9232	35205	24551	16275	35327	47002	837	1987	
13956	97288	24107	10274	37098	23798	16313	36195	53697	1021	1988	
14160	99306	27733	11987	38667	24939	17006	37948	59466	1052	1989	
14652	111269	34454	12229	45990	29621	19957	45365	69885	1283	1990	
4953	38619	6672	5309	45716	22214	16875	36698	21653	487	1980	Communauté Economique Européenne+ /2
10460	82794	15703	7785	32274	20606	13977	31210	39970	805	1987	
11840	83953	19677	8637	34063	20314	13836	31777	45613	988	1988	
11606	85168	23085	10145	35743	21797	14572	33824	50563	1022	1989	
11832	95908	28486	10149	42727	25706	16942	39878	58953	1241	1990	
821	5393	1509	819	3870	3260	2498	4653	3493	32	1980	Association Européenne de Libre Echange+
1745	12295	3359	1440	2877	3934	2293	4103	7016	32	1987	
2103	13222	4412	1636	2984	3472	2474	4388	8043	33	1988	
2553	14058	4629	1841	2880	3128	2432	4101	8868	29	1989	
2806	15310	5954	2078	3203	3903	3008	5462	10874	42	1990	
37400	78764	32098	11175	15378	50763	40786	25727	72302	1599	1980	Autres economies développées
63635	162983	42138	15402	10120	45873	33354	17665	108019	1732	1987	
74958	182263	55842	19133	11330	54306	41357	19363	143406	2139	1988	
82850	188819	63877	22821	11935	59534	45105	20971	156457	2389	1989	
86057	196933	67781	21993	12976	65609	50142	22663	168334	2433	1990	
.	41183	3726	660	932	3288	2396	681	2628	13	1980	Canada
.	71238	5124	635	679	3160	1683	730	4107	10	1987	
.	82951	7152	849	917	2367	1630	689	6284	26	1988	
.	85347	7430	1074	803	2169	1518	1068	5535	39	1989	
.	95217	7038	909	903	2325	1675	947	5911	27	1990	

118
World exports by commodity classes and by regions
Exportations mondiales par classes de marchandises et par régions

In million US dollars f.o.b.

Exports from	Year	World Monde /1,2,3	Developed economies Econ. dévelop- pées /2,3,4	Developing economies Economies en voie de développement /1,3,4		Eastern Europe and former USSR Europe de l'Est et l'ancienne URSS		Developed Economies			Developed Economies South Africa
				Total	OPEC+ OPEP+	Total /2,3	Former USSR an URSS	Europe			South Africa Afrique du Sud
								Total /2	EEC+ CEE+ /2	EFTA+ AELE+	Afrique du Sud

Total trade (SITC, Rev. 2 and Rev. 3, 0-9) /6 (continued)

Exports from	Year	World	Developed econ.	Developing Total	OPEC	E.Eur Total	Former USSR	Europe Total	EEC	EFTA	S.Africa
United States	1980	216592	128821	82805	17465	3850	1510	65064	57849	7185	2487
	1987	245421	157416	79379	10431	2193	1477	63487	57277	6109	1296
	1988	309600	194849	105594	13468	3641	2774	79633	71271	8207	1706
	1989	349356	223890	119793	12982	5278	4262	92594	82516	9988	1662
	1990	374449	242222	127368	13418	4226	3072	103832	93049	10666	1750
Japan	1980	129807	61602	64619	18482	3585	2778	21436	18120	3264	1809
	1987	229221	143514	82416	11384	3280	2563	45155	37945	7164	1871
	1988	264917	162643	98367	11441	3906	3130	55736	47168	8499	2083
	1989	275174	168064	103356	10772	3755	3082	56283	48191	8018	1789
	1990	286947	170164	113384	13575	3308	2563	62412	53846	8437	1518
Australia, New Zealand	1980	26673	15513	8291	1809	1513	1276	4334	4161	165	126
	1987	32325	19141	9700	1451	841	565	5715	5356	350	68
	1988	40413	25249	13497	1794	1100	720	6652	6219	429	118
	1989	43976	27244	14852	2145	1312	987	7164	6567	593	128
	1990	47735	29483	16701	2651	715	500	7863	6816	1041	105
Developing economies-Africa	1980	94942	78737	12962	1003	2450	733	46251	43864	2329	302
	1987	49665	40103	7293	1094	1680	699	29976	28821	1119	332
	1988	51830	40212	8427	1271	2191	887	30188	29428	744	347
	1989	56331	44744	8668	1526	2224	977	32735	31806	922	376
	1990	66548	55113	8651	1666	2168	1072	40125	39200	917	317
Developing economies-America	1980	107879	69452	29645	3601	6972	5196	26131	23991	2041	169
	1987	98190	61996	20476	3318	7194	4996	20783	18926	1739	187
	1988	110131	75966	24045	3586	7502	5268	27501	24764	2306	281
	1989	122876	85292	28680	3424	7804	5471	28832	26448	2355	342
	1990	133596	83892	28486	3971	6481	4224	30218	28088	2087	323
LAIA+	1980	79610	51871	23593	2851	3590	2116	20872	19222	1614	158
	1987	80032	51492	17823	3077	2399	1069	17557	16061	1382	185
	1988	92286	65813	21464	3407	2573	1259	24175	21822	1927	281
	1989	102185	73185	25412	3185	2625	1336	24981	23110	1849	342
	1990	112494	71239	24838	3685	1829	844	25838	24378	1433	322
Developing economies-Europe/ 5	1980	8977	3286	1659	851	4033	2489	2807	2385	417	0
	1987	11397	5560	1945	769	3892	2216	4674	3970	697	1
	1988	12597	6335	2033	913	4229	2354	5408	4631	773	0
	1989	13363	6705	2041	799	4617	2899	5916	4928	981	0
	1990	14391	8444	1880	747	4059	2681	7572	6587	978	1
Developing economies-Middle East	1980	210976	150989	53525	6014	3330	988	82802	76742	6059	72
	1987	84579	44670	36472	7985	2518	894	21210	19272	1933	7
	1988	87415	45366	38312	8373	2753	1085	20022	18310	1688	24
	1989	100233	55916	39450	8383	3702	2152	24941	22637	2266	33
	1990	105976	62556	38445	7861	4180	2240	30283	27400	2856	52
Developing economies-Other Asia	1980	161939	96881	57454	11376	6229	4020	27363	24821	2493	565
	1987	296031	183941	101405	10006	8575	5592	47559	42354	5101	918
	1988	363839	219296	132339	11971	9371	6235	59091	52740	6269	1002
	1989	405123	240936	150822	12744	10187	7178	64626	57635	6865	361
	1990	451273	257306	177354	14474	11124	8834	77982	69599	8252	441
Other Asia excluding China	1980	143669	88488	48918	10155	4889	3784	24590	22307	2241	565
	1987	256594	169355	79424	9020	5728	4336	43153	38456	4601	915
	1988	316324	201979	105351	10732	6275	4760	53854	48000	5781	998
	1989	352585	221916	121204	11613	6781	5329	59279	52756	6405	360
	1990	389182	236036	140913	13022	7976	6595	71835	63926	7787	437
Developing economies-Oceania	1980	2184	1930	232	0	0	0	764	761	3	0
	1987	2347	1768	568	2	0	0	696	686	9	
	1988	2858	2315	449	1	14	0	886	872	14	
	1989	3184	2571	534	6	46	0	909	889	20	
	1990	2821	2225	501	12	43	0	673	642	31	

En millions de dollars E.-U. f.o.b.

| Economies développées /2,3,4 | | | | Developing economies / Economies en voie de développement /1,3,4 | | | | | | ← Exportations vers | |
Canada	USA E-U	Japan Japon	Australia New Zealand Australie Nouvelle-Zélande	Africa Afrique	America Amerique Total	LAIA+ ALADI+	Asia Asie Mid.East Moyen Orient	Other Autres	Oceania Océanie	Année	Exportations en provenance de ↓

Commerce total (CTCI, Rev. 2 et Rev. 3, 0-9) /6 (suite)

Canada	USA E-U	Japan Japon	Australia NZ	Africa	America Total	LAIA+ ALADI+	Mid.East	Other	Oceania	Année	Provenance
34102	.	20574	4651	6357	38021	31668	10266	27224	184	1980	Etats-Unis
57513	.	26929	6125	4129	33763	26395	7061	33771	204	1987	
67440	.	36056	7573	5273	42278	33989	8830	48379	306	1988	
74971	.	42752	9215	5942	47475	37611	10267	55260	353	1989	
78212	.	46130	9406	6068	52280	41969	10124	58051	289	1990	
2437	31747	.	4065	5958	8537	5923	13114	'36480	407	1980	Japon
5611	84257	.	6274	3704	8086	4552	8336	61626	587	1987	
6424	90260	.	7719	3684	8672	4907	8222	77022	645	1988	
6807	93716	.	9151	3383	8855	5098	7775	82405	823	1989	
6726	90893	.	8106	3835	9712	5354	9478	89255	853	1990	
539	2804	6053	1627	611	358	267	1507	4754	991	1980	Australie, Nouvelle-Zélande
320	3346	7472	2177	378	405	333	1302	6638	924	1987	
729	4507	10408	2790	429	580	471	1414	9817	1158	1988	
616	4852	11289	3144	502	576	486	1577	10927	1169	1989	
763	5483	11877	3357	600	707	610	1754	12316	1258	1990	
180	29733	1998	72	2978	5879	1476	1801	1463	1	1980	Economies en voie de développement-Afrique
302	8100	1131	95	3007	1312	765	1157	1416	7	1987	
285	7653	1441	122	3330	1223	857	1399	2054	7	1988	
312	9428	1608	92	3865	756	516	1708	1893	3	1989	
705	12278	1444	60	3933	708	458	1717	1794	4	1990	
2804	34890	4541	189	2399	22985	11936	1592	2170	108	1980	Economies en voie de développement-Amérique
1501	33732	4912	537	1334	13943	8997	1706	3358	14	1987	
2096	38912	6176	620	1755	14783	10115	1827	5451	28	1988	
2340	46006	6697	706	1658	18626	11507	1978	6178	24	1989	
1587	43597	7128	630	1948	18174	12374	2109	5957	32	1990	
2430	23389	4154	144	1753	18408	10982	1263	1783	16	1980	ALADI+
1162	27229	4494	523	1248	11664	8566	1617	3174	13	1987	
1691	32912	5778	608	1682	12563	9758	1777	5233	27	1988	
1901	38681	6247	679	1520	15879	11164	1889	5903	24	1989	
1083	36211	6782	607	1605	15470	11854	2019	5474	32	1990	
28	393	31	12	735	65	36	628	228	2	1980	Economies en voie de développement-Europe /5
56	732	40	40	645	161	41	760	378	0	1987	
59	726	83	41	684	154	83	845	348	2	1988	
54	624	35	36	696	230	47	838	276	1	1989	
62	691	39	66	715	110	47	781	273	2	1990	
2546	20180	42724	2655	3524	11745	7167	11120	26020	112	1980	Economies en voie de développement-Moyen Orient
376	6634	15823	596	3021	6549	3346	11589	14538	33	1987	
348	7939	16345	645	3426	6714	3591	11674	15430	39	1988	
712	10565	18659	974	3131	6307	3647	12220	16953	44	1989	
798	9592	20860	920	5698	2452	261	12901	16711	54	1990	
1743	30648	32588	3848	5370	4019	1747	8463	38786	570	1980	Economies en voie de développement-Autres Pays d'Asie
5432	80129	43911	5642	5621	4188	1825	10557	80201	630	1987	
6192	90294	55206	7122	7123	4733	2177	11218	108256	760	1988	
6984	98632	61559	8339	6587	5810	2874	10627	126534	962	1989	
6825	99254	64415	7889	7145	7201	3679	11634	149761	1056	1990	
1619	29573	28402	3613	4264	3661	1641	7516	32748	539	1980	Autres pays d'Asie excluant la Chine
5024	77095	37518	5300	4199	3699	1477	8015	62736	612	1987	
5802	86910	47303	6722	5261	4347	2009	9248	85546	735	1988	
6574	94227	53165	7877	5953	5263	2627	9506	99275	933	1989	
6395	94079	55404	7386	6450	6426	3315	10521	115969	1030	1990	
24	296	570	275	0	6	5	0	143	83	1980	Economies en voie de développement-Océanie
6	389	445	233	1	1	0	1	423	141	1987	
23	321	815	270	1	2	1	3	347	96	1988	
18	411	841	392	0	3	0	2	363	165	1989	
15	429	634	475	0	3	1	2	325	171	1990	

118
World exports by commodity classes and by regions
Exportations mondiales par classes de marchandises et par régions

In million US dollars f.o.b.

Exports from / Year	World Monde /1,2,3	Developed economies Econ. développées /2,3,4	Developing economies Economies en voie de développement /1,3,4		Eastern Europe and former USSR Europe de l'Est et l'ancienne URSS		Europe			Developed Economies South Africa Afrique du Sud
			Total	OPEC+ OPEP+	Total /2,3	Former USSR an URSS	Total /2	EEC+ CEE+ /2	EFTA+ AELE+	

Food, beverages and tobacco (SITC, Rev. 2 and Rev. 3, 0 and 1)

Exports from / Year	World	Developed	Developing Total	OPEC+	Eastern Total	Former USSR	Europe Total	EEC+	EFTA+	South Africa
World /1,2,3 1980	200336	123059	53413	16026	21737	14263	87594	79670	7587	435
1987	225270	161544	45917	12638	16560	11491	111848	102491	8963	583
1988	253684	178790	54923	13700	18504	12774	123464	113296	9686	678
1989	266319	183219	61297	14290	20404	14482	125826	115287	9671	628
1990	294421	210833	63737	14226	17926	12531	147457	135732	10881	644
Developed economies /2,3,4 1980	128734	85109	33949	10501	8606	4967	66045	60022	5717	274
1987	148259	115771	27724	7996	3910	2679	89328	82058	6930	366
1988	168208	128348	33586	9169	5644	4084	98239	90366	7474	451
1989	178750	132716	37478	9651	7728	5624	100782	92949	7415	342
1990	204188	157061	38446	9292	7484	5187	120605	111322	8776	392
Developing economies /1,3,4 1980	62204	35320	17068	4973	8857	6588	19302	17974	1303	162
1987	67276	42893	15869	3942	8292	5660	20191	18619	1522	218
1988	76056	47448	19109	4013	8877	6080	22862	21148	1640	227
1989	77596	46559	21530	4272	9257	6602	21694	20032	1623	286
1990	80984	49591	23200	4540	7835	5552	23393	21849	1506	252
OPEC+ 1980	3722	2214	1095	290	406	95	1087	1054	32	0
1987	3860	2783	981	310	96	24	1029	977	51	1
1988	4290	2976	1188	335	125	21	1171	1141	26	2
1989	4319	2867	1292	341	159	69	1013	978	24	2
1990	5111	3533	1437	334	140	75	1530	1497	25	0
Eastern Europe and the former USSR /2,3 1980	9398	2629	2396	551	4274	2708	2246	1674	567	0
1987	9735	2880	2324	701	4359	3153	2329	1814	511	0
1988	9420	2994	2228	518	3983	2610	2363	1782	572	0
1989	9973	3943	2289	367	3420	2257	3351	2306	633	0
1990	9249	4180	2091	394	2608	1792	3459	2561	598	0
former USSR 1980	1278	154	963	1	161	.	96	71	24	0
1987	1557	351	983	9	222	.	140	118	22	0
1988	1786	420	1064	21	302	.	150	120	29	0
1989	1589	425	969	19	196	.	230	139	91	0
1990	...	...	...	...	...	.	...	...	...	...
Developed economies-Europe /2 1980	76971	59480	13647	6020	3747	1925	54142	49381	4484	147
1987	104536	89980	12186	4300	2163	1196	80347	73806	6213	255
1988	114406	98209	13381	4622	2461	1247	88233	81220	6633	320
1989	120362	101395	15076	5014	3523	1704	91037	84005	6633	244
1990	141605	120027	16965	5215	3925	1974	108679	100287	7908	238
European Economic Community+ /2 1980	72296	56008	12962	5644	3233	1598	51274	47307	3700	138
1987	97973	84369	11640	4098	1759	987	75950	70450	5188	249
1988	107525	92351	12785	4428	2043	1046	83542	77618	5565	314
1989	113320	95551	14395	4863	3008	1443	86362	80385	5600	239
1990	132813	112865	16025	5021	3247	1578	102709	95597	6656	234
European Free Trade Association+ 1980	4489	3295	681	372	514	327	2710	1925	776	9
1987	6268	5331	535	193	402	207	4161	3140	1004	7
1988	6559	5569	575	179	415	198	4454	3389	1044	6
1989	6710	5530	668	142	512	258	4414	3378	1013	5
1990	8394	6785	922	182	678	396	5624	4369	1227	4
Other developed economies 1980	51763	25630	20303	4482	4859	3042	11904	10641	1233	127
1987	43724	25791	15538	3696	1746	1484	8981	8252	717	110
1988	53802	30140	20206	4547	3183	2837	10005	9146	841	131
1989	58388	31322	22404	4637	4205	3920	9744	8945	781	97
1990	62583	37035	21482	4077	3559	3214	11926	11035	868	154
Canada 1980	7021	3640	2026	415	1355	1102	1190	1115	74	3
1987	8025	5391	2042	528	592	530	885	795	89	10
1988	9622	5897	2878	693	848	812	997	885	111	10
1989	8331	5863	1960	863	508	451	895	795	99	4
1990	9945	6801	2264	645	872	839	967	871	93	60

En millions de dollars E.-U. f.o.b.

← Exportations vers

| Economies développées /2,3,4 | | | | Developing economies / Economies en voie de développement /1,3,4 | | | | | | | |
Canada	USA E-U	Japan Japon	Australia New Zealand Australie Nouvelle-Zélande	Africa Afrique	America Amerique Total	LAIA+ ALADI+	Asia Asie Mid.East Moyen Orient	Other Autres	Oceania Océanie	Année	Exportations en provenance de ↓
Produits alimentaires, boisson et tabac (CTCI, Rev. 2 et Rev. 3, 0 et 1)											
3201	17595	12560	1144	12093	11826	8086	11557	16222	598	1980	Monde /1,2,3
3551	23749	19490	1694	9325	7867	4250	11100	16128	561	1987	
3865	23580	24533	1932	10493	9288	5127	11789	21706	670	1988	
4164	23586	26152	2141	10564	10963	6540	13215	24653	727	1989	
6204	26407	27152	2118	11209	12203	7638	12532	25226	731	1990	
2633	6984	8070	677	8980	8476	5940	6871	8688	511	1980	Economies dévelopées /2,3,4
2794	11425	10333	1021	6568	5369	2904	6641	8392	477	1987	
3046	11253	13653	1190	7442	6623	3831	7156	11474	574	1988	
3254	11116	15329	1340	7643	7297	4414	8394	13172	584	1989	
5241	12790	16007	1411	7944	7785	4689	7950	13158	602	1990	
556	10344	4399	459	2534	2851	2126	4178	7128	87	1980	Economies en voie de développement /1,3,4
740	12083	8876	664	2157	1992	1274	3975	7506	84	1987	
796	12090	10528	725	2602	2112	1235	4101	10000	95	1988	
885	12220	10529	784	2614	3074	2016	4356	11112	142	1989	
934	13269	10825	692	2708	3826	2920	4359	11884	128	1990	
10	763	295	57	44	259	165	402	340	0	1980	OPEP+
49	1101	561	39	166	100	77	397	311	0	1987	
14	1033	704	44	217	97	76	431	437	0	1988	
15	1063	726	46	148	162	107	440	519	0	1989	
24	1196	733	48	141	223	140	371	678	1	1990	
12	268	91	8	578	499	20	508	405	0	1980	Europe de l'Est et l'ancienne URSS /2,3
17	240	282	8	600	506	72	485	230	0	1987	
23	237	352	16	449	553	61	533	232	0	1988	
25	250	294	17	307	592	111	465	369	0	1989	
29	348	321	15	556	592	29	223	184	0	1990	
0	18	37	2	48	349	0	12	259	0	1980	l'ancienne URSS
1	23	185	3	69	353	5	10	145	0	1987	
4	22	238	6	112	420	7	16	169	0	1988	
5	20	161	8	98	444	18	9	128	0	1989	
...	...	...	...	...	...	...	...	...	...	1990	
556	3271	959	262	6028	2110	1178	4001	1224	75	1980	Economies développées-Europe /2
881	5784	2047	392	4417	1854	853	3701	1898	120	1987	
865	5441	2632	451	4780	1985	845	3925	2297	131	1988	
964	5424	2920	515	5116	2367	1140	4416	2672	146	1989	
1018	5986	3198	560	5631	2656	1283	4601	3062	164	1990	
513	2823	877	243	5683	1992	1097	3855	1179	74	1980	Communauté Economique Européenne+ /2
818	4952	1776	362	4251	1758	786	3538	1806	119	1987	
804	4753	2263	421	4635	1874	771	3751	2186	130	1988	
903	4703	2592	472	4981	2207	1026	4254	2531	146	1989	
944	5315	2808	520	5451	2476	1157	4395	2903	163	1990	
43	427	82	19	343	118	81	143	45	1	1980	Association Européenne de Libre Echange+
63	796	263	30	164	97	67	155	92	1	1987	
59	656	352	31	143	109	74	158	110	0	1988	
61	687	310	44	133	160	114	153	138	1	1989	
73	653	378	40	177	179	126	194	157	1	1990	
2077	3713	7111	415	2952	6366	4762	2870	7464	436	1980	Autres economies développées
1913	5641	8286	629	2151	3515	2051	2940	6494	358	1987	
2181	5812	11020	739	2663	4638	2986	3231	9177	443	1988	
2290	5692	12410	825	2527	4930	3274	3978	10500	443	1989	
4223	6803	12809	851	2313	5130	3406	3349	10096	438	1990	
·	1731	658	33	325	931	492	166	591	2	1980	Canada
·	3513	910	54	225	616	325	322	878	1	1987	
·	3622	1193	64	310	559	276	335	1671	1	1988	
·	3681	1204	76	223	427	210	649	660	2	1989	
·	4510	1196	63	235	519	311	413	1094	1	1990	

118
World exports by commodity classes and by regions
Exports mondiales par classes de marchandises et par régions

In million US dollars f.o.b.

Food, beverages and tobacco (SITC, Rev. 2 and Rev. 3, 0 and 1) (continued)

Exports from	Year	World Monde /1,2,3	Developed economies Econ. développées /2,3,4	Developing economies Economies en voie de développement /1,3,4 Total	OPEC+ OPEP+	Eastern Europe and former USSR Europe de l'Est et l'ancienne URSS Total /2,3	Former USSR an URSS	Europe Total /2	EEC+ CEE+ /2	EFTA+ AELE+	Developed Economies South Africa Afrique du Sud
United States	1980	30318	14058	12804	2372	2514	975	7598	7069	518	85
	1987	22687	12633	8936	1999	1047	860	5078	4702	371	72
	1988	30739	15818	12483	2622	2234	1946	5825	5347	466	88
	1989	35233	16847	14457	2430	3566	3345	5663	5188	464	67
	1990	36399	20561	13076	2061	2487	2205	6906	6385	508	75
Japan	1980	1590	519	1069	350	2	1	152	128	21	25
	1987	1549	685	861	59	2	2	164	154	9	13
	1988	1648	612	1034	36	3	2	159	151	8	11
	1989	1649	566	1076	52	8	7	159	151	8	13
	1990	1612	518	1081	60	10	10	144	136	8	10
Australia, New Zealand	1980	9673	4712	3967	1339	974	964	975	922	47	11
	1987	8970	5003	3342	1108	101	91	1227	1150	74	13
	1988	9359	5730	3510	1195	95	77	1268	1182	82	20
	1989	10548	5864	4527	1290	118	116	1301	1223	75	11
	1990	11165	6247	4584	1311	181	160	1527	1432	89	6
Developing economies-Africa	1980	9429	6975	1541	633	820	478	5516	5302	208	95
	1987	8281	6422	1315	423	489	359	5094	4902	190	122
	1988	9234	7118	1340	396	469	337	5594	5373	220	139
	1989	8153	6221	1380	380	496	359	4971	4747	222	150
	1990	8192	6397	1423	387	300	218	5127	4914	211	123
Developing economies-America	1980	28885	17531	5110	1308	5629	4482	8855	8024	820	44
	1987	27800	18570	3681	921	5525	3805	8848	7992	818	67
	1988	30670	20455	4318	998	5871	4135	10812	9784	968	67
	1989	31034	19767	5217	900	5969	4149	10116	9235	872	123
	1990	32268	21306	6227	1189	4695	3013	10724	9931	779	116
LAIA+	1980	19498	13053	3964	930	2480	1582	6983	6300	675	44
	1987	18694	14150	2971	819	1570	630	6649	5979	632	66
	1988	21621	16212	3594	942	1806	897	8496	7680	758	67
	1989	21448	15354	4342	813	1686	798	7726	7112	606	122
	1990	22774	16698	5042	1033	1003	398	8015	7550	453	116
Developing economies-Europe/ 5	1980	1023	517	132	72	375	260	450	387	63	0
	1987	992	677	127	40	187	110	607	513	90	0
	1988	1063	785	118	43	160	46	697	613	84	0
	1989	1089	754	129	41	206	110	645	570	72	0
	1990	1037	757	96	35	181	87	655	588	65	0
Developing economies-Middle East	1980	2709	1066	1212	672	369	138	890	789	101	
	1987	3671	1804	1672	1107	180	91	1418	1222	195	1
	1988	3975	1667	2006	1147	295	133	1281	1112	163	1
	1989	3960	1755	1958	1220	236	152	1239	1082	143	3
	1990	4555	2342	1902	1175	297	191	1870	1664	195	3
Developing economies-Other Asia	1980	19380	8572	8958	2288	1664	1231	3362	3243	111	22
	1987	25665	14693	8939	1451	1911	1294	3948	3713	229	28
	1988	30408	16840	11221	1430	2068	1429	4243	4031	204	19
	1989	32458	17373	12705	1730	2305	1831	4497	4173	314	10
	1990	34031	18036	13456	1750	2321	2043	4779	4516	254	9
Other Asia excluding China	1980	15040	7533	5988	2004	1335	1161	2878	2799	72	22
	1987	20709	12864	6613	1343	1131	814	3385	3220	161	28
	1988	24282	14614	8248	1309	1245	911	3685	3510	168	19
	1989	25880	14896	9391	1567	1517	1258	3914	3673	231	10
	1990	26954	15389	9842	1543	1546	1389	4146	3949	187	9
Developing economies-Oceania	1980	778	659	114	0	0	0	229	229	0	0
	1987	866	726	135	0	0	0	277	276	0	0
	1988	706	582	107	0	14	0	235	234	1	0
	1989	901	689	141	0	46	0	226	226	0	0
	1990	901	754	97	4	42	0	237	235	2	0

En millions de dollars E.-U. f.o.b.

Produits alimentaires, boisson et tabac (CTCI, Rev. 2 et Rev. 3, 0 et 1) (suite)

| Economies développées /2,3,4 | | | | Developing economies / Economies en voie de développement /1,3,4 | | | | | | Année | ← Exportations vers / Exportations en provenance de ↓ |
Canada	USA E-U	Japan Japon	Australia New Zealand Australie Nouvelle-Zélande	Africa Afrique	America Amerique Total	LAIA+ ALADI+	Asia Asie Mid.East Moyen Orient	Other Autres	Oceania Océanie		
1534	.	4489	131	1746	5188	4127	1260	4371	47	1980	Etats-Unis
1598	.	5517	188	1513	2619	1532	1524	3160	48	1987	
1810	.	7687	184	1945	3752	2485	1792	4888	63	1988	
1986	.	8671	216	1807	4193	2853	2124	6251	61	1989	
3891	.	9162	274	1496	4228	2821	1791	5367	60	1990	
45	258	.	37	177	38	10	213	579	62	1980	Japon
56	410	.	41	34	27	12	62	668	71	1987	
58	348	.	35	31	18	7	32	843	110	1988	
44	309	.	41	28	21	10	36	898	93	1989	
34	282	.	47	19	14	6	47	910	88	1990	
375	1583	1564	187	418	200	126	1225	1799	325	1980	Australie, Nouvelle-Zélande
219	1654	1562	326	276	244	180	1020	1564	238	1987	
261	1769	1974	437	291	302	217	1063	1585	269	1988	
211	1599	2263	477	361	279	197	1157	2447	283	1989	
250	1892	2114	457	434	347	257	1083	2431	289	1990	
23	1030	244	60	727	88	19	376	276	0	1980	Economies en voie de développement-Afrique
43	646	459	53	808	16	10	288	171	1	1987	
58	745	520	57	832	12	2	286	171	1	1988	
54	494	514	35	871	14	3	296	153	1	1989	
62	502	489	22	926	23	12	244	183	2	1990	
368	7328	828	36	849	2532	1981	888	663	3	1980	Economies en voie de développement-Amérique
351	8092	1007	125	363	1836	1199	768	626	4	1987	
383	7733	1194	115	540	1915	1163	876	878	11	1988	
427	7705	1163	130	432	2820	1916	858	980	13	1989	
476	8616	1160	112	610	3422	2680	986	1024	13	1990	
154	5181	589	33	591	2068	1879	661	487	3	1980	ALADI+
223	6218	796	122	324	1398	1158	691	482	4	1987	
230	6199	963	111	500	1439	1117	835	727	11	1988	
247	6122	920	122	383	2264	1863	785	797	13	1989	
222	7156	985	108	506	2819	2444	924	619	13	1990	
2	57	4	2	36	15	4	73	8	0	1980	Economies en voie de développement-Europe /5
5	58	5	3	32	13	8	76	6	0	1987	
5	75	3	4	25	16	12	63	14	0	1988	
6	84	5	6	31	2	2	87	9	0	1989	
9	82	3	8	43	2	1	50	2	0	1990	
11	126	23	7	87	13	8	1023	78	0	1980	Economies en voie de développement-Moyen Orient
53	243	65	12	147	14	12	1369	138	0	1987	
20	242	76	15	231	18	15	1437	311	0	1988	
21	351	103	23	190	24	22	1562	164	0	1989	
27	310	97	21	213	38	37	1376	226	0	1990	
128	1561	3247	240	834	199	110	1817	6036	42	1980	Economies en voie de développement-Autres Pays d'Asie
283	2692	7309	409	807	113	45	1474	6477	32	1987	
317	3059	8704	468	974	150	43	1438	8553	51	1988	
362	3260	8689	521	1090	214	72	1553	9723	71	1989	
346	3389	9014	459	916	341	190	1703	10415	53	1990	
110	1501	2791	218	627	146	106	1511	3647	30	1980	Autres pays d'Asie excluyant la Chine
246	2527	6263	390	713	56	24	1358	4421	31	1987	
280	2867	7281	450	855	83	30	1334	5873	50	1988	
313	2954	7175	496	916	111	51	1423	6823	64	1989	
290	3010	7464	431	729	195	128	1540	7303	47	1990	
24	241	52	113	0	5	5	0	67	42	1980	Economies en voie de développement-Océanie
4	351	31	63	0	0	0	1	87	46	1987	
13	236	32	66	0	0	0	1	73	32	1988	
15	325	54	69	0	0	0	1	84	57	1989	
14	371	62	71	0	0	0	1	35	61	1990	

118
World exports by commodity classes and by regions
Exportations mondiales par classes de marchandises et par régions

In million US dollars f.o.b.

Exports from / Year	World Monde /1,2,3	Developed economies Econ. développées /2,3,4	Developing economies Economies en voie de développement /1,3,4 Total	OPEC+ OPEP+	Eastern Europe and former USSR Europe de l'Est et l'ancienne URSS Total /2,3	Former USSR an URSS	Europe Total /2	EEC+ CEE+ /2	EFTA+ AELE+	Developed Economies South Africa Afrique du Sud

Crude materials (excluding fuels), oils, fats (SITC, Rev. 2 and Rev. 3, 2 and 4)

Exports from	Year	World	Dev. econ.	Developing Total	OPEC	E.Eur/USSR Total	Former USSR	Europe Total	EEC	EFTA	South Africa
World /1,2,3	1980	138324	95545	30434	4152	10214	3315	61531	55345	6149	633
	1987	142073	95161	35372	3710	9799	3484	60610	53863	6659	382
	1988	167447	113287	42990	4398	10204	3553	71889	63948	7822	443
	1989	177006	122097	43629	4646	10317	4175	76807	68301	8402	448
	1990	175342	123556	42764	4607	7909	3178	79798	71483	8192	426
Developed economies /2,3,4	1980	85024	64847	15608	2576	2705	953	44228	39639	4559	524
	1987	91968	69877	18157	1888	2313	1040	46811	41786	4997	270
	1988	106992	81811	21746	2288	2864	1358	55382	49331	6011	308
	1989	112536	86993	21899	2341	2940	1468	58115	51937	6132	325
	1990	114211	89309	22172	2460	2009	696	60191	53954	6182	299
Developing economies /1,3,4	1980	42787	26864	12130	1366	3626	1912	14361	13587	767	109
	1987	38730	21922	13389	1564	3354	2039	10888	10184	695	111
	1988	48576	27384	17643	1856	3280	1821	13096	12286	800	135
	1989	52284	30524	17827	1989	3828	2419	14852	13608	1234	123
	1990	50172	28865	17670	2048	3457	2286	15043	14029	1000	126
OPEC+	1980	5120	3301	1607	75	210	96	1224	1173	50	0
	1987	3021	1920	962	102	138	78	780	708	71	0
	1988	4068	2734	1225	107	109	34	1056	967	89	1
	1989	4566	3130	1318	177	117	53	1409	1074	335	2
	1990	4054	2575	1380	162	99	36	1202	1121	79	4
Eastern Europe and the former USSR /2,3	1980	10512	3835	2697	210	3883	449	2943	2119	824	0
	1987	11376	3362	3826	259	4132	406	2912	1894	967	0
	1988	11879	4092	3600	255	4060	374	3411	2331	1011	0
	1989	12186	4581	3903	317	3548	288	3839	2755	1036	0
	1990	10959	5382	2922	99	2443	196	4563	3501	1011	0
former USSR	1980	7047	2221	1991	85	2835	.	1381	1089	291	0
	1987	8004	1798	2985	7	3221	.	1392	1052	339	0
	1988	8486	2472	2851	17	3163	.	1844	1384	461	0
	1989	8360	2508	2999	20	2853	.	1833	1360	473	0
	1990	...	...	...	...	...	.	...	...	...	
Developed economies-Europe /2	1980	33950	28666	3678	1387	1585	476	27108	23777	3304	238
	1987	41541	35736	4387	832	1314	429	33298	28918	4352	144
	1988	48044	41312	4871	852	1401	400	38851	33787	5027	148
	1989	50408	43531	4882	804	1495	470	40851	35853	4954	171
	1990	53521	46420	5280	982	1274	321	43734	38582	5099	173
European Economic Community+ /2	1980	24349	20694	2622	931	1011	301	19469	17115	2332	149
	1987	31790	27455	3352	566	879	279	25427	22272	3133	137
	1988	37120	31987	3906	615	876	223	29935	26185	3720	140
	1989	38861	33589	3954	571	952	280	31305	27619	3650	155
	1990	41365	35953	4164	614	865	188	33608	29752	3807	166
European Free Trade Association+	1980	9587	7957	1055	455	574	175	7625	6652	969	88
	1987	9737	8269	1034	265	435	150	7859	6637	1216	8
	1988	10898	9303	962	235	525	178	8895	7587	1301	8
	1989	11526	9924	925	231	542	190	9528	8218	1302	16
	1990	12140	10452	1114	367	409	133	10112	8816	1291	7
Other developed economies	1980	51074	36183	11930	1189	1120	477	17120	15862	1255	286
	1987	50427	34143	13770	1056	999	611	13513	12867	645	126
	1988	58949	40500	16877	1436	1464	958	16530	15545	983	159
	1989	62128	43463	17019	1537	1445	998	17264	16085	1177	155
	1990	60690	42890	16893	1478	735	375	16458	15372	1083	126
Canada	1980	12577	11188	1136	175	86	43	3632	3402	230	74
	1987	15174	12909	2198	226	67	41	3113	2944	169	35
	1988	17215	15105	2026	225	84	42	4183	3726	457	55
	1989	18369	16351	1879	272	138	90	4698	4089	609	54
	1990	17697	15839	1784	250	74	36	4429	3884	546	29

En millions de dollars E.-U. f.o.b.

← Exportations vers

| Economies développées /2,3,4 | | | | Developing economies — Economies en voie de développement /1,3,4 | | | | | | Année | Exportations en provenance de ↓ |
Canada	USA E-U	Japan Japon	Australia New Zealand Australie Nouvelle-Zélande	Africa Afrique	America Amerique Total	LAIA+ ALADI+	Asia Asie Mid.East Moyen Orient	Other Autres	Oceania Océanie		
Matières brutes (sauf combustibles), huiles et graisses (CTCI, Rev. 2 et Rev. 3, 2 et 4)											
2497	10213	19357	1068	3612	4883	4033	2895	16432	64	1980	Monde /1,2,3
2391	12296	18083	1028	3337	5578	4212	2929	20589	57	1987	
2662	13832	22661	1367	3828	6309	5161	3376	26359	70	1988	
3127	14378	25499	1422	4058	6355	5091	3955	25905	72	1989	
3704	14213	23691	1288	4378	6121	4901	3872	25873	81	1990	
2159	6522	10507	696	2390	3048	2569	1417	8087	41	1980	Economies dévelopées /2,3,4
1996	8943	10807	728	2061	3547	2725	1418	10687	41	1987	
2188	9545	13075	938	2301	3727	3183	1596	13561	49	1988	
2664	9961	14669	923	2338	3700	3084	1961	13261	50	1989	
3263	9943	14409	848	2438	3354	2815	2011	13679	57	1990	
331	3643	8018	370	931	1577	1447	1278	7959	23	1980	Economies en voie de développement /1,3,4
390	3322	6870	300	964	1619	1444	1194	9381	16	1987	
468	4257	8950	430	1303	2188	1933	1441	12352	20	1988	
459	4383	10159	499	1395	2321	1970	1564	12168	22	1989	
437	4232	8516	440	1597	2417	2045	1661	11741	23	1990	
29	634	1411	2	50	48	45	95	1385	0	1980	OPEP+
49	555	522	2	54	27	27	143	718	0	1987	
47	710	889	16	94	36	35	144	927	0	1988	
19	593	1062	31	148	68	60	181	894	0	1989	
33	584	702	29	216	123	116	254	767	0	1990	
7	47	832	2	291	258	17	200	386	0	1980	Europe de l'Est et l'ancienne URSS /2,3
5	31	407	0	313	411	44	317	521	0	1987	
6	30	637	0	224	394	45	339	447	0	1988	
4	33	670	0	326	334	36	430	476	0	1989	
4	37	766	0	343	351	41	200	452	0	1990	
2	13	823	2	118	227	9	72	157	0	1980	l'ancienne URSS
3	16	387	0	114	343	0	21	369	0	1987	
3	16	609	0	28	321	0	51	381	0	1988	
1	14	634	0	92	280	0	54	381	0	1989	
...	...	...	...	...	...	...	...	...	...	1990	
129	737	275	91	1404	308	242	996	575	10	1980	Economies développées-Europe /2
224	1124	677	134	1254	381	290	810	1650	9	1987	
223	1162	658	139	1317	430	298	825	1957	13	1988	
254	1212	734	159	1355	454	331	901	1735	11	1989	
203	1276	710	163	1457	475	347	1065	1744	17	1990	
112	657	188	73	1054	260	205	653	403	10	1980	Communauté Economique Européenne+ /2
192	993	525	90	912	335	251	611	1278	9	1987	
204	1004	512	103	983	385	262	664	1635	12	1988	
230	1095	583	125	1021	420	308	750	1464	11	1989	
185	1159	594	136	1053	435	314	812	1504	17	1990	
17	80	87	19	350	48	37	343	172	0	1980	Association Européenne de Libre Echange+
32	131	151	43	342	46	39	198	372	0	1987	
19	157	146	36	334	45	37	159	321	1	1988	
24	117	151	34	334	34	24	149	271	0	1989	
18	116	116	26	404	40	34	252	239	0	1990	
2030	5786	10232	605	986	2741	2327	420	7512	31	1980	Autres economies développées
1772	7819	10130	594	807	3166	2434	608	9037	32	1987	
1964	8383	12417	799	984	3297	2885	771	11604	36	1988	
2411	8749	13936	764	983	3246	2753	1061	11526	39	1989	
3059	8668	13699	685	981	2879	2468	947	11936	40	1990	
.	5269	2014	192	217	265	218	41	584	0	1980	Canada
.	7100	2422	209	273	771	385	32	1112	0	1987	
.	7542	3052	246	343	371	349	38	1256	1	1988	
.	7883	3440	240	236	338	315	56	1224	0	1989	
.	7989	3165	187	275	277	258	82	1141	0	1990	

118
World exports by commodity classes and by regions
Exportations mondiales par classes de marchandises et par régions

In million US dollars f.o.b.

Exports to → Exports from	Year	World Monde /1,2,3	Developed economies Econ. dévelop- pées /2,3,4	Developing economies Economies en voie de développement /1,3,4		Eastern Europe and former USSR Europe de l'Est et l'ancienne URSS		Developed Economies Europe			Developed Economies South Africa
				Total	OPEC+ OPEP+	Total /2,3	Former USSR an URSS	Total /2	EEC+ CEE+ /2	EFTA+ AELE+	Afrique du Sud

Crude materials (excluding fuels), oils, fats (SITC, Rev. 2 and Rev. 3, 2 and 4) (continued)

Exports from	Year	World	Dev.econ	Total	OPEC	Total	Former USSR	Total Europe	EEC	EFTA	S.Africa
United States	1980	25688	17201	7882	735	430	84	9311	8798	512	159
	1987	21491	13298	7957	704	231	75	6219	5909	309	49
	1988	26796	15935	10435	1031	424	251	7085	6725	358	56
	1989	28296	17135	10850	1018	280	122	7089	6735	352	56
	1990	28176	17145	10738	914	288	103	6792	6470	319	65
Japan	1980	1580	395	1087	162	97	57	237	220	18	30
	1987	1649	431	1164	72	53	30	216	198	18	24
	1988	2033	554	1410	98	68	46	309	289	20	29
	1989	2040	509	1452	112	79	64	254	232	22	28
	1990	2044	529	1456	122	58	47	284	259	25	23
Australia, New Zealand	1980	7717	4474	1262	108	481	292	1716	1685	29	20
	1987	9295	5305	1995	47	611	465	2446	2421	25	12
	1988	9754	6269	2630	74	855	619	3066	2985	80	13
	1989	9672	6387	2358	130	926	722	2972	2900	72	12
	1990	8501	5884	2341	187	276	189	2391	2358	33	6
Developing economies-Africa	1980	6958	5386	982	117	484	95	4511	4353	155	64
	1987	4548	3240	910	200	363	86	2627	2531	95	62
	1988	5150	3442	1099	198	439	87	2757	2635	121	62
	1989	5797	3994	1335	245	444	106	3128	2994	133	68
	1990	5818	4047	1417	282	329	83	3348	3198	149	58
Developing economies-America	1980	11431	7849	2553	454	1004	610	4556	4192	362	35
	1987	10047	6430	2726	442	888	623	3547	3210	334	39
	1988	13095	8792	3571	549	729	474	4979	4547	424	62
	1989	14614	9640	4115	619	852	601	5550	5018	524	49
	1990	15788	10487	4273	749	1022	669	5991	5472	509	54
LAIA+	1980	9581	6613	2193	386	775	435	3883	3675	206	35
	1987	8740	5570	2585	420	583	338	3076	2885	190	39
	1988	11673	7801	3457	531	413	181	4448	4152	291	62
	1989	12947	8468	3936	601	539	309	4881	4550	323	49
	1990	13647	9182	4001	721	461	243	5237	5006	220	54
Developing economies-Europe/ 5	1980	685	364	142	28	179	76	353	312	41	0
	1987	573	377	90	13	106	29	361	311	50	0
	1988	723	405	132	29	186	61	388	326	62	0
	1989	885	421	149	19	315	199	406	332	75	0
	1990	1000	509	159	13	332	244	493	418	74	0
Developing economies-Middle East	1980	1510	667	487	151	349	138	586	491	95	0
	1987	1525	652	591	242	281	129	527	473	54	1
	1988	1892	787	774	263	331	152	621	539	83	2
	1989	2176	1039	822	276	313	148	882	546	336	2
	1990	2237	952	980	235	303	138	783	659	124	10
Developing economies-Other Asia	1980	21229	11716	7891	616	1610	993	4072	3959	112	10
	1987	21052	10543	8772	667	1717	1173	3556	3391	161	10
	1988	26362	12840	11838	816	1596	1047	3967	3856	109	10
	1989	27428	14252	11201	831	1904	1366	4441	4280	159	4
	1990	24310	12059	10637	766	1471	1152	4144	4018	123	4
Other Asia excluding China	1980	19201	10469	7389	593	1333	934	3457	3390	66	10
	1987	17318	8671	7466	543	1161	801	2790	2718	70	10
	1988	22031	11017	9908	710	1018	656	3358	3269	89	10
	1989	23130	11910	9736	731	1413	1034	3585	3467	116	4
	1990	20612	10013	9275	632	1182	925	3296	3205	87	4
Developing economies-Oceania	1980	973	882	74	0	0	0	282	279	3	0
	1987	984	681	300	0	0	0	269	268	1	0
	1988	1353	1118	230	0	0	0	384	383	1	0
	1989	1384	1177	204	0	0	0	446	439	7	0
	1990	1019	812	204	3	0	0	284	263	21	0

En millions de dollars E.-U. f.o.b.

| Economies développées /2,3,4 | | | | Developing economies / Economies en voie de développement /1,3,4 | | | | | | ← Exportations vers | |
Canada	USA E-U	Japan Japon	Australia New Zealand Australie Nouvelle-Zélande	Africa Afrique	America Amérique Total	LAIA+ ALADI+	Asia Asie Mid.East Moyen Orient	Other Autres	Oceania Océanie	Année	Exportations en provenance de ↓
Matières brutes (sauf combustibles), huiles et graisses (CTCI, Rev. 2 et Rev. 3, 2 et 4) (suite)											
1857	.	5561	214	566	2342	1987	180	4664	10	1980	Etats-Unis
1702	.	5024	176	397	2312	1970	463	4723	10	1987	
1850	.	6507	252	508	2811	2439	638	6387	13	1988	
2336	.	7300	231	592	2820	2356	881	6469	16	1989	
2981	.	6945	226	547	2521	2135	744	6852	14	1990	
40	44	.	38	29	46	35	79	930	1	1980	Japon
23	122	.	42	22	25	23	17	1098	1	1987	
25	140	.	49	22	28	26	28	1330	1	1988	
28	148	.	47	24	18	15	36	1374	1	1989	
32	148	.	40	19	15	13	36	1384	1	1990	
68	258	2254	149	47	33	32	98	997	18	1980	Australie, Nouvelle-Zélande
28	434	2207	160	46	27	27	85	1778	20	1987	
58	499	2366	244	56	60	45	59	2356	22	1988	
22	453	2670	241	56	38	37	74	2104	22	1989	
14	284	2955	222	49	33	32	71	2134	25	1990	
55	315	427	10	271	92	91	111	354	0	1980	Economies en voie de développement-Afrique
40	267	212	19	341	55	54	90	361	0	1987	
59	273	241	32	374	66	66	149	432	0	1988	
69	333	343	38	483	67	64	126	556	0	1989	
65	302	229	21	486	81	79	152	638	0	1990	
136	1600	1500	15	176	1171	1071	376	782	2	1980	Economies en voie de développement-Amérique
187	1484	1142	25	185	1268	1107	299	946	0	1987	
252	2009	1454	30	302	1687	1461	301	1195	0	1988	
241	2121	1627	45	299	1926	1597	426	1383	0	1989	
239	2222	1900	63	342	2003	1679	429	1442	0	1990	
98	1154	1431	5	131	1030	984	326	657	2	1980	ALADI+
93	1229	1104	24	181	1146	1054	295	934	0	1987	
134	1714	1406	30	295	1608	1433	300	1171	0	1988	
115	1791	1580	45	269	1813	1554	423	1353	0	1989	
133	1849	1834	58	283	1835	1606	428	1412	0	1990	
0	3	5	0	85	1	1	34	22	1	1980	Economies en voie de développement-Europe /5
0	3	5	0	52	1	1	21	16	0	1987	
0	4	9	0	80	11	2	26	16	0	1988	
0	3	6	0	101	1	0	35	12	0	1989	
0	6	3	0	91	6	6	26	36	0	1990	
0	34	44	2	57	1	1	174	221	0	1980	Economies en voie de développement-Moyen Orient
0	54	64	2	86	11	10	264	190	0	1987	
0	76	83	2	97	5	3	282	315	0	1988	
0	64	84	3	95	4	2	296	350	0	1989	
1	56	96	2	135	4	4	402	385	0	1990	
140	1679	5576	220	342	311	282	584	6525	2	1980	Economies en voie de développement-Autres Pays d'Asie
162	1504	5107	191	299	285	272	520	7578	5	1987	
156	1871	6518	301	450	418	401	682	10179	8	1988	
148	1858	7438	348	417	324	306	680	9676	9	1989	
132	1641	5823	297	542	324	278	649	9052	9	1990	
138	1544	5092	210	316	295	276	576	6083	2	1980	Autres pays d'Asie excluyant la Chine
127	1370	4181	180	283	266	258	458	6383	5	1987	
142	1793	5411	287	428	402	392	642	8341	6	1988	
130	1654	6190	331	411	311	299	654	8258	9	1989	
119	1462	4835	279	535	303	267	631	7738	9	1990	
0	10	466	124	0	0	0	0	57	17	1980	Economies en voie de développement-Océanie
0	9	340	63	0	0	0	0	290	10	1987	
1	24	645	64	0	0	0	1	215	12	1988	
1	5	662	64	0	0	0	2	190	12	1989	
0	5	466	58	0	0	0	2	189	14	1990	

118
World exports by commodity classes and by regions
Exportations mondiales par classes de marchandises et par régions

In million US dollars f.o.b.

Exports from / Exports to →	Year	World Monde /1,2,3	Developed economies Econ. développées /2,3,4	Developing economies Economies en voie de développement /1,3,4 Total	OPEC+ OPEP+	Eastern Europe and former USSR Europe de l'Est et l'ancienne URSS Total /2,3	Former USSR an URSS	Europe Total /2	EEC+ CEE+ /2	EFTA+ AELE+	Developed Economies South Africa Afrique du Sud
											Animal and vegetable oils, fats and waxes (SITC, Rev. 2 and Rev. 3, 4)
World /1,2,3	1980	11029	5208	5235	1229	580	352	4172	3893	270	54
	1987	10109	5048	4428	874	533	414	3961	3723	217	44
	1988	12406	6092	5744	1046	438	261	4561	4254	263	74
	1989	12748	6194	5772	1097	716	573	4801	4477	278	63
	1990	13493	7132	5770	938	511	366	5649	5298	299	76
Developed economies /2,3,4	1980	6036	2988	2868	737	177	84	2620	2416	196	46
	1987	5452	3357	1835	372	169	99	2940	2773	162	16
	1988	6351	3763	2366	450	145	74	3206	3003	197	24
	1989	6349	3816	2272	420	193	106	3240	3036	195	22
	1990	6921	4504	2180	404	166	72	3794	3571	208	29
Developing economies /1,3,4	1980	4726	2126	2247	469	351	268	1459	1447	11	8
	1987	4333	1598	2411	438	314	302	934	922	12	28
	1988	5700	2208	3230	533	208	181	1238	1220	18	49
	1989	6038	2242	3306	612	489	466	1431	1396	35	41
	1990	6280	2480	3480	516	314	291	1710	1669	38	47
OPEC+	1980	334	230	102	30	1	0	197	197	0	0
	1987	318	198	98	24	23	23	160	159	1	0
	1988	567	361	183	22	23	23	317	315	3	0
	1989	487	357	106	29	24	24	293	292	1	0
	1990	468	297	158	22	13	13	231	229	0	0
Eastern Europe and the former USSR /2,3	1980	267	95	120	23	52	1	93	29	64	0
	1987	324	93	182	64	50	12	87	28	43	0
	1988	354	120	148	63	85	5	116	31	48	0
	1989	362	135	193	65	33	2	130	46	48	0
	1990	291	149	111	17	31	2	145	58	53	0
former USSR	1980	92	20	66	0	7	.	20	10	10	0
	1987	85	6	79	0	0	.	6	2	5	0
	1988	103	12	54	0	38	.	12	2	10	0
	1989	95	9	85	0	0	.	9	1	8	0
	1990	...	...	...	...	...	.	...	...	...	...
Developed economies-Europe /2	1980	3515	2331	1056	483	124	48	2199	2003	189	11
	1987	3985	2951	832	201	144	77	2727	2565	157	11
	1988	4236	3111	936	215	111	44	2856	2664	186	11
	1989	4567	3321	1018	218	162	76	3023	2825	189	12
	1990	5289	3952	1127	232	138	46	3547	3330	202	17
European Economic Community+ /2	1980	3239	2139	1002	478	94	47	2017	1858	152	10
	1987	3769	2808	786	198	118	73	2596	2466	125	11
	1988	3980	2945	874	211	83	41	2701	2557	137	11
	1989	4338	3167	969	216	136	71	2882	2725	149	11
	1990	5043	3790	1068	224	114	44	3399	3223	162	16
European Free Trade Association+	1980	273	189	54	5	29	1	179	143	36	1
	1987	214	141	46	3	26	4	129	98	30	1
	1988	250	160	62	3	29	3	149	104	45	0
	1989	227	152	49	2	27	4	139	99	39	1
	1990	246	161	60	8	24	2	147	108	39	1
Other developed economies	1980	2521	656	1812	255	53	36	421	413	8	35
	1987	1467	407	1003	172	25	22	213	208	5	5
	1988	2115	652	1430	235	33	30	350	339	12	13
	1989	1781	495	1254	202	31	30	217	210	6	10
	1990	1632	552	1052	173	27	26	247	241	7	12
Canada	1980	200	88	112	0	0	0	53	51	2	0
	1987	154	77	77	7	0	0	21	21	1	0
	1988	253	152	100	33	1	0	38	37	1	2
	1989	180	127	52	14	0	0	30	30	0	0
	1990	183	144	38	12	1	0	28	27	0	0

En millions de dollars E.-U. f.o.b.

← Exportations vers

Huiles et graisses d'origine animale ou vegetale (CTCI, Rev. 2 et Rev. 3, 4)

Economies développées /2,3,4				Developing economies / Economies en voie de développement /1,3,4							
					America Amerique		Asia Asie				Exportations en provenance de ↓
Canada	USA E-U	Japan Japon	Australia New Zealand Australie Nouvelle-Zélande	Africa Afrique	Total	LAIA+ ALADI+	Mid.East Moyen Orient	Other Autres	Oceania Océanie	Année	
78	517	271	95	1173	815	538	835	2282	16	1980	Monde /1,2,3
77	621	240	85	867	613	348	778	2116	16	1987	
90	886	344	110	1162	912	590	886	2746	17	1988	
92	743	351	126	1128	954	617	1013	2574	17	1989	
110	819	323	128	1133	1032	695	886	2628	20	1990	
55	82	117	53	913	594	386	391	• 886	16	1980	Economies dévelopées /2,3,4
58	205	93	27	528	401	240	270	610	13	1987	
61	270	147	32	686	487	307	317	850	14	1988	
68	293	140	39	667	549	360	332	670	13	1989	
87	388	140	43	639	501	329	315	670	15	1990	
23	435	153	42	221	165	153	441	1393	1	1980	Economies en voie de développement /1,3,4
19	411	146	58	293	162	108	456	1494	3	1987	
29	613	197	77	437	372	283	518	1895	3	1988	
23	446	211	87	400	353	256	638	1903	5	1989	
23	429	182	84	477	473	365	566	1957	5	1990	
3	22	7	0	35	3	3	32	33	0	1980	OPEP+
0	31	6	0	18	1	0	26	53	0	1987	
1	30	10	2	47	2	1	27	107	0	1988	
3	45	7	9	18	3	3	29	55	0	1989	
8	34	9	15	19	16	15	34	89	0	1990	
0	1	1	0	39	56	0	3	4	0	1980	Europe de l'Est et l'ancienne URSS /2,3
0	5	1	0	46	50	0	52	12	0	1987	
0	4	0	0	39	53	0	51	1	0	1988	
0	5	0	0	61	53	1	43	1	0	1989	
1	2	1	0	17	58	1	6	1	0	1990	
0	0	0	0	0	52	0	2	2	0	1980	l'ancienne URSS
0	0	0	0	0	50	0	0	9	0	1987	
0	0	0	0	0	53	0	0	1	0	1988	
0	0	0	0	4	52	0	0	1	0	1989	
...	...	...	...	...	...	...	...	...	...	1990	
15	64	9	24	534	56	24	317	85	7	1980	Economies développées-Europe /2
21	144	13	21	343	69	29	182	221	7	1987	
22	163	16	27	398	111	3ᶠ	191	219	6	1988	
20	203	22	29	377	132	6ᵖ	224	238	7	1989	
28	275	29	37	434	185	114	205	254	9	1990	
14	59	6	23	513	46	22	313	71	7	1980	Communauté Economique Européenne+ /2
20	133	13	21	325	62	27	177	206	6	1987	
21	155	16	26	376	102	37	187	194	6	1988	
19	192	21	29	357	124	66	218	226	6	1989	
28	263	29	37	404	177	111	201	242	9	1990	
1	5	3	1	21	9	2	4	14	0	1980	Association Européenne de Libre Echange+
0	11	0	0	18	7	2	5	15	0	1987	
1	8	0	0	22	9	2	4	25	0	1988	
1	11	0	0	20	8	2	6	12	0	1989	
1	12	0	0	30	8	3	4	12	0	1990	
40	18	108	29	379	538	362	74	800	8	1980	Autres economies développées
37	61	80	6	186	332	211	89	389	7	1987	
39	106	131	6	289	376	268	126	631	7	1988	
48	90	118	10	290	417	292	108	432	6	1989	
59	112	111	6	205	315	216	110	416	6	1990	
.	10	21	3	10	10	3	0	92	0	1980	Canada
.	39	16	0	14	5	2	1	58	0	1987	
.	89	23	1	12	7	2	1	80	0	1988	
.	75	22	0	26	6	3	1	19	0	1989	
.	96	20	0	21	11	6	1	6	0	1990	

118
World exports by commodity classes and by regions
Exportations mondiales par classes de marchandises et par régions

In million US dollars f.o.b.

Exports from	Year	World Monde /1,2,3	Developed economies Econ. développées /2,3,4	Developing economies Economies en voie de développement /1,3,4		Eastern Europe and former USSR Europe de l'Est et l'ancienne URSS		Developed Economies Europe /2			South Africa Afrique du Sud
				Total	OPEC+ OPEP+	Total /2,3	Former USSR an URSS	Total /2	EEC+ CEE+ /2	EFTA+ AELE+	Afrique du Sud

Animal and vegetable oils, fats and waxes (SITC, Rev. 2 and Rev. 3, 4) (continued)

Exports from	Year	World	Dev.	Total	OPEC	E.Eur Total	Former USSR	Eur Total	EEC	EFTA	S.Africa
United States	1980	1975	465	1472	229	39	28	290	285	5	24
	1987	1037	243	775	157	20	19	151	147	4	4
	1988	1520	330	1163	196	26	26	202	194	7	10
	1989	1350	277	1046	183	27	26	150	146	3	6
	1990	1191	306	862	157	23	23	164	163	2	10
Japan	1980	114	66	41	6	6	0	59	58	0	0
	1987	85	42	38	4	5	4	21	21	0	1
	1988	154	105	43	7	6	4	85	82	3	1
	1989	81	38	38	4	4	4	19	17	2	4
	1990	93	54	35	3	3	3	35	32	4	2
Australia, New Zealand	1980	166	26	133	19	8	8	8	8	0	10
	1987	136	38	67	4	0	0	14	14	0	0
	1988	141	56	85	0	0	0	19	18	0	0
	1989	111	44	67	0	0	0	10	10	0	0
	1990	99	42	57	1	0	0	16	16	1	0
Developing economies-Africa	1980	347	319	27	4	0	0	311	309	1	1
	1987	224	177	27	1	11	11	167	165	2	1
	1988	263	164	45	1	0	0	154	153	1	1
	1989	294	252	42	1	0	0	246	245	1	2
	1990	410	372	39	3	0	0	364	358	6	0
Developing economies-America	1980	1296	445	670	207	182	123	371	365	6	7
	1987	1045	284	623	169	137	135	142	141	1	26
	1988	1442	493	872	172	78	73	190	187	3	48
	1989	1541	389	943	220	209	203	233	225	9	40
	1990	1731	450	1144	296	137	126	276	270	7	47
LAIA+	1980	1284	440	662	207	182	123	369	364	6	7
	1987	1026	276	613	169	137	135	135	134	1	26
	1988	1427	485	864	172	78	73	183	180	3	48
	1989	1523	387	927	220	209	203	233	225	9	40
	1990	1720	449	1135	295	137	126	276	269	7	47
Developing economies-Europe/ 5	1980	20	10	3	1	8	0	6	4	2	0
	1987	11	6	4	0	1	0	6	4	1	0
	1988	33	10	15	3	8	0	9	8	2	0
	1989	9	5	2	2	2	0	5	4	1	0
	1990	12	6	2	0	3	0	6	5	1	0
Developing economies-Middle East	1980	25	1	22	16	2	0	1	1	0	0
	1987	121	9	106	83	6	5	8	8	0	0
	1988	116	13	93	77	10	10	12	8	4	0
	1989	189	15	108	90	67	65	12	11	1	0
	1990	186	31	103	82	50	50	29	19	10	0
Developing economies-Other Asia	1980	2965	1279	1526	241	160	144	712	710	2	0
	1987	2870	1059	1651	185	159	151	564	556	8	0
	1988	3765	1449	2204	280	112	98	807	798	9	0
	1989	3918	1496	2211	299	211	197	861	838	23	0
	1990	3863	1542	2191	135	123	115	964	947	15	0
Other Asia excluding China	1980	2890	1252	1492	240	146	142	697	695	2	0
	1987	2788	1008	1629	185	152	146	525	519	6	0
	1988	3691	1416	2170	279	105	94	785	776	8	0
	1989	3832	1465	2163	299	204	191	840	817	23	0
	1990	3702	1477	2095	135	123	115	910	894	14	0
Developing economies-Oceania	1980	73	73	0	0	0	0	58	58	0	0
	1987	63	63	0	0	0	0	49	49	0	0
	1988	80	79	1	0	0	0	66	66	0	0
	1989	87	87	0	0	0	0	74	74	0	0
	1990	79	79	0	0	0	0	71	70	1	0

En millions de dollars E.-U. f.o.b.

| Economies développées /2,3,4 | | | | Developing economies Economies en voie de développement /1,3,4 | | | | | | ⟵ Exportations vers | |
Canada	USA E-U	Japan Japon	Australia New Zealand Australie Nouvelle-Zélande	Africa Afrique	America Amerique Total	LAIA+ ALADI+	Asia Asie Mid.East Moyen Orient	Other Autres	Oceania Océanie	Année	Exportations en provenance de ↓

Huiles et graisses d'origine animale ou vegetale (CTCI, Rev. 2 et Rev. 3, 4) (suite)

Canada	USA E-U	Japan Japon	Australia NZ	Africa	Total	LAIA ALADI	Mid.East	Other	Oceania	Année	Exportations en provenance de
39	.	82	24	345	525	355	49	540	0	1980	Etats-Unis
36	.	46	4	151	325	207	86	212	0	1987	
36	.	74	3	253	366	263	123	419	0	1988	
47	.	68	6	238	411	288	107	289	0	1989	
58	.	68	4	158	304	209	109	291	0	1990	
0	7	.	0	0	2	2	2	37	0	1980	Japon
1	18	.	1	0	2	2	0	36	0	1987	
2	16	.	0	0	3	3	0	40	0	1988	
1	13	.	0	0	0	0	0	38	0	1989	
1	15	.	0	0	0	0	0	35	0	1990	
0	1	5	1	11	0	0	18	96	7	1980	Australie, Nouvelle-Zélande
0	4	18	2	9	0	0	1	49	7	1987	
0	1	34	2	13	1	0	1	63	7	1988	
0	1	29	4	13	0	0	0	48	5	1989	
0	1	23	2	11	1	1	0	39	6	1990	
0	5	1	0	21	0	0	6	0	0	1980	Economies en voie de développement-Afrique
0	4	5	0	20	0	0	2	2	0	1987	
0	4	5	0	29	0	0	15	0	0	1988	
0	3	1	0	33	0	0	9	0	0	1989	
0	3	3	1	28	0	0	11	0	0	1990	
1	62	3	1	99	156	149	170	224	0	1980	Economies en voie de développement-Amérique
0	95	5	15	97	160	106	158	208	0	1987	
0	224	14	15	162	366	278	132	211	0	1988	
1	85	12	17	101	339	244	251	252	0	1989	
1	87	16	21	164	418	331	239	324	0	1990	
1	59	3	1	99	148	147	170	224	0	1980	ALADI+
0	93	5	15	97	149	105	158	208	0	1987	
0	223	14	15	162	359	278	132	211	0	1988	
1	82	12	17	101	324	244	251	252	0	1989	
1	86	16	21	164	408	328	239	324	0	1990	
0	0	4	0	0	0	0	3	0	0	1980	Economies en voie de développement-Europe /5
0	0	0	0	0	0	0	4	0	0	1987	
0	0	0	0	7	1	1	5	3	0	1988	
0	0	0	0	2	0	0	0	0	0	1989	
0	0	0	0	2	0	0	0	0	0	1990	
0	0	0	0	3	0	0	19	0	0	1980	Economies en voie de développement-Moyen Orient
0	0	0	0	55	0	0	51	0	0	1987	
0	0	0	0	46	0	0	46	0	0	1988	
0	2	0	0	62	0	0	45	0	0	1989	
0	1	0	0	50	0	0	51	1	0	1990	
23	358	145	35	98	9	4	244	1169	1	1980	Economies en voie de développement-Autres Pays d'Asie
19	305	136	34	121	2	2	241	1283	3	1987	
29	380	178	53	193	4	4	320	1681	3	1988	
22	352	198	61	202	13	13	333	1651	5	1989	
21	334	163	58	233	55	34	265	1632	5	1990	
23	356	136	34	96	5	4	244	1142	1	1980	Autres pays d'Asie excluyant la Chine
19	303	126	34	121	2	2	239	1263	3	1987	
29	380	167	53	193	4	4	319	1648	3	1988	
22	352	189	61	202	13	12	333	1603	5	1989	
21	333	154	57	233	54	34	265	1537	5	1990	
0	10	0	5	0	0	0	0	0	0	1980	Economies en voie de développement-Océanie
0	7	0	8	0	0	0	0	0	0	1987	
0	4	0	9	0	0	0	0	1	0	1988	
0	3	0	9	0	0	0	0	0	0	1989	
0	4	0	4	0	0	0	0	0	0	1990	

118
World exports by commodity classes and by regions
Exportations mondiales par classes de marchandises et par régions

In million US dollars f.o.b.

Exports to →		Developed economies	Developing economies Economies en voie de développement /1,3,4		Eastern Europe and former USSR Europe de l'Est et l'ancienne URSS		Developed Economies			
	World Monde /1,2,3	Econ. dévelop- pées /2,3,4					Europe /2			South Africa Afrique du Sud
Exports from / Year			Total	OPEC+ OPEP+	Total /2,3	Former USSR an URSS	Total /2	EEC+ CEE+ /2	EFTA+ AELE+	Afrique du Sud

Mineral fuels and related materials (SITC, Rev. 2 and Rev. 3, 3)

Exports from	Year	World	Dev.econ	Total	OPEC	Total E.Eur	Former USSR	Total Europe	EEC	EFTA	South Africa
World /1,2,3	1980	480789	362190	90913	5386	18846	1426	199825	178452	21049	411
	1987	280401	174920	58466	3647	33497	1707	98145	86013	11858	147
	1988	262984	167194	57824	3443	31228	1758	87737	78153	9362	67
	1989	292150	194407	63730	3616	27575	1552	99684	89019	10440	60
	1990	343414	234311	68066	3820	20383	1443	130004	116806	12867	71
Developed economies /2,3,4	1980	88066	75956	7208	2273	638	131	59156	51406	7563	108
	1987	77541	66201	7585	1015	631	166	45933	40300	5490	61
	1988	70855	60603	7244	835	498	156	39620	34989	4459	61
	1989	80150	67892	8678	1020	537	127	44975	39266	5511	54
	1990	104542	88418	11254	1137	875	181	60774	52582	7907	63
Developing economies /1,3,4	1980	350404	264533	78033	3101	3669	876	119387	111329	7924	303
	1987	147246	90596	43235	2476	3107	1167	34692	32815	1846	86
	1988	140517	89737	43442	2421	3198	1151	32200	30823	1366	7
	1989	162296	108624	47611	2438	3148	1167	37907	36838	1056	6
	1990	190384	121865	49165	2508	3222	1129	46353	45146	1183	8
OPEC+	1980	290878	222288	63696	1833	3051	637	109578	101930	7596	73
	1987	102104	62723	28833	1426	1990	422	29320	27618	1673	1
	1988	100673	66420	29253	1471	2207	448	28109	26866	1233	1
	1989	112941	78769	30534	1303	2215	516	32424	31525	899	1
	1990	133357	86858	29440	1196	2481	650	38209	37347	863	1
Eastern Europe and the former USSR /2,3	1980	42320	21701	5672	13	14539	418	21282	15717	5563	0
	1987	55614	18123	7646	155	29759	375	17520	12898	4522	0
	1988	51613	16854	7139	187	27532	452	15917	12341	3537	0
	1989	49704	17891	7440	159	23891	258	16803	12916	3873	0
	1990	48489	24028	7647	176	16287	133	22876	19078	3777	0
former USSR	1980	35949	17659	5127	5	13163	.	17363	13024	4339	0
	1987	50397	14607	7171	100	28619	.	14341	11044	3289	0
	1988	46654	13590	6766	130	26298	.	13066	10626	2434	0
	1989	43894	14486	6945	124	22463	.	13688	10809	2867	0
	1990	...	...	...	...	...	.	...	...	...	...
Developed economies-Europe /2	1980	66785	57733	4407	1915	437	88	55045	47652	7208	48
	1987	52599	47788	1945	665	364	87	42251	36797	5312	40
	1988	45070	41201	1615	518	243	87	35711	31243	4298	38
	1989	51434	46671	1993	577	277	73	40441	34935	5308	37
	1990	68687	62320	2461	615	605	127	54435	46455	7715	35
European Economic Community+ /2	1980	55498	46579	4345	1904	390	76	43900	37431	6293	48
	1987	41538	37018	1903	660	155	59	31916	27725	4064	40
	1988	34880	31232	1584	511	120	62	26149	22847	3152	38
	1989	37785	33279	1950	569	119	49	28255	24258	3826	36
	1990	49757	43709	2410	608	366	107	37634	32033	5374	35
European Free Trade Association+	1980	11263	11154	62	11	47	12	11146	10221	915	0
	1987	11021	10769	42	6	209	28	10334	9071	1247	0
	1988	10120	9966	31	8	123	25	9558	8393	1146	0
	1989	13593	13392	43	8	158	24	12187	10678	1482	1
	1990	18893	18611	44	7	239	20	16800	14423	2341	0
Other developed economies	1980	21281	18223	2801	358	201	43	4111	3754	355	60
	1987	24943	18413	5639	350	267	78	3682	3503	178	20
	1988	25785	19403	5629	316	255	69	3909	3747	161	22
	1989	28716	21221	6685	443	260	55	4533	4331	202	17
	1990	35855	26099	8792	521	270	54	6340	6127	192	27
Canada	1980	9392	9054	338	75	0	0	455	375	80	0
	1987	9482	9176	306	0	1	0	99	83	15	0
	1988	10187	9780	406	6	0	0	134	114	20	0
	1989	10624	10176	448	1	0	0	207	144	64	0
	1990	12670	12169	501	10	0	0	246	203	33	1

En millions de dollars E.-U. f.o.b.

Canada	USA E-U	Japan Japon	Australia New Zealand Australie Nouvelle-Zélande	Africa Afrique	America Amerique Total	America Amerique LAIA+ ALADI+	Asia Asie Mid.East Moyen Orient	Asia Asie Other Autres	Oceania Océanie	Année	Exportations en provenance de ↓

Combustibles minéraux et produits assimiles (CTCI, Rev. 2 et Rev. 3, 3)

Canada	USA E-U	Japon	Australia NZ	Afrique	Total	LAIA+ ALADI+	Mid.East	Other Autres	Océanie	Année	
6599	84939	65400	4162	7825	32414	12648	9781	36133	681	1980	Monde /1,2,3
3018	37978	33590	1512	4977	15983	7296	6495	27231	454	1987	
3465	40200	33827	1450	4699	14758	7278	6418	28071	491	1988	
4489	49936	37553	2139	5034	15261	7373	7302	31882	550	1989	
5267	51965	44316	2252	7760	10999	3622	8580	36634	742	1990	
1965	10547	3872	296	2646	1635	1069	1175	1499	125	1980	Economies dévelopées /2,3,4
2370	12892	4452	344	1312	2197	1433	643	3087	194	1987	
2690	12902	4909	294	1129	2003	1386	507	3228	235	1988	
3127	13627	5493	457	1336	2591	1832	558	3807	227	1989	
3914	17108	5792	576	1503	2858	1795	758	5676	287	1990	
4634	74301	61225	3866	4994	29686	11518	8170	33072	556	1980	Economies en voie de développement /1,3,4
593	24704	28980	1160	3253	11541	5635	5494	21745	260	1987	
775	26791	28489	1156	3251	10721	5746	5485	22565	257	1988	
1345	35746	31570	1672	3390	10990	5446	6126	25860	324	1989	
1339	34287	37958	1676	5945	6262	1677	6864	28826	455	1990	
4437	54896	50500	2804	3757	25329	9863	6649	26456	10	1980	OPEP+
399	12911	19317	776	2364	8603	4076	4476	12450	39	1987	
661	17125	19785	740	2434	8120	4404	4487	13028	45	1988	
1144	21813	22269	1118	2536	8195	4317	4924	13999	45	1989	
1214	18594	27801	1039	4854	3289	509	5422	15177	54	1990	
0	92	303	0	186	1093	61	436	1562	0	1980	Europe de l'Est et l'ancienne URSS /2,3
55	382	158	8	412	2245	228	358	2399	0	1987	
0	507	429	0	318	2033	146	426	2277	0	1988	
17	563	489	10	308	1680	94	618	2215	0	1989	
14	570	566	0	311	1879	151	958	2132	0	1990	
0	20	276	0	161	951	0	343	1469	0	1980	l'ancienne URSS
48	73	138	8	314	2079	80	285	2356	0	1987	
0	95	429	0	256	1885	19	353	2239	0	1988	
0	303	487	0	242	1560	0	559	2043	0	1989	
...	...	...	...	...	...	...	...	...	...	1990	
119	2428	32	52	2425	549	96	1034	324	2	1980	Economies développées-Europe /2
1006	4254	70	49	986	288	164	384	189	4	1987	
1290	3972	80	20	848	228	117	285	190	3	1988	
1440	4454	148	54	970	318	175	361	272	4	1989	
1758	5852	62	79	1118	291	137	463	485	5	1990	
118	2422	32	51	2418	546	96	1009	315	2	1980	Communauté Economique Européenne+ /2
954	3966	69	48	983	269	160	377	185	4	1987	
1188	3738	79	20	844	217	116	278	185	3	1988	
995	3763	147	54	967	309	174	350	259	4	1989	
1043	4809	60	79	1114	279	126	457	474	5	1990	
0	6	0	1	7	4	1	25	9	0	1980	Association Européenne de Libre Echange+
52	288	1	0	3	19	4	8	4	0	1987	
102	234	1	0	4	10	1	7	5	0	1988	
445	691	1	0	3	10	1	10	13	0	1989	
715	1043	2	0	3	12	11	6	4	0	1990	
1847	8119	3840	245	221	1086	973	141	1175	123	1980	Autres economies développées
1364	8638	4382	296	326	1909	1269	259	2899	190	1987	
1400	8930	4828	274	281	1776	1269	222	3038	231	1988	
1688	9173	5345	403	366	2273	1658	198	3536	223	1989	
2156	11256	5729	496	385	2567	1658	295	5192	282	1990	
.	8036	558	5	39	104	95	0	195	0	1980	Canada
.	8123	954	0	0	73	61	2	230	0	1987	
.	8429	1215	1	6	97	79	5	298	0	1988	
.	8697	1264	7	8	88	71	2	350	0	1989	
.	10647	1275	1	0	106	90	10	385	0	1990	

118
World exports by commodity classes and by regions
Exportations mondiales par classes de marchandises et par régions

In million US dollars f.o.b.

Mineral fuels and related materials (SITC, Rev. 2 and Rev. 3, 3) (continued)

Exports from	Year	World Monde /1,2,3	Developed economies Econ. développées /2,3,4	Developing economies Economies en voie de développement /1,3,4 Total	OPEC+ OPEP+	Eastern Europe and former USSR Europe de l'Est et l'ancienne URSS Total /2,3	Former USSR an URSS	Europe Total /2	EEC+ CEE+ /2	EFTA+ AELE+	South Africa Afrique du Sud
United States	1980	8017	6281	1621	207	115	26	2589	2444	142	43
	1987	7749	4901	2734	203	114	54	2074	1993	80	13
	1988	8229	5363	2734	224	131	52	2375	2286	90	14
	1989	9865	6288	3459	285	119	28	2822	2732	103	24
	1990	12175	7757	4278	270	140	29	3839	3724	103	24
Japan	1980	523	89	361	30	73	17	24	24	0	16
	1987	810	135	602	77	73	24	41	41	0	7
	1988	578	151	379	20	49	16	23	23	0	7
	1989	955	188	714	48	54	26	26	26	0	3
	1990	1274	137	1089	19	48	25	22	19	3	3
Australia, New Zealand	1980	2432	1882	481	47	14	0	301	299	2	0
	1987	5129	3140	1469	70	80	0	750	696	54	0
	1988	5090	3044	1675	67	75	0	548	509	39	0
	1989	5311	3435	1493	109	87	0	594	554	40	0
	1990	7011	4316	2235	222	80	0	801	779	22	0
Developing economies-Africa	1980	71778	61346	9124	47	920	0	32206	30296	1863	0
	1987	28583	24724	2968	27	466	26	17739	17023	687	1
	1988	26916	22678	3155	37	709	70	16246	15984	252	1
	1989	30435	26063	3021	70	843	192	17624	17243	381	1
	1990	40094	35640	2959	58	1106	391	24062	23652	411	0
Developing economies-America	1980	45728	32719	12443	285	13	12	7476	6973	420	9
	1987	26841	14116	4005	315	496	465	2580	2538	42	0
	1988	22476	16628	3098	123	502	470	2869	2715	154	0
	1989	27659	22564	3974	127	516	483	3427	3276	146	0
	1990	35389	17644	3330	140	281	272	3007	2935	58	0
LAIA+	1980	31749	22158	9311	49	13	12	5375	4948	402	1
	1987	21759	10529	3158	233	9	9	2378	2337	42	0
	1988	18184	13870	2208	54	3	2	2758	2604	154	0
	1989	22341	18928	2913	53	2	1	3245	3102	143	0
	1990	30035	13775	2226	127	21	20	2791	2740	50	0
Developing economies-Europe/ 5	1980	231	188	4	1	40	1	187	154	33	0
	1987	220	180	9	1	30	2	167	107	60	0
	1988	199	183	6	1	11	2	172	114	58	0
	1989	205	162	7	1	36	2	160	107	53	0
	1990	354	292	11	5	51	3	290	191	99	0
Developing economies-Middle East	1980	199479	146349	47751	2531	2461	628	79123	73522	5602	72
	1987	65669	35712	27434	2024	1692	398	14055	13011	1044	1
	1988	66749	36144	28118	2076	1597	378	12794	11897	897	1
	1989	75022	42967	29498	2006	1493	325	16180	15758	420	1
	1990	77650	48067	27403	2021	1537	260	18437	17919	518	1
Developing economies-Other Asia	1980	33158	23909	8703	236	235	234	391	382	6	221
	1987	25928	15864	8813	108	422	275	151	136	13	85
	1988	24169	14103	9061	183	378	229	117	111	6	5
	1989	28964	16862	11106	234	259	165	516	455	56	4
	1990	36824	20188	15449	285	247	203	557	450	98	7
Other Asia excluding China	1980	30175	21306	8324	236	235	234	238	233	2	221
	1987	21389	13045	7242	101	276	275	61	59	1	85
	1988	20219	12106	7270	180	229	229	40	39	0	5
	1989	24643	14113	9613	217	181	165	395	337	53	4
	1990	31587	16662	13762	257	224	203	430	333	88	7
Developing economies-Oceania	1980	31	23	8	0	0	0	4	4	0	0
	1987	6	0	6	0	0	0	0	0	0	0
	1988	8	1	4	0	0	0	1	1	0	0
	1989	12	6	5	0	0	0	0	0	0	0
	1990	72	34	15	0	0	0	0	0	0	0

En millions de dollars E.-U. f.o.b.

← Exportations vers

Economies développées /2,3,4				Developing economies Economies en voie de développement /1,3,4						Année	Exportations en provenance de ↓
Canada	USA E-U	Japan Japon	Australia New Zealand Australie Nouvelle-Zélande	Africa Afrique	America Amerique Total	LAIA+ ALADI+	Asia Asie Mid.East Moyen Orient	Other Autres	Oceania Océanie		

Combustibles minéraux et produits assimiles (CTCI, Rev. 2 et Rev. 3, 3) (suite)

Canada	USA E-U	Japan Japon	Australie N.-Z.	Afrique	Total	LAIA+ ALADI+	Moyen Orient	Autres	Océanie	Année	Exportations en provenance de
1844	.	1741	61	153	889	789	125	388	11	1980	Etats-Unis
1361	.	1298	144	112	1683	1058	93	787	13	1987	
1400	.	1425	125	89	1522	1034	142	893	21	1988	
1678	.	1510	214	127	2027	1430	138	1068	18	1989	
2155	.	1455	211	120	2342	1450	202	1530	19	1990	
0	44	.	5	11	90	86	4	255	1	1980	Japon
4	79	.	5	3	63	61	58	477	0	1987	
0	115	.	5	2	50	48	6	320	0	1988	
0	150	.	9	1	65	64	36	612	0	1989	
1	95	.	16	0	20	17	9	1059	0	1990	
3	18	1396	164	18	3	3	12	336	112	1980	Australie, Nouvelle-Zélande
0	410	1815	145	32	89	89	105	1059	174	1987	
0	361	1979	141	37	106	106	69	1241	208	1988	
0	302	2356	172	36	92	92	21	1132	202	1989	
0	488	2743	268	33	99	99	74	1765	258	1990	
89	27896	972	0	1219	5619	1321	1189	531	0	1980	Economies en voie de développement-Afrique
176	6540	123	0	850	1107	613	491	261	6	1987	
110	5939	233	2	888	1034	695	551	430	6	1988	
118	7961	192	4	1123	582	380	706	363	6	1989	
544	10766	263	4	1163	512	292	641	283	1	1990	
2013	21413	1149	24	698	11382	2915	52	177	96	1980	Economies en voie de développement-Amérique
170	9949	1175	5	213	3655	1471	2	130	4	1987	
414	12224	946	3	94	2866	1365	1	137	0	1988	
612	17225	1063	23	83	3778	1319	2	111	0	1989	
90	13040	1273	5	158	2964	1077	1	206	0	1990	
1911	13117	1117	0	363	8811	2274	0	96	4	1980	ALADI+
126	6696	1091	0	211	2828	1229	2	112	4	1987	
357	9683	902	0	89	1991	1154	1	127	0	1988	
564	13874	1010	20	47	2763	1164	0	102	0	1989	
44	9476	1233	2	23	1998	1037	1	204	0	1990	
0	0	0	0	1	1	0	1	2	0	1980	Economies en voie de développement-Europe /5
0	13	0	0	2	5	0	3	0	0	1987	
1	10	0	0	1	3	3	1	1	0	1988	
0	2	0	0	0	4	1	3	0	0	1989	
0	2	0	0	5	5	0	1	0	0	1990	
2531	19890	42140	2592	3037	11713	7144	6633	25304	112	1980	Economies en voie de développement-Moyen Orient
237	5555	15325	539	2143	6499	3306	4878	13200	33	1987	
242	6779	15756	572	2219	6662	3548	4804	13478	39	1988	
603	7437	17852	895	1994	6253	3617	5285	15258	44	1989	
693	8154	19968	803	4324	2361	193	6062	14163	52	1990	
0	5100	16952	1244	39	972	138	295	7053	345	1980	Economies en voie de développement-Autres Pays d'Asie
10	2647	12357	615	46	275	245	120	8154	212	1987	
8	1839	11554	578	49	156	134	128	8520	208	1988	
12	3115	12463	750	190	374	129	130	10128	274	1989	
12	2325	16454	831	296	420	115	157	14162	398	1990	
0	4965	14651	1230	33	963	135	293	6690	345	1980	Autres pays d'Asie excluyant la Chine
9	2153	10133	603	44	48	19	95	6843	211	1987	
8	1657	9826	569	48	55	34	114	6846	207	1988	
11	2508	10454	738	179	297	56	99	8752	274	1989	
12	1648	13766	797	280	322	18	156	12590	398	1990	
0	1	12	6	0	0	0	0	5	3	1980	Economies en voie de développement-Océanie
0	0	0	0	0	0	0	0	1	5	1987	
0	0	0	0	0	0	0	0	0	4	1988	
0	6	0	0	0	0	0	0	0	5	1989	
0	0	0	34	0	0	0	0	11	4	1990	

118
World exports by commodity classes and by regions
Exportations mondiales par classes de marchandises et par régions

In million US dollars f.o.b.

Chemicals (SITC, Rev. 2 and Rev. 3, 5)

Exports from	Year	World Monde /1,2,3	Developed economies Econ. développées /2,3,4	Developing economies Economies en voie de développement /1,3,4 Total	OPEC+ OPEP+	Eastern Europe and former USSR Europe de l'Est et l'ancienne URSS Total /2,3	Former USSR an URSS	Developed Economies Europe Total /2	EEC+ CEE+ /2	EFTA+ AELE+	South Africa Afrique du Sud
World /1,2,3	1980	140751	89739	40180	9035	10391	4877	69716	58758	10851	1418
	1987	214292	144998	55188	9264	12989	6133	108930	92876	15895	1568
	1988	253737	170737	66996	9949	13764	6517	126963	108199	18561	1896
	1989	263698	180429	68056	9855	13097	6878	133685	114388	19084	1731
	1990	298469	208283	75703	11310	11803	6269	157039	134928	21856	1796
Developed economies /2,3,4	1980	122564	83439	32948	7772	5999	2573	66003	55655	10249	1393
	1987	182603	134596	40923	7276	6251	2905	103211	88339	14716	1473
	1988	213215	157025	47793	7843	6570	2995	119535	102342	16994	1792
	1989	222042	163984	49230	7609	7076	3621	124470	106737	17528	1662
	1990	253724	190838	54347	8841	6395	3101	146937	126312	20379	1723
Developing economies /1,3,4	1980	10470	4205	5165	1029	896	601	1859	1725	126	25
	1987	20151	7833	11063	1691	1062	575	3474	3209	263	96
	1988	27647	10234	15792	1812	1310	740	4348	4009	335	104
	1989	30012	12317	15625	1907	1869	1387	5433	4881	547	69
	1990	34191	12907	18883	2337	2105	1761	5924	5423	496	72
OPEC+	1980	1419	875	532	93	5	0	407	404	3	2
	1987	1824	757	1047	132	4	0	510	465	44	0
	1988	2125	901	1177	136	34	29	601	542	59	0
	1989	2447	1129	1259	168	38	33	710	677	33	0
	1990	3602	1264	2233	334	95	93	815	699	114	0
Eastern Europe and the former USSR /2,3	1980	7718	2096	2067	233	3496	1703	1854	1377	476	0
	1987	11539	2569	3201	298	5676	2654	2245	1328	916	0
	1988	12874	3479	3412	294	5883	2782	3080	1847	1231	0
	1989	11644	4128	3201	339	4152	1871	3782	2771	1009	0
	1990	10555	4538	2474	132	3303	1407	4178	3193	981	0
former USSR	1980	1704	562	527	22	615	.	438	344	93	0
	1987	2945	593	929	5	1422	.	534	316	218	0
	1988	3536	1010	1099	10	1427	.	840	498	342	0
	1989	3569	1129	1122	11	1318	.	1004	668	336	...
	1990	...	...	...	...	...	.	...	...	...	...
Developed economies-Europe /2	1980	89278	65718	18108	5678	5405	2144	57407	48017	9293	960
	1987	137849	109171	22623	5378	5467	2310	92131	78123	13853	1106
	1988	158274	126410	24345	5566	5759	2363	106391	90256	15950	1315
	1989	160996	129174	24021	5489	6173	2919	109250	92788	16278	1180
	1990	188015	152428	27932	6394	5667	2535	130249	110920	19102	1293
European Economic Community+ /2	1980	78479	58118	15953	5048	4361	1769	50892	42896	7907	871
	1987	119675	95485	19550	4677	4268	1821	81025	69089	11793	996
	1988	137442	110389	21005	4816	4543	1899	93436	79700	13564	1187
	1989	139823	112795	20782	4788	4881	2373	95870	81827	13867	1068
	1990	162709	132601	24108	5576	4368	1937	114057	97502	16338	1156
European Free Trade Association+	1980	10794	7597	2153	629	1044	375	6512	5119	1386	89
	1987	18163	13679	3069	699	1198	489	11098	9028	2060	111
	1988	20819	16013	3336	747	1216	464	12948	10550	2385	128
	1989	21160	16370	3235	698	1293	545	13371	10953	2410	112
	1990	25290	19813	3820	815	1299	598	16181	13407	2764	137
Other developed economies	1980	33286	17721	14841	2095	594	430	8596	7638	956	432
	1987	44754	25426	18301	1897	784	594	11080	10216	863	366
	1988	54942	30616	23449	2277	811	633	13145	12087	1044	477
	1989	61046	34811	25211	2120	903	702	15220	13949	1249	482
	1990	65708	38411	26416	2448	729	567	16688	15392	1277	430
Canada	1980	3519	2844	597	46	78	76	633	600	33	17
	1987	4736	3931	796	57	8	2	614	582	33	6
	1988	6379	5097	1270	132	12	6	643	599	44	9
	1989	6141	5115	1014	102	11	9	605	579	26	9
	1990	6661	5607	1042	114	11	9	596	529	67	11

En millions de dollars E.-U. f.o.b.

← Exportations vers

Produits chimiques (CTCI, Rev. 2 et Rev. 3, 5)

Canada	USA E-U	Japan Japon	Australia New Zealand Australie Nouvelle-Zélande	Africa Afrique	America Amerique Total	LAIA+ ALADI+	Asia Mid.East Moyen Orient	Asia Other Autres	Oceania Océanie	Année	Exportations en provenance de ↓
2801	7895	5346	2154	6629	11229	8696	5979	14474	140	1980	Monde /1,2,3
4407	16240	9743	3266	6257	12278	9167	7424	27260	195	1987	
5149	19815	11979	4016	6724	13330	9984	7415	37173	225	1988	
5789	20811	12863	4565	6822	13928	10345	7737	37233	242	1989	
7756	22634	13361	4518	7298	14629	11216	8974	42269	253	1990	
2765	6828	4046	2000	5828	9437	7620	5057	11273	127	1980	Economies dévelopées /2,3,4
4287	13841	8033	2924	5349	9385	7401	5587	19159	166	1987	
4993	16703	9598	3511	5709	10140	8036	5622	24586	186	1988	
5628	16971	10295	4023	5640	10817	8452	5840	25189	191	1989	
7588	18600	10834	4009	6020	11703	9258	7135	27443	206	1990	
30	901	1243	144	481	1466	937	639	2541	12	1980	Economies en voie de développement /1,3,4
102	2241	1575	335	709	2008	1456	1309	6966	29	1987	
139	2899	2232	493	828	2236	1690	1267	11382	39	1988	
148	3667	2428	531	955	2384	1706	1284	10928	51	1989	
159	3837	2393	500	1038	2542	1867	1504	13720	48	1990	
0	29	394	43	28	67	51	110	327	0	1980	OPEP+
1	92	129	25	36	117	99	173	713	0	1987	
3	113	158	26	50	133	110	172	808	0	1988	
2	193	190	35	47	178	136	182	843	0	1989	
12	152	257	28	89	264	208	333	1528	1	1990	
6	166	57	10	320	326	139	283	660	0	1980	Europe de l'Est et l'ancienne URSS /2,3
18	158	135	7	200	885	310	528	1135	0	1987	
17	214	148	12	187	954	259	526	1205	0	1988	
13	173	139	12	227	728	187	613	1116	0	1989	
8	196	134	8	240	384	91	335	1106	0	1990	
0	116	5	3	38	135	26	46	113	0	1980	l'ancienne URSS
5	0	50	4	22	217	29	73	470	0	1987	
2	110	52	5	36	240	25	74	604	0	1988	
0	72	51	3	45	284	31	77	574	0	1989	
...	...	...	...	...	...	...	...	...	...	1990	
568	3836	1773	866	5107	3665	2822	4152	3994	52	1980	Economies développées-Europe /2
1089	8332	4311	1484	4799	4262	3167	4827	7423	93	1987	
1251	9921	5092	1685	5159	4180	3130	4790	8614	97	1988	
1229	9913	5031	1791	4959	4302	3193	4956	8227	98	1989	
1378	10992	5710	1864	5355	4850	3641	6062	9769	115	1990	
486	3383	1448	762	4710	3094	2376	3615	3544	51	1980	Communauté Economique Européenne+ /2
902	7243	3430	1240	4365	3441	2521	4185	6482	92	1987	
1029	8580	4069	1406	4689	3388	2483	4073	7540	94	1988	
1042	8566	4027	1517	4562	3505	2556	4246	7152	96	1989	
1140	9413	4459	1525	4936	3981	2937	5231	8395	114	1990	
82	453	326	104	397	571	445	536	451	1	1980	Association Européenne de Libre Echange+
187	1089	880	244	431	821	646	641	941	1	1987	
221	1340	1023	279	468	792	647	716	1074	4	1988	
187	1347	1003	274	395	796	637	709	1074	2	1989	
238	1579	1251	339	416	869	704	830	1373	2	1990	
2197	2993	2272	1134	720	5772	4798	905	7278	75	1980	Autres economies développées
3198	5509	3723	1440	550	5123	4234	760	11736	73	1987	
3742	6782	4506	1826	550	5960	4905	832	15972	89	1988	
4400	7058	5264	2232	680	6515	5259	884	16963	93	1989	
6210	7608	5124	2145	666	6853	5617	1073	17675	90	1990	
.	1983	148	59	20	197	159	21	358	0	1980	Canada
.	3027	198	82	15	166	105	19	595	0	1987	
.	4005	325	111	27	243	182	23	976	1	1988	
.	4039	318	141	38	173	117	26	776	1	1989	
.	4545	278	174	24	155	104	25	836	1	1990	

118
World exports by commodity classes and by regions
Exportations mondiales par classes de marchandises et par régions

In million US dollars f.o.b.

Exports from	Year	World Monde /1,2,3	Developed economies Econ. dévelop- pées /2,3,4	Developing economies Economies en voie de développement /1,3,4		Eastern Europe and former USSR Europe de l'Est et l'ancienne URSS		Europe			Developed Economies South Africa Afrique du Sud
				Total	OPEC+ OPEP+	Total /2,3	Former USSR an URSS	Total /2	EEC+ CEE+ /2	EFTA+ AELE+	

Chemicals (SITC, Rev. 2 and Rev. 3, 5) (continued)

Exports from	Year	World	Dev.econ	Total	OPEC	E.Eur Total	Former USSR	Eur Total	EEC	EFTA	S.Africa
United States	1980	20728	11266	9373	1375	89	31	6155	5487	667	300
	1987	25631	15091	10197	1299	343	264	7525	6941	584	242
	1988	31343	17762	13245	1521	336	268	8775	8072	691	327
	1989	36485	21356	14720	1376	409	332	10647	9758	869	348
	1990	38983	23721	14918	1509	339	280	11366	10510	838	311
Japan	1980	6619	2048	4160	631	412	322	865	759	106	72
	1987	11545	4585	6536	505	424	329	1978	1776	201	77
	1988	13904	5424	8029	576	450	357	2520	2258	261	83
	1989	14692	5831	8398	587	463	361	2676	2384	292	73
	1990	15778	6214	9206	759	354	271	3162	2836	326	69
Australia, New Zealand	1980	669	427	202	24	0	0	70	62	8	6
	1987	797	497	233	22	2	0	60	57	3	6
	1988	1063	676	384	37	2	1	87	84	3	6
	1989	1186	731	452	44	0	0	128	124	4	5
	1990	1320	830	482	55	6	5	168	161	7	6
Developing economies-Africa	1980	801	317	417	90	59	37	304	301	3	5
	1987	1582	646	802	147	129	74	604	590	13	8
	1988	2473	1111	1164	206	174	87	1035	1008	27	8
	1989	2266	1184	933	286	141	98	1140	1113	27	8
	1990	2461	1130	1139	362	186	171	1090	1075	15	3
Developing economies-America	1980	2854	1336	1486	169	28	10	526	479	46	14
	1987	4400	2161	2196	230	38	14	744	690	54	26
	1988	5945	3126	2783	288	34	7	1003	930	72	52
	1989	6517	3645	2824	237	41	18	1256	1183	70	39
	1990	6680	3723	2859	296	92	64	1281	1178	103	44
LAIA+	1980	2145	1056	1060	150	28	10	375	345	30	12
	1987	3540	1719	1797	211	22	14	614	564	50	26
	1988	4969	2470	2480	280	18	7	870	814	54	51
	1989	5224	2710	2485	226	26	18	1067	1014	51	39
	1990	5531	2956	2546	281	23	14	1181	1082	99	44
Developing economies-Europe/ 5	1980	1010	187	245	135	578	384	172	143	29	0
	1987	1288	429	458	235	401	151	390	323	67	0
	1988	1143	462	243	86	438	152	414	352	62	0
	1989	1824	744	208	93	873	613	671	383	288	0
	1990	1402	597	169	75	634	432	551	464	86	0
Developing economies-Middle East	1980	1437	825	566	185	33	16	395	391	4	0
	1987	2176	633	1409	602	114	103	417	372	44	1
	1988	2672	839	1637	662	183	165	552	498	54	8
	1989	2864	892	1429	665	524	514	587	543	44	5
	1990	3648	1023	2123	706	488	474	685	545	138	7
Developing economies-Other Asia	1980	4364	1538	2449	451	198	154	460	409	45	6
	1987	10693	3961	6191	476	379	233	1318	1233	84	61
	1988	15403	4690	9960	569	481	329	1342	1220	121	37
	1989	16528	5849	10222	626	290	144	1778	1659	119	17
	1990	19985	6432	12580	895	705	620	2316	2159	153	18
Other Asia excluding China	1980	3274	945	1985	386	165	154	190	166	18	6
	1987	8463	2804	5274	438	222	205	730	669	61	61
	1988	12506	3696	8220	499	320	299	929	852	76	37
	1989	13310	4279	8752	550	113	98	1023	933	89	17
	1990	16235	4761	10634	758	580	555	1482	1354	124	18
Developing economies-Oceania	1980	3	1	2	0	0	0	1	1	0	0
	1987	11	5	7	0	0	0	1	1	0	0
	1988	11	6	4	0	0	0	2	2	0	0
	1989	13	4	9	0	0	0	1	1	0	0
	1990	16	3	12	2	0	0	1	1	0	0

En millions de dollars E.-U. f.o.b.

← Exportations vers

Produits chimiques (CTCI, Rev. 2 et Rev. 3, 5) (suite)

Canada	USA E-U	Japan Japon	Australia New Zealand Australie Nouvelle-Zélande	Africa Afrique	America Amerique Total	LAIA+ ALADI+	Mid.East Moyen Orient	Other Autres	Oceania Océanie	Année	Exportations en provenance de ↓
2126	.	1972	638	413	5118	4223	602	3158	7	1980	Etats-Unis
3064	.	3366	809	248	4545	3770	516	4833	6	1987	
3567	.	3937	1039	258	5309	4370	595	7037	10	1988	
4210	.	4664	1354	354	5918	4762	616	7759	11	1989	
6050	.	4582	1239	322	6232	5090	690	7622	9	1990	
50	769	.	279	65	263	227	242	3564	14	1980	Japon
107	2086	.	316	79	252	205	178	6009	12	1987	
122	2303	.	381	86	252	208	160	7516	9	1988	
122	2535	.	409	85	252	216	171	7873	10	1989	
109	2479	.	373	68	260	226	257	8604	10	1990	
8	134	80	128	1	6	5	4	137	54	1980	Australie, Nouvelle-Zélande
10	151	80	189	1	7	7	4	166	54	1987	
12	185	141	244	1	11	10	5	298	69	1988	
12	162	152	272	2	10	9	8	360	72	1989	
10	208	141	297	12	12	11	11	376	71	1990	
1	6	1	0	131	35	27	46	194	0	1980	Economies en voie de développement-Afrique
1	24	3	7	237	106	68	156	284	0	1987	
9	40	6	15	283	69	64	183	609	0	1988	
7	20	4	5	333	37	30	178	375	0	1989	
10	22	2	3	312	57	51	324	432	0	1990	
10	602	167	16	45	1351	874	24	60	0	1980	Economies en voie de développement-Amérique
26	1082	221	58	61	1787	1310	39	303	1	1987	
42	1631	287	102	76	2056	1566	47	601	1	1988	
36	1986	232	86	105	2192	1595	44	481	2	1989	
40	2041	227	82	114	2284	1696	67	390	2	1990	
3	505	151	9	39	933	830	23	58	0	1980	ALADI+
20	848	154	53	34	1422	1235	39	300	0	1987	
24	1186	236	97	66	1771	1523	47	592	0	1988	
27	1294	200	79	98	1859	1555	44	480	2	1989	
26	1428	202	71	103	1997	1655	56	387	2	1990	
1	10	4	0	85	12	7	127	21	0	1980	Economies en voie de développement-Europe /5
3	29	3	2	109	6	6	202	141	0	1987	
4	34	7	3	78	17	5	57	90	2	1988	
2	40	5	5	76	7	5	62	62	0	1989	
5	32	7	3	71	7	5	44	46	1	1990	
0	3	386	41	26	14	10	238	285	0	1980	Economies en voie de développement-Moyen Orient
2	75	120	19	100	11	8	650	642	0	1987	
1	111	144	21	120	6	3	744	757	0	1988	
2	113	155	26	153	8	3	679	582	0	1989	
6	76	230	17	212	24	19	708	1167	0	1990	
18	281	685	87	194	52	20	204	1981	11	1980	Economies en voie de développement-Autres Pays d'Asie
71	1029	1228	249	201	97	65	263	5595	21	1987	
83	1081	1788	350	270	89	52	236	9326	34	1988	
100	1507	2031	406	287	140	73	321	9428	40	1989	
99	1666	1927	396	329	168	96	361	11681	36	1990	
13	178	494	64	117	21	12	157	1681	9	1980	Autres pays d'Asie excluyant la Chine
58	841	890	218	167	68	44	157	4860	19	1987	
73	989	1332	328	229	66	41	162	7730	31	1988	
85	1235	1529	381	239	78	44	175	8220	38	1989	
81	1357	1451	361	283	110	72	252	9952	34	1990	
0	0	0	0	0	0	0	0	0	2	1980	Economies en voie de développement-Océanie
0	3	0	0	0	0	0	0	0	7	1987	
0	2	0	2	0	0	0	0	1	4	1988	
0	0	0	2	0	0	0	0	0	9	1989	
0	1	0	1	0	0	0	0	3	9	1990	

Economies développées /2,3,4

Developing economies Economies en voie de développement /1,3,4

118
World exports by commodity classes and by regions
Exportations mondiales par classes de marchandises et par régions

In million US dollars f.o.b.

Machinery and transport equipment (SITC, Rev. 2 and Rev. 3, 7)

Exports from	Year	World Monde /1,2,3	Developed economies Econ. dévelop-pées /2,3,4	Developing economies Economies en voie de développement /1,3,4 Total	OPEC+ OPEP+	Eastern Europe and former USSR Europe de l'Est et l'ancienne URSS Total /2,3	Former USSR an URSS	Europe Total /2	EEC+ CEE+ /2	EFTA+ AELE+	South Africa Afrique du Sud
World /1,2,3	1980	513089	305291	160556	51780	44406	21034	195303	161616	33325	7526
	1987	850668	602830	187798	33794	54609	27993	344570	285584	58051	5706
	1988	990055	699667	223279	37058	60549	31351	407968	341450	65175	7283
	1989	1057500	754171	247214	35912	51835	25620	445484	376377	67546	6553
	1990	1211950	871468	289367	43690	44955	23351	538194	458372	77970	6375
Developed economies /2,3,4	1980	436318	287483	136421	45744	10982	5387	187709	154928	32428	7402
	1987	710375	544459	149644	28077	11961	5834	326230	270001	55427	5424
	1988	817139	624909	172498	30330	14490	7753	383785	320946	61907	6842
	1989	875891	669635	186788	29396	16768	8999	417289	352091	64224	6372
	1990	1021286	777039	220513	36703	19478	10440	502326	427034	73868	6164
Developing economies /1,3,4	1980	30650	14860	14241	3896	1368	851	5056	4704	345	123
	1987	84975	55401	26817	3735	2483	1654	15684	13854	1714	282
	1988	113470	71482	38669	4617	2899	1943	21265	18626	2245	441
	1989	131752	79141	49331	4592	2641	1691	23181	20714	2393	181
	1990	151796	88978	58238	5474	3236	2223	30793	27620	3111	211
OPEC+	1980	1539	245	1277	697	1	0	176	169	6	0
	1987	1584	378	1197	988	1	0	337	331	6	0
	1988	1738	424	1229	940	77	72	353	350	3	0
	1989	1952	527	1411	1094	1	0	337	322	15	0
	1990	2565	942	1554	862	56	0	672	660	12	0
Eastern Europe and the former USSR /2,3	1980	46121	2948	9894	2139	32055	14796	2538	1984	551	1
	1987	55318	2970	11337	1982	40165	20505	2657	1729	910	0
	1988	59446	3276	12112	2111	43160	21655	2919	1878	1024	0
	1989	49858	5396	11095	1924	32427	14930	5015	3573	930	1
	1990	38868	5452	10617	1513	22241	10688	5075	3719	991	1
former USSR	1980	12227	453	5015	854	6758	.	406	285	122	0
	1987	16524	751	7540	821	8233	.	710	444	266	0
	1988	18251	806	8049	837	9396	.	750	457	288	0
	1989	17871	852	7567	699	9452	.	780	505	275	0
	1990	...	...	...	...	...	.	...	...	...	
Developed economies-Europe /2	1980	256722	182104	65398	26000	9137	4215	150400	122593	27495	4715
	1987	399104	324252	64203	15794	10419	4906	260310	213441	46194	3300
	1988	448371	366840	67126	16848	12562	6515	299542	248307	50460	4353
	1989	483357	395436	71744	16942	14409	7413	326104	272858	52371	4151
	1990	592320	481632	90661	21231	16954	8808	401376	338573	61535	4137
European Economic Community+ /2	1980	225301	160088	58230	23618	6913	3107	132374	109379	22701	4383
	1987	346416	283248	55777	14252	7164	3059	227353	188558	38243	3068
	1988	390189	320981	58876	15015	8488	4157	262568	219735	42132	4157
	1989	423389	347752	63598	15494	10298	4971	287778	243252	43731	3932
	1990	517958	422806	79688	19054	12433	6232	353177	300351	51740	3867
European Free Trade Association+	1980	31349	21987	7139	2361	2223	1107	17998	13187	4793	333
	1987	52454	40826	8373	1513	3251	1847	32803	24733	7948	232
	1988	57849	45598	8180	1806	4071	2358	36787	28404	8309	196
	1989	59538	47354	8082	1426	4101	2434	38023	29314	8629	219
	1990	73706	58290	10891	2148	4515	2576	47690	37727	9781	270
Other developed economies	1980	179599	105380	71025	19745	1845	1173	37309	32335	4934	2687
	1987	311273	220208	85442	12284	1542	928	65921	56561	9233	2124
	1988	368769	258072	105373	13482	1927	1238	84245	72640	11447	2489
	1989	392535	274201	115045	12454	2358	1586	91187	79234	11852	2220
	1990	428968	295409	129853	15472	2524	1632	100952	88462	12333	2027
Canada	1980	16675	14573	1889	840	213	72	887	731	155	56
	1987	35925	34448	1403	249	73	29	1351	1120	231	18
	1988	42992	41384	1494	170	114	59	2011	1681	330	27
	1989	44591	42792	1723	238	76	20	2069	1764	305	24
	1990	47229	44982	2192	327	54	26	2347	2009	337	18

En millions de dollars E.-U. f.o.b.

← Exportations vers

Economies développées /2,3,4				Developing economies / Economies en voie de développement /1,3,4							
Canada	USA E-U	Japan Japon	Australia New Zealand / Australie Nouvelle-Zélande	Africa Afrique	America Amerique Total	LAIA+ ALADI+	Asia Asie Mid.East Moyen Orient	Other Autres	Oceania Océanie	Année	Exportations en provenance de ↓

Machines et matériél de transport (CTCI, Rev. 2 et Rev. 3, 7)

Canada	USA E-U	Japon	Australie N-Z	Afrique	Total	LAIA+	Moyen Orient	Autres	Océanie	Année	Provenance
22301	62553	7303	8702	31592	40075	31782	35235	48488	756	1980	Monde /1,2,3
40615	178239	16273	13782	21826	39829	27599	28312	92683	1072	1987	
45906	195413	22647	16814	23643	44849	32351	29928	119006	1349	1988	
46863	203502	27656	20879	25207	47331	33352	29780	139153	1618	1989	
56131	211003	35556	20157	30901	53939	38536	35305	162068	1672	1990	
21865	54558	6159	8198	27539	34750	28633	30437	39409	721	1980	Economies dévelopées /2,3,4
38962	145298	12600	12422	18492	33746	25008	23081	70155	948	1987	
43750	155224	16929	14864	19623	37805	29197	24268	86269	1219	1988	
44315	159955	20119	18483	20773	39183	29803	24720	97186	1439	1989	
53849	166444	26487	17892	26144	45826	34789	29927	112325	1516	1990	
362	7737	1099	475	2228	3333	2668	2513	6127	36	1980	Economies en voie de développement /1,3,4
1608	32752	3630	1327	1742	3630	2257	3188	18115	124	1987	
2102	39982	5682	1900	2517	4361	2899	3592	28039	130	1988	
2499	43320	7495	2336	3043	5232	3152	3451	37363	180	1989	
2241	44318	9025	2221	3442	5302	3498	4037	45070	155	1990	
1	63	6	0	145	47	43	944	141	0	1980	OPEP+
3	34	2	3	94	33	20	973	89	8	1987	
3	53	11	4	130	53	26	913	129	2	1988	
8	149	26	6	134	63	46	1067	143	2	1989	
6	220	35	8	268	106	72	970	206	4	1990	
74	258	45	29	1824	1992	480	2285	2952	0	1980	Europe de l'Est et l'ancienne URSS /2,3
45	189	43	33	1592	2453	335	2044	4413	0	1987	
53	208	36	50	1503	2683	255	2068	4699	0	1988	
49	227	42	60	1391	2917	397	1609	4604	0	1989	
42	241	43	44	1315	2811	249	1341	4673	0	1990	
27	7	9	4	710	1305	87	814	1859	0	1980	l'ancienne URSS
7	9	14	11	925	1991	112	925	3343	0	1987	
8	12	13	23	806	2177	21	834	3572	0	1988	
7	18	19	27	681	2419	159	545	3720	0	1989	
...	...	...	...	...	...	...	...	...	...	1990	
2331	18783	2250	2818	20448	12094	9988	17554	11884	196	1980	Economies développées-Europe /2
4792	44472	5304	4009	13963	11839	8221	14068	21016	344	1987	
5769	44114	6825	4425	14650	11492	8602	14557	22969	493	1988	
5654	44132	8319	5516	15492	11551	8464	15133	25892	482	1989	
5633	51081	11750	5648	19811	14550	10287	18661	32668	602	1990	
1930	16380	1855	2439	18708	10451	8746	15630	10512	174	1980	Communauté Economique Européenne+ /2
3983	38976	4647	3356	12690	9607	7187	12470	18147	326	1987	
4728	38240	5955	3702	13388	9746	7437	12804	19969	483	1988	
4524	38163	7292	4629	14204	10144	7341	13607	22383	472	1989	
4492	44583	10252	4605	18346	12471	8700	16473	27982	586	1990	
400	2403	395	379	1716	1639	1242	1923	1372	23	1980	Association Européenne de Libre Echange+
809	5478	657	647	1243	2226	1030	1596	2853	18	1987	
1032	5810	870	722	1239	1743	1162	1747	2960	10	1988	
1129	5945	1027	886	1268	1399	1122	1522	3476	10	1989	
1127	6486	1497	1041	1437	2071	1581	2183	4644	15	1990	
19534	35775	3909	5380	7092	22657	18645	12883	27525	525	1980	Autres economies développées
34170	100826	7296	8413	4530	21907	16787	9013	49139	604	1987	
37981	111109	10104	10439	4973	26313	20596	9711	63301	726	1988	
38661	115823	11800	12967	5281	27632	21340	9587	71295	956	1989	
48216	115363	14738	12244	6333	31276	24503	11266	79657	915	1990	
.	13387	58	165	204	1048	926	333	282	8	1980	Canada
.	32798	115	135	97	636	511	238	425	5	1987	
.	38966	145	218	154	561	426	186	562	14	1988	
.	40224	164	298	166	584	448	187	740	30	1989	
.	42118	208	271	252	704	541	177	1017	13	1990	

118
World exports by commodity classes and by regions
Exportations mondiales par classes de marchandises et par régions

In million US dollars f.o.b.

Machinery and transport equipment (SITC, Rev. 2 and Rev. 3, 7) (continued)

Exports from	Year	World Monde /1,2,3	Developed economies Econ. développées /2,3,4	Developing economies Total /1,3,4	OPEC+ OPEP+	Eastern Europe and former USSR Total /2,3	Former USSR an URSS	Europe Total /2	EEC+ CEE+ /2	EFTA+ AELE+	South Africa Afrique du Sud
United States	1980	84512	48579	34292	8763	466	269	22285	19739	2540	1334
	1987	110219	71534	34567	4634	235	87	29536	26319	3130	622
	1988	135247	87884	44538	6072	259	118	38660	34389	4176	819
	1989	148800	97194	51042	5703	564	242	45341	40194	5118	780
	1990	172522	114899	56624	6171	661	331	50324	44902	5387	855
Japan	1980	75870	41058	33650	10020	1162	832	13811	11582	2196	1236
	1987	161615	111777	48599	7277	1228	809	34317	28469	5815	1450
	1988	185786	125963	58277	7144	1545	1058	42648	35705	6883	1589
	1989	193669	130874	61094	6389	1701	1318	42702	36320	6316	1343
	1990	202934	131559	69548	8732	1777	1267	46974	40355	6500	1113
Australia, New Zealand	1980	1336	676	656	74	1	0	100	87	13	42
	1987	1530	1128	398	53	3	3	271	252	14	19
	1988	1844	1252	588	55	4	3	336	324	12	29
	1989	2115	1481	625	58	7	5	351	329	22	23
	1990	2676	1840	828	133	8	5	410	374	36	21
Developing economies-Africa	1980	337	231	99	23	1	0	211	210	1	11
	1987	554	329	220	70	2	0	286	282	2	11
	1988	792	390	315	120	81	72	332	328	3	11
	1989	784	467	307	97	5	2	428	421	6	11
	1990	1073	699	362	104	11	3	658	632	25	16
Developing economies-America	1980	5184	1992	3188	594	3	1	686	659	26	36
	1987	10518	7789	2701	565	18	10	1467	1221	174	25
	1988	12348	8998	3329	565	18	10	2100	1620	120	43
	1989	13861	9908	3927	452	20	13	1887	1731	155	59
	1990	14228	10883	3256	405	55	41	2068	1954	113	48
LAIA+	1980	4963	1880	3080	593	3	1	666	640	25	36
	1987	10250	7626	2612	564	5	1	1419	1179	168	25
	1988	12048	8798	3244	562	5	1	2049	1574	116	42
	1989	13494	9658	3824	449	7	4	1819	1667	150	59
	1990	13637	10472	3118	398	14	5	1810	1707	102	48
Developing economies-Europe/ 5	1980	2546	682	624	291	1240	777	637	594	42	0
	1987	3467	1227	515	176	1725	1190	1036	937	97	0
	1988	3875	1384	601	233	1889	1302	1247	1071	173	0
	1989	3715	1337	666	213	1712	1181	1223	1074	146	0
	1990	4322	2107	805	341	1409	1012	1965	1769	194	0
Developing economies-Middle East	1980	1750	258	1426	933	1	0	195	187	8	0
	1987	2847	1146	1680	1345	10	7	1110	865	245	2
	1988	2470	719	1724	1391	19	7	677	535	141	1
	1989	2410	698	1659	1381	43	33	653	607	41	8
	1990	3031	1170	1712	1154	127	55	1111	1065	39	6
Developing economies-Other Asia	1980	20823	11691	8901	2055	123	73	3322	3049	270	75
	1987	67557	44903	21674	1578	729	448	11782	10546	1196	244
	1988	93906	59934	32677	2309	892	552	16889	15053	1808	386
	1989	110892	66684	42731	2445	860	462	18980	16871	2045	102
	1990	129061	74099	52057	3470	1633	1113	24975	22184	2741	141
Other Asia excluding China	1980	20296	11652	8429	1954	107	73	3299	3026	269	75
	1987	65821	44689	20354	1534	527	403	11691	10460	1191	244
	1988	91137	58885	31199	2232	651	471	16462	14651	1783	386
	1989	102816	65787	35936	2309	477	271	18752	16656	2034	102
	1990	118228	72797	42960	3207	1206	810	24561	21802	2708	140
Developing economies-Oceania	1980	9	6	3	0	0	0	5	5	0	0
	1987	33	7	26	0	0	0	3	3	0	0
	1988	79	57	21	0	0	0	20	20	0	0
	1989	90	47	41	3	1	0	10	10	0	0
	1990	82	20	45	1	1	0	16	16	0	0

En millions de dollars E.-U. f.o.b.

Machines et matériél de transport (CTCI, Rev. 2 et Rev. 3, 7) (suite)

Economies développées /2,3,4				Developing economies — Economies en voie de développement /1,3,4							← Exportations vers
Canada	USA E-U	Japan Japon	Australia New Zealand Australie Nouvelle-Zélande	Africa Afrique	America Amerique Total	LAIA+ ALADI+	Asia Asie Mid.East Moyen Orient	Other Autres	Oceania Océanie	Année	Exportations en provenance de ↓
18009	.	3786	2452	2320	15928	14046	5427	10295	66	1980	Etats-Unis
29739	.	7067	3395	1395	14591	12774	3367	14947	69	1987	
32970	.	9822	4259	1965	18604	16469	4229	19365	118	1988	
33194	.	11460	5120	2309	19839	17153	4623	23906	162	1989	
42746	.	14302	5220	2718	22488	19892	4732	26410	111	1990	
1500	22008	.	2450	4127	5513	3528	7077	16615	246	1980	Japon
4400	67055	.	4320	2791	6499	3357	5369	33498	394	1987	
4962	71004	.	5437	2638	7009	3582	5262	42901	395	1988	
5421	74295	.	6890	2537	7042	3606	4748	46120	577	1989	
5422	71601	.	6066	3051	7872	3870	6310	51499	608	1990	
15	163	61	294	79	40	29	40	290	205	1980	Australie, Nouvelle-Zélande
15	207	71	542	14	13	8	26	212	133	1987	
17	285	86	496	19	14	11	25	330	199	1988	
16	348	119	622	19	16	13	17	381	187	1989	
25	555	174	655	22	18	14	31	573	183	1990	
1	8	0	0	87	4	3	6	3	0	1980	Economies en voie de développement-Afrique
3	29	0	1	202	7	6	6	5	0	1987	
13	33	0	1	270	9	8	27	8	0	1988	
4	18	1	0	274	13	11	13	6	0	1989	
3	21	0	1	243	7	7	30	80	0	1990	
67	1004	148	51	427	2550	2249	110	94	6	1980	Economies en voie de développement-Amérique
298	5807	91	99	271	2031	1588	212	185	3	1987	
272	6325	138	113	353	2478	2004	165	316	14	1988	
398	7300	142	110	321	3011	1933	225	362	3	1989	
234	8268	145	108	279	2617	2015	82	266	11	1990	
63	915	148	51	427	2444	2245	110	93	6	1980	ALADI+
277	5712	91	99	268	1947	1585	212	182	3	1987	
264	6185	137	113	353	2397	1999	165	312	14	1988	
382	7135	141	110	321	2909	1926	225	362	3	1989	
228	8124	141	108	274	2492	1977	82	258	11	1990	
3	36	0	6	325	20	16	148	131	0	1980	Economies en voie de développement-Europe /5
6	177	0	7	220	100	11	137	59	0	1987	
8	119	0	10	286	38	12	198	79	0	1988	
7	97	1	8	273	169	12	157	67	0	1989	
9	97	1	34	354	67	17	300	85	0	1990	
0	55	5	1	147	2	2	1244	33	0	1980	Economies en voie de développement-Moyen Orient
4	13	1	14	131	2	0	1468	77	0	1987	
5	17	1	17	194	1	1	1417	112	0	1988	
6	26	2	2	181	6	1	1380	88	0	1989	
6	38	2	4	333	15	2	1293	57	2	1990	
291	6633	945	416	1243	757	398	1005	5866	26	1980	Economies en voie de développement-Autres Pays d'Asie
1297	26727	3537	1202	919	1490	651	1364	17788	95	1987	
1799	33472	5542	1743	1414	1835	874	1785	27520	98	1988	
2083	35868	7349	2191	1995	2033	1195	1676	36830	144	1989	
1989	35893	8877	2070	2233	2596	1458	2333	44566	114	1990	
289	6627	941	413	1128	713	381	946	5614	24	1980	Autres pays d'Asie excluyant la Chine
1293	26648	3514	1186	861	1464	636	1212	16708	94	1987	
1760	33026	5439	1710	1344	1809	865	1681	26245	97	1988	
2065	35504	7093	2159	1849	1935	1145	1574	30385	142	1989	
1965	35400	8536	2040	2068	2430	1391	2207	35931	112	1990	
0	1	0	0	0	0	0	0	0	3	1980	Economies en voie de développement-Océanie
0	0	0	3	0	0	0	0	1	26	1987	
4	16	1	16	0	0	0	0	4	18	1988	
0	11	0	26	0	0	0	0	9	32	1989	
0	0	0	4	0	0	0	0	17	28	1990	

118
World exports by commodity classes and by regions
Exportations mondiales par classes de marchandises et par régions

In million US dollars f.o.b.

Other manufactured goods (SITC, Rev. 2 and Rev. 3, 6 and 8)

Exports from	Year	World Monde /1,2,3	Developed economies Econ. développées /2,3,4	Developing economies Economies en voie de développement /1,3,4 Total	OPEC+ OPEP+	Eastern Europe and former USSR Europe de l'Est et l'ancienne URSS Total /2,3	Former USSR an URSS	Developed Economies Europe Total /2	EEC+ CEE+ /2	EFTA+ AELE+	South Africa Afrique du Sud
World /1,2,3	1980	481145	332323	115972	38708	30971	15953	243909	196742	46525	3002
	1987	680957	507580	135632	25487	35090	19669	328554	269369	58275	2605
	1988	790330	586714	162398	28782	37577	21500	377074	312032	64031	3129
	1989	861751	634496	188501	29845	35406	19939	408158	338661	67993	2849
	1990	970845	728887	207619	31719	30199	16921	490096	408822	79663	3072
Developed economies /2,3,4	1980	370695	273238	83691	29299	12922	7358	213483	170297	42582	2617
	1987	478521	383633	81183	15306	12140	7500	282566	229777	52011	2010
	1988	548402	438525	95006	16909	12774	7843	321206	263516	56829	2432
	1989	590659	469174	106307	17390	13111	7569	344154	283546	59729	2447
	1990	683805	552736	116179	19589	12583	6099	414212	343027	70123	2622
Developing economies /1,3,4	1980	85075	52421	27459	8374	4469	2534	24538	21838	2668	385
	1987	170397	115602	48717	9401	5355	3238	38740	34292	4392	595
	1988	206343	138640	60255	11046	6273	4021	47668	42628	4982	697
	1989	238884	153569	76484	11597	7806	5391	53611	47287	6260	402
	1990	259108	162854	86541	11732	8145	6065	63801	56297	7443	450
OPEC+	1980	3701	1966	1686	971	43	27	1430	1250	180	0
	1987	7239	3852	3352	1341	15	7	1111	1001	109	0
	1988	8954	4739	4124	1436	63	49	1528	1406	121	1
	1989	12295	7425	4809	1740	25	7	2909	1855	1052	0
	1990	15117	9418	5470	1624	182	6	4530	3149	1379	1
Eastern Europe and the former USSR /2,3	1980	25375	6663	4822	1036	13581	6062	5889	4608	1275	0
	1987	32040	8345	5732	780	17596	8930	7248	5299	1871	0
	1988	35586	9549	7137	827	18530	9636	8200	5887	2221	0
	1989	32208	11753	5710	858	14489	6978	10394	7828	2004	0
	1990	27933	13297	4899	398	9471	4757	12084	9497	2097	0
former USSR	1980	4989	308	1451	96	3230	.	286	231	55	0
	1987	6828	460	2317	31	4051	.	372	297	74	0
	1988	8961	1057	3883	19	4020	.	913	773	138	0
	1989	7713	1229	2468	29	4015	.	1051	842	204	0
	1990	...	...	...	...	...	.	...	...	...	...
Developed economies-Europe /2	1980	264145	211401	41667	18164	10904	5772	187445	148196	38668	1642
	1987	358036	306546	40686	10285	10504	6088	258928	208980	49187	1427
	1988	403813	345114	46184	11304	10790	6138	291449	237446	53177	1710
	1989	431009	367188	50579	11700	11456	6185	309808	253497	55465	1737
	1990	508870	437469	58072	13254	11387	5177	375200	308747	65424	1919
European Economic Community+ /2	1980	220490	176168	36394	16218	7755	3917	156133	124392	31189	1441
	1987	296249	253555	34973	8856	7463	4143	213505	173423	39352	1293
	1988	332722	284436	39068	9654	7712	4255	239848	196548	42511	1593
	1989	356695	303714	43183	10190	8246	4250	255906	210475	44618	1619
	1990	422357	362990	49587	11471	8129	3266	311018	257172	52851	1791
European Free Trade Association+	1980	43301	34914	5254	1931	3132	1855	31006	23515	7461	200
	1987	61422	52670	5685	1409	3024	1931	45130	35271	9827	134
	1988	70704	60332	7081	1625	3073	1883	51274	40580	10657	117
	1989	73937	63142	7363	1487	3199	1926	53592	42719	10841	118
	1990	86038	74064	8444	1753	3240	1897	63790	51192	12563	127
Other developed economies	1980	106553	61839	42027	11135	2018	1585	26038	22101	3914	976
	1987	120488	77089	40499	5021	1635	1413	23639	20797	2825	583
	1988	144594	93412	48823	5605	1985	1705	29757	26071	3652	722
	1989	159654	101989	55730	5690	1655	1385	34348	30051	4263	709
	1990	174938	115270	58109	6336	1196	922	39015	34283	4699	704
Canada	1980	13694	11983	1565	313	41	21	2208	2023	184	25
	1987	19144	17339	1796	127	9	3	1399	1277	121	14
	1988	22818	21238	1534	182	46	23	1924	1718	206	19
	1989	23733	21955	1752	171	26	11	2320	2110	209	27
	1990	24942	23265	1644	259	32	18	2218	2022	196	25

En millions de dollars E.-U. f.o.b.

← Exportations vers

Articles manufacturés divers (CTCI, Rev. 2 et Rev. 3, 6 et 8)

| Economies développées /2,3,4 | | | | Developing economies — Economies en voie de développement /1,3,4 | | | | | | Année | Exportations en provenance de ↓ |
Canada	USA E-U	Japan Japon	Australia New Zealand Australie Nouvelle-Zélande	Africa Afrique	America Amerique — Total	LAIA+ ALADI+	Asia Asie — Mid.East Moyen Orient	Other Autres	Oceania Océanie		
10517	52906	13095	6804	20866	23352	16749	28888	38520	660	1980	Monde /1,2,3
17548	115828	29354	9895	15277	19668	11457	23960	72036	875	1987	
20321	127626	42286	11838	16681	22037	13801	26349	90744	1061	1988	
21247	134033	49895	13314	17785	26080	16318	27680	111095	1185	1989	
28068	138500	50928	12636	19223	29397	18941	29303	122349	1288	1990	
9012	34130	6905	5126	16545	17293	13056	20849	26334	511	1980	Economies dévelopées /2,3,4
13223	62270	13346	6670	10977	13822	8628	13832	39402	594	1987	
15520	68988	18379	7865	11711	15477	10448	15292	49001	686	1988	
16065	70634	22507	8710	12354	17567	11996	16136	55967	735	1989	
23121	74113	25137	8436	14150	20079	14101	17995	58198	808	1990	
1380	18294	6086	1627	3468	5492	3587	6768	11354	148	1980	Economies en voie de développement /1,3,4
4157	52970	15744	3170	3512	5010	2786	9184	30655	280	1987	
4630	57957	23507	3901	4156	5683	3284	10036	39897	375	1988	
4976	62834	26917	4527	4466	7636	4280	10491	53284	450	1989	
4789	63844	25405	4147	4638	8190	4779	10612	62283	478	1990	
6	130	383	18	83	130	108	1074	397	0	1980	OPEP +
79	1365	1236	61	160	235	127	1496	1458	2	1987	
100	1416	1604	89	229	310	146	1573	2007	3	1988	
85	1841	2482	107	253	473	202	1838	2236	6	1989	
122	2340	2286	138	226	606	301	1890	2733	10	1990	
125	481	104	51	852	567	106	1271	832	0	1980	Europe de l'Est et l'ancienne URSS /2,3
168	588	264	54	787	837	43	944	1978	0	1987	
171	681	400	72	814	877	69	1021	1846	0	1988	
206	565	471	77	965	877	42	1052	1844	0	1989	
158	543	387	53	436	1128	62	696	1868	1	1990	
6	9	4	2	42	301	2	144	248	0	1980	l'ancienne URSS
3	24	59	2	68	616	1	77	914	0	1987	
6	23	112	3	128	636	9	88	1002	0	1988	
29	20	128	2	75	698	2	79	1270	0	1989	
...	...	...	...	...	...	...	...	...	...	1990	
1904	14192	2784	1869	13463	6230	4609	13097	6651	173	1980	Economies développées-Europe /2
3970	29882	6394	2969	9371	5206	3067	9935	13429	253	1987	
4408	31939	8708	3410	10058	5177	3117	11044	16882	272	1988	
4410	33308	10350	3811	10425	5372	3213	11487	19543	300	1989	
4431	35175	12759	3764	12096	6422	3981	13182	21178	369	1990	
1624	12168	2168	1576	12423	5354	3922	11426	5328	165	1980	Communauté Economique Européenne + /2
3396	25411	4994	2495	8662	4498	2579	8460	10948	241	1987	
3738	26910	6688	2843	9241	4401	2565	9438	13322	254	1988	
3704	28051	8234	3211	9678	4651	2690	9947	15688	283	1989	
3805	29790	10074	3139	11323	5703	3443	11220	16767	345	1990	
278	2015	616	292	1025	875	687	1669	1323	8	1980	Association Européenne de Libre Echange +
573	4444	1399	473	689	704	488	1472	2481	12	1987	
669	5012	2019	567	793	770	552	1600	3560	18	1988	
705	5236	2116	599	726	716	523	1534	3854	17	1989	
626	5365	2685	625	747	715	538	1957	4405	24	1990	
7108	19939	4121	3257	3082	11063	8446	7753	19684	338	1980	Autres economies développées
9253	32388	6952	3702	1606	8616	5561	3897	25975	341	1987	
11112	37049	9671	4455	1654	10300	7331	4249	32119	414	1988	
11655	37326	12157	4900	1929	12195	8783	4649	36425	435	1989	
18689	38938	12378	4671	2053	13658	10119	4813	37021	439	1990	
.	9222	288	206	124	722	496	115	598	2	1980	Canada
.	15512	239	153	62	855	267	111	764	2	1987	
.	18459	590	194	69	496	300	95	864	6	1988	
.	18607	655	297	116	479	313	140	1007	6	1989	
.	20161	620	188	90	462	308	227	854	7	1990	

118
World exports by commodity classes and by regions
Exportations mondiales par classes de marchandises et par régions

In million US dollars f.o.b.

Other manufactured goods (SITC, Rev. 2 and Rev. 3, 6 and 8) (continued)

Exports from	Year	World Monde /1,2,3	Developed economies Econ. développées /2,3,4	Developing economies Total	Developing economies OPEC/OPEP	Eastern Europe Total /2,3	Eastern Europe Former USSR	Europe Total /2	Europe EEC/CEE /2	Europe EFTA/AELE	South Africa Afrique du Sud
United States	1980	38854	24657	13864	3350	229	123	13801	11682	2113	481
	1987	37255	24960	11875	1390	186	135	10803	9522	1275	238
	1988	48190	31801	16012	1737	209	136	13877	12167	1683	310
	1989	59880	38818	20792	1798	270	188	17395	15198	2173	319
	1990	70956	48373	22350	1932	227	107	19436	17068	2349	331
Japan	1980	42012	16605	23683	7223	1724	1441	5924	5056	852	426
	1987	49653	24518	23734	3338	1401	1275	8042	7002	1031	292
	1988	57380	27817	27879	3482	1684	1546	9400	8246	1147	358
	1989	58229	27782	29130	3490	1318	1178	9559	8418	1134	324
	1990	58727	28470	29313	3793	908	794	10723	9438	1276	295
Australia, New Zealand	1980	3707	2056	1473	190	16	0	837	817	19	24
	1987	4749	2945	1609	144	35	0	538	519	18	16
	1988	5995	4017	1938	178	40	0	633	612	21	16
	1989	6587	4275	2283	204	27	1	713	678	34	17
	1990	6820	4189	2619	325	11	2	810	767	43	18
Developing economies-Africa	1980	4968	4030	629	87	166	123	3078	2979	97	126
	1987	5861	4544	1031	223	224	153	3504	3374	127	84
	1988	6911	5216	1282	321	312	233	4044	3925	115	86
	1989	8491	6467	1646	444	287	219	5176	5026	147	95
	1990	8484	6832	1308	469	226	205	5572	5466	102	105
Developing economies-America	1980	12521	7472	4739	746	291	77	3701	3425	274	31
	1987	17764	12474	5043	772	228	77	3412	3143	265	29
	1988	24592	17337	6897	1052	342	167	5402	5055	345	58
	1989	28164	19165	8573	1088	403	206	6292	5873	417	72
	1990	28071	19218	8487	1187	334	163	6832	6401	428	60
LAIA+	1980	11285	7032	3960	731	291	77	3546	3277	268	31
	1987	16426	11607	4595	757	210	77	3279	3027	248	29
	1988	23010	16223	6456	1028	324	167	5276	4941	333	58
	1989	25864	17600	7883	1042	365	206	5981	5573	406	72
	1990	25879	17678	7870	1120	304	163	6509	6094	411	60
Developing economies-Europe/ 5	1980	3444	1316	510	322	1618	989	1004	792	209	0
	1987	4827	2642	743	303	1442	733	2110	1777	333	0
	1988	5570	3096	931	520	1543	788	2489	2152	335	0
	1989	5619	3268	879	432	1472	792	2792	2455	335	0
	1990	6230	4143	638	277	1447	900	3583	3126	455	0
Developing economies-Middle East	1980	3856	1620	2055	1534	116	69	1521	1272	249	0
	1987	8392	4461	3650	2661	240	166	3526	3181	342	1
	1988	9383	4972	4020	2828	328	249	3970	3611	341	11
	1989	11747	6539	4056	2830	1093	981	5279	3995	1268	14
	1990	14334	8527	4281	2568	1427	1122	7124	5455	1661	25
Developing economies-Other Asia	1980	59946	37669	19501	5685	2278	1276	15010	13147	1838	228
	1987	133346	91338	38186	5441	3219	2109	26114	22751	3318	480
	1988	159487	107677	47070	6324	3748	2584	31572	27705	3834	542
	1989	184263	117635	61225	6800	4551	3194	33849	29725	4082	221
	1990	201537	123782	71729	7230	4710	3675	40556	35723	4788	259
Other Asia excluding China	1980	52724	34829	15768	4936	1629	1170	13804	12082	1701	228
	1987	118522	84937	30651	4988	2331	1777	23888	20795	3053	477
	1988	140730	98602	38411	5733	2724	2128	28453	24947	3478	538
	1989	158719	108656	46147	6161	3064	2490	31048	27220	3792	220
	1990	171212	113712	52998	6548	3200	2685	37268	32733	4496	255
Developing economies-Oceania	1980	340	314	25	0	0	0	223	223	0	0
	1987	207	143	64	1	0	0	73	65	8	0
	1988	400	342	55	1	0	0	192	181	12	0
	1989	600	495	105	3	0	0	223	213	11	0
	1990	451	352	99	0	0	0	134	126	8	0

En millions de dollars E.-U. f.o.b.

| Economies développées /2,3,4 | | | | Developing economies / Economies en voie de développement /1,3,4 | | | | | | ← Exportations vers | |
Canada	USA E-U	Japan Japon	Australia New Zealand Australie Nouvelle-Zélande	Africa Afrique	America Amerique Total	LAIA+ ALADI+	Asia Asie Mid.East Moyen Orient	Other Autres	Oceania Océanie	Année	Exportations en provenance de ↓

Articles manufacturés divers (CTCI, Rev. 2 et Rev. 3, 6 et 8) (suite)

Canada	USA E-U	Japan Japon	Australia NZ	Africa	Total	ALADI+	Mid.East	Other	Oceania	Année	Exportations en provenance de
6203	.	2831	1077	894	7599	5758	2006	3285	31	1980	Etats-Unis
8104	.	4247	1159	357	6458	4336	941	4055	33	1987	
9652	.	6103	1401	413	8412	5929	1240	5859	48	1988	
10191	.	8495	1672	642	10238	7272	1535	8286	45	1989	
17331	.	8910	1689	709	11571	8495	1461	8495	36	1990	
796	8184		1242	1530	2548	2011	5452	14043	75	1980	Japon
1009	13606		1501	750	1187	870	2627	19049	100	1987	
1226	15027		1745	811	1276	1008	2712	22916	123	1988	
1166	15000		1680	696	1364	1105	2709	24200	125	1989	
1104	14783		1497	668	1460	1174	2786	24234	134	1990	
25	226	285	657	27	15	10	92	1110	229	1980	Australie, Nouvelle-Zélande
43	420	1134	794	8	17	13	60	1317	206	1987	
49	499	1795	1023	7	12	7	68	1615	236	1988	
49	519	1841	1132	10	12	8	83	1919	258	1989	
63	546	1551	1199	17	23	21	116	2202	261	1990	
11	459	346	2	397	27	14	64	103	0	1980	Economies en voie de développement-Afrique
37	573	332	12	530	20	12	125	328	0	1987	
35	596	440	12	620	31	22	200	397	0	1988	
61	578	547	7	743	42	27	385	438	0	1989	
22	590	450	5	763	28	16	324	174	0	1990	
205	2746	733	48	203	3877	2777	140	394	1	1980	Economies en voie de développement-Amérique
456	7064	1270	225	240	3247	2227	384	1168	1	1987	
710	8736	2144	256	388	3736	2536	437	2323	2	1988	
606	9408	2451	311	417	4861	3129	423	2852	5	1989	
468	9165	2393	261	445	4839	3204	543	2621	7	1990	
197	2487	716	47	203	3101	2751	140	392	1	1980	ALADI+
421	6379	1258	224	230	2822	2209	376	1164	1	1987	
673	7804	2126	255	378	3333	2514	428	2304	2	1988	
566	8275	2379	303	400	4250	3086	412	2802	5	1989	
413	8039	2358	260	415	4304	3113	527	2587	7	1990	
21	258	18	4	203	15	8	245	45	1	1980	Economies en voie de développement-Europe /5
42	426	27	27	231	36	15	322	155	0	1987	
40	466	65	23	215	69	50	499	149	0	1988	
38	398	18	16	213	47	28	494	125	0	1989	
39	468	26	21	150	24	17	360	104	1	1990	
2	55	29	11	166	1	1	1795	88	0	1980	Economies en voie de développement-Moyen Orient
75	628	216	7	408	11	10	2943	279	0	1987	
73	642	255	15	559	22	20	2972	449	0	1988	
49	753	411	25	514	12	2	3000	504	0	1989	
64	855	424	17	480	6	4	3037	695	0	1990	
1140	14734	4920	1555	2500	1571	787	4523	10713	132	1980	Economies en voie de développement-Autres Pays d'Asie
3546	44257	13870	2883	2103	1696	522	5411	28686	255	1987	
3768	47477	20523	3570	2373	1826	657	5928	36545	353	1988	
4220	51635	23385	4066	2579	2674	1094	6189	49290	414	1989	
4194	52716	22022	3766	2799	3292	1536	6348	58623	438	1990	
1046	14101	4176	1392	1826	1366	718	3999	8434	117	1980	Autres pays d'Asie excluyant la Chine
3235	42383	12133	2633	1952	1578	479	4627	22225	242	1987	
3479	45133	17499	3275	2199	1685	632	5196	28964	334	1988	
3910	48984	20523	3713	2330	2480	1025	5501	35380	396	1989	
3874	49578	19060	3408	2524	3007	1433	5654	41198	422	1990	
0	43	40	8	0	1	1	0	10	14	1980	Economies en voie de développement-Océanie
1	22	30	16	1	0	0	0	39	24	1987	
4	41	80	25	0	0	0	0	35	20	1988	
2	63	105	102	0	0	0	0	73	31	1989	
1	50	90	78	0	1	1	0	66	32	1990	

118
World exports by commodity classes and by regions
Exportations mondiales par classes de marchandises et par régions

Source:
International Trade statistics database of the Statistical Division of the United Nations Secretariat.

+/ For member states of this grouping, see Annex I:- Other groupings.

1/ Excluding the inter-trade between China, the Democratic People's Republic of Korea, Mongolia and Viet Nam in 1980. The figures shown for other years are based on the import statistics of China.

2/ Excluding the trade conducted in accordance with the supplementary protocol to the treaty on the basis of relations between the Federal Republic of Germany (FRG) and the former German Democratic Republic (GDR). Data reported by FRG are as follows:

Value in million United States dollars

	1980	1987	1988	1989	1990
FRG to GDR	2911	4140	4114	4351	13379
FRG from GDR	3072	3712	3868	3854	5129

3/ Exports of the former USSR for which country of destination is not available are included in the totals for the World, the 'Developed economies', the 'Developing economies' and 'Eastern Europe and former USSR' but are excluded from the regional components of these groupings.

4/ This classification is intended for statistical convenience and does not, necessarily, express a judgement about the stage reached by a particular country in the development process.

5/ Includes Yugoslavia only.

6/ Section 9 of the SITC, which comprises commodities and transactions not classified elsewhere, is included in the total trade but is not shown separately in this table.

Source:
Base de données pour les statistiques du commerce international de la Division de statistique du Secrétariat de l'ONU.

+/ Les etats membres de ce groupement, voir annexe I:- Autres groupements.

1/ Non compris le commerce entre la Chine, la République populaire démocratique de Corée, la Mongolie et le Viet Nam en 1980. Les chiffres indiqués pour les autres années sont bases sur les statistiques d'importations de la Chine.

2/ Non compris le commerce effectué en accord avec le protocole additionnel au traité définissant la base des relations entre la République fédérale d'Allemagne (RfA) et l'ancienne République démocratique allemande (Rda). Les données fournies par RfA sont les suivantes:

Valeur en millions de dollars des E.-U.

	1980	1987	1988	1989	1990
de RfA vers Rda	2911	4140	4114	4351	13379
de RfA en prov. de Rda	3072	3712	3868	3854	5129

3/ Les exportations en provenance de l'ancienne URSS dont les pays de destination ne sont pas disponibles sont comprises dans les totaux du Monde, des 'Economies développées', des 'Economies en voie de développement' et de 'Europe de l'Est et l'ancienne URSS', mais ils ne sont pas comprises dans chaque partie composant ces régions.

4/ Cette classification est utilisée pour plus de commodité dans la présentation des statistiques et n'implique pas nécessairement un jugement quant au stade de développement auquel est parvenu un pays donné.

5/ Y compris la Yougoslavie seulement.

6/ Section 9 de la CTCI, qui représente les articles et transactions non classes ailleurs est comprise dans le commerce total mais n'est pas présentée séparément dans ce tableau.

119
Total imports and exports: index numbers
Importations et exportations totales: indices

1980 = 100

Country or area Pays ou zone	1981	1982	1983	1984	1985	1986	1987	1988	1989	1990
A. Imports: Quantum index • Importations: Indice du quantum										
Australia Australie	114	121	106	139	134	128	131	154	187	173
Austria Autriche	96	95	101	109	116	120	127	137	152	168
Belgium-Luxembourg Belgique-Luxembourg	96	97	96	100	104	114	125	123	131	138
Bolivia Bolivie	...	...	...	...	...	...	...	...	...	...
Brazil Brésil	87	80	67	64	60	...	...	...	...	...
Bulgaria Bulgarie	109	113	119	121	134	139	137	...	...	...
Burkina Faso Burkina Faso	109	110	93	89	111	166	...	...	...	...
Canada Canada	103	147	170	205	118	131	146	137	...	...
Colombia Colombie	105	...	...	...	...	...	...	...	...	...
Cyprus Chypre	102	120	131	155	139	145	165	...	...	...
Czechoslovakia Tchécoslovaquie	93	96	98	98	103	105	110	114	117	...
Denmark Danemark	95	98	103	110	118	126	125	123	123	127
Dominica Dominique	103	87	84	114	110	...	...	...	...	...
Dominican Republic Rép. dominicaine	...	...	...	...	...	...	...	...	...	...
Ecuador Equateur	92	72	39	39	80	55	56	51	59	...
El Salvador El Salvador	96	82	89	...	...	...	...	...	...	...
Ethiopia Ethiopie	84	92	...	...	...	...	...	...	...	...
Faeroe Islands Iles Féroé	101	108	110	116	122	148	159	159	...	...
Fiji Fidji	121	100	113	109	101	99	92	...	...	...
Finland Finlande	94	95	98	98	104	110	119	130	143	138
France France	95	97	94	96	101	108	116	126	134	140
Germany† • Allemagne† F. R. Germany R. f. Allemagne	95	96	100	105	110	116	123	131	140	156
former German D. R. anc. R. d. allemande	99	94	99	103	105	110	108	...	...	...

Country or area Pays ou zone	1981	1982	1983	1984	1985	1986	1987	1988	1989	1990

A. Exports: Quantum index • Exportations: Indice du quantum

Country or area Pays ou zone	1981	1982	1983	1984	1985	1986	1987	1988	1989	1990
Australia Australie	94	100	97	112	126	131	146	138	147	156
Austria Autriche	105	106	111	122	134	134	138	151	172	190
Belgium-Luxembourg Belgique-Luxembourg	100	102	106	111	116	125	134	119	128	133
Bolivia Bolivie	108	99	86	89	74	83	72	...	...	...
Brazil Brésil	120	110	126	154	159	133	...	...	...	...
Bulgaria Bulgarie	108	121	126	132	142	136	139	...	...	...
Burkina Faso Burkina Faso	94	70	66	95	99	152	...	...	...	...
Canada Canada	103	154	169	206	137	144	159	124	...	...
Colombia Colombie	91	...	...	...	...	...	...	...	...	...
Cyprus Chypre	106	111	103	121	107	100	121	...	...	...
Czechoslovakia Tchécoslovaquie	101	106	113	123	127	128	132	130	127	...
Denmark Danemark	103	105	114	119	123	126	128	135	142	147
Dominica Dominique	197	209	215	216	229	...	...	...	...	...
Dominican Republic Rép. dominicaine	113	92	107	106	100	87	99	115	92	...
Ecuador Equateur	102	102	112	124	31	28	33	22	21	...
El Salvador El Salvador	73	64	91	...	...	...	...	...	...	...
Ethiopia Ethiopie	102	109	118	117	93	107	101	...	...	...
Faeroe Islands Iles Féroé	100	105	127	131	139	139	133	144	...	...
Fiji Fidji	94	98	84	92	101	92	94	...	...	...
Finland Finlande	103	100	104	114	115	116	118	121	121	130
France France	103	100	103	110	113	112	116	127	136	143
Germany† • Allemagne† F. R. Germany R. f. Allemagne	107	110	110	120	127	129	132	142	153	155
former German D. R. anc. R. d. allemande	108	114	126	129	132	134	130	...	...	...

119
Total imports and exports: index numbers
[cont.]

Importations et exportations totales: indices
1980 = 100 [suite]

Country or area Pays ou zone	1981	1982	1983	1984	1985	1986	1987	1988	1989	1990
Ghana Ghana	70	...	...	...	...	...	...	...	...	...
Greece Grèce	99	111	115	119	139	141	157	129	173	195
Guatemala Guatemala	99	79	70	55	80	65	100	71	104	...
Guyana Guyana	96	61	...	...	...	...	...	...	...	...
Hong Kong Hong-kong	111	109	119	137	146	164	217	275	299	334
Hungary Hongrie	101	98	99	100	107	110	113	112	114	108
Iceland Islande	108	109	101	109	118	125	156	150	134	134
India Inde	114	121	144	124	140	184	194	...	...	...
Indonesia Indonésie	...	...	...	...	...	...	...	...	...	...
Ireland Irlande	102	99	102	112	115	120	127	132	149	159
Israel Israël	102	113	124	121	126	147	165	166	159	172
Italy Italie	93	93	93	102	111	116	128	137	148	155
Japan Japon	99	98	99	109	110	120	131	153	165	175
Jordan Jordanie	112	123	124	116	119	125	130	137	114	...
Kenya Kenya	78	66	52	62	57	67	70	79	83	79
Korea, Republic of Corée, République de	111	111	126	146	155	168	131	149	174	303
Kuwait Koweït	101	...	...	...	...	...	...	...	...	...
Malawi Malawi	84	80	81	62	102	75	98	109	...	...
Malaysia Malaisie	...	...	...	...	...	...	...	...	...	...
Malta Malte	94	92	88	93	101	103	116	131	146	
Morocco Maroc	87	97	83	92	89	88	...	...	...	...
Myanmar Myanmar	...	...	...	...	...	...	...	...	...	...
Netherlands Pays-Bas	93	93	97	104	111	116	123	130	137	144
Netherlands Antilles Antilles néerlandaises	...	...	...	...	...	...	...	...	...	...
New Zealand Nouvelle-Zélande	104	110	102	123	123	121	137	124	150	161

Country or area Pays ou zone	1981	1982	1983	1984	1985	1986	1987	1988	1989	1990
Ghana Ghana	74	...	...	...	...	...	...	...	...	...
Greece Grèce	88	84	100	113	117	137	139	104	143	136
Guatemala Guatemala	95	104	106	133	88	123	98	99	109	...
Guyana Guyana	92	75	...	...	...	...	...	...	...	...
Hong Kong [4] Hong-kong [4]	114	111	128	155	164	189	249	315	348	378
Hungary Hongrie	104	109	116	125	131	128	132	141	142	136
Iceland Islande	99	82	93	96	106	117	122	119	122	120
India Inde	58	63	64	72	68	77	93	...	...	...
Indonesia Indonésie	82	78	94	89	99	107	103	...	...	...
Ireland Irlande	100	108	121	144	149	158	180	190	211	229
Israel Israël	106	104	106	121	132	146	162	154	161	161
Italy Italie	104	105	108	116	124	127	129	137	149	154
Japan Japon	111	108	117	135	142	141	141	147	153	162
Jordan Jordanie	122	125	124	174	181	186	221	256	260	...
Kenya Kenya	95	92	88	87	91	105	101	106	106	112
Korea, Republic of Corée, République de	118	125	146	168	181	205	139	156	147	277
Kuwait Koweït	...	...	...	...	...	...	...	...	...	...
Malawi [4] Malawi [4]	79	82	109	78	99	113	108	108	...	...
Malaysia Malaisie	97	103	114	111	117	126	132	220	...	...
Malta Malte	97	91	91	104	107	108	113	116	139	...
Morocco Maroc	107	106	118	124	125	130	...	...	...	...
Myanmar Myanmar	105	112	127	121	94	112	82	...	...	...
Netherlands Pays-Bas	101	99	105	112	117	122	127	138	144	152
Netherlands Antilles Antilles néerlandaises	99	...	...	...	...	...	...	...	...	...
New Zealand Nouvelle-Zélande	101	104	110	115	128	125	129	134	130	138

119
Total imports and exports: index numbers
[cont.]

Importations et exportations totales: indices
1980 = 100 [suite]

Country or area Pays ou zone	1981	1982	1983	1984	1985	1986	1987	1988	1989	1990
Niger Niger	87	78	65	...	...	...	...	...	...	...
Norway[1] Norvège[1]	99	100	96	109	122	140	138	124	118	130
Pakistan Pakistan	107	99	99	111	116	112	115	80	83	150
Panama Panama	...	...	...	...	...	...	...	...	...	...
Peru Pérou	...	...	...	...	...	...	...	...	...	...
Philippines Philippines	92	105	101	72	62	80	98	118	152	...
Poland Pologne	83	72	76	82	88	93	97	106	108	79
Saint Lucia Sainte-Lucie	102	92	...	...	...	...	...	...	...	...
Samoa Samoa	...	...	...	...	...	...	...	...	...	...
Singapore Singapour	112	121	122	127	123	134	153	196	215	...
Solomon Islands Iles Salomon	...	...	...	...	...	...	...	...	...	...
South Africa Afrique du Sud	113	92	78	94	69	72	72	84	84	76
Spain Espagne	94	97	98	97	103	121	148	...	...	...
Sri Lanka Sri Lanka	90	90	99	103	98	112	114	117	114	111
Sweden Suède	93	97	100	104	114	118	128	134	143	143
Switzerland Suisse	95	94	98	105	110	120	129	135	143	...
Syrian Arab Republic Rép. arabe syrienne	131	113	123	115	121	114	86	74	...	...
Thailand Thaïlande	97	86	110	113	109	106	130	168	206	265
Trinidad and Tobago Trinité-et-Tobago	93	92	103	117	103	106	83	70	86	...
Tunisia Tunisie	101	106	100	109	84	100	74	92	109	...
Turkey Turquie	121	124	141	249	...	...	...	...	...	...
former USSR ancienne URSS	107	118	123	128	134	126	124	129	162	159
United Kingdom Royaume-Uni	96	102	111	122	127	135	144	164	177	181
United Rep.Tanzania Rép. Unie de Tanzanie	93	70	56	85	87	79	...	...	...	...
United States Etats-Unis	103	97	108	133	145	161	165	171	...	...

Country or area Pays ou zone	1981	1982	1983	1984	1985	1986	1987	1988	1989	1990
Niger Niger	60	54	48	...	...	...	...	...	...	...
Norway[1] Norvège[1]	99	97	110	119	123	127	144	150	173	185
Pakistan Pakistan	119	95	110	86	110	146	160	109	114	195
Panama Panama	85	97	86	65	78	72	72	...	...	...
Peru Pérou	98	112	101	96	98	88	80	...	...	...
Philippines Philippines	101	107	101	99	94	115	116	131	145	156
Poland Pologne	81	88	97	106	108	113	118	129	129	140
Saint Lucia Sainte-Lucie	125	114	...	...	...	...	...	...	...	...
Samoa Samoa	52	218	...	...	...	...	...	...	...	...
Singapore Singapour	103	107	116	135	134	153	181	242	268	...
Solomon Islands Iles Salomon	109	104	126	139	127	156	114	126	...	...
South Africa Afrique du Sud	98	92	92	93	111	115	107	108	117	109
Spain Espagne	107	112	121	137	144	137	147	...	...	...
Sri Lanka Sri Lanka	105	111	104	122	126	135	137	141	153	170
Sweden Suède	102	105	118	127	131	135	140	144	147	146
Switzerland Suisse	101	98	98	110	117	121	121	129	136	...
Syrian Arab Republic Rép. arabe syrienne	91	97	95	89	83	80	84	93	...	...
Thailand Thaïlande	112	125	113	137	142	117	137	163	219	241
Trinidad and Tobago Trinité-et-Tobago	89	91	89	103	108	107	105	108	107	...
Tunisia Tunisie	98	97	90	89	80	90	93	94	108	...
Turkey Turquie	161	195	202	235	...	...	...	...	...	...
former USSR ancienne URSS	102	107	110	112	110	121	125	131	139	123
United Kingdom Royaume-Uni	99	102	104	113	120	124	131	132	141	151
United Rep.Tanzania Rép. Unie de Tanzanie	95	94	76	79	81	77	...	...	...	...
United States[5] Etats-Unis[5]	97	87	81	86	85	85	96	112	...	...

119
Total imports and exports: index numbers
[cont.]

Importations et exportations totales: indices
1980 = 100 [suite]

Country or area Pays ou zone	1981	1982	1983	1984	1985	1986	1987	1988	1989	1990
Yugoslavia[2] Yougoslavie[2]	95	79	72	67	67	70	78	...		...
Zimbabwe Zimbabwe	124	133	112	111	77	97	...	...		...

B. Imports: Unit value index • Importations: Indice du valeur unitaire

Country or area Pays ou zone	1981	1982	1983	1984	1985	1986	1987	1988	1989	1990
Australia[3] Australie[3]	102	109	118	121	144	157	166	162	161	167
Austria Autriche	111	111	110	114	117	107	102	104	106	104
Bangladesh Bangladesh	111	124	156	157	177	198	220	205	207	231
Belgium-Luxembourg Belgique-Luxembourg	115	131	141	152	152	127	119	84	90	88
Bolivia Bolivie	...	...	...	...	...	...	...	...	...	...
Brazil[2] Brésil[2]	111	107	102	97	93	...	...	...	...	...
Burkina Faso Burkina Faso	111	136	155	166	174	140	...	...	...	...
Canada Canada	112	116	116	124	122	122	118	114	111	112
Colombia[2] Colombie[2]	106	...	...	...	...	...	...	...	...	...
Cyprus Chypre	113	113	115	121	125	107	102	...	...	...
Czechoslovakia Tchécoslovaquie	114	121	129	142	144	146	142	130	130	...
Denmark Danemark	117	129	133	145	145	134	124	126	136	132
Dominica Dominique	101	100	99	102	101	...	...	...	...	...
Dominican Republic Rép. dominicaine	...	...	...	...	...	...	...	...	...	...
Ecuador Equateur	114	135	161	178	144	157	144	170	156	...
Egypt Egypte	106	104	102	101	102	125	194	280	...	...
El Salvador El Salvador	111	119	111	...	...	...	...	...	...	...
Ethiopia Ethiopie	110	110	...	...	...	...	...	...	...	...
Faeroe Islands Iles Féroé	115	127	133	144	148	138	159	136	...	...
Fiji Fidji	107	109	107	110	109	101	123	...	...	...
Finland Finlande	112	117	125	131	135	121	119	122	126	128
France France	119	135	146	160	162	137	135	138	148	146

Country or area Pays ou zone	1981	1982	1983	1984	1985	1986	1987	1988	1989	1990
Yugoslavia [2] Yougoslavie[2]	112	97	96	105	110	107	112	...	...	...
Zimbabwe Zimbabwe	95	98	102	100	96	119	...	...	...	...

B. Exports: Unit value index • Exportations: Indice du valeur unitaire

Country or area Pays ou zone	1981	1982	1983	1984	1985	1986	1987	1988	1989	1990
Australia [3] Australie[3]	102	106	114	115	129	131	136	152	160	162
Austria Autriche	106	111	111	115	119	114	112	112	109	109
Bangladesh Bangladesh	95	90	104	125	180	154	143	167	166	172
Belgium-Luxembourg Belgique-Luxembourg	109	124	132	142	145	130	123	90	97	94
Bolivia Bolivie	97	92	95	93	89	74	59	...	...	...
Brazil[2] Brésil[2]	94	88	84	85	78	84	...	...	...	...
Burkina Faso Burkina Faso	109	136	173	193	166	116	...	...	...	...
Canada Canada	106	107	108	113	112	108	107	106	108	108
Colombia[2] Colombie[2]	82	...	...	...	...	...	...	...	...	...
Cyprus Chypre	123	120	118	131	126	119	123	...	...	...
Czechoslovakia Tchécoslovaquie	109	112	115	116	118	119	119	118	123	...
Denmark Danemark	114	127	132	143	148	142	140	139	149	147
Dominica Dominique	98	105	115	102	111	...	...	...	...	...
Dominican Republic Rép. dominicaine	123	82	81	77	75	83	47	77	77	...
Ecuador Equateur	99	91	83	84	231	260	276	327	351	...
Egypt Egypte	106	100	94	94	101	92	102	141	...	...
El Salvador El Salvador	93	88	71	...	...	...	...	...	...	...
Ethiopia Ethiopie	91	96	93	96	95	127	93	...	...	...
Faeroe Islands Iles Féroé	117	128	134	136	144	144	152	148	...	...
Fiji[4] Fidji[4]	99	97	103	101	91	122	167	...	...	...
Finland Finlande	111	119	127	134	138	135	138	144	156	154
France France	113	127	140	152	157	150	149	154	163	160

119

Total imports and exports: index numbers
[*cont.*]

Importations et exportations totales: indices
1980 = 100 [*suite*]

Country or area Pays ou zone	1981	1982	1983	1984	1985	1986	1987	1988	1989	1990
Germany† · Allemagne† F. R. Germany R. f. Allemagne	114	115	114	121	124	104	98	98	106	103
Ghana Ghana	88	107	420	730	270	853	...	...	...	...
Greece Grèce	116	144	176	209	243	270	269	324	363	395
Guatemala Guatemala	106	110	102	126	92	93	91	117	107	...
Guyana Guyana	127	144	...	...	...	...	...	...	...	...
Hong Kong Hong-kong	111	117	131	146	142	149	155	162	168	172
Hungary Hongrie	105	108	117	125	130	136	139	145	163	178
Iceland Islande	146	228	439	526	682	788	840	989	294	569
India Inde	105	105	91	116	124	103	114	...	...	...
Indonesia Indonésie	...	...	...	...	...	...	...	...	...	...
Ireland Irlande	119	128	134	147	150	133	133	142	151	144
Israel[2] Israël [2]	97	89	85	85	83	82	91	98	104	112
Italy Italie	129	146	153	170	183	150	148	154	166	165
Japan Japon	100	104	95	93	89	56	52	49	55	61
Jordan Jordanie	111	113	103	108	106	80	83	87	126	...
Kenya Kenya	128	147	188	193	228	216	219	241	291	350
Korea, Republic of[2] Corée, République de [2]	105	98	93	94	90	84	101	111	114	103
Kuwait Koweït	109	...	...	...	...	...	...	...	...	...
Liberia Libéria	90	95	90	99	92	96	102	...	...	...
Malawi Malawi	115	124	138	157	212	250	362	385	...	...
Malaysia Malaisie	114	114	109	106	105	92	93	...	...	...
Malta Malte	110	110	112	110	108	104	105	106	109	...
Mauritius Maurice	107	132	129	145	160	129	134	146	174	...
Myanmar Myanmar	...	...	...	...	...	...	...	...	...	...

Country or area Pays ou zone	1981	1982	1983	1984	1985	1986	1987	1988	1989	1990
Germany† · Allemagne† F. R. Germany R. f. Allemagne	106	111	112	116	121	117	114	114	120	119
Ghana Ghana	64	49	310	829	115	490	692	...	...	...
Greece Grèce	119	148	168	207	235	252	275	326	376	408
Guatemala Guatemala	85	71	72	64	76	57	66	68	74	...
Guyana Guyana	107	104	...	...	...	...	...	...	...	...
Hong Kong[4] Hong-kong[4]	109	117	128	145	146	149	154	159	167	171
Hungary Hongrie	104	105	111	116	119	120	124	132	153	168
Iceland Islande	148	232	453	552	718	868	977	165	483	741
India Inde	120	130	145	167	176	171	191	...	...	...
Indonesia Indonésie	100	98	87	85	80	56	56	63	...	...
Ireland Irlande	115	128	140	151	155	144	144	154	164	149
Israel[2] Israël[2]	96	91	88	87	86	89	94	107	111	120
Italy Italie	123	142	153	167	181	172	174	183	195	199
Japan Japon	103	109	102	101	101	85	80	79	84	88
Jordan Jordanie	111	121	111	121	117	101	94	108	170	...
Kenya Kenya	110	121	145	174	171	183	152	175	188	204
Korea, Republic of[2] Corée, République de[2]	103	100	96	99	96	97	112	128	140	134
Kuwait Koweït	...	...	...	...	...	...	...	...	...	...
Liberia Libéria	83	90	84	83	84	81	77	...	...	...
Malawi[4] Malawi[4]	140	154	157	193	194	189	246	290	...	...
Malaysia Malaisie	97	91	87	91	82	61	71	77	...	...
Malta Malte	107	111	104	106	106	109	114	126	132	...
Mauritius Maurice	108	120	129	143	163	173	194	207	228	...
Myanmar Myanmar	114	93	88	88	94	75	58	...	...	...

119
Total imports and exports: index numbers
[cont.]

Importations et exportations totales: indices
1980 = 100 [suite]

Country or area Pays ou zone	1981	1982	1983	1984	1985	1986	1987	1988	1989	1990
Netherlands Pays-Bas	116	118	119	126	127	105	98	98	105	104
Netherlands Antilles Antilles néerlandaises	...	...	...	...	...	...	...	...	...	...
New Zealand Nouvelle-Zélande	115	128	139	158	175	170	163	162	174	176
Niger Niger	98	140	84	...	...	...	...	...	...	...
Norway[1] Norvège[1]	104	112	116	120	127	127	131	135	143	144
Pakistan Pakistan	115	132	140	165	171	153	165	125	147	236
Panama Panama	...	...	...	...	...	...	...	...	...	...
Peru[2] Pérou[2]	101	101	103	...	...	...	...	...	...	...
Philippines[2] Philippines[2]	111	95	96	108	104	83	85	95	93	...
Poland[3] Pologne[3]	110	115	119	137	167	195	263	443	1 175	8 527
Saint Lucia Sainte-Lucie	102	104	...	...	...	...	...	...	...	...
Samoa Samoa	...	...	...	...	...	...	...	...	...	...
Singapore Singapour	111	106	104	103	91	81	87	88	...	...
Solomon Islands Iles Salomon	...	...	...	...	...	...	...	...	...	...
South Africa Afrique du Sud	113	139	141	169	230	258	276	324	366	401
Spain[3] Espagne[3]	131	147	178	198	204	168	161	162	166	161
Sri Lanka Sri Lanka	116	121	127	133	143	133	150	183	272	377
Sweden[3] Suède[3]	111	126	142	148	151	139	143	148	155	157
Switzerland Suisse	97	96	97	101	105	95	90	94	101	...
Syrian Arab Republic Rép. arabe syrienne	130	122	141	132	110	132	133	164	...	...
Thailand Thaïlande	118	121	114	115	127	94	100	111	128	124
Trinidad and Tobago Trinité-et-Tobago	113	128	129	129	128	174	199	220	253	...
Tunisia Tunisie	111	123	126	136	146	145	160	168	185	...
Turkey Turquie	140	203	253	368	...	...	...	...	...	...
United Kingdom Royaume-Uni	108	117	127	138	144	138	141	142	149	155

Country or area Pays ou zone	1981	1982	1983	1984	1985	1986	1987	1988	1989	1990
Netherlands Pays-Bas	116	121	120	129	133	112	102	102	109	109
Netherlands Antilles Antilles néerlandaises	99	...	...	...	...	...	...	...	...	...
New Zealand Nouvelle-Zélande	113	125	132	150	163	159	168	179	203	201
Niger Niger	...	...	...	...	...	...	...	...	...	...
Norway[1] Norvège[1]	116	124	128	141	147	110	106	107	119	125
Pakistan Pakistan	95	111	110	121	123	129	155	90	98	197
Panama Panama	105	84	94	92	...	...	...	...	...	...
Peru[2] Pérou[2]	84	72	77	68	62	52	...	...	...	...
Philippines[2] Philippines[2]	98	81	85	94	85	73	84	100	98	87
Poland[3] Pologne[3]	107	111	110	124	155	185	268	459	1 410	9 084
Saint Lucia Sainte-Lucie	107	117	...	...	...	...	...	...	...	...
Samoa Samoa	84	79	...	...	...	...	...	...	...	...
Singapore Singapour	115	111	106	102	90	77	80	79	...	...
Solomon Islands Iles Salomon	87	89	94	140	135	120	181	217	...	...
South Africa Afrique du Sud	103	117	118	148	191	225	245	292	351	367
Spain[3] Espagne[3]	118	133	156	178	190	187	192	196	205	201
Sri Lanka Sri Lanka	109	107	136	174	154	137	159	180	290	411
Sweden[3] Suède[3]	109	122	136	146	151	150	153	162	172	175
Switzerland Suisse	103	107	108	109	113	109	109	110	117	...
Syrian Arab Republic Rép. arabe syrienne	111	97	91	93	91	67	62	60	...	...
Thailand Thaïlande	103	96	97	97	103	104	111	130	127	127
Trinidad and Tobago Trinité-et-Tobago	110	106	105	104	100	97	102	97	123	...
Tunisia Tunisie	115	126	130	142	143	128	146	156	178	...
Turkey Turquie	132	178	220	359	...	...	...	...	...	...
United Kingdom Royaume-Uni	109	116	126	136	144	132	137	140	146	154

119
Total imports and exports: index numbers
[*cont.*]

Importations et exportations totales: indices
1980 = 100 [*suite*]

Country or area Pays ou zone	1981	1982	1983	1984	1985	1986	1987	1988	1989	1990
United Rep.Tanzania Rép. Unie de Tanzanie	108	141	139	143	184	325	...	...	...	...
United States Etats-Unis	106	104	100	102	99	95	102	107	74	121
Yugoslavia[2] Yougoslavie[2]	110	112	112	115	117	106	111	...	...	...
Zimbabwe Zimbabwe	100	99	116	131	153	160	...	...	...	...

C. Terms of trade · Termes de l'échange

Country or area Pays ou zone	1981	1982	1983	1984	1985	1986	1987	1988	1989	1990
Australia Australie	100	97	97	95	90	83	81	94	99	97
Austria Autriche	96	100	101	101	101	107	109	108	102	105
Bangladesh Bangladesh	85	73	67	80	102	78	65	81	80	75
Belgium-Luxembourg Belgique-Luxembourg	...	...	...	...	...	...	...	...	...	...
Brazil Brésil	85	82	82	88	84	...	...	...	...	...
Burkina Faso Burkina Faso	98	100	112	116	95	83	...	...	...	...
Canada Canada	95	93	92	91	92	89	91	93	97	96
Colombia Colombie	77	...	...	...	...	...	...	...	...	...
Cyprus Chypre	108	106	102	108	101	111	121	...	...	...
Czechoslovakia Tchécoslovaquie	96	93	89	81	82	81	83	91	95	...
Denmark Danemark	97	98	99	99	102	106	113	110	109	112
Dominica Dominique	97	105	116	100	110	...	...	...	...	...
Ecuador Equateur	86	68	52	47	160	166	191	192	225	...
Egypt Egypte	100	97	92	93	99	74	53	51	...	...
El Salvador El Salvador	84	74	64	...	...	...	...	...	...	...
Ethiopia Ethiopie	83	87	...	...	...	...	...	...	...	...
Faeroe Islands Iles Féroé	102	101	101	94	97	104	96	109	...	...
Fiji Fidji	92	88	96	92	84	121	136	...	...	...
Finland Finlande	99	102	102	102	102	112	116	118	124	120
France France	94	94	96	95	97	110	110	111	110	109

Country or area Pays ou zone	1981	1982	1983	1984	1985	1986	1987	1988	1989	1990
United Rep.Tanzania Rép. Unie de Tanzanie	106	94	118	161	154	286	...	...	...	...
United States[5] Etats-Unis[5]	109	110	112	114	112	113	115	123	82	128
Yugoslavia [2] Yougoslavie[2]	109	117	113	107	106	103	109	...	...	...
Zimbabwe Zimbabwe	111	107	121	158	189	192	...	...	...	...

D. Purchasing power of exports • Pouvoir d'achat des exportations

	1981	1982	1983	1984	1985	1986	1987	1988	1989	1990
Australia Australie	94	97	94	106	112	108	119	129	145	151
Austria Autriche	101	107	112	123	135	144	150	163	176	199
Bangladesh Bangladesh	...	...	...	...	...	...	...	...	...	...
Belgium-Luxembourg Belgique-Luxembourg	...	...	...	...	...	...	...	...	...	...
Brazil Brésil	102	90	103	136	134	...	...	...	...	...
Burkina Faso Burkina Faso	92	70	73	110	94	126	...	...	...	...
Canada Canada	98	143	156	187	125	127	145	115	...	...
Colombia Colombie	70	...	...	...	...	...	...	...	...	...
Cyprus Chypre	115	117	105	131	108	111	146	...	...	...
Czechoslovakia Tchécoslovaquie	96	99	100	100	104	104	111	118	120	...
Denmark Danemark	100	103	113	118	126	133	145	148	155	164
Dominica Dominique	191	221	249	217	252	...	...	...	...	...
Ecuador Equateur	88	69	58	58	50	47	64	42	48	...
Egypt Egypte	...	...	...	...	...	...	...	...	...	...
El Salvador El Salvador	61	48	58	...	...	...	...	...	...	...
Ethiopia Ethiopie	85	94	...	...	...	...	...	...	...	...
Faeroe Islands Iles Féroé	102	106	128	124	135	145	127	157	...	...
Fiji Fidji	87	87	80	85	84	112	128	...	...	...
Finland Finlande	102	102	106	117	118	129	136	143	150	156
France France	97	94	99	104	110	123	128	141	150	157

119
Total imports and exports: index numbers
[*cont.*]

Importations et exportations totales: indices
1980 = 100 [*suite*]

Country or area Pays ou zone	1981	1982	1983	1984	1985	1986	1987	1988	1989	1990
Germany†·Allemagne† F. R. Germany R. f. Allemagne	93	97	99	96	98	112	116	116	113	115
Ghana Ghana	73	46	74	114	43	57	...	...	...	...
Greece Grèce	103	103	96	99	97	93	102	101	104	103
Guatemala Guatemala	80	65	71	51	83	61	72	58	69	...
Guyana Guyana	84	72	...	...	...	...	...	...	...	...
Hong Kong Hong-kong	98	99	98	99	103	100	99	98	99	100
Hungary Hongrie	99	97	95	93	92	88	89	91	94	94
Iceland Islande	101	102	103	105	105	110	116	17	164	130
India Inde	114	123	159	144	142	166	167	...	...	...
Ireland Irlande	97	100	105	103	104	108	108	109	109	104
Israel Israël	99	103	104	102	104	109	104	109	107	108
Italy Italie	95	98	100	98	99	115	118	119	117	121
Japan Japon	103	104	107	109	114	152	155	160	153	144
Jordan Jordanie	100	108	107	111	111	127	113	125	135	...
Kenya Kenya	86	82	77	90	75	85	69	72	65	58
Korea, Republic of Corée, République de	98	102	103	106	106	115	112	115	123	130
Liberia Libéria	92	95	94	84	91	84	76	...	...	...
Malawi Malawi	122	123	113	123	92	75	68	75	...	...
Malaysia Malaisie	85	80	80	86	78	66	76	...	...	...
Malta Malte	97	101	93	96	98	104	109	119	121	...
Mauritius Maurice	101	91	99	98	102	134	145	141	131	...
Netherlands Pays-Bas	100	103	101	102	105	107	104	105	104	105
New Zealand Nouvelle-Zélande	98	98	95	95	93	93	103	111	116	114
Norway Norvège	111	111	110	118	116	86	81	79	84	86

Country or area Pays ou zone	1981	1982	1983	1984	1985	1986	1987	1988	1989	1990
Germany† · Allemagne† F. R. Germany R. f. Allemagne	100	107	108	115	124	144	154	165	173	178
Ghana Ghana	54	...	...	...	...	...	...	...	...	...
Greece Grèce	90	87	96	112	113	128	142	104	148	140
Guatemala Guatemala	76	67	75	68	73	75	71	57	76	...
Guyana Guyana	77	54	...	...	...	...	...	...	...	...
Hong Kong Hong-kong	112	110	125	154	168	189	248	310	346	376
Hungary Hongrie	103	106	110	115	120	113	118	129	133	128
Iceland Islande	101	84	96	101	111	128	142	20	200	156
India Inde	67	78	102	103	97	128	155	...	...	...
Ireland Irlande	97	108	126	148	155	171	195	206	230	238
Israel Israël	105	107	109	124	137	158	168	169	171	173
Italy Italie	99	102	108	114	123	145	152	162	174	185
Japan Japon	114	113	125	148	161	214	219	234	234	233
Jordan Jordanie	122	135	133	194	201	236	251	320	351	...
Kenya Kenya	81	75	68	79	68	89	70	77	68	65
Korea, Republic of Corée, République de	115	128	150	178	192	234	155	179	181	360
Liberia Libéria	...	...	...	...	...	...	...	...	...	...
Malawi Malawi	96	102	124	95	91	85	73	81	...	...
Malaysia Malaisie	83	82	91	95	91	84	101	...	...	...
Malta Malte	94	92	85	100	105	113	122	138	168	...
Mauritius Maurice	...	...	...	...	...	...	...	...	...	...
Netherlands Pays-Bas	101	102	106	115	123	130	132	145	149	159
New Zealand Nouvelle-Zélande	100	102	105	109	119	117	133	148	151	158
Norway Norvège	110	108	121	140	142	110	117	119	144	160

119
Total imports and exports: index numbers
[*cont.*]

Importations et exportations totales: indices
1980 = 100 [*suite*]

Country or area Pays ou zone	1981	1982	1983	1984	1985	1986	1987	1988	1989	1990
Pakistan Pakistan	83	84	79	73	72	84	94	72	66	84
Peru Pérou	83	72	75	...	...	...	...	...	...	...
Philippines Philippines	88	86	89	87	82	88	99	106	105	...
Poland Pologne	97	96	92	91	93	95	102	104	235	16
Saint Lucia Sainte-Lucie	106	112	...	...	...	...	...	...	...	...
Singapore Singapour	103	105	103	99	99	96	92	90	...	...
South Africa Afrique du Sud	91	84	84	88	83	87	89	90	96	92
Spain Espagne	90	90	88	90	93	111	119	121	124	125
Sri Lanka Sri Lanka	93	88	107	131	108	103	106	99	107	109
Sweden Suède	98	97	96	99	100	108	108	109	111	112
Switzerland Suisse	106	111	112	108	108	115	121	118	116	...
Syrian Arab Republic Rép. arabe syrienne	85	79	65	71	83	51	47	36	...	...
Thailand Thaïlande	87	79	85	84	81	110	111	118	99	102
Trinidad and Tobago Trinité-et-Tobago	98	83	82	81	78	56	51	44	48	...
Tunisia Tunisie	103	102	103	105	97	89	91	93	96	...
Turkey Turquie	94	88	87	98	...	...	...	...	...	...
United Kingdom Royaume-Uni	101	100	99	98	100	96	97	98	98	99
United Rep.Tanzania Rép. Unie de Tanzanie	98	67	85	113	84	88	...	...	...	...
United States Etats-Unis	104	106	112	112	114	118	113	115	110	105
Yugoslavia Yougoslavie	99	104	101	93	91	97	98	...	...	...
Zimbabwe Zimbabwe	111	108	104	121	123	120	...	...	...	...

Source:
Trade statistics database of the Statistical Division of the
United Nations Secretariat.

† All data shown which pertain to Germany prior to 3 October
1990 are indicated separately for the Federal Republic of
Germany and the former German Democratic Republic based on
their respective territories at the time indicated. Where
data for united Germany (3 October 1990 and thereafter) are
not available, available data are shown separately under the

Source:
Base de données pour les statistiques du commerce extérieur
de la Division de statistique du Secrétariat de l'ONU.

† Toutes les données se rapportant à l'Allemagne avant le 3
octobre 1990 figurent dans deux rubriques séparées basées
sur les territoires respectifs de la République fédérale
d'Allemagne et l'ancienne République démocratique allemande
selon la période indiquée. En l'absence de données pour
l'Allemagne unifiée (à compter du 3 octobre 1990), les

Country or area Pays ou zone	1981	1982	1983	1984	1985	1986	1987	1988	1989	1990
Pakistan Pakistan	98	80	86	63	79	122	150	79	76	163
Peru Pérou	81	80	75	...	...	...	...	...	...	...
Philippines Philippines	89	91	91	86	77	101	116	139	153	...
Poland Pologne	79	85	90	97	100	107	120	134	303	22
Saint Lucia Sainte-Lucie	132	128	...	...	...	...	...	...	...	...
Singapore Singapour	107	112	119	134	133	147	166	218	...	...
South Africa Afrique du Sud	89	77	77	81	92	100	95	98	112	100
Spain Espagne	96	101	106	124	134	152	175	...	...	...
Sri Lanka Sri Lanka	98	97	112	160	136	139	145	139	163	185
Sweden Suède	100	102	113	125	131	146	150	158	163	163
Switzerland Suisse	107	108	110	118	125	140	147	152	157	...
Syrian Arab Republic Rép. arabe syrienne	77	76	61	63	69	41	39	34	...	...
Thailand Thaïlande	97	99	96	115	115	129	152	192	217	246
Trinidad and Tobago Trinité-et-Tobago	87	75	73	83	84	60	54	48	52	...
Tunisia Tunisie	100	100	92	93	78	80	85	87	103	...
Turkey Turquie	152	171	176	229	...	...	...	...	...	...
United Kingdom Royaume-Uni	100	102	103	111	120	119	127	130	138	150
United Rep.Tanzania Rép. Unie de Tanzanie	93	63	64	89	67	68	...	...	...	...
United States Etats-Unis	100	92	91	97	97	100	108	129	...	...
Yugoslavia Yougoslavie	111	101	97	98	100	104	110	...	...	...
Zimbabwe Zimbabwe	106	106	106	121	118	144	...	...	...	...

designations Federal Republic of Germany and former German Democratic Republic and pertain to the territorial boundaries prior to 3 October 1990. For detailed explanatory notes on data pertaining to Germany, see Annex I - Country Nomenclature.

1 Excluding ships.
2 Calculated in terms of US dollars.
3 Price index numbers.
4 Domestic exports only.
5 Excludes military exports.

données disponisbles sont fournies séparément sous les rubriques République fédérale d'Allemagne et ancienne République démocratique allemande et se rapportent aux limites territoriales antérieures au 3 octobre 1990. Pour les notes explicatives en détail sur les données concernant l'Allemagne, voir Annexe I – Nomenclature des pays.

1 Non compris les navires.
2 Calculés en dollars des Etats-Unis.
3 Indices des prix.
4 Seulement exportations domestiques.
5 Non compris les exportations militaires.

120
Manufactured goods exports
Exportations des produits manufacturés

1980 = 100

Region, country or area Region, pays ou zone	1981	1982	1983	1984	1985	1986	1987	1988	1989	1990
	Unit value indices in US dollars • Indices de valeur unitaire en dollars des E-U									
Total [1]	95	92	89	87	87	101	113	121	121	132
Developed economies Econ. développées	94	92	89	86	86	103	116	124	123	136
America, North Amérique du Nord	111	116	117	118	118	120	124	132	137	139
Canada Canada	105	107	109	110	105	105	108	117	122	121
United States [2] Etats-Unis [2]	112	118	120	121	123	127	130	138	142	145
Europe Europe	87	85	80	76	77	97	113	120	117	136
EEC+ CEE+	87	85	81	76	78	97	113	119	117	137
Belgium-Luxembourg [3] Belgique-Luxembourg [3]	85	79	76	71	73	93	106	113	112	126
Denmark Danemark	91	86	84	80	83	105	124	129	124	147
France France	87	83	79	78	81	95	110	114	112	130
Germany† • Allemagne† F. R. Germany R. f. Allemagne	86	85	82	73	75	100	118	121	119	137
Greece Grèce	93	96	83	78	74	83	93	99	103	118
Ireland [4] Irlande [4]	109	74	76	78	96	116	124	134	139	153
Italy [3] Italie [3]	81	87	83	81	81	102	119	126	122	160
Netherlands [3] Pays-Bas [3]	87	89	80	76	76	95	111	117	115	134
Portugal [4] Portugal [4]	89	76	70	71	73	80	100	107	103	121
Spain [4] Espagne [4]	84	86	76	77	75	97	93	113	113	136
United Kingdom Royaume-Uni	93	86	81	77	79	92	107	120	116	132
EFTA+ AELE+	90	86	80	76	77	97	114	121	118	135
Austria [5] Autriche [5]	87	87	81	76	78	99	117	123	110	130
Finland Finlande	96	93	85	83	83	100	117	130	137	153
Iceland [4] Islande [4]	88	74	76	78	70	81	92	121	126	122
Norway Norvège	92	85	77	79	78	89	104	129	133	134

120
Manufactured goods exports
[cont.]

Exportations des produits manufacturés
1980 = 100 [suite]

Region, country or area Region, pays ou zone	1981	1982	1983	1984	1985	1986	1987	1988	1989	1990
Sweden Suède	90	82	75	74	75	93	108	117	118	131
Switzerland[3] Suisse[3]	88	87	84	76	75	100	120	120	114	136
Other Europe · Autres pays d'Europe Malta[3] 　Malte[3]	89	100	82	80	79	96	115	132	165	...
Other **Autres**	**105**	**98**	**95**	**95**	**93**	**110**	**121**	**135**	**134**	**133**
Australia Australie	94	84	83	83	73	73	83	105	110	109
Israel Israël	96	93	89	87	86	89	94	106	114	125
Japan Japon	106	99	97	97	95	115	126	139	138	136
New Zealand Nouvelle-Zélande	101	99	97	95	91	98	115	146	143	137
South Africa Afrique du Sud	83	71	71	62	54	60	74	78	...	...
Developing economies **Econ. en dévelop.**	**100**	**94**	**89**	**93**	**92**	**91**	**101**	**110**	**112**	**114**

Unit value indices in 'SDR' · Indices de valeur unitaire en 'DTS'

	1981	1982	1983	1984	1985	1986	1987	1988	1989	1990
Total[1]	**105**	**109**	**108**	**110**	**112**	**112**	**114**	**118**	**122**	**126**
Developed economies **Econ. développées**	**104**	**109**	**108**	**109**	**111**	**114**	**117**	**120**	**125**	**131**
Developing economies **Econ. en dévelop.**	**110**	**111**	**108**	**118**	**118**	**101**	**102**	**107**	**113**	**109**

Unit value indices in national currency · Indices de valeur unitaire en monnaie nationale

America, North · Amérique du Nord Canada 　Canada	1981	1982	1983	1984	1985	1986	1987	1988	1989	1990
Canada 　Canada	108	113	115	122	123	125	123	122	123	120
United States[2] Etats-Unis[2]	112	118	120	121	123	127	130	138	142	145
Europe · Europe **EEC+ · CEE+** Belgium-Luxembourg[3] 　Belgique-Luxembourg[3]	107	123	132	141	146	141	136	...	...	...
Denmark 　Danemark	115	127	136	147	154	151	151	154	161	161
France 　France	112	129	143	161	170	155	156	161	169	167
Germany† · Allemagne† F. R. Germany 　R. f. Allemagne	107	113	115	116	121	119	116	117	122	121
Greece 　Grèce	120	149	169	204	238	271	291	337	392	435

120
Manufactured goods exports
[*cont.*]

Exportations des produits manufacturés
1980 = 100 [*suite*]

Region, country or area Region, pays ou zone	1981	1982	1983	1984	1985	1986	1987	1988	1989	1990
Italy[3] Italie[3]	107	137	147	166	179	176	179	191	194	...
Netherlands[3] Pays-Bas[3]	109	119	115	122	126	117	112	116	122	122
United Kingdom Royaume-Uni	107	114	124	134	143	146	151	157	166	172
EFTA+ · AELE+ Austria[5] Autriche[5]	107	115	113	117	123	116	114	117	112	...
Finland Finlande	111	120	128	134	137	135	137	146	157	156
Norway Norvège	107	110	116	130	134	134	142	170	186	169
Sweden Suède	108	121	136	144	152	156	160	169	180	183
Switzerland[3] Suisse[3]	103	106	106	106	109	106	...	...	...	...
Other Europe · Autres pays d'Europe Malta[3] Malte[3]	101	119	103	107	107	109	115	126	164	...
Other · Autres Australia Australie	93	95	105	108	119	125	136	153	160	...
Japan Japon	103	109	102	101	101	85	81	79	84	87
New Zealand Nouvelle-Zélande	114	129	141	165	179	181	189	217	233	224
South Africa Afrique du Sud	93	98	101	116	152	176	194	229	...	...

Quantum indices · Indices de volume

Region, country or area Region, pays ou zone	1981	1982	1983	1984	1985	1986	1987	1988	1989	1990
Total [1]	105	103	108	119	124	129	139	151	164	172
Developed economies **Econ. développées**	104	101	104	114	120	122	128	138	149	156
America, North **Amérique du Nord**	97	86	84	93	95	96	106	122	131	146
Canada Canada	105	105	115	139	151	157	163	182	181	193
United States Etats-Unis	96	82	77	82	82	81	92	108	120	135
Europe **Europe**	104	103	107	116	122	125	131	141	152	156
EEC+ **CEE+**	104	103	106	115	121	124	130	140	152	156
Belgium-Luxembourg **Belgique-Luxembourg**	99	99	104	109	114	118	126	137	150	158

120
Manufactured goods exports
[*cont.*]

Exportations des produits manufacturés
1980 = 100 [*suite*]

Region, country or area Region, pays ou zone	1981	1982	1983	1984	1985	1986	1987	1988	1989	1990
Denmark Danemark	106	106	115	122	125	127	130	139	144	155
France France	103	100	103	108	109	116	121	132	143	152
Germany† · Allemagne† F. R. Germany R. f. Allemagne	105	108	107	122	129	131	135	145	156	159
Greece Grèce	98	90	107	123	123	145	152	126	153	144
Ireland Irlande	90	144	153	167	151	153	180	199	218	234
Italy Italie	119	107	114	119	127	128	133	139	157	143
Netherlands Pays-Bas	102	97	108	116	122	129	134	142	149	156
Portugal Portugal	100	123	145	169	180	215	225	249	299	332
Spain Espagne	114	111	119	139	150	133	171	170	189	201
United Kingdom Royaume-Uni	93	94	93	99	106	108	116	122	134	142
EFTA+ AELE+	**103**	**104**	**110**	**121**	**128**	**133**	**136**	**144**	**153**	**160**
Austria Autriche	105	105	111	122	130	137	139	153	178	196
Finland Finlande	105	108	110	120	128	132	137	136	140	146
Iceland Islande	95	96	135	132	117	125	139	139	131	122
Norway Norvège	100	105	115	115	120	124	130	118	127	142
Sweden Suède	103	107	117	128	134	138	141	149	151	152
Switzerland Suisse	103	100	102	115	123	129	130	145	154	160
Other Europe · Autres pays d'Europe Malta Malte	103	85	91	102	106	112	116	118	115	...
Other Autres	**110**	**107**	**117**	**134**	**143**	**142**	**144**	**149**	**156**	**165**
Australia Australie	98	96	101	107	114	118	140	141	152	171
Israel Israël	104	100	101	119	133	148	168	173	180	185
Japan Japon	111	108	118	137	144	143	143	148	156	164
New Zealand Nouvelle-Zélande	101	103	106	120	130	127	128	126	136	148
South Africa Afrique du Sud	94	102	107	116	149	159	150	160	...	...

120
Manufactured goods exports
[*cont.*]

Exportations des produits manufacturés
1980 = 100 [*suite*]

Region, country or area Region, pays ou zone	1981	1982	1983	1984	1985	1986	1987	1988	1989	1990
Developing economies **Econ. en dévelop.**	112	117	136	151	157	178	219	250	277	293

Value (thousand million US $) · Valeur (millards de dollars E-U)

Region, country or area Region, pays ou zone	1981	1982	1983	1984	1985	1986	1987	1988	1989	1990
Total[1]	1 041.10	994.03	1 002.00	1 083.40	1 138.30	1 364.00	1 650.00	1 927.80	2 078.10	2 377.30
Developed economies **Econ. développées**	900.14	855.15	849.37	906.60	955.64	1 159.00	1 370.80	1 579.50	1 687.80	1 956.40
America, North **Amérique du Nord**	192.04	177.89	175.05	195.36	199.19	205.06	232.91	286.97	319.63	361.29
Canada Canada	37.64	38.01	42.47	52.04	53.86	56.19	59.81	72.19	74.46	78.83
United States Etats-Unis	154.40	139.88	132.58	143.32	145.34	148.87	173.11	214.78	245.17	282.46
Europe **Europe**	547.29	530.20	519.06	532.42	571.06	734.12	894.89	1 011.50	1 075.30	1 288.90
EEC+ **CEE+**	467.49	453.19	443.15	453.46	486.97	624.17	762.34	861.55	919.91	1 103.00
Belgium-Luxembourg Belgique-Luxembourg	40.45	37.41	37.59	37.21	39.75	52.29	64.09	73.83	80.42	95.18
Denmark Danemark	8.90	8.46	8.95	9.00	9.61	12.35	14.92	16.63	16.61	21.14
France France	75.50	69.98	68.66	70.48	73.71	92.19	111.69	126.05	134.56	165.29
Germany† · Allemagne† F. R. Germany R. f. Allemagne	150.82	152.61	146.45	148.63	161.58	217.32	264.48	293.10	309.59	362.01
Greece Grèce	2.39	2.27	2.31	2.51	2.37	3.15	3.69	3.25	4.12	4.45
Ireland Irlande	4.54	4.94	5.38	6.04	6.66	8.16	10.26	12.31	13.95	16.50
Italy Italie	63.26	61.49	62.34	63.06	67.77	86.09	104.04	115.83	125.92	150.63
Netherlands Pays-Bas	34.01	32.86	33.23	33.79	35.54	46.82	56.91	63.55	65.51	80.23
Portugal Portugal	2.92	3.06	3.34	3.92	4.32	5.62	7.34	8.72	10.09	13.20
Spain Espagne	14.84	14.89	14.17	16.68	17.57	20.11	24.62	29.90	33.18	42.63
United Kingdom Royaume-Uni	70.26	65.22	60.71	62.15	68.08	80.09	100.29	118.37	125.96	151.77
EFTA+ **AELE+**	79.43	76.67	75.60	78.63	83.75	109.51	132.01	149.37	154.64	185.03
Austria Autriche	13.64	13.66	13.48	13.73	15.06	20.17	24.28	27.94	29.07	37.89
Finland Finlande	10.46	10.39	9.70	10.36	10.88	13.57	16.56	18.17	19.77	22.98
Iceland Islande	0.16	0.14	0.20	0.20	0.16	0.20	0.25	0.33	0.32	0.29

120
Manufactured goods exports
[*cont.*]

Exportations des produits manufacturés
1980 = 100 [*suite*]

Region, country or area Region, pays ou zone	1981	1982	1983	1984	1985	1986	1987	1988	1989	1990
Norway Norvège	6.87	6.64	6.60	6.77	6.96	8.30	10.08	11.40	12.55	14.22
Sweden Suède	22.96	21.53	21.68	23.22	24.80	31.49	37.54	43.03	43.92	48.90
Switzerland Suisse	25.34	24.30	23.94	24.34	25.89	35.79	43.30	48.50	49.01	60.76
Other Europe · Autres pays d'Europe Malta Malte	0.37	0.34	0.30	0.33	0.34	0.43	0.54	0.63	0.77	...
Other Autres	**160.81**	**147.07**	**155.27**	**178.82**	**185.39**	**219.85**	**243.03**	**280.94**	**292.85**	**306.20**
Australia Australie	4.12	3.61	3.77	3.94	3.69	3.83	5.23	6.60	7.44	8.28
Israel Israël	4.61	4.26	4.15	4.77	5.26	6.10	7.29	8.44	9.42	10.52
Japan Japon	146.73	134.13	142.08	164.91	170.78	203.36	222.81	257.07	266.59	277.44
New Zealand Nouvelle-Zélande	1.28	1.28	1.29	1.43	1.48	1.56	1.85	2.31	2.45	2.54
South Africa Afrique du Sud	4.07	3.79	3.98	3.76	4.19	4.99	5.85	6.52	...	...
Developing economies Econ. en dévelop.	**140.91**	**138.88**	**152.58**	**176.84**	**182.69**	**204.98**	**279.13**	**348.34**	**390.30**	**420.88**

Source:
Trade statistics database of the Statistical Division of the
United Nations Secretariat.
+   For Member States of this grouping, see
     Annex I - Other groupings.

† All data shown which pertain to Germany prior to 3 October
  1990 are indicated separately for the Federal Republic of
  Germany and the former German Democratic Republic based on
  their respective territories at the time indicated. Where
  data for united Germany (3 October 1990 and thereafter) are
  not available, available data are shown separately under the
  designations Federal Republic of Germany and former German
  Democratic Republic and pertain to the territorial
  boundaries prior to 3 October 1990. For detailed
  explanatory notes on data pertaining to Germany, see Annex I
  - Country Nomenclature.

1 Excludes trade of the countries of Eastern Europe and the
  former USSR.
2 Beginning third quarter 1989 derived from price indices;
  national unit value index discontinued.
3 Derived from sub-indices using current weights. Indices for
  Belgium, beginning 1988, for Italy, beginning 1989, and for
  Switzerland, beginning 1987, are calculated by the United
  Nations Statistical Division.
4 Indices are calculated by the United Nations Statistical
  Division.
5 Series linked at 1988 by a factor calculated by the United
  Nations Statistical Division.
6 Special drawing right.

Source:
Base de données pour les statistiques du commerce extérieur
de la Division de statistique de la secrétariat de l'ONU.
+   Les Etats membres de ce groupement, voir
     annexe I - Autres groupements.

† Toutes les données se rapportant à l'Allemagne avant le 3
  octobre 1990 figurent dans deux rubriques séparées basées
  sur les territoires respectifs de la République fédérale
  d'Allemagne et l'ancienne République démocratique allemande
  selon la période indiquée. En l'absence de données pour
  l'Allemagne unifiée (à compter du 3 octobre 1990), les
  données disponibles sont fournies séparément sous les
  rubriques République fédérale d'Allemagne et ancienne
  République démocratique allemande et se rapportent aux
  limites territoriales antérieures au 3 octobre 1990. Pour
  les notes explicatives en détail sur les données concernant
  l'Allemagne, voir Annexe I – Nomenclature des pays.

1 Non compris le commerce des pays de l'Europe de l'Est et
  l'ancienne URSS.
2 A partir du troisième trimestre de l'année 1989 calculés à
  partir des indices des prix; l'indice de la valeur unitaire
  nationale est discontinué.
3 Calculé à partir de sous-indices à coéfficients de
  pondération correspondant à la période en cours. Les indices
  pour la Belgique, à partir de 1988, pour l'Italie, à partir
  de 1989, et pour la Suisse, à partir de 1987, sont calculés
  par la Division de statistique des Nations Unies.
4 Les indices sont calculés par la Division de statistique des
  Nations Unies.
5 Les séries sont enchaînés à 1988 par un facteur calculé par
  la Division de statistique des Nations Unies.
6 Le droit de tirage spécial.

121
Structure of world exports by commodity classes and regions
Structure des exportations mondiales par catégories de marchandises et par régions

In per cent

SITC Commodity classes	Year	World Monde /1,2,3	Developed economies Econ. dévelop-pées /2,3,4	Developing economies Total /1,3,4	OPEC+ OPEP+	Eastern Europe and former USSR Total /2,3	Former USSR an URSS	Europe Total /2	EEC+ CEE+ /2	EFTA+ AELE+	South Africa Afrique du Sud
Origin of exports of major commodity classes											
0-9 Total all commodities /5	1980	100.0	62.9	29.3	15.3	7.8	3.8	40.1	34.5	5.6	0.7
	1987	100.0	70.1	21.9	4.9	8.1	4.4	44.9	38.4	6.5	0.5
	1988	100.0	70.4	22.3	4.4	7.3	3.9	43.7	37.4	6.3	0.5
	1989	100.0	70.4	23.2	4.7	6.4	3.6	43.6	37.4	6.2	0.5
	1990	100.0	72.1	22.8	4.9	5.1	3.1	46.5	39.8	6.6	0.6
0&1 Food, live animals, beverages and tobacco	1980	100.0	64.3	31.0	1.9	4.7	0.6	38.4	36.1	2.2	1.3
	1987	100.0	65.8	29.9	1.7	4.3	0.7	46.4	43.5	2.8	0.8
	1988	100.0	66.3	30.0	1.7	3.7	0.7	45.1	42.4	2.6	0.6
	1989	100.0	67.1	29.1	1.6	3.7	0.6	45.2	42.6	2.5	0.7
	1990	100.0	69.4	27.5	1.7	3.1	0.5	48.1	45.1	2.9	0.8
2&4 Crude materials, oils and fats, (fuels excluded)	1980	100.0	61.5	30.9	3.7	7.6	5.1	24.5	17.6	6.9	2.3
	1987	100.0	64.7	27.3	2.1	8.0	5.6	29.2	22.4	6.9	1.7
	1988	100.0	63.9	29.0	2.4	7.1	5.1	28.7	22.2	6.5	1.6
	1989	100.0	63.6	29.6	2.6	6.9	4.7	28.5	21.9	6.5	1.9
	1990	100.0	65.1	28.6	2.3	6.3	4.4	30.5	23.6	6.9	2.2
3 Mineral fuels, lubricants and related material	1980	100.0	18.3	72.9	60.5	8.8	7.5	13.9	11.5	2.3	0.2
	1987	100.0	27.7	52.5	36.4	19.8	18.0	18.8	14.8	3.9	0.6
	1988	100.0	26.9	53.4	38.3	19.6	17.7	17.1	13.3	3.8	0.6
	1989	100.0	27.4	55.6	38.7	17.0	15.0	17.6	12.9	4.7	0.7
	1990	100.0	30.4	55.4	38.8	14.1	12.9	20.0	14.5	5.5	0.8
5 Chemicals	1980	100.0	87.1	7.4	1.0	5.5	1.2	63.4	55.8	7.7	0.6
	1987	100.0	85.2	9.4	0.9	5.4	1.4	64.3	55.8	8.5	0.4
	1988	100.0	84.0	10.9	0.8	5.1	1.4	62.4	54.2	8.2	0.3
	1989	100.0	84.2	11.4	0.9	4.4	1.4	61.0	53.0	8.0	0.4
	1990	100.0	85.0	11.5	1.2	3.5	1.2	63.0	54.5	8.5	0.4
7 Machinery and transport equipment	1980	100.0	85.0	6.0	0.3	9.0	2.4	50.0	43.9	6.1	0.1
	1987	100.0	83.5	10.0	0.2	6.5	1.9	46.9	40.7	6.2	0.1
	1988	100.0	82.5	11.5	0.2	6.0	1.8	45.3	39.4	5.8	0.0
	1989	100.0	82.8	12.5	0.2	4.7	1.7	45.7	40.0	5.6	0.0
	1990	100.0	84.3	12.5	0.2	3.2	1.3	48.9	42.7	6.1	0.1
6&8 Other manufactured goods	1980	100.0	77.0	17.7	0.8	5.3	1.0	54.9	45.8	9.0	1.1
	1987	100.0	70.3	25.0	1.1	4.7	1.0	52.6	43.5	9.0	0.8
	1988	100.0	69.4	26.1	1.1	4.5	1.1	51.1	42.1	8.9	0.7
	1989	100.0	68.6	27.7	1.4	3.7	0.9	50.0	41.4	8.6	0.7
	1990	100.0	70.4	26.7	1.6	2.9	0.7	52.4	43.5	8.9	0.8
Destination of exports of major commodity classes											
0-9 Total all commodities /5	1980	100.0	66.8	25.2	6.4	7.2	3.1	43.8	37.2	6.5	0.7
	1987	100.0	69.8	22.0	3.7	7.0	2.9	43.3	36.5	6.6	0.5
	1988	100.0	69.7	22.8	3.5	6.5	2.8	43.0	36.5	6.4	0.5
	1989	100.0	70.4	23.1	3.3	5.7	2.5	43.4	37.0	6.2	0.4
	1990	100.0	71.6	22.8	3.3	4.2	1.9	46.4	39.8	6.4	0.4
0&1 Food, live animals, beverages and tobacco	1980	100.0	61.4	26.7	8.0	10.9	7.1	43.7	39.8	3.8	0.2
	1987	100.0	71.7	20.4	5.6	7.4	5.1	49.7	45.5	4.0	0.3
	1988	100.0	70.6	21.7	5.4	7.3	5.0	48.8	44.8	3.8	0.3
	1989	100.0	68.8	23.0	5.4	7.7	5.4	47.2	43.3	3.6	0.2
	1990	100.0	71.6	21.7	4.8	6.1	4.3	50.1	46.1	3.7	0.2
2&4 Crude materials, oils and fats, (fuels excluded)	1980	100.0	69.1	22.0	3.0	7.4	2.4	44.5	40.0	4.4	0.5
	1987	100.0	67.0	24.9	2.6	6.9	2.5	42.7	37.9	4.7	0.3
	1988	100.0	67.7	25.7	2.6	6.1	2.1	43.0	38.3	4.7	0.3
	1989	100.0	69.0	24.7	2.6	5.8	2.4	43.4	38.6	4.7	0.3
	1990	100.0	70.5	24.4	2.6	4.5	1.8	45.5	40.8	4.7	0.2
3 Mineral fuels, lubricants and related material	1980	100.0	75.3	18.9	1.1	3.9	0.3	41.6	37.1	4.4	0.1
	1987	100.0	62.4	20.9	1.3	11.9	0.6	35.0	30.7	4.2	0.1
	1988	100.0	63.6	22.0	1.3	11.9	0.7	33.4	29.7	3.6	0.0
	1989	100.0	66.5	21.8	1.2	9.4	0.5	34.1	30.5	3.6	0.0
	1990	100.0	68.2	19.8	1.1	5.9	0.4	37.9	34.0	3.7	0.0
5 Chemicals	1980	100.0	63.8	28.5	6.4	7.4	3.5	49.5	41.7	7.7	1.0
	1987	100.0	67.7	25.8	4.3	6.1	2.9	50.8	43.3	7.4	0.7
	1988	100.0	67.3	26.5	3.9	5.4	2.6	50.0	42.6	7.3	0.7
	1989	100.0	68.4	25.9	3.7	5.0	2.6	50.7	43.4	7.2	0.7
	1990	100.0	69.8	25.4	3.8	4.0	2.1	52.6	45.2	7.3	0.6
7 Machinery and transport equipment	1980	100.0	59.5	31.3	10.1	8.7	4.1	38.1	31.5	6.5	1.5
	1987	100.0	70.9	22.1	4.0	6.4	3.3	40.5	33.6	6.8	0.7
	1988	100.0	70.7	22.6	3.7	6.1	3.2	41.2	34.5	6.6	0.7
	1989	100.0	71.3	23.4	3.4	4.9	2.4	42.1	35.6	6.4	0.6
	1990	100.0	71.9	23.9	3.6	3.7	1.9	44.4	37.8	6.4	0.5
6&8 Other manufactured goods	1980	100.0	69.1	24.1	8.0	6.4	3.3	50.7	40.9	9.7	0.6
	1987	100.0	74.5	19.9	3.7	5.2	2.9	48.2	39.6	8.6	0.4
	1988	100.0	74.3	20.6	3.6	4.8	2.7	47.8	39.5	8.1	0.4
	1989	100.0	73.6	21.9	3.5	4.1	2.3	47.4	39.3	7.9	0.3
	1990	100.0	75.1	21.4	3.3	3.1	1.7	50.5	42.1	8.2	0.3

En pourcentage

Economies développées /2,3,4				Developing economies — Economies en voie de développement /1,3,4						← En provenance ou vers	
Canada	U.S.A. E.-U.	Japan Japon	Australia New Zealand Australie Nouvelle-Zélande	Africa Afrique	America Amerique Total	LAIA+ ALADI+	Mid.East Moyen Orient	Other Autres	Oceania Océanie	Année	CTCI:Classes de marchandises ↓

Provenance des exportations de grandes catégories de marchandises

3.2	10.8	6.5	1.3	4.7	5.4	4.0	10.5	8.1	0.1	1980	0-9 Total tous produits /5
3.8	9.9	9.3	1.3	2.0	4.0	3.2	3.4	12.0	0.1	1987	
4.0	11.0	9.4	1.4	1.8	3.9	3.3	3.1	12.9	0.1	1988	
3.8	11.5	9.1	1.5	1.9	4.1	3.4	3.3	13.4	0.1	1989	
3.7	11.0	8.5	1.4	2.0	3.9	3.3	3.1	13.3	0.1	1990	
3.5	15.1	0.8	4.8	4.7	14.4	9.7	1.4	9.7	0.4	1980	0&1 Produits alimentaires, boissons et tabacs.
3.6	10.1	0.7	4.0	3.7	12.3	8.3	1.6	11.4	0.4	1987	
3.8	12.1	0.6	3.7	3.6	12.1	8.5	1.6	12.0	0.3	1988	
3.1	13.2	0.6	4.0	3.1	11.7	8.1	1.5	12.2	0.3	1989	
3.4	12.4	0.5	3.8	2.8	11.0	7.7	1.5	11.6	0.3	1990	
9.1	18.6	1.1	5.6	5.0	8.3	6.9	1.1	15.3	0.7	1980	2&4 Matières premières huiles & graisses (combust. exclu.)
10.7	15.1	1.2	6.5	3.2	7.1	6.2	1.1	14.8	0.7	1987	
10.3	16.0	1.2	5.8	3.1	7.8	7.0	1.1	15.7	0.8	1988	
10.4	16.0	1.2	5.5	3.3	8.3	7.3	1.2	15.5	0.8	1989	
10.1	16.1	1.2	4.8	3.3	9.0	7.8	1.3	13.9	0.6	1990	
2.0	1.7	0.1	0.5	14.9	9.5	6.6	41.5	6.9	0.0	1980	3 Combustibles minéraux et produits
3.4	2.8	0.3	1.8	10.2	9.6	7.8	23.4	9.2	0.0	1987	
3.9	3.1	0.2	1.9	10.2	8.5	6.9	25.4	9.2	0.0	1988	
3.6	3.4	0.3	1.8	10.4	9.5	7.6	25.7	9.9	0.0	1989	
3.7	3.5	0.4	2.0	11.7	10.3	8.7	22.6	10.7	0.0	1990	
2.5	14.7	4.7	0.5	0.6	2.0	1.5	1.0	3.1	0.0	1980	5 Produits chimiques
2.2	12.0	5.4	0.4	0.7	2.1	1.7	1.0	5.0	0.0	1987	
2.5	12.4	5.5	0.4	1.0	2.3	2.0	1.1	6.1	0.0	1988	
2.3	13.8	5.6	0.4	0.9	2.5	2.0	1.1	6.3	0.0	1989	
2.2	13.1	5.3	0.4	0.8	2.2	1.9	1.2	6.7	0.0	1990	
3.2	16.5	14.8	0.3	0.1	1.0	1.0	0.3	4.1	0.0	1980	7 Machines et matériels de transports
4.2	13.0	19.0	0.2	0.1	1.2	1.2	0.3	7.9	0.0	1987	
4.3	13.7	18.8	0.2	0.1	1.2	1.2	0.2	9.5	0.0	1988	
4.2	14.1	18.3	0.2	0.1	1.3	1.3	0.2	10.5	0.0	1989	
3.9	14.2	16.7	0.2	0.1	1.2	1.1	0.3	10.6	0.0	1990	
2.8	8.1	8.7	0.8	1.0	2.6	2.3	0.8	12.5	0.1	1980	6&8 Articles manufacturés divers
2.8	5.5	7.3	0.7	0.9	2.6	2.4	1.2	19.6	0.0	1987	
2.9	6.1	7.3	0.8	0.9	3.1	2.9	1.2	20.2	0.1	1988	
2.8	7.0	6.8	0.8	1.0	3.3	3.0	1.4	21.4	0.1	1989	
2.6	7.3	6.0	0.7	0.9	2.9	2.7	1.5	20.8	0.0	1990	

Destination des exportations de grandes catégories de marchandises

2.5	12.0	6.2	1.2	4.2	6.3	4.2	4.8	8.7	0.1	1980	0-9 Total tous produits /5
3.4	15.7	5.2	1.3	2.5	4.2	2.7	3.4	10.8	0.1	1987	
3.5	15.2	5.8	1.4	2.4	4.1	2.7	3.1	12.2	0.1	1988	
3.6	15.1	6.1	1.5	2.3	4.1	2.7	3.0	12.6	0.2	1989	
3.3	14.0	5.9	1.3	2.4	3.9	2.6	3.0	12.6	0.1	1990	
1.6	8.8	6.3	0.6	6.0	5.9	4.0	5.8	8.1	0.3	1980	0&1 Produits alimentaires, boissons et tabacs.
1.6	10.5	8.7	0.8	4.1	3.5	1.9	4.9	7.2	0.2	1987	
1.5	9.3	9.7	0.8	4.1	3.7	2.0	4.6	8.6	0.3	1988	
1.6	8.9	9.8	0.8	4.0	4.1	2.5	5.0	9.3	0.3	1989	
2.1	9.0	9.2	0.7	3.8	4.1	2.6	4.3	8.6	0.2	1990	
1.8	7.4	14.0	0.8	2.6	3.5	2.9	2.1	11.9	0.0	1980	2&4 Matières premières huiles & graisses (combust. exclu.)
1.7	8.7	12.7	0.7	2.3	3.9	3.0	2.1	14.5	0.0	1987	
1.6	8.3	13.5	0.8	2.3	3.8	3.1	2.0	15.8	0.0	1988	
1.8	8.1	14.4	0.8	2.3	3.6	2.9	2.2	14.6	0.0	1989	
2.1	8.1	13.5	0.7	2.5	3.5	2.8	2.2	14.7	0.0	1990	
1.4	17.7	13.6	0.9	1.6	6.7	2.6	2.0	7.5	0.1	1980	3 Combustibles minéraux et produits
1.1	13.5	12.0	0.5	1.8	5.7	2.6	2.3	9.7	0.2	1987	
1.3	15.3	12.9	0.6	1.8	5.6	2.8	2.4	10.7	0.2	1988	
1.5	17.1	12.9	0.7	1.7	5.2	2.5	2.5	10.9	0.2	1989	
1.5	15.1	12.9	0.7	2.3	3.2	1.1	2.5	10.7	0.2	1990	
2.0	5.6	3.8	1.5	4.7	8.0	6.2	4.2	10.3	0.1	1980	5 Produits chimiques
2.1	7.6	4.5	1.5	2.9	5.7	4.3	3.5	12.7	0.1	1987	
2.0	7.8	4.7	1.6	2.6	5.3	3.9	2.9	14.7	0.1	1988	
2.2	7.9	4.9	1.7	2.6	5.3	3.9	2.9	14.2	0.1	1989	
2.6	7.6	4.5	1.5	2.4	4.9	3.8	3.0	14.2	0.1	1990	
4.3	12.2	1.4	1.7	6.2	7.8	6.2	6.9	9.5	0.1	1980	7 Machines et matériels de transports
4.8	21.0	1.9	1.6	2.6	4.7	3.2	3.3	10.9	0.1	1987	
4.6	19.7	2.3	1.7	2.4	4.5	3.3	3.0	12.0	0.1	1988	
4.4	19.2	2.6	2.0	2.4	4.5	3.2	2.8	13.2	0.2	1989	
4.6	17.4	2.9	1.7	2.5	4.5	3.2	2.9	13.4	0.1	1990	
2.2	11.0	2.7	1.4	4.3	4.9	3.5	6.0	8.0	0.1	1980	6&8 Articles manufacturés divers
2.6	17.0	4.3	1.5	2.2	2.9	1.7	3.5	10.6	0.1	1987	
2.6	16.1	5.4	1.5	2.1	2.8	1.7	3.3	11.5	0.1	1988	
2.5	15.6	5.8	1.5	2.1	3.0	1.9	3.2	12.9	0.1	1989	
2.9	14.3	5.2	1.3	2.0	3.0	2.0	3.0	12.6	0.1	1990	

121
Structure of world exports by commodity classes and regions (cont.)
Structure des exportations mondiales par catégories de marchandises et par régions (suite)

In per cent

SITC Commodity classes	Year	World Monde /1,2,3	Developed economies Econ. développées /2,3,4	Developing economies Economies en voie de développement /1,3,4 Total	OPEC+ OPEP+	Eastern Europe and former USSR Europe de l'Est et l'ancienne URSS Total /2,3	Former USSR an URSS	Europe Total /2	EEC+ CEE+ /2	EFTA+ AELE+	Developed Economies South Africa Afrique du Sud
Commodity composition of the total exports of selected regions											
0-9 Total all commodities /5	1980	100.0	100.0	100.0	100.0	100.0	100.0	100.0	100.0	100.0	100.0
	1987	100.0	100.0	100.0	100.0	100.0	100.0	100.0	100.0	100.0	100.0
	1988	100.0	100.0	100.0	100.0	100.0	100.0	100.0	100.0	100.0	100.0
	1989	100.0	100.0	100.0	100.0	100.0	100.0	100.0	100.0	100.0	100.0
	1990	100.0	100.0	100.0	100.0	100.0	100.0	100.0	100.0	100.0	100.0
0&1 Food, live animals, beverages and tobacco	1980	10.0	10.2	10.6	1.2	6.1	1.7	9.6	10.5	4.0	18.8
	1987	9.1	8.5	12.4	3.2	4.9	1.4	9.4	10.3	3.9	13.4
	1988	9.0	8.5	12.1	3.5	4.6	1.6	9.3	10.2	3.7	12.2
	1989	8.8	8.4	11.1	3.1	5.1	1.5	9.1	10.0	3.6	12.0
	1990	8.7	8.4	10.5	3.1	5.4	1.5	9.0	9.8	3.7	13.0
2&4 Crude materials, oils and fats, (fuels excluded)	1980	6.9	6.8	7.3	1.7	6.8	9.2	4.2	3.5	8.6	23.8
	1987	5.7	5.3	7.1	2.5	5.7	7.4	3.7	3.3	6.1	19.1
	1988	5.9	5.4	7.7	3.3	5.8	7.7	3.9	3.5	6.1	20.7
	1989	5.9	5.3	7.5	3.2	6.3	7.7	3.8	3.4	6.2	22.0
	1990	5.2	4.7	6.5	2.5	6.4	7.4	3.4	3.1	5.4	20.2
3 Mineral fuels, lubricants and related material	1980	24.0	7.0	59.7	94.8	27.3	47.0	8.3	8.0	10.1	6.8
	1987	11.3	4.5	27.2	84.9	27.9	46.7	4.7	4.4	6.9	13.9
	1988	9.3	3.6	22.4	82.0	25.0	42.2	3.7	3.3	5.7	12.4
	1989	9.7	3.8	23.1	80.1	25.6	40.2	3.9	3.3	7.3	12.6
	1990	10.1	4.3	24.6	81.1	28.2	42.6	4.4	3.7	8.4	13.9
5 Chemicals	1980	7.0	9.7	1.8	0.5	5.0	2.2	11.1	11.4	9.7	6.7
	1987	8.7	10.5	3.7	1.5	5.8	2.7	12.4	12.6	11.3	6.6
	1988	9.0	10.7	4.4	1.7	6.2	3.2	12.8	13.0	11.7	6.7
	1989	8.7	10.4	4.3	1.7	6.0	3.3	12.2	12.4	11.3	6.6
	1990	8.8	10.4	4.4	2.2	6.1	3.6	11.9	12.0	11.2	6.5
7 Machinery and transport equipment	1980	25.6	34.7	5.2	0.5	29.7	16.0	32.0	32.7	28.1	3.6
	1987	34.3	40.9	15.7	1.3	27.7	15.3	35.9	36.4	32.8	3.7
	1988	35.1	41.2	18.0	1.4	28.8	16.5	36.4	37.0	32.6	3.2
	1989	35.0	41.1	18.8	1.4	25.7	16.4	36.6	37.5	31.8	3.3
	1990	35.7	41.8	19.6	1.6	22.6	15.2	37.5	38.3	32.7	3.6
6&8 Other manufactured goods	1980	24.0	29.4	14.5	1.2	16.4	6.5	32.9	32.0	38.8	38.8
	1987	27.5	27.6	31.4	6.0	16.1	6.3	32.2	31.1	38.4	40.2
	1988	28.0	27.6	32.8	7.3	17.3	8.1	32.7	31.5	39.9	42.3
	1989	28.5	27.7	34.0	8.7	16.6	7.1	32.7	31.6	39.5	41.0
	1990	28.6	28.0	33.5	9.2	16.2	7.0	32.2	31.3	38.2	40.3
Commodity composition of the world exports to selected regions											
0-9 Total all commodities /5	1980	100.0	100.0	100.0	100.0	100.0	100.0	100.0	100.0	100.0	100.0
	1987	100.0	100.0	100.0	100.0	100.0	100.0	100.0	100.0	100.0	100.0
	1988	100.0	100.0	100.0	100.0	100.0	100.0	100.0	100.0	100.0	100.0
	1989	100.0	100.0	100.0	100.0	100.0	100.0	100.0	100.0	100.0	100.0
	1990	100.0	100.0	100.0	100.0	100.0	100.0	100.0	100.0	100.0	100.0
0&1 Food, live animals, beverages and tobacco	1980	10.0	9.2	10.6	12.6	15.1	23.0	10.0	10.7	5.9	3.2
	1987	9.1	9.3	8.4	13.9	9.5	16.0	10.4	11.3	5.5	5.2
	1988	9.0	9.1	8.6	13.8	10.0	16.2	10.2	11.0	5.4	4.9
	1989	8.8	8.6	8.8	14.3	11.8	19.3	9.6	10.3	5.1	5.0
	1990	8.7	8.7	8.3	12.7	12.5	19.1	9.4	10.1	5.0	5.1
2&4 Crude materials, oils and fats, (fuels excluded)	1980	6.9	7.2	6.0	3.3	7.1	5.4	7.0	7.4	4.7	4.6
	1987	5.7	5.5	6.5	4.1	5.6	4.9	5.7	6.0	4.1	3.4
	1988	5.9	5.8	6.7	4.4	5.5	4.5	5.9	6.2	4.4	3.2
	1989	5.9	5.7	6.3	4.6	6.0	5.6	5.9	6.1	4.5	3.6
	1990	5.2	5.1	5.5	4.1	5.5	4.9	5.1	5.3	3.8	3.4
3 Mineral fuels, lubricants and related material	1980	24.0	27.1	18.0	4.2	13.1	2.3	22.8	24.0	16.2	2.9
	1987	11.3	10.1	10.7	4.0	19.3	2.4	9.2	9.5	7.2	1.3
	1988	9.3	8.5	9.0	3.5	16.9	2.2	7.2	7.6	5.2	0.5
	1989	9.7	9.1	9.1	3.6	16.0	2.1	7.6	8.0	5.5	0.5
	1990	10.1	9.6	8.8	3.4	14.3	2.2	8.3	8.7	5.9	0.6
5 Chemicals	1980	7.0	6.7	8.0	7.1	7.2	7.9	8.0	7.9	8.4	10.4
	1987	8.7	8.4	10.1	10.2	7.5	8.6	10.2	10.3	9.7	14.0
	1988	9.0	8.7	10.5	10.0	7.5	8.2	10.5	10.5	10.3	13.8
	1989	8.7	8.5	9.8	9.9	7.6	9.2	10.2	10.2	10.1	13.8
	1990	8.8	8.6	9.8	10.1	8.3	9.6	10.0	10.0	10.0	14.2
7 Machinery and transport equipment	1980	25.6	22.9	31.9	40.6	30.8	34.0	22.3	21.7	25.7	55.1
	1987	34.3	34.9	34.4	37.3	31.4	39.1	32.2	31.6	35.5	50.8
	1988	35.1	35.6	34.8	37.4	32.8	39.7	33.6	33.1	36.3	53.1
	1989	35.0	35.4	35.4	35.9	30.1	34.2	34.0	33.7	35.7	52.2
	1990	35.7	35.9	37.4	39.1	31.5	35.7	34.2	34.0	35.7	50.4
6&8 Other manufactured goods	1980	24.0	24.9	23.0	30.3	21.5	25.8	27.8	26.4	35.9	22.0
	1987	27.5	29.4	24.8	28.1	20.2	27.5	30.7	29.8	35.6	23.2
	1988	28.0	29.9	25.3	29.1	20.4	27.2	31.1	31.0	35.7	22.8
	1989	28.5	29.8	27.0	29.8	20.5	26.6	31.1	30.3	36.0	22.7
	1990	28.6	30.0	26.9	28.4	21.1	25.8	31.2	30.3	36.5	24.3

En pourcentage

Economies développées /2,3,4				Developing economies / Economies en voie de développement /1,3,4						← En provenance ou vers	
Canada	U.S.A. E.-U.	Japan Japon	Australia New Zealand Australie Nouvelle-Zélande	Africa Afrique	America Amerique Total	LAIA+ ALADI+	Asia Asie Mid.East Moyen Orient	Other Autres	Oceania Océanie	Année	CTCI:Classes de marchandises ↓

Composition par marchandises des exportations totales des régions selectionnées

100.0	100.0	100.0	100.0	100.0	100.0	100.0	100.0	100.0	100.0	1980	0-9 Total tous produits /5
100.0	100.0	100.0	100.0	100.0	100.0	100.0	100.0	100.0	100.0	1987	
100.0	100.0	100.0	100.0	100.0	100.0	100.0	100.0	100.0	100.0	1988	
100.0	100.0	100.0	100.0	100.0	100.0	100.0	100.0	100.0	100.0	1989	
100.0	100.0	100.0	100.0	100.0	100.0	100.0	100.0	100.0	100.0	1990	
10.8	14.0	1.2	36.3	9.9	26.3	24.5	1.3	12.0	35.6	1980	0&1 Produits alimentaires, boissons et tabacs.
8.5	9.2	0.7	27.7	16.7	28.3	23.4	4.3	8.7	36.9	1987	
8.5	9.9	0.6	23.2	17.8	27.8	23.4	4.5	8.4	24.7	1988	
7.2	10.1	0.6	24.0	14.5	25.3	21.0	4.0	8.0	28.3	1989	
7.8	9.7	0.6	23.4	12.3	24.2	20.2	4.3	7.5	31.9	1990	
19.4	11.9	1.2	28.9	7.3	10.6	12.0	0.7	13.1	44.6	1980	2&4 Matières premières huiles & graisses (combust. exclu.)
16.1	8.8	0.7	28.8	9.2	10.2	10.9	1.8	7.1	41.9	1987	
15.2	8.7	0.8	24.1	9.9	11.9	12.6	2.2	7.2	47.3	1988	
15.8	8.1	0.7	22.0	10.3	11.9	12.7	2.2	6.8	43.5	1989	
13.9	7.5	0.7	17.8	8.7	11.8	12.1	2.1	5.4	36.1	1990	
14.5	3.7	0.4	9.1	75.6	42.4	39.9	94.6	20.5	1.4	1980	3 Combustibles minéraux et produits
10.0	3.2	0.4	15.9	57.6	27.3	27.2	77.6	8.8	0.3	1987	
9.0	2.7	0.2	12.6	51.9	20.4	19.7	76.4	6.6	0.3	1988	
9.2	2.8	0.3	12.1	54.0	22.5	21.9	74.8	7.1	0.4	1989	
10.0	3.3	0.4	14.7	60.2	26.5	26.7	73.3	8.2	2.6	1990	
5.4	9.6	5.1	2.5	0.8	2.6	2.7	0.7	2.7	0.1	1980	5 Produits chimiques
5.0	10.4	5.0	2.5	3.2	4.5	4.4	2.6	3.6	0.5	1987	
5.6	10.1	5.2	2.6	4.8	5.4	5.4	3.1	4.2	0.4	1988	
5.3	10.4	5.3	2.7	4.0	5.3	5.1	2.9	4.1	0.4	1989	
5.2	10.4	5.5	2.8	3.7	5.0	4.9	3.4	4.4	0.6	1990	
25.7	39.0	58.4	5.0	0.4	4.8	6.2	0.8	12.9	0.4	1980	7 Machines et matériels de transports
38.1	44.9	70.5	4.7	1.1	10.7	12.8	3.4	22.8	1.4	1987	
38.0	43.7	70.1	4.6	1.5	11.2	13.1	2.8	25.8	2.8	1988	
38.4	42.6	70.4	4.8	1.4	11.3	13.2	2.4	27.4	2.8	1989	
37.2	46.1	70.7	5.6	1.6	10.7	12.1	2.9	28.6	2.9	1990	
21.1	17.9	32.4	13.9	5.2	11.6	14.2	1.8	37.0	15.6	1980	6&8 Articles manufacturés divers
20.3	15.2	21.7	14.7	11.8	18.1	20.5	9.9	45.0	8.8	1987	
20.2	15.6	21.7	14.8	13.3	22.3	24.9	10.7	43.8	14.0	1988	
20.5	17.1	21.2	15.0	15.1	22.9	25.3	11.7	45.4	18.8	1989	
19.7	18.9	20.5	14.3	12.7	21.0	23.0	13.5	44.7	16.0	1990	

Composition par marchandises des exportations mondiales vers des régions selectionnées

100.0	100.0	100.0	100.0	100.0	100.0	100.0	100.0	100.0	100.0	1980	0-9 Total tous produits /5
100.0	100.0	100.0	100.0	100.0	100.0	100.0	100.0	100.0	100.0	1987	
100.0	100.0	100.0	100.0	100.0	100.0	100.0	100.0	100.0	100.0	1988	
100.0	100.0	100.0	100.0	100.0	100.0	100.0	100.0	100.0	100.0	1989	
100.0	100.0	100.0	100.0	100.0	100.0	100.0	100.0	100.0	100.0	1990	
6.3	7.3	10.1	4.7	14.3	9.4	9.7	12.0	9.3	20.0	1980	0&1 Produits alimentaires, boissons et tabacs.
4.2	6.1	15.1	5.3	14.8	7.5	6.5	13.2	6.0	16.5	1987	
3.9	5.5	15.1	5.0	15.3	8.1	6.8	13.4	6.3	16.4	1988	
3.9	5.2	14.2	4.7	15.0	8.8	8.0	14.4	6.5	15.7	1989	
5.6	5.5	13.6	4.8	13.7	9.2	8.7	12.3	5.9	14.5	1990	
4.9	4.2	15.5	4.4	4.3	3.9	4.8	3.0	9.5	2.1	1980	2&4 Matières premières huiles & graisses (combust. exclu.)
2.9	3.2	14.0	3.2	5.3	5.3	6.4	3.5	7.7	1.7	1987	
2.7	3.2	13.9	3.6	5.6	5.5	6.8	3.8	7.7	1.7	1988	
2.9	3.2	13.9	3.1	5.8	5.1	6.2	4.3	6.8	1.6	1989	
3.3	3.0	11.9	2.9	5.3	4.6	5.6	3.8	6.1	1.6	1990	
13.0	35.3	52.5	17.0	9.3	25.7	15.2	10.1	20.8	22.7	1980	3 Combustibles minéraux et produits
3.6	9.7	25.9	4.7	7.9	15.2	11.1	7.8	10.2	13.4	1987	
3.5	9.4	20.8	3.8	6.8	12.9	9.7	7.3	8.2	12.0	1988	
4.2	11.0	20.4	4.7	7.1	12.3	9.0	8.0	8.3	11.9	1989	
4.7	10.9	22.2	5.1	9.5	8.3	4.1	8.5	8.6	14.7	1990	
5.5	3.3	4.3	8.8	7.9	8.9	10.4	6.2	8.3	4.7	1980	5 Produits chimiques
5.3	4.2	7.5	10.2	9.9	11.7	14.0	8.9	10.2	5.7	1987	
5.2	4.6	7.4	10.5	9.8	11.7	13.2	8.4	10.9	5.5	1988	
5.4	4.6	7.0	10.0	9.7	11.2	12.7	8.5	9.8	5.2	1989	
7.0	4.8	6.7	10.2	8.9	11.1	12.8	8.8	9.9	5.0	1990	
44.0	26.0	5.9	35.6	37.4	31.8	38.1	36.5	27.9	25.3	1980	7 Machines et matériels de transports
48.4	45.7	12.6	43.2	34.6	38.0	42.0	33.8	34.6	31.6	1987	
46.8	45.7	13.9	43.8	34.4	39.3	42.9	34.1	34.7	33.0	1988	
43.5	44.6	15.0	45.8	35.8	38.1	40.9	32.6	36.5	34.9	1989	
50.6	44.3	17.8	45.4	37.7	40.8	44.0	34.8	38.0	33.2	1990	
20.7	22.0	10.5	27.8	24.7	18.5	20.1	29.9	22.2	22.0	1980	6&8 Articles manufacturés divers
20.9	29.7	22.7	31.0	24.2	18.8	17.5	28.6	26.9	25.8	1987	
20.7	29.9	26.0	30.9	24.3	19.3	18.3	30.0	26.5	25.9	1988	
19.7	29.4	27.1	29.2	25.2	21.0	20.0	30.3	29.1	25.5	1989	
25.3	29.1	25.5	28.5	23.5	22.3	21.6	28.9	28.7	25.6	1990	

121
Structure of world exports by commodity classes and regions (cont.)
Structure des exportations mondiales par catégories de marchandises et par régions (suite)

Source:
International Trade Statistics database of the Statistical Division of the United Nations Secretariat.

+/ For member states of this grouping, see Annex I:- Other groupings.

1/ Excluding the inter-trade between China, the Democratic People's Republic of Korea, Mongolia and Viet Nam in 1980. The figures shown for other years are based on the import statistics of China.

2/ Excluding the trade conducted in accordance with the supplementary protocol to the treaty on the basis of relations between the Federal Republic of Germany and the former German Democratic Republic.

3/ Exports of the former USSR for which country of destination is not available are included in the totals for the World, the 'Developed economies', the 'Developing economies' and 'Eastern Europe and the former USSR' but are excluded from the regional components of these groupings.

4/ This classification is intended for statistical convenience and does not, necessarily, express a judgement about the stage reached by a particular country in the development process.

5/ Section 9 of the SITC, which comprises commodities and transactions not classified elsewhere, is included in the total trade but is not shown separately in this table.

Source:
Base de données pour les statistiques du commerce international de la Division de statistique du Secrétariat de l'ONU.

+/ Les etats membres de ce groupement, voir annexe I:- Autres groupements.

1/ Non compris le commerce entre la Chine, la République populaire démocratique de Corée, la Mongolie et le Viet Nam en 1980. Les chiffres indiqués pour les autres années sont bases sur les statistiques d'importations de la Chine.

2/ Non compris le commerce effectué en accord avec le protocole additionnel au traité définissant la base des relations entre la République fédérale d'Allemagne et l'ancienne République démocratique allemande.

3/ Les exportations en provenance de l'ancienne URSS dont les pays de destination ne sont pas disponibles sont comprises dans les totaux du Monde, des 'Economies développées', des 'Economies en voie de développement' et de 'Europe de l'Est et l'ancienne URSS', mais ils ne sont pas comprises dans chaque partie composant ces régions.

4/ Cette classification est utilisée pour plus de commodité dans la présentation des statistiques et n'implique pas nécessairement un jugement quant au stade de développement auquel est parvenu un pays donné.

5/ Section 9 de la CTCI, qui représente les articles et transactions non classes ailleurs est comprise dans le commerce total mais n'est pas présentée séparément dans ce tableau.

Technical notes, tables 116-121

Tables 116-121: Current data (annual, monthly and/or quarterly) for most of the series are published regularly by the Statistical Division in the United Nations *Monthly Bulletin of Statistics*. More detailed descriptions of the tables and notes on methodology appear in the United Nations 1977 *Supplement to the Statistical Yearbook and Monthly Bulletin of Statistics, International Trade Statistics: Concepts and Definitions* [47, 49] and the *International Trade Statistics Yearbook*.[22] More detailed data including series for individual countries showing the value in national currencies for imports and exports and notes on these series are to be found in the *International Trade Statistics Yearbook* and in the *Monthly Bulletin of Statistics*.

Data are obtained from national published sources, from data supplied by the Governments for use in the following United Nations publications: *Commodity Trade Statistics* [17], *Monthly Bulletin of Statistics* and *Statistical Yearbook*; and from publications of other United Nations agencies.

Territory

The statistics reported by a country refer to the customs area of the country. In most cases, this coincides with the geographical area of the country.

System of trade

Two systems of recording trade are in common use, differing mainly in the way warehoused and re-exported goods are recorded:

(a) Special trade (S): special imports are the combined total of imports for direct domestic consumption (including transformation and repair) and withdrawals from bonded warehouses or free zones for domestic consumption. Special exports comprise exports of national merchandise, namely, goods wholly or partly produced or manufactured in the country, together with exports of nationalized goods. (Nationalized goods are goods which, having been included in special imports, are then exported without transformation);

(b) General trade (G): general imports are the combined total of imports for direct domestic consumption and imports into bonded warehouses or free zones. General exports are the combined total of national exports and re-exports. Re-exports, in the general trade system, consist of the outward movement of nationalized goods plus goods which, after importation, move outward from bonded warehouses or free zones without having been transformed;

(c) Semi-special trade (SI): semi-special imports are general imports less all re-exports; semi-special exports are exports of domestic produce.

Direct transit trade, i.e., goods merely being trans-shipped or moving through the country for the purpose of transport only, is excluded from the statistics of both special and general trade.

Notes techniques, tableaux 116-121

Tableaux 116-121 : La Division de statistique des Nations Unies publie régulièrement des données courantes (annuelles, mensuelles et/ou trimestrielles) pour la plupart des séries de ces tableaux dans le *Bulletin mensuel de statistique* des Nations Unies. Des descriptions plus détaillées des tableaux et des notes méthodologiques figurent dans *1977 Supplément à l'Annuaire statistique et au Bulletin mensuel de statistique* des Nations Unies, dans la publication *Statistiques du Commerce international, Concepts et définitions* [47, 49] et dans l'*Annuaire statistique du Commerce international* [22]. Des données plus détaillées, comprenant des séries indiquant la valeur en monnaie nationale des importations et des exportations des divers pays et les notes accompagnant ces séries figurent dans l'*Annuaire statistique du Commerce international* et dans le *Bulletin mensuel de statistique*.

Les données proviennent de publications nationales et des informations fournies par les gouvernements pour les publications suivantes des Nations Unies : "Commodity Trade Statistics" [17], Bulletin mensuel de statistique et Annuaire statistique", ainsi que de publications d'autres institutions des Nations Unies.

Territoire

Les statistiques fournies par pays se rapportent au territoire douanier de ce pays. Le plus souvent, ce territoire coïncide avec l'étendue géographique du pays.

Système de commerce

Deux systèmes d'enregistrement du commerce sont couramment utilisés, qui ne diffèrent que par la façon dont sont enregistrées les marchandises entreposées et les marchandises réexportées :

(a) Commerce spécial (S) : les importations spéciales représentent le total combiné des importations destinées directement à la consommation intérieure (transformations et réparations comprises) et les marchandises retirées des entrepôts douaniers ou des zones franches pour la consommation intérieure. Les exportations spéciales comprennent les exportations de marchandises nationales, c'est-à-dire des biens produits ou fabriqués en totalité ou en partie dans le pays, ainsi que les exportations de biens nationalisés. (Les biens nationalisés sont des biens qui, ayant été inclus dans les importations spéciales, sont ensuite réexportés tels quels.)

(b) Commerce général (G) : les importations générales sont le total combiné des importations destinées directement à la consommation intérieure et des importations placées en entrepôt douanier ou destinées aux zones franches. Les exportations générales sont le total combiné des exportations de biens nationaux et des réexportations. Ces dernières, dans le système du commerce général, comprennent les exportations de biens nationalisés et de biens qui, après avoir été importés, sortent des entrepôts de douane ou des zones franches sans avoir été transformés.

Valuation

Goods are, in general, valued according to the transaction value. In the case of imports, the transaction value is the value at which the goods were purchased by the importer plus the cost of transportation and insurance to the frontier of the importing country (a c.i.f. valuation). In the case of exports, the transaction value is the value at which the goods were sold by the exporter, including the cost of transportation and insurance, to bring the goods onto the transporting vehicle at the frontier of the exporting country (a f.o.b. valuation).

Currency conversion

Conversion of values from national currencies into United States dollars is done by means of external trade conversion factors which are generally weighted averages of exchange rates, the weight being the corresponding monthly or quarterly value of imports or exports.

Coverage

The statistics relate to merchandise trade. Merchandise trade is defined to include, as far as possible, all goods which add to or subtract from the material resources of a country as a result of their movement into or out of the country. Thus, ordinary commercial transactions, government trade (including foreign aid, war reparations and trade in military goods), postal trade and all kinds of silver (except silver coin after its issue), are included in the statistics. Since their movement affects monetary rather than material resources, monetary gold, together with currency and titles of ownership after their issue into circulation, are excluded.

Commodity classification

The commodity classification of trade is in accordance with the United Nations *Standard International Trade Classification* (SITC).[52]

World and regional totals

The regional, economic and world totals have been adjusted: (a) to include estimates for countries or areas for which full data are not available; (b) to include insurance and freight for imports valued f.o.b.; (c) to include countries or areas not listed separately; (d) to approximate special trade; (e) to approximate calendar years; and (f) where possible, to eliminate incomparabilities owing to geographical changes, by adjusting the figures for periods before the change to be comparable to those for periods after the change.

Quantum and unit value index numbers

These index numbers show the changes in the volume of imports or exports (quantum index) and the average price of imports or exports (unit value index).

(c) Commerce semi-spécial (SI) : les importations semi-spéciales sont les importations générales moins l'ensemble des réexportations; les exportations semi-spéciales sont les exportations de produits du pays.

Le transit direct, c'est-à-dire les marchandises uniquement transbordées ou transportées à travers le pays, est exclu aussi bien du commerce général que du commerce spécial.

Evaluation

En général, les marchandises sont évaluées à la valeur de la transaction. Dans le cas des importations, cette valeur est celle à laquelle les marchandises ont été achetées par l'importateur plus le coût de leur transport et de leur assurance jusqu'à la frontière du pays importateur (valeur c.a.f.). Dans le cas des exportations, la valeur de la transaction est celle à laquelle les marchandises ont été vendues par l'exportateur, y compris le coût de transport et d'assurance des marchandises jusqu'à leur chargement sur le véhicule de transport à la frontière du pays exportateur (valeur f.o.b.).

Conversion des monnaies

Le conversion en dollars des Etats-Unis de valeurs exprimées en monnaie nationale se fait par application de coefficients de conversion du commerce extérieur, qui sont généralement les moyennes pondérées des taux de change, le poids étant la valeur mensuelle ou trimestrielle correspondante des importations ou des exportations.

Couverture

Les statistiques se rapportent au commerce des marchandises. Le commerce des marchandises se définit comme comprenant, dans toute la mesure du possible, toutes les marchandises qui ajoutent ou retranchent aux ressources matérielles d'un pays par suite de leur importation ou de de leur exportation par ce pays. Ainsi, les transactions commerciales ordinaires, le commerce pour le compte de l'Etat (y compris l'aide extérieure, les réparations pour dommages de guerre et le commerce des fournitures militaires), le commerce par voie postale et les transactions de toutes sortes sur l'argent (à l'exception des transactions sur les pièces d'argent après leur émission) sont inclus dans ces statistiques. La monnaie or ainsi que la monnaie et les titres de propriété après leur mise en circulation sont exclus, car leurs mouvements influent sur les ressources monétaires plutôt que sur les ressources matérielles.

Classification par marchandise

La classification par marchandise du commerce extérieur est celle adoptée dans la *Classification type por le commerce international* des Nations Unies (CTCI) [52].

Description of tables

Table 116: Trade conversion factors are weighted averages of exchange rates. The weights being the corresponding monthly or quarterly values of imports or exports. The exchange rates are the rates provided by the country concerned or compiled by the International Monetary Fund. For countries that provide trade statistics in terms of both national currencies and US dollars. These factors are obtained by dividing the values in US dollars by the values in national currencies. The United Nations Statistical Division uses the conversion factors show in this table to convert all the trade data expressed in national currencies into United States dollars.

The following countries which reported their external trade data in US dollars are not shown in table 116: Afghanistan, American Samoa, Argentina, Bermuda, Bolivia, Brazil, Chile, Colombia, Costa Rica, Dominican Rep., Ecuador, Guatemala, Indonesia, Israel, Korea, Republic of, Lao People's Democratic Republic, Maldives, Mexico, Nicaragua, Peru, Philippines, Poland, United States, Uruguay, Yugoslavia.

Table 118: The purpose of this table is to provide data on the network of flows of broad groups of commodities within and between important economic and geographic areas of the world. The regional analysis in this table is in accordance with that of table 117.

Export data in this table are largely comparable to data shown in table 117 except that table 117 contains revised data for total exports which may not be available at the commodity/destination level needed for this table. Also, the regional totals shown in table 117 have been adjusted to exclude the re-exports of countries comprising each region. This adjustment is not made in this table since re-exports are often not available by commodity and by destination.

The commodity classification is in accordance with the United Nations *Standard International Trade Classification* (SITC), Revision 2 for 1980 through 1987 except for countries which report trade data only in terms of the SITC, Revised. Beginning in 1988 the commodity classification is in accordance with SITC, Revision 3 where data are available from countries. [52]

The data approximate total exports of all countries and areas of the world with the exception of the inter-trade of the centrally planned economies of Asia in 1980 and trade conducted in accordance with the supplementary protocol to the treaty on the basis of relations between the Federal Republic of Germany and the former German Democratic Republic. They are based on official export figures converted, where necessary, to US dollars according to conversion factors published in table 116 in this volume. Where official figures are not available estimates based on the imports reported by partner countries and on other subsidiary data are used. Some official national data have been adjusted (a) to approximate the commodity groupings of SITC; and (b) to approximate calendar years.

Totaux mondiaux et régionaux

Les totaux économiques régionaux et mondiaux ont été ajustés de manière : (a) à inclure les estimations pour les pays ou régions pour lesquels on ne disposait pas de données complètes; (b) à inclure l'assurance et le fret dans la valeur f.o.b. des importations; (c) à inclure les pays ou régions non indiqués séparément; (d) à donner une approximation du commerce spécial; (e) à les ramener à des années civiles; et (f) à éliminer, dans la mesure du possible, les données non comparables par suite de changements géographiques, en ajustant les chiffres correspondant aux périodes avant le changement de manière à les rendre comparables à ceux des périodes après le changement.

Indices de quantum et de valeur unitaire

Ces indices indiquent les variations du volume des importations ou des exportations (indice de quantum) et du prix moyen des importations ou des exportations (indice de valeur unitaire).

Description des tableaux

Tableau 116 : Les facteurs de conversion pour le commerce extérieur sont des moyennes pondérées de taux de change, les pondérations étant les valeurs mensuelles ou trimestrielles correspondantes des importations ou des exportations. Les taux de change sont ceux fournis par le pays en question ou compilés par le Fonds monétaire international. Pour les pays qui fournissent des statistiques commerciales à la fois dans leur monnaie nationale et en dollars des Etats-Unis, ces facteurs s'obtiennent en divisant les valeurs en dollars des Etats-Unis par les valeurs en monnaie nationale. La Division de statistique des Nations Unies utilise les facteurs de conversion indiqués dans ce tableau pour convertir toutes les données sur le commerce exprimées en monnaie nationale en dollars des Etats-Unis.

Les pays suivants, qui ont fourni des données sur leur commerce extérieur en dollars des Etats-Unis, ne figurent pas au tableau 116 : l'Afghanistan, le Samoa américain, l'Argentine, les Bermudes, la Bolivie, le Brésil, le Chili, la Colombie, le Costa Rica, la Rép. dominicaine, l'Equateur, le Guatemala, l'Indonésie, Israël, la République de Corée, la République démocratique populaire lao, les Maldives, le Mexique, le Nicaragua, le Pérou, les Philippines, la Pologne, les Etats-Unis, l'Uruguay et la Yougoslavie.

Tableau 118 : Ce tableau a pour but de fournir des données sur l'ensemble des flux de grandes catégories de marchandises à l'intérieur des grandes régions économiques et géographiques du monde et entre ces régions. L'analyse régionale de ce tableau est conforme à celle du tableau 117.

Les données de ce tableau sur les exportations sont en grande partie comparables aux données fournies au tableau 117; toutefois, celui-ci présente des données révisées pour les totaux des exportations qui ne sont pas nécessairement disponibles au niveau des marchandises/destinations présentées au tableau 118. En outre, les totaux régionaux indiqués au tableau 117 ont été ajustés de manière à exclure les réexportations effectuées par

The data include special category (confidential) exports, ships' stores and bunders and exports of minor importance, the destination of which cannot be determined. These data are included in the world totals for each commodity group and in total exports, but are excluded from all regions of destination. Approximately 1 ½ percent of total exports are not distributed. All data are generally in accordance with the special trade system.

Table 119: These index numbers show the changes in the volume (quantum index) and the average price (unit value index) of total imports and exports. The terms of trade figures are calculated by dividing export unit value indices by the corresponding import unit value indices. The product of the net terms of trade and the quantum index of exports is called the index of the purchasing power of exports. The footnotes to countries appearing in table 117 also apply to the index numbers in this table.

Table 120: Manufactured goods are here defined to comprise sections 5 through 8 of the Standard International Trade Classification. These sections are: chemicals and related products, manufactured goods classified chiefly by material, machinery and transport equipment and miscellaneous manufactured articles. The economic and geographic groupings in this table are in accordance with those of table 117, although table 117 includes more detailed geographical sub-groups which make up the groupings "other developed market economies" and "developing market economies" of this table. In 1980 the exports of manufactured goods by countries included in the indices for "total market economies" accounted for approximately 91 per cent of world exports of manufactured goods.

The unit value indices are obtained from national sources, except those of a few countries which the United Nations Statistical Division compiles using their quantity and value figures. For countries that do not compiles indices for manufactured goods exports conforming to the above definition, sub-indices are aggregated to approximate an index of SITC sections 5-8. Unit value indices obtained from national indices are rebased, where necessary, so that 1980=100. Indices in national currency are converted into US dollars using conversion factors obtained by dividing the weighted average exchange rate of a given currency in the current period by the weighted average exchange rate in the base period. All aggregate unit value indices are current period weighted.

The indices in SDRs are calculated by multiplying the equivalent aggregate indices in United States dollars by conversion factors obtained by dividing the SDR/$US exchange rate in the current period by the rate in the base period.

The quantum indices are derived from the value data and the unit value indices. All aggregate quantum indices are base period weighted.

Table 121: The regional analysis in this table is in accordance with table 117. The figures in this table are derived from the data in table 118.

les pays composant chaque région. Cet ajustement n'apparaît pas sur ce tableau, car il est fréquent que l'on ne dispose pas des chiffres des réexportations par marchandise et par destination.

La classification des marchandises est conforme à la *Classification type pour le commerce international* (CTCI), Révision 2 des Nations Unies pour les années 1980 à 1987, sauf pour les pays qui ne fournissent de statistiques commerciales que sur la base de la CTCI révisée. A partir de 1988, la classification des marchandises est conforme à la CTCI, Révision 3, pour les pays qui ont fourni des données [52].

Les données fournissent une approximation des exportations totales de tous les pays et régions du monde à l'exception du commerce intrarégional des économies à planification centrale d'Asie en 1980 et des échanges commerciaux effectués selon le protocole supplémentaire du traité sur la base des relations entre la République fédérale d'Allemagne et l'ancienne République démocratique allemande. Elles sont fondées sur les chiffres officiels des exportations convertis, le cas échéant, en dollars des Etats-Unis, selon les facteurs de conversion publiés au tableau 116 du présent volume. En l'absence de chiffres officiels, on a utilisé des estimations fondées sur les importations notifiées par les pays partenaires et sur d'autres données subsidiaires. Certaines données nationales officielles ont été ajustées : a) sur la base des groupements de marchandises de la CTCI; et b) sur la base des années civiles.

Ces données englobent les catégories spéciales (confidentielles) d'exportation, les marchandises à bord des navires et les exportations d'importance mineure, dont la destination ne peut être déterminée. Ces données sont comprises dans les totaux mondiaux pour chaque classe de marchandises et dans les exportations totales, mais sont exclues de ceux des régions de destination. Environ 1,5 % des exportations totales ne sont pas distribuées. Dans l'ensemble, les données sont conformes au système du commerce spécial.

Tableau 119 : Ces indices indiquent les variations du volume (indice de quantum) et du prix moyen (indice de valeur unitaire) des importations et des exportations totales. Les chiffres relatifs aux termes de l'échange se calculent en divisant les indices de valeur unitaire des exportations par les indices correspondants de valeur unitaire des importations. Le produit de la valeur nette des termes de l'échange et de l'indice du quantum des exportations est appelé indice du pouvoir d'achat des exportations. Les notes figurant au bas du tableau 117 concernant certains pays s'appliquent également aux indices du présent tableau.

Tableau 120 : Les produits manufacturés se définissent comme correspondant aux sections 5 à 8 de la Classification type pour le commerce international. Ces sections sont : produits chimiques et produits connexes, biens manufacturés classés principalement par matière première, machines et équipements de transport et articles divers manufacturés. Les groupements économiques et géographiques de ce tableau sont conformes à ceux du tableau 117; toutefois, le tableau 117 comprend des subdivisions géographiques plus détaillées qui composent les groupements "autres pays

The commodity classification is in accordance with the United Nations *Standard International Trade Classification* (SITC), Revision 2, for 1980 through 1987 except for countries which report trade data only in terms of the SITC, Revised. Beginning 1988 the commodity classification is in accordance with the SITC, Revision 3, where data are available from countries. [52]

The data approximate total exports of all countries and areas of the world with the exception of the inter-trade of the centrally planned economies of Asia in 1980 and trade conducted in accordance with the supplementary protocol to the treaty on the basis of relations between the Federal Republic of Germany and the former German Democratic Republic. They are based on official export figures converted, where necessary, to United States dollars according to conversion factors published in table 116. Where official figures are not available estimates based on imports reported by partner countries and on other subsidiary data are used. Some official national data have been adjusted (a) to approximate the commodity groupings of SITC and (b) to approximate calendar years.

The data include special category (confidential) exports, ships' stores and bunkers and exports of minor importance, the destination of which cannot be determined. These data are included in the world totals for each commodity group and in total exports, but are excluded from all regions of destination. For "all commodities", approximately 1 ½ per cent of total exports are not distributed. All data are generally in accordance with the special trade system.

développés à économie de marché" et "pays en développement à économie de marché" du présent tableau. En 1980, les exportations de produits manufacturés des pays inclus dans les indices correspondant au "total économies de marché" représentaient environ 91 % des exportations mondiales de produits manufacturés.

Les indices de valeur unitaire sont obtenus de sources nationales, à l'exception de ceux de certains pays que la Division de statistique des Nations Unies compile en utilisant les chiffres de ces pays relatifs aux quantités et aux valeurs. Pour les pays qui n'établissent pas d'indices conformes à la définition ci-dessus pour leurs exportations de produits manufacturés, on fait la synthèse de sous-indices de manière à établir un indice proche de celui des sections 5-8 de la CTCI. Le cas échéant, les indices de valeur unitaire obtenus à partir des indices nationaux sont ajustés sur la base 1980=100. On convertit les indices en monnaie nationale en indices en dollars des Etats-Unis en utilisant des facteurs de conversion obtenus en divisant la moyenne pondérée des taux de change d'une monnaie donnée pendant la période courante par la moyenne pondérée des taux de change de la période de base. Tous les indices globaux de valeur unitaire sont pondérés pour la période courante.

On calcule les indices en DTS en multipliant les indices globaux équivalents en dollars des Etats-Unis par les facteurs de conversion obtenus en divisant le taux de change DTS/dollars EU de la période courante par le taux correspondant de la période de base.

On détermine les indices de quantum à partir des données de valeur et des indices de valeur unitaire. Tous les indices globaux de quantum sont pondérés par rapport à la période de base.

Tableau 121 : L'analyse régionale de ce tableau est conforme à celle du tableau 117. Les chiffres de ce tableau sont tirés des données du tableau 118.

La classification des marchandises est conforme à la *Classification type pour le commerce international* (CTCI), Révision 2 des Nations Unies pour les années 1980 à 1987, sauf pour les pays qui ne fournissent de statistiques commerciales que sur la base de la CTCI révisée. A partir de 1988, la classification des marchandises est conforme à la CTCI, Révision 3, pour les pays qui ont fourni des données [52].

Les données fournissent une approximation des exportations totales de tous les pays et régions du monde à l'exception du commerce intrarégional des économies à planification centrale d'Asie en 1980 et des échanges commerciaux effectués selon le protocole supplémentaire du traité sur la base des relations entre la République fédérale d'Allemagne et l'ancienne République démocratique allemande. Elles sont fondées sur les chiffres officiels des exportations convertis, le cas échéant, en dollars des Etats-Unis, selon les facteurs de conversion publiés au tableau 116 du présent volume. En l'absence de chiffres officiels, on a utilisé des estimations fondées sur les importations notifiées par les pays partenaires et sur d'autres données subsidiaires. Certaines données nationales officielles ont été ajustées :

a) sur la base des groupements de marchandises de la CTCI; et b) sur la base des années civiles.

Ces données englobent les catégories spéciales (confidentielles) d'exportation, les marchandises à bord des navires et les exportations d'importance mineure, dont la destination ne peut être déterminée. Ces données sont comprises dans les totaux mondiaux pour chaque classe de marchandises et dans les exportations totales, mais sont exclues de ceux des régions de destination. Environ 1,5 % des exportations totales ne sont pas distribuées. Dans l'ensemble, les données sont conformes au système du commerce spécial.

122
Tourist arrivals by region of origin
Arrivées de touristes par région de provenance

Country or area of destination and region of origin	1986	1987	1988	1989	1990	Pays ou zone de destination et région de provenance
Algeria[12]	**347 725**	**273 036**	**446 906**	**661 159**	**685 815**	**Algérie**[12]
Africa	151 886	90 000	256 741	449 346	431 164	Afrique
Americas	5 160	4 495	3 502	3 788	5 764	Amériques
Europe	181 071	163 541	172 640	193 834	226 277	Europe
Asia, East and South East/Oceania	4 963	6 000	2 357	2 578	4 402	Asie, Est et Sud-Est et Océanie
Western Asia	4 645	9 000	11 666	11 613	18 208	Asie occidentale
American Samoa[3]	**34 600**	**38 232**	**38 395**	**46 896**	**47 178**	**Samoa américaines**[3]
Americas	6 043	7 233	8 791	13 470	10 623	Amériques
Europe	3 546	1 066	1 068	3 778	1 331	Europe
Asia, East and South East/Oceania	25 011	29 933	28 536	29 648	35 224	Asie, Est et Sud-Est et Océanie
Anguilla	**16 476**	**20 649**	**24 138**	**26 744**	**29 712**	**Anguilla**
Americas	15 458	19 329	22 885	25 295	26 480	Amériques
Europe	1 018	1 320	1 253	1 449	3 232	Europe
Antigua and Barbuda[4]	**158 655**	**173 219**	**187 153**	**186 694**	**194 023**	**Antigua-et-Barbuda**[4]
Americas	124 955	131 564	134 577	134 573	136 918	Amériques
Europe	33 700	41 655	52 576	52 121	57 105	Europe
Argentina[2]	**1 737 829**	**1 433 277**	**2 042 563**	**2 405 047**	**2 621 474**	**Argentine**[2]
Americas	1 565 470	1 321 311	1 794 783	2 181 955	2 360 626	Amériques
Europe	172 359	111 966	247 780	223 092	260 848	Europe
Aruba	**180 548**	**231 013**	**277 288**	**342 243**	**427 948**	**Aruba**
Americas	171 256	216 505	252 848	305 204	387 253	Amériques
Europe	9 292	14 508	24 440	37 039	40 695	Europe
Australia[25]	**1 424 500**	**1 779 200**	**2 242 800**	**2 078 700**	**2 212 500**	**Australie**[25]
Africa	19 600	17 500	20 300	17 300	18 400	Afrique
Americas	301 800	373 100	404 400	326 600	317 100	Amériques
Europe	347 500	411 900	529 600	530 700	548 400	Europe
Asia, East and South East/Oceania	608 600	798 200	1 076 900	998 700	1 098 600	Asie, Est et Sud-Est et Océanie
Southern Asia	11 300	12 400	18 100	19 100	19 200	Asie du Sud
Western Asia	135 700	166 100	193 500	186 300	210 800	Asie occidentale
Austria[6]	**14 773 101**	**15 398 284**	**16 181 685**	**17 766 914**	**18 155 998**	**Autriche**[6]
Africa	22 041	25 192	25 576	25 628	24 955	Afrique
Americas	689 404	836 144	774 161	843 895	1 067 420	Amériques
Europe	13 851 806	14 273 367	15 107 282	16 554 592	16 685 230	Europe
Asia, East and South East/Oceania	186 950	239 733	245 634	293 976	350 433	Asie, Est et Sud-Est et Océanie
Western Asia	22 900	23 848	29 032	48 823	27 960	Asie occidentale
Bahamas	**1 359 205**	**1 465 455**	**1 464 860**	**1 562 445**	**1 545 025**	**Bahamas**
Americas	1 310 050	1 393 695	1 374 425	1 463 510	1 437 765	Amériques
Europe	46 450	67 950	85 135	91 320	96 625	Europe
Asia, East and South East/Oceania	2 705	3 810	5 300	7 615	10 635	Asie, Est et Sud-Est et Océanie
Bahrain[1]	**384 604**	...	...	...	...	**Bahreïn**[1]
Africa	15 956	...	...	...	...	Afrique
Americas	32 864	...	...	...	...	Amériques
Europe	51 659	...	...	...	...	Europe
Asia, East and South East/Oceania	103 424	...	...	...	...	Asie, Est et Sud-Est et Océanie
Southern Asia	169 510	...	...	...	...	Asie du Sud
Western Asia	11 191	...	...	...	...	Asie occidentale
Bangladesh[2]	**125 574**	**102 901**	**117 251**	**119 176**	**84 438**	**Bangladesh**[2]
Africa	76	83	90	64	49	Afrique
Americas	7 766	8 039	7 823	9 282	8 616	Amériques
Europe	16 274	16 541	18 915	20 004	14 889	Europe

122
Tourist arrivals by region of origin [*cont.*]
Arrivées de touristes par région de provenance [*suite*]

Country or area of destination and region of origin	1986	1987	1988	1989	1990	Pays ou zone de destination et région de provenance
Asia, East and South East/Oceania	14 326	14 877	23 692	20 433	14 565	Asie, Est et Sud-Est et Océanie
Southern Asia	84 750	60 678	64 193	66 784	44 294	Asie du Sud
Western Asia	2 382	2 683	2 538	2 609	2 025	Asie occidentale
Barbados	**365 490**	**417 798**	**446 965**	**457 078**	**427 950**	**Barbade**
Americas	298 244	314 183	310 240	293 441	274 021	Amériques
Europe	66 536	102 750	135 635	162 480	152 803	Europe
Asia, East and South East/Oceania	710	865	1 090	1 157	1 126	Asie, Est et Sud-Est et Océanie
Belgium * [6]	**2 453 467**	**2 515 253**	**2 642 959**	**3 039 710**	**3 220 783**	**Belgique * [6]**
Africa	60 007	57 499	58 715	70 670	71 861	Afrique
Americas	248 324	227 598	186 393	199 559	213 293	Amériques
Europe	2 059 893	2 141 917	2 306 843	2 657 563	2 825 259	Europe
Asia, East and South East/Oceania	81 020	83 605	86 345	106 389	103 353	Asie, Est et Sud-Est et Océanie
Western Asia	4 223	4 634	4 663	5 529	7 017	Asie occidentale
Belize	**60 468**	**66 871**	**82 055**	**101 194**	**111 732**	**Belize**
Americas	43 257	49 555	61 481	81 047	89 867	Amériques
Europe	17 211	17 316	20 574	20 147	21 865	Europe
Benin	**78 396**	**80 500**	**75 000**	**75 000**	...	**Bénin**
Africa	33 260	34 030	30 846	31 825	...	Afrique
Americas	3 514	4 373	1 494	1 496	...	Amériques
Europe	41 622	42 097	42 660	41 679	...	Europe
Bermuda [3]	**447 114**	**465 164**	**416 451**	**408 047**	**421 677**	**Bermudes [3]**
Americas	428 833	444 809	392 862	383 103	396 650	Amériques
Europe	18 281	20 355	23 589	24 944	25 027	Europe
Bhutan	**2 024**	**2 383**	**2 011**	**1 435**	...	**Bhoutan**
Americas	776	746	644	479	...	Amériques
Europe	740	1 151	982	621	...	Europe
Asia, East and South East/Oceania	508	486	385	335	...	Asie, Est et Sud-Est et Océanie
Bolivia [7]	**133 169**	**147 005**	**166 512**	**193 557**	**217 071**	**Bolivie [7]**
Africa	619	464	571	668	650	Afrique
Americas	74 663	79 862	84 601	98 858	118 190	Amériques
Europe	48 530	55 838	67 982	74 655	78 676	Europe
Asia, East and South East/Oceania	7 301	8 112	7 190	9 278	10 094	Asie, Est et Sud-Est et Océanie
Western Asia	2 056	2 729	6 168	10 098	9 461	Asie occidentale
Botswana [1]	**381 205**	**432 323**	**384 335**	**691 041**	**844 295**	**Botswana [1]**
Africa	333 167	382 908	333 783	612 457	759 197	Afrique
Americas	6 547	7 194	7 611	11 206	10 854	Amériques
Europe	37 473	36 918	37 737	60 075	65 473	Europe
Asia, East and South East/Oceania	4 018	5 303	5 204	7 303	8 771	Asie, Est et Sud-Est et Océanie
Brazil	**1 933 763**	**1 920 547**	**1 737 823**	**1 400 806**	**1 075 567**	**Brésil**
Africa	28 774	30 072	37 784	27 453	27 799	Afrique
Americas	1 476 890	1 344 992	1 115 362	979 991	673 578	Amériques
Europe	371 307	471 960	505 503	344 070	323 573	Europe
Asia, East and South East/Oceania	42 077	55 126	66 134	40 991	43 058	Asie, Est et Sud-Est et Océanie
Western Asia	14 715	18 397	13 040	7 753	7 559	Asie occidentale
British Virgin Islands	**140 074**	**169 631**	**172 778**	**173 091**	**152 970**	**Iles Vierges britanniques**
Americas	132 164	159 208	160 225	159 567	141 164	Amériques
Europe	7 910	10 423	12 553	13 524	11 806	Europe
Brunei Darussalam [1]	**408 691**	**520 310**	...	...	...	**Brunéi Darussalam [1]**
Africa	82	87	...	...	...	Afrique

122
Tourist arrivals by region of origin [cont.]
Arrivées de touristes par région de provenance [suite]

Country or area of destination and region of origin	1986	1987	1988	1989	1990	Pays ou zone de destination et région de provenance
Americas	6 475	2 852	...	...	...	Amériques
Europe	13 540	15 345	...	...	...	Europe
Asia, East and South East/Oceania	385 581	498 875	...	...	...	Asie, Est et Sud-Est et Océanie
Southern Asia	3 013	3 151	...	...	...	Asie du Sud
Bulgaria[1]	**7 309 300**	**7 483 274**	**8 228 836**	**8 145 330**	**10 305 202**	**Bulgarie**[1]
Africa	31 169	31 278	34 167	24 906	21 777	Afrique
Americas	9 702	17 349	22 583	17 136	17 195	Amériques
Europe	4 327 299	4 424 815	4 841 218	5 052 623	6 200 707	Europe
Asia, East and South East/Oceania	...	8 516	9 656	9 703	16 065	Asie, Est et Sud-Est et Océanie
Southern Asia	36 255	42 887	43 197	37 118	55 862	Asie du Sud
Western Asia	2 904 875	2 958 429	3 278 015	3 003 844	3 993 596	Asie occidentale
Burkina Faso[8]	**25 000**	**30 000**	**30 000**	**47 959**	**44 181**	**Burkina Faso**[8]
Africa	...	...	...	20 350	19 161	Afrique
Americas	3 000	3 000	3 000	2 960	2 728	Amériques
Europe	22 000	27 000	27 000	23 807	21 634	Europe
Asia, East and South East/Oceania	...	...	...	72	87	Asie, Est et Sud-Est et Océanie
Western Asia	...	...	...	770	571	Asie occidentale
Burundi[9]	**66 463**	**79 756**	**98 883**	**81 714**	**109 418**	**Burundi**[9]
Africa	31 238	36 741	46 530	38 406	51 426	Afrique
Americas	9 304	11 112	13 860	11 439	15 317	Amériques
Europe	25 256	29 811	37 620	31 052	41 579	Europe
Asia, East and South East/Oceania	665	2 092	873	817	1 096	Asie, Est et Sud-Est et Océanie
Cameroon[8]	**130 088**	**116 856**	**99 312**	**85 942**	**99 157**	**Cameroun**[8]
Africa	31 365	25 309	20 868	22 519	22 049	Afrique
Americas	8 292	9 032	7 875	8 171	8 727	Amériques
Europe	86 662	78 598	67 020	52 219	64 903	Europe
Western Asia	3 769	3 917	3 549	3 053	3 478	Asie occidentale
Canada	**15 615 200**	**14 970 500**	**15 475 500**	**15 102 900**	**15 257 800**	**Canada**
Africa	43 400	41 000	45 500	46 900	49 700	Afrique
Americas	13 810 000	12 941 100	13 047 000	12 457 400	12 548 400	Amériques
Europe	1 155 100	1 322 000	1 551 200	1 616 300	1 642 700	Europe
Asia, East and South East/Oceania	489 000	543 200	683 400	840 500	898 300	Asie, Est et Sud-Est et Océanie
Southern Asia	53 400	56 300	73 100	70 100	50 400	Asie du Sud
Western Asia	64 300	66 900	75 300	71 700	68 300	Asie occidentale
Cayman Islands	**163 092**	**205 724**	**214 560**	**209 360**	**247 445**	**Iles Caïmanes**
Americas	157 058	196 387	203 041	197 321	233 216	Amériques
Europe	6 034	9 070	11 122	11 008	13 134	Europe
Asia, East and South East/Oceania	...	267	397	1 031	1 095	Asie, Est et Sud-Est et Océanie
Central African Rep.[1]	**103 685**	**...**	**...**	**...**	**...**	**Rép. centrafricaine**[1]
Africa	53 956	...	...	...	...	Afrique
Americas	9 973	...	...	...	...	Amériques
Europe	35 333	...	...	...	...	Europe
Asia, East and South East/Oceania	2 968	...	...	...	...	Asie, Est et Sud-Est et Océanie
Southern Asia	203	...	...	...	...	Asie du Sud
Western Asia	1 252	...	...	...	...	Asie occidentale
Chad[1]	**3 813**	**26 555**	**20 080**	**12 332**	**29 079**	**Tchad**[1]
Africa	1 303	17 914	7 087	10 767	14 308	Afrique
Americas	378	1 138	2 547	246	1 156	Amériques
Europe	2 056	7 381	10 087	1 319	13 615	Europe
Asia, East and South East/Oceania	33	...	...	...	...	Asie, Est et Sud-Est et Océanie
Western Asia	43	122	359	...	...	Asie occidentale

122
Tourist arrivals by region of origin [*cont.*]
Arrivées de touristes par région de provenance [*suite*]

Country or area of destination and region of origin	1986	1987	1988	1989	1990	Pays ou zone de destination et région de provenance
Chile	**556 776**	**545 884**	**594 530**	**762 500**	**901 545**	**Chili**
Americas	507 672	490 664	531 521	678 900	814 700	Amériques
Europe	40 928	44 866	51 780	70 800	79 565	Europe
Asia, East and South East/Oceania	8 176	10 354	11 229	12 800	7 280	Asie, Est et Sud-Est et Océanie
China [10]	**1 479 599**	**1 722 039**	**1 836 919**	**1 454 996**	**1 741 053**	**Chine** [10]
Africa	5 812	6 811	9 093	14 349	12 582	Afrique
Americas	348 413	385 054	385 573	288 953	303 542	Amériques
Europe	289 742	366 817	421 892	394 484	446 260	Europe
Asia, East and South East/Oceania	811 412	929 279	945 810	704 259	884 319	Asie, Est et Sud-Est et Océanie
Southern Asia	24 220	34 078	74 551	52 951	94 350	Asie du Sud
Colombia [9]	**722 089**	**530 476**	**821 447**	**732 982**	**812 796**	**Colombie** [9]
Americas	659 547	468 733	772 578	682 780	766 271	Amériques
Europe	62 542	61 743	48 869	45 117	42 838	Europe
Asia, East and South East/Oceania	...	...	...	5 085	3 687	Asie, Est et Sud-Est et Océanie
Comoros	**5 560**	**7 663**	**7 390**	**12 959**	**7 519**	**Comores**
Africa	1 222	1 834	2 037	6 148	1 526	Afrique
Americas				835	448	Amériques
Europe	4 338	5 829	5 353	5 935	5 520	Europe
Asia, East and South East/Oceania	...	...	...	41	25	Asie, Est et Sud-Est et Océanie
Congo [11]	**38 401**	**38 635**	**38 369**	**...**	**...**	**Congo** [11]
Africa	14 347	14 199	14 122	...	...	Afrique
Americas	1 236	1 415	2 044	...	...	Amériques
Europe	22 818	23 021	22 203	...	...	Europe
Cook Islands [12]	**30 865**	**31 666**	**33 520**	**32 518**	**32 019**	**Iles Cook** [12]
Americas	6 205	5 182	5 305	4 988	6 754	Amériques
Europe	3 185	5 287	5 493	5 649	6 003	Europe
Asia, East and South East/Oceania	21 475	21 197	22 722	21 881	19 262	Asie, Est et Sud-Est et Océanie
Costa Rica	**260 598**	**277 566**	**328 663**	**373 949**	**433 657**	**Costa Rica**
Africa	152	145	246	171	244	Afrique
Americas	225 159	238 590	279 532	321 519	367 964	Amériques
Europe	29 026	32 354	41 396	45 355	57 654	Europe
Asia, East and South East/Oceania	4 735	4 786	5 406	4 871	5 641	Asie, Est et Sud-Est et Océanie
Western Asia	1 526	1 691	2 083	2 033	2 154	Asie occidentale
Côte d'Ivoire [3]	**183 304**	**171 662**	**176 241**	**...**	**...**	**Côte d'Ivoire** [3]
Africa	91 410	81 013	88 578	...	...	Afrique
Americas	9 575	10 478	9 689	...	...	Amériques
Europe	75 183	76 403	71 657	...	...	Europe
Asia, East and South East/Oceania	3 548	3 768	3 466	...	...	Asie, Est et Sud-Est et Océanie
Western Asia	3 588	...	2 851	...	...	Asie occidentale
Cuba [1]	**214 219**	**284 412**	**308 850**	**324 340**	**...**	**Cuba** [1]
Africa	1 998	2 973	1 369	1 904	...	Afrique
Americas	100 813	123 140	133 634	159 925	...	Amériques
Europe	109 793	157 949	173 399	161 443	...	Europe
Asia, East and South East/Oceania	1 615	350	448	1 068	...	Asie, Est et Sud-Est et Océanie
Cyprus	**787 945**	**902 942**	**1 083 835**	**1 303 008**	**1 535 533**	**Chypre**
Americas	12 144	11 031	12 524	13 836	22 197	Amériques
Europe	655 221	790 641	991 401	1 144 968	1 354 342	Europe
Western Asia	120 580	101 270	79 910	144 204	158 994	Asie occidentale

122
Tourist arrivals by region of origin [cont.]
Arrivées de touristes par région de provenance [suite]

Country or area of destination and region of origin	1986	1987	1988	1989	1990	Pays ou zone de destination et région de provenance
Czechoslovakia[1]	**18 779 731**	**21 470 405**	**24 176 420**	**29 221 568**	**...**	**Tchécoslovaquie**[1]
Europe	18 779 731	21 470 405	24 176 420	29 221 568	...	Europe
Denmark * [1]	**1 116 334**	**1 146 002**	**1 130 584**	**1 323 333**	**1 401 778**	**Danemark** * [1]
Americas	115 056	123 389	105 056	118 111	110 972	Amériques
Europe	981 806	999 585	995 611	1 172 250	1 260 806	Europe
Asia, East and South East/Oceania	19 472	23 028	29 917	32 972	30 000	Asie, Est et Sud-Est et Océanie
Dominica	**24 410**	**26 580**	**33 857**	**35 428**	**44 248**	**Dominique**
Africa	73	66	...	...	...	Afrique
Americas	19 092	20 610	24 058	28 434	34 919	Amériques
Europe	5 066	5 649	9 799	6 994	9 329	Europe
Southern Asia	179	255	...	...	...	Asie du Sud
Dominican Republic[2]	**518 775**	**419 571**	**411 153**	**...**	**...**	**Rép. dominicaine**[2]
Americas	470 667	339 266	373 911	...	...	Amériques
Europe	48 108	80 305	37 242	...	...	Europe
Ecuador[1 2]	**266 714**	**273 922**	**347 453**	**334 517**	**331 469**	**Equateur**[1 2]
Africa	270	233	283	311	357	Afrique
Americas	211 289	213 930	281 442	266 011	265 858	Amériques
Europe	44 957	48 607	56 321	59 044	55 500	Europe
Asia, East and South East/Oceania	10 198	11 152	9 407	9 151	9 754	Asie, Est et Sud-Est et Océanie
Egypt[1]	**1 310 937**	**1 794 652**	**1 968 841**	**2 502 603**	**2 599 287**	**Egypte**[1]
Africa	162 759	203 447	218 918	466 364	623 720	Afrique
Americas	94 763	148 999	164 141	200 479	179 144	Amériques
Europe	530 580	788 593	943 735	1 062 081	1 044 213	Europe
Asia, East and South East/Oceania	63 490	90 858	98 529	118 770	125 905	Asie, Est et Sud-Est et Océanie
Southern Asia	20 290	21 241	20 638	24 651	18 275	Asie du Sud
Western Asia	439 055	541 514	522 880	630 258	608 030	Asie occidentale
El Salvador[2]	**133 865**	**124 578**	**133 877**	**130 448**	**194 011**	**El Salvador**[2]
Americas	123 493	113 558	121 031	119 097	182 546	Amériques
Europe	8 858	9 457	11 118	9 862	10 002	Europe
Asia, East and South East/Oceania	1 514	1 563	1 728	1 489	1 463	Asie, Est et Sud-Est et Océanie
Ethiopia[2 13]	**49 745**	**56 801**	**62 573**	**64 396**	**59 084**	**Ethiopie**[2 13]
Africa	16 840	20 283	21 885	25 569	20 097	Afrique
Americas	3 520	4 640	5 470	4 881	5 636	Amériques
Europe	18 902	20 464	22 057	20 085	20 431	Europe
Asia, East and South East/Oceania	486	568	625	584	653	Asie, Est et Sud-Est et Océanie
Southern Asia	1 410	1 770	2 340	2 252	2 509	Asie du Sud
Western Asia	8 587	9 076	10 196	11 025	9 758	Asie occidentale
Fiji[2]	**252 076**	**185 376**	**202 404**	**245 480**	**277 719**	**Fidji**[2]
Americas	93 383	63 856	59 027	50 961	55 366	Amériques
Europe	25 060	23 237	28 962	35 320	43 984	Europe
Asia, East and South East/Oceania	133 633	98 283	114 415	159 199	178 369	Asie, Est et Sud-Est et Océanie
Finland * [1]	**523 079**	**568 490**	**580 651**	**643 683**	**621 806**	**Finlande** * [1]
Africa	4 500	4 650	4 839	4 900	5 100	Afrique
Americas	52 359	63 966	65 473	66 070	64 562	Amériques
Europe	449 215	479 399	488 388	549 409	527 639	Europe
Asia, East and South East/Oceania	16 645	20 105	21 570	22 914	24 055	Asie, Est et Sud-Est et Océanie
Western Asia	360	370	381	390	450	Asie occidentale
France	**36 080 000**	**36 974 000**	**38 288 000**	**49 949 000**	**47 135 000**	**France**
Africa	1 120 000	1 036 000	1 066 000	1 583 000	...	Afrique

122
Tourist arrivals by region of origin [cont.]
Arrivées de touristes par région de provenance [suite]

Country or area of destination and region of origin	1986	1987	1988	1989	1990	Pays ou zone de destination et région de provenance
Americas	2 750 000	2 931 000	3 045 000	3 005 000	2 072 000	Amériques
Europe	30 959 000	31 749 000	32 778 000	44 257 000	44 313 000	Europe
Asia, East and South East/Oceania	800 000	853 000	950 000	802 000	750 000	Asie, Est et Sud-Est et Océanie
Southern Asia	135 000	156 000	171 000	...	...	Asie du Sud
Western Asia	316 000	249 000	278 000	302 000	...	Asie occidentale
French Polynesia[2]	**160 985**	**142 647**	**135 228**	**139 512**	**132 152**	**Polynésie française**[2]
Africa	235	258	208	468	150	Afrique
Americas	108 326	79 970	71 336	61 795	51 592	Amériques
Europe	29 775	41 537	41 587	47 596	46 615	Europe
Asia, East and South East/Oceania	22 408	20 647	21 862	29 462	33 655	Asie, Est et Sud-Est et Océanie
Southern Asia	42	100	64	47	37	Asie du Sud
Western Asia	199	135	171	144	103	Asie occidentale
Gabon	**30 000**	**20 000**	**20 000**	**113 000**	**108 000**	**Gabon**
Africa	6 000	4 000	9 000	14 700	14 000	Afrique
Americas	3 000	2 000	2 000	13 400	12 600	Amériques
Europe	21 000	14 000	9 000	84 900	81 400	Europe
Gambia	**44 640**	**46 989**	**52 814**	**47 115**	**55 429**	**Gambie**
Africa	1 246	998	1 000	729	335	Afrique
Americas	478	536	400	433	659	Amériques
Europe	42 827	45 455	51 414	45 953	54 435	Europe
Southern Asia	73	...	...	...	...	Asie du Sud
Western Asia	16	...	...	...	...	Asie occidentale
Germany†						**Allemagne†**
F. R. Germany[6][14]	**12 077 172**	**12 626 160**	**12 943 118**	**14 474 084**	**15 398 011**[15]	**R. f. Allemagne**[6][14]
Africa	142 822	154 269	169 604	191 610	180 461	Afrique
Americas	2 244 162	2 432 837	2 229 285	2 474 179	2 871 511	Amériques
Europe	8 422 522	8 744 690	9 225 004	10 283 982	10 712 672	Europe
Asia, East and South East/Oceania	1 050 467	1 081 094	1 118 261	1 310 294	1 411 521	Asie, Est et Sud-Est et Océanie
Western Asia	217 199	213 074	200 964	214 019	221 846	Asie occidentale
former German D. R.	**604 409**	**655 867**	**683 459**	**639 630**	**...**	**anc. R. d. allemande**
Europe	604 409	655 867	683 459	639 630	...	Europe
Ghana	**55 111**	**41 226**	**70 193**	**...**	**...**	**Ghana**
Africa	31 567	19 146	31 256	...	...	Afrique
Americas	4 542	4 328	8 192	...	...	Amériques
Europe	14 994	15 541	26 967	...	...	Europe
Asia, East and South East/Oceania	1 527	1 442	2 037	...	...	Asie, Est et Sud-Est et Océanie
Southern Asia	1 273	403	1 183	...	...	Asie du Sud
Western Asia	1 208	366	558	...	...	Asie occidentale
Greece[16]	**6 885 452**	**7 475 582**	**7 910 298**	**8 057 855**	**8 832 354**	**Grèce**[16]
Africa	85 433	80 844	90 267	80 531	67 056	Afrique
Americas	321 455	398 000	383 339	399 802	382 623	Amériques
Europe	6 060 684	6 556 000	6 955 165	7 040 724	7 890 231	Europe
Asia, East and South East/Oceania	201 347	199 800	205 586	231 169	221 784	Asie, Est et Sud-Est et Océanie
Western Asia	216 533	240 938	275 941	305 629	270 660	Asie occidentale
Grenada	**41 184**	**41 760**	**45 849**	**50 947**	**64 266**	**Grenade**
Americas	33 896	31 684	31 983	34 080	44 671	Amériques
Europe	6 820	9 611	13 370	16 424	18 793	Europe
Western Asia	468	465	496	443	802	Asie occidentale
Guadeloupe[8]	**147 191**	**151 942**	**157 995**	**122 504**	**125 172**	**Guadeloupe**[8]
Americas	51 677	32 810	29 719	18 962	16 608	Amériques
Europe	95 514	119 132	128 276	103 542	108 564	Europe

122
Tourist arrivals by region of origin [*cont.*]
Arrivées de touristes par région de provenance [*suite*]

Country or area of destination and region of origin	1986	1987	1988	1989	1990	Pays ou zone de destination et région de provenance
Guam[4]	**393 644**	**475 173**	**570 487**	**666 288**	**776 147**	**Guam**[4]
Americas	28 688	31 513	32 362	43 393	52 199	Amériques
Europe	1 052	1 491	2 125	2 793	2 793	Europe
Asia, East and South East/Oceania	363 904	442 169	536 000	620 102	721 155	Asie, Est et Sud-Est et Océanie
Guatemala	**285 405**	**350 033**	**401 905**	**433 792**	**505 185**	**Guatemala**
Americas	239 221	289 337	328 257	352 187	411 871	Amériques
Europe	39 012	53 211	65 469	72 629	83 166	Europe
Asia, East and South East/Oceania	5 453	5 685	6 232	7 079	8 202	Asie, Est et Sud-Est et Océanie
Southern Asia	141	146	178	176	246	Asie du Sud
Western Asia	1 578	1 654	1 769	1 721	1 700	Asie occidentale
Haiti	**110 813**	**120 802**	**131 302**	**117 000**	...	**Haïti**
Americas	101 889	110 536	118 875	107 000	...	Amériques
Europe	8 924	10 266	12 427	10 000	...	Europe
Honduras[1]	**284 250**	**314 702**	**358 529**	**274 146**	**305 025**	**Honduras**[1]
Africa	58	37	3	94	101	Afrique
Americas	274 784	304 258	358 182	257 004	282 962	Amériques
Europe	9 408	10 407	340	13 919	18 094	Europe
Asia, East and South East/Oceania	...	...	4	3 129	3 868	Asie, Est et Sud-Est et Océanie
Hong Kong[1]	**3 713 829**	**4 475 186**	**5 583 232**	**5 356 361**	**5 928 188**	**Hong-kong**[1]
Africa	41 162	40 167	45 181	43 317	48 686	Afrique
Americas	861 112	981 316	952 097	812 919	807 649	Amériques
Europe	575 034	690 312	782 201	714 756	742 087	Europe
Asia, East and South East/Oceania	2 098 776	2 621 975	3 649 363	3 635 154	4 171 935	Asie, Est et Sud-Est et Océanie
Southern Asia	109 874	114 846	128 105	126 034	132 838	Asie du Sud
Western Asia	27 871	26 570	26 285	24 181	24 993	Asie occidentale
Hungary[2 17]	**10 513 000**	**11 697 000**	**10 426 000**	**14 479 000**	**20 510 000**	**Hongrie**[2 17]
Africa				26 000	23 000	Afrique
Americas	108 000	147 000	161 000	198 000	239 000	Amériques
Europe	10 405 000	11 550 000	10 265 000	14 106 000	20 088 000	Europe
Asia, East and South East/Oceania	...	...	...	50 000	58 000	Asie, Est et Sud-Est et Océanie
Western Asia	...	...	...	99 000	102 000	Asie occidentale
Iceland	**113 528**	**129 281**	**128 938**	**130 247**	**141 325**	**Islande**
Africa	283	334	341	382	349	Afrique
Americas	34 249	37 262	30 432	24 594	24 138	Amériques
Europe	76 737	88 768	95 534	102 538	114 097	Europe
Asia, East and South East/Oceania	1 869	2 374	2 300	2 242	2 343	Asie, Est et Sud-Est et Océanie
Southern Asia	95	145	130	127	116	Asie du Sud
Western Asia	295	398	201	364	282	Asie occidentale
India[2]	**1 450 473**	**1 483 793**	**1 590 074**	**1 735 252**	**1 705 546**	**Inde**[2]
Africa	50 607	51 898	58 014	61 667	60 750	Afrique
Americas	172 116	180 461	169 178	183 656	176 995	Amériques
Europe	465 925	503 698	575 093	631 995	627 540	Europe
Asia, East and South East/Oceania	154 244	174 720	179 927	200 161	207 630	Asie, Est et Sud-Est et Océanie
Southern Asia	491 885	451 064	484 871	535 981	520 248	Asie du Sud
Western Asia	115 696	121 952	122 991	121 792	112 383	Asie occidentale
Indonesia[1]	**823 590**	**1 060 347**	**1 300 189**	**1 616 990**	**2 177 566**	**Indonésie**[1]
Africa	1 451	774	1 607	1 687	2 380	Afrique
Americas	76 852	76 822	81 930	93 400	127 278	Amériques
Europe	209 031	265 767	321 612	379 625	484 383	Europe

122

Tourist arrivals by region of origin [cont.]

Arrivées de touristes par région de provenance [suite]

Country or area of destination and region of origin	1986	1987	1988	1989	1990	Pays ou zone de destination et région de provenance
Asia, East and South East/Oceania	517 300	697 343	879 110	1 116 955	1 526 306	Asie, Est et Sud-Est et Océanie
Southern Asia	9 132	8 111	7 386	9 240	13 454	Asie du Sud
Western Asia	9 824	11 530	8 544	16 083	23 765	Asie occidentale
Iran, Islamic Rep. of	**92 792**	**68 706**	**66 444**	**89 140**	**153 602**	**Iran, Rép. islamique d'**
Africa	1 206	837	931	1 316	1 442	Afrique
Americas	588	436	569	863	1 392	Amériques
Europe	20 812	13 229	12 708	16 186	27 774	Europe
Asia, East and South East/Oceania	8 676	6 318	5 816	8 717	11 767	Asie, Est et Sud-Est et Océanie
Southern Asia	44 161	33 839	32 820	44 323	83 308	Asie du Sud
Western Asia	17 349	14 047	13 600	17 735	27 919	Asie occidentale
Iraq[1]	**918 662**	**679 216**	**224 383**	**965 533**	**713 808**	**Iraq**[1]
Africa	425 457	387 354	...	340 486	95 940	Afrique
Americas	7 698	6 853	5 373	10 080	6 955	Amériques
Europe	89 544	66 132	94 363	85 369	55 104	Europe
Asia, East and South East/Oceania	14 998	7 381	3 983	11 016	11 159	Asie, Est et Sud-Est et Océanie
Southern Asia	21 993	20 022	11 422	45 651	33 107	Asie du Sud
Western Asia	358 972	191 474	109 242	472 931	511 543	Asie occidentale
Ireland[2]	**2 432 000**	**2 627 000**	**2 963 000**	**2 963 000**	**3 611 000**	**Irlande**[2]
Americas	343 000	398 000	419 000	427 000	443 000	Amériques
Europe	2 053 000	2 192 000	2 498 000	2 943 000	3 099 000	Europe
Asia, East and South East/Oceania	36 000	37 000	46 000	62 000	69 000	Asie, Est et Sud-Est et Océanie
Israel[17]	**1 097 989**	**1 371 824**	**1 163 676**	**1 172 972**	**1 058 898**	**Israël**[17]
Africa	32 837	36 102	32 137	30 071	29 919	Afrique
Americas	296 960	380 074	325 433	343 568	314 216	Amériques
Europe	665 921	839 776	711 493	695 777	614 547	Europe
Asia, East and South East/Oceania	30 699	41 438	37 939	39 653	35 399	Asie, Est et Sud-Est et Océanie
Southern Asia	6 735	5 780	4 864	4 708	4 018	Asie du Sud
Western Asia	64 837	68 654	51 810	59 195	60 799	Asie occidentale
Italy[18]	**18 324 722**	**20 367 038**	**20 836 410**	**19 543 504**	**19 762 923**	**Italie**[18]
Africa	57 542	64 544	70 425	68 743	75 986	Afrique
Americas	2 121 654	2 668 167	2 636 674	2 734 833	2 916 008	Amériques
Europe	15 352 014	16 660 875	17 072 762	15 499 743	15 456 722	Europe
Asia, East and South East/Oceania	595 230	741 072	835 833	1 016 072	1 084 923	Asie, Est et Sud-Est et Océanie
Western Asia	198 282	232 380	220 716	224 113	229 284	Asie occidentale
Jamaica	**662 211**	**737 675**	**647 753**	**713 440**	**838 878**	**Jamaïque**
Africa	619	380	275	511	461	Afrique
Americas	617 360	677 202	575 130	612 477	709 918	Amériques
Europe	42 193	58 119	69 985	96 550	121 049	Europe
Asia, East and South East/Oceania	1 832	1 851	2 215	3 649	7 091	Asie, Est et Sud-Est et Océanie
Southern Asia	207	123	148	253	359	Asie du Sud
Japan[2]	**2 057 149**	**2 151 328**	**2 352 316**	**2 832 308**	**3 233 098**	**Japon**[2]
Africa	11 165	9 981	10 510	11 883	12 135	Afrique
Americas	648 507	649 978	621 562	653 793	711 410	Amériques
Europe	360 691	380 939	400 796	455 228	516 125	Europe
Asia, East and South East/Oceania	950 988	1 023 172	1 219 461	1 625 453	1 897 829	Asie, Est et Sud-Est et Océanie
Southern Asia	59 514	63 197	68 046	61 964	71 384	Asie du Sud
Western Asia	26 284	24 061	31 941	23 987	24 215	Asie occidentale
Jordan[1]	**1 890 815**	**1 876 930**	**2 368 347**	**2 257 660**	**2 629 500**	**Jordanie**[1]
Africa	782 821	619 353	800 344	682 744	798 912	Afrique
Americas	37 078	47 235	47 590	48 257	38 538	Amériques
Europe	90 341	99 825	107 994	127 148	117 366	Europe

122
Tourist arrivals by region of origin [*cont.*]
Arrivées de touristes par région de provenance [*suite*]

Country or area of destination and region of origin	1986	1987	1988	1989	1990	Pays ou zone de destination et région de provenance
Asia, East and South East/Oceania	28 381	28 850	27 194	24 835	55 389	Asie, Est et Sud-Est et Océanie
Southern Asia	19 686	17 177	19 388	18 515	253 319	Asie du Sud
Western Asia	932 508	1 064 490	1 365 837	1 356 161	1 365 976	Asie occidentale
Kenya [17]	**595 000**	**656 700**	**671 500**	**709 200**	**691 100**	**Kenya** [17]
Africa	127 500	129 900	132 700	171 600	142 100	Afrique
Americas	68 900	77 400	79 400	97 700	77 700	Amériques
Europe	354 500	394 200	403 400	391 500	423 400	Europe
Asia, East and South East/Oceania	29 000	37 900	38 600	32 600	31 400	Asie, Est et Sud-Est et Océanie
Southern Asia	10 900	12 400	12 600	11 300	12 500	Asie du Sud
Western Asia	4 200	4 900	4 800	4 500	4 000	Asie occidentale
Kiribati [1]	**3 026**	**3 458**	**3 284**	**2 629**	**2 913**	**Kiribati** [1]
Americas	1 223	1 234	892	939	642	Amériques
Europe	154	165	315	107	10	Europe
Asia, East and South East/Oceania	1 649	2 059	2 077	1 583	2 261	Asie, Est et Sud-Est et Océanie
Korea, Republic of [1]	**1 454 415**	**1 630 496**	**2 040 393**	**2 407 296**	**2 635 598**	**Corée, République de** [1]
Africa	4 010	4 460	4 893	7 137	6 925	Afrique
Americas	312 247	357 600	388 450	359 790	370 143	Amériques
Europe	95 982	106 499	163 584	172 086	192 639	Europe
Asia, East and South East/Oceania	1 017 983	1 130 312	1 437 495	1 785 910	1 949 680	Asie, Est et Sud-Est et Océanie
Southern Asia	14 732	23 382	35 532	41 642	48 210	Asie du Sud
Western Asia	9 461	8 243	10 439	40 731	68 001	Asie occidentale
Kuwait [1]	**1 267 425**	**1 127 101**	**1 254 428**	**...**	**...**	**Koweït** [1]
Africa	1 931	1 774	2 527	...	...	Afrique
Americas	17 685	14 202	14 919	...	...	Amériques
Europe	78 246	53 818	55 806	...	...	Europe
Asia, East and South East/Oceania	352 664	289 831	280 309	...	...	Asie, Est et Sud-Est et Océanie
Western Asia	816 899	767 476	900 867	...	...	Asie occidentale
Lesotho [1]	**212 548**	**220 087**	**165 025**	**216 108**	**242 456**	**Lesotho** [1]
Africa	207 590	214 822	158 704	210 234	235 845	Afrique
Americas	1 730	1 738	2 392	1 185	1 570	Amériques
Europe	3 228	3 527	3 929	4 014	3 970	Europe
Southern Asia	...	...	...	675	1 071	Asie du Sud
Liechtenstein [8]	**75 793**	**75 246**	**71 538**	**76 982**	**77 339**	**Liechtenstein** [8]
Africa	174	218	189	292	249	Afrique
Americas	11 506	12 270	9 894	11 014	10 984	Amériques
Europe	60 983	59 685	59 021	62 238	63 287	Europe
Asia, East and South East/Oceania	2 815	2 770	1 891	2 857	2 312	Asie, Est et Sud-Est et Océanie
Western Asia	315	303	543	581	507	Asie occidentale
Luxembourg [1]	**614 058**	**643 817**	**760 063**	**872 928**	**820 367**	**Luxembourg** [1]
Africa	4 410	4 560	5 531	6 099	6 547	Afrique
Americas	52 274	53 537	48 677	47 138	50 026	Amériques
Europe	547 868	578 649	693 225	805 939	747 515	Europe
Asia, East and South East/Oceania	9 506	7 071	12 630	13 752	10 071	Asie, Est et Sud-Est et Océanie
Macau [1]	**767 841**	**849 149**	**908 520**	**997 177**	**1 126 053**	**Macao** [1]
Africa	1 905	2 307	3 102	3 444	2 896	Afrique
Americas	140 201	153 871	133 914	113 926	104 104	Amériques
Europe	232 519	195 887	190 664	187 279	181 147	Europe
Asia, East and South East/Oceania	380 785	484 424	566 344	675 637	821 963	Asie, Est et Sud-Est et Océanie
Southern Asia	12 431	12 660	14 496	16 891	15 943	Asie du Sud
Madagascar [3]	**26 723**	**27 417**	**35 351**	**38 980**	**52 923**	**Madagascar** [3]
Africa	5 517	5 375	5 838	7 974	10 149	Afrique

122
Tourist arrivals by region of origin [*cont.*]
Arrivées de touristes par région de provenance [*suite*]

Country or area of destination and region of origin	1986	1987	1988	1989	1990	Pays ou zone de destination et région de provenance
Americas	3 121	2 998	1 066	2 372	3 867	Amériques
Europe	16 928	17 286	27 714	27 401	36 702	Europe
Asia, East and South East/Oceania	1 076	1 664	733	1 208	2 035	Asie, Est et Sud-Est et Océanie
Western Asia	81	94	...	25	170	Asie occidentale
Malawi [17]	**68 409**	**76 134**	**96 634**	**115 676**	**...**	**Malawi** [17]
Africa	55 404	61 513	81 247	95 564	...	Afrique
Americas	2 651	2 775	3 398	3 810	...	Amériques
Europe	9 116	10 154	11 989	13 164	...	Europe
Asia, East and South East/Oceania	...	1 134	...	3 138	...	Asie, Est et Sud-Est et Océanie
Southern Asia	947	303	...	...	...	Asie du Sud
Western Asia	291	255	...	...	...	Asie occidentale
Malaysia	**3 014 418**	**3 142 629**	**3 383 493**	**4 696 854**	**7 276 395**	**Malaisie**
Americas	46 116	49 127	56 246	103 330	175 171	Amériques
Europe	123 742	131 259	150 539	282 744	453 458	Europe
Asia, East and South East/Oceania	2 808 028	2 924 435	3 135 878	4 253 953	6 502 936	Asie, Est et Sud-Est et Océanie
Southern Asia	36 532	37 808	40 830	39 396	110 360	Asie du Sud
Western Asia	...	...	...	17 431	34 470	Asie occidentale
Maldives	**113 787**	**131 261**	**155 685**	**158 450**	**195 147**	**Maldives**
Africa	285	322	280	372	296	Afrique
Americas	1 749	1 539	1 614	1 702	2 167	Amériques
Europe	82 720	98 809	119 496	123 006	151 560	Europe
Asia, East and South East/Oceania	15 172	15 780	15 459	16 867	21 006	Asie, Est et Sud-Est et Océanie
Southern Asia	13 861	14 811	18 836	12 153	15 529	Asie du Sud
Western Asia	...	...	...	4 350	4 589	Asie occidentale
Mali [8]	**30 990**	**31 810**	**33 503**	**29 984**	**41 027**	**Mali** [8]
Africa	6 776	6 548	7 269	9 140	13 669	Afrique
Americas	3 287	3 532	3 291	3 387	3 895	Amériques
Europe	19 141	19 634	19 784	15 316	22 330	Europe
Asia, East and South East/Oceania	499	691	717	558	954	Asie, Est et Sud-Est et Océanie
Western Asia	1 287	1 405	2 442	1 583	179	Asie occidentale
Malta [17]	**558 260**	**730 254**	**770 276**	**817 328**	**858 049**	**Malte** [17]
Africa	28 235	47 775	40 464	34 108	38 881	Afrique
Americas	8 761	11 052	13 280	15 160	15 456	Amériques
Europe	515 534	665 418	709 599	757 540	792 476	Europe
Asia, East and South East/Oceania	5 477	5 639	6 631	7 569	8 388	Asie, Est et Sud-Est et Océanie
Southern Asia	...	...	...	1 824	1 589	Asie du Sud
Western Asia	253	370	302	1 127	1 259	Asie occidentale
Martinique [3]	**154 204**	**203 259**	**232 708**	**287 280**	**258 432**	**Martinique** [3]
Americas	66 995	53 329	37 850	28 043	41 664	Amériques
Europe	87 209	149 930	194 858	259 237	216 768	Europe
Mauritius	**155 510**	**193 870**	**224 790**	**247 270**	**272 230**	**Maurice**
Africa	78 250	91 460	108 260	117 980	131 820	Afrique
Americas	1 780	1 880	1 470	1 890	2 220	Amériques
Europe	67 630	91 590	105 000	115 210	121 610	Europe
Asia, East and South East/Oceania	5 580	5 960	5 840	7 400	8 390	Asie, Est et Sud-Est et Océanie
Southern Asia	2 270	2 980	4 220	4 790	8 190	Asie du Sud
Mexico [9]	**4 624 958**	**5 407 000**	**5 692 000**	**6 186 000**	**6 357 000**	**Mexique** [9]
Africa	535	900	849	718	...	Afrique
Americas	4 459 519	5 160 500	5 554 000	6 007 000	6 168 000	Amériques

122
Tourist arrivals by region of origin [cont.]
Arrivées de touristes par région de provenance [suite]

Country or area of destination and region of origin	1986	1987	1988	1989	1990	Pays ou zone de destination et région de provenance
Europe	149 084	218 933	112 000	157 000	189 000	Europe
Asia, East and South East/Oceania	11 687	19 700	18 580	15 723	...	Asie, Est et Sud-Est et Océanie
Western Asia	4 133	6 967	6 571	5 559	...	Asie occidentale
Monaco[1]	**200 128**	**203 959**	**218 417**	**234 574**	**233 093**	**Monaco** [1]
Africa	510	390	340	455	637	Afrique
Americas	35 428	36 616	38 326	37 763	38 133	Amériques
Europe	156 756	160 045	172 621	187 152	181 601	Europe
Asia, East and South East/Oceania	3 958	4 409	4 839	6 660	10 048	Asie, Est et Sud-Est et Océanie
Western Asia	3 476	2 499	2 291	2 544	2 674	Asie occidentale
Mongolia	**199 194**	**185 899**	**239 738**	**236 537**	**147 138**	**Mongolie**
Africa	53	80	32	13	21	Afrique
Americas	842	1 100	949	782	1 117	Amériques
Europe	197 006	182 896	236 474	231 909	137 496	Europe
Asia, East and South East/Oceania	1 114	1 551	1 983	3 558	8 218	Asie, Est et Sud-Est et Océanie
Southern Asia	141	243	288	188	153	Asie du Sud
Western Asia	38	29	12	87	133	Asie occidentale
Montserrat[2]	**15 496**	**16 924**	**17 806**	**20 160**	**17 280**	**Montserrat**[2]
Americas	14 064	14 954	15 490	17 430	14 940	Amériques
Europe	1 432	1 970	2 316	2 730	2 340	Europe
Morocco	**1 451 349**	**1 545 657**	**1 956 202**	**2 485 868**	**2 945 361**	**Maroc**
Africa	102 091	42 183	466 703	1 033 829	1 585 238	Afrique
Americas	67 614	84 358	91 074	116 456	115 408	Amériques
Europe	1 229 830	1 365 287	1 341 348	1 272 058	1 189 981	Europe
Asia, East and South East/Oceania	12 756	15 016	16 716	17 790	18 315	Asie, Est et Sud-Est et Océanie
Western Asia	39 058	38 813	40 361	45 735	36 419	Asie occidentale
Myanmar	**40 605**	**41 904**	**22 251**	**5 044**	**...**	**Myanmar**
Americas	7 275	7 962	3 764	1 022	...	Amériques
Europe	25 447	26 592	14 566	3 340	...	Europe
Asia, East and South East/Oceania	6 649	6 128	3 260	622	...	Asie, Est et Sud-Est et Océanie
Southern Asia	1 234	1 222	661	60	...	Asie du Sud
Nepal	**216 720**	**240 742**	**262 828**	**238 922**	**...**	**Népal**
Africa	532	549	604	605	...	Afrique
Americas	29 700	33 344	32 253	30 013	...	Amériques
Europe	88 149	94 503	102 305	107 966	...	Europe
Asia, East and South East/Oceania	38 949	36 894	38 479	38 884	...	Asie, Est et Sud-Est et Océanie
Southern Asia	57 205	64 828	78 006	48 122	...	Asie du Sud
Western Asia	2 185	10 624	11 181	13 332	...	Asie occidentale
Netherlands[6]	**4 810 852**	**4 843 432**	**4 876 000**	**5 206 000**	**5 795 100**	**Pays-Bas**[6]
Africa	80 102	74 951	69 800	60 200	57 700	Afrique
Americas	609 208	619 405	629 600	646 400	690 400	Amériques
Europe	3 801 635	3 815 122	3 828 600	4 168 300	4 690 900	Europe
Asia, East and South East/Oceania	319 907	333 954	348 000	331 100	356 100	Asie, Est et Sud-Est et Océanie
New Caledonia[39]	**58 648**	**60 344**	**60 764**	**81 833**	**86 870**	**Nouvelle-Calédonie**[39]
Africa	...	...	...	...	165	Afrique
Americas	1 587	1 721	1 566	2 004	1 663	Amériques
Europe	20 432	14 376	16 674	17 185	29 855	Europe
Asia, East and South East/Oceania	35 846	43 369	41 704	61 601	55 187	Asie, Est et Sud-Est et Océanie
Southern Asia	783	878	820	1 043	...	Asie du Sud
New Zealand[19]	**704 476**	**823 252**	**839 881**	**866 071**	**938 414**	**Nouvelle-Zélande** [19]
Africa	2 870	3 894	3 457	3 017	3 834	Afrique
Americas	189 029	222 778	210 711	173 489	179 287	Amériques
Europe	93 105	118 853	137 535	144 959	169 683	Europe

122
Tourist arrivals by region of origin [cont.]
Arrivées de touristes par région de provenance [suite]

Country or area of destination and region of origin	1986	1987	1988	1989	1990	Pays ou zone de destination et région de provenance
Asia, East and South East/Oceania	417 861	471 852	481 732	537 013	578 543	Asie, Est et Sud-Est et Océanie
Southern Asia	1 576	2 537	3 391	4 687	3 670	Asie du Sud
Western Asia	35	3 338	3 055	2 906	3 397	Asie occidentale
Nicaragua	...	...	...	**66 878**	**99 892**	**Nicaragua**
Americas	...	...	...	52 816	80 333	Amériques
Europe	...	...	...	13 717	17 758	Europe
Asia, East and South East/Oceania	...	...	...	345	995	Asie, Est et Sud-Est et Océanie
Southern Asia	...	...	...	...	300	Asie du Sud
Western Asia	...	...	...	...	366	Asie occidentale
Niger	**27 086**	**29 733**	**31 028**	**23 501**	**20 621**	**Niger**
Africa	8 703	9 552	10 255	7 905	6 552	Afrique
Americas	2 293	2 478	2 480	19 151	1 814	Amériques
Europe	15 516	16 754	17 229	12 901	11 557	Europe
Asia, East and South East/Oceania	574	949	450	366	319	Asie, Est et Sud-Est et Océanie
Western Asia	...	...	614	414	379	Asie occidentale
Nigeria	**86 364**	**170 880**	**306 166**	**160 719**	...	**Nigéria**
Africa	65 296	146 893	253 801	137 053	...	Afrique
Americas	4 429	8 034	8 891	1 299	...	Amériques
Europe	10 812	9 462	29 684	15 427	...	Europe
Asia, East and South East/Oceania	1 594	2 001	3 712	2 883	...	Asie, Est et Sud-Est et Océanie
Southern Asia	2 280	2 568	4 510	2 349	...	Asie du Sud
Western Asia	1 953	1 922	5 568	1 708	...	Asie occidentale
Niue[3]	**1 734**	**1 503**	**361**	**481**	**1 044**	**Nioué**[3]
Americas	76	86	5	16	23	Amériques
Europe	53	10	7	7	17	Europe
Asia, East and South East/Oceania	1 605	1 407	349	458	1 004	Asie, Est et Sud-Est et Océanie
Norway *[1]	**1 532 373**	**1 606 769**	**1 566 991**	**1 599 206**	**1 665 813**	**Norvège ***[1]
Americas	240 592	262 707	206 024	205 629	215 272	Amériques
Europe	1 268 758	1 306 496	1 327 610	1 353 530	1 408 293	Europe
Asia, East and South East/Oceania	23 023	37 566	33 357	40 047	42 248	Asie, Est et Sud-Est et Océanie
Oman[1]	**51 000**	**51 000**	**98 159**	**108 366**	**121 508**	**Oman**[1]
Africa	...	...	1 198	5 119	5 524	Afrique
Americas	4 000	5 000	5 944	6 771	7 956	Amériques
Europe	47 000	46 000	45 835	53 343	58 170	Europe
Western Asia	...	...	45 182	43 133	49 858	Asie occidentale
Pakistan	**432 325**	**424 949**	**460 091**	**494 600**	**423 800**	**Pakistan**
Africa	9 003	11 458	13 009	9 700	5 200	Afrique
Americas	35 021	36 044	40 408	46 700	42 300	Amériques
Europe	103 839	109 000	132 449	141 900	146 200	Europe
Asia, East and South East/Oceania	23 878	24 427	30 095	31 900	33 000	Asie, Est et Sud-Est et Océanie
Southern Asia	237 213	219 910	217 599	238 800	177 500	Asie du Sud
Western Asia	23 371	24 110	26 531	25 600	19 600	Asie occidentale
Panama[1]	**339 025**	**299 769**	**227 859**	**216 632**	**230 960**	**Panama**[1]
Africa	398	356	238	312	256	Afrique
Americas	308 073	269 834	200 552	187 565	206 198	Amériques
Europe	19 994	19 290	16 610	17 365	16 359	Europe
Asia, East and South East/Oceania	5 585	5 219	4 339	4 452	3 684	Asie, Est et Sud-Est et Océanie
Western Asia	4 975	5 070	6 120	6 938	4 463	Asie occidentale

122
Tourist arrivals by region of origin [cont.]
Arrivées de touristes par région de provenance [suite]

Country or area of destination and region of origin	1986	1987	1988	1989	1990	Pays ou zone de destination et région de provenance
Papua New Guinea	**31 900**	**34 970**	**40 529**	**48 918**	**40 742**	**Papouasie-Nvl-Guinée**
Africa	89	80	97	133	161	Afrique
Americas	4 609	4 842	5 225	7 203	5 286	Amériques
Europe	5 093	4 131	5 499	6 530	7 497	Europe
Asia, East and South East/Oceania	20 733	24 675	28 132	33 376	26 124	Asie, Est et Sud-Est et Océanie
Southern Asia	1 376	1 242	1 576	1 676	1 674	Asie du Sud
Paraguay[2 5 19]	**363 661**	**295 810**	**275 058**	**271 061**	**273 247**	**Paraguay**[2 5 19]
Americas	334 188	250 810	234 473	234 941	230 786	Amériques
Europe	21 799	34 417	30 723	15 886	30 317	Europe
Asia, East and South East/Oceania	7 674	10 583	9 862	20 234	12 144	Asie, Est et Sud-Est et Océanie
Peru	**303 601**	**330 110**	**359 281**	**333 499**	**316 798**	**Pérou**
Africa	958	942	995	652	631	Afrique
Americas	171 131	178 793	192 450	172 278	159 689	Amériques
Europe	113 830	129 948	143 115	135 567	129 720	Europe
Asia, East and South East/Oceania	14 630	16 593	18 678	21 742	23 431	Asie, Est et Sud-Est et Océanie
Southern Asia	...	...	...	781	904	Asie du Sud
Western Asia	3 052	3 834	4 043	2 479	2 423	Asie occidentale
Philippines[1 2]	**690 492**	**699 276**	**903 320**	**1 042 400**	**854 686**	**Philippines**[1 2]
Africa	1 190	1 246	1 062	917	909	Afrique
Americas	210 649	219 501	259 116	298 665	245 496	Amériques
Europe	151 649	154 292	214 571	220 372	165 289	Europe
Asia, East and South East/Oceania	286 163	283 536	387 312	488 228	412 389	Asie, Est et Sud-Est et Océanie
Southern Asia	20 189	19 222	22 184	17 532	16 039	Asie du Sud
Western Asia	20 652	21 479	19 075	16 686	14 564	Asie occidentale
Poland[1]	**3 771 687**	**4 696 861**	**6 024 500**	**7 992 900**	**17 582 928**	**Pologne**[1]
Africa	5 029	8 147				Afrique
Americas	40 555	65 931	69 700	114 700	...	Amériques
Europe	3 705 351	4 597 548	5 936 800	7 853 300	17 582 928	Europe
Asia, East and South East/Oceania	11 177	15 545	18 000	24 900	...	Asie, Est et Sud-Est et Océanie
Southern Asia	5 610	6 358	...	...	...	Asie du Sud
Western Asia	3 965	3 332	...	...	...	Asie occidentale
Portugal[2]	**5 282 168**	**5 987 566**	**6 486 618**	**6 954 955**	**7 846 771**	**Portugal**[2]
Africa	73 978	68 527	78 355	93 118	102 030	Afrique
Americas	186 602	224 227	270 215	269 806	267 124	Amériques
Europe	4 982 925	5 649 269	6 091 795	6 541 886	7 423 471	Europe
Asia, East and South East/Oceania	38 663	45 543	46 253	50 145	54 146	Asie, Est et Sud-Est et Océanie
Puerto Rico	**1 342 290**	**1 575 223**	**1 651 968**	**1 742 247**	**1 801 309**	**Porto Rico**
Americas	1 342 290	1 575 223	1 651 968	1 742 247	1 801 309	Amériques
Réunion	...	...	...	**169 809**	**187 173**	**Réunion**
Africa	...	...	...	30 436	33 927	Afrique
Europe	...	...	...	139 373	153 246	Europe
Romania[1]	**4 520 600**	**5 126 600**	**5 493 300**	**4 831 500**	**6 533 190**	**Roumanie**[1]
Africa	...	...	...	...	20 952	Afrique
Americas	20 700	31 300	32 800	34 100	52 777	Amériques
Europe	4 366 600	4 959 700	5 322 300	4 639 200	6 222 547	Europe
Asia, East and South East/Oceania	91 900	89 300	84 500	86 900	29 902	Asie, Est et Sud-Est et Océanie
Southern Asia	...	...	...	...	22 397	Asie du Sud
Western Asia	41 400	46 300	53 700	71 300	184 615	Asie occidentale

122
Tourist arrivals by region of origin [cont.]
Arrivées de touristes par région de provenance [suite]

Country or area of destination and region of origin	1986	1987	1988	1989	1990	Pays ou zone de destination et région de provenance
Rwanda	31 806	...	...	...	...	Rwanda
Africa	4 650	...	...	...	...	Afrique
Americas	4 749	...	...	...	...	Amériques
Europe	21 461	...	...	...	...	Europe
Asia, East and South East/Oceania	654	...	...	...	...	Asie, Est et Sud-Est et Océanie
Southern Asia	292	...	...	...	...	Asie du Sud
Saint Kitts and Nevis	49 781	60 083	60 303	63 165	69 290	Saint-Kitts-et-Nevis
Americas	45 412	54 948	54 525	57 016	61 934	Amériques
Europe	4 369	5 135	5 778	6 149	7 356	Europe
Saint Lucia[23]	109 791	115 749	131 012	133 189	144 301	Sainte-Lucie[23]
Americas	75 191	76 024	82 381	85 730	100 875	Amériques
Europe	34 600	39 725	48 631	47 459	43 426	Europe
St. Vincent-Grenadines	34 346	45 081	45 811	49 158	53 416	St. Vincent-Grenadines
Americas	28 491	34 484	33 327	35 299	39 139	Amériques
Europe	5 855	10 597	12 484	13 859	14 277	Europe
Samoa[3]	48 577	47 518	47 839	52 754	44 501	Samoa[3]
Americas	5 854	6 029	5 486	5 700	4 686	Amériques
Europe	2 294	2 605	3 018	3 244	5 936	Europe
Asia, East and South East/Oceania	40 429	38 884	39 335	43 810	33 879	Asie, Est et Sud-Est et Océanie
San Marino[1]	...	2 832 747	2 917 061	2 822 132	2 330 290	Saint-Marin[1]
Africa	...	16	...	131	...	Afrique
Americas	...	3 394	2 568	2 300	...	Amériques
Europe	...	2 829 289	2 914 264	2 819 669	2 330 290	Europe
Asia, East and South East/Oceania	...	16	4	32	...	Asie, Est et Sud-Est et Océanie
Southern Asia	...	4	169	...	...	Asie du Sud
Western Asia	...	28	56	...	...	Asie occidentale
Sao Tome and Principe	...	840	...	...	...	Sao Tomé-et-Principe
Africa	...	67	...	...	...	Afrique
Americas	...	42	...	...	...	Amériques
Europe	...	673	...	...	...	Europe
Asia, East and South East/Oceania	...	56	...	...	...	Asie, Est et Sud-Est et Océanie
Western Asia	...	2	...	...	...	Asie occidentale
Saudi Arabia[1]	1 263 775	1 999 748	2 088 103	1 178 038	1 982 149	Arabie saoudite[1]
Africa	328 382	313 853	382 357	342 183	390 889	Afrique
Americas	870 989	1 636 971	1 652 790	789 212	1 540 576	Amériques
Europe	64 404	48 924	52 956	46 643	50 684	Europe
Senegal[1]	234 406	234 700	255 144	258 790	245 617	Sénégal[1]
Africa	43 481	43 181	43 177	39 847	37 127	Afrique
Americas	13 197	12 101	12 143	13 845	11 005	Amériques
Europe	174 623	175 771	196 033	201 397	193 694	Europe
Asia, East and South East/Oceania	2 139	2 355	2 191	2 713	2 824	Asie, Est et Sud-Est et Océanie
Western Asia	966	1 292	1 600	988	967	Asie occidentale
Seychelles	66 734[3]	71 626[3]	77 401[3]	86 095[3]	103 770[4]	Seychelles
Africa	5 956	5 640	7 957	11 174	15 871	Afrique
Americas	2 064	2 473	2 197	2 597	2 614	Amériques
Europe	55 078	59 621	64 184	68 613	81 236	Europe
Asia, East and South East/Oceania	1 031	1 776	1 022	1 390	1 837	Asie, Est et Sud-Est et Océanie
Southern Asia	260	203	266	416	339	Asie du Sud
Western Asia	2 345	1 913	1 775	1 905	1 873	Asie occidentale

122
Tourist arrivals by region of origin [cont.]
Arrivées de touristes par région de provenance [suite]

Country or area of destination and region of origin	1986	1987	1988	1989	1990	Pays ou zone de destination et région de provenance
Singapore[20]	**3 153 516**	**3 621 805**	**4 115 977**	**4 746 093**	**5 228 153**	**Singapour**[20]
Americas	260 305	286 636	307 783	344 564	370 551	Amériques
Europe	699 736	820 054	1 019 162	1 132 735	1 204 795	Europe
Asia, East and South East/Oceania	1 844 287	2 147 822	2 446 848	2 914 478	3 275 177	Asie, Est et Sud-Est et Océanie
Southern Asia	349 188	367 293	342 184	354 316	377 630	Asie du Sud
Solomon Islands	**10 710**	**11 655**	**9 792**	**8 970**	**8 478**	**Iles Salomon**
Americas	1 176	1 193	909	999	758	Amériques
Europe	1 057	1 139	772	657	791	Europe
Asia, East and South East/Oceania	8 477	9 323	8 111	7 314	6 929	Asie, Est et Sud-Est et Océanie
South Africa[2]	**644 502**	**703 351**	**801 470**	**926 894**	**1 027 621**	**Afrique du Sud**[2]
Africa	351 456	368 555	418 484	460 633	535 501	Afrique
Americas	49 721	50 610	55 010	63 863	66 972	Amériques
Europe	203 368	237 709	274 681	332 279	349 685	Europe
Asia, East and South East/Oceania	12 707	14 869	16 263	20 527	22 744	Asie, Est et Sud-Est et Océanie
Southern Asia	27 250	31 608	37 032	49 592	52 719	Asie du Sud
Spain[12]	**27 705 026**	**30 546 708**	**32 120 861**	**31 425 353**	**30 459 791**	**Espagne**[12]
Africa	259 335	266 300	294 169	310 716	303 812	Afrique
Americas	1 526 365	1 625 551	1 553 767	1 676 759	1 497 189	Amériques
Europe	25 520 941	28 240 429	29 819 780	28 924 388	28 105 198	Europe
Asia, East and South East/Oceania	259 166	274 235	334 394	389 918	433 700	Asie, Est et Sud-Est et Océanie
Southern Asia	37 725	39 043	36 075	39 796	42 653	Asie du Sud
Western Asia	101 494	101 150	82 676	83 776	77 239	Asie occidentale
Sri Lanka	**224 258**	**180 006**	**179 872**	**181 254**	**295 222**	**Sri Lanka**
Africa	454	416	530	362	556	Afrique
Americas	10 022	8 100	7 754	6 844	8 802	Amériques
Europe	145 670	124 516	116 000	120 064	198 132	Europe
Asia, East and South East/Oceania	32 122	24 528	27 114	31 672	50 706	Asie, Est et Sud-Est et Océanie
Southern Asia	35 990	22 058	28 046	21 914	29 158	Asie du Sud
Western Asia	...	388	428	398	7 868	Asie occidentale
Sudan[3]	**21 057**	**33 724**	**32 956**	**20 354**	**24 032**	**Soudan**[3]
Africa	4 338	7 941	6 574	4 365	4 206	Afrique
Americas	2 342	3 161	3 083	1 882	2 585	Amériques
Europe	9 342	13 206	15 900	9 350	10 678	Europe
Asia, East and South East/Oceania	1 581	1 855	1 614	1 282	1 858	Asie, Est et Sud-Est et Océanie
Southern Asia	800	1 820	1 478	965	1 174	Asie du Sud
Western Asia	2 654	5 741	4 307	2 510	3 531	Asie occidentale
Suriname[2]	**58 884**	**42 583**	**39 880**	**19 965**	**...**	**Suriname**[2]
Americas	31 486	24 215	10 030	3 204	...	Amériques
Europe	25 994	17 261	28 150	16 761	...	Europe
Asia, East and South East/Oceania	1 404	1 107	1 700	...	...	Asie, Est et Sud-Est et Océanie
Swaziland[6]	**229 360**	**209 477**	**211 586**	**267 706**	**...**	**Swaziland**[6]
Africa	191 206	182 458	184 365	232 842	...	Afrique
Americas	10 738	3 817	3 850	4 931	...	Amériques
Europe	25 864	21 146	21 329	27 318	...	Europe
Asia, East and South East/Oceania	1 552	2 056	2 042	2 615	...	Asie, Est et Sud-Est et Océanie
Sweden *[6]	**745 775**	**738 862**	**740 188**	**790 703**	**681 299**	**Suède** *[6]
Americas	40 667	43 899	40 634	41 747	41 264	Amériques
Europe	699 031	687 820	691 047	739 643	629 243	Europe
Asia, East and South East/Oceania	6 077	7 143	8 507	9 313	10 792	Asie, Est et Sud-Est et Océanie

122
Tourist arrivals by region of origin [*cont.*]
Arrivées de touristes par région de provenance [*suite*]

Country or area of destination and region of origin	1986	1987	1988	1989	1990	Pays ou zone de destination et région de provenance
Switzerland[6]	**8 970 200**	**9 121 200**	**9 152 100**	**9 856 100**	**10 264 032**	**Suisse**[6]
Africa	144 100	150 200	147 900	151 200	155 072	Afrique
Americas	1 305 900	1 402 900	1 264 800	1 397 200	1 580 099	Amériques
Europe	6 794 200	6 810 500	6 978 300	7 447 900	7 638 588	Europe
Asia, East and South East/Oceania	477 300	514 200	539 100	637 400	670 757	Asie, Est et Sud-Est et Océanie
Southern Asia	60 400	54 700	48 900	48 100	52 001	Asie du Sud
Western Asia	188 300	188 700	173 100	174 300	167 515	Asie occidentale
Syrian Arab Republic[19]	**1 124 378**	**1 194 134**	**1 246 211**	**1 327 763**	**1 402 990**	**Rép. arabe syrienne**[19]
Africa	40 543	45 385	45 990	47 854	52 210	Afrique
Americas	12 685	10 666	9 227	14 976	14 034	Amériques
Europe	64 322	52 890	66 686	69 776	88 397	Europe
Asia, East and South East/Oceania	5 780	5 423	5 434	9 264	7 315	Asie, Est et Sud-Est et Océanie
Southern Asia	182 538	135 965	173 429	215 880	200 878	Asie du Sud
Western Asia	818 510	943 805	945 445	970 013	1 040 156	Asie occidentale
Thailand[2]	**2 818 092**	**3 482 958**	**4 230 737**	**4 809 508**	**5 298 860**	**Thaïlande**[2]
Africa	13 683	18 873	24 922	27 484	31 943	Afrique
Americas	244 308	292 448	327 727	366 016	381 894	Amériques
Europe	621 514	794 320	1 068 043	1 207 332	1 322 752	Europe
Asia, East and South East/Oceania	1 617 404	2 042 437	2 410 043	2 844 662	3 214 779	Asie, Est et Sud-Est et Océanie
Southern Asia	212 819	219 442	277 247	251 149	270 568	Asie du Sud
Western Asia	108 364	115 438	122 755	112 865	76 924	Asie occidentale
Togo[1]	**98 839**	**97 887**	**103 446**	**115 083**	**103 185**	**Togo**[1]
Africa	39 316	38 554	41 745	51 238	48 399	Afrique
Americas	6 613	6 458	6 583	8 540	6 953	Amériques
Europe	48 879	48 076	50 844	51 374	44 660	Europe
Asia, East and South East/Oceania	4 031	4 799	4 274	3 931	3 173	Asie, Est et Sud-Est et Océanie
Tonga[3]	**15 562**	**16 621**	**18 428**	**20 994**	**19 020**	**Tonga**[3]
Americas	3 579	4 546	4 575	4 553	4 881	Amériques
Europe	2 002	2 648	3 310	3 708	3 175	Europe
Asia, East and South East/Oceania	9 981	9 427	10 543	12 442	10 964	Asie, Est et Sud-Est et Océanie
Southern Asia	...	...	...	291	...	Asie du Sud
Trinidad and Tobago	**181 820**	**192 372**	**179 305**	**184 805**	**184 691**	**Trinité-et-Tobago**
Americas	159 730	162 870	149 998	151 940	146 915	Amériques
Europe	22 090	29 502	29 307	32 865	37 776	Europe
Tunisia[2]	**1 480 208**	**1 850 233**	**3 436 844**	**3 184 465**	**3 150 504**	**Tunisie**[2]
Africa	313 916	273 187	1 710 374	1 454 814	1 386 714	Afrique
Americas	8 614	9 513	10 342	12 523	12 608	Amériques
Europe	1 135 328	1 547 664	1 681 882	1 670 538	1 705 451	Europe
Western Asia	22 350	19 869	34 246	46 590	45 731	Asie occidentale
Turkey[1]	**2 389 908**	**2 853 401**	**4 169 606**	**4 455 302**	**5 385 643**	**Turquie**[1]
Africa	69 107	139 725	202 008	133 912	89 924	Afrique
Americas	108 687	172 887	216 773	266 763	273 436	Amériques
Europe	1 744 266	2 033 777	3 148 002	3 388 274	4 346 119	Europe
Asia, East and South East/Oceania	43 651	49 576	66 713	75 588	85 342	Asie, Est et Sud-Est et Océanie
Southern Asia	196 666	181 012	250 950	272 286	280 425	Asie du Sud
Western Asia	227 531	276 424	285 160	318 479	310 397	Asie occidentale
Turks and Caicos Islands	**30 944**	**32 525**	**44 272**	**40 710**	**37 008**	**Iles Turques et Caïques**
Americas	27 367	29 707	40 975	37 393	31 991	Amériques
Europe	3 577	2 818	3 297	3 317	5 017	Europe

122
Tourist arrivals by region of origin [cont.]
Arrivées de touristes par région de provenance [suite]

Country or area of destination and region of origin	1986	1987	1988	1989	1990	Pays ou zone de destination et région de provenance
Tuvalu	**632**	**525**	**658**	**565**	**638**	**Tuvalu**
Americas	73	118	113	84	76	Amériques
Europe	107	110	142	110	129	Europe
Asia, East and South East/Oceania	452	297	403	371	433	Asie, Est et Sud-Est et Océanie
former USSR[1,21]	**3 934 400**	**4 797 700**	**5 407 000**	**2 541 557**	**3 054 029**	**ancienne URSS**[1,21]
Africa	...	...	...	69 000	37 000	Afrique
Americas	95 600	141 300	166 800	253 658	282 658	Amériques
Europe	3 769 800	4 581 300	5 164 300	1 634 277	1 968 867	Europe
Asia, East and South East/Oceania	51 700	54 700	57 100	412 000	570 000	Asie, Est et Sud-Est et Océanie
Southern Asia	17 300	20 400	18 800	80 622	88 504	Asie du Sud
Western Asia	...	...	...	92 000	107 000	Asie occidentale
United Kingdom[1,17]	**13 415 000**	**15 037 000**	**15 276 000**	**16 752 000**	**17 321 000**	**Royaume-Uni**[1,17]
Africa	613 000	542 000	522 000	548 000	563 000	Afrique
Americas	3 081 000	3 624 000	3 490 000	3 730 000	4 005 000	Amériques
Europe	8 387 000	9 368 000	9 749 000	10 803 000	10 889 000	Europe
Asia, East and South East/Oceania	764 000	927 000	999 000	1 163 000	1 326 000	Asie, Est et Sud-Est et Océanie
Western Asia	570 000	576 000	516 000	508 000	538 000	Asie occidentale
United Rep.Tanzania[1]	**103 363**	**130 851**	**130 343**	**137 889**	**153 000**	**Rép. Unie de Tanzanie**[1]
Africa	38 693	51 374	51 175	54 138	59 691	Afrique
Americas	18 206	29 267	29 153	30 841	34 005	Amériques
Europe	31 023	46 566	46 385	49 071	55 072	Europe
Asia, East and South East/Oceania	15 441	3 644	3 630	3 839	4 232	Asie, Est et Sud-Est et Océanie
United States	**25 358 501**	**29 556 810**	**34 245 000**	**36 604 000**	**39 772 000**	**Etats-Unis**
Africa	165 232	146 200	158 915	158 557	159 159	Afrique
Americas	18 547 283	21 298 613	24 108 895	25 231 832	27 589 609	Amériques
Europe	3 721 570	4 663 576	5 761 685	6 251 036	6 658 796	Europe
Asia, East and South East/Oceania	2 506 817	3 022 225	3 727 747	4 469 976	4 853 109	Asie, Est et Sud-Est et Océanie
Southern Asia	140 490	139 738	155 251	153 897	165 808	Asie du Sud
Western Asia	277 109	286 458	332 507	338 702	345 519	Asie occidentale
United States Virgin Is.[8]	**275 797**	**554 071**	**667 403**	**587 432**	**436 702**	**Iles Vierges américaines**[8]
Americas	270 988	543 982	653 570	575 958	426 451	Amériques
Europe	4 809	10 089	13 833	11 474	10 251	Europe
Uruguay[1]	**1 094 000**	**986 535**	**836 831**	**1 038 919**	**...**	**Uruguay**[1]
Americas	1 080 000	953 000	795 167	993 643	...	Amériques
Europe	14 000	33 535	40 177	42 372	...	Europe
Asia, East and South East/Oceania	...	...	1 487	1 452	...	Asie, Est et Sud-Est et Océanie
Western Asia	...	...	...	1 452	...	Asie occidentale
Vanuatu	**16 440**	**13 364**	**16 389**	**22 304**	**32 527**	**Vanuatu**
Americas	600	729	648	693	904	Amériques
Europe	731	666	676	716	1 311	Europe
Asia, East and South East/Oceania	15 109	11 969	15 065	20 895	30 312	Asie, Est et Sud-Est et Océanie
Venezuela	**310 311**	**337 612**	**372 392**	**411 050**	**523 132**	**Venezuela**
Africa	239	260	149	214	571	Afrique
Americas	187 759	204 279	213 296	236 104	296 412	Amériques
Europe	116 895	127 180	154 208	168 155	217 265	Europe
Asia, East and South East/Oceania	3 055	3 323	2 685	3 324	3 623	Asie, Est et Sud-Est et Océanie
Southern Asia	294	319	190	398	610	Asie du Sud
Western Asia	2 069	2 251	1 864	2 855	4 651	Asie occidentale

122
Tourist arrivals by region of origin [cont.]
Arrivées de touristes par région de provenance [suite]

Country or area of destination and region of origin	1986	1987	1988	1989	1990	Pays ou zone de destination et région de provenance
Yemen [1][22]	**52 125**	**50 723**	**60 174**	**65 188**	**51 849**	**Yémen** [1][22]
Africa	3 860	3 543	37 014	3 816	3 911	Afrique
Americas	4 246	2 949	2 463	2 826	3 370	Amériques
Europe	26 157	28 863	38 706	40 494	30 081	Europe
Asia, East and South East/Oceania	5 845	4 756	5 284	6 617	4 366	Asie, Est et Sud-Est et Océanie
Western Asia	12 017	10 612	10 020	11 435	10 121	Asie occidentale
Yugoslavia [6]	**8 340 917**	**8 784 275**	**8 887 230**	**8 513 554**	**7 743 473**	**Yougoslavie** [6]
Americas	181 553	275 072	342 766	299 342	279 455	Amériques
Europe	7 997 584	8 320 424	8 338 325	8 015 485	7 270 231	Europe
Asia, East and South East/Oceania	45 548	47 444	53 668	56 639	57 834	Asie, Est et Sud-Est et Océanie
Western Asia	116 232	141 335	152 471	142 088	135 953	Asie occidentale
Zaire [2][19]	**31 236**	**36 167**	**39 444**	**51 422**	...	**Zaïre** [2][19]
Africa	4 200	11 003	13 107	21 719	...	Afrique
Americas	3 409	3 377	4 598	5 365	...	Amériques
Europe	22 430	20 290	19 645	21 835	...	Europe
Asia, East and South East/Oceania	1 197	1 497	2 094	2 503	...	Asie, Est et Sud-Est et Océanie
Zambia	**122 890**	**120 589**	**108 290**	**113 156**	**139 988**	**Zambie**
Africa	88 089	83 515	83 265	79 194	105 405	Afrique
Americas	5 492	6 092	3 804	5 596	5 441	Amériques
Europe	24 214	25 323	16 845	22 297	22 497	Europe
Asia, East and South East/Oceania	1 876	2 717	1 855	2 612	3 781	Asie, Est et Sud-Est et Océanie
Southern Asia	2 760	2 394	2 116	2 899	2 679	Asie du Sud
Western Asia	459	548	405	558	185	Asie occidentale
Zimbabwe [23]	**318 963**	**339 890**	**412 543**	**436 248**	**552 686** [18]	**Zimbabwe** [23]
Africa	246 259	261 768	331 560	337 974	436 220	Afrique
Americas	17 833	17 728	17 173	19 849	21 539	Amériques
Europe	49 326	53 107	55 235	67 758	81 827	Europe
Asia, East and South East/Oceania	5 545	7 287	8 575	10 667	13 100	Asie, Est et Sud-Est et Océanie

Source:
World Tourism Organization (Madrid).

† All data shown which pertain to Germany prior to 3 October 1990 are indicated separately for the Federal Republic of Germany and the former German Democratic Republic based on their respective territories at the time indicated. Where data for united Germany (3 October 1990 and thereafter) are not available, available data are shown separately under the designations Federal Republic of Germany and former German Democratic Republic and pertain to the territorial boundaries prior to 3 October 1990. For detailed explanatory notes on data pertaining to Germany, see Annex I - Country Nomenclature.

1 International visitor arrivals at frontiers (including tourists and same-day visitors).
2 Excluding nationals of the country residing abroad.
3 Air arrivals.
4 Air and sea arrivals.
5 Excluding crew members.
6 International tourist arrivals at all means of accommodation.
7 International tourist arrivals in hotels of regional capitals.

Source:
Organisation mondiale du tourisme (Madrid).

† Toutes les données se rapportant à l'Allemagne avant le 3 octobre 1990 figurent dans deux rubriques séparées basées sur les territoires respectifs de la République fédérale d'Allemagne et l'ancienne République démocratique allemande selon la période indiquée. En l'absence de données pour l'Allemagne unifiée (à compter du 3 octobre 1990), les données disponibles sont fournies séparément sous les rubriques République fédérale d'Allemagne et ancienne République démocratique allemande et se rapportent aux limites territoriales antérieures au 3 octobre 1990. Pour les notes explicatives en détail sur les données concernant l'Allemagne, voir Annexe I - Nomenclature des pays.

1 Arrivées de visiteurs internationaux aux frontières (y compris touristes et visiteurs de la journée).
2 A l'exclusion des nationaux du pays résidant à l'étranger.
3 Arrivées par voie aérienne.
4 Arrivées par voie aérienne et maritime.
5 A l'exclusion des membres des équipages.
6 Arrivées de touristes internationaux dans l'ensemble des moyens d'hébergement.
7 Arrivées de touristes internationaux dans les hôtels des capitales de département.

122
Tourist arrivals by region of origin
[*cont.*]

Arrivées de touristes par région de provenance
[*suite*]

8 International tourist arrivals in hotels and similar establishments.
9 Including nationals of the country residing abroad.
10 Excluding ethnic Chinese arriving from Hong Kong, Macau and Taiwan.
11 International visitor arrivals at all means of accommodation in Brazzaville, Pointe Noire, Loubomo, Owando and Sibiti.

12 Air arrivals at Rarotonga.
13 International tourist arrivals at Addis Ababa, Asmara and Assab.
14 Excluding camping sites.
15 As of 1990, tourists from the former German Democratic Republic will be regarded as domestic tourists.

16 Data based on surveys.
17 Departures.
18 Travellers.
19 Arrivals by air and land.
20 Excluding Malaysian citizens arriving by land and cruise passengers, but including excursionists (same-day visitors).

21 Excluding arrivals from eastern and central European countries.
22 Excluding former Democratic Yemen.
23 Excluding transit passengers.

8 Arrivées de touristes internationaux dans les hôtels et établissements assimilés.
9 Y compris les nationaux du pays résidant à l'étranger.
10 A l'exclusion des arrivées de personnes d'ethnie chinoise en provenance de Hong-kong, Macao et Taïwan.
11 Arrivées des touristes internationaux dans l'ensemble des moyens d'hébergement de Brazzaville, Pointe Noire, Loubomo, Owando et Sibiti.
12 Arrivées par voie aérienne à Rarotonga.
13 Arrivées de touristes internationaux à Addis Abeba, Asmara et Assab.
14 A l'exclusion des terrains de camping.
15 A partir de 1990, les touristes en provenance de l'ancienne République démocratique allemande seront considérés comme des touristes nationaux.
16 Données obtenues au moyen d'enquêtes.
17 Départs.
18 Voyageurs.
19 Arrivées par voies aérienne et terrestre.
20 A l'exclusion des arrivées de malaisiens par voie terrestre et de passagers en croisière, mais y compris les excursionnistes (visiteurs de la journée).
21 A l'exclusion des arrivées en provenance des pays d'Europe centrale et de l'est.
22 Non compris l'ancien Yémen démocratique.
23 A l'exclusion des passagers en transit.

123
Tourist arrivals and international tourism receipts
Arrivées de touristes et recettes touristiques internationales

Country or area Pays ou zone	Number of tourist arrivals (000) Nombre d'arrivées de touristes (000)					Tourist receipts (million US dollars) Recettes touristiques (millions de dollars E-U)				
	1986	1987	1988	1989	1990	1986	1987	1988	1989	1990
World *Monde*	340 891	367 402	392 813	427 660	454 875	140 023	171 352	197 712	211 436	255 006
Africa **Afrique**	10 789	11 717	14 479	16 406	17 728	3 757	5 369	6 347	5 999	6 801
Algeria Algérie	849	778	967	1 207	1 137	104	100	85	64	64
Angola Angola	...	...	39	40	46	...	...	...	...	...
Benin Bénin	46	47	45	43	50	33	39	40	41	47
Botswana Botswana	381	432	384	691	844	40	49	38	54	65
Burkina Faso Burkina Faso	45	51	55	51	46	7	7	6	5	9
Burundi Burundi	66	80	99	82	109	1	1	2	3	4
Cameroon Cameroun	131	118	100	87	100	47	25	21	17	21
Central African Rep. Rép. centrafricaine	4	5	5	5	6	5	9	8	8	9
Chad Tchad	25	27	21	12	29	5	6	7	9	12
Comoros Comores	6	8	8	13	8	3	3	3	3	2
Congo Congo	39	39	39	40	46	6	7	6	6	7
Côte d'Ivoire Côte d'Ivoire	184	175	181	192	196	52	59	57	60	48
Djibouti Djibouti	19	22	29	41	47	6	6	5	5	6
Egypt Egypte	1 236	1 671	1 833	2 351	2 411	785	1 586	1 784	1 646	1 994
Ethiopia Ethiopie	59	73	76	74	73	5	15	19	21	25
Gabon Gabon	31	21	20	113	108	13	5	7	4	4
Gambia Gambie	74	96	100	86	101	23	14	23	23	26
Ghana Ghana	92	103	114	125	146	27	36	55	72	81
Kenya Kenya	604	662	677	714	801	313	344	410	375	443
Lesotho Lesotho	131	135	110	169	171	7	10	8	8	9
Libyan Arab Jamah. Jamah. arabe libyenne	95	97	98	95	96	2	3	7	6	6
Madagascar Madagascar	27	28	35	39	53	6	10	23	29	43
Malawi Malawi	69	76	99	117	130	8	6	6	7	8

123
Tourist arrivals and international tourism receipts
[cont.]

Arrivées de touristes et recettes touristiques internationales
[suite]

Country or area Pays ou zone	Number of tourist arrivals (000) Nombre d'arrivées de touristes (000)					Tourist receipts (million US dollars) Recettes touristiques (millions de dollars E-U)				
	1986	1987	1988	1989	1990	1986	1987	1988	1989	1990
Mali Mali	33	34	36	32	44	31	37	38	28	32
Mauritania Mauritanie	...	...	...	...	...	8	14	12	13	15
Mauritius Maurice	165	208	239	263	292	88	138	172	183	264
Morocco Maroc	2 128	2 248	2 841	3 468	4 024	739	936	1 110	1 146	1 259
Niger Niger	29	31	33	24	21	7	9	11	13	15
Nigeria Nigéria	311	177	143	161	190	147	77	53	21	25
Réunion Réunion	165	170	175	182	200	...	169	...	...	...
Rwanda Rwanda	35	34	36	37	43	7	7	7	9	10
Sao Tome and Principe Sao Tomé-et-Principe	1	1	1	1	1	1	1	1	1	1
Senegal Sénégal	235	235	256	259	246	115	133	143	138	152
Seychelles Seychelles	67	72	77	86	104	55	67	81	91	120
Sierra Leone Sierra Leone	194	88	75	86	98	10	14	15	17	19
Somalia Somalie	39	39	40	40	46	...	...	...	...	...
South Africa Afrique du Sud	645	703	805	930	1 029	414	589	673	709	815
Sudan Soudan	42	52	37	23	33	21	14	29	45	5
Swaziland Swaziland	240	211	213	270	300	14	20	22	25	25
Togo Togo	99	98	104	115	103	34	40	42	41	23
Tunisia Tunisie	1 502	1 875	3 468	3 222	3 204	488	672	1 234	933	953
Uganda Ouganda	32	37	40	44	50	3	5	8	9	10
United Rep.Tanzania Rép. Unie de Tanzanie	103	131	130	138	153	27	31	40	60	65
Zaire Zaïre	31	36	39	51	46	14	18	7	6	7
Zambia Zambie	123	121	108	113	141	7	6	5	5	6
Zimbabwe Zimbabwe	357	372	449	474	606	29	32	24	40	47
America, North Amérique du Nord	63 340	69 398	75 534	79 221	84 104	33 521	38 228	45 459	52 944	61 824
Anguilla Anguilla	17	21	26	28	31	13	17	21	25	29

123
Tourist arrivals and international tourism receipts
[*cont.*]

Arrivées de touristes et recettes touristiques internationales
[*suite*]

Country or area Pays ou zone	Number of tourist arrivals (000) Nombre d'arrivées de touristes (000)					Tourist receipts (million US dollars) Recettes touristiques (millions de dollars E-U)				
	1986	1987	1988	1989	1990	1986	1987	1988	1989	1990
Antigua and Barbuda Antigua-et-Barbuda	159	173	187	189	197	156	187	214	232	250
Aruba Aruba	181	232	278	344	433	158	204	279	310	353
Bahamas Bahamas	1 375	1 480	1 475	1 575	1 562	1 105	1 146	1 150	1 310	1 333
Barbados Barbade	370	422	451	461	432	327	379	459	505	494
Belize Belize	94	103	164	220	222	41	47	48	79	91
Bermuda Bermudes	460	478	427	417	434	422	468	441	451	490
British Virgin Islands Iles Vierges britanniques	141	173	176	176	160	89	111	121	125	132
Canada Canada	15 621	14 975	15 485	15 111	15 258	3 860	3 961	4 603	5 014	6 374
Cayman Islands Iles Caïmanes	166	209	219	210	253	94	146	244	236	326
Costa Rica Costa Rica	261	278	329	376	435	133	136	165	207	275
Cuba Cuba	276	282	298	314	300	150	185	223	230	268
Dominica Dominique	24	27	34	37	45	11	13	14	19	25
Dominican Republic Rép. dominicaine	747	902	1 116	1 400	1 533	420	545	616	750	750
El Salvador El Salvador	134	125	134	131	194	41	43	61	63	69
Grenada Grenade	57	57	62	69	82	29	30	29	31	38
Guadeloupe Guadeloupe	284	327	354	323	331	163	188	220	183	231
Guatemala Guatemala	287	353	405	437	508	77	103	124	152	185
Haiti Haïti	112	122	133	122	120	63	69	75	75	75
Honduras Honduras	156	158	162	176	202	26	27	28	28	29
Jamaica Jamaïque	663	739	649	715	841	516	595	525	593	740
Martinique Martinique	183	234	280	312	282	108	210	230	272	240
Mexico Mexique	12 659	13 831	14 142	14 964	15 695	2 984	3 499	4 000	4 766	5 324
Montserrat Montserrat	16	17	18	21	19	8	10	11	11	14
Nicaragua Nicaragua	44	49	55	77	106	8	9	5	4	12
Panama Panama	308	271	199	192	214	205	188	168	157	167

123
Tourist arrivals and international tourism receipts
[*cont.*]

Arrivées de touristes et recettes touristiques internationales
[*suite*]

Country or area	Number of tourist arrivals (000) Nombre d'arrivées de touristes (000)					Tourist receipts (million US dollars) Recettes touristiques (millions de dollars E-U)				
Pays ou zone	1986	1987	1988	1989	1990	1986	1987	1988	1989	1990
Puerto Rico Porto Rico	1 696	2 035	2 281	2 444	2 554	793	955	1 121	1 254	1 367
Saint Kitts and Nevis Saint-Kitts-et-Nevis	57	66	70	72	76	38	47	54	60	63
Saint Lucia Sainte-Lucie	112	118	133	133	147	118	126	134	145	154
St. Vincent-Grenadines St. Vincent-Grenadines	42	46	47	50	54	29	35	39	43	54
Trinidad and Tobago Trinité-et-Tobago	191	202	188	194	194	83	92	81	85	95
Turks and Caicos Islands Iles Turques et Caiques	35	37	47	48	42	22	24	32	34	31
United States Etats-Unis	25 359	29 658	34 245	36 604	39 772	20 454	23 505	28 935	34 432	40 579
United States Virgin Is. Iles Vierges américaines	463	542	553	505	523	510	639	657	682	707
America, South **Amérique du Sud**	**7 632**	**7 043**	**7 972**	**8 307**	**8 634**	**4 131**	**3 976**	**4 275**	**4 188**	**5 781**
Argentina Argentine	1 774	1 472	2 119	2 492	2 728	564	615	634	790	1 975
Bolivia Bolivie	133	147	167	194	217	35	40	65	75	93
Brazil Brésil	1 934	1 929	1 743	1 403	1 079	1 527	1 502	1 643	1 225	1 444
Chile Chili	581	575	624	797	950	183	182	205	407	548
Colombia Colombie	732	541	829	733	813	418	349	461	335	362
Ecuador Equateur	267	274	347	335	362	170	167	173	187	188
Guyana Guyana	47	60	71	67	67	19	24	30	30	30
Paraguay Paraguay	371	303	284	279	280	148	121	114	112	112
Peru Pérou	304	330	359	334	317	357	341	448	402	398
Suriname Suriname	29	27	21	21	29	8	11	8	8	11
Uruguay Uruguay	1 149	1 047	1 035	1 240	1 267	258	208	203	228	261
Venezuela Venezuela	311	338	373	412	525	444	416	291	389	359
Asia **Asie**	**41 396**	**46 499**	**53 224**	**54 939**	**62 495**	**21 293**	**27 190**	**33 840**	**37 525**	**41 044**
Afghanistan Afghanistan	9	9	9	8	8	1	1	1	1	1
Bahrain Bahreïn	96	176	192	193	195	83	101	114	115	110

123
Tourist arrivals and international tourism receipts
[cont.]

Arrivées de touristes et recettes touristiques internationales
[suite]

Country or area Pays ou zone	Number of tourist arrivals (000) Nombre d'arrivées de touristes (000)					Tourist receipts (million US dollars) Recettes touristiques (millions de dollars E-U)				
	1986	1987	1988	1989	1990	1986	1987	1988	1989	1990
Bangladesh Bangladesh	129	107	121	128	115	14	12	13	18	11
Bhutan Bhoutan	3	3	2	2	2	1	1	1	2	2
Brunei Darussalam Brunéi Darussalam	411	523	573	600	625	22	25	30	32	35
China Chine	9 000	10 760	12 361	9 361	10 484	1 531	1 862	2 247	1 861	2 218
Cyprus Chypre	828	949	1 112	1 378	1 561	497	666	782	990	1 258
Hong Kong Hong-kong	3 733	4 502	5 589	5 361	5 933	2 287[1]	3 261[1]	4 273[1]	4 731[1]	5 032[1]
India Inde	1 451	1 484	1 591	1 736	1 707	1 260	1 430	1 500	1 535	1 437
Indonesia Indonésie	823	1 060	1 301	1 626	2 178	647	924	1 283	1 628	2 105
Iran, Islamic Rep. of Iran, Rép. islamique d'	93	69	67	89	154	28	27	25	37	62
Iraq Iraq	1 004	739	227	1 025	747	84	62	61	59	55
Israel Israël	1 102	1 379	1 170	1 177	1 063	971	1 342	1 347	1 468	1 382
Japan Japon	1 055	1 069	1 116	1 499	1 879	1 463	2 097	2 893	3 143	3 578
Jordan Jordanie	1 912	1 898	2 391	2 278	2 633	535	582	617	551	500
Korea, Republic of Corée, République de	1 660	1 875	2 340	2 728	2 959	1 550	2 299	3 265	3 556	3 559
Kuwait Koweït	84	78	80	89	50	86	75	108	123	80
Lao People's Dem. Rep. Rép. dém. pop. lao	25	25	25	25	25	...	...	...	...	...
Macau Macao	772	908	1 005	1 008	1 138	...	...	...	...	139
Malaysia Malaisie	3 217	3 359	3 624	4 846	7 446	642	718	745	1 038	1 667
Maldives Maldives	114	131	156	158	195	43	45	53	70	85
Mongolia Mongolie	199	186	240	237	147	...	...	...	...	...
Myanmar Myanmar	47	48	26	14	21	10	12	8	3	5
Nepal Népal	223	248	265	240	255	51	60	64	68	57
Oman Oman	88	112	126	136	149	47	47	49	56	69
Pakistan Pakistan	432	425	460	495	424	180	173	143	159	156
Philippines Philippines	764	781	1 023	1 076	893	1 006[2]	1 029[2]	1 301[2]	1 465[2]	1 306[2]

123
Tourist arrivals and international tourism receipts
[*cont.*]

Arrivées de touristes et recettes touristiques internationales
[*suite*]

Country or area Pays ou zone	Number of tourist arrivals (000) Nombre d'arrivées de touristes (000)					Tourist receipts (million US dollars) Recettes touristiques (millions de dollars E-U)				
	1986	1987	1988	1989	1990	1986	1987	1988	1989	1990
Qatar Qatar	99	101	103	104	100	...	...	...	...	...
Saudi Arabia Arabie saoudite	857	960	763	769	827	2 000	2 600	2 066	2 050	1 884
Singapore Singapour	2 902	3 373	3 833	4 397	4 842	1 767	2 088	2 622	3 307	4 362
Sri Lanka Sri Lanka	230	183	183	185	298	82	82	77	76	125
Syrian Arab Republic Rép. arabe syrienne	618	493	421	411	562	363	204	332	290	244
Thailand Thaïlande	2 818	3 483	4 231	4 810	5 299	1 421	1 947	3 120	3 753	4 326
Turkey Turquie	2 079	2 468	3 715	3 921	4 799	1 215	1 721	2 355	2 557	3 349
United Arab Emirates Emirats arabes unis	730	584	590	593	616	...	...	...	...	...
Viet Nam Viet Nam	127	139	148	167	180	26	30	35	59	85
Yemen[3] Yémen[3]	52	51	60	65	52	47	48	21	26	20
Europe Europe	**214 311**	**228 827**	**237 023**	**264 072**	**276 755**	**74 345**	**92 699**	**102 628**	**105 002**	**132 977**
Austria Autriche	15 092	15 761	16 571	18 202	19 011	6 954[4]	8 863[4]	10 090[4]	10 717[4]	13 017[4]
Belgium Belgique	2 454	2 516	2 700	3 007	3 163	2 271	2 980	3 438	3 064	3 575
Bulgaria Bulgarie	3 506	3 604	3 967	4 316	4 500	356	357	359	362	394
Czechoslovakia Tchécoslovaquie	5 330	6 126	6 886	8 036	8 100	383	493	608	581	470
Denmark Danemark	1 216	1 171	1 150	1 218	1 275	1 759	2 219	2 423	2 313	3 322
Finland Finlande	598	823	877	882	866	598	823	983	1 013	1 169
France France	36 080	36 974	38 288	49 544	53 157	9 724	11 870	13 786	16 245	20 185
Germany† · Allemagne† F. R. Germany R. f. Allemagne	13 458	14 045	14 501	16 115	17 045	6 294	7 678	8 449	8 658	10 683
former German D. R. anc. R. d. allemande	1 950	2 102	2 231	3 102	...	...	...	...	...	...
Gibraltar Gibraltar	90	123	156	162	132	41	53	66	91	112
Greece Grèce	7 025	7 564	7 923	8 082	8 873	1 834	2 268	2 396	1 976	2 587
Hungary Hongrie	10 613	11 826	10 563	14 490	20 510	592	784	758	798	1 000
Iceland Islande	114	129	129	131	142	60	86	108	108	122

123
Tourist arrivals and international tourism receipts
[cont.]

Arrivées de touristes et recettes touristiques internationales
[suite]

Country or area Pays ou zone	Number of tourist arrivals (000) Nombre d'arrivées de touristes (000)					Tourist receipts (million US dollars) Recettes touristiques (millions de dollars E-U)				
	1986	1987	1988	1989	1990	1986	1987	1988	1989	1990
Ireland Irlande	2 467	2 664	3 007	3 484	3 666	639	839[5]	997[5]	1 070[5]	1 447[5]
Italy Italie	24 672	25 749	26 155	25 935	26 679	9 855	12 174	12 403	11 984	19 742
Liechtenstein Liechtenstein	76	75	72	77	78	...	...	...	...	...
Luxembourg Luxembourg	616	645	760	875	820	193	201	238	286	290
Malta Malte	574	746	784	828	872	203	327	382	372	496
Monaco Monaco	211	214	232	245	245	...	...	...	...	...
Netherlands Pays-Bas	4 829	4 922	4 876	5 206	5 795	2 219	2 695	2 899	3 052	3 615
Norway Norvège	1 638	1 782	1 704	1 867	1 955	1 059	1 255	1 466	1 336	1 517
Poland Pologne	2 500	2 484	2 495	3 293	3 400	136	184	206	202	266
Portugal Portugal	5 409	6 102	6 624	7 116	8 020	1 533	2 145	2 402	2 685	3 556
Romania Roumanie	4 535	5 142	5 514	4 901	6 533	141	176	171	167	106
San Marino[6] Saint-Marin[6]	454	505	504	437	582	...	...	...	...	...
Spain Espagne	29 910	32 900	35 000	35 350	34 300	12 058	14 760	16 686	16 174	18 593
Sweden Suède	824	814	830	837	731	1 553	2 033	2 346	2 543	2 895
Switzerland Suisse	11 400	11 600	11 700	12 600	13 200	4 227	5 345	5 720	5 543	6 839
former USSR ancienne URSS	4 309	5 246	6 007	7 752	7 204	163	198	216	250	270
United Kingdom Royaume-Uni	13 897	15 566	15 799	17 338	18 021	8 163	10 225	11 008	11 182	13 935
Yugoslavia Yougoslavie	8 464	8 907	9 018	8 644	7 880	1 337	1 668	2 024	2 230	2 774
Oceania Océanie	**3 423**	**3 918**	**4 581**	**4 715**	**5 159**	**2 976**	**3 890**	**5 163**	**5 778**	**6 579**
American Samoa Samoa américaines	37	38	39	47	47	6	7	8	9	10
Australia Australie	1 429	1 785	2 249	2 080	2 215	1 300	1 789	2 801	3 156	3 635
Cook Islands Iles Cook	31	32	34	33	34	13	13	15	15	16
Fiji Fidji	258	190	208	251	279	163	121	128	198	230
French Polynesia Polynésie française	161	143	135	140	132	136	150	157	136	142

123
Tourist arrivals and international tourism receipts
[cont.]

Arrivées de touristes et recettes touristiques internationales
[suite]

Country or area Pays ou zone	Number of tourist arrivals (000) Nombre d'arrivées de touristes (000)					Tourist receipts (million US dollars) Recettes touristiques (millions de dollars E-U)				
	1986	1987	1988	1989	1990	1986	1987	1988	1989	1990
Guam Guam	407	484	586	669	780	488	580	706	804	936
Kiribati Kiribati	3	4	3	3	3	1	1	1	1	2
New Caledonia Nouvelle-Calédonie	59	61	61	82	87	50	64	65	112	150
New Zealand Nouvelle-Zélande	734	844	865	901	976	622	935	1 014	1 005	1 072
Niue Nioué	2	2	1	1	1	...	...	...	...	...
Papua New Guinea Papouasie-Nvl-Guinée	32	35	41	49	41	18	21	25	32	28
Samoa Samoa	50	49	49	54	48	17	18	18	19	20
Solomon Islands Iles Salomon	12	13	11	10	9	5	6	5	5	4
Tonga Tonga	16	17	19	21	21	8	10	7	9	9
Tuvalu Tuvalu	1	1	1	1	1	...	...	...	...	...
Vanuatu Vanuatu	18	15	18	24	35	9	8	11	16	25

Source:
World Tourism Organization (Madrid).

† All data shown which pertain to Germany prior to 3 October 1990 are indicated separately for the Federal Republic of Germany and the former German Democratic Republic based on their respective territories at the time indicated. Where data for united Germany (3 October 1990 and thereafter) are not available, available data are shown separately under the designations Federal Republic of Germany and former German Democratic Republic and pertain to the territorial boundaries prior to 3 October 1990. For detailed explanatory notes on data pertaining to Germany, see Annex I - Country Nomenclature.

1 Receipts from visitors, excluding service men, air crew members and transit passengers.

2 Receipts from visitors.
3 Excluding former Democratic Yemen.
4 Including international fare receipts.

5 Including excursionist (same-day visitors) revenues.

6 Excluding Italian visitors.

Source:
Organisation mondiale du tourisme (Madrid).

† Toutes les données se rapportant à l'Allemagne avant le 3 octobre 1990 figurent dans deux rubriques séparées basées sur les territoires respectifs de la République fédérale d'Allemagne et l'ancienne République démocratique allemande selon la période indiquée. En l'absence de données pour l'Allemagne unifiée (à compter du 3 octobre 1990), les données disponibles sont fournies séparément sous les rubriques République fédérale d'Allemagne et ancienne République démocratique allemande et se rapportent aux limites territoriales antérieures au 3 octobre 1990. Pour les notes explicatives en détail sur les données concernant l'Allemagne, voir Annexe I – Nomenclature des pays.

1 Recettes des visiteurs à l'exclusion des recettes provenant des membres des forces armées, des membres des équipages et des passagers en transit.
2 Recettes provenant des visiteurs.
3 Non compris l'ancien Yémen démocratique.
4 Y compris les recettes au titre des transports internationaux.
5 Y compris les recettes des excursionnistes (visiteurs de la journée).
6 Non compris les visiteurs italiens.

124
International tourism expenditures
Dépenses provenant du tourisme international
Million US dollars
Millions de dollars E-U

Country or area Pays ou zone	1981	1982	1983	1984	1985	1986	1987	1988	1989	1990
World Monde	99 399	91 729	86 468	95 727	100 707	124 971	155 989	186 186	198 352	237 916
Africa Afrique	3 859	3 753	3 573	3 156	2 852	2 752	3 207	3 826	3 808	3 932
Algeria Algérie	405	452	427	503	606	440	239	294	212	149
Benin Bénin	4	4	4	3	8	10	12	13	12	12
Botswana Botswana	18	16	17	20	17	19	26	32	34	39
Burkina Faso Burkina Faso	32	32	24	20	20	19	28	30	30	35
Burundi Burundi	18	18	18	9	9	14	17	15	14	16
Cameroon Cameroun	71	70	99	111	130	205	244	283	283	283
Central African Rep. Rép. centrafricaine	19	26	28	23	24	31	41	41	41	41
Chad Tchad	8	10	16	18	20	30	47	32	33	36
Comoros Comores	5	7	7	7	7	13	5	5	5	6
Congo Congo	37	54	65	55	63	67	79	126	75	75
Côte d'Ivoire Côte d'Ivoire	239	146	116	112	106	195	233	235	216	246
Egypt Egypte	174	179	151	146	106	52	78	43	112	166
Ethiopia Ethiopie	3	4	4	4	4	5	6	6	10	10
Gabon Gabon	107	92	88	83	83	131	132	134	124	143
Gambia Gambie	2	26	2	3	2	2	3	5	5	8
Ghana Ghana	33	6	10	10	10	11	12	12	13	13
Kenya Kenya	11	14	12	14	15	22	24	23	27	38
Lesotho Lesotho	7	8	7	5	5	8	9	11	10	15
Libyan Arab Jamah. Jamah. arabe libyenne	358	293	608	503	409	149	322	551	513	424
Madagascar Madagascar	38	24	25	21	24	31	26	30	34	34
Malawi Malawi	8	7	7	7	8	7	6	3	3	3
Mali Mali	16	19	18	17	37	49	57	58	56	67

124
International tourism expenditures
Million US dollars [cont.]
Dépenses provenant du tourisme international
Millions de dollars E-U [suite]

Country or area Pays ou zone	1981	1982	1983	1984	1985	1986	1987	1988	1989	1990
Mauritania Mauritanie	17	22	21	18	17	18	25	23	31	31
Mauritius Maurice	20	20	20	18	19	26	51	64	79	94
Morocco Maroc	94	83	83	70	88	100	132	164	153	184
Niger Niger	17	14	13	10	9	9	34	35	32	32
Nigeria Nigéria	774	816	476	229	200	90	55	41	51	57
Rwanda Rwanda	12	12	11	11	11	14	15	16	17	23
Sao Tome and Principe Sao Tomé-et-Principe	1	1	1	1	1	1	1	1	2	2
Senegal Sénégal	48	46	45	42	38	49	55	91	88	105
Seychelles Seychelles	7	7	6	7	9	10	12	13	17	20
Sierra Leone Sierra Leone	6	4	5	5	1	6	17	7	7	7
South Africa Afrique du Sud	802	744	729	651	413	603	838	958	936	1 065
Sudan Soudan	74	60	72	48	43	32	35	99	144	15
Swaziland Swaziland	24	8	26	19	11	12	14	18	10	14
Togo Togo	22	18	22	21	22	27	30	38	37	46
Tunisia Tunisie	59	147	136	142	126	107	95	119	134	179
Uganda Ouganda	10	10	...	...	...	9	13	11	10	8
United Rep.Tanzania Rép. Unie de Tanzanie	14	11	12	15	16	15	23	23	25	19
Zaire Zaïre	38	33	48	64	35	45	22	16	17	16
Zambia Zambie	57	86	21	24	22	31	46	49	98	98
Zimbabwe Zimbabwe	150	104	73	66	58	38	48	58	58	58
America, North Amérique du Nord	22 110	20 085	20 897	30 298	32 315	33 846	38 419	44 301	48 415	54 420
Antigua and Barbuda Antigua-et-Barbuda	4	6	6	13	13	14	15	16	16	17
Aruba Aruba	...	...	23	23	17	12	22	23	28	40
Bahamas Bahamas	92	104	96	106	123	132	153	172	184	196

124
International tourism expenditures
Million US dollars [cont.]
Dépenses provenant du tourisme international
Millions de dollars E-U [suite]

Country or area Pays ou zone	1981	1982	1983	1984	1985	1986	1987	1988	1989	1990
Barbados Barbade	22	26	22	23	23	29	36	37	45	45
Belize Belize	...	...	6	6	5	5	5	7	8	8
Bermuda Bermudes	...	...	64	64	76	81	77	93	104	104
Canada Canada	3 200	3 188	3 916	3 985	4 130	4 294	5 304	6 316	7 370	8 390
Costa Rica Costa Rica	40	38	48	50	53	60	71	72	114	148
Dominica Dominique	1	2	2	2	3	2	2	2	4	4
Dominican Republic Rép. dominicaine	128	87	88	89	84	89	95	127	136	144
El Salvador El Salvador	69	60	74	74	89	74	76	75	104	104
Grenada Grenade	4	3	3	3	3	4	4	5	7	7
Guatemala Guatemala	133	101	90	61	61	82	95	109	126	139
Haiti Haïti	30	41	39	39	43	37	42	34	33	32
Honduras Honduras	27	23	21	25	27	30	35	37	38	38
Jamaica Jamaïque	14	30	25	21	32	35	44	57	54	54
Mexico Mexique	6 155	3 205	1 583	2 166	2 262	2 177	2 364	3 202	4 247	5 379
Nicaragua Nicaragua	...	...	...	...	6	4	6	2	2	2
Panama Panama	65	81	72	67	65	83	90	88	81	99
Puerto Rico Porto Rico	447	431	437	446	411	416	488	533	589	647
Saint Kitts-Nevis Saint-Kitts-et-Nevis	2	3	3	5	2	2	3	3	3	4
Saint Lucia Sainte-Lucie	26	29	35	39	44	15	15	21	21	21
St. Vincent-Grenadines St. Vincent-Grenadines	5	8	6	6	7	4	4	4	5	5
Trinidad and Tobago Trinité-et-Tobago	167	225	261	276	219	165	158	168	119	122
United States Etats-Unis	11 479	12 394	13 977	22 709	24 517	26 000	29 215	33 098	34 977	38 671
America, South **Amérique du Sud**	**6 427**	**6 464**	**3 779**	**3 879**	**3 551**	**4 562**	**4 416**	**4 302**	**4 200**	**5 651**
Argentina Argentine	1 471	565	508	682	671	888	890	975	1 014	1 171

124
International tourism expenditures
Million US dollars [cont.]
Dépenses provenant du tourisme international
Millions de dollars E-U [suite]

Country or area Pays ou zone	1981	1982	1983	1984	1985	1986	1987	1988	1989	1990
Bolivia Bolivie	50	40	20	30	38	27	56	55	59	60
Brazil Brésil	1 230	1 411	839	939	1 145	1 464	1 249	1 084	751	1 559
Chile Chili	260	210	242	327	269	319	353	423	397	426
Colombia Colombie	391	524	445	308	169	611	666	538	494	515
Ecuador Equateur	250	250	152	155	156	156	170	167	169	176
Paraguay Paraguay	38	42	44	44	47	48	51	59	75	105
Peru Pérou	133	176	162	160	244	320	335	344	424	571
Suriname Suriname	29	17	35	17	13	12	8	10	10	12
Uruguay Uruguay	203	304	259	154	162	174	129	138	167	111
Venezuela Venezuela	2 372	2 925	1 073	1 063	597	543	509	509	640	945
Asia Asie	13 611	10 939	11 779	12 529	13 840	16 560	21 529	31 598	38 192	43 746
Afghanistan Afghanistan	2	2	3	3	2	1	1	1	1	1
Bahrain Bahreïn	144	202	104	121	125	59	80	77	77	77
Bangladesh Bangladesh	23	18	23	33	45	54	53	99	123	78
China Chine	...	66	53	150	314	308	387	633	429	470
Cyprus Chypre	57	68	66	71	39	54	68	81	133	162
India Inde	146	190	227	306	354	302	352	397	425	425
Indonesia Indonésie	644	577	523	516	591	570	511	592	722	886
Iran, Islamic Rep. of Iran, Rép. islamique d'	681	400	404	488	508	247	246	396	396	396
Israel Israël	621	673	803	726	549	801	1 043	1 161	1 288	1 485
Japan Japon	4 616	4 116	4 428	4 607	4 814	7 229	10 760	18 682	22 490	24 928
Jordan Jordanie	367	367	370	381	424	445	444	480	419	419
Korea, Republic of Corée, République de	439	632	555	576	606	613	704	1 354	2 602	3 166
Kuwait Koweït	1 098	1 306	1 362	1 540	1 588	1 941	2 257	2 358	2 315	2 315

124
International tourism expenditures
Million US dollars [cont.]
Dépenses provenant du tourisme international
Millions de dollars E-U [suite]

Country or area Pays ou zone	1981	1982	1983	1984	1985	1986	1987	1988	1989	1990
Macau Macao	...	...	...	...	399	419	518	591	708	708
Malaysia Malaisie	678	850	1 047	1 141	1 158	1 669	1 795	2 012	2 803	4 473
Maldives Maldives	3	3	3	4	5	6	7	8	15	19
Myanmar Myanmar	7	2	2	1	19	1	1	1	2	1
Nepal Népal	24	24	21	28	29	29	35	43	48	45
Oman Oman	...	...	...	...	...	47	47	47	47	47
Pakistan Pakistan	116	123	156	198	202	221	248	334	326	429
Philippines Philippines	127	148	220	19	37	56	88	76	77	111
Saudi Arabia Arabie saoudite	2 761	...	...	...	...	...	...	...	...	...
Singapore Singapour	374	466	531	601	613	645	795	930	1 088	1 381
Sri Lanka Sri Lanka	40	39	39	47	46	55	63	68	68	79
Syrian Arab Republic Rép. arabe syrienne	208	213	282	311	302	160	160	170	194	190
Thailand Thaïlande	276	267	343	305	280	296	381	602	750	854
Turkey Turquie	103	109	127	277	324	314	448	358	565	520
Yemen Yémen	56	78	87	79	67	18[1]	37[1]	47[1]	81[1]	81[1]
Europe Europe	50 977	48 089	44 029	43 008	45 566	64 193	84 749	97 239	97 938	124 034
Austria Autriche	2 788	2 744	2 898	2 624	2 723	4 026	5 592	6 307	6 266	6 212
Belgium Belgique	2 644	2 191	2 095	1 953	2 050	2 889	3 881	4 428	4 272	5 445
Czechoslovakia Tchécoslovaquie	190	193	229	280	300	349	409	399	431	636
Denmark Danemark	1 269	1 330	1 212	1 227	1 410	2 119	2 860	3 087	2 932	3 676
Finland Finlande	592	630	622	681	777	1 060	1 506	1 842	2 047	2 765
France France	5 752	5 157	4 281	4 271	4 557	6 513	8 493	9 715	10 031	12 424
Germany†·Allemagne† F. R. Germany R. f. Allemagne	17 528	16 223	13 274	12 423	12 809	18 000	23 341	25 036	23 727	29 836

124
International tourism expenditures
Million US dollars [cont.]
Dépenses provenant du tourisme international
Millions de dollars E-U [suite]

Country or area Pays ou zone	1981	1982	1983	1984	1985	1986	1987	1988	1989	1990
Greece Grèce	249	231	362	339	368	494	508	735	816	1 088
Hungary Hongrie	164	142	152	157	208	225	250	647	1 008	600
Iceland Islande	53	54	66	85	94	129	213	200	176	218
Ireland Irlande	513	454	453	411	429	685	839	961	989	1 159
Italy Italie	1 664	1 731	1 822	2 098	2 283	2 910	4 536	5 989	6 772	13 826
Malta Malte	50	57	53	50	50	69	102	120	107	134
Netherlands Pays-Bas	3 516	3 406	3 289	3 277	3 448	4 901	6 408	6 750	6 481	7 340
Norway Norvège	1 475	1 641	1 587	1 488	1 722	2 511	3 067	3 442	2 855	3 413
Poland Pologne	289	111	195	225	184	186	203	251	215	220
Portugal Portugal	246	247	229	222	235	329	421	533	583	867
Romania Roumanie	67	84	92	85	64	22	30	33	35	103
Spain Espagne	1 008	1 008	894	835	1 010	1 513	1 938	2 440	3 080	4 254
Sweden Suède	2 193	1 895	1 619	1 713	1 967	2 821	3 784	4 572	4 966	6 066
Switzerland Suisse	2 122	2 216	2 296	2 282	2 399	3 368	4 339	5 019	4 907	5 989
United Kingdom Royaume-Uni	6 478	6 237	6 223	6 197	6 369	8 942	11 939	14 624	15 111	17 614
Yugoslavia Yougoslavie	127	107	86	85	110	132	90	109	131	149
Oceania **Océanie**	**2 415**	**2 399**	**2 411**	**2 857**	**2 583**	**3 058**	**3 669**	**4 920**	**5 799**	**6 133**
Australia Australie	1 852	1 850	1 721	2 125	1 918	2 058	2 351	2 965	3 859	4 120
Fiji Fidji	19	18	17	18	18	24	53	35	41	32
Guam Guam	...	...	210	221	227	304	304	589	589	589
New Zealand Nouvelle-Zélande	523	504	432	455	389	637	917	1 268	1 252	1 335
Papua New Guinea Papouasie-Nvl-Guinée	19	20	23	29	21	22	32	48	42	42
Samoa Samoa	1	1	1	1	1	2	2	1	2	2
Solomon Islands Iles Salomon	...	3	3	3	4	6	6	9	9	11

124
International tourism expenditures
Million US dollars [cont.]
       Dépenses provenant du tourisme international
       Millions de dollars E-U [suite]

Country or area Pays ou zone	1981	1982	1983	1984	1985	1986	1987	1988	1989	1990
Tonga Tonga	1	1	2	3	3	3	3	3	3	1
Vanuatu Vanuatu	...	2	2	2	2	2	1	2	2	1

Source:
World Tourism Organization (Madrid).

† All data shown which pertain to Germany prior to 3 October
 1990 are indicated separately for the Federal Republic of
 Germany and the former German Democratic Republic based on
 their respective territories at the time indicated. Where
 data for united Germany (3 October 1990 and thereafter) are
 not available, available data are shown separately under the
 designations Federal Republic of Germany and former German
 Democratic Republic and pertain to the territorial
 boundaries prior to 3 October 1990. For detailed
 explanatory notes on data pertaining to Germany, see Annex I
 - Country Nomenclature.

1 Excluding former Democratic Yemen.

Source:
Organisation mondiale du tourisme (Madrid).

† Toutes les données se rapportant à l'Allemagne avant le 3
 octobre 1990 figurent dans deux rubriques séparées basées
 sur les territoires respectifs de la République fédérale
 d'Allemagne et l'ancienne République démocratique allemande
 selon la période indiquée. En l'absence de données pour
 l'Allemagne unifiée (à compter du 3 octobre 1990), les
 données disponisbles sont fournies séparément sous les
 rubriques Réublique fédérale d'Allemagne et ancienne
 République démocratique allemande et se rapportent aux
 limites territoriales antérieures au 3 octobre 1990. Pour
 les notes explicatives en détail sur les données concernant
 l'Allemagne, voir Annexe I – Nomenclature des pays.

1 Non compris l'ancien Yémen démocratique.

Technical notes, tables 122-124

Tables 122 and 123: For statistical purposes, the term "international visitor" describes "any person who travels to a country other than that in which he/she has his/her usual residence but outside his/her usual environment for a period not exceeding 12 months and whose main purpose of visit is other than the exercise of an activity remunerated from within the country visited".

International visitors include:

(a) *Tourists*, (overnight visitor): "a visitor who stays at least one night in a collective or private accommodation in the country visited"; and

(b) *Same-day visitors*: "a visitor who does not spend the night in a collective or private accommodation in the country visited".

Unless otherwise stated, same-day visitors are not included in these tables.

The figures do not therefore include immigrants, residents in a frontier zone, persons domiciled in one country or area and working in an adjoining country or area, members of the armed forces and diplomats and consular representatives when they travel from their country of origin to the country in which they are stationed and vice-versa.

The figures also exclude persons in transit who do not formally enter the country through passport control, such as air transit passengers who remain for a short period in a designated area of the air terminal or ship passengers who are not permitted to disembark. This category would include passengers transferred directly between airports or other terminals. Other passengers in transit through a country are classified as visitors.

These data are based generally on a frontier check. In the absence of frontier check figures, data based on arrivals at accommodation establishments are given but these are not strictly comparable with frontier check data as they exclude certain types of tourists such as campers and tourists staying in private houses, while on the other hand they may contain some duplication when a tourist moves from one establishment to another.

Unless otherwise stated, table 122 shows the number of tourist arrivals at frontiers classified by their region of origin. Totals correspond to the total number of arrivals from the regions indicated in the table. However, these totals may not correspond to the number of tourist arrivals shown in table 123. The later excludes same-day visitors whereas they may be included in table 122. Moreover, data in table 122 exclude arrivals from "rest of the world" and, in general, exclude arrivals of nationals residing abroad. More detailed information will be found in *Yearbook of Tourism Statistics*, published by the World Tourism Organization. [36]

Unless otherwise stated the data on tourist receipts have been supplied by the World Tourism Organization. Tourist receipts are defined as "expenditure of international inbound visitors including their payments to national

Notes techniques, tableaux 122-124

Tableaux 122 et 123 : A des fins statistiques, l'expression "*visiteur international*" désigne "toute personne qui se rend dans un pays autre que celui où elle a son lieu de résidence habituelle, mais différent de son environnement habituel, pour une période de 12 mois au maximum, dans un but principal autre que celui d'y exercer une profession rémunérée".

Entrent dans cette catégorie :

(a) Les *touristes* (visiteurs passant la nuit), c'est à dire "les visiteurs qui passent une nuit au moins en logement collectif ou privé dans le pays visité";

(b) Les *visiteurs ne restant que la journée*, c'est à dire "les visiteurs qui ne passent pas la nuit en logement collectif ou privé dans le pays visité".

Sauf indication contraire, les visiteurs ne restant que la journée ne sont pas inclus dans ces tableaux.

Par conséquent, ces chiffres ne comprennent pas les immigrants, les résidents frontaliers, les personnes domiciliées dans une zone ou un pays donné et travaillant dans une zone ou pays limitrophe, les membres des forces armées et les membres des corps diplomatique et consulaire lorsqu'ils se rendent de leur pays d'origine au pays où ils sont en poste, et vice versa.

Ne sont pas non plus inclus les voyageurs en transit, qui ne pénètrent pas officiellement dans le pays en faisant contrôler leurs passeports, tels que les passagers d'un vol en escale, qui demeurent pendant un court laps de temps dans une aire distincte de l'aérogare, ou les passagers d'un navire qui ne sont pas autorisés à débarquer. Cette catégorie comprend également les passagers transportés directement d'une aérogare à l'autre ou à un autre terminal. Les autres passagers en transit dans un pays sont classés parmi les visiteurs.

Ces données reposent en général sur un contrôle à la frontière. A défaut, les données sont tirées des établissements d'hébergement, mais elles ne sont alors pas strictement comparables à celles du contrôle à la frontière, en ce qu'elles excluent, d'une part, certaines catégories de touristes telles que les campeurs et les touristes séjournant dans des maisons privées, et que, d'autre part, elles peuvent compter deux fois le même touriste si celui-ci change d'établissement d'hébergement.

Sauf indication contraire, le tableau 122 indique le nombre d'arrivées de touristes par région de provenance. Les totaux correspondent au nombre total d'arrivées de touristes des régions indiquées sur le tableau. Les chiffres totaux peuvent néanmoins, ne pas coïncider avec le nombre des arrivées de touristes indiqué dans le tableau 123, qui ne comprend pas les visiteurs ne restant que la journée, lesquels peuvent au contraire être inclus dans les chiffres du tableau 122. En outre, les chiffres du tableau 122 ne comprennent pas les arrivées depuis le "reste du monde", ni, en règle générale, les arrivées de ressortissants du pays considéré qui résident à l'étranger. Pour plus de renseignements, consulter l'*Annuaire des statistiques du tourisme* publié par l'Organisation mondiale du tourisme [36].

carriers for international transport. They should also include any other prepayments made for goods/services received in the destination country. They should in practice also include receipts from same-day visitors, except in cases when these are so important as to justify a separate classification. It is also recommended that, for the sake of consistancy with the Balance of Payments recommendations of the International Monetary Fund, international fare receipts be classified separately.

Table 124: International tourism expenditures are defined as consumption expenditures, i.e. payments for goods and services, made by residents of a country visiting abroad. They may, in practice, include expenditures of residents visiting abroad as excursionists except where these are so important as to justify separate classification. They should, however, exclude all forms of remuneration resulting from employment as well as international fare expenditures. This category corresponds to "travel debts" in the standard reporting form of the International Monetary Fund.

For more detailed statistics on the number of tourists and expenditures, see *Yearbook of Tourism Statistics* published by the World Tourism Organization.[36] For detailed definitions of tourist receipts, see *Balance of Payments Yearbook* published by the International Monetary Fund.[8]

For detailed information on methods of collection for frontier statistics, accommodation statistics and foreign exchange statistics see *Methodological Supplement to World Travel and Tourism Statistics* published by the World Tourism Organization.[55; see also 51]

Sauf indication contraire, les données sur les recettes du tourisme ont été fournies par l'Organisation mondiale du tourisme et sont définies comme "les sommes dépensées par les visiteurs internationaux arrivant dans le pays, y compris les sommes versées aux transporteurs nationaux en paiement de transports internationaux. Il faut également y inclure tout autre versement effectué à l'avance pour des biens ou services à recevoir dans le pays de destination. Dans la pratique, il faut également y inclure les recettes provenant de visiteurs ne restant que la journée, sauf dans les cas où elles sont suffisamment importantes pour justifier une classification distincte. Il est recommandé aussi, dans un souci de cohérence avec les recommandations du Fonds monétaire international visant la balance des paiements, de classer à part les recettes au titre des transports internationaux".

Tableau 124 : Dépenses du tourisme international sont définies comme étant les dépenses de consommation (achats de biens et de services) effectuées par les résidents du pays en visite à l'étranger. En pratique, ce poste devrait également inclure les dépenses des résidents qui voyagent à l'étranger en qualité d'excursionnistes, sauf dans les cas où ces dépenses sont suffisamment importantes pour justifier une classification distincte. Elles doivent néanmoins exclure toutes les formes de rémunération relatives à l'exercice d'un emploi et les paiements au titre des transports internationaux. Cette catégorie de paiements correspond à la rubrique "débits-voyages" du formulaire type du Fonds monétaire international.

Pour des statistiques plus détaillées sur le nombre de touristes et les dépenses, voir l'*Annuaire des statistiques du tourisme* publié par l'Organisation mondiale du tourisme [36]. Pour des définitions détaillées des recettes touristiques, voir *"Balance of Payments Yearbook"* publié par le Fonds monétaire international [8].

Pour plus de renseignements sur les méthodes de collecte de données statistiques sur les contrôles aux frontières, l'hébergement et les mouvements de devises, voir *"Supplément méthodologique aux statistiques des voyages et du tourisme mondiaux"* publié par l'Organisation mondiale du tourisme [55; voir aussi 51].

125
Summary of balance of payments
Résumé des balances des paiements
Millions of US dollars
Millions de dollars des E-U

Country or area	1984	1985	1986	1987	1988	1989	1990	Pays ou zone
Afghanistan								**Afghanistan**
Merchandise: Exp. fob	787.7	628.2	497.0	538.7	453.8	252.3	...	Marchandises : exp. fob
Merchandise: Imp. fob	-1204.7	-921.6	-1138.8	-904.5	-731.8	-623.5	...	Marchandises : imp. fob
Serv. & Income: Credit	53.8	69.2	53.1	54.8	92.9	28.3	...	Serv. & revenu : crédit
Serv. & Income: Debit	-214.7	-162.7	-215.9	-167.6	-131.5	-111.3	...	Serv. & revenu : débit
Private Unrequited Transfers	0.0	0.0	0.0	0.0	0.0	0.0	...	Transf. priv.sans contrep.
Offic. Unrequited Trans.,nie	127.3	143.7	267.4	311.7	342.8	312.1	...	Tr. offic. sans contrep.nia
Direct Investment, nie	0.0	0.0	0.0	0.0	0.0	0.0	...	Investissements direct, nia
Portfolio Investments, nie	0.0	0.0	0.0	0.0	0.0	0.0	...	Invest. portefeuille, nia
Other Capital, nie	313.2	100.8	302.1	-33.9	-4.1	-59.6	...	Autres capitaux, nia
Net Errors and Omissions	202.6	168.4	216.4	211.6	-47.9	181.5	...	Erreurs et omissions nettes
Reserves and Related Items	-65.2	-26.0	18.7	-10.8	25.8	20.2	...	Rés. et postes assimilables
Algeria								**Algérie**
Merchandise: Exp. fob	12792.0	13034.0	8065.0	9029.0	7620.0	9534.0	12964.0	Marchandises : exp. fob
Merchandise: Imp. fob	-9235.0	-8811.0	-7879.0	-6616.0	-6675.0	-8372.0	-8777.0	Marchandises : imp. fob
Serv. & Income: Credit	778.0	722.0	721.0	675.0	542.0	607.0	571.0	Serv. & revenu : crédit
Serv. & Income: Debit	-4442.0	-4309.0	-3900.0	-3464.0	-3917.0	-3390.0	-3671.0	Serv. & revenu : débit
Private Unrequited Transfers	186.0	367.0	765.0	522.0	385.0	535.0	332.0	Transf. priv.sans contrep.
Offic. Unrequited Trans.,nie	-5.0	11.0	-1.0	-5.0	5.0	6.0	1.0	Tr. offic. sans contrep.nia
Direct Investment, nie	-14.0	-2.0	11.0	-11.0	8.0	4.0	-4.0	Investissements direct, nia
Portfolio Investments, nie	0.0	0.0	0.0	0.0	2.0	0.0	0.0	Invest. portefeuille, nia
Other Capital, nie	-197.0	-119.0	579.0	321.0	734.0	751.0	-996.0	Autres capitaux, nia
Net Errors and Omissions	-197.0	127.0	142.0	-802.0	335.0	-448.0	-336.0	Erreurs et omissions nettes
Reserves and Related Items	333.0	-1020.0	1498.0	352.0	960.0	774.0	-84.0	Rés. et postes assimilables
Antigua and Barbuda								**Antigua-et-Barbuda**
Merchandise: Exp. fob	35.2	28.3	30.9	28.5	30.1	31.6	33.2	Marchandises : exp. fob
Merchandise: Imp. fob	-150.2	-174.8	-284.2	-258.3	-274.8	-317.0	-325.9	Marchandises : imp. fob
Serv. & Income: Credit	137.2	152.7	172.2	194.3	266.9	302.9	325.7	Serv. & revenu : crédit
Serv. & Income: Debit	-36.1	-43.5	-62.8	-66.6	-123.5	-136.6	-149.1	Serv. & revenu : débit
Private Unrequited Transfers	12.5	13.5	14.5	17.0	18.9	22.8	19.8	Transf. priv.sans contrep.
Offic. Unrequited Trans.,nie	2.0	0.7	2.7	1.7	1.8	0.7	0.8	Tr. offic. sans contrep.nia
Direct Investment, nie	4.4	15.6	17.7	29.2	59.5	60.8	84.7	Investissements direct, nia
Portfolio Investments, nie	0.0	0.0	0.0	0.0	0.0	0.0	0.0	Invest. portefeuille, nia
Other Capital, nie	-5.3	4.6	94.3	41.0	5.1	-1.4	-15.8	Autres capitaux, nia
Net Errors and Omissions	-0.1	0.0	18.0	8.8	-0.1	-0.1	-0.1	Erreurs et omissions nettes
Reserves and Related Items	0.4	2.9	-3.4	4.3	16.1	36.0	26.7	Rés. et postes assimilables
Argentina								**Argentine**
Merchandise: Exp. fob	8100.0	8396.0	6852.0	6360.0	9134.0	9573.0	12354.0	Marchandises : exp. fob
Merchandise: Imp. fob	-4118.0	-3518.0	-4406.0	-5343.0	-4892.0	-3864.0	-3726.0	Marchandises : imp. fob
Serv. & Income: Credit	1809.0	1933.0	1989.0	2046.0	2226.0	2469.0	2721.0	Serv. & revenu : crédit
Serv. & Income: Debit	-8288.0	-7763.0	-7296.0	-7290.0	-8040.0	-9491.0	-9631.0	Serv. & revenu : débit
Private Unrequited Transfers	2.0	0.0	2.0	-8.0	0.0	8.0	71.0	Transf. priv.sans contrep.
Offic. Unrequited Trans.,nie	0.0	0.0	0.0	0.0	0.0	0.0	0.0	Tr. offic. sans contrep.nia
Direct Investment, nie	268.0	919.0	574.0	-19.0	1147.0	1028.0	2036.0	Investissements direct, nia
Portfolio Investments, nie	372.0	-617.0	-542.0	-572.0	-718.0	-1098.0	-1614.0	Invest. portefeuille, nia
Other Capital, nie	-577.0	555.0	386.0	565.0	2.0	-7938.0	-2736.0	Autres capitaux, nia
Net Errors and Omissions	-55.0	-532.0	302.0	-112.0	-165.0	-249.0	715.0	Erreurs et omissions nettes
Reserves and Related Items	2487.0	627.0	2139.0	4373.0	1306.0	9562.0	-190.0	Rés. et postes assimilables
Aruba								**Aruba**
Merchandise: Exp. fob	...	...	29.6	45.1	87.4	107.5	157.0	Marchandises : exp. fob
Merchandise: Imp. fob	...	...	-210.4	-249.8	-354.6	-397.4	-556.7	Marchandises : imp. fob
Serv. & Income: Credit	...	...	224.2	269.0	330.0	364.7	425.7	Serv. & revenu : crédit
Serv. & Income: Debit	...	...	-64.2	-83.4	-105.2	-124.9	-154.3	Serv. & revenu : débit
Private Unrequited Transfers	...	...	2.3	-2.9	-0.9	3.4	-10.3	Transf. priv.sans contrep.
Offic. Unrequited Trans.,nie	...	...	0.0	0.0	0.0	0.0	0.0	Tr. offic. sans contrep.nia
Direct Investment, nie	...	...	0.0	0.0	0.0	0.0	0.0	Investissements direct, nia
Portfolio Investments, nie	...	...	0.0	0.0	0.0	0.0	0.0	Invest. portefeuille, nia
Other Capital, nie	...	...	61.8	24.6	56.8	47.9	149.9	Autres capitaux, nia
Net Errors and Omissions	...	...	13.7	10.4	-13.7	20.3	11.6	Erreurs et omissions nettes
Reserves and Related Items	...	...	-56.9	-12.9	0.4	-21.5	-22.8	Rés. et postes assimilables
Australia								**Australie**
Merchandise: Exp. fob	22769.0	22275.0	22187.0	26270.0	32780.0	36339.0	38935.0	Marchandises : exp. fob
Merchandise: Imp. fob	-23653.0	-23592.0	-24264.0	-26749.0	-33892.0	-40329.0	-38966.0	Marchandises : imp. fob
Serv. & Income: Credit	6271.0	5683.0	6426.0	8707.0	11384.0	12341.0	13455.0	Serv. & revenu : crédit
Serv. & Income: Debit	-14083.0	-13805.0	-14386.0	-17317.0	-22270.0	-28146.0	-30479.0	Serv. & revenu : débit

125
Summary of balance of payments
Millions of US dollars [*cont.*]
Résumé des balances des paiements
Millions de dollars des E-U [*suite*]

Country or area	1984	1985	1986	1987	1988	1989	1990	Pays ou zone
Private Unrequited Transfers	451.0	686.0	817.0	1196.0	1706.0	2124.0	1939.0	Transf. priv.sans contrep.
Offic. Unrequited Trans.,nie	-458.0	-263.0	-225.0	-178.0	-232.0	-182.0	-98.0	Tr. offic. sans contrep.nia
Direct Investment, nie	-1032.0	183.0	157.0	-1099.0	2864.0	4426.0	5365.0	Investissements direct, nia
Portfolio Investments, nie	736.0	2144.0	1188.0	4325.0	5834.0	377.0	1970.0	Invest. portefeuille, nia
Other Capital, nie	6323.0	4484.0	8141.0	4995.0	9592.0	10631.0	3361.0	Autres capitaux, nia
Net Errors and Omissions	1369.0	-79.0	664.0	220.0	-2516.0	3048.0	6241.0	Erreurs et omissions nettes
Reserves and Related Items	1307.0	2282.0	-705.0	-371.0	-5251.0	-628.0	-1725.0	Rés. et postes assimilables
Austria								**Autriche**
Merchandise: Exp. fob	15471.0	16683.0	21565.0	26558.0	30056.0	31832.0	40252.0	Marchandises : exp. fob
Merchandise: Imp. fob	-19485.0	-21089.0	-26176.0	-31702.0	-36358.0	-38437.0	-48234.0	Marchandises : imp. fob
Serv. & Income: Credit	12448.0	13423.0	16908.0	20059.0	23349.0	25690.0	33028.0	Serv. & revenu : crédit
Serv. & Income: Debit	-8637.0	-9198.0	-12168.0	-15263.0	-17469.0	-18897.0	-24088.0	Serv. & revenu : débit
Private Unrequited Transfers	-22.0	-55.0	2.0	-10.0	38.0	-57.0	108.0	Transf. priv.sans contrep.
Offic. Unrequited Trans.,nie	-39.0	-39.0	-44.0	-71.0	-74.0	-72.0	-109.0	Tr. offic. sans contrep.nia
Direct Investment, nie	92.0	195.0	-39.0	145.0	190.0	-66.0	-746.0	Investissements direct, nia
Portfolio Investments, nie	648.0	-58.0	539.0	929.0	2423.0	1197.0	1512.0	Invest. portefeuille, nia
Other Capital, nie	-345.0	-488.0	742.0	-597.0	-1464.0	-300.0	-984.0	Autres capitaux, nia
Net Errors and Omissions	-61.0	713.0	-600.0	356.0	-297.0	106.0	-775.0	Erreurs et omissions nettes
Reserves and Related Items	-70.0	-88.0	-728.0	-404.0	-395.0	-996.0	36.0	Rés. et postes assimilables
Bahamas								**Bahamas**
Merchandise: Exp. fob	261.9	295.8	293.4	273.1	273.6	259.2	307.6	Marchandises : exp. fob
Merchandise: Imp. fob	-866.2	-1096.2	-1012.8	-1154.7	-1058.9	-1203.5	-1228.8	Marchandises : imp. fob
Serv. & Income: Credit	1065.9	1240.5	1313.8	1389.4	1354.8	1506.7	1514.4	Serv. & revenu : crédit
Serv. & Income: Debit	-507.9	-592.1	-620.8	-670.2	-683.3	-721.2	-784.0	Serv. & revenu : débit
Private Unrequited Transfers	-14.6	-14.6	-14.0	-17.8	-28.9	-17.9	-10.7	Transf. priv.sans contrep.
Offic. Unrequited Trans.,nie	15.4	17.9	13.7	14.2	14.4	18.9	21.3	Tr. offic. sans contrep.nia
Direct Investment, nie	-4.9	-30.2	-13.2	10.8	36.7	25.0	-16.3	Investissements direct, nia
Portfolio Investments, nie	0.0	0.0	0.0	0.0	0.0	0.0	0.0	Invest. portefeuille, nia
Other Capital, nie	-28.7	17.2	35.4	-29.6	28.0	71.5	85.1	Autres capitaux, nia
Net Errors and Omissions	118.2	181.1	52.1	125.2	64.4	48.1	123.1	Erreurs et omissions nettes
Reserves and Related Items	-39.1	-19.4	-47.6	59.6	-0.8	13.2	-11.7	Rés. et postes assimilables
Bahrain								**Bahreïn**
Merchandise: Exp. fob	3204.0	2896.8	2199.5	2429.5	2411.4	2831.1	...	Marchandises : exp. fob
Merchandise: Imp. fob	-3131.6	-2796.0	-2164.6	-2442.3	-2334.0	-2820.2	...	Marchandises : imp. fob
Serv. & Income: Credit	1149.5	1234.0	1042.3	1146.3	1162.5	1367.0	...	Serv. & revenu : crédit
Serv. & Income: Debit	-1002.4	-1181.4	-1002.1	-1203.5	-1223.4	-1303.5	...	Serv. & revenu : débit
Private Unrequited Transfers	-125.5	-234.8	-264.6	-243.6	-193.1	-195.7	...	Transf. priv.sans contrep.
Offic. Unrequited Trans.,nie	124.5	120.2	120.7	113.3	366.5	102.1	...	Tr. offic. sans contrep.nia
Direct Investment, nie	140.7	101.3	-31.9	-35.9	222.1	180.9	...	Investissements direct, nia
Portfolio Investments, nie	0.0	0.0	0.0	0.0	0.0	0.0	...	Invest. portefeuille, nia
Other Capital, nie	-176.3	-548.9	-41.8	-17.8	-448.7	-363.0	...	Autres capitaux, nia
Net Errors and Omissions	-192.8	766.2	-36.9	-95.6	136.4	12.9	...	Erreurs et omissions nettes
Reserves and Related Items	10.1	-357.4	179.4	349.6	-99.7	188.5	...	Rés. et postes assimilables
Bangladesh								**Bangladesh**
Merchandise: Exp. fob	931.7	999.5	880.0	1076.9	1291.0	1304.8	1672.4	Marchandises : exp. fob
Merchandise: Imp. fob	-2340.0	-2286.4	-2300.7	-2445.6	-2734.5	-3300.1	-3259.4	Marchandises : imp. fob
Serv. & Income: Credit	275.6	279.5	246.7	295.4	332.5	423.1	455.8	Serv. & revenu : crédit
Serv. & Income: Debit	-613.8	-631.6	-670.2	-666.7	-793.9	-923.4	-880.2	Serv. & revenu : débit
Private Unrequited Transfers	535.4	531.0	605.8	788.3	827.2	806.8	828.3	Transf. priv.sans contrep.
Offic. Unrequited Trans.,nie	730.0	650.2	611.4	713.9	804.4	589.1	785.8	Tr. offic. sans contrep.nia
Direct Investment, nie	-0.6	0.0	2.4	3.2	1.8	0.2	3.2	Investissements direct, nia
Portfolio Investments, nie	1.6	-7.2	0.0	-0.1	0.0	1.7	0.3	Invest. portefeuille, nia
Other Capital, nie	569.5	453.3	739.9	555.0	396.8	831.3	694.3	Autres capitaux, nia
Net Errors and Omissions	-95.2	-68.0	8.7	-123.8	6.6	-43.1	-76.3	Erreurs et omissions nettes
Reserves and Related Items	5.7	79.7	-124.2	-196.5	-132.0	309.5	-224.2	Rés. et postes assimilables
Barbados								**Barbade**
Merchandise: Exp. fob	339.7	300.5	244.3	131.4	144.8	146.9	...	Marchandises : exp. fob
Merchandise: Imp. fob	-606.2	-559.2	-522.6	-458.4	-517.9	-599.3	...	Marchandises : imp. fob
Serv. & Income: Credit	487.5	494.3	516.1	556.9	658.4	774.9	...	Serv. & revenu : crédit
Serv. & Income: Debit	-216.6	-200.8	-265.2	-289.9	-300.9	-331.1	...	Serv. & revenu : débit
Private Unrequited Transfers	16.5	14.7	18.6	19.0	29.7	31.9	...	Transf. priv.sans contrep.
Offic. Unrequited Trans.,nie	-1.8	-9.1	-7.0	-12.5	-12.0	-26.1	...	Tr. offic. sans contrep.nia
Direct Investment, nie	-1.5	2.6	5.0	4.6	10.5	5.4	...	Investissements direct, nia
Portfolio Investments, nie	-0.2	-2.8	-5.4	-0.6	-0.1	-5.1	...	Invest. portefeuille, nia
Other Capital, nie	-15.7	-5.7	19.0	97.1	31.5	-0.3	...	Autres capitaux, nia
Net Errors and Omissions	-15.4	-12.1	17.4	-41.4	-6.1	-39.6	...	Erreurs et omissions nettes

125
Summary of balance of payments
Millions of US dollars [*cont.*]
Résumé des balances des paiements
Millions de dollars des E-U [*suite*]

Country or area	1984	1985	1986	1987	1988	1989	1990	Pays ou zone
Reserves and Related Items	13.6	-22.3	-20.1	-6.2	-38.0	42.4	...	Rés. et postes assimilables
Belgium-Luxembourg								**Belgique-Luxembourg**
Merchandise: Exp. fob	46213.0	47336.0	59955.0	76088.0	85496.0	89988.0	108762.0	Marchandises : exp. fob
Merchandise: Imp. fob	-47431.0	-47808.0	-59399.0	-76268.0	-84273.0	-89020.0	-108132.0	Marchandises : imp. fob
Serv. & Income: Credit	31153.0	33467.0	41241.0	48264.0	56112.0	71850.0	95959.0	Serv. & revenu : crédit
Serv. & Income: Debit	-29159.0	-31670.0	-37796.0	-43864.0	-51986.0	-67902.0	-90026.0	Serv. & revenu : débit
Private Unrequited Transfers	-176.0	-129.0	-212.0	-114.0	41.0	47.0	-595.0	Transf. priv.sans contrep.
Offic. Unrequited Trans.,nie	-656.0	-527.0	-734.0	-1313.0	-1796.0	-1765.0	-1418.0	Tr. offic. sans contrep.nia
Direct Investment, nie	96.0	755.0	-993.0	-427.0	1428.0	245.0	1831.0	Investissements direct, nia
Portfolio Investments, nie	-3981.0	-6001.0	-6592.0	-2429.0	-4572.0	-2900.0	-7203.0	Invest. portefeuille, nia
Other Capital, nie	4398.0	5067.0	4700.0	2346.0	-614.0	-2549.0	4141.0	Autres capitaux, nia
Net Errors and Omissions	152.0	-262.0	222.0	-11.0	59.0	-311.0	-2276.0	Erreurs et omissions nettes
Reserves and Related Items	-611.0	-228.0	-392.0	-2273.0	104.0	2319.0	-1043.0	Rés. et postes assimilables
Belize								**Belize**
Merchandise: Exp. fob	93.2	90.1	92.6	102.8	119.4	124.4	...	Marchandises : exp. fob
Merchandise: Imp. fob	-116.3	-113.8	-108.3	-126.9	-161.3	-188.5	...	Marchandises : imp. fob
Serv. & Income: Credit	36.7	41.0	52.0	61.8	82.4	95.5	...	Serv. & revenu : crédit
Serv. & Income: Debit	-44.2	-40.7	-51.7	-59.3	-69.0	-81.6	...	Serv. & revenu : débit
Private Unrequited Transfers	15.9	19.6	15.3	15.2	15.2	20.7	...	Transf. priv.sans contrep.
Offic. Unrequited Trans.,nie	9.4	12.8	12.1	15.7	10.6	10.4	...	Tr. offic. sans contrep.nia
Direct Investment, nie	-3.7	3.7	4.6	6.8	14.0	18.6	...	Investissements direct, nia
Portfolio Investments, nie	0.7	0.7	0.0	0.0	0.0	0.0	...	Invest. portefeuille, nia
Other Capital, nie	-0.6	5.0	-3.9	-4.2	13.3	6.8	...	Autres capitaux, nia
Net Errors and Omissions	3.6	-16.1	-0.9	-0.2	-2.9	9.1	...	Erreurs et omissions nettes
Reserves and Related Items	5.3	-2.3	-11.8	-11.9	-21.8	-15.5	...	Rés. et postes assimilables
Benin								**Bénin**
Merchandise: Exp. fob	223.8	314.1	303.2	363.4	379.1	...	...	Marchandises : exp. fob
Merchandise: Imp. fob	-290.9	-337.7	-412.4	-483.8	-511.0	...	...	Marchandises : imp. fob
Serv. & Income: Credit	65.7	77.2	99.3	111.8	118.9	...	...	Serv. & revenu : crédit
Serv. & Income: Debit	-149.3	-152.7	-196.1	-228.9	-238.7	...	...	Serv. & revenu : débit
Private Unrequited Transfers	33.2	31.6	43.0	50.9	51.7	...	...	Transf. priv.sans contrep.
Offic. Unrequited Trans.,nie	60.4	72.1	97.3	112.8	117.2	...	...	Tr. offic. sans contrep.nia
Direct Investment, nie	-0.5	0.0	0.0	0.0	0.0	...	...	Investissements direct, nia
Portfolio Investments, nie	0.0	0.0	0.0	0.0	0.0	...	...	Invest. portefeuille, nia
Other Capital, nie	-1.4	9.1	-16.5	44.3	-7.7	...	...	Autres capitaux, nia
Net Errors and Omissions	35.6	-43.3	10.8	-41.6	-7.8	...	...	Erreurs et omissions nettes
Reserves and Related Items	23.3	29.5	71.2	71.3	98.4	...	...	Rés. et postes assimilables
Bhutan								**Bhoutan**
Merchandise: Exp. fob	18.6	22.1	33.9	55.0	74.9	73.4	...	Marchandises : exp. fob
Merchandise: Imp. fob	-70.8	-80.3	-97.3	-88.7	-125.1	-105.1	...	Marchandises : imp. fob
Serv. & Income: Credit	7.9	9.5	9.9	11.5	13.2	15.9	...	Serv. & revenu : crédit
Serv. & Income: Debit	-41.1	-48.9	-44.9	-41.4	-45.8	-37.1	...	Serv. & revenu : débit
Private Unrequited Transfers	2.5	2.7	3.0	3.4	4.5	4.4	...	Transf. priv.sans contrep.
Offic. Unrequited Trans.,nie	60.8	78.5	81.5	75.2	63.4	57.9	...	Tr. offic. sans contrep.nia
Direct Investment, nie	0.0	0.0	0.0	0.0	0.0	0.0	...	Investissements direct, nia
Portfolio Investments, nie	0.0	0.0	0.0	0.0	0.0	0.0	...	Invest. portefeuille, nia
Other Capital, nie	26.7	26.9	21.9	18.5	32.4	3.7	...	Autres capitaux, nia
Net Errors and Omissions	6.5	-5.1	6.5	-9.0	5.5	-23.2	...	Erreurs et omissions nettes
Reserves and Related Items	-11.2	-5.4	-14.5	-24.5	-23.0	10.1	...	Rés. et postes assimilables
Bolivia								**Bolivie**
Merchandise: Exp. fob	724.5	623.4	545.5	518.7	542.5	723.5	830.8	Marchandises : exp. fob
Merchandise: Imp. fob	-412.3	-462.8	-596.5	-646.3	-590.9	-729.5	-775.6	Marchandises : imp. fob
Serv. & Income: Credit	123.5	114.0	138.7	147.7	146.5	167.2	164.7	Serv. & revenu : crédit
Serv. & Income: Debit	-698.4	-636.5	-571.9	-564.1	-537.8	-581.2	-580.7	Serv. & revenu : débit
Private Unrequited Transfers	21.8	19.7	18.5	18.0	12.7	20.6	21.6	Transf. priv.sans contrep.
Offic. Unrequited Trans.,nie	66.0	59.8	81.5	99.2	123.9	135.2	138.4	Tr. offic. sans contrep.nia
Direct Investment, nie	7.0	10.0	10.0	36.4	-12.0	-25.4	...	Investissements direct, nia
Portfolio Investments, nie	-0.9	-0.9	0.0	0.0	0.0	0.0	7.3	Invest. portefeuille, nia
Other Capital, nie	54.6	-241.1	-56.5	-178.1	-12.7	-78.0	45.6	Autres capitaux, nia
Net Errors and Omissions	-12.1	190.0	136.3	174.6	46.6	-32.1	75.5	Erreurs et omissions nettes
Reserves and Related Items	126.3	324.4	294.4	393.9	281.2	399.7	28.5	Rés. et postes assimilables
Botswana								**Botswana**
Merchandise: Exp. fob	677.7	727.6	852.5	1586.6	1468.9	1819.7	1753.2	Marchandises : exp. fob
Merchandise: Imp. fob	-583.4	-493.8	-608.4	-803.9	-986.9	-1185.1	-1606.2	Marchandises : imp. fob
Serv. & Income: Credit	199.9	161.2	216.6	296.9	330.4	355.3	497.3	Serv. & revenu : crédit
Serv. & Income: Debit	-399.3	-352.7	-400.9	-589.6	-791.9	-711.5	-782.1	Serv. & revenu : débit

125
Summary of balance of payments
Millions of US dollars [*cont.*]
Résumé des balances des paiements
Millions de dollars des E-U [*suite*]

Country or area	1984	1985	1986	1987	1988	1989	1990	Pays ou zone
Private Unrequited Transfers	-7.3	-3.2	-2.6	6.8	-17.5	-30.6	-40.8	Transf. priv.sans contrep.
Offic. Unrequited Trans.,nie	42.6	42.1	53.8	166.7	184.6	250.5	316.2	Tr. offic. sans contrep.nia
Direct Investment, nie	62.4	52.1	70.4	113.6	39.9	42.2	38.2	Investissements direct, nia
Portfolio Investments, nie	0.0	0.0	0.0	0.0	0.0	0.0	0.0	Invest. portefeuille, nia
Other Capital, nie	50.1	71.0	35.2	-203.3	-65.2	70.8	153.3	Autres capitaux, nia
Net Errors and Omissions	81.7	49.9	90.5	-12.2	220.0	-34.8	-21.7	Erreurs et omissions nettes
Reserves and Related Items	-124.3	-254.3	-306.9	-561.5	-382.3	-576.5	-307.2	Rés. et postes assimilables
Brazil								**Brésil**
Merchandise: Exp. fob	27002.0	25634.0	22348.0	26210.0	33773.0	34375.0	...	Marchandises : exp. fob
Merchandise: Imp. fob	-13916.0	-13168.0	-14044.0	-15052.0	-14605.0	-18263.0	...	Marchandises : imp. fob
Serv. & Income: Credit	3203.0	3675.0	2783.0	2520.0	3050.0	4442.0	...	Serv. & revenu : crédit
Serv. & Income: Debit	-16418.0	-16569.0	-16478.0	-15198.0	-18153.0	-19773.0	...	Serv. & revenu : débit
Private Unrequited Transfers	161.0	139.0	89.0	113.0	107.0	226.0	...	Transf. priv.sans contrep.
Offic. Unrequited Trans.,nie	10.0	16.0	-2.0	-43.0	-13.0	18.0	...	Tr. offic. sans contrep.nia
Direct Investment, nie	1556.0	1267.0	177.0	1087.0	2794.0	744.0	...	Investissements direct, nia
Portfolio Investments, nie	-272.0	-237.0	-450.0	-428.0	-498.0	-421.0	...	Invest. portefeuille, nia
Other Capital, nie	-7281.0	-9321.0	-8312.0	-10941.0	-11506.0	-12848.0	...	Autres capitaux, nia
Net Errors and Omissions	399.0	-530.0	66.0	-802.0	-827.0	-819.0	...	Erreurs et omissions nettes
Reserves and Related Items	5556.0	9094.0	13823.0	12534.0	5878.0	12319.0	...	Rés. et postes assimilables
Burkina Faso								**Burkina Faso**
Merchandise: Exp. fob	140.9	135.6	149.0	229.9	249.1	215.7	304.2	Marchandises : exp. fob
Merchandise: Imp. fob	-270.1	-352.8	-437.5	-475.2	-486.8	-501.6	-593.2	Marchandises : imp. fob
Serv. & Income: Credit	34.1	37.0	51.8	48.9	49.0	52.7	65.0	Serv. & revenu : crédit
Serv. & Income: Debit	-138.7	-145.9	-181.4	-213.3	-235.4	-237.3	-272.9	Serv. & revenu : débit
Private Unrequited Transfers	71.5	102.1	159.6	123.1	113.4	97.5	113.9	Transf. priv.sans contrep.
Offic. Unrequited Trans.,nie	158.9	163.9	237.7	235.6	261.4	444.3	272.5	Tr. offic. sans contrep.nia
Direct Investment, nie	1.7	-1.4	3.1	0.0	0.0	0.0	0.0	Investissements direct, nia
Portfolio Investments, nie	0.0	1.2	1.3	0.0	0.0	0.0	0.0	Invest. portefeuille, nia
Other Capital, nie	31.4	66.4	55.2	71.4	73.9	-217.3	77.5	Autres capitaux, nia
Net Errors and Omissions	6.2	-0.4	3.0	-3.0	-0.4	-0.8	-1.2	Erreurs et omissions nettes
Reserves and Related Items	-35.9	-5.7	-41.9	-17.5	-24.3	146.8	34.3	Rés. et postes assimilables
Burundi								**Burundi**
Merchandise: Exp. fob	...	113.6	129.1	98.3	124.4	93.2	72.5	Marchandises : exp. fob
Merchandise: Imp. fob	...	-149.7	-165.3	-159.2	-166.1	-151.4	-188.6	Marchandises : imp. fob
Serv. & Income: Credit	...	14.7	13.8	14.8	14.8	24.2	24.8	Serv. & revenu : crédit
Serv. & Income: Debit	...	-109.3	-125.6	-163.4	-140.7	-118.6	-123.5	Serv. & revenu : débit
Private Unrequited Transfers	...	10.0	7.7	7.2	9.9	8.6	10.0	Transf. priv.sans contrep.
Offic. Unrequited Trans.,nie	...	78.5	102.7	105.7	87.2	132.4	148.3	Tr. offic. sans contrep.nia
Direct Investment, nie	...	0.5	1.5	1.4	1.2	0.5	1.2	Investissements direct, nia
Portfolio Investments, nie	...	0.0	0.0	0.0	0.0	0.0	0.0	Invest. portefeuille, nia
Other Capital, nie	...	65.5	83.4	129.7	83.1	63.9	38.0	Autres capitaux, nia
Net Errors and Omissions	...	-8.0	-18.9	-37.3	-6.6	-14.9	13.9	Erreurs et omissions nettes
Reserves and Related Items	...	-15.9	-28.5	2.6	-7.1	-38.0	3.2	Rés. et postes assimilables
Cameroon								**Cameroun**
Merchandise: Exp. fob	1589.0	1626.3	2077.0	1688.7	1841.2	...	...	Marchandises : exp. fob
Merchandise: Imp. fob	-1064.7	-1135.9	-1634.5	-1434.8	-1220.8	...	...	Marchandises : imp. fob
Serv. & Income: Credit	450.3	536.8	518.5	423.8	473.0	...	...	Serv. & revenu : crédit
Serv. & Income: Debit	-1118.9	-1570.1	-1419.3	-1471.4	-1416.4	...	...	Serv. & revenu : débit
Private Unrequited Transfers	-77.3	-99.9	-122.8	-125.7	-134.9	...	...	Transf. priv.sans contrep.
Offic. Unrequited Trans.,nie	52.8	81.3	30.1	26.6	29.0	...	...	Tr. offic. sans contrep.nia
Direct Investment, nie	7.6	305.7	3.3	0.4	38.6	...	...	Investissements direct, nia
Portfolio Investments, nie	0.0	0.0	0.0	0.0	0.0	...	...	Invest. portefeuille, nia
Other Capital, nie	196.6	205.6	487.8	333.3	13.1	...	...	Autres capitaux, nia
Net Errors and Omissions	-121.6	108.9	-21.7	89.3	166.2	...	...	Erreurs et omissions nettes
Reserves and Related Items	86.4	-58.7	81.5	469.7	210.8	...	...	Rés. et postes assimilables
Canada								**Canada**
Merchandise: Exp. fob	88623.0	89648.0	89027.0	98052.0	116058.0	123016.0	129046.0	Marchandises : exp. fob
Merchandise: Imp. fob	-72656.0	-77074.0	-81350.0	-89092.0	-106637.0	-116284.0	-119093.0	Marchandises : imp. fob
Serv. & Income: Credit	13539.0	14681.0	15949.0	17709.0	22994.0	23201.0	23697.0	Serv. & revenu : crédit
Serv. & Income: Debit	-28109.0	-29316.0	-31999.0	-35368.0	-43865.0	-47680.0	-52363.0	Serv. & revenu : débit
Private Unrequited Transfers	24.0	7.0	41.0	394.0	663.0	767.0	759.0	Transf. priv.sans contrep.
Offic. Unrequited Trans.,nie	-203.0	-225.0	143.0	-449.0	-449.0	-501.0	-861.0	Tr. offic. sans contrep.nia
Direct Investment, nie	-836.0	-4845.0	-2646.0	-3548.0	-2260.0	-735.0	4901.0	Investissements direct, nia
Portfolio Investments, nie	4143.0	6379.0	14002.0	8615.0	9663.0	15946.0	9004.0	Invest. portefeuille, nia
Other Capital, nie	-1623.0	2054.0	-1798.0	8868.0	14944.0	6288.0	10948.0	Autres capitaux, nia
Net Errors and Omissions	-4724.0	-4588.0	-1945.0	-2403.0	-3552.0	-3726.0	-5413.0	Erreurs et omissions nettes

125
Summary of balance of payments
Millions of US dollars [*cont.*]
Résumé des balances des paiements
Millions de dollars des E-U [*suite*]

Country or area	1984	1985	1986	1987	1988	1989	1990	Pays ou zone
Reserves and Related Items	1822.0	3280.0	576.0	-2778.0	-7558.0	-293.0	-625.0	Rés. et postes assimilables
Cape Verde								**Cap-Vert**
Merchandise: Exp. fob	6.9	6.1	5.1	12.2	5.0	11.1	...	Marchandises : exp. fob
Merchandise: Imp. fob	-82.5	-86.7	-85.9	-92.8	-101.8	-110.2	...	Marchandises : imp. fob
Serv. & Income: Credit	25.4	26.7	35.8	44.9	43.6	55.9	...	Serv. & revenu : crédit
Serv. & Income: Debit	-16.6	-14.6	-23.3	-26.1	-20.8	-25.9	...	Serv. & revenu : débit
Private Unrequited Transfers	21.6	22.2	28.7	34.0	39.6	43.3	...	Transf. priv.sans contrep.
Offic. Unrequited Trans.,nie	38.7	37.3	42.2	46.9	38.7	35.7	...	Tr. offic. sans contrep.nia
Direct Investment, nie	0.0	0.0	0.0	2.8	0.4	-0.6	...	Investissements direct, nia
Portfolio Investments, nie	0.0	0.0	0.0	0.0	0.0	0.0	...	Invest. portefeuille, nia
Other Capital, nie	7.0	6.4	14.3	0.6	-6.5	6.4	...	Autres capitaux, nia
Net Errors and Omissions	-8.7	20.9	-38.6	10.2	1.9	-35.8	...	Erreurs et omissions nettes
Reserves and Related Items	8.1	-18.3	21.9	-32.7	0.0	20.1	...	Rés. et postes assimilables
Central African Rep.								**Rép. centrafricaine**
Merchandise: Exp. fob	114.4	131.0	129.5	127.1	...	...	...	Marchandises : exp. fob
Merchandise: Imp. fob	-140.1	-167.7	-201.0	-197.7	...	...	...	Marchandises : imp. fob
Serv. & Income: Credit	37.5	53.4	58.8	70.6	...	...	...	Serv. & revenu : crédit
Serv. & Income: Debit	-98.8	-122.1	-156.8	-176.4	...	...	...	Serv. & revenu : débit
Private Unrequited Transfers	-11.2	-11.8	-19.5	-23.7	...	...	...	Transf. priv.sans contrep.
Offic. Unrequited Trans.,nie	64.6	68.5	102.4	124.8	...	...	...	Tr. offic. sans contrep.nia
Direct Investment, nie	4.9	2.4	6.9	9.3	...	...	...	Investissements direct, nia
Portfolio Investments, nie	0.0	0.0	0.0	0.0	...	...	...	Invest. portefeuille, nia
Other Capital, nie	22.4	29.2	69.5	54.1	...	...	...	Autres capitaux, nia
Net Errors and Omissions	6.7	-8.1	7.1	0.1	...	...	...	Erreurs et omissions nettes
Reserves and Related Items	-0.6	25.1	3.0	11.6	...	...	...	Rés. et postes assimilables
Chad								**Tchad**
Merchandise: Exp. fob	109.7	61.8	98.6	109.4	145.9	155.4	...	Marchandises : exp. fob
Merchandise: Imp. fob	-128.3	-166.3	-212.1	-225.9	-228.4	-240.3	...	Marchandises : imp. fob
Serv. & Income: Credit	38.2	37.6	48.0	73.3	80.9	43.6	...	Serv. & revenu : crédit
Serv. & Income: Debit	-96.4	-161.2	-178.1	-211.0	-233.4	-220.7	...	Serv. & revenu : débit
Private Unrequited Transfers	-1.7	6.8	-5.4	-9.8	-17.1	-20.2	...	Transf. priv.sans contrep.
Offic. Unrequited Trans.,nie	87.6	133.9	189.5	238.5	277.7	226.3	...	Tr. offic. sans contrep.nia
Direct Investment, nie	9.2	53.4	27.8	0.2	-12.6	6.2	...	Investissements direct, nia
Portfolio Investments, nie	0.0	0.0	0.0	0.0	0.0	0.0	...	Invest. portefeuille, nia
Other Capital, nie	-11.3	16.2	5.1	8.5	36.9	55.8	...	Autres capitaux, nia
Net Errors and Omissions	13.7	-5.5	9.7	16.5	-83.7	23.7	...	Erreurs et omissions nettes
Reserves and Related Items	-20.6	23.2	16.8	0.3	33.8	-29.7	...	Rés. et postes assimilables
Chile								**Chili**
Merchandise: Exp. fob	3650.0	3804.0	4199.0	5224.0	7052.0	8080.0	8310.0	Marchandises : exp. fob
Merchandise: Imp. fob	-3288.0	-2954.0	-3099.0	-3994.0	-4833.0	-6502.0	-7037.0	Marchandises : imp. fob
Serv. & Income: Credit	986.0	865.0	1150.0	1268.0	1399.0	1777.0	2232.0	Serv. & revenu : crédit
Serv. & Income: Debit	-3566.0	-3104.0	-3471.0	-3432.0	-3962.0	-4337.0	-4494.0	Serv. & revenu : débit
Private Unrequited Transfers	47.0	47.0	40.0	65.0	63.0	58.0	54.0	Transf. priv.sans contrep.
Offic. Unrequited Trans.,nie	60.0	14.0	44.0	61.0	114.0	157.0	145.0	Tr. offic. sans contrep.nia
Direct Investment, nie	67.0	62.0	57.0	97.0	109.0	259.0	587.0	Investissements direct, nia
Portfolio Investments, nie	0.0	50.0	262.0	826.0	902.0	1324.0	428.0	Invest. portefeuille, nia
Other Capital, nie	-221.0	-1390.0	-2077.0	-1959.0	110.0	-317.0	1635.0	Autres capitaux, nia
Net Errors and Omissions	190.0	-3.0	88.0	-78.0	-109.0	-72.0	508.0	Erreurs et omissions nettes
Reserves and Related Items	2075.0	2609.0	2807.0	1922.0	-845.0	-427.0	-2368.0	Rés. et postes assimilables
China								**Chine**
Merchandise: Exp. fob	23905.0	25108.0	25756.0	34734.0	41054.0	43220.0	51519.0	Marchandises : exp. fob
Merchandise: Imp. fob	-23891.0	-38231.0	-34896.0	-36395.0	-46369.0	-48840.0	-42354.0	Marchandises : imp. fob
Serv. & Income: Credit	4819.0	4533.0	4927.0	5413.0	6327.0	6497.0	8873.0	Serv. & revenu : crédit
Serv. & Income: Debit	-2766.0	-3070.0	-3200.0	-3676.0	-5233.0	-5575.0	-6314.0	Serv. & revenu : débit
Private Unrequited Transfers	305.0	171.0	255.0	249.0	416.0	238.0	222.0	Transf. priv.sans contrep.
Offic. Unrequited Trans.,nie	137.0	72.0	124.0	-25.0	3.0	143.0	52.0	Tr. offic. sans contrep.nia
Direct Investment, nie	1124.0	1031.0	1425.0	1669.0	2344.0	2613.0	2659.0	Investissements direct, nia
Portfolio Investments, nie	83.0	742.0	1567.0	1051.0	876.0	-180.0	-241.0	Invest. portefeuille, nia
Other Capital, nie	-968.0	4914.0	2271.0	3361.0	3996.0	1289.0	5787.0	Autres capitaux, nia
Net Errors and Omissions	-889.0	5.0	-277.0	-1598.0	-1040.0	117.0	-8161.0	Erreurs et omissions nettes
Reserves and Related Items	-1859.0	4725.0	2048.0	-4783.0	-2374.0	478.0	-12042.0	Rés. et postes assimilables
Colombia								**Colombie**
Merchandise: Exp. fob	4273.0	3650.0	5331.0	5661.0	5343.0	6031.0	7105.0	Marchandises : exp. fob
Merchandise: Imp. fob	-4027.0	-3673.0	-3409.0	-3793.0	-4516.0	-4557.0	-5088.0	Marchandises : imp. fob
Serv. & Income: Credit	1055.0	966.0	1283.0	1368.0	1665.0	1589.0	1806.0	Serv. & revenu : crédit
Serv. & Income: Debit	-3001.0	-3213.0	-3607.0	-3901.0	-3672.0	-4156.0	-4431.0	Serv. & revenu : débit

125
Summary of balance of payments
Millions of US dollars [cont.]
Résumé des balances des paiements
Millions de dollars des E-U [suite]

Country or area	1984	1985	1986	1987	1988	1989	1990	Pays ou zone
Private Unrequited Transfers	289.0	455.0	801.0	1009.0	975.0	912.0	1014.0	Transf. priv.sans contrep.
Offic. Unrequited Trans.,nie	10.0	6.0	-16.0	-8.0	-11.0	-14.0	-15.0	Tr. offic. sans contrep.nia
Direct Investment, nie	561.0	1016.0	642.0	293.0	159.0	547.0	471.0	Investissements direct, nia
Portfolio Investments, nie	-3.0	-1.0	30.0	48.0	0.0	179.0	-4.0	Invest. portefeuille, nia
Other Capital, nie	382.0	1222.0	492.0	-353.0	781.0	-321.0	-449.0	Autres capitaux, nia
Net Errors and Omissions	76.0	-273.0	-251.0	67.0	-530.0	153.0	231.0	Erreurs et omissions nettes
Reserves and Related Items	385.0	-155.0	-1296.0	-391.0	-194.0	-363.0	-640.0	Rés. et postes assimilables
Comoros								**Comores**
Merchandise: Exp. fob	7.1	15.7	20.4	11.6	21.5	18.1	17.9	Marchandises : exp. fob
Merchandise: Imp. fob	-32.7	-28.2	-28.5	-44.2	-44.3	-35.7	-45.2	Marchandises : imp. fob
Serv. & Income: Credit	3.7	4.8	8.3	16.2	18.6	21.8	20.2	Serv. & revenu : crédit
Serv. & Income: Debit	-39.3	-37.9	-45.4	-44.7	-46.2	-42.6	-46.9	Serv. & revenu : débit
Private Unrequited Transfers	-3.0	-0.7	-2.0	0.9	3.1	2.7	5.5	Transf. priv.sans contrep.
Offic. Unrequited Trans.,nie	31.4	32.0	31.7	38.7	40.8	41.1	39.2	Tr. offic. sans contrep.nia
Direct Investment, nie	0.0	0.0	0.0	7.6	3.8	3.3	-0.7	Investissements direct, nia
Portfolio Investments, nie	0.0	-0.2	0.0	0.0	0.0	0.0	0.0	Invest. portefeuille, nia
Other Capital, nie	27.4	18.8	21.2	22.0	0.4	4.3	14.4	Autres capitaux, nia
Net Errors and Omissions	-0.5	1.8	-2.0	0.7	5.9	-7.6	-9.2	Erreurs et omissions nettes
Reserves and Related Items	5.9	-6.1	-3.5	-8.8	-3.6	-5.4	4.9	Rés. et postes assimilables
Congo								**Congo**
Merchandise: Exp. fob	1268.4	1144.8	672.6	876.7	843.2	1138.0	...	Marchandises : exp. fob
Merchandise: Imp. fob	-617.6	-630.1	-512.4	-419.9	-522.7	-534.0	...	Marchandises : imp. fob
Serv. & Income: Credit	86.8	83.8	111.2	127.7	99.7	90.0	...	Serv. & revenu : crédit
Serv. & Income: Debit	-524.8	-762.7	-891.0	-818.3	-873.5	-796.9	...	Serv. & revenu : débit
Private Unrequited Transfers	-45.0	-37.5	-41.0	-56.8	-56.6	-29.8	...	Transf. priv.sans contrep.
Offic. Unrequited Trans.,nie	42.3	40.4	59.9	67.9	64.4	51.7	...	Tr. offic. sans contrep.nia
Direct Investment, nie	34.9	12.7	22.4	43.4	9.1	0.0	...	Investissements direct, nia
Portfolio Investments, nie	0.0	0.0	0.0	0.0	0.0	0.0	...	Invest. portefeuille, nia
Other Capital, nie	-327.0	26.2	132.0	-336.9	-71.1	-374.4	...	Autres capitaux, nia
Net Errors and Omissions	20.5	41.7	48.1	27.6	40.6	100.3	...	Erreurs et omissions nettes
Reserves and Related Items	61.4	80.7	398.2	488.5	466.8	355.0	...	Rés. et postes assimilables
Costa Rica								**Costa Rica**
Merchandise: Exp. fob	997.5	939.1	1084.8	1106.7	1180.7	1333.4	1365.6	Marchandises : exp. fob
Merchandise: Imp. fob	-992.9	-1001.0	-1045.2	-1245.2	-1278.6	-1572.0	-1833.3	Marchandises : imp. fob
Serv. & Income: Credit	316.2	331.1	355.4	385.6	478.1	617.8	681.7	Serv. & revenu : crédit
Serv. & Income: Debit	-612.8	-613.9	-626.9	-729.5	-814.1	-985.5	-940.5	Serv. & revenu : débit
Private Unrequited Transfers	31.9	42.6	37.4	38.7	40.0	39.2	47.8	Transf. priv.sans contrep.
Offic. Unrequited Trans.,nie	9.0	11.0	33.9	67.3	90.4	87.2	94.9	Tr. offic. sans contrep.nia
Direct Investment, nie	52.0	65.2	57.4	75.8	121.4	95.2	108.7	Investissements direct, nia
Portfolio Investments, nie	-0.2	-13.5	-2.5	0.0	-6.0	-13.2	-28.2	Invest. portefeuille, nia
Other Capital, nie	-331.0	-284.2	-358.1	-517.0	-387.3	-264.5	-199.7	Autres capitaux, nia
Net Errors and Omissions	104.4	142.9	97.5	131.2	224.6	208.9	127.5	Erreurs et omissions nettes
Reserves and Related Items	426.0	380.7	366.3	686.4	350.8	453.5	575.4	Rés. et postes assimilables
Côte d'Ivoire								**Côte d'Ivoire**
Merchandise: Exp. fob	2624.8	2761.0	3187.4	2949.7	2774.2	2807.8	3120.1	Marchandises : exp. fob
Merchandise: Imp. fob	-1487.3	-1409.9	-1639.9	-1863.3	-1696.2	-1720.3	-1701.3	Marchandises : imp. fob
Serv. & Income: Credit	408.1	438.5	538.8	613.2	621.5	529.5	449.6	Serv. & revenu : crédit
Serv. & Income: Debit	-1355.3	-1468.9	-2027.7	-2309.9	-2373.7	-2447.6	-2702.9	Serv. & revenu : débit
Private Unrequited Transfers	-291.1	-278.9	-427.4	-501.1	-514.4	-469.6	-539.9	Transf. priv.sans contrep.
Offic. Unrequited Trans.,nie	24.1	21.8	68.3	141.4	58.9	90.3	91.8	Tr. offic. sans contrep.nia
Direct Investment, nie	21.7	29.2	70.7	87.5	56.1	78.4	47.7	Investissements direct, nia
Portfolio Investments, nie	-1.6	-1.3	-0.3	-8.0	-14.1	0.0	0.0	Invest. portefeuille, nia
Other Capital, nie	-302.8	-351.9	-94.1	-87.8	95.4	-316.6	-376.1	Autres capitaux, nia
Net Errors and Omissions	-119.1	84.2	-55.3	45.2	-410.8	144.9	45.4	Erreurs et omissions nettes
Reserves and Related Items	478.6	176.2	379.5	933.1	1403.1	1303.3	1565.6	Rés. et postes assimilables
Cyprus								**Chypre**
Merchandise: Exp. fob	522.6	416.9	451.6	566.4	645.5	717.3	846.5	Marchandises : exp. fob
Merchandise: Imp. fob	-1224.9	-1121.9	-1141.6	-1326.8	-1666.8	-2072.1	-2304.7	Marchandises : imp. fob
Serv. & Income: Credit	862.0	911.9	1112.6	1375.8	1628.3	1878.6	2321.1	Serv. & revenu : crédit
Serv. & Income: Debit	-406.0	-408.9	-478.6	-567.5	-650.6	-696.6	-857.8	Serv. & revenu : débit
Private Unrequited Transfers	21.6	22.2	23.7	26.4	24.6	22.9	21.9	Transf. priv.sans contrep.
Offic. Unrequited Trans.,nie	21.9	18.6	21.6	17.5	27.4	17.0	18.7	Tr. offic. sans contrep.nia
Direct Investment, nie	52.7	58.0	46.3	52.0	62.1	69.9	129.9	Investissements direct, nia
Portfolio Investments, nie	0.0	0.0	0.0	0.0	0.0	92.8	-37.4	Invest. portefeuille, nia
Other Capital, nie	216.9	49.6	150.9	28.9	86.8	247.6	371.8	Autres capitaux, nia
Net Errors and Omissions	29.1	23.9	-27.1	-109.0	-86.6	-48.9	-212.4	Erreurs et omissions nettes

125
Summary of balance of payments
Millions of US dollars [*cont.*]
Résumé des balances des paiements
Millions de dollars des E-U [*suite*]

Country or area	1984	1985	1986	1987	1988	1989	1990	Pays ou zone
Reserves and Related Items	-95.9	29.6	-159.5	-63.7	-70.7	-228.5	-297.6	Rés. et postes assimilables
Czechoslovakia								**Tchécoslovaquie**
Merchandise: Exp. fob	11564.0	11662.0	13503.0	15183.0	15027.0	14217.0	11635.0	Marchandises : exp. fob
Merchandise: Imp. fob	-11211.0	-11385.0	-13890.0	-15454.0	-14642.0	-14074.0	-13057.0	Marchandises : imp. fob
Serv. & Income: Credit	2643.0	2583.0	2983.0	3220.0	3301.0	3363.0	3171.0	Serv. & revenu : crédit
Serv. & Income: Debit	-2295.0	-2174.0	-2440.0	-2647.0	-2659.0	-2678.0	-3182.0	Serv. & revenu : débit
Private Unrequited Transfers	29.0	34.0	53.0	95.0	94.0	130.0	258.0	Transf. priv.sans contrep.
Offic. Unrequited Trans.,nie	-20.0	-30.0	-41.0	-27.0	-29.0	-22.0	-52.0	Tr. offic. sans contrep.nia
Direct Investment, nie	0.0	0.0	0.0	0.0	0.0	257.0	187.0	Investissements direct, nia
Portfolio Investments, nie	0.0	0.0	0.0	0.0	0.0	0.0	0.0	Invest. portefeuille, nia
Other Capital, nie	-438.0	-959.0	72.0	-108.0	-893.0	-548.0	455.0	Autres capitaux, nia
Net Errors and Omissions	-105.0	156.0	19.0	-3.0	6.0	-81.0	-543.0	Erreurs et omissions nettes
Reserves and Related Items	-166.0	112.0	-260.0	-260.0	-207.0	-563.0	1127.0	Rés. et postes assimilables
Denmark								**Danemark**
Merchandise: Exp. fob	16079.0	17123.0	21307.0	25695.0	27537.0	28728.0	35740.0	Marchandises : exp. fob
Merchandise: Imp. fob	-16285.0	-17887.0	-22357.0	-24900.0	-25654.0	-26304.0	-31174.0	Marchandises : imp. fob
Serv. & Income: Credit	6365.0	6896.0	8367.0	10475.0	13303.0	14547.0	19740.0	Serv. & revenu : crédit
Serv. & Income: Debit	-7864.0	-8764.0	-11529.0	-14051.0	-16219.0	-17720.0	-22710.0	Serv. & revenu : débit
Private Unrequited Transfers	-7.0	-55.0	-112.0	-56.0	-88.0	80.0	-46.0	Transf. priv.sans contrep.
Offic. Unrequited Trans.,nie	76.0	-80.0	-166.0	-164.0	-131.0	-223.0	-10.0	Tr. offic. sans contrep.nia
Direct Investment, nie	-88.0	-195.0	-491.0	-534.0	-217.0	-976.0	-271.0	Investissements direct, nia
Portfolio Investments, nie	89.0	1233.0	-2079.0	3683.0	1231.0	-2749.0	2900.0	Invest. portefeuille, nia
Other Capital, nie	1879.0	3555.0	5367.0	4210.0	2174.0	1142.0	1861.0	Autres capitaux, nia
Net Errors and Omissions	-611.0	-304.0	-285.0	85.0	-619.0	-347.0	-2657.0	Erreurs et omissions nettes
Reserves and Related Items	367.0	-1522.0	1979.0	-4443.0	-1316.0	3821.0	-3374.0	Rés. et postes assimilables
Dominica								**Dominique**
Merchandise: Exp. fob	25.6	28.4	44.6	49.3	57.0	46.3	59.9	Marchandises : exp. fob
Merchandise: Imp. fob	-50.7	-52.0	-49.0	-58.7	-77.0	-94.2	-103.9	Marchandises : imp. fob
Serv. & Income: Credit	11.8	10.2	15.3	18.1	23.8	26.1	37.4	Serv. & revenu : crédit
Serv. & Income: Debit	-11.4	-13.7	-19.7	-22.0	-24.3	-33.9	-40.9	Serv. & revenu : débit
Private Unrequited Transfers	6.3	6.5	6.7	7.6	8.4	11.1	13.0	Transf. priv.sans contrep.
Offic. Unrequited Trans.,nie	11.2	14.2	9.4	10.6	9.4	7.7	8.8	Tr. offic. sans contrep.nia
Direct Investment, nie	2.3	3.0	2.7	8.6	6.9	8.9	8.4	Investissements direct, nia
Portfolio Investments, nie	0.0	0.0	0.0	0.0	0.0	0.0	0.0	Invest. portefeuille, nia
Other Capital, nie	9.1	3.5	4.7	-3.2	-1.4	27.4	24.9	Autres capitaux, nia
Net Errors and Omissions	1.4	-0.7	-8.6	-1.7	-3.9	0.7	-2.8	Erreurs et omissions nettes
Reserves and Related Items	-5.6	0.6	-6.0	-8.6	1.1	-0.2	-4.7	Rés. et postes assimilables
Dominican Republic								**Rép. dominicaine**
Merchandise: Exp. fob	868.1	738.5	722.1	711.3	889.7	924.4	734.7	Marchandises : exp. fob
Merchandise: Imp. fob	-1257.1	-1285.9	-1351.7	-1591.5	-1608.0	-1963.8	-1792.9	Marchandises : imp. fob
Serv. & Income: Credit	507.3	605.9	702.9	859.9	1014.8	1228.9	1282.3	Serv. & revenu : crédit
Serv. & Income: Debit	-546.7	-522.4	-549.8	-678.1	-676.1	-706.4	-653.4	Serv. & revenu : débit
Private Unrequited Transfers	205.0	242.0	230.3	277.4	292.9	305.8	314.9	Transf. priv.sans contrep.
Offic. Unrequited Trans.,nie	60.0	114.3	60.7	57.5	64.8	83.9	55.8	Tr. offic. sans contrep.nia
Direct Investment, nie	68.5	36.2	50.0	89.0	106.1	110.0	132.8	Investissements direct, nia
Portfolio Investments, nie	0.0	0.0	0.0	0.0	0.0	0.0	0.0	Invest. portefeuille, nia
Other Capital, nie	155.6	39.3	41.2	103.4	-86.3	83.6	-0.7	Autres capitaux, nia
Net Errors and Omissions	29.7	155.7	147.6	28.4	11.8	-101.6	-71.6	Erreurs et omissions nettes
Reserves and Related Items	-90.4	-123.6	-53.3	142.8	-9.7	35.2	-1.8	Rés. et postes assimilables
Ecuador								**Equateur**
Merchandise: Exp. fob	2621.0	2905.0	2186.0	2021.0	2202.0	2354.0	2714.0	Marchandises : exp. fob
Merchandise: Imp. fob	-1567.0	-1611.0	-1643.0	-2054.0	-1583.0	-1693.0	-1711.0	Marchandises : imp. fob
Serv. & Income: Credit	350.0	422.0	468.0	444.0	455.0	541.0	542.0	Serv. & revenu : crédit
Serv. & Income: Debit	-1688.0	-1682.0	-1609.0	-1674.0	-1676.0	-1771.0	-1781.0	Serv. & revenu : débit
Private Unrequited Transfers	0.0	0.0	0.0	0.0	0.0	0.0	0.0	Transf. priv.sans contrep.
Offic. Unrequited Trans.,nie	20.0	80.0	45.0	132.0	97.0	97.0	100.0	Tr. offic. sans contrep.nia
Direct Investment, nie	50.0	62.0	70.0	75.0	80.0	80.0	82.0	Investissements direct, nia
Portfolio Investments, nie	0.0	0.0	0.0	0.0	0.0	0.0	0.0	Invest. portefeuille, nia
Other Capital, nie	-1163.0	-1040.0	-462.0	-309.0	-594.0	-503.0	-730.0	Autres capitaux, nia
Net Errors and Omissions	-150.1	36.9	-1160.5	-262.6	-165.2	-69.4	83.1	Erreurs et omissions nettes
Reserves and Related Items	1527.1	827.1	2105.5	1627.6	1184.2	964.4	700.9	Rés. et postes assimilables
Egypt								**Egypte**
Merchandise: Exp. fob	3864.0	3836.0	2632.0	3115.0	2770.0	2907.0	3604.0	Marchandises : exp. fob
Merchandise: Imp. fob	-10080.0	-9050.0	-7170.0	-8095.0	-9378.0	8841.0	-10303.0	Marchandises : imp. fob
Serv. & Income: Credit	3512.0	3442.0	3764.0	4130.0	4982.0	5123.0	7147.0	Serv. & revenu : crédit
Serv. & Income: Debit	-4188.0	-4400.0	-4139.0	-3725.0	-3859.0	-4672.0	-5667.0	Serv. & revenu : débit

125
Summary of balance of payments
Millions of US dollars [*cont.*]
Résumé des balances des paiements
Millions de dollars des E-U [*suite*]

Country or area	1984	1985	1986	1987	1988	1989	1990	Pays ou zone
Private Unrequited Transfers	3981.0	3216.0	2515.0	3604.0	3770.0	3293.0	4284.0	Transf. priv.sans contrep.
Offic. Unrequited Trans.,nie	923.0	791.0	586.0	725.0	666.0	880.0	1119.0	Tr. offic. sans contrep.nia
Direct Investment, nie	713.0	1175.0	1211.0	929.0	1178.0	1228.0	722.0	Investissements direct, nia
Portfolio Investments, nie	1.0	20.0	0.0	2.0	0.0	0.0	15.0	Invest. portefeuille, nia
Other Capital, nie	1004.0	186.0	724.0	-1263.0	130.0	-867.0	-11776.0	Autres capitaux, nia
Net Errors and Omissions	23.0	585.0	-155.0	893.0	-360.0	415.0	631.0	Erreurs et omissions nettes
Reserves and Related Items	247.0	200.0	31.0	-315.0	101.0	533.0	10224.0	Rés. et postes assimilables
El Salvador								**El Salvador**
Merchandise: Exp. fob	725.9	679.0	777.9	589.6	610.6	557.5	...	Marchandises : exp. fob
Merchandise: Imp. fob	-914.5	-895.0	-902.3	-938.7	-966.5	-1220.2	...	Marchandises : imp. fob
Serv. & Income: Credit	228.3	272.4	277.9	361.1	352.2	377.1	...	Serv. & revenu : crédit
Serv. & Income: Debit	-400.8	-428.7	-420.0	-415.0	-471.0	-519.5	...	Serv. & revenu : débit
Private Unrequited Transfers	118.0	129.4	149.6	180.5	202.1	232.8	...	Transf. priv.sans contrep.
Offic. Unrequited Trans.,nie	54.6	54.2	99.7	154.4	143.5	202.7	...	Tr. offic. sans contrep.nia
Direct Investment, nie	12.4	12.4	24.1	18.3	17.0	14.4	...	Investissements direct, nia
Portfolio Investments, nie	0.0	0.0	-3.1	0.0	0.0	0.0	...	Invest. portefeuille, nia
Other Capital, nie	6.7	-15.6	24.7	-77.4	35.4	103.7	...	Autres capitaux, nia
Net Errors and Omissions	-52.1	23.0	-141.8	7.0	-107.2	140.9	...	Erreurs et omissions nettes
Reserves and Related Items	221.5	169.0	113.2	120.3	184.1	110.7	...	Rés. et postes assimilables
Equatorial Guinea								**Guinée équatoriale**
Merchandise: Exp. fob	...	...	...	38.5	44.7	32.8	...	Marchandises : exp. fob
Merchandise: Imp. fob	...	...	...	-47.9	-56.5	-43.9	...	Marchandises : imp. fob
Serv. & Income: Credit	...	...	...	6.1	5.9	6.1	...	Serv. & revenu : crédit
Serv. & Income: Debit	...	...	...	-47.2	-56.6	-50.2	...	Serv. & revenu : débit
Private Unrequited Transfers	...	...	...	-2.5	-3.8	-2.0	...	Transf. priv.sans contrep.
Offic. Unrequited Trans.,nie	...	...	...	27.6	45.8	37.9	...	Tr. offic. sans contrep.nia
Direct Investment, nie	...	...	...	0.0	0.0	0.0	...	Investissements direct, nia
Portfolio Investments, nie	...	...	...	0.0	0.0	0.0	...	Invest. portefeuille, nia
Other Capital, nie	...	...	...	-1.0	4.9	5.4	...	Autres capitaux, nia
Net Errors and Omissions	...	...	...	0.8	-1.7	-3.9	...	Erreurs et omissions nettes
Reserves and Related Items	...	...	...	25.6	17.4	17.8	...	Rés. et postes assimilables
Ethiopia								**Ethiopie**
Merchandise: Exp. fob	416.8	332.9	477.1	355.2	400.0	440.4	297.5	Marchandises : exp. fob
Merchandise: Imp. fob	-798.4	-840.5	-932.6	-932.7	-956.0	-802.8	-914.8	Marchandises : imp. fob
Serv. & Income: Credit	210.3	313.6	277.9	318.7	288.9	286.3	...	Serv. & revenu : crédit
Serv. & Income: Debit	-290.1	-310.6	-332.9	-366.4	-402.9	-378.9	...	Serv. & revenu : débit
Private Unrequited Transfers	144.9	212.9	69.4	129.6	180.5	136.2	...	Transf. priv.sans contrep.
Offic. Unrequited Trans.,nie	186.1	397.7	113.7	278.0	261.6	175.2	171.2	Tr. offic. sans contrep.nia
Direct Investment, nie	0.0	0.0	0.0	0.0	0.0	0.0	0.0	Investissements direct, nia
Portfolio Investments, nie	0.0	0.0	0.0	0.0	0.0	0.0	0.0	Invest. portefeuille, nia
Other Capital, nie	224.2	225.0	239.6	292.8	299.6	85.1	214.5	Autres capitaux, nia
Net Errors and Omissions	-150.6	-168.8	201.6	-182.8	-94.0	76.3	...	Erreurs et omissions nettes
Reserves and Related Items	56.8	-162.2	-113.9	107.4	22.3	-17.9	21.5	Rés. et postes assimilables
Fiji								**Fidji**
Merchandise: Exp. fob	227.9	207.7	245.7	303.6	345.6	399.4	494.3	Marchandises : exp. fob
Merchandise: Imp. fob	-390.7	-382.6	-368.3	-329.0	-395.3	-495.2	-591.1	Marchandises : imp. fob
Serv. & Income: Credit	298.4	315.7	309.2	243.4	272.9	384.2	437.9	Serv. & revenu : crédit
Serv. & Income: Debit	-177.4	-176.8	-205.4	-225.5	-222.6	-274.5	-236.8	Serv. & revenu : débit
Private Unrequited Transfers	-3.8	-9.3	-5.3	-20.8	-3.6	-13.1	-22.2	Transf. priv.sans contrep.
Offic. Unrequited Trans.,nie	18.8	33.0	14.8	10.3	33.7	28.4	24.9	Tr. offic. sans contrep.nia
Direct Investment, nie	23.0	42.8	30.0	6.3	48.6	41.2	100.9	Investissements direct, nia
Portfolio Investments, nie	0.0	0.0	0.0	0.0	0.0	0.0	0.0	Invest. portefeuille, nia
Other Capital, nie	11.6	-10.7	-8.9	-54.3	16.3	-39.3	-39.7	Autres capitaux, nia
Net Errors and Omissions	-0.3	-24.6	16.1	3.1	16.5	-44.7	-32.3	Erreurs et omissions nettes
Reserves and Related Items	-7.5	4.8	-27.9	62.9	-112.1	-13.7	35.9	Rés. et postes assimilables
Finland								**Finlande**
Merchandise: Exp. fob	13087.0	13351.0	16005.0	19079.0	21826.0	22882.0	26089.0	Marchandises : exp. fob
Merchandise: Imp. fob	-11596.0	-12473.0	-14363.0	-17700.0	-20686.0	-23101.0	-25322.0	Marchandises : imp. fob
Serv. & Income: Credit	3310.0	3428.0	3784.0	5057.0	6363.0	6611.0	8784.0	Serv. & revenu : crédit
Serv. & Income: Debit	-4651.0	-4941.0	-5724.0	-7675.0	-9686.0	-11389.0	-15157.0	Serv. & revenu : débit
Private Unrequited Transfers	-22.0	-7.0	-157.0	-162.0	-87.0	-252.0	-342.0	Transf. priv.sans contrep.
Offic. Unrequited Trans.,nie	-153.0	-169.0	-236.0	-328.0	-425.0	-510.0	-735.0	Tr. offic. sans contrep.nia
Direct Investment, nie	-356.0	-235.0	-470.0	-885.0	-2092.0	-2620.0	-2606.0	Investissements direct, nia
Portfolio Investments, nie	1273.0	1343.0	1356.0	1298.0	3186.0	3401.0	5814.0	Invest. portefeuille, nia
Other Capital, nie	1421.0	329.0	-2906.0	6860.0	1112.0	2720.0	8816.0	Autres capitaux, nia
Net Errors and Omissions	-496.0	-38.0	431.0	-1523.0	745.0	1238.0	-1407.0	Erreurs et omissions nettes

125
Summary of balance of payments
Millions of US dollars [*cont.*]
Résumé des balances des paiements
Millions de dollars des E-U [*suite*]

Country or area	1984	1985	1986	1987	1988	1989	1990	Pays ou zone
Reserves and Related Items	-1817.0	-586.0	2280.0	-4022.0	-255.0	1058.0	-3935.0	Rés. et postes assimilables
France								**France**
Merchandise: Exp. fob	92214.0	95927.0	119360.0	141658.0	160188.0	170761.0	206672.0	Marchandises : exp. fob
Merchandise: Imp. fob	-96865.0	-101203.0	-121441.0	-150325.0	-168726.0	-181415.0	-220339.0	Marchandises : imp. fob
Serv. & Income: Credit	55491.0	58356.0	68441.0	79344.0	89530.0	102721.0	134726.0	Serv. & revenu : crédit
Serv. & Income: Debit	-48818.0	-50488.0	-59340.0	-69713.0	-79114.0	-89158.0	-121357.0	Serv. & revenu : débit
Private Unrequited Transfers	-1013.0	-1312.0	-1739.0	-2297.0	-2437.0	-2668.0	-3957.0	Transf. priv.sans contrep.
Offic. Unrequited Trans.,nie	-1884.0	-1315.0	-2850.0	-3114.0	-4237.0	-5863.0	-9517.0	Tr. offic. sans contrep.nia
Direct Investment, nie	279.0	353.0	-2147.0	-4071.0	-6010.0	-9113.0	-21793.0	Investissements direct, nia
Portfolio Investments, nie	6985.0	6478.0	1874.0	5442.0	7798.0	21642.0	28795.0	Invest. portefeuille, nia
Other Capital, nie	-4252.0	-4707.0	-1545.0	-6104.0	1999.0	-7573.0	12035.0	Autres capitaux, nia
Net Errors and Omissions	650.0	290.0	807.0	850.0	912.0	-1687.0	6555.0	Erreurs et omissions nettes
Reserves and Related Items	-2786.0	-2380.0	-1418.0	8329.0	95.0	2354.0	-11821.0	Rés. et postes assimilables
Gabon								**Gabon**
Merchandise: Exp. fob	2017.8	1951.4	1074.2	1286.4	1195.6	1626.0	2471.1	Marchandises : exp. fob
Merchandise: Imp. fob	-733.2	-854.7	-979.1	-731.8	-791.2	-751.7	-760.3	Marchandises : imp. fob
Serv. & Income: Credit	185.5	167.7	141.7	130.8	228.4	308.0	163.8	Serv. & revenu : crédit
Serv. & Income: Debit	-1290.4	-1331.1	-1147.4	-1142.6	-1090.3	-1248.6	-1491.2	Serv. & revenu : débit
Private Unrequited Transfers	-92.5	-109.2	-169.1	-147.8	-155.5	-135.2	-146.9	Transf. priv.sans contrep.
Offic. Unrequited Trans.,nie	25.3	13.5	22.3	24.4	11.2	9.3	-12.1	Tr. offic. sans contrep.nia
Direct Investment, nie	4.8	11.1	103.7	82.2	121.4	-38.6	-77.1	Investissements direct, nia
Portfolio Investments, nie	0.0	0.0	0.0	0.0	0.0	0.0	...	Invest. portefeuille, nia
Other Capital, nie	-20.6	153.1	807.3	283.2	513.3	99.7	-139.2	Autres capitaux, nia
Net Errors and Omissions	-57.6	-50.6	-48.1	80.0	-34.3	34.3	34.9	Erreurs et omissions nettes
Reserves and Related Items	-39.2	48.9	194.6	135.2	1.4	96.7	-42.9	Rés. et postes assimilables
Gambia								**Gambie**
Merchandise: Exp. fob	90.7	62.8	64.9	74.5	83.1	100.2	110.6	Marchandises : exp. fob
Merchandise: Imp. fob	-99.4	-74.8	-84.6	-95.0	-105.9	-125.4	-140.5	Marchandises : imp. fob
Serv. & Income: Credit	27.7	24.4	28.7	50.2	63.7	67.7	71.4	Serv. & revenu : crédit
Serv. & Income: Debit	-25.9	-19.3	-22.4	-61.9	-61.4	-66.4	-64.9	Serv. & revenu : débit
Private Unrequited Transfers	4.9	6.3	10.7	13.5	12.8	6.7	14.1	Transf. priv.sans contrep.
Offic. Unrequited Trans.,nie	10.0	8.0	6.9	24.8	34.3	32.2	43.2	Tr. offic. sans contrep.nia
Direct Investment, nie	0.0	0.0	0.0	1.5	1.2	14.8	0.0	Investissements direct, nia
Portfolio Investments, nie	0.0	0.0	0.0	0.0	0.0	0.0	0.0	Invest. portefeuille, nia
Other Capital, nie	-21.5	-3.9	-9.3	-2.1	8.5	-5.3	-6.1	Autres capitaux, nia
Net Errors and Omissions	-6.5	-9.2	-26.7	5.9	-19.0	-23.4	-19.9	Erreurs et omissions nettes
Reserves and Related Items	19.9	5.8	31.8	-11.4	-17.1	-1.1	-7.9	Rés. et postes assimilables
Germany ß								**Allemagne ß**
F. R. Germany[1]								**R. f. Allemagne[1]**
Merchandise: Exp. fob	161.4	173.7	231.0	278.5	308.6	324.9	391.6	Marchandises : exp. fob
Merchandise: Imp. fob	-139.2	-145.1	-175.3	-208.3	-228.8	-247.2	-320.0	Marchandises : imp. fob
Serv. & Income: Credit	48.1	50.1	67.5	82.4	87.2	100.3	132.6	Serv. & revenu : crédit
Serv. & Income: Debit	-50.3	-51.7	-70.6	-90.1	-98.2	-102.6	-134.6	Serv. & revenu : débit
Private Unrequited Transfers	-4.0	-3.5	-4.7	-5.5	-6.2	-5.5	-6.9	Transf. priv.sans contrep.
Offic. Unrequited Trans.,nie	-6.5	-6.5	-8.0	-11.1	-12.3	-12.9	-15.3	Tr. offic. sans contrep.nia
Direct Investment, nie	-3.8	-4.5	-9.0	-7.3	-10.2	-7.5	-21.1	Investissements direct, nia
Portfolio Investments, nie	1.3	1.8	23.6	-1.9	-43.6	-5.0	-2.3	Invest. portefeuille, nia
Other Capital, nie	-9.9	-15.4	-53.7	-14.4	-16.4	-58.3	-35.5	Autres capitaux, nia
Net Errors and Omissions	1.7	2.0	0.8	-2.2	1.5	3.2	17.1	Erreurs et omissions nettes
Reserves and Related Items	1.1	-0.9	-1.5	-20.4	18.4	10.6	-5.5	Rés. et postes assimilables
Ghana								**Ghana**
Merchandise: Exp. fob	565.9	632.4	773.4	826.8	881.0	807.2	890.6	Marchandises : exp. fob
Merchandise: Imp. fob	-533.0	-668.7	-712.5	-951.5	-993.4	-1002.2	-1198.9	Marchandises : imp. fob
Serv. & Income: Credit	45.8	43.6	45.1	79.3	77.7	81.9	93.1	Serv. & revenu : crédit
Serv. & Income: Debit	-279.8	-283.6	-343.6	-376.3	-399.6	-407.8	-429.0	Serv. & revenu : débit
Private Unrequited Transfers	21.1	32.6	72.1	201.6	172.4	202.1	201.9	Transf. priv.sans contrep.
Offic. Unrequited Trans.,nie	141.2	109.5	122.5	123.2	196.1	220.2	213.8	Tr. offic. sans contrep.nia
Direct Investment, nie	2.0	5.6	4.3	4.7	5.0	15.0	14.8	Investissements direct, nia
Portfolio Investments, nie	0.0	0.0	0.0	0.0	0.0	0.0	0.0	Invest. portefeuille, nia
Other Capital, nie	204.9	79.3	59.1	251.1	204.0	198.6	310.2	Autres capitaux, nia
Net Errors and Omissions	-132.5	63.4	-81.2	-18.7	37.9	40.6	8.8	Erreurs et omissions nettes
Reserves and Related Items	-35.6	-14.1	60.8	-140.2	-181.1	-155.6	-105.3	Rés. et postes assimilables
Greece								**Grèce**
Merchandise: Exp. fob	4394.0	4293.0	4513.0	5612.0	5933.0	5994.0	6365.0	Marchandises : exp. fob
Merchandise: Imp. fob	-8624.0	-9346.0	-8936.0	-11112.0	-12005.0	-13377.0	-16543.0	Marchandises : imp. fob
Serv. & Income: Credit	2939.0	2818.0	3392.0	4604.0	5445.0	5191.0	6968.0	Serv. & revenu : crédit

125
Summary of balance of payments
Millions of US dollars [*cont.*]
Résumé des balances des paiements
Millions de dollars des E-U [*suite*]

Country or area	1984	1985	1986	1987	1988	1989	1990	Pays ou zone
Serv. & Income: Debit	-2473.0	-2707.0	-3012.0	-3362.0	-3980.0	-4352.0	-5045.0	Serv. & revenu : débit
Private Unrequited Transfers	917.0	797.0	975.0	1370.0	1713.0	1381.0	1817.0	Transf. priv.sans contrep.
Offic. Unrequited Trans.,nie	715.0	869.0	1392.0	1665.0	1936.0	2602.0	2901.0	Tr. offic. sans contrep.nia
Direct Investment, nie	485.0	447.0	471.0	683.0	907.0	752.0	1005.0	Investissements direct, nia
Portfolio Investments, nie	0.0	0.0	0.0	0.0	0.0	0.0	0.0	Invest. portefeuille, nia
Other Capital, nie	1687.0	2477.0	1937.0	1291.0	947.0	1999.0	2997.0	Autres capitaux, nia
Net Errors and Omissions	-242.0	-44.0	-82.0	223.0	41.0	-538.0	-185.0	Erreurs et omissions nettes
Reserves and Related Items	202.0	396.0	-650.0	-974.0	-937.0	348.0	-280.0	Rés. et postes assimilables
Grenada								**Grenade**
Merchandise: Exp. fob	18.2	22.3	28.8	31.6	32.8	27.9	...	Marchandises : exp. fob
Merchandise: Imp. fob	-51.1	-65.6	-86.0	-92.8	-94.5	-92.5	...	Marchandises : imp. fob
Serv. & Income: Credit	23.0	31.8	40.5	48.4	52.7	58.1	...	Serv. & revenu : crédit
Serv. & Income: Debit	-21.3	-24.2	-29.5	-31.6	-31.6	-36.9	...	Serv. & revenu : débit
Private Unrequited Transfers	10.6	10.3	10.4	11.8	15.3	17.0	...	Transf. priv.sans contrep.
Offic. Unrequited Trans.,nie	22.4	27.3	25.5	8.2	5.2	12.2	...	Tr. offic. sans contrep.nia
Direct Investment, nie	2.8	4.1	5.0	12.7	13.0	10.0	...	Investissements direct, nia
Portfolio Investments, nie	0.0	0.0	0.0	0.0	0.0	0.0	...	Invest. portefeuille, nia
Other Capital, nie	4.8	1.9	7.1	11.2	1.0	19.7	...	Autres capitaux, nia
Net Errors and Omissions	-11.7	-1.8	-1.8	1.6	-1.9	-21.2	...	Erreurs et omissions nettes
Reserves and Related Items	2.3	-6.2	-0.1	-1.0	8.0	-5.7	...	Rés. et postes assimilables
Guatemala								**Guatemala**
Merchandise: Exp. fob	1132.2	1059.7	1043.8	977.9	1073.3	1126.1	...	Marchandises : exp. fob
Merchandise: Imp. fob	-1182.2	-1076.7	-875.7	-1333.2	-1413.2	-1484.4	...	Marchandises : imp. fob
Serv. & Income: Credit	129.0	131.6	159.4	189.4	227.4	328.7	...	Serv. & revenu : crédit
Serv. & Income: Debit	-485.1	-380.6	-420.2	-469.9	-525.8	-587.3	...	Serv. & revenu : débit
Private Unrequited Transfers	28.0	18.9	50.6	101.0	141.7	178.8	...	Transf. priv.sans contrep.
Offic. Unrequited Trans.,nie	0.7	0.8	24.5	92.3	82.6	71.0	...	Tr. offic. sans contrep.nia
Direct Investment, nie	38.0	61.8	68.8	150.2	329.7	76.2	...	Investissements direct, nia
Portfolio Investments, nie	-9.6	-27.6	-11.6	-16.0	-372.2	-63.9	...	Invest. portefeuille, nia
Other Capital, nie	-137.8	-162.5	-390.2	52.5	123.0	213.0	...	Autres capitaux, nia
Net Errors and Omissions	15.5	43.6	67.3	-72.7	-2.4	54.7	...	Erreurs et omissions nettes
Reserves and Related Items	471.3	331.0	283.3	328.5	336.0	87.1	...	Rés. et postes assimilables
Guinea-Bissau								**Guinée-Bissau**
Merchandise: Exp. fob	17.4	11.6	9.7	15.4	15.9	14.2	19.3	Marchandises : exp. fob
Merchandise: Imp. fob	-60.1	-59.5	-51.2	-44.7	-58.9	-68.9	-68.1	Marchandises : imp. fob
Serv. & Income: Credit	8.0	6.5	0.0	0.0	0.0	0.0	0.0	Serv. & revenu : crédit
Serv. & Income: Debit	-25.9	-31.0	-30.9	-33.2	-37.0	-49.0	-35.4	Serv. & revenu : débit
Private Unrequited Transfers	-4.9	-3.4	-1.5	-2.0	1.5	1.2	-2.1	Transf. priv.sans contrep.
Offic. Unrequited Trans.,nie	29.3	30.5	38.3	40.2	37.0	51.3	51.8	Tr. offic. sans contrep.nia
Direct Investment, nie	0.0	0.0	0.0	0.0	0.0	0.0	0.0	Investissements direct, nia
Portfolio Investments, nie	0.0	0.0	0.0	0.0	0.0	0.0	0.0	Invest. portefeuille, nia
Other Capital, nie	36.6	63.0	6.7	1.5	-3.4	-7.0	1.2	Autres capitaux, nia
Net Errors and Omissions	-12.9	-9.6	-3.7	-3.6	3.4	-10.6	1.3	Erreurs et omissions nettes
Reserves and Related Items	12.4	-8.0	32.6	26.4	41.5	68.8	32.1	Rés. et postes assimilables
Guyana								**Guyana**
Merchandise: Exp. fob	216.9	214.0	...	...	...	...	...	Marchandises : exp. fob
Merchandise: Imp. fob	-201.6	-209.1	...	...	...	...	...	Marchandises : imp. fob
Serv. & Income: Credit	29.5	48.0	...	...	...	...	...	Serv. & revenu : crédit
Serv. & Income: Debit	-144.0	-144.3	...	...	...	...	...	Serv. & revenu : débit
Private Unrequited Transfers	1.0	-2.0	...	...	...	...	...	Transf. priv.sans contrep.
Offic. Unrequited Trans.,nie	1.8	-3.2	...	...	...	...	...	Tr. offic. sans contrep.nia
Direct Investment, nie	4.5	1.8	...	...	...	...	...	Investissements direct, nia
Portfolio Investments, nie	0.0	0.0	...	...	...	...	...	Invest. portefeuille, nia
Other Capital, nie	-33.3	-39.5	...	...	...	...	...	Autres capitaux, nia
Net Errors and Omissions	-12.7	-4.3	...	...	...	...	...	Erreurs et omissions nettes
Reserves and Related Items	138.0	138.6	...	...	...	...	...	Rés. et postes assimilables
Haiti								**Haïti**
Merchandise: Exp. fob	214.6	223.0	190.8	210.1	180.4	148.3	139.0	Marchandises : exp. fob
Merchandise: Imp. fob	-337.9	-344.7	-303.2	-311.2	-283.9	-259.3	-224.8	Marchandises : imp. fob
Serv. & Income: Credit	109.0	119.3	105.8	115.5	100.7	83.1	88.2	Serv. & revenu : crédit
Serv. & Income: Debit	-211.7	-237.3	-190.2	-216.6	-230.4	-218.9	-207.3	Serv. & revenu : débit
Private Unrequited Transfers	45.0	48.5	52.0	56.2	63.4	59.3	46.8	Transf. priv.sans contrep.
Offic. Unrequited Trans.,nie	78.0	96.5	99.8	114.8	129.5	114.9	103.0	Tr. offic. sans contrep.nia
Direct Investment, nie	4.5	4.9	4.8	4.7	10.1	9.4	8.2	Investissements direct, nia
Portfolio Investments, nie	0.0	0.0	0.0	0.0	0.0	0.0	0.0	Invest. portefeuille, nia
Other Capital, nie	108.3	41.1	29.6	48.7	15.9	54.0	7.1	Autres capitaux, nia

125
Summary of balance of payments
Millions of US dollars [*cont.*]
Résumé des balances des paiements
Millions de dollars des E-U [*suite*]

Country or area	1984	1985	1986	1987	1988	1989	1990	Pays ou zone
Net Errors and Omissions	-29.9	44.2	17.6	-17.6	10.7	-8.5	62.4	Erreurs et omissions nettes
Reserves and Related Items	20.2	4.5	-7.1	-4.6	3.7	7.7	-22.5	Rés. et postes assimilables
Honduras								**Honduras**
Merchandise: Exp. fob	737.0	789.6	891.2	844.3	893.0	966.7	...	Marchandises : exp. fob
Merchandise: Imp. fob	-884.8	-879.2	-874.0	-893.8	-916.6	-964.0	...	Marchandises : imp. fob
Serv. & Income: Credit	126.2	128.6	130.9	137.5	135.7	138.6	...	Serv. & revenu : crédit
Serv. & Income: Debit	-374.8	-388.8	-453.6	-444.2	-476.1	-488.0	...	Serv. & revenu : débit
Private Unrequited Transfers	10.3	12.4	13.0	16.0	17.5	16.0	...	Transf. priv.sans contrep.
Offic. Unrequited Trans.,nie	11.9	44.7	37.8	34.0	27.5	28.5	...	Tr. offic. sans contrep.nia
Direct Investment, nie	20.5	27.5	30.0	38.7	46.8	37.3	...	Investissements direct, nia
Portfolio Investments, nie	-1.9	1.2	-1.0	0.6	-0.3	0.1	...	Invest. portefeuille, nia
Other Capital, nie	240.4	165.7	13.5	80.0	15.2	-57.4	...	Autres capitaux, nia
Net Errors and Omissions	-8.5	-54.6	33.3	27.1	22.4	-3.7	...	Erreurs et omissions nettes
Reserves and Related Items	123.8	152.8	178.8	159.7	234.8	325.8	...	Rés. et postes assimilables
Hungary								**Hongrie**
Merchandise: Exp. fob	9090.0	8578.0	9198.0	9967.0	9989.0	10493.0	9151.0	Marchandises : exp. fob
Merchandise: Imp. fob	-8310.0	-8130.0	-9663.0	-9887.0	-9406.0	-9450.0	-8617.0	Marchandises : imp. fob
Serv. & Income: Credit	734.0	817.0	990.0	1227.0	1287.0	1522.0	3164.0	Serv. & revenu : crédit
Serv. & Income: Debit	-1541.0	-1785.0	-1965.0	-2088.0	-2559.0	-3283.0	-4107.0	Serv. & revenu : débit
Private Unrequited Transfers	66.0	65.0	75.0	105.0	117.0	130.0	794.0	Transf. priv.sans contrep.
Offic. Unrequited Trans.,nie	0.0	0.0	0.0	0.0	0.0	0.0	-7.0	Tr. offic. sans contrep.nia
Direct Investment, nie	0.0	0.0	0.0	0.0	0.0	0.0	0.0	Investissements direct, nia
Portfolio Investments, nie	0.0	0.0	0.0	0.0	0.0	0.0	0.0	Invest. portefeuille, nia
Other Capital, nie	267.0	1066.0	1388.0	235.0	680.0	901.0	-1220.0	Autres capitaux, nia
Net Errors and Omissions	-240.0	-75.0	109.0	160.0	50.0	-141.0	659.0	Erreurs et omissions nettes
Reserves and Related Items	-66.0	-536.0	-132.0	281.0	-158.0	-172.0	182.0	Rés. et postes assimilables
Iceland								**Islande**
Merchandise: Exp. fob	743.2	814.0	1096.8	1376.1	1425.4	1401.5	1588.6	Marchandises : exp. fob
Merchandise: Imp. fob	-756.6	-814.3	-1000.0	-1428.2	-1439.4	-1267.3	-1509.1	Marchandises : imp. fob
Serv. & Income: Credit	362.8	410.4	472.9	569.7	557.8	549.4	624.1	Serv. & revenu : crédit
Serv. & Income: Debit	-481.5	-525.4	-556.5	-707.8	-764.1	-764.5	-863.7	Serv. & revenu : débit
Private Unrequited Transfers	2.1	1.3	5.7	0.9	0.7	0.9	6.1	Transf. priv.sans contrep.
Offic. Unrequited Trans.,nie	-1.3	-1.1	-1.7	-1.7	-1.9	-3.9	-6.3	Tr. offic. sans contrep.nia
Direct Investment, nie	13.7	23.6	6.4	1.7	-15.9	-35.6	-0.6	Investissements direct, nia
Portfolio Investments, nie	0.0	0.0	0.0	0.0	0.0	0.6	0.3	Invest. portefeuille, nia
Other Capital, nie	130.9	208.9	93.8	230.0	242.7	160.0	209.0	Autres capitaux, nia
Net Errors and Omissions	-28.5	-53.3	-18.5	-58.8	-4.1	13.5	26.0	Erreurs et omissions nettes
Reserves and Related Items	15.2	-64.1	-98.9	18.1	-1.2	-54.6	-74.4	Rés. et postes assimilables
India								**Inde**
Merchandise: Exp. fob	10192.0	9465.0	10248.0	11884.0	13510.0			Marchandises : exp. fob
Merchandise: Imp. fob	-14216.0	-15081.0	-15686.0	-17661.0	-20091.0	...	...	Marchandises : imp. fob
Serv. & Income: Credit	3719.0	3913.0	3746.0	3813.0	4218.0	...	...	Serv. & revenu : crédit
Serv. & Income: Debit	-4808.0	-5250.0	-5526.0	-6235.0	-7537.0	...	...	Serv. & revenu : débit
Private Unrequited Transfers	2278.0	2456.0	2223.0	2636.0	2295.0	...	...	Transf. priv.sans contrep.
Offic. Unrequited Trans.,nie	492.0	320.0	399.0	370.0	457.0	...	...	Tr. offic. sans contrep.nia
Direct Investment, nie	0.0	0.0	0.0	0.0	0.0	...	...	Investissements direct, nia
Portfolio Investments, nie	0.0	0.0	0.0	0.0	0.0	...	...	Invest. portefeuille, nia
Other Capital, nie	3044.0	3281.0	3992.0	5734.0	7243.0	...	...	Autres capitaux, nia
Net Errors and Omissions	368.0	500.0	197.0	-409.0	-112.0	...	...	Erreurs et omissions nettes
Reserves and Related Items	-1069.0	397.0	409.0	-133.0	16.0	...	...	Rés. et postes assimilables
Indonesia								**Indonésie**
Merchandise: Exp. fob	20754.0	18527.0	14396.0	17206.0	19509.0	22974.0	26832.0	Marchandises : exp. fob
Merchandise: Imp. fob	-15047.0	-12705.0	-11938.0	-12532.0	-13831.0	-16310.0	-20734.0	Marchandises : imp. fob
Serv. & Income: Credit	1398.0	1612.0	1576.0	1626.0	1861.0	2437.0	2623.0	Serv. & revenu : crédit
Serv. & Income: Debit	-9128.0	-9445.0	-8204.0	-8655.0	-9190.0	-10548.0	-11304.0	Serv. & revenu : débit
Private Unrequited Transfers	53.0	61.0	71.0	86.0	99.0	167.0	153.0	Transf. priv.sans contrep.
Offic. Unrequited Trans.,nie	114.0	27.0	188.0	171.0	155.0	172.0	61.0	Tr. offic. sans contrep.nia
Direct Investment, nie	222.0	310.0	258.0	385.0	576.0	682.0	964.0	Investissements direct, nia
Portfolio Investments, nie	-10.0	-35.0	268.0	-88.0	-98.0	-173.0	0.0	Invest. portefeuille, nia
Other Capital, nie	3245.0	1507.0	3651.0	3184.0	1739.0	2409.0	2810.0	Autres capitaux, nia
Net Errors and Omissions	-620.0	651.0	-1269.0	-753.0	-933.0	-1361.0	868.0	Erreurs et omissions nettes
Reserves and Related Items	-981.0	-510.0	1003.0	-630.0	113.0	-449.0	-2273.0	Rés. et postes assimilables
Iran, Islamic Rep. of								**Iran, Rép. islamique d'**
Merchandise: Exp. fob	17087.0	14175.0	7171.0	11916.0	10709.0	...	...	Marchandises : exp. fob
Merchandise: Imp. fob	-14729.0	-12006.0	-10585.0	-12005.0	-10608.0	...	...	Marchandises : imp. fob
Serv. & Income: Credit	1069.0	763.0	607.0	437.0	467.0	...	...	Serv. & revenu : crédit

125
Summary of balance of payments
Millions of US dollars [*cont.*]
Résumé des balances des paiements
Millions de dollars des E-U [*suite*]

Country or area	1984	1985	1986	1987	1988	1989	1990	Pays ou zone
Serv. & Income: Debit	-3841.0	-3408.0	-2348.0	-2438.0	-2436.0	...	...	Serv. & revenu : débit
Private Unrequited Transfers	0.0	0.0	0.0	0.0	0.0	...	...	Transf. priv.sans contrep.
Offic. Unrequited Trans.,nie	0.0	0.0	0.0	0.0	0.0	...	...	Tr. offic. sans contrep.
Direct Investment, nie	0.0	0.0	0.0	0.0	0.0	...	...	Investissements direct, nia
Portfolio Investments, nie	0.0	0.0	0.0	0.0	0.0	...	...	Invest. portefeuille, nia
Other Capital, nie	-2818.0	544.0	3127.0	1711.0	435.0	...	...	Autres capitaux, nia
Net Errors and Omissions	-904.0	487.0	814.0	155.0	421.0	...	...	Erreurs et omissions nettes
Reserves and Related Items	4136.0	-555.0	1214.0	224.0	1012.0	...	...	Rés. et postes assimilables
Ireland								**Irlande**
Merchandise: Exp. fob	9421.0	10131.0	12365.0	15569.0	18392.0	20355.0	23359.0	Marchandises : exp. fob
Merchandise: Imp. fob	-9183.0	-9500.0	-11224.0	-12948.0	-14563.0	-16348.0	-19382.0	Marchandises : imp. fob
Serv. & Income: Credit	1829.0	2058.0	2498.0	3048.0	3534.0	4021.0	5570.0	Serv. & revenu : crédit
Serv. & Income: Debit	-3903.0	-4425.0	-5612.0	-6604.0	-8258.0	-9081.0	-10731.0	Serv. & revenu : débit
Private Unrequited Transfers	-33.0	-21.0	-61.0	-162.0	-121.0	-94.0	-66.0	Transf. priv.sans contrep.
Offic. Unrequited Trans.,nie	833.0	1068.0	1352.0	1471.0	1659.0	1671.0	2683.0	Tr. offic. sans contrep.nia
Direct Investment, nie	121.0	164.0	-43.0	89.0	92.0	85.0	99.0	Investissements direct, nia
Portfolio Investments, nie	1177.0	970.0	1749.0	-208.0	990.0	651.0	-201.0	Invest. portefeuille, nia
Other Capital, nie	-89.0	-6.0	79.0	748.0	-882.0	-2306.0	-1803.0	Autres capitaux, nia
Net Errors and Omissions	-214.0	-389.0	-1198.0	-118.0	-251.0	109.0	1222.0	Erreurs et omissions nettes
Reserves and Related Items	43.0	-48.0	96.0	-885.0	-593.0	937.0	-750.0	Rés. et postes assimilables
Israel								**Israël**
Merchandise: Exp. fob	6319.0	6751.0	7839.0	9310.0	10355.0	11169.0	12260.0	Marchandises : exp. fob
Merchandise: Imp. fob	-8820.0	-9115.0	-9711.0	-13030.0	-13424.0	-12933.0	-15150.0	Marchandises : imp. fob
Serv. & Income: Credit	4325.0	4209.0	4081.0	4783.0	5212.0	5711.0	6235.0	Serv. & revenu : crédit
Serv. & Income: Debit	-6386.0	-5723.0	-5944.0	-6695.0	-7315.0	-7706.0	-8432.0	Serv. & revenu : débit
Private Unrequited Transfers	767.0	826.0	1183.0	1369.0	1145.0	1577.0	1982.0	Transf. priv.sans contrep.
Offic. Unrequited Trans.,nie	2514.0	4172.0	4197.0	3400.0	3370.0	3286.0	3807.0	Tr. offic. sans contrep.nia
Direct Investment, nie	19.0	48.0	46.0	173.0	133.0	105.0	-137.0	Investissements direct, nia
Portfolio Investments, nie	213.0	301.0	392.0	169.0	4173.0	1022.0	-204.0	Invest. portefeuille, nia
Other Capital, nie	1001.0	-664.0	-920.0	884.0	-4970.0	-2110.0	-285.0	Autres capitaux, nia
Net Errors and Omissions	-460.0	-414.0	-171.0	298.0	159.0	1275.0	483.0	Erreurs et omissions nettes
Reserves and Related Items	508.0	-391.0	-992.0	-661.0	1162.0	-1396.0	-559.0	Rés. et postes assimilables
Italy								**Italie**
Merchandise: Exp. fob	73836.0	76073.0	96719.0	116178.0	127416.0	140118.0	169940.0	Marchandises : exp. fob
Merchandise: Imp. fob	-79654.0	-82157.0	-92194.0	-116516.0	-128778.0	-142285.0	-169216.0	Marchandises : imp. fob
Serv. & Income: Credit	25714.0	27284.0	32188.0	39963.0	42430.0	48730.0	57334.0	Serv. & revenu : crédit
Serv. & Income: Debit	-24018.0	-25768.0	-32170.0	-40162.0	-46049.0	-55219.0	-68411.0	Serv. & revenu : débit
Private Unrequited Transfers	1454.0	1325.0	1465.0	1278.0	1449.0	1300.0	866.0	Transf. priv.sans contrep.
Offic. Unrequited Trans.,nie	274.0	-166.0	-2876.0	-2013.0	-2657.0	-3530.0	-3246.0	Tr. offic. sans contrep.nia
Direct Investment, nie	-687.0	-873.0	-2848.0	1753.0	1270.0	529.0	-667.0	Investissements direct, nia
Portfolio Investments, nie	86.0	272.0	-1081.0	-7346.0	395.0	3256.0	-394.0	Invest. portefeuille, nia
Other Capital, nie	2495.0	566.0	4148.0	14811.0	13756.0	20766.0	39946.0	Autres capitaux, nia
Net Errors and Omissions	2860.0	-4256.0	-1793.0	-2546.0	-1798.0	-2552.0	-15764.0	Erreurs et omissions nettes
Reserves and Related Items	-2361.0	7699.0	-1558.0	-5401.0	-7434.0	-11114.0	-10388.0	Rés. et postes assimilables
Jamaica								**Jamaïque**
Merchandise: Exp. fob	702.3	568.6	589.5	708.4	883.0	1000.4	1156.9	Marchandises : exp. fob
Merchandise: Imp. fob	-1037.0	-1004.2	-837.4	-1065.1	-1240.3	-1606.4	-1624.8	Marchandises : imp. fob
Serv. & Income: Credit	632.9	697.6	821.5	925.0	890.8	1008.6	1167.2	Serv. & revenu : crédit
Serv. & Income: Debit	-750.9	-784.8	-760.1	-876.0	-1008.9	-1187.2	-1237.7	Serv. & revenu : débit
Private Unrequited Transfers	80.4	153.2	111.6	117.2	436.5	299.5	159.0	Transf. priv.sans contrep.
Offic. Unrequited Trans.,nie	37.1	65.2	34.7	53.5	69.1	187.5	116.0	Tr. offic. sans contrep.nia
Direct Investment, nie	12.2	-9.0	-4.6	53.4	-12.0	57.1	93.8	Investissements direct, nia
Portfolio Investments, nie	0.0	0.0	0.0	0.0	0.0	0.0	0.0	Invest. portefeuille, nia
Other Capital, nie	538.9	236.3	-101.0	299.8	100.0	41.1	216.8	Autres capitaux, nia
Net Errors and Omissions	-64.0	17.0	80.0	82.8	-44.0	4.6	46.9	Erreurs et omissions nettes
Reserves and Related Items	-151.8	60.0	65.8	-299.0	-74.2	194.8	-94.1	Rés. et postes assimilables
Japan[1]								**Japon**[1]
Merchandise: Exp. fob	168.3	174.0	205.6	224.6	259.8	269.6	280.4	Marchandises : exp. fob
Merchandise: Imp. fob	-124.0	-118.0	-112.8	-128.2	-164.8	-192.7	-216.8	Marchandises : imp. fob
Serv. & Income: Credit	42.2	45.5	53.7	79.7	111.8	143.9	166.0	Serv. & revenu : crédit
Serv. & Income: Debit	-49.9	-50.7	-58.6	-85.4	-123.1	-159.5	-188.2	Serv. & revenu : débit
Private Unrequited Transfers	-0.1	-0.3	-0.6	-1.0	-1.1	-1.0	-1.0	Transf. priv.sans contrep.
Offic. Unrequited Trans.,nie	-1.4	-1.4	-1.5	-2.7	-3.0	-3.3	-4.5	Tr. offic. sans contrep.nia
Direct Investment, nie	-6.0	-5.8	-14.3	-18.4	-34.7	-45.2	-46.3	Investissements direct, nia
Portfolio Investments, nie	-24.0	-41.8	-102.0	-91.3	-52.8	-32.5	-14.5	Invest. portefeuille, nia
Other Capital, nie	-6.6	-6.0	42.8	64.3	21.3	29.8	39.2	Autres capitaux, nia

125
Summary of balance of payments
Millions of US dollars [*cont.*]
Résumé des balances des paiements
Millions de dollars des E-U [*suite*]

Country or area	1984	1985	1986	1987	1988	1989	1990	Pays ou zone
Net Errors and Omissions	3.7	3.8	2.5	-3.7	3.1	-21.8	-20.9	Erreurs et omissions nettes
Reserves and Related Items	-2.1	0.6	-14.8	-37.9	-16.5	12.8	6.6	Rés. et postes assimilables
Jordan								**Jordanie**
Merchandise: Exp. fob	751.9	788.9	732.0	933.1	1007.4	1109.4	...	Marchandises : exp. fob
Merchandise: Imp. fob	-2472.6	-2426.7	-2158.4	-2400.1	-2418.7	-1882.5	...	Marchandises : imp. fob
Serv. & Income: Credit	1231.7	1268.2	1158.9	1350.0	1461.3	1278.2	...	Serv. & revenu : crédit
Serv. & Income: Debit	-1491.6	-1476.6	-1390.9	-1576.9	-1695.2	-1298.9	...	Serv. & revenu : débit
Private Unrequited Transfers	1028.1	846.2	984.4	742.9	799.8	565.4	...	Transf. priv.sans contrep.
Offic. Unrequited Trans.,nie	687.8	739.6	634.2	599.1	551.7	613.2	...	Tr. offic. sans contrep.nia
Direct Investment, nie	74.8	25.7	18.8	38.3	23.8	-18.1	...	Investissements direct, nia
Portfolio Investments, nie	0.0	0.0	0.0	0.0	0.0	0.0	...	Invest. portefeuille, nia
Other Capital, nie	48.8	222.5	80.7	426.9	350.4	97.6	...	Autres capitaux, nia
Net Errors and Omissions	-47.9	-29.6	-17.2	27.9	123.4	0.3	...	Erreurs et omissions nettes
Reserves and Related Items	189.0	41.9	-42.5	-141.3	-203.9	-464.7	...	Rés. et postes assimilables
Kenya								**Kenya**
Merchandise: Exp. fob	1034.5	943.2	1170.2	908.7	1017.5	926.1	1010.5	Marchandises : exp. fob
Merchandise: Imp. fob	-1348.2	-1269.8	-1454.6	-1622.6	-1802.2	-1963.4	-2008.7	Marchandises : imp. fob
Serv. & Income: Credit	628.6	663.4	732.0	829.6	874.3	1008.5	1222.8	Serv. & revenu : crédit
Serv. & Income: Debit	-617.8	-641.4	-693.0	-824.6	-895.5	-941.2	-1076.6	Serv. & revenu : débit
Private Unrequited Transfers	60.1	81.5	58.2	72.0	89.0	101.5	167.8	Transf. priv.sans contrep.
Offic. Unrequited Trans.,nie	116.6	110.2	149.0	141.9	256.4	280.9	206.9	Tr. offic. sans contrep.nia
Direct Investment, nie	3.9	12.7	27.8	45.0	-1.8	68.8	23.0	Investissements direct, nia
Portfolio Investments, nie	0.0	0.0	0.0	0.0	0.0	0.0	0.0	Invest. portefeuille, nia
Other Capital, nie	175.2	8.6	104.8	316.8	384.1	573.0	291.9	Autres capitaux, nia
Net Errors and Omissions	8.6	39.6	43.2	107.9	34.7	67.6	70.0	Erreurs et omissions nettes
Reserves and Related Items	-61.6	52.1	-137.7	25.1	43.4	-122.0	92.5	Rés. et postes assimilables
Kiribati								**Kiribati**
Merchandise: Exp. fob	11.0	5.0	2.0	...	...	...	...	Marchandises : exp. fob
Merchandise: Imp. fob	-18.0	-15.0	-15.0	...	...	...	...	Marchandises : imp. fob
Serv. & Income: Credit	13.0	12.0	14.0	...	...	...	...	Serv. & revenu : crédit
Serv. & Income: Debit	-9.0	-7.0	-8.0	...	...	...	...	Serv. & revenu : débit
Private Unrequited Transfers	-1.0	-1.0	-1.0	...	...	...	...	Transf. priv.sans contrep.
Offic. Unrequited Trans.,nie	14.0	12.0	15.0	...	...	...	...	Tr. offic. sans contrep.nia
Direct Investment, nie	0.0	0.0	0.0	...	...	...	...	Investissements direct, nia
Portfolio Investments, nie	0.0	0.0	0.0	...	...	...	...	Invest. portefeuille, nia
Other Capital, nie	-11.0	-7.0	-2.0	...	...	...	...	Autres capitaux, nia
Net Errors and Omissions	-3.0	1.0	-3.0	...	...	...	...	Erreurs et omissions nettes
Reserves and Related Items	4.0	0.0	-2.0	...	...	...	...	Rés. et postes assimilables
Korea, Republic of								**Corée, République de**
Merchandise: Exp. fob	26335.0	26442.0	33913.0	46244.0	59648.0	61408.0	63123.0	Marchandises : exp. fob
Merchandise: Imp. fob	-27371.0	-26461.0	-29707.0	-38585.0	-48203.0	-56811.0	-65127.0	Marchandises : imp. fob
Serv. & Income: Credit	7317.0	6664.0	8052.0	10011.0	11252.0	12643.0	14269.0	Serv. & revenu : crédit
Serv. & Income: Debit	-8194.0	-8110.0	-8680.0	-9034.0	-9984.0	-12432.0	-14712.0	Serv. & revenu : débit
Private Unrequited Transfers	516.0	555.0	1028.0	1199.0	1404.0	200.0	266.0	Transf. priv.sans contrep.
Offic. Unrequited Trans.,nie	25.0	23.0	11.0	19.0	44.0	48.0	9.0	Tr. offic. sans contrep.nia
Direct Investment, nie	73.0	200.0	325.0	418.0	720.0	453.0	-105.0	Investissements direct, nia
Portfolio Investments, nie	333.0	982.0	301.0	-113.0	-482.0	-29.0	811.0	Invest. portefeuille, nia
Other Capital, nie	2417.0	780.0	-4619.0	-9239.0	-4492.0	-3050.0	2263.0	Autres capitaux, nia
Net Errors and Omissions	-891.0	-883.0	-547.0	1184.0	-591.0	690.0	-2005.0	Erreurs et omissions nettes
Reserves and Related Items	-560.0	-192.0	-77.0	-2104.0	-9316.0	-3120.0	1208.0	Rés. et postes assimilables
Kuwait								**Koweït**
Merchandise: Exp. fob	12156.0	10374.0	7216.0	8221.0	7709.0	11396.0	...	Marchandises : exp. fob
Merchandise: Imp. fob	-6117.0	-5327.0	-5007.0	-4773.0	-5447.0	-5525.0	...	Marchandises : imp. fob
Serv. & Income: Credit	6691.0	6417.0	9164.0	6897.0	8784.0	10164.0	...	Serv. & revenu : crédit
Serv. & Income: Debit	-4543.0	-4741.0	-4463.0	-4543.0	-4698.0	-4953.0	...	Serv. & revenu : débit
Private Unrequited Transfers	-963.0	-1044.0	-1084.0	-1102.0	-1179.0	-1283.0	...	Transf. priv.sans contrep.
Offic. Unrequited Trans.,nie	-415.0	-529.0	-182.0	-158.0	-140.0	-211.0	...	Tr. offic. sans contrep.nia
Direct Investment, nie	-95.0	-70.0	-248.0	-115.0	-254.0	-507.0	...	Investissements direct, nia
Portfolio Investments, nie	209.0	-346.0	-485.0	219.0	-487.0	-330.0	...	Invest. portefeuille, nia
Other Capital, nie	-7566.0	-1919.0	-6772.0	-4913.0	-6028.0	-6859.0	...	Autres capitaux, nia
Net Errors and Omissions	610.0	-2271.0	1778.0	-1581.0	-254.0	-638.0	...	Erreurs et omissions nettes
Reserves and Related Items	32.0	-545.0	83.0	1847.0	1996.0	-1255.0	...	Rés. et postes assimilables
Lao People's Dem. Rep.								**Rép. dém. pop. lao**
Merchandise: Exp. fob	43.8	53.6	55.0	64.3	57.8	...	...	Marchandises : exp. fob
Merchandise: Imp. fob	-161.9	-193.2	-185.7	-216.2	-188.0	...	...	Marchandises : imp. fob
Serv. & Income: Credit	15.0	21.2	24.3	26.2	23.2	...	...	Serv. & revenu : crédit

125
Summary of balance of payments
Millions of US dollars [*cont.*]
Résumé des balances des paiements
Millions de dollars des E-U [*suite*]

Country or area	1984	1985	1986	1987	1988	1989	1990	Pays ou zone
Serv. & Income: Debit	-24.8	-28.5	-17.8	-19.0	-17.6	...	...	Serv. & revenu : débit
Private Unrequited Transfers	2.8	3.5	3.7	3.5	6.7	...	...	Transf. priv.sans contrep.
Offic. Unrequited Trans.,nie	42.1	49.6	30.5	27.0	25.5	...	...	Tr. offic. sans contrep.nia
Direct Investment, nie	0.0	0.0	0.0	0.0	0.0	...	...	Investissements direct, nia
Portfolio Investments, nie	0.0	0.0	0.0	0.0	0.0	...	...	Invest. portefeuille, nia
Other Capital, nie	83.7	97.9	104.1	111.4	96.7	...	...	Autres capitaux, nia
Net Errors and Omissions	-3.2	19.2	-2.5	-5.3	-5.5	...	...	Erreurs et omissions nettes
Reserves and Related Items	2.5	-23.4	-11.6	8.1	1.1	...	...	Rés. et postes assimilables
Lesotho								**Lesotho**
Merchandise: Exp. fob	28.0	22.0	25.0	47.0	64.0	66.0	59.0	Marchandises : exp. fob
Merchandise: Imp. fob	-433.0	-324.0	-342.0	-452.0	-559.0	-593.0	-673.0	Marchandises : imp. fob
Serv. & Income: Credit	368.0	260.0	289.0	392.0	417.0	413.0	496.0	Serv. & revenu : crédit
Serv. & Income: Debit	-56.0	-45.0	-57.0	-76.0	-84.0	-91.0	-103.0	Serv. & revenu : débit
Private Unrequited Transfers	3.0	2.0	2.0	0.0	4.0	4.0	5.0	Transf. priv.sans contrep.
Offic. Unrequited Trans.,nie	96.0	72.0	79.0	112.0	134.0	211.0	281.0	Tr. offic. sans contrep.nia
Direct Investment, nie	2.0	5.0	2.0	6.0	21.0	13.0	17.0	Investissements direct, nia
Portfolio Investments, nie	0.0	0.0	0.0	0.0	0.0	0.0	0.0	Invest. portefeuille, nia
Other Capital, nie	-31.0	16.0	15.0	-2.0	-29.0	-34.0	-62.0	Autres capitaux, nia
Net Errors and Omissions	30.0	-3.0	-1.0	-26.0	27.0	2.0	-3.0	Erreurs et omissions nettes
Reserves and Related Items	-9.0	-6.0	-13.0	-1.0	6.0	8.0	-17.0	Rés. et postes assimilables
Liberia								**Libéria**
Merchandise: Exp. fob	447.0	430.0	408.0	375.0	...	...	...	Marchandises : exp. fob
Merchandise: Imp. fob	-325.0	-264.0	-259.0	-312.0	...	...	...	Marchandises : imp. fob
Serv. & Income: Credit	40.0	38.0	59.0	58.0	...	...	...	Serv. & revenu : crédit
Serv. & Income: Debit	-222.0	-211.0	-264.0	-262.0	...	...	...	Serv. & revenu : débit
Private Unrequited Transfers	-44.0	-28.0	-25.0	-21.0	...	...	...	Transf. priv.sans contrep.
Offic. Unrequited Trans.,nie	103.0	90.0	96.0	45.0	...	...	...	Tr. offic. sans contrep.nia
Direct Investment, nie	36.0	-16.0	-17.0	39.0	...	...	...	Investissements direct, nia
Portfolio Investments, nie	7.0	4.0	6.0	0.0	...	...	...	Invest. portefeuille, nia
Other Capital, nie	-59.0	-139.0	-192.0	-224.0	...	...	...	Autres capitaux, nia
Net Errors and Omissions	-134.0	-109.0	-74.0	30.0	...	...	...	Erreurs et omissions nettes
Reserves and Related Items	153.0	203.0	261.0	273.0	...	...	...	Rés. et postes assimilables
Libyan Arab Jamah.								**Jamah. arabe libyenne**
Merchandise: Exp. fob	11028.0	10353.0	5814.0	5828.0	5644.0	7283.0	11362.0	Marchandises : exp. fob
Merchandise: Imp. fob	-8464.0	-5754.0	-4434.0	-5391.0	-5753.0	-6517.0	-7582.0	Marchandises : imp. fob
Serv. & Income: Credit	720.0	526.0	629.0	845.0	889.0	566.0	784.0	Serv. & revenu : crédit
Serv. & Income: Debit	-3420.0	-2314.0	-1638.0	-1801.0	-2071.0	-1871.0	-1880.0	Serv. & revenu : débit
Private Unrequited Transfers	-1240.0	-859.0	-490.0	-470.0	-496.0	-472.0	-446.0	Transf. priv.sans contrep.
Offic. Unrequited Trans.,nie	-81.0	-45.0	-36.0	-56.0	-37.0	-16.0	-35.0	Tr. offic. sans contrep.nia
Direct Investment, nie	-17.0	119.0	-177.0	-213.0	42.0	90.0	54.0	Investissements direct, nia
Portfolio Investments, nie	47.0	55.0	-67.0	-2976.0	-221.0	-52.0	-115.0	Invest. portefeuille, nia
Other Capital, nie	801.0	610.0	-186.0	3061.0	342.0	1152.0	-946.0	Autres capitaux, nia
Net Errors and Omissions	-1096.0	-328.0	798.0	172.0	270.0	130.0	-37.0	Erreurs et omissions nettes
Reserves and Related Items	1721.0	-2362.0	-212.0	1000.0	1390.0	-292.0	-1159.0	Rés. et postes assimilables
Madagascar								**Madagascar**
Merchandise: Exp. fob	337.0	291.0	323.0	327.0	284.0	313.0	...	Marchandises : exp. fob
Merchandise: Imp. fob	-360.0	-336.0	-331.0	-315.0	-319.0	-314.0	...	Marchandises : imp. fob
Serv. & Income: Credit	58.0	63.0	80.0	105.0	132.0	146.0	...	Serv. & revenu : crédit
Serv. & Income: Debit	-306.0	-300.0	-366.0	-410.0	-443.0	-475.0	...	Serv. & revenu : débit
Private Unrequited Transfers	-1.0	24.0	21.0	34.0	38.0	47.0	...	Transf. priv.sans contrep.
Offic. Unrequited Trans.,nie	78.0	74.0	132.0	120.0	158.0	155.0	...	Tr. offic. sans contrep.nia
Direct Investment, nie	0.0	0.0	0.0	0.0	0.0	6.0	...	Investissements direct, nia
Portfolio Investments, nie	0.0	0.0	0.0	0.0	0.0	0.0	...	Invest. portefeuille, nia
Other Capital, nie	-23.0	6.0	22.0	-11.0	-22.0	-63.0	...	Autres capitaux, nia
Net Errors and Omissions	13.0	10.0	4.0	-12.0	53.0	6.0	...	Erreurs et omissions nettes
Reserves and Related Items	203.0	167.0	116.0	163.0	118.0	178.0	...	Rés. et postes assimilables
Malawi								**Malawi**
Merchandise: Exp. fob	311.8	245.5	248.4	278.5	297.0	...	...	Marchandises : exp. fob
Merchandise: Imp. fob	-162.0	-176.7	-154.1	-177.6	-253.0	...	...	Marchandises : imp. fob
Serv. & Income: Credit	38.0	37.7	27.8	43.7	37.8	...	...	Serv. & revenu : crédit
Serv. & Income: Debit	-272.6	-266.8	-254.3	-243.8	-230.3	...	...	Serv. & revenu : débit
Private Unrequited Transfers	11.5	11.1	13.1	13.8	15.1	...	...	Transf. priv.sans contrep.
Offic. Unrequited Trans.,nie	24.4	24.5	29.3	30.1	80.4	...	...	Tr. offic. sans contrep.nia
Direct Investment, nie	0.0	0.5	0.0	0.1	0.0	...	...	Investissements direct, nia
Portfolio Investments, nie	1.0	0.4	1.3	4.2	0.8	...	...	Invest. portefeuille, nia
Other Capital, nie	44.3	-4.6	47.3	72.3	130.6	...	...	Autres capitaux, nia

125
Summary of balance of payments
Millions of US dollars [*cont.*]
Résumé des balances des paiements
Millions de dollars des E-U [*suite*]

Country or area	1984	1985	1986	1987	1988	1989	1990	Pays ou zone
Net Errors and Omissions	7.1	102.8	40.7	24.2	-18.0	...	...	Erreurs et omissions nettes
Reserves and Related Items	-3.5	25.6	0.6	-45.4	-60.2	...	...	Rés. et postes assimilables
Malaysia								**Malaisie**
Merchandise: Exp. fob	16407.0	15133.0	13547.0	17754.0	20852.0	24667.0	28956.0	Marchandises : exp. fob
Merchandise: Imp. fob	-13426.0	-11556.0	-10302.0	-11918.0	-15306.0	-20754.0	-27032.0	Marchandises : imp. fob
Serv. & Income: Credit	2660.0	2643.0	2638.0	3229.0	3597.0	4185.0	5907.0	Serv. & revenu : crédit
Serv. & Income: Debit	-7274.0	-6828.0	-6043.0	-6567.0	-7484.0	-8390.0	-9580.0	Serv. & revenu : débit
Private Unrequited Transfers	-63.0	-46.0	-19.0	69.0	70.0	-17.0	16.0	Transf. priv.sans contrep.
Offic. Unrequited Trans.,nie	24.0	40.0	56.0	69.0	81.0	97.0	61.0	Tr. offic. sans contrep.nia
Direct Investment, nie	797.0	695.0	489.0	423.0	719.0	...	...	Investissements direct, nia
Portfolio Investments, nie	1108.0	1942.0	30.0	140.0	-448.0	-107.0	-255.0	Invest. portefeuille, nia
Other Capital, nie	1115.0	-704.0	583.0	-2100.0	-2608.0	-17.0	546.0	Autres capitaux, nia
Net Errors and Omissions	-863.0	-168.0	476.0	20.0	97.0	-101.0	431.0	Erreurs et omissions nettes
Reserves and Related Items	-486.0	-1151.0	-1455.0	-1119.0	430.0	-1230.0	-1953.0	Rés. et postes assimilables
Maldives								**Maldives**
Merchandise: Exp. fob	23.1	25.5	26.9	34.9	44.6	51.3	58.1	Marchandises : exp. fob
Merchandise: Imp. fob	-61.0	-58.0	-59.1	-62.9	-87.3	-108.0	-117.3	Marchandises : imp. fob
Serv. & Income: Credit	61.8	65.9	55.4	57.9	72.0	89.5	112.5	Serv. & revenu : crédit
Serv. & Income: Debit	-46.8	-40.5	-40.0	-42.3	-39.5	-43.2	-52.8	Serv. & revenu : débit
Private Unrequited Transfers	-0.9	-2.0	-1.0	-2.4	-5.0	-5.1	-10.0	Transf. priv.sans contrep.
Offic. Unrequited Trans.,nie	7.5	3.6	9.3	9.7	11.5	18.3	11.0	Tr. offic. sans contrep.nia
Direct Investment, nie	0.0	0.0	2.2	3.2	3.8	3.1	3.2	Investissements direct, nia
Portfolio Investments, nie	0.0	0.0	0.0	0.0	0.0	0.0	0.0	Invest. portefeuille, nia
Other Capital, nie	1.5	-4.3	-9.0	-10.8	-2.9	5.0	5.7	Autres capitaux, nia
Net Errors and Omissions	9.5	11.3	11.0	15.0	16.7	-8.5	-10.7	Erreurs et omissions nettes
Reserves and Related Items	5.3	-1.5	4.3	-2.3	-13.9	-2.4	0.3	Rés. et postes assimilables
Mali								**Mali**
Merchandise: Exp. fob	192.0	176.1	205.6	255.9	251.5	269.3	344.5	Marchandises : exp. fob
Merchandise: Imp. fob	-257.8	-328.5	-339.0	-335.4	-359.1	-338.8	-432.4	Marchandises : imp. fob
Serv. & Income: Credit	40.7	58.6	71.3	86.8	88.0	77.7	95.5	Serv. & revenu : crédit
Serv. & Income: Debit	-192.2	-282.4	-331.5	-355.1	-372.1	-353.0	-434.4	Serv. & revenu : débit
Private Unrequited Transfers	20.8	46.7	46.5	36.6	45.3	51.4	63.2	Transf. priv.sans contrep.
Offic. Unrequited Trans.,nie	127.9	197.0	185.7	210.0	254.8	213.5	270.0	Tr. offic. sans contrep.nia
Direct Investment, nie	10.1	2.9	-8.4	-6.0	0.7	15.0	-6.2	Investissements direct, nia
Portfolio Investments, nie	0.0	0.0	0.0	0.0	0.0	0.0	0.0	Invest. portefeuille, nia
Other Capital, nie	58.3	117.7	135.8	92.6	134.3	125.8	154.8	Autres capitaux, nia
Net Errors and Omissions	0.6	-14.8	-15.6	0.2	4.4	0.0	-22.4	Erreurs et omissions nettes
Reserves and Related Items	-0.5	26.6	49.5	14.3	-47.7	-60.9	-32.5	Rés. et postes assimilables
Malta								**Malte**
Merchandise: Exp. fob	410.8	421.6	521.6	631.3	758.3	866.3	...	Marchandises : exp. fob
Merchandise: Imp. fob	-639.2	-672.1	-786.5	-1024.6	-1222.3	-1327.7	...	Marchandises : imp. fob
Serv. & Income: Credit	404.7	414.1	529.3	718.4	831.5	860.0	...	Serv. & revenu : crédit
Serv. & Income: Debit	-229.7	-241.3	-324.4	-384.7	-465.8	-506.2	...	Serv. & revenu : débit
Private Unrequited Transfers	31.2	36.0	48.1	63.1	97.6	54.8	...	Transf. priv.sans contrep.
Offic. Unrequited Trans.,nie	33.4	18.8	23.2	23.4	68.6	49.6	...	Tr. offic. sans contrep.nia
Direct Investment, nie	26.2	19.0	21.9	19.4	40.8	51.7	...	Investissements direct, nia
Portfolio Investments, nie	0.9	-32.8	44.0	-7.5	-38.4	...	...	Invest. portefeuille, nia
Other Capital, nie	-5.0	-10.9	-84.2	-14.5	14.8	-41.3	...	Autres capitaux, nia
Net Errors and Omissions	-1.5	-19.6	2.6	-28.8	-50.4	64.7	...	Erreurs et omissions nettes
Reserves and Related Items	-31.9	67.1	4.6	4.5	-34.8	-14.2	...	Rés. et postes assimilables
Mauritania								**Mauritanie**
Merchandise: Exp. fob	293.8	371.5	418.8	402.4	437.6	447.9	...	Marchandises : exp. fob
Merchandise: Imp. fob	-302.1	-333.9	-401.2	-359.2	-348.9	-349.3	...	Marchandises : imp. fob
Serv. & Income: Credit	36.5	31.0	26.0	37.0	39.4	39.2	...	Serv. & revenu : crédit
Serv. & Income: Debit	-225.3	-297.9	-329.7	-306.3	-306.7	-251.7	...	Serv. & revenu : débit
Private Unrequited Transfers	-20.4	-20.8	-23.0	-20.6	-22.1	-25.0	...	Transf. priv.sans contrep.
Offic. Unrequited Trans.,nie	106.4	133.6	114.6	99.2	104.6	120.3	...	Tr. offic. sans contrep.nia
Direct Investment, nie	8.5	7.0	3.1	1.4	1.0	3.5	...	Investissements direct, nia
Portfolio Investments, nie	0.0	0.0	0.0	0.0	0.0	0.0	...	Invest. portefeuille, nia
Other Capital, nie	75.2	88.0	181.3	109.5	98.9	36.9	...	Autres capitaux, nia
Net Errors and Omissions	-0.8	-5.6	-5.7	-101.5	-16.0	-3.6	...	Erreurs et omissions nettes
Reserves and Related Items	28.2	27.1	15.8	137.9	12.1	-18.2	...	Rés. et postes assimilables
Mauritius								**Maurice**
Merchandise: Exp. fob	374.1	433.0	675.3	891.4	997.9	993.1	1206.6	Marchandises : exp. fob
Merchandise: Imp. fob	-415.5	-458.8	-617.1	-907.6	-1165.7	-1204.2	-1477.9	Marchandises : imp. fob
Serv. & Income: Credit	132.6	147.8	217.5	332.4	404.2	454.4	573.6	Serv. & revenu : crédit

125
Summary of balance of payments
Millions of US dollars [*cont.*]
Résumé des balances des paiements
Millions de dollars des E-U [*suite*]

Country or area	1984	1985	1986	1987	1988	1989	1990	Pays ou zone
Serv. & Income: Debit	-174.6	-186.7	-230.6	-319.0	-392.9	-423.3	-519.0	Serv. & revenu : débit
Private Unrequited Transfers	19.3	21.8	29.9	40.6	71.5	68.2	82.4	Transf. priv.sans contrep.
Offic. Unrequited Trans.,nie	8.0	13.3	19.2	25.2	21.3	7.3	14.5	Tr. offic. sans contrep.nia
Direct Investment, nie	4.9	8.0	7.4	17.1	23.6	35.1	40.5	Investissements direct, nia
Portfolio Investments, nie	0.0	0.0	0.0	0.0	0.0	0.0	-2.2	Invest. portefeuille, nia
Other Capital, nie	-12.8	-31.8	-11.6	42.4	103.8	15.5	100.5	Autres capitaux, nia
Net Errors and Omissions	23.8	51.3	32.6	96.3	121.6	198.9	213.2	Erreurs et omissions nettes
Reserves and Related Items	40.2	2.0	-122.5	-218.8	-185.4	-145.1	-232.2	Rés. et postes assimilables
Mexico								**Mexique**
Merchandise: Exp. fob	24196.0	21663.0	16031.0	20655.0	20566.0	22765.0	26773.0	Marchandises : exp. fob
Merchandise: Imp. fob	-11255.0	-13212.0	-11432.0	-12222.0	-18898.0	-23410.0	-29799.0	Marchandises : imp. fob
Serv. & Income: Credit	8182.0	7881.0	7653.0	9244.0	11441.0	13204.0	14817.0	Serv. & revenu : crédit
Serv. & Income: Debit	-17339.0	-16202.0	-14389.0	-14357.0	-16119.0	-18592.0	20519.0	Serv. & revenu : débit
Private Unrequited Transfers	325.0	327.0	345.0	384.0	397.0	1922.0	2207.0	Transf. priv.sans contrep.
Offic. Unrequited Trans.,nie	85.0	673.0	119.0	264.0	170.0	153.0	1266.0	Tr. offic. sans contrep.nia
Direct Investment, nie	390.0	491.0	1160.0	1796.0	635.0	2648.0	2548.0	Investissements direct, nia
Portfolio Investments, nie	-756.0	-984.0	-816.0	-397.0	-880.0	438.0	-5359.0	Invest. portefeuille, nia
Other Capital, nie	-3532.0	-2578.0	420.0	-3852.0	-5625.0	-2113.0	11942.0	Autres capitaux, nia
Net Errors and Omissions	-973.0	-1765.0	458.0	2605.0	-2840.0	2775.0	-1657.0	Erreurs et omissions nettes
Reserves and Related Items	677.0	3706.0	451.0	-4120.0	11153.0	210.0	-2219.0	Rés. et postes assimilables
Morocco								**Maroc**
Merchandise: Exp. fob	2161.0	2145.0	2410.0	2781.0	3608.0	3312.0	4210.0	Marchandises : exp. fob
Merchandise: Imp. fob	-3569.0	-3513.0	-3477.0	-3850.0	-4360.0	-4991.0	-6282.0	Marchandises : imp. fob
Serv. & Income: Credit	855.0	1015.0	1165.0	1411.0	1798.0	1700.0	2111.0	Serv. & revenu : crédit
Serv. & Income: Debit	-1374.0	-1612.0	-1864.0	-1938.0	-2185.0	-2431.0	-2572.0	Serv. & revenu : débit
Private Unrequited Transfers	847.0	965.0	1394.0	1579.0	1303.0	1356.0	2012.0	Transf. priv.sans contrep.
Offic. Unrequited Trans.,nie	91.0	109.0	159.0	191.0	303.0	265.0	320.0	Tr. offic. sans contrep.nia
Direct Investment, nie	47.0	20.0	1.0	60.0	85.0	167.0	165.0	Investissements direct, nia
Portfolio Investments, nie	0.0	0.0	0.0	0.0	0.0	0.0	0.0	Invest. portefeuille, nia
Other Capital, nie	752.0	843.0	622.0	104.0	-236.0	708.0	1962.0	Autres capitaux, nia
Net Errors and Omissions	97.0	-24.0	-67.0	33.0	-56.0	-76.0	-193.0	Erreurs et omissions nettes
Reserves and Related Items	93.0	52.0	-344.0	-372.0	-260.0	-9.0	-1734.0	Rés. et postes assimilables
Mozambique								**Mozambique**
Merchandise: Exp. fob	96.0	77.0	79.0	...	...	...	...	Marchandises : exp. fob
Merchandise: Imp. fob	-486.0	-381.0	-488.0	...	...	...	...	Marchandises : imp. fob
Serv. & Income: Credit	118.0	107.0	113.0	...	...	...	...	Serv. & revenu : crédit
Serv. & Income: Debit	-178.0	-217.0	-259.0	...	...	...	...	Serv. & revenu : débit
Private Unrequited Transfers	-26.0	-25.0	-23.0	...	...	...	...	Transf. priv.sans contrep.
Offic. Unrequited Trans.,nie	168.0	139.0	219.0	...	...	...	...	Tr. offic. sans contrep.nia
Direct Investment, nie	0.0	0.0	0.0	...	...	...	...	Investissements direct, nia
Portfolio Investments, nie	0.0	0.0	0.0	...	...	...	...	Invest. portefeuille, nia
Other Capital, nie	-113.0	-52.0	-22.0	...	...	...	...	Autres capitaux, nia
Net Errors and Omissions	26.0	-13.0	-77.0	...	...	...	...	Erreurs et omissions nettes
Reserves and Related Items	396.0	366.0	459.0	...	...	...	...	Rés. et postes assimilables
Myanmar								**Myanmar**
Merchandise: Exp. fob	364.1	310.8	330.7	...	...	...	...	Marchandises : exp. fob
Merchandise: Imp. fob	-564.5	-512.9	-620.9	...	...	...	...	Marchandises : imp. fob
Serv. & Income: Credit	66.4	69.0	70.6	...	...	...	...	Serv. & revenu : crédit
Serv. & Income: Debit	-151.7	-152.9	-167.6	...	...	...	...	Serv. & revenu : débit
Private Unrequited Transfers	7.2	5.8	5.8	...	...	...	...	Transf. priv.sans contrep.
Offic. Unrequited Trans.,nie	60.7	74.7	87.5	...	...	...	...	Tr. offic. sans contrep.nia
Direct Investment, nie	0.0	0.0	0.0	...	...	...	...	Investissements direct, nia
Portfolio Investments, nie	0.0	0.0	0.0	...	...	...	...	Invest. portefeuille, nia
Other Capital, nie	193.8	148.7	262.2	...	...	...	...	Autres capitaux, nia
Net Errors and Omissions	15.3	48.6	81.1	...	...	...	...	Erreurs et omissions nettes
Reserves and Related Items	8.8	8.2	-49.3	...	...	...	...	Rés. et postes assimilables
Nepal								**Népal**
Merchandise: Exp. fob	130.1	161.3	142.6	162.2	195.5	156.2	217.9	Marchandises : exp. fob
Merchandise: Imp. fob	-402.9	-444.0	-436.5	-512.4	-672.8	-571.4	-666.6	Marchandises : imp. fob
Serv. & Income: Credit	164.4	162.1	181.0	224.5	232.3	226.5	229.5	Serv. & revenu : crédit
Serv. & Income: Debit	-104.5	-119.6	-117.3	-137.9	-154.3	-154.6	-178.6	Serv. & revenu : débit
Private Unrequited Transfers	34.8	37.7	41.3	67.2	59.1	52.0	60.4	Transf. priv.sans contrep.
Offic. Unrequited Trans.,nie	82.9	81.0	69.8	73.1	60.0	48.0	48.2	Tr. offic. sans contrep.nia
Direct Investment, nie	0.0	0.0	0.0	0.0	0.0	0.0	0.0	Investissements direct, nia
Portfolio Investments, nie	0.0	0.0	0.0	0.0	0.0	0.0	0.0	Invest. portefeuille, nia
Other Capital, nie	61.5	25.8	87.4	190.7	253.7	194.4	304.5	Autres capitaux, nia

125
Summary of balance of payments
Millions of US dollars [*cont.*]
Résumé des balances des paiements
Millions de dollars des E-U [*suite*]

Country or area	1984	1985	1986	1987	1988	1989	1990	Pays ou zone
Net Errors and Omissions	12.8	2.3	-0.8	-3.6	1.1	4.8	4.9	Erreurs et omissions nettes
Reserves and Related Items	20.9	93.5	32.5	-63.8	25.4	44.2	-20.2	Rés. et postes assimilables
Netherlands								**Pays-Bas**
Merchandise: Exp. fob	59890.0	62286.0	73100.0	86158.0	97442.0	101274.0	122253.0	Marchandises : exp. fob
Merchandise: Imp. fob	-54335.0	-56893.0	-66051.0	-80991.0	-88966.0	-93135.0	-111737.0	Marchandises : imp. fob
Serv. & Income: Credit	26568.0	24669.0	30395.0	37246.0	41669.0	49238.0	55326.0	Serv. & revenu : crédit
Serv. & Income: Debit	-24660.0	-24755.0	-31521.0	-36302.0	-41204.0	-45495.0	-52422.0	Serv. & revenu : débit
Private Unrequited Transfers	-468.0	-453.0	-753.0	-1050.0	-909.0	-944.0	-1127.0	Transf. priv.sans contrep.
Offic. Unrequited Trans.,nie	-665.0	-663.0	-1130.0	-1233.0	-1159.0	-1309.0	-1981.0	Tr. offic. sans contrep.nia
Direct Investment, nie	-3284.0	-1323.0	-721.0	-5796.0	-1768.0	-6231.0	-3814.0	Investissements direct, nia
Portfolio Investments, nie	-33.0	272.0	-4729.0	2470.0	3972.0	8037.0	-4716.0	Invest. portefeuille, nia
Other Capital, nie	-2414.0	-1525.0	3146.0	2764.0	-3493.0	-7207.0	2424.0	Autres capitaux, nia
Net Errors and Omissions	-559.0	-863.0	-2121.0	-404.0	-3957.0	-3778.0	-3929.0	Erreurs et omissions nettes
Reserves and Related Items	-38.0	-751.0	386.0	-2861.0	-1626.0	-450.0	-277.0	Rés. et postes assimilables
Netherlands Antilles								**Antilles néerlandaises**
Merchandise: Exp. fob	3760.6	1679.4	64.4	77.2	...	...	...	Marchandises : exp. fob
Merchandise: Imp. fob	-4248.8	-2132.3	-665.0	-767.2	...	...	...	Marchandises : imp. fob
Serv. & Income: Credit	1137.3	1057.1	1103.3	1007.2	...	...	...	Serv. & revenu : crédit
Serv. & Income: Debit	-450.9	-471.8	-342.2	-340.0	...	...	...	Serv. & revenu : débit
Private Unrequited Transfers	-92.4	-89.0	-96.7	-62.2	...	...	...	Transf. priv.sans contrep.
Offic. Unrequited Trans.,nie	30.0	326.1	38.3	35.0	...	...	...	Tr. offic. sans contrep.nia
Direct Investment, nie	1.7	-271.5	-0.6	2.2	...	...	...	Investissements direct, nia
Portfolio Investments, nie	-36.3	-42.6	-49.4	13.9	...	...	...	Invest. portefeuille, nia
Other Capital, nie	-158.6	-3.2	21.1	-28.3	...	...	...	Autres capitaux, nia
Net Errors and Omissions	18.7	19.6	23.9	32.8	...	...	...	Erreurs et omissions nettes
Reserves and Related Items	38.7	-71.9	-97.2	29.4	...	...	...	Rés. et postes assimilables
New Zealand								**Nouvelle-Zélande**
Merchandise: Exp. fob	5434.0	5595.0	5836.0	7245.0	8817.0	8853.0	9283.0	Marchandises : exp. fob
Merchandise: Imp. fob	-5846.0	-5654.0	-5741.0	-6719.0	-6675.0	-7873.0	-8452.0	Marchandises : imp. fob
Serv. & Income: Credit	1688.0	1771.0	2153.0	3019.0	3145.0	3231.0	3607.0	Serv. & revenu : crédit
Serv. & Income: Debit	-3233.0	-3538.0	-4166.0	-5358.0	-6062.0	-5935.0	-6455.0	Serv. & revenu : débit
Private Unrequited Transfers	197.0	144.0	190.0	220.0	309.0	501.0	664.0	Transf. priv.sans contrep.
Offic. Unrequited Trans.,nie	-73.0	-60.0	-39.0	-40.0	-43.0	-29.0	-39.0	Tr. offic. sans contrep.nia
Direct Investment, nie	110.0	319.0	-100.0	-298.0	178.0	107.0	9.0	Investissements direct, nia
Portfolio Investments, nie	0.0	0.0	0.0	0.0	0.0	88.0	30.0	Invest. portefeuille, nia
Other Capital, nie	-468.0	-789.0	-1126.0	-3086.0	-1637.0	1063.0	-130.0	Autres capitaux, nia
Net Errors and Omissions	1123.0	893.0	434.0	881.0	-337.0	-1233.0	1661.0	Erreurs et omissions nettes
Reserves and Related Items	1068.0	1319.0	2558.0	4136.0	2305.0	1227.0	-179.0	Rés. et postes assimilables
Nicaragua								**Nicaragua**
Merchandise: Exp. fob	412.4	305.1	257.8	295.1	235.7	...	...	Marchandises : exp. fob
Merchandise: Imp. fob	-735.3	-794.1	-677.4	-734.4	-718.3	...	...	Marchandises : imp. fob
Serv. & Income: Credit	53.0	40.8	30.2	30.9	39.7	...	...	Serv. & revenu : crédit
Serv. & Income: Debit	-417.0	-404.4	-413.4	-405.8	-402.3	...	...	Serv. & revenu : débit
Private Unrequited Transfers	2.0	13.8	0.0	0.0	0.0	...	...	Transf. priv.sans contrep.
Offic. Unrequited Trans.,nie	87.8	113.1	115.1	135.4	130.0	...	...	Tr. offic. sans contrep.nia
Direct Investment, nie	0.0	0.0	0.0	0.0	0.0	...	...	Investissements direct, nia
Portfolio Investments, nie	0.0	0.0	0.0	0.0	0.0	...	...	Invest. portefeuille, nia
Other Capital, nie	240.8	353.9	-127.6	125.9	303.5	...	...	Autres capitaux, nia
Net Errors and Omissions	-38.7	-186.8	-183.6	-78.9	51.7	...	...	Erreurs et omissions nettes
Reserves and Related Items	395.1	558.6	998.9	631.8	360.0	...	...	Rés. et postes assimilables
Niger								**Niger**
Merchandise: Exp. fob	303.3	259.4	331.5	411.9	369.0	311.0	...	Marchandises : exp. fob
Merchandise: Imp. fob	-269.9	-345.6	-309.8	-409.6	-392.5	-368.6	...	Marchandises : imp. fob
Serv. & Income: Credit	44.8	57.0	0.0	60.6	50.0	57.7	...	Serv. & revenu : crédit
Serv. & Income: Debit	-179.4	-188.1	-101.7	-245.2	-218.6	-202.5	...	Serv. & revenu : débit
Private Unrequited Transfers	-43.3	-58.7	-43.3	-49.9	-34.9	-40.4	...	Transf. priv.sans contrep.
Offic. Unrequited Trans.,nie	145.8	212.1	148.4	142.1	143.7	132.0	...	Tr. offic. sans contrep.nia
Direct Investment, nie	3.9	-9.2	0.0	0.0	0.0	0.0	...	Investissements direct, nia
Portfolio Investments, nie	0.0	0.0	0.0	0.0	0.0	0.0	...	Invest. portefeuille, nia
Other Capital, nie	-42.7	1.0	-7.9	71.7	26.1	35.7	...	Autres capitaux, nia
Net Errors and Omissions	14.2	26.6	-65.6	-15.4	43.4	-4.1	...	Erreurs et omissions nettes
Reserves and Related Items	23.4	45.6	48.4	33.9	13.7	79.3	...	Rés. et postes assimilables
Nigeria								**Nigéria**
Merchandise: Exp. fob	11859.0	13115.0	6015.0	7545.0	6897.0	7870.0	13585.0	Marchandises : exp. fob
Merchandise: Imp. fob	-8867.0	-7499.0	-3702.0	-4097.0	-4271.0	-3692.0	-4932.0	Marchandises : imp. fob
Serv. & Income: Credit	491.0	396.0	298.0	270.0	404.0	704.0	1265.0	Serv. & revenu : crédit

125
Summary of balance of payments
Millions of US dollars [*cont.*]
Résumé des balances des paiements
Millions de dollars des E-U [*suite*]

Country or area	1984	1985	1986	1987	1988	1989	1990	Pays ou zone
Serv. & Income: Debit	-3042.0	-3190.0	-2132.0	-3762.0	-3212.0	-3918.0	-4877.0	Serv. & revenu : débit
Private Unrequited Transfers	-300.0	-254.0	-108.0	-19.0	-33.0	-19.0	-14.0	Transf. priv.sans contrep.
Offic. Unrequited Trans.,nie	-26.0	-2.0	-5.0	-5.0	21.0	145.0	99.0	Tr. offic. sans contrep.nia
Direct Investment, nie	200.0	478.0	167.0	603.0	377.0	1882.0	588.0	Investissements direct, nia
Portfolio Investments, nie	0.0	0.0	0.0	-535.0	-65.0	-220.0	-137.0	Invest. portefeuille, nia
Other Capital, nie	-1554.0	-4142.0	-1358.0	-4188.0	-4857.0	-5307.0	-4752.0	Autres capitaux, nia
Net Errors and Omissions	272.0	-134.0	-161.0	-306.0	-215.0	-110.0	82.0	Erreurs et omissions nettes
Reserves and Related Items	967.0	1232.0	986.0	4495.0	4954.0	2664.0	-906.0	Rés. et postes assimilables
Norway								**Norvège**
Merchandise: Exp. fob	19115.0	20059.0	18143.0	21191.0	23075.0	27171.0	34315.0	Marchandises : exp. fob
Merchandise: Imp. fob	-13957.0	-15331.0	-20257.0	-21951.0	-23284.0	-23401.0	-26545.0	Marchandises : imp. fob
Serv. & Income: Credit	8713.0	9555.0	10866.0	11585.0	12994.0	14195.0	16626.0	Serv. & revenu : crédit
Serv. & Income: Debit	-10446.0	-10666.0	-12491.0	-13948.0	-15544.0	-16619.0	-19080.0	Serv. & revenu : débit
Private Unrequited Transfers	-36.0	-66.0	-140.0	-194.0	-168.0	-222.0	-222.0	Transf. priv.sans contrep.
Offic. Unrequited Trans.,nie	-469.0	-499.0	-665.0	-788.0	-962.0	-910.0	-1208.0	Tr. offic. sans contrep.nia
Direct Investment, nie	-781.0	-1731.0	-582.0	-687.0	-699.0	161.0	-539.0	Investissements direct, nia
Portfolio Investments, nie	810.0	1776.0	4283.0	2283.0	4226.0	3043.0	457.0	Invest. portefeuille, nia
Other Capital, nie	460.0	1414.0	-739.0	3637.0	1373.0	-1148.0	-400.0	Autres capitaux, nia
Net Errors and Omissions	-340.0	-1059.0	-1626.0	-1349.0	-1149.0	-1305.0	-2990.0	Erreurs et omissions nettes
Reserves and Related Items	-3068.0	-3452.0	3211.0	220.0	138.0	-965.0	-414.0	Rés. et postes assimilables
Oman								**Oman**
Merchandise: Exp. fob	4421.0	4971.0	2861.0	3805.0	3342.0	4047.0	5488.0	Marchandises : exp. fob
Merchandise: Imp. fob	-2640.0	-3028.0	-2309.0	-1769.0	-2107.0	-2130.0	-2519.0	Marchandises : imp. fob
Serv. & Income: Credit	359.0	376.0	610.0	533.0	270.0	333.0	367.0	Serv. & revenu : crédit
Serv. & Income: Debit	-1232.0	-1400.0	-1357.0	-1091.0	-1095.0	-1151.0	-1338.0	Serv. & revenu : débit
Private Unrequited Transfers	-819.0	-906.0	-846.0	-702.0	-762.0	-791.0	-845.0	Transf. priv.sans contrep.
Offic. Unrequited Trans.,nie	211.0	-26.0	0.0	8.0	42.0	16.0	-57.0	Tr. offic. sans contrep.nia
Direct Investment, nie	158.0	161.0	140.0	35.0	92.0	112.0	144.0	Investissements direct, nia
Portfolio Investments, nie	0.0	0.0	0.0	0.0	0.0	0.0	0.0	Invest. portefeuille, nia
Other Capital, nie	281.0	296.0	875.0	-225.0	130.0	-90.0	-653.0	Autres capitaux, nia
Net Errors and Omissions	-421.0	-323.0	-588.0	-486.0	-379.0	-64.0	-282.0	Erreurs et omissions nettes
Reserves and Related Items	-319.0	-122.0	613.0	-108.0	467.0	-282.0	-304.0	Rés. et postes assimilables
Pakistan								**Pakistan**
Merchandise: Exp. fob	2480.0	2648.0	3191.0	3938.0	4405.0	4796.0	5351.0	Marchandises : exp. fob
Merchandise: Imp. fob	-6234.0	-5878.0	-5971.0	-6254.0	-7097.0	-7366.0	-8050.0	Marchandises : imp. fob
Serv. & Income: Credit	982.0	956.0	933.0	1083.0	946.0	1323.0	1476.0	Serv. & revenu : crédit
Serv. & Income: Debit	-1750.0	-1870.0	-1949.0	-2187.0	-2393.0	-2808.0	-3154.0	Serv. & revenu : débit
Private Unrequited Transfers	2942.0	2710.0	2676.0	2440.0	2101.0	2207.0	2275.0	Transf. priv.sans contrep.
Offic. Unrequited Trans.,nie	384.0	354.0	475.0	418.0	615.0	513.0	523.0	Tr. offic. sans contrep.nia
Direct Investment, nie	60.0	139.0	106.0	110.0	173.0	167.0	242.0	Investissements direct, nia
Portfolio Investments, nie	9.0	110.0	83.0	132.0	126.0	15.0	92.0	Invest. portefeuille, nia
Other Capital, nie	265.0	382.0	341.0	127.0	1248.0	957.0	705.0	Autres capitaux, nia
Net Errors and Omissions	-96.0	33.0	-42.0	17.0	23.0	-210.0	-46.0	Erreurs et omissions nettes
Reserves and Related Items	958.0	417.0	157.0	177.0	-147.0	406.0	585.0	Rés. et postes assimilables
Panama								**Panama**
Merchandise: Exp. fob	1686.0	1974.0	2366.0	2492.0	2346.0	2568.0	3195.0	Marchandises : exp. fob
Merchandise: Imp. fob	-2509.0	-2731.0	-2907.0	-3058.0	-2531.0	-3084.0	-3804.0	Marchandises : imp. fob
Serv. & Income: Credit	4852.0	4333.0	3769.0	3235.0	2059.0	2103.0	2159.0	Serv. & revenu : crédit
Serv. & Income: Debit	-3923.0	-3399.0	-2968.0	-2527.0	-1179.0	-1366.0	-1505.0	Serv. & revenu : débit
Private Unrequited Transfers	-32.0	-31.0	-27.0	-51.0	-40.0	-36.0	-35.0	Transf. priv.sans contrep.
Offic. Unrequited Trans.,nie	144.0	139.0	122.0	114.0	111.0	105.0	117.0	Tr. offic. sans contrep.nia
Direct Investment, nie	10.0	59.0	-62.0	57.0	-52.0	56.0	-30.0	Investissements direct, nia
Portfolio Investments, nie	59.0	-183.0	66.0	-71.0	259.0	-61.0	-29.0	Invest. portefeuille, nia
Other Capital, nie	-29.0	-37.0	9.0	277.0	140.0	-73.0	-514.0	Autres capitaux, nia
Net Errors and Omissions	-374.0	-253.0	-374.0	-519.0	-1237.0	-252.0	706.0	Erreurs et omissions nettes
Reserves and Related Items	115.0	129.0	6.0	53.0	123.0	39.0	-259.0	Rés. et postes assimilables
Papua New Guinea								**Papouasie-Nvl-Guinée**
Merchandise: Exp. fob	914.0	921.4	1030.6	1243.9	1475.3	1318.5	...	Marchandises : exp. fob
Merchandise: Imp. fob	-962.8	-874.8	-928.8	-1129.6	-1384.5	-1341.3	...	Marchandises : imp. fob
Serv. & Income: Credit	115.0	119.6	164.0	155.2	240.9	254.9	...	Serv. & revenu : crédit
Serv. & Income: Debit	-555.2	-450.3	-505.2	-583.9	-761.3	-674.0	...	Serv. & revenu : débit
Private Unrequited Transfers	-95.0	-88.8	-78.7	-104.5	-124.6	-130.7	...	Transf. priv.sans contrep.
Offic. Unrequited Trans.,nie	261.9	218.4	212.7	204.3	217.7	217.3	...	Tr. offic. sans contrep.nia
Direct Investment, nie	113.4	82.4	99.5	115.4	119.7	221.3	...	Investissements direct, nia
Portfolio Investments, nie	0.0	0.0	0.0	0.0	0.0	0.0	...	Invest. portefeuille, nia
Other Capital, nie	144.6	41.6	35.0	61.7	125.4	43.7	...	Autres capitaux, nia

125
Summary of balance of payments
Millions of US dollars [*cont.*]
Résumé des balances des paiements
Millions de dollars des E-U [*suite*]

Country or area	1984	1985	1986	1987	1988	1989	1990	Pays ou zone
Net Errors and Omissions	112.9	29.1	-25.9	39.5	37.9	31.6	...	Erreurs et omissions nettes
Reserves and Related Items	-48.8	1.4	-3.2	-1.9	53.5	58.7	...	Rés. et postes assimilables
Paraguay								**Paraguay**
Merchandise: Exp. fob	361.3	465.6	575.8	597.4	871.0	1166.5	1392.3	Marchandises : exp. fob
Merchandise: Imp. fob	-649.1	-659.3	-864.2	-918.7	-1030.1	-1001.3	-1353.6	Marchandises : imp. fob
Serv. & Income: Credit	304.7	221.2	254.3	222.4	352.6	465.9	547.0	Serv. & revenu : crédit
Serv. & Income: Debit	-343.6	-287.1	-341.9	-417.8	-438.8	-418.9	-527.4	Serv. & revenu : débit
Private Unrequited Transfers	2.1	1.8	0.8	2.0	2.0	1.6	2.4	Transf. priv.sans contrep.
Offic. Unrequited Trans.,nie	7.2	5.7	10.3	25.0	33.2	22.3	30.5	Tr. offic. sans contrep.nia
Direct Investment, nie	5.2	0.7	0.6	5.3	8.4	12.8	79.0	Investissements direct, nia
Portfolio Investments, nie	0.0	0.0	0.0	0.0	0.0	0.0	0.0	Invest. portefeuille, nia
Other Capital, nie	279.0	31.7	94.8	86.7	-207.7	-238.0	-215.8	Autres capitaux, nia
Net Errors and Omissions	17.6	73.7	43.8	338.3	198.3	-90.0	61.7	Erreurs et omissions nettes
Reserves and Related Items	15.6	146.0	225.7	59.4	211.1	79.1	-16.1	Rés. et postes assimilables
Peru								**Pérou**
Merchandise: Exp. fob	3147.0	2978.0	2531.0	2661.0	2691.0	3533.0	3276.0	Marchandises : exp. fob
Merchandise: Imp. fob	-2140.0	-1806.0	-2596.0	-3182.0	-2790.0	-2291.0	-2885.0	Marchandises : imp. fob
Serv. & Income: Credit	827.0	947.0	929.0	998.0	1038.0	1063.0	1046.0	Serv. & revenu : crédit
Serv. & Income: Debit	-2213.0	-2116.0	-2091.0	-2138.0	-2187.0	-2136.0	-2358.0	Serv. & revenu : débit
Private Unrequited Transfers	0.0	0.0	0.0	0.0	0.0	0.0	0.0	Transf. priv.sans contrep.
Offic. Unrequited Trans.,nie	156.0	132.0	149.0	180.0	157.0	155.0	247.0	Tr. offic. sans contrep.nia
Direct Investment, nie	-89.0	1.0	22.0	32.0	26.0	59.0	34.0	Investissements direct, nia
Portfolio Investments, nie	0.0	0.0	0.0	0.0	0.0	0.0	0.0	Invest. portefeuille, nia
Other Capital, nie	-681.0	-1165.0	-1162.0	-1180.0	-909.0	-1178.0	-805.0	Autres capitaux, nia
Net Errors and Omissions	-566.0	-298.0	45.0	-54.0	-114.0	-183.0	234.0	Erreurs et omissions nettes
Reserves and Related Items	1559.0	1326.0	2172.0	2683.0	2089.0	978.0	1211.0	Rés. et postes assimilables
Philippines								**Philippines**
Merchandise: Exp. fob	5391.0	4629.0	4842.0	5720.0	7074.0	7821.0	8186.0	Marchandises : exp. fob
Merchandise: Imp. fob	-6070.0	-5111.0	-5044.0	-6737.0	-8159.0	-10419.0	-12206.0	Marchandises : imp. fob
Serv. & Income: Credit	2626.0	3288.0	3791.0	3454.0	3592.0	4586.0	4842.0	Serv. & revenu : crédit
Serv. & Income: Debit	-3627.0	-3220.0	-3076.0	-3454.0	-3672.0	-4274.0	-4231.0	Serv. & revenu : débit
Private Unrequited Transfers	118.0	172.0	235.0	376.0	500.0	473.0	357.0	Transf. priv.sans contrep.
Offic. Unrequited Trans.,nie	268.0	207.0	206.0	197.0	275.0	357.0	357.0	Tr. offic. sans contrep.nia
Direct Investment, nie	9.0	12.0	127.0	307.0	936.0	563.0	530.0	Investissements direct, nia
Portfolio Investments, nie	-3.0	5.0	13.0	19.0	50.0	280.0	-50.0	Invest. portefeuille, nia
Other Capital, nie	775.0	311.0	6.0	-8.0	-415.0	511.0	1577.0	Autres capitaux, nia
Net Errors and Omissions	65.0	539.0	34.0	68.0	479.0	402.0	593.0	Erreurs et omissions nettes
Reserves and Related Items	448.0	-832.0	-1134.0	58.0	-660.0	-300.0	45.0	Rés. et postes assimilables
Poland								**Pologne**
Merchandise: Exp. fob	5528.0	5164.0	5589.0	6444.0	7722.0	8297.0	11329.0	Marchandises : exp. fob
Merchandise: Imp. fob	-4328.0	-4591.0	-4687.0	-5423.0	-6875.0	-8411.0	-9919.0	Marchandises : imp. fob
Serv. & Income: Credit	1295.0	1700.0	1621.0	1871.0	2110.0	2981.0	3383.0	Serv. & revenu : crédit
Serv. & Income: Debit	-3791.0	-4016.0	-4295.0	-4574.0	-4916.0	-6056.0	-6459.0	Serv. & revenu : débit
Private Unrequited Transfers	675.0	897.0	1022.0	1550.0	1684.0	1515.0	2196.0	Transf. priv.sans contrep.
Offic. Unrequited Trans.,nie	0.0	0.0	0.0	0.0	0.0	88.0	305.0	Tr. offic. sans contrep.nia
Direct Investment, nie	2.0	1.0	-9.0	2.0	-18.0	-12.0	88.0	Investissements direct, nia
Portfolio Investments, nie	0.0	0.0	0.0	0.0	0.0	0.0	0.0	Invest. portefeuille, nia
Other Capital, nie	-3365.0	-1680.0	-3843.0	-2364.0	-2868.0	-1552.0	5457.0	Autres capitaux, nia
Net Errors and Omissions	486.0	185.0	457.0	79.0	-141.0	-165.0	-63.0	Erreurs et omissions nettes
Reserves and Related Items	3498.0	2340.0	4145.0	2415.0	3302.0	3315.0	-6317.0	Rés. et postes assimilables
Portugal								**Portugal**
Merchandise: Exp. fob	5172.0	5674.0	7200.0	9266.0	10874.0	12720.0	16427.0	Marchandises : exp. fob
Merchandise: Imp. fob	-7297.0	-7179.0	-8876.0	-12847.0	-16392.0	-17585.0	-23007.0	Marchandises : imp. fob
Serv. & Income: Credit	1926.0	2283.0	2789.0	3646.0	4037.0	4630.0	6371.0	Serv. & revenu : crédit
Serv. & Income: Debit	-2599.0	-2642.0	-2862.0	-3401.0	-3906.0	-4152.0	-5413.0	Serv. & revenu : débit
Private Unrequited Transfers	2140.0	2118.0	2650.0	3418.0	3597.0	3726.0	4504.0	Transf. priv.sans contrep.
Offic. Unrequited Trans.,nie	35.0	126.0	263.0	368.0	725.0	814.0	980.0	Tr. offic. sans contrep.nia
Direct Investment, nie	187.0	252.0	238.0	476.0	842.0	1653.0	1984.0	Investissements direct, nia
Portfolio Investments, nie	149.0	124.0	404.0	816.0	1814.0	1050.0	725.0	Invest. portefeuille, nia
Other Capital, nie	323.0	405.0	-2074.0	-604.0	-2363.0	1302.0	-1003.0	Autres capitaux, nia
Net Errors and Omissions	-31.0	-253.0	156.0	639.0	1640.0	497.0	1974.0	Erreurs et omissions nettes
Reserves and Related Items	-6.0	-908.0	111.0	-1777.0	-867.0	-4654.0	-3542.0	Rés. et postes assimilables
Romania								**Roumanie**
Merchandise: Exp. fob	12646.0	10174.0	9763.0	10491.0	11392.0	10487.0	5770.0	Marchandises : exp. fob
Merchandise: Imp. fob	-10334.0	-8402.0	-8083.0	-8313.0	-7642.0	-8437.0	-9114.0	Marchandises : imp. fob
Serv. & Income: Credit	957.0	862.0	801.0	908.0	1023.0	1015.0	923.0	Serv. & revenu : crédit

125
Summary of balance of payments
Millions of US dollars [*cont.*]

Résumé des balances des paiements
Millions de dollars des E-U [*suite*]

Country or area	1984	1985	1986	1987	1988	1989	1990	Pays ou zone
Serv. & Income: Debit	-1550.0	-1253.0	-1086.0	-1043.0	-851.0	-551.0	-833.0	Serv. & revenu : débit
Private Unrequited Transfers	0.0	0.0	0.0	0.0	0.0	0.0	0.0	Transf. priv.sans contrep.
Offic. Unrequited Trans.,nie	0.0	0.0	0.0	0.0	0.0	0.0	0.0	Tr. offic. sans contrep.nia
Direct Investment, nie	0.0	0.0	0.0	0.0	0.0	0.0	-18.0	Investissements direct, nia
Portfolio Investments, nie	0.0	0.0	0.0	0.0	0.0	0.0	0.0	Invest. portefeuille, nia
Other Capital, nie	-1691.0	-1580.0	-791.0	-1083.0	-4223.0	-1376.0	1631.0	Autres capitaux, nia
Net Errors and Omissions	100.0	-118.0	8.0	81.0	16.0	114.0	-3.0	Erreurs et omissions nettes
Reserves and Related Items	-128.0	317.0	-612.0	-1041.0	285.0	-1252.0	1644.0	Rés. et postes assimilables
Rwanda								**Rwanda**
Merchandise: Exp. fob	142.6	126.1	184.1	121.4	117.9	104.7	102.6	Marchandises : exp. fob
Merchandise: Imp. fob	-197.5	-219.3	-259.2	-267.0	-278.6	-254.1	-227.7	Marchandises : imp. fob
Serv. & Income: Credit	40.1	44.1	52.7	56.4	56.9	52.3	46.6	Serv. & revenu : crédit
Serv. & Income: Debit	-127.9	-131.3	-170.8	-171.3	-165.0	-142.0	-151.1	Serv. & revenu : débit
Private Unrequited Transfers	1.9	4.3	7.0	7.5	10.5	7.9	5.8	Transf. priv.sans contrep.
Offic. Unrequited Trans.,nie	99.2	112.1	117.0	118.6	139.3	129.2	115.8	Tr. offic. sans contrep.nia
Direct Investment, nie	15.1	14.6	17.6	17.5	21.0	15.6	7.7	Investissements direct, nia
Portfolio Investments, nie	0.1	0.0	0.0	0.0	0.0	0.0	-0.3	Invest. portefeuille, nia
Other Capital, nie	37.9	54.7	86.3	104.7	72.7	58.5	78.1	Autres capitaux, nia
Net Errors and Omissions	-3.6	-3.6	-4.5	1.5	0.4	-18.2	-0.5	Erreurs et omissions nettes
Reserves and Related Items	-7.8	-1.7	-30.2	10.6	24.9	46.2	22.9	Rés. et postes assimilables
Saint Kitts-Nevis								**Saint-Kitts-et-Nevis**
Merchandise: Exp. fob	20.2	20.4	27.2	29.8	29.1	32.8	24.6	Marchandises : exp. fob
Merchandise: Imp. fob	-47.2	-46.7	-55.4	-70.0	-81.6	-89.7	-103.2	Marchandises : imp. fob
Serv. & Income: Credit	21.2	23.1	33.2	40.9	46.8	49.2	54.3	Serv. & revenu : crédit
Serv. & Income: Debit	-10.0	-11.9	-17.6	-21.6	-25.5	-35.7	-38.1	Serv. & revenu : débit
Private Unrequited Transfers	9.8	7.1	8.5	9.2	9.9	10.5	11.5	Transf. priv.sans contrep.
Offic. Unrequited Trans.,nie	1.7	1.4	2.4	3.1	3.7	4.7	0.7	Tr. offic. sans contrep.nia
Direct Investment, nie	6.0	8.0	6.0	8.7	14.7	29.6	13.1	Investissements direct, nia
Portfolio Investments, nie	0.0	0.0	0.0	0.0	0.0	0.0	0.0	Invest. portefeuille, nia
Other Capital, nie	1.4	1.2	-2.8	-1.7	3.2	13.0	39.6	Autres capitaux, nia
Net Errors and Omissions	-1.2	-0.5	1.7	2.3	-0.2	-8.0	-2.4	Erreurs et omissions nettes
Reserves and Related Items	-1.8	-1.9	-3.1	-0.6	-0.1	-6.4	-0.1	Rés. et postes assimilables
Saint Lucia								**Sainte-Lucie**
Merchandise: Exp. fob	47.8	52.0	82.9	77.3	119.1	111.9	...	Marchandises : exp. fob
Merchandise: Imp. fob	-107.7	-113.6	-140.7	-161.9	-201.0	-235.2	...	Marchandises : imp. fob
Serv. & Income: Credit	64.0	69.5	83.3	94.5	124.4	135.8	...	Serv. & revenu : crédit
Serv. & Income: Debit	-41.5	-41.3	-45.0	-47.5	-59.9	-66.5	...	Serv. & revenu : débit
Private Unrequited Transfers	13.7	14.8	14.7	15.0	10.8	8.7	...	Transf. priv.sans contrep.
Offic. Unrequited Trans.,nie	10.3	6.1	6.5	9.7	4.0	3.7	...	Tr. offic. sans contrep.nia
Direct Investment, nie	12.0	17.0	18.5	22.0	16.0	18.0	...	Investissements direct, nia
Portfolio Investments, nie	0.0	0.0	0.0	0.0	0.0	0.0	...	Invest. portefeuille, nia
Other Capital, nie	2.2	-4.8	-9.1	-1.7	-7.5	29.9	...	Autres capitaux, nia
Net Errors and Omissions	-1.1	-0.3	1.3	1.5	0.4	-0.1	...	Erreurs et omissions nettes
Reserves and Related Items	0.4	0.6	-12.5	-8.9	-6.3	-6.2	...	Rés. et postes assimilables
St. Vincent-Grenadines								**St. Vincent-Grenadines**
Merchandise: Exp. fob	53.6	63.2	68.0	52.3	85.3	74.6	...	Marchandises : exp. fob
Merchandise: Imp. fob	-68.9	-71.3	-78.3	-89.5	-110.0	-114.8	...	Marchandises : imp. fob
Serv. & Income: Credit	19.1	19.4	27.5	31.3	34.4	38.7	...	Serv. & revenu : crédit
Serv. & Income: Debit	-20.3	-21.0	-25.9	-30.9	-35.2	-39.2	...	Serv. & revenu : débit
Private Unrequited Transfers	13.0	10.0	13.2	17.8	20.0	21.7	...	Transf. priv.sans contrep.
Offic. Unrequited Trans.,nie	2.5	3.4	6.8	6.6	6.3	8.7	...	Tr. offic. sans contrep.nia
Direct Investment, nie	1.4	1.8	4.4	5.5	9.5	5.7	...	Investissements direct, nia
Portfolio Investments, nie	0.0	0.0	0.0	0.0	0.0	0.0	...	Invest. portefeuille, nia
Other Capital, nie	4.1	-1.0	1.0	3.9	-5.8	7.1	...	Autres capitaux, nia
Net Errors and Omissions	0.0	1.9	-5.2	-2.2	-2.2	-0.7	...	Erreurs et omissions nettes
Reserves and Related Items	-4.5	-6.4	-11.5	5.2	-2.3	-1.9	...	Rés. et postes assimilables
Samoa								**Samoa**
Merchandise: Exp. fob	18.3	16.1	10.5	11.8	15.1	12.9	8.9	Marchandises : exp. fob
Merchandise: Imp. fob	-45.6	-46.6	-42.8	-55.8	-68.5	-68.5	-73.2	Marchandises : imp. fob
Serv. & Income: Credit	8.9	11.0	14.6	19.2	28.4	32.3	35.7	Serv. & revenu : crédit
Serv. & Income: Debit	-14.5	-13.7	-14.7	-17.2	-18.7	-20.2	-23.1	Serv. & revenu : débit
Private Unrequited Transfers	20.3	23.6	28.4	36.4	35.5	38.2	39.7	Transf. priv.sans contrep.
Offic. Unrequited Trans.,nie	13.2	11.3	11.3	12.9	14.5	14.6	12.7	Tr. offic. sans contrep.nia
Direct Investment, nie	0.0	0.0	0.0	0.0	0.0	0.0	0.0	Investissements direct, nia
Portfolio Investments, nie	0.0	0.0	0.0	0.0	0.0	0.0	0.0	Invest. portefeuille, nia
Other Capital, nie	4.6	-0.5	-0.7	3.2	0.5	0.5	9.4	Autres capitaux, nia

125
Summary of balance of payments
Millions of US dollars [*cont.*]
Résumé des balances des paiements
Millions de dollars des E-U [*suite*]

Country or area	1984	1985	1986	1987	1988	1989	1990	Pays ou zone
Net Errors and Omissions	2.0	3.9	0.9	-1.9	3.2	1.0	4.3	Erreurs et omissions nettes
Reserves and Related Items	-7.2	-5.2	-7.4	-8.6	-10.1	-10.7	-14.5	Rés. et postes assimilables
Sao Tome and Principe								**Sao Tomé-et-Principe**
Merchandise: Exp. fob	12.2	7.1	9.9	6.5	9.5	4.9	4.2	Marchandises : exp. fob
Merchandise: Imp. fob	-23.6	-19.9	-23.2	-13.6	-14.1	-13.3	-13.0	Marchandises : imp. fob
Serv. & Income: Credit	2.3	2.3	3.7	1.7	2.0	4.6	3.9	Serv. & revenu : crédit
Serv. & Income: Debit	-10.9	-10.4	-14.2	-9.3	-9.0	-8.8	-9.3	Serv. & revenu : débit
Private Unrequited Transfers	3.2	-0.2	-0.8	-0.3	0.0	-0.2	0.1	Transf. priv.sans contrep.
Offic. Unrequited Trans.,nie	5.9	2.9	5.7	1.9	0.9	1.5	-0.2	Tr. offic. sans contrep.nia
Direct Investment, nie	0.0	0.0	0.0	0.0	0.2	0.0	0.0	Investissements direct, nia
Portfolio Investments, nie	0.0	0.0	0.0	0.0	0.0	0.0	0.0	Invest. portefeuille, nia
Other Capital, nie	11.0	0.8	5.4	8.9	6.1	6.7	7.6	Autres capitaux, nia
Net Errors and Omissions	-7.3	6.2	2.2	0.0	0.0	-1.0	-2.7	Erreurs et omissions nettes
Reserves and Related Items	7.3	11.2	11.3	4.1	4.4	5.6	9.4	Rés. et postes assimilables
Saudi Arabia								**Arabie saoudite**
Merchandise: Exp. fob	37423.0	27393.0	20125.0	23138.0	24315.0	28299.0	44283.0	Marchandises : exp. fob
Merchandise: Imp. fob	-28557.0	-20364.0	-17066.0	-18283.0	-19805.0	-19231.0	-21490.0	Marchandises : imp. fob
Serv. & Income: Credit	17600.0	16065.0	13944.0	13113.0	12809.0	13012.0	12502.0	Serv. & revenu : crédit
Serv. & Income: Debit	-35984.0	-27578.0	-20995.0	-19506.0	-15650.0	-20766.0	-23364.0	Serv. & revenu : débit
Private Unrequited Transfers	-5284.0	-5199.0	-4804.0	-4935.0	-6510.0	-8342.0	-11637.0	Transf. priv.sans contrep.
Offic. Unrequited Trans.,nie	-3598.0	-3249.0	-3000.0	-3300.0	-2499.0	-2200.0	-4401.0	Tr. offic. sans contrep.nia
Direct Investment, nie	4850.0	491.0	967.0	-1175.0	-328.0	-312.0	0.0	Investissements direct, nia
Portfolio Investments, nie	13406.0	8412.0	3451.0	6150.0	3057.0	-1786.0	0.0	Invest. portefeuille, nia
Other Capital, nie	-1335.0	3319.0	-242.0	7438.0	3092.0	7819.0	-1270.0	Autres capitaux, nia
Net Errors and Omissions	0.0	0.0	0.0	0.0	0.0	0.0	0.0	Erreurs et omissions nettes
Reserves and Related Items	1480.0	709.0	7619.0	-2640.0	1519.0	3508.0	5376.0	Rés. et postes assimilables
Senegal								**Sénégal**
Merchandise: Exp. fob	597.8	514.7	657.0	670.9	762.8	776.2	911.6	Marchandises : exp. fob
Merchandise: Imp. fob	-818.9	-795.8	-883.3	-955.8	-1009.6	-1004.1	-1176.1	Marchandises : imp. fob
Serv. & Income: Credit	318.1	340.5	419.7	453.2	505.0	495.3	585.8	Serv. & revenu : crédit
Serv. & Income: Debit	-508.3	-500.6	-658.4	-708.9	-761.4	-720.7	-831.5	Serv. & revenu : débit
Private Unrequited Transfers	0.3	7.6	10.7	6.3	31.9	27.3	29.4	Transf. priv.sans contrep.
Offic. Unrequited Trans.,nie	137.0	160.6	186.1	227.7	204.6	230.9	355.9	Tr. offic. sans contrep.nia
Direct Investment, nie	27.2	-18.9	-3.8	-2.0	0.0	0.0	0.0	Investissements direct, nia
Portfolio Investments, nie	1.7	0.6	2.1	0.7	0.0	0.0	0.0	Invest. portefeuille, nia
Other Capital, nie	161.1	188.3	202.9	210.6	153.3	26.8	4.2	Autres capitaux, nia
Net Errors and Omissions	-14.2	13.5	30.0	-0.2	-1.4	5.0	-16.7	Erreurs et omissions nettes
Reserves and Related Items	98.2	89.5	37.0	97.4	114.8	163.3	137.4	Rés. et postes assimilables
Seychelles								**Seychelles**
Merchandise: Exp. fob	5.0	4.6	4.4	8.1	17.3	14.5	14.5	Marchandises : exp. fob
Merchandise: Imp. fob	-73.9	-84.1	-89.3	-96.2	-135.0	-139.6	-158.2	Marchandises : imp. fob
Serv. & Income: Credit	97.2	114.2	125.3	147.6	167.5	189.9	229.8	Serv. & revenu : crédit
Serv. & Income: Debit	-54.2	-67.4	-85.1	-101.7	-102.1	-114.1	-130.0	Serv. & revenu : débit
Private Unrequited Transfers	-2.3	-0.8	-5.1	-2.3	-4.9	-5.2	-2.5	Transf. priv.sans contrep.
Offic. Unrequited Trans.,nie	15.0	14.3	16.5	23.5	28.9	31.7	29.3	Tr. offic. sans contrep.nia
Direct Investment, nie	5.9	1.1	8.4	14.0	17.5	14.9	19.4	Investissements direct, nia
Portfolio Investments, nie	0.0	0.0	0.0	0.0	0.0	0.0	0.0	Invest. portefeuille, nia
Other Capital, nie	10.1	15.4	25.2	4.9	2.4	17.9	-7.0	Autres capitaux, nia
Net Errors and Omissions	-4.3	2.7	-1.6	6.2	4.2	-6.5	8.7	Erreurs et omissions nettes
Reserves and Related Items	1.5	0.1	1.4	-4.0	4.3	-3.5	-4.0	Rés. et postes assimilables
Sierra Leone								**Sierra Leone**
Merchandise: Exp. fob	132.6	131.9	126.0	138.9	104.5	139.5	...	Marchandises : exp. fob
Merchandise: Imp. fob	-149.7	-141.2	-111.4	-114.8	-138.2	-160.3	...	Marchandises : imp. fob
Serv. & Income: Credit	41.3	28.2	26.7	44.1	52.1	38.5	...	Serv. & revenu : crédit
Serv. & Income: Debit	-80.2	-34.3	93.3	-105.2	-29.9	-43.6	...	Serv. & revenu : débit
Private Unrequited Transfers	7.6	2.3	1.5	0.0	0.3	0.1	...	Transf. priv.sans contrep.
Offic. Unrequited Trans.,nie	25.4	16.9	4.6	6.8	8.5	7.2	...	Tr. offic. sans contrep.nia
Direct Investment, nie	5.9	-31.0	-140.3	39.4	-23.1	21.2	...	Investissements direct, nia
Portfolio Investments, nie	0.0	0.0	0.0	0.0	0.0	0.0	...	Invest. portefeuille, nia
Other Capital, nie	-66.6	-37.6	-139.0	-40.6	16.4	-40.3	...	Autres capitaux, nia
Net Errors and Omissions	-12.4	-9.3	42.0	-21.9	-62.5	19.9	...	Erreurs et omissions nettes
Reserves and Related Items	96.1	74.1	96.6	53.4	71.9	17.8	...	Rés. et postes assimilables
Singapore								**Singapour**
Merchandise: Exp. fob	22662.0	21533.0	21336.0	27464.0	37993.0	43239.0	50683.0	Marchandises : exp. fob
Merchandise: Imp. fob	-26734.0	-24362.0	-23402.0	-29910.0	-40338.0	-45713.0	-55802.0	Marchandises : imp. fob
Serv. & Income: Credit	9184.0	8197.0	8439.0	10596.0	13070.0	15861.0	20693.0	Serv. & revenu : crédit

125
Summary of balance of payments
Millions of US dollars [*cont.*]
Résumé des balances des paiements
Millions de dollars des E-U [*suite*]

Country or area	1984	1985	1986	1987	1988	1989	1990	Pays ou zone
Serv. & Income: Debit	-5275.0	-5159.0	-5872.0	-7692.0	-9174.0	-10595.0	12864.0	Serv. & revenu : débit
Private Unrequited Transfers	-214.0	-205.0	-172.0	-170.0	-209.0	-254.0	-265.0	Transf. priv.sans contrep.
Offic. Unrequited Trans.,nie	-9.0	-8.0	-11.0	-64.0	-88.0	-90.0	-95.0	Tr. offic. sans contrep.nia
Direct Investment, nie	1210.0	809.0	1529.0	2630.0	3493.0	3915.0	4489.0	Investissements direct, nia
Portfolio Investments, nie	-151.0	175.0	-549.0	252.0	-359.0	-116.0	-131.0	Invest. portefeuille, nia
Other Capital, nie	522.0	-285.0	-1425.0	-2412.0	-2311.0	-1897.0	1964.0	Autres capitaux, nia
Net Errors and Omissions	329.0	638.0	664.0	400.0	-418.0	-1712.0	-3240.0	Erreurs et omissions nettes
Reserves and Related Items	-1524.0	-1333.0	-538.0	-1095.0	-1659.0	-2738.0	-5431.0	Rés. et postes assimilables
Solomon Islands								**Iles Salomon**
Merchandise: Exp. fob	93.1	71.0	63.9	63.2	81.9	74.7	70.4	Marchandises : exp. fob
Merchandise: Imp. fob	-67.6	-71.9	-67.5	-69.4	-105.1	-97.5	-80.4	Marchandises : imp. fob
Serv. & Income: Credit	15.2	16.3	20.3	25.9	29.8	33.9	30.7	Serv. & revenu : crédit
Serv. & Income: Debit	-44.1	-45.2	-55.9	-59.2	-69.9	-82.9	-81.2	Serv. & revenu : débit
Private Unrequited Transfers	-2.5	-2.7	0.3	-0.1	-1.4	1.2	1.4	Transf. priv.sans contrep.
Offic. Unrequited Trans.,nie	14.6	11.8	29.1	19.4	36.3	37.3	33.5	Tr. offic. sans contrep.nia
Direct Investment, nie	2.0	0.9	-1.0	8.4	3.4	5.7	12.9	Investissements direct, nia
Portfolio Investments, nie	0.0	0.0	0.0	0.0	0.0	0.0	0.0	Invest. portefeuille, nia
Other Capital, nie	-13.1	7.5	-1.0	1.3	34.8	13.3	4.0	Autres capitaux, nia
Net Errors and Omissions	-6.9	-2.6	-6.5	3.8	-14.8	0.9	-4.9	Erreurs et omissions nettes
Reserves and Related Items	9.4	14.8	18.3	6.6	5.1	13.5	13.7	Rés. et postes assimilables
Somalia								**Somalie**
Merchandise: Exp. fob	54.8	90.6	94.7	94.0	58.4	67.7	...	Marchandises : exp. fob
Merchandise: Imp. fob	-465.7	-330.7	-342.1	-358.5	-216.0	-346.3	...	Marchandises : imp. fob
Serv. & Income: Credit	52.0	37.0	0.0	0.0	0.0	0.0	...	Serv. & revenu : crédit
Serv. & Income: Debit	-137.1	-123.4	-188.5	-179.7	-164.6	-206.4	...	Serv. & revenu : débit
Private Unrequited Transfers	162.9	19.4	5.3	-13.1	6.4	-2.9	...	Transf. priv.sans contrep.
Offic. Unrequited Trans.,nie	194.0	204.3	304.9	343.3	217.3	331.2	...	Tr. offic. sans contrep.nia
Direct Investment, nie	-14.9	-0.7	0.0	0.0	0.0	0.0	...	Investissements direct, nia
Portfolio Investments, nie	0.0	0.0	0.0	0.0	0.0	0.0	...	Invest. portefeuille, nia
Other Capital, nie	102.5	76.7	8.7	-25.1	-102.8	-31.9	...	Autres capitaux, nia
Net Errors and Omissions	23.5	15.5	19.2	40.7	21.1	-0.8	...	Erreurs et omissions nettes
Reserves and Related Items	28.1	11.3	97.7	98.4	180.2	189.4	...	Rés. et postes assimilables
South Africa								**Afrique du Sud**
Merchandise: Exp. fob	16948.0	16244.0	18330.0	21088.0	22432.0	22399.0	23383.0	Marchandises : exp. fob
Merchandise: Imp. fob	-14774.0	-10402.0	-11130.0	-13925.0	-17210.0	-16810.0	-17045.0	Marchandises : imp. fob
Serv. & Income: Credit	3049.0	2606.0	2726.0	3110.0	3244.0	3544.0	4195.0	Serv. & revenu : crédit
Serv. & Income: Debit	-6910.0	-5899.0	-6861.0	-7549.0	-7424.0	-7755.0	-8397.0	Serv. & revenu : débit
Private Unrequited Transfers	125.0	84.0	76.0	97.0	90.0	129.0	107.0	Transf. priv.sans contrep.
Offic. Unrequited Trans.,nie	-25.0	-10.0	11.0	116.0	86.0	72.0	10.0	Tr. offic. sans contrep.nia
Direct Investment, nie	251.0	-497.0	-116.0	-163.0	98.0	10.0	-5.0	Investissements direct, nia
Portfolio Investments, nie	1115.0	309.0	-711.0	-181.0	-54.0	-138.0	-324.0	Invest. portefeuille, nia
Other Capital, nie	790.0	-1798.0	-1698.0	-1011.0	-1683.0	-989.0	-109.0	Autres capitaux, nia
Net Errors and Omissions	-1149.0	-1145.0	-134.0	-220.0	-965.0	-575.0	-416.0	Erreurs et omissions nettes
Reserves and Related Items	581.0	507.0	-493.0	-1361.0	1386.0	113.0	-1399.0	Rés. et postes assimilables
Spain								**Espagne**
Merchandise: Exp. fob	22714.0	23665.0	26754.0	33561.0	39652.0	43301.0	53888.0	Marchandises : exp. fob
Merchandise: Imp. fob	-26985.0	-27836.0	-33278.0	-46547.0	-57650.0	-67797.0	-83454.0	Marchandises : imp. fob
Serv. & Income: Credit	14064.0	14834.0	19624.0	23867.0	27440.0	28875.0	34496.0	Serv. & revenu : crédit
Serv. & Income: Debit	-8870.0	-8940.0	-10266.0	-13749.0	-17729.0	-19919.0	-26006.0	Serv. & revenu : débit
Private Unrequited Transfers	1191.0	1395.0	1492.0	2277.0	3018.0	3163.0	3053.0	Transf. priv.sans contrep.
Offic. Unrequited Trans.,nie	-96.0	-268.0	-360.0	358.0	1485.0	1444.0	1204.0	Tr. offic. sans contrep.nia
Direct Investment, nie	1523.0	1718.0	3073.0	3825.0	5786.0	6955.0	10904.0	Investissements direct, nia
Portfolio Investments, nie	54.0	232.0	1228.0	3799.0	2291.0	7989.0	5361.0	Invest. portefeuille, nia
Other Capital, nie	3345.0	-5167.0	-5942.0	6605.0	6538.0	3397.0	12037.0	Autres capitaux, nia
Net Errors and Omissions	-2123.0	-1908.0	20.0	-1291.0	-2414.0	-2692.0	-4521.0	Erreurs et omissions nettes
Reserves and Related Items	-4817.0	2275.0	-2344.0	-12706.0	-8416.0	-4716.0	-6962.0	Rés. et postes assimilables
Sri Lanka								**Sri Lanka**
Merchandise: Exp. fob	1461.6	1315.8	1208.5	1393.9	1477.1	1505.1	1853.0	Marchandises : exp. fob
Merchandise: Imp. fob	-1698.7	-1838.5	-1764.3	-1866.0	-2017.5	-2055.1	-2325.6	Marchandises : imp. fob
Serv. & Income: Credit	334.3	328.7	373.3	397.5	407.9	404.2	534.7	Serv. & revenu : crédit
Serv. & Income: Debit	-575.4	-667.2	-705.7	-744.0	-787.9	-787.1	-901.1	Serv. & revenu : débit
Private Unrequited Transfers	276.5	265.5	294.1	312.8	320.0	330.7	364.7	Transf. priv.sans contrep.
Offic. Unrequited Trans.,nie	202.6	177.2	176.9	179.9	206.1	188.6	178.1	Tr. offic. sans contrep.nia
Direct Investment, nie	32.6	24.8	29.2	58.2	43.6	17.7	29.6	Investissements direct, nia
Portfolio Investments, nie	0.0	0.0	0.0	0.0	0.0	0.0	0.0	Invest. portefeuille, nia
Other Capital, nie	326.5	338.2	332.6	336.7	212.4	559.3	407.0	Autres capitaux, nia

125
Summary of balance of payments
Millions of US dollars [cont.]
Résumé des balances des paiements
Millions de dollars des E-U [suite]

Country or area	1984	1985	1986	1987	1988	1989	1990	Pays ou zone
Net Errors and Omissions	-39.7	-43.0	-34.4	-122.5	37.4	-115.0	-75.7	Erreurs et omissions nettes
Reserves and Related Items	-320.3	98.5	89.6	53.6	101.2	-48.3	-64.8	Rés. et postes assimilables
Sudan								**Soudan**
Merchandise: Exp. fob	519.0	444.3	326.8	252.1	427.0	544.4	326.5	Marchandises : exp. fob
Merchandise: Imp. fob	-599.8	-583.4	-633.7	-696.6	-948.5	-1051.0	-648.8	Marchandises : imp. fob
Serv. & Income: Credit	268.2	386.3	228.8	192.2	171.6	279.7	184.0	Serv. & revenu : crédit
Serv. & Income: Debit	-459.2	-456.1	-272.8	-321.2	-341.3	-495.7	-373.0	Serv. & revenu : débit
Private Unrequited Transfers	276.8	248.6	89.3	133.7	216.3	412.4	60.7	Transf. priv.sans contrep.
Offic. Unrequited Trans.,nie	20.3	114.9	244.2	180.7	117.0	160.0	81.4	Tr. offic. sans contrep.nia
Direct Investment, nie	9.1	-3.0	0.0	0.0	0.0	3.5	0.0	Investissements direct, nia
Portfolio Investments, nie	0.0	0.0	0.0	0.0	0.0	0.0	0.0	Invest. portefeuille, nia
Other Capital, nie	-151.8	-455.9	-95.5	90.7	67.6	114.7	112.9	Autres capitaux, nia
Net Errors and Omissions	-1.5	-84.8	-89.4	-165.5	7.5	-136.4	32.7	Erreurs et omissions nettes
Reserves and Related Items	118.9	389.1	202.3	334.0	282.8	168.4	223.6	Rés. et postes assimilables
Suriname								**Suriname**
Merchandise: Exp. fob	374.1	336.1	337.1	338.8	358.4	549.2	465.9	Marchandises : exp. fob
Merchandise: Imp. fob	-391.6	-309.5	-304.1	-274.3	-239.4	-330.9	-374.4	Marchandises : imp. fob
Serv. & Income: Credit	60.2	47.0	27.3	82.1	24.0	24.7	22.9	Serv. & revenu : crédit
Serv. & Income: Debit	-116.6	-82.5	-81.6	-73.8	-86.1	-98.0	-105.7	Serv. & revenu : débit
Private Unrequited Transfers	-7.4	-4.0	-1.9	-0.4	-4.6	-5.6	-7.5	Transf. priv.sans contrep.
Offic. Unrequited Trans.,nie	2.6	1.3	0.9	2.4	10.3	23.5	31.7	Tr. offic. sans contrep.nia
Direct Investment, nie	-39.7	11.9	-33.8	-72.6	-95.8	-167.9	-43.0	Investissements direct, nia
Portfolio Investments, nie	0.0	-0.2	-0.2	-0.1	0.0	0.0	0.5	Invest. portefeuille, nia
Other Capital, nie	29.3	2.8	35.1	22.1	29.9	-5.0	27.6	Autres capitaux, nia
Net Errors and Omissions	36.6	-11.4	-18.7	-33.6	-1.8	9.9	-7.8	Erreurs et omissions nettes
Reserves and Related Items	52.5	8.5	40.0	9.4	5.2	0.1	-10.3	Rés. et postes assimilables
Swaziland								**Swaziland**
Merchandise: Exp. fob	230.8	176.7	278.1	423.5	466.2	486.5	557.4	Marchandises : exp. fob
Merchandise: Imp. fob	-371.2	-271.7	-295.9	-369.4	-441.0	-515.4	-632.2	Marchandises : imp. fob
Serv. & Income: Credit	114.6	95.3	101.9	148.3	179.3	198.7	257.9	Serv. & revenu : crédit
Serv. & Income: Debit	-119.3	-94.6	-120.3	-172.3	-195.2	-282.8	-277.3	Serv. & revenu : débit
Private Unrequited Transfers	0.1	2.5	2.5	3.6	2.1	0.5	-4.2	Transf. priv.sans contrep.
Offic. Unrequited Trans.,nie	65.2	49.3	40.8	46.6	83.5	105.4	110.0	Tr. offic. sans contrep.nia
Direct Investment, nie	0.0	10.1	28.1	39.3	38.5	61.5	40.4	Investissements direct, nia
Portfolio Investments, nie	-0.2	1.9	-0.6	1.1	6.6	8.0	1.0	Invest. portefeuille, nia
Other Capital, nie	28.7	17.6	-14.9	-62.0	-103.6	-67.6	-48.4	Autres capitaux, nia
Net Errors and Omissions	42.4	8.6	-12.3	-37.1	-21.9	56.5	6.4	Erreurs et omissions nettes
Reserves and Related Items	9.0	4.2	-7.5	-21.5	-14.6	-51.3	-11.1	Rés. et postes assimilables
Sweden								**Suède**
Merchandise: Exp. fob	29123.0	30173.0	36845.0	44011.0	49369.0	51071.0		Marchandises : exp. fob
Merchandise: Imp. fob	-25701.0	-27788.0	-31811.0	-39528.0	-44487.0	-47056.0	56835.0	Marchandises : imp. fob
Serv. & Income: Credit	8605.0	8705.0	10431.0	13291.0	15477.0	17848.0	-53433.0	Serv. & revenu : crédit
Serv. & Income: Debit	-10535.0	-11242.0	-13534.0	-16520.0	-19435.0	-23083.0	23622.0	Serv. & revenu : débit
Private Unrequited Transfers	-310.0	-397.0	-483.0	-379.0	-472.0	-709.0	-30196.0	Transf. priv.sans contrep.
Offic. Unrequited Trans.,nie	-617.0	-682.0	-858.0	-1016.0	-1140.0	-1338.0	-645.0	Tr. offic. sans contrep.nia
Direct Investment, nie	-1207.0	-1412.0	-2781.0	-3911.0	-5719.0	-8172.0	-1645.0	Investissements direct, nia
Portfolio Investments, nie	222.0	562.0	-155.0	-1065.0	-1352.0	-1267.0	-11828.0	Invest. portefeuille, nia
Other Capital, nie	-4827.0	-2042.0	170.0	4197.0	7209.0	19336.0	2270.0	Autres capitaux, nia
Net Errors and Omissions	-586.0	-527.0	-793.0	108.0	-1142.0	-5654.0	28785.0	Erreurs et omissions nettes
Reserves and Related Items	5834.0	4651.0	2969.0	810.0	1693.0	-976.0	-6175.0	Rés. et postes assimilables
Switzerland							-7587.0	**Suisse**
Merchandise: Exp. fob	36658.0	37057.0	48453.0	55219.0	62725.0	65366.0		Marchandises : exp. fob
Merchandise: Imp. fob	-37747.0	-38618.0	-53412.0	-60647.0	-67301.0	-69690.0	77488.0	Marchandises : imp. fob
Serv. & Income: Credit	19090.0	19555.0	25816.0	32483.0	35886.0	38204.0	-83865.0	Serv. & revenu : crédit
Serv. & Income: Debit	-11024.0	-11148.0	-15112.0	-19266.0	-20754.0	-24155.0	47417.0	Serv. & revenu : débit
Private Unrequited Transfers	-826.0	-841.0	-1206.0	-1550.0	-1711.0	-1665.0	-31744.0	Transf. priv.sans contrep.
Offic. Unrequited Trans.,nie	-9.0	35.0	116.0	45.0	-2.0	-18.0	-2185.0	Tr. offic. sans contrep.nia
Direct Investment, nie	-362.0	-3305.0	662.0	1047.0	-8289.0	-5023.0	-170.0	Investissements direct, nia
Portfolio Investments, nie	-2876.0	-1104.0	1401.0	-1733.0	-7415.0	-3023.0	-1449.0	Invest. portefeuille, nia
Other Capital, nie	-2589.0	-2866.0	-5223.0	-7377.0	963.0	429.0	2493.0	Autres capitaux, nia
Net Errors and Omissions	1164.0	2461.0	-402.0	4984.0	3473.0	995.0	-10548.0	Erreurs et omissions nettes
Reserves and Related Items	-1480.0	-1225.0	-1093.0	-3205.0	2426.0	-1420.0	3695.0	Rés. et postes assimilables
Syrian Arab Republic							-1132.0	**Rép. arabe syrienne**
Merchandise: Exp. fob	1834.0	1856.0	1037.0	1357.0	1348.0	3013.0		Marchandises : exp. fob
Merchandise: Imp. fob	-3687.0	-3946.0	-2363.0	-2226.0	-1986.0	-1821.0	4221.0	Marchandises : imp. fob
Serv. & Income: Credit	542.0	685.0	576.0	625.0	689.0	837.0	-2062.0	Serv. & revenu : crédit
							864.0	Serv. & revenu : crédit

125
Summary of balance of payments
Millions of US dollars [*cont.*]
Résumé des balances des paiements
Millions de dollars des E-U [*suite*]

Country or area	1984	1985	1986	1987	1988	1989	1990	Pays ou zone
Serv. & Income: Debit	-1033.0	-1113.0	-836.0	-1148.0	-1097.0	-1476.0	-1650.0	Serv. & revenu : débit
Private Unrequited Transfers	321.0	350.0	323.0	334.0	360.0	395.0	375.0	Transf. priv.sans contrep.
Offic. Unrequited Trans.,nie	1229.0	1212.0	759.0	760.0	536.0	223.0	80.0	Tr. offic. sans contrep.nia
Direct Investment, nie	0.0	0.0	0.0	0.0	0.0	0.0	0.0	Investissements direct, nia
Portfolio Investments, nie	0.0	0.0	0.0	0.0	0.0	0.0	0.0	Invest. portefeuille, nia
Other Capital, nie	1035.0	789.0	591.0	399.0	85.0	-453.0	-758.0	Autres capitaux, nia
Net Errors and Omissions	-25.0	-17.0	-26.0	-23.0	34.0	71.0	-348.0	Erreurs et omissions nettes
Reserves and Related Items	-216.0	186.0	-61.0	-79.0	32.0	-789.0	-721.0	Rés. et postes assimilables
Thailand								**Thaïlande**
Merchandise: Exp. fob	7338.0	7059.0	8803.0	11595.0	15781.0	19834.0	22811.0	Marchandises : exp. fob
Merchandise: Imp. fob	-9236.0	-8391.0	-8415.0	-12019.0	-17856.0	-22750.0	-29561.0	Marchandises : imp. fob
Serv. & Income: Credit	3077.0	3163.0	3333.0	4168.0	5944.0	7046.0	8478.0	Serv. & revenu : crédit
Serv. & Income: Debit	-3463.0	-3533.0	-3699.0	-4334.0	-5760.0	-6874.0	-9222.0	Serv. & revenu : débit
Private Unrequited Transfers	59.0	47.0	64.0	100.0	47.0	47.0	26.0	Transf. priv.sans contrep.
Offic. Unrequited Trans.,nie	115.0	118.0	161.0	125.0	189.0	199.0	187.0	Tr. offic. sans contrep.nia
Direct Investment, nie	400.0	162.0	261.0	182.0	1081.0	1726.0	2303.0	Investissements direct, nia
Portfolio Investments, nie	155.0	895.0	-29.0	346.0	530.0	1486.0	-38.0	Invest. portefeuille, nia
Other Capital, nie	2011.0	481.0	-363.0	534.0	2228.0	3387.0	6833.0	Autres capitaux, nia
Net Errors and Omissions	71.0	103.0	598.0	248.0	411.0	928.0	1419.0	Erreurs et omissions nettes
Reserves and Related Items	-529.0	-105.0	-714.0	-945.0	-2596.0	-5029.0	-3235.0	Rés. et postes assimilables
Togo								**Togo**
Merchandise: Exp. fob	291.0	282.0	362.4	402.7	325.0	331.0	396.3	Marchandises : exp. fob
Merchandise: Imp. fob	-263.2	-303.7	-418.6	-444.0	-352.5	-343.9	-501.7	Marchandises : imp. fob
Serv. & Income: Credit	93.9	110.5	131.2	133.2	161.8	159.2	197.3	Serv. & revenu : crédit
Serv. & Income: Debit	-168.6	-193.3	-241.8	-257.7	-271.3	-266.4	-312.7	Serv. & revenu : débit
Private Unrequited Transfers	-1.9	7.7	9.8	7.0	8.4	11.9	13.6	Transf. priv.sans contrep.
Offic. Unrequited Trans.,nie	65.1	63.4	91.4	97.1	69.5	73.7	108.7	Tr. offic. sans contrep.nia
Direct Investment, nie	-9.9	16.6	6.5	7.4	0.0	0.0	0.0	Investissements direct, nia
Portfolio Investments, nie	-0.2	0.4	-2.6	1.4	0.0	0.0	0.0	Invest. portefeuille, nia
Other Capital, nie	-26.9	14.1	19.5	-63.0	-28.2	4.7	72.5	Autres capitaux, nia
Net Errors and Omissions	-0.1	-0.7	-9.7	-4.6	-36.6	2.0	-2.5	Erreurs et omissions nettes
Reserves and Related Items	20.8	3.1	51.8	120.4	124.0	27.8	29.5	Rés. et postes assimilables
Tonga								**Tonga**
Merchandise: Exp. fob	7.2	7.9	5.9	7.0	6.4	9.8	8.9	Marchandises : exp. fob
Merchandise: Imp. fob	-33.4	-31.8	-32.3	-35.1	-44.1	-48.0	-47.0	Marchandises : imp. fob
Serv. & Income: Credit	15.2	18.0	17.8	22.1	24.7	25.3	31.6	Serv. & revenu : crédit
Serv. & Income: Debit	-11.0	-14.8	-16.7	-17.0	-25.4	-24.0	-23.8	Serv. & revenu : débit
Private Unrequited Transfers	15.7	16.2	23.2	22.4	21.3	24.9	28.8	Transf. priv.sans contrep.
Offic. Unrequited Trans.,nie	6.6	3.2	3.1	6.4	5.7	10.0	9.9	Tr. offic. sans contrep.nia
Direct Investment, nie	0.0	0.0	0.1	0.2	0.1	0.1	0.1	Investissements direct, nia
Portfolio Investments, nie	0.0	0.0	0.0	0.0	0.0	-2.2	-9.4	Invest. portefeuille, nia
Other Capital, nie	4.3	4.7	-0.8	0.1	4.8	3.2	-1.5	Autres capitaux, nia
Net Errors and Omissions	1.5	0.4	0.3	-4.9	7.2	-0.4	2.9	Erreurs et omissions nettes
Reserves and Related Items	-6.1	-3.8	-0.5	-1.1	-0.7	1.2	-0.4	Rés. et postes assimilables
Trinidad and Tobago								**Trinité-et-Tobago**
Merchandise: Exp. fob	2110.8	2110.7	1363.1	1396.9	1453.3	1534.6	1935.2	Marchandises : exp. fob
Merchandise: Imp. fob	-1704.9	-1354.6	-1209.4	-1057.6	-1064.2	-1045.2	-947.6	Marchandises : imp. fob
Serv. & Income: Credit	470.5	483.6	375.6	235.4	299.0	318.7	383.2	Serv. & revenu : crédit
Serv. & Income: Debit	-1315.6	-1270.7	-920.9	-785.2	-776.2	-849.9	-915.6	Serv. & revenu : débit
Private Unrequited Transfers	-71.5	-55.4	-30.8	-20.2	-23.0	-19.2	-21.0	Transf. priv.sans contrep.
Offic. Unrequited Trans.,nie	-11.7	-4.0	-19.1	-16.6	-6.6	-5.5	-4.4	Tr. offic. sans contrep.nia
Direct Investment, nie	109.8	-7.0	-21.8	35.0	62.9	148.9	109.4	Investissements direct, nia
Portfolio Investments, nie	0.0	0.0	0.0	0.0	0.0	0.0	0.0	Invest. portefeuille, nia
Other Capital, nie	-232.6	27.4	-174.5	43.4	-204.4	-315.4	-610.1	Autres capitaux, nia
Net Errors and Omissions	-47.6	-231.4	-83.5	-86.6	29.8	56.2	-107.8	Erreurs et omissions nettes
Reserves and Related Items	692.9	301.3	721.6	255.6	229.4	176.8	178.5	Rés. et postes assimilables
Tunisia								**Tunisie**
Merchandise: Exp. fob	1776.0	1700.0	1763.0	2101.0	2399.0	2931.0	3515.0	Marchandises : exp. fob
Merchandise: Imp. fob	-2929.0	-2567.0	-2698.0	-2829.0	-3496.0	-4139.0	-5193.0	Marchandises : imp. fob
Serv. & Income: Credit	1020.0	1011.0	984.0	1267.0	1901.0	1629.0	1785.0	Serv. & revenu : crédit
Serv. & Income: Debit	-983.0	-1039.0	-1067.0	-1119.0	-1256.0	-1280.0	-1415.0	Serv. & revenu : débit
Private Unrequited Transfers	304.0	259.0	354.0	481.0	547.0	485.0	593.0	Transf. priv.sans contrep.
Offic. Unrequited Trans.,nie	42.0	48.0	45.0	37.0	118.0	215.0	215.0	Tr. offic. sans contrep.nia
Direct Investment, nie	115.0	114.0	62.0	91.0	62.0	74.0	59.0	Investissements direct, nia
Portfolio Investments, nie	91.0	30.0	33.0	8.0	3.0	18.0	22.0	Invest. portefeuille, nia
Other Capital, nie	416.0	237.0	346.0	41.0	93.0	94.0	258.0	Autres capitaux, nia

125
Summary of balance of payments
Millions of US dollars [*cont.*]
Résumé des balances des paiements
Millions de dollars des E-U [*suite*]

Country or area	1984	1985	1986	1987	1988	1989	1990	Pays ou zone
Net Errors and Omissions	48.0	-19.0	94.0	48.0	69.0	40.0	38.0	Erreurs et omissions nettes
Reserves and Related Items	100.0	225.0	84.0	-128.0	-441.0	-65.0	123.0	Rés. et postes assimilables
Turkey								**Turquie**
Merchandise: Exp. fob	7389.0	8255.0	7583.0	10322.0	11929.0	11780.0	13026.0	Marchandises : exp. fob
Merchandise: Imp. fob	-10331.0	-11230.0	-10664.0	-13551.0	-13706.0	-15999.0	-22580.0	Marchandises : imp. fob
Serv. & Income: Credit	2366.0	3162.0	3338.0	4195.0	6026.0	7098.0	8933.0	Serv. & revenu : crédit
Serv. & Income: Debit	-2945.0	-3184.0	-3646.0	-4162.0	-4812.0	-5476.0	-6506.0	Serv. & revenu : débit
Private Unrequited Transfers	1885.0	1762.0	1703.0	2066.0	1827.0	3135.0	3349.0	Transf. priv.sans contrep.
Offic. Unrequited Trans.,nie	229.0	222.0	221.0	324.0	332.0	423.0	1162.0	Tr. offic. sans contrep.nia
Direct Investment, nie	113.0	99.0	125.0	106.0	354.0	663.0	713.0	Investissements direct, nia
Portfolio Investments, nie	0.0	0.0	146.0	282.0	1178.0	1586.0	547.0	Invest. portefeuille, nia
Other Capital, nie	80.0	966.0	1853.0	1503.0	-2490.0	-1413.0	2701.0	Autres capitaux, nia
Net Errors and Omissions	317.0	-836.0	-119.0	-505.0	515.0	915.0	-402.0	Erreurs et omissions nettes
Reserves and Related Items	897.0	784.0	-540.0	-580.0	-1153.0	-2712.0	-943.0	Rés. et postes assimilables
Uganda								**Ouganda**
Merchandise: Exp. fob	407.3	347.8	394.9	333.6	266.3	277.7	177.8	Marchandises : exp. fob
Merchandise: Imp. fob	-286.8	-238.3	-309.5	-475.6	-523.3	-588.3	-491.0	Marchandises : imp. fob
Serv. & Income: Credit	16.8	24.6	14.7	0.0	0.0	0.0	0.0	Serv. & revenu : crédit
Serv. & Income: Debit	-119.2	-150.9	-169.4	-236.2	-260.4	-260.5	-243.1	Serv. & revenu : débit
Private Unrequited Transfers	0.0	-0.3	-1.1	0.0	0.0	0.0	0.0	Transf. priv.sans contrep.
Offic. Unrequited Trans.,nie	85.4	21.7	66.6	266.2	322.4	311.6	293.0	Tr. offic. sans contrep.nia
Direct Investment, nie	0.0	0.0	0.0	0.0	0.0	0.0	0.0	Investissements direct, nia
Portfolio Investments, nie	0.0	0.0	0.0	0.0	0.0	0.0	0.0	Invest. portefeuille, nia
Other Capital, nie	-58.2	80.8	12.4	25.4	0.0	210.2	211.6	Autres capitaux, nia
Net Errors and Omissions	21.6	-52.2	-35.4	32.2	158.3	-35.2	9.8	Erreurs et omissions nettes
Reserves and Related Items	-66.9	-33.2	26.8	54.4	36.7	84.5	41.9	Rés. et postes assimilables
United Kingdom								**Royaume-Uni**
Merchandise: Exp. fob	93487.0	100966.0	106590.0	130039.0	143078.0	151080.0	182299.0	Marchandises : exp. fob
Merchandise: Imp. fob	-100601.0	-104816.0	-120488.0	-148866.0	-181494.0	-191485.0	-214800.0	Marchandises : imp. fob
Serv. & Income: Credit	97597.0	99159.0	107936.0	124753.0	149722.0	170997.0	201521.0	Serv. & revenu : crédit
Serv. & Income: Debit	-85804.0	-87208.0	-90654.0	-107339.0	-132586.0	-156542.0	-186291.0	Serv. & revenu : débit
Private Unrequited Transfers	493.0	405.0	112.0	-205.0	-480.0	-491.0	-535.0	Transf. priv.sans contrep.
Offic. Unrequited Trans.,nie	-2875.0	-4369.0	-3284.0	-5340.0	-5857.0	-6964.0	-8281.0	Tr. offic. sans contrep.nia
Direct Investment, nie	-8312.0	-5912.0	-10269.0	-17341.0	-19050.0	-7352.0	16143.0	Investissements direct, nia
Portfolio Investments, nie	-11474.0	-11448.0	-17234.0	41943.0	9111.0	-27674.0	-13910.0	Invest. portefeuille, nia
Other Capital, nie	-2903.0	7467.0	20481.0	-21567.0	25441.0	42409.0	20993.0	Autres capitaux, nia
Net Errors and Omissions	8054.0	445.0	10378.0	-2833.0	10348.0	11726.0	1143.0	Erreurs et omissions nettes
Reserves and Related Items	12339.0	5312.0	-3568.0	6754.0	1768.0	14298.0	1717.0	Rés. et postes assimilables
United Rep.Tanzania								**Rép. Unie de Tanzanie**
Merchandise: Exp. fob	398.5	328.5	335.9	346.8	386.5	415.1	407.8	Marchandises : exp. fob
Merchandise: Imp. fob	-760.3	-869.2	-913.3	-1000.5	-1033.0	-1070.1	-1186.3	Marchandises : imp. fob
Serv. & Income: Credit	107.4	108.1	110.2	101.6	120.6	123.2	140.1	Serv. & revenu : crédit
Serv. & Income: Debit	-264.2	-309.2	-327.9	-420.2	-471.1	-479.2	-481.1	Serv. & revenu : débit
Private Unrequited Transfers	63.0	236.3	250.6	230.0	231.9	182.4	164.5	Transf. priv.sans contrep.
Offic. Unrequited Trans.,nie	96.0	130.4	222.8	476.9	389.3	469.8	529.0	Tr. offic. sans contrep.nia
Direct Investment, nie	0.0	0.0	0.0	0.0	0.0	0.0	0.0	Investissements direct, nia
Portfolio Investments, nie	0.0	0.0	0.0	0.0	0.0	0.0	0.0	Invest. portefeuille, nia
Other Capital, nie	11.8	-72.2	-86.5	-31.7	33.9	21.7	126.5	Autres capitaux, nia
Net Errors and Omissions	125.1	-39.6	-43.5	-74.2	-42.5	-114.9	133.1	Erreurs et omissions nettes
Reserves and Related Items	222.6	487.0	451.8	371.3	384.4	452.0	166.3	Rés. et postes assimilables
United States[1]								**Etats-Unis**[1]
Merchandise: Exp. fob	219.9	215.9	223.4	250.3	320.3	361.5	389.5	Marchandises : exp. fob
Merchandise: Imp. fob	-332.4	-338.1	-368.4	-409.8	-447.3	-477.4	-497.7	Marchandises : imp. fob
Serv. & Income: Credit	159.5	150.2	160.6	181.5	213.1	244.1	263.4	Serv. & revenu : crédit
Serv. & Income: Debit	-133.3	-134.8	-145.1	-167.9	-197.5	-220.0	-225.1	Serv. & revenu : débit
Private Unrequited Transfers	-1.8	-2.1	-1.9	-1.8	-1.8	-1.9	-1.9	Transf. priv.sans contrep.
Offic. Unrequited Trans.,nie	-10.8	-13.4	-14.0	-12.5	-13.3	-13.6	-20.4	Tr. offic. sans contrep.nia
Direct Investment, nie	14.0	5.9	15.4	27.1	41.5	37.2	3.8	Investissements direct, nia
Portfolio Investments, nie	28.8	64.4	71.6	31.1	40.3	43.2	-31.2	Invest. portefeuille, nia
Other Capital, nie	29.8	37.9	8.8	51.9	17.4	24.6	27.4	Autres capitaux, nia
Net Errors and Omissions	27.2	19.9	15.9	-6.7	-9.1	18.4	63.6	Erreurs et omissions nettes
Reserves and Related Items	-0.7	-5.8	33.8	56.9	36.3	-16.9	28.5	Rés. et postes assimilables
Uruguay								**Uruguay**
Merchandise: Exp. fob	924.6	853.6	1087.8	1182.3	1404.5	1599.0	1692.9	Marchandises : exp. fob
Merchandise: Imp. fob	-732.2	-675.4	-814.5	-1079.9	-1112.2	-1136.2	-1266.9	Marchandises : imp. fob
Serv. & Income: Credit	452.0	476.6	504.5	467.6	472.9	598.9	684.9	Serv. & revenu : crédit

125
Summary of balance of payments
Millions of US dollars [*cont.*]
Résumé des balances des paiements
Millions de dollars des E-U [*suite*]

Country or area	1984	1985	1986	1987	1988	1989	1990	Pays ou zone
Serv. & Income: Debit	-783.5	-785.7	-758.5	-735.9	-777.6	-916.5	-895.1	Serv. & revenu : débit
Private Unrequited Transfers	0.0	0.0	0.0	0.0	0.0	0.0	0.0	Transf. priv.sans contrep.
Offic. Unrequited Trans.,nie	10.0	10.8	25.3	8.0	21.3	8.0	8.1	Tr. offic. sans contrep.nia
Direct Investment, nie	3.4	-7.9	32.5	55.0	44.5	0.0	0.0	Investissements direct, nia
Portfolio Investments, nie	18.5	96.5	86.4	13.1	37.4	49.9	17.5	Invest. portefeuille, nia
Other Capital, nie	144.0	-162.9	-100.3	221.1	104.6	-56.0	-103.4	Autres capitaux, nia
Net Errors and Omissions	-121.2	260.8	224.3	-93.1	-231.5	-60.3	127.3	Erreurs et omissions nettes
Reserves and Related Items	84.4	-66.4	-287.5	-38.2	36.1	-86.8	-265.3	Rés. et postes assimilables
Vanuatu								**Vanuatu**
Merchandise: Exp. fob	32.5	18.7	8.8	13.7	15.4	13.7	13.8	Marchandises : exp. fob
Merchandise: Imp. fob	-51.5	-52.3	-46.8	-57.1	-57.9	-57.9	-79.6	Marchandises : imp. fob
Serv. & Income: Credit	51.2	64.9	71.4	67.3	63.3	63.7	95.7	Serv. & revenu : crédit
Serv. & Income: Debit	-48.0	-60.6	-64.8	-67.4	-58.8	-48.3	-57.9	Serv. & revenu : débit
Private Unrequited Transfers	6.9	6.8	7.2	5.8	9.8	4.6	6.7	Transf. priv.sans contrep.
Offic. Unrequited Trans.,nie	28.1	23.8	21.3	32.3	30.8	20.7	29.4	Tr. offic. sans contrep.nia
Direct Investment, nie	7.4	4.6	2.0	12.9	10.8	9.2	13.2	Investissements direct, nia
Portfolio Investments, nie	0.0	0.0	0.0	0.0	0.0	0.0	0.0	Invest. portefeuille, nia
Other Capital, nie	-35.0	0.4	10.9	6.5	-0.9	15.0	8.9	Autres capitaux, nia
Net Errors and Omissions	2.1	-6.6	-5.3	-15.1	-17.3	-13.0	-34.2	Erreurs et omissions nettes
Reserves and Related Items	6.2	0.3	-4.8	1.1	4.7	-7.8	4.1	Rés. et postes assimilables
Venezuela								**Venezuela**
Merchandise: Exp. fob	15878.0	14283.0	8535.0	10437.0	10082.0	12915.0	17411.0	Marchandises : exp. fob
Merchandise: Imp. fob	-7246.0	-7501.0	-7866.0	-8870.0	-12080.0	-7283.0	-6543.0	Marchandises : imp. fob
Serv. & Income: Credit	2982.0	2906.0	2718.0	2446.0	2623.0	2695.0	3577.0	Serv. & revenu : crédit
Serv. & Income: Debit	-6791.0	-6190.0	-5511.0	-5312.0	-6287.0	-5979.0	-5989.0	Serv. & revenu : débit
Private Unrequited Transfers	-144.0	-147.0	-102.0	-73.0	-123.0	-171.0	-235.0	Transf. priv.sans contrep.
Offic. Unrequited Trans.,nie	-28.0	-24.0	-19.0	-18.0	-24.0	-16.0	-23.0	Tr. offic. sans contrep.nia
Direct Investment, nie	-3.0	57.0	-444.0	-16.0	21.0	77.0	96.0	Investissements direct, nia
Portfolio Investments, nie	0.0	0.0	0.0	0.0	0.0	-158.0	13579.0	Invest. portefeuille, nia
Other Capital, nie	-2009.0	-1113.0	-705.0	283.0	-1923.0	-5231.0	-17296.0	Autres capitaux, nia
Net Errors and Omissions	-996.0	-978.0	-930.0	-505.0	3065.0	1418.0	-2173.0	Erreurs et omissions nettes
Reserves and Related Items	-1643.0	-1293.0	4324.0	1628.0	4646.0	1733.0	-2404.0	Rés. et postes assimilables
Yemen[2]								**Yémen**[2]
Merchandise: Exp. fob	8.5	8.2	16.1	48.2	447.0	606.0	...	Marchandises : exp. fob
Merchandise: Imp. fob	-1401.1	-1078.9	-796.6	-1189.4	-1309.4	-1282.7	...	Marchandises : imp. fob
Serv. & Income: Credit	243.0	190.7	151.1	172.1	200.3	247.6	...	Serv. & revenu : crédit
Serv. & Income: Debit	-288.1	-259.5	-222.0	-332.1	-436.2	-488.5	...	Serv. & revenu : débit
Private Unrequited Transfers	995.5	763.2	527.4	707.7	313.7	242.3	...	Transf. priv.sans contrep.
Offic. Unrequited Trans.,nie	141.6	89.0	198.5	141.2	90.4	96.2	...	Tr. offic. sans contrep.nia
Direct Investment, nie	6.6	2.8	5.1	1.1	0.0	0.0	...	Investissements direct, nia
Portfolio Investments, nie	-1.2	-0.7	0.0	1.1	0.0	0.0	...	Invest. portefeuille, nia
Other Capital, nie	88.0	219.6	115.3	481.3	476.1	510.9	...	Autres capitaux, nia
Net Errors and Omissions	112.1	21.8	77.9	36.6	-58.8	57.9	...	Erreurs et omissions nettes
Reserves and Related Items	95.2	43.7	-72.9	-67.7	277.0	10.3	...	Rés. et postes assimilables
Yugoslavia								**Yougoslavie**
Merchandise: Exp. fob	10136.0	10622.0	11084.0	11425.0	12779.0	13560.0	14308.0	Marchandises : exp. fob
Merchandise: Imp. fob	-10925.0	-11210.0	-11786.0	-11343.0	-12000.0	-13502.0	-16984.0	Marchandises : imp. fob
Serv. & Income: Credit	3311.0	3458.0	4301.0	4465.0	4902.0	5844.0	7163.0	Serv. & revenu : crédit
Serv. & Income: Debit	-5385.0	-5316.0	-6430.0	-7578.0	-8063.0	-10117.0	-16679.0	Serv. & revenu : débit
Private Unrequited Transfers	3343.0	3281.0	3933.0	4281.0	4871.0	6645.0	9830.0	Transf. priv.sans contrep.
Offic. Unrequited Trans.,nie	-2.0	-2.0	-2.0	-2.0	-2.0	-3.0	-2.0	Tr. offic. sans contrep.nia
Direct Investment, nie	0.0	0.0	0.0	0.0	0.0	0.0	0.0	Investissements direct, nia
Portfolio Investments, nie	0.0	0.0	0.0	0.0	0.0	0.0	0.0	Invest. portefeuille, nia
Other Capital, nie	-586.0	-757.0	-613.0	-758.0	-683.0	-697.0	3504.0	Autres capitaux, nia
Net Errors and Omissions	211.0	83.0	349.0	-252.0	149.0	201.0	228.0	Erreurs et omissions nettes
Reserves and Related Items	-103.0	-159.0	-836.0	-238.0	-1953.0	-1931.0	-1368.0	Rés. et postes assimilables
Zaire								**Zaïre**
Merchandise: Exp. fob	1919.0	1853.0	1844.0	1731.0	2178.0	2201.0	2138.0	Marchandises : exp. fob
Merchandise: Imp. fob	-1176.0	-1247.0	-1283.0	-1376.0	-1645.0	-1683.0	-1539.0	Marchandises : imp. fob
Serv. & Income: Credit	141.0	153.0	189.0	262.0	185.0	165.0	171.0	Serv. & revenu : crédit
Serv. & Income: Debit	-1292.0	-1193.0	-1286.0	-1412.0	-1458.0	-1461.0	-1549.0	Serv. & revenu : débit
Private Unrequited Transfers	-91.0	-55.0	-62.0	-70.0	-67.0	-109.0	-81.0	Transf. priv.sans contrep.
Offic. Unrequited Trans.,nie	174.0	199.0	184.0	220.0	226.0	276.0	217.0	Tr. offic. sans contrep.nia
Direct Investment, nie	0.0	0.0	0.0	0.0	0.0	0.0	0.0	Investissements direct, nia
Portfolio Investments, nie	0.0	0.0	0.0	0.0	0.0	0.0	0.0	Invest. portefeuille, nia
Other Capital, nie	-201.0	-92.0	-88.0	70.0	-8.0	-60.0	-222.0	Autres capitaux, nia

125
Summary of balance of payments
Millions of US dollars [*cont.*]
Résumé des balances des paiements
Millions de dollars des E-U [*suite*]

Country or area	1984	1985	1986	1987	1988	1989	1990	Pays ou zone
Net Errors and Omissions	-23.0	-1.0	-17.0	13.0	-134.0	113.0	105.0	Erreurs et omissions nettes
Reserves and Related Items	548.0	383.0	519.0	561.0	723.0	558.0	761.0	Rés. et postes assimilables
Zambia								**Zambie**
Merchandise: Exp. fob	893.0	797.0	692.0	852.0	1189.0	1340.0	...	Marchandises : exp. fob
Merchandise: Imp. fob	-612.0	-571.0	-518.0	-585.0	-687.0	-774.0	...	Marchandises : imp. fob
Serv. & Income: Credit	80.0	71.0	48.0	49.0	61.0	87.0	...	Serv. & revenu : crédit
Serv. & Income: Debit	-478.0	-667.0	-553.0	-553.0	-893.0	-915.0	...	Serv. & revenu : débit
Private Unrequited Transfers	-45.0	-33.0	-42.0	-20.0	-25.0	-30.0	...	Transf. priv.sans contrep.
Offic. Unrequited Trans.,nie	10.0	6.0	22.0	8.0	59.0	109.0	...	Tr. offic. sans contrep.nia
Direct Investment, nie	17.0	52.0	28.0	75.0	93.0	0.0	...	Investissements direct, nia
Portfolio Investments, nie	0.0	0.0	0.0	0.0	0.0	0.0	...	Invest. portefeuille, nia
Other Capital, nie	134.0	323.0	-37.0	32.0	-13.0	37.0	...	Autres capitaux, nia
Net Errors and Omissions	-87.0	-145.0	258.0	-116.0	-17.0	41.0	...	Erreurs et omissions nettes
Reserves and Related Items	89.0	169.0	101.0	258.0	232.0	105.0	...	Rés. et postes assimilables
Zimbabwe								**Zimbabwe**
Merchandise: Exp. fob	1173.6	1119.6	1322.7	1452.0	1664.9	...	...	Marchandises : exp. fob
Merchandise: Imp. fob	-989.3	-918.9	-1011.6	-1071.0	-1163.6	...	...	Marchandises : imp. fob
Serv. & Income: Credit	194.6	333.3	205.4	197.6	207.8	...	...	Serv. & revenu : crédit
Serv. & Income: Debit	-528.5	-629.2	-541.5	-578.2	-644.9	...	...	Serv. & revenu : débit
Private Unrequited Transfers	-38.5	-45.2	-26.6	-32.1	-13.2	...	...	Transf. priv.sans contrep.
Offic. Unrequited Trans.,nie	88.2	64.8	58.3	79.7	65.6	...	...	Tr. offic. sans contrep.nia
Direct Investment, nie	-2.5	2.9	7.5	-30.5	4.1	...	...	Investissements direct, nia
Portfolio Investments, nie	2.8	9.1	1.3	0.9	-60.9	...	...	Invest. portefeuille, nia
Other Capital, nie	8.4	108.8	109.4	89.5	104.8	...	...	Autres capitaux, nia
Net Errors and Omissions	44.9	37.2	-69.4	16.6	-62.9	...	...	Erreurs et omissions nettes
Reserves and Related Items	46.3	-82.2	-55.5	-124.5	-101.6	...	...	Rés. et postes assimilables

Source:
International Monetary Fund (Washington, DC).

₰ All data shown which pertain to Germany prior to 3 October 1990 are indicated separately for the Federal Republic of Germany and the former German Democratic Republic based on their respective territories at the time indicated. Where data for united Germany (3 October 1990 and thereafter) are not available, available data are shown separately under the designations Federal Republic of Germany and former German Democratic Republic and pertain to the territorial boundaries prior to 3 October 1990. For detailed explanatory notes on data pertaining to Germany, see Annex I - Country Nomenclature.

1 Billions of US dollars.
2 Excluding former Democratic Yemen.

Source:
Fonds monétaire international (Washington, DC).

₰ Toutes les données se rapportant à l'Allemagne avant le 3 octobre 1990 figurent dans deux rubriques séparées basées sur les territoires respectifs de la République fédérale d'Allemagne et d'Allemagne et l'ancienne République démocratique allemande selon la période indiquée. En l'absence de données pour l'Allemagne unifiée (à compter du 3 octobre 1990), les données disponibles sont fournies séparément sous les rubriques République fédérale d'Allemagne et ancienne République démocratique allemande et se rapportent aux limites territoriales antérieures au 3 octobre 1990. Pour les notes explicatives en détail sur les données concernant l'Allemagne, voir Annexe I - Nomenclature des pays.

1 Milliards de dollars des E-U.
2 Non compris l'ancien Yémen démocratique.

Technical notes, table 125

A balance of payments can be broadly described as the record of an economy's international economic transactions. It shows (a) transactions in goods, services and income between an economy and the rest of the world, (b) changes of ownership and other changes in that economy's monetary gold, special drawing rights (SDRs) and claims on and liabilities to the rest of the world, and (c) unrequited transfers and counterpart entries needed to balance in the accounting sense any entries for the foregoing transactions and changes which are not mutually offsetting.

The detailed definitions concerning the content of the basic categories of the balance of payments are given in the *Balance of Payments Manual (Fourth Edition)*.[39] Brief explanatory notes are given below to clarify the scope of the major items.

Merchandise covers exports and imports of goods and related distributive services at the customs frontiers of the exporting economy, i.e. f.o.b. value. International transactions in non-monetary gold are included in this item. Most merchandise exports and imports data are based on customs records.

Services and income covers transactions in real resources between residents and non-residents other than those classified as merchandise, including (a) shipment and other transportation services, including freight, insurance and other distributive services in connection with the movement of commodities, (b) travel, i.e. goods and services acquired by non-resident travellers in a given country and similar acquisitions by resident travellers abroad, and (c) investment income which covers income of non-residents from their financial assets invested in the compiling economy (debit) and similar income of residents from their financial assets invested abroad (credit).

Private unrequited transfers covers unrequited transfers between residents other than the general government and central bank of the compiling economy on the one hand, and non-residents other than general government and central banks on the other hand.

Official unrequited transfers covers inter-official transfers together with other transfers in which either the transferor or the transferee is the general government or central bank whether of the compiling country or of a foreign country.

Direct investment covers all capital transactions between direct investment enterprises and the direct investors themselves or any of those investors' other direct investment enterprises.

Portfolio investment covers investment in long-term bonds and corporate equities other than "direct investment" and "reserves".

Notes techniques, tableau 125

La balance des paiements peut se définir d'une façon générale comme le relevé des transactions économiques internationales d'une économie. Elle indique (a) les transactions sur biens, services et revenus entre une économie et le reste du monde, (b) les transferts de propriété et autres variations intervenues dans les avoirs en or monétaire de cette économie, dans ses avoirs en droits de tirages spéciaux (DTS) ainsi que dans ses créances financières sur le reste du monde ou dans ses engagements financiers envers lui et (c) les "inscriptions de transferts sans contrepartie" et de "contrepartie" destinées à équilibrer, d'un point de vue comptable, les transactions et changements précités qui ne se compensent pas réciproquement.

Les définitions détaillées relatives au contenu des postes fondamentaux de la balance des paiements figurent dans le *Manuel de la balance des paiements (Quatrième édition)* [39]. De brèves notes explicatives sont présentées ci-après pour clarifier la portée de ces principales rubriques.

Marchandise : exportations et importations de biens et de services de distribution y afférents à la frontière douanière de l'économie exportatrice, c'est-à-dire leur valeur f.o.b. Les transactions internationales sur or non monétaire sont comprises sous ce poste. La plupart des données relatives aux exportations et importations de marchandises sont tirées des registres douaniers.

Services et revenus : transactions en ressources effectuées entre résidents et non résidents, autres que celles qui sont considérées comme des marchandises, notamment : (a) expéditions et autres services de transport, y compris le fret, l'assurance et les autres services de distribution liés aux mouvements de marchandises; (b) voyages, à savoir les biens et services acquis par des voyageurs non résidents dans un pays donné et achats similaires faits par des résidents voyageant à l'étranger; et (c) revenus des investissements, qui correspondent aux revenus que les non-résidents tirent de leurs avoirs financiers placés dans l'économie déclarante (débit) et les revenus similaires que les résidents tirent de leurs avoirs financiers placés à l'étranger (crédit).

Transferts privés sans contrepartie : transferts sans contrepartie entre les résidents de l'économie déclarante autres que le gouvernement et la banque centrale, d'une part, et les non-résidents autres que les gouvernements et les banques centrales, d'autre part.

Transferts officiels sans contrepartie : transferts entre secteurs officiels, ainsi que les transferts où l'une des parties est le gouvernement ou la banque centrale de l'économie déclarante ou d'une économie étrangère.

Investissements directs : toutes les transactions sur capitaux entre, d'une part, les entreprises d'investissement direct et, d'autre part, les investisseurs directs eux-mêmes ou toute entreprise où ils ont effectué des investissements directs.

Other capital (long-term and short-term) covers transactions in all financial assets and liabilities other than those specified in the table.

Reserves and related items covers those assets (together with the use of Fund credit) that are conceived of as available for use by an economy's central authorities in meeting balance of payments needs.

Investissements de portefeuille : investissements en obligations à long terme et actions de société autres que les "investissements directs" et les "réserves".

Autres capitaux (à long et à cour terme): transactions sur tous les avoirs et engagements financiers autres que ceux spécifiés dans le tableau.

Réserves et postes assimilables : variations de ces actifs (ainsi que l'utilisation des crédits du Fonds) qui sont considérés comme disponibles par les responsables de l'économie pour répondre aux besoins de la balance des paiements.

126
Exchange rates
Cours des changes
National currency per US dollar
Valeur du dollar des Etats-Unis en monnaie nationale

Country (monetary unit) Pays (unité monétaire)	1982	1983	1984	1985	1986	1987	1988	1989	1990	1991
Afghanistan: afghani Afghanistan : afghani										
End of period										
Fin de période	50.600	50.600	50.600	50.600	50.600	50.600	50.600	50.600	50.600	50.600
Period average										
Moyenne sur période	50.600	50.600	50.600	50.600	50.600	50.600	50.600	50.600	50.600	50.600
Albania: lek Albanie : lek										
End of period[1]										
Fin de période[1]	7.000	7.000	7.000	7.000	7.000	7.000	6.000	6.400	15.000	...
Algeria: dinar Algérie : dinar										
End of period										
Fin de période	4.635	4.916	5.123	4.773	4.823	4.936	6.731	8.032	12.191	21.392
Period average										
Moyenne sur période	4.592	4.789	4.983	5.028	4.702	4.850	5.915	7.609	8.957	18.473
Antigua and Barbuda: EC dollar Antigua-et-Barbuda : EC dollar										
End of period										
Fin de période	2.700	2.700	2.700	2.700	2.700	2.700	2.700	2.700	2.700	2.700
Argentina: peso Argentine : peso										
End of period										
Fin de période	0.000	0.000	0.000	0.000	0.000	0.000	0.001	0.179	0.558	0.998
Period average										
Moyenne sur période	0.000	0.000	0.000	0.000	0.000	0.000	0.001	0.042	0.488	0.954
Aruba: florin Aruba : florin										
End of period										
Fin de période	...	...	...	...	1.790	1.790	1.790	1.790	1.790	1.790
Period average										
Moyenne sur période	...	...	...	...	1.790	1.790	1.790	1.790	1.790	1.790
Australia: dollar Australie : dollar										
End of period										
Fin de période	1.020	1.109	1.208	1.469	1.504	1.384	1.169	1.262	1.293	1.316
Period average										
Moyenne sur période	0.986	1.110	1.140	1.432	1.496	1.428	1.280	1.265	1.281	1.284
Austria: schilling Autriche : schilling										
End of period										
Fin de période	16.687	19.341	22.050	17.280	13.710	11.250	12.565	11.815	10.677	10.689
Period average										
Moyenne sur période	17.059	17.963	20.009	20.689	15.267	12.642	12.348	13.231	11.370	11.676
Bahamas: dollar Bahamas : dollar										
End of period[2]										
Fin de période[2]	1.000	1.000	1.000	1.000	1.000	1.000	1.000	1.000	1.000	1.000
Period average[2]										
Moyenne sur période[2]	...	...	...	1.180	1.175	1.199	1.225	1.225	1.225	1.225
Bahrain: dinar Bahreïn : dinar										
End of period										
Fin de période	0.376	0.376	0.376	0.376	0.376	0.376	0.376	0.376	0.376	0.376
Period average										
Moyenne sur période	0.376	0.376	0.376	0.376	0.376	0.376	0.376	0.376	0.376	0.376
Bangladesh: taka Bangladesh : taka										
End of period										
Fin de période	24.074	25.000	26.000	31.000	30.800	31.200	32.270	32.270	35.790	38.580
Period average										
Moyenne sur période	22.118	24.615	25.354	27.995	30.407	30.950	31.733	32.270	34.569	36.596
Barbados: dollar Barbade : dollar										
End of period										
Fin de période	2.011	2.011	2.011	2.011	2.011	2.011	2.011	2.011	2.011	2.011

126
Exchange rates
National currency per US dollar [cont.]
Cours des changes
Valeur du dollar des Etats-Unis en monnaie nationale [suite]

Country (monetary unit) Pays (unité monétaire)	1982	1983	1984	1985	1986	1987	1988	1989	1990	1991
Belgium: franc Belgique : franc										
End of period Fin de période	46.920	55.640	63.080	50.360	40.410	33.153	37.345	35.760	30.982	31.270
Period average Moyenne sur période	45.691	51.132	57.784	59.378	44.672	37.334	36.768	39.404	33.418	34.148
Belize: dollar Belize : dollar										
End of period Fin de période	2.000	2.000	2.000	2.000	2.000	2.000	2.000	2.000	2.000	2.000
Benin: CFA franc Bénin : franc CFA										
End of period Fin de période	336.250	417.370	479.600	378.050	322.750	267.000	302.950	289.400	256.450	259.000
Period average Moyenne sur période	328.600	381.060	436.950	449.260	346.300	300.540	297.850	319.010	272.260	282.110
Bhutan: ngultrum Bhoutan : ngultrum										
End of period Fin de période	...	...	...	12.166	13.122	12.877	14.949	17.035	18.073	25.834
Period average Moyenne sur période	...	...	...	12.369	12.611	12.962	13.917	16.226	17.504	22.742
Bolivia: boliviano Bolivie : boliviano										
End of period Fin de période	0.000	0.000	0.009	1.692	1.923	2.210	2.470	2.980	3.400	3.745
Period average Moyenne sur période	0.000	0.000	0.003	0.440	1.922	2.055	2.350	2.692	3.173	3.581
Botswana: pula Botswana : pula										
End of period Fin de période	1.061	1.155	1.560	2.101	1.838	1.566	1.936	1.872	1.871	2.072
Period average Moyenne sur période	1.030	1.097	1.298	1.903	1.879	1.679	1.829	2.015	1.860	2.022
Brazil: cruzeiro Brésil : cruzeiro										
End of period[3] Fin de période[3]	0.000	-	0.003	0.010	0.015	0.072	0.765	11.358	177.060	1 068.800
Period average[3] Moyenne sur période[3]	0.000	-	0.002	0.006	0.014	0.039	0.262	2.834	68.300	406.609
Bulgaria: lev Bulgarie : lev										
End of period[1] Fin de période[1]	0.800	1.000	1.000	1.000	1.200	1.300	1.700	2.100	2.800	...
Burkina Faso: CFA franc Burkina Faso : franc CFA										
End of period Fin de période	336.250	417.370	479.600	378.050	322.750	267.000	302.950	289.400	256.450	259.000
Period average Moyenne sur période	328.600	381.060	436.950	449.260	346.300	300.540	297.850	319.010	272.260	282.110
Burundi: franc Burundi : franc										
End of period Fin de période	90.000	117.410	124.950	111.965	124.165	114.470	149.940	175.430	165.350	190.970
Period average Moyenne sur période	90.000	92.948	119.709	120.691	114.171	123.564	140.395	158.667	171.255	181.513
Cameroon: CFA franc Cameroun : franc CFA										
End of period Fin de période	336.250	417.370	479.600	378.050	322.750	267.000	302.950	289.400	256.450	259.000
Period average Moyenne sur période	328.600	381.060	436.950	449.260	346.300	300.540	297.850	319.010	272.260	282.110
Canada: dollar Canada : dollar										
End of period Fin de période	1.229	1.244	1.321	1.397	1.380	1.300	1.193	1.158	1.160	1.156
Period average Moyenne sur période	1.234	1.232	1.295	1.365	1.389	1.326	1.231	1.184	1.167	1.146

126
Exchange rates
National currency per US dollar [*cont.*]
Cours des changes
Valeur du dollar des Etats-Unis en monnaie nationale [*suite*]

Country (monetary unit) Pays (unité monétaire)	1982	1983	1984	1985	1986	1987	1988	1989	1990	1991
Cape Verde: escudo Cap-Vert : escudo										
End of period										
Fin de période	63.040	79.975	93.015	85.375	76.565	65.775	73.665	73.045	66.085	66.470
Period average										
Moyenne sur période	58.293	71.686	84.878	91.632	80.145	72.466	72.067	77.978	70.031	71.408
Central African Rep.: CFA franc Rép. centrafricaine : franc CFA										
End of period										
Fin de période	336.250	417.370	479.600	378.050	322.750	267.000	302.950	289.400	256.450	259.000
Period average										
Moyenne sur période	328.600	381.060	436.950	449.260	346.300	300.540	297.850	319.010	272.260	282.110
Chad: CFA franc Tchad : franc CFA										
End of period										
Fin de période	336.250	417.370	479.600	378.050	322.750	267.000	302.950	289.400	256.450	259.000
Period average										
Moyenne sur période	328.600	381.060	436.950	449.260	346.300	300.540	297.850	319.010	272.260	282.110
Chile: peso Chili : peso										
End of period										
Fin de période	73.430	87.530	128.240	183.860	204.730	238.140	247.200	297.370	337.090	374.510
Period average										
Moyenne sur période	50.909	78.842	98.656	161.081	193.016	219.540	245.047	267.155	305.062	349.372
China: yuan renminbi Chine : yuan renminbi										
End of period										
Fin de période	1.923	1.981	2.796	3.201	3.722	3.722	3.722	4.722	5.222	5.434
Period average										
Moyenne sur période	1.892	1.976	2.320	2.937	3.453	3.722	3.722	3.765	4.783	5.323
Colombia: peso Colombie : peso										
End of period[3]										
Fin de période[3]	70.290	88.770	113.890	172.200	219.000	263.700	335.860	433.920	568.730	706.860
Period average[3]										
Moyenne sur période[3]	64.085	78.854	100.817	142.312	194.261	242.607	299.174	382.568	502.259	633.045
Comoros: franc Comores : franc										
End of period										
Fin de période	336.250	417.375	479.600	378.050	322.750	267.000	302.950	289.400	256.450	259.000
Period average										
Moyenne sur période	328.605	381.065	436.955	449.261	346.305	300.535	297.847	319.007	272.264	282.106
Congo: CFA franc Congo : franc CFA										
End of period										
Fin de période	336.250	417.370	479.600	378.050	322.750	267.000	302.950	289.400	256.450	259.000
Period average										
Moyenne sur période	328.600	381.060	436.950	449.260	346.300	300.540	297.850	319.010	272.260	282.110
Costa Rica: colon Costa Rica : colon										
End of period										
Fin de période	40.250	43.400	47.750	53.700	58.875	69.250	79.500	84.350	103.550	135.425
Period average										
Moyenne sur période	37.407	41.094	44.533	50.453	55.986	62.776	75.805	81.504	91.579	122.432
Côte d'Ivoire: CFA franc Côte d'Ivoire : franc CFA										
End of period										
Fin de période	336.250	417.370	479.600	378.050	322.750	267.000	302.950	289.400	256.450	259.000
Period average										
Moyenne sur période	328.600	381.060	436.950	449.260	346.300	300.540	297.850	319.010	272.260	282.110
Cuba: peso Cuba : peso										
End of period[1]										
Fin de période[1]	0.800	0.900	0.900	0.900	0.800	0.800	0.800	0.800	0.700	...
Cyprus: pound Chypre : livre										
End of period										
Fin de période	0.488	0.556	0.644	0.543	0.512	0.439	0.466	0.479	0.435	0.439

126
Exchange rates
National currency per US dollar [cont.]
Cours des changes
Valeur du dollar des Etats-Unis en monnaie nationale [suite]

Country (monetary unit) Pays (unité monétaire)	1982	1983	1984	1985	1986	1987	1988	1989	1990	1991
Period average Moyenne sur période	0.475	0.527	0.588	0.613	0.518	0.481	0.467	0.494	0.458	0.458
Czechoslovakia: koruna Tchécoslovaquie : couronne										
End of period Fin de période	14.040	14.630	17.130	16.000	14.380	13.000	14.310	14.290	28.000	27.840
Period average Moyenne sur période	13.710	14.160	16.610	17.140	14.990	13.690	14.360	15.050	17.950	29.480
Denmark: krone Danemark : couronne										
End of period Fin de période	8.384	9.875	11.260	8.969	7.342	6.096	6.874	6.607	5.776	5.913
Period average Moyenne sur période	8.332	9.145	10.357	10.596	8.091	6.840	6.731	7.310	6.189	6.396
Djibouti: franc Djibouti : franc										
End of period Fin de période	177.721	177.721	177.721	177.721	177.721	177.721	177.721	177.721	177.721	177.721
Period average Moyenne sur période	177.721	177.721	177.721	177.721	177.721	177.721	177.721	177.721	177.721	177.721
Dominica: EC dollar Dominique : EC dollar										
End of period Fin de période	2.700	2.700	2.700	2.700	2.700	2.700	2.700	2.700	2.700	2.700
Dominican Republic: peso Rép. dominicaine : peso										
End of period Fin de période	1.000[2]	1.000[2]	1.000[2]	2.940	3.077	4.960	6.340	6.340	11.350	12.660
Period average Moyenne sur période	1.000[2]	1.000[2]	1.000[2]	3.113	2.904	3.845	6.112	6.340	8.525	12.692
Ecuador: sucre Equateur : sucre										
End of period Fin de période	33.150	54.100	67.170	95.750	146.500	221.500	432.510	648.420	878.200	1 270.580
Period average Moyenne sur période	30.030	44.120	62.540	69.560	122.780	170.460	301.610	526.350	767.750	1 046.250
Egypt: £ Egypte : £										
End of period Fin de période	0.700	0.700	0.700	0.700	0.700	0.700	0.700	1.100	2.000	3.330
Period average Moyenne sur période	...	...	...	1.300	1.350	1.518	2.223	2.517	2.708	...
El Salvador: colon El Salvador : colon										
End of period Fin de période	2.500	2.500	2.500	2.500	5.000	5.000	5.000	5.000	8.030	8.080
Period average Moyenne sur période	2.500	2.500	2.500	2.500	4.852	5.000	5.000	5.000	6.848	8.017
Equatorial Guinea: CFA franc Guinée équatoriale : franc CFA										
End of period[4] Fin de période[4]	251.200	313.400	347.660	378.050	322.750	267.000	302.950	289.400	256.450	259.000
Period average[4] Moyenne sur période[4]	219.720	286.860	321.520	449.260	346.300	300.540	297.850	319.010	272.260	282.110
Ethiopia: birr Ethiopie : birr										
End of period Fin de période	2.070	2.070	2.070	2.070	2.070	2.070	2.070	2.070	2.070	2.070
Fiji: dollar Fidji : dollar										
End of period Fin de période	0.947	1.046	1.143	1.120	1.145	1.441	1.405	1.494	1.459	1.473
Period average Moyenne sur période	0.932	1.017	1.083	1.154	1.133	1.244	1.430	1.483	1.481	1.476
Finland: markka Finlande : markka										
End of period Fin de période	5.291	5.810	6.530	5.417	4.794	3.946	4.169	4.059	3.634	4.133

126
Exchange rates
National currency per US dollar [cont.]
    Cours des changes
    Valeur du dollar des Etats-Unis en monnaie nationale [suite]

Country (monetary unit) Pays (unité monétaire)	1982	1983	1984	1985	1986	1987	1988	1989	1990	1991
Period average Moyenne sur période	4.820	5.570	6.010	6.198	5.069	4.396	4.183	4.291	3.823	4.044
France: franc France : franc										
End of period Fin de période	6.725	8.347	9.592	7.561	6.455	5.340	6.059	5.788	5.129	5.180
Period average Moyenne sur période	6.572	7.621	8.739	8.985	6.926	6.011	5.957	6.380	5.445	5.642
Gabon: CFA franc Gabon : franc CFA										
End of period Fin de période	336.250	417.370	479.600	378.050	322.750	267.000	302.950	289.400	256.450	259.000
Period average Moyenne sur période	328.600	381.060	436.950	449.260	346.300	300.540	297.850	319.010	272.260	282.110
Gambia: dalasi Gambie : dalasi										
End of period Fin de période	2.478	2.757	4.323	3.461	7.426	6.439	6.659	8.315	7.495	8.957
Period average Moyenne sur période	2.290	2.639	3.584	3.894	6.938	7.074	6.709	7.585	7.883	8.803
Germany† · Allemagne†										
F. R. Germany: deutsche mark R. f. Allemagne : deutsche mark										
End of period Fin de période	2.376	2.724	3.148	2.461	1.941	1.581	1.780	1.698	1.494	1.516
Period average Moyenne sur période	2.427	2.553	2.846	2.944	2.171	1.797	1.756	1.880	1.616	1.660
former German D. R.: mark anc. R. d. allemande : mark										
End of period¹ Fin de période¹	2.500	2.700	3.000	2.600	2.000	1.600	1.900	1.800	1.500	...
Ghana: cedi Ghana : cedi										
End of period Fin de période	2.750	30.003	50.000	59.988	90.009	176.056	229.885	303.030	344.828	...
Period average Moyenne sur période	2.750	8.830	35.986	54.365	89.204	153.733	202.346	270.000	326.332	...
Greece: drachma Grèce : drachme										
End of period Fin de période	70.570	98.670	128.480	147.760	138.760	125.925	148.100	157.790	157.625	175.280
Period average Moyenne sur période	66.803	88.064	112.717	138.119	139.981	135.429	141.860	162.417	158.514	182.266
Grenada: EC dollar Grenade : EC dollar										
End of period Fin de période	2.700	2.700	2.700	2.700	2.700	2.700	2.700	2.700	2.700	2.700
Guatemala: quetzal Guatemala : quetzal										
End of period Fin de période	1.000	1.000	1.000	1.000	2.500	2.500	2.705	3.400	5.015	5.043
Period average Moyenne sur période	1.000	1.000	1.000	1.000	1.875	2.500	2.620	2.816	4.486	5.029
Guinea: franc Guinée : franc										
End of period Fin de période	22.378	23.578	25.184	22.473	235.630	440.000	550.000	620.000	680.000	802.947
Period average Moyenne sur période	22.366	23.095	24.090	24.333	333.452	428.402	474.396	591.646	660.167	753.858
Guinea-Bissau: peso Guinée-Bissau : peso										
End of period Fin de période	39.890	84.160	128.580	173.610	238.980	851.650	1 363.580	1 987.200	2 508.620	...
Period average Moyenne sur période	39.870	42.100	105.290	159.620	203.950	559.330	1 111.060	1 811.420	2 185.450	...

126
Exchange rates
National currency per US dollar [cont.]
Cours des changes
Valeur du dollar des Etats-Unis en monnaie nationale [suite]

Country (monetary unit) Pays (unité monétaire)	1982	1983	1984	1985	1986	1987	1988	1989	1990	1991
Guyana: dollar Guyana : dollar										
End of period Fin de période	3.000	3.000	4.150	4.150	4.400	10.000	10.000	33.000	45.000	120.000
Period average Moyenne sur période	3.000	3.000	3.832	4.252	4.272	9.756	10.000	27.159	39.533	111.811
Haïti: gourde Haïti : gourde										
End of period[2] Fin de période[2]	5.000	5.000	5.000	5.000	5.000	5.000	5.000	5.000	5.000	...
Honduras: lempira Honduras : lempira										
End of period[2] Fin de période[2]	2.000	2.000	2.000	2.000	2.000	2.000	2.000	2.000	2.000	5.400
Period average[2] Moyenne sur période[2]	2.000	2.000	2.000	2.000	2.000	2.000	2.000	2.000	4.112	5.317
Hong Kong: dollar Hong-kong : dollar										
End of period Fin de période	6.495	7.780	7.823	7.811	7.795	7.760	7.808	7.807	7.801	7.781
Period average Moyenne sur période	6.070	7.265	7.818	7.791	7.803	7.798	7.806	7.800	7.790	7.771
Hungary: forint Hongrie : forint										
End of period Fin de période	39.610	45.193	51.199	47.347	45.927	46.387	52.537	62.543	61.449	75.620
Period average Moyenne sur période	36.631	42.671	48.042	50.119	45.832	46.971	50.413	59.066	63.206	74.735
Iceland: krona Islande : couronne										
End of period Fin de période	16.625	28.670	40.545	42.060	40.240	35.660	46.220	61.170	55.390	55.620
Period average Moyenne sur période	12.351	24.843	31.694	41.508	41.104	38.677	43.014	57.042	58.284	58.996
India: rupee Inde : roupie										
End of period Fin de période	9.634	10.493	12.451	12.165	13.122	12.877	14.949	17.035	18.073	25.834
Period average Moyenne sur période	9.455	10.099	11.363	12.369	12.611	12.961	13.917	16.225	17.503	22.742
Indonesia: rupiah Indonésie : rupiah										
End of period Fin de période	692.500	994.000	1 074.000	1 125.000	1 641.000	1 650.000	1 731.000	1 797.000	1 901.000	1 992.000
Period average Moyenne sur période	661.420	909.260	1 025.940	1 110.580	1 282.560	1 643.850	1 685.700	1 770.060	1 842.810	1 950.300
Iran, Islamic Rep. of: rial Iran, Rép. islamique d' : rial										
End of period Fin de période	83.433	88.161	93.993	84.228	75.644	65.622	68.589	70.235	65.307	64.591
Period average Moyenne sur période	83.602	86.358	90.030	91.052	78.760	71.460	68.683	72.015	68.096	67.505
Iraq: dinar Iraq : dinar										
End of period Fin de période	0.311	0.311	0.311	0.311	0.311	0.311	0.311	0.311	0.311	0.311
Period average Moyenne sur période	0.299	0.311	0.311	0.311	0.311	0.311	0.311	0.311	0.311	0.311
Ireland: pound Irlande : livre										
End of period Fin de période	0.716	0.881	1.009	0.804	0.715	0.597	0.663	0.643	0.563	0.571
Period average Moyenne sur période	0.705	0.805	0.923	0.946	0.743	0.673	0.656	0.706	0.605	0.621
Israel: new shekel Israël : nouveau shekel										
End of period Fin de période	0.034	0.108	0.639	1.499	1.486	1.539	1.685	1.963	2.048	2.283

126
Exchange rates
National currency per US dollar [cont.]
Cours des changes
Valeur du dollar des Etats-Unis en monnaie nationale [suite]

Country (monetary unit) Pays (unité monétaire)	1982	1983	1984	1985	1986	1987	1988	1989	1990	1991
Period average Moyenne sur période	0.024	0.056	0.293	1.179	1.488	1.595	1.599	1.916	2.016	2.279
Italy: lira Italie : lire										
End of period Fin de période	1 370.000	1 659.500	1 935.880	1 678.500	1 358.130	1 169.250	1 305.770	1 270.500	1 130.150	1 151.060
Period average Moyenne sur période	1 352.510	1 518.850	1 756.960	1 909.440	1 490.810	1 296.070	1 301.630	1 372.090	1 198.100	1 240.610
Jamaica: dollar Jamaïque : dollar										
End of period Fin de période	1.781	3.278	4.930	5.480	5.480	5.500	5.480	6.480	8.038	21.492
Period average Moyenne sur période	1.781	1.932	3.943	5.559	5.478	5.487	5.489	5.745	7.184	12.116
Japan: yen Japon : yen										
End of period Fin de période	235.000	232.200	251.100	200.500	159.100	123.500	125.850	143.450	134.400	125.200
Period average Moyenne sur période	249.080	237.510	237.520	238.540	168.520	144.640	128.150	137.960	144.790	134.710
Jordan: dinar Jordanie : dinar										
End of period Fin de période	0.351	0.371	0.405	0.368	0.344	0.329	0.477	0.648	0.665	0.675
Period average Moyenne sur période	0.352	0.363	0.384	0.395	0.350	0.338	0.374	0.575	0.664	0.681
Kenya: shilling Kenya : shilling										
End of period Fin de période	12.725	13.796	15.781	16.284	16.042	16.515	18.599	21.601	24.084	28.074
Period average Moyenne sur période	10.922	13.311	14.414	16.432	16.226	16.454	17.747	20.572	22.915	27.508
Kiribati: dollar Kiribati : dollar										
End of period Fin de période	1.020	1.109	1.208	1.469	1.504	1.384	1.169	1.261	1.293	1.316
Period average Moyenne sur période	0.986	1.110	1.140	1.432	1.496	1.428	1.280	1.265	1.281	1.284
Korea, Republic of: won Corée, République de : won										
End of period Fin de période	748.800	795.500	827.400	890.200	861.400	792.300	684.100	679.600	716.400	760.800
Period average Moyenne sur période	731.080	775.750	805.980	870.020	881.450	822.570	731.470	671.460	707.760	733.350
Kuwait: dinar Koweït : dinar										
End of period Fin de période	0.289	0.293	0.304	0.289	0.292	0.270	0.283	0.292	...	0.284
Period average Moyenne sur période	0.288	0.291	0.296	0.301	0.291	0.279	0.279	0.294	...	0.284
Lao People's Dem. Rep.: new kip Rép. dém. pop. lao : nouveau kip										
End of period Fin de période	35.000	35.000	35.000	95.000	95.000	367.500	452.500	713.500	695.500	711.500
Period average Moyenne sur période	35.000	35.000	35.000	45.000	95.000	175.117	392.012	583.015	708.570	702.500
Lebanon: pound Liban : livre										
End of period Fin de période	3.810	5.490	8.890	18.100	87.000	455.000	530.000	505.000	842.000	879.000
Period average Moyenne sur période	4.744	4.528	6.511	16.417	38.370	224.596	409.230	496.689	695.089	928.227
Lesotho: loti Lesotho : maloti										
End of period Fin de période	1.076	1.222	1.985	2.558	2.183	1.930	2.378	2.536	2.563	2.743
Period average Moyenne sur période	1.086	1.114	1.475	2.229	2.285	2.036	2.273	2.623	2.587	2.761

126
Exchange rates
National currency per US dollar [cont.]
    Cours des changes
    Valeur du dollar des Etats-Unis en monnaie nationale [suite]

Country (monetary unit) Pays (unité monétaire)	1982	1983	1984	1985	1986	1987	1988	1989	1990	1991
Liberia: dollar Libéria : dollar										
End of period [2] Fin de période[2]	1.000	1.000	1.000	1.000	1.000	1.000	1.000	1.000	1.000	1.000
Libyan Arab Jamah.: dinar Jamah. arabe libyenne : dinar										
End of period Fin de période	0.296	0.296	0.296	0.296	0.314	0.271	0.285	0.292	0.270	0.268
Period average Moyenne sur période	0.296	0.296	0.296	0.296	0.315	0.295	0.286	0.299	0.283	0.281
Luxembourg: franc Luxembourg : franc										
End of period Fin de période	46.920	55.640	63.080	50.360	40.410	33.153	37.345	35.760	30.982	31.270
Period average Moyenne sur période	45.691	51.132	57.784	59.378	44.672	37.334	36.768	39.404	33.418	34.148
Madagascar: franc Madagascar : franc										
End of period Fin de période	367.710	492.160	658.020	635.790	769.810	1 234.270	1 526.430	1 532.540	1 465.830	1 832.660
Period average Moyenne sur période	349.740	430.450	576.640	662.480	676.340	1 069.210	1 407.110	1 603.440	1 494.150	1 835.360
Malawi: kwacha Malawi : kwacha										
End of period Fin de période	1.097	1.299	1.565	1.679	1.952	2.054	2.535	2.679	2.647	2.664
Period average Moyenne sur période	1.056	1.175	1.413	1.719	1.861	2.209	2.561	2.760	2.729	2.803
Malaysia: ringgit Malaisie : ringgit										
End of period Fin de période	2.321	2.338	2.425	2.426	2.603	2.493	2.715	2.703	2.701	2.724
Period average Moyenne sur période	2.335	2.321	2.344	2.483	2.581	2.520	2.619	2.709	2.705	2.750
Maldives: rufiyaa Maldives : rufiyaa										
End of period Fin de période	7.050	7.050	7.050	7.129	7.244	9.395	8.525	9.205	9.620	10.320
Period average Moyenne sur période	7.174	7.050	7.050	7.098	7.151	9.223	8.785	9.041	9.552	10.253
Mali: CFA franc Mali : franc CFA										
End of period Fin de période	336.250	417.370	479.600	378.050	322.750	267.000	302.950	289.400	256.450	259.000
Period average Moyenne sur période	328.600	381.060	436.950	449.260	346.300	300.540	297.850	319.010	272.260	282.110
Malta: lira Malte : lire										
End of period Fin de période	0.415	0.445	0.492	0.424	0.369	0.312	0.332	0.337	0.301	0.306
Period average Moyenne sur période	0.412	0.432	0.461	0.469	0.393	0.345	0.331	0.348	0.318	0.323
Mauritania: ouguiya Mauritanie : ouguiya										
End of period Fin de période	52.960	57.030	67.290	77.070	74.080	71.600	75.730	83.550	77.840	77.820
Period average Moyenne sur période	51.769	54.812	63.803	77.085	74.375	73.878	75.261	83.051	80.609	81.946
Mauritius: rupee Maurice : roupie										
End of period Fin de période	10.860	12.723	15.603	14.310	13.137	12.175	13.834	14.996	14.322	14.794
Period average Moyenne sur période	10.872	11.706	13.800	15.442	13.466	12.878	13.438	15.250	14.863	15.652
Mexico: peso Mexique : peso										
End of period Fin de période	96.480	143.930	192.560	371.700	923.500	2 209.700	2 281.000	2 641.000	2 945.400	3 071.000

126
Exchange rates
National currency per US dollar [cont.]
Cours des changes
Valeur du dollar des Etats-Unis en monnaie nationale [suite]

Country (monetary unit) Pays (unité monétaire)	1982	1983	1984	1985	1986	1987	1988	1989	1990	1991
Period average Moyenne sur période	56.400	120.090	167.830	256.870	611.770	1 378.180	2 273.100	2 461.470	2 812.600	3 018.430
Mongolia: tugrik Mongolie : tughrik										
End of period Fin de période[1]	3.300	3.400	3.800	3.600	3.000	2.800	3.000	3.000	5.600	...
Morocco: dirham Maroc: dirham										
End of period Fin de période	6.267	8.061	9.551	9.621	8.712	7.800	8.211	8.122	8.043	8.150
Period average Moyenne sur période	6.023	7.111	8.810	10.062	9.104	8.359	8.209	8.488	8.242	8.706
Myanmar: kyat Myanmar: kyat										
End of period Fin de période	7.777	8.223	8.751	7.842	7.039	6.110	6.410	6.494	6.080	6.014
Period average Moyenne sur période	7.790	8.035	8.386	8.475	7.330	6.653	6.395	6.705	6.339	6.284
Nepal: rupee Népal : roupie										
End of period Fin de période	14.300	15.200	18.000	20.700	22.000	21.600	25.200	28.600	30.400	42.700
Period average Moyenne sur période	13.244	14.545	16.459	18.246	21.230	21.819	23.289	27.189	29.369	37.255
Netherlands: guilder Pays-Bas : florin										
End of period Fin de période	2.624	3.064	3.549	2.772	2.192	1.777	1.999	1.915	1.690	1.710
Period average Moyenne sur période	2.670	2.854	3.209	3.321	2.450	2.026	1.977	2.121	1.821	1.870
Netherlands Antilles: guilder Antilles néerlandaises : florin										
End of period Fin de période	1.800	1.800	1.800	1.800	1.800	1.800	1.800	1.790	1.790	1.790
New Zealand: dollar Nouvelle-Zélande : dollar										
End of period Fin de période	1.365	1.528	2.094	2.006	1.910	1.521	1.592	1.674	1.701	1.848
Period average Moyenne sur période	1.333	1.497	1.764	2.023	1.913	1.695	1.526	1.672	1.676	1.734
Nicaragua: córdoba Nicaragua : córdoba										
End of period Fin de période	...	...	...	28.000[5]	70.000[5]	70.000[5]	920.000	38 150.000	3 000.000[5]	25 000.000[5]
Period average Moyenne sur période	...	...	...	27.000[5]	67.000[5]	70.000[5]	270.000	15 655.000	705.000[5]	21 354.000[5]
Niger: CFA franc Niger : franc CFA										
End of period Fin de période[2]	336.250	417.370	479.600	378.050	322.750	267.000	302.950	289.400	256.450	259.000
Period average Moyenne sur période[2]	328.600	381.060	436.950	449.260	346.300	300.540	297.850	319.010	272.260	282.110
Nigeria: naira Nigéria : naira										
End of period Fin de période	0.670	0.749	0.808	1.000	3.317	4.141	5.353	7.651	9.001	9.862
Period average Moyenne sur période	0.673	0.724	0.766	0.894	1.754	4.016	4.537	7.365	8.038	9.909
Norway: krone Norvège : couronne										
End of period Fin de période	7.054	7.722	9.087	7.582	7.400	6.232	6.570	6.615	5.907	5.973
Period average Moyenne sur période	6.454	7.296	8.161	8.597	7.395	6.737	6.517	6.904	6.260	6.483
Oman: rial Omani Oman : rial omani										
End of period Fin de période[2]	0.345	0.345	0.345	0.345	0.384	0.384	0.384	0.384	0.384	0.384

126
Exchange rates
National currency per US dollar [*cont.*]
Cours des changes
Valeur du dollar des Etats-Unis en monnaie nationale [*suite*]

Country (monetary unit) Pays (unité monétaire)	1982	1983	1984	1985	1986	1987	1988	1989	1990	1991
Pakistan: rupee Pakistan : rouple										
End of period[6] Fin de période[6]	12.840	13.500	15.360	15.980	17.250	17.450	18.650	21.420	21.900	24.720
Period average[6] Moyenne sur période[6]	11.847	13.117	14.046	15.928	16.647	17.399	18.003	20.541	21.707	23.801
Panama: balboa Panama : balboa										
End of period[2] Fin de période[2]	1.000	1.000	1.000	1.000	1.000	1.000	1.000	1.000	1.000	1.000
Papua New Guinea: kina Papouasie-Nvl-Guinée : kina										
End of period Fin de période	0.748	0.875	0.941	1.012	0.961	0.878	0.826	0.860	0.953	0.953
Period average Moyenne sur période	0.738	0.836	0.899	1.000	0.971	0.908	0.867	0.859	0.955	0.952
Paraguay: guarani Paraguay : guaraní										
End of period Fin de période	126.000[3]	126.000[3]	240.000[3]	320.000[3]	550.000[3]	550.000[3]	550.000[3]	1 218.000	1 258.000	1 352.300
Period average Moyenne sur période	126.000[3]	126.000[3]	201.000[3]	306.670[3]	339.170[3]	550.000[3]	550.000[3]	1 056.220	1 229.810	1 325.180
Peru: new sol[7] Pérou : nouveau sol										
End of period[7] Fin de période[7]	0.990	2.271	5.696	14.000	14.000	33.000	500.000	5 261.000	# 516.900	960.000
Period average[7] Moyenne sur période[7]	0.698	1.629	3.467	11.000	14.000	17.000	129.000	2 666.000	# 187.900	772.500
Philippines: peso Philippines : peso										
End of period Fin de période	9.171	14.002	19.760	19.032	20.530	20.800	21.335	22.440	28.000	26.650
Period average Moyenne sur période	8.540	11.113	16.699	18.607	20.386	20.568	21.095	21.737	24.310	27.479
Poland: zloty Pologne : zloty										
End of period Fin de période	...	...	...	148.000	198.000	316.000	503.000	6 500.000	9 500.000	10 957.000
Period average Moyenne sur période	...	...	...	147.000	175.000	265.000	431.000	1 439.000	9 500.000	10 576.000
Portugal: escudo Portugal : escudo										
End of period Fin de période	89.064	131.450	169.280	157.487	146.117	129.865	146.371	149.841	133.600	134.184
Period average Moyenne sur période	79.473	110.780	146.390	170.395	149.587	140.882	143.954	157.458	142.555	144.482
Qatar: riyal Qatar : riyal										
End of period Fin de période	3.640	3.640	3.640	3.640	3.640	3.640	3.640	3.640	3.640	3.640
Period average Moyenne sur période	3.640	3.640	3.640	3.640	3.640	3.640	3.640	3.640	3.640	3.640
Romania: leu Roumanie : leu										
End of period Fin de période	15.000	18.330	17.790	15.730	15.280	13.740	14.370	14.440	34.710	189.000
Period average Moyenne sur période	15.000	17.178	21.280	17.141	16.153	14.557	14.277	14.922	22.432	76.387
Rwanda: franc Rwanda : franc										
End of period Fin de période	92.840	98.540	104.360	93.490	84.180	73.020	76.710	77.620	121.120	119.790
Period average Moyenne sur période	92.840	94.342	100.172	101.262	87.640	79.672	76.445	79.977	82.597	125.140
Saint Kitts-Nevis: EC dollar Saint-Kitts-et-Nevis : EC dollar										
End of period[2] Fin de période[2]	2.700	2.700	2.700	2.700	2.700	2.700	2.700	2.700	2.700	2.700

126
Exchange rates
National currency per US dollar [cont.]
Cours des changes
Valeur du dollar des Etats-Unis en monnaie nationale [suite]

Country (monetary unit) Pays (unité monétaire)	1982	1983	1984	1985	1986	1987	1988	1989	1990	1991
Period average[2] Moyenne sur période[2]	2.700	2.700	2.700	2.700	2.700	2.700	2.700	2.700	2.700	2.700
Saint Lucia: EC dollar Sainte-Lucie : EC dollar										
End of period[2] Fin de période[2]	2.700	2.700	2.700	2.700	2.700	2.700	2.700	2.700	2.700	2.700
St. Vincent-Grenadines: EC dollar St. Vincent-Grenadines : EC dollar										
End of period[2] Fin de période[2]	2.700	2.700	2.700	2.700	2.700	2.700	2.700	2.700	2.700	2.700
Samoa: tala Samoa : tala										
End of period Fin de période	1.237	1.620	2.183	2.306	2.198	2.011	2.148	2.290	2.333	2.449
Period average Moyenne sur période	1.207	1.549	1.862	2.245	2.236	2.122	2.080	2.270	2.315	2.398
Sao Tome and Principe: dobra Sao Tomé-et-Principe : dobra										
End of period Fin de période	41.020	43.221	46.164	41.196	36.993	72.827	98.176	140.366	140.982	280.021
Period average Moyenne sur période	40.999	42.335	44.159	44.604	38.589	54.211	86.343	124.672	143.331	201.816
Saudi Arabia: riyal Arabie saoudite : riyal										
End of period Fin de période	3.435	3.495	3.575	3.645	3.745	3.745	3.745	3.745	3.745	3.745
Period average Moyenne sur période	3.428	3.455	3.524	3.622	3.703	3.745	3.745	3.745	3.745	3.745
Senegal: CFA franc Sénégal : franc CFA										
End of period[2] Fin de période[2]	336.250	417.370	479.600	378.050	322.750	267.000	302.950	289.400	256.450	259.000
Period average[2] Moyenne sur période[2]	328.600	381.060	436.950	449.260	346.300	300.540	297.850	319.010	272.260	282.110
Seychelles: rupee Seychelles : roupie										
End of period[2] Fin de période[2]	6.547	6.923	7.358	6.602	5.929	5.143	5.397	5.467	5.119	5.063
Period average[2] Moyenne sur période[2]	6.553	6.768	7.059	7.134	6.177	5.600	5.384	5.646	5.337	5.289
Sierra Leone: leone Sierra Leone : leone										
End of period Fin de période	1.233	2.510	2.510	5.208	35.587	23.041	39.063	65.359	188.679	434.783
Period average Moyenne sur période	1.239	1.885	2.510	5.094	16.092	34.043	32.514	59.813	151.446	295.344
Singapore: dollar Singapour : dollar										
End of period Fin de période	2.108	2.127	2.178	2.105	2.175	1.998	1.946	1.894	1.744	1.630
Period average Moyenne sur période	2.140	2.113	2.133	2.200	2.177	2.106	2.012	1.950	1.813	1.728
Solomon Islands: dollar Iles Salomon : dollar										
End of period Fin de période	1.045	1.221	1.344	1.613	1.986	1.974	2.118	2.397	2.614	2.795
Period average Moyenne sur période	0.971	1.149	1.274	1.481	1.741	2.003	2.083	2.293	2.529	2.715
Somalia: shilling Somalie : shilling										
End of period Fin de période	15.206	17.556	26.000	42.500	90.500	100.000	270.000	929.500	...	...
Period average Moyenne sur période	10.750	15.788	20.019	39.487	72.000	105.177	170.453	490.675	...	...
South Africa: rand Afrique du Sud : rand										
End of period Fin de période	1.076	1.222	1.985	2.558	2.183	1.930	2.378	2.536	2.563	2.743

126

Exchange rates
National currency per US dollar [*cont.*]
Cours des changes
Valeur du dollar des Etats-Unis en monnaie nationale [*suite*]

Country (monetary unit) Pays (unité monétaire)	1982	1983	1984	1985	1986	1987	1988	1989	1990	1991
Period average Moyenne sur période	1.086	1.114	1.475	2.229	2.285	2.036	2.273	2.623	2.587	2.761
Spain: peseta Espagne : peseta										
End of period Fin de période	125.601	156.700	173.400	154.150	132.395	109.000	113.450	109.720	96.909	96.688
Period average Moyenne sur période	109.859	143.430	160.761	170.044	140.048	123.478	116.487	118.378	101.934	103.912
Sri Lanka: rupee Sri Lanka : roupie										
End of period Fin de période	21.320	25.000	26.280	27.408	28.520	30.763	33.033	40.000	40.240	42.580
Period average Moyenne sur période	20.812	23.529	25.438	27.163	28.017	29.445	31.807	36.047	40.063	41.371
Sudan: pound Soudan : livre										
End of period Fin de période	1.300	1.300	1.300	2.500	2.500	4.500	4.500	4.500	4.500	14.993
Period average Moyenne sur période	0.952	1.300	1.300	2.304	2.500	3.000	4.500	4.500	4.500	6.956
Suriname: guilder Suriname : florin										
End of period Fin de période	1.785	1.785	1.785	1.785	1.785	1.785	1.785	1.785	1.785	1.785
Swaziland: lilangeni Swaziland : lilangeni										
End of period Fin de période	1.076	1.222	1.985	2.558	2.183	1.930	2.378	2.536	2.563	2.743
Period average Moyenne sur période	1.086	1.114	1.475	2.223	2.285	2.036	2.273	2.623	2.587	2.761
Sweden: krona Suède : couronne										
End of period Fin de période	7.294	8.001	8.989	7.615	6.819	5.848	6.157	6.227	5.698	5.529
Period average Moyenne sur période	6.283	7.667	8.272	8.604	7.124	6.340	6.127	6.447	5.919	6.047
Switzerland: franc Suisse : franc										
End of period Fin de période	1.994	2.179	2.585	2.076	1.623	1.278	1.504	1.546	1.295	1.355
Period average Moyenne sur période	2.030	2.099	2.350	2.457	1.799	1.491	1.463	1.636	1.389	1.434
Syrian Arab Republic: pound Rép. arabe syrienne : livre										
End of period Fin de période	3.925	3.925	3.925	3.925	3.925	3.925	11.225	11.225	11.225	11.225
Thailand: baht Thaïlande : baht										
End of period Fin de période	23.000	23.000	27.150	26.650	26.130	25.070	25.240	25.690	25.290	25.280
Period average Moyenne sur période	23.000	23.000	23.639	27.159	26.299	25.723	25.294	25.702	25.585	25.517
Togo: CFA franc Togo : franc CFA										
End of period [2] Fin de période [2]	336.250	417.370	479.600	378.050	322.750	267.000	302.950	289.400	256.450	259.000
Period average [2] Moyenne sur période [2]	328.600	381.060	436.950	449.260	346.300	300.540	297.850	319.010	272.260	282.110
Tonga: pa'anga Tonga : pa'anga										
End of period [2] Fin de période [2]	1.020	1.109	1.208	1.469	1.504	1.384	1.170	1.258	1.296	1.332
Period average [2] Moyenne sur période [2]	0.986	1.110	1.140	1.432	1.496	1.428	1.280	1.264	1.281	1.296
Trinidad and Tobago: dollar Trinité-et-Tobago : dollar										
End of period [2] Fin de période [2]	2.400	2.400	2.400	3.600	3.600	3.600	4.250	4.250	4.250	4.250

126
Exchange rates
National currency per US dollar [*cont.*]
Cours des changes
Valeur du dollar des Etats-Unis en monnaie nationale [*suite*]

Country (monetary unit) Pays (unité monétaire)	1982	1983	1984	1985	1986	1987	1988	1989	1990	1991
Period average[2] Moyenne sur période[2]	2.400	2.400	2.400	2.450	3.600	3.600	3.844	4.250	4.250	4.250
Tunisia: dinar Tunisie : dinar										
End of period Fin de période	0.616	0.727	0.867	0.757	0.840	0.778	0.898	0.905	0.837	0.864
Period average Moyenne sur période	0.591	0.679	0.777	0.834	0.794	0.829	0.858	0.949	0.878	0.925
Turkey: lira Turquie : livre										
End of period Fin de période	186.750	282.800	444.730	576.860	757.790	1 020.900	1 814.840	2 313.690	2 930.070	5 079.918
Period average Moyenne sur période	162.550	225.460	366.680	521.980	674.510	857.210	1 422.350	2 121.680	2 608.640	4 171.816
Uganda: shilling Ouganda : shilling										
End of period Fin de période	1.060	2.400	5.200	14.000	14.000	60.000	165.000	370.000	540.000	915.000
Period average Moyenne sur période	0.940	1.540	3.600	6.720	14.000	42.840	106.140	223.090	428.850	734.010
former USSR: rouble ancienne URSS : rouble										
End of period[1] Fin de période[1]	0.700	0.800	0.800	0.800	0.700	0.600	0.600	0.600	1.600	...
United Arab Emirates: dirham Emirats arabes unis : dirham										
End of period Fin de période	3.671	3.671	3.671	3.671	3.671	3.671	3.671	3.671	3.671	3.671
Period average Moyenne sur période	3.671	3.671	3.671	3.671	3.671	3.671	3.671	3.671	3.671	3.671
United Kingdom: pound Royaume-Uni : livre										
End of period Fin de période	0.619	0.689	0.865	0.692	0.678	0.534	0.553	0.623	0.519	0.535
Period average Moyenne sur période	0.572	0.660	0.752	0.779	0.682	0.612	0.562	0.611	0.563	0.567
United Rep.Tanzania: shilling Rép. Unie de Tanzanie : shilling										
End of period Fin de période	9.567	12.456	18.105	16.499	51.719	83.717	125.000	192.300	196.600	233.900
Period average Moyenne sur période	9.283	11.143	15.292	17.472	32.698	64.260	99.292	143.377	195.056	219.157
United States: dollar Etats-Unis : dollar										
End of period Fin de période	1.000	1.000	1.000	1.000	1.000	1.000	1.000	1.000	1.000	1.000
Period average Moyenne sur période	1.000	1.000	1.000	1.000	1.000	1.000	1.000	1.000	1.000	1.000
Uruguay: new peso Uruguay : nouveau peso										
End of period[3] Fin de période[3]	33.750	43.250	74.250	125.000	181.000	281.000	451.000	805.000	1 594.000	2 489.000
Period average[3] Moyenne sur période[3]	13.910	34.540	56.120	101.430	151.990	226.670	359.440	605.510	1 171.050	2 018.820
Vanuatu: vatu Vanuatu : vatu										
End of period Fin de période	96.150	101.770	102.580	100.250	116.240	100.560	105.050	110.700	109.250	110.790
Period average Moyenne sur période	96.207	99.368	99.233	106.032	106.076	109.849	104.426	116.042	116.565	111.675
Venezuela: bolívar Venezuela : bolívar										
End of period Fin de période	4.292[3]	4.300	7.500	7.500	14.500	14.500	14.500	43.079	50.380	61.554
Period average Moyenne sur période	4.292[3]	4.297	7.017	7.500	8.083	14.500	14.500	34.681	46.900	56.816

126
Exchange rates
National currency per US dollar [cont.]
Cours des changes
Valeur du dollar des Etats-Unis en monnaie nationale [suite]

Country (monetary unit) Pays (unité monétaire)	1982	1983	1984	1985	1986	1987	1988	1989	1990	1991
Viet Nam: dong Viet Nam : dong										
End of period Fin de période	0.980	1.030	1.050	18.000	18.000	225.000	900.000	4 300.000	6 500.000	9 200.000
Period average Moyenne sur période	0.940	1.000	1.030	6.700	18.000	61.960	480.000	3 532.780	5 130.500	7 943.328
Yemen: rial Yémen : rial										
End of period' Fin de période'	4.562	4.676	5.860	8.100	12.250	9.900	9.760	9.760	...	...
Period average' Moyenne sur période'	4.562	4.579	5.353	7.363	9.639	10.342	9.772	9.760	...	...
Yugoslavia: new dinar Yougoslavie : nouveau dinar										
End of period Fin de période	0.006	0.013	0.021	0.031	0.046	0.124	0.521	11.816	10.657	19.735
Period average Moyenne sur période	0.005	0.009	0.015	0.027	0.038	0.074	0.252	2.876	11.318	19.638
Zaire: zaire Zaïre : zaïre										
End of period Fin de période	6.000	30.000	40.000	56.000	71.000	132.000	274.000	455.000	2 000.000	63 673.000
Period average Moyenne sur période	6.000	13.000	36.000	50.000	60.000	112.000	187.000	381.000	719.000	15 587.000
Zambia: kwacha Zambie : kwacha										
End of period Fin de période	0.930	1.511	2.201	5.700	12.710	8.000	10.004	21.650	42.753	88.968
Period average Moyenne sur période	0.929	1.259	1.813	3.140	7.788	9.519	8.266	13.814	30.289	64.640
Zimbabwe: dollar Zimbabwe : dollar										
End of period Fin de période	0.919	1.105	1.502	1.641	1.678	1.663	1.943	2.270	2.636	5.051
Period average Moyenne sur période	0.759	1.013	1.258	1.614	1.667	1.662	1.806	2.119	2.452	3.621

Source:
International Monetary Fund (Washington, DC).

† All data shown which pertain to Germany prior to 3 October 1990 are indicated separately for the Federal Republic of Germany and the former German Democratic Republic based on their respective territories at the time indicated. Where data for united Germany (3 October 1990 and thereafter) are not available, available data are shown separately under the designations Federal Republic of Germany and former German Democratic Republic and pertain to the territorial boundaries prior to 3 October 1990. For detailed explanatory notes on data pertaining to Germany, see Annex I - Country Nomenclature.

1 Data refer to non-commercial rates derived from the Operational Rates of Exchange for United Nations Programmes.

2 Fixed rate (Rwanda: beginning 1983).
3 Selling rate.
4 Prior to January 1985, national currency relates to bipkwele.
5 Per thousand US dollars.
6 Buying rate.
7 Per million US dollars through Dec.1989 and per thousand US dollars thereafter.
8 Excluding former Democratic Yemen.

Sources:
Fonds monétaire international (Washington, DC).

† Toutes les données se rapportant à l'Allemagne avant le 3 octobre 1990 figurent dans deux rubriques séparées basées sur les territoires respectifs de la République fédérale d'Allemagne et l'ancienne République démocratique allemande selon la période indiquée. En l'absence de données pour l'Allemagne unifiée (à compter du 3 octobre 1990), les données disponibles sont fournies séparément sous les rubriques République fédérale d'Allemagne et ancienne République démocratique allemande et se rapportent aux limites territoriales antérieures au 3 octobre 1990. Pour les notes explicatives en détail sur les données concernant l'Allemagne, voir Annexe I – Nomenclature des pays.

1 Les données se rapportent aux taux non-commerciaux tirées de "Operational Rates of Exchange for United Nations Progammes".

2 Taux fixe (Rwanda: à partir de 1983).
3 Cours de vente.
4 Avant janvier 1985, la monnaie nationale se rapporte à bipkwele.
5 Pour mille $EU.
6 Cours d'achat.
7 Pour un million de $EU jusqu'à la fin de 1989 et pour mille $EU par la suite.
8 Non compris l'ancien Yémen démocratique.

127
International reserves excluding gold
Réserves internationales, l'or non compris
Special drawing rights(SDR) + Reserve position in IMF + Foreign exchange: million $US (end of period)
Droits de tirage spéciaux(DTS) + Réserve disponible au FMI + Devises étrangères : $E-U millions (fin de la période)

Country or area and reserves Pays ou zone et réserves	1982	1983	1984	1985	1986	1987	1988	1989	1990	1991
Afghanistan **Afghanistan**	**257.8**	**214.2**	**228.7**	**295.2**	**258.5**	**279.7**	**261.1**	**243.7**	**266.4**	**234.9**
SDR DTS	17.2	10.4	13.2	13.6	14.0	14.9	12.8	10.6	9.0	6.7
Position in IMF Disponibilité au FMI	16.7	5.0	4.7	5.3	5.9	6.9	6.5	6.4	7.0	7.0
Foreign exchange Devises étrangères	223.9	198.7	210.8	276.3	238.6	257.9	241.8	226.7	250.4	221.2
Algeria **Algérie**	**2 422.0**	**1 880.4**	**1 464.2**	**2 819.0**	**1 660.2**	**1 640.5**	**900.2**	**847.0**	**724.8**	**1 485.9**
SDR DTS	154.5	106.9	110.9	138.0	167.1	202.4	2.2	4.0	2.8	1.9
Position in IMF Disponibilité au FMI	138.5	180.5	162.2	168.0	181.1	153.1	0.0	0.0	0.0	0.0
Foreign exchange Devises étrangères	2 129.0	1 593.0	1 191.0	2 513.0	1 312.0	1 285.0	898.0	843.0	722.0	1 484.0
Antigua and Barbuda **Antigua-et-Barbuda**	**8.5**	**9.9**	**15.4**	**16.6**	**28.3**	**25.6**	**28.0**	**28.1**	**27.5**	**...**
SDR DTS	0.0	0.0	0.0	0.0	0.0	0.0	0.0	0.0	0.0	0.0
Position in IMF Disponibilité au FMI	0.0	0.0	0.0	0.0	0.0	0.0	0.0	0.0	0.0	0.0
Foreign exchange Devises étrangères	8.5	9.9	15.4	16.6	28.3	25.6	28.0	28.1	27.5	...
Argentina **Argentine**	**2 506.0**	**1 172.0**	**1 243.0**	**3 273.0**	**2 718.0**	**1 617.0**	**3 363.0**	**1 463.0**	**4 592.0**	**6 615.0**
SDR DTS	0.0	0.4	0.6	0.0	0.0	0.0	0.5	0.3	297.3	193.0
Position in IMF Disponibilité au FMI	100.4	0.0	0.0	0.0	0.0	0.0	0.0	0.0	0.0	0.0
Foreign exchange Devises étrangères	2 406.0	1 172.0	1 242.0	3 273.0	2 718.0	1 617.0	3 363.0	1 463.0	4 295.0	6 422.0
Aruba[1] **Aruba**[1]	**...**	**...**	**...**	**...**	**73.6**	**78.7**	**90.9**	**87.1**	**97.9**	**119.6**
Foreign exchange Devises étrangères	...	...	...	...	73.6	78.7	90.9	87.1	97.9	119.6
Australia **Australie**	**6 371.0**	**8 962.0**	**7 441.0**	**5 768.0**	**7 246.0**	**8 744.0**	**13 598.0**	**13 780.0**	**16 264.0**	**16 534.0**
SDR DTS	85.7	80.6	209.2	310.3	331.5	369.0	334.1	307.5	310.5	289.5
Position in IMF Disponibilité au FMI	0.1	113.7	183.0	207.3	230.9	267.9	275.3	322.1	348.8	350.8
Foreign exchange Devises étrangères	6 285.0	8 768.0	7 049.0	5 250.0	6 684.0	8 107.0	12 989.0	13 150.0	15 605.0	15 894.0
Austria **Autriche**	**5 300.0**	**4 515.0**	**4 244.0**	**4 767.0**	**6 162.0**	**7 532.0**	**7 368.0**	**8 598.0**	**9 376.0**	**10 332.0**
SDR DTS	249.8	161.3	220.0	210.0	185.7	291.9	268.5	298.4	278.4	282.5
Position in IMF Disponibilité au FMI	285.5	468.1	438.3	445.1	442.1	468.2	388.7	360.6	344.0	394.7
Foreign exchange Devises étrangères	4 765.0	3 886.0	3 586.0	4 112.0	5 534.0	6 772.0	6 711.0	7 939.0	8 754.0	9 655.0
Bahamas **Bahamas**	**113.5**	**122.0**	**161.1**	**182.5**	**231.5**	**170.1**	**172.0**	**146.8**	**158.2**	**181.3**
SDR DTS	6.3	1.2	0.7	0.4	0.1	0.4	0.5	0.2	0.5	0.2
Position in IMF Disponibilité au FMI	7.3	11.4	10.7	12.0	13.4	15.2	12.2	11.4	11.2	10.2
Foreign exchange Devises étrangères	99.8	109.4	149.8	170.1	218.1	154.5	159.4	135.3	146.4	170.9

127
International reserves excluding gold
Special drawing rights(SDR) + Reserve position in IMF + Foreign exchange: million $US (end of period) [cont.]
Réserves internationales, l'or non compris
Droits de tirage spéciaux(DTS) + Réserve disponible au FMI + Devises étrangères : $E-U millions (fin de la période) [suite]

Country or area and reserves Pays ou zone et réserves	1982	1983	1984	1985	1986	1987	1988	1989	1990	1991
Bahrain **Bahreïn**	**1 534.8**	**1 426.4**	**1 302.4**	**1 659.7**	**1 489.4**	**1 148.5**	**1 251.7**	**1 050.0**	**1 234.9**	**1 514.6**
SDR DTS	15.4	12.8	12.8	15.0	17.3	20.7	20.4	20.9	24.0	25.5
Position in IMF Disponibilité au FMI	13.8	22.4	22.6	27.0	31.8	38.7	37.3	38.9	39.8	42.4
Foreign exchange Devises étrangères	1 505.6	1 391.2	1 267.0	1 617.7	1 440.3	1 089.1	1 194.0	990.3	1 171.1	1 446.7
Bangladesh **Bangladesh**	**182.6**	**524.1**	**389.9**	**336.5**	**409.1**	**843.1**	**1 046.1**	**501.5**	**628.7**	**1 278.2**
SDR DTS	0.8	13.5	0.3	13.1	10.3	53.3	54.0	3.0	25.8	71.3
Position in IMF Disponibilité au FMI	8.3	23.5	22.0	24.6	27.4	31.8	30.1	29.4	0.0	0.0
Foreign exchange Devises étrangères	173.5	487.1	367.7	298.8	371.4	758.1	961.9	469.0	602.9	1 206.9
Barbados **Barbade**	**121.6**	**123.3**	**132.5**	**139.8**	**151.7**	**145.2**	**135.5**	**109.5**	**117.5**	**86.4**
SDR DTS	0.9	0.3	0.0	0.0	0.0	0.9	0.6	0.0	0.0	0.7
Position in IMF Disponibilité au FMI	0.0	2.3	2.1	2.4	2.6	3.1	2.9	2.9	3.1	0.0
Foreign exchange Devises étrangères	120.7	120.7	130.4	137.4	149.1	141.2	131.9	106.6	114.4	85.6
Belgium **Belgique**	**3 927.0**	**4 714.0**	**4 564.0**	**4 849.0**	**5 538.0**	**9 620.0**	**9 333.0**	**10 766.0**	**12 151.0**	**12 180.0**
SDR DTS	741.3	417.7	445.3	360.7	342.2	700.4	562.5	556.2	566.2	587.9
Position in IMF Disponibilité au FMI	362.0	519.8	511.0	518.9	565.5	556.7	464.0	449.3	464.3	524.4
Foreign exchange Devises étrangères	2 824.0	3 776.0	3 608.0	3 969.0	4 630.0	8 363.0	8 306.0	9 760.0	11 121.0	11 068.0
Belize **Belize**	**9.8**	**9.3**	**6.1**	**14.8**	**26.9**	**36.4**	**51.7**	**59.9**	**69.8**	**53.0**
SDR DTS	0.0	0.0	0.0	0.0	0.0	0.1	0.0	0.0	0.0	0.1
Position in IMF Disponibilité au FMI	1.5	2.0	1.9	2.1	2.3	2.7	2.6	2.5	2.7	2.7
Foreign exchange Devises étrangères	8.4	7.3	4.2	12.7	24.6	33.6	49.1	57.4	67.1	50.2
Benin **Bénin**	**4.9**	**3.7**	**2.5**	**4.1**	**3.9**	**3.6**	**4.2**	**3.4**	**64.9**	**191.6**
SDR DTS	2.2	1.2	0.2	0.0	0.0	0.1	0.1	0.0	0.1	0.2
Position in IMF Disponibilité au FMI	2.2	2.1	2.0	2.2	2.5	2.9	2.7	2.7	2.9	2.9
Foreign exchange Devises étrangères	0.5	0.4	0.3	1.9	1.4	0.6	1.4	0.7	61.9	188.5
Bhutan **Bhoutan**	**...**	**...**	**...**	**50.3**	**61.0**	**74.9**	**94.1**	**98.5**	**86.0**	**...**
SDR DTS	...	...	...	0.1	0.1	0.2	0.2	0.3	0.4	0.4
Position in IMF Disponibilité au FMI	...	...	...	0.6	0.7	0.8	0.8	0.8	0.8	0.8
Foreign exchange Devises étrangères	...	...	...	49.6	60.2	73.9	93.1	97.5	84.8	...
Bolivia **Bolivie**	**155.9**	**160.1**	**251.6**	**200.0**	**163.7**	**97.3**	**105.8**	**204.9**	**166.8**	**106.4**
SDR DTS	0.0	0.1	0.0	0.0	2.5	0.0	0.0	0.0	1.0	0.1

127
International reserves excluding gold
Special drawing rights(SDR) + Reserve position in IMF + Foreign exchange: million $US (end of period) [cont.]
Réserves internationales, l'or non compris
Droits de tirage spéciaux(DTS) + Réserve disponible au FMI + Devises étrangères : $E-U millions (fin de la période) [suite]

Country or area and reserves Pays ou zone et réserves	1982	1983	1984	1985	1986	1987	1988	1989	1990	1991
Position in IMF Disponibilité au FMI	0.0	0.0	0.0	0.0	0.0	0.0	0.0	0.0	0.0	0.0
Foreign exchange Devises étrangères	155.9	160.0	251.6	200.0	161.2	97.3	105.8	204.9	165.8	106.3
Botswana **Botswana**	**293.0**	**395.7**	**474.3**	**783.2**	**1 197.7**	**2 057.1**	**2 258.1**	**2 841.1**	**3 385.3**	**3 772.4**
SDR DTS	6.8	7.6	8.4	10.7	17.1	21.8	22.6	24.9	30.9	34.5
Position in IMF Disponibilité au FMI	10.1	11.8	12.1	14.2	19.1	22.3	18.4	25.3	23.0	19.2
Foreign exchange Devises étrangères	276.1	376.2	453.8	758.4	1 161.5	2 013.0	2 217.1	2 791.0	3 331.5	3 718.7
Brazil **Brésil**	**3 928.0**	**4 355.0**	**11 508.0**	**10 605.0**	**# # 5 803.0**	**6 299.0**	**6 972.0**	**7 535.0**	**7 441.0**	**8 033.0**
SDR DTS	0.3	0.1	0.9	0.6	# # 0.0	0.2	0.4	0.0	10.9	12.8
Position in IMF Disponibilité au FMI	286.6	0.0	0.0	0.0	# # 0.0	0.0	0.0	0.0	0.0	0.0
Foreign exchange Devises étrangères	3 641.0	4 355.0	11 507.0	10 604.0	# # 5 803.0	6 299.0	6 971.0	7 535.0	7 430.0	8 020.0
Burkina Faso **Burkina Faso**	**61.8**	**85.0**	**106.3**	**139.5**	**233.5**	**322.6**	**320.9**	**265.5**	**300.5**	**346.1**
SDR DTS	8.3	7.9	5.5	6.2	6.9	8.0	7.6	7.4	8.0	8.0
Position in IMF Disponibilité au FMI	6.2	5.9	7.4	8.3	9.2	10.7	10.2	9.9	10.2	10.3
Foreign exchange Devises étrangères	47.3	71.2	93.4	125.1	217.4	303.9	303.1	248.2	282.2	327.8
Burundi **Burundi**	**29.5**	**26.9**	**19.7**	**29.5**	**69.1**	**60.7**	**69.4**	**99.6**	**105.0**	**141.4**
SDR DTS	4.6	1.0	0.1	0.1	0.7	0.1	0.1	0.0	0.1	3.8
Position in IMF Disponibilité au FMI	8.1	9.9	9.2	10.1	11.2	13.0	12.3	12.0	10.7	10.4
Foreign exchange Devises étrangères	16.8	16.0	10.4	19.3	57.2	47.7	57.0	87.6	94.2	127.3
Cameroon **Cameroun**	**67.2**	**159.1**	**53.9**	**132.5**	**59.0**	**63.8**	**175.9**	**79.9**	**25.5**	**...**
SDR DTS	1.8	0.7	6.1	4.6	3.5	0.3	0.0	0.3	0.6	5.6
Position in IMF Disponibilité au FMI	15.3	7.5	0.2	0.2	0.2	0.3	0.3	0.3	0.3	0.3
Foreign exchange Devises étrangères	50.1	150.9	47.6	127.6	55.3	63.2	175.5	79.3	24.6	...
Canada **Canada**	**2 999.5**	**3 465.5**	**2 491.3**	**2 502.7**	**3 251.2**	**7 277.4**	**15 390.6**	**16 055.3**	**17 845.4**	**16 252.5**
SDR DTS	70.8	21.0	72.2	217.9	247.4	398.9	1 369.2	1 377.4	1 525.8	1 581.6
Position in IMF Disponibilité au FMI	365.2	703.4	678.4	710.8	686.4	660.7	504.9	527.8	517.4	592.3
Foreign exchange Devises étrangères	2 563.5	2 741.0	1 740.7	1 574.0	2 317.5	6 217.8	13 516.6	14 150.2	15 802.2	14 078.6
Cape Verde **Cap-Vert**	**43.0**	**46.0**	**41.0**	**55.0**	**56.4**	**81.0**	**81.3**	**75.0**	**77.0**	**...**
SDR DTS	0.2	0.1	0.1	0.1	0.1	0.1	0.0	0.1	0.0	0.1
Position in IMF Disponibilité au FMI	0.6	1.0	0.9	1.0	0.0	0.0	0.0	0.0	0.0	0.0
Foreign exchange Devises étrangères	41.9	44.8	40.0	54.2	56.2	80.6	81.3	74.7	76.9	...

127
International reserves excluding gold
Special drawing rights(SDR) + Reserve position in IMF + Foreign exchange: million $US (end of period) [cont.]
Réserves internationales, l'or non compris
Droits de tirage spéciaux(DTS) + Réserve disponible au FMI + Devises étrangères : $E-U millions (fin de la période) [suite]

Country or area and reserves Pays ou zone et réserves	1982	1983	1984	1985	1986	1987	1988	1989	1990	1991
Central African Rep. **Rép. centrafricaine**	**46.4**	**46.8**	**52.7**	**49.6**	**65.3**	**96.7**	**108.5**	**113.1**	**118.6**	**...**
SDR DTS	0.2	0.7	2.5	1.7	0.5	7.0	12.3	0.0	4.9	0.7
Position in IMF Disponibilité au FMI	1.3	1.7	0.1	0.1	0.1	0.2	0.2	0.1	0.1	0.1
Foreign exchange Devises étrangères	44.9	44.4	50.0	47.8	64.7	89.6	96.0	112.9	113.6	...
Chad **Tchad**	**12.4**	**28.0**	**44.2**	**33.5**	**15.9**	**52.1**	**63.1**	**111.7**	**127.8**	**...**
SDR DTS	0.3	1.5	0.4	3.9	2.1	9.0	7.7	1.8	0.1	0.2
Position in IMF Disponibilité au FMI	5.6	3.6	0.3	0.3	0.3	0.4	0.4	0.3	0.4	0.4
Foreign exchange Devises étrangères	6.5	22.8	43.5	29.3	13.5	42.7	55.0	109.6	127.3	...
Chile **Chili**	**1 815.0**	**2 036.3**	**2 303.2**	**2 449.9**	**2 351.3**	**2 504.2**	**3 160.5**	**3 628.6**	**6 068.5**	**7 041.3**
SDR DTS	19.5	5.4	11.5	0.3	0.2	40.8	44.3	24.4	1.0	0.8
Position in IMF Disponibilité au FMI	77.8	0.0	0.0	0.0	0.0	0.0	0.0	0.0	0.0	0.0
Foreign exchange Devises étrangères	1 717.7	2 030.9	2 291.7	2 449.6	2 351.1	2 463.4	3 116.2	3 604.2	6 067.5	7 040.5
China **Chine**	**11 349.0**	**14 987.0**	**17 366.0**	**12 728.0**	**11 453.0**	**16 305.0**	**18 541.0**	**17 960.0**	**29 586.0**	**43 674.0**
SDR DTS	213.9	335.0	405.6	482.8	568.8	639.7	586.1	540.2	561.7	577.5
Position in IMF Disponibilité au FMI	0.0	175.6	255.4	332.4	370.2	429.3	407.2	397.7	430.5	432.9
Foreign exchange Devises étrangères	11 135.0	14 476.0	16 705.0	11 913.0	10 514.0	15 236.0	17 548.0	17 022.0	28 594.0	42 664.0
Colombia **Colombie**	**3 861.0**	**1 901.0**	**1 364.0**	**1 595.0**	**2 696.0**	**3 086.0**	**3 248.0**	**3 616.0**	**4 212.0**	**6 029.0**
SDR DTS	179.1	197.5	0.1	0.0	139.8	162.1	153.8	150.1	162.5	163.4
Position in IMF Disponibilité au FMI	193.3	274.4	0.0	0.0	0.0	0.0	0.0	0.0	0.0	0.0
Foreign exchange Devises étrangères	3 489.0	1 429.0	1 364.0	1 595.0	2 556.0	2 924.0	3 094.0	3 466.0	4 049.0	5 866.0
Comoros **Comores**	**10.8**	**10.8**	**3.5**	**11.8**	**17.6**	**30.7**	**23.5**	**30.8**	**29.7**	**...**
SDR DTS	0.0	0.0	0.2	0.2	0.2	0.2	0.2	0.1	0.1	0.0
Position in IMF Disponibilité au FMI	0.0	0.0	0.0	0.0	0.0	0.0	0.0	0.0	0.0	0.0
Foreign exchange Devises étrangères	10.8	10.8	3.3	11.5	17.3	30.5	23.4	30.7	29.6	...
Congo **Congo**	**37.0**	**7.4**	**4.1**	**4.0**	**6.8**	**3.4**	**14.6**	**15.8**	**22.0**	**...**
SDR DTS	1.1	0.2	2.1	1.7	4.7	2.7	1.1	1.6	1.7	0.1
Position in IMF Disponibilité au FMI	3.6	3.1	0.5	0.5	0.6	0.7	0.7	0.6	0.7	0.7
Foreign exchange Devises étrangères	32.3	4.0	1.6	1.8	1.6	0.0	12.8	13.6	19.7	...
Costa Rica **Costa Rica**	**226.1**	**311.3**	**405.0**	**506.4**	**523.4**	**488.9**	**668.0**	**742.6**	**520.6**	**919.8**
SDR DTS	0.1	3.0	0.1	0.0	0.0	0.0	0.0	0.1	1.6	0.3

127
International reserves excluding gold
Special drawing rights(SDR) + Reserve position in IMF + Foreign exchange: million $US (end of period) [cont.]
Réserves internationales, l'or non compris
Droits de tirage spéciaux(DTS) + Réserve disponible au FMI + Devises étrangères : $E-U millions (fin de la période) [suite]

Country or area and reserves Pays ou zone et réserves	1982	1983	1984	1985	1986	1987	1988	1989	1990	1991
Position in IMF										
Disponibilité au FMI	0.0	0.0	0.0	0.0	0.0	0.0	0.0	0.0	0.0	0.0
Foreign exchange										
Devises étrangères	226.0	308.3	404.9	506.4	523.4	488.9	668.0	**742.5	519.0	919.5
Côte d'Ivoire										
Côte d'Ivoire	**2.2**	**19.7**	**5.4**	**4.7**	**19.6**	**8.9**	**10.4**	**15.0**	**4.0**	**13.4**
SDR										
DTS	0.1	16.2	0.2	0.1	8.5	0.2	0.7	5.1	1.2	2.0
Position in IMF										
Disponibilité au FMI	0.0	0.0	0.0	0.0	0.0	0.0	0.0	0.0	0.0	0.0
Foreign exchange										
Devises étrangères	2.1	3.5	5.2	4.7	11.0	8.7	9.7	9.8	2.8	11.4
Cyprus										
Chypre	**523.2**	**519.1**	**540.5**	**595.3**	**752.7**	**873.5**	**927.9**	**1 124.0**	**1 506.9**	**1 390.2**
SDR										
DTS	0.2	0.2	0.1	0.1	0.0	0.4	0.1	0.1	0.1	0.1
Position in IMF										
Disponibilité au FMI	0.0	4.9	4.6	5.1	5.7	6.6	15.7	23.7	21.5	25.6
Foreign exchange										
Devises étrangères	523.0	514.0	535.9	590.0	747.0	866.5	912.1	1 100.2	1 485.3	1 364.5
Czechoslovakia										
Tchécoslovaquie	**776.0**	**803.0**	**966.3**	**854.3**	**1 115.2**	**1 382.0**	**1 583.0**	**2 156.5**	**1 102.0**	**3 189.7**
SDR										
DTS	0.0	0.0	0.0	0.0	0.0	0.0	0.0	0.0	0.0	139.6
Position in IMF										
Disponibilité au FMI	0.0	0.0	0.0	0.0	0.0	0.0	0.0	0.0	0.0	0.0
Foreign exchange										
Devises étrangères	776.0	803.0	966.3	854.3	1 115.2	1 382.0	1 583.0	2 156.5	1 102.0	3 050.1
Denmark										
Danemark	**2 265.7**	**3 620.7**	**3 008.9**	**5 428.6**	**4 964.5**	**10 066.4**	**10 765.0**	**6 396.8**	**10 591.4**	**7 404.3**
SDR										
DTS	194.5	124.2	155.0	196.3	253.5	304.4	224.8	280.0	215.7	241.8
Position in IMF										
Disponibilité au FMI	110.0	214.7	210.1	227.9	155.6	174.8	315.8	335.0	312.7	355.6
Foreign exchange										
Devises étrangères	1 961.1	3 281.8	2 643.8	5 004.4	4 555.4	9 587.2	10 224.4	5 781.8	10 063.0	6 806.9
Djibouti										
Djibouti	...	...	**44.9**	**50.9**	**53.6**	**63.5**	**64.4**	**59.2**	**93.6**	**100.0**
SDR										
DTS	0.5	0.5	0.4	0.4	0.4	0.5	0.4	0.4	0.3	0.3
Position in IMF										
Disponibilité au FMI	1.4	1.3	1.2	1.4	1.5	1.8	1.7	1.6	1.8	1.8
Foreign exchange										
Devises étrangères	...	...	43.3	49.2	51.7	61.3	62.3	57.2	91.5	98.0
Dominica										
Dominique	**4.3**	**1.5**	**5.3**	**3.3**	**9.6**	**18.4**	**14.1**	**11.7**	**14.5**	...
SDR										
DTS	0.3	0.4	0.0	0.0	1.0	1.0	0.7	0.4	0.3	0.0
Position in IMF										
Disponibilité au FMI	0.0	0.0	0.0	0.0	0.0	0.0	0.0	0.0	0.0	0.0
Foreign exchange										
Devises étrangères	4.0	1.1	5.2	3.3	8.6	17.4	13.3	11.3	14.2	...
Dominican Republic										
Rép. dominicaine	**129.0**	**171.3**	**253.5**	**340.1**	**376.3**	**182.2**	**254.0**	**164.0**	**61.6**	**441.9**
SDR										
DTS	0.6	0.2	0.4	31.6	0.0	0.0	0.0	0.0	0.0	0.1
Position in IMF										
Disponibilité au FMI	0.0	7.8	0.0	0.0	0.0	0.0	0.0	0.0	0.0	0.0
Foreign exchange										
Devises étrangères	128.4	163.3	253.1	308.5	376.3	182.2	254.0	164.0	61.6	441.8

127
International reserves excluding gold
Special drawing rights(SDR) + Reserve position in IMF + Foreign exchange: million $US (end of period) [cont.]
Réserves internationales, l'or non compris
Droits de tirage spéciaux(DTS) + Réserve disponible au FMI + Devises étrangères : $E-U millions (fin de la période) [suite]

Country or area and reserves Pays ou zone et réserves	1982	1983	1984	19.5	1986	1987	1988	1989	1990	1991
Ecuador **Equateur**	**304.2**	**644.5**	**611.2**	**718.2**	**644.1**	**491.1**	**397.6**	**540.4**	**838.5**	**924.3**
SDR DTS	0.0	0.1	0.5	28.8	56.0	0.9	1.3	0.9	14.7	41.4
Position in IMF Disponibilité au FMI	0.0	12.0	0.0	0.0	0.0	0.0	0.0	0.0	0.0	0.0
Foreign exchange Devises étrangères	304.2	632.4	610.7	689.4	588.2	490.2	396.3	539.5	823.8	882.9
Egypt **Egypte**	**698.0**	**771.0**	**736.0**	**792.0**	**829.0**	**1 378.0**	**1 263.0**	**1 520.0**	**2 684.0**	**5 325.0**
SDR DTS	0.1	0.3	0.2	0.1	0.0	0.3	0.4	0.1	0.6	1.3
Position in IMF Disponibilité au FMI	0.0	31.8	0.0	0.0	0.0	0.0	0.0	0.0	0.0	0.0
Foreign exchange Devises étrangères	698.0	739.0	736.0	792.0	829.0	1 378.0	1 263.0	1 520.0	2 683.0	5 324.0
El Salvador **El Salvador**	**108.5**	**160.2**	**165.8**	**179.6**	**169.7**	**186.1**	**161.6**	**265.9**	**414.8**	**287.2**
SDR DTS	1.8	0.1	0.0	0.0	0.0	0.0	0.0	0.0	0.0	0.0
Position in IMF Disponibilité au FMI	0.0	0.0	0.0	0.0	0.0	0.0	0.0	0.0	0.0	0.0
Foreign exchange Devises étrangères	106.7	160.1	165.8	179.6	169.7	186.1	161.6	265.9	414.8	287.2
Equatorial Guinea **Guinée équatoriale**	**2.8**	**1.3**	**1.4**	**3.5**	**2.7**	**0.6**	**5.5**	**0.8**	**0.7**	**...**
SDR DTS	0.0	0.0	0.0	3.4	0.8	0.2	0.0	0.1	0.1	8.0
Position in IMF Disponibilité au FMI	0.6	0.0	0.0	0.0	0.0	0.0	0.0	0.0	0.0	0.0
Foreign exchange Devises étrangères	2.2	1.3	1.4	0.1	1.9	0.4	5.5	0.7	0.6	...
Ethiopia **Ethiopie**	**181.8**	**125.9**	**44.3**	**148.0**	**250.5**	**122.7**	**64.2**	**46.1**	**20.2**	**54.5**
SDR DTS	3.5	2.4	2.9	0.2	0.0	1.7	0.0	0.1	0.3	0.2
Position in IMF Disponibilité au FMI	0.0	4.4	0.0	0.0	0.0	0.0	0.0	0.0	0.0	0.0
Foreign exchange Devises étrangères	178.3	119.1	41.4	147.8	250.5	121.0	64.2	46.0	19.9	54.3
Fiji **Fidji**	**126.9**	**115.8**	**117.4**	**130.8**	**171.1**	**132.2**	**233.4**	**211.6**	**260.8**	**271.4**
SDR DTS	4.1	0.3	6.2	5.7	7.1	14.0	20.3	20.9	23.5	13.3
Position in IMF Disponibilité au FMI	6.0	8.2	7.7	8.6	9.6	11.2	10.6	10.4	10.1	9.7
Foreign exchange Devises étrangères	116.9	107.4	103.6	116.6	154.3	107.0	202.5	180.3	227.2	248.5
Finland **Finlande**	**1 517.5**	**1 237.7**	**2 754.3**	**3 749.9**	**1 787.1**	**6 417.5**	**6 369.2**	**5 111.2**	**9 644.1**	**7 608.7**
SDR DTS	115.0	38.7	142.8	171.8	204.6	227.7	268.9	239.2	216.9	225.6
Position in IMF Disponibilité au FMI	85.5	128.7	131.1	143.1	165.2	200.8	225.6	235.0	214.8	275.1
Foreign exchange Devises étrangères	1 317.0	1 070.3	2 480.4	3 435.0	1 417.4	5 989.0	5 874.7	4 636.9	9 212.4	7 108.0
France **France**	**16 531.0**	**19 851.0**	**20 940.0**	**26 589.0**	**31 454.0**	**33 049.0**	**25 364.0**	**24 611.0**	**36 778.0**	**31 284.0**
SDR DTS	978.8	441.7	572.1	900.1	1 289.8	1 501.8	1 390.0	1 328.7	1 283.0	1 325.7

127
International reserves excluding gold
Special drawing rights(SDR) + Reserve position in IMF + Foreign exchange: million $US (end of period) [cont.]
Réserves internationales, l'or non compris
Droits de tirage spéciaux(DTS) + Réserve disponible au FMI + Devises étrangères : $E-U millions (fin de la période) [suite]

Country or area and reserves / Pays ou zone et réserves	1982	1983	1984	1985	1986	1987	1988	1989	1990	1991
Position in IMF / Disponibilité au FMI	957.9	1 352.3	1 265.4	1 369.5	1 736.1	1 913.7	1 615.5	1 414.3	1 427.8	1 666.3
Foreign exchange / Devises étrangères	14 594.0	18 057.0	19 102.0	24 319.0	28 428.0	29 634.0	22 359.0	21 868.0	34 067.0	28 292.0
Gabon / Gabon	311.9	186.9	199.5	192.6	126.4	12.0	67.4	34.4	273.8	...
SDR / DTS	0.8	0.4	5.7	2.3	12.3	11.6	8.8	0.2	0.3	6.4
Position in IMF / Disponibilité au FMI	0.0	7.4	0.0	0.0	0.0	0.0	0.0	0.1	0.1	0.1
Foreign exchange / Devises étrangères	311.0	179.1	193.7	190.2	114.0	0.3	58.6	34.2	273.4	...
Gambia / Gambie	8.4	2.9	2.3	1.7	13.6	25.8	19.1	20.6	55.4	67.6
SDR / DTS	0.1	0.1	0.0	0.0	0.7	4.6	1.3	1.4	1.8	0.8
Position in IMF / Disponibilité au FMI	0.0	0.0	0.0	0.0	0.1	0.1	0.1	0.1	0.0	0.0
Foreign exchange / Devises étrangères	8.2	2.8	2.2	1.7	12.8	21.1	17.7	19.2	53.6	66.8
Germany† · Allemagne† / F. R. Germany / R. f. Allemagne	44 762.0	42 674.0	40 141.0	44 380.0	51 734.0	78 756.0	58 528.0	60 709.0	67 902.0	63 001.0
SDR / DTS	2 054.1	1 613.2	1 362.2	1 546.6	2 019.7	1 963.7	1 857.4	1 804.0	1 879.7	1 916.8
Position in IMF / Disponibilité au FMI	3 087.5	3 748.0	3 750.4	3 808.3	3 848.3	3 899.6	3 346.2	3 042.7	3 055.6	3 567.1
Foreign exchange / Devises étrangères	39 620.0	37 313.0	35 028.0	39 025.0	45 866.0	72 893.0	53 324.0	55 862.0	62 967.0	57 517.0
Ghana / Ghana	138.9	144.8	301.6	478.5	513.0	195.1	221.3	347.3	218.8	...
SDR / DTS	0.2	2.2	0.1	18.9	2.0	15.9	0.3	29.9	4.4	12.5
Position in IMF / Disponibilité au FMI	0.0	0.0	0.0	0.0	0.0	0.0	0.0	0.0	0.0	0.0
Foreign exchange / Devises étrangères	138.7	142.6	301.5	459.6	511.0	179.2	221.0	317.4	214.4	...
Greece / Grèce	861.0	901.0	954.0	868.0	## 1 519.0	2 681.0	3 619.0	3 223.0	3 412.0	5 189.0
SDR / DTS	0.0	0.6	1.1	0.0	0.0	0.0	0.3	0.4	0.3	0.4
Position in IMF / Disponibilité au FMI	61.3	90.2	79.8	82.4	85.8	99.5	95.7	117.3	106.3	106.9
Foreign exchange / Devises étrangères	799.8	809.7	873.3	785.6	## 1 432.9	2 581.9	3 523.4	3 105.8	3 305.5	5 081.6
Grenada / Grenade	9.2	14.1	14.2	20.8	20.6	22.7	16.9	15.4	17.6	...
SDR / DTS	0.0	0.2	0.0	0.0	0.0	0.0	0.0	0.0	0.0	0.0
Position in IMF / Disponibilité au FMI	0.0	0.0	0.0	0.0	0.0	0.0	0.0	0.0	0.0	0.0
Foreign exchange / Devises étrangères	9.2	14.0	14.2	20.8	20.6	22.7	16.9	15.4	17.6	...
Guatemala / Guatemala	112.2	210.0	274.4	300.9	362.1	287.8	201.2	306.0	282.0	807.3
SDR / DTS	0.0	0.6	2.0	0.0	0.0	1.7	0.2	0.7	0.0	0.0
Position in IMF / Disponibilité au FMI	0.0	8.3	0.0	0.0	0.0	0.0	0.0	0.0	0.0	0.0
Foreign exchange / Devises étrangères	112.2	201.2	272.4	300.9	362.1	286.1	201.0	305.3	282.0	807.3

127
International reserves excluding gold
Special drawing rights(SDR) + Reserve position in IMF + Foreign exchange: million $US (end of period) [cont.]
Réserves internationales, l'or non compris
Droits de tirage spéciaux(DTS) + Réserve disponible au FMI + Devises étrangères : $E-U millions (fin de la période) [suite]

Country or area and reserves Pays ou zone et réserves	1982	1983	1984	1985	1986	1987	1988	1989	1990	1991
Guyana **Guyana**	**10.6**	**6.5**	**5.9**	**6.5**	**9.0**	**8.4**	**4.0**	**13.4**	**28.7**	**124.4**
SDR DTS	2.9	0.0	0.0	0.0	0.0	0.0	0.0	0.0	2.1	1.4
Position in IMF Disponibilité au FMI	0.0	0.0	0.0	0.0	0.0	0.0	0.0	0.0	0.0	0.0
Foreign exchange Devises étrangères	7.7	6.5	5.9	6.5	9.0	8.4	4.0	13.4	26.6	123.0
Haiti **Haïti**	**4.2**	**9.0**	**13.0**	**6.4**	**15.9**	**17.0**	**13.0**	**12.6**	**3.2**	**17.3**
SDR DTS	1.1	1.0	0.0	0.0	6.6	0.0	0.0	0.1	0.0	0.0
Position in IMF Disponibilité au FMI	0.1	0.1	0.1	0.1	0.1	0.1	0.1	0.1	0.1	0.1
Foreign exchange Devises étrangères	3.1	7.9	12.9	6.3	9.2	16.9	12.9	12.4	3.1	17.2
Honduras **Honduras**	**112.2**	**113.6**	**128.2**	**105.8**	**111.3**	**106.0**	**50.0**	**21.1**	**40.4**	**104.9**
SDR DTS	1.8	2.2	0.2	0.0	0.0	0.0	0.0	0.0	0.0	0.0
Position in IMF Disponibilité au FMI	0.0	4.4	0.0	0.0	0.0	0.0	0.0	0.0	0.0	0.0
Foreign exchange Devises étrangères	110.4	107.0	128.0	105.8	111.3	106.0	50.0	21.1	40.4	104.9
Hungary **Hongrie**	...	**1 231.0**	**1 560.0**	**2 153.0**	**2 302.0**	**1 634.0**	**1 467.0**	**1 246.0**	**1 070.0**	...
SDR DTS	2.5	46.5	0.0	0.2	0.0	0.3	0.2	0.0	1.0	1.3
Position in IMF Disponibilité au FMI	0.0	40.8	0.0	0.0	0.0	0.0	0.0	0.0	0.0	0.0
Foreign exchange Devises étrangères	...	1 144.0	1 560.0	2 153.0	2 302.0	1 634.0	1 467.0	1 246.0	1 069.0	...
Iceland **Islande**	**145.2**	**149.3**	**127.6**	**205.5**	**309.8**	**311.3**	**290.7**	**337.3**	**436.1**	**449.5**
SDR DTS	2.2	0.2	0.4	0.4	0.2	2.7	1.3	0.0	0.4	0.1
Position in IMF Disponibilité au FMI	0.0	4.2	4.0	4.4	4.9	5.7	5.4	5.3	5.7	5.8
Foreign exchange Devises étrangères	143.0	144.9	123.3	200.7	304.7	302.9	284.0	332.0	430.0	443.6
India **Inde**	**4 315.0**	**4 937.0**	**5 842.0**	**6 420.0**	**6 396.0**	**6 454.0**	**4 899.0**	**3 859.0**	**1 521.0**	**3 627.0**
SDR DTS	374.3	109.5	331.0	336.3	355.9	159.4	95.7	113.2	316.0	46.3
Position in IMF Disponibilité au FMI	401.8	509.7	477.3	535.0	595.9	691.1	655.6	640.3	0.0	0.3
Foreign exchange Devises étrangères	3 539.0	4 318.0	5 034.0	5 549.0	5 444.0	5 603.0	4 148.0	3 105.0	1 205.0	3 580.0
Indonesia **Indonésie**	**3 144.0**	**3 718.0**	**4 773.0**	**4 974.0**	**4 051.0**	**5 592.0**	**5 048.0**	**5 454.0**	**7 459.0**	**9 258.0**
SDR DTS	311.0	4.1	0.5	56.3	43.5	6.2	2.7	1.0	3.3	3.7
Position in IMF Disponibilité au FMI	240.9	75.8	71.0	79.6	88.6	102.8	97.5	95.2	103.0	103.6
Foreign exchange Devises étrangères	2 592.6	3 638.5	4 701.5	4 838.4	3 919.2	5 483.4	4 948.2	5 357.4	7 352.7	9 150.7
Iran, Islamic Rep. of **Iran, Rép. islamique d'**	**5 701.0**	...	...	...	...	...	...	...	...	...

127
International reserves excluding gold
Special drawing rights(SDR) + Reserve position in IMF + Foreign exchange: million $US (end of period) [cont.]
Réserves internationales, l'or non compris
Droits de tirage spéciaux(DTS) + Réserve disponible au FMI + Devises étrangères : $E-U millions (fin de la période) [suite]

Country or area and reserves Pays ou zone et réserves	1982	1983	1984	1985	1986	1987	1988	1989	1990	1991
SDR DTS	330.7	323.7	313.9	360.6	410.3	485.6	155.7	400.5	441.6	308.7
Position in IMF Disponibilité au FMI	83.5	74.1	69.4	77.7	86.6	100.4	0.0	0.0	0.0	0.0
Foreign exchange Devises étrangères	5 287.0	...	...	...	...	...	...	...	...	...
Iraq[2] **Iraq[2]**	**205.4**	**8.9**	**0.1**	**0.0**	**0.0**	**7.3**	**0.0**	**0.0**	**0.0**	**0.0**
SDR DTS	81.9	8.9	0.1	0.0	0.0	7.3	0.0	0.0	0.0	0.0
Position in IMF Disponibilité au FMI	123.5	0.0	0.0	0.0	0.0	0.0	0.0	0.0	0.0	0.0
Ireland **Irlande**	**2 622.0**	**2 640.0**	**2 352.0**	**2 940.0**	**3 236.0**	**4 796.0**	**5 087.0**	**4 057.0**	**5 223.0**	**5 740.0**
SDR DTS	106.4	68.5	87.6	108.7	138.6	179.4	180.7	191.1	225.1	243.2
Position in IMF Disponibilité au FMI	82.9	121.1	121.9	132.7	159.9	185.7	180.7	164.8	148.8	177.2
Foreign exchange Devises étrangères	2 433.0	2 450.0	2 143.0	2 698.0	2 938.0	4 431.0	4 725.0	3 702.0	4 849.0	5 320.0
Israel **Israël**	**3 839.3**	**3 651.2**	**3 060.3**	**3 680.2**	**4 659.6**	**5 876.1**	**4 015.6**	**5 276.2**	**6 275.1**	**6 279.1**
SDR DTS	0.6	1.7	0.1	0.1	0.0	0.1	0.1	0.1	0.2	0.4
Position in IMF Disponibilité au FMI	0.0	36.4	0.0	0.0	0.0	0.0	0.0	0.0	0.0	0.0
Foreign exchange Devises étrangères	3 838.7	3 613.1	3 060.2	3 680.1	4 659.6	5 876.0	4 015.5	5 276.1	6 274.9	6 278.7
Italy **Italie**	**14 091.0**	**20 105.0**	**20 795.0**	**15 595.0**	**19 987.0**	**30 214.0**	**34 715.0**	**46 720.0**	**62 927.0**	**48 679.0**
SDR DTS	784.7	591.4	632.6	326.1	587.3	947.9	948.8	998.1	1 037.5	929.8
Position in IMF Disponibilité au FMI	695.6	989.7	1 074.1	1 160.1	1 267.9	1 446.6	1 266.5	1 443.6	1 713.7	2 254.6
Foreign exchange Devises étrangères	12 611.0	18 524.0	19 088.0	14 109.0	18 132.0	27 819.0	32 500.0	44 278.0	60 176.0	45 495.0
Jamaica **Jamaïque**	**109.0**	**63.2**	**96.9**	**161.3**	**98.4**	**174.3**	**147.2**	**107.5**	**168.2**	**...**
SDR DTS	0.1	0.0	0.0	0.0	0.4	1.4	0.0	0.0	0.5	0.1
Position in IMF Disponibilité au FMI	4.1	0.0	0.0	0.0	0.0	0.0	0.0	0.0	0.0	0.0
Foreign exchange Devises étrangères	104.7	63.2	96.9	** 161.3	98.0	172.9	147.2	107.5	** 167.7	...
Japan **Japon**	**23 334.0**	**24 602.0**	**26 429.0**	**26 719.0**	**42 257.0**	**80 973.0**	**96 728.0**	**83 957.0**	**78 501.0**	**72 059.0**
SDR DTS	2 090.9	1 935.0	1 927.1	2 115.7	2 217.7	2 463.3	2 936.4	2 447.3	3 042.2	2 579.1
Position in IMF Disponibilité au FMI	2 071.1	2 302.6	2 219.0	2 274.9	2 381.9	2 852.6	3 277.8	3 518.0	5 971.4	7 721.8
Foreign exchange Devises étrangères	19 172.0	20 364.0	22 283.0	22 328.0	37 657.0	75 657.0	90 514.0	77 992.0	69 487.0	61 758.0
Jordan **Jordanie**	**884.1**	**824.2**	**515.0**	**422.8**	**437.1**	**424.7**	**109.6**	**470.7**	**848.8**	**825.8**
SDR DTS	18.2	18.2	15.5	24.1	23.9	12.1	0.1	11.0	1.0	1.1
Position in IMF Disponibilité au FMI	18.3	7.6	0.0	0.0	0.0	0.0	0.0	0.0	0.0	0.0
Foreign exchange Devises étrangères	847.6	798.4	499.5	398.7	413.2	412.6	109.5	459.7	847.8	824.7

127
International reserves excluding gold
Special drawing rights(SDR) + Reserve position in IMF + Foreign exchange: million $US (end of period) [cont.]
Réserves internationales, l'or non compris
Droits de tirage spéciaux(DTS) + Réserve disponible au FMI + Devises étrangères : $E-U millions (fin de la période) [suite]

Country or area and reserves Pays ou zone et réserves	1982	1983	1984	1985	1986	1987	1988	1989	1990	1991
Kenya **Kenya**	**211.7**	**376.0**	**389.8**	**390.6**	**413.3**	**255.8**	**263.7**	**284.6**	**205.4**	**116.9**
SDR DTS	15.4	17.4	2.1	0.9	12.1	16.2	0.6	11.5	3.9	1.4
Position in IMF Disponibilité au FMI	1.6	10.1	10.7	13.4	15.0	17.3	16.4	16.1	17.4	17.5
Foreign exchange Devises étrangères	194.6	348.5	377.0	376.4	386.3	222.3	246.7	257.1	184.1	98.1
Korea, Republic of **Corée, République de**	**2 807.3**	**2 346.7**	**2 753.6**	**2 869.3**	**3 319.6**	**3 583.7**	**12 346.7**	**15 213.6**	**14 793.0**	**13 701.1**
SDR DTS	63.7	63.1	30.3	39.8	17.7	16.4	5.7	1.6	14.4	29.8
Position in IMF Disponibilité au FMI	0.0	54.2	0.0	0.8	0.8	1.0	0.9	234.2	319.4	365.3
Foreign exchange Devises étrangères	2 743.6	2 229.5	2 723.3	2 828.8	3 301.1	3 566.3	12 340.1	14 977.8	14 459.2	13 306.0
Kuwait **Koweït**	**5 913.2**	**5 192.1**	**4 590.2**	**5 470.7**	**5 501.1**	**4 141.6**	**1 923.5**	**3 101.9**	**1 951.7**	**3 409.0**
SDR DTS	68.8	37.3	74.5	114.6	157.1	211.1	224.4	128.6	162.1	183.5
Position in IMF Disponibilité au FMI	508.6	729.4	702.3	702.3	631.0	536.2	332.8	207.9	175.7	159.0
Foreign exchange Devises étrangères	5 335.9	4 425.4	3 813.4	4 653.8	4 7 1	3 394.3	1 366.3	2 765.4	...	3 066.6
Lebanon **Liban**	**2 608.1**	**1 902.5**	**671.6**	**1 073.8**	**488.0**	**367.9**	**977.8**	**938.2**	**659.9**	**1 275.5**
SDR DTS	2.3	0.0	0.8	1.9	3.2	4.8	5.7	7.1	9.8	11.8
Position in IMF Disponibilité au FMI	6.8	19.7	18.5	20.7	23.0	26.7	25.3	24.8	26.8	26.9
Foreign exchange Devises étrangères	2 599.1	1 882.8	652.3	1 051.2	461.8	336.4	946.8	906.3	623.3	1 236.7
Lesotho **Lesotho**	**47.5**	**66.7**	**48.6**	**43.5**	**60.3**	**67.5**	**56.3**	**49.0**	**72.4**	**115.0**
SDR DTS	1.2	1.1	1.0	1.1	1.0	0.9	1.3	1.0	0.7	0.3
Position in IMF Disponibilité au FMI	0.1	1.3	1.2	1.4	1.5	1.8	1.7	1.7	1.9	1.9
Foreign exchange Devises étrangères	46.2	64.3	46.4	41.1	57.7	64.8	53.2	46.3	69.9	112.9
Liberia **Libéria**	**6.5**	**20.4**	**3.5**	**1.5**	**2.7**	**0.5**	**0.4**	**7.9**	**...**	**...**
SDR DTS	0.0	0.0	0.0	0.0	0.0	0.0	0.0	0.0	0.0	0.0
Position in IMF Disponibilité au FMI	0.0	0.0	0.0	0.0	0.0	0.0	0.0	0.0	0.0	0.0
Foreign exchange Devises étrangères	6.5	20.4	3.5	1.5	2.6	0.5	0.4	7.8	...	...
Libyan Arab Jamah. **Jamah. arabe libyenne**	**7 059.5**	**5 218.7**	**3 634.2**	**5 903.9**	**5 953.2**	**5 838.0**	**4 321.6**	**4 332.7**	**5 839.2**	**5 695.0**
SDR DTS	142.7	165.1	130.0	171.5	217.2	280.5	294.5	327.8	408.7	461.4
Position in IMF Disponibilité au FMI	208.7	198.1	238.7	267.5	297.9	345.5	327.7	320.0	346.4	348.3
Foreign exchange Devises étrangères	6 708.1	4 855.6	3 265.5	5 464.9	* * 5 438.1	* * 5 212.1	* * 3 699.4	* * 3 684.9	* * 5 084.1	* * 4 885.3
Luxembourg [2] **Luxembourg [2]**	**30.5**	**29.5**	**28.3**	**32.2**	**36.5**	**42.9**	**41.3**	**41.1**	**45.7**	**47.1**
SDR DTS	17.0	16.7	16.3	18.8	21.5	25.5	24.8	25.0	28.3	29.5

127
International reserves excluding gold
Special drawing rights(SDR) + Reserve position in IMF + Foreign exchange: million $US (end of period) [cont.]
Réserves internationales, l'or non compris
Droits de tirage spéciaux(DTS) + Réserve disponible au FMI + Devises étrangères : $E-U millions (fin de la période) [suite]

Country or area and reserves Pays ou zone et réserves	1982	1983	1984	1985	1986	1987	1988	1989	1990	1991
Position in IMF Disponibilité au FMI	13.5	12.8	12.0	13.4	15.0	17.4	16.5	16.1	17.4	17.6
Madagascar **Madagascar**	**20.0**	**29.2**	**58.9**	**48.4**	**114.5**	**185.2**	**223.7**	**245.3**	**92.1**	**...**
SDR DTS	1.3	0.1	1.5	0.0	0.0	0.1	0.1	0.1	0.2	0.1
Position in IMF Disponibilité au FMI	1.3	0.0	0.0	0.0	0.0	0.0	0.0	0.0	0.0	0.0
Foreign exchange Devises étrangères	17.4	29.1	57.4	48.4	114.5	185.1	223.6	245.2	91.9	...
Malawi **Malawi**	**22.7**	**15.4**	**56.6**	**45.0**	**24.6**	**51.8**	**145.6**	**100.3**	**137.2**	**153.2**
SDR DTS	4.0	0.9	2.8	0.0	0.5	0.0	3.2	0.4	3.2	0.3
Position in IMF Disponibilité au FMI	0.0	2.3	2.2	2.4	2.7	3.1	3.0	2.9	3.2	3.2
Foreign exchange Devises étrangères	18.7	12.3	51.7	42.5	21.5	48.7	139.4	97.0	130.8	149.8
Malaysia **Malaisie**	**3 768.0**	**3 784.0**	**3 723.0**	**4 912.0**	**6 027.0**	**7 435.0**	**6 527.0**	**7 783.0**	**9 754.0**	**...**
SDR DTS	130.2	107.9	97.1	115.8	135.4	163.3	161.1	167.0	193.9	207.4
Position in IMF Disponibilité au FMI	128.5	166.8	156.2	175.0	195.0	217.0	231.5	223.0	233.2	257.2
Foreign exchange Devises étrangères	3 509.0	3 509.0	3 470.0	4 621.0	5 697.0	7 055.0	6 134.0	7 393.0	9 327.0	...
Maldives **Maldives**	**8.4**	**4.5**	**5.1**	**4.6**	**6.9**	**8.2**	**21.6**	**24.8**	**24.4**	**23.5**
SDR DTS	0.2	0.0	0.0	0.0	0.0	0.0	0.0	0.0	0.0	0.0
Position in IMF Disponibilité au FMI	0.3	0.5	0.0	0.0	0.0	0.0	0.0	0.0	0.0	0.0
Foreign exchange Devises étrangères	7.9	4.1	5.1	4.6	6.9	8.2	21.6	24.7	24.4	23.4
Mali **Mali**	**16.7**	**16.2**	**26.6**	**22.5**	**23.3**	**15.8**	**36.0**	**115.8**	**190.5**	**319.4**
SDR DTS	0.6	0.2	1.7	1.8	0.0	0.3	0.2	0.2	0.5	0.4
Position in IMF Disponibilité au FMI	9.6	9.1	8.5	9.5	10.6	12.3	11.7	11.4	12.4	12.4
Foreign exchange Devises étrangères	6.6	7.0	16.5	11.2	12.7	3.3	24.1	104.2	177.7	306.5
Malta **Malte**	**1 083.6**	**1 112.3**	**990.2**	**986.9**	**1 145.4**	**1 414.6**	**1 364.7**	**1 355.1**	**1 431.8**	**1 333.3**
SDR DTS	23.4	32.4	35.0	43.5	52.8	66.1	67.5	70.4	84.1	91.8
Position in IMF Disponibilité au FMI	28.4	30.2	30.3	32.6	39.6	44.3	36.7	30.9	27.9	29.8
Foreign exchange Devises étrangères	1 031.8	1 049.7	925.0	910.9	1 053.0	1 304.2	1 260.5	1 253.8	1 319.8	1 211.7
Mauritania **Mauritanie**	**139.1**	**105.9**	**77.5**	**59.2**	**48.2**	**71.8**	**55.6**	**82.4**	**54.1**	**...**
SDR DTS	0.0	0.7	0.0	4.3	3.1	17.2	0.1	0.1	0.8	0.1
Position in IMF Disponibilité au FMI	0.0	0.0	0.0	0.0	0.0	0.0	0.0	0.0	0.0	0.0
Foreign exchange Devises étrangères	139.1	105.2	77.5	54.9	45.1	54.6	55.5	82.3	53.3	...

127
International reserves excluding gold
Special drawing rights(SDR) + Reserve position in IMF + Foreign exchange: million $US (end of period) [cont.]
Réserves internationales, l'or non compris
Droits de tirage spéciaux(DTS) + Réserve disponible au FMI + Devises étrangères : $E-U millions (fin de la période) [suite]

Country or area and reserves Pays ou zone et réserves	1982	1983	1984	1985	1986	1987	1988	1989	1990	1991
Mauritius **Maurice**	**38.0**	**17.9**	**23.6**	**29.9**	**136.0**	**343.5**	**442.0**	**517.8**	**737.6**	**893.2**
SDR DTS	1.9	0.1	0.1	0.0	0.6	5.9	5.0	6.6	14.6	25.7
Position in IMF Disponibilité au FMI	0.0	0.0	0.0	0.0	0.0	0.1	0.1	0.1	0.1	1.8
Foreign exchange Devises étrangères	36.1	17.8	23.5	29.9	135.4	337.5	437.0	511.2	722.9	865.8
Mexico **Mexique**	**834.0**	**3 913.0**	**7 272.0**	**4 906.0**	**5 670.0**	**12 464.0**	**5 279.0**	**6 329.0**	**9 863.0**	**17 726.0**
SDR DTS	5.9	22.9	3.0	0.4	8.8	706.1	393.7	383.1	416.9	585.5
Position in IMF Disponibilité au FMI	0.0	95.0	0.0	0.0	0.0	0.0	0.0	0.0	0.0	0.0
Foreign exchange Devises étrangères	828.0	3 795.0	7 269.0	4 906.0	5 661.0	11 758.0	4 885.0	5 946.0	9 446.0	17 140.0
Morocco **Maroc**	**218.0**	**107.0**	**49.0**	**115.0**	**211.0**	**411.0**	**547.0**	**488.0**	**2 066.0**	**3 100.0**
SDR DTS	0.6	0.7	0.7	0.1	19.3	4.1	0.4	0.5	1.5	147.1
Position in IMF Disponibilité au FMI	0.0	0.0	0.0	0.0	0.0	0.0	0.0	0.0	0.0	0.0
Foreign exchange Devises étrangères	217.0	106.0	48.0	115.0	192.0	407.0	547.0	488.0	2 065.0	2 953.0
Myanmar **Myanmar**	**104.3**	**89.4**	**62.1**	**33.9**	**33.1**	**27.2**	**77.4**	**263.4**	**312.8**	**258.4**
SDR DTS	1.2	0.2	0.1	0.0	0.0	0.1	0.1	0.6	0.8	0.2
Position in IMF Disponibilité au FMI	14.3	7.2	6.7	0.0	0.0	0.0	0.0	0.0	0.0	0.0
Foreign exchange Devises étrangères	88.8	82.0	55.3	33.9	33.1	27.1	77.3	262.8	312.0	258.2
Nepal **Népal**	**199.2**	**133.3**	**82.0**	**56.0**	**86.7**	**178.2**	**220.3**	**211.6**	**295.3**	**385.7**
SDR DTS	0.9	0.2	0.1	0.0	0.1	0.1	0.1	0.2	0.2	0.1
Position in IMF Disponibilité au FMI	6.3	6.0	5.6	6.3	7.0	8.1	7.7	7.5	8.1	8.2
Foreign exchange Devises étrangères	192.1	127.2	76.3	49.7	79.7	170.0	212.5	203.9	287.0	377.4
Netherlands **Pays-Bas**	**10 132.0**	**10 171.0**	**9 237.0**	**10 782.0**	**11 191.0**	**16 003.0**	**16 075.0**	**16 508.0**	**17 484.0**	**17 798.0**
SDR DTS	851.3	525.4	514.6	625.4	731.2	903.4	775.8	775.9	717.6	757.9
Position in IMF Disponibilité au FMI	619.2	943.3	944.2	986.6	877.1	925.6	757.1	705.6	738.0	799.6
Foreign exchange Devises étrangères	8 662.0	8 702.0	7 778.0	9 170.0	9 583.0	14 174.0	14 542.0	15 027.0	16 028.0	16 240.0
Netherlands Antilles[1] **Antilles néerlandaises**[1]	**187.0**	**164.0**	**118.0**	**176.0**	**239.0**	**217.0**	**263.0**	**207.0**	**215.0**	**...**
Foreign exchange Devises étrangères	187.0	164.0	118.0	176.0	239.0	217.0	263.0	207.0	215.0	...
New Zealand **Nouvelle-Zélande**	**636.0**	**778.0**	**1 787.0**	**1 596.0**	**3 771.0**	**3 260.0**	**2 836.0**	**3 027.0**	**4 129.0**	**3 203.0**
SDR DTS	1.9	2.8	6.7	6.6	11.0	1.4	0.9	0.6	0.6	0.5
Position in IMF Disponibilité au FMI	0.0	29.8	0.0	0.1	0.2	0.2	11.0	52.3	57.2	77.3
Foreign exchange Devises étrangères	634.0	745.0	1 780.0	1 589.0	3 760.0	3 258.0	2 824.0	2 974.0	4 071.0	3 125.0

127
International reserves excluding gold
Special drawing rights(SDR) + Reserve position in IMF + Foreign exchange: million $US (end of period) [cont.]
Réserves internationales, l'or non compris
Droits de tirage spéciaux(DTS) + Réserve disponible au FMI + Devises étrangères : $E-U millions (fin de la période) [suite]

Country or area and reserves Pays ou zone et réserves	1982	1983	1984	1985	1986	1987	1988	1989	1990	1991
Niger Niger	29.6	53.2	88.7	136.4	189.2	248.5	232.1	212.3	222.2	202.8
SDR DTS	8.2	4.8	2.2	0.0	1.2	0.2	0.2	1.2	0.1	0.4
Position in IMF Disponibilité au FMI	6.8	9.0	8.4	9.4	10.5	12.1	11.5	11.3	12.2	12.2
Foreign exchange Devises étrangères	14.6	39.4	78.1	127.0	177.6	236.1	220.4	199.8	210.0	190.1
Nigeria Nigéria	1 613.0	990.0	1 462.0	1 667.0	1 081.0	1 165.0	651.0	1 766.0	3 864.0	4 435.0
SDR DTS	44.5	26.9	10.3	1.1	0.3	0.2	0.1	0.5	1.2	0.0
Position in IMF Disponibilité au FMI	0.0	0.0	0.0	0.1	0.1	0.1	0.1	0.1	0.1	0.1
Foreign exchange Devises étrangères	1 568.0	963.0	1 452.0	1 666.0	1 081.0	1 165.0	651.0	1 765.0	3 863.0	4 435.0
Norway Norvège	6 873.5	6 629.2	9 365.0	13 916.7	12 524.6	14 276.5	13 267.7	13 784.8	15 332.3	13 232.0
SDR DTS	313.7	269.3	257.0	283.4	389.2	441.5	487.5	453.7	448.9	451.9
Position in IMF Disponibilité au FMI	271.9	430.3	461.1	509.3	589.4	706.6	607.4	580.7	580.2	570.8
Foreign exchange Devises étrangères	6 288.0	5 929.5	8 647.0	13 124.0	11 546.0	13 128.4	12 172.9	12 750.5	14 303.2	12 209.4
Oman Oman	872.4	762.6	900.2	1 090.2	968.0	1 402.2	1 054.2	1 354.3	1 672.4	1 663.3
SDR DTS	8.7	11.6	9.2	12.0	13.9	10.2	11.8	14.3	19.0	22.2
Position in IMF Disponibilité au FMI	22.9	31.9	32.5	35.2	39.2	45.5	38.7	36.4	35.8	32.5
Foreign exchange Devises étrangères	840.8	719.1	858.4	1 043.0	914.8	1 346.5	1 003.8	1 303.6	1 617.6	1 608.6
Pakistan Pakistan	969.0	1 973.0	1 035.0	807.0	709.0	502.0	395.0	521.0	296.0	527.0
SDR DTS	50.6	0.9	36.7	26.4	13.0	15.8	6.5	1.5	0.8	7.4
Position in IMF Disponibilité au FMI	64.9	92.7	86.8	0.0	0.0	0.1	0.1	0.1	0.1	0.1
Foreign exchange Devises étrangères	853.0	1 879.0	912.0	781.0	696.0	486.0	388.0	519.0	295.0	519.0
Panama Panama	101.0	206.7	215.6	98.0	170.2	77.8	72.2	119.4	406.4	498.7
SDR DTS	4.2	0.4	0.0	12.9	1.8	0.0	0.0	0.0	27.6	11.6
Position in IMF Disponibilité au FMI	0.0	9.1	0.0	0.0	0.0	0.0	0.0	0.0	0.0	0.0
Foreign exchange Devises étrangères	96.8	197.2	215.6	85.1	168.4	77.8	72.2	119.4	378.7	487.1
Papua New Guinea Papouasie-Nvl-Guinée	452.9	440.1	435.2	442.6	425.5	436.8	393.5	384.4	403.0	323.1
SDR DTS	34.2	17.8	4.9	6.5	3.2	4.7	4.1	3.5	0.0	0.0
Position in IMF Disponibilité au FMI	0.1	5.6	5.2	5.9	6.6	9.9	9.4	9.2	0.0	0.0
Foreign exchange Devises étrangères	418.6	416.8	425.1	430.2	415.6	422.3	380.1	371.7	403.0	323.0
Paraguay Paraguay	739.0	680.2	666.3	533.6	446.7	497.0	323.7	432.6	686.4	562.6
SDR DTS	26.1	31.8	34.3	42.6	51.5	63.5	63.8	66.8	78.2	84.1

127
International reserves excluding gold
Special drawing rights(SDR) + Reserve position in IMF + Foreign exchange: million $US (end of period) [cont.]
Réserves internationales, l'or non compris
Droits de tirage spéciaux(DTS) + Réserve disponible au FMI + Devises étrangères : $E-U millions (fin de la période) [suite]

Country or area and reserves Pays ou zone et réserves	1982	1983	1984	1985	1986	1987	1988	1989	1990	1991
Position in IMF Disponibilité au FMI	30.5	33.8	31.6	34.7	30.4	27.8	20.2	16.4	15.7	15.8
Foreign exchange Devises étrangères	682.4	614.6	600.3	456.3	364.8	405.7	239.8	349.4	592.6	462.8
Peru **Pérou**	**1 349.6**	**1 365.7**	**1 630.5**	**1 842.0**	**1 407.2**	**645.8**	**511.0**	**808.4**	**1 039.8**	**2 443.0**
SDR DTS	33.0	0.6	22.4	0.0	0.0	0.0	0.0	0.0	0.0	0.0
Position in IMF Disponibilité au FMI	0.0	0.0	0.0	0.0	0.0	0.0	0.0	0.0	0.0	0.0
Foreign exchange Devises étrangères	1 316.7	1 365.1	1 608.1	1 842.0	1 407.2	645.8	511.0	808.4	1 039.8	2 443.0
Philippines **Philippines**	**888.0**	**747.0**	**602.0**	**615.0**	**1 728.0**	**968.0**	**1 003.0**	**1 417.0**	**924.0**	**3 246.0**
SDR DTS	2.8	0.9	19.4	38.7	5.7	0.2	0.1	0.9	1.1	4.4
Position in IMF Disponibilité au FMI	0.0	0.0	8.7	26.2	47.5	55.1	52.3	51.0	55.2	55.5
Foreign exchange Devises étrangères	885.0	746.0	574.0	550.0	1 675.0	913.0	951.0	1 365.0	868.0	3 186.0
Poland **Pologne**	**646.8**	**765.2**	**1 106.0**	**870.4**	**697.8**	**1 494.7**	**2 055.3**	**2 314.3**	**4 492.1**	**3 632.6**
SDR DTS	0.0	0.0	0.0	0.0	0.1	0.1	0.1	0.1	0.8	7.7
Position in IMF Disponibilité au FMI	0.0	0.0	0.0	0.0	0.0	0.0	0.0	0.0	0.0	0.0
Foreign exchange Devises étrangères	646.8	765.2	1 106.0	870.4	697.7	1 494.6	2 055.2	2 314.2	76.9	3 624.9
Portugal **Portugal**	**447.0**	**385.0**	**516.0**	**1 395.0**	**1 456.0**	**3 327.0**	**5 127.0**	**9 952.0**	**14 485.0**	**20 629.0**
SDR DTS	2.5	0.9	12.3	17.1	66.1	79.5	3.8	1.8	57.3	97.6
Position in IMF Disponibilité au FMI	53.8	31.1	29.1	32.6	36.3	42.1	40.0	124.7	176.0	270.2
Foreign exchange Devises étrangères	391.0	353.0	475.0	1 345.0	1 354.0	3 205.0	5 083.0	9 826.0	14 252.0	20 261.0
Qatar **Qatar**	**386.6**	**384.1**	**380.0**	**446.1**	**571.9**	**618.4**	**474.5**	**533.4**	**631.1**	**...**
SDR DTS	15.4	9.1	15.9	20.7	25.8	35.0	35.6	37.7	44.6	48.3
Position in IMF Disponibilité au FMI	21.4	38.9	37.6	39.3	37.6	39.5	28.4	25.7	24.6	26.8
Foreign exchange Devises étrangères	349.8	336.1	326.5	386.2	508.5	543.9	410.5	470.0	561.9	...
Romania **Roumanie**	**450.0**	**525.0**	**709.0**	**199.0**	**582.0**	**1 402.0**	**780.0**	**1 859.0**	**373.0**	**380.0**
SDR DTS	13.0	0.2	0.3	0.0	0.2	0.0	0.0	99.8	0.1	57.7
Position in IMF Disponibilité au FMI	0.0	0.0	0.0	0.0	0.0	0.0	0.0	0.0	0.0	0.0
Foreign exchange Devises étrangères	437.0	525.0	709.0	199.0	582.0	1 402.0	780.0	1 759.0	373.0	323.0
Rwanda **Rwanda**	**128.4**	**110.9**	**106.9**	**113.3**	**162.3**	**164.2**	**118.3**	**70.4**	**44.4**	**110.1**
SDR DTS	12.0	8.8	8.1	9.0	9.9	11.4	10.6	9.9	10.2	9.6
Position in IMF Disponibilité au FMI	8.1	10.1	9.5	10.2	11.4	13.2	9.5	9.3	9.2	9.2
Foreign exchange Devises étrangères	108.4	92.0	89.3	94.1	141.0	139.6	98.2	51.1	25.0	91.3

127
International reserves excluding gold
Special drawing rights(SDR) + Reserve position in IMF + Foreign exchange: million $US (end of period) [cont.]
Réserves internationales, l'or non compris
Droits de tirage spéciaux(DTS) + Réserve disponible au FMI + Devises étrangères : $E-U millions (fin de la période) [suite]

Country or area and reserves Pays ou zone et réserves	1982	1983	1984	1985	1986	1987	1988	1989	1990	1991
Saint Kitts-Nevis **Saint-Kitts-et-Nevis**	**3.3**	**3.2**	**5.7**	**7.4**	**10.2**	**10.6**	**10.3**	**16.4**	**16.3**	...
SDR DTS	0.0	0.0	0.0	0.0	0.0	0.0	0.0	0.0	0.0	0.0
Position in IMF Disponibilité au FMI	0.0	0.0	0.0	0.0	0.0	0.0	0.0	0.0	0.0	0.0
Foreign exchange Devises étrangères	3.3	3.2	5.7	7.4	10.2	10.6	10.3	16.4	16.3	...
Saint Lucia **Sainte-Lucie**	**8.2**	**8.9**	**12.4**	**12.7**	**25.1**	**30.8**	**32.7**	**38.2**	**44.6**	...
SDR DTS	0.0	0.0	0.0	0.0	0.0	0.0	0.0	0.0	1.8	1.8
Position in IMF Disponibilité au FMI	0.0	0.0	0.0	0.0	0.0	0.0	0.0	0.0	0.0	0.0
Foreign exchange Devises étrangères	8.2	8.9	12.4	12.7	25.1	30.8	32.7	38.2	42.8	...
St. Vincent-Grenadines **St. Vincent-Grenadines**	**4.8**	**5.7**	**12.8**	**13.8**	**25.8**	**20.2**	**21.8**	**22.8**	**26.5**	...
SDR DTS	0.0	0.0	0.0	0.0	0.0	0.0	0.0	0.0	0.0	0.0
Position in IMF Disponibilité au FMI	0.0	0.4	0.0	0.0	0.0	0.0	0.0	0.0	0.0	0.0
Foreign exchange Devises étrangères	4.8	5.3	12.8	13.8	25.8	20.2	21.8	22.8	26.5	...
Samoa **Samoa**	**3.5**	**7.2**	**10.6**	**14.0**	**23.8**	**37.2**	**49.2**	**55.1**	**69.1**	**67.8**
SDR DTS	0.0	0.4	0.2	0.0	1.0	1.9	3.3	0.9	4.2	3.7
Position in IMF Disponibilité au FMI	0.0	0.0	0.0	0.0	0.0	0.0	0.0	0.0	0.0	0.0
Foreign exchange Devises étrangères	3.5	6.8	10.4	14.0	22.8	35.3	45.9	54.1	64.8	64.1
Saudi Arabia **Arabie saoudite**	**29 549.0**	**27 287.0**	**24 748.0**	**25 004.0**	**18 324.0**	**22 684.0**	**20 553.0**	**16 748.0**	**11 668.0**	**11 673.0**
SDR DTS	638.4	509.6	574.4	580.8	410.8	526.1	532.3	613.9	99.4	88.7
Position in IMF Disponibilité au FMI	5 097.8	9 320.7	9 986.0	10 662.0	10 811.0	11 372.0	8 801.5	4 846.6	2 986.3	1 847.6
Foreign exchange Devises étrangères	23 813.0	17 457.0	14 188.0	13 761.0	7 102.0	10 786.0	11 219.0	11 287.0	8 582.0	9 737.0
Senegal **Sénégal**	**11.4**	**12.2**	**3.7**	**5.1**	**9.4**	**9.2**	**10.5**	**19.0**	**11.0**	**13.2**
SDR DTS	5.4	3.9	0.1	0.1	3.0	0.0	0.0	4.7	0.2	0.4
Position in IMF Disponibilité au FMI	1.0	1.0	0.9	1.1	1.2	1.4	1.3	1.3	1.4	1.5
Foreign exchange Devises étrangères	5.0	7.3	2.6	3.9	5.2	7.8	9.1	13.0	9.3	11.3
Seychelles **Seychelles**	**13.1**	**10.0**	**5.4**	**8.5**	**7.8**	**13.7**	**8.7**	**12.1**	**16.6**	**27.6**
SDR DTS	0.2	0.1	0.0	0.0	0.0	0.0	0.0	0.0	0.0	0.1
Position in IMF Disponibilité au FMI	0.5	0.0	0.0	0.0	0.0	0.0	0.1	0.1	0.1	0.1
Foreign exchange Devises étrangères	12.4	9.9	5.4	8.5	7.7	13.7	8.6	12.0	16.5	27.5
Sierra Leone **Sierra Leone**	**8.4**	**16.2**	**7.7**	**10.8**	**13.7**	**6.3**	**7.4**	**3.7**	**5.4**	**9.6**
SDR DTS	0.5	0.1	0.0	0.0	0.3	0.0	0.0	0.0	0.0	0.0

127

International reserves excluding gold
Special drawing rights(SDR) + Reserve position in IMF + Foreign exchange: million $US (end of period) [cont.]
Réserves internationales, l'or non compris
Droits de tirage spéciaux(DTS) + Réserve disponible au FMI + Devises étrangères : $E-U millions (fin de la période) [suite]

Country or area and reserves Pays ou zone et réserves	1982	1983	1984	1985	1986	1987	1988	1989	1990	1991
Position in IMF Disponibilité au FMI	0.0	0.0	0.0	0.0	0.0	0.0	0.0	0.0	0.0	0.0
Foreign exchange Devises étrangères	7.9	16.1	7.7	10.8	13.3	6.3	7.4	3.7	5.4	9.6
Singapore **Singapour**	**8 479.8**	**9 264.2**	**10 416.0**	**12 846.6**	**12 939.0**	**15 227.0**	**17 072.5**	**20 345.1**	**27 748.4**	**34 128.2**
SDR DTS	54.5	62.5	56.9	72.5	90.3	115.2	106.3	104.4	115.9	116.3
Position in IMF Disponibilité au FMI	74.7	71.8	67.8	88.6	97.8	112.2	104.8	105.3	98.0	86.0
Foreign exchange Devises étrangères	8 350.6	9 130.0	10 291.3	12 685.5	12 750.9	14 999.6	16 861.4	20 135.5	27 534.5	33 926.0
Solomon Islands **Iles Salomon**	**37.2**	**47.3**	**44.7**	**35.6**	**29.6**	**36.8**	**39.6**	**26.2**	**17.6**	**8.5**
SDR DTS	1.4	1.7	1.3	0.9	1.6	0.2	0.0	0.1	0.4	0.1
Position in IMF Disponibilité au FMI	0.0	0.5	0.5	0.6	0.6	0.7	0.7	0.7	0.8	0.8
Foreign exchange Devises étrangères	35.8	45.1	42.9	34.2	27.4	35.8	38.9	25.4	16.5	7.7
Somalia **Somalie**	**6.5**	**9.2**	**1.0**	**2.5**	**12.8**	**7.3**	**15.3**	**15.4**	**...**	**...**
SDR DTS	0.8	0.4	0.1	0.0	0.0	0.0	0.0	0.0	0.0	0.0
Position in IMF Disponibilité au FMI	0.0	0.0	0.0	0.0	0.0	0.0	0.0	0.0	0.0	0.0
Foreign exchange Devises étrangères	5.7	8.8	1.0	2.5	12.8	7.3	15.3	15.4	...	...
South Africa **Afrique du Sud**	**485.0**	**823.0**	**242.0**	**315.0**	**370.0**	**641.0**	**780.0**	**960.0**	**1 008.0**	**899.0**
SDR DTS	109.3	28.6	2.3	0.6	0.0	1.6	1.1	1.7	2.3	1.7
Position in IMF Disponibilité au FMI	0.0	73.3	68.6	0.0	0.0	0.0	0.0	0.1	0.1	0.1
Foreign exchange Devises étrangères	376.0	721.0	171.0	314.0	370.0	639.0	779.0	958.0	1 006.0	897.0
Spain **Espagne**	**7 655.0**	**7 402.0**	**11 955.0**	**11 175.0**	**14 755.0**	**30 669.0**	**37 074.0**	**41 467.0**	**51 228.0**	**65 822.0**
SDR DTS	205.0	68.5	151.5	279.5	432.0	596.4	614.8	687.2	696.3	456.3
Position in IMF Disponibilité au FMI	226.8	336.6	355.9	405.4	517.5	785.4	1 057.8	1 221.8	1 133.6	1 071.1
Foreign exchange Devises étrangères	7 223.0	6 997.0	11 448.0	10 490.0	13 805.0	29 287.0	35 401.0	39 558.0	49 398.0	64 295.0
Sri Lanka **Sri Lanka**	**351.0**	**297.0**	**511.0**	**451.0**	**353.0**	**279.0**	**222.0**	**244.0**	**423.0**	**685.0**
SDR DTS	7.0	0.9	0.2	0.1	0.0	0.1	0.1	13.5	0.4	0.3
Position in IMF Disponibilité au FMI	6.5	17.9	5.9	6.6	0.0	0.0	0.0	0.0	0.1	0.1
Foreign exchange Devises étrangères	338.0	278.0	505.0	445.0	353.0	279.0	222.0	231.0	422.0	685.0
Sudan **Soudan**	**20.5**	**16.6**	**17.2**	**12.2**	**58.5**	**11.7**	**12.1**	**15.9**	**11.4**	**7.6**
SDR DTS	0.0	0.1	0.0	0.0	0.0	0.0	0.0	0.0	0.0	0.0
Position in IMF Disponibilité au FMI	0.0	0.0	0.0	0.0	0.0	0.0	0.0	0.0	0.0	0.0
Foreign exchange Devises étrangères	20.5	16.5	17.2	12.2	58.5	11.7	12.1	15.9	11.4	7.6

127
International reserves excluding gold
Special drawing rights(SDR) + Reserve position in IMF + Foreign exchange: million $US (end of period) [cont.]
Réserves internationales, l'or non compris
Droits de tirage spéciaux(DTS) + Réserve disponible au FMI + Devises étrangères : $E-U millions (fin de la période) [suite]

Country or area and reserves Pays ou zone et réserves	1982	1983	1984	1985	1986	1987	1988	1989	1990	1991
Suriname Suriname	**175.8**	**59.2**	**24.9**	**23.4**	**20.9**	**15.1**	**12.6**	**9.3**	**21.1**	**0.0**
SDR DTS	9.4	1.8	1.3	0.8	0.3	0.0	0.0	0.0	0.0	0.0
Position in IMF Disponibilité au FMI	8.7	3.1	0.0	0.0	0.0	0.0	0.0	0.0	0.0	0.0
Foreign exchange Devises étrangères	157.7	54.2	23.6	22.6	20.6	15.1	12.6	9.3	21.1	0.0
Swaziland Swaziland	**76.1**	**92.5**	**80.1**	**83.4**	**96.5**	**127.2**	**140.0**	**180.6**	**216.5**	**171.9**
SDR DTS	5.5	1.5	2.2	0.3	2.7	4.0	1.7	1.1	12.1	12.4
Position in IMF Disponibilité au FMI	0.0	1.8	1.7	1.9	0.0	0.0	0.0	0.0	0.0	0.0
Foreign exchange Devises étrangères	70.6	89.2	76.2	81.2	93.7	123.2	138.3	179.6	204.4	159.5
Sweden Suède	**3 513.0**	**4 034.0**	**3 845.0**	**5 793.0**	**6 551.0**	**8 174.0**	**8 492.0**	**9 559.0**	**17 988.0**	**18 331.0**
SDR DTS	257.3	128.8	177.8	246.3	319.6	295.7	402.5	341.1	290.2	414.3
Position in IMF Disponibilité au FMI	164.2	252.0	253.2	274.1	310.0	393.6	337.5	333.3	332.7	441.2
Foreign exchange Devises étrangères	3 091.0	3 653.0	3 414.0	5 273.0	5 921.0	7 485.0	7 752.0	8 885.0	17 365.0	17 476.0
Switzerland Suisse	**15 460.0**	**15 034.0**	**15 296.0**	**18 016.0**	**21 786.0**	**27 476.0**	**24 203.0**	**25 276.0**	**29 223.0**	**29 004.0**
SDR DTS	4.0	13.5	9.1	3.3	0.0	14.3	20.5	4.8	1.9	2.3
Position in IMF Disponibilité au FMI	515.4	665.7	581.1	550.0	452.4	300.1	137.3	41.5	0.0	0.0
Foreign exchange Devises étrangères	14 941.0	14 355.0	14 706.0	17 463.0	21 334.0	27 162.0	24 045.0	25 230.0	29 221.0	29 002.0
Syrian Arab Republic Rép. arabe syrienne	**198.0**	**52.0**	**268.0**	**83.0**	**144.0**	**223.0**	**193.0**	**...**	**...**	**...**
SDR DTS	13.1	9.2	5.2	3.1	0.2	0.0	0.0	0.0	0.1	0.1
Position in IMF Disponibilité au FMI	0.0	0.0	0.0	0.0	0.0	0.0	0.0	0.0	0.0	0.0
Foreign exchange Devises étrangères	185.0	43.0	263.0	80.0	144.0	223.0	193.0	...	...	...
Thailand Thaïlande	**1 538.0**	**1 607.0**	**1 921.0**	**2 190.0**	**2 804.0**	**4 007.0**	**6 097.0**	**9 515.0**	**13 305.0**	**17 517.0**
SDR DTS	24.5	15.9	2.4	1.5	33.2	60.2	60.8	16.5	12.9	8.2
Position in IMF Disponibilité au FMI	0.0	30.1	28.2	31.6	35.2	40.9	38.8	37.9	45.2	222.0
Foreign exchange Devises étrangères	1 513.0	1 561.0	1 890.0	2 157.0	2 736.0	3 906.0	5 997.0	9 461.0	13 247.0	17 287.0
Togo Togo	**167.7**	**172.8**	**203.3**	**296.6**	**332.7**	**354.9**	**232.1**	**285.3**	**353.2**	**364.9**
SDR DTS	4.3	1.3	2.0	0.1	0.6	0.1	0.1	1.7	0.2	0.4
Position in IMF Disponibilité au FMI	0.2	0.2	0.2	0.2	0.3	0.3	0.3	0.3	0.3	0.4
Foreign exchange Devises étrangères	163.2	171.4	201.1	296.3	331.8	354.5	231.7	283.3	352.7	364.1
Tonga Tonga	**15.6**	**21.0**	**26.0**	**27.5**	**22.5**	**28.9**	**30.5**	**24.9**	**31.3**	**32.3**
SDR DTS	0.0	0.0	0.0	0.0	0.0	0.1	0.1	0.2	0.2	1.1

127
International reserves excluding gold
Special drawing rights(SDR) + Reserve position in IMF + Foreign exchange: million $US (end of period) [cont.]
Réserves internationales, l'or non compris
Droits de tirage spéciaux(DTS) + Réserve disponible au FMI + Devises étrangères : $E-U millions (fin de la période) [suite]

Country or area and reserves Pays ou zone et réserves	1982	1983	1984	1985	1986	1987	1988	1989	1990	1991
Position in IMF Disponibilité au FMI	0.0	0.0	0.0	0.8	0.9	1.1	1.0	1.0	1.1	1.1
Foreign exchange Devises étrangères	15.6	21.0	26.0	26.7	21.5	27.8	29.4	23.7	30.0	30.2
Trinidad and Tobago **Trinité-et-Tobago**	**3 080.5**	**2 104.5**	**1 356.7**	**1 128.5**	**474.1**	**187.8**	**127.1**	**246.5**	**492.0**	**338.6**
SDR DTS	81.0	98.8	101.1	118.4	137.0	0.0	0.0	9.0	1.1	2.1
Position in IMF Disponibilité au FMI	106.7	124.2	124.3	136.6	94.4	75.2	0.0	0.0	0.0	0.0
Foreign exchange Devises étrangères	2 892.8	1 881.5	1 131.3	873.5	242.8	112.6	127.1	237.5	490.9	336.5
Tunisia **Tunisie**	**606.5**	**567.3**	**406.3**	**232.7**	**305.3**	**525.5**	**899.3**	**961.9**	**794.8**	**789.9**
SDR DTS	17.7	3.8	1.9	0.5	27.6	54.1	28.3	8.5	2.5	32.9
Position in IMF Disponibilité au FMI	21.2	31.5	28.9	29.0	0.0	0.0	0.0	0.0	0.0	0.0
Foreign exchange Devises étrangères	567.6	532.0	375.6	203.2	277.7	471.4	871.0	953.4	792.3	757.0
Turkey **Turquie**	**1 080.0**	**1 288.0**	**1 271.0**	**1 056.0**	**1 412.0**	**1 776.0**	**2 344.0**	**4 780.0**	**6 050.0**	**5 144.0**
SDR DTS	0.1	1.3	0.1	0.5	0.1	0.0	0.1	0.0	0.6	0.0
Position in IMF Disponibilité au FMI	0.0	33.8	31.6	35.5	39.5	45.8	43.4	42.4	45.9	46.2
Foreign exchange Devises étrangères	1 080.0	1 253.0	1 239.0	1 020.0	1 372.0	1 730.0	2 301.0	4 738.0	6 003.0	5 098.0
Uganda **Ouganda**	**78.3**	**106.5**	**67.9**	**27.3**	**29.2**	**54.6**	**49.3**	**14.1**	**44.0**	**58.9**
SDR DTS	11.1	0.9	0.2	0.0	0.0	0.0	0.0	0.0	6.8	10.3
Position in IMF Disponibilité au FMI	3.9	3.7	3.5	3.9	4.3	0.0	0.0	0.0	0.0	0.0
Foreign exchange Devises étrangères	63.3	101.9	64.3	23.4	24.9	54.6	49.3	14.1	37.2	48.6
United Arab Emirates **Emirats arabes unis**	**2 215.5**	**2 072.4**	**2 286.9**	**3 204.3**	**3 369.9**	**4 725.3**	**4 433.5**	**4 456.6**	**4 583.9**	**5 365.4**
SDR DTS	55.5	64.4	64.7	75.0	93.5	113.2	110.8	112.9	129.1	136.6
Position in IMF Disponibilité au FMI	222.2	224.7	217.9	221.5	221.5	225.6	182.0	181.9	179.7	180.4
Foreign exchange Devises étrangères	1 937.8	1 783.3	2 004.3	2 907.8	3 054.9	4 386.5	4 140.7	4 161.8	4 275.1	5 048.4
United Kingdom **Royaume-Uni**	**12 397.0**	**11 339.0**	**9 440.0**	**12 859.0**	**18 422.0**	**41 715.0**	**44 103.0**	**34 768.0**	**35 853.0**	**41 892.0**
SDR DTS	1 170.1	517.2	496.9	1 131.1	1 553.0	1 382.2	1 319.5	1 143.4	1 249.1	1 314.9
Position in IMF Disponibilité au FMI	1 553.3	2 104.6	1 972.0	1 987.8	1 982.8	1 777.0	1 666.7	1 637.0	1 677.3	1 849.1
Foreign exchange Devises étrangères	9 674.0	8 718.0	6 971.0	9 740.0	14 886.0	38 556.0	41 117.0	31 987.0	32 926.0	38 728.0
United Rep.Tanzania **Rép. Unie de Tanzanie**	**4.8**	**19.4**	**26.9**	**16.0**	**61.1**	**31.8**	**77.7**	**54.2**	**192.8**	**203.9**
SDR DTS	0.0	0.1	0.1	0.0	5.6	0.1	0.0	0.0	0.0	0.0
Position in IMF Disponibilité au FMI	0.0	0.0	0.0	0.0	0.0	0.0	0.0	0.0	0.0	0.0
Foreign exchange Devises étrangères	4.8	19.3	26.8	16.0	55.5	31.7	77.7	54.2	192.8	203.9

127
International reserves excluding gold
Special drawing rights(SDR) + Reserve position in IMF + Foreign exchange: million $US (end of period) [cont.]
Réserves internationales, l'or non compris
Droits de tirage spéciaux(DTS) + Réserve disponible au FMI + Devises étrangères : $E-U millions (fin de la période) [suite]

Country or area and reserves Pays ou zone et réserves	1982	1983	1984	1985	1986	1987	1988	1989	1990	1991
United States **Etats-Unis**	**22 809.4**	**22 626.7**	**23 837.8**	**32 095.5**	**37 452.4**	**34 720.3**	**36 744.7**	**63 549.6**	**72 258.3**	**66 661.5**
SDR DTS	5 249.8	5 025.4	5 640.7	7 292.8	8 394.6	10 280.0	9 636.6	9 950.8	10 990.0	11 240.0
Position in IMF Disponibilité au FMI	7 347.6	11 310.0	11 540.0	11 950.0	11 730.0	11 350.0	9 745.2	9 047.8	9 076.2	9 487.8
Foreign exchange Devises étrangères	10 212.0	6 289.0	6 656.0	12 856.0	17 328.0	13 088.0	17 363.0	44 551.0	52 193.0	45 934.0
Uruguay **Uruguay**	**116.3**	**206.9**	**134.4**	**174.2**	**481.6**	**530.0**	**532.0**	**501.3**	**544.4**	**345.8**
SDR DTS	1.9	3.8	5.0	14.6	11.9	68.2	29.8	23.0	11.2	4.7
Position in IMF Disponibilité au FMI	0.0	9.9	0.0	0.0	0.0	0.0	0.0	0.0	0.0	0.0
Foreign exchange Devises étrangères	114.4	193.2	129.4	159.6	469.7	461.8	502.2	478.4	533.2	341.1
Vanuatu **Vanuatu**	**5.7**	**6.6**	**8.1**	**10.6**	**21.4**	**40.2**	**40.7**	**35.1**	**37.7**	**39.8**
SDR DTS	0.0	0.0	0.1	0.2	0.2	0.4	0.4	0.5	0.7	0.8
Position in IMF Disponibilité au FMI	1.2	1.6	1.6	1.7	1.9	2.3	2.2	2.1	2.3	2.3
Foreign exchange Devises étrangères	4.5	4.9	6.5	8.7	19.2	37.6	38.1	32.4	34.7	36.7
Venezuela **Venezuela**	**6 579.0**	**7 643.0**	**8 901.0**	**10 251.0**	**6 437.0**	**5 963.0**	**3 092.0**	**4 106.0**	**8 321.0**	**10 665.0**
SDR DTS	440.4	352.6	374.5	495.9	609.1	757.0	75.9	47.3	9.5	269.3
Position in IMF Disponibilité au FMI	752.1	918.3	810.6	817.9	802.3	670.7	40.7	4.1	0.0	0.0
Foreign exchange Devises étrangères	5 386.0	6 372.0	7 716.0	8 937.0	5 026.0	4 535.0	2 975.0	4 055.0	8 311.0	10 396.0
Yemen[3] **Yémen**[3]	**554.2**	**366.0**	**318.5**	**296.8**	**431.7**	**539.5**	**285.1**	**279.2**	**...**	**...**
SDR DTS	15.5	9.1	8.6	25.4	28.2	26.9	32.2	20.8	0.0	0.0
Position in IMF Disponibilité au FMI	0.0	6.2	5.8	0.0	0.0	0.0	0.0	0.0	0.0	0.0
Foreign exchange Devises étrangères	538.6	350.7	304.0	271.4	403.5	512.6	252.9	258.3	...	...
Yugoslavia **Yougoslavie**	**775.3**	**976.4**	**1 158.1**	**1 094.7**	**1 459.6**	**697.7**	**2 297.9**	**4 135.7**	**5 473.9**	**2 681.8**
SDR DTS	0.0	0.0	0.1	0.2	0.0	0.0	0.1	0.0	13.1	0.3
Position in IMF Disponibilité au FMI	3.6	55.1	0.0	0.0	0.0	0.0	0.0	0.0	0.0	0.0
Foreign exchange Devises étrangères	771.6	921.2	1 158.0	1 094.5	1 459.6	697.7	2 297.8	4 135.7	5 460.8	2 681.5
Zaire **Zaïre**	**38.9**	**101.6**	**137.4**	**189.7**	**268.6**	**180.8**	**186.9**	**195.1**	**219.1**	**182.9**
SDR DTS	0.0	22.0	0.0	0.2	0.0	0.1	0.0	4.9	0.0	0.0
Position in IMF Disponibilité au FMI	0.0	0.0	0.0	0.0	0.0	0.0	0.0	0.0	0.0	0.0
Foreign exchange Devises étrangères	38.9	79.6	137.4	189.5	268.6	180.7	186.9	190.2	219.1	182.9
Zambia **Zambie**	**58.2**	**54.5**	**54.2**	**200.1**	**70.3**	**108.8**	**134.0**	**116.2**	**193.1**	**184.6**
SDR DTS	16.0	0.0	0.0	0.0	0.0	0.0	0.0	0.0	0.0	0.0

127
International reserves excluding gold
Special drawing rights(SDR) + Reserve position in IMF + Foreign exchange: million $US (end of period) [*cont.*]
Réserves internationales, l'or non compris
Droits de tirage spéciaux(DTS) + Réserve disponible au FMI + Devises étrangères : $E-U millions (fin de la période) [*suite*]

Country or area and reserves Pays ou zone et réserves	1982	1983	1984	1985	1986	1987	1988	1989	1990	1991
Position in IMF Disponibilité au FMI	0.0	0.0	0.0	0.0	0.0	0.0	0.0	0.0	0.0	0.0
Foreign exchange Devises étrangères	42.2	54.5	54.2	200.1	70.3	108.8	134.0	116.2	193.1	184.6
Zimbabwe **Zimbabwe**	**140.5**	**75.4**	**45.4**	**93.3**	**106.4**	**166.2**	**178.6**	**94.6**	**149.2**	**149.7**
SDR DTS	6.9	6.4	2.3	14.5	6.2	23.1	0.6	0.7	0.3	0.1
Position in IMF Disponibilité au FMI	0.0	0.0	0.0	0.0	0.0	0.1	0.1	0.1	0.1	0.1
Foreign exchange Devises étrangères	133.5	68.9	43.1	78.8	100.2	143.0	178.0	93.8	148.9	149.5

Source:
International Monetary Fund (Washington, DC).

† All data shown which pertain to Germany prior to 3 October 1990 are indicated separately for the Federal Republic of Germany and the former German Democratic Republic based on their respective territories at the time indicated. Where data for united Germany (3 October 1990 and thereafter) are not available, available data are shown separately under the designations Federal Republic of Germany and former German Democratic Republic and pertain to the territorial boundaries prior to 3 October 1990. For detailed explanatory notes on data pertaining to Germany, see Annex I - Country Nomenclature.

1 Foreign exchange data only.
2 SDR and Position in IMF data only.
3 Excluding former Democratic Yemen.

Source:
Fonds monétaire international (Washington, DC).

† Toutes les données se rapportant à l'Allemagne avant le 3 octobre 1990 figurent dans deux rubriques séparées basées sur les territoires respectifs de la République fédérale d'Allemagne et l'ancienne République démocratique allemande selon la période indiquée. En l'absence de données pour l'Allemagne unifiée (à compter du 3 octobre 1990), les données disponibles sont fournies séparément sous les rubriques République fédérale d'Allemagne et ancienne République démocratique allemande et se rapportent aux limites territoriales antérieures au 3 octobre 1990. Pour les notes explicatives en détail sur les données concernant l'Allemagne, voir Annexe I – Nomenclature des pays.

1 Devises étrangères seulement.
2 DTS et Disponibilité au FMI seulement.
3 Non compris l'ancien Yémen démocratique.

128
Gold reserves
Réserves d'or
Millions of fine troy ounces, end of period
Millions d'onces troy de fin, fin de la période

Country or area Pays ou zone	1982	1983	1984	1985	1986	1987	1988	1989	1990	1991
Afghanistan Afghanistan	0.97	0.97	0.97	0.97	0.97	0.97	0.97	0.97	0.97	0.97
Algeria Algérie	5.58	5.58	5.58	5.58	5.58	5.58	5.58	5.58	5.14	5.58
Argentina Argentine	4.37	4.37	4.37	4.37	4.37	4.37	4.37	4.37	4.23	4.12
Aruba Aruba	...	...	...	...	-	-	-	0.10	0.10	0.10
Australia Australie	7.93	7.93	7.93	7.93	7.93	7.93	7.93	7.93	7.93	7.93
Austria Autriche	21.12	21.13	21.13	21.14	21.14	21.15	21.15	20.66	20.39	20.03
Bahrain Bahreïn	0.15	0.15	0.15	0.15	0.15	0.15	0.15	0.15	0.15	0.15
Bangladesh Bangladesh	0.06	0.06	0.06	0.06	0.07	0.07	0.07	0.08	0.08	0.08
Barbados Barbade	0.01	0.01	0.01	0.01	0.01	0.01	0.01	0.01	0.00	0.00
Belgium Belgique	34.18	34.18	34.18	34.18	34.18	33.63	33.67	30.23	30.23	30.23
Benin Bénin	0.01	0.01	0.01	0.01	0.01	0.01	0.01	0.01	0.01	0.01
Bolivia Bolivie	0.89	0.91	0.91	0.89	0.89	0.89	0.89	0.89	0.89	0.89
Brazil Brésil	0.15	0.54	1.47	3.10	2.43	2.43	2.73	2.98	4.57	2.02
Burkina Faso Burkina Faso	0.01	0.01	0.01	0.01	0.01	0.01	0.01	0.01	0.01	0.01
Burundi Burundi	0.02	0.02	0.02	0.02	0.02	0.02	0.02	0.02	0.02	0.02
Cameroon Cameroun	0.03	0.03	0.03	0.03	0.03	0.03	0.03	0.03	0.03	...
Canada Canada	20.26	20.17	20.14	20.11	19.72	18.52	17.14	16.10	14.76	12.96
Central African Rep. Rép. centrafricaine	0.01	0.01	0.01	0.01	0.01	0.01	0.01	0.01	0.01	...
Chad Tchad	0.01	0.01	0.01	0.01	0.01	0.01	0.01	0.01	0.01	...
Chile Chili	1.71	1.53	1.53	1.53	1.53	1.81	1.53	1.75	1.86	1.86
China Chine	12.70	12.70	12.70	12.70	12.70	12.70	12.70	12.70	12.70	12.70
Colombia Colombie	3.82	4.22	1.37	1.84	2.01	0.68	1.10	0.61	0.63	0.86
Congo Congo	0.01	0.01	0.01	0.01	0.01	0.01	0.01	0.01	0.01	...
Costa Rica Costa Rica	0.05	0.09	0.02	0.06	0.07	0.06	0.02	0.01	0.01	0.03
Côte d'Ivoire Côte d'Ivoire	0.04	0.04	0.04	0.04	0.04	0.04	0.04	0.04	0.04	0.04

128
Gold reserves
Millions of fine troy ounces, end of period [*cont.*]
    Réserves d'or
    Millions d'onces troy de fin, fin de la période [*suite*]

Country or area Pays ou zone	1982	1983	1984	1985	1986	1987	1988	1989	1990	1991
Czechoslovakia Tchécoslovaquie	3.65	3.85	3.91	3.81	3.77	3.66	3.73	3.62	2.49	2.79
Cyprus Chypre	0.46	0.46	0.46	0.46	0.46	0.46	0.46	0.46	0.46	0.46
Denmark Danemark	1.63	1.63	1.63	1.63	1.63	1.63	1.63	1.64	1.65	1.66
Dominican Republic Rép. dominicaine	0.09	0.08	0.02	0.02	0.02	0.02	0.02	0.02	0.02	0.02
Ecuador Equateur	0.41	0.41	0.41	0.41	0.41	0.41	0.41	0.41	0.44	0.44
Egypt Egypte	2.43	2.43	2.43	2.43	2.43	2.43	2.43	2.43	2.43	2.43
El Salvador El Salvador	0.52	0.47	0.47	0.47	0.47	0.47	0.47	0.47	0.47	0.47
Ethiopia Ethiopie	0.21	0.21	0.21	0.21	0.21	0.21	0.21	0.19	0.09	0.15
Fiji Fidji	0.01	0.01	0.01	0.01	0.01	0.01	0.00	0.00	0.00	0.00
Finland Finlande	1.27	1.27	1.27	1.91	1.91	1.96	1.96	2.00	2.00	2.00
France France	81.85	81.85	81.85	81.85	81.85	81.85	81.85	81.85	81.85	81.85
Gabon Gabon	0.01	0.01	0.01	0.01	0.01	0.01	0.01	0.01	0.01	...
Germany†·Allemagne† F. R. Germany R. f. Allemagne	95.18	95.18	95.18	95.18	95.18	95.18	95.18	95.18	95.18	95.18
Ghana Ghana	0.38	0.38	0.44	0.23	0.28	0.28	0.22	0.22	0.24	...
Greece Grèce	3.87	3.88	4.11	4.12	3.31	3.34	3.40	3.40	3.40	3.43
Guatemala Guatemala	0.52	0.52	0.52	0.52	0.52	0.52	0.52	0.54	0.21	0.21
Haiti Haïti	0.02	0.02	0.02	0.02	0.02	0.02	0.02	0.02	0.02	0.02
Honduras Honduras	0.02	0.02	0.02	0.02	0.02	0.02	0.02	0.02	0.02	0.02
Hungary Hongrie	0.65	1.53	2.06	2.33	2.35	1.64	1.59	1.50	0.30	0.26
Iceland Islande	0.05	0.05	0.05	0.05	0.05	0.05	0.05	0.05	0.05	0.05
India Inde	8.59	8.59	8.74	9.40	10.45	10.45	10.45	10.45	10.69	11.28
Indonesia Indonésie	3.10	3.10	3.10	3.10	3.10	3.10	3.10	3.11	3.11	3.11
Iran, Islamic Rep. of Iran, Rép. islamique d'	5.92	...	...	...	...	...	...	...	...	...
Ireland Irlande	0.36	0.36	0.36	0.36	0.36	0.36	0.36	0.36	0.36	0.36

128
Gold reserves
Millions of fine troy ounces, end of period [cont.]
Réserves d'or
Millions d'onces troy de fin, fin de la période [suite]

Country or area Pays ou zone	1982	1983	1984	1985	1986	1987	1988	1989	1990	1991
Israel Israël	1.08	1.02	1.02	1.02	1.02	1.02	1.02	1.02	0.84	0.42
Italy Italie	66.67	66.67	66.67	66.67	66.67	66.67	66.67	66.67	66.67	66.67
Japan Japon	24.23	24.23	24.23	24.23	24.23	24.23	24.23	24.23	24.23	24.23
Jordan Jordanie	1.08	1.09	1.06	1.06	1.06	1.00	0.74	0.75	0.75	0.79
Kenya Kenya	0.08	0.08	0.08	0.08	0.08	0.08	0.08	0.08	* 0.08	0.08
Korea, Republic of Corée, République de	0.30	0.30	0.31	0.31	0.32	0.32	0.32	0.32	0.32	0.32
Kuwait Koweït	2.54	2.54	2.54	2.54	2.54	2.54	2.54	2.54	2.54	2.54
Lebanon Liban	9.22	9.22	9.22	9.22	9.22	9.22	9.22	9.22	9.22	9.22
Libyan Arab Jamah. Jamah. arabe libyenne	3.58	3.58	3.65	3.60	3.60	3.60	3.60	3.60	3.60	3.60
Luxembourg Luxembourg	0.46	0.46	0.43	0.43	0.43	0.43	0.43	0.43	0.43	0.43
Malawi Malawi	0.01	0.01	0.01	0.01	0.01	0.01	0.01	0.01	0.01	0.01
Malaysia Malaisie	2.33	2.33	2.33	2.34	2.34	2.35	2.35	2.37	2.35	...
Mali Mali	0.02	0.02	0.02	0.02	0.19	0.02	0.02	0.02	0.02	0.02
Malta Malte	0.46	0.47	0.47	0.47	0.47	0.47	0.47	0.23	0.16	0.12
Mauritania Mauritanie	0.01	0.01	0.01	0.01	0.01	0.01	0.01	0.01	0.01	...
Mauritius Maurice	0.04	0.04	0.04	0.04	0.04	0.04	0.05	0.06	0.06	0.06
Mexico Mexique	2.07	2.31	2.42	2.36	2.57	2.54	2.56	1.03	0.92	0.92
Morocco Maroc	0.70	0.70	0.70	0.70	0.70	0.70	0.70	0.70	0.70	0.70
Myanmar Myanmar	0.25	0.25	0.25	0.25	0.25	0.25	0.25	0.25	0.25	0.25
Nepal Népal	0.15	0.15	0.15	0.15	0.15	0.15	0.15	0.15	0.15	0.15
Netherlands Pays-Bas	43.94	43.94	43.94	43.94	43.94	43.94	43.94	43.94	43.94	43.94
Netherlands Antilles Antilles néerlandaises	0.55	0.55	0.55	0.55	0.55	0.55	0.55	0.55	0.55	...
New Zealand Nouvelle-Zélande	0.02	0.02	0.02	0.02	0.02	0.02	0.02	-	-	-
Nicaragua Nicaragua	0.02	0.12	...	...	...	...	...	...	...	...
Niger Niger	0.01	0.01	0.01	0.01	0.01	0.01	0.01	0.01	0.01	0.01

128
Gold reserves
Millions of fine troy ounces, end of period [*cont.*]
Réserves d'or
Millions d'onces troy de fin, fin de la période [*suite*]

Country or area Pays ou zone	1982	1983	1984	1985	1986	1987	1988	1989	1990	1991
Nigeria Nigéria	0.69	0.69	0.69	0.69	0.69	0.69	0.69	0.69	0.69	0.69
Norway Norvège	1.18	1.18	1.18	1.18	1.18	1.18	1.18	1.18	1.18	1.18
Oman Oman	0.28	0.29	0.29	0.29	0.29	0.29	0.29	0.29	0.29	0.29
Pakistan Pakistan	1.85	1.86	1.87	1.90	1.93	1.94	1.95	1.95	1.95	1.95
Papua New Guinea Papouasie-Nvl-Guinée	0.06	0.06	0.06	0.06	0.06	0.06	0.06	0.06	0.06	0.06
Paraguay Paraguay	0.04	0.04	0.04	0.04	0.04	0.04	0.04	0.04	0.04	0.04
Peru Pérou	1.40	1.40	1.40	1.95	2.14	1.50	1.71	1.97	2.21	1.83
Philippines Philippines	1.87	0.29	0.79	1.48	2.26	2.78	2.84	2.45	2.89	3.37
Poland Pologne	...	...	...	0.47	0.47	0.47	0.47	0.47	0.47	0.47
Portugal Portugal	22.09	20.43	20.30	20.23	20.16	20.06	16.07	16.05	15.83	15.87
Qatar Qatar	0.90	1.07	1.21	1.08	0.97	0.83	0.88	0.90	0.84	...
Romania Roumanie	3.55	3.62	3.73	3.82	3.25	1.36	1.45	2.17	2.21	2.25
Saudi Arabia Arabie saoudite	4.60	4.60	4.60	4.60	4.60	4.60	4.60	4.60	4.60	4.60
Senegal Sénégal	0.03	0.03	0.03	0.03	0.03	0.03	0.03	0.03	0.03	0.03
Somalia Somalie	0.02	0.02	0.02	0.02	0.02	0.02	0.02	0.02	...	...
South Africa Afrique du Sud	7.57	7.79	7.36	4.84	4.82	5.83	3.47	3.08	4.09	6.47
Spain Espagne	14.61	14.61	14.63	14.65	14.82	11.92	14.04	15.72	15.61	15.62
Sri Lanka Sri Lanka	0.06	0.06	0.06	0.06	0.06	0.06	0.06	0.06	0.06	0.11
Suriname Suriname	0.05	0.05	0.05	0.05	0.05	0.05	0.05	0.05	0.05	0.05
Sweden Suède	6.07	6.07	6.07	6.07	6.07	6.07	6.07	6.07	6.07	6.07
Switzerland Suisse	83.28	83.28	83.28	83.28	83.28	83.28	83.28	83.28	83.28	83.28
Syrian Arab Republic Rép. arabe syrienne	0.83	0.83	0.83	0.83	0.83	0.83	0.83	0.83	0.83	...
Thailand Thaïlande	2.49	2.49	2.49	2.49	2.49	2.48	2.48	2.48	2.48	2.48
Togo Togo	0.01	0.01	0.01	0.01	0.01	0.01	0.01	0.01	0.01	0.01
Trinidad and Tobago Trinité-et-Tobago	0.05	0.05	0.05	0.05	0.05	0.05	0.05	0.05	0.05	0.05

128
Gold reserves
Millions of fine troy ounces, end of period [*cont.*]
Réserves d'or
Millions d'onces troy de fin, fin de la période [*suite*]

Country or area Pays ou zone	1982	1983	1984	1985	1986	1987	1988	1989	1990	1991
Tunisia Tunisie	0.19	0.19	0.19	0.19	0.19	0.19	0.19	0.19	0.19	0.19
Turkey Turquie	3.77	3.78	3.80	3.86	3.84	3.83	3.82	3.79	4.10	4.16
United Arab Emirates Emirats arabes unis	0.82	0.82	0.82	0.82	0.82	0.82	0.82	0.80	0.80	0.80
United Kingdom Royaume-Uni	19.01	19.01	19.03	19.03	19.01	19.01	19.00	18.99	18.94	18.89
United States Etats-Unis	264.03	263.39	262.79	262.65	262.04	262.38	261.87	261.93	261.91	261.91
Uruguay Uruguay	2.86	2.60	2.62	2.62	2.61	2.61	2.61	2.61	2.40	2.26
Venezuela Venezuela	11.46	11.46	11.46	11.46	11.46	11.46	11.46	11.46	11.46	11.46
Yemen[1] Yémen[1]	0.04	0.04	0.04	0.04	0.04	0.04	0.04	...	...	...
Yugoslavia Yougoslavie	1.86	1.86	1.86	1.86	1.87	1.87	1.89	1.90	1.91	1.92
Zaire Zaïre	0.41	0.44	0.47	0.45	0.47	0.49	0.45	0.22	0.11	0.03
Zambia Zambie	0.22	0.22	0.00	0.00	...	0.00	0.01	0.02	0.02	0.01
Zimbabwe Zimbabwe	0.39	0.59	0.70	0.77	0.54	0.42	0.40	0.45	0.38	0.41

Source:
International Monetary Fund (Washington, DC).

Source:
Fonds monétaire international (Washington, DC).

† All data shown which pertain to Germany prior to 3 October 1990 are indicated separately for the Federal Republic of Germany and the former German Democratic Republic based on their respective territories at the time indicated. Where data for united Germany (3 October 1990 and thereafter) are not available, available data are shown separately under the designations Federal Republic of Germany and former German Democratic Republic and pertain to the territorial boundaries prior to 3 October 1990. For detailed explanatory notes on data pertaining to Germany, see Annex I - Country Nomenclature.

1 Former Democratic Yemen only.

† Toutes les données se rapportant à l'Allemagne avant le 3 octobre 1990 figurent dans deux rubriques séparées basées sur les territoires respectifs de la République fédérale d'Allemagne et l'ancienne République démocratique allemande selon la période indiquée. En l'absence de données pour l'Allemagne unifiée (à compter du 3 octobre 1990), les données disponisbles sont fournies séparément sous les rubriques République fédérale d'Allemagne et ancienne République démocratique allemande et se rapportent aux limites territoriales antérieures au 3 octobre 1990. Pour les notes explicatives en détail sur les données concernant l'Allemagne, voir Annexe I – Nomenclature des pays.

1 L'ancien Yémen démocratique seulement.

129
External debt of developing countries
Dette extérieure des pays en développement
Million US dollars
Millions de dollars E-U

A. Total external debt • Total de la dette extérieure

Country or area[1]	1980	1984	1985	1986	1987	1988	1989	1990	Pays ou zone[1]
Total long-term debt (LDOD)	420448	666237	789091	893439	1012555	996247	999687	1047041	**Total de la dette à long terme (LDOD)**
Public/publicly guaranteed	350347	565068	696042	809490	936396	933565	943545	982419	**Dette publique ou garantie par l'Etat**
Official creditors	157322	232305	304112	368607	449242	452549	474243	524638	Créanciers publics
Multilateral	49109	83687	107926	136949	171802	172438	181241	208647	Multilatéraux
IBRD	22356	36624	50676	68305	89048	84225	84684	95855	BIRD
IDA	11886	20863	24213	28007	33305	36119	39284	45035	IDA
Bilateral	108213	148618	196186	231658	277441	280111	293002	315991	Bilatéraux
Private creditors	193025	332763	391931	440883	487154	481016	469302	457781	Créanciers privés
Bonds	13091	22405	31516	35902	40558	45543	52573	110809	Obligations
Commercial banks	127013	241422	271760	301475	328698	318502	302388	227896	Banques commerciales
Other private	52920	68935	88655	103506	117898	116970	114341	119076	Autres institutions privées
Private non-guaranteed	70101	101169	93049	83949	76158	62682	56142	64622	**Dette privé non garantie**
Undisbursed debt[2]	133741	149476	172442	180668	203079	197011	195603	205469	**Dette (montants non versés)**[2]
Official creditors	91366	106549	117801	126319	142918	142913	147275	157149	Créanciers publics
Multilateral	47395	66888	72733	79468	87185	86839	93182	101157	Multilatéraux
IBRD	23091	33369	35206	38090	39505	38027	42599	43306	BIRD
IDA	9484	10726	12513	13714	15929	15394	16290	19183	IDA
Bilateral	43971	39661	45068	46852	55733	56074	54093	55992	Bilatéraux
Private creditors	42375	42926	54640	54349	60161	54098	48328	48320	Créanciers privés
Bonds	43	479	184	169	129	91	327	748	Obligations
Commercial banks	14285	12664	20570	18815	24479	18649	16916	18495	Banques commerciales
Other private	28047	29784	33887	35365	35553	35357	31086	29077	Autres institutions privées
Commitments[2]	93384	82097	91142	84880	99680	105292	95775	92753	**Engagements**[2]
Official creditors	42112	38687	40494	43780	51589	52785	56972	55627	Créanciers publics
Multilateral	18330	20961	23016	27562	28837	28796	32773	34878	Multilatéraux
IBRD	7968	9258	11667	14569	14112	12472	16820	15430	BIRD
IDA	4375	3261	3686	3221	4567	4552	4837	6316	IDA
Bilateral	23781	17726	17478	16218	22751	23990	24199	20750	Bilatéraux
Private creditors	51272	43409	50648	41100	48091	52507	38803	37126	Créanciers privés
Bonds	1518	2127	6067	3456	3278	8518	6361	6958	Obligations
Commercial banks	33587	26526	28548	21309	29718	21929	16927	14824	Banques commerciales
Other private	16168	14757	16033	16335	15095	22060	15516	15343	Autres institutions privées
Disbursements[2]	78207	76723	82408	84432	86072	96012	84492	84360	**Versements**[2]
Official creditors	27201	32155	33831	38745	40954	41980	43850	47254	Créanciers publics
Multilateral	9473	16139	16924	20226	21933	23792	23406	27556	Multilatéraux
IBRD	4545	8579	8405	10199	11329	12190	10845	13636	BIRD
IDA	1586	2566	2884	3189	3919	3834	3570	4369	IDA
Bilateral	17728	16016	16907	18518	19021	18187	20445	19699	Bilatéraux
Private creditors	51006	44568	48577	45687	45119	54033	40642	37106	Créanciers privés
Bonds	1601	1750	6359	3492	3333	8553	6097	6545	Obligations
Commercial banks	33920	27531	25619	23947	25556	26087	16974	13747	Banques commerciales
Other private	15484	15288	16600	18248	16230	19392	17571	16814	Autres institutions privées
Principal repayments[2]	29021	36699	45108	51794	60420	63456	57962	61074	**Remboursements du principal**[2]
Official creditors	6981	11527	12896	16804	19959	22737	22762	24490	Créanciers publics
Multilateral	1704	3826	4699	6733	9218	12431	11582	12777	Multilatéraux
IBRD	1079	2596	3131	4620	6572	9308	8022	8547	BIRD
IDA	31	91	119	136	151	173	209	251	IDA
Bilateral	5277	7702	8197	10071	10741	10307	11180	11714	Bilatéraux
Private creditors	22039	25172	32212	34989	40461	40719	35200	36584	Créanciers privés
Bonds	515	1425	1432	1992	2995	5378	3322	4930	Obligations
Commercial banks	13858	13001	18995	20254	24420	20333	15571	16231	Banques commerciales
Other private	7667	10746	11785	12743	13046	15008	16307	15423	Autres institutions privées

129
External debt of developing countries
Million US dollars [cont.]
Dette extérieure des pays en développement
Millions de dollars E-U [suite]

A. Total external debt · Total de la dette extérieure

Country or area	1980	1984	1985	1986	1987	1988	1989	1990	Pays ou zone
Net flows[2]	49186	40024	37300	32638	25653	32556	26530	23285	**Apports nets[2]**
Official creditors	20220	20627	20935	21940	20994	19242	21088	22764	Créanciers publics
Multilateral	7769	12313	12225	13493	12714	11362	11824	14779	Multilatéraux
IBRD	3466	5983	5274	5579	4758	2882	2823	5089	BIRD
IDA	1555	2474	2765	3053	3768	3661	3361	4118	IDA
Bilateral	12451	8314	8710	8447	8280	7881	9265	7985	Bilatéraux
Private creditors	28967	19396	16365	10698	4658	13314	5442	522	Créanciers privés
Bonds	1086	325	4927	1499	338	3175	2775	1615	Obligations
Commercial banks	20063	14530	6624	3693	1136	5754	1403	-2485	Banques commerciales
Other private	7818	4542	4814	5506	3184	4385	1264	1391	Autres institutions privées
Interest payments (LINT)[2]	24547	39702	45082	44504	45876	53667	48440	46897	**Paiements d'intérets (LINT)[2]**
Official creditors	5716	9167	11299	13770	15178	17041	16931	18579	Créanciers publics
Multilateral	2593	4524	5285	7547	9112	10023	9536	10965	Multilatéraux
IBRD	1830	3129	3561	5191	6247	6980	6382	7155	BIRD
IDA	79	163	188	237	276	295	267	303	IDA
Bilateral	3123	4644	6014	6223	6066	7018	7396	7614	Bilatéraux
Private creditors	18831	30535	33784	30734	30698	36626	31509	28319	Créanciers privés
Bonds	929	1375	1744	2071	2304	2603	3686	5108	Obligations
Commercial banks	14293	23802	26308	23120	21991	26599	20431	15490	Banques commerciales
Other private	3609	5358	5732	5542	6403	7424	7393	7721	Autres institutions privées
Net transfers[2]	24640	322	-7782	-11866	-20223	-21111	-21910	-23612	**Transferts nets[2]**
Official creditors	14504	11460	9636	8170	5816	2201	4157	4185	Créanciers publics
Multilateral	5176	7790	6940	5947	3603	1339	2288	3814	Multilatéraux
IBRD	1636	2854	1713	388	-1490	-4098	-3559	-2066	BIRD
IDA	1476	2311	2577	2816	3492	3366	3094	3815	IDA
Bilateral	9328	3671	2696	2224	2214	863	1869	371	Bilatéraux
Private creditors	10135	-11139	-17418	-20036	-26040	-23312	-26068	-27797	Créanciers privés
Bonds	157	-1050	3183	-572	-1966	573	-911	-3493	Obligations
Commercial banks	5770	-9273	-19684	-19427	-20855	-20846	-19028	-17974	Banques commerciales
Other private	4209	-816	-918	-36	-3219	-3039	-6129	-6330	Autres institutions privées
Total debt service (LTDS)[2]	53568	76401	90191	96298	106296	117123	106403	107972	**Total du service de la dette (LTDS)[2]**
Official creditors	12697	20694	24195	30575	35137	39778	39693	43069	Créanciers publics
Multilateral	4297	8349	9984	14280	18330	22454	21117	23742	Multilatéraux
IBRD	2909	5724	6692	9811	12819	16289	14404	15701	BIRD
IDA	110	255	307	373	427	468	476	553	IDA
Bilateral	8400	12345	14211	16295	16807	17325	18576	19327	Bilatéraux
Private creditors	40870	55707	65996	65723	71159	77345	66710	64902	Créanciers privés
Bonds	1444	2800	3176	4064	5299	7981	7008	10038	Obligations
Commercial banks	28151	36803	45303	43375	46411	46932	36002	31721	Banques commerciales
Other private	11276	16104	17517	18285	19449	22431	23700	23143	Autres institutions privées

129
External debt of developing countries
Million US dollars [*cont.*]
Dette extérieure des pays en développement
Millions de dollars E-U [*suite*]

B. Public and publicly guaranteed long-term debt • Dette publique extérieure à long terme garantie par l'Etat

Country or area Pays ou zone	1981	1982	1983	1984	1985	1986	1987	1988	1989	1990
Algeria Algérie	16089.8	14977.1	14328.1	14185.0	16512.2	19644.0	23432.7	23453.9	23604.2	24315.9
Angola Angola	...	...	...	...	1821.6	2547.8	3966.1	4649.0	6148.2	7151.9
Argentina Argentine	10570.2	15886.2	25439.9	26700.1	37327.3	40958.9	49221.1	47543.2	51460.3	46146.1
Bangladesh Bangladesh	3762.7	4263.0	4699.7	5036.8	5974.4	7368.3	8975.8	9497.8	9922.6	11463.6
Belize Belize	56.1	62.2	74.4	74.7	94.8	97.1	111.0	118.6	128.0	140.7
Benin Bénin	386.7	573.1	611.0	576.0	657.5	763.8	927.0	908.8	1046.4	1261.7
Bhutan Bhoutan	0.3	1.1	1.8	2.7	8.8	21.0	40.2	66.2	73.6	80.3
Bolivia Bolivie	2722.4	2837.4	3254.8	3372.0	3511.3	4070.2	4620.8	4139.9	3428.8	3683.3
Botswana Botswana	149.6	200.2	221.6	255.7	331.7	383.8	508.9	497.7	509.3	509.8
Brazil Brésil	45770.6	51700.3	59757.4	70638.5	74556.9	84723.2	91596.3	89903.1	84299.6	82098.1
Bulgaria Bulgarie	...	...	...	...	3943.7	5183.7	6721.2	8499.1	9359.1	9564.1
Burkina Faso Burkina Faso	279.7	307.1	358.0	369.1	456.2	576.7	744.0	767.4	648.2	749.9
Burundi Burundi	135.9	180.5	274.0	320.0	409.0	524.2	711.6	755.6	832.2	849.6
Cameroon Cameroun	2009.0	1920.9	1843.1	1700.2	2002.5	2399.1	2781.0	2934.6	3748.4	4784.4
Cape Verde Cap-Vert	39.4	59.0	71.7	77.0	98.7	112.4	127.2	129.0	132.1	144.2
Central African Rep. Rép. centrafricaine	176.3	195.1	203.3	215.0	289.8	396.6	543.9	582.2	641.6	815.3
Chad Tchad	175.8	146.8	150.4	146.7	153.8	195.3	262.2	303.4	317.1	430.4
Chile Chili	4488.1	5244.4	6598.0	10616.9	12896.8	14689.2	15523.9	13694.3	10850.6	10338.7
China Chine	4913.5	5220.3	5301.3	6178.5	9963.3	16597.9	25926.9	32587.7	37032.4	45319.2
Colombia Colombie	5076.3	5990.4	6874.4	7733.7	9573.1	12180.6	13827.7	13846.0	13989.0	14680.4
Comoros Comores	52.5	67.7	83.6	101.4	129.4	158.5	187.6	187.5	162.4	177.1
Congo Congo	1356.7	1753.1	1829.9	1850.2	2341.7	2761.6	3431.5	3550.1	3544.7	4380.2
Costa Rica Costa Rica	2206.4	2391.4	3141.9	3180.3	3531.9	3623.6	3707.4	3560.1	3559.3	3076.4
Côte d'Ivoire Côte d'Ivoire	4453.5	5062.5	4895.6	4841.5	5778.4	6668.5	8309.2	7946.2	8349.5	10050.0
Cyprus Chypre	486.3	609.3	654.4	761.5	940.3	1197.6	1418.7	1323.8	1433.2	1541.8
Czechoslovakia Tchécoslovaquie	...	...	...	...	2689.1	2876.9	3581.6	3968.9	4330.3	5346.1
Djibouti Djibouti	19.9	24.5	33.5	61.6	96.0	119.2	154.3	158.3	132.9	144.5
Dominica Dominique	...	...	1.2	3.1	42.0	44.5	55.6	57.3	63.9	76.0
Dominican Republic Rép. dominicaine	1400.3	1666.7	2203.1	2363.6	2690.9	2929.7	3220.2	3282.3	3334.7	3440.3
Ecuador Equateur	4192.1	3885.1	5546.1	6555.2	7198.7	8259.6	8991.1	9027.7	9424.6	9853.6
Egypt Egypte	19623.8	23606.2	26413.4	28957.4	34676.6	37742.4	43099.9	44005.5	42271.5	34242.0

129
External debt of developing countries
Million US dollars [*cont.*]
Dette extérieure des pays en développement
Millions de dollars E-U [*suite*]

B. Public and publicly guaranteed long-term debt · Dette publique extérieure à long terme garantie par l'Etat

Country or area Pays ou zone	1981	1982	1983	1984	1985	1986	1987	1988	1989	1990
El Salvador El Salvador	703.9	973.6	1385.5	1476.2	1556.1	1577.9	1669.0	1677.7	1818.1	1897.8
Equatorial Guinea Guinée équatoriale	62.5	85.1	90.0	80.3	112.8	139.5	173.2	184.5	203.1	206.0
Ethiopia Ethiopie	934.7	1010.1	1190.4	1358.5	1720.9	2047.6	2535.1	2790.7	2886.1	3116.2
Fiji Fidji	237.5	265.4	292.2	279.3	302.3	311.5	334.2	330.3	292.5	292.4
Gabon Gabon	974.6	816.7	678.6	669.0	944.7	1427.3	2077.1	2264.3	2483.3	2944.9
Gambia Gambie	132.2	147.4	151.7	151.8	176.6	212.1	265.5	272.9	290.5	304.1
Ghana Ghana	1164.5	1176.0	1205.1	1152.0	1299.7	1721.7	2255.6	2192.3	2334.3	2669.9
Grenada Grenade	19.6	31.0	43.1	42.1	46.6	51.8	65.6	69.7	72.2	91.2
Guatemala Guatemala	806.5	1144.1	1386.0	1947.3	2166.4	2268.3	2311.6	2108.5	2078.3	2178.6
Guinea Guinée	1239.3	1208.7	1188.0	1104.2	1275.5	1603.1	1859.1	1999.5	1966.5	2229.7
Guinea-Bissau Guinée-Bissau	132.6	140.0	146.4	189.2	263.9	310.6	402.3	416.7	460.5	544.3
Guyana Guyana	664.7	699.6	717.8	697.4	779.5	858.3	955.4	978.7	1484.1	1663.2
Haiti Haïti	340.6	395.7	428.9	475.1	522.3	575.7	673.5	683.3	684.4	745.4
Honduras Honduras	1235.6	1412.2	1627.1	1773.1	2137.0	2377.9	2703.3	2757.6	2822.2	3159.3
Hungary Hongrie	6932.8	6718.5	6268.9	7053.4	9964.9	12381.9	15673.1	15605.0	16627.5	18045.8
India Inde	19003.1	21095.8	23491.0	25394.8	31752.3	39084.8	46662.2	50705.7	56567.3	61096.7
Indonesia Indonésie	15908.4	18317.5	21493.9	22239.8	26767.4	32628.4	40899.2	41248.9	41204.4	44973.6
Iran, Islamic Rep. of Iran, Rép. islamique d'	3793.1	3468.1	2967.6	2456.7	2390.0	2412.8	2279.9	2055.1	1861.6	1797.0
Jamaica Jamaïque	1713.0	2115.3	2459.5	2616.8	3122.1	3261.4	3728.6	3710.7	3726.7	3873.5
Jordan Jordanie	1762.7	2137.8	2634.7	2831.6	3397.6	4190.7	4969.3	5341.8	6403.6	6486.2
Kenya Kenya	2209.6	2352.3	2389.0	2337.4	2682.6	3466.3	4322.5	4166.0	4117.4	4809.8
Korea, Republic of Corée, République de	18360.6	20190.7	22175.5	23833.0	28279.1	29350.9	23889.7	20024.9	17035.3	17814.1
Lao People's Dem. Rep. Rép. dém. pop. lao	306.5	340.5	377.3	422.2	466.1	588.8	719.4	815.8	937.0	1053.0
Lebanon Liban	320.1	442.2	498.4	477.6	527.8	488.8	552.9	499.4	517.3	545.0
Lesotho Lesotho	71.5	112.1	126.2	128.0	165.1	187.2	249.7	274.0	312.6	371.8
Liberia Libéria	598.3	628.2	714.0	773.8	886.8	990.1	1117.8	1083.6	1073.0	1126.4
Madagascar Madagascar	1405.8	1654.6	1753.6	1869.3	2187.0	2666.3	3265.7	3310.1	3361.3	3676.6
Malawi Malawi	657.7	673.8	695.2	708.6	791.7	947.5	1160.9	1185.6	1241.9	1366.4
Malaysia Malaisie	5743.2	8201.1	11664.7	13574.1	15072.2	17106.3	19246.8	17101.4	15515.9	16106.8
Maldives Maldives	36.8	42.1	48.1	49.9	49.0	58.7	61.9	59.3	54.4	63.9
Mali Mali	729.1	802.5	902.0	1092.0	1301.2	1578.5	1905.9	1910.5	2043.5	2306.0

129
External debt of developing countries
Million US dollars [cont.]
Dette extérieure des pays en développement
Millions de dollars E-U [suite]

B. Public and publicly guaranteed long-term debt · Dette publique extérieure à long terme garantie par l'Etat

Country or area Pays ou zone	1981	1982	1983	1984	1985	1986	1987	1988	1989	1990
Malta Malte	82.0	76.6	93.8	90.8	99.4	95.2	95.6	84.8	80.1	121.8
Mauritania Mauritanie	814.6	986.0	1125.6	1163.8	1349.0	1599.7	1845.6	1819.6	1776.8	1897.8
Mauritius Maurice	323.3	357.9	316.7	336.4	398.0	449.2	582.5	643.1	633.8	738.7
Mexico Mexique	43032.2	51550.7	66766.7	69726.1	72702.9	75817.9	84349.2	80589.7	76059.3	76204.2
Morocco Maroc	9117.9	10247.0	10975.6	11812.4	13823.4	15836.5	18615.4	19445.9	20310.5	22097.2
Mozambique Mozambique	...	...	...	...	2545.7	3025.9	3683.2	3739.4	3884.7	4053.2
Myanmar Myanmar	1579.7	1897.7	2166.2	2196.5	2896.5	3617.8	4245.1	4220.2	4044.9	4446.5
Nepal Népal	218.7	283.3	348.8	430.5	544.7	707.4	933.3	1095.9	1289.7	1556.8
Nicaragua Nicaragua	2076.0	2473.0	3366.3	4059.4	4892.4	5726.3	6349.5	6884.5	7508.1	8066.6
Niger Niger	589.8	586.1	656.3	674.4	832.5	995.2	1244.0	1286.0	1126.1	1326.3
Nigeria Nigéria	6361.9	9105.8	12180.6	11392.6	13139.2	19261.1	28696.9	29320.5	31661.3	33709.1
Oman Oman	537.3	723.9	1131.7	1339.8	1908.1	2463.0	2442.7	2486.2	2626.5	2205.3
Pakistan Pakistan	8578.3	9455.8	9483.8	9724.5	10563.8	11809.3	13440.7	13890.0	14467.3	16532.3
Panama Panama	2429.5	2917.2	3145.4	3180.7	3319.9	3533.1	4026.5	4005.3	3935.4	3986.8
Papua New Guinea Papouasie-Nv.-Guinée	609.5	721.1	912.9	960.5	1069.3	1233.0	1432.1	1259.4	1319.0	1509.0
Paraguay Paraguay	707.3	938.4	1142.9	1247.5	1533.8	1825.6	2223.9	2093.0	2095.7	1736.2
Peru Pérou	6047.7	6967.7	8248.9	9209.6	10338.8	11334.3	12732.6	12505.5	12669.4	13343.7
Philippines Philippines	7468.2	8857.3	10566.7	11298.7	14020.2	19577.6	23231.2	23022.4	22596.8	24107.7
Poland Pologne	...	...	...	...	29762.7	31932.2	36055.4	33670.3	34452.4	39282.0
Portugal Portugal	7818.5	8982.3	10090.4	10770.4	12825.5	13928.1	14979.4	13907.3	14433.7	14431.7
Romania Roumanie	8071.0	7796.9	7585.3	6254.8	5805.1	5652.5	5342.7	1680.0	...	19.0
Rwanda Rwanda	164.1	185.5	215.2	243.8	332.5	420.0	564.1	613.0	602.8	692.3
Sao Tome and Principe Sao Tomé-et-Principe	29.4	36.8	42.7	53.8	61.6	74.8	85.8	95.6	109.3	129.1
Samoa Samoa	54.8	58.5	58.6	63.0	63.7	64.6	71.4	71.1	71.9	91.9
Senegal Sénégal	1234.0	1425.5	1643.8	1691.1	2058.3	2622.8	3328.7	3264.1	2670.5	2953.6
Seychelles Seychelles	27.9	38.7	46.5	51.5	74.0	108.2	138.7	134.3	132.8	151.2
Sierra Leone Sierra Leone	327.2	374.2	367.2	326.5	392.5	472.2	551.5	540.2	532.0	605.4
Solomon Islands Iles Salomon	19.5	22.3	27.9	37.4	54.3	71.3	94.7	101.4	99.4	105.3
Somalia Somalie	975.1	1089.5	1217.4	1282.7	1411.9	1554.3	1743.3	1779.2	1813.3	1922.0
Sri Lanka Sri Lanka	1511.6	1864.8	2115.5	2348.3	2836.4	3449.3	4078.6	4149.0	4269.1	4911.0
St. Kitts and Nevis St. Kitts-et-Nevis	...	...	...	...	12.7	17.8	25.1	28.5	30.7	36.2

129
External debt of developing countries
Million US dollars [*cont.*]
Dette extérieure des pays en développement
Millions de dollars E-U [*suite*]

B. Public and publicly guaranteed long-term debt · Dette publique extérieure à long terme garantie par l'Etat

Country or area Pays ou zone	1981	1982	1983	1984	1985	1986	1987	1988	1989	1990
Saint Lucia Sainte-Lucie	...	...	...	...	20.6	25.9	35.9	44.9	51.5	60.3
St. Vincent - Grenadines St.-Vincent-Grenadines	17.1	19.7	22.6	22.7	24.7	28.3	38.7	45.4	51.3	57.3
Sudan Soudan	4505.2	5424.3	5837.5	6191.0	6601.7	7121.6	8043.1	8002.3	8468.8	9156.3
Swaziland Swaziland	173.8	179.8	182.8	170.5	209.2	250.8	285.3	255.3	248.5	251.1
Syrian Arab Republic Rép. arabe syrienne	4000.5	5253.6	7594.7	7406.3	9501.6	11536.8	14478.4	15165.6	15704.0	14958.8
Thailand Thaïlande	5017.0	6033.8	6901.6	7186.4	9860.3	11488.0	13831.8	13186.1	12407.9	12571.6
Togo Togo	817.3	792.2	795.6	681.4	791.7	889.7	1052.3	1064.8	947.1	1095.8
Tonga Tonga	...	...	...	...	...	32.3	44.3	43.8	44.1	49.2
Trinidad and Tobago Trinité-et-Tobago	794.8	906.9	1026.0	1062.9	1299.2	1584.6	1638.8	1897.0	1811.6	1807.8
Tunisia Tunisie	3312.6	3498.5	3815.6	3705.0	4452.6	5236.3	6007.2	5881.9	5988.7	6505.8
Turkey Turquie	15271.6	16095.0	16069.4	16569.5	19565.9	24905.4	31540.1	33576.6	34832.2	38595.4
Uganda Ouganda	580.7	654.3	698.4	779.5	962.4	1166.9	1656.0	1688.1	1975.9	2300.7
United Rep. of Tanzania Rép.-Unie de Tanzanie	2094.4	2291.3	2461.6	2424.0	2786.9	3622.8	4292.0	4470.9	4610.9	5294.1
Uruguay Uruguay	1348.2	1700.2	2510.3	2527.8	2695.0	2895.3	3098.4	2950.6	2977.9	3044.0
Vanuatu Vanuatu	3.0	4.0	3.5	4.9	6.9	8.1	13.7	15.3	20.9	30.6
Venezuela Venezuela	11540.8	12449.6	14828.3	18778.2	17737.7	25329.1	25118.9	25307.9	25280.4	24643.1
Yemen[3] Yémen[3]	1812.4	2144.4	2490.8	2582.8	2968.1	3421.0	4012.4	4287.7	4545.8	5039.8
Yugoslavia Yougoslavie	5197.8	5460.5	7234.3	8484.7	11200.8	12242.3	14278.5	13952.0	14650.0	13492.3
Zaire Zaïre	4223.9	4071.8	4433.4	4284.4	4957.8	5917.1	7204.6	6911.5	7912.3	8850.9
Zambia Zambie	2191.4	2335.2	2554.9	2613.9	3159.4	3792.5	4428.3	4415.2	4196.2	4784.4
Zimbabwe Zimbabwe	804.5	1207.7	1422.2	1495.3	1765.6	2030.7	2355.8	2206.7	2260.2	2448.4

Source:
World Debt Tables 1991-92, The World Bank (Washington, DC).

1   The following abbreviations have been used in the table:
    LDOD: Long-term debt outstanding and disbursed
    IBRD: International Bank for Reconstruction and Development
    IDA: International Development Association
    LINT: Loan interest
    LTDS: Long-term debt service
2   Public and publicly guaranteed debt.
3   Excluding former Democratic Yemen.

Source:
"World Debt Tables 1991-92", La Banque mondiale
(Washington, DC).
1   Les abréviations ci-après ont été utilisées dans le tableau:
    LDOD : Dette à long terme
    BIRD : Banque internationale pour la réconstruction
        et le développement
    IDA : Association internationale de développement
    LINT : Paiement des intérêts
    LTDS : Service de la dette
2   Dette publique extérieure garantie par l'Etat.
3   Non compris l'ancien Yémen démocratique.

Technical notes, tables 126-129

Table 126: Foreign exchange rates are shown in units of national currency per US dollar and refer to end-of-period and period-average quotations. Unless otherwise stated, the table refers to the midpoint (average of selling and buying rates) market rates for all dates so far as data are available. For dates prior to those for which market rates are available, the series represent fixed par value of central rates agreed with the International Monetary Fund. For further information see *International Financial Statistics*.[9]

Table 127: The international reserves series reports special drawing rights (SDR), reserve positions in the Fund, foreign exchange and corresponding totals of these three items for individual country members.

Special drawing rights (SDR) are unconditional international reserve assets created by the Fund.

Reserve positions in the Fund are unconditional assets that arise from countries' gold subscriptions to the Fund, from the Fund's use of member currencies to finance drawings of others and from Fund Borrowings.

Foreign exchange is defined as holdings by monetary authorities (central banks, currency boards, exchange stabilization funds and Treasuries to the extent that they perform similar functions) of claims on foreigners in the form of bank deposits, treasury bills, short and long-terms government securities, and other claims usable in the event of a balance of payments deficit including nonmarketable claims arising from inter-central bank and inter-governmental arrangements without regard to whether the claim is denominated in the currency of the debtor or the creditor.

For further details and current figures, see *International Financial Statistics* and the United Nations *Monthly Bulletin of Statistics*.[9, 23]

Table 128 reports data on official holdings of gold expressed in physical terms (millions of fine troy ounces) for all countries.

Table 129: Data are extracted from *World Debt Tables 1990-1991, External Debt of Developing Countries*, published by the World Bank.[33]

External debt is defined as debt that has an original or extended maturity of more than one year and is owed to non-residents and repayable in foreign currency, goods, or services. A distinction is made between:

— Public debt which is an external obligation of a public debtor, which could be a national government, a political sub-division, an agency of either of the above or, in fact, any autonomous public body;

— Publicly guaranteed debt, which is an external obligation of a private debtor that is guaranteed for repayment by a public entity;

— Private non-guaranteed external debt, which is an external obligation of a private debtor that is not guaranteed for repayment by a public entity.

Notes techniques, tableaux 126-129

Tableau 126 : Les cours des changes sont exprimés par nombre d'unités de monnaie nationale pour un dollar des Etats-Unis et se rapportent aux cotations en fin de période et pour les moyennes sur la période. Sauf indication contraire, le tableau indique les cours moyens (moyenne des cours à la vente et à l'achat) pour toutes les dates pour lesquelles des données sont disponibles. En ce qui concerne les dates antérieures à celles pour lesquelles on dispose de données, les séries portent sur les parités fixes des taux centraux convenus avec le Fonds monétaire international. Pour plus de renseignements, voir *Satistiques financières internationales* [9].

Tableau 127 : Le tableau des réserves internationales rend compte des droits de tirage spéciaux (DTS), des réserves disponibles au Fonds monétaire international, des devises étrangères et des totaux correspondants de ces trois éléments pour les pays membres.

Les droits de tirages spéciaux (DTS) sont des instruments inconditionnels de réserves créés par le FMI.

Les réserves disponibles au FMI sont des avoirs inconditionnels découlant des souscriptions-or des pays membres au capital du Fonds, de l'utilisation par le Fonds des monnaies de ses pays membres pour financer des tirages d'autres membres et des emprunts du Fonds.

Les devises sont définies comme les avoirs détenus par les autorités monétaires (banques centrales, offices des changes, fonds de stabilisation des changes et Trésor, dans la mesure où ils exercent des fonctions similaires) à titre de créances sur des égrangers sous la forme de dépôts bancaires, bons du Trésor, obligations d'Etat à court et à long terme et autres créances utilisables au cas où le déficit de la balance des paiements comprendrait des créances non négociables découlant d'arrangements entre banques centrales et entre gouvernements, que la créance soit libellée dans la monnaie du débiteur ou du créancier.

Pour plus de détails et pour les chiffres courants, voir *Statistiques financières internationales* et *Bulletin mensuel de statistique* des Nations Unies [9, 23].

Le *Tableau 128* présente les données sur les avoirs officiels en or exprimés en unités physiques (million d'onces troy de fin) pour tous les pays.

Tableau 129 : les données sont extraites des *Tableaux de la dette internationale 1990-1991, Dette extérieure des pays en développement*, publiés par la Banque mondiale [33].

La dette extérieure désigne la dette dont l'échéance initiale ou reportée est de plus d'un an, due à des non-résidents et remboursable en devises, biens ou services. On établit les distinctions suivantes :

— La dette publique, qui est une obligation extérieure d'un débiteur public, pouvant être un gouvernement, un organe politique, une institution de l'un ou l'autre ou, en fait, tout organisme public autonome.

— La dette garantie par l'Etat, qui est une obligation extérieure d'un débiteur privé, dont le remboursement est garanti par un organisme public.

The data referring to public and publicly guaranteed debt do not include data for (a) transactions with International Monetaty Fund, (b) debt repayable in local currency, (c) direct investment and (d) short-term debt (that is, debt with an original maturity of less than a year).

The data referring to private non-guaranteed debt also exclude the above items but include contractual obligations on loans to direct-investment enterprises by foreign parent companies or their affiliates.

Data are aggregated by type of creditor. The breakdown is as follows:

Official creditors:

(a) Loans from international organizations (multilateral loans), excluding loans from funds administered by an international organization on behalf of a single donor government. The latter are classified as loans from governments;

(b) Loans from governments (bilateral loans) and from autonomous public bodies;

Private creditors:

(a) Suppliers: Credits from manufacturers, exporters, or other suppliers of goods;

(b) Financial markets: Loans from private banks and other private financial institutions as well as publicly issued and privately placed bonds;

(c) Other: External liabilities on account of nationalized properties and unclassified debts to private creditors.

A distinction is made between the following categories of external public debt:

— Debt outstanding (including undisbursed) is the sum of disbursed and undisbursed debt and represents the total outstanding external obligations of the borrower at year-end;

— Debt outstanding (disbursed only) is total outstanding debt drawn by the borrower at year end;

— Commitments are the total of loans for which contracts are signed in the year specified;

— Disbursements are drawings on outstanding loan commitments during the year specified;

— Service payments are actual repayments of principal amortization and interest payments made in foreign currencies, goods or services in the year specified;

— Net flows (or net lending) are disbursements minus principal repayments;

— Net transfers are net flows minus interest payments or disbursements minus total debt-service payments.

The countries included in the table are those for which data are sufficiently reliable to provide a meaningful presentation of debt outstanding and future service payments as of the end of 1990.

— La dette extérieure privée non garantie, qui est une obligation extérieure d'un débiteur privé, dont le remboursement n'est pas garanti par un organisme public.

Les statistiques relatives à la dette publique ou à la dette garantie par l'Etat ne comprennent pas les données concernant : (a) les transactions avec le Fonds monétaire international; (b) la dette remboursable en monnaie nationale; (c) les investissements directs; et (d) la dette à court terme (c'est-à-dire la dette dont l'échéance initiale est inférieure à un an).

Les statistiques relatives à la dette privée non garantie ne comprennent pas non plus les éléments précités, mais comprennent les obligations contractuelles au titre des prêts consentis par des sociétés mères étrangères ou leurs filiales à des entreprises créées dans le cadre d'investissements directs.

Les données sont groupées par type de créancier, comme suit :

Créanciers publics :

(a)   Les prêts obtenus auprès d'organisations internationales (prêts multilatéraux), à l'exclusion des prêts au titre de fonds administrés par une organisation internationale pour le compte d'un gouvernement donateur précis, qui sont classés comme prêts consentis par des gouvernements.

(b)   Les prêts consentis par des gouvernements (prêts bilatéraux) et par des organisations publiques autonomes.

Créanciers privés :

(a)   Fournisseurs : Crédits consentis par des fabricants exportateurs et autre fournisseurs de biens;

(b)   Marchés financiers : prêts consentis par des banques privées et autres institutions financières privées, et émissions publiques d'obligations placées auprès d'investisseurs privés;

(c)   Autres créanciers : engagements vis-à-vis de l'extérieur au titre des biens nationalisés et dettes diverses à l'égard de créanciers privés.

On fait une distinction entre les catégories suivantes de dette publique extérieure :

— L'encours de la dette (y compris les fonds non décaissés) est la somme des fonds décaissés et non décaissés et représente le total des obligations extérieures en cours de l'emprunteur à la fin de l'année;

— L'encours de la dette (fonds décaissés seulement) est le montant total des tirages effectués par l'emprunteur sur sa dette en cours à la fin de l'année;

— Les engagements représentent le total des prêts dont les contrats ont été signés au cours de l'année considérée;

— Les décaissements sont les sommes tirées sur l'encours des prêts pendant l'année considérée;

— Les paiements au titre du service de la dette sont les remboursements effectifs du principal et les paiements d'intérêts effectués en devises, biens ou services pendant l'année considérée;

— Les flux nets (ou prêts nets) sont les décaissements moins les remboursements de principal;

— Les transferts nets désignent les flux nets moins les paiements d'intérêts, ou les décaissements moins le total des paiements au titre du service de la dette.

Les pays figurant sur ce tableau sont ceux pour lesquels les données sont suffisamment fiables pour permettre une présentation significative de l'encours de la dette et des paiements futurs au titre du service de la dette à la fin de 1990.

130
Disbursements to developing countries or areas of bilateral and multilateral official development assistance
Versements mises à la disposition des pays ou zones en développement au titre de l'aide publique bilatérale et multilatérale au développement

Region, country or area Région, pays ou zone	Year Année	Disbursements ($US) Versements ($E-U)			
		Bilateral Bilatérale (millions)	Multilateral[1] Multilatérale[1] (millions)	Total (millions)	Per capita[2] Par habitant[2]
Africa	1988	12532.1	4924.6	17456.7	28.9
Afrique	1989	12654.0	5528.8	18182.8	29.2
	1990	16561.9	6104.2	22666.1	35.3
Algeria	1988	120.0	24.1	144.1	6.1
Algérie	1989	89.4	40.7	130.1	5.4
	1990	185.7	36.1	221.8	8.9
Angola	1988	106.1	50.3	156.4	16.5
Angola	1989	87.9	57.8	145.7	14.9
	1990	142.6	69.1	211.7	21.1
Benin	1988	93.2	67.4	160.6	36.9
Bénin	1989	149.4	108.4	257.8	57.4
	1990	125.6	135.9	261.5	56.5
Botswana	1988	125.4	24.7	150.1	123.8
Botswana	1989	120.0	42.1	162.1	128.9
	1990	121.2	27.9	149.1	114.3
Burkina Faso	1988	218.8	76.4	295.2	34.7
Burkina Faso	1989	199.0	72.4	271.4	31.0
	1990	238.3	73.3	311.6	34.6
Burundi	1988	83.3	102.1	185.4	36.0
Burundi	1989	89.5	105.2	194.7	36.7
	1990	156.8	109.4	266.2	48.6
Cameroon	1988	240.0	48.5	288.5	26.1
Cameroun	1989	300.6	157.8	458.4	40.1
	1990	321.7	163.2	484.9	41.0
Cape Verde	1988	59.4	26.0	85.4	244.0
Cape Verde	1989	48.5	26.8	75.3	209.7
	1990	54.2	28.4	82.6	223.2
Central African Rep.	1988	107.4	86.5	193.9	67.5
Rép. centrafricaine	1989	99.3	92.7	192.0	65.0
	1990	99.9	131.3	231.2	76.1
Chad	1988	146.0	118.4	264.4	49.0
Tchad	1989	128.1	113.4	241.5	43.6
	1990	178.1	134.7	312.8	55.1
Comoros	1988	34.5	17.3	51.8	101.0
Comores	1989	31.8	12.6	44.4	83.6
	1990	30.6	11.8	42.4	77.1
Congo	1988	76.6	12.2	88.8	41.7
Congo	1989	79.1	12.1	91.2	41.5
	1990	197.7	11.7	209.4	92.2
Côte d'Ivoire	1988	226.3	212.7	439.0	39.5
Côte d'Ivoire	1989	260.1	142.9	403.0	34.9
	1990	530.2	159.3	689.5	57.5
Djibouti	1988	71.2	18.4	89.6	232.1
Djibouti	1989	63.9	10.2	74.1	186.6
	1990	88.3	20.1	108.4	265.0
Egypt	1988	1432.8	121.3	1554.1	31.1
Egypte	1989	1408.9	174.0	1582.9	30.9
	1990	3169.8	110.6	3280.4	62.6
Equatorial Guinea	1988	23.8	19.5	43.3	127.7
Guinée équatoriale	1989	19.6	22.2	41.8	121.2
	1990	20.2	18.8	39.0	110.8
Ethiopia	1988	559.9	411.3	971.2	20.9
Ethiopie	1989	377.8	375.8	753.6	15.7
	1990	503.1	385.1	888.2	18.0

130
**Disbursements to developing countries or areas of bilateral and multilateral
official development assistance [cont.]**
**Versements mises à la disposition des pays ou zones en développement au titre de l'aide publique bilatérale
et multilatérale au développement [suite]**

Region, country or area Région, pays ou zone	Year Année	Disbursements ($US) Versements ($E-U)			
		Bilateral Bilatérale (millions)	Multilateral[1] Multilatérale[1] (millions)	Total (millions)	Per capita[2] Par habitant[2]
Gabon Gabon	1988	98.7	8.9	107.6	98.3
	1989	121.0	12.6	133.6	117.9
	1990	126.8	13.1	139.9	119.4
Gambia Gambie	1988	54.7	29.5	84.2	103.6
	1989	56.1	37.6	93.7	111.9
	1990	56.1	37.2	93.3	108.4
Ghana Ghana	1988	235.8	225.3	461.1	32.6
	1989	350.4	199.1	549.5	37.7
	1990	261.7	234.4	496.1	33.0
Guinea Guinée	1988	159.5	88.7	248.2	45.8
	1989	191.7	136.6	328.3	58.8
	1990	138.9	147.8	286.7	49.8
Guinea-Bissau Guinée-Bissau	1988	47.7	45.0	92.7	100.2
	1989	53.4	45.6	99.0	104.9
	1990	62.1	60.2	122.3	126.9
Kenya Kenya	1988	609.9	195.3	805.2	36.0
	1989	620.6	346.8	967.4	41.7
	1990	735.2	344.0	1079.2	44.9
Lesotho Lesotho	1988	70.1	38.5	108.6	64.8
	1989	68.8	58.8	127.6	74.1
	1990	85.1	53.1	138.2	77.9
Liberia Libéria	1988	48.3	16.5	64.8	26.8
	1989	38.5	20.1	58.6	23.5
	1990	42.2	73.0	115.2	44.7
Libyan Arab Jamahiriya Jamahiriya arabe libyenne	1988	1.9	3.7	5.6	1.3
	1989	6.9	10.0	16.9	3.9
	1990	7.6	12.6	20.2	4.4
Madagascar Madagascar	1988	213.5	92.2	305.7	27.2
	1989	175.4	147.0	322.4	27.7
	1990	268.2	132.8	401.0	33.4
Malawi Malawi	1988	181.4	184.6	366.0	44.9
	1989	181.7	230.0	411.7	48.7
	1990	216.0	262.7	478.7	54.7
Mali Mali	1988	260.0	162.0	422.0	48.7
	1989	300.6	150.6	451.2	50.5
	1990	312.2	146.2	458.4	49.8
Mauritania Mauritanie	1988	111.0	77.0	188.0	98.2
	1989	160.3	86.3	246.6	125.2
	1990	106.1	104.2	210.3	103.9
Mauritius Maurice	1988	44.9	11.2	56.1	53.1
	1989	50.0	9.5	59.5	55.7
	1990	75.5	12.6	88.1	81.4
Morocco Maroc	1988	402.2	57.2	459.4	19.3
	1989	389.1	63.6	452.7	18.5
	1990	562.1	49.1	611.2	24.4
Mozambique Mozambique	1988	731.1	158.8	889.9	60.0
	1989	546.1	222.3	768.4	50.4
	1990	690.6	255.5	946.1	60.4
Namibia Namibie	1988	17.2	5.4	22.6	13.5
	1989	36.1	22.8	58.9	34.1
	1990	39.4	17.6	57.0	32.0
Niger Niger	1988	241.9	123.6	365.5	50.3
	1989	199.7	94.3	294.0	39.2
	1990	254.3	100.9	355.2	45.9
Nigeria Nigéria	1988	97.0	23.0	120.0	1.2
	1989	309.8	35.6	345.4	3.3
	1990	172.6	61.8	234.4	2.2
Réunion Réunion	1988	570.9	37.2	608.1	1053.9
	1989	678.4	20.2	698.6	1188.1
	1990	920.7	20.2	940.9	1573.4

130
Disbursements to developing countries or areas of bilateral and multilateral
official development assistance [*cont.*]
Versements mises à la disposition des pays ou zones en développement au titre de l'aide publique bilatérale
et multilatérale au développement [*suite*]

| Region, country or area
Région, pays ou zone | Year
Année | Disbursements ($US)
Versements ($E-U) | | | |
		Bilateral Bilatérale (millions)	Multilateral [1] Multilatérale [1] (millions)	Total (millions)	Per capita [2] Par habitant [2]
Rwanda	1988	137.2	112.0	249.2	36.9
Rwanda	1989	131.6	95.0	226.6	32.4
	1990	182.2	98.9	281.1	38.8
St. Helena	1988	27.0	0.1	27.1	4516.7
Sainte-Hélène	1989	47.2	0.0	47.2	6742.9
	1990	23.3	0.0	23.3	3328.6
Sao Tome and Principe	1988	8.0	16.0	24.0	208.7
Sao Tomé-et-Principe	1989	11.0	21.8	32.8	278.0
	1990	15.0	23.4	38.4	317.4
Senegal	1988	367.8	176.8	544.6	78.6
Sénégal	1989	536.4	101.7	638.1	89.5
	1990	581.7	203.3	785.0	107.1
Seychelles	1988	17.8	2.5	20.3	303.0
Seychelles	1989	15.1	4.8	19.9	292.6
	1990	31.6	3.5	35.1	508.7
Sierra Leone	1988	52.5	39.8	92.3	23.4
Sierra Leone	1989	72.2	27.8	100.0	24.7
	1990	39.9	30.3	70.2	16.9
Somalia	1988	311.3	117.9	429.2	60.8
Somalia	1989	266.9	159.2	426.1	58.5
	1990	259.8	137.3	397.1	53.0
Sudan	1988	500.6	332.8	833.4	35.0
Soudan	1989	434.3	312.2	746.5	30.5
	1990	418.7	371.6	790.3	31.4
Swaziland	1988	22.6	15.6	38.2	52.0
Swaziland	1989	11.6	17.8	29.4	38.6
	1990	36.1	19.0	55.1	69.9
Togo	1988	128.2	72.6	200.8	60.5
Togo	1989	108.0	75.5	183.5	53.6
	1990	154.9	55.1	210.0	59.5
Tunisia	1988	235.8	77.1	312.9	40.1
Tunisie	1989	177.7	60.8	238.5	29.8
	1990	213.1	51.1	264.2	32.3
Uganda	1988	188.0	175.4	363.4	20.8
Ouganda	1989	158.9	212.5	371.4	20.5
	1990	243.4	311.8	555.2	29.5
United Rep. Tanzania	1988	785.6	195.9	981.5	38.7
Rép. unie de Tanzanie	1989	687.6	231.1	918.7	34.9
	1990	837.5	313.8	1151.3	42.1
Zaire	1988	400.3	175.4	575.7	17.3
Zaïre	1989	432.9	201.5	634.4	18.4
	1990	628.5	194.4	822.9	23.1
Zambia	1988	407.4	70.8	478.2	61.0
Zambie	1989	314.2	77.7	391.9	48.1
	1990	408.9	82.7	491.6	58.2
Zimbabwe	1988	233.1	42.8	275.9	30.3
Zimbabwe	1989	227.5	38.9	266.4	28.3
	1990	295.8	45.7	341.5	35.2
Other & unallocated	1988	742.6	156.1	898.7	...
Autres et non-ventilés	1989	892.2	190.3	1082.5	...
	1990	835.4	212.8	1048.2	...
Americas	1988	4138.9	980.7	5119.6	7.3
Ameriques	1989	4599.0	1008.7	5607.7	7.9
	1990	5312.0	1087.6	6399.6	8.8
Argentina	1988	113.6	38.2	151.8	4.8
Argentine	1989	179.3	31.5	210.8	6.6
	1990	153.7	18.2	171.9	5.3

130
Disbursements to developing countries or areas of bilateral and multilateral
official development assistance [*cont.*]
Versements mises à la disposition des pays ou zones en développement au titre de l'aide publique bilatérale
et multilatérale au développement [*suite*]

Region, country or area Région, pays ou zone	Year Année	Disbursements ($US) Versements ($E-U)			
		Bilateral Bilatérale (millions)	Multilateral[1] Multilatérale[1] (millions)	Total (millions)	Per capita[2] Par habitant[2]
Aruba Aruba	1988 1989 1990	19.3 24.2 28.9	0.0 0.2 0.1	19.3 24.4 29.0	321.7 406.7 483.3
Bahamas Bahamas	1988 1989 1990	0.1 0.2 0.4	4.1 3.9 4.1	4.2 4.1 4.5	17.1 16.5 17.8
Barbados Barbade	1988 1989 1990	2.3 0.5 1.4	0.9 1.7 1.5	3.2 2.2 2.9	12.6 8.7 11.4
Belize Belize	1988 1989 1990	16.6 18.4 18.8	8.6 10.2 10.0	25.2 28.6 28.8	140.8 156.3 154.0
Bermuda Bermudes	1988 1989 1990	0.0 0.1 42.1	0.0 0.0 0.0	0.0 0.1 42.1	0.0 1.7 725.9
Bolivia Bolivie	1988 1989 1990	227.8 302.6 344.8	166.1 137.6 145.9	393.9 440.2 490.7	56.9 61.9 67.1
Brazil Brésil	1988 1989 1990	192.2 193.3 141.4	18.4 13.5 22.8	210.6 206.8 164.2	1.5 1.4 1.1
Chile Chili	1988 1989 1990	46.1 50.3 76.9	-2.1 10.8 17.4	44.0 61.1 94.3	3.5 4.7 7.2
Colombia Colombie	1988 1989 1990	60.3 49.0 78.8	1.3 18.1 8.2	61.6 67.1 87.0	1.9 2.1 2.6
Costa Rica Costa Rica	1988 1989 1990	163.7 205.6 203.8	23.1 19.9 23.8	186.8 225.5 227.6	65.2 76.7 75.5
Cuba Cuba	1988 1989 1990	2.9 10.6 16.1	16.6 13.2 17.3	19.5 23.8 33.4	1.9 2.3 3.1
Dominican Rep. Rép. dominicaine	1988 1989 1990	99.3 120.8 71.5	18.4 21.7 21.6	117.7 142.5 93.1	17.1 20.3 13.0
Ecuador Equateur	1988 1989 1990	99.9 125.9 114.6	36.5 33.7 39.0	136.4 159.6 153.6	13.5 15.5 14.5
El Salvador El Salvador	1988 1989 1990	380.9 365.5 308.7	38.6 78.0 38.0	419.5 443.5 346.7	83.5 86.4 66.0
French Guyana Guyane française	1988 1989 1990	130.4 110.3 113.2	14.9 8.6 8.7	145.3 118.9 121.9	1579.3 1251.6 1243.9
Guadeloupe Guadeloupe	1988 1989 1990	253.1 219.4 310.5	13.2 12.5 12.5	266.3 231.9 323.0	785.5 680.1 941.7
Guatemala Guatemala	1988 1989 1990	193.0 209.6 149.2	42.0 51.0 49.6	235.0 260.6 198.8	27.1 29.2 21.6
Guyana Guyana	1988 1989 1990	15.6 26.8 35.8	11.5 17.4 73.1	27.1 44.2 108.9	34.2 55.7 136.8
Haiti Haïti	1988 1989 1990	101.4 138.6 116.4	45.3 61.3 66.8	146.7 199.9 183.2	23.5 31.3 28.1
Honduras Honduras	1988 1989 1990	252.1 201.7 377.5	69.4 40.9 70.7	321.5 242.6 448.2	66.6 48.7 87.2

130
Disbursements to developing countries or areas of bilateral and multilateral
official development assistance [cont.]
Versements mises à la disposition des pays ou zones en développement au titre de l'aide publique bilatérale
et multilatérale au développement [suite]

Region, country or area Région, pays ou zone	Year Année	Disbursements ($US) Versements ($E-U)			
		Bilateral Bilatérale (millions)	Multilateral[1] Multilatérale[1] (millions)	Total (millions)	Per capita[2] Par habitant[2]
Jamaica	1988	173.0	19.2	192.2	80.1
Jamaïque	1989	225.9	35.4	261.3	107.7
	1990	251.9	28.5	280.4	114.2
Martinique	1988	447.8	13.7	461.5	1369.4
Martinique	1989	606.9	11.7	618.6	1824.8
	1990	841.6	11.7	853.3	2502.3
Mexico	1988	140.1	32.8	172.9	2.0
Mexique	1989	77.2	8.9	86.1	1.0
	1990	125.2	14.7	139.9	1.6
Netherlands Antilles	1988	51.0	2.5	53.5	289.2
Antilles néerlandaises	1989	57.0	4.3	61.3	327.8
	1990	53.0	4.4	57.4	305.3
Nicaragua	1988	166.6	46.6	213.2	58.9
Nicaragua	1989	182.8	42.2	225.0	60.1
	1990	273.7	50.1	323.8	83.6
Panama	1988	19.2	2.8	22.0	9.5
Panama	1989	14.2	3.4	17.6	7.4
	1990	90.1	2.4	92.5	38.3
Paraguay	1988	63.0	12.4	75.4	18.7
Paraguay	1989	88.4	3.8	92.2	22.2
	1990	46.6	9.9	56.5	13.2
Peru	1988	244.1	28.0	272.1	13.2
Pérou	1989	263.4	41.5	304.9	14.4
	1990	343.9	48.4	392.3	18.2
St. Pierre and Miquelon	1988	35.6	0.0	35.6	5933.3
Saint-Pierre-et-Miquelon	1989	32.9	0.0	32.9	5483.3
	1990	30.5	0.0	30.5	5083.3
Suriname	1988	14.7	6.6	21.3	52.5
Suriname	1989	45.0	5.9	50.9	122.9
	1990	51.2	6.4	57.6	136.5
Trinidad and Tobago	1988	2.7	5.9	8.6	6.9
Trinité-et-Tobago	1989	1.6	4.4	6.0	4.8
	1990	6.1	4.1	10.2	8.0
Uruguay	1988	29.8	11.0	40.8	13.3
Uruguay	1989	25.7	12.7	38.4	12.5
	1990	34.9	11.8	46.7	15.1
Venezuela	1988	17.6	0.1	17.7	0.9
Venezuela	1989	20.0	0.8	20.8	1.1
	1990	75.3	3.7	79.0	4.0
West Indies[3]	1988	54.2	23.9	78.1	126.6
Antilles[3]	1989	53.9	23.8	77.7	125.1
	1990	53.5	23.8	77.3	124.7
Other and unallocated	1988	228.3	96.6	324.9	...
Autres et non-ventilés	1989	274.5	125.9	400.4	...
	1990	269.1	121.9	391.0	...
Asia	1988	10345.6	4341.0	14686.6	4.8
Asie	1989	10626.5	4347.5	14974.0	4.8
	1990	11241.7	5084.1	16325.8	5.2
Afghanistan	1988	66.8	6.2	73.0	4.8
Afghanistan	1989	82.4	86.3	168.7	10.7
	1990	100.1	43.4	143.5	8.7
Bahrain	1988	1.1	0.9	2.0	4.2
Bahreïn	1989	3.2	2.7	5.9	11.8
	1990	1.9	0.7	2.6	5.0
Bangladesh	1988	930.9	674.5	1605.4	14.7
Bangladesh	1989	972.3	843.7	1816.0	16.1
	1990	1103.0	1004.6	2107.6	18.2

130
Disbursements to developing countries or areas of bilateral and multilateral
official development assistance [cont.]
Versements mises à la disposition des pays ou zones en développement au titre de l'aide publique bilatérale
et multilatérale au développement [suite]

Region, country or area Région, pays ou zone	Year Année	Disbursements ($US) Versements ($E-U)		Total (millions)	Per capita[2] Par habitant[2]
		Bilateral Bilatérale (millions)	Multilateral[1] Multilatérale[1] (millions)		
Bhutan	1988	18.7	19.7	38.4	26.5
Bhoutan	1989	19.6	22.6	42.2	28.5
	1990	20.1	27.0	47.1	31.1
Brunei Darussalam	1988	4.5	0.1	4.6	18.5
Brunéi Darussalam	1989	4.5	0.1	4.6	17.8
	1990	3.7	0.0	3.7	13.9
Cambodia	1988	9.8	8.7	18.5	2.4
Cambodge	1989	17.8	12.5	30.3	3.8
	1990	28.3	13.3	41.6	5.0
China	1988	1196.2	783.9	1980.1	1.8
Chine	1989	1494.9	656.3	2151.2	2.0
	1990	1416.4	659.8	2076.2	1.8
Hong Kong	1988	13.6	8.5	22.1	3.9
Hong-kong	1989	11.7	29.0	40.7	7.0
	1990	19.4	18.2	37.6	6.4
India	1988	949.5	1168.1	2117.6	2.6
Inde	1989	1133.7	760.2	1893.9	2.3
	1990	727.0	851.7	1578.7	1.9
Indonesia	1988	1497.9	124.8	1622.7	9.1
Indonésie	1989	1703.7	133.6	1837.3	10.2
	1990	1517.9	186.4	1704.3	9.2
Iran, Islamic Rep. of	1988	52.1	29.4	81.5	1.6
Iran, Rép. islamique d'	1989	60.0	36.3	96.3	1.8
	1990	34.3	34.2	68.5	1.3
Iraq	1988	-0.6	1.1	0.5	0.0
Iraq	1989	-5.3	16.5	11.2	0.6
	1990	-8.6	5.1	-3.5	-0.2
Israel	1988	1239.4	1.7	1241.1	278.7
Israël	1989	1188.0	3.7	1191.7	263.2
	1990	1370.6	3.7	1374.3	298.8
Jordan	1988	122.9	10.8	133.7	35.6
Jordanie	1989	130.5	17.3	147.8	38.1
	1990	431.2	23.9	455.1	113.5
Korea, Dem. People's Rep.	1988	4.5	5.9	10.4	0.5
Corée, Rép. pop. dém. de	1989	3.3	5.3	8.6	0.4
	1990	0.9	7.5	8.4	0.4
Korea, Rep. of	1988	9.3	2.3	11.6	0.3
Corée, Rep. de	1989	48.9	3.7	52.6	1.2
	1990	54.7	3.3	58.0	1.4
Kuwait	1988	4.4	1.6	6.0	3.1
Koweït	1989	2.1	1.5	3.6	1.8
	1990	2.2	1.2	3.4	1.7
Lao People's Dem.Rep.	1988	36.2	40.7	76.9	19.7
Rép. dém. pop. lao	1989	42.9	96.4	139.3	34.7
	1990	51.2	100.5	151.7	36.7
Lebanon	1988	99.1	30.4	129.5	48.6
Liban	1989	83.2	32.3	115.5	43.2
	1990	71.6	28.7	100.3	37.1
Malaysia	1988	96.7	12.2	108.9	6.4
Malaisie	1989	132.0	14.1	146.1	8.4
	1990	458.6	13.9	472.5	26.4
Maldives	1988	21.6	6.6	28.2	139.6
Maldives	1989	23.7	6.0	29.7	142.8
	1990	11.6	10.8	22.4	104.2
Mongolia	1988	0.7	2.4	3.1	1.5
Mongolie	1989	1.6	4.9	6.5	3.1
	1990	6.3	6.8	13.1	6.0
Myanmar	1988	332.7	118.2	450.9	11.3
Myanmar	1989	89.9	94.0	183.9	4.5
	1990	83.1	87.4	170.5	4.1

130
Disbursements to developing countries or areas of bilateral and multilateral official development assistance [*cont.*]
Versements mises à la disposition des pays ou zones en développement au titre de l'aide publique bilatérale et multilatérale au développement [*suite*]

Region, country or area Région, pays ou zone	Year Année	Disbursements ($US) Versements ($E-U)			
		Bilateral Bilatérale (millions)	Multilateral[1] Multilatérale[1] (millions)	Total (millions)	Per capita[2] Par habitant[2]
Nepal	1988	224.9	169.9	394.8	21.7
Népal	1989	248.9	242.4	491.3	26.3
	1990	238.8	187.6	426.4	22.3
Oman	1988	13.9	1.5	15.4	11.0
Oman	1989	16.9	6.1	23.0	15.9
	1990	11.3	6.4	17.7	11.8
Pakistan	1988	1003.7	422.5	1426.2	12.4
Pakistan	1989	682.2	471.7	1153.9	9.7
	1990	653.5	512.1	1165.6	9.5
Philippines	1988	789.3	65.1	854.4	14.4
Philippines	1989	757.3	86.9	844.2	13.9
	1990	1100.3	176.4	1276.7	20.5
Qatar	1988	1.1	0.8	1.9	5.6
Qatar	1989	3.0	0.9	3.9	11.0
	1990	1.3	0.8	2.1	5.7
Saudi Arabia	1988	12.8	6.5	19.3	1.5
Arabie saoudite	1989	9.2	26.9	36.1	2.7
	1990	12.8	30.8	43.6	3.1
Singapore	1988	20.7	1.2	21.9	8.2
Singapour	1989	93.7	1.1	94.8	35.2
	1990	-3.2	0.2	-3.0	-1.1
Sri Lanka	1988	436.0	156.0	592.0	35.3
Sri Lanka	1989	397.3	152.3	549.6	32.3
	1990	403.8	265.3	669.1	38.9
Syrian Arab Republic	1988	168.3	36.4	204.7	17.6
Rép. arabe syrienne	1989	108.8	33.3	142.1	11.8
	1990	69.4	30.3	99.7	8.0
Thailand	1988	514.1	55.9	570.0	10.5
Thaïlande	1989	657.4	88.7	746.1	13.6
	1990	731.5	76.9	808.4	14.5
United Arab Emirates	1988	-13.4	1.4	-12.0	-8.0
Emirats arabes unis	1989	-7.3	1.5	-5.8	-3.8
	1990	2.8	2.4	5.2	3.3
Viet Nam	1988	84.0	64.4	148.4	2.3
Viet Nam	1989	64.7	57.1	121.8	1.9
	1990	107.7	82.6	190.3	2.9
Yemen	1988	168.6	109.9	278.5	32.7
Yémen	1989	195.1	129.1	324.2	36.6
	1990	168.8	91.0	259.8	28.3
Other and unallocated	1988	211.7	140.7	352.4	...
Autres et non-ventilés	1989	142.6	81.7	224.3	...
	1990	230.8	155.2	386.0	...
Oceania	1988	1291.3	144.9	1436.2	55.9
Oceanie	1989	1273.6	87.8	1361.4	52.2
	1990	1214.7	133.5	1348.2	50.9
Cook Islands	1988	10.9	1.1	12.0	666.7
Iles Cook	1989	11.1	1.4	12.5	694.4
	1990	10.1	2.2	12.3	683.3
Fiji	1988	47.4	6.9	54.3	73.5
Fidji	1989	36.5	6.6	43.1	57.4
	1990	43.5	5.7	49.2	64.4
French Polynesia	1988	326.6	4.6	331.2	1716.1
Polynésie française	1989	286.0	2.7	288.7	1450.8
	1990	258.0	2.6	260.6	1265.0
Kiribati	1988	12.1	4.1	16.2	253.1
Kiribati	1989	15.1	2.4	17.5	269.2
	1990	17.7	2.8	20.5	310.6

130

Disbursements to developing countries or areas of bilateral and multilateral official development assistance [*cont.*]

Versements mises à la disposition des pays ou zones en développement au titre de l'aide publique bilatérale et multilatérale au développement [*suite*]

Region, country or area Région, pays ou zone	Year Année	Disbursements ($US) Versements ($E-U)			
		Bilateral Bilatérale (millions)	Multilateral[1] Multilatérale[1] (millions)	Total (millions)	Per capita[2] Par habitant[2]
New Caledonia	1988	260.9	0.2	261.1	1621.7
Nouvelle-Calédonie	1989	280.9	0.8	281.7	1717.7
	1990	300.2	0.7	300.9	1801.8
Niue	1988	5.1	0.2	5.3	1766.7
Nioue	1989	5.5	0.1	5.6	1866.7
	1990	7.0	0.2	7.2	2400.0
Pacific Islands (Palau)[4]	1988	151.5	0.9	152.4	...
Iles du Pacifique (Palaos)[4]	1989	157.9	1.1	159.0	...
	1990	61.9	1.1	63.0	...
Papua New Guinea	1988	307.4	72.4	379.8	102.6
Papouasie-Nouvelle-Guinée	1989	314.3	25.3	339.6	89.7
	1990	320.0	56.4	376.4	97.2
Samoa	1988	22.0	8.2	30.2	181.9
Samoa	1989	20.5	10.0	30.5	182.6
	1990	27.7	22.8	50.5	300.6
Solomon Islands	1988	34.8	23.4	58.2	194.0
Iles Salomon	1989	37.9	11.7	49.6	160.0
	1990	31.1	13.1	44.2	138.1
Tokelau	1988	3.4	0.3	3.7	...
Tokelaou	1989	4.3	0.3	4.6	...
	1990	4.4	0.4	4.8	...
Tonga	1988	13.6	5.2	18.8	197.9
Tonga	1989	19.7	4.8	24.5	257.9
	1990	24.2	7.0	31.2	328.4
Tuvalu	1988	13.3	0.7	14.0	1555.6
Tuvalu	1989	5.7	1.2	6.9	766.7
	1990	4.8	0.2	5.0	555.6
Vanuatu	1988	29.0	10.3	39.3	263.8
Vanuatu	1989	31.8	8.0	39.8	258.4
	1990	42.1	6.8	48.9	309.5
Wallis and Futuna Islands	1988	0.0	1.1	1.1	68.7
Iles Wallis et Futuna	1989	0.0	0.1	0.1	5.9
	1990	0.0	0.1	0.1	5.9
Other and unallocated	1988	53.0	5.4	58.4	...
Autres et non-ventilés	1989	46.4	11.5	57.9	...
	1990	61.9	11.5	73.4	...
Unspecified	1988	4370.9	849.8	5220.7	...
Non-specifiés	1989	4716.5	723.8	5440.3	...
	1990	5111.9	996.1	6108.0	...
Total developing countries	1988	33155.9	11326.9	44482.8	...
Total des pays en voie	1989	34228.1	11736.3	45964.4	...
de développement	1990	40225.7	13447.1	53672.8	...

Source:
Organization for Economic Co-operation and Development (Paris).

1   As reported by OECD/DAC, IDA, agencies of the United Nations family and the European Development Fund. Excluding non-concessional flows (i.e., less than 25% grant element).
2   Population based on estimates of midyear population.
3   Anguilla, Antigua and Barbuda, British Virgin Islands, Cayman Islands, Dominica, Grenada, Montserrat, St. Kitts and Nevis, St. Lucia, St Vincent and the Grenadines, Turks and Caicos Islands.
4   Including data for Federated States of Micronesia, Marshall Is. and Northern Mariana Is.

Source:
Organisation de coopération et de développement economiques (Paris).

1   Chiffre communiqués par le Comité d'aide au développement de l'OCDE, l'IDA, les agences des Nations Unies et le Fonds Européen de Développement. Non compris les apports non libéraux (c'est-à-dire, dont l'élément de libéralité est intérieur à 25 p.100).
2   Population d'après des estimations de la population au milieu de l'année.
3   Anguilla, Antigua-et-Barbuda, Iles Vierges britanniques, Iles Caïmanes, Dominique, Grenade, Montserrat, Saint-Kitts-et-Nevis, Sainte-Lucie, Saint-Vincent-et-Grenadines, Iles Turques et Caïmanes.
4   Y compris les données pour les Etats fédéré de Micronésie, les îles Marshall et les îles Mariannes du Nord.

131
Socio-economic development assistance through the United Nations system
Assistance en matière de développement socio-économique fournie par le système des Nations Unies

Thousand US dollars
Milliers de dollars E-U

A. Development grant expenditures [1] · Aide au développement [1]

Country or area / Pays ou zone	Year / Année	UNDP PNUD — UNDP programme / Programme de PNUD	Special funds / Fonds gérés	UNFPA / FNUAP	UNICEF / FISE	WFP / PAM	Other UN system / Autres organis. (ONU) — Regular budget / Budget ordinaire	Extra-budgetary / Extra-budgétaire	Total	Gov't self-supporting Auto-assistance gouvernementale
Total	1990	1035683	97960	179535	584332	956391	233930	686146	3773977	80833
Total	1991	1123595	114205	171778	591091	1337882	287508	663356	4289415	68938
Regional programmes	1990	209092	20859	52319	101388	6535	72395	392695	855283	2136
Totaux régionaux	1991	65739	23479	55069	97450	5450	75804	404014	727005	6238
Africa	1990	73891	4658	10219	6055	0	17550	91826	204199	669
Afrique	1991	0	5070	11842	5637	0	19315	74693	116557	2622
Asia	1990	45644	1721	8209	8969	0	11837	39747	116127	47
Asie	1991	0	1786	7773	9294	0	13163	46361	78377	2
Latin America	1990	21012	2020	5034	11761	0	11514	25074	76415	35
Amérique latine	1991	0	1943	4999	13585	0	15423	24981	60931	56
Middle East	1990	13716	145	3910	80	0	8680	10213	36744	1385
Moyen orient	1991	0	193	2886	53	0	5073	4188	12393	65
Interregional	1990	37599	10564	24947	0	0	17533	85926	176569	0
Interrégional	1991	46470	9541	27565	0	0	16507	116555	216638	88
Global	1990	17230	1751	0	74523	6535	5281	139909	245229	0
Global	1991	19269	4946	4	68881	5450	6323	137236	242109	3405
Country programmes	1990	823339	77101	126864	481119	949856	154660	288912	2901851	77518
Programmes, pays	1991	865332	88107	115601	492245	1332432	199617	248814	3342148	61485
Afghanistan	1990	7098	6131	261	14	12989	1259	9033	36785	3818
Afghanistan	1991	16319	3011	274	0	6840	2541	1097	30082	112
Albania	1990	1498	0	286	0	0	186	51	2021	0
Albanie	1991	1289	0	388	282	0	268	169	2396	26
Algeria	1990	3978	1	512	446	2278	1519	326	9060	57
Algérie	1991	3665	0	308	433	4190	1364	373	10333	86
Angola	1990	8370	154	674	7274	10176	1249	2477	30374	274
Angola	1991	14522	303	1023	11352	26024	1693	1185	56102	158
Anguilla	1990	222	13	64	0	0	0	0	299	0
Anguilla	1991	323	30	1	0	0	0	0	354	0
Antigua and Barbuda	1990	293	4	28	0	0	203	37	565	0
Antigua-et-Barbuda	1991	187	17	62	0	25	214	45	550	0
Argentina	1990	14891	137	72	552	0	723	1160	17535	606
Argentine	1991	23655	47	9	1685	0	1533	1424	28353	799
Aruba	1990	49	0	0	0	0	5	19	73	0
Aruba	1991	157	0	0	0	0	5	0	162	0
Bahamas	1990	340	11	0	0	0	247	329	927	0
Bahamas	1991	302	0	0	0	0	442	235	979	0
Bahrain	1990	141	0	-4	51	0	106	64	358	5
Bahreïn	1991	147	0	66	0	0	196	35	444	0
Bangladesh	1990	27595	1840	4506	33778	51193	3669	2290	124871	11
Bangladesh	1991	23564	5746	2474	31974	47578	5944	2371	119651	14
Barbados	1990	294	0	9	0	0	142	262	707	0
Barbade	1991	328	58	8	0	0	309	242	945	0
Belize	1990	380	1	5	96	0	220	123	825	0
Belize	1991	602	39	7	96	0	279	136	1159	0
Benin	1990	5736	994	1124	2220	4130	928	1093	16225	105
Bénin	1991	8447	2013	1218	2571	2048	1033	801	18131	3
Bermuda										
Bermudes	1991	1	0	0	0	0	0	0	1	0
Bhutan	1990	10110	834	477	3031	2543	730	930	18655	0
Bhoutan	1991	7761	1188	181	2678	4028	983	662	17481	0
Bolivia	1990	7991	1597	1202	4136	6555	1421	1879	24781	7
Bolivie	1991	10084	3068	776	4973	6376	1504	3029	29810	54
Botswana	1990	3885	524	511	818	1021	663	1596	9018	139
Botswana	1991	2941	1129	926	813	2513	1255	996	10573	-2
Brazil	1990	16671	111	1699	6574	12567	1992	4400	44014	2483
Brazil	1991	21735	172	1720	3563	12459	2555	2834	45038	1395

131
Socio-economic development assistance through the United Nations system
Thousand US dollars [*cont.*]
Assistance en matière de développement socio-économique fournie par le système des Nations Unies
Milliers de dollars E-U [*suite*]

A. Development grant expenditures [1] · Aide au développement [1]

Country or area Pays ou zone	Year Année	UNDP PNUD — UNDP programme Programme de PNUD	Special funds Fonds gérés	UNFPA FNUAP	UNICEF FISE	WFP PAM	Other UN system Autres organis. -ONU — Regular budget Budget ordinaire	Extra-budgetary Extra-budgétaire	Total	Gov't self-supporting Auto-assistance gouverne-mentale
British Virgin Islands	1990	30	5	17	0	0	47	0	99	0
Iles vierges britanniques	1991	29	7	6	0	0	48	0	90	0
Brunei Darussalam	1990	5	0	0	0	0	32	0	37	0
Brunéi Darussalam	1991	1	0	0	0	0	50	0	51	0
Bulgaria	1990	1003	0	184	0	0	856	1753	3796	0
Bulgarie	1991	652	0	101	2	1459	1086	519	3819	1
Burkina Faso	1990	11164	3258	1104	5375	7257	1254	4487	33899	0
Burkina Faso	1991	11133	3503	1212	3291	6518	1736	652	28045	0
Burundi	1990	8743	828	1207	3253	2952	947	1588	19518	41
Burundi	1991	13138	714	1122	2781	1594	1330	1397	22076	227
Cambodia	1990	388	0	0	5311	4780	280	304	11063	0
Cambodge	1991	342	12	0	7348	9050	292	1269	18313	0
Cameroon	1990	5633	201	955	1140	1614	1076	1302	11921	24
Cameroun	1991	4622	256	680	1967	6901	1659	909	16994	8
Cape Verde	1990	1605	2285	851	943	9558	906	2588	18736	30
Cap Vert	1991	1285	1018	605	726	6767	1096	2909	14406	5
Cayman Islands	1990	183	0	0	0	0	0	0	183	0
Iles Caïmanes	1991	286	0	0	0	0	0	-2	284	0
Central African Rep.	1990	6558	556	490	2939	3749	1408	1464	17164	0
Rép. centrafricaine	1991	6561	578	687	2689	3285	1094	909	15803	0
Chad	1990	13145	807	486	4295	10481	1268	3185	33667	193
Tchad	1991	11254	1070	643	3638	13309	1650	1812	33376	-2
Chile	1990	8992	4	63	450	0	1392	382	11283	30
Chili	1991	6909	135	213	862	1332	1556	1278	12285	57
China	1990	48034	462	4579	14639	8652	6589	7255	90210	28
Chine	1991	48927	967	13560	16081	21624	4728	5190	111077	93
Colombia	1990	10105	17	441	2850	1953	1364	975	17705	13
Colombie	1991	10906	13	799	2224	2614	1507	1293	19356	68
Comoros	1990	2171	895	786	499	1213	760	472	6796	22
Comores	1991	1818	1283	765	844	935	1265	345	7255	0
Congo	1990	1542	21	758	1091	987	680	923	6002	9
Congo	1991	1730	-476	570	1476	1224	946	962	6432	0
Cook Islands	1990	348	81	18	0	0	346	41	834	0
Iles Cook	1991	205	166	38	0	0	319	31	759	0
Costa Rica	1990	2400	172	375	239	2437	795	2555	8973	-8
Costa Rica	1991	2547	202	218	91	459	1209	2317	7043	25
Côte d'Ivoire	1990	6829	25	335	1616	6560	1438	1305	18108	119
Côte d'Ivoire	1991	5625	4	478	2618	5612	1166	1486	16989	125
Cuba	1990	1335	0	222	164	13191	1322	189	16423	0
Cuba	1991	2099	13	514	522	12478	1843	657	18126	37
Cyprus	1990	557	15	36	0	0	780	355	1743	34
Chypre	1991	583	46	9	0	-97	760	428	1729	17
Czechoslovakia	1990	51	0	0	0	0	119	71	241	0
Tchécoslovaquie	1991	98	0	0	0	0	369	449	916	1
Djibouti	1990	1575	654	85	1014	1971	610	106	6015	0
Djibouti	1991	1326	654	158	1268	2052	971	190	6619	0
Dominica	1990	499	31	14	0	-17	355	74	956	0
Dominique	1991	261	34	52	0	275	324	49	995	0
Dominican Republic	1990	3208	-6	249	1038	201	672	317	5679	0
Rép. dominicaine	1991	4833	130	379	861	282	941	342	7768	0
Ecuador	1990	6235	6	833	1574	7543	1488	1592	19271	21
Equateur	1991	4682	19	512	2534	8639	1766	1636	19788	0
Egypt	1990	9709	33	1697	3973	10141	1723	3777	31053	1548
Egypte	1991	6498	55	1186	4088	13732	2691	7199	35449	1857
El Salvador	1990	2491	18	484	2006	8105	780	379	14263	105
El Salvador	1991	3446	10	529	1350	12469	932	353	19089	101
Equatorial Guinea	1990	2981	972	592	784	2632	590	527	9078	134
Guinée équatoriale	1991	2919	630	435	836	2186	851	235	8092	23

131
Socio-economic development assistance through the United Nations system
Thousand US dollars [*cont.*]

Assistance en matière de développement socio-économique fournie par le système des Nations Unies
Milliers de dollars E-U [*suite*]

A. Development grant expenditures [1] · Aide au développement [1]

Country or area Pays ou zone	Year Année	UNDP PNUD UNDP programme Programme de PNUD	Special funds Fonds gérés	UNFPA FNUAP	UNICEF FISE	WFP PAM	Other UN system Autres organis. -ONU Regular budget Budget ordinaire	Extra-budgetary Extra-budgétaire	Total	Gov't self-supporting Auto-assistance gouverne-mentale
Ethiopia	1990	14829	3327	2433	22393	88465	2548	6407	140402	605
Ethiopie	1991	13628	2336	1737	17668	248804	2458	3533	290164	263
Fiji	1990	1242	197	177	0	0	1162	877	3655	143
Fidji	1991	1359	125	158	0	0	1418	699	3759	41
French Guiana	1990	0	0	0	0	0	22	0	22	0
Guyane française	1991	0	0	0	0	0	29	0	29	0
French Polynesia	1990	0	0	0	0	0	8	0	8	0
Polynésie française	1991	0	0	0	0	0	46	0	46	0
Gabon	1990	1013	29	297	0	0	634	1119	3092	438
Gabon	1991	941	9	517	0	0	786	589	2842	209
Gambia	1990	3603	558	506	563	2769	919	1858	10776	22
Gambie	1991	4497	1752	675	911	5648	1015	1328	15826	3
Ghana	1990	8051	197	1330	1997	10360	1473	2886	26294	135
Ghana	1991	11102	415	1454	2929	14067	1611	2855	34433	153
Greece	1990	0	0	0	0	0	241	185	426	41
Grèce	1991	0	0	0	0	0	131	179	310	33
Grenada	1990	134	21	31	0	-13	129	38	340	0
Grenade	1991	97	44	90	0	285	208	95	819	0
Guam	1990	0	0	0	0	0	6	0	6	0
Guam	1991	0	0	0	0	0	40	0	40	0
Guatemala	1990	2465	917	318	2723	18049	530	753	25755	0
Guatemala	1991	4967	616	305	1670	15564	846	656	24624	0
Guinea	1990	10497	1979	544	2799	8614	1138	1246	26817	93
Guinée	1991	10332	3446	981	5023	4218	2208	1151	27359	9
Guinea-Bissau	1990	4200	980	423	1647	3692	890	2261	14093	290
Guinée-Bissau	1991	4426	832	909	1753	1855	1094	1933	12802	576
Guyana	1990	3635	97	120	439	1298	515	451	6555	0
Guyana	1991	4791	155	45	609	701	685	303	7289	0
Haiti	1990	10931	1001	875	2158	1717	524	1091	18297	0
Haïti	1991	5474	1753	705	2595	629	927	2665	14748	0
Honduras	1990	3781	467	893	1455	6201	386	720	13903	0
Honduras	1991	4766	-62	696	1741	5482	624	1076	14323	12
Hong Kong	1990	65	70	0	0	0	106	0	241	0
Hong-kong	1991	0	56	204	0	0	128	0	388	0
Hungary	1990	353	0	260	0	0	170	362	1145	45
Hongrie	1991	202	0	354	0	0	245	1123	1924	20
Iceland	1990	0	0	0	0	0	25	0	25	0
Islande	1991	0	0	0	0	0	23	12	35	12
India	1990	22870	132	16558	79148	44092	2172	7495	172467	1837
Inde	1991	21612	83	11217	40848	45473	6859	6717	132809	1986
Indonesia	1990	16673	302	2148	9984	7956	4306	6261	47630	2471
Indonésie	1991	17082	153	3618	11268	7634	6651	5655	52061	2562
Iran, Islamic Rep. of	1990	5263	0	73	2461	11477	2264	356	21894	35
Iran, Rép. islamique d'	1991	6256	64	1057	5510	29270	2505	624	45286	392
Iraq	1990	2001	18	132	1453	0	680	275	4559	231
Iraq	1991	-175	101	-23	23646	30562	1081	940	56132	7
Jamaica	1990	2391	115	185	689	6371	739	379	10869	78
Jamaïque	1991	2656	282	179	1080	3343	927	569	9036	9
Jordan	1990	3059	28	895	2467	9806	818	1742	18815	87
Jordanie	1991	2835	55	849	4574	16177	1379	910	26779	42
Kenya	1990	12338	244	155	5213	3516	1165	4545	27176	0
Kenya	1991	14304	662	1774	8182	10645	1532	4037	41136	1
Kiribati	1990	680	287	20	0	0	284	15	1286	0
Kiribati	1991	646	281	12	0	0	351	71	1361	0
Korea, Dem. People's Rep.	1990	3752	44	757	463	0	1968	496	7480	0
Corée, Rép. pop. dém.	1991	3343	5	1298	210	0	2297	421	7574	0
Korea, Rep. of	1990	2631	58	513	445	0	1245	379	5271	59
Corée, Rép. de	1991	1664	48	388	427	0	1586	318	4431	30

131
Socio-economic development assistance through the United Nations system
Thousand US dollars [cont.]
Assistance en matière de développement socio-économique fournie par le système des Nations Unies
Milliers de dollars E-U [suite]

A. Development grant expenditures [1] · Aide au développement [1]

Country or area / Pays ou zone	Year / Année	UNDP PNUD — UNDP programme / Programme de PNUD	UNDP PNUD — Special funds / Fonds gérés	UNFPA FNUAP	UNICEF FISE	WFP PAM	Other UN system / Autres organis. -ONU — Regular budget / Budget ordinaire	Other UN system / Autres organis. -ONU — Extra-budgetary / Extra-budgétaire	Total	Gov't self-supporting / Auto-assistance gouvernementale
Kuwait	1990	1086	0	0	0	0	28	82	1196	73
Koweït	1991	243	0	20	0	0	259	95	617	37
Lao People's Dem. Rep.	1990	10894	1305	135	1933	-5	1525	1540	17327	0
Rép. dém. pop. lao	1991	8633	1189	154	2354	-36	1825	1522	15641	0
Lebanon	1990	430	0	0	3398	4842	511	342	9523	0
Liban	1991	730	0	9	3423	4485	1063	814	10524	0
Lesotho	1990	3641	715	252	1553	8378	1190	2571	18300	-1
Lesotho	1991	4167	1080	704	1133	10437	1307	2003	20831	134
Liberia	1990	2683	20	448	528	22930	749	500	27858	124
Libéria	1991	2288	9	119	4173	78745	1006	138	86478	75
Libyan Arab Jamahiriya	1990	737	0	0	0	0	678	11148	12563	8872
Jamah. arabe libyenne	1991	1156	1	2	0	2	851	17724	19736	13406
Madagascar	1990	14912	184	801	4008	779	1518	1481	23683	39
Madagascar	1991	14319	38	917	3828	8008	1573	903	29586	114
Malawi	1990	13832	2297	1255	3351	43337	1336	2997	68405	-6
Malawi	1991	14971	1433	1479	4867	59281	1655	730	84416	-3
Malaysia	1990	1833	2	415	443	0	1461	323	4477	18
Malaisie	1991	2950	25	426	401	0	1617	350	5769	49
Maldives	1990	1631	410	192	560	0	663	315	3771	0
Maldives	1991	1926	2746	227	731	0	833	283	6746	0
Mali	1990	9589	4474	775	6143	8459	1586	2676	33702	20
Mali	1991	11470	4875	833	6186	3887	1650	2541	31442	0
Malta	1990	236	0	0	0	0	213	65	514	0
Malte	1991	150	0	0	0	0	257	91	498	3
Marshall Islands	1990	0	0	129	0	0	54	16	199	16
Iles Marshall	1991	0	0	102	0	0	58	-6	154	0
Mauritania	1990	3532	1893	980	2286	6060	1353	1076	17180	111
Mauritanie	1991	3889	1550	883	1369	11529	1486	757	21463	0
Mauritius	1990	1021	0	456	359	2484	459	363	5142	0
Maurice	1991	857	176	242	378	2404	762	101	4920	0
Mexico	1990	3416	150	2488	3682	10172	1694	3028	24630	1502
Mexique	1991	4125	0	1827	3783	5253	1919	2792	19699	1261
Mongolia	1990	4860	32	602	51	0	1185	90	6820	0
Mongolie	1991	3448	7	335	410	2464	2107	158	8929	0
Montserrat	1990	365	27	16	0	0	0	0	408	0
Montserrat	1991	303	45	1	0	0	2	0	351	0
Morocco	1990	6913	30	1850	2589	25756	1521	857	39516	34
Maroc	1991	7286	54	1347	2583	16987	2109	591	30957	16
Mozambique	1990	16169	3888	959	15599	33879	1327	5048	76869	118
Mozambique	1991	21950	1109	816	16460	35691	1609	4309	81944	114
Myanmar	1990	13292	12	383	7870	10	1725	2723	26015	353
Myanmar	1991	17852	36	237	6443	13	3326	2466	30373	375
Namibia	1990	2482	1914	79	3487	2018	793	1618	12391	0
Namibie	1991	2774	2633	352	2846	2514	1435	1502	14056	19
Nepal	1990	20721	2111	2258	7918	6570	2035	3200	44813	69
Népal	1991	18871	1826	3145	7386	1103	3421	2814	38566	36
Netherlands Antilles	1990	346	0	0	0	0	89	148	583	107
Antilles néerlandaises	1991	246	0	0	0	0	56	75	377	65
Nicaragua	1990	2765	1754	476	3578	7587	839	2818	19817	0
Nicaragua	1991	5628	2577	811	2631	7130	1254	3247	23278	0
Niger	1990	10441	1576	1291	2443	8773	1321	8877	34722	444
Niger	1991	14395	3842	1194	3017	10190	1603	3803	38044	31
Nigeria	1990	11657	3	2253	25491	0	2511	3158	45073	549
Nigéria	1991	14216	11	3300	13605	0	2991	3307	37430	1199
Niue	1990	216	20	0	0	0	0	0	236	0
Nioue	1991	491	24	0	0	0	0	0	515	0
Oman	1990	5028	62	144	451	0	582	154	6421	0
Oman	1991	4005	89	85	543	0	657	168	5547	5

131
Socio-economic development assistance through the United Nations system
Thousand US dollars [*cont.*]
Assistance en matière de développement socio-économique fournie par le système des Nations Unies
Milliers de dollars E-U [*suite*]

A. Development grant expenditures [1] · Aide au développement [1]

Country or area Pays ou zone	Year Année	UNDP PNUD — UNDP programme Programme de PNUD	Special funds Fonds gérés	UNFPA FNUAP	UNICEF FISE	WFP PAM	Other UN system Autres organis. -ONU — Regular budget Budget ordinaire	Extra-budgetary Extra-budgétaire	Total	Gov't self-supporting Auto-assistance gouverne-mentale
Pacific Islands (Palau)[2]	1990	218	264	15	1326	0	24	0	1847	0
Iles du Pacifique (Palaos)[2]	1991	884	20	32	1351	0	0	81	2368	0
Pakistan	1990	23446	129	2470	13385	62746	1868	2628	106672	878
Pakistan	1991	23457	49	894	15520	97269	3597	2225	143011	92
Panama	1990	1087	0	321	0	590	588	229	2815	0
Panama	1991	1704	0	240	1	422	1069	246	3682	48
Papua New Guinea	1990	4779	132	199	1036	0	1362	749	8257	112
Papouasie-Nvl-Guinée	1991	6082	245	271	745	0	1628	1303	10274	669
Paraguay	1990	1163	113	893	691	2110	435	314	5719	0
Paraguay	1991	4721	6	480	904	2241	624	260	9236	0
Peru	1990	3623	104	1252	4479	10656	1176	3115	24405	38
Pérou	1991	3336	281	780	4778	18300	1792	2777	32044	208
Philippines	1990	8066	698	4416	7366	2736	1877	3646	28805	223
Philippines	1991	7718	539	3274	9308	3977	2149	5050	32015	114
Poland	1990	1032	0	33	0	23814	1630	835	27344	36
Pologne	1991	1457	19	45	0	172	1029	1414	4136	27
Portugal	1990	228	0	147	0	0	205	81	661	0
Portugal	1991	915	0	237	0	0	359	56	1567	18
Qatar	1990	582	0	14	0	0	69	151	816	151
Qatar	1991	833	1	-2	0	0	88	122	1042	122
Réunion	1990	0	0	0	0	0	15	0	15	0
Réunion	1991	0	0	0	0	0	33	0	33	0
Romania	1990	1643	0	323	210	0	624	443	3243	9
Roumanie	1991	748	0	420	2480	1416	1057	654	6775	50
Rwanda	1990	11007	1294	718	2814	1948	936	3244	21961	14
Rwanda	1991	11222	1547	1640	3813	3170	1223	2433	25048	0
Saint Kitts and Nevis	1990	289	33	5	0	0	186	13	526	0
Saint Kitts et Nevis	1991	84	15	0	0	32	218	15	364	0
Saint Lucia	1990	281	16	24	0	-13	148	93	549	0
Sainte-Lucie	1991	144	21	16	0	0	206	99	486	0
Saint Vincent/Grenadines	1990	362	62	44	0	57	61	66	652	0
St. Vincent/Grenadines	1991	269	39	69	0	30	72	51	530	0
Samoa	1990	1628	222	168	0	1035	1064	135	4252	0
Samoa	1991	1062	199	184	0	74	1093	28	2640	0
Sao Tome and Principe	1990	1288	261	506	316	3237	641	561	6810	0
Sao Tomé-et-Principe	1991	1084	374	555	343	2107	695	333	5491	315
Saudi Arabia	1990	4911	1	0	24	0	503	25369	30808	24695
Arabie saoudite	1991	5452	2	0	-5	0	741	25893	32083	20092
Senegal	1990	7962	3612	1689	5072	11673	1385	6465	37858	60
Sénégal	1991	9132	3665	1413	3007	8977	1364	7347	34905	0
Seychelles	1990	296	3	39	0	471	553	376	1738	0
Seychelles	1991	380	9	35	0	138	716	260	1538	0
Sierra Leone	1990	4833	1696	229	2698	1281	692	1544	12973	340
Sierra Leone	1991	4060	888	353	3947	2258	1173	836	13515	289
Singapore	1990	449	0	0	0	0	683	21	1153	0
Singapour	1991	126	0	0	0	0	485	9	620	0
Solomon Islands	1990	711	41	98	0	0	801	90	1741	0
Iles Salomon	1991	395	28	125	0	0	888	79	1515	0
Somalia	1990	8293	516	814	3644	21523	1279	2150	38219	68
Somalie	1991	3535	223	-4	4899	11789	1426	660	22528	-4
Sri Lanka	1990	7678	1	1000	4542	3699	1748	1486	20154	135
Sri Lanka	1991	9967	19	773	5022	1997	2948	1763	22489	39
Sudan	1990	11254	2271	2344	21606	37637	2833	9162	87107	998
Soudan	1991	25122	3458	1129	29896	115015	3284	6809	184713	545
Suriname	1990	645	-4	57	0	0	198	534	1430	0
Suriname	1991	604	0	18	0	0	336	343	1301	0
Swaziland	1990	1534	220	212	938	3608	578	365	7455	9
Souaziland	1991	1961	154	613	812	2065	791	434	6830	16

131
Socio-economic development assistance through the United Nations system
Thousand US dollars [*cont.*]

Assistance en matière de développement socio-économique fournie par le système des Nations Unies
Milliers de dollars E-U [*suite*]

A. Development grant expenditures [1] · Aide au développement [1]

Country or area Pays ou zone	Year Année	UNDP PNUD — UNDP programme Programme de PNUD	UNDP PNUD — Special funds Fonds gérés	UNFPA FNUAP	UNICEF FISE	WFP PAM	Other UN system Autres organis. -ONU — Regular budget Budget ordinaire	Other UN system Autres organis. -ONU — Extra-budgetary Extra-budgétaire	Total	Gov't self-supporting Auto-assistance gouverne-mentale
Syrian Arab Republic	1990	3621	35	477	660	22649	1366	577	29385	39
Arabe rép. syrienne	1991	2154	74	1041	1183	13451	2104	809	20816	0
Thailand	1990	6800	10	1332	3396	27615	2627	3843	45623	513
Thaïlande	1991	8158	201	444	4271	27638	3881	3444	48037	306
Togo	1990	6889	1128	894	1168	1762	1131	1748	14720	0
Togo	1991	5642	2119	411	1817	666	1098	776	12529	0
Tokelau	1990	359	11	0	0	0	0	2	372	0
Tokélaou	1991	399	0	0	0	0	0	0	399	0
Tonga	1990	420	24	165	0	0	646	38	1293	0
Tonga	1991	443	32	148	0	0	587	55	1265	0
Trinidad and Tobago	1990	955	12	19	0	0	513	410	1909	0
Trinité-et-Tobago	1991	937	21	45	0	0	687	387	2077	44
Tunisia	1990	2711	0	965	418	13562	1389	391	19436	241
Tunisie	1991	1736	0	818	893	3402	1529	1186	9564	121
Turkey	1990	3794	0	566	1279	2726	904	5538	14807	4704
Turquie	1991	2800	25	795	4028	1427	1014	3514	13603	2405
Turks and Caicos Islands	1990	896	39	90	0	0	0	206	1231	0
Iles Turques et Caïques	1991	197	0	1	0	0	2	-2	198	0
Tuvalu	1990	21	3	3	0	0	10	38	75	0
Tuvalu	1991	386	424	95	0	0	28	0	933	0
Uganda	1990	14411	1736	3033	11769	16535	1300	4952	53736	639
Ouganda	1991	15548	1498	3197	14931	18060	1866	2310	57410	565
former USSR ancienne URSS	1991	0	0	0	293	0	305	352	950	83
United Arab Emirates	1990	1757	7	97	0	0	87	416	2364	381
Emirats arabes unis	1991	1482	9	0	0	0	169	169	1829	165
United Rep. of Tanzania	1990	16090	812	2539	15034	7785	2323	10122	54705	2324
Rép. unie de Tanzanie	1991	16094	2282	3106	15720	6881	2575	7568	54226	2828
Uruguay	1990	2722	9	89	152	0	613	654	4239	174
Uruguay	1991	3058	9	95	57	0	506	469	4194	17
Vanuatu	1990	766	52	221	0	0	495	250	1784	0
Vanuatu	1991	1027	149	124	0	0	666	240	2206	0
Venezuela	1990	3042	13	133	0	0	1109	1163	5460	811
Venezuela	1991	4362	5	54	727	0	1074	654	6876	172
Viet Nam	1990	33960	91	4578	10176	12218	4244	8339	73606	7081
Viet Nam	1991	33913	740	3528	9650	16700	4541	1788	70860	260
Yemen	1990	16256	949	2180	4311	20957	2694	4804	52151	1279
Yémen	1991	11886	1577	1720	2582	15139	3705	4710	41319	2402
Yugoslavia	1990	1288	0	120	0	0	1085	571	3064	28
Yougoslavie	1991	1041	10	118	0	0	629	409	2207	13
Zaire	1990	12166	27	1231	6955	3231	1479	1279	26368	153
Zaïre	1991	11506	27	814	6830	2187	1499	1179	24042	70
Zambia	1990	6843	442	2426	2826	2606	1460	5326	21929	596
Zambie	1991	7113	176	425	2739	3201	1922	5352	20928	337
Zimbabwe	1990	5497	565	787	2427	3380	1249	5340	19245	0
Zimbabwe	1991	4197	478	1374	1972	4192	1504	4115	17832	0
Not elsewhere classified	1990	3252	0	352	1825	0	6875	4539	16843	1179
Non-classé ailleurs	1991	192524	2619	1108	1396	0	12087	10528	220262	1215

131
Socio-economic development assistance through the United Nations system
Thousand US dollars [*cont.*]

Assistance en matière de développement socio-économique fournie par le système des Nations Unies
Milliers de dollars E-U [*suite*]

B. Development loan and relief expenditures [1] · Prêts au développement et secours [1]

Country or area Pays ou zone	Year Année	IFAD FIDA	IDA IDA	IBRD BIRD	Training, consultants Formation, consultants	IFC SFI	Total develop. grants Aide totale au dévelop- pement	Grand total Total général	Relief and related grants Secours et aide connexe
Total	1990	139322	3698512	-1947773	1053490	1878460	3773977	7542497	873686
Total	1991	114936	4016000	-4048000	1096914	985000	4289415	5357351	1406035
Regional programmes	1990	6200	0	0	0	0	855283	861483	15545
Totaux régionaux	1991	9660	0	0	0	0	727005	736665	86091
Africa	1990	2100	0	0	0	0	204199	206299	300
Afrique	1991	4947	0	0	0	0	116557	121504	229
Asia	1990	1900	0	0	0	0	116127	118027	2357
Asie	1991	1516	0	0	0	0	78377	79893	0
Latin America	1990	500	0	0	0	0	76415	76915	1123
Amérique latine	1991	880	0	0	0	0	60931	61811	484
Middle East	1990	1700	0	0	0	0	36744	38444	11765
Moyen orient	1991	2317	0	0	0	0	12393	14710	85378
Interregional	1990	0	0	0	0	0	176569	176569	0
Interrégional	1991	0	0	0	0	0	216638	216638	0
Global	1990	0	0	0	0	0	245229	245229	0
Global	1991	0	0	0	0	0	242109	242109	0
Country progammes	1990	112822	3698512	-2020957	1052750	1878460	2901851	6570687	578948
Programmes, pays	1991	79476	4016000	-4044000	1096914	985000	3342148	4378624	745143
Afghanistan	1990	0	-1677	0	0	0	36785	35108	909
Afghanistan	1991	0	0	0	0	0	30082	30082	1929
Albania	1990	0	0	0	0	0	2021	2021	0
Albanie	1991	0	0	0	0	0	2396	2396	0
Algeria	1990	161	0	96918	4022	0	9060	106139	5896
Algérie	1991	2981	0	64000	7121	0	10333	77314	9596
Angola	1990	0	0	0	0	0	30374	30374	4062
Angola	1991	0	0	0	544	0	56102	56102	2782
Anguilla	1990	0	0	0	0	0	299	299	0
Anguilla	1991	0	0	0	0	0	354	354	0
Antigua and Barbuda	1990	0	0	0	0	0	565	565	0
Antigua-et-Barbuda	1991	0	0	0	0	0	550	550	0
Argentina	1990	0	0	-12129	34567	161930	17535	167336	1310
Argentine	1991	750	0	-107000	34435	36000	28353	-41897	1282
Aruba	1990	0	0	0	0	0	73	73	0
Aruba	1991	0	0	0	0	0	162	162	0
Bahamas	1990	0	0	-2310	207	0	927	-1383	0
Bahamas	1991	0	0	0	591	0	979	979	0
Bahrain	1990	0	0	0	0	0	358	358	430
Bahreïn	1991	0	0	0	0	0	444	444	0
Bangladesh	1990	13217	421446	-5349	17199	1190	124871	555375	238
Bangladesh	1991	3695	207000	0	17712	0	119651	330346	2940
Barbados	1990	0	0	-4667	1172	40	707	-3920	0
Barbade	1991	0	0	-7000	779	0	945	-6055	0
Belize	1990	278	0	1661	1635	1000	825	3763	526
Belize	1991	133	0	0	438	0	1159	1292	759
Benin	1990	3051	52684	0	8615	0	16225	71960	157
Bénin	1991	264	39000	0	7169	0	18131	57395	27
Bermuda									
Bermudes	1991	0	0	0	0	0	1	1	0
Bhutan	1990	761	1868	0	102	0	18655	21284	0
Bhoutan	1991	659	1000	0	0	0	17481	19140	0
Bolivia	1990	1659	44560	-36175	4496	4240	24781	39065	63
Bolivie	1991	4178	43000	0	10865	8000	29810	84988	150
Botswana	1990	-205	-382	-26323	1256	0	9018	-17892	1024
Botswana	1991	58	0	-19000	2962	0	10573	-8369	958
Brazil	1990	-1055	0	-1200906	33539	205540	44014	-952407	412
Brazil	1991	-1051	0	-1075000	46908	48000	45038	-983013	349
British Virgin Islands	1990	0	0	0	0	0	99	99	0
Iles vierges britanniques	1991	0	0	0	0	0	90	90	0

131
Socio-economic development assistance through the United Nations system
Thousand US dollars [cont.]

Assistance en matière de développement socio-économique fournie par le système des Nations Unies
Milliers de dollars E-U [suite]

B. Development loan and relief expenditures [1] · Prêts au développement et secours [1]

Country or area Pays ou zone	Year Année	Loans Prêts IFAD FIDA	IBRD/IDA IDA IDA	BIRD/IDA IBRD BIRD	Training, consultants Formation, consultants	IFC SFI	Total develop. grants Aide totale au dévelop- pement	Grand total Total général	Relief and related grants Secours et aide connexe
Brunei Darussalam	1990	0	0	0	0	0	37	37	10
Brunéi Darussalam	1991	0	0	0	0	0	51	51	-10
Bulgaria	1990	0	0	0	0	0	3796	3796	0
Bulgarie	1991	0	0	58000	0	0	3819	61819	0
Burkina Faso	1990	584	12231	0	2737	0	33899	46713	250
Burkina Faso	1991	144	39000	0	2846	0	28045	67189	97
Burundi	1990	524	45147	0	6575	0	19518	65189	803
Burundi	1991	-116	35000	0	7878	0	22076	56960	1108
Cambodia	1990	0	0	0	0	0	11063	11063	1190
Cambodge	1991	0	0	0	0	0	18313	18313	19200
Cameroon	1990	5635	-4229	-16236	12088	970	11921	-1940	1674
Cameroun	1991	1385	0	-3000	13631	0	16994	15379	1410
Cape Verde	1990	-67	1850	0	384	0	18736	20519	0
Cap Vert	1991	184	3000	0	262	0	14406	17590	0
Cayman Islands	1990	0	0	0	0	0	183	183	0
Iles Caïmanes	1991	0	0	0	0	0	284	284	0
Central African Rep.	1990	1005	69554	0	9322	0	17164	87723	588
Rép. centrafricaine	1991	2253	21000	0	8460	0	15803	39056	1595
Chad	1990	0	46215	0	13546	0	33667	79882	0
Tchad	1991	0	45000	0	14443	0	33376	78376	0
Chile	1990	0	-798	2183	8886	170840	11283	183507	213
Chili	1991	0	0	-99000	11600	77000	12285	-9715	463
China	1990	8213	486467	175893	45304	2710	90210	763493	4446
Chine	1991	12489	587000	310000	49935	0	111077	1020566	8595
Colombia	1990	-534	-819	-539434	16904	68420	17705	-454661	38
Colombie	1991	-544	0	-498000	16349	16000	19356	-463188	38
Comoros	1990	190	349	0	285	0	6796	7335	0
Comores	1991	74	1000	0	539	0	7255	8329	0
Congo	1990	137	-947	-16259	1678	2340	6002	-8727	573
Congo	1991	828	0	0	1548	0	6432	7260	556
Cook Islands	1990	0	0	0	0	0	834	834	0
Iles Cook	1991	0	0	0	0	0	759	759	0
Costa Rica	1990	-687	-192	-73300	617	370	8973	-64836	5507
Costa Rica	1991	-461	0	-49000	277	5000	7043	-37418	7133
Côte d'Ivoire	1990	219	-128	-1953	5504	7710	18108	23956	5948
Côte d'Ivoire	1991	411	31000	-130000	6504	1000	16989	-80600	9649
Cuba	1990	0	0	0	0	0	16423	16423	0
Cuba	1991	0	0	0	0	0	18126	18126	265
Cyprus	1990	0	0	608	532	0	1743	2351	12827
Chypre	1991	0	0	-2000	609	6000	1729	5729	10447
Czechoslovakia	1990	0	0	0	0	0	241	241	0
Tchécoslovaquie	1991	0	0	200000	0	0	916	200916	0
Djibouti	1990	58	2318	0	191	0	6015	8391	915
Djibouti	1991	294	3000	0	698	0	6619	9913	4679
Dominica	1990	389	790	0	728	0	956	2135	0
Dominique	1991	361	0	0	438	0	995	1356	0
Dominican Republic	1990	918	-340	5931	4751	2660	5679	14848	4
Rép. dominicaine	1991	-2207	0	-11000	1411	10000	7768	4561	3
Ecuador	1990	797	-938	-56451	5468	0	19271	-37322	107
Equateur	1991	358	0	-90000	3906	0	19788	-69854	137
Egypt	1990	6635	-5467	-221641	5905	16410	31053	-173010	1073
Egypte	1991	2350	-14000	-257000	6246	6000	35449	-227201	1271
El Salvador	1990	828	-837	-26916	61	0	14263	-12662	965
El Salvador	1991	-467	0	33000	496	0	19089	51622	1683
Equatorial Guinea	1990	269	1820	0	1236	0	9078	11167	0
Guinée équatoriale	1991	746	2000	0	1314	0	8092	10838	0
Ethiopia	1990	923	63374	-8876	10422	0	140402	195822	78046
Ethiopie	1991	1807	48000	0	6222	0	290164	339971	85094

131
Socio-economic development assistance through the United Nations system
Thousand US dollars [*cont.*]
Assistance en matière de développement socio-économique fournie par le système des Nations Unies
Milliers de dollars E-U [*suite*]

B. Development loan and relief expenditures [1] · Prêts au développement et secours [1]

Country or area Pays ou zone	Year Année	Loans Prêts IFAD FIDA	IBRD/IDA BIRD/IDA IDA IDA	IBRD BIRD	Training, consultants Formation, consultants	IFC SFI	Total develop. grants Aide totale au dévelop- pement	Grand total Total général	Relief and related grants Secours et aide connexe
Fiji	1990	0	0	-9647	1263	1590	3655	-4402	0
Fidji	1991	0	0	-8000	934	0	3759	-4241	0
French Guiana	1990	0	0	0	0	0	22	22	295
Guyane française	1991	0	0	0	0	0	29	29	260
French Polynesia	1990	0	0	0	0	0	8	8	0
Polynésie française	1991	0	0	0	0	0	46	46	0
Gabon	1990	0	0	4207	2752	27000	3092	34299	74
Gabon	1991	0	0	4000	3897	9000	2842	15842	99
Gambia	1990	774	8805	0	2261	0	10776	20355	0
Gambie	1991	1160	5000	0	2980	2000	15826	23986	0
Ghana	1990	214	194550	-19612	26354	102070	26294	303516	206
Ghana	1991	1648	186000	0	25565	22000	34433	244081	572
Greece	1990	0	0	-49334	0	0	426	-48908	1653
Grèce	1991	0	0	0	0	0	310	310	1537
Grenada	1990	0	933	0	151	280	340	1553	0
Grenade	1991	0	0	0	278	0	819	819	0
Guam	1990	0	0	0	0	0	6	6	0
Guam	1991	0	0	0	0	0	40	40	0
Guatemala	1990	73	0	7702	1289	0	25755	33531	935
Guatemala	1991	1211	0	0	388	0	24624	25835	891
Guinea	1990	519	48677	-14166	22327	3760	26817	65608	9927
Guinée	1991	1056	79000	0	22288	0	27359	107415	15712
Guinea-Bissau	1990	1319	14083	0	8352	0	14093	29495	0
Guinée-Bissau	1991	914	13000	0	6845	0	12802	26716	250
Guyana	1990	277	54358	-61863	167	0	6555	-673	0
Guyana	1991	474	38000	0	888	0	7289	45763	0
Haiti	1990	2418	9938	0	4232	90	18297	30743	133
Haïti	1991	903	6000	0	3174	0	14748	21651	137
Honduras	1990	51	-3455	-108351	27	1370	13903	-96482	11986
Honduras	1991	0	46000	-58000	138	0	14323	2323	2833
Hong Kong	1990	0	0	0	0	0	241	241	15896
Hong-kong	1991	0	0	0	0	0	388	388	20802
Hungary	1990	0	0	39286	5234	17600	1145	58031	2773
Hongrie	1991	0	0	113000	11868	18000	1924	132924	4010
Iceland	1990	0	0	-7520	0	0	25	-7495	0
Islande	1991	0	0	0	0	0	35	35	0
India	1990	4059	446234	374966	54493	90980	172467	1088707	4692
Inde	1991	2970	714000	92000	59416	138000	132809	1079779	3775
Indonesia	1990	2531	-17540	-294353	243522	3410	47630	-258321	4273
Indonésie	1991	3834	0	-15000	281222	62000	52061	102895	6882
Iran, Islamic Rep. of	1990	0	0	-76998	0	10	21894	-55094	13286
Iran, Rép. islamique d'	1991	0	0	-45000	0	0	45286	286	57474
Iraq	1990	0	0	-8781	0	0	4559	-4222	523
Iraq	1991	0	0	0	0	0	56132	56132	68898
Jamaica	1990	-498	0	-84805	5998	5200	10869	-69235	0
Jamaïque	1991	-674	0	-78000	5794	3000	9036	-66638	0
Jordan	1990	353	-1580	33443	3294	5440	18815	56472	22742
Jordanie	1991	1130	0	-57000	1897	0	26779	-29091	5850
Kenya	1990	2553	223297	-169154	12460	20210	27176	104083	2675
Kenya	1991	2061	169000	0	11601	0	41136	212197	13437
Kiribati	1990	0	0	0	0	0	1286	1286	0
Kiribati	1991	0	0	0	0	0	1361	1361	0
Korea, Dem. People's Rep.	1990	0	0	0	0	0	7480	7480	0
Corée, Rép. pop. dém.	1991	0	0	0	0	0	7574	7574	0
Korea, Rep. of	1990	0	-3136	-662366	980	630	5271	-659600	348
Corée, Rép. de	1991	0	0	-570000	831	0	4431	-565569	130
Kuwait	1990	0	0	0	0	· 0	1196	1196	0
Koweït	1991	0	0	0	0	0	617	617	1000

131
Socio-economic development assistance through the United Nations system
Thousand US dollars [cont.]

Assistance en matière de développement socio-économique fournie par le système des Nations Unies
Milliers de dollars E-U [suite]

B. Development loan and relief expenditures [1] · Prêts au développement et secours [1]

Country or area Pays ou zone	Year Année	Loans Prêts IFAD FIDA	IBRD/IDA BIRD/IDA IDA IDA	IBRD BIRD	Training, consultants Formation, consultants	IFC SFI	Total develop. grants Aide totale au dévelop- pement	Grand total Total général	Relief and related grants Secours et aide connexe
Lao People's Dem. Rep.	1990	3044	31606	0	1774	0	17327	51978	983
Rép. dém. pop. lao	1991	560	13000	0	1727	0	15641	29201	1345
Lebanon	1990	0	0	-10702	0	0	9523	-1179	1287
Liban	1991	0	0	0	0	0	10524	10524	1633
Lesotho	1990	543	7899	0	2254	0	18300	26742	283
Lesotho	1991	358	7000	0	1493	0	20831	28189	540
Liberia	1990	0	0	41	0	650	27858	28549	638
Libéria	1991	0	0	0	0	0	86478	86478	418
Libyan Arab Jamahiriya	1990	0	0	0	0	0	12563	12563	0
Jamah. arabe libyenne	1991	0	0	0	0	0	19736	19736	267
Madagascar	1990	1119	55673	-5137	9818	7250	23683	82588	0
Madagascar	1991	2118	97000	0	8125	0	29586	128704	0
Malawi	1990	2521	90070	-13208	13315	3200	68405	150987	36698
Malawi	1991	5496	90000	0	12150	0	84416	179912	49823
Malaysia	1990	0	0	-45713	10092	4700	4477	-36536	7351
Malaisie	1991	0	0	-62000	7188	2000	5769	-54231	9862
Maldives	1990	97	1087	0	588	0	3771	4955	0
Maldives	1991	27	4000	0	1332	0	6746	10773	81
Mali	1990	2132	38141	0	9571	0	33702	73975	42
Mali	1991	366	64000	0	10681	0	31442	95808	707
Malta	1990	0	0	0	0	0	514	514	40
Malte	1991	0	0	0	0	0	498	498	40
Marshall Islands	1990	0	0	0	0	0	199	199	0
Iles Marshall	1991	0	0	0	0	0	154	154	0
Mauritania	1990	2394	37275	-15836	5544	1050	17180	42064	1146
Mauritanie	1991	-44	11000	0	4842	3000	21463	35419	748
Mauritius	1990	808	-354	-24344	312	5900	5142	-12848	0
Maurice	1991	65	0	-25000	570	6000	4920	-14015	0
Mexico	1990	-1212	0	1772972	26307	254360	24630	2050750	9235
Mexique	1991	-1248	0	-235000	38966	218000	19699	1451	9815
Mongolia	1990	0	0	0	0	0	6820	6820	0
Mongolie	1991	0	0	0	0	0	8929	8929	0
Montserrat	1990	0	0	0	0	0	408	408	0
Montserrat	1991	0	0	0	0	0	351	351	0
Morocco	1990	4027	-1166	-2033	40897	113980	39516	154324	135
Maroc	1991	3708	0	-113000	8706	1000	30957	-77335	157
Mozambique	1990	0	73614	0	6411	500	76869	150983	7090
Mozambique	1991	0	54000	0	10640	0	81944	135944	6335
Myanmar	1990	0	49550	0	2793	0	26015	75566	0
Myanmar	1991	0	33000	0	1088	0	30373	63373	124
Namibia	1990	0	0	0	0	0	12391	12391	1660
Namibie	1991	0	0	0	0	0	14056	14056	187
Nauru	1990	0	0	0	0	0	0	0	0
Nauru	1991	0	0	0	0	0	0	0	0
Nepal	1990	6871	60492	0	14049	4720	44813	116896	331
Népal	1991	1880	41000	0	11167	0	38566	81446	504
Netherlands Antilles	1990	0	0	0	0	0	583	583	0
Antilles néerlandaises	1991	0	0	0	0	0	377	377	0
Nicaragua	1990	1368	-174	-1469	0	0	19817	19541	7698
Nicaragua	1991	68	49000	0	96	0	23278	72346	12367
Niger	1990	-311	48759	0	12595	0	34722	83170	45
Niger	1991	-304	12000	0	6852	0	38044	49740	170
Nigeria	1990	2938	6076	-99644	24326	21490	45073	-24066	672
Nigéria	1991	2386	22000	-355000	27818	19000	37430	-274184	912
Niue	1990	0	0	0	0	0	236	236	0
Nioue	1991	0	0	0	0	0	515	515	0
Oman	1990	0	0	-10652	370	0	6421	-4230	0
Oman	1991	0	0	0	577	0	5547	5547	0

131
Socio-economic development assistance through the United Nations system
Thousand US dollars [cont.]
Assistance en matière de développement socio-économique fournie par le système des Nations Unies
Milliers de dollars E-U [suite]

B. Development loan and relief expenditures [1] · Prêts au développement et secours [1]

Country or area Pays ou zone	Year Année	Loans Prêts — IFAD FIDA	IBRD/IDA IDA IDA	BIRD/IDA IBRD BIRD	Training, consultants Formation, consultants	IFC SFI	Total develop. grants Aide totale au développement	Grand total Total général	Relief and related grants Secours et aide connexe
Pacific Islands (Palau)[2]	1990	0	0	0	0	0	1847	1847	0
Iles du Pacifique (Palaos)[2]	1991	0	0	0	0	0	2368	2368	0
Pakistan	1990	850	100928	156972	23457	14870	106672	380292	53327
Pakistan	1991	2450	173000	159000	26931	37000	143011	514461	43505
Panama	1990	-1804	0	-68362	0	9500	2815	-57851	222
Panama	1991	-1945	0	0	0	0	3682	1737	193
Papua New Guinea	1990	596	-1725	35248	9147	0	8257	42376	1569
Papouasie-Nvl-Guinée	1991	-316	0	16000	4991	0	10274	25958	1593
Paraguay	1990	4331	-1313	-48861	1200	650	5719	-39475	18
Paraguay	1991	-1481	0	-54000	1891	0	9236	-46245	18
Peru	1990	2166	0	-54	161	2930	24405	29447	518
Pérou	1991	749	0	0	290	0	32044	32793	66
Philippines	1990	125	-1817	-90365	10268	99920	28805	36667	13286
Philippines	1991	674	32000	-269000	13063	26000	32015	-178311	12745
Poland	1990	0	0	53549	100	1060	27344	81953	0
Pologne	1991	0	0	341000	383	11000	4136	356136	228
Portugal	1990	0	0	-35876	1593	6180	661	-29036	470
Portugal	1991	0	0	0	1892	3000	1567	4567	297
Qatar	1990	0	0	0	0	0	816	816	0
Qatar	1991	0	0	0	0	0	1042	1042	0
Réunion	1990	0	0	0	0	0	15	15	0
Réunion	1991	0	0	0	0	0	33	33	0
Romania	1990	0	0	-14316	0	0	3243	-11073	1
Roumanie	1991	0	0	3000	0	0	6775	9775	32
Rwanda	1990	2321	18246	0	6982	50	21961	42578	1730
Rwanda	1991	2118	45000	0	5429	0	25048	72166	1497
Saint Kitts and Nevis	1990	0	0	0	0	0	526	526	0
Saint Kitts et Nevis	1991	0	0	0	4	0	364	364	0
Saint Lucia	1990	-42	0	0	0	0	549	508	0
Sainte-Lucie	1991	280	0	0	85	4000	486	4766	0
Saint Vincent/Grenadines	1990	0	815	-4	260	0	652	1463	8
St. Vincent/Grenadines	1991	0	0	0	141	0	530	530	-8
Samoa	1990	111	3339	0	273	0	4252	7702	0
Samoa	1991	83	0	0	1383	0	2640	2723	0
Sao Tome and Principe	1990	437	5222	0	2513	0	6810	12470	0
Sao Tomé-et-Principe	1991	108	5000	0	1746	0	5491	10599	0
Saudi Arabia	1990	0	0	0	0	0	30808	30808	0
Arabie saoudite	1991	0	0	0	0	0	32083	32083	0
Senegal	1990	865	108016	-19368	7331	4650	37858	132021	5744
Sénégal	1991	-221	42000	0	6710	0	34905	76684	4779
Seychelles	1990	0	0	389	18	130	1738	2257	0
Seychelles	1991	80	0	0	14	0	1538	1618	0
Sierra Leone	1990	0	12	-11	12	0	12973	12974	2656
Sierra Leone	1991	0	0	0	13	0	13515	13515	2190
Singapore	1990	0	0	-26819	0	0	1153	-25666	462
Singapour	1991	0	0	0	0	0	620	620	576
Solomon Islands	1990	217	2420	0	182	0	1741	4378	0
Iles Salomon	1991	221	2000	0	79	0	1515	3736	0
Somalia	1990	-372	28276	0	8678	30	38219	66153	12516
Somalie	1991	27	0	0	410	0	22528	22555	8176
Sri Lanka	1990	1119	117856	-13395	10336	-1750	20154	123983	2401
Sri Lanka	1991	2397	176000	0	7883	0	22489	200886	1438
Sudan	1990	2628	110370	-11268	19575	90	87107	188926	34053
Soudan	1991	2003	63000	0	20215	0	184713	249716	41777
Suriname	1990	0	0	0	0	0	1430	1430	87
Suriname	1991	0	0	0	0	0	1301	1301	212
Swaziland	1990	683	-203	-23039	117	770	7455	-14333	1528
Souaziland	1991	740	0	-7000	0	0	6830	570	1926

131
Socio-economic development assistance through the United Nations system
Thousand US dollars [*cont.*]

Assistance en matière de développement socio-économique fournie par le système des Nations Unies
Milliers de dollars E-U [*suite*]

B. Development loan and relief expenditures [1] · Prêts au développement et secours [1]

Country or area Pays ou zone	Year Année	Loans Prêts IFAD FIDA	IBRD/IDA IDA IDA	BIRD/IDA IBRD BIRD	Training, consultants Formation, consultants	IFC SFI	Total develop. grants Aide totale au dévelop- pement	Grand total Total général	Relief and related grants Secours et aide connexe
Syrian Arab Republic	1990	-3	-2100	-1983	0	0	29385	25299	200
Arabe rép. syrienne	1991	-3871	0	0	0	0	20816	16945	1110
Thailand	1990	-1093	-1977	-222295	7867	9110	45623	-170631	23454
Thaïlande	1991	-2150	0	-233000	14648	30000	48037	-157113	33274
Togo	1990	2746	27629	-697	8711	340	14720	44737	267
Togo	1991	1118	44000	0	7707	0	12529	57647	221
Tokelau	1990	0	0	0	0	0	372	372	0
Tokélaou	1991	0	0	0	0	0	399	399	0
Tonga	1990	534	981	0	0	0	1293	2807	0
Tonga	1991	546	0	0	0	0	1265	1811	0
Trinidad and Tobago	1990	0	0	11017	928	17630	1909	30556	0
Trinité-et-Tobago	1991	0	0	-7000	995	0	2077	-4923	0
Tunisia	1990	-941	-1819	-1407	3095	4150	19436	19419	519
Tunisie	1991	-986	0	61000	4078	4000	9564	73578	90
Turkey	1990	1020	-5214	-505245	20611	132240	14807	-362392	2386
Turquie	1991	-761	0	-721000	22258	63000	13603	-645158	12042
Turks and Caicos Islands	1991	0	0	0	0	0	198	198	0
Iles Turques et Caïques	1990	0	0	0	0	0	1231	1231	0
Tuvalu	1990	0	0	0	0	0	75	75	0
Tuvalu	1991	0	0	0	0	0	933	933	0
Uganda	1990	3504	199074	-6392	21873	2000	53736	251922	3059
Ouganda	1991	1613	127000	0	19780	0	57410	186023	4740
former USSR ancienne URSS	1991	0	0	0	0	0	950	950	-69
United Arab Emirates	1990	0	0	0	0	0	2364	2364	24
Emirats arabes unis	1991	0	0	0	0	0	1829	1829	45
United Rep. of Tanzania	1990	-39	172689	-43490	18654	1550	54705	185415	1860
Rép. unie de Tanzanie	1991	850	164000	0	15878	1000	54226	220076	1885
Uruguay	1990	0	0	-19147	897	15260	4239	352	69
Uruguay	1991	0	0	11000	1632	1000	4194	16194	120
Vanuatu	1990	0	776	0	23	0	1784	2560	0
Vanuatu	1991	0	4000	0	551	0	2206	6206	0
Venezuela	1990	1479	0	821178	1495	19560	5460	847676	416
Venezuela	1991	2391	0	268000	541	38000	6876	315267	837
Viet Nam	1990	0	-1039	0	0	0	73606	72567	8665
Viet Nam	1991	0	-1000	0	0	0	70860	69860	9528
Yemen	1990	1859	26335	0	5360	3670	52151	84015	387
Yémen	1991	865	40000	0	10360	0	41319	82184	1083
Yugoslavia	1990	0	0	-326893	2620	100960	3064	-222868	3410
Yougoslavie	1991	0	0	-501000	2446	12000	2207	-486793	5895
Zaire	1990	1195	79621	14197	15055	7090	26368	128471	4254
Zaïre	1991	2085	54000	0	19036	3000	24042	83127	8309
Zambia	1990	3179	2861	-3988	2905	7390	21929	31371	4018
Zambie	1991	2475	202000	0	2378	0	20928	225403	2619
Zimbabwe	1990	265	-486	-26413	4345	74470	19245	67081	4147
Zimbabwe	1991	3655	0	-16000	783	35000	17832	40487	3391
Not elsewhere classified	1990	20300	0	73184	740	0	16843	110327	279193
Non-classé ailleurs	1991	25800	0	-4000	0	0	220262	242062	574801

Source:
1991 Annual Report on Operational Activities for Development of the United Nations system (A/46/206); Comprehensive statistical data on operational activities for development for the year 1990 (A/46/206/Add.4); 1992 Report for the Triennial comprehensive policy review of operational activities of the United Nations system (A/47/419) and Comprehensive statistical data on operational activities for development for the year 1991 (A/47/419/Add.2).

Source:
Rapport annuel de 1991 sur les Activités opérationelles pour le développement du système des Nations Unies (A/46/206); Information statistique détail concernant les activités opérationelles du développement, 1990 (A/46/206/Add.4); Rapport de 1992 pour l'examen triennal d'ensemble des orientations des activités opérationelles du système des Nations Unies (A/47/419) et Information statistique détail concernant les activités opérationelles du développement, 1991 (A/47/419/Add.2).

131
Socio-economic development assistance through the United Nations system
Thousand US dollars [*cont.*]
Assistance en matière de développement socio-économique fournie par le système des Nations Unies
Milliers de dollars E-U [*suite*]

1 The following abbreviations have been used in the table: UNDP: United Nations Development Programme UNFPA: United Nations Population Fund UNICEF: United Nations Children's Fund WFP: World Food Programme IDA: International Development Association IFAD: International Fund for Agricultural Development IBRD: International Bank for Reconstruction and Development IFC: International Finance Corporation 2 Including data for Federated States of Micronesia and Northern Mariana Is.	1 Les abbréviations ci-après ont été utilisées dans le tableau: PNUD : Programme des Nations Unies pour le développement FNUAP : Fonds des Nations Unies pour la population FISE : Fonds des Nations Unies pour l'enfance PAM : Programme alimentaire mondiale IDA : Association internationale de développement FIDA : Fonds international de développement agricole BIRD : Banque internationale pour la réconstruction et le développement SFI : Société financière internationale 2 Y compris les données pour les Etats fédérés de Micronésie et les îles Mariannes du Nord.

Technical notes, tables 130 and 131

Table 130 presents estimates of flows of financial resources to countries in the developing regions either directly (bilaterally) or through multilateral institutions (multilaterally).

For the purpose of this table, the developing regions include all Africa, except South Africa; Latin America and the Caribbean, including the United States Virgin Islands; Asia, except Japan; and Oceania, except Australia, New Zealand and United States possessions and territories (however, the Trust Territory of the Pacific Islands is included) "Developed market economies" comprise 20 members of the Development Assistance Committee of the Organisation for Economic Co-operation and Development (OECD): Australia, Austria, Belgium, Canada, Denmark, Finland, the Federal Republic of Germany, France, Ireland, Italy, Japan, Netherlands, New Zealand, Norway, Portugal (since December 1991), Spain (since December 1991), Sweden, Switzerland, the United Kingdom of Great Britain and Northern Ireland and the United States of America.

The multilateral institutions include the World Bank Group, regional banks, financial institutions of the European Community and a number of United Nations institutions, programmes and trust funds.

The main source of data is the Development Assistance Committee of OECD to which member countries reported data on their flow of resources to developing countries and multilateral institutions. Data in the *Statistical Yearbook* do not include the less developed countries in Europe as recipients.

Additional information on definitions, methods and sources can be found in OECD's *Geographical Distribution of Financial Flows to Developing Countries*.[15]

Table 131: includes data on expenditures on operational activities for development undertaken by the organizations of the United Nations system. Operational activities encompass, in general, those activities of a development cooperation character that seek to mobilize or increase the potential and capacity of countries to promote economic and social development and welfare, including the transfer of resources to developing countries or regions in a tangible or intangible form. The table also covers, as a memo item, expenditures on activities of an emergency character, the purpose of which is immediate relief in crisis situations, such as assistance to refugees, humanitarian work and activities in respect of disasters.

Expenditures on operational activities for development are financed from contributions from governments and other official and non-official sources to a variety of funding channels in the United Nations system. These include United Nations funds and programmes such as contributions to the United Nations Development

Notes techniques, tableaux 130 et 131

Le *Tableau 130* présente les estimations des flux de ressources financières mises à la disposition des pays des régions en développement soit directement (aide bilatérale) soit par l'intermédiaire d'institutions multilatérales (aide multilatérale).

Aux fins de ce tableau, les régions en développement comprennent l'Afrique, sauf l'Afrique du Sud; l'Amérique latine et les Caraïbes, y compris les Iles vierges américaines; l'Asie, sauf le Japon; et l'Océanie, sauf l'Australie, la Nouvelle-Zélande et les possessions et territoires des Etats-Unis (le Territoire sous tutelle des Iles du Pacifique étant toutefois inclus dans les pays en développement). Les "pays développés à économie de marché" comprennent les 20 membres du Comité d'aide au développement de l'Organisation de coopération et de développement économiques (OCDE), à savoir : l'Allemagne (République fédérale d'), l'Australie, l'Autriche, la Belgique, le Canada, le Danemark, l'Espagne (à partir de décembre 1991), les Etats-Unis d'Amérique, la Finlande, la France, l'Irelande, l'Italie, le Japon, la Norvège, la Nouvelle-Zélande, les Pays-Bas, le Portugal (à partir de décembre 1991), le Royaume-Uni de Grande-Bretagne et d'Irlande du Nord, la Suède et la Suisse.

Les institutions multilatérales comprennent le Groupe de la Banque mondiale, les banques régionales, les institutions financières de la Communauté européenne et un certain nombre d'institutions, de programmes et de fonds d'affectation spéciale des Nations Unies.

La principale source de données est le Comité d'aide au développement de l'OCDE, auquel les pays membres ont communiqué des données sur les flux de ressources qu'ils mettent à la disposition des pays en développement et des institutions multilatérales. Les données présentées dans l'*Annuaire statistique* ne comprennent pas l'aide fournie aux pays moins développés d'Europe.

Pour plus de renseignements sur les définitions, méthodes et sources, se reporter à la *Répartition géographique des ressources financières de l'OCDE* [15].

Le *Tableau 131* présente des données sur les dépenses consacrées à des activités opérationnelles pour le développement par les organisations du système des Nations Unies. Par "activités opérationnelles", on entend en général les activités ayant trait à la coopération au développement, qui visent à mobiliser ou à accroître les potentialités et aptitudes que présentent les pays pour promouvoir le développement et le bien-être économiques et sociaux, y compris les transferts de ressources vers les pays ou régions en développement sous forme tangible ou non. Ce tableau indique également, pour mémoire, les dépenses liées à des activités revêtant un caractère d'urgence, qui ont pour but d'apporter un secours immédiat dans les situations de crise, telles que l'aide aux réfugiés, l'assistance humanitaire et les secours en cas de catastrophe.

Programme; contributions to funds administered by United Nations Development Programme; regular (assessed) and other extrabudgetary contributions to specialized agencies, International Atomic Energy Agency, and other organizations for the purposes of operational activities; contributions to the International Development Association (IDA) and to the International Fund for Agricultural Development (IFAD).

Data have been taken from the 1991 and 1992 reports of the Secretary-General to the forty-seventh Assembly on operational activities for development. [18]

Les dépenses consacrées aux activités opérationnelles pour le développement sont financées au moyen de contributions que les gouvernements et d'autres sources officielles et non officielles apportent à divers organes de financement, tels que fonds et programmes, du système des Nations Unies. On peut citer notamment les contributions au Programme des Nations Unies pour le développement; les contributions aux fonds gérés par le Programme des Nations Unies pour le développement; les contributions régulières (budgétaires) et les contributions extrabudgétaires aux institutions spécialisées, à l'Agence internationale de l'énergie atomique et à d'autres organismes aux fins d'activités opérationnelles; les contributions à l'Association internationale de développement (IDA) et au Fonds international de développement agricole (FIDA).

Les données sont extraites des rapports annuels de 1991 et 1992 du Secrétaire général à la quarante-septième session de l'Assemblée générale sur les activités opérationnelles pour le développement [18].

Annex I

Country and area nomenclature, regional and other groupings

A. Changes in country or area names

In the periods covered by the statistics in the present issue of the *Statistical Yearbook* (in general, 1981-1990, 1982-1991), the following changes in designation have taken place:

Belize was formerly listed as British Honduras;

Brunei Darussalam was formerly listed as Brunei;

Burkina Faso was formerly listed as Upper Volta;

Cambodia was formerly listed as Democratic Kampuchea;

Cameroon was formerly listed as United Republic of Cameroon;

Côte d'Ivoire was formerly listed as Ivory Coast;

Equatorial Guinea was formerly listed as Spanish Guinea;

Germany: Through the accession of the German Democratic Republic to the Federal Republic of Germany with effect from 3 October 1990, the two German States have united to form one sovereign State. As from the date of unification, the Federal Republic of Germany acts in the United Nations under the designation "Germany". All data shown which pertain to Germany prior to 3 October 1990 are indicated separately for the Federal Republic of Germany and the former German Democratic Republic based on their respective territories at the time indicated;

Myanmar was formerly listed as Burma;

Pacific Island (Palau) was formerly listed as Pacific Islands and includes data for Federated States of Micronesia, Marshall Islands and Northern Mariana Islands;

Saint Kitts and Nevis was formerly listed as Saint Christopher and Nevis;

St. Vincent and the Grenadines was formerly listed as St. Vincent;

Vanuatu was formerly listed as New Hebrides;

Yemen comprises the former Republic of Yemen and the former Democratic Yemen;

Yugoslavia: Data provided for Yugoslavia prior to 1 January 1992 refer to the Socialist Federal Republic of Yugoslavia which was composed of six republics. Data provided for Yugoslavia after that date refer to the Federal Republic of Yugoslavia which is composed of two republics (Serbia and Montenegro);

Zimbabwe was formerly listed as Southern Rhodesia.

Data relating to the People's Republic of China generally include those for Taiwan Province in the field of statistics relating to population, area, natural resources, natural conditions such as climate. In other fields of statistics, they do not include Taiwan Province unless otherwise stated.

Annexe I

Nomenclature des pays ou zones, groupements régionaux et autres groupements

A. Changements dans le nom des pays ou zones

Au cours des périodes sur lesquelles portent les statistiques, dans cette édition de l'*Annuaire Statistique* (1981-1990 et 1982-1991, en générale), les changements de désignation suivants ont eu lieu:

Le *Belize* apparaissait antérieurement sous le nom de Honduras britannique;

Le *Brunéi Darussalam* apparaissait antérieurement sous le nom de Brunéi;

Le *Burkino Faso* apparaissait antérieurement sous le nom de l'Haute-Volta;

Le *Cambodge* apparaissait antérieurement sous le nom de la Kampuchea démocratique;

Le *Cameroun* apparaissait antérieurement sous le nom de République-Unie du Cameroun;

La *Guinée équatoriale* appariassait antérieurement sous le nom de la Guinée espagnol.

Allemagne: En vertu de l'adhésion de la République démocratique allemande à la République fédérale d'Allemagne, prenant effet le 3 octobre 1990, les deux Etats allemands se sont unis pour former un seul Etat souverain. A compter de la date de l'unification, la République fédérale d'Allemagne est désigné à l'ONU sous le nom d'"Allemagne". Toutes les données se rapportant à l'Allemagne avant le 3 octobre figurent dans deux rubriques séparées basées sur les territoires respectifs de la République fédérale d'Allemagne et l'ancienne République démocratique allemande selon la période indiquée;

Myanmar apparaissait antérieurement sous le nom de Birmanie;

Iles du Pacifique (Palaos) apparaissait antétieurement sous le nom de Iles du Pacifique y compris les données pour les Etats fédérés de Micronésie, les îles Marshall et îles Mariannes du Nord;

Saint-Kitts-et-Nevis apparaissait antérieurement sous le nom de Saint-Christophe-et-Nevis;

Saint-Vincent-et-Grenadines apparaissait antérieurement sous le nom de Saint-Vincent;

Vanuatu apparaissait antérieurement sous le nom des Nouvelles-Hébrides;

Le *Yémen* comprend l'ancienne République de Yémen et l'ancien Yémen démocratique;

Yougoslavie : Les données fournies pour la Yougoslavie avant le 1er janvier 1992 se rapportent à la République fédérative socialiste de Yougoslavie, qui était composée de six républiques. Les données fournies pour la Yougoslavie après cette date se rapportent à la République fédérative de Yougoslavie, qui est composée de deux républiques (Serbie et Monténégro);

Le *Zimbabwe* apparaissait antérieurement sous le nom de la Rhodésie du Sud.

Les données relatives à la République populaire de Chine comprennent en général les données relatives à la province de Taïwan lorqu'il s'agit de statistiques concernant la population, la superficie, les ressources naturelles, les conditions naturelles

Some tables contain data, shown in parentheses, for Belarus and for Ukraine; such data are also included in the total comprising the former USSR.

B. *Regional groupings*

The scheme of regional groupings given below presents seven regions based mainly on continents. Five of the seven continental regions are further subdivided into 20 regions that are so drawn as to obtain greater homogeneity in sizes of population, demographic circumstances and accuracy of demographic statistics.[19, 25] This nomenclature is widely used in international statistics and is followed to the greatest extent possible in the present *Yearbook* in order to promote consistency and facilitate comparability and analysis. However, it is by no means universal in international statistical compilation, even at the level of continental regions, and variations in international statistical sources and methods dictate many unavoidable differences in particular fields in the present *Yearbook*. General differences are indicated in the footnotes to the classification presented below. More detailed differences are given in the footnotes and technical notes to individual tables.

Neither is there international standardization in the use of the terms "developed" and "developing" countries, areas or regions. These terms are used in the present publication to refer to regional groupings generally considered as "developed": these are Europe and former USSR, the United States of America and Canada in northern America, and Australia, Japan and New Zealand in Asia and Oceania. These designations are intended for statistical convenience and do not necessarily express a judgement about the stage reached by a parlicular country or area in the development process. Differences from this usage are indicated in the notes to individual tables.

Africa
Northern Africa
Algeria
Egypt
Libyan Arab Jamahiriya
Morocco
Sudan
Tunisia
Western Sahara

telles que le climat, etc. Dans les statistiques relatives à d'autres domaines, la province de Taïwan n'est pas comprise, sauf indication contraire.

Certains tableaux présentent, entre parenthèses, les données du Bélarus et d'Ukraine, qui font partie également de l'estimation totale relative à l'ancienne URSS.

B. *Groupements régionaux*

Le système de groupements régionaux présenté ce-dessous comporte sept régions basés principalement sur les continents. Cinq des sept régions continentales sont elles-mêmes subdivisées, formant ainsi 20 régions délimitées de manière à obtenir une homogénéité accrue dans les effectifs de population, les situations démographiques et la précision des statistiques démographiques [19, 25]. Cette nomenclature est couramment utilisée aux fins des statistiques internationales et a été appliquée autant qu'il a été possible dans le présent *Annuaire* en vue de renforcer la cohérence et de faciliter la comparaison et l'analyse. Son utilisation pour l'établissement des statistiques internationales n'est cependant rien moins qu'universelle, même au niveau des régions continentales, et les variations que présentent les sources et méthodes statistiques internationales entraînent inévitablement de nombreuses différences dans certains domaines de cet *Annuaire*. Les différences d'ordre général sont indiquées dans les notes figurant au bas de la classification présentée ci-dessous. Les différences plus spécifiques sont mentionées dans les notes techiques et notes infrapaginales accompagnant les divers tableaux.

L'application des expressions "développés" et "en développement" aux pays, zones ou régions n'est pas non plus normalisée à l'échelle internationale. Ces expressions sont utilisées dans la présente publication en référence aux groupements régionaux généralement considérés comme "développés", à savoir l'Europe et l'ancienne URSS, les Etats-Unis d'Amérique et le Canada en Amérique du Nord, et l'Australie, le Japon et la Nouvelle-Zélande dans la région de l'Asie et du Pacifique. Ces appellations sont employées pour des raisons de commodité statistique et n'expriment pas nécessairement un jugement sur le stade de développement atteint par tel ou tel pays ou zone. Les cas différant de cet usage sont signalés dans les notes accompagnant les tableaux concernés.

Afrique
Afrique septentrionale
Algérie
Egypte
Jam. arabe libyenne
Maroc
Soudan
Tunisie
Sahara occidental

Sub-Saharan Africa	*Afrique subsaharienne*
Eastern Africa	Afrique orientale
British Indian Ocean Territory	Territoire britannique de l'océan indien
Burundi	Burundi
Comoros	Comores
Djibouti	Djibouti
Ethiopia	Ethiopie
Kenya	Kenya
Madagascar	Madagascar
Malawi	Malawi
Mauritius	Maurice
Mozambique	Mozambique
Reunion	Réunion
Rwanda	Rwanda
Seychelles	Seychelles
Somalia	Somalie
Uganda	Ouganda
United Republic of Tanzania	République-Unie de Tanzanie
Zambia	Zambie
Zimbabwe	Zimbabwe
Middle Africa	Afrique centrale
Angola	Angola
Cameroon	Cameroun
Central African Republic	République centrafricaine
Chad	Tchad
Congo	Congo
Equatorial Guinea	Guinée équatoriale
Gabon	Gabon
Sao Tome and Principe	Sao Tomé-et-Principe
Zaire	Zaïre
Southern Africa	Afrique méridionale
Botswana	Botswana
Lesotho	Lesotho
Namibia	Namibie
South Africa	Afrique du Sud
Swaziland	Swaziland
Western Africa	Afrique occidentale
Benin	Bénin
Burkina Faso	Burkina Faso
Cape Verde	Cap-Vert
Côte d'Ivoire	Côte d'Ivoire
Gambia	Gambie
Ghana	Ghana
Guinea	Guinée
Guinea-Bissau	Guinée-Bissau
Liberia	Libéria
Mali	Mali
Mauritania	Mauritanie
Niger	Niger
Nigeria	Nigéria
St. Helena	Sainte-Hélène
Senegal	Sénégal
Sierra Leone	Sierra Leone
Togo	Togo

Northern America	**Amérique septentrionale**
Bermuda	Bermudes
Canada	Canada
Greenland	Groenland
St. Pierre and Miquelon	Saint-Pierre-et-Miquelon
United States of America	Etats-Unis d'Amérique
Latin America and Caribbean	**Amérique latine et Caraïbes**
Caribbean	*Caraïbes*
Anguilla	Anguilla
Antigua and Barbuda	Antigua-et-Barbuda
Aruba	Aruba
Bahamas	Bahamas
Barbados	Barbade
British Virgin Islands	Iles Vierges britanniques
Cayman Islands	Iles Caïmanes
Cuba	Cuba
Dominica	Dominique
Dominican Republic	République dominicaine
Grenada	Grenade
Guadeloupe	Guadeloupe
Haiti	Haïti
Jamaica	Jamaïque
Martinique	Martinique
Montserrat	Montserrat
Netherlands Antilles	Antilles néerlandaises
Puerto Rico	Porto Rico
St. Kitts and Nevis	St. Christophe/Nevis
St. Lucia	Sainte-Lucie
St. Vincent/Grenadines	St. Vincent/Grenadines
Trinidad and Tobago	Trinité-et-Tobago
Turks and Caicos Islands	Iles Turques et Caiques
US Virgin Islands	Iles Vierges américaines
Central America	*Amérique centrale*
Belize	Belize
Costa Rica	Costa Rica
El Salvador	El Salvador
Guatemala	Guatemala
Honduras	Honduras
Mexico	Mexique
Nicaragua	Nicaragua
Panama	Panama
South America	*Amérique du Sud*
Argentina	Argentine
Bolivia	Bolivie
Brazil	Brésil
Chile	Chili
Colombia	Colombie
Ecuador	Equateur
Falkland Islands (Malvinas)	Iles Falkland (Malvinas)
French Guiana	Guyane française
Guyana	Guyana
Paraguay	Paraguay
Peru	Pérou

Suriname	Suriname
Uruguay	Uruguay
Venezuela	Venezuela
Asia	**Asie**
Eastern Asia	*Asie orientale*
China	Chine
Hong Kong	Hong-kong
Japan	Japon
Korea, Democratic People's Republic	Corée, république populaire démocratique de
Korea, Republic of	Corée, République de
Macau	Macao
Mongolia	Mongolie
South-eastern Asia	*Asie méridionale orientale*
Brunei Darussalam	Brunéi Darussalam
Cambodia	Cambodge
East Timor	Timor oriental
Indonesia	Indonésie
Lao People's Democratic Republic	République démocratique populaire Lao
Malaysia	Malaisie
Myanmar	Myanmar
Philippines	Philippines
Singapore	Singapour
Thailand	Thaïlande
Viet Nam	Viet Nam
Southern Asia	*Asie méridionale*
Afghanistan	Afghanistan
Bangladesh	Bangladesh
Bhutan	Bhoutan
India	Inde
Iran (Islamic Republic of)	Iran, République islamique d'
Maldives	Maldives
Nepal	Népal
Pakistan	Pakistan
Sri Lanka	Sri Lanka
Western Asia	*Asie occidentale*
Bahrain	Bahreïn
Cyprus	Chypre
Gaza Strip (Palestine)	Zone de Gaza (Palestine)
Iraq	Iraq
Israel	Israël
Jordan	Jordanie
Kuwait	Koweït
Lebanon	Liban
Oman	Oman
Qatar	Qatar
Saudi Arabia	Arabie saoudite
Syrian Arab Republic	République arabe syrienne
Turkey	Turquie
United Arab Emirates	Emirats arabes unis
Yemen	Yémen

Europe
Eastern Europe
Bulgaria
Czechoslovakia
Germany: [a]
  former German Democratic Republic
Hungary
Poland
Romania

Western Europe [b]
Austria
Belgium
France
Germany: [a]
  Federal Republic of Germany
Liechtenstein
Luxembourg
Monaco
Netherlands
Switzerland

Northern Europe
Channel Islands
Denmark
Faeroe Islands
Finland
Iceland
Ireland
Isle of Man
Norway
Sweden
United Kingdom

Southern Europe
Albania
Andorra
Gibraltar
Greece
Holy See
Italy
Malta
Portugal
San Marino
Spain
Yugoslavia

Oceania
Australia and New Zealand
Australia
Christmas Islands
Cocos (Keeling) Islands
New Zealand
Norfolk Island

Europe
Europe orientale
Bulgarie
Tchécoslovaquie
Allemgne: [a]
  ancienne République démocratique allemande
Hongrie
Pologne
Roumanie

Europe occidentale [b]
Autriche
Belgique
France
Allemagne: [a]
  République fédérale d'Allemagne
Liechtenstein
Luxembourg
Monaco
Pays-Bas
Suisse

Europe septentrionale
Iles Anglo-Normandes
Danemark
Iles Féroé
Finlande
Islande
Irlande
Iles de Man
Norvège
Suède
Royaume-Uni

Europe méridionale
Albanie
Andorre
Gibraltar
Grèce
Saint-Siège
Italie
Malte
Portugal
Saint-Marin
Espagne
Yougoslavie

Océanie
Australie et Nouvelle Zélande
Australie
Iles Christmas
Iles des Cocos (Keeling)
Nouvelle-Zélande
Ile Norfolk

Pacific
   Melanesia
Fiji
New Caledonia
Papua New Guinea
Solomon Islands
Vanuatu

   Micronesia
Guam
Kiribati
Marshall Islands
Micronesia, Federated States of
Nauru
Northern Marianna Islands
Pacific Islands (Palau)
Wake Island

   Polynesia
American Samoa
Cook Islands
French Polynesia
Johnston Island
Midway Islands
Niue
Pitcairn
Samoa
Tokelau
Tonga
Tuvalu
Wallis and Futuna Islands

former **Union of Soviet Socialist Republics**
former USSR

Pacifique
   Melenésie
Fidji
Nouvelle-Calédonie
Papouasie-Nouv.-Guinée
Iles Salomon
Vanuatu

   Micronésie
Guam
Kiribati
Iles Marshall
Micronésie, Etats fédératives de
Nauru
Iles Mariannes du Nord
Iles du Pacifique (Palaos)
Ile de Wake

   Polynésie
Samoa américaine
Iles Cook
Polynésie française
Ile Johnston
Iles Midway
Nioué
Pitcairn
Samoa
Tokélau
Tonga
Tuvalu
Iles Wallis et Futuna

ancienne **Union des républiques Socialistes Soviétiques**
ancienne URSS

---

[a] Through the accession of the German Democratic Republic to the Federal Republic of Germany with effect from 3 October 1990, the two German States have united to form one sovereign State. As from the date of unification, the Federal Republic of Germany acts in the United Nations under the designation of "Germany". All data shown which pertain to Germany prior to 3 October 1990 are indicated separately for the Federal Republic of Germany and the former German Democratic Republic based on their respective territories at the time indicated.

[b] Where the term "western Europe" is used in the present publication in distinction to "eastern Europe", it refers to all regions of Europe except eastern Europe (that is, it is comprised of northern and southern as well as western Europe).

[a] En vertu de l'adhésion de la République démocratique allemande à la République fédérale d'Allemagne, prenant effet le 3 octobre 1990, les deux Etats allemands se sont unis pour former un seul Etat souverain. A compter de la date de l'unification, la République fédérale d'Allemagne est désigné à l'ONU sous le nom d'Allemagne'. Toutes les données se rapportant à l'Allemagne avant le 3 octobre figurent dans deux rubriques séparées basées sur les territoires respectifs de la République fédérale d'Allemagne et l'ancienne République démocratique allemande selon la période indiquée.

[b] Lorsque l'expression "Europe occidentale" est utilisée dans la présente publication par opposition à l'expression "Europe orientale", elle s'applique à toutes les régions de l'Europe à l'exception de l'Europe orientale (c'est-à-dire qu'elle englobe l'Europe septentrionale et l'Europe méridionale aussi bien que l'Europe occidentale proprement dite).

C. *Other groupings*

Following is a list of other groupings and their compositions presented in the *Yearbook*. These groupings are organized mainly around economic and trade interests in regional associations.

Central American Common Market (CACM)
   Costa Rica
   El Salvador
   Guatemala
   Honduras
   Nicaragua

Customs and Economic Union of Central Africa (CEUCA)
   Cameroon
   Central African Republic
   Chad
   Congo
   Equatorial Guinea
   Gabon

Economic Community of West African States (ECOWAS)
   Benin
   Burkina Faso
   Cape Verde
   Côte d'Ivoire
   Gambia
   Ghana
   Guinea
   Guinea-Bissau
   Liberia
   Mali
   Mauritania
   Niger
   Nigeria
   Senegal
   Sierra Leone
   Togo

European Economic Communities (EEC)
   Belgium
   Denmark
   France
   Germany
   Greece
   Ireland
   Italy
   Luxembourg
   Netherlands
   Portugal
   Spain
   United Kingdom

C. *Autres groupements*

On trouvera ci-après une liste des autres groupements et de leur composition, présentée dans l'*Annuaire*. Ces groupements correspondent essentiellement à des intérêts économiques, et commerciaux d'après les associations régionales.

Marché commun de l'Amérique centrale (MCAC)
   Costa Rica
   El Salvador
   Guatemala
   Honduras
   Nicaragua

Union douanière et économique de l'Afrique centrale (UDEAC)
   Cameroun
   République centrafricaine
   Tchad
   Congo
   Guinée équatoriale
   Gabon

Communauté économique des états de l'Afrique de l'Ouest (CEDEAO)
   Bénin
   Burkina Faso
   Cap-Vert
   Côte d'Ivoire
   Gambie
   Ghana
   Guinée
   Guinée-Bissau
   Libéria
   Mali
   Mauritanie
   Niger
   Nigéria
   Sénégal
   Sierra Leone
   Togo

Communauté économique européenne (CEE)
   Belgique
   Danemark
   France
   Allemagne
   Grèce
   Irlande
   Italie
   Luxembourg
   Pays-Bas
   Portugal
   Espagne
   Poyaume-Uni

European Free Trade Association (EFTA)
   Austria
   Finland
   Iceland
   Norway
   Sweden
   Switzerland

Latin American Integration Association (LAIA)
   Argentina
   Bolivia
   Brazil
   Chile
   Colombia
   Ecuador
   Mexico
   Paraguay
   Peru
   Uruguay
   Venezuela

Least developed countries (LDC)*
   Afghanistan
   Bangladesh
   Benin
   Bhutan
   Botswana
   Burkina Faso
   Burundi
   Cambodia
   Cape Verde
   Central African Republic
   Chad
   Comoros
   Djibouti
   Equatorial Guinea
   Ethiopia
   Gambia
   Guinea
   Guinea-Bissau
   Haiti
   Kiribati
   Lao Pelple's Democratic Republic
   Lesotho
   Liberia
   Madagascar
   Malawi
   Maldives
   Mali
   Mauritania
   Mozambique
   Myanmar
   Nepal
   Niger

Association européenne de libre échange (AELE)
   Autriche
   Finlande
   Islande
   Norvège
   Suède
   Suisse

Association Latino-américaine d'intégration (LAIA)
   Argentine
   Bolivie
   Brésil
   Chili
   Colombie
   Equateur
   Mexique
   Paraguay
   Pérou
   Uruguay
   Venezuela

Les pays moins avancés (PMA)*
   Afghanistan
   Bangladesh
   Bénin
   Bhoutan
   Botswana
   Burkina Faso
   Burundi
   Cambodge
   Cap-Vert
   République centrafricaine
   Tchad
   Comores
   Djibouti
   Guinée équatoriale
   Ethiopie
   Gambie
   Guinée
   Guinée-Bissau
   Haïti
   Kiribati
   République démocratique populaire lao
   Lesotho
   Libéria
   Madagascar
   Malawi
   Maldives
   Mali
   Mauritanie
   Mozambique
   Myanmar
   Népal
   Niger

---

*   As determined by the General Assembly in its resolution 46/206.

*   Comme déterminés par l'Assemblée générale dans sa résolution 46/206.

Rwanda	Rwanda
Samoa	Samoa
Sao Tomé and Principe	Sao Tomé-et-Principe
Sierra Leone	Sierra Leone
Solomon Islands	Iles Salomon
Somalia	Somalie
Sudan	Soudan
Togo	Togo
Tuvalu	Tuvalu
Uganda	Ouganda
United Republic of Tanzania	République-Unie de Tanzanie
Vanuatu	Vanuatu
Yemen	Yémen
Zaire	Zaïre
Zambia	Zambie

Organization of Petroleum Exporting Countries (OPEC)	*Organisation des pays exportateurs de pétrole* (OPEP)
Algeria	Algérie
Ecuador	Equateur
Gabon	Gabon
Indonesia	Indonésie
Iran, Islamic Republic of	Iran, République islamique d'
Iraq	Iraq
Kuwait	Koweït
Libyan Arab Jamahiriya	Jamahiriya arabe libyenne
Nigeria	Nigéria
Qatar	Qatar
Saudi Arabia	Arabie saoudite
United Arab Emirates	Emirats arabes unis
Venezuela	Venezuela

Annex II
Conversion coefficients and factors

The metric system of weights and measures is employed in the *Statistical Yearbook*. In this system, the relationship between units of volume and capacity is: 1 litre = 1 cubic decimetre exactly (as decided by the 12th International Conference of Weights and Measures, New Delhi, November 1964).

Section A shows the equivalents of the basic metric, British imperial and United States units of measurements. According to an agreement between the national standards institutions of English-speaking nations, the British and United States units of length, area and volume are now identical, and based on the yard = 0.9144 metre exactly. The weight measures in both systems are based on the pound = 0.45359237 kilogram exactly (Weights and Measures Act 1963 (London), and *Federal Register* announcement of 1 July 1959: *Refinement of Values for the Yard and Pound* (Washington D.C.)).

Section B shows various derived or conventional conversion coefficients and equivalents.

Section C shows other conversion coefficients or factors which have been utilized in the compilation of certain tables in the *Statistical Yearbook*. Some of these are only of an approximate character and have been employed solely to obtain a reasonable measure of international comparability in the tables.

For a comprehensive survey of international and national systems of weights and measures and of units weights for a large number of commodities in different countries, see *World Weights and Measures* (United Nations publication, Sales No. E.66.XVII.3).

Annexe II
Coefficients et facteurs de conversion

L'*Annuaire statistique* utilise le système métrique pour les poids et mesures. La relation entre unités métriques de volume et de capacité est: 1 litre = 1 décimètre cube (dm³) exactement (comme fut décidé à la Conférence internationale des poids et mesures, New Delhi, novembre 1964).

La section A fournit les équivalents principaux des systèmes de mesure métrique, britannique et américain. Suivant un accord entre les institutions de normalisation nationales des pays de langue anglaise, les mesures britanniques et américaines de longueur, superficie et volume sont désormais identiques, et sont basées sur le yard = 0:9144 mètre exactement. Les mesures de poids se rapportent, dans les deux systèmes, à la livre (pound) = 0.45359237 kilogramme exactement ("Weights and Measures Act 1963" (Londres), et "*Federal Register Announcement of 1 July 1959: Refinement of Values for the Yard and Pound*" (Washington, D.C.)).

La section B fournit divers coeficients et facteurs de conversion conventionnels ou dérivés.

La section C fournit d'autres coefficients ou facteurs de conversion utilisés dans l'élaboration de certains tableaux de l'*Annuaire statistique*. D'aucuns ne sont que des approximations et n'ont été utilisés que pour obtenir un degré raisonnable de comparabilité sur le plan international.

Pour une étude d'ensemble des systèmes internationaux et nationaux de poids et mesures, et d'unités de poids pour un grand nombre de produits dans différents pays, voir "*World Weights and Measures*" (publication des Nations Unies, No de vente E.66.XVII.3).

A. Equivalents of metric, British imperial and United States units of measure
A. Equivalents des unités métriques, britanniques et des Etats-Unis

Metric units Unités métriques	British imperial and US equivalents Equivalents en mesures britanniques et des Etats-Unis	British imperial and US units Unités britannniques et des Etats-Unis	Metric equivalents Equivalents en mesures métriques	
Length–Longeur				
1 centimetre–centimètre (cm)	0.3937008 inch	1 inch.	2.540	cm
1 metre–mètre (m)	(3.280840 feet	1 foot	30.480	cm
	(1.093613 yard	1 yard	0.9144	m
1 kilometre – kilomètre (km)	(0.6213712 mile	1 mile	1609.344	m
	(0.5399568 int. naut. mile	1 international nautical mile	1852.000	m
Area – Superficie				
1 square centimetre – cm²	(0.1550003 square inch	1 square inch	6.45160	cm²
1 square metre – m²	(10.763910 square feet	1 square foot	9.290304	dm²
	(1.195990 square yards	1 square yard ·	0.83612736	m²
1 hectare – ha	2.471054 acres	1 acre	0.4046856	ha²
1 square kilometre – km²	0.3861022 square mile	1 square mile	2.589988	km²
Volume				
1 cubic centimetre – cm³	0.06102374 cubic inch	1 cubic inch	16.38706	cm³
1 cubic metre – m³	(35.31467 cubic feet	1 cubic foot	28.316847	dm³
	(1.307951 cubic yards	1 cubic yard	0.76455486	m³
Capacity – Capacité				
1 litre (l)	(0.8798766 imp. quart	1 British imperial quart	1.136523	l
	(1.056688 U.S. liq. quart	1 U.S. liquid quart	0.9463529	l
	(0.908083 U.S. dry quart	1 U.S. dry quart	1.1012208	l
1 hectolitre (hl)	(21.99692 imp. gallons	1 imperial gallon	4.546092	l
	(26.417200 U.S. gallons	1 U.S. gallon	3.785412	l
	(2.749614 imp. bushels	1 imperial bushel	36.368735	l
	(2.837760 U.S. bushels	1 U.S. bushel	35.239067	l
Weight or mass – Poids				
1 kilogram (kg)	(35.27396 av. ounces	1 av. ounce	28.349523	g
	(32.15075 troy ounces	1 troy ounce	31.10348	g
	(2.204623 av. pounds	1 av. pound	453.59237	g
		1 cental (100 lb.)	45.359237	kg
		1 hundredweight (112 lb.)	50.802345	kg
1 ton – tonne (t)	(1.1023113 short tons	1 short ton (2 000 lb.)	0.9071847	t
	(0.9842065 long tons	1 long ton (2 240 lb.)	1.0160469	t

B. Various conventional or derived coefficients

Railway and air transport

1 passenger-mile = 1.609344 voyageur (passager) - kilomètre
1 short ton-mile = 1.459972 tonne-kilomètre
1 long ton-mile = 1.635169 tonne kilomètre

Ship tonnage

1 register ton (100 cubic feet) – tonne de jauge = 2.83m³
1 British shipping ton (42 cubic feet) = 1.19m³
1 U.S. shipping ton (40 cubic feet) = 1.13m³
1 deadweight ton (dwt ton = long ton) = 1.016047 metric ton – tonne métrique

Electric energy

1 Kilowatt (kW) = (1.34102 British horsepower (hp)
(1.35962 cheval vapeur (cv)

B. Divers coefficients conventionnels ou dérivés

Transport ferroviaire et aérien

1 voyageur (passager) - kilomètre = 0.621371) passenger-mile
1 tonne-kilomètre = (0.684945 short ton-mile
(0.611558 long ton-mile

Tonnage de navire

1 cubic metre – m³ = (0.353 register ton – tonne de jauge
(0.841 British shipping ton
(0.885 US shipping ton
1 metric ton – tonne métrique – 0.984 dwt ton

Energie électrique

1 British horsepower (hp) = 0.7457 kW
1 cheval vapeur (cv) = 0.735499 kW

Agricultural and forest products, etc. – Produits agricoles et forestiers, etc.

Commodity Produit	1 Bushel [1] 1 Boisseau [1]	lb.	Metric tons Tonnes métr.	Bushels Boisseaux per – par m.t – t.m	Commodity Produit	1 Bushel [1] 1 Boisseau [1]	lb.	Metric tons Tonnes métr.	Bushels Boisseaux per – par m.t – t.m
Wheat – Froment....	U.S. Brit.	60	0.027216	36.744	Maize (corn) – Maïs.	U.S. Brit.	56	0.025401	39.368
Barley – Orge.	U.S.	48	0.021772	45.930	Rice–Riz (Paddy). . .	U.S.	45	0.020412	48.992
	Brit.	50	0.022680	44.092		Australia	42	0.019051	52.491
Oats–Avoine.	U.S.	32	0.014515	68.894	Potatoes.	U.S. Brit.	60	0.027216	36.744
	Canada	34	0.015422	64.842	Pommes de terre				
	Australia	40	0.018144	55.115					
	New Zealand				Soybeans–Soya.	U.S.	60	0.027216	36.743

Coffee–Café 1 bag–sac		kg	lb.	bags–sacs per m.t–t.m.	Cotton–Coton 1 bale–balle		lb.	metric tons tonnes métr.	bales–balles per m.t.–t.m.
Brazil–Brésil.		60	132.28	16.667	United States–Etats-Unis:				
Colombia–Colombie..............		60	132.28	16.667	conventional: gross–brute.		500	0.226796	4.409
El Salvador.		69	152.12	14.493	conventional: net		480	0.217724	4.593
Haiti–Haïti.		60	132.28	16.667	Brazil–Brésil.		397	0.180076	5.553
Mexico–Mexique		60	132.28	16.667	India-Inde		392	0.177808	5.624
Venezuela.		60	132.28	16.667	1 Kantar Egypt–Egypte.		110.23	0.05	kantar per m.t–t.m. 20.0

C. Other coefficients or conversion factors employed in *Statistical Yearbook* tables

Cotton

Where cotton production was reported in terms of raw (unginned) cotton, the ginned cotton equivalent was estimated by applying a coefficient of one-third.

Roundwood

Equivalent in solid volume without bark.

Sugar

1 metric ton raw sugar = 0.9 metric ton refined sugar.
For the United States and its possessions:
1 metric ton refined sugar = 1.07 metric tons raw sugar

C. Autres coefficients ou facteurs de conversion utilisés dans les tableaux de l'*Annuaire statistique*

Coton

Lorsque la production a été indiquée en poids brut (coton non égrené), un coefficient de conversion d'un tiers a été utilisé pour estimer l'équivalent en coton égrené.

Bois rond

Equivalences en volume solide sans écorce.

Sucre

1 tonne métrique de sucre brut = 0.9 tonne métrique de sucre raffiné.
Pour les Etats-Unis et leurs possessions:
1 tonne métrique de sucre raffiné = 1.07 t.m. de sucre brut.

---

[1] US: United States bushel; unless otherwise indicated, applies also to Canada. Brit.: British imperial bushel; unless otherwise indicated, applies also to Australia and New Zealand.

[1] US: boisseau des Etats-Unis; sauf indication contraire, s'applique aussi au Canada. Brit.: boisseau britannique; sauf indication contraire, s'applique aussi à l'Australie et à la Nouvelle-Zélande.

<table>
<tr><td>

D. Selected energy conversion factors

Crude petroleum

1 barrel = 42 U.S. gallons = 34.97 imperial gallons = 158.99 litres = 0.15899 cubic metres.
1 cubic metre = 6.2898 barrels.

The equivalent of barrels in metric tons depends on the specific gravity of the petroleum which varies from country to country. The average specific gravity for each producing country is indicated in the table on the production of crude petroleum in the *Energy Statistics Yearbook*.

Coal equivalent[1] (metric tons unless otherwise indicated):

Coal, anthracite and bituminous	1.0
Coal briquettes	1.0
Cokes of coal	0.9
Lignite	0.385
Cokes of brown coal or lignite	0.67
Lignite briquettes	0.67
Peat for fuel	0.325
Peat briquettes	0.5
Crude petroleum	1.429
Natural gas liquids (weighted average)	1.542
Liquefied petroleum gases	1.554
Natural gasolene	1.532
Condensate and other	1.512
Motor spirit	1.5
Kerosene and jet fuel	1.474
Gas-diesel oils	1.45
Residual fuel oils	1.416
Natural gas (terajoules[2])	34.121
Manufactured gas (terajoules[2])	34.121
Fuelwood ($10^3 m^3$)	0.333

Coal equivalent (metric tons of hydro, nuclear and geothermal electricity:
1000 kWh = 0.123

---

[1]   It should be noted that the base used for coal equivalency comprises 7000 calories/gramme.
[2]   Under standard conditions of l5°C, 1013.25 mbar, dry.

</td><td>

D. Facteurs de conversion pour certains produits en matière d'énergie

Pétrole brut

1 baril = 42 gallons E.U. = 34.97 gallons britanniques = 158.99 litres = 0.15899 m³.
1m³ = 6.2898 barils.

L'équivalent du baril en tonnes métriques dépend du poids spécifique du pétrole qui varie d'un pays à l'autre. Le poids spécifique moyen utilisé pour chaque pays producteur se trouve dans l'Annuaire des statistiques de l'énergie dans le tableau relatif à la production de pétrole brut.

Equivalent en houille[1] (tonnes métriques sauf indication contraire):

Charbon, anthracite et la houille bitumineuse . .	1.0
Briquettes de charbon	1.0
Cokes de charbon	0.9
Lignite	0.385
Cokes de charbon brun ou de lignite	0.67
Briquettes de lignite	0.67
Tourbe pour combustible	0.325
Briquettes de tourbe	0.5
Pétrole brut	1.429
Condensats provenant du gaz naturel	
(moyenne pondérée)	1.542
Gaz de pétrole liquéfié	1.554
Gazoline naturelle	1.532
Condensat et autres	1.512
Essence	1.5
Pétrole lampant et carburéacteur	1.474
Gaz oil fuel oil fluides	1.45
Huile lourde	1.416
Gaz naturel (terajoules[2])	34.121
Gaz d'usine (terajoules[2]	34.121
Bois de chauffage ($10^3 m^3$)	0.333

Equivalent en houille (tonnes métriques) d'électricité, hydraulique, nucléaire et géothermique:
1000 kWh = 0.123

---

[1]   Veuillez noter que l'équivalence en houille est faite sur la base de 7000 calories/gramme.
[2]   En volume standard (à 15°C, 1013.25 mbar, gaz sec).

</td></tr>
</table>

Annex III
Tables added and omitted

A. *Tables added*

In the 37th issue of the *Statistical Yearbook* (1988/89), the following tables were added:

Table 14: Population in urban and urual areas, rates of growth and largest city popultion;

Table 22: Selected indicators of life expectancy, child-bearing and mortality;

Table 23: Numbers of reported AIDS cases and current assessment of total cases and infected persons;

Table 120: Surface and land area and land use.

In the present issue of the *Statistical Yearbook* (1990/91), the following tables have been added:

Table 81: Wood-based panels;

Table 108: Selected indicators of natural resources;

Table 109: Indicators of environmental pollution and management;

Table 124: International tourism expenditures.

B. *Tables omitted*

The following tables which appeared in the *Statistical Yearbook 1987* were deleted from the 37th issue of the *Statistical Yearbook* (1988/89):

Table 3: Growth of gross domestic product or net material product, by regions: This table was merged with the previous table 4. The merged table appeared as table 3 in the 1988/89 edition;

Table 36: Health personnel: These data are no longer compiled by the World Health Organization;

Table 121: Steel, total and per capita, industrial consumption: These data are no longer compiled by the United Nations Economic Commission for Europe.

A cumulative listing of tables omitted as of the *1987* edition, relative to all previous issues of the *Yearbook*, is provided in the *1987* edition.

The following tables have been deleted from the present issue:

Table 3 in the *1988/89* edition: Growth of gross domestic product by kind of economic activity;

Table 4 in the *1988/89* edition: Growth of gross domestic product by type of expenditure;

Table 41 in the *1988/89* edition: Hours of work per week: manufacturing;

Table 49 in the *1988/89* edition: Barley;

Table 50 in the *1988/89* edition: Oats;

Table 53 in the *1988/89* edition: Potatoes;

Table 60 in the *1988/89* edition: Milk;

Table 61 in the *1988/89* edition: Eggs (hen);

Table 70 in the *1988/89* edition: Output and employment in mining and quarrying;

Annexe III
Tableaux ajoutés et supprimés

A. *Tableaux ajoutés*

Dans la trente-septième édition de l'*Annuire statistique* (1988/89), les tableaux suivants ont été ajoutés :

Tableau 14 : Population urbaine, population urbaine, taux d'accroissement et population de la ville la plus peuplée;

Tableau 22 : Choix d'indicateurs de l'expérance de vie, de maternité et de la mortalité;

Tableau 23 : Nombre de cas de SIDA déclarés et evaluation actuelle du nombre de cas et de personnes infectées;

Tableau 120 : Superficie totale, superficie des terres et utilisation des terres.

Dans ce numéro de l'*Annuaire statistique* (1990/91), les tableaux suivants ont été ajoutés:

Tableau 81 : Panneaux à base de bois;

Tableau 108 : Indicateurs concernant certaines ressources naturelles;

Tableau 109 : Indicateurs de la pollution et de la gestion de l'environnement;

Tableau 124 : Dépenses provenant du tourisme international.

B. *Tableaux supprimés*

Les tableaux suivants, qui figuraient dans l'*Annuaire statistique 1987*, ont été supprimés de la 37ème édition de l'*Annuaire statistique* (1988/89):

Tableau 3 : Accroissement du produit intérieur brut et produit matériel net, par régions : ce tableau a été incorporé à l'ancien tableau 4. Le tableau combiné constitue le tableau 3 dans l'édition 1988/89;

Tableau 36 : Personnel de santé : l'Organisation mondiale de la santé a cessé de rassembler les données correspondantes;

Tableau 121 : Acier, totale et par habitant, consommation industrielle : la Commission économique des Nations Unies pour l'Europe a cessé de rassembler les données correspondantes.

L'édition de *1987* contient une liste récapitulative des tableaux supprimés dans toutes les éditions précédentes de l'*Annuaire* jusqu'à l'édition de *1987* incluse.

Les tableaux suivants ont été supprimés de la présente édition :

Tableau 3 de l'édition de *1988/89* : Accroissement du produit intérieur brut par genre d'activité économique;

Tableau 4 de l'édition de *1988/89* : Accroissement du produit intérieur brut par type de dépense;

Tableau 41 de l'édition de *1988/89* : Durée du travail par semaine : industries manufacturières;

Tableau 49 de l'édition de *1988/89* : Orge;